AF744296

ENCYCLOPEDIA OF

BIOPROCESS TECHNOLOGY: FERMENTATION, BIOCATALYSIS, AND BIOSEPARATION

VOLUME 1

WILEY BIOTECHNOLOGY ENCYCLOPEDIAS

Encyclopedia of Bioprocess Technology: Fermentation, Biocatalysis, and Bioseparation
Edited by Michael C. Flickinger and Stephen W. Drew

Encyclopedia of Molecular Biology
Edited by Thomas E. Creighton

Encyclopedia of Cell Technology
Edited by Raymond E. Spier

Encyclopedia of Ethical, Legal, and Policy Issues in Biotechnology
Edited by Thomas J. Murray and Maxwell J. Mehlman

ENCYCLOPEDIA OF BIOPROCESS TECHNOLOGY: FERMENTATION, BIOCATALYSIS, AND BIOSEPARATION

EDITORIAL BOARD

Chairman
Elmer Gaden, Jr.
University of Virginia, Charlottesville

Associate Editors
H.W. Blanch
University of California, Berkeley

Yusuf Chisti
University of Almería

Arnold Demain
Massachusetts Institute of Technology

Peter Dunnill
Advanced Centre for Biochemical Engineering

David Estell
Khepri Pharmaceuticals

Csaba Horvath
Yale University

Arthur E. Humphrey
Pennsylvania State University

Bjorn K. Lydersen
Irvine Scientific

Poul B. Poulson
Novo Nordisk

Dane Zabriskie
Biogen, Inc.

Editorial Board
Stuart E. Builder
Strategic Biodevelopment

John R. Birch
Lonza Biologics

Charles L. Cooney
Massachusetts Institute of Technology

Edward L. Cussler
University of Minnesota

Jonathan S. Dordick
Rensselaer Polytechnic Institute

Bryan Griffiths
Centre for Applied Microbiology and Research

Lars Hagel
Amersham Pharmacia

Zhao Kai
National Vaccine and Serum Institute

Subash B. Karkare
AMGEN

Murry Moo-Young
University of Waterloo

Tetsuo Oka
Kyowa Hakko Kogyo Co., Ltd.

Karl Schugerl
University of Hannover

Atsuo Tanaka
Kyoto University

Kathryn Zoon
U.S. Food and Drug Administration

Series Editor
Leroy Hood
University of Washington

Editorial Staff
Publisher: **Jacqueline I. Kroschwitz**

Managing Editor: **Camille Pecoul Carter**

Editor: **Glenn Collins**

Editorial Assistant: **Hugh Kelly**

ENCYCLOPEDIA OF

BIOPROCESS TECHNOLOGY: FERMENTATION, BIOCATALYSIS, AND BIOSEPARATION

VOLUME 1

Michael C. Flickinger
University of Minnesota
St. Paul, Minnesota

Stephen W. Drew
Merck and Co., Inc.
Rahway, New Jersey

A Wiley-Interscience Publication
John Wiley & Sons, Inc.
New York / Chichester / Weinheim / Brisbane / Singapore / Toronto

21434735

This book is printed on acid-free paper. ♾

Copyright © 1999 by John Wiley & Sons, Inc. All rights reserved.

Published simultaneously in Canada.

No part of this publication may be reproduced, stored in a retrieval system or transmitted in any form or by any means, electronic, mechanical, photocopying, recording, scanning or otherwise, except as permitted under Sections 107 or 108 of the 1976 United States Copyright Act, without either the prior written permission of the Publisher, or authorization through payment of the appropriate per-copy fee to the Copyright Clearance Center, 222 Rosewood Drive, Danvers, MA 01923, (978) 750-8400, fax (978) 750-4744. Requests to the Publisher for permission should be addressed to the Permissions Department, John Wiley & Sons, Inc., 605 Third Avenue, New York, NY 10158-0012, (212) 850-6011, fax (212) 850-6008, E-Mail: PERMREQ @ WILEY.COM.

For ordering and customer service, call 1-800-CALL-WILEY.

Library of Congress Cataloging-in-Publication Data:

M0168020BU

Flickinger, Michael C.
The encyclopedia of bioprocess technology : fermentation, biocatalysis, and bioseparation / Michael C. Flickinger, Stephen W. Drew.
p. cm.
Includes index.
ISBN 0-471-13822-3 (alk. paper)
1. Biochemical engineering--Encyclopedias. I. Drew, Stephen W., 1945- . II. Title.
TP248.3.F57 1999
660.6′03--dc21 99-11576
CIP

Printed in the United States of America.

10 9 8 7 6 5 4 3 2 1

PREFACE

The Wiley Biotechnology Encyclopedias, composed of the *Encyclopedia of Molecular Biology*; the *Encyclopedia of Bioprocess Technology: Fermentation, Biocatalysis, and Bioseparation*; the *Encyclopedia of Cell Technology*; and the *Encyclopedia of Ethical, Legal, and Policy Issues in Biotechnology* cover very broadly four major contemporary themes in biotechnology. The series comes at a fascinating time in that, as we move into the twenty-first century, the discipline of biotechnology is undergoing striking paradigm changes.

Biotechnology is now beginning to be viewed as an informational science. In a simplistic sense there are three types of biological information. First, there is the digital or linear information of our chromosomes and genes with the four-letter alphabet composed of G, C, A, and T (the bases guanine, cytosine, adenine, and thymine). Variation in the order of these letters in the digital strings of our chromosomes or our expressed genes (or mRNAs) generates information of several distinct types: genes, regulatory machinery, and information that enables chromosomes to carry out their tasks as informational organelles (e.g., centromeric and telomeric sequences).

Second, there is the three-dimensional information of proteins, the molecular machines of life. Proteins are strings of amino acids employing a 20-letter alphabet. Proteins pose four technical challenges: (*1*) Proteins are synthesized as linear strings and fold into precise three-dimensional structures as dictated by the order of amino acid residues in the string. Can we formulate the rules for protein folding to predict three-dimensional structure from primary amino acid sequence? The identification and comparative analysis of all human and model organism (bacteria, yeast, nematode, fly, mouse, etc.) genes and proteins will eventually lead to a lexicon of motifs that are the building block components of genes and proteins. These motifs will greatly constrain the shape space that computational algorithms must search to successfully correlate primary amino acid sequence with the correct three-dimensional shapes. The protein-folding problem will probably be solved within the next 10–15 years. (*2*) Can we predict protein function from knowledge of the three-dimensional structure? Once again the lexicon of motifs with their functional as well as structural correlations will play a critical role in solving this problem. (*3*) How do the myriad of chemical modifications of proteins (e.g., phosphorylation, acetylation, etc.) alter their structures and modify their functions? The mass spectrometer will play a key role in identifying secondary modifications. (*4*) How do proteins interact with one another and/or with other macromolecules to form complex molecular machines (e.g., the ribosomal subunits)? If these functional complexes can be isolated, the mass spectrometer, coupled with a knowledge of all protein sequences that can be derived from the complete genomic sequence of the organism, will serve as a powerful tool for identifying all the components of complex molecular machines.

The third type of biological information arises from complex biological systems and networks. Systems information is four dimensional because it varies with time. For example, the human brain has 1,012 neurons making approximately 1,015 connections. From this network arise systems properties such as memory, consciousness, and the ability to learn. The important point is that systems properties cannot be understood from studying the network elements (e.g., neurons) one at a time; rather the collective behavior of the elements needs to be studied. To study most biological systems, three issues need to be stressed. First, most biological systems are too complex to study directly, therefore they must be divided into tractable subsystems whose properties in part reflect those of the system. These subsystems must be sufficiently small to analyze all their elements and connections. Second, high-throughput analytic or global tools are required for studying many systems elements at one time (see later). Finally, the systems information needs to be modeled mathematically before systems properties can be predicted and ultimately understood. This will require recruiting computer scientists and applied mathematicians into biology—just as the attempts to decipher the information of complete genomes and the protein folding and structure/function problems have required the recruitment of computational scientists.

I would be remiss not to point out that there are many other molecules that generate biological information: amino acids, carbohydrates, lipids, and so forth. These too must be studied in the context of their specific structures and specific functions.

The deciphering and manipulation of these various types of biological information represent an enormous technical challenge for biotechnology. Yet major new and powerful tools for doing so are emerging.

One class of tools for deciphering biological information is termed high-throughput analytic or global tools. These tools can be used to study many genes or chromosome features (genomics), many proteins (protcomics), or many cells rapidly: large-scale DNA sequencing, genomewide genetic mapping, cDNA or oligonucleotide arrays, two-dimensional gel electrophoresis and other global protein separation technologies, mass spectrometric analysis of proteins and protein fragments, multiparameter, high-throughput cell and chromosome sorting, and high-throughput phenotypic assays.

A second approach to the deciphering and manipulation of biological information centers around combinatorial strategies. The basic idea is to synthesize an informational string (DNA fragments, RNA fragments, protein fragments, antibody combining sites, etc.) using all combinations of the basic letters of the corresponding alphabet, thus creating many different shapes that can be used to activate, inhibit, or complement the biological functions of designated three-dimensional shapes (e.g., a molecule in a signal transduction pathway). The power of combinational chemistry is just beginning to be appreciated.

A critical approach to deciphering biological information will ultimately be the ability to visualize the functioning of genes, proteins, cells, and other informational elements within living organisms (in vivo informational imaging).

Finally, there are the computational tools required to collect, store, analyze, model, and ultimately distribute the various types of biological information. The creation presents a challenge comparable to that of developing new instrumentation and new chemistries. Once again this means recruiting computer scientists and applied mathematicians to biology. The biggest challenge in this regard is the language barriers that separate different scientific disciplines. Teaching biology as an informational science has been a very effective means for breeching these barriers.

The challenge is, of course, to decipher various types of biological information and then be able to use this information to manipulate genes, proteins, cells, and informational pathways in living organisms to eliminate or prevent disease, produce higher-yield crops, or increase the productivity of animals for meat and other foods.

Biotechnology and its applications raise a host of social, ethical, and legal questions, for example, genetic privacy, germline genetic engineering, cloning of animals, genes that influence behavior, cost of therapeutic drugs generated by biotechnology, animal rights, and the nature and control of intellectual property.

Clearly, the challenge is to educate society so that each citizen can thoughtfully and rationally deal with these issues, for ultimately society dictates the resources and regulations that circumscribe the development and practice of biotechnology. Ultimately, I feel enormous responsibility rests with scientists to inform and educate society about the challenges as well as the opportunities arising from biotechnology. These are critical issues for biotechnology that are developed in detail in the *Encyclopedia of Ethical, Legal, and Policy Issues in Biotechnology*.

The view that biotechnology is an informational science pervades virtually every aspect of this science, including discovery, reduction to practice, and societal concerns. These Encyclopedias of Biotechnology reinforce the emerging informational paradigm change that is powerfully positioning science as we move into the twenty-first century to more effectively decipher and manipulate for humankind's benefit the biological information of relevant living organisms.

Leroy Hood
University of Washington

CONTRIBUTORS

Nicholas R. Abu-Absi, *University of Minnesota, St. Paul, Minnesota,* Cell Cycle, Eukaryotes

Luis A. Actis, *Miami University, Oxford, Ohio,* Plasmid DNA Replication

Michael W.W. Adams, *University of Georgia, Athens, Georgia,* Enzymes, Extremely Thermostable

P.S. Adams, *Trudeau Institute, Saranac Lake, New York,* Professional Societies, Association of Biomolecular Resource Facilities

William R. Adams, *Merck Manufacturing Division, West Point, Pennsylvania,* Adsorption, Protein, Batch

Eduardo Agosin, *Pontificia Universidad Católica de Chile, Santiago, Chile,* Solid Substrate Fermentation, Automation

Eric Ailor, *Johns Hopkins University, Baltimore, Maryland,* Insect Cells and Larvae, Gene Expression Systems

Kazuo Aisaka, *Kyowa Hakko Kogyo Co., Ltd., Tokyo, Japan,* Cholesterol Oxidase

Hiroyuki Akatsuka, *Tanabe Seiyaku Co., Ltd., Osaka, Japan,* Diltiazem Synthesis

Mohamed Al-Rubeai, *University of Birmingham, Birmingham, England,* Apoptosis; Cell Cycle

Seigo Amachi, *Hokkaido University, Sapporo, Japan,* Pyruvate, Production Using Defective ATPase Activity

Teruo Amachi, *Kyoto University, Kyoto, Japan,* β-Galactosidase, Enzymology and Applications

Graham Andrews, *Idaho National Engineering and Environmental Laboratory, Idaho Falls, Idaho,* Bioreactors, Gas Treatment

Ruth Hogue Angeletti, *Albert Einstein College of Medicine, Bronx, New York,* Professional Societies, Association of Biomolecular Resource Facilities

William Apel, *Idaho National Engineering and Environmental Laboratory, Idaho Falls, Idaho,* Bioreactors, Gas Treatment

Frances H. Arnold, *California Institute of Technology, Pasadena, California,* Enzymes, Directed Evolution

David L. Aronson, *Consultant, Bethesda, Maryland,* Coagulation Factors, Therapeutic

Yasuhisa Asano, *Toyama Prefectural University, Toyama, Japan,* D-Aminopeptidase, Alkaline D-Peptidase; Opine Dehydrogenase, Secondary Amine Dicarboxylic Acids; Phenylalanine Dehydrogenase; *Mitsubishi Chemical Co., Yokohama, Japan,* Malate, D-Malate

Yoshiro Ashina, *Nitto Chemical Industry Co., Ltd., Tokyo, Japan,* Nitrile Hydratase

John Aunins, *Merck and Company, West Point, Pennsylvania,* Roller Bottle Culture, Mixing

Hans Axelsson, *Alfa Laval Separation AB, Tumba, Sweden,* Cell Separation, Centrifugation

Irina Bagyan, *University of Connecticut Health Center, Farmington, Connecticut,* Gene Transfer, Gram-Positive Bacteria

Frederick S. Baker, *Westvaco Corporation, Charleston, South Carolina,* Activated Carbon, Decoloration of Pharmaceutical Products

Kym N. Baker, *University of Kent, Canterbury, U.K.,* Glycosylation of Recombinant Proteins

Richard H. Baltz, *CognoGen Enterprises, Indianapolis, Indiana,* Mutagenesis

A. Baradarajan, *Indian Institute of Technology–Madras, Chennai, India,* Dextran, Microbial Production Methods

Ana Paulina Barba de la Rosa, *University of California, Irvine, Irvine, California,* Crystallization, Bulk, Macromolecules

Claudia Bardouille, *ECACC CAMR Porton Down, Salisbury, Wiltshire, United Kingdom,* Animal Cells Used in Manufacturing

Ann-Kristin Barnfield Frej, *Amersham Pharmacia Biotech, Uppsala, Sweden,* Adsorption, Expanded Bed

Jonathan Basch, *Bristol-Myers Squibb Company, Syracuse, New York,* Cephalosporins

J.W. Bennett, *Tulane University, New Orleans, Louisiana,* Aspergillus

I.S. Bentley, *ABM Brewing and Enzymes Group / Rhone-Poulenc, Stockport, United Kingdom,* Enzymes, Starch Conversion

A.R. Bernard, *Serono Pharmaceutical Research Institute, Geneva, Switzerland,* Transient Expression Systems

Michael J. Betenbaugh, *Johns Hopkins University, Baltimore, Maryland,* Insect Cells and Larvae, Gene Expression Systems

Kami Beyzavi, *Bioprocessing Limited, Consett, County Durham, United Kingdom,* Adsorbents, Inorganic

John R. Birch, *Lonza Biologics PLC, Berkshire, United Kingdom,* Suspension Culture of Animal Cells

H.D. Blasey, *Serono Pharmaceutical Research Institute, Geneva, Switzerland,* Transient Expression Systems

Thomas Bley, *Dresden University of Technology, Dresden, Germany,* Flow Cytometry

Lynda F. Bonewald, *University of Texas Health Science Center, San Antonio, Texas,* Professional Societies, Association of Biomolecular Resource Facilities

C.J.-P. Boonaert, *Université Catholique de Louvain, Louvain-la-Neuve, Belgium,* Cell Separation, Flocculation

Michelle K. Bothwell, *Oregon State University, Corvallis, Oregon,* Adsorption, Proteins with Synthetic Materials

Joye Bramble, *Merck and Company, West Point, Pennsylvania,* Roller Bottle Culture, Mixing

R. Bruttini, *Criofarma Freeze-drying Equipment, Turin, Italy,* Freeze-drying, Pharmaceuticals

James D. Bryers, *University of Connecticut, Farmington, Connecticut,* Biofilms, Microbial

R. Buccholz, *Technical University of Berlin, Berlin, Germany,* Microencapsulation

Michael Butler, *University of Manitoba, Winnipeg, Canada,* Energy Metabolism, Microbial and Animal Cells

Eleanor Canova-Davis, *Genentech, San Francisco, California,* Professional Societies, Association of Biomolecular Resource Facilities

Ningjun Cao, *Purdue University, West Lafayette, Indiana,* Hemicellulose Conversion; Organic Compounds, Cellulose Conversion

Christine L. Case, *Skyline College, San Bruno, California,* Professional Societies, Society for Industrial Microbiology (SIM)

S.A. Casnocha, *Monsanto Co., St. Louis, Missouri,* Media, Animal Cell Culture

Francisco J. Castillo, *Berlex Biosciences, Richmond, California,* Hybridoma, Antibody Production

Marvin Charles, *Lehigh University, Bethlehem, Pennsylvania,* Fermenter Design

Peter S. J. Cheetham, *Zylepsis, Ltd., Ashford, United Kingdom,* Enzymes, for Flavor Production

Shu-Jen D. Chiang, *Bristol-Myers Squibb Company, Syracuse, New York,* Cephalosporins

Yusuf Chisti, *University of Almería, Almería, Spain,* Mass Transfer; Shear Sensitivity; Solid Substrate Fermentations, Enzyme Production, Food Enrichment

Jeffrey L. Cleland, *Genentech, Inc., South San Francisco, California,* Formulation and Delivery, Protein Pharmaceuticals

O. Colagrande, *Università Cattolica Sacro Cuore, Piacenza, Italy,* Wine Production

Attilio Converti, *Genoa University, Genoa, Italy,* Biofilters

Charles L. Cooney, *Massachusetts Institute of Technology, Cambridge, Massachusetts,* Bioreactors, Continuous Stirred-Tank Reactors

Athel Cornish-Bowden, *National Center for Scientific Research, Marseilles, France,* Kinetics, Enzymes

Rosalie J. Cote, *Becton Dickinson Microbiology Systems, Sparks, Maryland,* Media Composition, Microbial, Laboratory Scale; Medium Formulation and Design, *E. coli* and *Bacillus* spp.

John W. Crabb, *Cleveland Clinic Foundation, Cleveland, Ohio,* Professional Societies, Association of Biomolecular Resource Facilities

Steven M. Cramer, *Rensselaer Polytechnic Institute, Troy, New York,* Chromatography, Ion Exchange

Ronald L. Crawford, *University of Idaho, Moscow, Idaho,* Bioremediation

Jorge H. Crosa, *Oregon Health Sciences University, Portland, Oregon,* Plasmid DNA Replication

John Cullum, *LB Genetik, University of Kaiserslautern, Kaiserslautern, Germany,* Genetic Instability

Simon Cutting, *Royal Holloway University of London, Egham, Surrey, United Kingdom,* Gene Transfer, Gram-Positive Bacteria

Hans von Döhren, *Technical University Berlin, Berlin, Germany,* Peptide

Alessandro D'Aprano, *University of Rome, La Sapienza, Rome, Italy,* Conductivity

W.-D. Deckwer, *GBF-Gesellschaft für Biotechnologische Forschung GmbH, Braunschweig, Germany,* Xanthan Gum

Michael L. Dekleva, *Merck and Co., Inc., West Point, Pennsylvania,* Vaccine Technology

Arnold L. Demain, *Massachusetts Institute of Technology, Cambridge, Massachusetts,* Metabolites, Primary and Secondary

P.B. Dengis, *Université Catholique de Louvain, Louvain-la-Neuve, Belgium,* Cell Separation, Flocculation

C.P. Dillon, *C.P. Dillon & Associates, St. Albans, West Virginia,* Stainless Steels

Dennis Dobie, *Fluor Daniel, Marlton, New Jersey,* Heating, Ventilating, and Air Conditioning

D. Dochain, *Université Catholique de Louvain, Louvain-La-Neuve, Belgium,* Process Control, Strategy and Optimization

Ed Domanico, *Tri-Clover, Valencia, California,* Pumps, Industrial

John Dougherty, *Lilly Research Laboratories, Indianapolis, Indiana,* Professional Societies, Association of Biomolecular Resource Facilities

Alan Doyle, *Centre for Applied Microbiology and Research, Salisbury, Wilts, United Kingdom,* Culture Collections

Y.F. Dufrêne, *Université Catholique de Louvain, Louvain-la-Neuve, Belgium,* Cell Separation, Flocculation

C.C. Dupont-Gillain, *Université Catholique de Louvain, Louvain-la-Neuve, Belgium,* Cell Separation, Flocculation

H. Dziallas, *Technische Universität Braunschweig, Braunschweig, Germany,* Scale-Up, Stirred-Tank Reactors

Heinrich Ebner, *Linz, Austria,* Vinegar, Acetic Acid Production

G. Larry Eitel, *Jacobs Engineering Group, Denver, Colorado,* ASTM Standards for Biotechnology

Lynda B.M. Ellis, *University of Minnesota, Minneapolis, Minnesota,* Biocatalysis Databases

Takakazu Endo, *Nitto Chemical Industry Co., Ltd., Tokyo, Japan,* Nitrile Hydratase

Larry E. Erickson, *Kansas State University, Manhattan, Kansas,* Anaerobes

Bruno Fabiano, *University of Genoa, Genoa, Italy,* Corrosion, Microbial

Pierre Fauquenberque, *Gist-brocades, Seclin Cedex, France,* Enzymes, Fruit Juice Processing

Joseph Fernandez, *Rockefeller University, New York, New York,* Professional Societies, Association of Biomolecular Resource Facilities

J.-L. Flores Candia, *GBF-Gesellschaft für Biotechnologische Forschung GmbH, Braunschweig, Germany,* Xanthan Gum

Ian G. Fotheringham, *NSC Technologies, Mount Prospect, Illinois,* Phenylalanine

Beth Fowler, *Autoimmune, Inc., Lexington, Massachusetts,* Professional Societies, Association of Biomolecular Resource Facilities

Cornelius G. Friedrich, *University of Dortmund, Dortmund, Germany,* Waste Gas Cleaning, Biological

Roger Fuentes-Granados, *Iowa State University, Ames, Iowa,* Soybean (Fermentation, Meal Oil)

Daniel Y.C. Fung, *Kansas State University, Manhattan, Kansas,* Anaerobes

Sean R. Gallagher, *Motorola Phoenix Corporate Research Laboratories, Tempe, Arizona,* Electrophoresis of Proteins and Nucleic Acids

Subinay Ganguly, *Smith Kline Beecham Pharmaceuticals, King of Prussia, Pennsylvania,* Expression Systems, Mammalian Cells

F.A.P. Garcia, *University of Coimbra, Coimbra, Portugal,* Cell Disruption and Lysis

Alan R. Gardner, *Smith Kline Beecham, King of Prussia, Pennsylvania,* Process Validation

Maria Gavrilescu, *Research Centre for Antibiotics, Iaşi, Romania,* Static Mixing, in Fermentation Process

Craig J.L. Gershater, *S. B. Pharmaceuticals, Harlow, Essex, England,* Inoculum Preparation

Robert L. Gherna, *American Type Culture Collection, Rockville, Maryland,* Culture Preservation, Bacteria, Fungi, Yeast, and Cell Lines; Medium Formulation and Design, *E. coli* and *Bacillus* spp.

Siddartha Ghose, *Aston University, Birmingham, United Kingdom,* Protein Adsorption, Expanded Bed

Barbara Ghrist, *Genentech, San Francisco, California,* Professional Societies, Association of Biomolecular Resource Facilities

M. Gluz, *Ben-Gurion University of the Negev, Beer-Sheva, Israel,* Bioreactors, Airlift Reactors

Francesc Gòdia, *Universitat Autònoma de Barcelona, Barcelona, Spain,* Bioreactors, Fluidized-Bed

Victor Goetz, *Merck Research Labs, West Point, Pennsylvania,* Adsorption, Protein, Batch

C.S. Gong, *Purdue University, West Lafayette, Indiana,* Hemicellulose Conversion; Organic Compounds, Cellulose Conversion

Catherine Grassin, *Gist-brocades, Seclin Cedex, France,* Enzymes, Fruit Juice Processing

Thomas P. Graycar, *Genencor International Inc., Palo Alto, California,* Proteolytic Cleavage, Reaction Mechanisms

Lasse Greiner, *Forschungszentrum Jülich GmbH, Jülich, Germany,* Enzymes, Immobilized, Reactors

J. Bryan Griffiths, *Scientific Consultancy & Publishing, Salisbury, United Kingdom,* Mammalian Cell Culture Reactors, Scale-Up

E. Molina Grima, *University of Almería, Almería, Spain,* Microalgae, Mass Culture Methods

Tingyue Gu, *Ohio University, Athens, Ohio,* Chromatography, Radial Flow

Maria J. Guardia, *University of Minnesota, Minneapolis, Minnesota,* Mammalian Cell Bioreactors

S. Guillouet, *Massachusetts Institute of Technology, Cambridge, Massachusetts,* Corynebacteria, Brevibacteria

H. Håkanson, *Lund University, Lund, Sweden,* Sampling Methods (Reactors, Contamination)

Daniel D. Hanle, *Kendro Laboratory Products, Newtown, Connecticut,* Centrifuges, Animal Cells

Satoshi Hanzawa, *TOSOH Corp., Ayase-shi, Japan,* Aspartame; *Tokyo Research Laboratory, Hayakawa, Japan,* Thermolysin

Colin R. Harwood, *University of Newcastle upon Tyne, Newcastle upon Tyne, United Kingdom, Bacillus*

Junzo Hasegawa, *Kaneka Corporation, Hyogo, Japan,* Optically Active 1,2-Diols, Microbial Production by Stereoinversion

Gary Hathaway, *California Institute of Technology, Pasadena, California,* Professional Societies, Association of Biomolecular Resource Facilities

Holly Haughney, *Pall Corporation, Long Island, New York,* Filtration, Air; Filtration, Cartridge

J.J. Heijnen, *Delft University of Technology, Delft, The Netherlands,* Bioenergetics of Microbial Growth

Martin Held, *Swiss Federal Institute of Technology, Zürich, Switzerland, Pseudomonas,* Process Applications

Ulf Hellman, *Ludwig Institute for Cancer Research, Uppsala, Sweden,* Professional Societies, Association of Biomolecular Resource Facilities

D.C. Hempel, *Universität Gesamthochschule Paderborn, Paderborn, Germany,* Scale-Up, Stirred-Tank Reactors

Diane L. Hevehan, *Northwestern University, Evanston, Illinois,* Hypoxia, Effects on Animal Cells

S.M. Heydarian, *University College London, London, United Kingdom,* Rheology of Filamentous Microorganisms, Submerged Culture

Paula M. Hicks, *North Carolina State University, Raleigh, North Carolina,* Enzymes, Extremely Thermostable; Thermophilic Microorganisms

Sophia Hober, *Royal Institute of Technology, Stockholm, Sweden,* Affinity Fusions, Gene Expression

Ann C. Horan, *Schering Plough Research Institute, Kenilworth, New Jersey,* Secondary Metabolite Production, Actinomycetes, Other Than Streptomyces

Daslav Hranueli, *PLIVA d.d., Research Institute, Zagreb, Croatia,* Genetic Instability

Tsu-An Hsu, *Vaccine Pharmaceutical R&D, Merck & Co., Inc., West Point, Pennsylvania,* Insect Cells and Larvae, Gene Expression Systems

Wei-Shou Hu, *University of Minnesota, Minneapolis, Minnesota,* Mammalian Cell Bioreactors; Microcarrier Culture

Holger Huebner, *Technical University of Berlin, Berlin, Germany,* Microencapsulation

Tony Hunt, *Advanced Minerals Corporation, Santa Barbara, California,* Filter Aids

Takamitsu Iida, *Niigata University, Niigata, Japan,* Wastewater Treatment, Immobilized Cells

Yasuhiro Ikenaka, *Kaneka Corporation, Hyogo, Japan,* Aminohydrolases, for Production of D-Amino Acids

Masayuki Inui, *Mitsubishi Chemical Corporation, Ibaraki, Japan,* L-Isoleucine

A.P. Ison, *University College London, London, United Kingdom,* Rheology of Filamentous Microorganisms, Submerged Culture

Kathryn M. Ivanetich, *University of California–San Francisco, San Francisco, California,* Professional Societies, Association of Biomolecular Resource Facilities

Cornelius F. Ivory, *Washington State University, Pullman, Washington,* Electrophoresis, Proteins, Batch and Continuous

Mahendra K. Jain, *MBI International, Lansing, Michigan,* Anaerobes, Industrial Uses

David C. James, *University of Kent, Canterbury, United Kingdom,* Glycosylation of Recombinant Proteins

Nigel Jenkins, *Eli Lilly, Inc., Indianapolis, Indiana,* Protein Glycosylation

Steen Weber Jensen, *Novo Nordisk A/S, Bagsvaerd, Denmark,* Insulin, Purification

Michael R. Johns, *University of Queensland, Brisbane, Australia,* Crystallization, Proteins, Kinetics

Alois Jungbauer, *Institute of Applied Microbiology, Vienna, Austria,* Chromatography, Computer-Aided Design

Beth H. Junker, *Merck Research Laboratories, Rahway, New Jersey,* Good Manufacturing Practice (GMP) and Good Industrial Large Scale Practice (GLSP); Pilot Plants, Design and Operation

Oliver Kaltenbrunner, *Institute of Applied Microbiology, Vienna, Austria,* Chromatography, Computer-Aided Design

Manohar Kalyanpur, *Consultant, Bioseparations and Pharmaceutical Validation, Plaisir, France,* Membrane Separations

Per Kårsnäs, *Percreative, Göteborg, Sweden,* Chromatography, Hydrophobic Interaction

R.S. Karthikeyan, *Indian Institute of Technology–Madras, Chennai, India,* Dextran, Microbial Production Methods

Michihiko Kataoka, *Kyoto University, Kyoto, Japan,* Aldehyde Reductase; Lactonohydrolase; Pantothenic Acid and Related Compounds

Yasuo Kato, *Toyama Prefectural University, Toyama, Japan,* Opine Dehydrogenase, Secondary Amine Dicarboxylic Acids

Randal J. Kaufman, *University of Michigan Medical School, Ann Arbor, Michigan,* Secretion from Animal Cells

Tetsuya Kawakita, *Ajinomoto Co., Inc., Kanazawa, Japan,* Amino Acids, Glutamate

Takuo Kawamoto, *Kyoto University, Kyoto, Japan,* Cell Immobilization; Organosilicon Compounds

Robert M. Kelly, *North Carolina State University, Raleigh, North Carolina,* Enzymes, Extremely Thermostable; Thermophilic Microorganisms

Birgit Kessler, *Institute of Biotechnology, Zürich, Switzerland,* Poly(3-Hydroxyalkanoates)

Shun-ichi Kidokoro, *Sagami Chemical Research Center, Kanagawa, Japan,* Thermolysin

Shukuo Kinoshita, *Kyowa Hakko Kogyo Co., Ltd., Tokyo, Japan,* Glutamic Acid Producing Microorganisms; Production of L-Glutamic Acid

Horst Kleinkauf, *Technical University Berlin, Berlin, Germany,* Peptide

M.A. Klich, *U.S. Department of Agriculture, New Orleans, Louisiana, Aspergillus*

Miki Kobayashi, *Mitsubishi Chemical Corporation, Ibaraki, Japan,* L-Aspartic Acid

Arthur L. Koch, *Indiana University, Bloomington, Indiana,* Microbial Growth Measurement, Methods

Fragiskos N. Kolisis, *National Technical University of Athens, Athens, Greece,* Reverse Micelles, Enzymes

Hidenobu Komeda, *Toyama Prefectural University, Toyama, Japan,* D-Aminopeptidase, Alkaline D-Peptidase

Tadashi Kometani, *Toyama College of Technology, Toyama-shi, Japan,* Bioreduction; Yeast, Baker's

Udo Kragl, *Forschungszentrum Jülich GmbH, Jülich, Germany,* Biotransformation, Engineering Aspects; Enzymes, Immobilized, Reactors

M.R.V. Krishnan, *Anna University, Chennai, India,* Citric Acid, Processes

Maria-Regina Kula, *Heinrich Heine University Düsseldorf, Jülich, Germany,* Protein Purification, Aqueous Liquid Extraction

Hidehiko Kumagai, *Kyoto University, Kyoto, Japan,* Dihydroxyphenylalanine (4), Produced by Microorganisms; Tyrosine Phenol-lyase

Hidetoshi Kutsuki, *Kaneka Corporation, Hyogo, Japan,* Optically Active 1,2-Diols, Microbial Production by Stereoinversion

Andreas Lübbert, *Martin Luther University, Wittenberg, Germany,* Gas Hold-Up

David C. LaPorte, *University of Minnesota, Minneapolis, Minnesota,* Glyoxylate Bypass, Regulation

Xanthe M. Lam, *Genentech, Inc., South San Francisco, California,* Formulation and Delivery, Protein Pharmaceuticals

Per Larsen, *Novo Nordisk A/S, Bagsvaerd, Denmark,* Insulin, Purification

Michael E. Laska, *Massachusetts Institute of Technology, Cambridge, Massachusetts,* Bioreactors, Continuous Stirred-Tank Reactors

Anh LeDuy, *Laval University, Sainte-Foy, Canada,* Pullulan, Microbial Production Methods

David Leak, *Imperial College of Science, Technology, and Medicine, London, United Kingdom,* Methylotrophs, Industrial Applications

Ann L. Lee, *Merck Research Labs, West Point, Pennsylvania,* Adsorption, Protein, Batch

Gyun Min Lee, *Korea Advanced Institute of Science and Technology, Taejon, Korea,* Osmotic Stress, Secretion Rate

P.A. Lessard, *Massachusetts Institute of Technology, Cambridge, Massachusetts,* Corynebacteria, Brevibacteria

Andreas Liese, *Forschungszentrum Jülich, Jülich, Germany,* Biotransformations, Engineering Aspects

Carsten Lindemann, *University of Hannover, Hannover, Germany,* Fluorescence Techniques for Bioprocess Monitoring

M. Liu, *Rutgers University, Piscataway, New Jersey,* Roller Bottle Culture, Mixing

David R. Lloyd, *University of Birmingham, Birmingham, England,* Cell Cycle

Deryk T. Loo, *Bristol-Myers Squibb Pharmaceutical Research Institute, Seattle, Washington,* Attachment Factors

Shun-ichi Maemoto, *Kaneka Corporation, Hyogo, Japan,* Optically Active 1,2-Diols, Microbial Production by Stereoinversion

Savvas C. Makrides, *PRAECIS Pharmaceuticals, Inc., Cambridge, Massachusetts,* Expression Systems, *E. coli*

Stefan Marose, *University of Hannover, Hannover, Germany,* Fluorescence Techniques for Bioprocess Monitoring

R.E. Marquis, *University of Rochester, Rochester, New York,* Transport, Microbial Solute Uptake

Seiji Masuda, *Kyoto University, Kyoto, Japan,* Erythropoietin

Jennie P. Mather, *Rauen Biotechnologies, Inc., Mountain View, California,* Culture Media, Animal Cells, Large Scale Production

Hiroaki Matsumae, *Tanabe Seiyaku Co., Ltd., Osaka, Japan,* Diltiazem Synthesis

Ryuichi Matsumo, *Kyoto University, Kyoto, Japan,* Glucosidases

Tadashi Matsunaga, *Tokyo University of Agriculture and Technology, Tokyo, Japan,* Algal Culture

Ryuichi Matsuno, *Kyoto University, Kyoto, Japan,* Yeast, Baker's

B. Mattiasson, *Lund University, Lund, Sweden,* Sampling Methods (Reactors, Contamination)

Ian R. McDonald, *University of Warwick, Coventry, United Kingdom,* Methanotrophs

Joseph McGuire, *Oregon State University, Corvallis, Oregon,* Adsorption, Proteins with Synthetic Materials

Alexander McPherson, *University of California, Irvine, Irvine, California,* Crystallization, Bulk, Macromolecules

J.C. Merchuk, *Ben-Gurion University of the Negev, Beer-Sheva, Israel,* Bioreactors, Air-lift Reactors

Charles E. Miller, *Westvaco Corporation, Charleston, South Carolina,* Activated Carbon, Decoloration of Pharmaceutical Products

Stephen P. Miller, *University of Minnesota, Minneapolis, Minnesota,* Glyoxylate Bypass, Regulation

William M. Miller, *Northwestern University, Evanston, Illinois,* Hypoxia, Effects on Animal Cells

Sheenah Mische, *Rockefeller University, New York, New York,* Professional Societies, Association of Biomolecular Resource Facilities

David A. Mitchell, *Universidade Federal do Parana, Curitiba, Brazil,* Solid-State Fermentation, Microbial Growth Kinetics

Inger Mollerup, *Novo Nordisk A / S, Bagsvaerd, Denmark,* Insulin, Purification

Alison Moore, *Amgen, Inc., Newbury Park, California,* Culture Media, Animal Cells, Large Scale Production

Vadim V. Mozhaev, *University of Iowa, Iowa City, Iowa,* Denaturation, Proteins, Solvent Mediated

Susann Müller, *University of Leipzig, Leipzig, Germany,* Flow Cytometry

David W. Murhammer, *University of Iowa, Iowa City, Iowa,* Pluronic Polyols, Cell Protection

J. Colin Murrell, *University of Warwick, Coventry, United Kingdom,* Methanotrophs

F.J. Muzzio, *Rutgers University, Piscataway, New Jersey,* Roller Bottle Culture, Mixing

Clayton Naeve, *St. Jude Children's Research Hospital, Memphis, Tennessee,* Professional Societies, Association of Biomolecular Resource Facilities

Kazuhiro Nakanishi, *Okayama University, Okayama, Japan,* Membrane Surface Liquid Culture, Microorganisms, Fungi

Toru Nakayama, *Tohoku University, Sendai, Japan,* β-Galactosidase, Enzymology and Applications

M.P. Nandakumar, *Lund University, Lund, Sweden,* Sampling Methods (Reactors, Contamination)

Venkatesh Natarajan, *Rensselaer Polytechnic Institute, Troy, New York,* Chromatography, Ion Exchange

Robert Newburgh, *The Protein Society, Bethesda, Maryland,* Professional Societies, The Protein Society

Ronald L. Niece, *University of California–Irvine, Irvine, California,* Professional Societies, Association of Biomolecular Resource Facilities

Hans Ole Nielsen, *DELTA Light&Optics, Lyngby, Denmark,* Fluorescence Techniques for Bioprocess Monitoring

Jens Nielsen, *Technical University of Denmark, Lyngby, Denmark,* Fermentation Monitoring, Design and Optimization

Zivko L. Nikolov, *Iowa State University, Ames, Iowa,* Soybean (Fermentation, Meal Oil)

Joakim Nilsson, *Royal Institute of Technology, Stockholm, Sweden,* Affinity Fusions, Gene Expression

Yasuo Ninomiya, *Nitto Denko Company, Osaka, Japan,* Insecticides, Microbial Production; Optical Resolution, Biocatalysis

Takuo Nishida, *Tanabe Seiyaku Co., Ltd., Osaka, Japan,* Diltiazem Synthesis, Microbial Asymmetric Reduction

P.T. Noble, *Fluor Daniel GmbH, Wiesbaden, Germany,* Sterilization-in-Place

Per-Åke Nygren, *Royal Institute of Technology, Stockholm, Sweden,* Affinity Fusions, Gene Expression

Jun Ogawa, *Kyoto University, Kyoto, Japan,* Oils, Microbial Production

Masahiro Ogura, *Kaneka Corporation, Hyogo, Japan,* Optically Active 1,2-Diols, Microbial Production by Stereoinversion

Takehisa Ohashi, *Kaneka Corporation, Hyogo, Japan,* Optically Active 1,2-Diols, Microbial Production by Stereoinversion

Hiromichi Ohta, *Keio University, Yokohama, Japan,* Lactones, Biocatalytic Synthesis

Tetsuo Oka, *Kyowa Hakko Kogyo, Co., Ltd., Tokyo, Japan,* Amino Acids, Production Processes

Hans Sejr Olsen, *Novo Nordisk, Bagsvaerd, Denmark,* Enzymes, Protein Hydrolysis

Gustaf Olsson, *Lund Institute of Technology, Lund, Sweden,* Waste Treatment, Activated Sludge, Control Strategies

Tetsuo Omata, *Nitto Denko Company, Osaka, Japan,* Insecticides, Microbial Production; Optical Resolution, Biocatalysis

Sadettin S. Ozturk, *Bayer Corporation, Berkeley, California,* Ammonia Toxicity, Animal Cells

J. Ricardo Pérez-Correa, *Pontificia Universidad Católica de Chile, Santiago, Chile,* Solid Substrate Fermentation, Automation

Philip Packer, *Centre for Applied Microbiology and Research, Salisbury, Wilts, United Kingdom,* Culture Collections

Nicolai S. Panikov, *Institute of Microbiology, Russian Academy of Sciences, Moscow, Russian Federation,* Kinetics, Microbial Growth

Patrizia Perego, *University of Genoa, Genoa, Italy,* Corrosion, Microbial

Madhusudan V. Peshwa, *Dendreon Corp., Mountain View, California,* Microcarrier Culture

Joerg Peters, *Bayer AG, Wuppertal, Germany,* Cofactor Regeneration, Nicotinamide Coenzymes

David Pollard, *Merck and Co., Inc., Rahway, New Jersey,* Vent Gas Analysis

Kaisa Poutanen, *VTT Biotechnology and Food Research, Espoo, Finland,* Hemicellulases

Joan Qi Si, *Novo Nordisk Ferment Ltd., Dittingen, Switzerland,* Enzymes, Baking, Bread Making

Bruce Ramsay, *Queen's University, Kingston, Ontario, Canada,* Polyhydroxyalkanoates, Separation, Purification, and Manufacturing Methods

Juliana Ramsay, *Queen's University, Kingston, Ontario, Canada,* Polyhydroxyalkanoates, Separation, Purification, and Manufacturing Methods

Douglas W. Rea, *Cytogen Corp., Princeton, New Jersey,* Antibody Purification

Kenneth F. Reardon, *Colorado State University, Fort Collins, Colorado,* Fluorescence Techniques for Bioprocess Monitoring

Matthias Redenbach, *LB Genetik, University of Kaiserslautern, Kaiserslautern, Germany,* Genetic Instability

Hans Reichenbach, *Gesellschaft für Biotechnologische Forschung, Braunschweig, Germany,* Myxobacteria

Harold Reisman, *Biotechnology Results, Weston, Connecticut,* Economics

Albert J. Repik, *Westvaco Corporation, Charleston, South Carolina,* Activated Carbon, Decoloration of Pharmaceutical Products

Nadine Ritter, *Abbott Laboratories, Abbott Park, Illinois,* Professional Societies, Association of Biomolecular Resource Facilities

Palmer Rogers, *University of Minnesota, Minneapolis, Minnesota,* Clostridia, Solvent Formation

Rodica-Viorica Roman, *Chemical Pharmaceutical Research Institute, Iaşi, Romania,* Transfer Phenomena in Multiphase Systems in Mixing Vessels

Charles M. Roth, *Massachusetts General Hospital, Shriners Burns Hospital, and Harvard Medical School, Boston, Massachusetts,* Chromatography, Size Exclusion

P.G. Rouxhet, *Université Catholique de Louvain, Louvain-la-Neuve, Belgium,* Cell Separation, Flocculation

Joon Soo Ryu, *Korea Advanced Institute of Science and Technology, Taejon, Korea,* Osmotic Stress, Secretion Rate

Peter Salmon, *Merck and Co., Inc., Rahway, New Jersey,* Vent Gas Analysis

Gautam Sanyal, *Astra Research Center Boston, Cambridge, Massachusetts,* Thermal Unfolding, Proteins

Ryuzo Sasaki, *Kyoto University, Kyoto, Japan,* Erythropoietin

Tadashi Sato, *Tanabe Seiyaku Co., Ltd., Osaka, Japan,* Enzymes, Immobilization Methods; Malic Acid, Production by Fumarase; Production of L-Amino Acids by Aminoacylase

Catherine H. Schein, *University of Texas Medical Branch, Galveston, Texas,* Protein Aggregation, Denaturation; Protein Expression, Soluble

Thomas Scheper, *University of Hannover, Hannover, Germany,* Fluorescence Techniques for Bioprocess Monitoring

S. Schiewer, *McGill University, Montreal, Canada,* Biosorption, Metals

Andrew Schmid, *Swiss Federal Institute of Technology, Zurich, Switzerland, Pseudomonas*, Process Applications

Ole Schou, *Novo Nordisk A / S, Bagsvaerd, Denmark,* Insulin, Purification

Robert Schwartz, *Abbott Laboratories, North Chicago, Illinois,* Professional Societies, Society for Industrial Microbiology (SIM)

Henry Schwartzberg, *University of Massachusetts, Amherst, Massachusetts,* Food Process Engineering

James Searles, *Merck and Company, West Point, Pennsylvania,* Roller Bottle Culture, Mixing

Katsuya Seguro, *Food Research and Development Lab, Kawaski, Japan,* Transglutaminase

Klaus Selber, *Heinrich Heine University Düsseldorf, Jülich, Germany,* Protein Purification, Aqueous Liquid Extraction

Sylvia Sellmer-Wilsberg, *Bad Honnef, Germany,* Vinegar, Acetic Acid Production

Shuji Senda, *Nitto Denko Company, Osaka, Japan,* Insecticides, Microbial Production; Optical Resolution, Biocatalysis

Pir M. Shah, *De Montfort University, Leicester, United Kingdom,* Protein Glycosylation

P. Ayazi Shamlou, *University College London, London, United Kingdom,* Rheology of Filamentous Microorganisms, Submerged Culture

Bhav P. Sharma, *CV Therapeutics, Palo Alto, California,* Cell Separation, Sedimentation

Allan R. Shatzman, *SmithKline Beecham Pharmaceuticals, King of Prussia, Pennsylvania,* Expression Systems, Mammalian Cells

Li Shi, *Merck Research Laboratories, West Point, Pennsylvania,* Thermal Unfolding, Proteins

Takeji Shibatani, *Tanabe Seiyaku Co., Ltd., Osaka, Japan,* Diltiazem Synthesis; Diltiazem Synthesis, Microbial Asymmetric Reduction

Sakayu Shimizu, *Kyoto University, Kyoto, Japan,* Aldehyde Reductase; Lactonohydrolase; Oils, Microbial Production; Pantothenic Acid and Related Compounds

Suteaki Shioya, *Osaka University, Osaka, Japan,* Mixed Culture

M. S. Showell, *Procter and Gamble, Cincinnati, Ohio,* Enzymes, Detergent

Rabinder P. Singh, *University of Birmingham, Birmingham, England,* Apoptosis

Satinder K. Singh, *University of Minnesota, Minneapolis, Minnesota,* Glyoxylate Bypass, Regulation

A.J. Sinskey, *Massachusetts Institute of Technology, Cambridge, Massachusetts,* Corynebacteria, Brevibacteria

Augustine Smith, *Abbott Laboratories, Abbott Park, Illinois,* Professional Societies, Association of Biomolecular Resource Facilities

Thomas M. Smith, *Smith Kline Beecham, King of Prussia, Pennsylvania,* Process Validation

Leo Snel, *Novo Nordisk A / S, Bagsvaerd, Denmark,* Insulin, Purification

Gail Sofer, *BioReliance, Rockville, Maryland,* Cleaning, Cleaning Validation

Carles Solà, *Universitat Autònoma de Barcelona, Barcelona, Spain,* Bioreactors, Fluidized-Bed

Richard Sparling, *University of Manitoba, Winnipeg, Canada,* Energy Metabolism, Microbial and Animal Cells

Dave Speicher, *Wistar Institute, Philadelphia, Pennsylvania,* Professional Societies, Association of Biomolecular Resource Facilities

Koti Sreekrishna, *Procter and Gamble Co., Ross, Ohio,* *Pichia,* Optimization of Protein Expression

Friedrich Srienc, *University of Minnesota, St. Paul, Minnesota,* Cell Cycle, Eukaryotes

Stefan Ståhl, *Royal Institute of Technology, Stockholm, Sweden,* Affinity Fusions, Gene Expression

Keith Stephenson, *University of Newcastle upon Tyne, Newcastle upon Tyne, United Kingdom, Bacillus*

Kathryn Stone, *Yale University, New Haven, Connecticut,* Professional Societies, Association of Biomolecular Resource Facilities

Bob Stover, *Tri-Clover, Valencia, California,* Pumps, Industrial

William R. Strohl, *Merck Research Laboratories, Rahway, New Jersey,* Secondary Metabolites, Antibiotics

Daniel J. Strydom, *Bio Nebraska, Inc., Lincoln, Nebraska,* Professional Societies, Association of Biomolecular Resource Facilities

Deidre M. Stuart, *Queensland University of Technology, Brisbane, Australia,* Solid-State Fermentation, Microbial Growth Kinetics

John Stults, *Genentech, San Francisco, California,* Professional Societies, Association of Biomolecular Resource Facilities

Takeshi Sugai, *Keio University, Yokohama, Japan,* Lactones, Biocatalytic Synthesis

Masaru Suto, *Nitto Chemical Industry Co., Ltd., Tokyo, Japan,* Nitrile Hydratase

T. Swaminathan, *Indian Institute of Technology–Madras, Chennai, India,* Dextran, Microbial Production Methods

P.D. Swanson, *Rutgers University, Piscataway, New Jersey,* Roller Bottle Culture, Mixing

Kristine M. Swiderek, *ZymoGenetics, Seattle, Washington,* Professional Societies, Association of Biomolecular Resource Facilities

Satomi Takahashi, *Kaneka Corporation, Hyogo, Japan,* Aminohydrolases, for Production of D-Amino Acids; Phenylglycines, D-Phenylglycines

Hiroyuki Takano, *Chichibu Onoda Cement Corporation, Chiba-prefecture, Japan,* Algal Culture

Haruko Takeyama, *Tokyo University of Agriculture and Technology, Tokyo, Japan,* Algal Culture

Atsuo Tanaka, *Kyoto University, Kyoto, Japan,* Cell Immobilization

Robert D. Tanner, *Vanderbilt University, Nashville, Tennessee,* Solid-State Fermentation, Microbial Growth Kinetics

Maija Tenkanen, *VTT Biotechnology and Food Research, Espoo, Finland,* Hemicellulases

Masato Terasawa, *Mitsubishi Chemical Corporation, Ibaraki, Japan,* L-Aspartic Acid; L-Isoleucine

Jörg Thömmes, *Heinrich-Heine Universität Düsseldorf, Jülich, Germany,* Membrane Chromatography

Jules Thibault, *Laval University, Sainte-Foy, Canada,* Pullulan, Microbial Production Methods

J.S. Tolan, *Iogen Corporation, Ottawa, Canada,* Enzymes, Pulp and Paper Processing

E. Donald Tolles, *Westvaco Corporation, Charleston, South Carolina,* Activated Carbon, Decoloration of Pharmaceutical Products

Marcelo E. Tolmasky, *California State University, Fullerton, Fullerton, California,* Plasmid DNA Replication

Fusao Tomita, *Hokkaido University, Sapporo, Japan,* Pyruvate, Production Using Defective ATPase Activity

Tetsuya Tosa, *Tanabe Seiyaku Co. Ltd., Osaka, Japan,* Enzymes, Immobilization Methods; Malic Acid, Production by Fumarase; Production of L-Amino Acids by Aminoacylase

R. Reid Townsend, *University of California–San Francisco, San Francisco, California,* Professional Societies, Association of Biomolecular Resource Facilities

Mario R. Tredici, *University of Florence, Florence, Italy,* Bioreactors, Photo

G.T. Tsao, *Purdue University, West Lafayette, Indiana,* Hemicellulose Conversion; Organic Compounds, Cellulose Conversion

Satoru Tsuda, *Kaneka Corporation, Hyogo, Japan,* Optically Active 1,2-Diols, Microbial Production by Stereoinversion

Radu Z. Tudose, *Technical University Gh.Asachi Iaşi, Iaşi, Romania,* Static Mixing, in Fermentation Process

A.P.F. Turner, *Cranfield Biotechnology Centre, Bedfordshire, United Kingdom,* Process Monitoring

Makoto Ueda, *Mitsubishi Chemical Co., Yokohama, Japan,* Malate, D-Malate

Mathias Uhlén, *Royal Institute of Technology, Stockholm, Sweden,* Affinity Fusions, Gene Expression

Michiel E. Ultee, *Cytogen Corp., Princeton, New Jersey,* Antibody Purification

D.R. Unger, *Rutgers University, Piscataway, New Jersey,* Roller Bottle Culture, Mixing

Mahesh Uttamlal, *Glasgow Caledonian University, Glasgow, Scotland,* Optical Sensors

Robert van Reis, *Genentech, Inc., South San Francisco, California,* Protein Ultrafiltration

James Vaughn, *ARS, USDA, Beltsville, Maryland,* Insect Cell Culture, Protein Expression

Liisa Viikari, *VTT Biotechnology and Food Research, Espoo, Finland,* Hemicellulases

B. Volesky, *McGill University, Montreal, Canada,* Biosorption, Metals

David B. Volkin, *Merck Research Laboratories, West Point, Pennsylvania,* Thermal Unfolding, Proteins

Lawrence P. Wackett, *University of Minnesota, Minneapolis, Minnesota,* Biocatalysis Databases

David R. Walt, *Tufts University, Medford, Massachusetts,* Optical Sensors

Christian Wandrey, *Forschungszentrum Jülich GmbH, Jülich, Germany,* Enzymes, Immobilized, Reactors

Sally Warburton, *ECACC, Wiltshire, United Kingdom,* Human and Primate Cell Lines

Udo Werner, *University of Dortmund, Dortmund, Germany,* Waste Gas Cleaning, Biological

David White, *Indiana University, Bloomington, Indiana,* Electron Transport

S.F. White, *Cranfield Biotechnology Centre, Bedfordshire, United Kingdom,* Process Monitoring

Erik M. Whiteley, *Johns Hopkins University, Baltimore, Maryland,* Insect Cells and Larvae, Gene Expression Systems

Ken Williams, *Yale University, New Haven, Connecticut,* Professional Societies, Association of Biomolecular Resource Facilities

L.B. Willis, *Massachusetts Institute of Technology, Cambridge, Massachusetts,* Corynebacteria, Brevibacteria

Jack Wilson, *ABEC, Inc., Allentown, Pennsylvania,* Fermenter Design

Patrick L. Wintrode, *California Institute of Technology, Pasadena, California,* Enzymes, Directed Evolution

Bernard Witholt, *Swiss Federal Institute of Technology, Zürich, Switzerland,* Poly(3-Hydroxyalkanoates); *Pseudomonas*, Process Applications

K.D. Wittrup, *Kyowa Hakko Kogyo Co., Ltd., Tokyo, Japan,* Protein Secretion, *Saccharomyces cerevisiae*

R.A. Wolfe, *Monsanto Co., St. Louis, Missouri,* Media, Animal Cell Culture

Kathy Wong, *University of Minnesota, Minneapolis, Minnesota,* Microcarrier Culture

Marcel Wubbolts, *Swiss Federal Institute of Technology, Zürich, Switzerland, Pseudomonas*, Process Applications

Florian M. Wurm, *Swiss Federal Institute of Technology, Lausanne, Switzerland,* Chinese Hamster Ovary Cells, Recombinant Protein Production

Xinfa Xiao, *Apotex Fermentation, Inc., Winnipeg, Canada,* Energy Metabolism, Microbial and Animal Cells

Feng Xu, *Novo Nordisk Biotech, Davis, California,* Laccase

Hideaki Yamada, *Mitsubishi Chemical Co., Yokohama, Japan,* Malate, D-Malate

Tsuneo Yamane, *Nagoya University, Nagoya, Japan,* Monoacylglycerols

Martin L. Yarmush, *Massachusetts General Hospital, Shriners Burns Hospital, and Harvard Medical School, Boston, Massachusetts,* Chromatography, Size Exclusion

Atsushi Yokota, *Hokkaido University, Sapporo, Japan,* Pyruvate, Production Using Defective ATPase Activity

Hideaki Yukawa, *Mitsubishi Chemical Corporation, Ibaraki, Japan,* L-Aspartic Acid; L-Isoleucine

Dane W. Zabriskie, *Smith Kline Beecham Pharmaceuticals, King of Prussia, Pennsylvania,* Process Validation

J. Gregory Zeikus, *MBI International, Lansing, Michigan,* Anaerobes, Industrial Uses

Mario Zilli, *Genoa University, Genoa, Italy,* Biofilters

Marko Zlokarnik, *Graz, Austria,* Dimensional Analysis, Scale-Up

Andrew L. Zydney, *University of Delaware, Newark, Delaware,* Protein Ultrafiltration

COMMONLY USED ACRONYMS AND ABBREVIATIONS

ABS	acrylonitrile-butadiene-styrene copolymer
AISI	American Iron and Steel Institute
ANSI	American National Standards Institute
APHIS	Animal and Plant Health Inspection Service (of the USDA)
API	American Petroleum Institute
ASME	American Society of Mechanical Engineers
AS/RS	automated storage and retrieval system
ASTM	American Society for Testing and Materials
ATCC	American Type Culture Collection
BLx	biosafety level x
BLx-Ls	biosafety level x, large-scale
BOD	Biological oxygen demand
BPVC	Boiler and Pressure Vessel Code
BSCC	Biotechnology Science Coordinating Committee
Btu	British thermal unit (unit of energy)
CADD	computer-aided design and drafting
CBER	Center for Biologics Evaluation & Research (of FDA)
CDER	Center for Drug Evaluation & Research
CDW	cell dry weight
CFF	cross-flow filtration
cfm	cubic feet per minute
CFR	Code of Federal Regulations (U.S.)
CFU	colony-forming unit
cGMPs	current good manufacturing practices
CIP	clean-in-place
COD	chemical oxygen demand
cP	centipoise
CPAC	Central Pharmaceutical Affairs Council (Japan)
DCS	distributed control system
DI	deionized (water)
DIN	Deutsches Institut fur Normung
DNA	deoxyribonucleic acid
DOP	dioctyl phthalate
DOT	dissolved oxygen tension
dP	differential pressure
DW	dry weight
DCW	dry cell weight
E&C	engineering and construction
EDTA	ethylenediaminetetraacetic acid
EEC	European Economic Community
ELA	Establishment License Application
EPA	Environmental Protection Agency
EPDM	ethylene-propylene-diene monomer
EU	endotoin unit
FDA	Food and Drug Administration
fpm	feet per minute
GILSP	good industrial large-scale practice
GLSP	good large-scale practice
GMP	good manufacturing practice
gpm	gallons per minute
h	hour(s)
HACCP	hazard analysis of critical control points
HEPA	high-efficiency particulate air (filter)
HEPES	name of a buffer
HMSO	Her Majesty's Stationery Office
hp	horsepower; unit of power, 0.75 kW
HVAC	heating, ventilating, and air conditioning
ID	inside diameter
IDF	International Dairy Federation
INO	Investigating new drug application (FDA)
IQ	installation qualification
ISS	International Sanitary Standard
Kla	volumetric gas-liquid mass transfer coefficient
L	liter(s)
LAL	Limulus amoebocyte lysate
MAb	monoclonal antibody
MAWP	maximum allowable working pressure
MF	microfiltration
MAW	Ministry of Health & Welfare (Japan)
MWCO	molecular weight cutoff
NAD, NADH, NADP, NADPH	nicotine adenine dinucleotides
NBS	National Bureau of Standards
NDA	new drug application (FDA)
NEMA	National Electrical Manufacturers Association
NIH	National Institutes of Health
NIH/RAC	National Institutes of Health Recombinant DNA Advisory Committee
NIST	National Institute of Standards and Testing
NPSH	net positive suction head
NPT	National Pipe Taper (pipe thread)
NSF	National Science Foundation
NWP	normalized water permeability
O&M	operation and maintenance
OD	outside diameter
OECD	Organization for Economic Cooperation and Development
OQ	operational qualification
OSHA	Occupational Safety and Health Administration
PAB	Pharmaceutical Affairs Bureau (Japan)
P&ID	piping (or process) and instrumentation diagram
PFD	process flow diagram
PI	proportional/integral
PID	proportional/integral/derivative
PIEL	pharmacologically insignificant exposure level
PLA	Product License Application
PLC	programmable logic controller
PMN	premanufacture notification
ppb	parts per billion

ppm	parts per million
PQ	performance (or process) qualification
psi	pounds per square inch
QA	quality assurance
QC	quality control
R&D	research & development
Ra	arithmetic average surface roughness
rDNA	recombinant DNA
RH	relative humidity
RO	reverse osmosis
RODI	reverse osmosis deionized water
ROI	return on investment
rpm, -s	revolutions per minute, second
RTD	resistance temperature device
s	second(s)
SAL	sterility assurance level
SAMA	Scientific Apparatus Makers Association
scfm	standard cubic feet per minute
SIP	sterilization in place
SLPM	standard liters per minute
STA	slit-to-agar
SVP	small-volume parenteral
T1/2	half-life
TDS	total dissolved solids
TFE	= PTFE, polytetraflouroethylene
TFF	tangential flow filtration
TEL	toxicologically insignificant exposure level
TIG	tungsten-inert-gas (welding)
TMP	transmembrane pressure
TOC	total organic carbon
TSCA	Toxic Substances Control Act
UF	ultrafiltration
UHPW	ultra high purity water
UK	United Kingdom
UL	Underwriters Laboratories
USDA	United States Department of Agriculture
USP/NF	United States Pharmacopoeia/National Formulary
VVM	volume per volume per minute
WFI	Water for Injection
WHO	World Health Organization

CONVERSION FACTORS, ABBREVIATIONS, AND UNIT SYMBOLS

SI UNITS (Adopted 1960)

The International System of Units (abbreviated SI), is being implemented throughout the world. This measurement system is a modernized version of the MKSA (meter, kilogram, second, ampere) system, and its details are published and controlled by an international treaty organization (The International Bureau of Weights and Measures).

SI units are divided into three classes:

BASE UNITS	
length	meter† (m)
mass	solid angle
time	second (s)
electric current	ampere (A)
thermodynamic temperature‡	kelvin (K)
amount of substance	mole (mol)
luminous intensity	candela (cd)

SUPPLEMENTARY UNITS	
plane angle	radian (rad)
steradian (sr)	kilogram (kg)

Quantity	Unit	Symbol	Acceptable equivalent
volume	cubic meter	m^3	
	cubic diameter	dm^3	L (liter) (5)
	cubic centimeter	cm^3	mL
wave number	1 per meter	m^{-1}	
	1 per centimeter	cm^{-1}	

In addition, there are 16 prefixes used to indicate order of magnitude, as follows:

Multiplication factor	Prefix	Symbol
10^{18}	exa	E
10^{15}	peta	P
10^{12}	tera	T
10^{9}	giga	G
10^{6}	mega	M
10^{3}	kilo	k
10^{2}	hecto	h[a]
10	deka	da[a]
10^{-1}	deci	d[a]
10^{-2}	centi	c[a]
10^{-3}	milli	m
10^{-6}	micro	€
10^{-9}	nano	n
10^{-12}	pico	p
10^{-15}	femto	f
10^{-18}	atto	a

[a]Although hecto, deka, deci, and centi are SI prefixes, their use should be avoided except for SI unit-multiples for area and volume and nontechnical use of centimeter, as for body and clothing measurement.

For a complete description of SI and its use the reader is referred to ASTM E380.

A representative list of conversion factors from non-SI to SI units is presented herewith. Factors are given to four significant figures. Exact relationships are followed by a dagger. A more complete list is given in the latest editions of ASTM E380 and ANSI Z210.1.

†The spellings "metre" and "litre" are preferred by ASTM; however, "-er" is used in the *Encyclopedia*.

‡Wide use is made of Celsius temperature (t) defined by

$$t = T - T_0$$

where T is the thermodynamic temperature, expressed in kelvin, and $T_0 = 273.15$ K by definition. A temperature interval may be expressed in degrees Celsius as well as in kelvin.

CONVERSION FACTORS TO SI UNITS

To convert from	To	Multiply by
acre	square meter (m^2)	4.047×10^{3}
angstrom	meter (m)	1.0×10^{-10}†
are	square meter (m^2)	1.0×10^{2}†
astronomical unit	meter (m)	1.496×10^{11}
atmosphere, standard	pascal (Pa)	1.013×10^{5}
bar	pascal (Pa)	1.0×10^{5}†
barn	square meter (m^2)	1.0×10^{-28}†
barrel (42 U.S. liquid gallons)	cubic meter (m^3)	0.1590
Bohr magneton (€$_B$)	J/T	9.274×10^{-24}
Btu (International Table)	joule (J)	1.055×10^{3}
Btu (mean)	joule (J)	1.056×10^{3}
Btu (thermochemical)	joule (J)	1.054×10^{3}
bushel	cubic meter (m^3)	3.524×10^{-2}
calorie (International Table)	joule (J)	4.187
calorie (mean)	joule (J)	4.190
calorie (thermochemical)	joule (J)	4.184†
centipoise	pascal second (Pa s)	1.0×10^{-3}†
centistokes	square millimeter per second (mm^2/s)	1.0†
cfm (cubic foot per minute)	cubic meter per second (m^3/s)	4.72×10^{-4}
cubic inch	cubic meter (m^3)	1.639×10^{-5}
cubic foot	cubic meter (m^3)	2.832×10^{-2}
cubic yard	cubic meter (m^3)	0.7646
curie	becquerel (Bq)	3.70×10^{10}†
debye	coulomb meter (C m)	3.336×10^{-30}
degree (angle)	radian (rad)	1.745×10^{-2}
denier (international)	kilogram per meter (kg/m)	1.111×10^{-7}
	tex‡	0.1111
dram (apothecaries')	kilogram (kg)	3.888×10^{-3}
dram (avoirdupois)	kilogram (kg)	1.772×10^{-3}
dram (U.S. fluid)	cubic meter (m^3)	3.697×10^{-6}
dyne	newton (N)	1.0×10^{-5}†
dyne/cm	newton per meter (N/m)	1.0×10^{-3}†
electronvolt	joule (J)	1.602×10^{-19}
erg	joule (J)	1.0×10^{-7}†
fathom	meter (m)	1.829
fluid ounce (U.S.)	cubic meter (m^3)	2.957×10^{-5}
foot	meter (m)	0.3048†
footcandle	lux (lx)	10.76
furlong	meter (m)	2.012×10^{-2}
gal	meter per second squared (m/s^2)	1.0×10^{-2}†
gallon (U.S. dry)	cubic meter (m^3)	4.405×10^{-3}
gallon (U.S. liquid)	cubic meter (m^3)	3.785×10^{-3}
gallon per minute (gpm)	cubic meter per second (m^3/s)	6.309×10^{-5}
	cubic meter per hour (m^3/h)	0.2271
gauss	tesla (T)	1.0×10^{-4}
gilbert	ampere (A)	0.7958
gill (U.S.)	cubic meter (m^3)	1.183×10^{-4}
grade	radian	1.571×10^{-2}
grain	kilogram (kg)	6.480×10^{-5}
gram force per denier	newton per tex (N/tex)	8.826×10^{-2}
hectare	square meter (m^2)	1.0×10^{4}†
horsepower (550 ft · lbf/s)	watt (W)	7.457×10^{2}
horsepower (boiler)	watt (W)	9.810×10^{3}
horsepower (electric)	watt (W)	7.46×10^{2}†
hundredweight (long)	kilogram (kg)	50.80
hundredweight (short)	kilogram (kg)	45.36
inch	meter (m)	2.54×10^{-2}†
inch of mercury (32°F)	pascal (Pa)	3.386×10^{3}
inch of water (39.2°F)	pascal (Pa)	2.491×10^{2}

CONVERSION FACTORS TO SI UNITS

To convert from	To	Multiply by
kilogram-force	newton (N)	9.807
kilowatt hour	megajoule (MJ)	3.6†
kip	newton (N)	4.448×10^3
knot (international)	meter per second (m/S)	0.5144
lambert	candela per square meter (cd/m^3)	3.183×10^3
league (British nautical)	meter (m)	5.559×10^3
league (statute)	meter (m)	4.828×10^3
light year	meter (m)	9.461×10^{15}
liter (for fluids only)	cubic meter (m^3)	1.0×10^{-3}†
maxwell	weber (Wb)	1.0×10^{-8}†
micron	meter (m)	1.0×10^{-6}†
mil	meter (m)	2.54×10^{-5}†
mile (statute)	meter (m)	1.609×10^3
mile (U.S. nautical)	meter (m)	1.852×10^3†
mile per hour	meter per second (m/s)	0.4470
millibar	pascal (Pa)	1.0×10^2
millimeter of mercury (0°C)	pascal (Pa)	1.333×10^2†
minute (angular)	radian	2.909×10^{-4}
myriagram	kilogram (kg)	10
myriameter	kilometer (km)	10
oersted	ampere per meter (A/m)	79.58
ounce (avoirdupois)	kilogram (kg)	2.835×10^{-2}
ounce (troy)	kilogram (kg)	3.110×10^{-2}
ounce (U.S. fluid)	cubic meter (m^3)	2.957×10^{-5}
ounce-force	newton (N)	0.2780
peck (U.S.)	cubic meter (m^3)	8.810×10^{-3}
pennyweight	kilogram (kg)	1.555×10^{-3}
pint (U.S. dry)	cubic meter (m^3)	5.506×10^{-4}
pint (U.S. liquid)	cubic meter (m^3)	4.732×10^{-4}
poise (absolute viscosity)	pascal second (Pa · s)	0.10†
pound (avoirdupois)	kilogram (kg)	0.4536
pound (troy)	kilogram (kg)	0.3732
poundal	newton (N)	0.1383
pound-force	newton (N)	4.448
pound force per square inch (psi)	pascal (Pa)	6.895×10^3
quart (U.S. dry)	cubic meter (m^3)	1.101×10^{-3}
quart (U.S. liquid)	cubic meter (m^3)	9.464×10^{-4}
quintal	kilogram (kg)	1.0×10^2†
rad	gray (Gy)	1.0×10^{-2}†
rod	meter (m)	5.029
roentgen	coulomb per kilogram (C/kg)	2.58×10^{-4}
second (angle)	radian (rad)	4.848×10^{-6}†
section	square meter (m^2)	2.590×10^6
slug	kilogram (kg)	14.59
spherical candle power	lumen (lm)	12.57
square inch	square meter (m^2)	6.452×10^{-4}
square foot	square meter (m^2)	9.290×10^{-2}
square mile	square meter (m^2)	2.590×10^6
square yard	square meter (m^2)	0.8361
stere	cubic meter (m^3)	1.0†
stokes (kinematic viscosity)	square meter per second (m^2/s)	1.0×10^{-4}†
tex	kilogram per meter (kg/m)	1.0×10^{-6}†
ton (long, 2240 pounds)	kilogram (kg)	1.016×10^3
ton (metric) (tonne)	kilogram (kg)	1.0×10^3†
ton (short, 2000 pounds)	kilogram (kg)	9.072×10^2
torr	pascal (Pa)	1.333×10^2
unit pole	weber (Wb)	1.257×10^{-7}
yard	meter (m)	0.9144†

†Exact.

A

ACTIVATED CARBON, DECOLORATION OF PHARMACEUTICAL PRODUCTS

FREDERICK S. BAKER
CHARLES E. MILLER
ALBERT J. REPIK
E. DONALD TOLLES
Westvaco Corporation
Charleston, South Carolina

KEY WORDS

Activated carbon
Adsorption isotherms
Adsorption of radionuclides
BET surface area
Emissions, motor vehicles
Food, beverage, and cooking oil
Pharmaceuticals
Pore volume
Powdered carbon regeneration
Solvent recovery

OUTLINE

INTRODUCTION

Activated carbon is a predominantly amorphous solid that has an extraordinarily large internal surface area and pore volume. These unique characteristics are responsible for its adsorptive properties, which are exploited in many different liquid- and gas-phase applications. Activated carbon is an exceptionally versatile adsorbent because the size and distribution of the pores within the carbon matrix can be controlled to meet the needs of current and emerging markets (1). Engineering requirements of specific applications are satisfied by producing activated carbons in the form of powders, granules, and shaped products. Through choice of precursor, method of activation, and control of processing conditions, the adsorptive properties of products are tailored for applications as diverse as the purification of potable water and the control of emissions from bioproduct recovery processes.

In 1900, two very significant processes in the development and manufacture of activated carbon products were patented (2). The first commercial products were produced in Europe under these patents: Eponite, from wood in 1909, and Norit, from peat in 1911. Activated carbon was first produced in the United States in 1913 by Westvaco Corp. under the name Filtchar, using a by-product of the papermaking process (3). Further milestones in development were reached as a result of World War I. In response to the need for protective gas masks, a hard, granular activated carbon was produced from coconut shell in 1915. Following the war, large-scale commercial use of activated carbon was extended to refining of beet sugar and corn syrup and to purification of municipal water supplies (4). The termination of the supply of coconut char from the Philippines and India during World War II forced the domestic development of granular activated carbon products from coal in 1940 (5). More recent innovations in the manufacture and use of activated carbon products have been driven by the need to recycle resources and to prevent environmental pollution.

PHYSICAL AND CHEMICAL PROPERTIES

The structure of activated carbon is best described as a twisted network of defective carbon layer planes, cross-linked by aliphatic bridging groups (6). X-ray diffraction patterns of activated carbon reveal that it is nongraphitic, remaining amorphous because the randomly cross-linked network inhibits reordering of the structure even when heated to 3,000 °C (7). This property of activated carbon contributes to its most unique feature, namely, the highly developed and accessible internal pore structure. The surface area, dimensions, and distribution of the pores depend on the precursor and on the conditions of carbonization and activation. Pore sizes are classified (8) by the International Union of Pure and Applied Chemistry (IUPAC) as micropores (pore width <2 nm), mesopores (pore width 2–50 nm), and macropores (pore width >50 nm).

The surface area of activated carbon is usually determined by application of the Brunauer-Emmett-Teller (BET) model of physical adsorption (9,10) using nitrogen as the adsorptive (8). Typical commercial products have specific surface areas in the range 500–2,000 m^2/g, but values as high as 3,500–5,000 m^2/g have been reported for some activated carbons (11,12). In general, however, the effective surface area of a microporous activated carbon is far smaller because the adsorption of nitrogen in micropores does not occur according to the process assumed in the BET model, which results in unrealistically high values for surface area (10,13). Adsorption isotherms are usually determined for the appropriate adsorptives to assess the effective surface area of a product in a specific application. Adsorption capacity and rate of adsorption depend on the internal surface area and distribution of pore size and shape but are also influenced by the surface chemistry of the activated carbon (14). The macroporosity of the carbon is important for the transfer of adsorbate molecules to adsorption sites within the particle.

Functional groups are formed during activation by interaction of free radicals on the carbon surface with atoms such as oxygen and nitrogen, both from within the precursor and from the atmosphere (15). The functional groups render the surface of activated carbon chemically reactive and influence its adsorptive properties (6). Activated carbon is generally considered to exhibit a low affinity for water, which is an important property with respect to the adsorption of gases in the presence of moisture (16). However, the functional groups on the carbon surface can interact with water, rendering the carbon surface more hydrophilic (15). Surface oxidation, which is an inherent feature of activated carbon production, results in hydroxyl, carbonyl, and carboxylic groups that impart an amphoteric character to the carbon, so that it can be either acidic or basic. The electrokinetic properties of an activated carbon product are, therefore, important with respect to its use as a catalyst support (17). As well as influencing the adsorption of many molecules, surface oxide groups contribute to the reactivity of activated carbons toward certain solvents in solvent recovery applications (18).

In addition to surface area, pore size distribution, and surface chemistry, other important properties of commercial activated carbon products include pore volume, particle size distribution, apparent or bulk density, particle density, abrasion resistance, hardness, and ash content. The range of these and other properties is illustrated in Table 1 together with specific values for selected commercial grades of powdered, granular, and shaped activated carbon products used in liquid- or gas-phase applications (19).

MANUFACTURE AND PROCESSING

Commercial activated carbon products are produced from organic materials that are rich in carbon, particularly coal, lignite, wood, nut shells, peat, pitches, and cokes. The choice of precursor is largely dependent on its availability, cost, and purity, but the manufacturing process and intended application of the product are also important considerations. Manufacturing processes fall into two categories, thermal activation and chemical activation. The effective porosity of activated carbon produced by thermal activation is the result of gasification of the carbon at relatively high temperatures (20), but the porosity of chemically activated products is generally created by chemical dehydration reactions occurring at significantly lower temperatures (1,21).

Thermal Activation Processes

Thermal activation occurs in two stages: thermal decomposition or carbonization of the precursor and controlled gasification or activation of the crude char. During carbonization, elements such as hydrogen and oxygen are eliminated from the precursor to produce a carbon skeleton possessing a latent pore structure. During gasification, the char is exposed to an oxidizing atmosphere that greatly increases the pore volume and surface area of the product through elimination of volatile pyrolysis products and from carbon burn-off. Carbonization and activation of the char are generally carried out in direct-fired rotary kilns or multiple-hearth furnaces, but fluidized-bed reactors have also been used (22). Materials of construction, notably steel and refractories, are designed to withstand the high-temperature conditions (i.e., >1,000 °C) inherent in activation processes. The thermal activation process is illustrated in Figure 1 for the production of activated carbon from bituminous coal (23,24).

Bituminous coal is pulverized and passed to a briquette press. Binders may be added at this stage before compression of the coal into briquettes. The briquetted coal is then crushed and passed through a screen, from which the on-size material passes to an oxidizing kiln. Here, the coking properties of the coal particles are destroyed by oxidation at moderate temperatures in air. The oxidized coal is then devolatilized in a second rotary kiln at higher temperatures under steam. To comply with environmental pollution regulations, the kiln off-gases containing dust and volatile matter pass through an incinerator before discharge to the atmosphere.

The devolatilized coal particles are transported to a direct-fired multihearth furnace where they are activated by holding the temperature of the furnace at about 1,000 °C. Product quality is maintained by controlling coal feed rate and bed temperature. As before, dust particles in the furnace off-gas are combusted in an afterburner before discharge of the gas to the atmosphere. Finally, the granular product is screened to provide the desired particle size. A

Table 1. Properties of Selected U.S. Activated Carbon Products

	Manufacturer	Gas-phase carbons			Liquid-phase carbons		
	Precursor	Calgon coal	Norit peat	Westvaco wood	Calgon coal	Norit peat	Westvaco wood
	Product grade	BPL	B4	WV-A 1100	SGL	SA 3	SA-20
	Product form	granular	extruded	granular	granular	powdered	powdered
Property	Typical range						
Particle size, U.S. mesh[a,b]	<4	12 × 30	3.8 mm	10 × 25	8 × 30	60% <325	65–85% <325
Apparent density, g/cm^3	0.2–0.6	>0.48	0.43	0.27	0.52	0.46	0.34–0.37
Particle density, g/cm^3	0.4–0.9	0.80		0.50	0.80		
Hardness number	50–100	>90	99				
Abrasion number					>75		
Ash, wt %	1–20	<8	6		<10	6	3–5
BET surface area, N_2 m^2/g	500–2,500	1,050–1,150	1,100–1,200	1750	900–1,000	750	1,400–1,800
Total pore volume, cm^3/g	0.5–2.5	0.8	0.9	1.2	0.85		2.2–2.5
CCl_4 activity, wt%	35–125	>60					
Butane working capacity, $g/100\ cm^3$	4–14			>11.0			
Iodine number	500–1,200	>1,050			>900	800	>1,000
Decolorizing index							
Westvaco	15–25						>20
Molasses number							
Calgon	50–250				>200		
Norit	300–1,500					440	
Heat Capacity at 100 °C, J/(g K)[c]	0.84–1.3	1.05			1.05		
Thermal conductivity, W/(m·K)	0.05–0.10						

Note: Specific values shown are those cited in manufacturers' product literature (19). Typical ranges shown are based on values reported in the open literature.
[a]Unless otherwise noted.
[b]Approximate mm corresponding to cited meshes are (mesh:mm 4:4.76; 8:2.38; 10:2; 12:1.68; 25:0.72; 30:0.59; 325:0.04.
[c]To convert J to cal, divide by 4.184.

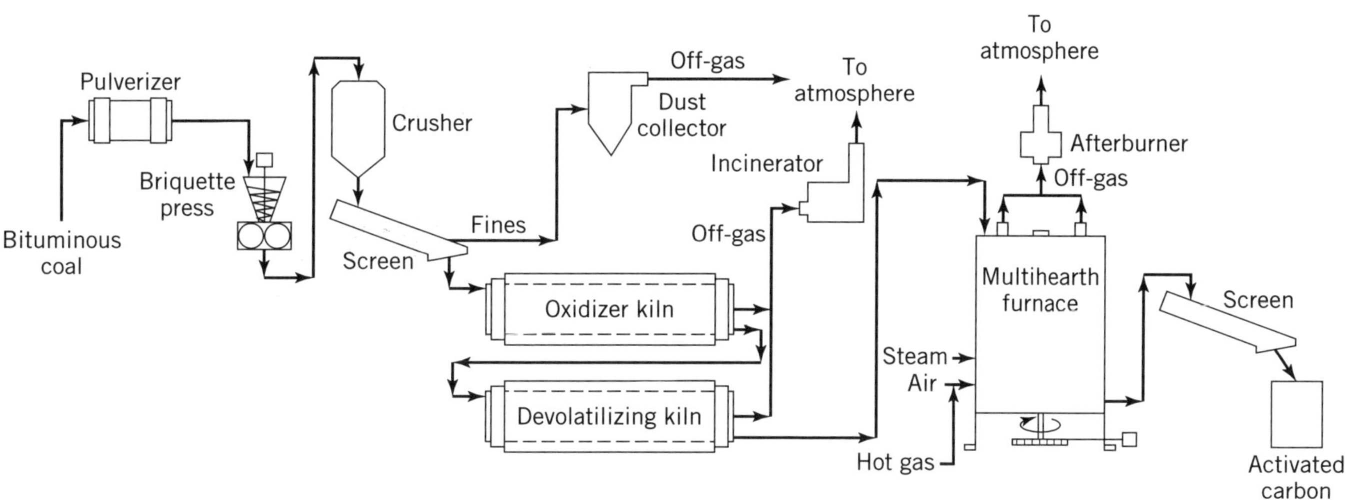

Figure 1. Thermal activation of bituminous coal.

typical yield of activated carbon is about 30–35% by weight based on the raw coal.

The process for the thermal activation of other carbonaceous materials is modified according to the precursor. For example, the production of activated carbon from coconut shell does not require the stages involving briquetting, oxidation, and devolatilization. To obtain a high-activity product, however, it is important that the coconut shell is charred slowly prior to activation of the char. In some processes, the precursor or product is acid-washed to obtain a final product with a low ash content (23,25).

Chemical Activation Processes

In contrast to the thermal activation of coal, chemical activation is generally carried out commercially in a single kiln. The precursor, usually wood, is impregnated with a chemical activation agent, typically phosphoric acid, and the blend is heated to a temperature of 450–700 °C (26). Chemical activation agents reduce the formation of tar and other by-products, thereby increasing carbon yield. The chemical activation process is illustrated in Figure 2 for the production of granular activated carbon from wood (23,27).

Sawdust is impregnated with concentrated phosphoric acid and fed to a rotary kiln, where it is dried, carbonized, and activated at a moderate temperature. To comply with environmental pollution regulations, the kiln off-gases are treated before discharge to the atmosphere. The char is washed with water to remove the acid from the carbon, and the carbon is separated from the slurry. The filtrate is then passed to an acid-recovery unit. Some manufacturing plants do not recycle all the acid but use a part of it to manufacture fertilizer in an allied plant. If necessary, the pH of the activated carbon is adjusted, and the product is dried. The dry product is screened and classified into the size range required for specific granular carbon applications. Carbon yields as high as 50% by weight of the wood precursor have been reported (26).

Novel Manufacturing Processes. Different chemical activation processes have been used to produce carbons with enhanced adsorption characteristics. Activated carbons of exceptionally high surface area (>3,000 m^2/g) have been produced by the chemical activation of carbonaceous materials with potassium hydroxide (28,29). Activated carbons are also produced commercially in the form of cloths (30), fibers (31), and foams (32) generally by chemical activation of the precursor with a Lewis acid such as aluminum chloride, ferric chloride, or zinc chloride.

Forms of Activated Carbon Products

To meet the engineering requirements of specific applications, activated carbons are produced and classified as granular, powdered, or shaped products. Granular activated carbons are produced directly from granular precursors, such as sawdust and crushed and sized coconut char or coal. The granular product is screened and sized for specific applications. Powdered activated carbons are obtained by grinding granular products. Shaped activated carbon products are generally produced as cylindrical pellets by extrusion of the precursor with a suitable binder before activation of the precursor.

Shipping and Storage

Activated carbon products are shipped in bags, drums, and boxes in weights ranging from about 10 to 35 kg. Containers can be lined or covered with plastic and should be stored in a protected area both to prevent weather damage and to minimize contact with organic vapors that could reduce the adsorption performance of the product. Bulk quantities of activated carbon products are shipped in metal bins and bulk bags, typically 1–2 m^3 in volume, and in railcars and tank trucks. Bulk carbon shipments are generally transferred by pneumatic conveyors and stored in tanks. However, in applications such as water treatment where water adsorption does not impact product performance, bulk carbon may be transferred and stored as a slurry in water.

Specifications

Activated carbon producers furnish product bulletins that list specifications, usually expressed as a maximum or minimum value, and typical properties for each grade produced. Standards helpful in setting purchasing specifications for granular and powdered activated carbon products have been published (33,34).

ECONOMIC ASPECTS

Excluding eastern European countries and China, where production figures have not been published, the world pro-

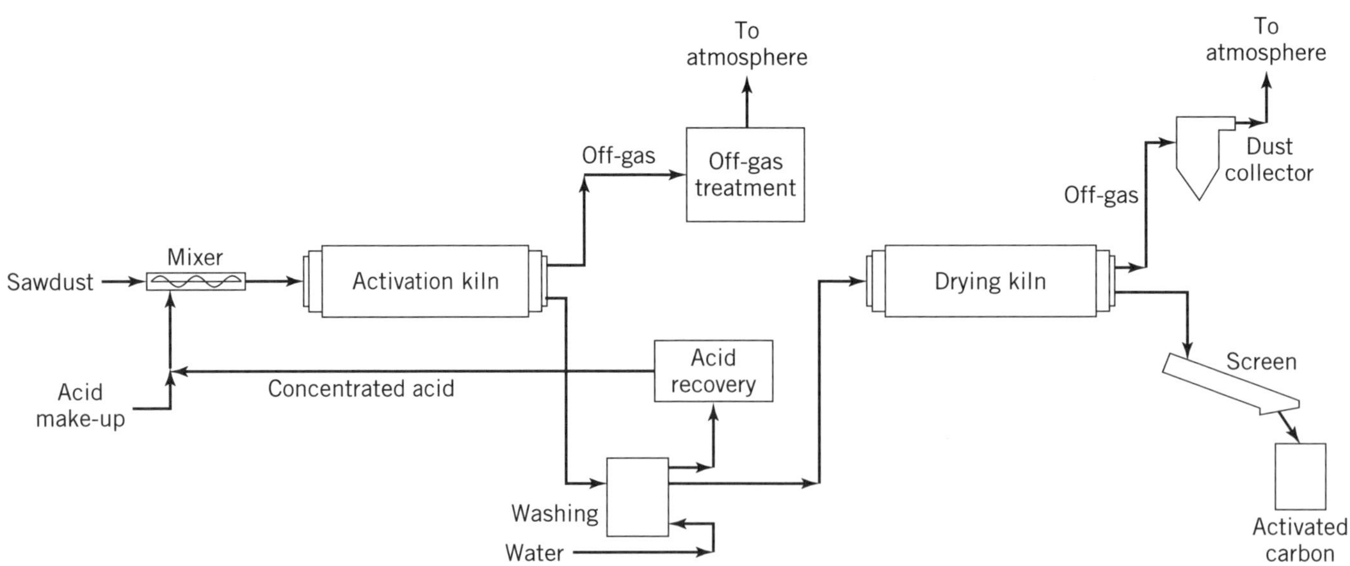

Figure 2. Chemical activation of wood.

duction capacity of activated carbon was estimated to be 375,000 metric tons in 1990 (35). The price of most products was 0.70 to 5.50 $/kg, but some specialty carbons were more expensive (36, pp. 731.2001P, 731.2001Q). Forty percent of the production capacity was in the United States, 30% in western Europe, 20% in Japan, and 10% in other Pacific Rim countries (Table 2).

Production capacity was almost equally split between powdered and nonpowdered activated carbon products. Powdered activated carbon, a less expensive form used in liquid-phase applications, is generally used once and then disposed of. In some cases, however, granular and shaped products are regenerated and reused (35). In 1990 production capacity for granular and shaped products was split, with about two-thirds for liquid-phase and one-third for gas-phase applications (37).

Over the last decade, production capacity in the United States remained essentially unchanged, but minor fluctuations occurred in response to changes in environmental regulations (36pp.731.2000S–731.2000Y). A similar reaction was noted worldwide (35). The current demand for activated carbon is estimated at 93% of production capacity. The near-term growth in demand is projected to be approximately 5.5%/year (37pp.4,5).

In 1970 the U.S. Congress enacted the Clean Air Act, the Clean Water Act, and the Safe Drinking Water Act. Because activated carbon can often be used to help meet Environmental Protection Agency (EPA) regulations, the U.S. activated carbon industry reacted by increasing its production capacity. A proposed amendment to the Safe Drinking Water Act in 1979 required the use of granular activated carbon systems, but the amendment was not enacted. In response to the projected increase in demand for activated carbon, production capacity remained high until the late 1980s, but when the anticipated need did not materialize, some production facilities were shut down. Currently, because of stricter EPA regulations implementing all three acts in 1990, the industry will increase production capacity by 25% during the next several years (35,38).

The estimated production capacity of activated carbon in the United States is shown in Table 3 for seven manufacturers (35pp.54–65). The principal producers are Calgon Carbon (37%), American-Norit (26%), Westvaco (19%), and Atochem (10%). Several other companies purchase activated carbon for resale but do not manufacture products.

Western Europe has seven manufacturers of activated carbon. The two largest, Norit and Chemviron (a subsidiary of Calgon), account for 70% of western European production capacity, and Ceca accounts for 13% (35p.13), Japan is the third-largest producer of activated carbon, having 18 manufacturers, but 4 companies share over 50% of the total Japanese capacity (35p.25–32). Six Pacific Rim countries account for the balance of the world production capacity of activated carbon, 90% of which is in the Philippines and Sri Lanka (35p.13). As is the case with other businesses, regional markets for activated carbon products have become international, leading to consolidation of manufacturers. Calgon, Norit, Ceca, and Sutcliffe-Speakman are examples of multinational companies.

Table 2. World Production Capacity, Estimated 1990

Country	Capacity (10^3 t)
United States	146
Western Europe	108
Japan	72
Pacific Rim, other	49
Total	375

Note: Excluding eastern Europe and China.

Activated carbon is a recyclable material that can be regenerated. Thus the economics, especially the market growth of activated carbon, particularly granular and shaped products, is affected by regeneration and industry regeneration capacity. The decision to regenerate an activated carbon product is dependent on the cost, size of the carbon system, type of adsorbate, and the environmental issue involved. Large carbon systems, such as those used in potable and wastewater treatment, generally require a high-temperature treatment, which is typically carried out in rotary or multihearth furnaces. During regeneration, carbon losses of 1 to 15% typically occur from the treatment and movement of the carbon (39). However, material loss is compensated for by the addition of new carbon to the adsorber system. In general, regeneration of spent carbon is considerably less expensive than the purchase of new activated carbon. For example, fluidized-bed furnace regeneration of activated carbon used in a 94,600-m^3/day water treatment system cost only 35% of new material (40). For this system, regeneration using either infrared or multihearth furnaces was estimated to be more expensive but still significantly less so than the cost of new carbon.

Because powdered activated carbon is generally used in relatively small quantities, the spent carbon has often been disposed of in landfills. However, landfill disposal is becoming more restrictive environmentally and more costly. Thus large consumers of powdered carbon find that regeneration is an attractive alternative. Examples of regeneration systems for powdered activated carbon include the Zimpro/Passavant wet air oxidation process (41), the multihearth furnace as used in the DuPont PACT process (41pp.389–447,42), and the Shirco infrared furnace (40p.51,43).

Other types of regenerators designed for specific adsorption systems may use solvents and chemicals to remove susceptible adsorbates (44), steam or heated inert gas to recover volatile organic solvents (45), and biological systems in which organics adsorbed on the activated carbon during water treatment are continuously degraded (46).

ANALYTICAL TEST PROCEDURES AND STANDARDS

Source references for frequently used test procedures for determining properties of activated carbon are shown in Table 4. A primary source is the *Annual Book of American Society for Testing and Materials (ASTM) Standards* (54). Other useful sources of standards and test procedures include manufacturers of activated carbon products, the American Water Works Association (AWWA) (33,34), and the U.S. Department of Defense (54).

Table 3. Production Capacity in the United States, Estimated 1990

Company	Location	Capacity (10^3 t)
Acticarb Division, Royal Oak Enterprises	Romeo, Fla.	6.8
American Norit Co.	Marshall, Tex.	38.6
Barneby and Sutcliffe	Columbus, Ohio	3.0
Calgon Carbon Corp.	Catlettsburg, Ky. and Pittsburgh, Pa.	53.5
Ceca Division, Atochem NA	Pryor, Okla.	15.0
Trans-Pacific Carbon	Blue Lake, Calif.	2.3
Westvaco Corp.	Covington, Va.	27.2
Total		*146.4*

Table 4. Source References for Activated Carbon Test Procedures and Standards

Title of procedure or standard	Source
Standard Definitions of Terms Relating to Activated Carbon	ASTM D2652
Apparent Density of Activated Carbon	ASTM D2854
Particle Size Distribution of Granular Activated Carbon	ASTM D2862
Total Ash Content of Activated Carbon	ASTM D2866
Moisture in Activated Carbon	ASTM D2867
Ignition Temperature of Granular Activated Carbon	ASTM D3466
Carbon Tetrachloride Activity of Activated Carbon	ASTM D3467
Ball-Pan Hardness of Activated Carbon	ASTM D3802
Radioiodine Testing of Nuclear-Grade Gas-Phase Adsorbents	ASTM D3803
pH of Activated Carbon	ASTM D3838
Determination of Adsorptive Capacity of Carbon by Isotherm Technique	ASTM D3860
Determining Operating Performance of Granular Activated Carbon	ASTM D3922
Impregnated Activated Carbon Used to Remove Gaseous RadioIodines from Gas Streams	ASTM D4069
Determination of Iodine Number of Activated Carbon	ASTM D4607
Military Specification, Charcoal, Activated, Impregnated	Ref. 47
Military Specification, Charcoal, Activated, Unimpregnated	Ref. 47
AWWA Standard for Granular Activated Carbon	Ref. 33
AWWA Standard for Powdered Activated Carbon	Ref. 34
BET Surface Area by Nitrogen Adsorption	Refs. 6, 8, 9, 48
Pore Volume by Nitrogen Adsorption or Mercury Penetration	Refs. 10, 49–52
Particle Density	Ref. 53

HEALTH AND SAFETY

Activated carbon generally presents no particular health hazard as defined by NIOSH (55). However, it is a nuisance and mild irritant with respect to inhalation, skin contact, eye exposure, and ingestion. On the other hand, special consideration must be given to the handling of spent carbon that may contain a concentration of toxic compounds.

Activated carbon products used for decolorizing food products in liquid form must meet the requirements of the *Food Chemical Codex* as prepared by the Food and Nutrition Board of the National Research Council (56).

According to the National Board of Fire Underwriters, activated carbons normally used for water treatment pose no dust explosion hazard and are not subject to spontaneous combustion when confined to bags, drums, or storage bins (67). However, activated carbon burns when sufficient heat is applied; the ignition point varies between about 300 and 600 °C (58p.353).

Dust-tight electrical systems should be used in areas where activated carbon is present, particularly powdered products (58pp.84,85). When partially wet activated carbon comes into contact with unprotected metal, galvanic currents can be set up; these result in metal corrosion (59).

Manufacturer material safety data sheets (MSDS) indicate that the oxygen concentration in bulk storage bins or other enclosed vessels can be reduced by wet activated carbon to a level that will not support life. Therefore, self-contained air packs should be used by personnel entering enclosed vessels where activated carbon is present (60).

LIQUID-PHASE APPLICATIONS

Activated carbons for use in liquid-phase applications differ from gas-phase carbons primarily in pore size distribution. Liquid-phase carbons have significantly more pore volume in the macropore range, which permits liquids to diffuse more rapidly into the mesopores and micropores (61). The larger pores also promote greater adsorption of large molecules, either impurities or products, in many liquid-phase applications. Specific-grade choice is based on the isotherm (62,63) and, in some cases, bench or pilot-scale evaluations of candidate carbons.

Liquid-phase activated carbon can be applied either as a powder, granular, or shaped form. The average size of powdered carbon particles is 15–25 μm (62). Granular or shaped carbon particle size is usually 0.3–3.0 mm. A significant factor in choosing between powdered and nonpowdered carbon is the degree of purification required in the adsorption application. Granular and shaped carbons are usually used in continuous flow through deep beds to remove essentially all contaminants from the liquid being treated. Granular and shaped carbon systems are preferred when a large carbon buffer is needed to withstand significant variations in adsorption conditions, such as in cases where large contaminant spikes may occur. A wider range of impurity removal can be attained by batch application of powdered carbon, and the powdered carbon dose per batch can be controlled to achieve the degree of purification desired (61).

Batch-stirred vessels are most often used in treating material with powdered activated carbon (64). The type of carbon, contact time, and amount of carbon vary with the desired degree of purification. The efficiency of activated carbon may be improved by applying continuous, countercurrent carbon-liquid flow with multiple stages (Fig. 3). Carbon is separated from the liquid at each stage by settling or filtration. Filter aids such as diatomaceous earth are sometimes used to improve filtration.

Granular and shaped carbons are generally used in continuous systems where the liquid to be treated is passed through a fixed bed (41pp.8–19,64). Compounds are adsorbed by the carbon bed in the adsorption zone (Fig. 4). As carbon in the bed becomes saturated with adsorbates, the adsorption zone moves in the direction of flow, and breakthrough occurs when the leading edge of the adsorption zone reaches the end of the column. Normally at least two columns in series are on line at any given time. When the first column becomes saturated, it is removed from service, and a column containing fresh carbon is added at the discharge end of the series. An alternative approach is the moving bed column (41pp.8–19). In this design the adsorption zone is contained within a single column by passing liquid upward while continuously or intermittently withdrawing spent carbon at the bottom and adding fresh carbon at the top.

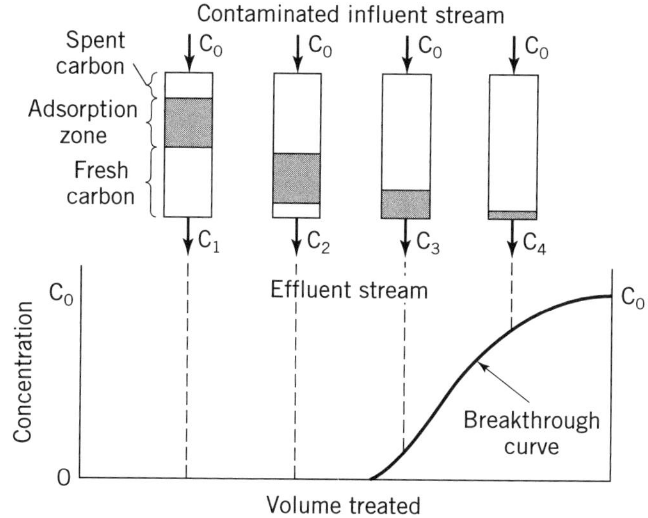

Figure 4. Adsorption and breakthrough curve for fixed bed of granular or shaped activated carbon.

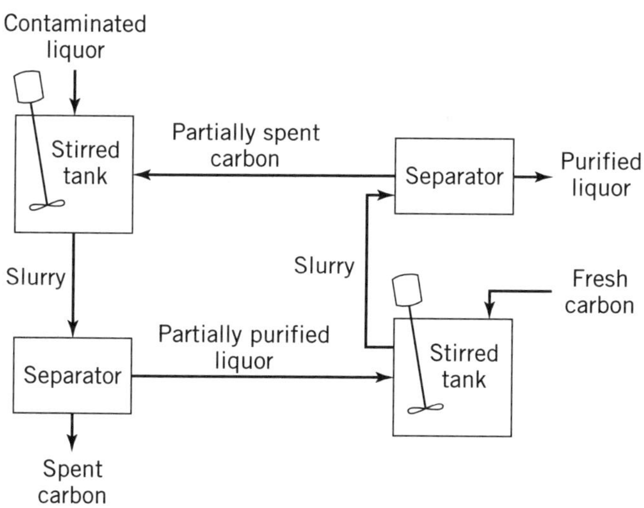

Figure 3. Multistage countercurrent application of powdered activated carbon.

The total activated carbon consumption for liquid-phase applications in the United States in 1987 was estimated to be about 76,700 t, which accounted for nearly 80% of the total activated carbon use. The consumption by application is summarized in Table 5 (36pp.731.2000V–731.2001L)

Potable Water Treatment

Treatment of drinking water accounts for about 24% of the total activated carbon used in liquid-phase applications (36pp.731.2000V–731.2001L). Rivers, lakes, and groundwater from wells, the most common drinking water sources, are often contaminated with bacteria, viruses, natural vegetation decay products, halogenated materials,

Table 5. Liquid-Phase Activated Carbon Consumption (10^3 t) in the United States in 1987.

	Granular-shaped	Powdered	Total
Potable water	4.5	13.6	18.1
Wastewater			
Industrial	6.4	6.6	13.0
Municipal	0.9	2.0	2.9
Sweetener decolorization	6.8	9.1	15.9
Chemical processing and misc.	4.1	2.3	6.4
Food, beverage, and oils	0.9	3.9	4.8
Pharmaceuticals	2.0	2.3	4.3
Mining	1.6	2.5	4.1
Groundwater	0.9	2.3	3.2
Household uses	1.4	0.9	2.3
Dry cleaning	0.7	0.4	1.1
Electroplating	0.2	0.4	0.6
Total	*30.4*	*46.3*	*76.7*

and volatile organic compounds. Normal water disinfection and filtration treatment steps remove or destroy the bulk of these materials (65). However, treatment by activated carbon is an important additional step in many plants to remove toxic and other organic materials (66–68) for safety and palatability.

Groundwater Remediation

Concern over contaminated groundwater sources increased in the 1980s, and in 1984 an Office of Groundwater Protection was created by the EPA (36pp.731.2000V–731.2001L). This led to an increase in activated carbon consumption in 1987 for groundwater treatment to about 4% of the total liquid-phase usage, and further growth is expected in the 1990s. There are two ways to apply carbon in groundwater cleanup. One is the conventional method of applying powdered, granular, or shaped carbon to adsorb contaminants directly from the water. The other method utilizes air stripping to transfer the volatile compounds from water to air. The compounds are then recovered by passing the contaminated air through a bed of carbon (69,70).

Industrial and Municipal Wastewater Treatment

Wastewater treatment consumes about 21% of the total U.S. liquid-phase activated carbon (36pp.731.2000V–731.2001L), and governmental regulations are expected to increase demand over the next several years. Wastewater may contain suspended solids, hazardous microorganisms, and toxic organic and inorganic contaminants that must be removed or destroyed before discharge to the environment. In tertiary treatment systems, powdered, granular, or shaped carbon can be used to remove residual toxic and other organic compounds after the primary filtration and secondary biological treatment (71). Powdered carbon is also used in the PACT process by direct addition of the carbon to the secondary biological step (41pp.389–447).

Sweetener Decolorization

About 21% of the liquid-phase activated carbon is used for purification of sugar (qv) and corn syrup (36pp.731.2000V–731.2001L). White sucrose sugar is made from raw juice squeezed from sugarcane or sugar beets. The clarified liquor is decolorized using activated carbon or ion exchange resins (58pp.87–125,274–292). High-fructose corn sweeteners (HFCS) are produced by hydrolysis of corn starch and are then treated with activated carbon to remove undesirable taste and odor compounds and to improve storage life. The demand for HFCS rose sharply in the 1980s primarily because of the switch by soft drink producers away from sucrose (83).

Chemical Processing

Activated carbon consumption in a variety of chemical processing applications is about 8% of the total (36pp.731.2000V–731.2001L). The activated carbon removes impurities to achieve high quality. For example, organic contaminants are removed from solution in the production of alum, soda ash, and potassium hydroxide (58pp.87–125,274–292). Other applications include the manufacture of dyestuffs, glycols, amines, organic acids, urea, hydrochloric acid, and phosphoric acid (35pp.92–135).

Food, Beverage, and Cooking Oil

Approximately 6% of the liquid-phase activated carbon is used in food, beverage, and cooking oil production (36pp.731.2000V–731.2000L). Before being incorporated into edible products, vegetable oils and animal fats are refined to remove particulates, inorganics, and organic contaminants. Activated carbon is one of several agents used in food purification processes. In the production of alcoholic beverages, activated carbon removes haze-causing compounds from beer, taste and odor from vodka, and fusel oil from whiskey (58pp.87–125,274–292). The feed water for soft drink production is often treated with carbon to capture undesirable taste and odor compounds and to remove free chlorine remaining from disinfection treatment. Caffeine is removed from coffee beans by extraction with organic solvents, water, or supercritical carbon dioxide prior to roasting. Activated carbon is used to remove the caffeine from the recovered solvents (35pp.92–135).

Pharmaceuticals

Pharmaceuticals account for 6% of the liquid-phase activated carbon consumption (36pp.731.2000V–731.2000L). Many antibiotics, vitamins, and steroids are isolated from fermentation broths by adsorption onto carbon followed by solvent extraction and distillation (58pp.87–125,274–292). Other uses in pharmaceutical production include removal of chromogenic impurities, process water purification, and removal of impurities from intravenous solutions prior to packaging (35pp.92–135).

Mining

The mining industry accounts for only 4% of liquid-phase activated carbon use, but this figure may grow as low-grade ores become more common (36pp.731.2000V–731.2000L). Gold, for example, is recovered on activated carbon as a cyanide complex in the carbon-in-pulp extraction process (58pp.87–125, 274–292). Activated carbon serves as a catalyst in the detoxification of cyanides contained in wastewater from cyanide stripping operations (46pp.8–19). Problems caused by excess flotation agent concentrations in flotation baths are commonly cured by adding powdered activated carbon (58pp.87–125,274–292).

Miscellaneous Uses

Several relatively low-volume activated carbon uses comprise the remaining 6% of liquid-phase carbon consumption (36pp.731.2000V–731.2000L). Small carbon filters are used in households for purification of tap water. Oils, dyes, and other organics are adsorbed on activated carbon in dry cleaning recovery and recycling systems. Electroplating solutions are treated with carbon to remove organics that can produce imperfections when the thin metal layer is deposited on the substrate (58pp.87–127,274–292). Medical applications include removal of toxins from the blood of patients with artificial kidneys (35pp.92–135), and oral ingestion into the stomach to recover poisons or toxic materials (58pp.87–125,274–292;72). Activated carbon also is used as a support for metal catalysts in low-volume production of high-value specialty products such as pharmaceuticals, fragrance chemicals, and pesticides (73).

GAS-PHASE APPLICATIONS

Gas-phase applications of activated carbon include separation, gas storage, and catalysis. Although only 20% of activated carbon production is used for gas-phase applications, these products are generally more expensive than liquid-phase carbons and account for about 40% of the total dollar value of shipments. Most of the activated carbon used in gas-phase applications is granular or shaped. Activated carbon use by application is shown in Table 6 (74).

Separation processes comprise the main gas-phase applications of activated carbon. These usually exploit the differences in the adsorptive behavior of gases and vapors on activated carbon on the basis of molecular weight and size. For example, organic molecules with a molecular weight greater than about 40 are readily removed from air by activated carbon.

Solvent Recovery

Most of the activated carbon used in gas-phase applications is employed to prevent the release of volatile organic compounds into the atmosphere. Much of this use has been in response to environmental regulations, but recovery and recycling of solvents from a range of industrial processes such as printing, coating, and extrusion of fibers also provides substantial economic benefits.

The structure of activated carbons used for solvent recovery has been predominantly microporous. Micropores provide the strong adsorption forces needed to capture small vapor molecules such as acetone at low concentrations in process air (75). In recent years, however, more mesoporous carbons, specifically made for solvent recovery, have become available and are giving good service, especially for the adsorption of heavier vapors such as cumene and cyclohexanone that are difficult to remove from micropores during regeneration (75). Regeneration of the carbon is performed on a cyclic basis by purging it with steam or heated nitrogen.

Table 6. Gas-Phase Activated Carbon Uses in the United States in 1987

Application	Consumption (10^3 t)
Solvent recovery	4.5
Automotive/gasoline recovery	4.1
Industrial off-gas control	3.2
Catalysis	2.7
Pressure swing separation	1.1
Air conditioning	0.5
Gas mask	0.5
Cigarette filters	0.5
Nuclear	0.3
Total	*17.4*

Gasoline Emission Control

A principal application of activated carbon is in the capture of gasoline vapors that escape from vents in automotive fuel systems (76). Under EPA regulations, all U.S. motor vehicles produced since the early 1970s have been equipped with evaporative emission control systems. Most other auto-producing countries now have similar controls. Fuel vapors vented when the fuel tank or carburetor are heated are captured in a canister containing 0.5–2 L of activated carbon. Regeneration of the carbon is then accomplished by using intake manifold vacuum to draw air through the canister. The air carries desorbed vapor into the engine where it is burned during normal operation. Activated carbon systems have also been proposed for capturing vapors emitted during vehicle refueling, and activated carbon is used at many gasoline terminals to capture vapor displaced when tank trucks are filled (77). Typically, the adsorption vessels contain around 15 m^3 of activated carbon and are regenerated by application of a vacuum. The vapor that is pumped off is recovered in an absorber by contact with liquid gasoline. Similar equipment is used in the transfer of fuel from barges (78). The type of carbon pore structure required for these applications is substantially different from that used in solvent recovery. Because the regeneration conditions are very mild, only the weaker adsorption forces can be overcome, and therefore the most effective pores are in the mesopore size range (79). A large adsorption capacity in these pores is possible because vapor concentrations are high, typically 10–60%.

Adsorption of Radionuclides

Other applications that depend on physical adsorption include the control of krypton and xenon radionuclides from nuclear power plants (80). The gases are not captured entirely, but their passage is delayed long enough to allow radioactive decay of the short-lived species. Highly microporous coconut-based activated carbon is used for this service.

Control by Chemical Reaction

Pick-up of gases to prevent emissions can also depend on the chemical properties of activated carbon or of impregnants. Emergency protection against radioiodine emissions from nuclear power reactors is provided by isotope exchange over activated carbon impregnated with potassium iodide (81). Oxidation reactions catalyzed by the carbon surface are the basis for several emission control strategies. Sulfur dioxide can be removed from industrial off-gases and power plant flue gas because it is oxidized to sulfur trioxide, which reacts with water to form nonvolatile sulfuric acid (82,83). Hydrogen sulfide can be removed from such sources as Claus plant tail gas because it is converted to sulfur in the presence of oxygen (84). Nitric oxide can be removed from flue gas because it is oxidized to nitrogen dioxide. Ammonia is added and reacts catalytically on the carbon surface with the nitrogen dioxide to form nitrogen (85).

Protection against Atmospheric Contaminants

Activated carbon is widely used to filter breathing air to protect against a variety of toxic or noxious vapors, including war gases, industrial chemicals, solvents, and odorous compounds. Activated carbons for this purpose are highly microporous and thus maximize the adsorption forces that hold adsorbate molecules on the surface. Although activated carbon can give protection against most organic gases, it is especially effective against high molecular weight vapors, including chemical warfare agents such as mustard gas or the nerve agents that are toxic at parts per million concentrations. The activated carbon is employed in individual canisters or pads, as in gas masks, or in large filters in forced air ventilation systems. In air-conditioning systems, adsorption on activated carbon can be used to control the buildup of odors or toxic gases such as radon in recirculated air (86).

Inorganic vapors are usually not strongly adsorbed on activated carbon by physical forces, but protection against many toxic agents is achieved by using activated carbon impregnated with specific reactants or decomposition catalysts. For example, a combination of chromium and copper impregnants is used against hydrogen cyanide, cyanogen, and cyanogen chloride, whereas silver assists in the removal of arsine. All of these are potential chemical warfare agents; the Whetlerite carbon, which was developed in the early 1940s and is still used in military protective filters, contains these impregnants (87). Recent work has shown that chromium, which loses effectiveness with age and is itself toxic, can be replaced with a combination of molybdenum and triethylenediamine (88). Oxides of iron and zinc on activated carbon have been used in cigarette filters to absorb hydrogen cyanide and hydrogen sulfide (89). Mercury vapor in air can be removed by activated carbon impregnated with sulfur (90). Activated carbon impregnated with sodium or potassium hydroxide has long been used to control odors of hydrogen sulfide and organic mercaptans in sewage treatment plants (91). Alkali-impregnated carbon is also effective against sulfur dioxide, hydrogen sulfide, and chlorine at low concentrations. Such impregnated carbon is used extensively to protect sensitive electronic equipment against corrosion by these gases in industrial environments (92).

Process Stream Separations

Differences in adsorptivity between gases provides a means for separating components in industrial process gas streams. Activated carbon in fixed beds has been used to separate aromatic compounds from lighter vapors in petroleum refining process streams (93) and to recover gasoline components from natural and manufactured gas (94,95).

Molecular sieve activated carbons are specially made with restricted openings leading to micropores. These adsorbents are finding increasing use in separations utilizing pressure swing adsorption, in which adsorption is enhanced by operation at high pressure and desorption occurs upon depressurization (96). Larger molecules are restricted from entrance into the pores of these carbons and, therefore, are not retained as strongly as smaller molecules. The target product can be either the adsorbed or unadsorbed gases. Examples include separation of oxygen from air and recovery of methane from inorganic gases in biogas production. Hydrogen can be removed from gases produced in the catalytic cracking of gasoline, and carbon monoxide can be separated from fuel gases. Use of pressure swing techniques for gas separation is an area of growing interest in engineering research.

The hypersorption process developed in the late 1940s used a bed of activated carbon moving countercurrent to gas flow to separate light hydrocarbons from each other and from hydrogen in refinery operations. The application is of interest because of its scale, treating up to 20,000 m^3/h of gas, but the plants were shut down within a few years, probably because of problems related to attrition of the rapidly circulating activated carbon (97). It should be noted, however, that in recent years moving-bed and fluid-bed adsorption equipment using activated carbon has been successfully employed for solvent recovery (98).

Gas Storage

Adsorption forces acting on gas molecules held in micropores significantly densify the adsorbed material. As a result, activated carbon has long been considered a medium for lowering the pressure required to store weakly adsorbed compressed gases (99). Recent work with modern high-capacity carbons has been directed toward fueling passenger cars with natural gas, but storage volume targets have not yet been attained (100). Natural gas storage on activated carbon is now used commercially in portable welding cylinders (101). These can be refilled easily at about 2,000 kPa and hold as much gas as a conventional cylinder pressurized to 6,000 kPa (59 atm).

Catalysis

Catalytic properties of the activated carbon surface are useful in both inorganic and organic synthesis. For example, the fumigant sulfuryl fluoride is made by reaction of sulfur dioxide with hydrogen fluoride and fluorine over activated carbon (102). Activated carbon also catalyzes the addition of halogens across a carbon-carbon double bond in the production of a variety of organic halides (73) and is used in the production of phosgene from carbon monoxide and chlorine (103,104).

BIBLIOGRAPHY

1. H. Jüntgen, *Carbon*, **15**, 273–283 (1977).
2. Brit. Pat. 14,224 (1900), R. von Ostrejko; Fr. Pat. 304,867 (1900); Ger. Pat. 136,792 (1901); U.S. Pat. 739,104 (1903).
3. J.W. Hassler, *Forest Products J.* **8**, 25A–27A (1958).
4. J.W. Hassler, *Activated Carbon*, Chemical Publishing Co., Inc., New York, 1963, pp. 1–14. A comprehensive account of the development and use of activated carbon products to about 1960.
5. R.V. Carrubba, J.E. Urbanic, N.J. Wagner, and R.H. Zanitsch, *AIChE Symp. Ser.* **80**, 76–83 (1984).
6. B. McEnaney and T.J. Mays, in H. Marsh ed., *Introduction to Carbon Science*, Butterworths, London, 1989, pp. 153–196. A good introduction to carbon science in general.

7. H. Marsh and J. Butler, in K.K. Unger, J. Rouquerol, K.S.W. Sing, and H. Kral eds., *Characterization of Porous Solids, Proceedings of the IUPAC Symposium (COPS I)*, Bad Soden a.Ts., FRG, Apr. 26–29, 1987, Elsevier, Amsterdam, The Netherlands, 1988, pp. 139–149.
8. K.S.W. Sing et al., *Pure Appl. Chem.* **57**, 603–619 (1985).
9. S. Brunauer, P.H. Emmett, and E. Teller, *J. Am. Chem. Soc.* **60**, 309–319 (1938).
10. S.J. Gregg and K.S.W. Sing, *Adsorption, Surface Area, and Porosity*, 2nd ed., Academic Press, London, 1982. An indispensable text on the interpretation and significance of adsorption data.
11. H. Marsh, D. Crawford, T.M. O'Grady, and A. Wennerberg, *Carbon* **20**, 419–426 (1982).
12. *Jpn. Chem. Week* **30**, 5 (Mar. 16, 1989).
13. M.M. Dubinin, *J. Colloid Interface Sci.* **46**, 351–356 (1974).
14. K.S.W. Sing, *Carbon* **27**, 5–11 (1989).
15. J. Zawadzki, in P.A. Thrower ed., *Chemistry and Physics of Carbon*, vol. 21, Marcel Dekker, New York, 1989, pp. 147–380. *Chemistry and Physics of Carbon*, published in 23 volumes through 1991, is a primary source of excellent review articles on carbon, many relevant to activated carbon.
16. D. Atkinson, A.I. McLeod, K.S.W. Sing, and A. Capon, *Carbon* **20**, 339–343 (1982).
17. J.M. Solar, C.A. Leon y Leon, K. Osseo-Asare, and L.R. Radovic, *Carbon* **28**, 369–375 (1990).
18. K.D. Henning, W. Bongartz, and J. Degel, *19th Biennial Conference on Carbon*, Penn State University, Pa., June 25–30, 1990, extended abstracts, pp. 94–95.
19. Product data bulletins from activated carbon manufacturers, Calgon Carbon Corp., 1990, American Norit Co., 1990, and Westvaco Corp., 1988.
20. T. Wigmans, *Carbon* **27**, 13–22 (1989).
21. F. Derbyshire and M. Thwaites, *Proc. 4th Australian Coal Science Conf.*, Brisbane, Australia, Dec. 3–5, 1990, pp. 372–379.
22. U.S. Pat. 3,976,597 (Aug. 24, 1976), A.J. Repik, C.E. Miller, and H.R. Johnson (to Westvaco Corp.).
23. W. Gerhartz, Y.S. Yamamoto, and F. Thomas Campbell eds., *Ullmann's Encyclopedia of Industrial Chemistry*, 5th ed., vol. A5, VCH, New York, 1986, pp. 124–140. Good descriptions of activation processes.
24. Product literature on Pittsburgh activated carbon, Pittsburgh Coke & Chemical Co. (now Calgon Carbon Corp.), Pittsburgh, Pa., ca 1960.
25. U.S. Pat. 4,014,817 (Mar. 29, 1977), B.C. Johnson, R.K. Sinha, and J.E. Urbanic (to Calgon Corp.).
26. A. Cameron and J.D. MacDowall, in J.M. Haynes and P. Rossi-Doria eds., *Principles and Applications of Pore Structural Characterization, Proceedings of the RILEM / CNR International Symposium*, Milan, Italy, Apr. 26–29, 1983, J.W. Arrowsmith, Ltd., Bristol, U.K., 1985, pp. 251–275.
27. R.C. Bansal, J.-B. Donnet, and F. Stoeckli, *Active Carbon*, Marcel Dekker, New York, 1988, p. 8. A modern treatise on activated carbon based on a comprehensive review of the literature.
28. U.S. Pat. 4,082,694 (Apr. 4, 1978), A.N. Wennerberg and T.M. O'Grady (to Standard Oil Co.).
29. T. Kasuh, D.A. Scott, and M. Mori, *Proceedings of an International Conference on Carbon*, The University of Newcastle upon Tyne, U.K., Sept. 18–23, 1988, pp. 146–148.
30. Product literature on activated carbon cloth, Charcoal Cloth Ltd., U.K., 1985, and on C-tex products, Siebe Gorman & Co., Ltd., U.K., 1985.
31. Product literature on KYNOL activated carbon fibers and cloths, GUN EI Chemical Industry Co., Ltd., Japan, 1987; product literature on AD'ALL activated carbon fibers, Unitika, Ltd., Japan, 1989.
32. Product literature on KURASHEET activated carbon foam sheets, Kuraray Chemical Co., Ltd., Japan, 1987.
33. *AWWA Standard for Granular Activated Carbon*, ANSI/AWWA B604, American Water Works Association, Denver, Colo., 1991.
34. *AWWA Standard for Powdered Activated Carbon*, ANSI/AWWA B600, American Water Works Association, Denver, Colo., 1990.
35. *The Economics of Activated Carbon*, 3rd ed., Roskill Information Services, London, 1990.
36. J. Goin, V. von Schuller-Goetzburg, and Y. Sakuma, *Chemical Economics Handbook—SRI International*, Menlo Park, Calif., 1989.
37. Wertheim Schroder, *Calgon Carbon Corp.*, Company Report No. 955346, Mar. 19, 1990.
38. S. Irving-Monshaw, *Chem. Eng.* **97(2)**, 43–46 (1990).
39. W.G.P. Schuliger, *Waterworld News* **4(1)**, 15–17 (1988).
40. R.M. Clark and B.W. Lykins, Jr., *Granular Activated Carbon—Design, Operation, and Cost*, Lewis Publishers, Chelsea, Mich., 1989, pp. 295–338.
41. P.N. Cheremisinoff and F. Ellerbusch, *Carbon Adsorption Handbook*, Ann Arbor Science Publishers, Ann Arbor, Mich., 1978, pp. 539–626. An excellent reference book on activated carbon, ranging from theoretical to applied aspects.
42. Product literature on PACT systems, Zimpro/Passavant, Inc., Rothschild, Wis., 1990.
43. W.E. Koffskey and B.W. Lykins, Jr., *J. Am. Water Works Assoc.* **82(1)**, 48–56 (1990).
44. A. Yehaskel, *Activated Carbon—Manufacture and Regeneration*, Noyes Data Corporation, Park Ridge, N.J., 1978, pp. 202–217. A dated, but still useful summary of key patent literature.
45. P.N. Cheremisinoff, *Pollut. Eng.* **17(3)**, 29–38 (1985).
46. R.G. Rice and C.M. Robson, *Biological Activated Carbon—Enhanced Aerobic Biological Activity in GAC Systems*, Ann Arbor Science Publishers, Ann Arbor, Mich., 1982, 611 pp.
47. *Department of Defense Military Specifications*, MIL-C-0013724D(EA), Sept. 22, 1983; MIL-C-0013724D(EA) Amendment 1, Mar. 5, 1986; and MIL-C-17605C(SH), Mar. 22, 1989.
48. S.J. Gregg and K.S.W. Sing, *Adsorption, Surface Area, and Porosity*, 1st ed., Academic Press (London) London, 1967, pp. 308–355.
49. H.M. Rootare, *Advanced Experimental Techniques in Powder Metallurgy*, Plenum, New York, 1970, pp. 225–252. A comprehensive review of the use of mercury penetration to measure porosity.
50. G. Horvath and K. Kawazoe, *J. Chem. Eng. Jpn.* **16**, 470–475 (1983).
51. D. Dollimore and G.R. Heal, *J. Appl. Chem.* **14**, 109–114 (1964).
52. M.M. Dubinin and H.F. Stoeckli, *J. Colloid Interface Sci.* **56**, 34–42 (1980).
53. C. Orr, Jr., *Powder Technol.* **3**, 117–123 (1970).

54. *Annual Book of ASTM Standards*, 15.01, Section 15, American Society for Testing and Materials, Philadelphia, Pa., 1989.
55. *1985–1986 Registry of Toxic Effects of Chemical Substances*, vol. 2, National Institute for Occupational Safety and Health, U.S. Department of Health and Human Services, Washington, D.C., 1987, p. 1475.
56. National Research Council, Assembly of Life Sciences, Division of Biological Sciences, Food and Nutrition Board, and Committee on Codex Specifications, *Food Chemicals Codex*, 3rd ed., National Academy Press, Washington, D.C., 1981, pp. 70, 71.
57. American Society of Civil Engineers, American Water Works Association and Conference of State Sanitary Engineers, *Water Treatment Plant Design*, American Water Works Association, New York, 1969, p. 297.
58. J.W. Hassler, *Purification with Activated Carbon*, 3rd ed., Chemical Publishing, New York, 1974. Contains much of the information given in Ref. 4 but with more emphasis on the commercial uses of activated carbon.
59. U.S. Environmental Protection Agency, *Process Design Manual for Carbon Adsorption*, Swindell-Dressler, Pittsburgh, Pa., 1971, pp. 3–68.
60. Material safety data sheets on activated carbon products, available from the manufacturers, 1991.
61. R.A. Hutchins, *Chem. Eng.* **87(2)**, 101–110 (1980). A particularly useful paper on liquid-phase adsorption.
62. M. Suzuki, *Adsorption Engineering*, Kodansha Ltd., Tokyo and Elsevier Science Publishers B.V., Amsterdam, The Netherlands, 1990, pp. 11, 35–62.
63. T.F. Speth and R.J. Miltner, *J. Am. Water Works Assoc.* **82(2)**, 72–75 (1990).
64. F.L. Slejko ed., *Adsorption Technology*, Marcel Dekker, New York, 1985, pp. 23–32. A good account of the theory, design, and application of adsorption systems.
65. American Water Works Association, *Water Quality and Treatment*, 3rd ed., McGraw-Hill, New York, 1971, pp. 1–216.
66. W.J. Weber, Jr. and B.M. Van Vliet, in I.H. Suffet and M.J. McGuire eds., *Activated Carbon Adsorption of Organics from the Aqueous Phase*, vol. 1, Ann Arbor Science Publishers, Ann Arbor, Mich., 1980, pp. 15–41. A comprehensive, two-volume treatise with many key references.
67. J.L. Oxenford and B.W. Lykins, Jr., *J. Am. Water Works Assoc.* **83(1)**, 58–64 (1991).
68. I.N. Najm and co-workers, *J. Am. Water Works Assoc.* **83(1)**, 65–76 (1991).
69. L.W. Canter and R.C. Knox, *Ground Water Pollution Control*, Lewis Publishers, Chelsea, Mich., 1985, pp. 89–125.
70. Environmental Science and Engineering, Inc., *Removal of Volatile Organic Chemicals from Potable Water—Technologies and Costs*, Noyes Data Corp., Park Ridge, N.J., 1986, pp. 23–40.
71. G. Culp, G. Wesner, R. Williams, and M.V. Hughes, *Wastewater Reuse and Recycling Technology*, Noyes Data Corp., Park Ridge, N.J., 1980, pp. 343–432. A useful review of wastewater treatment with activated carbon.
72. M. Smisek and S. Cerny, *Active Carbon—Manufacture, Properties, and Applications*, Elsevier Publishing, New York, 1970, pp. 290–294.
73. A.J. Bird, in A.B. Stiles ed., *Catalyst Supports and Supported Catalysts*, Butterworths, Stoneham, Mass., 1987, pp. 107–137.
74. pp. 731.2000W, 731.2001M-731.2001P.
75. P.J. Luft and P.C. Speers, Paper 52c, *AIChE Summer National Mtg.*, Aug. 19–22, 1990.
76. P.J. Clarke and co-workers, *SAE Trans.* **76**, 824–837 (1967).
77. Product literature on hydrocarbon vapor recovery systems, John Zink Co., Tulsa, Okla., 1990.
78. J. Hill, *Chem. Eng.* **97**, 133–143 (1990).
79. H.R. Johnson and R.S. Williams, S.A.E. Technical Paper No. 902119, International Fuels and Lubricants Exposition, Tulsa, Okla., Oct. 23, 1990.
80. D.W. Moeller and D.W. Underhill, *Nucl. Saf.* **22**, 599–611 (1981).
81. M.L. Hyder, *Comm. Eur. Communities [Rep.] EUR 1986, EUR 10580, Gaseous Effluent Treat. Nucl. Install.*, pp. 451–462 (1986).
82. F.J. Ball, S.L. Torrence, and A.J. Repik, *APCA J.* **22**, 20–26 (1972).
83. P. Ellwood, *Chem. Eng.* **76**, 62–64 (1969).
84. J. Klein and K.-D. Henning, *Fuel* **63**, 1064–1067 (1984).
85. E. Richter, *Catal. Today* **7**, 93–112 (1990).
86. M.A. Brisk and A. Turk, *Proc. APCA Ann. Meetg., 77th* **2**, 84–93 (1984).
87. U.S. Pat. 2,920,050 (Jan. 5, 1960), R.J. Grabenstetter and F.E. Blacet (to U.S. Dept. of Army).
88. U.S. Pat. 4,801,311 (Jan. 31, 1989), E.-D. Tolles (to Westvaco Corp.).
89. U.S. Pat. 3,460,543 (Aug. 12, 1969), C.H. Kieth, V. Norman, and W.W. Bates, Jr. (to Ligget & Meyers Corp.).
90. R.K. Sinah and P.L. Walker, *Carbon* **10**, 754–756 (1972).
91. W.D. Lovett and R.L. Poltorak, *Water and Sewage Works* **121**, 74–75 (1974).
92. G.N. Brown, M.A. Lunn, C.E. Miller, and C.D. Shelor, *Tappi J.* **66**, 33–36 (1983).
93. S. Dunlop and R. Banks, *Hydrocarbon Process.* **56**, 147–152 (1977).
94. G.F. Russell, *Petrol. Refiner.* **40**, 103–106 (1961).
95. T. Scott, *Gas. J.* **303**, 300–307 (1960).
96. E. Richter, *Erdol Kohle, Erdgas, Petrochem.* **40**, 432–438 (1987).
97. C. Berg, *Chem. Eng. Prog.* **47**, 585–590 (1951).
98. *Gastak Solvent Recovery System*, product literature, Kureha Chemical Industry Co., Ltd., New York, 1990.
99. H. Briggs and W. Cooper, *Proc. R. Soc. Edinburgh* **41**, 119–127 (1920–1921).
100. J. Braslaw, J. Nasea, and A. Golovoy, *Alternative Energy Sources: Proceedings of the Miami Int. Conf. on Alternative Energy Sources*, 4th ed., Ann Arbor Science Publishers, Ann Arbor, Mich., pp. 261–270, 1980.
101. U.S. Pat. 4,817,684 (Apr. 4, 1989), J.W. Turko and K.S. Czerwinski (to Michigan Consolidated Gas Co.).
102. U.S. Pat. 4,102,987 (July 25, 1978), D.M. Cook and D.C. Gustafson (to The Dow Chemical Company).
103. H. Jüntgen, *Fuel* **65**, 1436–1446 (1986).
104. H. Jüntgen, *Erdol Kohle, Erdgas, Petrochem.* **39(12)**, 546–551 (1986).

ADDITIONAL READING

"Active Carbon" under "Carbon" in *Encyclopedia of Chemical Technology (ECT)* 1st ed., Vol. 2, pp. 881–899, by J.W. Hassler, Nuchar Active Carbon Division, West Virginia Pulp and Paper Co., and J.W. Goetz, Carbide and Carbon Chemicals Corp.

"Activated Carbon" under "Carbon" in *ECT* 2nd ed., Vol. 4, pp. 149–158, by E.G. Doying, Union Carbide Corp., Carbon Products Division.
"Activated Carbon" under "Carbon (Carbon and Artificial Graphite)" in *ECT* 3rd ed., Vol. 4, pp. 561–570, by R.W. Soffel, Union Carbide Corp.

ADSORBENTS, INORGANIC

KAMI BEYZAVI
Bioprocessing Limited
Consett, County Durham, United Kingdom

KEY WORDS

Alumina
Chromatography
CPG (controlled pore glass)
Hydroxyapatite
Rigid matrices
Separation
Silica
Titania
Zirconia

OUTLINE

Introduction
Choice of Matrices for Bioprocess Separations
Inorganic Matrices
- Controlled Pore Glass
- Silica
- Alumina
- Zirconia and Titania
- Hydroxyapatite

Controlled Pore Glass
- Manufacturing Process
- Physical Characteristics
- Surface Modification
- Applications

Future Trends
List of Suppliers
Bibliography

INTRODUCTION

The design of a process for the large-scale purification of biological molecules is one of the several complex tasks that must be successfully completed to bring a product to market. Most downstream processes include some chromatography steps as well as other conventional separation methods. Chromatography is generally recognized as the technique that enables the highest degree of purification of biomolecules. For example, the U.S. Food and Drug Administration (FDA) recommends that at least two chromatography steps be used in the purification of therapeutic-grade monoclonal antibodies.

Chromatography separations are sophisticated and can be expensive processes, which explains the continuous development of effective approaches by scientists and engineers to improve the productivity of chromatography technologies. Fluidized bed adsorption, various column technologies, computerized optimization strategies, and continuous and semicontinuous processes are some of the recent approaches that have been proposed. However, one of the most important aspects of chromatography technology is the choice of matrix.

In this article, we examine the use of inorganic rigid matrices in chromatography separation processes of biological molecules, with emphasis on the preparation, properties, and applications of controlled pore glass (CPG).

CHOICE OF MATRICES FOR BIOPROCESS SEPARATIONS

The subject of matrices for process chromatography and the requirements of an ideal matrix have been reviewed by many authors (1–3). In process-scale purification, the important characteristics of a matrix are chemical and physical stability, rigidity of the matrix to withstand high flow rates and differing hydrodynamic pressure, uniform porosity, and high surface area. In addition, for most chromatography applications, surface modification of the matrix will be necessary to facilitate ligand immobilization, which requires the presence of suitable functional groups on the matrix. Finally, the cost and commercial availability of the matrix are also important considerations in selecting a matrix.

The majority of matrices used in chromatography processes can be divided into two categories: natural and synthetic. Natural matrices include glass, silica, ceramic, alumina, agarose, chitosan, and cellulose. Synthetic matrices are prepared by polymerization of functional monomers. The incorporation of monomers with suitable functional groups can provide activation sites on these supports for ligand attachments. There are also the composite matrices, which are mixed networks constituted of at least two components. Generally, one component acts as a rigid skeleton, and the other component has active groups that interact with the biological molecule to be separated. Dextran-coated silica and polyacrylamide-agarose are examples of composite matrices that have been applied to the separation of biological molecules.

INORGANIC MATRICES

A large selection of inorganic matrices are commercially available for incorporating homogenous catalysts, and some have been applied to chromatography separations. These matrices include silica, alumina, glass, zeolite, clay, celite, zerconia, titania, magnesium oxide, and carbon. Only a limited number of these matrices with suitable pore structure, particle size, and mechanical stability can be used for chromatography separation of biomolecules. Silica, alumina, and glass are available with a large variety of surface areas, pore structure, and particle sizes, and some have been used extensively in both analytical and preparative chromatography of biological molecules.

Controlled Pore Glass

Controlled pore glass (CPG) has excellent mechanical stability and can be manufactured with a wide range of pore diameters, a high surface area, and a range of particle sizes. The material is almost pure silica and has a very narrow pore size distribution. Surface-modified CPG with the trade name PROSEP manufactured by Bioprocessing Limited has been extensively used in large-scale purification of biomolecules. (See later sections on CPG for a more detailed discussion of this material.)

Silica

Silica is extensively used in the analytical field (HPLC) and to a much lesser extent for preparative chromatography of biomolecules. Silica matrices for chromatography applications are generally produced by a wet process, that is, by acidification of sodium silicate in water (4,5), and are available in particle sizes of up to 40 μm and pore sizes of up to 30 nm. In most cases, these materials are unsuitable for large-scale purification processes. The small particle size causes excessive backpressure in preparative chromatography columns, and the pore size of 30 nm is not large enough for the separation of most proteins, especially in the field of affinity chromatography, where both the ligand and the ligate could be large proteins.

In our laboratories, we have used Matrex™ silica (Amicon) for the large-scale purification of biomolecules. This material is available in a range of pore diameters and particle sizes (Table 1) and is manufactured by first forming a hydrosol from organosilicic acid. Hydrosol is then allowed to set to a jelly-like mass called hydrogel. After setting, the hydrogel is thoroughly washed to remove salts resulting from the reaction. Temperature and pH of the washing medium and the washing time influence the specific surface area, pore volume, and purity of the hydrogel. The hydrogel is subsequently dried to give a structurally stable xerogel. This material is 99.5% silica, with impurities including Na^+, Ca^{2+}, Fe^{3+}, SO_4^{2-}, and Cl^- ranging from 600 to 60 ppm.

Alumina

Alumina-based materials have a much wider pH stability than silica matrices and therefore are of great interest in chemical separations and industrial-scale chromatography. Commercially available alumina matrices are all aluminium oxides prepared by thermal dehydration of aluminum hydroxide (6,7). The origin of the starting material together with other factors, such as the heating rate, drying time, and temperature, influence the properties of the final product.

Alumina matrices can be obtained in a wide range of particles and pore sizes (Table 2). The physical stability of alumina particles in chromatography columns, even under moderate flow rates and pressure, has been a major problem in our laboratories. Generally, the particles break up and cause a substantial increase in the backpressure generated by the column. Furthermore, any risk of contamination of the purified product with aluminum may not be acceptable to the pharmaceutical and biotechnology industries, particularly when the reports relating to Alzheimer's disease to aluminum are considered (8).

Zirconia and Titania

Recently, there has been an increasing interest in zirconia- and titania-based matrices in chromatography applications, mainly because of their chemical stability over a relatively wide pH range and their physical characteristics. Zirconia or titania supports with properties suitable for preparative chromatography or HPLC are not commercially available, although zirconia and titania powders can be obtained commercially from a number of sources and have been used by researchers to prepare matrices with a small particle size (1–10 μm) and pore sizes of up to 100 nm, suitable for HPLC applications (9). More recently, larger porous zirconia particles (30–150 μm) have been prepared and used in packed and fluidized beds for protein purification (10–13).

Zirconia and titania as base stable matrices not only enable separation of biomolecules at high pH values, but they can also be sanitized with alkaline media, which is the standard sanitization procedure in downstream processing of biological molecules. Various adsorbents such as affinity, ion-exchange, and reversed phase can be prepared from these matrices for protein purification by either polymer coating or reaction with organosilanes (14–17).

Table 1. Properties of Matrex™ Silica

Nominal pore diameter (nm)	Surface area (m^2/g)	Pore volume (ml/g)	Particle size range (μm)
6	540	0.93	20–500
10	320	1.20	45–300
25	160	1.10	45–70
50	90	1.10	50
100	45	1.10	50–130
150	25	1.10	50

Table 2. Properties of Alumina Matrix (Selecto Scientific)

Nominal pore diameter (nm)	Surface area (m^2/g)	Particle size range (μm)
6	120–150	60–120
10	100	60–120
15	75	60–120
30	80	40
50	50	40
100	8	40
180	6–8	20

Hydroxyapatite

Hydroxyapatite is a special crystalline form of calcium phosphate that has been used in the separation of biological molecules. The adsorbent consists of a mosaic of crystals that result from a combination of phosphate and calcium salts after a number of steps of wet alkaline treatment and heat aging. The application of hydroxyapatite to the field of protein separation was first introduced by Tiselius et al. (18). The standard procedures for the operation of hydroxyapatite columns were developed by Bernardi and Gorbunoff (19,20) using a number of different proteins in several solvent systems, Bernardi also proposed a mechanism for protein adsorption to, and desorption from, hydroxyapatite.

Proteins bind reversibly to hydroxyapatite by a complex ionic mechanism different from the principle of ion-exchange chromatography (19,20). Both the amino and carboxyl groups on the protein molecules act in the adsorption mechanism. The adsorption of proteins to hydroxyapatite via the amino groups is the result of their positive charges and the negative charge on the adsorbent when the column is equilibrated in phosphate buffer. The surface of hydroxyapatite crystals consists of sites with both positive (calcium) and negative (phosphate) charges. When the hydroxyapatite columns are equilibrated in phosphate buffers, normally at pH 6.8, the surface of the adsorbent can be regarded as negative because of partial neutralization of the positive calcium loci by phosphate ions. The carboxyl groups act in two ways. First, they are repelled electrostatically from the negative charge of the adsorbent; second, they bind specifically by complex formation with the calcium sites.

The elution of basic proteins from hydroxyapatite is generally carried out by specific displacement with Ca^{2+} and Mg^{2+} ions that neutralize the negative charges of the hydroxyapatite phosphate sites by complex formation. Acidic proteins are eluted by displacement of their carboxyl groups from hydroxyapatite calcium sites by ions, such as fluoride or phosphate, which form stronger complexes.

The difficulty with using hydroxyapatite in medium- and large-scale purification of biological molecules is the physical nature of the material. The crystals are highly fragile, generating small particles resulting in column clogging and precluding any long-term usage of the column. An improved form of hydroxyapatite crystals has recently been prepared and is commercially available from a number of sources, including Bio-Rad. The new ceramic material from Bio-Rad is spherical and has been reported to be stable both chemically and physically.

CONTROLLED PORE GLASS

The method used for the preparation of CPG was originally invented by Hood and Nordberg (21). They discovered that by heat treating a certain sodium borosilicate glass composition, the glass separated into two intermingled and continuous glassy phases. One phase, rich in boric oxide, was soluble in acids; the other phase was high in silica and stable toward acid solutions. This process was used by the inventors to prepare nonporous silica particles by first removing the boron phase with an acid solution to obtain porous particles and then revitrifying this material by heating to 900 °C or above to yield transparent homogenous nonporous particles having a composition of greater than 95% silica.

The porous glass material obtained by this method typically had an average pore diameter of about 4 to 5 nm. A procedure for enlarging the pore size to approximately 20 nm was invented by Chapman and Elmer (22). This was carried out by impregnating a porous glass body with an aqueous solution of a weakly reactive fluorine-containing compound, reacting the compound in situ with a mineral acid to release hydrofluoric acid at a temperature sufficient to dissolve a portion of the glass body and washing the body to remove the soluble constituents. Ammonium bifluoride and nitric acid at 95 °C were applied to a porous glass disc to enlarge the pore size from 4 to 24 nm.

The process for making porous glass was modified by Haller (23) to produce the material with a range of controllable pore sizes. This was achieved by varying the temperature and duration of the heat treatment to obtain porous glass with pore sizes from 12 to 250 nm. Larger pore sizes were also obtained by the addition of a sodium hydroxide leaching process.

Manufacturing Process

CPG can be commercially obtained from a number of sources, and the manufacturing processes are all based on the process invented by Haller (23). The composition of the starting metal oxides mixture and the exact conditions used for the heat treatment and leaching of the glass, which influence the quality and physical properties of the porous glass, are the propriety technology of the manufacturing companies. An outline of the general procedure used for preparing CPG is shown in Figure 1.

CPG is manufactured in a range of particle sizes, 40 to 200 μm, and different pore sizes, 7 to 300 nm (Table 3). The manufacturing batch sizes are generally much smaller than silica particles, possibly because of the heat treatment process. The largest batch size commercially available is in the region of 50 L (20 kg).

Physical Characteristics

Most commercially available CPG granules are irregularly shaped and highly porous (Figs. 2 and 3). CPG has a narrow pore size distribution, with 80 to 90% of the pores showing a deviation less than $\pm 10\%$ from the nominal pore diameter (Fig. 4). CPG has a very high surface area that varies according to the pore diameter and pore volume. The surface area increases with increasing pore volume and decreases with increasing pore diameter, as is shown by the data in Table 3. CPG is a rigid insoluble matrix, compatible with most reagents used in the purification of biological molecules. Because CPG is essentially pure silica, it generally exhibits a high degree of solubility in alkali solutions. The solubility of CPG is a function of temperature, time, and pH of the solution.

The surface of native CPG is slightly hydrophobic and contains acidic hydroxyl groups that could denature bio-

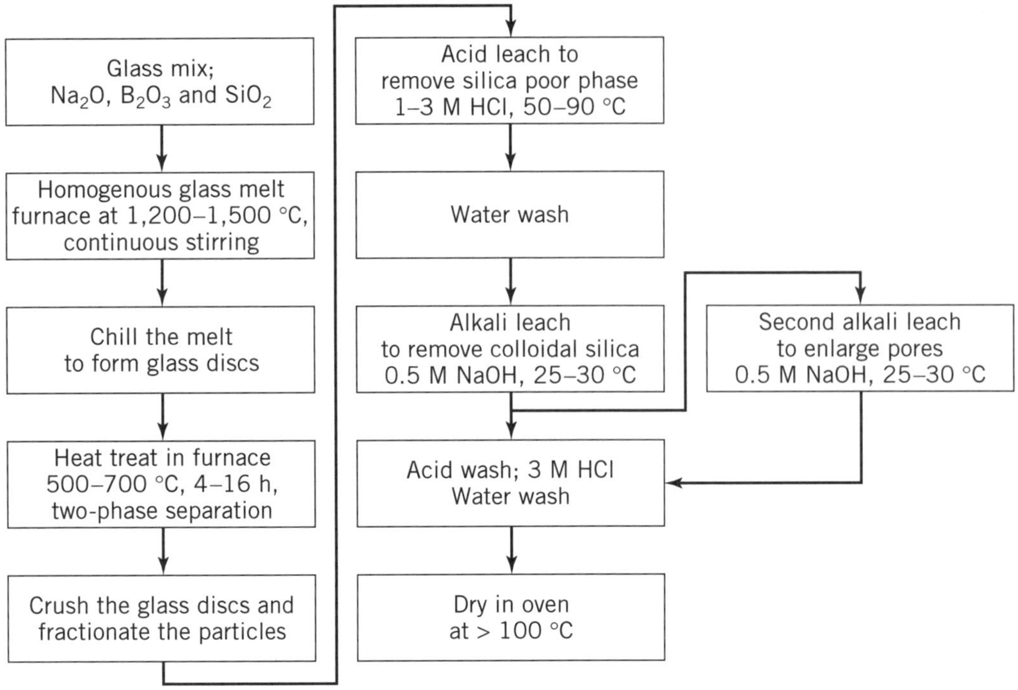

Figure 1. Process for the preparation of porous glass.

Table 3. Properties of Controlled Pore Glass (CPG, Inc.)

Nominal pore diameter (nm)	Surface area (m^2/g); pore volume = 1.0 cm^3/g	Surface area (m^2/g); pore volume = 0.9 cm^3/g
7.5	340	240
12	210	150
17	150	100
24	110	75
35	75	50
50	50	35
70	36	25
100	25	18
140	18	13
200	13	9
300	9	7

logical molecules and cause strong, often irreversible binding of proteins. For most separation applications, surface modification of CPG with organosilanes is necessary, although native CPG has been used for the purification of a number of biological molecules, such as factor VIII (24). Surface modification also enhances the stability and durability of CPG.

Surface Modification

Siliceous materials such as CPG can be reacted with the organosilane compounds in a number of different ways to obtain the desired bonding through hydrolyzable groups of the organosilane. Depending on the type of organosilane, the reaction can be performed in organic, aqueous, or vapour phase, resulting in matrices suitable for different applications. The majority of the work cited on this subject relates to the surface modification of silica material (25), which is not always applicable to CPG because of the difference in the surface characteristics of the two matrices.

Surface modification of CPG is carried out by treating the material with an organosilane containing an organic functional group at one end and a silylalkoxy group at the other. Attachment of the silane compound to the matrix is through the surface silanol or oxide groups to the silylakoxy groups with polymerization between the adjacent silanes. The result is an inorganic matrix with available organic functional groups that can be used in the preparation of chromatography adsorbents. A large number of different organosilanes are commercially available with functional groups such as epoxy, amine, and sulfhydryl (Table 4).

The first step in the surface modification of porous glass is to clean the surface by washing the matrix in nitric acid at elevated temperature. The matrix is then washed with water and dried in the oven at temperatures not exceeding 180 °C. The dry matrix is reacted with the silane compound dissolved in water or an organic solvent at temperatures ranging from 40 °C to 120 °C, depending on the solvent. After removing the excess silane by washing the matrix with suitable solvents or water, the functional groups on the matrix can be further modified to obtain various adsorbents. The exact conditions used in the surface modification of CPG will depend on the nature of the silane compound and the application of the resulting adsorbent.

Extensive optimization of the surface modification procedure is usually required to obtain a suitable adsorbent for a specific application. At Bioprocessing, special tech-

Figure 2. The irregularly shaped granules of CPG matrix.

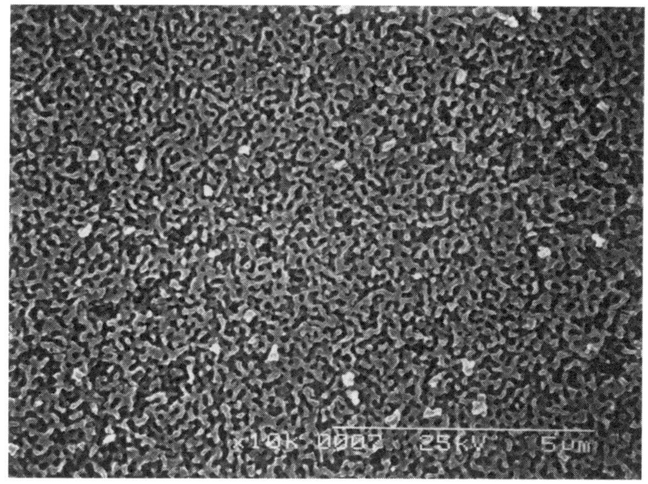

Figure 3. Scanning electron micrograph of CPG matrix.

niques for preparing surface modified CPG (PROSEP®) have been developed for applications in affinity and other types of separation processes. PROSEP® has been designed for effective immobilization of both large and small ligands, resulting in affinity adsorbents with a high capacity for the target molecules, minimum nonspecific binding of contaminating proteins, and low ligand leakage (26).

Applications

CPG has been effective in the separation of a range of biological molecules, using a variety of chromatography techniques that include adsorption, size exclusion, affinity, and ion exchange. It has also been used extensively in the field of enzyme immobilization for the modification of biological molecules. The rigidity and incompressibility of the matrix allows it to be used in large columns at very high flow rates generating low backpressure (Fig. 5), which makes this matrix especially useful for processing large volumes of feedstock in short cycle times.

Application of native CPG in adsorption chromatography has been generally limited because of its denaturing surface characteristics and irreversible binding of proteins. A number of proteins, such as factor VIII (24), have been purified from plasma by first blocking the surface of CPG with albumin. The use of CPG in ion exchange chromatography applications has also been very limited because of the instability of the matrix in sodium hydroxide solution, which is the standard cleaning and sanitization

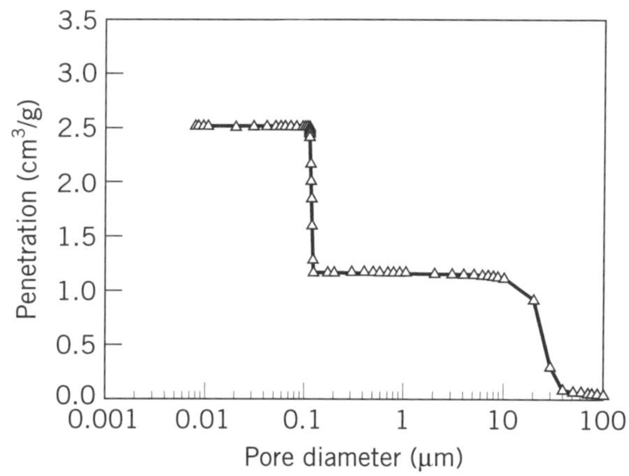

Figure 4. Pore size distribution of CPG matrix as determined by mercury porosimetery method. Average pore diameter was determined to be 0.1125 μ.

solution for ion exchange adsorbents. Furthermore, ion exchange adsorbents prepared with CPG as the matrix have generally shown low functional capacity when compared to the polymetric matrices. Nevertheless, some applications of surface-modified CPG in ion exchange chromatography have been reported (27–29).

The use of CPG in size-exclusion chromatography has generally been for the separation of macromolecules such

Table 4. Surface Modification of CPG

Organosilane	Reactant	Activated CPG
3-Aminopropyltriethoxysilane	—	CPG-amine
3-Aminopropyltriethoxysilane	Glutaraldehyde	CPG-aldehyde
3-Aminopropyltriethoxysilane	Succinic anhydride	CPG-carboxyl
3-Glycidoxypropyltrimethoxysilane	—	CPG-epoxy
3-Glycidoxypropyltrimethoxysilane	Acid solution	CPG-diol
3-Glycidoxypropyltrimethoxysilane	Sodium periodate	CPG-aldehyde
Mercaptopropyltrimethoxysilane	—	CPG-sulfyhyryl

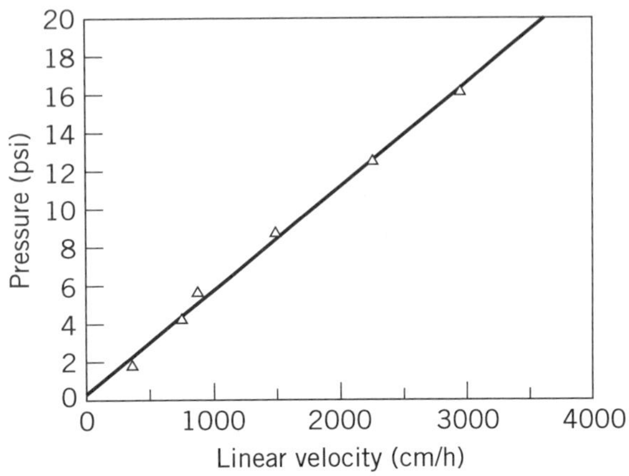

Figure 5. Backpressure generated from a column (11.3 × 10.0 cm) packed with PROSEP®.

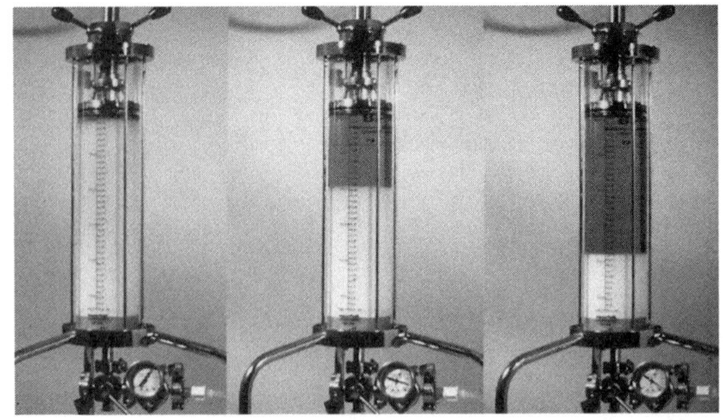

Figure 6. Principle of fluidized-bed adsorption. From left to right: PROSEP® adsorbent is packed into a column specially adopted for fluidized-bed operation. The bed is gradually expanded by an upward flow of solution through the column until full expansion is reached.

as viruses, cell components, and polymers (30,31). Separation of these species from low molecular weight contaminants has been achieved by taking advantage of the very sharp exclusion limits and narrow pore size distribution of CPG. Generally, surface-modified CPG has been used for size-exclusion chromatography to eliminate the problem of nonspecific and irreversible binding of biological molecules to native CPG. The surface of the matrix is modified with glycidoxypropyltrimethoxysilane, converting the solid phase to diol-CPG (Table 4). This is a nonionic, hydrophilic coating compatible with aqueous and most solvent systems (32).

Surface-modified CPG has been widely applied in the area of affinity chromatography for the purification of biological molecules. Biologically active molecules will often bind to other molecules in a reversible and highly specific manner. The formation of these stable, specific, and dissociable complexes forms the basis of this powerful separation technique.

Affinity adsorbents prepared using PROSEP® and other surface-modified CPG have been used in many applications for the purification of biological molecules, such as antibodies (33,34) and enzymes (35). PROSEP® adsorbents have been used in packed-bed chromatography columns and in fluidized or expanded bed mode (Fig. 6) for the direct capture of biological molecules from feedstocks containing particulate or colloidal solids (36,37). Table 5 shows the application of PROSEP® adsorbents in the field of affinity purification.

FUTURE TRENDS

The progress of inorganic matrices such as silica and CPG as adsorbents in the downstream processing of biological molecules is hampered by their instability toward alkaline solutions. Sodium hydroxide solution up to 0.5 M concentration is the standard cleaning and sanitizing reagent for chromatography adsorbents used in the separation processes of biomolecules. Furthermore, certain chemistries used in the surface modification of matrices and some purification protocols require alkaline conditions.

Alkaline-resistant inorganic matrices are under development using different approaches. Development of zirconia and titania matrices by sol-gel techniques have already been mentioned. Surface coating of silica and CPG with an inert metal oxide such as zirconium oxide and preparation of glass–ceramic matrices may result in suitable products that satisfy all the requirements of an ideal matrix for the purification of biomolecules.

Methods for coating silica matrices with zirconium oxide were studied by Meijers et al. (38) and applied to CPG at Bioprocessing. Great improvement in the stability of the

Table 5. Application of Surface-Modified CPG (PROSEP®) in Separation Processes

Product	Trade name	Application
Protein A adsorbent	PROSEP®-A High Capacity	Purification of IgG from all species and subclasses except rat IgG
Protein G adsorbent	PROSEP®-G	Purification of IgG from all species and subclasses
Heparin adsorbent	PROSEP®-Heparin	Purification of antithrombin III and blood clothing factors
Lysine adsorbent	PROSEP®-Lysine	Purification of plasminogen and tPA
Metal-chelating adsorbent	PROSEP®-Chelating	A metal-binding adsorbent used in purification of metal-binding proteins and metal removals from dilute solutions
Thiophilic adsorbent	PROSEP®-Thiosorb	Purification of antibodies
Immobilized antibodies	—	Purification of specific antigens and second antibodies

Note: IgG, immunoglobulin G; tPA, tissue-type plasminogen activator.

coated CPG to alkaline solutions were observed; however, total stability has not yet been achieved.

A preferred approach for making alkaline-stable inorganic matrices is the glass–ceramic route. Porous glass–ceramic with a skeleton of $CaTi_4(PO_4)_6$ crystals has been prepared by controlled crystallization of glasses in the $CaO\text{-}TiO_2\text{-}P_2O_5\text{-}Na_2O$ system and subsequent acid leaching of the resulting dense glass–ceramic (39). The medium pore diameter has been reported to be controllable in the range of 40 to 150 nm by changing the temperature of the heat treatment. Such an approach is being examined at Bioprocessing in collaboration with a glass–ceramic research center for developing alkaline stable inorganic matrices suitable for downstream processing of biological molecules.

LIST OF SUPPLIERS

Bioprocessing Limited, Medomsley Road, Consett, Co Durham, DH8 6SZ, UK

Amicon Products–Millipore Corporation, 80 Ashby Road, Bedford, MA 01730-2271, USA

Selecto Scientific, 5933 Peachtree Industrial Blvd., Bldg. B, Norcross, GA 30092, USA

Bio-Rad Laboratories, 2200 Right Avenue, Richmond, CA 94804, USA

CPG Inc., 3 Borinski Road, Lincoln Park, NJ 07035, USA

BIBLIOGRAPHY

1. E. Boschetti, *J. Chromatogr. A* **658**, 207–236 (1994).
2. S.R. Narayanan, *J. Chromatogr. A* **658**, 237–258 (1994).
3. G.B. Wisdom, *Peptide Antigens: A Practical Approach*, IRL Press at Oxford Univ. Press, London, 1993, pp. 117–136.
4. R.K. Iler, *The Chemistry of Silica*, John Wiley, New York, 1979.
5. K.K. Unger, *Porous Silica*, Elsevier, Amsterdam, 1979.
6. M.A. Coletti-Previero and A. Previero, *Anal. Biochem.* **180**, 1–10 (1989).
7. L.K. Huston, C. Misra, and K. Weters, *Ulmann Encyclopedia of Industrial Chemistry*, Volume A1, VCH, Weinheim, 1985, p. 557.
8. *Lancet* **343**, 993 (1994).
9. J. Nawrocki, M.P. Rigney, A. McCormick, and P.W. Carr, *J. Chromatogr. A* **657**, 229–282 (1993).
10. M.J. Robichaud, A.N. Sathyagal, P.W. Carr, A.V. McCormick, and M.C. Flickinger, *Sep. Sci. Technol.* **32**, 2547–2559 (1997).
11. C.M. Griffith, J. Morris, M. Robichaud, M.J. Annen, A.V. McCormick, and M.C. Flickinger, *J. Chromatogr. A* **776**, 179–195 (1997).
12. A. Mullick, C.M. Griffith, and M.C. Flickinger, *Biotech. Bioeng.* (1998), **63**, 333–340.
13. A. Mullick and M.C. Flickinger, *Prep. Biochem. Biotechnol.* **28**, 1–21 (1998).
14. A.J. Alpert and F.E. Regnier, *J. Chromatogr.* **185**, 375–392 (1979).
15. J.J. Pesek and V.H. Tang, *Chromatographia*, **39**, 649–654 (1994).
16. J. Yu and Z. El Rassi, *J. High Resolut. Chromatogr.* **17**, 705–712 (1994).
17. H.J. Wirth and M.T.W. Hearn, *J. Chromatogr.* **646**, 143–151 (1993).
18. A. Tiselius, S. Hjerten, and O. Levin, *Arch. Biochem. Biophys.* **65**, 132–137 (1956).
19. G. Bernardi, *Methods Enzymol.* **22**, 325–330 (1971).
20. M.J. Gorbunoff, *Methods Enzymol.* **182**, 329–339 (1990).
21. U.S. Pat. 2,106,744 (February 1, 1938), H.P. Hood and M.E. Nordberg (to Corning Glass Works, Corning, N.Y.).
22. U.S. Pat. 3,485,687 (December 23, 1969), I.D. Chapman and T.H. Elmer.
23. U.S. Pat. 3,549,524 (December 22, 1970), W. Haller.
24. J. Margolis and P.H. Rhoades, *Lancet* 446–448 (August 29, 1981).
25. P. Van Der Voot and E.F. Vansant, *J. Liq. Chromatogr. Relat. Tech.* **19**, 2723–2752 (1996).
26. M.A.J. Godfrey, P. Kwasowski, R. Clift, and V. Marks, *J. Immunol. Methods* **160**, 97–105 (1993).
27. S.H. Chang, K.M. Gooding, and F.E. Regnier, *J. Chromatogr.* **120**, 321–333 (1976).
28. A.J. Albert and F.E. Regnier, *J. Chromatogr.* **185**, 375–392 (1979).
29. S.H. Chang, R. Noel, and F.E. Regnier, *Anal. Chem.* **48**, 1833–1845 (1976).
30. T. Darling, J. Albert, P. Russel, D.M. Albert, and T.W. Reid, *J. Chromatogr.* **131**, 383–390 (1977).
31. R.C. Collins and W. Haller, *Anal. Biochem.* **54**, 47–53 (1973).
32. F.E. Regnier and R. Noel, *J. Chromatogr. Sci.* **14**, 316–320 (1976).
33. K. Beyzavi and H.C. Wood, *Separation for Biotechnology*, Elsevier Applied Science, London, 1990, pp. 452–461.
34. M.A.J. Godfrey, *Affinity Separations: A Practical Approach*, IRL Press at Oxford Univ. Press, London, 1997, pp. 141–195.
35. P.J. Robinson, P. Dunnill, and M.D. Lilly, *Biochim. Biophys. Acta* **242**, 659–661 (1971).
36. K. Beyzavi and R. Clift, *Genetic Engineering News* **14** (March 1, 1994).
37. C. Spence, C.A. Schaffer, S. Kessler, and P. Bailon, *Biomed. Chromatogr.* **8**, 236–241 (1994).
38. A.C.Q.M. Meijers, A.M. de Jong, L.M.P. van Gruijthuijsen, and J.W. Niemantsverdriet, *Appl. Catal.* **70**, 53–71 (1991).
39. H. Hosono, Y. Sakai, and Y. Abe, *J. Non-Cryst. Solids* **139**, 90–92 (1992).

See also ADSORPTION, EXPANDED BED; ADSORPTION, PROTEIN, BATCH; ADSORPTION, PROTEINS WITH SYNTHETIC MATERIALS.

ADSORPTION, EXPANDED BED

ANN-KRISTIN BARNFIELD FREJ
Amersham Pharmacia Biotech
Uppsala, Sweden

KEY WORDS

Capture
Expanded bed
Chromatography
Downstream process
Fluidized bed
Protein adsorption
Protein purification
Recovery

OUTLINE

INTRODUCTION

Expanded-bed adsorption is a technique based on fluidization for recovery of proteins directly from unclarified feedstocks such as fermentation broths, cell homogenates, and other solutions containing suspended particulates. An expanded bed is a chromatography bed with an increased void volume between the adsorbent particles, as compared to a packed chromatography bed. An expanded bed is formed when high-density adsorbent particles are lifted by an upward liquid flow through a column specially designed for expanded bed adsorption. When the bed has expanded and is stable, the particulate-containing feedstock is applied, and the protein of interest can bind to the adsorbent particles while the particulates in the feedstock can pass unhindered through the void of the bed. Expanded-bed adsorption is a one-passage technique designed for large-scale recovery processes. Traditionally, when conventional packed-bed chromatography is used to recover proteins, the feedstock has to be clarified by centrifugation or filtration prior to application or the particulates can block the bed. Expanded-bed adsorption can handle particulate-containing feedstocks; as a result, clarification, concentration, and initial purification are accomplished in one unit operation. Reducing the number of unit operations in a downstream processing scheme decreases the risk of product loss, decreases the processing time, and lowers the overall production costs. The basic principles of expanded-bed adsorption and some applications are described in this article.

BACKGROUND

The biotechnology industry produces proteins for therapeutic and diagnostic use by using biological systems such as bacteria, yeast, and mammalian cells. A common bottleneck in the production of these proteins is the initial purification, during which the proteins are recovered from large volumes of cell- and debris-containing feedstocks, such as fermentation broths and homogenates. The purification processes are usually based on packed-bed chromatography techniques in which the protein is bound to an adsorbent packed in a column. A packed bed (fixed bed) cannot handle particulates very well and would eventually become totally blocked if an unclarified feedstock was passed through it. Another concern with these types of feedstocks is that because of the sometimes low production level of the protein of interest, the feedstock volume is large. A large volume takes a long time to pass through a chromatography column, and a long processing time increases the risk of proteolytic breakdown of the protein. Unit operations to clarify and reduce the volume of these feedstocks are therefore vital for a successful purification process. It is not uncommon for a purification process to consist of a sequence of unit operations just to achieve initial recovery. The techniques traditionally used for feedstock clarification are centrifugation and filtration (1).

Centrifugation is well suited to industrial processes because the throughput is high and it can be performed in a contained environment (a requirement when handling recombinant host organisms). The efficiency of the centrifugation process will depend on the properties of the feedstock, but it is usually difficult to remove all particulates by centrifugation, and the resulting fluid may therefore not be sufficiently free of particulates for direct application to a packed chromatography bed (2). Whole cells are relatively easy to remove by centrifugation but disrupted cells are not, and cells (especially mammalian) may even be damaged by the centrifugation process (3) because of shear stress and centrifugal forces. New and improved centrifuges that are more gentle can be used for mammalian cells (4), but often the clarification problem remains because at harvest, culture fluids usually contain relatively large amounts of subcellular particulates, such as debris. When the feedstock is secreting bacteria or a bacterial homogenate, for example, clarification is even more difficult and centrifugation is not very efficient.

When filtration is used for clarification of feedstocks, it is usually performed in cross-flow mode (5,6). The fluxes that can be used will depend on the properties of the feedstock, but it is often difficult to reach an acceptable economic limit (100 $L/m^2/h$) (7) for a cross-flow filtration process because fluxes are typically less than 50 $L/m^2/h$. Cross-flow filtration can easily be operated as a completely contained system, and it yields a particulate-free solution. The fouling of the filtration membranes is a major concern, and the resulting decrease in flux per filter area often leads to long processing times. In addition, the cost of membranes must be taken into account in a process where membrane fouling is severe because the membranes have to be replaced frequently. Because of the drawbacks associated with both the centrifugation and the filtration techniques, they are often used in combination in a purification process to achieve the best clarification of the feedstock.

The volume of the feedstock is not significantly reduced by either centrifugation or filtration, which is why product concentration does not usually increase until the protein has been adsorbed to a chromatography adsorbent in a packed-bed column. Because the feedstock volume is not reduced before application to the packed bed, processing time will be long. Thus, for a number of reasons, the best option for successful initial recovery of a protein from a large-volume cell- or debris-containing feedstock is direct adsorption of the protein.

A variety of methods for direct recovery have been tried. One such method is to use a batch procedure in which an adsorbent is added directly to the unclarified feedstock in a stirred tank (8,9). The advantage of using a batch procedure is that the feedstock does not have to be clarified. The disadvantage is that adsorption of protein to the chromatography adsorbent is inefficient, because the stirred tank acts as one single theoretical plate in a separation process (10). Several attempts, however, have been made to improve batch adsorption, for example, continuous-batch adsorption processes (11).

Another alternative for direct recovery is fluidized bed adsorption. This technique uses high-density adsorbent particles in an upflow fluidized bed column (12,13). As with a batch procedure, a fluidized bed behaves like a column with a low number of theoretical plates. The adsorbent particles and the feedstock are constantly "back-mixed," which usually results in poor adsorption of the protein by the adsorbent particles, making it necessary to recirculate the feedstock through the column, leading to long processing times. A number of solutions to improve the efficiency of fluidized beds have been suggested, such as columns fitted with perforated horizontal plates (14). This creates a column with a number of smaller, each completely back-mixed, fluidized beds. The number of fluidized beds depends on the number of plates inserted. The end result is a column with more than one theoretical plate and consequently a more efficient process. Another way to improve the efficiency of a fluidized bed is to use magnetically stabilized fluidized beds (15). The adsorbent particles are magnetically susceptible and are fluidized while being stabilized by a magnetic field. Stabilizing the particles in this way prevents back-mixing and gives the column several theoretical plates. There are, however, drawbacks that have prevented this technique from being used at industrial scale, such as the heat generated by the magnetic field.

The concept of expanded-bed adsorption (or expanded-bed chromatography) was introduced after another attempt to stabilize and prevent back-mixing in a fluidized bed. An expanded bed is a chromatography bed in which the bed voidage is larger than in a settled or packed bed. Furthermore, it displays no or very low back-mixing of adsorbent and liquid. An expanded bed can be also be described as a stable fluidized bed because the liquid flow shows a constant velocity profile over the whole cross-section of the bed (i.e., plug flow) (Worth mentioning is that the terms *fluidized bed* and *expanded bed* are used differently in the literature, and sometimes authors do not differ between the expressions.) The early expanded bed experiments were performed using Sepharose™ Fast Flow (Amersham Pharmacia Biotech, Uppsala, Sweden) as the adsorbent and a sintered glass filter as the liquid flow distributor at the bottom of the column (16). The results showed that if the bed expansion was limited to approximately twice the settled bed height, the liquid flow through the column was close to plug flow and the protein adsorption efficiency was similar to that of a packed bed. These promising results showed that in an expanded bed it is possible to combine the fluidized bed's ability to handle unclarified feedstocks with a packed bed's characteristic of efficient protein adsorption. The adsorbent used in these early experiments is intended for use in packed beds. The density of the particles is relatively low, which meant that the flow through the expanded bed column also had to be low, otherwise the bed would expand to an extent that adsorbent would start to pack at the top of the column. For this technique to be attractive at industrial scale, the flow velocities must be significantly increased. The sintered glass filter that was used as a flow distributor is not appropriate for large-scale operations for a number of reasons: particulates may get trapped in the filter, the filter is difficult to clean, and it is difficult to produce large homogenous sinters. Hence, it soon became apparent that for expanded-bed adsorption to be successful at industrial scale, new types of adsorbent particles and new column designs were needed.

Many different types of adsorbents have been tested: agarose (17,18), zirconia, silica (19), glass (20,21), hydrophilized perfluorocarbon (22), and composites such as kieselguhr/agarose (23), cellulose/titanium dioxide (24), and dextran/silica (25). These adsorbents are suitable for certain applications, but they lack the combination of properties needed for an efficient expanded-bed adsorption process. Agarose adsorbents are well suited for protein recovery but they are not dense enough to allow high flow velocities. Silica cannot withstand the harsh cleaning and sanitization conditions that are used in early recovery processes (26). Zirconia particles are the most dense, are stable at high pH, and are sterilizable. Porous glass and the described perfluorocarbons have promising sedimentation properties, but porous glass sometimes shows quite low protein-binding capacities (21), and the perfluorocarbons have low protein capacities because of the low particle porosity (22).

With the recent development of STREAMLINE™ (Amersham Pharmacia Biotech) adsorbents (agarose/crystalline quartz composite, agarose-dextran/crystalline quartz composite, and agarose/metal alloy composite) and STREAMLINE™ columns, it was possible to secure a recovery system that has the properties needed for an efficient expanded-bed process. The columns have a flow distributor in the bottom that generates an even flow distribution over the whole cross-section of the bed. The density of the adsorbent permits process flow velocities (300–500 cm/h), and the size distribution creates a classified bed that enhances the stability of the expanded bed. The result is a bed with protein adsorption characteristics similar to those of a packed chromatography bed. Expanded-bed adsorption is a scalable, one-passage technique that is intended for industrial-scale operations (27).

There have been a number of applications reported that use expanded bed adsorption with various feedstocks: secreting bacteria (28), periplasmic preparations of bacteria (29), bacterial homogenates (30,31), secreting yeast (32–34), yeast homogenates (35,36), mammalian cells (37–45), milk/whey (46–48), and inclusion body preparations (49).

THEORY

As mentioned earlier, an expanded bed is a fluidized bed that is devoid of or shows low back-mixing of liquid and adsorbent. The stability of the expanded bed makes it resemble a packed bed in that it shows an almost constant flow velocity profile and yields a bed with an increased number of theoretical plates (relative to a fully mixed fluidized bed). These characteristics are achieved by a combination of column and adsorbent design.

Column

The liquid flow in a column used for expanded-bed adsorption must be uniform over the entire cross-section of the bed. A conventional packed-bed chromatography column with one or a few inlets cannot meet this criterion. In a packed bed, the bed itself creates a pressure drop that achieves the desired plug flow. In an expanded bed, there is a very low pressure drop over the column, thus requiring a specially designed flow distributor at the bottom of the column (50). This distributor must ensure that the flow is directed in a vertical direction only, because radial flow inside the column will cause disturbances in the expanded bed. It must also prevent the adsorbent particles from leaving the column while allowing cells and debris to pass through. Other requirements are that it must not create high shear stresses, or cells and shear sensitive molecules may break, and it should be designed to facilitate cleaning. There should be no stagnant zones in the column where cells and other contaminants can accumulate and, in the worst case, escape cleaning.

Adsorbent

The adsorbents used for expanded-bed adsorption must exhibit additional properties to those required of an adsorbent used in packed-bed chromatography. The density of the particles must be relatively high to enable high liquid flow velocities through the expanded bed without the adsorbent particles packing against the adaptor. The adsorbent particles in a fluidized/expanded bed are in constant motion; however, an efficient protein adsorption process requires that the solid dispersion (the movement of the adsorbent particles) is reduced to a minimum. This can be achieved if the adsorbent is polydispersed with respect to size, that is, the particles exhibit a size distribution. When such a polydispersed adsorbent with a small variation in particle density is expanded, each particle will find its equilibrium position in the upward flow. The result is called a classified bed (51), and at a closer look the larger particles will be found at the bottom of the bed and the smaller ones at the top. A distribution of density with size (a large particle exhibiting a larger density than a small particle) will further enhance the classification of the bed. One way of describing the behavior of such an adsorbent in an expanded bed is the Richardson-Zaki (52) model for monodispersed particles:

$$u_0/u_t = \epsilon^n$$

where u_0 is the superficial velocity of the fluid, u_t is the terminal falling velocity of a particle in infinite dilution, ϵ is bed voidage, and n is an index. The model describes expansion for a bed that has reached equilibrium (the expansion height is constant). The terminal falling velocity u_t is described by Stoke's law:

$$u_t = ((d_p)^2 g(\rho_p - \rho))/18\mu$$

where d_p is particle diameter, g is acceleration due to gravity, ρ_p is particle density, ρ is fluid density, and μ is fluid viscosity. Stoke's law can only be used for particles with a Reynold number lower than 0.2. The Reynold number Re_p for a particle in a fluid is defined by:

$$Re_p = ud_p\rho/\epsilon\mu$$

The Richardson-Zaki model is corrected for the particle size distribution of the adsorbent by using the perfectly classified bed model (51):

$$\Delta H(\epsilon) = H_0(1 - \epsilon_0)\Delta F_v/(1 - \epsilon)$$

where H_0 is sedimented bed height, ϵ_0 is sedimented bed voidage, ΔF_v is the proportion of the total volume of particles having the diameter within the interval $d - \Delta d/2 < d < d + \Delta d/2$, and ϵ is bed voidage for the particles within that interval. This means that the particle size group between d and $(d - \Delta d)$ is found within the expanded bed between heights h and $(h + \Delta h)$. Thus, each segment of height is considered a bed with a specific voidage that can be calculated using the Richardson-Zaki equation. The total bed height is obtained by adding up the heights of each segment.

EQUIPMENT AND METHODOLOGY

Column

The major difference between a standard packed-bed column and a column used for expanded-bed adsorption is the liquid distribution system. STREAMLINE™ columns are specially designed for expanded bed adsorption and have a fixed bottom distributor and a movable top distributor (adaptor). The bottom and top distributors have essentially the same design: a perforated plate with a single-weave screen or net facing the adsorbent. The distributor plate in the adaptor has a lower pressure drop than the bottom distributor. The purpose of the top distributor plate is to create a more even flow (compared with a column with one or a few inlets in the adaptor) when the column is run in downward mode during elution in settled bed mode. The steel screens help to stabilize the flow and confine the adsorbent particles to the column. These columns do not cause cell damage, and it has been shown that shear-sensitive cells pass through the column without breaking (40,42).

Expanded-bed adsorption handles unclarified feedstocks, and for this reason, harsh cleaning procedures are necessary. The columns are constructed of materials that withstand the chemicals used in such procedures and are also designed to have the minimum of stagnant zones where cells could accumulate and risk contaminating the next run. It has been shown that the columns can be efficiently cleaned and sanitized after passage of unclarified feedstocks such as whole cultures of bacteria and yeast (27).

Columns range in size from small lab-scale columns with a 25-mm inner diameter to industrial-scale columns, presently the largest with a 1,200-mm inner diameter. They all have essentially the same design, but differ in the material used for the column tube: the smaller columns have tubes made of glass and industrial columns have tubes made of stainless steel. The larger columns also have a greater number of inlets and a higher pressure drop over the distributors than the smaller columns. One method of verifying successful scale-up and comparing the performance of different sizes of expanded-bed columns is to determine the axial dispersion of a tracer substance (the method is described in "Operation of an Expanded Bed"). Another method is to determine protein breakthrough capacity using a model protein in a buffer system. The results obtained from these methods give a good measure of bed stability, a factor that is crucial for a successful expanded-bed process. The hydrodynamic properties of the columns are tested using the axial-dispersion method, and the "function"—the protein adsorption properties—is evaluated using the protein breakthrough method. The columns have been tested using both methods, and the results demonstrate that the performance of an industrial-sized column is very similar to that of a small lab-scale column (27). Table 1 shows that the number of theoretical plates, and the axial dispersion varies little between the different column sizes. Figure 1 compares protein breakthrough curves (bovine serum albumin [BSA] at 300 cm/h using 15-cm settled bed height), illustrating the maintained protein adsorption characteristics from lab scale to industrial scale.

Table 1. Number of Theoretical Plates (N and Axial Dispersion (D_a) for Expanded-Bed Columns with Inner Diameters from 25 to 600 mm

Column	N	D_a (m^2/s)
STREAMLINE™ 25	50–70	3–5 10^{-6}
STREAMLINE™ 50	40–60	4–6 10^{-6}
STREAMLINE™ 200	35–50	5–7 10^{-6}
STREAMLINE™ 600	35–50	5–7 10^{-6}

Note: The experiments were performed at 300 cm/h flow velocity and 15 cm settled bed height adsorbent using acetone as tracer substance.

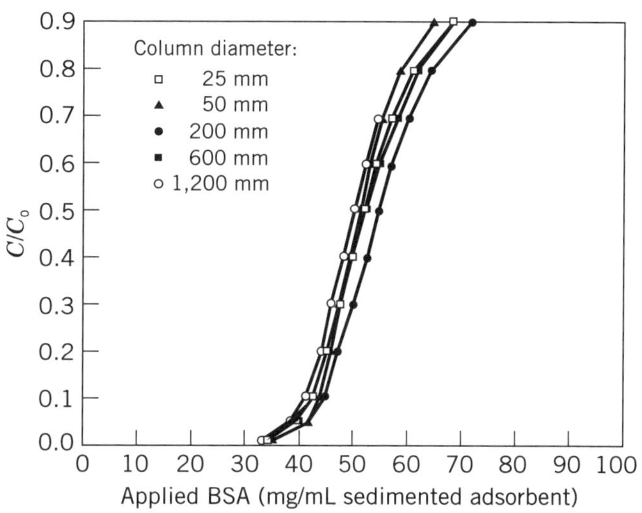

Figure 1. Protein breakthrough capacity. Breakthrough curves for bovine serum albumin (BSA) on STREAMLINE™ DEAE (15 cm settled bed height at 300 cm/h flow velocity) using STREAMLINE™ expanded-bed columns with inner diameters ranging from 25 to 1,200 mm. C_0 is the concentration of BSA in the applied feedstock, and C is the concentration in the flowthrough. *Source*: Copyright Amersham Pharmacia Biotech, Uppsala, Sweden, modified and reproduced with permission.

Because industrial-scale columns have to be automated, these columns are equipped with an adsorbent sensor that locates the position of the expanded bed surface. In response to the signal from the sensor, the adaptor is automatically adjusted to its correct position. Figure 2 shows different scales of expanded-bed columns.

UpFront™ Columns (UpFront Chromatography A/S, Denmark) offer another approach to expanded-bed column design (53). The columns are without screens and without distribution plates, and the feedstock is introduced at the bottom of the column, right below a stirring device that distributes the feedstock over the cross-section of the column. The stirrer causes a mixed zone, but gradually above that zone plug flow is obtained. Columns of 20-, 50-, and 200-mm inner diameter are available.

BioProcessing Ltd. (England) uses modified standard chromatography columns for expanded bed adsorption (54). The standard column bottom support net or screen is replaced with a 300-μm nylon mesh, and then the column

Figure 2. Different sizes of STREAMLINE™ expanded bed columns: 25-mm, 600-mm, and 1,200-mm inner diameters. *Source*: Courtesy of Amersham Pharmacia Biotech, Uppsala, Sweden.

is packed with a layer (approximately 3 cm in height) of surface-modified solid glass beads (0.4–0.52 or 0.85–1.23 mm in diameter). This bed of beads serves as a distributor of fluid to the bed. The adsorbent particles are then added directly on top of the distributor beads. During operation of the column, the bed of the relatively heavy distributor beads remains static, and the bed of adsorbent particles expands.

Adsorbent

STREAMLINE™ adsorbents are specially designed for expanded bed adsorption. To ensure that the bed does not expand too much at industrial chromatography flow velocities, a dense core material has been incorporated into the agarose particles to increase the apparent density. The adsorbent has a defined Gaussian-like particle size distribution, which yields a stable classified bed when the adsorbent is expanded. The influence of particle size distribution on protein adsorption performance has been studied using three different particle fractions (120–160 μm, 250–300 μm, and 120–300 μm) (55). The results showed that a wide particle size distribution gives an expanded bed with low axial dispersion and that small particles give high dynamic protein capacity (due to reduced diffusion length and large surface area). The results also showed that with increasing (settled) bed heights, the dynamic binding capacity increases over a wide range of flow velocities (due to reduced axial dispersion and increased numbers of transfer units for pore and film diffusion).

The STREAMLINE™ DEAE, Q XL, SP, and SP XL, chelating and heparin adsorbents are spherical, macroporous, cross-linked 6% agarose particles with crystalline quartz as core material. The Q XL and SP XL adsorbents have dextran chains coupled to the agarose, and the Q and SP charged groups are attached to the dextran through chemically stable ether bonds. This increases the exposure of the Q and SP groups and results in very high protein binding capacities. The adsorbents can be cleaned with 1 M NaOH. The matrix used for the protein A adsorbent contains highly cross-linked 4% agarose with a metal alloy core. The alloy is stable and composed of 77% Ni, 15.5% Cr, and 7.5% Fe. No leakage of metal was detected in 0.1 M glycine, pH 3.0, a solution commonly used in protein A chromatography. The metal leakage was below the detection limit of the analysis method (0.02 μg/mL eluent using induced coupled plasma atomic-emission spectroscopy). The protein A adsorbent is stable in all aqueous buffers commonly used in protein A chromatography and cleaning. No significant change was observed in chromatographic performance after either 1 week's storage or 100 cycles, normal use, at room temperature with 10 mM HCl (pH 2), 0.1 M sodium citrate (pH 3), 1 mM NaOH (pH 11), 6 M GuHCl, or 20% ethanol. Figure 3 shows micrographs of the adsorbent particles, and Table 2 summarizes the properties of the various STREAMLINE™ adsorbents.

Apart from the purpose-designed column and adsorbent, the equipment and the experimental set-up used for expanded bed adsorption is very similar to that used for packed bed chromatography. An example of an industrial-scale set-up with a 600-mm-diameter column is seen in Figure 4.

Method Development

The strategy for developing an expanded-bed process is in general analogous to that of a conventional packed-bed process. In order to save time, it is advisable to do the first experiments with clarified feedstock, using the expanded-bed adsorbent in a small, packed-bed column, for example, 15 mm in diameter and with 15-cm bed/height. (A bed height below 10 cm of these relatively large particles usually results in a decreased dynamic binding capacity. The increase in dynamic binding capacity with increased bed height is an effect of longer residence time for the sample.) From the packed-bed experiments, the most suitable adsorbent is selected (e.g., ion exchanger or chelating adsorbent), and the conditions for protein binding and elution are determined. It is important to establish that the conditions selected for protein binding are compatible with the properties of the unclarified, cell- or debris-containing feedstock (e.g., low pH might cause aggregation). Such problems are not apparent with clarified feedstock and

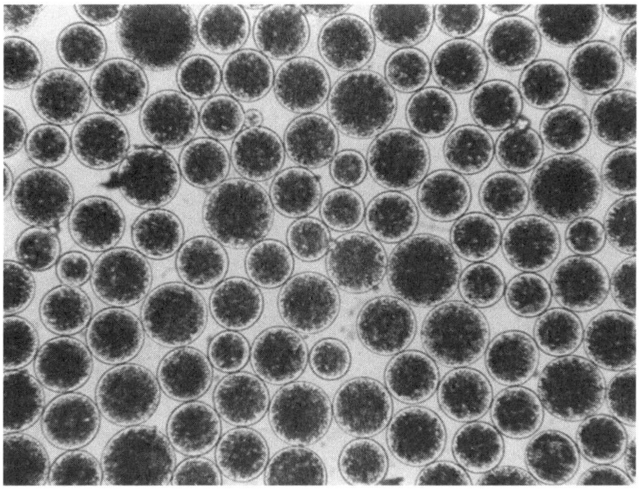

(a)

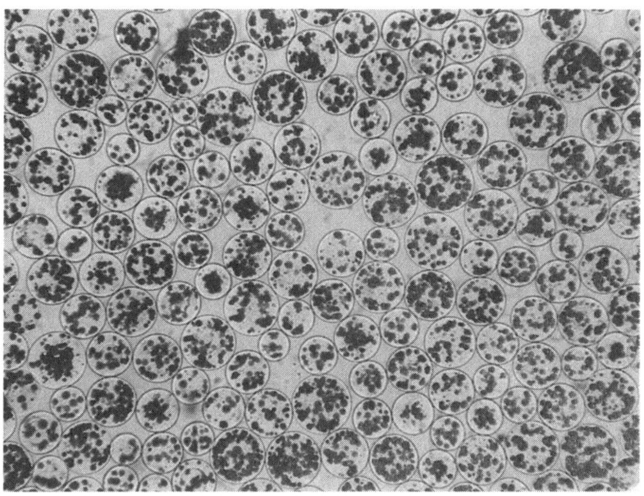

(b)

Figure 3. STREAMLINE™ adsorbents. Micrographs of STREAMLINE™ DEAE (**a**) and STREAMLINE™ rProtein A (**b**). *Source*: From *Bioseparation and Bioprocessing, Vol. 1 Biochromatography: Expanded Bed Adsorption Chromatography*, Wiley-VCH Verlagsgesellschaft, Weinheim, Germany, 1998.

should be investigated before the process has progressed too far. Once the most promising method has been selected from the packed-bed experiments, it is then optimized in a lab-scale expanded-bed column using unclarified feedstock. At this stage, the effect of the now present cells or debris on protein binding is evaluated. In theory, if the net surface-charge of the cells and debris is negative, the protein and the cells or debris may compete for binding sites on an anion exchanger. A consequence of this may be that the feedstock load per milliliter adsorbent has to be decreased. When STREAMLINE™ DEAE was used to recover insulinase from whole yeast culture, the presence of cells did not cause any reduction in protein capacity (33). However, it has been reported that when the same adsorbent was used to recover glucose-6-phosphate dehydrogenase from a yeast homogenate, the presence of cell debris did cause a slight reduction in capacity of the adsorbent (36). Feedstock properties and their effects on the expanded-bed process are discussed further in "Feedstocks." Before the process is transferred into larger columns, appropriate cleaning procedures should be determined. The whole process is then verified at pilot scale before scaling up to the final production scale.

Operation of an Expanded Bed

Figure 5 is a schematic representation of how an expanded bed is operated—essentially as a packed bed. Avoid getting air into the column; air trapped in the bottom distributor may disturb the bed. If air does enter the column, it can usually be removed with high-flow velocity pulses combined with back-flushes (reversal of flow direction). Before start-up always check that the column tube is positioned vertically; otherwise, flow will be turbulent when the liquid hits the inside walls of the column tube. The height of the settled bed (H_0) is noted before start-up.

Bed Expansion and Equilibration. The bed is expanded and equilibrated with buffer. It usually takes 20–30 min at a flow velocity of 300 cm/h before the bed is stable and has stopped expanding. The height of the expanded bed (H) is then noted. At this stage, the bed can be checked empirically, before application of the feedstock, by determining the axial dispersion of a tracer substance. Figure 6 shows the UV signal recording from such a test procedure. Before applying the tracer substance, the adaptor is lowered to a position where the adaptor screen is 0.5–1.0 cm above the expanded-bed surface. When the signal from the UV monitor is stable, the solution is changed to buffer-tracer substance mixture (e.g., 0.25% acetone v/v), that is, positive step input signal. The solution is changed to buffer when the UV signal is stable at maximum absorbance (100%), that is, negative step input signal. The change is marked on the chart recorder paper, and the UV signal is allowed to stabilize at the baseline level (0%). The number of theoretical plates (N) is calculated from the negative step input signal (57):

$$N = t^2/\sigma^2$$

where t is the mean residence time, the distance from the mark on the chart recorder paper to 50% of the maximum absorbance; and σ is the standard deviation, half the distance between the points 15.85% and 84.15% of maximum absorbance.

The procedure gives the number of theoretical plates not only in the column but also the contribution from pumps, valves, tubing, and so forth. Because different experimental set-ups may vary in their contribution to N, the results should only be compared between runs performed using the same set-up. The negative input signal is recommended for evaluation (at least when acetone is used as tracer) because the reproducibility of the results is higher than that obtained from the positive step input signal (from 0% to 100%).

The relation between N and the axial dispersion coefficient (D_a) is

Table 2. Properties of STREAMLINE™ Adsorbents Used in Expanded-Bed Adsorption

Adsorbent name and base matrix	Ligand and functionality	Particle size distribution (μm)	Mean density (g/mL)	Ligand density	Breakthrough capacity (mg/mL)	Total binding capacity (mg/mL)
STREAMLINE™ DEAE Agarose/quarts	DEAE Anion exchange	100–300	1.15	0.13–0.21	35	55
STREAMLINE™ Q XL Agarose-dextran/quartz	Q Anion exchange	100–300	1.2	0.23–0.33	>110	Not done
STREAMLINE™ SP Agarose/quartz	SP Cation exchange	100–300	1.15	0.17–0.24	65	75
STREAMLINE™ SP XL Agarose-dextran/quartz	SP Cation exchange	100–300	1.2	0.18–0.24	>140	Not done
STREAMLINE™ Chelating Agarose/quartz	IDA Metal affinity	100–300	1.15	40	8	40
STREAMLINE™ Heparin Agarose/quartz	Heparin Affinity	100–300	1.15	4	Not done	Not done
STREAMLINE™ rProtein A Agarose/metal alloy	rProtein A Affinity	80–165	1.3	6	20	50

Note: Adsorbent name: XL stands for extreme load and consists adsorbents with very high loading capacities. *Ligand*: DEAE, diethyl aminoethyl; Q, quatenary amine; SP, sulphopropyl; IDA, iminodiacetic acid; rProtein A, recombinant protein A. *Functionality*: Functionality of STREAMLINE Heparin was evaluated by its ability to bind antithrombin from bovine plasma. *Ligand density*: Ligand densities for STREAMLINE DEAE, Q XL, SP, and SP XL were determined as millimole charged groups per milliliter of adsorbent. For STREAMLINE Chelating, ligand density is expressed as Cu^{2+} binding capacity in μmole per milliliter adsorbent. Ligand density for STREAMLINE Heparin is milligram of heparin bound per milliliter adsorbent. For STREAMLINE rProtein A, ligand density is expressed as milligram protein A per milliliter adsorbent. *Breakthrough capacity*: For STREAMLINE DEAE, SP, and rProtein A, breakthrough capacity was determined at 300 cm/h linear flow velocity in expanded-bed mode (15 cm settled bed height) at 1% breakthrough of 2 mg/mL bovine serum albumin (DEAE) or lysozyme (SP) or human immunoglobulin G (IgG) (rProtein A). For STREAMLINE Q XL AND SP XL, the breakthrough capacity was determined at 300 cm/h in packed bed mode (10 cm bed height) at 10% breakthrough of 2 mg/mL bovine serum albumin (Q XL) or lysozyme (SP XL). Breakthrough capacity for STREAMLINE Chelating was determined in packed bed mode at 300 cm/h flow velocity (5 cm sedimented bed height) at 1% breakthrough of 2 mg/mL bovine serum albumin. *Total binding capacity*: Total binding capacity was determined in packed bed mode at 50 cm/h flow velocity (15 cm bed height) at 100% breakthrough of 2 mg/mL bovine serum albumin (DEAE, Chelating), lysozyme (SP), or human IgG (rProtein A).

$$D_a = (u_0 \cdot H)/(2 \cdot N \cdot \epsilon)$$

where u_0 is the superficial velocity, H is the expanded bed height, and ϵ is the bed voidage in expanded mode. ϵ is calculated from $H/H_0 = (1 - \epsilon_0)/(1 - \epsilon)$. In a settled bed, the bed voidage (ϵ_0) is approximately 0.4. The higher the value for N, the lower the value for D_a, indicating a low degree of back-mixing (a flow profile approaching plug flow). The relation between N and D_a is derived from the coupling between the tanks in series and the axial dispersion models by comparing the variances between the models' step-response curves. It is not completely accurate to make this comparison because the responses of the two different models are never identical (56). But for a system with a low degree of back-mixing, this way of calculating the axial dispersion is accurate.

Feedstock Application. The feedstock is applied in expanded bed mode, usually at a constant flow velocity (e.g., 300 cm/h). The adaptor is kept at its uppermost position, or at a position 5–10 cm above the expanded-bed surface. The uppermost position is usually the choice when a new (unknown properties) feedstock is applied or when there are feedstock batch variations, making it difficult to predict how the feedstock will behave in the expanded bed from run to run. If feedstock properties, such as viscosity or biomass or others that affect expansion, are known to be consistent and do not vary, then the adaptor can be left at a position closer to the bed surface. The feedstock is always stirred during application to prevent cells and debris from settling. If there is a build-up of cells and debris on the adaptor, the column is intermittently back-flushed. The duration of a back-flush is about 5–10 seconds, and the interval between back-flushes is usually 10 to 30 minutes, depending on the properties of the feedstock. Another method of applying the feedstock is to change the flow velocity (normally lowering it) during the run, so that the expansion height is kept constant. It has been reported that this approach, under certain conditions, may give a higher productivity than the constant flow velocity method does (58).

Wash. The wash is performed in expanded-bed mode at the highest possible flow velocity. This means that the flow velocity at the beginning of the wash is often the same as during feedstock application, but as the bulk of the cells and debris are washed out, the flow velocity is increased. At the beginning of the wash, the column is usually back-flushed as described in the previous section. During the wash, when the bulk of the particulates has been washed out, the adaptor is lowered closer to the bed surface. Lowering the adaptor significantly reduces the volume of wash buffer and shortens the time required for the wash step. If, instead, the wash is performed with a wide gap between the bed surface and the adaptor, mixing of the liquids is usually severe (a low density wash buffer being introduced after the high density/viscosity feedstock), resulting in a large wash volume. After the wash, when the UV curve has levelled out, the column is back-flushed to remove any trapped cells or debris from the distribution system. The bed is allowed to settle and the adaptor is lowered to the bed surface. The wash is continued in settled-bed mode, in

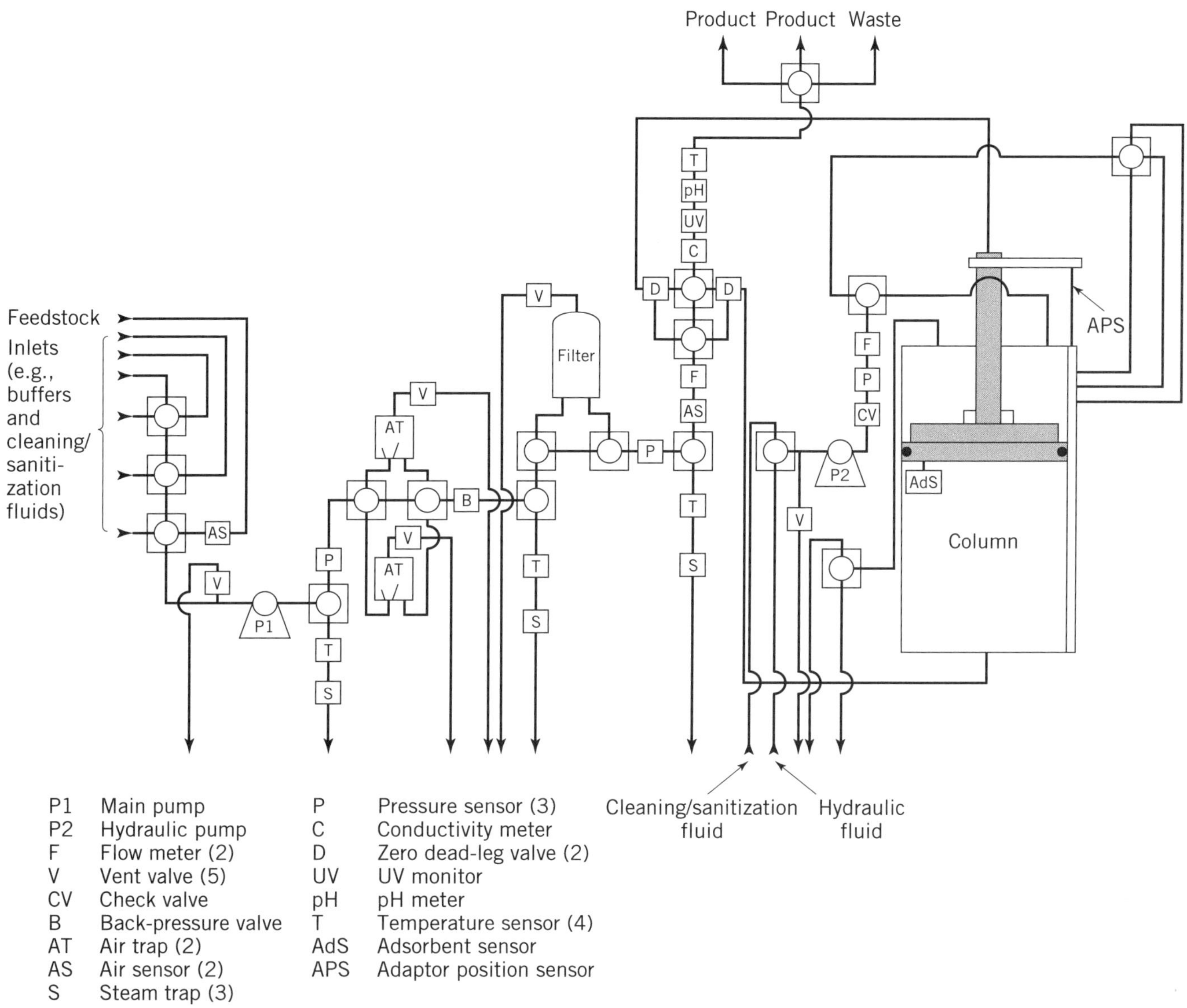

Figure 4. Experimental set-up for production-scale expanded-bed adsorption using a 600-mm inner diameter column. *Source*: From Ref. 27.

a downward flow direction at 300 cm/h, for at least one settled bed volume or until the UV curve levels out. When the wash step has been optimized approximately, it then requires 15 settled bed volumes of wash buffer to flush out particulates and unbound proteins (59).

Elution. Elution is usually performed in settled bed mode with downward flow. The flow velocity is lower than during the previous steps, usually 100 cm/h, in order to keep the elution volume as small as possible. A high flow velocity during elution will increase the elution volume. It has been shown that elution in expanded-bed mode may increase the volume by approximately 40% (58).

Cleaning. Cleaning is performed in expanded bed mode to allow any trapped particulates in the bed or column to be flushed out. The adaptor is typically placed at a position corresponding to twice the settled bed height, and the flow velocity is about 100 cm/h. Cleaning is facilitated if it is performed directly after each purification cycle. As with any protocol for cleaning, it will depend on the adsorbent and on the nature of the feedstock, etc. A general cleaning protocol suggested by the manufacturer for the different types of adsorbents (e.g., ion exchanger) is always a good starting point, as is a cleaning protocol used for a corresponding packed-bed process (if such exists). As previously mentioned, the cleaning is facilitated if as many of the particulates as possible have been washed out during the previous wash step. Back-flushes disturb the build-up of particulates, and it has been shown that back-flushes at a high flow velocity were effective in removing the build-up of sticky yeast from the distributor plate in an industrial-scale expanded-bed column (27).

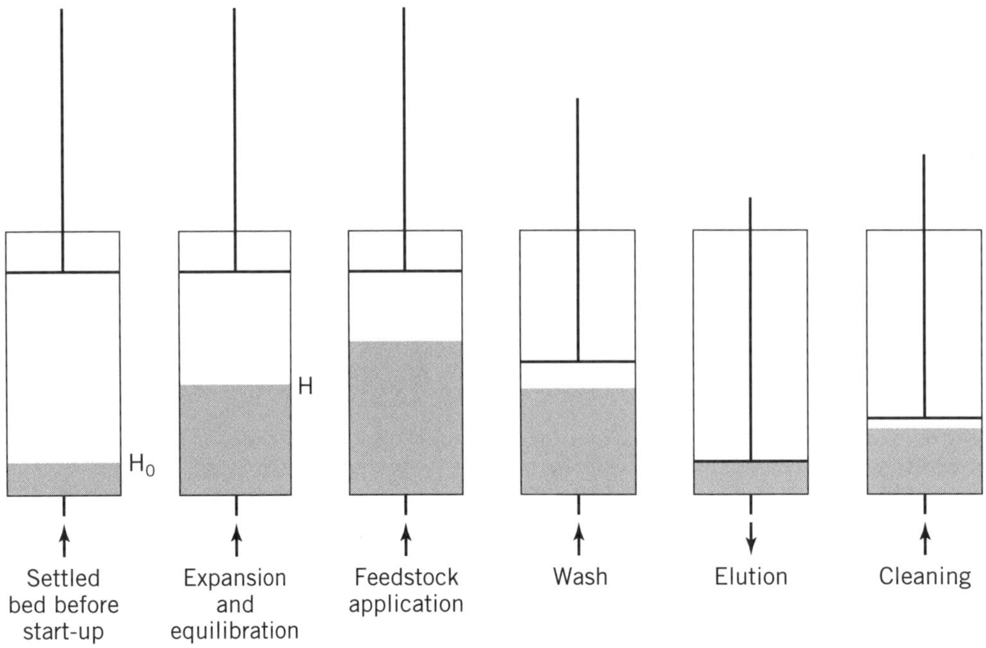

Figure 5. Operational stages in an expanded-bed adsorption process. The arrows indicate the liquid flow direction. H_0 is the settled bed height and H is the expanded bed height.

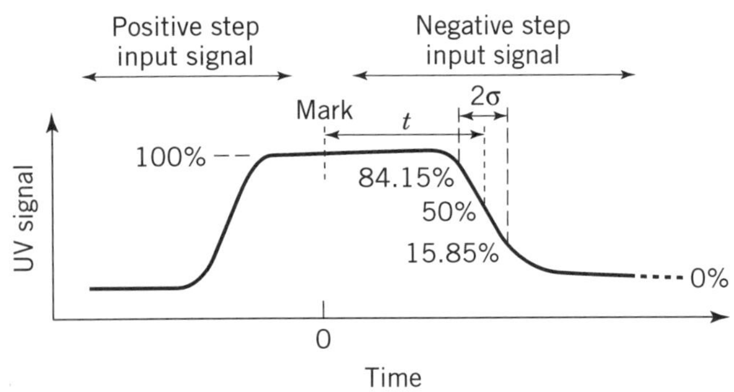

Figure 6. UV signal recording from the test procedure for determination of the number of theoretical plates (N). $N = t^2/\sigma^2$, where t is the mean residence time and σ is the standard deviation. *Source*: Courtesy of Amersham Pharmacia Biotech, Uppsala, Sweden.

After cleaning, the adsorbent can be stored in the column in an appropriate storage solution.

Feedstocks

Particulate-containing feedstocks can be applied to an expanded bed without prior clarification steps such as centrifugation and filtration; however, there are certain physical properties a feedstock must have to make it suitable for an expanded-bed process. The size and shape of the particulates must be such that they can pass through the screen or net in the column, that is, the particulates must be smaller than approximately 60 μm in diameter. Feedstocks containing small, round single cells or fine precipitates (e.g., renatured inclusion body preparations) will easily pass through the screens, but large aggregates of cells will not. If the particulates are too large, they will build up under the bottom screen and eventually block the column. When the feedstock contains particulates that are too large, they can be removed by centrifugation or by filtering through a simple mesh screen. It is of course desirable for these steps to be avoided completely, but they are relatively easy to perform and should not be compared to the complete clarification procedures needed for a packed-bed process. It is important to stir all feedstocks during application, otherwise the particulates may settle and disturb the expanded bed. Other parameters that will influence the performance of an expanded bed are the viscosity and biomass content of the feedstock. High values for these parameters will cause the bed to overexpand (relative to a buffer); if the bed expands up to the adaptor, the adsorbent particles will start to pack against the adaptor. This build-up will act as a filter and eventually cause the column to block. Furthermore, if the viscosity or the biomass content is too high, flow may become turbulent and channels may appear in the bed. If the turbulence is severe, the feedstock will pass through the channels in the bed, resulting in poor adsorption efficiency. The upper limits for feedstock viscosity and biomass content will vary from case to case, but for STREAMLINE™ with homogenates of *Escherichia coli*, there are a few guidelines (30). A dry weight of 3 to 4% gives the best result, 5% gives an acceptable result, and 7 to 8% is the upper limit. Viscosities up to 10 mPa s (measured at a shear rate of 1/s) give the best results; the upper limit is about 50 mPa s (at 1/s). In cases where the viscosity or biomass content of the feedstock does not cause the bed to expand too much, build-up on the adaptor can be prevented if the column is intermittently back-flushed. The only way to lower the biomass content of the feedstock is to dilute it. The disadvantage is the resulting increase in feedstock volume. Dilution is only one of way of lowering viscosity. When the viscosity is caused by nucleic acids, the feedstock may be treated with a nuclease (30). When such a feedstock is a homogenate, further homogenization can be used to shear the nucleic acids and thereby lower the viscosity. The viscosity of a feedstock decreases with increased temperature, making it advantageous to apply the feedstock directly from the fermentor without prior cooling. Since temperature also has a positive effect on protein adsorption (59), expanded-bed adsorption should always be performed at room temperature (or without feedstock cooling), unless the protein of interest is, for example, sensitive to proteases.

To keep aggregation of cells and debris to a minimum, any pH adjustment of the feedstock should be performed immediately before application to the expanded-bed column. Aggregation of cells usually occurs at low pH and is therefore a problem associated with cation exchange, where pH has to be below the isoelectric point of the protein for it to bind to the adsorbent. If aggregation of cells and debris occurs within the expanded bed, when the feedstock is in contact with the adsorbent, and blocks the screen in the adaptor, the expanded bed may be operated without the adaptor screen (provided that the bed has not expanded all the way up to the adaptor) (49). As mentioned previously, cells or debris may, under certain conditions, interact with the adsorbent particles. If this occurs, it is usually with anion exchangers because they are positively charged and may attract the negatively charged cells or debris. A slight change in the binding conditions can help, for example, the addition of some salt and/or lowering the pH.

Alternatively, the feedstock load per milliliter adsorbent may have to be decreased, or it might be necessary to change to an adsorbent with another type of ligand, such as a cation exchanger.

INDUSTRIAL-SCALE APPLICATIONS

Expanded-bed adsorption is a technique with great potential for initial purification. The number of published industrial scale applications is limited because until recently columns and adsorbents suitable for use in expanded beds were not available. In addition to the three industrial-scale applications described below, an IND (Investigational New Drug) application has been filed with the Food and Drug Administration for a therapeutic recombinant protein in which expanded-bed adsorption is central to the purification process (60).

Purification of Recombinant Human Serum Albumin from Whole Yeast Culture

The number of unit operations in the process for purification of recombinant human serum albumin (rHSA) from yeast was reduced by three when expanded-bed adsorption replaced the conventional packed-bed method (32). In the conventional packed-bed method, three filtration steps were necessary to prepare the yeast culture for the first cation exchange column; when using expanded-bed adsorption, the unclarified culture could be applied directly to the expanded cation exchanger. By reducing the number of unit operations, the total processing time was deceased and the overall yield was increased. In contrast to the conventional packed-bed process, the expanded bed process could be performed in a totally closed system, and the quality (determined as degree of coloring) of the rHSA was reported to be higher using the expanded-bed process. The process was successfully scaled up (determined by comparing yield of rHSA) from a 50-mm diameter column with 300 mL adsorbent to a 1,000-mm diameter process scale column with 150 L adsorbent. Approximately 2,000 L unclarified feedstock (1,000 L culture diluted with 1,000 L distilled water) was applied to the process scale column, and the average yield of rHSA from four different runs was 87% (ranging from 82 to 91%). Using a specially made column with a 4-mm inner diameter, the lifetime of the resin was investigated. The results showed that the adsorbent could be reused up to 1,000 times without compromising performance (61).

Purification of Monoclonal Antibodies from Whole Mammalian Cell Culture

A process for recovery of recombinant monoclonal antibodies was directly scaled up from a small lab-scale expanded-bed column containing 70 mL cation exchange adsorbent to a process-scale expanded-bed column to which 12,000 L of cell culture was applied (45). In this single process step, all the cells were removed, the monoclonal antibody was concentrated fivefold, and the yield of antibody obtained was 95%.

Purification of Bovine Serum Albumin from Whole Yeast Culture

In a mimicked purification process, a 4.5% dry weight yeast culture was spiked with bovine serum albumin (2 mg/mL) and used as feedstock. The bovine serum albumin was then recovered from the feedstock using expanded-bed adsorption with an anion exchange adsorbent (27). The process was scaled up from a 25-mm diameter column to a pilot scale column with a 200-mm diameter and finally run in a fully automated system with a production-size column with a 600-mm diameter. The settled bed height was 15 cm throughout all scales, and the feedstock volume was approximately 750 mL in the 25-mm column and 430 L in the 600 mm column. The results from the different scales showed good agreement. The yield of bovine serum albumin was approximately 89% (ranging from 87 to 92%). The wash volume was approximately 11 settled bed volumes, and the volume of the elution peak was approximately 2 settled bed volumes.

Potential Industrial-Scale Applications

The successful use of expanded beds at laboratory and pilot scales has increased significantly during the past few years, giving rise to a number of potential industrial-scale applications. One such example is the purification of the protease inhibitor aprotinin (34), which has good potential as a therapeutic and diagnostic compound. Aprotinin was expressed and secreted from the methylotrophic yeast *Hanensula polymorpha*, and the initial recovery was performed using 300 mL cation exchange adsorbent in a 50-mm diameter expanded-bed column. The only pretreatment of the yeast culture was a 1 + 1 dilution with water and an adjustment of pH. The final volume of the feedstock was 6.4 L. The dilution achieved two goals: it lowered the conductivity to favor binding to the ion exchanger, and it reduced the biomass and viscosity of the feedstock so that the bed did not expand too much. The yield of aprotinin was 76%, the purification factor was 4, and the concentration factor was 7.

Another example is the recovery of a recombinant mitotoxin fusion protein, fibroblast growth factor–saporin (rFGF-2-SAP) (31), which is a possible therapeutic agent for the treatment of diseases characterized by cellular proliferation (e.g., cancer). The fusion protein was expressed in *E. coli*. After homogenization, it was recovered from the diluted homogenate using 300 mL cation exchanger in a 50-mm diameter expanded-bed column. Homogenate from 300 to 400 g (wet weight) cells was applied to the expanded-bed column. The yield of fusion protein was 65%, and the purification factor was 20. It was also reported that a somewhat modified process was scaled up to handle 8,000 g (wet weight) cells. This process was carried out under GLP (good laboratory practice) conditions, and one such process yielded an average of 1.6 g of protein. The yield and quality of the fusion protein were comparable between the different scales.

Recombinant human nerve growth factor (rhuNGF) produced in Chinese hamster ovary cells is another example of a potential industrial scale application (38). Here, the rhuNGF was recovered using a 25-mm diameter expanded-bed column with cation exchange adsorbent using flow velocities up to 375 cm/h. The yield of rhuNGF was approximately 95% and concentrated approximately 45 to 50 times. During method development, it was found that the dynamic binding capacity increased when the feedstock was applied at 37 °C (normally at 25 °C). The optimal conditions for recovery were determined in this small scale before the process was scaled up 65-fold.

In another example, *E. coli* was used to express modified exotoxin A from *Pseudomonas aeruginosa* (29). The exotoxin lacks the enzymatic activity but has retained binding activity. It accumulates in the periplasm of *E. coli*. After release of the exotoxin by osmotic shock, it was recovered using expanded-bed adsorption with an anion exchange adsorbent. To reduce the viscosity of the extract, it was treated with a DNase prior to application to a pilot-scale expanded-bed column with a 200-mm diameter. Cell extract from 4.5 kg of *E. coli* was applied to 4.7 L adsorbent at 400 cm/h. Recovery by expanded-bed adsorption gave an exotoxin with a concentration three times greater than that obtained using the conventional packed-bed process and with a slightly higher yield. Furthermore, the entire process was completed in one third of the time.

Humanized immunoglobulin G_4 was affinity purified from myeloma cell culture using expanded-bed adsorption at pilot scale (41). A 200-mm diameter expanded-bed column containing 4.7 L protein A adsorbent was used. The cell reactor was connected directly to the expanded bed, and approximately 100 L of cell culture was applied to the column at 37 °C. The purification process (application, wash, and elution) took 2.5 h and 40 g of the antibody was quantitatively recovered. The purity of the antibody was comparable to that obtained by the corresponding packed-bed process.

Other potential large-scale processes include the purification of proteins from milk or whey (46–48).

BIBLIOGRAPHY

1. S.-M. Lee, *J. Biotechnol.* **11**, 103–118 (1989).
2. W. Berthold and R. Kempken, *Cytotechnology* **15**, 229–242 (1994).
3. R. Kempken, A. Preißmann, and W. Berthold, *J. Indust. Microbiol.* **14**, 52–57 (1995).
4. H. Tebbe, D. Lütkemeyer, F. Gudermann, R. Heidemann, and J. Lehmann, *Cytotechnology* **22**, 119–127 (1996).
5. F.J. Bailey, R.T. Warf, and R.Z. Maigetter, *Enzyme Microb. Technol.* **12**, 647–652 (1990).
6. A.G. Fane, and J.M. Radovich, in J.A. Asenjo ed. *Separation Processes in Biotechnology*, Marcel Dekker, New York, 1990, pp. 209–262.
7. K.H. Kroner, H. Schütte, H. Hustedt, and M.-R. Kula, *Process Biochem.* **19**, 67–74 (1984).
8. H.G.J. Brummelhuis, in J. Curling ed. *Methods in Plasma Protein Fractionation*, Academic Press, London, 1980, pp. 117–128.
9. M. Better, R. Richieri, A. Tran, and W.C. McGregor, *Abstract from Recovery of Biological Products VIII*, Tucson, Ariz., Oct. 1996, p. 11.
10. M.J. Slater, *The Principles of Ion Exchange Technology*, Butterworth-Heinemann, Oxford, 1991, pp. 41–49.
11. E. Pungor Jr., N.B. Afeyan, N.F. Gordon, and C.L. Cooney, *Biotechnology* **5**, 604–608 (1987).
12. C.R. Bartels, G. Kleiman, J.N. Korzun, and D.B. Irish, *Chem. Eng. Prog.* **54**, 49–51 (1958).
13. F.P. Gailliot, C. Gleason, J.J. Wilson, and J. Zwarick, *Biotechnol. Prog.* **6**, 370–375, (1990).
14. A. Buijs and J.A. Wesselingh, *J. Chromatogr.* **201**, 319–327 (1980).
15. M.A. Burns and D.J. Graves, *Biotechnol. Prog.* **1**, 95–103 (1985).
16. N.M. Draeger and H.A. Chase, *Inst. Chem. Eng. Symposium Series* **118**, 161–172 (1990).
17. H.A. Chase and N.M. Draeger, *J. Chromatogr.* **597**, 129–145 (1992).
18. N.M. Draeger and H.A. Chase, *Bioseparation* **2**, 67–80 (1991).
19. G. Dasari, I. Prince, and M.T.W. Hearn, *J. Chromatogr.* **631**, 115–124 (1993).
20. C. Spence, C.A. Schaffer, S. Kessler, and P. Bailon, *Biomed. Chromatogr.* **8**, 236–241 (1994).
21. J. Thömmes, M. Weiher, A. Karau, and M.-R. Kula, *Biotechnol. Bioeng.* **48**, 367–374 (1995).
22. G.E. McCreath, H.A. Chase, R.O. Owen, and C.R. Lowe, *Biotechnol. Bioeng.* **48**, 341–354 (1995).
23. C.M. Wells, A. Lyddiatt, and K. Patel, in M.S. Verrall and M.J. Hudson eds., *Separations for Biotechnology*, Ellis-Horwood, Chichester, UK, 1987, pp. 217–224.

24. G.R. Gilchrist, M.T. Burns, and A. Lyddiatt, in D.L. Pyle ed., *Separations for Biotechnology 3*, The Royal Society of Chemistry, London, 1994, pp. 184–190.
25. P.H. Morton and A. Lyddiatt, in M.J. Slater ed., *Ion Exchange Advances*, Elsevier, London, 1992, pp. 237–244.
26. K.K. Unger, *Journal of Chromatography,* vol. 16, Elsevier, Amsterdam, 1979, pp. 12–14.
27. A.-K. Barnfield Frej, H.J. Johansson, S. Johansson, and P. Leijon, *Bioprocess Eng.* **16**, 57–63 (1997).
28. M. Hansson, S. Ståhl, R. Hjorth, M. Uhlén, and T. Moks, *Biotechnology* **12**, 285–288 (1994).
29. H.J. Johansson, C. Jägersten, and J. Shiloach, *J. Biotechnol.* **48**, 9–14 (1996).
30. A.-K. Barnfield Frej, R. Hjorth, and Å. Hammarström, *Biotechnol. Bioeng.* **44**, 922–929 (1994).
31. J.R. McDonald, M. Ong, C. Shen, Z. Parandoosh, B. Sosnowski, S. Bussell, and L.L. Houston, *Protein Express. Pur.* **8**, 97–108 (1996).
32. European Patent Application EP 0 699 687 A2 (Published 1996), M. Noda, A. Sumi, T. Ohmura, and K. Yokoyama (to The Green Cross Corporation, Osaka, Japan).
33. A. Pessoa Jr., R. Hartmann, M. Vitolo, and H. Hustedt, *J. Biotechnol.* **51**, 89–95 (1996).
34. C. Zurek, E. Kubis, P. Keup, D. Hörlein, J. Beunink, J. Thömmes, M.-R. Kula, C.P. Hollenberg, and G. Gellissen, *Process Biochem.* **31**, 679–689 (1996).
35. Y.K. Chang, G.E. McCreath, and H.A. Chase, *Biotechnol. Bioeng.* **48**, 355–366 (1995).
36. Y.K. Chang and H.A. Chase, *Biotechnol. Bioeng.* **49**, 204–216 (1996).
37. B.C. Batt, V.M. Yabannavar, and V. Singh, *Bioseparation* **5**, 41–52 (1995).
38. J.T. Beck, A. Liten, S. Viswanathan, A. Lindgren, J.C. Emery, and S.E. Builder, *Abstract from Recovery of Biological Products VIII*, Tucson, Ari., Oct. 1996, p. 74.
39. C. Born, J. Thömmes, M. Biselli, C. Wandrey, and M.-R. Kula, *Bioprocess Eng.* **15**, 21–29 (1996).
40. J.C. Erickson, J.D. Finch, and D.C. Greene, in B. Griffiths, R.E. Spier, and W. Berthold eds., *Animal Cell Technology: Products for Today, Perspectives for Tomorrow*, Butterworth-Heinemann, Oxford, 1994, pp. 557–560.
41. C. Jägersten, S. Johansson, R. Pardon, and J. Bonnerjea, *Abstract from Recovery of Biological Products VIII*, Tucson, Ariz., Oct. 1996, p. 46.
42. D. Lütkemeyer, N. Ameskamp, H. Tebbe, J. Wittler, and J. Lehmann, *Abstract from Recovery of Biological Products VIII*, Tucson, Ariz., Oct. 1996, p. 78.
43. J. Thömmes, M. Halfar, S. Lenz, and M.-R. Kula, *Biotechnol. Bioeng.* **45**, 205–211 (1995).
44. J. Thömmes, A. Bader, M. Halfar, A. Karau, and M.-R. Kula, *J. Chromatogr., A* **752**, 111–122 (1996).
45. G. Zapata, A. Lindgren, A.-K. Barnfield Frej, P. Leijon, A.D. Liten, T.L. Mayes, G. Blank, D. Narindray, B. Wagner, W. Galan, J. Beck, M. Press, and S. Builder, *Abstract from Recovery of Biological Products VIII*, Tucson, Ariz., Oct. 1996, p. 16.
46. A. Degener, M. Belew, and W.H. Velander, *Abstract from Recovery of Biological Products VIII*, Tucson, Ariz., Oct. 1996, p. 15.
47. W. Noppe, I. Hanssens, and M. De Cuyper, *J. Chromatogr., A.* **719**, 327–331 (1996).
48. W. Noppe, I. Hanssens, F. Geeraerts, and M. De Cuyper, *Abstract from First International Conference on Expanded Bed Adsorption*, Cambridge, U.K., Dec. 1996, p. 15.
49. A.-K. Barnfield Frej, *Bioseparation* **6**, 265–271 (1996).
50. M. Asif, N. Kalogerakis, and L.A. Behie, *AIChE J.* **37**, 1825–1832 (1991).
51. M.R. Al-Dibouni and J. Garside, *Trans. Inst. Chem. Eng.* **57**, 94–103 (1979).
52. J.F. Richardson and W.N. Zaki, *Trans. Inst. Chem. Eng.* **32**, 35–53 (1954).
53. E. Zafirakos and A. Lihme, *Abstract from Second International Conference on Expanded Bed Adsorption*, Napa Valley, Calif., June 21–23, 1998, p. 15.
54. K. Beyzavi and V. Sarantschin, *Abstract from Second International Conference on Expanded Bed Adsorption*, Napa Valley, Calif., June 21–23, 1998, p. 42.
55. A. Karau, C. Benken, J. Thömmes, and M.-R. Kula, *Biotechnol. Bioeng.* **55**, 54–64 (1997).
56. *Bioseparation and Bioprocessing, Vol. 1 Biochromatography: Expanded Bed Adsorption Chromatography*, Wiley-VCH Verlagsgesellschaft, Weinheim, Germany, 1998.
57. O. Levenspiel, *Chemical Reaction Engineering*, 2nd ed., Wiley, New York, 1972, pp. 253–295.
58. Y.K. Chang and H.A. Chase, *Biotechnol. Bioeng.* **49**, 512–526 (1996).
59. R. Hjorth, S. Kämpe, and M. Carlsson, *Bioseparation* **5**, 217–223 (1995).
60. J.A. Purvis, A. Binieda, C.A. Lewis, M. Pearce-Higgins, and P.G. Varley, *Abstract from First International Conference on Expanded Bed Adsorption*, Cambridge, U.K., Dec. 1996, p. 27.
61. A. Sumi, K. Okuyama, K. Kobayashi, W. Ohtani, T. Ohmura, and K. Yokoyama, *Abstract from Second International Conference on Expanded Bed Adsorption*, Napa Valley, Calif., June 21–23, 1998, p. 12.

ADDITIONAL READING

A. Mullick, C.M. Griffith, and M.C. Flickinger, *Biotechnol. Bioeng.* **60**, 333–340 (1998).

M.C. Flickinger, M.J. Robichaud, J.E. Morris, C.M. Griffith, M.J. Annen, P.W. Carr, and C. Dunlap, *U.S. Patent 5,837,826*, November 17, 1998.

See also ADSORBENTS, INORGANIC; ADSORPTION, PROTEIN, BATCH; ADSORPTION, PROTEINS WITH SYNTHETIC MATERIALS.

ADSORPTION, PROTEIN, BATCH

WILLIAM R. ADAMS
Merck Manufacturing Division
West Point, Pennsylvania

VICTOR GOETZ
ANN L. LEE
Merck Research Labs
West Point, Pennsylvania

KEY WORDS

Adsorption
Batch

Modeling
On-off chromatography
Protein
Separation

OUTLINE

INTRODUCTION

Batch adsorption technology has been successfully applied in a number of important protein separation processes. The principles of batch adsorption are used in a wide range of diverse applications, including several commercial-scale processes for the purification of protein products, a large number of biochemical assays (e.g., enzyme-linked immunosorbent assays), and design of biosensors and biomedical devices, where the aim could be to promote or retard protein adsorption. The focus of this article is on methods for recovery and purification of proteins, although the principles and considerations can be readily applied to the full range of applications.

Batch adsorption is considered here as a separation technique in which an entire protein-containing solution is contacted simultaneously with a fixed quantity of adsorbent. This is in contrast to adsorption in a fixed-bed column mode, where the feed stream flows through the column and is differentially contacted with the adsorbent. After adsorption, the wash and desorption procedures can be performed in a number of different ways. For example, elution may be performed in the original mixing vessel by changing the surrounding solution, in a different vessel after adsorbent recovery through an additional step such as centrifugation, or after transfer of the adsorbent into a column for ease of flowthrough operation. For desorption to be considered a batch operation, however, the elution must be performed in a manner consistent with single-stage solid–liquid contact. This implies that the eluting solution will have uniform product concentration, which is in contrast to the differential elution characteristic of chromatography.

Protein batch adsorption is often regarded as an outdated unit operation now superseded by modern column chromatography techniques incorporating a new generation of chromatographic supports. Before the introduction of rigid resins and high-performance liquid chromatography techniques, however, stirred tank batch contactors were the preferred configuration for scale-up of adsorptive separation steps, because the highly compressible resins could not even support their own weight when packed into process-scale chromatography columns. Nevertheless, an appropriate train of such stirred tanks could provide a moderate number of theoretical separation stages.

Although it is true that modern chromatographic techniques in many applications offer efficiencies and purification factors far superior to the stirred tank systems, there remain a number of applications where creative implementations of batch adsorption offer decided advantages. Because of the immediate and uniform contact of solution and adsorbent, batch adsorption can be performed faster than column-mode adsorption. Thus, it could offer an advantage in applications where a large volume of feed solution must be processed quickly to achieve efficient cycle times or where the limited stability or proteolytic susceptibility of protein products makes rapid processing essential. Batch adsorption may also be favored for handling crude feedstreams having high viscosity or debris that cannot be processed readily through column operations.

This review summarizes the physicochemical principles that define and control batch adsorption, describes a variety of classical and novel applications of these principles, and attempts to demonstrate that a widely overlooked unit operation still has significant utility in bioprocessing.

PRINCIPLES GOVERNING PROTEIN BATCH ADSORPTION

The overall effectiveness of batch adsorption processes will depend on both the equilibrium adsorbent capacities as well as the mass transfer and kinetic limitations. This section reviews the fundamental principles of protein adsorption phenomena, through which the suitability of batch adsorption as a separation or recovery process may be more readily understood or predicted.

Models for Protein Adsorption: Capacity and Competition

Single Component Adsorption. The specific mechanisms by which large macromolecular proteins contact and adsorb to a solid surface are varied and generally highly complex. Depending on the choice of protein, adsorbent, and solution properties, the surface attachment mechanism may be driven by several different general types of interactions, such as ion exchange, affinity (protein–ligand reaction), less-specific hydrophilic, hydrophobic, van der Waals, or hydrogen bonding. In principle, the thermodynamic treatment of protein adsorption from solution could be directly analogous to that for adsorption of small nonpolar molecules from the vapor phase for which there are several well-developed surface equations of state. For such simple molecule adsorption, the adsorbent surface geometric and energetic irregularities along with specific sorbate–surface interactions are generally the chief complexities (1). For protein adsorption in practice, however, thermodynamic treatment of equilibria is severely limited by the additional complexities arising from specific solvent–protein interactions, protein conformational variations both in solution and on the adsorbent surface, and frequent multipoint attachment configurations.

As a consequence of the somewhat intractable experimental and mathematical complexities, most models used for protein adsorption equilibria are simple empirical or semitheoretical models that are useful according to their goodness-of-fit to the experimental data. The semitheoretically based Langmuir adsorption isotherm given by

$$q = q_m C/(K_D + C) \quad (1)$$

where q is quantity adsorbed per unit of adsorbent, C is concentration of protein in solution, q_m is maximum quantity adsorbed at high C, and K_D is the disassociation or binding constant has been frequently fitted to protein adsorption data with adequate correlation. Although the physical assumptions underlying the development of this model are not followed by protein adsorption via ion exchange (2), the model nonetheless has provided a good fit if salt concentration-dependent parameters are used. For example, Weaver and Carta (3) fitted data for lysozyme on POROS 50 HS (a macroporous resin based on styrene-divinyl benzene copolymer) and on S-Hyper D-M (polystyrene-coated silica filled with functionalized polyacrylamide hydrogel), both of which possess strong cationic functionality. In their studies, the fitted values for q_m ranged from 109 to 262 mg/cm^3, and disassociation constants ranged from 5.3×10^{-3} to 2.6×10^{-1} mg/cm^3. The isotherms displayed classical protein adsorption characteristics; the nearly rectangular-shaped isotherms indicative of highly favorable adsorption at low salt concentration moderated to an almost linear-shaped isotherm indicative of weak adsorption at high salt concentration. Numerous other ion exchange protein adsorption studies have been performed with similar relative measures of adherence to the Langmuirian behavior.

Affinity adsorption may be highly selective, depending on the specific ligand–protein interaction, which may often be driven by a combination of electrostatic, hydrophobic, or hydrogen bonding type forces. Some affinity adsorption data has been successfully correlated with simple equilibrium models, based on the assumed simple reaction

$$[P] + [L] \Leftrightarrow [P - L] \quad (2)$$

where $[P]$ and $[L]$ denote concentration of protein and ligand, respectively, and from which an equilibrium constant may be defined such as

$$K_{eq} = [P][L]/[P - L_{complex}] \quad (3)$$

Generally, K_{eq} is small, often ranging between 10^{-4} M to 10^{-10} M, implying that binding is nearly irreversible or that the rate constants are related by $k_{adsorp} \gg k_{desorp}$. Adsorption data for many affinity systems can also be approximated through the fitted Langmuir model, although the range of fitted constants vary widely, depending on substrate–enzyme, antigen–antibody, carrier protein–hormone, or base sequence nucleic acid interactions. As an example, Chase describes the fit of data to equation 1 above for β-galactosidase adsorbing onto monoclonal antibodies immobilized on agarose gels, where it was found that K_D and q_m typically had values of 1.5×10^{-8} M and 6×10^{-9} gmol/mL, respectively (4).

Protein adsorption onto dye–ligand modified supports generally is controlled by a combination of electrostatic and hydrophobic protein–dye interactions. The Langmuir equation provides a reasonable fit to the data for many systems (5), although some data are better correlated through other models, such as the Freundlich isotherm. Typically, q_m for such systems is between those observed for high capacity ion exchange and the lower capacity affinity adsorbents. There are many other protein–adsorbent systems in which adsorption is driven predominantly by nonspecific mechanisms of interaction that could be used for either batch adsorption or chromatographic separations. Some of these systems are adequately fitted by simple models, whereas others, such as those displaying sigmoidal isotherms, clearly follow more elaborate adsorption mechanisms. A listing of representative adsorption capacities for several typical types of protein–adsorbent systems is shown in Table 1.

As indicated in Table 1, the choice of adsorbent and solution conditions provide a wide range of possible adsorbent capacities and relative strengths of adsorption. The suitability of any particular adsorbent–solution system to accomplish an effective separation process, however, will depend significantly on the adsorption capacity for the protein of interest relative to other competing proteins as well as on the relative rates of protein component adsorption and desorption.

Multicomponent Adsorption. Application of adsorption models to protein mixtures has been much more limited because of several factors:

1. Limitations of applicability of the empirical or semitheoretical models when extended to multiple adsorbing protein components
2. Limited adsorption database available for even model protein mixtures
3. The fact that industrially important separations, whether by batch adsorption or chromatographic adsorption processes, typically involve mixtures containing numerous protein components (many of which may be poorly characterized) or additional impurities that substantially confound the analysis

As an example model binary system, adsorption of mixtures of bovine serum albumin (BSA) and lysozyme on the strong cation exchanger S Sepharose FF has been studied (9). For this system, the single component isotherms were well correlated by the Langmuir equation. For the mixture data, the authors evaluated both competitive and

Table 1. Representative Capacities and Binding Constants for Protein Adsorption

Protein	Adsorbent	Mode	q_m (mg/mL)	K_D (mg/mL)	Ref. no.
BSA	DEAE Sepharose CL-6B	Ion exchange	82.1	6.0×10^{-2}	6
BSA	Q Sepharose (CL-4B) FF	Ion exchange	41.2	7.0×10^{-2}	6
α-Chymotrypsinogen	SynChroprep Propyl HIC	Hydrophobic	70.7	3.7×10^{-2}	7
β-galactosidase	Monoclonal antibody on agararose	Affinity (antibody)	3.2×10^{-3}	8.1×10^{-6}	4
IgG (human)	Avid AL gel	Affinity	22.7	5.8×10^{-3}	8

Note: BSA, bovine serum albumin; IgG, immunoglobulin G.

noncompetitive multicomponent extensions of the Langmuir equation. Neither model gave an accurate correlation to the data, although from this comparison and from additional chromatographic data it was concluded that competitive adsorption was apparent but that additional factors such as interprotein interactions were likely important as well. More recently, Lassen and Malmsten (10) reported on a study of multicomponent adsorption from ternary mixtures of human serum albumin (HSA), human immunoglobulin G (IgG) and human fibrinogen (Fgn) in 0.01 M phosphate buffer on three different polymer surfaces. The three surfaces were prepared to yield a spectrum of overall surface properties covering hydrophobic and hydrophilic with either positive or negative overall charge, respectively. For all surfaces, the single component isotherms were approximately Langmuirian in shape, with the relative maximum quantities adsorbed falling in the pattern Fgn > IgG > HSA. Adsorption from the multicomponent mixtures was notable in that (*1*) the total quantity adsorbed relative to the single component isotherms was reduced, and (*2*) the emergent dominant protein varied depending on the specific surface evaluated. Overall, the observed ternary behavior could not be readily predicted from the single component data.

Although not strictly of concern with respect to equilibrium loadings, the so-called Vroman effect, through which some protein mixtures exhibit sequential adsorption with smaller molecular weight proteins generally adsorbing early only to be later displaced by larger molecular weight proteins, has been noted for several protein–adsorbent systems (11). This rate-based competitive effect could occur within the time scale of practical application in separations, and thereby provide a contributing factor to adsorptive process performance.

The complex adsorption behavior observed even in relatively simple binary or ternary mixtures amply highlights the limitations of adsorption models and weakness of extensions thereof to multicomponent mixtures. It is therefore clear that the development of batch adsorption separation processes must be firmly based on experimental data.

Implications of Adsorption Performance to Separations. Equilibrium adsorption capacities (or q vs. C isotherms) define the first criteria for adsorption process feasibility. If capacity appears sufficient and if competitive adsorption is not a major problem, then the next concern would be fractional recovery. As discussed by Scopes (12), for practical applications that use small or modest volumes of adsorbent relative to solution (e.g., a ratio of volume adsorbent V_a to volume solution V_s of ~0.1), the separation factor α, defined by $\alpha = q/(C + q)$, should be greater than approximately 0.98. This in turn implies that K_{eq} values less than about 10^{-6} M are needed to achieve high fractional recoveries (e.g., a recovery of >90% in one adsorptive stage).

Assuming loading capacity and fractional recoveries from solution would be practical, the next major challenge in applying batch adsorption is to identify appropriate combinations of adsorbent and solution properties to provide a useful selectivity. Selectivity among a mixture of n components may be defined by

$$S = [(q_1/C_1)/(q_i/C_i)] \tag{4}$$

where subscript *1* denotes the desired product and i denotes some unwanted component, with i ranging from 2 to n. The required value for S depends of course on the objective of the separation. To achieve a 10-fold purification from an equimolar solution, for example, would require that S be greater than 10. High selectivity may be achievable through a particular combination of adsorption and desorption conditions; however, it is uncommon for most general adsorbents to provide high selectivity between many proteins. Because of the broad range of proteins generally present in bioseparation feedstreams (such as clarified broths or cell lysates originating from fermentation operations), the adsorbent candidates more likely to enable a successful application would employ specific interactions. Affinity adsorbents, for example, may be able to achieve such selectivity, and immunoadsorbents should be especially capable in light of the highly specific antigen–antibody complexation. If the selectivities of interest are low, highly effective separations could be more readily achieved through non–batch operations such as column chromatography, which can exploit even modest selectivity differences in equilibrium loadings or in relative rates of adsorption.

Dynamics of Protein Adsorption

Kinetic and Mass Transfer Considerations. Separations of practical industrial application will generally require substantial adsorbent capacity. The large surface areas consequently required will usually be provided through extensive macroporous structures. A general schematic of the paths available for a protein molecule in the bulk liquid phase to adsorb to the surface of such an adsorbent is indicated in Figure 1. The overall process will likely involve several diffusional steps as well as multiple interactions with the adsorbent surface moities.

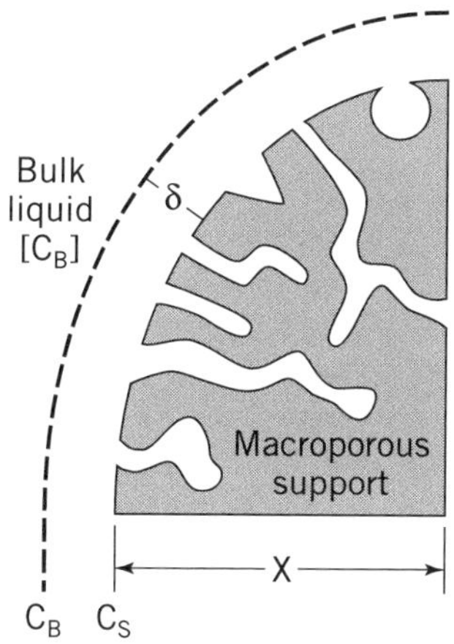

Figure 1. Schematic representation of available paths for protein adsorption on macroporous adsorbents.

The bulk phase in batch adsorption applications or experiments will normally be well mixed through appropriate design of mixing conditions in the contacting vessel. As such, the concentration of protein throughout the bulk phase will be uniform, as denoted by C_B. Surrounding each adsorbent particle will be a stagnant fluid layer, the thickness of which (δ) will depend on the hydrodynamic conditions. Diffusion through this film is normally approximated using a linear driving force in concentration

$$J = -D \times (C_B - C_S)/\delta \qquad (5)$$

where J is the protein mass flux, D is the free solution diffusivity, and C_S is the concentration at the external surface of the adsorbent. Diffusion within the macroporous structure is generally approximated by an equation of form

$$J = -\epsilon_p \times D_e \times dC/dx \qquad (6)$$

where ϵ_p is the adsorbent void fraction and D_e is the effective diffusivity, related to the free solution diffusivity by the approximation

$$D_e = D/\tau \qquad (7)$$

where τ is the tortuosity factor. There are a number of means to measure or estimate values for D and δ (13,14) as well as τ (15). Although the approximation of pore diffusional mass transfer with such simplistic equations allows for practical modeling and data correlation, the mechanistic inaccuracies are quite apparent. For example, because of the irregular geometry of the pores, there will be a large distribution of actual mass transfer trajectories (or random walks). Furthermore, because the proteins may be of comparable size to some pore passageways in certain adsorbents, there may be frequent opportunities for pore blocking, diffusional hindrances, or substantial variation in the number of contacts with the surface.

The kinetics of interaction between the protein in solution and the adsorbent surface will depend on many factors. Because the transitions in three-dimensional conformational states of the protein during the adsorption process are not well known, simplistic mechanisms are usually postulated in order to construct a mathematical representation. Normally, the adsorption kinetics are modeled using first- or second-order reaction rate equations to describe rates of transition between a few different states of protein or protein–adsorbent complexes. The kinetic constants thereby serve as adjustable model parameters. The rationale for the use and the accuracy of such representations is discussed in the next section.

Kinetic Studies: Ideal Surfaces and Industrial Adsorbents. Recent progress has been made in the study of protein–adsorption kinetics using well-characterized surfaces. In contrast to macroporous industrial adsorbents, in which diffusional limitations will usually be significant, planar surfaces within small experimental reactor volumes often will not be diffusion limited and hence can allow intrinsic adsorption rate measurement. Adsorption onto planar optical waveguide surfaces as monitored through reflectance techniques can be measured to within approximately ± 10 molecules per μm^{-2} at high sample frequencies (14). Kurrat et al. (16), for example, evaluated the kinetics of HSA and BSA adsorption onto a hydroxylated silica-titania surface. Approach to half saturation was achieved within 5 to 10 min, although adsorption toward equilibrium loadings continued out beyond 50 min. In developing and fitting kinetic equations, based on an elementary mechanistic model, they found that the model required at least two states of protein (distinguished by reversibly and irreversibly bound, respectively) to provide an acceptable fit to the data. Within the context of their equations, the adsorption kinetic constants k_a were on the order of 10^{-6} cm/s.

Wahlgren and Eloffson (17) studied adsorption kinetics of β-lactoglobulins A and B to hydrophobic methylated silica waveguide surfaces. Adsorption of these proteins, which show complicating self-associating behavior in solution, proceeded to completion over the time course of approximately 50 min. In applying their kinetic model, which allowed for both a first-order surface conformational change as well as an exchange reaction between adsorbed and solubilized protein, a reasonable fit to the data was obtained. It appears that an increase in model accuracy, however, would have to come through an allowance of additional nonidealities, such as from the range of protein orientations and conformations, surface heterogeneity, and probability factors for interactions, steric hindrances, etc. Along these lines, Jin et al. (18) attempted to account for steric hindrance effects by applying random sequential adsorption (RSA) principles to the case where adsorption may proceed reversibly. The resulting formalism provided an explanation for why the apparent adsorption kinetic constants can show a functional dependence on surface coverage.

The state-of-the-art adsorption rate modeling exemplified in the works cited earlier are useful for correlation of data required for practical applications and for aiding the development of theoretical understanding of viable biophysical mechanisms. However, the data analysis and modeling efforts also highlight a fundamental limitation:

the inability to experimentally distinguish or measure protein conformational variations and the distribution thereof on adsorbent surfaces. Until further progress in technique is achieved, modeling and understanding will remain limited to semitheoretical models based on an indirectly assessed distribution of proteins between only a few postulated states of conformation and surface binding.

The kinetics of protein adsorption to industrial macroporous adsorbents has been evaluated for many systems, of course, since the definition of adsorption rates is a prerequisite to application of any batch adsorption process. Basic rate data have generally been obtained through classical methods using either batch- or column-based experiments. As mentioned above, the adsorption dynamics will often have some significant diffusional resistances, and hence the kinetic data will reflect combined mass transfer and intrinsic adsorption reaction rates.

Modeling of Adsorption Process Dynamics. The ultimate success of potential applications of batch adsorption separation depends on the adroit manipulation of the thermodynamic and kinetic properties of the adsorbent-solute system to achieve high process efficiency. This aim can best be realized through development and application of a mathematical model for the process dynamics, which necessarily would be based on the underpinning factors governing intrinsic adsorption kinetics, equilibrium phase distribution, and mass transfer effects. The model used to simulate the process dynamics would facilitate the following:

1. *Feasibility analysis.* To assess the performance, whether for batch or column mode, for approximate performance in terms of yield and purification achievable
2. *Process design.* To enable detailed design, including scale-up effects, as well as to perform operational analyses to ascertain cycle times, economics of operation, etc.
3. *Optimization.* To readily calculate effects resulting from changes to the myriad of variables, such as adsorbent properties, relative volumes, initial concentration, contact times, regeneration procedures, etc., all of which could have substantial impact on process viability

Modeling of adsorption process dynamics is straightforward in principle but requires numerous simplifications at the detail level in order to retain mathematical tractability as well as biophysical reality in the adjustable model parameters. The general approach follows directly from the differential equations for mass balance, mass transfer, and kinetics for adsorption and any additional reactions. The process model, which is the resulting system of differential equations, will vary in complexity depending on the simplifying assumptions used to approximate the collective molecular behavior, which is usually highly complex because of the distribution of interaction rates and pathways occurring throughout an irregular three-dimensional geometry.

A number of models for protein adsorption within a finite batch have been published. As an example, Mao et al. (19) have presented a model for batch protein adsorption applicable specifically for nonporous adsorbents along with a simplified extension for porous adsorbents. By combining variants of equations 1 and 2, along with the differential form of the overall mass balance and an assumed second-order adsorption kinetic equation, the resulting equations could then be integrated to provide C and q as functions of time. The parameters in their model were first fitted to adsorption data for single component proteins HSA, lysozyme, and ferritin, as adsorbed onto either dye-affinity or ion-exchange adsorbents. In assessing the model fits, it was observed that the sensitivity of the models with respect to the kinetic constants may not be significant enough relative to the accuracy of the adsorption data to accurately ascertain true values of parameters. This conclusion may generally be extended to more elaborate models as well, where the accuracy of the fitted model may be improved by including additional kinetic parameters to account for additional reactions (states of conformation, etc.), but where the resulting understanding and the accuracy of extrapolation using the models is not improved. Mao et al. further used the models to evaluate relationships between various operational parameters and system performance. For example, calculations showed that the initial concentration of HSA could significantly affect both the final concentrations as well as the time required to approach equilibrium (e.g., time to 90% of final C/C_o), where a C_o varied from 0.8 to 0.05 mg/mL increased the required adsorption time from about 10 to more than 30 minutes. Of particular note was the extreme sensitivity of both the rate of adsorption and the equilibrium loadings to the ratio of adsorbent volume to liquid volume, $R_v(V_a/V_s)$. This would clearly be an important parameter to routinely evaluate in process design and operational optimization.

These sample calculations highlight the real utility of such models, which is that once proven capable of accurately representing the mass transfer and kinetic reaction rates of adsorption, they can be readily applied to adsorptive system design and optimization for either batch or column modes of operation.

Implications and Applications to Batch Separations. Knowledge of the mass transfer and kinetic limitation principles provides several general guidelines. In terms of process timing, for many systems it can be appreciated that equilibrium loadings may take hours to attain, but practical process-type loadings may be achieved on the order of 10 minutes. Analogously, adsorbent regeneration will often require hours, depending on the conditions employed to induce complete desorption and the requirements for regenerated adsorbent site occupancy. Although some approximate predictions of capacity and adsorption rates could be made by considering the basic features of the adsorbent (particle diameter, average pore size, primary mode of interaction with the protein, etc.) and of the protein in solution (molecular weight, size and shape, average surface potential, etc.), it is clear that acquisition of adsorption rate data for the specific system under consideration will always be an important prerequisite to successful process applications. Along these lines, relative to column-based experiments, batch experiments will generally be easier to perform and will provide more accurate information on the fundamentals of equilibrium capacity and limitations for mass transfer and kinetics of adsorption for particular protein–adsorbent systems.

The rationale for successful batch adsorption applications reviewed later in this article can be readily interpreted from the performance characteristics of relative adsorption rates and capacities. For example, it can be seen that if fast adsorption is required because of product lability, then a batch mode may be preferred to the column mode of operation. If speed and overall cycle time is a main driver, it has been shown (20) that a batch system may be more efficient than a column-based system under some conditions. Batch-contacting systems may also be suggested if there are other hydrodynamic factors, such as high solids/debris content or high viscosity which would negate other column operational advantages. Batch operations will also allow greater flexibility in choice of adsorbent, since even highly compressible gels can readily be contacted with the solution. On the other hand, if a high purification factor is desired from the adsorptive separation process, then a column mode of differential protein-adsorbent contact will generally be required other than for highly specific affinity adsorbents.

BATCH ADSORPTION AS A PRIMARY SEPARATION STEP

Except for highly specific affinity methods, batch adsorption techniques are not generally applicable to high-resolution separations. Instead, batch adsorption is ideal for so-called capture applications, where entire classes of impurities are to be eliminated in advance of high-resolution purification techniques, such as differential elution chromatography. Such bulk adsorptions also serve to reduce large volumes of dilute cell culture fluids or microbial lysates to more manageable volumes for further downstream processing.

Whereas solvent extraction is a popular first step in recovery of antibiotics from fermentation broths, adsorption is generally more gentle to proteins, which can be adsorbed without the denaturation often produced by organic extraction (21). Ion exchange adsorbents are preferred for highly polar molecules that, even in neutral form, exhibit low distribution coefficients for extraction.

Selectivity in batch ion exchange adsorption can be enhanced by judicious choice of adsorption buffer conditions (e.g., moderate salt concentration) to exploit differences in properties between the desired product and contaminants. This may be especially true in the case of recombinant proteins, which may be fundamentally different from the host organism's natural proteins. Adsorption equilibria that rely on weak physical interactions such as hydrophobic interaction, hydrogen bonding, and ion exchange are desirable for separations because these low-energy interactions can be readily reversed. Additional selectivity can be introduced during the desorption step so that adsorption processes are not without resolving power. Nonetheless, high selectivity is not the primary requirement for such an early process step.

With the introduction of the more specific nature of affinity interactions, well-thought-out batch adsorptions could accomplish clarification, concentration, and at least partial purification in one integrated step. Creative new configurations such as expanded-bed adsorption to approach this one-step purification ideal are discussed in more detail later ("Expanded Bed Chromatography").

CLASSICAL BATCH ADSORPTION PROCESSES

Antibiotic Purification

Much of the technology for batch adsorption of proteins and other biomolecules was originally developed for antibiotic purification from microbial fermentation sources. Both ion-exchange and hydrophobic adsorbents have been used for antibiotic purifications. Like other packed-bed applications, column adsorption is extremely susceptible to plugging by suspended insoluble materials in a feedstream. Therefore, either efficient precolumn filtration or stirred tank adsorber configurations predominated.

In the late 1950s, Bartels et al. (22) developed the technology for sorption of the basic antibiotic streptomycin on cation exchange resins directly from fermentation broths in well-agitated tanks (actually in the form of upflow columns fitted with agitators to keep solids dispersed) without need for costly and loss-associated prefiltration. This breakthrough helped make adsorption processes competitive with traditional solvent extraction methods. Belter et al. published a similar anion exchange process to isolate the acidic antibiotic novobiocin in a series of specially designed, well-mixed columns with screens to pass mycelia but retain the resin (23). Periodically, the lead column was removed from the train, washed free of insolubles, and eluted in fixed-bed mode. Periodic countercurrent operation was obtained in the sorption sequence by advancing each of the uneluted columns one position and placing a freshly eluted column in the trail position of the train.

Modifications to surface chemistry of nonpolar adsorbent resins such as XAD-2 allowed specific capture of known fermentation by-products, such as removal of desacetylcephalosporin C from cephalosporin C, to produce a more potent antibiotic (24).

Blood Plasma Fractionation

The well-known Cohn fractionation process, originally developed in the 1940s to isolate albumin as a blood volume expander during World War II, relies primarily on a series of ethanol fractionation steps to separate the various proteins of human blood plasma from one another according to their solubility behavior in the presence of ethanol (25). As the role of other plasma components became recognized, such as factors VIII and IX for the treatment of clotting disorders and IgG for use in passive immunization, higher yield and higher purity large-scale purification methods for these components were sought, and the importance of adsorption techniques grew.

With the development of cellulose ion exchangers in the mid-1950s, followed by the introduction of agarose-based resins (Sephadex®) a few years later, new chromatographic techniques specifically designed for protein separations became available for use in purification of blood plasma components. Despite the advances made, large-scale application of these methods in the biological industry developed slowly. One reason was the aseptic processing requirements often specified for biological raw materials. Chromatography using saline solutions and biodegradable supports could result in bacterial contamination if the appropriate precautions were not strictly followed. An inherent advantage of the Cohn fractionation process is that ethanol is bacteriostatic. But the main cause for the slow acceptance of chromatography at the large scale (i.e., fixed-bed operations rather than stirred-tank adsorption) was the poor mechanical qualities of the available macroporous supports designed for protein separations. However, these soft gels were used for some batch adsorption processes in stirred tank reactors for recovery of various blood products. Continual advances in resin technology through the 1970s provided supports that were mechanically strong and had surfaces and porosities well suited for protein chromatographic separations.

Although there are a number of advantages associated with column operations, packed beds of adsorptive media tended to act as depth filters for many of the biological feedstocks that were commonly used. This problem could be avoided by carrying out the adsorption in a stirred tank. This method has been used for many years for factor IX isolation from plasma, one of the first major applications of an adsorption method in plasma fractionation.

Factor IX complex, also known as prothrombin complex, can be prepared by a number of methods, but one of the most popular is to use batch adsorption with anion exchangers based on diethylaminoethyl (DEAE) groups attached to a stable support matrix such as cellulose, dextran, or agarose. Suomela et al. of the Finnish Red Cross Blood Transfusion Centre in Helsinki reported on a method developed to purify coagulation factor IX concentrate from human plasma (26). The production scale consisted of starting with 150 L of plasma, which was combined with pretreated and autoclaved DEAE-Sephadex A-50 and mixed for 30 minutes to complete adsorption. About 95% of the factor IX along with 5 to 6% of the other proteins were adsorbed from the plasma. The weakly bound proteins were removed by a low ionic strength wash. Adsorption was carried out in stainless steel tanks with mixing, followed by settling the gel to the bottom of the tank while the clear plasma was pumped off. The product containing DEAE-Sephadex gel was then pumped into a steel cylinder equipped with a bottom screen to retain the resin. Elution of the factor IX was achieved by increasing the buffer ionic strength. The resulting factor IX concentrate was purified about 100-fold, although it still contained relatively smaller amounts of factors II, VII, and X. The overall yield of the process was 60%.

A semicontinuous method for purification of factor IX complex from human plasma was described by Tharakan et al. (27). In the semicontinuous process, cryosupernatant plasma is pumped through a stirred column containing resin. Plasma was pumped out of the bottom of the column through the flow adapter screen at the same flow rate so that the volume in the column remained constant. After all the plasma had contacted the resin, the column was packed by gravity settling, and an upper flow adapter was installed. The column was washed with low ionic strength buffer until the absorbance of the effluent was negligible. The factor IX complex was then eluted with a higher ionic strength buffer. It was found that a residence time of 15 minutes was sufficient to capture 95% of the factor IX in the starting plasma. In the pilot plant, 550 L of plasma was passed through a 50-L column containing 8.5 L of resin, yielding a 68% recovery. Comparative studies involving batch, semicontinuous, and packed bed adsorption methods showed that although the amount of factor IX adsorbed from the plasma remained the same regardless of contacting mode, the mode of operation did affect the percentage of factor IX recovered, with increasing recoveries encountered as the resin density or ratio of resin to reactor volume increased (recoveries were 32%, 82%, and 100% batch, semicontinuous, and packed-bed operations, respectively). These results indicate that the adsorption and desorption process is not simple, and it was speculated that multipoint attachment resulting in denaturation or competitive binding of other proteins may have played a role. The semicontinuous method was developed to facilitate contacting plasma with inexpensive soft resins. It eliminates the need for handling a resin–plasma slurry, and it reduces total process time and labor requirements in comparison to batch adsorption. Furthermore, the resin is contained within one vessel (a modified column) throughout its use, thus minimizing handling and overall equipment requirements.

More recently, highly selective chromatography steps have been combined with the traditional batch adsorption step to separate factor IX from the other clotting factors (28) in human plasma. The factor IX concentrate from batch adsorption is passed through a more efficient anion exchanger, and factor VII is removed by stepwise elution. Separation of factor IX from factors II and X is carried out using a highly selective affinity column using heparin-derivatized agarose. Both factors II and X have a lower binding affinity for heparin than does factor IX. The resulting factor IX is purified more than 30-fold, resulting in an overall purification factor from plasma of about 10,000-fold.

The DEAE-ion exchangers have also been applied to the preparation of IgG, which, unlike other plasma proteins, is not bound by the DEAE-group at a neutral pH and low ionic strength. Several methods have been described that use ion-exchange materials alone or in combination with ethanol fractionation to purify IgG from plasma. A DEAE-Sephadex in combination with the ethanol fractionation technique was reported for the fractionation of IgG from blood plasma (29). The plasma proteins were first separated into three main fractions by increasing ethanol fractionations (8%, 25%, and 40% ethanol). The γ-globulin was recovered in the 25% ethanol fractionation precipitate. The paste was then dissolved and the pH adjusted to 6.5 in low ionic strength. DEAE-Sephadex (A50 coarse) was added at a level of about 1 g per gram of protein to be adsorbed, and the mixture was stirred for 30 minutes. The Sephadex gel was filtered off on a stainless steel Buchner-type funnel, and the γ-globulin, which was nonbound and in the filtrate, was subsequently dispensed for direct lyophilization.

Hepatitis B Purification

Fused silica, or Aerosil, was first used in a batch adsorption mode to adsorb lipoproteins from human serum (30).

The methodology was then applied to the adsorption of hepatitis B surface antigen (HBsAg) to remove hepatitis B virus from donated plasma (31). In the mid-70s, these methods were refined with respect to optimizing HBsAg adsorption selectivity and elution conditions for the purpose of purifying HBsAg from human plasma (32,33).

The classical methods for purification of HBsAg from sera were by isopycnic banding in a CsCl density gradient, followed by rate zonal centrifugation on a sucrose density gradient. It was cumbersome, however, to process large volumes of plasma or serum containing HBsAg by ultracentrifugation as the first step of purification. Duimel et al. (32) described a method whereby large quantities of antigen contained in serum could be partially purified by means of batch adsorption to colloidal silica (Aerosil) and then eluted with low ionic strength buffer (0.01 M borax, pH 9.3). Human positive serum was first extracted with Freon, which extracted part of the lipoids from serum, and then adsorbed with 2% Aerosil at 37 °C for 4 hours. After centrifugation for 15 min at 3,000 rpm, the Aerosil was washed four times with physiological saline. The Aerosil was eluted batchwise twice with borax, pH 9.3, for 30 min at 37 °C. Recoveries of HBsAg across this step were reported to be about 60%, providing a 15-fold purification factor. Pillot et al. (33) reported improving the recovery of the desorption step to 100% by eluting with 0.25% sodium deoxycholate in 0.01 M borax, pH 9.3, at 56 °C. Furthermore, elution at the higher temperature favored HBsAg desorption relative to some of the serum proteins, particularly albumin, which often represented the major contaminating protein eluted along with HBsAg.

The French vaccine manufacturer, Pasteur Merieux, used Aerosil adsorption with elution conditions similar to those described above for the production of a commercial plasma-derived HBsAg vaccine (34). In the mid-1980s, new second-generation hepatitis B vaccines based on recombinant DNA technology became available commercially. Many of the general purification principles that were used for the plasma-derived products were incorporated into purification schemes for HBsAg from recombinant sources (35). In the purification process for RECOMBIVAX HB® (Merck & Co., Inc.), batch adsorption of HBsAg onto colloidal silica is employed. The HBsAg is isolated from the clarified lysate by adsorption onto colloidal silica (Aerosil). The silica suspension is then collected by centrifugation, washed multiple times through resuspension and recentrifugation, and finally eluted from the silica by treatment with warm borate buffer. Sanford et al. reported that the antigen adsorbs to bare silica by a mixed mode and is specifically eluted by a chemical interaction between the silica surface and borate ion (36). More recently, this traditional batch operation has been evaluated for application in a fixed-bed column mode using a macroporous silica packing (37). The elution uses an extended warm borate buffer recycle through the column to achieve an equilibrium distribution between the entire bed volume before product collection. Although the adsorption, washing, and elution steps all take place within a column, it is still fundamentally a batch operation because the separation is achieved in a truly single stage contactor. The column mode of operation is able to be scaled and is more efficient in terms of overall cycle time as a result of the elimination of multiple centrifugation steps, despite the more lengthy column adsorption loading time.

NOVEL CONFIGURATIONS OF BATCH ADSORPTION

Fluidized Bed Contactors

In practice, batch adsorption generally requires large amounts of adsorbent because of relatively low product adsorption. Thus, a continuous or semicontinuous operation would be more efficient, but until recently, this has been severely constrained by the difficulties in processing viscous or debris-laden streams through fixed beds that act as depth filters and quickly clog. The use of a fluidized bed for batch adsorption of biomolecules is not entirely new, having been demonstrated in hybrid form for antibiotic recovery as early as 1958 (22). Because of the limitation of a well-mixed fluidized system to one theoretical plate, there was little inherent advantage of fluidized bed contactors over stirred tank systems. However, the introduction of segmented beds, magnetic stabilization, and, most recently, Pharmacia's expanded bed chromatography system has allowed multistage contacting in fluidized systems that exhibit comparable dispersion to packed beds of similarly sized particles (38).

Expanded-Bed Chromatography. In process-scale chromatography of proteins, most separations rely more on selectivity than on a high number of theoretical plates because achieving the latter necessarily requires high pressures (small diameter particles) and costly equipment. Given the accepted preference for low-pressure operations, a step from conventional packed-bed systems to recently developed expanded bed systems is often appropriate for unclarified feeds. These stable fluidized beds show axial dispersion and dynamic capacities comparable to those measured in packed bed.

An analogous technique for stabilizing fluidized beds incorporating the application of uniform magnetic fields to beds of magnetically susceptible adsorbent particles was extensively investigated by Rosensweig (39) and subsequently applied to biological separations by Burns and Graves (40). Although these systems have not been commercialized, they exhibit similar hydrodynamic characteristics to expanded-bed adsorption, facilitating processing of debris-laden streams while maintaining dispersion characteristics not dramatically greater than corresponding fixed beds. In this case, a uniform magnetic field prevents tumbling of the fluidized bed, creating a multistage contactor for potentially improved separation efficiency.

Whole Broth Adsorptive Extraction

Integrated fermentation and recovery may enable reduction of feedback inhibition resulting from product accumulation in the bioreactor, thus providing for an overall more productive system (41). Whereas integrated solvent extraction requires intermittent phase separation and culture recycle, integrated adsorption can be accomplished either by inclusion of adsorbents within the bioreactor or by passing filter-clarified broth through an external contactor with continuous recycle to the bioreactor.

An improved system for direct-contact broth extraction was developed by immobilizing finely divided affinity adsorbent particles in a large hydrogel bead (42). The large bead size facilitates separation from broth components, and the reversible Ca^{2+} hydrogel facilitates efficient recovery of the costly affinity ligand for recycle and reuse. Hydrogels, by virtue of their extremely high water content (>90%), offer limited diffusional resistance to the desired product while protecting the affinity ligand from fouling components of the broth. Addition of adsorptive particles to a cell culture or fermentation broth can also remove those components from the liquid medium that are responsible for low filtration fluxes frequently encountered in clarification of untreated streams (43). Even in those cases where flocculants thus formed raise the viscosity of the medium, the resulting flux is still less significantly impaired than that decreased by the presence of a significant concentration polarization layer.

Incorporation of an external fluidized bed contactor minimizes attrition to both adsorbent particles and cells encountered in direct inclusion systems while allowing circulation of whole broth to avoid frequent membrane fouling associated with broth clarification. Single-stage recirculating fluidized beds are less susceptible to time-based changes in biomass and viscosity over the course of a fermentation cycle than expanded beds, which are best suited to single-pass batch treatments. The greatest limitation on implementation of such integrated fermentation and recovery systems, as in the case of fluidized-bed purification systems, has been the lack of available supports. Because of the strict sterility requirements in fermentation or cell culture applications, any such adsorbents must also withstand sterilization.

Continuous Affinity Recycle Extraction

A variant of batch adsorption referred to as continuous affinity-recycle extraction (CARE) was developed in the late 1980s. It combines adsorption purification with membrane filtration for the continuous separation of proteins (44). The method employs two continuous stirred-tank reactors (CSTRs), one for adsorption and the other for desorption, between which the adsorbent is recirculated. The feed solution is continuously added to the adsorption tank, where the desired product is contacted with the selectively adsorbing resin. Simultaneously, contaminants are diluted with the addition of wash buffer and removed by passage through a screen filter, which allows even crude lysates to be processed. A stream of the product-containing adsorbent is pumped into the second CSTR along with the desorbing buffer. The product is eluted in the tank and recovered through an ultrafiltration membrane in the permeate stream. The regenerated adsorbent is recycled to the adsorption tank. By controlling the relative rates of the feed, product, and recycle streams, which are dictated by the rates of adsorption in CSTR-1 and desorption in CSTR-2, the system can be operated continuously at a steady state.

This scheme was demonstrated for both ion exchange and affinity modes of purification for the recovery of β-galactosidase from a partially clarified *Escherichia coli* lysate. A recovery yield of 70% was achieved with a 35-fold purification factor, although the product was substantially diluted (45).

Ultrafiltration-Coupled Adsorption Systems

Mattiasson et al. (46) described a method of ultrafiltration affinity purification to combine purification based on affinity interactions with membrane separation. By selecting membranes with pore sizes large enough to freely pass the protein of interest but then introducing recirculating (membrane-rejected) macromolecular ligands with specific affinity for this protein, all molecules except the target protein are rapidly washed from the feed pool. When all proteins not bound to the ligand are removed, the affinity complex may be dissociated, and the liberated material can pass through the membrane for collection. In batch experiments capturing concanavalin A using yeast cells as ligands, yields up to 70% of electrophoretically homogeneous product have been obtained. As in column-affinity chromatography, the adsorption, wash, and desorption steps can be optimized independently.

Fletcher and Deley (47) applied a similar concept using recirculating fine adsorbent particles, such as Biocryl bioprocessing aids, to purification of a peptide from highly colored cell culture streams. After pH adjustment for optimal binding to the adsorbent resin and a short incubation, the bound peptide was concentrated twofold and then diafiltered to remove remaining color bodies. The permeate contained colored material but minimal peptide. The loaded particles were then diafiltered against elution buffer, and product peptide was collected in the permeate. This preliminary removal of load-limiting contaminant allowed a subsequent reverse phase chromatography step to be used for polishing purposes, with a 100-fold increase in column capacity.

OUTLOOK FOR PROTEIN BATCH ADSORPTION

As described in the section on protein adsorption principles and evidenced within the context of the examples discussed in the preceding text, the application of batch adsorption operations has been limited by the modest protein selectivities generally found for most commercially available adsorbent systems. Development of more sophisticated high-affinity adsorbents through combinatorial chemistry or phage display approaches, which use high throughout screening to tailor and design specific ligands, will certainly enable additional batch adsorption applications by providing high selectivity separations.

Important future applications of batch adsorption for biomolecule recovery and purification will likely be developed for several niche bioseparations. For example, adsorbent-based purification of very large biomolecules including plasmid DNA, viruses, and recombinant viruslike particles are often characterized by very slow diffusion through pores in even the most macroporous resins available. This can make chromatographic-based operations problematic, because the overall mass transfer can require long contact times or make the effective surface area appear quite small. The development of new adsorbent materials with megapores may provide improvements in separation performance. Moreover, limitations in column operation because of slow adsorption can be overcome in some cases if the separation is performed in a batch operation mode. Analogously, slow desorption may be handled optimally through batch operation using either a well-mixed contactor or a column with column effluent-recycle configuration. In such cases, the recycle mode may result in minimal volumes for product elution relative to the large volume if carried out in a single-pass flow mode.

Improvements in adsorbents and processing equipment, especially for product-laden adsorbent handling to be consistent with aseptic operations, along with greater understanding and modeling of protein transport and adsorptive reactions will allow invention of additional adsorbent-based process technology schemes. The coupling of purification technologies, such as through adsorptive membranes or more specific fluidized bed batch adsorbers, will likely be increasingly applied as enhanced variants of traditional batch adsorption.

BIBLIOGRAPHY

1. D.M. Ruthven, *Principles of Adsorption and Adsorption Processes*, Wiley, New York, 1984, pp. 29–121.
2. A. Velayudhan and C. Horvath, *J. Chromatogr.* **443**, 13–29 (1988).
3. L. Weaver and G. Carta, *Biotechnol. Prog.* **12**, 342–355 (1996).
4. H.A. Chase, *Chem. Eng. Sci.* **38**, 1099–1125 (1984).
5. P.M. Boyer and J.T. Hsu, *Adv. Biochem. Eng. Biotechnol.* **49**, 1–44 (1993).
6. G.L. Skidmore and H.A. Chase, in M. Streat ed. *Ion Exchange for Industry*, Ellis Horwood Limited, Chichester, UK, 1988, pp. 520–532.
7. A. Tongta, A.I. Liapis, and D.J. Siehr, *J. Chromatogr., A* **686**, 21–29 1994.
8. J.Y. Shi and R.A. Goffe, *J. Chromatogr. A* **686**, 61–71 1994.
9. G.L. Skidmore and H.A. Chase, in D.L. Pyle ed., *Separations for Biotechnology 2*, Elsevier, New York, 1990. pp. 112–128.
10. B. Lassen and M. Malmsten, *J. Colloid Interface Sci.* **186**, 9–16 (1997).
11. S.M. Slack and T.A. Horbett, in T.A. Horbett and J.L. Brash eds. *Proteins at Interfaces II*, ACS Symp. Series No. 602, American Chemical Society, Washington, D.C., 1995, pp. 112–128.
12. R.K. Scopes, *Protein Purification: Principles and Practice*, 3rd ed., Springer, New York, 1994, pp. 121–126.
13. M. Wahlgren, T. Amebrant, and I. Lundstrom, *J. Colloid Interface Sci.* **175**, 506–514 (1995).
14. J.J. Ramsden, *Chem. Soc. Rev.* **24**, 73–78 (1995).
15. D. Farnan, D.D. Frey, and C. Horvath, *Biotechnol. Prog.* **13**, 429–439 (1997).
16. R. Kurrat, J.E. Prenosil, and J.J. Ramsden, *J. Colloid Interface Sci.* **185**, 1–8 (1997).
17. M. Wahlgren and U. Elofsson, *J. Colloid Interface Sci.* **188**, 121–129 (1997).
18. X. Jin, J. Talbot, and N.-H.L. Wang, *AIChE J.* **40**, 1685–1696 (1994).
19. Q. Mao, R. Stockman, I. Prince, M. Hearn, *J. Chromatogr.* **646**, 67–80 (1993).
20. B. Yang, M. Goto, and S. Goto, *Sep. Sci. Technol.* **24**, 741–754 (1989).
21. P.A. Belter, E.L. Cussler, and W.-S. Hu, *Bioseparations*, Wiley, New York, 1988, pp. 145–177.
22. C.R. Bartels, G. Kleiman, J.N. Korzun, and D.B. Irish, *Chem. Eng. Prog.* **54**, 49–51 (1958).
23. P.A. Belter, F.L. Cunningham, and J.W. Chem, *Biotech. Bioeng.* **15**, 533–549 (1973).
24. J.L. Casillas, J.L. Garrido, J. Aracil, M. Martinez, F. Adde-Yobo, in D.L. Pyle ed. *Separations for Biotechnology*, Volume 2, Elsevier, New York, 1990, pp. 285–294.
25. E.J. Cohn, L.E. Strong, W.L. Hughes Jr., D.J. Mulford, J.N. Ashworth, M. Melin, and H.L. Taylor, *J. Am. Chem. Soc.* **68**, 459–475 (1946).
26. H. Suomela, G. Myllyla, and E. Raaska, *Vox Sang.* **33**, 37–50 (1977).
27. J.P. Tharakan, D.M. Gee, and D.B. Clark, *Vox Sang.* **57**, 233–239 (1989).
28. T. Burnouf, in C.V. Prowse ed. *Plasma and Recombinant Blood Products in Medical Therapy*, Wiley, New York, 1992, pp. 67–87.
29. L.-G. Falksveden and G. Lundblad, in J.M. Curling ed. *Methods of Plasma Protein Fractionation*, Academic Press, New York, 1980, pp. 93–103.
30. W. Stephan and L. Roka, *Z. Clin. Chem. Clin. Biochem.* **6**, 186–190 (1968).
31. J.C. Siebke, E. Kjeldsberg, and T. Traavik, *Acta Pathol. Microbiol. Scand., B*, **80**, 935–936 (1972).
32. W.J.M. Duimel, H.G.J. Brummelhuis, and H.W. Krijnen, *Vox Sang.* **23**, 249–255 (1972).
33. J. Pillot, S. Goueffon, and R.G. Keros, *J. Clin. Microbiol.* **4**, 205–207 (1976).
34. F. Barin, M. Andre, A. Goudeau, P. Coursager, and P. Maupas, *Ann. Microbiol.* **129B**, 87–100 (1978).
35. R.D. Sitrin, D.E. Wampler, and R.W. Ellis, in R. Ellis ed. *Hepatitis B Vaccines in Clinical Practice*, Marcel Dekker, New York, 1993, pp. 83–101.
36. W. Sanford, D. Kubek, and R.D. Sitrin, *Am. Inst. Chem. Eng.*, Fall Meeting, Abstract 2751, Los Angeles, Calif., November 17–22, 1991.
37. M. Kosinski, D. Krips, W. Adams, W. Sanford, D. Kubek, and R. Sitrin, *Recovery of Biological Products VIII Meeting*, co-sponsored by ACS and Engineering Foundation, Abstract D.9, Tucson, AZ, October 20–25, 1996.
38. V. Goetz and D.J. Graves, *Powder Technol.* **64**, 81–92 (1991).
39. R.E. Rosensweig, *Science* **204**, 57–60 (1979).
40. M.A. Burns and D.J. Graves, *Biotechnol. Prog.* **1**, 95–103 (1985).
41. P. Morton and A. Lyddiatt, in D.L. Pyle ed. *Separations for Biotechnology*, Volume 3, Elsevier, New York, 1994, pp. 329–335.
42. S.C. Nigam and H.Y. Wang, in J.A. Asenjo and J. Hong eds. *Separation, Recovery, and Purification in Biotechnology*, ACS Symp. Series 314, American Chemical Society, Washington, D.C., 1986, pp. 153–168.
43. U.S. Pat. 4,830,753 (May 16, 1989), J.M.S. Cabral, E.M. Robinson, and C.L. Cooney (to Rohm and Haas Company, Philadelphia, Penn.).

44. N.F. Gordon and C.L. Cooney, in M.R. Ladisch ed., *Protein Purification: From Molecular Mechanisms to Large-Scale Processes*, ACS Symp. Series 427, American Chemical Society, Washington, D.C., 1990, pp. 118–138.
45. E. Pungor, N.B. Afeyan Jr., N.F. Gordon, C.L. Cooney, *Biotechnology* **5**, 604 (1987).
46. B. Mattiasson, T.G.I. Ling, and J.L. Nilsson, in I.M. Chaiken, M. Wilcheck, and I. Parikh eds. *Affinity Chromatography and Biological Recognition*, Academic Press, New York, 1983, pp. 223–227.
47. K. Fletcher and S. Deley, in D.L. Pyle ed. *Separations for Biotechnology*, Volume 2, Elsevier, New York, 1990, pp. 142–294.

See also ADSORBENTS, INORGANIC; ADSORPTION, EXPANDED BED; ADSORPTION, PROTEINS WITH SYNTHETIC MATERIALS.

ADSORPTION, PROTEINS WITH SYNTHETIC MATERIALS

JOSEPH MCGUIRE
MICHELLE K. BOTHWELL
Oregon State University
Corvallis, Oregon

KEY WORDS

Adhesion
Adsorption
Adsorption kinetics
Competitive adsorption
Conformational change
Exchange reaction
Interface
Kinetic modeling
Surface energetics
Synthetic materials

OUTLINE

INTRODUCTION

In general, the surfaces of materials of almost any type that come into contact with protein mixtures tend to become quickly occupied by proteins. These protein films can and often do lead to profound alterations in the properties of the material–fluid interface, affecting material performance in a number of bioprocess and biomedical applications. Performance of immobilized enzyme bioreactors, purification of therapeutic proteins in the biotech industry, drug formulation strategies, bacterial adhesion and contamination of industrial process equipment, and cell adhesion to natural and synthetic materials in the body are only a few examples of processes impacted by protein behavior at interfaces. The key impediment to needed progress in all these areas continues to be our inability to predict what the eventual make-up and nature of these films will be under a given set of conditions. Protein interactions with solid surfaces have been studied for decades, and several reviews are available (1–5). In this treatment, we describe some generally well-understood, important results that have provided a basis for meeting the challenges facing engineers and scientists in this area.

INTERFACES

Interfaces, as well as the interactions that take place in interfacial regions, can be complex. In fact, the interface has been described as a fourth state of matter (6). The properties of atoms or atomic groups at a material surface are different than those of the bulk material. The first layer of atoms, in contact with the fluid phase, is particularly unique. Chemical composition, molecular orientation, and properties relevant to crystallinity differ at the surface. In addition, surfaces have different electrical and optical properties and can be characterized by atomic- or molecular-level textures and roughnesses. Surfaces have wettabilities or hydrophobic/hydrophilic balances related to the factors named above. Further, surfaces are generally energetically heterogeneous. For example, although a surface may be assigned a particular wettability, it would most likely be the result of a distribution of surface regions of varying wettabilities.

In spite of this complexity, many researchers have met with success in describing some aspect of protein adsorption in terms of one or several surface properties. The effects of charge distribution, surface energy (i.e., whether it is high or low), and surface hydrophobicity, for example, have received much attention (1–5). From a purely thermodynamic standpoint, the extent of protein adsorption or biological adhesion in general could be determined purely by surface energetics, that is, the surface energies of the synthetic material, liquid medium, and adsorbates involved. Such an approach would imply that the free energy of adsorption is minimized at equilibrium. Adsorption would be favored if it caused the free energy function to decrease and would not be favored if it caused the function to increase. In the absence of electrostatic and specific receptor–ligand interactions, the change in free energy upon adsorption could be written

$$\Delta F_{ads} = \gamma_{AS} - \gamma_{AL} - \gamma_{SL} \tag{1}$$

where F_{ads}(J/m^2) is the free energy of adsorption per unit of surface area, and γ_{AS}, γ_{AL}, and γ_{SL} (J/m^2) are the adsorbate-solid, adsorbate-liquid, and solid-liquid interfacial energies, respectively.

If all the required interfacial energies of equation 1 could be estimated, one could predict the relative extent of adsorption among different surfaces. This would lead to a distinction between two situations (7,8), depending on whether adsorbate surface energy is greater than or less than the surface energy of the suspending liquid. Concerning protein adsorption from aqueous media, equation 1 would predict increasing adsorption with decreasing surface energy. In other words, a given protein would be expected to adsorb with greater affinity to hydrophobic as opposed to hydrophilic surfaces.

The importance of hydrophobic/hydrophilic balance in protein adsorption has prompted numerous investigators to develop techniques for measurement of this property at solid surfaces. Contact angle methods have been prominent in this regard (9). Contact angle analysis is inexpensive, rapid, and fairly sensitive. However, contact angle data can be difficult to interpret, and the technique is subject to artifacts caused by macroscopic, energetic heterogeneities in the surface, hysteresis, and drop-volume effects, among others. Still, useful conclusions regarding biological interactions with surfaces have been based on the results of contact angle analysis in areas of red blood cell adhesion, platelet adhesion, bacterial adhesion, and protein adsorption (9,10). Surface properties have been correlated to biological responses using other methods as well, including electron spectroscopy for chemical analysis (ESCA), secondary ion mass spectroscopy (SIMS), infrared and vibrational methods, and scanning probe microscopies. These methods and their relevance to biomedical technology were reviewed by Ratner and Porter (9).

The properties of a synthetic material's surface play a large role in dictating any biological response the material may evoke. But although much is known about selected surface property effects on protein adsorption, in a quantitative sense we know very little about how the molecular properties of protein influence its adsorption. Interfacial behavior is a cumulative property of a protein, influenced by many factors; among these are its size, shape, charge, and thermodynamic (thermal, structural, or conformational) stability. Experimentally observed differences in interfacial behavior among different protein molecules have been very difficult to quantify in terms of these factors because proteins usually vary substantially from one another in each category. The following discussion is an attempt to summarize the salient results from a wide range of experiments, focused on study of surface, solution, and protein effects on adsorption. Note that many observations have been explained in terms of a protein's charge, its tendency to unfold, and contact surface hydrophobicity.

PROTEINS AT INTERFACES

General Features of Adsorption from Single-Protein Solutions

Solution Chemistry Effects on Adsorption. The net charge of a protein in solution is dependent on the difference between pH of the solution and the isoelectric point (pI) of the protein. If the pH of the solution is greater than the pI, the net charge of the protein would be negative, whereas if the pH is less than the pI, the net charge of the protein would be positive. It is generally accepted that maximum adsorption occurs at the isoelectric point. As the out-of-balance charge of a protein increases, it will be in a more extended form than when the net charge is zero (11).

Norde and Lyklema (12) suggested that the degree to which pH affects the adsorption of a protein is determined by its conformational stability. They found that plateau values of adsorbed mass were independent of pH for structurally stable proteins, whereas those of less stable proteins varied considerably, apparently because less stable proteins were able to change structure with solution conditions. The effect of pH on protein adsorption and desorption can depend on solution history as well (13). Kondo and Higashitani (14) studied the adsorption of model proteins with wide variation in molecular properties. They explained the pH dependence of adsorbed mass in terms of lateral interactions. In particular, they suggested that lateral interactions between large protein molecules are stronger than those between small molecules. Large proteins would thus be expected to show maximum adsorption around their pIs, whereas the effect of pH on smaller proteins would be less pronounced.

The degree to which ionic strength affects protein adsorption is a function of the role electrostatic plays in the adsorption driving force. At low ionic strength, protein surface charge fully contributes to the total electrostatic interaction (11). At high ionic strength, the surface charges of proteins are shielded, reducing electrostatic interactions between proteins, whether attractive or repulsive (13). Luey et al. (15) showed that ionic strength effects on adsorbed mass are very much related to solid surface properties. They observed that increased ionic strength reduced the electrostatic repulsion between negatively charged β-lactoglobulin molecules and the hydrophilic, negatively charged surface they studied, increasing adsorbed mass. By contrast, increased ionic strength resulted in little change in adsorbed mass at hydrophobic surfaces.

Surface-Induced Conformational Changes. It is well accepted that a given protein can exist in multiple adsorbed conformational states on a surface (16–20). These states can be distinguished by differences in occupied area, binding strength, propensity to undergo exchange events with other proteins, and catalytic activity or function. All these features of adsorbed protein are interrelated and can be time-dependent. For example, decreases in surfactant-mediated elution of proteins from an adsorbed layer (an indirect measure of binding strength) are observed as protein–surface contact time increases (21). This time-dependence is illustrated in Figure 1. As conformational change proceeds, the likelihood of desorption decreases.

It has been observed that the extent of conformational change experienced by adsorbed fibrinogen increases with contact surface hydrophobicity (22). This is consistent with findings of Elwing et al. (23), who used ellipsometry to make inferences regarding conformational changes experienced by complement factor III, a plasma protein, on hydrophilic and hydrophobic silica surfaces. The results of

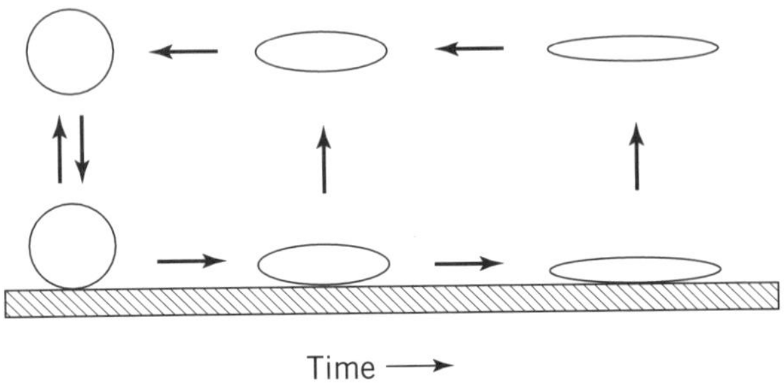

Figure 1. Surface-induced conformational changes undergone by adsorbed protein, resulting in multiple noncovalent bonds with the surface and coverage of greater interfacial area per molecule.

Elwing et al. also showed that greater values of adsorbed mass were found on hydrophobic as opposed to hydrophilic surfaces. Protein molecules are assumed, in general, to change conformation to a greater extent on hydrophobic surfaces. This is due to the effect of hydrophobic interactions between the solid surface and hydrophobic regions in the protein molecule. In fact, surface-induced unfolding is often characterized as entropically driven, because the hydrophobic protein interior associates with hydrophobic regions of the surface during unfolding. These interactions can give the molecule an extended structure, covering a relatively large area of the surface. If the repulsive force normally acting between native protein molecules is decreased for such structurally altered molecules, one would expect to measure a greater adsorbed mass on hydrophobic as opposed to hydrophilic surfaces. On the other hand, adsorption of positively charged protein to hydrophilic (negatively charged) silica can result in greater conformational change than adsorption of the same protein to hydrophobic silica, even with a greater extent of adsorption being observed at the hydrophobic silica surface (24). It is thus important to recognize that multiple factors affect the extents of protein adsorption and conformational change.

The concept that adsorbed proteins can exist in multiple states on a surface plays a role in interpretation of most if not all experiments in protein adsorption. Biophysicists rather routinely gain information relevant to protein structure in solution with circular dichroism (CD). It would be attractive to use CD in structural study of adsorbed protein as well. A recent innovation has made CD more applicable to study of structural changes during adsorption, and that is the use of colloidal silica particles, or nanoparticles (16–20). In these tests, particles have ranged from less than 10 to about 30 nm and are small enough not to interfere with the CD spectra. Individual molecules are allowed to adsorb to nanoparticles, resulting in a stable suspension of adsorbed protein. In this way, structural changes on adsorption have been unambiguously measured.

Work by Billsten et al. (19) and Tian et al. (20) have provided the most direct illustration of the effect of stability on structural rearrangements at a solid surface. Using site-directed mutants of bacteriophage T4 lysozyme, these investigators showed that both the rate and extent of secondary structure loss upon adsorption to colloidal silica were clearly related to protein thermal stability. With the same mutants, Fröberg et al. (25) used the interferometric surface force technique to study structural characteristics of adsorbed layers of T4 lysozyme. The results demonstrated that less-stable mutants lose their tertiary structure upon adsorption, whereas more stable mutants retain their globular shape.

Steady-State Adsorption Behavior. A great deal is known about how various conditions affect the steady-state adsorbed mass of protein. Numerous protein adsorption isotherms have been constructed and compared on the basis of temperature, pH, ionic strength, conformational stability of the protein in solution, and solid surface charge and hydrophobicity. The effects of protein conformational stability and solid surface properties are perhaps best revealed with reference to effects of pH and ionic strength.

In general, the effect of pH and ionic strength on protein adsorption is dependent on which type of interactions predominate (e.g., electrostatic, hydrophobic, or van der Waals interactions). At a negatively charged surface, if electrostatic interactions predominate, adsorbed mass should be greater at pH values below the isoelectric point relative to pH values above it. Below the isoelectric point, the protein and surface would be of opposite charge, whereas both the protein and surface would be negatively charged at pH values greater than the isoelectric point. As ionic strength increases, the electrostatic interaction would be reduced because of shielding of the protein by counterions; consequently, increasing the ionic strength should decrease adsorbed mass at pH values less than the isoelectric point and increase the adsorbed mass at greater values of pH. The relationship between adsorbed mass and changes in pH and ionic strength becomes inextricably linked to protein conformational stability. In general, pH and ionic strength conditions that lead to a less stable conformation for the protein in solution will lead to an increased adsorbed mass, assuming that the protein molecule would be more stable on the solid surface (15).

Another observation of importance is that protein adsorption is often an apparently irreversible process, at least in the sense that is often irreversible to dilution or buffer elution. The adsorbed mass remains constant or decreases very little when the solution in contact with the solid surface is depleted of protein. This irreversibility is more pronounced as protein–surface contact time increases. However, although spontaneous desorption is not generally observed, adsorbed protein can undergo exchange reactions with similar or dissimilar protein molecules adsorbing from solution (26). Such exchange reactions are shown schematically in Figure 2. Adsorbed protein exchange rates are likely state-dependent, being slower for more conformationally altered protein.

Some workers have reported that protein adsorbs onto a solid surface in more than one layer. Arnebrant et al. (27) studied adsorption of β-lactoglobulin and ovalbumin on hydrophilic and hydrophobic chromium surfaces using ellipsometry and electrical potential measurements. On hydrophilic surfaces, their results showed that a highly hydrated layer is obtained that can be partially removed by rinsing. They suggested that the protein adopts a bilayer formation

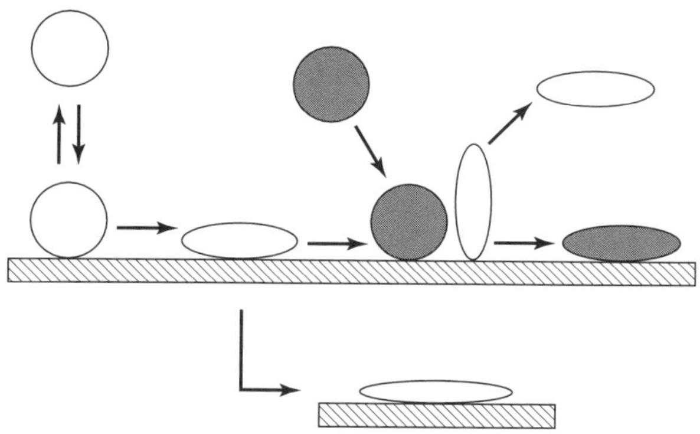

Figure 2. Exchange reaction between a conformationally altered, adsorbed protein and a dissimilar protein adsorbing from solution.

on the surface, with the layer in direct contact with the surface being unfolded and attached by strong polar bonds. Rinsing showed that the outer protein layer is loosely attached, which would imply that molecules in the outer layer have a structure closer to that of their native state. This adsorption behavior was described in terms of surface-induced conformational changes and charge interactions between the protein and surface. In particular, there are always polar amino acid side chains that can interact strongly with a surface, even if both the protein and surface are negatively charged. Such binding might be expected to result in unfolding of the protein. A consequence of this might be exposure of hydrophobic regions into aqueous solution; adsorption of a second protein layer would therefore reduce the interfacial free energy. In the case of protein in contact with a hydrophobic metal surface, values of adsorbed mass were found to be consistent with formation of a monolayer. Arnebrant and Nylander (28) reported possible bilayer formation upon adsorption of oligomeric units of insulin as well.

Adsorption Kinetics. In considering the kinetics of any interfacial process, the question of transport versus reaction control must be addressed. Protein adsorption at an interface is dependent not only on the intrinsic kinetic rate, which is a function of protein, solution, and surface properties, but also by the rate of protein transport from the bulk solution through the concentration boundary layer near the interface.

Proteins are macromolecules, and they can possess domains that differ chemically and physically. Diffusion coefficients may vary widely among proteins, depending on their concentration and the electrostatic condition of the solution (29). The initial adsorption rate of protein molecules at an interface can be transport-limited at either low or high concentration. The diffusion limitation exists as long as there is a significant concentration gradient near the solid surface. With careful design of an experimental system to minimize the transport-limited period, however, an intrinsic adsorption kinetic rate can be estimated. Still, relatively little is known about the nature of the adsorbed layer, and predictive models to describe any aspect of adsorption as a function of protein and interfacial properties are lacking. Protein adsorption is characterized by the likely presence of a time dependence in the development of bonds with the surface, a time dependence in the lateral mobility and exchangeability of the protein molecules, and time-dependent conformational changes; it is thus very difficult to describe mathematically.

Many experimental observations have indicated that a major portion of the final adsorbed amount had been adsorbed within the first few minutes of contact. Soderquist and Walton (30) proposed the existence of three distinct processes contributing to protein adsorption kinetics on polymeric surfaces. First, rapid and reversible adsorption of the proteins occurs during a short period of time. At up to 50 to 60% of surface coverage there is a random arrangement of adsorbed molecules, but then some form of surface transition occurs that is probably in the direction of surface ordering, thereby allowing further protein adsorption. Second, molecules on the surface undergo structural transitions as a function of time that occur in the direction of optimizing the protein-surface interaction. Third, as time increases the probability of desorption decreases, and the adsorption becomes irreversible to dilution.

Competitive Adsorption

Molecular Structure and Interfacial Behavior. Study of molecular influences on protein adsorption has received much attention because of its relevance to better understanding of adsorption competition in multiprotein mixtures. Important contributions to current understanding of molecular influences on protein adsorption have evolved from several comparative studies of protein interfacial behavior, in which similar or otherwise very well-characterized proteins (31–34), genetic variants (35,36), or site-directed mutants (19,20,24,25,37–39) of a single protein had been selected for study. A number of factors are known to affect protein adsorption, and these studies have stressed the importance of protein charge, hydrophobicity, and structural stability in interfacial behavior.

Shirahama et al. (33) studied hen lysozyme, ribonuclease A, and α-lactalbumin adsorption to hydrophilic and hydrophobic, polystyrene-coated silica (both negatively charged surfaces). At hydrophilic silica, they found that adsorbed mass increased with increasing charge contrast between the surface and protein. At the hydrophobic surface they found electrostatic interaction to have a lesser effect, in that the adsorbed mass was not clearly related to charge contrast between the surface and protein. Arai and Norde (31) described adsorption from single-protein solutions of hen lysozyme, ribonuclease A, myoglobin, and α-lactalbumin to synthetic materials varying in surface charge density and hydrophobicity. They concluded that at a given surface, adsorption of a globular protein is related to its structural stability. That is, proteins of high stability behave like hard particles at a surface, with the interactions governed by surface hydrophobicity and electrostatics, whereas adsorption of proteins of low stability (soft proteins) may be influenced by structural rearrangement, allowing adsorption to occur even under conditions of electrostatic repulsion.

Horsley et al. (35) compared isotherms constructed for hen and human lysozymes at silica derivatized to exhibit negatively charged, positively charged, or hydrophobic surfaces. Differences in adsorptive behavior observed between the two lysozyme variants were largely explained with reference to the fact that human lysozyme contains one less disulfide bond and is less thermally stable than hen lysozyme. Xu and Damodaran (36) compared adsorption kinetic data measured for native and denatured hen, human, and bacteriophage T4 lysozymes at the air–water interface. Their results showed substantial differences in adsorption dynamics among the three variants, as influenced by their structural state and the physical and chemical nature of the protein and surface.

Kato and Yutani (37) evaluated the interfacial behavior of six site-directed mutants of tryptophan synthase α-subunits, produced by amino acid substitution in the protein's interior, using surface tension, foaming, and emulsifying property measurements. The stability of these subunits, as measured by their free energy of denaturation in water, varied from about 5 to 17 kcal/mol. They were able to attribute differences in interfacial behavior to protein stability with good success. In particular, they observed that less stable mutants were most surface active, that is, they more rapidly adsorbed or more readily unfolded at the hydrophobic interfaces studied in that work.

Multicomponent Systems. Shirahama et al. (33) used hen lysozyme, ribonuclease A, and α-lactalbumin to study sequential and competitive adsorption at hydrophilic and hydrophobic, polystyrene-coated silica. At hydrophilic silica, they found that once an adsorbed layer of a given protein was formed, almost total displacement of that protein would occur upon introduction of a second protein to solution if the second protein had a more favorable capacity for electrostatic attraction with the surface (otherwise, sequential adsorption was not observed). In addition, when adsorbed from a mixture, the protein capable of more favorable electrostatic attraction with the surface preferentially adsorbed, essentially to the exclusion of the other protein. At the hydrophobic surface, they found that once an adsorbed layer of a given protein was formed, only partial displacement of that protein would occur upon introduction of a second protein to solution, even if the second protein had a more favorable capacity for electrostatic attraction with the surface. Moreover, the eventual make-up of a film adsorbed from a mixture was not related to charge contrast between the protein and surface. Other experimental observations (40,41) indicated that adsorbed protein molecules undergo exchange with protein from solution more readily on hydrophilic than on hydrophobic regions of a surface. Arai and Norde (32) studied the sequential and competitive adsorption behavior exhibited by hen lysozyme, ribonuclease A, myoglobin, and α-lactalbumin and concluded that whether introduced in sequence or in a mixture, adsorption of a globular protein is related to its structural stability. In particular, interfacial behavior of proteins of high stability is governed by surface hydrophobicity and electrostatics, whereas that of proteins of low stability are more influenced by structural rearrangement.

The elution of adsorbed protein by surfactant has been used to provide a measure of protein binding strength (21,24,42–51). The essential steps of this type of experiment are illustrated in Figure 3. Adsorption is allowed to occur, followed by rinsing with protein-free buffer. A surfactant solution is then introduced, after which adsorbed protein is displaced or solubilized (45). This is followed by rinsing and comparison of the amounts of protein present before surfactant addition and after the final rinse. Elution by dissimilar protein has been used as a measure as binding strength as well. Slack and Horbett (52) evaluated the strength of attachment of fibrinogen to solid surfaces by measuring its time-dependent elution in plasma and modeled fibrinogen adsorption, with reference to its rate of conversion from a weakly bound (exchangeable) to a tightly bound (nonexchangeable) state. Wahlgren and Arnebrant (45,46) used in situ ellipsometry to continuously monitor the different effects of cationic and anionic surfactants on the elution of β-lactoglobulin and lysozyme from hydrophilic and hydrophobic surfaces as well as adsorption from protein and surfactant mixtures. The elution studies allowed postulation of four mechanisms for surfactant-mediated elution of adsorbed protein. With the aim of relating elutability to protein molecular properties, Wahlgren et al. (47) studied removal of well-characterized proteins from silica surfaces using dodecyltrimethylammonium bromide. Some general trends regarding molecular property effects on elutability emerged from that work, but clear correlations between molecular properties and elutability remained difficult to quantify. By contrast, similar tests conducted with synthetic mutants of bacteriophage T4 lysozyme showed a clear correlation between protein stability and elutability (24). In particular, less stable proteins are more resistant to elution, presumably because they are more able to alter their conformation at the surfaces.

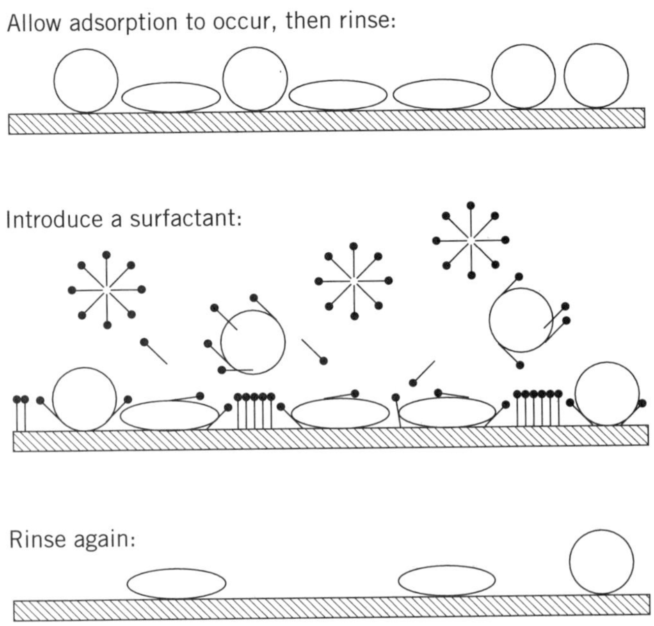

Figure 3. Experimental approach to evaluating adsorbed protein-binding strength using surfactant-mediated elution.

Modeling the Process. A number of mathematical models of protein adsorption at air–water and solid–water interfaces have been constructed (38,53–57). The problem is generally approached as an issue of molecular diffusion through a potential gradient, a reaction-diffusion problem involving interactions between diffusive-convective protein transport from the bulk solution and competitive adsorption and exchange kinetics on the surface, or as a kinetically controlled phenomenon, involving adsorption, unfolding, and exchange. Such models and the kinetic simulations they allow provide a framework with which the complexity of protein adsorption can be better understood and quantified. In addition, they can be used to provide direction for further experiments, particularly involving surface modification.

Lundström (58) presented an equilibrium model of protein adsorption on solid surfaces. The model described the fractional surface coverage of adsorbed molecules as a function of equilibrium concentration and allowed for reversible adsorption and conformation change. Later, Lundström and Elwing (26) described a model that allowed for bulk-surface exchange reactions among proteins in single-component and binary mixtures. The work featured manipulation of the equations describing the fractional surface coverage of protein in specific states and simulations of total surface coverage as a function of equilibrium concentration and as a function of time. Although no experimental data were presented, the shapes of the curves were in qualitative agreement with experimental observations. Currently, there is no adequate method to directly monitor changes in fractional surface coverages of protein in different adsorbed states. A less-complex model that can be statistically compared with the available data would be useful, because it would enable individual rate constants to be related to surface, solution, and protein properties.

Past work with synthetic mutants of bacteriophage T4 lysozyme have involved in situ ellipsometry and surfactant-mediated elution (24,51), ring tensiometry (38), the interferometric surface force technique (25), CD (19,20), and spectrophotometric assays of bound enzyme activity (39). These studies have shown that structural alterations definitely occur upon adsorption, with the extent and rate of structural change being related to thermal stability. In addition, mutants exhibited resistances to surfactant-mediated elution that were proportional to thermal stability and consequently related to extent of structural change. Finally, concerning a number of T4 lysozyme variants, results could be explained by modeling adsorption as occurring such that molecules adopt one of only two states, each varying in binding strength and occupied area, with differences in behavior among the molecules attributable to the relative populations in each state.

The simplest adsorption mechanism consistent with the fact that adsorbed proteins can exist in multiple states would include two adsorbed states. Figure 4 shows such a mechanism. Rate constants k_1 and k_2 govern adsorption into states 1 and 2, respectively. Although the mechanism is drawn to depict molecules adsorbing directly into states 1 and 2 from solution, a more accurate and detailed mechanism might include a multistep path to state 2. However, for modeling purposes, the actual path to state 2 is not consequential; we need only account for the different rates of generation of two functionally dissimilar forms of adsorbed protein. If adsorption of practically relevant proteins can be adequately described in this way, extension to the case of competitive protein adsorption would be straightforward.

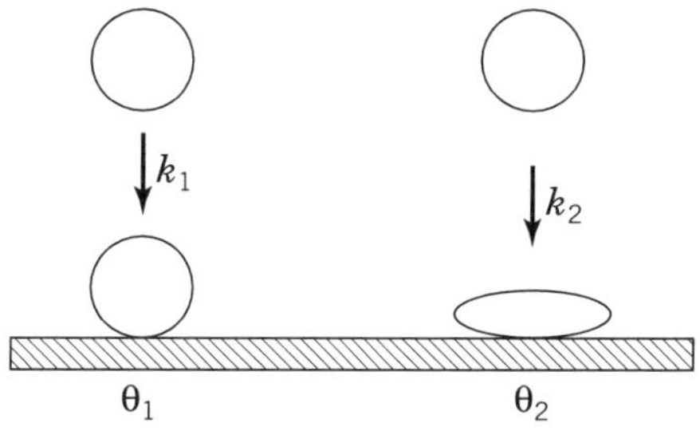

Figure 4. A simple mechanism for adsorption from a single-protein solution, allowing adsorption into one of two states exhibiting different resistances to elution and occupying different interfacial areas.

Figure 5 shows a mechanism for competitive adsorption (between two proteins, A and B) based on Figure 4. In each case, all associated rate constants can be determined a priori. The protein-specific k_1C and k_2C of Figure 4 can be obtained from single-component kinetic data, and the various exchange constants can be determined through sequential adsorption experiments (59). Figure 5 can be easily redrawn to depict competitive adsorption of three proteins, A, B, and C or more. As long as sequential adsorption data are available for each pair permutation of A, B, and C, for example, an a priori estimate of all rate constants can be made, and the adsorption competition can be simulated. Such comparisons would provide a basis for design of further experiments to better resolve molecular ef-

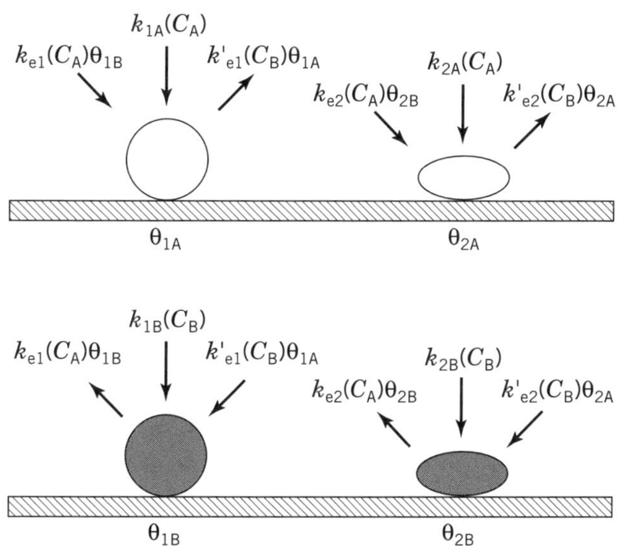

Figure 5. A mechanism for competitive adsorption between two proteins A and B, based on the single-component mechanism of Figure 4.

fects on competition in complex mixtures, subsequently enabling more quantitative prediction of adsorbed layer effects on practically relevant phenomena.

BIBLIOGRAPHY

1. J.D. Andrade, in J.D. Andrade ed. *Surface and Interfacial Aspects of Biomedical Polymers. Volume 2; Protein Adsorption*, Plenum Press, New York, 1985, pp. 1–80.
2. W. Norde, *Adv. Colloid Interface Sci.* **25**, 267–340 (1986).
3. T.A. Horbett and J.L. Brash, in J.L. Brash and T.A. Horbett eds. *Proteins at Interfaces: Physicochemical and Biochemical Studies*, ACS Symp. Series 343, American Chemical Society, Washington, D.C., 1987, pp. 1–35.
4. J.L. Brash and T.A. Horbett, in T.A. Horbett and J.L. Brash eds. *Proteins at Interfaces II: Fundamentals and Applications*, ACS Symp. Series 602, American Chemical Society, Washington, D.C., 1995, pp. 1–23.
5. J.D. Andrade, V. Hlady, L. Feng, and K. Tingey. in J.L. Brash and P.W. Wojciechowski eds. *Interfacial Phenomena and Bioproducts*, Dekker, New York, 1996, pp. 19–56.
6. C.B. Duke, *J. Vac. Sci. Technol., A* **2**, 139–143 (1984).
7. D. Absolom, F. Lamberti, Z. Policova, W. Zingg, C. Van Oss, and A. Neumann, *Appl. Environ. Microbiol.* **46**, 90–97 (1983).
8. D.R. Absolom, L.A. Hawthorne, G. Chang, *J. Biomed. Mater. Res.* **22**, 271–285 (1988).
9. B.D. Ratner, S.C. Porter, in J.L. Brash and P.W. Wojciechowski eds. *Interfacial Phenomena and Bioproducts*, Dekker, New York, 1996, pp. 57–84.
10. R.E. Baier and A.E. Meyer, in J.L. Brash and P.W. Wojciechowski eds. *Interfacial Phenomena and Bioproducts*, Dekker, New York, 1996, pp. 85–122.
11. S. Lee and E. Ruckenstein, *J. Colloid Interfac. Sci.* **125**, 365–379 (1988).
12. W. Norde and J. Lyklema, *J. Colloid Interfac. Sci.* **66**, 257–265 (1978).
13. P. Bagchi and S. Birmbaum, *J. Colloid Interfac. Sci.* **83**, 460–478 (1981).
14. A. Kondo and K. Higashitani, *J. Colloid Interfac. Sci.* **150**, 344–351 (1992).
15. J. Luey, J. McGuire, and R.D. Sproull, *J. Colloid Interface Sci.* **143**, 489–500 (1991).
16. A. Kondo, S. Oku, and K. Higashitani, *J. Colloid Interface Sci.* **143**, 214–221 (1991).
17. A. Kondo, F. Murakami, and K. Higashitani, *Biotechnol. Bioeng.* **40**, 889–894 (1992).
18. W. Norde and J.P. Favier, *Colloids Surf.* **64**, 87–93 (1992).
19. P. Billsten, M. Wahlgren, T. Arnebrant, J. McGuire, and H. Elwing, *J. Colloid Interface Sci.* **175**, 77–82 (1995).
20. M. Tian, W.-K. Lee, M.K. Bothwell, and J. McGuire, *J. Colloid Interface Sci.*, **200**, 146–154 (1998).
21. J.L. Bohnert and T.A. Horbett, *J. Colloid Interface Sci.* **111**, 363–377 (1986).
22. D.R. Lu and K. Park, *J. Colloid Interface Sci.* **144**, 271–281 (1991).
23. H. Elwing, S. Welin, A. Askendal, and I. Lundström, *J. Colloid Interface Sci.* **123**, 306–308 (1988).
24. J. McGuire, M.C. Wahlgren, and T. Arnebrant, *J. Colloid Interface Sci.* **170**, 182–192 (1995).
25. J.C. Fröberg, T. Arnebrant, J. McGuire, and P.M. Claesson, *Langmuir*, **14**, 456–462 (1998).
26. I. Lundström and H. Elwing, *J. Colloid Interface Sci.* **136**, 68–84 (1990).
27. T. Arnebrant, B. Ivarsson, K. Larsson, I. Lundström, and T. Nylander, *Prog. Colloid Polym. Sci.* **70**, 62–66 (1985).
28. T. Arnebrant and T. Nylander, *J. Colloid Interface Sci.* **122**, 557–565 (1988).
29. E.L. Cussler, *Diffusion: Mass Transfer in Fluid Systems*, Cambridge Univ. Press, Cambridge, Mass., 1989.
30. M.E. Soderquist and A.G. Walton, *J. Colloid Interface Sci.* **75**, 386–397 (1980).
31. T. Arai and W. Norde, *Colloids Surf.* **51**, 1–15 (1990).
32. T. Arai and W. Norde, *Colloids Surf.* **51**, 17–28 (1990).
33. H. Shirahama, J. Lyklema, and W. Norde, *J. Colloid Interface Sci.* **139**, 177–187 (1990).
34. A.-P. Wei, J.N. Herron, and J.D. Andrade, in D.J.A. Crommelin and H. Schellekens eds. *From Clone to Clinic*, Kluwer, Amsterdam, 1990, pp. 305–313.
35. D. Horsley, J. Herron, V. Hlady, and J.D. Andrade, in J.L. Brash and T.A. Horbett eds. *Proteins at Interfaces: Physicochemical and Biochemical Studies*, ACS Symp. Series 343, American Chemical Society, Washington, D.C., 1987, p. 290.
36. S. Xu and S. Damodaran, *J. Colloid Interface Sci.* **159**, 124–133 (1993).
37. A. Kato and K. Yutani, *Protein Eng.* **2**, 153–156 (1988).
38. J. Wang and J. McGuire, *J. Colloid Interface Sci.* **185**, 317–323 (1997).
39. C.K. Bower, Q. Xu, and J. McGuire, *Biotechnol. Bioeng.*, **58**, 658–662 (1998).
40. H. Elwing, S. Welin, A. Askendal, U. Nilsson, and I. Lundström, *J. Colloid Interface Sci.* **119**, 203–210 (1987).
41. H. Elwing, A. Askendal, and I. Lundström, *Prog. Colloid Polym. Sci.* **74**, 103–107 (1987).
42. R.J. Rapoza and T.A. Horbett, *J. Colloid Interface Sci.* **136**, 480–493 (1990).
43. R.J. Rapoza and T.A. Horbett, *J. Biomed. Mater. Res.* **24**, 1263–1282 (1990).
44. S.I. Ertel, B.D. Ratner, and T.A. Horbett, *J. Colloid Interface Sci.* **147**, 433–442 (1991).
45. M.C. Wahlgren and T. Arnebrant, *J. Colloid Interface Sci.* **142**, 503–511 (1991).
46. M.C. Wahlgren and T. Arnebrant, *J. Colloid Interface Sci.* **148**, 201–206 (1992).
47. M.C. Wahlgren, M.A. Paulsson, and T. Arnebrant, *Colloids Surf., A* **70**, 139–149 (1993).
48. M.C. Wahlgren, T. Arnebrant, A. Askendal, and S. Welin-Klinström, *Colloids Surf., A* **70**, 151–158 (1993).
49. V. Krisdhasima, P. Vinaraphong, and J. McGuire, *J. Colloid Interface Sci.* **161**, 325–334 (1993).
50. P. Vinaraphong, V. Krisdhasima, and J. McGuire, *J. Colloid Interface Sci.* **174**, 351–360 (1995).
51. J. McGuire, M.C. Wahlgren, and T. Arnebrant, *J. Colloid Interface Sci.* **170**, 193–202 (1995).
52. S.M. Slack and T.A. Horbett, *J. Colloid Interface Sci.* **133**, 148–165 (1989).
53. G. Narsimhan and F. Uraizee, *Biotechnol. Prog.* **8**, 187–196 (992).
54. V. Krisdhasima, J. McGuire, and R. Sproull, *J. Colloid Interface Sci.* **154**, 337–350 (1992).

55. C.F. Lu, A. Nadarajah, and K.K. Chitter, *J. Colloid Interface Sci.* **168**, 152–161 (1994).
56. M.C. Wahlgren, T. Arnebrant, and I. Lundström, *J. Colloid Interface Sci.* **175**, 506–514 (1995).
57. P. Dejardin, I. Cottin, *Colloids Surf., B* **4**, 111–120 (1995).
58. I. Lundström, *Prog. Colloid Polym. Sci.* **70**, 76–82 (1985).
59. M.A. Daeschel and J. McGuire, *Biotechnol. Genet. Eng. Rev.*, **15**, 413–438 (1998).

See also ADSORBENTS, INORGANIC; ADSORPTION, EXPANDED BED; ADSORPTION, PROTEIN, BATCH; CHROMATOGRAPHY, ION EXCHANGE; CHROMATOGRAPHY, SIZE EXCLUSION.

AFFINITY FUSIONS, GENE EXPRESSION

STEFAN STÅHL
JOAKIM NILSSON
SOPHIA HOBER
MATHIAS UHLÉN
PER-ÅKE NYGREN
Royal Institute of Technology
Stockholm, Sweden

KEY WORDS

Affibody
Affinity purification
Affinity tag
Downstream processing
Expression system
Functional cDNA analysis
IgG binding
Immunodetection
In vivo folding
Phage display
Renaturation
Serum albumin
Stabilization
Staphylococcal protein A
Streptococcal protein G
Vaccine development

OUTLINE

INTRODUCTION

The strong, specific, and well-characterized interaction between the Fc part of immunoglobulin G (IgG) and *Staphylococcus aureus* protein A (SpA) was the first biointeraction exploited for the creation of an affinity fusion system allowing affinity chromatography purification of expressed recombinant gene products (1). To date, hundreds of proteins have been produced as SpA fusions in a wide variety of host cells (2–5). During the past decade a large number of additional affinity fusion systems have been presented (for reviews, see 2,6–9). Affinity fusions are primarily used to simplify recovery of a fused target gene product but are also widely employed for detection and immobilization purposes (10). Most affinity tags are easy to use as detection moieties to monitor the production or distribution of a produced fusion protein, and certain affinity tags are particularly suited to obtain efficient and directed immobilization of a target protein.

This article describes the use of two affinity tags, the IgG-binding SpA domains and the serum albumin–binding region of streptococcal protein G (SpG), which both carry a number of features that make them highly suited as fusion partners in different applications. The wide spectrum of applications for affinity fusions in biotechnology is illustrated by selected examples in which SpA or SpG fusion proteins have been used to *(1)* improve the stability to proteolysis of produced recombinant proteins, *(2)* allow integrated downstream processing schemes by careful genetic design, *(3)* improve in vitro renaturation schemes for expressed gene products, *(4)* enable affinity-assisted folding in vivo of target proteins, *(5)* prolong the in vivo half-life of therapeutic proteins, *(6)* facilitate subunit vaccine development and functional cDNA analysis, and *(7)* select novel receptor variants with new specificities by the use of phage display technology.

COMMONLY USED AFFINITY TAGS

A large number of different affinity fusion systems have been described involving fusion partners that range in size from one amino acid to large multisubunit proteins capable of selective interaction with a ligand immobilized onto a chromatography matrix. Different types of interactions have been utilized, such as enzymes–substrates, bacterial receptors–serum proteins, polyhistidines–metal ions, and antibodies–antigens (2). The degree of specificity in the interaction and conditions for the affinity purification differ from system to system. The environment tolerable by the target protein is an important factor for deciding which affinity fusion system to choose. Also, other factors, including costs for the affinity matrix and buffers and the possibilities for column sanitation and reuse, are important to consider. Numerous gene fusion systems for affinity purification have been described in the literature, all with different characteristics (2,6–9). Some of the most commonly used systems are listed in Table 1, and some relevant references are given for each system. These different affinity fusion systems, with their advantages and disadvantages, were described in a recent review (10). The polyhistidyl tags, which suffer from a somewhat lower specificity, have the advantage of allowing affinity purification under denaturing conditions. The in vivo biotinylated "Bio-tag" is particularly suited for immobilization purposes because of the strong interaction (association constant; $K_A \approx 10^{15}$ M^{-1}) between biotin and streptavidin, but is inconvenient to use for affinity purification purposes because efficient elution is difficult to accomplish. One of the most extensively used affinity tags is the glutathione S-transferase (GST) from the parasitic helminth *Shistosoma japonicum* (17). GST fusion proteins can be purified using glutathione as immobilized ligand, also following renaturation from inclusion bodies after solubilization with 6 M guanidinium hydrochloride (18). A possible complication associated with the GST fusion system is the use of reduced glutathione (a reducing agent) for elution, which can affect target proteins containing disulfides (10).

Although a multitude of systems have been described, no single affinity fusion strategy is ideal for all expression or purification situations. For example, if secretion of the product into the periplasm or culture medium is desired, fusion to affinity tails derived from normally intracellular proteins (e.g., the β-galactosidase system) is not applicable, and if the purification must be performed under denaturating conditions, protein ligands such as monoclonal antibodies (mAbs; e.g., anti-FLAG M1 and M2 mAbs) are unsuitable. Therefore, to obtain general expression vectors for affinity gene fusion strategies applicable in several situations, a combination of affinity fusion domains could be introduced into a single fusion partner. Such composite fusion partners, consisting of several independent affinity domains, could potentially be used in different detection, purification, and immobilization situations, employing the affinity function that is most suitable for the situation. However, the included affinity domains should be carefully chosen not to interfere functionally with each other, and each moiety should be able to withstand the purification conditions dictated by the affinity domain requiring the harshest affinity chromatography conditions. A further advantage using combined affinity tags is that different methods for the detection and immobilization of the fusion protein also can be envisioned.

Recently, a new affinity fusion partner was described in which this strategy was utilized (26). An affinity-tag combination was designed consisting of the in vivo biotinylated peptide (Bio), a hexahistidyl peptide (His_6), and the SpG-derived albumin-binding protein (ABP) that could allow for flexible use in different detection, purification, and immobilization situations (26). Fusions to the Bio–His_6–ABP affinity tag could be used for easy detection using commercial streptavidin conjugates, for purification under both native and denaturating conditions, and for directed immobilization purposes with an option to choose among three strong affinity interactions. In addition, the ABP part of the affinity tag is highly soluble and effectively refolds after denaturation, which can be an advantage during refolding of the target protein fused to the Bio–His_6–ABP affinity tag. Such composite affinity tags have the potential of becoming widely used in the future because they offer flexibility.

This article does not focus on the description of different characteristics of various affinity-tag systems, as up-to-date reviews are available (4,9,10), but instead describes various application areas in which affinity fusions to the

Table 1. Commonly Used Affinity Fusion Systems

Fusion partner	Size	Ligand	Supplier	Refs.
Protein A	31 kDa	hIgG	Pharmacia Biotech	5,11
Z	7 kDa	hIgG	Pharmacia Biotech	12,13
ABP	5–25 kDa	HSA		5,14
Poly His	≈1 kDa	Me^{2+} chelator	Novagen, Qiagen	15,16
GST	25 kDa	Glutathione	Pharmacia Biotech	17,18
MBP	41 kDa	Amylose	New England Biolabs	19,20
FLAG	1 kDa	MAbs M1/M2	Kodak	21,22
PinPoint[a]	13 kDa	Streptavidin/avidin	Promega	23,24
Bio[b]	13 aa	Streptavidin/avidin		25,26

Note: aa, amino acids; ABP, albumin-binding protein; GST, glutathione S-transferase; hIgG, human immunoglobulin G; HSA, human serum albumin; MAb, monoclonal antibody; MBP, maltose-binding protein.

[a] A 13-kDA subunit of the transcarboxylase complex from *Propionibacterium shermanii*, which is biotinylated in vivo by the *E. coli* cells.

[b] A 13-amino acid peptide that is selected from a combinatorial library and found to be biotinylated in vivo.

IgG-binding domains of SpA and the serum albumin-binding regions of SpG have been used.

PROTEIN A AND PROTEIN G AFFINITY FUSION SYSTEMS

SpA (Fig. 1a) is an immunoglobulin-binding surface receptor found in the Gram-positive bacterium *Staphylococcus aureus*. SpA has found extensive use in immunological and biotechnological research (2–4,11,27,28). SpA binds to the constant (Fc) part of certain immunoglobulins, but the exact biological significance of this property is not clarified (29). Functional and structural analysis of SpA has revealed the presence of a signal peptide, S, which is processed during secretion, five highly homologous domains (E, D, A, B and C) all capable of binding to IgG (30), followed by a cell wall–attaching structure designated XM (31–35) (Fig. 1a). Here, X represents a charged and repetitive region, postulated to interact with the peptidoglycan cell wall (31), whereas M is a region common for Gram-positive cell surface-bound receptors containing an $LPX_{aa}TG$ motif, linked to a C-terminal hydrophobic region ending with a charged tail (32–35). It has been demonstrated that the complete M region is required for cell surface anchoring and that the cell wall sorting is accompanied by proteolytic cleavage at the SpA C-terminus and subsequent covalent linking of the surface receptor to the cell wall (32–35). SpA binds to IgG from most mammalian species, including humans. Of the four subclasses of human IgG, SpA binds to IgG1, IgG2, and IgG4 but shows only weak interaction with IgG3 (36).

Several characteristics of the IgG-binding domains of SpA have made them suitable as fusion partners for the production and purification of recombinant proteins: *(1)* SpA is proteolytically stable. *(2)* The structure of the IgG-binding domains, each being a three-helix bundle (37–39), appears to be favorable for independent folding of the fusion partner and the fused product because the N- and C-termini of each IgG-binding domain are solvent exposed (38,40). *(3)* It has been demonstrated to be feasible to introduce sequences accessible for site-specific cleavage of SpA fusion proteins so as to release the target gene product (41). *(4)* The high solubility of SpA enables the production

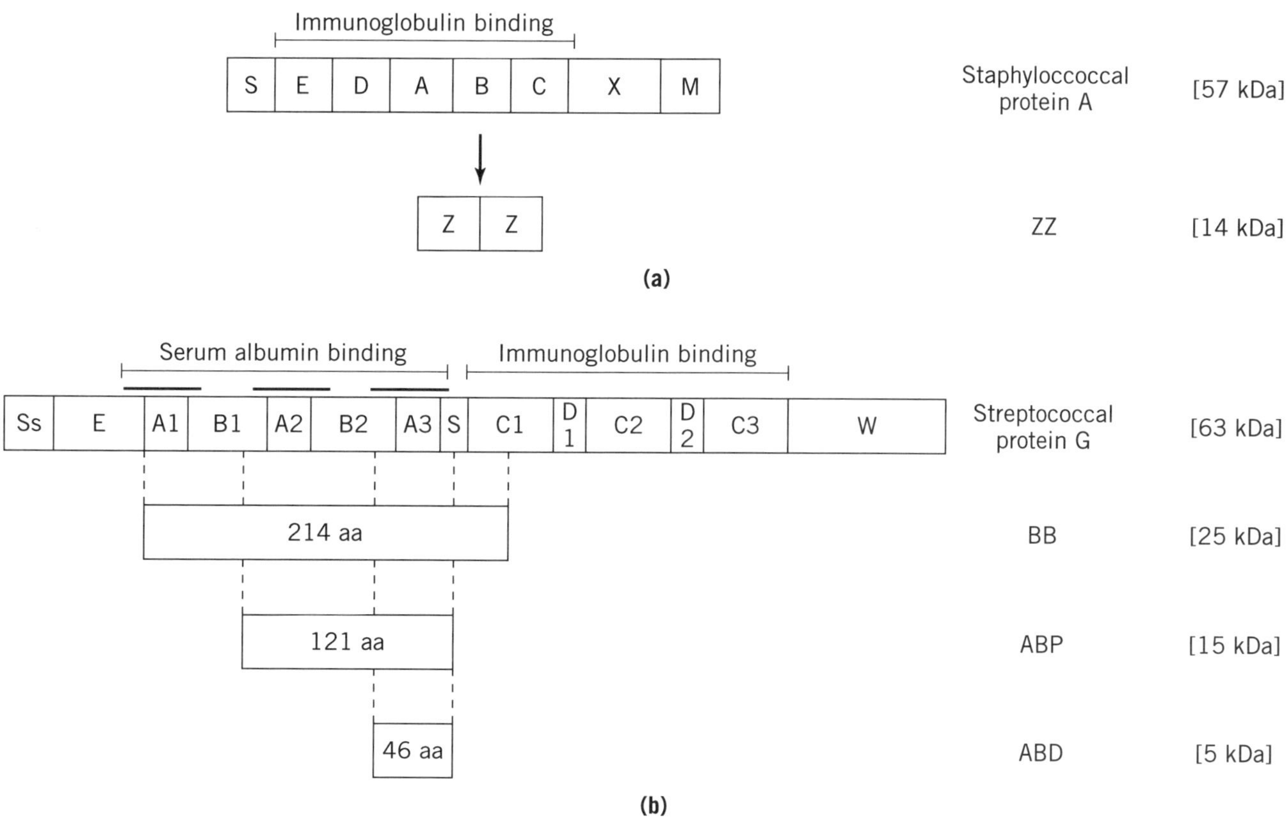

Figure 1. Schematic presentation of staphylococcal protein A (SpA) and streptococcal protein G (SpG). **(a)** The complete SpA gene product and the 7-kDa B-domain analog Z, represented in the most widely used divalent form. **(b)** The SpG gene product, with the separate albumin- and immunoglobulin-binding regions indicated, and a presentation of some of the most commonly used albumin-binding affinity tags. BB (25 kDa) was the first isolated serum albumin–binding affinity tag (14). More recently, the new tags ABP (15 kDa) and ABD (5 kDa), containing two or one serum albumin–binding motifs, respectively, were designed on the basis of postulated borders for the albumin-binding protein domains (13,47). The three postulated minimal albumin-binding motifs are indicated as *bars* above the SpG.

of soluble fusion proteins to very high concentrations within the *Escherichia coli* cell (42,43).

To create a "second-generation" affinity tag, an engineered IgG-binding domain, Z, was designed based on domain B of SpA (Fig. 1a) (12). This domain naturally lacks methionines, making it resistant to CNBr cleavage. In addition, an NG dipeptide sequence present in all native domains was changed to NA by altering the glycine codon to a codon for alanine, rendering the Z-domain also resistant to cleavage by hydroxylamine (12). Analysis of the interaction between IgG and Z domains, polymerized to different multiplicities, demonstrated that a dimeric form, ZZ (Fig. 1a), was the optimal fusion partner by its strong IgG-binding (44) and efficient secretion (12).

A different affinity-tag system, which has a number of features in common with the SpA system but also some additional advantageous properties, is based on the serum albumin-binding region of streptococcal protein G (SpG; Fig. 1b) (14,45). The regions of SpG mediating binding activities to serum albumin and IgG have been shown to be structurally separated (Fig. 1b) (14). Figure 1b illustrates SpG schematically and presents some existing variants of albumin-binding affinity tags. SpG binds to serum albumins of different mammalian species, including humans, mouse, and rat (45). The complete region comprises three serum albumin-binding motifs (46), each being approximately 46 amino acids in size. One of these postulated minimal motifs has been produced and tested for its serum albumin binding (47) and is considered to be a suitable affinity tag in a mono- or divalent fashion. However, tags comprising one, two, two-and-a-half, or three serum albumin-binding motifs (Fig. 1b) have all been successfully produced and used in different applications (4,13,27,47–50). The affinity tags, denoted BB, ABP (for albumin-binding protein), and ABD (for albumin-binding domain) (Fig. 1b), are proteolytically stable, highly soluble, and possible to produce at high yields (13,26,48,51). The 46–amino acid albumin-binding domain (ABD) was recently subjected to a heteronuclear nuclear magnetic resonance (NMR) analysis, concluding that it constitutes a three-helix bundle (52), remarkably similar to that of the IgG-binding domains of SpA.

Host Cell Systems, Affinity Resins, and Detection Systems

To date, a multitude of proteins have been produced as fusions to the IgG-binding domains of SpA (native domains or the engineered Z-domains; see following), in host cells from all five kingdoms of living organisms: *(1)* Gram-negative bacteria, such as *E. coli* (more than 100 published examples, including Refs. 13,28,53–55) and *Salmonella typhimurium* (56,57), as well as Gram-positive bacteria such as *Staphylococcus aureus* (58,59), *Streptococcus sanguis* (60), and *Bacillus subtilis* (61); *(2)* the yeast *Saccharomyces cerevisiae* (62–66); *(3)* plants (67); *(4)* insect cells (68,69); and *(5)* mammalian cells (70–75).

The SpG-derived expression systems have found several applications (as are described later), some completely new, and some as a complement to the SpA/ZZ systems. The albumin-binding tags have not yet been evaluated for production in all host organisms but have been successfully expressed for different purposes in Gram-negative bacteria such as *E. coli* (13,14,26,47,51), Gram-positive bacteria such as *Staphylococcus xylosus* and *Staphylococcus carnosus* (48,76–80), and mammalian cells such as Chinese hamster ovary (CHO) cells (81).

The SpA fusion proteins, which bind to the Fc region of IgG, are commonly affinity purified using IgG-Sepharose (50,82) and the serum albumin-binding SpG fusions are normally purified on human serum albumin (HSA) Sepharose (14,50). However, other matrices, such as protein-coated paramagnetic beads, have also been utilized (83). SpA (ZZ) and SpG (BB, ABP, or ABD) fusion proteins can be easily detected during expression, without the need of specific antisera, by blotting techniques after polyacrylamide gel electrophoresis (PAGE) by taking advantage of their IgG- (Fc) and HSA-binding properties, respectively. SpA fusions are efficiently stained using a commercially available complex of rabbit anti-horseradish peroxidase–IgG and horseradish peroxidase (13). SpG fusions are stained using biotinylated HSA and conjugated streptavidin–alkaline phosphatase (13).

Expression Vector Systems for *E. coli*

A number of expression vectors based on gene fusions to the SpA gene, often in the form of the divalent SpA analogue ZZ (12), or all five native IgG-binding domains, have been developed for recombinant protein production in *E. coli* (11,82,84). The promoter and secretion signals of SpA are functional also in the Gram-negative *E. coli* (1). Protein A fusions expressed in *E. coli* can thus be efficiently secreted to the periplasm of the bacteria, and in many cases also to the culture medium (50,53,55,85,86) from which they can be easily purified by IgG affinity chromatography (86,87). In the vectors designed for secretion, the expressed SpA-fusion products are in most examples transcribed from the SpA promoter, which is constitutive and efficiently recognized in *E. coli* (27,50).

Analogous expression vectors for production of secreted fusions also to the serum albumin–binding regions of SpG have been developed (14,50), taking advantage of the same SpA promoter and secretion signals. Fusion proteins produced in this way can be efficiently affinity purified using human serum albumin (HSA) columns (14).

There exists no general strategy applicable to all recombinant gene products to guarantee secretion. Instead, the inherent properties of the target protein largely dictate if a secretion route is possible to use. However, a large fraction of the proteins produced as SpA fusions are soluble and efficiently secreted to the periplasmic space and culture medium of *E. coli* (2,55,86). This may at least be partially caused by the high solubility of the ZZ domains, which most likely increases the overall solubility of the fusion proteins. There are a number of advantages connected with a secretion strategy. First, the gene product will be less exposed to cytoplasmic proteases, which might enable production of also labile proteins (88). Second, disulfide bond formation, which is occurring in the nonreducing environment outside the cytoplasm, could improve on the folding of certain gene products (86). Third, the recovery of the recombinant protein is simplified because a large degree of the purification from host cell proteins has been achieved through the secretion (55,86). Hansson and coworkers (55) described the *E. coli* production of a fusion protein ZZ-M5, a malaria subunit vaccine candidate, using a secretion strategy. More than 65% of the recombinant gene product was found to be secreted to the culture medium, from which it could be recovered in a single step by expanded-bed anion exchange adsorption. As the ZZ-M5 protein was to be used in a preclinical malaria vaccine trial in *Aotus* monkeys (89), a polishing step was included in which the IgG-binding capacity of the fusion protein was employed. After affinity chromatography on IgG-Sepharose, contaminating DNA and endotoxin levels were well below the demands set by regulatory authorities. The overall yield of the process, performed in pilot scale, exceeded 90%, resulting in 550 mg product per liter of culture (55).

As mentioned earlier, there is no guarantee that a certain protein can be secreted even if an expression system with secretion signals is chosen for the production. Proteins with a strong tendency to precipitate intracellularly and proteins containing hydrophobic transmembrane regions may be extremely difficult to secrete (27,80). Intracellular expression of such proteins could be a more attractive alternative because of recent advances in in vitro renaturation of recombinant proteins from intracellular precipitates, that is, inclusion bodies (90). The mechanism for inclusion body formation is not fully elucidated, but expression driven from exceptionally strong promoters such as the T7 promoter (91), leading to very high expression levels, seems to increase the tendency for intracellular aggregation of the produced recombinant protein (90). Production by the inclusion body strategy has the main advantages that the recombinant product normally is protected from proteolysis and that it can be produced in large quantities. Levels up to 50% of total cell protein content have been reported (90).

Vectors for intracellular production of SpA fusions have thus been constructed in which the transcription is under the control of inducible promoters, such as *lac*UV5 (11), λ_{PR} (11), *trp* (92,93), *trc* (94), or T7 (13), enabling a more controlled expression. Also, for intracellular production of BB and ABP fusion proteins, inducible vector systems have been used, employing *trc* (94), *trp* (51,93), or T7 (13) promoters, respectively. The most robust and in addition the most tightly controlled of these systems seems to be the T7 RNA polymerase–regulated promoter system (91). Using this system, Larsson and coworkers (13) recently described the expression of five different mouse cDNAs as fusions to ZZ and ABP, respectively. Expression levels ranged from 4 to 500 mg gene product per liter of culture.

Intracellularly produced and still soluble fusion proteins can be released by sonication or high-pressure homogenization of cells before purification by IgG or HSA affinity. However, precipitated gene products need to be solubilized before the affinity purification procedure. It was recently demonstrated that affinity purification indeed can be useful also for the recovery of proteins with a strong tendency to aggregate during renaturation from inclusion bodies following standard protocols (93). Murby and coworkers (93) devised a recovery scheme for the production of such ZZ and BB fusions, applied on various fragments of the fusion glycoprotein (F) from the human respiratory syncytial virus (RSV). Earlier attempts to produce these labile and precipitation-prone polypeptides in *E. coli* had failed, but several different F fusion proteins could by this novel strategy be produced and recovered as full-length products with substantial yields (20–50 mg/L). Because it was demonstrated that the IgG and HSA affinity ligands both were resistant to 0.5 M guanidine hydrochloride, efficient recovery from inclusion bodies of the ZZ-F and BB-F fusions could be achieved by affinity chromatography in the presence of the chaotropic agent throughout the purification process (93). The described strategy has so far been successfully evaluated for a number of proteins of low solubility and should be of interest for efficient recovery of other heterologous proteins that form inclusion bodies when expressed in a bacterial host.

It is, of course, almost impossible to give general information concerning the production levels for the described expression systems because production levels are strongly dependent on inherent properties of the target gene products. It is however evident that both the secretion strategy and intracellular protein production in *E. coli* using the SpA and SpG expression systems can give high expression levels. For the secreted production, expression levels of 0.5 to 0.8 g product per liter of culture have been reported (55,95), and for intracellular production levels up to 3 g gene product per liter of culture have been obtained (96). The SpA and SpG expression systems are thus for many gene products a good choice because high expression levels can be obtained and efficient affinity purification of produced fusion proteins can be easily achieved.

Mild Elution of Protein A and Protein G Fusions by Competition

One possible drawback associated with the SpA and SpG affinity purification systems has been the need for elution by low pH from the affinity column. The elution of fusion proteins at pH 3, used routinely, has for some products been destructive, yielding biologically inactive products. By a recently presented concept, the low-pH elution can be circumvented by competitive elution based on an engineered competitor protein that can be efficiently removed from the eluate mix after elution. The principle for the competitive elution strategy was described by Nilsson and coworkers (47) (Fig. 2). The target protein in the presented examples was produced as a fusion to a monovalent Z-domain of SpA. A sample (e.g., crude cell lysate) containing the recombinant fusion Z-target is passed through an IgG-Sepharose column. The recombinant fusion protein is thus captured, enabling extensive washing to remove contaminants. A stoichiometric excess of an engineered bifunctional competitor fusion protein ZZ-BB (Fig. 2) was added in which BB constitutes a second affinity tail capable of selective binding (association constant; $K_A \approx 1.4 \times 10^9$ M^{-1}) to HSA. The competitor ZZ-BB, having an affinity to human IgG more than 10-fold higher than that to the monovalent Z, is used for competitive elution of the Z-fusion at neutral pH. Specific removal of ZZ-BB from the effluent mix is accomplished by a passage through a second affinity column (HSA) to which the ZZ-BB fusion is bound.

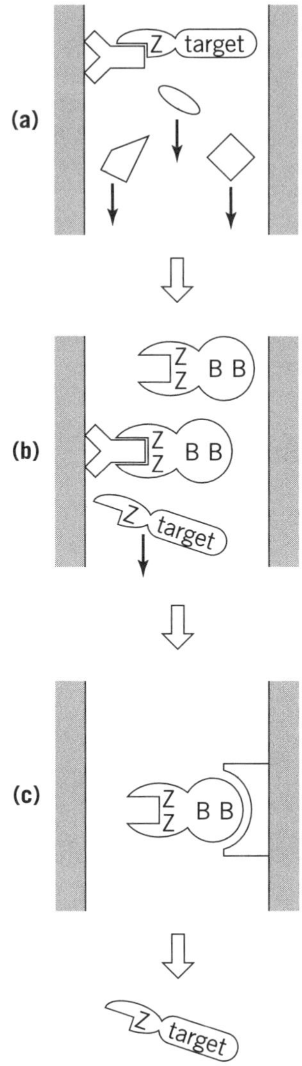

Figure 2. The concept of competitive elution, here used for elution of a monovalent Z-fusion by the competitor ZZ-BB. *Source*: Modified from Nisson and coworkers (47).

The described strategy has been used to produce the Klenow fragment of *E. coli* DNA polymerase I as a Z-Klenow fusion, and the recovered product retained full enzymatic activity (47).

Stabilization to Proteolysis of Labile Gene Products

When expressing mammalian proteins in a heterologous host such as *E. coli*, proteolytic degradation is one of the more common problems encountered. A review describing the different strategies available to minimize proteolytic degradation of recombinant gene products was recently presented (97). Fusion of a proteolytically labile target protein to stable fusion partners, such as the ZZ and ABP tags, has been shown to have a certain, but limited, stabilizing effect (88,97). However, a much more pronounced stabilizing effect has been obtained when employing a dual affinity fusion concept (Fig. 3) (98), where the protein of interest, X, is fused between the two different affinity tag proteins, ZZ and BB. The concept was applied to the production of proteolytically sensitive human insulin-like growth factor II (IGF-II) in *E. coli*. The concept first of all allows the use of two consecutive affinity chromatography steps, but in addition, dramatic stabilization effects can be obtained as compared with more conventional expression routes. The dual affinity fusion approach was found to have a stabilizing effect on three small and labile recombinant proteins, namely human proinsulin, a domain from rat protein disulfide isomerase, and a fragment from a human T-cell receptor (88), when expressed in *E. coli*. The concept of dual affinity fusions has thereafter been employed also with affinity tags other than the SpA and SpG tags (99–102).

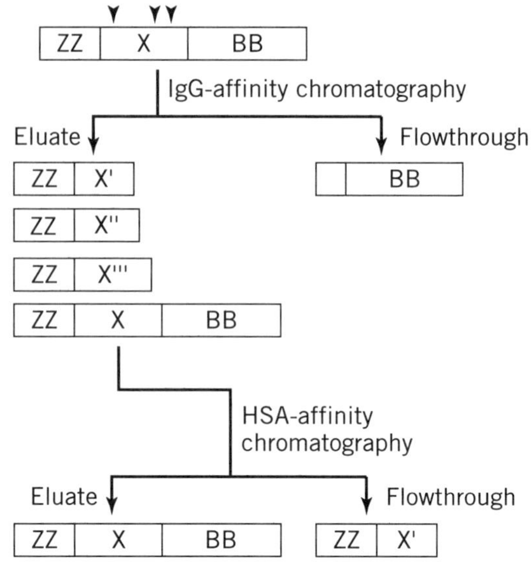

Figure 3. The concept of dual-affinity fusions for recovery of proteolytically labile target proteins. The gene encoding the target protein X is fused between the genes encoding the ZZ and BB tags, respectively. A first passage over an IgG column results in recovery of the full-length product as well as potential degradation products. A second passage over an HSA column results in recovery of only the full-length product.

Strategies for Integrated Downstream Processing

An important part of modern biotechnology is to develop simplified schemes for fermentation and downstream processing of recombinant proteins by the integration of unit operations. Genetic product design can be used in several ways to obtain such schemes, that is, by influencing the yield and localization of the product as well as to adapt the gene product by fusion technology to specific unit operations suitable for large-scale downstream processing.

The recently presented expanded-bed adsorption procedure (mentioned earlier) constitutes a nice example of process integration (55). Efficient recovery of a secreted recombinant fusion protein ZZ-M5 was achieved directly from a crude fermentor broth without prior cell removal. The fusion protein had a relatively low isoelectric point (pI), which allowed anionic exchange adsorption at pH 5.5 at which most *E. coli* host proteins are not adsorbed. This strategy allowed an integration of the cell separation step with ion exchange adsorption of the gene product with simultaneous volume reduction, which resulted in a highly condensed but still efficient recovery process (55). The two-step purification process, ion exchange chromatography in an expanded-bed format followed by IgG affinity chromatography for polishing, demonstrated an overall yield of more than 90%.

An alternative strategy to simplify downstream processing was investigated by Köhler and coworkers (92), where recombinant technology was applied to alter the partitioning properties of a ZZ-fusion protein, which thereby could be efficiently recovered by aqueous two-phase extraction as the primary purification step. An attractive approach to simplify recovery of products, which may be produced with modified primary sequences, e.g., industrial enzymes, would be to alter the partitioning properties by substitution of one or more surface amino acid residues, which are not involved in the biological activity, for tryptophans, which are known to influence partitioning characteristics. This type of protein engineering will most probably be used in the future to reduce production costs for bulk proteins.

A third example of how unit operations can be integrated by careful design of the gene product and the recovery process was recently demonstrated for the production of native human insulin and its C-peptide in *E. coli* (96,103). Human proinsulin was genetically fused to ZZ, and recovered from inclusion bodies via solubilization and refolding of the intact fusion protein, followed by IgG affinity chromatography. Native insulin, consisting of cystein-bridged A- and B-chains, as well as the clinically relevant C-peptide (104), could be released by a single-step trypsin treatment, cleaving off the ZZ-tail simultaneously to the processing of the C-peptide (96,103). The careful genetic design of a fusion protein as an intermediate product, which could be efficiently converted to the final products, allowed the design of a pilot-scale process involving few unit operations (96) that thus might result in a cost-efficient bioprocess. A further improved bioprocess for production of the proinsulin C-peptide was presented (105) in which the C-peptide-encoding gene fragment was heptamerized to increase yields. A fusion protein containing seven copies of the C-peptide was thus produced, and after affinity purification, enzymatic digestion was employed to release native C-peptide (105).

Improved Renaturation Schemes for Recombinant Gene Products

The recovery of biologically active or native proteins by in vitro refolding from insoluble inclusion bodies can in some cases be hampered by the aggregation of the product during the procedure, leading to low overall yields. To make the procedure more efficient, several improved protocols have been developed, including the addition of different "folding enhancers" (4). Alternatively, the protein itself can be engineered to facilitate the refolding. The presence of hydrophilic peptide extensions during the refolding can dramatically improve the folding yield, probably by conferring a higher overall solubility of the protein (90). Samuelsson and coworkers (42) showed that the reshuffling of misfolded disulfides in recombinant insulin-like growth factor I (IGF-I) was greatly facilitated by fusion to the highly soluble ZZ fusion partner (42). Compared to unfused IGF-I, the fusion ZZ-IGF-I could be successfully refolded at a 100-fold higher concentration (1–2 mg/mL), without forming precipitates (43).

Affinity-Assisted In Vivo Folding

A completely new concept to improve the fraction of correctly folded recombinant IGF-I was recently presented (106). It was demonstrated that coexpression of a specific binding protein, IGF binding protein 1 (IGFBP-1), significantly increased the relative yield of IGF-I having native disulfide bridges when expressed in a secreted form in *E. coli*. In addition, a glutathione redox buffer was added to the growth medium to enhance formation and breakage of disulfide bonds in the periplasm of the bacteria. In the presented example, both IGF-I and IGFBP-1 were produced as affinity fusions, to Z and BB (Fig. 4a), respectively, which facilitated the purification of *in vivo*–assembled heterodimers by alternative purification methods (106). A further development of the strategy would be to express the target protein as a nonfused gene product and the specific binding protein in an affinity-tagged configuration (Fig. 4b). Correctly folded target protein would thus be affinity captured as a heterodimer via the tagged binding protein. This would employ the benefits of high specificity for affinity chromatography without introducing requirements of proteolytic processing (see following) to recover the native target protein.

Affinity-Tagged Proteases for Site-Specific Cleavage of Fusion Proteins

When gene fusion strategies are used for the production of native proteins, efficient means for site-specific cleavage of the fusion protein and subsequent removal of the affinity fusion partner are needed. In addition, if employing enzymatic strategies for cleavage, strategies for the removal of the proteolytic enzyme itself must be developed. Special considerations must be taken if the target protein itself also has to be further processed to give the desired final product (96,103). Careful upstream design of the fusion protein construct using genetic strategies can greatly facilitate the subsequent purification of the target protein and also allow for integrated systems involving coprocessing of the protein and efficient removal of the affinity fusion partner as well as the protease used for cleavage (96,102).

Enzymes for use in biotechnological applications should preferably be highly specific proteases, easy to produce by recombinant means in large scale. Interesting candidate enzymes can be found in human picorna viruses, whose maturation relies on the site-specific cleavage of a large polyprotein precursor to yield the viral components. Some of these proteases are functionally produced at high levels in bacterial expression systems (107,108), and this could facilitate the production of variants constructed by protein engineering. Recently, a new general strategy was described where the 3C protease of rhinovirus was fused to

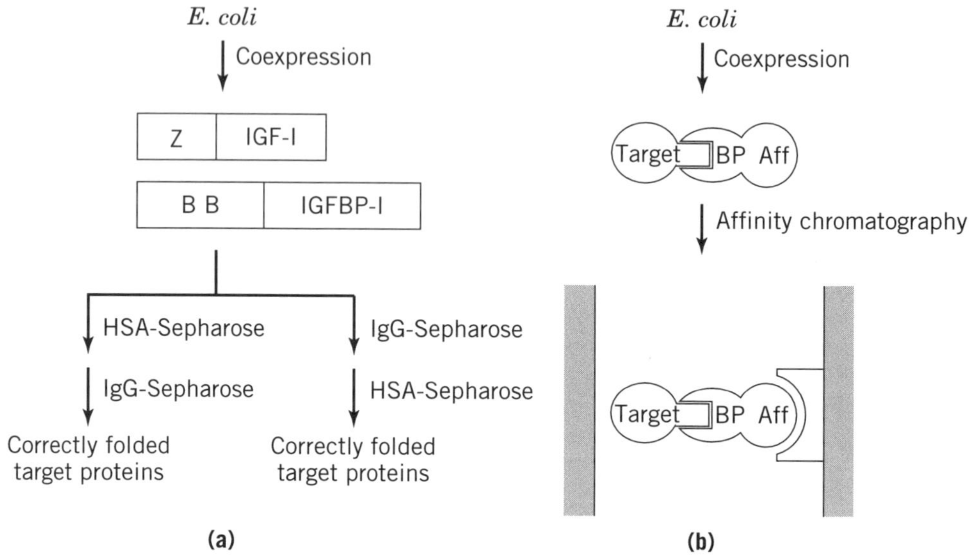

Figure 4. Affinity-assisted in vivo folding and its use for recovery of correctly folded gene products. **(a)** Schematic description of the example presented by Samuelsson and coworkers (106) where a fusion protein Z-IGF-I, with correctly folded IGF-I, could be recovered using alternative affinity chromatography strategies, taking advantage of the interaction between IGF-I and IGFBP-I. **(b)** The basic principle might be suitable for general use both to increase the yield of correctly folded target protein and to achieve recovery of only correctly folded target protein via the interaction of a specific binding protein (*BP*), which binds only to the correctly folded form of the target protein. If the binding protein is expressed as an affinity fusion (*Aff*), this fusion protein should be a suitable ligand for recovery of correctly folded target protein.

an affinity tag and subsequently used for specific cleavage of a target protein (108). This enabled removal of the affinity-tagged protease and the cleaved affinity fusion partner by simple affinity chromatography steps. It was in addition possible to add any desired amount of protease to ensure efficient cleavage because the protease could be easily removed after cleavage (108). Recently, a different affinity-tagged protease, consisting of coxsackievirus 3C ($3C^{pro}$) fused to the serum albumin–binding ABP at the N-terminus and His_6 at the C-terminus, was produced in *E. coli* and used for the production of a truncated *Taq* DNA polymerase (*ΔTaq*) (102). The heat-stable polymerase was produced as an ABP-*ΔTaq* fusion protein having a $3C^{pro}$ cleavage site introduced between the two protein moieties. After affinity purification of ABP-*ΔTaq*, the fusion was efficiently cleaved using the affinity-tagged protease ABP-$3C^{pro}$, which allowed for subsequent recovery of nonfused, fully active *ΔTaq* via passage of the cleavage mixture over an HSA-column (102). Note that the affinity-tagged protease, ABP-$3C^{pro}$, is removed together with eventual remains of unprocessed ABP-*ΔTaq* fusion protein.

The use of affinity fusions for recombinant production of therapeutic target gene products has probably been hampered by the difficulties with efficiently cleaving off the affinity tag and thereafter removing not only the tag but also the used protease. The increasing availability of affinity-tagged, highly specific proteases for cleavage of fusion proteins to enable straightforward methods for recovery of unfused authentic gene products (102,108) will probably influence researchers to further increase their use of fusion strategies for production of target proteins.

Increased In Vivo Half-Lives for Therapeutic Gene Products

Gene fusion technology might also prove to be useful in prolonging the in vivo half-life of pharmaceutical proteins by administering the therapeutic protein as fused to a protein with extended half-life. The principle was demonstrated to be efficacious by Capon and coworkers (109), who showed that by fusing a soluble portion of CD4 (the target receptor for the human immunodeficiency virus particle) to the Fc part of IgG, the serum half-life of CD4 in rabbits increased 200-fold. Nygren and coworkers (110) described an alternative strategy where in vivo stabilization was achieved by fusion of the unstable CD4 protein to BB, leading to the formation of complexes between the fusion protein and serum albumin. The serum half-lives of different fusion proteins containing the albumin-binding receptor were tested in mice. In these studies, the half-life of the CD4 fusion in vivo was shown to be comparable to the CD4-IgG molecule described earlier. Interestingly, the albumin-binding receptor in macaque monkeys was found to have a serum half-life similar to that of serum albumin itself (110). Recently, the same strategy was employed to increase the serum half-life of human soluble complement receptor type 1 (sCR1) in rats, by fusion to different serum albumin–binding fragments of SpG (81). These examples suggest that gene fusions to BB, ABP, or the minimal binding motif ABD of SpG might constitute useful strategies to improve the in vivo stability of pharmaceutical proteins. It may also be possible to combine these approaches with new methods for drug targeting or slow drug release.

Protein A Fusions in Immunodetection Techniques

Bifunctional SpA fusion proteins have found extensive use for various analytical purposes, for example, in enzyme immunoassays. Protein A fusion proteins can after expression be efficiently purified using the IgG-binding activity, and in most assays the protein A moiety binds antibodies to be quantified and the activity of the fused gene product is monitored. Thus, the fusion protein can replace more traditional reagents, such as secondary antibodies conjugated to an enzyme. The concept of genetically engineered conjugates in immunoassays was recently reviewed (111). The fusion between catechol dioxygenase and protein A produced in *E. coli* was used in an enzyme immunoassay to quantify antibodies (112). The principle of a bioluminescent immunoassay using the fusion between protein A and bacterial luciferase has also been demonstrated (113), and recently a bifunctional fusion between domain D of SpA and luciferase from sea-firefly was used in a sandwich enzyme-linked immunosorbent assay (ELISA) (72). Furthermore, an SpA-streptokinase fusion was shown to be equally useful as a commercial SpA-horseradish peroxidase conjugate for use in immunoassays (60). A fusion between SpA and LacG, 6-phospho-β-galactosidase from *S. aureus*, was shown to be able to replace secondary antibodies in an ELISA (114). In a similar approach, an SpA–alkaline phosphatase fusion was used in ELISA (115) and a fusion between maltose-binding protein and protein A was used for a solid-phase immunoassay, in which the fusion protein was immobilized to determine IgG concentrations (116).

Recently, the use of SpA-LacI and SpA-streptavidin fusion proteins in various immunodetection techniques was reviewed (3). Using sandwich techniques, DNA fragments also can be involved in various purification and detection strategies (Fig. 5). Taking advantage of the lac repressor (LacI) from *E. coli*, binding to the *lac* operator sequence, a LacI-SpA fusion protein was produced (117) for simplified analysis of polymerase chain reaction (PCR) amplified DNA fragments generated for diagnosis of pathogens (118) (Fig. 5a). In addition, this LacI-SpA fusion protein has been used for mild reversible recovery of protein antigens or whole cells using paramagnetic beads with a coupled DNA fragment containing the *lacO* recognition sequence (119) (Fig. 5b).

An antigen detection system, termed immuno-PCR, has been described (120) in which the principle of immunodetection was combined with the sensitivity of PCR. A streptavidin-SpA chimera was used to attach a biotinylated DNA fragment to antigen–IgG complexes immobilized on microtiter plate wells (Fig. 5c). Although the method gives a sensitivity greater than any existing antigen–antibody detection system, the sensitivity of the PCR assay could have the drawback of causing false positives resulting from nonspecific binding of IgG, fusion protein, or DNA. Nevertheless, SpA fusion proteins have through the IgG Fc binding activity found extensive use in a vast number of immunodetection techniques.

Protein G Fusion Used to Accomplish "Hot-Start PCR"

Immobilization of enzymes has proven to be a useful procedure for many applications because of the altered characteristics of the immobilized enzyme as compared with the enzyme's soluble counterpart (121). One practical application utilizing such differential activities was recently described in which a heat-mediated release of an affinity-immobilized recombinant *Taq* DNA polymerase (TagTaq) was used to create a hot-start PCR procedure. *Taq* DNA polymerase was produced in *E. coli* as fused to the multifunctional Bio–His_6–ABP tag (see foregoing text; 122) for affinity purification and immobilization. HSA-affinity immobilization of the fusion employing the ABP moiety resulted in an apparent deactivation of the *Taq* DNA polymerase. However, the ABP–HSA interaction was shown to

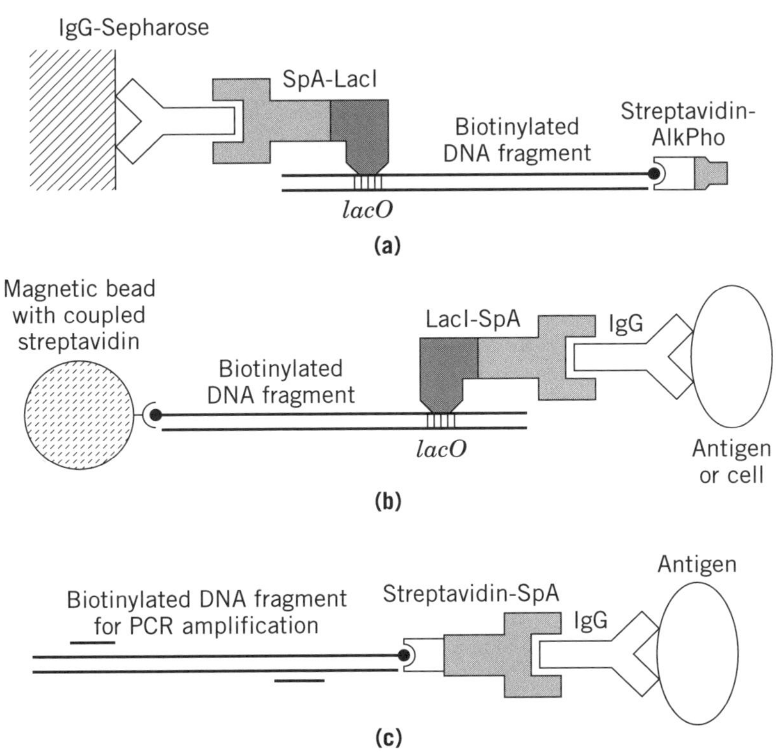

Figure 5. Schematic presentation of some sandwich concepts utilizing SpA fusion proteins and interacting DNA fragments for detection and purification purposes. **(a)** Colorimetric detection of DNA fragments containing the *lac* operator (*lacO*) sequence, using a lac repressor (LacI)-SpA fusion protein immobilized to IgG-Sepharose. The colorimetric signal was obtained using a streptavidin–alkaline phosphatate conjugate. The DNA fragments, generated by the PCR technique, introduce biotin and the *lac* operator by the PCR primers. This assay has been used to detect *Plasmodium falciparum* DNA in clinical samples (118). **(b)** This system takes advantage of the same interactions as described in **a**, with the only difference that paramagnetic beads with covalently coupled streptavidin are used as solid support. The technique has been used to purify different proteins and also whole cells using specific antibodies reversibly immobilized to the paramagnetic beads via the LacI-SpA fusion and a biotinylated DNA fragment. Mild recovery of the antigen–antibody complexes can be achieved using deoxyribonuclease (DNase) or restriction endonucleases (119). **(c)** In this system, the antigen, immobilized on a microtiter plate, is allowed to react with a monoclonal antibody, linked to a biotinylated DNA fragment via an SpA-streptavidin chimera. The detection is accomplished by PCR amplification using primers specific for the bound DNA. The generated fragment is analyzed by agarose gel electrophoresis (120).

be disrupted at elevated temperatures (>70 °C), resulting in a restored DNA polymerase activity of released *Taq* DNA polymerase fusion protein. This concept for controlling the activity of the *Taq* DNA polymerase was demonstrated to be of practical use to achieve a so-called hot-start PCR procedure, where it is desired to suppress the DNA polymerase activity until temperatures are reached at which nonspecific annealing of PCR primers is eliminated. The affinity immobilization concept for hot-start PCR was successfully used in full multiple-cycle PCR amplification and was shown to eliminate artefactual primer-dimer products in the amplification of the human oncogene K-*ras* gene in contrast to standard amplification protocols. Noteworthy, the ABP affinity fusion partner was still functional after 30 PCR cycles, allowing post-PCR reimmobilization of the *Taq* DNA polymerase fusion protein onto fresh HSA-Sepharose, facilitating the removal of the DNA polymerase from reaction mixtures after PCR (122). Further studies are needed to investigate if this concept for modulating biological activity can also be applied to other widely used enzymes, for example, reverse transcriptase.

Protein A and Protein G Fusion Proteins in Subunit Vaccine Development

Immunogenic proteins and peptides that are produced as fusion proteins have been frequently used for immunization purposes and represent a strategy to construct subunit vaccines against infectious diseases. An important aspect regarding production of antigens by recombinant DNA technology is the purification of the gene product from contaminating host components. The approach to express peptides or proteins as one part of a fusion protein, where the other part constitutes an affinity handle, allows rapid and efficient affinity purification of the gene product.

As a powerful strategy in subunit vaccine development, a dual expression system was devised, combining the two similar affinity-fusion expression systems (ZZ and BB) for the parallel expression of immunogenic peptides and proteins (Fig. 6) (50). After affinity purification, one fusion protein is used for immunization while the corresponding second fusion protein is used to analyze the induced antibody response to the fused peptide. This strategy eliminates any background originating from antibodies reactive with the affinity tag of the fusion used for immunization. In addition, the second fusion protein can, after immobilization, be used as a ligand for affinity purification of peptide-specific antibodies (Fig. 6). The dual expression system has been used to produce immunogenic structures derived from different *Plasmodium falciparum* malaria blood-stage antigens, and the expressed fusion proteins have been shown to be immunogenic in mice, rabbits, and monkeys and to induce antigen-specific antibody and T-cell responses (for review, see Ref. 27). The dual expression strategy has thus shown to be of significant value in subunit vaccine research.

Besides the use of BB and ABP fusion proteins in the dual expression concept for analysis of elicited immune responses, the serum albumin–binding region of SpG has been shown to have inherent immunopotentiating properties when used as a carrier protein genetically fused to

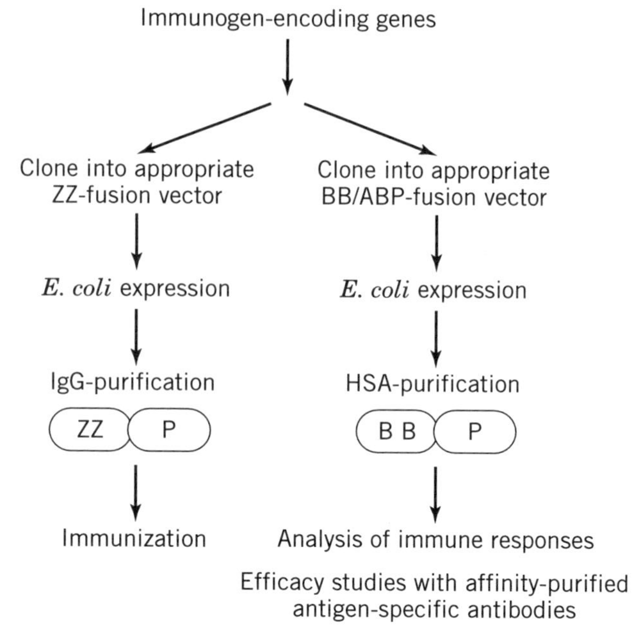

Figure 6. A flow chart representation of how the dual expression concept has been used in subunit vaccine research. Note that the BB-P fusion protein can be used both as coating antigen in various assays to analyze elicited immune responses and also in an immobilized form for enrichment of antigen-specific antibodies from the immune sera. The use of this concept has been thoroughly reviewed (27).

the immunogen used for immunization (27,49,123,124). Recently, it was demonstrated that a BB-fusion protein (BB-M3), containing a malaria peptide, induced significant antibody responses in mice strains that were nonresponders to the malaria peptide (M3) alone, suggesting that BB has the ability to provide T-cell help for antibody production (49). Furthermore, a fusion protein BB-G2N, comprising a 101–amino acid sequence from human RSV, was shown to induce protective immunity in mice to RSV challenge (123,124). It was shown that, by inclusion of the BB part, a more potent G2N-specific B-cell memory response was evoked (124). This indicates that the SpG-derived BB can function both as an affinity tag to facilitate purification and as a carrier protein with immunopotentiating properties. To date, it is not fully elucidated whether this capacity results from strong T-cell epitopes (49) or is related to the serum albumin–binding activity resulting in a prolonged exposure of the immunogen to the immune system, or is due to a combination of both.

Protein A and Protein G Fusion Proteins in Functional cDNA Analysis

The principle of dual expression systems was recently also implemented for functional analysis of cDNA-encoded proteins (Fig. 7) (13). Selected cDNAs were then expressed as ZZ fusion proteins and used for immunization to elicit antibodies, whereas the corresponding ABP fusion protein, having the cDNA-encoded portion in common, was used for purification of target protein–specific antibodies. Such

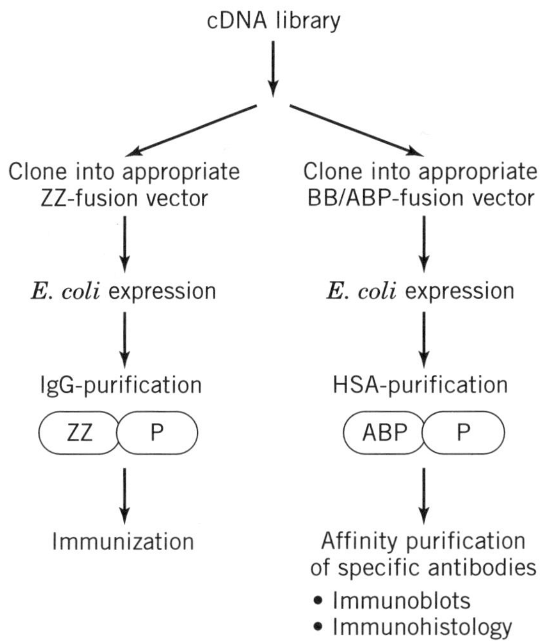

Figure 7. A flow chart representation showing how the dual expression concept has been used in functional cDNA research. The concept is thus used for the generation of highly specific antibodies employed in various analyses to assign function to cDNA-encoded proteins. Such sera have indeed shown to be useful to localize gene products to a certain cell type within a tissue section and also further down to subcellular levels (13).

highly enriched antisera should be suitable for functional analysis of the cDNA-encoded protein, for example, in Western blots and immunohistology on tissue sections or within cells. The principle was evaluated by expressing five cDNAs, isolated from prepubertal mouse testis by a differential cDNA library screening strategy. All five clones could be expressed intracellularly in *E. coli* as fusion proteins with high production levels, and affinity purification yielded essentially full-length products. Characterization of affinity-purified antibodies revealed that there existed no cross-reactivity between the two fusion systems, and that such antibodies indeed were valuable for various analyses to elucidate the functions of the cDNA-encoded proteins (13). For example, in the presented study a previously unknown cDNA, with no sequence homology to genes in the public databases, could be connected with a gene product found in nuclear bodies within spermatogonial cells (13).

To make it possible to use the dual expression system for functional analysis of cDNA, where robustness of the expression systems is crucial, completely new expression vectors were constructed. Intracellular expression was used and transcription was tightly controlled by the T7 promoter system (13). This method should ensure that a higher fraction of investigated cDNA-encoded proteins are successfully produced. The different ongoing genome projects have generated enormous amounts of partially sequenced cDNAs, and there exists a significant need for various alternative systems to take care of this information to elucidate the function of the corresponding proteins. By automation of some of the unit operations, the system described by Larsson and coworkers (13) might have the potential to make large-scale functional analysis of cDNA libraries possible.

Bacterial Surface Display

The display of heterologous proteins on the outer surface of bacteria has become an emerging topic in different fields of research within biotechnology, microbiology, and immunology (125,126). Genetic insertion of a target sequence into the genes for various outer membrane proteins has constituted the general strategy to enable secretion and subsequent surface anchoring of the recombinant target gene products. For Gram-negative bacteria, represented by *E. coli* and *Salmonella* spp., a number of different types of heterologous proteins have been surface displayed including antigenic determinants for the purpose of vaccine development, enzymes, metal-chelating histidyl peptides, and single-chain antibodies, as well as entire peptide libraries (125–128). Recently, surface display on Gram-positive bacteria has gained interest in the context of vaccine development. Approaches based on the attenuated mycobacteria (129), commensal streptococci (130), and nonpathogenic food-fermenting staphylococci have been presented (48,76–78).

To achieve surface display of various target proteins, the staphylococcal systems take advantage of the cell surface anchoring regions (XM) of SpA (see Fig. 1). The surface-exposed proteins are thus genetically fused between a functional signal peptide, responsible for the translocation from the cell, and the XM-regions of SpA. Heterologous antigenic determinants of various origin (bacteria, protozoa, and human virus) have been expressed as anchored on recombinant staphylococci, which are being investigated as a bacterial vaccine delivery system (48,76–78,80,131–133). Even active single-chain antibody fragments have been surface displayed on recombinant staphylococci (79). Such recombinant bacteria could be envisioned as inexpensive tools in diagnostic tests.

Protein A Domains and Phage Technology for Selection of "Affibodies"

An example of emerging interest in which gene fusions have made new technology possible is the display of libraries of peptides or proteins on filamentous phage surfaces. Phage particles decorated in such manner can be enriched in vitro from a background of irrelevant phages by biopanning techniques, employing an immobilized ligand capable of binding to the displayed protein. Moreover, inside its protein coat the captured phage carries the genetic information of the displayed protein, which allows later identification by DNA sequencing. Thus, the technology provides a link between genotype and phenotype and allows the selection of peptides or proteins of desired functions to be expressed on the surface of the filamentous phage. This technology, as pioneered by Smith (134), has emerged into a powerful technology to select, rather than screen, for rare phages capable of binding to a defined target molecule from libraries comprising as many as 10^{10}

phage species. Such phage libraries are obtained by the fusion to phage pIII or pVIII genes, of cassettes encoding randomized peptides (usually 6- to 10-mers), antibody repertoires, or multiple variants of a protein (135). Thus, it is possible to rapidly and efficiently select high-affinity ligands or effector molecules that can be used for analytical or preparative applications.

The Z-domain from SpA has several features including proteolytic stability, high solubility, small size, and compact and robust structure, which makes it ideal as a fusion partner for the production of recombinant proteins. The residues responsible for the IgG Fc–binding activity have been identified from the crystal structure of the complex between its parent B-domain and human IgG Fc, showing that these residues are situated on the outer surfaces of two of the helices making up the domain structure and do not contribute to the packing of the core (37). The Fc-binding surface covers an area of approximately 600 Å^2, similar in size to the surfaces involved in many antigen–antibody interactions. Taken together, these data suggest that if random mutagenesis of this binding surface was performed, novel domains could be obtained with the possibility of finding variants capable of binding molecules other than Fc of IgG. In addition, these domains would theoretically share the overall stability and α-helical structure of the wild-type domain.

Recently, an attempt to obtain such ligands with completely new affinities was initiated, employing phage display technology (136) (Fig. 8a). A genetic library was created in which 13 surface residues (involved in the IgG Fc binding) of the Z-domain were randomly and simultaneously substituted to any of the 20 possible amino acids. This Z-library has, after genetic fusion to the gene for coat protein 3 of filamentous phage M13, resulted in a phage library adapted for selection of novel specificities by biopanning (137). The library has been subjected to such affinity selection against different molecules for investigation as a source of novel binders (Fig. 8b). So far, novel Z-variants, so-called affibodies, have successfully been selected to diverse targets, such as *Taq* DNA polymerase, human insulin, and a human apolipoprotein variant (137). This concept suggests that the described strategy might prove to be beneficial in generating a "next generation" of ligands or artificial antibodies to various target molecules,

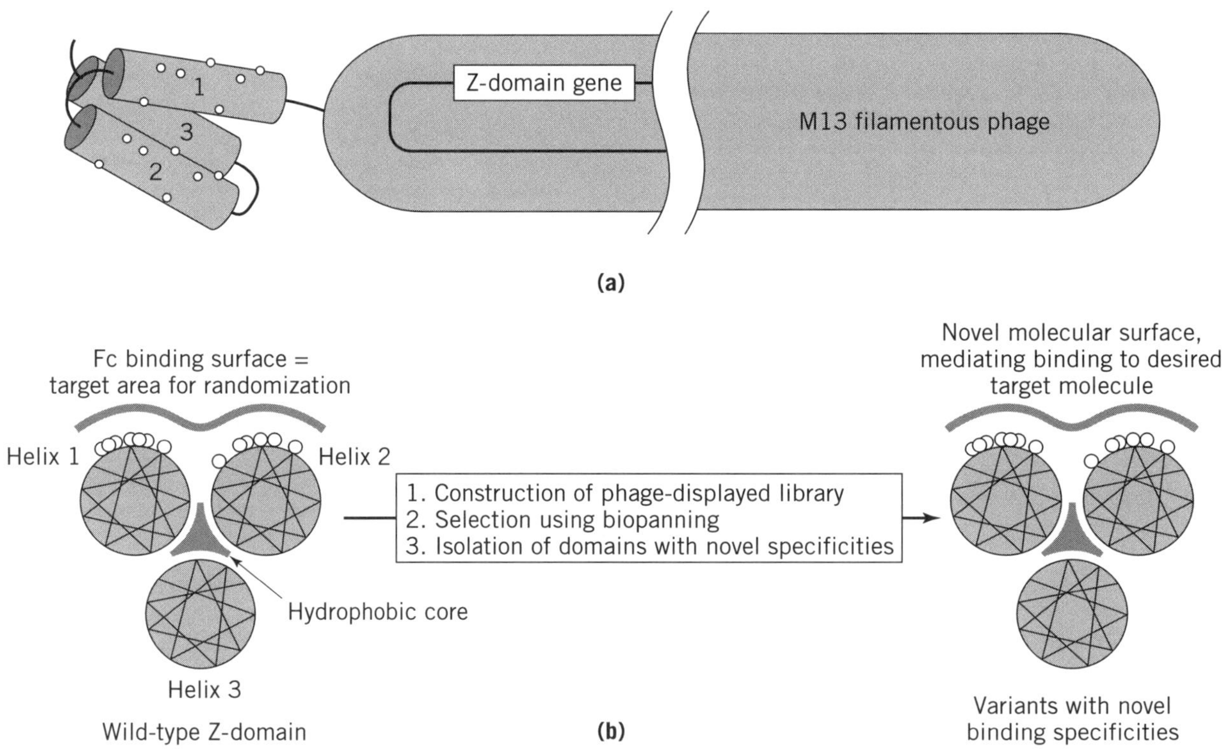

Figure 8. Schematic description showing how the Z-domain has been used as molecular scaffold in search of domain variants with novel binding specificities. **(a)** Using phage display technology, Z-variants can be expressed for presentation on the surface of filamentous phage particles. Phage libraries containing large numbers of phage clones, each displaying an individual Z-variant, can be subjected to rounds of biopannings in which Z-variants capable of binding to a desired target molecule can be selected. Identification of binding variants is accomplished through DNA sequencing of the packed phage DNA, containing the gene of the displayed Z-variant. **(b)** Top view of the Z-domain consisting of a triple α-helix. The library is constructed through randomization of amino acids present at the domain surface and involved in the Fc binding of the wild-type domain. Using the selection procedure (biopanning), variants with new molecular surfaces are isolated that are capable of binding the desired target.

to be used for purification, detection, immobilization, and perhaps even therapeutic purposes.

CONCLUSIONS

In this article, we have illustrated how fusions to the immunoglobulin-binding domains of SpA and the serum albumin-binding regions of SpG, respectively, have become important tools in many fields of research. Several strategies for improved heterologous expression and simplified recovery are based on such gene fusions. We focused on some current trends in certain fields of applications for SpA and SpG fusion proteins. It is apparent that SpA and SpG fusion proteins will soon find extensive use in novel applications in medicine, immunology, and biotechnology.

ACKNOWLEDGMENTS

All former and present members at the Department of Biochemistry and Biotechnology who have been involved in the development and evaluations of the SpA and SpG expression systems are gratefully acknowledged. We express our gratitude to Drs. Björn Nilsson, Tomas Moks, Lars Abrahmsén, and Maria Murby. In particular we thank Dr. Birger Jansson for help with literature searches concerning the use of SpA fusions in immunodetection techniques.

BIBLIOGRAPHY

1. M. Uhlén, B. Nilsson, B. Guss, M. Lindberg, S. Gatenbeck, and L. Philipson, *Gene* **23**, 369–378 (1983).
2. M. Uhlén, G. Forsberg, T. Moks, M. Hartmanis, and B. Nilsson, *Curr. Opin. Biotechnol.* **3**, 363–369 (1992).
3. S. Ståhl, P.-Å. Nygren, A. Sjölander, and M. Uhlén, *Curr. Opin. Immunol.* **5**, 272–277 (1993).
4. P.-Å. Nygren, S. Ståhl, and M. Uhlén, *Trends Biotechnol.* **12**, 184–188 (1994).
5. S. Ståhl, P.-Å. Nygren, and M. Uhlén, in R.S. Tuan ed., *Methods in Molecular Biology*, vol. 63, *Recombinant Proteins: Detection and Isolation Protocols*, Humana Press, Totowa, N.J., 1997, pp. 103–118.
6. E. Flaschel and K. Friehs, in M. Moo-Young and B.R. Glick, eds., *Biotechnology Advances*, Pergamon Press, Oxford, 1993, pp. 31–78.
7. C.F. Ford, I. Suominen, and C.E. Glatz, *Prot. Expr. Purif.* **2**, 95–107 (1991).
8. E.R. Lavallie and J.M. McCoy, *Curr. Opin. Biotechnol.* **6**, 501–505 (1995).
9. S.C. Makrides, *Microbiol. Rev.* **60**, 512–538 (1996).
10. J. Nilsson, S. Ståhl, J. Lundeberg, M. Uhlén, and P.-Å. Nygren, *Prot. Expr. Purif.* **11**, 1–16 (1997).
11. B. Nilsson and L. Abrahmsén, *Methods Enzymol.* **185**, 144–161 (1990).
12. B. Nilsson, T. Moks, B. Jansson, L. Abrahmsén, A. Elmblad, E. Holmgren, C. Henrichson, T.A. Jones, and M. Uhlén, *Prot. Eng.* **1**, 107–113 (1987).
13. M. Larsson, E. Brundell, L. Nordfors, C. Höög, M. Uhlén, and S. Ståhl, *Prot. Expr. Purif.* **7**, 447–457 (1996).
14. P.-Å. Nygren, M. Eliasson, E. Palmcrantz, L. Abrahmsén, and M. Uhlén, *J. Mol. Recognit.* **1**, 69–74 (1988).
15. J. Porath, J. Carlsson, I. Olsson, and G. Belfrage, *Nature* **258**, 598–599 (1975).
16. E. Hochuli, W. Bannwarth, H. Döbeli, R. Gentz, and D. Stüber, *Biotechnology* **6**, 1321–1325 (1988).
17. D.B. Smith and K.S. Johnson, *Gene* **67**, 31–40 (1988).
18. X. Zhao and C.H. Siu, *J. Biol. Chem.* **270**, 29413–29421 (1995).
19. C. di Guan, P. Li, P.D. Riggs, and H. Inouye, *Gene* **67**, 21–30 (1988).
20. R. Aitken, J. Gilchrist, and M.C. Sinclair, *Gene* **144**, 69–73 (1994).
21. T.H. Hopp, K.S. Prickett, V.L. Price, R.T. Libby, C.J. March, D.P. Cerretti, D.L. Urdal, and P.J. Conlon, *Biotechnology* **6**, 1204–1210 (1988).
22. B.L. Brizzard, R.G. Chubet, and D.L. Vizard, *Biotechniques* **16**, 730–734 (1994).
23. J.E. Cronan, *J. Biol. Chem.* **265**, 10327–10333 (1990).
24. J.K. Choi, F. Yu, E.S. Wurtele, and B.J. Nikolau, *Plant Physiol.* **109**, 619–625 (1995).
25. P.J. Schatz, *Biotechnology* **11**, 1138–1143 (1993).
26. J. Nilsson, M. Larsson, S. Ståhl, P.-Å. Nygren, and M. Uhlén, *J. Mol. Recognit.* **9**, 585–594 (1996).
27. A. Sjölander, S. Ståhl, and P. Perlmann, *ImmunoMethods* **2**, 79–92 (1993).
28. M. Uhlén and T. Moks, *Methods Enzymol.* **185**, 129–143 (1990).
29. J.J. Langone, *Adv. Immunol.* **32**, 157–252 (1982).
30. T. Moks, L. Abrahmsén, B. Nilsson, U. Hellman, J. Sjöquist, and M. Uhlén, *Eur. J. Biochem.* **156**, 637–643 (1986).
31. B. Guss, M. Uhlén, B. Nilsson, M. Lindberg, J. Sjöquist, and J. Sjödahl, *Eur. J. Biochem.* **138**, 413–420 (1984).
32. O. Schneewind, P. Model, and V.A. Fischetti, *Cell* **70**, 267–281 (1992).
33. O. Schneewind, D. Mihaylova-Petkov, and P. Model, *EMBO J.* **12**, 4803–4811 (1993).
34. W.W. Navarre and O. Schneewind, *Mol. Microbiol.* **14**, 115–121 (1994).
35. O. Schneewind, A. Fowler, and K.F. Faull, *Science* **268**, 103–106 (1995).
36. M. Eliasson, R. Andersson, A. Olsson, H. Wigzell, and M. Uhlén, *J. Immunol.* **142**, 575–581 (1989).
37. J. Deisenhofer, *Biochemistry* **20**, 2361–2370 (1981).
38. H. Gouda, H. Torigoe, A. Saito, M. Sato, Y. Arata, and I. Shimada, *Biochemistry* **31**, 9665–9672 (1992).
39. M. Tashiro and G.T. Montelione, *Curr. Opin. Struct. Biol.* **5**, 471–481 (1995).
40. B.A. Lyons, M. Tashiro, L. Cedergren, B. Nilsson, and G. Montelione, *Biochemistry* **32**, 7839–7845 (1993).
41. G. Forsberg, B. Baastrup, H. Rondahl, E. Holmgren, G. Pohl, M. Hartmanis, and M. Lake, *J. Prot. Chem.* **11**, 201–211 (1992).
42. E. Samuelsson, H. Wadensten, M. Hartmanis, T. Moks, and M. Uhlén, *Biotechnology* **9**, 363–366 (1991).
43. E. Samuelsson, T. Moks, B. Nilsson, and M. Uhlén, *Biochemistry* **33**, 4207–4211 (1994).
44. C. Ljungquist, B. Jansson, T. Moks, and M. Uhlén, *Eur. J. Biochem.* **186**, 557–561 (1989)
45. P.-Å. Nygren, C. Ljungquist, H. Trömborg, K. Nustad, and M. Uhlén, *Eur. J. Biochem.* **193**, 143–148 (1990).
46. C. Falkenberg, L. Björck, and B. Åkerström, *Biochemistry* **31**, 1451–1457 (1992).
47. J. Nilsson, P. Nilsson, Y. Williams, L. Pettersson, M. Uhlén, and P.-Å. Nyren, *Eur. J. Biochem.* **224**, 103–108 (1994).

48. P. Samuelson, M. Hansson, N. Ahlborg, C. Andréoni, F. Götz, T. Bächi, T.N. Nguyen, H. Binz, M. Uhlén, and S. Ståhl, *J. Bacteriol.* **177**, 1470–1476 (1995).
49. A. Sjölander, P.-Å. Nygren, S. Ståhl, K. Berzins, M. Uhlén, P. Perlmann, and R. Andersson, *J. Immunol. Methods* **201**, 115–123 (1997).
50. S. Ståhl, A. Sjölander, P.-Å. Nygren, K. Berzins, P. Perlmann, and M. Uhlén, *J. Immunol. Methods* **124**, 43–52 (1989).
51. M. Murby, E. Samuelsson, T.N. Nguyen, L. Mignard, U. Power, H. Binz, M. Uhlén, and S. Ståhl, *Eur. J. Biochem.* **230**, 38–44 (1995).
52. P.J. Kraulis, P. Jonasson, P.-Å. Nygren, M. Uhlén, L. Jendeberg, B. Nilsson, and J. Kördel, *FEBS Lett.* **378**, 190–194 (1996).
53. L. Abrahmsén, T. Moks, B. Nilsson, and M. Uhlén, *Nucleic Acids Res.* **14**, 7487–7500 (1986).
54. F. Ducancel, J.-C. Boulain, O. Trémeau, and A. Ménez, *Prot. Eng.* **3**, 139–143 (1990).
55. M. Hansson, S. Ståhl, R. Hjorth, M. Uhlén, and T. Moks, *Biotechnology* **12**, 285–288 (1994).
56. D. Haddad, S. Liljeqvist, S. Kumar, M. Hansson, S. Ståhl, H. Perlmann, P. Perlmann, and K. Berzins, *FEMS Immunol. Med. Microbiol.* **12**, 175–186 (1995).
57. S. Liljeqvist, D. Haddad, K. Berzins, M. Uhlén, and S. Ståhl, *Biochem. Biophys. Res. Commun.* **218**, 356–359 (1996).
58. X. Cao, T. Preiss, E.P. Slater, H.M. Westphal, and M. Beato, *J. Steroid Biochem. Mol. Biol.* **44**, 1–11 (1993).
59. B. Hammarberg, T. Moks, M. Tally, A. Elmblad, E. Holmgren, M. Murby, B. Nilsson, S. Josephson, and M. Uhlén, *J. Bacteriol.* **14**, 423–438 (1990).
60. K.H. Schmidt, C. Klessen, W. Kohler, and H. Malke, *J. Immunol. Methods* **143**, 111–117 (1991).
61. V. Nagarajan, H. Albertson, M. Chen, and J. Ribbe, *Gene* **114**, 121–126 (1992).
62. B. Antonsson, S. Montessuit, L. Friedli, M.A. Payton, and G. Paravicini, *J. Biol. Chem.* **269**, 16821–16828 (1994).
63. B. Jansson, Ph.D. Thesis, Royal Institute of Technology, Stockholm, Sweden, 1996.
64. R.L. McKown and G.J. Warren, *Cryobiology* **28**, 474–482 (1991).
65. D.A. Stirling, A. Petrie, D.J. Pulford, D.T.W. Paterson, and M.J.R. Stark, *Mol. Microbiol* **6**, 703–713 (1992).
66. J. Zueco and A. Boyd, *Anal. Biochem.* **207**, 348–349 (1992).
67. R. Hightower, C. Baden, E. Penzes, P. Lund, and P. Dunsmuir, *Plant Mol. Biol.* **17**, 1013–1021 (1991).
68. D. Andersons, A. Engström, S. Josephson, L. Hansson, and H. Steiner, *Biochem. J.* **280**, 219–224 (1991).
69. C. Oker-Blom, A.-M. Soumalainen, K. Åkerman, Z. Qi, C. Lindqvist, A. Kuusisto, and M. Karp, *Biotechniques* **14**, 800–809 (1993).
70. H. Dorai, J.E. McCartney, R.M. Hudziak, M.-S. Tai, A.A. Laminet, L.L. Houston, J.S. Huston, and H. Oppermann, *Biotechnology* **12**, 890–897 (1994).
71. F. Liu and B. Roizman, *J. Virol.* **65**, 5149–5156 (1991).
72. Y. Maeda, H. Ueda, T. Hara, J. Kazami, G. Kawano, E. Suzuki, and T. Nagamune, *Biotechniques* **20**, 116–121 (1996).
73. R. Sanchez-Lopez, R. Nicholson, M.-C. Gesnel, L.M. Matrisian, and R. Breathnach, *J. Biol. Chem.* **263**, 11892–11899 (1988).
74. T. de Vries, C. Srnka, M.M. Palcic, S.J. Swiedler, D.H. van den Eijnden, and B.A. Macher, *J. Biol. Chem.* **270**, 8712–8722 (1995).
75. L. Zhang and J.D. Esko, *J. Biol. Chem.* **269**, 19295–19299 (1994).
76. M. Hansson, S. Ståhl, T.N. Nguyen, T. Bächi, A. Robert, H. Binz, A. Sjölander, and M. Uhlén, *J. Bacteriol.* **174**, 4239–4245 (1992).
77. T.N. Nguyen, M. Hansson, S. Ståhl, T. Bächi, A. Robert, W. Domzig, H. Binz, and M. Uhlén, *Gene* **128**, 89–94 (1993).
78. T.N. Nguyen, M.-H. Gourdon, M. Hansson, A. Robert, P. Samuelson C. Libon, C. Andréoni, P.-Å. Nygren, H. Binz, M. Uhlén, and S. Ståhl, *J. Biotechnol.* **47**, 207–219 (1995).
79. E. Gunneriusson, P. Samuelson, M. Uhlén, P.-Å. Nygren, and S. Ståhl, *J. Bacteriol.* **178**, 1341–1346 (1996).
80. A. Robert, P. Samuelson, C. Andréoni, T. Bächi, M. Uhlén, H. Binz, T.N. Nguyen, and S. Ståhl, *FEBS Lett.* **390**, 327–333 (1996).
81. S.C. Makrides, P.-Å. Nygren, S.B. Andrews, P.J. Ford, K.S. Evans, E.G. Hayman, H. Adari, J. Levin, M. Uhlén, and C.A. Toth, *J. Pharmacol. Exp. Ther.* **277**, 534–542 (1996).
82. Amersham Pharmacia Biotech, Uppsala, Sweden.
83. A. Hedrum, J. Lundeberg, C. Påhlson, and M. Uhlén, *PCR Methods Appl.* **2**, 167–171 (1992).
84. B. Löwenadler, B. Jansson, S. Paleus, E. Holmgren, B. Nilsson, T. Moks, G. Palm, S. Josephson, L. Philipson, and M. Uhlén, *Gene* **58**, 87–97 (1987).
85. L. Abrahmsén, T. Moks, B. Nilsson, U. Hellman, and M. Uhlén, *EMBO J.* **4**, 3901–3906 (1985).
86. T. Moks, L. Abrahmsén, B. Österlöf, S. Josephson, M. Östling, S.-O. Enfors, I. Persson, B. Nilsson, and M. Uhlén, *Biotechnology* **5**, 379–382 (1987).
87. Nilsson, L. Abrahmsén, and M. Uhlén, *EMBO J.* **4**, 1075–1080 (1985).
88. M. Murby, L. Cedergren, J. Nilsson, P.-Å. Nygren, B. Hammarberg, B. Nilsson, S.-O. Enfors, and M. Uhlén, *Biotechnol. Appl. Biochem.* **14**, 336–346 (1991).
89. K. Berzins, S. Adams, J.R. Broderson, C. Chizzolini, M. Hansson, K. Lövgren, P. Millet, C.L. Morris, H. Perlmann, P. Perlmann, A. Sjölander, S. Ståhl, J.S. Sullivan, M. Troye-Blomberg, B. Wåhlin-Flyg, and W.E. Collins, *Vaccine Res.* **4**, 121–133 (1995).
90. R. Rudolph, in J.L. Cleland and C.S. Craik, eds., *Principles and Practice of Protein Folding*, Wiley, New York, 1995, pp. 283–298.
91. F.W. Studier, A.H. Rosenberg, J.J. Dunn, and J.W. Dubendorff, *Methods Enzymol.* **185**, 60–89 (1990).
92. K. Köhler, C. Ljungquist, A. Kondo, A. Veide, and B. Nilsson, *Biotechnology* **9**, 642–646 (1991).
93. M. Murby, T.N. Nguyen, H. Binz, M. Uhlén, and S. Ståhl, in D.L. Pyle, ed., *Separations for Biotechnology*, vol. 3, Bookcraft, Bath, UK, 1994, pp. 336–344.
94. A. Sjölander, S. Ståhl, K. Lövgren, M. Hansson, L. Cavelier, A. Walles, H. Helmby, B. Wåhlin, B. Morein, M. Uhlén, K. Berzins, P. Perlmann, and M. Wahlgren, *Exp. Parasitol.* **76**, 134–145 (1993).
95. S. Josephson and R. Bishop, *Trends Biotechnol.* **6**, 218–224 (1988).
96. J. Nilsson, P. Jonasson, E. Samuelsson, S. Ståhl, and M. Uhlén, *J. Biotechnol.* **248**, 241–250 (1996).
97. M. Murby, M. Uhlén, and S. Ståhl, *Prot. Expr. Purif.* **7**, 129–136 (1996).
98. Hammarberg, P.-Å. Nygren, E. Holmgren, A. Elmblad, M. Tally, U. Hellman, T. Moks, and M. Uhlén, *Proc. Natl. Acad. Sci. USA* **86**, 4367–4371 (1989).

99. B. Jansson, C. Palmcrantz, M. Uhlén, and B. Nilsson, *Prot. Eng.* **2**, 555–561 (1990).
100. J.S. Kim and R.T. Raines, *Anal. Biochem.* **219**, 165–166 (1994).
101. C.A. Panagiotidis and S.J. Silverstein, *Gene* **164**, 45–47 (1995).
102. T. Gräslund, J. Nilsson, M. Lindberg, M. Uhlén, and P.-Å. Nygren, *Prot. Expr. Purif.* **9**, 125–132 (1997).
103. P. Jonasson, J. Nilsson, E. Samuelsson, T. Moks, S. Ståhl, and M. Uhlén, *Eur. J. Biochem.* **236**, 656–661 (1996).
104. B.-L. Johansson, A. Kernell, S., Sjöberg, and J. Wahren, *J. Clin. Endocrinol. Metab.* **77**, 976–981 (1993).
105. P. Jonasson, P.-Å. Nygren, B.-L. Johansson, J. Wahren, M. Uhlén, and S. Ståhl, *Gene* **210**, 203–210 (1998).
106. E. Samuelsson, P. Jonasson, F. Viklund, B. Nilsson, and M. Uhlén, *Nat. Biotechnol.* **14**, 751–755 (1996).
107. K. Miyashita, M. Kusumi, R. Utsumi, T. Komano, and N. Satoh, *Biosci. Biotechnol. Biochem.* **56**, 746–750 (1992).
108. P.A. Walker, L.E.-C. Leong, P.W.P. Ng, S. Han Tan, S. Waller, D. Murphy, and A.G. Porter, *Biotechnology* **12**, 601–605 (1994).
109. D.J. Capon, S.M. Chamow, J. Mordenti, S.A. Marsters, T. Gregory, H. Mitsuya, R.A. Byrn, C. Lucas, F.M. Wurm, J.E. Groopman, S. Broder, and D.H. Smith, *Nature* **337**, 525–531 (1989).
110. P.-Å. Nygren, P. Flodby, R. Andersson, H. Wigzell, and M. Uhlén, in R.M. Chanock, H.S. Ginsberg, F. Brown, and R.A. Lerner eds., *Vaccines 91*, Cold Spring Harbor Laboratory Press, New York, 1991, pp. 363–368.
111. C. Lindbladh, K. Mosbach, and L. Bülow, *Trends Biochem. Sci.* **18**, 279–283 (1993).
112. E. Kobatake, Y. Ikariyama, M. Aizawa, K. Miwa, and S. Kato, *J. Biotechnol.* **16**, 87–96 (1990).
113. C. Lindbladh, K. Mosbach, and L. Bülow, *J. Immunol. Methods* **137**, 199–207 (1991).
114. E. Witt, G.C. Stewart, and W. Hengstenberg, *Prot. Eng.* **5**, 267–271 (1992).
115. P.S. Chowdhury, A. Kushwaha, S. Abrol, and V.K. Chaudhary, *Prot. Expr. Purif.* **5**, 89–95 (1994).
116. E. Kobatake, Y. Ikariyama, and M. Aizawa, *J. Biotechnol.* **38**, 263–268 (1995).
117. J. Lundeberg, J. Wahlberg, and M. Uhlén, *Gene Anal. Technol.* **7**, 47–53 (1990).
118. J. Lundeberg, J. Wahlberg, M. Holmberg, U. Pettersson, and M. Uhlén, *DNA Cell Biol.* **9**, 287–292 (1990).
119. C. Ljungquist, J. Lundeberg, A.M. Rasmussen, E. Hornes, and M. Uhlén, *DNA Cell Biol.* **12**, 191–197 (1993).
120. T. Sano, C. Smith, and C.R. Cantor, *Science* **258**, 120–122 (1992).
121. D.S. Clark, *Trends Biotechnol.* **12**, 439–443 (1994).
122. J. Nilsson, M. Bosnes, F. Larsen, P.-Å. Nygren, M. Uhlén, and J. Lundeberg, *Biotechniques* **22**, 744–751 (1997).
123. U.F. Power, H. Plotnicky-Gilquin, T. Huss, T.N. Nguyen, A. Robert, M. Trudel, S. Ståhl, M. Uhlén, and H. Binz, *Virology* **230**, 155–166 (1997).
124. C. Libon, N. Corvaïa, J.-F. Haeuw, T.N. Nguyen, S. Ståhl, J.-Y. Bonnefoy, and C. Andréoni, *Vaccine* **17**, 406–414 (1999).
125. G. Georgious, C. Stathopoulos, P.S. Daugherty, A.R. Nayak, B.L. Iverson, and R. Curtiss III, *Nat. Biotechnol.* **15**, 29–34 (1997).
126. S. Ståhl and M. Uhlén, *Trends Biotechnol.* **15**, 185–192 (1997).
127. Z. Lu, K.S. Murray, V. van Cleave, E.R. Lavallie, M.L. Stahl, and J.M. McCoy, *Biotechnology* **13**, 366–372 (1995).
128. C. Sousa, A. Cebolla, and V. de Lorenzo, *Nat. Biotechnol.* **14**, 1017–1020 (1996).
129. C.J. Stover, G.P. Bansal, M.S. Hanson, J.E. Burlein, S.R. Palaszynski, J.F. Young, S. Koenig, D.B. Young, A. Sadziene, and A.G. Barbour, *J. Exp. Med.* **178**, 197–209 (1993).
130. V. A. Fischetti, D. Medaglini, and G. Pozzi, *Curr. Opin. Biotechnol.* **7**, 659–666 (1996).
131. S. Ståhl, P. Samuelson, M. Hansson, C. Andréoni, L. Goetsch, C. Libon, S. Liljeqvist, E. Gunneriusson, H. Binz, T.N. Nguyen, and Uhlén M., in J. Wells and G. Pozzi, eds., *Recombinant Gram-Positive Bacteria as Vaccine Vehicles for Mucosal Immunization*, Landes, Springer-Verlag, New York, 1997, pp. 61–81.
132. S. Liljeqvist, P. Samuelson, M. Hansson, T.N. Nguyen, H. Binz, and S. Ståhl, *Appl. Environ. Microbiol.* **63**, 2481–2488 (1997).
133. C. Andréoni, L. Goetsch, C. Libon, P. Samuelson, A. Robert, T.N. Nguyen, M. Uhlén, H. Binz, and S. Ståhl, *Biotechniques* **23**, 696–704 (1997).
134. G.P. Smith, *Science* **228**, 1315–1317 (1985).
135. P.-Å. Nygren, and M. Uhlén, *Curr. Opin. Struct. Biol.* **7**, 463–469 (1997).
136. K. Nord, J. Nilsson, B. Nilsson, M. Uhlén, and P.-Å. Nygren, *Prot. Eng.* **6**, 601–608 (1995).
137. K. Nord, E. Gunneriusson, J. Ringdahl, S. Ståhl, M. Uhlén, and P.-Å. Nygren, *Nat. Biotechnol.* **15**, 772–777 (1997).

See also ANTIBODY PURIFICATION; HYBRIDOMA, ANTIBODY PRODUCTION.

AIR HANDLING. See FILTRATION, AIR; HEATING, VENTILATION AND AIR CONDITIONING.

ALANINE. See AMINO ACIDS, PRODUCTION PROCESSES.

ALDEHYDE REDUCTASE

SAKAYU SHIMIZU
MICHIHIKO KATAOKA
Kyoto University
Kyoto, Japan

KEY WORDS

Aldehyde reductase
Aldo-keto reductase superfamily
Chiral alcohol
4-Chloro-3-hydroxybutanoate ester
Glucose dehydrogenase
$NADP^+$
NADPH
Optically active alcohol

Oxidoreductase
Stereospecific reduction

OUTLINE

INTRODUCTION

There is a group of reductases that catalyze the reduction of various aldehydes and ketones (1,2). They occur widely in living organisms and are suggested to be involved in the metabolism of biogenic and xenobiotic carbonyl compounds. From the viewpoint of the practical application of these reductases, the stereospecific reduction of carbonyl compounds with enzymes exhibiting stereospecificity is an effective method for the production of various optically active alcohols (3–6).

Among these enzymes catalyzing the reduction of various carbonyl compounds, a class of monomeric NADPH-dependent oxidoreductases with molecular masses of about 35 kDa, called the aldo-keto reductase superfamily (7), has been well characterized. This superfamily contains many reductases, such as aldehyde reductase (EC 1.1.1.2), aldose reductase (EC 1.1.1.21), and so forth, and these proteins show high similarity in their substrate specificities and primary amino acid sequences (7). Here we describe aldehyde reductase and the application of microbial aldehyde reductase to the production of chiral alcohols.

ALDEHYDE REDUCTASE AND ALDO-KETO REDUCTASE SUPERFAMILY ENZYMES

The aldo-keto reductase superfamily is a class of monomeric NADPH-dependent oxidoreductases with molecular masses of about 35 kDa (7). This group includes aldehyde and aldose reductases (8–14), prostaglandin F synthase (EC 1.1.1.188) (15), 2,5-diketogluconate reductase (16,17), morphine dehydrogenase (EC 1.1.1.218) (18), chlordecone reductase (EC 1.1.1.225) (19), a number of hydroxysteroid dehydrogenases (EC 1.1.1.62, 1.1.1.149, 1.1.1.213) (20–25), plant chalcone reductases (26,27), and so on. From the strong similarity in their amino acid sequences, ρ-crystallin (28) and a yeast protein encoded by the *GCY* gene (29), which do not exhibit oxidoreductase activity, are also included in this superfamily. (A list of current members of the aldo-keto reductase superfamily can be seen on the AKR Web page at *http: / /pharme26.med.upenn.edu. /*.) Among them, aldehyde reductase and aldose reductase catalyze the NADPH-dependent reduction of a wide range of aromatic and aliphatic aldehydes, such as *p*-nitrobenzaldehyde, pyridine-3-aldehyde, and glyceraldehyde, to the corresponding alcohols (8–14). These two enzymes are widely distributed in various kinds of mammals including humans, cows, pigs, rats, and mice.

Because aldose and aldehyde reductases cause complications in diabetes (30), the crystal structures of human and pig aldose reductases and aldehyde reductases have been determined by X-ray diffraction methods (31–34). As oxidoreductases, the enzymes are unique in that they have a β/α–triosephosphate isomerase (TIM) barrel structure, and are the first oxidoreductases known to possess such a structure instead of a dinucleotide, or Rossmann, binding fold. Elucidation of the tertiary structures and reaction mechanisms of these enzymes will facilitate the design of specific inhibitors that may be used as drugs for the treatment of diabetic complications.

NOVEL MICROBIAL ALDEHYDE REDUCTASE

Yamada et al. found a novel aldehyde reductase in a red yeast, *Sporobolomyces salmonicolor* (35–37), and succeeded in cloning the aldehyde reductase gene from this strain (38).

Aldehyde reductase was isolated in a crystalline form from cells of *S. salmonicolor* and was characterized in some detail (35–37). It should be noted that the cellular content of the enzyme comprised more than 4.5% of the total extractable protein. The enzyme has a monomeric form of about 35 kDa. The enzyme absolutely requires NADPH as a cofactor. The substrate specificity of the enzyme is shown in Table 1. The enzyme did not catalyze the reduction of any keto acids or monoketones tested. Several aldehydes such as *p*-nitrobenzaldehyde, *p*-chlorobenzaldehyde, pyridine-3-aldehyde, or D-glyceraldehyde were readily reduced. These aldehydes are typical substrates for aldo-keto reductase family members, as already described. Quinones, such as *p*-benzoquinone, α-naphthoquinone, β-naphthoquinone, and menadione, which are good substrates for carbonyl reductases (14), were not reduced. Although no detectable oxidation of NADPH was observed with D-glucuronate, D-glucose, D-xylose, or D-galactose as the substrate at a low concentration (0.2 mM), the enzyme effectively catalyzed the reduction of these compounds when they were present at a high concentration (100 mM). The K_m values for these substrates are quite high in comparison with those for other substrates, which suggests that the affinity of the enzyme for these substrates is very low. Furthermore, the enzyme effectively catalyzed the reduction of 4-haloacetoacetate esters to the corresponding (*R*)-4-halo-3-hydroxybutanoate esters (100% ee) (35,36). Because the reduction products derived from 4-haloacetoacetate esters are optically active alcohols, the enzyme can recognize the stereopositions of these substrates.

2-Chloroacetoacetate esters were also effective substrates. Initial velocity and inhibition studies indicated that the enzyme reaction sequence is ordered, with NADPH binding to the free enzyme, and $NADP^+$ being the last product to be released. Quercetin, dicoumarol, and some SH-reagents inhibited the enzyme activity. These results of enzymological studies suggested that the enzyme of *S. salmonicolor* belongs to the aldo-keto reductase superfamily.

cDNA coding the aldehyde reductase was cloned from cells of *S. salmonicolor* and sequenced (38). The aldehyde reductase gene comprises 969 bp and encodes a polypeptide of 35,232 Da. The deduced amino acid sequence shows a high degree of similarity to that of other members of the aldo-keto reductase superfamily. The identities between *S. salmonicolor* aldehyde reductase and the bovine (39), rat (40), human (7), rabbit (41), and barley (42) aldose reductases, human aldehyde reductase (7), and xylose reductase from *Pichia stipitis* (43) are 45.1, 45.0, 43.7, 42.9, 39.5, 43.9, and 37.7%, respectively. In general, aldo-keto reductases exhibit greater similarity in their N-terminal regions than in their C-terminal regions (Fig. 1). Analysis of the genomic DNA sequence indicated that the aldehyde reductase gene was interrupted by six introns (two in the 5′ noncoding region and four in the coding region).

Table 1. Substrate Specificity of the Aldehyde Reductase from *S. salmonicolor*

Substrate	Relative activity (%)	Substrate	Relative activity (%)
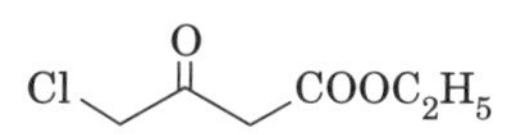	100	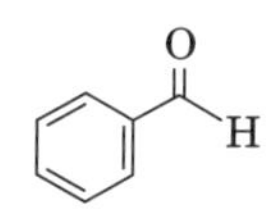	14
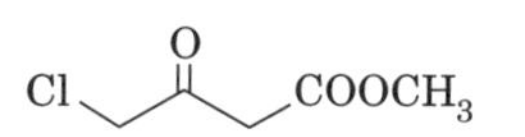	25	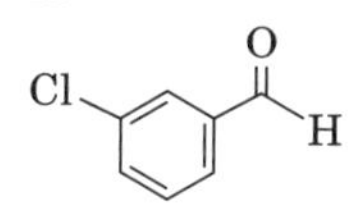	56
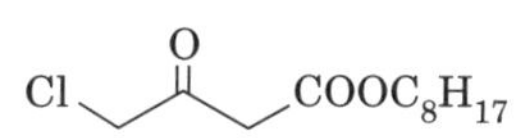	240	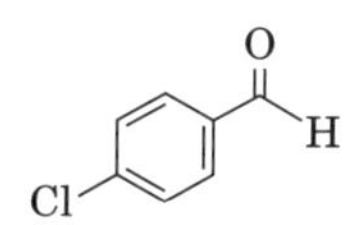	52
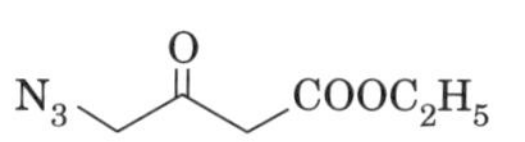	65	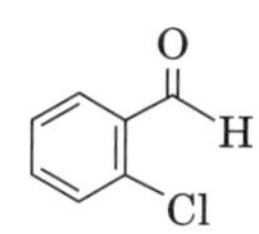	58
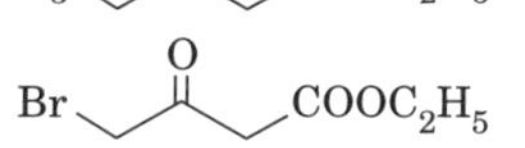	75	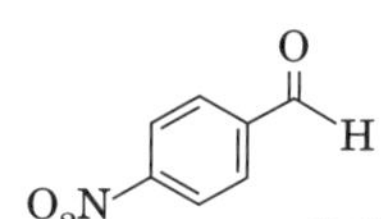	468
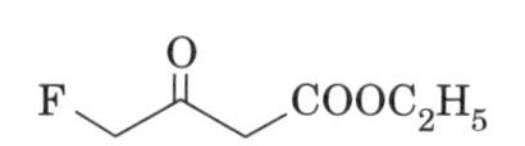	153	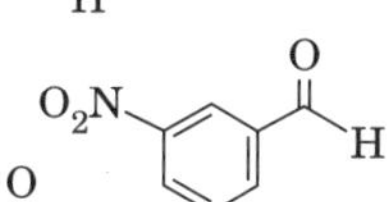	63
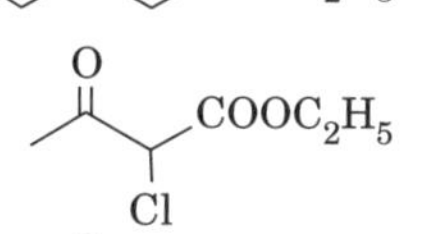	330	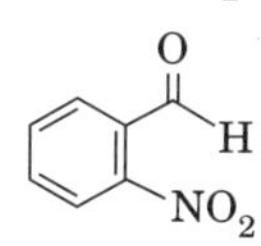	14
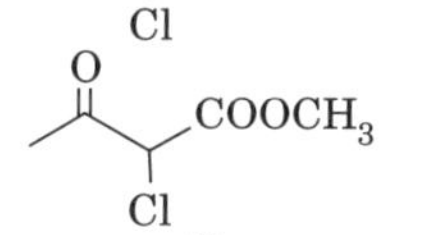	74	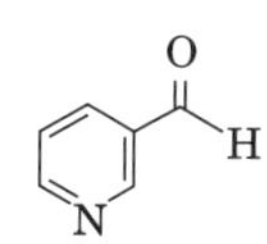	228
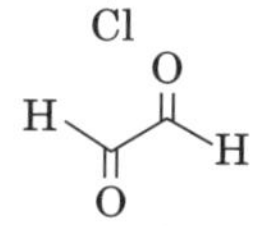	74	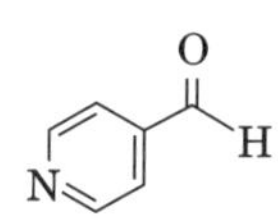	54
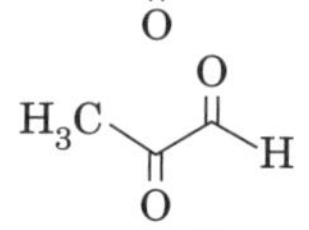	219	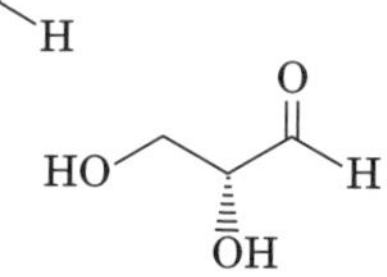	64
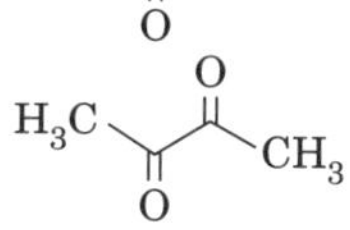	75	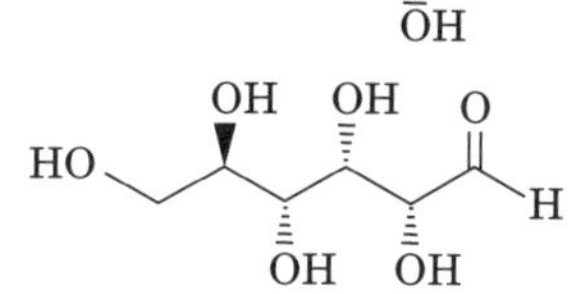	81
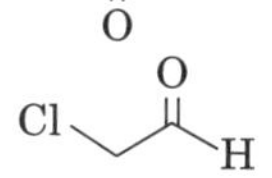	17	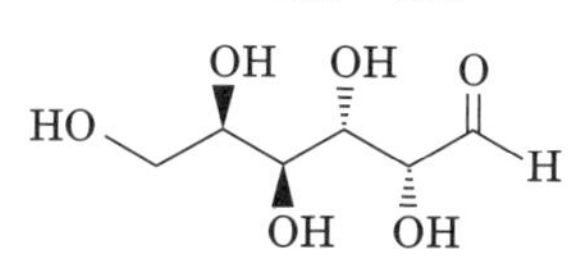	24
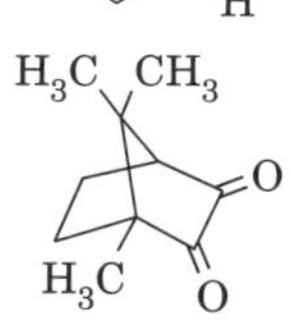	16	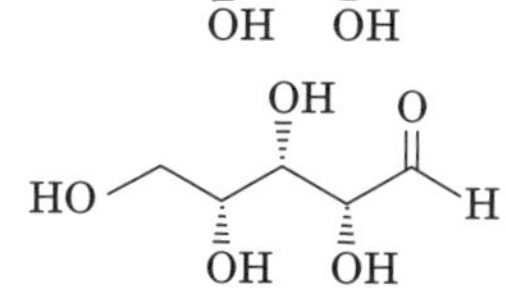	472
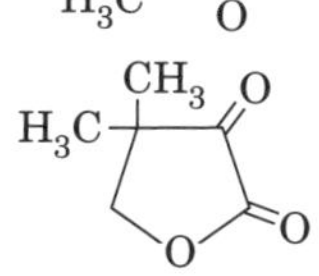	0	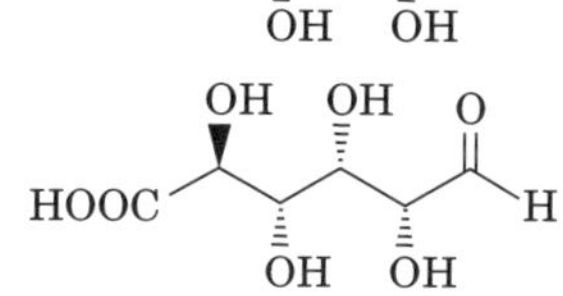	173

Note: See Refs. 35 and 36 for details.

```
sALR         MVGTTTLNTGASLELVGYGTWQA------APGEVGQGVKVAIETGYRHLDLAK
hALR        MAASCVLLHTGQKMPLIGLGTW------KSEPGQVKAAVKYALSVGYRHIDCAA
hADR         MASRLLLNNGAKMPILGLGTW------KSPPGQVTEAVKVAIDVGYRHIDCAH
rADR         MASHLELNNGTKMPTLGLGTW------KSPPGQVTEAVKVAIDMGYRHIDCAQ
bPGF      MDPKSQRVKLNDGHFIPVLGFGTYAPEEVPKSEA---LEATKFAIEVGFRHVDSAH
fRHO     MTLTKETRVTLNDGNMMPILGLGTYAAPDVPKSLA---EEAVKTAIDVGYRHIDCAF
sGCY  MPATLHDSTKILSLNTGAQIPQIGLGTWQS----KENDA--YKAVLTALKDGYRHIDTAA
cDKG        MTVPSIVLNDGNSIPQLGYGVF------KVPPADTQRAVEEALEVGYRHIDTAA

sALR  VYSNQPEVGAAIKEA-----GVKREDLFITSKLWNNSHRPEQVEPALDDTLKELGLEYLD
hALR  IYGNEPEIGEALKEDVGPGKAVPREELFVTSKLWNTKHHPEDVEPALRKTLADLQLEYLD
hADR  VYQNENEVGVAIQEKLREQ-VVKREELFIVSKLWCTYHEKGLVKGACQKTLSDLKLDYLD
rADR  VYQNEKEVGVALQEKLKEQ-VVKRQDLFIVSKLWCTFHDQSMVKGACQKTLSDLQLDYLD
bPGF  LYQNEEQVGQAIRSKIADG-TVKREDIFYTSKLWCNSLQPELVRPALEKSLQNLQLDYVD
fRHO  ITGNEMHIGNGIRSKISDG-TVEREDIFYTGKLWCTYFSPDMVRKGLERSLRDVGMDYLD
sGCY  IYRNEDQVGQAIK-----DSGVPREEIFVTTKLWCTQHHEPEV--ALDQSLKRLGLDYVD
cDKG  IYGNEEGVGAAIA-----ASGIARDDLFITTKLWNDRHDGDEPAAAIAESLAKLALDQVD

sALR  LYLIHWPVAFPPEG----DITQNLFPKANDKEVKLDLEVSLVDTWKAMVKLLDTGKVKAI
hALR  LYLMHWPYAFERG--------DNPFPKNADGTICYDSTH-YKETWKALEALVAKGLVQAL
hADR  LYLIHWPTGFKPG--------KEFFPLDESGNVVPSDTN-ILDTWAAMEELVDEGLVKAI
rADR  LYLIHWPTGFKPG--------PDYFPLDASGNVIPSDTD-FVDTWTAMEQLVDEGLVKAI
bPGF  LYIIHSPVSLKPG--------NKFVPKDESGKLIFDSVD-LCHTWEALEKCKDAGLTKSI
fRHO  LFLMHWPVSLKPS--------GASDPSDKDKPFIYDNVD-LCATWEALEACKDAGLVRSL
sGCY  LYLMHWPARLDPAYIKNEDILSVPTKKDGSRAVDITNWN-FIKTWELMQELPKTGKTKAV
cDKG  LYLVHWP--------------------------TPAADN-YVHAWEKMIELRAAGLTRSI

sALR  GVSNFDAKMVDAIIEATGV--TPSVNQIERHPLLLQPELIAHHKAKNIHITAYSPLG---
hALR  GLSNFNSRQIDDILSVASV--RPAVLQVECHPYLAQNELIAHCQARGLEVTAYSPLGSS-
hADR  GISNFNHLQVEMILNKPGLKYKPAVNQIECHPYLTQEKLIQYCQSKGIVVTAYSPLGSP-
rADR  GVSNFNPLQIERILNKPGLKYKPAVNQIECHPYLTQEKLIEYCHCKGIVVTAYSPLGSP-
bPGF  GVSNFNHKQLEKILNKPGLKYKPVCNQVECHPYLNQSKLLEFCKSHDIVLVAYAALGAQL
fRHO  GVSNFNRRQLERILNKPGLKYKPVCNQVECHVYLNQNKLHAYCKSKDIVLVAYSVLGSHR
sGCY  GVSNFSINNLKDLLASQGNKLTPAANQVEIHPLLPQDELINFCKSKGIVVEAYSPLGS--
cDKG  GVSNHLVPHLERIVAATGV--VPAVNQIELHPAYQQREITDWAAAHDVKIESWGPLG---

sALR  ---NNTVGAPLLVQHPEIKRLAEKNGCTPAQVLIAWAIVGGHSVIPKSVTPSRIGENFKQ
hALR  DRAWRDPDEPVLLEEPVVLAIAEKYGRSPAQILLRWQVQRKVICIPKSITPSRILQNIKV
hADR  DRPWAKPEDPSLLEDPRIKAIAAKHNKTTAQVLIRFPMQRNLVVIPKSVTPERIAENFKV
rADR  DRPWAKPEDPSLLEDPRIKEIAAKYNKTTAQVLIRFPIQRNLVVIPKSVTPARIAENFKV
bPGF  LSEWVNSNNPVLLEDPVLCAIAKKHKQTPALVALRYQVQRGVVVLAKSFNKKRIKENMQV
fRHO  DRNWVDLSLAVLLDDPILNKIAAKYNRSSAEVAMRFILQKGIVVLAKSFTPARIKQNLGV
sGCY  -------TDAPLLKEPVILEIAKKNNVQPGHVVISWHVQRGYVVLPKSVNPDRIKTNRKI
cDKG  -----QGKYDLFGAEPVTAAAAA-HGKTPAQAVLRWHLQKGFVVFPKSVRRERLEENLDV

sALR  V--SLSQEDVDAVSKLGEGSGRRRYNIPCTYSPKWDINVFGEEDEKSCKNAVKIK
hALR  FDFTFSPEEMKQLNALNKNWRYIVPMLTVDGKRVPRDAGHPLYPFNDPY
hADR  FDFELSSQDMTTLLSYNRNWRVCALLSC---------TSHKDYPFHEEF
rADR  FDFELSNEDMATLLSYNRNWRVCALMSC---------AKHKDYPFHAEV
bPGF  FDFELTPEDMKAIDGLNRNIRYYDFQKG---------IGHPEYPFSEEY
fRHO  FEFELKPEDMKTLESLDRNLHYGPFREV---------KQHPEYPFHDEY
sGCY  F--TLSTEDFEAINNISKE-------------KGEKRVVHPNW---SPFEVFK
cDKG  FDFDLTDTEIAAIDAMDPG------------DGSGRVSAHPD----EVD
```

Figure 1. Comparison of the primary structures of the aldehyde reductase from *S. salmonicolor* and aldo-keto reductase superfamily proteins. The sequences of the aldehyde reductase (sALR), human aldehyde reductase (hALR), human aldose reductase (hADR), rat aldose reductase (rADR), bovine prostaglandin F synthase (bPGF), frog ρ-crystallin (fRHO), the yeast nuclear gene product (sGCY), and 2,5-diketogluconic acid reductase from *Corynebacterium* sp. (cDKG) are aligned. Gaps in the aligned sequences are indicated (–). Identical amino acid residues are enclosed in boxes. See Ref. 38 for details.

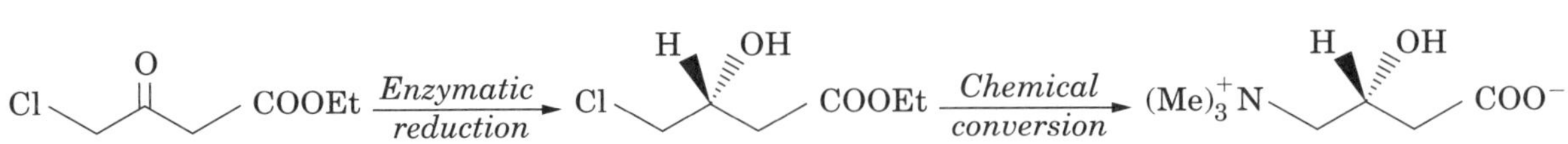

CAAE (*R*)-CHBE L-Carnitine

Figure 2. Chemicoenzymatic route for the synthesis of L-carnitine. CAAE, ethyl 4-chloroacetoacetate; CHBE, ethyl 4-chloro-3-hydroxybutanoate.

APPLICATION OF MICROBIAL ALDEHYDE REDUCTASE TO THE PRODUCTION OF OPTICALLY ACTIVE ALCOHOLS

The asymmetric reduction of carbonyl compounds with microorganisms or enzymes is a useful method for the synthesis of optically active compounds (3–6). The aldehyde reductase produced by *S. salmonicolor* effectively catalyzes the asymmetric reduction of 4-haloacetoacetate esters to the corresponding (*R*)-isomers, which are promising chiral building blocks for the chemical synthesis of L-carnitine (Fig. 2) and so forth (44–47).

Stereospecific Reduction of Ethyl 4-Chloroacetoacetate

Ethyl (*R*)-4-chloro-3-hydroxybutanoate formed through the reduction of ethyl 4-chloroacetoacetate with cells of *S. salmonicolor* showed low optical purity (8.5% ee) because the cells contained a significant amount of an enzyme(s) catalyzing the reduction of ethyl 4-chloroacetoacetate to ethyl (*S*)-4-chloro-3-hydroxybutanoate in addition to the aldehyde reductase forming ethyl (*R*)-4-chloro-3-hydroxybutanoate. Because heat treatment and acetone fractionation of a cell-free extract of *S. salmonicolor* remove the (*S*)-isomer-forming enzyme(s) from the cell extract, the cell extract after these treatments was used as the source of aldehyde reductase. Commercially available glucose dehydrogenase, NADP$^+$, and glucose were added as a cofactor regenerator. A practical preparation can be carried out in an organic solvent–water two-phase system, as shown in Fig. 3a, because the substrate is unstable in an aqueous solution, and both the substrate and product strongly inhibit the enzyme reaction. In this system, most of the substrate is present in the organic phase; consequently, the ethyl 4-chloroacetoacetate is quite stable and cannot in-

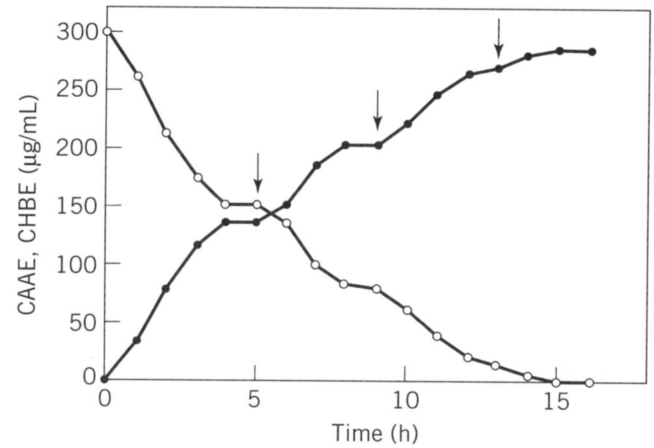

Figure 4. Conversion of ethyl 4-chloroacetoacetate to ethyl (*R*)-4-chloro-3-hydroxybutanoate in a two-phase reaction system by *E. coli* JM109 [pKAR, pKKGDH] cells. Glucose (100 mg/mL) was added to the aqueous phase periodically (arrows). Symbols: ○, ethyl 4-chloroacetoacetate (CAAE); ●, ethyl (*R*)-4-chloro-3-hydroxybutanoate (CHBE). See Refs. 49–52 for details.

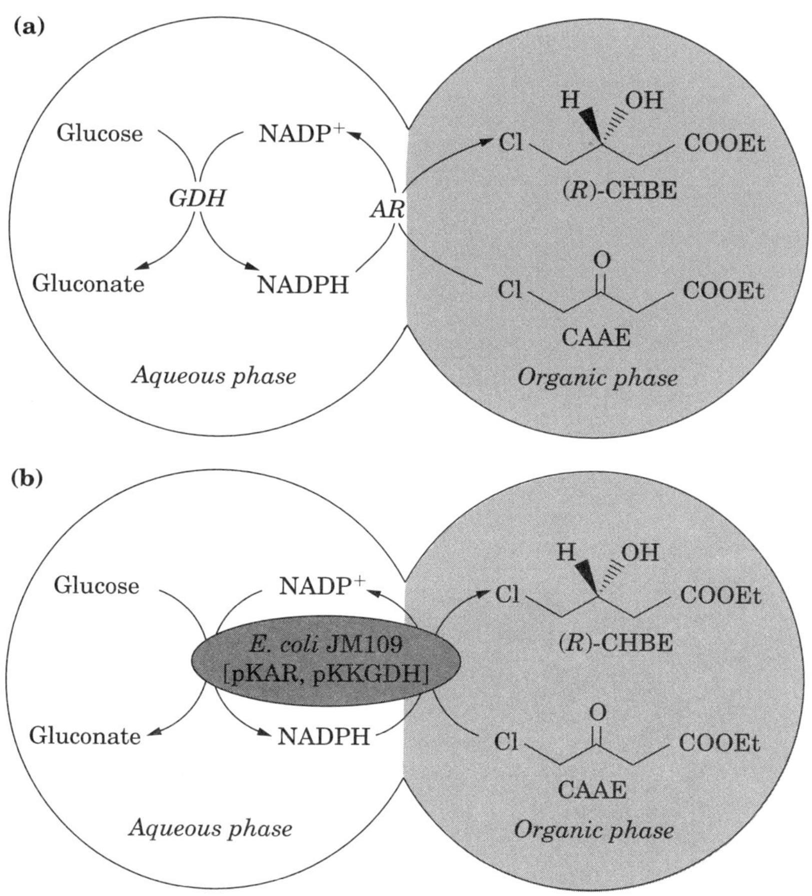

Figure 3. Outline of the stereospecific reduction of ethyl 4-chloroacetoacetate by aldehyde reductase (AR) and glucose dehydrogenase (GDH) (**a**), and washed cells of *E. coli* JM109 [pKAR, pKKGDH] (**b**), as catalysts, in an organic solvent–water two-phase system. CAAE, ethyl 4-chloroacetoacetate; CHBE, ethyl 4-chloro-3-hydroxybutanoate. See Refs. 48–52 for details.

hibit the enzyme reaction occurring in the aqueous phase. n-Butyl acetate is regarded as the most suitable organic solvent for such a two-phase system because it shows high partition efficiencies with regard to both the substrate and product, and both enzymes are stable in the presence of this organic solvent. In a bench-scale two-phase reaction, 83.8 g/L of ethyl (R)-4-chloro-3-hydroxybutanoate (86% ee) was produced from ethyl 4-chloroacetoacetate, with a molar yield of 95.4% (48,53).

Enzymatic Production of Ethyl (*R*)-4-Chloro-3-Hydroxybutanoate

The two-phase reaction was improved by using an *Escherichia coli* transformant. The aldehyde reductase gene from *S. salmonicolor* (38) and the glucose dehydrogenase gene from *Bacillus megaterium* (54) were transformed into the same *E. coli* strain; consequently, a transformant overproducing both aldehyde reductase and glucose dehydrogenase, *E. coli* JM109 [pKAR, pKKGDH], was obtained (49–52). The reduction reaction was carried out in an organic solvent–water two-phase system containing ethyl 4-chloroacetoacetate, glucose, and $NADP^+$, and *E. coli* JM109 [pKAR, pKKGDH] cells as the catalyst (Fig. 3b). When the *E. coli* cells were incubated in the n-butyl acetate–water two-phase system, 300 mg/mL of ethyl 4-chloroacetoacetate was almost stoichiometrically converted to ethyl (R)-4-chloro-3-hydroxybutanoate (92% ee) in 16 h (Fig. 4). Because the use of *E. coli* transformant cells as the catalyst is simple and does not require isolation of the aldehyde reductase, which is necessary for *S. salmonicolor* cells, it is highly advantageous for the practical synthesis of ethyl (R)-4-chloro-3-hydroxybutanoate.

BIBLIOGRAPHY

1. H. Weiner, D.W. Crabb, and T.G. Flynn eds., *Enzymology and Molecular Biology of Carbonyl Metabolism 4, Advances in Experimental Medicine and Biology*, Vol. 328, Plenum, New York, 1993.
2. H. Weiner, R.S. Holmes, and B. Wermuth eds., *Enzymology and Molecular Biology of Carbonyl Metabolism 5, Advances in Experimental Medicine and Biology*, Vol. 372, Plenum, New York, 1995.
3. W. Hummel and M.-R. Kula, *Eur. J. Biochem.* **184**, 1–13 (1989).
4. J.B. Jones, *Tetrahedron* **42**, 3351–3403 (1986).
5. G.M. Whitesides and C.-H. Wong, *Angew. Chem. Int. Ed. Engl.* **24**, 617–638, (1985).
6. H. Yamada and S. Shimizu, *Angew. Chem. Int. Ed. Engl.* **27**, 622–642 (1988).
7. K.M. Bohren, B. Bullock, B. Wermuth, and K.H. Gabbay, *J. Biol. Chem.* **264**, 9547–9551 (1989).
8. T.G. Flynn, *Biochem. Pharmacol.* **31**, 2705–2712 (1982).
9. B. Wermuth, J.D.B. Munch, and J.-P. von Wartburg, *J. Biol. Chem.* **252**, 3821–3828 (1977).
10. T.G. Flynn, J.A. Cromlish, and W.S. Davidson, *Methods Enzymol.* **89**, 501–506 (1982).
11. J.-P. von Wartburg and B. Wermuth, *Methods Enzymol.* **89**, 506–513 (1982).
12. B. Wermuth, H. Burgisser, K. Bohren, and J.-P. von Wartburg, *Eur. J. Biochem.* **127**, 279–284 (1982).
13. B. Wermuth and J.-P. von Wartburg, *Methods Enzymol.* **89**, 181–186 (1982).
14. B. Wermuth, *J. Biol. Chem.* **256**, 1206–1213 (1981).
15. K. Watanabe, Y. Fujii, K. Nakayama, H. Ohkubo, S. Kuramitsu, H. Kagamiyama, S. Nakanishi, and O. Hayaishi, *Proc. Natl. Acad. Sci. U.S.A.* **85**, 11–15 (1988).
16. S. Anderson, C.B. Marks, R. Lazarus, J. Miller, K. Stafford, J. Seymour, D. Light, W. Rastetter, and D. Estell, *Science* **230**, 144–149 (1985).
17. J.F. Grindley, M.A. Payton, H. van de Pol, and K.G. Hardy, *Appl. Environ. Microbiol.* **54**, 1770–1775 (1988).
18. D.L. Willey, D.A. Caswell, C.R. Lowe, and N.C. Bruce, *Biochem. J.* **290**, 539–544 (1993).
19. C.J. Winters, D.T. Molowa, and P.S. Guzelian, *Biochemistry* **29**, 1080–1087, (1990).
20. Y. Deyashiki, K. Ohshima, M. Nakanishi, K. Sato, K. Matsuura, and A. Hara, *J. Biol. Chem.* **270**, 10461–10467 (1995).
21. A. Hara, K. Matsuura, Y. Tamada, K. Sato, Y. Miyabe, Y. Deyashiki, and N. Ishida, *Biochem. J.* **313**, 373–376 (1996).
22. R. Miura, K. Shiota, K. Noda, S. Yagi, T. Ogawa, and M. Takahashi, *Biochem. J.* **299**, 561–567 (1994).
23. W.R. Lacy, K.J. Washenick, R.G. Cook, and B.S. Dunbar, *Mol. Endocrinol.* **7**, 58–66 (1993).
24. J.E. Pawlowski, M. Huizinga, and T.M. Penning, *J. Biol. Chem.* **266**, 8820–8825 (1991).
25. M. Khanna, K.N. Qin, R.W. Wang, and K.C. Cheng, *J. Biol. Chem.* **270**, 20162–20168 (1995).
26. C. Sallaud, J. El-Turk, L. Bigarré, H. Sevin, R. Welle, and R. Esnault, *Plant Physiol.* **108**, 869–870 (1995).
27. R. Welle, G. Schröder, E. Schilts, H. Grisebach, and J. Schröder, *Eur. J. Biochem.* **196**, 423–430 (1991).
28. S.I. Tomarev, R.D. Zinovieva, S.M. Dolgilevich, S.V. Luchin, A.S. Krayev, K.G. Skryabin, and G. G. Gause, Jr., *FEBS Lett.* **171**, 297–302 (1984).
29. U. Oechsner, V. Magdolen, and W. Bandlow, *FEBS Lett.* **238**, 123–128 (1988).
30. J.H. Kinoshita and C. Nishimura, *Diabetes Metab. Rev.* **4**, 323–337 (1988).
31. O. El-Kabbani, N.C. Green, G. Lin, M. Carson, S.V.L. Narayana, K.M. Moore, T.G. Flynn, and L.J. DeLucas, *Acta Cryst.* **D50**, 859–868 (1994).
32. O. El-Kabbani, S.V.L. Narayana, Y.S. Babu, K.M. Moore, T.G. Flynn, J.M. Petrash, E.M. Westbrook, L.J. DeLucas, and C.E. Bugg, *J. Mol. Biol.* **218**, 695–698 (1991).
33. J.-M. Rondeau, F. Tête-Favier, A. Podjarny, J.-M. Reymann, P. Barth, J.-F. Biellmann, and D. Moras, *Nature* **355**, 469–472 (1992).
34. D.K. Wilson, K.M. Bohren, K.H. Gabbay, and F.A. Quiocho, *Science* **257**, 81–84 (1992).
35. H. Yamada, S. Shimizu, M. Kataoka, H. Sakai, and T. Miyoshi, *FEMS Microbiol. Lett.* **70**, 45–48 (1990).
36. M. Kataoka, H. Sakai, T. Morikawa, M. Katoh, T. Miyoshi, S. Shimizu, and H. Yamada, *Biochim. Biophys. Acta* **1122**, 57–62 (1992).
37. M. Kataoka, K. Yamamoto, S. Shimizu, K. Kita, T. Hashimoto, H. Yanase, K. Kita, and K. Miki, *Acta Cryst.* **D52**, 405–406 (1996).
38. K. Kita, K. Matsuzaki, T. Hashimoto, H. Yanase, N. Kato, M.C.M. Chung, M. Kataoka, and S. Shimizu, *Appl. Environ. Microbiol.* **62**, 2303–2310 (1996).

39. S.Z. Schade, S.L. Early, T.R. Williams, F.J. Kezdy, R.L. Heinrikson, C.E. Grimshaw, and C.C. Doughty, *J. Biol. Chem.* **265**, 3628–3635 (1990).
40. D. Carper, C. Nishimura, T. Shinihara, B. Dietzchold, G. Wistow, C. Craft, P. Kador, and J.H. Kinoshita, *FEBS Lett.* **220**, 209–213 (1987).
41. A. Garcia-Perez, B. Martin, H.R. Murphy, S. Uchida, H. Murer, B.D. Cowley, Jr., J.S. Handler, and M.B. Burg, *J. Biol. Chem.* **264**, 16815–16821 (1989).
42. D. Bartels, K. Engelhardt, R. Roncarati, K. Schneider, M. Rotter, and F. Salamani, *EMBO J.* **10**, 1037–1043 (1991).
43. R. Amore, P. Kotter, C. Kuster, M. Ciriacy, and C.P. Hollenberg, *Gene* **109**, 89–97 (1991).
44. J. Grunwald, B. Wirz, M.P. Scollar, and A.M. Klibanov, *J. Am. Chem. Soc.* **108**, 6732–6734 (1986).
45. W-R. Shieh, A.S. Gopalan, and C.J. Sih, *J. Am. Chem. Soc.* **107**, 2993–2994 (1985).
46. C-H. Wong, D.G. Drueckhammer, and H.M. Sweers, *J. Am. Chem. Soc.* **107**, 4028–4031 (1985).
47. B. Zhou, A.S. Gopalan, F. VanMiddlesworth, W-R. Shieh, and C.J. Sih, *J. Am. Chem. Soc.* **105**, 5925–5926 (1983).
48. S. Shimizu, M. Kataoka, M. Katoh, T. Morikawa, T. Miyoshi, and H. Yamada, *Appl. Environ. Microbiol.* **56**, 2374–2377 (1990).
49. M. Kataoka, L.P.S. Rohani, K. Yamamoto, M. Wada, H. Kawabata, K. Kita, H. Yanase, and S. Shimizu, *Appl. Microbiol. Biotechnol.* **48**, 699–703 (1997).
50. M. Kataoka, L.P.S. Rohani, M. Wada, K. Kita, H. Yanase, I. Urabe, and S. Shimizu, *Biosci. Biotechnol. Biochem.* **62**, 167–169 (1998).
51. S. Shimizu, M. Kataoka, and K. Kita, *J. Mol. Catal. B: Enzymatic* **5**, 321–325 (1998).
52. S. Shimizu, M. Kataoka, and K. Kita, in A. Laskin, G.-X. Li, and Y.-T. Yu eds., *Enzyme Engineering XIV, Annals of the New York Academy of Sciences*, Vol. 864, The New York Academy of Sciences, New York, 1998, in press.
53. S. Shimizu, M. Kataoka, A. Morishita, M. Katoh, T. Morikawa, T. Miyoshi, and H. Yamada, *Biotechnol. Lett.* **12**, 593–596 (1990).
54. Y. Makino, S. Negoro, I. Urabe, and H. Okada, *J. Biol. Chem.* **264**, 6381–6385 (1989).

See also BIOCATALYSIS DATABASES.

ALGAL CULTURE

TADASHI MATSUNAGA
HARUKO TAKEYAMA
Tokyo University of Agriculture and Technology
Tokyo, Japan
HIROYUKI TAKANO
Chichibu Onoda Cement Corporation
Chiba-prefecture, Japan

KEY WORDS

Algal biomass
CO_2 fixation
Cyanobacteria
Large-scale biomass culture systems
Microalgae
On-site CO_2 removal systems
Open culture systems
Photobioreactor
Useful material production

OUTLINE

INTRODUCTION

Photosynthetic microorganisms play an important role in the conversion of solar energy into chemical energy through their photosynthetic activities. Photosynthetic conversion can be expected to be an efficient and alternative process in several industrial fields; for example, microalgae have been studied because they produce a wide variety of metabolites.

Attempts to develop large-scale methodologies for the cultivation of microalgae have been summarized in a document produced by the Carnegie Institute of Washington (1), in which algae were predicted to provide an alternative to fermentation and agriculture products. Algal biomass has historically served as fertilizer, and as a food source for both humans and animals (2–4). They are also used for secondary wastewater treatment (5). This use of algal biomass is an important industrial application of microalgal cultures. With advances in processing technology, algal biomass has come to be seen as a possible source of fuel, as well as of fine chemicals and pharmaceuticals (6). Several microalgae that produce useful chemicals, such as amino acids, vitamins, carotenoids (β-carotene), fatty acids (DHA, EPA, γ-LA), polysaccharides, and antibiotics have been reported (Table 1). Several microalgal products have already been commercialized (7).

It is estimated that there are roughly over 10,000 species of microalgae in the natural environment. Unique algal strains may be found in highly diversified marine environments. We have discovered several marine microalgal strains that produce useful substances, including novel

Table 1. Commercial Products from Microalgae

Products	Uses	Value (dollars/kg)	Market (millions of dollars)
Phycobiliproteins	Research	10,000	1–10
	Food coloring	100	
β-Carotene	Provitamin A	500	1–10
	Food coloring	300	
Xanthophylls	Feed	200–500	10–100
Vitamin C	Vitamin	10	10–100
Vitamin E	Vitamin	50	10–100
Polysaccharides	Viscosifiers	5–10	10–1,000
Amino acids			
Proline		5–50	1–10
Arginine		50–100	1–10
aspartic acid	Gums	2–5	100–1,000
Single cell protein	Animal feeds	10–20	100–2,000

plant-growth regulators that promote redifferentiation, germination, and plantlet formation; tyrosinase inhibitors; a sulfated polysaccharide showing anti-HIV activity; novel antibiotics with light-regulated activity; and a UV-A absorbing compound (Table 2).

The development of an efficient culture system is necessary for algal mass production and the industrial application of microalgae. Hence, we describe several technologies for microalgal mass cultivation under open and closed systems. Furthermore, recent studies of applied microalgal culture investigating the possible use of biological CO_2 fixation are presented. Also discussed are examples of the technical designs for on-site CO_2 removal systems combined with algal cultures, which promise to play a role in developing technologies for environmental protection.

MICROALGAL MASS CULTIVATION TECHNOLOGIES

The growth rate and maximum biomass yield of microalgal strains are affected by culture parameters (light, temperature, and pH) and nutritional status (CO_2, nitrogen, and phosphate concentrations). On the other hand, increasing the density of cultures decreases photon availability to individual cells. Light penetration of microalgal cultures is poor, especially at high cell densities (Fig. 1), and such poor photon availability decreases specific growth rates. Higher biomass yields could be expected if sufficient photons were provided to high-density cultures of microalgae.

Large-scale culture systems have been constructed (classified as open and closed systems) with the greatest attention paid to the light supply (Fig. 2) (28). Strains such as *Chlorella, Scenedesmus, Dunaliella, Spirulina, Porphyridium*, and *Haematococcus* have been cultured using these systems to provide for several useful materials (6,29–32).

Open Systems

Several different types of open culture systems have been proposed (Fig. 2a–d). These open culture systems are the simplest method of algal cultivation and offer advantages in construction cost and ease of operation. Open culture systems require a large surface area and shallow depth (ca. 12–15 cm) to improve light utilization. Furthermore, mixing the culture prevents the cells from sinking to the bottom and distributes efficient cell growth with sunlight. The raceway pound proposed by Oswald (33) has been developed into various types, including raceway pounds employing a paddle wheel, which have most generally been used for outdoor production of microalgae (34,35). The raceway pounds for commercial production of microalgae require an area of 1,000–5,000 m^2.

Contamination by different algal species and other organisms is a serious problem in open culture systems. *Chlorella, Dunaliella*, and *Spirulina*, which are tolerant to extreme conditions (high nutrients, high salinity, and high pH), are adequate strains for open culture systems. Vonshak et al. (36) demonstrated that contamination by *Chlorella* in outdoor *Spirulina* culture was prevented by maintaining the culture medium at a high bicarbonate concentration (0.2 M). Grazers sometimes found in *Spirulina* culture were arrested by addition of ammonia (2 mM).

The open culture system is easily affected by weather conditions. Rain dilutes salinity, causing contamination.

Table 2. Production of Useful Materials by Marine Microalgae

Product	Strain	Amount (mg or unit*/g dry wt)	Ref.
Coccolith	*Emiliania huxleyi*	740	8–10
	Pleurochrysis carterae		
γ-Linolenic acid	*Chlorella* sp. NKG042401	9.5	11, 12
Palmitoleic acid	*Phormidium* sp. NKBG 041105	47	13
Docosahexaenoic acid (DHA)	*Isochrysis galbana* UTEX LB 2307	15.7	14
Polysaccharide	*Aphanocapsa halophytia*	45	15
Glutamate	*Synechococcus* sp. NKBG040607	15.4	16, 17
Phycocyanin	*Synechococcus* sp. NKBG042902	150	18
UV-A absorbing biopterin glucoside	*Oscillatoria* sp. NKBG 091600	0.2	19, 20
Antimicrobial compound	*Chlorella* sp. NKG 0111	—	21
Plant growth regulator	*Synechococcus* sp. NKBG042902	—	22–24
Lactic acid bacteria growth regulator	*Synechococcus* sp. NKBG040607	—	
Tyrosinase inhibitor	*Synechococcus* sp. NKBG15041c	107*	25, 26
SOD	*Synechococcus* sp. NKBG042902	—	27

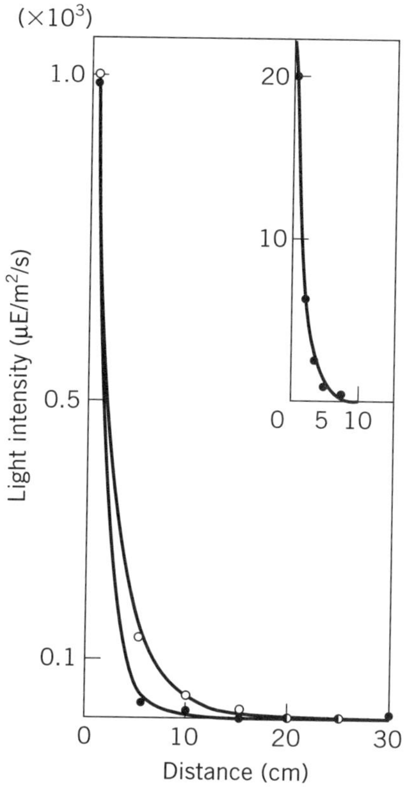

Figure 1. The distribution of light intensity as a function of distance from the surface at various cell concentrations; solid circles, 2.3×10^8 cells/ml; open circles, 7.9×10^7 cells/ml.

Furthermore, several algae that produce useful chemicals require more restricted conditions for efficient growth and for metabolite production.

Photobioreactor (Closed System)

Closed systems have been expected to overcome such open culture system disadvantages, and several photobioreactors have been devised. Figure 2e–i shows some large-scale closed photobioreactors. These photobioreactors offer several advantages: easier maintenance of monoalgal cultures because of protection from contamination; avoidance of water loss and subsequent increase of salinity in the culture medium; higher productivity with greater cell densities, reducing harvesting costs; and applicability to varying microalgal species under favorable culture conditions. However, for an efficient and reliable large-scale culture system, several criteria need to be considered (37), such as light-utilization efficiency, homogeneous mixing (turbulence), low-shear environment, temperature control, and efficient gas transfer.

Yield of algal biomass depends on the light path to each cell, and therefore, surface-to-volume ratio is an important factor for efficient light utilization in a photobioreactor. Tubular reactors have been well studied since their development by Tamiya et al. (38). The diameter of a tubular reactor is now less than 40 cm. Richmond et al. (39) demonstrated that reduction of the tube diameter from 5.0 cm to 2.8 cm enhanced the biomass yield 1.8 times. Narrower tube diameters may allow efficient light utilization as well as a faster flow rate, enhancing the algal productivity and reducing fouling on the inside walls of the tubes. In tubular reactors, flow rates of 30–50 cm/s are generally used, and air-lift is the most effective circulation method of culture, rather than a centrifugal pump, a rotary positive displacement pump, and a peristaltic pump. The main advantages of air-lift systems are their low shear and relative simplicity of construction. Furthermore, the advantage of the tubular system in biomass production per volume using *Anabaena siamensis* was shown where productivities of an open raceway (300-L capacity, height of 15 cm) and a tubular reactor with a 2.8-cm-diameter tube were 86 and 550 mg dry wt/L/day, respectively. Several modifications in tube arrangement have been carried out for optimizing light penetration. Most of the tubular reactor is laid on the ground, and the lower parts of the tubes receive less light than the upper parts. Torzillo et al. (40) have constructed a two-plane tubular bioreactor for optimizing light availability, in which each tube in the lower plane was placed in the vacant space between two tubes in the upper plane. They showed that this two-plane tubular bioreactor (145 L culture volume) has an effective surface area (ratio of surface area [m^2] to culture volume [m^3] = 49/m) and gave a net volumetric productivity of 1.5 g dry wt/L/day using *Spirulina platensis*. Furthermore, a pilot plant called the Biocoil has been set up (Fig. 2e) (41). This helical tubular reactor is constructed with a vertical tower coiled up with a tube, increasing the available efficient land surface area. When biomass productivity of various tubular reactors is compared on area basis (surface area), a similar production range of 15–30 g dry wt/m^2/day is obtained for all reactor types.

The effect of temperature on biomass yield is significant in algal culture. The culture in the tubular reactor is often maintained at higher temperature than that in the open raceway. Richmond et al. (39) demonstrated that the *Spirulina* culture using the tubular reactor could be warmed faster than in the open raceway. The culture could be kept at 35–37 °C, the optimal range for *Spirulina* growth during most of the light period. However, the closed reactors are sometimes overheated and thus are more suitable for the thermophilic or thermotolerant strains. Temperature control using a heat exchanger and/or evaporative cooling by spraying water onto the surface is sometimes required for the cultivation of general strains (40,42).

A problem in the closed system is photooxidative damage to the cells caused by accumulation of dissolved oxygen produced by photosynthesis during the light period. In the open system, evolved oxygen is diffused to the atmosphere. By contrast, because oxygen cannot escape from the closed reactors, degasser systems are sometimes required.

The culture part of closed photobioreactors has been constructed with materials such as glass, methyl polymetacrylate, polyethylene, polypropylene, polyvinylchloride, silicone, and stainless steel. These photobioreactors have been designed to utilize external illumination, such as sunlight. Fiber optics also have been employed as internal light sources (43). Photobioreactors employing fiber optics have the merit of controlling the illumination and duration of the light period. The authors have demonstrated growth

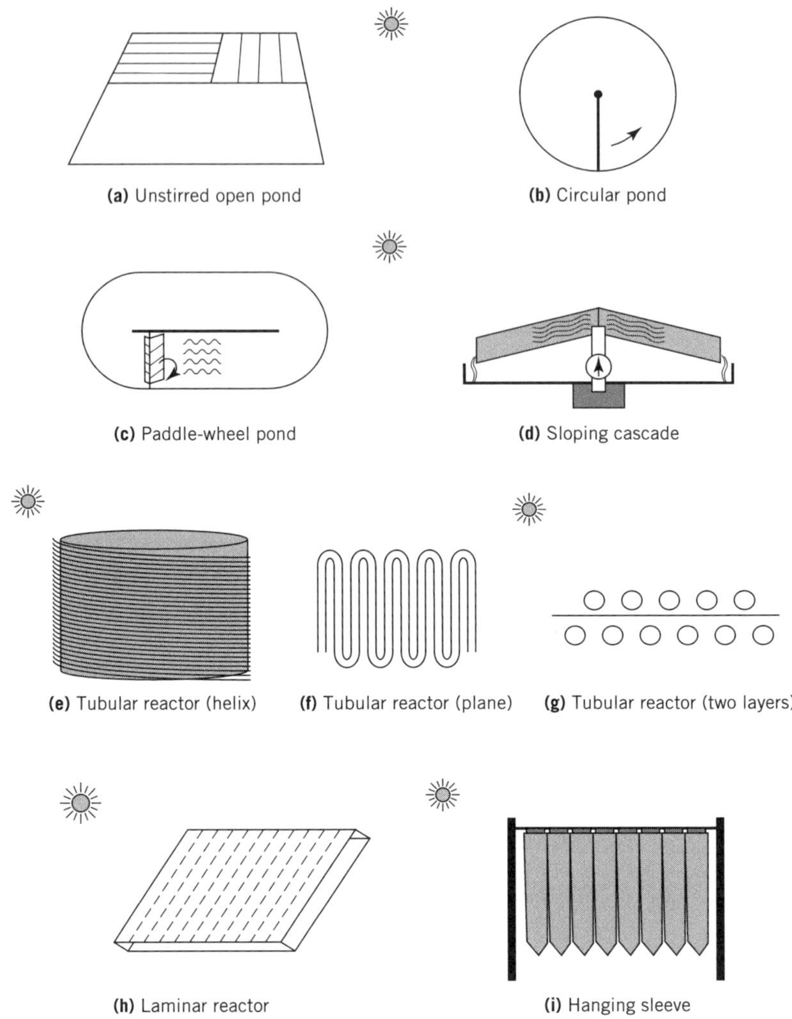

Figure 2. Illustration of algal mass culture systems.

of a high-density culture using a photobioreactor employing novel light-diffusing optical fibers (LDOF). Efficient light distribution was achieved as a result of a bundle of LDOFs passing through the cell culture (17). The ratio of surface area to volume was very high (692/m), and uniform illumination in the cell culture was obtained. Takano et al. (44) cultured a marine cyanobacterium *Synechococcus* sp. in the LDOF photobioreactor, achieving a biomass yield of 2g/L/day.

Photobioreactors are less economical but have many superior aspects to the open culture systems mentioned earlier. Thus, they are limited to high-value products. Suitable culture systems have to be adopted according to the target products and available environmental conditions.

PRODUCTION OF USEFUL MATERIALS BY RECOMBINANT CYANOBACTERIA

Higher productivity of valuable primary or secondary metabolites in microalgae will be possible by high-density cultivation and/or genetically engineered microalgae. Genetic engineering of cyanobacteria (blue–green algae) has been developed in which gene-transfer systems have been established in several cyanobacterial strains (45,46), and these have been used as hosts for production of foreign proteins (47).

Recent pharmaceutical interest in unsaturated fatty acids has triggered a search for sources of these valuable compounds. Monounsaturated fatty acids such as palmitoleic acid (C16:1), undecylenic acid (C11:1), and tridecenic acid (C13:1) have the potential to prevent several diseases.

Several eukaryotic microalgae are known to produce highly unsaturated fatty acids such as eicosapentaenoic acid (EPA; C20:5) and docosahexaenoic acid (DHA; C22:6), which are valuable dietary components (14,48). They may be involved in preventing several human diseases (49,50). However, wild-type cyanobacteria do not possess the biosynthetic pathway for these fatty acids. The EPA synthesis gene cluster (ca. 38 kbp) from the marine bacterium *Shewanella putrefaciens* SCRC-2738 was cloned into the marine cyanobacterium *Synechococcus* sp. using a broad-host cosmid vector, pJRD215 (10.2 kbp, Sm^r Km^r). The constructed plasmid (pJRDEPA; ca. 48 kbp) could be transferred by transconjugation to the cyanobacterial host at a frequency of 2.2×10^{-7}. Cyanobacterial transconjugants grown at 29 °C produced only 0.12 mg EPA/g dry wt, whereas those grown at 23 °C produced 0.56 mg/g dry wt.

Furthermore, the content of EPA in cells grown at 23 °C increased by 0.64 mg/g dry wt after preincubation for one day at 17 °C (51).

Spirulina spp. are commercially valuable cyanobacteria. Several *Spirulina* genes have been characterized (52). The genetic study of *Spirulina* has been focused on the construction of mutants and the cloning of genes. Gene-transfer systems have not been developed because *Spirulina* has high restriction endonuclease activity and is resistant to most antibiotics. Establishment of gene-transfer systems in *Spirulina* will contribute to their industrial application.

CO_2 FIXATION USING MICROALGAL CULTURES

Recently, the possible use of biological CO_2 fixation to reduce anthropogenic CO_2 emission has been investigated. For development of on-site CO_2 fixation systems using microalgae, efficient photobioreactors and strains that can fix large quantities of CO_2 are required. Recently, several applied studies aimed at the direct biological utilization of CO_2 in emission gas from power plants have been carried out. In Japan, coal-fired power plants and the steel and cement industries produce large quantities of CO_2. These industries also emit NO_x and SO_x. It is known that NO_x and SO_x reduce biological CO_2 fixation activity, primarily by inhibiting photosynthesis. Therefore, microalgae that can grow under such extreme conditions will be required for direct CO_2 fixation.

Identification of microalgal strains capable of growing rapidly in water sparged with emitted gas and under other extreme conditions, such as at high pH, has been carried out. Their ability to grow in saline water (e.g., seawater) and their resistance to other gas constituents (e.g., SO_2 and NO_x) are additional considerations. Desirable properties of microalgae for the direct biological utilization of CO_2 in gas emitted from a power plant are summarized in Table 3.Several strains have been isolated from environments that have some of the properties described above. Among those strains, the marine alga *Chlorococcum littorale* (53) is characterized by high CO_2 tolerance and high growth rate in the linear growth phase.

A Designed System for the Reduction of CO_2 Emission: Case I—Coal-Fired Power Plants

A designed system for the reduction of CO_2 emission from coal-fired power plants is presented (28). Microalgal cultivation using CO_2 in gas emitted from power plants has been described. In this system, CO_2-fixed product by microalgal culture is used as biomass fuel, which will substitute for fossil fuel (Fig. 3). CO_2 recycling in coal-fired power plants will achieve the reduction of CO_2 emission from power plants.

Table 3. Desired Properties of Microalgae for CO_2 Fixation

1. High pCO_2 tolerance
2. Low pH tolerance
3. Stable and high growth rate in the linear growth phase
4. Capacity to grow at high cell densities
5. Acidic gas (NO_x, SO_x) tolerance
6. Thermotolerance

Cost of microalgae CO_2 mitigation using the designed system has been estimated based on several designed specifications:

1. Plant size: emission gas from a 500 MW power plant
2. Gas condition: 30.85 kg C/S (CO_2) concentration in dry flue gas: 14.2% (vol)
3. Biofixation conditions
   a. One hundred percent of the CO_2 produced by the power plant is fed to biological systems, and 90% of the CO_2 fed to the system is utilized during daylight summer hours.
   b. Direct biofixation of CO_2 is applied where CO_2 in emission gas from the power plant is directly fed to algal ponds for photosynthesis.
   c. Both 10% (case 1) and 20% (case 2) of photosynthetic efficiency based on visible light are studied as are 4.5% (case 1) and 9.0% (case 2) based on total light.
   d. The average solar radiation should be greater than 200 W/m^2.
4. Operation of the plant: The power plant is operated at 70% operation factor. That is to say, the plant is operated at 100% of rated capacity for 16.8 h (70% of 24 h) during the day and is shut down for 7.2 h (30% of 24 h) during the night.
5. CO_2 production rate
   a. Hourly CO_2 production: 407 tCO_2/h
   b. Annual CO_2 production: 2,495,724 tCO_2/year
6. Algal Strain: NANNP2 (*Nannochloropsis* sp. from the DOE/NREL/ASP culture collection) The expected productivities of algal biomass: 42 g/m^2/day (case 1) and 84 g/m^2/day (case 2)

Table 4 presents preliminary cost estimates for a large-scale (over 1,000 ha) microalgal system for biomass fuel production. Two different productivities, 42 and 84 g/m^2/day, were assumed to correspond to about 10 and 20% solar conversion efficiencies, based on visible light, and 4.5 and

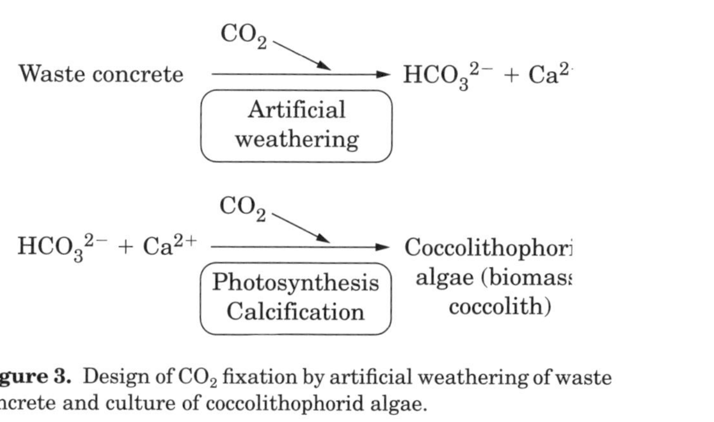

Figure 3. Design of CO_2 fixation by artificial weathering of waste concrete and culture of coccolithophorid algae.

Table 4. Capital and Operating Costs

Productivity assumption	Case 1	Case 2
Avg. ash-free dry weight (g/m^2/day) @20 Mj/m^2/day	42	84
Annual productivity (mt/ha/year)	101.9	203.8
Solar conversion efficiency (%)	10	20
CO_2 fixation (tC/ha/year)	48.4	96.8
Average annual utilization of CO_2 (%)	54	54
Growth pond area required (ha)	7,390	3,700
Land area required (500 MW power plant, growth pond × 1.2 ha)	8,900	4,500
Capital costs ($/ha)		
Ponds (earthworks, CO_2 sumps, mixing)	31,000	45,000
Harvesting (settling ponds, centrifuge)	10,500	15,000
Systemwide costs (water, CO_2 supply, etc.)	25,100	36,900
Drying	9,800	19,600
Engineering, contingency, fees	16,700	25,700
Land-related costs	3,800	5,800
Total capital cost ($/ha)	96,900	148,000
Capital costs ($tC/year)	2,002	1,529
Operation costs ($/ha/year)		
Power, nutrient, maintenance, labor	15,400	27,000
Annualized capital cost (10% DCF)	10,048	15,346
Credit for biomass fuel (82.4 $/mt)	−8,400	−16,793
Net operating cost	17,048	25,553
CO_2 mitigation cost ($/tC)	352	264

Note: Solar conversion efficiencies are for visible light.
The credit for biomass fuel is $25/barrel of oil.

9%, based on total light. The lower productivity reflects what could be expected based on present experience. The higher productivity is the theoretical limit. CO_2 mitigation costs greatly depend on the productivity of algae and solar radiation. Microalgal strains that can achieve higher photosynthetic efficiency at higher solar radiation are necessary. A photobioreactor, in which microalgae can convert from CO_2 to biomass at higher photosynthetic efficiency, is also required.

Further research into microalgae, photobioreactors, the downstream process of microalgal culture, and the burning process of microalgal biomass is still needed. If such research is successful and the economic projections shown in Table 4 are verified, this system will represent one of the most promising technologies for future CO_2 mitigation processes.

A Designed System for the Reduction of CO_2 Emission: Case II—Cement Plants

Coccolithophorids are unicellular planktonic marine algae that produce elaborate structures called coccoliths, comprising scales or plates of $CaCO_3$. In the oceans, huge blooms of coccolithophorid algae occur. These blooms have been recognized as contributing to ocean floor sediment formation, and algae play an important role in the global carbon cycle by recycling $CaCO_3$.

We have chosen coccolithophorid algae as model organism to investigate biomineralization and have focused on the ecological significance of $CaCO_3$ recycling. We have proposed CO_2 fixation by artificial weathering of waste concrete and coccolithophorid algae cultures (Fig. 4). During artificial weathering of waste concrete suspended in seawater, atmospheric CO_2 can be absorbed and dissolved as bicarbonate ions, which are a major source of coccolith particles. Coccolithophorid algae can use bicarbonate ions to form $CaCO_3$ particles. As a result, CO_2 absorbed by artificial weathering can be mineralized and fixed permanently. Artificial weathering of waste concrete is also a useful method to supply bicarbonate ions to cells of the coccolithophorid alga *Emiliania huxleyi*.

The CO_2 fixation method by artificial weathering of waste concrete and coccolithophorid algae cultures can be applied to the reduction of CO_2 emission in cement plants. A designed system for the reduction of CO_2 emission in cement plant is shown in Fig. 5. Coccoliths can be used as an alternative to limestone, which is a carbonate source for cement production. In the cement industry, CO_2 is mainly produced by the decomposition of limestone during the burning of cement clinker. If $CaCO_3$ recycling can be achieved by artificial weathering of waste concrete and coccolithophorid culture, CO_2 emission by the cement industry might be reduced (Fig. 5).

Moreover, it is proposed that microalgal biomass with fixed CO_2 products may be stored in concrete. If microalgal biomass can be stored in concrete without the decomposition of the biomass back to CO_2, removal of anthropogenic CO_2 may be achieved.

CO_2 absorption by artificial weathering of concrete has been determined. When concrete (calcium content was 15% wt/wt) pieces were weathered by CO_2 and seawater, over 310 mg CO_2/g concrete could be absorbed and dissolved in the form of HCO_3^- (54). The amount of CO_2 absorbed is directly related to calcium content in concrete, 310 mg CO_2/g concrete is equal to almost 1 g CO_2/g Portland cement (calcium content is almost 46%).

On the other hand, amount of emitted CO_2 during a cement production is 708 kg CO_2/ton cement. The amount of absorbed CO_2 by the weathering of waste concrete is greater than that of emitted CO_2 during a cement production. Moreover, CO_2 is absorbed by following coccolithophorid algae cultures.

CONCLUSION

The authors reviewed large-scale algal culture systems. Open pond systems have lower productivity of algal biomass, require larger land areas, and involve larger land costs. However, their operation cost is lower. On the other hand, closed systems can achieve high-density culture, and the volume of algal culture can be reduced. The reduction of culture volume decreases the land costs. However, a certain land area is still required for the collection of solar energy, and the operating cost is larger than that of open systems. Moreover, solar radiation, temperature, and other factors regulating algal productivity are significantly affected by location. Suitable culture systems have to be

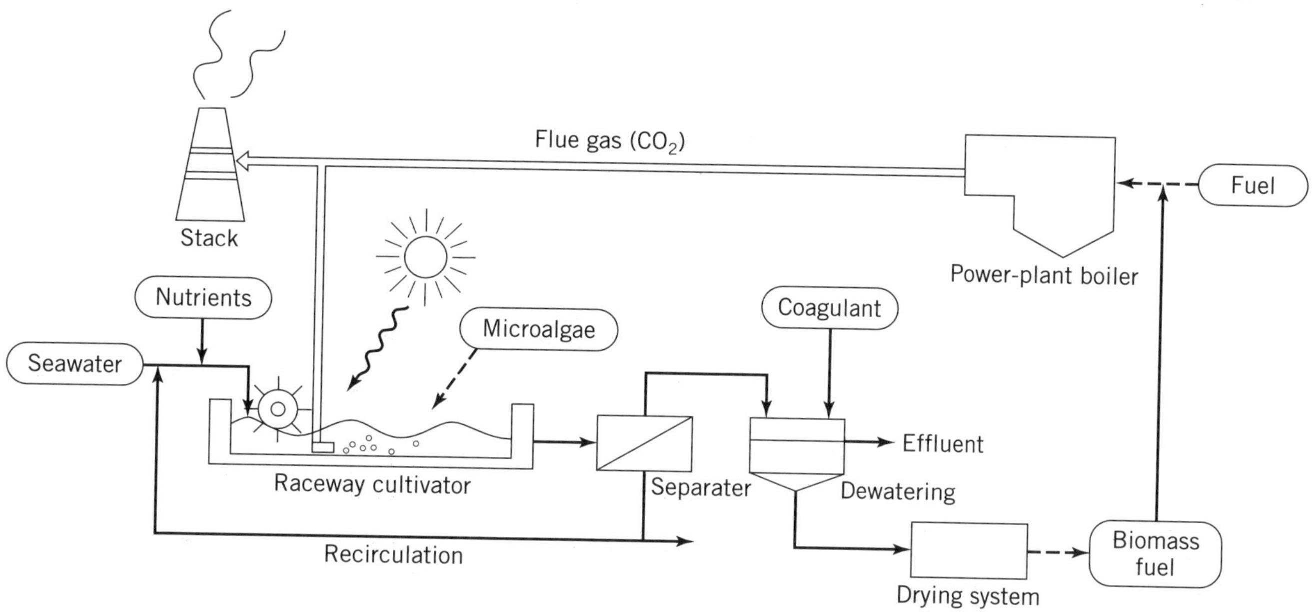

Figure 4. Conceptional system of CO_2 fixation by microalgae.

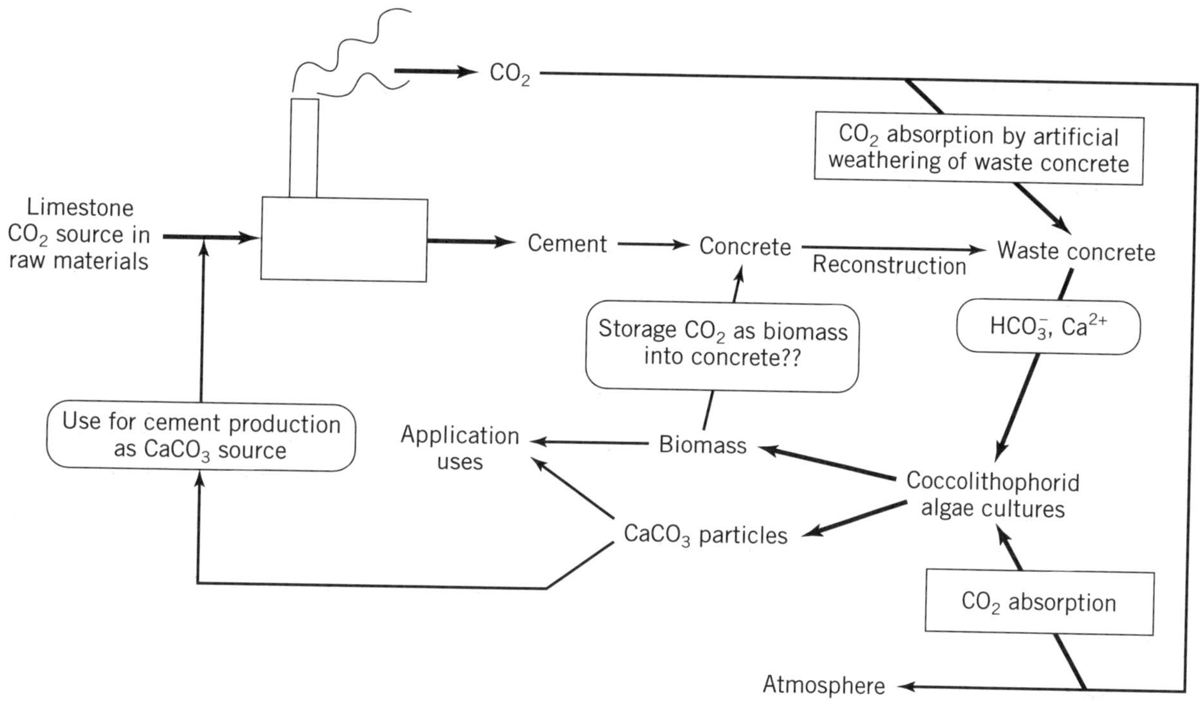

Figure 5. Design of CO_2 removal system in cement industry.

adopted according to the target products and available environmental conditions.

Extensive studies of biological CO_2 fixation using microalgal cultures have been pursued for the past 10 years. A primary goal is the complete removal of CO_2 in discharged gas emitted by such an on-site system. Because of the land-area requirements and the current CO_2 mitigation cost of $264/ton as carbon, it is to be difficult at present to apply microalgal cultures to CO_2 removal. Most nations are seriously considering the increase of atmospheric CO_2 concentration, and intensive efforts to reduce the anthropogenic CO_2 emission will be made. Microalgae culture may be one of the important processes used for such efforts (55).

Regarding technological use of microalgal biomass, several examples, including historical findings, have been presented. Microalgal production of useful chemicals and energy resources have been extensively investigated. Processes that utilize the majority of the resulting microalgal biomass as energy sources would be desired. Such processes may allow the recycling of evolved CO_2 from human energy consumption rather than a one-way emission, as is the case with fossil fuels. The following six products can be produced from microalgal biomass for use as fuels: hydrogen (through biophotolysis), methane (through anaerobic digestion), ethanol (through yeast or other alcohol fermentation), triglycerides (through extraction of lipids), methyl ester fuels (through transesterification of lipids), and liquid hydrocarbon (from *Botryococcus braunii*).

Increasing attention has been paid to resource sustainability in all industries. It must be considered whether resources used for manufacturing products can be sustainable. Developing technologies of microalgal culture will be expected to provide sustainable resources.

BIBLIOGRAPHY

1. J.S. Burlew ed., *Algal Culture from Laboratory to Pilot Plant*, Carnegie Institute of Washington. Washington, D.C., 1953.
2. B. Metting, *Micro-algal Biotechnology*, Cambridge Univ. Press, Cambridge, UK, 1988, pp. 288–304.
3. E.W. Backer, *Handbook of Microalgal Mass Culture*, CRC Press, Boca Raton, FL, 1986, pp. 339–420.
4. J.R. Benemann, *J. Appl. Phycol.* **4**, 233–245 (1992).
5. V.M. Kaya, J. de la Noüe, and G. Picard, *J. Appl. Phycol.* **7**, 85–95 (1995).
6. M.A. Borowitzka, *J. Appl. Phycol.* **7**, 3–15 (1995).
7. J.R. Benemann, D.M. Tillett, and J.C. Weissman, *Trends Biotechnol.* **5**, 47–53 (1987).
8. H. Takano, H. Furu-une, J.G. Burgess, E. Manabe, M. Hirano, M. Okazaki, and T. Matsunaga, *Appl. Biochem. Biotechnol.* **39/40**, 159–167 (1993).
9. H. Takano, J. Jeon, J.G. Burgess, E. Manabe, Y. Izumi, M. Okazaki, and T. Matsunaga, *Appl. Microbiol. Biotechnol.* **40**, 946–950 (1994).
10. H. Takano, R. Takei, E. Manabe, J.G. Burgess, M. Hirano, and T. Matsunaga, *Appl. Microbiol. Biotechnol.* **43**, 460–465 (1995).
11. M. Hirano, H. Mori, Y. Miura, N. Matsunaga, N. Nakamura, and T. Matsunaga, *Appl. Biochem. Biotechnol.* **24/25**, 183–191 (1990).
12. Y. Miura, K. Sode, Y. Narasaki, and T. Matsunaga, *FEMS Microbiol. Lett.* **107**, 163–168 (1993).
13. T. Matsunaga, H. Takeyama, Y. Miura, T. Yamazaki, H. Furuya, and K. Sode, *FEMS Microbiol. Lett.* **133**, 137–141 (1995).
14. J.G. Burgess, K. Iwamoto, Y. Miura, H. Takano, and T. Matsunaga, *Appl. Microbiol. Biotechnol.* **39**, 456–459 (1993).
15. H. Sudo, J.G. Burgess, H. Takemasa, N. Nakamura, and T. Matsunaga, *Curr. Microbiol.* **30**, 1–4 (1994).
16. T. Matsunaga, N. Nakamura, N. Tsuzaki, and N. Takeda, *Appl. Microbiol. Biotechnol.* **28**, 373–376 (1988).
17. T. Matsunaga, H. Takeyama, H. Sudo, N. Oyama, S. Ariura, H. Takano, M. Hirano, J.G. Burgess, K. Sode, and N. Nakamura, *Appl. Biochem. Biotechnol.* **28/29**, 157–167 (1991).
18. H. Takano, T. Arai, M. Hirano, and T. Matsunaga, *Appl. Microbiol. Biotechnol.* **43**, 1014–1018 (1995).
19. T. Matsunaga, J.G. Burgess, N. Yamada, K. Komatsu, S. Yoshida, and Y. Wachi, *Appl. Microbiol. Biotechnol.* **39**, 250–253 (1993).
20. Y. Wachi, J.G. Burgess, K. Iwamoto, N. Yamada, N. Nakamura, and T. Matsunaga, *Biochim. Biophys. Acta* **1244**, 165–168 (1995).
21. Y. Miura, K. Sode, Y. Narasaki, and T. Matsunaga, *J. Mar. Biotechnol.* **1**, 143–146 (1993).
22. H. Wake, H. Umetsu, K. Shimomura, and T. Matsunaga, *Plant Cell Rep.* **9**, 655–658 (1991).
23. H. Wake, A. Akasaka, H. Umetsu, Y. Ozeki, K. Shimomura, and T. Matsunaga, *Plant Cell Rep.* **11**, 62–65 (1992).
24. H. Wake, A. Akasaka, H. Umetsu, Y. Ozeki, K. Shimomura, and T. Matsunaga, *Appl. Microbiol. Biotechnol.* **36**, 684–688 (1992).
25. Y. Wachi, J.G. Burgess, J. Takahashi, N. Nakamura, and T. Matsunaga, *J. Mar. Biotechnol.* **2**, 210–213 (1995).
26. Y. Wachi, K. Sode, K. Horikoshi, H. Takeyama, and T. Matsunaga, *Biotechnol. Techniques* **9**, 633–636 (1995).
27. Y. Wachi, J.G. Burgess, J. Takahashi, T. Matsunaga, and N. Nakamura, *J. Mar. Biotechnol.* **3**, 258–261 (1995).
28. The IEA Greenhouse Gas R&D Programme, Report G/93/OE16B, Chemical Society of Japan, 1994.
29. A. Ben-Amotz, and M. Avron, *Trends Biotechnol.* **8**, 121–126 (1990).
30. A. Richmond, *Micro-algal Biotechnology*, Cambridge Univ. Press, Cambridge, UK, 1988, pp. 85–121.
31. A. Vonshak, *Micro-algal Biotechnology*, Cambridge Univ. Press, Cambridge, UK, 1988, pp. 122–134.
32. P. Bubrick, *Bioresource Technol.* **38**, 237–240 (1991).
33. W.J. Oswald, *Chem. Eng. Symp. Ser.* **65**, 87–92 (1969).
34. J.C. Dodd, *Handbook of Microalgal Mass Culture*, CRC Press, Boca Raton, FL, 1986, pp. 265–283.
35. A. Richmond, *J. Appl. Phycol.* **4**, 281–286 (1992).
36. A. Vonshak, S. Boussiba, A. Abeliovich, and A. Richmond, *Biotechnol. Bioeng.* **25**, 341–351 (1983).
37. M.A. Borowitzka, *J. Mar. Biotechnol.* **4**, 185–191 (1996).
38. H. Tamiya, E. Hase, K. Shibata, K. Mitsuya, T. Iwamura, T. Nihei, and T. Sasa, *Algal Culture from Laboratory to Pilot Plant*, Carnegie Institute of Washington, Washington, DC, 1953, pp. 204–232.
39. A. Richmond, S. Boussiba, A. Vonshak, and R. Kopel, *J. Appl. Phycol.* **5**, 327–32 (1993).
40. G. Torzillo, P. Carlozzi, B. Pushparaj, E. Montaini, and R. Materassi, *Biotechnol. Bioeng.* **42**, 891–898 (1993).
41. L.J. Borowitzka and M.A. Borowitzka, *Algal and Cyanobacterial Biotechnology*, Longman Scientific, London, 1989, pp. 294–316.
42. Y.K. Lee and C.S. Low, *Biotechnol. Bioeng.* **40**, 1119–1122 (1992).
43. K. Mori, *Biotechnol. Bioeng.* **15**, 331–345 (1985).
44. H. Takano, H. Takeyama, N. Nakamura, K. Sode, J.G. Burgess, E. Manabe, M. Hirano, and T. Matsunaga, *Appl. Biochem. Biotechnol.* **34/35**, 449–458 (1992).
45. T. Matsunaga and H. Takeyama, *J. Appl. Phycol.* **7**, 77–84 (1995).
46. J. Elhai, *J. Appl. Phycol.* **6**, 177–186 (1994).
47. R.C. Murphy and S.E. Stevens, Jr, *Appl. Environ. Microbiol.* **58**, 1650–1655 (1992).
48. H. Takeyama, K. Iwamoto, S. Hata, H. Takano, and T. Matsunaga, *J. Mar. Biotechnol.* **3**, 244–247 (1996).

49. T. Terano, J.A. Salmon, and S. Moncada, *Prostaglandins* **27**, 217–232 (1984).
50. M. Abbey, P. Clifton, M. Kestin, B. Belling, and P. Nestel, *Arteriosclerosis* **10**, 85–94 (1990).
51. H. Takeyama, D. Takeda, K. Yazawa, A. Yamada, and T. Matsunaga, *Microbiology* **143**, 2725–2731 (1997)
52. A.K. Vachhani and A. Vonshak, *Spirulina platensis (Arthrospira): Physiology, Cell-Biology and Biotechnology*, Taylor and Francis, London, 1997, pp. 67–77.
53. M. Kodama, H. Ikemoto, and S. Miyachi, *J. Mar. Biotechnol.* **1**, 21–25 (1993).
54. H. Takano and T. Matsunaga, *Energy Conversion Manage.* **36**, 697–700 (1994).
55. M. Murakami and M. Ikenouchi, *Energy Conversion Manage.* **38S**, 493–497 (1997).

AMINO ACIDS, GLUTAMATE

TETSUYA KAWAKITA
Ajinomoto Co., Inc.
Kanazawa, Japan

KEY WORDS

Acceptable daily intake (ADI)
Brevibacterium
Corynebacterium
Fermentation process
Glutamic acid
Polymorph of crystals of glutamic acid α
Pyrroridon carboxylic acid (PCA)
Synergy effect
Umami seasoning

OUTLINE

INTRODUCTION

Glutamic acid was isolated from wheat gluten hydralyzate by Ritthousen in 1866. Wolff carried out the first chemical synthesis of glutamic acid in 1890. In 1908, Ikeda (1) isolated the specific taste fraction from the kelplike seaweed *Laminariaceae*. Kelp, called konbu in Japan, is traditionally used as a soup-stock component with dried bonito, and has been identified as the salt of glutamate, used to enhance the flavor of foods. He also recognized it as "umami" taste, which does not belong to the four traditional taste categories; bitter, sour, salty, and sweet. Immediately after, Ikeda filed a patent application and began the commercial production of monosodium L-glutamate monohydrate (MSG) with his partner, Suzuki, in 1909.

With the increasing demand for MSG is use as a seasoning, biochemical procedures were developed for its production. For example, α-ketoglutaric acid was first fermented from glucose by using a species of *Pseudomonas* (2), followed by the conversion with amination to L-glutamate in a high yield. The vigorous study on the accumulation of L-glutamate in bacterial cultures by Kinoshita et al. (3) opened the door to the microbial production of L-glutamate and other amino acids.

The annual amount of production was estimated at more than 1,000,000 tons in 1997, an amount that increases by about 5% globally every year. Production is increased because of the demand due to changes in eating habits, and its use in commercially prepared, packaged foods; ready-made sours; and table seasonings in both Western and Eastern countries.

METABOLISM

L-glutamic acid is split into α-ketoglutaric acid and ammonia by glutamate dehydrogenase. By the reverse reaction, L-glutamic acid is synthesized from α-ketoglutaric acid, a member of the tricarboxylic acid (TCA) cycle of glycolysis. Whereas glutamate transaminase is nonspecific for the pairs of keto acids and amino acids, L-glutamate is the only amino acid in mammalian tissues that undergoes oxidative deamination at an appropriable rate. The formation of ammonia from α-amino groups requires their conversion to the α-amino nitrogen of L-glutamate. Thus, L-glutamic acid is a key substance in the nitrogen metabolism of amino acids. Glutamate dehydrogenase is widely distributed in microorganisms and higher plants as a catalyst in the synthesis of L-glutamic acid from α-ketoglutaric acid and free ammonia. Transaminase is contained in a wide variety of microorganisms.

MICROBIAL PRODUCTION

Although various kinds of microorganisms accumulate amino acids in culture medium, only bacteria have sufficient productivity to warrant the commercial production of amino acids. Since amino acids are essential components of microbial cells, and their biosyntheses are teleologically regulated to maintain an optimal level, they are normally synthesized in limited quantities and subject to negative feedback regulation. It is necessary to overcome this regulation to achieve overproduction of amino acids. Microbial amino acid overproduction can be achieved using the following procedures and has been attained by introducing mutation techniques (4):

1. Stimulation of cellular uptake of the starting materials
2. Hindrance of the side reaction
3. Stimulation of the formation and the activity of the enzymes for biosynthesis
4. Inhibition or reduction of the enzyme concerned with the degradation of the amino acids produced
5. Stimulation of the excretion of the product into the extracellular space

L-Glutamate production by a wild-type strain of *Corynebacterium* was explained by the release of L-glutamate regulation and efflux of L-glutamate through the cell membrane by the limitation of biotin or the addition of penicillin ester of fatty acids to the culture medium (5).

Fermentation Method

Excretion of L-amino acids by *Escherichia coli* when an excess of ammonium salt was added was reported in 1950 (6). In 1957, a fermentation process was successfully commercialized (3) that used a strain of bacteria isolated from soil, later found to be *Corynebacterium glutamicum*, which is able to excrete considerable amounts of L-glutamic acid.

Numerous microorganisms have been isolated and found to be able to overproduce L-glutamic acid, including *Brevibacterium flavum, B. lactofermentum*, and *Microbacterium ammoniaphlum*. Because of minor differences in the character of these bacteria, which are all Gram positive, non–spore forming, nonmotile, and require biotin for growth, the name of genus *Corynebacterium* was suggested for these coryneform bacteria (7).

For production of L-glutamic acid, the key factors in controlling the fermentation are as follows:

1. The presence of biotin in the range of 5 to 10 μg/L, which is optimal for the excretion of L-glutamic acid through cell walls
2. A sufficient supply of oxygen to reduce the accumulation of lactic acid and succinic acid as by-products

When a biotin-rich medium, such as molasses, is used as the carbon source, the addition of penicillin or cephalosporin C favors the overproduction of L-glutamic acid, supposingly due to the repression of peptide glycan synthesis on the cell membrane. The supplementation of a fatty acid or surfactant also results in an increased permeability of the cell wall, thus enhancing L-glutamic acid excretion. Kramer (8) reports the existence of specific carrier mechanisms that are responsible for active L-glutamic acid secretion in *C. glutamicum*.

Generally, the intracellular accumulation of L-glutamic acid does not reach levels sufficient for feedback control in glutamate overproducers due to rapid excretion of glutamate. However, the regulatory mechanisms of L-glutamic acid biosynthesis have been studied intensively to obtain mutants with increased productivity, as shown in Figure 1.

Shiio (9) discussed two enzymes that have played key roles in the biosynthesis of L-glutamic acid. The first, phosphoenolpyruvate carboxylase, catalyzes carboxylation of phosphoenolpyruvate to yield oxaloacetate; it is inhibited by L-aspartic acid and repressed by both L-aspartic acid and L-glutamic acid. Second, α-ketoglutaric acid dehydrogenase converts α-ketoglutaric acid to succinyl-CoA; in L-glutamic acid–overproducing strains, α-ketoglutaric acid dehydrogenase limits further oxidation of α-ketoglutaric acid to carbon dioxide and succinic acid, thus favoring the formation of L-glutamic acid.

In L-glutamic acid–overproducing strains, the K_m value of α-ketoglutaric acid dehydrogenase was nearly two magnitudes lower than that of L-glutamic acid dehydrogenase, which catalyzes the last step to L-glutamic acid. Consequently, v_{max} of L-glutamic acid dehydrogenase proved to be about 150 times higher than that of α-ketoglutaric acid dehydrogenase. Borman (10) isolated and characterized the *C. glutamicum* glutamic acid dehydrogenase.

A strain of *Microbacterium ammoniaphlum* cultured under biotin-deficient conditions produced 58% of L-glutamic acid formed from glucose via phosphoenolpyrvate, citrate, and α-ketoglutaric acid; the other 42% was formed via the TCA cycle or glyoxylate cycle (11).

The mutants are reported as either having sensitivity in cell permeability (12), having the capability for increased carbon dioxide fixation (13), or having a too-low activity level of pyruvate dehydrogenase (14).

Another approach focused on the development of thermophilic mutants. A strain of *Corynebacterium thermoaminogenes* is reported to accumulate L-glutamic acid at temperatures above 43 °C (15).

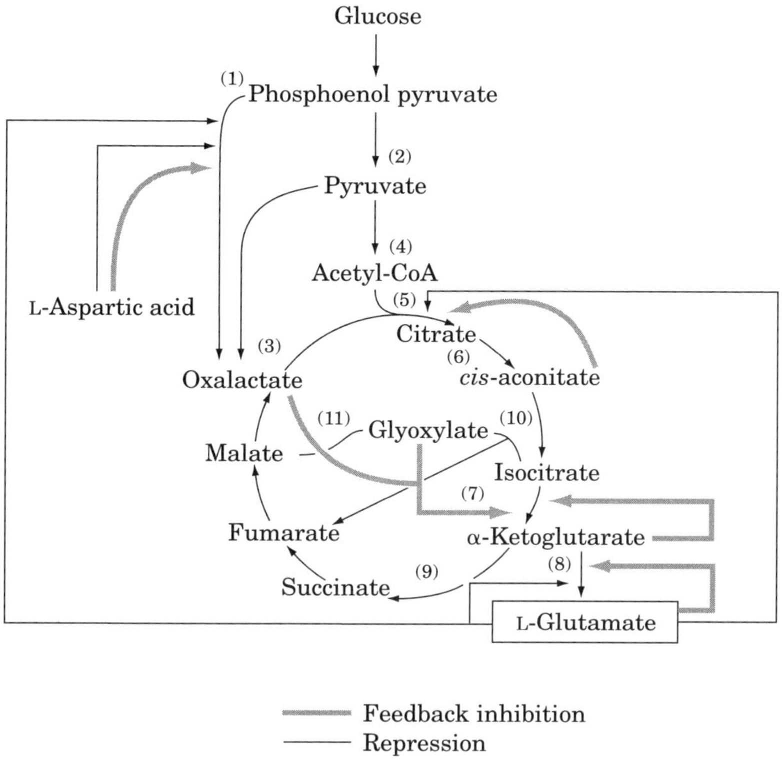

Figure 1. Regulation of L-glutamic acid biosynthesis in *Corynebacterium glutamicum* (5). Regulatory enzymes: 1, phosphoenolpyruvate carboxylase; 2, pyruvate kinase; 3, pyruvate carboxylase; 4, pyruvate dehydrogenase; 5, citrate synthetase; 6, aconitase; 7, isocitrate dehydrogenase; 8, L-glutamate dehydrogenase; 9, α-ketoglutarate dehydrogenase; 10, isocitrate lyase; 11, malate synthetase.

CHEMICAL PROPERTIES

Dissociation Constant

Amino acids exit aqueous solutions as dipolar ions (zwitterions), and glutamic acid may be represented as follows (16). Following Brönsted, an acid may be defined as a substance capable of combining with a proton. When an amino acid is titrated with HCl and NaOH, a curve exhibiting two points of inflection is obtained. Thus amino acid can react either as an acid or as a base, that is, as an ampholyte.

The dissociation constants for glutamic acid are pK_1 2.19 (COOH), pK_2 4.25 (COOH), and pK_3 9.67 (NH_3^+). (It is customary to designate the pK' values of amino acids in order of decreasing acidity.) The isoelectric point (pI) of glutamic acid is the pH 3.2 calculated from the relation pI = $pK_1 + pK2/_2$, at which there is no net charge on the amino acid molecule. At this pH, no migration will occur in an electric field.

Racemization

Racemization proceeds, in general, in alkali solutions as shown in Figure 2, such that α-H liberated from α-C by a base-catalyzed reaction yields carboion (C^-—C═O) on α-C, which equilibrates with α,β unsaturated from C═C—O^-. In the case of acidic solutions, racemization proceeds the same as in alkali, via carboion, C^-—C═OH^-, which equilibrates with α,β unsaturated from C═C—O^-.

Racemization in neutral solution occurs at 190 °C with the loss of taste characteristics after the formation of the lactam and pyroglutamic acid. This reaction is very slow, but the equilibrium shifts toward the lactam rather than

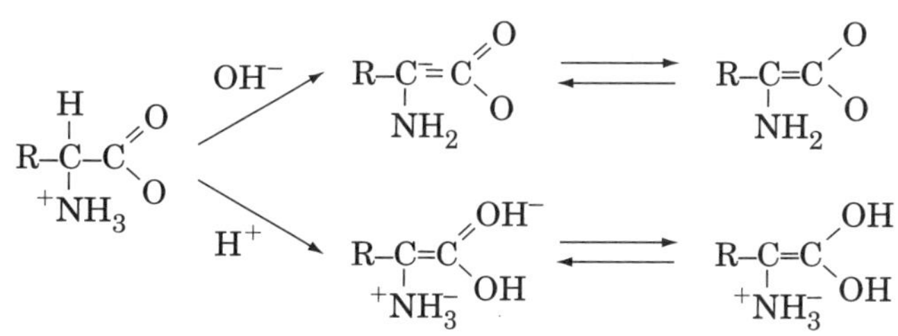

Figure 2. Mechanism of racemization of amino acid.

glutamic acid. Under strongly acidic or alkaline conditions, the ring-opening reaction requires a very short time; therefore, neutralization of L-glutamic acid should be performed cautiously to avoid intramolecular dehydration even below 190 °C (17).

Racemization is also brought about by a microbial racemase. As in other amino acids, the racemase of glutamic acid is found in *Lactobacillus fermenti*. This enzyme acts specifically on glutamic acid but cannot act on its simplest derivatives (18).

PROPERTIES OF L-GLUTAMIC ACID

Salt Formation

An amino acid can react with a copper ion to form a copper complex of the following type:

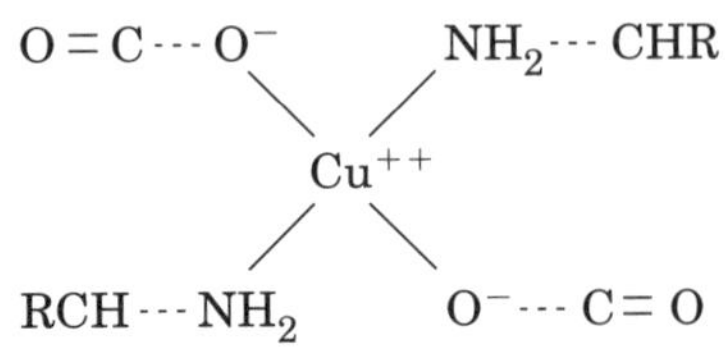

Formation of the copper complex is the basis of a quantitative procedure for the determination of the amino acid.

In alkaline solution, amino acids react with carbon dioxide to form carbamino acids, which may be precipitated as salts:

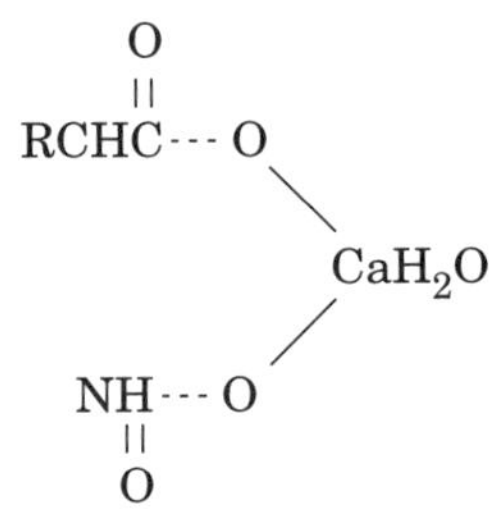

Some of these are quite insoluble and have been used for isolation of amino acids. Decomposition of carbamino acid occurs by boiling in aqueous solution.

L-Glu is able to react with various kind of cations, especially with metallic ions to form their corresponding specific complexes. The complex constant (log K_s) for various complexes of L-Glu is 10.3 for Ni^{2+}, 8.5 for Zn^{2+}, 8.1 for Co^{2+}, 4.6 for Fe^{2+}, and 3.0 for Mn^{2+} (19). Glutamic acid reacted with sodium ion to form monosodium L-glutamate, on the other hand, reacts as a base with sulfuric acid to form diglutamate sulfate crystals.

Infrared Spectra

Comparison of the infrared (IR) spectra of L-Glu, L-GluHCl, and L-GluNa H_2O are shown in Figure 3 (20). The IR spectra of L-Glu shows strong bands at 1,620 and 1,420 cm^{-1}, which correspond to the COO; these are 1,660 cm^{-1} for COOH and 2,740 cm^{-1} for NH_3^+, respectively. The spectra of L-Glu HCl, on the other hand, does not show bands at 1,620 and 1,420 cm^{-1} for COO^- and that of L-Glu NaH_2O does not show the band at 1,670 cm^{-1} for COOH, but does show a strong band at 1,620 and 1,420 cm^{-1} for COO^-.

Ionic Forms

The four ionic forms of L-Glu exist depending on the pH of the solution. The dipolar nature of amino acids in solution has been shown by Raman spectroscopy. The Raman spectra of L-Glu shows a strong line at 1730 cm^{-1}, which corresponds to the C═O vibration. The spectrum of sodium L-Glu^-, on the other hand, does not show this frequency because the salt is completely dissociated and its structure represented is by the resonant forms

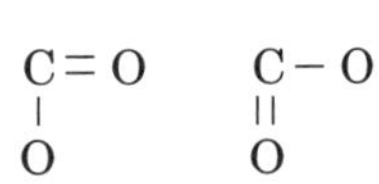

This does not have a C═O group, but instead intermediate

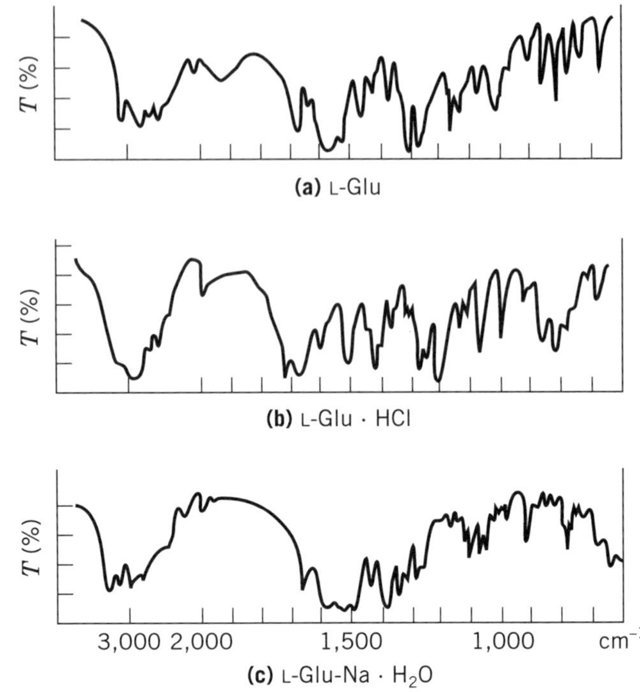

Figure 3. IR spectra of L-glutamic acid– related substances.

some between C—O and C═O. The spectra of the amino acids at the isoelectric point of pH do not show the existence of C═O, which indicates that the —COOH group is ionized. In a similar way, the ionization of the —NH_2 is indicated.

The optical rotation changes with changing pH in solution at 25 °C; glutamic acid $[\alpha]_D(G^{\pm}; H_2O)$ is +12.0, $[\alpha]_D(G^+; 5N\ HCl)$ is +31.8, $[\alpha]_D(G^-; NaOH)$ is −4.2, and $[\alpha]_D(G^=; NaOH)$ is +10.9, which corresponds to the ratio of ionic form in solution (21).

Solubility

For the production of the umami seasoning MSG, the characteristic solubility changes in L-Glu HCl, L-Glu, and L-GluNa are used to separate it from other impurities to meet the category of food additives. The change in solubility of L-Glu for α- and β-forms, respectively, with temperature are expressed by the following (22):

$$\log S = 0.0174t - 0.377 \text{ (0–30 °C) for } \alpha\text{-form}$$

$$\log S = 0.0153t - 0.328 \text{ (30–70 °C) for } \alpha\text{-form}$$

$$\log S = 0.0159t - 0.461 \quad \text{for } \beta\text{-form}$$

The solubility of the β-form is lower than that of the α-form throughout the temperature range measured. It follows that the β-form is a stable form from the aspect of thermodynamic kinetics. In Table 1, the solubility for various salts of L-Glu versus temperature are summarized (23).

The solubility changes with change in the concentration of both hydrochloric acid and sodium hydroxide in Figure 4, which corresponds to the existence of solid crystals, in

Table 1. Solubility of Various Salts of L-Glu (g/L)

Temperature (°C)	L-Glu	DL-Glu	L-GluNa	DL-GluNa	L-GluHCl	DL-GluHCl
0	3.41	8.55	514	158	298	471
25	8.64	20.5	627	243	479	698
50	21.9	49.3	765	372	769	1,030
75	55.3	119	933	570	1,240	1,540
100	140	285	1,140	875	1,990	2,280

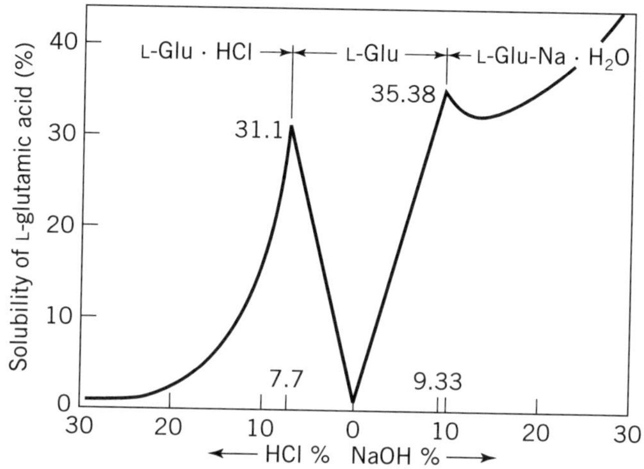

Figure 4. The change in the solubility of L-glutamic acid with change in the concentration of both hydrochloric acid and sodium hydroxide.

equilibrium solution, of the hydrochloride, free, and sodium salts of L-Glu, respectively (i.e., in the range of 0 to 30% for HCl and of 0 to 20% for NaOH). In the range of 0 to 7.7% of HCl, the solubility increases linearly so that L-Glu is soluble until the equivalent amount of HCl in solution in which L-Glu exists as solid crystals. Above the concentration of 7.7% HCl, with an increase in the HCl concentration, the solubility decreases steeply. It flattens in the vicinity 20% and then becomes a constant of about 1% above 25% HCl. Above 7.7% HCl, the hydrochloride is a solid crystal in solution. At the invariant point where equilibrium is attained between L-Glu and L-Glu HCl, the maximum solubility of 31.1% is obtained.

On the other hand, in the range of 0 to 9.33% of NaOH, L-Glu solvates an equal amount of NaOH in solution and exists as a solid crystal. At the invariant point (NaOH 9.33%), the solubility of L-Glu reaches the maximum of 35.38%. Above the concentration of 9.33% NaOH, L-GluNa H_2O exists as solid crystals. A slight decrease in solubility is observed in the range from 9.33 to about 13%, and then increases with increasing concentration of NaOH. The minimum value of the solubility of L-Glu is obtained from the range of pH 2 to 4, in the vicinity of the pI.

L-GLUTAMIC ACID (L-GLU)

$$\begin{array}{c}HOOCCH_2CH_2\underset{\displaystyle NH_2}{\underset{|}{C}}HCOOH\text{; L-Glu}\end{array}$$

Chemical formula: $C_6H_9O_4N$ Molecular weight: 147.14

C = 40.82, H = 6.17, O = 43.50, N = 9.52 (%)

Crystal: L-Glu is a polymorph, having two crystal forms, α and β. Both are anhydrous.

Crystals of the α-form are obtained by crystallization from the saturated solution of L-Glu at 60 to 70 °C by cooling rapidly with agitation or from the acidic or alkalic solution of L-Glu to which alkali or acid was added to rapidly neutralize it until the isoelectric point of L-Glu was reached (pH 3.2).

The crystal shape is a column or pyramid. The α-form crystals transform gradually to β-form when kept in solution for a long time at room temperature. β-form crystals are obtained from either the saturated solution of L-Glu at 80 to 90 °C, which is cooled gradually with agitation, or the relatively higher concentration of L-Glu hydrochloride or sodium salt, to which alkali or acid was added to neutralize it slowly until pH 3.2 was reached. The crystal shape is a needle or thin plate (24).

Crystal Structure

The α- and β-forms of the crystals belong to the orthorhombic crystal group, and the space group is $P2_12_12_1$ with the crystal-lattice constant for both crystals as follows (25): a = 7.06, b = 10.3, and c = 8.75 Å for the α-form and a = 5.17, b = 17.34, and c = 6.95 Å for the β-form. The molecular arrangement obtained from an X-ray diffraction for two crystals are shown in Figure 5. From the crystal-structure analysis (25), L-glutamic acid in both forms exists as a zwitterion; a proton transfers from the α-carboxyl group to the amino group. The C(1)-O(1) and C(1)-O(2) distances of 1.24 and 1.27 å, for both crystal forms, are the values of the ionized carboxyl group. This discrepancy of 0.03 å found in both crystal forms can be attributed to differences in their intermolecular hydrogen bonds. O(2) makes a very short hydrogen bond with the OH group and a NH—O hydrogen bond with the NH_3^+ group, while O(1) makes one NH—O bond with NH_3^+.

In the γ-carboxyl group, the length of C(5)-O(4) and C(5)-O(3), 1.21 and 1.37 Å for the α-form, and 1.19, and 1.24 Å for the β-form, respectively, indicate that it is un-ionized in both crystal forms.

The bond angles around C(2), C(3), and C(4) in the α-form are significantly different from the corresponding ones in the β-form. This discrepancy may be a result of the steric repulsions that are due to the conformational differences. The OH—O bond seems to be most predominant of the four hydrogen bonds. It is reasonable to assume that considerable amounts of molecules are connected by OH—O hydrogen bonds to form chains in solution. The neigh-

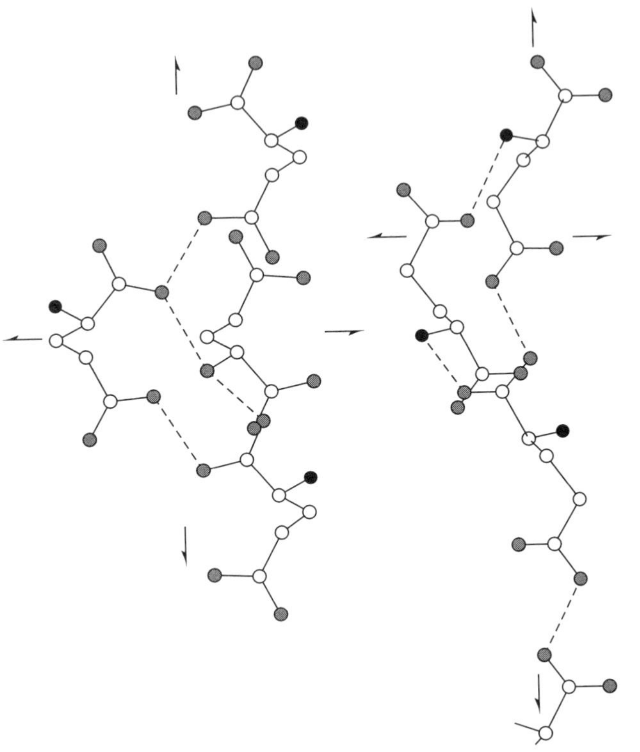
(a) (b)

Figure 5. The molecular arrangement of L-glutamic acid crystals: a, α-form; b, β form; Dashed line shows the hydrogen bond systems in the crystals, ○, carbon; ●, oxygen, ●, nitrogen.

boring chains can be connected by the NH—O hydrogen bonds, and the conformation and the arrangement of the molecules are stabilized by these hydrogen bonds.

Stability

When heated as a dried crystal to 150 °C, L-Glu undergoes cyclization and is converted to an interdehydrated compound of pyrrolidone carboxylic acid. Moreover, heated to 180–200 °C, the optically inactive DL-pyrrolidone carboxylic acid is obtained. At greater than 200 °C, glutamic acid begins to decompose (26). At pH 4 or 10 when the solution is heated at 100 °C for about 50 h, more than 98% of L-Glu undergoes cyclization. At pH 3 and when heated at 120 °C for 3 h, 100% cyclization is observed.

The melting point of L-pyrrolidone carboxylic acid is 160 to 161 °C, and the specific optical rotation in water is −11.35 at 20 °C.

This cyclization is a reversible reaction when treated with strong acid to yield glutamic acid, or when heated with 2N HCl or 0.5N NaOH at 100 °C for 1 to 2 h.

L-GLUTAMIC ACID HYDROCHLORIDE

Chemical formula: $C_5H_{10}O_4NCl$ Molecular weight: 183.61

C = 32.71, H = 5.49, O = 34.86, N = 7.53, Cl = 19.33 (%)

L-Glu HCl forms transparent crystals that crystallize from the high concentration of hydrochloric acid of L-Glu by cooling. The melting point is 160 °C and optical rotation calculated from the ionic form (G^+) is +25.65.

The relationship between the solubility of L-GluHCl and the temperature is (27):

$$\log S = 1.474 + 0.00825t$$

The solubility of L-glutamic acid hydrochloride crystals at 20 and 30% HCl with changing temperature is shown in Table 2.

As mentioned earlier, with an increase in the concentration of HCl above 17%, the solubility of hydrochloride decreases and at more than 30% becomes constant. Conversely, when the temperature is lowered, the solubility decreases sharply. For example, at 3 °C in 30% HCl, the solubility is only 0.38%.

It is profitable when crystallizing L-GluHCl crystals to allow the solution to cool to as low a temperature as possible. The crystal of DL-GluHCl is racemic compound, and it is possible to resolute by the seed method. The solubility of DL-Glu HCl is $\log S = 1.673 + 0.00685t$.

SODIUM SALTS OF L-GLUTAMIC ACID

Monosodium L-glutamate (MSG), L-GluNa

$$\begin{array}{c} HOOCCH_2CH_2CHCOONa\ H_2O \\ | \\ NH_2 \end{array}$$

Chemical formula: $C_5H_8O_4N\ Na\ H_2O$ Molecular weight: 187.13

C = 32.09, H = 4.31, O = 34.20, N = 7.49, Na = 12.29, H_2O = 9.63(%)

The monosodium salt of L-Glu crystallizes generally as a monohydrate. The melting point is 195 °C and dehydration temperature is 150 °C.

Crystal Structure

This crystal belongs to the orthorhombic crystal group $P2_12_12_1$. The crystal lattice constant is a = 15.30, b = 17.86, and c = 5.54 å, and the crystal density 1.635 g/cm^3. The packed specific volume is 1.20 to 1.25 cm^3/g.

Table 2. The Solubility of L-GluHCl for 20% and 30% of HCl

	Hydrochloric acid	
Temperature (°C)	20%	30%
3	1.15	0.38
7	1.31	0.44
10	1.48	0.47
15	1.38	0.58
25	2.60	1.16
35	5.29	1.19
50	8.42	2.68

The crystal shape is an octahedral column and has bilateral planes at both termini where the atomic coordination is unstable, causing piezoelectricity. The crystal shows weak hygroscopic character with increasing temperature, as summarized in Table 3, which shows the relationship between the temperature and critical humidity, although the degree of hygroscopicity was relatively weak, when compared with that of NaCl. It is easily soluble in water, and the solubility S is designated by the following relation (28):

$$S = 35.30 + 0.098t + 0.012t^2$$

where S is g/100 g H_2O and t is in degrees centigrade.

L-Glutamic Acid Monosodium Salt Pentahydrate

$HOOCH_2CH_2CHCOONa\ 5H_2O$ (with NH_2 on the CH)

Chemical formula of pentahydrate: $C_5H_8O_4Na\ 5H_20$
Molecular weight: 259.5

The condensed solution of L-Glu Na was cooled between −0.8 and −8.5 °C, from which pentahydrate crystals were crystallized (29). The crystal structure belongs to the triclinic and space group P1, and lattice constant is (30)

$$a = 6.12,\ b = 16.06,\ \text{and}\ c = 6.01\ \text{Å}$$
$$\alpha = 99.3,\ \beta = 100.9,\ \gamma = 99.1$$

The crystal density is 1.480 g/cm^3.

As shown in Figure 6, the transition point form monohydrate to pentahydrate is −0.8 °C. Above −0.8 °C monohydrate crystals exist as the stable form, between −0.8 and −8.5 °C, pentahydrate exists as the stable form, and below −8.5 °C the solution itself turns into ice. The solubility of pentahydrate is changeable with temperature compared to monohydrate, and is designated as follows:

$$S = 35.72 + 0.630\ t + 0.0016\ t^2$$

Table 3. Critical Humidity of L-GluNa and NaCl

	Critical humidity (%)	
Temperature (°C)	MSG	NaCl
10	96.1	74.9
20	96.0	75.3
30	94.8	75.1
40	92.6	75.2
50	90.0	75.2

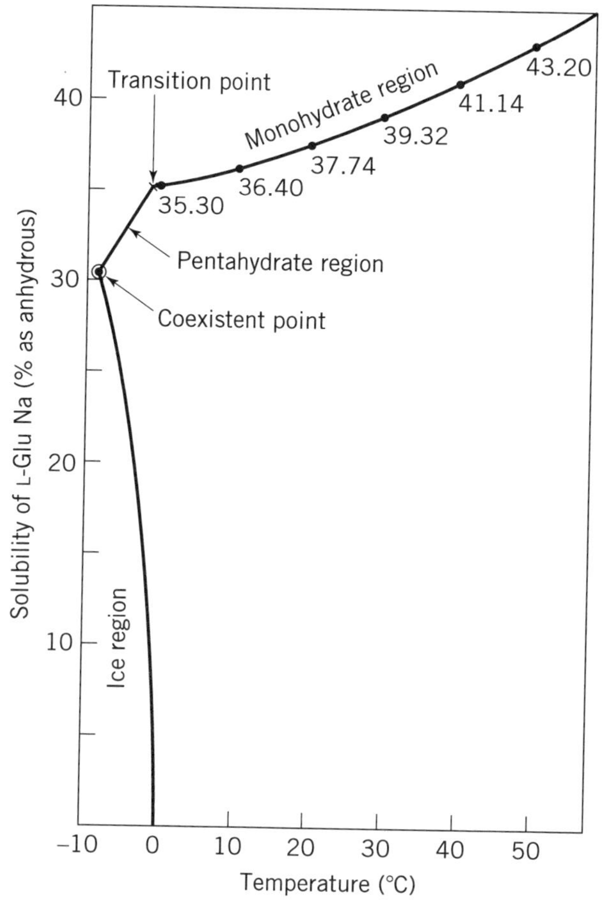

Figure 6. The solubility of sodium L-glutamate with change in temperature.

CALCIUM SALT OF L-GLUTAMIC ACID

Several forms of calcium salt are reported.

$(Glu)_2$-CaH_2O: An equimolar amount of $Ca(OH)_2$ was added to L-Glu solution to crystallize. This is soluble in water.

(Glu)-CaH_2O: L-Glu HCl in solution reacts with $Ca(OH)_2$ at 80 °C to form (Glu)-CaH_2O. This is sparely soluble in water.

$(Glu)_2$-$CaCaCl_2H_2O$: $Ca(OH)_2$ was added to L-Glu-HCl in solution, saturated with CO_2, and boiled. Concentration yielded the complex salt.

INDUSTRIAL PRODUCTION

Extraction Method

From 1909, when MSG production started, until 1965, when the extraction process ended, wheat gluten and defatted soybeans were used as the main raw materials, containing as much as 15 to 25% L-glutamaic acid. Crude raw materials were hydrolyzed by heating with hydrochloric acid. After concentration under reduced pressure, the hydrolyzate was cooled to crystallize L-glutamic acid hydrochloride. L-Glutamic acid can be easily separated from other amino acids in the form of its hydrochloride because of its low solubility in concentrated hydrochloric acid. The crude hydrochloride is dissolved in hot water and filtered to separate the humic substances formed in the hydrolyzation process. The pH was adjusted with caustic soda to 3.2, the isoelectric point of L-glutamic acid needed to crystallize L-glutamic acid crystals at a yield of more than 90%. The crude L-glutamic acid crystals are suspended in water and neutralized with sodium hydroxide. The solution is decolonized with activated carbon and crystallized MSG (monosodium L-glutamate monohydrate) under reduced pressure. Commercially preferable crystals are grown with a small amount of amino acids (31). The crystals are separated by centrifuge and then dried for packaging.

Beet, sugarcane molasses or Stefan's waste (which contained 2-pyrrolidone carboxylic acid) has been used as raw material for MSG production in the United States and Europe. The pyrrolidone carboxylic acid was hydrolyzed at 85 °C at pH 10.5 to 11.5 for about 2 h to avoid racemization (32).

Industrial Fermentation Method

For industrial production of MSG, molasses or starch hydrolyzate is generally used as raw material. Sugar molasses usually contains excess amounts of biotin, so when used, it is necessary to repress the activity of biotin by adding penicillin or certain surface-active substances at an early stage of the production to retard microorganism propagation. Furthermore, sugar concentration in culture media at below about 10% is necessary so as not to retard propagation. Additional sugar was fed during fermentation to obtain a higher concentration of excreted L-glutamate. This batch-fed fermentation process made possible a concentration of more than 120 g/L of L-glutamic acid in media.

Microorganisms used for L-glutamate fermentation are usually preserved under lyophilization below −80 °C or, for a short period, by keeping the stock culture below 10 to 15 °C. To refresh the microorganisms, stocked in either form, they are inoculated on an agar medium composed of 1% yeast extract and polypeptone, 0.5% sodium chloride, and 2% agar, at an optimum temperature of microorganisms. The refreshed microorganisms are then cultivated in liquid medium, shaken vigorously, and transferred to a small fermenter to allow them to propagate to about several kiloliters for seed culture. Industrial-scale fermenters are pressure tight, stainless steel containers, built to hold up to several hundred kiloliters of cultivating medium. They are equipped with aeration and stirring devices, as well as other automatic controls. Fermentation takes from 30 to 45 h.

Gaseous ammonia or a solution of urea and ammonium salts are convenient nitrogen sources for fermentation, not only as the initial medium, but also to maintain the pH of the medium at 7 to 7.5 for microbial growth and product formation. The culture medium becomes acidic because of assimilation of ammonium ions and the formation of L-glutamate. Gaseous ammonia can be used advantageously to maintain neutral pH and avoid dilution of culture medium, resulting in the high accumulation of L-glutamate in the fermentation broth, because it does not contain OH^- ions or water.

Microorganisms required several minerals, such as ferrous and potassium ions, which play important roles in L-glutamate fermentation. Other important conditions include regulating the aeration stirring. The biosynthesis of glutamate is performed under regulated aerobic conditions. When oxygen is not sufficiently dissolved, lactic acid and succinic acids accumulate and reduce the accumulation of glutamate. On the other hand, an excess of dissolved oxygen results in the formation of α-ketoglutaric acid. The optimal oxygen transfer rate was determined by measuring the rate of consumption of sodium sulfite, which is considered to depend on the characteristics of microorganisms used (33). The pressure of dissolved oxygen was usually kept above 1 kPa (0.01 atm) by aeration and agitation in the fermenter. The sterilized air is fed through a filter. Carbon dioxide is formed during fermentation and causes the medium to become heavy. Mechanical and chemical defoaming systems are provided.

The optimum temperature for fermentation is dependant on the character of microorganisms used. Regulation of the temperature using heat exchangers installed inside of the fermenter is indispensable. Fermentation is an exothermic reaction, and the temperature critically affects not only the propagation of microorganisms, but also the formation of glutamate.

In large-scale production the formation of trehalose very often reduces the product yield of glutamate. Trehalose consists of two α-1,1-bond glucose molecules and is excreted by the bacteria as an osmoprotectant. A process has been developed and successfully industrialized in which trehalose formation is controlled and decreased by culturing the overproducing mutant in media with inverted sucrose from molasses (34).

When fermentation is complete and after sterilization of the broth, the glutamate-recovery process is performed.

Separation Process

The separation process consists mainly of the following two processes (35):

1. Recovery of L-glutamic acid crystals from the fermentation broth
2. The purification of L-monosodium glutamate monohydrate from crude L-glutamic acid crystals

The typical outline of a production flow sheet of L-MSG from the fermentation broth is shown in Figure 7.

Hydrochloric acid at 35%% is added to the fermentation broth with agitation to crystallize the L-glutamic acid crystals at pH 3.2, the isoelectric point of L-glutamic acid. The crude L-glutamic acid crystals, retaining some mother liquor when removed from the slurry, are dissolved in water with equimolar sodium hydroxide; then the solution is applied to the activated carbon column to obtain the decolonized L-MSG solution, on which most of the color-related and hydrophobic substances are adsorbed. This purified L-MSG solution is fed continuously to a vacuum crystallizer to crystallize L-MSG crystals. L-MSG crystals separated by centrifugation are dried, sieved, and finally packaged to deliver to the consumers.

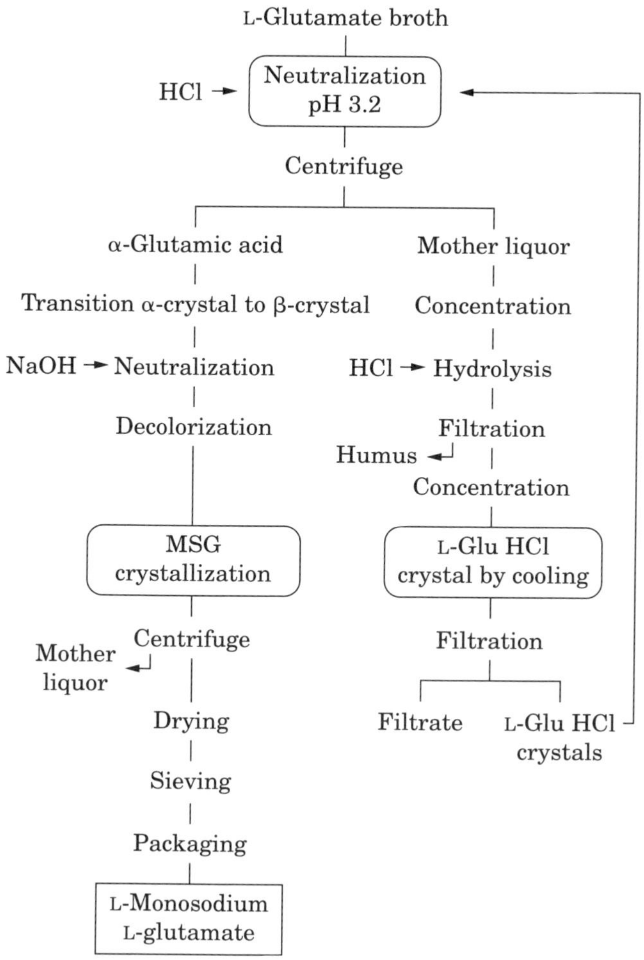

Figure 7. Flow diagram of production of MSG from the fermentation broth.

Crystallization of L-Glutamic Acid

In the fermentation broth, L-glutamic acid exists as an ammonium salt in the vicinity of pH 7. For obtaining L-MSG directly from the fermentation broth, it is necessary to exchange ammonium ion for sodium ions. This process is rather complicated, taking into account an environmental approach. To avoid this complexity, as a first step, crystallization of the α-form of L-glutamic acid crystals, which show lower solubility as compared to L-MSG and fairly good sedimentation as compared to β-form crystals, is employed in actual industrial production. The presence of salts can affect solubility of L-glutamic acid in solution. For example, the chloride-related inorganic substances NaCl, $CaCl_2$, NH_4Cl, and KCl increase the solubility of L-glutamic acid.

When crystallized under ordinary conditions, pyramid-type α-form crystals are obtained by the effect of the impurities contained in the fermentation broth.

Pyramid Type

In the growing of crystals, only those faces having the lowest translation velocities survive. The face of [001] of the α-form are fast-growing (high translation velocities) and this face tends to disappear, as overlapped by slower [111] faces, and finally results in the pyramid type.

If Ostwald's step theory is obeyed, as α-form crystals are unstable from the thermodynamic point of view, α-form crystals first appear and then gradually transform to β-form. But this theory is not based on the thermodynamics, but on the kinetics, so exceptions frequently occur.

The conditions to obtain α-form crystals are summarized as follows (36):

1. When neutralized, 35% hydrochloric acid is added as quickly as possible to the fermentation broth and then cooled to below 10 °C instantly with vigorous agitation.
2. A small amount of a third substance is added to affect the stability of α-form, such as a peptide, active surfactant, protein hydrolyzate, or amino acid, especially L-phenylalanine and L-cysteine.

Centrifugation

The α-form of L-glutamic acid crystals are separated by a centrifuge. Residual mother liquor after centrifugation amounts to 5 to 15% of the weight of the crystals, depending on the viscosity of the mother liquor and the degree of uniformity of crystal size. It is more effective to wash the crystals on the centrifuge with fresh water to reduce impurities than recrystallization. In this process, the bacterial cells remain effectively in the mother liquor, and without another bacterial cell separation process, L-glutamic acid crystals remain effectively separated from the bacterial cells.

Transformation of α-Crystals to β-Crystals

When the α-form crystals are poured into the water and the temperature is elevated to 60 to 80 °C, the transformation of α-form to β-form occurs within about 1 h. During this solvent-mediated transformation, a large amount of adhering impurities are rejected out of the crystals, and higher purified β-form miceous plate-shape crystals are obtained (37). This transformation process has so marked an efficiency for removing impurities, especially color-related substances, that when higher purity is required, a small amount of activated carbon is used to obtain a transparent MSG solution.

For another purification of L-glutamic acid solution, the application of UF membrane or NF membrane is investigated: When these membranes are applied to permeate the fermentation broth, the permeate solution obtained is neutralized followed by cooling. Purified β-form crystals are obtained directly, without α-form crystallization.

Crystallization of L-MSG

Neutralized L-glutamic acid solution (L-MSG solution) is applied to the activated carbon column to obtain decolonized solution, which is fed to the continuous evaporator to concentrate until the saturation of L-MSG (50% at 60 °C). To meet the requirements for L-MSG monohydrate crystals—crystal size, crystal shape (aspect ratio), and size distribution—it is necessary to strictly control the operational conditions of the crystallization process, in which nucleation and crystal growth occur simultaneously. The result of the combined processes on the mentioned requirement for crystals depends on the relative rates of nucleation and growth. To obtain a more uniform product, nucleation is controlled by adding the desired number of crystals, usually crushed, to the crystallizer when the solution is either saturated or supersaturated. In a continuous crystallizer, the number of nuclei formed per unit time will be continuous and will equal the number of crystals withdrawn from the crystallizer, so that all nuclei can receive the same time of growth. Otherwise, there must be a classifying action in the crystallizer that will retain the small crystals under treatment until they have grown to the proper size, when they can be removed from the crystallizer.

For practical control, it is necessary to reduce as much as possible the local variation in the temperature or concentration of the solution that causes excess nucleation. Usually, the objective of the crystallizer operator is to achieve the maximum growth rate, consistent with low nucleation rate. More importantly, impurities in solution may inhibit the formation of new nuclei or retard crystal growth. The degree of the effect of a given impurity cannot be predicted; in general, higher molecular weight materials seem to be more effective inhibitors.

Crystal Shape (Aspect Ratio)

When no impurity is contained in the solution the aspect ratio (referred to as L/D in crystal) is 7 to 8 which is not convenient for umami seasoning for cooking. To overcome this problem, the appropriate amount of amino acids is added. The crystal shape is determined by the change in the difference in the relative crystal growth of constituted crystal faces. For L-MSG crystals, a small amount of amino acids changes the crystal shape, the amino acids such as L-alanine, L-lysine, and L-arginine are easily adsorbed on the [110] face of the L-MSG crystal and inhibit the growth of this face, those having the lowest translation velocities survive, and the aspect ratio is 4 to 5, which satisfies the demand of users (38).

Recovery of L-Glutamic Acid from Mother Liquor

In mother liquor–separated L-glutamic acid crystals other than L-glutamic acid, many impurities derived from the raw material used as the carbon source, as well as ammonium hydrochloride and bacterial cells, are found. For recovery of L-glutamic acid from this mother liquor, the following process is proposed: The mother liquor is condensed and ammonium chloride crystallized by cooling for a day and night followed by filtration to obtain the ammonium chloride crystals that are used as fertilizer. The separated mother liquor is hydrolyzed with excess hydrochloric acid to convert the L-glutamic acid to hydrochloride and the bacterial cells and other debris to insoluble matters (so-called humic acid). The humic acid is separated by filtration. The filtrate is condensed, then cooled to crystallize the hydrochloride crystals at lower than 10 °C for more than several days. This is recycled in the process of L-glutamic acid crystallization from the fermentation broth to curtail the consumption of hydrochloric acid. The residual mother liquor combined with humic acid is used as raw material for the organic fertilizer.

USES

Seasoning

MSG is widely used as a long-established seasoning or flavor enhancer to improve the palatability of foods. The effect of MSG is due to its characteristic taste, umami. The development of food science, including electrophysiology, psychophysics, and nutrition, have provided evidence that umami should be regarded as a basic taste, independent from the traditional four basic tastes—sweet, salty, sour, and bitter. Glutamate is ubiquitous in foods such as tomato, cheese, human milk, and seaweed. Other umami substances, that is, the 5′-ribonucleotides disodium 5′-inosinic acid and disodium 5′-guanylic acid, are found in meats, fish, vegetables, and mushrooms.

When glutamate and nucleotides coexist, a very strong synergistic effect occurs, and even a small amount of glutamate added to a food containing nucleotide causes umami to be dramatically enhanced—seven to eight times as much as the original umami of the food (39).

The taste threshold of MSG is about 0.03% in aqueous solution. The intensity of umami increases linearly with a logarithmic increase in the concentration of MSG. The synergistic effect of MSG with 5′-ribonucleotides is expressed by the following relation (41):

$$y = u + A\,u\,v$$

where y is the intensity equivalent to the concentration of MSG alone, and u and v are the concentrations of MSG and 5′-ribonucleotides, respectively. A is a constant that is 1,200 for disodium 5′-inosinate (IMP) and 2,800 for disodium 5′-guanylate (GMP). In some commercial umami seasonings, 5′-ribonucleotides are mixed in to take advantage of the synergistic effect with MSG.

Although MSG increases the palatability of food, it is not always palatable by itself. When MSG is tested in crystal form or in simple solutions, it does not cause a pleasant sensation. It gives a pleasant taste only in flavored solution or in actual foods. Another important property of MSG is the concentration dependence of its pleasantness rating. An excess of MSG causes an unpleasant sensation, and thus the amount of MSG that is added to foods should be limited (41).

The desirable usage level of MSG depends on the properties of the food, including the amount of original umami substances. Results of extensive taste studies indicate that the optimum MSG level in food ranges from 0.1 to 0.8 wt %, varying from food to food. In most cases, MSG is used in combination with salt. A relationship between the levels of MSG and salt, and palatability of foods, has been clarified. The lower the salt concentration, the higher the MSG concentration required to maintain the palatability, and vice versa. When umami substances are added to meals, the amount of salt can be reduced without a decrease in the palatability or degree of satisfaction.

Umami taste is considered to be a universal taste sensation for humankind. An ethnic comparison of taste sensitivity has shown that there is no difference in the thresholds of MSG and IMP between Japanese and European-American Caucasians (42).

Industrial Use

Depending on the pH, glutamic acid can be used to neutralize acidic or basic compounds, that is, arginine glutamate is used as a pharmaceutical and as a raw material for cosmetics. The reaction of glutamic acid with hydrochloric acid affords glutamic acid hydrochloride, which is used as a gastric acidifier. The property of glutamic acid as an amphoteric electrolyte is exploited in chelating agents, buffers, and builders for detergents.

The sodium salt of pyroglutamic acid (43), which is obtained by dehydration of glutamic acid followed by neutralization with sodium hydroxide, is very hydroscopic and used as a component of a natural moisturizing factor for human skin and as a humectant for cosmetics. The *N*-acylated compound (44) produced by reaction with a long-chain fatty acid is an anionic surfactant and is popularly used for skin and hair cleansers due to its mildness and high safety. The dibutylamine of the *N*-acyl-glutamate gelatinizes nonpolar oils and can be used as a recovering agent for marine-oil spills (45).

SAFETY

Monosodium L-glutamate is metabolized in the same way as glutamic acid from any digested protein. After oral ingestion, glutamate is absorbed from the intestine by an active-transport system. During this process, a large portion is metabolized to alanine and α-ketoglutarate, which enters the tricarboxylic acid cycle. Glutamate is further metabolized in the liver to give glucose, lactate, glutamine, and other amino acids. Consequently, blood glutamate levels do not rise significantly unless very large doses are administered. Blood glutamate levels transiently rise when large doses are ingested on their own, but the coingestion of foods that contain metabolizable carbohydrate increases the metabolism of monosodium L-glutamate and eliminates or greatly attenuates this rise. Human infants, including premature infants, metabolize glutamate similar to adults (46,47).

The acute toxicity of MSG is low (48), the oral LD_{50} for humans, calculated on the basis of doses administered in different ways to various animals, would represent a single dose greater than 1 kg for a person weighing 70 kg. In contrast, the oral LD_{50} for sodium chloride in rats is 3.75 kg/kg body weight (49).

The main use of MSG is as a umami seasoning or a food ingredient, and so its safety when used in the diet is the most important aspect of its use. Both short-term and chronic toxicity studies on MSG in the diet of several species at doses of up to 4% (approximately 6 to 8 g/kg body weight/day) in the diet showed no specific toxic effects and no evidence for carcinogenicity or mutagenecity. Reproduction studies of up to three generations did not reveal any adverse effects of dietary MSG ingestion. Fertility, gestation, viability, and lactation indexes, or the pre- and post-weaning performance of offspring, were unaffected. In primates, it was found that the placenta serves as an effective barrier to the transfer of glutamate from the oral ingestion of large doses of MSG.

The concentration of glutamate is higher in the brain than in the blood. The demonstration that injected or force-fed neonatal rodents given extremely high doses of MSG showed evidence of brain leisons has led to much additional research to determine any possible link between neurotoxicity and human use of MSG (50). However, from animal tests on monkeys, guinea pigs, rat, and mice, glutamate levels in the brain remain unchanged after the administration of large oral doses of MSG (51). Numerous experiments on rodents, as well as dogs and monkeys, using dosage levels up to 43 g MSG/kg body weight have failed to show any link between dietary use of MSG and brain damage. In the case of dogs and monkeys, even experiments involving injection of MSG have not shown any effects on the brain.

An anecdotal report in 1968 triggered interest in Chinese restaurant syndrome (CRS), a reaction associated with transient subjective symptoms of burning, numbness, and a tight sensation in the upper part of the body. Possible association with food ingredients such as MSG was suggested. No objective changes in skin temperature, heart rate, ECG, or tone were observed, and no correlation was seen between plasma glutamate levels and symptoms. In 1979 a questionnaire survey of more than 3,000 people in the United States was conducted for CRS. Only 1.8% of those surveyed acknowledged that they experienced possible CRS feelings (52), and 0.19% of those with symptoms attributed them to eating in Chinese restaurants. These feelings were also found to occur after consumption of spicy tomato juice, orange juice, coffee, and tea. In 1986 self-identified MSG responders were challenged in a properly controlled double-blind study, and it was concluded that the link between MSG and CRS was not supportable (53). In 1987, the Joint FAO/WHO Expert Committee on Food Additives (JECFA), a scientific advisory body to the World Health Organization and Food and Agriculture Organization of the United Nations, concluded that properly conducted double-blind studies among individuals who claimed to suffer from the syndrome did not confirm MSG as the causal agent (54).

In 1981 Chinese restaurant asthma was reported following capsule administration of MSG to several asthmatics (55), but the researchers failed to account for other allergens to which the subjects could have been exposed and did not utilize the scientific practice of a "control" substance that would have helped to determine if glutamate triggered this response. In a double-blind crossover study in which chronic asthmatics were challenged with MSG or a placebo, no decrease in pulmonary function was observed (56).

JECFA reviewed the safety studies of glutamate and endorsed its safety by allocation of an acceptable daily intake (ADI) for L-glutamic acid and its monosodium salt, potassium salt, ammonium salt, calcium salt, and magnesium salt as being "not specified." This result is ascertained by the scientific committee for food of EC (57).

BIBLIOGRAPHY

1. K. Ikeda, *J. Tokyo Chem. Soc.* **30**, 820–836 (1908).
2. L.B. Lockwood and F.H. Stodola, *J. Biol. Chem.* **164**, 81–93 (1946).

3. S. Kinoshita, K. Nakayama, and S. Udaka, *J. Gen. Appl. Microbiol.* **3**, 193–205 (1957).
4. Y. Hirose and H. Okada, *Microbial Technology*, 2nd ed., vol 1, Academic Press, New York, 1979, p. 211.
5. I. Shiio, *Seikagaku* **50**, 1–16 (1978).
6. S. Dagley, E.A. Davis, and G.A. Morrison, *Nature* **165**, 437–438 (1950).
7. W. Liebl, U. Ehrman, W. Ludwig, and K.H. Schleifer, *Int. J. Syst. Bacteriol.* **41**, 225–260 (1991).
8. R. Kramer, *FEMS Microbiol. Rev.* **13**, 75–93 (1994).
9. I. Shiio and K. Ujikawa, *Agric. Biol. Chem.* **42**, 1897–1904 (1980).
10. E.R. Boermann, B.J. Eikmanns, and H. Sahm, *Mol. Microbiol.* **6**, 317–326 (1992).
11. T.E. Walker, C.H. Han, V.H. Kollman, R.E. London, and N.A. Matwiyoff, *J. Biol. Chem.* **257**, 1189 (1982).
12. H. Momose and T. Takagi, *Agric. Biol. Chem.* **42**, 1911–1917 (1978).
13. Japanese Pat. 50-92795 (1980), O. Tosaka, Y. Murakami, K. Akashi, and S. Ikeda, Ajinomoto Co.
14. Japanese Pat. 50-521762 (1980), E. Ono, O. Tasako, and K. Takinami, Ajinomoto Co.
15. French Pat. 2612937 (March 24, 1988), Y. Yamada and A. Seto, Ajinomoto Co.
16. H. Wilson and R.K. Cannan, *J. Biol. Chem.* **119**, 309–331 (1937).
17. C.B. Anfensen, J.T. Edsall, F.M. Richards, and M.L. Anson, *Advances in Protein Chemistry*, vol. 4, Academic Press, New York, 1948, p. 339.
18. W.F. Diven, *Biochim. Biophys. Acta* **191**, 702–706 (1969).
19. P. Pfeiffer and J. Wurgler *Z. Physiol Chem.* **97**, 128–147 (1916); P. Pfeiffer and O. Angern, *Z. Physiol. Chem.* **133**, 180–192 (1924).
20. T. Takenishi, *Journal of the Chemical Society of Japan* **81**, 1380–1382 (1960); **82**, 15 (1961).
21. M.C. Otey, J.P. Greenstein, M. Winitz, and S.M. Birnbaum, *J. Am. Chem. Soc.* **77**, 3112–3114 (1955).
22. Y. Sakata, *Agric. Biol. Chem.* **25**, 829–834 (1961).
23. T. Ogawa, *Journal of the Chemical Society of Japan, Ind. Chem. Section* **52**, 69 (1949).
24. Y. Sakata, *Agric. Biol. Chem.* **25**, 835–837 (1961); **26**, 355–361 (1962).
25. M.S. Lehman, T.F. Koetzle, and W.C. Hamilton, *J. Cryst. Mol. Struct.* **2**, 225–233 (1972); M.S. Lehman and A.C. Nenes, *Acta Crystallogr.*, **B36**, 1621–1625 (1980); N. Hiroyama, K. Shirahata, Y. Ohashi, and Y. Sasada, *Bull. Chem. Soc. Jpn.* **53**, 30–35 (1980).
26. S. Budavari ed., *Merck Index*, 12th ed., Merck Co., Inc. Rahway, N.J., 1996, p. 760.
27. A. Albert, *Biochem. J.* **50**, 690–698 (1952).
28. T. Ogawa, *Journal of the Chemical Society of Japan, Ind. Chem. Section* **52**, 102–105 (1949).
29. Japanese Pat. 23-3312 (Dec. 17, 1949), T. Ogawa, Ajinomoto Co.
30. T. Kashiwagi, N. Nagashima, C. Sano, and T. Kawakita, *Acta Crystallogr.*, **C51**, 1053–1056 (1995).
31. U.S. Pat. 2834805 (May 13, 1958) J.L. Purvis and V. Bassel.
32. H. Olbrich ed., *Die Melasse*, 4th ed., Nippon Seito Kogyokai, Tokyo, Japan, 1961, p. 227.
33. Y. Hirose, H. Sonoda, K. Kinoshita, and H. Otsuka, *Agric. Biol. Chem.* **30**, 585–593 (1966).
34. H. Yoshii, M. Yoshimura, S. Nakamura, and S. Inoue, *Nippon Nogei Kagakukaishi* **67**, 955–960 (1993).
35. T. Kawakita and M. Saeki, *Chem. Eng. Jpn.* **50**, 196–203 (1986).
36. S. Hiramatsu, Ph.D. Thesis, University of Tokyo, Tokyo, Japan (1976).
37. Japanese Pat. 45-4730 (Feb. 17, 1970), K. Ito, N. Mizoguchi, and M. Dazai, Ajinomoto Co.
38. H. Tasuke, *Nippon Nogei Kagakukaishi* **50**, 175–179 (1976).
39. A. Kuninaka, *Nippon Nogei Kagakukaishi* **34**, 489–492 (1960).
40. S. Yamaguchi, *J. Food. Sci.* **36**, 846–849 (1971).
41. S. Yamaguchi and C. Takahashi, *Agric. Biol. Chem.* **48**, 1077–1081 (1984).
42. S. Yamaguchi, H. Kiminami, and R. Ishii, *Proceedings of the 22nd Japanese Symposium on Taste and Smell, Japanese Association for Safety of Taste and Smell*, Gifu, 1988, p. 73.
43. K. Laden and R. Spitzer, *J. Soc. Cosmet. Chem.* **19**, 351–360 (1967).
44. M. Takehara, R. Yoshida, and M. Yoshikawa, *Cosmet. Toiletries* **94**, 31 (1979).
45. M. Honma and T. Ikeda, *Region of Chemistry* (in Japanese), **36**, 697–701 (1982).
46. L.J. Filer, Jr., S. Garattini, M.R. Kare, W.A. Reynolds, and R.J. Wurtman eds., *Glutamic Acid*, Raven Press, New York, 1979.
47. Joint FAO/WHO Expert Committee on Food Additives, Toxicological Evaluation of Certain Food Additives, World Health Organization, 1988.
48. H. Moriyuli and M. Ichimura, *Appl. Pharmaceuticals* (in Japanese) **15**, 433 (1978).
49. S. Budavari, M.J. O'Neill, A. Smith, P.E. Heckelman, and J.F. Kinneary eds., *Merck Index*, 12th ed., Merck Co., Rahway, N.J., 1996, p. 1474.
50. J.W. Olney, *Science* **164**, 719–721 (1969).
51. Y. Takasaki, *Toxicology* **9**, 293–305, 307–318 (1978).
52. G.R. Kerr, *J. Am. Diet. Assoc.* **75**, 29–33 (1979).
53. R.A. Kenny, *Food Chem. Toxicol.* **24**, 351–354 (1986).
54. D.H. Allen and G.J. Baker, *New Engl. J. Med.* **305**, 1154–1155 (1981).
55. R.E. Gosselin, R.P. Smith, H.C. Hodge, and J.E. Braddock, *Clinical Toxiocology of Commercial Products*; 5th ed., Williams & Wilkins, London, 1989, p. 407.
56. J.M. Drazen, *J. Asthma* **167**, 167–172 (1987).
57. Commission of the EC Reports of the Scientific Committee for Food, **25**, 16, 1990.

ADDITIONAL READING

T. Kawakita, in J.I. Kroschwitz ed., *Encyclopedia of Chemical Technology*, 4th ed. vol. 2, 1982, p. 410.

T. Kawakita, C. Sano, S. Shioya, M. Takehara, and S. Yamaguchi, in B. Elvers, S. Hawkins, and G. Schulz eds., *Ullman's Encyclopedia of Industrial Chemistry*, 5th ed., vol. A16, VCH, Weinheim, Germany, 1991, p. 711.

W. Leuchtenberger, in H.-J. Rehm and G. Reed eds., *Biochemistry*, 2nd ed., vol. 6, VCH, Weinheim, Germany, 1996, p. 466.

K. Soda, H. Tanaka, and N. Esaki, in H.-J. Rehm and G. Reed eds., *Biotechnology*, vol. 3, Verlag Chemie, Weinheim Germany 1983, p. 481.

T. Yoshida, in J.I. Kroschwitz ed., *Encyclopedia of Chemical Technology*, 3rd ed., vol. 2, 1982, p. 410.

AMINO ACIDS, PRODUCTION PROCESSES

TETSUO OKA
Kyowa Hakko Kogyo, Co., Ltd.
Tokyo, Japan

KEY WORDS

Amino acids
Corynebacterium
Corynebacterium glutamicum
Fermentation
Lysine
Mutant
Recombinant

OUTLINE

INTRODUCTION

Monosodium L-glutamate (MSG) was found by Ikeda in 1908 as a flavoring component of *konbu*, or sea tangle. This was the start of the later development of the amino acid industry. Initially, MSG was produced industrially by extraction from a protein hydrolysate, more specifically, first from wheat gluten hydrolysate with hydrochloric acid, and then from defatted soybean hydrolysate or by conversion from pyrrolidone-5-carboxylic acid in beet molasses, for use as a seasoning agent. This latter process was not sufficient to meet the increasing demand because it involved the problem of by-products disposal, requiring the development of new production processes. Patent applications filed around that time related to chemical synthesis of MSG from succinate semialdehyde or cyclopentane. In 1956, Kinoshita et al. discovered a microbial process for L-glutamic acid production by direct fermentation of a microorganism (*Corynebacterium glutamicum*) from sugar and ammonia (1). Consequently, industrial production of MSG as a seasoning was rapidly enlarged, and at the same time, a new industry of amino acid production by fermentation of microorganisms emerged. In 1958, Kinoshita, Nakayama, and Kitada found that an auxotrophic mutant of *C. glutamicum* requiring homoserine accumulated L-lysine in a medium (2), which enabled industrial production of L-lysine by fermentation. Establishing the basis for the development of the amino acid fermentation industry, this technology suggested the possibility of producing various other amino acids by auxotrophic mutants and also the importance of research in fermentative production of biological components with regulatory mutants. Since then, direct fermentation of various amino acids has been broadly studied.

Reduction of the cost of amino acids, owing to the development of fermentation techniques together with the establishment of their production processes, facilitated extended application of amino acids to uses other than as a seasoning. Particularly, L-lysine, an essential amino acid that is lacking in several food proteins, has been enjoying remarkably increased demand as a feed additive. Table 1 shows the production of the major amino acids. Most of the amino acids produced in large quantities are still manu-

Table 1. Estimated Worldwide Production of Amino Acids (1996)

Amino acids	Estimated production (ton/year)	Process[a]
Glycine	22,000	C
L-Aspartic acid	7,000	E
L-Arginine	1,200	F
L-Cysteine	1,500	E
Monosodium L-Glutamate	1,000,000	F
L-Glutamine	1,300	F
L-Histidine	400	F
L-Isoleucine	400	F
L-Leucine	500	Ex, F
L-Lysine HCl salt	250,000	F
DL-Methionine	350,000	C
L-Phenylalanine	8,000	F, C
L-Proline	350	F
L-Threonine	4,000	F
L-Tryptophan	500	F, E
L-Tyrosine	120	Ex

Source: Ref. 3.
[a]C, chemical synthesis; E, enzymatic method; Ex, extraction; F, fermentation.

factured by fermentation, except for glycine, which does not have optical isomers, and methionine, which has a similar effect as a feed additive in both L- and DL-forms.

In addition to being used as a seasoning agent and a feed additive, amino acids are currently used widely as a raw material for the sweetener Aspartame (*N*-L-α-aspartyl-L-phenylalanine 1-methyl ester), for pharmaceuticals and agrichemicals, for infusion and oral nutrition, for surfactants, and so forth, based on their reactivities and nutritional, pharmacological, and flavoring effects.

In connection with amino acid fermentation, many reports have been published on the physiological properties of amino acid–producing microorganisms. The following discussions stress problems from the standpoint of practically manufacturing amino acids on an industrial scale.

FERMENTATION PROCESSES

Amino acid fermentation may be defined in two ways. One is a definition in the narrow sense and refers to direct accumulation of amino acids in a medium containing a sugar, ammonia, and other nutrients, and the other is a broad definition that covers the fermentation process involving the addition of specific precursors and amino acid production by reaction using enzymatic functions of microorganisms. Here, mainly direct fermentation of the former type is discussed.

Fermentation processes generally have a number of advantages and benefits that accrue to the user of the process. On the other hand, they also have drawbacks. Both are summarized as follows:

Benefits

- Mild conditions are used both in fermentation and in product recovery; hence, little product degradation takes place.
- Fermentation requires relatively less complex operation.
- Once the plant is built and operation has begun, there are relatively low maintenance costs.
- Only L-form amino acids are obtained.

Problems

- Operations provide low product concentrations compared with chemical synthesis processes and require large volumes of water, large fermenter capacity, and comparatively high capital investment.
- The requirements for strict sterility add to capital costs and operation costs.
- Large amounts of energy for oxygen transfer and mixing are required.
- Product recovery may be complex, difficult, and expensive.
- The process time necessary to reach maximum concentrations of the desired product is usually comparatively long.

The aforementioned narrow-sense definition of direct fermentation of amino acids includes three types of fermentation: batch-type, fed-batch-type, and continuous.

Batch Fermentation

Industrial fermentation is mostly performed using batch processes. In a batch process, a large volume of a medium containing nutrients and substrate material is inoculated with a viable culture of one or more appropriate microorganisms. Microbial growth and biochemical synthesis are allowed to proceed until an optimum yield of metabolite or a desired biochemical transformation has been obtained. Although this process is the basic form of fermentation, it is not frequently practiced on an industrial scale because productivity is limited by the amount and nature of the nutrients present at the time of inoculation. In most amino acid fermentation, enhanced product concentration is important to improve production efficiency and requires more advanced fermentation processes.

Fed-Batch Fermentation

Fed-batch fermentation types are classified in Table 2. This fermentation method is aimed at efficiently carrying out fermentation and is characterized by a low concentration of components in the initial medium to minimize metabolic regulation. At the time of inoculation the medium promotes initial growth of microbes; subsequent supplies of more raw materials drive the desirable increase in metabolite biosynthesis. Feeding is effected either intermittently or continuously. Industrial fermentation of most amino acids is accomplished with this method. Fed-batch fermentation requires feeding equipment in addition to the equipment required for batch fermentation and therefore leads to higher fixed costs (Fig. 1). However, the process can provide improved productivity as a whole because of the enhanced yield and reduced fermentation time. Actual product cost depends on the cost of the carbon source or specific precursors that are used for feeding. The nitrogen source is supplied generally through pH control with ammonia. This process is of particular importance in industrial operations where the reaction of the microbial catalyst does not last long; hence, continuous fermentation cannot be practiced, unlike in the case of L-glutamic acid fermentation from cane molasses using penicillin, described next.

Table 2. Classification of Fed-Batch Microbial Reaction Processes

Case 1 (nonfeedback regulation)
Constant feeding-rate method
Exponential feeding method
Optimized feeding method
Others (intermittent feeding, etc.)
Case 2 (feedback regulation)
Indirect control
Direct control
Constant control
Program control
Others

Source: Ref. 4.

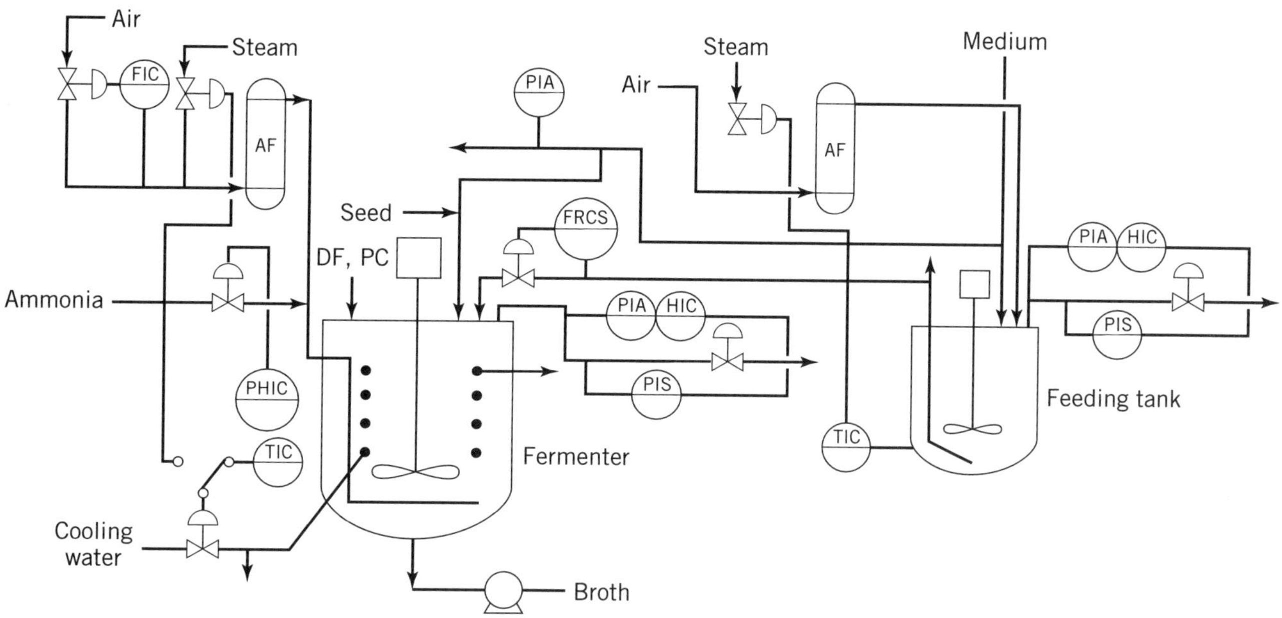

Figure 1. Typical system for fed-batch fermentation. AF, air filter; DF, defoamer; PC, penicillin; FIC, flow indication control; PHIC, pH indication control; TIC, temperature indication control; PIS, pressure indication sum; PIA, pressure indication alarm; HIC, highest indication control; FRCS, flow record control sum.

Continuous Fermentation

In continuous fermentation, a complete medium is fed to a fermenter after an appropriate period of batch fermentation, and the same quantity of broth is continuously taken from the fermenter to maintain the fermentation broth at a fixed volume. This may be performed either by the chemostat method using a substrate or limiting substance, or by the turbidostat method in which the cell level is adjusted to maintain constant cell mass. Because the continuous fermentation process allows improvement of productivity compared with the ordinary fermenter, the initial investment in equipment is small relative to the production volume, and operation cost is low. However, one drawback is that it is not suitable for small-scale production, and the challenges of sterile operation and equipment maintenance are more necessary than they are for batch and fed-batch fermentation.

This process shifts from batch fermentation to continuous fermentation when productivity per unit time in the former is relatively high. There are many reports analyzing the steady-state condition in continuous fermentation. Most of them relate to cell culture, but a few reports are available specifically on amino acid fermentation (5). One of the reasons may be that studies on amino acid fermentation have been mainly directed to the influence of metabolic regulation, as in the case of the penicillin addition method in glutamic acid fermentation, or because some amino acid processes have a distinct growth phase and production phase, and continuous culture cannot be used. Many of the microbial strains used in amino acid fermentation are released from metabolic regulation to a remarkable extent. This makes it easier to analyze continuous processes and thereby optimize them. It is necessary to study optimum conditions for each process and to optimize their industrial application.

Unlike processes of chemical synthesis, continuous fermentation processes have their own restrictions and duration. This is because microbes undergo spontaneous mutation within the system, and an increase in the fraction of microbes with decreased productivity may lead to rapid reduction in productivity (Fig. 2). Hence, it is necessary to breed a strain with high genetic stability.

Enzymatic Method

Of the amino acid production processes using direct enzymatic biotransformation, those for L-alanine (6), L-aspartic acid (7), L-lysine (8), and L-tryptophan (9,10) have been the most extensively studied, and some of the results have led to standard industrial processes. Although the enzymatic production process of L-lysine from DL-aminocaprolactam did not result in practical application, this technology is interesting because of its use of petrochemical products for fermentation raw materials. The outline of the process is shown in Figure 3.

MICROORGANISMS

In amino acid fermentation, screening/breeding of microorganisms to be used for the process is the most important step. These microorganisms are classified into (*1*) wild-type strains isolated from nature to meet a specific purpose, (*2*) mutants that provide the desired properties that are developed by spontaneous or artificial mutation of

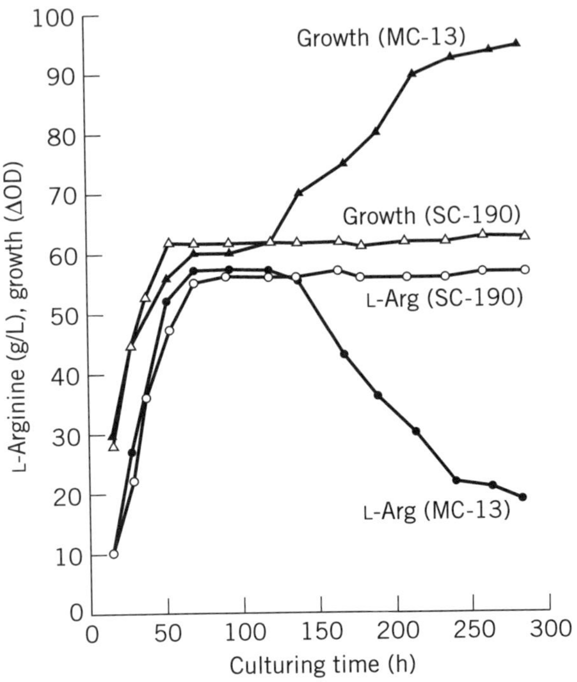

Figure 2. Comparison of L-arginine continuous culture profile of MC-13 and SC-190. *Source*: Ref. 5.

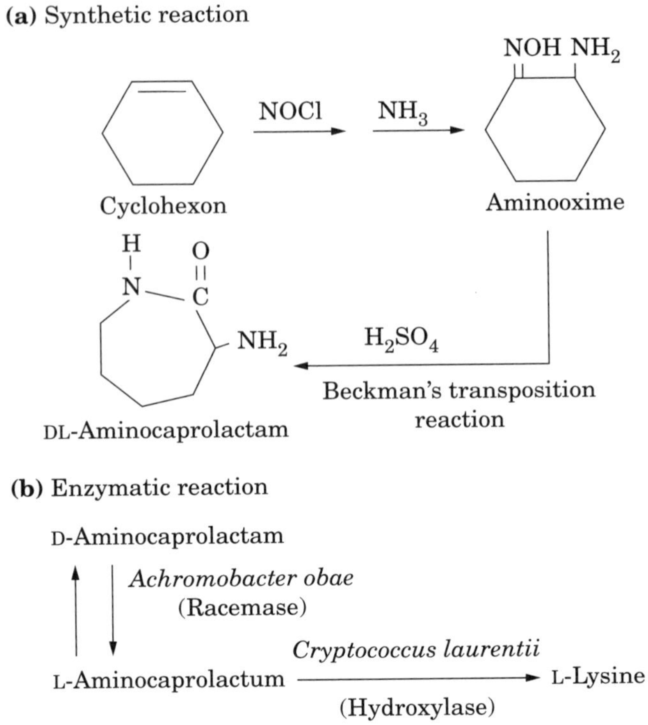

Figure 3. L-Lysine production by a combination of synthetic and enzymatic reactions. *Source*: Ref. 8.

wild-type strains, and (*3*) recombinant strains bred or constructed through genetic recombination technologies.

Wild-Type Strains

Microorganisms suitable for the purpose are selected from nature. They are cultured under adjusted fermentation conditions so as to produce excess amounts of the desired products. Typical examples are *C. glutamicum* and *Brevibacterium flavum* used for L-glutamic acid fermentation.

Mutant Strains

Auxotrophic strains and amino acid analog–resistant strains fall into the mutant strain category. Most amino acids can be produced by these types of microorganisms. Currently, most fermentative production of amino acids is accomplished in this fashion. Tables 3 and 4 and Figure 4 show L-lysine producers and L-tryptophan producers as typical examples of strain improvement.

Table 3. L-Lysine-Producing Microorganisms

Microorganisms[a]	L-Lysine HCl productivities Conc. (g/L)	Yield (%)	Ref.
C. glutamicum (Hse^-)	13	13	2
B. flavum (Thr^-, Met^-)	34	34	11
B. flavum (AEC^r)	32	32	12
C. glutemicum (Hse^-, Leu^-, $Pant^-$, ACE^r)	42	42	13

Source: Refs. 2, 11–13.
[a]AEC, *S*-(2-aminoethyl)-L-cysteine; Pant, pantothenate.

Table 4. Isolation and Productivities of Tryptophan-Producing *C. glutamicum*

Strains	Phenotype[a]	Production of L-Trp (g/L)
KY 9456	Phe^-, Tyr^-	0.15
MT-11	Phe^-, Tyr^-, $5MT^r$, $TrpHx^r$, $6FT^r$, $4MT^r$	4.9
PFP-2-32	Phe^-, Tyr^-, $5MT^r$, $TrpHx^r$, $6FT^r$, $4MT^r$, PFP^r	5.7
PAP-136-50	Phe^-, Tyr^-, $5MT^r$, $TrpHx^r$, $6FT^r$, $4MT^r$, PFP^r, PAP^r	7.1
TX-49	Phe^-, Tyr^-, $5MT^r$, $TrpHx^r$, $6FT^r$, $4MT^r$, PFP^r, PAP^r, $TyrHx^r$	10.0
Px-115-L-67	Phe^-, Tyr^-, $5MT^r$, $TrpHx^r$, $6FT^r$, $4MT^r$, PFP^r, PAP^r, $TyrHx^r$, $PheHx^r$	12.0

Source: Ref. 14.
[a]5Mt, 5-methyltryptophan; TrpHx, tryptophanhydroxamate; 4MT; 4-methyltryptophan; 6FT, 6-fluorotryptophan; PFP, *p*-fluorophenylalanine; PAP, *p*-aminophenylalanine; TyrHx, tyrosinehydroxamate; PheHx, phenylalaninehydroxamate.

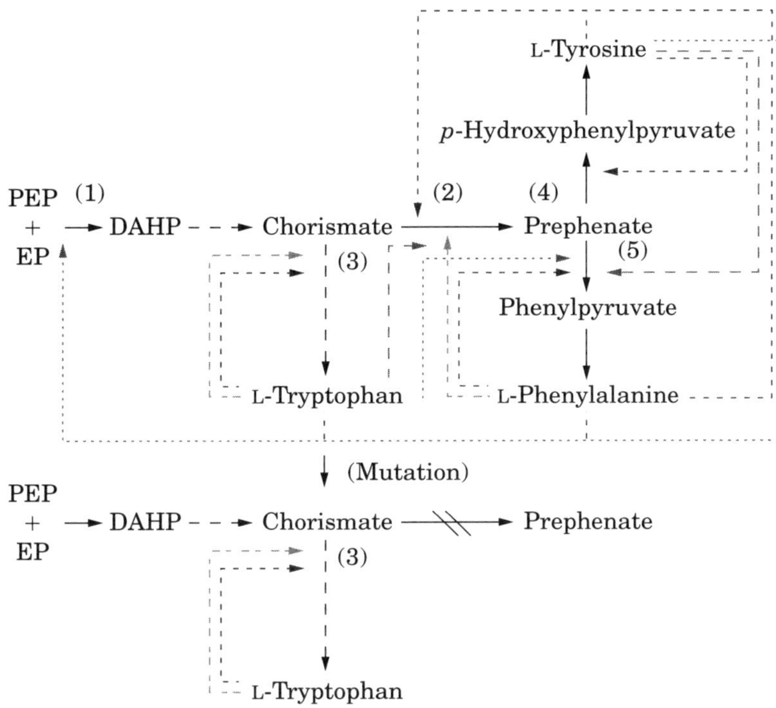

Figure 4. Metabolic regulation of aromatic amino acids in *C. glutamicum*. 1, DAHP synthetase; 2, chorismate mutase; 3, anthranilate synthetase; 4, prephenate dehydrogenase; 5, prephenate dehydratase. PEP, phosphoenolpyruvate; EP, erythrose 4-phosphate; DAHP, 3-deoxy-D-arabinoheptulosonic acid 7-phosphate; ·····➤, repression; ·····➤, inhibition; ·····➤, activation. *Source*: Ref. 14.

Recombinant Strains

There are many reports on the strains prepared by recombination of the various genes required for amino acid biosynthetic enzymes, as well as their possible applications to amino acid fermentation. Self-cloning recombinants are often reported in the literature because they are highly practical from the viewpoint of regulation. They will be put to use in the future as a primary technology for amino acid fermentation. Table 5 shows two L-threonine-producing strains. Whichever types of strains are constructed, conditions such as raw materials, temperature, process, and so on, as well as the by-products to be produced must be determined.

MEDIA

Components

Components of a medium are decided upon with regard to the microorganism, cost, purification process, waste treatment, and so forth. Except for particular cases, the cost of the carbon source, a main raw material, accounts for most of the cost of raw materials. Hence, its selection is of primary importance.

Cane molasses, beet molasses, glucose, methanol, *n*-paraffin, and other carbon source materials have been studied. In enzymatic processes and fermentation processes with addition of precursors, petrochemical products may be used as a precursor for the desired product. Carbon sources currently used for the industrial production of amino acids are mostly sugars, including cane molasses, beet molasses, and corn syrup (glucose), because of their relatively low cost and availability. When a natural substance such as molasses is used as a carbon source, composition of the medium must be properly adjusted and empirically optimized because natural substances contain additional components.

Typically, the concentration of the nitrogen source is relatively low in an initial medium to avoid overregulation of the biosynthetic pathways, and nitrogen is supplemented by way of pH control with ammonia. Because amino acid fermentation is a process of converting ammonia into amino groups, the importance of the nitrogen source is second only to that of the carbon source.

In addition, phosphate and other minerals are added in the required amounts. When an auxotrophic strain is used for fermentation, the required substance is added at an optimum concentration. Furthermore, an antifoam agent must be added to prevent foaming of the broth at the sterilization and fermentation stages.

Table 5. Accumulation of Amino Acids by *E. coli* Strain No. 29-4 and β-IM4

	Amino acids produced (g/L)						
Strain	Thr	Lys	Pro	Glu[a]	Ile	Asp	Phe
β-IM4[b]	3.82	0.31	0.25	0.20	0.14	0.09	0.06
No. 29-4[c]	13.40	0.03	0.00	0.13	0.12	0.00	0.02

Source: Ref. 15.
[a]Or homoserine.
[b]Threonine-overproducing mutant.
[c]Recombinant (self) or β-IM4 with pBR322-thr.

Sterilization

When the fermenter has a capacity of approximately 10,000 L or less, the preparation of raw materials and steam sterilization are often both carried out within the fermenter. On the other hand, when a large fermenter (40,000-L capacity or more) is used, as in the case of amino acid fermentation, each piece of equipment is separately sterilized with steam and cooled with sterile air, and raw materials other than sugar are prepared in a make-up tank and subjected to steam sterilization and cooling. Sugar is sterilized separately and sent to the fermenter with other nutrients. The flow diagram is shown in Figure 5.

Similar procedures apply to preparation of the feed tank and the feeding solution in fed-batch fermentation. Feeding is carried out from the feed tank through a sterilized line while controlling the flow.

FACTORY FERMENTATION OPERATION

Figures 1 and 6, respectively, show a process flow diagram of amino acid fermentation. The process comprises the steps of laboratory seed culture, factory seed culture (Fig. 6), and fermentation (Fig. 1), and the size of culturing is scaled up in this order.

Laboratory Seed Culture and Factory Seed Culture

The most important feature in the steps of laboratory and factory seed culture is to ensure active microbes. This promotes initial growth of microbes at the fermentation step and stabilizes fermentation. Although the cost required for these steps accounts for a relatively large portion of that of the whole process, the steps are of much importance to properly operate the process; it is thus necessary to carry out the steps only under optimum conditions. Laboratory seed culture requires an exceedingly high level of sterility because all subsequent steps for all of the lots may be lost as a result of contamination at this step. Likewise, transfer from laboratory seed culture to factory seed culture must be performed under highly sterile conditions. Typically, all transfer steps after laboratory seed culture are closed, and the whole fermentation train of equipment must be sterilized carefully before the initiation of transfer. Because amino acid fermentation is generally carried out under aerobic conditions at a neutral pH, the process will easily support the growth of a contaminant.

Fermentation Equipment

Figure 7 illustrates a typical fermenter. Amino acid fermentation is performed aerobically, so an efficient oxygen supply is required. The effect of oxygen supply on amino acid fermentation is shown in Table 6, which indicates that the sufficiency rate of oxygen required (r_{ab}/KrM) by microbes is controlled by the oxygen transfer efficiency of the fermenter. Oxygen transfer efficiency correlates with aeration and agitation conditions. It strongly affects the results of fermentation and at the same time constitutes the main factor of the cost of utilities.

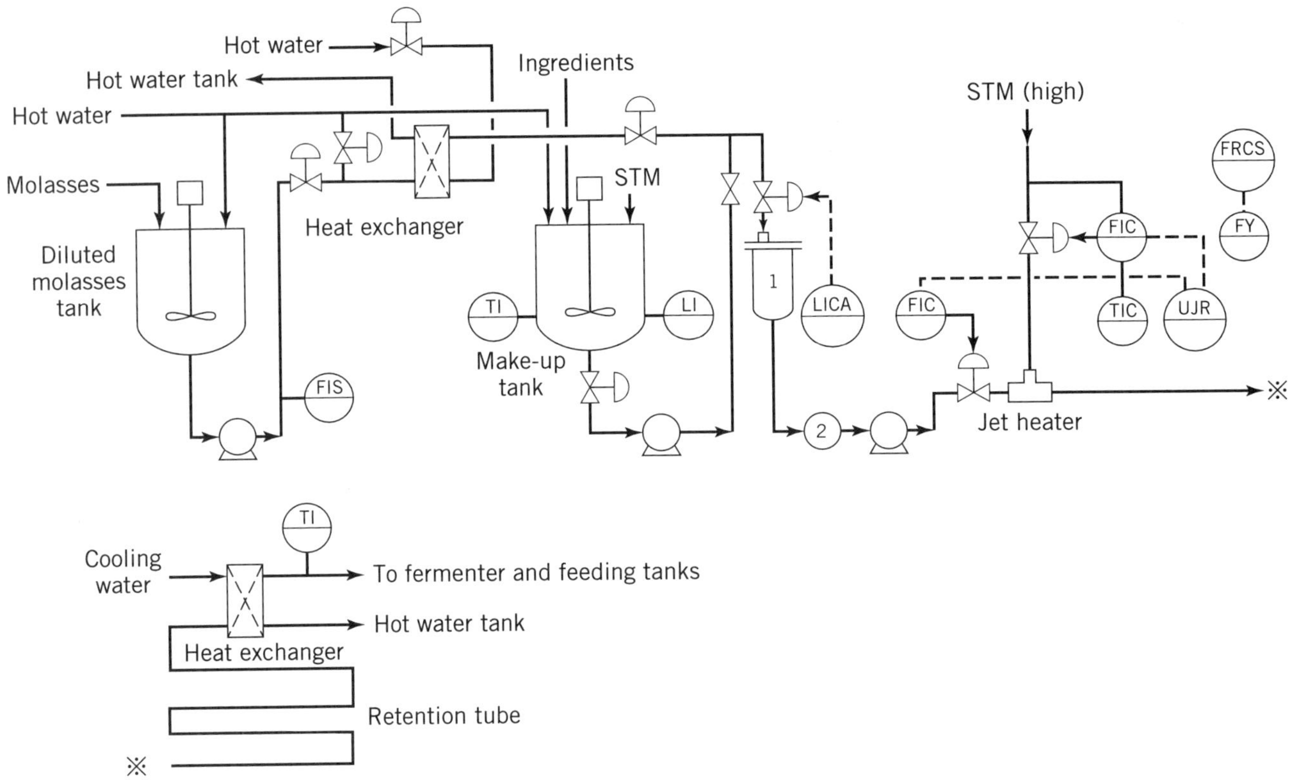

Figure 5. Piping and instrumentation drawing of the sterilizing section of a continuous sterilizer. 1, Cushion tank, 2, strainer; FIS, flow indication sum; TI, temperature indication; STM, steam LI, lowest indication; LICA, lowest indication control alarm; FIC, flow indication control; TIC, temperature indication control; FRCS, flow record control sum; FY, flow indication converter; UJR, point recorder. *Source*: Ref. 16.

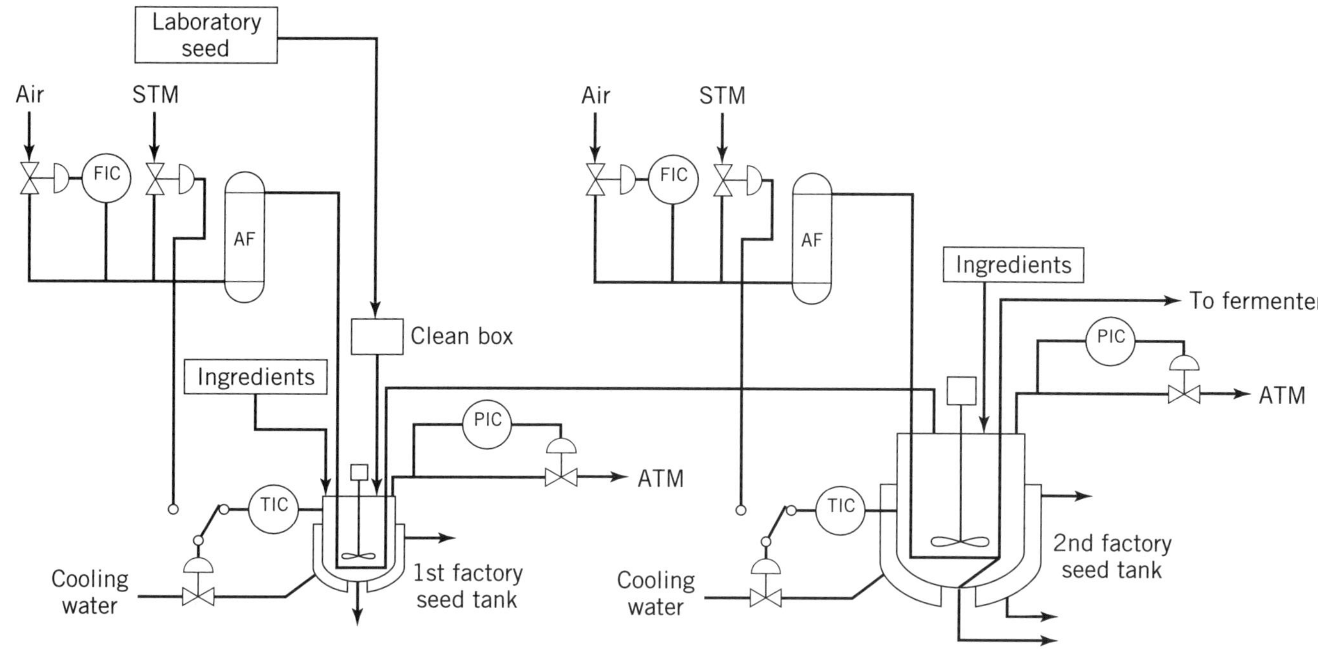

Figure 6. Factory seed culture equipment. AF, air filter; FIC, flow indication control; STM, steam; TIC, temperature indication control; PIC, pressure indication control; ATM, atmosphere.

Process Measurement and Control

In a fermentation process, aeration, pH, feed rate, dissolved oxygen, temperature, and foaming level must be measured and controlled. Parameters automatically measured during fermentation generally include fermenter temperature, volume and temperature of cooling water, volume of air supplied into the fermenter, concentrations of O_2 and CO_2 in the exhaust gas, pressure within the tank, feed volume, agitation rate, agitation energy, pH, dissolved oxygen, and degree of foaming. Sampled broth is analyzed mainly for optical density, sugar concentration, amino acids, and organic acids. Fermentation temperature, pH, aeration, feed volume, and foaming level are automatically controlled. In continuous fermentation, equipment capable of controlling the volume of both discharged and the liquid surface is provided.

Downstream Processing

Amino acids are typically recovered and purified either by chromatographic methods or by concentration–direct crystallization methods. Although most amino acids may be isolated and purified using an ion exchange resin, this tends to increase the cost of waste-liquor treatment because of the large volume of diluted waste liquor. On the other hand, with the concentration–direct crystallization method, the discharged mother liquors have a high biological oxygen demand and contain a large amount of plant-growth-promoting factors and may thus be used as fertilizer. This makes the cost of waste-liquor treatment less burdensome; however, this method cannot be applied to all amino acids because the chemical composition of fermentation broth strongly affects the purification steps.

Wastewater Treatment

In recent years, treatment of fermentation waste liquor has become more and more important in the planning process. When chromatographic purification is employed, the low concentration and large volume of waste liquor is burdensome, and the cost of treatment by the active sludge method is high, exerting much influence on the cost of the final product. In this respect, the direct crystallization method is advantageous, though it may not be applicable to many products due to characteristics peculiar to those products. It is necessary to construct a process that takes waste treatment into consideration as soon as the raw materials have been selected.

ECONOMY OF THE FERMENTATION PROCESS

As already mentioned, economy of the fermentation processes must consider the cost of waste-liquor treatment. It is difficult to accurately estimate the economic contribution of waste treatment because this factor is affected by many factors other than fermentation, such as environmental regulation around the location of the plant and so forth. In this discussion, process economics is limited to the steps that begin with the selection of raw materials and end with the desired products. A primary part of the cost of fermentation processes is the cost of the carbon source, which is decided by the *(1)* unit price of the carbon source, *(2)* fermentation yield, and *(3)* purification yield. In L-glutamic acid fermentation, because the carbon source to be used directly affects these three points, selection of the carbon source is particularly important. On the other hand, in the case of fermentation of other amino acids in

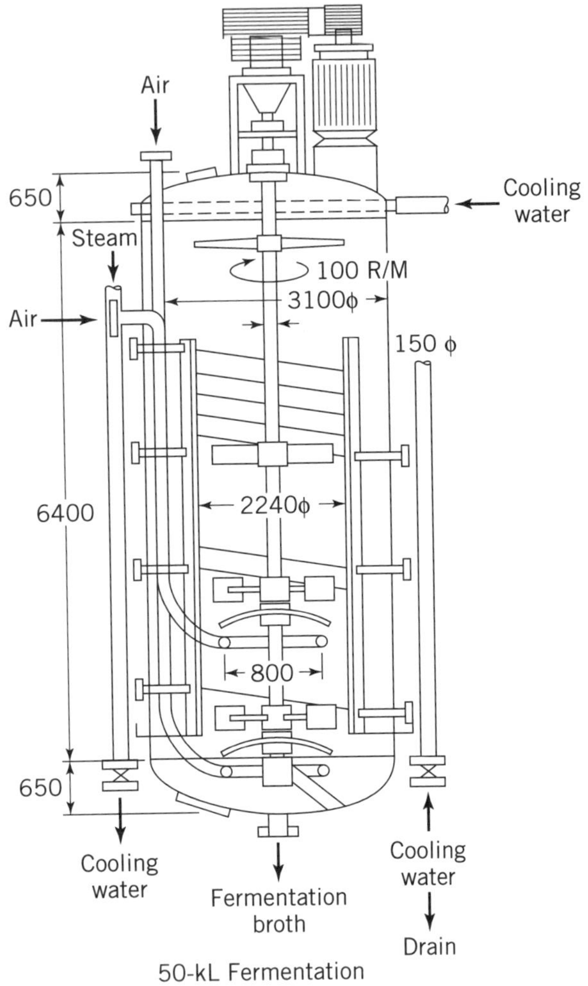

Figure 7. Example of an industrial fermenter. R/M, revolutions per minute; *φ*, diameter; all distances in millimeters. *Source*: From Y. Su and K. Yamada, *Bull. Agric. Chem.*, 1960, cited in Ref. 17.

Table 6. Rate of Oxygen Sufficiency for L-Glutamic Acid–Family Amino Acids

Amino acid	Rate of oxygen sufficiency[a]	Rate of productivity
Glutamic acid	0.2	0.1
	0.43	0.41
	1	1
Arginine	0.22	0.2
	0.4	0.44
	0.6	0.64
	1	1
Proline	0.22	0.2
	1	1
Glutamine	0.55	0.39
	0.25	0.1
	0.4	0.2
	0.9	0.8
	1	1

Source: Ref. 18.
[a] r_{ab}/KrM, where r_{ab} is the respiratory rate of the microbe (mol of O_2 mL^{-1} min^{-1}), and KrM is the maximum respiratory rate of the microbe (mol of O_2 mL^{-1} min^{-1}).

which auxotrophic strains are used, the cost of the required nutrient also is a major factor. Thus, it is necessary to use a strain requiring a low concentration of such a nutrient. Production of pharmaceutical-grade neutral amino acids requires difficult separation of analogous amino acids by-produced to the level at which they contribute to increased cost. To overcome this problem, it is necessary to minimize by-produced amino acids to a level at which purification of the desired product is not affected, and to this end, breeding of suitable strains and improvement of fermentation processes that result in few by-products are important. The cost of utilities supporting fermentation is generally high because although the fermentation is carried out mostly at normal temperature and atmospheric pressure, the following factors create additional demands:

1. Sterility must be secured.
2. Aerobic fermentation is involved.
3. It is necessary to more intensely concentrate the fermentation broth because of low product concentration compared to other processes.

AMINO ACID PRODUCTION

Monosodium Glutamate

Corynebacterium glutamicum, which produces L-glutamate, requires biotin for growth but excessively forms L-glutamic acid at a biotin concentration suboptimal for growth. Under the condition of excess biotin, L-glutamic acid can be produced by *(1)* adding penicillin (19) or a surface-active agent (20), or *(2)* culturing an oleic acid–requiring strain at growth-limiting oleic acid concentrations (20,21). Cane molasses, beet molasses, and corn syrup are mainly used as the carbon source because of their low cost. By way of example, L-glutamic acid fermentation by the fed-batch method using cane molasses is described here. Although cane molasses contains most of the

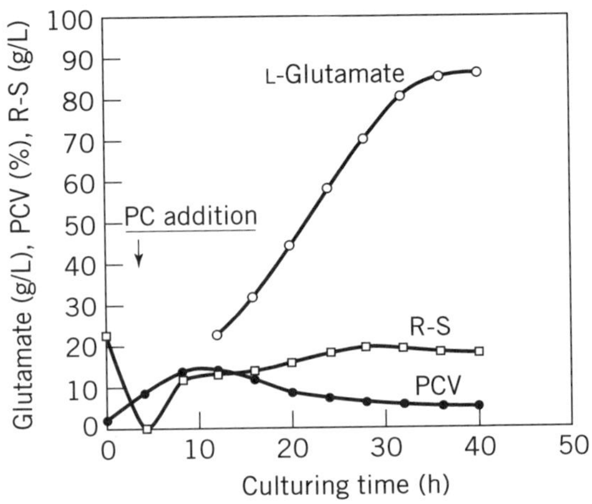

Figure 8. Typical fermentation profile (L-glutamate): *C. glutamicum*, 120-kL fermenter, fed-batch culture, cane molasses medium, penicillin method. R-S, residual sugar; PCV, packed cell volume.

carbon, nitrogen, phosphate, and minerals for microbial growth, its use as a carbon source is basic, and the growth medium is optimized by supplementing with other nutrient sources. Such medium contains an excess concentration of biotin. Potassium salt of penicillin G (8 units/L) is added to the culture broth late in the logarithmic growth phase. Upon addition of penicillin, the microorganism produces L-glutamic acid—mainly due to a change of the permeating system and/or the enzyme activities in TCA cycle (22)—and L-glutamic acid is excreted into the broth. Fermentation is continued while feeding molasses and controlling the pH to neutrality with liquid ammonia. Molasses is fed until L-glutamic acid is no longer produced in the broth. The final L-glutamic acid concentration is around 95 g/L. Figure 8 shows a typical fermentation profile in a commercial plant. The method of penicillin addition provides remarkably high productivity of this amino acid compared with fermentation under suboptimal biotin conditions.

For purification of L-glutamic acid, two methods are available: the direct crystallization method and the chromatographic ion exchange resin method. Because it is well known that the resin method can be used to purify all amino acids produced by fermentation, only the direct crystallization method is discussed here.

An outline of the recovery process (monosodium L-glutamate from L-glutamic acid fermentation broth) in a commercial plant is shown in Figure 9. The purification process may comprise an L-glutamic acid (GA) extraction process and an MSG process to obtain MSG from L-glutamic acid. In Figure 9 the GA extraction process discharges impurities out of the system as a GA mother liquor (containing microbial cells) to recover high-purity GA from the fermentation broth and an MSG mother liquor produced from the MSG process. High purity can be secured by efficiently washing crystals with water at a pH range near the isoelectric point. To this end, the crystal type (there are α- and β-types) must be in the α-form, which can be easily separated by centrifugation.

In the MSG process it is essential to maintain the purity of the product while securing a high yield. These may be achieved by enhancing the purity of L-glutamic acid in the GA process and efficiently using the carbon source to achieve minimum levels of residues. Because the MSG mother liquor contains a high concentration of dissolved MSG, it is sent back into the GA process to augment the total yield of L-glutamic acid.

The GA mother liquor (containing spent cells) and a decolorized active carbon cake are the main fractions dis-

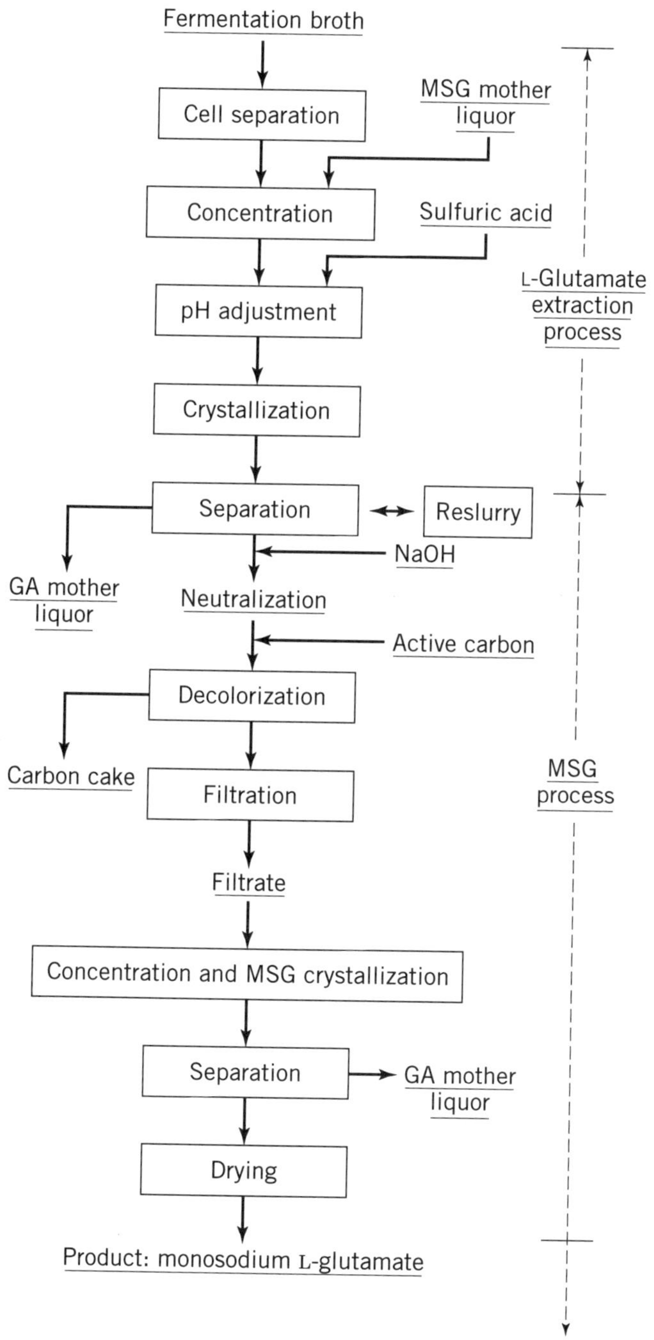

Figure 9. Recovery process flow of monosodium L-glutamate.

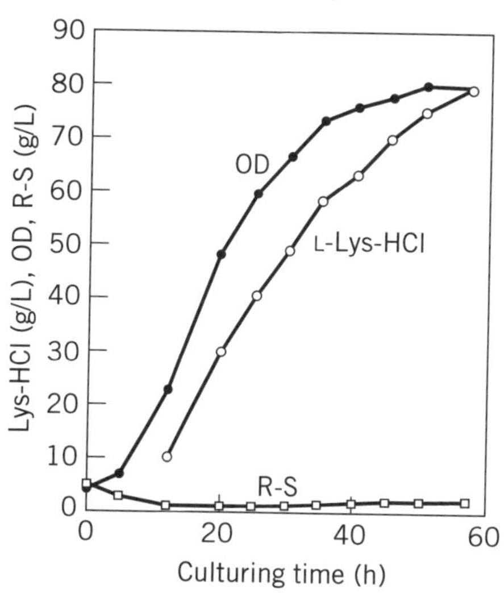

Figure 10. Typical fermentation profile (L-lysine): *C. glutamicum* (Homoser$^-$, Leu^L, AEC^r), 120-kL fermenter, fed-batch culture, molasses medium. OD, indicated cell concentration; R-S, residual sugar.

charged out of the system. Therefore, it is clear that the properties of the fermentation broth strongly affect the total yield of MSG. How the mother liquor of the GA process is used is also important. Hence, with the direct crystallization method of MSG production, there is unification of the whole process from raw materials to final product.

The yield of MSG from the fermentation broth after purification is 65–70 wt %. This is an excellent and effective production system with an attractive process economy if the mother liquor by-product can be used as a fertilizer. As already mentioned, yield is strongly affected by the ratio of L-glutamic acid concentration to total solids in the fermentation broth. This ratio varies, depending upon the purity of raw material sugar and the fermentation efficiency. Whereas the latter depends on the strain of bacteria used, purity of the sugar depends on the quality of the raw material.

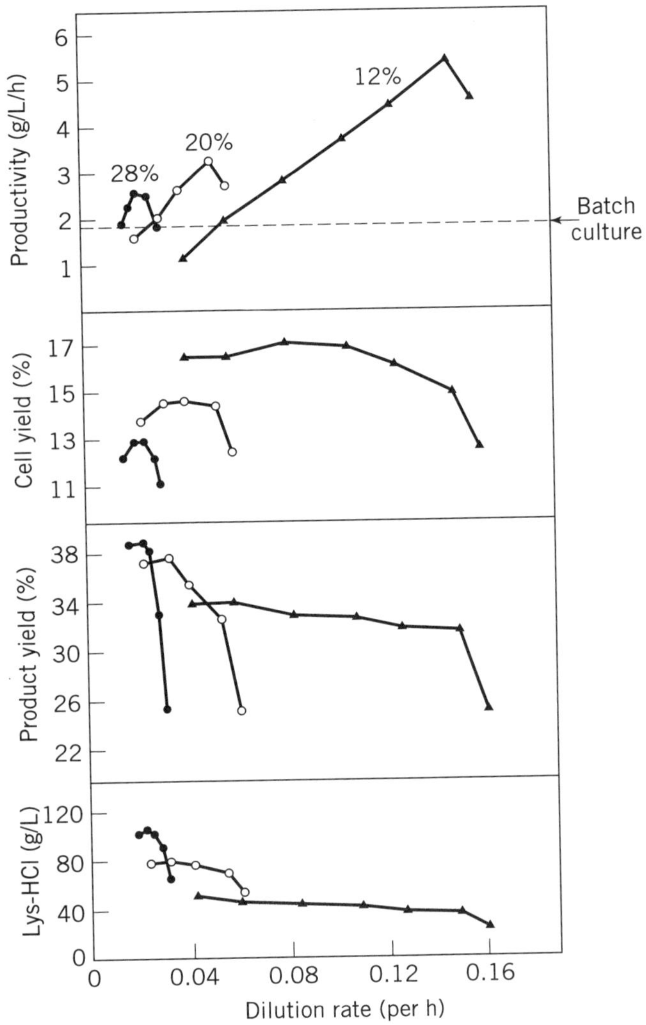

Figure 11. Effect of dilution rate on steady-state parameters for *C. glutamicum* B-6; *S*-(2-aminoethyl)-L-cysteine^r; rifampicin^r, streptomycin^r, 6-azauracil^r. Culture conditions: 5-L jar fermenter, molasses medium, pH 7.0, 32 °C, 600 rpm, air at 3 L/min, working volume of 2.5 L. Stable L-lysine production for over 300 h. Feeding sugar concentration: ▲, 12%; ○, 20%; ●, 28%. *Source*: Ref. 23.

L-Lysine HCl

L-Lysine HCl is demanded in large quantities for animal feed and must be available at low cost. Reduction of cost through the improvement of productivity is very important, as is the quality of the product.

Mutants of *C. glutamicum* (regulatory mutants such as homoserine and/or leucine leaky mutants, which are revertants from auxotrophic mutants and/or lysine analog–resistant mutants) are used. The main raw material for L-lysine fermentation is either molasses or corn syrup. Addition of a biotin source may be required, depending on the raw materials. Molasses contains a sufficient amount of biotin. Because the strains used for L-lysine fermentation are released from metabolic regulation to a remarkable extent, propagation and product formation proceed simultaneously. This is one of the characteristic features of L-lysine fermentation.

In L-lysine fermentation carried out by the fed-batch method, more nitrogen must be supplied compared to the amount required for L-glutamic acid production because L-lysine is a basic amino acid. It is necessary to add ammonium sulfate to the feed solution to supplement the NH_3 supplied for pH control and to neutralize the L-lysine formed. Furthermore, the feeding rate of the solution is controlled to maintain the sugar in the broth at a low concentration because a rapid increase causes catabolite repression. Fermentation is usually complete in about 60 h at a concentration of 95 g/L (as L-lysine HCl). Figure 10 shows a typical profile of the L-lysine fermentation process in a commercial plant. Continuous fermentation of L-lysine is interesting from the viewpoints of productivity and the price required by the market. There has been a report on laboratory-scale steady-state production of L-lysine HCl at a concentration of 80–100 g/L with a productivity of more than 3 g/L per h (Fig. 11). To be successful, however, the following conditions must be satisfied:

1. The equipment for industrial fermentation must ensure a pure culture of *C. glutamicum*.
2. The strain's production of L-lysine must not be decreased in the presence of a high concentration of L-lysine.
3. The strain must be genetically stable.

If these conditions are met, high productivity is secured by determining the optimal dilution rate (feeding volume/working volume) for stable operation (Fig. 11). Continuous fermentation is superior to batch fermentation in that the productivity per fermenter is higher. However, it is difficult to operate a continuous fermenter for a long period, unlike synthetic chemical processes, due to the three requirements just listed. All factors must be taken into consideration when deciding whether to employ this process.

Figure 12 shows a flow diagram of the purification process for feed-grade L-lysine HCl according to the adsorption-desorption method using a cation exchange resin. After separation of the cells, sulfuric acid–acidified broth is passed through the resin for adsorption, followed by washing with water and elution with ammonia. (Lysine is removed from the broth and spent water is sent to the waste treatment step.) Fractions containing L-lysine are collected and concentrated by removing ammonia (ammonia recovery). The concentrate is neutralized, further concentrated in a vacuum pan, and subjected to crystallization. The resulting mother liquor is spent on the resin-adsorption step and recovered. The yield of the final product from the broth is 80–82 wt %. This process is readily operable and provides stable product quality. However, the relatively high cost of wastewater treatment is a problem because there is a large volume of wastewater with a low-concentration chemical oxygen demand.

L-Threonine

L-Threonine is used in pharmaceuticals and animal feed. In recent years, there has been a growing demand for L-threonine as an additive to low-protein feeds.

The strains used include mutants of *C. flavum* and *Escherichia coli* (mutants requiring diaminopimeric acid, methionine, or isoleucine, or their partial revertants) and recombinants of genes involved in the biosynthetic pathway of threonine. The fermentation process is basically similar to that of L-lysine. Because L-threonine is a neutral amino acid, less nitrogen is required compared with L-lysine fermentation. The purification process differs depending on whether the product is of feed grade or for pharmaceutical use. In the case of feed-grade L-threonine, the resin adsorption method or the concentration–direct crystallization methods may be selected depending upon the composition of the fermentation broth.

L-Tryptophan

L-Tryptophan is used in pharmaceuticals and animal feed. As with L-threonine, feed-grade L-tryptophan has been in increasing demand. L-Tryptophan is produced by direct fermentation, fermentation with a precursor, or an enzymatic method using indole and L-serine as raw materials. In the direct fermentation method, strains of the genus *Corynebacterium* requiring aromatic amino acids and having a resistance to aromatic amino acid analogues or re-

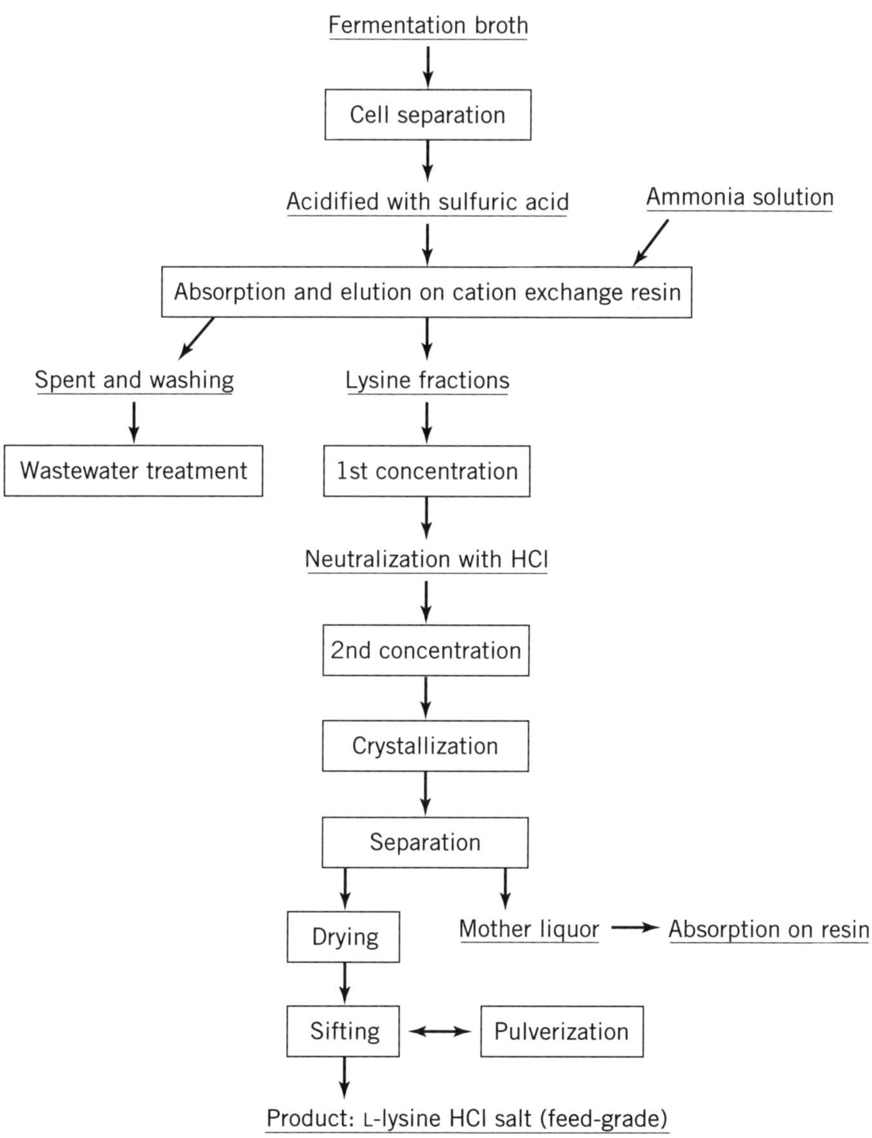

Figure 12. Purification process flow of L-lysine HCl salt (feed-grade).

combinants of aromatic amino acids biosynthetic genes in *E. coli* are used. The direct fermentation method using mutants and recombinants and the enzymatic method are currently practiced. Because L-tryptophan is not highly soluble in the aqueous media, the purification methods have been developed to take this into consideration.

CONCLUSION

Further technical improvement of amino acid fermentation is expected to greatly reduce the production cost of amino acids. Development of L-glutamic acid fermentation resulted in the growth of the seasoning market, and that of L-lysine promoted development of the market for amino acids used in animal feed. Likewise, it is expected that low-cost amino acid production will lead to the development of new uses of amino acids.

BIBLIOGRAPHY

1. S. Kinoshita, S. Udaka, and M. Shimono, *J. Gen. Appl. Microbiol.* **3**, 193–205 (1957).
2. K. Nakayama, S. Kitada, and S. Kinoshita, *J. Gen. Appl. Microbiol.*, **7**, 145–154 (1961).
3. Association for Essential Amino Acids (Japan), Tokyo, Japan.
4. T. Yamane, in T. Yamane ed., Bio-Reaction Engineering, Sangyo-tosho, Tokyo, 1979, pp 167–180.
5. T. Nakano, T. Hirao, T. Azuma, and T. Nakanishi, *Kagakukogakkai* (Yamaguchi) **42** (1990).
6. I. Chibata, T. Kakimoto, and J. Kato, *Appl. Microbiol.* **13**, 638–645 (1965).
7. M. Kisumi, Y. Ashikaga, and I. Chibata, *Bull. Agric. Chem. Soc. Jpn.* **24**, 296–305 (1966).
8. T. Hukumura, *Sekiyu-to-Biseibutsu* **15**, 10–17 (1976).
9. P. Zaffaroni, V. Vitobello, F. Cercere, E. Glacomozzi, and F. Morisi, *Agric. Biol. Chem.* **38**, 1335–1342 (1974).
10. H. Yoshida, H. Kumagai, and H. Yamad, *Agric. Biol. Chem.* **38**, 463–464 (1974).
11. K. Sano and I. Shilo, *J. Gen. Appl. Microbiol.* **13**, 349–358 (1967).
12. K. Sano and I. Shilo, *J. Gen. Appl. Microbiol.* **16**, 373–391 (1970).
13. Jpn. Pat. 1445888 (June 30, 1988), K. Nakayama and K. Araki to Kyowa Hakko Kogyo Co., Ltd.
14. H. Hagino, *Nogei-kagakukai-shi* **50**, R79–R81 (1976).
15. K. Miwa, T. Tsuchiya, O. Kurahashi, S. Nakamori, K. Sano, and H. Momose, *Agric. Biol. Chem.* **47**, 2329–2334 (1983).
16. H.C. Vogel, in H.C. Vogel ed., Fermentation and Biochemical Engineering Handbook, Noyes Publication, New Jersey, 1983.
17. S. Kinoshita, in S. Kinoshita ed., Fermentation Technology, Dainippon-tosho, Tokyo, 1975, pp 84–85.
18. K. Akashi, H. Shibai, and Hirose, *J. Ferment. Technol.* **57**, 321–327 (1979).
19. Jpn. Pat. 509952 (January 31, 1968), N.L. Somerson and T. Phillips (to Merck & Co., Inc.).
20. S. Okumura, R. Tsugawa, T. Tsunoda, and S. Kitai, *Nogei-kagakukai-shi* **36**, 197–211 (1962).
21. K. Takinami, H. Okada, and T. Tsunoda, *Agric. Biol. Chem.* **28**, 114–119 (1964).
22. Y. Kawahara, K. Takahashi, E. Shimizu, T. Nakamatsu, and S. Nakamori, *Biosci. Biotech. Biochem.* **61**, 1109–1112 (1997).
23. T. Hirao, T. Nakano, T. Azuma, and T. Nakanishi, *Int. Conf. Ferment. Technol.* Wellington, New Zealand, February 12–15, 1990.

See also AMINOHYDROLASES, FOR PRODUCTION OF D-AMINO ACIDS; FERMENTATION MONITORING, DESIGN AND OPTIMIZATION; GLUTAMIC ACID PRODUCING MICROORGANISMS; METABOLITES, PRIMARY AND SECONDARY; PHENYLALANINE; PRODUCTION OF L-AMINO ACIDS BY AMINOACYLASE.

AMINOHYDROLASES, FOR PRODUCTION OF D-AMINO ACIDS

YASUHIRO IKENAKA
SATOMI TAKAHASHI
Kaneka Corporation
Hyogo, Japan

KEY WORDS

D-Amino acid
N-Carbamoyl-D-amino acid amidohydrolase
Thermotolerant enzyme

OUTLINE

INTRODUCTION

N-Carbamoyl-D-amino acid amidohydrolase (hereinafter abbreviated as DCase) is the enzyme that catalyzes the following reaction:

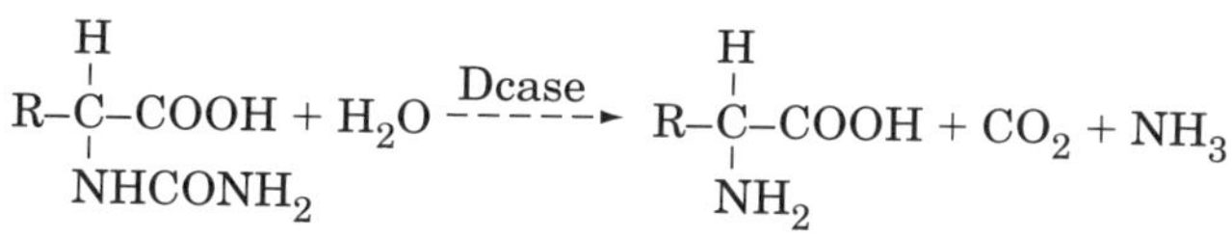

The DCase converts *N*-carbamoyl-D-amino acids to D-amino acids with strict D-form specificity and is expected to be applicable to the industrial D-amino acid production process instead of chemical diazonation, in which *N*-carbamoyl-D-amino acids are converted by DCase after hydrolysis of the corresponding 5-substituted hydantoins by the hydrolyzing enzyme, D-hydantoinase. These reactions can be performed separately as two serial reactions or as one batch reaction with microorganisms possessing both enzymes.

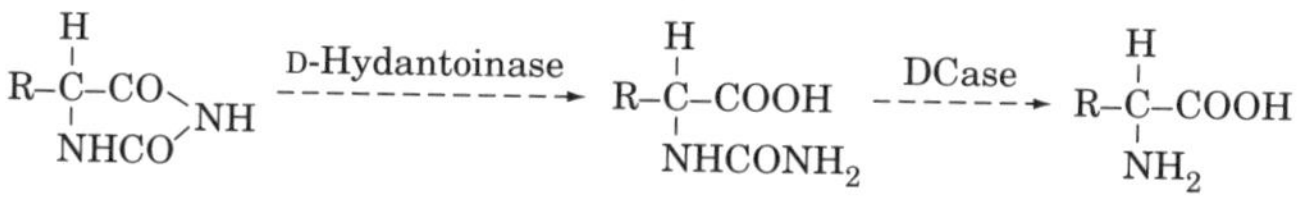

D-Amino acids are useful compounds for the preparation of physiologically active peptides and β-lactam antibiotics such as semisynthetic penicillins and cephalosporins. DCase activity was first found in rat liver (1) and microbial cells (2–5) about 40 years ago. These enzymes were examined as to their properties and substrate specificities, and their reaction mechanisms and biological functions were deduced. The enzymes from rat liver and *Clostridium uracilicum* (2) were found to be involved in the degradation of pyrimidine, and the enzyme from *Pseudo-*

Table 1. Substrate Specificities of Screened Enzymes

	Mesophile (30 °C)		Thermophile (45 °C)	
	KNK 712	KNK 1415	KNK 003A	KNK 505
c-D-Ala	+ + + +	+ + +	+ + + +	+ +
c-D-Val	+ + + +	+ + +	+ + +	+ +
c-D-Leu	+ + + +	+ + + +	+ + + +	+ + +
c-D-Phe	+ + + +	+ + +	+ + +	+ +
c-D-PEG	+ + +	+ + +	+ +	+
c-D-PG	+ + +	+ + +	+ −	+ −
c-D-HPG	+ + +	+ + +	+ +	+ −

Note: With 1 mL culture broth of each isolated strain, the resting cell reaction was performed under standard conditions at 30 °C (mesophile, 3-h reaction) or 45 °C (thermophile, 24-h reaction), followed by analysis by TLC. Seven *N*-carbamoyl-D-amino acids (c-D-amino acids) were examined, and the amount of D-amino acid produced in the reaction is represented approximately as follows: + −, 0.1 g/L; +, 0.2 g/L; + +, 0.5 g/L; + + +, 1–2 g/L; + + + +, 4 g/L. PEG and PG represent phenetylglycine and phenylglycine, respectively.

Table 2. Properties of DCases from *Agrobacterium* sp. KNK712 and *Pseudomonas* sp. KNK003A

	KNK712	KNK003A
Molecular weight (monomer)		
SDS-PAGE	37,000	38,000
DNA sequence	34,300	35,400
Thermal stability (60 °C, 20 min) (residual activity)	5%	95%
Optimal temperature	67 °C	
Optimal pH	6.8–7.3	7.0–7.5
Number of amino acids (deduced)		
Total	304	312
Hydrophobic	162	154
Neutral	50	51
Hydrophilic	92	107
Cystein	5	5
G + C Content	60.3%	62.9%
Homology		
DNA	62%	
Amino acid	60%	

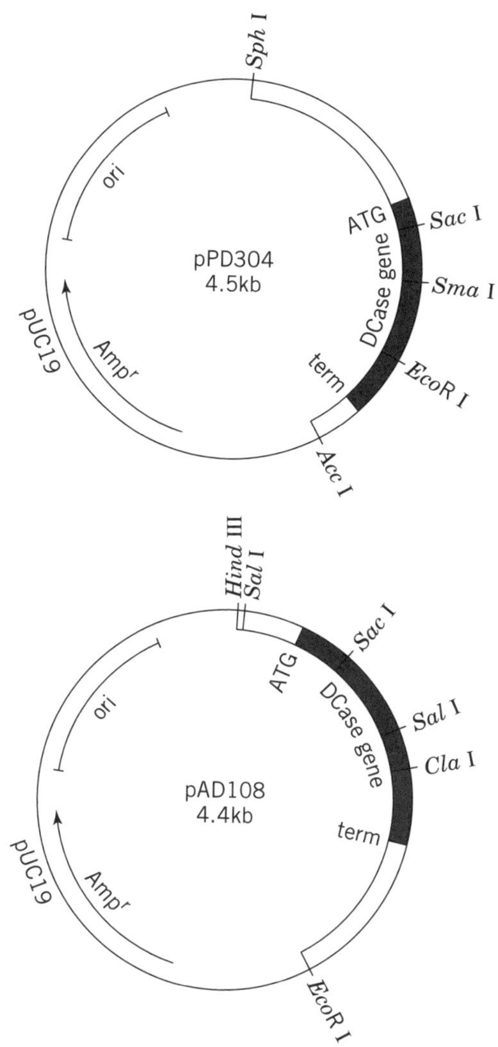

Figure 1. Restriction maps of pAD108 and pPD304. The *thin line* indicates the vector DNA and the *boxed line* the inserted DNA fragment. The open reading frame is represented from ATG to term (the *black-boxed* line).

(a)

```
GTCGACGGCGGGCTCGCGCGAGAGCTTGTCAAGCAGCGCAAATTCCGGTTCCGCTCCGGTTGACAGATCAAAAATTTTACGCCTGTTATTGTCGTGCTGC  100

ATGTAATATTTCGTACTTTATGTAGAATTTGCATTGCGCCGCGAGTCACAAAGCCGGTTTTCGGCGATGTGTTTCACAACGTTTTCCCGGCCGCTGGGCC  200

GGACATCACCTAGGAAGGAGCAGAGGTTCATGACACGTCAGATGATACTTGCAGTGGGACAACAAGGTCCGATCGCGCGCGCGGAGACACGCGAACAGGT  300
                             M  T  R  Q  M  I  L  A  V  G  Q  Q  G  P  I  A  R  A  E  T  R  E  Q  V

CGTCGTTCGTCTTCTCGACATGCTGACGAAAGCCGCGAGCCGGGGCGCGAATTTCATTGTCTTCCCCGAACTCGCGCTTACGACCTTCTTCCCGCGCTGG  400
 V  V  R  L  L  D  M  L  T  K  A  A  S  R  G  A  N  F  I  V  F  P  E  L  A  L  T  T  F  F  P  R  W

CATTTCACCGACGAGGCCGAGCTCGATAGCTTCTATGAGACCGAAATGCCCGGCCCGGTGGTCCGTCCACTCTTTGAGAAGGCCGCGGAACTCGGGATCG  500
H  F  T  D  E  A  E  L  D  S  F  Y  E  T  E  M  P  G  P  V  V  R  P  L  F  E  K  A  A  E  L  G  I  G

GCTTCAATCTGGGCTACGCTGAACTCGTCGTCGAAGGCGGCGTCAAGCGTCGCTTCAACACGTCCATTTTGGTGGATAAGTCAGGCAAGATCGTCGGCAA  600
  F  N  L  G  Y  A  E  L  V  V  E  G  G  V  K  R  R  F  N  T  S  I  L  V  D  K  S  G  K  I  V  G  K

GTATCGTAAGATCCATTTGCCGGGTCACAAGGAGTACGAGGCCTACCGGCCGTTCCAGCATCTTGAAAAGCGTTATTTCGAGCCGGGCGATCTCGGCTTC  700
 Y  R  K  I  H  L  P  G  H  K  E  Y  E  A  Y  R  P  F  Q  H  L  E  K  R  Y  F  E  P  G  D  L  G  F

CCGGTCTATGACGTCGACGCCGCGAAAATGGGGATGTTCATCTGCAACGATCGCCGCTGGCCTGAAGCCTGGCGGGTGATGGGCCTCAGGGGCGCCGAGA  800
P  V  Y  D  V  D  A  A  K  M  G  M  F  I  C  N  D  R  R  W  P  E  A  W  R  V  M  G  L  R  G  A  E  I

TCATCTDCGGCGGCTACAACACGCCGACCCACAATCCCCCTGTTCCCCAGCACGACCACCTGACGTCCTTCCACCATCTCCTATCGATGCAGGCCGGGTC  900
  I  C  G  G  Y  N  T  P  T  H  N  P  P  V  P  Q  H  D  H  L  T  S  F  H  H  L  L  S  M  Q  A  G  S

TTATCAGAACGGGGCCTGGTCCGCGGCCGCGGGCAAGGTGGGCATGGAGGAGAACTGCATGCTGCTCGGCCACTCCTGCATCGTGGCGCCGACCGGGGAA  1000
 Y  Q  N  G  A  W  S  A  A  A  G  K  V  G  M  E  E  N  C  M  L  L  G  H  S  C  I  V  A  P  T  G  E

ATCGTCGCTCTCACTACGACGCTGGAAGACGAGGTGATCACCGCCGCCGTCGATCTCGATCGCTGCCGGGAACTGCGTGAACACATCTTCAACTTCAAGC  1100
I  V  A  L  T  T  T  L  E  D  E  V  I  T  A  A  V  D  L  D  R  C  R  E  L  R  E  H  I  F  N  F  K  Q

AGCATCGTCAGCCCCAGCACTATGGTCTGATCGCGGAACTCTGAGGTTGCCGAAAAGCATGTGTGTCGTTGTTCTCGGCGCCTGGGTCACATCCAGGCGC  1200
  H  R  Q  P  Q  H  Y  G  L  I  A  E  L

GCCAGGGTGACGCTGGTGGAATAGTACCACGACCGCTTCAGGGCGATCCGCAAGGAGATGCGGGTCGCCGGAGCGGCAAAGCCCGACATTCGTTTCGCAC  1300

CGACGGCCGTCGTGAACTCGACAGTCCGCGAGAAGGGCGTATTGCGCGGCCTGGACCTGTACGTGGAACTGTAGCCCATATATAGATTTCCAAAGAGTTT  1400

CGGCGAGGCGCGGCGCGCCTAGCCCCATGTGAGCGAGAACCGTGCCCAGATCAAAGAATGAGACCGACGCGCCGGCCGCGGCAAAGGATGATCCTCAGGG  1500

TCGGATCTATCGCTCCGCCCTGAAGCAGGAGGGCGCACGCTGGCTGCTGACGGCGGAGGAAGGGTTGCTGGCAAAGCCCAAGCCGCCCGGCCTTGTTCCG  1600

GCACTTGAGAATGCGATCGCCATCGTCGATTACATCAACGGTACACCGCCCCATATCGCGTCCCTGGCGGAGCTTTCAACGACGCTCGGGATATCCAAGA  1700

GCCACTGTCACTCCATCCTCAAGACGCTGACGCATTTCGGCTGGCTGAAATTCGACAATCGCTCAAAGAGCTACGAGCTGAATTC
```

Figure 2. Nucleotide sequences of the insert DNAs of pAD108 (**a**) and pPD304 (**b**) and the deduced amino acid sequences of these DCases. The numbers at the *right* are nucleotide positions. Sequences similar to the −10 and −35 consensus sequences for the *E. coli* promoter are *double-underlined*, and the putative ribosome-binding site (Shine–Dalgarno) sequence and initiation and termination codons are also *double-underlined*. The open reading frame of the DCase is *underlined*.

monas putida 77 (3), which is *N*-carbamoyl-sarcosine amidohydrolase, is involved in creatinine metabolism. Olivieri et al. found DCase activity of *Agrobacterium radiobacter* NRRL B11291 in intact cells or a cell-free extract, together with D-hydantoinase activity (4). Yokozeki and Kubota attempted the partial purification of DCase from *Pseudomonas* sp. AJ11220 (5). These two enzymes were examined as to their substrate specificities and reaction profiles. They showed relatively broad substrate specificities such as for aliphatic and aromatic compounds, with strict D-form specificity, and almost the same optimal conditions, that is, pH 7 and 55–60 °C. Ogawa et al. screened enzyme producers by means of enrichment cultures with the *N*-carbamoyl-D-amino acids citrulline and ornithine as sole nitrogen or carbon sources. They isolated alkaliphilic and thermotolerant enzymes, which were produced by strains classified as *Comamonas* sp. and *Blastobacter* sp., respectively, purified them to homogeneity, and determined their properties (6,7). These enzymes also showed strict specificity toward *N*-carbamoyl-D-amino acids and hydrolyzed *N*-carbamoyl-amino acids having a hydrophobic side chain very efficiently; on the other hand, those having a polar group or short-chain alkyl group were only weakly hydrolyzed.

(b)

```
GCATGCGCGGGGAACTGAAGAACTTGCAAGACGAACTCGGCATTACCTTCGTGCATGTAACCCATACCCAGCCTGAGGCGATCGCGCTCGCCGACATGGT  100

GGTTGTGATGGATACGGGCCGCATAGAGCAGGCAGCGAGCGCCAACGAAATCTACAACCGGCCCGCGACGCCCTATGTGGCGCGCTTCATGGGCGGCCAA  200

AACGTGTTGACGGGGAGGGTGGAGAGCATCTCGCCCACCGGCATGGTGCTGAAAAGCGAAAAGGGCGAGATCTTCAATGCGCCTCTTACGGGTGCTGCGC  300

CGAAGCTGGGCGAACCCGTATCGATATCCATGCGCCGCGACCGCATCAGCATCAGCAAGCCGCAAAACGGCAAGGGCGCGCAGCAGGCTGACGCGGTAAC  400

GGGTGTGGTCGATTCCACGGAATACCAGGGCAGCTTCGTGAAGGTCAGCATAGTGCTCGACGGTGGCGAGACCTTCGTCGCAAACATGCCCGACCATGAA  500

TTTTTCGCGGAACCGGTGGATCACGGCGTCCCGGTGGTCGCCCGCTGGAAACCGGAGCATGTGCATGTCCTGTCCAAGTCTGACCGGGGCGCCGACCACA  600

CCGAAATCTACCGCTTCCCTGCAGGCGAAAATACCGTTTCAATGGGCAAGGGGCGGCAAACGGGGTTGAGACGACCCGGTTTATCGAGGAGGACGAGATG  700
                                                                                                  M

ACACGCATCGTCAATGCAGCCGCCGCGCAGATGGGGCCCATCAGCCGGTCCGAAACGCGCAAGGATACGGTCCGGCGCCTGATCGCGCTCATGCGCGAGG  800
T  R  I  V  N  A  A  A  A  Q  M  G  P  I  S  R  S  E  T  R  K  D  T  V  R  R  L  I  A  L  M  R  E  A

CGAAGGCCCGCGGTTCCGACCTTGTCGTCTTTACCGAACTCGCGCTCACCACCTTCTTTCCCCGCTGGGTGATCGAGGACGAAGCTGAGCTCGACAGCTT  900
 K  A  R  G  S  D  L  V  V  F  T  E  L  A  L  T  T  F  F  P  R  W  V  I  E  D  E  A  E  L  D  S  F

CTACGAGAAGGAGATGCCAGGGCCCGAAACCCAGCCGCTCTTCGATGAGGCGAAGCGCTTGGAGATCGGCTTCTATCTCGGTTATGCCGAGCTGGCGGAG  1000
 Y  E  K  E  M  P  G  P  E  T  Q  P  L  F  D  E  A  K  R  L  E  I  G  F  Y  L  G  Y  A  E  L  A  E

GAGGGCGGCAGGAAGCGGCGCTTCAACACCTCTATCCTTGTGGACCGCAGCGGCCGGATCGTCGGCAAGTACCGCAAGGTGCACCTGCCCGGGCACAAAG  1100
E  G  G  R  K  R  R  F  N  T  S  I  L  V  D  R  S  G  R  I  V  G  K  Y  R  K  V  H  L  P  G  H  K  E

AGCCGCAGCCCGGCAGGAAACACCAGCATCTCGAGAAACGCTATTTCGAGCCCGGCGATCTCGGCTTCGGTGTCTGGCGCGCCTTCGACGGCGTAATGGG  1200
 P  Q  P  G  R  K  H  Q  H  L  E  K  R  Y  F  E  P  G  D  L  G  F  G  V  W  R  A  F  D  G  V  M  G

CATGTGCATTTGCAACGACCGCCGCTGGCCGGAGACCTACCGGGTCATGGGCTTGCAGGGAGTGGAGATGGTCATGCTGGGCTACAACACGCCGTATGAC  1300
 M  C  I  C  N  D  R  R  W  P  E  T  Y  R  V  M  G  L  Q  G  V  E  M  V  M  L  G  Y  N  T  P  Y  D

CATACCGGTCACGACGACATCGATTCACTCACCCAGTTTCACAATCATCTCTCCATGCAGGCGGGCGCCTACCAGAATTCGACCTGGGTGATCGGCACCG  1400
H  T  G  H  D  D  I  D  S  L  T  Q  F  H  N  H  L  S  M  Q  A  G  A  Y  Q  N  S  T  W  V  I  G  T  A

CCAAATGCGGCACCGAGGAGGGCTCCAAAATGGTGGGGCAGAGCGTGATCGTTGCGCCCTCCGGCGAGATCGTCGCTATGGCCTGCACGATCGAGGACGA  1500
 K  C  G  T  E  E  G  S  K  M  V  G  Q  S  V  I  V  A  P  S  G  E  I  V  A  M  A  C  T  I  E  D  E

GATCATCACCGCACGCTGCGATCTCGACATGGGCAAGCGCTACCGCGAGACCATCTTCGATTTCGCCCGCCATCGCGAGCCCGACGCCTATCGCCTGATC  1600
 I  I  T  A  R  C  D  L  D  M  G  K  R  Y  R  E  T  I  F  D  F  A  R  H  R  E  P  D  A  Y  R  L  I

GTCGAACGCAAAGGGGCTGTGCCGCCGCCGCAGTGATCGGAACCTGAAAACGAAATATCCCGCCGGACGGTGGGAAGGTGAAAGGAGGAGTCTCCATGAC  1700
V  E  R  K  G  A  V  P  P  P  Q

AACAGTTATCAAGGGTGGAACGATCGTCGCCGCCGATCGCAGCTATGAAGCCGATATCCTGATCGAAGGCGAAAAGATCGCCCAGATCGGCAGGGATCTG  1800

CAGGGCGACAAGATTGTCGAC
```

Figure 2. Continued.

However, these enzymes showed that the *N*-carbamoyl compounds involved in the metabolism of pyrimidine and purine are not hydrolyzed at all. On characterization of these enzymes, the reactive molecular weights of the native enzymes were found to be about 120,000 and those of the subunits to be about 40,000. Enzymochemical analysis was also performed. Recently, Louwrier and Knowles purified a DCase from *Agrobacterium* sp. which was composed of genetically engineered self-cloning cells, and characterized it (8). This enzyme was able to cleave a variety of *N*-carbamoyl substrates but was strictly D-form specific. The active enzyme was suggested to be present as a dimer with a subunit molecular weight of 38,000 Da, differing from the trimer enzymes of Ogawa et al. (6,7).

Concerning the enzyme genes, some DCase genes have been cloned and analyzed. The DCase gene from *Agrobacterium radiobacter* NRRL B11291 was cloned, analyzed, and expressed in *Escherichia coli*, and the recombinant enzyme was then characterized (9,10). By means of a site-directed mutagenesis experiment, the relationship between activity and amino acid substitutions was examined, and some mutations concerning enzyme stability were found. Neal et al. also isolated the DCase gene of *Agrobacterium* sp. and expressed the gene in *E. coli* and *Agrobacterium* sp. (11).

We also screened some strains producing the enzyme from mesophile (12) and thermotolerant strains (13) and cloned two enzyme genes. We tried to improve the native DCase to obtain a practical DCase that exhibits both high reactivity and sufficient stability for repeated use in a bioreactor system by means of amino acid substitutions using recombinant DNA technology (14,15). We succeeded in creating a practical DCase by the substitution of three amino acids (16), and applied it to an industrial production process as an immobilized enzyme (17,18).

In this section, we report our attempts to establish a new D-amino acid production process, such as screening of DCase-producing bacteria from soil, purification of the enzymes, cloning of the enzyme genes, mutagenesis to obtain thermostabilized enzymes, and immobilization of the enzymes for a bioreactor system to produce D-amino acids.

SCREENING OF *N*-CARBAMOYL-D-AMINO ACID AMIDOHYDROLASE

Screening of DCase-Producing Microorganisms

The screening and isolation methods used for soil microorganisms were as follows (12). For the isolation of microorganisms that can hydrolyze *N*-carbamoyl-D-amino acids to D-amino acids, soil samples were suspended in saline and the supernatants were inoculated into growth medium containing *N*-carbamoyl-D-amino acids as sole nitrogen sources for enrichment culture. After aerobic cultivation, the reduction of *N*-carbamoyl-D-amino acids and the production of D-amino acids were detected by silica gel thin-layer chromatography (TLC) with a solvent system of *n*-butanol:acetic acid:water (4:1:1). *N*-Carbamoyl-D-amino acids and D-amino acids were detected with *p*-dimethylaminobenzaldehyde in a 6 M HCl solution and ninhydrin, respectively. From the culture broth from which *N*-carbamoyl-D-amino acids disappeared or in which D-amino acids accumulated, microorganisms were isolated on agar plates.

The substrate specificities of these DCases were examined by means of the resting cell reaction with detection by the TLC method (Table 1). We screened two good mesophile strains producing a lot of enzymes that were classified as *Agrobacterium* sp. or *Rhizobium* sp. (12). A similar enzyme-producing strain, classified as *Agrobacterium radiobacter*, was also found by Olivieri et al. (4) and characterized. We also screened thermotolerant strains by means of enrichment culture at 45 °C and obtained some *Pseudomonas* sp. strains (Table 1) (13). The strain isolated by Yokozeki and Kubota was classified as *Pseudomonas* sp. (5) but was not a thermotolerant strain.

Purification and Characterization of the Enzymes

The DCases from *Agrobacterium* sp. KNK712 and *Pseudomonas* sp. KNK003A were purified after large-scale incubation by ammonium sulfate precipitation and chromatography such as ion exchange and gel filtration.

DCase activity was assayed by measurement of D-*p*-hydroxyphenyl glycine (D-HPG) production from *N*-carbamoyl-D-HPG (c-D-HPG) (12). The reaction was started by the addition of 100 μL of an enzyme solution diluted with 100 mM potassium phosphate buffer, pH 7.0 (hereinafter abbreviated as KP buffer), containing 5 mM dithiothreitol (DTT) to the assay mixture containing 47.6 μmol c-D-HPG and 100 μmol KP buffer in a total volume of 1 mL. After 20 min incubation at 40 °C, the reaction was stopped by the addition of 0.25 μL of 20% (w/v) trichloroacetic acid. The D-HPG produced was analyzed by means of the following high-performance liquid chromatography (HPLC) method. *N*-Carbamoyl-amino acids and amino acids were detected and quantified by HPLC at 210 nm on a reverse-phase HPLC column (e.g., Finepack SIL C18-5 column; Nipponbunko), using 36.7 mM KH_2PO_4 containing 15% methanol adjusted to pH 2.5 with phosphoric acid. One unit of the enzyme was defined as the amount of enzyme that catalyzed the formation of D-HPG at the rate of 1 μmol/min under the assay conditions already mentioned.

The reactivity of the DCase of *Agrobacterium* sp. KNK712 is about 20 times higher than that of *Pseudomonas* sp. KNK003A with c-D-HPG as the substrate, but as concerns heat stability, in contrast, the *Pseudomonas* sp. KNK003A enzyme is more stable, with an about 10 °C increase in view of its denaturation temperature.

The molecular weights of these DCases were found to be about 37,000 and 38,000 Da, respectively, on SDS-PAGE. The properties of these two enzymes are shown in Table 2.

ISOLATION OF ENZYME GENES

Cloning of Enzyme Genes

The DCase genes from *Agrobacterium* sp. KNK712 and *Pseudomonas* sp. KNK003A were isolated and their DNA sequences analyzed as follows (12). Chromosome DNA was partially digested with *Sau*3AI, and the fractionated DNA fragment of 4–9 kb was inserted into pUC18 and then transformed into *E. coli* JM109. Recombinant colonies were collected, inoculated into enrichment culture medium containing *N*-carbamoyl-D-amino acids or 5-substituted hydantoins, and then incubated at 37 °C. After repeating the enrichment cultures, the recombinant clones were isolated. Plasmid DNAs were prepared from these clones and analyzed with several restriction endonucleases. The DCase gene of *Agrobacterium* sp. KNK712 was located in a 1.8-kb *Sal*I–*Eco*RI fragment, and the gene of *Pseudomonas* sp. KNK003A was found in a 1.8-kb *Sph*I–*Acc*I fragment. These fragments were inserted into pUC19, the resultant plasmids being designated as pAD108 and pPD304, respectively. The restriction maps are shown in Figure 1. The recombinant strains containing these plasmids showed increased DCase expression compared with the native strains.

Analysis of the Enzyme Genes

The DNA sequences of the genes in the 1.8-kb fragments were analyzed (Fig. 2), there being one open reading frame in each case, consisting of 912 and 936 bases with a starting triplet, ATG, and predicted to encode polypeptides of 304 and 312 amino acids with calculated molecular weights of 34,285 and 35,438, respectively. Upstream of the open reading frame, sequences similar to the −35 and −10 consensus sequences for *E. coli* promoters and a putative ribosomal-binding site (Shine–Dalgarno sequence) were found. The G+C contents of the open reading frames were 60.3% and 62.9%, respectively (see Table 2).

The homologies of the DNA sequences and the deduced amino acid sequences between these two DCases were 62% and 60%, respectively. An 8–amino acid length of the C-terminal region was different in these two enzymes. The homologies among those of *Agrobacterium* sp. KNK712 and those of two other reported *Agrobacterium* strains (9,10) which had completely the same sequences, were 97.0% (amino acid sequences) and 93.0% (DNA sequences). The homologies of the N-terminal regions of 30 amino acids in five DCases (6,7,9,12,13) were about 50%, the homology between the enzyme of *Pseudomonas* sp. KNK003A and the *Comamonas* sp. E222c DCase being the highest (56.7%), besides the homology among *Agrobacterium* strains. Multiple alignment of the amino acid sequences of these enzymes is shown in Figure 3.

Similarity research using a database (PIR release, 45.0) showed that the aliphatic amidase of *Pseudomonas aeruginosa* (19) showed the highest similarity (30%) to DCase, but a closely related enzyme was not found. Multiple alignment with Clustal W (1.4) software of the DCase and 10 known related enzymes (PIR accession numbers [PIR Ac. No.] are given in the legend to Fig. 4) showed that 6 amino acids, that is the 119th amino acid of DCase, glycine (hereinafter abbreviated as 119th Gly), 126th Arg, 127th Lys, 146th Glu, 153rd Gly, and 172nd Cys, were all at corresponding positions.

IMPROVEMENT OF THE ENZYME IN STABILITY

For its use in a bioreactor, DCase was immobilized on a resin and used repeatedly. Although a DCase requires both high reactivity and high enzyme stability for such use, the known DCases do not have both properties. Thus, the DCase gene of *Agrobacterium* sp. KNK712, which exhibits high reactivity, was mutagenized to increase the enzyme's thermostability as an index of enzyme stability.

Mutagenesis

The DCase gene of *Agrobacterium* sp. KNK 712 was separated from recombinant plasmid pAD108 and then inserted into M13mp18. Random mutation was performed using two mutagens, hydroxylamine hydrochloride and $NaNO_2$ (14,15). For the mutation with hydroxylamine, recombinant phage particles containing the DCase gene were treated with 0.25 M mutagen (pH 6.0) for 1–8 h at 37 °C. For the mutation with $NaNO_2$, single-strand recombinant phage DNA was treated with 0.9 M mutagen (pH 4.3) for 30 min at 25 °C. After preparation of double-strand phage DNA, the 1.8-kb *Eco*RI–*Hin*dIII fragment corresponding to the DCase gene was prepared, inserted into pUC19, and then transformed into *E. coli* JM109.

The recombinant clones were screened by means of a newly developed colorimetric enzyme assay to select thermostabilized enzyme-producing colonies. The method is as follows. For first screening for colony assaying, recombinant *E. coli* colonies on the plate medium (the plates being called the master plates) were transferred to sterilized filter paper and then lysed for 30 min at 37 °C by adding a lysis solution containing lysozyme and Triton X-100. After inactivation of the heat-sensitive enzymes (65 °C, 5 min), a color reagent (1 mL), a mixture of two solutions (Sol. A, 30 mM potassium phosphate buffer, pH 7.4, 3 mg/mL *N*-carbamoyl-D-phenylglycine, 10 mM phenol, and 0.8 u/mL D-amino acid oxidase; Sol. B, 70 u/mL horseradish peroxidase and 50 mM 4-aminoantipyrine: Sol. A:Sol. B = 100:1, mixed just before use) was added to the filter paper, followed by incubation at 37 °C. The candidate colonies, which were colored red, were separated from the master plates. These clones were then subjected to the following second screening for cell-free assaying to confirm the mutant clones.

Recombinant *E. coli* cells were disrupted by sonication and the heat-sensitive enzymes were then inactivated (65 °C, 10 min). The crude enzyme sample was dispensed in 150-μL aliquots onto a 96-well microtiter plate; these aliquots were serially diluted twofold and 50 μL of the modified color reagent (Sol. A:Sol. B = 19:1) was added. The plate was incubated at 37 °C, and then the enzyme activity was measured as absorbance at 505 nm.

The candidate clones that exhibited high absorbance were separated and the enzyme genes were analyzed.

Nucleotide Sequence Analysis of the Mutagenized DCase Genes

The DNA sequences of the 1.8-kb *Eco*RI–*Hin*dIII fragments that contained the mutagenized DCase genes from these representative clones were analyzed (Table 3). One or some nucleotides of the DCase gene were changed. The amino acid changes related to the thermostability increase proved that the 57th amino acid of DCase, histidine, was changed to tyrosine, the 203rd proline was changed to ser-

```
                                                  50                                                 100
1. MTRQMILAVGQQGPIARAETREQVVVRLLDMLTKAASRGANFIVFPELALTTFFPRWHFTDEAELDSFYETEMPGPVVRPLFEKAAELGIGFNLGYAELV
2. MTRQMILAVGQQGPIARAETREQVVGRLLDMLTNAASRGVNFIVFPELALTTFFPRWHFTDEAELDSFYETEMPGPVVRPLFETAAELGIGFNLGYAELV
3. MTRIVNAAAAQMGPISRSETRKDTVRRLIALMREAKARGSDLVVFTELALTTFFPRWVIEDEAELDSFYEKEMPGPETQPLFDEAKRLEIGFYLGYAELA
4.  SRIVNYAAAQLGPIQRADSRADVMERLLAH
5.  ARKLNLAVAQLGPIARAETRDQVVARLMEM

                                                  150                                                200
1. VEGGVKRRFNTSILVDKSGKIVGKYRKIHLPGHKEYEAYRPFQHLEKRYFEPGDLGFPVYDVDAAKMGMFICNDRRWPEAWRVMGLRGAEIICGGYNTPT
2. VEGGVKRRFNTSILVDKSGKIVGKYRKIHLPGHKEYEAYRPFQHLEKRYFEPGDLGFPVYDVDAAKMGMFICNDRRWPETWRVMGLKGAEIICGGYNTPT
3. EEGGRKRRFNTSILVDRSGRIVGKYRKVHLPGHKEPQPGRKHQHLEKRYFEPGDLGFGVWRAFDGVMGMCICNDRRWPETYRVMGLQGVEMVMLGYNTPY

                                                  250                                                300
1. HNPPVPQHDHLTSFHHLLSMQAGSYQNGAWSAAAGKVGMEENCMLLGHSCIVAPTGEIVALTTTLEDEVITAAVDLDRCRELREHIFNFKQHRQPQHYGL
2. HNPPVPQHDHLTSFHHLLSMQAGSYQNGAWSAAAGKVGMEEGCMLLGHSCIVAPTGEIVALTTTLEDEVITAAVDLDRCRELREHIFNFKAHRQPQHYGL
3. DHTGHDDIDSLTQFHNHLSMQAGAYQNSTWVIGTAKCGTEEGSKMVGQSVIVAPSGEIVAMACTIEDEIITARCDLDMGKRYRETIFDFARHREPDAYRL

1. IAEL
2. IAEF
3. IVDRKGAVPPPQ
```

Figure 3. Multiple alignment of the deduced amino acid sequences of DCases. The amino acid sequences of the DCases of *Agrobacterium* sp. KNK712 and homologous amino acid residues in other DCases are shown in *black boxes. 1, Agrobacterium* sp. KNK712; *2, Agrobacterium radiobacter* NRRL B11291; *3, Pseudomonas* sp. KNK003A; *4, Comamonas* sp. E222C; *5, Blastobacter* sp. A17p-4.

ine or leucine, and the 236th valine was changed to alanine.

Amino Acid Substitution of Thermostability-Related Sites by PCR

The three thermostability-related sites of the DCase were substituted (15) through site-directed random amino acid changes by use of the PCR technique according to Ito et al. (20), and then thermostabilized DCase-producing strains were screened by means of the colorimetric colony assay method. As a result of DNA analysis of the mutant thermostabilized DCases, it was proved that the following amino acid changes at the sites increased the thermostability in addition to the known amino acid changes; the 57th amino acid of DCase was changed to leucine; the 203rd amino acid to alanine, asparagine, glutamate, histidine, isoleucine, or threonine; and the 236th amino acid to serine or threonine. These results are shown in Table 4.

Combination of Thermostability-Related Mutations

To produce multiple mutants with combinations of two or three thermostability-related amino acid mutations, restriction fragments containing the mutations were used to replace the corresponding mutation-free DNA fragments (16). In the DCase of *Agrobacterium* sp. KNK712, the mutation of the 57th amino acid is located in a *Nde*I–*Sac*I DNA fragment of about 190 bp; the 203rd amino acid mutation is in a *Sal*I–*Cla*I DNA fragment of about 170 bp (hereinafter this fragment is referred to as fragment A); and the 236th amino acid mutation is in a *Cla*I–*Sph*I DNA fragment of about 75 bp. These fragments were replaced and the thermostabilities of the enzymes were then mea-

```
                   *      **                  *      *                    *
1.  GVKRRFNTSILVDKSGKIVGKYRKIHLPGHKEYEAYRPFQHLEKRYFEPGD--LGFPVYDVDAAKMGMFICNDRRWPEAWRVMGLRGAEII
2.  GVKRRFNTSILVDKSGKIVGKYRKIHLPGHKEYEAYRPFQHLEKRYFEPGD--LGFPVYDVDAAKMGMFICNDRRWPETWRVMGLKGAEII
3.  GRKRRFNTSILVDRSGRIVGKYRKVHLPGHKEPQPGRKHQHLEKRYFEPGD--LGFGVWRAFDGVMGMCICNDRRWPETYRVMGLQGVEMV
4.  PNKPPYNTLILIDNKGEIVQRYRKI-LPWCP----------IEGWY--PGD--TTYVTEGPKGLKISLIICDDGNYPEIWRDCAMKGAELI
5.  PRKAPYNTLVLIDNNGEIVQKYRKI-IPWCP----------IEGWY--PGG--QTYVSEGPKGMKISLIICDDGNYPEIWRDCAMKGAELI
6.  GV--LWNTAVVISNSGLVMGKTRKNHIPRVG--------DFNESTYYMEGN--LGHPVFQTQFGRIAVNICYGRHHPLNWLMYSVNGAEII
7.  YT--LYCTALFFSPQGQFLGKHRKL-MP-----------TSLERCIWGQGDG-STIPVYDTPIGKLGAAICWENRMPLYRTALYAKGIELY
8.  YT--LYCTVLFFSPQGQFLGKHRKL-MP-----------TSLERCIWGQGDG-STIPVYDTPIGKLGAAICWENRMPLYRTALYAKGIELY
9.  AS--RYLSQVFIDQNGDIVANRRKL-KP-----------THVERTIYGEGNG-TDFLTHDFGFGRVGGLNCWEHFQPLSKYMMYSLNEQIH
10. GS--LYMTQLVIDADGQLVARRRKL-KP-----------THVERSVYGEGNG-SDISVYDMPFARLGALNCWEHFQTLTKYAMYSMHEQVH
11. RT--LYMSQMLIDADGITKIRRRKL-KP-----------TRFERELFGEGDG-SDLQVAQTSVGRVGALNCAENLQSLNKFALAAEGEQIH
12. GS--LYLGQCLIDDKGQMLWSRRKL-KP-----------THVERTVFGEGYA-RDLIVSDTELGRVGALCCWEHLSPLSKYALYSQHEAIH
13. AT--LYLTQVLISPLGDVINHRRKI-KP-----------THVEKLVYGDGSGDSFEPVTQTEIGRLGQLNCWENMNPFLKSLAVARGEQIH
```

Figure 4. Optimal alignment with the CLUSTAL W program of the high-scoring segments in the predicted amino acid sequences of DCases and related enzymes. Identical amino acid residues in all enzymes are indicated by *asterisks. 1, Agrobacterium* sp. KNK712 DCase (104–192); *2, Agrobacterium radiobacter* NRRL B11291 DCase (104–192); *3, Pseudomonas* sp. KNK003A DCase (104–192); *4, Brevibacterium* sp. R312 aliphatic amidase, PIR Ac. NO. JC1174 (104–186); *5, Pseudomonas aeruginosa* aliphatic amidase, PIR Ac. No. A26741 (115–186); *6,* rat beta-alanine synthetase, PIR Ac. No. S27881 (176–252); *7, Arabidopsis thaliana* nitrilase, PIR Ac. No. S31969 (126–199); *8, Arabidopsis thaliana* nitrilase, PIR Ac. No. S22398 (133–206); *9, Rhodococcus rhodochrous* aliphatic nitrilase, PIR Ac. No. A43470 (117–190); *10, Rhodococcus rhodochrous* nitrilase, PIR Ac. No. A45070 (112–185); *11, Klebsiela ozaenae* nitrilase, PIR Ac. No. A28658 (108–181); *12, Alcaligenes faecalis* nitrilase, PIR Ac. No. A47181 (110–183); *13, Gloeocercospora sorghi* cyanide hydratase, PIR Ac. No. JQ1613 (109–183).

Table 3. Nucleotide Analysis of Mutagenized DCase Genes

Mutant	Locations of nucleotide changes	Amino acid substitutions
401M	840 C → T	203 Pro → Leu
402M	401 C → T	57 His → Tyr
	715 C → T	(161 Val → Val)
403M	177 C → T	(noncoding)
	499 C → T	(89 Ile → Ile)
	839 C → T	203 Pro → Ser
404M	840 C → T	203 Pro → Leu
406M	839 C → T	203 Pro → Ser
414M	939 T → C	236 Val → Ala

Note: The mutagenized DCase genes of thermostable enzyme-producing strains were analyzed and the locations of the nucleotide changes were determined. The amino acid substitutions resulting from the mutations were deduced.

sured (Table 5). The thermostabilities of the enzymes increased cumulatively with the accumulation of the individual mutations. Among the mutant enzymes with two amino acid changes, a mutant with mutations of the 203rd amino acid, proline, to glutamate, and the 236th, valine, to alanine, showed a thermostability increase of about 17 °C. Among the mutant enzymes with three amino acid changes, a mutant enzyme from KNK455M that had mutations of the 57th amino acid to tyrosine, the 203rd to glutamate, and the 236th to alanine, showed about 19 °C increase in thermostability.

CHARACTERIZATION OF THE THERMOSTABILIZED ENZYMES

Effects of Temperature and pH on Enzyme Stability

To evaluate the thermostability of the DCases, a cell-free extract (100 μL) was incubated at various temperatures (55–85 °C) for 10 min, the denatured protein was removed by centrifugation, and then the residual activity was measured by the standard HPLC method. The heat denaturation profiles of the four mutant enzymes from KNK402M, KNK404M, KNK416M, and KNK455M and the native enzyme were investigated. The relative activities in comparison with the activities of the non-heat-treated enzymes were plotted against temperature (Fig. 5). The thermostabilities of the mutant enzymes increased about 5–20 °C under these conditions compared to that of the native enzyme.

To examine the effect of pH on DCase stability, a cell-free extract (200 μL) was mixed with 800 μL of a pH-adjusted buffer (pH 6–10, 0.1 M each), followed by incu-

Table 4. Thermostability Analysis of Mutant DCases

Location of mutation	Substituted amino acid	Thermostable temperature (°C)
57 His	Tyr*	67.3
	Leu	67.5
203 Pro	Leu*	68.0
	Ser*	66.5
	Ala	67.7
	Asn	67.0
	Glu	70.0
	His	65.2
	Ile	67.2
	Thr	67.5
236 Val	Ala*	71.4
	Ser	72.0
	Thr	69.5
Native type	—	61.8

Note: The mutagenized DCase genes were analyzed and the amino acid substitutions resulting from the mutations were deduced. To compare the thermostability of the mutant enzymes easily, the index of thermostability, thermostable temperature, was defined as the temperature on heat treatment for 10 min that caused a decrease in DCase activity of 50%. The substituted amino acids, on random mutagenesis, in these mutants are indicated by asterisks.

Table 5. Thermostability of Mutant DCases Having Two or Three Mutations

Strain	Thermostable temperature[a] (°C)	Amino acid mutation		
		57 His	203 Pro	236 Val
KNK712	61.8	—	—	—
KNK421M	71.5	Tyr	Leu	—
KNK422M	70.0	Tyr	Ser	—
KNK423M	75.4	Tyr	—	Ala
KNK424M	77.5	—	Leu	Ala
KNK425M	75.9	—	Ser	Ala
KNK426M	78.8	Tyr	Leu	Ala
KNK427M	77.8	Tyr	Ser	Ala
KNK451M	74.0	Tyr	Glu	—
KNK452M	75.3	Tyr	—	Ser
KNK453M	76.0	—	Glu	Ser
KNK461M	79.0	—	Glu	Ala
KNK454M	80.4	Tyr	Glu	Ser
KNK455M	80.8	Tyr	Glu	Ala
KNK456M	78.5	Tyr	Leu	Ser

[a]Thermostable temperature is defined as the temperature on heat treatment for 10 min that caused a decrease in the DCase activity of 50%.

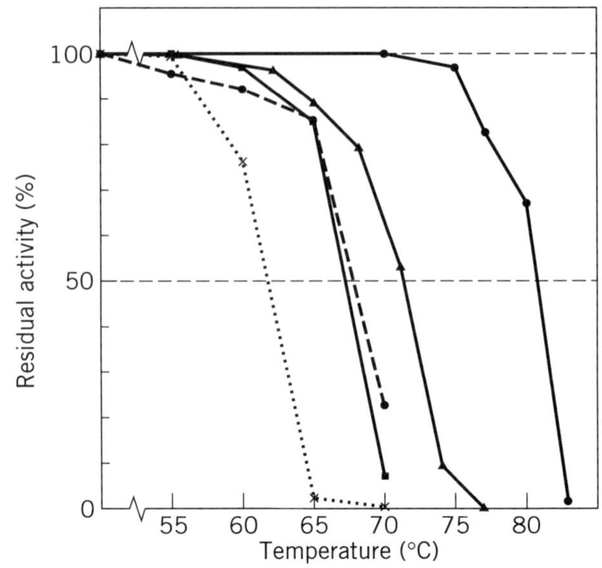

Figure 5. Effect of temperature on the stability of the mutagenized enzymes. A cell-free extract of each mutant DCase-producing strain was kept at various temperatures and the remaining activity was assayed under standard conditions. The remaining activities of the mutant enzymes from KNK402M (*solid line, squares*), KNK404M (*dotted line, circles*), KNK416M (*solid line, triangles*), and KNK455M (*solid line, circles*), and the native enzyme (*asterisks*), are expressed as a percentage of the activity of each nonheated enzyme.

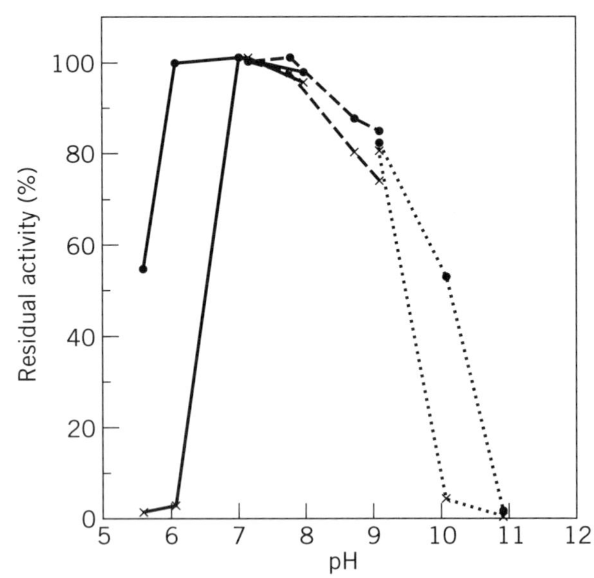

Figure 6. Effect of pH on the enzyme stability of the mutagenized DCases. The mutant enzyme from KNK455M was incubated at 40 °C for 4 h at the indicated pH in 0.1 M K_2HPO_4-KH_2PO_4 (*solid line*), 0.1 M Tris-HCl (*dashed line*), or Na_2CO_3-$NaHCO_3$ (*dotted line*) buffer. The remaining activities were then assayed under standard conditions and expressed as a percentage of the activity of each nontreated enzyme. As a control, the native enzyme (*asterisk*) was treated in the same manner.

bation at 40 °C for 4 h, and then the residual activity was measured under standard conditions. The relative activity compared with the activity of each nontreated sample was plotted against the incubation pH (Fig. 6). The stability of the mutant enzyme from KNK455M was increased in the lower- and higher-pH regions in comparison with those of the native enzyme.

Substrate Specificities of the DCases

The substrate specificity of the enzyme from KNK455M was investigated using a broad range of *N*-carbamoyl-D-

Table 6. Comparison of Substrate Specificities of Native and Mutant DCases

	Relative activity (%)[a]	
Substrate	Native DCase	DCase of KNK455M
c-D-HPG	100	100
c-D-Phenylglycine	110	130
c-D-Ala	86	75
c-D-Val	42	47
c-D-Leu	120	110
c-D-Met	180	160
c-DL-Ile	79	76
c-DL-Phe	55	46
c-DL-Ser	58	39
c-DL-Asp	7.2	3.4
c-DL-Norvaline	82	69
c-DL-Norleucine	120	99
c-DL-α-Amino butyrate	100	87

[a]Relative activity is expressed as a percentage of the DCase activity in comparison with the activity with c-D-HPG as a substrate.

Table 7. Comparison of the Properties of Native and Mutant DCases

	Native DCase	DCase of KNK455M
Specific activity (U/mg)	6.8	6.6
Optimum pH	7.0	6.4
pH stability	6.5–7.5	6.0–8.0
Optimum temperature (°C)	65	75
Thermostable temperature (°C)	61.8	80.8
Substrate specificity	c-D-amino acid	c-D-amino acid
K_m (mM, pH 7.0 c-D-HPG)	0.89	1.3
V_{max} (μmol/min/mg, c-D-HPG)	9.6	7.6

amino acids and related compounds. Like the native enzyme, the enzyme showed strict specificity toward *N*-carbamoyl-D-amino acids, *N*-carbamoyl-L-amino acids not being hydrolyzed. *N*-Carbamoyl-D-amino acids having a hydrophobic side chain, as well as *N*-carbamoyl-amino acids having a polar group or short-chain alkyl group, were hydrolyzed. The substrate specificity of the mutant enzyme was almost the same as that of the native enzyme (Table 6). It seemed that these three thermostability-related amino acid changes little affected the substrate specificity. The other known DCases from *Agrobacterium* showed relatively similar tendencies as to substrate specificity.

Kinetic Properties of the Enzymes

The kinetic properties of the enzymes from *Agrobacterium* sp. KNK712 and mutant KNK455M were measured using c-D-HPG as a substrate. The K_m and V_{max} values obtained on Lineweaver–Burk plotting are shown in Table 7.

As the affinity toward substrates and the reactivity were slightly decreased by the mutations, the mutant enzyme was thought to be utilized under industrial conditions.

PRODUCTION OF AN IMPROVED ENZYME

Construction of an Expression Vector

To express the DCase gene effectively, the mutant DCase gene was inserted into a newly constructed vehicle, pUCNT. This vehicle was constructed as follows (21).

To insert the improved DCase gene just after the lac promoter of pUC19, a *Nde*I cleavage site was generated in the initiation codon of the *lacZ* gene, through PCR amplification of the *Hin*dIII–*Cfr*10I 1.3-kb fragment. The amplified fragment was used to replace the corresponding fragment of pUC19, and the plasmid constructed was designated as pUCNde. After generating a *Nde*I site instead of the *Nco*I site of pTrc99A in the same manner, a 0.6-kb *Nde*I–*Ssp*I fragment was ligated into the 2.0-kb *Nde*I-*Ssp*I fragment, pUCNT being constructed.

The 455M mutant DCase gene was inserted into pUCNT between the *Nde*I and *Pst*I cleavage sites, the recombinant plasmid pNT4553 being constructed (Fig. 7).

Expression of the Thermostabilized Enzyme

The expression plasmid was transformed into *E. coli* HB101, and then the recombinant clone was cultivated overnight in 2 × YT medium at 37 °C. In the recombinant *E. coli*, the plasmid was stably maintained without the addition of ampicillin and was expressed efficiently without IPTG, which revealed that the DCase accounted for nearly 50% of all the soluble protein in a cell.

PREPARATION OF AN IMMOBILIZED ENZYME

Immobilization of the Improved DCase

Escherichia coli JM109 (pAD108) and JM109 (pAD404) cells were harvested by centrifugation, suspended in 50 mL of 0.1 M KP buffer, pH 7.0, containing 5 mM DTT, and then disrupted by sonication. The cell debris was removed and the supernatant was obtained, as was the enzyme solution. Duolite A-568 (Rohm and Haas) was washed with 1 M NaCl, water, and then 0.1 M KP buffer, pH 7.0, and then equilibrated with the same buffer for 18 h at room temperature. The wet resin was subjected to filtration, and then DTT (final concentration of 5 mM) was added to the enzyme solution, followed by stirring at 4 °C for 20 h under nitrogen sealing. For adsorption, the weight of the wet resin and that of protein in the enzyme solution were in the ratio of 1 to 0.04. After adsorption, the resin was washed three times with a fivefold volume of 0.1 M KP buffer, pH 7.0, containing 10 mM DTT, cross-linked with a fivefold volume of 0.2% (w/w) of glutaraldehyde at 4 °C for 10 min, washed three times with a fivefold volume of the same buffer at 4 °C, and then filtered (17,18).

Stability of the Immobilized Enzyme

For repeated batch reactions with the immobilized DCase, a substrate solution (3%) of c-D-HPG (pH 7.0) was prepared, with nitrogen gas bubbling at 40 °C. The DCase immobilized on Duolite A-568 and the prepared substrate

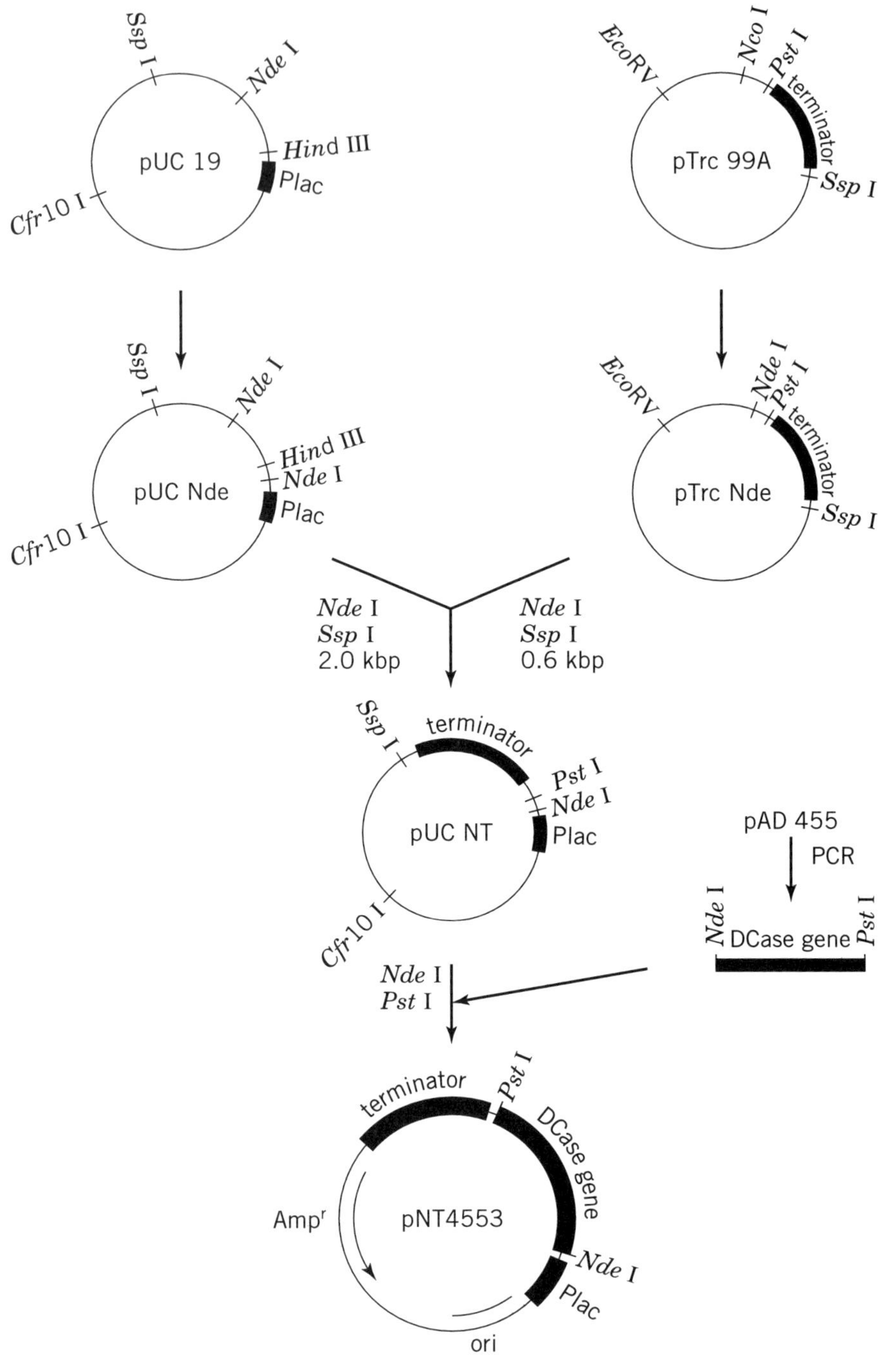

Figure 7. Construction profile of pNT4553.

solution were stirred at 40 °C under a stream of nitrogen gas while the pH was controlled at pH 7.0 with 2 M HCl. Samples were taken for activity measurement at 10 and 60 min, the reaction being continued for 23.5 h in total. The reaction mixture was removed by suction, after which a fresh substrate solution was introduced and allowed to react in the same manner as already described. The residual activity was measured as the relative activity compared with the initial activity and plotted against the reaction times (Fig. 8). The stability of the mutant enzyme increased in comparison with that of the native enzyme.

UTILIZATION OF THE IMPROVED DCASE FOR INDUSTRIAL PRODUCTION

An enzymatic conversion process involving the improved DCase in an immobilized form was introduced for D-HPG production instead of chemical diazonation. As a result of the introduction of the new process, the efficiency of the

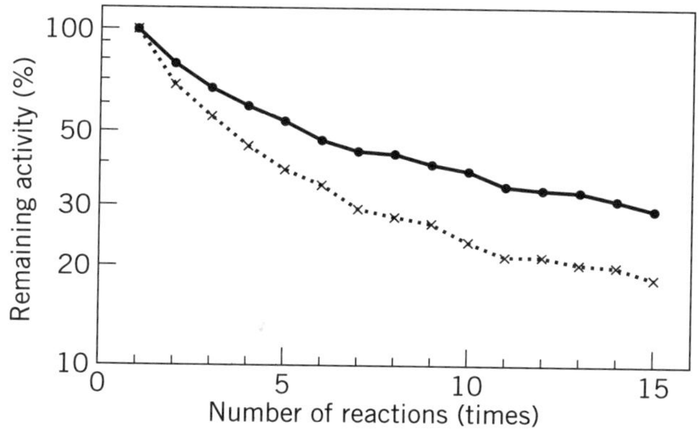

Figure 8. Stability of the immobilized DCase on repeated batch reactions. Repeated batch reactions were carried out using the immobilized mutant DCase from KNK404M (*circles*) and the native enzyme (*asterisks*).

reaction was greatly increased, the purification process became simpler because of the decrease in the undesirable color formation of the product, and the energy required for production was reduced. In addition to these advantages, by-products such as inorganic salts were greatly reduced by the change in the reaction mechanism. By means of these advantages, the new process decreases the burden on the environment.

BIBLIOGRAPHY

1. J. Caravaca and S. Grisolia, *J. Biol. Chem.* **231**, 357–365 (1958).
2. L.L. Campbell, *J. Biol. Chem.* **235**, 2375–2378 (1960).
3. J.M. Kim, S. Shimizu, and H. Yamada, *J. Biol. Chem.* **261**, 11832–11839 (1986).
4. R. Olivieri, E. Fascetti, L. Angelini, and L. Degan, *Biotechnol. Bioeng.* **23**, 2173–2183 (1981).
5. K. Yokozeki and K. Kubota, *Agric. Biol. Chem.* **51**, 721–728 (1987).
6. J. Ogawa, S. Shimizu, and H. Yamada, *Eur. J. Biochem.* **212**, 685–691 (1993).
7. J. Ogawa, M.C.M. Chung, S. Hida, H. Yamada, and S. Shimizu, *J. Biotechnol.* **38**, 11–19 (1994).
8. A. Louwrier and C.J. Knowles, *Enzyme Microb. Technol.* **19**, 562–571 (1996).
9. R. Grifantini, C. Pratesi, G. Galli, and G. Grandi, *J. Biol. Chem.* **271**, 9326–9331 (1996).
10. A. Buson, A. Negro, L. Grassato, M. Tagliaro, M. Basaglia, C. Grandi, A. Fontana, and M.P. Nuti, *"FEMS Microbiol. Lett.* **145**, 55–62 (1996).
11. World Pat. WO94/00577 (January 6, 1994), R.J. Neal, A.M. Griffin, M.O. Scott, A.R. Shatzman, and H.C. Gorham (to Smith-Kline Beecham P.L.C.).
12. H. Nanba, Y. Ikenaka, Y. Yamada, K. Yajima, M. Takano, and S. Takahashi, *Biosci. Biotechnol. Biochem.* **62**, 875–881 (1998).
13. Y. Ikenaka, H. Nanba, Y. Yamada, K. Yajima, M. Takano, and S. Takahashi, *Biosci. Biotechnol. Biochem.* **62**, 882–886 (1998).
14. Y. Ikenaka, H. Nanba, K. Yajima, Y. Yamada, M. Takano, and S. Takahashi, *Biosci. Biotechnol. Biochem.* **62**, 1668–1671 (1998).
15. Y. Ikenaka, H. Nanba, K. Yajima, Y. Yamada, M. Takano, and S. Takahashi, *Biosci. Biotechnol. Biochem.* **62**, 1672–1675 (1998).
16. Y. Ikenaka, H. Nanba, K. Yajima, Y. Yamada, M. Takano, and S. Takahashi, *Biosci. Biotechnol. Biochem.* **63**, in press.
17. H. Nanba, Y. Ikenaka, Y. Yamada, K. Yajima, M. Takano, K. Ohkubo, Y. Hiraishi, K. Yamada, and S. Takahashi, *J. Mol. Catal. B: Enzymatic*, in press.
18. H. Nanba, Y. Ikenaka, Y. Yamada, K. Yajima, M. Takano, and S. Takahashi, *Biosci. Biotechnol. Biochem.* **62**, 1839–1844 (1998).
19. R.P. Ambler, A.D. Auffret, and P.H. Clarke, *FEBS Lett.* **215**, 285–290 (1987).
20. W. Ito, H. Ishiguro, and Y. Kurosawa, *Gene* **102**, 67–70 (1991).
21. H. Nanba, Y. Ikenaka, Y. Yamada, H. Yajima, M. Takano, and S. Takahashi, *J. Ferment. Bioeng.*, in press.

See also AMINO ACIDS, PRODUCTION PROCESSES.

D-AMINOPEPTIDASE, ALKALINE D-PEPTIDASE

YASUHISA ASANO
HIDENOBU KOMEDA
Toyama Prefectural University
Toyama, Japan

KEY WORDS

Alkaline D-peptidase
D-Amino acid–containing peptide
D-Aminopeptidase
Bacillus cereus
Ochrobactrum anthropi
Organic solvent
Penicillin-recognizing enzyme

OUTLINE

INTRODUCTION

Enzymatic reaction has increasingly been applied to organic synthesis (1–3), and there are several reports of the attempted in vitro synthesis of the D-amino acid–containing peptides and amides (4–10). However, because these enzymatic reactions are not stereospecific for substrates with D-configurations, they are at an innate disadvantage. Some peptidases and amidases act on peptides and amides containing D-amino acids. Soluble *Streptomyces* carboxypeptidase DD catalyzes not only the transpeptidation reaction on the peptide intermediate in peptidoglycan biosynthesis but also the hydrolysis of N_α,N_ϵ-diacetyl-L-lysyl-(D-Ala)$_2$ in water (11). A D-peptidase has been purified and characterized from an actinomycete, although it is not strictly specific toward peptides containing D-amino acids (12). In *Enterococcus*, the *vanX* gene product (D-Ala)$_2$ hydrolase plays a role in vancomycin resistance (13). The chemically synthesized D-enzyme of an HIV-1, in which all the amino acids were replaced with the corresponding D-amino acids, displays D-stereospecificity (14). D-Stereospecific amino acid amidases (amidohydrolases) were isolated from some microorganisms (15–17). Of the enzymes, a D-alaninamide–specific amidohydrolase from *Arthrobacter* has been used in the industrial manufacture of D- and L-alanine (17,18). Production of D-amino acids by use of D- or L-specific enzymes such as D-amino acylase, D-hydantoin hydrolase, *N*-carbamoyl-D-amino acid hydrolase, D-amidase, D-transaminase, and amino acid racemase was recently reviewed (19,20).

A new enzyme, D-aminopeptidase, was isolated from *Ochrobactrum anthropi* and characterized. We describe its

Table 1. Substrate Specificity of D-Aminopeptidase from *O. anthropi* SCRC C1-38

Substrate	Relative activity (%)	K_m (mM)	V_{max} (U/mg)
D-Alanine amide	100	0.65	600
Glycine amide	44	22.3	365
D-α-Aminobutyric acid amide	30	18.3	576
D-Serine amide	29	27.0	22.0
D-Threonine amide	9	100	60.3
D-Methionine amide	2		
D-Norvaline amide	1.8		
D-Norleucine amide	0.8		
D-Phenylglycine amide	0.7		
D-Alanylglycine	95	0.98	1,000
D-Alanylglycylglycine	45	0.37	799
D-Alanyl-D-alanine	21	10.2	326
D-Alanyl-D-alanyl-D-alanine	92	0.57	866
D-Alanyl-D-alanyl-D-alanyl-D-alanine	89	0.32	702
D-Alanyl-L-alanine	46	1.03	312
D-Alanyl-L-alanyl-L-alanine	100	0.65	730
DL-Alanyl-DL-serine	27		
DL-Alanyl-DL-methionine	20		
DL-Alanyl-DL-phenylalanine	9		
DL-Alanyl-DL-asparagine	7		
DL-Alanyl-DL-leucine	1		
DL-Alanyl-DL-valine	0.5		
Glycine methyl ester	229		
D-Alanine methyl ester	75		
D-Alanine-β-naphthylamide	32		
D-Alanine benzylamide	72	0.51	768
D-Alanine anilide	73		
D-Alanine-*p*-nitroanilide	96	0.51	696
D-Alanine *n*-butylamide	66	0.73	670
D-Alanine-3-aminopentane amide	32	2.27	288
D-Alanine *n*-laurylamide	19		
D-Threonine benzyl ester	3.2		

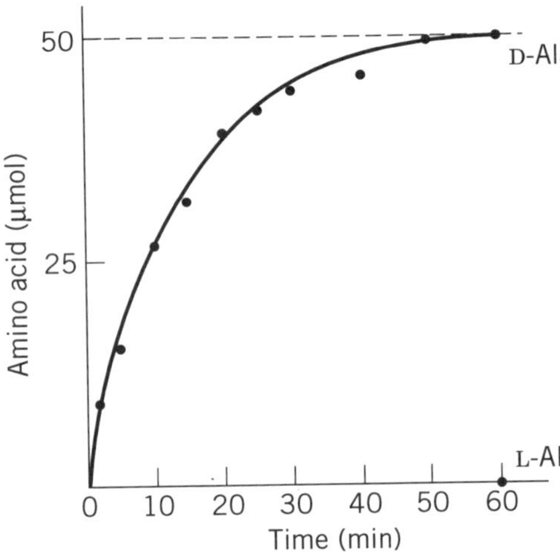

Figure 1. Kinetic resolution of DL-alanine amide by *O. anthropi* D-aminopeptidase. A reaction mixture containing 0.5 mmol tris(hydrochloride), pH 8.0, 0.1 mmol DL-alanine amide hydrochloride, and 5 units of the purified enzyme in a total volume of 5.0 mL was incubated at 30 °C. The content of L-alanine was measured by L-alanine dehydrogenase.

Table 2. D-Aminopeptidase-Catalyzed Synthesis of D-Amino Acid Derivatives in Organic Solvents

Acyl donor	Acyl acceptor	Yield (%)
In 1,1,1-trichloroethane		
D-Alanine methyl ester HCl	3-Aminopentane	99
D-Alanine methyl ester HCl	Neopentylamine	98
D-Alanine methyl ester HCl	*n*-Butylamine	66
D-Alanine methyl ester HCl	Benzylamine	15
In butyl acetate		
D-Alanine methyl ester HCl	3-Aminopentane	99
Glycine methyl ester HCl	3-Aminopentane	80
D-α-Aminobutyric acid methyl ester HCl	3-Aminopentane	44
In toluene		
D-Alanine methyl ester HCl	D-Alanine methyl ester HCl	58[a]

[a]Product: D-alanine dimer (plus D-alanine trimer, yield 6%).

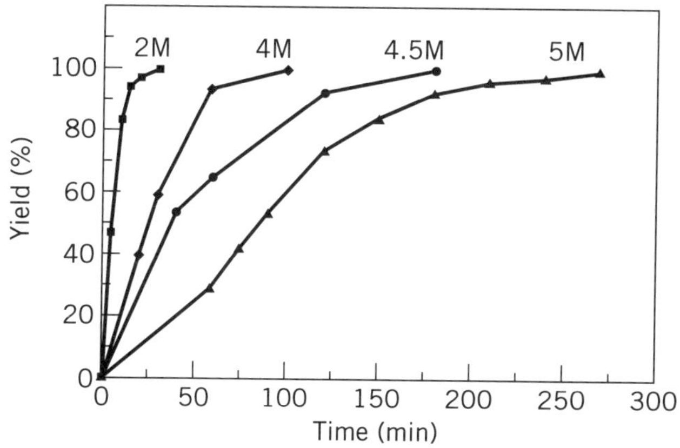

Figure 2. Synthesis of D-alanine from alanine amide HCl by the cells of *E. coli* JM109/pC138DP. The reaction mixture contained various amounts of racemic alanine amide HCl, washed *E. coli* cells harvested from 2.0 mL of culture broth, and 50 μmol of tris(HCl), pH 8.0, in a total volume of 2.0 mL. The enzyme activity shown by the *E. coli* transformant was 288,000 Units per liter of culture. The mixtures were incubated at 30 °C. Formation of D-alanine from D-alanine amide is shown as yield (%).

function, structure, and application to the D-stereospecific hydrolysis of racemic amino acid amides and the formation of peptide bonds. A new enzyme, alkaline D-peptidase, acting on a synthetic peptide, (D-Phe)$_4$, was also isolated from *Bacillus cereus.* We describe its structure and function, as well as its application to the synthesis of D-phenylalanine oligomers. We propose that these two enzymes are new members of the group of penicillin-recognizing enzymes.

D-AMINOPEPTIDASE

During the course of studies on the enzyme-catalyzed organic synthesis of D-amino acid derivatives, a D-stereospecific aminopeptidase was needed. It was speculated that microorganisms might be a likely source for such an enzyme, given their important role in the environment as synthesizers and degraders of a wide variety of substances.

Isolation and Properties of the D-Aminopeptidase

An enrichment culture in a medium containing a synthetic substrate (D-Ala-NH_2) as the sole nitrogen source led to isolation of a bacterial strain, *Ochrobactrum anthropi* SCRC C1-38. The enzyme hydrolyzing D-Ala-NH_2, named D-aminopeptidase (EC 3.4.11.19), was purified to homogeneity from the cell-free extract of the strain and characterized in detail (21). The molecular weight of the native enzyme was estimated to be approximately 122,000, with two identical subunits of molecular weight of about 59,000.

A typical time course for the stereoselective hydrolysis of D-alanine amide is shown in Figure 1, with racemic alanine amide as a substrate. D-Alanine amide was completely hydrolyzed by the action of the enzyme, whereas the L-enantiomer remained unhydrolyzed in the reaction mixture. The rate of the hydrolysis of L-alanine amide was less than 0.01% that of D-alanine amide. The enzyme showed strict chemo- and stereospecificities toward D-amino acid amides, peptides, and esters, as shown in Table 1. Each substrate (100 mM) was incubated under the standard enzyme assay condition (21). The substrates include D-alanine amide (relative velocity: 100%, K_m value: 0.65 mM); glycine amide (44%, 22.3 mM); D-α-aminobutyramide (30%, 18.3 mM); D-serine amide (29%, 27.0 mM); D-alanine 3-aminopentane amide (32%, 2.27 mM); D-alanine-*p*-nitroanilide (96%, 0.51 mM); D-alanine methyl ester (75%), dimer (21%, 10.2 mM), trimer (92%, 0.57 mM), and tetramer (89%, 0.32 mM) of D-alanine; D-alanylglycine (95%, 0.98 mM), D-alanylglycylglycine (45%, 0.37 mM); and D-alanyl-L-alanyl-L-alanine (100%, 0.65 mM). These results show that the enzyme has higher affinity toward peptide substrates than amino acid amides. The enzyme showed neither endopeptidase nor carboxypeptidase activity. The mode of action of the enzyme toward a peptide substrate was studied with D-alanylglycylglycine. The reaction was followed over time, and it was observed that alanine and glycylglycine were released until the substrate tripeptide was nearly completely consumed, whereupon glycine release began, and finally alanine and glycine were produced. This result shows that the enzyme catalyzes the hydrolysis of a single amino acid from the N-terminus of a peptide. Because the mode of action is typical of an aminopeptidase, we named the enzyme D-aminopeptidase.

Application to Organic Synthesis

The D-aminopeptidase from *O. anthropi* SCRC C1-38 was utilized for a stereoselective synthesis of D-alanine *N*-alkylamide from an amine and D-amino acid amide or D-amino acid methyl ester (22). In water, D-alanine *N*-alkylamide once formed by the enzyme was successively hydrolyzed to yield D-alanine. The enzyme was immobilized by urethane prepolymer PU-6. When the aminolysis reaction was performed in water-saturated organic solvents such as butylacetate, benzene, and 1,1,1-trichloroethane with the immobilized enzyme as a catalyst, it progressed well quantitatively in a highly D-stereoselective manner, to give optically pure D-alanine *N*-alkylamides. The acyl donor specificity of this reaction was rather limited, with D-alanine, glycine, and D-α-amino butyric acid as substrates, while the acyl acceptor specificity was relatively wide, with bulky, aromatic-containing, and straight-chain amines (Table 2). The catalytic center activity (k_{cat}) of this reaction, 7,700 (min^{-1}), was much higher than in the case of the nonstereoselective synthesis of D-amino acid–containing peptides (4,5).

Alkaline conditions are usually essential for peptide bond formation in kinetically controlled systems (23,24). Several attempts to oligomerize D-alanine methyl ester with D-aminopeptidase in an alkaline aqueous medium were unsuccessful, probably because the substrate is unstable under alkaline conditions. The immobilized D-aminopeptidase catalyzed the synthesis of D-alanine oligomers from D-alanine methylester in organic solvents (25). Dimer and trimer of D-alanine were obtained in 58% and 6% yield, respectively, when urethane prepolymer PU-6 immobilized D-aminopeptidase (1.5 U/mL) was incubated

```
dd     1  MVSGTVGRGTALGAVLLALLAVPAQAGTAAAADLPAPDDTGLQAVLHTALSQGAPGAMVRVDDNGTIHQL
dap    1                                 MSKFDTSALEAFVRHIPQNYKGPGGVVAVVKDGEVVLQH
bla    1              MFKTTLCALLITASCSTFAAPQQINDIVHRTITPLIEQQKIPGMAVAVIYQGKPYY-F

dd    71  SEGVADRATGRAITTTDRFRVGSVTKSFSAVVLLQLVDEGKLDLDASVNTYLPGLLPDDRITVRQVMSHR
dap   40  AWGFADLRTRTPMTLDTRMPICSVSKQFTCAVLLDAVGEPEL-LDDALEAYL-DKFEDERPAVRDLCNNQ
bla   58  TWGYADIAKKQPVTQQTLFELGSVSKTFTGVLGGDAIARGEIKLSDPTTKYW-PELTAKQWNGITL----

dd   141  SGLYDYTNDMFAQTVPGFESVRNKVFSYQDLITLSLKHGVTNAPGAAYSYSNTNFVVAGMLIEKLTGHSV
dap  107  SGLRDY----WALSVLCGADPEGVFLPAQAQSLLRRLKTTHFEPGSHYSYCNGNFRILADLIEAHTGRTL
bla  123  LHLATY-----TAGGLPLQVPDEVKSSSDLLRFYQNWQPA-WAPGTQRLYANSSIGLFGALAVKPSGLSF

dd   211  ATEYQNRIFTPLNLTDTFYVHPDTVIPGTHANGYLTPDEAGGALVDSTEQTVSW-AQSAGA-VISSTQDL
dap  174  VDILSERIFAPAGMKRAELI-SDTALFD-ECTGY--EGDTVRGFLPATNR-IQWMGDAGICASLNDMIAW
bla  187  EQAMQTRVFQPLKLNHTWINVPPAEEKN-YAWGY--REGKAVHVSPGALDAEAYGVKSTIEDMARWVQSN

dd   279  DTFFSALMSGQLMSAAQLAQMQQWTTVNSTQGYGLGLRRRDLSCGISVYGHTGTVQGYYTYAFASKDGKR
dap  239  EQFIDATRDDESGLYRRLSGPQTFKD-GVAAPYGFGLNLHETG-GKRLTGHGGALRGWRCQRWHCADERL
bla  254  LKPLDINEKTLQQGIQLAQSRYWQTGDMYQGLGWEMLDWPVNPDSIINGSDNKIALAARPVKAITPPTRA

dd   349  SVTALANTSNNVNVLNTMARTLESAFCGKPTTAKLRSATSSATTVERHEDIAPGIARD
dap  307  STIAMFNFEGGASEVAFKLMNIALGVSSSEVSRVEADSAWFGSWLDDETGLVLSLEDAGHGRMKARFGTS
bla  324  VRASWVHKTGATGGFGSYVAFIPEKELGIVMLANKNYPNPARVDAAWQILNALQ

dap  377  PEMMDVVSANEARSAVTTIRRDGETIELVRASENLRLSMKRVKGEAKHDIIGRYHSDELDADLLLVSEGG
dap  447  AIYGAFEGFLGKSDMYPLYSVGSDVWLLPVQRSMDAPSPGEWKLVFRRDDKGEITGLSVGCWLARGVEYR
dap  517  RVQP
```

Figure 3. Comparison of the amino acid sequences of the *dap* gene from *O. anthropi* SCRC C1-38, carboxypeptidase DD from *Streptomyces* R61, and class C β-lactamase from *E. coli*: asterisk (*), identical residue; single dot (·) similarity of functional group.The first amino acid sequence (dd) is the sequence of carboxypeptidase DD from *Streptomyces* R61 (sequence size: 406) (30). The second amino acid sequence (dap) is that of D-aminopeptidase from *O. anthropi* (sequence size: 521) (28). The third amino acid sequence (bla) is that of class C β-lactamase from *E. coli* K12 (sequence size: 377) (32).

in water-saturated toluene with 250 mM of D-alanine methyl ester HCl and 750 mM of triethylamine (Table 2). The k_{cat} of the reaction was calculated to be 19,500 (min^{-1}), which is several tens of thousands times greater than that of the known enzymatic syntheses of amino acid oligomers (26,27).

The gene for the D-aminopeptidase was cloned in *Escherichia coli* JM109 to overproduce the enzyme (28). An expression plasmid pC138DP was constructed. The amount of the enzyme in a cell-free extract of *E. coli* JM109/pC138DP was elevated to 288,000 units per liter of culture, which is about 3,600-fold over that of the wild strain (29). The intact cells of *E. coli* transformant were used as a catalyst for the D-stereospecific hydrolysis of several racemic amino acid amides HCl. Complete hydrolysis of D-alanine amide was achieved in a short time (4.5 h) from 5.0 M of racemic alanine amide HCl using the cells of *E. coli* transformant (Fig. 2). The concentration of D-alanine reached up to 220 g/L. The cells or the cell-free extract catalyzed the synthesis of D-α-aminobutyric acid, D-methionine, D-norvaline, and D-norleucine from their amides in a similar manner.

Enzyme		Organism	Sequence	Reference
DAP		*O. anthropi*	D L R T R T P M T L D T R M P I C **S** V S **K** Q F T C A V L L D A V G E P E L L	28
BLA	Class A	*E. coli* RTEM	S G K I L E S F R P E E R F P M M **S** T F **K** V L L C G A V L S R V D A G Q E Q	33
		B. licheniformis	G T N R T V A Y R P D E R F A F A **S** T I **K** A L T V G V L L Q Q K S I E D L N	34
	Class C	*C. freundii*	D I A N N H P V T Q Q T L F E L G **S** V S **K** T F N G V L G G D R I A R G E I K	31
		E. coli K12	L I A K K Q P V T Q Q T L F E L G **S** V S **K** T F T G V L G G D A I A R G E I K	32
		E. cloacae	D I A A N K P V T P Q T L F E L G **S** I S **K** T F T G V L G G D A I A R G E I S	52
		S. marcescens	Q T G K P I T E Q T L F E V G **S** L S **K**	51
	Class D	*Staphylococcus*	A M L V F D P V R S K K R Y S P A **S** T F **K** I P H T L F A L D A G A V R D E F	41
		S. marcescens	S C A T N D L A R A S K E Y L P A **S** T F **K** I P N A I I G L E T G V I K N E H	42
PBP 1A		*E. coli*	F N Q S K F N R A T Q A L R Q V G **S** N I **K** P F L Y T A A M D K G L T L A S M	36
1B			P Q F A G Y N R A M Q A R R S I G **S** L A **K** P A T Y L T A L S Q P K I Y R L N	36
2			K S G K E V K F N S D K R F A Y A **S** T S **K** A I N S A I L L E Q V P Y N K L N	37
3			P K E A M R N R T I T D V F E P G **S** T V **K** P M V V M T A L Q R G V V R E N S	35
5			S G K V L A E Q N A D V R R D P A **S** L T **K** M M T S Y V I G Q A M K A G K F K	36
6			S G K V L A E G N A D E K L D P A **S** L T **K** I M T S Y V V G Q A L K A D K I K	36
PBP		*S. aureus*	D K K E P L L N K F Q I T T S P G **S** T Q **K** I L T A M I G L N N K T L D D K T	38
PBP 2		*N. gonorrhoeae*	D S E Q R R N R A V T D M I E P G **S** A I **K** P F V I A K A L D A G K T D L N E	40
CPase		*Streptomyces*	D R A T G R A I T T T D R F R V G **S** V T **K** S F S A V V L L Q L V D E G K L D	30

Figure 4. Partial alignments of the amino acid sequences of D-aminopeptidase and active sites of the penicillin-recognizing enzymes: BLA, β-lactamase; Cpase, carboxypeptidase; DAP, D-aminopeptidase; PBP, penicillin-binding protein.

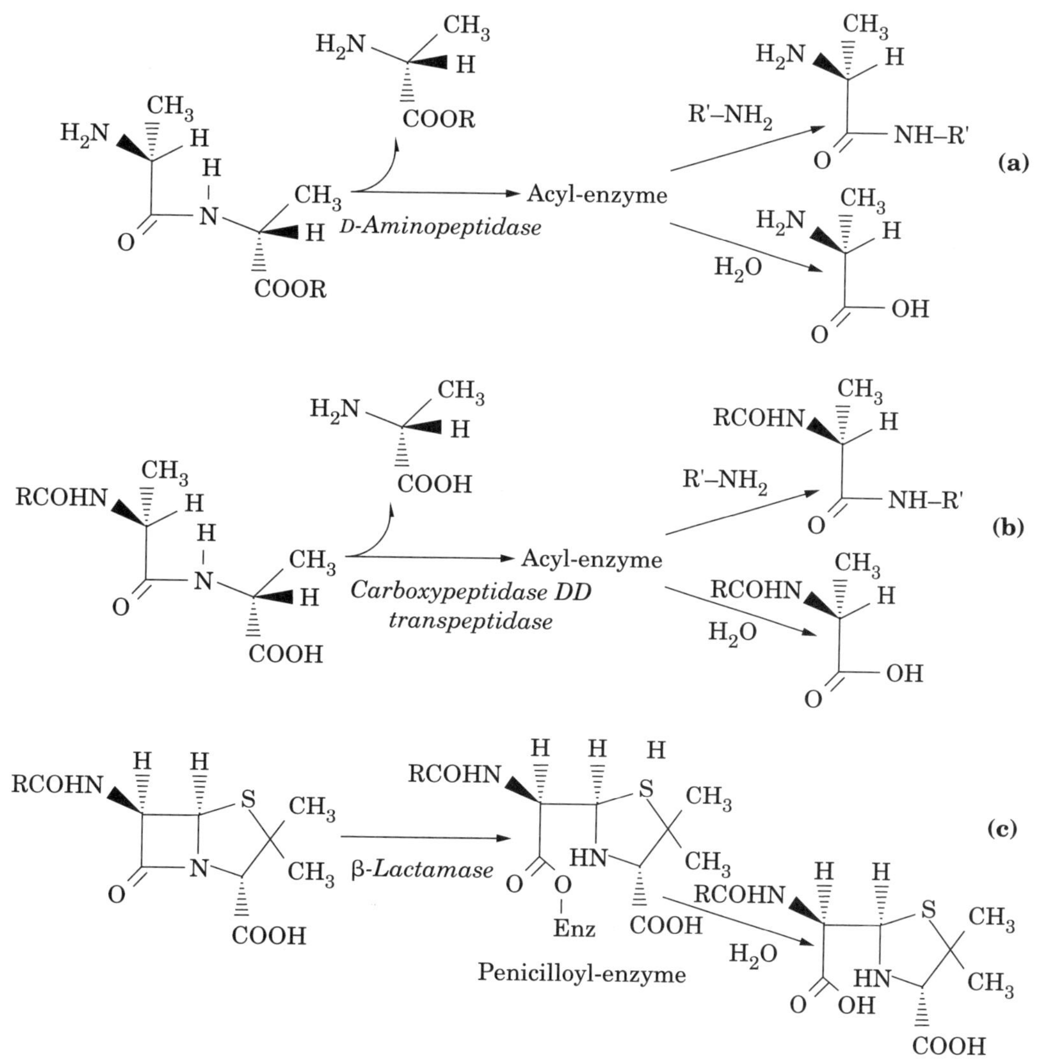

Figure 5. Reactions catalyzed by (**a**) D-aminopeptidase, (**b**) carboxypeptidase DD, and (**c**) β-lactamase.

Table 3. Substrate Specificity of Alkaline D-Peptidase from *Bacillus cereus* DF4-B

Substrate[a]	Relative activity (%)[b]	K_m (mM)	V_{max} (U/mg)	V_{max}/K_m (U/mg/mM)
$(D\text{-Phe})_6$	1.8[c]			
$(D\text{-Phe})_4$	100[c]	0.398	199	500
$(D\text{-Phe})_3$	90[c]	0.127	130	1020
$(D\text{-Phe})_2$	0.2[d]	50.1	13.7	0.270
D-Phe-L-Phe	<0.1[d]			
$(D\text{-Phe})_2$-L-Phe	14.9[c]	0.522	30.6	59.0
L-Phe-$(D\text{-Phe})_2$	119[e]	0.455	154	346
L-Phe-D-Phe-L-Phe	28.1[e]	1.63	66.0	41.0
D-Tyr$(D\text{-Phe})_2$	83.6[d]			
$(D\text{-Phe})_2$-D-Tyr	83.6[c]			
D-Phe-OMe	15[c] 1.8[d]			
D-Phe-NH_2	0.1[c] 0.1[d]			
D-Phe-*p*-Nitroanilide	4.2[d]			
Boc-$(D\text{-Phe})_4$	1.8[f] 0.8[g]			
Boc-$(D\text{-Phe})_3$	3.2[f] 1.1[g]			
Boc-$(D\text{-Phe})_2$	7.0[f]			
Boc-$(D\text{-Phe})_3$-O^{t}Bu	1.2[f] 0.3[g]			
Boc-$(D\text{-Phe})_4$-OMe	0.5[f] 0.2[g]			
Boc-$(D\text{-Phe})_3$-OMe	0.7[f] 0.3[g]			
Boc-$(D\text{-Phe})_2$-OMe	1.4[f]			
Ampicillin	8.9[h]	73.1	262	3.58
Penicillin G	9.7[h]	48.9	250	5.11

[a] O^{t}Bu, *tert*-butyl ester; OMe, methyl ester.
[b] Formation of five D-Phe compounds and consumption of a β-lactam compound were measured.
[c] Formation of $(D\text{-Phe})_2$.
[d] Formation of D-Phe.
[e] Formation of L-Phe-D-Phe.
[f] Formation of Boc-D-Phe.
[g] Formation of Boc-$(D\text{-Phe})_2$.
[h] Consumption of a β-lactam compound.

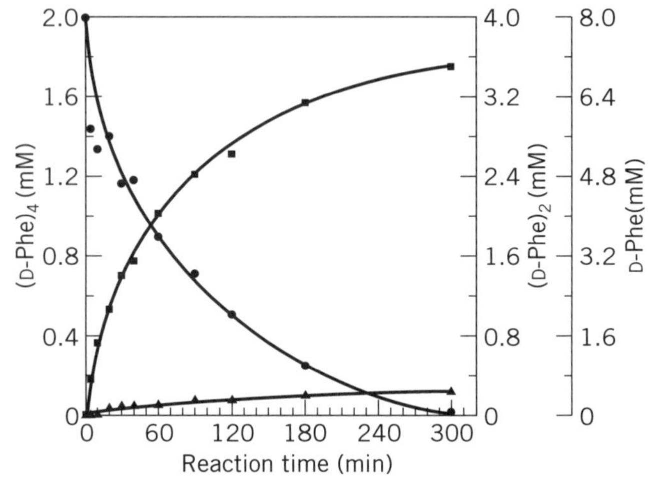

Figure 6. Time course of $(D\text{-Phe})_4$ hydrolysis by D-aminopeptidase. The reaction mixture consisted of 2 mM of $(D\text{-Phe})_4$, 100 mM of tris (HCl), pH 9.0, 2 mM of $MgSO_4$, 2% (v/v) of DMSO, and 0.038 unit of the enzyme solution in a total volume of 500 μL. The reaction was carried out at 30 °C and terminated with $HClO_4$. The amounts of $(D\text{-Phe})_4$ (circles), $(D\text{-Phe})_2$ (squares), and D-Phe (triangles) in the supernatant were measured by a Waters HPLC equipped with a Cosmosil 5C18-MS at a flow rate of 1.0 mL/min with a mixture of methanol (55%) and 5 mmol KH_2PO_4/H_3PO_4, pH 2.9 (45%).

Structural Similarities of D-Aminopeptidase to Carboxypeptidase DD and β-Lactamases

The nucleotide sequence of the D-aminopeptidase gene was determined and analyzed to study the structural relationship to other proteins. The gene consisted of an open reading frame of 1,560 nucleotides, which specifies a protein of M_r of 57,257 (28). The deduced primary structure is considerably similar to that of *Streptomyces* R61 carboxypeptidase DD (30) (27% identical over 287 amino acids) and class C β-lactamase from *E. coli* (22% identical over 234 amino acids) (31,32) (Fig. 3). The enzyme is also structurally related to class A β-lactamases (33,34), penicillin-binding proteins (35–40), and class D β-lactamases (41,42). The sequence Ser-X_{aa}-X_{aa}-Lys is perfectly conserved among this class of enzymes (Fig. 4). With carboxypeptidase DD and class C β-lactamase, the sequences at the highly similar area, Ser-Val-X_{aa}-Lys-X_{aa}-Phe-X_{aa}-Ala-X_{aa}-Val-Leu-Leu (30) and Ser-Val-Ser-Lys-X_{aa}-Phe-Tyr (31), respectively, are well conserved. Furthermore, some of the residues scattered around the conserved region are similar among the enzymes. The Met-57 located four residues upstream from the Ser-61 is found similarly in the carboxypeptidase DD and β-lactamases as a hydrophobic residue Phe. A common observation regarding D-aminopeptidase and the β-lactamase is that the consensus sequence is located around 60 from the N-terminus of the enzymes (43).

The inhibition by *p*-chloromercuribenzoate (PCMB) and other sulfhydryl reagents suggested that the enzyme would be a thiol peptidase (44). However, the findings of the site-specific mutagenesis study showed that the amino acid sequences Ser-X_{aa}-X_{aa}-Lys, which had been conserved in the penicillin-recognizing enzymes, are essential in exerting the D-aminopeptidase activity (28). The sites Ser-61 and Lys-64 are essential in the catalysis, because the V_{max}/K_m value measured with the mutants modified to Cys-61 and Asn-64 have been reduced to 0.26% and 0.008%, respectively, of those of the natural enzyme, although these K_m values were hardly lowered. The mutant with the inert Gly residue in place of the likely active center Ser-61 had a lower activity (10^{-5}-fold that of the native enzyme). On the other hand, the mutations at other sites (Met-57 and Cys-68) did not greatly affect the enzyme activity. The mutations at Cys-60, which is adjacent to the likely active center Ser-61, gave notable effects on the kinetic profiles, indicating that the residue is important in the binding of the substrate. The substitutions at Cys-60 to Ser and Gly produced mutants with slightly altered V_{max}/K_m values. They were tolerant to inhibition by PCMB, which suggests that the inhibition of the native enzyme by PCMB would have been due to the steric hindrance by a mercaptide bond formation between Cys-60 of the enzyme and PCMB.

A mutant of D-aminopeptidase with increased thermostability was obtained. The enhancement of the thermal stability of mutant enzyme was attributed to the substitution of Gly-155 to Ser (45).

Figure 5 shows similarities of the reactions catalyzed by D-aminopeptidase, carboxypeptidase DD, and β-lacta-

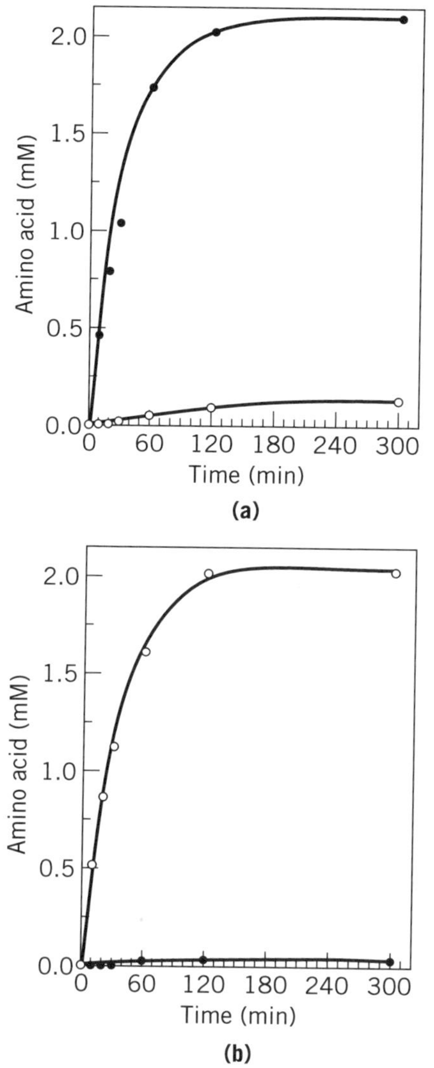

Figure 7. Mode of action toward (**a**) (D-Phe)$_2$-D-Tyr and (**b**) D-Tyr-(D-Phe)$_2$: solid circles, D-Tyr; open circles, D-Phe. The reaction mixture (500 μL) containing 2 mM of the substrate (D-Phe)$_2$-D-Tyr or D-Tyr-(D-Phe)$_2$, 50 mM of tris (HCl), pH 9.0, 2 mM of $MgSO_4$, 2% (v/v) of DMSO, and 0.038 unit of the enzyme was incubated at 30 °C. After termination of the reaction, the reaction mixture was analyzed with a Hitachi L-8500 amino acid analyzer.

mase. All three enzymes catalyze the hydrolytic cleavage of the amide bond between D-amino acids of the same configuration. The former two catalyze the transpeptidation reaction; carboxypeptidase DD catalyzes the cross-linkage of the peptideglycan, and D-aminopeptidase catalyzes the transpeptidation and aminolysis of D-amino acid ester and amides, respectively, in water and organic solvents. The aminolysis reaction catalyzed by D-aminopeptidase in butylacetate is efficient (k_{cat}: 7,700/min) (22). The transpeptidation reaction can also be catalyzed in water, although the product D-alanine-3-aminopentane amide was easily hydrolyzed in water (22). Inheritance of the carboxypeptidase-like tertiary structure in the D-aminopeptidase would have made the transpeptidation reaction in organic solvents possible. The enzyme is actually inhibited by β-lactam compounds, such as 6-APA, 7-ACA, benzylpenicillin, and ampicillin, although none of these is the substrate for the enzyme.

The D-aminopeptidase is a new member of the penicillin-recognizing enzymes by virtue of the following characteristics: similarities in the primary structure by gene sequencing, similarities in the reactions catalyzed in water and organic solvents, and the findings obtained by kinetic studies of the mutants generated by site-directed mutagenesis and the inhibition by β-lactam compounds. The mutants generated by inhibition of β-lactam compounds may constitute evidence to rebut the contention that the β-lactamases evolve from the penicillin-binding proteins (46). The existence of a third enzyme with a similar structure, which does not appear to be a selective target of the β-lactam compounds, was shown. To our knowledge, this enzyme is the first example of an aminopeptidase with Ser at the active site.

ALKALINE D-PEPTIDASE

D-Aminopeptidase (EC 3.4.11.19) from *Ochrobactrum anthropi* was discovered, and its primary structure has been found to have similarity to the β-lactamases and penicillin-binding proteins (21,28). The enzyme acts mostly on peptides with D-Ala at the N-terminus to yield D-amino acids; it does not act on D-amino acid derivatives with bulkier substituents. We proposed that D-aminopeptidase is a new penicillin-recognizing enzyme (11) based on its primary structure, inhibition by β-lactam compounds, and the ability to catalyze peptide bond formation in organic solvents, although the enzyme does not show β-lactamase activity (21,47).

Here, we describe the screening of soil microorganisms for D-stereospecific endopeptidases using a synthetic peptide (D-Phe)$_4$, characterization of the new enzyme alkaline D-peptidase (ADP), as well as cloning and sequencing of the *adp* gene from *B. cereus* strain DF4-B.

Isolation and Properties of the Alkaline D-Peptidase

An enrichment culture in LB medium containing a synthetic substrate (D-Phe)$_4$ led to the isolation of a bacterial strain *Bacillus cereus* DF4-B. The extracellular enzyme hydrolyzing (D-Phe)$_4$ was purified and characterized (48). The M_r of the subunit calculated was about 36,000 by gel electrophoresis (SDS-PAGE), 37,000 by HPLC, and 37,952 by mass spectrophotometry. The absorption of the purified enzyme in 0.01 M potassium phosphate buffer, pH 7.0, was maximal at 281 nm. The enzyme is an endopeptidase that acts D-stereospecifically on peptides composed of aromatic D-amino acids, recognizing the D-configuration of the amino acid whose carboxy-terminal peptide bond is hydrolyzed. The enzyme had an optimum pH at around 10.3. Thus, the enzyme was named alkaline D-peptidase (D-stereospecific peptide hydrolase [EC 3.4.11.–]). The substrate specificity of the enzyme was examined as shown in Table 3. Each substrate (2 mM) was incubated under standard enzyme assay conditions (48). The enzyme was active

```
                 9            18            27            36            45            54            63            72            81
5' GAA TTC AAT GTT GAC TTT TTT TTT TGG TGG AGA TGA TAG CTG TGT GGT TAA TAC GAA ATA GCA AGA AAC CAC CTA TAC AAT

                90            99           108           117           126           135           144           153           162
   ATG AAT AGA GGA GAG AAA TGC CCT ATG AAA ACA CGT AGT CAA ATT ACA TGT GCA AGT CTT GCC CTT TTA ATA GCT GGA AGT
                                   Met Lys Thr Arg Ser Gln Ile Thr Cys Ala Ser Leu Ala Leu Leu Ile Ala GGA AGT

               171           180           189           198           207           216           225           234           243
   TCC CTG TTA TAC GAG ACG CAA ACC TTA ATT GTA AAA GCA GAA CCT ACT CAA AGT GTA TCT AGT TCG GTA CAA ACA AGT ACT
   Ser Leu Leu Tyr Thr Thr Gln Thr Leu Ile Val Lys Ala Glu Pro Thr Gln Ser Val Leu Ser Ser Val Gln Thr Ser Thr

               252           261           270           279           288           297           306           315           324
   CAA CGA GAT CGT AAT TCT GTC AAG CAA GCA GTG CGG GAT ACA TTG CAA CTT GGA TTC CCG GGG ATA CTT GCT AAA ACT TCT
   Gln Arg Asp Arg Asn Ser Val Lys Gln Ala Val Arg Asp Thr Leu Gln Leu Gly Phe Pro Gly Ile Leu Ala Lys Thr Ser

               333           342           351           360           369           378           387           396           405
   GAG GGT GGA AAA ACA TGG AGT TAT GCC GCT GGG GTA GCG AAT CTG AGC AGC AAG AAA CCC ATG AAA ACA GAT TTT CGC TTT
   Glu Gly Gly Lys Thr Trp Ser Tyr Ala Ala Gly Val Ala Asn Leu Ser Ser Lys Lys Pro Met Lys Thr Asp Phe Arg Phe

               414           423           432           441           450           459           468           477           486
   CGC ATT GGT AGC GTG ACG AAG ACG TTC ACG GCA ACA GTT GTA CTT CAa TTA GCC GAA GAG AAC CGC TTG AAT CTA GAC GAC
   Arg Ile Gly Ser Val Thr Lys Thr Phe Thr Ala Thr Val Val Leu Gln Leu Ala Glu GLu Asn Arg Leu Asn Leu Asp Asp

               495           504           513           522           531           540           549           558           567
   TCT ATT GAA AAA TCG TTG CCT GGT GTC ATT CAA GGA AAT GGG TAT GAT GAT AAA CAG ATT ACT ATC CGG CAA TTA TTG AAC
   Ser Ile Glu Lys Trp Leu Pro Gly Val Ile Gln Gly Asn Gly Tyr Asp Asp Lys Gln Ile Thr Ile Arg Gln Leu Leu Asn

               576           585           594           603           612           621           630           639           648
   CAC ACA AGT GGT ATC GCT GAA TAC ACA AGG TCA AAA AGT TTT GAT CTT ATG GAT ACT AAA AAA TCG TAT AGG GCT GAA GAA
   His Thr Ser Gly Ile Ala Glu Tyr Thr Arg Ser Lys Ser Phe Asp Leu Met Asp Thr Lys Lys Ser Tyr Arg Ala Glu Glu

               657           666           675           684           693           702           711           720           729
   TTA GTA AAG ATG GGG ATT TCG ATG CCC CCA GAT TTT GCT CCA GGA AAG AGC TGG TCT TAT TCA AAT ACA GGA TAC GTA TTA
   Leu Val Lys Met Gly Ile Ser Met Pro Pro Asp Phe Ala Pro Gly Lys Ser Trp Ser Tyr Ser Asn Thr Gly Tyr Val Leu

               738           747           756           765           774           783           792           801           810
   CTT GGG ATC CTT ATT GAA ACA GTA ACC GGG AAC AGC TAT GCG GAA GAG ATT GAA AAT CGG ATT ATT GAA CCG CTT GAA TTA
   Leu Gly Ile Lys Ile Glu Thr Val THr Gly Asn Ser Tyr Ala Glu Glu Ile Glu Asn Arg Ile Ile Glu Pro Leu Glu Leu

               819           828           837           846           855           864           873           882           891
   TCG AAT ACA TTC TTA CCT GGC AAT TCA AGC GTG ATT CCA GGA ACC AAG CAT GCC CGT GGA TAT ATC CAA CTA GAC GGA GCA
   Ser Asn Thr Phe Leu Pro Gly Asn Ser Ser Val Ile Pro Gly Thr Lys His Ala Arg Gly Tyr Ile Gln Leu Asp Gly Ala

               900           909           918           927           936           945           954           963           972
   AGT GAG CCA AAA GAT GTT ACT TAT TAT AAC CCA AGT ATG GGG AGC TCG GCT GGA GAT ATG ATT TCT ACT GCT GAT GAT TTA
   Ser Glu Pro Lys Asp Val Thr Tyr Tyr Asn Pro Ser Met Gly Ser Ser Ala Gly Asp Met Ile Ser Thr Ala Asp Asp Leu

               981           990           999          1008          1017          1026          1035          1044          1053
   AAC AAA TTC TTC TCT TAC TTA CTT GGG GGT AAA TTA CTA AAA GAA CAG CAA CTA AAA CAG ATG CTT ACT ACA GTT CCA ACA
   Asn Lys Phe Phe Ser Tyr Leu Leu Gly Gly Lys Leu Leu Lys Glu Gln Gln Leu Lys Gln Met Leu Thr Thr Val Pro Thr

              1062          1071          1080          1089          1098          1107          1116          1125          1134
   GGA GAA GCT GCA CTT GGC AGA TAT GGT CTT GGA ATC TAT GAA ACT AAG CTT CCA AAC GGT GTC TCA ATA TGG GGA CAC GGA
   Gly Thr Ala Ala Leu Gly Arg Tyr Gly Leu Gly Ile Tyr Glu Thr Lys Leu Pro Asn Gly Val Ser Ile Trp Gly His Gly

              1143          1152          1161          1170          1179          1188          1197          1206          1215
   GGT AGC ATT CCA GGG TTT GTT ACT TTT GCT GGA GGC ACA CTT GGA GGC AAG CAT ACA TTA GCT GTC AAT TTG AAC AGC CTT
   Gly Ser Ile Pro Gly Phe Val Thr Phe Ala Gly Gly Thr Leu Gly Gly Lys His Thr Leu Ala Val Asn Leu Asn Ser Leu

              1224          1233          1242          1251          1260          1269          1278          1287          1296
   AAT GCT GAG AGT CCT GAT CCT TTT AAA AAT ATT TTA CTT GCT GAA TTT AGC AAG TAG GCC AAA AGG AAG AAC GGA TAA ATT
   Asn Ala Glu Ser Pro Asp Pro Phe Lys Asn Ile Leu Leu Ala Glu Phe Ser Lys ***

              1305          1314          1323          1332          1341          1350          1359          1368          1377
   TTA TCA TAT TTT CAT AGT TAA ACA AAT AAA TGA TGA AAT TGA TCA CTT ATC TAA GGA GGC AAT TCC CTC ATG AAA AAT AAA

              1386          1395          1404          1413          1422          1431          1440          1449          1458
   CTA ACT GGT GTA TTA GCC GCT ACA TTA GCG CTA ACT ATG ATT CTA CCA ACA GGG GCG AAA GCA TTT TCT GAC GGT AAG ACT

              1467          1476          1485          1494          1503          1512          1521          1530          1539
   ACA GTT GTA TCT AAC GCA GAA GTA GCT AGC CAA GAA CTG AAG AAG ATT GCA GCA GAA AAG GCT GCG CTA CTC ACG AAG TCC

              1548          1557          1566          1575          1584          1593          1602          1611          1620
   CAT GGG ACG AGT AGC GTG CAG TAT GCG CTA ATT GAT AAT GGC AAG CTT ACT TTG TCG GGG CAG GCT GGA AAG AAC GAT ATG

              1629          1638          1647          1656          1665          1674          1683          1692          1701
   GAA GGG GAA CAG CCG CTA ACC AAA GAT ACT CTA TAC GGG ATT GGT TCA GTC AGT AAA ATG TAT GCA ACA GCG GCT GTA ATG

              1710
   AAA TTG GTC GAC 3'
```

Figure 8. Nucleotide sequence of the *adp* gene and the predicted amino acid sequence. In this open reading frame of 1,164 bp (388 amino acids), shown with the deduced amino acid sequence, sequences confirmed by the protein sequencer are underlined.

Bacillus cereus ADP (48)	55	M K T D F R F R I G **S** V T **K** T F T A T V V	75
Streptomyces Cpase DD (30)	83	I T T T D R F R V G **S** V T **K** S F S A V V L	103
Streptomyces lactamdurans PBP (49)	50	V T T D S V F Q S G **S** V A **K** V Y T A T L V	70
Bacillus subtilis PBP 4 (50)	50	L K T N S L F E L A **S** L S **K** P F T A L G I	70
Serratia marcescens BLA (51)	69	I T E Q T L F E V G **S** L S **K** T F T A T L A	89
Enterobacter cloacae BLA (52)	74	V T P Q T L F E L G **S** I S **K** T F T G V L G	94
Dichelobacter nodosus Protein D (53)	103	M R V N S R F R Y A **S** V T **K** V L T S A L V	123
Ochrobactrum anthropi DAP (28)	55	M T L D T R M P I C **S** V S **K** Q F T C A V L	75
Pseudomonas sp. Esterase (31)	65	W H S D T I V N L F **S** C T **K** T F T A V T A	85

Figure 9. Partial alignments of the amino acid sequences of ADP and other similar enzymes. Met^{55} found in the partial amino acid sequence of ADP corresponds to ^{387}ATG in Figure 8; abbreviations as in Figure 4.

Table 4. Comparison of Properties of the D-Amino Acid–Containing Peptide Hydrolases

Property	D-Aminopeptidase	Alkaline D-peptidase
Origin	*Ochrobactrum anthropi*	*Bacillus cereus*
Molecular weight	114,514 (57,252 × 2)	37,952
Number of subunits	2	1
Optimum pH	8.5	10.5–11
Substrates (V_{max}/K_m)		
(D-Ala)$_4$	2,190	
(D-Phe)$_4$		364
Ampicillin	0.04	3.58
Penicillin G	0.05	5.11
Mode of action	D-Stereospecific aminopeptidase	D-Stereospecific endopeptidase
Similarity	Cpase DD (27%)	Cpase DD (35%)
	Class C BLA (22%)	Class C BLA (25%)
	PBP 4 (29%), DAP (27%)	

Note: BLA, β-lactamase; C Base, carboxy peptides; DAP, D-aminopeptidase; PBP, penicillin-binding protein.

toward (D-Phe)$_3$ and (D-Phe)$_4$, forming (D-Phe)$_2$ and D-Phe. The enzyme is also active toward tripeptides with D-Tyr at the C- or N-terminus and on Boc-(D-Phe)n (n = 2–4), forming Boc-D-Phe, (D-Phe)$_2$, and D-Phe. The enzyme had esterase activity toward D-Phe methyl ester and (D-Phe)$_2$ methyl ester. The products from Boc-(D-Phe)$_3$ *tert*-butyl ester were Boc-D-Phe, D-Phe, and D-Phe *tert*-butyl ester. The enzyme was not active toward L-Phe methyl ester, (L-Phe)$_2$ methyl ester, (L-Phe)$_4$, Boc-(L-Phe)$_4$, Boc-(L-Phe)$_4$ methyl ester, (D-Val)$_3$, (D-Leu)$_2$, and (D-Ala)n (n = 2–5). These properties indicated that the enzyme is an endopeptidase that acts D-stereospecifically on peptides composed of aromatic D-amino acids. On the other hand, a dimer was formed when D-Phe methyl ester and D-Phe amide were the substrates. Since the enzyme is found to be a serine peptidase as described it is anticipated that later, the enzyme will be useful in the kinetically controlled synthesis of peptides.

Eight stereoisomers of phenylalanine trimer were synthesized, and their effectiveness as substrates for the enzyme was tested. The enzyme recognized the configuration of the second D-Phe of tripeptides and catalyzes the hydrolysis of the second peptide bond from the N-terminus. The calculated V_{max}/K_m values for the peptides containing L-Phe were lower than that for (D-Phe)$_3$, affected by the neighboring L-Phe. The enzyme also showed β-lactamase activity toward ampicillin and penicillin G. The calculated V_{max} values of the enzyme for β-lactam compounds were about the same as those for (D-Phe)$_3$ and (D-Phe)$_4$, whereas the K_m values were several hundred times larger. On the other hand, carboxypeptidase DD (30) and D-aminopeptidase (28) activities were undetectable.

The time course of the (D-Phe)$_4$ degradation was measured. As shown in Figure 6, (D-Phe)$_4$ was hydrolyzed to (D-Phe)$_2$ and D-Phe. No (D-Phe)$_3$ was detected. These results coincide with the kinetic properties of the enzyme described earlier. The mode of action of the enzyme was examined with the synthetic substrates D-Tyr-(D-Phe)$_2$ and (D-Phe)$_2$-D-Tyr, as shown in Figure 7. When D-Tyr-(D-Phe)$_2$ was the substrate, D-Phe was released first, then D-Tyr was slowly formed. When (D-Phe)$_2$-D-Tyr was used as a substrate, D-Tyr was released first, then D-Phe was slowly formed. In both reactions, the second peptide bond from the N-terminus of the substrate was hydrolyzed first. These results showed that the enzyme acts as a D-stereospecific dipeptidylendopeptidase.

The enzyme activity was maximal at 45 °C. About 60% activity remained after an incubation at 43 °C in 0.1 M potassium phosphate buffer, pH 8.0, for 10 min. No activity was lost between pH 5.0 and 10.0 after an incubation at 30 °C for 1 h in 0.05 M buffers at various pH values (48).

The enzyme activity was enhanced by Mg^{2+} (138%), Mo^{3+} (130%), and Ba^{2+} (123%). When measured after incubating at 30 °C for 30 min, the activity was inhibited to 94% by phenylmethylsulfonyl fluoride (PMSF), 76% by Ag^{+}, 74% by Fe^{2+}, and 32% by Hg^{2+} at 5 mM. These results, together with the information about its primary structure as described later, indicated that the enzyme is a serine peptidase (48).

Cloning of the Gene for the Alkaline D-Peptidase

The gene coding for alkaline D-peptidase (*adp*) was cloned into plasmid pUC118, and a 1,164 bp open reading frame consisting of 388 codons with an M_r of 42,033 was identified as the *adp* gene (Fig. 8) (48). The enzyme would be synthesized with a signal peptide.

The deduced primary structure of the enzyme is similar to carboxypeptidase DD from *Streptomyces* R61 (35.0% identical over 346 amino acids [a.a.]) (14), penicillin-binding proteins from *Streptomyces (Nocardia) lactamdurans* (28.1% over 263 a.a.) (49) and that of *B. subtilis* (28.5% over 309 a.a.) (50), and class C β-lactamases of *Serratia marcescens* (24.9% over 217 a.a.) (51), class C β-lactamases of *Enterobacter cloacae* (25.1% over 191 a.a.) (52), fimbrial protein D from *Dichelobacter nodosus* (24.1% over 261 a.a.) (53), D-aminopeptidase from *O. anthropi* (27.5% over 182 a.a.) (28), and esterase from *Pseudomonas* sp. (30.5% over 154 a.a.) (54). As shown in Figure 9, the sequence of Ser-X_{aa}-X_{aa}-Lys is perfectly conserved among this class of enzymes and the consensus sequence is located around 60 residues from the N-termini of most of the enzymes. Thus, we propose that alkaline D-peptidase from *B. cereus* be categorized as a new member of the penicillin-recognizing enzymes, which include penicillin-binding proteins, β-lactamases, and D-aminopeptidase. Table 4 summarizes the properties of the D-amino acid–containing peptide hydrolases, D-aminopeptidase, and alkaline D-peptidase.

Synthesis of D-Phe Oligopeptides by the Alkaline D-Peptidase (55)

The alkaline D-peptidase not only hydrolyzed $(\text{D-Phe})_3$ and $(\text{D-Phe})_4$ to form $(\text{D-Phe})_2$ and D-phenylalanine but also acted on D-Phe methyl ester and D-Phe amide to form $(\text{D-Phe})_2$. This finding gave us the opportunity to use the enzyme to further investigate the synthesis of D-Phe oligopeptides from D-Phe methyl ester. Because the alkaline D-peptidase was found to be a serine peptidase (48), we attempted to use it for kinetically controlled peptide synthesis. An expression plasmid pKADP was constructed by placing the alkaline D-peptidase gene (*adp*), amplified by means of the polymerase chain reaction, under the *tac* promoter of pKK223-3. Oligomerization of D-phenylalanine methyl ester by means of the purified enzyme from the transformant *E. coli* was investigated under several conditions. D-Phe dimer, $(\text{D-Phe})_2$, and trimer, $(\text{D-Phe})_3$, were produced in 25.4% and 8.6% yield, respectively, when 50 mM of the substrate was incubated for 8 h with ADP (2.0 and 0.4 U/mL, respectively) in 100 mM triethylamine HCl (pH 11.5). Addition of dimethyl sulfoxide to the reaction mixture resulted in the production of tetramer $(\text{D-Phe})_4$ in 6.7% yield with the decrease of the $(\text{D-Phe})_2$ and $(\text{D-Phe})_3$ production. This is the first example of the synthesis of D-phenylalanine oligomers by means of a D-stereospecific endopeptidase.

BIBLIOGRAPHY

1. G.M. Whitesides and C.-H. Wong, *Angew. Chem. Int. Ed. Engl.* **24**, 617–638 (1985).
2. C. Laane, J. Tramper, and M.D. Lilly, eds., *Biocatalysts in Organic Media*, Elsevier, Amsterdam, 1987.
3. H. Yamada and S. Shimizu, *Angew. Chem. Int. Ed. Engl.* **27**, 622–642 (1988).
4. J.B. West and C.-H. Wong, *J. Org. Chem.* **51**, 2728–2735 (1986).
5. A.L. Margolin, D.-F. Tai, and A.M. Klibanov, *J. Am. Chem. Soc.* **109**, 7885–7887 (1987).
6. K. Kato, K. Kawahara, T. Takahashi, and S. Igarashi, *Agric. Biol. Chem.* **44**, 821–825 (1980).
7. I.B. Stoineva and D.D. Petkov, *FEBS Lett.* **183**, 103–106 (1985).
8. H. Nakajima, S. Kitabatake, R. Tsurutani, K. Yamamoto, I. Tomioka, and K. Imahori, *Int. J. Pept. Protein Res.* **28**, 179–185 (1986).
9. A.L. Margolin and A.M. Klibanov, *J. Am. Chem. Soc.* **109**, 3802–3804 (1987).
10. C.F. Barbas, III and C.-H. Wong, *J. Chem. Soc., Chem. Commun.* **1987**, 533–534.
11. J.-M. Frère and B. Joris, *CRC Crit. Rev. Microbiol.* **11**, 299–396 (1985).
12. M. Sugie and H. Suzuki, *Agric. Biol. Chem.* **50**, 1397–1402 (1986).
13. P.E. Reynolds, F. Depardieu, S. Dutka-Malen, M. Arthur, and P. Courvalin, *Mol. Microbiol.* **13**, 1065–1070 (1994). GenBank database accession number M97297.
14. R.C. Del Milton, S.C.F. Milton, and S.B.H. Kent, *Science* **256**, 1445–1448 (1987).
15. Y. Asano, T. Mori, S. Hanamoto, Y. Kato, and A. Nakazawa, *Biochem. Biophys. Res. Commun.* **162**, 470–474 (1989).
16. Japanese Pat. 88-87998 (1988), M. Dotani, H. Igarashi, and U. Uragami.
17. A. Ozaki, H. Kawasaki, M. Yagasaki, and Y. Hashimoto, *Biosci. Biotechnol. Biochem.* **56**, 1980–1984 (1992).
18. A. Ozaki, H. Kawasaki, M. Yagasaki, and Y. Hashimoto, *Biosci. Biotechnol. Biochem.* **57**, 520–521 (1993).
19. J. Ogawa and S. Shimizu, *J. Mol. Catal.* **B2**, 163–176 (1997).
20. M. Yagasaki and A. Ozaki, *J. Mol. Catal.* **B4**, 1–11 (1998).
21. Y. Asano, A. Nakazawa, Y. Kato, and K. Kondo, *J. Biol. Chem.* **264**, 14233–14239 (1989).
22. Y. Kato, Y. Asano, A. Nakazawa, and K. Kondo, *Tetrahedron* **45**, 5743–5754 (1989).
23. I.M. Chaiken, A. Komoriya, M. Ohno, and F. Widmer, *Appl. Biochem. Biotechnol.* **7**, 385–399 (1982).
24. K. Morihara, *Trends Biotechnol.* **5**, 164–170 (1987).
25. Y. Kato, Y. Asano, A. Nakazawa, and K. Kondo, *Biocatalysis* **3**, 207–215 (1990).
26. L. Sluyterman and J. Wijdenes, *Biochim. Biophys. Acta* **289**, 194–202 (1972).
27. K. Aso, T. Uemura, and Y. Shiokawa, *Agric. Biol. Chem.* **52**, 2443–2449 (1988).
28. Y. Asano, Y. Kato, A. Yamada, and K. Kondo, *Biochemistry* **31**, 2316–2328 (1992). GenBank database accession number M84523.
29. Y. Asano, K. Kishino, A. Yamada, S. Hanamoto, and K. Kondo, *Recl. Trav. Chim. Pays-Bas* **110**, 206–208 (1991).

30. C. Duez, C. Piron-Fraipont, B. Joris, J. Dusart, M.S. Urdea, J.A. Martial, J.-M. Frère, and J.-M. Ghuysen, *Eur. J. Biochem.* **162**, 509–518 (1987). GenBank database accession number X05109.
31. F. Lindberg and S. Normark, *Eur. J. Biochem.* **156**, 441–445 (1986). GenBank database accession number X03866.
32. B. Jaurin and T. Grundström, *Proc. Natl. Acad. Sci. U.S.A.* **78**, 4897–4901 (1981).
33. R.P. Ambler and G.K. Scott, *Proc. Natl. Acad. Sci. U.S.A.* **75**, 3732–3736 (1978).
34. K. Neugebauer, R. Sprengel, and H. Schaller, *Nucleic Acids Res.* **9**, 2577–2588 (1981). GenBank database accession number V00093.
35. M. Nakamura, I.N. Maruyama, M. Soma, J. Kato, H. Suzuki, and Y. Horota, *Mol. Gen. Genet.* **191**, 1–9 (1983). GenBank database accession number K00137.
36. J.K. Broome-Smith, A. Edelman, S. Yousif, and B.G. Spratt, *Eur. J. Biochem.* **147**, 437–446 (1985). GenBank database accession number X02163.
37. S. Asoh, H. Matsuzawa, F. Ishino, J.L. Strominger, M. Matsuhashi, and T. Ohta, *Eur. J. Biochem.* **160**, 231–238 (1986). GenBank database accession number X04516.
38. M.D. Song, M. Wachi, M. Doi, F. Ishino, and M. Matsuhashi, *FEBS Lett.* **221**, 167–171 (1987). GenBank database accession number Y00688.
39. J.K. Broom-Smith, I. Ioannidis, A. Edelman, and B.G. Spratt, *Nucleic Acids Res.* **16**, 1617 (1988). GenBank database accession number X06479.
40. B.G. Spratt, *Nature* **332**, 173–176 (1988). GenBank database accession number X59632.
41. J.W. Dale, D. Godwin, D. Mossakowska, P. Stephenson, and S. Wall, *FEBS Lett.* **191**, 39–44 (1985). GenBank database accession number X03037.
42. P. Huovinen, S. Huovinen, and G.A. Jacoby, *Antimicrob. Agents Chemother.* **32**, 134–136 (1988). GenBank database accession number J03427
43. B. Joris, F. De Meester, M. Galleni, S. Masson, J. Dusart, J.-M. Frère, J. Van Beeumen, K. Bush, and R. Sykes, *Biochem. J.* **239**, 581–586 (1986).
44. Y. Asano, A. Nakazawa, Y. Kato, and K. Kondo, *Angew. Chem.* **101**, 511–512 (1989).
45. Y. Asano and K. Yamaguchi, *J. Ferment. Bioeng.* **79**, 614–616 (1995).
46. R.A. Nicholas and J.L. Strominger, *Rev. Infect. Dis.* **10**, 733–738 (1988).
47. Y. Asano, in K. Takai ed., *Frontiers and New Horizons in Amino Acid Research*, Elsevier, Amsterdam, 1992, pp. 333–336.
48. Y. Asano, H. Ito, T. Dairi, and Y. Kato, *J. Biol. Chem.* **271**, 30256–30262 (1996). GenBank database accession number D86380.
49. J.J.R. Coque, P. Liras, and J.F. Martin, *EMBO J.* **12**, 631–639 (1993). EMBL database accession number Z13971.
50. D.L. Popham and P. Setlow, *J. Bacteriol.* **175**, 2917–2925 (1993). GenBank database accession number L10629.
51. B. Joris, F. De Meester, M. Galleni, G. Reckinger, J. Coyette, and J.-M. Frère, *Biochem. J.* **228**, 241–248 (1985).
52. M. Galleni, F. Lindberg, S. Normark, S. Cole, N. Honore, B. Joris, and J.-M. Frère, *Biochem. J.* **250**, 753–760 (1988).
53. M. Hobbs, B.P. Dalrymple, P.T. Cox, S.P. Livingstone, S.F. Delaney, and J.S. Mattick, *Mol. Microbiol.* **5**, 543–560 (1991). GenBank database accession number M32230.
54. D.B. McKay, M.P. Jennings, E.A. Godfrey, I.C. MacRae, P.J. Rogers, and I.R. Beacham, *J. Gen. Microbiol.* **138**, 701–708 (1992). GenBank database accession number M68491.
55. H. Komeda and Y. Asano, *J. Mol. Catal. B.* In press.

AMMONIA TOXICITY, ANIMAL CELLS

SADETTIN S. OZTURK
Bayer Corporation
Berkeley, California

KEY WORDS

Ammonia
Ammonium
Cell culture
Fermentation
Inhibition
Ion exchange
Toxicity

OUTLINE

INTRODUCTION

Mammalian cells are used for the production of several important biologicals including interferons, growth factors, vaccines, hormones, and monoclonal antibodies (1–4).

The process efficiency of cell culture fermentation should be increased in most cases for economically viable production (4). Maintaining high numbers of viable cells in the bioreactor for extended periods of time results in higher product concentration and higher reactor productivity, thus increasing process efficiency.

The cell growth rate and the number of cells attained in the bioreactor are determined, in part, by the concentration of nutrients available to the cells. Even though fed-batch or perfusion systems can eliminate nutrient depletion, they are still prone to inhibitory metabolite accumulation. Several metabolic by-products have been reported to inhibit cell growth. Ammonia* and lactate are major metabolic by-products of glucose and glutamine metabolism (5,6), and their presence has been demonstrated to diminish growth of mammalian cells and to influence protein production. Lactate is generated mainly from incomplete oxidation of glucose, and it affects the cells through alteration of culture pH and osmolarity (6). The effects of ammonia on cells are more direct and occur at much lower concentrations. Thus, ammonia is generally more important as an inhibitory metabolic by-product in cell culture.

Ammonia is produced mainly from glutamine, an essential component of cell culture media. Glutamine gives rise to ammonia through chemical degradation and cellular metabolism. Concentration of ammonia in cell culture is influenced by many factors, including the mode of reactor operation, cell concentration, glutamine concentration, and cellular metabolism. Typically, ammonia concentrations reach 2 to 5 mM in batch cultures. This concentration can be still higher in fed-batch reactors where glutamine is the limiting substrate and therefore supplemented continually, leading to increased ammonia accumulation.

Although different cells exhibit different tolerance to ammonia, ammonia concentrations as low as 2 mM have been shown to adversely affect mammalian cells. Numerous studies have reported the effect of ammonia on mammalian cells in different cell culture systems. It has been established that ammonia not only affects cell growth but also alters cell metabolism, product expression, and product quality (7). The action mechanism for ammonia is complicated and not well characterized. Ammonia was shown to alter intracellular pH, membrane potential, and intracellular composition and to result in futile cycles.

Minimization of ammonia inhibition should yield better cell growth and increase the longevity of the cultures. Several techniques were used to minimize ammonia accumulation in cell culture systems. Media modification involves replacement or elimination of glutamine from cell culture medium. The success of this method depends on the cell line, and in most cases the cells should be adapted or genetically altered. Alternatively, the ammonia generated in the media can be removed by physical and chemical means.

In this article, we review the current knowledge on ammonia inhibition in cell culture. The sources of ammonia accumulation; the effects of ammonia on cell growth, metabolism, product secretion, and product quality; the mechanism of ammonia inhibition; and the methods of reducing ammonia concentration in cell culture are discussed.

*In this text we use ammonia in reference to both gaseous (ammonia) and ammonium ion. The ammonia and ammonium ion are present simultaneously in the culture. The relative concentration of ammonium ion is determined by the pH-dependent equilibria.

SOURCES OF AMMONIA ACCUMULATION

There are two major sources for ammonia accumulation, and in both cases glutamine is involved. Glutamine is one of the important energy sources for the mammalian cells in vitro (8,9). Addition of glutamine has been shown to stimulate cell growth and antibody production (10–15). Because almost all conventional cell culture media contain glutamine, the problem of ammonia accumulation is of universal importance (16,17). However, depending on the cell line and media composition, the effects can be different from one case to another.

Generation of Ammonia Caused by Glutamine Degradation

Glutamine is not stable in aqueous solutions and is degraded to yield a five-member ring structure, pyrrolidone carboxylic acid, and ammonia (18–20):

$$\text{Glutamine} \rightarrow \text{Pyrrolidone carboxylic acid} + NH_4^+ \quad (1)$$

This reaction is irreversible, and the extent of the reaction is affected by the chemical environment of the medium. Glutamine degradation follows first-order kinetics, the half-life of glutamine is reported to be between 6 and 20 days, depending on the conditions (16,21–23).

Glutamine degradation is pH (18,21,22,24) and temperature (20) dependent. Alkaline pH increases the degradation rate, and altering pH from 7.2 to 7.6 was reported to elevate the degradation rate by 300% (21,22). The presence of serum, especially if not heat inactivated, can elevate degradation of glutamine and subsequent ammonia accumulation caused by glutaminase and asparaginase activity (25). The media composition, especially the phosphate levels in the media, affects glutamine degradation significantly (22).

Glutamine degradation is an important problem for cell culture. The degradation not only generates inhibitory ammonia, but also decreases the level of glutamine available to the cells. The cell performance is affected negatively because of these two factors.

Generation of Ammonia Caused by Cell Metabolism

Although ammonia can be produced from other amino acids, glutamine metabolism is established to be the most important pathway. The pathways involving ammonia are summarized in Figure 1 and have been studied extensively in the literature (9,26–46).

Glutamine metabolism primarily takes place in mitochondria. Glutamine is initially deaminated by glutaminase, with ammonia as a byproduct (47,48). Ammonia can also be removed from glutamine via biosynthetic pathways, where the ammonia group is used in the formation of a biosynthetic product, particularly for pyrimidine and purine base synthesis (Fig. 1). The product formed by ei-

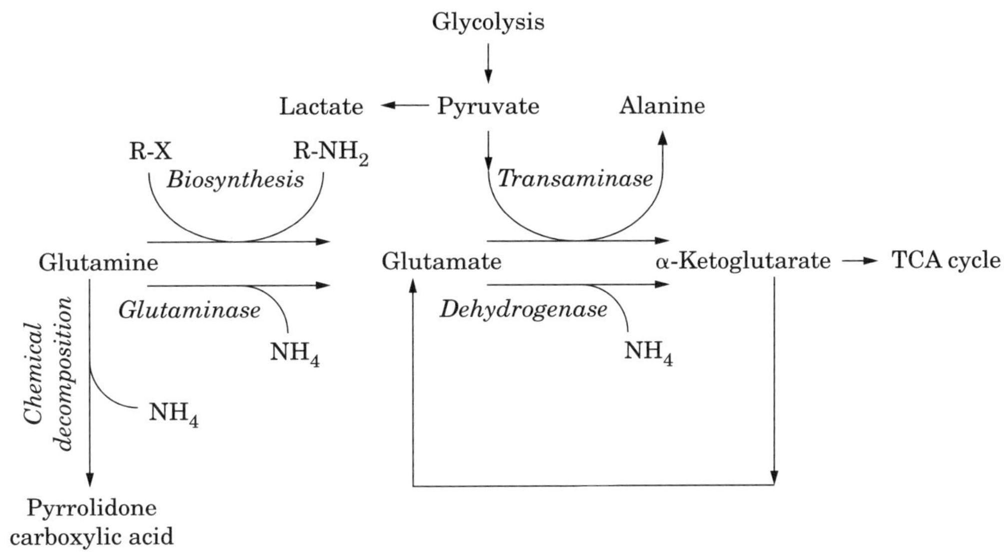

Figure 1. Ammonia generation by glutamine degradation and cell metabolism.

ther route is glutamate. The second step of glutaminolysis is the conversion of glutamate to α-ketoglutarate. Again, this conversion may be accomplished by more than one enzyme. An active transaminase can remove the ammonia group from glutamate. Less-active transaminases appear to deliver the ammonia group to the glycolytic intermediate 3-phosphoglycerate to form serine, which subsequently can be transformed to glycine. Glutamate can also be transformed by glutamate dehydrogenase to form α-ketoglutarate, which liberates a second ammonium ion. α-Ketoglutarate is then metabolized via the tricarboxylic acid (TCA) cycle.

The amount of ammonia generated is dependent on the metabolic pathway for glutamine. Depending on the pathway and the fate of glutamine, the ammonia yield on glutamine varies. Based on stoichiometry, Glacken (49) reported possible scenarios for glutamine metabolism and ammonia yield, between 1 and 2 mol of ammonia are generated per mole of glutamine consumed (Table 1). These yield coefficients, however, are very high and are not typically observed in cell culture.* For hybridoma cells, the ammonia yield was reported to be between 0.5 to 1 mol of ammonia generated per mole of glutamine consumed (6,33,50–52). Glutamine concentrations in typical cell culture media range between 1 and 7 mM (17). Thus, the ammonia concentrations in cell culture can reach 5 mM (6,50–53). Consequently, the higher glutamine concentrations found in special media formulations yield much more ammonia levels. Glutamine concentrations used in fed-batch cultures, for instance, can vary between 8 and 40 mM, and very high ammonia concentrations are obtained (14,54,55).

*It should be noted that the determination of ammonia yield from glutamine requires inclusion of the glutamine degradation rate in the analysis. When the degradation rate is omitted, the analysis results in erroneously high "apparent yields" (22).

Table 1. Amount of ATP Produced and Ammonia Generated in Different Metabolic Pathways

End product	ATP produced (mol/mol Gln)	NH_4 produced (mol/mol Gln)	NH_4/ATP
Alanine	9	1	0.11
Aspartate	9	1	0.11
Acetyl-CoA	15	2	0.13
Pyruvate	12	2	0.17
Lactate	9	2	0.22

Source: Glacken (49).

EFFECTS OF AMMONIA ON CELL PHYSIOLOGY

The influence of ammonia concentration on mammalian cell growth and product expression has been reported for a variety of cell lines and culture conditions (56). Most of these studies emphasized the influence of ammonia on cell growth. The effects of elevated ammonia concentrations on metabolic rates and cell productivity have been studied less extensively. Although it is not discussed in detail in this article, the effect of ammonia on virus replication and virus production from cell culture was demonstrated clearly in the literature (57–63).

Effects of Ammonia on Cell Growth

Although ammonia inhibited cell growth in all these studies, the sensitivity of the growth rate to ammonia concentration varied among the cell lines used. The extent of ammonia inhibition seems to be more severe for primary cells, whereas transformed cells tolerate higher ammonia levels. For 3T3 cells, ammonia levels as low as 0.6 mM reduced cell growth by 60%. However, when the cells are transformed, the reduction was reduced to 15% (64). Addition of 2 mM ammonia reduced growth by 30% for BSC-1 monkey epithelial cells (65). For established cell lines, inhibitory effects of ammonia were observed at higher ammonia concentrations. Hassell et al. (66) reported reduction of cell growth at 2 mM for a variety of cell lines. Several reports indicate ammonia inhibition for hybridoma cells with varying severity (6,41,50,52,67–82). Inhibitory concentrations of ammonia were observed to be in the range of 2 to 10 mM in these studies. For BHK cells, even lower concentrations of ammonia resulted in growth inhibition. Wentz and Schügerl (83) reported a 80% reduction at 1 mM ammonia, and Butler and Spier (84) reported a 75% decrease in growth at 3 mM of ammonia in BHK cells. For CHO cells, the inhibitory ammonia levels seems to be higher. Kurano et al. (85) obtained a 50% reduction in growth at 8 mM ammonia. Some cell lines are reported to be more resistant to ammonia. For instance, Schneider (81) observed only a 20% reduction in growth at 6 mM ammonia for HeLa cells, and Hansen and Emborg (86) did not observe any inhibition at up to 8 mM for a CHO clone. Other factors that can influence the inhibitory effects of ammonia on growth are the serum level and cell adaptation. Holley et al. (65) for instance, observed that increasing the level of serum from 0.1 to 10% reduced the ammonia inhibition. Miller et al. (52) observed adaptation of hybridoma cells to 8 mM ammonia. Maiorella et al. (54) used serial passaging in the presence of 5 mM ammonia and demonstrated that the resulting cells survived better in 10 mM ammonia. Adaptation to high ammonia concentrations can be used for selection of ammonia-resistant cell lines.

The effects of ammonia on cell growth in most of these studies reported were quantified as the reduction in cell density, and only a few studies evaluated the specific growth rates as a function of ammonia concentration. Ammonia inhibition on the specific growth rate can be described by a second-order inhibition model:

$$\mu = \frac{\mu_0}{1 + [NH_4^+]^2/K_a} \tag{2}$$

where μ and μ_0 are the growth rates in the presence and absence of ammonia, respectively; $[NH_4^+]$] is the initial ammonia concentration; and K_a is the inhibition constant. The square root of K_a corresponds to the ammonia concentration at which the specific growth rate decreases by 50%. Figure 2 presents the effect of ammonia on cell growth rate and cell density for 163.4.G5.3 murine hybridoma cells. Equation 2 describes the cell growth rate accurately with parameters $\mu_0 = 0.037$/h and $K_a = 24$ mM^2. For comparison, K_a value of 26 mM^2 was reported for the CRL-1606 hybridoma cell line (87), however, the value reported for the SB-4082 hybridoma cell line was almost one order of magnitude lower ($K_a = 3.2$ mM^2) (65).

Effects of Ammonia on Cell Death

Even though ammonia affects the cell growth rate and maximum cell density, effect on cell death is not well documented. Ozturk et al. (6) reported no effect of ammonia on the cell-specific death rate in the range of 0 to 5 mM; their data on the death rate are presented in Figure 2. When the ammonia concentration was increased from 0 to 3.75 mM, the specific death rate did not change. Ozturk and Palsson (88) also observed that elevated ammonia concentrations did not significantly influence the cell viability.

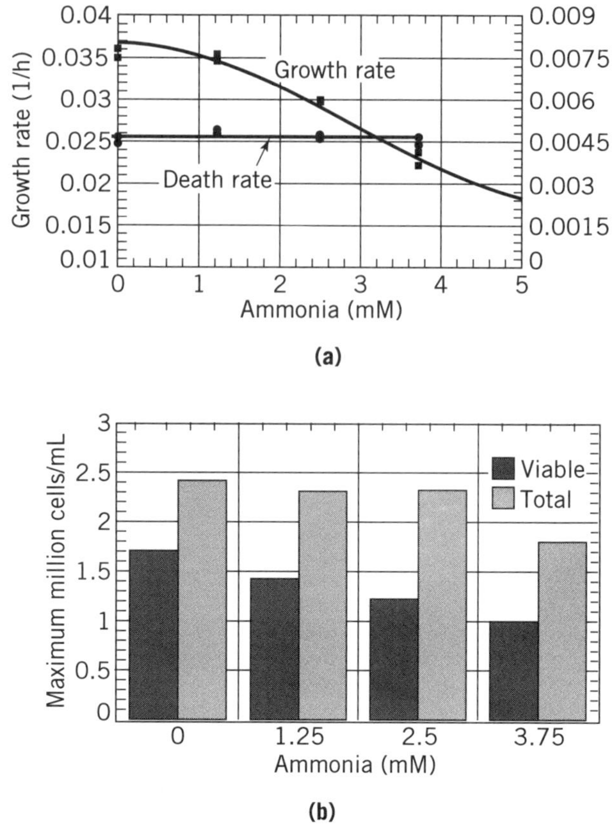

Figure 2. Effect of ammonia on murine hybridoma culture. (**a**) Cell growth and death rates, and (**b**) maximum viable and total cell concentrations in batch culture. *Source*: Data obtained from Ozturk and Palsson (88).

Thus, the reduction in cell density at higher ammonia concentrations was simply the result of lower cell specific growth rates. On the other hand, Goergen et al. (89) used much higher concentrations of ammonia (up to 17 mM) and observed an increase in the death rate. Newland et al. (80) also reported the influence on death rates by ammonia at high concentrations. More quantitative studies are needed to verify the effects of ammonia on specific death rates and the mechanism(s) involved.

Effects of Ammonia on Cell Metabolism

A number of studies reported the effects of ammonia on cell metabolism. For a murine hybridoma cell line, Ozturk et al. (6) observed a roughly twofold increase in glucose and glutamine consumption rates in cultures exposed to 3.75 mM of ammonia, compared to control conditions (Fig. 3). The production rates of lactate and ammonia were also enhanced by the presence of ammonia, indicating an elevated metabolic state of the cells. The yield coefficients of ammonia from glutamine and lactate yield from glucose decreased about 15% with 3.75 mM ammonia (Fig. 3). A similar elevation in metabolic rates was reported by Miller et al. (52) for AB2-143.2 hybridoma cells and by Alex et al. (90) for C127 cells. Glacken (87) has also shown an increase in the glutamine consumption rate by added ammonia.

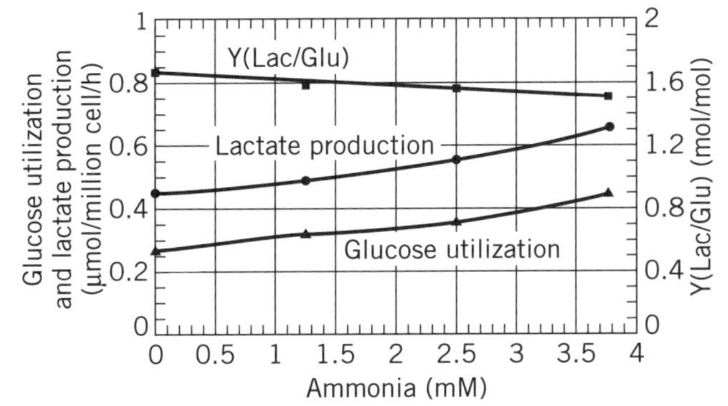

(a)

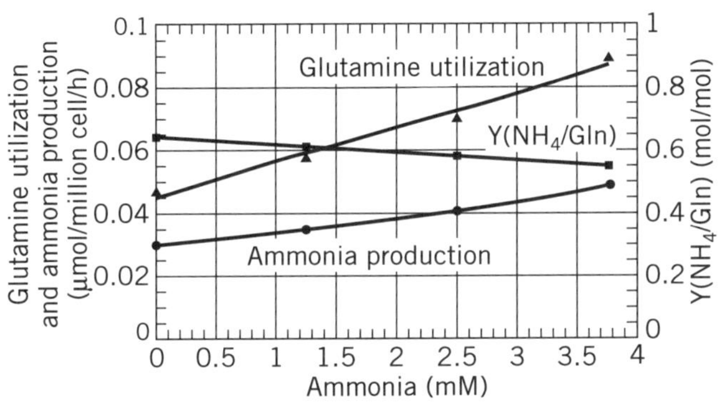

(b)

Figure 3. Effect of ammonia on murine hybridoma culture. (**a**) Glucose consumption, lactate production, and lactate yield on glucose; (**b**) Glutamine consumption, ammonia production, and ammonia yield on glutamine. Y, yield; Lac, lactate; Gln, glutamine; Glu, glucose. *Source*: Data obtained from Ozturk and Palsson (88).

McQueen and Bailey (91) indicated an increase in hybridoma cell yield from glucose and glutamine that resulted from a decrease in growth rate and an increase in metabolic rate.

Miller et al. (52) and Ozturk et al. (6) studied the effects of ammonia on amino acid metabolism. Table 2 summarizes the effects of ammonia on amino acid metabolic rates for murine hybridoma 163.4G5.3 (6). For this cell line, glutamate, serine, glycine, and alanine were produced and all other amino acids were consumed. The consumption or production rates of these amino acids were doubled at 3.75 mM ammonia. The increase in the amino acid consumption or production rates were parallel to the glutamine consumption rate because they increased at higher ammonia concentrations by the same magnitude. Only serine, arginine, alanine, and valine showed a relatively different response to ammonia concentration. The consumption rate for serine decreased. The rate of increase for arganine (1.5-fold increase at 3.75 mM ammonia) was lower compared to glutamine (1.8-fold increase). On the other hand, the increase in alanine production (2.1-fold) and valine consumption (2.6-fold) was higher (6).

The yield coefficients of lactate from glucose and ammonia from glutamine decreased, whereas alanine yield

Table 2. Amino Acid Consumption and Production Rates (Parentheses) as a Function of the Ammonia Concentrations

	$[NH_4^+]$			
Amino acid	0 mM	1.25 mM	2.5 mM	3.75 mM
Aspartate	1.74	2.21	2.58	3.75
Glutamate	(2.92)	(1.69)	(2.03)	(3.25)
Asparagine	2.13	2.26	2.67	4.18
Serine	(2.45)	(2.07)	(2.03)	(0.47)
Glutamine	48.50	58.00	67.50	89.00
Histidine	0.12	0.23	0.35	0.20
Glycine	(0.58)	(0.32)	(0.66)	(0.66)
Threonine	3.10	3.01	3.86	4.14
Arginine	3.18	3.24	3.61	4.89
Alanine	(34.92)	(42.78)	(55.33)	(72.14)
Tyrosine	3.75	4.51	5.08	6.84
Methionine	2.14	2.98	2.54	3.51
Valine	5.68	8.46	9.61	14.70
Phenylalanine	2.21	3.48	3.86	4.57
Isoleucine	7.06	10.84	11.26	14.16
Leucine	8.64	11.14	13.35	18.57
Lysine	5.39	6.94	8.17	11.73

Source: Ozturk et al. (6).
Note: Rates are in nmol/106/cell/h.

from glutamine increased as a result of ammonia addition. Miller et al. (52) observed similar data for ammonia yield from glutamine. Hansen and Emborg (86) also observed a decrease in ammonia yield from glutamine at high ammonia concentrations for CHO cells. A decrease in the lactate yield coefficient has been reported at elevated ammonia concentrations for other mammalian cells (90). Similarly, McQueen and Bailey (71) reported a decrease in lactate yield from glucose and in ammonia yield from glutamine. An increase in the alanine yield from glutamine was reported by Miller et al. (52). An increase in amino acid metabolic rates was observed for the C127 cell line (90).

The effects of ammonia on amino acid metabolism can be interpreted in the view of Figure 1. Ozturk et al. (6) showed a 12% increase in the alanine yield from glutamine at elevated ammonia concentrations, whereas the ammonia yield decreases by the same amount. These data indicate that at elevated ammonia concentrations, relatively more of the glutaminolytic flux is via the alanine transaminase pathway than that of glutamate dehydrogenase. This shift could be due to the fact that ammonia is a byproduct of the dehydrogenase reaction, which is believed to operate nearly at equilibrium; thus, elevated ammonia concentrations would shift the equilibrium toward glutamate (Fig. 1). Both Miller et al. (52) and Ozturk et al. (6) showed an increase in the specific glutamine consumption rate at elevated ammonia concentrations. This increase in the consumption rate suggests that inhibition of glutaminase by ammonia is not operative in the hybridoma cell line investigated.

Effects of Ammonia on Energy Metabolism

Mammalian cells use various pathways to produce energy in the form of adenosine triphosphate (ATP). The production rate of ATP can be estimated using the lactate production and oxygen consumption rates (87,92,93):

$$q_{ATP} = q_{Lac} + 2(P/O) \cdot q_{O_2} = q_{Lac} + 6 \cdot q_{O_2} \quad (3)$$

where q_{ATP}, q_{Lac}, and q_{O_2} are the production rates of ATP, lactate, and oxygen, respectively, and P/O is the phosphorylation ratio. The term q_{Lac} corresponds to the contribution of glycolysis, and the term $6 \cdot q_{O_2}$ corresponds to the contribution of oxidative phosphorylation.

Miller et al. (52) observed a decrease in oxygen uptake at high ammonia concentrations in a continuous reactor operation. On the other hand, Kimura et al. (94) and Ozturk et al. (6) reported no significant effect of ammonia on oxygen consumption rates for a human leukemia line and a hybridoma line, respectively. The data of Ozturk et al. (6) on the effect of ammonia on oxygen consumption rates are presented in Figure 4A.

Miller et al. (52) and Ozturk et al. (6) reported an increase in the ATP production rate at high ammonia concentrations. The dependence of the ATP production rate on ammonia is presented in Figure 4a. Although cells generated more ATP at higher ammonia levels, this increased energy production was not, however, used for growth, because the growth rate was inhibited (6). It appears that the cells produce more energy under stressful conditions and use it for maintenance. The relative contribution of oxidative phosphorylation was observed to decrease, and

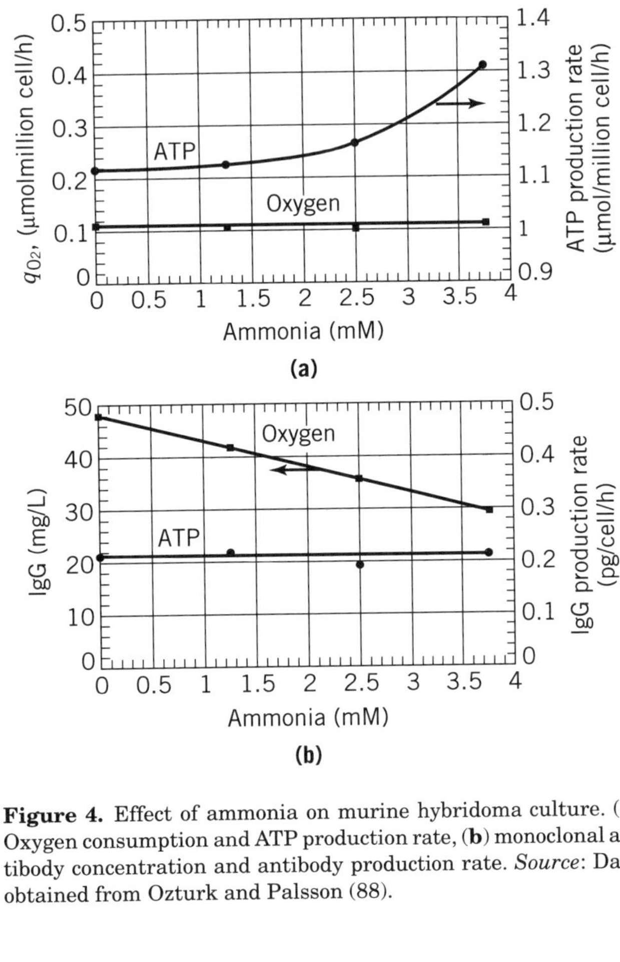

Figure 4. Effect of ammonia on murine hybridoma culture. (**a**) Oxygen consumption and ATP production rate, (**b**) monoclonal antibody concentration and antibody production rate. *Source*: Data obtained from Ozturk and Palsson (88).

the contribution of glycolysis increased with elevated ammonia levels (6,52).

Effects of Ammonia on Product Expression

Because ammonia affects cell growth and metabolism, it is conceivable that it also affects the product expression. Either directly or indirectly, ammonia can affect the product concentration obtained in the culture and thus the process efficiency.

The indirect effect of ammonia on production results from the reduced cell growth rate and subsequent cell density achieved in the culture. Even though the specific productivity may not have been altered, the cell density and longevity of the culture can be restricted by high ammonia levels. Thus, cultures with high ammonia levels result in lower product concentrations. Several studies have reported a decrease in monoclonal antibody production at high ammonia concentrations, mainly because of lower cell densities obtained (6,70,71). Figure 4b presents the effect of ammonia on antibody titer and specific productivity for a murine hybridoma line in batch culture. Although the titer was reduced, the specific antibody production rate remained constant under these conditions.

The direct effect of ammonia on specific productivity varies and the mechanism(s) are poorly understood. Dyken and Sambanis (95) observed a slight increase in protein secretion from AtT-20 cells. Ammonia was reported to alter the secretion rate of interferon (INF) γ (96) and synthesis of INF-β; (97); however, the specific production rates were not provided in these studies. Hansen and Emborg (86) reported a decrease in tissue-type plasminogen activator (t-PA) production from CHO cells at high ammonia concentrations. Specific antibody productivity was evaluated for a number of hybridoma cell lines in the literature. Glacken (87) reported a decrease in antibody productivity at elevated ammonia levels for the C-1606 hybridoma line. Other investigators, however, reported an unaltered specific antibody production rate (6,70,71).

Effect of Ammonia on Product Quality

Ammonia was reported to affect the conditions in intracellular compartments and enzymatic reactions for protein processing. High ammonia concentrations were reported to alter protein after translations, glycosylation, and secretion. Ammonia can alter the intracellular or intracompartmental pH, alter membrane potentials, and directly interact with enzymes. Thorens and Vassalli (98) reported inhibition of sialic acid transferase for Immunoglobulin (Ig) M at 10 mM ammonia. The results indicate a pH increase in Golgi because of ammonia load. Similarly, Andersen and Goochee (99) reported significant reduction in the terminal sialylation of O-linked glycosylation of recombinant granulocyte colony-stimulating factor. Borys et al. (100) studied the N-linked glycosylation of recombinant mouse placental lactogen-I from CHO cells and observed inhibition of glycosylation by ammonia. The effect of ammonia was also dependent on extracellular pH. Gawlitzek et al. (101,102) studied the glycosylation pattern of recombinant proteins expressed by BHK-21 cells. For the production of HuIL-2 variant in BHK cells, ammonia was observed to increase the intracellular uridine diphosphate-N-actylglucosamine (UDP-GlcNAc) pool. UDP-GlcNAc is a precursor substrate for the glycosylation process in the cytosol and Golgi. High ammonia levels led to a decrease in terminal sialyation and to an increase in branching. To test the hypothesis that UDP-GlcNAc is involved, the authors used glucosamine, which is the precursor for UDP-GlcNAc. Kopp et al. (103) studied product consistency and glycosylation patterns in recombinant CHO-expressed glycoproteins. High ammonia levels influenced glycosylation patterns significantly for INF-ω and t-PA. The alteration of product glycosylation by ammonia was described in monoclonal antibody production (104). To minimize the effects of ammonia, a multilevel pH control was proposed for commercial production. Genetic and process engineering strategies for control of ammonia in cell cultures were recently described to yield enhanced glycosylation (104).

MECHANISM FOR AMMONIA INHIBITION

Although a significant number of articles have been published on the affect of ammonia, the mechanism of action is not fully understood. Several mechanisms were hypothesized and tested experimentally for ammonia inhibition. Alteration of intracompartmental pH and membrane potential, futile cycles, metabolic inhibition, and alteration of critical ribonucleotides are identified as potential mechanisms for ammonia inhibition.

Intracellular pH

A possible explanation for the effect of ammonia is simply the alteration of intracellular and intracompartmental pH (6,72,106,107). Ammonia in cell culture is present in both gaseous and ionic form, and these species are in equilibrium at a ratio determined by the medium pH. At normal pH values, almost all the ammonia present is in the form of ammonium ion, and the concentration of gaseous ammonia is very low ($pK_a = 9.2$).

The regulation of intracellular (cytosolic) and intracompartmental pH in response to added ammonia is complex (Fig. 5). Both gaseous ammonia and ammonia ion can permeate the membranes. This permeation involves both passive (diffusion) and active transport.

The permeability of gaseous ammonia is much greater than that of the ammonium ion (6,49,72,107,108). Thus, initially, gaseous ammonia permeates rapidly, raising the intracellular pH. Then, the slower penetration of ammonium ion decreases the intracellular pH. McQueen and Bailey (72,107) and Ozturk et al. (6) studied the response of hybridoma cells to added ammonia and verified this mechanism. Figure 6 shows the response of intracellular pH to ammonia addition, followed by fluorescent dye. After the addition, there is an immediate increase in intracellular pH. This is due to rapid gaseous ammonium diffusion into the cells. The intracellular pH decays to steady-state intracellular pH values that are lower than those observed prior to the exposure to ammonia because of the transport of ammonium ion. The steady-state intracellular pH values for this experiment are presented in Table 3.

The rate of diffusion for ammonium ion through the cell membrane is four to five orders of magnitude lower than the rate of diffusion for gaseous ammonia (109). However, ammonium ion can be actively transported across the cell membrane via [Na,K]ATPase, [Na,K,Cl]-cotransporter, and [Na,H]-exchanger, as indicated in Figure 5 (109–114). The dynamics of ammonia transport for mammalian cells were analyzed by mathematical models (107).

The transport of ammonia and ammonium ion to the cells were studied in detail by Martinelle and Haggsrom (115–117) for the murine myeloma cell (Sp2/O-Ag14). In addition to diffusional transport, these authors identified the active transport of ammonium ion via [Na,K]ATPase and [Na,K,C]-cotransporter in hybridoma cells. The presence of K^+ could inhibit the [Na,K]ATPase active transport and alter the transport of ammonium ion. When K^+ (10 mM) was added, the change in intracellular pH due to ammonia addition was observed to be negligible. The authors postulate that one of the reasons for ammonia inhibition is the increased energy demand resulting from energy wasted because NH_4^+ is actively transported to the cells via [Na,K]ATPase, and they proposed the use of K^+ for minimization of ammonia inhibition. Use of KOH instead of NaOH for pH control is proposed to increase K^+ concentration in the culture.

A change in intracellular or intracompartmental pH can alter the activity of numerous enzymes and, depending on the location, different results can be obtained. The *net* result of ammonium ion transport to the cells is a decrease in intracellular pH. However, the ammonia transported to the cells further diffuses to compartments in the cells. Gaseous ammonia easily permeates intracellular membranes and the pH in mitochondria, Golgi, endoplasmic reticulum, and lysosymes are elevated by its presence. The pH in vesicles is normally lower than in cytosol (118,119), and the pH can be altered by ammonia. The rise in pH by the transport of ammonia gas into the lysosomes has been demonstrated (120,121).

The generation of ammonia inside the cell caused by cellular metabolism alters intracompartmental pH. Ammonia is generated mainly in mitochondria because of glutamine metabolism. Ammonia gas readily diffuses out of the mitochondria, and the pH decreases. Ammonia diffuses not only to the cytoplasm of cells but also to the compartments. In all cases, ammonia that diffuses from mitochondria increases the pH in other intracellular compartments (i.e., Golgi, endoplasmic reticulum, and lysosymes).

McQueen and Bailey (72,107) have shown that the net result of ammonia addition is a decrease in intracellular pH for the hybridoma line ATCC TIB 131. When internal pH was altered by external pH, McQueen and Bailey associated ammonia effects to the variations in intracellular pH, because both ammonia addition and low external pH resulted in lower cell yields on glucose and glutamine. On the other hand, Ozturk et al. (6) individually evaluated the cell growth rate and the metabolic rates and could not relate the effects of ammonia to the effects of lowering external pH. There is a decrease in glucose consumption and an increase in glutamine uptake rates when pH was controlled below 7.2 (93). However, a decrease in intracellular pH as a result of ammonia addition increased both rates.

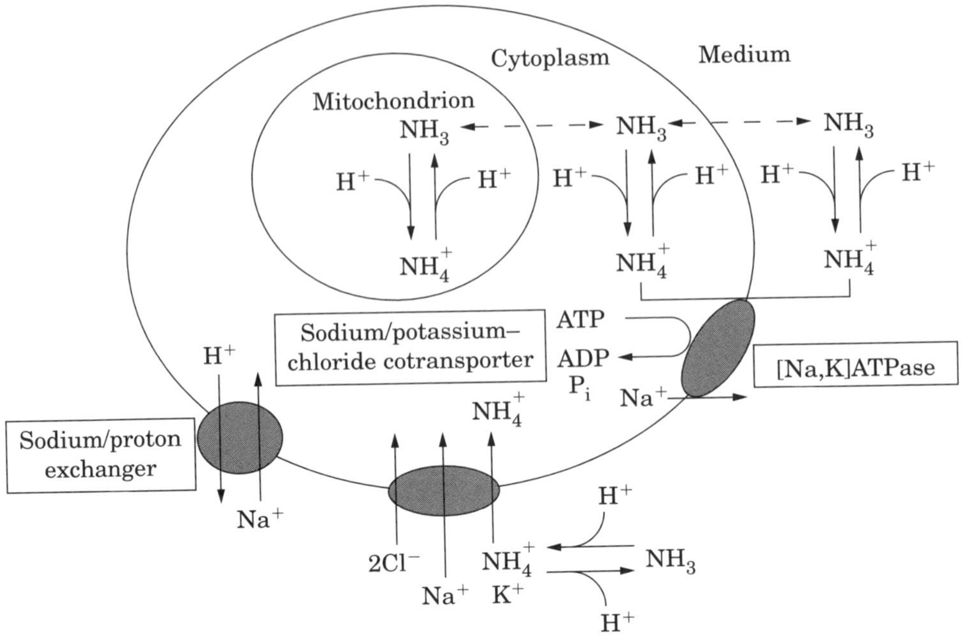

Figure 5. Ammonia transport and alteration of intracellular pH.

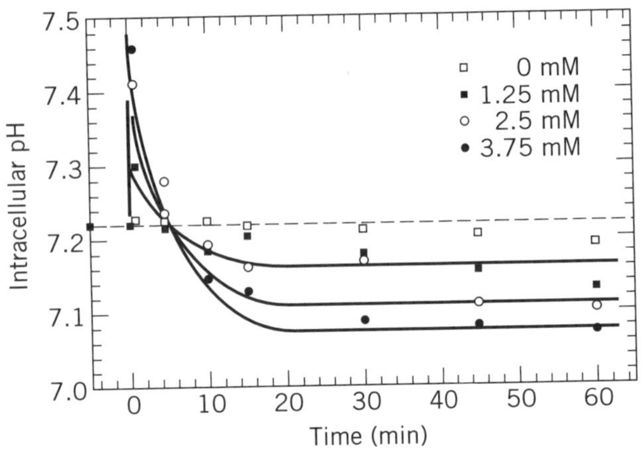

Figure 6. Alteration of intracellular (cytosolic) pH in response to ammonia addition. *Source*: Data obtained from Ozturk and Palsson (88).

Table 3. Effect of Ammonia on Intracellular pH at External pH of 7.20

$[NH_4^+]$ mM	Peak pH_i	Steady-state pH_i	Delta pH
0	7.23	7.19	0
1.25	7.3	7.14	−0.05
2.5	7.42	7.1	−0.09
3.75	7.43	7.07	−0.13

Source: Ozturk et al. (6).

ATP production was also influenced differently from ammonia addition (increasing ATP production) and from lowering external pH (decreasing ATP production). Miller et al. (52) observed an increase in glucose consumption rate as a result of ammonia addition, although the decrease in extracellular pH led to a decrease in glucose consumption rate. Thus, it can be concluded that the mechanisms of the ammonia effects may not be as simple as they were originally thought to be. The alteration of intracellular pH by ammonia cannot alone explain the observations.

Membrane Potential

As previously mentioned, ammonium ions can be transported across cellular membranes via [Na,K]ATPase and [Na,K,Cl]-cotransporters. Ammonium ions compete with K^+ ions and disturb the membrane potential. The [Na,K]ATPase transport system has a high energy demand (55,122,123), and the competition of ammonium ions with K^+ ions decreases the efficiency. Martinella and Hagstrom (116) reported futile cycles originating from ammonia diffusion and ammonium ion transport. These futile cycles increase the maintenance energy required by the cells. When ammonia is generated in the mitochondria, it diffuses out. However, ammonia is transported back in the form of ammonium ion, and cells use [Na,K]ATPase for this transfer. Thus, the alteration of membrane potential and elevation of maintenance energy are possible mechanisms for ammonia inhibition.

Alteration of Metabolic Pathways

Ammonia can also interact with the enzymes and alter the metabolic pathways. Phosphofructokinase (PFK) can be activated by ammonia (124,125). PFK is a highly regulated key enzyme for glucose metabolism, and its activation by ammonia can explain the increase in glucose consumption at high ammonia levels. Ammonia can also affect the metabolic pathways in mitochondria. The glutamine metabolism pathway, for instance, involves glutaminase, transaminase, and glutamate dehydrogenase reactions. The glutaminase enzyme is known to be inhibited by ammonia. However, Ozturk and Palsson (88) did not observe a decrease in the specific glutamine consumption rate at elevated ammonia concentrations. The glutamate dehydrogenase reaction is known to operate near equilibrium, and elevated ammonia levels would shift the equilibrium toward glutamate. At high ammonia levels, Ozturk and Palsson (88) observed a shift to the glutaminolytic flux via the alanine transaminase reaction from glutamate dehydrogenase, confirming this change. Thus, at high ammonia concentrations, the ammonia production by the cells decreased and alanine yield increased. This finding was also supported by other investigators (52,87,126).

The involvement of ammonia in metabolic pathways can explain the effects on cell metabolism. Growth inhibition by ammonia, on the other hand, can be explained by the presence of futile cycles (49,127). Futile cycles are cyclic reactions in which the hydrolysis of ATP is the only result (49); thus, cells have to increase ATP production to maintain the energy required for their survival.

Tagler et al. (127) studied the glutamate dehydrogenase and transhydrogenase enzymes in isolated mitochondria and liver cells. This enzyme system catalyzes the following reactions:

$$\text{Glutamate} + \text{NAD}^+ \leftrightarrow \alpha\text{-Ketoglutarate} + \text{NH}_4^+ + \text{NADH} \quad (4)$$

$$\text{NADH} + \text{NADP}^+ + \text{ATP} \leftrightarrow \text{NADPH} + \text{NAD}^+ + \text{P}_i \quad (5)$$

$$\alpha\text{-Ketoglutarate} + \text{NADPH} + \text{NH}_4^+ \leftrightarrow \text{Glutamate} + \text{NADP}^+ \quad (6)$$

Ammonia can also drive the glutaminase–glutamine synthetase enzyme system in a futile cycle mode. Glutaminase enzyme catalyzes the deamidation of glutamine to glutamate and ammonia. Glutamine synthetase, on the other hand, uses ATP to convert the glutamate to glutamine:

$$\text{Glutamine} \leftrightarrow \text{Glutamate} + \text{NH}_4^+ \quad (\text{Glutaminase}) \quad (7)$$

$$\text{Glutamate} + \text{NH}_4^+ + \text{ATP} \leftrightarrow \text{Glutamine} + \text{ADP} \quad (\text{Glutamine synthetase}) \quad (8)$$

Thus, for both glutamate dehydrogenase–transhydrogenase and glutaminase–glutamine synthetase cycles, the ATP is dissipated and the reactions are driven by ammonia. This increases the maintenance energy of the cells and can explain the inhibitory effects of ammonia (49). Although these futile cycles have been studied for isolated mitochondria, their functionality has not been demonstrated for mammalian cells in culture.

Alteration of Ribonucleotide Pools

Another explanation for ammonia inhibition comes from detailed studies on various ribonucleotides by Ryll and Wagner (128) and Ryll et al. (129). The intracellular pools of UDP-*N*-actylglucosamine (UDP-GlcNAc) and UDP-*N*-actylgalactosamine (UDP-GalNAc) have been shown to be responsible for inhibition of protein, DNA, and RNA synthesis (129–132). The pool of UDP-GNAc refers to both UDP-GlcNAc and UDP-GalNAc collectively, and it was demonstrated to be elevated in response to increased ammonia levels for a variety of cell lines. Ammonia is incorporated into the glycolytic pathway in cytosol (Fig. 7). Fructose-6-phosphate and ammonia combine to form glucosamine-6-phosphate, a direct precursor for the UDP-GNAc pool in the cytosol. In the mitochondria, ammonium leads to formation of carbamoylphosphate. When the ammonia concentration is high in the mitochondria, then subsequently up-regulated carbamoylphosphate can enter the cytoplasm. Carbamoylphosphate in the cytosol stimulates de novo synthesis of UMP and UTP. Finally, UTP and glucosamine-6-phosphate are combined to supplement the UDP-GNAc pool.

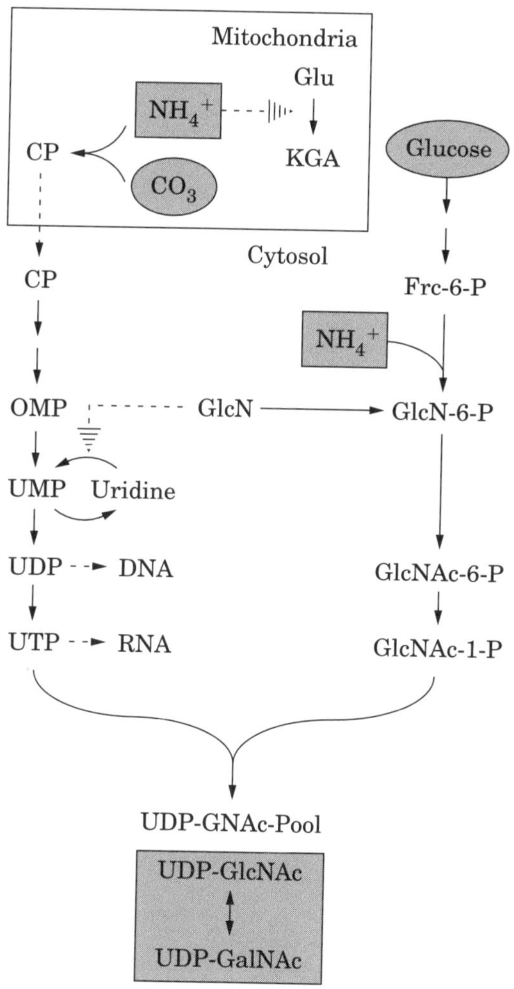

Figure 7. Biosynthesis of UDP-GNAc pool after ammonia application. *Source*: Adapted from Ryll et al. (129).

Ryll et al. (129) studied the biochemistry of the UDP-GNAc pool formation and regulation of the enzymes involved. The formation of the UDP-GNAc pool was observed to be dependent on both glucose and ammonia. Ammonia increases the size of the UDP-GNAc pool, which has been correlated with growth inhibition. It is also important to note that the UDP-GlcNAc is also known as a precursor substrate for glycosylation processes in the endoplasmic reticulum and Golgi. Thus, the effects of ammonia on glycosylation could also be explained by this mechanism (101,102).

CONTROL OF AMMONIA INHIBITION

Several techniques were used to minimize ammonia accumulation in cell culture. The techniques can be classified into two groups: minimization of ammonia generation and removal of ammonia.

Minimization of Ammonia Generation

As mentioned before, ammonia is generated mainly by the spontaneous degradation and cellular metabolism of glutamine. Ammonia generation can be minimized by controlling these processes.

Minimization of Glutamine Degradation. As previously mentioned, the glutamine degradation rate is dependent on several factors, and it can be minimized by several strategies. Temperature is the most obvious factor, and the media should be kept cold when stored. In the reactor, however, the temperature cannot be lowered much. Other factors, such as media pH, can be used to minimize the degradation. The pH optimum for cells varies between pH 7.0 and 7.4 (22,23). Cultivation at lower pH values should decrease the degradation rate and ammonia generation. Serum, if used, should be inactivated to minimize any glutaminase activity. Finally, glutamine degradation is dependent on several ions, such as phosphate. Various media should be evaluated for a given cell line, and glutamine degradation rates for each should be compared.

Attempts have been made to replace glutamine by glutamine-containing peptides (133,134). Peptides such as alanyl-glutamine or glycyl-glutamine are more stable, and the media prepared with these peptides can be autoclaved. These peptides were also shown to generate less ammonia. Although the cells exhibited a considerable lag phase, the cell yields from peptide-substituted media were comparable to those obtained from standard media. Holmlund et al. (135) used glycyl-glutamine and *N*-acetyl-glutamine for CHO cultures. Although cells did not perform well in the presence of *N*-acetyl-glutamine (low growth and t-PA production), glycyl-glutamine provided good culture performance.

Control of Cellular Metabolism. The specific ammonia production rate is affected by the cellular environment, and variables such as pH, dissolved oxygen (DO), temperature, and the level of metabolites can be manipulated to minimize the production rate.

There seems to be an optimal pH for reducing the ammonia production rate. For instance, Ozturk and Palsson (93) observed the ammonia production rates to be minimal at pH 7.2. The cell growth rates were optimal, and glutamine consumption was also observed to be minimal at this pH. The ammonia yield from glutamine, on the other hand, was reduced at low pH values (92,136). Cells seem to produce less ammonia when they are growing under optimal conditions. At very low (1% air saturation) and very high (100% air saturation) DO levels, the ammonia generation rates increased (93,136,137).

In most cases, the ammonia production follows closely with the glutamine consumption with a relatively constant yield coefficient. Reducing the glutamine concentration ([Gln]) decreases the glutamine consumption (q_{Gln}), and Monod-type saturation kinetics can be used to describe the data:

$$q_{\mathrm{Gln}} = q^{0}_{\mathrm{Gln}} \frac{[\mathrm{Gln}]}{K_{\mathrm{m}} + [\mathrm{Gln}]} \tag{9}$$

where q^{0}_{Gln} and K_{m} are constants. The value of K_{m} is in the order of 0.2 to 1 mM (10,87).

Reducing glutamine not only decreases the glutamine consumption and ammonia production rates, but also decreases the yield of ammonia from glutamine. When glutamine was fed in a controlled fashion below 1 mM levels, Glacken et al. (50) observed a substantial decrease in ammonia production in Madin-Darby bovine kidney (MDCK) and human fibroblast cells. A similar strategy was successfully used to reduce ammonia production by 50% in hybridoma cells (36,38).

Although a glucose limitation alone did not cause a change in ammonia secretion, Ljunggren and Haggstrom (38) observed an enhancement in ammonia reduction when glucose *and* glutamine were limited. The replacement of glucose by fructose, mannose, and galactose virtually eliminated the ammonia-induced generation of UDP-GNAc.

Altering the amino acid composition of the media seems to affect ammonia generation. Hiller at al. (138) increased the concentrations of leucine, isoleucine, valine, and lysine and observed a decrease in ammonia secretion in a hybridoma line. In these experiments, alanine production rates increased, indicating an elevation in the activity of alanine transaminase.

Replacement of Glutamine. The replacement of glutamine in cell culture media can eliminate most of the ammonia generation in culture. Although the idea of replacement is a good one, it is difficult to implement in most cases because the cells seem to be strongly dependent on glutamine for energy and biomass production. Cells need to adapted or genetically altered to grow in the absence of supplemented glutamine.

Mammalian cells can use other amino acids as a substitute for glutamine. Studies were conducted to investigate the replacement of glutamine by glutamate, α-ketoglutarate, and asparagine. The efficiency of glutamate for supporting cell growth is very low, and high concentrations of glutamate (up to 20 mM) are required (139,140). The replacement of glutamine by glutamate was possible for mouse LS cells (141) and for the McCoy cells (142). On the other hand, MDCK cells could not be adapted to glutamine-free media (143). Cells can convert glutamate to glutamine via the glutamine synthetase reaction, and the success of growing the cells on glutamate can depend on the concentration and activity of this enzyme. However, McDermott and Butler (143) observed that the key factor in cell adaptation to glutamine-free media is not the glutamine synthetase, but the uptake rate of glutamate by the cells.

Asparagine is another amino acid used to replace glutamine. Although asparagine is also unstable in media (25), the degradation rate is much lower (half-life, 87 days). Kurano et al. (85) observed the growth of CHO cells on asparagine after an initial lag phase. The asparagine-containing cultures generated 40% less ammonia than cultures in standard media. The use of α-ketoglutarate instead of glutamine was also effective in reducing ammonia generation (142).

The success of growing cells in the absence of glutamine can be increased by genetic engineering. Scientists at Celltech Ltd. developed a vector containing glutamine synthetase and infected NS0 myeloma and CHO cells with plasmid containing this vector (144–146). The construction of this vector is presented in Figure 8. The cells use glutamate as a substrate in glutamine-free media. The genes for protein expression were also integrated into the vector; thus, the glutamine synthetase gene was used as an amplifiable, selectable marker; in the glutamine-free media only the cells containing plasmid could grow. This resulted in the selection of high-producing clones. The system works best for NS0 myeloma cells because these cells lack any endogenous glutamine synthetase activity. CHO cells, on the other hand, contain endogenous glutamine synthetase genes, and the selective pressure induced by the absence of glutamine does not work effectively. The use of a specific inhibitor, methionine sulphoximine (MSX), increases the efficiency of selection and amplification.

Adaptation of Cells to High Ammonia. The adaptation of cells to grow at high ammonia concentrations can result in more ammonia-tolerant cultures. Adaptation is a complex process, and it is not clear whether the cells alter themselves or a particular clone is selected as a result of this process (147,148). Regardless of the mechanism, the cells can tolerate higher ammonia levels after the adaptation, and this method of ammonia adaptation can be used to minimize the ammonia inhibition. The adaptation of hybridoma cells to ammonia has been demonstrated by several investigators (52,54,149).

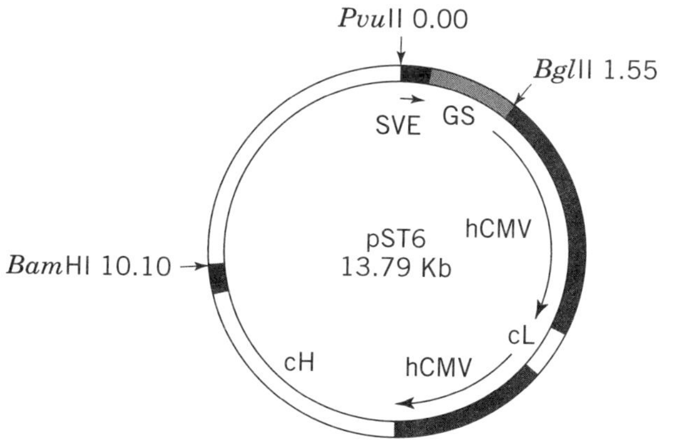

Figure 8. Expression plasmid for cB72.3 antibody containing glutamine synthetase. Expression vectors: GS, glutamine synthetase; cH, heavy chain; cL, light chain; hCMV, CMV promoter. *Source*: From Bebbington et al. (144).

Removal of Ammonia

Several techniques were investigated to remove ammonia during cell culture and to improve culture performance. These techniques involve the use of adsorbents, gas exchange and ion exchange membranes, and electrodialysis. Although the systems can be placed in the bioreactor, they are normally used as a loop system (Fig. 9). Cell culture fluid is recycled through a column, where ammonia is removed. In the case of adsorbents, the recycle is halted routinely to allow the regeneration of the column. The column is regenerated by stripping the ammonia from the adsorbents, using a stripping solution. The methods using gas and ion exchange membranes and electrodialysis, on the other hand, can be run continuously. In these cases, the ammonia fixing–stripping solution is continuously recycled on the other side of the membrane.

Use of Adsorbents. Several resins and natural adsorbents were investigated to selectively remove ammonia from cell culture media. Carbonell et al. (150) and Capiaumont et al. (151) used natural adsorbents such as Clinoptilolite to adsorb ammonia in a column where cells culture media is cycled through. The operation of the adsorbent had to be interrupted periodically to regenerate the column. Although these columns removed ammonia from the culture, the cell culture performance did not change significantly. Similar results were obtained when Zeolite (Phillipsite-Gismondine) and ion exchange resins were used as adsorbents (77). On the other hand, Nayve et al. (152,153) and Matsumura and Nayve (154) obtained good results in hybridoma cultures by combining the adsorption method and other removal systems. In fact, Matsumura and Nayve (154) obtained a viable cell concentration of 25 million cells/mL in a perfusion system with ammonia removal.

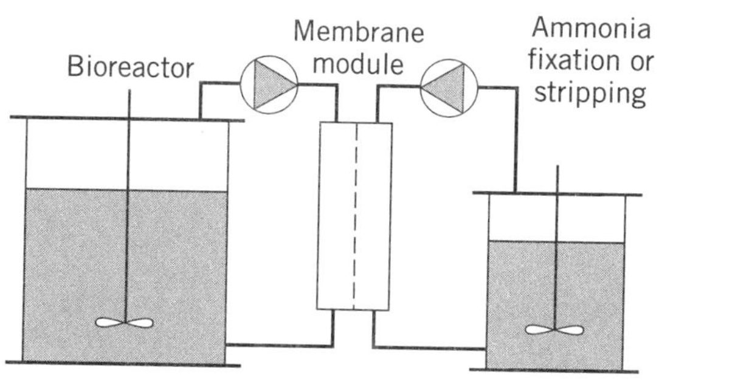

Figure 9. Ammonia removal from cell culture using adsorbents, gas or ion exchange membranes, or electrodialysis.

The use of adsorbents to remove ammonia requires complicated systems. The regeneration of the column prohibits continuous operation and requires complicated process control. These systems have to be proven to be reliable and effective for commercial scale operation.

Gas Exchange Membranes. Another method of ammonia removal from cell culture uses gas exchange membranes (such as polypropylene) where ammonia is stripped from the solution because of its gaseous properties (155–158). These systems were first developed for microbial systems and for chemical processes (159–162). The efficiency of ammonia removal in these systems depends on the concentration gradient for gaseous ammonia across the membrane (Fig. 10). The ammonia removal rate (R) is given by:

$$R = k \cdot A \cdot ([NH_3]_c - [NH_3]_s) \quad (10)$$

where k is the mass transfer coefficient, A is the surface area, and $[NH_3]_c$ and $[NH_3]_s$ are the concentrations of gaseous ammonia, in the cell culture and receiving side, respectively. The concentration gradient can be maximized by maintaining a very low gaseous ammonia concentration on the receiving side of the membrane ($[NH_3]_s \rightarrow 0$). In most cases, the transported ammonia is irreversibly converted by either enzymatic or chemical reactions (e.g., protonization). On the cell culture side, the concentration of gaseous ammonia ($[NH_3]_c$) is determined by the *total* ammonia concentration. Gaseous ammonia is at equilibrium with ammonium ion, and the culture pH determines the concentration of ammonia in gaseous form:

$$[NH_3]_c = [NH_4^+]_T/(1 + 10^{(pK_a - pH)}) \quad (11)$$

where $[NH_4^+]_T$ is the total ammonia concentration (ionized and gas) and pK_a is the ionization constant. At typical culture conditions (pH below 7.5), gaseous ammonia constitutes less than 1% of the total ammonia concentration (pK_a = 9.2 at 37 °C). Thus, only a minute fraction of ammonia is available for its removal. The efficiency of ammonia removal can be increased by increasing the pH on the cell culture side. Higher pH shifts the equilibrium to ammonia gas and increases the concentration gradient. However, this method cannot be utilized fully because the cells require a tight pH range for growth and maintenance. Increasing pH also affects the media components, and high pH can precipitate proteins, including the product.

The removal of ammonia using gas exchange membranes was tested in a variety of hybridoma lines (7,81,151,163) in an effort to increase antibody yield. Although the final cell density was reported to have increased, the effect on production was not significant. Ammonia removal, however, did result in an alteration in cell metabolism.

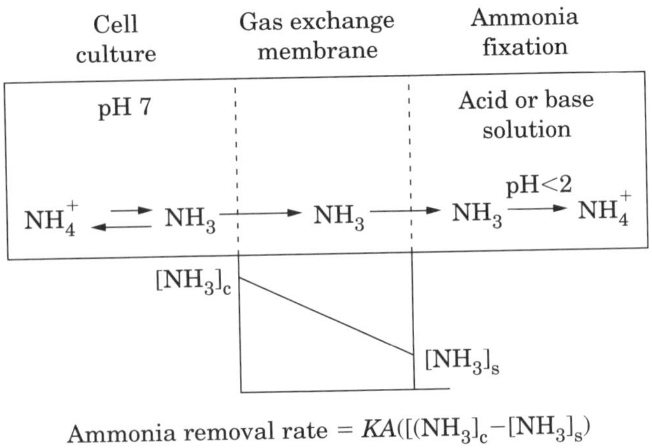

Figure 10. Ammonia removal from cell culture using gas exchange membranes. The rate of removal depends on the ammonia gas concentration difference.

Ion Exchange Membranes. Ammonia can be selectively removed by ion (cation) exchange membranes. Contrary to gas exchange membranes, ion exchange membranes remove ammonium ion instead of gaseous ammonia. As mentioned before, ammonia in the cell culture is almost 99% in ammonium ion form, so the removal via ion exchange membranes should be more effective.

The use of autoclavable cation exchange membranes (DuPont, Wilmington, Del., Nafion 417) was described by Thommes et al. (164). Ammonium ions bind to the fixed anions in the membrane and pass into the strip–fixation solution, where they are deprotonized or stripped from the solution by pervaporation (Fig. 11). The elimination of ammonia on the strip–fixation side and a high concentration of ammonium ions on the cell culture side establishes an efficient concentration gradient so the removal of ammonia is very effective. Thommes et al. (164) reported an increase in cell density and antibody production in a murine hybridoma culture. The problem with the ion exchange membrane is the simultaneous removal of some other cations from the culture. Some cations such as calcium and magnesium can also precipitate as bicarbonates as a result of ion exchange (56).

Electrodialysis. Another technique for ammonia removal involves the use of an electrokinetic mechanism utilizing electrophoresis. Chang and Wang (82,165) applied a continuous DC electrical field to remove charged ammonia and lactate from the culture of ATCC CRL 1606 hybridoma cells. At a current density of 50 A/m^2, almost all ammonia in the culture could be removed. This ammonia removal system allowed the use of 4× concentrated media. The system allowed enhancement of cell density and antibody production significantly. The applied current did not cause any detrimental effect on the cells. Removal of ammonia increased the glutamine consumption rate. The electrodialysis system also removed lactate and minimized lactate related inhibition.

CONCLUSIONS

In this article, we reviewed several aspects of ammonia inhibition. Ammonia accumulation in cell cultures can be a serious problem for cell growth, productivity, and product quality.

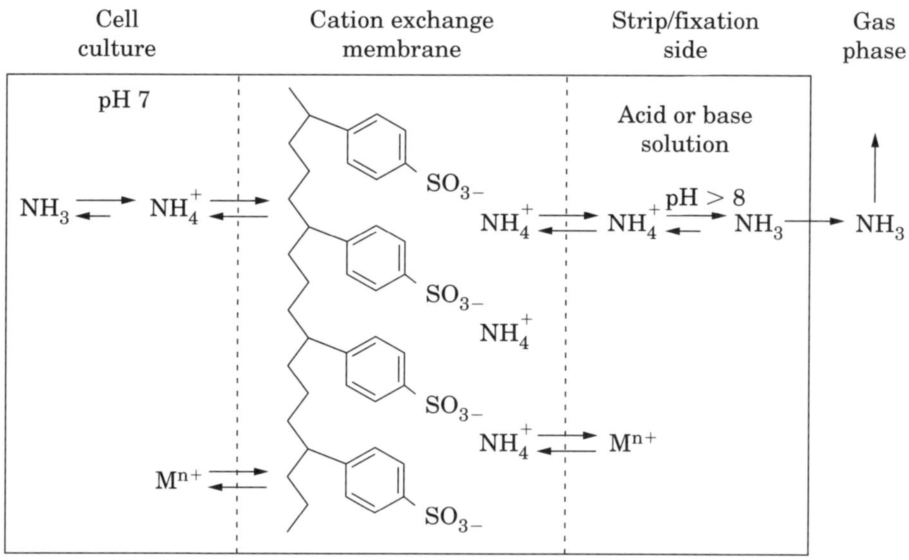

Figure 11. Ammonia removal from cell culture using cation exchange membranes. *Source*: Adapted from Thommes et al. (164).

Ammonia is generated mainly by glutamine degradation and glutamine metabolism. Cell physiology can be affected by ammonia concentrations between 2 and 10 mM. Different cell lines display varying tolerance to ammonia inhibition. Cell growth rate and final cell density in batch and fed-batch cultures decrease at high ammonia levels. Under normal conditions, ammonia does not *directly* influence the death rate or the decrease in viability of the cultures.

Ammonia is involved in several metabolic pathways and can influence the metabolic rates. Specific metabolic rates for glucose, glutamine, ammonia, lactate, amino acids are accelerated at higher ammonia concentrations. The ammonia yield from glutamine decreases, and more alanine is produced at higher ammonia levels. Although ammonia does not seem to alter the specific oxygen consumption rate, the ATP production rate increases at elevated ammonia concentrations. Cells become more active metabolically and generate more ATP for maintenance when their growth is suppressed by ammonia.

Specific rates for product secretion are not influenced by ammonia concentration. Cultures at higher ammonia levels result in lower product concentrations because of the low cell concentrations achieved. Ammonia can influence product quality, glycosylation, and sialyation; this was demonstrated for a number of cell lines and products.

Several mechanisms have been hypothesized and examined experimentally in an attempt to understand ammonia inhibition. Alteration of intracompartmental pH and membrane potential, futile cycles, metabolic inhibition, and alteration of critical ribonucleotides are identified as potential mechanisms for ammonia inhibition. Although the dynamics of ammonia transport to and from the cells were studied and intracellular pHs were measured, the data on metabolic rates could not be explained by internal pH hypothesis. Energy dissipation caused by futile cycles and disturbance of membrane potentials are viable explanations for ammonia inhibition. Ammonia was shown to elevate levels in several ribonucleotide pools. Ammonia increases the size of UDP-GNAc pool and results in growth inhibition.

Several techniques were investigated to minimize ammonia accumulation in cell culture. Generation of ammonia resulting from glutamine degradation and metabolism can be reduced by controlling the environment of the cells. By maintaining glutamine at low levels, the ammonia generation can be minimized. Glutamine can also be replaced by several other amino acids; however, the cells have to be adapted or genetically altered for this strategy to be successful. Cells can also be adapted to high ammonia concentrations and can thus tolerate higher ammonia levels.

Methods for ammonia removal in cell culture were investigated to improve culture performance. These techniques involve use of adsorbents, gas exchange and ion exchange membranes, and electrodialysis. These systems are fairly complicated, and more work has to be done to improve the effectiveness and reliability for commercial production.

BIBLIOGRAPHY

1. M.P. Backer, L.S. Metzger, P.L. Slaber, K.L. Nevitt, and G.B. Boder, *Biotechnol. Bioeng.* **32**, 993–1000 (1988).
2. C.H. Leist, H.-P. Meyer, and A. Fiechter, *J. Biotechnol.* **15**, 1–46 (1990).
3. A.S. Lubiniecki, *Large Scale Mammalian Cell Culture Technology*, Dekker, New York, 1990.
4. S.S. Ozturk, *Cytotechnology* **23**, 1–20 (1996).
5. K.W. Lanks and P.-W. Li, *J. Cell. Physiol.* **135**, 151–155 (1986).
6. S.S. Ozturk, M.R. Riley and B.O. Palsson, *Biotechnol. Bioeng.* **39**, 418–431 (1992).
7. M. Schneider, M. El Alaoui, U. Von Stockar, and I.W. Marison, *Enzyme Microb. Technol.* **20**, 268–276 (1997).
8. J. Crawford and H.J. Cohen, *J. Cell. Physiol.* **124**, 275–282 (1985).
9. H.R. Zielke, C.L. Zielke, and P.T. Ozand. *Fed. Proc.* **43**, 121–125 (1984).

10. M. Dalili, G.D. Sayles, and D.F. Ollis, *Biotechnol. Bioeng.* **36**, 74–82 (1990).
11. C. Dyring, H.A. Hansen, and C. Emborg, *Cytotechnology* **16**, 37–42 (1994).
12. D. Duval, C. Demangel, K. Munier-Jolain, S. Miossec, and I. Geahel, *Biotechnol. Bioeng.* **38**, 561–570 (1991).
13. O.T. Ramirez and R. Mutharasan, *Biotechnol. Bioeng.* **36**, 839–848 (1990).
14. M.C. Flickinger, N.K. Goebel, T. Bibila, and S. Boyce-Jacino, *J. Biotechnol.* **22**, 201–226 (1992).
15. T. Omasa, M. Ishimoto, K. Higashiyama, S. Shioya, and K. Suga, *Cytotechnology* **8**, 75–84 (1992).
16. S. Heeneman, N.E.P. Deutz, and W.A. Buurman, *J. Immunolog. Methods* **166**, 85–91 (1993).
17. R.I. Freshney, *Culture of Animal Cells: A Manual of Basic Technique*, 2nd ed., Liss, New York, 1987.
18. H.G. Bray, S.P. James, I.M. Raffan, and W.V. Thorpe, *Biochem. J.* **44**, 625–627 (1949).
19. J.B. Gilbert, V.E. Price, and J.P. Greenstein, *J. Biol. Chem.* **180**, 209–218 (1949).
20. G.L. Tritsch and G.E. Moore, *Exp. Cell. Res.* **28**, 360–364 (1962).
21. A.A. Lin, P. Agrawal, *Biotechnol. Lett.* **10**, 695–698 (1988).
22. S.S. Ozturk and B.O. Palsson, *Biotechnol. Prog.* **6**, 121–128 (1990).
23. S.S. Ozturk, M.E. Meyerhoff, and B.O. Palsson, *Biotechnol. Tech.* **3**, 217–222 (1989).
24. S.S. Seaver, J.L. Rudolph, and J.E. Gabriels Jr., *Biotechnology* **2**, 254–260 (1984).
25. J. Wein and I.E. Goetz, *In Vitro* **9**, 186–193 (1973).
26. M.S.M. Ardavi and E.A. Newsholme, *Biochem. J.* **212**, 835–842 (1993).
27. L.G. Baggetto, *Biochimie* **74**, 959–974 (1992).
28. K. Brand, J. von Hintzenstarn, K. Langer, and W. Fekl, *J. Cell Physiol.* **132**, 559–564 (1987).
29. K. Brand, W. Fekl, and J. von Hintzenstem, *Metabolism* **38**, 29–33 (1989).
30. M. Butler and H. Jenkins, *J. Biotechnol.* **12**, 97–110 (1989).
31. M. Donnely and I.E. Schefffler, *J. Cell Physiol.* **89**, 39–52 (1976).
32. L. Fitzpatrick, H.A. Jenkins, and M. Butler, *Appl. Biochem. Biotechnol.* **43**, 93–116 (1993).
33. H.A. Jenkins, M. Butler, and A.J. Dickson, *J. Biotechnol.* **23**, 167–182 (1992).
34. Z. Kovacevic and J.D. McGivan, *Physiol. Rev.* **63**, 547–605 (1983).
35. K.W. Lanks and P.-W. Li, *J. Cell Physiol.* **135**, 151–155 (1988).
36. J. Ljunggren and L. Haggstrom, *Biotechnol. Lett.* **12**, 705–710 (1990).
37. J. Ljunggren and L. Haggstrom, *Cytotechnology* **8**, 45–56 (1992).
38. J. Ljunggren and L. Haggstrom, *Biotechnol. Bioeng.* **44**, 808–818 (1994).
39. W.L. McKeehan, in M.J. Morgan ed., *Carbohydrate Metabolism in Cultured Animal Cells*. Plenum, New York, 1986, pp. 111–150.
40. R.W. Moreadith and A.L. Lehninger, *J. Biol. Chem.* **259**, 6215–6221 (1984).
41. M. Newland, P.F. Greenfield, and S. Reid, *Cytotechnology* **3**, 215–229 (1990).
42. L.J. Reitzer, PhD Thesis, Washington University, St. Louis, Mo., 1978.
43. L.J. Reitzer, B.M. Wice, D. Kennell, *J. Biol. Chem.* **254**, 2669–2676 (1979).
44. L.J. Reitzer, B.M. Wice, D. Kennell, *J. Biol. Chem.* **255**, 5616–5626 (1980).
45. S.T. Sharfstein, S.N. Tucker, A. Manusco, H.W. Blanch, and D.S. Clark, *Biotechnol. Bioeng.* **43**, 1059–1074 (1994).
46. M. Watford, *Biochim. Biophys. Acta* **1200**, 73–78 (1994).
47. E. Kvamme ed., *Glutamine and Glutamate in Mammals*, Vol. 1. CRC, Boca Raton, Fla., 1988.
48. J. Swierczynski, Z. Bereznowski, and W. Makarewicz, *Biochim. Biophys. Acta* **1157**, 55–62 (1993).
49. M.W. Glacken, *Biotechnology.* **6**, 941–1050 (1988).
50. M.W. Glacken, R.J. Fleischaker, and A.J. Sinskey, *Biotechnol. Bioeng.* **28**, 1376–1389 (1986).
51. P.M. Hayter, E.M.A. Curling, A.J. Baines, N. Jenkins, L. Salmon, P.G. Strange, and A.T. Bull, *Appl. Microbiol. Biotechnol.* **34**, 559–564 (1991).
52. W.M. Miller, C.R. Wilke, and H.W. Blanch, *Bioproc. Eng.* **3**, 113–122 (1988).
53. H. Buntemeyer, C. Wallerius, and J. Lehmann, *Cytotechnology* **9**, 59–67 (1992).
54. B.L. Maiorella, A. Shauger, M. Poulhazan, B. Howarth, and D. Inlow, *203th ACS Natl. Mtg.*, 1992.
55. L. Xie and D.I.C. Wang, *Biotechnol. Bioeng.* **43**, 1175–1189 (1994).
56. M. Schneider, I.W. Marison, and U. Von Stockar, *J. Biotechnol.* **46**, 161–181 (1996).
57. W.M. Canning and B.N. Fields, *Science* **219**, 987–988 (1983).
58. M.D. Eaton and A.R. Scala, *Virology* **13**, 300–307 (1961).
59. E. Furusawa and W. Cutting, *Proc. Soc. Exp. Biol. Med.* **111**, 71–75 (1962).
60. G. Farias, E. Navarrete, J. Kiss, and J. Kuznar, *Arch. Virol.* **98**, 155–162 (1988).
61. E.M. Jensen and O.C. Liu, *Proc. Soc. Exp. Biol. Med.* **107**, 834–838 (1961).
62. A.H. Koyama and T. Uchida, *Virus Res.* **13**, 271–282 (1989).
63. C. Scholtissek, R. Rott, and H.-D. Klenk, *Virology* **63**, 191–200 (1975).
64. W.J. Visek, G.M. Konoldny, and P.R. Gross, *J. Cell. Physiol.* **80**, 373–382 (1980).
65. R.W. Holley, R. Armour, and J.H. Baldwin, *Proc. Natl. Acad. Sci. USA* **75**, 1864–1866 (1978).
66. T. Hassell, S. Gleave, and M. Butler, *Appl. Biochem. Biotechnol.* **30**, 29–41 (1991).
67. G.A. Truskey, D.P. Nicolakis, D. Dimasi, A. Haberman, and R.W. Swartz, *Biotechnol. Bioeng.* **36**, 797–812 (1990).
68. T.C. Dodge, G.-Y. Ji, and W.-S. Hu, *Enzyme Microb. Technol.* **9**, 607–611 (1987).
69. J.G. Gaertner and P. Dhurjati, *Biotechnol. Prog.* **9**, 309–316 (1993).
70. S. Reuveny, D. Velez, L. Miller, and J.D. Macmillan, *J. Immunol. Methods* **86**, 53–59 (1986).
71. A. McQueen and E. Bailey, *Biotechnol. Bioeng.* **35**, 1067–1077 (1990).
72. A. McQueen and E. Bailey, *Biotechnol. Bioeng.* **35**, 897–906 (1990).
73. D. Doyle and M. Butler, *J. Biotechnol.* **15**, 91–100 (1990).
74. A.J. Racher, D. Looby, and J.B. Griffiths, *J. Biotechnol.* **15**, 129–146 (1990).
75. A.J. Racher, D. Looby, and J.B. Griffiths, *J. Biotechnol.*, **29**, 145–156 (1993).
76. O.W. Ronning, M. Schartum, A. Winsnes, and G. Lindberg, *Cytotechnology* **7**, 15–24 (1991).

77. Y-H. Jeong and S.S. Wang, *Biotechnol. Tech.* **6**, 341–346 (1992).
78. Y.-H. Jeong and S.S. Wang, *Enzyme Microb. Technol.* **17**, 47–55 (1995).
79. I. Ludemann, R. Portner, and H. Markl, *Cytotechnology* **14**, 11–20 (1994).
80. M. Newland, M.N. Kamal, P.F. Greenfield, and L.K. Neilsen, *Biotechnol. Bioeng.* **43**, 434–438 (1994).
81. M. Schneider, PhD Thesis No. 1412, Ecole Polytechnique Federal de Lausanne (EPFL), Switzerland, 1995.
82. Y.H.D. Chang, A.J. Grodzinsky, and D.I.C. Wang, *Biotech. Bioeng.* **47**, 308–318 (1995).
83. D. Wentz and K. Schügerl, *Enzyme Microb. Technol.* **14**, 68–75 (1992).
84. M. Butler and R.E. Spier, *J. Biotechnol.* **1**, 187–196 (1984).
85. N. Kurano, C. Leist, F. Messi, S. Kurano, and A. Fiechter, *J. Biotechnol.* **15**, 113–128 (1990).
86. H. Hansen and C. Emborg, *Biotechnol. Prog.* **10**, 121–124 (1994).
87. M.W. Glacken, Ph.D. Thesis, MIT, Cambridge, Mass., 1987.
88. S.S. Ozturk and B.O. Palsson, *Biotechnol. Bioeng.* **37**, 989–993 (1991).
89. J.L. Goergen, A. Marc, and J.M. Engasser, in R.E. Spier, J.B. Griffiths, W. Berthold eds. *Animal Cell Technology: Products of Today, Prospects of Tomorrow*, Butterworth-Heinemann, London, 1994, pp 161–164.
90. P.C. Alex, D. Barngrover, and R.W. Swartz, *ACS Natl. Mtg.*, Miami, Fla., 1989.
91. A. McQueen and J. Bailey, *Biotechnol. Bioeng.* **35**, 1067–1077 (1989).
92. S.S. Ozturk and B.O. Palsson, *Biotechnol. Prog.* **7**, 471–480 (1991).
93. S.S. Ozturk and B.O. Palsson, *Biotechnol. Prog.* **7**, 481–494 (1991).
94. T. Kimura, S. Iijima, and T. Kobayashi, *J. Ferment. Technol.* **65**, 341–344 (1987).
95. J.J. Dyken and A. Sambanis, *Enzyme Microb. Technol.* **16**, 90–98 (1994).
96. C.L. Atanassov, C.D. Muller, S. Sarhan, B. Knodgen, G. Rebel, and N. Seiler, *Res. Immunol.* **145**, 277–288 (1994).
97. M. Ito and W.F. McLimans, *Cell. Biol. Int. Rep.* **5**, 661–666 (1981).
98. B. Thorens and P. Vassalli, *Nature* **321**, 618–620 (1986).
99. D.C. Andersen and C.F. Goochee, *Biotech. Bioeng.* **47**, 96–105 (1995).
100. M.C. Borys, D.L.H. Linzer, and E.T. Papoutsakis, *Biotechnology* **11**, 720–724 (1993).
101. M. Gawlitzek, U. Valley, R. Wagner, and H.S. Conradt, *Cytotechnology* **14**, 3–5 (1994).
102. M. Gawlitzek, U. Valley, M. Nimtz, R. Wagner, and H.S. Conradt, in E.C. Beuvery, J.B. Griffiths, P.W. Zeijlemaker eds., *Animal Cell Technology: Developments Towards the 21st Century*. Kluwer, Norwell, Mass., 1995, pp. 379–384.
103. K. Kopp, W. Noe, M. Schlueter, W. Werner, and F. Goetz, in E.C. Beuvery, J.B. Griffiths, P.W. Zeijlemaker eds., *Animal Cell Technology: Developments Towards the 21st Century* Kluwer, Norwell, Mass., 1995, pp 403–407.
104. U.S. Pat. 5,156,964 (1992) Cetus Corp.
105. T.S. Stoll, M.B. Sliwkoeski, and Jacobson, *ACS Mtg.*, San Francisco, April 1977.
106. Y. Fukami, *Brain Res.* **463**, 140–143 (1988).
107. A. McQueen and J.E. Bailey, *Bioproc. Eng.* **6**, 49–61 (1991).
108. W.F. Boron and P. De Weer, *J. Gen. Physiol.* **67**, 91–112 (1976).
109. M.A. Knepper, R. Packer, and D.W. Good, *Physiol. Rev.* **69**, 179–249 (1989).
110. D. Kikeri, A. Sun, M.L. Zeidel, and S.C. Heben, *Nature*, **339**, 478–480 (1989).
111. R.M. Lynch and R.S. Balaban, *Am J. Physiol.* **252**, C225–C231 (1987).
112. R.M. Lynch and R.S. Balaban, *Am. J. Physiol.* **253**, C269–C276 (1987).
113. W.H. Moolenaar, R.Y. Tsien, P.T. van der Saag, and S.W. de Laat, *Nature* **304**, 645–648 (1983).
114. R.L. Post and P.C. Jolly, *Biochim. Biophys. Acta* **25**, 118–128 (1957).
115. M. Martinelle and L. Haggsrom, in J.B. Spier, C. Griffiths, C. MacDonald eds., *Animal Cell Technology: Developments, Processes, and Products*, Butterworth-Heinemann, London, 1992, p. 163.
116. K. Martinelle and L. Haggstrom, *J. Biotechnol.* **30**, 339–350 (1993).
117. K. Martinelle and L. Haggstrom, *Cytotechnology* **14**, 2–17 (1994).
118. R.G.W. Anderson and R.K. Pathak, *Cell*, **40**, 635–643 (1985).
119. D.D. Wagner, T. Mayadas, and V.J. Marder, *J. Cell Biol.* **102**, 1320–1324 (1986).
120. B. Poole and S. Ohkuma, *J. Cell. Biol.* **90**, 665–669 (1981).
121. F. Van Leuven, J.-J. Cassiman, and H. Van Den Berghe, *Cell* **20**, 37 (1980).
122. S.I. Harris, R.S. Balban, L. Barrett, and L.J. Mandel, *J. Biol. Chem.* **256**, 319–328 (1981).
123. S.M. Wall and L.M. Koger, *Am. J. Physiol. Ren. Fluid Electrol. Physiol.* **36**, F660–F670 (1994).
124. K. Uyeda and E. Racker, *J. Biol. Chem.* **240**, 4682–4688 (1965).
125. A. Parmeggiani, J.H. Luft, D.S. Love, and E.G. Krebs, *J. Biol. Chem.* **241**, 4625–4637 (1966).
126. J.C. Street, A.-M. Delort, P.S. Braddock, and K.M. Brindle, *Biochem. J.* **291**, 485–492 (1993).
127. J.M. Tagler, T.P.M. Akerboom, J.B. Hoek, A.J. Meijer, W. Vaartjes, L. Ernster, and J.R. Williamson, in F.A. Hommes, C.J. Van der Berg, eds., *Normal and Pathological Development of Energy Metabolism*. Academic Press, New York, 1975, pp. 63–75.
128. T. Ryll and R. Wagner, *Biotechnol. Bioeng.* **40**, 934–946 (1992).
129. T. Ryll, U. Valley, and R. Wagner, *Biotechnol. Bioeng.* **44**, 184–193 (1994).
130. I.G. Bekesi, E. Bekesi, and R.J. Einzler, *J. Biol. Chem.* **244**, 3766–3772 (1969).
131. J.G. Bekesi and R.J. Winzler, *Cancer Res.* **30**, 2905–2912 (1970).
132. I.-N. Chou, J. Zeigher, and E. Rapaport, *Proc. Natl. Acad. Sci. USA* **81**, 2401–2405 (1984).
133. E. Roth, G. Ollenschlager, A. Hamilton, K. Langer, W. Fekl, and R. Jakesz, *In Vitro Cell Dev. Biol.* **24**, 696–698 (1988).
134. A. Christie and M. Butler, in J.B. Spier, C. Griffiths, W. Berthold, eds., *Animal Cell Technology: Products of Today, Prospects for Tomorrow*, Butterworth-Heinemann, London, 1994, p. 164.
135. A.-C. Holmlund, S.L. Chatzisavido, S.L. Bell, and E. Lindner-Olsson, *Proc. of the 11th ESACT Mtg.*, Brighton, UK, 1992, pp. 176–179.

136. W. Miller, C. Wilke, and H. Blanch, *J. Cell Physiol.* **132**, 524–530 (1987).
137. S.S. Ozturk and B.O. Palsson, *Biotechnol. Prog.* **6**, 437–446 (1990).
138. G.W. Hiller, D.S. Clark, and H.W. Blanch, *Biotechnol. Bioeng.* **44**, 303–321 (1994).
139. J.E. Darnell Jr. and H. Eagle, *Virology* **6**, 556–566 (1958).
140. S.C. Nagle and B.L. Brown, *J. Cell. Physiol.* **77**, 259–264 (1971).
141. J.B. Griffiths and S.J. Pirt, *Proc. R. Soc. Lond., Ser. B* **168**, 421–438 (1967).
142. T. Hassell and M. Butler, *J. Cell Sci.* **96**, 501–508 (1990).
143. R.C. McDermott and M. Butler, *J. Cell Sci.* **104**, 51–58 (1993).
144. C.R. Bebbington, G. Renner, S. Thomson, D. King, D. Abrams, and G.T. Yarranton, *Biotechnology* **10**, 169–175 (1992).
145. S.L. Bell, M.E. Bushell, M.F. Scott, J.N. Wardell, R.E. Spier, and P.G. Sanders, *Proc. of the 11th ESACT Mtg.* Brighton, UK, 1992, pp. 180–182.
146. S.L. Bell, C. Bebbington, M.F. Scott, J.N. Wardell, R.E. Spier, M.E. Bushell, and P.G. Sanders, *Enzyme Microb. Technol.* **17**, 98–106 (1995).
147. S.S. Ozturk and B.O. Palsson, *Hybridoma* **9**, 167–175 (1990).
148. S.S. Ozturk and B.O. Palsson, *Biotechnology Bioeng.* **37**, 35–46 (1991).
149. M. Matsumura, M. Shimoda, T. Arii, and H. Kataoka, *Cytotechnology* **7**, 103–112 (1991).
150. D. Carbonell, B. Besnainou, J. Capiaumont, P. Lessart, and P. Nabet, in J.B. Spier, C. Griffiths, C. MacDonald, eds., *Animal Cell Technology: Developments, Processes, and Products*, Butterworth-Heinemann, London, 1992, p. 166.
151. J. Capiaumont, C. Legrand, D. Carbonell, B. Dousset, F. Belleville, and P. Nabet, *J. Biotechnol.* **39**, 49–58 (1995).
152. F.R.P. Nayve Jr., M. Motoki, M. Matsumura, and H. Kataoka, *Cytotechnology* **6**, 121–130 (1991).
153. F.R.P. Nayve Jr., T. Misato, M. Matsumura, and H. Kataoka, *J. Biotechnol.* **34**, 217–225 (1994).
154. M. Matsumura and F.R.P. Nayve Jr., *Cytotechnology* **18**, 35–50 (1995).
155. European Pat. Appl. No. 90301630.1 (1990), Bend Research Inc.
156. U.S. Pat. 5,071,561 (1991), Bend Research Inc.
157. V. Blet, M.-N. Pons, and J.L. Greffe, *Anal. Chim. Acta* **219**, 309–311 (1989).
158. D.J. Brose and P. Van Eikeren, *Appl. Biochem. Biotech.* **24/25**, 457–458 (1990).
159. V. Hecht, L. Bischoff, and K. Gerth, *Biotechnol. Bioeng.* **35**, 1042–1050 (1990).
160. M. Imai, S. Furusaki, and T. Miyauchi, *Ind. Eng. Chem. Proc. Des. Dev.* **21**, 421–426 (1982).
161. L. Qin and J.M.S. Cabral, *J. Chem. Technol. Biotechnol.* **67**, 323–328 (1996).
162. M.J. Semmens, D.M. Foster, and E.L. Cussler, *J. Membr. Sci.* **51**, 127–140 (1990).
163. M. Schneider, I.W. Marison, and U. Von Stockar, *Enzyme Microb. Technol.* **16**, 957–963 (1994).
164. J. Thommes, U. Garske, M. Biselli, and W. Wandrey, in J.B. Spier, C. Griffiths, MacDonald, eds., *Animal Cell Technology: Developments, Processes, and Products*, Butterworth-Heinemann, London, 1992, p. 171.
165. Y.-H.D. Chang, A.J. Grodzinsky, and D.I.C. Wang, *Biotechnol. Bioeng.* **47**, 319–326 (1995).

See also CULTURE MEDIA, ANIMAL CELL, LARGE SCALE PRODUCTION; ENERGY METABOLISM, MICROBIAL AND ANIMAL CELLS; FERMENTATION MONITORING, DESIGN AND OPTIMIZATION; INSECT CELL CULTURE, PROTEIN EXPRESSION; PROCESS CONTROL, STRATEGY AND OPTIMIZATION.

ANAEROBES

LARRY E. ERICKSON
DANIEL Y.C. FUNG
Kansas State University
Manhattan, Kansas

KEY WORDS

Anaerobes
Biodegradation
Bioenergetics
Cultivation
Mixed cultures
Products

OUTLINE

INTRODUCTION

There are both aerobic and anaerobic microorganisms that are important to the advancement of biotechnology. Anaerobic microorganisms include obligate anaerobes that require oxidation–reduction potentials of −150 to −420 mV and facultative anaerobes that can grow at oxidation–reduction potentials between +300 and −420 mV (1). Oxygen must be excluded in working with obligate anaerobes.

Many important fermentation products are produced under anaerobic conditions (1–69). These include food products such as bread, yogurt, cheeses, wine, beer, sufu, and sauerkraut. Acetone, butanol, and ethanol are examples of industrial chemicals produced by anaerobes. Specialty chemicals produced by anaerobes include vitamins and pharmaceutical products. Major references describing the microbial products and the microorganisms that produce them include those by Erickson and Fung (2), Reed (3), Steinkraus (4), and Zeikus and Johnson (20).

Anaerobic digestion and biodegradation processes are applied widely to treat process waste products and in environmental restoration. Environmental microbiology is significant because of the beneficial impact of natural and nurtured processes and because of the isolation and identification of useful microorganisms that have evolved in natural environments. Anaerobic processes occur in production agriculture; cattle and sheep are still the greatest commercial success in harvesting and utilizing cellulosic plant products because of the anaerobic fermentations that occur in the rumen of these animals.

The taxonomy of anaerobes has evolved as new information has become available (64–66). Based on 16/18 S rRNA sequence comparisons, a universal phylogenetic tree has been described that includes bacteria, archaea, and eukarya (64,65). Anaerobes are found in all three branches of this tree.

METHODS OF CULTIVATION

The microbial world consists of aerobic and anaerobic organisms, defined by the ability to tolerate molecular oxygen. Sonnewirth (63) traced the evolution of anaerobic methodology and recorded that Leeuwenhoek demonstrated that some life forms could exist in the presence of gases other than oxygen and that Pasteur discovered anaerobiosis in 1861, coining the terms *aerobes* and *anaerobes*. The separation between these groups is not clearly defined. A convenient way to categorize these organisms is by observing their ability to grow at different oxidation–reduction potentials (E_h). E_h can be measured and expressed + or − millivolts (mV). Thus, aerobes can grow between +300 and −50 mV; facultative anaerobes, +300 to −420 mV; aerotolerant anaerobes, +180 to −350 mV; and obligate anaerobes, −150 to −420 mV (28).

The microbiological procedures for studying obligate anaerobes are similar to those used for the more familiar aerobic organisms, with the exception of the need to exclude oxygen from the working environment during manipulation, inoculation, and incubation of cultures and samples. As a class of organisms, anaerobes are very diversified in their behavior in terms of oxygen tolerance and nutritional requirements for growth. There is no single set of procedures for isolation of all anaerobes. Clinically important anaerobes constitute the most studied class. Specific methods were developed for isolating anaerobes from various sites in patients (57).

A detailed treatment of clinically important anaerobic organisms is recorded in the *Manual for the Determination of the Clinical Role of Anaerobic Microbiology* by Gall and Riely (43). This manual contains a definition of clinically important anaerobic bacteria methods for the collection and transportation of anaerobic specimens, culturing of anaerobic specimens, and identification of anaerobes.

The book *Isolation of Anaerobes*, by Shapton and Board (56), contains chapters describing the isolation of clostridia from animal tissues, feces, soil, and foods. Some chapters deal with the isolation of anaerobes from rumen as well as with sulfate-reducing bacteria and photosynthetic bacteria. Zeikus (62) describes the biology of methanogenic bacteria and methods of studying these organisms. In the area of food microbiology, very little attention has been given to the anaerobic microbiology. The *Compendium of Methods for the Microbiological Examination of Foods* (59) contains some chapters on anaerobes, mainly on the spore formers and the detection of toxins in foods. Anderson and Fung (30) reviewed the status of anaerobic methods, techniques, and principles for food bacteriology in detail. The history of cultivation of anaerobes was detailed by Hall (44) and more recently summarized by Fung (1).

The general conclusion on isolation procedures is that samples must be free from contact with air during collection time. The best method of collecting samples is by aspiration with needle and syringe. Sample should be placed in media that are designed to handle anaerobes. A variety of liquid and solid media are marketed for anaerobic cultivation by such companies as DIFCO, BBL, UNIPATH/OXOID, and others (69).

Currently, anaerobic cultivation systems include anaerobic jars, anaerobic glove boxes, roll tubes, anaerobic tube systems, and biological membrane systems.

Anaerobic Jars

Anaerobic jars that use both H_2 and CO_2 were developed by Brewer and Allgeier (33). The chemical generator consisted of a sodium borohydride tablet and a citric acid–sodium bicarbonate tablet in a convenient package. These two tablets, when activated by the addition of water to the package, produce H_2 and CO_2 gas, respectively. The system required no vacuum or electric current because the H_2 + O_2 reaction was catalyzed by a cold catalyst (alumina pellets coated with 0.5% palladium). CO_2 is used in the system for the promotion of growth of some anaerobic bacteria. Some companies are marketing these anaerobic jars (e.g., BBL, B.T.L., and UNIPATH/OXOID). The advantages of the anaerobic jar include (*1*) ease of operation, especially for the small laboratory; (*2*) economy of space; (*3*) convenience of prepackaged materials for the generation of gases;

and (*4*) safety of the units. The disadvantages include (*1*) the culture plates are inoculated under aerobic conditions; (*2*) the jar remains aerobic for 15 to 20 minutes even while the Gas Paks are activated, which may affect some strict anaerobes; and (*3*) occasionally the jar fails to achieve anaerobiosis.

Anaerobic Glove Boxes

To eliminate the chance of exposure of anaerobes to the natural environment during bacteriological manipulation, the anaerobic chamber, or glove box, was developed. The basic principles for achieving anaerobiosis in the glove box and for promoting binding of residual O_2 and H_2 in the presence of a catalyst (Laidlaw principle) are to remove air first and then flush the chamber with inert gas. Reducing agents in the medium also help to reduce the E_h to a desired low level. Companies marketing anaerobic glove boxes include Format Scientific, Don Whitney Scientific, and Microflow MDH Ltd. The advantages of the glove box include (*1*) the bacteriological procedure can be operated under anaerobic conditions from the sampling stage to the incubation stage; (*2*) the area for manipulation is large; and (*3*) in combination with temperature control, the culture can be incubated in the chamber without further disturbance of the cultural environment. The disadvantages are (*1*) high initial cost; (*2*) high operating expenses (gas and electricity); (*3*) high maintenance cost; (*4*) requirement for operator training; (*5*) oxygen contamination through leakage, operator error, or mechanical malfunction; (*6*) occasional high labor requirements; and (*7*) moisture buildup in the box.

Roll Tubes

One of the most widely used anaerobic cultivation systems is the Hungate roll tube, developed by Hungate (46–48). Sterile tubes were flushed with CO_2 while the medium was aseptically added to the tubes to prevent air contamination. The tubes were then inoculated, sealed with rubber stoppers, cooled (48 °C), and hand spun so that the agar with cultures forms a thin film in the inner wall of the test tube. After incubation the colonies could be examined with ease. Isolation of colonies was also feasible with minimum effort by flushing CO_2 into the cavity to allow maintenance of reduced condition while performing microbiological operations. A popular version of the roll tube system is the VPI Anaerobic Culture System, developed by Moore (53). This apparatus has a platform so that the tubes can stand upright with cannulas positioned above each tube. The system also allows for streaking cultures into the roll tube anaerobically. The system is continuously flushed by CO_2 to exclude O_2 from the system.

The advantages of the Hungate roll tube system include (*1*) good anaerobiosis with minimum investment, (*2*) unit operation for convenience of small sample numbers; and (*3*) ease of observation of colonies and isolation of cultures for further studies. Disadvantages include (*1*) time requirement for large number of samples; (*2*) need for manual dexterity; and (*3*) occasional failure of some tubes to achieve anaerobiosis.

Anaerobiosis Tube Systems

Ogg et al. (54) designed a tube with double wells so that agar with cultures can be introduced into the tube and form a thin film. Anaerobiosis is initially achieved by boiling and autoclaving of the medium. By the action of reducing agent in the medium and minimum reabsorption of O_2 due to the design of the tube (commercially called the Lee tube), good anaerobiosis is maintained in the system. The glass cylinder forming the inner wall (the inverted smaller tube) makes it easy for enumeration and observation of colony morphologies. The Lee tube has the advantage of ease of operation and low operating cost (no gas or special equipment needed). The disadvantages include (*1*) difficulty in picking isolates from the agar; (*2*) difficulty in cleaning the tubes; and (*3*) fragility of the system.

Fung and Lee (41) developed a double-tube system for anaerobic cultivation of bacteria from foods. It consists of a glass tube (15×1.0 cm o.d.) placed into a larger glass tube (15×1.5 cm o.d.) with anaerobic agar (designed for specific organisms) and culture sandwiched between the two tubes. The screw cap of the larger tube when closed seals the system and makes it anaerobic. Anderson and Fung (29) were successful in cultivating strict anaerobic rumen microbes, such as *Megasphaera elsdenii, Butyrivibro fibrisolvens, Clostridium perfringens, Eubacterium limosum*, and anaerobes from beef and poultry. Ali et al. (31) successfully used this system to isolate *C. perfringens* from meat. Aramouni et al. (32) used the double-tube system to study *C. sporogenes* in home-style canned quick bread. The advantages of the system include (*1*) ease of operation; (*2*) inexpensive and ease for cleanup; (*3*) ease of enumeration and observation of cultures; (*4*) ease of obtaining colonies (by removing the inner tube and pick colonies); and (*5*) flexibility and ease of adaptation by other laboratories. Disadvantages include (*1*) operation of samples in the open environment and (*2*) occasional breakage of agar by gas-producing organisms.

Biological Membrane Fragments

A novel method using biological materials to achieve anaerobiosis was developed by Adler and Crow (26). Sterile suspensions of membrane fragments derived from *Escherichia coli* were able to remove oxygen rapidly and efficiently as a result of the presence of a cytochrome-based electron transport system that transfers hydrogen from suitable donors in the bacteriological media to oxygen and produces water as an end product. The membranes are stable, nontoxic, and active in both liquid and solid media. The history of the development of this concept and manufacturing of membrane fragments were described recently by Adler and Spady (27). The commercial product is named Oxyrase and marketed by Oxyrase Inc. (Mansfield, Ohio). Fung et al. (42) described the use of Oxyrase to stimulate the growth of *Listeria monocytogenes, Campylobacter jejuni, E. coli* O157:H7, etc. to high numbers (10^6 CFU/mL) so that secondary detection methods such as polymerase chain reaction (PCR), enzyme-linked immunosorbent assay, and DNA/RNA probes can detect the presence of these facultative anaerobic pathogens much faster for food safety concerns. Fung et al. (42) also detailed the use of Oxyrase and membranes derived from *Acetobacter* and *Gluconobacter* to stimulate the growth of starter cultures to produce fermented food faster. Claerbout (34) described the applications for bacterial membrane fragments in sterile pharmaceutical quality control. Thurston and Gannon (58) made a detailed comparison between Oxyrase anaerobic agar plates and conventional anaerobic chambers for the isolation and identification of anaerobic bacteria from clinical infections and concluded that Oxyrase is a valuable compound for anaerobic cultivation of bacteria.

The future of membrane fragments for anaerobic cultivation of microorganisms from food, clinical, industrial, and environmental samples is very bright.

METHODS OF IDENTIFICATION

The identification of anaerobes is a concern for applied microbiologists. Classic methods for identification can be found in *Bergey's Manual of Systematic Bacteriology*, Volume 1 (50). A variety of rapid and automated methods in microbiology have been developed for the isolation, enumeration, and identification of microorganisms (37), and many of these new methods can be adapted to anaerobic bacteriology. Identification of anaerobes can be achieved by the following methods and procedures: conventional methods, miniaturized methods, commercial diagnostic kits, physical and biochemical methods, immunological methods, and PCR and related methods.

Conventional Methods

Identification of microorganisms using conventional procedures is summarized in Table 1 (38). For anaerobes, the most important aspect is to have all the liquid and solid media incubated in anaerobic environments, as described in the previous section. After the reactions are obtained, microbiologists will determine the identities of the cultures by a variety of schemes, charts, and flow diagrams. The conventional method is time consuming and highly dependent on the knowledge and skill of the microbiologist involved.

The morphology is one of the important considerations in identifying microorganisms by conventional methods. Under the microscope, the morphology of anaerobic microorganisms are similar to the aerobic counterparts. The morphology of bacteria are rod-shaped, coccal-shaped, and spiral-shaped organisms. An example of a sporulating anaerobic rod-shaped bacterium, *C. perfringens*, is presented in Figure 1. There are also other bacteria that are pleomorphic (variable shapes depending on growth conditions). The morphology of yeast and mold are similar to the aerobic counterparts. Morphology of an anaerobic protozoa, *Isotricha prostoma*, is presented in Figure 2.

Table 1. Needed Information for Identification of Microorganisms

Morphology under magnification and on agar plates
Gram reaction and special staining properties
Biochemical activity profile and special enzyme systems
Pigment production, bioluminescence, chemiluminescence, and fluorescent compound production
Nutritional and growth-factor requirements
Temperature and pH requirements and tolerance
Fermentation products, metabolites, and toxin production
Antibiotic sensitivity pattern (antibiogram)
Gas requirements and tolerance
Genetic profile; DNA/RNA sequences and fingerprinting
Pathogenicity to animals and humans
Serology and phage typing
Cell wall, cell membrane, and cellular components
Growth rate and generation time
Ecological niche and survival ability
Motility and spore formation
Extracellular and intracellular products
Response to electromagnetic fields, light, sound, and radiation
Resistance to organic dyes and special compounds
Impedence, conductance, and capacitance characteristics

Source: Adapted from Fung (38).

Miniaturized Methods

Fung and Hartman (40) developed a variety of miniaturized microbiological methods for the identification and enumeration of aerobic bacteria. The entire concept and system were adapted for miniaturized anaerobic bacteriology simply by operating the procedures in an anaerobic chamber (60,61). Guy Miller and T.G. Nagaraja at Kansas State University (personal communication, 1987) found that miniaturized techniques could be used for studying rumen bacteria effectively. The advantages of miniaturized techniques are savings of material, space, time of incubation and operation, and labor. Another advantage is the flexibility of the system to be able to adapt to many different laboratories, especially for environmental and industrial settings where commercial diagnostic kits are not available for specialized organisms.

Commercial Diagnostic Kits

Around the beginning of 1970s with the advancement of miniaturized microbiological techniques and concepts de-

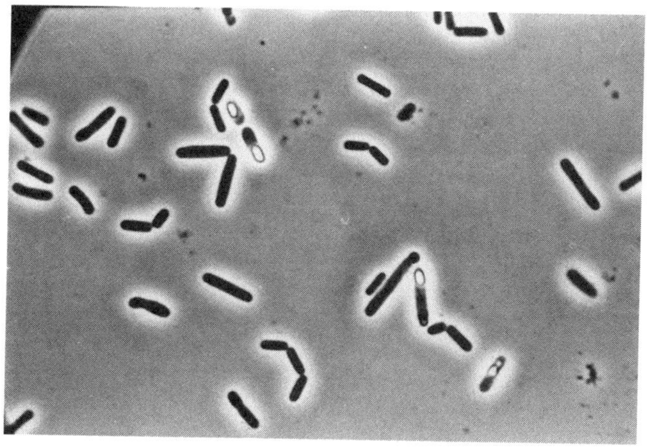

Figure 1. Morphology of *Clostridium perfringens* under 1,000X light-phase-contrast microscopy. This is a gram-positive, rod-shaped, anaerobic, spore-forming bacterium that causes large numbers of foodborne outbreaks and cases in the world. The dark rod-shaped cells are vegetative bacteria, and the bright refractile oval objects are spores developed in the vegetative cells.

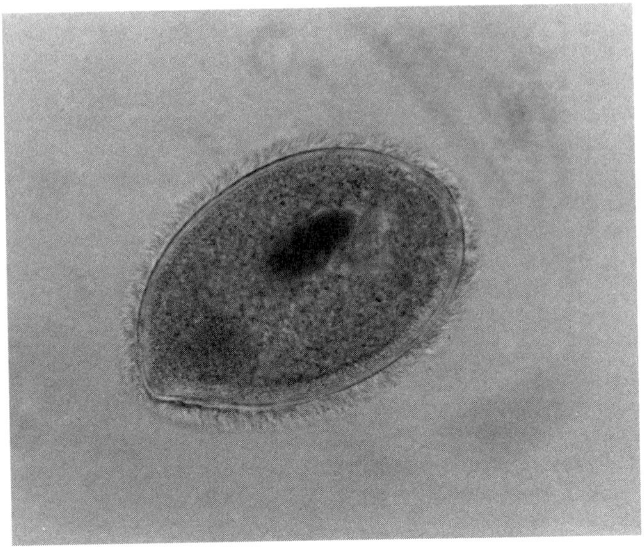

Figure 2. Morphology of *Isotricha prostoma*, under 100X light-phase-contrast microscopy. This is an anaerobic protozoan an important member of the ruminal flora in cattle. *Source*: Courtesy of T. G. Nagaraja of Kansas State University.

veloped by Fung and colleagues and other scientists, many diagnostic kits appeared on the market mainly for clinical microbiology. These methods gradually found their ways into food microbiology laboratories and sometime into industrial and environmental laboratories. Currently commercially successful diagnostic kits are marketed toward clinical settings. This may change if the need arises in the near and far future. There are two major types of diagnostic kits. One type depends on *growth* of the pure culture and biochemical reactions obtained in the media. This type of test is typically slow, requiring overnight incubation and anaerobic incubation of cultures. Another type is by monitoring *enzyme systems* of the pure culture. In this system, anaerobiosis is not necessary and the results can be obtained in about 4 hours. Kelly and Schipp (49) made a detailed analysis of the approaches and designs of anaerobic diagnostic kits, including discussions on substrate concentration, organism concentration, pH of reaction, buffering capacity, additives, incubation time, and growth medium for the identification of anaerobes. In that article, important variables affecting system performance and issues concerning comparative analysis of data reported in journal articles were also studied. To compare data reported in the literature, one needs to address the issues such as source of strains, database version, the "gold standard," number of species/strains tested, need of supplementary tests, and accuracy. In general, a diagnostic kit providing 90 to 95% accuracy compared with the conventional method is considered good. Any system with less than 85% agreement with the conventional method is considered unacceptable.

The major growth-dependent commercial kits for anaerobes are API 20A, marketed by bioMerieux Vitek Inc., Hazelwood, Mo., and Minitek Anaerobe system, marketed by BBL Microbiology Systems, Cockeysville, Md. The API 20A kit consists of 20 microtubes with dehydrated substrates. A liquid suspension of a pure culture is made and then introduced into each microtube. The kit is then incubated anaerobically, and after 1 to 2 days the reactions are read and data recorded in a convenient chart. The results are converted into a number system. From a code book the identity of the culture can be ascertained. These types of results provide a statistical probability of the unknown culture by comparing the attributes in the computerized database. Thus, an organism may be identified as *C. perfringens* at 0.9500 level. Sometimes the printout may include requests for additional tests, and in some cases the organism may be given two identifications with probability of each listed. This general format of identification is used by all commercial diagnostic kits.

Minitek Anaerobe II uses paper disks saturated with substrates. Thirty-five disks are available. After a suspension of the pure culture is made, the liquid is dispensed into wells of the units containing the appropriate disks. The assemblage is incubated anaerobically for 2 days. Reactions are read by color changes and recorded. The identity of the unknown can be obtained by following the procedure outlined for the API system.

Use of enzyme test systems include Rapid ID 32A (bioMerieux), AN-Ident (bioMerieux), RapID ANA II (IDS Inc., Norcross, Ga.), Vitek ANI Card (bioMerieux), and MicroScan Rapid Anaerobe Panel (Baxter Diagnostic, Sacramento, Calif.). All these methods are incubated aerobically because enzyme activities will not be influenced by the presence of air. Rapid ID 32A and RapID ANA II are similar in that the cultures are made into a liquid suspension, and the liquid is inoculated into microwells. After 4 hours, the color reactions are read and the identities are ascertained from a code book. Vitek ANI Card involves injection of liquid sample into a plastic card that is the size of a credit card with 30 wells. After 4 hours of incubation, the results are read and data are entered into the ANI program on the VITEK computer. Identification is made by the computer. In the MicroScan Rapid Anaerobe Panel system, the culture is inoculated into 24 substrates and two color interpretation controls. After 4 hours of aerobic incubation and addition of reagents, the panels are read either manually or with the aid of a Touchscan light box. Data are then interpreted by computer for identification.

It should be emphasized that the aforementioned systems are designed for clinical bacteriology. Identifications of unknowns from environmental, food, and industrial sources may not be as accurate because the databases of these systems are from clinical isolates.

Physical and Biochemical Methods

Physical and biochemical analysis of components of cells and metabolites of cells have also been used for identification of anaerobes. Fingerprinting techniques have been explored to identify anaerobes by matching profiles from unknown cultures with known profiles in the data base. One such approach is the gas–liquid chromatography method popularized by the VPI Anaerobic Laboratory (45). Mitruka (51) provided a detailed discussion on this method. The Hewlett-Packard company developed a procedure to fingerprint the fatty acid profiles of microorganisms, including anaerobes, using a high-performance liquid chromatography technique. The AMBIS system involves computer-integrated fingerprinting of electrophoretically separated labeled proteins of bacteria to match profiles of unknowns versus knowns. In all these systems, these cultures must be pure and grown anaerobically before accurate identification can be made. Again, the database must be from the appropriate sources for the identification to be accurate.

Immunological Methods

Microorganisms have specific antigen properties, and many of these antigens can elicit the production of antibodies in animals. With the appropriate antibodies, one can then use the antibodies to identify the unknown by antigen–antibody reactions. Historically, polyclonal antibodies have been used for detection and characterization of bacteria. More recently, monoclonal antibodies have been used for the reactions. Both types of antibodies are useful for anaerobic microbiology. The reactions can be agglutination of cells directly with antibodies or indirect agglutination of antigens reacting with antibodies fixed on a variety of support systems, including blood cells, latex beads, plastic steel balls, etc. The most popular format of antigen–antibody reactions used is ELISA. In this test, the antibody is fixed on a support surface and the antigen is captured. A second antibody then reacts with another site of the antigen. The second antibody is conjugated with an enzyme. A substrate is then applied, and if the enzyme is present a color reaction will occur. The intensity of the color reaction will indicate the presence of the antigen. A cutoff color intensity is established to indicate a positive or negative test of the presence of the antigen. At first, this test was done manually, but recently, it has been completely automated. The analyst presents the instrument with a suspect sample, and the instrument will automatically complete the test with a computer printout.

An exciting new development in immunological analysis is the field of immunomagnetic capture technology. In this technology, beads are magnetized and then antibodies or other capture particles are attached on the surface of the beads. Antibodies against whole cells, antigens, and other target particles are fixed on the beads. The charged beads are then introduced to a broth or food sample to interact with target bacteria or other molecules. In this phase, the sample is mixed thoroughly with the charged beads tumbling in the liquid food to interact with target organisms. After the interaction, a powerful magnet is applied to the side of the reaction tube. The magnet holds all the beads with or without captured organisms. The rest of the debris is discarded, and the beads are released and washed. The captured organisms can be subject to growth in a special broth or agar, or these may be used for further immunological reaction, PCR, DNA/RNA reactions, or other tests.

PCR and Related Methods

DNA and RNA hybridization tests have been developed to detect organisms whose specific gene sequence of interest is known. At first, radioactive compounds were used to report the hybridization, but recently, the reaction has been reported by enzyme and color reactions. The truly revolutionary development in genetic type of identification of microorganisms is in the field of PCR technology. In this technology, the gene sequence of the organism of interest (an anaerobe, for example) is known at the DNA level. A piece of the DNA can be unfolded by heat, and after cooling of the unfolded DNA, two primers will interact with specific regions of the single DNA strands. A heat-stable polymerase will complete the complementary strand in the presence of nucleotides from the 3′ end and the 5′ end. Thus, in one cycle, one DNA molecule will become two, and after another cycle, two will become four, etc. In about 2 hours, one molecule of DNA can be amplified to 10^6. To use this technology, the genetic sequence of the target organism must be known, and the appropriate primers must be developed for successful use of this powerful tool. There are many new variations of this technology, such as the search for a cold amplification method to bypass the heating cycle and the development of an ELISA-type detection scheme to bypass the need to do electrophoresis of PCR products.

Another technology is called ribotyping. In this technology, the DNA of a target organism is extracted from the cells, an appropriate enzyme or a collection of enzymes is then used to cut the DNA. The DNA fragments are then separated by electrophoresis, and the pattern of the bands are photographed and quantified, resulting in special fingerprints.

The fingerprints can be matched with fingerprints of known cultures for identification of unknowns. One added feature is that for certain organisms (e.g., *E. coli*) there are different fingerprints, even for the same organism. This becomes very important when one wants to trace the occurrence of an organism in the case of an outbreak of foodborne disease. For example, finding *E. coli* O157:H7 in several foods and the environment in an outbreak is not good enough to trace the etiology of the outbreak. With ribotyping, the culprit *E. coli* O157:H7 can be traced to a particular food by matching the ribotyping pattern of the strain that caused the outbreak with the origin of the strain in certain foods or in the environment. This will help pinpoint the source of the infection. Dupont's Qualicon Division is leading the way for the development and commercialization of this technology.

There are, of course, many other methods for identification of unknowns, such as impedance, conductance, capacitance, pyrolysis pattern, microcalorimetry, flow cytometry, etc. *Automated Microbial Identification and Quantitation: Technologies for the 2000s*, an excellent recent reference book edited by Olson (55), highlights many of these developments. There will be an explosion of technologies in applied microbiology in the near future that will directly and indirectly influence the enumeration, characterization, isolation, and identification of anaerobes.

INDUSTRIALLY IMPORTANT STRAINS AND PRODUCTS

Many products are or have been produced by fermentation in commercial quantities (2–4). Food products, industrial bulk chemicals, solvents, enzymes, and amino acids are produced with the help of anaerobic microorganisms. In this work, the emphasis is on those processes carried out by anaerobes, which generally means that important pathways yield useful products under conditions where oxygen is not required. However, oxygen is present in modest quantities to enhance microbial growth in some cases. Alternative manufacturing processes have been developed for some products; they can be produced by fermentation and by chemical processes that do not require microorganisms. Often these products are made by a microbial process in some countries and a chemical process in other countries; both processes may be used in the same country as well. The choice may depend on local raw materials, taxes, and scale of operation; often local economics determines which process is used.

Table 2 lists several compounds produced by anaerobes and some of the microorganisms that can produce significant quantities of the product. The pathways by which these products are produced have been investigated; many of the well-known industrially important pathways such as the Embden-Meyerhof-Parnas pathway are described in detail elsewhere (2,6,7). Ethanol is produced in large quantities for use as a motor fuel in several countries (2,68). Methane is often produced by mixed cultures as a product of anaerobic digestion of waste materials. Lactic acid is widely produced in food fermentations where milk, meat, or vegetables are fermented. Table 3 lists some of the many food fermentations and microorganisms that are present and active in these fermentations. Detailed descriptions of food fermentations are given by Reed (3) and Steinkraus (4). Flavors are introduced into fermented foods as a result of the fermentation process (8); however, preservation is often as important as flavor in food fermentations.

Knowledge of the biochemical pathways and the enzymes that facilitate the biochemical transformations is needed to commercialize and optimize many anaerobic fermentations. The control systems are often altered to overproduce desired products. In the case of biodegradation, it may be necessary to add plasmids that provide genetic material necessary for a biodegradation pathway to function. As new products such as gasoline oxygenates enter the soil and water environment, microorganisms capable of biodegradation of methyl-*tert*-butyl ether (MTBE) have evolved (67).

Table 2. Anaerobic Fermentation Products and Microorganisms That Can Produce Significant Quantities of the Product

Product	Microorganism
Methane	*Methanobacterium*
	Methanococcus
	Methanosarcina
	Methanobrevibacter
	Methanospirillum
Acetate	*Clostridium thermoaceticum*
Ethanol	*Saccharomyces cerevisiae*
	Saccharomyces uvarum
	Saccharomyces carlsbergensis
	Zymomonas mobilis
	Clostridium thermocellum
Lactic acid	*Lactobacillus*
	Streptococcus (Lactococcus)
	Leuconostoc
	Pediococcus
Propionate	*Clostridium propionicum*
	Clostridium tetanomorphum
Acetone	*Clostridium acetobutylicum*
Isopropanol	*Clostridium beijerisckii*
Acetoin	*Clostridium acetobutylicum*
Butanediol	*Enterobacter aerogenes*
	Bacillus
	Serratia
	Erwinia
	Aerobacter
Butyrate	*Clostridium butyricum*
	Clostridium tyrobutyricum
Butanol	*Clostridium acetobutylicum*
	Clostridium beijerinskii
	Clostridium tetranomorphum

Source: Linden et al. (5).

GROWTH AND PRODUCT FORMATION

Anaerobic growth and product formation are often investigated together when the product is growth associated. In anaerobic fermentations, adenosine triphosphate (ATP) needed for growth is often formed as a result of substrate phosphorylation. Products such as ethanol are produced because the organism is growing and needs ATP for synthesis of microbial cell mass. Mass balances and yield parameters that describe growth and product formation can be written in terms of the mass of substrate, biomass, and product or in terms of the equivalents of available electrons of each of these.

Consider the anaerobic production of a simple product, such as ethanol, by anaerobic fermentation. The chemical balance equation is as follows:

$$CH_mO_l + aNH_3 = y_cCH_pO_nN_q + zCH_rO_sN_t + cH_2O + dCO_2 \quad (1)$$

where CH_mO_l, $CH_pO_nN_q$, and $CH_rO_sN_t$ give the elemental compositions of the carbon, hydrogen, oxygen, and nitrogen in the substrate, biomass, and extracellular product, respectively. For glucose as substrate, $m = 2$ and $l = 1$; for ethanol as product, $r = 3$, $s = 0.5$, and $t = 0$. For the valences C = 4, H = 1, O = −2, and N = −3, the available electron balance (2,9) is

$$y_c \frac{\gamma_b}{\gamma_s} + z \frac{\gamma_p}{\gamma_s} = 1.0 \quad (2)$$

or

$$\eta + \xi_p = 1.0 \quad (3)$$

where γ is the reductance degree (2,9) and the subscripts s, b, and p refer to substrate, biomass, and product, respectively.

Luedeking and Piret (10,11) reported that product formation kinetics and growth kinetics were related; product

Table 3. Fermented Food Products and Associated Microorganisms

Product	Microorganisms	Reference
Yogurt	*Streptococus thermophilus*	2
	Lactobacillus bulgaricus	
Sauerkraut	*Enterobacter cloacae*	2,3
	Erwinia herbicola	
	Leuconostoc mesenteroides	
	Lactobacillus brevis	
	Lactobacillus plantarum	
	Pediococcus cerevisiae	
Cucumber fermentation (pickles)	*Lactobacillus brevis*	3
	Lactobacillus plantarum	
	Pediococcus cerevisiae	
	Leuconostoc mesenteroides	
Sausage	*Lactobacillus*	3
	Leuconostoc	
	Pediococcus	
Bread and sourdough bread	*Saccharomyces cerevisiae*	21
	Lactobacillus	
	Candida milleri	
	Leuconostoc mesenteroides	
	Saccharomyces cerevisiae	
Cheddar cheese	*Streptococcus (Lactococcus) cremoris*	3
	Streptococcus (Lactococcus) lactis	
Swiss cheese	*Lactobacillus bulgaricus*	3
	Streptococcus thermophilus	
	Propionibacterium shermanii	
Blue cheese	*Streptococcus (Lactococcus) lactis*	3
	Streptococcus (Lactococcus) cremoris	
	Penicillium roquiforti	
Camembert cheese	*Streptococcus (Lactococcus) lactis*	3
	Streptococcus (Lactococcus) cremoris	
	Penicillium camemberti	
Cottage cheese	*Streptococcus (Lactococcus) lactis*	3
	Streptococcus (Lactococcus) cremoris	
Acidophilus milk	*Lactobacillus acidophilus*	3
Kefir	*Streptococcus (Lactococcus) lactic*	3
	Streptococcus (Lactococcus) cremoris	
	Torula or *Candida yeast*	
Buttermilk and sour cream	*Streptococcus (Lactococcus) lactis*, (including subspecies diacetylactis)	3
	Streptococcus (Lactococcus) cremoris	
	Leuconostoc cremoris	
Wine	*Saccharomyces cerevisiae*	3
	Saccharomyces uvarum	
	Other *Saccharomyces*	
Beer	*Saccharomyces cerevisiae*	3
	Succharomyces uvarum	
Soy sauce	*Asperigillus oryzae*	3
	Saccharomyces rouxii	
	Torulopsis versatilis	
	Pediococcus soyae	
Miso	*Pediococcus halophilus*	3
	Saccaromyces rouxii	
	Torulopsis	
	Streptococcus faecalis	
Tempeh	*Rhizopus oligosporus*	3

formation was modeled using a growth-associated term and a maintenance-associated term; that is,

$$\frac{dP}{dt} = \alpha \frac{dX}{dt} + \beta X \tag{4}$$

where X is biomass concentration and P is product concentration. Growth is dependent upon the availability of substrate according to the Monod model

$$\frac{1}{X}\frac{dX}{dt} = \mu = \frac{\mu_{max}S}{K_s + S} \tag{5}$$

Here S is substrate concentration, μ is specific growth rate, μ_{max} is maximum specific growth rate, and K_s is the saturation constant. The organism uses substrate for growth and maintenance as follows:

$$-\frac{dS}{dt} = \frac{1}{Y_{x/s}^{max}}\frac{dX}{dt} + m_s X \tag{6}$$

where $Y_{x/s}^{max}$ is the true growth yield corrected for maintenance and m_s is the maintenance coefficient. In this process, the product formation kinetic parameters and the bioenergetic parameters are related. Equation 6 may be written as (2)

$$\frac{\mu}{\eta} = \frac{\mu}{\eta_{max}} + m_e \tag{7}$$

and equation 4 becomes

$$\frac{\mu}{\eta}\xi_p = \frac{\mu}{\eta}(1 - \eta) = \alpha_e\mu + \beta_e \tag{8}$$

if the following relationships are used (2)

$$m_e = \beta_e = m_s\frac{\sigma_s\gamma_s}{\sigma_b\gamma_b} = \beta\frac{\sigma_p\gamma_p}{\sigma_b\gamma_b} \tag{9}$$

$$\frac{1}{\eta_{max}} = \alpha_e + 1 = \frac{\sigma_s\gamma_s}{\sigma_b\gamma_b Y_{x/s}^{max}} = \alpha\frac{\sigma_p\gamma_p}{\sigma_b\gamma_b} + 1 \tag{10}$$

The true growth yield in equation 7 must satisfy the relationship (2)

$$\eta_{max} = \frac{(\sigma_b\gamma_b/12)Y_{ATP}^{max}}{(\sigma_b\gamma_b/12)Y_{ATP}^{max} + \delta} \tag{11}$$

where Y_{ATP}^{max} is the true growth yield based on ATP, δ is the equivalents of available electrons transferred to products to produce 1 mol of ATP from ADP, and σ_b is the weight fraction carbon in biomass. Information on the theoretical maximum yield with respect to ATP and the value of δ for a particular fermentation allows one to estimate the range of values of the growth-associated kinetic term in the product formation model. For $\delta = 12$ equivalents of available electrons per mole of ATP generated, $Y_{ATP}^{max} = 28.8$ g cells per mole ATP, $\sigma_b = 0.462$, and $\gamma_b = 4.291$; $\eta_{max} = 0.283$ and $\alpha_e \geq 2.52$.

Product formation is directly linked to the specific growth rate, true growth yield corrected for maintenance, and the maintenance coefficient as shown in equations 1 to 11. Equation 3 shows that the product yield is related to the biomass yield; the largest product yields are expected when the cells are using carbon and energy primarily for maintenance and nearly all the available electrons are being converted to product. When oxygen is present, some of the available electrons in the carbon substrate are transferred to oxygen, and the product yield is reduced accordingly.

Growth kinetics, product formation rates, and product yields are affected by temperature, pressure, water activity, osmolality, oxidation–reduction potential, bioenergetics, and the concentrations of organic substrate, inorganic nutrients, biomass, hydrogen ions (pH), carbon dioxide, products, and electron acceptors such as oxygen, nitrate, and sulfate (2).

BIOENERGETICS AND PRODUCT YIELDS

The product yields under anaerobic conditions depend on the operating conditions and the fermentation pathway for the selected strain of microorganisms. Equations 2 and 7 can be rearranged as follows:

$$\xi_p = 1 - \eta \tag{12}$$

$$\eta = \frac{\mu}{\mu/\eta_{max} + m_e} \tag{13}$$

If equation 13 is substituted into equation 12, one obtains

$$\xi_p = 1 - \frac{\mu}{\mu/\eta_{max} + m_e} \tag{14}$$

Equation 14 indicates that the product yield is a function of the biomass true growth yield, which depends on the growth yield with respect to ATP and the efficiency with which ATP is formed through substrate phosphorylation. The product yield also depends on the values of the maintenance coefficient and specific growth rate. Table 4 shows

Table 4. Effect of ATP Yield, True Growth Yield, Maintenance Coefficient, and Specific Growth Rate on Biomass Yield and Product Yield

Y_{ATP}^{max}	δ	η_{max}[a]	m_e	μ	η	ξ_p[b]
28.8	12	0.283	0.1	0	0	1.0
28.8	12	0.283	0.1	0.1	0.221	0.779
28.8	12	0.283	0.1	0.2	0.248	0.752
28.8	12	0.283	1.0	0.1	0.0739	0.926
10.5	12	0.126	0.1	0.1	0.112	0.888
10.5	12	0.126	1.0	0.1	0.056	0.944
10.5	24	0.0674	0.1	0.1	0.063	0.937
10.5	24	0.0674	1.0	0.1	0.040	0.960

[a]Values estimated from equation 11.
[b]Values estimated from equation 14.

how the true growth yield varies with the growth yield with respect to ATP and the number of moles of available electrons of substrate required to produce 1 mol of ATP according to equation 11. It also illustrates how the product yield varies as a function of specific growth rate and maintenance coefficient. The product yield decreases as the specific growth rate increases. As values of the maintenance coefficient increase, the product yield increases also.

Estimated values of the true growth yield and maintenance coefficient are presented in the book by Erickson and Fung (2) for several sets of experimental data reported by a variety of investigators. Selected values are presented in Table 5. These values are within the expected range of values based on theoretical yields and known pathways; however, not all of the estimated values reported by Erickson and Fung (2) are below the estimated theoretical yield. It is common to have experimental error because of the difficulty of making measurements. When data consistency is checked by making carbon and available electron balances and yields are compared with theoretical yields, it is possible to identify experiments where errors are present; however, it is not always possible to determine which particular measurements have the largest errors associated with them. When errors are present in the data, the covariate adjustment method (2) should be used in data analysis.

The results in Tables 4 and 5 may be related to metabolic pathways. The Embden-Meyerhof-Parnas pathway commonly produces 2 mol of ATP per mole of glucose (δ = 12 equivalents of available electrons per mole of ATP generated), whereas the Entner-Doudoroff pathway produces 1 mol of ATP per mole of glucose ($\delta = 24$). *Zymomonas mobilis* has the Entner-Doudoroff pathway; thus, the lower values of true growth yield in Table 5 are expected. As shown in Table 4, the product yield is predicted to be larger for *Z. mobilis*.

ANAEROBIC DIGESTION AND BIODEGRADATION

Anaerobic digestion is widely used in wastewater treatment to degrade the solids from primary and secondary treatment. Anaerobic processes are also applied in the treatment of industrial effluents and in bioremediation of contaminated soil and ground water. Although environmental microorganisms have been functioning and serving the needs of humans for a very long time, the commercialization and engineering of this aspect of bioprocess technology was started after the discovery of microorganisms. Wastewater treatment is the second oldest bioprocess technology after food processing.

Anaerobic degradation processes depend on the electron acceptors that are present. Stouthamer (13) points out that when several electron acceptors are present, the microorganism selects the electron acceptor that yields the largest amount of energy by repressing the formation of reductase enzymes for the other electron acceptors. Table 6 gives the standard free energy changes for oxidation of acetate with oxygen, nitrate, sulfate, and water as electron acceptors, respectively. When oxygen is present, aerobic processes are observed and anaerobic processes are generally absent unless there are spatial regions or time periods where oxygen is absent. Nitrate is the electron acceptor of choice when oxygen is absent, and when nitrate and oxygen are both absent, sulfate is the favored electron acceptor. Methane production, which occurs when water is the electron acceptor, is generally inhibited if sulfate is present. Stouthamer points out that other electron acceptors may participate, including other oxidized forms of nitrogen and fumarate.

Many bacteria have been found that can use nitrate as the terminal electron acceptor. The first product of nitrate respiration or nitrate reduction is formation of nitrite, which can be further used as an electron acceptor by many microorganisms. In denitrification, the end product of further reduction is nitrogen gas; however, in other cases the end product is ammonia. Stouthamer (13) reviewed the enzymology and bioenergetics of nitrate reduction.

Sulfate-reducing bacteria include *Desulfovibrio, Desulfobacterium, Desulfotomaculum, Desulfococcus, Desulfosarcina, Desulfomonile, Desulfonema, Desulfoarculus, Desulfobulbus, Desulfobotulus*, and *Desulfobacula* (19). For these genera, many species have been described in the literature (13,14,19). These organisms have great diversity in biochemical properties, and they can completely oxidize a variety of organic substrates.

Methanogenic bacteria are widely found under anaerobic conditions in nature. Methane is produced in soil, landfills, anaerobic digesters, rumens of animals, and many other places. In anaerobic digestion, methanogenic bacteria are found together with hydrolytic fermentative microorganisms and syntrophic acetogenic bacteria. The hydrolytic fermentative organisms, which include eubac-

Table 5. Estimated Values of m_e and η_{max} for Selected Experiments

m_e	η_{max}	Organism	Substrate	Product
0.896	0.164	*Saccharomyces cerevisiae*	Glucose	Ethanol
1.791	0.0835	*Zymomonas mobilis*	Glucose	Ethanol
3.264	0.078	*Zymomonas mobilis*	Glucose	Ethanol
0.078	0.281	*Lactobacillus delbruekii*	Glucose and yeast extract	Lactic acid
0.112	0.308	*Lactobacillus delbruekii*	Glucose and yeast extract	Lactic acid
0.491	0.267	*Lactobacillus delbruekii*	Glucose	Lactic acid
0.396	0.206	*Lactobacillus bulgaricus*	3% Nonfat dry milk	Lactic acid
0.286	0.230	*Clostridium acetobutylicum*	Glucose	Acetone, butanol, ethanol, butyrate

Source: Data reported by Oner et al. (12).

Table 6. Standard Free Energy Changes per Equivalent of Available Electrons for Oxidation of Acetate with Various Electron Acceptors

Electron acceptor	Free energy change (kJ/equiv)
Oxygen	−106
Nitrate	−99
Sulfate	−6
Water	−4

Source: Stouthamer (13).

teria, fungi, and often protozoa, produce fermentation products such as ethanol, lactate, butyrate, succinate, and acetate. The acetogenic bacteria (*Pellobacter, Desulfovibrio, Syntrophobacter, Clostridium, Syntrophomonas*, and others) (16) produce acetate, formate, bicarbonate and hydrogen from the ethanol, lactate, butyrate, and other organic acids and alcohols with more than two carbons. The methanogenic bacteria use the acetate, formate, and hydrogen as substrates. In anaerobic digesters, about 75% of the available electrons in the complex organics are first transformed to fermentation products other than acetate, formate, and hydrogen, which make up the remaining 25%. More than 60% and as much as 72% of the available electrons is converted to acetate, which is then used by methanogens to form methane. More than 25% of the available electrons are converted to hydrogen before being incorporated into methane (2). Thiele and Zeikus (15) and Thiele (16) reviewed mixed culture interactions in methanogenesis and provided information on selected anaerobic microorganisms that are active in anaerobic digestion.

Many microorganisms that are active in environmental microbiology have been studied and described; however, each year many new organisms are described for the first time (17). New anaerobic treatment processes for specific applications, such as nitroaromatic munitions compounds, are being developed (18). Although anaerobic processes are often slower than aerobic processes, they have been found to be economically attractive in a wide variety of applications.

MIXED CULTURES

Mixed cultures are used extensively in food fermentations and in biodegradation of complex substances. Mutualism, commensalism, amensalism, competition, and prey–predator relationships are the main interactions observed in mixed cultures (16). Mutualism occurs when both species benefit from the interaction, whereas commensalism describes processes where only one species benefits and the other is not affected. Amensalism occurs when one species restricts another, whereas competition affects both species because of nutrient limitations. The predator consumes the prey in prey–predator interactions. In actual situations, more than two species are often present, and several types of interaction may be taking place.

In the recent book on mixed cultures by Zeikus and Johnson (20), most of the contributions are related to either food fermentations or environmental microbiology. There are food fermentation chapters on bread, milk, oriental foods, wine, and vegetables and environmental microbiology chapters related to degradation of polysaccharides, methanogenesis, detoxification of hazardous waste, corrosion, and leaching processes in mineral biotechnology.

In San Francisco sourdough bread, *Lactobacillus sanfrancisco* ferments only the maltose, whereas *Candida milleri* uses all the other free sugars but not maltose. Lactic acid, acetic acid, and carbon dioxide are the main products of the fermentation; however, small amounts of propionic, isobutyric, butyric, α-methyl-*n*-butyric, isovaleric, and valeric acids have been found in commercial San Francisco sourdough breads (21).

Anaerobic cellulose degradation is accomplished by a mixed culture of *Clostridium* cellulolytic bacteria that produce cellobiose, glucose, and cellodextrins; fermentative bacteria that produce propionate, butyrate and other fermentation products from glucose and other intermediates; syntrophic acetogenic bacteria that convert the fermentation products to acetate, hydrogen, and carbon dioxide; homoacetogens that convert hydrogen and carbon dioxide to acetate; and methanogens that produce methane from acetate, formate, and hydrogen (22). Leschine (22) reports that *Clostridium papyrosolvens* grow mutualistically in coculture with a noncellulytic *Klebsiella; C. papyrosolvens* hydrolyzes cellulose for the *Klebsiella,* whereas *Klebsiella* excretes vitamins required by the *C. papyrosolvens.* The soluble sugar products of cellulose hydrolysis provide substrates for many noncelluloytic commensal organisms that depend on the cellulolytic bacteria, but do not provide known compounds in return.

In the natural environment, the human environment, and even in the industrial environment, mixed cultures of bacteria, yeast, mold, viruses, protozoa, nematodes, and higher organisms live in antagonistic, cooperative, or inert coexistence. Pure culture systems are the result of understanding of the fundamental role of microbes in a particular process and the isolation and cultivation of such cultures for specific reactions. This section deals with single-culture processes mixed pure-culture processes, and mixed natural-culture processes, using food systems as examples. This information certainly can be applied to other industrial and fermentation processes.

Other publications deal adequately with the subject of single-cell process. The key success of single-culture process is to provide the culture with a sterile substrate and environment with no contamination during the process. Single-cell process is a manmade situation classified as a controlled process because the substrate is prepared and processed in such a way as to minimize contamination. Examples of this type of process are wine making, beer making, bread making, single-culture dairy product fermentation, and vinegar production. The kinetics of growth and product formation are easier to control and monitor.

Mixed Pure-Culture Process

For some processes, it is desirable to have more than one pure culture for proper product development. The mixed pure cultures can be a controlled mixture of bacterium with bacterium or with a combination of yeast or mold, or both. The relationships among these pure cultures during growth become very complex.

Mixed Natural-Culture Process

In many situations, such as in nature and noncontrolled fermentation, the flora in the process are mixed natural cultures. The interactions of these microbes are even more complex than those of mixed pure-culture process. The microbial successions in these processes follow a special sequence. If the products are treated properly, the desirable cultures provide the characteristic end results. On the other hand, if the conditions are not treated properly, the process will fail.

Mixed Bacteria Interactions

An example of mixed pure bacterial culture fermentation is yogurt fermentation, which typically uses *Streptococcus thermophilus* and *Lactobacillus bulgaricus*. The interactions are commensal. Figure 3 illustrates the development of acidity of pure and mixed cultures of *S. thermophilus* and *L. bulgaricus* in liquid. In mixed cultures, *S. thermophilus* grows faster at the beginning, reduces the pH by lactic acid production, and also produces formic acid, which is a stimulatory compound for *L. bulgaricus*. *L. bulgaricus* produces amino acids that stimulate the growth of *S. thermophilus*. Higher acid production with mixed cultures was obtained and compared to pure cultures, as can be seen in Figure 3 (39,52). This is an example of commensalism in mixed cultures. Details of the kinetics and mathematical relationships of interactions between *S. thermophilus* and *L. bulgaricus* are presented in an article by Fung et al. (39).

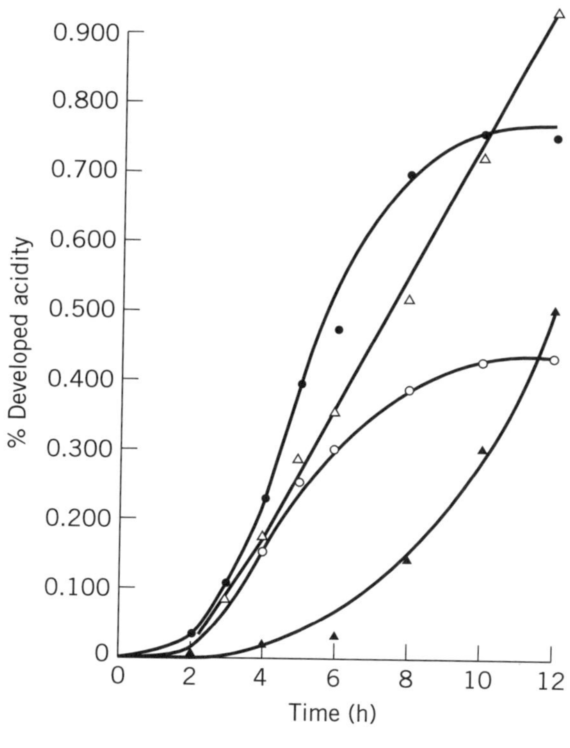

Figure 3. Developed acidity of pure and mixed cultures of *L. bulgaricus* and *S. thermophilus*. *Open circles*, *S. thermophilus*; *closed triangles*, *L. bulgaricus*; *open triangles*, sum of developed acidity of pure cultures; *closed circles*, mixed culture. *Source*: Reprinted with permission from *Journal of Food Protection* and Marcel Dekker, Inc. Reprinted from Reference 39, page 511, courtesy of Marcel Dekker, Inc.

Mixed Bacterium and Yeast Interaction

Yeast is one of the most important organisms for industrial use in alcoholic fermentation. Bacterial contamination may occasionally occur during yeast fermentation, which may pose serious problems for the fermentation industry. This section deals with the interaction of *Saccharomyces cerevisiae* with *Acetobacter suboxydans* or *E. coli* in mixed-culture growth. *S. cerevisiae* was grown in mixed cultures with either *A. suboxydans* or *E. coli* in a culture medium and glucose at 28 °C. Growth was monitored by viable cell counts of yeast and bacteria using selective agars, direct count by microscopic reading of cells in a Petroff-Hauser counting chamber, and electronic counts by the Coulter counter. By selecting the appropriate threshold windows, the electronic instrument can differentiate yeast count and bacterial count in the same liquid. During the growth period, pH values were monitored by a conventional pH meter and glucose was determined by the Glucostat, a commercial kit designed to detect glucose in solution. A vinometer was used to measure alcohol level of the solution.

Figure 4 illustrates the interaction between *S. cerevisiae* and *A. suboxydans* as reported by Fung et al. (39). For the yeast, both direct count and electronic count increased with time. The direct count registered about 10 times more cells than the electronic count. Because *A. suboxydans* does not grow well in solid agar medium, viable cell count data were not obtained. For *A. suboxydans*, the direct count increased with time, but the electronic count increased to about log 7.8 mL^{-1} and then started to decrease. The decrease in number is due to the adhesion of many of the bacterial cells to yeast cells. The electronic count cannot differentiate these two populations in this condition. However, microscopic observations can ascertain the different counts. The concentration of glucose completely disappeared in the first 10 h of yeast and bacterial interactions. The pH of the medium first decreased and then stabilized at pH 3. This is because of oxidation of alcohol (produced by *S. cerevisiae*) by *A. suboxydans* to acetic acid, making the reaction mixture acidic.

The interactions of yeast and *E. coli* (39) in Figure 5 show interesting contrasts compared with the yeast–*Acetobacter* interaction. Growth of yeast as monitored by viable cell count, direct count, and electronic counts all increased with time. After reaching a stationary phase, viable yeast count decreased, but direct count and electronic count registered no reduction in numbers. A likely explanation is that the former two biomass measurements counted both live and dead cells, but the viable cell count only registered living cells. After the stationary phase, some of the yeast cells went through the death cycle. The growth curves of *E. coli* showed an increase in the viable

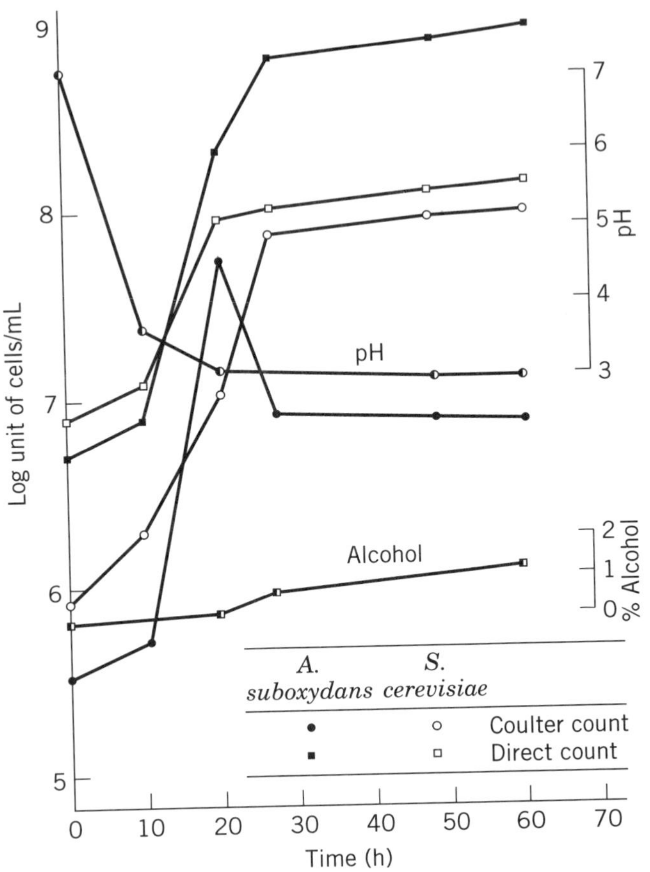

Figure 4. Interaction of mixed cultures of *S. cerevisiae* and *A. suboxydans* in terms of cell numbers monitored by direct count and electronic count as well as product formation (alcohol and acid production. *Source*: Reprinted with permission from reference 39, page 527, by courtesy of Marcel Dekker, Inc.

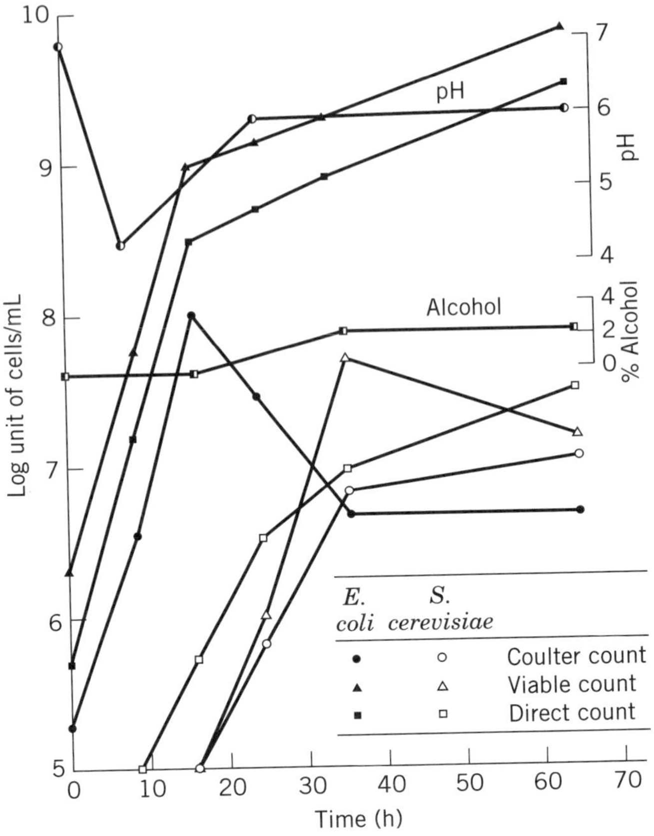

Figure 5. Interaction of mixed culture of *S. cerevisiae* and *E. coli* in terms of cell numbers monitored by direct count, electronic count, and viable cell count as well as product formation (alcohol and acid production). *Source*: Reprinted with permission from reference 39, page 528, by courtesy of Marcel Dekker, Inc.

cell count and direct cell count. The electronic count reached log 8 mL^{-1} and then declined, exhibiting a trend similar to that observed in the yeast–*Acetobacter* interactions.

Alcohol contents increased to about 2%, and the pH first dropped to around 4 and then rose to 6. This pattern differs from that obtained from the interaction for yeast and *Acetobacter. E. coli* cannot oxidize alcohol and thus did not create large amounts of acid to counteract the basic metabolites in the reaction vessel. This resulted in a medium that reverted to a more alkaline state.

Mixed Bacteria and Mold Interaction

Another example of mixed culture interaction is ensilage of crops. Dalamacio and Fung (35) and Dalmacio et al. (36) studied the influence of ammonia on bacteria and mold during ensilage of high-moisture corn. Bacteria were not affected by ammonia treatment as much as mold at the onset of the treatment. Bacterial populations remained detectable after treatment of 1.0, 1.5, and 2% ammonia. However, no mold was detected for corn treated with 1.0, 1.5, 2.0% ammonia. Corn treated with 0.5% ammonia had high bacterial and mold counts. After 2 months of storage, the bacterial population increased to more than 10^7 g^{-1}. Mold count increased to around 10^4 g^{-1} for the corn treated with 0.5% ammonia. After 4 months of storage, the bacterial count reached saturation level (higher than 10^{10} g^{-1}) but the mold population remained at around 10^4 or 10^5 g^{-1}. Not only did the bacteria and mold counts change in the fermentation, but the genera of bacteria and mold also changed concomitant to the fermentation stages. At the beginning, most of the bacteria isolated were *Bacillus*, but as the fermentation progressed most of the bacterial isolates belonged to *Lactobacillus*.

At the beginning of the storage of ammonia-treated corn (in the fall season), the predominant mold was *Mucor*, which is considered a field mold. As the storage continued into the cold winter, *Penicillium* isolates increased, and at the end of the storage in spring, *Scopulariopsis* predominated. These successions occurred for a variety of reasons. For example, the winter months favored the development of *Penicillium*, which is cold tolerant. As temperature increased in the spring, *Penicillium* population was overtaken by *Scopulariopsis*, which can use complex nitrogenous foods better than *Penicillium*.

The aforementioned examples of microbial interactions are just some of the typical interaction patterns in food and in nature. Countless other interactions can occur. The main point to emphasize is that mixed-culture interactions are very complex, but scientists can use this knowledge to optimize process control in obtaining desirable products in food and industrial microbiological processes.

ACKNOWLEDGMENTS

This research was partially supported by the U.S. Environmental Protection Agency (EPA) under assistance agreement R-819653 to the Great Plains–Rocky Mountain Hazardous Substance Research Center. It has not been submitted to the EPA for peer review and, therefore, may not necessarily reflect the views of the agency; no official endorsement should be inferred. The Center for Hazardous Substance Research and the Agricultural Experiment Station also provided partial funding.

BIBLIOGRAPHY

1. D.Y.C. Fung, in L.E. Erickson and D.Y.C. Fung eds., *Handbook on Anaerobic Fermentations*, Dekker, New York, 1988, pp. 3–25.
2. L.E. Erickson and D.Y.C. Fung eds., *Handbook on Anaerobic Fermentations*, Dekker, New York, 1988.
3. Gerald Reed ed. *Prescott and Dunn's Industrial Microbiology*, 4th ed, AVI Publishing, Westport, Conn., 1982.
4. K.H. Steinkraus, *Handbook of Indigenous Fermented Foods*, Dekker, New York, 1983.
5. J.C. Linden, in L.E. Erickson and D.Y.C. Fung eds., *Handbook on Anaerobic Fermentations*, Dekker, New York, 1988, pp. 59–80.
6. J.E. Bailey and D.F. Ollis, *Biochemical Engineering Fundamentals*, 2nd ed. McGraw-Hill, New York, 1986.
7. C.K. Matthews and K.E. van Holde, *Biochemistry*, Benjamin/Cummings, New York, 1990.
8. E.R. Vedamuthu, in L.E. Erickson and D.Y.C. Fung eds., *Handbook on Anaerobic Fermentations*, Dekker, New York, 1988, pp. 641–692.
9. L.E. Erickson, I.G. Minkevich, and V.K. Eroshin, *Biotechnol. Bioeng.* **20**, 1595–1621 (1978).
10. R. Luedeking and E.L. Piret, *Biotechnol. Bioeng.* **1**, 393–412 (1959).
11. R. Luedeking and E.L. Piret, *Biotechnol. Bioeng.* **1**, 431–459 (1959).
12. M.D. Oner, L.E. Erickson, and S.S. Yang, in L.E. Erickson and D.Y.C. Fung eds., *Handbook on Anaerobic Fermentations*, Dekker, New York, 1988, pp. 293–323.
13. A.H. Stouthamer, in L.E. Erickson and D.Y.C. Fund eds., *Handbook on Anaerobic Fermentations*, Dekker, New York, 1988, pp. 345–437.
14. A.J.B. Zehnder, *Biology of Anaerobic Microorganisms*, Wiley, New York, 1988.
15. J.H. Thiele and J.G. Zeikus, in L.E. Erickson and D.Y.C. Fung eds., *Handbook on Anaerobic Fermentations*, Dekker, New York, 1988, pp. 537–595.
16. J.H. Thiele, in J.G. Zeikus and E.A. Johnson, eds., *Mixed Culture Biotechnology*, McGraw-Hill, New York, 1991, pp. 261–292.
17. S.G. Pavlostathis, G. Misra, M. Prytula, and D. Yeh, *Water Environ. Res.* **68**, 479–497 (1996).
18. J.C. Spain ed. *Biodegradation of Nitroaromatic Compounds*, Plenum, New York, 1995.
19. L.L. Barton ed. *Sulfate-Reducing Bacteria*, Plenum, New York, 1995.
20. J.G. Zeikus and E.A. Johnson eds., *Mixed Culture Biotechnology*, McGraw-Hill, New York, 1991.
21. H.J. Peppler, in J.G. Zeikus and E.A. Johnson eds., *Mixed Culture Biotechnology*, McGraw-Hill, New York, 1991, pp. 17–36.
22. S.B. Leschine, in L.N. Ornston, A. Balows, and E.P. Greenbert eds., *Ann. Rev. Microbiol.* **49**, 399–426 (1995).
23. J.C. Spain, in L.N. Ornston, A. Balows, and E.P. Greenberg eds., *Ann. Rev. Microbiol.* **49**, 523–555 (1995).
24. G.S. Omenn ed. *Environmental Biotechnology*, Plenum, New York, 1988.
25. M.L. Rochkind-Dubinsky, G.S. Sayler, and J.W. Blackburn, *Microbial Decomposition of Chlorinated Aromatic Compounds*, Dekker, New York, 1987.
26. H.I. Adler and W.D. Crow, *Biotechnol. Bioeng. Symp.* **11**, 533–540 (1981).
27. H.I. Adler and G. Spady, *J. Rapid Methods Automat. Microbiol.* **5**, 1–12 (1997).
28. American Society for Microbiology, *Fundamentals of Anaerobic Bacteriology as Related to the Clinical Laboratory*, American Society for Microbiology, Washington, D.C., 1980.
29. K.L. Anderson and D.Y.C. Fung, *Proc. of the 42nd Ann. Mtg.*, Inst. of Food Technologists, Abstract 163, Las Vegas, June 22–25, 1982.
30. K.L. Anderson and D.Y.C. Fung, *J. Food Protect.* **40**, 811–827 (1983).
31. M.S. Ali, D.Y.C. Fung, and C.L. Kastner, *J. Food Sci.* **56**, 367–370 (1991).
32. F.M. Aramouni, K. Kone, J.A. Craig, and D.Y.C. Fung, *J. Food Protect.* **57**, 882–886 (1994).
33. J.H. Brewer and D.L. Allgeier, *Appl. Microbiol.* **14**, 985–988 (1966).
34. M.E. Claerbout, *J. Rapid Methods Automat. Microbiol.* **5**, 21–28 (1997).
35. I.F. Dalmacio and D.Y.C. Fung, Kalikasan, Philippines. *Journal of Biology* **6**, 203–216 (1977).
36. I.F. Dalmacio, D.Y.C. Fung, and J.H. McNeil, *Philippine Phytopathol.* **18**, 102–107 (1982).
37. D.Y.C. Fung, in C.M. Bourgeois and J.V. Leveau eds., *Microbiology Control for Food and Agricultural Products*, VCH Publishers, New York, 1995, pp. 1–25.
38. D.Y.C. Fung, *Food Technol.* **49**, 64–67 (1995).
39. D.Y.C. Fung, M.A. Buono, and L.E. Erickson, in L.E. Erickson and D.Y.C. Fung eds., *Handbook on Anaerobic Fermentations*, Dekker, New York, 1988, pp. 501–536.
40. D.Y.C. Fung and P.A. Hartman, in C.G. Heden and T. Illeni eds., *New Approaches to the Identification of Microorganisms*, Wiley, New York, 1975.
41. D.Y.C. Fung and C.M. Lee, *Food Sci. (Republic of China)* **7**, 209–213 (1981).
42. D.Y.C. Fung, L.S.L. Yu, F. Niroomand, and K. Tuitemwong, in R.C. Spencer, E.P. Wright, and S.W.B. Newsom eds., *Rapid Methods and Automation in Microbiology*, Inercept Limited, Andover, UK, 1994, pp. 313–326.

43. L.S. Gall and P.E. Riely, *Manual for the Determination of the Clinical Role of Anaerobic Microbiology*, CRC, Boca Raton, Fla., 1981.
44. I.C. Hall, *J. Bacteriol.* **17**, 255–301 (1929).
45. L.V. Holdeman, E.P. Cato, and W.E.C. Moore, *Anaerobe Laboratory Manual*, 4th ed., Virginia Polytechnic Institute and State University, Blacksburg, Vir., 1977.
46. R.E. Hungate, *J. Bacteriol.* **53**, 631–645 (1947).
47. R.L. Hungate, *Bacteriol. Rev.* **14**, 1–49 (1950).
48. R.E. Hungate, in J.R. Norris and D.W. Ribbons eds., *Methods in Microbiology*, Vol. 3B, Academic Press, New York, 1969, pp. 117–132.
49. R.W. Kelly and T.C. Schipp, in W.P. Olson ed., *Automated Microbial Identification and Quantitation: Technologies for the 2000s*, Interpharm Press, Buffalo Grove, Ill., 1996.
50. N.R. Krieg and J.G. Holt, *Bergey's Manual of Systematic Bacteriology*, Vol. 1, Williams and Wilkins, Baltimore, 1984.
51. B.M. Mitruka, *Methods of Detection and Identification of Bacteria*, CTC, Cleveland, Ohio, 1976.
52. N.J. Moon and G.W. Reinbold, *J. Milk Food Technol.* **39**, 337–341 (1976).
53. W.E.C. Moore, *Inst. J. Syst. Bacteriol.* **16**, 173–190 (1966).
54. J.E. Ogg, S.Y. Lee, and B.J. Ogg, *Can. J. Microbiolol.* **25**, 987–990 (1979).
55. W.P. Olson, *Automated Microbial Identification and Quantitation: Technologies for the 2000s*. Interpharm Press, Buffalo Grove, Ill., 1996.
56. D.A. Shapton and R.G. Board, *Isolation of Anaerobes*, Academic Press, New York, 1971.
57. V.L. Sutter, D.M. Citron, and S.M. Finegold, *Anaerobic Bacteriology Manual*, C.V. Mosby, St. Louis, Mo., 1980.
58. M. Thurston and C. Gannon, *J. Rapid Methods Automat. Microbiol.* **5**, 13–20 (1997).
59. C. Venderzant and D. Splittoesser, *Compendium of Methods for the Microbiological Examination of Foods*, Amer. Public Health Assoc., Washington, D.C., 1992.
60. T.D. Wilkins and C.B. Walker, *Appl. Microbiol.* **30**, 825–830 (1975).
61. T.D. Wilkins, C.B. Walker, and W.E.C. Moore, *Appl. Microbiol.* **30**, 831–837 (1975).
62. J.C. Zeikus, *Bacteriol. Rev.* **41**, 514–541 (1977).
63. A.C. Sonnewirth, *Am. J. Clin. Nutr.* **25**, 1295–1298 (1972).
64. K. Horikoshi and W.D. Grant eds., *Extremophiles: Microbial Life in Extreme Environments*, Wiley, New York, 1998.
65. C.R. Woese, O. Kandler, and M.L. Wheelis, *Proc. Natl. Acad. Sci. USA* **87**, 457–466 (1990).
66. C.R. Woese, *Microbial. Rev.* **51**, 221–271 (1987).
67. M.R. Mormile, S.H. Lin, and J.M. Sulfita, *Environ. Sci. Technol.* **28**, 1727–1732 (1994).
68. R.J. Bothast and B.C. Saha, *Adv. Appl. Microbiol.* **44**, 261–286 (1997).
69. R.M. Atlas, *Handbook of Media for Environmental Microbiology*, CRC, Boca Raton, Fla., 1995.

See also BIOENERGETICS OF MICROBIAL GROWTH; CLOSTRIDIA, SOLVENT FERMENTATION; PROTEIN EXPRESSION, SOLUBLE; YEAST, BAKER'S.

ANAEROBES, INDUSTRIAL USES

MAHENDRA K. JAIN
J. GREGORY ZEIKUS
MBI International
Lansing, Michigan

KEY WORDS

Anaerobes
Enzymes
Extremophiles
Metabolic diversity
Methanogenesis
Organic acids
Solvents
Waste treatment

OUTLINE

INTRODUCTION

Anaerobes, in the simplest form, are microorganisms that are unable to grow at the surface of a medium exposed to air, where the oxygen may either be bacteriocidal or bacteriostatic. These organisms do not use oxygen for catabolic reactions, and oxygen is not required for their survival. Anaerobes, with the exception of autotrophic methanogenic and acetogenic bacteria, sulfur-metabolizing archaea, and sulfate-reducing bacteria, obtain their energy by fermentation, a redox process in which organic substances serve as electron donors and terminal electron acceptors. Obligate anaerobes share two common characters: an extreme sensitivity to molecular oxygen, and a capacity to produce energy and perform essential biosyntheses without molecular oxygen. The sensitivity to O_2 is extremely variable from one group to another, and evaluation of O_2 sensitivity requires that all organisms be cultivated under identical conditions, both for medium composition (including reducing substances) and gaseous environment. The physiological, ecological, and genetic diversity of anaerobic bacteria makes them very interesting organisms.

Unlike aerobic microorganisms whose anabolic features are largely exploited for production of single-cell proteins, antibiotics, and amino acids, anaerobes offer the potential for utilizing catabolic features in the production of specialty and bulk chemicals, fuels, and unique enzymes that are stable and active under extreme environmental conditions. These properties of anaerobes make them highly desirable for certain industrial uses. Today, a large body of information is available on the metabolic diversity of anaerobes. Anaerobes seem to have adapted from the primordial atmospheric conditions to inhabit both moderate and extreme environments with metabolic machinery that is active from about 0 °C to over 100 °C, in very little salt to saturated salt conditions, in highly acidic (pH < 2) to alkaline (pH > 9) conditions, and even in the presence of highly toxic chemicals. In this chapter, we will only cover procaryotes and archaea, not anaerobic eucaryotes (i.e., yeasts and fungi). We will include examples of pure cultures used to produce specific products and mixed cultures or consortia used in environmental processes.

PHYSIOLOGY, BIOCHEMISTRY, AND MOLECULAR BIOLOGY

Species Diversity

Knowledge about microbial diversity is essential in understanding the relationship between environmental factors and metabolic functions. Such knowledge can be used to assess the effects of environmental conditions on production of microbial products. However, relatively little is known about the entire spectrum of anaerobic bacteria that may be of importance to various industries. What is known about anaerobic species diversity is presented in *Bergey's Manual* (1), where over 170 genera and over 800 species of anaerobic bacteria and archaea are described.

Cultivation of microorganisms greatly helps in determining the full complement of genetic and physiological diversity of newly isolated species. Strains available in pure and mixed culture can be examined to determine what novel features they have, including those that might be of special interest to industry. Thus, this aspect of microbial diversity is particularly important. Pure cultures are suitable for determination of morphological, physiological, and metabolic characteristics and commercial potential. The 16S rRNA sequence can provide a powerful means to organize the phylogenetic relationships among different microbial species in a meaningful way, but this analysis is often considered too conservative to provide species-level distinctions because any two different species can show 97–100% 16S rRNA sequence homology.

Initially, only mesophilic anaerobic species were isolated and characterized from moderate environments. A good example is the rumen, which provides a great variety of anaerobic species with different metabolic properties. However, in recent years, studies on microbial diversity in extreme ecosystems, such as thermal springs and deep-sea vents, have offered an unparalleled diversity in extremophile communities. The Yellowstone National Park ecosystem holds the world's greatest diversity of accessible, extreme microbial habitats. Not only are there temperature and pH extremes, but also areas with high levels of sulfur, heavy metals, radionuclides, and UV radiation, as well as a range of organic nutrients. Since oxygen solubility in water decreases as the temperature increases, anaerobes are common at the highest temperatures. Interest in novel microorganisms thriving under environmental extremes reflects a growing awareness of their value in biotechnological and industrial processes. Not only are the enzymes and metabolites often more stable, but in some cases, novel enzymes, not known in organisms from moderate habitats, have been found in extremophiles (2,3).

The range of anaerobe diversity has proven to be extraordinary. For example, following the discovery of the deep-sea hydrothermal vents, hyperthermophilic anaerobes from hot springs and volcanic fields, as well as from shallow and abyssal marine hydrothermal vents were identified (4,5). Numerous *Archaea* have been identified and classified (6), and some species produce enzymes that have, or are expected to have, commercial utility (2,7).

Table 1 provides some examples of representative anaerobic bacteria and archaea that illustrate the spectrum of anaerobic bacteria that live under extreme environmental conditions. The different ecophysiologies of these species explain, in part, the great diversity among anaerobic microorganisms, as it is the function of the organism or its enzyme(s) and gene(s) that is practiced by industry.

Metabolic Diversity

Microbial diversity at the biochemical level is more important from the industrial perspective. During the past decade, molecular biological and genetic approaches have come to dominate the study of microorganisms. Although fruitful, exclusive reliance on such approaches poses both practical and philosophical problems for microbiology. After an early phase of discounting the importance of physiology, people in industry are learning from the "biotechnology revolution" that gene cloning and biochemical manipulations do not solve all the product-development problems that may arise. Some of the more valued intellectual benefits anticipated in the field will emerge from understanding the physiology of peculiar and diverse microorganisms from unusual environments. As long as oxygen is present as the electron acceptor, denitrification, sulfate reduction, and methanogenesis are inhibited. Denitrification begins after consumption of oxygen, and sulfate reduction begins after consumption of nitrate. Finally, methanogenesis occurs, which is partly a dissimilatory carbon dioxide reduction. The sequence of these reactions agrees with the sequence of their free-energy changes, which decrease from respiration to methanogenesis.

Table 2 lists some representative anaerobic bacteria that display different metabolic types based on their catabolic pathways. These different biochemistries explain, in part, the great diversity of anaerobic microorganisms.

Methanogenesis. Methanogenic bacteria are strictly anaerobic archaea with a unique form of energy metabolism involving the generation of methane. Biological methanogenesis is an important component of the carbon cycle in a variety of anaerobic habitats. It represents the terminal

Table 1. Representative Bacterial and Archaeal Species of Anaerobic Extremophiles

Extremophiles	Subgroup	Optimal growth condition	Representative species
Thermophile	Moderate	60 °C < T_{opt} > 80 °C	*Clostridium thermocellum*
		60 °C < T_{opt} > 70 °C	*Methanobacterium thermoautotrophicum*
	Hypher	T_{opt} ≥ 80 °C	*Thermotoga maritima*
			Pyrococcus furiosus
Halophile	Moderate salinity	$[NaCl]_{opt}$ ≥ 1.8 M	*Desulfohalobium retbaense*
	High salinity	$[NaCl]_{opt}$ = 3.0 M	*Halomethanococcus doii*
Acidophile		pH_{opt} ≤ 4.0	*Sarcina ventriculi*
Alkaliphile		pH_{opt} ≥ 9.2	*Methanohalphilus zhilinae*
Syntrophile		Grows with a metabolic partner (e.g., a methanogen or a sulfate reducer)	*Syntrophospora bryantii*
Toxophile		Carbon monoxide as a substrate	*Clostridium ljungdahlii*

Table 2. Potential Industrial Fermentation Products from Anaerobic Bacteria

Fermentation product	Substrate	Typical species	Concentration (g/L)	Status	Reference
Ethanol	Glucose	*Zymomonas mobilis*	84.5	Bench-scale	Kesava et al. (8)
	Paddystraw	*Clostridium thermocellum*	23.6	Bench-scale	Sudha Rani et al. (9)
	CO	*Clostridium ljungdahlii*	48	Bench-scale	Klasson et al. (10)
n-Butanol	Starch/glucose	*Clostridium acetobutylicum*	20	Bench-scale	Glassner et al. (11)
	CO	*Butyribacterium methylotrophicum*	2.7	Bench-scale	Grethlein et al. (12)
1,3-Propanediol	Glycerol	*Clostridium butyricum*	58	Pilot-scale	Günzel et al. (13)
Lactic acid	Glucose	*Lactobacillus acidophilus*	82	Industrial	Datta (14)
Acetic acid	Glucose	*Clostridium thermoaceticum*	83	Bench-scale	Parekh and Cheryan (15)
Butyric acid	Glucose	*Clostridium tyrobutyricum*	62.8	Bench-scale	Fayolle et al. (16)
Propionic acid	Glucose	*Propionibacterium acidipropionici*	57	Bench-scale	Paik and Glatz (17)
	Glycerol	*Propionibacterium acidipropionici*	42	Bench-scale	Barbirato et al. (18)
Succinic acid	Glucose	*Actinobacillus succinogenes*	105	Pilot-scale	Guettler et al. (19)
Caproic acid	Cellulose & ethanol	*Clostridium kluyveri* plus *Fibrobacter succinogenes*	4.6	Bench-scale	Kenealy et al. (20)

step in the anaerobic breakdown of organic matter under sulfate-limiting conditions. The substrate range for methanogenesis is limited to carbon dioxide, formate, carbon monoxide, methanol, methylamines, and acetate, but no single isolate is able to utilize all these carbon sources (21). The metabolic pathways of methane formation are unique and involve a number of enzymes and coenzymes that occur only in methanogens. A primary Na^+ gradient and H^+ gradients are formed by interesting and unique membrane-bound enzymes, and the reactions by which these ion-motive forces are generated are also novel (22). The organisms and the methanogenic reactions (21,23), and the pathways of energy conservation, have been extensively reviewed (22).

In recent years, bioconversion of agricultural and industrial wastes to methane has been considered important from two perspectives, as a waste-disposal method for residual organic matter and as a source of energy. Anaerobic digestors for treatment of sewage, agricultural and animal waste, and industrial effluent are being used worldwide. More recently, methanogenic microbial communities have been found to be very efficient in treating toxic organic wastes. For example, transformation of C_1 and C_2 halocarbons was observed in the early 1980s and, as one of the mechanisms responsible for these transformations, was proposed as a biologically mediated reductive dechlorination (24–26). Anaerobic dechlorination and degradation of chlorinated aromatic compounds such as chlorophenols, chlorobenzenes, and polychlorinated biphenyls (PCBs) occurs in a wide variety of environments. Methanogenic microbial consortia that dechlorinate tetrachloroethylenes (PCE) and trichloroethylenes (TCE), and dechlorinate, degrade, and mineralize chlorophenols and PCBs have been developed (27–31). The potential of these consortia in bioremediating the contaminated soil and sediment systems is being examined. DDE [1,1-dichloro-2,2-bis(*p*-chlorophenyl)ethane], a metabolite of the pesticide DDT, has been observed to reductively dechlorinate to DDMU at a relatively high rate under methanogenic conditions as compared to sulfidogenic conditions (32).

The catabolic reactions carried out by methanogens yield very little energy compared to those of aerobes. For example, use of H_2 to reduce CO_2 to CH_4 has a $\Delta G^{0'}$ of -34 kJ/mol H_2, whereas the corresponding energy value for an aerobe oxidizing H_2 using O_2 as an electron acceptors is -237 kJ/mol H_2. Most of the known species of methanogens can use H_2 to reduce CO_2 to CH_4. Many of the hydrogenotrophic methanogens also utilize formate. In contrast, only species belonging to the genera *Methanosarcina* and *Methanosaeta* use acetate (acetotrophic) and convert it to methane. Acetate is an important end product of many fermentative anaerobes, and is a primary methanogenic substrate in anaerobic digestors. The methylotrophic methanogens can utilize several simple methylated compounds. These methanogens include *Methanosarcina*, which can also use acetate and usually H_2—CO_2, and *Methanolobus* and *Methanococcoides*, which are only known to grow on methylated compounds. Carbon monoxide is also converted to methane by methanogenic species. Methanogenesis has been observed at low, moderate, and high temperatures, and in neutral to mildly acidic or alkaline conditions. In summary, methane is produced under a variety of environmental conditions and is considered a byproduct of the anaerobic treatment of industrial wastes.

Solventogenesis. Ethanol production is a very well-known fermentation process since it has been an outgrowth of the alcoholic beverages industry. During World War I, shortages of acetone in the manufacture of munitions led to the development of an acetone–butanol–ethanol process involving the fermentation of starch by *Clostridium acetobutylicum*. This fermentation gave approximately 2 parts butanol to 1 part acetone and 1 part ethanol. During the 1940s and 1950s, lower costs of petrochemical processes stopped butanol production by fermentation routes in the U.S. and Europe. The process was also stopped in South Africa, but continued in China into the early 1990s.

The present market for ethanol is met by yeast fermentation, but the industrial market for isopropanol, acetone, and butanol could be met by other anaerobic fermentations. Biomass-derived solvents produced by fermentation can enter into the current petrochemical synthetic pathways through a number of reactions. The most important of these is the dehydration of alkanols to alkene to form ethylene, propylene, butylene, and butadiene. Therefore, production of chemical feedstocks from biomass via fermentation is becoming increasingly attractive because biomass production costs are not as tightly bound to energy costs.

A great diversity exists among the solvent-producing microorganisms. These organisms span several groups of yeasts and a broad range of both mesophilic and thermophilic, and aerobic and anaerobic, bacteria. By far, the ethanol-producing organisms are the most abundant solventogens. However, industrial use is mainly limited to strains of yeasts, but great potential exists for recombinant *Zymomonas mobilis*. Several clostridia are capable of producing butanol and either acetone or isopropanol with ethanol as a byproduct. The *C. acetobutylicum* and *Clostridium beijerinckii* strains are of prime importance for these industrial fermentations.

Clostridium thermocellum, Thermoanaerobacter ethanolicus, Clostridium thermosaccharolyticum, Thermoanaerobacterium thermosulfurigenes, Thermoanaerobacter brockii, and *Thermobacteroides acetoethylicus*, use a wide range of substrates, from polymeric carbohydrates such as cellulose, pectin, xylan, and starch, to mono- and disaccharides such as glucose, cellobiose, xylose, and xylobiose. The primary fermentation product is ethanol, with the production of acetate, lactate, carbon dioxide, and hydrogen in various ratios. *T. acetoethylicus* has not been shown to produce lactic acid. At present, one of the major limitations in the use of thermoanaerobes is the variability of end-product ratios, yield, and low ethanol concentration. These are affected by species, enzyme complement, and environmental conditions. Certain strains of *T. ethanolicus* have the best conversion of carbohydrates to ethanol, forming 1.6–1.9 mol/mol of glucose fermented (33). High ethanol and hydrogen concentrations also reduce the yield of ethanol (33) owing to the flexibility of the carbon and electron pathways, which may possess many reversible enzyme systems (33,34).

In addition to ethanol, several saccharolytic clostridia produce butanol and acetone besides volatile fatty acids and gaseous products from carbohydrate fermentation. In some cases acetone is further reduced to isopropanol. In most species, the production of solvents only occurs late in the fermentation cycle, following a shift from the pathways leading to acetate and butyrate production. The production of butanol and ethanol is usually associated with the uptake and reutilization of acids, and the production of acetone or butanol. The ability to produce solvents is influenced by the type and concentration of substrate, the pH and the buffering capacity of the culture medium, and the environmental conditions. Several strains that have the ability to produce solvents undergo degenerative changes, resulting in the loss of their ability to produce solvents. The solvent-producing clostridia are found in a wide variety of natural habitats (35); however, the strains of *C. acetobutylicum* and *Clostridium beijerinckii* are the most frequently studied species. *C. acetobutylicum*, which characteristically produces butanol, acetone, and ethanol in the ratio of 6:3:1, has been used extensively for the industrial production of solvents. Most butanol producers are mesophiles with a temperature optima for fermentation between 30 and 37 °C. Recently, solvent yield was shown to increase at lower temperatures and acetone but not butanol yield to decrease at elevated temperatures (36; B.K. Soni and M.K. Jain, unpublished data 1991).

At present, many ethanol- and butanol-producing strains of *Clostridium* remain poorly classified, and there is still no accepted standard classification for the clostridia group as a whole. Strains of *C. beijerinckii* (formerly classified as *C. butylicum*) constitute a second group of solvent producers that do not show DNA homology with the *C. acetobutylicum* group. This group contains both high- and low-solvent-producing strains that produce either acetone or isopropanol in addition to butanol. The strains of *Clostridium aurantibutyricum* group also include butanol-producing strains that produce both acetone and isopropanol (37). *Clostridium tetanomorphum* produces butanol and ethanol, but not acetone or isopropanol. As obligate anaerobes, butanol producers require anaerobic conditions. However, vegetative cells of *C. acetobutylicum*, for example, survived several hours exposure to oxygen and formed fruiting-body-like structures on agar plates when exposed to oxygen (38).

Acetogenesis. The term acetogen is commonly applied to a bacterium that forms acetate, whether the acetate is produced by a catabolic process such as fermentation or by an autotrophic-type synthesis. Acetogenic bacteria are ubiquitous in anaerobic ecological systems. Homoacetogens have the ability to grow well on H_2 plus CO_2 and poorly on CO, and to form acetyl CoA by a CO-dependent pathway involving CO dehydrogenase (39). *Clostridium thermoaceticum* and *Clostridium thermoautotrophicum* are homoacetogens and synthesize acetate from C_1 compounds by the recently established Wood pathway of acetyl CoA synthesis (40). This pathway was established by studies with *C. thermoaceticum* that can grow on CO or CO_2—H_2, utilizing them as carbon and energy sources (41).

In industrial fermentations, the formation of a single product is advantageous because its recovery would be simplified. In this respect, homoacetogenic bacteria essentially form only acetate in the fermentations of hexoses and pentoses. They appear to be ideal for the microbial production of acetate. Energy for cell growth is obtained by the reduction of CO_2 to acetate via the Wood pathway. The homoacetogens also grow on other one-carbon compounds, including CO, formate, and methanol with acetate as the product. In the acetyl CoA pathway for synthesis of acetate, acetyl CoA is the first two-carbon intermediate of the autotrophic fixation of CO_2, and is either used for the synthesis of cell carbon or converted to acetate.

Butyribacterium methylotrophicum is an acetogenic anaerobe that can grow on multicarbon compounds as well as on one-carbon compounds (e.g., CO or methanol). It produces acetate or butyrate using the Wood pathway, but acetyl CoA can be condensed and reduced to butyrate. Higher levels of NADH were found in butyrate-producing than in acetate-producing cells. In *B. methylotrophicum*, butyrate production is regulated by the carbon source and is dependent on cellular NADH/NAD ratios, and the levels and direction of ferredoxin- and NAD-linked oxidoreductases (42). Also, the growth pH regulates both hydrogenase and FD-NAD oxidoreductase activities such that, at acid pH, more intermediary electron flow was directed towards butyrate synthesis than H_2 production.

Acidogenesis. The acidogenic bacteria include those bacteria that only form acetate by fermentation. These bacteria invariably produce other acids in addition to acetate. The acidogenic bacteria grow heterotrophically on a variety of carbohydrates, producing a mixture of acids such as lactic, propionic, butyric, succinic, and caproic. The pathways for production of these acids have been reviewed (43). A generalized scheme for anaerobic production of fermentative organic acids and alcohols is presented in Figure 1. The acidogens differ from the acetogens in that they do not synthesize acetate from CO_2 or other C_1 compounds by the Wood autotrophic pathway. Many of the species can utilize hexoses such as glucose, fructose, galactose, and mannose; disaccharides including cellobiose, lactose, and maltose; pentoses represented by xylose; and other substrates such as glycerol, mannitol, sorbitol, inositol, and polymeric carbohydrates. For example, *Clostridium butyricum*, the type strain of the genus *Clostridium*, is saccharolytic in nature and metabolizes glucose, producing butyrate, acetate, CO_2, and H_2 as fermentation products. Glucose is converted to pyruvate by the Embden–Meyerhof–Parnas glycolytic pathway. Pyruvate is then simultaneously decarboxylated and oxidized by the enzyme complex pyruvate–ferredoxin oxidoreductase to yield acetyl CoA, CO_2, and reduced ferredoxin. The reduced ferredoxin is reoxidized in several reactions, the most important of which involves H_2 evolution catalyzed by hydrogenase. Acetyl CoA is condensed and reduced to butyrate.

In some other acidogens, phosphoenolpyruvate (PEP) serves as a branching point in the pathway producing pyruvate and oxaloacetate. The PEP–oxaloacetate step involves CO_2 fixation mediated by PEP carboxykinase. Depending upon the species, products such as succinate and propionate are produced as fermentation products with acetate, formate, or H_2. For example, oxaloacetate is converted to propionate in species like *Propionispira arboris* (44), whereas, oxaloacetate is further converted to succinate in species like *Anaerobiospirillum succiniciproducens* (45). Lactic acid bacteria such as *Lactobacillus acidophilus* convert pyruvate to lactic acid, whereas *Clostridium kluvyri* produces caproic acid from ethanol and acetate (43).

Sulfidogenesis. Sulfate, a chemically inert, nonvolatile, and nontoxic compound, is widespread in rocks, soil, and water. In contrast, hydrogen sulfide, because of its chemical properties and physiological effects, is a far more conspicuous substance and is chemically reactive. The presence of black-colored sediments due to the formation of ferrous sulfide from iron-containing materials, accompanied by a distinctive smell, is indicative of H_2S production. Hydrogen sulfide acts as a reductant, and is toxic to plants, animals, and humans. It is produced by reduction of sulfate by sulfate-reducing bacteria, which are present in ecological niches where oxygen has no access. In the absence of oxygen, the oxidized form of sulfur (SO_4^{2-}) is used as an electron acceptor by sulfate-reducing bacteria (sulfate reducers). Since reduction of the inorganic compound serves for energy conservation, the process is distinguished as dissimilatory reduction from the assimilatory reduction of sulfate in plants and bacteria.

Under anaerobic, reduced conditions, hydrogen sulfide is the energetically stable form of sulfur, as sulfate is the stable form under aerobic conditions. The most important reducers of elemental sulfur in nature are probably special anaerobes that may be designated as true sulfur-reducing bacteria; these carry out the dissimilatory reduction of sulfur as their primary or even obligate metabolic reaction during oxidation of organic substrates. The true sulfur-reducing bacteria do not reduce sulfate, and only some reduce other oxygen-containing sulfur compounds. The majority of the eubacterial sulfur reducers are mesophilic, as are most sulfate reducers. In contrast, all described archaebacterial sulfur reducers are hyperthermophiles with temperature optima (up to 110 °C) the highest known in the living world.

Sulfate-reducing bacteria include a rather heterogenous assemblage of microorganisms having in common merely dissimilatory sulfate metabolism and obligate anaerobiosis. The genus *Desulfovibrio* includes species that are still the best-studied sulfate reducers. These are nonsporing, having curved motile cells and growing on a relatively limited range of organic substrates, preferably lactate or pyruvate, which are incompletely oxidized to acetate and H_2S. Spore-forming sulfate-reducing bacteria with a similar metabolism were classified within the genus

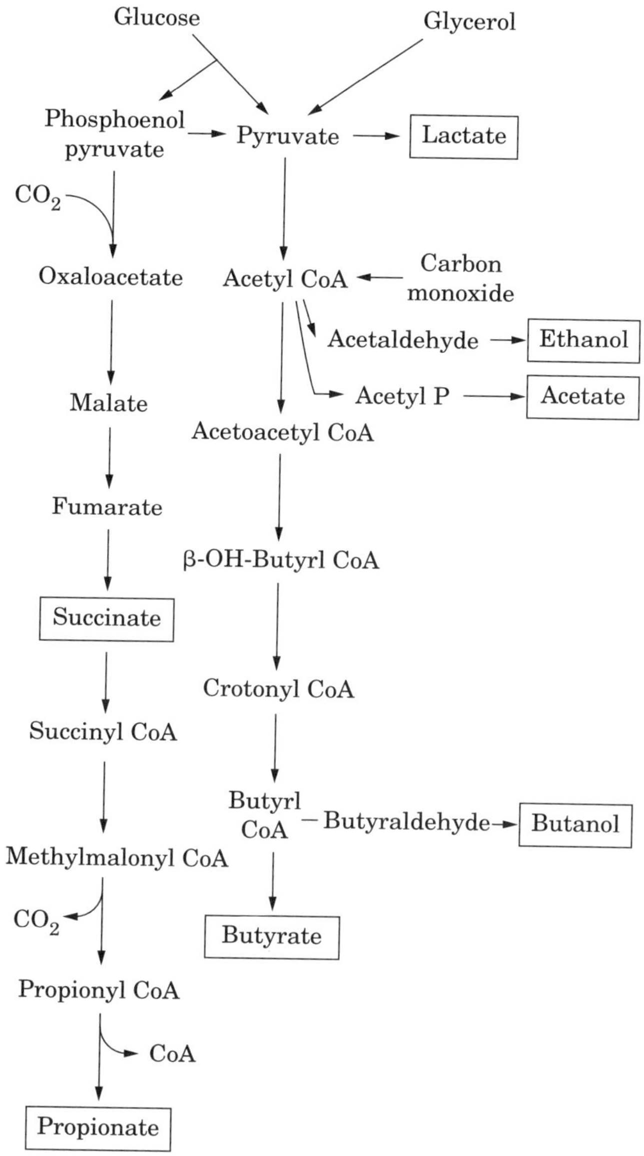

Figure 1. Generalized scheme for anaerobic fermentative production of common organic acids and alcohols.

Desulfotomaculum. Other types of sulfate-reducing bacteria differ markedly physiologically and morphologically from the known *Desulfovibrio* and *Desulfotomaculum* species. The reactions by which sulfate-reducing bacteria are involved in anaerobic degradation are indicated by their metabolic capacities. As far as tested, these bacteria use low molecular weight compounds as electron donors and therefore depend on fermentative bacteria that degrade the original polymers from biomass. Thus sulfate reducers are terminal degraders and their role is analogous to that of methanogenic bacteria that form methane and carbon dioxide as final anaerobic products. The recognition of dissimilatory sulfate reduction and methanogenesis as two alternative terminal carbon degradation processes has contributed a great deal to our understanding of anaerobic mineralization.

Genes and Genetic System

The field of genetics has been restricted for many years to *Escherichia coli* and few other genera of aerobic or facultative anaerobic bacteria such as *Pseudomonas, Bacillus*, and *Salmonella*. Anaerobic bacteria were known and studied since before 1900, but work on the genetics and molecular biology of anaerobic bacteria began to emerge only in the 1970s. The volume by Sebald is the most recent comprehensive review on genetics of anaerobes (46). Improvements in basic techniques for culturing anaerobes and recent advances in molecular biology techniques have helped in making rapid progress in understanding genes and genetic systems in anaerobic bacteria. Thermoanaerobic bacteria produce many thermostable enzymes but the yields are low. However, their enzymes can be overproduced by cloning the genes into mesophilic, industrially important aerobic bacteria. The genes can be overexpressed in both gram-positive and gram-negative hosts such as *Bacillus subtilis* and *E. coli* (47). One major advantage of cloning genes for thermostable enzymes in mesophiles is in the recovery of enzymes in greater than 95% purity by a single-step high-temperature treatment that inactivates host proteins, including protease enzyme (47). Aerobic bacteria produce about 10-fold higher biomass yields than the anaerobic bacteria. Therefore, there is no advantage, except for biotransformation, in producing the enzymes in an anaerobic system.

Several species of the genus *Clostridium* are of biotechnological interest because of the end products of their fermentative metabolism, the stereospecific reductions that they undertake, and the potentially important enzymes that they produce. Other species, such as *C. botulinum*, produce some of the most powerful toxins known today, which have been commercialized for medical treatments such as the correction of crossed eyes.

Various elements of gene-transfer technology have been developed in several species. Two gene-transfer procedures should prove widely applicable throughout the genus—electroporation and conjugal plasmid mobilization. In clostridia, mutants defective in a variety of functions, including purine, pyrimidine, vitamin, and amino acid biosynthesis, pathways of fermentative metabolism, and sporulation, have been isolated with ethyl methane sulfonate, *N*-methyl-*N*′-nitro-*N*-nitrosoguanidine, and ultraviolet light, and have been characterized (48–52). These mutant strains provide valuable information about gene regulation and function. Conjugal transfer of plasmids and transposons as well as phase transfection/plasmid transformation has now been documented in several species. In addition, recombinant DNA technology has also been employed to gain insights into the gene-transfer systems in *Clostridium*. Protoplasts of *C. acetobutylicum* can be transfected with phage DNA (53). The whole cells of *C. acetobutylicum* can be transformed with plasmid DNA using an electroporation procedure. Permealized cells of *Thermoanaerobacter thermohydrosulfuricus* have been transformed with plasmid pUB110 (encoding resistance to kanamycin, Km^R) and a derivative, pGS13, carrying a chloramphenicol resistance (Cm^R) marker (54).

A variety of cloning vectors have been developed for use in *C. acetobutylicum*. To date only pMTL500E has been used to introduce cloned heterologous genes into *C. acetobutylicum*. These genes included the *Clostridium pasteurianum leu*B gene (55) and the *C. thermocellum cel*A gene (see Ref. 56). Acquisition of the *cel*A gene was demonstrated by an in situ plate assay using Congo red. In most cases, screening of clostridial gene banks has relied on expression of the heterologous genes in *E. coli*. Appropriate *E. coli* mutants have been employed for the isolation of *C. acetobutylicum* genes involved in the acetone–butanol fermentation. The butyraldeyde dehydrogenase gene of *C. acetobutylicum* has been isolated by complementation of an *E. coli* aldehyde dehydrogenase–deficient mutant. Endo-β-glucanases and xylanse of *C. thermocellum* and other cellulolytic clostridia are easily detectable on plates containing the appropriate substrate by the Congo red assay. Upon staining with Congo red, positive clones are surrounded by a yellow hydrolysis zone on a red background.

BIOTECHNOLOGICAL APPLICATIONS

Industrial biotechnology has developed into a more than $25 billion business of enzymes and of chemically defined compounds produced by fermentation, microbial transformation, and enzymatic conversion. These include organic acids, amino acids, alcohols, vitamins, microbial polymers, antibiotics, industrial enzymes, biopesticides, and pharmaceutical specialities. Of these, ethanol, citric acid, L-lactic acid, gluconic acid, monosodium L-glutamate (MSG), and L-lysine are examples of bulk fermentation products.

Solvents

Solvents are compounds that are essentially neutral in character and comprise organic alcohols and ketones. The major solvent products of microbial carbohydrate catabolism include ethanol, acetone, isopropanol, butanol, acetoin, 1,3-propanediol, and butanediol. These are the most useful oxychemicals produced by fermentation. Solvents have various uses in the chemical industry, including the direct use as organic chemicals and indirect use as fuel additives, extenders, or feedstocks for further chemical synthesis. Butanol is primarily used as a feedstock chemical in the manufacture of lacquers, rayon, plasticizers, coatings, detergents, and brake fluids. It can also be used as a solvent for fats, waxes, resins, shellac, and varnish. In addition, butanol may also be used as an extractant and solvent in the food industry.

Butanol. The acetone–butanol fermentation currently has potential because butanol has many characteristics that makes it better than the currently used liquid-fuel extender, ethanol. Higher alcohols have several characteristics that are favorable for motor fuel use, either in gasoline blends or, in some cases, when used directly as fuel. They are miscible in gasoline; their heats of combustion per gallon are greater than those of methanol, and they have good octane-enhancing properties. Butanol, for example, has a heat of combustion 54% higher than methanol and 83% of that for gasoline (57,58). Although the octane numbers are less than that for methanol (RON 100 vs. 110), the research octane number (RON) value of 100 is still above that of gasoline (RON 92), and therefore useful for fuel blends. In addition, butanol has low miscibility with water and exhibits high miscibility with both diesel and gasoline. Owing to its high heat of combustion butanol solutions containing as much as 20% (v/v) water have the same combustion value as anhydrous ethanol.

In the early 1900s, butanol was used to produce butadiene, the most desirable raw material for synthetic rubber. The annual production of fermentation-derived butanol was over 45 million pounds during 1945. But the fermentation-derived butanol process declined after World War II in the U.S. due to both changes in availability of renewable feedstocks (molasses, sugarcane) and the increase in availability of inexpensive petrochemical feedstocks. Butanol fermentation of beet molasses continued through the 1970s in the Soviet Union, and fermentation of sugarcane molasses continued through the 1980s in South Africa. In China, butanol was still produced fermentatively until recently, but the last viable industrial acetone–butanol–ethanol (ABE) fermentation in the Western world was carried out by National Chemical Products in Germiston, South Africa, using *C. acetobutylicum*.

Butanol is now synthesized chemically from petroleum-derived ethylene, propylene, and triethylaluminium, or carbon monoxide and hydrogen. The major domestic producers of butanol and its derivatives are BASF, Chem Service Inc., Dow Chemical, Eastman Chemical, Hoechst Celanese, Shell, Union Carbide, and Vista. The current U.S. production of butanol is more than 1.2 billion pounds per year and is experiencing a growth of 3–4% annually.

Butanol pricing has been rising steadily in recent years due to an increase in demand. The bulk butanol pricing in the United States is $0.40–$0.50 per pound. The price of butanol in Europe and the Far East is higher than $0.50 per pound. If the production cost of fermentation-derived butanol could be in the range of $0.25–$0.30 per pound, the market for butanol will likely expand: The market penetration of fermentation-based butanol will drive the utilization of corn from 7.2 million bushels in 1995 to 153 million bushels in 2010. In addition, the supply of other alternative feedstock, such as biomass, is more available than corn. However, biomass requires effective pretreatment technology that has not yet been developed. Biomass-derived solvents produced by fermentation can enter into the current petrochemical synthetic pathways through a number of reactions, the most important of which is the dehydration of alkanols to alkenes to form ethylene, propylene, butylene, and butadiene.

The ABE fermentation in batch culture is a sequential process characterized by a primary acid-producing or acidogenic phase that is coupled to growth, and a secondary solvent-producing or solventogenic phase. Thus, solventogenesis appears as a secondary metabolic process caused by uncoupling of growth. Acetate and butyrate are produced as soluble fermentation products during the acidogenic phase. As the acids accumulate, the growth rate and the culture pH decreases, and solventogenesis begins. The acids and H_2 are consumed and are reduced to solvents in the solventogenic phase. The "switch" from acidogenesis to solventogenesis is accompanied by a large number of physiological and biochemical changes, and is clearly a multifactorial process. The exact mechanism that regulates this switch has not been clearly elucidated, but the concentration of undissociated butyric acid within the cytoplasm functions as a "bioregulator." The end products of the ABE fermentation alter and inhibit growth and metabolism. Acids are more toxic than solvents, and the more hydrophobic a product, the higher is its toxicity. It should be noted that toxicity of acids depends upon the pH. At neutral pH, high concentrations of organic anions (i.e., butyrate, acetate) are not toxic because they are dissociated. In normal batch-fermentation conditions, organic acids are produced, and then consumed during solventogenesis to prevent dissipation of the cellular proton-motive force. Butanol inhibits the fermentation at a relatively low concentration.

Butanol producers use a wide variety of carbohydrates, including hexoses, pentoses, oligosaccharides, and polysaccharides as fermentation substrates, and including starch, molasses, whey, wood hydrolysates, pentosans, inulin, and sulfite liquor. Cellulosic materials are one of the most attractive substrates for chemicals, as they are the most abundant and least expensive raw material. Few attempts have been made to use cellulosic substrates for the production of butanol. A simultaneous saccharification fermentation process was studied to produce butanol from alkali-pretreated wheat straw using *Trichoderma reesi* cellulase and *C. acetobutylicum* (59). The final butanol concentration was 10.7 g/L from about 140 g of straw. Pentoses did not accumulate during the fermentation, suggesting simultaneous use of both hemicellulose and cellulose components. Different substrates alter fermentation parameters of *C. acetobutylicum,* such as growth rate and the solvent production ratio. Higher growth and substrate consumption rates were obtained when glucose, cellobiose, or mannose were used than when xylose, arabinose, or galactose were used. The solvent production ratio with the first sugar group was 1:4:10 (ethanol:acetone:butanol), whereas ratios of 1:2:5 were obtained with the latter.

Starch is at present the most useful raw material for fermentation because it is abundant, has high density, and requires limited pretreatments to be made from biomass. Starch is produced from corn, cereals (wheat, oats, rice), and starchy roots (sweet potatoes, cassava). Typically, *C. acetobutylicum* produces 12–13 g/L butanol with about 20–22 g/L total solvents. In recent years, the majority of investigators have focused their interests on the type culture strain *C. acetobutylicum* ATCC 824. Other cultures included DSM 792 strain and NCIMB 8052. The *C. acetobutylicum* ATCC 4259 strain and the equivalent DSM 1731, NCIMB 619, and NRRL B-530 strains are all derived from the original Weizmann industrial strain. An asporogenous mutant ATCC 39236 derived from this strain has been patented by scientists at the Moffett Technical Center CPC International (60).

Recently, a limited sporulating *C. acetobutylicum* E604 mutant strain has been developed by treating the parent *C. acetobutylicum* strain (ATCC 4259) with ethane methane sulfonate (61). This mutant strain has been used to ferment a high-carbohydrate substrate concentration in a multistage, continuous-temperature programmed-fermentation process followed by batch fermentation (11), to yield an extraordinarily high concentration of butanol (>20 g/L) and total solvent (>30 g/L) production (61). This mutant strain has higher butyrate uptake rate (0.33 g/L) in comparison to the parent strain (0.26 g/L) at a lower temperature of 30 °C (62). Butyrate up to an externally added concentration of 11.4 g/L did not inhibit butyrate uptake. Optimization studies for butyrate uptake by *C. acetobutylicum* suggested a direct correlation between minimum pH and butyrate concentration or temperature (63). These developments make it possible to commercially produce butanol fermentatively with no residual butyrate.

The saccharolytic strain *C. acetobutylicum* P262, one of the NCP production strains, has been studied by groups working in South Africa and New Zealand. The various solvent-producing strains now tend to be regarded as varieties of *C. acetobutylicum* or *C. beijerinckii*. For batch ABE fermentation to be industrially viable, and to compete with the chemical process, concentrations of between 22 and 28 g/L must be obtained in 40–60 h (64). A large number of publications on butanol-producing microbial cultures and the process demonstrate the extensive research work that has been done in past years (33,60,64–68).

Simultaneous recovery of butanol from the broth during fermentation can eliminate product inhibition. This will allow complete utilization of sugar and continual fermentation of the substrate at relatively higher rates. In contrast to the traditional distillation processes, removal of butanol can be accomplished by integrating a pervaporation process with the multistage fermentation process. In this process, the solvents will pass through the membranes preferentially and the organic acids and cells will remain on the liquid side and return to the fermentor. The permeate stream that is enriched in solvent relative to the feed will separate into a low-butanol aqueous phase and a butanol-rich solvent phase. This aqueous phase could be recycled back to the feed tank. Pervaporation also provides for the retention of cells in the fermentor, which improves volumetric productivity.

Ethanol. In the United States, use of ethanol as a fuel is almost wholly confined to its use as an octane enhancer in the higher grade of gasoline. Currently, this fuel ethanol is obtained from both fermentation and chemical synthesis. All of the fermentation ethanol produced in the United States is made by yeast fermentation.

Interestingly, the substrate range for both yeasts and *Zymomonas* for ethanol production is similar. Traditionally, yeasts have been used to produce alcoholic beverages and, more recently, to produce a considerable portion of industrial ethanol all over the world. Among yeasts, *Saccharomyces* is the most important and most commonly used organism. *Saccharomyces cerevisiae* converts glucose to ethanol and carbon dioxide. Other yeasts include *Pachysolen tannophilus* and *Kluyveromyces marxianus*. Many facultative and anaerobic bacteria produce varying amounts of ethanol as one of the metabolic end products. Bacterial species such as *Zymomonas mobilis*, *Sarcina ventriculi*, and *Erwinia amylovorans* utilize pyruvate decarboxylase to convert pyruvate to CO_2 and acetaldehyde, which is then reduced to ethanol. *Zymomonas* grows on glucose, fructose, and sucrose, but is unable to use starch, maltose, or pentose sugars. *Z. mobilis* is a promising bacterium with some potential to replace yeast as the major organism for production of industrial ethanol.

Z. mobilis was metabolically engineered to broaden its range of fermentable substrates to include the pentose sugar xylose (69). Two operons encoding xylose assimilation and pentose-phosphate-pathway enzymes were constructed and transformed into *Z. mobilis* in order to generate a strain that grew on xylose and efficiently fermented both glucose and xylose to ethanol. Subsequent metabolic-pathway engineering further expanded the fermentation substrate range of the ethanologenic bacterium *Z. mobilis* to include the pentose sugar L-arabinose (70). Direct fermentation of cassava starch to ethanol by *Z. mobilis* was obtained in the medium containing amylase-rich culture filtrate of *Endomycopsis fibuligera* (71). The ethanol concentration of 105 g/L could be increased to 132 g/L by further addition of glucoamylase enzyme at 0.01%.

A number of different groups of bacteria, including clostridia, produce ethanol via the reduction of acetyl CoA, which is generated by the cleavage of pyruvate produced during glycolysis. At least 30 species of *Clostridium* have been reported to produce ethanol in amounts varying from trace to close to the theoretical maximum of 2 mol ethanol/mol glucose fermented. In most cases, the extent of ethanol production is dependent upon the nature and concentration of fermentation substrate and the fermentation conditions, such as pH and temperature. For a number of mesophilic clostridial species, ethanol yields ranging from 1.7–1.9 mol/mol hexose fermented have been observed. In a single-step biomass to ethanol conversion process, *C. thermocellum* has been shown to produce 23.6 g/L ethanol from an alkali-treated paddy straw (9). The bacterium *C. ljungdahlii*, a fast-growing bacteria, has been shown to produce ethanol from carbon monoxide in concentrations up to 48 g/L from synthesis gas in a CSTR with cell recycling (10).

In recent years, thermophilic bacteria have been shown to produce solvents and are considered potential candidates for use in process development because of certain advantages over mesophiles. Thermophiles have the ability to use complex plant polymers such as cellulose or starch, and offer high growth rates, fast fermentation, reduced contamination, increased process stability, and eliminate high energy demand in cooling the media and product recovery. The major disadvantage is production of end products in low concentrations and yield due to lower end-product tolerance. However, this can be partially overcome if the solvents are removed simultaneously.

There are increasing number of patented processes concerning the use of thermophilic microorganisms, including ethanol fermentation (72–75). The thermophilic anaerobic bacteria have provided a great deal of information on ethanol-fermentation pathways via acetyl-CoA (34). There are two different ethanol pathways in thermophiles (76). The type I pathway uses NADH-linked alcohol dehydrogenase and produces ethanol from acetaldehyde. The Type II pathway uses NADPH-linked alcohol dehydrogenase to produce ethanol from both acetyl CoA directly or acetaldehyde. In Type II ethanologens, the NAD-linked alcohol dehydrogenase functions to consume ethanol. The Type I pathway is applicable in anaerobic bacteria such as *C. thermocellum*. The Type II pathway is present in *Thermoanaerobacter thermohydrosulfuricus, Thermoanaerobacter brockii*, and *Thermoanaerobacter ethanolicus*. Ethanol yields of heterofermentations vary considerably with the specific growth conditions used. The biochemical basis for different reduced end-product ratios of thermophilic ethanol producers that contain the same glycolytic pathways is related to subtle differences in the specific activities and regulatory properties of the enzymes that control electron flow during fermentation. Similarly, specific changes in culture conditions such as temperature and pH influence the rate and direction of the enzymatic machinery responsible for end-product formation.

T. ethanolicus 39E has a low tolerance for ethanol, with growth inhibition occurring at 2% (wt/vol) ethanol. An ethanol-tolerant strain (39EA) that was tolerant to 4% (wt/vol) ethanol at 60 °C, produced ethanol under these conditions (77), and lacked the NAD-linked alcohol dehydrogenase was selected. Another strain (H8) can grow at 8% ethanol but produces lactic acid at high solvent concentrations. The mechanism of high (i.e., 8%) ethanol tolerance in *T. ethanolicus* is related to its ability to produce unique transmembrane lipids (C_{30} to C_{34} fatty acids) that provide solvent tolerance (78). Alcohol increases membrane fluidity, and these transmembrane lipids may serve to reduce the fluidity and maintain membrane integrity.

1,3-Propanediol. 1,3-Propanediol (PD) is a versatile intermediate compound for the synthesis of heterocycles and as a monomer for the production of polymers such as polyethers, polyesters, and polyurethanes. PD can also be used as a solvent and an additive for lubricants. The microbial conversion of glycerol to PD is simple in comparison to the chemical conversion of acrolein. It has been shown that some *Clostridium* species are able to convert glycerol to PD with an appreciable yield (79). 1,3-Propanediol from glycerol was produced by *C. butyricum* DSM 5431 at 50–58 g/L with productivities of 2.3–2.9 g/L per hour (13). The fermentations were conducted in 300- and 3000-L fermentors with almost similar results, indicating that the scale-up of this microbial fermentation process should not cause a major problem. The substrate glycerol is relatively cheap and its conversion to PD could help reduce glycerol surpluses in the market.

Organic Acids. Currently, organic acids such as acetic, propionic, butyric, fumaric, succinic, malic, and lactic acids can be produced by anaerobic fermentation technology. The key technical problems blocking the rapid advances in developing the bioprocess technology for organic-acid fermentations have been low product concentration, low specificity to desired product(s), low productivity or rate of fermentation, and inefficient or energy-intensive recovery processes. However, recently, significant technological advances have occurred in anaerobic fermentations for production of organic acids by fermentation of carbohydrates (80). Acidogenic fermentation of carbohydrates to volatile organic acids is well known (43). In general, anaerobic bacterial fermentations are of interest because of the wide range of products formed and substrate fermented, the high substrate-to-product conversion yields and rates, and the potential for enhanced process stability and product recovery due to physiological diversity of species (e.g., extreme thermophily, acidophily, halophily). For acetate fermentation, homacetogenic fermentation producing 3 mol of acetate from 1 mol of dextrose (e.g., by using *C. thermoaceticum*) has been described by many authors. Currently, the available organisms produce salts of organic acids in low concentrations (400–800 mM for acetate, 200–400 mM for propionate and 200–300 mM for butyrate) with low productivity of 0.2–2 g/L per hour (81). New fermentation processes with better yield and productivities have been developed, making it feasible to produce volatile as well as nonvolatile organic acids economically.

Lactic Acid. Lactic acid is a natural organic acid with a long history of use in the food industry. Lactic acid, because of its flavoring and preservation properties, is used as an acidulant, particularly in dairy products, confectionery, beverages, pickles, bread, and meat products. It is also used in the pharmaceutical and cosmetic industries. The cyclic dimers of lactic acid are used as raw materials in the synthesis of biodegradable polymers for a variety of uses, including absorbable surgical sutures, slow-release drugs, and prostheses. Ethyl lactate is the active ingredient in many anti-acne preparations. Crude grades of lactic acid are used for the deliming of hides in the leather industry, and it is used for fabric treatment in the textile and laundry industries. Its ability to form polymeric polylactic acids finds application in production of various resins. Because of its structure and its two functionally reactive groups, hydroxyl and carboxyl, lactic acid can be used to make numerous value-added functional chemicals (82). These include biodegradable polymers and copolymers, propylene glycol, propylene oxide, and acrylates. Lactic acid–based polymers and copolymers have potentially large-volume uses as biodegradable thermoplastics, pesticide formulation, and environmentally benign polymers, with a potential market exceeding several hundred million dollars per year.

The market for fermentation lactic acid is growing every year because it is now feasible to produce large-volume chemicals from lactic acid. Homofermentative lactic-acid bacteria produce none or only trace amounts of end products other than lactic acid, and are used for industrial processes. Numerous organisms are known to produce lactic acid with high (>90%) yield and productivity (>2 g/L per hour) from carbohydrate fermentation. Some of the common homolactic acid bacteria are: *Lactobacillus casei, L. pentosus, L. leichmannii, L. acidophilus, L. delbrueckii, L. bulgaricus, Streptococcus cremoris, S. lactis, S. diacetylactis, Sporolactobacillus* sp. and *Pediococcus* sp. Homolactic acid–producing bacteria ferment starch, hexoses, pentoses, and cellobiose. Depending upon the cultural conditions, the yield of lactic acid by the homolactic acid bacterium *L. casei* and heterolactic acid bacterium *T. brockii* are nearly equivalent. In non-energy-limited batch culture, *L. casei* produces lactic acid as the sole end product, whereas in energy-limited, continuous culture, nearly equivalent amounts of lactic, acetic, and formic acids, and as well as ethanol, are produced (43). Lactate is the major fermentation product of *T. brockii* grown in high-yeast extract, batch fermentation. It should be noted that lactate yields in homolactic-acid bacteria depend upon specific growth conditions and that additional products (e.g., formic acid, acetic acid, and ethanol) can also be formed. Lactic acid yields are generally highest during glycolysis via the homolactic-acid-fermentation pathway. Theoretically, 2 mol of lactate and 2 mol of ATP are formed per 1 mol of glucose fermented.

Anaerobic fermentations for production of organic acids operate optimally at pHs where salts of organic acids, rather than free acids, are produced. The free acids and their derivatives are required for the manufacture of functional chemicals. The development of new technologies to recover and purify the fermentation acids from broth have reduced the costs associated with product recovery and purification, thus making the commodity production of organic acids such as lactic acid practical and economically attractive. One of the approaches that has been traditionally applied is simultaneous fermentation and neutralization with calcium salt to obtain calcium lactate precipitate. The precipitate can be separated easily, and free lactic acid can be obtained after treating the salt with acids such as sulfuric acid. The disadvantage of such a system is in handling the solids and slurry. Another emerging technology for separation and recovery of lactic acid is electrodialysis (ED). ED is a rapid process for altering the composition and/or concentration of electrolytes in aqueous solutions by transferring ions across ion-exchange membranes under the influence of a direct electric current. The recovery and purification of lactic acid using ED has been successfully demonstrated (14). In this process, sodium lactate is first selectively separated from proteins, amino acids, and cells, then, using water-splitting membrane, the lactic acid salt is converted to free lactic acid and base. The traditional process using lime for neutralization followed by cell filtration, carbon adsorption, evaporation, acidulation, gypsum filtration, carbon adsorption, evaporation, filtration, esterification, and final purification is uneconomical (83). However, separation and recovery of lactic acid by ED seems a cost-effective and attractive approach.

Succinic Acid. Succinic acid is a relatively new nonhygroscopic acidulant. Succinic acid is listed by the FDA as a GRAS additive for miscellaneous and/or general-purpose uses. It readily combines with proteins in modifying the plasticity of bread doughs. It is also an dibasic acid for producing edible synthetic fats with desirable thermal properties. Succinic acid is manufactured by the catalytic hydrogenation of maleic or fumaric acid. It has also been produced commercially by aqueous alkali or acid hydrolysis of succinonitrile, which is derived from ethylene bromide and potassium cyanide.

Succinic acid is a common intermediate in the metabolic pathway of several anaerobic microorganisms. For example, succinate is a key intermediate for anaerobic fermentations by propionate-producing bacteria, but it is only produced in low yields and in low concentrations. Succinate is also produced by anaerobic rumen bacteria. These bacteria include *Ruminobacter amylophilus* and *Prevotella ruminicola*. As such, accumulation of succinic acid in fermentation broth has been observed in a number of rumen microorganisms. Although the rumen bacteria give higher yields of succinate than do the propionate-producing bacteria, the reported fermentations were generally run in very dilute solutions and gave a variety of products in generally low yields. Moreover, the rumen organisms tend to lyse after a comparatively short fermentation time, thereby leading to unstable fermentations.

Anaerobiospirillum succiniciproducens, an anaerobic non-rumen bacterium, has been demonstrated to produce succinic acid in higher concentrations and yield (84,85). The pathway for succinic acid production in *A. succiniciproducens* has been established (45). Two processes to produce succinic acid as sodium succinate (86) and calcium succinate (87) were developed using *A. suciniciproducens*, as this organism produced succinate in a relatively higher concentration and yield with fewer cofermentation products, had a faster growth rate, and could grow in an industrial grade medium.

In order to develop a commercially attractive process to produce succinic acid by fermentation, the product should be produced in a high concentration and in a high yield (wt%) using inexpensive raw materials and nutrients. *A. succiniciproducens* produces succinate and acetate in a ratio of about 4:1, thus a good amount of carbon is lost as acetate and the separation of succinate and acetate becomes very critical. Therefore, a fluoroacetate-resistant variant FA-10 of *A. succiniciproducens* was developed that produced very minute amounts, if any, of acetic acid (88). The organism produces enough succinate to allow simultaneous fermentation and calcium succinate precipitation under strictly controlled conditions and when the correct calcium salt is used for neutralization. However, the process seems to have a major disadvantage in handling gypsum at a commercial scale. The succinic acid fermentation in *A. succiniciproducens* is regulated by fermentation pH and the amount of carbon dioxide (45). For example, lactate is the major fermentation product at pH 6.5 and higher with limiting levels of CO_2, whereas succinate is the major product at a pH of 6.0–6.2 with an abundance of CO_2. Phosphoenolpyruvate carboxykinase has been purified and described for *A. succiniciproducens* (89), a nonruminal anaerobic bacterium, as well as from a ruminal anaerobic bacterium, *Ruminococcus flavefaciens* (90). In *E. coli*, overexpression of PEP carboxikinase had no effect on succinic acid production, but succinic acid was produced as the major fermentation product by weight (3.5-fold increase in the concentration) by overexpression of PEP carboxylase (91).

Recently, a rumen-facultative anaerobic bacteria, *Actinobacillus* sp. strain 130Z, has been shown to produce succinic acid in much higher concentration (60–80 g/L) than any of the previously known succinic acid–producing microorganisms (92). Acetate and small amounts of pyruvate and formate are produced as cofermentation products. The variants of *Actinobacillus* sp. strain 130Z have been developed to produce succinic acid in a concentration of >105 g/L and yield of 85–95 wt % (19) under neutralization with magnesium. The organism is reported to be robust with a wide pH range and tolerance to high substrate and product concentrations. The levels of key succinic acid–producing pathway enzymes, such as phosphoenolpyruvate (PEP) carboxykinase, a key CO_2-fixing enzyme, malate dehydrogenase, and fumarase were significantly higher in *Actinobacillus* sp. strain 130Z than in *E. coli* K-12 (93). The key enzymes in end-product formation in *Actinobacillus* sp. 130Z were regulated by the energy substrates.

Butyric Acid. Butyric acid, butyrate esters, and other butyrate derivatives are important flavor ingredients in many natural and processed foods. For example, butyric acid is present in butter as an ester to the extent of 4–5%. Some of its esters serve as bases of artificial flavoring ingredients of certain liqueurs, soda water syrups, candies, and so on. Development of a bioprocess for production of natural butyric acid by anaerobic fermentation thus is likely to satisfy the potential market of natural fermentation products. In addition, ethyl and methyl esters of butyric acid can also be used as octane enhancers when added to gasoline (94). Some common anaerobic bacterial genera producing butyrate are *Butyribacterium, Butyrivibrio,*

Clostridium, Eubacterium, Fusobacterium, Haloanaerobium, and *Sarcina*. Most butyric-acid producing bacteria ferment starch, hexose, pentose, and cellobiose, and form acetic acid, in addition to butyric acid, as the major (i.e., percentage of substrate weight converted to product) fermentation product. Butyric acid production is also related to cultural conditions of the specific species. For example, *C. thermosaccharolyticum* produces butyric acid as the fermentation product during exponential growth (95). *B. methylotrophicum* can produce either acetate or butyrate as the sole end product, or it can form a mixture of butyrate and acetate. *B. methylotrophicum* ferments glucose to acetate with small amounts of butyrate at near neutral pH; it produces high levels of butyrate at acidic pH (42,96). In addition, it will produce butyrate from methanol in the presence of carbon dioxide and acetate under the same pH conditions. Products of CO fermentation by this organism are acetate and butyrate in a ratio of approximately 30:1. However, the fermentation can be shifted towards butyrate by decreasing the pH at the onset of the stationary phase (97). Theoretically, yields of 1 mol of butyrate and 2 mol of hydrogen and CO_2 are obtainable from 1 mol of hexose via the butyrate fermentation path (43). *B. methylotrophicum* or *C. butyricum* under cell-recycle can achieve this scenario.

In general, the most studied butyrate-producing microorganisms belonged to the genus *Clostridium*. A large number of fermentation substrates, including hydrolysates of waste cellulosic material, lactose from whey, molasses, and cellulosic materials can be utilized to produce butyric acid. However, the low concentration in the fermentation broth did not make the process commercially attractive as a commodity product. In addition, a major characteristic of this fermentation is the concomitant production of acetate, which is observed in butyrate-producing *Clostridium* and other bacterial species. Recently, fed-batch fermentation of glucose *with Clostridium tyrobutyricum* have yielded butyrate in a concentration of 42.5 g/L with a selectivity of 0.90, a productivity of 0.82 g/L per hour and a weight yield of 36% (98). In glucose-limited, fed-batch cultures, initially produced acetate was reutilized, resulting in exclusive production of butyrate. Because the ratio of butyrate to total acids was strongly influenced by the growth rate of bacteria, acetate being produced along with butyrate at higher growth rates (16) achieved increases in butyrate concentrations, productivity, selectivity, and yield by controlling the substrate feeding by the rate of gas production. In a fermentation of wheat flour hydrolysate (380 g/L of glucose) with *C. tyrobutyricum*, a butyrate concentration of 62.8 g/L was obtained with a productivity of 1.25 g/L per hour, a selectivity of 91.5%, and a weight yield of 45% (16). Thus, production of butyrate for specialty uses by fermentation is feasible.

Propionic Acid. Propionic acid is a value-added specialty chemical used in various chemical and food-processing industries. It has wide-ranging applications, such as an antifungal agent in foods and feeds, and as an ingredient in thermoplastics, antiarthritic drugs, perfumes, flavors, and solvents. Currently, propionic acid is produced by chemical synthesis from petroleum feedstocks, but small amounts of natural propionic acid are produced by fermentation. Propionic acid is a major end product of fermentations carried out by a variety of anaerobic bacteria, many of which ferment glucose to propionate, acetate, and CO_2. Lactate can also be fermented to propionate either via the acrylate pathway in a stepwise reduction to propionate, or via the succinate pathway, where lactate is converted to propionate via pyruvate and succinate.

Although several microorganisms can produce propionic acid, fermentations using propionibacteria have been studied most extensively. The slow growth rate of propionibacteria usually results in propionate production at a slow rate. *Propionispira arboris* ferments glucose to propionate, acetate, and CO_2 via the succinate pathway. The ratio of propionate to acetate is low, resulting in a carbon loss to cofermentation products. However, the ratio of propionate to acetate can be changed from 2 to 16:1, approaching a homofermentation yield from glucose, by use of H_2 as a cosubstrate because this species consumes H_2 (44). Various fermentation systems have been examined to increase the propionate fermentation rate, including batch, fed-batch, continuous, continuous with cell recycle, extractive, and immobilized cell fermentations. Also, various substrates such as glucose, xylose, lactate, glycerol, food-processing waste, and whey have been examined to produce propionate not only with increased productivity, but also economically. Using *Propionibacterium acidipropionici*, propionate has been produced at 42 g/L from glycerol (18) and at 57 g/L from glucose (17) in batch and fed-batch fermentation systems, respectively. It seems fermentative production of propionate is feasible if an appropriate fermentation system and recovery technology can be integrated to obtain the acid in high concentration, yield, and productivity.

Enzymes

In recent years, enzymes have gained wider applications in the biotechnology industry. The global industrial market has recently been valued at $1.4 billion, with an expected market increase of 4–5% (99). Enzymes are grouped into six major classes: oxidoreductases, transferases, hydrolases, lyases, ligases, and isomerases. The majority of commercialized enzymes are hydrolases such as amylase, cellulase, xylanases, pectinase, proteases, lipase, and collagenase. Other important enzymes are glucose isomerase and glucose oxidase. Anaerobic bacteria are very diverse and thus produce a wide variety of enzymes, excluding oxygenases. Enzymes are widely used in pulp and paper, textile, detergent, and food- and feed-additive industries. In addition, enzymes also find applications in pharmaceutical synthesis, therapeutic contexts, and clinical and chemical analysis. The rumen anaerobic microbial population presents a rich and, until recently, underutilized source of novel enzymes with tremendous potential for industrial applications. The enzymes from these microorganisms include cellulases, xylanases, β-glucanases, pectinases, amylases, proteases, phytases, and tannases. High-molecular-mass complexes containing numerous cellulases have been identified in a number of rumen bacteria, including *Butyrivibrio fibrisolvens, Ruminococcus albus*, and *Fibrobacter succinogenes*.

The most common enzyme-producing anaerobic bacteria are mesophilic and thermophilic clostridia, and moderate- and hyper-thermophilic nonclostridial species (100,101). Among the amylolytic enzymes, α-amylase hydrolyzes internal α-1,4 linkages of starch at random in an endo fashion, producing oligosaccharides of varying chain lengths. Generally, it cannot act on α-1,6 linkages of starch. *C. butyricum, C. acetobutylicum, T. ethanolicus, C. thermoamylolyticum* have been reported to produce this amylase enzyme. β-Amylase hydrolyzes alternate α-1,4-glycosidic linkages of starch in an exo fashion from the nonreducing end, producing β-maltose. The β-amylase has been produced in high yield as a primary product during growth of *Thermoanaerobacterium thermosulfurigenes* (102). A hyperproductive mutant was isolated that produced eightfold more β-amylase than the wild type. Synthesis of the enzyme was both constitutive and resistant to catabolite repression. The β-amylase has also displayed industrial potential for the production of high-maltose syrups from raw or soluble starch at 75 °C. Various maltose-containing syrups are used in the brewing, baking, canning, and confectionery industries.

Glucoamylase is an exoacting carbohydrase that cleaves glucose units consecutively from the nonreducing end of starch molecules. High levels of a thermostable glucoamylase activity has been reported in crude extracts of *T. ethanolicus* 39E (103), although it was purified and later described as an α-glucosidase activity. Glucoamylase is widely used in alcoholic fermentation of starchy materials and in the commercial production of glucose and high-glucose corn syrups. However, the current source of this enzyme is fungal. Pullulanase is a debranching enzyme that specifically cleaves α-1,6 linkages in starch, amylopectin, pullulan, and related oligosaccharides. It is generally used in combination with saccharifying amylases such as glucoamylase, fungal α-amylase, or fungal β-amylase for the production of various sugar syrups because it improves saccharification and yield (104). Hyun and Zeikus (103) found that *T. ethanolicus* produces highly thermoactive and thermostable cell-bound pullulanase. They also developed a hyperproductive mutant that displayed improved starch metabolism features. Pullulanase has also been reported to be produced by *C. thermosaccharolyticum*. Taking advantage of the high thermoactivity and acidoactivity of these pullulanases, it may be assumed that these enzymes might effectively replace α-amylase and pullulanase in both starch liquefaction and saccharification processes. α-Glucosidase hydrolyzes terminal nonreducing α-1,4-linked glucose residues of various substrates, releasing α-D-glucose. It is generally considered to be maltase, but has a wide specificity, being able to cleave glucosides of non-sugars in addition to maltose, maltotriose, and other maltooligosaccharides, and to transfer α-D-glycosyl residues of maltose and α-D-glucosides to suitable acceptors.

α-Amylase, which is used in starch liquefaction, solubilizes α-1-4 linkages in starch-forming maltodextrin syrups. α-Amylase has been recently described in thermoanaerobes. For example, the *Pyrococcus furiosus* α-amylase gene has been cloned and expressed in *E. coli* (105). The *P. furiosus* α-amylase is twice as active as *Bacillus litchenformis* commercial α-amylase, but does not require Ca^{2+} for stability at high temperature (i.e., >100 °C).

The cellulase system in both bacteria and fungi comprises three different classes of enzymes: (*1*) endo-1,4-β-glucanases; (*2*) exo-1,4-β-D-glucanases, including both 1,4-β-D-glucan cellobiohydrolases and 1,4-β-D-glucan glucohydrolases; and (*3*) 1,4-β-D-glucosidases, also referred as cellobiases. Several species of cellulolytic clostridia have been described in the literature. These include *Clostridium cellobioparum, C. acetobutylicum, Clostridium cellulovorans, Clostridium stercorarium*, and *Clostridium thermocellum*. Most of the work on industrial cellulases has been accomplished using aerobic fungal systems. However, due to the high specific-enzyme activity on the one hand and the high thermostability on the other, the cellulases from *C. thermocellum* have been considered for potential industrial utilization in direct alcohol fermentations, but not for saccharification *per se* (i.e., glucose production). Recently, endoglucanases have been used in laundry detergents.

Collagenases are endopeptidases that hydrolyze native, insoluble fibrous collagen. One of the anaerobic bacteria producing collagenase that has been extensively studied is *C. histolyticum*. The collagenase of *C. histolyticum* is available commercially. Other collagenase-producing anaerobic bacterial species include *C. collagenovorans* and *C. proteolyticum* (106). Pure collagenase can be applied as a sensitive probe for biosynthetic studies and sequence determinations. It is useful in prevention or cure of keloids. Collagenase is useful for the dispersal and dissociation of animal tissues in the laboratory. It is routinely used to separate cells from their parent tissues. *C. histolyticum* also produces an extracellular sulfhydryl proteolytic enzyme called clostripain (clostridiopeptidase). This enzyme possesses amidase, esterase, and proteolytic activity, which is directed toward the carboxyl peptide linkage of arginine (107).

Pectin-degrading enzymes are produced by a variety of microorganisms, including clostridia. These have been produced by *C. aurantibutyricum, C. felsineum, C. multifermentas, C. roseum*, and *T. thermosulfurigenes*. An active thermostable polygalactunonate hydrolase and pectin methylesterase have been produced by *T. thermosulfurigenes* (108). These thermostable pectinolytic activities may have application in fruit juice clarification and for processing food or agricultural/forestry products.

Several xylose (glucose) isomerases have been isolated from thermoanaerobes including from *Thermotoga neapolitana* (109). This enzyme is more stable and active than commercial enzymes. An alkaline phosphatase that is more stable and active than commercial calf enzyme has been purified and characterized from *T. neapolitana* (110). Table 3 provides examples of some selected enzymes of industrial importance from anaerobic extremophiles. This serves to illustrate the diversity of enzymes from anaerobes.

Specialty Chemicals and Biotransformations

Diverse species of acetogenic and methanogenic bacteria grow on CO, H_2—CO_2, or methanol. These bacteria have been studied for conversion of synthesis gas (i.e., CO + H_2) or CO to methane by pure cultures and consortia, or into organic acids and alcohols by pure culture. Some of these

Table 3. Enzymes of Potential Industrial Importance from Anaerobic Extremophiles

Enzyme	Optimal temperature (°C)	Localization	Organism	Reference
Endoglucanase	65	extracellular	*Clostridium thermocellum*	Ng et al. (111)
α-Amylase	100	extracellular	*Pyrococcus furiosus*	Dong et al. (105)
β-Amylase	75	extracellular	*Thermoanaerobacterium thermosulfurigenes*	Hyun and Zeikus (102)
Xylose/glucose isomerase	95	intracellular	*Thermotoga neopolitana*	Vielle et al. (109)
Amylopullulunase	85	extracellular	*Thermoanaerobacter ethanolicus* 39E	Mathupala and Zeikus (112)
	105	extracellular	*Pyrococcus furiosus*	Dong et al. (113)
Endo-xylanase	70	extracellular	*Thermoanaerobacterium saccharolyticum* B6A-RI	Lee et al. (114)
1° Alcohol dehydrogenase	80	intracellular	*Thermococcus litoralis*	Ma et al. (115)
2° Alcohol dehydrogenase	90	intracellular	*Thermoanaerobacter ehtanolicus* 39E	Burdette et al. (116)
β-Xylosidase	70	intracellular	*Thermoanaerobacterium saccharolyticum* B6A-RI	Lee and Zeikus (117)
Alkaline phosphatase	85	intracellular	*Thermotoga neopolitana*	Dong and Zeikus (110)

bacteria include *B. methylotrophicum, Eubacterium limosum, Peptostreptococcus productus, C. thermoaceticum, C. ljungdahlii*, and *Methanosarcina barkeri*. Synthesis gas represents a cheap feedstock for microbial conversion to higher-value commodity products. Recent work has focused at syngas fermentation to liquid-fuel additives. The feasibility of ethanol production from syngas fermentation by *C. ljungdahlii* (118) and ethanol plus butanol production by *B. methylotrophicum* (119,120) has been established. However, these processes need to be developed further for their commercialization.

Recently, anaerobes have also been examined for a wide range of specialty fermentation products including antimicrobials, bioflavors, biopigments, biopesticides, and anticancer agents. Antimicrobial compounds such as antibiotics and bacteriocins are produced by some of the anaerobic species. The most common examples are that of nisin and pediocin production by *Lactococcus lactis* and *Pediococcus acidilactici*, respectively (121). Some strains of lactobacilli catalyze the decarboxylation of glutamate to γ-aminobutyrate (122).

Many chiral compounds can be synthesized by microbial hydrogenation using hydrogen (or formate) and hydrogenase-containing microorganisms. Oxidoreductase enzymes are involved in electron-transfer reactions and can be applied in a stereoselective catalysis. The best known alcohol dehydrogenase is that present in both *T. ethanolicus* and other thermoanaerobes, the most notable being *T. brockii*. This NADP-linked secondary alcohol (aldehyde/ketone) dehydrogenase, found in thermoanaerobes, exhibits a wide substrate specificity toward linear and cyclic secondary alcohols and thioesters (116,123,124). Table 4 summarizes novel oxidoreductases present in some clostridial species that are of commercial interest (125). The biotransformation reactions can be carried out either by whole-cell fermentation (which eliminates coenzyme regeneration but may be subject to interference by competing enzymes), or by crude or purified enzyme in batch or continuous-flow systems using an immobilized enzyme column. *C. sporogenes* performs the stickland reaction, in which pairs of amino acids are fermented, one amino acid acting as an electron donor (e.g., valine, leucine, isoleucine), and the other acting as an electron acceptor (e.g., proline, glycine). Synthesis of pyruvate and other 2-oxacids from acyl phosphate derivatives have been shown with permeabilized cells of *C. sporogenes* (125). In amino acid–fermenting anaerobic bacteria, a set of unusual dehydratases is found that use 2-hydroxyacetyl-CoA, 4-hydroxybutyryl-CoA, or 5-hydroxyvaleryl-CoA as substrates. These anaerobic bacteria include *C. propionicum, C. aminobutyricum*, and *C. aminovalericum* (126).

Waste Treatment

Residual Organics. Microorganisms excel in using organic substances as sources of nutrients and energy. The challenge of wastewater treatment is to remove (*1*) compounds with a high biochemical oxygen demand, (*2*) pathogenic organisms and viruses, and (*3*) a multitude of human-made chemicals. Anaerobic digestion (AD) is commonly used to treat materials with a high content of insoluble organic matter, such as cellulose, and to degrade concentrated industrial wastes, such as those from the food-processing industry. AD is a complex biological process that utilizes a consortium of anaerobic microorganisms to act in concert to hydrolyze complex organics to simple monomers and then to volatile fatty acids (VFAs). These VFAs are ultimately converted to methane and carbon dioxide in the final step in the anaerobic food chain. The degradative and fermentative reactions in the anaerobic treatment processes can be divided into two stages: acid-forming and methane-forming, which involve at least three groups of anaerobic bacteria—acidogenic, syntrophic, and methanogenic. In the acid-forming stage, complex organic polymers, including carbohydrates, fats, and proteins, are hydrolyzed and fermented to VFAs, alcohols, and ketones by selective anaerobic bacteria. Organic acids and alcohols such as lactic acid, propionic acid, butyric acid, and ethanol are converted to acetic acid and carbon dioxide by acetogenic bacteria in the syntrophic association of, generally, methanogenic bacteria. The acetate and CO_2 are finally converted to CH_4 by methanogenic bacteria that

Table 4. Novel Oxidoreductase in *Clostridium* Species

Reduction	Electron donor	Organism
Aldehyde/ketone dehydrogenase		
Steroids	Unknown	*C. paraputrificum*
		C. bifermentans
Methyl ketones	NADPH	*C. thermohydrosulfuricum*
Ketones	Unknown	*C. pasteurianum*
	Unknown	*C. tyrobutyricum*
2-Oxoacid synthase		
Fatty acids	Ferredoxin	*C. sporogenes*
Acetate	Ferredoxin	*C. kluyveri*
Linoleic reductase		
Linoleic acid	Unknown	*C. sporogenes*
Enoate reductase		
Cinnamic acid	NADH	*C. sporogenes*
Crotonic acid	NADH	*C. tyrobutyricum*
2-Oxoacid reductase		
Phenylpyruvic acid	NADH	*C. sporogenes*
Nitroaryl reductase		
Chloramphenicol	Ferredoxin/flavodoxin	*C. acetobutylicum*
Metronidazole	Ferredoxin/flavodoxin	
Paro-nitrobenzoate	Ferredoxin/flavodoxin	
2-Nitrobenzene	Ferredoxin/flavodoxin	
Lipoamide hydrogenase		
NAD/lipoamide	Lipoamide/NADH	*C. kluyveri*

Source: From Lovitt et al. (125).

may include acetate- and H_2-CO_2-consuming methanogens. Anaerobic microorganisms in general show a high degree of metabolic specialization. The success of the anaerobic digestion process therefore depends upon cooperative interactions between microorganisms with different metabolic capabilities. Figure 2 illustrates the anaerobic food chain that converts complex organic matter into methane. Since the rate of VFA production can be significantly higher than that of VFA conversion to methane, an imbalance between these two rates can occur in a single-stage digestor, resulting in VFA accumulation, a concomitant pH drop, cessation of methane fermentation, and ultimate process failure. A two-stage AD process that separates the acid-forming and methane-forming stages has been practiced with a variety of wastes. Alternatively, a high-rate syntrophic methanogenic microbial consortium can be developed (127,128). The leach-bed two-phase digestion process, consisting of an acid-phase, solid-bed reactor operated in tandem with a separate, packed-bed, methane-phase digestor (anaerobic filter), for high solids waste has been described (129). Low-strength wastewaters can now be treated, even under psychrophilic conditions, by using specific rheologic conditions in the expanded granular sludge bed (EGSB) reactor (130). Solid wastes can be treated anaerobically using the thermophilic "high-solids" fermentation technology (131). New reactor designs permit S° recovery from SO_2-rich waste gases (132).

Many human-made organic compounds are degraded during sewage treatment. Since 1986, full-scale UASB reactors have been used to treat municipal wastewater throughout the world. Treatment costs are halved when anaerobic treatment (e.g., UASB reactor) is applied instead of just aerobic processes (133). In warm climates, a simple upflow sludge blanket (USB) septic tank with an aerobic posttreatment (e.g., a trickling filter) can be combined to remove the bulk of its suspended solids (134). AD is expected to reduce most of the pathogenic bacteria. Thus, the treated water may be considered reclaimed for use in crop production in the next decade. Organic slurries containing particulate organic matter, such as animal manures and primary or secondary sewage sludges, are normally digested in completely mixed reactors. Because the hydrolysis rates of certain solids is slow, separation of hydraulic retention time (HRT) and solids retention time (SRT) may improve performance of digestors treating slurries. AD does not remove NH_4^+. In a new NH_4^+-removal pathway observed in the methanogenic reactors, NH_4^+ was oxidized anaerobically to N_2 in the presence of NO_2^-, with a lab-scale reactor achieving a removal rate of 0.7 kg NH_4^+-N/m^3 per day (135).

The UASB technology is now widely used to anaerobically digest industrial wastewaters with concentrations in the range 2–20 g COD/L. UASB reactors are usually implemented when wastewater is rich in carbohydrates and relatively poor in other contaminants. Biomass retention through adequate granulation is of utmost importance in UASB technology, first to obtain a good effluent quality and, second, in order to ensure a minimal cell residence time of 7–12 day that is required to avoid the wash-out of the slowest-growing anaerobic bacteria. The onset of sludge granulation after start-up of an anaerobic-sequencing batch reactor was enhanced by adding a cationic polymer, divalent cations, and granulation nuclei such as clay minerals (136). Consideration of surface thermodynamics is important in granulation (137). Some key methanogenic species have been identified and implicated in the granu-

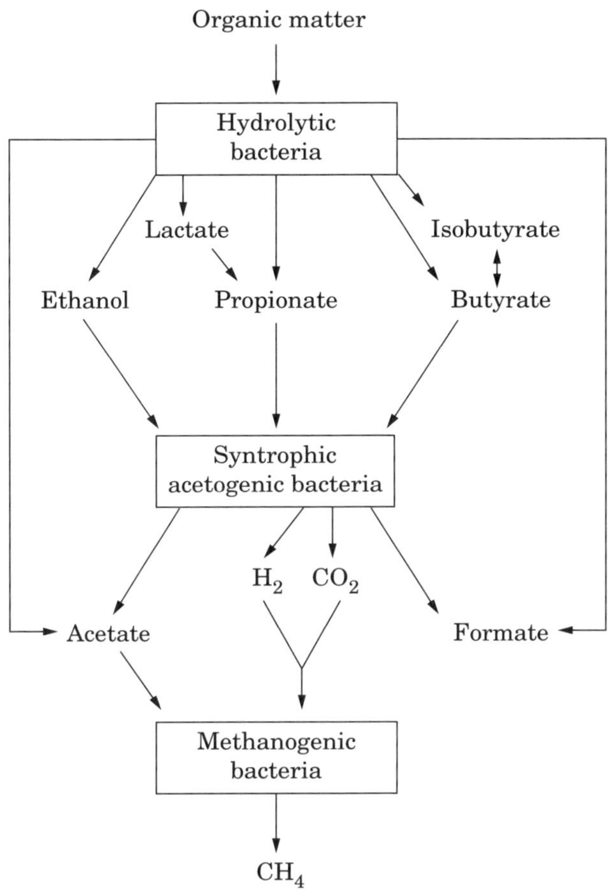

Figure 2. Anaerobic degradation of organic matter to methane in relation to microbial trophic groups.

lation process (138), and exopolysaccharides produced by these methanogens, specifically *Methanobacterium formicicum*, *M. mazei*, and *Methanosaeta* sp. seem to be responsible for the stability of granular structure (139). Industrial wastes are often unreliable in terms of their composition. Inhibition of methanogens caused by toxic compounds present in industrial wastewaters can be alleviated by adding activated carbon directly in the anaerobic reactors (140). Addition of GAC to UASB reactors treating textile wastewater prevented the chronic intoxication of anaerobic sludge that takes place in the absence of the GAC (134).

Anaerobic treatment systems are gaining in popularity and finding new applications. As revealed in papers presented at the 8th International Conference on Anaerobic Digestion, Sendai, Japan, in May 1997, the databank of anaerobic reactors presently includes 1066 operating anaerobic reactors worldwide. Of these reactors, 956 (89.7%) are for the treatment of industrial effluents, while 78 (7.3%) treat domestic sewage, and only 32 (3.0%) of the total are anaerobic-treatment systems for organic solid waste (excluding the vast number of biogas plants for manure treatment that have been installed worldwide). It is estimated that there are about two million biogas plants for manure treatment just in India alone. The largest number of municipal anaerobic treatment plants presently exist on a moderate, but remarkably growing, scale for sewage treatment in countries such as Brazil, Colombia, Mexico, India, and China. In industrial countries, only sewage sludge is treated anaerobically on a large scale due to climatic factors. For industrial wastewater treatment, countries such as India, China, Thailand, Brazil, and Mexico have a growing interest in anaerobic treatment for the local industry; in industrialized regions the leading countries, especially The Netherlands and Belgium, have almost already saturated their home markets. For anaerobic solid-waste treatment (and cofermentation), Germany and Denmark have significant numbers of plants so far. In the near future, anaerobic treatment of solid wastes will become a viable option provided that appropriate collection systems are introduced.

Toxic Organics. In recent years, AD has been used to degrade xenobiotics present in wastewater, soils, and river sediments. Complex xenobiotics often require more than one species to be completely mineralized. Microorganisms need to adapt, maybe for several months, before the maximal conversion rate of a new xenobiotic compound can be achieved. Anaerobic tunnel reactors are used to treat soils and sediments polluted with chloroethene, BTEX, or TNT (141). Efficiency of removal of adsorbable organic halogens by a UASB treating kraft-mill bleach wastewater varied between 27–65%, depending on the residence time (142). *Desulfomonile tiedjei*, which can rapidly transform 3-chlorobenzoate (3-CB), has been shown to become established in an UASB reactor. This indicates that a specific strain or desired species can be incorporated in microbial granular culture of an UASB reactor. Incorporation of an adapted microbial sludge culture into syntrophic biomethanation granules has been demonstrated to treat PCP (14). These studies were further validated by developing PCE/TCE-dechlorinating methanogenic granules (143) using a similar protocol. In this, the anaerobic granules, using methanol as an electron donor for reductive dehalogenation, completely converts highly oxidized PCE to ethylene. In contrast, PCBs biodegrade very slowly and persist in the environment for decades. An effective microbial enrichment that was shown to remove *ortho*-, *meta*-, and *para*-chlorine was incorporated into the PCP-degrading methanogenic microbial granules to produce PCB-dechlorinating anaerobic microbial granules in a manner similar to achieving PCP-granules in the first place (31). The PCB-dechlorinating anaerobic methanogenic granules have shown PCB-dechlorination of defined PCB-congeners (29) and Aroclor-contaminated river sediments (30). These granules are stable, culturable, and can be produced on a mass scale for bioaugmentation for application in both in situ and ex situ processes. These granules have also demonstrated degradation and mineralization of biphenyl, a non-chlorinated end-product of PCB-dechlorination process reductively (Natarajan, Wu, and Jain, unpublished data). Since PCBs are a complex mixture of various PCB compounds, and each PCB compound varies in its degree of dechlorination, it is expected that a mixture of bacteria capable of removing chlorines from different positions on the PCB molecule must be involved in achieving a complete dechlorination of PCB. A similar assumption, however, is no longer true for dechlorination of perchloroethylene (PCE). Recently, Zinder and his group isolated a pure culture of an anaerobic bacteria, tentatively named *Dehalococcoides* strain 195, that dechlorinates PCE and TCE in a stepwise manner to ethene (144). This novel eubacterium is not methanogenic, acetogenic, or sulfidogenic.

2,4,6-Trinitrotoluene (TNT), one of the most widely used explosives, occurs as a pollutant of soil and ground water, especially at sites of ammunition factories. There have been several reports of degradation or transformation of nitro and aminoaromatic compounds under anaerobic conditions. A nonspecific reduction of the nitro group to the corresponding amine was assigned to a variety of methanogenic bacteria, sulfate-reducing bacteria, and clostridia (145). In contrast, it was reported that 2,4-dinitrophenol, 2,4- and 2,6-dinitrotoluene (146), and TNT (147) were used as sources of nitrogen by sulfate-reducing bacteria. Thus, it is clear that many nitroaromatic compounds can serve as growth substrates for anaerobic bacteria. Anaerobic treatment systems have been proposed as a means of avoiding the accumulation of partially reduced intermediates during degradation of TNT. Under strictly anaerobic conditions, TNT can be completely reduced to triaminotoluene (147). *C. acetobutylicum* transformed 2,4,6-TNT to undetermined end products via monohydroxylamino derivatives (148). In contrast to solventogenic cells, acidogenic cultures showed rapid transformation rates and the ability to transform TNT and its primary reduction products to below detection limits. Anaerobic treatment of explosive (primarily TNT)-contaminated soil was demonstrated in open bulk containers containing soil, phosphate buffer, and potato starch (149). The potato starch served as a rapidly degradable carbon source that allowed the rapid establishment of anaerobic conditions. Cresols and small organic acids were observed as end products. Similarly, Dinoseb, a very persistent herbicide that is highly toxic to virtually all living systems, has also been shown to be degraded by anaerobic microbial consortia using potato starch as a carbon source (150).

Inorganics. Microorganisms degrade certain toxic constituents used in mineral processing, and concentrate and immobilize soluble heavy metals released as a result of mining and mineral-processing activities. The initial mechanism of metal binding by microorganisms is electrostatic attraction between charged metal ions in solution and charged functional groups on microbial cell walls. The cell walls are composed of macromolecules with functional groups (principally carboxylate, amine, imidazole, phosphate, sulfhydryl, and sulfate) that contribute a net negative charge to the surface of the microorganism. These functional groups remain active even when the microorganism is not viable. Sulfate-reducing bacteria (SRBs) are used in highly controlled reactor systems and in constructed anaerobic wetlands for removal of sulfate and heavy metals from acid-rock drainage and other aqueous, metal-contaminated, streams. SRBs such as *Desulfovibrio* and *Desulfatomaculum* oxidize organic matter or H_2 by using sulfate as an electron acceptor to produce hydrogen sulfide and bicarbonates. The sulfide immediately reacts with soluble heavy metal ions to form highly insoluble metal sulfides. A system to evaluate anaerobic and aerobic treatment of mine effluents containing excessive sulfate and heavy metals has been described (151). In the continuous system, the H_2S is produced by the SRB biofilm on the dolomite pebbles in the anaerobic stage, which precipitates metal sulfides from a waste stream amended with an organic energy source. In the aerobic stage, a completely mixed reactor and a settling tank facilitate oxidation of residual H_2S and biodegradation of residual organics from the primary anaerobic column. Sulfate reduction has been considered as a method for permanent stabilization of sulfidic mine tailings.

High sulfate concentrations are common in wastewaters from paperboard industries, molasses-based fermentation industries, edible oil refineries, and in acidic leachates of pyritic waste rock and tailings (152). Sulfidogenic UASB reactors have attracted some attention for treating such polluted streams, but the biological processes to remove high SO_4^{2-} contents have not yet been optimized. A new bioprocess in which sulfate and heavy metals are removed simultaneously from groundwater has been developed (153). In this two-step process, SO_4^{2-} bacteria produce sulfides that precipitate with the heavy metals after ethanol, an electron donor, is added to the first reactor (BIOPAQ UASB). In the second reactor (THIOPAQ-submerged fixed-film reactor), sulfur bacteria reoxidize the excess sulfide selectively to solid S° under controlled dosage of O_2. With HRT of 4 h in the BIOPAQ reactor and 30 min in THIOPAQ reactor, removal efficiencies of 99% for heavy metals and of 85% for sulfate have been achieved. This technology has also been adapted for desulfurization of flue gas from power plants (132). In the modified process for flue-gas desulfurization, H_2 can be used as an electron source in lieu of ethanol.

Minewaters and industrial effluents containing high sulfate concentrations create a disposal problem that requires an urgent solution in order to avoid excess mineralization of surface waters. Sulfate can be converted quantitatively to hydrogen sulfide by *D. desulfuricans* (154). Further conversion to elemental sulfur can be effected by the photosynthetic bacteria *Chlorobium limicola* and *Chromatium cinosum*. Two separate reactors were used for hydrogen sulfide and sulfur production. A possible way to increase the sulfate reduction rate is by making use of a packed-bed, instead of a completely mixed, reactor.

CONCLUSIONS AND FUTURE RESEARCH

The use of anaerobes in industrial processes has grown dramatically over the past 20 years. Traditionally, anaerobes were used routinely in the processing of fermented foods such as lactic acid and propionic acid bacteria in cheeses, yogurts, sausages, and pickles. In waste treatment, mixed methanogenic cultures were used in both industrial and municipal anaerobic digestors for removal of residual organic matter.

Now, *C. botulinum* toxins are used as medicines. Numerous organic alcohols and acids are produced by industry as natural fermentation chemicals. These natural chemicals are used in flavors, fragrances, preservatives, and other specialty products. Fermentation-derived lactic acid is now produced by industry as a commodity chemical and is used as an acidulant, disinfectant, green solvent, and precursor for polylactide-based biodegradable plastics. Anaerobic composting systems are used to treat food and municipal solid wastes because they produce less end product that must be used or land spread. Sulfite is also removed from coal-bearing flue gases in bioreactor systems comprising mixed sulfidogenic cultures.

With the rapid growth in isolation and characterization of new anaerobic species from normal and extreme environments, the diversity of anaerobic species with unique biochemical attributes far surpasses that of aerobic microbes. This great diversity of anaerobic microbes should be expected since anaerobes were the first organisms to evolve on earth, yet they were the last large natural group to be studied in detail by biologists. The vast array of biochemical diversity shown by anaerobes is related in part to the fact that they are not limited to use of O_2 as an electron acceptor, they can use fermentative metabolism or CO_2, SO_4, or other electron acceptors for anaerobic respiration. This biochemical diversity will undoubtably be exploited in the future since many of the fermentation products and enzymes of anaerobes are of interest to industry (i.e., food-feed, chemical, pharmaceutical, energy, and environmental companies).

Several research areas on anaerobes showing special industrial promise include succinate fermentation, thermozymes, and dechlorinating methanogenic granules. Ethanol fermentations are limited in part because two moles of CO_2 are lost from the product per mole of glucose fermented. On the other hand, one can derive more than one pound of succinate per pound of glucose fermented because the theoretical chemical yield is 1 glucose $+ 2\ CO_2 + 2\ H_2 \rightarrow 2$ succinate. Succinate has a wide variety of uses as both a specialty chemical and commodity intermediate chemical. Perhaps the largest markets for succinate includes their use as a feedstock to produce stronger-than-steel engineered plastics and polyesters, and as the chelator EDDS (ethylene diamine disuccinate) to replace nonbiodegradable EDTA.

The saccharolytic enzymes of thermoanaerobes are very active and stable. These thermozymes could be used to develop the next generation of enzymes used in the starch-processing industry (i.e., α-amylase, glucose isomerase, glucoamylase, and pullulanase) or to initiate a cellulose-processing industry for enhanced biomass utilization based on very active and stable cellulases and hemicellulases. Anaerobic sediments and soils are contaminated with a wide variety of chlorinated compounds. The use of dechlorinating methanogenic granules offers an alternative to expensive dredging and landfilling. Dechlorinating methanogenic granules can degrade a wide variety of toxicants (DDT, DIOXINS, PCE/TCE, PCBs, PCP, etc.) by in situ bioagumentation technology.

BIBLIOGRAPHY

1. N.R. Krieg and J.G. Holt eds., *Bergey's Manual*, Williams and Wilkins, Baltimore, Md., 1984.
2. M.W.W. Adams, *Annu. Rev. Microbiol.* **47**, 627–658 (1993).
3. J.G. Zeikus, in K.H. Park, J.F. Robyt, and Y.-D. Choi, eds., *Enzymes for Carbohydrate Engineering*, Elsevier Science, Amsterdam, 1996, pp. 145–161.
4. K.O. Stetter, A. Hoffman, and R. Huber, in R. Guerrero and C. Pedros-Alio eds., *Trends in Microbial Ecology*, Spanish Society for Microbiology, Barcelona, Spain, 1993, pp. 25–28.
5. L.D. Mermelstein and J.G. Zeikus. *Extremophiles: Microbial Life in Extreme Environments*, Wiley-Liss, New York, 1998, pp. 255–286.
6. C.R. Woese, *PNAS* **91**, 1601–1603 (1994).
7. M. Watanabe, *Scientist* **8**, 14–15 (1994).
8. S.S. Kesava, T. Panda, and S.K. Rakshit, *Proc. Biochem.* **31**, 449–456 (1996).
9. K.S. Rani, M.V. Swamy, and G. Seenayya, *Proc. Biochem.* **33**, 435–440 (1998).
10. K.T.A. Klasson, M.D. Ackerson, E.C. Clausen, and J.L. Gaddy, *Fuel* **72**, 1673 (1993).
11. U.S. Patent 5,063,156 (November 5, 1991), D.A. Glassner, M.K. Jain, and R. Datta (to Michigan Biotechnology Institute, Lansing, Mich.).
12. A.J. Grethlein, R.M. Worden, M.K. Jain, and R. Datta, *J. Fermen. Bioeng.* **72**, 58–60 (1991).
13. B. Günzel, S. Yonsel, and W.-D. Deckwer, *Appl. Microbiol. Biotechnol.* **36**, 288–294 (1991).
14. U.S. Patent 4,885,247 (December 5, 1989), R. Datta (to Michigan Biotechnology Institute, Lansing, Mich.).
15. S.R. Parekh and M. Cheryan, *Biotech. Lett.* **16**, 139–142 (1994).
16. F. Fayolle, R. Marchal, and D. Ballerini, *J. Ind. Microbiol.* **6**, 179–183 (1990).
17. H.-D. Paik and B.A. Glatz, *Appl. Microbiol. Biotechnol.* **42**, 22–47 (1994).
18. F. Barbirato, D. Chedaille, and A. Bories, *Appl. Microbiol. Biotechnol.* **47**, 441–446 (1997).
19. U.S. Patent 5,573,931 (November 12, 1996), M.V. Guettler, M.K. Jain, and D. Rumler (to Michigan Biotechnology Institute, Lansing, Mich.).
20. W.R. Kenealy, Y. Cao, and P.J. Weimer, *Appl. Microbiol. Biotechnol.* **44**, 507–513 (1995).
21. L. Bhatnagar, M.K. Jain, and J.G. Zeikus, in J.M. Shively and L.L. Barton eds., *Variations in Autotrophic Life*, Academic, London, 1991, pp. 251–270.
22. U. Deppenmeier, V. Müller, and G. Gottschalk, *Arch. Microbiol.* **165**, 149–163 (1996).
23. M. Blaut, *Antonie van Leewenhoek* **66**, 187–208 (1994).
24. E.J. Bouwer, B.E. Rittman, and P.L. McCarty, *Environ. Sci. Technol.* **15**, 596–599 (1981).
25. E.J. Bouwer and P.L. McCarty, *Appl. Environ. Microbiol.* **45**, 1286–1294 (1983).
26. E.J. Bouwer and P.L. McCarty, *Appl. Environ. Microbiol.* **45**, 1295–1299 (1983).
27. L. Bhatnagar, W.-M. Wu, M.K. Jain, and J.G. Zeikus, in *Proceedings of the International Symposium on Environmental Biotechnology*, Royal Flemish Society of Engineers, Ostend, Belgium, 1991, pp. 1–10.
28. W.-M. Wu, L. Bhatnagar, and J.G. Zeikus, *Applied Environ. Microbiol.* **59**, 389–397 (1993).
29. M.R. Natarajan, W.-M. Wu, J. Nye, H.-Y. Wang, L. Bhatnagar, and M.K. Jain, *Appl. Microbiol. Biotechnol.* **46**, 673–677 (1996).

30. M.R. Natarajan, J. Nye, W.-M. Wu, H.-Y. Wang, and M.K. Jain, *Biotech. Bioeng.* **55**, 182–190 (1997).
31. U.S. Patent 5,635,393 (June 3, 1997), L. Bhatnagar, W.-M. Wu, M.R. Natarajan, H.Y. Wang, and M.K. Jain (to Michigan Biotechnology Institute, Lansing, Mich.).
32. J.F. Quensen, III, S.A. Mueller, M.K. Jain, and J.M. Tiedje, *Science* **280**, 435–440 (1998).
33. R.W. Lovitt, B.-H. Kim, and J.G. Zeikus, *CRC Crit. Rev. Biotechnol.* **7**, 107–186 (1988).
34. R. Lamed and J.G. Zeikus, *J. Bacteriol.* **144**, 569–578 (1980).
35. H. Hippe, J.R. Andreesen, and G. Gottschall, in A. Balows, H.G. Truper, M. Dworkin, W. Harder, and K.H. Schleifer eds., *The Prokaryotes, a Handbook on the Biology of Bacteria: Ecophysiology, Isolation, Identification, Applications,* Vol. 2, Springer-Verlag, New York, 1992, pp. 1800–1866.
36. B. McNeil and B. Kristensen, *Biotechnol. Lett.* **7**, 449 (1985).
37. H.A. George, J.L., Johnson, W.E.C. Moore, L.V. Holdeman, and J.S. Chen, *Appl. Environ. Microbiol.* **45**(3), 1160–1163 (1983).
38. D.T. Jones, J.R. Webster, and D.R. Woods, *J. Gen. Microbiol.* **116**, 195 (1980).
39. H.G. Wood and L.G. Ljungdahl, in J.M. Shively and L.L. Barton eds., *Variations in Autotrophic Life*, Academic Press, London, 1991, pp. 201–205.
40. H.G. Wood, *Annu. Rev. Biochem.* **54**, 1–41 (1985).
41. R. Kerby and J.G. Zeikus, *Curr. Microbiol.* **8**, 27–30 (1983).
42. G.J. Shen, B.A. Annous, R.W. Lovitt, M.K. Jain, and J.G. Zeikus, *Appl. Microbiol. Biotechnol.* **45**, 355–362 (1996).
43. J.G. Zeikus, *Ann. Rev. Microbiol.* **34**, 423–464 (1980).
44. T.E. Thompson, R. Conrad, and J.G. Zeikus, *FEMS Microbiol. Lett.* **22**, 265–271 (1984).
45. N. Samuelov, R. Lamed, S. Lowe, and J.G. Zeikus, *Appl. Environ. Microbiol.* **57**, 3013–3019 (1991).
46. M. Sebald, *Genetics and Molecular Biology of Anaerobic Bacteria*, Springer-Verlag, New York, 1993, pp. 1–703.
47. C. Lee, L. Bhatnagar, B.C. Saha, Y.-E. Lee, M. Takagi, T. Imanaka, M. Bagdasarian, and J.G. Zeikus, *Appl. Environ. Microbiol.* **56**, 2638–2643 (1990).
48. R.L. Robson, R.M. Robson, and J.G. Morris, *Biochem J.* **144**, 503–511 (1974).
49. M. Sebald and R.N. Costilow, *Appl. Microbiol.* **29**, 1–16 (1975).
50. I.R. Booth and J.G. Morris, *Biosci. Rep.* **2**, 47–53 (1982).
51. B.S. Mendez and R.F. Gomez, *Appl. Environ. Microbiol.* **43**, 495–496 (1982).
52. S.N. Bowring and J.G. Morris, *J. Appl. Bacteriol.* **58**, 577–584 (1985).
53. S.J. Reid, E.R. Allcock, D.T. Jones, and D.R. Woods, *Appl. Environ. Microbiol.* **45**, 305–307 (1983).
54. E. Stoutschek-Bauer, L. Hartl, and W.L. Staudenbauer, *Biotechnol. Lett.* **7**, 705–710 (1985).
55. J.D. Oultram, M. Loughlin, T.J. Swinfield, J.K. Brehm, D.E. Thompson, and N. Minton, *FEMS Microbiol. Lett.* **56**, 83–88 (1988).
56. M. Young, W.L. Staudenbauer, and N.P. Minton, in N.P. Minton and D.J. Clarke eds., Clostridia, Plenum Press, New York, 1989, pp. 63–103.
57. P.F. Levy, J.E. Sanderson, E. Ashare, and S.R. de Riel, *Enzyme Microbiol. Technol.* **3**, 207–215 (1981).
58. P.E. Liley, R.C. Reid, and E. Buck, in R.H. Perry, D.W. Green, and J.O. Maloney eds., *Perry's Chemical Engineers' Handbook*, 6th ed. McGraw-Hill, New York, 1950, pp. 3-1–3-291.
59. R. Marchal, M. Rebeller, and J.P. Vandecasteele, *Biotechnol. Lett.* **6**, 523–528 (1984).
60. J.A. Marlatt and R. Datta, *Biotech. Prog.* **2**, 23–28 (1986).
61. U. S. Patent 5,192,673 (March 9, 1993), M.K. Jain, D. Beacom, and R. Datta (to Michigan Biotechnology Institute, Lansing, Mich.).
62. B.K. Soni and M.K. Jain, *Bioprocess Eng.* **17**, 261–267 (1997).
63. B.K. Soni and M.K. Jain, *Bioprocess Eng.* **17**, 329–334 (1997).
64. D.R. Woods, *Trends Biotechnol.* **13**, 259–264 (1995).
65. D.T. Jones and D.R. Woods, *Microbiol. Rev.* **50**, 484–524 (1986).
66. H.P. Blaschek, *Food Technol.* **(October)**, 84–87 (1986).
67. M.R. Laddisch, *Enzyme Microbiol. Technol.* **13**, 280–283 (1991).
68. I.S. Maddox, N. Qureshi, and K. Roberts-Thomson, *Proc. Biochem.* **30**, 209–215 (1995).
69. M. Zhang, C. Eddy, K. Deanda, M. Finkelstein, and S. Picataggio, *Science* **267**, 240–243 (1995).
70. K. Deanda, M. Zhang, C. Eddy, and S. Picataggio, *Appl. Environ. Microbiol.* **62**, 4465–4470 (1996).
71. O.V.S. Reddy and S.C. Bassapa, *Biotechnol. Lett.* **18**, 1315–1318 (1996).
72. U.S. Patent 4,292,406 (September 29, 1981), L.G. Ljungdahl and J.K.W. Wiegel (to U.S. Department of Energy, Washington, D.C.).
73. U.S. Patent 4,292,407 (September 29, 1981), L.G. Ljungdahl and J.K.W. Wiegel (to U.S. Department of Energy, Washington, D.C.).
74. U.S. Patent 4,400,470 (August 23, 1983), J.G. Zeikus, T.K. Ng, A. Ben-Bassat, and R.J. Lamed (to Wisconsin Alumni Research Foundation, Madison, Wisc.).
75. U.S. Patent 4,385,117 (May 24, 1983), L.G. Ljungdahl and L.H. Carriera (to University of Georgia Research Foundation, Inc., Athens, Ga.).
76. D. Burdette and J.G. Zeikus, *Biochem. J.* **302**, 163–170 (1994).
77. R.W. Lovitt, R. Longin, and J.G. Zeikus, *Appl. Environ. Microbiol.* **48**, 764–770 (1984).
78. S. Jung, J.G. Zeikus, and R. Hollingsworth, *J. Lipid Research* **35**, 1057–1065 (1994).
79. H. Biebl, S. Marten, H. Hippe, and W.-D. Deckwer, *Appl. Microbiol. Biotechnol.* **36**, 592–597 (1992).
80. M.K. Jain, R. Datta, and J.G. Zeikus, in T.K. Ghose ed., *Bioprocess Engineering: The First Generation*, Ellis Harwood, Chichester, England, 1989, pp. 366–398.
81. L.G. Ljungdahl, L.H. Carreira, R.J. Garrison, N.E. Rabek, and J. Wiegel, *Biotech. Bioeng. Symp.* **15**, 207–223 (1985).
82. E.S. Lipinsky and R.G. Sinclair, *Chem. Eng. Prog.* **82**, 26–32 (1986).
83. Anonymous, *Biomass Process Handbook*, Technical Insights Inc., Fort Lee, N.J., 1981, pp. 72–110.
84. U.S. Patent 5,143,833 (September 1, 1992), R. Datta (to Michigan Biotechnology Institute, Lansing, Mich.).
85. N.P. Nghiem, B.H. Davison, B.E. Suttle, and G.R. Richardson, *Appl. Biochem. Biotechnol.*, **63–65**, 565–576 (1997).
86. U.S. Patent 5,143,834 (September 1, 1992), D. Glassner and R. Datta (to Michigan Biotechnology, Lansing, Mich.).
87. U.S. Patent 5,168,055 (December 1, 1992), R. Datta, D.A. Glassner, M.K. Jain and J.R. Vick Roy (to Michigan Biotechnology Institute, Lansing, Mich.).

88. U.S. Patent 5,521,075 (May 28, 1996), M.V. Guettler and M.K. Jain (to Michigan Biotechnology Institute, Lansing, Mich.).
89. S. Podkovyrov and J.G. Zeikus, *J. Gen. Microbiol.* **139**, 223–228 (1993).
90. L. Schöcke and P.J. Weimer, *Arch. Microbiol.* **167**, 289–294 (1997).
91. C.S. Millard, Y.P. Chao, J.C. Liao, and M.I. Donnelly, *Appl. Environ. Microbiol.* **62**, 1808–1810 (1996).
92. U.S. Patent 5,504,004 (April 2, 1996), M.V. Guettler, M.K. Jain, and B.K. Soni (to Michigan Biotechnology Institute, Lansing, Mich.).
93. M.J. van der Werf, M.V. Guettler, M.K. Jain, and J.G. Zeikus, *Arch. Microbiol.* **167**, 332–342 (1997).
94. R. Datta, *Biotech Bioeng. Symp.* **11**, 521–532 (1981).
95. E. Hsu and J. Ordal, *J. Bacteriol.* **102**, 369–376 (1970).
96. B.A. Annous, J.-S. Shieh, G.-J. Shen, M.K. Jain, and J.G. Zeikus, *Appl. Microbiol. Biotechnol.* **45**, 804–810 (1996).
97. R.M. Worden, A.J. Grethlein, J.G. Zeikus, and R. Datta, *App. Biochem. Biotech.* **20**, 687–698 (1989).
98. D. Michel-Savin, R. Marchal, and J.P. Vandecasteele, *Appl. Microbiol. Biotechnol.* **32**, 387–392 (1990).
99. D. Cowan, *Trends Biotechnol.* **14**, 177–178 (1996).
100. B.C. Saha, R. Lamed, and J.G. Zeikus, in N.P. Minton and D.J. Clarke eds., Clostridia, Plenum Publishing, New York, 1989, pp. 227–263.
101. C. Vieille, D.S. Burdette, and J.G. Zeikus, *Ann. Rev. Biotech.* **2**, 1–83 (1996).
102. H.H. Hyun and J.G. Zeikus, *Appl. Environ. Microbiol.* **49**, 1162–1167 (1985).
103. H.H. Hyun and J.G. Zeikus, *Appl. Environ. Microbiol.* **49**, 1168–1173 (1985).
104. B.E. Norman, *Starch / Starke* **34**, 340–346 (1982).
105. G. Dong, C. Vieille, A. Savchenko, and J.G. Zeikus, *Appl. Environ. Microbiol.* **63**, 3569–3576 (1997).
106. M.K. Jain and J.G. Zeikus, *Sys. Appl. Microbiol.* **10**, 134–141 (1988).
107. W.M. Mitchell and W.F. Harrington, in P.D. Boyer ed., *The Enzymes 3*, Academic, New York, 1971, pp. 699–719.
108. B. Schink and J.G. Zeikus, *FEMS Microbiol. Lett.* **17**, 295–298 (1983).
109. C. Vieille, J.M. Hess, R.M. Kelly, and J.G. Zeikus, *Appl. Environ. Microbiol.* **61**, 1867–1875 (1995).
110. G. Dong and J.G. Zeikus, *Enzyme Microbial Technol.* **21**, 335–340 (1997).
111. T.K. Ng, P.J. Weimer, and J.G. Zeikus, *Arch. Microbiol.* **114**, 1–7 (1977).
112. S.P. Mathupala and J.G. Zeikus, *Appl. Microbiol. Biotechnol.* **39**, 487–493 (1993).
113. G. Dong, C. Vieille, and J.G. Zeikus, *Appl. Environ. Microbiol.* **63**, 3577–3584 (1997).
114. Y.-E. Lee, S.E. Lowe, and J.G. Zeikus, *Appl. Environ. Microbiol.* **59**, 763–771 (1993).
115. K. Ma, F.T. Robb, and M.W.W. Adams, *Appl. Environ. Microbiol.* **60**, 562–568 (1994).
116. D.S. Burdette, C. Vieille, and J.G. Zeikus, *Biochem. J.* **316**, 115–122 (1996).
117. Y.E. Lee and J.G. Zeikus, *J. Gen. Microbiol.* **139**, 1235–1243 (1993).
118. S. Barik, S. Prieto, S.B. Harrison, E.C. Clausen, and J.L. Gaddy, *Appl. Biochem. Biotechnol.* **18**, 363–378 (1988).
119. A.J. Grethlein, R.M. Worden, M.K. Jain, and R. Datta, *J. Appl. Biochem.* **24/25**, 875–884, (1990).
120. A.J. Grethlein and M.K. Jain, *Trends Biotechnol.* **10**, 418–423 (1992).
121. R.W. Jack, J.R. Tagg, and B. Ray, *Microbiol. Rev.* **59**, 171–200 (1995).
122. Y. Hanaoka, *Hakkokogaku* **45**, 312–319 (1967).
123. R.J. Lamed and J.G. Zeikus, *Biochem. J.* **195**, 183–190 (1981).
124. R.J. Lamed, E. Keinan, and J.G. Zeikus, *Enzyme Microbiol. Technol.* **3**, 144–148 (1981).
125. R.W. Lovitt, E.W. James, D.B. Kell, and J.G. Morris, in G.W. Moody and P.B. Baker eds., *Bioreactors and Biotransformations*, Elsevier Applied Science, London, 1987, pp. 263–276.
126. W. Buckel, *FEMS Microbiol. Rev.* **88**, 211–232 (1992).
127. J. Thiele, M. Chartrain, and J.G. Zeikus, *Appl. Environ. Microbiol.* **54**, 10–19 (1988).
128. J.H. Thiele and J.G. Ziekus, in L.E. Erickson and D.Y.-C. Fung eds., *Handbook on Anaerobic Fermentations*, Marcel Dekker, New York, 1988, pp. 537–595.
129. S. Ghosh, in T. Noike, A. Tilche, and K. Hanaki eds., *Proceedings of the 8th International Conference on Anaerobic Digestion*, vol. 1, Sendai, Japan, 1997, pp. 9–16.
130. S. Rebac, J. Ruskova, S. Gerbens, J.B. van Lier, A.J.M. Stams, and G. Lettinga, *J. Ferm. Bioeng.* **80**, 499–506 (1995).
131. V. Gellens, J. Boelens, W. Verstraete, *Antonie van Leewenhoek* **67**, 79–89 (1995).
132. C.J.N. Buisman, in *Wider Application and Diffusion of Bioremediation Technologies: The Amsterdam '95 Workshop*, Organization for Economic Cooperation and Development, Paris, 1996, pp. 103–114.
133. P.N. Lens and W. Verstraete, in T.G. Villa and J. Abalde eds., *Profiles on Biotechnology*, Universidade de Santiago, Spain, 1992, pp. 333–356.
134. W. Verstraete and P. Vandevivere, in T. Noike, A. Tilche, and K. Hanaki eds., *Proceedings of the 8th International Conference on Anaerobic Digestion*, vol. 1, Sendai, Japan, 1997, pp. 67–74.
135. M. Strous, E. VanGerven, P. Zheng, J.G. Kuenen, and M.S.M. Jetten, *Water Res.* **31**, 1955–1962 (1997).
136. R.A. Wirtz and R.R. Dague, *Water Environ. Res.* **68**, 883–892 (1996).
137. J. Thaveesri, D. Daffonchio, B. Liessens, P. Vandermeren, and W. Verstraete, *Appl. Environ. Microbiol.* **61**, 3681–3686 (1995).
138. W.-M. Wu, M.K. Jain, and J.G. Zeikus, *Appl. Environ. Microbiol.* **62**, 2037–2044 (1996).
139. M.C. Veiga, M.K. Jain, W.-M. Wu, R.I. Hollingsworth, and J.G. Zeikus, *Appl. Environ. Microbiol.* **63**, 403–407 (1997).
140. S.R. Berchtold, S.L. Vanderloop, M.T. Suidan, and S.W. Maloney, *Water Environ. Res.* **67**, 1081–1091 (1995).
141. O. Meyer, R.I. Refae, J. Warrelmann, and H. von Reis, *Microb. Releases* **2**, 11–22 (1993).
142. W.H. Parker, E.R. Hall, and G.J. Farquhar, *Water Environ. Res.* **65**, 264–270 (1993).
143. W.-M. Wu, J. Nye, R.F. Hickey, M.K. Jain, and J.G. Zeikus, in R.E. Hinchee, A. Leeson and L. Semprini eds., *Bioremediation of Chlorinated Solvents*, Battelle Press, Columbus, Ohio, 1995, pp. 45–52.
144. X. Maym-Gatell, Y.T. Chien, J.M. Gossett, and S.H. Zinder, *Science* **276**, 1568–1571 (1997).
145. T. Gorontzy, J. Küver, and K.-H. Blotevogel, *J. Gen. Microbiol.* **139**, 1331–1336 (1993).

146. R. Boopathy and C.F. Kulpa, *Can. J. Microbiol.* **39**, 430–433 (1993).
147. A. Preuss, J. Fimpel, and G. Diekert, *Arch. Microbiol.* **159**, 345–353 (1993).
148. T.A. Khan, R. Bhadra, and J. Hughes, *J. Ind. Microbiol. Biotechnol.* **18**, 198–203 (1997).
149. S.B. Funk, D.J. Roberts, D.L. Crawford, and R.L. Crawford, *Appl. Environ. Microbiol.* **59**, 2171–2177 (1993).
150. R.H. Kaake, D.L. Crawford, and R.L. Crawford, in R.E. Hinchee, D.B. Richardson, F.B. Metting, Jr., and G. Sayles eds., *Applied Biotechnology for Site Remediation*, Lewis Publishers, Boca Raton, Fla., 1995, pp. 337–341.
151. J.P. Maree, A. Gerber, and E. Hill, *J. Water Pollut. Control Fed.* **59**, 1069–1074 (1987).
152. E. Colleran, S. Finnegan, and P. Lens, *Antonie van Leeuwenhoek*, **67**, 29–46 (1995).
153. P. Yspeert and C. Buisman, in Y. Ando and S. Wald eds., *Bioremediation: The Tokyo '94 Workshop*, Organization for Economic Cooperation and Development, Paris, 1995, pp. 335–344.
154. D.J. Cork and M.A. Cusanovich, in L.E. Murr, A.E. Torma, and J.A. Brierly eds., *Metallurgical Applications of Bacterial Leaching and Related Microbiological Phenomena, Proceedings of International Symposium*, Academic Press, New York, 1978, pp. 207–221.

ANIMAL CELLS USED IN MANUFACTURING

CLAUDIA BARDOUILLE
ECACC CAMR Porton Down
Salisbury, Wiltshire
United Kingdom

KEY WORDS

Animal cell line
Biopharmaceutical production
Posttranslational modifications
Protein
Recombinant
Three-dimensional culture

OUTLINE

INTRODUCTION

Animal cell lines are a resource in biotechnological processes. They are used increasingly in molecular biology developments, fermentation technology, the production of diverse health care products, and cell-based screening systems in toxicology and pharmacology. Large-scale technology began with the use of primary monkey kidney cells for the production of poliomyelitis vaccines in the 1950s. After intense debates on safety issues, human diploid cell strains (e.g., WI-38) were accepted for the production of mumps, measles, and rubella vaccines in the 1960s. The next step was taken in 1964 with the commercial production of foot-and-mouth disease virus (FMDV) from baby hamster kidney (BHK 21) cells for veterinary purposes. After the discovery of interferons and their clinical importance, the growing demand for these compounds led to the search for alternative, easily accessible sources. Interferon was the first licensed product derived from Namalwa cells, a human lymphoblastoid heteroploid cell line, in the late 1970s. Soon after this, hybridoma and monoclonal antibodies (mAb) took over the fast-developing market with their diverse applications in diagnostic and therapy.

Although strict safety standards have to be met, the value of animal cell lines is undeniable, and they are indispensable for the industrial production of eukaryotic proteins.

Since the acceptance of cell lines for the production of vaccines, cytokines, mAb, and therapeutic proteins or even the cells as product themselves, constant progress in a broad range of areas has contributed to the improvement of products. Research fields having major impact are analysis and adaptation of suitable cell lines, high-level expression systems, genetically engineered cell lines and products (e.g., protein modifications), scale-up methodology, bioreactor development, and downstream techniques (e.g., purification, quality control).

This chapter presents an overview of cell lines used for various applications and assays and gives an insight into different technologies and products.

ANIMAL CELL TYPES

The purification of therapeutic products from natural sources was and still is restricted because of limited supply, and contamination of known or unknown origin presents a serious danger (e.g., HIV in blood products). In addition, bacterial systems are not capable of the post-translational modifications eukaryotic proteins undergo. Animal cell technology has been able to fill this gap and provide proteins of high quality on an industrial scale. Cell lines used most frequently for different production systems are described in the following sections.

Chinese Hamster Ovary

The anchorage-dependent cell line from Chinese hamster ovary (CHO) was derived from a biopsy of an ovary of an adult Chinese hamster *(Cricetulus griseus)* in 1957. Various derivatives were developed, including nutritionally deficient mutants and transfected cells. The most popular variants are CHO-K1, a proline-requiring mutant, and CHO/dhFr$_{-}$, which is deficient in dihydrofolate reductase and therefore useful for the selection of transfected cells. Other variants have been adapted to grow in suspension and in serum-free media.

Advantages of CHO are their high capacity of amplification and expression of recombinant genes and proteins, the ability to synthesize proteins with similar glycosylation patterns as observed in native proteins, and the ability to grow in large-scale bioreactors. Therefore, they are very suitable for the expression of heterologous proteins by recombinant DNA technology. Table 1 presents examples of proteins produced by transfected CHO cells in a suspension or attached culture system. Because CHO cells are the most important substrate for the production of therapeutic proteins, new products are developed and licensed continuously (1). Product status, either in research or as industrial process, is indicated in the tables under Application.

Generally, CHO cells are kept in attached culture on tissue culture material or on microcarriers (see "Three-Dimensional Culture Systems"), but they can readily be adapted to grow as single cells or aggregates in suspension. This allows the usage of highly efficient bioreactors such as fluidized bed reactors (16). Because aggregates can lead to nonhomogenous product formation and affect system monitoring, single-cell suspension cultures are often preferred. To achieve this status, media formulations have been developed maintaining single-cell suspension during the production process (17). Cell lines that have been adapted to grow in suspension do not maintain all characteristics seen in attached cultivation. In a study by Brand et al. (18), it was shown that specific production rates of cells in attached cultures are not necessarily connected to similar rates after adaptation to suspension. This observation illustrates the difficulties of scale-up procedures.

Serum-containing media impede the purification of pharmaceutical products to a great extent, and therefore efforts are directed to establish cultures in serum-reduced, serum-free, or even protein-free media (19). To facilitate the scale-up of transfected cells, CHO cells preadapted to serum-free and suspension culture have been developed by Sinacore et al. (13). Other aspects important for efficient production are the study and regulation of inhibitory substances such as ammonia (20) that can exert an influence on the glycosylation pattern (21).

A reproducible product consistency is of course a major aim in all animal cell systems used for the manufacture of proteins to be administered as therapeutic agents. The effect of genetic and environmental factors on purity and posttranslational modifications are discussed in "Posttranslational Modifications of Recombinant Proteins in Animal Cell Lines."

Baby Hamster Kidney Cells

The attached BHK 21 cell line was derived from five unsexed 1-day-old Syrian hamster (*Mesocricetus auratus*) kidneys in 1961. BHK 21 clone 13 was obtained after single-cell isolation and is now widely used, very often only addressed as BHK or BHK 21. Several variants have been developed that are being adapted to suspension culture or suitable for genetic engineering techniques.

Because of their susceptibility to various viruses, they have been extensively used for the production of vaccines. Since the 1960s, BHK cells have played the major role in the production of FMDV vaccine in industrial-scale fermenters (with a capacity of several thousand liters) for veterinary applications. The experience and know-how gained with this manufacturing system led to the increased use of animal cells and the improvement of process parameters. Like CHO, BHK cells show spontaneous aggregation that can be advantageous for cell retention and recycling systems (22). Efforts have also been directed toward high-cell-density culture and serum- or protein-free media formulations (23). The first product approved by the Food and Drug Administration (FDA) from BHK cells cultured in perfusion reactors was the recombinant human factor VIII for the therapy of coagulation disorders. Since then, many other proteins have been produced on laboratory- and industrial-scales using this versatile cell line, ranging from human erythropoietin and antithrombin III to enzymes and a range of vaccines for veterinary or human health applications (Table 2).

Hybridoma and Myeloma Cell Lines

Since Köhler and Milstein (30) discovered the technique to produce immortalized, antibody-secreting cells—hybridomas—the development in this area has had a major impact on the production and application of the secreted proteins (mAb). Therefore, it is not surprising that this tool occupies an important and still growing place in industrial biotechnology because the range of applications is constantly expanding. The advantages of hybridomas and myelomas (i.e., their natural secretory ability and, as lymphocytes, growth in suspension) renders these cell lines ideal for exploitation in biotechnical processes.

To optimize the production of mAb, efforts have been undertaken to develop efficient and economic media, including serum- and protein-free formulations (31), generating high cell densities by using perfusion reactors instead of batch processes (32), and selection of suitable clones with higher resistance against shear stress in agitated bioreactors (14). In addition to the ready-to-use serum-free media that are already commercially available, cultivation of hybridomas is constantly improved by research activities related to specific needs. Examples are cholesterol-free media or the use of genetically engineered cell lines that are glutamine-independent, which reduces the accumulation of inhibitory substances such as ammonia (33).

The ability to produce recombinant mAb contributed to the development of chimeric mAb containing several therapeutic advantages. It is now possible to construct heterohybridoma cell lines where different species can contrib-

Table 1. Recombinant Proteins Expressed in CHO cells

Product	Application[a]	Attached/suspension	Serum-free	Ref.
γ-Glutamyl transferase (hu)	Reference enzyme for clinical tests (diagnosis of hepatic cancer, alcohol abuse) (RA)	Attached	No	2
Humanized IgG1 binding CD52 (CAMPATH-IH®)	Transplant rejection, rheumatoid arthritis, non-Hodgkin's lymphoma (IP)	Suspension	Yes	3
Chimeric human/murine mAb reacting with tumor-associated glycoprotein TAG72	Tumor imaging and therapy (IP)	Suspension	Yes	4
Thyroid peroxidase (hu)	Testing of antibodies in autoimmune thyroid diseases (RA)	Attached	Yes	5
Interferon-γ	Immunostimulation; tumor therapy; antiviral therapeutic agent (IP)	Suspension	No	6
HIV 1 gp120	HIV research (RA)	Attached	No	7
Tissue plasminogen activator (t-PA) (hu); trade name Activase	Thrombolytic agent, acute myocardial infarction (IP)	Suspension	Yes	8,9
Antithrombin III (hu)	Antithrombin III deficiencies (IP)	n.k.	Yes	10
von Willebrand's factor (hu)	Hemostatic factor (IP)	Attached	Yes	11
Erythropoietin receptor (mu)	Study of physiological function detection assays (RA)	Attached	No	12
Erythropoietin; trade name Epogen/Procrit	Myocardial infarction (IP)	n.k.	n.k.	1
Macrophage colony-stimulating factor (hu)	Cytokine study (IP)	Suspension	Yes	13
Factor IX (hu)	Blood clotting factor (IP)	Suspension	Yes	13
Nerve growth factor (hu)	Bioactivity study (IP)	Suspension	Yes	14
Peptidylglycine α-amidating enzyme (rat)	Catalytic reagent for calcitonin production (IP)	Attached/Suspension	Yes	15

Note: hu, human; n.k., not known; mu, murine.
[a]RA, research application; IP, industrial process.

ute to the variant (e.g., mouse) and constant (e.g., human) region of the secreted mAb. These antibodies are less immunogenic and more effective in therapeutical applications. The constant region can also be replaced with enzymes or toxins, which is useful in tumor targeting. Bispecific mAb can react with two antigens (e.g., tumor markers and specific toxic drugs) and thereby bind and destroy specific cells only (34). Details of techniques for the construction of these new generations of hybridomas, bioreactor systems, purification, and applications are reviewed in James (35), Larrick and Balint (36), and Merten (37).

Because of the rapid development of new cell lines in this area, only a few examples were selected for inclusion in Table 3. Table 4 overviews the various applications for mAb and the potential for future developments in animal cell biotechnology.

Myeloma cell lines such as J558 L, NS0, SP2/0, or P3X63Ag.653 are not only malignant fusion partners used for the generation of hybridomas, but are also suitable for the production of recombinant proteins (40,41). Once again, it can be attributed to the tremendous new techniques of genetic engineering that opened the door for myeloma-based expression systems leading to a new branch in animal cell technology. These naturally secretory cells can be grown to high cell densities and are robust and relatively easy to transfect (42). With the use of recombinant myeloma cell lines, raising of antigen-specific B cells as prerequisite for the production of hybridomas and finally mAb can be avoided, and time- as well as cost-consuming optimization procedures can be limited.

Insect Cell Lines

In addition to mammalian cells, insect cell lines for the production of recombinant proteins have been established in the past decade. The techniques used most frequently can be distinguished by the application of different expression systems: the baculovirus expression vector system (BEVS) and the *Drosophila* metallothionein (Mt) promotor system.

Baculovirus Expression Systems. Baculoviruses replicate exclusively in invertebrates and are often highly species-specific. The BEVS most commonly used for the expression of heterologous proteins is the *Autographa californica* (alfalfa looper) nuclear polyhedrosis virus (AcNPV), where the gene of choice is inserted into the baculovirus genome under the control of a strong late viral transcription promoter. The recombinant virus is then used to infect lepidopteran insect cells, mainly from *Spodoptera frugiperda* (fall armyworm). These cells express the recombinant protein during the late stages of the infection process. Several human, animal, and plant proteins have been expressed using this system (43).

The insect cell line used most frequently in combination with BEVS is SF 9; the parental cell line, SF 21, was obtained from pupal ovarian tissue of *S. frugiperda*. SF 9 cells are highly susceptible to infection with various baculoviruses. A main advantage of the baculovirus/insect cell system is its potential to yield milligram quantities of recombinant product. To obtain mammalian cell lines producing reasonable amounts of heterologous proteins, it is

Table 2. Recombinant Proteins Expressed in BHK Cells

Product	Application[a]	Attached/suspension	Serum-free	Ref.
FMDV	Vaccine (cattle and pigs)	Suspension	No	24
Rabies	Vaccine	Suspension	No	25
Antithrombin III (hu)	Antithrombin III deficiencies	Suspension/attached	Yes	26
Erythropoietin (hu)	Regulation of red blood cell formation	Attached	No	27
Interleukin-2 (hu)	Immunodeficiencies, leukemia, cancer	Attached	Protein-free	28
Transferrin (hu)	Iron-transporting protein	Attached	No	29
Placental-secreted alkaline phosphatase	Aggregation studies	Suspension	No	22

Note: hu, human.
[a]All industrial processes.

Table 3. Hybridoma Cell Lines

Cell Line	MAb Specificity	Application[a]	Ref.
Orthoclone OKT®3 (mouse)	Anti-CD3	Suppression of rejection of transplanted kidneys	1
N12-16.63 (mouse-hu-hu, heterohybridoma)	Anti-tetanus toxoid	Tetanus therapy, diagnosis	14
HBW-4.16 (hu-hu; heterohybridoma)	Anti-hepatitis B virus	Hepatitis B	14
CB-STL-1 (hu-hu-mouse; heterohybridoma)	Anti-α-hemolysin	Sepsis, septic shock	38
2F5 (heterohybridoma)	HIV-1 neutralizing	HIV-1 therapy	39

[a]All industrial processes.

Table 4. Applications for mAb

Diagnostic
- Clinical: virus, bacteria, parasite detection, tumor marker, allergy
- Self-testing: pregnancy and fertility tests

Therapeutic (human and veterinary)
- Vaccines
- Growth factors, hormones
- Disease treatment, e.g., autoimmune disorders, cancer, AIDS, infections, toxifications

Industrial purifications
Quality control laboratories

not only necessary to invest valuable time on transfection, selection, and optimization, but the resulting cell line might also show a basic constitutive expression activity and is constantly under selective pressure.

***Drosophila* Metallothionein Expression System.** In the *Drosophila* Mt. system, the inducible Mt. promotor is used to control the expression of recombinant proteins. This system offers the benefit of a well-known host and genomic structure with a tightly regulated promotor. In contrast to the viral system BEVS, where the production peak is found when the cells already die from the viral infection, *Drosophila* cells and the Mt. expression system can be used continuously.

Despite the different expression systems for insect cell lines, development of efficient fermentation techniques and process optimization are similar to the ones already described for mammalian cell types. High-cell-density reactors to achieve a maximum yield of desired product have been developed by various groups. Cavegn et al. established a system monitoring the production of insulin-receptor tyrosine kinase domain and the soluble part of endothelial leukocyte adhesion molecule as model proteins (44) and achieved up to 3×10^7 cells/mL in a serum-free perfusion bioreactor. Other developments are immobilized systems, such as cultivation on porous microcarrier (45), or the use of simulated microgravity to reduce hydrodynamic forces, as described by O'Connor et al. applying the rotating-wall vessel (46).

Examples of heterologous proteins expressed in insect cell lines can be found in Table 5. Investigations at the laboratory level indicate interesting areas still to be exploited for biotechnological purposes. The high level of expression of measles virus protein in SF 9 facilitates research into structure, function, and immunogenicity (53). Expression of recombinant bovine β-lactoglobulin for the processing of milk proteins to inhibit allergic reactions is another potential application (54). Because of the ability to express two or more proteins simultaneously in the baculovirus system, analysis of multisubunit formation (e.g., for bluetongue viruslike particles) is possible (49).

Research activities are also concentrating on other insect host cells to improve the production process and, as an important requirement for clinically administered compounds, obtain a correctly glycosylated and biological active protein. SF 9 cells display only a very simple glycosylation capacity, whereas *Estigmena acrea* cells are able to produce complex glycoforms very similar to their authentic counterparts (55). Other insect cell lines investigated for the manufacture of recombinant proteins are Schneider 2 *(Drosophila melanogaster)* and Tr 5 *(Trichoplusia ni)*. Apart from using insect cell lines to facilitate production, they can also be applied in screening or cytotoxicity tests.

Other Cell Lines

Although the cell lines described in the previous section probably contribute to the majority of industrial processes,

Table 5. Recombinant Proteins Expressed in Insect Cell Lines

Cell Line	Product	Application[a]	Expression system	Ref.
SF 9	Soluble human TNF receptor	Biochemical studies, structure-function analysis	BEVS	47
SF 9	Soluble IFN-γ receptor	Structure determination, identification of antagonists, rational drug design	BEVS	48
SF 9	Bluetongue viruslike particles	Vaccine in sheep and cattle	BEVS	49
SF 9/Schneider 2	VCAM	Model system	BEVS/Mt	50
SF 9/Tn 5	Osteoblast-specific factor 2	Model system	BEVS	51
SF 9	Prostate specific antigen	Tumor marker	BEVS	52

Note: TNF, tumor necrosis factor; IFN, interferon; VCAM, vascular cell adhesion molecule.
[a]All research applications

various other cell types are used as substrate for the production of biologicals.

On the one hand, this might be of historical reason. Certain cell lines are well established; the production of vaccines and systems for manufacture have been optimized already. On the other hand, cell lines synthesizing proteins naturally without transformation procedures have clear advantages considering the production process (where a selective pressure to maintain expression is hence not necessary) and acceptance of the final product. Other cell lines have been shown to possess valuable properties such as glycosylation characteristics. This section describes some of these cell lines and their biotechnological usage; although not great in number, they are still significant in the pharmaceutical industry.

VERO cells, derived from the kidney of a normal adult African green monkey, are susceptible to a wide range of viruses, including measles, polioviruses, and rubella. These properties led to the early establishment of vaccine production in VERO cells (for polio) (Table 6). This now rather classical cell line is still used as capable substrate for virus production. VERO cells replaced primary monkey kidney cells for the production of oral polio vaccine and are now cultured on microcarriers in large-scale bioreactors (68). They were also successfully applied for the study of infection with Aujeszky virus in the VERAX system, a perfused fluidized bed with macroporous collagen carriers as microspheres to evaluate the possibility of vaccine production in a high-cell-density system (67). Baijot et al. (65) established the manufacture of an Aujeszky vaccine using swine testicular cells (NLST) on microcarriers. Although VERO cells are mostly grown as attached cells, they can be adapted to grow in suspension in serum-free medium (70). The inducible expression of HIV-1 viruslike particles in transfected VERO and COS 7 cells (a SV 40 transformed monkey kidney cell line) was described by Haynes et al. (69). This important development offers an exciting alternative to the use of live virus vectors for the production and evaluation of AIDS vaccines based on noninfectious particles.

The human lymphoblastoid cell line Namalwa has been used as substrate for biologicals since the 1970s. The secretion of INF-α being naturally induced in Namalwa cells by treatment with sodium butyrate and addition of sendai virus is well established for industrial manufacturing (61). Process optimization and increase of productivity are still continuing and achieved by various modifications, such as treatment with tetramethyl urea (62) or sequential induction with sendai virus (63). Namalwa KJM 1 was derived from Namalwa and adapted to grow in serum-free medium. Miyaji et al. (71) describe the efficient expression of recombinant human INF-β in these cells in high-cell-density perfusion culture. The comparison of recombinant pro-urokinase expression in Namalwa KJM 1 and CHO cells revealed a partly cleaved and thereby inactive form secreted in CHO cells due to proteases present in the supernatant (64). Because the protein secreted by Namalwa KJM 1 showed a single-chain form, this indicates that

Table 6. Examples of Cell lines and Their Use in Biotechnological Processes

Cell Line	Product	Application[a]	Ref.
293 (human embryonal kidney, Ad5 transformed)	Acetylcholinesterase	Structural and functional analysis (RA)	56
293S (Adapted to suspension)	Tyrosine phosphatase 1C	Model system for scale-up (RA)	57
HuCCl (Human colon adenocarcinoma)	Carcinoembryonic antigen	Tumor-associated marker (IP)	58
HUH-6 (Human hepatoblastoma)	Fibronectin	Attachment factor (RA)	59
MDBK (Madin-Darby bovine kidney)	BVD virus (bovine viral diarrhea)	Vaccine (IP)	60
Namalwa (human lymphoblastoid)	IFN-α	Pharmaceutical agent (IP)	61,62,63
Namalwa KJM 1	Pro-urokinase	Model system (RA)	64
NLST (swine testicular)	Aujeszky virus	Vaccine (IP)	65
RPMI 8866 (human B cells)	CD 23 fragment	Structural and functional analysis (RA)	66
VERO (monkey kidney)	Aujeszky virus	Vaccine (IP)	67
VERO	Oral polio virus	Vaccine (IP)	68
VERO and COS-7 (monkey kidney)	HIV-1 viruslike particles	Vaccine (IP)	69

[a]RA, research applications; IP, industrial process.

these cells are more suitable for the production of protease-susceptible heterologous proteins.

The carcinoembryonic antigen (CEA) is a high molecular weight glycoprotein identified as tumor-associated marker. For the purpose of early clinical diagnosis, therapy response, and recurrences, it is important to be able to detect CEA in monitored patients. Several cell lines synthesize CEA and have been used in various large-scale production systems (58). mAb raised against this antigen can improve clinical detection assays and cancer specificity. Other examples of cell lines used for the production of naturally secreted proteins are HUH 6 (fibronectin) and RPMI 8866 (CD 23) (see Table 6 for details).

A list of all cell lines currently used in biotechnology is clearly beyond the scope of this article. The constant improvements in this field, such as transformation of cell types without losing specific properties or differentiation characteristics (e.g., hepatocytes, neurons, keratinocytes) and optimization of cell lines for fermentation processes, will certainly lead to a variety of options for animal cell biotechnologists.

THREE-DIMENSIONAL CULTURE SYSTEMS

Animal cell lines have to be classified according to their ability to grow in suspension or be anchorage-dependent. For the large-scale production of biologicals, suspension cells are clearly the substrate of choice. With the development of microcarriers initiated by van Wezel 1967 (72), it is now possible to culture attached cells in bioreactors very similar to suspension cells. Growth of animal cells on microcarriers or immobilized systems facilitates production processes and allows high yield of cells, heterologous proteins, or viruses (73).

Microcarrier techniques are especially important for the handling of primary cell types or cell strains because these are, with only a few exceptions, mostly anchorage-dependent. The microcarrier technology offers several advantages: providing a high surface-to-volume ratio; allowing manipulation during culture without interrupting the fermentation process, and facilitating scale-up possibilities (74). A further advantage is the stabilization of cells in culture, which applies to both attached and suspension cells, thereby permitting prolonged culture periods. Macroporous microcarriers have proved to be suitable for the protection of cells from shear forces and have been applied successfully for the immobilization of adherent and suspension cells (75–77).

The ability of some cell lines to form aggregates is another feature used for increased productivity in perfusion bioreactors (16). However, aggregation has to be evaluated in a case-to-case study because it can result in inhomogeneous expression of proteins and might impair monitoring (17).

Finally, three-dimensional culture conditions can simulate the natural environment of cells, including cell–cell interaction and differentiation processes that are not always possible to achieve in two-dimensional systems. An example is the establishment of a bioreactor that provides high, almost tissuelike cell densities and thereby simulates cell–cell interactions (78).

POSTTRANSLATIONAL MODIFICATIONS OF RECOMBINANT PROTEINS IN ANIMAL CELL LINES

Recombinant DNA technology opened the field for the production of an unlimited variety of clinically relevant therapeutics in various host cell lines. The majority of proteins applied undergo more or less extensive posttranslational modifications in their natural counterparts, including mainly glycosylation, but also phosporylation, carboxylation, and signal-peptide processing. These modifications affect the biological activity of the heterologous protein by being involved in solubility, pharmacokinetics, antigeneicity, circulating half-life, secretion, protein folding, oligomer assembly, and susceptibility to proteolytic attack (79,80).

Choosing a host cell line able to fulfill these criteria is, therefore, important. Regulatory authorities such as the FDA demand a comprehensive analysis of carbohydrate structures and consistency in the production process. Because prokaryotes are not capable of performing posttranslational modifications observed in eukaryotic cells, they cannot be utilized for the production of recombinant proteins, where bioactivity relies heavily on correct glycosylation, for example. These unique advantages of animal cell technology inspired research directed toward overcoming process or regulatory difficulties.

Research has increasingly revealed already substantial differences in the glycosylation of recombinant proteins using several host cell lines or even transgenic animals. Variations can also stem from different environmental factors, such as media or bioreactor configuration, or the internal cell status at the time of production (80).

A few examples illustrate the importance of choosing a host system able to produce heterologous proteins with appropriate characteristics for clinical applications. Although CHO and BHK lack certain functional glycosyltransferases found in human cells (80), they still represent the most favorable host cell lines for large-scale manufacture of protein-based pharmaceuticals such as erythropoietin or tissue-type plasminogen activator (t-PA). Studies on product consistency revealed an independence of glycosylation pattern by applying varied process parameters (8) and showed that glycosylation was identical to natural sources (81). Constant improvement in the methods available for the analysis of carbohydrate structures allow a detailed pattern identification (79) and have been used to detect species-specific oligosaccharide variations of recombinant human IFN-γ produced in CHO and SF cells and transgenic mice (82).

The glycosylation status of recombinant proteins synthesized in baculovirus-infected insect cells can vary quite considerably, as shown by Ogonah et al. (55). *Estigmena* cells are capable of producing complex oligosaccharides, whereas the still more frequently used SF 9 cell line is restricted to the performance of only simple glycosylation reactions.

These findings demonstrate the importance of investigating and identifying an animal cell system best suited for the production of therapeutics. Even shortcomings, such as the lack of glycosyltransferases in CHO cells, can be overcome by transfection of appropriate enzymes (80), indicating the potential for future processes.

TRENDS

Several examples given in the preceding sections already point toward activities undertaken to optimize animal cell technology and bioproducts of interest.

Cell lines other than the well-established hamster lines CHO and BHK present valuable alternatives, for example, the efficient production of recombinant proteins. Concomitant with an increasing understanding of media formulations, bioreactor design, and processes, efficiency and quality are still subject to improvement. Tailored cell lines can be selected by appropriate cloning strategies, using either naturally occurring variants or conferring specific characteristics by recombinant DNA methods (33).

The development of vectors with a predictable integration site for foreign DNA into the host genomic structure, including efficient expression systems, can overcome low-productivity problems. An example for enhanced productivity is the oncogene-activated system applied by Ternya et al. (83), where cells are cotransfected with the *ras* oncogene.

Recombinant DNA technology gives new directions for the design of therapeutics with improved or even novel characteristics compared to their natural counterparts (e.g., pharmacokinetic properties). Ahern et al. (84) investigated recombinant variants of tPA produced by site-directed mutagenesis and identified proteins with increased fibrinolytic activity in vitro and decreased clearance in vivo.

New technologies, such as the encapsulation of cells, genetically engineered or as primary isolates, and transplantation into patients with various disorders or illnesses, could provide a cellular therapy tool for long-term treatment or even cure (85).

Other approaches attempt to simulate the three-dimensional microenvironment for constructing organotypical systems, such as artificial liver support devices (86) or drug metabolism studies (87). Investigations into the involvement of cytokines, extracellular matrixes, and cell–cell interactions are required to optimize organ models and lead to successful applications.

CONCLUSION

Animal cell lines are an established tool in biotechnology and will be used extensively in a variety of emerging areas, such as tissue engineering, organ replacement, or drug discovery. A vast amount of knowledge and experience has accumulated concerning the cell lines described in this article, and research into improved industrial large-scale use is still on going. Safety concerns using continuous cell lines with a tumorigenic potential are addressed frequently. As a result of numerous efforts made in response to recommendations by regulatory bodies and the World Health Organization, cell lines and derived products are now a prosperous factor in biotechnology.

Selecting a cell line for the manufacture of pharmaceuticals is dependent on the protein that has to be produced. Post-translational modifications, secretion, and purification are aspects that have to be considered, but species (e.g., human or rodent origin) and regulatory issues are equally important. The examples described in this article give some indications on cell line suitability, but research as well as industrial production is improving rapidly in this area. New sources (i.e., cell lines and methods) are being established continuously and should therefore be closely studied before choosing an appropriate cell substrate.

BIBLIOGRAPHY

1. A.S. Lubiniecki and S.A. Vargo, in A.S. Lubiniecki and S.A. Vargo eds., *Regulatory Practice for Biopharmaceutical Production*, Wiley-Liss, New York, 1994, pp. 1–12.
2. J.L. Goergen, I. Chevalot, A. Visvikis, T. Oster, K. Hess, J.M. Engasser, and A. Marc, in R.E. Spier, J.B. Griffiths, and C. MacDonald eds., *Animal Cell Technology: Developments, Processes and Products*, European Society for Animal Cell Technology, Butterworth-Heinemann, Oxford, U.K., 1992, pp. 593–595.
3. M.J. Keen and N.T. Rapson, *Cytotechnology* **17**, 153–163 (1995).
4. R.P. Field, H. Brand, G.L. Renner, H.A. Robertson, and R. Boraston, in R.E. Spier, J.B. Griffiths, and B. Meignier eds., *Production of Biologicals from Animal Cells in Culture*, European Society for Animal Cell Technology, Butterworth-Heinemann, Oxford, U.K., 1991, pp. 742–744.
5. T. Ii, M. Murakami, M. Mizuguchi, K. Matsumoto, and K. Onodera, in E.C. Beuvery, J.B. Griffiths, and W.P. Zeijlemaker eds., *Animal Cell Technology: Developments towards the 21st Century*, Proc. of the Veldhoven Mtg., Kluwer, Dordrecht, The Netherlands, 1995, pp. 385–389.
6. V. Leelavatcharamas, A.N. Emery, and M. Al-Rubeai, *Cytotechnology* **15**, 65–71 (1994).
7. S.J. Froud, G.J. Clements, M.E. Doyle, E.L.V. Harris, C. Lloyd, P. Murray, A. Preneta, P.E. Stephens, S. Thompson, and G.T. Yarranton, in R.E. Spier, J.B. Griffiths, and B. Meignier eds., *Production of Biologicals from Animal Cells in Culture*, European Society for Animal Cell Technology, Butterworth-Heinemann, Oxford, U.K., 1991, pp. 110–115.
8. K. Kopp, W. Noé, M. Schlüter, F. Walz, and R. Werner, in R.E. Spier, J.B. Griffiths, and W. Berthold eds., *Animal Cell Technology: Products of Today, Prospects for Tomorrow*, European Society for Animal Cell Technology, Butterworth-Heinemann, Oxford, U.K., 1994, pp. 661–666.
9. D. Beebe and G. Murano, in A.S. Lubinicck i and S.A. Vargo eds., *Regulatory Practice for Biopharmaceutical Production*, Wiley-Liss, New York, 1994, pp. 281–293.
10. G. Zettlmeissl, H.S. Conradt, M. Nimtz, and H.E. Karges, *J. Biol. Chem.* **264**, 21153– 21159 (1989).
11. G. Mignot, T. Faure, V. Ganne, B. Arbeille, A. Pavirani, and J.L. Romet-Lemonne, *Cytotechnology* **4**, 163–171 (1990).
12. M. Nagao, S. Masuda, S. Abe, M. Ueda, and R. Sasaki, in S. Kaminogawa, A. Ametani, and S. Hachimura eds., *Animal Cell Technology: Basic and Applied Aspects*, Proc. of the Fifth Int. Mtg. of the Japanese Assoc. for Animal Cell Technology, Kluwer, Dordrecht, The Netherlands, 1993, pp. 71–77.
13. M.S. Sinacore, T.S. Charlebois, S. Harrison, S. Brennan, T. Richards, M. Hamilton, S. Scott, S. Brodeur, P. Oakes, M.

Leonard, M. Switzer, A. Anagnostopoulos, B. Foster, A. Harris, M. Jankowski, M. Bond, S. Martin, and S.R. Adamson, *Biotechnol. Bioeng.* **52**, 518–528 (1996).

14. K. Kitano, in R.E. Spier, J.B. Griffiths, and W. Berthold eds., *Animal Cell Technology: Products of Today, Prospects for Tomorrow*, European Society for Animal Cell Technology, Butterworth-Heinemann, Oxford, U.K., 1994, pp. 35–42.
15. D.E. Matthews, K.E. Piparo, V.H. Burkett, and C.R. Pray, in R.E. Spier, J.B. Griffiths, and W. Berthold eds., *Animal Cell Technology: Products of Today, Prospects for Tomorrow*, European Society for Animal Cell Technology, Butterworth-Heinemann, Oxford, U.K., 1994, pp. 315–319.
16. M. Reiter, G. Blüml, T. Gaida, N. Zach, C. Schmatz, N. Borth, O. Hohenwarter, and H. Katinger, in R.E. Spier, J.B. Griffiths, and C. MacDonald eds., *Animal Cell Technology: Developments, Processes and Products*, European Society for Animal Cell Technology, Butterworth-Heinemann, Oxford, U.K., 1992, pp. 421–423.
17. R. Boraston, C. Marshall, P. Norman, G. Renner, and J. Warner, in R.E. Spier, J.B. Griffiths, and C. MacDonald eds., *Animal Cell Technology: Developments, Processes and Products*, European Society for Animal Cell Technology, Butterworth-Heinemann, Oxford, U.K., 1992, pp. 424–426.
18. H.N. Brand, S.J. Froud, H.K. Metcalfe, A.O. Onadipe, A. Shaw, and A.J. Westlake, in R.E. Spier, J.B. Griffiths, and W. Berthold eds., *Animal Cell Technology: Products of Today, Prospects for Tomorrow*, European Society for Animal Cell Technology, Butterworth-Heinemann, Oxford, U.K., 1994, pp. 55–59.
19. M. Zang, H. Trautmann, C. Gandor, F. Messi, F. Asselbergs, C. Leist, A. Fiechter, and J. Reiser, *Biotechnology* **13**, 389–392 (1995).
20. C. Dyring, H.A. Hansen, and C. Emborg, in R.E. Spier, J.B. Griffiths, and W. Berthold eds., *Animal Cell Technology: Products of Today, Prospects for Tomorrow*, European Society for Animal Cell Technology, Butterworth-Heinemann, Oxford, U.K., 1994, pp. 155–157.
21. E.T. Papoutsakis, D.I.H. Linzer, and M.C. Borys, in R.E. Spier, J.B. Griffiths, and W. Berthold eds., *Animal Cell Technology: Products of Today, Prospects for Tomorrow*, European Society for Animal Cell Technology, Butterworth-Heinemann, Oxford, 1994, U.K., pp. 658–660.
22. J.L. Moreira, P.M. Alves, J.G. Aunins, and M.J.T. Carrondo, in R.E. Spier, J.B. Griffiths, and C. MacDonald eds., *Animal Cell Technology: Developments, Processes and Products*, European Society for Animal Cell Technology, Butterworth-Heinemann, Oxford, U.K., 1992, pp. 411–413.
23. M. Lucki-Lange and R. Wagner, in R.E. Spier, J.B. Griffiths, and B. Meignier eds., *Production of Biologicals from Animal Cells in Culture*, European Society for Animal Cell Technology, Butterworth-Heinemann, Oxford, U.K., 1991, pp. 180–185.
24. P.J. Radlett, *Adv. Biochem. Eng. Biotechnol.* **34**, 129–145 (1987).
25. T.W.F. Pay, A. Boge, F.J.R.R. Menard, and P.J. Radlett, *Develop. Biol. Standard.* **60**, 171–174 (1985).
26. S.-Y. Li, B. Röder, and M. Wirth, in R.E. Spier, J.B. Griffiths, and B. Meignier eds., *Production of Biologicals from Animal Cells in Culture*, European Society for Animal Cell Technology, Butterworth-Heinemann, Oxford, U.K., 1991, pp. 445–450.
27. Y. Shirai, K. Hashimoto, H. Kawahara, R. Sasaki, K. Hitomi, and H. Chiba, *Cytotechnology* **2**, 141–145 (1989).
28. R. Kratje and R. Wagner, in R.E. Spier, J.B. Griffiths, and C. MacDonald eds., *Animal Cell Technology: Developments, Processes and Products*, European Society for Animal Cell Technology, Butterworth-Heinemann, Oxford, U.K., 1992, pp. 488–493.
29. D.W. Lee, D. Gregory, D.J. Haddow, J.M. Piret, and D.G. Kilburn, in R.E. Spier, J.B. Griffiths, and C. MacDonald eds., *Animal Cell Technology: Developments, Processes and Products*, European Society for Animal Cell Technology, Butterworth-Heinemann, Oxford, U.K., 1992, pp. 480–486.
30. G. Köhler and C. Milstein, *Nature* **256**, 495–497 (1975).
31. E.-J. Schlaeger, T. Klier, and K. Christensen, R.E. Spier, J.B. Griffiths, and W. Berthold eds., *Animal Cell Technology: Products of Today, Prospects for Tomorrow*, European Society for Animal Cell Technology, Butterworth-Heinemann, Oxford, U.K., 1994, pp. 130–133.
32. H. Pinton, A. Lourenco da Silva, J.L. Goergen, A. Marc, J.M. Engasser, J.N. Rabaud, and G. Pierry, in R.E. Spier, J.B. Griffiths, and W. Berthold eds., *Animal Cell Technology: Products of Today, Prospects for Tomorrow*, European Society for Animal Cell Technology, Butterworth-Heinemann, Oxford, U.K., 1994, pp. 470–475.
33. J.R. Birch, R.C. Boraston, H. Metcalfe, M.E. Brown, C.R. Bebbington, and R.P. Field, *Cytotechnology* **15**, 11–16 (1994).
34. S.A. Clark, *Animal Cell Biotechnol.* **4**, 318–341 (1990).
35. K. James, *Animal Cell Biotechnol.* **4**, 206–255 (1990).
36. J.W. Larrick and R. Balint, in S. Kaminogawa, A. Ametani, and S. Hachimura eds., *Animal Cell Technology: Basic and Applied Aspects*, Proc. of the Fifth Int. Mtg. of the Japanese Association for Animal Cell Technology, Kluwer, Dordrecht, The Netherlands, 1993, pp. 625–639.
37. O.-W. Merten, *Animal Cell Biotechnol.* **4**, 258–315 (1990).
38. R. Waterstrat, R.W. Glaser, H. Tanzmann, C. Riese, and U. Marx, in E.C. Beuvery, J.B. Griffiths, and W.P. Zeijlemaker eds., *Animal Cell Technology: Developments towards the 21st century*, Proc. of the Veldhoven Mtg., Kluwer, Dordrecht, The Netherlands, 1995, pp. 647–651.
39. M. Reiter, A. Buchacher, G. Blüml, N. Zach, W. Steinfellner, C. Schmatz, T. Gaida, A. Assadian, and H. Katinger, in R.E. Spier, J.B. Griffiths, and W. Berthold eds., *Animal Cell Technology: Products of Today, Prospects for Tomorrow*, European Society for Animal Cell Technology, Butterworth-Heinemann, Oxford, U.K., 1994, pp. 333–335.
40. C.A. Mitchell and P.P. Gray, in R.E. Spier, J.B. Griffiths, and C. MacDonald eds., *Animal Cell Technology: Developments, Processes and Products*, European Society for Animal Cell Technology, Butterworth-Heinemann, Oxford, U.K., 1992, pp. 583–585.
41. C.W. Buser, R. Beaudet, N. Soohoo, and G.G. Pugh, in R.E. Spier, J.B. Griffiths, and W. Berthold eds., *Animal Cell Technology: Products of Today Prospects for Tomorrow*, European Society for Animal Cell Technology, Butterworth-Heinemann, Oxford, U.K., 1994, pp. 121–123.
42. A. Traunecker, F. Oliveri, and K. Karjalainen, *TibTech* **9**, 109–113 (1991).
43. V.A. Luckow and M.D. Summers, Biotechnology **6**, 47–55 (1988).
44. C. Cavegn, H.D. Blasey, M.A. Payton, B. Allet, J. Li, and A.R. Bernard, in R.E. Spier, J.B. Griffiths, and C. MacDonald eds., *Animal Cell Technology: Developments, Processes and Products*, European Society for Animal Cell Technology, Butterworth-Heinemann, Oxford, U.K., 1992, pp. 569–576.
45. E.-J. Schlaeger, H. Loetscher, and R. Gentz, in R.E. Spier, J.B. Griffiths, and C. MacDonald eds., *Animal Cell Technology: De-*

velopments, Processes and Products, European Society for Animal Cell Technology, Butterworth-Heinemann, Oxford, U.K., 1992, pp. 562–567.

46. K.C. O'Connor, T.L. Prewett, T.J. Goodwin, K.M. Francis, A.D. Andrews, and G. F. Spaulding, in R.E. Spier, J.B. Griffiths, and W. Berthold eds., *Animal Cell Technology: Products of Today, Prospects for Tomorrow*, European Society for Animal Cell Technology, Butterworth-Heinemann, Oxford, U.K., 1994, pp. 293–295.
47. G.R. Pettman and C.J. Mannix, in R.E. Spier, J.B. Griffiths, and W. Berthold eds., *Animal Cell Technology: Products of Today, Prospects for Tomorrow*, European Society for Animal Cell Technology, Butterworth-Heinemann, Oxford, U.K., 1994, pp. 264–266.
48. G. Schmid, N. Wild, M. Fountoulakis, H. Gallati, R. Gentz, L. Ozmen, and G. Garotta, in R.E. Spier, J.B. Griffiths, and W. Berthold eds., *Animal Cell Technology: Products of Today, Prospects for Tomorrow*, European Society for Animal Cell Technology, Butterworth-Heinemann, Oxford, U.K., 1994, pp. 625–631.
49. K. Baker, Y.-Z. Zheng, S. Reid, and P.F. Greenfield, in S. Kaminogawa, A. Ametani, and S. Hachimura eds., *Animal Cell Technology: Basic and Applied Aspects*, Proc. of the Fifth Int. Mtg. of the Japanese Assoc. for Animal Cell Technology, Kluwer, Dordrecht, The Netherlands, 1993, pp. 521–528.
50. A.R. Bernard, T.A. Kost, L. Overton, C. Cavegn, J. Young, M. Bertrand, Z. Yahia-Cherif, C. Chabert, and A. Mills, *Cytotechnology* **15**, 139–144 (1994).
51. T. Sugiura and E. Amann, *Biotechnol. Bioeng.* **51**, 494–499 (1996).
52. R. Kurkela, A. Herrala, P. Henttu, H. Nal, and P. Vihko, *Biotechnology* **13**, 1230–1234 (1995).
53. A.R. Fooks, J.R. Stephenson, A. Warnes, B. Dowsett, B.K. Rima, and G.W. Wilkinson, in R.E. Spier, J.B. Griffiths, and W. Berthold eds., *Animal Cell Technology: Products of Today, Prospects for Tomorrow*, European Society for Animal Cell Technology, Butterworth-Heinemann, Oxford, U.K., 1994, pp. 605–609.
54. K. Mizumachi, J. Kurisaki, and N.M. Tsuji, in S. Kaminogawa, A. Ametani, and S. Hachimura eds., *Animal Cell Technology: Basic and Applied Aspects*, Proc. of the Fifth Int. Mtg. of the Japanese Assoc. for Animal Cell Technology, Kluwer, Dordrecht, The Netherlands, 1993, pp. 115–121.
55. O.W. Ogonah, R.B. Freedman, N. Jenkins, and B.C. Rooney, in E.C. Beuvery, J.B. Griffiths, and W.P. Zeijlemaker eds., *Animal Cell Technology: Developments towards the 21st century*, Proc. of the Veldhoven Mtg., Kluwer, Dordrecht, The Netherlands, 1995, pp. 427–430.
56. A. Lazar, S. Reuveny, C. Kronman, B. Velan, and A. Shafferman, in R.E. Spier, J.B. Griffiths, and W. Berthold eds., *Animal Cell Technology: Products of Today, Prospects for Tomorrow*, European Society for Animal Cell Technology, Butterworth-Heinemann, Oxford, U.K., 1994, pp. 324–328.
57. A. Garnier, J. Côté, I. Nadeau, A. Kamen, and B. Massie, *Cytotechnology* **15**, 145–155 (1994).
58. N. Epstein, S. Reuveny, J. Friedman, and N. Ariel, *Animal Cell Biotechnol.* **4**, 445–473 (1990).
59. K. Nagamine, M. Shiraishi, Z. Kong, K. Shinohara, and H. Murakami, in R.E. Spier, J.B. Griffiths, and B. Meignier eds., *Production of Biologicals from Animal Cells in Culture*, European Society for Animal Cell Technology, Butterworth-Heinemann, Oxford, U.K., 1991, pp. 167–169.
60. W. Noe, R. Bux, W. Berthold, and W. Werz, *Cytotechnology* **15**, 169–176 (1994).
61. S.C. Musgrave, Y. Douglas, G. Layton, J. Merrett, M.F. Scott, and C.A. Caulcott, S. Kaminogawa, A. Ametani, and S. Hachimura eds., *Animal Cell Technology: Basic and Applied Aspects*, Proc. of the Fifth Int. Mtg. of the Japanese Assoc. for Animal Cell Technology, Kluwer, Dordrecht, The Netherlands, 1993, pp. 223–230.
62. S.C. Musgrave, N.J. Bangerter, C.M. Bentley, C.D. Brown, Y.E. Douglas, N.J. Furminger, S.T. McGowan, and M.F. Scott, in R.E. Spier, J.B. Griffiths, and C. MacDonald eds., *Animal Cell Technology: Developments, Processes and Products*, European Society for Animal Cell Technology, Butterworth-Heinemann, Oxford, U.K., 1992, pp. 89–91.
63. S.C. Musgrave, C.M. Bentley, C.D. Brown, G.G. Feeney, N.J. Furminger, and T.W. Steward, in R.E. Spier, J.B. Griffiths, and C. MacDonald eds., *Animal Cell Technology: Developments, Processes and Products*, European Society for Animal Cell Technology, Butterworth-Heinemann, Oxford, U.K., 1992, pp. 92–94.
64. S. Hosoi, M. Satoh, H. Miyaji, T. Nishi, T. Mizukami, M. Hasegawa, S. Itoh, and T. Tamaoki, in S. Kaminogawa, A. Ametani, and S. Hachimura eds., *Animal Cell Technology: Basic and Applied Aspects*, Proc. of the Fifth Int. Mtg. of the Japanese Assoc. for Animal Cell Technology, Kluwer, Dordrecht, The Netherlands, 1993, pp. 63–70.
65. B. Baijot, M. Duchene, and J. Stephenne, *Develop. Biol. Standard.* **66**, 523–530 (1987).
66. C. Isch, A. Bernard, J.Y. Bonnefoy, C. Cavegn, P. Graber, T. Jamotte, and H.D. Blasey, in R.E. Spier, J.B. Griffiths, and C. MacDonald eds., *Animal Cell Technology: Developments, Processes and Products*, European Society for Animal Cell Technology, Butterworth-Heinemann, Oxford, U.K., 1992, pp. 530–533.
67. T. Marique, D. Malarme, P. Stragier, and J. Wérenne, in R.E. Spier, J.B. Griffiths, and W. Berthold eds., *Animal Cell Technology: Products of Today, Prospects for Tomorrow*, European Society for Animal Cell Technology, Butterworth-Heinemann, Oxford, U.K., 1994, pp. 369–371.
68. B. Montagnon, B. Fanget, R. Vinas, L. Peyron, J.-C. Vincent-Falquet, and P. Caudrelier, in R.E. Spier, J.B. Griffiths, and B. Meignier eds., *Production of Biologicals from Animal Cells in Culture*, European Society for Animal Cell Technology, Butterworth-Heinemann, Oxford, U.K., 1991, pp. 695–704.
69. J.R. Haynes, S.X. Cao, B. Rovinski, C. Sia, O. James, G. A. Dekaban, and M. H. Klein, *AIDS Res. Hum. Retroviruses* **7**, 17–27 (1991).
70. J. Litwin, in R.E. Spier, J.B. Griffiths, and C. MacDonald eds., *Animal Cell Technology: Developments, Processes and Products*, European Society for Animal Cell Technology, Butterworth-Heinemann, Oxford, 1992, pp. 414–417.
71. H. Miyaji, N. Harada, T. Mizukami, S. Sato, N. Fujiyoshi, and S. Itoh, *Cytotechnology* **4**, 173–180 (1990).
72. A.L. van Wezel, *Nature* **216**, 64–65 (1967).
73. S. Reuveny, in A.S. Lubiniecki ed., *Large Scale Mammalian Cell Culture Technology*, Dekker, New York, 1990, pp. 271–341.
74. C.A.M. van der Velden-de Groot, in E.C. Beuvery, J.B. Griffiths, and W.P. Zeijlemaker eds., *Animal Cell Technology: Developments towards the 21st century*, Proc. of the Veldhoven Mtg., Kluwer, Dordrecht, The Netherlands, 1995, pp. 899–905.
75. B. Griffiths, *Animal Cell Biotechnol.* **4**, 149–166 (1990).
76. P.M. Alves, J.L. Moreira, J.M. Rodrigues, J.G. Aunins, and M.J.T. Carrondo, *Biotechnol. Bioeng.* **52**, 429–432 (1996).
77. F. Cahn, *Tib Tech* **8**, 131–136 (1990).

78. M. Schläfke, W. Döcke, and U. Marx, in E.C. Beuvery, J.B. Griffiths, and W.P. Zeijlemaker eds., *Animal Cell Technology: Developments towards the 21st century*, Proc. of the Veldhoven Mtg., Kluwer, Dordrecht, The Netherlands, 1995, pp. 489–493.
79. N. Jenkins and E.M.A. Curling, *Enzyme Microb. Technol.* **16**, 354–364 (1994).
80. N. Jenkins, R.B. Parekh, and D.C. James, *Nat. Biotechnol.* **14**, 975–981 (1996).
81. H.S. Conradt, M. Nimtz, K.E.J. Dittmar, W. Lindenmaier, J. Hoppe, and H. Hauser, *J. Biol. Chem.* **264**, 17368–17373 (1989).
82. D.C. James, R.B. Freedman, M. Hoare, O.W. Ogonah, B.C. Rooney, O.A. Larionov, V.N. Dobrovolsky, O.V. Lagutin, and N. Jenkins, *Biotechnology* **13**, 592–596 (1995).
83. K. Teruya, S. Shirahata, T. Yano, J. Watanabe, K. Seki, K. Osada, H. Tachibana, H. Ohashi, and H. Murakami, in E.C. Beuvery, J.B. Griffiths, and W.P. Zeijlemaker eds., *Animal Cell Technology: Developments towards the 21st century*, Proc. of the Veldhoven Mtg., Kluwer, Dordrecht, The Netherlands, 1995, pp. 91–95.
84. T.J. Ahern, G.E. Morris, K.M. Barone, P.G. Horgan, G.A. Timony, L.B. Angus, K.S. Henson, J.B. Stoudemire, P.R. Langer-Safer, and G.R. Larsen, *J. Biol. Chem.* **265**, 5540–5545 (1990).
85. R.P. Lanza, J.L. Hayes, W.L. Chick, *Nat. Biotechnol.* 14, 1107–1111, (1996).
86. A. Bader, K. Bork, G. Schumann, and K.F. Sewing, in E.C. Beuvery, J.B. Griffiths, and W.P. Zeijlemaker eds., *Animal Cell Technology: Developments towards the 21st century*, Proc. of the Veldhoven Mtg., Kluwer, Dordrecht, The Netherlands, 1995, pp. 939–943.
87. A. Lazar, F.J. Wu, H.J. Mann, R.P. Remmel, F.B. Cerra, and W.-S. Hu, in E.C. Beuvery, J.B. Griffiths, and W.P. Zeijlemaker eds., *Animal Cell Technology: Developments towards the 21st century*, Proc. of the Veldhoven Mtg. Kluwer, Dordrecht, The Netherlands, 1995, pp. 1017–1023.

ANTIBIOTICS. See SECONDARY METABOLITES, ANTIBIOTICS.

ANTIBODY PURIFICATION

MICHIEL E. ULTEE
DOUGLAS W. REA
Cytogen Corp.
Princeton, New Jersey

KEY WORDS

Antibody
Chromatography
Isolation
Monoclonal antibody
Polyclonal antibody
Purification
Separations
Ultrafiltration

OUTLINE

INTRODUCTION

Scope

This review focuses on the fullest extent of purification: that needed for an injectable or parenteral product for human use. This application requires the use of multistep procedures, with steps of high resolution (1,2). Many uses of antibodies do not require such stringent purity. Our objective is to present methods that when combined are capable of achieving parts-per-million levels of contaminants, with the understanding that such criteria are not needed for many uses of antibodies. In such cases, single steps of many of the methods presented will prove to be adequate.

Our focus is also primarily on monoclonal antibodies, which now dominate the antibody field, but we include a special section on polyclonal antibodies. Similarly, we describe purification processes for the IgG and IgM classes only, because these account for almost all uses of antibodies. Occasional reference is made to purification of the IgA subclass, whose use is rare.

Our discussion is limited to methods of purification. The choice of equipment needed to carry out these methods is a very broad topic. It depends on whether process development or production is being performed. It also depends on the scale, on whether the method is good manufacturing process (GMP) or not, and on the nature of the antibody being purified. Thus, details of this selection process are beyond the scope of this article.

Properties of Antibodies Compared to Typical Feedstock Contaminants

Typical feedstock contains serum proteins such as albumin and transferrin as well as cell-degradation products such as DNA and cytosol proteins. In addition, if the antibody is in ascites fluid, the host animal's (typically mouse) antibodies would be present as well. By a combination of chromatographic or precipitation methods, one seeks to purify the desired antibody away from these contaminants. More recently, cell-culture media have become available that are described as "protein-free" and usually contain only low molecular weight digests of proteins or amino acids. If the antibody-secreting cells can grow and produce well in such media, the purification process becomes much easier, because the IgG molecules at a molecular weight (MW) of 150 kDa are readily separated by size from the components of such media.

As described in other articles, IgG antibody molecules are multichain glycoproteins that have been divided into various subclasses based on structure and properties. Their carbohydrate content is only 2% and is invariably located in the CH_2 domain of the Fc portion. The carbohydrate is sequestered, in that the carbohydrate chains are directed inward and are the actual contact residues between the heavy chains in this region (3), with the two heavy chains bowed outward to accommodate the carbohydrate. Thus, lectins generally do not bind to IgG, and purification must be directed toward the polypeptide portion. It should be noted that there are occasional reports of a carbohydrate site in the variable region of the IgG (3), in which case a purification method could be directed toward the carbohydrate moiety.

IgM molecules, on the other hand, contain about 9 to 12% carbohydrate, attached at five locations throughout the constant domains (4,5). Therefore, carbohydrate-directed methods have been developed for IgM and will be discussed later.

Specific affinity interactions for IgG and to some extent for IgM have been exploited in their purification. These affinity methods fall into three categories. First are those that rely on the key functional activity of the antibody, that is, its binding to antigen. Thus, a classic method of purification of antibody is binding on a solid-phase absorbent to which antigen has been immobilized. The second category relies on certain proteins that specifically bind to IgG. In this category are the well-known bacterial proteins A and G as well as second antibodies raised to bind against the desired antibody. Third, less-specific but often considered affinity chromatographic techniques include those relying on the interaction of the antibody with immobilized metals (IMAC), dye-ligands, hydroxyapatite, and thiophilic adsorbents.

The isoelectric points vary widely among different antibodies, mostly between about pH 6–9, but within different subclasses the range is generally much tighter. The isoelectric point directly affects the ability to use ion exchange, hydroxyapatite, chromatofocusing, and preparative electrofocusing techniques to purify antibodies. For example, the isoelectric point of serum albumin at 4.5 to 5.0 (6,7) indicates that its removal by ion exchange should be feasible for most antibody-purification schemes. Transferrin, at pI = 5.5–6.0 (8), can be more challenging.

The hydrophobicity and solubility of antibodies are important in their purification by such methods as hydrophobic interaction chromatography and precipitation methods. In general, antibodies are more hydrophobic and less soluble in the presence of precipitating salts than feedstock contaminants.

Finally, the size of antibodies, 150 kDa for IgG and 900 kDa for IgM, makes this a key property to exploit in their purification. Size-exclusion chromatography is widely used, as are ultrafiltration methods. Most contaminants are of lower molecular weight than antibodies. The nucleic acids DNA and RNA, which vary widely in size, can be digested using nucleases (9–11) to convert them to small oligonucleotides that are readily removed by sizing methods.

PURIFICATION METHODS

Introduction

The choice of a purification method is based on a number of different factors: the antibody, the feedstock, scale, economics, timing, and desired purity. As mentioned above, this article will concentrate on the most demanding type of antibody purification, that of multigram lots of monoclonal antibodies produced under GMP and for use in parenteral formulations.

The purification process can be divided into four parts. Figure 1 presents a flow chart of these steps, along with the most common techniques used for each step. The purpose of the first part, *sample preparation*, is to ensure the readiness of the sample for the first purification step. Sample preparation is sometimes not necessary, depending on the nature of the crude feedstock and on the choice of the first purification step.

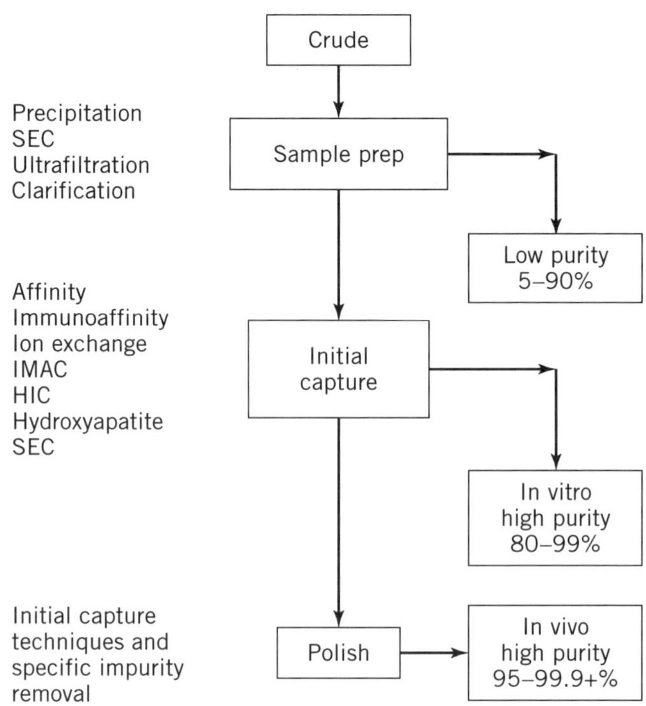

Figure 1. A general flow path for antibody purification, showing (on left) techniques typically used at each step.

Sample preparation is followed by the first purification step, often known as *initial product capture*. Most typically, the methodology for this step is chosen to give a high purification factor. Indeed, for antibody products with less stringent purification requirements (e.g., for in vitro use), this may be the only purification step required, if properly chosen.

The third part of the procedure is any *additional purification step(s)* necessary to reach the purity goal. These steps are sometimes referred to as polishing steps.

The final part of the procedure is to put the product into its *final formulation*, that is, the desired final concentration and physical state. This may sometimes be accomplished by the final purification step itself, eliminating the need for separate additional steps.

Sample Preparation

The purpose of sample preparation is to modify the chemical and physical state of the crude product into a state compatible with the initial capture step. From a chemical standpoint, this could involve changing the ionic strength, pH, or type of ion in the product solution. In the simplest cases, this may involve dilution or the addition of salt as a solid or concentrated solution. In a current GMP (cGMP) manufacturing setting, such methods are not particularly desirable. Addition of volume means increased costs. Solids that may be added will generally not be sterile, thus requiring an additional sterile filtration of the modified sample.

A more typical process at manufacturing scale would be a buffer exchange by size-exclusion chromatography (SEC) or by ultrafiltration or diafiltration. Because of its time- and volume-intensive nature, the related technique—dialysis—is typically used only on a laboratory scale.

From a physical standpoint, there are two typical processes carried out. One is concentration by ultrafiltration. The second is removal of particulate matter, by microfiltration or centrifugation. A very dirty feedstock (e.g., a batch fermentation) would clog a filter rapidly, making centrifugation the choice. However, for a relatively clean feedstock, the simplicity of filtration makes it the economic choice.

On a small scale, precipitative techniques such as ammonium sulfate, capyrilic acid, and polyethylene glycol (PEG) are also in common use. These techniques actually have at least modest purification power and generally have a concentrating effect as well (12–15). These techniques are less common at large scale, especially sterile operations, due to engineering considerations. Also, recovery from these operations can be low (15–17).

In addition to the above very common sample preparations, there are a number of other possibilities. It is sometimes desirable to remove certain contaminants early in the process, even though their concentration is low. Their removal is needed because their presence could affect subsequent purifications step(s). Some examples of these contaminants are DNA and RNA, endotoxins, and cell culture media dyes such as phenol red, which binds tightly to certain chromatography resins and could diminish their capacity. Bound dyes may also change the selectivity of a chromatographic resin (18,19). Many of these are removed when antibody is prepared by the precipitative methods noted earlier. Other techniques are possible, but these are generally used as secondary purification steps and will be described as such later. Another example of a troublesome contaminant is lipid, which can clog columns, affecting their performance and lifetime. Bulk lipids, especially as found in small volumes of ascites, may be partially removed by high-speed centrifugation. As before, the precipitative methods for antibody preparation will generally leave lipids in solution, thus effecting their removal. In addition, there are a number of extractive and chromatographic techniques that can remove lipids (16,20–21).

Initial Capture

The first major purification step in a process typically involves binding the antibody to some chromatography matrix, while impurities either flow through or are differentially eluted from the antibody. Alternately, conditions can be chosen where contaminants bind, and the antibody does not.

General Affinity. The most common matrix used to bind antibodies is immobilized protein A. This *Staphylococcus aureus* cell surface protein binds to many, although not all, antibodies and is widely available (17,22–31). Another bacterial cell surface protein, known as protein G, has been introduced more recently. Protein G binds more types of antibodies than does protein A, and it binds them more tightly (28,29,32,33). Although more versatile than protein A due to its wider specificity, the tighter binding of protein G can cause problems, as will be described later. Both of these are widely used because of their broad specificity, wide availability, and the high degree of purification. Indeed, purities of more than 95% are readily obtained, and antibodies of this purity are usable in many applications.

Historically, there have been three problems related to the use of protein A. First, it is expensive. However, given proper care, it is possible to reuse protein A resins many times, decreasing the long-term cost (34–37). Second, early protein A resins tended to leach protein A into the product (35). This is no longer a common problem, although leaching should be tested for, especially for products intended for in vivo use (38). Protein A assays are now readily available for such testing (39). The third problem is that many antibodies require a fairly low (pH 3–4) pH to elute from protein A. There are some antibodies that cannot withstand exposure to these conditions, even if the condition is neutralized as soon as possible. Needless to say, this is an even greater problem with the tighter binding of protein G (15,17,40).

In those instances where an antibody is acid sensitive, but it is still desirable to use protein A or G, a number of potential alternative elution conditions have been reported. These include the use of basic pH conditions (41), chaotropic ions or basic pH followed by ethylene glycol (42), and peptides that mimic the protein A or G binding site (43–45).

In addition to purifying bulk antibodies, proteins A and G have other applications in the field. For instance, they can be used to remove residual whole antibody and Fc fragments in preparations of Fab and $F(ab')_2$ fragments (17), since both proteins A and G bind to the Fc region of antibodies. Another use is subclass separation from a polyclonal antibody feedstock (e.g., serum). Although this is not always the case, often different subclasses of antibodies have different affinities for proteins A and G, which allows chromatographic fractionation of the subclasses (15,17,22–23,40,46–48). Table 1 illustrates this effect. The differential binding of protein A for different subclasses is not only a blessing, but also a curse and a need for caution. One subclass that does not generally bind well to protein A under physiological conditions is murine IgG1. This is currently the most common monoclonal antibody type seen. However, this problem has been overcome, in that a number of different regimens, in particular the use of slightly alkaline binding buffers of high ionic strength, have been developed to efficiently purify murine IgG1 on protein A (49–52). The need for caution lies in that not even antibodies of the same subclass will behave exactly the same (Table 1; 40,46). Thus, although general guidelines for the binding and elution of different classes of antibodies are helpful, the protein A or G purification of each antibody must be individually optimized.

Similar natural affinity ligands exist with other specificities. Some of these are based on the fact that antibodies are glycoproteins. Examples are mannan binding protein, which targets the mannan sugars of IgM, and the lectin jacalin, which binds human IgA1 (26,49). Many others have been described in the literature (26).

Immunoaffinity. The most selective technique is that of immunoaffinity, which can take a number of forms. In the most specific, the antigen for the desired antibody, or an anti-antibody against it, is used to make a chromatographic resin. In a somewhat less-specific fashion, antibodies against specific types of light or heavy chains can be used. However, immunoaffinity suffers from a number of problems. First, the antigen or anti-antibody must be produced and purified. Second, binding is often so strong that very vigorous and potentially harmful elution conditions must be used (12,53–56). On a production scale, the economics of producing the antigen or anti-antibody often make this technique undesirable. Finally, for those antibodies intended for in vivo use, assay techniques must be available to show that the product is free of antigen or anti-antibody. Furthermore, unless process steps downstream of the immunoaffinity column provide adequate viral clearance, as shown by validation studies, the ligand used to make the immunoaffinity support will need to have been purified from a process validated to remove viruses (1). Also, it must be demonstrated that either the product is free of any leached ligand (as noted above) or the ligand does not affect product safety (1). These additional requirements would add significantly to the cost of immunoaffinity chromatography.

A renaissance of this technique may come from the advent of chemically and genetically engineered pseudoantigens. Developed for specificity and reasonable binding affinity and produced by recombinant or chemical methods, these compounds (often small peptides) could overcome most, if not all, of the earlier objections (57). A similar case is the development of mimics of protein A and of generic anti-antibodies (58,59). Generic anti-antibodies are those targeted against constant regions of the antibody, making

Table 1. Percentage of Given Antibody Eluted at Given pH

Clonality	Subclass	pH 8.5	pH 8.0	pH 6.0	pH 5.8	pH 5.5	pH 5.0	pH 4.5	pH 4.0	pH 3.5	pH 3.0	Ref.
Poly	G1			>90			2					23
Mono	G1	24.3			68							47
Mono	G1	73.2			3.1			1				47
Mono	G1			66				33.3				46
Mono	G2	1			44			25		0.6		47
Poly	G2a						90		10			23
Mono	G2a							100				46
Mono	G2a							56		44		46
Mono	G2a			45				55				46
Mono	G2a			61				39				46
Mono	G2a			38				62				46
Mono	G2a			60				40				46
Mono	G2a			7.5				92.5				46
Mono	G2a			29				71				46
Mono	G2a			12				88				46
Poly	G2b										>90	23
Mono	G2b			20.5				71.5				46
Mono	G2b			6				28		66		46
Mono	G2b			23				77				46
Mono	G2b							85.2		14.8		46
Mono	G3					6		103			5	48
Mono	G3			26				74				46
Poly	M		>80									23
Poly	A		>90									23

Note: mono, monoclonal; poly, polyclonal.

their usage more general. A number of these are available commercially.

Ion Exchange. Another very widely used technique is ion exchange chromatography (IEC). Although less specific than proteins A and G, it can often provide purities of more than 80% in one step, which is adequate for many purposes (18,60–64). Both cationic and anionic IEC can be used, in either binding or flowthrough modes. The choice of method and its utility depend on the relative ionic characteristics of the antibody and its impurities. Care must be taken in selecting the resin and conditions to be used, not only from the standpoint of resolution, but also of capacity. If conditions are chosen where both the antibody and impurities bind, the capacity of the resin for antibody could be dramatically reduced.

The pI of a protein is not an exact predictor of a protein's behavior on IEC. The charge distribution on the surface of the protein and a number of other minor effects also play a role. Nonetheless, the pI is a good first approximation. In one study of 15 murine monoclonal antibodies, pIs of 4.9 to 8.3 were reported (65). Bovine albumin has been reported to have a pI of about 4.5 to 5.0 (4,5). The pI of another common impurity, transferrin, ranges from 5.5 to 6.0, depending on the species and degree of iron saturation (66). Thus, it is quite possible to have an antibody that is acidic enough to be inseparable from transferrin and even albumin by IEC. One of the first steps in developing an antibody purification procedure is to determine the pI of the antibody and the contaminants. This may be done quickly and easily and will give a good indication of how useful IEC might be.

Immobilized-Metal Affinity Chromatography. In addition to the widely used techniques described in the previous section, a number of others may be used for the initial purification of antibody. One technique that is growing in use is immobilized metal affinity chromatography (IMAC). Metal ions, usually Ni^{2+}, are held to a chromatography resin by ligands. In turn, these metal ions can interact with proteins, primarily with histidine groups. Many engineered antibodies and fragments are being produced with a polyhistidine tail, which gives this technique a high specificity. IMAC can sometimes also be used for nonengineered antibodies that have naturally occurring surface histidines, although histidine is an uncommon amino acid (67–69).

Hydrophobic Interaction Chromatography. Hydrophobic interaction chromatography (HIC) is another technique that can be used for initial capture. Antibodies tend to be among the most hydrophobic of the proteins in the crude feedstock (70,71). Thus, this technique can be quite powerful (6,8,72,73). An example is given in Figure 2 (72), where bovine serum proteins are separated from antibody. On the down side, experience shows that the loading buffer must be carefully chosen (67). As with IEC, conditions where most of the contaminating proteins bind along with the antibody should be avoided for the initial capture. In such a situation, capacity is decreased and selectivity diminished. Also, selection of an appropriate media can be somewhat more tedious than for IEC, because the choice of potential binding strengths is very broad. Not only is there supplier-to-supplier variation in ligand density and surface chemistry, variabilities also seen in IEC, but there are a large number of binding groups, such as ethyl, propyl, butyl, phenyl, ether, phenylether, and others. Even reversed-phase media, such as octyl, can be used in the aqueous HIC mode with weakly hydrophobic proteins.

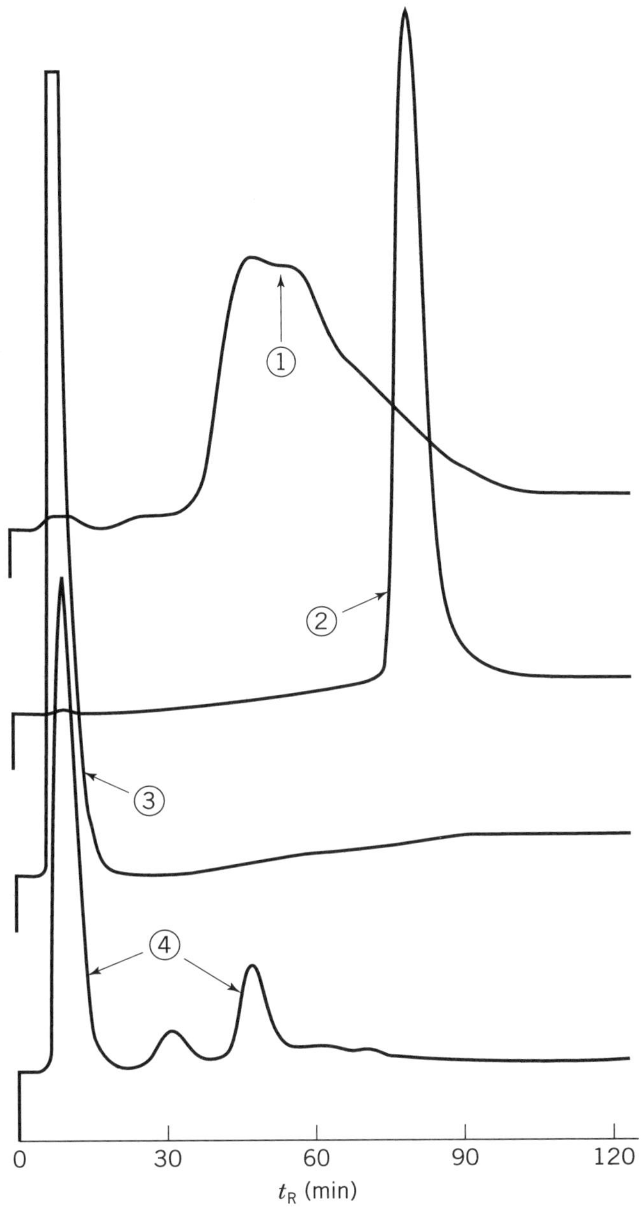

Figure 2. HIC separation of mouse polyclonal IgG from fetal bovine serum proteins and bovine polyclonal IgG. Chromatogram 1, bovine polyclonal IgG; 2, mouse polyclonal IgG; 3, partially purified (on ABx™, J.T. Baker) fetal bovine serum; 4, neat fetal bovine serum. *Source*: Reprinted from Ref. 72 with kind permission of Eaton Publishing.

Because of the variation in manufacturing, very little can be said in a general fashion about the strengths of the various resins. Within an individual manufacturer's line, usually the longer the alkyl chain length, the stronger the binding. In practice, however, tests should be made. It is important to make these tests, because the selection of a resin is an important way to manipulate selectivity.

Another means of altering selectivity is through the choice of buffer. In general, the stronger the lyotropic effect, the stronger the induction of hydrophobic binding by a salt (Fig. 3) (71). An example of this is seen in Figure 4 (74). However, strong salt solutions can have a detrimental effect on the integrity of proteins, so the choice should be limited, if possible, to those with strong salting out effects. Ammonium sulfate is the most commonly used salt, but it has the disadvantage of outgassing ammonia above pH 7.5 (71,75). Sodium sulfate and potassium phosphate are also good choices, but are somewhat limited in solubility (71,75). Because of the limited choice of buffers, the proper choice of resin becomes important.

In addition to salt, the selection of pH can have an effect on retention. In general, above pH 9 and below pH 4, retention increases, probably a result of denaturation of the protein. Between these pHs some proteins show an effect, but others do not. It is not possible to make generalizations; rather, the effect must be tested (71,75). Solution modifiers such as sugars, alcohols, PEG, and urea can also affect both binding and elution and can play a role in special situations (71,75).

HIC is perhaps deserving of more attention than it has received. Not only is it potentially a powerful technique itself, but for multistep procedures, it can be an ideal complement to IEC. IEC uses a low-ionic-strength load and a high-ionic-strength elution. HIC used a high-ionic-strength load and a low-ionic-strength elution. Thus, it is quite possible that conditions can be developed to allow one column to flow directly to the next, without intermediate sample handling.

Hydroxyapatite. Hydroxyapatite chromatography (HAP) has seen considerable use on a small basis, but little use on the large scale. The main reason for this is that until very recently, all versions of this material were mechanically fragile. Low flow rates were typical, and reuse of the columns was limited (53,76,77).

The mechanism of separation is not fully understood, but interaction of the protein with the calcium and phosphate in the resin matrix plays an important role, because a phosphate buffer gradient is most typically used for elution (53,76,77). As with other nonspecific techniques such as HIC and IEC, antibodies exhibit a wide variety of behavior on HAP, and generalizations are somewhat unsafe. However, HAP has been used not only for purifying antibodies (Fig. 5) (78), but also for separating different antibody subclasses (Fig. 6) (77). It is well known for its ability

Increasing salting out characteristics (lyotropic) →

$SCN^- < I^- < ClO_4^- < NO_3^- < Br^- < Cl^- < CH_3COO^- < SO_4^{2-} < PO_4^{3-} <$ Citrate

$Ba^{2+} < Ca^{2+} < Mg^{+2} < Li^+ < Cs^+ < Na^+ < K^+ < Rb^+ < NH_4^+ < (CH_3)_4N^+$

Decreasing salting in characteristics (chaotropic) →

Figure 3. Lyotropic/chaotropic characteristics of salts.

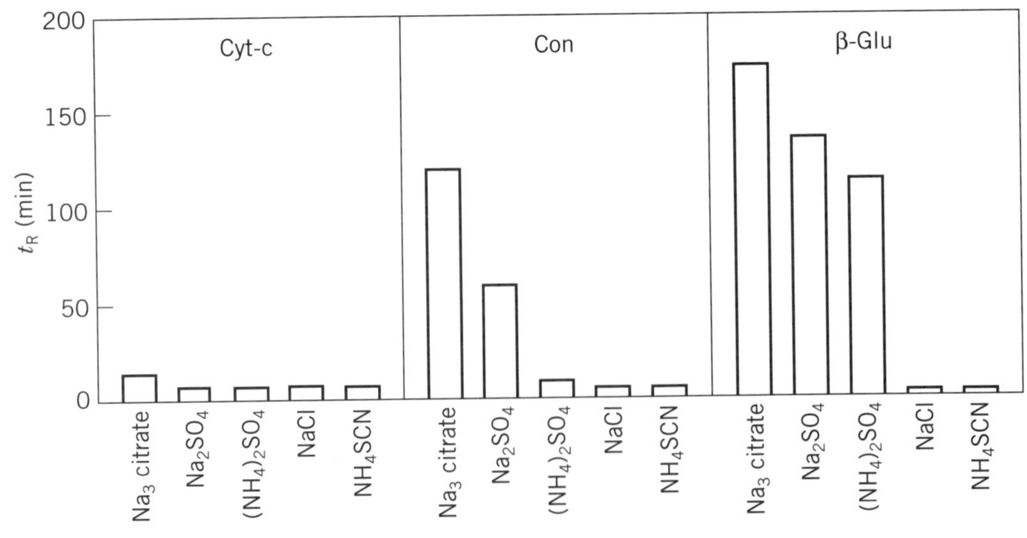

Figure 4. The effect of salt type on protein retention on a butyrate HIC column. The proteins were separated in a 20-min linear gradient from 1.0 M salt in 10 mM potassium phosphate buffer (pH 7.0) to 10 mM potassium phosphate buffer (pH 7.0). Cyt-c, cytochrome c; Con, conalbumin; β-Glu, β-glucosidase; t_R, retention time. *Source*: Reprinted from Ref. 74 with kind permission of Elsevier Science-NL, Sara Burgerharstraat 25, 1055 KV Amsterdam, The Netherlands.

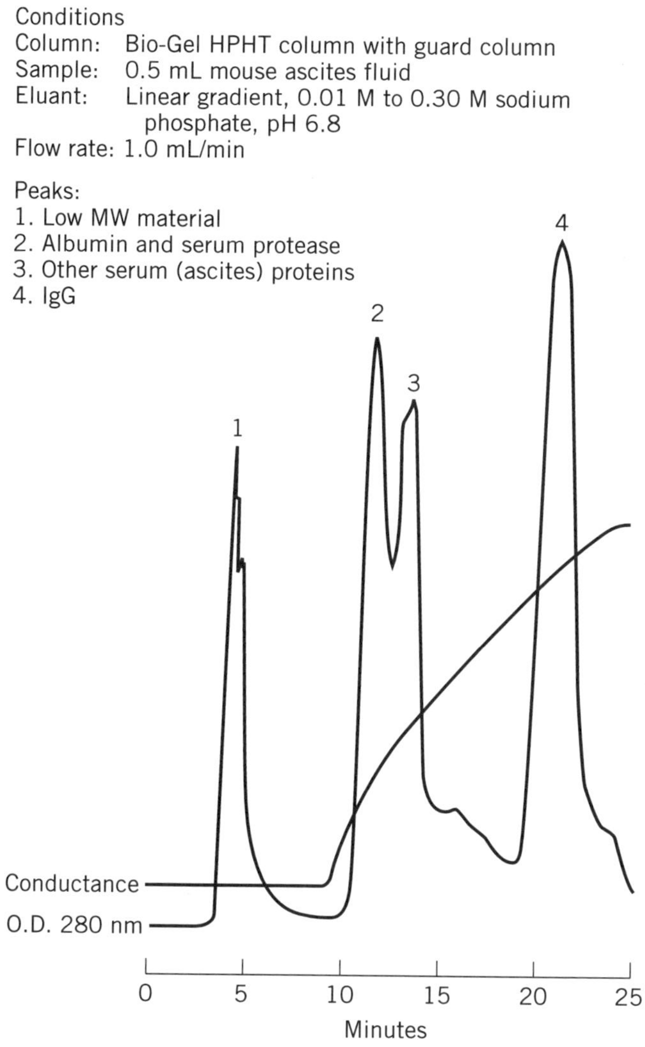

Figure 5. HPHT chromatography of unpurified ascites fluid containing monoclonal antibody. *Source*: Reprinted from reference 78 with kind permission of Eaton Publishing.

to bind DNA and to separate idiotypes, which will be discussed in "Secondary Purification Steps." With the recent development by several companies of more robust versions of this resin, its use for large-scale production of antibodies is bound to expand.

Size-Exclusion Chromatography. SEC is typically a low-resolution, low-throughput technique. Thus, in general, it is more often used for buffer exchange or polishing steps. However, there are instances when it can be used for initial purification. This is the case when dealing with very large antibodies such as IgM and IgA. IgM is typically a pentamer, with a weight nearing 1 million Da. IgA is usually found as a dimer, with a weight of about 300,000 Da. Thus, these are much larger than most contaminating proteins and can be purified by SEC.

Other Techniques. Another technique that has seen some use is thiophilic chromatography. In this technique, the sulfur ligand of the resin is thought to interact with the aromatic groups of the proteins. It has been shown that this technique can be used to effectively purify proteins (22,79). Like HIC, this technique generally loads proteins in high-salt buffers. However, unlike HIC, high salt concentrations also promote elution. Needless to say, different salts have different effects (79).

In addition to chromatographic techniques, there are a number of other techniques that can substantially purify antibodies. The precipitative techniques were mentioned earlier in "Sample Preparation." Ultrafiltration and diafiltration may also be used. Certainly, low molecular weight impurities may be removed by this technique. However, in addition, by careful selection of the membrane and buffer used, some contaminant proteins may be removed as well, such as albumin from IgG (80). The technique of preparative electrophoresis has gradually been advancing during the past few years. Currently it is can be considered useful for small-scale (milligram) bench preparations, especially non-GMP. Techniques for true production-scale electrophoresis that can be performed aseptically are being developed (81). Given the resolving power of analytical electrophoresis, success at developing production-scale techniques could add a valuable new tool to the development of purification methods.

In purification of antibodies on a large scale, process time is very important for economic reasons. One way to increase the efficiency of a process is to combine steps. In the past few years, a number of approaches to combining the concentration and clarification aspects of the sample prep step and the initial capture step have been commercialized. These approaches are generally either a fluidized bed, large-diameter bead resins, or wide-channel tangential flow filters. They are all designed so that cells and cell debris will flow through without clogging the setup, thus achieving clarification. The antibody is to bind to the resin or membrane, and after clarification is complete, it can then be eluted, providing both concentration and purification. Although simple in theory, a number of technical challenges had to be overcome before these approaches became practical.

Secondary Purification Steps

Depending on the feedstock, the technique selected, and the optimization, the initial capture step can often give purities in the range of 80 to 95%. This is often adequate for in vitro usages. However, for in vivo use, purities in excess of 99% are usually required. In addition to protein contaminants, other impurities such as DNA, endotoxins, viruses, and aggregates need to be removed. In such cases, a multistep procedure is almost inevitable.

All the same techniques used for initial capture can be used for secondary purification. Indeed, many of the techniques, such as HIC and HAP, are used more often as polishing steps than as initial capture steps.

The orientation of these secondary steps tends to be different from the initial capture step. With initial capture, the goal is to remove as many and as much contamination as possible. The secondary purification steps tend to be planned to remove individual contaminants. For instance,

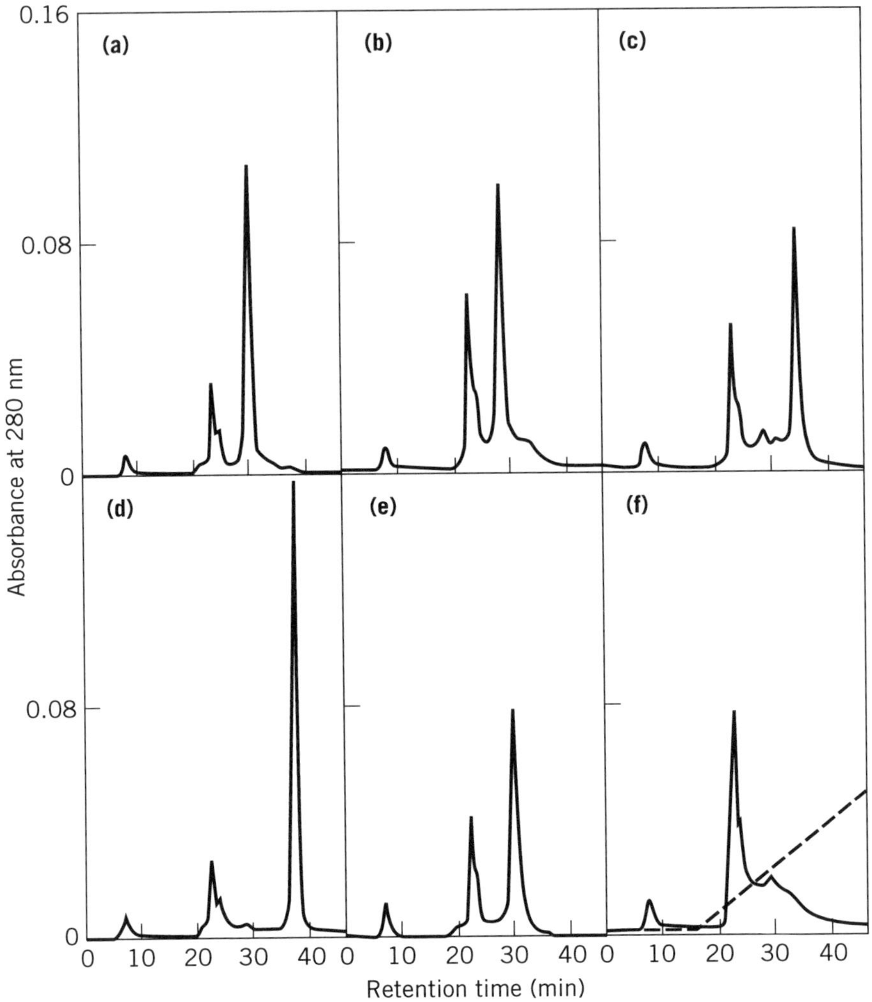

Figure 6. HPLC of mouse monoclonal antibodies in ascites on hydroxyapatite gel. (**a**), IgG1; (**b**), IgG2a; (**c**), IgG2b; (**d**), IgG3; (**e**), IgM; (**f**), antibody-free ascites. The *dashed line* on **f** indicates the gradient profile of phosphate ion concentration. *Source*: Reprinted from Ref. 77, p. 670, by courtesy of Marcel Dekker, Inc.

DNA and RNA tend to be highly negatively charged at just about any physiological pH. Thus, anion exchange resins are often used to remove these nucleic acids from antibodies, with conditions usually selected so that the nucleic acids bind while the antibody flows through.

In addition to the techniques mentioned in the initial capture step, there are a number of other techniques aimed at removing specific contaminants. For example, materials are available commercially that are intended specifically for the removal of impurities such as endotoxins, viruses, DNA and RNA, and albumin. In one way or another, these products take advantage of specific attributes of the impurity to remove them.

Some general examples of secondary purifications are discussed in this section. These include both techniques covered in the initial capture section and special polishing techniques.

The removal or inactivation of viruses must be validated for any animal-sourced or animal cell–sourced products, including antibodies, intended for in vivo use (1–2,82,83). All the techniques mentioned to date have the potential for removing virus. A specific attribute that is applicable to all viruses is their size. Being significantly larger than antibodies, they can be removed by a size basis. Although this can be done by SEC, a higher degree of virus removal is typically seen by ultrafiltration, with a membrane specifically selected for a molecular weight cutoff appropriate to retain virus and pass antibody (82,84,85).

In addition to virus clearance, virus inactivation is often performed. The most common techniques used are detergent treatment, low pH, or high temperature (82). Needless to say, the antibody must be tested to be resistant to these conditions before their routine use.

As mentioned earlier, one characteristic of DNA and RNA is the high negative charge. As well as anion IEC, HAP is well known to be effective at binding nucleic acids because of this high charge (76,86,87). HIC also may remove nucleic acids, because nucleic acids do not bind, whereas antibodies do bind to HIC resins (72).

Other properties of nucleic acids are the long, thin nature of the polymer, and, of course, the nucleotide sequence itself. Various precipitation techniques take advantage of this to remove DNA and RNA. Also, complementary nucleotides bound to chromatography supports can be used as an affinity adsorbent. Finally, as alluded to in the "Introduction," DNA and RNA can be enzymatically digested using nuclease enzymes (e.g., Benzonase®) to generate very small fragments that can be readily removed from antibodies by sizing techniques (9,10).

Endotoxins are lipopolysaccharides. Many can be removed by anion exchange chromatography (6,53,88). As with the nucleic acids, HIC has been reported to be able to remove endotoxins for antibodies (72). In addition, both precipitative and chromatography techniques focusing in on the lipid portion of the endotoxins have been developed.

Albumin is probably the most typical and highest concentration contaminant in most feedstocks, whether serum, ascites, or cell culture media. As with the other contaminants, the techniques already mentioned have the potential for removing albumin. However, if additional specific removal is required, antialbumin affinity chromatography is possible. Also, many serum proteins, including albumin, can be removed using dye-ligand chromatography. A well-known technique in laboratory chromatography, it has not been much used in preparative-scale work. The dyes used were originally taken from the textile industry and were not particularly pure. In addition, there was concern about ligand leakage. However, modern dye-ligand resins are made with high-purity dyes prepared especially for this purpose; coupling procedures have improved, minimizing leakage; and assays are often available to track what leakage there may be (89–93).

Perhaps the greatest challenge for the polishing steps is the removal of form variants. The easiest form variant to remove is aggregate, which can be removed by SEC, a step that may be needed for final formulation (see next section) anyway. In addition, there may be light or heavy chain variants if a cell-line producing antibody is not stable and specific. There is the possibility of host antibody from serum, ascites, or serum-supplemented cell culture. And there can be intramolecule variants caused by variations in glycosylation, deamidation, proteolytic digestion, and a number of other possible causes. It is difficult to make a blanket statement in regard to how or even if such variants need be removed. If these variants prove to be extremely difficult to remove, and the product can be proved to be safe and effective with their presence, the economics of the situation may dictate that they not be removed.

Nonetheless, variants may need to be removed, and some examples of possible techniques follow. For chain variants and other source antibodies, if the physical characteristics are significantly different compared to the target antibody, they probably can be removed by IEC, HIC, etc. But, if not, one possibility is HAP. This technique has long been known to sometimes be able to separate different idiotypes (76,78) (Fig. 7). Dye-ligand chromatography has also been shown to be capable of separating subclasses (93). In addition, the different affinity of different antibodies for protein A and protein G allows these resins to be used for variant removal at times. Immunoaffinity directed against the undesirable variant heavy or light chain(s) is a powerful technique. Glycosylation variants may sometimes be removed by lectin or immunoaffinity chromatography directed against specific sugars. If physical changes such as deamidation do not change the antibody's charge enough for separation by IEC, preparative electrophoresis or chromatofocusing may sometimes be successful (94). But, basically, this is a struggle that must be developed on a case-by-case basis and often requires a unique solution.

In general, the difference between initial capture and additional purification steps may be summarized in one word: resolution. In initial capture, the feedstock is a moderately to highly complex mixture. The feedstock may be dirty, that is, contain particulates, if sample preparation is not performed. On the production scale, large volumes are the norm. The goals in initial capture are to remove most of the impurities, with good yield, and to concentrate the product. With secondary purification, the feedstock is relatively clean, most of the protein is the target antibody, and some residual impurities may be physically very similar to the target antibody. Thus, the goal in secondary purification is to achieve high-resolution separation.

Final Formulation

Final formulation can be considered as a part of purification in that it removes conditions that would impair the stability or utility of the antibody in its intended use. One goal of a purification strategy should be to limit the additional processing required to put the antibody into the final formulation. For example, ideally the antibody should be recovered from the final purification step in the same buffer needed for its final formulation. Often this can be done if this goal is kept in mind when the purification process is being developed.

Final formulation may be as simple as a straightforward sterilization by membrane filtration through a sterile filter with pores 0.2 μm or less. Another relatively simple formulation step is adjusting the antibody concentration, either by dilution with buffer or by concentration on ultrafiltration. In other circumstances, the buffer composition may need to be changed to achieve optimal stability of the antibody. For this purpose, SEC or diafiltration is widely used. A more complex formulation step would be the addition of excipients to confer stability (95). Finally, the antibody solution may need to be lyophilized to confer stability, and the liquid formulation may be changed to be compatible with the lyophilization process. Such complex formulation strategies are beyond the scope of this article, so the reader is referred to reviews on this subject (95–97).

PURIFICATION OF POLYCLONAL AND PROBLEMATIC ANTIBODIES

Polyclonal Antibodies

Purification of polyclonal antibodies is actually more of a challenge than the purification of monoclonal antibodies. This is the case for two reasons.

The first is related to the method of production. Polyclonal antibodies are produced naturally, by the immune response of an immunized animal. The antibody is purified from serum, with the necessity of removing all the serum impurities, including host antibodies. Monoclonal antibodies, however, are produced in incubators. Historically, for small batches, the incubator has been a mouse, with the resulting ascites fluid containing the desired antibody. This ascites fluid is related to serum and can be an equally

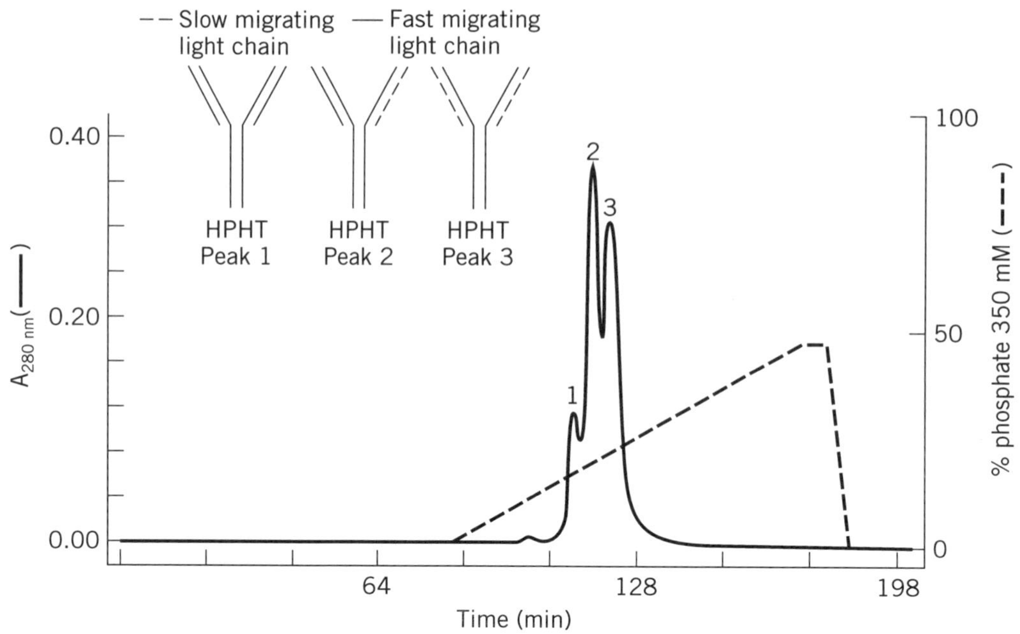

Figure 7. Preparative HPHT chromatography of IgG. *Source*: Reprinted from Ref. 76 with kind permission of Eaton Publishing.

complicated mixture. For large lots, and increasingly for small lots, monoclonal antibodies are produced in an artificial reactor, using at least a semidefined culture media. Again, historically, culture media could be considered only semidefined because it was supplemented with serum. However, even in this case, the concentration and number of macromolecular contaminants to be removed are lower than those in serum. Also, the current trend for monoclonal antibody production is toward use of serum-free or even protein-free media.

The second reason that polyclonal antibodies may be more challenging to purify is intrinsic to the antibody itself. Monoclonal antibodies are not necessarily absolutely homogenous. Often, there are differences in glycosylation. And there will be aging differences, such as in degree of deamidation. They will, however, all be of the same isotype, directed against the same epitope. This is not the case for polyclonal antibodies. Polyclonal antibodies tend to have a broader pI range than monoclonal antibodies, making IEC and related techniques such as HAP more complicated (77,86,94). They also tend to give broader peaks on hydrophobic interaction (72).

The same techniques used for purification of monoclonal antibodies are used for the purification of polyclonal antibodies. Indeed, most of the techniques were developed before the advent of monoclonal antibodies. As with monoclonal antibodies, the key to selecting the right purification techniques lies in knowing the desired yield, the desired purity, and the nature of the contaminants. Success at producing high-purity polyclonal antibodies centers around two points: extensive preparation and the use of affinity techniques.

In regard to preparation, characterizing all the individual impurities in a crude polyclonal mixture would be outlandishly time consuming. However, a valuable approach is to run standard IEC, HIC, and HAP conditions, where most if not all the proteins initially bind. Analysis of the elution fractions by an analytical technique such as two-dimensional electrophoresis will indicate which column technique can remove which impurities. The limitation to this approach is the well-known fact that the binding conditions used in chromatography affect the resolution of elution.

The affinity and pseudoaffinity techniques such as dye-ligand, IMAC, and thiophilic chromatography can be important in several roles. These techniques can be selected not only to bind to the target antibody, but also to remove specific impurities. For instance, one of the most vexing problems in antibody purification is removal of host antibody. If a specific isotype is desired, affinity techniques that bind to the other isotypes can be used to remove these unwanted host antibodies. Generic affinity techniques such as protein A and G are still very powerful tools. And, if practical to use, immunoaffinity has excellent selectivity.

In summary, there is nothing unique in purifying polyclonal antibodies compared to monoclonal antibodies. To reach a given degree of purity, the polyclonal antibody will probably take more planning; more preparation; and for high purity, use of more powerful techniques, especially affinity techniques.

Problematic Antibodies

As has been seen, antibodies can be remarkably diverse. One problem seen is when an antibody does not behave in a fashion similar to others of the same isotype. The solution of this type of problem is just good old-fashioned process optimization.

A much more serious problem is the purification of an antibody with limited solubility. A wide variety of additives can potentially be used for solubilization of antibodies, including urea, guanidine HCl, detergents, and sugars. Needless to say, there is the potential for these agents to interfere with purification steps as well. Thus, solution of this problem tends to require rather extensive development (48).

A similar problem is lack of stability. This may be intrinsic to the antibody, that is, it occurs even under physiological conditions, or it may be caused by a harsh purification condition, such as a low pH protein A elution or a virus-killing step. In the second case, generally alternate conditions are possible. Examples of alternates for both the examples are given earlier in this article. The solution of an intrinsic stability problem is similar to that of a solubility problem. These problems can be quite complicated and are beyond the scope of this article, but good references are available (98,99).

PURITY CONSIDERATIONS FOR USE OF ANTIBODIES AS PHARMACEUTICALS

Although purities of 95% or greater are often suitable for antibodies intended for nonpharmacological use, such as in vitro diagnostic reagents, much greater levels of purity are required for those intended for use as pharmaceuticals. In general, there are three types of contaminants of concern to regulatory agencies: microbes, viruses, and their products, such as endotoxins, proteins, and DNA (1,2). These contaminants must be removed to trace levels.

The need for sterility of a biological product intended for use in humans is a given. This is primarily assured by strict adherence to both aseptic technique and cGMP (100). Less apparent is the need for a validated process to remove any *potential* viruses that may be present in the crude material, including murine retroviruses, which are present in almost all hybridoma cell lines. These issues have been addressed in some detail by the regulatory authorities (1,2,83,101,102).

These regulatory authorities, such as the Center for Biologics Evaluation and Research division of the FDA, have relied on a three-tiered approach to controlling any infectious agents that may be present in the crude product or acquired during the downstream processing.

1. Careful selection of raw materials, including media components, for the absence of biological contaminants
2. Validating that the purification process can remove or inactivate viruses
3. Full testing of the product at appropriate stages in the process for the presence of viruses and infectious agents

Each of these three tiers of control must be adhered to by those engaged in the purification of antibodies for human use. A full treatment of these regulatory considerations is beyond the scope of this article. The reader is encouraged to refer to the cited references and keep abreast of current regulations issued by the various governmental agencies.

SUMMARY

The antibody molecule is one of the most versatile biological molecules created. Advances in immunology, biochemistry, and molecular biology have yielded a wealth of information about the structure and function of this class of proteins. The huge diversity of their binding sites allows one to acquire exquisitely specific binding reagents for almost any molecular target. Yet, the conserved nature of their constant domains as well as their characteristic three-dimensional structure allows the purification specialist to approach their recovery with much understanding of the behavior of these biomolecules. In spite of this, the adage that each monoclonal antibody is unique must be remembered, because even monoclonal antibodies from the same species, class, and subclass may behave very differently in purification procedures. With this caveat in mind, it is our hope that this review will provide a helpful guide to the purification of these fascinating biomolecules.

BIBLIOGRAPHY

1. Center for Biologics Evaluation and Research, *Points to Consider in the Manufacture and Testing of Monoclonal Antibody Products for Human Use*, 3rd ed., U.S. Food & Drug Administration, 1997.
2. European Regulatory Guidance Document, *Production and Quality Control of Monoclonal Antibodies*, vol. 3, addendum 3, Committee on Proprietary Medical Products, 1995.
3. T.W. Rademacher, S.W. Homans, R.B. Parekh, and R.A. Dwek, *Biochemical Society Symposium* **51**, 131–148 (1986).
4. D.R. Anderson, P.H. Atkinson, and W.J. Grimes, *Arch. Biochem. Biophys.* **243**, 605–618 (1985).
5. A. Nisonoff, J.E. Hopper, and S.B. Spring eds., *The Antibody Molecule*, Academic Press, New York, 1975, pp. 87, 314.
6. L.S. Hanna, P. Pine, G. Reuzinsky, S. Nigam, and D.R. Omstead, *BioPharm* 33–37 (Oct. 1991).
7. P.G. Righetti and T. Caravaggio, *J. Chromatogr.* **127**, 1–28, (1976).
8. K. Morimoto and K. Inouye, *J. Biochem. Biophys. Methods* **24**, 107–117 (1992).
9. A.J. Hagen, R.A. Aboud, P.A. DePhillips, C.N. Oliver, C.J. Orella, and R.D. Sitrin, *Biotechnol. Appl. Biochem.* **23**, 209–215 (1996).
10. A-K.B. Frej, R. Hjorth, and Å. Hammarström, *Biotechnol. Bioeng.* **44**, 922–929 (1994)
11. G.B. Dove, G. Mitra, M.A. Shearer, and M.S. Cho, in M.R. Ladisch, R.C. Willson, and C.-D.C. Painton eds., *Protein Purification: From Molecular Mechanisms to Large-Scale Processes*, American Chemical Society, Symposium 427, Washington, D.C., 1989, pp 194–209.
12. E. Harlow and D. Lane, *Antibodies, A Laboratory Manual*, Cold Spring Harbor Laboratory, Cold Spring Harbor, New York, 1988, pp. 283–318.
13. A. Polson, M.B. von Wechmar, and M.H. V. van Regenmortel, *Immunol. Commun.* **9**, 475–493 (1980).
14. S.H. Neoh, C. Gordon, A. Potter, and H. Zola, *J. Immunol. Methods* **91**, 231–235 (1986).

15. A.P. Phillips, K.L. Martin, and W.H. Horton, *J. Immunol. Methods* **74**, 385–393 (1984).
16. P.W. Thompson, A.C. Kenney, P. Moulding, and D. Wormald, *Ann. N.Y. Acad. Sci.* **529**, 529–539 (1990).
17. C.M. Fraser and J. Lindstrom in J. C. Venter and L.C. Harrison eds., *Receptor Biochemistry and Methodology*, Liss, New York, 1984, pp. 1–30.
18. M.J. Gemski, B.P. Doctor, M.K. Gentry, M.G. Pluskal, and M.P. Strickler, *Biotechniques* **3**, 378–384 (1985).
19. F. Chen, G.S. Naeve, and A.L. Epstein, *J. Chromatogr.* **444**, 153–164 (1988).
20. T.J. Smith, in D.J. Asai ed., *Methods in Cell Biology*, Vol. 37, Academic Press, San Diego, 1993, pp. 75–93.
21. U.S. Pat. 4,565,652 (Jan. 21, 1986), R. Schmidtberger (to Behringwerke Aktiengesellschaft).
22. L. Jacob, E. Schmitt, and W. Brümmer, *Am. Biotechnol. Lab.* 44–45 (Nov. 1994).
23. P.L. Ey, S.J. Prowse, and C.R. Jenkin, *Immunochemistry* **15**, 429–436 (1978).
24. J. Groudswaard, J.A. van der Donk, A. Noordzij, R.H. van Dam, and J.P. Vaerman, *Scand. J. Immunol.* **8**, 21–28 (1978).
25. J. Goding, *J. Immunol. Methods* **20**, 241–253 (1978).
26. M.D. P. Boyle, in M.D. P. Boyle ed., *Bacterial Immunoglobulin Binding Proteins*, Vol. 2, Academic Press, San Diego, 1990, pp. 1–21.
27. L. Schwartz, in M.D. P. Boyle ed., *Bacterial Immunoglobulin Binding Proteins*, Vol. 2, Academic Press, San Diego, 1990, pp. 309–340.
28. L. Björck and G. Kronvall, *J. Immunol.* **133**, 969–974 (1984).
29. A. Surolia, D. Pain, and M.I. Khan, *TIBS* 74–76 (Feb. 1982).
30. R. Lindmark, K. Thorén-Tolling, and J. Sjöquist, *J. Immunol. Methods* **62**, 1–13 (1983).
31. J.J. Langone, in F.J. Dixon ed., *Advances in Immunology*, Vol. 32, Academic Press, San Diego, 1982, pp. 157–252.
32. B.W. Walker, in M.D.P. Boyle ed., *Bacterial Immunoglobulin Binding Proteins*, Vol. 2, Academic Press, San Diego, 1990, pp. 255–368.
33. B. Äkerstrom and L. Bjorck, *J. Biol. Chem.* **261**, 10240–10247 (1966).
34. A.C. Kenney and H.A. Chase, *J. Chem. Technol.* **39**, 173–182 (1987)
35. P. Füglistaller, *J. Immunol. Methods* **124**, 171–177 (1989).
36. G. Hale, A. Drumm, P. Harrison, and J. Phillips, *J. Immunol. Methods* **171**, 15–21 (1994).
37. R. Francis, J. Bonnerjea, and C.R. Hill, in D.L. Pyle ed., *Separations for Biotechnology 2*, Elsevier, New York, 1990, pp. 491–498.
38. M.A. J. Godfrey, P. Kwasowski, R. Clift, and V. Marks, *J. Immunol. Methods* **160**, 97–105 (1993).
39. M.T. Dertzbaugh, M.C. Flickinger, and W.B. Lebherz III., *J. Immunol. Methods* **83**, 169–177 (1985).
40. C.L. Villemez, M.A. Russell, and P.L. Carlo, *Mol. Immunol.* **21**, 993–998 (1984).
41. Eur. Pat. Appl. 453,767 (Oct. 30, 1991), E.M. Croze (to E.R. Squibb and Sons, now Bristol-Myers Squibb).
42. N. Fornstedt, *FEBS Lett.* **177**, 195–199 (1984).
43. R. Bywater, G. Eriksson, and T. Ottosson, *J. Immunol. Methods* **64**, 1–6 (1983).
44. R. Bywater, *Chromatogr. Synth. Biol. Polym.* **2**, 337–340 (1978).
45. M. Yarnell and M.D.P. Boyle, *Biochem. Biophys. Res. Commun.* **135**, 1105–1111 (1986).
46. J.R. Stephenson, J.M. Lee, and P.D. Wilton-Smith, *Analyt. Biochem.* **142**, 189–195 (1984).
47. L. Manil, P. Motté, P. Pernas, F. Troalen, C. Bohuon, and D. Bellet, *J. Immunol Methods* **90**, 25–37 (1986).
48. W. Jiskoot, A-M.V. Hoven, A.A.M. DeKonig, M.F. Leerling, C.H.K. Reubsaet, D.J.A. Crommelin, and E.C. Beuvery, *J. Immunol. Methods* **138**, 181–189 (1991).
49. G.T. Hermanson, G.R. Mattson, and R.I. Krohn, *J. Chromatogr., A* **691**, 113–122 (1995).
50. G. Schuler and M. Reinacher, *J. Chromatogr.* **587**, 61–70 (1991).
51. A.P.G. van Sommeren, P.A.G.M. Machielsen, and T.C.J. Gribnau, *Prep. Biochem.* **22**, 135–149 (1992).
52. U.K. Pat. Appl. 2,160,530A (June 22, 1984), H. Juarez-Salinas and G.S. Ott (to Bio-Rad Laboratories).
53. D.R. Nau, *Biochromatography* **4**, 4–18 (1989).
54. Pharmacia, in Pharmacia ed., *Monoclonal Antibody Purification*, Pharmacia Biotech., Piscataway, N.J., 1993, pp. 21–25.
55. J.W. Eveleigh and D.E. Levy, *J. Solid Phase Biochem.* **2**, 45–78 (1977).
56. L.O. Narhi, D.J. Caughey, T. Horan, Y. Kita, D. Chang, and T. Arakawa, *Analyt. Biochem.* **253**, 236–245 (1997).
57. G.A. Baubach and D.J. Hammond, *Biopharm* 24–35 (May 1992).
58. N. Khatter, R.S. Matson, and T. Ngo, *Am. Lab.* 32LL– RR (May 1991).
59. J.Y. Shi and R.A. Goffe, *J. Chromatogr., A* **686**, 61–71 (1994).
60. J.R. Deschamps, J.E.K. Hildreth, D. Derr, and J.T. August, *Analyt. Biochem.* **147**, 451–454 (1985).
61. P. Clezardin, J.L. McGregor, M. Manach, H. Boukerche, and M. Dechanvanne, *J. Chromatogr.* **319**, 67–77 (1985).
62. B. Pavlu, U. Johansson, C. Nyhlén, and A. Wichman, *J. Chromatogr.* **359**, 449–460 (1986).
63. Y. Yang and K. Harrison, *J. Chromatogr., A* **743**, 171–180 (1996).
64. E. McCarthy, G. Vella, R. Mhatre, and Y. Lim, *J. Chromatogr., A* **743**, 163–170 (1996).
65. M. Carlsson, A. Hedin, M. Inganäs, B. Härfast, and F. Blomberg, *J. Immunol. Methods* **79**, 89–98 (1985).
66. E.N. Baker, *Perspect. Bioinorg. Chem.* **2**, 161–205 (1993).
67. V. Boden, J.J. Winzerling, M. Vijayalakshmi, and J. Porath, *J. Immunol. Methods* **181**, 225–232 (1995).
68. F.H. Arnold, *Biotechnology* **9**, 151–156 (1991).
69. J.E. Hale and D.E. Beidler, *Analyt. Biochem.* **222**, 29–33 (1994).
70. P. Gagnon, P.G. Cartier, J.J. Maikner, R. Eksteen, and M. Kraus, *LC-GC* **11**, 26–34 (1993).
71. P. Gagnon and E. Grund, *Biopharm*, 54–64 (May 1996).
72. D.R. Nau, *Biochromatography* **5**, 62–73 (1990).
73. O. Manzke, H. Tesch, V. Diehl, and H. Bohler, *J. Immunol. Methods* **208**, 65–73 (1997).
74. J.L. Fausnaugh, L.A. Kennedy, and F.E. Regnier, *J. Chromatogr.* **317**, 141–155 (1984).
75. T. Arakawa and L.O. Narhi, *Biotechnol. Appl. Biochem.* **13**, 151–172 (1991).
76. R. Vola, A. Lombardi, and M. Mariani, *Biotechniques* **14**, 650–655 (1993).

77. Y. Yamakawa and J. Chiba, *J. Liq. Chromatogr.* **11**, 665–681 (1988).
78. H. Juarez-Salina, S.C. Engelhorn, W.L. Bigbee, M.A. Lowry, and L.H. Stanker, *Biotechniques* 164–169 (May 1984).
79. T.W. Hutchens and J. Porath, *Analyt. Biochem.* **159**, 217–226 (1986).
80. S. Gadam, S. Orlando, S. Mundt, R. Kuriyel, S. Pearl, and M.E. Ultee, in S. Minden ed., *Monoclonal Antibody Purification*, International Business Communications, Southborough, Mass., 1996, pp. 1.3.1–1.3.4
81. C.W. Wrigley and H.P. Manusu, *Am. Lab.* **28**, 20K–20O (1996).
82. J.B. Grun, E.M. White, and A.F. Sito, *BioPharm* 22–30 (Nov.-Dec. 1992).
83. F. Brown and A.S. Lukiniecki eds., *Developments in Biological Standardization*, vol. 88, Karger, New York, 1996.
84. H. Maerz, S.O. Hahn, A. Maassen, H. Meisel, D. Roggenbuck, T. Sato, H. Tanzmann, F. Emmrich, and U. Marx, *Nat. Biotechnol.* **14**, 651–652 (1996).
85. L. Hoffer, H. Schwinn, L. Biesert, and D. Josic, *J. Chromatogr., B* **669**, 187–196 (1995).
86. T.L. Brooks and A. Stevens, *Am. Lab.* 54–64 (Oct. 1985).
87. R.J. von Wedel, in S.S. Seaver ed., *The Commercial Production of Monoclonal Antibodies*, Dekker, New York, 1987, pp. 159–173.
88. E.A. Neidhardt, M.A. Luther, and M.A. Recny, *J. Chromatogr.* **590**, 255–261 (1992).
89. C.R. Lowe, S.J. Burton, J.C. Pearson, and Y.D. Clonis, *J. Chromatogr.* **376**, 121–130 (1986).
90. C.R. Lowe, N. Burton, S. Dimaghanian, S. McLoughlin, J. Pearson, D. Stewart, and Y.D. Clonis, in M.A. Vijayalakshmi and O. Bertrand eds., *Protein-Dye Interactions: Developments and Applications*, Elsevier Applied Science, London, 1989, pp. 11–20.
91. G.E. McCreath, H.A. Chase, D.R. Purvis, and C.R. Lowe, *J. Chromatogr.* **629**, 201–213 (1993).
92. D.J. Stewart, D.R. Purvis, J.M. Pitts, and C.R. Lowe, *J. Chromatogr.* **623**, 1–14 (1992).
93. M. Hasnaoui, M. Debbia, S. Cochet, J.P. Cantron, P. Lambin, and O. Bertrand, *J. Chromatogr., A* **766**, 49–60 (1997).
94. J.C. Waldrep and J.R. Schulte, *J. Immunol. Methods* **118**, 273–277 (1988).
95. A. Wade and P.J. Weller eds., *Handbook of Pharmaceutical Excipients*, 2nd ed., American Pharmaceutical Assoc., Washington, D.C., and the Pharmaceutical Press, London, 1994.
96. J.L. Cleland and R. Langer, eds., *Formulation and Delivery of Proteins and Peptides*, American Chemical Society, Symposium 567, Washington, D.C., 1994.
97. A.P. MacKenzie, in D. Marshak and D. Liu eds., *Therapeutic Peptides and Proteins: Formulation, Delivery and Targeting*, Cold Spring Harbor Laboratory, Cold Spring Harbor, N.Y., 1989, pp. 17–21.
98. S. Li, C. Schöneich, and R.T. Borchardt, *Biotech. Bioeng.* **48**, 490–500 (1995).
99. T.J. Arern and M.C. Manning eds., *Stability of Protein Pharmaceuticals*, Plenum Press, New York, 1992.
100. Code of Federal Regulations 21:CFR 211, 600, 610.
101. Center for Biologics Evaluation and Research, *Q5 Viral Safety Evaluation of Biotechnology Products Derived from Cell Lines of Human or Animal Origin*, ICH Guidance Document, Washington, D.C., U.S. Food and Drug Administration, 1998.
102. F. Horaud and F. Brown eds., *Developments in Biological Standardization*, vol. 75, International Symposium on Virological Aspects of the Safety of Biological Products, Karger, New York, 1990.

See also ADSORPTION, PROTEINS WITH SYNTHETIC MATERIALS; AFFINITY FUSIONS, GENE EXPRESSION; CENTRIFUGES, ANIMAL CELLS; GOOD MANUFACTURING PRACTICE (GMP) AND GOOD INDUSTRIAL LARGE SCALE PRACTICE (GLSP); HYBRIDOMA, ANTIBODY PRODUCTION.

APOPTOSIS

RABINDER P. SINGH
MOHAMED AL-RUBEAI
University of Birmingham
Birmingham, England

KEY WORDS

Apoptosis
bcl-2
Bioreactor
Cell culture
Cell death
Flow cytometry
Necrosis

OUTLINE

INTRODUCTION

In the past it has generally been assumed that certain cell-culture conditions, such as interactions with specific hormones and deprivation of basic nutrients and growth factors, result in physical damage or metabolic collapse of the cell, leading to a passive, or "necrotic," death. It was thought that the cell was at the complete mercy of its environment and had no control over its fate. However, it is now clear that the cell does not always respond in such a simple and passive manner. Instead, a range of factors can trigger a highly complex and genetically regulated cellular response during which specific "death proteins" are activated, a phenomenon that has been named apoptosis (1). It is these proteins that are responsible, ultimately, for the death and destruction of the cell. Indeed, in many cases, the cell may have sustained only low levels of damage. Clearly, under these conditions apoptotic death will be premature and is often referred to as *cellular suicide*.

Apoptosis is now acknowledged as a fundamentally important process that plays an essential role in embryogenesis and in the maintenance and functionality of the highly ordered cell populations that constitute higher organisms. Indeed, there are now few aspects of biomedical research upon which apoptosis has not had an impact. It is as a consequence of its fundamental nature that any failure in its regulatory mechanisms leads to many of the diseases that pose the greatest challenges to medical science. This has resulted in an explosion of research into the genetic basis of apoptosis, the objective of which is to develop novel therapeutics for a wide range of disorders, including cancer, AIDS, and ischemic injury. Consequently a growing number of proteins have been identified that are involved in the induction suppression and execution of the apoptotic program. It is now clear that many of the cell lines used in the biotechnology industry for the production of therapeutics undergo apoptotic death. Obviously, the advances made in the characterization of the apoptotic pathway may be applied to the suppression of the apoptotic response in industrial culture processes. This should provide the biotechnologist with a new route to the optimization of culture performance.

We begin by describing the biochemical basis of apoptosis, the morphological changes that accompany it, and some of the techniques that have been used to identify apoptotic cells. This is followed by a general discussion of the regulation and induction of apoptosis. We then describe studies that have investigated this phenomenon from a biotechnological perspective. Particular reference is made to the conditions that elicit an apoptotic response, the cell lines that are susceptible, and the effect of *bcl-2* overexpression on cell survival and productivity in the bioreactor environment.

BACKGROUND

Morphology of Apoptosis

Apoptosis is defined by its highly characteristic morphology, which is illustrated in Figure 1. Early changes include a reduction in cell volume (2,3) and loss of surface microvilli. Cytoskeletal changes result in the formation of protrusions on the surface of the cell, which are referred to as blebbs. These may break away as intact vesicular structures, giving rise to "apoptotic bodies" (4).

One of the most striking changes during apoptosis occurs within the nucleus. Fluorescence microscopy of a typical viable cell following staining with a DNA stain such as acridine orange, reveals a large spherical nucleus that may constitute most of the cellular volume. Often, the chromatin is highly diffuse, although occasionally it may be possible to see condensed chromosomes in mitotic cells. During apoptosis, the nucleus undergoes major changes. Initially, the chromatin condenses and marginates to the nuclear membrane, forming crescent or ring-shaped structures that exhibit intense fluorescence. Eventually, the chromatin collapses into two or more particles, which are often highly spherical. Again, the remainder of the cell will be devoid of chromatin and will be almost transparent in appearance. The shape of the cell also undergoes a highly characteristic change. While viable cells tend to be highly irregular in shape, on entry into apoptosis they become smooth-surfaced and in many cases almost spherical.

The changes in nuclear morphology coincide with the activation of a nuclease enzyme that cleaves chromatin first into 300 and/or 50-kbp fragments (5), and then ultimately into multiples of 200 bp (6). The first stage is believed to be responsible for the morphological changes described above. The second stage represents the cleavage of chromatin at the internucleosomal linker regions. This generates a striking ladder like pattern when DNA samples from apoptotic cells are subjected to DNA gel electrophoresis. Together with condensation of chromatin, this so-called DNA ladder has become one of the hallmarks of apoptotic death.

Many cell types express a further enzymatic activity—that of the transglutaminase enzyme (7). This crosslinks proteins within the cell, generating a protein scaffold that is believed to hold the dead cell together, thus explaining the relative robustness of dead apoptotic cells as compared with cells that have undergone necrotic death.

These highly controlled changes appear to have a specific task: to provide a very clean and rapid method of eliminating dead cells, which is a vital consideration when one considers the extent of apoptotic cell death during, for example, embryogenesis. The cleavage of chromatin does not appear to be responsible, in itself, for the death of the cell. Instead, it has been argued that this ensures the complete destruction of the genetic material. It is suggested that this reduces the possibility of malignant transformation of surrounding cells by DNA from dying cells. The stabilization of the dead cell by the transglutaminase enzyme minimizes the probability of leakage of its contents onto surrounding cells. In vivo, surrounding cells and phagocytes engulf the dead cell before it can cause damage to surrounding tissue. Thus, inflammation of tissue as usually results from necrotic death is avoided. In vitro, in the absence of phagocytes, the apoptotic cell will eventually enter a degenerative phase called secondary necrosis.

Identification of Apoptosis

The search for simple techniques that allow for the identification of apoptotic cells has attracted a considerable

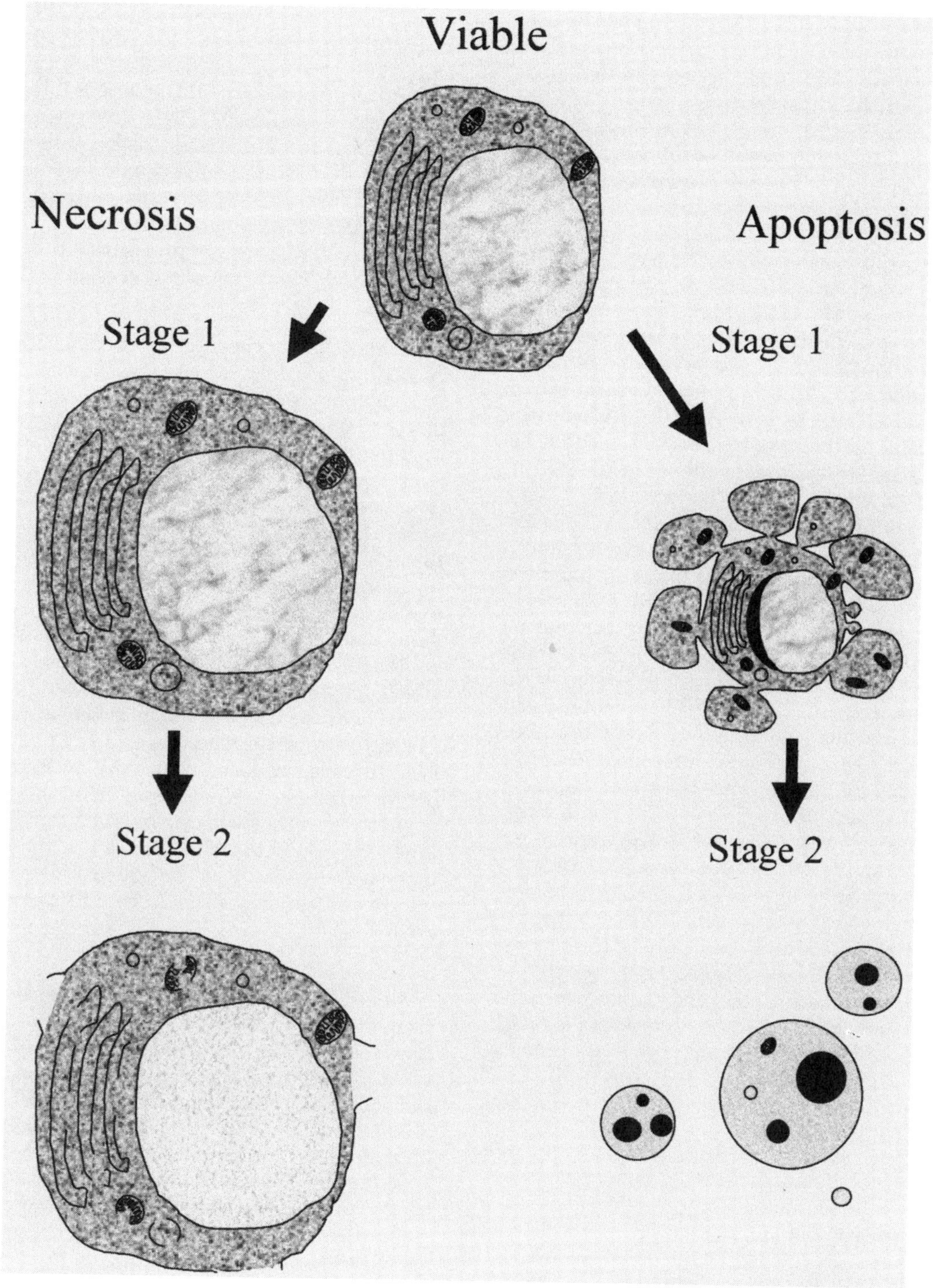

Figure 1. Schematic diagram showing the morphological features of apoptotic and necrotic cell death. The cell shown at stage 1 of necrosis exhibits clear swelling and dilation of mitochondria and Golgi. At this stage, nuclear morphology remains unchanged. By stage 2 of necrosis, disruption of organelles and plasma membrane is apparent, nuclear morphology is altered, and complete destruction of the cell quickly follows. During stage 1 of apoptosis, on the other hand, the cell exhibits extensive boiling or blebbing of the plasma membrane and a reduction in cellular volume. Condensation of the nuclear morphology may also be apparent. By stage 2, the chromatin has condensed and fragmented into several spherical particles and the cell is smooth surfaced in appearance. Apoptotic bodies containing chromatin and organelles are also shown. Eventually, plasma-membrane integrity is lost and degraded chromatin escapes from the dead cells, leaving nuclear-free ghosts. These structures may persist in the culture for many days.

amount of interest. As already stated, visualization of nuclease-mediated cleavage of DNA has been widely used to identify the presence of apoptotic cells. However, the technique can produce variable quality results that are of a qualitative, rather than a quantitative, nature.

Perhaps the simplest techniques for the study of apoptosis are based on the identification of the morphological features of cell death. For example, fluorescence microscopic analysis of nuclear morphology is a highly effective method for identification and quantification of apoptosis. There are two ways in which this may be done. If samples cannot be analyzed immediately, cells may be fixed in formaldehyde and stored at 4 °C. Analysis involves staining with acridine orange, which reveals the condensation of chromatin in apoptotic cells (7). However, this technique has one major drawback—an inexperienced operator may confuse early necrotic cells with viable cells, which have a very similar nuclear morphology. In order to avoid this difficulty, cells may be analyzed immediately while still in their culture medium by using the acridine orange–propidium iodide dual-staining technique (8). All cells are permeable to acridine orange, which stains chromatin green. Only membrane-damaged cells take up propidium iodide, and as a result exhibit red fluorescence. This technique therefore provides information regarding plasma membrane integrity, as well as nuclear morphology, and can consequently be used to simplify identification of necrotic cells. Additionally, the classification of apoptotic cells into early apoptotic and membrane-damaged apoptotic (sometimes referred to as secondary necrotic) cells gives an indication of the cell growth and death kinetics under the influence of a variety of environmental conditions during the cultivation process. The early-membrane-intact phase of death is relatively brief and, therefore, the presence of a large proportion of cells at this stage indicates that the rate of cell death is very high.

Microscopic techniques have two major drawbacks—they are subjective and time consuming. In theory, the development of flow cytometric methods should overcome these difficulties and provide a powerful tool for the study of the biochemical features of apoptosis in heterogenous cell populations. At present, there are a number of techniques that are available, and some of the most commonly used are discussed further.

Light-scattering Properties. The simplest flow cytometric method is based on the changes in light scattering properties that accompany cell death. Cells can be studied without the need for pretreatment, with the decrease in cell size producing a decrease in forward-scattered light. The increased granularity caused by nuclear condensation produces an increase in orthogonal light scatter (9,10). However, the latter is a transient stage, and eventually, a reduction in orthoganol light scatter is observed. The technique can also be used for the identification of necrotic cells (at least in their early stages). When necrosis is induced by a permeablizing agent such as saponin, a reduction in forward scatter is observed, but the increase in side scatter that occurs during apoptosis is not seen.

Changes in DNA Content. When stained with a DNA-specific stain such as propidium iodide (PI), apoptotic cells exhibit a characteristically low DNA content that appears as a sub-G_1 peak (i.e., it appears below the position of the G_1 peak of the cell cycle of viable cells). This is believed to result from the leakage of cleaved DNA from apoptotic cells (11–14). The technique provides a rather good correlation with the fluorescence microscopic technique already described. However, necrotic cells undergoing degradation may also exhibit a reduced DNA content, although the passive and asynchronous nature of this process means that a clear "peak" is not usually observed.

Annexin V. In viable cells, phosphatidylserine (PS) is located on the inner leaflet of the plasma membrane (15). During apoptosis, one of the earliest changes is the loss of this asymmetrical distribution (15,16). Annexin V has a very high affinity for PS. Conjugation of annexin V to a fluorescent tag, such as FITC, enables the use of this interaction to identify cells that have lost PS asymmetry. By combining this method with PI staining, it is also possible to classify apoptotic cells into two subpopulations: early (membrane intact and therefore PI negative) and late (membrane damaged and therefore PI-positive). The technique has been found to give a good correlation with levels of apoptosis during hybridoma batch cultures, during which apoptosis accounts for around 90% of cell deaths, as revealed by the fluorescence microscopic method described earlier. However, as with other flow cytometric techniques, necrotic cells can also give a false-positive result. This is because damage to the plasma membrane allows the annexin to enter the dead cell and bind to PS residues on the inner surface of the plasma membrane. Thus, the technique is reliable only when used to analyze early apoptotic cells, which can be seen as Annexin V positive and PI negative (16,17).

General Comments Regarding Identification of Apoptosis. Each of the techniques we described has advantages and drawbacks. When designing experiments to identify and quantify apoptosis, a number of points should be taken into consideration. First, there can be significant variations in the morphology of apoptosis from one cell type to the next, including, for example, absence of chromatin condensation. It is therefore recommended that several methods be used simultaneously to identify the mechanism of cell death. However, quantification of actual levels of apoptosis can vary significantly depending upon the technique used. Thus, in order to draw valid conclusions, it is essential that comparison of quantitative data collected using one technique is not made with data collected using a second technique.

Inducers of Apoptosis: Physiological Factors

The list of factors that induce apoptosis has grown steadily over the last few years. In the present section factors that have been the center of purely biological studies will be considered, although where necessary, implications for animal cell technology will be highlighted. Factors that are specifically of interest to the process biotechnologist will be explored under "Apoptosis and Animal Cell Biotechnology."

Presence/Absence of Receptor–Ligand Interactions. The absence of certain hormones can result in the induction of apoptosis in certain cell types. This has led to the suggestion that apoptosis may play a pivotal role in the maintenance of tissue organisation in vivo. It is thought that all cells are primed to undergo apoptosis and are prevented from doing so through constant stimulation by parakrine survival factors (18). If a cell is removed from its physiologically correct location, the absence of the appropriate survival signal will lead to the induction of apoptosis. This role of survival factors as regulators of cellular distribution in vivo may also have an impact on the development of serum-free media preparations for industrial-scale cell-culture processes, as described later.

The presence of receptor–ligand interactions can also lead to the induction of apoptosis. For instance, in the Fas–FasL system, binding of the Fas ligand to the Fas receptor can trigger apoptosis (19). This mechanism is responsible for the regulation of the immune system. Autoreactive B cells undergoing maturation (20) and autoreactive mature T cells (21) are eliminated by the Fas-mediated induction of apoptosis. Fas also acts as the "off" switch for the immune system by inducing apoptosis in antigen-activated B and T cells (22). Molecular dissection of the Fas–FasL system has provided important insights into the early stages of the signaling cascade that leads to the expression of the death pathway (see "Genetics of Apoptosis").

Induction by Viruses. A number of viruses have been shown to interact with the cellular apoptotic machinery. An apoptotic response to viral infection would appear to act as a protective mechanism that prevents viral replication by triggering the suicide of the infected cell. However, a number of virally encoded antiapoptotic genes have now been identified that suppress the expression of the death program, thereby providing the virus with the opportunity to propagate itself. Perhaps one of the most interesting examples of this anti-death mechanism from a biotechnology perspective is that of baculovirus, which has been synthesized at production scale by infection of insect cell lines, and has applications as a biological pesticide and, more recently, for the expression of recombinant proteins. A mutant was identified that induced high levels of apoptosis during infection of Sf21 insect cells. This was attributed to a mutation in the p35 viral gene that, in its wild-type form, acts as an antiapoptosis gene (23).

Arguably, the most important example of virus-induced apoptosis is that mediated by HIV (24–27). A number of reports have indicated that the binding of the viral gp 120 protein to the $CD4^+$ receptor of T cells triggers the induction of apoptosis, thus leading to the depletion of this class of cells during HIV infection (28–30). Interestingly, HIV infection of $CD4^+$ cells appears to provide protection from apoptosis. The viral Nef protein downregulates the expression of $CD4^+$, thus preventing the induction of apoptosis. As a result, virus propagation in the infected cell is not prevented (31).

There have also been reports of virus-induced suicide of bacterial cells. Until recently, it was assumed that altruistic cell death was not possible in single-celled organisms, simply because the genes involved in such a phenomenon would be lost when the cell concerned dies. However, bacteriophage infection in bacterial colonies has been reported to induce a suicide response, thus preventing the spread of the infection to the remainder of the colony (32). The genes that mediate such a response are propagated by other clones in the colony, thus preserving the altruistic nature of cellular suicide. Bacteria and viral expression systems have been used for the production of recombinant proteins. Clearly, the possibility that bacterial cells in such systems exhibit an apoptosis-like response needs to be investigated.

Free Radicals and Apoptosis. Free-radical-mediated cellular damage has become an important area of study. In recent years there have been numerous reports that cell death following oxidative stress occur by apoptosis. Moreover, generation of free radicals has been postulated as being a universal triggering event in the induction of apoptosis (33–38). Indeed, for a period in the early 1990s, it was suggested that the widely studied antiapoptosis gene *bcl-2* functioned as an antioxidant (39). However, this theory has been challenged by the demonstration that anoxia-induced apoptosis can also be suppressed by Bcl-2 in the absence of measurable levels of free radicals (8,40,41).

Induction by Toxins and Therapeutic Agents. Exposure to high levels of toxic chemicals results in necrotic death of the cell. However, long-term exposure at a low level can trigger an apoptotic response. Many of the agents used in chemotherapy exert their affect by inducing apoptosis in tumor and normal cells. Overexpression of genes such as *bcl-2* has been linked to resistance to chemotherapy. Clearly, establishing the mechanism of induction of apoptosis in response to such agents and providing strategies that minimize the affect of antiapoptotic genes should provide novel and more effective chemotherapeutic strategies (42).

Exposure of cells to ionizing radiation induces high levels of apoptosis in many normal tissues. Particularly susceptible are those cells from tissues that undergo rapid proliferation, such as spermatogonia (43) and lymphocytes (44). Such tissue would be especially prone to malignant transformation, and consequently the induction of apoptosis following DNA damage minimizes the likelihood of such an event. Irradiation of tumors can also lead to the induction of apoptosis (45–47). Notably, tumors that respond least favorably to irradiation exhibit the lowest level of apoptosis under such conditions (45). The induction of apoptosis following DNA damage is regulated by the product of the p53 gene, which has been referred to as the gaurdian of the genome because of its central role in preventing the propagation of cells that may have sustained genetic damage and thus, potentially, malignant transformation (48).

Genetics of Apoptosis

The study of the genetic basis of apoptosis has become a highly complex and fast-moving area of research. Consequently, this section will only deal with some of the most important aspects of this subject in order to provide a basis for the later discussion on genetic manipulation of the apoptotic pathway in commercially important cell lines.

The genes involved in apoptotic death may be classified into three groups. The first group consists of the modulators of the apoptotic pathway that suppress or induce death. The second group comprises the components of the cell death pathway that mediate the cell death signal. The final component is the group of effector enzymes that is responsible for the death and destruction of the cell. Some examples of these genes are given in Table 1.

Of the modulators of the apoptotic pathway, one group of closely related proteins, the Bcl-2 family, has attracted particular attention. The *bcl-2* gene, the first and best characterized member, was identified at the t(14;18)(q32↗1) breakpoint found in human follicular lymphoma (54). It encodes a 24-kDa protein that is located on the outer mitochondrial membrane, the cytosolic face of the nuclear membrane, and endoplasmic reticulum. Numerous studies have demonstrated the ability of this protein to suppress apoptosis in response to a wide variety of inducers. Table 1 lists other members of this family, some of which are functionally similar to Bcl-2, whereas others act as antagonists of Bcl-2 and thereby trigger apoptosis.

Until recently, studies of the molecular basis of cancer were directed at the identification and characterization of genes involved in the regulation of cellular proliferation. However, it is now evident that the failure of cells to undergo apoptosis at the correct time and location is also an important step in malignant transformation. Indeed, studies of *bcl-2* in this context have been instrumental in establishing the role of apoptosis in cancer. Mutations that result in the overexpression of *bcl-2* lead to the accumulation of cells due to life span extension. These cells then undergo further mutation, most notably involving the *c-myc* gene, which leads to the formation of high-grade tumors that combine the characteristics of high cellular survival with a high rate of proliferation (61–63). Clearly, these are characteristics that would be desirable in the ideal candidate host for the expression of recombinant proteins.

One of the most common mutations identified in tumors is that of the p53 gene, mentioned earlier. In response to DNA damage, p53 causes cell cycle arrest, allowing the cell to repair damaged DNA, thus ensuring that any potentially carcinogenic mutation is not propagated. However, in some cell types, p53 triggers the induction of apoptosis, again in order to minimize the risk of transformation (48,64).

The well-characterized pattern of development of the nematode *Caenorhabtis elegans*, which includes the induction of apoptosis in specific cell types, has provided an important insight into the genetic basis of apoptosis. Most notable among the genes identified are *ced 9* and *ced 3*. The former is a *bcl-2* homologue that blocks the ability of *ced 3* to induce cell death (65). The *ced 3* gene is a cysteine protease that shares extensive homology with the mammalian protein called interleukin 1β converting enzyme (ICE) (66,67). As the name suggests, this enzyme cleaves the active precursor pro–interleukin 1β to generate the active molecule interleukin 1β, and a number of studies have now implicated it in the induction of apoptosis. However, apoptosis can also be induced in macrophages and thymocytes of ICE-negative mice, indicating that ICE is not a universal mediator of apoptosis (68). Indeed, several ICE-related enzymes have been identified in recent years (see Table 2).

The molecular mechanism of apoptosis induced by the interaction of the Fas-ligand with its receptor centers on the activity of ICE-related proteases. Indeed, important progress has now been made in deciphering the earliest events in the signal cascade that transduces the initial death stimulus to the death machinery of the cell. Activation of the Fas receptor by its ligand or agonist antibodies leads to the binding of the adapter protein MORT1/FADD (Fas-associating protein with death domain) (75,76). This, in turn, binds to the ICE homologue FLICE (FADD-like ICE) or MACH (MORT1-associated ced-3 homologue) (77,78). The targets of this enzyme are still under investigation.

APOPTOSIS AND ANIMAL CELL BIOTECHNOLOGY

At present, there are around 100 new biopharmaceuticals in phase I, II, and III clinical trials (79). Production of a biologically active therapeutic will, in many cases, necessitate expression of the protein in mammalian cell lines. As a result of intense commercial pressure, tried and tested production processes are often adopted in order to ensure that the product reaches the market in the shortest possible time. Initially, such production processes were relatively inefficient and there was tremendous scope for process optimization. Advances in cell biology, in addition

Table 1. Examples of Genes Involved in Apoptosis

Inducers/ Promoters	Ref.	Suppressors	Refs.	Mediators	Refs.
bax	49	*bcl-2*	54	ICE family	50
bak	50,51	*bcl-xL*	53	Trans-glutaminase	7
bad	52	*mcl-l*	55	NUC-18	58
bcl-xS	53	*ced-9*	56	DNase-I	59
FasL / Fas	19	*p35*	23	DNase-II	60
		bhfrI	57		

Table 2. Proteases Involved in the Induction of Apoptosis

Protease	Refs.
Ced-3	55
Interleukin 1β-converting enzyme (ICE)	69
Nedd2	70
CPP32	71
$ICE_{rel\ II}$/TX	72,73
$ICE_{rel\ III}$	73
mch2	74

to improvements in process monitoring, control, and optimization, and the development of novel bioreactor designs, are making the transition from the research laboratory to industry standard practice. Clearly, this will lead to a reduction in the cost and complexity of the process of recombination-protein production using mammalian cell lines by placing the technology on a firmer scientific footing.

In order to optimize the viable cell number and protein productivity, a detailed understanding of the factors that lead to cell death in the bioreactor is required. These factors may be classified into three groups:

1. The hydrodynamic environment of the cell
2. The accumulation of toxic metabolites
3. The exhaustion or local limitation of nutrients and oxygen

Although each of these areas has now been investigated to varying degrees, recent studies have demonstrated that at least some of the cell lines used in bioreactors undergo apoptotic death rather than necrosis, indicating that a reassessment of the subject is required. As described later, the greater understanding of the mechanism of cell death in commercial cultures should provide new routes to culture optimization. The resultant enhancement in culture efficiency would be expected to manifest itself in three forms:

1. First, the nutrients and culture time invested in generating a viable cell will be wasted if that cell should die prematurely. If the survival time of the cell can be enhanced, the proportion of culture resources utilized for production of the biopharmaceutical of interest can be increased by eliminating the need for regeneration of cellular biomass.
2. Second, as a cell dies, it releases proteolytic enzymes into the culture medium, which can lead to degradation of the product. Thus, product stability should be enhanced by minimization of cell death.
3. Finally, high levels of cellular debris in the culture medium can complicate the recovery of the target protein. This will add to the cost of downstream processing and lead to a reduction in the efficiency of target protein recovery.

A Brief History

The first suggestion that apoptosis may account for cell death during the cultivation of hybridoma cells came from an electron microscopic study conducted by Al-Rubeai et al. (73). Further studies by Franek and Dolnikova (74) demonstrated the accumulation of nucleosomal DNA fragments in culture medium during the death phase of batch hybridoma cultures. Based on this observation, they estimated that around 30% of cells had undergone apoptotic death. Further evidence of apoptosis during hybridoma cultures was provided by the studies of Mercille and Massie (80) and Singh et al. (81). Upon DNA gel electrophoresis, both studies revealed the laddering pattern that, as mentioned earlier, is characteristic of apoptosis. Morphological analysis of the nuclei of the cells indicated that apoptosis accounted for 90% of the dead cells. Both groups also found high levels of apoptosis during the cultivation of murine plasmacytoma cell lines (sometimes incorrectly referred to as myeloma cells). Furthermore, Singh et al. (82) reported an absence of apoptosis during the death phase of CHO and Sf-9 batch cultures. However, studies by Moore et al. (77) indicated significant levels of apoptosis during the death phase of serum-free batch cultures of CHO cells. Recent studies in our laboratory on a CHO 320 cell line overexpressing interferon indicate that this cell line may also be susceptible to apoptosis, although the morphology was not typical, and the frequency was much lower than that seen during hybridoma cultures under comparable conditions.

Studies are now required to give an indication of variability in susceptibility to apoptosis between clones of the same cell type, and between different unrelated cell lines. The objective will be to produce a correlation between susceptibility to apoptosis and general cell robustness. Clearly, such a correlation would provide a simple and easily identifiable predictor of robustness following exposure of cells to a range of stresses, thus simplifying the process of cell-line selection.

Induction and Suppression of Apoptosis in the Bioreactor Environment

The features of the bioreactor environment that may result in cell death were outlined in the introduction. In the present section, studies that have considered these factors in terms of their ability to induce apoptosis are described and the implications of suppression of apoptosis are discussed.

Death during Batch Cultivation. The nutrient limitations encountered by cells in the bioreactor may be classed into two groups: cycling or terminal limitation. The former may be encountered at all stages of large-scale or intensive culture systems. For example, in large-scale stirred tank reactors, the cells may be exposed to fluctuating nutrient levels because of inhomogeneity due to poor mixing. In intensive culture systems, the high cell densities reached will result in low local-nutrient concentrations. As a result, the level of cell death in such systems is relatively high. Terminal nutrient limitations will occur at the end of batch cultivation and, as a result, will become more extreme with time, invariably leading to cell death. In the case of hybridoma batch cultures, the first nutrient to become limiting is glutamine, and its exhaustion coincides with the onset of the death phase of the culture. As stated above, cell death under these conditions is almost exclusively by apoptosis. Two studies have reported on the effect of *bcl-2* overexpression on cell survival during the death phase of hybridoma cultures. Itoh et al. (78) found that bcl-2 overexpression significantly extended the duration of the culture by reducing the rate of cell death. Moreover, they reported a fourfold increase in the Mab (monoclonal antibody) titre in the culture medium. Simpson et al. (8) have also reported an extension in culture duration, although there was only a 40% improvement in Mab titre. Necrosis became the predominant mechanism of cell death in the *bcl-2*-transfected cell line, indicating near-complete suppression of apoptosis under batch culture conditions.

More recently, Suzuki et al. (79) have reported that transfection of COS-1 cells with *bcl-2* and then with the vector pcDNA-λ carrying the immunoglobulin λ gene for transient expression of λ protein have resulted in higher expression of the protein when compared to the control transfectant (i.e., *bcl-2* negative). In the same study, the mouse plasmacytoma p3-X63-Ag.8.653, which is used as a fusion partner in the generation of hybridomas, and the hybridoma cell line 2E3 were transfected with the human *bcl-2* gene. In both cases an extension of batch culture duration was reported. The *bcl-2*-transfected 2E3 cells survived 2 to 4 days longer in culture, producing a 1.5- to 4-fold larger amount of antibody in comparison with the control vector transfectants. A further enhancement in survival and antibody production in hybridoma 2E3 cultures was observed when cells were cotransfected with *bcl-2* and *bag-1*.

In contrast with these promising studies, Murray et al. (83) found that *bcl-2* transfection of the murine plasmacytoma NS0 failed to provide any protection from apoptosis. Although there was no endogenous Bcl-2 expression, they did report expression of Bcl-xL, a functional homologue of bcl-2. Thus, they suggested that Bcl-2 may be functionally redundant in this cell line.

Nutrient Limitation

Amino Acids and Glucose. The link between the onset of apoptosis and exhaustion of glutamine during batch cultures of hybridoma cells has prompted systematic studies of the role of the various nutrients used in culture medium. Initial studies indicated that deprivation of glucose, serum, glutamine, cysteine, and methionine (84–86) could all individually induce high levels of apoptosis. Moreover, recent studies in our laboratory (87) indicate that this is not a feature of these particular nutrients alone. Deprivation of each amino acid individually from the commonly used RPMI 1640 culture medium was found to result in the induction of apoptosis, with particularly high levels observed following deprivation of essential amino acids.

How might the deprivation of nutrients trigger an apoptotic response? Perreault and Lemieux (88) have found that hybridoma cells undergo apoptosis when their protein biosynthetic machinery is compromised. It may be that deprivation of nutrients such as amino acids has the same effect, possibly resulting in the failure of the synthesis of a critical regulatory protein required to keep the apoptotic pathway in check. Vaux and Strasser (89) suggest that some agents or treatments lead to a reduction in the cellular ATP pool, and that this may be a trigger of apoptosis. They propose that such changes may be interpreted by the cell as being a consequence of viral infection, and the cell responds by inducing the apoptotic pathway. Overexpression of Bcl-2 was found to offer a high degree of protection following deprivation of each individual amino acid, with two exceptions, glutamine and threonine, which exhibited relatively less protection (87). This may indicate that these two amino acids either play a particularly important role in cellular metabolism and biosynthesis, or that they are essential components of the mechanism by which Bcl-2 protects the cell.

The survival of *bcl-2*-transfected cells, even in the absence of supposedly essential amino acids, suggests a reduction in amino acid utilization due to downregulation of nonessential cellular functions. Indeed, metabolic arrest has been reported following interleukin 3 (IL-3) withdrawal from the IL-3-dependent murine cell line Bo, which consequently underwent apoptosis. This state was stabilized by *bcl-2* transfection of the cells, thus extending survival time by 300% (90). Presumably, Bcl-2 also stabilizes the metabolic arrest caused by amino acid starvation in murine hybridoma cells, possibly by maintaining the ATP pool above a threshold level. Moreover, studies by Simpson et al. (8) would suggest that this state may be reversible by feeding the cells with fresh medium. Clearly, such a characteristic is far more desirable than a rapid, and obviously nonreversible, entry into apoptosis that occurs in apoptosis-susceptible cell lines.

Serum. The role of serum in the suppression of apoptosis has been well documented, and consequently, it was of no surprise when it was reported that commercially important hybridoma and plasmacytoma cell lines also undergo apoptosis on withdrawal of serum (85). This would be expected to have important consequences for the development of new serum-free media formulations. Previously, the rational behind the design of such media was not particularly scientific and often involved the inclusion of chemicals that were identified by empirical studies. However, demonstration of the role of serum in the regulation of apoptosis may provide a new avenue of research for the development of novel, and perhaps cheaper, serum-free media. The Bcl-2-mediated suppression of apoptosis following serum withdrawal was the earliest demonstration of the antiapoptosis activity of this gene. A Burkitt's lymphoma cell line transfected with *bcl-2* has been reported to grow better than control vector–transfected cells in commercially available serum-free media without the need for adaptation (91). Similar results were obtained using a *bcl-2* murine hybridoma cell line.

Oxygen Limitation

In large-scale and intensive culture systems, effective aeration of the culture is a major difficulty. Thus, oxygen limitation is often the major limiting factor determining maximum cell number. Studies by Mercille and Massie (92) demonstrated that deprivation of oxygen can induce apoptosis. Subsequently, it has been shown that Bcl-2 overexpression protects hybridoma (8) and Burkitt's lymphoma cells (80) from apoptosis. Clearly this will provide the cells with a considerable advantage in oxygen-limited intensive culture systems.

Hydrodynamic Stress

During the very early days of mammalian cell culture, it was generally assumed that cell lines would be far more sensitive to shear damage due to the absence of a cell wall. Considerable progress has been made in our understanding of the exact interactions that result in the greatest damage to the cells. It is now clear that it is not the shear forces generated at the impeller tip that are responsible for cell death induced by the hydrodynamic environment of the reactor. Far more damage is caused by the events that take place at the gas head space–liquid interface during bubble disengagement (95).

Despite the importance of the hydrodynamic environment, the biochemical response of the cell to this component of the bioreactor environment has been rather neglected. To address this issue Al-Rubeai et al. (91) investigated the mechanism of cell death following exposure of cells to very high agitation levels. Flow cytometric and morphological analysis indicated that cell death occurred mostly by apoptosis, although levels of necrosis were also significant. Singh et al. (82) found that a Burkitt's lymphoma cell line that had been routinely passaged in stationary cultures underwent apoptosis when attempts were made to grow the cells in suspension. *Bcl-2* transfection of this cell line was found to allow much better cell growth in suspension without the need for adaptation. Simpson et al. (8) reported similar behavior of a hybridoma cell line.

Physiological studies have also investigated the relationship between apoptosis and shear stress. Dimmeler et al. (89) have found that shear stress actually prevents induction of apoptosis in endothelial cells in the presence of the inducer tumor necrosis factor α or following growth factor withdrawal. Clearly, the possibility that a sublethal level of shear stress in the bioreactor protects the cells from apoptosis-inducing agents needs to be investigated.

Growth Arrest, Specific Productivity, and Apoptosis

Studies of p53 and c-*myc* have revealed a close relationship between cell proliferation and apoptosis. These studies may have important implications for process optimization strategies that have centered on the control of cellular proliferation. Such an approach allows for improvements in specific recombinant protein productivity during cultures that exhibit a negative correlation between growth rate and productivity. Furthermore, by controlling maximum cell number at an optimal level, cell death due to limitation of nutrients and oxygen should be minimized, thus simplifying medium clarification during downstream processing. However, such studies have had one major drawback: when attempts were made to control cellular proliferation, the cultures rapidly lost viability (93,94), and it would appear that, at least in the case of hybridoma cultures, this was due to the induction of apoptosis (85). Thus strategies designed to control cellular proliferation must also incorporate methods that minimize apoptosis. Initial results would appear to suggest that such an approach does work. Simpson et al. (8) found that Bcl-2 overexpression delays hybridoma cell death following cell cycle arrest induced by thymidine treatment. Similar results have been reported during Burkitt's lymphoma cultures (81).

FUTURE PROSPECTS

The study of apoptosis from an animal-cell-technology standpoint is still in its infancy. However, even at this early stage, significant progress has been made, and it would appear that suppression of apoptosis can lead to improvements in culture viability and, indeed, productivity. It is now necessary to study the prevalence of apoptosis in other culture systems, and then to examine the effect of the suppression of bcl-2 overexpression on cell survival. Studies of a variety of high-density perfusion systems are currently underway. It is also necessary to establish the incidence of apoptosis in other industrially important mammalian cell lines, and to investigate the level of protection afforded by antiapoptosis genes. If the results presented here are found to be reproducible in many different industrial culture systems and cell lines, the study of apoptosis will provide the process biotechnologist with a further strategy in the quest for the optimal production process.

BIBLIOGRAPHY

1. J.F.R. Kerr, A.H. Wyllie, and A.R. Currie, *Br. J. Cancer* **26**, 239–257 (1972).
2. H. Ohyama, T. Yamada, and I. Watanabe, *Radiat. Res.* **85**, 333–339 (1981).
3. N. Thomas and P.A. Bell, *Mol. Cell. Endocrinol.* **22**, 71–84 (1981).
4. C.J. Sanderson, in W.R. Clark and P. Golstein eds., *Mechanisms of Cell Mediated Cytotoxity*, Plenum, New York, 1982, pp. 3–21.
5. F. Oberhammer, J.W. Wilson, C. Dive, I.D. Morris, J.A. Hickman, A.E. Wakeling, P.R. Walker, and M. Sikorska, *EMBO J.* **12**, 3679–3684 (1983).
6. A.H. Wyllie, *Nature* **284**, 555–556 (1980).
7. M. Piacentini, L. Fesus, M.G. Farrace, L. Ghibelli, L. Piredda, and G. Melino, *Eur. J. Cell Biol.* **54**, 246–254 (1991).
8. N. Simpson, A.N. Milner, and M. Al-Rubeai, *Biotechnol. Bioeng.* **54**, 1–16 (1997).
9. K.D. Steck, T.J. Mcdonnell, and A.K. Elnaggar, *Cytometry* **20**, 54–161 (1995).
10. M.G. Ormerod, X.M. Sun, R.T. Snowden, R. Davies, H. Fearnhead, and G.M. Cohen, *Cytometry* **14**, 595–602 (1993).
11. K.H. Elstein and R.M. Zucker, *Exp. Cell Res.* **211**, 322–331 (1994).
12. K.H. Elstein, D.J. Thomas, and R.M. Zucker, *Cytometry* **21**, 170–176 (1995).
13. I. Nicoletti, G. Migliorati, M.C. Pagliacci, F. Grignani, and C. Riccardi, *J. Immunol. Methods* **139**, 271–279 (1991).
14. Y. Tsujimoto, N. Ikegaki, And C.M. Croce, *Oncogene* **2**, 3–7 (1987).
15. V.A. Fadok, D.R. Voelker, P.A. Campbell, J.J. Cohen, D.L. Bratton, and P.M. Henson, *J. Immunol.* **148**, 2207–2216 (1992).
16. G. Koopman, C.P.M. Reutelingsperger, G.A.M. Kuijten, R.M.J. Keehnen, S.T. Pals, and M.H.J. Vanoers, *Blood* **8**, 1415–1420 (1994)
17. M. Castedo, T. Hirsch, S.A. Susin, N. Zamzami, P. Marchetti, A. Macho, and G. Kroemer, *J. Immunol.* **157**, 512–521 (1996).
18. M.C. Raff, *Nature* **356**, 397–400 (1992).
19. S. Nagata, *Adv. Immunol.* **57**, 129–144 (1994).
20. J.C. Rathwell and C.C. Goodnow, *J. Immunol.* **153**, 2831–2842 (1994).
21. G.G. Singer and A.K. Abbas, *Immunity* **1**, 365–371 (1994).
22. P.T. Daniel and P.H. Krammer, *Immunology* **152**, 5624–5632 (1994).
23. R.J. Clem, M. Fechheimer, and L.K. Miller, *Science* **254**, 1388–1390 (1991).

24. J.C. Ameisen and A. Capron, *Immunol. Today* **12**, 102–391 (1991).
25. C. Terai, R.S. Kornbluth, C.D. Pauza, D.D. Richman, and D.A. Carson, *J. Clin. Invest.* **87**, 1710–1715 (1991).
26. H. Groux, G. Torpier, D. Monte, Y. Mouton, J.M. Bourez, A. Capron, and J.C. Ameisen, *J. Exp. Med.* **175**, 331–340 (1992).
27. L. Meyaard, S.A. Otto, R.R. Jonker, M.J. Mijinster, R.P.M. Keet, and F. Miedema, *Science* **257**, 217–219 (1992).
28. D.I. Cohen, Y. Tani, H. Tian, E. Boone, L. Samelson, and H.C. Lane, *Science* **256**, 542–545 (1992).
29. A.G. Laurent-Crawford, B. Krust, Y. Riviere, C. Desgranges, S. Muller, M.P. Kieny, C. Dauguet, and A.G. Hovanessian, *AIDS Res. Human Retroviruses* **9**, 761–773 (1993).
30. N.K. Banda, J. Bernier, D.K. Kurahara, R. Kurrle, N. Haigwood, R.P. Sekaly, and T.H. Finkel, *J. Exp. Med.* **174**, 1099–1106 (1992).
31. T.H. Finkel, G. Tudor-Williams, N.K. Banda, M.F. Cotton, T. Curiel, C. Monks, T.W. Baba, R.M. Ruprecht, and A. Kupfer, *Nature Med.* **1**, 129–134 (1995).
32. D.A. Shub, *Curr. Biol.* **4**, 55–556 (1994).
33. S.V. Lennon, S.J. Martin, and T.G. Cotter, *Cell Prolif.* **24**, 203–214 (1991).
34. L.T. Zhong, T. Sarafian, D.J. Kane, A.C. Charles, S.P. Mah, R.H. Edwards, and D.E. Bredesen, *Proc. Natl. Acad. Sci. U.S.A.* **90**, 4533–4537 (1993).
35. J.W. Larrick and S.C. Wright, *FASEB J.* **4**, 3215–3223.
36. M. Iwata, M. Mukai, Y. Nakia, and R. Iseki, *J. Immunol.* **148**, 3302–3308 (1992).
37. N. Ramakrishnan and G. Catravas *J. Immunol.* **148**, 1817–1821 (1992).
38. B. Brune, P. Hartzell, P. Nicotera, and S. Orrenius, *Exp. Cell Res.* **195**, 323–329 (1991).
39. D.M. Hockenbery, Z.N. Oltvai, X.M. Yin, C.L. Milliman, and S.J. Korsmyer, *Cell* **75**, 241–251 (1993).
40. S. Shimizu, Y. Eguchi, H. Kosaka, W. Kamiike, H. Matsuda, and Y. Tsujimoto, *Nature* **374**, 811–813, (1995).
41. M.D. Jacobson, and M.C. Raff, *Nature* **374**, 814–816 (1995).
42. J.F.R. Kerr, C.M. Winterford, and B.V. Harmon, *Cancer* **73**, 2013–2026 (1994).
43. D.J. Allen, B.V. Harmon, and J.F.R. Kerr, in C.S. Potten ed., *Perspectives on Mammalian Cell Death*, Oxford University Press, Oxford, 1987, pp. 229–258.
44. K.S. Sellins, and J.J. Cohen, *Immunol* **139**, 3199–3206 (1987).
45. L.C. Stephens, K.K. Ang, T.E. Schutheiss, L. Milas, and R.E. Meyn, *Rad. Res.* **127**, 308–316 (1991).
46. K.H. Falkvoll, *Int. J. Radiat. Oncol. Biol. Phys.* **21**, 989–94, 1991.
47. T.H. Forster, D.J. Allan, G.C. Gobe, B.V. Harmon, T.P. Walsh, and J.F.R. Kerr, *Int. J. Radiat. Biol.* **61**, 365–367 (1992).
48. D.P. Lane, *Nature* **362**, 786–787 (1993).
49. M.C. Kiefer, M.J. Brauer, V.C. Powers, J.J. Wu, S.R. Umansky, L.D. Tomei, and P.J. Barr, *Nature* **374**, 736–739 (1995).
50. M.L. Gaido and J.A. Cidlowski, *J. Biol. Chem.* **266**, 580–585 (1991).
51. M.C. Peitsch, B. Polzar, H. Stephan, T. Crompton, H.R. Macdonald, H.G. Mannherz, J. Tschopp, *EMBO J.* **12**, 371–377 (1993).
52. M.A. Barry and A. Eastman, *Arch. Biochem. Biophys.* **300**, 440–450 (1993).
53. Y. Tsujimoto, J. Cossman, E. Jaff, and C.M. Croce, *Science* **228**, 1440–1443 (1985).
54. T.J. McDonnell, N. Deane, F.M. Platt, G. Nunez, U. Jaeger, J.P. Mckearn, and S.J. Korsemeyer, *Cell* **57**, 79–88 (1989).
55. T.J. McDonnell, G. Nunez, F.M. Platt, D.M. Hockenbery, L. London, J.P. Mckearn, and S.J. Korsemeyer, *Mol. Cell. Biol.* **10**, 1901–1907 (1990).
56. A. Strasser, A.W. Harris, M.L. Bath, and S. Cory, *Nature* **348**, 331–333 (1990).
57. S.W. Lowe, E.M. Schmitt, S.W. Smith, B.A. Osborne, and T. Jacks, *Nature* **362**, 847–849 (1993).
58. M.O. Hengartner, and H.R. Horvitz, *Cell* **76**, 665–676 (1994).
59. J. Yuan, S. Shaham, S. Ledoux, H.M. Ellis, and H.R. Horvitz, *Cell* **75**, 641–652 (1993).
60. M. Miura, H. Zhu, R. Rotello, E. Hartwig, and J. Yuan, *Cell* **75**, 653–660 (1993).
61. K. Kiuda, J.A. Lippke, G. Ku, M.W. Harding, D.S. Livingstone, M.S.S. Su, and R.A. Flavell, *Science* **267**, 2000–2003 (1995).
62. M. Miura, H. Zhu, R. Rotello, E.A. Hartwieg, and J.Y. Yuan, *Cell* **75**, 653–660 (1993).
63. S. Kumar, M. Kinoshita, M. Noda, N.G. Copeland, and N.A. Jenkins, *Genes Dev.* **8**, 1613–1626 (1994).
64. A.T. Fernandes, G. Litwack, and E.S. Alnemri, *J. Biol. Chem.* **269**, 30761–30764 (1994).
65. C. Faucheu, A. Diu, A.W.E. Chan, A.M. Blanchet, C. Miossec, F. Herve, V. Collarddutilleul, Y. Gu, R.A. Aldape, J.A. Lippke, C. Rocher, M.S.S. Su, D.J. Livingston, T. Hercend, and J.L. Lalanne, *EMBO J.* **14**, 1914–1922 (1995).
66. N.A. Munday, J.P. Vaillancourt, A. Ali, F.J. Casano, D.K. Miller, S.M. Molineaux, T.T. Yamin, V.L. Yu, and D.W. Nicholson, *J. Biol. Chem.* **270**, 15870–15876 (1995).
67. A.T. Fernandes, G. Litwack, and E.S. Alnemri, *Cancer Res.* **55**, 2737–2742 (1995).
68. M.P. Boldin, E.E. Varfolomeev, Z. Pancer, I.L. Mett, J.H. Camonis, and D. Wallach, *J. Biol. Chem.* **270**, 7795–7798 (1995).
69. A.M. Chinnaiyen, K. O'Rourke, M. Tewari, and V.M. Dixit, *Cell* **81**, 505–512 (1995).
70. M.P. Boldin, T.M. Goncharov, Y.V. Goltsev, and D. Wallach, *Cell* **85**, 803–815 (1996).
71. M. Muzio, A.M. Chinnaiyen, F.C. Kischkel, K. O'Rourke, A. Shevchenko, J. Ni, C. Scaffidi, J.D. Bretz, M. Zhang, R. Gentz, P.H. Krammer, M.E. Peter, and V.M. Dixit, *Cell* **85**, 817–827 (1996).
72. L.C. Cooney, in E.C. Beuvery, J.B. Giffiths, and W.P. Zeijlemaker, eds., *Animal Cell Technology: Developments towards the 21st Century*, Kluwer Academic, Netherlands, 1995, pp. xxv–xxx.
73. M. Al-Rubeai, D. Mills, and A.N. Emery, *Cytotechnology* **4**, 13–28 (1990).
74. F. Franek and J. Dolnikova *FEBS Lett.* **248**, 285–287 (1991).
75. S. Mercille and B. Massie, Induction of apoptosis in nutrient-limited cultures of hybridoma cells. *Biotechnol. Bioeng.* **44**, 1140–1154 (1994).
76. R.P. Singh, M. Al-Rubeai, C.D. Gregory, and A.N. Emery, *Biotechnol. Bioeng.* **44**, 720–726 (1994).
77. A. Moore, C.J. Donahue, J. Hooley, D.L. Stocks, K.D. Bauer, and J.P. Mather, *Cytotechnology* **17**, 1–11 (1995).
78. Y. Itoh, H. Ueda, and E. Suzuki, *Biotechnol. Bioeng.* **48**, 118–122 (1995).
79. E. Suzuki, S. Terada, H. Ueda, T. Fujita, T. Komatsu, S. Takayama, and J.C. Reed, *Cytotechnology* **23**, 55–59 (1997).

80. S. Mercille and B. Massie, *Cytotechnology* **15**, 117–128 (1994).
81. R.P. Singh, A.N. Emery, and M. Al-Rubeai, *Biotechnol. Bioeng.* **52**, 166–175 (1996).
82. R.P. Singh, G. Finka, A.N. Emery, and M. Al-Rubeai, *Cytotechnology* **23**, 87–93 (1997).
83. K. Murray, C.E. Ang, K. Gull, J.A. Hickman, and A.J. Dickson, *Biotechnol. Bioeng.* **55**, 298–304 (1996).
84. N.H. Simpson, R.P. Singh, A. Perani, C. Goldenzon, and M. Al-Rubeai, *Biotechnol. Bioeng.* **59**, 90–98 (1997).
85. D.L. Vaux and A. Strasser, *Proc. Natl. Acad. Sci. U.S.A.* **93**, 2239–2244 (1996).
86. A. Handa-Corrigan, A.N. Emery, and R.E. Spier, *Enzyme Microb. Technol.* **11**, 230–235 (1989).
87. M. Al-Rubeai, R.P. Singh, M.H. Goldman, and A.N. Emery, *Biotechnol. Bioeng.* **45**, 463–472 (1994).
88. J. Perreault and R. Lemieux, *Cytotechnology* **13**, 99–105 (1994).
89. S. Dimmeler, J. Haendeler, V. Rippmann, M. Nehls, and A.M. Zieher, *FEBS Lett.* **399**, 71–74 (1996).
90. M. Al-Rubeai and A.N. Emery, *J. Biotechnol.* **16**, 67–86 (1990).
91. M. Al-Rubeai, A.N. Emery, S. Chalder, and D.C. Jan, *Cytotechnology* **9**, 85–97 (1992).
92. Z. Oltvai, C. Milliman, and S.J. Korsmeyer, *Cell* **74**, 609–619 (1993).
93. M.O. Hengartner, R.E. Ellis, and H.R. Horvitz, *Nature* **356**, 494–499 (1992).
94. S. Henderson, D. Huen, M. Rowe, C. Dawson, G. Johnson, and A.B. Rickinson, *Proc. Natl. Acad. Sci. U.S.A.* **90**, 8479–8483 (1993).

ADDITIONAL READING

E. Yang, J.P. Zha, J. Jockel, L.H. Boise, C.B. Thompson, and S.J. Korsmeyer, *Cell* **80**, 285–291 (1995).

S.N. Farrow, J.H.M. White, I. Martinou, T. Raven, K.T. Pun, C.J. Grinham, J.C. Martinou, and R. Brown, *Nature* **374**, 731–733 (1995).

L.H. Boise, M. Gonzalezgarcia, C.E. Postema, L.Y. Ding, T. Lindsten, L.A. Turka, X.H. Mao, G. Nunez, and C. Thompson, *Cell* **74**, 597–608 (1993).

K.M. Kozopas, T. Yang, H.I. Buchan, P. Zhou, and R.W. Craig, *Proc. Natl. Acad. Sci. U.S.A.* **90**, 3516–3520 (1993).

ASPARTAME

SATOSHI HANZAWA
TOSOH Corp.
Ayase-shi, Japan

KEY WORDS

Aspartame
Enzymatic synthesis
Protease
Protein engineering
Sweetener

OUTLINE

INTRODUCTION

Aspartame, α-L-aspartyl-L-phenylalanine methyl ester (APM) (Fig. 1), is an artificial sweetener whose sweetness is about 200 times stronger than sucrose by weight. Its sweetness was found accidentally by a researcher of Searle Research Laboratories, where APM was produced as an intermediate of a C-terminal tetrapeptide of gastrin (1). Its strong sweetness allows reduction of the amount of calorie sweetener added to foods to achieve a sweetness equivalent to that from sucrose. Although other intense sweeteners often exhibit a slightly bitter aftertaste, APM has none (2). Its sweet taste is close to that of sucrose, a traditional sweetener and the standard of sweetness. Therefore, the demand for APM has grown rapidly since its approval by the Food and Drug Administration (FDA) in 1981. Worldwide APM consumption is currently over 10,000 tons per year, and it is the most common intense sweetener for beverages, tabletop sweetener, ice cream, chewing gum, and so forth. APM is produced by two methods: a chemical method, used by Nutrasweet in the United States and Ajinomoto in Japan, and an enzymatic method, used by Holland Sweetener Company, a joint venture of TOSOH (formerly Toyo Soda Manufacturing) in Japan and DSM in the Netherlands. In the enzymatic method, an APM precursor, *N*-benzyloxycarbonyl-APM (Z-APM) is produced by a protease-catalyzed condensation reaction of *N*-benzyloxycarbonyl-L-aspartic acid (Z-L-Asp) and L-phenylalanine methyl ester (L-PheOMe). The advantages of the enzymatic method are that a racemic substrate can be used and there is no production of β-aspartame (β-APM) (Fig. 1). The reduction of enzyme consumption was, however, desirable for further cost reduction. In the present article, the APM production process with the protease reaction and various efforts to reduce the enzyme consumption therein are described.

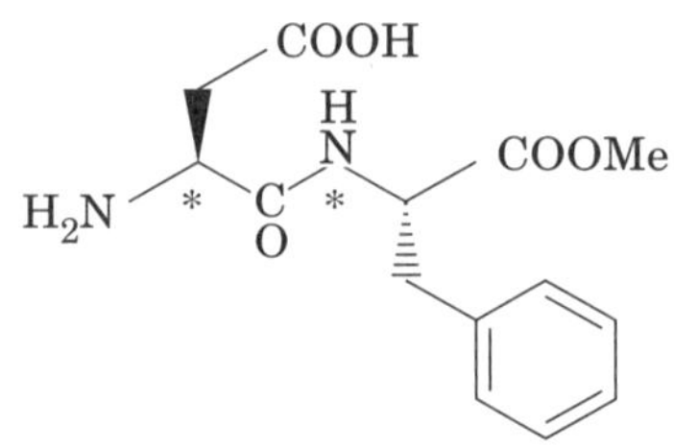

Aspartame

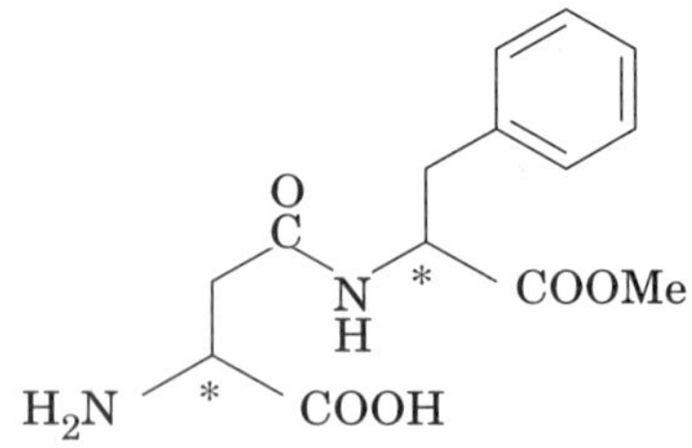

β-Aspartame

Figure 1. Structures of aspartame and its regioisomer. Asymmetric carbon atoms are indicated by an asterisk. The configuration of aspartame is L-α-L.

ASPARTAME PRODUCTION BY PROTEASE

Advantages of Aspartame Production by Protease

The structures of APM and its regioisomer are shown in Figure 1. In APM production by conventional organic synthesis, the side-chain carboxylic group (β-carboxylic group) of aspartic acid must be protected to avoid the formation of unwanted β-APM; additionally, optically pure substrates are required to avoid the production of stereoisomers of APM. These APM-related compounds, such as β-APM and stereoisomers, exhibit a bitter taste (1). Use of a protease as catalyst does not give formation of these undesirable compounds, even if racemic substrates are used as starting materials and without protection of the β-carboxylic group of aspartic acid. The use of racemic phenylalanine in the enzyme process is very advantageous, because L-phenylalanine is more expensive than racemic phenylalanine.

Proteases such as thermolysin, subtilisin, or papain were found to be useful tools for the peptide synthesis because of their high stereo and regio selectivity, and they have been applied in the synthesis of various biologically active peptides, such as peptide hormones, neuropeptides, and insulin (3). APM synthesis catalyzed by a protease was also investigated by Isowa et al. (4). They reported that N-protected aspartic acid and Phe-alkyl ester are condensed to N-protected L-α-Asp-L-Phe-alkyl ester by thermolysin in high yield (83–96%) as a precipitate consisting of the addition compound of the condensation product and one molecule of Phe-alkyl ester, as shown in Scheme 1.

In this scheme, R indicates N-protective group, such as Z or PMZ, and R1 indicates lower alkyl ester, such as methyl or ethyl ester. Proteases usually catalyze the hydrolysis of peptides in water as solvent; the condensation reaction is, in other words, the reverse reaction. Therefore, it is usually difficult to obtain products in high yield because of the occurrence of hydrolysis of the products by the proteases. The synthesis of APM-related peptides can, however, be performed in high yield when the product is removed from the reaction system as a precipitate and is not hydrolyzed back by the protease, for example, by thermolysin.

Although this method requires an expensive catalyst, such as thermolysin, the high stereo and regio selectivity is a great advantage in APM production. The industrial APM production process based on the proposal of Isowa et al. was developed by Toyo Soda Manufacturing and was adopted for APM production by Holland Sweetener Company.

In contrast to the enzymatic method, the chemical method produces an APM precursor where formyl-APM (f-APM) is derived from *N*-f-Asp anhydride and L-PheOMe, (5) as is shown in Scheme 2.

Although β-APM is produced as by-product in this method, α-selectivity of the reaction can be enhanced by selection of the optimal solvent for the reaction, such as alkaline aqueous medium (6) or acetic acid partially replaced with an alkyl ester with a secondary or tertiary alcohol (7). It was reported that the production of β-APM was reduced to 20% of total peptide by selection of a proper solvent. The β-APM reportedly can be converted to APM by alkaline treatment in alcohol in the presence of metal ions such as zinc or copper as catalysts (8). Productivity of APM seems to be enhanced by these methods in the chemical procedure.

As described earlier, the enzymatic and chemical methods both have individual advantages. An advantage of the enzymatic method is the high selectivity of the reaction, where APM is produced from cheap raw materials by a simple procedure in high yield. Not requiring expensive catalysts, such as enzymes, in the condensation step is an advantage of the chemical method; however, expensive starting materials must be used and by-product is produced. Attempting cost reduction in APM production and improving the condensation step are of great importance for APM production by Holland Sweetener Company, and many efforts have been made by both TOSOH and DSM. Hereafter, I describe the enzymatic APM production process and the efforts to improve the condensation reaction.

Aspartame Production by Thermolysin

The principles of the thermolysin process were introduced by Oyama et al. (9) in 1987. At first, L-PheOMe from racemic PheOMe is coupled with Z-Asp stereo and regio selectively by thermolysin. The reaction product, Z-APM, forms an insoluble addition compound with one molecule of remaining D-PheOMe immediately and precipitates. When the reaction is completed, the Z-APM·D-PheOMe addition compound is collected as a solid product, and thermolysin is recovered from the solution. D-PheOMe is separated from Z-APM as its hydrochloride by acid washing (with hydrochloric acid, for example). The Z-group is removed from the Z-APM by hydrogenolysis. Finally, APM is obtained by crystallization and drying. D-PheOMe is recycled after racemization by alkaline treatment and reesterification in methanol. The process is summarized in Figure 2.

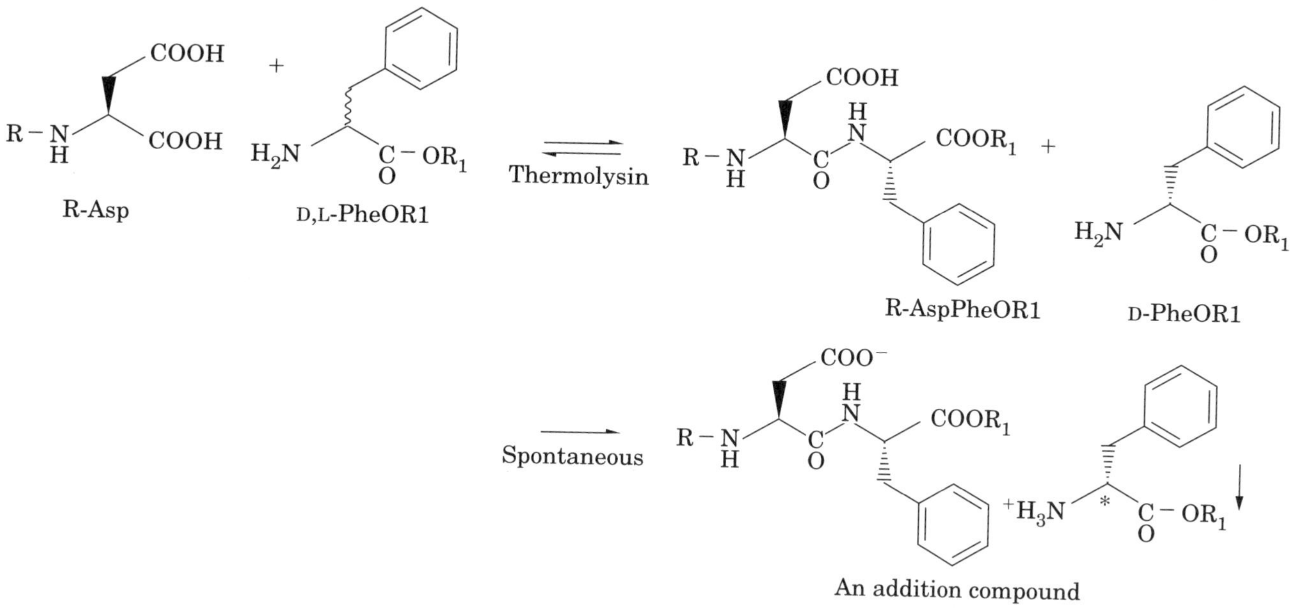

Scheme 1.

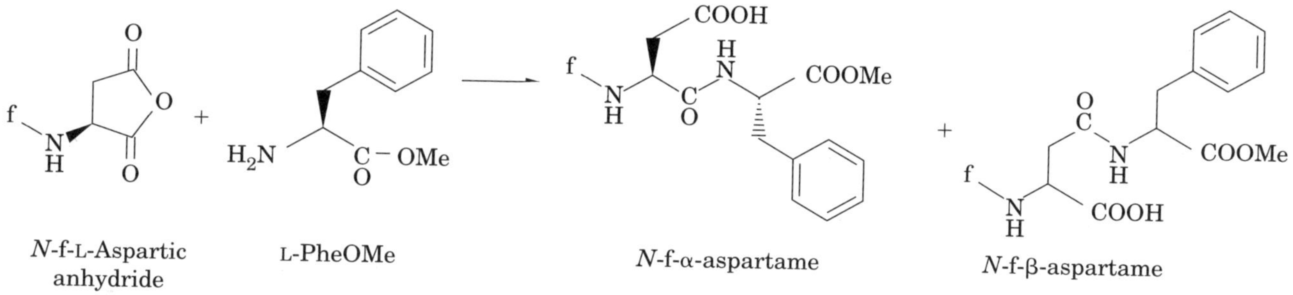

Scheme 2.

Although the β-carboxylic group of aspartic acid does not need to be protected, the α-amino group of aspartic acid and the carboxylic group of phenylalanine must be protected. This plays an important role not only for the selective reaction of the α-carboxylic group of L-aspartic acid and the amino group of L-phenylalanine, but also for being suitable to be used by thermolysin as substrates. Thermolysin can not catalyze the condensation reaction without such protections (9,10). The methyl ester group is used preferably for the protection of the carboxyl group of phenylalanine because this is an essential group expressing the sweetness of APM. The amino group of aspartic acid is protected by the Z-group because it is easily removed by hydrogenolysis without removal of the methyl ester group at the C-terminal of APM. Additionally, a higher reaction rate is obtained when a hydrophobic and bulky group exists at this position. Although formyl or acetyl groups can also be used for the protection, the Z-group, which is hydrophobic and bulky, is more suitable for the reaction rate of the condensation.

Thermolysin and similar microbial metalloproteases (thermolysin-like metalloproteases) were selected from among a number of industrially available proteases. Although other serine or cysteine proteases, such as subtilisin or papain, have been more widely used in various industrial fields, they are not very suitable for APM production because they tend to hydrolyze lower alkyl esters at the C-terminal of peptides (11–13) and also the methyl ester group of L-PheOMe and Z-APM. The cleaving sites of proteases in Z-APM are summarized in Table 1, where residues of Z-APM are numbered as P1, P2, and P3, respectively, for those toward the N-terminal from the cleavage site, and as P′1 and P′2, respectively, for those toward the C-terminal from the cleavage site. Specificity of thermolysin and thermolysin-like metalloproteases is restricted to hydrophobic and bulky amino acids, such as phenylalanine or leucine (14) at the P′1 position. The metalloproteases also have esterase activity, but this activity is restricted in peptide mimic esters like Bz-Gly-OPhe-Ala or furylacryloyl-Gly-OLeu-NH_2 (15). The methyl ester of L-PheOMe or Z-APM is not hydrolyzed by them at all, because the methyl ester group is too small at P′1 position. Z-APM is, therefore, produced by thermolysin or thermolysin-like metalloproteases without by-products.

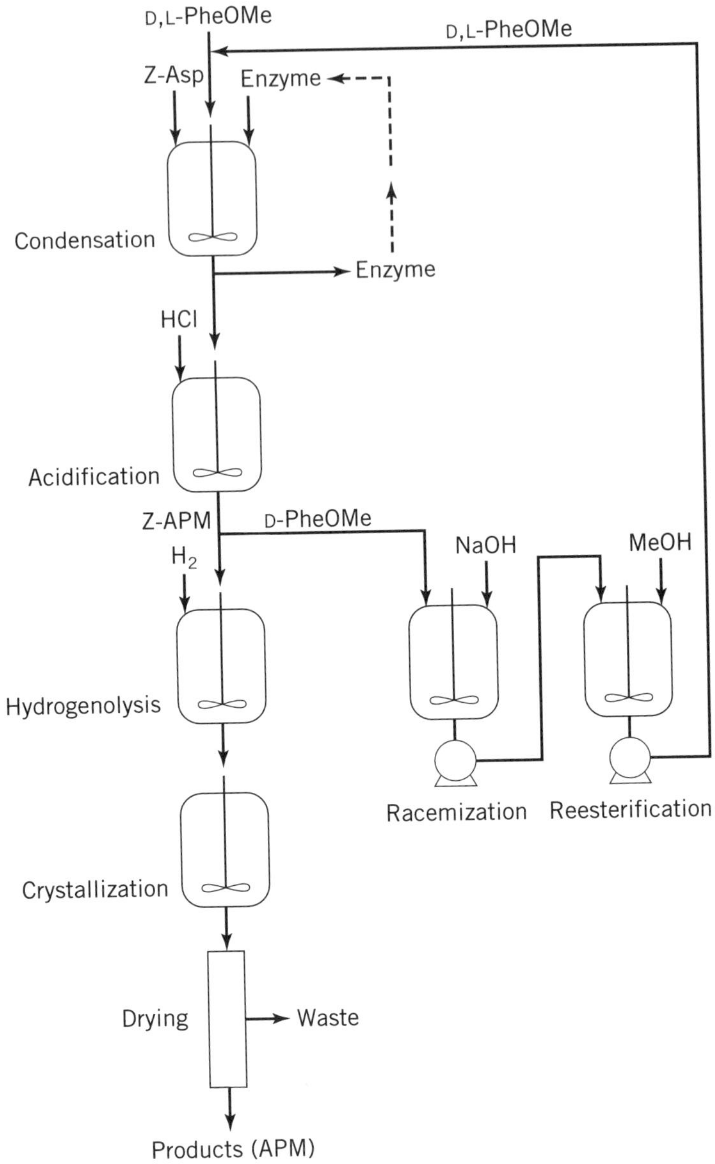

Figure 2. Process of aspartame production via thermolysin catalyzed reaction.

Table 1. Cleavage Sites on Z-APM

Cleavage site	P3 - P2 - P1 - P′1 - P′2 ↑
Thermolysin	Z - Asp - Phe- OMe
Subtilisin	Z - Asp-Phe - OMe
Papain	Z - Asp - Phe- OMe
	Z - Asp-Phe - OMe
V8 protease	Z - Asp - Phe- OMe

IMPROVEMENT OF THE CONDENSATION REACTION

Stability of Enzymes

To improve the condensation reaction in the enzymatic process, reduction of enzyme consumption may be of great importance. Although thermolysin is currently the most stable of the proteases industrially available (16), some portion of it is inactivated during the condensation reaction. This enzyme inactivation, of course, influences production costs. Stabilization of the enzyme during the condensation reaction thus was one of the most important aspects of cost reduction.

It is well known that calcium ion stabilizes many microbial proteases, and thermolysin is also stabilized by it (14). If the concentration of calcium ions in the reaction mixture is increased, it will diminish the inactivation of thermolysin. However, this would not be suitable to improve this process, because high concentration of calcium will frequently cause troubles caused by scaling.

If thermolysin can be replaced by a more stable protease, the enzyme loss can be reduced further. Extremely stable proteases such as archaelysin (17) and a serine protease from *Desulfurococcus* strain SY (18) were recently found in cultures of hyperthermophilic archaea (formerly called archaeobacteria). The archaea, such as *Desulfurococcus, Pyrococcus*, or *Thermococcus*, showed an optimum temperature for growth around 90 °C and even can grow above 100 °C. Their enzymes exhibited extremely high stability against extreme conditions; archaelysin and the protease from *D.* strain SY retained 50% of their activity when they were heated at 95 °C for 80 min and 450 min, respectively (17,18). These proteases are expected to have applicability in various industrial fields (19). Unfortunately, no metalloprotease has yet been found from hyperthermophilic archaea. The proteases separated from the archaea up to now belong to serine proteases (17,18) or cysteine proteases (20). Because their specificity toward Z-APM is similar to that of subtilisin or papain, they are not appropriate for APM production. They will hydrolyze the ester bond of PheOMe or APM in the same way as subtilisn or papain. Although finding a metalloprotease from such archaea would be desirable, thermolysin remains the most stable protease currently available for APM production. Recently a metalloprotease was found from a culture broth of a hyperthermophilic aerobic archaeon "Aeropyrum pernix K1" that also exhibited extremely high stability against thermal denaturation with a half-life of 70 min at 125 °C (21), although its specificity for APM relating peptides has not been solved.

Immobilized Thermolysin

Immobilization of thermolysin was also expected to reduce enzyme inactivation because it is protected from autolysis. However, an application of immobilized thermolysin in the current production process seems difficult, because the reaction for Z-APM production takes place in the presence of a dense solid precipitate, that is, an insoluble addition compound of Z-APM and PheOMe. If both the immobilized enzyme and the reaction product occur in the reaction mixture as solids, they cannot be separated by a simple procedure. Additionally, if the reaction would be performed in a continuous process equipped with an immobilized thermolysin-packed column, the precipitate would block the flow of the reaction solution in the column. Z-APM forms its insoluble addition compound with PheOMe quite easily in water.

To overcome this problem, Oyama et al. (9,22) attempted to perform the reaction in ethyl acetate. Because ethyl acetate dissolves Z-APM without formation of the insoluble addition compound, it was expected that Z-APM could be recovered in solution and that the immobilized thermolysin could be recovered as a solid, thus enabling complete recycling of enzyme (9). Moreover, because the equilibrium state of the synthesis and hydrolysis of Z-APM shifts toward the synthesis side in organic solvents, it was expected that Z-APM could be obtained in high yield in ethyl acetate even though the product is not removed from reaction system in situ. Unfortunately, significant inactivation of immobilized thermolysin was observed in both batchwise operation and continuous operation. Additionally, Z-APM synthesis was disturbed by channeling of organic solvent and water in the immobilized thermolysin column when it was operated continuously (9); thus, high production yield could not be obtained by this system. Only if these problems can be solved, immobilized thermolysin can be applied in the APM production. Studies for APM production by immobilized thermolysin are being continued by various researchers (23,24).

Enzyme Improvement by Site-Directed Mutagenesis

Improvement of enzymes by site-directed amino acid substitutions at selected sites has recently become possible by the progress in genetic engineering and of computer simulation of the three-dimensional structure of the protein molecules. These so-called protein engineering techniques are becoming an important tool for the improvement of enzyme character. A first example of improved thermolysin-like metalloproteases was shown by Imanaka et al. (24) using a metalloprotease from *Bacillus stearothermophilus* CU21. This protease exhibited 85% homology in amino acid sequence with thermolysin. They substituted the 141st glycine of this protease to alanine and expected to find enhancement of the internal hydrophobicity and stabilization of the internal α-helix. The resulting mutant protease indeed exhibited higher thermostability than did wild-type protease (25). Another example of stabilization of this metalloprotease was reported by Hardy et al. (26), who introduced proline at the 65th or 69th site. Although the stability of this protease was lower than that of thermolysin even with introduction of these mutations, it may be expected that thermolysin can be improved by a similar procedure. Thus, studies for the improvement of thermolysin by protein engineering were started by a collaboration of TOSOH and Sagami Chemical Research Center.

In this research, attention was focused to enhancing the activity of thermolysin rather than its stability. Even when the stability of thermolysin is enhanced until the enzyme inactivation during the condensation reaction is at a negligible level, the enzyme has to be recycled for reduction of enzyme costs. This requires additional equipment or procedures to recover the enzyme from the process solution, such as concentration of enzyme by ultrafiltration, salting out, or immobilization of enzyme (9). This will increase the fixed costs at an APM plant. On the other hand, if the activity is enhanced adequately and if the enzyme amount in the APM production can be reduced to a negligible level by application of highly active mutant thermolysins, it would be possible to use the enzyme without recycling. A highly active mutant thermolysin would be more effective than a highly stable one. Additionally, useful knowledge would be gained about the mechanism determining the relationships among the proteins' structures, properties, and functions from the improvement of thermolysin activity.

Thus, studies for improvements of thermolysin were started, and highly active mutant thermolysins were successfully constructed. One of these mutants exhibited a five times higher activity toward Z-APM synthesis when it is evaluated by the initial reaction rates (27). Details of the improvements of thermolysin are described in the article entitled THERMOLYSIN. The stability of these mutant thermolysins was the same as that of wild-type thermolysins when it was determined by calorimetry (28). It is therefore expected that mutant thermolysin can be used for APM production in the same way as wild-type thermolysin, except that the enzyme amount to obtain the same reaction rate is lower with mutant thermolysin.

Evaluation of Mutant Thermolysins in APM Production

The highly active thermolysin exhibiting a five times higher activity than wild-type thermolysin was called TZ-1. TZ-1 has substituted amino acid residues at three sites, namely the 144th leucine to serine (L144S), the 150th aspartic acid to histidine (D150H) and the 227th asparagine to histidine (N227H). Although this mutant exhibits five times higher activity toward Z-APM synthesis, it should be noticed that its activity was evaluated at conditions different from real APM production conditions as described in the article on thermolysin referred to earlier. The evaluation conditions were chosen to reduce the viscosity of the reaction solution; the concentration of substrates was lower than that under conventional production conditions, and the ratio of Z-Asp and PheOMe was set at about 1:1. These conditions enabled to obtain exact data in a milliliter scale of reaction mixture. These conditions are referred to as "model conditions." The model conditions enabled evaluating several candidates during molecular engineering of mutants. Because TZ-1 showed very high activity under the model conditions, its efficiency in small-scale APM production under the conventional production conditions was also examined.

The reactions were performed in 200 mL of a mixture containing 0.2 M of Z-L-Asp, 0.5 M of D,L-PheOMe, 5.1 mM $CaCl_2$, 0.17 M NaCl and 50–150 mg of the thermolysin at 40 °C and in a range of pH between 5 and 7. Mutant thermolysins used here were produced by fermentation of *Bacillus subtilis* MT-2, harboring recombinant plasmid p-UBTZ2 in a jar fermentor equipped with a 16-L vessel (New Brunswick Scientific Co. Ltd., U.S.). Other conditions for the fermentation were the same as described in the article on thermolysin. Enzyme amounts in the reaction mixture were determined by casein hydrolytic activity according to the method of Endo (29) as are also described in the article on thermolysin. The condensation reactions were monitored by HPLC equipped with TSK gel G2000SW (TOSOH, Japan) with 5% acetonitrile in 0.05% trifluoroacetic acid as solvent for determining the amount of the product and the remaining substrates. Before HPLC, the samples were diluted by appropriate solution, such as 10 mM calcium acetate, and the insoluble addition compound of Z-APM and PheOMe was removed by centrifugation. The conversion rates were calculated from remaining Z-Asp using the following formula:

$$CV(\%) = ([Z\text{-}Asp]_0 - [Z\text{-}Asp]_t)/[Z\text{-}Asp]_0$$

Herein, $[Z\text{-}Asp]_0$ and $[Z\text{-}Asp]_t$ refer to the concentration of Z-Asp at the initial and at an appropriate later time, respectively.

Effect of pH on the initial reaction rates by 50 mg of TZ-1 or wild type is shown in Figure 3. The optimum pH of TZ-1 was 6.0, and that of wild type was 6.5. At pH 6.0, the activity of the TZ-1 was about fivefold that of wild type. The optimum pH shift of TZ-1 seems to result from the introduction of a plus charge of histidine instead of the 150th aspartic acid and the 227th asparagine. Because the 150th residue and the 227th residue exist at vicinity of the Zn ion and the 231st histidine residue in the active center of thermolysin, respectively, the plus charge of histidine may influence the electrostatic interactions in this center. Both Zn ion and the 231st histidine are regarded as important to express thermolysin activity (30,31). The mutation N227H especially seems to influence the pH profile of thermolysin activity, because the pK_a value of the 231st histidine is likely to be shifted toward the acidic side by introduction of a plus charge of histidine at the 227th site. This shift in optimum pH is not observed, however, when the activity is evaluated under the model conditions where the optimum pH of both wild type and TZ-1 are pH 7.0, as is mentioned in THERMOLYSIN. An explanation for the difference of the optimum pH between the model conditions described in there and the production conditions described herein have not yet been found, although important factors may be *(1)* concentration of substrates, *(2)* ratio of the two substrates, and *(3)* optical purity of PheOMe.

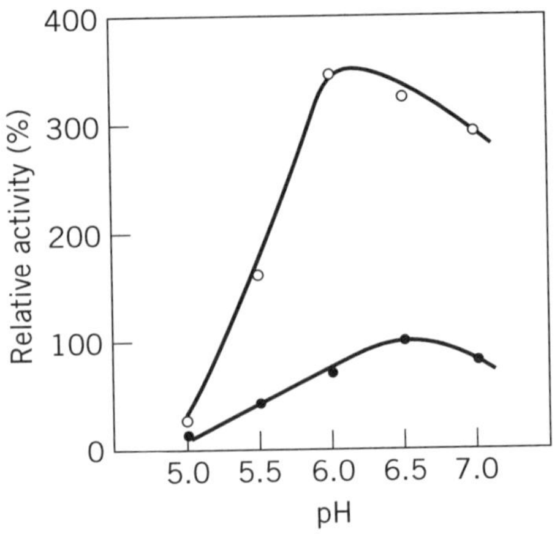

Figure 3. Effect of pH on the initial reaction rate in thermolysin catalyzed condensation reaction. Curves are shown for TZ-1 (open circle) and wild type (closed circle), respectively, for 0.2 M of Z-L-Asp and 0.5 M of D,L-PheOMe and containing 5.1 mM $CaCl_2$ and 0.17 M NaCl at 40 °C.

Typical time courses of the Z-APM synthesis by TZ-1 and wild type at pH 6.0 are shown in Figure 4. Reaction rate and production yield by 50 mg of TZ-1 are the same as those by 150 mg of wild type. When the reaction is performed with 150 mg of TZ-1, the initial reaction rate is three times higher than that by the same amount of wild type. It seems that activity of TZ-1 is three times higher than that of wild type under this condition. Remaining activities of both wild type and TZ-1 after the condensation reaction were 70%. This indicates that the stability of TZ-1 is the same as that of wild type. Moreover, there are no differences in thermostability between TZ-1 and wild type as has been demonstrated by heat capacity analysis with the calorimeter as described earlier. From these results, it follows that mutant thermolysins can be useful for cost reduction in APM production.

Thus, studies for improvement of thermolysin were continued. Another important amino acid substitution, the 150th aspartic acid to tryptophan (D150W) (32) was recently found. The activity of the single site mutant D150W was five times higher than that of wild type. This is the same as the effect of the combination of three mutations, L144S, D150H, and N227H in TZ-1. Combination of D150W with L144S and N227H resulted in a new mutant, L144S-D150W-N227H, which exhibits about 10 times higher activity than wild type under the model conditions

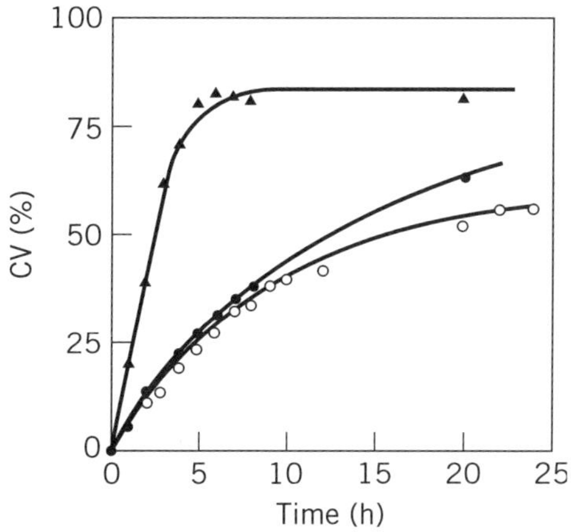

Figure 4. Time course of the condensation reaction of Z-L-Asp and L-PheOMe catalyzed by 150 mg of wild type (closed circle) and 150 mg (closed triangle) and 50 mg (open circle) of TZ-1, respectively, in 200 mL of a reaction mixture containing 0.2 M of Z-L-Asp 0.5 M of D,L-PheOMe, 5.1 mM $CaCl_2$, and 0.17 M NaCl at pH 6.0 at 40 °C.

(32). This mutant is called TZ-5 and will be useful for further reducing of costs in APM production.

Stabilization of Thermolysin by Change of Reaction Conditions

Studies on stabilization of thermolysin by controlling the reaction conditions were also continued separately by a collaboration of TOSOH and DSM. Harada et al. reported such stabilization by an N-protected amino acid, such as Z-Asp, Z-Glu or Z-Gly, during storage of thermolysin solution (33). Z-amino acids seem to act as inhibitors against the autolysis of thermolysin herein. The inhibitory effect of Z-Ala was reported earlier by Morihara and Tsuzuki (10) after they had found that Z-Ala inhibited hydrolysis of Z-Ala-Leu-Ala catalyzed by thermolysin at 30 mM of Ki value. Although these results for Z-Ala are not conclusive for determining whether inhibition by protected amino acids is competitive, it now appears that Z-amino acids are likely to bind to the active site of thermolysin instead of substrates.

Moreover, Harada et al. demonstrated that thermolysin in the reaction mixture was kept stable when Z-Asp concentration is kept at a high level during the condensation reaction. They studied effects of the ratio of Z-Asp and D,L-PheOMe at the start of the condensation reaction. An example of the results is summarized in Table 2. When the reaction was started at a 1:2 ratio of Z-Asp to PheOMe, the enzyme recovery was 92%, whereas it was 44% at a ratio of 1:2.5. Remaining Z-Asp after the reactions were 55.9 and 10.5 mmol/kg, respectively. At the ratio of 1:2, much more enzyme and Z-Asp were remaining than at the ratio of 1:2.5. The remaining Z-Asp apparently protects the enzyme from inactivation. Productivity of Z-APM at the ratio of 1:2 was 0.225 mol/kg; at 1:2.5 it was 0.238. Thus, significant decrease of productivity was not observed by higher concentration of Z-Asp. This indicates that the synthesis is not inhibited by Z-Asp, whereas Z-Asp is an inhibitor against autolysis or peptide hydrolysis. Nagayasu

Table 2. Condensation Reaction at Different Z-Aspartic Acid-to-PheOMe Ratio

Z-Asp/ PheOMe ratio	Initial [Z-Asp] (mol/kg)	Conversion rate (%)	Produced Z-APM (mol/kg)	Remained [Z-Asp] (mmol/kg)	Enzyme recovery (%)
1:2	0.28	80.5	0.225	55.9	92
1:2.5	0.25	95.2	0.238	10.5	44

et al., however, reported that Z-Asp inhibited the condensation reaction by immobilized thermolysin (24). The inhibition effect toward the condensation reaction can be avoided by choosing the reaction conditions, as suggested by Harada et al. According to Harada et al. (33), enzyme inactivation may be caused by adsorption of thermolysin on the surface of crystals of the addition compound of Z-APM and PheOMe. Apparently, Z-Asp is likely to prevent such adsorption.

In contrast to the protection by Z-Asp, thermolysin is not protected from inactivation by L-PheOMe, although the latter is another substrate of the condensation reaction; this may be because of much higher K_m value of thermolysin toward L-PheOMe. According to results of earlier kinetic studies of the condensation reaction by Oyama et al. (34), a double reciprocal plot toward L-PheOMe and D,L-PheOMe showed no apparent K_m value, whereas that toward Z-Asp is 10.3 mM. These authors presented a reaction model where thermolysin at first forms an enzyme-Z-Asp complex, and then Z-APM is produced by the attack of L-PheOMe on the complex, as shown in Scheme 3. In this scheme, TLN indicates thermolysin. This model suggests that thermolysin cannot form an enzyme–L-PheOMe complex to prevent autolysis and adsorption on crystals of the addition compound. Therefore, thermolysin does not seem to be stabilized by L-PheOMe.

The stabilization by Z-Asp is quite effective for thermolysin stabilization during APM production because the effect can be obtained quite simply by adjusting the concentration of the substrates. Moreover, Harada et al. mentioned (33) that Z-Asp stabilized thermolysin without presence of calcium ions. This is another advantage of this stabilization method because various problems caused by calcium scaling can be avoided by this method. As described earlier, enzyme inactivation during the condensation reaction can be prevented by the adjusting the ratio of the two substrates. Recently another method to stabilize thermolysin during condensation was revealed by Harada et al. (35): an addition of toluene into the reaction mixture at about 30% (w/w) of low materials. It prevented adsorption of thermolysin onto the crystal surface of the addition compound of Z-APM and PheOMe and resulted in an increase of thermolysin recovery after the reaction.

$$\text{Z-Asp+PheOMe+TLN} \rightleftharpoons \text{Z-Asp}-\text{TLN+PheOMe} \longrightarrow \text{Z-APM+TLN}$$

Scheme 3.

ALTERNATIVE PROCESSES

Condensation without N-Protection

Various strategies have been used in the attempt to improve the APM production process. The possibility of reducing enzyme consumption was indicated by mutant thermolysins and by adjusting of the concentration of substrates. On the other hand, an alternative production system is desired to take long strides in reducing costs. Much of this research has been focused on avoiding N-protection of aspartic acid. If APM can be produced without N-protection, it will not only reduce raw materials costs but will also simplify the production process by eliminating the protection and de-protection steps. Although low yields have prevented industry adoption of a new method, the production costs of APM would be further reduced by N-protection-free methods if productivity could be improved. Hereafter, I summarize such attempts as appeared in patent applications.

Direct production of APM by condensation of aspartic acid and PheOMe was proposed by TOSOH and Ajinomoto, respectively, where both starting materials are condensed by microbial cells belonging to *Pseudomonas, Alkaligenes*, and so forth (36,37), as shown in Scheme 4. In this process, the reaction seems to be catalyzed by aminopeptidases in the bacterial cells. Although this is a very simple production method, its yield was very low. According to the TOSOH patent application (36), only 0.06 g of APM was produced from 1.0 g of aspartic acid after 16 h of reaction. Because APM is not removed from the reaction mixture as a precipitate under this condition, it seems to be hydrolyzed back by the aminopeptidases. If the product could be separated from the reaction mixture in situ, the production yield will be increased. It is, however, difficult to remove APM from the reaction mixture as a precipitate, because APM is not only highly soluble (similar to the substrates) but also does not form an insoluble addition compound with remaining substrate. No other additive that would form an insoluble addition compound with APM has yet been found. Although separation of APM by electrodialysis was attempted by Snedecor and Hsu (38), sufficient separation yields do not seem to be obtained. Development of an effective method to separate APM from its starting materials and enzymes will be of importance in establishing a feasible production process based on those patent applications.

Another APM production method without N-protection is the coupling of α-AspOR (R indicates a lower alkyl group) and PheOMe by an ester–amino exchange reaction catalyzed by a protease from *Staphylococcus aureus* strain V8 (V8 protease) (39) as shown in Scheme 5. Because this reaction is a dealcoholation reaction, the reaction rate is higher than that of a dehydration reaction, such as the reaction shown in Scheme 4. Additionally, by-products due to misconfiguration, such as PheAspOR or PhePheOMe, are not produced because V8 protease is specific for a peptide bond or ester bond that involves the carbonyl group of amino acids with an acidic side chain (40), for example, aspartic acid at the P1 position (Table 1). However, removal of APM from the reaction system encounters the same difficulties as in the first example of the APM direct production; therefore, a higher production yield cannot be expected. It will also be difficult to produce α-AspOR industrially without formation of β-AspOR and $Asp(OR)_2$, because the reactivities of the two carboxyl group of aspartic acid are not so different. Thus, various barriers exist to establishing a production process based on this method.

Production from Other Raw Materials

The methods described in the previous section are aimed at coupling of aspartic acid and PheOMe without protection; the methods described in this section do not depend on aspartic acid as the starting material. Because aspartic acid can be produced by ammonia addition to fumaric acid by an ammonia lyase, APM synthesis from fumaric acid, PheOMe, and ammonia as starting materials was proposed by TOSOH (41). *Pseudomonas putida* TS-15001 was found to be catalyzing the coupling of the three substrates simultaneously. Although all details of this reaction system have not been studied, this system seems to consist of two reactions in the cells. At first, aspartic acid is produced from fumarate by the ammonia addition reaction by an ammonia lyase; second, aspartic acid is coupled with PheOMe by an aminopeptidase. The low production yield seems to result from the hydrolysis of APM by the aminopeptidase in the same way as in the first example of the direct production.

Production of APM from similar substrates was proposed by Showa Denko Corporation (42), where APM was derived from fumaric-PheOMe (FPM) by ammonia addition reaction by certain bacterial cells that produced an ammonia lyase, as shown in Scheme 6. In this proposal, FPM was derived from its *cis-trans* isomer, maleic-PheOMe (MPM) which is more easily produced by the coupling reaction of maleic anhydride and PheOMe than by the direct production of FPM by the coupling reaction of fumaric acid and PheOMe. When fumaric acid and PheOMe are coupled directly, use of a coupling reagent is required, such as oxalyl chloride, thionyl chloride, or carbodiimide. However, the coupling of maleic anhydride and

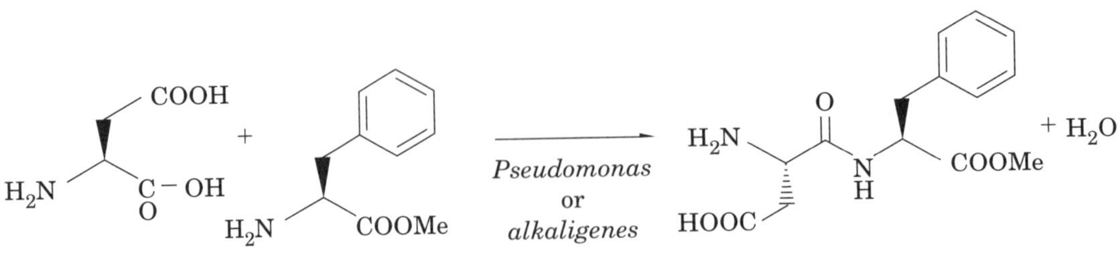

Scheme 4.

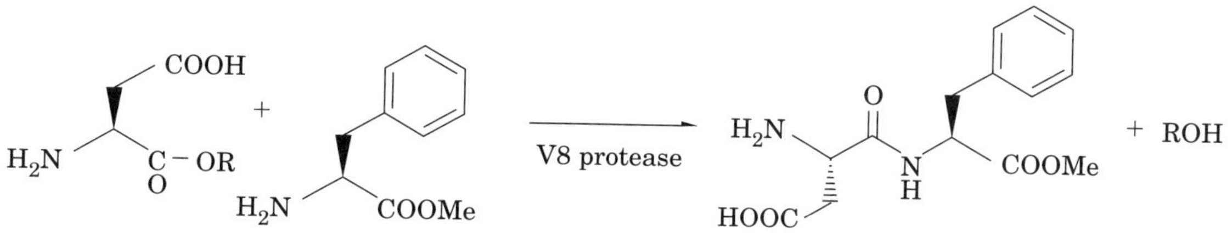

Scheme 5.

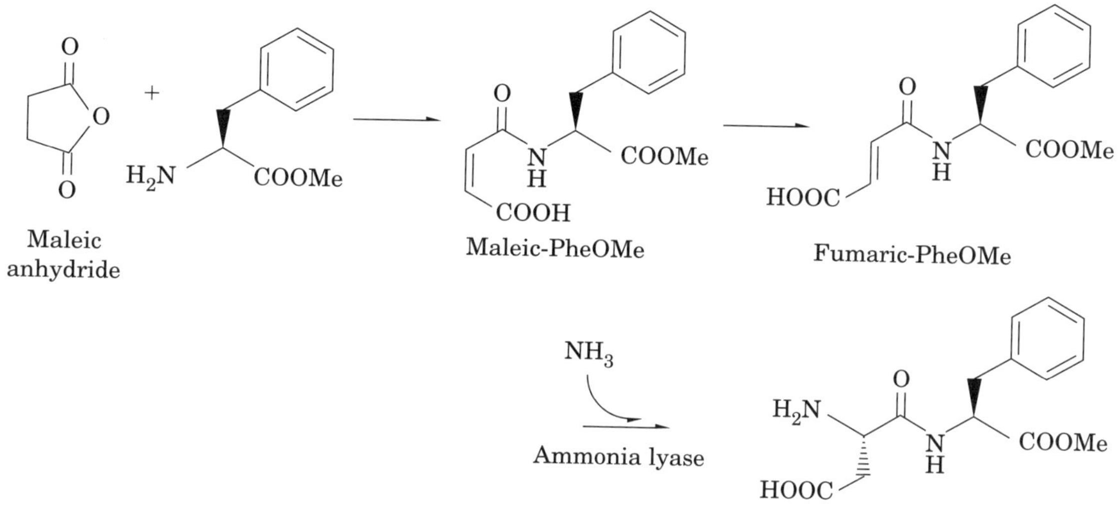

Scheme 6.

PheOMe proceeds spontaneously. Fumaric-PheOMe was reportedly produced in good yield by isomerization of MPM (40). Unfortunately, the yield at the step of ammonia addition is too low to be applicable in an industrial APM production process. A new microbial strain exhibiting higher lyase activity has recently been reported by Xu et al. (43), but even with this strain, productivity seems to be insufficient. Further screening of bacteria and improvement of ammonia lyase thus appears to be important.

CONCLUSION

Because APM is an excellent high-intensity sweetener, especially in its taste, demand for it will grow in the future, and APM will gradually become a commodity chemical. The drive for cost reductions will therefore increase. Although alternative production methods have been proposed, their yields are lower than the current, conventional production methods. The current processes will therefore continue to be used while attempts to improve them are ongoing. I focused on the reduction of enzyme costs to advance enzymatic APM production because the condensation by thermolysin is a key step of the production process used by Holland Sweetener Company. Enzymatic APM production processes will be further improved in many ways.

ACKNOWLEDGMENTS

Improvement of thermolysin was supported by Holland Sweetener Company. We are grateful for analysis of mutant thermolysin in production condition by Shin-ichiro Nakamura, Atsuo Aoyama, Yoshikazu Tanaka, and Akira Tokuda. We would like to give special thanks to Kiyotaka Oyama and Ichiro Hiraki for the organization of this work.

BIBLIOGRAPHY

1. R.H. Mazur, in L.D. Steging and L.J. Filer, Jr. ed. *Aspartame Physiology and Biochemistry*, Dekker, New York, 1984, pp. 3–9.
2. M. Tajima, in S. Tsumura and T. Itano ed. *Foods Bio Technology*, Sangyo Chosakai, Tokyo, 1988, pp. 132–134.
3. V. Kashe, in R.J. Beynone and J.S. Bond ed. *Proteolytic Enzymes*, Information Press, Oxford, U.K., 1989, pp. 125–143.
4. Y. Isowa, M. Ohmori, T. Ichikawa, K. Mori, Y. Nonaka, K. Kihara, K. Oyama, H. Saitoh, and S. Nishimura, *Tetrahedron Lett.* **28**, 2611–2612 (1979).
5. U.S. Pat. 3,786,039 (April 15, 1970), Y. Ariyoshi and M. Nagao (to the Ajinomoto Co. Inc.).
6. JP Pat. 2,513,159 (April 30, 1996), T. Takemoto, K. Hisamitsu, and T. Yukawa (to the Ajinomoto Co. Inc.).

7. EP Pat. 0468063B1 (June 5, 1996), J.B. Hill, R.A. Erickson, Y. Gelman, K. Kuang, H.L. Dryden Jr., and M.R. Hohnson (to the Nutrasweet Company).
8. JP Pat. 2,503,592 (June 21, 1988), S. Takahashi and T. Takemoto (to the Ajinomoto Co. Inc.).
9. K. Oyama, S. Irino, and N. Hagi, *Methods Enzymol.* **136**, 503–516 (1987).
10. K. Morihara and H. Tsuzuki, *Eur. J. Biochem.* **15**, 374–380 (1970).
11. M. Ottesen and I. Svendsen, *Methods Enzymol.* **19**, 199–215 (1970).
12. R. Arnon, *Methods Enzymol.* **19**, 226–244 (1970).
13. E.E. Ricks, M.C. Estrada-Valdes, T.L. McLean, and G.A. Iacobucci, *Biotechnol. Prog.* **8**, 197–203 (1992).
14. H. Matsubara, *Methods Enzymol.* **19**, 642–649 (1970).
15. B. Holmquist and B.L. Vallee, *Biochemistry* **15**, 101–107 (1976).
16. K. Aunstrup, in A.H. Rose ed. *Microbial Enzymes and Bioconversions*, Academic Press, New York, 1980, pp. 50–114.
17. D.A. Cowan, K.A. Smolenski, R.M. Daniel, and H.W. Morgan, *Biochem. J.* **247**, 121–133 (1987).
18. S. Hanzawa, T. Hoaki, H.W. Jannasch, and T. Maruyama, *J. Marine Biotechnol.* **4**, 121–126 (1996).
19. T. Coolbear, R.M. Daniel, and H.W. Morgan, *Adv. Biochem. Eng. Biotechnol.* **45**, 58–98 (1995).
20. M. Morikawa, Y. Izawa, N. Rashid, T. Hoaki, and T. Imanala, *Appl. Environ. Microbiol.* **60**, 4559–4566 (1994).
21. Y. Sako, P.C. Croocker, and Y. Ishida, *FEBS Lett.* **415**, 329–334 (1997).
22. K. Oyama, S. Nishimura, Y. Nonaka, K. Kihara, and T. Hashimono, *J. Org. Chem.* **46**, 5341–5242 (1981).
23. K. Nakanishi, A. Takeuchi, and R. Matsuno, *Appl. Microbiol. Biotechnol.* **32**, 633–636 (1990).
24. T. Nagayasu, M. Miyanaga, T. Tanaka, T. Sakiyama, and K. Nakanishi, *Biotechnol. Bioeng.* **43**, 1118–1123 (1994).
25. T. Imanaka, M. Shibazaki, and M. Takagi, *Nature* **324**, 695–697 (1986).
26. F. Hardy, G. Vriend, O.R. Veltman, B. van der Vinne, G. Venema and V.G.H. Eijsink, *FEBS Lett.* **317**, 89–92 (1993).
27. EP Pat. 0616033A1 (May 17, 1993), H. Nagao, T. Yoneya, T. Miyake, A. Aoyama, K. Kai, S. Kidokoro, Y. Miki, K. Endo, and A. Wada (to the Holland Sweetener Company and Sagami Chemical Research Center).
28. S. Kidokoro, K. Endo, Y. Miki, A. Wada, T. Yoneya, A. Aoyama, K. Kai, H. Nagao, T. Miyake, S. Hanzawa, K. Misawa, S. Ooe, T. Hanya, and H. Yagame, *Seibutu butsuri* **35**, supplement S19 (1995).
29. S. Endo, *J. Ferment. Technol.* **40**, 346–353 (1962).
30. S. Toma, S. Campagnol, E.D. Gregoriis, R. Gianna, I. Margrit, M. Zamai, and G. Grand, *Protein Eng.* **2**, 359–364 (1989).
31. W.R. Kester and B.W. Matthews, *J. Biol. Chem.* **252**, 7704–7710 (1977).
32. JP Pat. 9,402,050 (June 15, 1995), Y. Tanaka, T. Miyake, S. Hanzawa, S. Ohe, S. Kidokoro, Y. Miki, K. Endo, and A. Wada (to the Holland Sweetener Company and Sagami Chemical Research Center).
33. JP Pat. 0,8131,170 (May 28, 1996), T. Harada, Y. Kunizawa, K. Oyama, C.M.Y. Smith, S. Takasuga, and W.J.J. van den Tweel (to the Holland Sweetener Company).
34. K. Oyama, K. Kihara, and Y. Nonaka, *J. Chem. Soc., Perkins Trans.* 2 356–360 (1981).
35. EP Pat. 0768384A1 (September 23, 1996), T. Harada, S. Irino, Y. Kunisawa, and K. Oyama (to Holland Sweetener Company).
36. JP Pat. 9,014,040 (April 5, 1990), T. Harada, K. Takemoto, and T. Igarashi (to the TOSOH Corp.).
37. JP Pat. 9,015,196 (April 11, 1990), K. Yokoseki and K. Kubota (to the Ajinomoto Co. Ltd.).
38. U.S. Pat. 5,055,399 (Oct. 8, 1991), B.R. Snedecor and C.C. Hsu (to the Genentech Inc.).
39. EP Pat. 0269390B1 (Nov. 21, 1986), M.G. Boston and A.J. Poulose (to the Genencor Int. Inc.).
40. G.R. Drapeau, *Methods Enzymol.* **45**, 469–475 (1976).
41. JP Pat. 912,520 (Jan. 16, 1991), T. Harada, H. Takemoto, and I. Igarashi (to the TOSOH Corp.).
42. JP Pat. 63,185,395 (July 30, 1988), A. Nakayama and Y. Torigoe (to the Showa Denko Corp.).
43. J. Xu, W. Wu, S. Jin, and W. Yao, *J. China Pharm. Univ.* **25**, 53–55 (1994).

See also AMINO ACIDS, PRODUCTION PROCESSES.

ASPARTIC ACID. See AMINO ACIDS, PRODUCTION PROCESSES; L-ASPARTIC ACID.

L-ASPARTIC ACID

MIKI KOBAYASHI
MASATO TERASAWA
HIDEAKI YUKAWA
Mitsubishi Chemical Corporation
Ibaraki, Japan

KEY WORDS

Aspartase
L-Aspartic Acid
Coryneform bacteria
Isomerase
Ultrafiltration

OUTLINE

INTRODUCTION

L-Aspartic acid is widely used as a food additive as well as an ingredient of infusion solutions and medicines. In addition, a dipeptide, L-aspartyl-L-phenylalanyl methyl ester, is used as a synthetic low-calorie sweetener in many countries. Moreover, studies for production and application of L-aspartic acid derivatives have been carried out by many companies in Europe, America, and Japan because L-aspartic acid is also chemically reactive. L-Aspartic acid derivatives are hoped to function as a kind of biodegradable precursor for environment friendly products (Fig. 1). It is presumed that each of these would be produced at the 1,000-kiloton level in the near future. However, the high price of L-aspartic acid produced using the existing process has been a high hurdle in expansion of its use. Therefore, we tried to develop a novel process to produce L-aspartic acid economically.

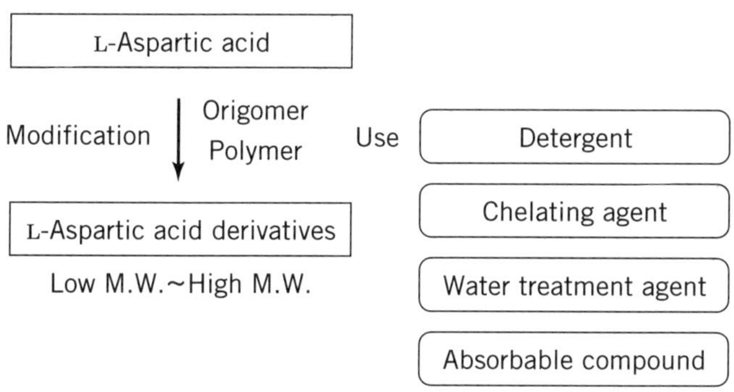

Figure 1. Uses of aspartic acid.

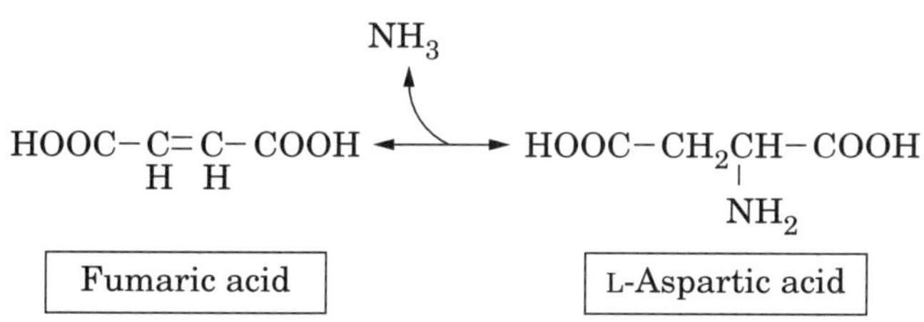

Figure 2. Pathway of L-aspartic acid formation from fumaric acid.

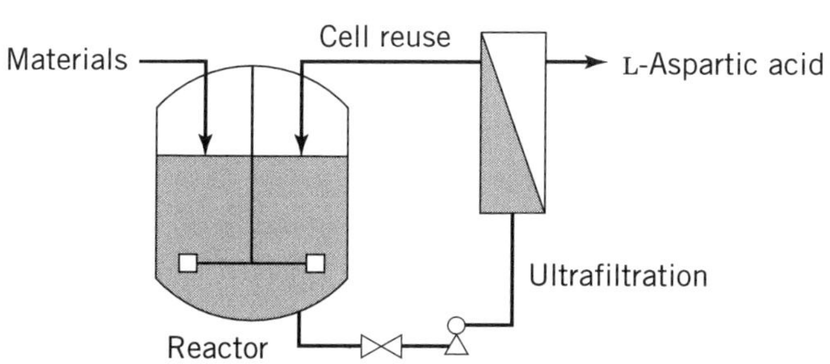

Figure 3. Membrane reactor system

CONVENTIONAL METHODS

Since Quastel and Woolf (1) observed the reversible formation of L-aspartic acid from fumaric acid and ammonium ion by a cell suspension of *Escherichia coli*, many reports have been published on aspartase (E.C.4.3.1.1), which catalyzes this reaction. This enzyme degrades L-aspartic acid into fumaric acid and ammonium ion under physiological conditions, but the reverse reaction proceeds in high concentrations of ammonia (Fig. 2).

High-yield conversion of L-aspartic acid from fumaric acid and ammonia was first reported by Sumiki et al. (2) in 1928. This method employed resting yeast cells to produce L-aspartic acid. Fermentation methods using medium containing fumaric acid, and enzymatic methods using bacterial aspartase activity have also been reported (3–5).

A process for industrial production of L-aspartic acid was first established in 1973; bacterial cells having aspartase activity were immobilized in a gel matrix to stabilize the enzyme activity. A continuous operation was also reported (6–9). In the immobilized-enzyme process, the enzyme cost was reduced compared with that of a batch process.

However, in the immobilized-cell process, cells had to be immobilized and packed into reactor columns under aseptic conditions. Contamination-free operations were needed, which resulted in additional costs. Also, it did not seem to be possible to use the reactor columns for conventional purposes because different specifications were required according to each individual reaction mode or the chemical properties of substrate and product.

PROPERTIES OF A NOVEL BIOREACTOR SYSTEM

As already mentioned, there are several biological and engineering disadvantages that need to be solved in conventional production methods. To overcome these disadvantages, we have developed a novel bioreactor process, the membrane reactor system, using microbial cells without artificial immobilization for the enzymatic reaction. Coryneform bacterium MJ-233, which is employed in this process, is characterized by nonlytic properties under nongrowing conditions. The cost of the enzyme is low because the bacterial cells can be reused repeatedly for enzymatic reactions without immobilization. Intracellular components such as protein and nucleic acid do not leak out, so the purity of the products is very high. From the engineering stand point, the ultrafiltration membrane system is employed for recycling bacterial cells (Fig. 3). The advantages of the ultrafiltration process are as follows:

1. Damage of bacterial cells by compression is reduced compared with centrifugation.
2. Capital outlay and running costs are relatively low.
3. Operation and maintenance are easier.

On the basis of these properties, we established an economical new bioprocess for the production of L-aspartic acid (10–13).

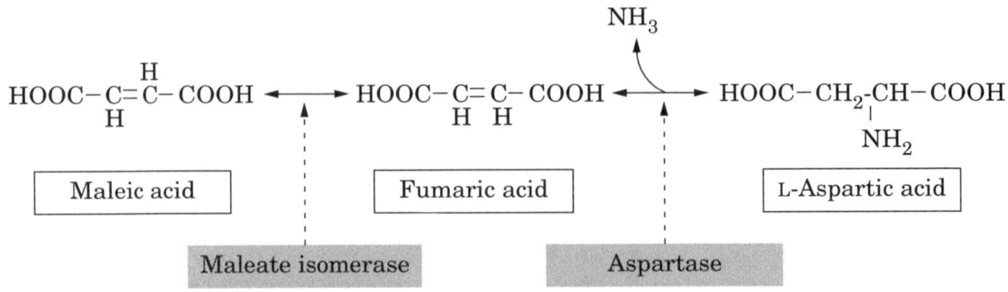

Figure 4. L-Aspartic acid formation from maleic acid

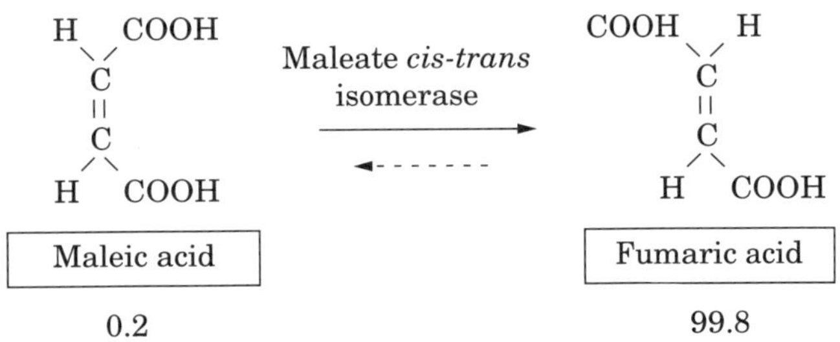

Figure 5. Equilibrium of the maleate isomerase reaction

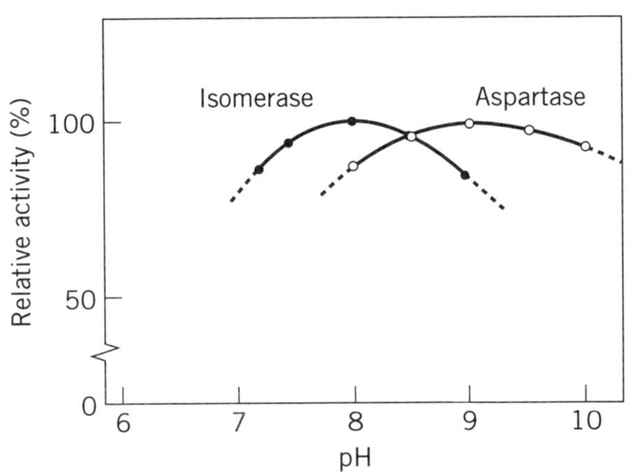

Figure 6. The relationship between pH and relative enzyme activity.

TECHNICAL COMPONENTS

Enzymes for L-Aspartic Acid Production

We tried to develop a new bioprocess for L-aspartic acid production using maleic acid as a starting material. Maleate isomerase (E.C.5.2.1.1), which isomerizes maleic acid to fumaric acid, and aspartase (E.C.4.3.1.1), which converts fumalic acid and ammonium ion to L-aspartic acid, were used for L-aspartic acid production from maleic acid (Fig. 4). The equilibrium of the isomerase reaction strongly favors fumaric acid formation (Fig. 5). This results in almost complete conversion of maleic acid to fumaric acid, so it is very useful for L-aspartic acid formation. On the other hand, the equilibrium of the aspartase reaction also favors L-aspartic acid formation under high concentrations of ammonium ion. Therefore, the conversion yield from maleic acid to L-aspartic acid can reach 99%.

Conditions for L-Aspartic Acid Production

In order to establish a one-pot reactor system using two enzymes, maleate isomerase and aspartase, we analyzed the properties of each enzyme, including pH profile (Fig. 6). Consequently, we could determine the potential for coexistence of two enzymes in a one-pot reactor even in the high concentrations of ammonium ion that are an essential condition for L-aspartic acid formation by aspartase.

Improvement of L-Aspartic Acid Productivity

In order to improve the productivity of the process, genetic engineering techniques were applied to coryneform bacterial strain MJ-233. An efficient transformation system in

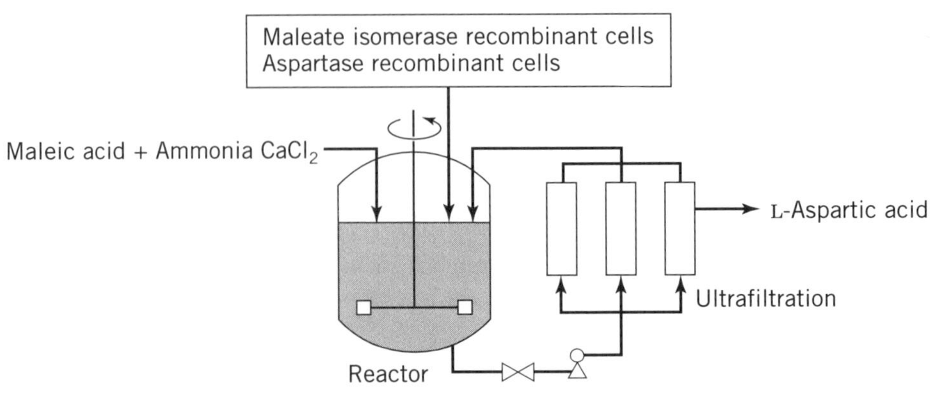

Figure 7. The L-Aspartic acid production process

coryneform bacteria using electroporation (14–16) and other techniques has also been developed (17,18). Both maleate isomerase and aspartase can be overproduced in coryneform bacterial strain MJ-233. These recombinant cells can therefore be employed for L-aspartic acid production from maleic acid as a starting raw material.

Development of a Stable Host–Vector System. One of the important factors in the application of genetic engineering techniques for industrial use is the development of a stable host–vector system. For this purpose, we isolated various plasmids from coryneform bacteria and identified a genetic element that is important for stable plasmid maintenance (19). We then constructed a highly stable host–vector system.

Cloning and Expression of the Maleate Isomerase and Aspartase Genes. Isolation of both the maleate isomerase gene and the aspartase gene from a genomic library was achieved by hybridization with DNA probes synthesized on the basis of the N-terminal amino acid sequences of each purified enzyme. Using this methodology, we cloned maleate isomerase (20) and aspartase (21) genes from the *Alcaligenas faecalis* and coryneform bacterial strain MJ-233, respectively, and achieved high-level expression in recombinant cells of coryneform bacterial strain MJ-233.

Membrane Reactor System

Coryneform bacterial strain MJ-233 exhibited nonlytic properties under nongrowing conditions and, therefore, showed no leakage of intracellular macromolecules, as already mentioned. Taking advantage of these physiological characteristics, an ultrafiltration system was adopted for the separation and recycling of cells from the reaction mixture. Employment of the ultrafiltration system greatly simplified the purification step (Fig. 7). In this ultrafiltration system, the filtration capacity can be easily modified by varying the number of modules. A hollow-fiber tubular-type membrane made of polysulfon, which can withstand high temperature and alkaline pH, was chosen for our process. We can operate this system at high temperature and alkaline pH, allowing us to carry out a contamination-free process.

L-Aspartic acid can be produced at high purity using this membrane reactor system and, therefore, more economically compared with conventional fermentation processes.

CONCLUSION

The host–vector system in coryneform bacterial strain MJ-233 was developed and L-aspartic acid can now be economically produced at high yield using recombinant intact cells in a membrane reactor system. We hope that our method can provide the quantities of L-aspartic acid needed to meet the increasing demands for biodegradable chemicals in an expanding market.

BIBLIOGRAPHY

1. J.H. Quastel and B. Woolf, *Biochem. J.* **20**, 545–555 (1926).
2. Y. Sumiki, *Bull. Jpn. Soc. Ferment.* **23**, 33–37 (1928).
3. M. Kisumi, Y. Ashikaga, and I. Chibata, *Bull. Agric. Chem. Society. Soc. Jpn.* **24**, 296–305 (1960).
4. S. Kinoshita, K. Nakayama, and S. Kitada. *Bull. Jpn. Soc. Ferment.* **16**, 517–520 (1958).
5. K. Kitahara, S. Fukui, and M. Misawa, *Bull. Agric. Chem. Soc. Jpn.* **34**, 44–48 (1960).
6. T. Tosa, T. Sato, T. Mori, Y. Matsuo, and I. Chibata, *Biotechnol. Bioeng.* **15**, 69–84 (1973).
7. I. Chibata, T. Tosa, and T. Sato, *Appl. Microbiol.* **27**, 878–885 (1974).
8. T. Sato, T. Mori, T. Tosa, I. Chibata, M. Furui, Y. Yamashita, and A. Sumi, *Biotechnol. Bioeng.* **17**, 1797–1804 (1975).
9. Y. Nishida, T. Sato, T. Tosa, and I. Chibata, *Enzyme Microb. Technol.* **1**, 95–99 (1979).
10. H. Yukawa, S. Yamada, T. Nara, M. Terasawa, and Y. Takayama, *Nippon Nogeikagaku Kaishi* **59**, 31–37 (1985).
11. H. Yukawa, T. Nara, M. Terasawa, and Y. Takayama, *Nippon Nogeikagaku Kaishi* **59**, 279–285 (1985).
12. H. Yukawa and M. Terasawa, *Nippon Nogeikagaku Kaishi*, **61**, 1279–1284 (1987).
13. M. Terasawa, H. Yukawa, and Y. Takayama, *Process Biochem.* **20**, 124–128 (1985).
14. Y. Satoh, K. Hatakeyama, K. Kohama, M. Kobayashi, Y. Kurusu, and H. Yukawa, *J. Ind. Microbiol.* **5**, 159–166 (1990).
15. Y. Kurusu, M. Kainuma, M. Inui, Y. Satoh, and H. Yukawa, *Agric. Biol. Chem.* **54**, 443–447 (1990).
16. A.A. Vertes, M. Inui, M. Kobayashi, Y. Kurusu, and H. Yukawa, *Res. Microbiol.* **144**, 181–185 (1993).
17. A.A. Vertes, M. Inui, M. Kobayashi, Y. Kurusu, and H. Yukawa, *Mol. Microbiol.* **11**, 739–746 (1994).
18. A.A. Vertes, K. Hatakeyama, M. Inui, M. Kobayashi, Y. Kurusu, and H. Yukawa, *Biosci. Biotechnol. Biochem.* **57**, 2036–2038 (1993).
19. Y. Kurusu, Y. Satoh, M. Inui, K. Kohama, M. Kobayashi, M. Terasawa, and H. Yukawa, *Appl. Environ. Microbiol.* **57**, 759–764 (1991).
20. K. Hatakeyama, Y. Asai, Y. Uchida, M. Kobayashi, M. Terasawa, and H. Yukawa, *Biochem. Biophys. Res. Commun.* **239**, 74–79 (1997).
21. Y. Asai, M. Inui, A.A. Vertes, M. Kobayashi, and H. Yukawa, *Gene* **158**, 87–90 (1995).

ASPERGILLUS

J.W. Bennett
Tulane University
New Orleans, Louisiana

M.A. Klich
U.S. Department of Agriculture
New Orleans, Louisiana

KEY WORDS

Aflatoxin
Amylases
Expression host
Filamentous fungi
Genetic model
Koji
Mycoses
Mycotoxins
Organic acids
Secondary metabolites

OUTLINE

INTRODUCTION

Biodiversity experts estimate that there may be 1.5 million species of fungi on earth, of which only approximately 100,000 have been described. Of these described species, the approximately 200 taxa that comprise the genus *Aspergillus* are among the most commonly encountered molds. Adapted to many habitats, exhibiting a wide range of metabolic activities, and capable of the production of astronomical numbers of spores, *Aspergillus* species are found throughout the biosphere. A few species are economically important as biotechnological resources, whereas others have gained notoriety for their negative impact on human health and commerce. For example, some aspergilli are used in industrial fermentations to produce enzymes, pharmaceuticals, and bulk chemicals; others are important components of Asian food fermentations; still others are major agents of biodeterioration; and a few common species cause allergenic, pathogenic, and toxigenic diseases in humans and other animals.

TAXONOMY

The filamentous fungi that comprise the genus *Aspergillus* reproduce mitotically through the production of spores borne on a specialized structure called the conidiophore. The morphology of this spore-bearing structure, with its distinctive foot cell, long stalk, swollen apex, and protuberant phialides, is the defining feature of the genus. It was first described in 1729 by Micheli, an Italian priest, who was reminded of a Roman Catholic device used to sprinkle holy water, the aspergillum.

All members of the genus *Aspergillus* produce an aspergillum-like conidiophore (Fig. 1). (The special vocabulary regularly encountered in *Aspergillus* taxonomy is summarized in Table 1.) The conidiophore wall is usually thicker and more hydrophobic than that of the other hyphal cells. The stipe of the conidiophore attaches to the vegetative mycelium in a T- or L-shaped base that has been traditionally called a foot cell, even though it is not a separate cell. The presence of the foot cell is an important diagnostic criterion of the genus. The apex of the stipe swells into the vesicle. Spore-producing phialides are borne directly (uniseriate) or indirectly on metulae (biseriate) on the surface of the vesicle, another important characteristic for distinguishing species within the genus. The phialides are sporogenous. Repeated mitotic divisions in the phialide nucleus lead to the formation of a chain of uninucleate or multinucleate conidiospores, also called conidia. The phialides also extrude primary wall layers and pigments that form part of the mature spore. Conidiospores are spherical or elliptical, smooth or echinulate, extremely hydrophobic, and easily airborne when mature.

Charles Thom (1872–1956), most famous for his work on penicillin and the other great mold genus, *Penicillium*, was also a major figure in the development of *Aspergillus* taxonomy. With Margaret Church in 1926 (1) and Kenneth Raper in 1945 (2) he produced the first two complete monographic treatments. Kenneth Raper went on to expand this work and with Dorothy Fennell published *The Genus Aspergillus* in 1965 (3), still considered by many to be the most authoritative source of *Aspergillus* taxonomy. Highly sensitive to substrate, temperature, moisture level, and other environmental variables, potential morphological plasticity is controlled in modern taxonomic schemes by the use of standard media and culture conditions. Pure cultures are grown on standardized agar media for specific lengths of time with defined temperature, light, and other environmental parameters. Species within *Aspergillus* are distinguished by cultural characters such as colony color, growth rate, and texture and by microscopic traits such as phialides and metulae, shape of the vesicle, spore size, and spore ornamentation. The ability to produce certain pigments, sclerotia, or cleistothecia is highly dependent on culture conditions, but can also be useful in distinguishing species. Modern taxonomists place increasing emphasis on secondary metabolite profiles, isoenzyme electrophoresis, and molecular genetic characters such as ribosomal DNA sequences to supplement the traditional characters.

After adequate characterization, identification of species depends on careful comparison with published descriptions. The last monograph of the entire genus was Raper and Fennell's *The Genus Aspergillus* (3), but several more recent books provide identification systems for the more common species (for example, see, Klich and Pitt [4]) Medically important species are described in Kwon-Chung and Bennett (5).

Some aspergilli possess a sexual stage. All the sexual stages result in the formation of ascospores inside of cleistothecia and are classified formally in the Ascomycete family Trichocomaceae. According to the rules of botanical nomenclature, which govern the naming of all fungi, the sexual (teleomorphic or perfect) state requires its own name. Because *Aspergillus* by definition refers only to the asexual (anamorphic or imperfect) state, many species within the genus have two names. For example, the asexual species *Aspergillus tetrazonus* produces a teleomorphic state called *Emericella quadrilineata*, and the teleomorph of the well-known genetic model *Aspergillus nidulans* is called *Emericella nidulans*. This dual nomenclature mystifies workers not trained in the rules of botanical nomen-

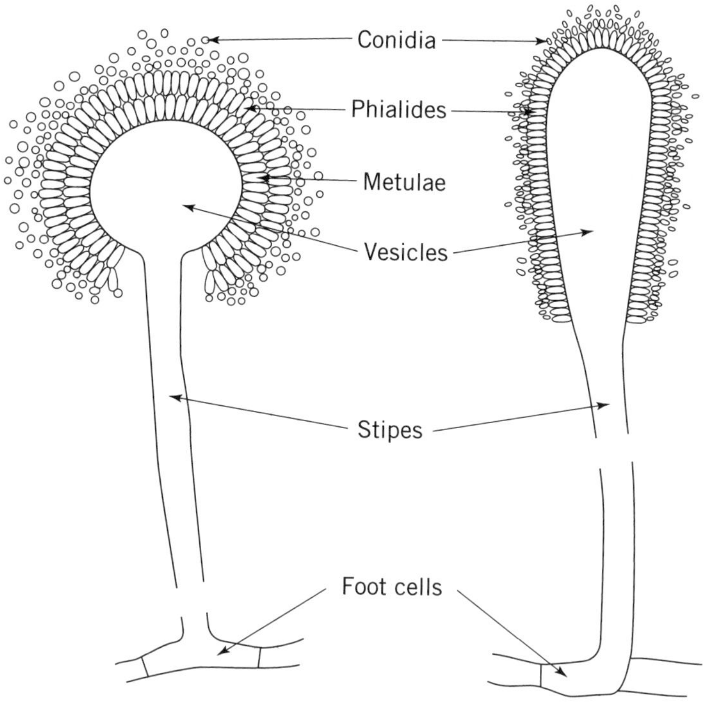

Figure 1. Conidiophores of *A. flavus* (left) and *A. clavatus* (right).

Table 1. Glossary of *Aspergillus* Taxonomy

Term	Definition
Anamorph	The nonsexual form of a fungus, loosely synonymous with *imperfect*
Ascus	A baglike sexual structure containing ascospores
Cleistothecium	A closed ascus-containing structure; the type of ascocarp (ascoma) found in the sexual stages of *Aspergillus*
Conidium (*pl.* conidia)	A thin-walled asexual spore borne on a conidiophore (also called *conidiospore*)
Foot cell	Lower portion of conidiophore, not a separate cell
Heterokaryon	A mycelium containing genetically dissimilar nuclei
Metullae	Sterile cells subtending the phialides in biseriate species; also called *secondary sterigmata*
Phialides	Conidiogenous cells; also called *primary sterigmata*
Sclerotium	Multicellular structure composed of hardened, thick-walled hyphae containing no spores inside or outside
Stipe	Stalk of the conidiophore
Teleomorph	The sexual stage of the fungal life cycle, loosely synonymous with *perfect stage*
Vesicle	The swollen apex of the conidiophore

clature. Moreover, Thom, Church, Raper, and Fennell all resisted these rules and applied the generic name *Aspergillus* to both the conidial and ascosporic stage of these organisms. Much common practice is also blind to the rules of botanical nomenclature.

Revisions of the Botanical Code in 1981 have added another layer of complexity by endangering many well-known specific epithets. Strict constructionists renamed the asexual state of *A. nidulans* as *Aspergillus nidulellus* (6). Rules of priority also endanger the name *Aspergillus niger*, one of the most widely used industrial species. Although the change to *A. nidulellus* is not followed, and although most workers continue to use the name *A. niger*, it demonstrates the potential for nomenclatural confusion in both the traditional literature and with on-line data retrieval. For a more detailed description of the nomenclatural conundrum, see Bennett (7); the contemporary rules of taxonomy are outlined in Samson and Pitt (6,8). A list of the Ascomycete genera (e.g., *Emericella, Eurotium*) with *Aspergillus* anamorphs are presented in Samson (9). Additionally, a list of *Aspergillus* species names in current use has been published in an attempt to help stabilize nomenclature in the genus (10).

DISTRIBUTION AND PHYSIOLOGY

Members of the genus *Aspergillus* are found almost everywhere. They are truly ubiquitous and have been isolated from penguin dung in the Antarctic, binocular lenses in the tropics, and spoiled foodstuffs all over the planet. Aspergilli can utilize an incredible variety of substrates, including plant debris, foods and feeds, fabrics, feathers, leather, and dung. Aspergilli have been associated with the biodeterioration of kerosene, paper, pesticides, plasticisers, and rubber and have been isolated from unlikely sources that range from biblical scrolls to sauna bath boards. They often grow on the surface of humid walls in cellars and stables, and their spores are found in high concentrations in attics, crawl spaces, air conditioning ducts, house dust, and insulation. Although the majority of species are saprophytic, the genus also includes opportunistic pathogens of mammals, insects, and plants.

The major natural habitats for aspergilli are soil and decaying vegetation. Water and temperature are the major factors influencing mycelial growth. Within the soil environment, aspergilli occur most frequently in subtropical/warm temperate zones; however, certain species such as *Aspergillus niveus, A. nidulans*, and *Aspergillus tamarii* are more common in tropical zones. Most aspergilli prefer moist habitats, although several species are xerophilic. Whatever the optimal conditions for growth, all species of *Aspergillus* form abundant conidiospores that are resistant to harsh environments. These propagules bridge survival until hospitable conditions for germination prevail.

Aspergillus colonies are visible macroscopically as black, green, yellow, or brown mold. The colony color reflects the spore color of the particular species. Pigment production is also influenced by trace elements. In fact, the color of *A. niger* spores was once used as a bioassay to detect copper.

As with all filamentous fungi, the hyphae grow by apical extension, allowing for rapid exploitation of new habitats and facilitating the colonization of solid and semisolid substrates. The ramifying hyphal filaments are profusely branched and have a high surface-to-volume ratio in intimate contact with their substrate. As mycelia colonize, they extrude enzymes that break down complex carbon sources such as cellulose, hemicellulose, pectin, and starch, not only changing their substrate, but literally becoming part of it. Simultaneously, mycelia influence the growth of other organisms, especially bacteria, and create complex degradative consortia.

Because of their robust physiology, aspergilli are easy to grow in the laboratory and can be cultured with simple carbon and nitrogen sources, mineral salts, water, and oxygen. Mutations blocked in various catabolic and anabolic pathways are also easily isolated. Biochemical and molecular genetic analyses in *A. nidulans* have revealed considerable detail about the physiology and genetics of these processes. In addition, the weedy nature of these molds means they often cause trouble as common contaminants in tissue culture and bacteriology laboratories. Techniques for handling aspergilli as well as common media for culturing species are given by Raper and Fennell (3) and Onions et al. (11).

GENETICS

Aspergillus nidulans is an excellent model organism for genetics. If the work by Raper and Fennell (3) is the bible of *Aspergillus* taxonomy, then the paper of Pontecorvo et al. (12) is often referred to as the old testament. This paper "contained not only the foundations of *Aspergillus* genetics, but most of the superstructure as well" (13, p. 3).

The parasexual cycle was first elucidated in *A. nidulans*. The first step of the parasexual cycle is the production of heterokaryons, a mycelium with genetically dissimilar nuclei, usually forced between different auxotrophic mutants. Somatic diploids are formed that are identified by their ability to grow on minimal media. Parasexual recombination occurs through mitotic crossing over and vegetative haploidization. After the parasexual cycle was elucidated in *A. nidulans*, it was found in a large number of other fungi, allowing formal genetic analysis in anamorphic species. Because most species of *Aspergillus* do not form a sexual stage, these artificially produced diploids provide "an alternative to sex." For example, early genetic studies on the aflatoxin-producing species *Aspergillus flavus* and *Aspergillus parasiticus*, both of which lack a sexual stage, were conducted by means of the parasexual cycle. Even more importantly, parasexual cycle genetics was adopted to mammalian tissue culture, where it was usually referred to as somatic cell genetics. For several decades before the recombinant DNA revolution, somatic cell genetics was the major method for mapping human genes.

Aspergillus nidulans remains an important genetic model in its own right. The steps of conidiation display morphological phenotypes amenable to genetic dissection and have been exploited as a model for developmental biology through molecular analysis of successive stages of the defined developmental sequence. The ability of the species to degrade all kinds of natural materials and to utilize a wide range of inorganic nitrogen sources has also been exploited for the genetic dissection of carbon metabolism, alcohol metabolism, purine degradation, proline utilization, sulfur metabolism, and many other metabolic pathways. The genetic regulation of nitrogen metabolite repression and carbon catabolite repression, both tightly controlled by induction of appropriate enzymes, has also received considerable attention. In addition, *A. nidulans* has been a valuable model for the study of mitochondrial and population genetics, for penicillin biosynthesis, and for the molecular biology of gene expression. The formal genetics of *Aspergillus* has been reviewed by Smith and Pateman (14) and Martinelli and Kinghorn (15).

The small size of fungal chromosomes, long a disadvantage for cytogenetic analysis, has become an advantage for doing electrophoretic karyotyping by means of pulsed gel electrophoresis. Electrophoretic karyotypes allow workers to bypass conventional genetic and cytological maps. Any DNA sequence for which a probe is available can be mapped to a chromosome by Southern analysis. The fungal genomics community recently selected *A. nidulans* as a premier target for genomic analysis (16).

The genetic versatility and well-elucidated genetic systems have made *Aspergillus* a suitable genus for industrial application of the latest recombinant DNA techniques to large-scale fermentation processes. These industrial processes are discussed below.

ECONOMICALLY IMPORTANT PRODUCTS

Aspergilli have been used since prehistoric times as components of several Asian food fermentations. Members of the genus have exceptional protein- and metabolite-secreting capacity, and these physiological activities have been adapted to human enterprise. From a contemporary perspective, the most important economic impact is the manufacture of fungal products such as enzymes, organic acids, and bioactive secondary metabolites.

Modern fermenters are sterilizable and operate under aseptic conditions. They are frequently controlled by computers that monitor temperature, agitation, pH, dissolved oxygen, and other parameters. In general, wild-type strains do not work well in industrial fermentations. Yield of the product is low, and the strain usually is not adapted to the conditions of large-scale fermentation. Genetic strain improvement has traditionally been conducted by means of brute strength mutagenesis and screening. Although repeated rounds of mutagenesis and screening are generally successful, they are also laborious. Recombinant DNA techniques have revolutionized strain improvement programs. *Aspergillus* gene expression systems out perform and are cheaper and simpler than analogous animal or insect systems.

Organic Acids

Citric acid is the most important acid produced by fungi and the only organic acid produced almost exclusively through fermentation. Widely used in the food and beverage industry, its pleasant acid taste enhances products such as soft drinks, fruit juices, jams, jellies, candies, prepared desserts, and frozen fruits. Citrates are also efficient buffering and chelating agents and are used by the cosmetics industry and in blood transfusion products, effervescent tablets, detergent manufacturing, electroplating, printing, inks, leather tanning, and a host of other applications. With increasing emphasis on nonpolluting chemical products, the market for nature's acidulant is increasing.

Many molds and yeasts accumulate large amounts of citric acid from glucose. The biochemical basis of this metabolic overproduction has been studied for a century. Names associated with early research include C. Wehmer in Germany, and J.N. Currie and C. Thom in the United States. The first commercial fermentation was perfected by Currie at Pfizer and Company in Brooklyn, New York, during the 1920s and involved surface fermentations of *A. niger*. This species is still used in many industrial processes worldwide, although submerged fermentations have generally replaced surface and solid-state systems. Production strains are selected not only for their high yield of citric acid, but also for their ability to use cheap raw materials as substrates. Optimal yields of citric acid are associated with low pH (below 2.5), high carbohydrate, and manganese deficiency.

Although a great deal is known about the environmental cues involved in maximizing citric acid production, the exact biochemical basis and the crucial regulatory control points are the subject of continuing research (17,18).

Mycelium is removed by filtering or centrifugation; citric acid is then precipitated. Subsequent recovery steps can include treatment with activated carbon, cation and anion exchanges, and final crystallization. The whole process is highly optimized, with *A. niger* capable of producing about 90% of the theoretical yield from a carbon source.

Other organic acids produced by *Aspergillus* fermentation are listed in Table 2. None of these is as economically important as citric acid.

Secondary Metabolites

The families of low molecular weight compounds that have no obvious role in the life cycle of producing species, often produced after active growth has ceased, are called secondary metabolites. Both biologically and chemically, secondary metabolites are extremely diverse; however, they arise from just a few biosynthetic pathways based in primary metabolism. Acetate-derived polyketides and amino acid–derived compounds are the most important biosynthetic routes in filamentous fungi, including aspergilli. Members of the genus *Aspergillus* are prolific producers of these natural products, and the structures of a large number of secondary metabolites from *Aspergillus* have been elucidated by chemists (19,20).

In general, the most famous natural products are bioactive and are classified on the basis of this activity as antibiotics, chemotherapeutic agents, toxins, and so forth (21). The single best-known fungal secondary metabolite is the antibiotic penicillin. Although several species of *Aspergillus* make penicillin, and although *A. nidulans* has been developed as a model system for studying its biosynthesis, high-yielding industrial fermentations all use strains of *Penicillium*.

Economically speaking, toxic secondary metabolites of *Aspergillus* are more important than the medically useful ones. Aflatoxins and sterigmatocystins, for example, cause millions of dollars of damage in agriculture each year (see "Mycotoxins"). Currently, the most important pharmacological agent is lovastatin (mevinolin) a potent inhibitor of cholesterol synthesis, derived from *Aspergillus terreus*. Lovastatin is a polyketide. Several amino acid–derived metabolites also show pharmacological activity. For example, echinocandin B (cilofungin), produced by *A. nidulans* and *Aspergillus rugulovalvus* (also called *E. rugulosis*), is under development as an antifungal agent (Table 2).

Food Production

Foods fermented with microorganisms have interesting flavors and improved digestibility. In Western culture, the majority of food fermentations involve yeasts (e.g., bread and wine) or bacteria (e.g., cheese and pickles). Aspergilli play only minor roles in Western cuisine (e.g., ripening salamis and country-cured hams). This is in sharp contrast to Asian food traditions, where filamentous fungi have been used for millennia, especially in complex two-stage fermentations. The best known of these processes include shoyu (soy sauce), sake (rice wine), and miso (fermented soybean paste). The initial step in each case is the production of a *koji*, a Japanese word that roughly translates as "bloom of mold." To produce a koji, *Aspergillus oryzae* or the closely related species *A. sojae* is grown on steamed rice or other cereal and incubated in warm, humid conditions. The hyphae infiltrate the grain, secreting hydrolytic enzymes that partially degrade the substrate, producing the koji, a fragrant, crumbly mass of mold and substrate. Perhaps because of the paucity of Western fungal fermentations, the English language does not have a word that similarly denotes a substance thus composed of living mycelia, extracellular enzymes, and partially degraded substrate. This mold–substrate mixture has different proper-

Table 2. Economically Useful Low Molecular Weight Products of *Aspergillus*

Product	Producing species	Use
Organic Acids		
Citric acid	*A. niger*	Flavoring agency, antioxidant, chelating and cleaning agent, component of effervescent powders
Gluconic acid	*A. niger*	Manufacturing of toothphaste
Itaconic acid	*A. terreus*	Copolymer in resin manufacture
Pharmacologically active secondary metabolites		
Asperillomarasmine	*A. oryzae*	Angiotensin-converting enzyme inhibitor
Asperlicins	*A. alliaceus*	Cholecystokinin antagonists
Lovastatin	*A. terreus*	Cholesterol inhibitor
Echinocandin	*A. nidulans* / *A. rugulovalvus*	Antifungal agent

ties than either alone and constitutes a form of matter not captured by the term solid-state fermentation.

Kojis are classified according to the degree of mycelial development and sporulation, their intended use, and the cereal substrate. Kojis are usually used as inocula for further fermentations. For soy sauce, the koji is mixed with soybeans and wheat. After fungal amylases and proteases have partially degraded the mixture, a salt solution is added. An anaerobic yeast–lactobacilli fermentation ensues; the final pressed liquid is clarified as soy sauce. For sake, a koji inoculum is mixed with steamed rice and water and incubated for 3 to 4 days in the moto stage. As the fungal enzymes partially saccharify the rice, a predictable change in the microbial flora takes place, with yeasts becoming the predominant microorganisms, to yield the moroni stage. After appropriate incubation, yeast metabolic activities yield alcohol. Miso is a fermented paste with the consistency of peanut butter, often made in Japanese homes. Widely used as a soup base, miso varies according to substrate (rice, barley, soybeans, mixtures), length of fermentation, amount of salt, etc. Similar products, based solely on fermented soybeans, are known as *chiang* in China, *jang* in Korea, *tao-jo* in Indonesian and Thailand, and *tao-tsi* in the Philippines (22,23).

Aspergillus oryzae and *A. sojae* are close relatives of the toxigenic species *A. flavus* and *A. parasiticus*. These domesticated species do not produce toxins; *A. oryzae* is on the GRAS list (Generally Regarded as Safe) by the U.S. Food and Drug Administration.

The first industrial enzyme patented in the United States was an amylase mixture purified from a koji by the early Japanese-American biotechnologist Jokichi Takamine in 1894 (7). Contemporary scientists have subsequently identified more than 50 different enzymes from koji.

Enzymes

Enzymes are organic catalysts. Long before they were purified by scientists, their properties had been used in baking, brewing, tanning, and other processes. Because of their high specificity, their biodegradability, and their ability to work in watery solutions at moderate temperatures, enzymes are becoming increasingly important in today's environmentally conscious industries. In many processes, enzymes have replaced the strong chemicals and high temperatures formerly used in commercial catalysis.

The major industrial enzymes from fungi are hydrolytic, and the major producing species are *A. oryzae* and *A. niger*. Amylases and glucoamylases are used for turning starches into sugars and oligosaccharides. Pectinases are applied in fruit juice clarification. *Aspergillus* proteases are used in bread making and in chill-proofing beer, and a thermostable phytase is used widely in the animal feed industry as an additive. Representative commercial enzymes produced by *Aspergillus* are listed in Table 3.

With the advent of modern molecular techniques, the genes for many native *Aspergillus* enzymes have been cloned, sequenced, amplified, and engineered to improve yields, stability, and other properties (24,25). Genetic manipulation also has allowed *Aspergillus* to become the host for the production of various nonnative (foreign) proteins, the best known of which is chymosin. Rennet is a substance extracted from the stomachs of slaughtered calves that is used in cheese making. The principle active ingredient of rennet is chymosin, a protease that coagulates milk. Structural genes for the calf stomach enzyme can be transferred to fungi. The mammalian enzyme is then expressed and secreted by the fungus in fermentation broths. The yeast *Kluyveromyces lactis* and *A. niger* have both been developed as production hosts for chymosin production. Given the economic advantages of fungal fermentations, it can be

Table 3. Commercially Important *Aspergillus* Enzymes

Enzyme	Applications
Amylases	Hydrolysis of starch; bread and beer production; removal of sizing from fabrics; high-fructose syrups
Invertase	Confections; soft center in chocolates
Lactase	Hydrolysis of lactose; production of syrups for sweetening agents
Pectinases	Pretreatment of fruit juices to remove turbidity, reduction cloudiness in wines
Phytase	Animal feed additive for liberation of phosphate from plant material
Proteases	Meat tenderizing; removal of bitter flavors and chill proofing of beer; reduction elasticity of gluten proteins in bread

predicted that *Aspergillus* will be the production host of choice for many other nonfungal enzymes that traditionally have been extracted from higher plants and animals.

Descriptions of the use of enzymes in fermentation, diagnostics, baking, brewing, cheese manufacturing, chemical biotransformation, detergents, effluent and waste treatment, leather, textile, starch conversion, and the like are presented in Godfrey and West's *Industrial Enzymology* (26). This work also gives detailed information on the legislation and regulation of commercial enzymes, toxicological and regulatory considerations, and the scientific and regulatory aspects surrounding the use of genetically engineered production strains (26). The long history of safe use and the GRAS status of *A. oryzae* and *A. niger* will ensure that they will remain the species of choice for industrial enzyme production, especially in the food industry.

DISEASE

Although *Aspergillus* species are very common in the human environment, only a limited number of taxa (approximately 17) are capable of initiating human disease. The usual reservoir for an *Aspergillus* infection is not another infected person but, rather, a nonliving site where the fungus is growing saprophytically. Air conditioning vents, attic insulation, cellar walls, stables, and even houseplants in hospitals have been implicated in various disease states. Because *Aspergillus* conidia are among the most common of airborne molds, the diseases they cause are usually pulmonary. The single species most commonly associated with human disease is *Aspergillus fumigatus*. In addition, *A. flavus, A. nidulans, A. niger*, and *A. terreus* are regularly implicated.

The term *aspergillosis* has been broadly used to describe any disease caused by an *Aspergillus* species, excluding mycotoxicoses. Because this term encompasses so many different clinical entities, with different researchers applying different emphases, and because some conditions defy simple categorization, the medical mycology literature can be confusing. The classification of Kwon-Chung and Bennett (5) and Dixon and Walsh (27) recognizes three broad categories of aspergillosis: *(1)* allergy, *(2)* saprophytic colonization, and *(3)* invasive mycosis.

Allergenic Responses

Like other fungal spores, *Aspergillus* conidia are potent allergens, especially associated with respiratory allergies. Spore inhalation may cause hypersensitivity reactions in susceptible individuals. For example, allergic bronchopulmonary aspergillosis occurs in patients with preexisting asthma who develop bronchial sensitivity reactions to inhaled conidia. Occupational exposure is a risk factor. Industrial-scale fermentations can cause sensitization of workers to high concentrations of airborne spores or to culture fluid materials, resulting in asthma or other adverse health effects. Malt worker's lung is a hypersensitivity pneumonitis caused by *Aspergillus clavatus*. Some medical mycologists classify respiratory allergies as extrinsic (inhalation of conidia) or intrinsic (*Aspergillus* growing in airways). Intrinsic conditions are an example of saprophytic growth.

Saprophytic Growth

In contrast to extrinsic respiratory allergies caused by spore inhalation, intrinsic aspergillosis is a localized hyphal growth associated with colonization of air spaces. Aspergilloma, or fungus ball, is formed when *Aspergillus* grows in a preexisting lung cavity. Fungus balls have also been reported to occur in the paranasal sinus.

Invasive Aspergillosis

Sometimes *Aspergillus* species are opportunistic pathogens causing frank mycosis. These mycoses are collectively called invasive aspergillosis and are usually classified by the portal of entry. Invasive pulmonary aspergillosis is typically manifested as acute pneumonia. Nonpulmonary portals of entry include the eye, paranasal sinus, or gastrointestinal tract or may be associated with vascular surgery and drug addiction. Invasive aspergillosis is largely a disease of a compromised host. Those at high risk include patients with diabetes, drug addicts, chronically ill patients, transplant patients receiving immunosuppressive drugs, and cancer patients receiving chemotherapy. Occupational exposure is also a factor. Historically, the earliest reports of aspergillosis were among wig cleaners and pigeon fanciers.

Aspergillus fumigatus and *A. flavus* are the most common species associated with invasive and noninvasive aspergillosis. It is hypothesized that these species possess special virulence factors that facilitate their role as opportunistic pathogens.

Invasive disease is treated with antifungal chemotherapy, but existing antimycotic drugs have toxic side effects, and the therapeutic outcome is often poor. As the use of immunosuppressive drugs increases in modern medicine, invasive aspergillosis is becoming more common. There is an important need for effective systemic antifungal drugs. Several pharmaceutical companies have active drug discovery programs in this area.

Mycotoxins

Filamentous fungal metabolites that cause human or veterinary diseases are called mycotoxins. The mycotoxins that have received the most attention are the aflatoxins, a family of highly toxic, mutagenic, and carcinogenic compounds produced by certain strains of *A. flavus, Aspergillus nomius*, and *A. parasiticus*. These species regularly contaminate peanuts, corn, cottonseed, rice, tree nuts, and other agricultural commodities (28). A large body of epidemiological and molecular biological data implicates aflatoxin with human liver cancer (29). Acute human aflatoxicosis, however, is rare and usually only occurs when famine or poverty forces people to subsist on moldy foodstuffs. On the other hand, veterinary aflatoxicosis is a major problem, with poultry and trout being particularly susceptible.

Other mycotoxins produced by members of the genus *Aspergillus* tend to be produced at toxicologically less important levels. These include cyclopiazonic acid (*A. flavus*), patulin (*A. clavatus*), and sterigmatocystin (*A. nidulans* and *Aspergillus versicolor*). The toxigenic potential of *Aspergillus ochraceus* is particularly varied, with different strains reported to make ochratoxin, citrinin, xanthomegnin, viomellein, and penicillic acid. Like aflatoxin, patulin and sterigmatocystin are carcinogenic; citrinin and ochratoxin A are nephroxic; cyclopiazonic acid is a neurotoxin; and xanthomegnin and viomellein have been associated with photosensitization and liver damage. Many of the compounds classified as cytotoxic or neurotoxic in vertebrates also show toxicity against insects. Sclerotia have been a particularly rich source of metabolites with insecticidal properties (30).

Mycotoxin contamination varies from year to year based on weather, agricultural practices, storage conditions, and other factors. Aflatoxin is usually worst in drought years. International agencies have set levels for permissible standards of contamination (31).

SUMMARY

Among the most commonplace and versatile of all fungi, the aspergilli play an important role in natural degradative cycles. Although a few species have gained notoriety as animal pathogens and mycotoxin producers, the vast majority are benign. In processes ranging from traditional Asian food preparation to massive contemporary industrial fermentations, the capacity of aspergilli to secrete enzymes and metabolites is exploited for human benefit. In basic science, *A. nidulans* is a model system for genomics and molecular biology as well as classical genetics. The increasing societal emphasis on environmentally safe processes and commodities guarantees that the metabolic capabilities of this genus will continue to be used in modern industry.

BIBLIOGRAPHY

1. C. Thom and M.B. Church, *The Aspergilli*, Williams & Wilkins, Baltimore, Md., 1926.
2. C. Thom and K.B. Raper, *A Manual of the Aspergilli*, Williams & Wilkins, Baltimore, Md., 1945.
3. K.B. Raper and D.I. Fennell, *The Genus Aspergillus*, Williams & Wilkins, Baltimore, Md., 1965.
4. M.A. Klich and J.I. Pitt, *A Laboratory Guide to the Common Aspergillus Species and Their Teleomorphs*, CSIRO Div. Food Processing, North Ryde, Australia, 1988.
5. K.J. Kwon-Chung and J.E. Bennett, in *Medical Mycology*, Lea & Febiger, Philadelphia, 1992, pp. 201–247.
6. R.S. Samson and J.I. Pitt, *Advances in Penicillium and Aspergillus Systematics*, Plenum Press, New York, 1985.
7. J.W. Bennett, in A.L. Demain and N.A. Soloman eds., *Biology of Industrial Microorganisms*, Benjamin/Cummings, Menlo Park, Calif., 1985, pp. 359–406.
8. R.A. Samson and J.I. Pitt, *Modern Concepts in Penicillium and Aspergillus Classification*, Plenum Press, New York, 1990.
9. R.A. Samson, in J.W. Bennett and M.A. Klich eds., *Aspergillus: Biology and Industrial Applications*, Butterworth-Heinemann, Boston, 1992, pp. 355–390.
10. J.I. Pitt and R.A. Samson, in W. Grauter ed. *NCU-2, Names in Current Use in the Families Trichocomaceae, Cladoniaceae, Pinaceae and Lemnaceae*, Koeltz Scientific Books, Konigstein, Germany, 1993, pp. 13–57.
11. A.H.S. Onions, D. Allsopp, and H.O.W. Eggins, *Smith's Introduction to Industrial Mycology*, E. Arnold, London, 1981.
12. G. Pontecorvo, J.A. Roper, L.M. Hemmons, K.D. MacDonald, and A.W.J. Bufton, *Adv. Genet.* **5**, 141–239 (1953).
13. O.A. Clutterbuck, in J.W. Bennett and M.A. Kich eds., *Aspergillus Biology and Industrial Applications*, Butterworth-Heinemann, Boston, 1992, pp. 3–18.
14. J.E. Smith and H.A. Pateman eds., *Genetics and Physiology of Aspergillus*, Academic Press, London, 1977.
15. S.D. Martinelli and J.R. Kinghorn eds., *Aspergillus: 50 Years On*, Elsevier, Amsterdam, 1994.
16. J.W. Bennett, *Fungal Gen. Biol.* **21**, 3–7 (1997).
17. M. Roehr, C.P. Kubicek, and J. Kominek, in J.W. Bennett and M.A. Klich eds., *Aspergillus: Biology and Industrial Applications*, Butterworth-Heinemann, Boston, 1992, pp. 91–131.
18. C.P. Kubicek, C.F.B. Witteveen, and J. Visser, in K.A. Powell, A. Renwick, and J.F. Peberdy eds., *The Genus Aspergillus: From Taxonomy and Genetics to Industrial Application*, Plenum, New York, 1994, pp. 135–145.
19. W.B. Turner, *Fungal Metabolites*, Academic Press, London, 1971.
20. W.B. Turner, *Fungal Metabolites II*, Academic Press, London, 1983.
21. G. Lancini and R. Lorenzetti, *Biotechnology of Antibiotics and Other Bioactive Microbial Metabolites*, Plenum, New York, 1993.
22. N.R. Reddy, M.D. Pierson, and D.K. Salunkhe eds., *Legume-Based Fermented Foods*, CRC, Boca Raton, Fla., 1986.
23. K.H. Steinkraus ed., *Handbook of Indigenous Fermented Foods*, Dekker, New York, 1983.
24. T. Christensen, in K.A. Powell, A. Renwick, and J.F. Peberdy eds., *The Genus Aspergillus: From Taxonomy and Genetics to Industrial Application*, Plenum, New York, 1994, pp. 251–259.
25. R.M. Berka, N. Dunn-Coleman, M. Ward, in J.W. Bennett and M.A. Klich eds., *Aspergillus: Biology and Industrial Applications*, Butterworth-Heinemann, Boston, 1992, pp. 155–202.
26. T. Godfrey and S. West eds., *Industrial Enzymology*, Stockton, New York, 1996.
27. D.M. Dixon and T.J. Walsh, in J.W. Bennett and M.A. Klich eds., *Aspergillus: Biology and Industrial Applications*, Butterworth-Heinemann, Boston, 1992, pp. 249–267.
28. L.A. Goldblatt, *Aflatoxin*, Academic Press, New York, 1969.
29. D.L. Eaton and J.D. Groopman eds., *The Toxicology of Aflatoxins*, Academic Press, New York, 1994.
30. V. Betina, *Mycotoxins—Production, Isolation, Separation, and Purification*, Elsevier, Amsterdam, 1984.
31. Food and Agricultural Organization of the United Nations, *World Wide Regulations for Mycotoxins: A Compendium*, FAO, Rome, 1995.

See also METABOLITES, PRIMARY AND SECONDARY; RHEOLOGY OF FILAMENTOUS MICROORGANISMS, SUBMERGED CULTURE; SECONDARY METABOLITE PRODUCTION, ACTINOMYCETES, OTHER THAN STREPTOMYCES.

ASTM STANDARDS FOR BIOTECHNOLOGY

G. Larry Eitel
Jacobs Engineering Group
Denver, Colorado

KEY WORDS

ASTM standards
Bioprocess conversion
Consensus standards
Cryogenic
Preservation
Standards
Validation
Voluntary standards

OUTLINE

INTRODUCTION

As the biotechnology products and processes industry has grown during the past several decades, so has the realization grown that more standard test methods are needed to keep pace with regulatory requirements to assure the public that these products and processes are safe and effective methods of producing pure products. In 1996, the federal government passed a law directing federal agencies to participate in and use technical standards developed by voluntary consensus standards organizations, with few exceptions, to carry out policy objectives and activities. The American Society for Testing and Materials (ASTM) was specifically mentioned as an example of a voluntary consensus standards organization.

The ASTM E-48 Committee for Biotechnology was organized in 1985. This was in response to urging from the biotechnology community to lead the development of voluntary consensus standards in biosafety, characterization and identification of biotech organisms, preservation and storage, bioprocess controls and processing equipment, environmental issues, process validation, equipment qualification, and terminology. Biomass conversion standards responsibility was transferred to E-48 in 1987. Although many of these topics have analytical procedures currently in use by individual companies or trade organizations, there is a well-known need to standardize these procedures for fully acceptable application in industry and governmental regulatory organizations.

By 1997, the E-48 committee had published and is still responsible for more than 30 standards. More standards, based on the ASTM structured manner using technically qualified volunteers from international resources, are currently being developed.

This article explains the characteristics of voluntary consensus standards, how ASTM develops these standards, what standards are currently available, and which ASTM E-48 subcommittees are responsible for these and future standards development.

ASTM BACKGROUND

A Century of Consensus Standards

The ASTM has developed during the past century into one of the largest voluntary standards development organizations in the world. Organized in 1898, ASTM is a not-for-profit organization that enables all interests to participate in a common effort to write, review, and approve standards. These standards are used by producers, ultimate consumers, users, and general interest groups such as government, academia, and others in domestic and international settings.

What Is the Purpose of ASTM?

As stated in the *ASTM Technical Committee Officer Handbook* (1), the charter purpose is "the development of standards on characteristics and performance of materials, products, systems and services; and the promotion of related knowledge."

Currently, ASTM has 132 standards-writing technical committees, publishes more than 9,000 standards annually, and has more than 35,000 technically qualified members who voluntarily develop and maintain current standards in their committee specialty (2). The more familiar standards developed by ASTM include the concrete, iron, and steel products standards used in the construction industry today. These standards have been developed for public safety. Other categories of ASTM standards include nonferrous metal products; metals; construction products (cement, road paving, roofing, masonry, insulation, sealants, soil, wood); petroleum fuels; lubricants; paints; coatings; textiles; plastics; rubber, electrical insulation; electronics; water; environmental technology; nuclear, solar, and geothermal energy; medical devices; emergency services; general laboratory analytical methods; hazardous substances; biotechnology, and others (2). These published standards are available for purchase from ASTM (100 Barr Harbor Drive, West Conshohocken, Penn., 19428-2959; phone (610) 832-9585).

What Are Consensus Standards?

Because ASTM standards are developed voluntarily and used voluntarily, the basic principle in their development is that a rigorous due process procedure is followed, resulting in a full consensus of all concerned parties prior to approval and publication of the standard by ASTM. ASTM technical committees are dedicated to this consensus process because these standards are often adopted by regulatory agencies in government regulations or referred to in business contracts between client's and contractors. This use of ASTM standards goes beyond the voluntary use and becomes legally binding.

Governmental agencies encourage the use of consensus standards, as demonstrated on March 7, 1996, when President Clinton signed Public Law 104-113, Technology Transfer Improvements Act of 1995 (3). This new law codified the existing Office of Management and Budget (OMB) Circular A-119 titled *Federal Participation in the Development and Use of Voluntary Standards*. This law directs all federal agencies and departments to use technical standards developed or adopted by voluntary consensus standards organizations, such as ASTM, to carry out policy objectives or activities. These federal agencies and departments consult with voluntary, private sector, and consensus standards bodies, when possible within budgetary resources and priorities, they should also participate in the development of these technical standards.

Senator John D. Rockefeller (D-W.Va.) proposed this legislation (4) and stated that

> Consensus standards are standards which are developed by voluntary, private sector, consensus standards bodies. These organizations are established explicitly for the purpose of developing such standards through a process having three characteristics. First, openness, defined as meaning that participation in the standards development process shall be open to all persons who are directly and materially affected by the activity in question. Second, balance of interest, which means that the consensus body responsible for the development of a standard shall be comprised of representatives of all categories of interest that relate to the subject. For example, manufacturer, user, regulatory, insurance, inspection, employee and union interest. Third, due process, which means a procedure by which any individual or organization who believes that an action or inaction of a third party causes unreasonable hardship or potential harm is provided the opportunity to have a fair hearing of their concerns. In short, a legitimate consensus standards organization provides open process in which all parties and experts have ample opportunity to participate in developing the consensus. Examples include traditional standards organizations such as the American Society for Testing and Materials.

How Is ASTM Organized to Produce Standards?

The basic organization of ASTM is illustrated in Figure 1. The board of directors is the governing body elected by the ASTM membership of more than 35,000. The standing committees maintain and enforce the procedures specified by the charter for developing standards. The Committee on Technical Committee Operations oversees the application of the regulations governing technical committee operations. It recommends changes to the *Regulations*, resolves jurisdictional disputes between committees, develops guidelines for long-range planning, and ensures that committee bylaws are consistent with ASTM regulations (5). The Committee on Standards maintains procedures for preparing standards and reviews all technical committee standards actions for consistency with the regulations before approval and publication by the society. The Committee on Publications advises the board on publication policy matters and oversees publication of all ASTM materials. The technical committees exist as semiautonomous groups to develop the standards. A more detailed description of the ASTM E-48 Committee for Biotechnology is presented later. Each technical committee is assigned a staff manager who assists the volunteer leadership in management functions and coordinates administrative services.

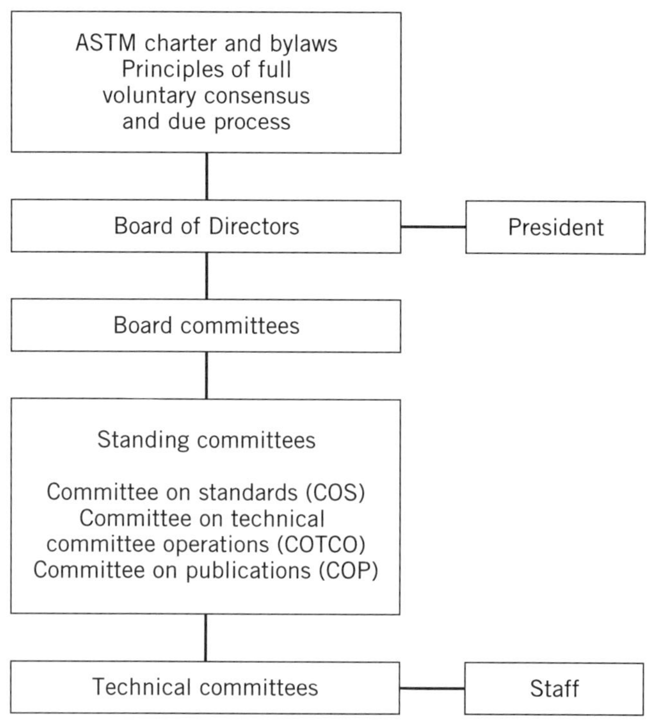

Figure 1. ASTM Organization. The president is a nonvoting member of the board of directors. For full descriptions of these committees, consult *Regulations Governing ASTM Technical Committees*.

TYPES OF ASTM STANDARDS

Categories of Standards

First, the word *standard* in ASTM is used primarily as an adjective with a specific document, such as a practice or test method, to denote consensus and approval (6). The technical committees, such as ASTM E-48 Committee on Biotechnology, determine the type of standard document to be developed and approved. There are two categories of documents within ASTM, as shown in the following (5):

- *Standard.* This document has been developed and approved by consensus within the principles of ASTM and meets the approval requirements of ASTM procedures and regulations.
- *Provisional Standard.* This document has been approved and published by ASTM for a limited period of time to meet urgent needs, such as an emergency situation, regulatory requirements, or other special circumstances. Provisional standards are not full consensus documents because they are approved by technical subcommittees only and not by full society consensus.

Types of Standards

The two categories of ASTM documents are found in the following seven types and forms, based on the technical content and intended use, not on the degree of consensus achieved (5). These different types of ASTM documents provide a flexibility of form, communication, and usage for the technical committees and the wide variety of users of these documents.

1. *Classification.* This is a systematic arrangement or division of products, systems, materials, or services into groups that have similar characteristics, such as composition, origin, properties, or use. An example is classifications of types of coal having similar specifications.
2. *Guide.* A guide is a documentation of general options or instructions that do not recommend a specific course of action. Normally, a guide suggests an approach based on consensus of viewpoints but does not establish a fixed procedure. A guide can be used to create increased awareness of the available techniques on a given subject and to provide information for subsequent evaluation and standardization. Note the difference between a guide and a practice: a guide suggests an approach, whereas a practice dictates a general usage principle.
3. *Practice.* This document is a definitive procedure for performing one or more specific operations or functions where a definitive test result is *not* produced (compare with test method). Examples of practices include procedures for conducting interlaboratory testing programs, statistical procedures, sampling procedures, testing equipment selection, operation, maintenance, installation, cleaning, training, application, assessment, decontamination, inspection, screening, and disposal.
4. *Specification.* This precise statement of requirements to be satisfied by a material, product, system, or service identifies the procedures for determining whether each of the requirements is satisfied. Examples include purity, composition, pH, or minimum or maximum ranges of acceptable values of the properties for a biochemical product. The specification could include physical, mechanical, or chemical properties of the material. Also applicable are quality, safety, or performance criteria with identification of the test methods to be used in evaluating these issues.
5. *Terminology.* This document contains definitions, descriptions of terms, and explanations of symbols, abbreviations, or acronyms used in the subject area. Each standard can have terminology specific to that standard, and then a consolidated standard of terminology can be approved for the general subject or committee area of responsibility. For example, ASTM Committee E-48 on Biotechnology has developed more than 30 individual standards with their own terminology sections. In addition, Standard E 1705-95 has been approved; it summarizes appropriate terminology specific to biotechnology and related ASTM standards.
6. *Test method.* A test method is a definitive procedure that produces a test result that can include identification, measurement, and evaluation of qualities, characteristics, or properties of the material, product, system, or service.
7. *Other types of ASTM documents.* Other documents include those less frequently used, such as charts, tables, matrixes, and reference photographs recommended by the sponsoring technical committees with approval from the ASTM Committee on Standards.

Who Uses ASTM Standards?

ASTM standards have a far-reaching international and domestic impact. Manufacturers of goods for the public use many ASTM procedures to specify raw materials used in the production of other products and services. The specifications for these produced goods are often tied into ASTM standards, where the producer and consumer are both testing the products for acceptability, identity, safety, and quality using the same test methods and specifications. Research and development interests use ASTM standards that are tried and proved to be accurate and reliable. In production of drugs by biotechnical or pharmaceutical routes, for example, it is critical that testing and analytical characterization of the chemical intermediates and final products be accurate, repeatable, and able to be validated. This enables regulatory agencies, such as the U.S. Food and Drug Administration (FDA) to monitor the quality of drug products using established, proved test methods and production protocols. Governmental agencies use consensus standards in concert with regulatory standards to assure the public that products being purchased are safe and proper. For example, in the mid-1980s the U.S. Department of Agriculture (USDA) asked ASTM to help prepare a checklist for USDA officials. This checklist would be used to screen applicants requesting financial support to install small ethanol fermentation plants on their farms. Some of these plants did not work effectively because of incomplete designs. The checklists were prepared and used by the USDA to improve the effectiveness of evaluating the technical success potential of the new ethanol projects being proposed. These checklists were developed by volunteers in the ASTM Biomass Subcommittee with participation from USDA employees.

Other users of these consensus standards include insurance and inspection organizations that use standard guides, practices, and test methods to monitor and check the quality of products being used by the public. These standards may also find their way into the court system, where successful litigation may depend on using test methods that are credible and used as standards within the industry.

ASTM standards are international; the Biotechnology Committee E-48 includes members from Japan, Europe, and Canada, among others, often representing international standard-setting organizations with common interests. An example of cooperative efforts includes development of the polymerase chain reaction (PCR) test methods. ASTM E-48 members cooperated with the Deutsches Institut fur Normung (DIN) standards committee in Germany to develop PCR standards.

ASTM COMMITTEE E-48 FOR BIOTECHNOLOGY

Technical Committee Organization

The structure of the workhorse organization to produce ASTM standards in biotechnology is shown in Figure 2.

In 1985, the ASTM Board of Directors approved establishing the E-48 Committee on Biotechnology to answer requests from industry and government and provide voluntary consensus standards that would meet the needs in this rapidly emerging new field. E-48 founding members developed new bylaws and organization structure to represent a balanced interest from producers, general interest groups, and consumers. Because E-48 develops standards for products, services, systems, and materials that are offered for sale, the committee is required to be classified and balanced; the number of voting members representing producers does not exceed the total membership in the remaining classifications of voting members (5). Currently, there are about 100 members of E-48 from international and domestic countries. Officers are elected by the E-48 membership, and those in key positions serve for 2 years.

The chairperson and two vice chairpersons direct and manage the activities of the E-48 administrative and technical subcommittees (Fig. 2). The Executive Subcommittee provides direction for the main committee, including long-range planning, symposium planning, and coordination of the technical and technical service subcommittees. The Main Committee meets semiannually to conduct business, review balloting, approve new members, and plan for the short- and long-range interests of developing standards for biotechnology. E-48 has sponsored several symposia (7) for biotechnology and has published papers through ASTM. Papers resulting from the symposia are subject to peer review and approval by the Committee on Publications before publication.

E-48 technical subcommittees have been organized to represent specific subjects within the scope of the E-48 charter. Each technical subcommittee is the primary resource to develop and approve ASTM standards for biotechnology as described in "Standards Development". Each technical subcommittee is normally subdivided into task groups that focus on specific standards development. The

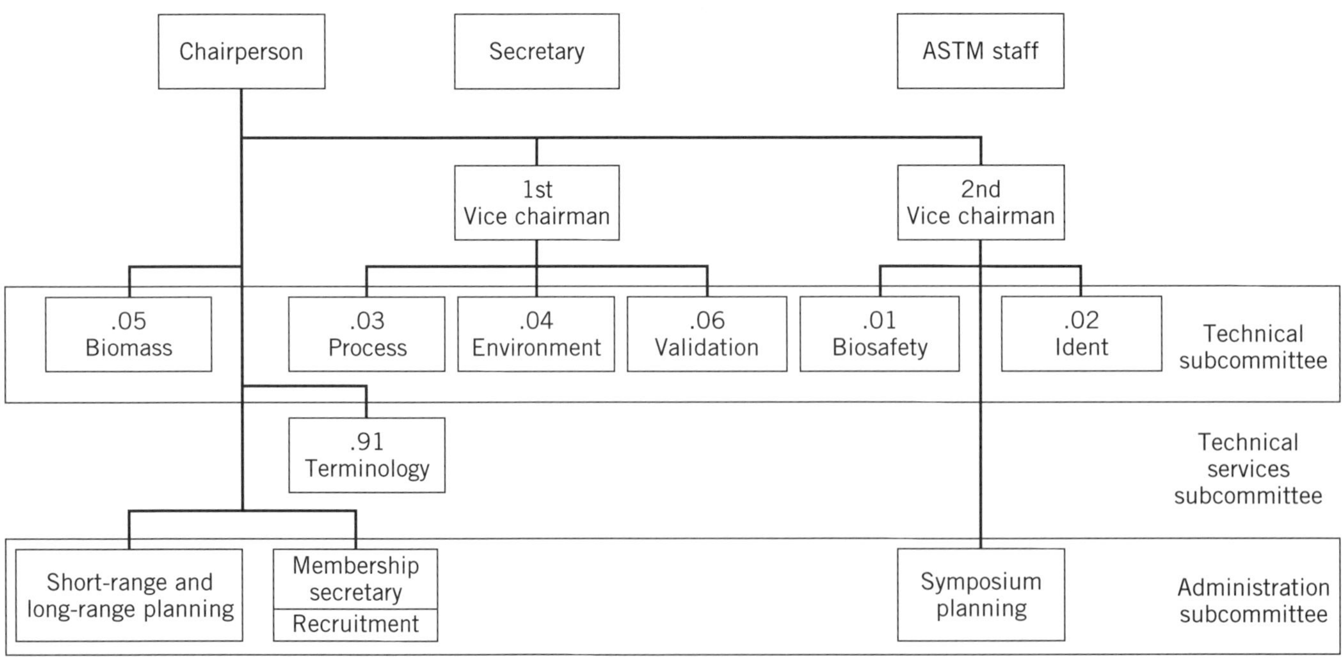

Figure 2. Main committee E-84 on biotechnology.

task group is composed of ASTM, affiliated, and non-ASTM members who are knowledgeable about the subject matter. The numbers of task groups vary, depending on the subjects being considered for standards. Minutes of all formal meetings are published for distribution to ASTM members to communicate the status and to correlate the overall committee work. For information on the E-48 subcommittee standards developed by specific subcommittees, refer to "E-48 Standards Summary."

Liaison assignments to committee members are made to coordinate E-48 activities and interests with other ASTM committees or other standards organizations (8). For example, the charter of Committee E-47 on Biological Effects and Environmental Fate includes standards on (*1*) the effects of physical and chemical stress on aquatic and terrestrial plants and animals (including humans); and (*2*) the properties of materials that affect and determine their fate, distribution, and persistence when introduced into the environment. The Committee D-22 on Sampling and Analysis of Atmospheres covers biological aerosols, mainly from sampling and analytical aspects. These and other ASTM main committees may have overlapping scopes of interest with E-48, and close coordination is needed to properly develop a standard of common interest. Sometimes, a joint task group is formed, and representatives from various committees develop the common standard. This is necessary because the charter of E-48 covers interests that include characterization of biological species, biosafety, processes using living organisms, environmental issues, biomass conversion, process validation, and terminology. Members of E-48 also serve as liaisons with other organizations, such as the American Society of Mechanical Engineers and professional, pharmaceutical, and drug manufacturing organizations.

ASTM STANDARDS DEVELOPMENT

Overview of Consensus Standard Development by E-48

The life cycle of an ASTM standard begins when a need is recognized and deemed appropriate for development by the E-48 technical subcommittee and executive subcommittee (1). Task groups are assigned by the subcommittee chair, and experts are assigned to develop a standard using proper format (6), which is then voted on by the subcommittee members.

- *Subcommittee ballots* require an affirmative vote of at least two-thirds of the votes cast by voting members, at least 60% of the voting members must return the letter ballots. If a negative vote is received with explanation of the reasons for the negative, then the subcommittee must resolve the negative by a meeting of the subcommittee or by another letter ballot. In a consensus standard, it is essential to properly resolve any negative ballot issues so that a full consensus is obtained.
- *Main committee ballots* are used to vote on the standard after completion of a subcommittee letter ballot. ASTM headquarters issues all letter ballots for the main committee. An affirmative vote of at least 90% of the voting members is needed, at least 60% of the voting members must return the letter ballots. The task of resolving all negative votes not previously considered belongs to the original subcommittee. If substantive changes to the standard are required, then the standard is reballoted normally at the subcommittee level.
- *Society review* of the standard is conducted simultaneously with the main committee ballot through publication in the monthly ASTM magazine, *Standardization News*. Each member of the society can then vote on the standard. Negatives again must be fully addressed by the appropriate subcommittee that developed the standard.
- The *Committee on Standards* then reviews the balloting and approves publication of the standard in the *Annual Book of ASTM Standards*.
- *Updating* the approved standard can be achieved whenever a significant change is needed by using the entire balloting sequence. Every 4 years, each standard is reviewed and balloted for either revision, reapproval, or withdrawal.

The E-48 Committee also seeks comments on new draft standards from other reviewers, even if they are not members of ASTM, to be sure that the new standard receives full consensus from biotechnical interests. Often, E-48 is requested by other biotechnology associations to discuss biotech standards and activities at ASTM.

E-48 STANDARDS SUMMARY

Since 1985 when E-48 was organized, the number of approved standards has increased to more than 40. Three symposia and seminars have been held on biotechnology subjects such as biosafety, validation, and general biotech interests (7,9,10). The abstracts from these documents are described in the next section, organized by subcommittee.

E-48.01 Subcommittee on Biosafety

Although the E-48.01 Subcommittee on Biosafety was established in 1996, biosafety was a main topic when E-48 was organized in 1985, and a symposium was conducted in 1987. This section summarizes the work of this subcommittee.

Scope and Background. Although multiple agencies regulate and provide guidelines for safety and transport of biomaterials within manufacturing facilities, a need for clarification of these standards was indicated in 1985. However, in the past 14 years, significant progress has been made in the industry to produce safety guidelines, especially at the laboratory-scale level. Possible augmentation of safety standards may be an area for ASTM to investigate. For example, the D-22 Committee on Biological Aerosols is considering standards to measure biocontaminants in indoor air. This is a very difficult area to develop effective test methods, but the industry is highly interested in this topic. E-48 has interest in developing test methods that can be applied to large-scale operations relative to employee protection. Standardization of guidelines for safety action plans that can be customized for biotech operations could help regulatory agencies in their monitoring of industry manufacturing.

Available ASTM Standards. No biosafety standards have been produced to date. However, the BioSafety Symposia in 1987–1988 included papers on the following topics, some of which may provide basic information and guidance.

1. The section on safety in the bioprocessing industry section discussed the NIH Recombinant DNA Advisory Committee, which is an example of the self-regulatory process. Concepts and concerns of biosafety during the past decade were presented, such as risk taking, opportunities, and roadblocks to minimizing hazards in bioprocessing.
2. The section on design criteria for safety discussed biological production facility containment issues, regulations that govern, and examples of design criteria for safety in the isolation and purification of antibiotics and biologically active compounds. The emphasis was on protecting the worker and community, and significant strides have been made by the industry without additional standards from ASTM.
3. The personnel safety planning section reviewed the general considerations for work practices and personal protective equipment for biological operations. Significant standards are in place today from industry associations, regulatory agencies, and specialty safety equipment suppliers.
4. Safe waste disposal issues are as fundamental today as they were 10 years ago, and the need to minimize community environmental impact is just as essential. Community awareness and emergency response issues were presented. International, national, and local regulatory aspects of bioprocessing were presented. Ten years ago, the fear of overregulation was evident. Today, concern and self-regulation by the industry members have encouraged regulatory agencies to continue monitoring this progress without dictating additional guidelines. However, if these industry members or regulatory agencies see a need for additional standard guides, test methods, or practices, then ASTM E-48 will assist in developing a consensus procedure.

New Standards Being Developed ASTM E-48 is currently looking to the industry membership and regulatory agencies to identify standards that are justified to help improve biosafety.

Future Trends. Surveys and discussions with the industry have been and will continue to be used by the E-48.01 subcommittee to identify new standards for development. Participation with the overall E-48 effort to identify and develop needed, effective, consensus standards is encouraged from industry organizations, manufacturing companies, research institutions, and regulatory agencies.

E-48.02 Subcommittee on Characterization and Identification of Biological Systems

The E-48.02 subcommittee, organized in 1985, is responsible for standard test methods and guides that identify, characterize, and preserve biological species. At this time, 13 standards are published for use.

Scope and Background. The primary charter of Subcommittee E-48.02 is to develop test methods and guides for identifying plasmids, cells, viruses, enzymes, bacteria, and fungi. This includes methods for determining contamination in biological products, detecting mycoplasma, detecting nucleic acids by PCR, and methods for preserving biological products. Guides to meet FDA in vitro diagnostic requirements are also being considered.

Available ASTM Standards. The following standards are published by ASTM (2):

1. Identification of viruses
   a. E 1285, *Guide for Identification of Bacteriophage Lambda or Its DNA*, describes laboratory characterization procedures that are sufficient to verify that a biological preparation believed to contain λ or λ-DNA for use in any step of a biotechnology process actually does contain this material. This procedure should be used if the bacteriophage λ is to be used to construct a recombinant molecule.
   b. E 1286, *Standard Guide for Identification of Herpes Simplex Virus or Its DNA*, describes laboratory characterization procedures that would be sufficient to verify that a biological preparation believed to contain primarily HIV types 1 and 2 or their DNA for use in any step of a biotechnology process actually contains this virus or its DNA. This guide does not cover the identification of HSV in HSV-infected host cells without first isolating the virus from the sample using standard techniques of HSV purification.
   c. E 1493, *Standard Guide for Identification of Bacteriophage M13 or Its DNA*, includes procedures to verify that a biological preparation believed to contain M13 or M13 DNA for use in any step of a biotechnology process actually does contain this bacteriophage or its DNA. This procedure is applicable where M13 is to be used to construct a recombinant molecule and characterization of double-stranded RF DNA is made by restriction enzyme analysis or characterization of single-stranded phage DNA is made by DNA sequencing.
2. Detection of contamination or impurities
   a. E 1298, *Standard Guide for Determination of Purity, Impurities, and Contaminants in Biological Drug Products*, provides definitions of types and levels of impurities, purity of drug product, and

contaminants. It suggests analytical methods generally applied within the pharmaceutical industry to identify and quantitate the level of impurities in a drug product developed through biotechnology. The general effects of these contaminants on these drugs are also described.

b. E 1531, *Standard Practice for Direct Detection of Mycoplasma in Cell Culture by Broth Enrichment and Agar Growth*, describes the procedure that can be used for detection of mycoplasma contamination in cell cultures by direct microbiological culture. Preparation of the broth medium is described. This detection technique can be used in conjunction with the indirect method that is included in E 1532 93.

c. E 1532, *Standard Practice for Indirect Detection of Mycoplasma in Cell Culture by DNA Binding with Bisbenzamide Fluorochrome Stain*, describes the procedure that can be used for mycoplasma contamination detection, based on DNA staining and intended for use in examining cultured mammalian cells. It is not intended for use in detection of mycoplasma in serum, culture media, or systems other than cultures of mammalian cells. Preparation of the bisbenzamide stain concentrate is included with the cell staining procedures.

d. E 1533, *Standard Practice for Indirect Detection of Mycoplasma in Cell Culture by 4′-6-Diamidino-2-2 Phenylindole (DAPI) Staining*, is intended for use in examining only cultured animal cells for mycoplasma contamination using DAPI fluorochrome staining. The procedure includes preparation of DAPI stain concentrate.

e. E 1536, *Standard Practice for Large Volume Testing of Serum for Mycoplasma Contamination*, describes the procedure used for detection of mycoplasma contamination in serum, based on direct microbiological culture of a large volume (50 to 100 mL) of serum as an inoculum in a broth medium. The procedure includes preparation of the broth medium.

f. E 1759, *Standard Test Method for Isoaspartic Acid in Proteins: Method for the Determination of Asparagine Deamidation Products*, measures the amount of isoaspartic acid residues in a protein or peptide solution by using the enzyme protein isoaspartyl methyl transferase and radioactive *S*-adenosyl-L-methionine. The storage of proteins in aqueous solutions often results in the formation of isoaspartic acid linkages within the polypeptide chain as a result of the deamidation of aspargine residues and rearrangement of aspartic acid linkages. The concentration range of isoaspartic acid residues in the sample should be 2.5 to 50 μmol/L. The basis of the procedure is the production of radioactive methanol equal to the amount of isoaspartic acid residues in the sample by action of the enzyme protein isoaspartyl methyl transferase and radiolabeled *S*-adenosyl-L-methionine.

3. Preservation of biological materials

a. E 1342, *Standard Practice for Preservation by Freezing, Freeze-Drying, and Low Temperature Maintenance of Bacteria, Fungi, Protista, Viruses, Genetic Elements, and Animal and Plant Tissues*, outlines procedures to minimize the adverse effects of handling biological materials during low-temperature preservation and maintenance. Although a limited number of methods of preservation can be used on biological systems, the low-temperature methods provide the only real assurance of genetic stability. Cryogenic temperatures are those at or below −70°C. Suggested cooling rates and use of cryoprotectants are presented, and freeze-drying optimum temperatures are discussed. Safety precautions are included for handling cryogenic materials.

b. E 1564, *Standard Guide for Design and Maintenance of Low-Temperature Storage Facilities for Maintaining Cryopreserved Biological Materials*, covers recommended procedures for developing and maintaining low-temperature storage facilities for freezers with mechanical refrigeration and freezers cooled with liquid nitrogen. Suggestions are presented for selecting the proper equipment for the intended service.

c. E 1565, *Standard Guide for Inventory Control and Handling of Biological Material Maintained at Low Temperatures*, suggests safety and security procedures to be used for materials stored at low temperatures to avoid problems during storage. Properly designed inventory control systems ensure the maximum use of freezer space and that all material is located and retrieved easily, without compromising the stability of other items in the freezer (mechanical or liquid nitrogen design). Safety precautions are included for handling vials of materials at cryogenic temperatures.

d. E 1566, *Standard Guide for Handling Hazardous Biological Materials in Liquid Nitrogen*, lists special handling and storage procedures that avoid penetration of vials by liquid nitrogen during storage, which could contaminate the liquid nitrogen bath or contaminate personnel if improperly sealed vials were to explode. Safe handling techniques are discussed for retrieval of cultures from liquid nitrogen, it is recommended that vials be filled to only two-thirds of capacity. A leak test for sealed vials is also described.

4. Collaborative development of DIN standards on PCR. Subcommittee E-48.02 is actively participating in the collaborative development of DIN Committee E9 for the Polymerase Chain Reaction.

a. DIN 58967-10, *Serodiagnosis of Infectious Diseases and Diseases of the Immune System—Polymerase Chain Reaction* (PCR). *Part 60: Terminology, General Method-Specific Requirements*, is published in English and German.

b. DIN 58969-61, *Serodiagnosis of Infectious Diseases. Part 61: Polymerase Chain Reaction*

(PCR)—Special Requirements for the Detection of Nucleic Acid Sequences of the Human Immunodeficiency Viruses HIV-1 and HIV-2, is published in English and German.

5. Collaborative standards with Japan Bioindustry Association (JBA). Currently, JBA and E-48.02 are actively participating in a collaborative development of a standard on the detection of mycoplasma by PCR (*Mycoplasma Detection Method; Nested PCR Assay*)

New Standards Being Planned and Developed

1. Detection of nucleic acids by PCR
2. Identification of cell lines
3. Characterization of industrial bacteria
4. Guidance to help meet FDA requirements
5. Identification of plasmids and vectors
6. Characterization of fungi

Future Trends. Focus will be on completing the test methods and guides needed to safely and economically meet the rapidly growing application of biotechnology to produce drugs, chemicals, services, and products. Subcommittee E-48.02 continues to actively participate in the collaborative development of DIN (German Institute for Standardization) Committee E-9 standards for the PCR.

E-48.03 Subcommittee on Unit Processes and Their Control

The E-48.03 subcommittee was organized in 1985 for the purpose of developing guides and test methods for applying engineered equipment to biological processes. New instrumentation to detect process variables, such as pH, requires uniform procedures to ensure consistent performance. Separation techniques such as liquid chromatography provide opportunities for developing standard testing methods. The E-48.03 subcommittee was chartered to develop standards for these and other bioprocesses.

Scope and Background. The initial focus of this subcommittee was to develop standards for the emerging biosensor instrumentation and the specialized processing equipment used to properly handle living organisms to meet a specific objective. Unit processes being evaluated include bioreactor design, fermenter performance, membrane ultrafilter characterization, biosensors, and bio-specific equipment. Supporting processes such as aseptic sampling, sterile cleaning, clean-in-place, sterilize-in-place, and issues such as solution compatibility and materials of construction to meet aseptic requirements are also in the E-48.03 charter.

Available ASTM Standards. More than six standards have been developed or are in the process of development at this time. These standards are highlighted in the following, based on ASTM published resources (2).

1. Ultrafiltration membrane, chromatographic media, and protein characterization
   a. E 1294, *Test Method for Pore Size Characteristics of Membrane Filters Using Automated Liquid Porosimeter*, describes a procedure for measuring the pore size characteristics of membrane filters in the range of 0.05 to 300 μm. This method uses the automated bubble point method described in ASTM Test Method F 316. This test uses the capillary rise created by surface tension and uses the Washburn equation to calculate the pore diameter. This method determines maximum pore size, minimum pore size, and mean flow pore size of the membrane.
   b. E 1343, *Standard Test Method for Molecular Weight Cutoff Evaluation of Flat Sheet Ultrafiltration Membranes*, evaluates the molecular weight cutoff of flat sheet membranes between 4,500 and 1 million Da. A nonadsorbing penetrant liquid (dextran T-fractions) is used and can be applied to a wide variety of membrane substrates, excluding those that strongly adsorb dextran, from highly hydrophilic to highly hydrophobic.
   c. E 1772, *Standard Test Method for Particle Size Distribution of Chromatography Media by Electric Sensing Zone Technique*, is valuable for measuring particle size and distribution of a dilute suspension of chromatography media in the overall size range of 1 to 450 μm, using the electric sensing zone apparatus. This apparatus uses an electrical current path of small dimensions that modulates individual particles passing through an aperture to produce individual pulses of amplitude proportional to the particle volume.
   c. E 1470, *Standard Test Method for Characterization of Proteins by Electrophoretic Mobility*, is for determining the electrophoretic mobility of proteins of molecular weight greater than 10,000 Da using automatic electrophoretic light-scattering principles to determine its mobility. The instrument measures the Doppler shifts of scattered light at four different angles to determine the mobility distribution of particles in a dispersed, dilute suspension of the protein particles. The test results are expressed as either electrophoretic mobility or ζ potential distribution of the protein particles.
2. Bioprocesses
   a. E 1287, *Standard Practice for Aseptic Sampling of Biological Materials*, summarizes principles, state-of-the-art concepts, and generally accepted methods for aseptic sampling of materials in biotechnical process where contamination of either the sample or the source is not acceptable. Suggested designs include the advantages and disadvantages being discussed. Fabrication and maintenance considerations are presented.
   b. E 1357, *Standard Test Method for Determining the Rate of Bioleaching of Iron from Pyrite by* Thiobacillus ferrooxidans, describes how cells of *T. ferrooxidans* grown on ferrous iron are added to finely ground iron pyrite in an inorganic salt

medium of 2% pulp density. The culture is incubated with agitation, and samples are periodically withdrawn for soluble iron analysis. The rate of pyrite leaching is determined from the linear portion of a curve-plotting soluble iron produced over time. A high leaching rate is evidence for potential degradability of the mineral in mining operations using biologically active cultures.

New Standards Being Developed

1. Guide for evaluating the performance of stirred tank and bioreactors
2. Upgrading ultrafiltration membrane performance measurements

Future Trends. Equipment used to process cell cultures, viruses, bacteria, fungi, tissue, and other biological materials can be quite different from the equipment used in pharmaceutical processes. Unit processes such as reaction, separation, concentration, and purification have industry support to establish evaluation test methods, practices, and guides for proper use, cleaning, sterilization, and application. These needs are to receive focus by E-48.03 in the near future as experts in this field come forward to assist in identifying and preparing these standards.

E-48.04 Subcommittee on Environmental Issues

Although the E-48.04 subcommittee was established in 1985, its primary activity has been to review environmental issues in standards prepared by other E-48 subcommittees. This subcommittee has participated in symposiums, workshops, and surveys to define standards needed within the biotech industry.

Scope and Background. Two categories and specific topics of interest have been identified by industry, government, and research.

1. Biomanufacturing processes
   a. Containment of processes and local manufacturing areas
   b. Biotechnology risk assessment guides that are not currently available from National Institutes of Health, Centers for Disease Control, and other existing organizations
   c. Standardized methods for determining occupational exposure
   d. Decontamination of waste streams. Significant challenges still exist to effectively sample waste streams before discharge to the local waste treatment facilities.
   e. Standardizing construction of and protocols for model ecosystem mycocosm tests. Significant technical problems exist and need resolution before writing standards.
2. Deliberate release or uncontained applications could encourage development of guides for bioremediation microorganisms and their efficacy and efficiency.

Available ASTM Standards

1. E 1287, *Standard Practice for Aseptic Sampling of Biological Materials*, discusses containment options to avoid environmental releases. See E-48.03 "Subcommittee" for more details.
2. E 1344, *Evaluation of Fuel Ethanol Manufacturing Facilities*, presents containment and sterility issues for environmental, safety, and process yield considerations. See "E-48.05 Subcommittee" for other information.

New Standards Being Developed. Surveys and discussions with biotech industry members continue today, and volunteers are needed to identify new standards that are needed and have value to the industry.

Future Trends. Because the objectives and charters of E-48.04 include topics that are of interest to E-48.01 Subcommittee on BioSafety, a close cooperation between the two is planned to identify and develop the standards that are needed by the industry. Volunteers are needed to participate in this consensus standard development activity.

E-48.05 Subcommittee on Biomass Conversion

Biomass conversion standards were first developed in the biomass subcommittee of E-44 Committee on Solar Energy in the early 1980s. In 1982, standards were published for the measurement of moisture, volatiles, ash, and bulk density of particulate wood fuels. Initially, the focus of these standards was to characterize wood for fuel as an energy source and not to characterize biomass materials for conversion to other products using biotechnology. In 1982, the USDA requested ASTM to develop standards for evaluating the design and performance of small grain-fermentation plants that produce ethanol for oxygenate-enhanced fuel blending. This was the entry into biotechnology leading to biomass conversion using living organisms.

In 1985, biotechnology interests in government and industry encouraged ASTM to establish the E-48 Committee on Biotechnology. In 1987, the E-48 Committee requested and received approval for the transfer of the biomass subcommittee from E-44 because the biotechnical similarities would be more efficiently handled in E-48.

Scope and Background. The fuel crises starting in 1973 spurred interest in alternative fuels manufacturing, including fuel-grade ethanol from fermenting grains and other renewable feedstocks. Because grain-based feedstocks have high value for animal and human feeds, interest grew in converting other biomass materials to fuels using emerging bioconversion technology. Although wood, grasses, bagasse, and agricultural residues all contain carbohydrates, as do grains, their exposure for bioconversions to sugars and fermentation to alcohols is more difficult because insoluble lignin surrounds and protects the carbohydrates. This protection limits economical conversion by microorganisms and enzymes. However, with genetic engineering, a new, more efficient source of biological materials was emerging that could be commercially viable for biomass conversions. With the intensified research by industry and U.S. Department of Energy–sponsored grants to organizations such as the National Renewable Energy Laboratory to conduct biomass conversion research, new test methods had to be developed to properly characterize biomass and measure the efficiency of its conversion to usable products. Existing standards from the wood products industry (TAPPI) describe characteristics of wood for use in the pulp and paper industry, but they have limited value for biotechnical use. Accordingly, the E-48.05 subcommittee has developed the following charter for developing standards for the biomass conversion industry.

1. Maintain current standards for wood materials used directly for fuel and energy generation
2. Develop standards for biomass feedstock characterization
3. Develop standards for biomass conversion to usable products through enzymatic saccharification, fermentation, and thermochemical processes
4. Develop terminology standards specific to the biomass feedstock characterization and conversion

Available ASTM Standards. The following standards are available for the wood and biomass conversion industries.

1. Wood fuels
   a. E 870, *Standard Test Methods for Analysis of Wood Fuels*, includes proximate and ultimate analysis of wood fuels. Lists of existing ASTM standards are referenced and include D 1102 (ash in wood), E 711 (gross calorific value of refuse-derived fuels [RDF] by bomb calorimeter), E 775 (total sulfur of RDF), E 777 (carbon and hydrogen in RDF), E 778 (nitrogen and oxygen in RDF), and other standards developed by E-44 and E-48.
   b. E 871, *Standard Method for Moisture Analysis of Particulate Wood Fuels*, determines total weight basis moisture in particulate wood fuel having a maximum particle volume of 16.39 cm^2 or 1 cu in. The method includes sample reduction techniques and oven-drying procedures for calculating moisture content, which can be used for establishing burning characteristics and process yields and as a basis for purchasing or selling wood fuels.
   c. E 872, *Standard Test Method for Volatile Matter in the Analysis of Particulate Wood Fuels*, determines the percentage of gaseous products, exclusive of moisture vapor, for wood particle sizes up to 16.39 cm^3. The technique includes sample preparation and volatilization in a vertical electric tube furnace at 950 °C for 7 minutes.
   d. E 873, *Standard Test Method for Bulk Density of Densified Particulate Biomass Fuels*, uses a fixed-volume open box with prescribed pouring height of cellulosic materials having up to 16.39-cm^3 particulate size. Repeated test values are averaged for final result.
   e. E 1288, *Standard Test Method for the Durability of Biomass Pellets*, describes a tumbling and sieve analysis test that evaluates the relative durability of a 45-kg sample of biomass fuel pellets. This sample is reduced to 1 kg, by a riffle or by coning and dividing. A 0.25-kg sample tumbles for 5 minutes at 50 rpm. A sieve test determines percentage fines passing through a no. 200 sieve.
   f. E 1358, *Standard Test Method for Determination of Moisture Content of Particulate Wood Fuels Using a Microwave Oven*, is a rapid test to determine moisture in wood fuels in several minutes using a microwave oven. The sample is stirred and reheated every 4 minutes or less, depending on values in the tables provided until a constant weight is reached and the percentage of moisture can be calculated.
   g. E 1534, *Standard Test Method for Determination of Ash Content of Particulate Wood Fuels*, discusses the ash content measurement of the noncombustible part of a particulate wood fuel, which uses an electric muffle furnace with an indicating pyrometer to oxidize a 2-g sample to ash.
2. Grain ethanol manufacturing plants
   a. E 869, *Standard Test Method for Performance Evaluation of Fuel Ethanol Manufacturing Facilities*, evaluates four main performance factors for a grain ethanol fermentation plant: conversion efficiency, energy for conversion, production rate, and mass balance. Procedures are contained to measure, interpret, and assign values for these performance parameters in a standard format. Data collection forms are also provided, and standard calculation equations are listed with examples.
   b. E 1117, *Standard Practice for Design of Fuel Alcohol Manufacturing Facilities*, presents minimum good engineering practices for use in designing, constructing, operating, and modifying fuel alcohol plants. Categories of equipment discussed include vessels, towers, heat exchangers, rotating equipment, electrical, instruments, safety, environmental, utilities, and piping.
   c. E 1344, *Standard Guide for Evaluation of Fuel Ethanol Manufacturing Facilities*, provides evaluation criteria to enable a prospective purchaser or lender to effectively review the plans, specifications, and plant operating concept of a fuel ethanol manufacturing facility and to determine whether its design meets the requirements of ASTM design practice standards. This guide can be used in conjunction with E 1117 practice. A design review checklist is included. The guide provides an explanation of the ethanol production process, followed by a discussion of design parameters and potential problem areas within the process.
3. Anaerobic digestion systems
   a. E 1535, *Standard Test Method for Performance Evaluation of Anaerobic Digestion Systems* pro-

vides for measuring the concentration and mass of the influent and effluent wastes and operational parameters, such as input energy, output gas production, and waste biomass, to provide a methodology for evaluation of the anaerobic digester operation. Data collection sheets, calculation procedures, and test report formats are presented.

4. Biomass and biotechnology terminology
   a. E 1126, *Standard Terminology Relating to Biomass Fuels*, presents and defines more than 70 terms used frequently in biomass fuels industries. These terms are directed to the wood fuels industry.
   b. E 1705, *Standard Terminology Relating to Biotechnology*, presents more than 160 terms used in biotechnology, with references to ASTM standards where they are used. As new E-48 standards are developed, new terms will be included.
5. Biomass feedstock characterization
   a. E 1690, *Standard Test Method for Determination of Ethanol Extractives in Biomass*, describes the determination of ethanol-soluble extractives as a percentage of the oven-dried biomass, which includes hard and soft woods, herbaceous materials (such as switchgrass and sericea), agricultural residues (corn stover, wheat straw, bagasse), and wastepaper (office waste, boxboard, newsprint). The extractive content of biomass is not considered to be part of the structural components of biomass and needs to be removed before any chemical analysis of the sample. Some of these extractives include waxes, fats, resins, tannins, gums, sugars, starches, and pigments. The procedure involves using a Soxhlet extraction apparatus and 190-proof ethanol to extract the sample for 24 hours. The sample slurry is filtered, and the filtrate is evaporated in a rotary vacuum evaporator. The resulting extractive weights are reported as a percentage of the original samples oven-dried weight.
   b. E 1721, *Standard Test Method for Determination of Acid-Insoluble Residue in Biomass*, can be used to determine the acid-insoluble residue of hard and soft woods, herbaceous materials, agricultural wastes, wastepaper, acid and alkaline pretreated biomass, and the solid fraction of fermentation residues. This acid-insoluble residue contains the acid-insoluble lignin and condensed proteins from the original oven-dried sample of biomass. Samples for this test can come from several sources and have separate calculation procedures; for example, prepared biomass samples are treated in accordance with practice E 1757, and extractives-free material is prepared in accordance with test method E 1690. The samples are reacted with 72% sulfuric acid (reagent grade) for hydrolysis and then diluted to 4% acid and autoclaved for an hour at 121 °C. The filtrate is removed by vacuum filtering, and the remaining insolubles are dried, weighed, and then ignited in a muffle furnace to determine the ash weight. The difference in weight of the dried insolubles minus the ash weight as a percentage of the original dry weight of the sample is defined as the acid-insoluble content of the biomass.
   c. E 1755, *Standard Test Method for Ash in Biomass*, expresses the ash content as the mass percentage of residue remaining after dry oxidation (575 °C in a muffle furnace) on the original sample dry weight. Samples of biomass can include hard and soft woods, herbaceous materials, agricultural residues, wastepaper, acid and alkaline pretreated biomass, and the solid fraction of fermentation residues.
   d. E 1756, *Standard Test Method for Determination of Total Solids in Biomass*, describes two test methods that depend on the nature of the sample. Because moisture is not considered a structural component of biomass and can change with storage and handling, it is important to analyze biomass samples and report the results on a moisture-free basis. Test method A uses a drying oven with controlled temperatures at 105 °C. Test method B uses an automated moisture analyzer with direct infrared heating elements; it should not be used for samples containing large amounts of free sugars or proteins that can degrade or caramelize.
   e. E 1757, *Standard Practice for Preparation of Biomass for Compositional Analysis*, describes a reproducible way to convert the many types of biomass feedstocks into a uniform material suitable for compositional analysis. Three methods are presented. Test method A uses field-collected samples and natural air drying to a constant weight. Test methods B and C produce dried feedstocks using convection oven and lyophilization, respectively. The samples are milled and sieved for additional analyses. Test methods B and C are used for very wet biomass samples that are at risk for mold growth during drying or that could degrade after standing for prolonged times.
   f. E 1758, *Standard Test Method for Determination of Carbohydrates in Biomass by High Performance Liquid Chromatography*, discusses the carbohydrates making up the major portion of dried biomass samples—polysaccharides constructed primarily of glucose, xylose, arabinose, galactose, and mannose subunits. These polysaccharides can be hydrolyzed to their component sugar monomers with 72% sulfuric acid in a two-stage hydrolysis process. These monosaccharides can then be quantified by ion-moderated partition HPLC using calibration sugar standards. These results can then be used with other biomass assays to determine the total composition of the biomass sample.

New Standards Being Developed. Now that the feedstock characterizations are almost complete, a new category of standards is being developed including biomass conversion standards. Just as standards were developed to evaluate the yields, efficiency, and quality of alcohols produced by cooking and fermenting grain feedstocks, a similar opportunity applies to biomass conversion by the following methods that are currently in the research stage of development:

1. Chemical pretreatment using acid prehydrolysis
2. Biochemical conversion of biomass using liquefaction, saccharification, fermentation, simultaneous saccharification and fermentation, enzymatic hydrolysis, and bacterial conversion
3. Thermochemical conversion processes using pyrolysis and gasification

Future Trends. As research continues to uncover new biological and engineered organisms that improve the efficiency and economics to convert renewable biomass feedstocks to fuels, chemicals, drugs, and other usable products, additional standards may be justified in this rapidly expanding industry. A focused, concerted effort from the biomass research community is needed to develop these new test methods and to standardize biomass conversion processes and product characterization techniques. The ASTM E-48.05 Subcommittee on Biomass Conversion will continue to be a significant force to support this progress.

E-48.06 Subcommittee on Biotechnical Equipment Qualification and Process Validation

For biological processes producing drugs and chemicals that are under the regulatory responsibility of the FDA, there is a need to validate the biological processes and qualify the equipment used in the process. Because many processes use common equipment, the protocols used to qualify this equipment can be standardized to the point where their specific application may not be confidential or specific to the process owner. Process validation protocols can also be standardized for uniform application and acceptance by regulatory agencies. The current *Good Manufacturing Practices* (cGMPs) published by FDA call for process validation, equipment qualification, and utility qualification (12). The work of E-48.06 will assist the biotech industry to use standard test methods comparable to quality standards.

Scope and Background. Subcommittee E-48.06 was organized in October 1988 to develop standards and protocols for equipment qualification and process validation. Its charter is to develop standards required for or used in the qualification and validation of facilities, for process-related utilities, production equipment, computer controls, sampling methods, and processes that include at least one step involving living organisms.

Available ASTM Standards

1. E 1567, *Standard Guide for Biopharmaceutical Facilities Architectural Design Considerations*, covers architectural design considerations for buildings and facilities that use biological processing to produce drugs, chemicals, and other products, especially when cGMP criteria apply under Code of Federal Regulations Title 21 (2). The guide is intended for use in designing, constructing, and operating pilot plant and commercial production facilities to meet federal regulations. References are provided to existing federal regulations that apply to all general phases of a bioprocess activity.
2. The symposium *Validation Practices for Biotechnology Products* (9) brought together representatives from industry, federal agencies, research, and academia to identify more than 100 standards for development.

New Standards Being Developed. Task groups have been established to develop standards for the following issues:

1. Multiuse cleaning of process equipment
2. Computer validation protocols
3. Master plan for validation
4. Standards for installation qualification, operational qualification, performance qualification, purchase, or construction specifications
5. Qualification of utilities, filtered air, plant steam, and clean steam

Future Trends. By considering industry needs obtained from workshops, symposia, surveys, and technical committee meetings, E-48.06 will focus on protocols for fermenters, cell recovery systems, purification systems, sterile filling operations, environmental systems, sterilization equipment, and other production equipment. Computer controls validation is a high-interest topic in industry and regulatory agencies. In recent workshops (9,10) more than 100 potential standards have been identified for prioritization and development into voluntary consensus standards.

E-48.91 Subcommittee on Terminology

The charter of this subcommittee is to maintain E 1705, *Terminology Relating to Biotechnology* (2), which contains more than 130 definitions pertaining to the E-48 standards and other words relating specifically to biotechnology. As new standards are developed and approved, any new biotechnical words are added automatically to E 1705. As discussed previously, biomass standards currently include a biomass terminology standard, E 1126, *Terminology Relating to Biomass Fuels*.

BIBLIOGRAPHY

1. *ASTM Technical Committee Officer Handbook*, ASTM, June 1996, pp. 1–5.
2. *Annual Book of ASTM Standards*, vol. 11.05, ASTM, Philadelphia, Pa., 1996, pp. iii–v.

3. Technology Transfer Improvements Act of 1995, H.R. 2196, Provision 12(d), Public Law 104-113.
4. *Congressional Record—Senate*, February 7, 1996, p. 1081.
5. *Regulations Governing ASTM Technical Committees*, ASTM, July 1995, pp. 5–13.
6. *Form and Style for ASTM Standards*, ASTM, West Conshohocken, Pa., January 1996, p. vi.
7. *Symposium on Bioprocessing Safety Standard Guidelines and Practices to Insure Work Safety in the Bioprocessing of Industrial Chemicals, Foods, and Waste Products*, ASTM E-48 Symposium, October 1987, STP no. 1051.
8. *ASTM 1996 Directory*, ASTM, Philadelphia, Pa., pp. 40, 114.
9. *Validation Practices for Biotechnology Products*, ASTM E-48 Symposium, April 1995, STP no. 1260.
10. Conference Report, *J. Res. Natl. Inst. Standards Technol.* **99** (Jan.–Feb. 1994).
11. N.A. Roscioli, C.A. Renshaw, A.A. Gilbert, C.F. Kerry, and P.G. Probst, *BioPharm* 32–40 (Sept. 1996).
12. *ASTM Standardization News*, August 1989, pp. 44–49.

See also GOOD MANUFACTURING PRACTICE (GMP) AND GOOD INDUSTRIAL LARGE SCALE PRACTICE (GLSP); PROCESS VALIDATION.

ATTACHMENT FACTORS

DERYK T. LOO
Bristol-Myers Squibb Pharmaceutical Research Institute
Seattle, Washington

KEY WORDS

Cadherins
Cell adhesion molecules
Collagen
Fibronectin
Glycosaminoglycan
Laminin
Selectins
Tenascin
Thrombospondin
Vitronectin

OUTLINE

INTRODUCTION

The development and maintenance of tissues in vivo is acutely dependent on the ability of cells to form connections with substrates and neighboring cells. The formation of intimate contacts with various substrates provides sites of attachment and the scaffolding necessary for cells to communicate and develop cooperative function. These substrates, or attachment factors, include extracellular matrix (ECM) and basement membrane molecules such as collagen, laminin, fibronectin, and vitronectin. Additional substrates include proteins expressed on the surface of cells that function as attachment factors and mediate cell–cell interactions. Cell culture provides a powerful tool with which to identify attachment factors involved in cellular adhesion and to define the molecular mechanisms regulating cell adhesion. Additionally, in vitro approaches play a key role in the identification and cloning of specific attachment factor receptors. The cloning and characterization of attachment factor receptors has forced us to redefine the role of attachment factors in cellular functions. Growing in vitro evidence, combined with in vivo studies, make it clear that attachment factors serve not only as molecular scaffolds to support cell attachment and movement, but also as signaling molecules that play pivotal roles in regulating numerous cellular functions including growth, activation, differentiation, and death.

I review here the ECM molecules that are widely used as attachment factors in culture and illustrate ways in which attachment factors regulate cellular function. Additionally, cell surface molecules that serve as attachment factors and facilitate cell–cell interactions in vivo are also reviewed.

EXTRACELLULAR MATRIX MOLECULES

Fibronectin

Biochemical Properties. Fibronectins are glycoproteins consisting of two disulfide-linked subunits with a molecular weight of 220,000–250,000 kDa. Fibronectin was originally purified to homogeneity in 1970 by Mosesson and Umfleet from plasma cryoprecipitate and termed cold-insoluble globulin (CIg) (1). Later, fibronectin was detected as a cell surface protein on fibroblasts whose expression was lost upon transformation (2,3). In addition to existing as a disulfide-linked dimer, cell surface–associated fibronectin also exists as higher molecular weight disulfide-linked aggregates that are secreted by cells and accumulate as a major component of the basement membrane and ECM in vivo.

The primary structure of fibronectin consists of a series of three different types of modular units that combine to generate the functional domains of the protein. The two subunits of fibronectin are connected by a pair of interchain disulfide linkages at the carboxy-terminal end. The intact dimer appears as an extended, flexible molecule consisting of two linear arrays of globular domains. The identification and characterization of the individual globular domains has resulted largely from examination of proteolytic fragments of fibronectin, many of which retain biological function. The N-terminal domain of fibronectin binds fibrin (4–6), heparin (6–8), *Staphylococcus aureus* (8–11), thrombospondin (12), factor XIIIa transglutaminase (4,11), and IgG (13). Adjacent to the amino-terminal domain of fibronectin is a domain that binds gelatin and collagen (7,14). This interaction is important for localization of fibronectin to basement membranes and ECMs. The collagen-binding domain also binds to the C1q component of complement via the collagenous tail of C1q (15). Adjacent to the gelatin-binding domain of fibronectin is the cell-binding domain. Pierschbacher et al. (16) showed that the cell-binding activity of fibronectin resided within a 12-kDa proteolytic fragment that contains the tetrapeptide sequence Arg-Gly-Asp-Ser (RGDS) (17). Peptides containing soluble synthetic Arg-Gly-Asp (RGD) were shown to detach cultured cells and platelets from fibronectin substrates (18,19), implicating the RGDS sequence in cell binding.

In addition to binding collagen, fibronectin has an affinity for several other components of the ECM, including the glycosaminoglycans heparin, heparan sulfate and chondroitin sulfate proteoglycan, and hyaluronic acid (6,20–25). The interaction of fibronectin with these various components of the ECM is thought to play a role in the formation and maintenance of the ECM.

The high-affinity cell surface receptors for fibronectin are the integrins, which consist of heterodimeric glycoprotein α- and β-subunits of approximately 140 and 120 kDa that bind to the RGD sequence of fibronectin (26). The extracellular domains mediate cell adhesion via binding to fibronectin and other ECM proteins. The cytoplasmic domains are responsible for transmitting signals to the cell interior following ligand engagement (outside-in signaling), and for modulating the affinity of the integrin for its ligand by interacting with intracellular signaling molecules (inside-out signaling). The cytoplasmic domains also interact with the actin-binding proteins filamin (27,28) and α-actinin (29), thereby connecting the cell membrane with the cytoskeleton. These interactions play a role in the formation of focal adhesions and in phagocytosis.

Fibronectin plays a central role in vertebrate development. Fibronectin matrices are utilized by migrating cells during embryogenesis (30–32). The migration of avian, amphibian, and mammalian neural crest cells is blocked by RGD-containing peptides and by antibodies to fibronectin or its cell surface receptor (33,34). Furthermore, inactivation of the fibronectin gene in mice leads to early embryonic lethality (35). Its role in development and in maintenance of the adult organism is due to its diverse influences on cell growth, adhesion, migration, and differentiation.

Role in Cell Adhesion. Fibronectin promotes the attachment of multiple cell types, including epithelial cells, follicular cells, ovary cells, neuronal cells, fibroblasts, keratinocytes, and adult liver cells. As previously mentioned, fibronectin is not solely an attachment factor for cells. In many cases, cell attachment to fibronectin is required for cell survival and promotion of cell growth. Fibronectin promotes cell division of numerous cell types, including follicular cells (36), primary chick fibroblasts (37), murine and human fibroblasts (38), neuroblastoma cells (39), keratinocytes (40), rat granulosa cells (41), and lymphocytes (42). Fibronectin also promotes DNA synthesis by serum-deprived quiescent normal hamster fibroblasts (43).

In addition to regulating cell division, fibronectin has been shown to modulate the differentiation of cells in culture. Fibronectin coating of culture surfaces inhibits morphological changes and biosynthesis of lipogenic enzymes associated with the differentiation of preadipocyte fibroblasts (44,45). Fibronectin also attenuates hormone-induced expression of cytoskeletal proteins associated with granulosa cell differentiation (46). Additionally, exogenously added fibronectin inhibits both spontaneous and chemically induced differentiation of cultured normal human keratinocytes in vitro (47,48). Alternatively, fibronectin induces the differentiation of neural crest cells, neuroepithelial embryonal carcinoma cells (F9), and neuroblastoma cells in vitro (49–51), consistent with observations showing that receptors for fibronectin are found on neural crest cell derivitives colocalized with fibronectin (52) in vivo. Fibronectin also promotes the differentiation of primary human hepatocytes, as measured by the expression of albumin and the liver-specific enzyme cytochrome P450 (53). It has been suggested that modulation of cell growth and differentiation by fibronectin is due in part to its role in regulating the cytoskeletal architecture of cells (36,50). The recent cloning of fibronectin receptors and identification of signaling pathways coupled to these receptors has provided evidence that specific signaling events play an important role in controlling cell growth and differentiation.

Collagen

Biochemical Properties. Collagen is the most abundant component of the ECM and has been used extensively as an attachment factor for cultured cells. Collagen is composed of three helical α-chains that are wrapped around each other in a triple helical arrangement. In some forms of collagen the three chains are identical, whereas in other forms the chains differ from one another. The α-chains are composed of a series of triplet Gly-X-Y residues, where X and Y may be any amino acid. Frequently, X is a proline and Y is a hydroxyproline. Alternatively, hydroxylysine may also occupy the X or Y position and contribute to interchain cross-linking of the collagen α-chains. Collagen is secreted from cells as procollagen, with each of the α-chains containing large amino- and carboxy-terminal nonhelical domains. Although some collagens exist in vivo as intact procollagen, many forms of collagen undergo proteolytic removal of the amino- and carboxy-terminal nonhelical domains.

To date, more than 14 collagen types have been described, which are generally subgrouped into six classes defined by the structural forms in which they assemble within the ECM. Several forms of collagen display ordered fibrous structures. For example types I, II, III, V, and XI collagen form fibrillar structures that contribute to connective tissues such as cartilage, skin, bone, and tendon, whereas type IX and XII collagens are fibril associated and interact with type I and II collagen fibrils. Type IV collagen is a nonfibrillar collagen and is the principle collagen type found in basement membrane. Type IV collagen differs from the fibrillar collagens by its existence in tissues in an unprocessed procollagen form (54,55). Other collagen types include filamentous collagen (type VI), short-chain collagens (types VIII and X), and long-chain collagen (type VII).

Collagen matrices are associated with many cell attachment proteins in vivo, including fibronectin and laminin. Thus the role of collagen as an attachment factor may be divided into two types: (*1*) a direct mechanism whereby collagen serves as the substrate for cell attachment (VLA-2; collagen receptor); and (*2*) an indirect mechanism whereby a secondary attachment factor, through its interaction with collagen, serves as the substrate for cell attachment.

Role in Cell Adhesion. The principle collagens used as attachment factors in cell culture are the fibrillar collagens and the nonfibrillar type IV collagen. Fibrillar collagen isolated from connective tissues was the first collagen used in cell culture. A mouse tumor source of type IV collagen was subsequently identified, which produced large amounts of this collagen type (56,57). Collagen substrates serve as an attachment and growth-promoting factor for a variety of cells in culture. Collagen promotes the sustained growth of epithelial cells from various tissues including mammary epithelial cells (58–61), endometrial epithelial cells (62), vaginal epithelial cells (63), esophageal epithelial cells (64), and liver epithelial cells (65). Collagen substrates also promote the extended viability of primary chondrocytes (66). Chondrocytes exhibit three-dimensional growth within collagen gels, deposit ECM, and incorporate or rearrange exogenous collagen into newly formed matrix, suggesting the formation of tissuelike structures.

In addition to providing attachment and growth-promoting activity, collagen also modulates the differentiation state of cells in culture. Collagen induces morphological and biochemical characteristics associated with the differentiated state of various epithelial cell types including human endometrial cells (62), trachial epithelial cells (67), uterine epithelial cells (68), mammary epithelial cells (69), and intestinal epithelial cells (70). Collagen also promotes the differentiation of adrenocortical cells (71,72), osteoblasts (73), keratinocytes (74), hepatocytes (75–79), liver dendritic cell progenitors (80), thyroid follicle cells (81), smooth muscle cells (82), and Sertoli cells (83). Regulation of growth and differentiation, which is accompanied by alterations in gene expression, implicate collagen as an active signal-transducing molecule. Candidate receptors for collagen that may play a role in cell signaling include the integrins VLA-1 (84,85), VLA-2 (86–88), and VLA-3 (86–88).

Laminin

Biochemical Properties Laminin is the first extracellular protein expressed during development and plays a critical role in cellular development and tissue organization through modulation of cellular attachment, motility, growth, and differentiation. Laminin is a large glycoprotein that is a major component of basement membranes. The prototypical laminin molecule is composed of three subunits: a 440-kDa A-chain, a 200-kDa B1-chain, and a 220-kDa B2-chain. These are joined together by disulfide linkages to form a cruciform-shaped structure possessing multiple globular and rodlike domains that are responsible for its cell-binding activity and its interaction with a variety of ECM-associated molecules. The terminal end of the long arm, which contains a large, multilobed globular domain and forms the base of the cruciform, is responsible for receptor-mediated cell attachment, promotion of neurite outgrowth, and heparin binding (89). Regions within the short B-chain arms provide a second cell attachment site and also play a role in cell signaling and binding of the glycoprotein nidogen/entactin. Similar to fibronectin, laminin also binds to bacteria (90) and glycosaminoglycans (91).

Laminin promotes cell attachment through interaction with specific cell surface receptors. The most extensively studied laminin receptors are the integrins. Laminin binds to at least four VLA integrins: $\alpha1\beta1$ (92,93), $\alpha2\beta1$ (87,94), $\alpha3\beta1$ (95), and $\alpha6\beta1$ (92,96). Laminin also binds to the $\alpha V\beta3$ integrin (97) and $\alpha6\beta4$ integrin (98,99). The interaction of laminin with integrins is believed to occur through a non-RGD-mediated mechanism, although RGD-containing peptides can inhibit cell adhesion to the second cell adhesion site located within the central portion of the laminin cruciform (96). Other recently identified receptors for laminin include the 67-kDa laminin receptor, which has been implicated in tumor metastasis (100,101), and a 110-kDa laminin-binding protein from brain, which appears to be immunologically and biologically related to the β-amyloid precursor protein (APP) family (102). The interaction of the 110-kDa laminin-binding protein with the Ile-Lys-Val-Ala-Val (IKVAV) residues of the laminin A-chain plays a role in neurite outgrowth (102). A distinct 110-kDa protein, termed nucleolin, was copurified with the APP-related 110-kDa protein and has also been identified as a laminin-binding protein (103) that may play a role in ECM signaling. Laminin also binds to the actin-binding cell surface protein connectin (104) and the 67-kDa elastin/laminin receptor (105).

Role in Adhesion. Laminin plays an important role in the attachment, growth, and differentiation of many cell types in culture, including neural and epithelial cells. When added to the culture medium, laminin promotes Schwann cell attachment, growth, and the formation of a stellate, process-bearing morphology (106) and regulates glycolipid synthesis (107). Laminin promotes attachment and neurite outgrowth of sensory neurons (108) and induces the differentiation of neuroblastoma cells (51). Antibodies directed against the heparin-binding domain of laminin reduce neuronal viability and inhibit neurite outgrowth in vitro, indicating that the heparin-binding domain is responsible for the effects of laminin on neurite outgrowth and neuronal survival (109). Laminin also modulates the proliferation of cultured astrocytes (110) and oligodendrocytes (111). Laminin promotes the differentiation of cultured endothelial cells to form capillary-like structures (112). Laminin also induces the differentiation of various epithelial cells, as measured by β-casein production by cultured mammary epithelial cells (113,114), the formation of cordlike structures by Sertoli cells (115), and the expression of alkaline phosphatase and lactase activity by intestinal epithelial cells (116). Laminin promotes the attachment and survival of F9 embryonal carcinoma cells (117), while it potentiates the differentiation of PCC4uva embryonal carcinoma cells to neurons following treatment with retinoic acid and dibutyryl cyclic AMP (118). Laminin also promotes the differentiation of granulosa cells (119) and hepatocytes (120,121).

Vitronectin

Biochemical Properties. Vitronectin, or serum-spreading factor, is a glycoprotein that is present in the blood and in other tissues. Vitronectin was initially described as a cell-spreading and growth-promoting α1-glycoprotein enriched in a fraction of human serum isolated by glass bead chromatography (122). The spreading–promoting activity of this fraction was further purified from this fraction by Barnes et al. (58) and was shown to be biochemically and immunologically distinct from fibronectin and laminin (123,124). Vitronectin exists in two distinct forms in human serum: a 75-kDa single-chain form and a 65-kDa two-chain form (125,126). The 65-kDa form is the product of proteolytic processing and lacks the carboxy-terminal 10-kDa fragment (127). Sensitivity to proteolytic processing may result from allelic differences in the vitronectin gene (128). The gene encoding vitronectin was identified by expression cloning (127), and sequence analysis revealed a 1,545-bp open reading frame corresponding to the full-length plasma vitronectin. Sequence analysis of an independently isolated serum protein associated with complement, termed S-protein, revealed it to be identical to vitronectin (129).

The major source of circulating vitronectin is the liver (130), although it is also produced by various hematopoetic cell types including platelets, monocytes, and macrophages (131,132). Vitronectin is also associated with tissues (124,133). In situ hybridization analysis indicates that vitronectin mRNA is expressed early in mammalian development primarily in the liver and central nervous system (134), suggesting that it plays an important role in mammalian development. In the mouse, vitronectin mRNA is detected by day 10 in the liver and central nervous system. Expression of vitronectin mRNA in the central nervous system is first detected in the floor plate; later, vitronectin mRNA can be observed in the meninges of the cortex and the spinal cord. Vitronectin mRNA is associated with the vasculature of the central nervous system but not with blood vessels of peripheral tissues, suggesting that vitronectin may play a specific role in vascular function in the central nervous system. Surprisingly, mice deficient in vitronectin gene expression demonstrate normal development and survival, indicating that the functional role of vitronectin in development may be compensated for by other molecules (135).

The vitronectin gene is composed of eight exons and seven introns (136), which define the domain structure of the mature protein. The amino-terminal cysteine-rich 44-residue domain is identical to somatomedin B (137). The second domain contains an RGD sequence that, analogous to fibronectin, is responsible for cell attachment via a number of integrins (137,138) and also contributes to collagen binding (139,140). The second domain is also implicated in binding to plasminogen-activator inhibitor-1, and consistent with its localization to platelets, this interaction plays an important role in blood clotting. Vitronectin also binds to heparin (125), and a domain located toward the carboxy-terminal region of vitronectin is responsible for the heparin-binding activity of vitronectin (137,141), which is dependent on the conformational state of vitronectin. Although native plasma vitronectin binds heparin weakly, vitronectin denatured by 8 M urea or other denaturing agents has an increased affinity for heparin (125,141). Denaturation is hypothesized to expose a cryptic heparin-binding domain and suggests that the biological functions of vitronectin are dependent on its conformation. Recently a form of vitronectin distinct from plasma vitronectin has been identified within platelet α-granules, which is able to bind heparin (142) and may play a role in platelet function at sites of vascular injury. Additionally, similar to collagen and fibronectin, vitronectin binds to bacteria, including *S. aureus, Escherichia coli, S. pneumoniae*, and *Neisseria gonorrhoeae* (143–147), implicating vitronectin in the adherence and phagocytosis of bacteria by host immune defense cells.

Role in Adhesion. Based on the tissue distribution of vitronectin, it is not surprising that vitronectin promotes the attachment and growth of a wide variety of cell types in vitro, including endothelial cells, lymphoid cells,and neural cells (58). Human endothelial cells attach to vitronectin, adopt a flattened morphology, and exhibit increased viability (148). Vitronectin promotes the attachment and proliferation of IL-3-dependent mast cells (149). However, vitronectin is unable to replace IL-3 as a survival factor, indicating that vitronectin is not itself a survival factor for bone marrow–derived mast cells but is able to augment the IL-3-dependent signal in mast cells. Vitronectin promotes the attachment of myeloblast cells through the αv-containing integrins (150) and stimulates megakaryocyte proplatelet formation through interaction with $\alpha v/\beta 3$ integrin (151). Vitronectin also promotes the attachment and long-term survival of cultured glioma cells (152), and this interaction may be mediated by gangliosides (153). In addition, vitronectin mediates attachment and nerve growth factor–dependent neurite outgrowth of rat peochromoccytoma cells via an RGD-dependent mechanism (154).

Other Attachment Factors

In addition to collagen, fibronectin, laminin, and vitronectin, other components of the ECM and basement membrane also contribute to cell attachment. Thrombospondin is a large trimeric glycoprotein consisting of three identical 140-kDa subunits joined together through interchain disulfide linkages (155). Thrombospodin is released by activated platelets and is found associated with the ECM proteins heparin, collagen, laminin, and fibrinogen. Thrombospondin is synthesized by human long-term bone marrow cells in vitro, promotes the adhesion of hematopoietic progenitor cells (156), and promotes the attachment and migration of monocytes and neutrophils (157). Receptors for thrombospondin that play a role in its function as an attachment factor include membrane-bound heparan sulphate (158), platelet glycoprotein IV (CD36; GPIIIb) (159), and $\alpha v/\beta 3$ integrin (160).

Tenascin is a very large, star-shaped ECM glycoprotein composed of six related subunits joined through interchain disulfide linkages. The individual subunits contain 13 epidermal growth factor–like repeats, 8 or more fibronectin type III repeats, and a globular carboxy-terminal fibronectin-like domain. Tenascin is associated with mesenchymal–epithelial interactions during development, tissue remodeling, and wound healing. Tenascin promotes the attachment of many cell types in vitro, including tumor cells, fibroblasts, and endothelial cells, in an RGD-dependent manner (161) through interaction with multiple integrins (162). Janusin, an ECM protein with structural homology to tenascin, promotes attachment and neurite outgrowth of hippocampal neurons in vitro (163).

Glycosaminoglycans are a major component of the ECM and, as already mentioned, bind to fibronectin (164), collagen (165), laminin (166), and vitronectin (127). Glycosaminoglycans consist of long polysaccharide chains made up of disaccharide subunits and, with the exception of hyaluronan (HA), are sulfated and covalently attach to core proteins through serine residues. In addition to playing a role in tissue formation during development and remodeling, glycosaminoglycans also contribute to the process of inflammation and tumorigenesis (167). One glycosaminoglycan that has been widely studied is HA. The role of HA in various cellular functions including cell proliferation, cell activation, and cell migration is dependent on its interaction with specific cell surface glycoprotein CD44 (168–170). Molecular cloning of the CD44 gene revealed the presence of multiple isoforms, which result from alternate splicing. Analysis of the expression pattern of the different isoforms suggests that regulation of expression of specific isoforms plays a critical role in cellular interactions during early development (171). The CD44–HA interaction is involved in endothelial cell adhesion and proliferation (172) and T-cell adhesion and activation (173). CD44 has also been implicated in the regulation of tumor growth and metastasis, and this activity is dependent on its interaction with HA (174).

CELL SURFACE ATTACHMENT MOLECULES

In addition to molecules associated with the ECM and the basement membrane, cell surface receptors also serve as attachment factors to promote cell–cell interactions. Cell surface receptors implicated in cell attachment include the cell adhesion molecule family, the selectin family, and the cadherin family.

Cell Adhesion Molecules

The cell adhesion molecules, or CAMs, comprise a family of cell surface receptors that are members of the immunoglobulin gene superfamily. Members of this gene superfamily are characterized by the presence of multiple tandem immunoglobulin fold domains. Vascular cell adhesion molecule 1 (VCAM-1) is present on most peripheral blood leukocytes, activated endothelial cells, macrophages, dendritic cells, bone marrow fibroblasts, and myoblasts (175–177) and plays an important role in the recruitment of leukocytes to sites of inflammation. At sites of inflammation, activated endothelial cells express VCAM-1, which facilitates the adhesion and subsequent transmigration of circulating leukocytes. The $\beta 1$ integrin VLA-4 (CD$\alpha 4/\beta 1$) (178) and the $\beta 7$ integrin (CD$\alpha 4/\beta 7$) (179) are ligands for VCAM-1. Similar to VCAM-1, intercellular adhesion molecules (ICAM-1, -2, -3) are expressed on most circulating blood leukocytes, activated endothelial cells, fibroblasts, keratinocytes, chondrocytes, and epithelial cells (180,181) and play a role in immune and inflammatory responses by mediating cell–cell interactions. The ligands for ICAM-1 and ICAM-2 include the integrins LFA-1 (CD18/CD11a) (182) and MAC-1 (CD18/CD11b) (183,184), and CD43 (185). Neural cell adhesion molecule (NCAM) is expressed on neurons, astrocytes, Schwann cells, myoblasts, natural killer (NK) cells, and activated T lymphocytes. NCAM is thought to play an important role in vertebrate development through modulation of cell–cell interations between neurons, astrocytes, oligodendrocytes, and myoblasts (186). In addition to homotypic interactions, NCAM binds to heparan sulfate and heparin glycosaminoglycans (187). Two related members, L1-CAM and Ng-CAM, are expressed on postmitotic, premigratory neurons and play a role in neuronal migration and neurite outgrowth. Ligands for L1-CAM include axonin 1 (188) and $\alpha V\beta 3$ integrin (189). Ng-CAM also binds to axonin 1 (190,191). Neurite outgrowth is stimulated by the interactions of L1-CAM and Ng-CAM with axonin I (188) and requires calcium influx through L- and N-type calcium channels (192). The condroitin sulfate proteoglycan neurocan also binds to NCAM, Ng-CAM, and L1-CAM and inhibits neuronal adhesion and neurite outgrowth (193). Platelet endothelial cell adhesion molecule (PECAM-1, CD31) is expressed on platelets, endothelial cells, monocytes, granulocytes, and a subset of T cells (194,195). Exogenous expression of PECAM-1 on mouse L cells induces cell aggregation both in a PECAM-1-dependent, homophilic manner (196), and a calcium-dependent heterophilic manner through interaction with heparin and chondroitin sulfate (197). Recently, the $\alpha v/\beta 3$ integrin has been demonstrated to be a ligand for PECAM-1 (198).

Selectins

The selectin family of cell surface adhesion molecules function specifically in leukocyte–endothelial cell adhesion. Selectins have a multidomain structure consisting of an amino-terminal carbohydrate-binding lectin domain, an epidermal growth factor–homologous domain, and a number of cysteine-rich domains homologous to the consensus repeats contained within complement regulatory proteins. Three members of the selectin family have been cloned and characterized, to date. L-selectin (CD62L) is expressed by leukocytes, E-selectin (CD62E) is found on endothelial cells activated by inflammatory mediators, and P-selectin (CD62P) is found on platelets and activated endothelial cells (199). The E- and P-selectins are implicated in the initial adhesion and rolling of leukocytes on activated vascular endothelial cells that line blood vessels (200,201). Studies showing that monoclonal antibodies directed against the lectin domain of selectins blocked cell adhesion have implicated carbohydrates as ligands for selectins. Currently, a large number of sialylated, sulfated, and/or fucosylated carbohydrates have been identified as ligands for selectins (199,202). In addition to their adhesive properties, selectins activate leukocytes, causing intracellular calcium release, induction of IL-8 and TNF-α mRNA transcription, and activation of β2 integrin–dependent strong adhesion (203).

Cadherins

The cadherin family of cell adhesion molecules are calcium-dependent membrane glycoproteins that are grouped into nine types based on structural and functional differences (204). Type I, or classical cadherins (E-, P-, N-, B-, R-, EP-cadherin), are the only cadherins that have been conclusively shown to promote cell adhesion. Type I cadherins contain five extracellular domains of approximately 110 amino acids each, a single transmembrane domain, and two cytoplasmic domains. The extracellular domains contain Asp-X-Gln-Asp-Gln-Asp and Asp-X-Asp amino acid motifs, which are responsible for calcium binding. The amino-terminal extracellular domain contains the cell adhesion recognition sequence His-Ala-Val (HAV), which mediates calcium-dependent homotypic and heterotypic interactions between cadherins. The cytoplasmic domains of type I cadherins interact with a family of intracellular proteins termed catenins (105). The catenins in turn bind to microfilaments and serve to connect the cadherins to the cytoskeleton. The interaction of catenins with the cytoplasmic domain of cadherins is thought to play an important role in cadherin adhesive function.

Cadherin interactions regulate the formation of adherens junctions and tight junctions between cells. In addition to mediating cell–cell contact, cadherins also regulate cell morphology and differentiation. E-cadherin, which has been classically associated with the formation, differentiation, and polarization of developing epithelia, has recently been shown to regulate erythropoietin-mediated differentiation of bone marrow mononuclear cells (205). Neural cadherin, or N-cadherin, promotes neurite outgrowth, which can be blocked by HAV-containing peptides (206). N-cadherin also regulates the adhesion of O-2A-lineage glial progenitor cells in culture (207).

FUTURE DIRECTIONS

It has become clear that cell–cell and cell–substrate interactions mediated through adhesion molecules and their corresponding receptors play critical roles in cellular functions including growth, activation, differentiation, and death. One of the challenges facing researchers will be to define the molecular mechanisms triggered by the interaction between attachment factors and cell surface receptors that regulate cellular function. Recent advances in molecular biology have provided powerful tools by which to identify both extracellular and intracellular components of these complex signaling systems. Soluble, chimeric proteins containing the extracellular regions of attachment factors and cell surface receptors have been used to identify and clone novel interacting partners, to define domains and individual residues that participate in binding, and to examine the biological role of the attachment factor–cell surface receptor interaction. In addition to being a tool for studying the biochemistry and biology of attachment factor–cell surface interactions, soluble chimeric proteins are also being pursued as novel drug candidates. The ability of soluble, chimeric proteins to mimic their native counterparts may allow for the generation of effective biological-based drugs that can modulate protein–protein, or cell–cell interactions and their downstream signaling events.

In addition to defining extracellular interations, the development of powerful molecular biology approaches has provided novel methods by which to identify protein–protein interactions that occur inside cells. The yeast two-hybrid system has been used to identify signaling proteins that interact with the cytoplasmic domains of various attachment factor cell surface receptors. This system has been used to identify several proteins that play important roles in integrin function, including the cytoskeletal protein filamin (27,28), cytohesin 1 (208), integrin-linked kinase (209), and β3 endonexin (210). This system will continue to provide a means of identifying molecules that interact with the cytoplasmic domain of attachment factor cell surface receptors and participate in attachment factor–cell surface receptor interactions. These approaches, together with the current explosion in the identification and cloning of novel attachment factor receptors through use of genomic-based bioinformatics, will provide an exciting opportunity for interface between attachment factor biologists, molecular and cell biologists investigating cell signaling, and bioinformaticians in the development of the biology of cell attachment.

BIBLIOGRAPHY

1. M.W. Mosesson and R.A. Umfleet, *J. Biol. Chem.* **245**, 5728–5736 (1970).
2. R.O. Hynes, *Proc. Natl. Acad. Sci. U.S.A.* **70**, 3170–3174 (1973).
3. C.G. Gahmberg and S. Hakomori, *Proc. Natl. Acad. Sci. U.S.A.* **70**, 3329–3333 (1973).
4. H. Hormann and M. Seidl, *Hoppe Seyler's Z. Physiol. Chem.* **361**, 1449–1452 (1980).

5. M. Seidl and H. Hormann, *Hoppe Seyler's Z. Physiol. Chem.* **364**, 83–92 (1983).
6. K. Sekiguchi and S. Hakomori, *Biochemistry* **22**, 1415–1422 (1983).
7. K. Sekiguchi and S. Hakomori, *Proc. Natl. Acad. Sci. U.S.A.* **77**, 2661–2665 (1980).
8. M. Hayashi and K.M. Yamada, *J. Biol. Chem.* **257**, 5263–5267 (1982).
9. P. Kuusela, *Nature* **276**, 718–720 (1978).
10. P. Kuusela, T. Vartio, M. Vuento, and E.B. Myhre, *Infect. Immun.* **50**, 77–81 (1985).
11. D.F. Mosher and R.A. Proctor, *Science* **209**, 927–929 (1980).
12. G.A. Homandberg and J. Kramer-Bjerke, *Thromb. Res.* **48**, 329–335 (1987).
13. A.A. Rostagno, B. Frangione, and L.I. Gold, *J. Immunol.* **143**, 3277–3282 (1989).
14. D.D. Wagner and R.O. Hynes, *J. Biol. Chem.* **254**, 6746–6754 (1979).
15. D.H. Bing, S. Almeda, H. Isliker, J. Lahav, and R.O. Hynes, *Proc. Natl. Acad. Sci. U.S.A.* **79**, 4198–4201 (1982).
16. M.D. Pierschbacher, E. Ruoslahti, J. Sundelin, P. Lind, and P.A. Peterson, *J. Biol. Chem.* **257**, 9593–9597 (1982).
17. M.D. Pierschbacher and E. Ruoslahti, *Nature* **309**, 30–33 (1984).
18. E.F. Plow, M.D. Pierschbacher, E. Ruoslahti, G.A. Marguerie, and M.H. Ginsberg, *Proc. Natl. Acad. Sci. U.S.A.* **82**, 8057–8061 (1985).
19. E.G. Hayman, M.D. Pierschbacher, and E. Ruoslahti, *J. Cell Biol.* **100**, 1948–1954 (1985).
20. M.E. Perkins, T.H. Ji, and R.O. Hynes, *Cell* **16**, 941–952 (1979).
21. K.M. Yamada, D.W. Kennedy, K. Kimata, and R.M. Pratt, *J. Biol. Chem.* **255**, 6055–6063 (1980).
22. M. Isemura, Z. Yosizawa, T. Kiode, and T. Ono, *J. Biochem.* **91**, 731–734 (1982).
23. A. Oldberg and E. Ruoslahti, *J. Biol. Chem.* **257**, 4859–4863 (1982).
24. D.E. Smith and L.T. Furcht, *J. Biol. Chem.* **257**, 6518–6523 (1982).
25. J. Calaycay et al., *J. Biol. Chem.* **260**, 12136–12141 (1985).
26. M.A. Schwartz, M.D. Schaller, and M.H. Ginsberg, *Annu. Rev. Cell Dev. Biol.* **11**, 549–599 (1995).
27. D.T. Loo, S.B. Kanner, and A. Aruffo, *J. Cell Biol.* **273**, 23304–23312 (1998).
28. C.P. Sharma, R.M. Ezzell, and M.A. Arnaout, *J. Immunol.* **154**, 3461–3470 (1995).
29. C.A. Otey, F.M. Pavalko, and K. Burridge, *J. Cell Biol.* **111**, 721–729 (1990).
30. D. Newgreen and J.P. Thiery, *Cell Tissues Res.* **211**, 269–291 (1980).
31. J.L. Duband, S. Rocher, W.T. Chen, K.M. Yamada, and J.P. Thiery, *J. Cell Biol.* **102**, 160–178 (1986).
32. B.W. Mayer, E.D. Hay, and R.O. Hynes, *Dev. Biol.* **82**, 267–286 (1981).
33. J.C. Boucaut, T. Darribere, H. Boulekbache, and J.P. Thiery, *Nature* **307**, 364–367 (1984).
34. M. Bronner-Fraser, *Dev. Biol.* **117**, 528–536 (1986).
35. E.L. George, E.N. Georges-Labouesse, R.S. Patel-King, H. Rayburn, and R.O. Hynes, *Development* **119**, 1079–1091 (1993).
36. J. Orly and G. Sato, *Cell* **17**, 295–305 (1979).
37. J.R. Couchman, D.A. Rees, M.R. Green, and C.G. Smith, *J. Cell Biol.* **93**, 402–410 (1982).
38. P.B. Bitterman, S.I. Rennard, S. Adelberg, and R.G. Crystal, *J. Cell Biol.* **97**, 1925–1932 (1983).
39. J.E. Bottenstein and G.H. Sato, *Exp. Cell Res.* **129**, 361–366 (1980).
40. B.A. Gilchrest, R.E. Nemore, and T. Maciag, *Cell. Biol. Int. Rep.* **4**, 1009–1016 (1982).
41. P. Morley, D.T. Armstrong, and R.E. Gore-Langton, *J. Cell Physiol.* **132**, 226–236 (1987).
42. H.G. Klingemann and F.R. Kohn, *J. Leukoc. Biol.* **50**, 464–470 (1991).
43. M.J. Humphries and S.R. Ayad, *Nature* **305**, 811–813 (1983).
44. T.E. Broad and R.G. Ham, *Eur. J. Biochem.* **135**, 33–39 (1983).
45. B.M. Spiegelman and C.A. Ginty, *Cell* **35**, 657–666 (1983).
46. A. Ben-Ze'ev and A. Amsterdam, *J. Biol. Chem.* **262**, 5366–5376 (1987).
47. L. Staiano-Coico and P.J. Higgins, *Exp. Cell Res.* **201**, 126–136 (1992).
48. F.M. Watt, M.D. Kubler, N.A. Hotchin, L.J. Nicholson, and J.C. Adams, *J. Cell Sci.* **106**, 175–182 (1993).
49. M. Sieber-Blum, F. Sieber, and K.M. Yamada, *Exp. Cell Res.* **133**, 285–295 (1981).
50. A. Rizzino, *Dev. Biol.* **95**, 126–136 (1983).
51. T. Kidowaki, C.J. Thiele, H.K. Kleinman, and M.A. Israel, *Pathobiology* **59**, 316–323 (1991).
52. A.M. Sheppard, M.D. Onken, G.D. Rosen, P.G. Noakes, and D.C. Dean, *Cell Adhes. Commun.* **2**, 27–43 (1994).
53. C. Trautwein, M. Davies, E. Elias, A. Strain, and M.P. Manns, *J. Hepatol.* **22**, 50–56 (1995).
54. J.G. Heathcote, C.H. Sear, and M.E. Grant, *Biochem. J.* **176**, 283–294 (1978).
55. J.G. Heathcote, A.J. Bailey, and M.E. Grant, *Biochem. J.* **190**, 229–237 (1980).
56. R. Timpl, H. Wiedemann, V.v. Delden, H. Furthmayer, and K. Kuhn, *Eur. J. Biochem.* **120**, 203–211 (1981).
57. I. Oberbaumer, H. Wiedemann, R. Timpl, and K. Kuhn, *EMBO J.* **1**, 805–810 (1982).
58. D. Barnes, R. Wolfe, G. Serrero, D. McClure, and G. Sato, *J. Supramol. Struct.* **14**, 47–63 (1980).
59. J. Yang et al., *Proc. Natl. Acad. Sci. U.S.A.* **76**, 3401–3405 (1979).
60. J. Yang, R. Guzman, J. Richards, and S. Nandi, *In Vitro* **16**, 502–506 (1980).
61. J. Richards, D. Pasco, J. Yang, R. Guzman, and S. Nandi, *Exp. Cell Res.*. **146**, 1–14 (1983).
62. A. Akoum, et al., *J. Reprod. Med.* **41**, 555–561 (1996).
63. T. Iguchi, F.D. Uchima, P.L. Ostrander, and H.A. Bern, *Proc. Natl. Acad. Sci. U.S.A.* **80**, 3743–3747 (1983).
64. M.S. Babcock, M.R. Marino, W.T. Gunning, and G.D. Stoner, *In Vitro* **19**, 403–415 (1983).
65. L. Yang, R.A. Faris, and D.C. Hixson, *Gastroenterology* **104**, 840–852 (1993).
66. N. Yasui, S. Osawa, T. Ochi, H. Nakashima, and K. Ono, *Exp. Cell Biol.* **50**, 92–100 (1982).
67. E.A. Davenport and P. Nettesheim, *Exp. Cell Res.* **223**, 155–162 (1996).
68. D. Ghosh, K.G. Danielson, J.T. Alston, and S. Heyner, *In Vitro Cell Dev. Biol.* **27A**, 713–719 (1991).
69. H. Satoh et al., *Cell Struct. Funct.* **18**, 315–321 (1993).

70. M.D. Basson, G. Turowski, and N.J. Emenaker, *Exp. Cell Res.* **225**, 301–305 (1996).
71. C. Fujiyama, Z. Masaki, and H. Sugihara, *Pathol. Res. Pract.* **189**, 1205–1214 (1993).
72. J.A. East, S.P. Langdon, K.M. Townsend, and J.A. Hickman, *Differentiation* **50**, 179–188 (1992).
73. J. Green, S. Schotland, D.J. Stauber, C.R. Kleeman, and T.L. Clemens, *Am. J. Physiol.* **268**, 1090–1103 (1995).
74. S.T. Boyce and J.F. Hansbrough, *Surgery* **103**, 421–431 (1988).
75. A. Ben-Ze'ev, G.S. Robinson, N.L. Bucher, and S.R. Farmer, *Proc. Natl. Acad. Sci. U.S.A.* **85**, 2161–2165 (1988).
76. B. Rana, D. Mischoulon, Y. Xie, N.L. Bucher, and S.R. Farmer, *Mol. Cell Biol.* **14**, 5858–5869 (1994).
77. Z. Gatmaitan et al., *J. Cell Biol.* **97**, 1179–1190 (1983).
78. M. Haramaki, *Acta Pathol. Jpn.* **43**, 490–499 (1993).
79. N. Sawada et al., *In Vitro Cell Dev. Biol.* **23**, 267–273 (1987).
80. A.W. Thomson et al., *Adv. Exp. Med. Biol.* **378**, 511–518 (1995).
81. S. Toda et al., *Cell Struct. Funct.* **20**, 345–354 (1995).
82. M. Yamamoto, K. Yamamoto, and T. Noumura, *Exp. Cell Res.* **204**, 121–129 (1993).
83. M.A. Hadley, S.W. Byers, C.A. Suarez-Quian, H.K. Kleinman, and M. Dym, *J. Cell Biol.* **101**, 1511–1522 (1985).
84. R.H. Kramer and N. Marks, *J. Biol. Chem.* **264**, 4684–4688 (1989).
85. V.M. Belkin, A.M. Belkin, and V.E. Koteliansky, *J. Cell Biol.* **111**, 2159–2170 (1990).
86. W.D. Staatz et al., *J. Biol. Chem.* **266**, 7363–7367 (1991).
87. M.J. Elices and M.E. Hemler, *Proc. Natl. Acad. Sci. U.S.A.* **86**, 9906–9910 (1989).
88. Y. Takada and M.E. Hemler, *J. Cell Biol.* **109**, 397–407 (1989).
89. S. Sakashita, E. Engvall, and E. Ruoslahti, *FEBS Lett.* **116**, 243–246 (1980).
90. P. Speziale, M. Hook, T. Wadstrom, and R. Timpl, *FEBS Lett.* **146**, 55–58 (1982).
91. M.D. Rosso, R. Cappelletti, M. Viti, S. Vannucchi, and V. Chiarugi, *Biochem. J.* **199**, 699–704 (1981).
92. D.E. Hall et al., *J. Cell Biol.* **110**, 2175–2184 (1990).
93. E. Forsberg, M. Paulsson, R. Timpl, and S. Johansson, *J. Biol. Chem.* **265**, 6376–6381 (1990).
94. D. Kirchhofer, L.R. Languino, E. Ruoslahti, and M.D. Pierschbacher, *J. Biol. Chem.* **265**, 615–618 (1990).
95. K.R. Gehlsen, K. Dickerson, W.S. Argraves, E. Engvall, and E. Ruoslahti, *J. Biol. Chem.* **264**, 19034–19038 (1989).
96. M. Aumailley, R. Timpl, and A. Sonnenberg, *Exp. Cell Res.* **188**, 55–60 (1990).
97. R.H. Kramer, Y.-G. Cheng, and R. Clyman, *J. Cell Biol.* **111**, 1233–1243 (1990).
98. A. Sonnenberg et al., *J. Cell Biol.* **110**, 2145–2155 (1990).
99. M.M. Lotz, C.A. Korzelius, and A.M. Mercurio, *Cell Regul.* **1**, 249–257 (1990).
100. N.N. Tandon et al., *Biochem. J.* **274**, 535–542 (1991).
101. K. Satoh et al., *Cancer Lett.* **62**, 199–203 (1992).
102. M.C. Kibbey et al., *Proc. Natl. Acad. Sci. U.S.A.* **90**, 10150–10153 (1993).
103. M.C. Kibbey, B. Johnson, R. Petryshyn, M. Jucker, and H.K. Kleinman, *J. Neurosci. Res.* **42**, 314–322 (1995).
104. S.S. Brown, H.L. Malinoff, and M.S. Wicha, *Proc. Natl. Acad. Sci. U.S.A.* **80**, 5927–5930 (1983).
105. L. Hinck, I.S. Nathke, J. Papkoff, and W.J. Nelson, *J. Cell Biol.* **125**, 1327–1340 (1994).
106. M.L. McGarvey, A.B.-V. Evercooren, H.K. Kleinman, and M. Bubois-Dalcq, *Dev. Biol.* **105**, 18–28 (1984).
107. R.G. Farrer and R.H. Quarles, *J. Neurosci. Res.* **45**, 248–257 (1996).
108. A.B.-V. Evercooren et al., *J. Neurosci. Res.* **8**, 179–193 (1982).
109. D. Edgar, R. Timpl, and H. Thoenen, *EMBO J.* **3**, 1463–1468 (1984).
110. N. Nagato, M. Aoyagi, and K. Hirakawa, *Glia* **8**, 71–76 (1993).
111. H. Ovadia et al., *Brain Res.* **322**, 93–100 (1984).
112. Y. Kubota, H.K. Kleinman, G.R. Martin, and T.J. Lawley, *J. Cell Biol.* **107**, 1589–1598 (1988).
113. D. Medina, M.L. Li, C.J. Oborn, and M.J. Bissell, *Exp. Cell Res.* **172**, 192–203 (1987).
114. C.H. Streuli et al., *J. Cell Biol.* **129**, 591–603 (1995).
115. M.A. Hadley, B.S. Weeks, H.K. Kleinman, and M. Dym, *Dev. Biol.* **140**, 318–327 (1990).
116. U. Hahn, A. Stallmach, E.G. Hahn, and E.O. Riecken, *Gastroenterology* **98**, 322–335 (1990).
117. A. Rizzino, V. Terranova, D. Rohrbach, C. Crowley, and H. Rizzino, *J. Supramol. Struct.* **13**, 243–253 (1980).
118. T.M. Sweeney, R.C. Ogle, and C.D. Little, *J. Cell Sci.* **97**, 23–31 (1990).
119. R.F. Aten, T.R. Kolodecik, and H.R. Behrman, *Endocrinology* **136**, 1753–1758 (1995).
120. K. Hirata, Y. Yoshida, K. Shiramatsu, A.E. Freeman, and H. Hayasaka, *Exp. Cell Biol.* **51**, 121–129 (1983).
121. D. Mooney et al., *J. Cell Physiol.* **151**, 497–505 (1992).
122. R. Holmes, G. Mercer, and N. Mohamed, *In Vitro* **15**, 522–531 (1979).
123. E.G. Hayman et al., *J. Cell Biol.* **95**, 20–23 (1982).
124. D.W. Barnes, J. Silnutzer, C. See, and M. Shaffer, *Proc. Natl. Acad. Sci. U.S.A.* **80**, 1362–1366 (1983).
125. D.W. Barnes, J.E. Reing, and B. Amos, *J. Biol. Chem.* **260**, 9117–9122 (1985).
126. M. Izumi, K.M. Yamada, and M. Hayashi, *Biochim. Biophys. Acta* **990**, 101–108 (1989).
127. S. Suzuki, A. Oldberg, E.G. Hayman, M.D. Pierschbacher, and E. Ruoslahti, *EMBO J.* **4**, 2519–2524 (1985).
128. M.G. Conlan, B.R. Tomasini, R.L. Schultz, and D.F. Mosher, *Blood* **72**, 185–190 (1988).
129. D. Jenne, and K.K. Stanley, *EMBO J.* **4**, 3153–3157 (1985).
130. D.W. Barnes, and J. Reing, *J. Cell Physiol.* **125**, 207–214 (1985).
131. C.J. Parker, O.L. Stone, V.F. White, and N.J. Bernshaw, *Br. J. Haematol.* **71**, 245–252 (1989).
132. G. Hetland, H.B. Pettersen, T.E. Mollnes, and E. Johnson, *Scand. J. Immunol.* **29**, 15–21 (1989).
133. E.G. Hayman, M.D. Pierschbacher, Y. Ohgren, and E. Ruoslahti, *Proc. Natl. Acad. Sci. U.S.A.* **80**, 4003–4007 (1983).
134. D. Seiffert, M.L. Iruela-Arispe, E.H. Sage, and D.J. Loskutoff, *Dev. Dyn.* **203**, 71–79 (1995).
135. X. Zheng, T.L. Saunders, S.A. Camper, L.C. Samuelson, and D. Ginsburg, *Proc. Natl. Acad. Sci. U.S.A.* **92**, 12426–12430 (1995).
136. D. Jenne and K.K. Stanley, *Biochemistry* **26**, 6735–6742 (1987).

137. S. Suzuki et al., *J. Biol. Chem.* **259**, 15307–15314 (1984).
138. E. Ruoslahti and M.D. Pierschbacher, *Cell* **44**, 517–518 (1986).
139. M. Izumi, T. Shimo-Oka, N. Morishita, I. Ii, and M. Hayashi, *Cell Struct. Funct.* **13**, 217–225 (1988).
140. C. Gebb, E.G. Hayman, E. Engvall, and E. Ruoslahti, *J. Biol. Chem.* **261**, 16698–16703 (1986).
141. K.T. Preissner and G. Muller-Berghaus, *J. Biol. Chem.* **262**, 12247–12253 (1987).
142. D. Seiffert and R.R. Schleef, *Blood* **88**, 552–560 (1996).
143. J.I. Fuquay, D.T. Loo, and D.W. Barnes, *Infect. Immun.* **52**, 714–717 (1986).
144. G.S. Chhatwal, K.T. Preissner, G. Muller-Berghaus, and H. Blobel, *Infect. Immun.* **55**, 1878–1883 (1987).
145. M. Paulsson and T. Wadstrom, *FEMS Microbiol. Immunol.* **2**, 55–62 (1990).
146. M. Kostrzynska and T. Wadstrom, *Int. J. Med. Microbiol. Virol. Parasitol. Infect. Dis.* **277**, 80–83 (1992).
147. R.J. Arko et al., *J. Clin. Microbiol.* **29**, 70–75 (1991).
148. F. Re et al., *J. Cell Biol.* **127**, 537–546 (1994).
149. P.J. Bianchine, P.R. Burd, and D.D. Metcalfe, *J. Immunol.* **149**, 3665–3671 (1992).
150. K.M. Neugebauer, K.A. Venstrom, and L.F. Reichardt, *J. Cell Biol.* **116**, 809–815 (1992).
151. R.M. Leven and F. Tablin, *Exp. Hematol.* **20**, 1316–1322 (1992).
152. R.A. Wolfe, G.H. Sato, and D.B. McClure, *J. Cell Biol.* **87**, 434–441 (1980).
153. A. Merzak, S. Koochekpour, and G.J. Pilkington, *Cell Adhes. Commun.* **3**, 27–43 (1995).
154. P.W. Grabham, P.H. Gallimore, and R.J. Grand, *Exp. Cell Res.* **202**, 337–344 (1992).
155. D.F. Mosher, *Annu. Rev. Med.* **41**, 85–97 (1990).
156. M.W. Long and V.M. Dixit, *Blood* **75**, 2311–2318 (1990).
157. P.J. Mansfield and S.J. Suchard, *J. Immunol.* **153**, 4219–4229 (1994).
158. D.D. Roberts and V. Ginsburg, *Arch. Biochem. Biophys.* **267**, 405–415 (1988).
159. A.S. Asch, S. Silbiger, E. Heimer, and R.L. Nachman, *Biochem. Biophys. Res. Commun.* **182**, 1208–1217 (1992).
160. R.O. Hynes, *Cell* **69**, 11–25 (1992).
161. M.A. Bourdon and E. Ruoslahti, *J. Cell Biol.* **108**, 1149–1155 (1989).
162. A.L. Prieto, G.M. Edelman, and K.L. Crossin, *Proc. Natl. Acad. Sci. U.S.A.* **90**, 10154–10158 (1993).
163. A. Lochter, J. Taylor, B. Fuss, and M. Schachner, *Eur. J. Neurosci.* **6**, 597–606 (1994).
164. E. Ruoslahti, *Annu. Rev. Cell Biol.* **4**, 229–255 (1988).
165. J.E. Koda and M. Bernfield, *J. Biol. Chem.* **259**, 11763–11770 (1984).
166. G.R. Martin and R. Timpl, *Annu. Rev. Cell Biol.* **3**, 57–85 (1987).
167. L. Sherman, J. Sleeman, P. Herrlich, and H. Ponta, *Curr. Opin. Cell Biol.* **6**, 726–733 (1994).
168. I. Stamenkovic, M. Amiot, J.M. Pesando, and B. Seed, *Cell* **56**, 1057–1062 (1989).
169. L.A. Goldstein et al., *Cell* **56**, 1063–1072 (1989).
170. A. Aruffo, I. Stamenkovic, M. Melnick, C.B. Underhill, and B. Seed, *Cell* **61**, 1303–1313 (1990).
171. S.C. Wheatley, C.M. Isacke, and P.H. Crossley, *Development* **119**, 295–306 (1993).
172. V.B. Lokeshwar, N. Iida, and L.Y. Bourguignon, *J. Biol. Chem.* **271**, 23853–23864 (1996).
173. B.F. Haynes, M.J. Telen, L.P. Hale, and S.M. Denning, *Immunol. Today* **10**, 423–428 (1989).
174. A. Bartolazzi, R. Peach, A. Aruffo, and I. Stamenkovic, *J. Exp. Med.* **180**, 53–66 (1994).
175. M.E. Hemler, *Annu. Rev. Immunol.* **8**, 365–400 (1990).
176. P.-Y. Chan and A. Aruffo, *J. Biol. Chem.* **268**, 24655–24664 (1993).
177. A.E. Koch et al., *Lab. Invest.* **64**, 313–320 (1991).
178. M.J. Elices et al., *Cell* **60**, 577–584 (1990).
179. C. Ruegg et al., *J. Cell Biol.* **117**, 179–189 (1992).
180. D. Simmons, M.W. Makgoba, and B. Seed, *Nature* **331**, 624–627 (1988).
181. J. Fawcett et al., *Nature* **360**, 481–484 (1992).
182. D.E. Staunton, M.L. Dustin, and T.A. Springer, *Nature* **339**, 61–64 (1989).
183. S.D. Marlin and T.A. Springer, *Cell* **51**, 813–819 (1987).
184. M.S. Diamond, D.E. Staunton, S.D. Marlin, and T.A. Springer, *Cell* **65**, 961–971 (1991).
185. Y. Rosenstein et al., *Nature* **354**, 233–235 (1991).
186. G.M. Edelman, *Annu. Rev. Cell Biol.* **2**, 81–116 (1986).
187. G.J. Cole and R. Akeson, *Neuron* **2**, 1157–1165 (1989).
188. T.B. Kuhn, E.T. Stoeckli, M.A. Condrau, F.G. Rathjen, and P. Sonderegger, *J. Cell Biol.* **115**, 1113–1126 (1991).
189. A.M. Montgomery et al., *J. Cell Biol.* **132**, 475–485 (1996).
190. M. Grumet, *J. Neurosci. Res.* **31**, 1–13 (1992).
191. C. Rader et al., *Eur. J. Biochem.* **215**, 133–141 (1993).
192. S.J. Harper, S.R. Bolsover, F.S. Walsh, and P. Doherty, *Cell Adhes. Commun.* **2**, 441–453 (1994).
193. D.R. Friedlander et al., *J. Cell Biol.* **125**, 669–680 (1994).
194. H. Stockinger et al., *J. Immunol.* **145**, 3889–3897 (1990).
195. P.J. Newman et al., *Science* **247**, 1219–1222 (1990).
196. Y. Xie and W.A. Muller, *Proc. Natl. Acad. Sci. U.S.A.* **90**, 5569–5573 (1993).
197. H.M. DeLisser et al., *J. Biol. Chem.* **268**, 16037–16046 (1993).
198. L. Piali et al., *J. Cell Biol.* **130**, 451–460 (1995).
199. L.A. Lasky, *Annu. Rev. Biochem.* **64**, 113–139 (1995).
200. G.S. Kansas, K. Ley, J.M. Munro, and T.F. Tedder, *J. Exp. Med.* **177**, 833–838 (1993).
201. O. Abbassi, T.K. Kishimoto, L.V. McIntire, D.C. Anderson, and C.W. Smith, *J. Clin. Invest.* **92**, 2719–2730 (1993).
202. A. Varki, *Proc. Natl. Acad. Sci. U.S.A.* **91**, 7390–7397 (1994).
203. S.I. Simon et al., *J. Immunol.* **155**, 1502–1514 (1995).
204. S.B. Munro and O.W. Blaschuk, in P. Brodt ed., *Cell Adhesion and Invasion in Cancer Metastasis*, Chapman & Hall, New York, 1996, pp 17–34.
205. S. Armeanu, H.J. Buhring, M. Reuss-Borst, C.A. Muller, and G. Klein, *J. Cell Biol.* **131**, 243–249 (1995).
206. P. Doherty, L.H. Rowett, S.E. Moore, D.A. Mann, and F.S. Walsh, *Neuron* **6**, 247–258 (1991).
207. H.R. Payne and V. Lemmon, *Dev. Biol.* **159**, 595–607 (1993).
208. W. Kolanus et al., *Cell* **86**, 233–242 (1996).
209. G.E. Hannigan et al., *Nature* **379**, 91–96 (1996).
210. S.J. Shattil et al., *J. Cell Biol.* **131**, 807–816 (1995).

B

BACILLUS

COLIN R. HARWOOD
KEITH STEPHENSON
University of Newcastle upon Tyne
Newcastle upon Tyne, UK

KEY WORDS

Batch culture
Continuous culture
Gene regulation
Genetic manipulation
Genetic transformation
Industrial enzymes
Insecticidal toxins
Large-scale fermentation
Protein secretion
Taxonomy of *Bacillus*

OUTLINE

INTRODUCTION

Members of the genus *Bacillus* are gram-positive, endospore-forming bacteria that include both mesophiles and extremophiles. They are widely distributed in nature but are most commonly found in soil and associated water sources. Their extraordinary metabolic diversity has been exploited by old and new biotechnologies—from the production of Natto, a traditional Japanese food obtained by fermenting soybeans with *Bacillus subtilis* var. *natto*, to engineered industrial enzymes used in the food and detergents industries.

DISTRIBUTION AND TAXONOMY

The genus *Bacillus* comprises gram-positive, endospore-forming, aerobic or facultatively anaerobic bacteria. Members of the genus show extraordinary metabolic diversity and lifestyles and include thermophiles, psychrophiles, alkalophiles, and acidophiles (1,2). They are metabolically chemoorganotrophs, being dependent on organic compounds as sources of carbon and energy. *B. subtilis*, the most extensively characterized species, is a prototroph capable of growing at mesophilic temperatures on chemically defined salts media with glucose or other simple carbon sources. In contrast, some insect pathogenic species are nutritionally fastidious and require highly specialized growth media (3).

Bacillus species are among the most widely distributed microorganisms, found commonly in soil and associated environments, including plants and water sources such as rivers, estuaries, and coastal waters (4). Most species are harmless to humans and animals, and only a few pathogens have been identified. The latter include *B. anthracis*, the causative agent of anthrax, *B. cereus*, which causes food poisoning, and several insect pathogens.

The primary reservoir for most *Bacillus* species is the soil, where they secrete a variety of carbohydrate- and protein-degrading enzymes that allow them to grow at the expense of plant material and other nutrients (4). These macromolecular hydrolases are the source of much of the commercial interest in these species (5). Their ability to form spores means that they can often be detected in an environment long after periods of metabolic activity. The major soil types are members of the *B. subtilis* and *B. sphaericus* groups. However, representatives of the more fastidious species in the *B. polymyxa* group tend to accumulate in soils with a high organic content.

Some species form a close relationship with plant roots, particularly representatives of the *B. polymyxa* group. Several species (e.g., *B. azotofixans, B. macerans*, and *B. polymyxa*) fix nitrogen under anaerobic conditions. Certain strains of *B. subtilis* are categorized as plant growth–promoting rhizobacteria (PGPR). In return for stimulating plant growth by increasing the availability of inorganic nutrients, or secreting plant growth hormones or compounds (often peptide antibiotics) that inhibit the growth of plant pathogens, PGPR receive nutrients from the plant in the form of plant exudates (6).

The metabolic diversity of the genus is matched by significant genetic diversity, and the %GC content of their genomic DNA varies from about 33 to 67 mol%. This variation is much larger than would be expected for a well-defined genus, and the 60 to 70 currently recognized species of *Bacillus* are likely to be reassigned to an increased number of more clearly defined genera. This process has already started with the reassignment of *B. acidocaldarius* and some other acidophilic thermophiles to the genus *Alicyclobacillus* (7). Numerical taxonomic and 16*S* ribosomal (r)RNA sequence analyses show good but not perfect congruence (8,9). These studies have resulted in *Bacillus* species being assigned to five groups. The evolutionary distance between these groups is large enough to affect the outcome of interspecies gene transfer and cloning experiments, particularly when there are large discrepancies between the %GC contents of the species involved.

The *B. subtilis* group includes many of the better-known bacilli. These include facultative anaerobes such as *B. licheniformis*, which grow fermentatively in the absence of exogenous electron acceptors, and aerobes such as *B. subtilis* that grow weakly in the absence of oxygen except in the presence of nitrate, which they can use as an alternative electron acceptor (10,11). Members of this group produce acid from several sugars, and several (e.g., *B. lentus, B. firmus*) are halotolerant. The group includes toxigenic species such as *B. anthracis, B. cereus*, and *B. thuringiensis*.

The *B. sphaericus* group is strictly oxidative, using acetate or amino acids rather than sugars as carbon and energy sources (12). Members of the *B. sphaericus* group are virtually unique among bacilli in having the *meso*-diaminopimelic acid that is usually present in cell walls replaced by lysine or ornithine.

The *B. polymyxa* group (renamed *Paenibacillus*) is made up of facultative anaerobes that use sugars and polysaccharides. *B. macerans* grows better in the absence of oxygen with an acidogenic type of fermentation that changes to an ethanolic fermentation at acidic pHs. *B. polymyxa* exhibits a butanediol type of fermentation. Members of the *B. brevis* group (renamed *Brevi bacillus*) are taxonomically heterologous. They are strict aerobes that generally do not produce acid from sugars. Some species, notably *B. azotoformans*, can grow anaerobically by using nitrate as an electron acceptor.

The thermophilic bacilli, which grow optimally at temperatures above 50 °C, represent a heterologous group with at least three lines of descent that range from anaerobes to strict aerobes. Some are able to grow fermentatively on lactic acid (*B. coagulans*) whereas others are chemolithotrophic autotrophs. The group includes *B. acidocaldarius* and the other acidophilic thermophiles that have been reassigned to the genus *Alicyclobacillus* (7).

Preservation of Strains

Although most strains of *Bacillus* survive well on agar plates, either at room temperature or at 4 °C, it is recommended that they are subcultured on a weekly basis. Because *Bacillus* species are able to sporulate, viable cells may even be recovered from severely dehydrated agar plates, particularly from minimal agar plates that encourage sporulation. Long-term stocks of *Bacillus* may be preserved as glycerol or lyophilized cultures, in the form of spores or vegetative cells (13). Strains that sporulate well may be preserved as spore suspensions in sterile water (14). Spore suspensions are generally stable for many years. Most *Bacillus* strains can be transported on freshly inoculated nutrient agar slopes or as spore suspensions spotted on sterilized filter paper disks encased in sterilized aluminium foil.

Culture Collections

The *Bacillus* Genetic Stock Center (BGSC, Department of Biochemistry, Ohio State University, 484 West 12th Avenue, Columbus, OH 43210-1292) has an extensive collection of mutant strains, bacteriophages, and plasmids. Although mainly devoted to *B. subtilis*, the collection also includes strains of *B. cereus, B. licheniformis, B. megaterium, B. pumilus, B. stearothermophilus, B. thuringiensis*, and a few others. BGSC produces an extensive catalog that can be requested by e-mail from *dzeigler @magnus.acs.ohio-state.edu*. Strains of *Bacillus* are also available from a variety of international culture collections and a comprehensive list has been published (15). The American Type Culture Collection (ATCC, 12301 Parklawn Drive, Rockville, MD 20852) has a World Wide Web (WWW) site at *http://www.atcc.org/*, and the Japan Collection of Micro-organisms (JCM; Riken, Wako-shi, Saitama 351, Japan) is available at *http://www.wdcm.riken.go.jp/wdcm/JCM*. More general links to culture collections are available at the World Data Collection for Micro-organisms at *http://www.wdcm.riken.go.jp/*.

The publication of the genome of *B. subtilis* 168 (16) has provided the opportunity to investigate the function of open reading frames (ORFs) with no known function (unknown reading frames (URFs) (17). A collection of isogenic mutants has been established in which the URFs of *B. subtilis* 168 have been systematically inactivated with a purpose-designed integrational vector, pMUTin. These mutants also permit the expression of the target genes to be monitored and any gene(s) downstream in the same transcriptional unit to be induced. The depository for the collection is accessible to account holders via the Micado WWW site at *http://locus.jouy.inra.fr/cgi-bin/genmic/madbase/progs/madbase.operl*.

GENETIC MANIPULATION

A wide range of genetic techniques has been developed for *B. subtilis*, making it among the most amenable bacteria to genetic manipulation. These techniques include classical in vitro and in vivo mutagenic methods and extremely efficient in vitro genome manipulation methods that are in advance of those available for other bacteria, including *Escherichia coli* (18). *Bacillus* vectors have been designed to allow recombinant or mutant genes to be maintained autonomously in the host or integrated into the chromosome by double-crossover (Fig. 1) or single-crossover (Fig. 2) recombinations. In the case of a single-crossover recombination, tandem copies of the gene, one wild type and one mutated, can be engineered into the chromosome, if required, each under the control of independently controlled promoters (Fig. 2). The techniques for the manipulation of other *Bacillus* species are limited by their transformability.

Mutagenesis

A number of DNA repair systems have been detected and studied in *B. subtilis*, including error-prone SOS-like translesion DNA synthesis but not photoreactivation (19). Consequently, classical in vivo mutagenic methods generate mutants efficiently. Limitations in whole-cell mutagenic techniques, including lack of specificity and the generation of multiple mutations that may influence the cell's phenotype in an unpredictable manner, can be overcome by use of an alternative approach in which cloned copies of the target DNA are subjected to in vitro mutagenesis and subsequently reintegrated into the chromosome (20). This approach is aided by the publication of the genome sequence of *B. subtilis* strain 168 (16).

Specific changes are currently introduced into DNA sequences by PCR techniques, with more extensive changes being engineered by splicing PCR methodologies (e.g., splicing by overlap extension [21]). The modified DNA can then be introduced into the host bacterium either on replicating or integration plasmids. The ability to generate insertion mutants is one of the great technical strengths of *B. subtilis*. The techniques rely on the ease with which *Bacillus* DNA can be cloned into plasmids that replicate in *E. coli* but not in *B. subtilis* and the high frequency with which homologous recombination occurs within *B. subtilis*.

The identification and analysis of *B. subtilis* genes involved in specific functions or responding to particular stimuli has been facilitated by the use of transposons and transpositional mutagenesis. Although the value of transposons as a tool for the analysis of *B. subtilis* has diminished with the publication of the genome sequence (16), nevertheless the technology remains important for the analysis of other *Bacillus* species. Because of the lack of suitable native *Bacillus* transposons, Tn*917* from *Enterococcus faecalis* and Tn*10* from *E. coli* have been used (22–24). In addition to their value for mutagenesis, transposons have been used to clone DNA adjacent to the site of integration, to generate transcriptional fusions to reporter genes, to control the expression of adjacent genes, and as phenotypic tags for mapping and cloning studies.

Transposon Tn*917* is a 5.3-kb Tn*3*-like transposon conferring inducible erythromycin resistance (MLS-type). This transposon has a number of properties that make it a valuable genetic tool, including *(1)* the ability to insert relatively randomly into bacterial chromosomes, *(2)* a relatively high transposition frequency, *(3)* the ability to accommodate at least 8 kb of DNA without markedly affecting transposition frequency or insertional stability, and *(4)* a host range that includes gram-positive and gram-negative bacteria. In addition to *B. subtilis*, Tn*917* has been used in *B. amyloliquefaciens, B. licheniformis*, and *B. megaterium* (23). Insertion of Tn*917* is not entirely random, and a limited preference for specific loci has been observed in the case of *B. subtilis* and *B. licheniformis* (25,26).

An alternative to Tn*917* is an engineered miniderivative of Tn*10* (24). This transposon consists of a *Staphylococcus aureus*–derived *cat* (chloramphenicol resistance) gene flanked by 307-bp fragments derived from IS*10*. The gene encoding the transposase is incorporated into a delivery vector rather than the minitransposon itself. Tn*10*-derived transposons have two potential advantages over Tn*917*-based transposons. They appear to transpose at a significantly higher frequency and insert randomly without bias to specific targets. A series of special-purpose delivery systems have been developed for gene inactivation, the recovery of adjacent chromosomal sequences, and transcription fusions.

Transformation

Bacillus subtilis was the first nonpathogenic bacterium to be successfully transformed with extracted DNA (27–29). The availability of methods for introducing DNA into cells is essential for recombinant DNA technology, and several transformation techniques have been developed for *B. subtilis*. The most frequently used method is transformation of naturally competent cells, although protoplasts of *B. subtilis* (30) and several other *Bacillus* species can be efficiently transformed. Current methods for electrotransformation of *Bacillus* species usually result in low yields of transformants.

Natural competence is one of several postexponential phase phenomena that are characteristic of *B. subtilis* strain 168. In glucose-minimal media, maximal competence develops shortly after the transition from exponential to stationary phase, and only about 10% of the cells are able to take up DNA. Natural competence has been observed with a limited number of *B. subtilis* strains, and few natural competence systems are known for *Bacillus* species. Unlike *Haemophilus influenzae*, competent *B. subtilis* cells do not discriminate between native and foreign DNA. In recombination-proficient strains, homologous chromosomal DNA is integrated in the host chromosome at the site of homology. If saturating amounts of DNA (>1 μg/mL of culture) are used, the cotransfer of unlinked genetic markers occurs. This phenomenon, called congression, can be used to introduce genes for which no selectable phenotype is known (31).

The frequency of plasmid-mediated transformation is usually between 0.001% and 0.1%, and this frequency is reduced still further with ligation mixtures. A major reason for these low efficiencies is that donor plasmid DNA becomes randomly fragmented and converted into single-stranded DNA before entry into the cell. Recirculation of the plasmid DNA requires a region of homology provided on fragments of opposite polarity, and consequently only plasmid multimers, present in most plasmid preparations, or monomers containing internal repeats are effective in transformation (32).

Bacillus protoplasts can incorporate DNA if treated with chaotropic agents such as polyethylene glycol. The

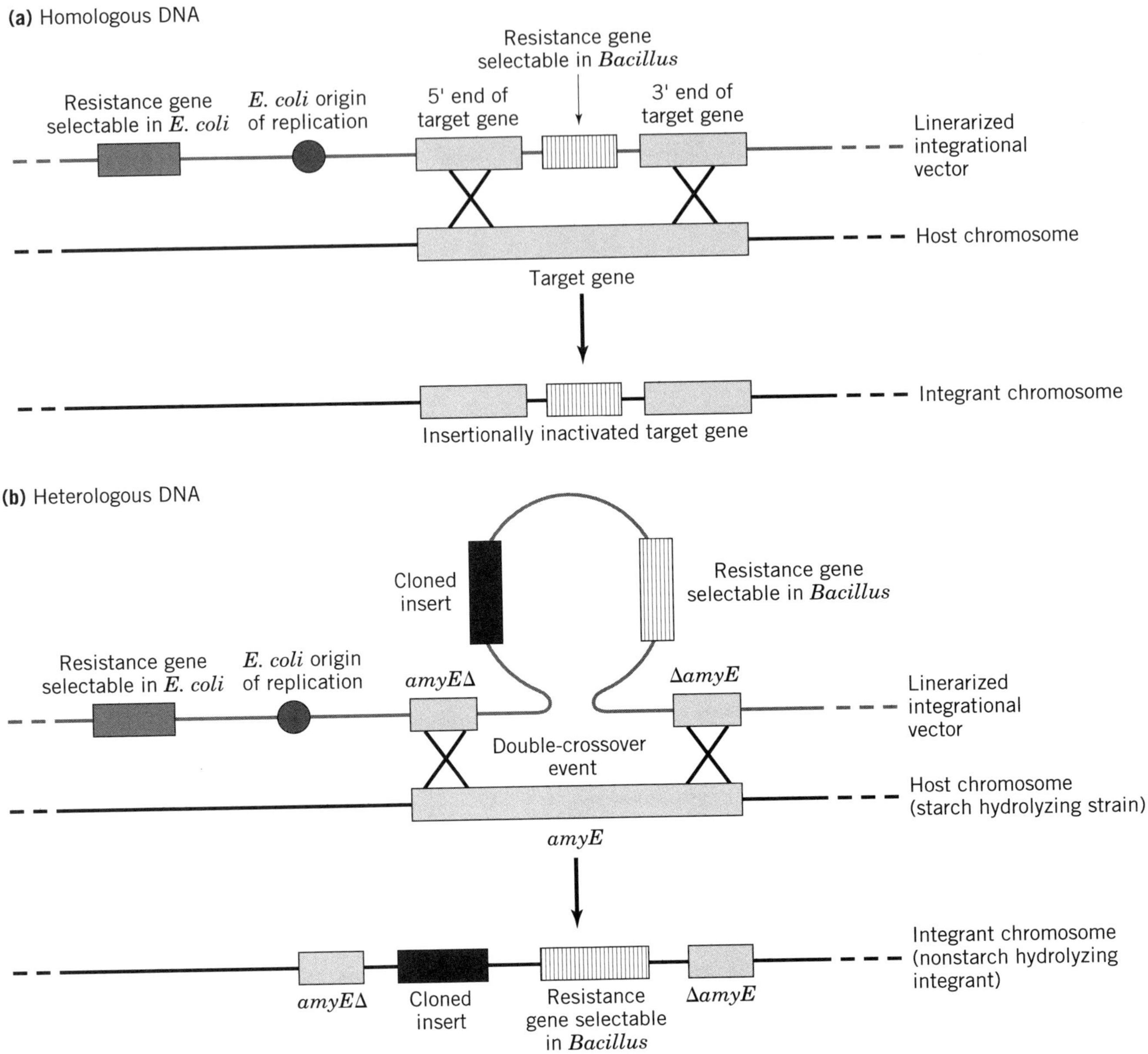

Figure 1. Double-crossover recombination shows the integration of a fragment of DNA from a linearized vector that is unable to replicate in *Bacillus* species. (**a**) Recombination event involving sequences homologous to the target DNA. (**b**) Recombination event involving heterologous DNA, in this case inserted between 5′ (*amyE*Δ) and 3′ (Δ*amyE*) fragments of the nonessential *amyE* gene encoding α-amylase. Integrants lose the ability to hydrolyze starch.

fragility of the wall-less protoplasts means that transformation needs to be carried out in an isosmotic medium; the cell walls are subsequently regenerated on complex regeneration media. Protoplast transformation was originally developed for *B. subtilis* by Chang and Cohen (30) but has subsequently been modified in many laboratories. A major advantage of protoplast transformation is that plasmid DNA remains double stranded throughout the procedure and consequently is less likely to undergo rearrangements. Despite its advantages, protoplast transformation methods are laborious and difficult to reproduce; transformation frequencies up to 10% per protoplast are claimed but in practice are usually a great deal lower. In addition, the complexity of the regeneration media precludes the selection for prototrophic markers. Protoplast transformation protocols have been developed for several other *Bacillus* species including *B. amyloliquefaciens, B. licheniformis, B. stearothermophilus, B. anthracis*, and *B. firmus*.

Genome Analysis and Manipulation

The genome of *B. subtilis* strain 168 is 4.2 Mbp in length, has a G+C content of 43.5%, and encodes slightly more

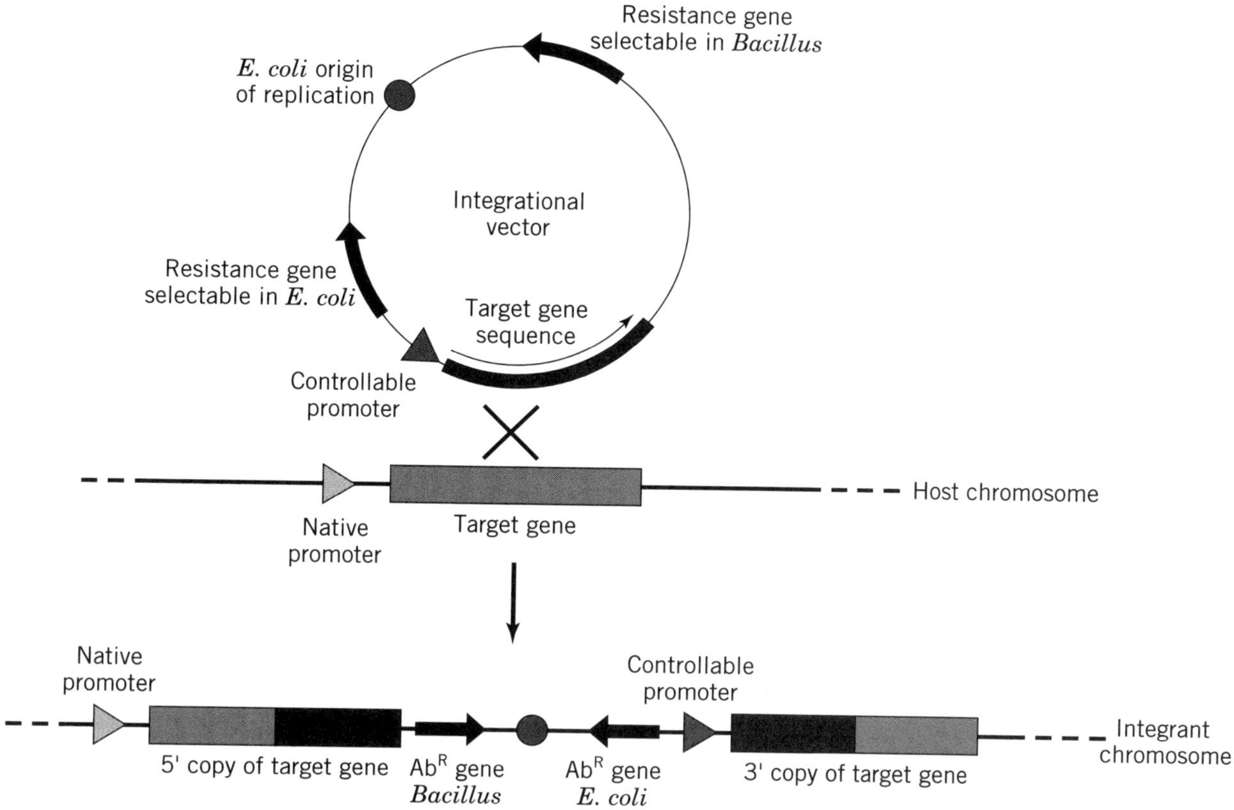

Figure 2. Single-crossover recombination showing integration of an intact integration vector into the host chromosome at the site of the target gene. Part or all of the target gene is duplicated as a result of the integration event. The inclusion of a controllable promoter upstream of the target gene fragment on the vector allows for controllable gene expression of the 3′ copy of the gene or any genes downstream in the same operon. Ab^R, selectable antibiotic resistance gene.

than 4,100 protein-coding genes (16). The genome shows evidence of 10 prophages or remnants of prophages, which may account for its susceptibility to lysis at the end of exponential growth in the absence of an energized membrane (33). Nearly half the protein-coding genes have paralogues in the genome, and several expanded gene families were detected, particularly for two-component signal transducers (34 genes) and ATP-binding transporters (77 genes).

Although analysis of a complete microbial genome sequence reveals many new ORFs, it is not possible to ascribe functions to all the detected ORFs on the basis of currently available sequence databases and bioinformatics tools. Even in well-characterized species such as *E. coli* and *B. subtilis*, URFs represent about 40% of the detected ORFs. Given the intensive in vitro studies on bacteria, it is likely that the function of a significant proportion of the URFs will relate to growth and survival in natural habitats.

The inability to ascribe function to more than a third of the ORFs in *B. subtilis* has required the development of new strategies for analyzing gene function. In the case of *B. subtilis*, a highly coordinated systematic functional analysis program involving groups with a wide range of technical expertise has been established. Approximately 2,000 unknown genes have been systematically disrupted with an integrational vector to generate a collection of isogenic mutants strains (34). The mutant construction strategy includes the introduction of a *lacZ* transcriptional reporter for monitoring the expression of the target gene and a controllable promoter that facilitates the expression of any genes downstream in the same operon (Fig. 2).

The availability of the *B. subtilis* genome sequence has revolutionized the methods of analysis and manipulation of this bacterium, and consequently traditional methods of genetic linkage mapping involving generalized transducing for long-range mapping and transformation for fine structure mapping have largely been rendered redundant (31). This is not the case for species such as *B. megaterium* for which generalized transducing phages such as MP13 and protoplast transformation remain important analytical tools (35).

Increasingly, integration systems are used for the analysis and manipulation of the *B. subtilis* genome. Transformation of cells with homologous DNA fragments is very efficient, and if heterologous sequences are flanked by sequences homologous to chromosomal DNA they can be inserted efficiently at the sites of homology via a double-crossover recombination event (Fig. 1). In the case of circular molecules, a single site of homology is all that is required and integration is via a single-crossover recombination event (Fig. 2).

Single or Campbell-type integration events generally use *E. coli*–based plasmid vectors that are not able to replicate in gram-positive bacteria but that include an antibiotic resistance gene that can be selected in *B. subtilis*. A fragment of *B. subtilis* DNA (≥0.4 kb) is cloned via *E. coli* into the integrational vector, which is then transformed into *B. subtilis*. Selection for an antibiotic-resistant gene results in transformants in which the vector has integrated into the host chromosome via a single-crossover recombination between homologous regions on the vector and bacterial chromosomes. Integration results in a duplication of the cloned fragment at the flanking ends of the integrated vector. These tandem repeats represent an amplifiable unit; increasing the amounts of antibiotic used for selection can result in the amplification of the intervening sequences up to 70 copies (36).

The integrational vector system can be used to generate different mutational outcomes. If the cloned fragment carries an entire operon, including promoter and transcription terminator, two fully functional copies of the gene are present in the integrant. If 5′- or 3′-end fragments of an operon are used, only one functional copy of the target gene is generated. If the 5′-end fragment includes a ribosome-binding site and start codon, and is placed downstream of a controllable promoter, then the expression of the single functional copy of the target gene is under the control of this promoter (Fig. 2). Finally, if regions internal to the target gene are used, no functional copies are formed after integration.

With double-crossover integrations, chromosomal DNA is replaced by either mutationally altered homologous DNA (Fig. 1a) or heterologous DNA (Fig. 1b). Only a single copy of the target gene is introduced. The incoming fragment, which must encode a selectable phenotype, is flanked by DNA that is homologous to sequences at the chromosomal target site. The *amyE* gene, encoding a nonessential α-amylase, is often used as a target of double-crossover integrations because its disruption provides a positive selection phenotype on starch plates (Fig. 1b).

Cloning and Expression Systems

The development of recombinant DNA techniques for *B. subtilis* is now well advanced, and previously encountered problems of vector stability have largely been overcome.

Plasmid Vectors. *B. subtilis* 168, the strain that has been sequenced and that is used for most genetic studies, does not contain endogenous plasmids, whereas most plasmids present in other *Bacillus* strains are cryptic. Consequently, plasmid vectors designed for use in *B. subtilis* have been taken from other gram-positive bacteria, notably *Enterococcus faecalis, Lactococcus lactis*, and *Staphylococcus aureus* (18,37). Staphylococcal plasmids such as pUB110 and pE194, which formed the basis of the first generation of *Bacillus* vectors, are still in common use for *B. subtilis* (38). The kanamycin resistance plasmid pUB110 is the most widely used replicon but suffers from problems of structural and segregational instability that result from its single-stranded (rolling circle) mode of replication (39). The second generation of *Bacillus* vectors involved the construction of chimeras between gram-positive and gram-negative replicons (40). These shuttle or bifunctional vectors enabled initial cloning experiments to be carried out in *E. coli*, exploiting this organism's highly efficient transformation system. In recent years, stable single-stranded replicating plasmids and theta-replicating plasmids have been identified and are being developed as a new generation of *Bacillus* vectors (18). The former includes derivatives of plasmids such as pWVO1 derived from *Lactococcus lactis* (41) and certain endogenous *Bacillus* plasmids, the latter plasmids based on pAMβ1, originally isolated from *Enterococcus faecalis* (42,43).

An alternative to using autonomously replicating plasmid vectors is to use plasmid integration vectors. Such integrants are usually stable; reversal of the process, by recombination in the duplicated sequences, occurs at a frequency of about 10^{-4} to 10^{-5} per cell generation (44). Integration vectors have been reviewed by Albertini and Galizzi (45) and Pérego (46).

Various special-purpose plasmid vectors have also been developed, including expression and secretion vectors. Although the expression of many genes has been studied in *B. subtilis*, relatively few promoter systems have been developed for controlled, high-level expression. The majority of extracellular enzymes synthesized for industrial purposes are expressed from native promoters induced at the end of exponential growth and for extended periods in the stationary phase (47). During this time they can be subjected to a variety of control pathways, including transition-state regulators and catabolite repression (48,49). Industry has developed promoter–expression systems (including growth regimes) that can direct the synthesis of extracellular proteins to concentrations of about 20 g/L, although for commercial reasons the details of these systems are not generally available. Three promoter systems that are widely used in research laboratories are discussed next.

***P_{spac}* Promoter.** The P_{spac} promoter was constructed by fusing the 5′ sequences of a promoter from SPO1, a lytic *B. subtilis* bacteriophage, and the 3′ sequences of the *lac* promoter, including the entire *lac* operator. The point of fusion is within the −10 region of the hybrid promoter (50). The controllability of P_{spac} is dependent on the presence of the *lac* repressor, expressed in *B. subtilis* by placing it downstream of a promoter and ribosome-binding site (RBS) derived from a *B. licheniformis* penicillinase gene. The P_{spac} promoter can be induced 50 fold with the lactose analog IPTG.

XylR-Controlled Promoters. The genes for the utilization of xylose polymers are controlled in *B. subtilis* at the level of transcription (51) through the activity of a repressor, the product of the *xylR* gene. The xylose-inducible promoter–operator elements have been used without modification to control expression from chromosomal and plasmid locations. A copy of the *xylR* gene is usually included on high-copy-number expression vectors to maintain a balance between the number of repressor molecules and operator sites. Although genes in the xylose regulon are subject to catabolite repression, the catabolite-responsive element is located well downstream of the promoter and operator, and is therefore not included in the expression vectors. In contrast to IPTG, xylose is a cheap and readily available substrate and can be used for the induction of large-scale fermentations.

SacR-Controlled Promoters. The inducible expression of *sacB*, the gene encoding extracellular levansucrase by sucrose, involves a number of regulatory mechanisms, not all of which are fully understood. The *sacB* gene is controlled positively by sucrose, the *sacY* antiterminator, and the products of *degQ* and *sacU* and negatively by *sacX*, a putative PTS enzyme $II^{sucrose}$. The *sacR* regulatory region located upstream of the *sacB* promoter contains a target sequence for DegQ and SacU, which together enhance transcription. Downstream of this promoter, but before the *sacB* RBS, a termination-like stem-loop structure is present, which is acted on by the SacY antiterminator to mediate sucrose-induced expression (52). Expression vectors based on the *sacB* promoter have the advantage that this promoter can be activated during exponential growth when the expression of most *Bacillus* extracellular proteases is repressed and is not subject to catabolite repression. The level of induction is also geared to the level of sucrose in the medium; the promoter is only weakly induced in the presence of 1 mM sucrose and is fully induced at 30 mM.

Phage Vectors. The development of phage vectors for *B. subtilis* has lagged well behind that of *E. coli*, and no phage vectors with the range and versatility of those of lambda exist for this organism. Libraries have been generated in derivatives of *Bacillus* phages φ105 and SPβ, although the former has been used more extensively (53). Attempts to develop a cosmid system for *B. subtilis* have failed because of the lack of an efficient in vitro packaging system for any of the candidate phages.

Bacteriophages φ105 and PBSX (a defective phage of *B. subtilis* strain 168) have both been developed for the production of proteins in *B. subtilis* (53). In both cases the phages have been modified to make them temperature inducible by the introduction of mutations in the genes encoding the immunity repressor proteins. Target genes are introduced downstream of a strong prophage promoter. In the case of φ105MU333, temperature induction can lead to the production of the target gene product to 0.5 mg/mL culture supernatant (54). Bacteriophage-based expression systems have the advantage of being relatively stable because they are maintained as a single copy during growth. Copy number increases at the time of induction due to the replication of the phage. The promoter is tightly controlled but is induced cheaply and in a manner favored in industrial fermentations by increasing temperature. However, vector development is still required to maximize the full potential of this system for high-level protein production.

PROTEIN SECRETION

Bacillus subtilis and its close relatives secrete proteins directly into the growth medium efficiently and to high concentrations. The productivity of *Bacillus* species for the production of extracellular proteins has been improved by a variety of techniques including random and directed mutagenesis, recombinant techniques, and growth regimes. Yields of tens of gram per liter are expected from industrial strains secreting native *Bacillus* proteins (55–58).

The secretion process is highly specific, involving functional sequences on the secretory protein and recognition between the secretory proteins and the host's protein secretion machinery (secretory preprotein translocase). For convenience, the process in *B. subtilis* can be divided into three stages: intracellular processes, involving the targeting of secretory proteins to the cytoplasmic membrane–secretory translocase; translocation across the cytoplasmic membrane; and posttranslocational stages, involving the processing of the secreted protein, its folding into the native conformation, and passage through the cell wall into the growth medium.

A major limiting step in the production of native secretory proteins is the level of transcription of their genes. The transcription of genes encoding native secretory proteins is tightly controlled and involves a number of complex, often interacting, regulatory networks, including transcriptional activators (e.g., DegQ, DegR, DegU, and SenS) and repressors (e.g., AbrB, Hpr, and Sin). For example, the nucleotide sequence upstream of *apr*, encoding the major alkaline protease of *B. subtilis*, includes binding sites for Hpr, Sin, SacU/Q, and AbrB (48).

Most *Bacillus* secretory proteins are synthesized as precursors with amino-terminal extensions known as signal peptides, or pre-peptides. The signal peptide is required to target the precursor to the cytoplasmic membrane and for initiating the interaction with the secretory translocase. During or shortly after translocation across the membrane, the signal peptide is removed through the action of signal peptidases. *Bacillus* signal peptides vary in length from 18 to 35 amino acid residues; their average length of about 30 residues is 5 to 7 residues longer than their counterparts in gram-negative bacteria and eukaryotes. The C-terminal end of the signal peptide contains the signal peptidase recognition sequence. Lipoproteins, which remain attached at the *trans* side of the membrane after translocation, are cleaved by a prolipoprotein-specific signal peptidase that has a different recognition sequence (47).

Many exported *Bacillus* proteins are synthesised as pre-pro-proteins. The pro-peptide, which is located between the signal peptide (pre-) and the mature protein, can vary in length from about 8 (e.g., *B. subtilis* α-amylase) to more than 200 amino acids (various neutral proteases). Propeptides have been implicated in the folding of the mature protein into its native conformation following translocation (47,59). Cleavage of the pro-peptide occurs after translocation, autocatalytically in the case of proteases or via extracellular proteases in the case of other secretory proteins (60). Long (8–20 residues) hydrophobic domains, flanked by positively charged residues, are usually absent from secretory proteins because such domains act as "stop transfer" or "membrane anchor" sequences. Similarly, few positively charged residues are found in the amino terminus of the mature protein as these reduce export efficiency.

Surprisingly little is known about early events in the *B. subtilis* secretory pathway. Secretion-specific molecular chaperones, required to maintain secretory precursor proteins in an unfolded conformation, to prevent their aggregation, and to target them to the secretory machinery in the membrane, have not been detected in *B. subtilis*. This could indicate that secretion is cotranslational, and indeed elements of the signal recognition particle–like (SRP-like) pathway have been found in this bacterium (47). The secretory translocase complex appears to be similar to that of *E. coli*. The SecAEY complex is primarily responsible for the ATP-dependent transport of proteins across the membrane, and homologs of other components of the *E. coli* translocase, such as SecD, F, and G, also appear to be present, in the former two cases as a SecDF fusion. A unique feature of *B. subtilis* strain 168 is the presence of five type I signal peptidases and a single type II lipoprotein signal peptidase; some industrial strains contain plasmids specifying additional type I signal peptidases.

The release of the translocated mature protein from the membrane is usually accompanied by its folding to the native conformation. For many *B. subtilis* secretory proteins, this requires folding factors such as Fe^{3+} and Ca^{2+} and/or the extracytoplasmic lipoprotein PrsA. Efficient folding of the mature protein on the *trans* side of the membrane appears to be important for avoiding or reducing electrostatic interactions with negatively charged components of the cell wall and degradation by extracellular proteases (61,62).

Although there are limits to the amounts of native proteins secreted by bacilli, such proteins are usually capable of being produced at commercially significant levels. In contrast, the extracellular production of heterologous proteins, especially from eukaryotic sources, is frequently inefficient. The reasons for this are not fully understood but are likely to include that the inherent incompatibilities with the secretory apparatus and their susceptibility to the many extracellular proteases, cell associated and released, produced by these bacteria (61,63).

GENE EXPRESSION

Most of the extracellular hydrolase enzymes produced by *Bacillus* species are induced during the transition from vegetative growth to stationary phase, where they appear to function to increase the range of substrates that may be used to sustain growth. The expression of the genes encoding some of these enzymes appears to be linked to, but not necessary for, sporulation itself (e.g., alkaline and neutral proteases), while others (e.g., α-amylases and alkaline phosphatases) are linked to the exhaustion of specific substrates.

Sporulation

Sporulation is a developmental process in which bacilli develop into heat-, chemical-, solvent-, radiation-, and desiccation-resistant spores (64,65). Their transition from dormancy to vegetative cells involves germination and outgrowth. Under laboratory conditions at 37 °C, sporulation takes approximately 8 h. The process has been divided into a series of stages that are defined primarily on the basis of their morphology. These stages, labeled 0, II, III, . . . through VII (stage I is no longer recognized), broadly correlate with the time (in hours) after the induction of sporulation (Fig. 3).

Sporulation is induced in response to nutrient deprivation and is only initiated after the failure of transition phase processes (e.g., motility and chemotaxis, production of macromolecular hydrolases, development of competence) to restore vegetative growth. Sporulation is induced by activated SpoOA (SpoOA–P) via a phosphorelay that facilitates multifactorial signal processing. The first morphologically distinct stage is stage II, which is defined by the formation of an asymmetric cell division event that divides the cell into two compartments, the larger sporangium (or mother cell) and the prespore, or forespore. During stage II, the prespore becomes completely engulfed by the sporangium so that, at stage III, the two compartments are separated by a double membrane system. During stage IV, peptidoglycan is laid down between the two membrane layers to form the spore cortex and the germ cell wall. During this stage the prespore becomes refractile to light. In stages V and VI, spore coat proteins are laid down on the surface of the inner and outer membrane of the prespore. As these mature the spore becomes increasingly resistant to chemicals and heat. In the final stage, stage VII, the by-now-dehydrated spore is released from the sporangium by cell lysis. The fully mature spore is metabolically inactive and capable of surviving as such for certainly tens, and possibly hundreds, of years.

Sporulation is a tightly controlled process requiring differential expression of genes on the chromosomes in each of the two compartments, as well as developmental checkpoints to ensure synchrony between their patterns of development. This is achieved through the differential activation of sporulation-specific sigma (σ) factors at or about stages II to III. σ^F is active in the prespore, and σ^E is active in the sporangium. These sigma factors are, in turn, responsible for the induction of sporangium-specific (σ^K) and prespore-specific (σ^G) sigma factors. Synchrony between the two compartments is maintained by signal communication processes.

Much less is known about germination and outgrowth. In the laboratory at 37 °C, the time between induction of germination and the first cell division event is about 90 min. Germination is triggered by a variety of chemical and physical stimuli; spores of *B. subtilis* germinate in response to L-alanine or to a mixture of asparagine, glucose, fructose, and KCl (66). Germination is accompanied by a series of rapid chemical and physiological changes including the degradation of the spore cortex, disruption of the spore coat, rehydration, and swelling. These changes are accompanied by a loss of heat and chemical resistance. It takes some 30 min before de novo RNA synthesis takes place. During outgrowth the cell begins to increase in mass and to replicate its DNA, forming an elongated cylindrical cell that is only subsequently reduced to a normal-sized cell by a delayed cell division.

Phosphate

Phosphorus is a major growth-limiting substrate in natural environments such as soil (67); consequently, under phosphate limitation, *B. subtilis* induces a programmed series of responses to reduce the cell's requirement for phosphorus, increase its affinity for phosphate uptake, and pro-

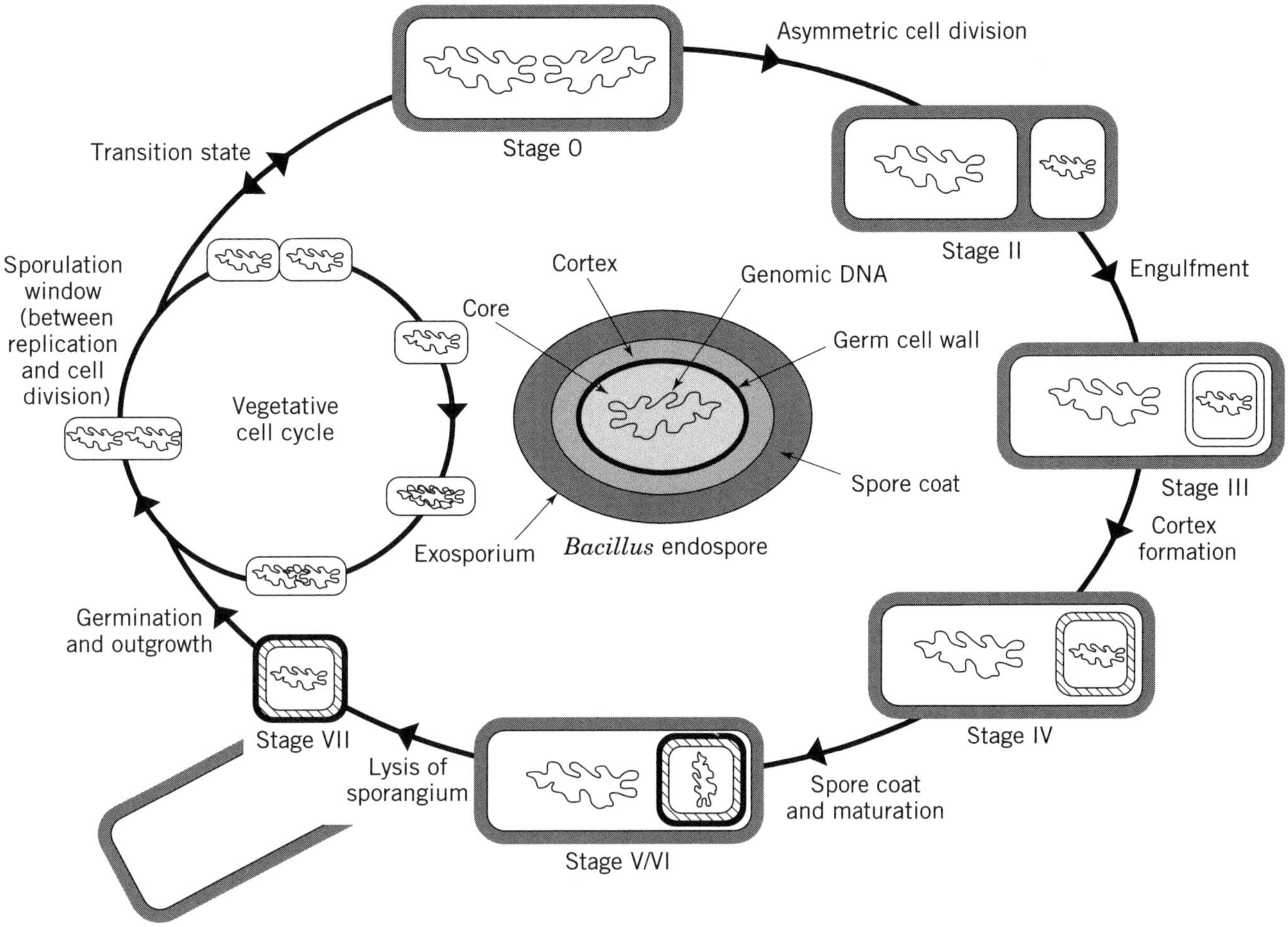

Figure 3. The sporulation pathway in *Bacillus subtilis* and its relationship to the vegetative cell cycle. Sporulation is only initiated toward the end of the vegetative cell cycle when the cell has two complete copies of the chromosome.

duce extracellular hydrolytic enzymes to recover inorganic phosphate from organic sources (68,69).

B. subtilis responds to phosphate stress by the coordinated induction of approximately 40 genes (69). Prominent among these are genes of the Pho regulon (67,70–72), whose products include alkaline phosphatases (APase), alkaline phosphodiesterase (APDase), a high-affinity phosphate transporter, and enzymes required for teichuronic acid synthesis.

The Pho regulon of *B. subtilis* is controlled by two *trans*-acting regulators, PhoP and PhoR, that together form the components of a two-component, environment-sensing, signal transduction system (70,73). PhoR is a membrane-spanning sensor protein with histidine protein kinase (HPKase) activity that is activated at low phosphate concentrations to PhoR–P. This in turn acts as a substrate for the activation of PhoP, a response regulator required for the induction of the Pho regulon. *B. subtilis* genes activated by PhoP–P, which include the bicistronic operon encoding PhoP and PhoR, constitute the Pho regulon. The response regulators of at least two other signal transduction systems, SpoOA–P and ResD, indirectly affect the expression of the Pho regulon by repressing the *phoP/R* operon (71,74,75).

The promoters of the Pho regulon genes/operons that have been studied to date have a conserved sequence, TTAACA (74), located around position −22. This sequence may be equivalent to the Pho box observed in *E. coli* (76), which also tends to be located toward the 5′ end of the promoter region of Pho regulon genes/operons.

Carbon

Bacillus species can grow on a wide range of simple carbon sources, including glucose, sucrose, and amino acids as well as a range of complex substrates such as proteins and polysaccharides, particularly plant-derived substrates such as starch and arabinans. In the case of more complex substrates, extracellular enzymes are required to reduce the size of these substrates before uptake and utilization. Carbon substrates are transported into the cell either in an unmodified form or via a PTS system (77).

In *B. subtilis* many genes/operons encoding enzymes involved in the catabolism of substrates as diverse as starch and histidine are repressed by readily metabolizable carbon sources such as fructose, glucose, glycerol, or mannitol—so-called PTS sugars. Catabolite repression functions at the level of transcription, but differs substantially from the cyclic AMP-mediated system observed in *E. coli*. Three components of the *B. subtilis* system have been identified: a *cis*-acting catabolite-responsive element (CRE) and two *trans*-acting factors, CcpA and Hpr. The 14-bp palindomic CRE sequence (78) is essential for catabolite repression. CREs are found at diverse locations in relation to the transcriptional units of catabolite-sensitive genes, in some cases overlapping the promoter (*amyE*), in others being located within the structural gene itself (*xylA*). The latter positioning makes it extremely difficult to engineer catabolite insensitivity in such genes.

CREs are likely to be binding sites for a catabolite repressor protein. Although it has not been directly shown to be the case, CcpA, which is related to *E. coli* LacI-type repressors, is the most likely candidate for this protein. Mutants in the *ccpA* gene abolish catabolite repression, including glucose repression of sporulation, without affecting substrate-specific induction of the individual genes/operons concerned. They also severely affect growth on substrates such as glucose, fructose, glucitol, and glycerol that are metabolized via glycolysis or the TCA cycle, but not on substrates such as xylose which are metabolized via the pentose phosphate pathway. In the former case, growth is restored if TCA cycle intermediates are added to the growth medium. Interestingly, a CRE appears to be responsible for glucose-mediated activation of the acetate kinase gene (*ackA*) and in genes encoding enzymes of central metabolism.

HPr is a component of the phosphotransferase system (PTS) which, when activated by phosphorylation to HPr(Ser–P), binds to CcpA. The key signaling component of the catabolite repression pathway appears to be the Embden–Meyerhof pathway (EMP) intermediate fructose-1,6-diphosphate (FDP). High concentrations of this compound activate the ATP-dependent HPr serine kinase and possibly stimulates an interaction between CcpA and HPr(Ser–P). This complex may then bind to the CREs of catabolite-responsive genes/operons (49).

Nitrogen

Bacillus spp. can use ammonium and a number of other nitrogen-containing compounds (notably certain amino acids) as sole sources of nitrogen (79). Ammonium is assimilated via glutamine synthetase (GS), glutamine:2-oxoglutarate amidotransferase (GOGAT), and glutamate dehydrogenase (GDH) (80). Most members of the genus are able to use all three enzymes, while a smaller group, including many (but possibly not all) strains of *B. subtilis*, use GS and GOGAT, while nitrogen-fixing *Bacillus* species use GDH. In *B. subtilis*, the relative levels of GS and GOGAT are dependent on the available sources of nitrogen. The levels of GS synthesis are dependent on the need for glutamine synthesis; they are lowest in the presence of glutamine, a nitrogen source favored by this bacterium, and highest in the presence of nonfavored nitrogen sources such as ammonium. No global regulatory network analogous to that of Ntr of *E. coli* (81) has been detected in *B. subtilis*, although GS synthesis is autoregulated via the GlnR repressor and has also been implicated as a key signal for the nitrogen status of the cell modulating the activity of a variety of other metabolic pathways.

GOGAT levels are low in the presence of glutamate and amino acids (e.g., glutamine, arginine, and histidine) that can be catabolized to glutamate, and high in the presence of ammonium and amino acids that are not broken down to glutamate and other sources of nitrogen such as nitrate.

In addition to its use as a nitrogen source, nitrate (and nitrite) can be used by *B. subtilis* in place of oxygen as an electron acceptor, allowing growth under anaerobic conditions. Nitrate respiration requires the products of the *narGHJI* operon (10,11).

PRODUCTS

Bacillus species are an important source of industrial enzymes, fine biochemicals, antibiotics, and insecticides (5). Additionally, the ability of *B. subtilis* and close relatives to secrete grams per liter quantities of protein directly into the growth medium, their ease of growth, and their well-proven safety have also made them prime candidates for the production of heterologous proteins.

Many commercially important enzymes have traditionally been produced in their native hosts. However, when organisms with novel enzymic activities are newly isolated, the costs and uncertain outcome associated with process optimization are often prohibitive. This even applies to well-established strains of *B. amyloliquefaciens* and *B. licheniformis* that have been used extensively in commercial enzyme fermentations. Both these organisms are generally refactory to genetic engineering, and relatively little information is available concerning their biochemistry and physiology. Instead, companies are increasingly seeking to develop well-characterized strains such as *B. subtilis* strain 168 as universal hosts for commercial applications.

Enzymes

The world annual sales for industrial enzymes was recently valued at $1 billion, with strong growth in the paper, textiles, and waste treatment markets. Three-quarters of the market is for enzymes involved in the hydrolysis of natural polymers, including proteolytic enzymes used in the detergents, dairy, and leather industries and carbohydrases used in the baking, brewing, distilling, starch, and textile industries. About two-thirds of these enzymes are produced by fermentation from *Bacillus* species, the main producers, together with their products and activities, are listed in Table 1.

Two *Bacillus* enzymes dominate the industrial enzymes markets alkaline (serine) protease and α-amylase. Although many species of *Bacillus* secrete enzymes of these types, their catalytic properties vary from one producer strain to the next. Moreover, commercially significant enzymes may be modified to improve their performance or stability in particular industrial processes.

Alkaline proteases are the single largest enzyme market and are used extensively as additives to detergents. They are relatively nonspecific endoproteases that convert substrate proteins into small, readily soluble fragments. The two main proprietary products, subtilisin Novo (BPN′) from *B. amyloliquefaciens* and subtilisin Carlsberg from *B. licheniformis*, have been the subject of extensive develop-

Table 1. Commercially Significant Products Produced by *Bacillus* species

Product	Producer organism	Properties/use
Glucose isomerase	*B. coagulans*	Conversion of glucose to fructose in manufacture of low calorie sweeteners
α-Amylase	*B. subtilis*	Limited hydrolysis of starch
	B. amyloliquefaciens	Liquefication of starch
	B. stearothermophilus	Thermophilic liquefaction of starch
	B. licheniformis	Thermostable liquefaction of starch
β-Amylase	*B. polymyxa*	Formation of low-dextrose-equivalent syrups from starch
β-Glucanase	*B. amyloliquefaciens*	Hydrolysis of barley β-glucans in brewing and animal feeds
	B. circulans	
	B. subtilis	
Pullulanase	*Bacillus* species	1,6-Bond starch debranching enzyme
Alkaline protease	*B. alkalophilus*	Detergents and leather industries
	B. licheniformis	
Neutral protease	*B. thermoproteolyticus*	Temperature-stable enzyme
Thermolysin	*B. thermoproteolyticus*	Condensation of L-aspartic acid and D,L-phenylalanine to produce aspartame
β-1,4-Galactosidase	*B. stearothermophilus*	Dairy industry
	B. subtilis	
Chymosin (rennin/pepsin)	*B. mesenterius*	Rennet substitutes
	B. cereus	
	B. polymyxa	
	B. megaterium	
	B. licheniformis	
Glucose dehydrogenase	*B. megaterium*	Glucose assay
Glycerol kinase	*B. stearothermophilus*	Glycerol/trigycerides assay
Phospholipase C	*B. cereus*	Phosphatidylcholine assay
Restriction endonucleases	*B. aneurinolyticus*	*Ban*I
	B. amyloliquefaciens	*Bam*HI
	B. subtilis	*Bsu*
	B. caldolyticus	*Bcl*
	B. globiii	*Bgl*
	B. stearothermophilus	*Bst*
	B. sphaericus	*Bsp*
	B. brevis	*Bbv*
δ-Endotoxin	*B. thuringiensis*	Biocontrol agent
Peptide antibiotics	*B. licheniformis*	Bacitracin
	B brevis	Edeines, gramicidin, tyrocidine
	B. subtilis	Fengycin, subtilin, surfactin
	B. polymyxa	Polymyxins

ment programs. Their three-dimensional structures have been solved (82,83), and functionality testing has resulted in the identification of the catalytic triad, an oxyanion-binding site, and substrate-binding determinants. Subtilisins have been engineered to improve their characteristics in washing detergents by increasing their stability in the presence of oxidizing chemicals, at temperatures above 60 °C and at pH values up to 12. They have pH optima of about 11.0 and half-lives of approximately 50 min at 40 °C. They are used at 0.4–1% in formulations that includes anionic and nonanionic surfactants, oxidants, soap, and optical brighteners. Although these were originally used simply in the form of a powder, hypersensitivity reactions among producers and users meant that the enzymes had to be added to formulations in a microencapsulated form. This process not only reduces dust formation but also protects the enzymes from other components of the formulation during storage.

α-Amylases are used extensively in the starch industry where, because of the physical properties of starch, they need to be used at high temperatures and good thermostability is an important property of these enzymes. Industry has sought to obtain thermostable amylases by screening for new sources and by improving the thermostability of existing enzymes by protein engineering. For example, the α-amylases from *B. amyloliquefaciens*, *B. stearothermophilus*, and *B. licheniformis* are closely related (3) but have very different thermostabilities of 2, 50, and 270 min, respectively, at 90 °C and pH 6.5 (84,85). Extra salt bridges caused by specific lysine residues in the more-thermostable *B. licheniformis* amylase are responsible for its increased stability, and the equivalent residues in the less-stable enzymes have been the target for protein engineering. Additionally, the extensive homology between the genes encoding these amylases has also been used for in vivo recombination experiments in which hybrid enzymes combining beneficial characteristics of these enzymes have been screened. The three-dimensional structure of the *B. licheniformis* α-amylase has recently been published (86), and the three amino acid residues that constitute the calcium-binding site (i.e., asparagine-104, aspartate-200, and histidine-235), have been identified.

The α-amylase from *B. acidocaldarius* strain A2 shows greater thermostability under acid conditions (87). Additionally, *B. subtilis* strain KP1064 (88) produces an enzyme that is active against starch and pullulan.

The industrial production of α-amylase was developed in 1917 using a strain subsequently reclassified as *B. amyloliquefaciens*. This α-amylase could be used to a maximum temperature of about 85 °C. Because starch needs to be cooked to 100 °C to rupture the starch granules, the introduction of a thermostable α-amylase from *B. licheniformis* (89) that can operate at about 105 °C was a distinct improvement.

B. coagulans is an important source of glucose isomerases, required for the conversion of glucose (0.75 times as sweet as sucrose) to fructose (twice as sweet as sucrose) in the production of high-fructose corn syrup. In contrast to the proteases and amylases already discussed, glucose isomerase is an intracellular enzyme used either in the form of immobilized cells or extracted and immobilized on a solid matrix.

Metabolites

Bacillus species are used for the production of a number of primary metabolites for the food and health care industries. Extensive knowledge of their biochemistry and regulation has allowed growth and rational mutant isolation strategies to be developed to the extent that this organism has been considered for the production of a wide range of metabolites at commercially viable levels.

B. subtilis has been used for the production of the nucleotides xanthanylic acid (XMP), inosinic acid (IMP), and guanylic acid (GMP), which are of commercial importance as flavor enhancers. The primary fermentation products are the nucleosides hypoxanthine, inosine, and guanosine because their greater membrane permeability allows them to accumulate to higher concentrations in the culture medium. These are then converted to their respective nucleotides by chemical phosphorylation (90).

Attempts have been made to develop strains of *Bacillus* for the production of amino acids such as tryptophan, histidine, and phenylalanine and of vitamins such as biotin, folic acid, and riboflavin. The use of analogs has led to the isolation of resistance mutants overproducing a number of amino acids, nucleosides, and vitamins (91). Knowledge of *B. subtilis* is such that rational approaches are now possible and attempts have been made to engineer this bacterium for the production of folic acid. Promising strains have been obtained, although their production levels cannot yet compete with those of *Brevibacterium*, *Corynebacterium*, or *Serratia* species.

Peptide Antibiotics

Under conditions of nutritional stress *Bacillus* species produce a variety of special metabolites that enhance their survival in natural environments (92). Prominent among these are peptide antibiotics, generally short peptides (2–20 amino acid residues) that are synthesized by large multienzyme complexes (peptide synthetases) rather than ribosomes. Individual amino acid residues are often extensively modified and may be linked to each other by peptide bonds or through esters or lactones. Gramicidin-S is a cyclic decapeptide from *B. brevis* with antibacterial and surfactant properties. Bacitracin is a branched cyclic dodecapeptide produced by *B. licheniformis* and *B. subtilis* (93) that is used as a topical antibiotic. Its antibacterial activity is directed against bacterial cell wall synthesis where it inhibits the recycling of lipid-P carriers. Surfactin, produced by most strains of *B. subtilis*, has both antibacterial activity and powerful surfactant properties. Polymyxins are membrane-active branched cyclic acylpeptides produced by *B. polymyxa*.

A minority of the peptide antibiotics produced in *Bacillus* species are synthesized on ribosomes and subsequently modified extensively by posttranslational processing (92). The resulting products are usually larger, but only marginally so, than those produced by the peptide synthetases. The best studied are the lantibiotics, so-called because they contain the unusual amino acids lanthionine and methyl-lanthionine. They include subtilin, a 32-residue lantibiotic produced by *B. subtilis* that shows antibacterial and antitumor activity.

Heterologous Proteins

The efficient large-scale production of heterologous enzymes and proteins by *B. subtilis* has been the focus of considerable research effort. The ability of this bacterium to secrete proteins directly into the culture medium in quantities of grams per liter offers considerable process advantages, in terms of product yield, structural integrity, and downstream processing costs, over production systems in which proteins are accumulated in the cytoplasm or periplasm, often as insoluble aggregates.

Nevertheless, attempts to use *B. subtilis* for the production of heterologous proteins has met with only limited success, and there are no reports of the use of this host in an industrial fermentation. Although extracellular proteins from close relatives can be produced in this bacterium at high concentrations, the yield of proteins from unrelated species remains disappointingly low. This is likely to be due, at least in part, to the production of at least seven extracellular proteases, which have been shown to cause extensive degradation of heterologous proteins, and to incompatibilities with the *Bacillus* protein secretion pathway. The isolation of strains defective in the identified proteases has helped in some but by no means all cases (61,63).

Other species of *Bacillus* may prove useful for development as hosts for the production of heterologous proteins, particularly those naturally producing low levels of extracellular proteases. Of interest are species such as *B. brevis* that produce a crystalline surface (S) layer at the outer surface of a thin cell wall. S-layer proteins have been observed at concentrations up to 35 g/L in the culture medium and, consequently, attempts have been made to develop the expression and signal sequences of their genes into components of a new generation of secretion vectors (94). However, it may ultimately be necessary to develop strains that are currently used in industrial processes as production hosts by improving our understanding of their genetics and physiology.

B. subtilis has also proved useful for the intracellular production of outer membrane proteins of gram-negative pathogens (95). These proteins, which have potential for use for immunodiagnostics and as vaccines, are produced in *B. subtilis* in preference to their native hosts to avoid contamination with endotoxins with which they often form tight associations.

Insecticides

The use of chemical insecticides, with a world market worth $5 billion, is increasingly seen as problematical because of the development of resistance in target insect populations, lack of specificity against the insect pest, persistence in the environment, and toxicity to man and other animals. Alternative biocontrol methods have been developed involving the use of viruses, fungi, and bacteria. *B. thuringiensis* (96) is the most successful insecticidal biocontrol agent and accounts for about 90% of the worldwide market for bioinsecticides. Although *B. thuringiensis* was originally isolated in 1901 from an infection of a Japanese silkworm farm and has been used as an insecticidal agent since 1920, it still only represents about 1–2% of the total insecticide market.

Strains of *B. thuringiensis* have been identified that are pathogenic for each of the main groups of insect pests, namely Lepidotera (moths, butterflies), Diptera (flies and mosquitoes), and Coleoptera (beetles). More recently strains of *B. thuringiensis* have been isolated that are active against noninsect pests, notably nematodes, mites, and protozoa (97). The toxicity of *B. thuringiensis* results from the production of proteinaceous δ-endotoxins during sporulation. The toxins, which may represent as much as 30% of the cell's weight, form crystals within the mother cell (98). The toxins are synthesized from plasmid-encoded *cry* genes, and a single strain may produce several toxins encoded by more than one plasmid. They are synthesized initially in the form of an inactive pro-toxin (70–140 kDa) that is considerably larger than the active form (60 kDa). The δ-endotoxins, which have the properties of chemical agents rather than infectious agents, are approximately 300 times more potent than pyrethroids and 8,000 times more potent than organophosphates. *B. thuringiensis* toxins therefore combine high toxicity and specificity for their target pests with little or no toxicity for nontarget insects (e.g., pollinators, predators) and other animals.

A large number of *cry* genes have been cloned, and the regions responsible for the pro- and mature forms, host specificity, and toxicity have been mapped. As different toxins exhibit a range of host specificities and levels of toxicity, chimeric toxins that combine the specificities and toxic regions of different natural toxins are currently being developed.

B. thuringiensis strains have been used extensively for the control of crop and forest insect pests and of insect disease vectors. However, as an alternative means of application, genes encoding the δ-endotoxin have been used directly to transform a variety of plant crops, including tomatoes, tobacco, potatoes, and cotton, in which field trials have confirmed their activity against target pests (99).

SAFETY

Bacillus species have been used extensively in the fermentation industry for the production of a range of products including enzymes, fine biochemicals, antibiotics, and insecticides. In addition, a strain of *B. subtilis* is used for the fermentation of soybeans to Natto, an extensively consumed ($\sim 10^8$ kg/annum) traditional Japanese food product. In a review of the safety of *B. subtilis* and *B. amyloliquefaciens* (these strains were considered to be synonymous up to 1973 [100]), no reports were found of infection or toxicity relating to *B. amyloliquefaciens*. In the case of *B. subtilis*, the situation was complicated by the fact that before about 1970 diagnostic laboratories did not distinguish between *B. cereus* and *B. subtilis*. Virtually all reports of putative *B. subtilis* infections since that time either were associated with drug abuse or occurred in patients whose immune system had been compromised by treatment with immunosuppressants or chemotherapeutic agents. This low incidence of pathogenicity and the widespread use of their products in the food, beverage, and detergents industries have resulted in GRAS (generally regarded as safe) status being granted to these species by the U.S. Food and Drug Administration (FDA).

GROWTH

Laboratory Culture Conditions

The majority of *Bacillus* species will grow at mesophilic temperatures on commercially prepared nutrient media, although in some cases it is necessary to modify the pH or salt concentration. Obligate thermophilic species such as *B. stearothermophilus* that do not grow satisfactorily at 37 °C are usually grown at 60 °C. Moderately thermophilic species, such as *B. coagulans*, are grown between 45 °C and 50 °C. The more fastidious insect pathogens, *B. larvae and B. popilliae*, require the addition of thiamine for growth and are usually grown at 25 °C to 30 °C. *B. stearothermophilus* requires additional calcium and iron, and *B. pasteuri* requires 0.5–1% urea.

B. subtilis and many other species are able to grow in simple salts media containing ammonium or amino acids as nitrogen sources and glucose or other simple sugars as carbon sources. The most commonly used general-purpose chemically defined medium is Spizizen's minimal medium (27). The growth rate in this medium can be improved by the replacement of ammonium, a poor nitrogen source for *B. subtilis*, with glutamine.

B. subtilis is able to use a number of amino acids as nitrogen sources (e.g., arginine, glutamine, glutamate, asparagine, and aspartate), and the respective catabolic pathways are induced in the presence of these compounds. Consequently, the concentration of many amino acids recommended to overcome auxotrophy are actually growth limiting. It should also be borne in mind that derivatives of the well-studied *B. subtilis* strain 168 often show tryptophan auxotrophy, and consequently complex media in which acid-hydrolyzed casein provides the main source of nitrogen need to be supplemented with this amino acid.

Although many *Bacillus* species sporulate readily, special media and growth protocols are required for efficient sporulation (14). Sporulation is induced in response to nutrient deprivation and occurs after the end of exponential growth. In *B. subtilis*, sporulation is repressed in the presence of glucose and at low cell densities. A widely used sporulation medium was described by Schaeffer and colleagues (101). In some species sporulation can also be induced by decoyinine, an analog of GTP (14).

Growth in Batch Culture. Although many species of *Bacillus* grow under anaerobic conditions, they are normally grown aerobically. Some strains (e.g., *B. subtilis*) have a tendency to lyse in the absence of an energized membrane (33). Growth is normally carried out in conical flasks in an orbital (reciprocating) incubator at >200 rpm, although most species can be grown in batch fermentors. Ideally, the culture volume should not exceed 20% of the flask volume. The media should be prewarmed as temperature shock can also induce autolysis.

Nutrients are usually supplied in excess and growth continues until a particular substrate is exhausted, the oxygen tension or pH falls to an inhibitory level, or an inhibitor accumulates to an inhibitory level. The various phases in the growth cycle have been described in detail elsewhere (102).

Growth in Continuous Culture. *Bacillus* species undergo marked metabolic, physiological, and morphological changes during batch culture. Many of these changes occur during transition to and during stationary phase and include morphologically distinct spores and physiologically distinct competent cells. To some extent these population variations can be reduced by growth in a chemostat under continuous culture conditions. This method allows the effect of individual growth parameters (e.g., pH, oxygen tension, growth rate, limiting substrate) on particular cellular characteristics to be identified. Chemostat cultures can be used to study cells under "steady-state" conditions, or during transitions from one steady state to another (103).

In principle any liquid culture medium that supports growth in batch culture can be used for continuous culture. However, because the pH is actively controlled during growth, it is not necessary to include components required only for buffering pH. The concentration of the limiting nutrient determines the culture density, and all other essential nutrients are supplied so that they are in excess. For aerobes, the concentration of the limiting nutrient is set so that the culture density is not so high as to impose an excessive oxygen demand. For *B. subtilis* this may mean limiting the culture density to about 1–2 g dry weight per liter. Complex media can be used if a defined limitation can be achieved (104). Several defined media have been described using, for example, phosphate, magnesium, potassium, glucose, and sulfate as limiting substrates.

Commercial Culture Conditions

Most of the products traditionally produced from *Bacillus* species, including enzymes, antibiotics, and insecticides, are synthesized during stationary phase using a traditional batch fermentor. However, insecticides require both a high growth yield and efficient sporulation because the δ-endotoxin accumulates in the sporangium. The production of α-amylase and of *B. thuringiensis* δ-endotoxin are given as examples of the factors that need to be taken into account during commercial *Bacillus* fermentations.

Most species of *Bacillus* grow well on commercial formulations of nutrient media, although in some cases the medium needs to be modified by adjustment of the pH or ionic strength or by the addition of specific nutrient supplements. This means that media can, to some extent, be based on cheap, readily available substrates. Despite the extensive expertise in the fermentation of *Bacillus* species, commercial considerations have limited the published literature on large-scale fermentations. Culture conditions and media formulations are highly geared to the particular strains and processes used and these have been selectively reviewed elsewhere (105–107).

Media components have been developed empirically over a number of years to avoid catabolite repression that prevents the synthesis of a number of the commercially important products from *Bacillus* species. For example, α-amylases are produced in media containing starch in various forms including corn, wheat malt, and potato that is noncatabolite repressing. When carbohydrates are used as a carbon source, phosphate is required at a concentration of about 100 mM. Peptones and amino acids are a better source of nitrogen than ammonium for *B. subtilis*, probably because of the efficiency with which nitrogen is assimilated into amido groups. In the case of *B. apiaries*, however, ammonium is the preferred nitrogen source. Glutamine is the nitrogen source that sustains the fastest growth rate. The use of amino acids as a nitrogen source represses the synthesis of proteases and the onset of sporulation, allowing the productive stationary phase to be extended.

The concentrations of divalent metal cations Mn^{2+}, Ca^{2+}, and Zn^{2+} are important to the final yield of α-amylase, even when they have little or no effect on the growth of the bacterium. For example, Ca^{2+} is required for both the enzymatic activity and structural integrity of the liquefying α-amylases such as those from *B. licheniformis*, *B. amyloliquefaciens*, and *B. stearothermophilus*. We have evidence that this results, at least in part, from the rate of folding of the enzyme on the *trans* side of the cytoplasmic membrane (62). Other metal ions, including Fe^{2+}, Fe^{3+}, Na^{+}, and Mg^{2+}, enhance enzyme production presumably in part as a result of their effect in increasing growth yield.

The growth kinetics in *Bacillus* cultures is complicated by the operation of the stringent response and sporulation, both of which are similarly induced by nutrient limitations and other negative growth factors. Commercial processes often use fully or partially (oligosporogenous) sporulation-deficient strains. The stringent response, modulated by variations in the levels of ppGpp(p), appears to be a cause of linear growth. Substrates with a low carbon to energy ratio gives rise to carbon-limited growth and consequential "energy spilling" and the production of overflow metabolites. Glucose excess chemostat cultures of *B. stearothermophilus* accumulate acetate up to 34% of the carbon consumed. Under these conditions, energy efficiency is sacrificed to obtain the ATP turnover required to sustain the high growth rates. Similar metabolite overflows have been observed for *B. licheniformis* (48%) and *B. subtilis* (22%).

Catabolite repression is one of the major factors limiting the production of many commercially important products. Although mutants (e.g., *ccpA*) have been isolated that lack elements of the catabolite repression pathway, these often grow poorly. An alternative approach in which CRE elements are deleted or inactivated may be frustrated by their occurrence within DNA sequences encoding the protein itself. Often, the most effective method for overcoming catabolite repression involves a combination of appropriate media formulations and growth regimes.

Many commercially significant products are produced after the end of exponential growth (transition phase) and during the early stages of sporulation. It is therefore essential that, when sporulation-deficient strains are used, they still retain the ability to synthesize the required product. For example, mutants deficient in Spo0A (the sporulation-specific response regulator) are unable to induce various extracellular proteases (108).

Although batch cultures are traditionally used for α-amylase production, because the main production phase is during stationary phase, Emanuilova and Toda (109) compared the production of α-amylase by *B. caldolyticus* in batch and continuous culture. Casitone and starch were required for optimal α-amylase production. The concentration of amylase was 10- to 20-fold higher in continuous rather than batch culture, although there was an inverse relationship between growth rate and amylase production.

Industrial fermentations are carried out at 30–40 °C and at pH 7.0 with $CaCO_3$ as the calcium source to stabilize the enzyme. The complex medium (110,111) allows growth to 10^{10} cell/ml within 20 h, after which time production continues for about 6 days (111). The fermentations also produce proteases that are important for the utilization of protein substrates, but which have to be separated from the α-amylase during subsequent downstream processing.

α-Amylase production can be increased by cloning the structural gene onto a multicopy plasmid, and production, possibly up to the limitation of the native secretory pathway, is achievable (112). However, because many of the plasmids used in *Bacillus* species are based on staphylococcal plasmids (e.g., pUB110), both segregational and structural instabilities are frequently observed.

Because δ-endotoxin production of *B. thuringiensis* occurs during sporulation, growth and process conditions need to be optimized to give cell densities $>5 \times 10^9$ per milliliter and sporulation rates greater than 90%. Although significant strain-to-strain variations are observed, certain basic requirements can be defined: these include good oxygenation, a growth temperature between 26 and 30 °C (higher temperatures can lead to plasmid curing and consequential loss of δ-endotoxin production [113]), and pH 6.5–7.5. The production of amylases and proteases by *B. thuringiensis* allows a wide range of natural raw materials to be used as growth substrates. Glucose is catabolized with the production of acetate, lactate, pyruvate, and acetoin (114,115), and pH control is required for some strains to prevent the acidic conditions restricting growth. The production of poly-β-hydroxybutyrate during early stationary phase under carbon excess can be mistaken for δ-endotoxin but disappears during sporulation.

Cultures are usually grown in submerged liquid cultures as large as 200 L using fermentors agitated with paddle blades and with static baffles on the side wall of the vessel (99). Occasionally vessels with impellers or airlift fermentors are used. Because the industrial media contain significant amounts of insoluble particles, sterilization is accompanied by agitation. Toward the end of the fermentation, the released spores attach to and stabilize the foam, and silicon- or polypropylene-based antifoam agents are required as foam suppressers. Antifoam agents can form aggregates that interfere with downstream processing, and consequently their use is kept to a minimum. During stationary phase, oxygen consumption is markedly reduced and, provided saturation levels do not fall below 20%, agitation can be reduced, with a concomitant reduction in foaming. At the end of fermentation, the fermentation vessel will be heavily contaminated with spores that need to be inactivated if different strains are used in the next fermentation run.

Bacillus thuringiensis will grow on glucose minimal medium only when supplemented with sources of amino acids (e.g., peptones, casamino acids) or proteins. Because *B. thuringiensis* produces amylases, glucose can be replaced with starch as the carbon source. Growth is stimulated by yeast extract and optimal concentrations of Fe^{3+}, Co^{+}, and Zn^{+}, and sporulation is improved by metal ions such as Ca^{2+} and Mn^{2+}. Phosphate is added both as a source of phosphorus and as a buffer.

Preparations of *B. thuringiensis* are applied at a rate of about 0.5 to 2 kg per hectare. Spores can be applied directly as a pest control agent, but are often processed to improve their physical characteristics. For example, granular or bait formulations are applied to the plants or soil; in other cases, concentrates are diluted in water and applied by spraying. Stickers, spreaders, or even chemical insecticides can be added to improve dispersion and activity (99).

BIBLIOGRAPHY

1. F.G. Priest, in A.L. Sonenshein, J.A. Hoch, and R. Losick eds., *Bacillus subtilis and Other Gram-Positive Bacteria: Biochemistry, Physiology, and Molecular Genetics*, American Society for Microbiology, Washington, D.C., 1993, pp. 1–16.
2. D. Claus and R.C.W. Berkeley, in P.H.A. Sneath ed., *Bergey's Manual of Systematic Bacteriology*, Williams & Wilkins, Baltimore, Md., 1986, pp. 1105–1141.
3. F.G. Priest and C.R. Harwood, in Y.H. Hui and G.G. Khachatourians eds., *Food Biotechnology Micro-organisms*, VCH, New York, 1994, pp. 377–421.
4. F.G. Priest, in C.R. Harwood ed., *Biotechnology Handbook 2: Bacillus*, Plenum Press, New York, 1989, pp. 27–56.
5. C.R. Harwood, *Trends Biotechnol.* **10**, 247–256 (1992).
6. J.W. Kloepper, R. Lifshitz, and R.M. Zablotowicz, *Trends Biotechnol.* **7**, 39–44 (1989).
7. J.D. Wisotzkey, P.J. Jurtshuck, G.E. Fox, G. Reinhard, and K. Poralla, *Int. J. Syst. Bacteriol.* **42**, 236–269 (1992).
8. C.H. Ash, A.E. Farrow, S. Wallbanks, and M.D. Collins, *Lett. Appl. Microbiol.* **13**, 202–206 (1991).

9. D. Rössler, W. Ludwig, K.H. Schleiffer, C. Lin, T.J. McGill, J. Wisotzkey, P.J. Jurtshuk, and G.E. Fox, *Syst. Appl. Microbiol.* **14**, 266–269 (1991).
10. T. Hoffman, B. Troup, A. Szabo, C. Hungerer and D. Jahn, *FEMS Microbiol. Lett.* **131**, 219–225 (1995).
11. M.M. Nakano, Y.P. Dailly, P. Zuber, and D.P. Clark, *J. Bacteriol.* **179**, 6749–6755 (1997).
12. B.L. Russell, S.A. Jelly, and A.A. Yousten, *Appl. Environ. Microbiol.* **55**, 294–297 (1989).
13. C.R. Harwood and A.R. Archibald, in C.R. Harwood and S.M. Cutting eds., *Molecular Biological Methods for* Bacillus, Wiley, Chichester, UK, 1990, pp. 1–26.
14. W.L. Nicholson and P. Setlow, in C.R. Harwood and S.M. Cutting eds., *Molecular Biological Methods for* Bacillus, Wiley, Chichester, UK, 1990, pp. 391–450.
15. D. Claus and D. Fritze, in C.R. Harwood ed., *Biotechnology Handbook 2:* Bacillus, Plenum Press, New York, 1989, pp. 5–26.
16. F. Kunst, N. Ogaswara, I. Moszer, et al. *Nature* **390**, 249–256 (1997).
17. R.F. Doolittle, Nature **392**, 339–342 (1998).
18. S. Bron, R. Meima, J.-M. van Dijl, A. Wipat, and C.R. Harwood, in G. Cohen ed., *Molecular Biology of Industrial Microorganisms*, American Society for Microbiology, Washington, D.C., 1998, in press.
19. R.E. Yasbin, W. Firshein, J. Laffan, and R.G. Wake, in C.R. Harwood and S.M. Cutting eds., *Molecular Biological Methods for* Bacillus, Wiley, Chichester, UK, 1990, pp. 296–320.
20. J.P. Müller, Z. An, T. Merad, I.C. Hancock, and C.R. Harwood, *Microbiology* **143**, 947–956 (1997).
21. R.M. Horton, H.D. Hunt, S.N. Ho, J.K. Pullen, and L.R. Pease, *Gene* **77**, 61–68 (1989).
22. P. Youngman, in C.R. Harwood and S.M. Cutting eds., *Molecular Biological Methods for* Bacillus, Wiley, Chichester, UK, 1990, pp. 221–266.
23. P. Youngman, in A.L. Sonenshein, J.A. Hoch, and R. Losick eds., Bacillus subtilis *and other Gram-Positive Bacteria: Biochemistry, Physiology, and Molecular Genetics*, American Society for Microbiology, Washington, D.C., 1993, pp. 585–596.
24. M.-A. Petit, C. Bruand, L. Jannière, and S.D. Ehrlich, *J. Bacteriol.* **172**, 6736–6740 (1990).
25. J. Perkins and P. Youngman, *Proc. Natl. Acad. Sci. U.S.A.* **83**, 140–144 (1986).
26. M.R. Wati, F.G. Priest, and W.J. Mitchell, *FEMS Microbiol. Lett.* **71**, 211–214 (1990).
27. J. Spizizen, *Proc. Natl. Acad. Sci. U.S.A.* **44**, 1072–1078 (1958).
28. D. Dubnau, *Microbiol. Rev.* **55**, 395–424 (1991).
29. D. Dubnau, in A.L. Sonenshein, J.A. Hoch, and R. Losick eds., Bacillus subtilis *and Other Gram-Positive Bacteria: Biochemistry, Physiology, and Molecular Genetics*, American Society for Microbiology, Washington, D.C., 1993, pp. 555–584.
30. S. Chang and S.N. Cohen, *Mol. Gen. Genet.* **168**, 111–115 (1979).
31. S.M. Cutting, and P.B. Vander Horn, in C.R. Harwood and S.M. Cutting eds., *Molecular Biological Methods for* Bacillus, Wiley, Chichester, UK, 1990, pp. 27–74.
32. B. Michel, B. Niaudet, and S.D. Ehrlich, *EMBO J.* **12**, 1565–1571 (1982).
33. L.K. Jolliffe, R.J. Doyle, and U.N. Streips, *Cell* **25**, 753–763 (1981).
34. C.R. Harwood and A. Wipat, *FEBS Lett.* **389**, 84–87 (1996).
35. P.S. Vary, *Microbiology* **140**, 1001–1013 (1994).
36. R. Meima, G. Venema, and S. Bron, *Plasmid* **35**, 14–30 (1996).
37. S. Bron, in C.R. Harwood and S.M. Cutting eds., *Molecular Biological Methods for* Bacillus, Wiley, Chichester, UK, 1990, pp. 75–175.
38. S.D. Ehrlich, *Proc. Natl. Acad. Sci. U.S.A.* **74**, 1680–1682 (1977).
39. A.D. Gruss and S.D. Ehrlich, *Microbiol. Rev.* **53**, 231–241 (1989).
40. S.D. Ehrlich, *Proc. Natl. Acad. Sci. U.S.A.* **75**, 1433–1436 (1978).
41. K.J. Leenhouts and G. Venema, in K.G. Hardy ed., *Plasmids, a Practical Approach*, Oxford Univ. Press, New York, 1993, pp. 65–94.
42. C. Bruand, S.D. Ehrlich, and L. Jannière, *EMBO J.* **10**, 2171–2177 (1991).
43. L. Jannière, A. Gruss, and S.D. Ehrlich, in A.L. Sonenshein, J.A. Hoch, and R. Losick eds., Bacillus subtilis *and Other Gram-Positive Bacteria: Biochemistry, Physiology, and Molecular Genetics*, American Society for Microbiology, Washington, D.C., 1993, pp. 625–644.
44. V. Vagner and S.D. Ehrlich, *J. Bacteriol.* **170**, 3978–3982 (1988).
45. A.M. Albertini and A. Galizzi, *J. Bacteriol.* **162**, 1203–1211 (1985).
46. M. Perego, in A.L. Sonenshein, J.A. Hoch, and R. Losick eds., Bacillus subtilis *and Other Gram-Positive Bacteria: Biochemistry, Physiology, and Molecular Genetics*, American Society for Microbiology, Washington, D.C., 1993, pp. 615–624.
47. M. Simonen and I. Palva, *Microbiol. Rev.* **57**, 109–137 (1993).
48. J. Pero and A. Sloma, in A.L. Sonenshein, J.A. Hoch, and R. Losick eds., Bacillus subtilis *and Other Gram-Positive Bacteria: Biochemistry, Physiology, and Molecular Genetics*, American Society for Microbiology, Washington, D.C., 1993, pp. 939–952.
49. G.H. Chamblis, in A.L. Sonenshein, J.A. Hoch, and R. Losick eds., Bacillus subtilis *and Other Gram-Positive Bacteria: Biochemistry, Physiology, and Molecular Genetics*, American Society for Microbiology, Washington, D.C., 1993, pp. 213–219.
50. D.G. Yansura and D.J. Henner, in A.T. Ganesan and J.A. Hoch eds., *Genetics and Biochemistry of* Bacilli, Academic Press, Orlando, Fla., 1984, pp. 249–263.
51. D.M. Gartner, M. Geissendorfer, and W. Hillen, *J. Bacteriol.* **170**, 3102–3109 (1988).
52. A.-M. Crutz, M. Steinmetz, S. Aymerich, R. Richter, and D. LeCoq, *J. Bacteriol.* **172**, 1043–1050 (1990).
53. J. Errington, in C.R. Harwood and S.M. Cutting eds., *Molecular Biological Methods for* Bacillus, Wiley, Chichester, UK, 1990, pp. 175–220.
54. S.J. Thornewell, A.K. East, and J. Errington, *Gene* **133**, 47–53 (1993).
55. D.E. Eveleigh and B.S. Montenencourt, *Adv. Appl. Microbiol.* **25**, 57–74 (1979).
56. V.P. Malik, *Adv. Genet.* **20**, 38–126 (1979).
57. K. Yamane and B. Maruo, in K. Sakaguchi and M. Ouishi eds., *Molecular Breeding and Genetics of Applied Microorganisms*, Academic Press, New York, 1980, pp. 117–123.
58. E.A. Ferrari, A.S. Jarnagin, and B.F. Schmidt, in A.L. Sonenshein, J.A. Hoch, and R. Losick eds., Bacillus subtilis *and*

Other Gram-Positive Bacteria: Biochemistry, Physiology, and Molecular Genetics, American Society for Microbiology, Washington, D.C., 1993, pp. 917–937.

59. U.P. Shinde, J.J. Liu, and M. Inouye, *Nature* **389**, 520–522 (1997).
60. V. Nagarajan, in A.L. Sonenshein, J.A. Hoch, and R. Losick eds., Bacillus subtilis *and Other Gram-Positive Bacteria: Biochemistry, Physiology, and Molecular Genetics*, American Society for Microbiology, Washington, D.C., 1993, pp. 713–726.
61. K. Stephenson and C.R. Harwood, *Appl. Environ. Microbiol.* **64**, 2875–2881 (1998).
62. K. Stephenson, C.R. Harwood, N.M. Carter, M.-F. Petit-Glatron, and R. Chambert, *FEBS Lett.* **432**, 385–389 (1998).
63. X. Wu, W. Lee, L. Tran, and S.L. Wong, *J. Bacteriol.* **173**, 4952–4958 (1991).
64. J. Errington, *Microbiol. Rev.* **57**, 1–33 (1993).
65. P. Stragier and R. Losick, *Annu. Rev. Genet.* **30**, 297–341 (1996).
66. A. Moir, I.M. Feavers, A.R. Zuberi, R. Sammons, I.S. Roberts, J.R. Yon, E.A. Wolff, and D. Smith, in J.A. Hock and P. Setlow eds., *Molecular Biology of Differentiation*, American Society for Microbiology, Washington, D.C., 1985, pp. 36–46.
67. F.M. Hulett, in A.L. Sonenshein, J.A. Hoch, and R. Losick eds., Bacillus subtilis *and Other Gram-Positive Bacteria: Biochemistry, Physiology, and Molecular Genetics*, American Society for Microbiology, Washington, D.C., 1993, pp. 229–235.
68. W.D. Grant, *J. Bacteriol.* **137**, 35–43 (1979).
69. C. Eymann, H. Mach, C.R. Harwood, and M. Hecker, *Microbiology* **142**, 3163–3170 (1996).
70. T. Seki, H. Yoshikawa, H. Takahashi, and H. Saito, *J. Bacteriol.* **169**, 2913–2916 (1987).
71. F.M. Hulett, J. Lee, L. Shi, G. Sun, G.R. Chestnut, E. Sharkova, M.F. Duggan, and N.V. Kapp, *J. Bacteriol.* **176**, 1348–1358 (1994).
72. F.M. Hulett, G. Sun, and W. Liu, in A. Torriani-Gorini, E. Yagil, and S. Silver eds., *Phosphate in Microorganisms: Cellular and Molecular Biology*, American Society for Microbiology, Washington, D.C., 1994, pp. 50–54.
73. T. Seki, H. Yoshikawa, H. Takahashi, and H. Saito, *J. Bacteriol.* **170**, 5935–5938 (1988).
74. F.M. Hulett, *Mol. Microbiol.* **19**, 933–939 (1996).
75. G.F. Sun, S.M. Birkey, and M.F. Hulett, *Mol. Microbiol.* **19**, 941–948 (1996).
76. B.L. Wanner, in F.C. Neidhardt ed., Escherichia coli *and* Salmonella: *Cellular and Molecular Biology*, American Society for Microbiology, Washington, D.C., 1996, pp. 1357–1381.
77. M.H. Saier, M.J. Fagan, C. Hoischen, and J. Reizer, in A.L. Sonenshein, J.A. Hoch, and R. Losick eds., Bacillus subtilis *and Other Gram-Positive Bacteria: Biochemistry, Physiology, and Molecular Genetics*, American Society for Microbiology, Washington, D.C., 1993, pp. 133–156.
78. C.J. Hueck, W. Hillen, and M.H. Saier, *Res. Microbiol.* **145**, 503–518 (1994).
79. S.H. Fisher, in A.L. Sonenshein, J.A. Hoch, and R. Losick eds., Bacillus subtilis *and Other Gram-Positive Bacteria: Biochemistry, Physiology, and Molecular Genetics*, American Society for Microbiology, Washington, D.C., 1993, pp. 221–228.
80. H.J. Schreier, in A.L. Sonenshein, J.A. Hoch, and R. Losick eds., Bacillus subtilis *and Other Gram-Positive Bacteria: Biochemistry, Physiology, and Molecular Genetics*, American Society for Microbiology, Washington, D.C., 1993, pp. 281–298.
81. L. Reitzer, in F.C. Neidhardt ed., Escherichia coli *and* Salmonella: *Cellular and Molecular Biology*, American Society for Microbiology, Washington, D.C., 1996, pp. 391–407.
82. C.S. Wright, R.A. Alden, and J. Kraut, *Nature* **221**, 235–242 (1969).
83. D.J. Neidhart and G.A. Petsko, *Prot. Eng.* **2**, 271–276 (1988).
84. S.J. Tomazic and A.M. Klibanov, *J. Biol. Chem.* **263**, 3086–3091 (1988).
85. S.J. Tomazic and A.M. Klibanov, *J. Biol. Chem.* **263**, 3092–3096 (1988).
86. M. Machius, G. Weigand, and R. Huber, *J. Mol. Biol.* **246**, 545–559 (1995).
87. M. Kanno, *Agric. Biol. Chem.* **50**, 23–31 (1986).
88. Y. Suzuki and T. Imai, *Appl. Microbiol. Biotechnol.* **21**, 20–31 (1986).
89. G.B. Madsen, B.E. Norman, and S. Slott, *Die Starkr.* **25**, 304–308 (1973).
90. F.G. Priest, in C.R. Harwood ed., *Biotechnology Handbook 2:* Bacillus, Plenum Press, New York, 1989, pp. 293–320.
91. S.W. Queener and D.H. Lively, in A.L. Domain and N.A. Solomon eds., *Manual of Industrial Microbiology and Biotechnology*, American Society for Microbiology, Washington, D.C., 1989, pp. 155–169.
92. P. Zuber, M.M. Nakano, and M.A. Marahiel, in A.L. Sonenshein, J.A. Hoch, and R. Losick eds., Bacillus subtilis *and Other Gram-Positive Bacteria: Biochemistry, Physiology, and Molecular Genetics*, American Society for Microbiology, Washington, D.C., 1993, pp. 897–916.
93. D. Konz, A. Klens, K. Schörgendorfer, and M.A. Marahiel, *Chem. Biol.* **4**, 927–937 (1997).
94. N. Tsukagoshi, S. Iritani, T. Sasaki, T. Takamura, H. Ihara, Y. Idota, H. Yamagata, and S. Udaka, *J. Bacteriol.* **164**, 1182–1187 (1985).
95. R. Puohiniemi, S. Butcher, E. Tarkka, and M. Sarvas, *FEMS Microbiol. Lett.* **83**, 29–34 (1991).
96. A.L. Aronson, in A.L. Sonenshein, J.A. Hoch, and R. Losick, eds., Bacillus subtilis *and Other Gram-Positive Bacteria: Biochemistry, Physiology, and Molecular Genetics*, American Society for Microbiology, Washington, D.C., 1993, pp. 953–963.
97. J.S. Feitelson, J. Payne, and L. Kim, *Bio/technology* **10**, 271–275 (1992).
98. J. Li, J. Carroll, and D.J. Ellar, *Nature* **353**, 815–821 (1991).
99. S. Ely, in P.F. Entwistle, J.S. Cory, M.J. Bailey, and S. Higgs eds., Bacillus thuringiensis, *an Environmental Biopesticide: Theory and Practice*, Wiley, Chichester, UK, 1993, pp. 105–124.
100. A. Sietske de Boar and B. Diderichsen, *Appl. Microbiol. Biotechnol.* **36**, 1–4 (1991).
101. P. Schaeffer, J. Millet, and P.J. Aubert, *Proc. Natl. Acad. Sci. U.S.A.* **54**, 704–711 (1965).
102. U. Wanner and T. Egli, *FEMS Microbiol. Rev.* **75**, 19–44 (1990).
103. T. Merad, A.R. Archibald, I.C. Hancock, C.R. Harwood, and J.A. Hobot, *J. Gen. Microbiol.* **135**, 645–655 (1989).
104. A.R. Archibald, in I.C. Hancock and I.R. Poxton eds., *Bacterial Cell Wall Techniques*, Wiley, Chichester, UK, 1988, pp. 37–54.

105. R.J. Sharp, M.D. Scawen, and T. Atkinson, in C.R. Harwood ed., *Biotechnology Handbook 2*: Bacillus, Plenum Press, New York, 1989, pp. 255–292.
106. F.G. Priest and R.J. Sharp, in J.O. Neway ed., *Fermentation Process Development of Industrial Organisms*, Dekker, New York, 1989, pp. 73–132.
107. M.V. Arbige, B.A. Bulthuis, J. Schultz, and D. Crabb, in A.L. Sonenshein, J.A. Hoch, and R. Losick eds., Bacillus subtilis *and Other Gram-Positive Bacteria: Biochemistry, Physiology, and Molecular Genetics*, American Society for Microbiology, Washington, D.C., 1993, pp. 871–895.
108. E. Ferrari, D.J. Henner, M. Perego, and J.A. Hoch, *J. Bacteriol.* **170**, 289–295 (1988).
109. E.I. Emanuilova and K. Toda, *Appl. Microbiol. Biotechnol.* **19**, 301–305 (1984).
110. L.A. Underkofler, in B.M. Miller and W. Litsky eds., *Industrial Microbiology*, McGraw-Hill, New York, 1976, pp. 128–146.
111. K. Aunstrup, *Appl. Biochem. Bioeng.* **2**, 27–69 (1976).
112. J. Vehmaanpera, P.N.A. Nybergh, R. Tanner, E. Pohjonen, R. Bergelin, and M. Korhola, *Enzyme Microb. Technol.* **9**, 546–549 (1987).
113. J.M. Gonzalez and B.C. Carlton, *Plasmid* **11**, 28–38 (1984).
114. K.W. Nickerson, G. St. Julian, and L.A. Bulla, *Appl. Environ. Microbiol.* **28**, 129–132 (1974).
115. T.G. Benoit, G.R. Wilson, and C.L. Baugh, *Lett. Appl. Microbiol.* **10**, 15–18 (1990).

BIOCATALYSIS DATABASES

LAWRENCE P. WACKETT
LYNDA B.M. ELLIS
University of Minnesota
Minneapolis, Minnesota

KEY WORDS

Biocatalysis
Biodegradation
Bioremediation
Microbial catabolism
World Wide Web

OUTLINE

INTRODUCTION TO MICROBIAL BIOCATALYSIS

The microbial transformation of organic compounds has been studied at the molecular level only during the last century of human history, driven by the development of techniques for determining the structure of organic compounds and cultivating microorganisms in pure culture. As information on these transformation reactions accrued, it became clear that microorganisms were, via their diverse metabolic pathways, principally responsible for the recycling of complex organic compounds on Earth. Almost simultaneously, it was recognized that the enormous diversity of microbial enzymes represents an invaluable resource for biosynthesizing desirable molecules. Historically, applications predate an understanding of the process, for example, the fermentation of sugars to yield ethanol. Although alcoholic beverage production was recorded in early written records, its biocatalytic basis was only proposed in the nineteenth century. Its enzymatic basis was first revealed in 1897 using a microbial cell-free system (1).

Now we appreciate that microorganisms collectively have evolved the greatest enzymatic diversity found on Earth to metabolically degrade disparate organic compounds, and thus obtain chemical energy for adenosine triphosphate (ATP) synthesis. This metabolism, called catabolism or biodegradation, is a major contribution to the Earth's carbon cycle (2). Microorganisms are increasingly used in engineered systems to biodegrade hazardous, xenobiotic compounds; this application is called bioremediation.

In the twentieth century, work on microbial biodegradation of organic molecules and on biocatalysis to yield desirable chemicals has gone hand in hand. The versatile enzymes found in biodegradation pathways are sometimes effective biocatalysts for specialty chemical synthesis. For example, recombinant *Escherichia coli* strains expressing naphthalene dioxygenase, the first enzyme in the biodegradation pathway of naphthalene and other polycyclic aromatic hydrocarbons, are used in an emerging commercial process for the environmentally benign production of the blue jean dye indigo (3). The process is based on the regiospecificity of naphthalene dioxygenase that dioxygenates indole at carbons 2 and 3 to yield an unstable dihydrodiol that spontaneously eliminates water. The resultant compound, indoxyl, spontaneously dimerizes to yield indigo. In a commercial biocatalytic process, indole is not supplied exogenously, because of its expense, but is generated in situ via overexpression of the enzymes functioning in the tryptophan biosynthetic pathway (4). A major impetus for bacterial indigo production is to preclude the formation of hazardous side products that are formed in the chemical process for indigo manufacture. Thus, biocatalysis and pollution prevention by biodegradation are intimately intertwined.

THE INFORMATION EXPLOSION AND THE WORLD WIDE WEB

New developments in commercial biosynthesis and environmental bioremediation are certain to emanate from a wider availability of information on microbial biocatalysis and biodegradation. Because the number of chemical substances subjected to microbial catabolism is large (10^6–10^7), a comparably large number of distinct enzymes and pathways must occur in nature. The vast majority have not yet been studied, in contrast to the better-characterized enzymes and pathways in common intermediary metabolism. Information on these reactions or pathways is not typically available in biochemistry or microbiology textbooks.

The increasing desire for this information has led to a virtual explosion of information published in journals, specialty books, and symposium proceedings over the past 2 years. There are also sources that contain extensive information but that are less readily available, for example, technical reports of government agencies such as the U.S. Environmental Protection Agency (EPA). The information retrieval problem is exacerbated by the different clientele for this information, including microbiologists, chemists, chemical engineers, and civil engineers. Most of these groups do not routinely survey the same publications and are accustomed to somewhat different presentations of the relevant data.

The rapid growth of the World Wide Web (WWW or Web) offers a way to overcome availability, outdatedness, and inconsistency problems because it can provide timely access to large amounts of appropriate, consistently formatted information. The Web builds on the success of the Internet, a collection of thousands of linked computer networks. Since the Internet's beginnings in the early 1970s, many scientists have used electronic mail as a very effective way to convey information. In January 1998, the Internet had approximately 30 million hosts (5), and its use is still growing rapidly.

In 1991, a client/server software system called Gopher was developed to allow individual client computers to access textual information on the computer networks of the Internet. Gopher, developed at the University of Minnesota, appears to be a hierarchical text database that allows the user to "go fer" (look for) Internet resources on various topics and display them on even a modest computer. The World Wide Web, developed shortly thereafter at the European Laboratory for Particle Physics, combines graphics with text and allows for links between related Internet resources. For example, clicking on a highlighted or underlined term transports the user to the designated Internet resource, which can be on a server housed anywhere in the world.

BIOCATALYSIS INFORMATION ON THE WORLD WIDE WEB

The Web information of most value to people involved in biocatalysis deals with chemical compounds, microorganisms, metabolic pathways, enzymes, and genes. Users can access each field of information separately, although in some cases there are hypertext links that facilitate finding related information in independent databases. There are commercial databases that cover biocatalysis, for example, focusing on the use of enzymes in organic synthesis. These databases are not freely available and thus are not linked to public Web databases and are not discussed here.

For each of the fields of chemical compounds, microorganisms, metabolic pathways, enzymes, and genes, Internet resources are listed with brief annotations in Table 1, along with their uniform resource locators (URLs). The enzyme and metabolism databases focus most directly on the biocatalysts that might be employed in organic synthesis or metabolic engineering. Most metabolism and enzyme databases, however, deal largely with the reactions common to most living things, often denoted as intermediary metabolism. Emerging applications in biocatalysis are less likely to involve such reactions, being more likely to be derived from metabolism that is carried out by a more limited range of organisms, typically bacteria. As an example, nitrile hydratase from *Rhodococcus* sp. is now used in the commercial synthesis of acrylamide from acrylonitrile. This is not a reaction carried out by animals, or even *E. coli*, but it nonetheless provides for clean catalysis with high product yield and purity. These more exotic enzymatic reactions are described in the first of the Web resources listed in Table 1, the University of Minnesota Biocatalysis/Biodegradation Database. It is described next in more detail.

THE UNIVERSITY OF MINNESOTA BIOCATALYSIS/BIODEGRADATION DATABASE

Introduction

The University of Minnesota Biocatalysis/Biodegradation Database (UM-BBD) first appeared on the World Wide Web (*http://www.labmed.umn.edu/umbbd/index.html*) on February 25, 1995, to meet the information needs of this distinct, diverse scientific community (6,7). It was designed to provide information on specialized nonintermediary microbial metabolism that is at the heart of scientific efforts in biocatalysis and biodegradation, in an easily accessible format, and to interface readily with Web-based databases of intermediary metabolism, such as the Kyoto Encyclopedia of Genes and Genomes (KEGG) (8). Given that there are 10^6–10^7 organic compounds subject to catabolism, and that their catabolism is largely nonintermediary and unique to selected microorganisms, the UM-BBD has a much larger information domain to cover than intermediary metabolism databases. At the same time, information contained in the UM-BBD has potentially the greatest impact. Hazardous waste treatment is a $200 billion industry worldwide, and bioremediation comprises 5–10% of that total. There is the potential for an expansion of this market with the development of new methods for metabolic and protein engineering. For example, researchers working on biodegradative pathways have begun to use metabolic engineering to develop new metabolic routes to handle particularly problematic pollutants (9). One example de-

Table 1. Useful Internet Resources for Microbial Biocatalysis

University of Minnesota Biocatalysis / Biodegradation Database

http: / / www.labmed.umn.edu / umbbd / index.html
The University of Minnesota Biocatalysis/Biodegradation Database (UM-BBD) provides well-organized information on microbial metabolism of primarily xenobiotic nonintermediary compounds. Its metabolic pathways include appropriate hypertext links to compound, reaction and pathway information in complementary Web databases. The UM-BBD also provides a list of useful Internet resources for microbial biotechnology, a few of which are described below.

Intermediary Metabolism

KEGG: Kyoto Encyclopedia of Genes and Genomes
http: / / www.genome.ad.jp / kegg / kegg.html
KEGG contains current knowledge of molecular and cellular biology in terms of information pathways of interacting genes or molecules and links individual components of the pathways with the gene catalogs produced by genome projects.

WIT
http: / / www.cme.msu.edu / WIT
WIT helps users reconstruct metabolism from complete bacterial genomic sequences.

Chemical Compounds

ChemFinder
http: / / chemfinder.camsoft.com /
ChemFinder is a database containing thousands of chemical compounds, including, for each compound, physical property data, chemical structures, and links to external information.

National Toxicology Program
http: / / ntp-server.niehs.nih.gov /
The NTP website contains information about the Program, Chemical Health & Safety Data, 2D and 3D chemical structures, and abstracts of Short-Term Toxicity and Long-Term Carcinogenesis studies.

Enzymes and Genes

Ligand Chemical Database for Enzyme Reactions
http: / / www.genome.ad.jp / dbget / ligand.html
LIGAND contains entries for enzymes and their substrates, products, and cofactors.

GenBank
http: / / www.ncbi.nlm.nih.gov / Web / Genbank / index.html
GenBank is the NIH genetic sequence database, an annotated collection of all publicly available DNA sequences.

Microorganisms

World Data Center for Microorganisms
http: / / wdcm.nig.ac.jp /
The WDCM provides a comprehensive directory of culture collections, databases on microbes and cell lines, and a getaway to biodiversity, molecular biology, and genome projects.

Bacterial Nomenclature Up-to-Date
http: / / www.gbf-braunschweig.de / DSMZ / bactnom / bactname.htm
This site, compiled by the DSMZ-Deutsche Sammlung von Mikroorganismen und Zellkuturen GmbH, includes all bacterial names that have been validly published since January 1, 1980, and nomenclatural changes which have been validly published since then.

Source: From Ref. 34.

scribes using a knowledge of existing biodegradative enzymes to design new pathways to metabolize polyhalogenated compounds (10). The UM-BBD could clearly stimulate similar projects by compiling and disseminating the crucial information needed to design such pathways de novo. We briefly describe the present status of the UM-BBD, including its format, use and users, peer acceptance, scientific advisory board, and future plans.

Format

In its first year of operation, the UM-BBD has evolved a standard format for reaction and compound pages. Both can link to many other internal and external information sources. One reaction page, for the first step in the biodegradation pathway for 1,2-dichloroethane, and a few of its many links, are shown in Figure 1. The Entrez Medline link (Fig. 1e) is especially interesting. The link back to the UM-BBD is prominent at the top. Also, when Entrez added a protein structure database, at our request they inserted the "Structure" link that is seen at the top of Figure 1e, so that our users would be only a click or two away from viewing and manipulating any available tertiary structure of the enzymes contained in the UM-BBD. Such a display for haloalkane dehalogenase is shown in Figure 1b. The

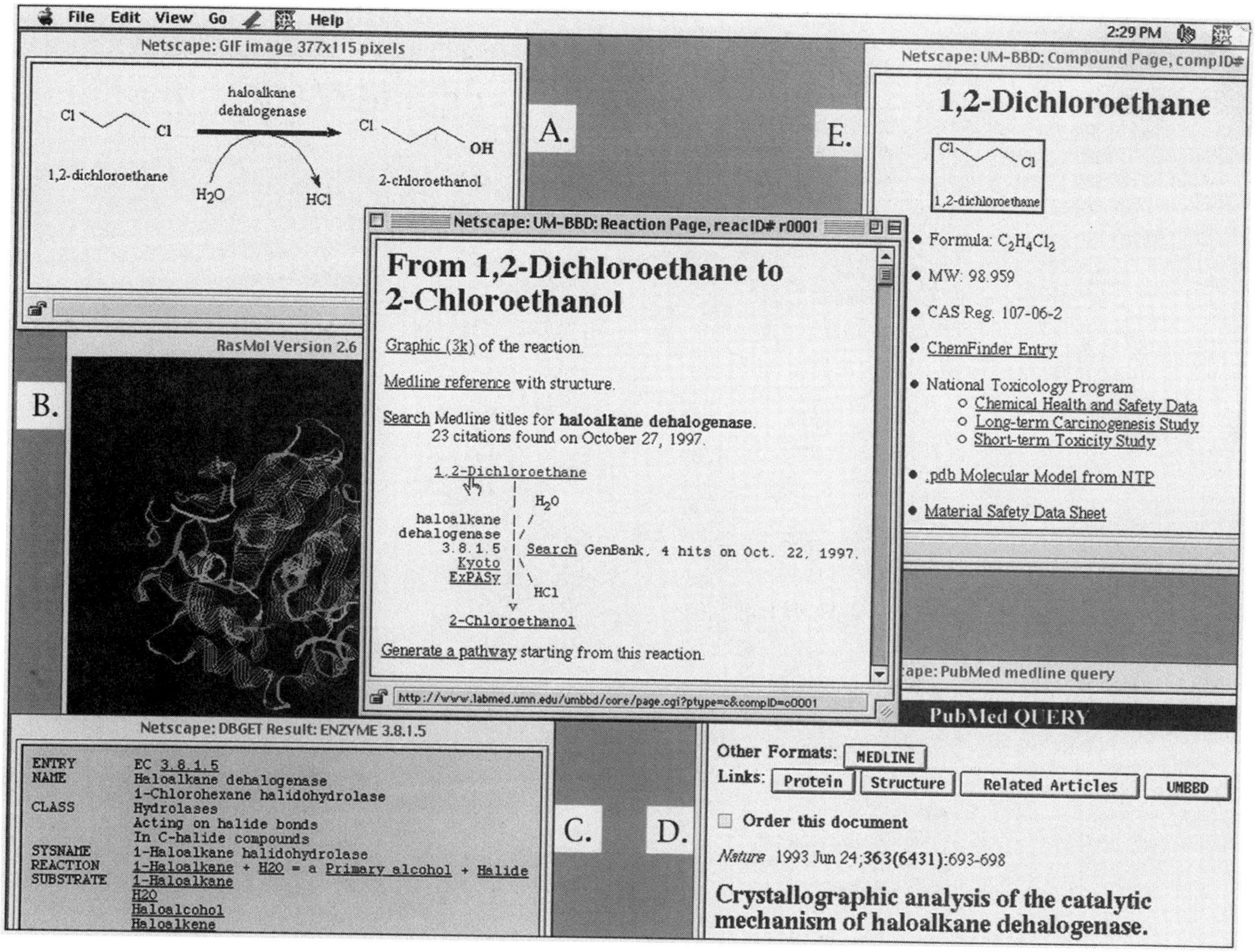

Figure 1. Selected windows of information from the UM-BBD, as viewed on a computer screen, representing the range of information available using the first reaction of 1,2-dichloroethane catabolism as an example. The respective windows are organized with the reaction page at the center and links from **a** to **e** at the periphery, starting with **a** in the upper-left corner and going counterclockwise. The six windows are (*center*) the haloalkane dehalogenase reaction page, (**a**) a reaction GIF showing the substrate and product structures, (**b**) a RasMol depiction of the haloalkane dehalogenase structure, (**c**) the enzyme page from the Kyoto Ligand Chemical Database, (**d**) the Entrez/Medline report on a selected paper on haloalkane dehalogenase, and (**e**) the 1,2-dichloroethane compound page.

UM-BBD now includes links to graphics of enzyme reaction mechanisms on selected reaction pages. An index to graphics helps users find them (11). The usefulness of a database is a function both of the information it contains and the access users have to this information.

Initially, the only way to access UM-BBD information was through browsing lists of pathways, reactions, and compounds. As the UM-BBD grew, browsing became an increasingly less convenient mode of access. The ability to search the database was then added (12). At present one can search for compounds by full or partial name or synonym (e.g., toluene or methyl benzene), CAS registry number (e.g., 108-88-3), or formula (e.g., C7H8), and for enzymes by full or partial name (e.g., urease) or full or partial EC number (e.g., 3.5.1.5 or 3.5.1). This search option will continue to be developed with input from the UM-BBD user community and may be expanded in the future to include chemical substructure searches.

Use and Users

Two months after the UM-BBD initially appeared on the Web, we began collecting use statistics. Usage has grown continuously; use statistics by Internet domain for 1997 are shown in Table 2. Currently, the UM-BBD has approximately 80,000 accesses per month (Fig. 2). U.S. educational (30%) and industrial (20%) users together make up half of all present use (Table 2). UM-BBD users are invited to “sign our guestbook” (i.e., voluntarily complete a seven-item electronic form). Those who sign are given the option

Table 2. UM-BBD Use Statistics by Domain from January 1 to December 31, 1997

#reqs	%bytes	Domain
187790	29.54%	.edu (USA Educational)
166052	19.66%	.com (USA Commercial)
87740	18.86%	[unresolved numerical addresses]
39050	8.13%	.net (Network)
12560	2.70%	.ca (Canada)
10008	2.40%	.de (Germany)
9527	2.23%	.uk (United Kingdom)
5944	1.56%	.gov (USA Government)
13285	1.45%	[Unknown]
4193	1.14%	.jp (Japan)
4960	1.07%	.au (Australia)
3950	0.97%	.fr (France)
3494	0.94%	.nl (Netherlands)
3718	0.68%	.se (Sweden)
2212	0.60%	.es (Spain)
2352	0.51%	.dk (Denmark)
1954	0.49%	.ch (Switzerland)
1948	0.46%	.us (United States)
2097	0.45%	.it (Italy)
1907	0.42%	.kr (South Korea)
1621	0.36%	.org (Non-Profit Making Organisations)
1473	0.34%	.mx (Mexico)
2210	0.33%	.il (Israel)
1332	0.31%	.be (Belgium)
1224	0.28%	.br (Brazil)
1014	0.26%	.fi (Finland)
1388	0.24%	.tw (Taiwan)
949	0.23%	.cz (Czech Republic)
925	0.23%	.no (Norway)
1022	0.22%	.mil (USA Military)
761	0.17%	.at (Austria)
672	0.14%	.sg (Singapore)
507	0.14%	.za (South Africa)
768	0.14%	.nz (New Zealand)
526	0.14%	.gr (Greece)
596	0.13%	.ar (Argentina)
467	0.11%	.pt (Portugal)
422	0.11%	.ru (Russian Federation)
468	0.11%	.my (Malaysia)
390	0.11%	.ie (Ireland)
524	0.11%	.co (Colombia)
406	0.10%	.th (Thailand)
436	0.09%	.pl (Poland)
381	0.09%	.su (Former USSR)
317	0.09%	.id (Indonesia)
284	0.08%	.cl (Chile)
331	0.08%	.hu (Hungary)
259	0.05%	.ee (Estonia)
131	0.05%	.hr (Croatia)
215	0.04%	.si (Slovenia)
112	0.04%	.sk (Slovak Republic)
258	0.04%	.tr (Turkey)
172	0.03%	.hk (Hong Kong)
185	0.03%	.in (India)
125	0.03%	.cn (China)
131	0.02%	.cr (Costa Rica)
78	0.02%	.ro (Romania)
124	0.02%	.lt (Lithuania)
70	0.02%	.bh (Bahrain)
64	0.02%	.pe (Peru)
104	0.01%	.ve (Venezuela)
71	0.01%	.uy (Uruguay)
69	0.01%	.eg (Egypt)
77	0.01%	.arpa (Old style Arpanet)
29	0.01%	.jo (Jordan)
57	0.01%	.ae (United Arab Emirates)
63	0.01%	.ph (Philippines)
31		.is (Iceland)
36		.yu (Yugoslavia)
18		.jm (Jamaica)
25		.pa (Panama)
23		.lb (Lebanon)
10		.do (Dominican Republic)
11		.lu (Luxembourg)
16		.pr (Puerto Rico)
11		.bm (Bermuda)
20		.ec (Ecuador)
10		.ua (Ukraine)
5		.zw (Zimbabwe)
5		.az (Azerbaidjan)
4		.md (Moldavia)
8		.gy (Guyana)
3		.am (Armenia)
8		.by (Belarus)
5		.mk (Macedonia)
9		.kw (Kuwait)
6		.bg (Bulgaria)
7		.gb (Great Britain)
9		.mt (Malta)
1		.dm (Dominica)
5		.gt (Guatemala)
4		.dz (Algeria)
9		.et (Ethiopia)
6		.cy (Cyprus)
7		.ge (Georgia)
6		.pg (Papua New Guinea)
2		.bo (Bolivia)
2		.pk (Pakistan)
2		.mu (Mauritius)
2		.lv (Latvia)
1		.om (Oman)

UM-BBD, University of Minnesota Biocatalysis/Biodegradation Database.

to regularly receive e-mail progress reports, and most choose to do so. This e-mail user list presently has about 400 members, who are surveyed on an annual basis. Users have recommended pathways and features to us, and we have acted on some of their suggestions: A pathway for degradation of a chlorinated aliphatic compound and graphical pathway maps were added in part because of user requests.

Connections

When the UM-BBD was first installed on a server in February 1995, it linked to several Internet resources. Links to others have been added since then. Most of the Web databases the UM-BBD links to have been contacted and collaborate on cross-linkage where possible. For example, the UM-BBD now has reciprocal links to Entrez PubMed (13), Entrez Nucleotide (14), Entrez Protein (15), ChemFinder (16), and Ligand (17), and others are under discussion. Each such interconnection is usually the result of individualized programming. The UM-BBD links to others

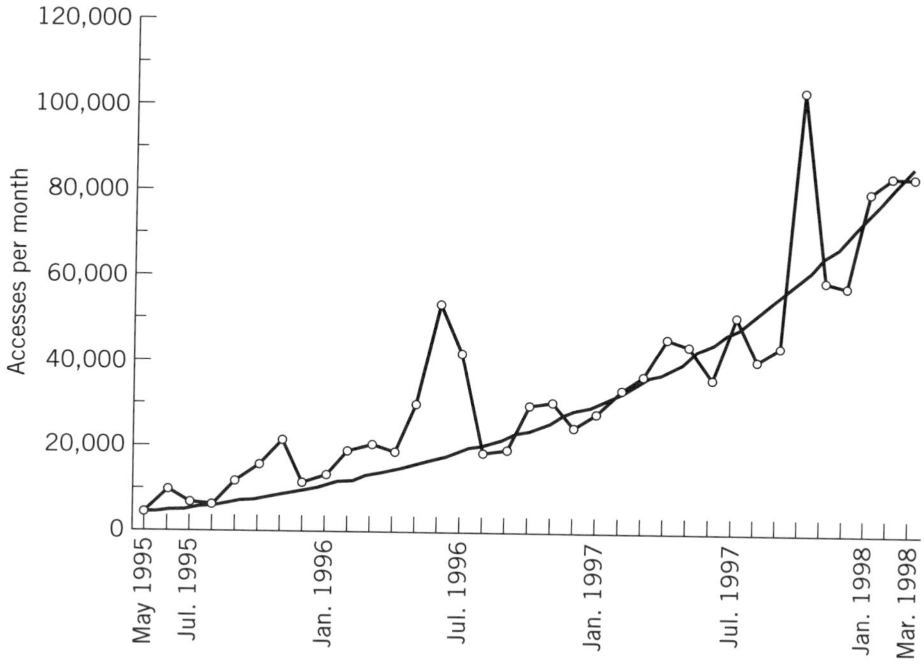

Figure 2. UM-BBD access statistics from May 1995 to March 1998.

in separate text lines, or, especially for reaction information, as links inserted in the text-formatted reaction (see Fig. 1, center).

Unique among present Web-based metabolic databases such as KEGG (8), the UM-BBD specializes in a more broadly based set of metabolic reactions. Because 10 million organic compounds are known and most are enzymically biodegraded, the potential range of the UM-BBD is enormous. These reactions are under active study now and will be in the foreseeable future, and the ability to search for current information in remote databases, as opposed to more static links to individual documents, is very important to UM-BBD users.

At present, the UM-BBD provides dynamic search links to EXTOXNET (18), Entrez/MEDLINE (13), and Entrez/GenBank (14). Although more difficult to construct and maintain than links to individual documents (in the first 6 months of 1996, each of these links was reconstructed at least once because the Web sites changed search format), they allow UM-BBD users to obtain the very latest information installed on these servers. Thus, a search is installed in preference to an individual link wherever possible, even if it returns only one document at present. Tomorrow or next week additional items may appear, and when they do UM-BBD users will have access to them.

UM-BBD search links include information on the number of hits that were returned when the search was last tested, the date of that search, and, if possible, the exact records being searched and terms that are being searched for. For example, Entrez/MEDLINE searches may be carried out for MEDLINE titles and abstracts, or if this produces too many hits, they may be restricted to titles alone. Entrez/MEDLINE and GenBank searches require custom formulation to return the best results because of the ambiguities and complexities of chemical and biochemical nomenclature and database search syntax.

Peer Acceptance

As mentioned earlier, the UM-BBD links to other metabolic databases. It also provides users with links to relevant information in several other biological and chemical Internet databases. The five prestigious international Web databases that it links to most closely—ChemFinder (16), Ligand (17), Entrez/Medline (13), Entrez/Protein (15), and Entrez/GenBank (14)—all link back to it. The UM-BBD has also been selected to appear on many peer-reviewed scientific Web referral lists, including Internet Chemical Resources (19), the American Society of Microbiology Biology Resources List (20), and BioTech: Biochemistry, Biophysics, and Molecular Biology (21). In addition, it was the subject of an article in the American Chemical Society's publication, *Chemical & Engineering News* (22), and in *Science* (23), and articles on it were solicited by the *Society of Industrial Microbiologists News* (6) and the *Journal of Microbial Methods* (7).

Scientific Advisory Board

During the first year of operation, a world-renowned scientific advisory board was established, including members in three countries from educational institutions, commercial companies, and government agencies: Peter Chapman (EPA, Florida); I. C. Gunsalus (EPA, Florida); Andreas Kiener (Lonza AG, Switzerland); Thomas Leisinger (ETH Zurich, Switzerland); and Jack T. Trevors (Guelph University, Canada). Most recently, Alfred Spormann (Stanford University) has joined the UM-BBD advisory board.

Growth and Future Developments

The UM-BBD began with only 4 pathways in February 25, 1995 and grew to 13 pathways by September 15, 1996, 25 pathways by February 20, 1997, and 59 pathways by April

21, 1998. However, with 10 million compounds in the biosphere, it will never, and is not meant to, contain information on the biodegradation of more than a representative fraction of these compounds. It was developed to contain reactions and pathways that span the diversity of organic functional groups and microbial metabolism. The next step in its development is to use this representative information to develop ways to predict microbial catabolic metabolism for compounds that the UM-BBD does not contain.

CONCLUSIONS

Biocatalysis will expand in scope as a knowledge of microbial catabolism increases. Most of the more than 10 million organic compounds are thought to be biodegradable, providing a virtually limitless potential for expansion of known microbial enzymatic reaction types. Additionally, information is rapidly emerging from the more than 60 bacterial genome sequencing projects completed or ongoing. This will lead to a new era of microbial metabolism studies, building on the extensive data on intermediary metabolism. These studies will increasingly reveal the vast catabolic capabilities of bacteria from diverse environments and taxonomic classes.

Research advances will create a need for more people trained in microbial genomics, metabolic reconstruction, and enzymology. The UM-BBD, and associated courses, can contribute to training the next generation of microbial biocatalysis specialists. The UM-BBD has been used instructionally at several colleges and universities, including the University of Minnesota (24–27), Iowa State University (28), Pomona College (29), the University of Waterloo (30), the University of Basel (31), and Hebrew University in Rehovot, Israel (32). A University of Minnesota course on biocatalysis and biodegradation is being offered completely over the Internet, using the UM-BBD as its focal point (24). The biocatalysis and biodegradation course instructs students on the general principles of microbial environmental biotechnology and microbial catabolic reactions. Using these fundamentals, students peruse the UM-BBD and related World Wide Web databases to research specific microbial transformation reactions. Student assignments are transferred to the instructors for assessment via electronic mail and the Web. This is a venue for advanced college-level instruction that is increasingly becoming available to students worldwide. It reflects a fundamental change in university education: for 6,000 years students were required to come to the scrolls, books, and personal contact with scholars that allowed learning to advance (33). Now, people can be at remote sites and increasingly access the accumulated knowledge of humanity via electronic means. The UM-BBD is one small part of this profound change in human information transfer and learning.

BIBLIOGRAPHY

1. E. Buchner, *Ber. Dtsch. Chem. Ges.* **30**, 117–124 (1897).
2. D.T. Gibson, *J. Ind. Microbiol.* **12**, 1–12 (1993).
3. B.D. Ensley, B.J. Ratzkin, T.D. Osslund, M.J. Simon, L.P. Wackett, and D.T. Gibson, *Science* **222**, 167–169 (1983).
4. H. Bialy, *Nat. Biotechnol.* **15**, 110 (1997).
5. R.H. Zakon, Robert H'obbes' "Hobbes' Internet Timeline" (May 1998): *http://info.isoc.org/guest/zakon/Internet/History/HIT.html*
6. L.B.M. Ellis and L.P. Wackett, *Soc. Ind. Microb. News* **45**, 167–173 (1995).
7. L.P. Wackett and L.B.M. Ellis, *J. Microbiol. Methods* **25**, 91–93 (1996).
8. KEGG: Kyoto Encyclopedia of Genes and Genomes: *http://www.genome.ad.jp/kegg/kegg.html*
9. K.N. Timmis, R.J. Steffan, and R. Unterman, *Annu. Rev. Microbiol.* **48**, 525–557 (1994).
10. L.P. Wackett, M.J. Sadowsky, L.M. Newman, H.-G. Hur, and S. Li, *Nature* **368**, 627–629 (1994).
11. S. Stephens, "UM-BBD Graphic Index" (May, 1998): *http://www.labmed.umn.edu/umbbd/search/graph_BBD.html*
12. R. McLeish and D. Hershberger, "UM-BBD Search Page" (May 1998): *http://www.labmed.umn.edu/umbbd/search/index.html*
13. National Center for Biotechnology Information (May 1998) Entrez/PubMed: *http://www.ncbi.nlm.nih.gov/PubMed/index.html*
14. National Center for Biotechnology Information (May 1998) Entrez/Nucleotide: *http://www.ncbi.nlm.nih.gov/Entrez/nucleotide.html*
15. National Center for Biotechnology Information (May 1998) Entrez/Protein: *http://www.ncbi.nlm.nih.gov/Entrez/protein.html*
16. CambridgeSoft (May 1998) ChemFinder: *http://chemfinder.camsoft.com/*
17. Kyoto University Ligand Chemical Database (May 1998): *http://www.genome.ad.jp/dbget-bin/www_bfind?ligand*
18. EXTOXNET: Extension Toxicology Network: *http://ace.ace.orst.edu/info/extoxnet/*
19. Internet Chemical Resources: *http://www.rpi.edu:80/dept/chem/cheminfo/chemres.html*
20. American Society for Microbiology—other biology sources: *http://www.asmusa.org/others.htm*
21. BioTech: Biochemistry, Biophysics, and Molecular Biology: *http://golgi.harvard.edu:80/biopages/biochem.html*
22. J. Kreiger, *Chem. & Eng. News*, p. 36–37, September 11 (1995).
23. D. Voss, *Science* **277**, 53 (1997).
24. Biocatalysis and Biodegradation, University of Minnesota: *http://biosci.cbs.umn.edu/class/bioc/5309/*
25. Advanced Biochemistry I, University of Minnesota: *http://biosci.cbs.umn.edu/class/bioc/8001/*
26. Ecological Biochemistry, University of Minnesota: *http://biosci.cbs.umn.edu/class/bioc/5301/*
27. M.C. Flickinger, *Applied Microbial Biochemistry*, Univ. of Minnesota, St. Paul, Minn. (1997).
28. Biotechnology Instructional Component, Iowa State University (1998): *http://www.lib.iastate.edu/biotech/biotech2/instr_bt.html*
29. Environmental Chemistry Sites, Pomona College, California (May 1998): *http://pages.pomona.edu/~wsteinmetz/envchem.htm*
30. Environmental Microbiology, University of Waterloo (May 1998): *http://bordeaux.uwaterloo.ca/biol447new/wwwsitesbiodegradation.htm*

31. Chemistry Sites, University of Basel, Switzerland (May 1998): *http://www.chemie.unibas.ch/Sites/Chemistry.html*

32. R. Mandelbaum, *Environmental Microbiology*, Hebrew University, Rehovot, Israel (1998).

33. E. Noam, *Science* **270**, 247–249 (1995).

34. L.B.M. Ellis and L.P. Wackett, *Nat. Biotechnol.* **15**, 1406 (1997).

See also ENZYMES, DIRECTED EVOLUTION; ENZYMES, EXTREMELY THERMOSTABLE; LACTONES, BIOCATALYTIC SYNTHESIS; LACTONOHYDROLASE; PHENYLALANINE DEHYDROGENASE.

BIOCATALYSIS, OPTICAL RESOLUTION. See OPTICALLY ACTIVE 1,2-DIOLS, MICROBIAL PRODUCTION BY STEREOINVERSION; OPTICAL RESOLUTION, BIOCATALYSIS.

BIOENERGETICS OF MICROBIAL GROWTH

J.J. HEIJNEN
Delft University of Technology
DelftThe Netherlands

KEY WORDS

Biomass yield
Conservation constraints
Heat production
Maintenance
Maximal growth rate
Thermodynamics

OUTLINE

INTRODUCTION

The growth of microorganisms occurs within a wide range of pHs and temperatures, and on a wide variety of nutrients. Figure 1 shows a typical batch experiment where a substrate (starting at concentration C_{so}) is converted by a microorganism (C_{xo} at $t = 0$). The microorganism grows exponentially at a specific growth rate μ^{max} and with yield Y^{m}_{DX}. After depletion of the substrate, which is characterized by the substrate affinity constant K_s and a threshold concentration, the biomass concentration reaches $C_x = C_{xo} + Y^{m}_{DX}C_{so}$. Subsequently the biomass concentration decreases due to maintenance and/or biomass decay, which is characterized by the maintenance coefficient m_D (or the decay coefficient k_d). The relevant substrate always acts as electron donor, and therefore it is proper to define the biomass yield and maintenance on donor (D).

In the design of processes with growing microorganisms (fermentation processes and biological waste-treatment processes) the key parameters that need to be considered are the maximal biomass yield on substrate (Y^{m}_{DX}), the substrate maintenance coefficient m_D, the maximal growth rate (μ_{max}), and the substrate affinity constant (K_s). These four key parameters are sufficient to describe growth of

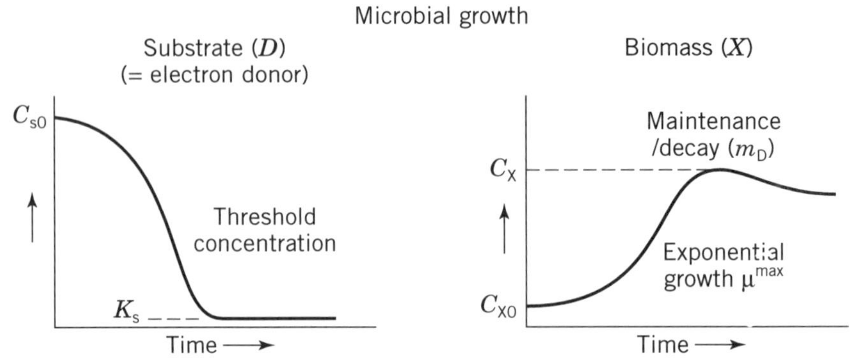

Four basic parameters

Units	Range
Stoichiometric	
Y^{m}_{DX} [C-mol X/(C)-mol D]	0.010–0.7
m_D [(C)-mol D/C-mol X/h]	0.01–3
Kinetic	
μ^{max} (h^{-1})	0.005–2
K_s (mol D/L)	10^{-6}–10^{-3}

Figure 1. A typical microbial growth batch profile of biomass concentration and substrate concentration.

microorganisms in a standard mathematical model (1–4). A practical problem is, however, that the values of these parameters can vary by more than two orders of magnitude for different electron donors or acceptors used by different microorganisms, as indicated in Figure 1. It should be realized that conventionally stoichiometric parameters are expressed with C-mol of biomass, C-mol of electron donor for organic, and mol of donor for inorganic donors (C-mol X/(C)-mol D). It is therefore of interest to provide a general method to estimate values of these parameters for any chemotrophic growth system. Such methods have been provided by, for example, Battley (5), Roels (1), and Westerhoff (6). Recently (2,9) these methods have been critically evaluated with respect to general applicability and internal consistency. It was concluded (2,7,8) that none of these methods was satisfying. However a new method was proposed that is generally applicable and lacks the mentioned problems (2,8). Further, it should be recognized that in growth processes not only biomass production (r_x in C-mol biomass per m^3 reactor/h) and electron donor (substrate) consumption r_D in C-mol substrate (for carbon compounds), or mol substrate (for noncarbon compounds) per m^3 reactor/h are important. Also, the other conversions, such as O_2 consumption, N source consumption, heat production, and CO_2 production are highly relevant for the process design to calculate, for example, the required O_2 and heat transfer. Clearly, the full stoichiometry of the growth process should also be calculated and methods to achieve this are of major interest.

A STANDARD DESCRIPTION OF MICROBIAL GROWTH STOICHIOMETRY

The stoichiometry of microbial growth is most easily understood from Figure 2. Figure 2a introduces the biomass composition of 1 C-mol biomass (the ash-free organic fraction). The composition shown is fairly typical and is taken from Roels (1). One C-mol ash-free organic biomass is the amount of organic dry biomass that contains 12 g of carbon. The indicated biomass organic fraction corresponds to an elemental composition of 48.8% carbon, 7.3% hydrogen, 32.5% oxygen, and 11.4% nitrogen (w/w).

In practice, total dry biomass, which includes the organic fraction and the ash fraction (S, P, K, Mg, etc.), is measured. In general, the organic and ash fraction are obtained by combusting the organic biomass at 500 to 600 °C and weighing the ashes. Recently Battley (9) has indicated that this simple procedure underestimates the real organic biomass weight by 5 to 6%. This is due to the formation of P, S, and metal oxides in the ash during combustion, whereas such oxides are not present in the dry biomass. The composition formula follows directly from the elemental analysis of the biomass. In Figure 2a only the four major elements (C, H, O, N) are shown; however, it is straightforward to include P, S, and metals such as K or Mg in this composition formula, and also in the stoichiometric/energetic calculation. Figure 2a also shows that in the formation of biomass for all chemotrophic growth systems a C source, N source, H_2O, CO_2, and H^+ are always involved. These five compounds provide the building elements for making biomass. This is also called anabolism. For heterotrophic organisms the C source is organic; for autotrophic organisms the C source is CO_2.

Although it is possible to establish a stoichiometrically correct description to make biomass from these five building compounds, it is easily shown that this is not acceptable from the point of view of the second law of thermodynamics. It has been calculated that the Gibbs energy of such a hypothetical reaction, depending on the C source used, is often positive (7), although sometimes small negative values can also be calculated (5). In addition, it is

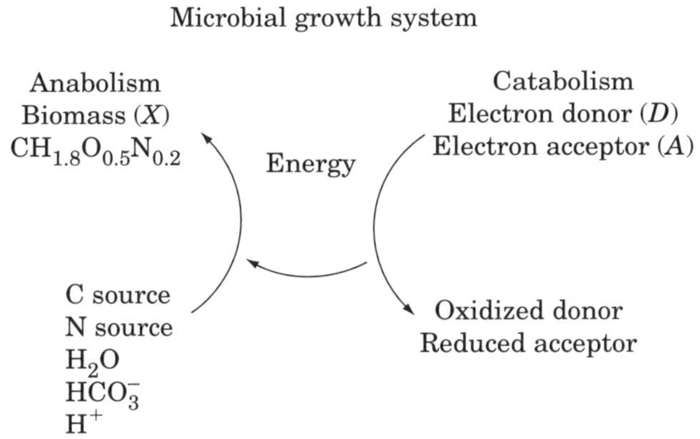

(a)

Relevant stoichiometry

$$-\frac{1}{Y_{DX}}\text{(C-)mol electron donor} - (\ldots)\text{ N-source}$$
$$-\frac{1}{Y_{AX}}\text{ mol electron acceptor} + 1\text{ C-mol biomass}$$
$$+\frac{1}{Y_{QX}}\text{(heat)} + \frac{1}{Y_{GX}}\text{ Gibbs energy}$$
$$+ (\ldots)\,H_2O + (\ldots)\,HCO_3^+ + (\ldots)\,H^+$$

Many different possibilities

Donor	Acceptor	C-source
Organic	Organic	Organic
Inorganic	Inorganic	Inorganic

Y_{DX}0.01 – 0.7 C-mol$X$/(C)mol donor

(b)

Figure 2. (**a**) System definition of microbial growth. (**b**) Macrochemical reaction equation of microbial growth.

known that to convert the five compounds into biomass, microorganisms use a large amount of biochemical energy in the form of ATP (10). Clearly the production of biomass from the five building compounds requires input of large quantities of Gibbs energy. The amount of energy needed to make biomass depends on the type of C source used. Intuitively, one expects that making 1 C-mol biomass from CO_2 requires more Gibbs energy than making 1 C-mol biomass from an organic compound. A quantitative relation for this energy need is presented later (equations 2, 3a, and 3b). The required energy, which must be taken as Gibbs energy and not as enthalpy, is delivered by a redox reaction between an electron donor and an electron acceptor. This redox reaction is called catabolism (Fig. 2a). Examples are the aerobic combustion of glucose ($C_6H_{12}O_6 + 6O_2 \rightarrow 6HCO_3^- + 6H^+$) and the anaerobic formation of ethanol from glucose ($C_6H_{12}O_6 + 2H_2O \rightarrow 2HCO_3^- + 2H^+ + 2C_2H_5OH$). Obtaining the required Gibbs energy is as essential for micro-organisms as it is for higher organisms, and even for human society. Therefore it should not be surprising that during evolution a wide diversity of microorganisms developed that are mainly different in the applied redox reaction for catabolism to obtain Gibbs energy (Fig. 2b). Electron donor or acceptor couples can be organic and inorganic compounds. This microbial variety in catabolic possibilities for generating Gibbs energy has led to the use of a classification system for naming microorganisms (Table 1). This system is understandably based on the source of Gibbs energy (light or chemical energy), the source of electron donor (inorganic or organic), and the source of biomass carbon (CO_2 or organic).

In addition, microorganisms may employ a wide variety of electron acceptors, as reflected in their class names. These class names are related to the electron acceptor used in catabolism (O_2, aerobic; NO_3^-, denitrification; SO_4^{2-}, sulphate reduction) fermentation (absence of external electron acceptor), or to the product of the catabolic reaction (CH_4, methanogenic; acetate, acetogenic; H_2S, sulphidogenic, etc.). The C source also functions often as electron donor, except in autotrophic microorganisms, where the C source is CO_2. For example, a microorganism growing aerobically in the dark on H_2S as the electron donor (inorganic compound) using CO_2 as the C source is called an aerobic chemolithoautotrophic organism. In summary, in each realistic chemotrophic microbial growth system there must be present the five compounds of anabolism and an electron donor/acceptor combination for catabolism.

These considerations bring us then to Figure 2b, which shows the macrochemical reaction equation containing all the stoichiometric information of the growth process. The macrochemical equation of Figure 2b should not be considered a mathematical equation but is a chemical reaction where substrates and products have negative and positive stoichiometric coefficients, respectively. Therefore for a = 0, sign is absent. In addition, the stoichiometric involvement of enthalpy and Gibbs energy is expressed in their respective stoichiometric coefficients (Y_{QX}, and Y_{GX}, which have units of C-mol X/kJ). The macrochemical reaction equation is therefore a compact, but exact, form of notation of the relevant stoichiometry of growth. This macrochemical equation shows that for the formation of +1 C-mol of biomass an amount of $-1/Y_{DX}$ of electron donor is required. The minus sign shows that the electron donor is consumed. Y_{DX} is in C-mol biomass per C-mol electron donor (in case of an organic donor) or per mol donor (in case of an inorganic donor). Its units are written as C-mol X/(C) mol D. An amount of $-1/Y_{AX}$ mol electron acceptor is consumed (minus sign) per 1 C-mol biomass produced and, in addition, $1/Y_{QX}$ kJ of heat and $1/Y_{GX}$ kJ of Gibbs energy are involved in the production of 1 C-mol of biomass. $1/Y_{QX}$ and $1/Y_{GX}$ are found as the conventionally calculated enthalpy of reaction and Gibbs energy of reaction of the macrochemical reaction equation, which produces 1 C-mol of biomass. Finally, certain amounts of H_2O, CO_2, (or HCO_3^-), H^+, and N source are involved. It is important that in each macrochemical reaction equation in which biomass is grown, HCO_3^-, H_2O, H^+, and N source be present. The differences between different organisms occur mostly in the electron acceptor/donor combinations used. The N source is often NH_4^+ and sometimes NO_3^-, N_2, or something else. The most important point in stoichiometry is to recognize that it is nearly always sufficient to measure one stoichio-

Table 1. Microbial Classification System

Source of Gibbs energy	Source of electron donor	C-source
Light (phototrophic)	Inorganic (lithotrophic)	CO_2 (autotrophic)
Chemical (chemotrophic)	Organic (organotrophic)	Organic (heterotrophic)

metric coefficient, that is, Y_{DX}, which is the traditional biomass yield on substrate (equal to carbon source and electron donor). All the other stoichiometric coefficients then follow from the so-called conservation equations (elements, electric charge, and enthalpy) (Example 1a) and the Gibbs energy balance (Example 1b).

EXAMPLE 1a

Calculation of stoichiometric coefficients in the macrochemical equation

Consider the aerobic growth of *Pseudomonas oxalaticus* on oxalate using NH_4^+ as the N source. The relevant chemical compounds in this growth system are the five compounds [biomass ($CH_{1.8}O_{0.5}N_{0.2}$), NH_4^+, HCO_3^-, H^+, H_2O], the electron donor oxalate ($C_2O_4^{2-}$), and the electron acceptor O_2. In total there are seven compounds and four elements (C, H, O, N). The conversion rates of these compounds are mathematically related by the conservation relations of C, H, O, N, and electric charge. In total there are five independent relations. This means that seven conversion rates are related by five conservation equations, and that the measurement of two rates (e.g., biomass production r_x and consumption of the electron donor oxalate r_D), which is equivalent to the measurement of $Y_{DX} = r_X/-r_D$, allows the calculation of all other yields.

Suppose that from measurement the biomass yield Y_{DX} is found to be +0.086 C-mol biomass produced per C-mol oxalate consumed. The proper macrochemical reaction equation can be written in a general form, without knowing all the stoichiometric coefficients but one (+1 for biomass), as

$$fC_2O_4^{2-} + aNH_4^+ + bH^+ + cO_2 + dH_2O + 1CH_{1.8}O_{0.5}N_{0.2} + eHCO_3^-$$

The following conservation equations can now be written:

C conservation	$2f + 1 + e = 0$
H conservation	$4a + b + 2d + 1.8 + e = 0$
O conservation	$4f + 2c + d + 0.5 + 3e = 0$
N conservation	$a + 0.2 = 0$
Charge conservation	$-2f + a + b - e = 0$

Clearly there are six unknown stoichiometric coefficients (a–f) that are related by five conservation equations. (Biomass has been assigned a convenient, yet arbitrary coefficient +1.) Having one measured coefficient allows the calculation of all other coefficients. Y_{DX} was measured as 0.086. This means that 1/0.086 = 11.63 C-mol oxalate are consumed to produce 1 C-mol biomass. The previously defined macrochemical equation contains f mol of oxalate, which was two carbon atoms. The stoichiometric coefficient f therefore has the value −11,63/2 = −5.815 (remember the minus sign). Using this f value and the five conservation equations, one can calculate the whole chemical growth stoichiometry. The result is

$$-5.815C_2O_4^{2-} - 0.2NH_4^+ - 0.8H^+ - 1.857O_2 - 5.42H_2O + 1CH_{1.8}O_{0.5}N_{0.2} + 10.63HCO_3^-$$

All the different biomass yields can be read from this reaction equation; thus, $Y_{AX} = 1/1.857 = 0.538$ C-mol biomass/mol O_2 or $Y_{CX} = 1/10.63 = 0.094$ C-mol biomass per mol CO_2.

In the Example 1a, only the chemical stoichiometry was calculated. However, there are two additional biomass yields of interest that relate the heat production and Gibbs energy dissipation occurring during the growth process to biomass production. These yields can be simply calculated if the full chemical stoichiometry is known by using tabulated ΔH_f^0 and ΔG_{f-}^{01} values (at pH = 7 and standard conditions) and calculating the enthalpy and Gibbs energy of reaction (Example 1b).

Table 2 contains all the required thermodynamic information as taken from Thauer et al. (11). The values for biomass are taken from Roels (1). Although there is some discussion about the value of ΔG_f^{01} for biomass, its value is not very important in thermodynamic calculations, as shown by Heijnen (3).

EXAMPLE 1b

Calculation of the yield of biomass on enthalpy and Gibbs energy (Y_{QX} and Y_{GX})

The chemical stoichiometry from Example 1a and the appropriate ΔH_f^0 and ΔG_f^{01} values from Table 2 can be used to obtain the heat (enthalpy) and Gibbs energy of reaction.

The enthalpy of reaction, using ΔH_f^0 from Table 2, is calculated as

$$(10.63)(-692) + 1(-91) - (5.42)(-286) - (1.857)(0) - (0.8)(0) - (0.2)(-133) - (5.815)(-824) = -1078.7 \text{ kJ}$$

For the Gibbs energy of reaction using ΔG_f^{01} values there follows a value of −1052.4 kJ. Because in the macrochemical reaction 1 C-mol of biomass is produced, this means that for each 1 C-mol biomass produced there is a heat production of 1078.7 kJ and a Gibbs energy dissipation of 1052.4 kJ, showing that $Y_{QX} = 1/1078.7 = 0.00093$ C-mol biomass produced per kilojoule heat produced and that $Y_{GX} = 1/1052.4 = 0.0095$ C-mol biomass produced per kilojoule of Gibbs energy dissipated. The complete chemical and energetic stoichiometry now can be written as

$$-5.815C_2O_4^{2-} - 0.2NH_4^+ - 0.8H^+ - 1.857O_2 - 5.42H_2O + 1CH_{1.8}O_{0.5}N_{0.2} + 10.63HCO_3^- + 1078.7 \text{ kJ heat} + 1052.4 \text{ kJ Gibbs energy}$$

Example 1 shows that the complete chemical and energetic stoichiometry of microbial growth can be calculated from one measured yield using conservation equations and the Gibbs energy and enthalpy balance (elements, charge,

Table 2. Standard Gibbs Energy and Ethalpy of Formation

Compound name	Composition	ΔG_f^{01} (kJ/mol)	ΔH_f (kJ/mol)
Biomass	$CH_{1.8}O_{0.5}N_{0.2}$	−67	−91
Water	H_2O	−237.18	−286
Bicarbonate	HCO_3^-	−586.85	−692
CO_2 (g)	CO_2	−394.359	−394.1
Ammonium	NH_4^+	−79.37	−133
Proton	H^+	−39.87	0
O_2 (g)	O_2	0	0
Oxalate^{2-}	$C_2O_4^{2-}$	−674.04	−824
Carbon monoxide	CO	−137.15	−111
Formate	CHO_2^-	−335	−410
Glyoxylate$^-$	$C_2O_3H^-$	−468.6	—
Tartrate^{2-}	$C_4H_4O_6^{2-}$	−1,010	—
Malonate^{2-}	$C_3H_2O_4^{2-}$	−700	—
Fumarate^{2-}	$C_4H_2O_4^{2-}$	−604.21	−777
Malate^{2-}	$C_4H_4O_5^{2-}$	−845.08	−843
Citrate^{3-}	$C_6H_5O_7^{3-}$	−1,168.34	−1,515
Pyruvate$^-$	$C_3H_3O_3^-$	−474.63	−596
Succinate^{2-}	$C_4H_4O_4^{2-}$	−690.23	−909
Gluconate$^-$	$C_6H_{11}O_7^-$	−1,154	—
Formaldehyde	CH_2O	−130.54	—
Acetate	$C_2H_3O_2^-$	−369.41	−486
Dihydroxyacetone	$C_3H_6O_3$	−445.18	—
Lactate	$C_3H_5O_3^-$	−517.18	−687
Glucose	$C_6H_{12}O_6$	−917.22	−1,264
Mannitol	$C_6H_{14}O_6$	−942.61	—
Glycerol	$C_3H_8O_3$	−488.52	−676
Propionate$^-$	$C_3H_5O_{2+}$	−361.08	—
Ethylene glycol	$C_2H_6O_2$	−330.50	—
Acetoine	$C_4H_8O_2$	−280	—
Butyrate	$C_4H_7O_2^-$	−352.63	−535
Propanediol	$C_3H_8O_2$	−327	—
Butanediol	$C_4H_{10}O_2$	−322	—
Methanol	CH_4O	−175.39	−246
Ethanol	C_2H_5O	−181.75	−288
Propanol	C_3H_8O	−175.81	−331
n-Alkane	$C_{15}H_{32}$	+60	−439
Propane	C_3H_8	−24	−104
Ethane	C_2H_6	−32.89	−85
Methane	CH_4	−50.75	−75
H_2 (g)	H_2	0	0
N_2 (g)	N_2	0	0
Nitrite ion	NO_2^-	−37.2	−107
Nitrate ion	NO_3^-	−111.34	−173
Iron II	Fe^{2+}	−78.87	−87
Iron III	Fe^{3-}	−4.6	−4
Hydrogen sulfide (g)	H_2S	−33.56	−20
Sulfide ion	HS^-	+12.05	−17
Sulfate ion	SO_4^{2-}	−744.63	−909
Thiosulfate ion	$S_2O_3^{2-}$	−513.2	−608

Note: pH = 7, 1 atm, 1 mol/L, 298 K.

enthalpy, and the Gibbs energy balance). This means also that there must exist mathematical relations between Y_{DX}, Y_{AX}, Y_{CX}, Y_{QX}, and Y_{GX} (see Fig. 2b). These relations are addressed in a later section (see equations 9a–9e). It is obvious that this knowledge of the complete growth stoichiometry provides essential engineering information with respect to reactor design on the amount of O_2 that must be transferred (aeration capacity), the amount of carbon dioxide that must be removed (ventilation), the amount of heat to be removed (cooling capacity), or the amount of fermentation products (in anaerobic growth). The amounts of the required N source and HCO_3^- (autotrophic growth) also follow from these stoichiometric calculations.

MEASUREMENT OF GROWTH STOICHIOMETRY

As shown earlier, the measurement of one stoichiometric coefficient suffices, in general, to calculate all the other stoichiometric coefficients using the conservation relations. This measured stoichiometric coefficient requires the measurement of two conversion rates because, by definition, a stoichiometric coefficient is the ratio of two conversion rates. For example, $Y_{DX} = r_X/-r_D$. The most simple growth system contains eight conversion rates (biomass, N source, H^+, H_2O, CO_2, electron donor, electron acceptor, heat production) and six conservation equations (C, H, O, N, enthalpy, charge). Measurement of two conversion rates is then sufficient to calculate all other rates and, hence, the complete growth stoichiometry. Currently, the most common measurements are biomass production and substrate (equal to electron donor) consumption. For aerobic growth the on-line measurement of O_2 consumption and CO_2 production by the analysis of O_2 and CO_2 in the off gas in air-sparged fermentors is becoming more and more routine. Especially for autotrophic growth, the on-line measurement of CO_2 consumption by off-gas analysis gives direct and highly accurate information on microbial growth (because all consumed CO_2 appears as biomass). This method was very successfully applied to study the growth stoichiometry and kinetics of solid pyrite oxidation by Fe^{2+}-oxidizing bacteria (12,13) and of *Methanobacterium thermoautotrophicum* on H_2/CO_2 (14).

Most recently, it was also shown that on-line measurement of heat production during microbial growth can be used to explore growth stoichiometry and kinetics (15–17).

However, such a simple approach of measuring only two conversion rates often makes certain assumptions:

- Each chosen pair of measured conversion rates will allow the complete calculation of all other conversion rates.
- All measurements are reliable within a certain statistical error but without a systematic deviation.
- The assumed description of the growth system is correct, which means that by-products or additional substrates are assumed to be absent.

All these assumptions are subject to critical considerations, which are dealt with extensively in a recent series of publications (18–21). Here, simple examples are provided to illustrate the points of interest. The reader is referred to Refs. 18–21 for a more elaborate introduction, including the full mathematical and statistical aspects.

Noncalculability of Stoichiometry

Suppose that in Example 1a the chosen two conversion rates to be measured are biomass production (r_X) and

NH_4^+ consumption (r_N). Measurement of these two rates would not lead to a calculation of the other rates because r_X and r_N occur in the nitrogen-conservation equation in such a way that r_X uniquely determines r_N, and vice versa. It is then said that r_X and r_N are redundant. The N balance gives a constraint for these two measured conversion rates that can be used to calculate the statistically best estimate of r_X and r_N, which also exactly satisfies the N balance.

Clearly the choice of the two measured rates must be such that calculability of all other conversion rates is assured. In example 1a, suitable combinations would be the oxygen consumption rate (r_O) and biomass production rate (r_X), r_O and the carbondioxide production rate (r_C), or r_X and the heat production rate (r_Q).

Ill-Conditioned Calculability of Stoichiometry (Error Propagation)

It is well known that measured conversion rates have a certain measurement error. The subsequently calculated conversion rates, from combining the conservation relations and the two measurements, have an error due to error propagation. It is obviously of great practical importance to choose two measured conversion rates where this error propagation is minimal. A simple example to illustrate this problem is the aerobic growth of biomass on the donor glucose. If oxygen consumption ($-r_O$) and carbon dioxide production (r_C) are the measured rates, then the following relations (using conservation relations and the standard biomass composition) to calculate r_X and $-r_D$ (in C-mol glucose/m^3 h) from the measured r_C and ($-r_O$) can be derived:

$$r_X = 20r_C - 20(-r_O)$$

$$(-r_D) = 20(-r_O) - 21r_C$$

Due to the large multiplication factors of 20 and 21 in these equations, the propagation of the measurement errors in r_C and r_O into r_X and $-r_D$ is enormous.

If the donor conversion rate ($-r_D$) and the carbon dioxide production rate (r_C) were chosen as the measured rates r_X and ($-r_O$) would be calculated as

$$r_X = (-r_D) - r_C$$

$$(-r_O) = -0.05(-r_D) + 1.05r_C$$

The error propagation now is much lower and, therefore, from the measured ($-r_D$) and (r_C), r_X and ($-r_O$) can be calculated, as can be the other conversion rates involved. Clearly the aspect of error propagation is of major importance, and this propagation can be significantly decreased by a proper choice of the conversion rates to be measured.

Redundancy of Measurements

As stated earlier, in general two well-chosen measured conversion rates are usually sufficient to reliably calculate the complete stoichiometry. However, it is advantageous (Example 2) to measure more conversion rates than the minimum requirement of two. This leads to so-called redundant measurements, which can be used for two purposes error diagnosis and data reconciliation.

Error Diagnosis

- To check the validity of the defined growth systems with respect to the absence of by-products or possible second substrates
- To check the measured conversion rates for systematic errors

Data Reconciliation

- To decrease the measurement error in the calculated and measured conversion rates, provided that the statistically based checks (error diagnosis) on the validity of the growth system and the systematic errors in the measured conversion rates are passed

---

EXAMPLE 2

Use of redundant measurements to establish the presence of errors in the definition of the growth system or in the measurements

Consider the microbial growth system of Example 1a, where the following four conversion rates have been measured. The biomass has the standard elemental composition.

Biomass production	$r_X = +1$ C-mol/h
O_2 consumption	$-r_O = 1.2$ mol/h
HCO_3^- production	$r_C = 10.5$ mol/h
Oxalic acid ($C_2O_4^{2-}$ consumption)	$-r_D = 5.8$ mol/h

We know that a minimum of two rate measurements are needed to calculate the full stoichiometry. Therefore there are two redundant measurements. We can now establish the conservation equations (with r_W, r_H, and r_N as the water, proton, and NH_4^+ conversion rates, respectively) based on conversion rates as

C conservation	$2r_D + r_X + r_C = 0$
H conservation	$4r_N + r_H + 2r_W + 1.8r_X + r_C = 0$
O conservation	$4r_D + 2r_O + r_W + 0.5r_X + 3r_C = 0$
N conservation	$r_N + 0.2r_X = 0$
Charge conservation	$-2r_D + r_N + r_H - r_C = 0$

By eliminating the three nonmeasured rates (r_W, r_H, r_N) from these five conservation equations, one obtains $5 - 3 = 2$ equations, which relate the measured conversion rates only. The result is as follows:

$$2r_D + r_X + r_C = 0$$

$$2r_D - 4r_O + 4.2r_X = 0$$

The first relation can be recognized as the carbon balance, and the second is so-called electron balance or the balance of degree of reduction (1) (see also a later section). With respect to the C balance, one finds from the measurements:

$$\text{C-in} = 2 \times 5.8(\text{oxalate}) = 11.6 \text{ C-mol/h}$$

$$\text{C-out} = 1(\text{biomass}) + 10.5(CO_2) = 11.5 \text{ C-mol/h}$$

Clearly the C balance seems satisfying (0.86% gap).

For the balance of degree of reduction one obtains

$$\text{electrons in} = 2 \times 5.8 = 11.6 \text{ mol electrons/h}$$

$$\text{electrons out} = -1.2(-4) + 1(4.2) = 9 \text{ mol electrons/h}$$

Clearly there is a large gap of 2.6 mol electrons/h.

Because the C balance fits, it is reasonable to assume that the measured values of r_D, r_X, and r_C are reliable. The balance of degree of reduction can therefore be wrong for two reasons:

1. A very inaccurate measurement of r_O.
2. If the measurement of r_O is found to be correct, then the only other possibility is the presence of an additional electron acceptor (e.g., NO_3^-). This would be an error in the defined growth system.

A Mathematically Complete Analysis of Calculability, Analysis of Redundancy, Error Diagnoses, and Data Reconciliation

In the preceding section simple examples were provided to highlight the problems in accurately establishing the full growth stoichiometry from measurements. Because all these calculations are based on linear conservation relations, it is highly appropriate to use matrix algebra. Basic to these calculations is the "elemental" matrix, which specifies the element, charge, and enthalpy information for each compound in the growth system. Recently, an extensive and coherent mathematical description has been provided for calculability, redundancy analysis, error diagnosis, statistical aspects, and data reconciliation using involved matrix algebra (18–21). The developed mathematical theory has been put in a user-friendly computer program called Macrobal (22).

THE EFFECT OF GROWTH RATE ON GROWTH STOICHIOMETRY

Maintenance Energy Concept

In his pioneering work, Monod (23) found that in exponential growth the amount of biomass formed increased in proportion to the amount of substrate consumed. This led to the definition of growth yield Y_{DX} [amount of biomass produced per amount of electron donor (substrate) consumed]. We have seen that Y_{DX} usually determines the complete growth stoichiometry. With the introduction of the chemostat in the early 1950s, microbial growth could be studied at a range of growth rates, and it became clear that Y_{DX} decreased at lower growth rates μ, as shown in Figure 3 (24). This phenomenon was explained by two different concepts (24–26):

- Endogenous respiration or microbial decay, determined by the parameter k_d
- Electron donor (substrate) requirements for maintenance, determined by the parameter m_D

The basic idea is, however, similar in recognizing that a microorganism is a complex structure where the polymers (proteins, etc.) are subject to slow thermal denaturation

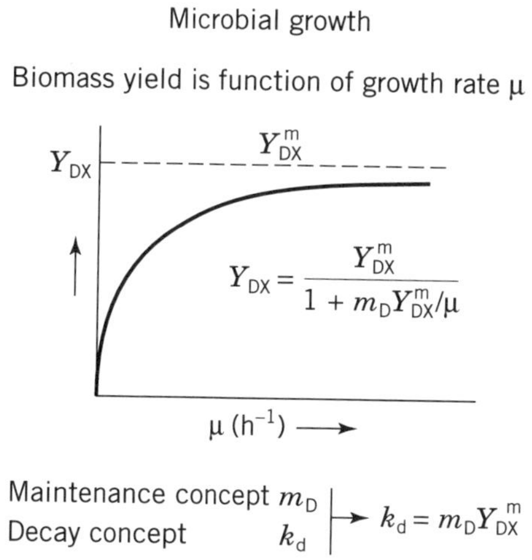

Figure 3. Dependence of Y_{DX} on growth rate μ.

and where there are numerous small leaks associated with the many transmembrane gradients (e.g., Na^+ leaking into the microorganism). These leaking substances must be pumped out, and the degraded polymers must be rebuilt at the expense of Gibbs energy. This results in a small, but finite, need of Gibbs energy to maintain the biomass structure and the transmembrane gradients (maintenance Gibbs energy). In the concept of endogeneous respiration or microbial decay, this energy is produced by catabolism of biomass itself. In the concept of maintenance this energy is produced by catabolism of a part of the substrate (electron donor).

Mathematically, the dependence of Y_{DX} on growth rate μ is described by equation 1a and shown in Figure 3.

$$1/Y_{DX} = 1/Y_{DX}^m + m_D/\mu \tag{1a}$$

This equation contains two model parameters, Y_{DX}^m and m_D, where m_D is the rate of consumption of electron donor (substrate) that is catabolized to generate the necessary Gibbs energy flow for maintenance in C-mol electron donor per C-mol biomass per hour. Y_{DX}^m is the maximal biomass yield. Figure 3 shows that Y_{DX}, using equation 1a, decreases with decreasing growth rate. Clearly, at higher growth rates Y_{DX} comes close to Y_{DX}^m. Using typical values for m_D it can be shown that only for $\mu < 0.01$ to $0.05\ h^{-1}$, Y_{DX} starts dropping significantly below Y_{DX}^m. This means that in exponential growth, as occurs in batch fermentation where μ is high, the stoichiometry is properly covered by Y_{DX}^m. However, in many industrial-fed batch-production processes, maintenance is extremely important due to the low growth rates applied. For example, in penicillin fermentation $\mu \approx 0.01\ h^{-1}$ and about 70% of all consumed glucose is spent for maintenance (27). Similarly, in wastewater-treatment processes, where low growth rates are also applied, the maintenance effects are very relevant. However, in this area one often uses the biomass decay coefficient k_d. This coefficient is however related to m_D according to $k_d = m_D\, Y_{DX}^m$. In general, it can be shown that all biomass yields, Y_{iX} as defined in Figure 2a, decrease with decreasing growth rate μ.

Measuring m_D

Equation 1a shows that Y^m_{DX} and m_D can be obtained directly by measuring Y_{DX} at different growth rates μ. Using equation 1a to plot $1/Y_{DX}$ versus $1/\mu$ as a straight line to obtain $1/Y^m_{DX}$ and m_D is, however, not desirable, because the error distribution of the measurements Y_{DX} and μ is completely distorted due to the use of $1/Y_{DX}$ and $1/\mu$.

It is more proper to directly use the nonlinear equation 1a in combination with the measured Y_{DX} and μ, and to use an algorithm for nonlinear parameter estimation. It is stressed that for accurate m_D values one should measure Y_{DX} at low growth rates (0.005–0.03 h^{-1}).

EXAMPLE 3a

Calculating Y^m_{DX} and m_D from measured Y_{DX} as function of growth rate μ

Consider aerobic growth on glucose and that there are two measurements available. At μ = 0.5 h^{-1}, Y_{DX} = 0.49 C-mol biomass per C-mol glucose, and at μ = 0.02 h^{-1}, Y_{DX} = 0.33.

Applying equation 1a will show that Y^m_{DX} = 0.50 and m_D = 0.02 C-mol glucose/C-mol biomass/h.

Other Maintenance Quantities

Microorganisms require Gibbs energy for maintenance. This is obtained by catabolizing the required amount of electron donor m_D. It is then obvious that other quantities, such as electron acceptor, heat, Gibbs energy, oxidized electron donor, and reduced electron acceptor, are also involved in maintenance, to catabolize the m_D electron donor. These maintenance-related quantities are directly obtained from the stoichiometry of the catabolic reaction (Example 3b).

EXAMPLE 3b

Calculating other maintenance rates using the catabolic reaction

In Example 3a it was found that m_D = 0.02 C-mol glucose/C-mol biomass/h.

In the growth system being considered, the catabolic reaction is the aerobic oxidation of glucose according to

$$C_6H_{12}O_6 - 6O_2 + 6HCO_3^- + 6H^+ + 2{,}814 \text{ kJ heat} + 2{,}843 \text{ kJ Gibbs energy}$$

Using the stoichiometry of this catabolic reaction and the known m_D, it is now easy to calculate the other maintenance rates:

$$\text{maintenance glucose } m_D = 0.02 \text{ C-mol glucose/C-mol Xh} = 0.02/6 \text{ mol glucose/C-mol biomass/h}$$

$$\text{maintenance oxygen } m_A = \frac{0.02}{6} \times 6 \text{ mol } O_2\text{/C-mol biomass/h}$$

$$\text{maintenance } HCO_3^- \; m_C = \frac{0.02}{6} \times 6 \text{ mol } CO_2\text{/C-mol biomass/h}$$

$$\text{maintenance heat } m_q = \frac{0.02}{6} \times 2{,}814 \text{ kJ/C-mol biomass/h}$$

$$\text{maintenance Gibbs energy } m_G = \frac{0.02}{6} \times 2{,}843 \text{ kJ/C-mol biomass/h}$$

Complete Growth Stoichiometry as a Function of Growth Rate

Equation 1a shows how $1/Y_{DX}$ depends on the growth rate μ and the two parameters Y^m_{DX} and m_D. Completely similar equations can be derived for growth yields on acceptor (A), carbon dioxide (C), heat (Q), and Gibbs energy (G) according to equation 1b:

$$1/Y_{iX} = 1/Y^m_{iX} + \frac{m_i}{\mu} \tag{1b}$$

Here i can be A, C, Q, or G. The maintenance coefficients for the different compounds are related to m_D according to Example 3b. The maximal yields Y^m_{AX}, Y^m_{CX}, Y^m_{QX} and Y^m_{GX} are related to Y^m_{DX}, and can be found by solving the macrochemical equation, using the available Y^m_{DX} value, according to Example 1.

A THERMODYNAMICALLY BASED METHOD TO ESTIMATE GROWTH STOICHIOMETRY

In the previous paragraphs the methods for accurate measurement of a growth stoichiometric coefficient, as, for example, the biomass yield on electron donor Y_{DX} and the subsequent calculation of all the nonmeasured stoichiometric coefficients of the macrochemical equation (using the conservation principles) have been provided. In past decades, the value of Y_{DX} for many different microorganisms, different electron donors, C sources, and electron acceptors has been measured under C- and energy-limited growth conditions. Many methods have been proposed to predict Y_{DX} because of its obvious importance. Recently, a critical evaluation of these methods has been performed (2). The following criteria were used for the evaluation:

- The method should be generally applicable to all chemotrophic growth systems.
- The method should relate directly to the second law of thermodynamics.
- No detailed knowledge of metabolism is required; only the identity of the electron donor, C source, and electron acceptor is known.
- Methodological problems are absent.

The conclusion of this evaluation was that none of the published methods satisfied these simple criteria. Therefore, an alternative method that satisfies the mentioned criteria has been proposed (2). This method is based on $1/Y_{GX}$, which is the amount of Gibbs energy (in kilojoules) that must be dissipated for the production of 1 C-mol biomass.

The Gibbs energy stoichiometric parameter $1/Y_{GX}$ has already been introduced as one of the stoichiometric coefficients in the macrochemical reaction equation (Fig. 2b). Therefore, it is obvious that this energetic parameter can be calculated directly if only one of the chemical stoichiometric coefficients has been measured and if the electron donor, electron acceptor, and C source are known (see Example 1b, where $1/Y_{GX} = 1{,}052$ kJ/C-mol biomass).

Furthermore it is well known that the value of growth yields depends on the growth rate (μ) due to the Gibbs energy that must be used for maintenance (see earlier section). This means that the Gibbs energy needed to produce biomass should be divided into two parts:

1. A growth-related part
2. A maintenance-related part

Mathematically this can be expressed as

$$\underbrace{1/Y_{GX}}_{\text{Total needed Gibbs energy kJ/C-mol biomass}} = \underbrace{\frac{1}{Y_{GX}^{m}}}_{\text{Gibbs energy for new biomass}} + \underbrace{\frac{m_G}{\mu}}_{\text{Maintenance Gibbs energy for existing biomass}} \qquad (1c)$$

where $1/Y_{GX}^{m}$ is the Gibbs energy needed to make 1 C-mol of biomass (kJ/C-mol X) and m_G is the Gibbs energy needed for biomass maintenance (kJ/C-mol biomass h). The biomass specific growth rate (h^{-1}) is μ.

Clearly, at high growth rate μ, the m_G/μ term becomes negligible and $1/Y_{GX}$ becomes practically equal to Y_{DX}^{m}. At low growth rates Y_{GX} becomes much lower than Y_{GX}^{m}.

Equation 1c shows that in order to calculate Y_{GX} as a function of growth rate μ we need information about Y_{GX}^{m} and m_G. In the past years two simple correlations have been found with which to estimate Y_{GX}^{m} and m_G (2,4). These correlations were established using a very large body of experimental growth yields, which covered carbon- and energy-limited growth for the following:

- Many different microorganisms (bacteria, fungi, plant cells)
- Many different C sources, including CO_2 and a wide variety of organic substrates
- Different electron acceptors (aerobic, anaerobic, denitrifying)
- Electron donors that need reversed electron transport (RET)

The resulting correlations are given in equations 2 and 3. Figure 4 shows the m_G data used to establish equation 2; Figure 5 shows the $1/Y_{GX}^{m}$ data used to establish the correlations 3a and 3b.

Maintenance Gibbs Energy Need m_G

The data for m_G as a function of temperature shown in Figure 4 can be correlated with an Arrhenius type of relation:

$$m_G = 4.5 \exp\left[\frac{-69{,}000}{R}\left(\frac{1}{T} - \frac{1}{298}\right)\right] \qquad (2)$$

This correlation was found to hold (with ± 40% accuracy) for a very wide variety of organisms, for different electron donors, for aerobic and anaerobic conditions, and for a temperature range of 5–75 °C (4). Obviously, the main influencing factor is the temperature, which behaves as an Arrhenius function with an activation energy of 69,000 J/mol. The type of electron donor (organic or inorganic), the microorganism, and the electron acceptor are of minor importance. This seems logical, because maintenance is a biomass-linked Gibbs-energy-requiring process that counteracts the biomass-deteriorating processes (protein degradation, leakage over cell membranes, etc.).

Gibbs Energy for Growth

The data for Y_{GX}^{m} shown in Figure 5a (heterotrophic growth) and Figure 5b (autotrophic growth) can be correlated by equations 3a and 3b as shown in Ref. 2. For autotrophic growth it was found to be important to distinguish electron donors for which reversed electron transport (RET) was necessary. Such electron donors (e.g., Fe^{2+}/Fe^{3+}, NH_4^+/NO_2^-) provide electrons that have insufficient Gibbs energy to reduce the C source CO_2 to biomass. Microorganisms using such electron donors first have to increase the Gibbs energy level of the donor electrons by the biochemical process RET.

Heterotrophic growth/autotrophic growth (−RET)

$$1/Y_{GX}^{m} = 200 + 18(6 - C)^{1.8} + \exp[((3.8 - \gamma)^2)^{0.16}(3.6 + 0.4C)] \qquad (3a)$$

Autotrophic growth (+RET)

$$1/Y_{GX}^{m} = 3{,}500 \qquad (3b)$$

It was found (4) that the Gibbs energy dissipation required for the production of 1 C-mol biomass mainly depends on the C source used (equations 3a and 3b) The type of microorganism and the type of electron acceptor have only minor effects, as shown in Figures 5a and 5b. The influence of the C source on Y_{GX}^{m} can be characterized by the following:

- Its number C of carbon atoms (e.g., for CO_2 $C = 1$ and for glucose $C = 6$) as shown in Figure 5a
- Its degree of reduction γ (1)

γ is a stoichiometric number of a chemical compound that represents the number of electrons in the compound. For organic compounds γ is per C-mol, for inorganic compounds γ is per mol. For example, for CO_2 $\gamma = 0$, for CH_4 $\gamma = 8$, and for glucose $\gamma = 4$. The concept of degree of reduction will be further elucidated extensively later in this article. For organic compounds (Fig. 5a), γ has a value between 0 and 8. For inorganic compounds (Fig. 5b), only a lower value of 0 holds; a maximal value does not exist because there is no normalization per atom. It is relevant to know that biomass has a degree of reduction of about 4.2. Equation 3a and Figure 5a show that, in the situation that RET is not required for both hetero- and autotrophic growth, the Gibbs energy needed to produce biomass

- Increases if the number of C atoms in the carbon source (the parameter C in equation 3a) decreases

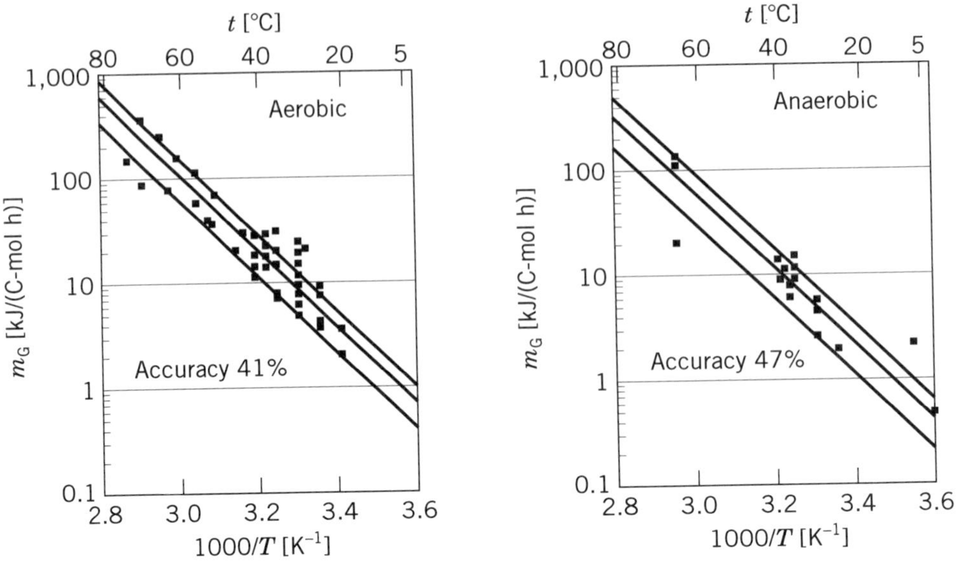

Figure 4. Maintenance Gibbs energy m_G (in kJ/C-mol biomass h) for aerobic (**a**) and anaerobic (**b**) growth, shown as an Arrhenius function of temperature. The lines represent equation 2.

- Increases if the degree of reduction of the carbon source (the parameter γ in equation 3a) is smaller or larger than about 3.8

Equation 3a further shows that for heterotrophic growth $1/Y^{m}_{GX}$ ranges between about 200 and 1,000 kJ/C-mol biomass, for the C sources explored, for which:

- The number of carbon atoms in the carbon source ranges between C = 1 (e.g., CO_2, formate, methane) and C = 6 (e.g., glucose, citrate).
- The degree of reduction of the C source γ ranges between 0 (for CO_2) and 8 (for CH_4).

The effect of the number of C atoms (C) and degree of reduction (γ) of the C source can be simply understood as follows:

- Biomass contains many polymers that contain monomers of four to six C atoms. If the C source contains fewer than four to six C atoms, the microorganism must perform extra biochemical reactions to achieve C–C couplings. This requires extra Gibbs energy, compared to a C source that has six C-atoms, Hence $1/Y^{m}_{GX}$ increases for C-sources with less carbon atoms.
- Biomass has $\gamma = 4.2$. If the C source is more reduced ($\gamma > 4.2$) or more oxidized ($\gamma < 4.2$), there is a need for additional oxidation reactions or reduction reactions, respectively, as compared to a carbon source (like glucose) with $\gamma = 4$. These additional reactions lead to extra Gibbs energy dissipation, leading to a higher value of $1/Y^{m}_{GX}$.

Simply stated, the more biochemical tinkering is needed to convert an organic C source into biomass, the more Gibbs energy is dissipated and the higher $1/Y^{m}_{GX}$ becomes. Obviously glucose (C = 6, $\gamma = 4$) is a nearly ideal C-source because it requires the least Gibbs energy dissipation for biomass production. According to equation 3a, for glucose $1/Y^{m}_{GX} = 200 + 0 + 36 = 236$ kJ/C-mol biomass. In contrast, CO_2 is a very poor C source, because it requires about four times as much Gibbs energy ($1/Y^{m}_{GX} = 200 + 236 + 460 = 986$ kJ/C-mol according to equation 3a). Equation (3b) shows that for autotrophic growth, in the situation where RET is needed (which occurs for many inorganic electron donors), $1/Y^{m}_{GX}$ has a very high value of 3,500 kJ/C-mol biomass. This value should be compared to autotrophic growth without RET as occurs with, for example, H_2 or CO as electron donor (for which $1/Y^{m}_{GX} \approx 1{,}000$ kJ/C-mol according to equation 3a).

Obviously, the use of RET increases the Gibbs energy dissipation needed for biomass production tremendously. The explanation is that, using the RET process, the electrons of the electron donor are increased in energy level, up to the energy level of electrons in NADH in order to make CO_2 reduction to biomass thermodynamically feasible. This "energy-pumping" process (RET) apparently requires a large amount of Gibbs energy, of about 3,500 − 1,000 = 2,500 kJ/C-mol biomass produced.

The effect that the type of the available C source has on the Gibbs energy needed for biomass synthesis is well known in biochemistry. Biochemists express the energy need in ATP. Figure 6 compares the calculated Gibbs energy dissipation needed for biomass synthesis ($1/Y^{m}_{GX}$, in kJ/C-mol$X$) with the theoretically calculated amount of ATP expenditure for biomass synthesis in mol ATP/C-mol X. The points shown are for different C sources, ranging from glucose (28.8) to CO_2 (2.5). The parenthetical numbers are the published (10) biomass yields on ATP in gram-X/mol ATP. It is clear that there is a close correspondence, which is logical. Equations 3a and 3b provide the energy needed for biomass synthesis in kilojoules, whereas the biochemists use mol ATP as the energy measure.

In conclusion, it should be realized that equations 2, 3a and 3b are completely sufficient to estimate a biomass yield and the full macrochemical equation for any arbitrary chemotrophic growth system (Example 4).

The predictive accuracy of this correlation for chemotrophic growth has been shown (2,4) to be ±10 to 20% rela-

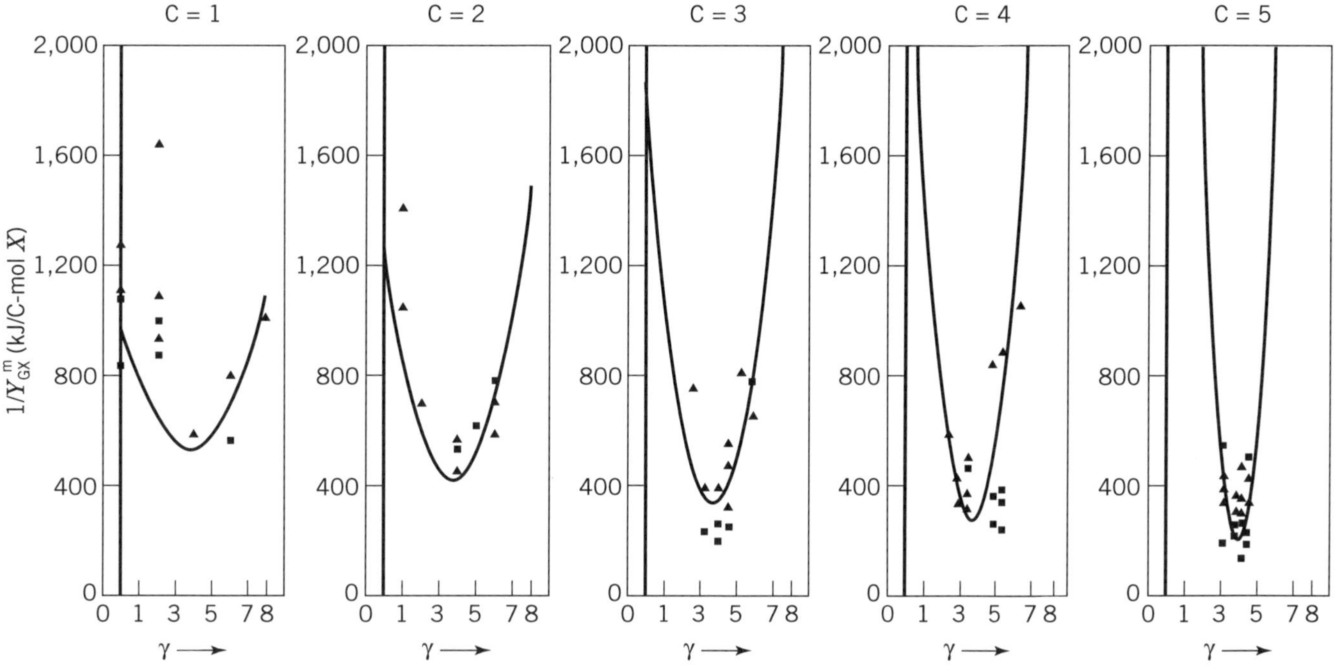

(a)

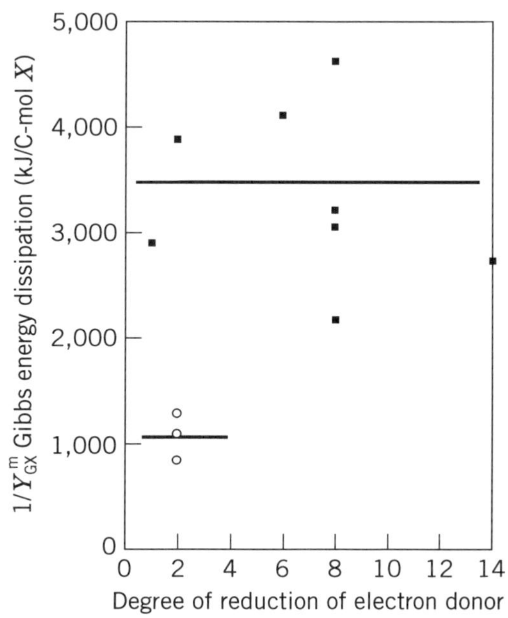

(b)

Figure 5. Reciprocal maximal yield of biomass on Gibbs energy, $1/Y_{GX}^{m}$, (kJ/C-mol X); (**a**) Heterotrophic growth (triangles, aerobic; squares, fermentation, X's denitrifying systems); C is the number of carbon atoms in the carbon source; γ is the degree of reduction of the carbon source. (**b**) Autotrophic growth (squares, electron donors where reversed electron transport [RET] is needed; circles, donors without RET). The lines represent equations 3a and 3b.

tive error in a yield range of nearly two orders of magnitudes of 0.01–0.70 C-mol biomass per C-mol organic electron donor or per mol inorganic donor while covering aerobic, anaerobic, denitrifying, autotrophic microbial systems with and without RET (Fig. 7). The measured yield data used were taken from Refs. 2 and 4.

EXAMPLE 4

Calculation of the full macrochemical reaction equation using the correlations of equations 2, 3a, and 3b

It is assumed that a microorganism grows anaerobically on methanol as C source and electron donor with NH_4^+ as the N source

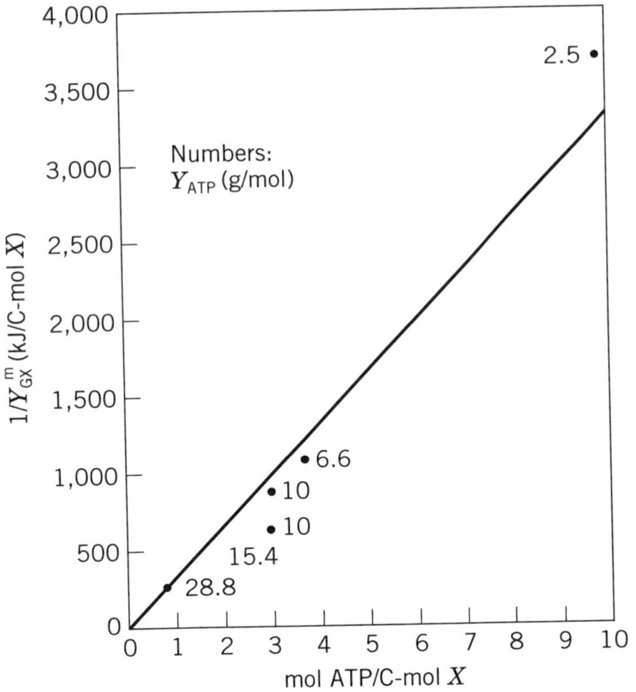

Figure 6. Comparison of energy needed for biomass synthesis on different carbon sources in mol ATP/C-mol biomass and in kJ/C-mol biomass ($1/Y_{GX}^{m}$). The numbers refer to the conventional biomass yield on ATP in gram biomass/mol ATP for different carbon sources (10).

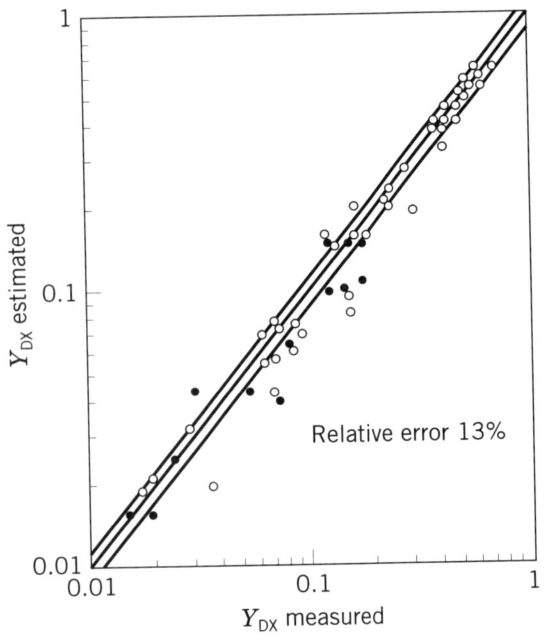

Figure 7. Comparison of measured and predicted biomass yield Y_{DX} (solid circles, fermentative; open circles, aerobic growth systems).

and acetate is produced. The growth system contains biomass, NH_4^+, H^+, HCO_3^-, H_2O, methanol, and acetate as the seven compounds. The general macrochemical reaction equation for the production of 1 C-mol biomass can be written as follows:

$$fCH_3OH + aNH_4^+ + bH^+ + cC_2H_3O_2^- + dH_2O + 1CH_{1.8}O_{0.5}N_{0.2} + eHCO_3^-$$

Clearly there are six unknown stoichiometric coefficients (a–f). However, using equation 3a we can calculate that (C = 1, γ = 6 for methanol, maintenance has been neglected) $1/Y_{GX}^{m} = 200 + 326 + 172 = 698$ kJ/C-molX. This means that we know that the Gibbs energy of reaction of the macrochemical reaction equation equals −698 kJ.

We can now write the conservation equations for C, H, O, N, electric charge, and the Gibbs energy balance (taking values of ΔG_f^{01} from Table 2):

C balance	$f + 2c + 1 + e = 0$
H balance	$4f + 4a + b + 3c + 2d + 1.8 + e = 0$
O balance	$f + 2c + d + 0.5 + 3e = 0$
N balance	$a + 0.2 = 0$
charge balance	$a + b - c - e = 0$

$$\text{Gibbs energy balance } (-175.39)f + (-79.37)a + (-39.87)b + (-369.41)c + (-237.18)d + (-67)1 + (-586.85)e + 698 = 0$$

Solving these six equations gives, for a to f,

$$a = -0{,}2;\ b = 2.866;\ c = 8.898;\ d = 12.964;\ e = -6.232;\ f = -12.564$$

This gives a biomass yield on methanol of 1/12.564 = 0.08 C-mol biomass/C-mol methanol. The stoichiometric result also shows that the acetate production is 8.898/12.564 = 0.70 mol acetate/mol methanol, showing a C yield of 1.4 acetate carbon/methanol-carbon. This is, of course, due to the CO_2 fixation that occurs (6.232 mol HCO_3^- per 12.564 mol methanol).

If maintenance is not allowed to be neglected m_G must be taken into account. For example, the temperature is assumed to be 50 °C. Equation 2 then shows that m_G = 38.8 kJ/C-mol biomass h. If the growth rate $\mu = 0.03\ h^{-1}$, then we can calculate, using equation 1c that $1/Y_{GX} = 698 + 38.8/(0.03) = 1{,}991$ kJ/C-mol biomass. Using this number one can solve the six equations to obtain the complete stoichiometry, which holds under these conditions.

Before ending this section, a final warning is relevant. The described thermodynamic method of predicting growth stoichiometry is based on a very wide database of experimentally measured growth systems. No detailed biochemical information is required, because intrinsically a kind of average biochemistry used by most organisms is assumed. This is an attractive feature, but in the end we should consider that, of course, the biochemistry used by microorganisms does have a significant influence. For example, for the anaerobic ethanol fermentation on glucose the mentioned method will give Y_{DX} = 0.15 C-mol biomass/C-mol glucose. This is indeed found for *Saccharomyces cerevisae*. However, another organism, *Zymomonas mobilis*, does the same glucose/ethanol process, but with Y_{DX} = 0.07. The explanation is that *Z. mobilis* uses a completely different biochemical pathway for glucose catabolism than *S. cerevisae*. From this example we can also learn that if the predicted biomass yield differs very substantially from the actually measured yield, it might be possible that the microorganism being studied uses a novel pathway for catabolism or anabolism.

A USEFUL REFERENCE SYSTEM TO SIMPLIFY GROWTH STOICHIOMETRIC AND ENERGETIC CALCULATIONS AND TO GAIN INSIGHT

The Growth Reference System

In the preceding sections, the stoichiometric coefficients for the macrochemical reaction equation of biomass formation have been solved by setting up the proper conservation equations (C, H, O, N, charge, enthalpy) and the Gibbs energy balance. Although this is a sufficient and straightforward method, solving these linear equations remains unattractive and does not provide insight. To simplify these calculations and to gain insight, a special reference system has been designed—the growth reference system. This reference system is based on the observation that, in all chemotrophic growth systems, H_2O, HCO_3^-, H^+, and N source (mostly NH_4^+) occur as chemical compounds (see earlier section on growth system definition). In this special reference system each chemical compound is assigned three new numbers.

- γ the degree of reduction, which represents the electron content per C-mol (for organic compounds) or per mol (for inorganic compounds)
- ΔG_e the Gibbs energy per electron present in the compound
- ΔH_e the enthalpy per electron present in the compound

Clearly γ is a stoichiometric quantity and ΔG_e and ΔH_e are energetic parameters.

The reference system is designed such that for H_2O, HCO_3^-, H^+ (pH = 7), N source for growth, HPO_4^{2-}, NO_3^-, SO_4^{2-}, and Fe^{3+}, the values of γ, ΔG_e, and ΔH_e are zero. For ΔG_e, the biochemical standard conditions (1 mol/L, 1 bar, pH = 7,298 K) are assumed, ΔH_e is calculated for CO_2 (gas) because of the large heat effect of HCO_3^- (liq) $\Leftrightarrow$ CO_2 (gas) transfer.

The calculation of γ, ΔG_e, and ΔH_e follows from the reference redox half reaction where 1 C-mol of organic or 1 mol of inorganic compound is converted into the reference chemicals and a number of electrons. The number of electrons is by definition equal to γ (Example 5). From the Gibbs energy and enthalpy of this reference reaction, called ΔG_{ref} and ΔH_{ref} (calculated with the usual thermodynamic ΔG_f^{01} and ΔH_f^{01} values, see Table 2), the values of ΔG_e and ΔH_e follow from equations 4a and 4b.

$$\Delta G_e = \frac{-\Delta G_{ref}}{\gamma} \tag{4a}$$

$$\Delta H_e = \frac{-\Delta H_{ref}}{\gamma} \tag{4b}$$

EXAMPLE 5

The reference redox half reaction and calculation of γ and ΔG_e for chemical compounds

For methanol the following reference redox half reaction can be set up according to the preceding definition by converting methanol to the reference compounds HCO_3^-, H_2O, and H^+

$$-1CH_4O - 2H_2O + HCO_3^- + 7H^+ + 6e^-$$

In this reference redox half reaction, 1 C-mol methanol is converted and six electrons are produced, hence $\gamma = +6$ for methanol. Using the ΔG_{f-}^{01} values from Table 2, the ΔG_{ref} for the methanol-reference redox half reaction follows as (standard conditions)

$$\Delta G_{ref}^{01} = (7)(-39.87) + 1(-586.85) - (2)(-237.18) - (1)(-181.75) = -216.192 \text{ kJ}$$

This gives for the ΔG_e^{01} value of methanol by equation 4a

$$\Delta G_e^{01} = -\left(\frac{-216.192}{6}\right) = +36.032 \text{ kJ/e-mol}$$

Obviously ΔH_e can be calculated in a similar way by calculation of ΔH_{ref}.

For biomass the following redox half reaction can be set up, assuming that NH_4^+ is the N source:

$$-1CH_{1.8}O_{0.8}N_{0.2} - 2.5H_2O + HCO_3^- + 0.2NH_4^+ + 5H^+ + 4.2e^-$$

Obviously, the degree of reduction for biomass is 4.2. The ΔG_{ref}^{01} value is obtained similarly as earlier for methanol. ΔG_{ref}^{01} can be calculated to be -142.128 kJ, giving

$$\Delta G_e = -(-142.128)/(4.2) = +33.840 \text{ kJ/e-mol}$$

In a similar way as shown in Example 5 for each chemical compound, the values of γ, ΔG_e, and ΔH_e can be calculated for a large number of relevant compounds. Table 3 contains all relevant stoichiometric and energetic information for growth systems, clearly shown in the following. A point of attention is the finding (Table 3) that for biomass the degree of reduction depends on the N source used in the growth system. For example $\gamma = 4.2$ for NH_4^+ and 5.8 for NO_3^- as N source. This is a consequence of the reference definition. The advantage is that the N source disappears from the stoichiometric calculations using γ, ΔG_e, and ΔH_e (Examples 7a, 7b, and 8b). The defined reference system is closely related to the generalized degree of reduction as defined by Roels (1) and Erickson et al. (28). It can be seen that for reduced organic compounds γ is between 0 and 8 (per C-mol). For inorganic compounds, such an upper limit does not exist (because there is not a normalization per atom). For O_2, γ is negative (-4), which is logical for an acceptor. ΔG_e is related to the conventional redox potential of redox half reactions ($\Delta G_e^{01} = -FE_0^1$). ΔG_e is calculated using HCO_3^- (the most abundant form of carbon dioxide at pH = 7); ΔH_e has been calculated using CO_2 (gas) as reference, to take the large heat effect of $HCO_3^- \rightarrow CO_2$ (gas) into account.

Table 3. Calculated γ, ΔG_e^{01}, and ΔH_e^0 Values for Chemical Compounds under Standard Conditions

	γ Degree of reduction per C-mole for organic and per mole for inorganic compounds in electrons/(C)-mole	ΔG_e^{01} (kJ/e-mol)	ΔH_e^0 (kJ/e-mol)
biomass/NH_4^+ – N source	+4.2	+33.840	−26.1
Biomass/NO_3 – N source	+5.8	+14.820	−44.2
Biomass/N_2 – N source	+4.8	+32.948	−26.3
N source for growth	0	0	0
HCO_3^-	0	0	0
Oxalate	+1	+52.522	−20
Formate	+2	+39.186	−15.50
Glyoxylate	+2	+48.229	—
Tartrate	+2.5	+39.577	—
Malonate	+2.67	+28.976	—
Fumarate	+3	+33.662	−31.60
Malate	+3	+33.354	−32.20
Citrate	+3	+32.282	−33.90
Pyruvate	+3.33	+34.129	−23.60
Succinate	+3.50	+28.405	−36.30
Gluconate	+3.67	+39.106	—
Formaldehyde	+4	+45.326	−0.10
Acetate	+4	+26.801	−33.50
Lactate	+4	+31.488	−28.90
Glucose	+4	+39.744	−25.75
Mannitol	+4.33	+38.777	—
Glycerol	+4.67	+37.625	−24.30
Propionate	+4.67	+26.939	−33.80
Ethylene glycol	+5	+37.292	—
Acetoin	+5	+32.625	—
Butyrate	+5	+27.000	−33.30
Propanediol	+5.33	+33.177	—
Acetone	+5.33	+28.718	−30.90
Butanediol	+5.50	+31.374	—
Methanol	+6	+36.032	−23
Ethanol	+6	+30.353	−28.90
Propanol	+6	+29.144	−32.50
n-Alkane	+6.13	+26.694	—
Propane	+6.66	+25.948	−31.90
Ethane	+7	+25.404	−31.40
Methane	+8	+22.925	−31.50
CO	+2	+47.477	−1.5
H_2	+2	+39.870	0
SO_4^{2-}	0	0	0
SO_3^{2-}	+2	+50.296	—
S^0	+6	+19.146	−55.2
$S_2O_3^{2-}$	+8	+23.584	−27.5
HS^-	+8	+20.850	−43.9
NO_3^-	0	0	0
NO_2^-	+2	−41.650	−108.5
NO(g)	+3	−96.701	—
N_2O(g)	+8	−57.540	−124.55
NH_4^+	+8	−35.109	−101.9
N_2	+10	−72.194	−136.4
Fe^{3+}	0	0	0
Fe^{2+}	+1	−74.270	−46.8
H_2O	0	0	0
O_2	−4	−78.719	−143

Note: pH = 7, 1 mol/L, 1 atm, 298 K.

Balance of Degree of Reduction, Atomic Degrees of Reduction, and the COD Balance

In the previously defined "growth reference" system, γ_i was introduced as the degree of reduction of compound *i*. This parameter is important in stoichiometric calculations, because due to the principle of electron conservation, an electron balance can be defined. This is the so-called balance of degree of reduction. This is not an additional conservation principle (in addition to C, H, O, N, and charge conservation). The balance of degree of reduction can be obtained from the usual C, H, O, N, and charge balances by eliminating (by suitable substitutions) H^+, H_2O, HCO_3^-, and the N source. Hence, the balance of degree of reduction is a suitable linear combination of already available conservation equations. The importance of the balance of degree of reduction is that, by definition, in this balance, only biomass formation, consumption of electron donor, and consumption of electron acceptor are related. Based on the previous definition of the reference set of compounds in the growth reference system it is also possible to calculate the degree of reduction of atoms and of electric charge (Table 4).

It should be noted that the atomic degree of reduction for the N atom in biomass depends on the applied N source as a consequence of the defined growth reference system as explained earlier. Using the γ values of atoms and electric charge in Table 4, it is straightforward to calculate the γ values for any chemical compound for which the elemental composition is known (Example 6a). This is an equivalent alternative to writing the reference redox half reaction to obtain γ (Example 5).

EXAMPLE 6a

Direct calculation of γ from elemental composition

Using the atomic degrees of reduction (Table 4) it can easily be checked that indeed for the reference chemicals $\gamma = 0$:

H_2O	$\gamma = 2 \times 1 + 1(-2) = 0$
CO_2	$\gamma = 1 \times 4 + 2(-2) = 0$
HCO_3^-	$\gamma = 1 \times 1 + 1 \times 4 + 3(-2) + 1 = 0$
H^+	$\gamma = 1 \times 1 + 1(-1) = 0$

Table 4. Degree of Reduction of Atoms and Electric Charge According to the Definition of the Growth Reference System

Atom or charge	Degree of reduction of atoms
H	+1
O	−2
C	+4
Charge +1	−1
Charge −1	+1
S	+6
P	+5
N	+5
N in N index in biomass	+3 for NH_4^+ or NH_3 as N source 0 for N_2 as N source +5 for NO_3^- or HNO_3 as N source

For the degree of reduction of biomass (γ_X) it is easy to show that this is a function of the N source used. Using the standard elemental biomass composition $CH_{1.8}O_{0.5}N_{0.2}$ and using the N degree of reduction for the different N sources (Table 4) one obtains

$$-NH_4^+ \text{ as N-source } \gamma_X = 1 \times 4 + 1.8 \times 1 + 0.5(-2) + 0.2(-3) = 4.2$$

$$-NO_3^- \text{ as N-source } \gamma_X = 1 \times 4 + 1.8 \times 1 + 0.5 \times (-2) + 0.2(5) = 5.8$$

For the electron content of an organic substrate (e.g., acetate ion, $C_2H_3O_2^-$) the amount of electrons is $2 \times 4 + 3 \times 1 + 2(-2) + 1(+1) = 8$. Because for organic compounds γ is defined as the number of electrons per C atom, we obtain for acetate with 2 carbon atoms $\gamma = 8/2 = 4$.

The degree of reduction balance is also called chemical oxygen demand (COD) balance in wastewater engineering (26). The COD balance is equivalent to the balance of degree of reduction. COD is a number assigned to each chemical and represents the consumed O_2 on total oxidation in g O_2/g compound. There is a direct link with degree of reduction. Each mole of electrons represents 8 g COD. This is easily understood, because the consumption of 1 mol O_2 represents the acceptance of 4 mol electrons ($\gamma = -4$; see Table 3). One mol O_2 represents −32 g COD and therefore 1 mol electrons ≡ 8 g COD.

EXAMPLE 6b

Calculation of COD values

Consider glucose, in which 1 mol (≡ 180 gram) represents (according to Table 3) a total of 6×4 electrons $\equiv 6 \times 4 \times 8 = 192$ g O_2. Clearly glucose has a COD value of 192/180 = 1.0667 g COD/g glucose.

The NO_3^-/N_2 acceptor couple has $\gamma_A = -5$ electrons. The COD value of NO_3^- − N is then $-5 \times 8/14 = -2.857$ g COD per gram nitrate-nitrogen.

The values of γ, ΔG_e, and ΔH_e from Table 3 can be used for very easy stoichiometric and energetic calculations as shown in Examples 7a and 7b.

EXAMPLE 7a

Calculation of the stoichiometry of example 1a using γ values

In Example 1a, five equations were solved to calculate the full macrochemical equation. Using the γ values of Table 3 we can first make the balance of degree of reduction. For the electron donor oxalate $\gamma_D = 1$ per carbon or 2 per mole oxalate; for biomass $\gamma_X = 4.2$ and for the electron acceptor $O_2 \gamma_A = -4$ (Table 3). For all the other chemicals (N source, H^+, H_2O, HCO_3^-) $\gamma = 0$ by definition. The γ balance is now

$$2f - 4c + 4.2 = 0$$

Because $f = -5.815$ we obtain $c = -1.857$ directly.

From the C balance we then obtain $e = +10.63$. From the N balance $a = -0.20$, from the charge balance $b = -0.8$, and from the O or H balance we finally find $d = -5.42$. This is, as expected, the same result as before in Example 1a.

EXAMPLE 7b

Calculation of the Gibbs energy of reaction in example 1b using ΔG_e values

Using the now-available full macrochemical stoichiometry, it is possible to calculate the Gibbs energy of reaction using the Gibbs energy balance.

For each chemical compound the Gibbs energy contribution follows from the product of its number of electrons and its ΔG_e number. For example (using Table 3), the Gibbs energy contribution for oxalate, O_2, and biomass in the growth reference system follows as

$$\text{oxalate} = 2 \times 1 \times 52.522 = +105.04 \text{ kJ}$$

$$O_2 = -4 \times (-78.719) = 314.876 \text{ kJ}$$

$$\text{biomass} = 4.2 \times 33.840 = 142.128 \text{ kJ}$$

For all other reactants the Gibbs energy contribution in the growth reference system is zero. For the Gibbs energy of the macrochemical-reaction equation we obtain then from the available full stoichiometry:

$$-5.815(105.04) - 1.857(314.876) + 142.128 = -1053 \text{ kJ}$$

This is the same as obtained before, but now the calculation has only three terms.

Energetics of Redox Couples, Catabolic Redox Reactions, RET, and Energetic Regularities

It was pointed out earlier that for each microbial growth system a catabolic redox reaction is needed where an electron donor couple reacts with an electron acceptor couple. For the generation of maintenance energy the catabolic reaction is also required. For example, in aerobic growth on glucose the electron donor couple is glucose/HCO_3^- (glucose is oxidized to HCO_3^-) and the electron acceptor couple is O_2/H_2O (O_2 is reduced to H_2O). However, for anaerobic growth on glucose, where the catabolic reaction is the conversion of glucose into ethanol, the electron donor couple is glucose/HCO_3^- and the electron acceptor is the HCO_3^-/ethanol couple. To be able to quickly calculate the catabolic energy production, it is relevant to define the ΔG_e, ΔH_e, and γ values of redox couples i/j. These can be calculated from γ, ΔG_e, and ΔH_e values (Table 3) using equations 5a–5c.

$$\gamma_{\text{couple}} = \gamma_i - \gamma_j \tag{5a}$$

$$(\Delta G_e)_{\text{couple}} = \frac{(\gamma \Delta G_e)_i - (\gamma \Delta G_e)_j}{\gamma_i - \gamma_j} \tag{5b}$$

$$(\Delta H_e)_{\text{couple}} = \frac{(\gamma \Delta H_e)_i - (\gamma \Delta H_e)_j}{\gamma_i - \gamma_j} \tag{5c}$$

Equations 5a–5c have the following properties:

- If we invert the redox couple, for example, i/j into j/i, we are in fact inverting the redox half reaction of the redox couple. The value of γ changes sign, which is logical because produced electrons become consumed electrons. The value of ΔG_e and ΔH_e does not change, which is logical because the energetics of the reaction do not change.
- If the redox couple i/j contains a biological reference compound (e.g., j = HCO_3^-, H_2O, H^+, NO_3^-, SO_4^{2-}) then for compound j the value of γ, ΔG_e, and ΔH_e is zero. Equations 5a–5c then show that the value of γ, ΔG_e, and ΔH_e of the redox couple i/j becomes equal to the tabulated values of the i component in Table 3. For example, if the redox couple is an organic compound/HCO_3^-, then the γ, ΔG_e-, and ΔH_e value of the organic compound follow directly from Table 3.
- If the redox couple does not contain a reference chemical then equations 5a–5c must be used to calculate γ, ΔG_e, and ΔH_e. Table 5 shows some examples.

It can be seen from Table 5 that electron donor couples are characterized by positive γ values and electron-acceptor couples by negative γ values (which is logical). As stated earlier, in each microbial growth system there functions a catabolic reaction between an electron donor and an electron acceptor, which generates the required (for anabolism and maintenance) Gibbs energy. It is noted (equation 6c) that in a full catabolic redox reaction, the redox couple with the highest ΔG_e value must be the electron donor; the redox couple with the lowest ΔG_e value is the acceptor (Example 8a).

EXAMPLE 8a

Recognizing electron donor and acceptor

Consider the following catabolic reaction: $C_6H_{12}O_6 + 6O_2 \rightarrow 6HCO_3^- + 6H^+$. The two redox couples are $C_6H_{12}O_6/HCO_3^-$ and O_2/H_2O. According to Table 3, the glucose couple has ΔG_e = +39.744 kJ/e-mol and the O_2 couple has ΔG_e = −78.719 kJ/e-mol. Clearly glucose is the electron donor and O_2 is the acceptor. Consider now the catabolic reaction $C_6H_{12}O_6 + 2H_2O \rightarrow 2HCO_3^- + 2H^+ + 2C_2H_5OH$. This is the catabolic reaction in the ethanol fermentation. The redox couples are $C_6H_{12}O_6/HCO_3^-$ and

Table 5. ΔG_e, and ΔH_e Values of Redox Couples under Standard Conditions

Redox couple	γ (per (C-)mol)	ΔG_e (kJ/e-mol)	ΔH_e (kJ/e-mol)
$NH_4^+/½N_2$	3	+26.703	−44.3
NH_4^+/NO_2^-	6	−32.928	−99.665
NO_2^-/NO_3^-	2	−41.647	−108.5
Lactate/pyruvate	2	+18.283	−55.4
Fumarate/succinate	−2	−3.137	−64.5
$NO_3^-/½N_2$	−5	−72.194	−136.4
$NO_2^-/½N_2$	−3	−92.559	−155
$NO(g)/½N_2$	−2	−35.434	—
$N_2O(g)/N_2$	−2	−130.809	−183.8
Glucose/HCO_3^-	4	+39.744	−25.75

HCO_3^-/C_2H_5OH. From Table 3 we read for the glucose couple that $\Delta G_e = +39.744$ and for the ethanol couple $\Delta G_e = +30.353$. Now glucose is the donor and HCO_3^-/ethanol is the acceptor.

Because the catabolic reaction liberates the Gibbs energy required for the anabolism, it is important to calculate this amount of energy. Using the ΔG_e approach, we can then write directly equations 6a and 6b to calculate the Gibbs energy of reaction (ΔG_{CAT}) of the catabolic reaction. ΔG_{ED} and ΔG_{EA} are the electron Gibbs energy of the acceptor and donor couples:

$$-\Delta G_{CAT} = \gamma_D(\Delta G_{ED} - \Delta G_{EA}) \tag{6a}$$

The enthalpy of reaction ΔH_{CAT} of the catabolic reaction can be calculated similarly:

$$-\Delta H_{CAT} = \gamma_D(\Delta H_{ED} - \Delta H_{EA}) \tag{6b}$$

ΔG_{CAT} and ΔH_{CAT} represent the Gibbs energy and enthalpy of reaction of the catabolic reaction consuming 1 C-mol organic or 1 mol inorganic compound of electron donor. Dimensions are kJ/(C)-mol donor. γ_D is the degree of reduction of the donor couple in mol electrons/(C)-mol donor, which is always positive. According to the second law of thermodynamics ΔG_{CAT} must be negative. Therefore equation 6c, the second law of thermodynamics, holds:

$$\Delta G_{ED} > \Delta G_{EA} \tag{6c}$$

This shows that indeed the electron donor couple always has the highest ΔG_e value.

EXAMPLE 8b

Calculation of catabolic Gibbs energy production using ΔG_e

Consider the catabolic reactions in Example 8a. For aerobic glucose oxidation we can calculate, using equation 6a and Table 3

$$\begin{aligned}(-\Delta G_{CAT}) &= 4 \times [39.644 - (-78.719)] \\ &= 473.852 \text{ kJ, so that } \Delta G_{CAT} = -473.852 \text{ kJ}\end{aligned}$$

This is the Gibbs energy released for the aerobic combustion of 1 C-mol glucose. For 1 mol of glucose (6 C-atoms) the Gibbs energy of the catabolic reaction is $6 \times (-473.852) = -2{,}843$ kJ.

For anaerobic ethanol fermentation of glucose we can calculate for the catabolic reaction $(-\Delta G_{CAT}) = 4 \times (39.744 - 30.353) = 37.564$ kJ per C-mol glucose. For 1 mol glucose the catabolic Gibbs energy of the catabolic reaction becomes $6 \times (-37.564) = -225.34$ kJ. This is the same as calculated in Example 3b.

The use of ΔG_e and ΔH_e now also reveals some interesting energetic regularities (4). Table 3 shows that for many organic donor compounds, the ΔH_e and ΔG_e values are rather close, with an average $\Delta G_e = +32 \pm 8$ kJ/e-mol and average $\Delta H_E = -28 \pm 5$ kJ/e-mol. These values are also close to the ΔG_E and ΔH_E value of biomass. Hence we can write for organic electron donors the important regularity

$$\Delta G_{ED} \approx \Delta G_{EX} \tag{7a}$$

$$\Delta H_{ED} \approx \Delta H_{EX} \tag{7b}$$

For autotrophic growth using inorganic electron donors, the ΔG_{ED} values are, in general, much lower than ΔG_{EX} (consider NH_4^+/NO_2^-, F_e^{2+}/F_e^{3+}, etc., in Table 4). For autotrophic growth (CO_2 as C source), this means that for these electron donors there is a need for Gibbs energy input in order to realize CO_2 reduction to biomass. This is achieved by RET. Knowing this we can write equation 7c to recognize RET.

$$\Delta G_{EX} > \Delta G_{ED} \tag{7c}$$

Finally, it is now easy to calculate the heat production and Gibbs energy dissipation in oxidative catabolism of organic compounds. As stated earlier, for organic compounds the average $\Delta G_{ED} = 32$ kJ/e-mol and for O_2 as acceptor $\Delta G_{EA} = -78.719$ kJ/e-mol. Hence, per mole of electron transferred between donor and acceptor, the available Gibbs energy is $32 - (-78.719) = 110.72$ kJ. Per mole of consumed O_2 (which accepts four electrons, $\gamma_A = -4$) in the combustion of any organic compound, the Gibbs energy made available by combustion of the organic compound is then $4 \times 110.72 = 443$ kJ per consumed mol O_2. Analogously, one can find for the produced heat per mole of O_2 in the combustion of organic compounds a value of 460 kJ per mole O_2. It is also obvious that the mentioned inaccuracy in the average ΔG_E or ΔH_E values for organic compounds only results in a minor error of 5 to 8% in the calculated Gibbs energy dissipation and heat production. These are very important rules of thumb for the fermentation industry (1).

MATHEMATICAL EQUATIONS TO CALCULATE THE GROWTH STOICHIOMETRY FROM KNOWN GIBBS ENERGY DISSIPATION

Deriving the Equations

In Example 1, it was shown that one suitably measured stoichiometric coefficient, for example, Y_{DX}, allows the calculation of all other stoichiometric coefficients, including the dissipated Gibbs energy $1/Y_{GX}$. This means that knowledge of $1/Y_{GX}$ should enable the calculation of all stoichiometric coefficients, as shown in Example 4. It has also been shown that Y_{GX} can be estimated for arbitrary growth systems under different growth rates and temperatures using the correlations (equations 2 and 3). Here we show that particularly simple equations to calculate all yields from Y_{GX} can be obtained by using the γ, ΔG_e, and ΔH_e parameters introduced in the previous section.

The general macrochemical reaction equation can be written as shown in Figure 2b. In this macrochemical equation, the electron donor and acceptor are written in C-mol (for organic compounds) or in mol (for inorganic compounds). The (. . .) stoichiometric coefficients are not given separate symbols. They follow easily from the charge balance (for H^+), N balance (for N source), from O or H balance (for H_2O), and the carbon balance (for HCO_3^-). For autotrophic growth, the HCO_3^- stoichiometric coefficient is -1. Three balances can be written based on the conservation principles of electrons and enthalpy and the Gibbs energy balance.

Balance of degree of reduction

$$-\gamma_D/Y_{DX} - \gamma_A/Y_{AX} + \gamma_X = 0 \qquad (8a)$$

Enthalpy balance

$$-\gamma_D \Delta H_{ED}/Y_{DX} - \gamma_A \Delta H_{EA}/Y_{AX} + \gamma_X \Delta H_{EX} + 1/Y_{QX} = 0 \qquad (8b)$$

Gibbs energy balance

$$-\gamma_D \Delta G_{ED}/Y_{DX} - \gamma_A \Delta G_{EA}/Y_{AX} + \gamma_X \Delta G_{EX} + 1/Y_{GX} = 0 \qquad (8c)$$

From equation 8a it follows that Y_{AX} is directly related to Y_{DX} (equation 9a)

$$\frac{1}{Y_{AX}} = \frac{\gamma_D}{(-\gamma_A)Y_{DX}} - \frac{\gamma_X}{(-\gamma_A)} \qquad (9a)$$

The degree of reduction of the acceptor γ_A is (by definition) a negative number. Using equations 8a–8c, it is also possible to calculate directly Y_{DX}, Y_{AX}, and Y_{QX} as function of Y_{GX}. If we further use the found energetic regularity that $\Delta G_{EX} \approx \Delta G_{ED}$ (equation 7a) and replace $(\Delta G_{ED} - \Delta G_{EA})$ by $(-\Delta G_{CAT})/\gamma_D$ using equation 6a, it is possible to derive the simple equations 9b–9e. It is noted that ΔG_{CAT} is the Gibbs energy of the catabolic reaction of 1 C-mol organic or 1 mol inorganic compound, and that therefore $(-\Delta G_{CAT})$ is the Gibbs energy released in the catabolic reaction of 1 C-mol of organic or 1 mol of inorganic electron donor. $(-\Delta G_{CAT})$ is then by definition >0 and its units are kJ/(C)-mol donor. ΔG_{CAT} and ΔH_{CAT} follow from equations 6a and 6b using ΔG_{EA} and ΔG_{ED} values (Tables 3 and 5).

$$Y_{DX} = \frac{(-\Delta G_{CAT})}{1/Y_{GX} + \gamma_X/\gamma_D(-\Delta G_{CAT})} \qquad (9b)$$

$$Y_{AX} = \frac{(-\gamma_A/\gamma_D)(-\Delta G_{CAT})}{1/Y_{GX}} \qquad (9c)$$

$$1/Y_{QX} = \frac{(-\Delta H_{CAT})}{(-\Delta G_{CAT})} 1/Y_{GX} \qquad (9d)$$

Furthermore it is often interesting to study Y_{DA}, which is the amount of electron acceptor couple consumed relative to the amount of electron donor consumed. For microorganisms growing aerobically on organic matter, this would be the mole of O_2 consumed per C-mole of organic compound consumed. For anaerobic growth this would be the amount of anaerobic products per amount of organic substrate in C-mole product per C-mole substrate. Because $Y_{DA} = Y_{DX}/Y_{AX}$ we obtain

$$Y_{DA} = [\gamma_D/(-\gamma_A)] \frac{1/Y_{GX}}{1/Y_{GX} + \gamma_X/\gamma_D(-\Delta G_{CAT})} \qquad (9e)$$

Application of the Mathematical Stoichiometry Relations

The obtained stoichiometric relations (equations 9a–9e) can now easily be applied. In this section their use is demonstrated with the following subjects:

- Calculation of the complete growth stoichiometry
- Calculation of maintenance coefficients and maximal growth yields
- Calculation of the limit to growth yield posed by the second law
- Calculation of COD-based growth yields
- Calculation of the relation between heat production and Gibbs energy dissipation
- Calculation of maximal product yields in anaerobic metabolism

Calculation of the Complete Growth Stoichiometry. If for a given growth system the C source, the electron donor (ΔG_{ED} to decide on RET using equation 7c), the temperature, and the growth rate are known, then equations 2 and 3 allow a direct calculation of the required Gibbs energy $1/Y_{GX}^m$ to produce 1 C-mol of biomass at high growth rates μ. Knowing the electron donor couple and acceptor couple, and using Table 3 and equations 5a–5c, the values of ΔG_{ED}, ΔG_{EA}, ΔH_{ED}, ΔH_{EA}, γ_D, and γ_A can be calculated and from this $(-\Delta G_{CAT})$ and $(-\Delta H_{CAT})$ using equations 7a and b. Using equations 9b–9e subsequently allows the complete stoichiometric calculation where H^+, N source, H_2O, and HCO_3^- must be calculated using the conservation equations of electric charge, N, O or H, and carbon.

EXAMPLE 9a

Calculation of stoichiometry using equations 9b–9e

Consider Example 4, where a microorganism is grown anaerobically on methanol, producing acetate. Assume first that the growth rate is high, such that maintenance can be neglected. Equation 3a then shows that $1/Y_{GX}^m = 698$ kJ Gibbs energy per C-mol biomass. Methanol is the C-source, methanol/HCO_3^- is the electron donor, and HCO_3^-/acetate is the electron acceptor. From Table 3 and using equations 5a–5c, we can then find that $\Delta G_{ED} = 36.032$ kJ/e-mol, $\Delta H_{ED} = -23$ kJ/e-mol; $\Delta G_{EA} = 26.801$ kJ/e-mol; $\Delta H_{EA} = -33.5$ kJ/e-mol. Also, $\gamma_D = 6$, $\gamma_A = -4$, and $\gamma_X = 4.2$.

This provides that $-\Delta G_{CAT} = 6(36.032 - 26.801) = 55.386$ and $-\Delta H_{CAT} = -23 - (-33.5) = 63$ kJ/C-mol methanol. Using equations 9b–9e, we obtain the maximal growth yields (maintenance neglected).

$$1/Y_{DX}^m = 13.3 \text{ mol methanol/C-mol biomass}$$

$$1/Y_{AX}^m = 18.9 \text{ C-mol acetate/C-mol biomass} = 9.45 \text{ mol acetate/C-mol biomass}$$

$$1/Y_{QX}^m = 794 \text{ kJ heat/C-mol biomass}$$

$$Y_{DA}^m = 1.42 \text{ C-mol acetate/C-mol methanol} = 0.71 \text{ mol acetate/mol methanol}$$

Using the C- balance $1/Y_{CX}^m$ is calculated as 6.6 mol HCO_3^- consumed/C-mol biomass produced. This overall stoichiometric result is very close to the exact solution obtained in Example 4.

The small deviation arises from the assumption that $\Delta G_{EX} \approx \Delta G_{ED}$ (as discussed before).

In general it can be shown that the simple set of equations 9b–9e seldom deviates more than 5% from the exact solution.

Maintenance Coefficients and Maximal Yield Coefficients. As indicated previously, relations between the maintenance coefficients follow from the catabolic reaction. The following relations can now be written to link the various maintenance coefficients to the maintenance Gibbs energy m_G using $-\Delta G_{CAT}$ (see also Example 9b). It is noted that m_G follows from the correlation (equation 2) and that ΔG_{CAT} is calculated for the catabolism of 1 (C)-mol of electron donor.

$$m_D = m_G/(-\Delta G_{CAT}) \quad (10a)$$

$$m_A = (\gamma_D/-\gamma_A)m_G/(-\Delta G_{CAT}) \quad (10b)$$

$$m_Q = m_G \frac{(-\Delta H_{CAT})}{(-\Delta G_{CAT})} \quad (10c)$$

$$m_C = m_G/(-\Delta G_{CAT}) \quad (10d)$$

Using equations 10a–10d, the value of the maintenance Gibbs energy requirement m_G can be calculated from either measured maintenance coefficients (for electron donor m_D, electron acceptor m_A, heat production m_Q, or carbon dioxide production m_C). Furthermore, it can easily be understood that the maximal biomass yields for electron donor, acceptor, and heat are found from equations 9b–9e by substitution of $1/Y^m_{GX}$ (instead of $1/Y_{GX}$) because the maintenance contribution is then neglected.

EXAMPLE 9b

Effect of maintenance on stoichiometry

In Example 9a the maintenance contribution was neglected. Assume that the microorganism is growing at 37 °C. Equation 2 then leads to m_G = 13 kJ/C-mol biomass h. Using equations 10a–10c and using ΔG_{CAT} and ΔH_{CAT} (Example 9) we obtain

$$m_D = 0.2347 \text{ mol methanol/C-mol biomass/h}$$

$$m_A = 0.3521 \text{ C-mol acetate/C-mol biomass/h}$$

$$m_Q = 14.787 \text{ kJ/C-mol biomass/h}$$

Further assume that the growth rate μ = 0.02 h^{-1}. Using equation 1b, the Y^m_{iX} values obtained in Example 9a, and the m_i values obtained here, one obtains for the stoichiometry:

$$\text{electron donor}(D)\,1/Y_{DX} = 13.3 + \frac{0.2347}{0.02} = 25.03 \text{ mol methanol/C-mol X}$$

$$\text{electron donor}(A)\,1/Y_{AX} = 18.9 + \frac{0.3521}{0.02} = 36.50 \text{ C-mol acetate/C-mol X}$$

$$\text{heat } 1/Y_{QX} = 794 + \frac{14.787}{0.02} = 1533 \text{ kJ heat/C-mol X}$$

$$Y_{DA} = 36.50/25.03 = 1.46 \text{ C-mol acetate/mol methanol}$$

These values can also be obtained directly from equations 9b–9e by substituting the complete Gibbs energy of growth and maintenance according to equation 1c:

$$1/Y_{GX} = 1/Y^m_{GX} + \frac{m_G}{\mu} = 698 + \frac{13}{0.02} = 1348 \text{ kJ Gibbs energy/C-mol X}$$

Clearly, comparing Examples 9a and 9b, one observes that the yield of biomass Y_{DX} drops from 0.077 to 0.04 due to maintenance, but the acetate/methanol yield Y_{DA} increases from 1.42 to 1.46 C-mol acetate/mol methanol.

Second Law Limit of Growth Yield. As for any chemical reaction, the microbial growth yield is also limited by the second law of thermodynamics. This limit is achieved if $1/Y_{GX} = 0$, because this defines equilibrium. Equations 9c–9f then show that for the thermodynamic limits we can write [see also Ref. 4] the following:

Thermodynamic limits for growth yields

$$Y_{DX} = \gamma_D/\gamma_X$$

$$1/Y_{AX} = 0$$

$$1/Y_{HX} = 0$$

Clearly, the more reduced electron donors (γ_D higher) have a higher Y_{DX} limit. This limit has already been determined (1).

COD-Based Yields. In wastewater treatment, the biomass yield is calculated on COD basis. Y_{COD} is the gram biomass COD over gram-substrate COD. Based on the COD definition we can write

$$Y_{COD} = \frac{\gamma_X}{\gamma_D} Y_{DX} \quad (11a)$$

This allows the following relation for Y_{COD} from equation 9b

$$Y_{COD} = \frac{(-\Delta G_{CAT})}{(-\Delta G_{CAT}) + (\gamma_D/\gamma_X)1/Y_{GX}} \quad (11b)$$

For aerobic growth on organic substrate $-\Delta G_{CAT} = \gamma_D[32-(-78.719)] \approx \gamma_D \times 110$ kJ/C-mol. Here the average value of ΔG_{ED} = +32 kJ/e-mol for organic matter was used as shown before. Substitution of $-\Delta G_{CAT}$ gives Y_{COD} for aerobic growth on organic substrate

$$Y_{COD} = \frac{110}{110 + (1/Y_{GX})/\gamma_x} \quad (11c)$$

This equation shows that Y_{COD} for aerobic growth is not a constant, as often assumed (26) with $Y_{COD} \approx 0.50$–0.67, but it depends also on the type of C source, because this determines $1/Y_{GX}$ (equations 3a and 3b). Also, to decrease the Y_{COD} leading to lower surplus-sludge production, one must, according to equations 11b and 11c:

- Decrease ($-\Delta G_{CAT}$), by, for example, using anaerobic metabolism producing CH_4
- Increase $1/Y_{GX}$ by increasing temperature or decreasing the growth rate according to equation 1

These predicted phenomena are well known and applied in waste-treatment processes.

Relation between Heat Production and Gibbs Energy Dissipation. According to equation 9d, the heat production ($1/Y_{QX}$) is related to Gibbs energy need ($1/Y_{GX}$) by the enthalpy and Gibbs energy of the catabolic reaction (ΔG_{CAT} and ΔH_{CAT}). Table 6 shows some examples of growth systems to illustrate the relation between heat production and dissipated Gibbs energy for growth (29). From Table 6 we can conclude the following rules of thumb:

- For aerobic (or denitrifying) growth systems on organic substrate, the Gibbs energy dissipation and heat production are nearly equal. The entropy contribution in the catabolic reaction is minimal (see glucose and acetate aerobic growth).
- For anaerobic growth, heat production and Gibbs energy dissipation can be substantially different, due to entropic effects.

Obviously, if in the catabolic reaction there is a net decrease of molecules or a consumption of gaseous molecules, then there is a strong negative entropy contribution (see H_2/CO_2 aerobic and anaerobic) and there is a much higher heat production than Gibbs energy dissipation. If, however, in the catabolic reaction there is a net production of the amount of molecules and/or production of gaseous molecules (e.g., the glucose/ethanol fermentation or the methane production from acetate), then there is a very large positive entropy contribution, leading to a much lower heat production than the Gibbs energy dissipation. The entropic effect can even be so large that there is a calculated heat uptake during growth (e.g., methanation of acetate). This is obviously endothermic growth. So, contrary to a common belief, growth of microorganisms is not necessarily related to heat production; there can be heat uptake as well. Experimental proof is, however, still lacking.

Maximal Product Yields in Anaerobic Metabolism. In many microbial processes the valuable product (e.g., ethanol or lactic acid) is related to catabolism. The relevant stoichiometric coefficient is then the yield of the electron acceptor couple to electron donor Y_{DA}.

Equation 9e shows how this coefficient is determined by various factors and it appears that Y_{DA} is maximized.

- For high Gibbs energy dissipation $1/Y_{GX}$. This means that high catabolic product yields are achieved for poor carbon sources, low growth rate, and high temperature, because $1/Y_{GX}$ is then maximized.
- For catabolic reactions with low ΔG_{CAT}. This is understandable, because then the growth yield is minimized, which leads directly to higher product yield.
- For highly reduced electron donors (γ_D high) and highly oxidized products (γ_A low). It is then even possible to achieve C yields larger than 1. An excellent example is the anaerobic production of acetate from methanol, where $Y_{DA} \approx 1.4$ C-mol acetate/C-mol methanol (Example 4).

KINETICS OF MICROBIAL GROWTH FROM A THERMODYNAMIC POINT OF VIEW

In the previous sections, the full stoichiometric description of growth has been given from a unifying thermodynamic point of view. However, for a complete description, growth kinetics are also required. In this section these kinetics are also presented from a thermodynamic point of view.

A Basic Kinetic Description of Microbial Growth

In Figure 1, the typically observed batch-growth curves of a microorganism growing on one substrate (electron donor) are shown. In this section we use the subscript S to denote substrate (electron donor, D). Usually these curves are described (for constant batch volume) with two mass balances for substrate (donor) and biomass, giving two differential equations (DE):

Table 6. Relation between Heat Production and Gibbs Energy Need

Microorganism	Growth condition	Y_{DX} (C-mol/(C)-mol)	$1/Y_{GX}$, Gibbs energy (kJ/C-mol biomass)	$1/Y_{QX}$, heat (kJ/C-mol biomass)	Entropy contribution (kJ/C-mol biomass)
Saccharomyces cerevisae	Glucose aerobic	0.57	332	+339	−7
Saccharomyces cerevisae	Glucose anaerobic	0.14	270	+95	+175
Hydrogenotroph	$H_2 + CO_2$ aerobic	0.13	1,265	+1686	−421
Methanobacterium arborophilus	$H_2 + CO_2$ anaerobic	0.015	1,035	+3,923	−2,888
Pseudomonas oxalaticus	Acetate aerobic	0.406	562	+593	−31
Methanobacterium soehngenii	Acetate anaerobic	0.024	597	−90	+687

Note: Heat production and Gibbs energy dissipation for (an)aerobic growth on glucose, H_2, and acetate; relative contribution of heat- and entropy-related dissipation.

$$\frac{dC_s}{dt} = q_s C_x \tag{12a}$$

$$\frac{dC_x}{dt} = \mu C_x \tag{12b}$$

where q_s and μ are the biomass-specific electron-donor (substrate) uptake and growth rate. To solve these DEs, we need kinetic expressions for q_s and μ. However, q_s and μ are stoichiometrically coupled to each other according to the Herbert–Pirt equation (equation 1a). This equation for substrate can be written as (realizing that $Y_{DX} \equiv Y_{SX} = \mu / -q_s$, and $m_S \equiv m_D$)

$$-q_s = \frac{1}{Y^{m}_{SX}} \mu + m_S \tag{13}$$

It is now only necessary to specify one kinetic equation, which can be for q_s of μ. In practice the Monod kinetic equation for μ ($\mu = \mu^m C_s/(C_s + K_s)$) is often chosen. This choice leads, however, to a very nasty inconsistency. According to Monod, $\mu = 0$ if the concentration of substrate C_s becomes 0. Substituting this result in the Herbert–Pirt equation shows, however, that (for $\mu = 0$) $q_s = m_s$. The inconsistency is then that in the absence of substrate ($C_s = 0$, $\mu = 0$) there is still consumption of substrate for maintenance ($v = -q_s = m_s$); this is very strange indeed. These problems can be avoided in a most simple way by introducing the kinetics of the substrate consumption rate in the most simplistic way (a Michaelis–Menten type of relation, see Figure 8):

$$-q_s = q_s^{max} \frac{C_s}{K_s + C_s} \tag{14a}$$

Combining equation 14a with equation 13 by eliminating q_s, leads to the kinetic relation of μ

$$\mu = Y^{m}_{SX} q_S^{max} \frac{C_s}{K_s + C_s} - Y^{m}_{SX} m_S \tag{14b}$$

The kinetic relations 14a and 14b contain four model parameters: Y^{m}_{SX}, K_s, m_s, and q_s^{max}. Figure 8 shows how μ and q_s depend on C_s.

Clearly, there is a maximal substrate uptake rate q_s^{max}, but also for μ a maximal value (μ^{max}) is seen at high C_s. The value of μ^{max} follows from equation 14b by taking the limit $C_s \gg K_s$, and the result is

$$\mu^{max} = Y^{m}_{SX}(q_s^{max} - m_s) \tag{15a}$$

In addition, it is clear from Figure 8 that there is a minimal substrate concentration ($C_{s,min}$) at which $\mu = 0$. For $C_s < C_{s,min}$, the growth rate μ becomes negative. An expression for $C_{s,min}$ can be found from equation 14b by putting $\mu = 0$, which results in

$$C_{s,min} = K_s\left(\frac{m_S}{q_s^{max} - m_s}\right) \tag{15b}$$

The occurrence of a minimal substrate concentration has indeed been observed (30,31). It can also be seen (Fig. 8) that at $C_s = 0$ the growth rate is negative.

$$(\mu)_{C_s=0} = -Y^{m}_{SX} m_S = -k_d \tag{15c}$$

Clearly, in the absence of substrate, the microorganism decreases its biomass; this phenomena is often called endogenous respiration, or biomass decay, where the kinetic parameter is k_d. Equation 14b can be rewritten by eliminating q_s^m and m_S using equations 15a and 15b leading to

$$\mu = \mu^{max} \frac{(C_s - C_{s,min})}{K_s + C_s} \tag{14c}$$

Alternatively equation 14b can be rewritten using the parameter k_d

$$\mu = \mu^{max} \frac{C_s}{K_s + C_s} - k_d \frac{K_s}{K_s + C_s} \tag{14d}$$

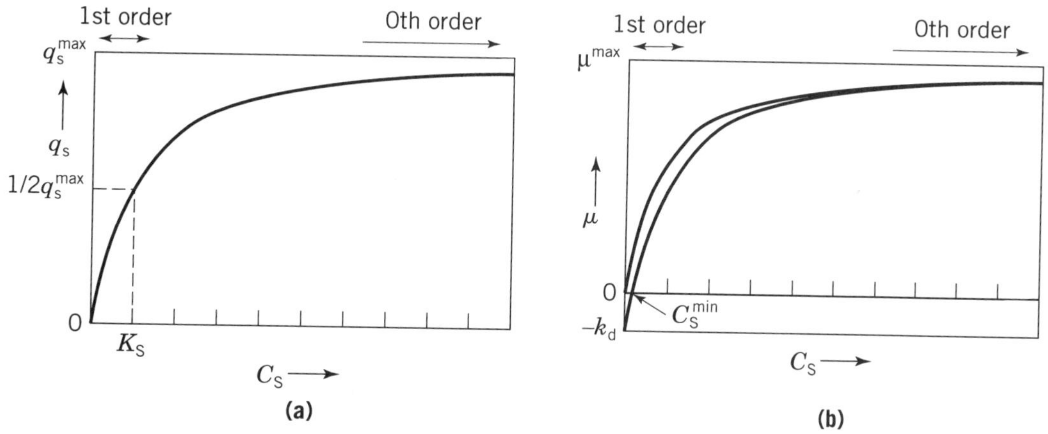

Figure 8. Dependence of q_s and μ as function of C_s for the Monod equation for μ (a) or the Michaelis–Menten equation for q_s (b).

It is clear that the obtained formulations for μ (equations 14b, 14c, and 14d) all become the familiar Monod equation if $m_S = 0$ (meaning $C_{s,min} = 0$ and $k_d = 0$; see equations 15b and 15c). The Monod equation and equations 14c and 14d are very close for $C_s \gg K_s$, but substantial differences arise for $C_s < K_s$ (see Fig. 8).

A final point to address is the implicit assumption (in equation 14a) that in the substrate-uptake process the substrate concentration can become zero. This is, in principle, not possible, because there always exists a substrate concentration where the catabolic reaction reaches a point where energy production in the form of ATP becomes impossible. This concentration is called the threshold concentration of substrate $C_{s,thresh}$.

From published data it is known that μ^{max} can vary in a wide range (0.001–1 h^{-1}) and that K_s is also covering an even wider range (10^{-6} to 10^{-3} mol/L). In the preceding sections m_S and Y_{SX}^{m} have been studied from a thermodynamic point of view. In the following, we present correlations and points of view that allow the estimation k_d, μ^{max}, K_s, and $C_{s,thresh}$, all based on a thermodynamic point of view. Such relations are very relevant in, for example, considerations of waste-treatment processes where the threshold concentration determines the effluent quality, where k_d determines the surplus sludge production, and where μ^{max} directly determines the reactor size or the type of reactor (suspended organisms or biomass retention systems). For batch-fermentation processes, μ^{max} is important because it determines the duration of a batch-growth process.

A Thermodynamic Relation for the Endogeneous/Decay Parameter k_d

According to equation 15c, k_d can be calculated from the relations for Y_{SX}^{m} and m_G found previously (equations 9c and 10a). Substitution gives

$$k_d = \frac{m_G}{1/Y_{GX}^{m} + (\gamma_X/\gamma_D)(-\Delta G_{CAT})} \qquad (16)$$

The values of m_G and Y_{GX}^{m} are obtained from the thermodynamic correlations equations 2 and 3. This equation shows that for aerobic heterotrophic growth (where $1/Y_{GX}^{m} \approx$ 300–600 kJ/C-mol biomass) the expected k_d will be a factor three times larger than for aerobic autotrophic growth (such as nitrification, for which $1/Y_{GX}^{m} \approx 3500$ kJ/C-mol x). This has indeed shown to be the case (4), indicating that equation 16 is useful.

A Thermodynamic Correlation for μ^{max}

The value μ^{max} is the result of a limiting factor in metabolism. This metabolism can be represented very schematically as shown in Figure 9. Three possible bottlenecks are identified (Fig. 9):

1. The uptake rate of substrate
2. The rate of synthesis of biomass, as related to the ribosomal capacity to synthesize biomass protein

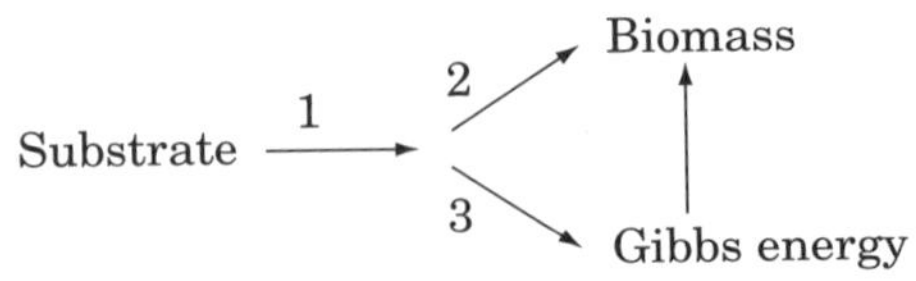

Figure 9. Possible bottlenecks in metabolism.

3. The rate of making Gibbs energy available to enable growth and maintenance

Here we assume as a simple working hypothesis that the rate of making Gibbs energy available for microorganisms is limited by a maximum rate of electron transport in the catabolic energy production. This hypothesis is supported by studies with *Escherichia coli*, where μ^{max} was measured aerobically (at the same temperature) for different growth substrates (32). It was observed that for all substrates the same maximal biomass specific O_2 uptake rate was achieved by *E. coli*. This supports the hypothesis of a constant maximal electron-transport capacity in microorganisms. A similar hypothesis was put forward by McCarthy (26). The following correlation is now proposed for the maximal electron-transport capacity in the microbial electron transport chain (in mol electrons/C-mol biomass/h) as a function of temperature:

$$\text{Maximal electron-transport capacity} = 3 \exp\left[\frac{-69000}{R}\left(\frac{1}{T} - \frac{1}{298}\right)\right]$$

This correlation is an Arrhenius type of relation with an energy of activation of 69,000 J/mol. R is the gas constant (8.314 J/mol K). The value of 69,000 J/mol is an activation energy that follows from observations of the effect of temperature on μ^{max} of microorganisms (1).

For a given electron donor/acceptor combination, the maximal rate of Gibbs energy made available per unit biomass (q_G^{max} in kJ Gibbs energy per C-mol biomass per hour) follows then as

$$q_G^{max} = 3[(-\Delta G_{CAT})/\gamma_D] \exp\left[\frac{-69,000}{R}\left(\frac{1}{T} - \frac{1}{298}\right)\right]$$

It should be noted that $(-\Delta G_{CAT})/\gamma_D$ is the Gibbs energy made available per mole electron transported between donor and acceptor by the electron transport chain.

Now, by definition, the Gibbs energy is spent for growth and maintenance and we can write under maximum growth rate condition

$$q_G^{max} = \frac{1}{Y_{GX}^{m}}\mu^{max} + m_G$$

Eliminating q_G^{max} leads to the final equation (equation 17), which gives μ^{max} as a function of temperature (after replacing m_G using equation 2), Y_{GX}^{m}, γ_D, and ΔG_{CAT}.

Table 7. Estimated μ^{max} Values at 25 °C Using Equation 17

Microbial system	$-\Delta G_{cat}/\gamma_D$ (kJ/mol electron)	$1/Y_{GX}^{m}$ (kJ/C-mol biomass)	μ^{max} (h^{-1}, 25 °C)
Aerobic/glucose	118.5	236	1.5
Aerobic/acetate	105.5	432	0.7
Anaerobic/CH_4 from acetate	3.87	432	0.015
Anaerobic/ethanol from glucose	9.39	236	0.10
Aerobic/Fe^{2+} oxidation (pH = 1.5)	38.6	3,500	0.03
Aerobic/sulfide oxidation to SO_4^{2-}	99.6	3,500	0.08
Aerobic/nitrification	45.8	3,500	0.04

$$\mu^{max} = \frac{[3(-\Delta G_{CAT})/\gamma_D - 4.5]}{1/Y_{GX}^{m}} \exp\left[\frac{-69{,}000}{R}\left(\frac{1}{T} - \frac{1}{298}\right)\right] \quad (17)$$

Table 7 shows the estimated μ^{max} values (25 °C = 298 K) for well-known growth systems. These estimated maximal growth rates are reasonable in the range of reported values. For the only prediction, which is rather wrong (system 4), it is indeed known that the Gibbs energy conversion is not due to electron transport, but due to substrate phosphorylation.

Affinity Constant of Electron Donor, K_s

The constant K_s is an affinity constant in the substrate-uptake kinetics. It is known that reported K_s values can vary over a wide range (10^{-3} to 10^{-6} mol/L). Even for the same organism, for example, *E. coli*, many different K_s values are found (30). Much of this wide spread is probably due to systematic errors in the determination of the K_s parameter. These errors can be due to ill-defined conditions of the microorganisms (adaptation), statistically unsound procedures to evaluate K_s from experimental data, neglect of mass-transfer limitation in the case of microorganisms growing as flocs, and analysis/sampling problems in determining the substrate concentration (see Ref. 30 for an extensive discussion). Moreover, K_s is a typical kinetic parameter, determined by the kinetic properties of the first step in substrate transport into the microorganism. For example, for the anaerobic acetate conversion into CH_4, the *Methanosarcina* type of organism has a high K_s value of 3 mM acetate and its substrate uptake is passive (non-energy linked). However, for the *Methanotrix* type of organism, where acetate is transported into the cell utilizing energy, $K_s \approx 0.5$mM (33).

Due to these factors, it is not possible to give, from a thermodynamic point of view, a generalization about the value of K_s for different microbial growth systems.

Threshold Concentration

In the previous section the substrate consumption kinetics q_s was assumed as an irreversible Michaelis–Menten kinetics equation (equation 14a). From this it follows that q_s becomes zero at substrate concentration $C_s = 0$. However, it is known that microbial metabolism stops at a certain concentration of the substrate. This is called the threshold concentration $C_{s,thresh}$.

This threshold concentration is thus the substrate concentration where substrate consumption q_s becomes zero. This should be distinguished from the minimal substrate concentration $C_{s,min}$, where the growth rate μ becomes zero.

This threshold situation is achieved when the coupled system of catabolic reaction and the energy generating system of the organism [most probably the proton motive force (pmf) process] are in equilibrium. Because the pmf requires at least 15–20 kJ for each proton (6), and assuming that a full catabolic reaction is minimally coupled to the extrusion of one proton, it appears that there must be a minimal catabolic Gibbs energy release of about 15–20 kJ. The following examples seem to support this idea where threshold behavior has been reported for a number for growth systems (Example 10):

- Anaerobic metabolism of CH_4 production from acetate (33)
- Aerobic metabolism of ferrous into ferric iron (12,13)
- Anaerobic production of H_2 converting 1 ethanol into 2 H_2 and 1 acetate and consumption of $4H_2$ to produce acetate, CH_4, or H_2S from $2HCO_3^-$, HCO_3^- or SO_4^{2-} (34,35)

In all these situations of observed threshold concentration the actual ΔG_{CAT} could be calculated to be in the range of -10 to -30 kJ per catabolic reaction. Example 10 shows how the actual $(\Delta G)_{CAT}$ follows from textbook thermodynamic calculations.

EXAMPLE 10

Threshold concentrations and minimally required catabolic Gibbs energy for aerobic Fe^{2+}-oxidizing bacteria (12,13)

Fe^{2+} can be aerobically oxidized by specific bacteria. The catabolic reaction is $Fe^{2+} + 1/4O_2 + H^+ \rightarrow Fe^{3+} + 1/2H_2O$. The ΔG^0_{CAT} can be calculated to be -44.32 kJ (standard conditions). This aerobic iron oxidation can be performed by two different microorganisms, *Thiobacillus ferro oxidans* (T.f.) and *Leptospirillum ferro oxidans* (L.f.). Recently (12,13) the Fe^{2+} concentration was observed where the O_2 consumption stopped (threshold concentration). The following results were found (30 °C):

T.f.: $P_{O2} = 0.12$ bar, pH = 1.85, $Fe^{3+} = 0.21$ M,
$Fe^{2+} = 3 \times 10^{-4}$ M (= threshold concentration)

L.f.: P_{02} = 0.12 bar, pH = 1.55, Fe^{3+} = 0.21 M,
Fe^{2+} = 7 × 10^{-6} M (= threshold concentration)

Using these concentrations the following ΔG_{CAT} can be calculated for *T. ferro oxidans*

$$\Delta G_{CAT} = -44.32 + 5.8 \log\left[\frac{0.21}{3 \times 10^{-4} \times 10^{-1.85}(0.12)^{0.25}}\right]$$
$$= -15.75 \text{ kJ}$$

Similarly we can calculate for *L. ferro oxidans* $\Delta G_{CAT} = -8.03$ kJ.

Threshold concentrations and minimally required Gibbs energy for anaerobic acetate consuming methanogens

CH_4 can be produced anaerobically from acetate according to the catabolic reaction Acetate + $H_2O \rightarrow CH_4 + HCO_3^-$. For this reaction ΔG^0_{CAT} and ΔH^0_{CAT} at 298 K follow as

$$\Delta G^{01}_{CAT} = -31.0 \text{ kJ}$$
$$\Delta H^0_{CAT} = +5 \text{ kJ}$$

This provides, using the Van 't Hoff equation, for $\Delta S^{01}_{CAT} = \Delta H_r - \Delta G_r/T$ = 0.121 kJ/mol K.

There are two organisms known to perform this reaction at 60 °C, both of which have threshold acetate concentrations. According to Ref. 32 *Methano sarcina* has a threshold concentration of 0.3 × 10^{-3} M and *Methano trix* of 16 × 10^{-5} M of acetate.

Using a CH_4 pressure of 0.6 bar, T = 333 K, and a bicarbonate concentration of 0.03 M, we can calculate the following ΔG_{CAT} of the catabolic reaction at the threshold acetate concentration (using the Van 't Hoff relation $\Delta G = \Delta H - T\Delta S$ to take the temperature effect into account):

$$\textit{Methanosarcina}\ \Delta G_{CAT} = -23 \text{ kJ}$$
$$\textit{Methanotrixe}\ \Delta G_{CAT} = -15.7 \text{ kJ}$$

It appears indeed that at the threshold concentration the coupled system of catabolic reaction and proton translocation may be in equilibrium. This leads to the statement that a threshold concentration can be estimated from the requirement that

$$(\Delta G_{CAT})_{thresh} \approx -20 \text{ kJ} \tag{18}$$

This explanation is further supported by the observation that for anaerobic H_2-consuming systems (36), the observed effect of temperature on H_2 threshold values can be explained from correlation equation 18. Example 11 shows how the expected threshold concentration can be estimated for such systems.

EXAMPLE 11

Calculation of the threshold concentration as a function of temperature

Consider the anaerobic catabolic system $4H_2 + HCO_3^- + H^+ \Leftrightarrow CH_4 + 3H_2O$. One can calculate that ΔG^{01}_{CAT} = −135.57 kJ, and ΔH^0_{CAT} = −241 kJ. Using the Van 't Hoff relation ($\Delta G = \Delta H - T\Delta S$) one calculates ΔS^0_{CAT} = −0.354 kJ/molK.

Assuming P_{CH_4} = 1 bar, HCO_3^- = 0.01 M, $H^+ = 10^{-8}$ M, T = 303 K, and ΔG_{CAT} equals the threshold minimum of −20 kJ, one can calculate at threshold conditions:

$$-20 = -241 - (-0.354) \times 303 + 0.008314 \times 303 \ln \frac{1}{(10^{-8}/10^{-7})(0.01)(P_{H_2})^4}$$

From this follows that the threshold hydrogen pressure is P_{H_2} = 7 × 10^{-5} bar at a temperature of 30 °C (303 K).

For a temperature of 75 °C (=348 K), one finds that $(P_{H_2})_{threshold}$ = 120 × 10^{-5} bar. Such an increase of hydrogen threshold partial pressure with increasing temperature has indeed been observed (36). The calculated threshold H_2 pressures are also in the correct range.

From the previous findings it is clear that, especially for systems with a low ΔG_{CAT} (e.g., anaerobic, inorganic electron donors), threshold values can be found. For aerobic growth, ΔG_{CAT} is so large that a measurable threshold value is not expected because the thermodynamically calculated $C_{s,thresh}$ would be extremely low.

The existence of threshold concentrations makes it desirable to change the kinetics of substrate uptake (q_s, equation 14a) from the irreversible to reversible form

$$q_s = \frac{q_s^m(C_s - C_{s,thresh})}{K_m + (C_s - C_{s,thresh})} \tag{19}$$

In conclusion it appears that threshold concentrations of electron donor (or substrate) do exist and their value can be estimated from the catabolic reaction using the value of $(\Delta G_{cat})_{thresh}$ in equation 18. In addition a more proper kinetic expression for q_s is then given by equation 19.

NOMENCLATURE

Symbols

C_i	Concentration of compound i, mol/m^3
Y_{ij}	Yield of compound j on compound i, mol$_j$/mol$_i$
μ^{max}	maximal specific growth rate of biomass, h^{-1}
k_d	Decay coefficient of biomass, h^{-1}
m_i	Biomass specific maintenance requirement of compound i, mol$_i$/C-mol biomass h
K_s	Affinity constant for substrate uptake, mol/m^3
r_i	Rate of conversion of compound i per reactor volume, (C)mol/m^{3}h
q_i	Biomass specific rate of uptake or secretion of compound i, (C) mol i/C-mol biomass h
H	Enthalpy, kJ
G	Gibbs energy, kJ
Q	Heat, kJ
γ	Degree of reduction: for organic compounds it is defined per C-mol, electron/C-mol; for inorganic compounds it is defined per mol, electron/mol

Subscripts

D	Electron donor
A	Electron acceptor
X	Biomass
e	Calculated per mol electron
E	Calculated per mol electron
O	Oxygen
S	Substrate
C	Carbon dioxide
N	N-source
Q	Heat
H	Proton
W	Water
CAT	Catabolic reaction

Superscripts

m	Value at infinite substrate concentration
max	Value under maximal conditions
o	Reference conditions (1 mol/L, 1 atm, 298 K)
01	Reference conditions, but with pH = 7 (H^+ = 10^{-7} mol/L)

ACKNOWLEDGMENTS

I would like to thank J. A. Roels, H. V. Westerhoff, U. von Stockar, and E. H. Battley for fruitful and pleasant discussions on thermodynamics in the past years.

BIBLIOGRAPHY

1. J.A. Roels, *Energetics and Kinetics in Biotechnology*, Elsevier, New York, 1983.
2. J.J. Heijnen and J.P. van Dijken, *Biotechnol. Bioeng.* **39**, 833–858 (1992).
3. J.J. Heijnen, M.C.M. van Loosdrecht, and L. Tijhuis, *Biotechnol. Bioeng.* **40**, 1139–1154 (1992).
4. L. Tijhuis, M.C.M. van Loosdrecht, and J.J. Heijnen, *Biotechnol. Bioeng.* **42**, 509–519 (1993).
5. E.H. Battley, *Energetics of Microbial Growth*, Wiley New York, 1987.
6. H.V. Westerhoff and K. van Dam, *Mosaic Non-Equilibrium Thermodynamics and the Control of Biological Free Energy Transduction*, Elsevier, Amsterdam, 1987.
7. J.J. Heijnen, *Antonie van Leeuwenhoek* **60**, 235–256 (1991).
8. J.J. Heijnen and J.P. van Dijken, *Biotechnol. Bioeng.* **42**, 1127–1130 (1993).
9. E.H. Battley, *Appl. Environ. Microbiol.* **61**, 1655–1657 (1995).
10. A.H. Stouthamer, *Antonie van Leeuwenhoek* **29**, 545–565 (1973).
11. R.K. Thauer, K. Jungermann, and K. Decker, *Bacterial. Rev.* **41**, 100–180 (1977).
12. M. Boon, J.J. Heijnen, and G.S. Hansford, in D.S. Holmes and R.W. Smith eds., *Minerals Bioprocessing II. Proceedings of the Engineering Foundations Conference: Minerals Processing II*, Snowbird, Utah, July 10–15, 1995; TMS, Warrendale, Pa., 1995, pp. 41–62.
13. M. Boon, J.J. Heijnen, and G.S. Hansford, in D.S. Holmes and R.W. Smith eds., *Minerals Bioprocessing II. Proceedings of the Engineering Foundations Conference: Minerals Processing II*, Snowbird, Utah, July 10–15 1995, TMS, Warrendale, Pa., 1995, pp. 63–82.
14. N. Schill, W.M. van Gulik, D. Voisard, and U. von Stockar, *Biotechnol. Bioeng.* **51**, 645–658 (1996).
15. B.H.A. van Kleeff, J.G. Kuenen, and J.J. Heijnen, Biotechnol. Bioeng. **12**, 510–518 (1996).
16. U. von Stockar, L. Gustafsson, C. Larsson, J. Marison, P. Tissot, and E. Gnaiger, *Biochim. Biophys. Acta* **1183**, 221–240 (1993).
17. U. von Stockar and I.W. Marison, *Adv. Biochem. Eng. Biotechnol.* **40**, 93–136 (1989).
18. H.J. Noorman, B. Romein, K.Ch.A.M. Luyben, and J.J. Heijnen, Biotechnol. Bioeng. **49**, 364–376 (1996).
19. R.T.J.M. van der Heijden, J.J. Heijnen, C. Hellinga, B. Romein, and K.Ch.A.M. Luyben, *Biotechnol. Bioeng.* **43**, 3–10 (1994).
20. R.T.J.M. van der Heijden, B. Romein, J.J. Heijnen, C. Hellinga, and K.Ch.A.M. Luyben, *Biotechnol. Bioeng.* **43**, 11–20 (1994).
21. R.T.J.M. van der Heijden, B. Romein, J.J. Heijnen, C. Hellinga, and K.Ch.A.M. Luyben, *Biotechnol. Bioeng.* **44**, 781–791 (1994).
22. C. Hellinga and B. Romein, in M.N. Karim and G. Stephanopoulos eds., *Modeling and Control of Biotechnical Processes 1992*, IFAC Symposia Series 10, 1992.
23. J. Monod, *Ann. Rev. Microbiology* **3**, 371–394 (1949).
24. S.J. Pirt, *Proc. R. Soc. London, Ser. B* **163**, 224–231 (1965).
25. O.M. Neijssel and D.W. Tempest, *Arch. Microbiol.* **107**, 215–230 (1976).
26. P.L. McCarty, in S.D. Faust and J.V. Hunter eds., *Organic Compounds in Aquatic Environments*, Marcel Dekker, New York, 1971, pp. 495–531.
27. J.J. Heijnen, J.A. Roels, and A.H. Stouthamer, *Biotechnol. Bioeng.* **21**, 2175–2201 (1979).
28. L.E. Erickson, I.G. Minkevich, and V.K. Eroshin, *Biotechnol. Bioeng.* **20**, 1595–1605 (1978).
29. J.J. Heijnen, *TIBTECH* **12**, 483–492 (1994).
30. K. Kovárová, A.J.B. Zehnder, and T. Egli, *J. Bacteriol.* **178**, 4530–4539 (1996).
31. M.E. Tros, G. Schraa, and A.J.B. Zehnder, *Appl. Environ. Microbiol.* **62**, 437–442 (1996).
32. K.B. Andersen and K. von Meyenburg, J. Bacteriol. **144**, 114–123 (1980).
33. S.H. Zinder, *FEMS Microbiol. Rev.* **75**, 125–138 (1990).
34. H.J. Seitz, B. Schink, N. Pfennig, and R. Conrad, *Arch. Microbiol.* **155**, 82–88 (1990).
35. H.J. Seitz, B. Schink, N. Pfennig, and R. Conrad, *Arch. Microbiol.* **155**, 89–93 (1990).
36. R. Conrad and B. Wetter, *Arch. Microbiol.* **155**, 94–98 (1990).

See also ANAEROBES; BIOREMEDIATION; CLOSTRIDIA, SOLVENT FORMATION; ELECTRON TRANSPORT; ENERGY METABOLISM, MICROBIAL AND ANIMAL CELLS; YEAST, BAKER'S.

BIOFILMS, MICROBIAL

JAMES D. BRYERS
University of Connecticut
Farmington, Connecticut

KEY WORDS

Bacterial adhesion
Biofilm detection
Biofilm eradication
Biofilm formation
Biofilms
Biofouling
Cell attachment
Sloughing

OUTLINE

THE IMPACT OF BIOFILM FORMATION

Biofilms are collections of microorganisms, predominantly bacteria, enmeshed within a three-dimensional gelatinous matrix of extracellular polymers secreted by the microorganisms (Fig. 1). In an aqueous environment, a support, termed a substratum, will be immediately biased by dissolved organic molecules and macromolecules that adsorb rapidly from the liquid phase. Bacterial cells present in the fluid contact the substratum by a variety of transport mechanisms. Once at the substratum, the cells can adsorb either reversibly or irreversibly. Provided the cells remain at the surface for a sufficient time, they will secrete extracellular polymers that serve to attach the cells tenaciously to the substratum. Attached cells metabolize prevailing energy and carbon sources (either dissolved within the surrounding fluid, adsorbed to the substratum surface, or existing as a constituent of the substratum itself), reduce terminal electron acceptors, grow, replicate, and produce insoluble extracellular polysaccharides, thus accumulating an initial viable biofilm community. Inert particles, bacterial cells of the same or different species, and higher life forms (e.g., algae, amoeba, protozoa) continue to be recruited from the fluid and incorporated into the biofilm community.

As the biofilm bacterial communities mature, the adherent populations will oxidize electron donors and reduce existing terminal electron acceptors in an order of decreasing redox potential. As the biofilm community becomes denser and thicker, the penetration depth of one electron acceptor will overlap that of another; thus, a succession of different microbial populations can be established. Biofilm layers near oxygen-saturated water will harbor aerobic heterotrophic or autotrophic activity, reducing the available oxygen. As the distance from the aerobic interface increases and oxygen is depleted, a sequence of terminal electron acceptors is utilized by specific microbial populations: facultative denitrifiers using nitrate and nitrite; anaerobic sulfate reducers using sulfate; fermentative microbes partially reducing organic carbon compounds; and finally, anaerobic methane producers. Depending upon the balance between mass transfer rates and microbial reaction rates, biofilms can develop stratified ecosystems, on a microscopic scale, similar to those observed in lake or marine sediments. In engineered systems, highly active aerobic activity within a biofilm can completely deplete oxygen concentrations over distances of 10–20 μm, creating anaerobic layers capable of supporting obligate, anaerobic sulfate-reducing bacteria. Consequently, it should come as no surprise to discover anaerobic sulfate reduction and hydrogen sulfide corrosion of the metallic surfaces in heat exchange systems that utilize highly aerated cooling water to condense steam.

As a result of hydrodynamic forces and stresses exerted by replication, there can be a continual erosion of cells and extracellular material from the biofilm back to the bulk fluid. A more random stochastic process known as "sloughing" can also occur where either large sections or the entire biofilm becomes displaced from the substratum and enters the liquid.

Biofilms are as versatile as they are ubiquitous. Intentional use of biofilms can serve many benefits, for example, in the water and waste treatment industry, in bioremediation applications, and in industrial biotechnology. Unintentionally formed biofilms can create such detriments as biofouling of heat exchange systems and marine structures; microbial-induced corrosion of metal surfaces or the deterioration of dental surfaces; contamination of household products, food preparations, and pharmaceuticals; and the infection of short- and long-term indwelling biomedical implants and devices. Such detriments can range in severity from being a mere nuisance to being life threatening. This chapter will first summarize the impact of biofilms, both beneficial and detrimental, on engineered and natural processes. A detail discussion of the fundamental rate processes governing biofilm formation and persistence is then given, followed by a summary of recent advances in biofilm research.

Detrimental Biofilms

Figure 2 illustrates, for an arbitrary analytical measure, the typical accumulation of biofilm at a surface as a func-

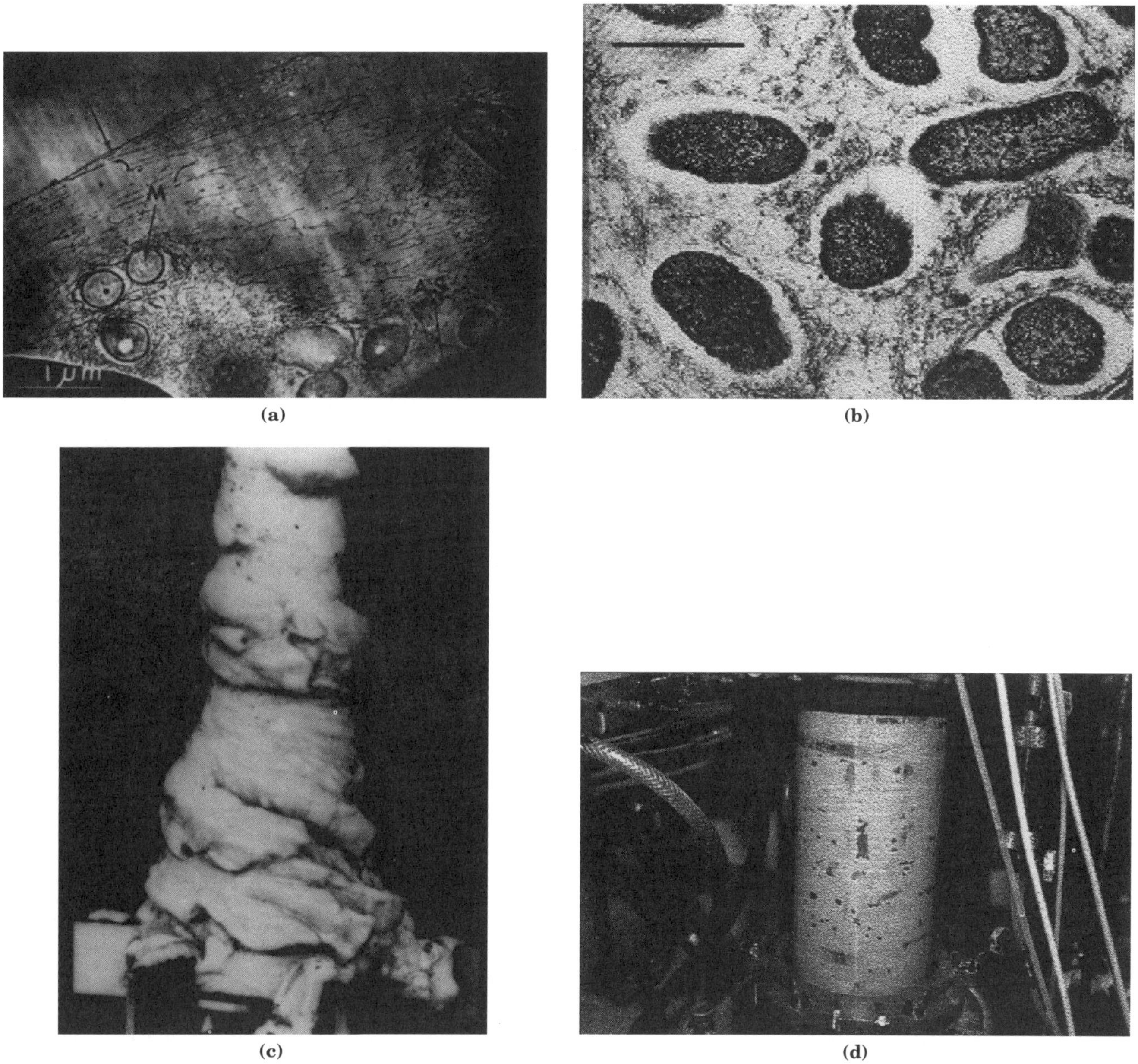

Figure 1. (**a**) Transmission electron micrograph of a sagital Section of a mixed-culture biofilm accumulating upon the surface of a rock in an alpine stream (1). (**b**) Transmission electron micrograph of a *Pseudomonas aeruginosa* biofilm isolated from a cystic fibrosis patient (2). (**c**) *Hyphomicrobium* spp. biofilm accumulating upon surfaces of a laboratory bioreactor vessel (3). (**d**) Accumulation of a yeast biofilm upon the impeller of a 200-L bioreactor (4).

tion of time. Also shown are the changes in frictional and heat transfer resistances observed in engineered systems experiencing biofilm formation.

Uncontrolled biofilm formation within natural, engineered, and biomedical systems can create numerous detriments as detailed in several definitive reviews (5–7). Detriments arise by way of a biofilm's influence on the transport of mass, momentum, and energy. Detriments attributed directly to bacterial biofilms include (*1*) material deterioration and corrosion (8–16), (*2*) increases in both frictional and heat transfer resistances (14,17), (*3*) attachment to and infection of biomedical implant devices (18–25), and (*4*) operational problems that plague lab-scale and full-scale bioreactors (6,7,26).

Beneficial Biofilms

Benefits afforded by biofilms in a continuous reactor situation arise chiefly because the cell population is immobilized, and thus the residence time of the cells in the reactor is independent of the fluid-phase residence time. In continuous suspended culture bioreactors (i.e., chemostats), the mean residence time of the system cannot be less than the generation time of the bacterial species, otherwise cells do not have sufficient time to replicate within the reactor and are eventually diluted from the system.

Immobilized and biofilm-bound cells remain in a continuous reactor system independent of the fluid phase, thus the mass loading of limiting substrate (or influent pollutant in the case of a wastewater treatment reactor) can be increased well beyond the growth rate limit imposed on suspended cultures. Consequently, immobilized-cell or biofilm reactors can provide (*1*) added volumetric reactivity, (*2*) more stable operating performance, (*3*) an inherent ease in biomass–fluid separation, and (*4*) the prospect of staging different bioconversion processes in sequential reactors. Due to these inherent advantages, the use of biofilm reactors is not confined to just bacterial cells, but also comprises plant and animal cell applications.

Bacterial biofilm reactors are employed either in commodities production or in wastewater treatment applications. Biofilm reactors have been reportedly used to produce acetic acid (27), L-lysine (28), gluconic acid (29), kojic acid (30,31), and ethanol (32,33), and in the epoxidation of propene (34). Such biofilm reactors are operated either as packed- or fluidized-bed reactor systems, with cells either attached to an inert support particle or artificially immobilized within a gel matrix.

One major application that relies on a microbial culture's ability to form biofilms is wastewater treatment. Biofilm reactor configurations, applied in both pilot- and full-scale wastewater treatment, include packed-bed trickling filters, high-rate plastic media filters, rotating biological contactors, fluidized-bed biofilm reactors, and membrane-immobilized cell reactors. Examples of biofilm reactors employed for wastewater treatment reported in the literature include polychlorinated hydrocarbon degradation (35), toluene degradation (36), denitrification (37,38), cadmium removal (39), anaerobic butyrate degradation (40), nitrification (41–43), glyphosphate degradation (44), anaerobic propionate degradation (45), phenolic wastewaters (46), uranium removal (47), and anaerobic carbon removal (48,49).

PROCESSES GOVERNING THE FORMATION AND PERSISTENCE OF BIOFILMS

Figure 2 illustrates, for an arbitrary analytical measure, the typical accumulation of biofilm at a surface as a function of time. Initially, the substratum is conditioned and cells attach reversibly, then irreversibly. Next, attached cells grow, reproduce, and secrete insoluble extracellular polysaccharidic material. As the biofilm matures, biofilm detachment and growth processes come into balance, such that the total amount of biomass on the surface remains approximately constant in time. Multiple physical, chemical, and biological processes operate in each phase (Fig. 3).

Processes governing biofilm formation and persistence include the following:

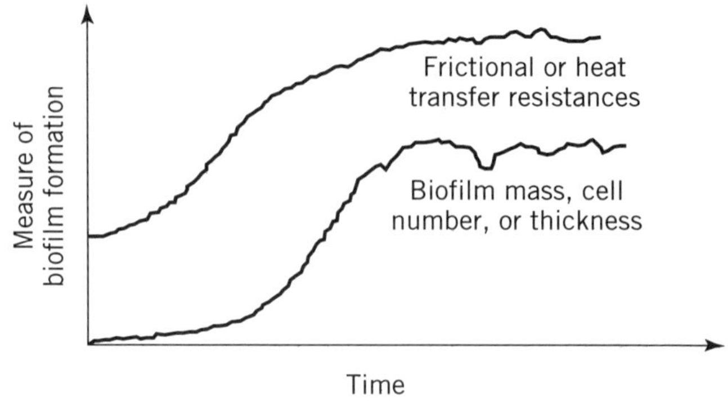

Figure 2. Net accumulation of biofilm and corresponding increases in frictional and heat transfer resistances.

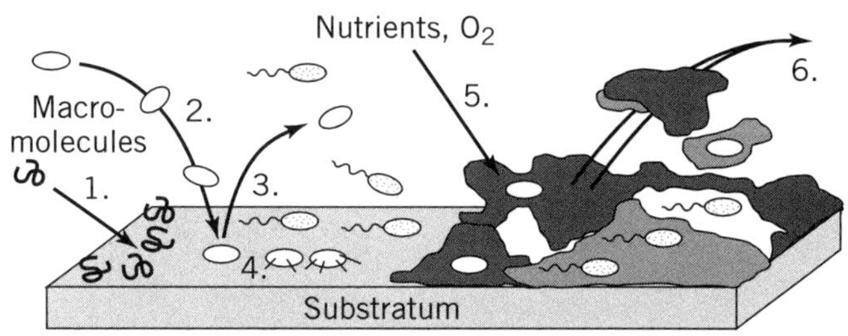

Figure 3. Schematic illustration of processes governing biofilm formation and persistence. (Processes are defined within the text.)

1. Biasing or preconditioning of the substratum either by macromolecules present in the bulk liquid or, intentionally, as in the case of precoating biomedical materials with adhesion molecules (e.g., fibronectin, vitronectin, fibrinogen, von Willebrand's factor).
2. Transport of planktonic cells from the bulk liquid to the substratum.
3. Adsorption of cells at the substratum for a finite time followed by desorption (release) of reversibly adsorbed cells.
4. Irreversible adsorption of bacterial cells at a surface.
5. Transport of substrates into the biofilm, substrate metabolism by the biofilm-bound cells, and transport of products out of the biofilm. These processes are accompanied by cellular growth, replication, and extracellular polymer production.
6. Biofilm removal can be either continuous, as in the case of biofilm detachment brought on by forces associated with either fluid flow or cell replication, or discrete, as in the case of biofilm sloughing.

Each of these processes is reviewed in some detail.

Initial Events

Processes 1–4 are often grouped in mathematical models as a single process termed deposition.

Preconditioning of the Substratum. In the case of intentional pretreatment of a surface (i.e., adhesion molecules for biomaterials), there is no rate to be considered; the substratum essentially enters the system with a biased surface chemistry. However, in cases where a clean substratum material is exposed to an aqueous environment, transport of dissolved organic molecules or macromolecules in laminar flow is basically by molecular diffusion; in turbulent flow, transport must also consider convective transport effects. Once at the surface, adsorption of macromolecules occurs almost instantaneously, immediately changing the surface chemistry of the exposed material. Loeb and Neihof (50) and Depalma et al. (51) measured adsorption rates of organic molecules to various solid substrata in seawater, and Bryers (52) reported analogous adsorption rates in a freshwater laboratory system. The net rate of adsorption in these studies can be described by equation 1.

$$R_{\text{adsorption}} = k_a S_i[1 - (S_i/k_s)] \quad (1)$$

where $R_{\text{adsorption}}$ is the rate of net adsorption ($ML^{-2}t^{-1}$); k_a is the adsorption rate constant (Lt^{-1}); k_S is the surface saturation coefficient (ML^{-2}); and S_i is the real concentration of adsorbed species (ML^{-2}). Although experiments indicate the maximum amount of adsorbed material may not exceed 0.10 μm in thickness, the surface properties resulting from adsorption of an organic film can significantly bias subsequent microbial events.

Cellular Adsorption and Desorption. The net deposition rate of cells upon a "conditioned" substratum is itself a composite of the individual rates of cell transport to the substratum, cell adsorption, and adherent cell desorption.

Many models exist to describe bacterial transport from the fluid phase to the target surface. Bowen et al. (53) and Beal (54) derive continuity equations describing the transport of suspended particles from a flowing fluid to the surfaces of a surrounding conduit that accounts for both convective and molecular transport mechanisms. From their work, the maximum flux of particles transported to the surface, L distance from the inlet, of a rectangular flow cell, with gap height $2h$, can be written as

$$J(\text{cells-}L^{-2}t^{-1}) = k_{tr}X\mathcal{D}_x X/h \quad (2)$$

where k_{tr} is the particle transport coefficient ($(2/9)K^{0.33}/\Gamma(4/3)$); $K = (1/Pe)(8L/3h)$; Pe is the Peclet number $= 4vh/\mathcal{D}_X$; v is the average velocity of the fluid; Γ is the gamma function ($\Gamma(4/3) = 0.89338$); and $\mathcal{D}_X$ is the diffusivity of cells in the liquid. For nonmotile cells, the Brownian diffusion coefficient can be estimated from the Stokes-Einstein equation,

$$\mathcal{D}_x X = k_b T/3\pi\mu d_c \quad (3)$$

where k_b is the Boltzmann constant (1.38×10^{-23} J/K); T is absolute temperature; μ is absolute viscosity; d_c is cell diameter. To compensate for cell motility, Jang and Yen (55) propose the following equation,

$$\mathcal{D}_x X = v_m d_r/3(1 - \cos\sigma) \quad (4)$$

where v_m is the velocity of motility, d_r is the free length of a random run, and σ is the angle turned by the motile cell.

Once at the substratum interface, cells stick. Observations have indicated that cells can adsorb to a surface for a period of time and then may desorb from the substratum, returning to the bulk liquid. If cells spend sufficient time at the surface, they become permanently bound, initiate within minutes the secretion of extracellular polymer production, and can be removed only by rather aggressive physical or chemical means. One can conceive of this overall process as much like two reaction steps in series, with the reversible adsorption process followed by an irreversible step pertaining to permanent adsorption,

$$X \leftrightarrow X_{rev} \rightarrow X_{irr} \quad (5)$$

where X is the suspended cells, X_{rev} is the reversibly adsorbed cells, and X_{irr} is the irreversibly adsorbed cells. Assuming no cell replication, the total amount of cells observed at a substratum, X_t, at any one time comprises both reversibly and permanently adsorbed cells, that is, $X_t = X_{rev} + X_{irr}$.

Cell Attachment. Attachment is the deposition of cells from the fluid phase onto an existing biofilm. This process differs from adsorption in that the target surface is now a developed biofilm. The biofilm itself poses a different type of substratum in that biofilm surface morphology is generally more irregular (perhaps filamentous), very porous, compliant, and gelatinous. Experimental results indicate that the increased attachment of cells to a biofilm-covered substratum may depend on far more subtle mechanisms than mere changes in surface morphology. The local chemistry of a biofilm surface will reflect the composition of the bacterial cell and that of the insoluble polymer matrix, both of which could affect subsequent chemical interactions with incoming cells.

Banks (56) and Banks and Bryers (57) report that the rates of cell deposition increased when cells were exposed to biofilm surfaces versus clean glass (Fig. 4). Two species were investigated, *Pseudomonas putida* and *Hyphomicrobium ZV620*, as to their deposition rates, under laminar flow conditions, onto both clean glass and biofilm-covered surfaces. *Pseudomonas putida* cells attached to a *P. putida* biofilm at a rate (zero order in attached cell concentration) that was 3.6 times higher than the cell attachment rate to clean glass. *Pseudomonas putida* cells attached to a *Hyphomicrobium* biofilm (with zero-order kinetics in attached cell concentration) that was 2.4 times higher than to clean glass. *Hyphomicrobium* illustrated an attachment rate to a *Hyphomicrobium* biofilm (again a zero dependency on attached cells) that was 3.3 times higher than the *Hyphomicrobium* attachment rate to clean glass. Attachment rate of *Hyphomicrobium* spp. cells to a *P. putida* biofilm was only 1.3 times that to clean glass. The attachment rate of *Hyphomicrobium* cells to a *Hyphomicrobium* biofilm was about half of the rate of *Pseudomonas* cells attaching to a *Pseudomonas* biofilm, even though suspended *Pseudomonas* cell concentrations were three times higher. Results insinuate that species-dependent enhancement of cell attachment may occur by specific rather than nonspecific adhesion mechanisms.

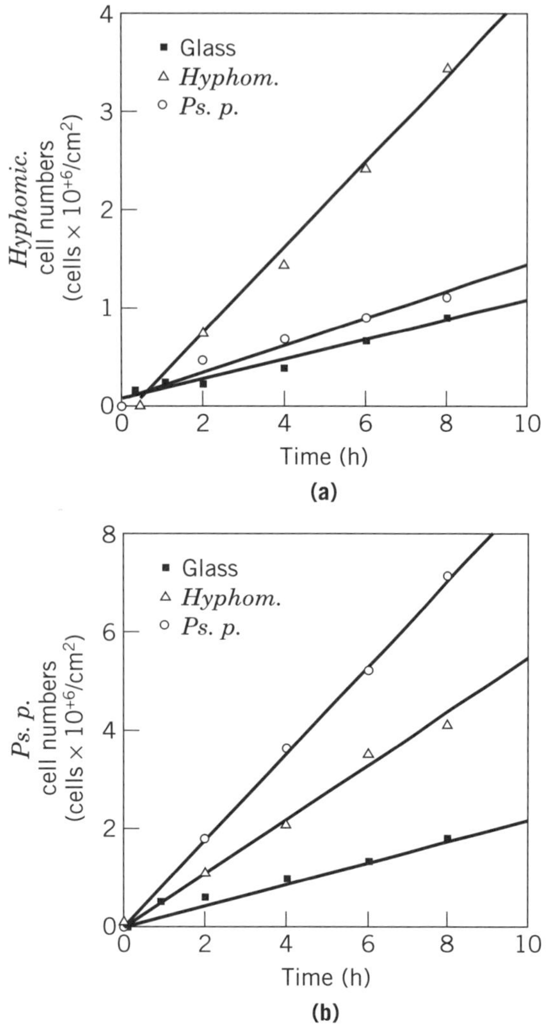

Figure 4. (**a**) Attachment of *Hyphomicrobium* spp. to various surfaces. Suspended cell concentration is 4.5E + 07 cells/mL. (**b**) Attachment of *Pseudomonas putida* to various surfaces. Suspended cell concentration is 1.5E + 08 cells/mL (57).

Biofilm Cell Metabolic Processes

Transformation collectively refers to those processes in which molecular rearrangements occur, that is, reactions. Once a cell adsorbs to a surface or becomes embedded within an existing biofilm, it will continue its metabolic processes in response to its immediate environment. Three fundamental rate processes can be identified: *(1)* alginate or exopolymer secretion, *(2)* cellular growth and replication, and *(3)* cell death and lysis. Prior to the mid-1990s; it was often easier to measure "per reactor area" performance using observed rates determined from bulk liquid parameters (e.g., electron donor removal rate or electron acceptor uptake rate) than to directly measure growth, replication, and death of cells in a biofilm. This experimental limitation led to the development of a number of unstructured mathematical models that were useful in estimating the flux of growth-rate-limiting substrate into a biofilm of fixed thickness, density, and reactivity, taking into account both external and internal molecular diffusion and biological reaction. It was not until the advent of various molecular methods and advances in computer-aided microscopy that it was possible to determine various biofilm processes directly within the biofilm.

Adhesion-Induced Alginate Synthesis. Reporter gene technology has been used to observe the regulation of the alginate biosynthesis gene, *algC*, in a mucoid strain of *P. aeruginosa* in developing and mature biofilms (58). In vivo detection of *algC* up-expression in developing biofilms was performed with a fluorogenic substrate for the plasmid-borne *lacZ* gene product (β-galactosidase) using microscopy coupled with image analysis. By this technique, cells were tracked over time and analyzed for *algC* activity. During the initial stages of biofilm development, cells attached to a glass surface for at least 15 min exhibited up-expression of *algC*, detectable as the development of whole-cell fluorescence. However, initial cell attachment to the substratum appeared to be independent of *algC* promoter activity. Furthermore, cells not exhibiting *algC* up-expression were shown to be less capable of remaining at a glass surface under flowing conditions than were cells in which *algC* up-expression was detected.

Cell Growth and Replication. The major transformation carried out by cells in the biofilm is the metabolism of both an electron donor and a terminal electron acceptor to produce soluble by-products, extracellular polymers, carbon dioxide, and water. Depending on the microbial population in question and the ambient concentration of electron donor and acceptor, a biofilm can be either aerobic, anoxic (denitrifying), anaerobic (sulfate-reducing bacteria, methane formers), or fermentative. Analysis of biofilm bacterial metabolic rates are frequently complicated by the effects of significant mass transfer resistances in both the liquid phase and within the developing biofilm. Steady-state mathematical models (59,60) were derived in the late 1970s to estimate the observed flux of growth-rate-limiting substrate into a biofilm of fixed thickness, density, and reactivity. These steady-state biofilm models are based on assumptions of a constant biofilm concentration (tacitly implying a spatially uniform reactivity) and a constant diffusion path (biofilm thickness). Such models allow one to predict *(1)* the concentration profile of limiting substrate (and, by stoichiometry, all other nutrients) with biofilm depth, and *(2)* the maximum substrate uptake or flux to the biofilm.

Biofilm cellular growth rates are typically simulated using a classical microbial growth rate expression, R_g, that is related by a stoichiometric coefficient, Y (in units of cell mass per substrate mass), to the rate of utilization of growth-limiting substrate, R_S ($M_{\text{substrate}}\,L^{-2}t^{-1}$), that is,

$$R_g(M_{\text{cells}}L^{-2}t^{-1}) = -R_S Y \tag{6}$$

where the rate of substrate utilization is typically a Monod-like dependency on the concentration of one or more growth-rate-limiting substrates. For the example of a single substrate limitation,

$$R_S = \mu^* SB/(K_S + S) \quad (7)$$

where μ^* is the species maximum growth rate constant (t^{-1}); S is the local dissolved substrate concentration ($M_{substrate}L^{-3}$); K_S is the saturation kinetics half-saturation constant ($M_{substrate}L^{-3}$); and B is the local biofilm cell concentration ($M_{cell}L^{-2}$).

Some differences in biofilm kinetics versus typical suspended cell microbial kinetics include the following:

- The concentration of active bacterial cell mass metabolizing the substrate, B, is a per area concentration and may also be a function of time and space, $B = B(t, z)$. In a mixed culture biofilm, each bacterial species could exhibit a different dependency on time and space.
- The concentration of substrate S in the biofilm will be affected by mass transfer resistances and biological reaction rates, such that S may also be a function of time and space.
- Endogenous decay, maintenance energy, or death terms for cells not receiving adequate nutrients are treated as simple first-order functions of the biofilm population, for example,

$$R_{death} = k_d B \quad (8)$$

where k_d is the cell death specific rate constant (t^{-1}). Thus the local, instantaneous net production of a species biomass in the biofilm can be modeled as

$$R_g(t, z) = [\mu^* S(t, z)B(t, z)/Y(K_s + S(t, z))] - k_d B(t, z) \quad (9)$$

Biofilm Removal Processes

Biofilm removal processes involve the actual loss of material from the biofilm—either cells or cell–biofilm matrix debris. Desorption (see "Deposition Processes") is the release of cells from the substratum and occurs during the early stages of substratum colonization. Biofilm removal represents loss of biomass from a developing or mature biofilm that arises by way of two distinctively different mechanisms: detachment and sloughing.

Detachment. Detachment is the continuous loss of biomass or individual cells from a biofilm. The rate of biofilm detachment has been mathematically modeled in a number of ways (61). In about the only definitive experimental study on biofilm detachment, Applegate and Bryers (62) reported that the growth conditions of the biofilm strongly affect biofilm removal processes. Carbon-substrate-limited growth conditions fostered biofilms that *(1)* contained less extracellular polymer per cell and per area, *(2)* bound less calcium per unit amount of biofilm, and *(3)* exhibited a higher detachment rate than oxygen-substrate-limited biofilms that accumulated more extracellular polymer per cell, bound more calcium per biofilm amount, and developed a rigid compact morphology that was very resistant to detachment prior to predictable sloughing. Carbon-limited biofilms did not exhibit sloughing for extended periods of cultivation or after prolonged (300 h) of carbon-nutrient deprivation.

Perhaps biofilm detachment processes may be controlled, not by fluid hydrodynamics as once imagined, but rather by cell physiology. Boyd and Chakrabarty (63) suggest biofilm detachment could be genetically controlled. Their work showed that the exopolysaccharide alginate of *P. aeruginosa* was important in determining the degree of cell detachment from an agar surface. Nonmucold strain 8822 gave rise to 50-fold more sloughed cells than mucoid strains 8821 and 8830. Alginate anchors the bacteria to the surface, thereby influencing the extent of detachment. The role of the *P. aeruginosa* alginate lyase in the process of cell detachment was investigated. Increased expression of the alginate lyase in mucoid strain 8830 led to alginate degradation and increased cell detachment. Similar effects were seen both when alginate lyase synthesis was induced at the initial stage of cell inoculation and when it was induced at a later stage of growth. Apparently, high molecular weight alginate polymers are required to efficiently retain the bacteria within the developing biofilm. When expressed from a regulated promoter, the alginate lyase did enhance detachment of cells through the degradation of the alginate. This suggests a possible role for the lyase in the removal process of bacterial growth films.

Sloughing. The catastrophic, apparently random, loss of large pieces or entire sections of biofilm is known as sloughing (Fig. 5). At the industrial scale, sloughing creates excursions of intolerably high biomass in bioreactor effluents, and periodic "blooms" of pathogenic bacteria in drinking water distribution systems. At the laboratory scale, all too often, random sloughing of a biofilm will terminate an experiment, resulting in very little quantitative data regarding the fundamental causes of sloughing. A number of situation-specific causes of sloughing have been identified: bubble formation in either anaerobic methane-producing biofilms or denitrifying biofilms, and artificially induced sloughing by way of EGTA addition, which chelates the bound calcium ion used in biofilm alginate cross-linking. Applegate and Bryers (62) reports that the growth conditions of the biofilm strongly affected the biofilm removal processes of detachment and sloughing. "Fluffier" biofilms, produced under carbon-limited growth, exhibited high detachment rates but never sloughed, even when subjected to over 300 h of nutrient starvation. Conversely, rigid biofilms cultivated under oxygen limitations showed little detachment, but the onset of catastrophic sloughing was repeatable between replicate experiments.

Net Accumulation of Biofilm

As shown in Figure 1, development of a bacterial biofilm comprises the following fundamental processes: *(1)* substratum preconditioning by adsorption of fluid-phase organic molecules; *(2)* bacterial cell deposition to the conditioned substratum (cell transport to the surface, cell adhesion to the substratum, cell desorption); *(3)* bacterial metabolism (cell substrate conversion; cell growth and replication; extracellular exopolymer production; cell starvation, death, lysis); *(4)* biofilm removal (cell and biofilm de-

Figure 5. Sequence of biofilm formation (**a**), onset of partial biofilm sloughing (**b**), and complete biofilm sloughing (**c**) of an anaerobic biofilm. *Source*: Ref. 3.

tachment, biofilm sloughing). Naturally, the relative influence of each process to the overall rate of accumulation is dependent upon the specific system, the prevailing environmental conditions, and biological changes throughout the lifetime of the biofilm.

Simple unstructured models of biofilm formation can be written. Biofilm formation is assumed to occur on surfaces exposed to a well-mixed bulk fluid phase, thus tacitly neglecting any concentration gradients in the bulk liquid. Any mass transfer limitations in the bulk liquid will only complicate the already difficult task of determining the kinetics of a heterogeneous reaction. Suspended biomass in the bulk liquid arises due to either cell growth and replication or to detachment of biofilm material. Planktonic cells leave the liquid phase by either the effluent liquid leaving the reactor volume or by cell deposition onto the reactor surfaces. A single, sterile growth-limiting substrate, S, enters the reaction volume, where it is consumed by either the suspended or biofilm-bound cells. Any reasonable measure of both biofilm and suspended cell concentrations is acceptable (cell number, biomass dry weight, biomass organic carbon). However, to complete material balances over the entire system, one most know the stoichiometric relationship between changes in the suspended and attached cell concentrations due to growth and the limiting substrate utilized.

Based on these assumptions, material balances for both the suspended cell biomass and single growth-limiting substrate can be written as equations 10 and 11:

Suspended biomass balance: $$VdX/dt = -FX + (R_gV)_{\text{suspended}} + R_{\text{det}}A - R_{\text{dep}}A \quad (10)$$

Limiting substrate balance: $$VdS/dt = F(S_{\text{in}} - S) - (R_gV/Y)_{\text{suspended}} - (R_gA/Y)_{\text{biofilm}} \quad (11)$$

where X is the suspended cell biomass (ML^{-3}); S is the limiting substrate concentration (ML^{-3}); V is the reaction volume (L^3); $(R_g)_{\text{suspended}}$ is the rate of suspended cell growth ($M_{\text{biomass}}L^{-3}t^{-1}$); R_{det} is the rate of biofilm detachment ($M_{\text{biomass}}L^{-2}t^{-1}$); R_{dep} is the net rate of bacterial cell deposition at a substratum ($M_{\text{biomass}}L^{-2}t^{-1}$); $(R_g)_{\text{biofilm}}$ is the rate of biofilm cell growth ($M_{\text{biomass}}L^{-2}t^{-1}$); Y is the yield coefficient ($M_{\text{biomass}}/M_{\text{substrate}}$); and A is the reaction area (L^2). Rate expressions for the process rates (R_g, R_{det}, R_{dep}) may depend directly on such changing parameters as biofilm amount, limiting substrate, and suspended biomass concentration.

Biofilm net accumulation can be described by equation 12:

Biofilm net accumulation: $$A(dB/dt = R_{\text{dep}} + R_{\text{gb}} - R_{\text{det}}) \quad (12)$$

Because rate expressions for both planktonic and biofilm growth are nonlinear saturation kinetic functions of the instantaneous substrate concentration and first-order functions of X and B, respectively, equations 10–12 are coupled, nonlinear ordinary differential equations that require either simultaneous numerical integration (for the transient situation) or simultaneous algebraic solution under steady-state conditions (i.e., dS/dt, dX/dt, and $dB/dt = 0$). Numerous biofilm accumulation studies have been carried out that can be mathematically described by this set of equations.

Should the reactor be operated at a hydraulic residence time shorter than the generation time of the cells, then one can assume the term ($R_{gs}V$) is negligible; however, the corresponding term in equation 11, representing suspended cell substrate metabolism ($R_{gs}V/Y_{X/S}$), cannot be disregarded unless in the specific system this term is determined to be small compared to the other terms, or substrate is not supplied.

In a simple bacterial adhesion study, experiments can be carried out without exogenous substrate, thus the R_g terms in equations 10–12 could be ignored. Thus, equation 12 becomes

$$dB/dt = R_{dep} - R_{det} \tag{13}$$

and is valid provided the adherent bacterial replication is zero.

RECENT ADVANCES IN BIOFILM RESEARCH

This section highlights recent research advances in understanding biofilm phenomena, which include advances in biofilm experimental techniques, new findings regarding mass transport mechanisms in biofilm, and genetic processes within biofilm communities.

Advances in Biofilm Experimental Techniques

All biofilm experimental systems should contain certain critical components: *(1)* the biofilm study reactor itself (where the surface accumulation of cells and biofilm will be studied) and *(2)* the environmental support system that controls the experimental conditions (e.g., temperature, flow velocity, nutrient concentrations, pH) of the bulk fluid. Under quiescent conditions, the rate of transport of dissolved nutrients and particulates (i.e., the bacteria) to the substratum may be the slowest in a series of rate processes. For example, particle transport may limit the rate of cell deposition in quiescent conditions. Thus, it is recommended that only experimental systems that create a gradient-less, well-mixed bulk fluid phase be employed. Otherwise, instead of determining the true kinetics of whatever biofilm process is of interest, the resultant data will erroneously reflect mass transfer limitations.

Invasive Biofilm Diagnosis. Detection of biofilms often requires direct sampling and removal of a finite quantity of biofilm from a reactor surface to be used in a number of destructive analytical procedures that may measure *(1)* overall biofilm amount (cell mass or biofilm thickness), *(2)* a component of the biofilm (biofilm carbon, cell number, biofilm nitrogen content), or *(3)* a biofilm cell component or cellular activity (ATP, cell protein, dehydrogenase activity, active DNA synthesis, species-specific 16S rRNA). Reactor designs that provide access for destructive sampling involve removable sections of substratum and are quite numerous. The reader is directed to a number of excellent reviews for further details (5,7). One drawback to destructive analyses is that the parameter determined represents an average over the entire sample area, which ignores any spatial variations in that parameter that may exist.

Almost any parameter determined by a molecular or chemical assay (ATP, DNA, total protein, biofilm polysaccharide content, 16S rRNA, cellular phospholipid signatures, specific enzyme levels, and oligonucleotide probes), cytochemical (acridine orange, DAPI stain, and Hoechst DNA stains) or immunofluorescent staining, or cellular activity (CTC stains, INT stains, and plate counting) techniques that has been applied to planktonic (freely suspended) cells can be used to quantify changes in biofilm on a substratum. Rodriguez et al. (64) reports on the potential of a fluorescent redox probe, 5-cyano-2,3-ditolyltetrazolium chloride (CTC), for direct observation of actively respiring bacteria. Oxidized CTC is colorless, but upon reduction by electron transport activity, the insoluble CTC-formazan fluoresces at 365 nm upon excitation. Respiring marine bacteria within thin (10–50 μm) biofilms attached to optically nontransparent polysulfone substrata were rapidly and easily enumerated. Respiring cell counts via CTC were also useful in directly assessing the efficacy of various biocidal control agents.

The advent of modern molecular biology has provided many useful diagnostic methods with which to analyze biofilm ecology. For example, Amann et al. (65) characterize the population structure of sulfidogenic biofilms, established in anaerobic bioreactors, by selective polymerase chain reaction (PCR) amplification and fluorescent microscopy. 16S rRNA common to the sulfate-reducing bacteria in the biofilm was selectively amplified by PCR and used to design both general and specific fluorescent hybridization probes. Biofilms, (5–10 μm thick) on glass cover slips immersed within laboratory anaerobic biofilm reactors, were fixed in formaldehyde and dehydrated prior to hybridization with the RNA probes. Hybridized biofilm samples were then viewed under epifluorescent microscopy.

Rogers and Keevil (66) report on the destructive sampling of a multispecies biofilm, intentionally inoculated with *Legionella pneumophilia*, followed by both immunogold and fluorescein immunolabeling for this species. Episcopic differential interference contrast microscopy was employed to simultaneously visualize the total biofilm community and the labeled *Legionella* species. This techniques was able to provide observations of variations in the biofilm community from one position to another (parallel to the substratum) but was unable to discern such variations with depth (perpendicular to the substratum).

McCarter et al. (67) report a very elegant series of genetic techniques to investigate bacterial responses to various stimuli, one being the adhesion to a surface. Using a transposon (mini-Mu*lux*), the authors were able to incorporate recombinant reporter gene insertions adjacent to target gene promoter sites that encode for light production as a function of target gene expression. Light production is conveniently measured by exposure of X-ray or photographic film, visual examination, chemiluminescence, or photometry. Such techniques also obviate the need to disrupt biofilms prior to quantification. McCarter et al. employed a *lux* reporter gene adjacent to the chromosomal gene encoding for swarmer cell differentiation that *Vibria parahaemolyticus* experiences upon association with a solid surface. Resultant mutants thus emit light upon growth at a surface but not in suspension. Dagostino et al.

(68) also employed transposon mutagenesis to insert into an appropriate recipient bacteria a marker gene that lacks its own promoter. The premise is that if a suitable target gene is "on" at the surface, then expression of the marker will be observed only in the presence of the surface. Dagostino et al. employed *Escherichia coli* C600 (pRK2013: mini-MuTetr *lacZ*) as donor and two marine bacteria as recipients. The authors were able to successfully isolate a transposon-generated mutant in which the *lacZ* gene was not expressed in either liquid or agar, but was expressed when grown on a polystyrene substratum.

Noninvasive Biofilm Diagnosis. The "holy grail" of biofilm research is a diagnostic method that creates as little disturbance to the biofilm and the reactor operation as possible. Regrettably, most parameters of interest in cell adhesion and biofilm accumulation require destructive sampling and analyses.

True noninvasive diagnostics of biofilm accumulation, amount, and biofilm-cell reactivity are limited to relatively few sophisticated techniques. The simplest of these techniques is on-line microscopy, which places the biofilm surface within the viewing field of a microscope. This technique requires a reactor dimension compatible with the microscope, and a flat reactor surface on which to focus. Microscopic observation of adherent cells can be quantified and preserved by use of video recorders or image analyzer systems that digitize observed images and save those images on a computer for future numerical interpretations.

For about 14 years, confocal scanning laser microscopy (CSLM) has been a tool in the biological sciences. CSLM uses confocal apertures, or pinholes, to create a thin (0.4 μm) depth of field, which eliminates out-of-focus light emitted from the laser-excited fluorophores. Laser light sources provide the intense, highly coherent, collimated light necessary to penetrate thick specimens. The laser light is used to excite fluorophores—either those intrinsic (e.g., chlorophyll) to the sample, or selected chemical or immunological stains intentionally applied to the sample. Resultant fluorescence that passes through the aperture is detected by a photomultiplier, and a digital image is collected. Using a computer-manipulated stage, optical sections can be collected from a specimen in three dimensions. This allows, for the first time, the ability to digitally locate cells on a substratum or within a biofilm, noninvasively (69). Numerous fluorochromes are available that can be used to localize and measure intracellular and extracellular conditions in three dimensions within living biofilms.

Camper et al. (70) detail image analysis software that can be used to collect images of bacterial adhesion patterns from the CSLM, digitize those images, then superimpose those images onto digitized images of the substratum itself, collected from any one of a number of other analytical instruments (e.g., X-ray photoelectron spectroscopy, atomic force microscopy, time-of-flight secondary ion mass spectroscopy); all images are collected at the same location on the substratum. Thus, one can now correlate the adhesion of a single cell to the local topographical or chemical properties of the substratum directly underneath the cell.

Lawrence et al. (71) illustrates the ability to optically section fully hydrated biofilms, both horizontally and sagitally, producing optical sections of undisturbed biofilms with a spatial thickness of 2 μm. Any fluorescent probe (cellular DNA stain; immunofluorescent stain; pH-, ion-, and redox-sensitive stains; viability redox stains) or any combination of probes can be simultaneously recorded by the CLSM with great clarity and little background interference, due to the confocal exclusion of any fluorescence originating from excited fluorochrome above and below the focal point (Fig. 6). CLSM is a critical tool in population analysis of biofilms consisting of mixed strains, mixed species, and mixed cell lines and will allow detailed examination of the relationships between biofilm structure, adaptation, reactivity, and response to external stress (72).

Another noninvasive technique that can be used to collect information on the molecular chemistry of cell adhesion at the substratum–fluid interface is Fourier transform infrared spectroscopy (FT-IR) coupled with attenuated total reflectance (ATR) waveguides integrated within flow cells. Atoms and groups of atoms within a molecule vibrate with a characteristic frequency and will adsorb light at those frequencies. Light that contains infrared frequencies can be focused on a molecule, and the amount of light adsorbed is then measured as a function of frequency. Specific IR adsorptions can thus be assigned to particular bonds, and alterations in these bonds due to changes in local environment can be assessed from the resultant spectral details. Recent increases in the capability of FT-IRs and the focusing of the IR wave within specific crystal wave guides (germanium or zinc selenide crystals) allows one to establish a standing IR evanescent wave at the surface of the crystal. Fourier transform signal processing and multiple scanning allow aqueous samples to be processed where the IR spectra of molecules directly adjacent to the waveguide can be collected. In biological systems, the IR absorbance of water must be subtracted via computer manipulations to provide the IR spectra of molecules accumulating at the crystal surface. The effective depth of penetration for the evanescent IR wave is a function of crystal material and the wavenumber but ranges around 0.3–0.7 μm, thus providing spectra that reflect those chemical species directly adjacent to the surface.

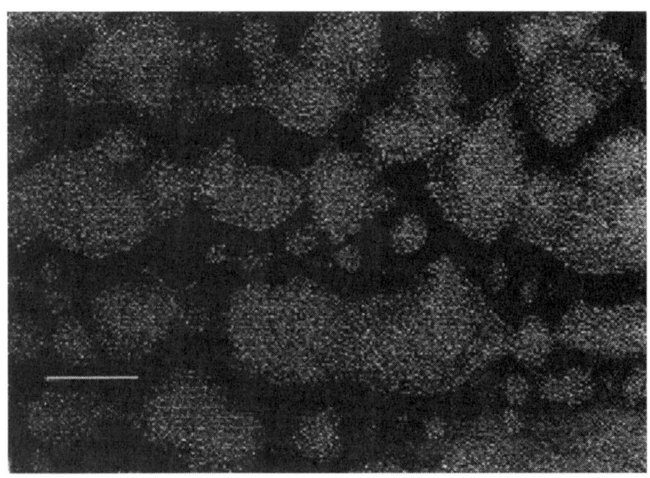

Figure 6. Confocal laser microscope optical section of a biofilm.

ATR/FT-IR has been used to study the effects of protein-conditioning films on alginate adsorption to the germanium surface (73). Alginate adsorbed in greater quantities to protein-coated ATR crystals than to uncoated ones. For the development of biofilms on ATR crystals, the observed spectrum is the integral of all biomolecules on the surface and over the entire surface. Bremer and Geesey (74) report the chemical changes as detected by FT-IR that occur upon inoculating with a microbial culture, at a germanium ATR crystal situated within a flow cell. Spectral intensities representing various bonds within proteins (amide I at 1,645/cm; amide II at 1,550/cm) and polysaccharide (C-O stretches at 1,058/cm) are seen to accumulate with time, indicating the feasibility of the ATR/FT-IR system to detect biofilm formation. Because the IR wave can only penetrate less than 1 μm into the biofilm, reported increases in absorbance intensities most likely represent increases in cell surface coverage of the crystal and not increases in biofilm thickness. A plot of attached cells per area versus the increases in area under the amide I peak was generated to estimate a lower limit of detection by FT-IR of $\sim 5 \times 10^5$ cells/cm^2. In a medical application, ATR/FT-IR has been used to follow the penetration of an antibiotic (100 μg/mL ciprofloxacin) into the IR wave region underneath a *P. aeruginosa* biofilm (75).

Several reports exist of the use of invasive but nondestructive analysis of solute concentration profiles, by microsensor chemical probes, as a function of spatial dimension within the developing biofilm. Microsensors (tip size $\leq$ 15 μm) exist that can detect various dissolved solutes including glucose, oxygen, pH, sulfide, and ammonia. DeBeer et al. (76) report using microoxygen sensors (tip size = 15 μm) to estimate oxygen profiles, local oxygen uptake rates, and oxygen diffusion coefficients in a mixed-culture biofilm community cultivated within a laboratory biofilm reactor. Schramm et al. (77) determined microprofiles of oxygen and nitrate in nitrifying biofilms from a trickling filter and correlated these solute profiles with bacterial population profiles for *Nitrosomonas* and *Nitrobacter* spp., as determined using fluorescence in situ hybridization of cells fixed with 16S rRNA oligonucleotide probes.

Transport Phenomena in Biofilms

Bacterial biofilms are heterogeneous reaction systems wherein limiting substrates, essential nutrients, cellular antagonists, and cellular by-products are exchanged between the surrounding fluid and the site of the reaction (i.e., the cell). In numerous circumstances, the performance of biofilm processes can be limited by both internal and external mass transfer rates (e.g., antibiotic treatment of a biofilm infection, biocide challenge to eliminate biofilm formation in heat exchange systems, biofilm reactor treatment of wastewaters or waste gases, and in situ hazardous waste decontamination). Recent advances in characterizing biofilm architecture have led to the need to reformulate concepts of the mass transport mechanism in biofilms.

For about 25 years, estimates of substrate consumption for biofilms of known thickness have been based on the steady-state solution of equation 14,

$$\mathcal{D}_{\text{eff}} d^2C/dx^2 = \upsilon R_c \quad (14)$$

where C is the solute concentration, R_c is all reactions transforming solute C, υ is the stoichiometric coefficient for the transformation, x is the coordinate of molecular transport, and $\mathcal{D}_{\text{eff}}$ is the effective diffusivity of solute in the biofilm. These various models are based on the assumption that transport of reactants (i.e., substrates) into the biofilm and export of products is by molecular diffusion alone and that bulk transport (convection) does not occur within the biofilm. These models also do not account for *(1)* any heterogeneities in the biofilm gel structure, *(2)* any locally dependent bacterial concentrations, or *(3)* any nonlinearity in diffusion coefficients.

Traditionally, either $\mathcal{D}_{\text{eff}}$ is estimated as the solute diffusivity in pure water, reduced by an arbitrary fraction, or $\mathcal{D}_{\text{eff}}$ is determined for the particular biofilm system in question from diffusion experiments. For small ions and uncharged small organic compounds, both Libicki et al. (78) and Siegrist and Gujer (79) indicate that transport in thin microbial layers is indeed by molecular diffusion. Both reports indicate that the decrease in observed diffusivity values relative to those in pure water is a function of the microbial aggregate cell-to-polymer composition. Siegrist and Gujer (79) provided the first report to insinuate that molecular diffusion may not be the only mass transport process acting within a biofilm and that convective mass transport may predominate in the upper layers of thick, filamentous biofilms.

Results of Drummond (80) and Bryers and Drummond (81) suggest these dense alginate clusters are themselves porous, with a wide distribution of pore sizes ranging from angstroms to tenths of micrometers. Their results have shown the transport of larger solutes (i.e., macromolecules) through the hydrated biofilm polysaccharide gel matrix is controlled by transport mechanisms far more complex than molecular diffusion. Overall effective diffusion coefficients for various molecules, increasing in molecular weight and differing in molecular size and stereochemistry, were determined using half-cell diffusion chambers. Transient diffusion studies were carried out at constant temperature and pH using bacterial biofilms of increasing thickness. Half-cell diffusion results were considered erroneous due to the "perforated" nature of the biofilms.

The existence of water channels in these biofilms was documented and found to significantly affected mass transfer coefficient estimates obtained from half-cell studies. Consequently, a local microtransport technique was developed based on the fluorescence recovery after photobleaching (FRAP) technique of Poitevin and Wahl (82). FRAP is a local, instantaneous, and noninvasive method that can be repeatedly carried out over a wide expanse of biofilm. Corrections of the FRAP "tracer" diffusion coefficients must be made in order to estimate "Fickian" diffusivities. Diffusion coefficients determined for large macromolecule solutes (dextrans, proteins, DNA fragments) moving in biofilms did not correlate simply with increasing molecular weight except within the homologous chemical groups. Transport of macromolecules through the biofilm was affected not only by molecular weight but by the size or length of the molecule in solution. Using a computer-actuated laser and motorized microscope stage, FRAP analysis across a large section of biofilm illustrates that local mass transfer coefficients can vary significantly in a biofilm over very short distances (Fig. 7).

Thus, research reported by Drummond (80) and Bryers and Drummond (81) and independently corroborated by Stoodley et al. (84) has shown biofilms are not architecturally spatially continuous. Rather, biofilms are structurally nonuniform, consisting of dense microbial clusters surrounded by tortuous water channels that perforate the biofilm matrix. Thus, theory and experiments addressing overall mass transport within a real biofilm system must account for both convective transport and molecular diffusion.

Molecular Processes in Biofilm Communities

Plasmid Expression and Retention in Biofilm Cultures. Any exposure of plasmid-recombinant microorganisms to an open system environment, either inadvertently or intentionally, mandates research into those fundamental organism–plasmid processes that influence plasmid retention, transfer, and expression. Use of plasmid-recombinant bacteria to impart a desired phenotypic ability to an existing ecosystem has been desired for the bioremediation of hazardous wastes in both controlled reactor systems and in waste-contaminated open environments. In open environmental systems, a majority of the microbial activity occurs associated with an interface, within biofilms. In wastewater or waste gas treatment, efficient reactor design often dictates the use of immobilized, biofilm-bound bacterial communities. Consequently, the fate and ability to manifest a plasmid-borne phenotype within a biofilm community requires basic information on the biofilm's genetic processes.

A plasmid is deemed unstable if it either undergoes molecular rearrangement (structural instability) or is not absolutely inherited by progeny (segregational instability).

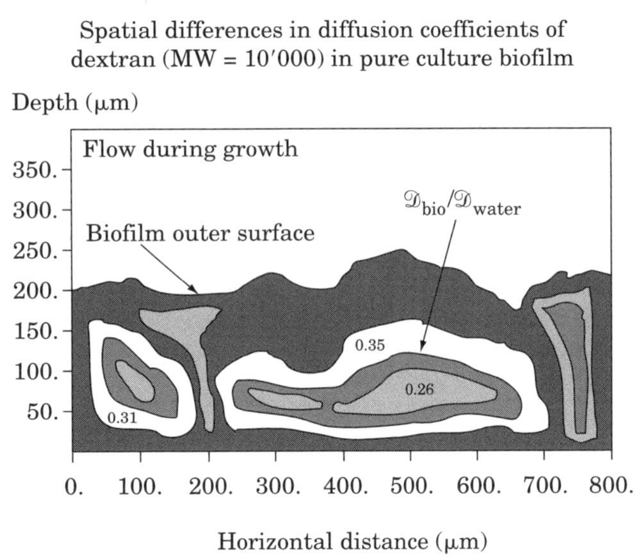

Figure 7. Spatial differences in diffusion coefficients of dextran (MW = 10,000) in pure culture biofilms. *Source*: Ref. 83.

Such phenomena can have significant effects on the outcome of cell cultivation processes. It has been established, both experimentally and theoretically, that plasmid maintenance and cloned gene expression can reduce the overall growth rate of the plasmid-bearing cell relative to the plasmid-free cell. Reduction of copy number and loss of plasmids from populations under continuous suspended culture have been reported in many cases, even in the presence of selective pressure.

That immobilization might stabilize a plasmid-bearing population can be shown mathematically (85,86). Plasmid persistence in suspended cultures has been observed in cases where the plasmid-bearing cell was at a growth rate disadvantage; this observation was directly attributed to biofilm formation and cell detachment from the biofilm (87). Improved plasmid stability upon artificial cell immobilization in κ-carrageenan gel beads has been reported by de Taxis du Poët et al. (88), Nasri et al. (89), and Sayadi et al. (90).

Huang et al. (91,92) cultivated biofilm cultures under controlled hydrodynamic conditions in a parallel-plate flow cell reactor using *E. coli* DH5α containing a recombinant plasmid either with (pTKW106) or without (pMJR1750) a segregational stability factor, *hok/sok*. Both plasmid constructs provided for the inducible expression of β-galactosidase. Results in these studies suggest, for these two plasmids, that the expression of the chromosomal gene for synthesis of the polysaccharide matrix of the biofilm commands a higher priority on cellular intermediates than the foreign genes. Consequently, a series of experiments were carried out to indirectly manipulate bacterial polysaccharide production in the biofilm by varying the nutrient carbon-to-nitrogen (C:N) ratio. Plasmid loss probability, total polysaccharide production, and ratios of total polysaccharide to total protein in biofilm cultures of *E. coli* DH5α (pMJR1750) all increased with increasing C:N ratios. Synthesis rates of total RNA, rRNA, and β-galactosidase-specific mRNA in *E. coli* DH5α (pTKW106) all increased after induction by IPTG, but levels of each parameter seen in biofilm cultures were 10–20 times lower than the same values observed in suspended cell cultures.

Cell–Cell Communication. Cell–cell communication involves a chemically unique molecule, produced biologically by one cell, that reacts with a set of specific receptors located at another cell's surface to subsequently alter cellular behavior. Two examples of cell–cell communication in higher life forms would be *(1)* the thousands of lectins that control embryonic development in plants and animals, and *(2)* signaling molecules, such as the pheromones, that control the behavior of adult insects.

Cell–cell communication in bacterial systems, both sessile and planktonic, has been recognized for years (93), but our understanding of the control mechanisms has entered a new era with the advent of various molecular diagnostic tools. Recent research has identified the specific chemical structures of various signaling molecules that manifest certain cellular responses. The majority of cell–cell communication studies have dealt with suspended bacteria (e.g., quorum sensing), but recent advances suggest that certain structural architecture in a biofilm arises due to molecular signaling.

Bacterial swarming, floc formation, and biofilm formation or aggregation are multicellular, cell-concentration-dependent behavior that can be induced in response to the recognition of surfaces of certain viscosity. Cells differentiate into multinucleated, elongated, and hyperflagellated form and orient themselves lengthwise across the surface. For a number of species, the ability to differentiate into swarmer cells is critical to virulence, surface adhesion, and colonization.

Acylated homoserine lactones serve as one class of signal molecule in bacterial communication. These molecules and their derivatives have been linked to control of bioluminescence, Ti plasmid transfer, production of virulence factors, antibiotic resistance, and swarming motility (94). Higher life forms have evolved mechanisms to interfere with such signaling processes. For example, certain seaweeds and sea grasses produce halogenated furanones, structurally similar to acylated homoserine lactones, that interfere with the swarming process and have been shown effective as antifoulants and antimicrobials (95). Givskov et al. (96) illustrated that two different furanones derived from the seaweed *Delisea pulchra* could progressively inhibit and eliminate the swarming behavior of *Serratia liquefaciens* as concentrations were increased from 0 μg/mL to 100 μg/mL (concentrations far too low to affect the growth of the bacteria).

Davies et al. (unpublished results) employed homoserine lactone mutants of the classic PAO-1 strain of *P. aeruginosa* to determine if cell–cell communication plays any role in bacterial biofilm polysaccharide matrix formation. Wild strains and mutants were separately used to inoculate flow cells to observe cell adhesion to surfaces and subsequent formation of biofilms. Cells of the wild-type strain (PAO-1), retaining the ability to secrete homoserine lactone, adhered to surfaces avidly and formed complex biofilms comprising discrete microcolonies and well-developed water channels.

Cells of the double homoserine lactone mutant (JP2), which lacked the ability to synthesize either the *Pseudomonas* autoinducer, PAI, or the factor U homoserine lactone, did adhere to surfaces avidly but failed to form the complexly structured biofilms. These mutant cells did not form detectable amounts of exopolysaccharide matrix material, did not aggregate into discrete microcolonies, and did not develop the water channels characteristic of the wild-type biofilms. Cells of the double homoserine lactone mutant simply formed unstructured masses of cells that were tightly packed together. These unstructured adherent masses of cells could be easily removed from the colonized surfaces by simple washing with surfactants. Chemical analyses proved that these biofilm contained no detectable biofilm exopolysaccharide matrix material. When a cell-free supernatant from a wild-type biofilm (PAO-1), which contained both types of homoserine lactone molecule, was added to the bulk fluid delivered to the flow cell, adherent JP2 mutant cells gradually began to produce matrix material and form complex biofilms; 12 h after the addition of the homoserine lactone molecules from the wild-type supernatant, the JP2 mutant strain biofilm was observed to be identical to the wild-type biofilm strain. Such research could potentially lead to nontoxic control strategies for disrupting detrimental biofilms, which would be economically and environmentally significant.

Novel Biofilm Removal Strategies

One traditional approach to eliminate biofilm formation within an engineered system is to prevent cell adhesion. One such anti–cell adhesion approach used in the biomedical materials sector focuses on modification of the surface chemical properties of the substratum by various methods including photochemical coupling of benzephenone derivatives of polyethylene glycol, polyacrylamide, and poly vinyl pyrrolidone (97); passive adsorption of pluronic surfactants (copolymers of polyethylene oxide and polypropylene oxide) to polystyrene (98); incorporation of polyethylene oxide into the upper layers of polyethylene terephthalate by solvent swelling (99); and a series of neutral, anionic, and cationic surfactants adsorbed onto stainless steel or glass (100). Rather than prevent adhesion, one alternative control approach for biomedical materials is to retard surface microbial activity by incorporating within the substratum a slowly released antibiotic agent (20,101,106).

Remedial approaches to eliminate or eradicate fouling biofilms existing in nonmedical engineered systems consist of mechanical cleaning, materials or unit replacement, or chemical biocide challenges (107). One novel alternative to chemical antagonism of a biofilm has been to eliminate an nutrient essential for microbial growth (108). In the petrochemical, refining, and power industries, either oxidants (chlorine, ozone, bromides) or biocides are employed regularly to suppress biofouling of water-cooled heat exchangers; biocides prove mostly ineffective in that they inactivate the bacterial cells but, unlike oxidants, are incapable of eliminating the source of the system inefficiencies, the biofilm matrix.

One of the most innovative technologies to control detrimental biofilm formation was reported by Wood et al. (109). Substrata containing layers of copper and cobalt phthalocyanine catalysts were constructed for bacterial adhesion trials. These immobilized catalysts serve to promote the formation of active oxygen species from various oxidizing agents such as peroxide or persulfates. Biofilms on the treated catalytic surfaces were exposed to various concentrations of hydrogen peroxide. Generation of antibacterial oxidant concentrations right at the interface of the substrata aided in the almost complete removal of biofilm.

BIBLIOGRAPHY

1. H.C. Jones, I.L. Roth, and W.M. Sanders, III, *J. Bacteriol.* **99**, 316–325 (1969).
2. J.W. Costerton, in L.D. Sabath ed., *Int. Symp.* Pseudomonas aeruginosa, Huber, Berne, 1979, pp. 1–3.
3. J.D. Bryers, *Biotechnol. Progr.* **3**, 57–68 (1987).
4. B. Atkinson, H.W. Fowler, in T.K. Ghose, A. Fiechter, and N. Blakebrough eds., *Advances in Biochemical Engineering*, vol. 3, Springer-Verlag, Berlin, 1974, pp. 221–277.
5. W.G. Characklis and K.C. Marshall, *Biofilms*, Wiley-Interscience, New York, 1990.

6. S.A. Blenkinsopp and J.W. Costerton, *Trends Biotechnol.* **9**, 138–144 (1991).
7. J.D. Bryers, *Trends Biotechnol.* **9**, 422–426 (1991).
8. S. Deckena and K.H. Blotevogel, *Biofouling* **5**, 287–294 (1992).
9. N.J.E. Dowling, J. Guezennec, J. Bullen, B. Little, and D.C. White, *Biofouling* **5**, 315–322 (1992).
10. R.D. Bryant, W. Jansen, J. Boivin, E.J. Laishley, and J.W. Costerton, *Appl. Environ. Microbiol.* **57**, 2804–2809 (1991).
11. P.J. Bremer and G.G. Geesey, *Appl. Environ. Microbiol.* **57**, 1956–1962 (1991).
12. I.B. Beech, C.C. Gaylarde, J.J. Smith, and G.G. Geesey, *Appl. Environ. Microbiol.* **35**, 65–71 (1991).
13. M.J. Franklin, and D.C. White, *Curr. Opinion Biotechnol.* **2**, 450–456 (1991).
14. N.D. Benbouzid-Rollet, M. Conte, J. Guezennec, and D. Prieur, *J. Applied Bacteriol.* **71**, 244–251 (1991).
15. R. Boopathy and L. Daniels, *Appl. Environ. Microbiol.* **57**, 2104–2108 (1991).
16. M.B. Deshmukh, I. Akhtar, R.B. Srivastava, and A.A. Karande, *Biofouling* **6**, 13–32 (1992).
17. P. Vandevivere and P. Baveye, *Appl. Environ. Microbiol.* **58**, 1690–1698 (1992).
18. G. Reid, H.S. Beg, C.A.K. Preston, and L.A. Hawthorn, *Biofouling* **4**, 171–176 (1991).
19. I.H. Pratt-Terpstra, J. Mulder, A.H. Weerkamp, J. Feijen, and H.J. Busscher, *J. Biomater. Sci. Polym. Ed.* **2**: 239–253 (1991).
20. C.C. Chang and K. Merritt, *J. Biomed. Mater. Res.* **26**, 197–207 (1992).
21. K. Merritt and C.C. Chang, *J. Biomat. Appl.* **5**, 185–203 (1991).
22. L. Passerini, K. Lam, J.W. Costerton, E.G. King, *Crit. Care Med.* **20**, 665–673 (1992).
23. C.P. Virden, M.K. Dobke, P. Stein, C.L. Parsons, and D.H. Frank, *Aesthetic Plant. Surg.* **16**, 173–179 (1992).
24. D.A. Jennings, M.J. Morykwas, W.W. Burns, M.E. Crook, W.P. Hudson, and L.C. Argenta, *Ann. Plastic Surg.* **27**, 216–220 (1991).
25. J.C. Nickel, J. Downey, and J.W. Costerton, *Urology* **39**, 93–96 (1992).
26. R.L.S. Whitehouse, E. Peters, J. Lizotte, and C. Lilge, *J. Dent.* **19**, 290–295 (1991).
27. Y.S. Park and K. Toda, *Biotechnol. Lett.* **14**, 609–612 (1992).
28. S.G. Velizarov, E.I. Rainina, A.P. Sinitsyn, S.D. Varfolomeyev, V.I. Lozinsky, and A.L. Zubov, *Biotechnol. Lett.* **14**, 291–296 (1992).
29. M. Moresi, E. Parente, and A. Mazzatura, *Appl. Microbiol. Biotechnol.* **36**, 320–323 (1991).
30. M.Y. Kwak and J.S. Rhee, *Appl. Microbiol. Biotechnol.* **36**, 578–583 (1992).
31. M.Y. Kwak and J.S. Rhee, *Biotechnol. Bioeng.* **39**, 903–906 (1992).
32. J.W. Tzeng, L.S. Fan, Y.R. Gan, and T.T. Hu, *Biotechnol. Bioeng.* **38**, 1253–1258 (1991).
33. B. Demuyakor and Y. Ohta, *Appl. Microbiol. Biotechnol.* **36**, 717–721 (1992).
34. G.A. Kovalenko and V.D. Sokolovskii, *Biotechnol. Bioeng.* **39**, 522–528 (1992).
35. B.Z. Eathepure and T.M. Vogel, *Appl. Environ. Microbiol.* **57**, 3418–3422 (1991).
36. J.P. Arcangeli and E. Arvin, *Appl. Microbiol. Biotechnol.* **37**, 510–517 (1992).
37. I. Coelhoso, R. Boaventura, and A. Rodrigues, *Biotechnol. Bioengrg.* **40**, 625–633 (1992).
38. D. Lemoine, T. Jouenne, and G.A. Junter, *Appl. Microbiol. Biotechnol.* **36**, 257–264 (1992).
39. J.A. Scott and A.M. Karanjkar, *Biotechnol. Lett.* **14**, 737–740 (1992).
40. G. Zellner, M. Gevcke, E.C. Demacario, and H. Dickmann, *Appl. Environ. Microbiol.* **36**, 404–409 (1991).
41. C.D. Gooijer, R.H. Wijffels, J. Tramper, *Biotechnol. Bioeng.* **38**, 224–231 (1991).
42. R.H. Wijffels, C.D. Gooijer, S. Kortekaas, J. Tramper. *Biotechnol. Bioeng.* **38**, 232–240 (1991).
43. T. Sumino, H. Nakamura, N. Mori, Y. Kawaguchi, and M. Tada, *Appl. Microbiol. Biotechnol.* **36**, 556–560 (1992).
44. L.E. Hallas, W.J. Adams, and M.A. Heitkamp, *Appl. Environ. Microbiol.* **58**, 1215–1219 (1992).
45. B. Heppner, G. Zellner, and H. Dickmann, *Appl. Environ. Microbiol.* **36**, 810–816 (1992).
46. A.M. Anselmo and J.M. Novais, *Biotechnol. Lett.* **14**, 239–244 (1992).
47. L.E. Macaskie, P.J. Clark, J.D. Gilbert, and M.R. Tolley, *Biotechnol. Lett.* **14**, 525–530 (1992).
48. G. Gonzalez, F. Ramirez, and O. Monroy, *Biotechnol. Lett.* **14**, 149–154 (1992).
49. D.E. Fennel, S.E. Underhill, and W.J. Jewell, *Biotechnol. Bioeng.* **40**, 1218–1232 (1992).
50. G.I. Loeb and R.A. Neihof, in R.E. Baier ed., *Applied Chemistry at Protein Surfaces*, Vol. 145, American Chemical Society, Washington, D.C. 1975, pp. 314–335.
51. V.A. DePalma, D.W. Goupil, and C.K. Akers, 6th Offshore Thermal Energy Conversion Conf., Washington, D.C., May 15–18, 1979.
52. J.D. Bryers, Ph.D. Dissertation, Wm. M. Rice University, Houston, Tex., 1980.
53. B.D. Bowen, S. Levine, and N. Epstein, *J. Coll. Interfac. Sci.* **54**, 375–390 (1978).
54. S.K. Beal, *Nucl. Sci. Eng.* **40**, 1–11 (1972).
55. L.K. Jang and T.F. Yen, in Zajic and Donaldson, eds., *Microbes and Oil Recovery*, Vol. 1, International Bioresources J., 1985, pp. 226–246.
56. M.K. Banks, Ph.D. Dissertation, Duke University, Durham, N.C., 1989.
57. M.K. Banks and J.D. Bryers, *Appl. Microbiol. Biotechnol.* **33**, 596–601 (1990).
58. D. Davies and G.G. Gessey, *Appl. Environ. Microbiol.* **61**, 860–867 (1995).
59. P. Harremoës, *Vatten* **33**, 122–143 (1977).
60. H. Tanaka and I.J. Dunn, *Biotechnol. Bioeng.* **24**, 669–689 (1982).
61. P.S. Stewart, *Biotechnol. Bioeng.* **41**, 111–117 (1993).
62. D.H. Applegate, and J.D. Bryers, *Biotechnol. Bioeng.* **37**, 17–25 (1991).
63. A. Boyd and A. Chakrabarty, *Appl. Environ. Microbiol.* **60**, 2355–2359 (1994).
64. G.G. Rodriguez, D. Phipps, K. Ishiguro, and H.F. Ridgway, *Appl. Environ. Microbiol.* **58**, 1801–1808 (1992).
65. R.I. Amann, J. Stromley, R. Devereux, R. Key, and D.A. Stahl, *Appl. Environ. Microbiol.* **58**, 614–623 (1992).
66. J. Rogers and C.W. Keevil, *Appl. Environ. Microbiol.* **58**, 2326–2330 (1992).

67. L.L. McCarter, R.E. Showalter, and M.R. Silverman, *Biofouling* **5**, 163–176 (1992).
68. L. Dagostino, A.E. Goodman, K.C. Marshall, *Biofouling* **4**, 113–120 (1991).
69. T.C. Brelje, M.W. Wessendorf, and R.L., Sorenson, *Methods Cell Biol.* **38**, 97–181 (1993).
70. A.C. Camper, M.A. Hamilton, K.R. Johnson, P. Stoodley, G.J. Harkin, and D.S. Daly, *Ultrapure Water* **11**, 26–35 (1994).
71. J.R. Lawrence, D.R. Korber, B.D. Hoyle, J.W. Costerton, and D.E. Caldwell, *J. Bacteriol.* **173**, 6558–6567 (1991).
72. S. Moller, A.R. Pedersen, L.K. Poulsen, E. Arvin, and S. Molin, *Appl. Environ. Microbiol.* **62**, 4632–4640 (1996).
73. K.P. Ishida and P.R. Griffiths, *J. Colloid Interface Sci.* **160**, 190–200 (1993).
74. P.J. Bremer and G.G. Geesey, *Biofouling* **3**, 89–100 (1991).
75. P.A. Suci, M.W. Mittleman, F.P. Yu, and G.G. Geesey, *Antimicrob. Agents Chemother.* **38**, 2127–2133 (1994).
76. Z. Lewandowski, G. Walser, W.G. Characklis, *Biotechnol. Bioeng.* **38**, 877–882 (1991).
77. A. Schramm, L.H. Larsen, N.P. Revsbech, N.B. Ramsing, R. Amann, and K.-H. Schleifer, *Appl. Environ. Microbiol.* **62**, 4641–4647 (1996).
78. S.B. Libicki, P.M. Salmon, and C.R. Robertson, *Biotechnol. Bioeng.* **32**, 68–80 (1988).
79. H. Siegrist and W. Gujer, *Water Res.* **19**, 1369–1378 (1985).
80. F.E. Drummond, M.S. Thesis, Duke University, Durham, N.C. (1993).
81. J.D. Bryers and F.E. Drummond, in R.H. Wijffels, R.M. Buitelaar, C. Bucke, and J. Tramper eds., *Progress in Biotechnology 11: Immobilized Cells*, Elsevier Science BV, Amsterdam, The Netherlands, 1996, pp. 196–204.
82. E. Poitevin and P. Wahl, *Biophys. Chem.* **31**, 247–258 (1988).
83. J.D. Bryers and F. Drummond, *Biotechnol. Bioeng.* **60(4)** (1998).
84. P. Stoodley, D. deBeer, and Z. Lewandowski, *Appl. Environ. Microbiol.* **60**, 2711–2716 (1994).
85. J.E. Bailey and D.F. Ollis, *Biochemical Engineering Fundamentals*, 2nd ed., McGraw-Hill, New York, 1986.
86. D.F. Ollis, *Phil. Trans. R. Soc. London* B, **297**, 617–629 (1982).
87. D.E. Dykhuizen and D.L. Hartl, *Microbiol. Rev.* **47**, 150–168 (1983).
88. P. de Taxis du Poët, P. Dhuster, J.-N. Barbotin, and D. Thomas, *J. Bacteriol.* **165**, 871–877 (1986).
89. M. Nasri, S. Sayadi, J.-N. Barbotin, P. Dhuster, and D. Thomas, *Appl. Environ. Microbiol.* **53**, 740–744 (1987).
90. S. Sayadi, M. Nasri, J.N. Barbotin, and D. Thomas, *Biotechnol. Bioeng.* **33**, 801–808 (1989).
91. C.-T. Huang, S.W. Peretti, and J.D. Bryers, *Biotechnol. Bioeng.* **41**, 211–220 (1993).
92. C.-T. Huang, S.W. Peretti, and J.D. Bryers, *Biotechnol. Bioeng.* **44**, 329–336 (1994).
93. W.C. Fuqua, S.C. Winans, and E.P. Greenberg, *J. Bacteriol.* **176**, 269–275 (1994).
94. G.P.C. Salmon, B.W. Bycroft, G.S.A.B. Stewart, and P. Williams, *Mol. Microbiol.* **16**, 615–624 (1995).
95. J.S. Todd, R.C. Zimmerman, P. Crews, and R.S. Alberte, *Phytochemistry* **34**, 401–404 (1993).
96. M. Givskov, R. de Nys, M. Manefield, L. Gram, R. Maximillien, L. Eberl, S. Molin, P.D. Steinberg, S. Kjelleberg, *J. Bacteriol.* **178**, 6618–6622 (1996).
97. S.G. Dunkirk, S.L. Gregg, L.W. Duran, J.D. Monfils, J.E. Haapala, J.A. Marcy, D.L. Clapper, R.A. Amos, and P.E. Guire, *J. Biomaterials. Appl.* **6**, 131–155 (1991).
98. M.J. Bridgett, M.C. Davies, and S.P. Denyer, *Biomaterials* **13**, 411–416 (1992).
99. N.P. Desai, S.F.A. Hossainy, and J.A. Hubbell, *Biomaterials* **13**, 417–420 (1992).
100. W.K. Whitekettle, *J. Industrial Microbiol.* **7**, 105–116 (1992).
101. C.C. Chang and K. Merritt, *J. Orthop. Res.* **9**, 284–288 (1991).
102. G. Golomb and A. Shpigelman, *J. Biomed. Mater. Res.* **25**, 937–952 (1991).
103. G.D. Kamal, M.A. Pfaller, L.E. Rempe, and P.J. Jebson, *JAMA* **265**, 2364–2368 (1991).
104. J.W. Leung, G.T. Lau, J.J. Sung, and J.W. Costertoni, *Gastrointest. Endosc.* **38**, 338–340 (1992).
105. D. Stickler and P. Hewett, *Eur. J. Clin. Microbiol. Infect. Dis.* **10**, 416–421 (1991).
106. D. Stickler and P. Hewett, *Eur. J. Clin. Microbiol. Infect. Dis.* **10**, 157–162 (1991).
107. C.L. Wiatr and O.X. Fedyniak, *J. Industrial Microbiol.* **7**, 7–13 (1991).
108. R. Bakke, B. Rivedal, and S. Mehan, *Biofouling* **6**, 53–60 (1992).
109. P. Wood, M. Jones, M. Bhakoo, and P. Gilbert, *Appl. Environ. Microbiol.* **62**, 2598–2602 (1996).

See also BIOREMEDIATION; DEXTRAN, MICROBIAL PRODUCTION METHODS; ENZYMES, IMMOBILIZATION METHODS.

BIOFILTERS

MARIO ZILLI
ATTILIO CONVERTI
Genoa University
Genoa, Italy

KEY WORDS

Air pollution control biotechnologies
Biofilter
Biofiltration
Biological methods
Degradation pathways
Microorganisms

OUTLINE

Introduction
Biological Technologies for Waste-Gas Treatment

INTRODUCTION

Biofiltration is the biological removal of air contaminants from waste gases by means of aerobic microorganisms that are immobilized on a porous solid support. In a biofilter, the waste-air stream is passed through a biologically active packing material; organic or inorganic air pollutants are degraded by microorganisms into harmless end products such as water, carbon dioxide, mineral salts, and new microbial mass. Clearly, biofiltration technology is limited to non-toxic gases with good biodegradability and water solubility, because these gases are used by microorganisms as their sole source of energy and carbon.

The idea of employing microorganisms to eliminate contaminants from waste gases is very old. Initially, biofiltration was used mainly for the treatment of odorous gases (such as hydrogen sulfide and ammonia) emanating from sewage-treatment plants. The first reports on actual applications of this technology, dating back to the late 1950s and the early 1960s, dealt with soil-bed installations in West Germany and the United States, where it was demonstrated that biodegradation, rather than sorption, was responsible for most of the odor abatement (1). After the first publication of Bohn's extensive investigation on soil-filter beds (2,3), several additional full-scale applications of biofilters in West Germany and the Netherlands followed. These biofilters were employed to control the odors from wastewater- and thermal-sludge-treatment plants, composting facilities, rendering plants, carcass incinerators, and others (4–6). Jäger and Jager, who compared several methods for off-gas purification from the composting plant of Heidelberg, first demonstrated the economic convenience of biofiltration (7).

Process and technology development was attempted by Thistlethwayte et al., who utilized a trickling filter column packed with river gravel or glass balls and seeded with activated sludge. Waste air contaminated by H_2S, C_2H_5SH, $(C_2H_5)_2NH$, and C_4H_9CHO was purified, in countercurrent with a nutrient solution, with removal yields ranging from 40% to almost 100% (8). In addition, applications in which organic impurities were eliminated from waste gas, or air in beds inoculated with biologically active material, have been reported together with the first criteria for correct process design and operation (9).

A further development of this technology was marked by a biofilter installation with continuous regeneration of the bed, which was obtained by removing active material at the bottom of the reactor and recharging it after regeneration at the top of the reactor (9). A regeneration process by microbial, thermal, and chemical methods was also reported by Helmer (10), who used dry- and wet-soil filter beds to extract products of anaerobic fermentation of organic materials (e.g., NH_3, H_2S, sulfides, alcohols, and aldehydes). The importance of a sufficient moisture content in the filter bed and of a good distribution of the gas supplied to the biofilter was demonstrated for the first time (11).

Leson and Winer estimated that in 1991 the total number of biofilter and soil-bed installations used in the United States and Canada mainly for odor control was less than 50 (1). There was greater confidence in biofiltration in Germany and the Netherlands, where over 500 biofilters were operating in the same period. In the last two decades, the number of installations has significantly increased in Japan, growing from about 40 in the 1980s to 90 in the 1990s (12). On the contrary, in other European countries, including Italy, Switzerland, and Austria, the number of biofilter installations has been quite limited (1).

Although biofiltration has developed mainly with the primary aim of eliminating odorous compounds from exhaust gases, since the early 1980s this technology has been of increasing importance in air pollution control, and the number of applications of biological filters has strongly increased due to an increasing concern about environmental quality and more stringent air-emission quality standards.

The application field of biofiltration has widened thanks to better knowledge of the biodegradation process, in particular the control of the process conditions, as well as to improved biofilter construction and packing material composition (13,14). In particular, special packing materials having optimal structural properties prevent aging phenomena, decrease the pressure drop over the packed bed, and increase the long-term operational stability of the filter, as well as its microbial activity.

Furthermore, the considerable research carried out on the isolation, selection, and construction of strains or mixed cultures of microorganisms (mainly bacteria) has led to the extension of biodegradation to both anthropogenic (i.e., manmade) and xenobiotic compounds (chemicals that do not occur naturally). Compounds that sufficiently resemble structures of biological origin are eliminated rather quickly, whereas biodegradation of xenobiotics is very slow (recalcitrant compounds) or even impossible (persistent compounds). Nevertheless, the continuous adaptation of microorganisms to new substrates is at the basis of the increasing number of microbes capable of degrading xenobiotics.

As a consequence, biofiltration is now a well-recognized technology that is successfully applied to a wide range of industrial facilities, waste-disposal and food-processing activities to control odors, volatile organic compounds (VOCs), and air toxics (both inorganic and organic), especially those that are readily biodegradable.

Biofiltration development is the result of the implementation of health and safety guidelines, enforced in many countries to protect people and the environment, combined with several advantages that this technique has demonstrated over traditional chemical and physicochemical treatment methods, such as water washing, chemical scrubbing (ozonation and chlorination), activated carbon adsorption, incineration, and catalytic oxidation.

Besides the overall low energy requirements (moderate temperature and pressure conditions), easy maintenance, simple control, and low capital and operating costs, it must be stressed that biological filters do not transfer the pollution problem to another environmental compartment, as occurs with many other alternative control technologies. Because the pollutants are converted into harmless oxidation products (which are part of a natural cycle), they do not create any secondary pollution and thus can be considered environmentally safe.

BIOLOGICAL TECHNOLOGIES FOR WASTE-GAS TREATMENT

Biological Systems

There are three basic types of biological reactor systems used to treat waste gases: bioscrubbers, biotrickling filters, and biofilters. These can be grouped into those employing microorganisms dispersed freely throughout the liquid phase (bioscrubbers), and those using microorganisms immobilized on a packing or carrier material (biotrickling filters and biofilters). Moreover, in bioscrubbers and biotrickling filters the water phase is continuously moving, whereas in biofilters it is stationary.

A bioscrubber consists of a scrubber unit and a regeneration unit. In the scrubber (absorption column), water-soluble gaseous pollutants are absorbed and partially oxidized in the liquid phase (the culture medium containing the microorganisms), which is distributed from the top of the unit. The contaminated water is subsequently transferred into an aerated stirred-tank reactor (regeneration unit), similar to an activated-sludge unit, where the contaminants are fully biodegraded. The regenerated suspension is continuously recirculated to the top of the scrubber section, thereby enhancing process efficiency (Fig. 1).

Biotrickling filters (Fig. 2) and biofilters (Fig. 3) are different from bioscrubbers in that gaseous pollutant absorption and biological degradation occur simultaneously in the same compartment, resulting in more compact systems. The polluted air flows through a biologically active bed, where microorganisms are attached in the form of a biofilm. As the gas diffuses through the packed bed, the pollutants are transferred to the biolayer and degraded.

As illustrated in Figure 3, in order to ensure optimal operation of a biofilter, the inlet gas usually requires pretreatment processes such as: (*1*) particulate removal in order to prevent possible clogging and/or sludge build up, (*2*) load equalization in case the waste-gas concentration is subject to strong fluctuations, (*3*) temperature control, and (*4*) humidification.

In biological trickling filters the packed bed consists only of inert materials (glass, ceramics, and plastics), while the liquid phase, containing inorganic nutrients, flows downward over the packaging material in countercurrent with the contaminated gaseous stream and is continuously recirculated through the bioreactor.

Bioscrubbers and biotrickling filters are applicable mainly to the treatment of waste gases containing good or moderately water-soluble compounds, whereas biofilters, due to the large surface area available for mass transfer, are also suited to treat poorly water-soluble compounds. Moreover, due to their high reaction selectivity, biofilters are particularly suitable for treating large volumes of air containing easily biodegradable pollutants with relatively low concentrations, typically below 1,000 ppm.

Compared with the other biological systems, biofilters have the widest application because they are easy to operate, simply structured, and imply low installation and operating/maintenance costs. Also, the reliability of biofilter operation is higher than that of bioscrubbers, where the risk exists of washing away the active microorganisms. Moreover, the presence of a large amount of packing material with a buffering capacity diminishes the sensitivity of biofilters to different kinds of fluctuations (15). Because the major disadvantage is the difficult control of parameters such as pH, temperature, and nutrient supply, biofilters may be unsuitable for degrading halogenated compounds (as acid metabolites are produced) and treating gas streams containing high concentrations of VOCs, unless long residence times or large bed volumes are applied (3). Biotrickling filters and biofilters are currently utilized mainly in compost-production plants, sewage-treatment plants, and agriculture, whereas biofilters and bioscrubbers are preferred in industrial applications.

Different Types of Biofilters

Among the different types of biofilters described in the literature for a variety of applications (16), open biofilters, consisting of single beds of compost or porous soils, commonly 1-m deep, are used mainly for odor and VOC abatement. Because they are in direct contact with the open air, their performance is strongly influenced by weather conditions (rain, frost, temperature fluctuations, etc.). The simple design and low cost of these systems are counterbalanced by difficult monitoring and control, as well as by large space requirements.

All the other real-scale configurations are closed biofilters containing mixtures of organic materials and bulking

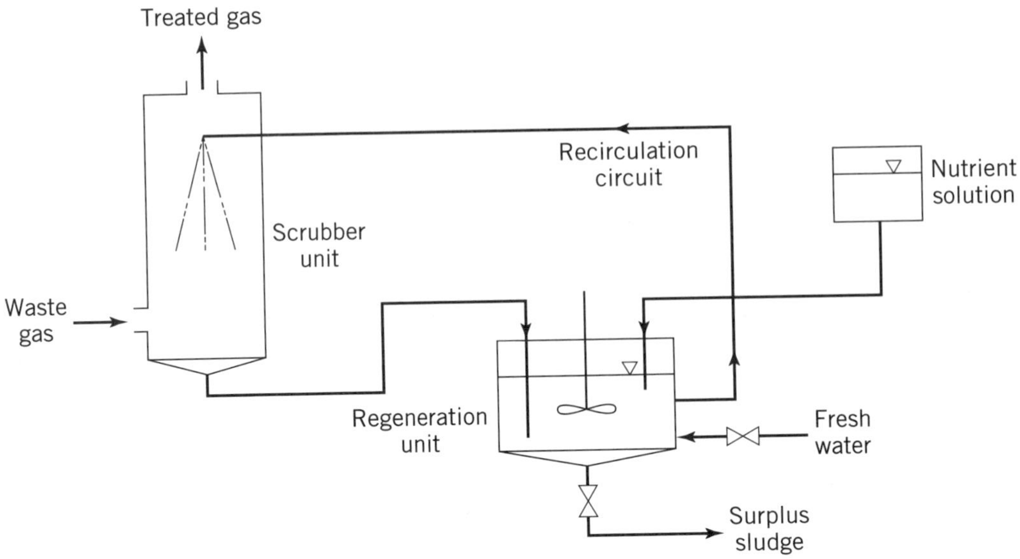

Figure 1. Scheme of a bioscrubber system.

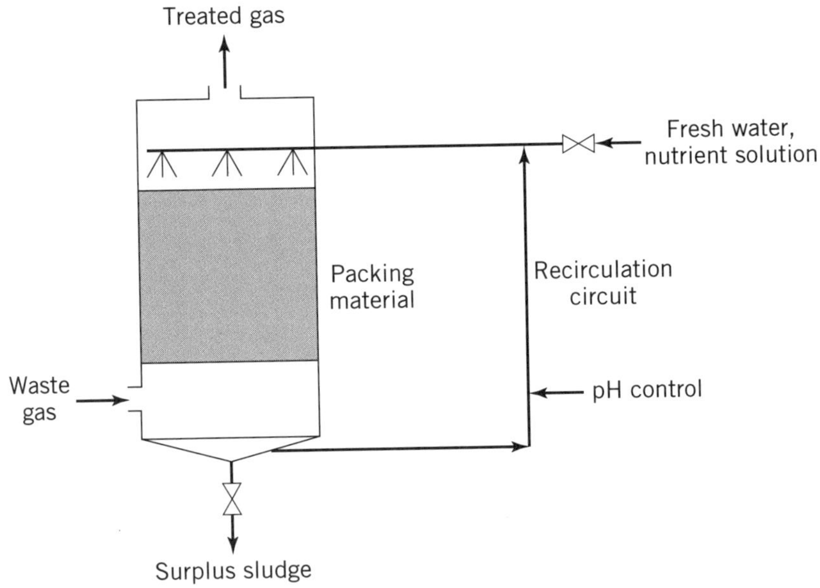

Figure 2. Scheme of a biotrickling filter system.

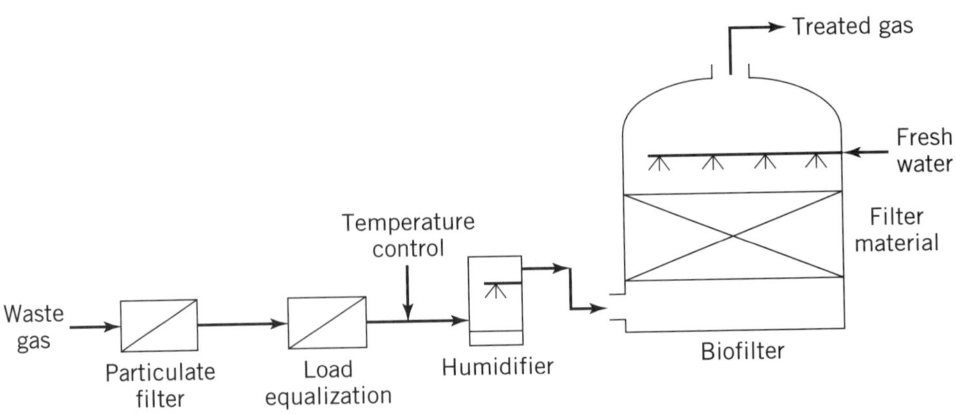

Figure 3. Scheme of a closed single-layer biofilter system.

agents. Among these, simple single-layer systems, often accurately controlled and monitored, but scarcely flexible and requiring large space, can be used for VOC treatment. Multiple-layer systems, characterized by increased complexity and cost but less space requirements, consist of separately supported layers that allow optimal growth conditions for different microbial populations in each stage; a better control of the bed environment during the treatment of pollutant mixtures is also achieved.

More sophisticated and very promising configurations have been successfully tested on bench- or pilot-scale projects for the treatment of VOC mixtures, BTEX (benzene, toluene, ethyl benzene and xylene), and reduced-sulfur compounds, among which there are multistage, modular, and step-feed systems, that in general are characterized by better control and flexibility, but higher costs and operative complexity.

PROCESS PARAMETERS

Biofiltration is a very complex process that involves mass transfer and reaction and is also influenced by the fluid dynamic of the gaseous stream through the reactor. Furthermore, it is necessary to maintain an environment as close as possible to the optimal conditions for the microorganisms. Therefore, several important process parameters, such as filter material structure, moisture content, temperature, and pH, must be kept under strict control and within an optimum range.

Filter Material

The choice of filter material is of great importance in order to maintain the efficiency of a biofilter, in that it must (*1*) guarantee optimum environmental living conditions for the microorganisms; (*2*) constitute at the same time a nutritious reserve, a humidity reservoir, and a mechanical odorless support; and (*3*) provide the structural stability of the bed.

The choice of the material is strongly influenced by the need to minimize the overall volume necessary for the reactor, optimize the removal efficiency, keep energy consumption to a minimum, and minimize maintenance. In addition, the characteristics of the carrier material impact directly on the microbial growth and activity, thus in turn affecting biofilter performance. Biofilter beds have the advantage of immobilizing the microflora on the packing material, as a result of which these organisms, forming a biolayer, are not drained from the system, as is often the case in freely dispersed systems.

Small particles of natural organic materials, such as compost, peat, soil, or mixtures of these materials with bark, leaves, wood chips, heather branches, humus earths, or brushwood (less than 10 mm in diameter), are widely used as packing media in biofilters because they provide a high specific surface area (from 300 to 1,000 m^{-1}), favorable living conditions for the resident microbial population (ensured by high retention capacities of water and nutrients), and favorable immobilization for the microflora involved. In practice these packing materials have shown the common disadvantage of being strongly subject to aging phenomena, resulting in bed shrinkage. This phenomenon disturbs the homogeneous flow distribution of the gas and may cause a considerable increase in the pressure drop, a decrease in the specific area, and shorter filter lifetime. Moreover, the natural unhomogeneity of the structure of these natural materials prevents the uniform distribution of the gas flow, provoking short-circuiting channel development in the filter bed, with an increase in the flow resistance and a decrease in the biological degradation capacity of the bed.

In order to prevent these problems, inert materials, such as polystyrene spheres, lava particles, glass beads, porous clay, and ceramic, are usually added to the natural filling materials. This combination, improving the uniform distribution of the gas flow, lowers the head losses up to 100–150 mm of water gauge (6), thus contributing to reducing the power requirement necessary to convey the gas through the filter and, consequently, the operating costs. Moreover, the addition of porous materials (e.g., granular activated carbon) with high internal porosity and hydrophilic properties, increases the buffering capacity of the filter, (which is very favorable, particularly in the treatment of gas streams with strongly fluctuating pollutant concentrations). The selection of the most suitable material for a specific application depends on the nature of the pollutants and on the magnitude of the concentration fluctuations in the gaseous stream. In order to buffer the fluctuations in the concentration so that a constant supply of contaminants to the biofilter can be achieved, the adsorbent should have both good adsorbing power and reasonable desorbing properties. This combination provides the adsorbent with the capacity of adsorbing at high concentrations and desorbing at low concentrations and allows a significant reduction in the required filter volume (15).

Compared to soil, compost has the advantage of providing lower resistance to the gas flow and consequently contained pressure drop, which should not, in general, exceed 250 mm of water gauge (17). Peat is the material with highest water-retention capacity and constitutes the optimal substrate for the microorganisms; nevertheless, as it implies a higher pressure drop, it is often mixed with other materials in order to improve the structural stability.

Moisture Content

The good performance of a biofilter depends greatly on an optimal moisture content of the filter bed. Moisture, the most critical operational parameter, is, in fact, essential for the survival and activity of the microbial flora (which can absorb and degrade substances only in the aqueous phase) and contributes to the filter buffer capacity.

A lack of moisture causes cracking of the bed mainly at the gas inlet side, where the microbial activity is stronger due to the higher air-pollutant concentration. This results in (*1*) a considerable decrease of the microbial activity, (*2*) development of short-circuiting channels that cause bypass of the gas flow, and (*3*) contraction of the filter bed with consequent breakthroughs. On the contrary, an excess of water content promotes the development of anaerobic zones in the bed, due to occlusion of the pores, and provokes the formation of odorous volatile metabolic products, which are released in these zones and transported by the gas flow. Moreover, due to the higher gas-flow resistance, these zones are bypassed by the gas flow, resulting in a decrease in the mean residence time of the gas and, hence, in the efficiency of the process, and causes a 10% increase in the power requirement (2). A moisture excess causes problems also in the oxygen transfer due to a reduction of the gas–water interface per unit biofilm volume, and in the drainage of the filter components, with production of low-pH and high-load leachate that requires disposal. The optimal moisture content of the filter bed should then range between 40 and 60% by weight, depending on the material. In order to preserve the microbial activity, it should exceed at least 40% (18).

Water consumption by a biofilter is generally low, depending on the temperature and the relative humidity of the input gas and, for open filters, on precipitation. Due to evaporation of the aqueous phase, freshwater must be continuously supplied to the system in order to ensure a degree of saturation of more than 95%, which is necessary for optimal living conditions. This is normally realized by humidifying the inlet gas stream in a water-spray scrubber upstream of the filter unit and by periodically spraying water onto the surface of the packing material by means of surface sprinklers, spray nozzles, and so on. This periodical water supply is necessary not only because it is not always possible to humidify the inlet waste gas up to the required level, but also because heat is produced by the microbial oxidation of the pollutants.

Moistening equipment must be operated in such a way that the moisture content stays within the indicated limits at any point of the medium. This operation must be realized in a very controlled way, otherwise the structure of the bed may be disturbed, which results in compaction of the filter bed in the long run and in increased gas-phase pressure drop.

pH

Maintaining an optimal pH in the filter material is a very important operational requirement for maximizing the biofiltration process because the microbial activity is strongly dependent on the pH. The effect of pH on biological systems is the indirect result of its action on the enzymatic reaction rate, which usually falls outside an optimal value. In biofiltration, as well as in most aerobic biological processes, the range of pH within which the biological systems can operate is typically between 5 and 9, with an optimum range of 6–8.

Because in biofilters the aqueous phase is stationary, problems of package acidification may arise during the degradation of pollutants such as sulfur- or nitrogen-containing compounds and halogenated hydrocarbons, which can generate acidic metabolites such as sulfuric, nitric, or hydrochloric acid. As a consequence of their accumulation, the microbial activity may decrease or even stop, seriously compromising the overall elimination capacity of the biofilter. These problems can be prevented by addition of buffering agents, such as lime, dolomite, limestone, marl, phosphates, or other water-insoluble alkaline materials, even if the neutralization products can sometimes reach inhibitory levels (13).

Besides the above adverse effect on the process, acidification of the biofilter, as well as the acidic nature of the waste gas, filter material, or leachate, can corrode the air distribution and leachate collection pipes. In such cases, the use of materials resistant to corrosion, together with a continuous pH monitoring, are suggested.

Temperature

Temperature remarkably influences the microbial growth through its effects on both bacterial enzyme activity and on various metabolic processes. Because the microorganisms generally applied in biofilters are mesophilic, temperature should be maintained between 20 and 40 °C, with an optimum of 35 °C for most aerobic microorganisms. Suboptimal temperatures can slow down the degradation capacity. The bed temperature should not rise above 45 °C, although the resident microflora can gradually adapt itself to temperatures as high as 50 °C (19). Microbial activity slows dramatically above 65 °C. The effect of low temperature on biofilters depends largely on the inlet gas temperature. In fact, although low bed temperatures, especially below 10 °C, decelerate biological oxidation, this effect can be partially counterbalanced by increased gas sorption by solids (3). However, the microbial activity will recover again upon preheating the influent exhaust gas. Fortunately, microorganisms have shown a tremendous ability to rapidly adapt themselves to drastic variations of temperature.

Reaction and diffusion rates typically increase with increasing temperature; nevertheless this effect is offset by a decrease of the solubility of the compounds to be removed, as well as a decrease in the physical adsorption capacity of the packing material at higher temperatures. Operation of the biofilter at high temperature shifts the microbial population towards thermophilic microorganisms.

In order to prevent the death of the microorganisms, it is often necessary to control the influent gas temperature by precooling through heat recovery, water injection, or wetting of the packing material, thus offering additional economic advantage and the possibility of raising the gas relative humidity. Moreover, the heat liberated by aerobic microbiological activity causes a significant increase in temperature within the filter material, which must be taken into consideration at the design phase, particularly in biofilters that treat waste gases containing high concentrations of volatile organic compounds. For good operation of the biofilter, it is obviously necessary to continuously monitor the filter-bed temperature.

PROCESS PRINCIPLES

Physical Model

In a biofilter, the microorganisms adhere to the surface of the porous solid material used as the support and grow, thus forming a biofilm. All models commonly used to describe the phenomena taking place in a biofilter under steady-state conditions are based on the principles of heterogeneous catalysis. According to the macrokinetic model proposed by Ottengraf (20), the support particle is surrounded by a biofilm; both the pollutant and oxygen molecules dispersed in the gaseous phase continuously transfer to the active biofilm region. Due to the metabolic activity of aerobic microorganisms, a concentration gradient forms that is responsible for the continuous mass transfer of the pollutant from the gaseous to the liquid phase. At the same time, the products of the aerobic degradation (carbon dioxide, water, inorganic compounds, etc.) continuously counterdiffuse toward the gaseous phase, where they are removed by the upward flow and are finally released from the biofilter.

The metabolic activity is made possible by the continuous diffusion of inorganic nutrilites (mainly nitrogenous and phosphoric compounds) from the support to the biofilm. Nutrilite consumption is only partially counterbalanced by the mineralization of dead cells, therefore periodic addition or support renewal is necessary.

Because the small particle size and the low solubility of the organic pollutants in water make the mass-transfer resistance in the gas phase negligible, pollutant concentrations at the biolayer interface and in the gas-phase bulk can reasonably be assumed in equilibrium and related by Henry's law.

Degradation Kinetics

The degradation macrokinetics in biofilters were reviewed by Ottengraf (20). Assuming gas flow through the biofilter to be plug flow, and that degradation follows typical Monod kinetics law (21), two different situations may occur: first-order or zero-order kinetics, respectively. Several experimental works (13) demonstrated that zero-order kinetics are typically observed in biofilters for most of the volatile compounds, even at very low concentration levels (13,20,22).

Reaction Limitation. The process is reaction-limited above a critical pollutant concentration (C_{crit} in the off-gas (Fig. 4a). Due to there being no diffusion limitation, the wet biolayer is fully active (e.g., the biofilter works at its maximum elimination capacity), and pollutant elimination is limited by the reaction (Fig. 5a). Under these conditions, the degree of conversion (η) is given by the ratio between

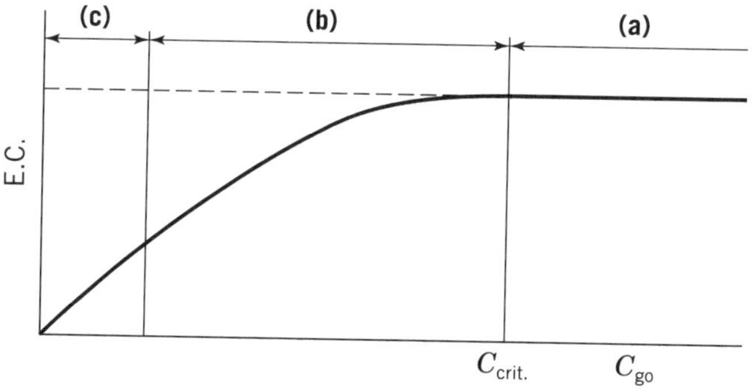

Figure 4. Dependence of the elimination capacity of a biofilter on the inlet off-gas concentration. (**a**) Reaction limitation; (**b**) Diffusion limitation; (**c**) 100% conversion.

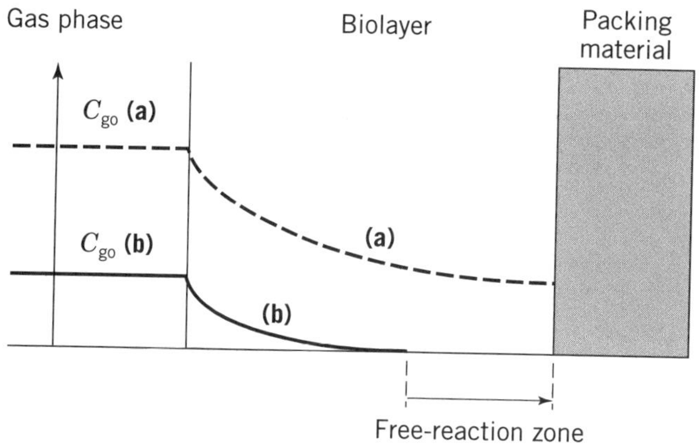

Figure 5. Biophysical substrate-penetration model.

the maximum elimination capacity, K_o which corresponds to an apparent zero-order kinetic constant, and the bed load per unit volume ($U_g C_{go}/H$) (20):

$$\eta = 1 - (C_{ge}/C_{go}) = HK_o/U_g C_{go}$$

So, at the macroscopic level, the pollutant concentration in the filter bed linearly decreases with the height, and the compound is completely consumed when $H\ K_o$ equals $U_g\ C_{go}$.

Diffusion Limitation. Below the critical concentration, diffusion limitation occurs and the rate of reaction tends to exceed that of diffusion (Fig. 4b). As a consequence, the depth of penetration becomes smaller than the layer thickness and the biolayer is not fully active (Fig. 5b). The pollutant is nearly completely consumed by the biofilm before having entirely crossed it, and the conversion rate, which decreases with decreasing pollutant concentration in the waste gas, is controlled by diffusion. Under these conditions, the elimination capacity of the biofilter is influenced by both the reaction and the diffusion rates and the compound is completely removed when $H\ K_o \geq 2\ U_g\ C_{go}$ (20).

The applicability of this model to multicomponent waste gases is limited by the theoretical complexity of the system because the various compounds are not biodegraded independently. Industrial application regarding mixture of pollutants requires pilot testing for accurate sizing of a full-scale system.

DESIGN AND MANAGEMENT CRITERIA

The main criteria for biofilter design and management are the degree of conversion or removal efficiency, the residence time, the superficial gas flow rate, the mass loading, and the elimination capacity (16).

Removal Efficiency

The removal efficiency (η), which provides information on the biofilter effectiveness, is given by the fraction of removed pollutant mass ($m_o - m_e$) with respect to total mass in the influent gas (m_o). The most common way of designing and managing biofilters consists of selecting the desired removal efficiency for one or more pollutants and deriving from this all the other criteria. The best removal efficiencies (95–99%) are usually observed for aromatic compounds, such as benzene, toluene, and benzoic acid (23).

Residence Time

The residence time of a gaseous pollutant in a biofilter (τ), given by the ratio of the empty bed volume of the filter (V_e) to the waste-gas flow rate (Q_g), is a fundamental parameter that cannot assume values smaller than a critical level if satisfactory removal efficiencies are desired. However, due to the difficulty in estimating empty-bed volume, residence time is often referred to total-bed volume. According to the pollutant degradability, this parameter typically ranges between 15 and 60 s. For liquid pollutants it is necessary to take into account their partition coefficients between liquid and gaseous phases.

Superficial Gas Flow Rate

The superficial gas-flow rate (U_g), defined as the ratio of the gas flow rate (Q_g) to the total cross-sectional area (A), is a parameter increasing with flow rate and decreasing with residence time; it usually ranges, depending on pollutant type, from 50 to 200 $m^3\ m^{-2}\ h^{-1}$. Since a superficial gas-flow rate increase obviously implies a reduction in removal efficiency, a maximum threshold of about 200 $m^3\ m^{-2}\ h^{-1}$ should not be exceeded (24).

Mass Loading

The mass loading (L_v) is defined as the ratio of the gas-flow rate (Q_g) to the bed volume (V) multiplied by the inlet pollutant concentration in the off gas (C_{ge}); for VOCs it should not exceed values of 3,000 ÷ 5,000 mg m^{-3}. In a biofilter, load increase could be ensured by either increasing waste-gas concentration or decreasing residence time. Although increasing gas concentration should, in general, accelerate biodegradation, excess concentrations above a certain threshold can inhibit microbial activity, so that decreasing residence time should be preferred. In addition, since the removal efficiency is affected by an increase in mass loading, and because excess organic load may obstruct the biofilter with the possible formation of toxic and/or acidic intermediates (25), a value between 10 and 160 g $m^{-3}\ h^{-1}$ should be employed to achieve satisfactory removal yields.

Elimination Capacity

The elimination capacity (EC) is defined as the ratio of gas flow rate (Q_g) to the bed volume (V), multiplied by the difference between the pollutant concentration in the influent and the effluent waste gases ($C_{go} - C_{go}$), it indicates a measure of the biofilter ability in removing the pollutants. From this definition it is evident that, because removal efficiency is expected to be close to 100%, the elimination capacity also ranges between 10 and 160 g $m^{-3}\ h^{-1}$ (1). In addition, this parameter also is higher close to the inlet side, as happens for mass loading. The elimination capacity increases with mass loading and inlet pollutant concentration and decreases with residence time, the limiting value depending not only on the pollutant degradability, but also on the support material and the operating conditions. Beyond this value it is nearly constant (20).

APPLICATIONS OF BIOFILTERS

Bench-Scale Results

Although many experimental works on biofiltration, both at laboratory and full-scale size, have been published during the last two decades, a critical comparison of results is difficult due to various reasons, among which are the use of German and Dutch as the most common languages in this research field, the use of different criteria to express the performances, the utilization of different volatile organic and inorganic compounds, the different types of biofilters used, and so on. Nevertheless, the work of Ottengraf (20) can be taken as a useful reference basis to provide a complete overview on biofilter performance, as well as on the various effects influencing the process macrokinetics.

For most organic pollutants, the steady-state continuous elimination follows zero-order kinetics, so that pollutant concentration decreases with the column height. In the presence of mixtures of pollutants characterized by different biodegradability, a nearly complete removal of the most easily degradable compounds takes place at the bottom of the filter, whereas the most recalcitrant are mainly removed at the top. This suggests that different microbial populations could be responsible in the filter bed for the degradation of different organic components.

The path of Figure 4 explains another typical behavior of biofilter performances besides the one illustrated earlier. At low organic loads (corresponding to low concentrations for a given residence time), the elimination capacity linearly increases with this parameter, which means that nearly complete pollutant conversion is achieved. Over a critical load, the elimination capacity reaches a maximum value (corresponding to the apparent zero-order kinetic rate constant, K_o) and stays nearly constant. Maximum thresholds of 21, 27, 32, and 32 g $m^{-3}\ h^{-1}$, calculated for a mixture of toluene, ethyl acetate, butyl acetate, and butanol in a five-stage system simulating a biofiltration column with variable height (20), seem to reflect a direct dependence of this parameter on the pollutant degradability.

Pilot-Scale and Real-Scale Applications

Biofilters of different sizes (areas ranging from 10 to 2,000 m^2) are currently used in many industrial sectors, mainly in Europe, to remove a variety of mixtures of volatile organic and inorganic compounds from off-gases characterized by variable flow rates (1,000–150,000 $m^3\ h^{-1}$) and low pollutant concentrations (usually less than 1,000 ppm). As one can see from Table 1, where the main pollutants removable by biofiltration are listed (3,13,16,26,27), a relatively variable degradability characterizes not only the different classes of pollutants, but also the various pollutants within the same class. While for easily biodegradable substances the microorganisms suited for their degradation are usually present in the natural packing materials, in-

Table 1. Classification of the Main Pollutants that can be Removed by Biofiltration, according to their Biodegradability

Excellent	Good	Minimum	No	Uncertain
Aliphatics	*Aliphatics*	*Alicyclics*	*Halogenated*	*Aliphatics*
Butadiene	Hexane	Cyclohexane	1,1,1-Trichloroethane	Acetylene
Aromatics	*Aromatics*	*Aliphatics*		*Nitrogenous*
Ethyl benzene, xylene Cresols	Benzene, styrene, toluene Phenols	Methane, pentane		Isocyanates
Nitrogenous	*Halogenated*	*Aromatics*		*Oxygenated*
Trimethylamine	Chlorophenols	PAH		Methyl methacrylate
Oxygenated	*Nitrogenous*	*Halogenated*		*Sulfuric*
Alcohols: butanol, ethanol, methanol Aldehydes: acetaldehyde, formaldehyde Esters: Ethyl acetate Ethers: Tetrahydrofurane Ketones: Acetone Organic acids: Butyric acid	Amides Heterocycles: Pyridine Isonitriles Nitriles: Acetonitrile	Carbon tetrachloride Dicloroethane Dichloromethane Pentachlorophenol Perchloroethylene Trichloroethane Trichloroethylene		Isothiocyanates
Sulfuric	*Oxygenated*	*Nitrogenous*		
Methyl mercaptan	Methylisobutylketone	Nitrocompounds		
Inorganics	*Sulfuric*	*Oxygenated*		
HCl, HF, H_2S, NH_3, NO_x (except NO_2), PH_3, SiH_4, SO_2	Heterocycles: Thiophene Thiocyanates Thioethers: Dimethylsulphide	Dioxane		
		Sulfuric		
		Carbon disulfide		

Source: From Refs. 3, 13, 16, 26, and 27.

oculation and enrichment with cultivated organisms is often necessary for the slowly biodegradable ones.

On the other hand, Table 2 provides a picture of the most significant industrial sectors where biofiltration is applied with success, with indication of the related pollutants (26).

Finally, from Table 3, where the results of the principal pilot and full-scale applications of biofilters are summarized, it is evident that conversion yields often higher than 90% are not uncommon in industrial practice (16,17,20,27–32).

MICROORGANISMS

Another factor strongly influencing biofiltration performance is the composition of the heterogeneous microflora, which primarily depends on the composition of the polluted gaseous stream. As a consequence of the progressive adaptation of microorganisms to the organic pollutants contained in the off-gas, which typically takes about 10 days (20), the population distribution shifts towards strains that naturally catabolize these pollutants.

Microbial Populations of Biofilters

Microbial population distribution along biofilters has recently been reviewed by Benedusi et al. (33). The purification of waste air containing a limited number of pollutants is effectively achieved by a population restricted to a few microbial species and generally requires, to shorten the start-up operation time, inoculation of the biofilter with pure cultures or activated sludges previously adapted to each pollutant. This practice is particularly suited to the degradation of very recalcitrant or complex organic pollutants, such as halogenated and aromatic hydrocarbons, which often require cometabolism with more easily degradable substances to accelerate the process (34,35). On the other hand, gaseous emissions simultaneously polluted by several compounds, such as those coming from water works, compost plants, and so on, require as inoculum a more heterogeneous population possessing a wider variety of metabolic pathways, which is available in the microorganisms naturally present in activated sludges and compost. Microbial loads of about 10^5, 10^4, and 10^3 colony-forming units per gram of support (CFU/g) have been reported for oligoheterotrophic, copioheterotrophic, and autotrophic metabolic types, respectively in heather-peat biofilters treating septic-tank emissions (33). Although Bohn (3) demonstrated that enriching compost-based biofilters with microorganisms is not advantageous in terms of removal yields, probably because of the unfavorable competition with the indigenous population, the addition of activated sludges to medium devoid of microorganisms

Table 2. Typical Pollutants Removed by Biofiltration, in Addition to Odor Control, in Different Industrial Applications

	Organics						Inorganics	
Type of Industry	Aliphatic	Aromatic	Oxygenated compounds	S-containing compounds	N-containing compounds	Halogenated compounds	H_2S	NH_3
			Food and Agriculture Industry					
Aroma extraction		X						
Yeast drying			X		X			
Blood and fish-meal production			X	X	X		X	X
Bristle and feather drying			X	X	X			X
Fish roasting			X	X	X		X	
Fat processing			X	X	X		X	X
Gelatin production			X	X				
Coffee and cacao roasting		X	X	X	X			
Bone processing			X	X	X		X	X
Meat and fish curing	X	X	X		X			X
Animal feed production			X	X	X		X	X
Slaughterhouses			X	X	X		X	X
Tobacco processing		X	X		X			X
Livestock farming			X	X	X		X	X
			Chemical Industry					
Foundries	X	X	X					X
Plastics processing	X	X	X		X			
Lacquer production and processing	X	X	X					
Adhesive production		X	X					
Oil and fat production	X	X		X		X		
Polyester manufacture				X	X			
Friction-linings production			X	X	X			X
Rendering plants			X	X	X	X	X	X
Glue production		X	X					
			Treatment Plants					
Used oil processing	X	X		X		X		
Dump gas removal			X	X			X	
Sewage treatment and sludge drying	X	X	X	X	X	X	X	X
Composting and waste processing	X		X	X	X	X	X	X
Manure drying	X		X	X	X		X	X

Source: Elaboration from Ref. 26.

is a common practice (1). Finally, the growth and activity of the microorganisms are strongly influenced by the availability of oxygen and nutrients, the eventual presence of toxic substances in the exit gas, the degree of moisture, temperature, pH, and so on.

A great advantage of the heterogeneous population present in the biofilter is the excellent ability to survive long periods (up to two months) without activity, provided that periodic aeration of the bed is ensured, even if microbial activity seems to be hardly affected (36).

As revealed by plate counting in nutrient agar, most of the microorganisms growing in biofilters treating organic pollutants are heterotrophic eubacteria, actinomycetes, and fungi (1,37), which utilize the organic compounds contained in the gaseous streams as carbon and energy sources. Even when the presence in the off-gas of inorganic substances, like thiosulfate or hydrogen sulfide, should favor the growth of chemio-litoautotrophic microorganisms, such as *Thiobacillus* sp., by utilizing these substances as the energy source and carbon dioxide as the carbon source, the heterotrophic population becomes prevalent and is probably responsible for most of the degradation activity (33,38,39).

The bacteria more frequently detected in biofilters are soil bacteria, such as *Bacillus cereus* var. *mycoides* and several strains and species of actinomycetes belonging to the genus *Streptomyces*. Helmer (40), who did a microbiological characterization of compost-filter population, observed abundant growth of eubacteria (10^9 CFU/g), actinomycetes (2×10^6 CFU/g), and fungi (10^6 CFU/g) along the whole filter height, with higher density of fungi at the bottom. The highest bacterial densities are usually detectable near the inlet side, where substrates are more concentrated and more easily degradable, while different strains well-adapted to low substrate levels, as well as to more recalcitrant compounds, can be found at higher depth (37). The growth of several bacteria and fungi, whose cell densities decreased according to the sequence: *Actinomyces*

Table 3. Management Data of Representative Pilot and Industrial-Scale Applications of Biofilters

Applications	U_g ($m\ h^{-1}$)	Q_g (m^3/h)	Pollutants	C_{go} (mg/m^3)	τ (s)	N of stages	η (%)	Reference
			Food industry					
Gelatin production		35,000	Odor		12–21	0.6–1	70–93	27
Cocoa and chocolate processing		10,000	Odor		22	2	99	27
	60		Odor	$4{\cdot}10^7$ o.u.		1	90	28
Fishmeal factory		40,000	Odor	230 (C)	20	1	50–90	27
Flavor and fragrance industry		25,400	Odor	10^5 o.u.	22	2	98	27
	550		Odor	$5{\cdot}10^4$ o.u.		3	98	28
Food processing industry		9,000	Odor (oil)	10^5 o.u.	20	2	93	27
	400		Odor	$2{\cdot}10^5$ o.u.		2	95	28
Meat rendering		80,000–120,000	Odor	$2{\cdot}10^4$ o.u.	60–90	1	99	16
Commercial bakery		90–200	VOC		25–60	2		
			Ethanol	1,000–4,000			80–100	
			Methane	700			10	16
			Aliphatics	200			10–30	
			Ethyl acetate				80–90	
Coffee industry	500		Odor	10^8 o.u.		2	90	28
Yeast production	100		Odor	7,500 o.u.		2	99	28
			Ethanol	500			99	
			Petrochemical industry					
Gasoline tank		222	BTEX	3–70		2	75–90	16
Oil storage	100		H_2S	240		2	99	28
Tank cleaning			Benzene	30			50	
	100		Acrylonitriie	16–300		2	50–98	28
			Alcohols	60			80	
Petroleum processing			BTEX				>95	
		8.5–70	C2–C5		45–360	1	≤50	17
			Methane				≤5	
			>C5				40–90	
			Other industries					
Tobacco industry		30,000	Odor		14	2	95	
			NH_3	1.5				27
			Nicotine	3.5				
Paint production		11,700	VOC	1,800	38	2	90	27
Pharmaceutical plant		75,000	VOC		108	3	80	27
Photo film production		140,000	VOC	400	30	2	75	27
Ceramics production		30,000	Ethanol		8	1	98	27
Metal foundry		40,000	Benzene	9	30	1	80	27
Industrial emission		17,000	Ethanol	1,000–4,000	60	2	80–90	16
Resin synthesis	65		Organics	25,000		2	40	28
			odor	10^6 o.u.			45	
Chemical industry		0.7–0.8	VOC	2,000–10,000		1	82	29
Printing industry	200		Esters, alcohols	1200		2	90	28
			Treatment plants					
Wastewater treatment		10,000	Odor		29	2	90–95	
			H_2S	10				27
			Acetone	8				
		30,000	Odor			2		
			H_2S	4.7			96	30
			NH_3	0.15			45–60	
	200		H_2S	300		1	80	28
Petrochemical sewage treatment	100		Odor	$1.3{-}10^5$ o.u.		2	75	28
Composting		154,000–240,000	Odor		30–47	4	>90	31
		16,000–22,000	Odor	45–230 (C)	40–60	1	93–96	20
	100		Odor	$1.3{\cdot}10^4$ o.u.		2	95	28
			Handicraft activities					
Wood paint and varnish	200		VOC	250		1	50	28
Metal paint and varnish	70		H_2S	4000 o.u.		2	90	28
Paint spraying			Styrene	60–600	15–60	1	95	32

globisporus, Penicillium sp., *Cephalosporium* sp., *Mucor* sp., *Micromonospora albus, Micrococcus albus*, and *Ovularia* sp., was also reported for similar supports used for mixtures of halogenated and aromatic hydrocarbons (9). An opposite bacteria distribution has been reported by Benedusi et al. (33) along a pilot heather-peat filter used to remove hydrogen sulfide and ethyl mercaptan from septic-tank emissions. They observed maximum bacterial densities just near the biofilter outlet side, with oligo-heterotrophic population (2.4×10^8 CFU/g) predominating over all the others: copiochemioheterotrophs (5.7×10^6 CFU/g), sulfur-oxidizing chemioautotrophs (2.6×10^5 CFU/g), and nitrifying chemioautotrophs (7.2×10^5 CFU/g). Bacterial densities at biofilter bottom vary between 1% and 3% of these values, probably due to the hard pH decrease (up to 2.5–3.5) consequent to hydrogen sulfide oxidation (to H_2SO_4), as well as to the inhibiting effect of high concentrations of this pollutant on microbial activity. Other bacteria detected in biofilters are *Micromonospora vulgaris* and *Proteus vulgaris* (41).

Among the main soil fungal taxa able to degrade polysaccharides naturally present as symbionts or saprophites in heather or peat (*Geomyces pannorum* var. *pannorum*, *Penicillium citrinum*, *Oidiodendron griseum*, *Penicillium frequentans*, *Pestalotiopsis guepinii*, and *Trichoderma koningii*), only *G. pannorum* var. *pannorum* and *T. koningii* seem to be able to develop in the bed under regular and irregular moisture conditions, respectively (33). Other taxa, initially absent in the support material but normally living in polluted waters and soil, appear as a consequence of the air flowing and progressive acclimation, under conditions either of irregular (*Mucor* sp.) or regular moisture (*Mortierella bainieri, Drechslera australiensis*, and *Trichotecium roseum*). Other fungi detected in filter beds are *Circinella* sp., *Cephalotecium* sp., and *Stemphilium* sp. (41). Although the contribution of fungi in biofiltration is not so clear, it is likely that an important role is played in the degradation of organic nitrogen, sulfur, and potassium, such as takes place in the soil.

As far as the influence of the support structure on microbial density inside the bed is concerned, it was demonstrated that compost-based filters usually contain much higher population densities of these microorganisms than does soil (43). On the other hand, it seems that concentrations of bacteria cells and fungi spores in the treated emissions are only slightly higher than those observed in open air (43,44). Nevertheless, Liebe and Herbig advised against the risk of relevant fungi-spore emissions during biofilter maintenance and support-replacement operations (45).

A great number of other bacterial strains have been inoculated and tested in bench-scale biofilters to treat waste gases polluted by specific compounds, most of which appear to be particularly promising for future development of this technology (see "New Developments and Potentials").

Metabolic Pathways for Organic Pollutant Degradation

Different metabolic pathways are employed by microorganisms to degrade a variety of organic pollutants.

Aliphatic Hydrocarbons. Several bacteria (mainly belonging to the genera *Nocardia*, *Pseudomonas*, and *Mycobacterium*), and some yeast and fungi, which are capable of utilizing saturated aliphatic hydrocarbons for their growth, are the main candidates for new applications in bioscrubbers and biofilters for the removal of such air pollutants. The oxidation of the terminal carbon to an alcohol, catalyzed by a monoxygenase, is responsible for the introduction in the hydrocarbon structure of only one of the atoms of molecular oxygen, which acts as a cosubstrate, while the other is reduced to water:

$$\text{R-CH}_3 + \text{NADH} + \text{H}^+ + \text{O:O} \xrightarrow{\text{monoxygenase}} \text{R-CH}_2\text{-OH} + \text{NAD}^+ + \text{H}_2\text{O}$$

Further oxidations by NAD^+-dependent alcohol and aldehyde dehydrogenases lead to a carboxylic acid that can be decomposed to acetyl-CoA by the β-oxidation route (46):

$$\text{R-CH}_2\text{-OH} + \text{NAD}^+ \xrightarrow{\text{alcohol dehydrogenase}} \text{R-CH=O} + \text{NADH} + \text{H}^+$$

$$\text{R-CH=O} + \text{NAD}^+ \xrightarrow{\text{aldehyde dehydrogenase}} \text{R-COOH} + \text{NADH} + \text{H}^+$$

$$\text{R-COOH} + \text{ATP} + \text{CoA} \xrightarrow{\beta\text{-oxidation}} \text{CH}_3\text{-CO-CoA} + \text{AMP} + \text{PP}_\text{i}$$

On the other hand, polychlorinated hydrocarbons can be cometabolized in the presence of isoprene-degrading bacteria (47).

Aromatic Compounds. The aerobic degradation of aromatics is only possible if the aromatic ring undergoes an enzymatic cleavage. Benzene can be degraded by several microorganisms by two divergent pathways after preliminary dioxygenation to catechol: the so-called *ortho*-pathway, utilizing the intradiol cleavage catalyzed by the catechol-1,2-dioxygenase, and the so-called *meta*-pathway, following an extradiol cleavage. Mixtures of benzene, toluene, and xylene are preliminarily dioxygenated, decomposed mainly via the *meta*-pathway, and finally metabolized through the citric acid cycle (47).

The presence of an alkyl group in the benzene ring gives the microbe the opportunity to attack the compounds either on the side chain or the ring. In arenes with short side chains (up to seven carbons), both alternative routes are possible, mainly due to the widespread presence of transformable plasmids. In particular, the plasmid TOL (48), containing catabolic operons for both alkyl-group oxidation and ring cleavage via *meta*-pathway, gives several strains of *Pseudomonas* the ability to degrade toluene, *m*- and *p*-xylenes, and other monoalkylbenzenes (49,50). On the contrary, if the side chain is sufficiently long (more than seven carbons), its metabolization by initial side-chain attack via ω- and β-oxidation (51), provides the cells of different species with enough energy for growth, and then the ring cleavage is not necessary.

Of the dialkylbenzenes, *m*- and *p*-xylenes can be biodegraded by certain strains of *Pseudomonas* containing the TOL plasmid by preliminary oxidation of one of the methyl groups to a carboxylic group and methylcatechol (52), followed by *meta*-cleavage. Members of *Nocardia* are able to cometabolize *m*- and *p*-xylenes via the *ortho*-pathway, and *o*-xylene via the *meta*-pathway in the presence of alkanes as carbon sources (53). More recently, it has also been demonstrated that *o*-xylene can be metabolized as the sole carbon and energy source, via 3,4-dimethylcatechol and subsequent *meta*-cleavage, by some strains of *Pseudomonas stutzeri* (54) and *Corynebacterium* (55).

There are reports of different bacteria (*Pseudomonas* sp., *Xanthobacter* sp., etc.) able to grow on styrene (56) and methylstyrenes (57) as sole carbon and energy sources, during which oxidation of the aromatic nucleus and subsequent *meta*-cleavage probably take place.

In several bacteria, after a preliminary monoxygenation to a catecholic structure, phenol is degraded (58) through the *meta*-pathway to directly give intermediates of the citric acid cycle (46). While for 4-chlorophenol degradation the chlorinated lactonic intermediate is directly dechlorinated before its introduction into the cycle, for 4-methylphenol a specific enzyme catalyzes the isomerization of the related intermediate into a more easily metabolizable lactone (47).

Biphenyl and Fused-Ring Hydrocarbons. Although the biodegradation of biphenyl, naphthalene, and polycyclic aromatic hydrocarbons (PAH) is a subject of less relevance for biofiltration applications given their solid state at ambient temperature, their high toxicity has brought lively interest in the actual possibility of purifying low contaminated off-gases by utilizing biofilters inoculated with liquid suspensions of strains specifically adapted.

Apart some subtle differences, biphenyl is biodegraded through a pathway common to many different bacteria belonging to the genera *Pseudomonas* (59) and *Nocardia* (60), including a preliminary dioxygenation of the ring in the C2 and C3 positions to give a catechol-type intermediate, which is subsequently *meta*-cleaved and catabolyzed via 2-oxopenta-4-enoate and benzoate (60).

Naphthalene is degraded mainly by members of the genus *Pseudomonas* via dioxygenation to 1,2-dihydroxynaphthalene, followed by extradiol cleavage of the ring between C1 and C9, whereas PAHS, which are potential carcinogens, mutagens, and tetragens, are dioxygenated and cleaved in two different ways, according to their phenanthrenic or anthracenic structures (53). Less detail is available for higher molecular weight PAHs.

ECONOMIC CONSIDERATIONS

Comparing cost data quoted in the literature for biofiltration systems is very difficult because of the use of different currencies, the fluctuations of their reciprocal rates, the cost variability with time, the different units employed, and so on. Nevertheless, there is no doubt of biofilter economic convenience when compared with other traditional abatement techniques, due to the following advantages (6): minimum operational input, reduced production and disposal of by-products, easy start-up procedures, operational stability at steady state, low operative temperature, low cost of materials, simple building technologies, simple control systems, and low power requirements.

From the comparison of different biofilters in use in Europe (1,37,61–63), total operating and maintenance costs ranging from \$0.2 to 0.5 per 1,000 m^3 of waste gas can be estimated, which includes depreciation and interest, water consumption, replacement of filter materials and personnel, wastewater treatment, and energy costs for compressors. Costs reductions showing a half of these amounts for United States installations do not include expenses for support replacement and reflect the lower cost of electricity (1).

Capital costs are often negatively influenced by several factors, among which high pollutant concentration in the waste gas, recalcitrant nature of the components, necessity of gas pretreatment, reduced waste-gas availability, lack of space near the polluting source, and transport. Since the transport costs are strongly dependent on the country and the nature of the site, a cost comparison is only possible if this factor is not considered. On these bases, capital costs for open single-bed filters are in the range between \$300 and \$1,000 per m^2 of filter material in Europe (64,65) and between \$600 and \$1,000 per m^2 in the United States (1). These costs would increase up to \$3,800–5,700 per m^2 if lack of space would force the build-up of the filter on the plant cover, because of the additional costs of personnel, piping, and compression (66), and up to \$1,000–5,000 per m^2 if particular meteorological situations (snow, frequent atmospheric precipitation, etc.) require an upper covering or enclosed systems (1). Multiple-bed systems seem to be twice as expensive as the single-bed ones.

The filling material may consist, according to the pollutants, of heather, peat, and earth, often mixed with other inert materials necessary to minimize head losses. Because of these variables and the market situation, costs may fluctuate remarkably from \$60 to \$700 per m^3 (65,66).

Taking odor abatement as a basis for comparison of different techniques, biofiltration is by far the most convenient process in terms of space availability, resulting in, as a function of waste-gas composition, total cost savings of 15–30%, 45–70%, 50–75%, and 80% with respect to bioscrubbing, chemical scrubbing, adsorption, and ozonation, respectively (17,65,67). The success of biofiltration appears to be the result of the very low incidence of running costs to total costs as compared to other techniques. Extending the above comparison also to incineration, it has been estimated that biofiltration allows total cost savings of 60–75%, and investment and operational cost savings of 40–75% and 70–80%, respectively (17,63).

NEW DEVELOPMENTS AND POTENTIALS

Pollutants' degradability often depends on their origin and source: although xenobiotics, which are characterized by unnatural structures, cannot be degraded (persistent compounds), or can be degraded only with difficulty (recalcitrant compounds), the biogenic ones are easily biodegradable. For this reason, much research effort is being spent on the isolation, selection, and construction of different microorganisms (27,69,70) able to grow on difficult carbon sources in pure cultures, at rates comparable with those usually encountered on common substrates (Table 4).

For these reasons, it is evident that future developments and new applications of this technology strongly depend on the actual possibility of utilizing these microbial strains in pure cultures or selected consortia under the peculiar conditions, as present in biofilters.

Table 4. Xenobiotic Compounds and Microorganisms Able to Degrade Them in Pure Cultures

Pollutant	Microorganism
Chlorinated hydrocarbons (27,68)	
Methyl chloride	*Hyphomicrobium* sp.
Dichloromethane	*Pseudomonas* DM1
	Methylobacter DM111
	Hyphomicrobium sp.
1,2-Dichloroethane	*Xanthobacter* GJ10
Vinylchloride	*Mycobacterium* L1
Epichlorohydrine	*Pseudomonas* AD1
Chlorobenzene	*Pseudomonas* WR 1306
1,2-Dichlorobenzene	*Pseudomonas* GJ 60
1,3-Dichlorobenzene	*Pseudomonas* sp.
1,4-Dichlorobenzene	*Alcaligenes* A175
Arenes (69)	
Benzene	*Pseudomonas* sp.
	Acinetobacter calcoaceticus RJE 74
Toluene	*Pseudomonas* sp.
	Pseudomonas putida
	Bacillus sp.
Monoalkylbenzenes	*Pseudomonas* sp. NCIB 10643
	Pseudomonas fluorescens
	Pseudomonas putida RE204
	Acinetobacter Iwoffi
m-,*p*-Xylene	*Pseudomonas* sp.
	Nocardia sp.
o-Xylene	*Pseudomonas stutzeri*
	Corynebacterium sp. C125
Styrene	*Xanthobacter* sp 124X
	Pseudomonas putida
α-Methylstyrene	*Pseudomonas* sp.
Biphenyl (69)	
	Pseudomonas sp. NCIB 10643
	Nocardia sp. NCIB 10503
	Pseudomonas cruciviae
	Pseudomonas putida
	Pseudomonas pseudoalcaligenes
	Pseudomonas aeruginosa
	Escherichia coli
(Fused-ring compounds) (69)	
Naphthalene	*Pseudomonas* sp.
	Pseudomonas fluorescens
	Pseudomonas putida
PAH	*Pseudomonas* sp.
	Pseudomonas paucimobilis
	Pseudomonas vesicularis
	Algaligenes denitrificans

CONCLUDING REMARKS

As pointed out here, biofiltration has become by now a well-recognized technology for the control of odors and the elimination of air contaminants from waste gases, and represents a new generation of air pollution control technologies.

Over the past two decades, thanks to the considerable progress made in the microbiological and process-engineering fields, biofiltration has gained the confidence of many of industries and has found increasingly widespread application. The suitability of this technique for the treatment of a wide range of inorganic and organic pollutants has been proven. Removal efficiencies as high as 90% or more are usually achieved for common pollutants, such as alcohols, ethers, aldehydes, ketones, and monoaromatic volatile chemicals.

The operating cost savings, the technological simplicity of the plants, the reduced maintenance and energy requirements, and the minimal generation of by-products that must be disposed of, make biofiltration, with respect to other air pollution control technologies, an attractive cost-effective alternative, particularly if applied to large volumes of gaseous-waste streams containing readily biodegradable contaminants in relatively low concentrations.

Although its application to different compounds has been demonstrated, considerable research is being undertaken to find appropriate strains able to degrade recalcitrant compounds, in particular, most of which are xenobiotics. In addition, new strategies should be developed to improve the control of essential operating parameters affecting biofilter performance and to optimize reactor design and performance, in order to widen the application field of biofilters.

BIBLIOGRAPHY

1. G. Leson and A.M. Winer, *J. Air Waste Manage. Assoc.* **41**, 1045–1054 (1991).
2. H.L. Bohn, *J. Air Pollut. Control. Assoc.* **25**, 953–955 (1975).
3. H.L. Bohn, *Chem. Eng. Progr.* **88**, 34–40 (1992).
4. W.H. Prokop and H.L. Bohn, *J. Air Pollut. Control. Assoc.* **35**, 1332–1338 (1985).
5. D.H. Kampbell, J.T. Wilson, H.W. Read, and T.T. Stocksdale, *J. Air Pollut. Control. Assoc.* **37**, 1236–1240 (1987).
6. F. Alfani, L. Cantarella, A. Gallifuoco, and M. Cantarella, *Acqua-Aria* **10**, 877–884 (1990).
7. B. Jäger and J. Jager, *Müll und Abfall* **2**, 48–54 (1978).
8. D.K.B. Thistlethwayte, B. Hardwick, and E.E. Goleb, *Chimie Ind.* **106**, 795–801 (1973).
9. M. Gust, H. Grochowski, and S. Schirz, *Staub-Reinh. Luft* **39**, 397–402 (1979).
10. R. Helmer, *Ges. Ing.* **94**, 21–30, (1974).
11. H. Hartmann, *Stuttg. Ber. Siedlungswasserwirtsch* **59**, 3–19 (1977).
12. S. Cernuschi and V. Torretta, *Ingegneria Ambientale* **25**, 248–264 (1996).
13. S.P.P. Ottengraf, J.J.P. Meesters, A.H.C. van den Oever, and H.R. Rozema. *Bioprocess Eng.* **1**, 61–69 (1986).
14. S.P.P. Ottengraf, A.H.C. van den Oever, and F.J.C.M. Kempenaars, in E.H. Houwink and R.R. van der Meer eds., *Inno-*

vations in Biotechnology, Elsevier, Amsterdam, 1984, pp. 157–167.

15. F.J. Weber and S. Hartmans, in A.J. Dragt and J. van Ham eds., *Biotechniques for Air Pollution Abatement and Odour Control Policies*, Elsevier, Amsterdam, 1992, pp. 125–130.
16. W.J. Swanson and R.C. Loeher, *J. Environ. Eng.* **123**, 538–546 (1997).
17. G. Leson and B.J. Smith, *J. Environ. Eng.* **123**, 556–563 (1997).
18. E. Eitner, in Verein Deutscher Ingenieure ed., *VDI Berichte 735: Biologische Abgasreinigung*, VDI Verlag, Düsseldorf, 1989, pp. 191–214.
19. H.L. Bohn, *86th Mtg. Air Waste Management Assoc.*, Denver, Col., June 14–18, 1993.
20. S.P.P. Ottengraf, in H.J. Rehm and G. Reed eds., *Biotechnology*, vol. 8, VCH, Weinheim, 1986, pp. 425–452.
21. J. Monod, *Ann. Rev. Microbiol.* **3**, 371–394 (1949).
22. S.P.P. Ottengraf and A. van den Oever, *Biotechnol. Bioeng.* **19**, 1411–1417 (1986).
23. R. Marsh, *Inst. Chem. Eng.* **3**, 13.1–13.14 (1992).
24. F. Sabo, U. Motz, and K. Fischer, *86th Mtg. Air Waste Management Assoc.*, Denver, Col., June 14–18, 1993.
25. J.S. Devinny and D.S. Hodge, *J. Air Waste Manage. Assoc.* **45**, 125–131 (1995).
26. Kommission Reinhaltung der Luft, in Verein Deutscher Ingenieure ed., *VDI 3477-Handbuch Reinhaltung der Luft: Biological Waste Gas / Waste Air Purification, Biofilters*, VDI Verlag, Düsseldorf, 1991, Band 6, pp. 1–29.
27. S.P.P. Ottengraf and R.M.M. Diks, in A.J. Dragt and J. van Ham eds., *Biotechniques for Air Pollution Abatement and Odour Control Policies*, Elsevier, Amsterdam, 1992, pp. 17–31.
28. A.J. Dragt and S.P.P. Ottengraf, *Proc. Conf. on Advanced Technologies and Biotechnology for the Pollution Prevention and Environmental Protection*, SIBA-ISER, Genoa, Italy, 1987, pp. 243–258.
29. M.A. Deshusses, *J. Environ. Eng.* **123**, 563–568 (1997).
30. M. Pescari and V. Pizzirani, *Microsymposium on Air Pollution Control*, Pavia, Italy, May 23, 1997.
31. I. Trentini and R. Vismara, *Biologi Italiani*, **28**, 16–26 (1998).
32. A.P. Togna and S. Frisch, *86th Mtg. Air Waste Management Assoc.*, Denver, Col., June 14–18, 1993.
33. L. Benedusi, G. Caretta, C. Collivignarelli, C. Dacarro, G. Del Frate, and V. Riganti, *Ingegneria Ambientale* **22**, 271–284 (1993).
34. S.J. Ergas, K. Kinney, M.E. Fuller, and K.M. Scow, *Biotechnol. Bioeng.* **44**, 1048–1054 (1994).
35. D.B. Janssen, R. Oldenhuis, and A.J. van den Wijngaard, in Verein Deutscher Ingenieure ed., *VDI Berichte 735: Biologische Abgasreinigung*, VDI Verlag, Düsseldorf, 1989, p. 25.
36. S.P.P. Ottengraf and A.H.C. van den Oever, *Biotechnol. Bioeng.* **25**, 3089–3102 (1983).
37. K.F. Kosky and C.R. Neff, *NWWA / API Conf. on Petroleum Hydrocarbons and Organic Chemicals in Ground Water*, Houston, Tex., November 9–11, 1988.
38. H. van Langenhove, E. Wuyts, and N. Schamp, *Water Res.* **20**, 1471–1476 (1986).
39. C. Kyeoung-Suk, H. Mitsuyo, and S. Makoto, *Appl. Environ. Microbiol.* **58**, 1183–1189 (1992).
40. R. Helmer, *Stuttgarter Berichte zur Siedlungswasserwirtschaft*, Band 49, Oldenbourg, München, 1972.
41. W. Ludwig, O. Fischer, and F.X. Kneer, *Chem. Ing. Tech.* **58**, 742 (1986).
42. M. Wainwright, *Trans. Br. Mycol. Soc.* **90**, 159–170 (1988).
43. S.P.P. Ottengraf and J.H.G. Konings, *Bioprocess Eng.* **7**, 89–96 (1991).
44. R.M.M. Diks and S.P.P. Ottengraf, in Verein Deutscher Ingenieure ed., *VDI Berichte 735: Biologische Abgasreinigung*, VDI Verlag, Düsseldorf, 1989, p. 7.
45. H.G. Liebe and V. Herbig, *Bericht II*, 1.1-50441-5/132, Umweltbundesamt, Berlin, May, 1986.
46. T.D. Brock, M.D. Madigan, J.M. Martinko, and J. Parker, *Microbiologia*, Città Studi Edizioni, Milan, 1995.
47. K.H. Engesser, in A.J. Dragt and J. Van Ham eds., *Biotechniques for Air Pollution Abatement and Odour Control Policies*, Elsevier Amsterdam, 1992, pp. 33–40.
48. R.C. Bayly and M.G. Barbour, in D.T. Gibson ed., *Microbial Degradation of Organic Compounds*, Marcel Dekker, New York, 1984, pp. 253–294.
49. S.F. Vecht, M.W. Platt, Z. Er-El, and I. Goldberg, *Appl. Microbiol. Biotechnol.* **27**, 587–592 (1988).
50. R.W. Eaton and K.N. Timmis, *J. Bacteriol.* **168**, 123–131 (1986).
51. O.O. Amund and I.J. Higgins, *Antonie van Leewenhoek* **51**, 45–56 (1985).
52. J.F. Davey and D.T. Gibson, *J. Bacteriol.* **119**, 923–929 (1974).
53. D.T. Gibson and V. Subramanian, in D.T. Gibson ed., *Microbial Degradation of Organic Compounds*, Marcel Dekker, New York, 1984, pp. 361–369.
54. G. Baggi, P. Barbieri, E. Galli, and S. Tollari, *Appl. Environ. Microbiol.* **53**, 2129–2132 (1987).
55. G. Schraa, B.M. Bethe, A.R.W. van Neerven, W.J.J. van den Tweels, E. van der Wende, and A.J.B. Zehnder, *Antonie van Leewenhoek* **53**, 159–170 (1987).
56. S. Hartmans, M.J. van der Werf, and J.A.M. de Bont, *Appl. Environ. Microbiol.* **56**, 1347–1351 (1990).
57. D.B. Dzhusupova, B.P. Baskunov, L.A. Golovleva, R.M. Alieva, and A.N. Ilyaletdinov, *Mikrobiologiya* **54**, 136–140 (1985).
58. M. Zilli, A. Converti, A. Lodi, M. Del Borghi, and G. Ferraiolo, *Biotechnol. Bioeng.* **41**, 693–699, (1993).
59. T. Omori, H. Ishigooka, and Y. Minoda, *Agric. Biol. Chem.* **50**, 1513–1518 (1986).
60. M.R. Smith and C. Ratledge, *Appl. Microbiol. Biotechnol.* **30**, 395–401 (1989).
61. J.A. Don, in Verein Deutscher Ingenieure ed., *VDI Berichte 561*, VDI Verlag, Düsseldorf, 1986, pp. 63–74.
62. B. Hippchen, *Mikrobiologische Untersuchungen zur Eliminierung Organischer Lösungmittel im Biofilter*, Stuttgarter Berichte zur Siedlungswasserwirtschaft, Band 94, Oldenbourg, München, 1985.
63. P.G. Maurer, *BMFT Forschungsbericht T*, 79–114 (1979).
64. L. Rodhe, L. Thyselius, and U. Berglung, Report N. 76 Jordbrukstekniska Intitutet, Uppsala, Sweden, (Joint JTI/AFRC Engineering, Silsoe, translation N. 1, New Series), 1988.
65. C.C. Pearson, V.R. Phillips, G. Green, and I.M. Scotford, in A.J. Dragt and J. van Ham eds., *Biotechniques for Air Pollution Abatement and Odour Control Policies*, Elsevier, Amsterdam, 1992, pp. 245–254.
66. V. Torretta, *Ingegneria Ambientale* **25**, 379–389 (1996).
67. J.L. Blitz and A.F. Elvidge, *Environ. Protect. Bull.* **16**, 3–6 (1992).
68. G. Stucki, R. Galli, H. Ebersold, and T. Leisinger, *Arch. Microbiol.* **130**, 366–371 (1981).
69. M.R. Smith, *Biodegradation* **1**, 191–206 (1990).

BIOREACTORS, AIR-LIFT REACTORS

J.C. MERCHUK
M. GLUZ
Ben-Gurion University of the Negev
Beer-Sheva, Israel

KEY WORDS

Bubble column
Fluid dynamics
Gas hold-up
Heat transfer
Liquid flow
Mass transfer
Scale-up
Three-phase airlift reactors

OUTLINE

INTRODUCTION

General

The term *airlift reactor* (ALR) covers a wide range of gas–liquid or gas–liquid–solid pneumatic contacting devices that are characterized by fluid circulation in a defined cyclic pattern through channels built specifically for this purpose. In ALRs, the content is pneumatically agitated by a stream of air or sometimes by other gases. In those cases, the name *gas lift reactors* has been used. In addition to agitation, the gas stream has the important function of facilitating exchange of material between the gas phase and the medium; oxygen is usually transferred to the liquid, and in some cases reaction products are removed through exchange with the gas phase.

The main difference between ALRs and bubble columns (which are also pneumatically agitated) lies in the type of fluid flow, which depends on the geometry of the system. The bubble column is a simple vessel into which gas is injected, usually at the bottom, and random mixing is produced by the ascending bubbles. In the ALR, the major patterns of fluid circulation are determined by the design of the reactor, which has a channel for gas–liquid upflow—the riser—and a separate channel for the downflow (Fig. 1). The two channels are linked at the bottom and at the top to form a closed loop. The gas is usually injected near the bottom of the riser. The extent to which the gas disengages at the top, in the section termed the *gas separator*, is determined by the design of this section and the operating conditions. The fraction of the gas that does not disengage, but is entrapped by the descending liquid and taken into the downcomer, has a significant influence on the fluid dynamics in the reactor and hence on the overall reactor performance.

Airlift Reactor Morphology

Airlift reactors can be divided into two main types of reactors on the basis of their structure (Fig. 1): (1) external-loop vessels, in which circulation takes place through separate and distinct conduits; and (2) baffled (or internal-loop) vessels, in which baffles placed strategically in a single vessel create the channels required for the circulation. The designs of both types can be modified further, leading to variations in the fluid dynamics, in the extent of bubble disengagement from the fluid, and in the flow rates of the various phases.

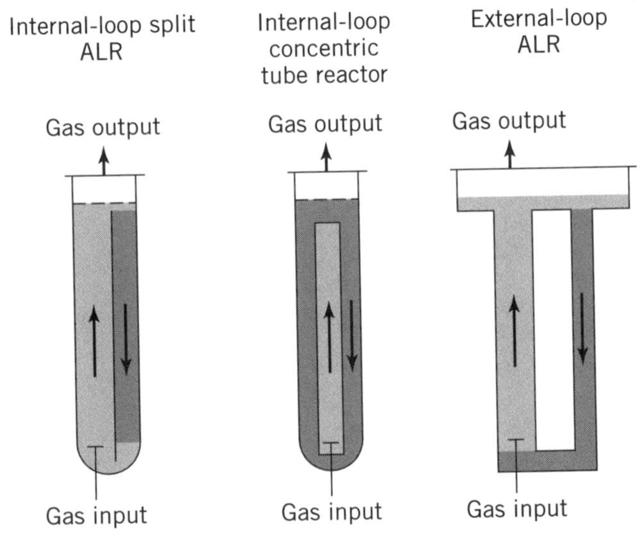

Figure 1. Different types of ALRs.

All ALRs, regardless of the basic configuration (external loop or baffled vessel), comprise four distinct sections with different flow characteristics:

- *Riser*. The gas is injected at the bottom of this section, and the flow of gas and liquid is predominantly upward.
- *Downcomer*. This section, which is parallel to the riser, is connected to the riser at the bottom and at the top. The flow of gas and liquid is predominantly downward. The driving force for recirculation is the difference in mean density between the downcomer and the riser; this difference generates the pressure gradient necessary for liquid recirculation.
- *Base*. In the vast majority of airlift designs, the bottom connection zone between the riser and downcomer is very simple. It is usually believed that the base does not significantly affect the overall behavior of the reactor, but the design of this section can influence gas holdup, liquid velocity, and solid phase flow (1,2).
- *Gas separator*. This section at the top of the reactor connects the riser to the downcomer, facilitating liquid recirculation and gas disengagement. Designs that allow for a gas residence time in the separator that is substantially longer than the time required for the bubbles to disengage will minimize the fraction of gas recirculating through the downcomer (Fig. 2).

Momentum, mass transfer, and heat transfer will be different in each section, but the design of each section may influence the performance and characteristics of each of the other sections, since the four regions are interconnected.

Advantages of Airlift Bioreactors

For the growth of microorganisms, ALRs are considered to be superior to traditional stirred-tank fermenters despite the fact that the conventional fermenters provide the major requirements for culturing microorganisms: gas–medium interface for the supply of oxygen and the removal of waste gases; means of agitation to ensure proper nutrient distribution and to minimize damage resulting from addition of concentrated acid or base (for pH control); means of heat transfer (for temperature control); and a contamination-free environment. Therefore, the reason for the more successful growth reported in ALRs (3,4) appears to lie in the difference in the fluid dynamics between ALRs and the more conventional fermenters. In conventional stirred tanks or bubble columns, the energy required for the movement of the fluids is introduced focally, at a single point in the reactor, via a stirrer or a sparger, respectively. Consequently, energy dissipation is very high in the immediate surroundings of the stirrer and decreases away from it toward the walls. Similarly, shear will be greatest near the stirrer (5), since the momentum is transferred directly to the fluid in that region (6), which, in turn, transfers this energy to the slower-moving, more distant elements of the fluid. This results in a wide variation of shear

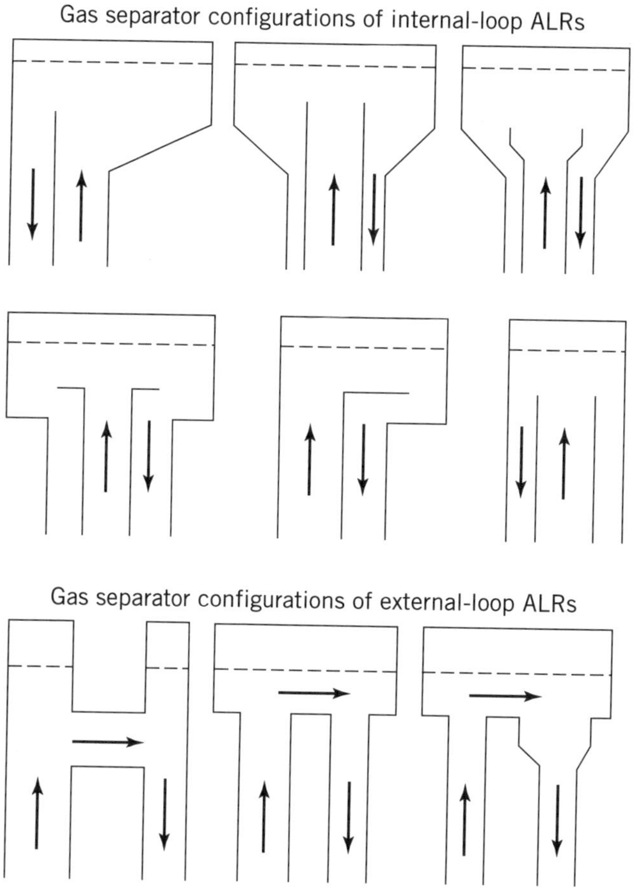

Figure 2. Different types of gas separators.

forces; for example, the maximum shear gradient in a stirred tank with a flat-blade turbine has been reported to be approximately 14 times the mean shear gradient (7).

Cells in culture may thus be exposed to contrasting environments in a mechanically stirred vessel, either to minimal shear forces that may generate potentially undesirable gradients in temperature and in substrate, metabolite, and electrolyte concentrations or, alternatively, to highly turbulent zones, with no problems of heat or mass transfer, but with very high shear gradients that may endanger cell integrity or exert some influence on cell morphology and metabolism (6). Changes in the morphology of microorganisms associated with high shear forces in the medium have frequently been observed (8–10). The nature of the relationship between such morphological changes and the rates of growth and metabolite production is still not properly understood, although it may be of great importance in the design and scale-up of bioreactors.

In ALRs, as in bubble columns, the gas is injected at a single point, but in ALRs the direct contribution of gas injection to the dynamics of the system is small; circulation of liquid and gas is facilitated by the difference in gas holdup between the riser and the downcomer, which creates a pressure difference at the bottom of the equipment:

$$\Delta P_{\mathrm{b}} = \rho_{\mathrm{L}} g(\varphi_{\mathrm{r}} - \varphi_{\mathrm{d}}) \quad (1)$$

where ΔP_{b} is the pressure difference, ρ_{L} is the density of the liquid (the density of the gas is considered to be negligible), g is the gravitational constant, and φ_r and φ_d are the fractional gas holdup of the riser and downcomer, respectively. The pressure difference forces the fluid from the bottom of the downcomer toward the riser, generating fluid circulation in the ALR. Since φ_r and φ_d are both average values integrated along the height of the reactor, it follows that there are no focal points of energy dissipation and that shear distribution is homogeneous throughout the ALR. There is thus a relatively constant environment, with minimization of sharp changes in the mechanical forces acting on suspended particles. Because good mixing is required, shear forces cannot be avoided completely. One of the most critical points is the bottom, where there is a sharp 180° turn.

Shear-sensitive mammalian and plant cells in culture should benefit from such an environment. Currently, the research and development of new bioreactors for mammalian cells is indeed focusing on the issue of shear-related damage to suspended cells (8,11–24).

Mammalian and plant cells in culture are more susceptible than microorganisms to the reactor conditions. Mammalian cells, which lack the rigid cell wall of microorganisms, have a larger size (one order of magnitude) than microorganisms and are very sensitive to mechanical stress. Plant cells have a rigid cellulose wall, but they are also much larger than microorganisms (usually by about an order of magnitude) and are therefore also sensitive to reactor conditions. Kolmogoroff's model of isotropic turbulence (25) indicates that serious damage may occur at relatively large values of the length scale. The last length is a parameter of the model and indicates the size of the eddy where energy starts to be dissipated by viscous resistance. Indeed, it has been observed that plant cells, in spite of their rigid wall, are shear-sensitive, and difficulties have been found in stirred-tank cultures. This is especially true when large-scale systems are considered. Although high agitation rates may be detrimental to cell growth, low agitation rates lead to an increase in the number and size of cell aggregates—also an undesirable phenomenon. The aggregates are formed as a result of daughter cells failing to separate after division and as a consequence of the stickiness of the polysaccharides excreted by the cells, especially at the end of the growth phase. An optimal shear rate between these two extremes must be found for each culture.

It has recently been shown experimentally that velocity fluctuations related to turbulent shear are relatively homogeneously distributed in an ALR (26,27). The measurements of fluctuating velocity made by Tan et al. (26) show that the liquid turbulence in ALRs is homogeneously distributed in both the riser and the downcomer. It thus seems reasonable to assume that the homogeneity of the stress forces is the main advantage offered by ALRs and that this homogeneity is responsible for the success of shear-sensitive cultures in the ALR type of fermenter (3,28–31).

Another advantage of the ALR is the mechanical simplicity of the device. The absence of a shaft and of the associated sealing, which is always a weak element from the point of view of sterility, confers on the ALR an obvious advantage over agitated tanks. This consideration is especially important in processes involving slow-growing cultures, such as animal and plant cells, for which the risk of contamination is large.

All the points mentioned above are particularly relevant for sophisticated processes in which the product is usually of high value. But ALRs may be used also for processes involving low-value products, in which case efficiency of energy use may well become the key point for design, as in the use of ALRs for wastewater treatment (32). The superiority of ALRs over mechanically agitated contactors in terms of mass transfer rates for a given energy input has been demonstrated by Legrys (33). Comparison of the efficiency of oxygen transfer, that is, the mass of oxygen absorbed per unit energy invested and unit time, showed that the efficiency of the ALR is among the highest in agitated systems (32). The ALRs are particularly suited to processes with changing oxygen requirements because aeration efficiency and performance are relatively insensitive to changes in operating conditions. Performance decreases markedly in mechanically stirred systems as the energy input (or oxygen transfer rate) increases, but it is quite constant in ALRs (34) (Fig. 3).

The efficiency of ALRs decreases relatively slowly as the energy input per unit volume of reactor is increased, as is shown in Figure 4 (32). In contrast, in the operation of stirred tanks, the mass transfer rate can be easily increased by increasing the power input, but this improvement is achieved at the cost of a considerable decrease in the efficiency of oxygen transfer. This decrease may constitute a crucial disadvantage in a process like wastewater treatment, where the energy input is an important element in the cost of the final product and flexibility of operating conditions is required because of the constant change of feed composition and flow rate.

Energy economy in the ALR may be improved by placing a second sparger in the upper part of the downcomer (32,35,36). If the liquid velocity is greater than the free rising velocity of the bubbles generated, the gas is carried down, resulting in a longer contact time between the bubble and the liquid. This diminishes the energy requirements, since part of the gas is injected against a lower hydrostatic pressure.

The advantages described above counterbalance the obvious disadvantage of ALRs, which is the requirement for a minimum liquid volume for proper operation. Indeed, the changes in liquid volume in these reactors are limited to the region of the gas separator, since the liquid height must always be sufficient to allow liquid recirculation in the reactor and must therefore be above the separation between the riser and the downcomer.

FLUID DYNAMICS

The interconnections between the design variables, the operating variables, and the observable hydrodynamic variables in an ALR are presented diagramatically in Figure 5 (37). The design variables are the reactor height, the riser-to-downcomer area ratio, the geometrical design of the gas separator, and the bottom clearance (C_b, the distance between the bottom of the reactor and the lower end of the draft tube, which is proportional to the free area for

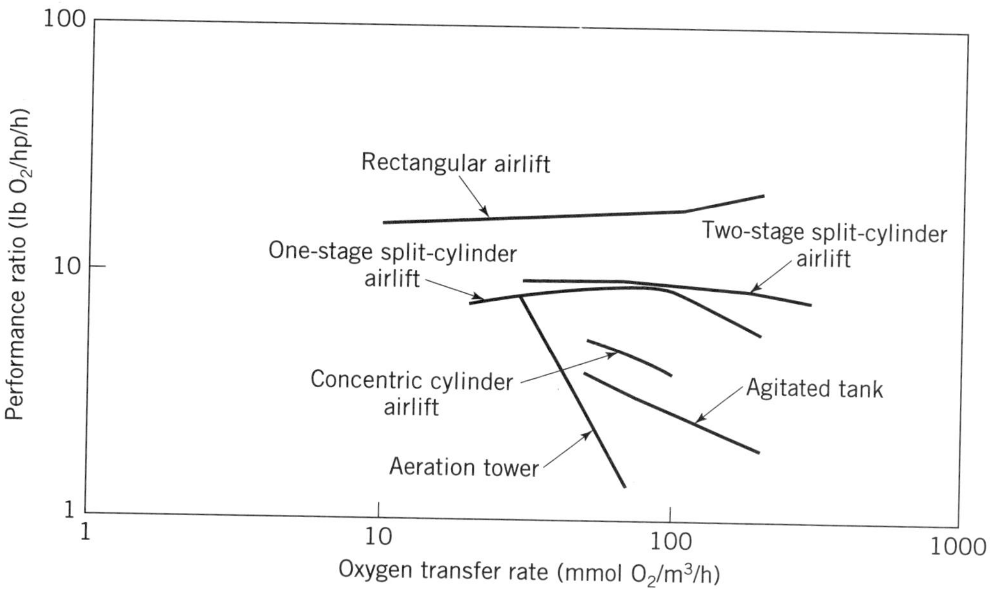

Figure 3. Performance ratio as a function of oxygen-transfer rate, showing that aeration efficiency and performance are relatively insensitive to changes in operating conditions in different types of ALRs (1–5) versus an agitation tank (6) and an aeration tower (7). Adapted from Orazem and Erickson (34).

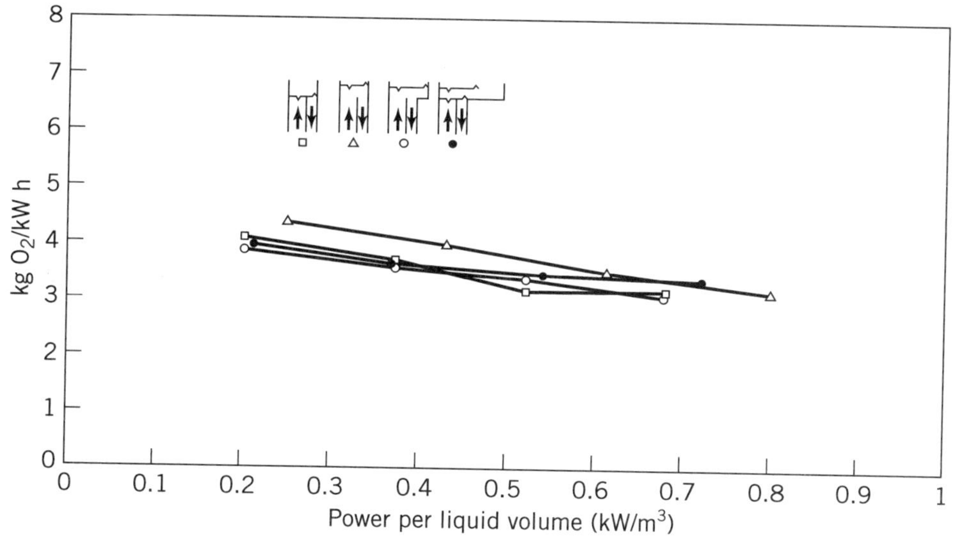

Figure 4. Aeration efficiency as a function of pneumatic power of gas input per unit volume in a straight-baffle ALR. The level indicated corresponds to no-aeration conditions. Adapted from Siegel et al. (32).

flow in the bottom and represents the resistance to flow in this part of the reactor). The main operating variables are primarily the gas input rate and, to a lesser extent, the top clearance (C_t, the distance between the upper part of the draft tube and the surface of the nonaerated liquid). These two independent variables set the conditions that determine the liquid velocity in the ALR via the mutual influences of pressure drops and holdups, as shown in Figure 5 (37). Viscosity is not shown in Figure 5 as an independent variable because in the case of gas–liquid mixtures, it is a function of the gas holdup (and of liquid velocity in the case of non-Newtonian liquids), and because in a real process, it will change with time due the changes in the composition of the liquid.

Flow Configuration

Riser. In the riser, the gas and liquid flow upward, and the gas velocity is usually larger than that of the liquid. The only exception is homogeneous flow, in which case both phases flow at the same velocity. This can happen only with very small bubbles, in which case the free-rising velocity of the bubbles is negligible with respect to the liquid velocity. Although about a dozen different gas–liquid flow configurations have been developed (38), only two of them are of interest in ALRs (39,40):

1. Homogeneous bubbly flow regime, in which the bubbles are relatively small and uniform in diameter and turbulence is low

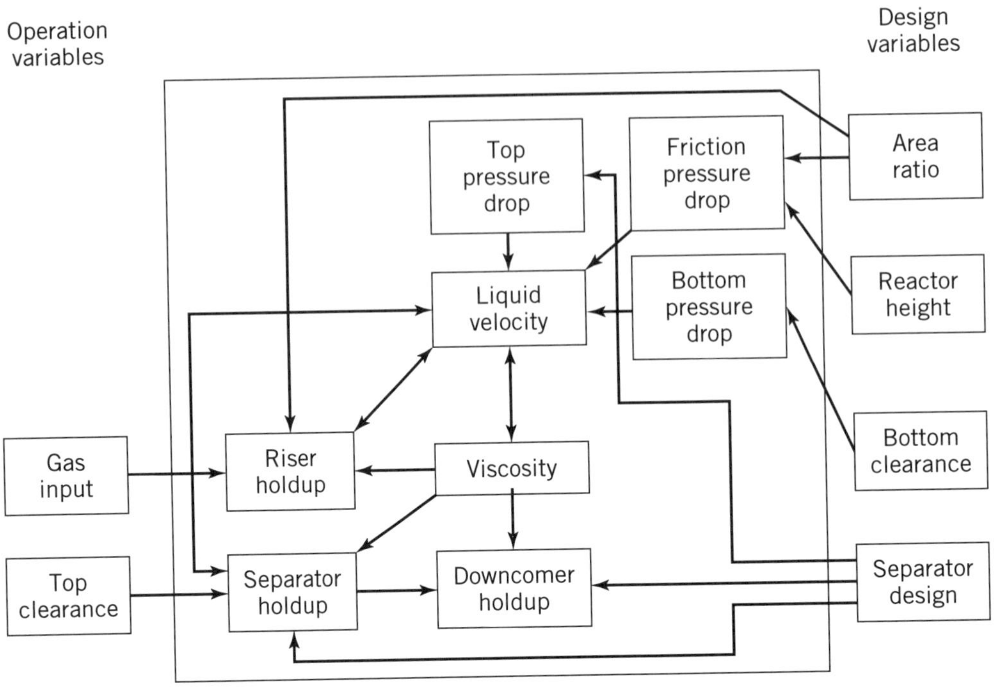

Figure 5. Interaction between geometric and fluid dynamic variables in an ALR. Adapted from Merchuk et al. (37).

2. Churn-turbulent regime, in which a wide range of bubble sizes coexist within a very turbulent liquid

The churn-turbulent regime can be produced from homogeneous bubbly flow by increasing the gas flow rate. Another way of obtaining a churn-turbulent flow zone is by starting from slug flow and increasing the liquid turbulence, by increasing either the flow rate or the diameter of the reactor, as can be seen in Figure 6 (41). The slug-flow configuration is important only as a situation to be avoided at all costs, because large bubbles bridging the entire tower cross-section offer very poor capacity for mass transfer.

Downcomer. In the downcomer, the liquid flows downward and may carry bubbles down with it. For bubbles to be entrapped and flow downward, the liquid velocity must be greater than the free-rise velocity of the bubbles. At very low gas flow input, the liquid superficial velocity is low, practically all the bubbles disengage, and clear liquid circulates in the downcomer. As the gas input is increased, the liquid velocity becomes sufficiently high to entrap the smallest bubbles. Upon a further increase in liquid velocity larger bubbles are also entrapped. Under these conditions the presence of bubbles reduces the cross-section available for liquid flow, and the liquid velocity increases in this section. Bubbles are thus entrapped and carried downward, until the number of bubbles in the cross-section decreases, the liquid velocity diminishes, and the drag forces are not sufficient to overcome the buoyancy. This feedback loop in the downcomer causes stratification of the bubbles, which is evident as a front of static bubbles, from which smaller bubbles occasionally escape downward and larger bubbles, produced by coalescence, escape upward. The bubble front descends, as the gas input to the system is increased, until the bubbles eventually reach the bottom and recirculate to the riser. When this point is reached, the bubble distribution in the downcomer becomes much more uniform. This is the most desirable flow configuration in the downcomer, unless a single pass of gas is required. The correct choice of cross-sectional area ratio of the riser to the downcomer will determine the type of flow.

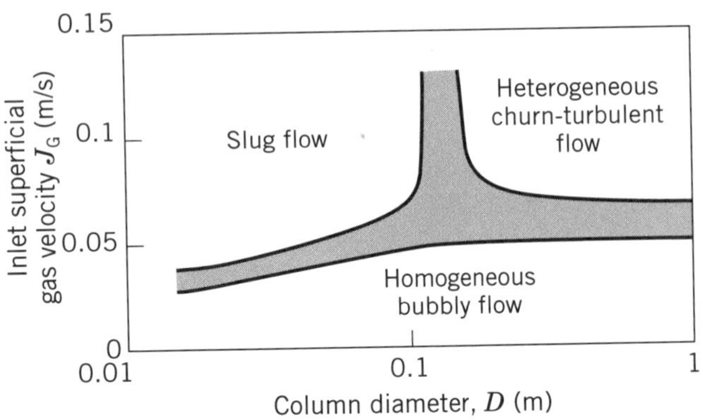

Figure 6. Map of flow configurations for gas–liquid concurrent flow in a vertical tube. Adapted with permission from Wiswanathan (40).

Gas Separator. The gas separator is often overlooked in descriptions of experimental ALR devices, although it has considerable influence on the fluid dynamics of the reactors. The geometric design of the gas separator will determine the extent of disengagement of the bubbles entering from the riser. In the case of complete disengagement, clear liquid will be the only phase entering the downcomer.

In the general case, a certain fraction of the gas will be entrapped and recirculated. Fresh gas may also be entrapped from the headspace if the fluid is very turbulent near the interface. The extent of this entrapment influences strongly gas holdup and liquid velocity in the whole reactor.

It is quite common to enlarge the separator section to reduce the liquid velocity and to facilitate better disengagement of spent bubbles. Experiments have been reported in which the liquid level in the gas separator was high enough to be represented as two mixed vessels in series (42,43). This point will be analyzed further in the section devoted to mixing.

Gas Holdup

Gas holdup is the volumetric fraction of the gas in the total volume of a gas–liquid–solid dispersion:

$$\varphi_i = \frac{V_G}{V_L + V_G + V_S} \qquad (2)$$

where the subindexes L, G, and S indicate liquid, gas, and solid, and i indicates the region in which the holdup is considered, that is, gas separator (s) the riser (r), the downcomer (d), or the total reactor (T).

The importance of the holdup is twofold: (1) the value of the holdup gives an indication of the potential for mass transfer, since for a given system a larger gas holdup indicates a larger gas–liquid interfacial area; and (2) the difference in holdup between the riser and the downcomer generates the driving force for liquid circulation. It should be stressed, however, that when referring to gas holdup as the driving force for liquid circulation, only the total volume of the gas is relevant. This is not the case for mass-transfer phenomena, in this case, the interfacial area is of paramount importance, and therefore some information on bubble size distribution is required for a complete understanding of the process.

Because gas holdup values vary within a reactor, average values, referring to the whole volume of the reactor, are usually reported. Values referring to a particular section, such as the riser or the downcomer, are much more valuable, since they provide a basis for determining liquid velocity and mixing. However, such values are less frequently reported.

The geometric design of the ALR has a significant influence on the gas holdup. Changes in the ratio A_d/A_r, the cross-sectional areas of the downcomer and the riser, respectively, will change the liquid and gas residence time in each part of the reactor and hence their contributions to the overall holdup. Gas holdup increases with decreasing A_d/A_r (44–47).

Gas Holdup in Internal Airlift Reactors. Correlations presented for internal-loop ALRs are shown in Table 1. These take into account liquid properties and geometric differences within a particular design. Most of the correlations take the form:

$$\varphi_r = a(J_G)^{\alpha}\left(\frac{A_d}{A_r}\right)^{\beta}(\mu_{ap})^{\gamma} \qquad (3)$$

where φ_r is the gas holdup in the riser, J_G is the superficial gas velocity (gas volumetric flow rate per unit of cross-sectional area), μ_{ap} is the effective viscosity of the liquid, and α, β, γ, and a are constants that depend on the geometry of the reactor and the properties of the liquid. The correlation can be used to predict the holdup in a system that is being designed or simulated as a function of the operating variables, the geometry of the system, or the liquid properties. Such correlations are effective for fitting data for the same type of reactor (e.g., a split-vessel reactor) with different area ratios or even different liquid viscosities, but they are mostly reactor-type specific.

The cyclic flow in the ALR complicates the analysis of the system. The riser gas holdup depends strongly on the geometric configuration of the gas–liquid separator and the water level in the gas separator. This has been shown experimentally in a split-vessel rectangular ALR (60), but the premise can essentially be extended to any internal-loop ALR. Analysis of the system revealed that these factors influence the gas disengagement and hence the gas recirculation in the downcomer. When this influence is taken into account and the holdup is plotted against the true gas superficial velocity, $J_{G,true}$, which is defined as the sum of the gas superficial velocity due to the freshly injected gas, Q_{in}, and to the recirculated gas, Q_d, that is,

$$J_{G,true} = \left(\frac{Q_{in} + Q_d}{A_r}\right) \qquad (4)$$

then all the data for the different gas separators may be represented by a single relationship, such as equation 3. In other words, if the actual gas flow is known, the influence of gas recirculation (which depends on A_d/A_r and the design of the gas separator) has been already taken into account and does not need to be considered again. Nevertheless, this simple approach has a drawback in that the true gas superficial velocity is difficult to measure because the gas recirculation rate is usually not known. A method for evaluation of the extent of the maximum gas recirculation has recently been developed and will be discussed later in this article.

Thus, correlations that take into account all the variables, which may be easily measured, remain the option of choice. Table 1 shows most of the correlations of this type that have been proposed for the riser holdup in internal-loop ALRs. Comparison of a number of these correlations shows that there is reasonable agreement between the predictions of the different sources (Fig. 7).

Figure 7 can be used as an example of the actual state-of-the-art in ALR design. A number of correlations have been proposed, and three variables (A_d/A_r, μ_{ap}, and J_G) have been tested by most researchers. The ranges in which these variables were studied varies from source to source. In addition, some other variables (such as bottom clearance, top clearance or gas separator design, and surface tension) have been used by some authors but ignored by others. One example is the disengagement ratio defined by Siegel and Merchuk (64), which represents the mean horizontal path of a recirculating bubble relative to the external diameter and is equivalent to the parameter obtained by dimensional analysis (1) as:

$$M = \frac{D_S}{4D}$$

where D is the diameter of column and D_s the diameter of

Table 1. Gas Hold-Up in Internal-Loop ALR

No.	Formula	Ref.
1	$\varphi_r = 0.441 J_{Gr}^{0.841} \mu_{ap}^{-0.135}$ $\varphi_d = 0.297 J_{Gr}^{0.935} \mu_{ap}^{-0.107}$	48
2	$\varphi_r = 2.47 J_{Gr}^{0.97}$	49
3	$\varphi_r = 0.465 J^{0.65}\left(1 + \frac{A_d}{A_r}\right)^{-1.06} \mu_{ap}^{-0.103}$	50
4	$\varphi_r = 0.65 J_{Gr}^{(0.603 + 0.078C_0)}\left(1 + \frac{A_d}{A_r}\right)^{-0.258}$ $\varphi_d = 0.46\varphi_r - 0.0244$	51
5	$\varphi_r = (0.491 - 0.498) J_{Gr}^{0.706}\left(\frac{A_d}{A_r}\right)^{-0.254} D_r \mu_{ap}^{-0.0684}$	52
6	$\varphi_r = 0.16\left(\frac{J_{Gr}}{J_{1r}}\right)^{0.57}\left(1 + \frac{A_d}{A_r}\right)$ $\varphi_d = 0.79\varphi_r - 0.057$	45
7	$\varphi_r = 0.364 J_{Gr}$	53
8	$\frac{\varphi_r}{1 - \varphi_r} = \frac{J_{Gr}^{n+2/2(n+1)}}{2^{3n+1/n+1} n^{n+2/2(n+1)}\left(\frac{K}{\rho_1}\right)^{1/2(n+1)} g^{n/2(n+1)}\left(1 + \frac{A_d}{A_r}\right)^{3(n+2)/4(n+1)}}$	54
9	$\frac{\varphi}{(1 - \varphi)^4} = \frac{0.124\left(\frac{J_G \mu_1}{\sigma_1}\right)^{0.996}\left(\frac{\rho_1 \sigma_1^3}{g \mu_1^4}\right)^{0.294}\left(\frac{D_r}{D}\right)^{0.114}}{1 - 0.276(1 - e^{-0.0368Ma})}$	55
10	$\varphi_r = \frac{Fr}{0.415 + 4.27\left(\frac{J_{Gr} + J_{1r}}{\sqrt{gD_r}}\right)\left(\frac{g\rho_1 D^2}{\sigma_1}\right)^{-0.188} + 1.13 Fr^{1.22} Mo^{0.0386}\left(\frac{\Delta\rho}{\rho_1}\right)^{0.0386}}$	56
11	$\frac{\varphi}{(1 - \varphi)^4} = 0.16\left(\frac{J_G \mu_1}{\sigma_1}\right) Mo^{-0.283}\left(\frac{D_r}{D}\right)^{-0.222}\left(\frac{\rho_1}{\Delta\rho}\right)^{0.283} * (1 - 1.61(1 - e^{-0.00565Ma}))^{-1}$	2
12	$\varphi_d = 4.51 \cdot 10^6 Mo^{0.115}\left(\frac{A_r}{A_d}\right)^{4.2} \varphi_r$ when $\varphi_r < 0.0133\left(\frac{A_d}{A_r}\right)^{-1.32}$ and $\varphi_d = 0.05 Mo^{-0.22}\left[\left(\frac{A_r}{A_d}\right)^{0.6} \varphi_r\right]^{0.31Mo^{-0.0273}}$ when $\varphi_r > 0.0133\left(\frac{A_d}{A_r}\right)^{-1.32}$	57
13	$\varphi_r = 0.0057\left[(\mu_1 - \mu_w)^{2.75} - 161\frac{73.3 - \sigma}{79.3 - \sigma}\right] \cdot J_{Gr}^{0.88}$	58
14	$\varphi_r = \frac{0.4Fr}{1 + 0.4Fr\left(1 + \frac{J_1}{J_{Gr}}\right)}$	57
15	$\varphi = 0.24 n^{-0.6} Fr^{0.84 - 0.14n} Ga$	59

gas separator. If this parameter is not taken into account, then studies of the influence of the top clearance (42,65) are incomplete and difficult to extrapolate to other designs. The same can be said about the filling factor (66) given by the ratio of the gas separator volume to the total volume.

The foregoing discussion thus explains why all the correlations coincide for some ranges of these secondary variables while in other ranges they may diverge. In addition, in some cases the number of experiments may not have been sufficient to provide correlations or they may have been ill-balanced from the statistical point of view. The obvious solution to this problem lies in the collection of a large and detailed bank of reliable data that will constitute the basis for correlations with greater accuracy and validity.

The safest procedure for the prediction of the gas holdup in an ALR under design is to take data provided by researchers who have made the measurements in that par-

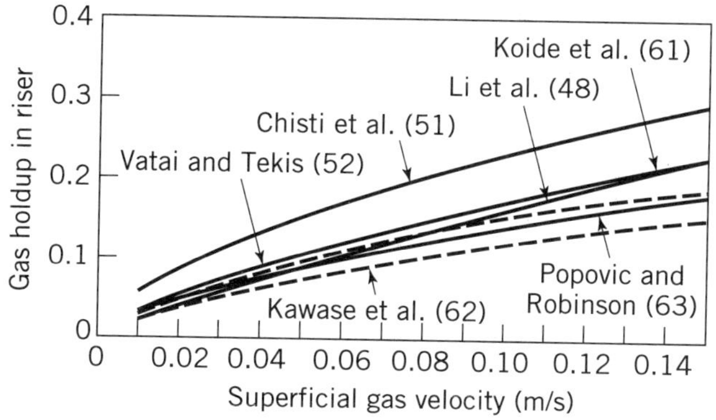

Figure 7. Some correlations proposed for prediction of gas holdup in the riser of internal-loop ALRs (Table 1). Gas holdup (φ_r) is presented as a function of superficial gas velocity (J_G). Other parameters related to geometry and physicochemical properties that were used in the calculations are shown on the figure.

ticular type of reactor with the same physicochemical properties of the system. If this option is not available, then correlation 9 in Table 1 (55) is recommended for prediction of the gas holdup in the riser.

Gas holdup in the downcomer is lower than that in the riser. The extent of this difference depends mainly on the design of the gas separator (67). The downcomer gas holdup is linearly dependent on the riser holdup, as a consequence of the continuity of liquid flow in the reactor. Many expressions of this type have been published (68). At low gas flow rates, φ_d is usually negligible, since most of the bubbles have enough time to disengage from the liquid in the gas separator. This usually happens at the low gas flow rates frequently used for animal cell cultures.

The gas holdup in the separator is very close to the mean gas holdup in the whole reactor (1) as long as the top clearance C_t is relatively small (one or two diameters). For larger top clearances, the behavior of the gas separator begins to resemble that of a bubble column, and the overall performance of the reactor is influenced by this change.

External-Loop Airlift Reactors. From the point of view of fluid dynamics, neither the external configuration (shape and architecture) nor the fact that both riser and downcomer are easily accessible is the most important difference between external- and internal-loop reactors. The most important point is that the gas separator of the external-loop ALR is built in such way that gas disengagement is usually much more effective in this type of reactor. This can be easily seen in Figure 2. In concentric tubes or split vessels, the shortest path that a bubble has to cover from the riser to the downcomer is a straight line across the baffle that separates the two sections. In the case of external-loop ALRs, there is usually a minimum horizontal distance to be covered, which increases the chances of disengagement of the bubbles. In this case, it is worth pointing out that if gas does appear in the downcomer, then most of it will be fresh air entrained in the reactor because of interfacial turbulence or vortices that appear in the gas separator above the entrance to the downcomer. In many of the studies reported in the literature on holdup in external-loop ALRs, total disengagement is attained. No such data are available for the concentric tubes of split-vessel ALRs, since total disengagement is possible only at very low gas flow rates.

Several authors (37,69–73) have presented their results of gas holdup as the gas velocity versus the superficial mixture velocity, based on the drift flux model of Zuber and Findlay (74). These authors derived general expressions for prediction of the gas holdup and for interpretation of experimental data applicable to nonuniform radial distributions of liquid velocity and gas fraction. The drift velocity is defined as the difference between the velocity of the particular phase (U) and the volumetric flux density of the mixture (J) where:

$$J = J_G + J_L \tag{5}$$

The drift velocities of the gas and liquid phases may thus be expressed as:

$$J_G = U_G - J \tag{6}$$

$$J_L = U_L - J \tag{7}$$

Zuber and Findlay (74) derived the relationship [8], which has been shown to be more than adequate to provide a correlation of gas holdup measurements in tower reactors with high liquid velocities, such as ALRs (71):

$$U_G = \frac{J_G}{\varphi} = C_0 J + \frac{\frac{1}{A}\int \varphi(U_G - J)dA}{\frac{1}{A}\int \varphi \cdot dA} \tag{8}$$

where A is cross-sectional area, C_0 is distribution parameter, J is superficial velocity, J_G is superficial gas velocity, U_G linear gas velocity, and φ is gas holdup.

Equation 7 describes the relationship between the gas velocity in a two-phase flow and the volumetric flow density of the mixture, J.

As stressed by Zuber and Findlay (74), J has the advantage of being independent on space coordinates for both one-dimensional flow and multidimensional irrotational flows. The distribution parameter C_0 is given by (75):

$$C_0 = \frac{\frac{1}{A}\int_A \varphi J \cdot dA}{\left[\frac{1}{A}\int_A J \cdot dA\right]\left[\frac{1}{A}\int_A \varphi \cdot dA\right]} \tag{9}$$

The value of C_0 depends mainly on the radial profile of the gas holdup. Zuber and Findlay (74) calculated $C_0 = 1$ for a flat profile and $C_0 = 1.5$ for a parabolic profile. Experimental values have been reported in the range of 1.03–1.2 for upflow (69–73,76) and 1.0–1.16 for downflow (70,73).

Equation 9 shows that this parameter is a function of the profiles of velocities and holdup. The last term of the right-hand side of equation 8 is the weighted mean value of the drift velocity:

$$U_{GI} = U_G - J \tag{10}$$

The drift velocity of a swarm of bubbles can be evaluated by using the expression given by Zuber and Findlay (74).

$$U_{2J} = 1.53 \cdot \left[\frac{\sigma g \Delta\rho}{\rho_L^2}\right]^{0.25} (1 - \varphi)^{1.5} \tag{11}$$

where U_{2J} is the velocity of the swarm of bubbles, g is gravitational acceleration, ρ_L is the density of liquid, $\Delta\rho$ is the density difference, σ is the surface tension, and φ is the gas holdup. This equation is valid for bubble diameters of the order of 0.1 to 2 cm, which covers the population usually observed in ALRs.

It has, however, been shown (71) that a plot of U_G versus J gives a straight line, suggesting that a constant value of the drift velocity satisfactorily represents the two-phase flow in the riser of an external-loop ALR. In this plot, the distribution parameter was $C_0 = 1.03$, and U_{GS}, the value of the slip velocity of a bubble, was taken as the mean drift velocity. Siegel et al. (35) applied the same model for the study of gas recirculation in a split-vessel ALR and obtained the values of $C_0 = 1.11$. The slip velocity that they obtained fitting their data to equation 8 was 0.238 m/s. It has been suggested (71) that this simplification holds as long as coalescence is not a predominant factor in the process.

It is very important to stress the difference between holdup, φ, and the flowing volumetric concentration (β), which is defined as:

$$\beta = \frac{Q_G}{Q_G + Q_L} = \frac{J_G}{J} \tag{12}$$

Zuber–Findlay's drift flux model allows us to derive the following equation, which establishes a connection between the gas holdup and β.

$$\frac{\beta}{\varphi} = C_0 + \frac{U_{b\infty}}{J} \tag{13}$$

where C_0 is the distribution parameter, J is the superficial velocity, $U_{b\infty}$ is the terminal gas velocity, β is the flowing volumetric concentration, and φ is the gas holdup.

Figure 8 gives a representation of the $\varphi - \beta$ plane. The 45° line indicates that $\varphi = \beta$, an equality that is true only for nonslip flow, where the velocity of the gas is equal to the velocity of the liquid. Such a situation can be visualized for the case of very small bubbles in a relatively fast liquid. In this case, there is no influence of one phase on the motion of the other. As indicated in Figure 8, all the points below the 45° line indicate operation situations in which the liquid is driven by the gas:

$$U_G > U_L;\ \varphi < \beta \tag{14}$$

This happens in the riser of ALRs. For all points above the line the opposite is true:

$$U_G < U_L;\ \varphi > \beta \tag{15}$$

This latter condition reflects the operation of the downcomer.

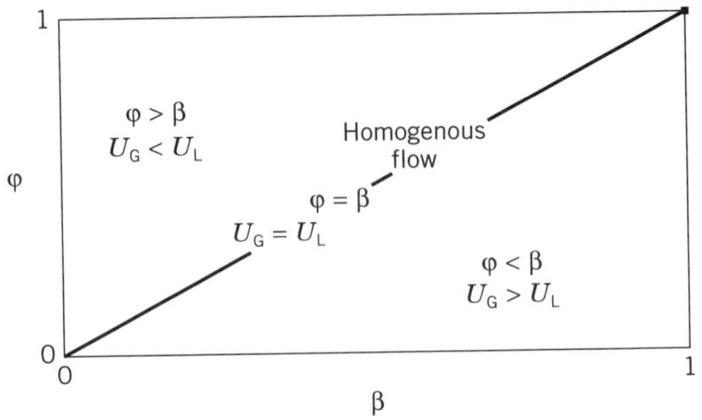

Figure 8. Gas flow holdup (φ) vs. flowing volumetric concentration (β). The different zones in the plane φ–β identify the two-phase flow. Adapted from Merchuk and Berzin (77).

A number of authors (71,76,78,79) have measured the local holdup profile along the riser of an external-loop ALR. In general, it was found that the holdup increases with height. This finding concurs with the expected expansion of gas bubbles as regions of lower pressure are reached. Common sense indicates that this situation must be limited to a certain range; an increase in bubble size will enhance turbulence and result in an increase in bubble encounters, leading eventually to bubble coalescence. The larger bubbles will rise much faster, resulting in a decrease in holdup. Such a scenario was indeed observed by Merchuk and Stein (71), as is illustrated in Figure 9. Merchuk and Stein (71) reported a maximum in the holdup profile for the case of a single-orifice gas distributor. For a multiple-orifice sparger, producing a more homogeneous bubble size distribution, a maximum was not observed within the studied length of the riser, which was 4 m.

Literature data from different sources for gas holdup in the riser under conditions of little or no carryover of gas from the separator into the downcomer for different A_d/A_r and top clearance C_t may be represented by the simple exponential:

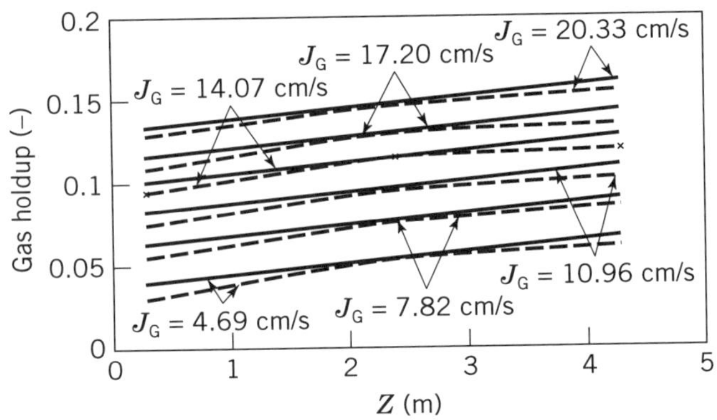

Figure 9. Dependence of the riser gas holdup in a 4-m high external-loop ALR with a multiple-orifice sparger (*solid lines*) and a single-orifice sparger (*broken lines*). Adapted from Merchuk and Stein (71).

$$\varphi_r = \alpha J_G^{\beta} \tag{16}$$

where the constant α depends on the friction losses in the loop, and β is usually a value between 0.6 and 0.7, as is illustrated in Figure 10 (65). The fact that neither the area ratio nor the top clearance affects the gas holdup demonstrates the role of the gas–liquid separator in determining the performance of the reactor in general. In the absence of gas recirculation, there is no effect on these variables. Moreover, this means that under conditions of no gas entrainment from the separator to the downcomer, it is possible to predict the riser gas holdup as a function of the riser superficial gas velocity alone, which is of great importance for design purposes.

It is accepted that liquid velocity has a mild negative effect on gas holdup in the riser. This effect is usually studied by reducing the liquid flow; this is achieved by adding resistance to the liquid loop by means of a valve or other controlled obstruction (71,78,82) under conditions of low or nil gas recirculation. Such experiments, which are relatively simple in external-loop ALRs, indicate that the holdup decreases as the liquid velocity is increased from zero (bubble column) to 0.3 m/s (which is close to the bubble free-rise velocity). For higher velocities, the effect of U_L is small. These findings add to our understanding of the fluid dynamics in the column. At liquid velocities that are smaller than the bubble free-rising velocity, the liquid transported in the wake of the bubbles, which must return downward to balance the mass flux, is the cause of the meandering and loops that typically appear in bubble column operation (83). As the overall liquid flux increases, the patterns straighten out, the bubbles begin to ascend in a straight pattern, and the holdup goes down. When the liquid velocity is higher than the free-rise velocity of the bubbles, piston flow of bubbles ensues in the tube, and the decrease in holdup for further increases in liquid velocity is due solely to the change in the ratio of gas–liquid volumetric flow rates.

When there is gas recirculation, the area ratio A_d/A_r becomes an important variable affecting gas holdup. The effect of A_d/A_r starts in the region in which gas entrainment from the separator to the downcomer occurs. If it is assumed that the riser cross-sectional area A_r remains unchanged and the downcomer cross-sectional area A_d is increased, then it can be expected that the liquid velocity in the riser will increase as a result of the smaller resistance to flow in the loop, which in turn leads to a decrease in the riser gas holdup. An increase in A_d/A_r will result in a decrease in the liquid velocity in the downcomer, which leads to a decrease in the gas recirculation, since fewer bubbles are entrapped in the downcomer. The final outcome of increasing A_d/A_r is thus a decrease is the riser gas holdup. A similar argument can be applied in the discussion of the effect of reactor height on the riser gas holdup, that is, an increase in the height of the downcomer will result in a higher liquid velocity, which will in turn lead to a decrease, as in the former case, in the holdup in the riser. In contrast, an increase in A_d/A_r will lead to an increase in the extent of bubble entrapment in the downcomer, which will serve to inject some additional gas into the riser. On the other hand, an increase of gas holdup in the downcomer diminishes the driving force for recirculation, as shown in equation 1, and this will moderate the increase of liquid velocity generated by the larger height. This feedback control of the liquid velocity is one of the characteristics particular to ALRs.

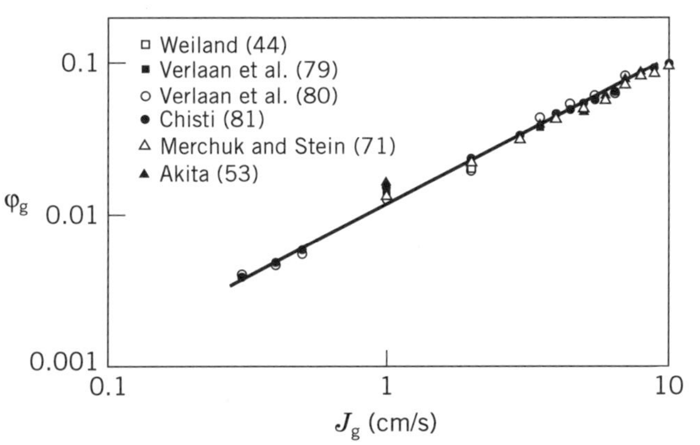

Figure 10. Gas holdup reported by various sources for the riser of airlift reactors under conditions of little or no gas recirculation. The data correspond to different A_d/A_r ratios.

Table 2 shows most of the expressions published for the correlation of experimental data obtained in external-loop ALRs. Some of these expressions are presented in Figure 11. The differences between the predictions obtained with the different correlations are probably due to the design of the gas separator. The equation given by Popovic and Robinson (63) seems to give an average of the proposed correlations.

Table 2. Gas Hold-up in External-Loop ALR

No.	Formula	Ref
1	$\varphi_r = \dfrac{0.6\rho_G^{0.062}\rho_1^{0.069}\mu_G^{0.107}}{\mu_1^{0.053}s_1^{0.185}} \cdot \dfrac{J_{Gr}^{0.936}}{(J_{Gr} + J_{1r})^{0.474}}$	84
2	$\varphi_r = 0.16\left(\dfrac{J_G^2}{J_{1r}}\right)^{0.56}\left(1 + \dfrac{A_d}{A_r}\right)$ $\varphi_d = 0.89\varphi r$	85
3	$\varphi_r = 1.07Fr^{0.333}$ $Fr = \dfrac{J_G^2}{gD_r}$	62
4	$\varphi = 0.55J_{Gr}^{0.78}F^{0.2}D_r^{0.42}$ $F = \dfrac{V_{1s}}{V_1}$	66
5	$\varphi_r = 0.203\,\dfrac{Fr_*^{0.31}}{Mo^{0.012}}\left(\dfrac{J_{Gr}}{J_{1r}} \cdot \dfrac{A_r}{A_d}\right)^{0.74}$ $Mo = \dfrac{g(\rho_1 - \rho_G)}{\sigma_1\rho_1^2} \cdot K^4\left(\dfrac{8J_{1r}}{D_r}\right)^{4(n-1)}\left(\dfrac{3n + 1}{4n}\right)^{4n}$ $Fr_* = \dfrac{(J_{1r} + J_{Gr})^2}{gD_r}$	86
6	$\varphi_d = 0.997\varphi_r$	81
7	$\varphi_r = 0.16\left(\dfrac{J_{Gr}}{J_1}\right)^{0.56}\left(1 + \dfrac{A_d}{A_r}\right)$	45

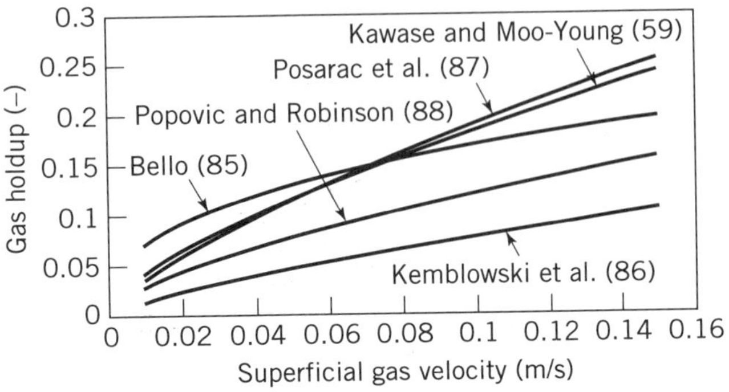

Figure 11. Some correlations proposed for prediction of gas holdup in the riser of external ALRs (Table 2). The gas holdup is presented as a function of the superficial gas velocity.

Effects of Liquid Rheology. The effect of rheology on the reactor behavior and performance is of great interest because in most biotechnological processes an increase in biomass provokes changes in the rheology of the fluid, especially in the case of mycelial growth. This effect is enhanced when in addition to the biomass growth, a product of the process is released into the medium in appreciable amounts. A good example of this scenario is the biosynthesis of polysaccharides, which cause an increase in the liquid viscosity.

The effect of viscosity on gas holdup in bubble columns has been studied by a number of authors. The main problem to be overcome is that of non-Newtonian flow. If the viscosity is not constant, but changes with changes in the shear rate, then the evaluation of shear rates becomes particularly relevant for the identification of the system. Several authors have confronted this issue. Nishikawa et al. (89,90) analyzed the problem of heat transfer in a bubble column with non-Newtonian liquids. They found a direct proportionality between the superficial gas velocity and the global shear rate:

$$\gamma = 5000 \cdot J_G \quad (J_G > 0.04 \text{m/s}) \tag{17}$$

This global shear rate was then used to calculate a global viscosity. In shear-sensitive cultures, the definition of a global shear rate in itself is of great importance.

A number of researchers, Henzler (91), Kawase and Moo-Young (59), Schumpe and Deckwer (92,93) have followed the approach of Nishikawa et al. (89) but have suggested different proportionality constants relating the global shear rate to the superficial gas velocity. This approach is questionable from the rheological point of view because it will predict the same shear rate for a certain superficial gas velocity, no matter which liquid is used. El-Tamtamy et al. (94) introduced an improvement by calculating the shear rate from the bubble velocity divided by the bubble diameter. However, accurate evaluation of the latter two parameters is difficult. Henzler and Kauling (95) suggested relating the shear rate to power input based on dimensional analysis by expressing the shear rate as a function of the power input per unit volume, $(P/[V\rho v])^{1/2}$. Their analysis gives different shear rates for liquids that are rheologically different.

The above-described relationships predict different shear rates that vary in up to three orders of magnitude. It is thus generally agreed that the correct solution is still to be found. Recently, a more general approach, known as a global approach, has been proposed by Merchuk and Ben-Zvi (Yona) (96). The shear stress in a bubble column was defined as being equal to the acting force, which can be calculated from the power input divided by the sum of the areas of all the bubbles:

$$\tau = \frac{P}{L_R} S_{ab} \tag{18}$$

where L_R is an effective length that represents the mean circulation path of a bubble in the system considered, P is the power input, S_{ab} is the total surface of all of bubbles, and τ is the shear stress.

Assuming ideal gas isothermal expansion, the power input P can be calculated. The interfacial area can be evaluated from correlations or can be obtained by direct measurement if available. A correlation taking into account other variables, like sparger configuration, surface tension, etc., will broaden the range of applications of this approach.

If a constitutive equation describing the rheology of the system is available (such as the power law, which has been reported to correspond to many biological systems), equation 17 facilitates the calculation of a global shear force acting on the liquid. The shear rate can be in this case expressed as:

$$\gamma = \left[\frac{\tau}{k}\right]^{1/n}$$

where γ is shear rate and κ is behavior coefficient, and equation 17 can be now used to express γ as:

$$\gamma = \left[\frac{p_1 J_G \ln\left(\frac{p_1}{p_2}\right)}{a L_R^2 \kappa}\right]^{1/n} \tag{19}$$

where the subindexes 1 and 2 represent the two extremes of the section considered.

Equation 19 thus gives a global shear rate that is a function of both fluid dynamics and rheology. This approach has been found to be useful for the presentation of results on mass transfer rates in bubble columns (96).

In contrast to the marked influence of rheology on gas holdup in bubble columns, the data available for ALRs show clearly that the effect of liquid viscosity is less dramatic, but not simpler. Figure 12 (65) illustrates the effect of the addition of glycerol to water in an internal-loop ALR. At low concentrations of glycerol, a moderate increase of the gas holdup is evident, particularly in the downcomer but also in the riser. These increases are caused by the lower free rise velocity of the bubbles, which increase the gas retention due to the longer residence time. In addition, the entrapment of the bubbles is increased, and this is reflected mainly in φ_d. When the concentration of glycerol becomes too high, a strong decrease of the gas holdup is

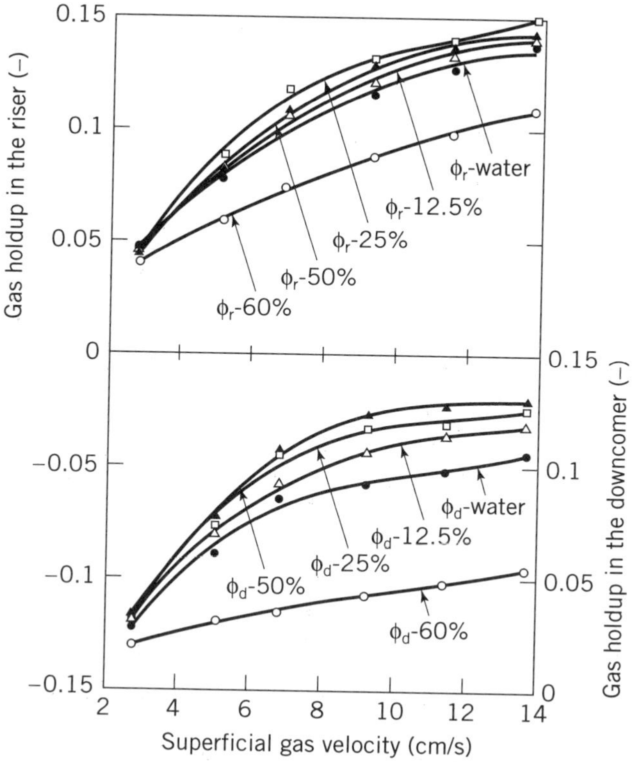

Figure 12. Effect of liquid viscosity on the gas holdup in the riser and in the downcomer of an internal-loop ALR. The viscosity corresponding to the solutions used was 4–14 mPa s. Percentages refer to percent glycerol in water (65).

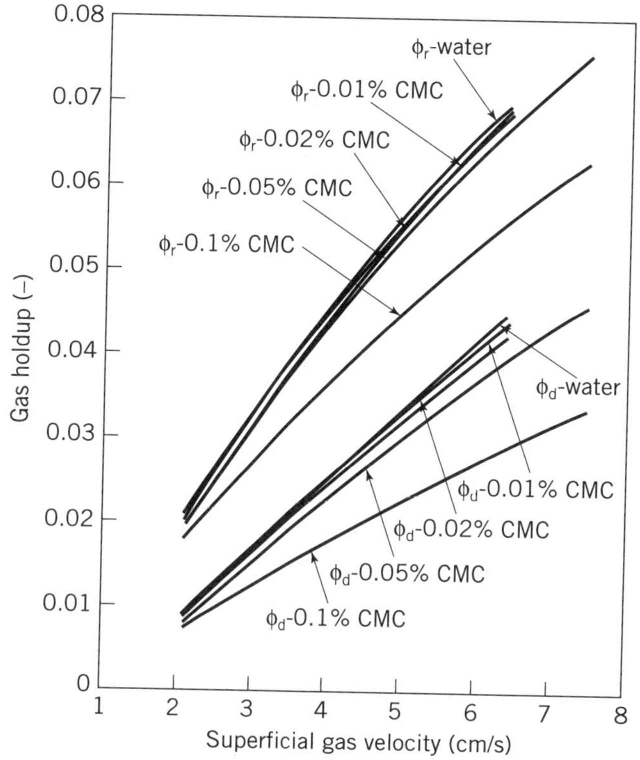

Figure 13. Effect of liquid rheology on the gas holdup in the riser and in the downcomer of an internal-loop ALR with a non-Newtonian liquid. The apparent viscosity corresponding to the solutions used, calculated for $\gamma = 50\ s^{-1}$, was 5–56 mPa s (65).

seen. This decrease is probably due to the onset of coalescence, which produces larger bubbles that ascend faster in the liquid and easily disengage in the gas separator. The viscosities corresponding to these solutions ranged from 4 mPa s to 14 mPa s. In Figure 13, the addition of carboxymethyl cellulose (CMC) to water is shown (65). The change in CMC concentration had only a slight effect on the gas holdup for additions in the range 0.01 to 0.05% CMC. Only for solutions with concentrations higher than 0.5% CMC was an appreciable decrease in holdup seen.

Effect of Liquid Level. The influence of the liquid level C_t on the gas holdup is exerted as a consequence of changes in the extent of disengagement of the bubbles in the gas separator. This influence is therefore dependent on the geometric design of this section. Whether a bubble will disengage or will be entrapped into the downcomer depends on the interrelationship of several parameters—the free rising velocity of the bubble $U_{b\infty}$ (a function of size and viscosity), the liquid velocity in the downcomer U_{Ld} (a function of the difference in gas holdup between the riser and the downcomer and frictional losses), and the residence time of the bubble in the gas separator (a function of geometric design and liquid height). For a given bubble size, if U_{Ld} is smaller than the $U_{b\infty}$ corresponding to the smallest bubble, then there is no carryover. For smaller bubbles, the balance between the time required to cover the path from the end of the riser to the zone near the entrance of the downcomer and the time needed for disengagement will give the fraction of bubbles recirculated. It should nevertheless be kept in mind that this is a feedback process. A higher bubble disengagement rate leads to a lower gas holdup in the downcomer, which in turn increases the liquid velocity, enabling larger bubbles to be trapped, until the system eventually reaches a steady state.

Due to the above-explained influence of the geometric design, the influence of the liquid height is completely different in internal- and external-loop ALRs. In internal-loop reactors, an increase in C_t increases the zone of the separator in which the gas holdup is higher, and as a result, the gas holdup increases. The extent of this increase depends, as said earlier, on the free-rise velocity of the bubbles. Figure 14 (1) shows the gas holdup in the riser and downcomer of a 30-liter ALR for two values of the top clearance, C_t = 0.178 m and C_t = 0.308 m, for two different liquids, water and a 0.5% CMC solution. It can be seen that although for water C_t has a small effect, this is not so for the more viscous solution. In the latter, the lower rising velocity of the bubbles causes a greater number of bubbles to be entrained and carried down by the liquid. Thus, the residence time in the disengagement section becomes very important in determining the fraction of bubbles that recirculate. A lower C_t will give a shorter residence time in the separator, a larger bubble recirculation, and, hence, a larger gas holdup.

In external-loop ALRs, the opposite effect is obtained, as may be seen in Figure 15 (65). The data in Figure 15,

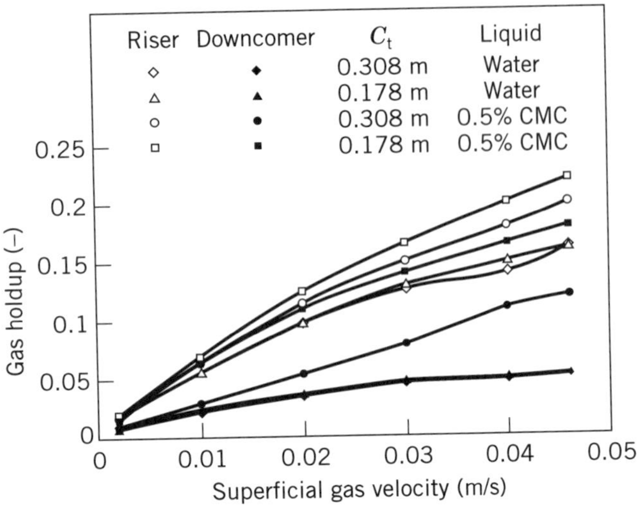

Figure 14. Riser and downcomer gas holdup in an internal-loop ALR for two different top clearances and two liquids. Adapted from Merchuk et al. (1).

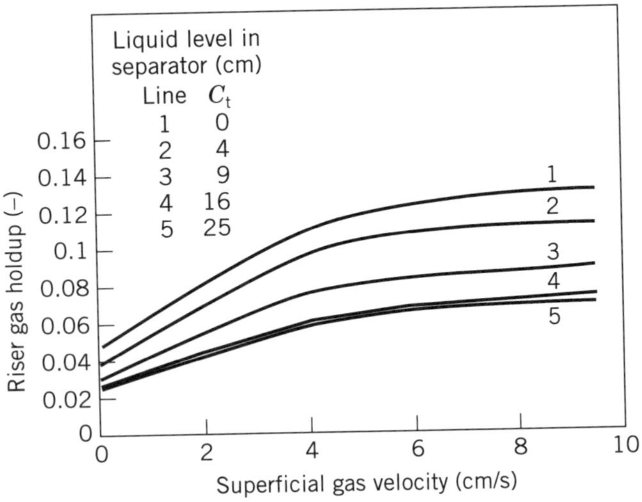

Figure 15. Gas holdup in the riser of an external-loop ALR for several top clearances. Adapted from Hallaile (65).

obtained for a 4-m high external-loop ALR, show that the holdup in the riser decreases as C_t increases. This is due to the construction of these reactors, in which much of the gas that enters into the downcomer is trapped from the headspace due to the turbulence in this zone. An increase in the liquid height serves to reduce the amount of gas trapped, so that less gas circulates in the downcomer and the liquid velocity increases. The final result is a reduction in the gas holdup, both in the riser and the downcomer (65).

Gas Recirculation

The degree to which gas flowing in the riser is entrapped and recirculated through the downcomer is an important variable, since it influences not only the flow configuration in the downcomer, but also the overall performance of the ALR. The liquid velocity depends mainly on the difference in holdup between the riser and the downcomer, and it in turn influences the gas holdup in the riser. Despite the importance of recirculation, very little quantitative data are available on this phenomenon. Siegel et al. (35) evaluated the gas recirculation in a split-vessel ALR by an indirect method based on holdup measurements. From their results, shown in Figure 16, it may be seen that the recirculation rate remains fairly constant for changing gas flow rates in the riser for high values of the last variable. Thus, the recirculation rate is determined largely by the geometric configuration of the gas–liquid separator and the liquid level in the separator.

Three zones are evident in Figure 16; they represent operating conditions giving oscillating, borderline, and straight bubble flow in the downcomer. Oscillating flow patterns produce much larger fractions of gas retained in the downcomer, but they are much more sensitive to J_G. At low superficial gas velocity, the recirculation increases very sharply with J_G. The bubbles exhibit an oscillating swirling flow pattern, with some larger bubbles escaping toward the top. The borderline condition is defined as oscillatory bubble flow at a low gas flow rate that shifts to straight bubble flow with increasing input gas flow rate. The straight flow operation zone is distinguished by bubble flow in a straight, well-defined flow pattern for all the input gas flow rates studied. If a straight bubble flow pattern is desired, the reactor should be operated at high riser gas flow rates, at which the reactor will shift toward stable operation.

Lubbert et al. (97) attempted to evaluate the recirculation of gas during the cultivation of yeast (*Saccharomyces cerevisiae*) on waste from a starch factory in a 4-m^3 pilot plant. They used microprocessor-aided pseudostochastic tracer input and cross-correlation techniques, which facilitated very reduced tracer feeds due to a high signal-to-noise ratio. The response to a pulse of helium was measured directly at the surface of the liquid in the separator by a quadruple mass spectrometer. The peak obtained showed pronounced shoulders (Fig. 17) which could be interpreted as superimposition of a second peak that represents the helium tracer one loop after. The fitting of such

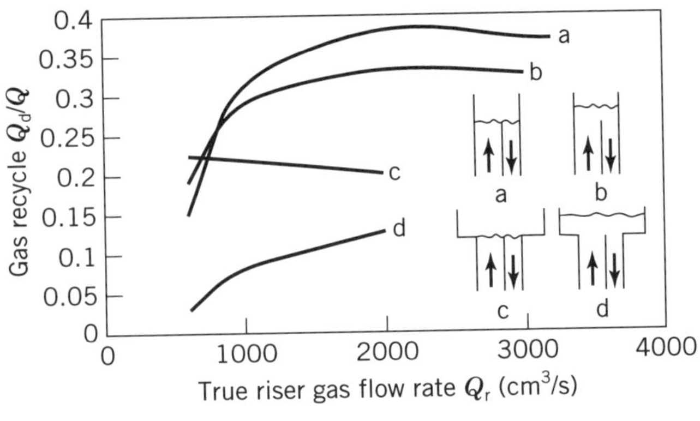

Figure 16. Gas recirculation in a split-cylinder ALR. The level indicated corresponds to no-aeration conditions. Adapted from Siegel et al. (35).

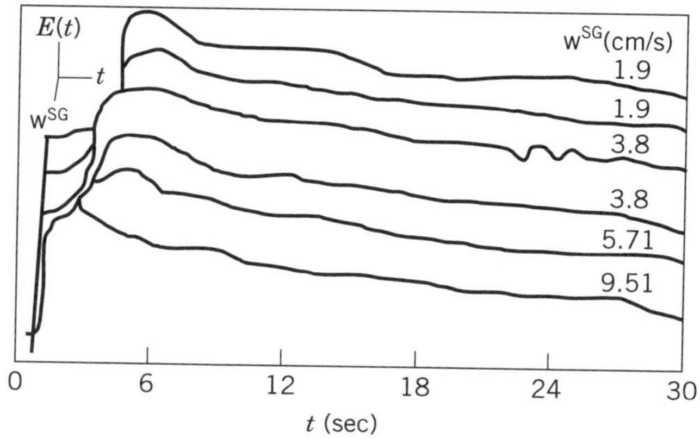

Figure 17. Response to a pulse of helium in an airlift reactor. The pronounced shoulders could be interpreted as the result of superimposing a second peak, which represents the He tracer one loop after. Reprinted with permission from Lubbert et al. (98).

a model to the experimental data suggested recirculation of 25% of the gas. This figure is in the range of the recirculation rates presented by Siegel et al. (35), considering the differences in the coalescing properties of the liquids used in these two works and the corresponding differences in the downcomer holdup.

Recently, Merchuk and Berzin (77) developed a mathematical model based on the application of the first law of the thermodynamics to each of the regions of an ALR. This model facilitates the evaluation of the maximum liquid recirculation possible in the system. The calculation is based on the premise that the gas recirculated must be compressed from the pressure at the top of the downcomer to the pressure at its bottom. The mathematical expression that gives this maximal gas recirculation is:

$$Q_d = \frac{Q_L(P_2 - P_3) + Q_L\rho_L gh - \frac{1}{2}\rho_L C_d A_d(1 - \varphi_d)U_L^3}{P_4 \ln\left(\frac{P_3}{P_2}\right)} \tag{20}$$

where Q_d is the gas flow rate in the downcomer, Q_L is the liquid circulation flow rate, P_i is pressure at point i of the reactor (1 is top of the riser, 2 is top of the downcomer, 3 is bottom of the downcomer, 4 is bottom of the riser), C_d is the hydraulic resistance coefficient, A_d is the downcomer cross-sectional area, U_L is the linear liquid velocity, g is the gravitational acceleration, ρ_L is the liquid density, and φ_d is the downcomer gas holdup.

The calculation of Q_d thus requires knowledge of the liquid flow rate, the pressures, and the geometry of the reactor. This equation represents the maximum recycling of gas in the downcomer, which will take place only if all the energy dissipated in the downcomer is invested in gas compression.

Liquid Velocity

The liquid velocity is one of the most important parameters in the design of ALRs. It affects the gas holdup in the riser and downcomer, the mixing time, the mean residence time of the gas phase, the interfacial area, and the mass and heat transfer coefficients.

Circulation in ALRs is induced by the difference in hydrostatic pressure between the riser and the downcomer as a consequence of a difference in gas holdup. Liquid velocity—like gas holdup—is not an independent variable, because (see Fig. 5) the gas flow rate is the only variable that can be manipulated. As shown in Figure 5, the geometric design of the reactor will also influence the liquid velocity, but this remains constant during operation. Experiments have been carried out in devices specially designed to artificially change the resistance to flow, with the aim of studying the effect of the velocity at a fixed rate of aeration (71,79). The information emerging from these experiments indicates that an increase in the liquid velocity leads to a decrease in the mean residence time of bubbles in the riser and hence of the gas holdup in the riser. In practice, when the gas flow rate is increased, the higher liquid velocity increases the carryover of bubbles from the gas separator into the downcomer; the carryover dampens the liquid flow by reducing the hydrostatic driving force. As a result, the overall change in liquid velocity is tempered.

Liquid Velocity Measurement. Several different methods can be used for measuring the liquid velocity. The most reliable ones are based on the use of tracers in the liquid. If a tracer is injected and two probes are installed in a section of the tube, the velocity of the liquid traveling the distance between probes can be taken directly from the recorded peaks, as the quotient of the distance between the two electrodes and the time required by the tracer to travel from the one to the other. The latter is obtained as the difference of between the first moments of the two peaks. A second method is to calculate the liquid velocity (U_L) from the circulation time (t_c) and holdup (φ) as:

$$U_L = \frac{\text{liquid volume}}{t_c \times A \times (1 - \varphi)} \tag{21}$$

where A is cross-sectional area.

In this case, only one electrode is necessary, φ is the holdup at the point at which the electrode is installed, and the circulation time is obtained from two successive peaks recorded by the electrode.

Modeling of Liquid Flow. A number of expressions are available for the estimation of the liquid velocity. Two main methods have been used for the modeling of two-phase flow in ALRs—energy balances and momentum balances. Chakravarty et al. (58) used the energy balance approach to obtain a relationship between superficial gas velocity, holdup, and liquid velocity. Lee et al. (99) calculated U_L by a similar type of model for a series of published data for concentric and external-loop ALRs and from their own results for split vessels. In both the above-mentioned models, constants accounting for friction losses were obtained by adjusting the models to the experimental data. Jones (100), on the other hand, managed to express the results of his energy balance (based on previous work of Niklin

[101] and Freedman and Davidson [73]) in a relationship free of empirical constants. His results, however, fit the experimental data only qualitatively, and the fit is satisfactory only for very small diameters. An improvement of this method was suggested by Clark and Jones (102), who took into account the radial distribution of the gas holdup through the drift flux model. However, the values of the distribution coefficient C_0 needed for satisfactory fitting of the experimental data for lager diameters is far from the range usual in this type of flow.

Chisti and Moo-Young (103) extended a model originally proposed by Bello (85), based on an energy balance over the airlift loop. Their expression for the average superficial liquid velocity is:

$$U_{\mathrm{Lr}} = \left(\frac{2gH_{\mathrm{d}}(\varphi_{\mathrm{r}} - \varphi_{\mathrm{d}})}{\dfrac{K_{\mathrm{t}}}{(1 - \varphi_{\mathrm{r}})^2} + K_{\mathrm{B}}\left(\dfrac{A_{\mathrm{r}}^2}{A_{\mathrm{d}}}\right) \dfrac{1}{(1 - \varphi_{\mathrm{d}})^2}} \right)^{0.5} \quad (22)$$

where U_{L} is the superficial liquid velocity, A_{r} is the riser cross-sectional area, A_{d} is the downcomer cross-sectional area, H_{d} is the downcomer height, K_{b} and K_{t} are the hydraulic pressure loss coefficients, φ_{r} is the riser gas holdup, and φ_{d} is the downcomer gas holdup.

By choosing suitable values for the friction coefficients in each case, the authors showed that much of the published data on liquid velocity for the different types of ALRs could be satisfactorily correlated by equation 22. Only one coefficient has to be adjusted, since the authors assume that K_{t}, the friction coefficient at the top of the loop, is negligible in concentric-tube type reactors and that in external-loop reactors K_{t} can be taken as equal to K_{b}, the friction coefficient for the bottom of the loop. Equation 22 has thus been adopted by many scientists. Wachi et al. (104) claimed that their derivation of the same equation gives a clearer physical meaning to the adjustable parameters. Equation 22 can also be also derived from a simple momentum balance (77).

Chisti et al. (51) presented an empirical correlation for K_{b} obtained by comparison of results obtained from several sources:

$$K_{\mathrm{b}} = 11.402 \cdot \left(\frac{A_{\mathrm{d}}}{A_{\mathrm{b}}}\right)^{0.789} \quad (23)$$

where A_{b} is the minimal cross section at the bottom of the airlift reactor and A_{d} is the downcomer cross-sectional area.

Equation 22 has the particularly that the gas flow rate, which is the main, and often the only, manipulable variable in the operation, is not present directly, but exerts its influence through the gas holdup. Therefore, either experimental data or a valid mathematical expression for the gas holdup in both the riser and the downcomer are required.

Chisti and Moo-Young (103) extended this model further in order to facilitate the prediction of liquid circulation in ALRs operating with pseudoplastic fluids, such as mold suspensions. This improvement is very important, since many commercial fermentation processes involve such non-Newtonian liquids. Kemblowski et al. (86) presented a method for the prediction of gas holdup and liquid circulation in external-loop ALRs. In their experiments there was almost no gas recirculation, because of the large size of the gas separators used.

Garcia Calvo (105) presented an ingenious model based on energy balances and on an idea originally proposed by Richardson and Higson (106), and Garcia Calvo and Leton (107) extended the model to bubble columns. The model is based on the assumption that the superficial gas velocity (J_{G}) in any region can be considered to be the sum of two streams (J' and J'') as follows. The J' stream has a velocity equal to that of the liquid and can therefore be treated by the laws of homogeneous two-phase flow (no slip between the bubbles and the liquid). The second stream (J'') is considered to be responsible for all the energy loses at the gas–liquid interface. The concept in itself is simple and elegant, and it is possible to envisage its application even to the flow in the downcomer, where $U_{\mathrm{G}} < U_{\mathrm{L}}$. In such a case, we would divide the gas flow rate into two parts as follows: One part would be larger than the actual flow rate, i.e., it would have the same velocity as the liquid. In order to arrive at the actual gas flow rate, the second flow rate must have the reverse direction. This type of gas flow can actually be seen under certain conditions, such as when there is coalescence of bubbles and larger bubbles ascend along the walls of the downcomer.

Another technique used by several researchers to predict liquid velocity is the momentum balance of the ALR. This method has been used by Blenke (108) in jet-loop reactors and by Hsu and Dudukovic (109), Kubota et al. (36), Bello (85), Koide et al. (47), and Merchuk and Stein (71). The latter authors presented a simple model for the prediction of the liquid velocity as a function of the gas input in an ALR. They assumed that the pressure drop between the bottom and the top of their external-loop reactor could be expressed as a continuation of the downcomer, using an equivalent length L_{E}. This length was set as an adjustable parameter describing the pressure loss in the loop. Kubota et al. (36) used a similar approach for the analysis of Imperial Chemical Industries' deep-shaft reactor. They were able to simulate the operation of the reactor and to predict the minimum air supply required to prevent flow reversal.

Verlaan et al. (76) used a similar model, in combination with the expression of Zuber and Findlay (74), to calculate the friction coefficients from experimental data reported by several authors for a wide range of reactor volumes. Koide et al. (47) presented an analysis of the liquid flow in a concentric-tube ARL that was also based on a momentum balance. The main difference between this model and that used by Merchuk and Stein (71) was that Koide et al. used a convergence–divergence flow model for the bottom and the top of the loop. At the bottom, the effect of flow reversal on the pressure drop was included in the effective width of the gas-liquid flow path under the lower end of the draft tube h_{l}, which was smaller than the actual gap. Miyahara et al. (57), who studied both the bubble size distribution in an internal-loop ALR and the pressure drop at the top and the bottom of the draft tube, also presented a model facilitating the prediction of the liquid velocity.

Other models use the drift-flux model (59) presented in equations 5–7, as:

$$U_{Lr} = \frac{J_{Gr}\left(\frac{1}{\varphi_r} - C_0\right) - U_{GJ}}{C_0(1 - \varphi_r)} \tag{24}$$

U_{GJ} can be taken from equation 10. The range of variation of C_0 is rather narrow, as shown in the previous section, and therefore it is not difficult to make a judicious guess as to the value of C_0 in an unknown system. The drift flux model has also been used together with energy balances (110) or with the momentum balance (111). Some studies on liquid measurement present the results in the form of empirical correlations (42,52,112). The usefulness of these correlations depends on the amount of data and the number of parameters taken into account.

Most of those correlations are shown on Table 3, and some of them are presented in Figure 18. In general, the superficial liquid velocity increases with an increase in the superficial gas velocity, but its rate of increase is much lower at high superficial gas velocities. From Table 3, it can be seen that the riser-to-downcomer cross-sectional area ratio and the reactor height are the main parameters that affect the superficial liquid velocity at constant superficial gas velocity. The superficial liquid velocity increases with an increase in A_d/A_r.

Table 3. Liquid Circulation Velocity in ALRs

No.		Formula	Ref.
1	ELR	$J_{Lr} = 1.55 J_{Gr}^{0.33}\left(\frac{A_d}{A_r}\right)^{0.74}$	45
	ILR	$J_{Lr} = 0.66 J_{Gr}^{0.33}\left(\frac{A_d}{A_r}\right)^{0.78}$	
2	Bubbly	$J_{Lr} = 0.024 J_{Gr}^{0.322}\left(\frac{A_d}{A_r}\right)^{0.794} \mu_{ap}^{-0.395}$	50
	Slug flow	$J_{Lr} = 0.052 J_{Gr}^{0.322}\left(\frac{A_d}{A_r}\right)^{0.794} \mu_{ap}^{-0.395}$	
3	Slug flow	$J_{Lr} = 0.23 J_{Gr}^{0.322}\left(\frac{A_d}{A_r}\right)^{0.97} \mu_{ap}^{-0.39}$	63
4		$J_{Lr} = 2.858 J_{Gr}^{0.482}\left(\frac{A_d}{A_r}\right)^{0.97} 416\mu_{ap}^{-0.0105}$	52
5		$J_{Lr} = \left[\frac{2gH_d(\varphi_r - \varphi_d)}{K_b\left(\frac{1}{(1-\varphi_r)^2} + \left(\frac{A_d}{A_r}\right)^2 \frac{1}{(1-\varphi_d)^2}\right)}\right]^{0.5}$	103
		$K_b = 11.4\left(\frac{A_d}{A_r}\right)^{0.79}$	

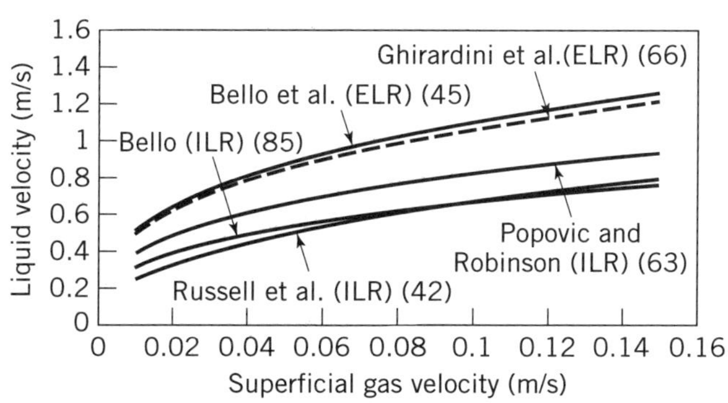

Figure 18. Liquid velocity predicted by some of the proposed correlations from Table 3. ILR, internal-loop ALR; ELR, external-loop ALR.

The effect of the properties of the liquid, such as viscosity, surface tension, and ionic strength, on the superficial liquid velocity are much milder in ALR than in bubble columns (113). It is expected that increasing the liquid viscosity will decrease the liquid velocity because of frictional losses, but this, in turn, will increase the gas holdup in the riser and consequently increase the driving force for liquid recirculation. Hence, it seems that these two effects balance each other partially and result in a milder effect on the superficial liquid velocity. The effects of the surface tension and the ionic strength are also exerted via their influence on the gas holdup, as analyzed above.

As a rule, it can be recommended that the Zuber-Findlay expression (equation 20) be used when the holdup is known and the liquid circulation velocity is high. For low liquid velocity, a correlation obtained in a piece of equipment as similar as possible to the one under design should be used.

Liquid Mixing

For the design, modeling, and operation of ALRs, a thorough knowledge of mixing behavior is necessary. This is of particular importance during the process of scale-up from laboratory-scale to industrial-scale reactors. The optimum growth rate of a microorganism or the optimum production rate of a specific secondary metabolite usually relates to well-defined environmental conditions, such as pH range, temperature, substrate level, limiting factors, dissolved oxygen, and inhibitor concentration in a specific well-mixed laboratory-scale vessel. Because of the compromises made during scale-up, it is difficult to keep, at different scales of operation, the same hydrodynamic conditions established in the laboratory; mixing on an industrial scale may not be as good as mixing on a laboratory scale (5). In smaller-scale reactors it is easier to maintain the optimal conditions of pH, temperature, and substrate concentration required for maximum productivity of metabolites in a fermenter.

Furthermore, in fermentation systems efficient mixing is required to keep the pH within the limited range, giving maximum growth rates or maximum production of the microorganism during addition of acid or alkali for pH control. Mixing time—or the degree of homogeneity—is also very important in fed-batch fermentation, where a required component, supplied either continuously or intermittently, inhibits the microorganisms or must be kept within a particular concentration range (114,115). A large number of commercially important biological systems are operated in batch or fed-batch mode. In this operation mode, fast distribution of the incoming fluid is required, and the necessity for understanding the dynamics of mixing behavior in these vessels is obvious. Even for batch systems, good control of the operating conditions, such as pH, temperature, and dissolved oxygen, require prior estimation of mixing so that the addition rates can be suitably adjusted. Deviation of the pH or temperature from the permitted range may cause a damage to the microorganism, in addition to its effect on the growth and production rates. Moreover, a knowledge of the mixing characteristics is required for modeling and interpreting mass and heat transfer data.

A parameter used frequently to represent mixing in reactors is the mixing time (t_m). It has the disadvantage that it is specific to the reactor design and scale, but it is easy to measure and understand. Mixing time is defined as the time required to achieve the desired degree of homogeneity (usually 90–95%) after the injection of an inert tracer pulse into the reactor. The so-called degree of homogeneity (I), is given by:

$$I = \frac{C - C_m}{C_m} \tag{25}$$

where C is the maximum local concentration and C_m is the mean concentration of tracer at complete mixing.

A more comprehensive way of analyzing mixing, applicable to continuous systems, is a study of the residence time distribution (RTD). Although ALRs are usually operated in a batch-wise manner, at least in the laboratory, advantage is taken of the fact that the liquid circulates on a definite path to characterize the mixing in the reactor. Hence, a single-pass RTD through the whole reactor or through a specific section is usually measured. Based on the observed RTD, several models have been proposed. These models have the advantage of reducing the information of the RTD to a small number of parameters, which can later be used in design and scale-up.

The axial dispersion model, which has the advantage of having a single parameter, is widely accepted for the representation of tower reactors. This model is based on visualization of the mixing process in the tower reactor as a random, diffusion-like eddy movement superimposed on a plug flow. The axial dispersion coefficient D_z is the only parameter in the formulation:

$$\frac{\partial C}{\partial \tau} = D_z \frac{\partial^2 C}{\partial z^2} + U_L \frac{\partial C}{\partial z} \tag{26}$$

where C is the concentration of a tracer. The boundary conditions depend on the specific type of tower reactor. This model is attractive, since it has a single parameter, the Bodenstein number (Bo), which is used to describe the mixing in the reactor:

$$Bo = \frac{U_L L}{D_z} \tag{27}$$

where L is the characteristic length. When the Bo number tends to infinity, the mixing conditions are similar to those of a plug-flow reactor, and the reactor can be considered as well-mixed for low Bo numbers.

The alternative approach of Buffham and Mason states that the mixing characteristics of a piece of equipment should be expressed as the variance σ^2 of the distribution obtained by injection of a pulse of tracer without adopting any mechanistic model (116). The relationship between Bo and σ^2 depends on the reactor configuration (117). The approach of Buffham and Mason facilitates the presentation of mixing characteristics free of any modeling assumptions. The variance σ^2 is the second moment of the distribution and carries information on the spread of the distribution around the mean value (first moment). Nevertheless, most of the data on mixing in bioreactors are presented either as t_m or as overall Bo numbers, which can be obtained by relatively simple experiments of pulse injection.

Single-pass mixing in the ALR is due to mixing in the individual and interrelated sections of the reactor—riser, separator, downcomer, and bottom. Repeated passage mixing is the sum of the mixing in the subsequent passages. The latter is usually reported as the mixing time (t_m), the former as Bo or σ^2. Indeed, these parameters are interrelated, and knowledge of Bo or t_m is sufficient for calculating, theoretically, the mixing time (108,118) based on the deviation of the envelope of the maxima in the response curve to a pulse, which is a measure of the degree of inhomogeneity. Verlaan et al. (80) and Lin et al. (119) correlated their results as follows:

$$t_m = MBo \tag{28}$$

where M is a constant equal to 0.093 (80) or to 0.089 (119). The coefficient M given by Verlaan et al. is in exact agreement with the theoretical relationship derived by Murakami et al. (118) for $Bo > 50$ and a degree of inhomogeneity, $I = 0.05$.

Equation 28 shows that the circulation path, which enters in the definition of Bo, has a linear effect on the mixing time. If the mean circulation time and the axial dispersion coefficients are known, it is possible to theoretically estimate the mixing time using equation 28. Experimental details must, however, be carefully planned to avoid complications. Note that in order to simplify data processing it is important to inject the signal and to measure the response at exactly the same point (120) (often the position of the injection point is not specified despite its effect on the mixing time). In a study of the effect of the injection point on the dynamics of the mixing time, Schügerl et al. (12) concluded that the gas–liquid separator is the best choice for tracer injection for short mixing times. Fields and Slater (114) reported a marked dependence of the respiratory quotients upon the injection point of methanol during unlimited fed-batch growth of *Methylophilus methylotropus* in a concentric-tube ALR. The lines in Figure 19 show experimental data for Bo (as overall values) reported for different types of ALR (in which the reactor is considered as a single unit). The dimensions of the reactors are given in Table 4. Because of the definition of the overall Bo, the values are specific to the reactor for which they were obtained and can be used only as indication of trends and orders of magnitude.

As explained above, the ALR is, in fact, a combination of several regions having quite different fluid dynamic characteristics. The overall mixing is the result of the contributions of each of them, and the overall Bo represents

Table 4. Bodenstein Number as a Function of the Superficial Gas Velocity (Key for Fig. 19)

Curve number	Authors	Type	A_d/A_r	H/D_r	Liquid	Ref.
1	Weiland	External loop	0.25	85	Water	44
2	Weiland	External loop	0.25	85	2-propanol 1.65%	44
3	Weiland	External loop	0.25	85	CMC 50%	44
4	Lin et al.	External loop, unbaffled	0.11	20	Water	119
5	Lin et al.	External loop, baffled	0.11	20	Water	119
6	Lin et al.	External loop, unbaffled	0.11	40	Water	119
7	Bello et al.	External loop	0.69	12	Water	45
8	Bello et al.	External loop	0.69	12	Water	45
9	Bello et al.	External loop	0.11	20	Water	45
10	Moor Nagar	External loop	1.0	30	Water	122
11	Fields and Slater	Concentric tube	1.56	10.5	Water	114
12	Fields and Slater	Concentric tube	1.56	21	Water	114
13	Fields and Slater	Concentric tube	1.56	10.5	Water, antifoam	114
14	Fields and Slater	Concentric tube	1.56	10.5	1% Ethanol	114
15	Bello et al.	Concentric tube	0.13	39.65	Water	112
16	Bello et al.	Concentric tube	0.56	55.6	Water	112
17	Verlaan et al.	External loop, total *Bo* number	0.25	16.5	50 mM KCl aqueous solution	79
18	Verlaan et al.	*Bo* number in the riser	0.25	16.5	50 mM KCl aqueous solution	79
19	Verlaan et al.	*Bo* number in the downcomer	0.25	16.5	50 mM KCl aqueous solution	79
20	Verlaan et al.	External loop, total *Bo* number	0.25	16.5	50 mM KCl aqueous solution	79

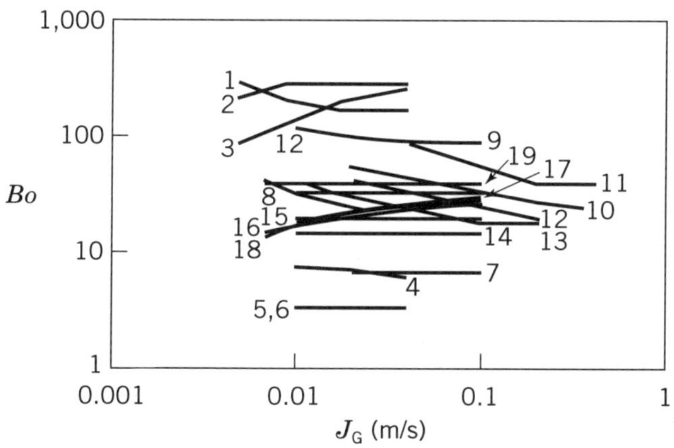

Figure 19. Global Bodenstein numbers reported for ALRs as a whole unit. See references in Table 4.

this combination for a particular reactor. This value thus has some limitations for extrapolation to other configurations or scales. Because each section has different mixing characteristics (it may have a different cross-section, flow configuration, etc.), from a strictly engineering point of view, the mixing in each of the sections of the ALR should be defined and considered separately.

One way of obtaining information on the mixing characteristics of each of the regions of the ALR is the simultaneous measurement of the response in the ALR at several points, so that after one single pulse injection the response of each section in the loop can be obtained (37,80). This method of measurement has the advantage that multiple measurements are made for the same tracer injection experiment. This enables us to check the consistency of the liquid velocity results obtained, since independent measurements can obtained in the same run, as can be seen in Figure 20.

The results for *Bo* obtained by Verlaan (80) in an external-loop ALR are shown in Figure 21 (the much higher *Bo* in the downcomer was explained by the fact that the data were obtained under conditions of complete gas disengagement so that only liquid flowed). The results indicated that most of the mixing took place in the gas separator, and both riser and downcomer could be considered as plug-flow conduits. The same conclusions were drawn for internal-loop ALRs by Merchuk et al. (37) by analysis of mixing times: the shift in results for mixing times obtained with different probes in response to the same impulse perturbation indicated that the riser and the downcomer were introducing a pure delay in the response, that is, that they behaved as plug-flow sections.

Indeed, the simplest model of an ALR, from the point of view of fluid dynamics, is a combination of two plug-flow reactors, representing the riser and the downcomer, and a perfectly stirred tank, representing the gas separator (Fig. 22) (67,123). Merchuk and Yunger (67) showed that this simple model could take into account the mixing in the ALR. The validity of using this simple model as a first approximation is supported by experimental evidence that shows that the mixing time decreases when the separator volume increases (31,124). When the volume of the separator is increased substantially without changing the reactor diameter, the gas separator becomes tall and slender and may depart from total mixing behavior. It has been reported that under such conditions the circulation time measured from two successive peaks of a probe is not influenced by changes in the top clearance (42,43). This indicates that the gas separator acts as two interconnected well-mixed regions, one of them (the lower one in Figure 23) being the link between riser and downcomer, as shown schematically in Figure 23.

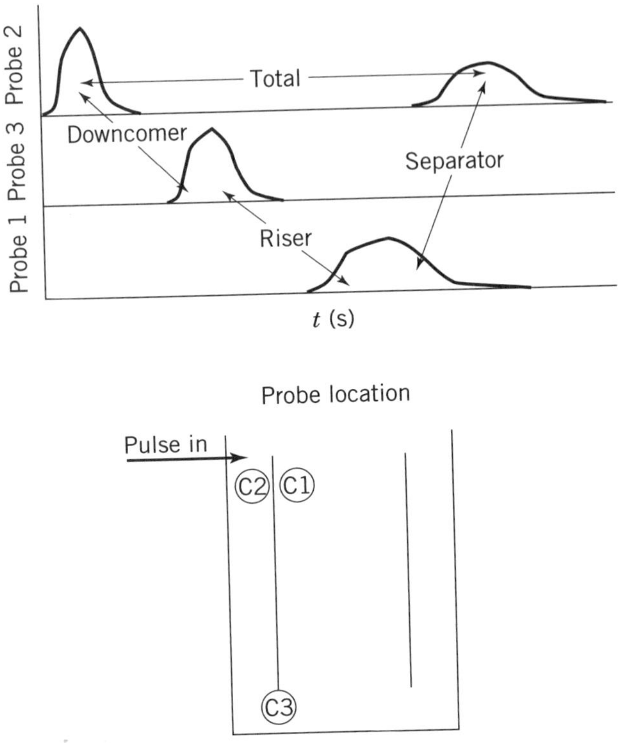

Figure 20. Response of three electrodes in the riser, separator, and downcomer to a pulse of electrolyte. The information that can be obtained from a single experiment is indicated by the *arrows*. From Merchuk et al. (37).

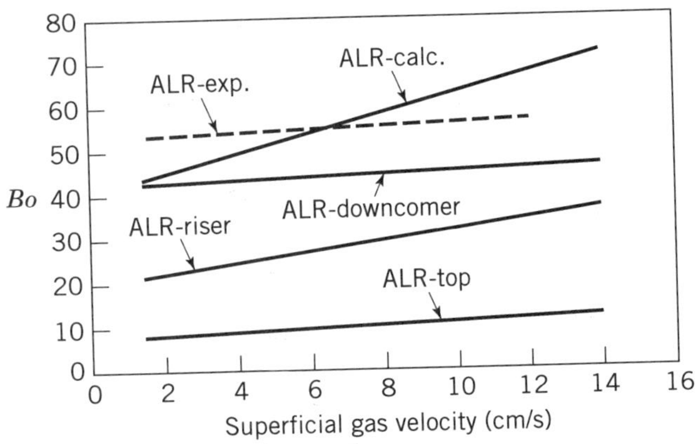

Figure 21. Bodenstein numbers as a function of the superficial gas velocity in an external-loop ALR. ALR-exp, experimental values for the whole reactor; ALR-calc, calculated values for the whole reactor. Adapted from Vorlaan et al. (80).

The extent of mixing in the gas separator of internal-loop ALRs can probably be compared to that in a bubble column of the same dimensions. In the case of external-loop ALRs, the extent of mixing depends on the geometric design. In special designs, like the channel-loop reactor (125), a different approach must be taken. However, in general, a rule of thumb may be recommended: consider that all the mixing takes place in the gas separator, and calculate the degree of mixing in it using published data for bubble columns. Correlations available for bubble columns can thus be used. Godboole and Shah (126) recommend the use of the correlation proposed by Deckwer et al. (127):

$$D_Z = 0.678 \cdot D^{1.4} \cdot J_G^{0.3} \tag{29}$$

where D_Z is the dispersion coefficient and D is the column diameter, or the relationship proposed by Joshi (128):

$$D_Z = 0.33(U_c + J_L)D \tag{30}$$

where J_L is the superficial liquid velocity and U_c is the cell circulation velocity given by:

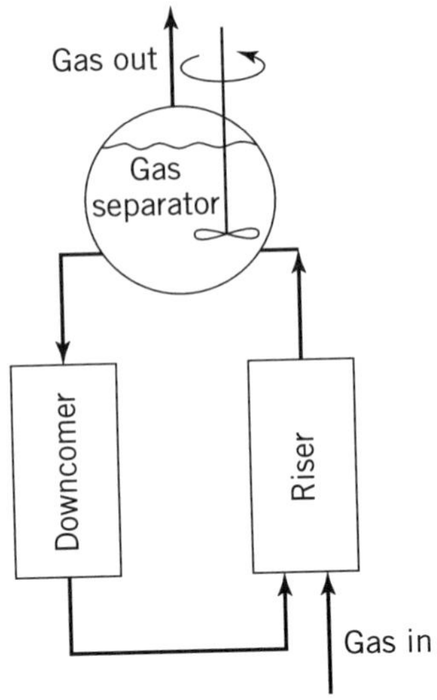

Figure 22. Simple model of an ALR representing the riser and the downcomer as plug-flow reactors and the gas separator as a perfectly mixed region. From Merchuk and Siegel (67).

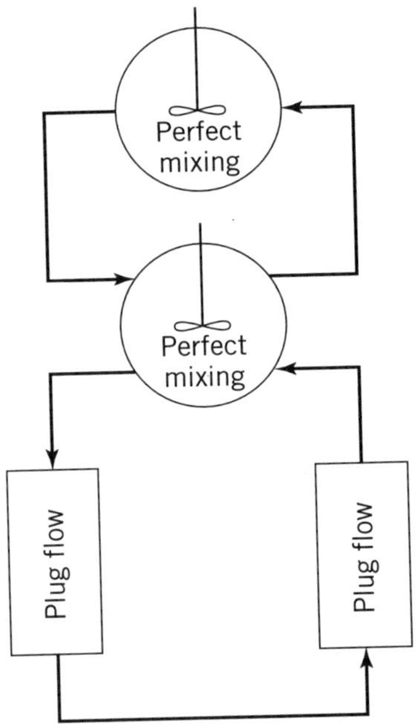

Figure 23. Schematic representation of the effect of a tall slender gas separation region in a concentric-tube ALR.

$$U_C = 1.31\left[gD\left(J_G - \frac{\varphi_G}{1 - \varphi_G} J_L - \varphi_G U_{b\infty}\right)\right] \quad (31)$$

where $U_{b\infty}$ is the terminal bubble velocity.

The liquid superficial velocity (J_L) can be easily calculated for external-loop reactors, and for internal loop reactors, J_L may be taken as half the liquid velocity in the riser. The viscosity may also play a role in the rate of mixing. An increase in liquid viscosity will increase the energy dissipation in the loop and result in an increase in mixing time and *Bo*. In non-Newtonian fluids, however, the behavior may be different, as shown in Figure 24. It has been reported that in the case of the addition of polymers that confer pseudoplastic behavior to the liquid, low concentrations produce a decrease of the mixing time (65,129). This can be explained in terms of the drag reduction due to the presence of polymers in the boundary layer near the walls. Fields et al. (129) found that the mixing time increased for concentrations of the natural polymer, xanthan gum, above the critical concentration. (The critical concentration is a theoretically calculated value at which appreciable overlapping of polymer molecules occurs and which marks the onset of a rapid increase of the apparent viscosity [130]).

Mixing in the Gas Phase

For all practical purposes, the gas in the riser of an ALR exhibits plug flow behavior. Only for extremely high J_G or hindered liquid circulation will the axial dispersion of the bubbles have some effect on the gas RTD. In the downcomer, the gas flow is almost plug-flow when the bubble recirculation is fully developed. But at the stage at which a stationary phase of suspended bubbles appears at the top of the downcomer, appreciable dispersion will occur. This zone has a large degree of mixing due to coalescence and consequent rise of larger bubbles amid smaller ones, with repeated events of breakup and coalescence. However, this type of operation has no relevance to practical applications. It is an operation mode to be avoided at all costs. Indeed, no data on mixing under these conditions have been reported. The main question related to the mixing of the gas phase is, in fact, related to gas recirculation. When a particular gas flow has developed in the downcomer, part of the gas is being recirculated (see "Liquid Velocity"). A pulse of gas tracer at the inlet would produce, as a response, a series of pulses, separated one from the other by the gas circulation time. In practice, not many of these pulses would be detected, due to dilution and disengagement of the tracer in each pass through the separator.

The only reported study on gas phase mixing in an ALR is that of Frehlich et al. (131). The distribution of the gas residence times in two reactors, one of 0.08 m^3 and the other of 4 m^3, was measured using pseudostochastic tracer signals and a mass spectrometer. The values of *Bo* were calculated from the first and second peaks, indicating the main gas stream and the recirculation. Figure 25 shows the *Bo* obtained for a laboratory-scale ALR. The values obtained in a pilot plant were of the same order.

Energy Dissipation and Shear Rate in Airlift Reactors

ALRs are being increasingly used in processes involving shear-sensitive cells (mammalian, insect, and plant cell cultures) (3,11,14,18,21,28,29). This situation has created the need for considering shear stress as one of the parameters relevant in the design of such reactors. Although the a priori evaluation of shear rates is a matter that has been studied for many years in stirred tanks, information on this subject is scanty for pneumatically stirred reactors. The first approach made in this direction in pneumatically agitated vessels was that of Nishikawa et al. (70), who were interested in the problem of heat transfer to a non-Newtonian liquid in a bubble column. This study was extended to mass transfer by Nakanoh and Yoshida (132), who proposed the expression:

$$\gamma = 5000\, J_G \quad (32)$$

This expression has been widely accepted despite the criticism sometimes leveled at it (62,68,123,133). Some modifications have thus been proposed (92,95,134), which, like the original approach, were based on data for bubble

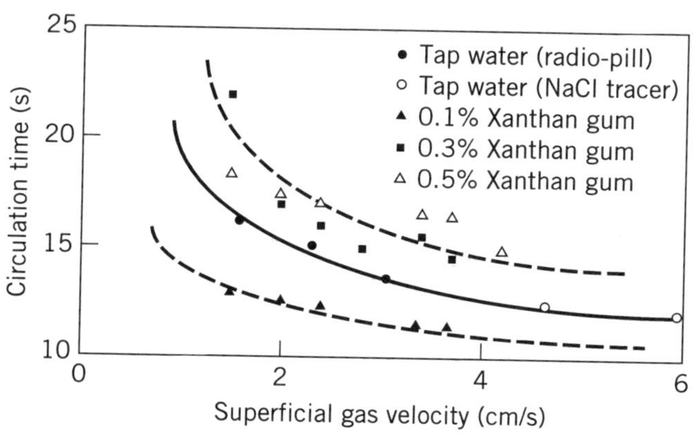

Figure 24. Variation of mean circulation times with J_G for several solutions. Adapted with permission from Fields and Slater (129).

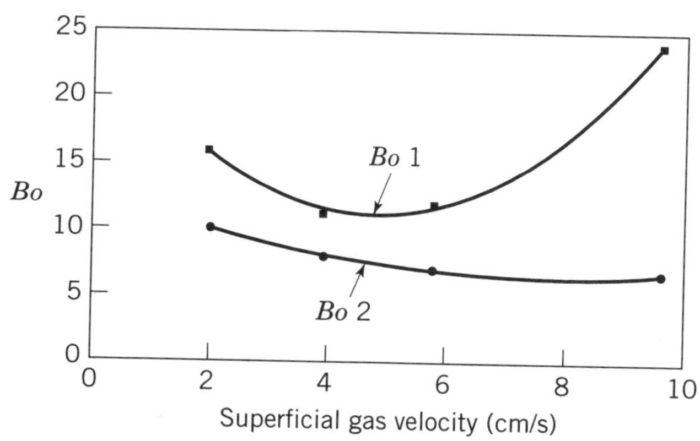

Figure 25. Bodenstein numbers of the first and second peaks of the gas residence time distribution in an 80-L internal-loop ALR, as a function of the gas superficial velocity. From Frehlich et al. (131).

column performance and therefore have limited application for ALRs. Recently, a method facilitating the prediction of the distribution of the energy dissipated in an ALR, based on a simple thermodynamic approach, has been developed (77). Energy dissipation was considered to occur in the ALR by two main mechanisms, wall friction and bubble-associated dissipation (ideal gas behavior was assumed). The work done by the gas on the liquid (and vice versa) was expressed assuming isothermal expansion of the bubbles. The energy dissipation inside the gas phase was considered negligible. The general energy balance (135) was written as:

$$\Delta(PQ) + \Delta E_p + E_D = W_S \tag{33}$$

In this equation, the first term represents the flow work lost by the system under consideration, E_D is the energy dissipated per unit of time, and W_S is the shaft work done by the surroundings on the system under consideration. The schematic representation of the concentric-tube ALR in Figure 26 indicates the different points in the reactor considered in the mathematical expressions. The expressions found for the energy dissipated in each zone were the following:

Riser

$$(E_d)_R = Q_L(P_4 - P_5) - Q_L\rho_L gh - Q_r P_4 \ln\left(\frac{P_5}{P_4}\right) \tag{34}$$

Gas separator

$$(E_d)_S = Q_L(P_5 - P_2) - Q_{in}P_4 \ln\left(\frac{P_1}{P_5}\right) - Q_d P_4 \ln\left(\frac{P_2}{P_5}\right) \tag{35}$$

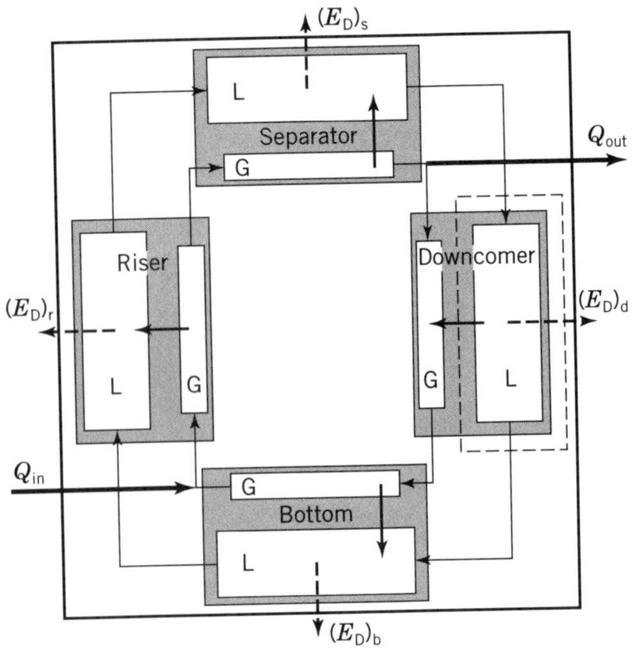

Figure 26. Schematic description of the variables in the thermodynamic model for energy dissipation distribution in an ALR. From Merchuk and Berzin (77).

Downcomer

$$(E_d)_d = Q_L(P_2 - P_3) - Q_L\rho_L gh - Q_d P_4 \ln\left(\frac{P_3}{P_2}\right) \tag{36}$$

Bottom

$$(E_d)_b = Q_L(P_3 - P_4) - Q_d P_4 \ln\left(\frac{P_4}{P_3}\right) \tag{37}$$

Results of this model of an ALR can also be used to estimate the global shear rate in each region of the reactor, according to the global approach presented Merchuk and Ben-Zvi (Yona) (96). The shear stress in the liquid of each region of the reactor can be defined as the energy dissipated divided by the mean path of circulation in the region and by the sum of the areas of all the bubbles. For the region i in the ALR

$$\tau_i = \frac{(E_d)_{\text{bulk},i} t_i}{a_i h_i^2 A_i} \tag{38}$$

where t_i is the residence time of the liquid, h_i is the effective length, and a_i is the specific interfacial area, in the region i.

A global shear rate γ_i can be calculated for each region i as

$$\gamma_i = \frac{\tau_i}{\mu} \tag{39}$$

where μ is the effective viscosity of the fluid.

For liquids exhibiting different types of rheological behavior, the corresponding constitutive equation must be used. Such an approach has been used for the interpretation of shear effects on mammalian cells (136) and algal growth (43).

MASS TRANSFER

The volumetric mass transfer coefficient (k_La) is the rate of gas transfer across the gas-liquid interface per unit of driving force (the driving force is the gas concentration gradient between the liquid and the gas). The mass transfer coefficient k_La can be seen as the product of two terms: the mass transfer coefficient k_L and the specific interfacial area a. Both terms depend on a series of variables that can be grouped into three categories: (*1*) static properties of the liquid, such a density, diffusivity, and surface tension; (*2*) dynamic properties of the liquid (related to liquid flow), such as rheological parameters; and (*3*) liquid dynamics. In general, the variables in group 1 do not change very drastically. The variables in groups 2 and 3, however, may span wide ranges.

Mass Transfer Rate Measurements

Methods for the determination of k_La in a reactor can be grossly classified as steady-state and nonsteady-state methods. In the steady-state methods, the rates of oxygen uptake in steady-state operation are evaluated, either by measurement of inlet and outlet rates of oxygen or by direct analysis of a compound that reacts with the oxygen, as in the case of the sulfite method. One of the problems associated with these procedures is that the changes in oxygen concentration in the gas streams are usually small, and the errors of measurement thus have a substantial influence. When a chemical is added to the system, there is the question of whether the addition has provoked changes in the physicochemical properties, which thus become different from the properties of the original system.

Transient methods may be applied to follow the response of the dissolved oxygen concentration in the system after a step-change of oxygen concentration in the inlet gas stream. These methods have the advantage that addition of an alien material is not required and that a single concentration is measured. The correct use of this method has been analyzed in depth by Linek et al. (137).

One important point to take into account is the dynamics of the oxygen electrode. The lag in the response of the oxygen electrode makes it necessary to discern between the electrode response and the real oxygen concentration, especially when close to a sharp change in concentration. A correct analysis should also include the model of the dynamic behavior of the electrode. In order to simplify these procedures, approximations based on truncation of parts of the response curve have been proposed (133,138). These methods are based on truncating the first part of the electrode response obtained in a transient experiment. Once the error included in the value of k_La is set, the extent of truncation is fixed, allowing simplification of the analysis of the remainder of the curve. It should be kept in mind that this simplification implies the loss of part of the information, and due care should be given to statistical analysis of the results.

Variations of the method have been proposed to minimize disturbances in the system by introduction of step variations of agitation or pressure (137,139). In this way, the method can be applied to bioreactors during real operation of the system.

One problem that may appear in the measurement of mass transfer rates, especially when viscous liquids are used, is related to the presence of very small bubbles that are depleted of oxygen very rapidly but do not disengage in the gas separator, thus constituting an inert volume of gas in the reactor (134). Kawase and Moo-Young (140) analyzed the use of transient absorption of CO_2 for the determination of k_La and concluded that the error due to small bubble retention was much smaller than that in the case of O_2.

Whatever the method used for the determination of k_La, the process of data elaboration is basically the same as is shown schematically in Figure 27 (141). The measured variable (usually oxygen concentration) is compared to the value predicted by a mathematical model of the process. The model includes k_La as a parameter, and the value of k_La that gives the best fit is chosen. It is thus obvious that the choice of the model is very important, and poor assumptions on flow characteristics of gas or liquid phases may lead to errors and deviations from the true values (142). All the models are a simplified (amenable to mathematical treatment) description of the system. It should be kept in mind, therefore, that k_La is not a property of the system, but a parameter of the model adopted. If total mixing is assumed in the model adopted, the mass transfer coefficient obtained will consequently be limited. The assumption of complete mixing is such a common practice that many reports do not even specify explicitly that this has been done, and in many texts the mass transfer coefficient k_La is defined by the equation:

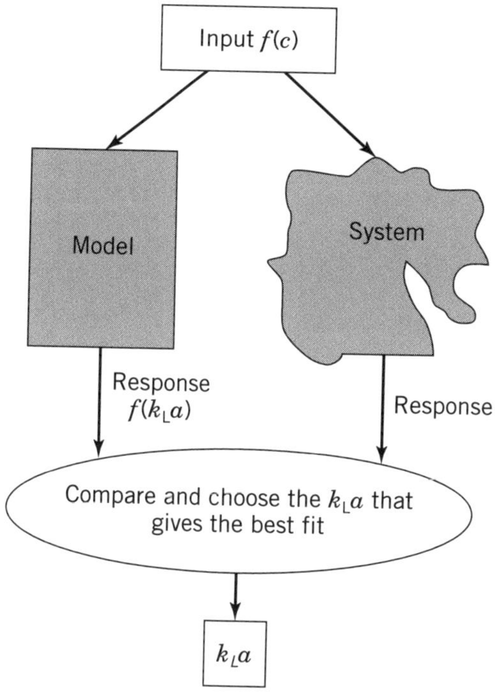

Figure 27. Steps in the determination of the mass transfer coefficient (k_La). From Merchuk et al. (141).

$$\text{OTR} = k_la(C^* - C) \quad (40)$$

It is often forgotten to state that this equation is valid only for perfectly mixed systems.

Strictly, the different sections of ALRs (riser, downcomer, and gas separator) have different flow characteristics, and the mass transfer coefficient may be expected to differ from one region to another. Some researchers (108) have assumed that the contribution of the downcomer to the overall mass transfer is negligible and have reported values of k_La that are, in fact, the entire mass transfer divided by the volume of the riser. The fact that the values of the mass transfer coefficient are calculated and reported in different ways introduces some ambiguity into the published data; special care should therefore be taken when comparing data or using published mass transfer coefficients for design purposes.

The assumption of perfect mixing in the liquid phase may be questionable in the case of tall reactors. One very simple method to verify this assumption is the simultaneous measurement of the response of the concentration to a step-change in the inlet. In a perfectly mixed system, the location of the probe should be irrelevant. Figure 28 (143) shows the response of three probes, one located at

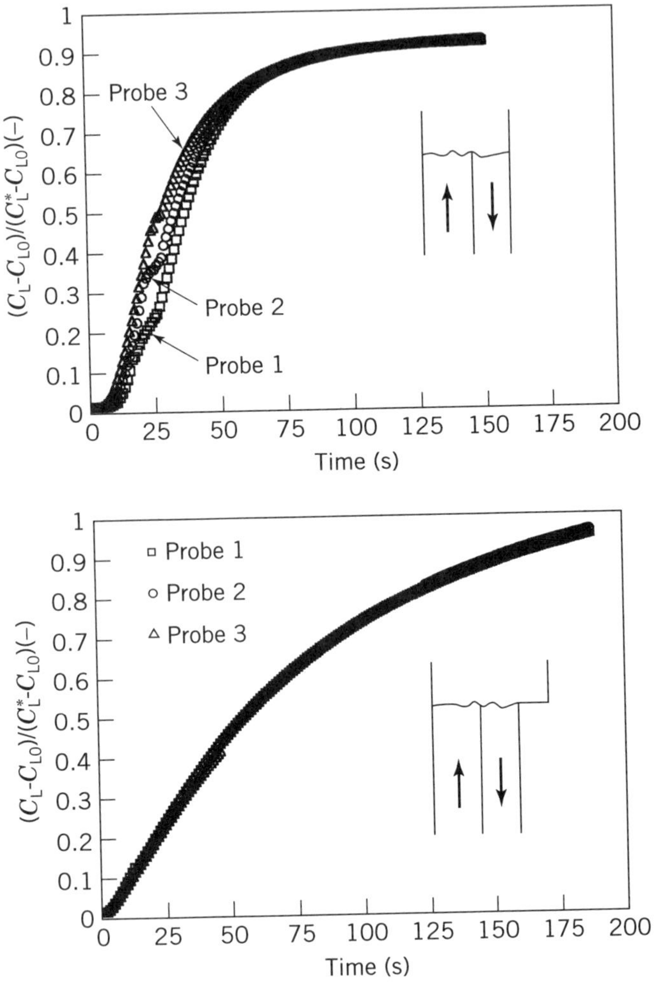

Figure 28. Response of three probes, one located at the end of the riser, one at the inlet of the downcomer, and one at the bottom of a split-cylinder ALR. From Siegel et al. (143).

the end of the riser, one at the inlet of the downcomer, and one at the bottom of a split-cylinder ALR. The difference between the figures stems from a different gas separator section, which changes the fluid dynamics in the system. In one case the system behaved as perfectly mixed, because of the large volume of the gas separator and the faster liquid circulation. In the other (closed system in the figure), the responses of the three electrodes were clearly different, calling for a different analysis. The validity of the criterion, originally proposed by Andre et al. (144), that compares the circulation time in the reactor (t_c) with the characteristic time for mass transfer, which is the inverse of the mass transfer coefficient, was thus confirmed (143):

$$t_c \cdot k_L a < 0.5 \tag{41}$$

If this condition is not fulfilled, the bioreactor cannot be considered as a perfectly mixed volume, and more sophisticated ways of analysis are required. In this case, the mathematical model, and, consequently, the analysis of the experimental data become much more complicated. Nevertheless, this approach has been used (48,122,145,146), and the mass transfer experiments render values of $k_L a$ for each of the regions of the ALR. Figure 29 shows the results reported by Hwang and Lu (146) in an internal-loop ALR. The graph shows that the mass transfer takes place at the highest rate in the riser. Values for the downcomer are 50% lower, and those for the gas separator are intermediate between the riser and the downcomer. The overall mass transfer rate is the result of the balance between the volumes and rates in the three sections.

Bubble Size and Interfacial Area

As said earlier, the interfacial area per unit volume is an important component of the volumetric mass transfer coefficient. In fact, it is the part of $k_L a$ that is most susceptible to changes in operation variables and fluid properties. The mass transfer coefficient k_L varies only within a limited range (147), but the interfacial area is the main component responsible for the changes in mass transfer rate due to variations in turbulence, initial bubble size, and liquid properties.

The methods for measurement of the interfacial area are based either on rapid chemical reactions or on direct measurement of bubble size. If a mean bubble size can be defined, then the interfacial area can be evaluated with the aid of the holdup measurement, since, in a population of homogeneous bubble size, it can be applied:

$$a = \frac{6\varphi}{d_s} \tag{42}$$

where the Sauter mean diameter (d_S) is given by

$$d_S = \frac{\sum n_i d_i^3}{\sum n_i d_i^2} \tag{43}$$

and n_i is the number of bubbles of diameter d_i

There are very few published data on bubble size applicable to ALRs. For the riser, the correlation presented by Miyahara et al. (57) for the volumetric mean diameter

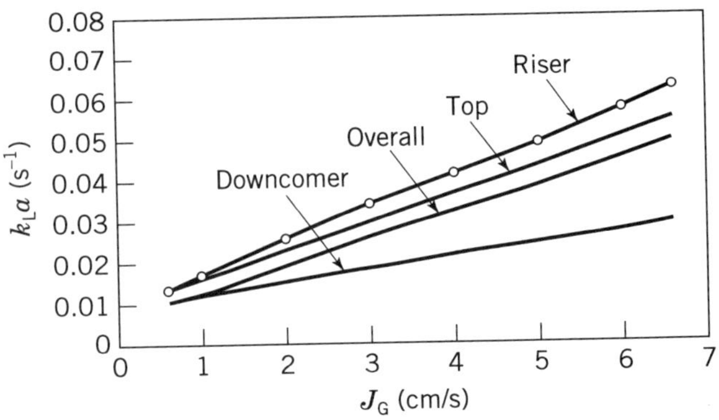

Figure 29. Influence of the superficial gas velocity on overall $k_L a$ and on the $k_L a$ in each of the regions of an ALR. Adapted from Hwang and Lu (146).

of the bubbles in the riser of a concentric-tube ALR can be used:

$$\bar{d}\left(\frac{g\rho_L}{\sigma d_0}\right)^{1/3} = f(N_W) \quad (44)$$

where d_0 is the diameter of the sparger orifice and σ the surface tension,

$$N_W = \frac{We}{Fr^{0.5}} \quad (45)$$

and the function $f(N_W)$ is different for each range of N_W:

$$\begin{aligned} f(N_W) &= 2.9 & N_W < 1 \\ f(N_W) &= 2.9N_W^{-0.188} & 1 < N_W < 2 \\ f(N_W) &= 2.9N_W^{-0.5} & 2 < N_W < 4 \\ f(N_W) &= 3.6 & 4 < N_W \end{aligned} \quad (46)$$

Data on bubble size in the downcomer of an ALR were published by Popovic and Robinson (63). No data on interfacial area measured by chemical methods have been published.

Data Correlations for Mass Transfer Rate

There are two ways of correlating experimental data from ALRs. First, the hydrodynamic point of view suggests that the movement of the fluid in the reactor determines its overall behavior; the gas superficial velocity is therefore the more appropriate independent variable. Second, the thermodynamic point of view is based on a consideration of energy balance, a more global approach to the system. This will lead to correlation of the phenomena in the system as a function of the energy input. Indeed, it is easier to compare mass transfer coefficients in ALRs with those in conventional reactors when the data are presented as a function of the total power input (both mechanical and pneumatic) per unit volume of the medium (34) (Fig. 30).

Figure 31 shows data for mass transfer coefficients as a function of specific power input (149). References and geometric characteristics can be seen in Table 5.

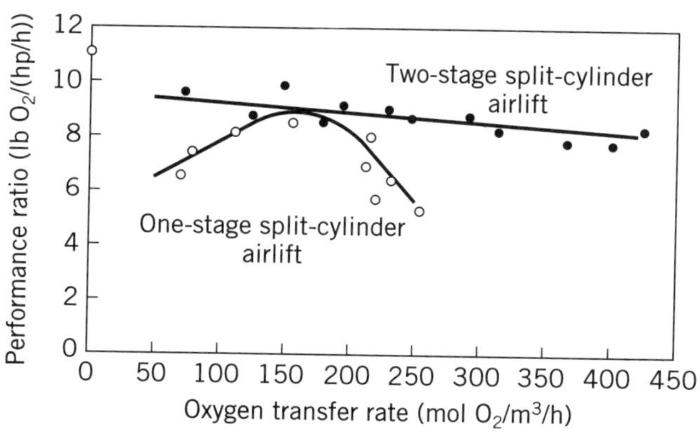

Figure 30. Performance of ALRs. Adapted from Orazem and Erickson (148).

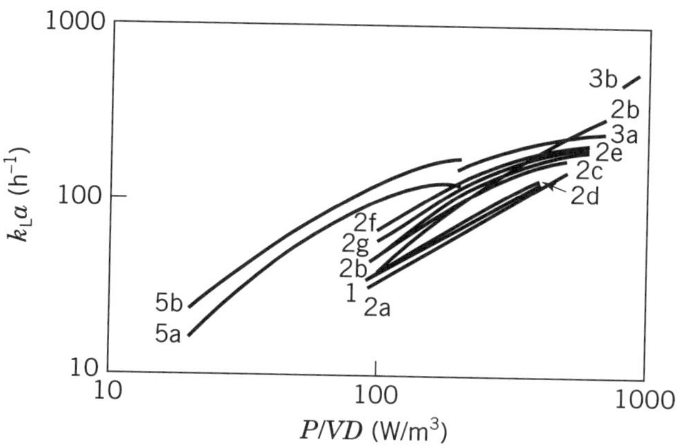

Figure 31. Mass transfer coefficients as a function of specific power input. With permission from Siegel et al. (123). See Table 5.

Selected of correlations proposed for the prediction of mass transfer coefficients are shown in Tables 6 and 7 for internal- and external-loop ALRs, respectively, and the predictions are compared in Figures 32 and 33. Among the internal-loop reactors, two correlations presented for reactors with a rectangular cross-section are shown (32,154). For external-loop ALRs, the correlation by Popovic and Robinson (88) is recommended. In the case of internal-loop ALRs, most of the correlations predict similar values. The correlation of Merchuk et al. (37) can be recommended on the basis that more geometric parameters have been taken into account, and this gives greater generality. Nevertheless, the general considerations already expressed when analyzing other correlations in this chapter are valid here as well.

HEAT TRANSFER

Because of the relatively low reaction rates of processes involving microorganisms and cells, it may—in a very general way—be said that heat-effect problems related to local variations of temperature are not common in bioreactors. Even in the case in which polymeric products are released into the medium and very high viscosity is reached, heat transfer is not the controlling step, because such viscous media will hinder the mass transfer, and heat generation will consequently be limited. In such cases the main point of focus is thus, mass, rather than heat, transfer. Reactions catalyzed by immobilized enzymes, however, may require different considerations, because of higher reaction rates.

There is a far greater body of published data on heat transfer in bubble columns than in ALRs, and some of the basic observations are valid for both types of reactor. The heat transfer rate in bubble columns is much larger than that expected from single-phase flow (155). This is a result of the bubble-driven turbulence and liquid recirculation, which are characteristic of the flow in pneumatically agitated reactors.

Several correlations have been proposed for the prediction of the heat transfer coefficient in these reactors. Recently, Kawase and Moo-Young (62) presented an expres-

Table 5. Mass Transfer in ALRs

Curve no.	Reactor type	Height (cm)	*D* Riser (cm) draught tube	*D* Downcomer (cm) reactor tube	A_d/A	Ref.
1	Concentric tube	143	21	30	1.04	150
2a	External loop	180	15.2	76 + 10.2	0.69	45
2b	External loop	180	15.2	7.6	0.25	45
2c	External loop	180	15.2	5.1	0.11	45
2d	Concentric tube	180	8.9	15.2	0.56	45
2e	Concentric tube	180	7.6	15.2	0.35	45
2f	Concentric tube	180	5.1	15.2	0.13	45
2g	Bubble column	180	0	15.2	0.00	45
3a	Rec. split vessel; all experiments, except high recirculation and two spargers	435–450	9 × 250	7 × 25	0.78	32
3b	Rec. split vessel; high recirculation and two spargers	435–450	9 × 25	7 × 25	0.78	32
4	External loop	850	10	5	0.25	78
5a	Concentric tube	170	11.7	20	1.96	44
5b	Concentric tube	170	17.6	20	0.26	44

Note: Curve numbers correspond to Figure 31.

Table 6. Mass Transfer in Internal-Loop ALRs

No.	Formula	Ref.
1	$Sh = 3 \cdot 10^4 Fr^{0.97} M^{-5.4} Ga^{0.045}\left(1 + \frac{A_d}{A_r}\right)^{-1}$	1
2	$Sh = 2.66 Sc^{0.5} Bo^{0.715} Ga^{0.25}\left(\frac{D_r}{D}\right)^{-0.429} \varphi^{1.34}$	151
3	$k_La = 0.0343 J_{Gr}^{0.524} \mu_{ap}^{-0.255}$ $\gamma = 5000 J_{Gr}$ for $J_{Gr} \geq 0.04\ \mathrm{ms^{-1}}$ $\gamma = 5000 J_{Gr}^{0.5}$ for $J_{Gr} \leq 0.04\ \mathrm{ms^{-1}}$	48
4	$Sh = 0.68 n^{-0.72} Fr^{0.38n+0.52} Sc^{0.38-0.14n}$	59

sion that satisfactorily fits most of the published data. The model is based on Levich's (156) three-zone concept and Kolmogoroff's (157) isotropic turbulence theory and has no empirical adjustable parameters. The model can take into consideration non-Newtonian behavior of the liquid and predicts the enhancement of the heat transfer due to the shear-thinning effect of the fluid. The dimensionless expression may be written as:

$$Nu = 0.075(10.3n^{-0.63})^{\beta} n^{1/3} (Pr^*) \frac{1}{3} Fr^{-\beta} (Re^*)^{\beta + 3(n+1)} \tag{47}$$

with

$$\beta = \frac{4 - n}{6(n + 1)};\ Pr^* = \frac{\kappa D^{1-n}}{\left(\frac{\kappa}{C_p}\right) J_G^{1-n}};\ Re^* = \frac{D^n J_G^{2-n}}{\left(\frac{k}{\rho}\right)} \tag{48}$$

where the symbols defined in "Nomenclature" have been used.

Table 7. Mass Transfer in External-Loop ALRs

No.	Formula	Ref.
1	$K_La = k_1 J_{Gr}^{0.8}\left(1 + \frac{A_d}{A_r}\right)^{-2}$ Water $K_L = 0.75\mu^{-0.8}\sigma^{-0.2}$ NaCl $K_L = 0.79\mu^{-0.8}\sigma^{-0.2}$	45
2	$k_La = 0.5 \cdot 10^{-2} J_{Gr}^{0.52}\left(1 + \frac{A_d}{A_r}\right)^{-0.85} D_1^{0.5} \rho_1^{1.03} \sigma_1^{-0.25} \mu_{ap}^{-0.89}$ or in simplified form: $k_La = 1.911 \cdot 10^{-4} J_{Gr}^{0.52}\left(1 + \frac{A_d}{A_r}\right)^{-0.85} \mu_{ap}^{-0.85}$ $k_La = 0.24 J_{Gr}^{0.837}\left(1 + \frac{A_d}{A_r}\right)^{-1}$	88
3	in case of suspended solids: $k_La = (0.349 - 0.102C_s) J_{Gr}^{0.837}\left(1 + \frac{A_d}{A_r}\right)^{-1}$	152
4	$k_La = 913\left(\frac{P}{VD}\right)^{1.04} (U_L)^{-0.15}$	32
5	$k_La = (0.349 - 0.102C_s\left(1 + \frac{A_d}{A_r}\right)^{-1} J_{Gr}^{0.837}$	152

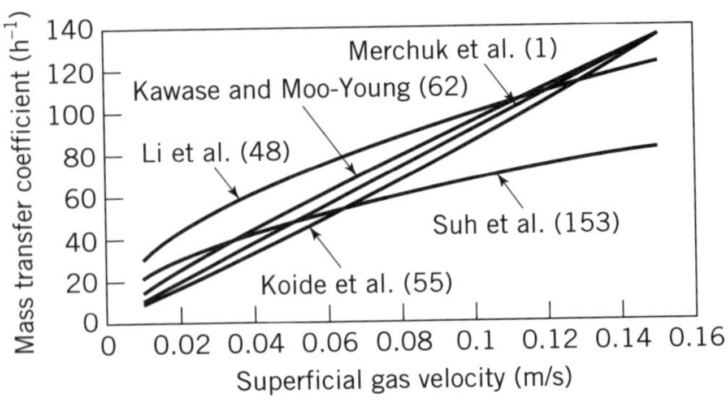

Figure 32. Mass transfer coefficient k_La as a function of gas superficial velocity for internal-loop ALRs. Correlations are presented in Table 6.

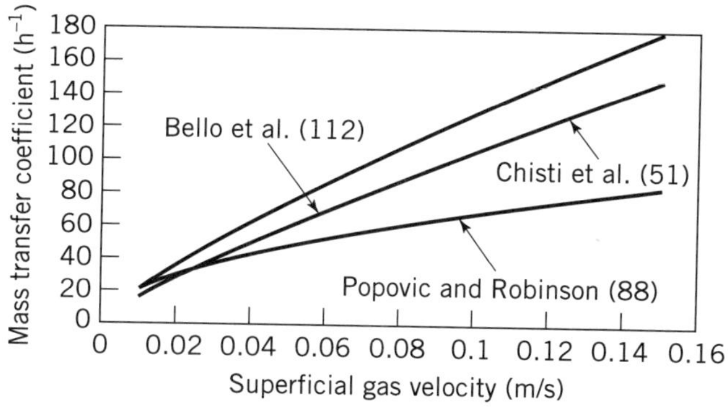

Figure 33. Mass transfer coefficient k_La as a function of gas superficial velocity for external-loop ALRs. Correlations are presented in Table 7.

For Newtonian liquids, this equation reduces to:

$$Nu = 0.134^{1/3}(Pr)^{1/3}Fr^{-1/4}(Re)^{3/4} \tag{49}$$

Kawase and Moo-Young (140) compared their model with data and correlations published by various investigators and found satisfactory agreement for both Newtonian and non-Newtonian liquids.

For ALRs (64), the flow in the reactor may be similar to that in bubble columns if the internal recirculation is high, or it may be even closer to net two-phase flow in pipes, in which case the equations for heat transfer in pipes can be used (126).

Blakebrough et al. (158) studied heat transfer rates in an external-loop airlift fermentor used for culturing *Aspergillus niger*. They concluded that the enhancement of the heat transfer rate could be explained by disturbances in the liquid layer near the surface caused by the presence of the microorganisms.

THREE-PHASE AIRLIFT REACTORS

The special qualities of the ALR stem, as stated before, from its fluid dynamic characteristics. One such characteristic is the directionality of the liquid flow. Independently of superimposed fluctuations, a clear net flow is present in the reactor, with exception of the gas separator in internal-loop designs. Therefore, it is to be expected that the fluidization capacity of the ALR will be markedly superior to that of a bubble column. Several studies have been conducted on the suspension of solids in ALRs, particularly on the use of this type of device for catalytic processes in the chemical industry, where the solid support is usually heavy (61,159–163). In this regard, a very important point is the minimum gas superficial velocity that leads to complete solid fluidization (61,87,159,162,164–166). Hysteresis has been observed in some cases; once total fluidization has been attained, the superficial velocity can be reduced to values lower than that required to reach this state. This is due to the high pressure drop related to passage of liquid through a bed of solids, before fluidization, as compared to the drag forces required to maintain the solid in suspension after all the solids are suspended. Contradictory data on the effect of the suspended solids on the reactor performance have been reported. Fan et al. (167) claimed that the overall gas holdup increased due to the presence of the solids, whereas Koide et al. (168,169) showed the opposite effect on the gas holdup and reported a small decrease in k_La as well. It is possible that these discrepancies are due to the use of different solids. One of the properties of solids that is often overlooked is wettability. Small bubbles may adhere on wettable solids, leading to a change in the apparent density of the particle and thus changing their solid circulation velocity.

In the case of suspended solids that take an active part in the process, the mass transfer rate from the liquid to the solid may become the limiting step. The dependence of particle size on the mass transfer to the suspended solids has been studied by several authors (19,170,171).

All the comments made above relate to heavy solids. This situation is not very frequent in biological processes, with the exception of biolixiviation (172) or the special case of microbial desulfuration of coal (173). In most of the biological processes that may take place in an ALR, the solids are either cells, clumps of cells, or supports that are not much more dense than the medium. Therefore, neither fluidization nor the distribution of solids in the reactor constitutes a problem. Assa and Bar (174) found very small variations in the axial distribution of animal and plant cells suspended in an internal-loop ALR. This is due to the small free-falling velocity of the solids, which is the reason for the difference in loading between the riser and the downcomer when heavy particles are used (160–162). Because of the small difference between solid and medium density, the movement of the particles usually present in biological processes is not as dependent on gravity forces as on liquid and bubble movement. In this case, the transport of liquid in the wake of the bubbles may be considered to be the prevailing transport mechanism (175). Snape and Thomas (176) proposed a Monte Carlo algorithm for modeling particle movement by this mechanism in bubbly flow to predict distribution of circulation times in the reactor. Koide et al. (169) conducted a broad study of gas holdup and mass transfer rates in an internal-loop ALR containing Ca-alginate beads, which are often used for cell immobilization. The authors found that solid loading had a negative effect on both parameters, but the particle diameter had no influence in the range studied (1.8 mm $< d_p <$ 3.98 mm). Chang et al. (31) studied the influence of suspended pellets (cylindrical pellets of immobilized penicillin acylase) on the mixing time in an internal-loop ALR. They found mixing time increased when the solid volumetric concentration was raised up to 15%. For higher concentrations the trend inverted, and the mixing time decreased. No such effects were observed with heavier solids. Increases in gas superficial velocity and in top clearance both lead to decreases in mixing time in all cases.

AIRLIFT REACTOR—SELECTION AND DESIGN

Scale-up of Airlift Bioreactors

The problems encountered in the scale-up of bioreactors can be concentrated into two groups. The first includes the cases in which a high power input per unit volume is used on a laboratory scale, but cannot be maintained on an industrial scale due to economic or mechanical limitations. This is not the case for plant or animal cultures, for which a very high specific power input cannot be used because of cell fragility. The second group of problems can be generalized as lack of knowledge about the hydrodynamics of large-volume vessels.

The methods available for scale-up of bioreactors have been reviewed by Oosterhuis (177), Kossen and Oosterhuis (178), Sweere et al. (179), and Sola and Godia (180), among others. Because, in general, design from first principles cannot be undertaken because of the lack of basic knowledge about the hydrodynamics of the bioreactors, one possible solution is the semifundamental method, which comprises using approximate simple models for fluid dynamics and integrating them with basic known kinetics and heat and mass transfer rates.

It does not happen very often that all this information is available with a degree of certainty that allows safe design of a large system. Thus, the designer must usually resort to dimensional analysis. This method requires a knowledge of all the variables affecting the process, which can be obtained from a qualitative, but realistic, model (181).

A simplified version of this method is to limit the number of variables to one or two and to follow rules of thumb (182), which, depending on the specific case, may be constant P/V, constant k_La, etc. The literature shows, however, many cases of inconsistency for this method. For example, design of a scaled-up bioreactor in which the oxygen transfer rate is kept constant can lead to a better performance than expected, as in the example reported by Taguchi et al. (183) for glucoamylase production by *Endomyces* sp., or worse than expected, as in the case of protease production by *Streptomyces* sp. (184). These two examples are shown in Figure 34.

The method of regime analysis and scale down, proposed by Kossen and Oosterhuis (178), combines two tools to overcome the problem posed by the complexity of biochemical reactors. Their method is based on considering the regime of the full-scale process as the objective and planning the strategy of process development from this point on. This method is therefore applicable only to conventional and well-studied bioreactors that are to be used in a new bioprocess.

Regime analysis is based on the consideration that, generally, biochemical processes involve a series of steps, some being mass or heat transfer by convection, some being diffusive mechanisms (activated or not), and others being chemical reaction steps. In the latter case, a mass-transfer mechanism is superimposed, since molecules must encounter one another in order to react, and usually a heat effect will accompany the reaction. Depending on whether these steps take place in parallel or in series and on the relaxation time for each step (185), the rate of the total process is often given as the rate of one single step. But the equilibrium between all the individual rates can be (and usually is) upset by a change in scale. This is to be expected, because a change in scale will not bring a change in the physicochemical or kinetic parameters (scale-insensitive variables), but will affect the overall convective mass and

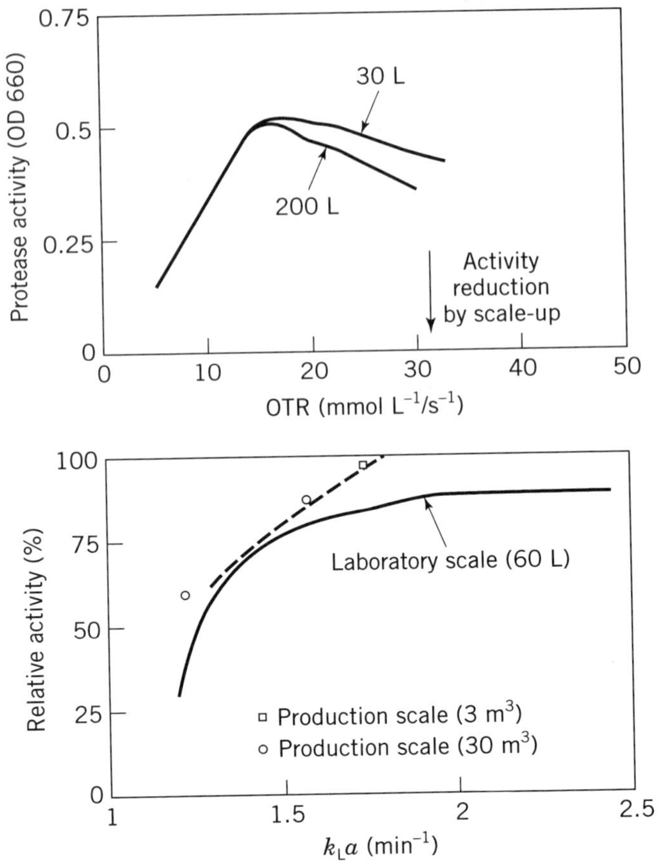

Figure 34. Results for a scaled-up bioreactor with a constant oxygen transfer rate (183,184). OD 660 is the optical density measured at wavelength of 660 nm.

heat rates (scale-sensitive variables). A new equilibrium will be established, and the interplay of all the parameters of the system may lead to a regime in which a different step becomes the step-controlling the process rate.

The method of Kossen and Oosterhuis (178) starts with an analysis of the operation of the large-scale system. Once the regime is clarified, a small-scale system is designed in such a way that it simulates the operation regime of the larger one. Optimization studies can be done on the smaller model, and conclusions will then be extrapolated to the full-scale process. This concept is depicted in Figure 35. An example of this method is given by Oosterhuis (177), in which a large-scale stirred tank reactor is simulated on a laboratory scale by two interconnected vessels, one with a small liquid volume, high agitation rate, and high rate of oxygen supply (representing the zone near the impeller-sparger in the reactor), and a second vessel, with a much larger liquid volume, oxygen consumption (simulated by nitrogen sparging), and gentle agitation. The success of this model suggests that large vessels must be carefully analyzed and their internal structure studied for proper modeling and design to be performed. The analysis starts with the definition of the characteristic time constants of the system.

In the case of low gas recirculation, the relationship between the total circulation time and the residence time in

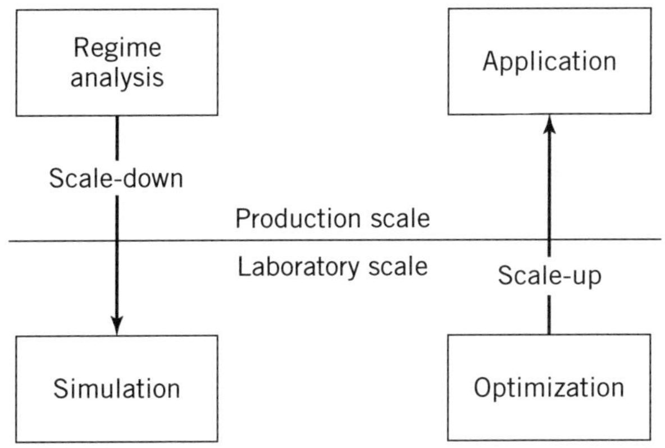

Figure 35. Schematic representation of the scale-down method. Adapted from Oosterhuis (177).

the downcomer may become very relevant. This was shown by MacNeil and Kristiansen (180,186) for the scale-up of an external-loop ALR for the production of *Aureobasidium pullulans*. The poor oxygen supply in the downcomer was reflected on the level of product synthesis, as shown in Figure 36. The smaller residence times lead to higher *A. pullulans* production.

An analysis of the time constants required for the design of an ALR for plant cell culture has recently been presented (187). Figure 37 shows the reported values of the time constants calculated for an external-loop ALR for mixing, mass transfer, and oxygen consumption at two different cell concentrations: 5 and 30 kg/m^3. These graphs exemplify the changes in the controlling mechanism that may take place as the physical properties of the broth change in the course of the bioprocess.

Details on bioreactor scale-up may be found in the recent review of Sola (180). It should be remembered, however, that the reliability of a scale-up will always be limited by the quality of the predictions of gas holdup, liquid velocity, and mass transfer rate.

Design Improvements

In the design of ALR, several modifications aiming at the improvement of some of the characteristics of the equipment have been proposed. One of the earliest modifications was the two-staged ALR, proposed by Orazem and Erickson (148). Their design was inspired by the improvement observed by them in the performance obtained with multistage sieve trays over single-stage bubble columns. They claimed that a substantially higher mass transfer coefficient was obtained, as was a better performance in terms of oxygen transferred per unit of energy invested.

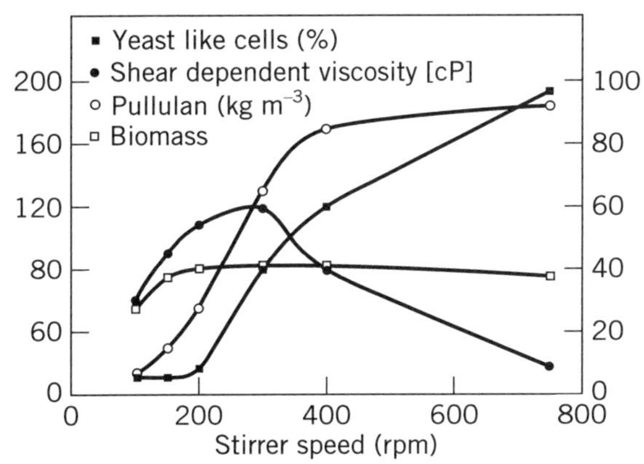

Figure 36. Influence of residence time in the poorly aerated downcomer on the production of *A. pullulans*. Adapted from McNeil and Kristiansen (186).

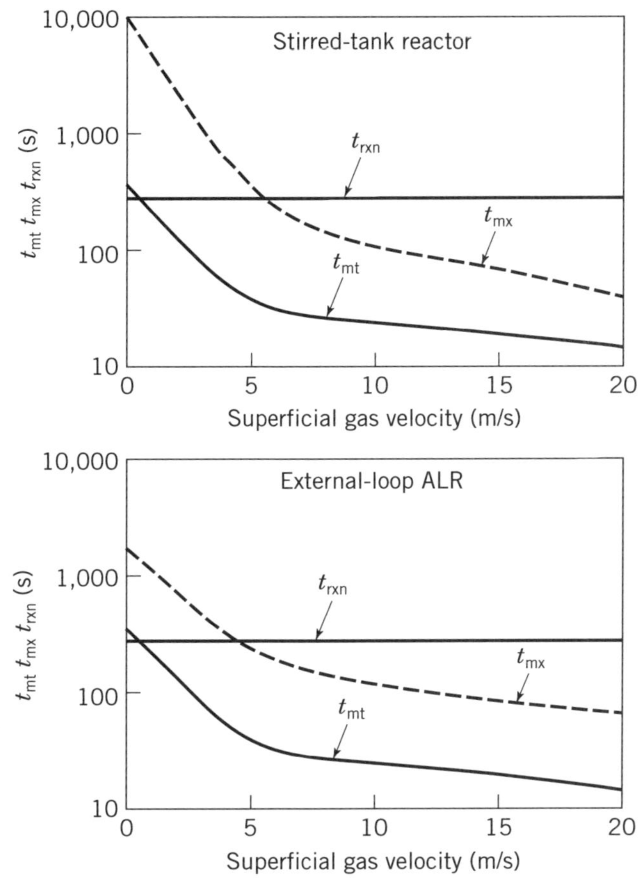

Figure 37. Time constants calculated in an external-loop ALR as a function of the superficial gas velocity (187).

The combination of a concentric-tube ALR with a marine propeller (1000 rpm) was studied by Pollard et al. (188) (Fig. 38). This modification, in which the stirrer was located near the bottom of the draft-tube, served to increase the circulation rate, which may be low for viscous liquids. This configuration showed enhanced oxygen transfer and more uniform distribution of the dissolved oxygen concentration throughout the reactor in comparison with that of the unmodified ALR. The liquid circulation was also intensified. However, the improvement was achieved at the cost of the introduction of a focus of energy input. As has been mentioned before, the absence of such a high shear region is one of the advantages of the ALR. This advantage is, in part at least, lost. It is interesting to note that a bakers' yeast culture (188) grown in the same type of modified reactor did not reveal any improvement. This may be related to the strong shear rate in the vicinity of the propeller.

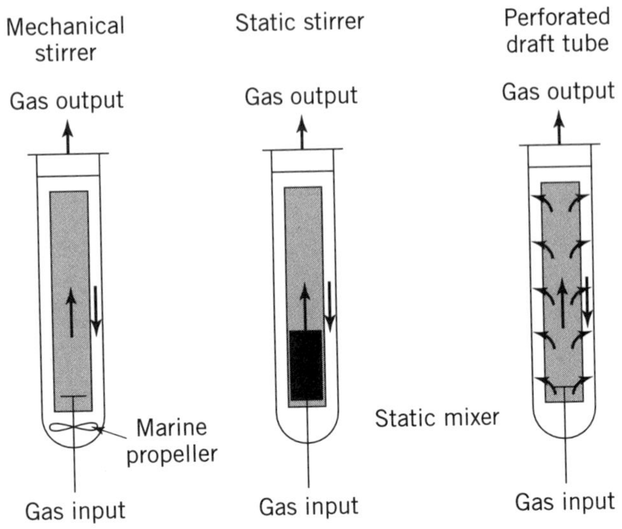

Figure 38. Modifications proposed in ALRs.

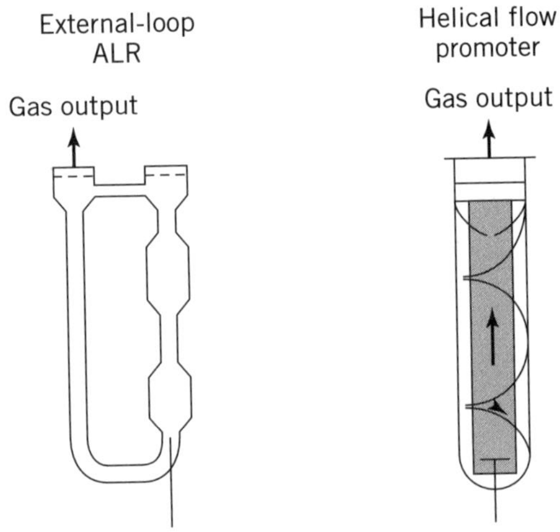

Figure 39. Comparison between a bubble column an ALR and an ALR with helical flow promotors. From Schlotelburg et al. (196).

Different types of static mixers, usually located in the riser, have also been used for enhancing the performance of ALRs (Fig. 38). Potucek (189) studied the influence of static mixers on the gas holdup and interfacial gas-liquid area in a concentric-tube ALR and found an improvement in interfacial area. Enhancement of k_La has been also obtained, especially in viscous liquids, as reported by Gaspillo and Goto (190) and Chisti et al. (191). Zhou et al. (192) showed better growth of *Cephalosporum acremonium* in an ALR with static mixers.

As mentioned above, it has been shown repeatedly that the mass transfer rate in a bubble column is usually higher than that in the conventional ALR. It therefore makes sense to try to bring into the ALR some of the characteristics of bubble columns in a controlled fashion. This was done by Bando et al. (193,194), who tested a perforated draft tube in a concentric-tube ALR. The perforations in the draft tube facilitated communication between the less-well aerated liquid in the downcomer and the better aerated riser (Fig. 38). The reported improvements in mass transfer rates were undoubtedly obtained at the cost of a reduction of circulation velocity.

Another variant tested is the converging-diverging tube ALR (195), which can be seen in Figure 38. The authors report that the changes in cross-section in the riser produced an increase in the gas holdup.

One of the advantages of the directionality of flow in ALRs is the improved fluidization capacity. The strengthening of this advantage was the aim of another modification, the helical flow promoter, proposed by Gluz and Merchuk (170). The helical flow promoter causes the fluid to flow down in the downcomer (Fig. 39) in a helical pattern. The device comprises several fins or baffles, which have the effect of modifying the flow paths; instead of going in straight lines along the axis, the flow paths move along an helix. The baffles may be installed in a small section at almost any place along the riser or the downcomer, and the effect is perceived throughout the reactor. One of the best positions for the helical flow promoter is the top of the riser. A helical flow is then generated in the downcomer to produce a swirl at the bottom and a corkscrew-like path in the riser. This has a strong potential for the culture of photosynthetic microorganisms. The helical movement causes secondary flow, which leads to enhanced radial mixing, and therefore more homogeneous distribution of light and heat among liquid elements and suspended particles. With the helical flow promoter, it is thus more likely that all the elements of the fluid get the same exposure to light and heat exchange.

One of the most important characteristics of the helical flow promoter is the enhanced capacity for fluidizing solid particles. This is due to the swirls that develop at the bottom of the reactor. Thus, this modification is especially suited to processes operating with cells immobilized on a solid. The minimal gas flow rate for complete fluidization of solids in an ALR may be reduced drastically by the use of a helical flow promoter (170). In addition, the mass transfer rate to suspended solids may be enhanced up to 50%, because of the higher relative velocity between the particles and solids.

Another interesting aspect of the performance of helical flow promoters can be seen in Figure 39 (196). Mass transfer coefficients obtained in a bubble column, an ALR, and an ALR with HFP were compared, for water and two solutions of CMC: 0.2% and 0.04%. The corresponding effective viscosities for $\gamma = 50\ s^{-1}$ are 16 and 38 mPa s respectively. It is clearly seen that although with water there is a consistent difference in favor of the bubble column, the differences diminish as the CMC concentration increases, and for the highest concentration tested the difference is negligible. This implies that under these conditions the disadvantage of the HFP in mass transfer vanishes, whereas all the advantages mentioned before subsist.

SUMMARY AND CONCLUSIONS

ALRs are popular in modern bioprocess research and development. These reactors are particularly suitable for biological processes in which a high mass transfer rate is required, but excessive power input may lead to damage of the cells due to shear effects. ALRs also have very appealing characteristics for bioprocesses for low-value products in which efficiency of energy utilization may become the key point for design. Such is the case for ALRs for wastewater treatment. The ALR is particularly effective for solid fluidization, which is important in many biological processes in which the biocatalyst is available in the form of pellets or is immobilized on a solid support.

The distinctive characteristics of ALRs are conferred by the fluid dynamics of the liquid–gas or liquid–gas–solid mixtures in it. These characteristics are expressed as gas holdup, liquid velocity, and mixing in each of the zones of the ALR. It is important to recognize the differences in the fluid dynamic characteristics of these zones: the riser, the gas separator, and the downcomer. Only a correct understanding of behavior and interconnection of these regions can make possible the correct design of a new reactor or the scale-up of a laboratory device up to pilot or industrial size.

The purpose of scale-up is to conserve and repeat on a larger scale the fluid dynamics of the reactor. Therefore, one of the most important factors in the design and scale-up of reactors is the influence of the geometric characteristics of the system on the flow of the different phases present.

The variables affecting the performance of the reactor are geometric design, operation variables, and fluid properties. Several correlations are available in the technical literature for the prediction of the fluid dynamic characteristics of the reactor (gas holdup, liquid velocity, mixing time of axial dispersion coefficient) and of the mass and heat transfer coefficients. Although attempts have been made to study all the aspects described, no single research group has managed to cover all the variables over a wide range. It is therefore extremely important that the engineer confronting scale-up or de novo design of an ALR analyzes in depth the range of validity of the correlations used for the calculations.

NOMENCLATURE

a	Interfacial area (m^{-1})
A	Cross-sectional area (m^{-2})
B	Heat transfer area (m^2)
Bo	Bond number or Bodenstein number (–)
C	Concentration (mg/L)
Co	Distribution parameter (–)
C	Clearance or fiction losses coefficient (–)
d	Mean diameter (mm)
D	Diameter (m), diffusivity, dispersion coefficient
D_L	Diffusivity, dispersion coefficient ($m^2\ sec^{-1}$)
E	Energy (J)
Fr	Froude number (–)
g	Gravitational constant ($m\ sec^{-2}$)
Ga	Galileo number (–)
h	Height (m)
h_b	Overall heat transfer coefficient ($W\ m^{-2}sec^{-1}K^{-1}$)
I	Degree of homogeneity (–)
J	Superficial velocity ($cm\ sec^{-1}$)
K	Saturation constant in Monod's equation mgL^{-1}
k_La	Volumetric mass transfer coefficient (hr^{-1})
K_b	Hydraulic coefficient bottom (–)
K_t	Hydraulic coefficient top (–)
L	Length (m)
M	Wetted surface parameter (–)
Ma	Bubble coalescence parameter (–)
Mo	Morton number (–)
n	Flow index (–)
N	Agitation speed (sec^{-1})
Nu	Nusselt number (–)
N_W	Parameter (–)
OTR	oxygen transfer rate
P	Pressure, power input (Pa)
Pe	Peclet number (–)
Pr	Prandtl number (–)
r	Reaction rate ($mol\ m^{-3}sec^{-1}$)
Re	Reynolds number (–)
S	Surface area (m^2)
Sc	Schmidt number (–)
Sh	Sherwood number (–)
t	Time (sec)
U	Velocity ($m\ sec^{-1}$)
V	Volume (m^3)
Q	Flow rate (m^3sec^{-1})
We	Weber number (–)
Wi	Weissenberg number (–)

Greek Letters

α	Coefficient (–)
β	Flowing volumetric concentration (m^3/m^3)
γ	Shear rate (sec^{-1})
Δ	Difference
φ	Gas holdup (–)
κ	Behavior coefficient ($Pa\ s^{n+2}$)
λ	Heat conductivity coefficient ($W\ m^{-1}K^{-1}$)
μ	Dynamic viscosity (Pa s), specific growth coefficient (h^{-1})
ρ	Density ($kg\ m^{-3}$)
σ	Surface tension ($N\ m^{-1}$)
τ	Shear stress (Pa)
ν	Kinematic viscosity ($m^2\ sec^{-1}$)

Indices

ab	All bubbles
ap	Apparent
b	Bottom
$b\infty$	Terminal
c	Circulation

d	Downcomer, dissipation
D	Dissipation
d	Downcomer
g	Growth
G	Gas
ht	Heat transfer
HM	Heat production by the microorganisms
HS	Heat production by stirring
i	Region of reactor
in	Input
L	Liquid
m	Mixing
max	Maximal
mt	Mass transfer
oc	Oxygen consumption
p	Potential
r	Riser
s	Gas separator
sc	Sugar consumption
sh	Shaft
sol	Solid
t	Top
T	Total
w	Water

BIBLIOGRAPHY

1. J.C. Merchuk, N. Ladwa, A. Cameron, M. Bulmer, and A. Pickett, *AIChE J.* **40**, 1105–1117 (1994).
2. K. Koide, K. Kurematsu, S. Iwamoto, Y. Iwata, and K. Horibe, *J. Chem. Eng. Jpn.* **16**, 413–418 (1983).
3. H. Tanaka, *Proc. Biochem.* **22**, 106–112 (1987).
4. W.R. Kessler, M.K. Popovic, C.W. Robinson, *Can. J. Chem. Eng.* **71**, 195–202 (1993).
5. J.Y. Oldshue, *Fluid Mixing Technology*, McGraw-Hill, New York, 1983.
6. J.C. Merchuk, *Adv. Biochem. Eng.* **44**, 65–95 (1991).
7. R.J. Weetman and J.Y. Oldshue, *Power, Flow and Shear Characteristics of Mixing Impellers, Proceedings of the 6th European Conference on Mixing*, Cranfield, 1988, p. 43.
8. H. Markl, R. Bronnenmeier, B. Wittek, *Int. Chem. Eng.* **31**, 185–197 (1991).
9. JD.a.P. Wase, Y.R., *J. Gen. Microbiol.* **131**, 725–731 (1985).
10. N. Edwards, S. Beeton, A.T. Bull, and J.C. Merchuk, *Appl. Microbiol. Biotechnol.* **30**, 190–196 (1989).
11. J. Tramper, D. Smit, J. Straatman, and J.M. Vlak, *Bioprocess Eng.* **3**, 37–41 (1988).
12. L. Yerushalmi and B. Volesky, *Biotechnol. Bioeng.* **27**, 1297–1305 (1985).
13. F.J. Petersen, V.L. McIntire and E.T. Papoutsakis, *J. Biotechnol.* **7**, 229–246 (1988).
14. D.E. Martens, C.D. de Gooijer, C.A.M. van der Velden-de Groot, E.C. Beuvery, and J. Tramper, *Biotechnol. Bioeng.* **41**, 429 (1993).
15. H.R. Millward, B.J. Bellhouse, A.M. Nicholson, S. Beeton, N. Jenkins, and C.J. Knowles, *Biotechnol. Bioeng.* **43**, 899–906 (1994).
16. H. N.a.U. Shiragami, *Bioprocess Eng.* **10**, 43–45 (1994).
17. N.A. Stathopoulos and J.D. Hellums, *Biotechnol. Bioeng.* **27**, 1021–1026 (1984).
18. H.J. Silva, T. Cortinas, and J.R. Ertola, *J. Chem. Technol. Biotechnol.* **40**, 41–49 (1987).
19. H. Tanaka, H. Semba, T. Jitsufuchi, and H. Harada, *Biotechnol. Lett.* **10**, 485–490 (1988).
20. C.G. Smith, P.F. Greenfield, and D.H. Randerson, *Biotechnol. Tech.* **1**, 39–44 (1987).
21. Z. Zhang, Y. Christi, and M. Moo-Young, *J. Biotechnol.* **43**, 33–40 (1995).
22. J.J. Meijer, H.J.G. ten Hoopen, Y.M. van Gameren, and K.Ch.A.M. Luyben, *Enzyme Microb. Technol.* **16**, 467–477 (1994).
23. G.Z. Lu, B.G. Thompson, and M.R. Gray, *Biotechnol. Bioeng.* **40**, 1277–1281 (1992).
24. H.E. Dunlop, K.N. Pradyumna, and Z.M. Rosenberg, *Chem. Eng. Sci.* **49**, 2263–2276 (1994).
25. L.D.a.L. Landau, E.M., *Fluid Mechanics*, Volume 6, Pergamon, Oxford, U.K., 1975, pp. 115–123.
26. W.S. Tan, G.C., Dai, W. Ye, and J.P. Shen, *Chem. Eng. J. Biochem. Eng. J.* **57**, B31–B36 (1995).
27. H.F. Svendsen, H.A. Jakobsen, and R. Torvik, *Chem. Eng. Sci.* **47**, 1297–1304 (1992).
28. A. Stafford, L. Smith, and M.W. Fowler, *Plant Cell Tissue Organ Culture* **4**, 83–94 (1985).
29. P.J. Wilkinson, in G.W. Moody and P.B. Baker eds., *Biotransformations*, Elsevier, London, 1992.
30. A. Gebauer, T. Scheper, K. Schugerl, *Bioprocess Eng.* **2**, 13–23 (1987).
31. C.M. Chang, W.J. Lu, K.S. Own, and S.J. Hwang, *Biotechnol. Tech.* **7**, 317–322 (1993).
32. M.H. Siegel and J.C. Merchuk, *Biotechnol. Bioeng.* **32**, 1128–1137 (1988).
33. G.A. Legrys, *Chem. Eng. Sci.* **33**, 83–86 (1978).
34. M.E. Orazem and L.E. Erickson, *Biotechnol. Bioeng.* **21**, 69–88 (1979).
35. M.H. Siegel, J.C. Merchuk, and K. Schugerl, *AIChE J.* **32**, 1585–1596 (1986).
36. H. Kubota, Y. Hosono, and K. Fujie, *J. Chem. Eng. Jpn.*, **11**, 319–325 (1978).
37. J.C. Merchuk, N. Ladwa, A. Cameron, M. Bulmer, A. Pickett, and I. Berzin, *Chem. Technol. Biol.* **66**, 174–182 (1996).
38. D. Barnea, Y. Taitel, in N.P. Cheremisinoff, ed. *Encyclopedia of Fluid Mechanics*, Gulf, Houston, Tex., 1986, pp. 403–491.
39. G.B. Wallis, *One Dimensional Two-Phase Flow*, McGraw-Hill, New York, 1969.
40. K. Wiswanathan, *Flow Patterns in Bubble Columns*, Gulf, Houston, Tex., 1986, pp. 291–308.
41. J.C. Merchuk, in K. Schügerl ed. *Biotechnology*, Springer, New York, 1990.
42. A.B. Russell, C.R. Thomas, and M.D. Lilly, *Biotechnol. Bioeng.* **43**, 69–76 (1994).
43. A.C. Gomez, Univ. of Almeria, Spain, 1996.
44. P. Weiland, *Ger. Chem. Eng.* **7**, 374–385 (1984).
45. R.A. Bello, C.W. Robinson, and M. Moo-Young, *Chem. Eng. Sci.* **40**, 53–58 (1985).
46. M.Y. Chisti and M. Moo-Young, *Chem. Eng. Commun.* **60**, 195–242 (1987).

47. K. Koide, S. Iwamoto, Y. Takasaka, S. Matsura, E. Takasashi, M. Kimura, and H. Kubota, *J. Chem. Eng. Jpn.* **17**, 611–618 (1984).
48. G.Q. Li, S.Z. Yang, Z.L. Cai, and J.Y. Chen, *Chem. Eng. J.* **56**, B101–B107 (1995).
49. J. Cai, T.J. Niewstad, and J.H. Kop, *Water Sci. Technol.* **26**, 2481–2484 (1992).
50. M. Popovic and C.W. Robinson, *Proc. 34th Can. Chem. Eng. Conf.* Quebec, Oct., 1984, pp. 258–264.
51. M.Y. Chisti, B. Halard, and M. Moo-Young, *Chem. Eng. Sci.* **43**, 451–457 (1988).
52. G. Vatai and M.N. Tekis, *Effect of Pseudoplasticity on Hydrodynamic Characteristics of Airlift Loop Contactors*, Prague, 1986.
53. K. Akita and M. Kawasaki, *Proc. 48th Meet. of Chem. Engrs. Engrs. Jpn.* Kyoto, 1983, pp. 122–126.
54. Y. Kawase, M. Tsujimura, and T. Yamaguchi, *Bioproc. Eng.* **12**, 21–27 (1995).
55. K. Koide, K. Horibe, H. Kawabata, and I. Shigetaka, *J. Chem. Eng. Jpn.* **18**, 248–254 (1995).
56. K. Koide, M. Kimura, H. Nitta, and H. Kawabata, *J. Chem. Eng. Jpn.* **21**, 393–399 (1988).
57. T. Miyahara, M. Hamaguchi, Y. Sukeda, and T. Takahashi, *Can. J. Chem. Eng.* **64**, 718–725 (1986).
58. M. Chakravarty, H.D. Singh, J.N. Baruah, and M.S. Lyengar, *Indian Chem. Eng.* **16**, 17–22 (1974).
59. Y. Kawase, M. Moo-Young, *Chem. Eng. Commun.* **40**, 67–83 (1986).
60. J.C. Merchuk, M. Bulmer, N. Ladwa, A.M. Pickett, and A. Cameron, in R. King, ed., *Bioreactor Fluid Dynamics* Elsevier, London, 1988, pp. 391–414.
61. K. Koide, K. Horibe, H. Kawabata, and S. Ito, *J. Chem. Eng. Jpn.* **17**, 368–374 (1984).
62. Y. Kawase, M. Moo-Young, *Chem. Eng. Res. Des.* **65**, 121–126 (1987).
63. M. Popovic and C.W. Robinson, *Biotechnol. Bioeng.* **32**, 301–312 (1988).
64. M. Siegel and J.C. Merchuk, *Can. J. Chem Eng.* **69**, 465–471 (1991).
65. M. Hallaile, *Biotechnology*, Ben-Gurion Univ. Beer Sheva, Israel, 1993.
66. M. Ghirardini, G. Donati, and F. Rivetti, *Chem. Eng. Sci.* **47**, 2209–2214 (1992).
67. J.C. Merchuk and R. Yunger, *Chem. Eng. Sci.* **45**, 2973–2976 (1990).
68. Y. Chisti and M. Moo-Young, *Biotechnol. Bioeng.* **34**, 1391–1392 (1989).
69. J.H. Hills, *Ind. Chem. Eng.* **52**, 1–9 (1974).
70. N.N. Clark, C.M. Atkinson, and R.L.C. Flemmer, *AIChE J.* **33**, 515–521 (1987).
71. J.C. Merchuk and Y. Stein, *AIChE J.* **27**, 377–381 (1981).
72. B. Glennon, W. Al-Masry, P.F. MacLoughlin, and D.M. Malone, *Chem. Eng. Commun.* **121**, 181–192 (1993).
73. W. Freedman and J.F. Davidson, *Trans. Inst. Chem. Eng.* **47**, 251–262 (1969).
74. N. Zuber and J.A. Findlay, *Trans. ASME J. Heat Transfer* **87**, 453–468 (1965).
75. S.G. Bankoff, *J. Heat Transfer* **82**, 252–258 (1960).
76. P. Verlaan, J. Tramper, K. van't Riet, and K. Luyben, *Chem. Eng. J.* **33**, B43–B48 (1986).
77. J.C. Merchuk and I. Berzin, *Chem. Eng. Sci.* **50**, 2225–2233 (1995).
78. P.a.O.U. Weiland, *Ger. Chem. Eng.* **4**, 174–181 (1981).
79. P. Verlaan, J. Tramper, K. van't Riet, and K.Ch.A.M. Luyben, *Hydrodynamics and Axial Dispersion in an Airlift Loop Reactor with Two and Three Phase Flow*, Cambridge, UK, 1986.
80. P. Verlaan, A.M.M. Van Eijs, J. Tramper, K. van't Riet, and K. Ch. A.M. Luyben, *Chem. Eng. Sci.* **44**, 1139–1144 (1989).
81. Y. Chisti, *Airlift Bioreactors*, Elsevier, London, 1989.
82. P. Verlaan, J.C. Vos, and K. van't Riet, in R. King ed., *Bioreactor Fluid Dynamics II*, Elsevier, London, 1988.
83. J.C. Merchuk, *Chem. Eng. Sci.* **41**, 11–16 (1986).
84. D. Posarac and D. Petrovic, *Chem. Eng. Sci.* **43**, 116–119 (1991).
85. R.A. Bello, Ph.D. Thesis, Univ. of Waterloo, Ontario, Canada, 1981.
86. Z. Kemblowski, J. Prywarski, and A. Diab, *Chem. Eng. Sci.* **48**, 4023–4035 (1993).
87. D. Posarac, D. Petrovic, A. Dudukovic, and D. Skala, *J. Serb. Chem. Soc.* **56**, 227–240 (1991).
88. M.K. Popovic and C.W. Robinson, *AIChE J.* **35**, 393–405 (1989).
89. M. Nishikawa, H. Kato, and K. Hashimoto, *Ind. Eng. Chem. Process Des. Dev*, **16**, 133–137 (1977).
90. M. Nishikawa, *Biotechnol. Bioeng.* **37**, 691–692 (1991).
91. H.J. Henzler, *Chem.-Ing.-Tech.* **52**, 643–652 (1980).
92. A. Schumpe and W.D. Deckwer, *Bioproc. Eng.* **2**, 79–94 (1987).
93. A. Zaidi, H. Bourziza, and L. Echihabi, *Chem.-Ing.-Tech.* **59**, 748–749 (1987).
94. S.A. El-Tamtamy, S.A., Khalil, A.A. Nour-El-Din, and A. Gaber, *Appl. Microbiol. Biotech.* **19**, 376–381 (1984).
95. H.J. Henzler, J. Kauling, *Scale-Up of Mass Transfer in Highly Viscous Liquids*, Wurzburg, Germany, 1985, pp. 303–312.
96. J.C. Merchuk and S. Ben-Zvi (Yona), *Chem. Eng. Sci.* **47**, 3517–3523 (1992).
97. A. Lubbert, S. Frohlich, B. Larson, and K. Schugerl, *Fluid Dynamics in Airlift Loop Reactors as Measured During Real Cultivation Processes*, London, 1988, pp. 379–394.
98. A. Lubbert, B. Larson, L.W. Wan, and S. Broring, *I. Chem. E. Symp. Ser.* **121**, 203–213 (1990).
99. C.H. Lee, L.A. Glasgow, L.E. Erickson, and S.A. Patel, *Liquid Circulation in Airlift Fermentors*, Miami Beach, 1986.
100. A.G. Jones, *Chem. Eng. Sci.* **40**, 449–462 (1985).
101. D.J. Niklin, *Chem. Eng. Sci.* **17**, 693–702 (1962).
102. M. Clark and A.G. Jones, *Chem. Eng. Sci.* **42**, 378–384 (1987).
103. Y. Chisti and M. Moo-Young, *J. Chem. Technol. Biotechnol.* **42**, 211–219 (1988).
104. S. Wachi, A.G. Jones, and T.P. Elson, *Chem. Eng. Sci.* **46**, 657–661 (1991).
105. E. Garcia Calvo, *Chem. Eng. Sci.* **44**, 321–323 (1989).
106. R.S. Richardson and D.J. Higson, *Trans. Inst. Chem. Eng.* **40**, 169–182 (1962).
107. E. Garcia Calvo and P. Leton, *Chem. Eng. Sci.* **46**, 2947–2951 (1989).
108. H. Blenke, *Adv. Biochem. Eng.* **13**, 121–127 (1979).
109. Y.C. Hsu and M.P. Dudukovic, *Chem. Eng. Sci.* **35**, 135–141 (1980).

110. P. Ayazi Shamlou, D.J. Pollard, A.P. Ison, and M.D. Lilly, *Chem. Eng. Sci.* **49**, 303–312 (1994).
111. W.J. Lu, S.J. Hwang, and C.M. Chang, *Chem. Eng. Sci.* **30**, 1301–1310 (1995).
112. R.A. Bello, C.W. Robinson, and M. Moo-Young, *Biotech. Bioeng.* **27**, 369–381 (1985).
113. J. Zahradnik, M. Fialova, F. Kastanek, K.D. Green, and N.H. Thomas, *Trans. IChE* **73A**, 341–346 (1995).
114. P.R. Fields and N.K.H. Slater, *Chem. Eng. Sci.* **38**, 647–651 (1983).
115. T. Yamane, in J.A. Asenjo and J.C. Merchuk eds., *Bioreactor System Design*, Dekker, New York, 1994.
116. G. B.A.a.M. Buffham and G. Mason, *Chem. Eng. Sci.* **48**, 3879–3887 (1993).
117. O. Levenspiel and K.B. Bishoff, *Adv. Chem. Eng.* **4**, 95–126 (1963).
118. Y. Murakami, T. Hiroshe, S. Ono, and T. Nishijima, *J. Chem. Eng. Jpn.* **15**, 121–126 (1982).
119. C.H. Lin, B.S. Fang, H.Y. Wu, T.F. Fang, T.F. Kuo, and C.Y. Hu, *Biotechnol. Bioeng.* **18**, 1557–1562 (1976).
120. J.C. Merchuk and J.A. Asenjo, *Bioreactor System Design*, Dekker, New York, 1994.
121. K. Schügerl, *Adv. Biochem. Eng.* **19**, 71–94 (1981).
122. Moor Nagar, Graduate Thesis, Ben Gurion University, Dept. of Chemical Engineering, 1992.
123. M.H. Siegel, M. Hallaile, and J.C. Merchuk, in A. Mizrahi ed., *Advances in Biotechnological Processes*, Liss, New York, 1988, pp. 79–124.
124. W.J. Lu, S.J. Hwang, and C.M. Chang, *Ind. Eng. Chem. Res.* **33**, 2180–2186 (1994).
125. M. Siegel and J.C. Merchuk, *Influence of the Gas-Liquid Separator on Air-Lift Reactor Design and Operation*, 1990.
126. S.P. Godboole and Y.T. Shah, *Design and Operation of Bubble Column Reactors*, Gulf, Houston, Tex., 1986.
127. W.D. Deckwer, R. Burckhart, and G. Zoll, *Chem. Eng. Sci.* **29**, 2177–2187 (1974).
128. J.B. Joshi, *Trans Inst. Chem. Eng.* **58**, 155–161 (1980).
129. P.R. Fields and N.K.H. Slater, *Biotechnol. Bioeng.* **26**, 719–723 (1984).
130. M.R. Makley and A. Keller, *Phylos. Trans. Roy. Soc. London*, **278**, 1276–1281 (1975).
131. S. Frehlich, M. Lotz, T. Korte, A. Lübbert, K. Schügerl, and M. Seekamp, *Biotech. Bioeng.* **38**, 43–48 (1991).
132. M. Nakanoh, F. Yoshida, *Ind. Eng. Chem. Process Des. Dev.* **19**, 190–195 (1980).
133. J.C. Merchuk, S. Yona, M.H. Siegel, and A. Ben-Zvi, *Biotech. Bioeng.* **35**, 1161–1163 (1990).
134. S.P. Godbole, A. Schumpe, Y.T. Shah, and N.L. Carr, *AIChE J.* **30**, 213–220 (1984).
135. R.B. Bird, *Chem. Eng.* **27**, 102–109 (1993).
136. E. Molina-Grima, Y. Chisty, and M. Moo-Young, *J. Biotechnol.* **52**, 195–201 (1997).
137. V. Linek, V. Vacek, and P. Benes, *Chem. Eng. J.* **34**, 11–34 (1987).
138. T. Philichi, L. Stenstrom, K. Michael, *J. Water Pollut. Control Fed.* **4**, 83–86 (1989).
139. C.F. Mignone and J. Ertola Rodolfo, *J. Chem. Technol. Biotechnol.* **34**, 121–126 (1984).
140. Y. Kawase and M. Moo-Young, *Biotechnol. Bioeng.* **30**, 345–347 (1987).
141. J.C. Merchuk, S. Ben-Zvi, and K. Niranjan, *TibTech* **12**, 501–511 (1994).
142. W.D. Deckwer, *Bubble Column Reactors*, Wiley, New York, 1991.
143. M. Siegel, S. Ben-Zvi, and J.C. Merchuk, *Measurement and Interpretation of Mass Transfer Data in Air-Lift Bioreactors*, Kuyngju, Korea, 1990, pp. 449–453.
144. G. Andre, C.W. Robinson, and M. Moo-Young, *Chem. Eng. Sci.* **38**, 1845–1854 (1983).
145. J.C. Merchuk, G. Osemberg, M. Siegel, and M. Shacham, *Chem. Eng. Sci.* **47**, 2221–2226 (1992).
146. S.-J. Hwang, W.-J. Lu, *Chem. Eng. Sci.* **52**, 853–857 (1997).
147. J.C. Merchuk, *TibTech*, **8**, 66–71 (1990).
148. M.E. Orazem and L.E. Erickson, *Biotechnol. Bioeng.* **21**, 69–88 (1979).
149. M. Siegel, M. Hallaile, and J.C. Merchuk, in A. Mizrahi ed. *Advances in Biotechnological Processes*, Liss, New York, 1988, pp. 51–75.
150. T.W. Barker and J.T. Worgan, *Eur. J. Appl. Microbiol. Biotechnol.* **13**, 77–82 (1981).
151. K. Koide, K. Horibe, H. Kawabata, and S. Ito, *J. Chem. Eng. Jpn.* **18**, 248–254 (1985).
152. Y. Chisti, K. Fujimoto, and M. Moo-Young, *Biotechnol. Prog.* **5**, 72–76 (1988).
153. I.S. Suh., A. Schumpe, and W.D. Deckwer, *Can. J. Chem. Eng.* **69**, 506–512 (1991).
154. M.Y. Chisti and M. Moo-Young, *Biotechnol. Bioeng.* **31**, 487–494 (1988).
155. Y.T. Shah, B.G. Kelkar, S.P. Godbole, and W.D. Deckwer, *AIChE J.* **38**, 353–359 (1982).
156. V.G. Levich, *Physicochemical Hydrodynamics*, Prentice Hall, Englewood Cliffs, N.J., 1962.
157. A.N. Kolmogoroff, *Dokl. Akad. Nauk SSSR* **66**, 1949.
158. N. Blakebrough, I.A. Fatile, W.J. McManamey, and G. Walker, *Chem. Eng. Res. Des.* **61**, 383–387 (1983).
159. K. Koide, M. Terasawa, and H. Takekawa, *J. Chem. Eng. Jpn.* **19**, 341–344 (1986).
160. R.S. Douek, G.F. Hewitt, and A.G. Livingston, *Trans. Inst. Chem. Eng.* **73A**, 336–340 (1995).
161. A.G. Livingstone and S.F. Zhang, *Chem. Eng. Sci.* **48**, 1641–1654 (1993).
162. J.C. Merchuk and M. Herskowitz, *Can. J. Chem. Eng.* **64**, 57–62 (1986).
163. L.S. Fan, S.J. Hwang, and A. Matsuura, *Chem. Eng. Sci.* **39**, 1677–1681 (1984).
164. D. Petrovic, D. Posarac, and D. Skala, *Chem. Eng. Sci.* **44**, 997–998 (1989).
165. D. Petrovic, D. Posarac, A. Dudukovic, and D. Skala, *Chem. Eng. Sci.* **45**, 2967–2970 (1990).
166. M. Immich and U. Onken, *Chem. Eng. Sci.* **47**, 3379–3386 (1992).
167. L.S. Fan, K. Fujie, and T.R. Long, *AIChE Symp. Ser.* **80C24T**, 102–107 (1984).
168. K. Koide, K. Horibe, H. Kawabata, and S. Ito, *J. Chem. Eng. Jpn.* **18**, 248–254 (1985).
169. K. Koide, K. Shibata, H. Ito, S.Y. Kim, and K. Ohtaguchi, *J. Chem. Eng. Jpn.* **25**, 11–16 (1992).
170. M.D. Gluz and J.C. Merchuk, *Chem. Eng. Sci.* **51**, 1920–1925 (1996).
171. K.B. Kushalkar and V.G. Pangarkar, *Chem. Eng. Sci.* **49**, 139–144 (1994).

172. G. Rossi, *Biohidrometallurgy*, McGraw-Hill, Hamburg, Germany, 1990.
173. B.C. Smith and D.R. Skidmore, *Biotechnol. Bioeng.* **35**, 483–491 (1990).
174. A. Assa and R. Bar, *Biotechnol. Bioeng.* **38**, 1325–1330 (1991).
175. J. Schmidt, R. Nassar, and A. Lubbert, *Chem. Eng. Sci.* **47**, 2295–2300 (1992).
176. J.B. Snape and N.H. Thomas, *Biotechnol. Bioeng.* **40**, 337–345 (1992).
177. N.M.G. Oosterhuis, Ph.D. Thesis, Dept. of Chemical Technology, Univ. of Technology, Delft, The Netherlands, 1984.
178. N.W.F. Kossen and N.M.G. Oosterhuis, in N.J. Rhem and J. Reed eds. *Biotechnology* VCH, Weinheim, Germany, 1985.
179. A.P.J. Sweere, K.C.A.M. Luyben, and N.W.F. Kossen, *Enzyme Microb. Technol.* **9**, 386–392 (1987).
180. C. Sola and F. Godia, in J.A. Asenjo, J.C. Merchuk eds., *Bioreactor System Design*, Dekker, New York, 1994.
181. F.H. Johnson, H. Eyring, and M.J. Polissar, *The Kinetic Basis of Molecular Biology*, Wiley, New York, 1954.
182. D.G. Jordan, *Chemical Process Development*, Volume 6, Wiley, New York, 1968.
183. H. Taguchi, T. Imanaka, S. Teramoto, M. Takatsu, and M. Sato, *J. Ferment. Technol.* **46**, 823–830 (1986).
184. H. Takei, K. Misusawa, and F. Yoshida, *Ferment. Technol.* **53**, 151–156 (1975).
185. J.A. Roels, *Energetics and Kinetics in Biotechnology*, Elsevier, Amsterdam, 1983.
186. B. McNeil and B. Kristiansen, *Biotechnol. Lett.* **9**, 101–107 (1987).
187. P.M. Doran, *Adv. Biochem. Eng. Biotechnol.* **48**, 115–168 (1993).
188. D.J. Pollard, A.P. Ison, P. Ayazi Shamlou, and M.D. Lilly, *The Examination of Bioreactor Heterogeneity with Rheological Different Fermentation Broths*, Kluwer, Dordrecht, The Netherlands, 1994, pp. 163–170.
189. F. Potucek, *Collect. Czech. Chem. Commun.* **55**, 981–986 (1990).
190. P.D. Gaspillo and S. Goto, *J. Chem. Eng. Jpn.* **24**, 680–682 (1991).
191. Y. Chisti, M. Kasper, and M. Moo-Young, *Can. J. Chem. Eng.* **68**, 45 (1990).
192. S.Q. Zhou, L.M. Tang, and K. Schugerl, *J. Biotechnol.* **28**, 165–177 (1993).
193. Y. Bando, M. Nishimura, H. Sota, S. Suzuki, and N. Kawase, *Chem. Eng. Sci.* **47**, 3371–3378 (1992).
194. Y. Bando, M. Nishimura, A. Hayaashi, S. Hiura, S. Indo, and A. Idota, *J. Chem. Eng. Jpn.* **28**, 225–227 (1995).
195. T.K. Ghosh, B.R. Maiti, and B.C. Bhattacharyya, *Biotechnol. Tech.* **7**, 301–307 (1993).
196. C. Schlotelburg, M. Gluz, M. Popovic, and J.C. Merchuk, *Int. Symp. on Bubble Columns*, Kyoto, Nov. 1997.

See also BIOREACTORS, CONTINUOUS STIRRED-TANK REACTORS; BIOREACTORS, FLUIDIZED-BED; BIOREMEDIATION; ENZYMES, IMMOBILIZATION METHODS; FERMENTATION MONITORING, DESIGN AND OPTIMIZATION; MASS TRANSFER; RHEOLOGY OF FILAMENTOUS MICROORGANISMS, SUBMERGED CULTURE; SCALE-UP, STIRRED TANK REACTORS.

BIOREACTORS, CONTINUOUS STIRRED-TANK REACTORS

MICHAEL E. LASKA
CHARLES L. COONEY
Massachusetts Institute of Technology
Cambridge, Massachusetts

KEY WORDS

Chemostat
Continuous culture
Continuous stirred-tank reactor (CSTR)

OUTLINE

INTRODUCTION

Definitions

A continuous stirred-tank reactor (CSTR) is defined as an agitated vessel with continuous addition and removal of material and energy. The CSTR is one of the basic continuous reactor types widely used in the chemical process industries because of its amenability to process control and scale-up, although in biotechnology applications the CSTR is used more often as a research tool than as a production technology. An idealized, well-mixed CSTR can be modeled as having no spatial variations in temperature, concentration, fluid properties, or reaction rates. Thus, the properties of the exit stream may be considered the same as those throughout the vessel. Although such ideal mixing is never observed, the vessel is designed to provide good mixing through selection of operating conditions and vessel, baffle, and impeller geometries. The stirred tank used in continuous bioprocesses is similar to that used in batch bioprocesses, with the exception that the CSTR likely has an overflow or other level control device. Oxygen can be introduced into the vessel by sparging through inlets at the base of the vessel, where impellers then disperse the bubbles. Vessel jacketing or internal cooling coils provide a means for heat transfer. Continuous systems that are not agitated vessels often are modeled as CSTRs when their behavior approximates that of the ideal CSTR. CSTRs are also known as backmix reactors, continuous-flow stirred-tank reactors (CFSTRs), or chemostats, when used for cell growth.

Strategy for CSTR Analysis

The performance and analysis of a CSTR is based on the material and energy conservation balances and the underlying processes governing the reaction kinetics. Because of the large variety of CSTR applications and limited space for discussion in this article, a brief outline of the steps in systems analysis will be beneficial in understanding any CSTR-based process.

The strategy for analyzing CSTR performance first requires defining the problem statement and goals. The second step is system identification, which includes defining the system boundaries and the interactions between the system and its environment across the system boundaries. The system could be a cell, the fluid in the reactor, or an entire bioprocessing plant. The third step is to identify the state variables that characterize the system. During the course of the analysis, new state variables may be identified and added to the original list. The fourth step is to characterize the state of the system using material and energy balances that account for the accumulation of mass and energy. In a general balance for a particular quantity, the rate of accumulation of that quantity in the system is equal to the net influx of the quantity across the system boundaries plus the rate at which the quantity is generated. Separate material balances are written for the reactants, products, and the catalyst (e.g., cells or enzymes). The final step is to calculate performance metrics and revisit the assumptions to determine the conditions under which they are valid.

REACTION KINETICS

This section discusses the theory governing ideal CSTR performance in two important bioprocess applications (cell growth and enzyme reactions), the underlying assumptions of the theory, and some practical aspects associated with CSTR operation.

Cell Growth

Introduction. Cell growth occurs in response to the environment. It is useful to classify growth in four mechanistic categories and identify those features relevant to the analysis of CSTRs. These classes are fission, budding, mycelial, and viral growth. The modes of growth serve to highlight the impact of morphology on bioprocessing considerations.

Bacteria grow by a process of binary fission yielding two identical daughter cells with doubling times typically between 0.5 and 3.0 h. The high specific growth rates of bacteria make them especially suitable for many CSTR applications; however, high O_2 requirements and metabolic heat generation become important considerations in bioprocess scale-up. Animal cells (5 to 20 μm) are much larger than bacteria ($\sim$1 μm) and yeast (5 to 10 μm) cells but grow more slowly, with typical doubling times between 18 and 48 h. Because of their larger size and lack of a protective cell wall, mammalian cells are particularly sensitive to the fluid shear in the vessel as well as to the osmotic pressure of the medium. Mammalian cells grow over a narrow range of osmotic/pressures and pH and typically have more complex nutritional requirements than bacteria and yeasts do.

Cell division in yeast occurs primarily by budding, with typical doubling times between 1 and 3 h. The budding process leads to a mother and a daughter cell, each having different growth rates and cell surface characteristics. Unlike bacterial cultures, yeast cell populations have a broad and time-varying distribution of ages and properties that may influence the formation of a desired product.

Mycelial growth occurs in molds, actinomycetes, and some yeasts by a process of hypha chain elongation and branching. Mycelial cultures also are characterized by a distribution of ages, with younger cells located at the hyphal tips. The hyphae form intertwined cellular strands, or mycelia, that increase the broth viscosity and lead to nonideal fluid mixing. High broth viscosity can be problematic for process monitoring and control, cell separation and recycle, and oxygen and heat transfer. Furthermore, fluid shear in the vessel can cause hyphal breakage and the formation of denser, more highly branched pellets or flocs. Cells in these pellets may be exposed to different microenvironments because of mass transfer limitations, which can vary with culture conditions and influence important cell properties.

Viral growth initially requires the infection of a host cell, which occurs by attachment of the virus to the cell surface and injection of viral nucleic acids into the cell interior. New viruses are constructed from biological molecules synthesized by the host cell under the direction of the viral genome. Viral nucleic acid is replicated many times (e.g., $>$500) and encapsulated in coat proteins to form a large number of new viral particles. Viral growth can proceed to one of two phases. In the lytic cycle, the host cell will lyse or break open and release infectious viral particles, whereas in the lysogenic cycle, the viral DNA will be integrated into the host cell DNA and the host cell will continue to reproduce normally.

CSTRs used for cell growth are commonly referred to as chemostats or turbidostats, depending on the strategy used to control the vessel environment. The most common arrangement is the chemostat (1–3), in which the medium fed to the vessel is designed so that all but a single nutrient essential for growth are present in excess of the cells' requirements. Any nutrient necessary for growth can be used to control the size of the cell population in the vessel, making the chemostat a flexible tool to study cellular behavior under different nutrient limitations. In a turbidostat, the cell concentration in the vessel is maintained constant by monitoring the optical density of the culture and modulating the medium feed rate to achieve a set point optical density. When the optical density rises above the set point, the feed rate is increased and, because the fluid volume is maintained constant by an overflow device, the well-mixed culture is diluted and the optical density approaches the set point value. The turbidostat is less commonly used because of difficulties in continuously monitoring the cell concentration. Its main utility is to control the growth rate near the maximum growth rate, an operating region in which the chemostat is less stable.

Material Balances. Cell mass is most often used to quantify microbial growth and usually is proportional to cell number under conditions of balanced growth, in which cellular chemical and physical properties are preserved in subsequent generations. Material balances based on cell number may have particular utility for some applications, such as mammalian cell culture, where number rather than mass is the conventional method of analysis. One may need to be wary of variations in cell size and morphology that may not be apparent from measurements of cell mass or number. The material balance for a uniform cell population in a CSTR can be written as shown in equation 1, in which $\mu[\text{h}^{-1}]$ represents the specific growth rate and $\alpha[\text{h}^{-1}]$ represents the specific rate of cell lysis and/or endogenous metabolism (i.e., resulting in a decrease in cell mass). The specific growth and death rates differ among organisms and are functions of the cell environment (e.g., pH, temperature, nutrients).

$$\underbrace{F \cdot X_0}_{\text{CELLS IN}} - \underbrace{F \cdot X}_{\text{CELLS OUT}} + \underbrace{V \cdot \mu \cdot X}_{\text{CELL GROWTH}} - \underbrace{V \cdot \alpha X}_{\text{CELL LYSIS}} = \underbrace{\frac{d(V \cdot X)}{dt}}_{\text{ACCUMULATION}} \qquad (1)$$

It may be necessary to reformulate the cell balance if the cell population is significantly differentiated; examples include mixed cultures, mammalian cell culture, and recombinant fermentations with plasmid instability (see "Selection/Mutation and Contamination"). If we assume the system is operating at steady state and there is no accumulation of fluid or cells in the vessel, then the time derivative can be set to zero. Under normal bioprocessing conditions, cell death is assumed to be negligible (i.e., $\mu \gg \alpha$). Bacterial and yeast cells maintain approximately complete viability except in suboptimal environments and at very low dilution rates, whereas in mammalian cell culture, viability and cell lysis may vary significantly with time.

Assuming that the feed to the reactor is sterile ($X_0 = O$), the material balance reduces to equation 2, in which the specific growth rate is equal to the liquid flow rate from the vessel divided by the liquid volume. This quantity is called the dilution rate D (the inverse residence time for the CSTR) and has units of h^{-1}.

$$\mu = \frac{F}{V} = D \qquad (2)$$

Equation 2 illustrates one of the most important attributes of the chemostat: that the specific growth rate can be controlled by manipulating the dilution rate. Control of the specific growth rate, combined with the ability to maintain a constant, defined cell environment, makes the chemostat a powerful experimental tool with which to investigate the many factors that influence cell growth, metabolism, and product formation. It is as an investigative tool, rather than as a production technology, that continuous culture has found its widest and most successful application.

A second material balance can be written for the growth-limiting substrate (S) using an *allocation model* for substrate utilization, in which substrate uptake is divided into cell growth, cell maintenance, and product formation components.

$$\underbrace{F \cdot S_0}_{\text{SUBSTRATE IN}} - \underbrace{F \cdot S}_{\text{SUBSTRATE OUT}} - V \cdot \left(\underbrace{\frac{\mu \cdot X}{Y_{X/S}}}_{\text{GROWTH}} + \underbrace{m \cdot X}_{\text{MAINTENANCE}} + \underbrace{\frac{q_P \cdot X}{Y_{P/S}}}_{\text{PRODUCT FORMATION}} \right) = \underbrace{\frac{d(V \cdot S)}{dt}}_{\text{ACCUMULATION}} \qquad (3)$$

In this equation, $Y_{X/S}$ is the cell yield (or dry cell weight, DCW) on substrate (g DCW/g substrate), m is the cell maintenance coefficient (g substrate/g DCW $\times$ h), q_P is the specific product formation rate (g product/g DCW $\times$ h), and $Y_{P/S}$ is the mass yield of product on substrate (g product/g substrate). Each specific yield coefficient describes the allocation of substrate to cells, product, or maintenance. Depending on whether the fermentation goal is to produce cells (biomass) or a metabolic product, it may be possible to simplify the substrate balance by assuming that some uptake terms are dominant. The following sections describe the analysis of biomass production and product formation.

Biomass Production. In the case of biomass production, in which the goal of the fermentation is to produce cells, the large majority of nutrient uptake goes toward cell growth. The rate of substrate uptake for growth is assumed to be much greater than that for maintenance (i.e., $\mu/Y_{X/S} \gg m$) and product formation (i.e., $\mu/Y_{X/S} \gg q_P/Y_{P/S}$). Assuming the system is at steady state, the substrate balance (equation 3) can be rewritten and solved for the cell concentration using equation 2.

$$X = Y_{X/S} \cdot (S_0 - S) \qquad (4)$$

To determine the relationship between the specific growth rate and the cell environment, a suitable growth model must be adopted. The simplest and most common relationship used is the unstructured Monod growth model, in which cell growth is a function of a single limiting substrate, usually the carbon source. Alternative unstructured growth models are given in Table 1. These unstructured growth models are empirically derived from observations of chemostat behavior, and their applicability to dynamic-batch or fed-batch processes should not be assumed. Some unstructured models (e.g., substrate inhibition) can have more than one solution, making different steady states possible depending on the starting conditions. Structured models, which typically account for either changes in cell composition, intracellular concentra-

Table 1. Common Unstructured Growth Models

Monod

$$\mu = \frac{\mu_{max} \cdot S}{K_S + S}$$

Modified monod

$$\mu = \frac{\mu_{max} \cdot S^{\lambda}}{K_S + S^{\lambda}}$$

where λ is an adjustable parameter

Inhibition models	Noncompetitive	Competitive
Substrate inhibition	$\mu = \frac{\mu_{max} \cdot S}{(K_S + S) \cdot \left(1 + \frac{S}{K_I}\right)}$	$\mu = \frac{\mu_{max} \cdot S}{K_S + S + \left(\frac{K_S}{K_I}\right) \cdot S}$
Product inhibition	$\mu = \frac{\mu_{max} \cdot S}{(K_S + S) \cdot \left(1 + \frac{P}{K_P}\right)}$	$\mu = \frac{\mu_{max} \cdot S}{K_S + S + \left(\frac{K_S}{K_P}\right) \cdot P}$

tions (4), or cell morphology (5), have been proposed in varying degrees of complexity (6,7). Structured models can have utility when such properties significantly influence the kinetics and are required to accurately describe behavior (such as in dynamic process modeling).

$$\mu = \frac{\mu_{max} \cdot S}{K_S + S} = D \tag{5}$$

In the Monod model, μ_{max} is the maximum specific growth rate of the organism, and K_S, called the saturation constant, is inversely proportional to the cell's affinity for the substrate. The value of K_S is typically quite low (1 to 5 mg/L for *Escherichia coli* on glucose), which means that $\mu \approx \mu_{max}$ when $S > 10K_S$ and that μ only becomes a strong function of the substrate concentration when $S < 10K_S$.

Note that the maximum specific growth rate of the organism limits the extent to which the dilution rate can be increased.

$$\mu_{max} = D_C \tag{6}$$

This threshold dilution rate is called the critical dilution rate (D_C), and increasing the dilution rate beyond D_C results in "washout"; cells are removed from the vessel at a rate faster than their growth rate. This can limit the productivity of the simple CSTR and motivates cell retention or recycle strategies that enable operation at higher throughputs.

Rearranging the growth model to solve for the substrate concentration in the vessel gives

$$S = \frac{D \cdot K_S}{\mu_{max} - D} \tag{7}$$

At dilution rates above D_C, cells have been washed out of the vessel, and the substrate concentration equals the inlet concentration (S_O). By substituting equation 7 into equation 4, the cell concentration becomes an explicit function of D and S_O.

$$X = Y_{X/S} \cdot \left(S_0 - \frac{D \cdot K_S}{\mu_{max} - D}\right) \tag{8}$$

The biomass productivity of the CSTR (R_{CSTR}), defined as the cell output per reactor volume, is calculated as

$$R_{CSTR} = \frac{F \cdot X}{V} = (D \cdot X) = D \cdot Y_{X/S} \cdot \left(S_0 - \frac{D \cdot K_S}{\mu_{max} - D}\right) \tag{9}$$

Figure 1 shows the steady-state cell and substrate concentrations and biomass productivity for the ideal chemostat. Note that the substrate is almost completely utilized

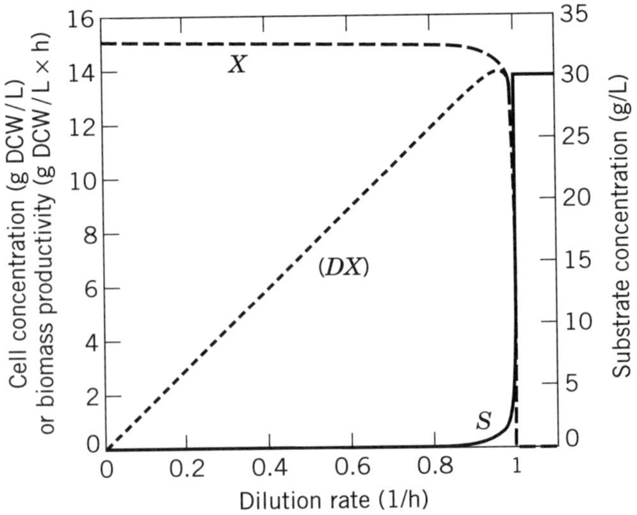

Figure 1. Ideal chemostat performance: $\mu_{max} = 1.0\ h^{-1}$, $K_S = 0.05$ g/L, $S_0 = 30$ g/L, $Y_{X/S} = 0.5$ g DCW/g substrate.

over most of the operating dilution rates. This high conversion of substrate is a key attribute of the ideal CSTR system, improving the economics of processes that depend on efficient substrate utilization and minimizing the effects of substrate inhibition. Also shown in Figure 1 is the high concentration of cells in the vessel until washout at the critical dilution rate. The maximum biomass productivity is located close to the washout point, making operation at the maximum productivity point very sensitive to deviations in dilution rate.

The dilution rate associated with maximum biomass productivity for a fixed limiting nutrient concentration in the feed (D_M) can be calculated by setting the derivative of (DX) with respect to D to zero and solving for the dilution rate. D_M is then given as

$$D_M = \mu_{max}\left(1 - \sqrt{\frac{K_S}{K_S + S_0}}\right) \tag{10}$$

By substituting D_M into equation 8, we can solve for X_M, the cell concentration corresponding to D_M.

$$X_M = Y_{X/S}\left[S_0 + K_S - \sqrt{K_S(S_0 + K_S)}\right] \tag{11}$$

Considering the limiting nutrient concentration (S_0) as an independent design variable, equation 11 suggests that the substrate concentration in the feed can be increased arbitrarily to achieve extraordinary cell densities and productivities. In reality, the productivity of an aerobic reactor system is ultimately limited by the rates of heat and/or mass transfer when the reaction kinetics are fast (i.e., high X and D) (see "Mass Transfer"). Oxygen transfer, and not the carbon substrate, is often growth limiting because oxygen is an essential nutrient for aerobic metabolism. It is poorly soluble in the medium (typically around 7 mg/L for air at 1 atm), and its transfer rate is restricted by the physical capabilities of the oxygenation system. Oxygen-limited growth may be expressed as a steady-state balance between the oxygen uptake rate (OUR) and the oxygen transfer rate (OTR). Oxygen transfer is usually limited by transfer from the gas to liquid phases, leading to the steady-state balance:

$$\left(\underbrace{\frac{\mu \cdot X}{Y_{X/O_2}} + m_{O_2} \cdot X}_{\text{OUR}}\right) = \underbrace{k_L a(C^* - C_L)}_{\text{OTR}} \tag{12}$$

in which Y_{X/O_2} is the cell yield on oxygen (g DCW/g O_2), m_{O_2} is the maintenance coefficient for oxygen (g O_2/g DCW × h), k_L is the liquid phase mass transfer coefficient (cm/h) a is the specific interfacial area for mass transfer (cm^2/cm^3), k_La is the mass transfer coefficient (h^{-1}), and ($C^* - C_L$) is the driving force for mass transfer where C^* is the equilibrium oxygen concentration (mmol/L) and C_L is the dissolved oxygen concentration (mmol/L). Typical values for Y_{X/O_2} are located in Table 2. The ability of the heat transfer system to remove heat generated during microbial growth can also limit R_{CSTR}, as discussed in "Energy Balance". These limitations will be important in evaluating and comparing CSTR performance.

Table 2. Yield Coefficients for Bacteria on Different Carbon Substrates

Substrate (i)	$Y_{X/S}$ [g DCW/g substrate]	Y_{X/O_2} [g DCW/g O_2]	Y_{HI} [g DCW/kcal]
Acetate	0.36	0.70	0.21
Glucose	0.51	1.5	0.42
Methanol	0.40	0.44	0.12
Ethanol	0.68	0.61	0.18
n-Paraffins	1.0	0.50	0.16
Methane	0.62	0.20	0.061

Source: From Ref. 16.

The biomass productivity of the CSTR can be compared to the productivity of a batch fermentor (R_{BATCH}) by defining a relevant batch productivity. The batch fermentation cycle consists of a lag phase, an exponential growth phase, cell harvest, and a batch turnaround time associated with cleaning, sterilizing, and filling the vessel. The lag, harvest, and turnaround activities can be grouped into a term $t_{TURNAROUND}$ in order to determine the batch cycle time (t_{CYCLE}) using equation 13, in which X_i is the concentration of cells in the vessel following inoculation (typically $X_i \sim 0.1 \times X$).

$$t_{CYCLE} = \frac{1}{\mu_{max}} \ln\left(\frac{X}{X_i}\right) + t_{TURNAROUND} \tag{13}$$

Cell growth can be calculated from the cell yield on the growth-limiting substrate and the initial concentration of substrate, again assuming cell maintenance and product formation are negligible.

$$X - X_i = Y_{Y/S}S_0 \tag{14}$$

Subsequently, the ratio of biomass productivities in the CSTR at D_M, $(R_{CSTR})_M$, and in the batch fermentor is given by equation 15, in which it is assumed that $S_0 \gg K_S$ (as it often is).

$$\frac{(R_{CSTR})_M}{R_{BATCH}} = \frac{(D_M \cdot X_M)}{\left(\dfrac{X - X_i}{t_{CYCLE}}\right)} = \ln\left(\frac{X}{X_i}\right) + \mu_{max} \cdot t_{TURNAROUD} \tag{15}$$

Equation 15 often appears as a measure of relative biomass productivity, in which the CSTR is favored over batch operation at high growth rates and long turnaround times. However, equations 10 and 11, associated with maximum productivity, do not reflect the ultimate limitation posed by heat and mass transfer in industrial aerobic processes. Given that productivity has a limit dictated by the system, the independent parameters, D and S_0, can be adjusted so that the maximum productivity is attained during operation. Recognizing that (R_{CSTR}) is ultimately limited by the maximum oxygen transfer rate, equation 12 can be rearranged to give the biomass productivity.

$$(D \cdot X) = \frac{D \cdot Y_{X/O_2} \cdot OTR_{MAX}}{(D + m_{O_2} \cdot Y_{X/O_2})} \tag{16}$$

Notice that the cell concentration is fixed once an operating dilution rate is specified. Solving equation 16 for X and substituting into equation 4 yields

$$S_0 = \frac{Y_{X/O_2} \cdot \mathrm{OTR_{MAX}}}{Y_{X/S} \cdot \delta_S \cdot (D + m_{O_2} \cdot Y_{X/O_2})} \quad (17)$$

where δ_S is the fractional substrate conversion. High substrate conversion leads to lower raw materials costs and reduces the burden of residual substrate on downstream purification and waste treatment operations. When operating near the maximum reactor productivity (i.e., maximum OTR), the dilution rate and the inlet substrate concentration can be adjusted independently to achieve target levels of cell concentration and/or substrate conversion, as shown in Figure 2.

For biomass production when O_2 transfer is not limiting productivity, the CSTR is favored over batch operation when the specific growth rate is high and the batch turnaround time is long. However, most industrial processes will be operating at or near the oxygen or heat transfer limitation. The desirability of enhanced O_2 transfer has motivated the development of novel bioreactor designs (e.g., bubble columns, loop or airlift reactors). For therapeutic products, however, the overwhelming majority of fermentations are batch or fed-batch processes. In these applications, the choice of operating mode is not based on biomass productivity. Performance metrics such as volumetric productivity of the product, product yield, and product concentration become more important in evaluating potential operating strategies than simply the biomass concentration.

Product Formation. When the goal of the fermentation is to produce a product other than biomass, the criteria used to evaluate alternative operational modes are less straightforward than biomass productivity. In order to evaluate process alternatives, a proper set of performance metrics must be identified that relate to overall process economics. When process economics are dominated by fermentation costs (e.g., fuel alcohol or gluconic acid production), volumetric productivity and conversion yields are especially relevant criteria as they relate to the size and cost of the reactor system and the cost of raw materials. In processes where recovery costs dominate (e.g., antibiotics), however, the size and operational costs of the recovery system are proportional to the fluid volume processed and inversely proportional to the product concentration (8). As a result, the final product concentration, or titer, is more important than biomass productivity.

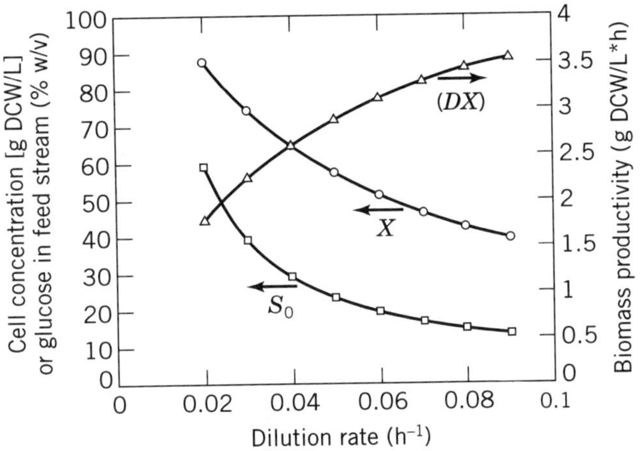

Figure 2. Chemostat operating at maximum oxygen transfer rate: OTR^{max} = 100 mmol O_2/L × h, μ_{max} = 0.09 h^{-1}, Y_{X/O_2} = 1.56 g DCW/g O_2, m_{O_2} = 0.024 g O_2/g DCW × h, $Y_{X/S}$ = 0.45 g DCW/g substrate, 95% substrate conversion.

The material balance on the product can be written as shown below, in which product formation is expressed using a specific product formation rate q_P (g product/g DCW × h), and product degradation by a specific rate constant k_P(h^{-1}).

$$\underbrace{q_P \cdot X}_{\text{PRODUCT FORMATION}} - \underbrace{k_P \cdot P \cdot D}_{\text{PRODUCT DEGRADATION}} - \underbrace{D \cdot P}_{\text{PRODUCT OUT}} = \underbrace{\frac{dP}{dt}}_{\text{PRODUCT ACCUMULATION}} \quad (18)$$

Product formation can be characterized in relation to growth, being growth (primary metabolites) or nongrowth (secondary metabolites) associated. Examples of growth-associated products are direct catabolic products of the carbon substrate, such as ethanol and citric or acetic acid. Nongrowth-associated products, comprising many antibiotics, are metabolites that are not necessary for cell growth and typically are only produced during slow or stationary growth phases. Some products, such as xanthan gum and lactic acid, are mixed growth associated in that they are produced during slow and stationary growth phases. Numerous models of product formation have been proposed, taking into account variables such as hyphal morphology, cell age, surface area, metabolic carbon flux, and plasmid copy number. A simple model expresses the growth dependence of q_P as (9)

$$q_P = \underbrace{\alpha \cdot \mu}_{\text{GROWTH ASSOCIATED}} + \underbrace{\beta}_{\text{NONGROWTH ASSOCIATED}} \quad (19)$$

The steady-state product balance can be rewritten and solved for P.

$$P = \frac{q_P \cdot X}{D} = \frac{(\alpha \cdot \mu + \beta)}{D} \cdot X = \left(\alpha + \frac{\beta}{D}\right) \cdot X \quad (20)$$

For growth-associated products (i.e., $\alpha \gg \beta$) product concentration is proportional to biomass and is independent of the dilution rate when X is approximately constant. Increasing the dilution rate results in increased product formation up to the region near D_C. The growth dependence of product formation and intracellular metabolic fluxes can be determined using a chemostat (10). An example of growth-associated product formation is shown in Figure 3 (11), in which product concentration is proportional to biomass and the growth dependence of q_P is evident.

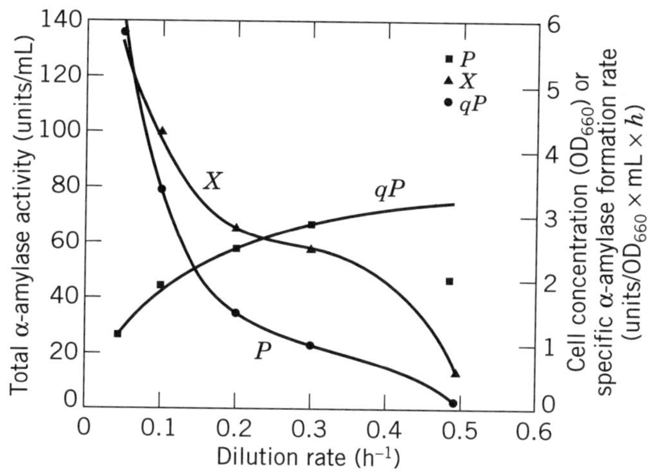

Figure 3. Production of α-amylase in a chemostat by recombinant *Escherichia coli.* The authors (11) used a modified Luedeking–Piret model, $q_P = (\alpha\mu + \beta)(1 + k\mu)^{-1}$, to describe α-amylase kinetics. The term $(1 + k\mu)^{-1}$ accounts for an observed increase in plasmid copy number with decreasing growth rate. Model parameters were regressed from data: α = 34.12 units/mL × OD_{660}, $\beta = 4.2 \times 10^{-10}$ U/mL × OD_{660} × h, and k = 8.63 h. Experimental data are depicted as points, model predictions as solid lines, and trend lines as dotted lines.

In the case of secondary metabolites (i.e., $\alpha \ll \beta$), product concentration is inversely proportional to the dilution rate, and the productivity ($D \times P$) is independent of dilution rate. The low dilution rates favorable for secondary metabolite production approach batch operation, which is generally favored over the CSTR in such instances.

Other issues can impact the decision between batch and continuous culture. The ability of the CSTR to maintain an ideal environment for product formation may offer a competitive advantage over the batch fermentation, with its time-varying environment and prolonged lag and stationary growth phases. Regulatory and market factors also play an important role in deciding on the operating mode. The CSTR is a dedicated manufacturing system used to produce a single product. Such a system may not be well suited for the production of specialty chemicals and pharmaceuticals because it can neither adapt to variable market demand nor satisfy demand for multiple products.

Energy Balance. Heat transfer is an important consideration in fermentor design, scale-up, operation, and sterilization. The energy balance is used to determine the time–temperature profile of the fermentation broth by accounting for the transfer and accumulation of energy. The heat transfer rate limits the ability to reduce the cycle time for sterilization (12). More importantly, the rate of heat removal from the broth during cell growth can constrain volumetric productivity, an issue in very large reactors with reduced area-to-volume ratios. Because considerable heat generation accompanies rapid cell growth, the high specific growth rates favored in industrial CSTR applications will exacerbate the problem of heat transfer in large reactors. The steady-state energy balance for the fluid in the CSTR is written as

$$Q_{AGIT} + Q_{MET} + \Delta Q_{SENS} - Q_{LOSS} - Q_{EVAP} - Q_{EXCH} = 0 \quad (21)$$

in which Q_{AGIT} is the mechanical energy imparted to the fluid through impeller agitation (equal to the gassed power input), Q_{MET} is the metabolic heat generated by cell growth, ΔQ_{SENS} is the net sensible heat added to the system by streams entering and leaving the system, Q_{LOSS} is the sum of the heat losses from the system to the surroundings, Q_{EVAP} is the latent heat removed by evaporation, and Q_{EXCH} is the heat removed from the system by an appropriate heat exchanger system (13). In some cases, the heats of solution and mixing must be accounted for, but in most cases they are negligible. The terms ΔQ_{SENS}, Q_{LOSS}, and Q_{EVAP} are comparatively small, leading to the simplified energy balance

$$Q_{EXCH} = Q_{MET} + Q_{AGIT} \quad (22)$$

For fast-growing microorganisms, the heat exchanger duty can be as high as 7.7 to 23.2 kW/m^3, of which Q_{MET} and Q_{AGIT} typically represent about 75% and 25% of the total, respectively (14).

Approximately 40 to 50% of the energy contained by a substrate is converted into useful chemical energy, whereas the balance is released as heat. If this metabolic heat is not removed from the fermentation broth, the temperature will rise and possibly hinder performance. Metabolic heat generation is a function of the growth rate of the organism, the cell concentration, the fluid volume, and the efficiency of cell growth on a particular substrate (i), which can be expressed as the metabolic heat released per gram of cell produced ($1/Y_{Hi}$) (kcal/g DCW).

$$Q_{MET} = V\frac{\mu \cdot X}{Y_{Hi}} \quad (23)$$

In aerobic fermentations, oxygen is the final electron acceptor in substrate metabolism, enabling a correlation between the rates of oxygen uptake and heat generation. The following empirical correlation (15) gives Q_{MET} (kcal/h) as a function of specific oxygen consumption, q_{O_2} (mmol O_2/L × h), and V(L):

$$Q_{MET} = 0.12 \cdot V \cdot q_{O_2} \quad (24)$$

where the oxygen demand during exponential growth can be expressed using the cellular yield on oxygen, Y_{X/O_2} (g DCW/g O_2).

$$Q_{O_2} = \frac{\mu \cdot X}{Y_{X/O_2}} \quad (25)$$

Table 2 gives values of Y_{Hi} and Y_{X/O_2} for bacterial growth that can be used to estimate oxygen demand and metabolic heat generation. Notice that the more reduced substrates result in greater O_2 demand and heat generation and subsequently a larger burden on the O_2 transfer and heat exchange system.

Heat exchange systems are chosen based on the expected heat exchanger duty, influence on fluid mixing, impact on cleaning and sterilization, utility economy, and maintenance, operating, and capital costs. Heat exchangers for fermentors commonly consist of a jacket or shell around the vessel, internal coolant coils, or occasionally an external heat exchanger. The rate of heat removal by a heat exchanger system can be described using a Fourier's law expression:

$$Q_{\mathrm{EXCH}} = U \cdot A \cdot (T - Tc) \tag{26}$$

in which A is the surface area available for heat transfer, U is an overall heat transfer coefficient accounting for all heat transfer resistances, and $(T - Tc)$ is the driving force for heat transfer, where T is the bulk temperature of the fermentation broth and T_C is the temperature of the cooling fluid used in the heat exchanger. Typical values of the heat transfer coefficient are 50 to 150 BTU/ft^2 × h × °F (280 to 850 W/m^2 × K). An important consequence of fermentor scale-up is a decreasing surface-area-to-volume ratio (A/V): at increasing scale the capacity for heat removal relative to heat generation diminishes and often becomes limiting at larger volumes.

The rate of heat removal can be increased by increasing the temperature driving force, increasing the heat transfer surface area, or reducing resistance to heat transfer (increasing U). Water is primarily used as the coolant fluid because of its availability and low cost relative to a refrigerated coolant system. The temperature of the cooling water increases as it passes through the heat exchanger, so an arithmetic or logarithmic mean Tc may give a more representative measure of the coolant temperature. Increasing the coolant flow rate can decrease the mean coolant temperature but with a subsequent increase in utility consumption. Furthermore, the pressure drop of the exchanger and pump capacity limit the extent to which coolant flow rate can be increased. The heat exchanger area is dictated by the size and type of heat exchange system chosen during process scale-up. The overall heat transfer coefficient and constituent resistances are thoroughly discussed elsewhere. In general, poor mixing contributes to decreased heat transfer (see "Nonideal Mixing"). It is sufficient to focus on the convective heat transfer coefficient on the fermentation broth side, h[W/m^2 * K], which is often the dominant heat transfer resistance. The convective heat transfer coefficient is a function of the Reynolds number (Re), Prandtl number (Pr), viscosity ratio (Vi), and the system geometry (F_{GEOMETRY}). In many cases $Vi \approx 1$, and the exponents a and b have typical values of 0.8 and 0.3, respectively.

$$Nu = C_1 \cdot Re^{\mathrm{a}} \cdot Pr^{\mathrm{b}} \cdot Vi^{\mathrm{c}} \cdot F_{\mathrm{GEOMETRY}}$$

$$\frac{h \cdot D_{\mathrm{T}}}{k} = C_1 \cdot \left(\frac{D_{\mathrm{i}}^2 N\rho}{\mu}\right)^{\mathrm{a}} \left(\frac{Cp\mu}{k}\right)^{\mathrm{b}} \left(\frac{\mu}{\mu_{\mathrm{w}}}\right)^{\mathrm{c}} F_{\mathrm{GEOMETRY}} \tag{27}$$

Economic Considerations. Operating costs are those costs associated with maintaining a given production level dictated by the scale and scheduling of the process, typically grouped as raw materials, direct expenses, and indirect expenses. Although the focus here will be on raw materials and direct expenses (e.g., utilities, maintenance, operating supplies, and labor) associated with the CSTR, it is important to view the reactor system in the context of the entire process. The process flow diagram will form the basis for determining not only the reactor operating costs but also the capital and operating costs of air compressors, pumps, holding tanks, and various downstream units. It is important to remember that the relative costs associated with the reactor system, batch or continuous, may not have a significant impact on the total process economics if product recovery costs are dominant (see "Product Formation"). Raw materials costs include the material and handling costs of components added to the system to satisfy metabolic (e.g., C, N, O sources) or process (e.g., acids or bases and antifoam agents) requirements. Direct expenses include the cost of utilities, maintenance, operating supplies, operating labor, direct supervision, laboratory charges, and patent royalties. Continuous systems are easier to automate and offer the potential of lower labor costs than batch production systems, with their labor intensive start-up and shut-down operations. Indirect expenses include taxes and depreciation, usually expressed as a percentage of the plant cost.

Media. The selection of bioprocess media typically involves a trade-off between media cost and the product yield and titer. It is an important stage of development that can influence the design and performance of the entire process. Selecting an apparently cheap media, for example, may result in more expensive downstream recovery and waste treatment operations. In general, media selection includes a number of technical and economic considerations: yield or titer of desired and undesired products; cost; variability in composition and price; availability; effect on downstream processes; need for pretreatment or supplements; shipment, storage, and handling, and need for testing and validation.

Media can be classified as defined or undefined with respect to chemical composition. Laboratory-scale chemostat investigations commonly use defined media to allow precise control over the growth-limiting nutrient and medium composition. Small bioreactors (1 to 4 L) are preferred for continuous operation in the laboratory because media preparation and storage are less of a burden. Industrial microbial fermentations predominantly use undefined media because they generally are less costly and perform better than defined media. Undefined microbial media often contain agricultural by-products (e.g., molasses or corn steep liquor) and thus are subject to source market fluctuations in quality and price. Processes that have been validated with a variety of media may change the production media to take advantage of market changes. The costs of raw materials as a percentage of operating costs for primary metabolites can range from 40% for citric acid to 70% for ethanol from sugar cane (17), whereas for secondary metabolites media costs can be around 10 to 20%. At small scale (less than 10 m^3), mammalian cells are often grown in undefined serum, an expensive and sometimes scarce media derived from mammalian plasma. Viral contamination and the presence of serum proteins can complicate cultivation, product recovery, and quality control and quality assurance (QA–QC) in industrial processes. These complications, combined with regulatory pressure to safeguard against viral contamination, are a driving force toward the development of defined, serum-free media in order to avoid use of animal-derived media components.

Utility Expenses. Steam, cooling water, and power requirements comprise the majority of utility expenses. Process demand for water-for-injection (WFI) must also be determined for clean-in-place (CIP) systems. Steam usage occurs predominantly during media and equipment sterilization and can be calculated using knowledge of the sterilization cycle. Continuous reactors readily lend themselves to the use of continuous, as opposed to batch, media sterilization systems, offering advantages in reduced thermal degradation of heat-sensitive media, reduced sterilization time, more efficient fermentor use, and greater steam economy (about 20 to 25% of the steam used by batch sterilization) (12). The cooling water requirement, w_{H_2O} (kg/h), can be calculated knowing the heat exchanger duty from the energy balance (equation 21):

$$w_{H_2O} = \frac{Q_{EXCH}}{C_p \cdot (T_{out} - T_{in})} \tag{28}$$

where C_p is the heat capacity of water [$\sim$1 kcal/kg $*$ K] and T_{out} and T_{in} are the outlet and inlet cooling water temperatures, respectively. The availability of abundant low-temperature water can reduce cooling water requirements and possibly allow the use of larger reactors because of the improved ability to remove metabolic heat.

Oxygen demand and mixing requirements drive power consumption by agitators and compressors, whereas larger volumes, higher cell densities, higher specific O_2 uptake rates, and higher broth viscosities result in increased power requirements. Total power consumption for agitated vessels is typically in the range of 2 to 10 kW/m^3. Centrifugal pumps are predominantly used in bioprocesses for their relatively low cost and ability to handle suspended solids. Although pumps represent only a small fraction of power consumption, they can be a significant percentage of overall maintenance costs. In addition, the performance of continuous processes is particularly susceptible to pump failure and may warrant additional capital investment toward the installation of backup pumps in parallel.

Actual vs. Ideal Behavior. In this section, the assumptions used in the development of the ideal CSTR theory are revisited in order to determine when they are invalid and to gauge the impact of nonidealities on performance.

Nonideal Mixing. The major assumption of the ideal CSTR is that there are no spatial variations of properties inside the vessel. Such ideal mixing is never observed in actual systems and even deteriorates on scale-up, although it can be a good approximation of behavior. The challenge is to determine when nonidealities can be expected and how they will influence performance.

Equipment design, operating conditions, and broth properties all influence the quality of fluid mixing in the vessel. Agitated tanks are designed to provide good mixing through selection of tank geometry, baffle placement, and impeller design, although mixing quality decreases with increasing scale. Agitator power input and gas sparging rates, although typically associated with oxygen transfer, are critical to mixing. High broth viscosities, typical of mycelial fermentations, contribute to poor mixing. Poor mixing can affect oxygen transfer (18), product formation (19), heat transfer, process monitoring and control, and the distribution of components added to the system (20,21).

Nonideal mixing in continuous systems is typically characterized by either the residence time distribution, the distribution of fluid residence times around the ideal residence time (τ) (22), or the mixing time (t_M), the time required for a system to respond to a feed disturbance. The ideal mixing time is equal to zero (instantaneous), and the mean value of the residence time distribution is τ (or D^{-1}). Residence time distributions are typically measured by pulse or step addition of a tracer into the reactor feed and tracking the appearance of the tracer in the exit stream as a function of time. Correlations obtained from dimensional analysis can be used to predict mixing times, which increase with scale and broth viscosity (23). Small vessels (<500 L) are generally well mixed ($t_M \sim$ s), but large fermentors (>5,000 L) typically have poor mixing ($t_M \sim$ min). Compartmental mixing models, in which the bulk fluid is modeled as discrete CSTRs and plug flow reactors (PFRs) with fluid interchange, have been used to describe heterogeneity in large vessels (24).

Wall growth is a special case of nonideal mixing in which cells adhere and proliferate on vessel surfaces (25), where they are hidden from cell mass measurements obtained from samples of the bulk fluid. The system, in addition to being heterogeneous, is no longer at steady state because cells are accumulating in the vessel. This can pose serious problems if the accumulating organism is a contaminant or otherwise undesirable organism. The metabolism and growth of wall-bound cells can be quite different from the suspended population because of mass transfer limitations. Wall growth can reduce heat transfer, create sterilization and cleaning problems, and corrupt measurements in experimental systems. It can be a significant factor when the surface-area-to-volume ratio (A/V) is high (e.g., laboratory-scale systems and vessels with internal cooling coils) and may require modifications to chemostat design or operation (26). For example, the glass walls of laboratory vessels are sometimes treated with organosilane compounds to minimize wall growth. Yeast and mycelial cells with a propensity to form pellets are prone to wall growth under the same conditions that favor flocculation. A discrete washout point does not exist with wall growth, because cells are effectively immobilized in the vessel even past D_C.

Substrate Assumptions. Deviations from ideal chemostat behavior may arise when assumptions regarding the magnitudes of nutrient uptake, the consistency of cell composition, or the identity of the growth-limiting nutrient become invalidated. Substrate uptake for growth in equation 3 was assumed to be much larger than that for maintenance and product formation. At low dilution rates (i.e., low growth rates) the maintenance term becomes significant ($m \sim \mu/Y_{X/S}$) and less substrate goes toward cell growth, causing the actual X to be less than predicted by the ideal theory at low D. Similarly, at high dilution rates, the production and accumulation of growth-related products and intermediates can become significant, leading to a reduced cell yield at higher D.

The ideal chemostat derivation assumes that cell composition does not change over the operating region. In actuality, cell composition varies with pH, temperature, growth rate (27), and medium composition. As cell composition changes, the demand for essential nutrients will change in ways that were not accounted for in the derived equations. Proper evaluation of experimental data from a chemostat may require consideration of the variation of nutrient uptake and cell composition with environment and growth rate.

When using complex and undefined media or an organism with complex nutritional requirements (e.g., mammalian cells), it is often difficult to identify the growth-limiting nutrient. In addition, the limiting nutrient itself may change because the nutrient demand, and subsequently the media composition, may change with operating conditions. The cell concentration profile in these situations likely would be constantly decreasing with dilution rate, unlike the ideal chemostat where X is approximately constant over the majority of $0 < D < D_C$. More complicated, structured growth models would be required to account for such behavior.

Non-Steady-State Behavior. The potential sources of process variability and non-steady-state behavior are perhaps too numerous to mention. However, typical instances include chemostat start-up, execution of control actions, induced disturbances (e.g., pulse and shift methods [28]), variations in feed composition, culture degeneration (e.g., plasmid loss or apoptosis), wall growth, or equipment failures. Steady state often is declared when the measurable process states are maintained constant for 3 to 5 residence times. Sustained oscillations are sometimes observed in continuous culture and often are the result of growth inhibition resulting from either an accumulated product (29) or the burden of product formation (30).

Selection/Mutation and Contamination. By controlling the culture conditions in the CSTR, a highly selective environment for the selection and proliferation of certain microorganisms can be created. Cells in this selective environment with growth rates less than the dilution rate will be washed out of the reactor, leaving only those cells with the properties that have been selected for. In this way, continuous culture can be used as a strain improvement tool to select organisms that possess a desirable trait, such as yeast with higher ethanol tolerance (31).

Because of the metabolic burden imposed by high levels of product formation, the production strain has a growth disadvantage relative to unproductive strains that are present in the reactor. Without selection pressure in favor of the production strain, a gradual decline in productivity will be observed over time as nonproductive cells dominate the culture. Examples include the reversion of specially selected antibiotic strains to low productivity mutants or the domination of recombinant protein processes by plasmid-free cells. The configuration of CSTRs in series can be used to circumvent this problem by providing separate environments for growth and product formation (see "CSTRs in Series"). Selective pressure in favor of the production strain (and against contaminants) may be exerted through the application of elevated temperatures, extremes of pH, the use of narrowly defined or modified media, and the use of specially selected cultures (e.g., antibiotic-resistant strains).

The prolonged operating periods of continuous culture increase the probability of contamination by a foreign organism. The threat posed by contamination depends on the ability of the undesirable microorganism to complete and thrive in the CSTR environment. Consider the case of two types of microorganisms with concentrations X (the desired strain) and Z (the contaminant) competing for the same limiting substrate in a CSTR. The material balances on cell mass, neglecting cell lysis, can be written as

$$\frac{dX}{dt} = \mu \cdot X - D \cdot X \tag{29}$$

$$\frac{dZ}{dt} = \mu_z \cdot Z - D \cdot Z \tag{30}$$

Subtracting equation 30 from equation 29 and rearranging leads to equation 31:

$$\frac{d \ln[X/Z]}{dt} = \mu - \mu_z \tag{31}$$

which shows that the growth rates and their dependence on the limiting substrate will determine the fate of the culture (32). The contaminant (Z) could be washed out ($\mu > \mu_z$), remain at a stable level ($\mu = \mu_z$), or dominate ($\mu < \mu_z$) the culture. This simple analysis of selection can be complicated if the contaminant has properties that prevent it from being washed out (e.g., adhesion to reactor surfaces), if the contaminant competes for a different substrate than the production strain, or if there are interactions from inhibitory cellular products. For example, lactobacilli are often a persistent contaminant of continuous ethanol production processes because of their intimate association with flocculant yeast aggregates and ability to adapt to high ethanol concentrations (33). Selective recycle of desirable organisms back to the vessel has been used to prevent domination of the culture by undesirable strains (34–36).

Enzymes

Introduction. Enzymes are biological catalysts with high selectivity toward reactants and products, making them attractive for use in a number of industrial applications. Enzyme activity is strongly influenced by the environment (e.g., pH, temperature, metal ions). Loss of activity or denaturation can be reversible or irreversible, depending on the type, strength, and duration of an unfavorable interaction. A benefit of using the CSTR for enzyme reactions is that the constant, controlled reactor environment can be designed for maximum enzyme activity and life.

Material Balances. Assuming that the inlet and outlet flow rates are approximately equal (i.e., solutions are dilute), the steady-state material balance on the substrate can be written as

$$F \cdot (S_0 - S) = v \cdot V \tag{32}$$

where F is the volumetric flow rate, v is the rate of substrate consumption by reaction, V is the fluid volume in the reactor, and S_0 and S are the substrate concentrations in the feed and vessel, respectively. Rewriting equation 32 in terms of the fluid residence time (τ) and the fractional substrate conversion (δ_S) yields the CSTR design equation

$$\tau = \frac{V}{F} = \frac{S_0 \cdot \delta_S}{v} \tag{33}$$

Using a valid rate expression for v, the design equation can be used to determine the reactor volume required to yield a given conversion rate ($S_0 \times \delta_S \times F$). Fast reaction kinetics are obviously favorable because reactor cost scales with reactor size. Enzyme loading in the reactor can be increased beyond the solubility limit by immobilization on inert support particles, which increases v, reduces the necessary reactor volume, facilitates enzyme retention and recycle, and may improve enzyme stability.

Enzyme Reaction Kinetics. Numerous mechanistic models have been developed to describe enzyme reaction rates as a function of enzyme and substrate concentrations. Some of the more common models appear in Table 3 accompanied by the corresponding solution to the design equation (equation 33).

Unlike a plug flow reactor (PFR), in which the substrate enters at a high concentration and leaves at a lower concentration, the substrate concentration in a CSTR is at a uniform, low concentration. The reduced substrate concentration leads to a slower reaction rate, so that the CSTR requires more of the active enzyme than the PFR to attain the same substrate conversion rate. Substrate inhibition is less problematic in a CSTR than in a PFR because of the lower substrate concentration in the bulk fluid, whereas product inhibition is generally more of a problem in CSTRs than PFRs. Arranging CSTRs in series can reduce the effects of product inhibition (approaching PFR behavior) while taking advantage of the good mixing characteristics of the CSTR to provide optimal pH control (37).

Temperature Effects. Like many chemical reactions, increasing the temperature enhances the rate of enzyme reactions. Higher temperatures also result in increased rates of thermal denaturation and loss of the active biocatalyst. Process economics often depend on optimal temperature control to maintain high substrate conversion and long catalyst life (38–40). The effects of temperature on the catalytic rate constant (k_2) can be described using an Arrhenius expression

$$k_2 = A \cdot e^{-\mathrm{Ea/RT}} \tag{34}$$

where A is the Arrhenius constant, E_a is the activation energy, R is the gas constant, and T is the absolute temperature. The activation energy of enzyme-catalyzed reactions ranges from 4 to 20 kcal/mol, with most reactions near 11 kcal/mol.

Thermal denaturation usually can be described as a first-order decay reaction:

$$\frac{dE}{dt} = -k_d \cdot E \text{ or } E = E_0 e^{-k_d t} \tag{35}$$

where k_d is the thermal denaturation constant, which also follows an Arrhenius temperature dependence. For thermal denaturation, E_a varies from 40 to 130 kcal/mol, with most in the vicinity of 70 kcal/mol. Increasing temperature has a greater effect on the rate of denaturation than catalysis. For a typical enzyme (i.e., 11 and 70 kcal/mol), an increase in temperature from 30 to 40 °C results in a 1.8-fold increase in the rate of catalysis, but a 41-fold increase in the denaturation rate.

Energy Balance. An optimal temperature control strategy requires good heat removal because most industrial enzyme reactions are exothermic. Although heat transfer is generally good for soluble enzymes in agitated tanks, the high enzyme concentrations attained with immobilization can result in fast reaction rates and appreciable heat generation. Heat transfer resistance within the catalyst pellet can reduce heat removal rates, resulting in higher pellet temperatures and shorter catalyst life. Catalyst degradation from insufficient heat removal is more of a concern in packed beds with high enzyme loading, where heat transfer resistances can be significant. The steady-state energy

Table 3. Some Common Enzyme Kinetic Expression

	Rate expression $v =$	Design equation $\tau =$
Michaelis–Menten	$\dfrac{v_{max} \cdot S}{K_m + S}$	$\dfrac{1}{v_{max}}\left[S_0 \cdot \delta_S + K_m \cdot \dfrac{\delta_S}{1 - \delta_S}\right]$
Substrate inhibition	$\dfrac{v_{max} \cdot S}{S + K'_m + \dfrac{S^2}{K'_S}}$	$\dfrac{1}{v_{max}}\left[S_0 \cdot \delta_S + K'_m \cdot \dfrac{\delta_S}{1 - \delta_S} + \dfrac{S_0^2}{K'_S} \cdot (\delta_S - \delta_S^2)\right]$
Competitive product inhibition	$\dfrac{v_{max} \cdot S}{K'_m\left(1 + \dfrac{P}{K_P}\right) + S}$	$\dfrac{1}{v_{max}}\left[S_0 \cdot \delta_S + K'_m \cdot \dfrac{\delta_S}{1 - \delta_S} + \dfrac{K'_m}{K_P} \cdot \left(\dfrac{S_0 \cdot \delta_S^2}{1 - \delta_S}\right)\right]$
	Where $v_{max} = k_2 \times E_0$	Where the product concentration (P) is related to converted substrate; $P = \delta_S \times S_0$

balance for an exothermic ($\Delta H_{RXN} < 0$) enzyme reaction in a well-mixed CSTR is

$$F\rho c_P(T_F - T) + v' \cdot E \cdot V(-\Delta H_{RXN}) - Q_{EXCH} = 0 \quad (36)$$

where F is the volumetric flowrate (m^3/h), ρ is the fluid density (kg/m^3), c_P is the fluid heat capacity (kJ/kg × K), v' is the specific reaction rate (kmol substrate/kg enzyme × h), E is the enzyme concentration (kg enzyme/m^3), ΔH_{RXN} is the heat of reaction (kJ/kmol substrate), Q_{EXCH} is the heat removed by the heat exchanger (kJ/h) from equation 26, and T_F and T are the feed and bulk fluid temperatures (K), respectively.

Economic Considerations. For enzyme CSTRs, the primary operating costs are associated with enzyme replacement. Prolonged catalyst activity and marked reductions in raw materials costs can be achieved by maintaining an optimal environment for the enzyme during operation. Preserving enzyme activity reduces the number and frequency of labor-intensive cleaning and changeovers, facilitating downstream operations by consistently providing a constant-quality product stream. With multiple reactors installed in parallel, changeovers can be scheduled to minimize production variations and downtime. Industrial enzymes are often sold as a crude mixture containing only a fraction of active enzyme. Selecting enzymes among different vendors may involve a trade-off between cost and purity (percent active enzyme) and a consideration of how the impurities may affect the process.

MASS TRANSFER

Introduction

The rate of mass transfer ultimately will limit the maximum aerobic reactor performance. Oxygen transfer to the fermentor broth, for example, can limit both the extent and rate of cell growth. Mass transfer limitations to microbial flocculants and immobilized catalyst pellets can result in reduced reaction rates and inefficient conversion. Liquid–liquid mass transfer rates from hydrocarbon substrates to suspended cells may limit productivity in two-phase systems (41). To determine the rate-controlling regime, it is useful to characterize the relative rates of mass transfer and reaction using the dimensionless Damkohler number (Da):

$$Da = \frac{\text{Maximum rate of reaction}}{\text{Maximum rate of diffusion}} \quad (37)$$

The observed reaction rate may be limited by the rate of diffusion depending on the value of the Damkohler number: if $Da \gg 1$ the diffusion rate is limiting, if $Da \ll 1$ the reaction rate is limiting, and if $Da \sim 1$ then the reaction and diffusion rates are comparable. As with all dimensionless numbers, the Damkohler number is only meaningful if it is calculated using the proper time and length scales for a given system. Consider spherical pellets ($r = 400\,\mu$m) of *Penicillium chrysogenum*, assuming a pellet cell density of 0.1 g/cm^3, an effective oxygen diffusivity ($D_{O_2}^{\text{eff}}$) of 1×10^{-6} cm^2/s, a particle size of 400 μm, an oxygen concentration (c_{O_2}) of 7 mg/L, and the following cell parameters (42): $\mu = 0.075\ h^{-1}$, $Y_{X/O_2} = 1.56$ g DCW/g O_2, and $m_{O_2} = 0.024$ g O_2/(g DCW × h). The Damkholer number shows that internal mass resistance is considerable and that cell growth in the pellet is likely to be limited by oxygen transfer.

$$Da = \frac{\overbrace{\left(\frac{\mu}{Y_{X/O_2}} + m_{O_2}\right)}^{\text{O}_2\text{ DEMAND}} \cdot \overbrace{X}^{\text{PELLET CELL DENSITY}} \cdot \overbrace{\left(\frac{4\pi r^3}{3}\right)}^{\text{PELLET VOLUME}} \cdot}{\underbrace{\left(D_{O_2} \cdot \frac{c_{O_2}}{r}\right)}_{\text{DIFFUSIVE FLUX}} \cdot \underbrace{(4 \cdot \pi \cdot r^2)}_{\text{PELLET SURFACE AREA}}} =$$

$$\frac{\left(\frac{0.075\ h^{-1}}{1.56\ \text{g DCW/g O}_2} + \frac{0.024\ \text{g O}_2}{\text{g DCW} \cdot \text{h}}\right) \cdot \frac{0.1\ \text{g DCW}}{\text{cm}^3} \cdot (4 \times 10^{-2}\ \text{cm})^2}{3 \cdot (1 \times 10^{-6}\ \text{cm}^2/\text{s}) \cdot (7 \times 10^{-6}\ \text{g/cm}^3)} \cong 5.5 \times 10^5 \quad (38)$$

Gas–liquid mass transfer is often rate-limiting for gases that are sparingly soluble in the broth, such as oxygen and methane. Although highly soluble, carbon dioxide exhibits pH-dependent partitioning between gaseous and dissolved forms (CO_2, H_2CO_3, HCO_3^-, CO_3^{2-}) that is influenced by the rates of both reaction and mass transfer. Proper interpretation of the respiratory coefficient (RQ) in fermentations operated at neutral pH requires consideration of CO_2 dynamics. The OTR has already been used to determine the productivity limit of a CSTR used for biomass production in equation 12. In general, the rate of mass transfer from the gas to the liquid phase is given as

$$N_A = k_L a(C^* - C_L) \quad (39)$$

where N_A is the rate of gas transfer (mmol/l × h) and the remaining terms have the same definitions as in equation 12. For sparged, agitated tanks, $k_L a$ has typical values in the range from 50 to 1,400 h^{-1}. Several correlations have been developed for $k_L a$ as a function of the gassed power input per unit volume and the superficial gas velocity for Newtonian broths in a variety of fermentors (43). The correlations can offer wide variability in mass transfer estimates and should be used in conjunction with knowledge from past experience or empirical measurements of $k_L a$ (e.g., dynamic or sulfite oxidation methods). Oxygen transfer to shear-sensitive mammalian cells requires gentle agitation combined with surface or membrane aeration, or light sparging, as opposed to the large power inputs and high rates of gas sparging in microbial fermentations. This limitation is somewhat offset by the fact that mammalian cells have lower O_2 requirements (0.05 to 0.5 mmol/10^9 cells × h) (44) and grow to lower cell densities (10^6 to 10^7 cells/mL) than microbial cultures.

VARIATIONS ON THE SINGLE CSTR

Single CSTR with Recycle

Volumetric productivity is related to the concentration of active catalyst. Cell or enzyme concentrations greater than the steady state obtained from the simple CSTR can be achieved by separating cells from the effluent stream and recycling them to the vessel (27,45) or by retaining them within reactor. Higher catalyst concentrations enhance substrate conversion and reduce the reactor size necessary to attain a given conversion. Recycle operation improves system stability in the face of feed disturbances by retaining cells in the vessel even under conditions that would cause washout in the simple CSTR. Recycle systems can be operated at dilution rates, or throughputs, greater than the specific growth rate of the organism. Productivity improvements achieved with cell recycle are demonstrated in Table 4 for *Saccharomyces cerevisiae* ATCC 4126 and *Zymomonas mobilis* ATCC 10988 at 100 g/L glucose feed.

Cell Recycle Methods. Cell recycle is implemented through a cell separation step, often by a unit operation commonly used in the initial stages of downstream processing. Typical methods of continuous cell separation include centrifugation, filtration, and sedimentation. Cell separation can be viewed as having two often equally important purposes: (*1*) the recovery or retention of cells for reuse and (*2*) the removal of potentially inhibitory byproducts or products from the culture environment. The separation step often has to satisfy additional performance requirements such as handling of shear- or temperature-sensitive materials, selectivity in rejection or recovery, containment, maintenance of asepsis, corrosion resistance, brief retention time, and ease of cleaning, sterilization, maintenance, and validation. Recycle operation is standard for reactors using stable enzymes, because discarding expensive active catalyst is economically unfeasible. Cell separation operations are discussed elsewhere in the context of downstream processing, although a brief description is presented here in relation to cell recycle.

Sedimentation. Sedimentation is the settling of particles in a gravitational field. With low energy requirements and simple equipment, sedimentation is a relatively inexpensive way of separating a dilute cell phase. Waste treatment is by far the largest application of sedimentation-based cell recycle, in which cells are typically separated in large sedimentation tanks using lime or clay to enhance flocculant formation. The settling velocity (u_0) for an isolated spherical particle can be described using Stoke's law:

$$u_0 = \frac{d_P^2(\rho_P - \rho_F)g}{18\eta} \quad (40)$$

in which d_P is the particle diameter, ρ_P is the particle density (the specific gravity of a typical cell is 1.05), ρ_F is the fluid density, g is the gravitational acceleration constant, and η is the fluid viscosity. The functional dependence of Stoke's law suggests ways of increasing the settling velocity. The easiest and most common method is to increase the effective cell size by promoting flocculation (cell aggregation) through physiological, chemical, and physical factors: selection of flocculant strains; modification of cell wall structure or surface charge; changing the pH, temperature, or shear stress; addition of inorganic salts (e.g., Ca^{2+} and Mg^{2+} salts) or clays; controlling the concentration of certain nutrients or products (e.g., extracellular polysaccharides); and controlling the cell age or growth phase. Selective cell recycle has been implemented using the differential sedimentation properties of a desired and unwanted microorganism (34–36). The properties of a particular broth are generally unchangeable and will probably only impede particle settling.

Although equation 40 holds for dilute suspensions of cells, the interactions among settling particles in concentrated slurries results in hindered settling. The hindered particle velocity (u_h) is influenced by the particle concentration and can be expressed with the following correlation (49):

$$\frac{u_h}{u_0} = \frac{1}{1 + \lambda \cdot \epsilon_P^{1/3}} \quad (41)$$

in which ϵ_P is the volume fraction of particles and λ is an empirical function of ϵ_P. For dilute suspensions, $\epsilon_P < 0.15$, whereas in slurries $0.15 < \epsilon_P < 0.50$.

The limiting settling velocity for a system has a strong influence on equipment design and operation. Consider the case of a continuous sedimentation tank with volumetric throughput (F) and constant cross-sectional area (A). The sedimentation tank performance can then be described by equation 42, in which throughput is directly proportional to A and independent of tank depth:

$$F = u_{lim} \cdot A \quad (42)$$

The throughput and limiting settling time will thus dictate equipment size and costs. Another design consideration is the residence time of cells in the settling tank, which must be considered in the context of nutrient depletion (particularly for oxygen) and its potential effects on performance.

Sedimentation at laboratory scale may be implemented with an external settling column (47,50,51). Similar devices may be used at bioprocessing scales, whereas large open-air tanks must be used in high-volume wastewater treatment. Internal sedimentation has been implemented

Table 4. Ethanol Productivity Enhancements for *S. cerevisiae* and *Z. mobilis*

		Dilution rate (h^{-1})	Ethanol productivity g/(L × h)	Cell density (g DCW/L)	Reference
S. cerevisiae	CSTR	0.17	7	12	46
	CSTR with recycle	0.68	29	50	47
Z. mobilis	CSTR	0.175	8	2.5	29
	CSTR with recycle	2.7	120	38	48

in tower fermenters, in which immobilized cells and enzymes or microbial flocculants are retained in the vessel by a sedimentation zone within the vessel. Unlike the ideal CSTR, tower fermenters may exhibit spatial variations in nutrient concentrations and broth properties along the height of the tower that can significantly influence reactor performance. In addition, the productivity in these reactors may be limited by the need to maintain low upward velocities (e.g., low aeration or CO_2 evolution) to allow adequate cell sedimentation.

Centrifugation. The operating principle behind centrifugation is the same as that of sedimentation; however, much higher settling velocities than in sedimentation may be obtained in the centrifugal field. Centrifugal separators enable high-volume continuous processing of fluids containing many particles, with short retention times and small space requirements. To determine the unhindered particle velocity in a centrifugal field (u_{0C}), equation 40 is multiplied by the centrifugal coefficient (C), also known as the G-value, which describes the increase in sedimentation rate due to centrifugation relative to gravitational settling:

$$u_{0C} = \frac{d_{\mathrm{P}}^2 g(\rho_{\mathrm{P}} - \rho_{\mathrm{F}})}{18\eta} \underbrace{\left(\frac{r\omega^2}{g}\right)}_{C} \tag{43}$$

where r is the radial distance from the axis of rotation and ω is the angular velocity.

Industrial centrifuges are most often classified by internal structure (e.g., disk stack, tubular bowl) and mode of operation (e.g., solids retaining, continuous or intermittent solids ejecting). The selection of sturdier construction and materials will enable higher rotation speeds for separation of smaller particles. The equation describing throughput in a centrifuge is analogous to equation 42, except that the centrifuge area is expressed using the Σ value, which is the area equivalent for a given centrifuge and rotation speed. Centrifuge manufacturers will often provide machine-specific Σ values, although the Σ value for simple disk-stack and tubular-bowl centrifuges can be calculated directly.

Filtration. Filtration is separation based on size, allowing retention of molecules larger than the pore size of the filter and passage of smaller molecules. Membrane filtration thus offers the twin benefits of cell retention and inhibitory by-product removal. In cell recycle systems, the most common arrangements are internal filters for cell and enzyme retention (52,53) or external membrane filters (54,55) in plate and frame, spiral cartridge, and hollow fiber configurations. In all these configurations, flow patterns tangential to the membrane surface can reduce fouling and improve the filtrate flux across the membrane. Compared to internal filters, external filters have higher surface-area-to-volume ratios and may be easier to maintain; however, they may be less easily sterilized (particularly for some polymer membranes) and could introduce problems of nutrient depletion in the external recycle loop. Membrane selection depends primarily on the critical particle size, with other criteria being cost, mechanical stability, and susceptibility to plugging and fouling. Because membranes have the potential for complete cell recycle, a purge or bleed stream is typically split from the recycle stream to prevent accumulation of inert particles and debris in the vessel.

Material Balances. A schematic of a CSTR with recycle of cells is shown in Figure 4. A material balance on cell mass for the CSTR with recycle system, neglecting cell death, may be written

$$\underbrace{F \cdot X_0}_{\text{CELLS IN FEED}} + \underbrace{\alpha \cdot F \cdot C \cdot X_1}_{\text{CELLS IN RECYCLE STREAM}} - \underbrace{(1+\alpha) \cdot F \cdot X_1}_{\text{CELLS OUT}} + \underbrace{V \cdot \mu \cdot X_1}_{\text{CELL GROWTH}} = \underbrace{\frac{d(V \cdot X_1)}{dt}}_{\text{CELL ACCUMULATION}} \tag{44}$$

where α is the recycle ratio equal to the recycle volume divided by the feed volume, C is the concentration factor (cell concentration in the recycle divided by the effluent concentration) related to the efficiency of the separation step, and X_0, X_1, and X_2 are the cell concentrations in the feed, recycle, and separator effluent streams, respectively. Note that the low substrate concentrations in waste treatment create suboptimal growth environments in which cell death cannot be neglected.

Assuming the system is at steady state ($dX_1/dt = 0$) and that the feed is sterile ($X_0 = 0$), equation 44 yields

$$\mu = (1 + \alpha - \alpha \cdot C) \cdot D \tag{45}$$

The dilution rate is no longer equal to the specific growth rate; in fact, because $C > 1$ and $\alpha < 1$, the dilution rate is greater than the specific growth rate.

A material balance on the limiting substrate, again neglecting maintenance and product formation, may be written

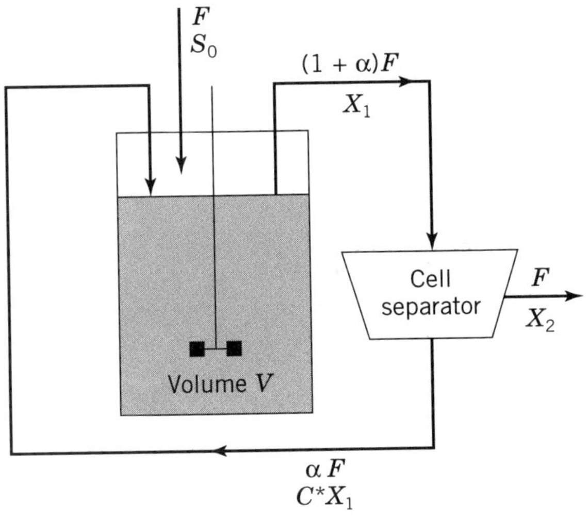

Figure 4. CSTR with recycle. Cell separation can be achieved through centrifugation, sedimentation, or filtration.

$$\underbrace{F \cdot S_0}_{\text{SUBSTRATE IN FEED}} + \underbrace{\alpha \cdot F \cdot S}_{\text{SUBSTRATE IN RECYCLE}} - \underbrace{(1 + \alpha) \cdot F \cdot S}_{\text{SUBSTRATE OUT}} - \underbrace{\frac{\mu \cdot X}{Y_{X/S}} \cdot V}_{\text{SUBSTRATE CONSUMED}} = \underbrace{V \cdot \frac{dS}{dt}}_{\text{SUBSTRATE ACCUMULATION}} \quad (46)$$

Solving equation 46 for the cell concentration, assuming steady-state ($dS/dt = 0$) operation, yields

$$X_1 = \frac{Y_{X/S} \cdot (S_0 - S)}{[1 + \alpha(1 - C)]} \quad (47)$$

in which the steady-state cell concentration with recycle is greater than that in the simple CSTR by a factor of $1/[1 + \alpha(1 - C)]$. By adopting a suitable expression for cell growth, the substrate concentration can be determined. Using the Monod expression, as before, and solving for the substrate concentration gives

$$S = K_S \frac{\mu}{\mu_{max} - \mu} = K_S \frac{D(1 + \alpha - \alpha C)}{\mu_{max} - D(1 + \alpha - \alpha C)} \quad (48)$$

Substituting equation 48 into equation 47 yields

$$X_1 = \frac{Y_{X/S}}{(1 + \alpha - \alpha C)} \left[S_0 - K_S \frac{D(1 + \alpha - \alpha C)}{\mu_{max} - D(1 + \alpha - \alpha C)} \right] \quad (49)$$

A material balance on cell mass around the separator gives the cell concentration in the outlet:

$$X_2 = (1 + \alpha - \alpha \cdot C) \cdot X_1 \quad (50)$$

Figure 5 shows the cell mass and biomass productivity of a CSTR with recycle compared to a simple CSTR. The higher biomass productivity of the recycle system results from a dilution rate higher than the specific growth rate and the increased cell concentration in the vessel.

Implementation of a recycle system is often critical to the economic viability of processes using expensive biocatalysts (e.g., enzymes). Typically, this is accomplished by immobilizing the enzymes on inert support particles to facilitate either internal or external recycle. The potential use of cell recycle in an industrial process involves weighing the effectiveness and economics associated with the cell separation step against the marginal improvement in process performance. It should also be noted that higher cell densities exacerbate the oxygen transfer and heat removal burden of the system.

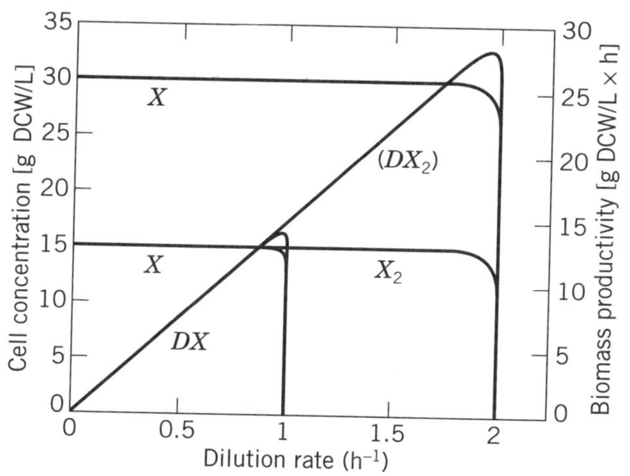

Figure 5. Comparison of steady-state behavior of a chemostat (*solid lines*) and a chemostat with recycle (*dotted lines*) using the following parameters: $Y_{X/S} = 0.5$ g DCW/g substrate, $\mu_{max} = 1.0$ h^{-1}, $K_S = 0.02$ g/L, $S_0 = 30$ g/L, $C = 2$, and $\alpha = 0.5$.

CSTRs in Series

In the single CSTR, the constant, controlled environment gives the advantage of being able to control the cellular–enzyme environment for maximum utility. Sometimes, however, a particular cell system will exhibit multiple properties of interest that can only be realized in different environments. The optimal environments for cell growth and product formation, for example, may be characterized by different temperatures, pH, and limiting nutrients. The configuration of CSTRs in series lends itself to those applications in which multiple environments are required.

Cell Growth. Bacterial growth in the presence of multiple carbon substrates often results in diauxic growth, in which cells preferentially metabolize a single substrate over all others. In a waste treatment application, the preferred substrate would be consumed by the microorganisms and the remaining substrates would pass through the system untreated. Configuring CSTRs in series provides a partitioning of cell metabolism so that less-favored substrates are consumed in subsequent stages.

CSTRs in series have been used to improve recombinant protein fermentations in which performance is threatened by plasmid instability (56,57) and lethal protein overproduction (30). Cells are grown to high density in the first stage without inducer so that plasmid-free cells have little growth advantage over plasmid-containing cells. Induction in the second stage results in higher productivity than the simple CSTR because the continuous introduction of plasmid-containing cells from the first stage reduces the ability of nonproductive cells to dominate the culture.

Consider the two-stage system for biomass production in Figure 6, in which a separate feedstream can be added to the second stage. The steady-state material balances for cells and substrate in the first reactor are identical to the single CSTR case (equations 1 and 3), with steady-state solutions as equations 5 and 8 for Monod growth. Considering the case without the second feedstream, the material balances on cell mass and growth-limiting substrate in the second stage can be written as

$$\underbrace{\frac{F \cdot X_1}{V_2}}_{\text{CELLS IN}} - \underbrace{\frac{F \cdot X_2}{V_2}}_{\text{CELLS OUT}} + \underbrace{\mu_2 \cdot X_2}_{\text{CELL GROWTH}} = \underbrace{\frac{dX_2}{dt}}_{\text{ACCUMULATION}} \quad (51)$$

and

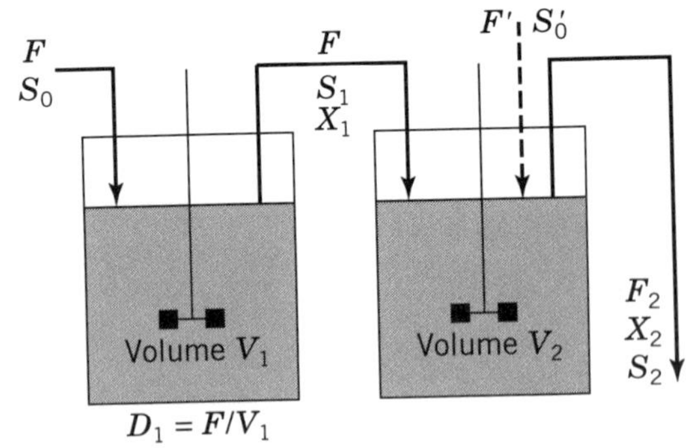

Figure 6. Two-stage chemostat system with possibility of a separate feed to second reactor.

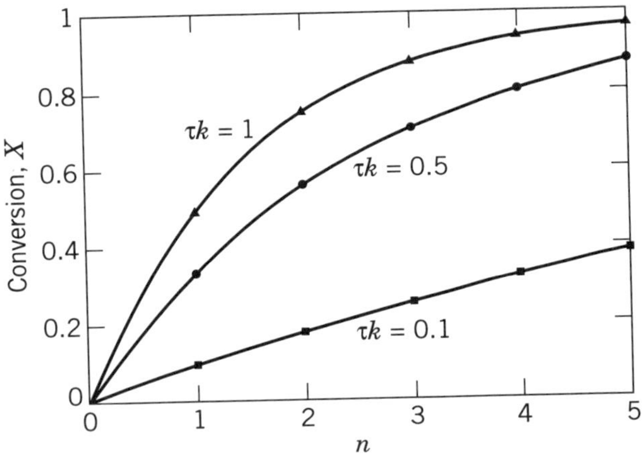

Figure 7. Substrate conversion for first order reaction in n CSTRs in series.

$$\underbrace{\frac{F \cdot S_1}{V_2}}_{\text{SUBSTRATE IN}} - \underbrace{\frac{F \cdot S_2}{V_2}}_{\text{SUBSTRATE OUT}} - \underbrace{\frac{\mu_2 \cdot X_2}{Y_{X/S}}}_{\text{CONSUMPTION FOR GROWTH}} = \underbrace{\frac{dS_2}{dt}}_{\text{ACCUMULATION}} \tag{52}$$

with steady-state solutions shown in Table 5. A growth model on the limiting substrate must be adopted to further complete the system description. Minimal cell growth will occur in the second stage if no additional substrate is added, because the majority of substrate is consumed in the first stage. Thus, non-growth-related cellular behavior and product formation could be studied in the second CSTR.

Adding an additional feedstream to the second stage (Fig. 6) provides the opportunity to introduce more of the limiting nutrient, other nutrients required for growth or product formation, inducers, or inhibitors. The material balances on cell mass and substrate on the second stage can be written as equations 53 and 54, respectively, with steady-state solutions given in Table 5. In equation 53, F' is the volumetric flow of the second feedstream, which is assumed to be sterile ($X' = 0$).

$$\underbrace{\frac{F_1}{V_2} X_1}_{\text{CELLS IN FROM STAGE 1}} - \underbrace{\frac{F_1 + F'}{V_2} X_2}_{\text{CELLS OUT}} + \underbrace{\mu_2 \cdot X_2}_{\text{CELL GROWTH}} = \underbrace{\frac{dX_2}{dt}}_{\text{ACCUMULATION}} \tag{53}$$

The dilution rate for the second stage is given by $D_2 = (F_1 + F')/V_2$, and the concentration of the limiting nutrient in the second feed is S_0'.

$$\underbrace{\frac{F_1}{V_2} S_1}_{\text{SUBSTRATE IN FROM STAGE 1}} + \underbrace{\frac{F'}{V_2} S_0'}_{\text{SUBSTRATE IN FROM SECOND FEED}} - \underbrace{\frac{F_1 + F'}{V_2} S_2}_{\text{SUBSTRATE OUT}} - \underbrace{\frac{\mu_2 \cdot X_2}{Y_{X/S}}}_{\text{CONSUMPTION FOR GROWTH}} = \underbrace{\frac{dS_2}{dt}}_{\text{ACCUMULATION}} \tag{54}$$

Feeding additional substrate to the second stage allows for more growth to occur. In addition, the dilution rate in the second stage is larger than the maximum specific growth rate of the organism because the second stage has a continuous feed of cells.

Enzyme Reaction. Enzyme reactions may also be carried out in multiple CSTRs. The performance of CSTRs in series approaches that of a single PFR while maintaining the good mixing characteristics of the stirred-tank reactor (37). Considering a first-order enzymatic reaction for substrate conversion ($v = k \times S$), the reactor design equation for a single CSTR can be written as

Table 5. Steady-State Solutions to Material Balances for a Two-Stage Chemostat

	Cell mass	Substrate
First stage	$\mu_1 = D_1$	$X_1 = Y_{X/S}(S_0 - S_1)$
Second stage	$\mu_2 = D_2\left(1 - \frac{X_1}{X_2}\right)$	$X_2 = \frac{D_2}{\mu_2} Y_{X/S}(S_1 - S_2)$
Second stage with additional feedstream	$\mu_2 = D_2 - \frac{F_1 \cdot X_1}{V_2 \cdot X_2}$	$X_2 = \frac{Y_{X/S}}{\mu_2}\left(\frac{F_1}{V_2} S_1 + \frac{F'}{V_2} S' - D_2 S_2\right)$
	where $D_2 = D_1 = F/V_1$	where $D_2 = (F_1 + F')/V_2$

$$\tau = \frac{S_0 - S}{k \cdot S} \quad (55)$$

Assuming that there is no volume change upon reaction, the conversion in the single CSTR is given by

$$X = \frac{\tau \cdot k}{1 + \tau \cdot k} \quad (56)$$

Then for a system of n CSTRs in series with equal volumes and reactor conditions (constant k) the conversion in the nth CSTR is given by equation 57 and depicted in Figure 7.

$$X_n = 1 - \frac{1}{(1 + \tau \cdot k)^n} \quad (57)$$

NOMENCLATURE

Abbreviations

CER	Carbon dioxide evolution rate
CIP	Clean-in-place
CSTR	Continuous stirred-tank reactor
OTR	Oxygen transfer rate
OUR	Oxygen uptake rate
PFR	Plug flow reactor
RQ	Respiratory quotient, RQ = CER/OUR
WFI	Water-for-injection

Symbols

ΔH_{RXN}	Heat of reaction, >0 endothermic, <0 exothermic, kJ/mol
ΔQ_{SENS}	Net sensible heat input, W
$(R_{CSTR})_M$	Biomass productivity at X_M and D_M, g DCW/m^3 × h
A	Area, m^2
C	Concentration factor in cell recycle system
C^*	Equilibrium dissolved oxygen concentration, g/m^3
C_L	Dissolved oxygen concentration, g/m^3
Cp	Heat capacity, J/kg × K
D	Dilution rate, h^{-1}
Da	Dimensionless Damkohler number
D_C	Critical dilution rate, h^{-1}
D_M	Dilution rate associated with maximum R_{CSTR} at fixed S_0, h^{-1}
d_P	Particle diameter, m
E	Enzyme concentration
E_0	Initial enzyme concentration
Ea	Activation energy, kJ/mol
F	Volumetric flow rate, m^3/h
g	Gravitational acceleration, 9.8 m/s^2
k_2	Catalytic rate constant, h^{-1}
k_d	Thermal denaturation constant, h^{-1}
k_La	Overall mass transfer coefficient, h^{-1}
k_P	Specific product degradation constant, h$_{-1}$
K_S, K_I, K_P	Model parameters for cell growth
K_m, K_S, K'_m, K'_S	Model parameters for enzyme kinetics
m	Maintenance coefficient g/g DCW × h
N_{discs}	Number of discs in a disk-stack centrifuge
OD_{660}	Optical density at 660 nm; measure of cell concentration
P	Product concentration, g/m^3
Q_{AGIT}	Agitation heat input, W
Q_{EVAP}	Heat loss by evaporation, W
Q_{EXCH}	Heat removal by heat exchanger, W
Q_{LOSS}	Heat loss to surroundings, W
Q_{MET}	Metabolic heat generation, W
q_{O_2}	Specific oxygen uptake rate, g O_2/g DCW * h
q_P	Specific product formation rate, g/g DCW * h
R	Ideal gas constant, J/mol * K
r_2, r_1	Outer and inner radii for centrifuge, m
R_{BATCH}	Biomass productivity for batch fermentation, g DCW/m^3 × h
R_{CSTR}	Biomass productivity for CSTR, g DCW/m^3 × h
S	Substrate concentration, g/m^3
S_0	Inlet substrate concentration, g/m^3
T	Bulk fluid temperature, K
T_C	Coolant temperature, K
t_{CYCLE}	Batch cycle time, h
t_M	Mixing time, h
$t_{TURNAROUND}$	Lumped batch turnaround time, h
U	Overall heat transfer coefficient, W/m^2 × K
u	Particle velocity, m/h
u_{lim}	Limiting particle velocity for a separator, m/h
V	Fluid volume, m^3
v	Reaction rate,
v_{max}	Maximum reaction rate
w_{H_20}	Mass flow rate of water, kg/h
X	Cell concentration, g DCW/m^3
X_0	Inlet cell concentration, g DCW/m^3
Xi	Inoculum cell concentration, g DCW/m^3
X_M	Cell concentration associated with D_M, g DCW/m^3
Y_{Hi}	Cell mass produced per heat evolved, g DCW/kcal
$Y_{P/S}$	Product yield on substrate, g/g
Y_{X/O_2}	Cell yield on oxygen, g DCW/g O_2

$Y_{X/S}$	Cell yield on substrate, g DCW/g
Z	Concentration of contaminant microorganism, g DCW/m^3

Greek Symbols

μ	Specific growth rate, h^{-1}
α	Specific rate of cell lysis or endogenous metabolism, h^{-1}
λ	Adjustable parameter
τ	Residence time, h
Σ	Σ Factor, area equivalent for a centrifuge
ω	Angular velocity
α	Recycle ratio
ρ	Density, kg/m^3
η	Viscosity, kg/m $\times$ s
λ	Empirical function of ϵ_P
α, β	Product formation parameters, g product/g DCW and g product/g DCW $\times$ h, respectively
μ_{max}	Maximum specific growth rate, h^{-1}
ϵ_P	Volume fraction of particles
δ_S	Fractional substrate conversion

BIBLIOGRAPHY

1. J. Monod, *Ann. Rev. Microbiol.* **3**, 371 (1949).
2. A. Novick and L. Szilard, *Proc. Nat. Acad. Sci. Wash.* **36**, 708 (1950).
3. D. Herbert, R. Elsworth, and R.C. Telling, *J. Gen. Microbiol.* **14**, 601–622 (1956).
4. M.L. Shuler and M.M. Domach, in H.W. Blanch, E.T. Papoutsakis, G. Stephanopoulos eds., *Foundations in Biochemical Engineering*, American Chemical Society, Washington, D.C., 1983, p. 93.
5. J. Nielsen, *Biotechnol. Bioeng.* **41**, 715–727 (1993).
6. A. Harder and J.A. Roels, *Adv. Biochem. Eng.* **21**, 55–107 (1982).
7. J. Nielsen and J. Villadsen, *Chem. Eng. Sci.* **47**, 4225–4270 (1992).
8. N.M. Fish and M.D. Lilly, *Biotechnology* **2**, 623–627 (1984).
9. R. Luedeking and E.L. Piret, *J. Biochem. Microbiol. Tech. Eng.* **1**, 431–459 (1959).
10. R.D. Kiss and G. Stephanopoulos, *Biotechnol. Bioeng.* **39**, 565–574 (1992).
11. P. Yu and K.-Y. San, *Biotechnol. Prog.* **9**, 587–593 (1993).
12. G.K. Raju and C.L. Cooney, in H.-J. Rehm, G. Reed, A. Puhler, P. Stadler eds., *Biotechnology*, 2nd ed., vol. 3, VCH, Weinheim, Germany, 1993, pp. 159–184.
13. M.F. Edwards and W.L. Wilkinson, *Chem. Eng.* **265**, 310 (1972).
14. M. Charles, *Trends Biotechnol.* **3**, 134–139 (1985).
15. C.L. Cooney, D.I.C. Wang, and R.I. Matales, *Biotechnol. Bioeng.* **11**, 269–281 (1968).
16. B.J. Abbott and A. Clamen, *Biotechnol. Bioeng.* **15**, 117–127 (1973).
17. B.L. Maiorella, H.W. Blanch, and C.R. Wilke, *Biotechnol. Bioeng.* **26**, 1003–1025 (1984).
18. A.P.J. Sweere, L. Mesters, L. Janse, K.Ch.A.M. Luyben, and N.W.F. Kossen, *Biotechnol. Bioeng.* **10**, 567–578 (1988).
19. F. Vardar and M. Lilly, *Eur. J. Appl. Microbiol. Biotechnol.* **14**, 203–211 (1982).
20. C.G. Sinclair and D.E. Brown, *Biotechnol. Bioeng.* **12**, 1001–1017 (1970).
21. P.J. Senior and J. Windass, *Biotechnol. Lett.* **2**, 205–210 (1980).
22. O. Levenspiel, *Chemical Reaction Engineering*, 2nd ed., Wiley, New York, 1972, pp. 253–325.
23. M. Charles, in T.K. Ghose, A. Fiechter, N. Blakebrough eds., *Advances in Biochemical Engineering*, vol. 8, Springer-Verlag, New York, 1978, pp. 1–62.
24. N.M.G. Oosterhuis and N.W.F. Kossen, *Biotechnol. Bioeng.* **26**, 546–550 (1984).
25. J.A. Howell, C.T. Chi, and U. Pawlowsky, *Biotechnol. Bioeng.* **14**, 253–265 (1972).
26. R.C. Righelato and S.J. Pirt, *J. Appl. Bacteriol.* **30**, 246–250 (1967).
27. D. Herbert, in *Continuous Culture of Microorganisms*, Monograph no. 12, Society of Chemistry and Industry, London, 1961, pp. 21–53.
28. H. Kuhn, U. Friederich, and A. Fiechter, *Eur. J. Appl. Microbiol.* **6**, 341–349 (1979).
29. K.J. Lee, D.E. Tribe, and P.L. Rogers, *Biotechnol. Lett.* **1**, 421–426 (1979).
30. J. Fu, D.B. Wilson, and M.L. Shuler, *Biotechnol. Bioeng.* **41**, 937–946 (1993).
31. S.W. Brown and S.G. Oliver, *Eur. J. Microbiol. Biotechnol.* **16**, 119–122 (1982).
32. D.E. Dykhuizen and D.L. Hartl, *Microbiol. Rev.* **47**, 150–168 (1983).
33. M. Nagashima, in H. Verachtert, and R.D. Mot eds., *Yeast: Biotechnology and Biocatalysis*, Dekker, New York, 1990, pp. 57–84.
34. R.H. Davis and C.S. Parnham, *Biotechnol. Bioeng.* **33**, 767–776 (1989).
35. K.L. Henry and R.H. Davis, *Biotechnol. Prog.* **6**, 7–12 (1990).
36. K.L. Ogden and R.H. Davis, *Biotechnol. Bioeng.* **37**, 325–333 (1991).
37. S.W. Carleysmith and M.D. Lilly, *Biotechnol. Bioeng.* **21**, 1057–1073 (1979).
38. L.Y. Ho and A.E. Humphrey, *Biotechnol. Bioeng.* **12**, 291–311 (1970).
39. S.H. Park, S.B. Lee, and D.Y. Ryu, *Biotechnol. Bioeng.* **23**, 1237–1254 (1981).
40. C. Kim, H.S. Kim, and D.D. Ryu, *Biotechnol. Bioeng.* **24**, 1889–1896 (1982).
41. M. Moo-Young, *Can. J. Chem. Eng.* **53**, 113–118 (1975).
42. R.C. Righelato, A.P.J. Trinci, S.J. Pirt, and A. Peat, *J. Gen. Microbiol.* **50**, 399–412 (1968).
43. M. Moo-Young and H.W. Blanch, *Adv. Biochem. Eng.* **19**, 1–69 (1981).
44. R.J. Fleischaker and A.J. Sinskey, *Eur. J. Appl. Microbiol. Biotechnol.* **12**, 193–197 (1981).
45. S.J. Pirt and W.M. Kurowski, *J. Gen. Microbiol.* **63**, 357–366 (1970).

46. G.R. Cysewski and C.R. Wilke, *Biotechnol. Bioeng.* **18**, 1297–1313 (1976).
47. G.R. Cysewski and C.R. Wilke, *Biotechnol. Bioeng.* **19**, 1125–1143 (1977).
48. P.L. Rogers, K.J. Lee, M.L. Skotnicki, and D.E. Tribe, *Adv. Biochem. Eng.* **23**, 37–84 (1982).
49. J.E. Bailey and D.F. Ollis, *Biochemical Engineering Fundamentals*, 2nd ed., McGraw-Hill, New York, 1986, pp. 733–734.
50. T.K. Ghose and R.D. Tyagi, *Biotechnol. Bioeng.* **21**, 1387–1400 (1979).
51. D. Gold, A. Mohagheghi, C.L. Cooney, and D.I.C. Wang, *Biotechnol. Bioeng.* **23**, 2105–2116 (1981).
52. A. Margaritis and C. Wilke, *Biotechnol. Bioeng.* **20**, 727–754 (1978).
53. H.N. Chang, W.G. Lee, and B.S. Kim, *Biotechnol. Bioeng.* **41**, 677–681 (1993).
54. Y.L. Lee and H.N. Chang, *Biotechnol. Bioeng.* **36**, 330–337 (1990).
55. H.N. Chang, I.-K. Yoo, and B.S. Kim, *Biotechnol. Adv.* **12**, 467–487 (1994).
56. R. Siegel and D.D.Y. Ryu, *Biotechnol. Bioeng.* **27**, 28–33 (1985).
57. S.B. Lee, D.D.Y. Ryu, R. Seigel, and S.H. Park, *Biotechnol. Bioeng.* **31**, 805–820 (1988).

See also BIOREACTORS, AIR-LIFT REACTORS; BIOREACTORS, FLUIDIZED-BED; FERMENTATION MONITORING, DESIGN AND OPTIMIZATION; MASS TRANSFER; SCALE-UP, STIRRED TANK REACTORS.

BIOREACTORS, FLUIDIZED-BED

FRANCESC GÒDIA
CARLES SOLÀ
Universitat Autònoma de Barcelona
Barcelona, Spain

OUTLINE

KEY WORDS

Bioreactors
Fluidized-bed
Hydrodynamics
Immobilized biocatalysts

INTRODUCTION

Fluidized-bed bioreactors are directly linked to the use of biocatalysts (cells or enzymes) for transformations in an immobilized form. The solid particles of the immobilized biocatalyst are maintained in fluidization by means of the circulation of a fluid phase (either liquid, gas, or a mixture of both) that compensates their weight. In this way, good liquid mixing and mass transfer between the solid and the liquid phases can be obtained with low attrition. Also, fluidized-bed bioreactors can accommodate a gas phase and can be used to feed solids in suspension. High productivities can be achieved in these systems, but their hydrodynamic complexity and operational stability have to be well defined for a proper operation.

THE FLUIDIZATION CONCEPT: GENERAL CONSIDERATIONS

The term *fluidized-bed* is used to define those physical systems composed of a solid phase in the form of individual particles that move within a fluid phase and are not in continuous contact with each other. Fluidization of the solid particles is reached when the flow of fluid through the bed is high enough to compensate their weight. On the other hand, in order to be kept in the fluidized-bed reactor and not be washed out (elutriated), the superficial velocity of the fluid in the bed (that is, the ratio between the flow rate and the bed cross-sectional area) has to be lower than the settling velocity of the particles. These two extreme situations are outlined in Figure 1. When the flow rate of a fluid through a packed bed of solid particles steadily increases, the pressure drop increases proportionally to the flow rate, as long as the bed height remains constant. When the drag force of the fluid equilibrates the weight of

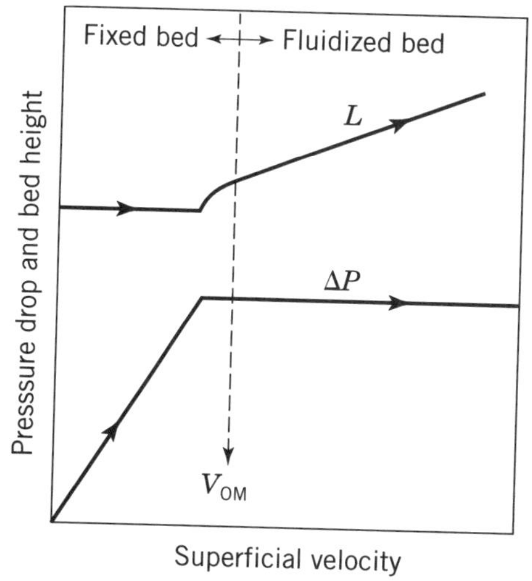

Figure 1. Illustration of the basic concept of fluidization. Variation of the pressure drop (ΔP) and the height of the solid particles in a bed (L), with increasing fluid superficial velocity. v_{OM}, minimum fluidization velocity.

the particles, the bed starts to expand and, after a transition period, reaches fully developed fluidization. At this point, further increments in the flow rate do not produce an increase in pressure drop, but instead lead to an increase of the height occupied by the solid particles in the reactor. If the flow rate is increased significantly, the elutriation of the solid particles occurs when the fluid's superficial velocity is higher than the solid's settling velocity. The fundamentals of the fluidization phenomena are discussed comprehensively in the chemical engineering literature (1,2).

Figure 2 represents the basic scheme of a fluidized-bed bioreactor. Although various configurations are possible (3), the most extensively used is the gas–liquid cocurrent up-flow reactor. In it, liquid usually comprises the continuous phase and is fed from the reactor bottom. Its flow upward in the reactor promotes fluidization of the solid particles. Usually, the reactor will have two or three phases. In addition to the liquid and solid phases, the occurrence of a gas phase is quite common in those systems using cells as biocatalysts, either for aeration requirements (in which case, an air or oxygen stream is fed to the reactor, as shown in Fig. 2) or because cell metabolism produces a gas product (for example, CO_2, CH_4). In systems using enzymes as biocatalysts, the most common situation is two-phase fluidization, without any gas phase. Very often, due to the low reaction rates of most biological transformations, long liquid residence times are needed for the completion of the reaction, and therefore the drag force created by the low liquid flow rate in a single pass reactor is not enough to promote fluidization of the solid particles. Fluidization is obtained either by external liquid recirculation or by the gas loaded to the reactor, as depicted in Figure 2. In systems where a gas is produced by cell metabolism, the gas can also be an additional factor contributing to solid particle fluidization, although other effects are also observed in this case, such as internal liquid recirculation patterns. Fluidization at relatively low liquid flow rates is also favored in tapered fluidized-bed configurations; the liquid superficial velocity at the bottom of the reactor is higher due to the reduced cross-sectional area. In general, one can distinguish three main sections in fluidized-bed bioreactors: *(1)* the bottom section, where feed (liquid, gas, or both) and recirculation are provided; *(2)* the central main section, where most of the reaction takes place; *(3)* and the top section, with a wider diameter that serves to decelerate the movement of the particles by decreasing the superficial velocity of the liquid, thus enhancing the retention of the solid phase and at the same time allowing gas disengagement from the liquid phase. It is a common trend for fluidized-bed bioreactors to use biocatalysts, either cells or enzymes, in the form of immobilized preparations. In general, the particles can be of three different types: *(1)* inert cores on which a biofilm is created by cell attachment, or in the case of enzymes, by adsorption or covalent binding immobilization; *(2)* porous particles in which the biocatalysts are entrapped; *(3)* cell aggregates obtained by self-immobilization caused by the ability of some cell strains to form flocs, pellets, or aggregates. Fluidized-bed bioreactors are usually differentiated from air-lift bioreactors by the fact that the latter do not specifically require the use of immobilized biocatalysts. Indeed, they were developed for free cell suspensions. In addition, air-lift bioreactors have different compartments, created by physical internal divisions, with different degrees of aeration.

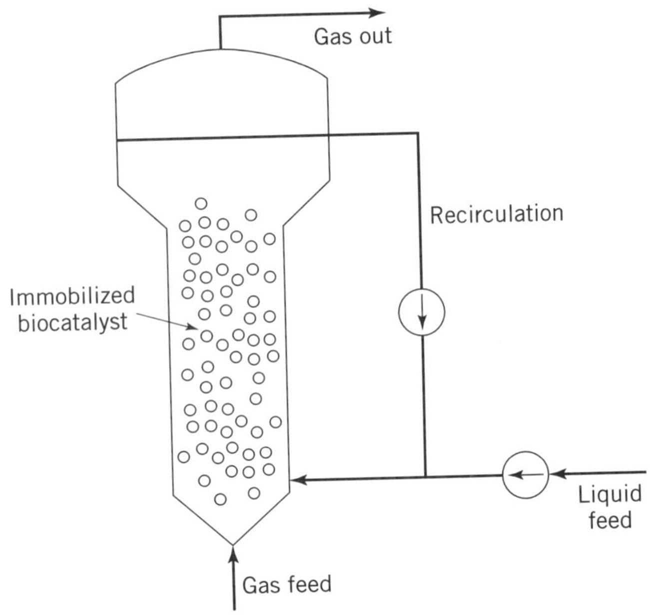

Figure 2. General scheme of a fluidized-bed reactor.

CHARACTERISTICS AND POTENTIAL OF FLUIDIZED-BED BIOREACTORS

The use of fluidized-bed bioreactors can provide a number of advantages that makes them an interesting alternative for bioprocesses, especially for continuous operation. In comparison with conventional mechanically stirred bioreactors, fluidized-bed bioreactors provide a much lower attrition of solid particles, and almost any kind of immobilized biocatalyst preparation can be used without physical disruption. Biocatalyst concentration can be significantly higher because of immobilization, and the typical wash-out limitations of continuous bioreactors operating with free cells are overcome because the solid particles are physically retained in the reactor vessel; operation at flow rates higher than the maximum growth rate of the cells can be achieved. As a consequence, the final productivity of the bioreactor can be increased substantially. Compared with packed-bed reactors, fluidized-bed bioreactors can be operated with smaller-size particles and without the drawbacks of clogging, creation of preferential flowpaths, or particle compression due to bed weight. Moreover, the smaller particle size minimizes the internal diffusional resistances, and the higher level of mixing enhances external mass and heat transfer from liquid to solid phase.

The degree of internal mixing in a fluidized bed can vary to a great extent, and it depends on various factors: density and diameter of the solid particles, liquid and gas flow rates at the reactor inlet, endogenous gas production by cell metabolism, and recirculation rate. For some operational conditions, high gas, liquid, or recirculation flow rates, the internal mixing will be very high, and the reactor will approach the behavior of a complete mixed tank. On the other hand, operation with low gas, liquid, or recirculation rates will provide a flow pattern close to plug flow, with some degree of axial mixing. This implies that reaction kinetics is an important factor to consider in the analysis and design of fluidized-bed bioreactors. Configurations favoring liquid mixing will be more appropriate for substrate-inhibited reactions, and configurations approaching plug flow will be indicated for product-inhibited reactions. Another advantage of fluidized-bed bioreactors is the ease of separation of the gas produced in most transformations involving cells (i.e., CO_2), or the feed of a gas stream to the reactor, for example, for aeration purposes. Also, fluidized-bed reactors make biocatalyst replacement easy, without disruption of the operation, enabling good control of the overall activity of the reactor. For example, one may replace particles with deactivated enzyme or removing an excess of biomass created by biofilms. On the other hand, solids attrition is higher in fluidized-bed than in packed-bed bioreactors. From the productivity point of view, the advantages of the fluidized bed, especially with respect to mass transfer rates, make it possible to obtain higher levels of overall productivity than in packed-bed reactors, in spite of the fact that the fraction of immobilized biocatalyst particles is lower for a fluidized bed.

Fluidized-bed bioreactors are complex with regard to hydrodynamic aspects, especially taking into account that the properties of the biocatalyst particles may change considerably during the operation time and in the presence of three different phases (solid, liquid, gas) in many cases. In fact, the nature of the particles (for example, their density and size, or their evolution with time, which are especially important with respect to certain kinds of immobilized cells), the liquid and gas flow rates employed, the type of reaction kinetics, and the kinetics of cell growth or enzyme deactivation influence each other and have a direct effect on the reactor design and performance (4).

Using the biocatalyst in immobilized form also contributes to the complexity of a fluidized-bed bioreactor. The behavior of the immobilized biocatalyst, especially when cells are used, can be substantially different than that of free suspensions (5). The behavior must be determined at the kinetic level, the physiological level, and the genetic level, and the biocatalyst's relationship with the diffusional restrictions in the particles and the possible direct effects associated with the immobilization itself must be well understood and correctly described in order to build appropriate and reliable models of for reactor design, control, and scaling up.

The potential of fluidized-bed bioreactors can be further exploited by considering multistage units and using two solid particles with different properties. Figure 3 gives an example of the concept of multistage operation, in particular, a countercurrent multistage fluidized bed working with immobilized enzymes (6). The main characteristic of this bioreactor is the continuous transport of the solid particles of immobilized enzymes from one stage to another in a downward direction. The overall catalytic activity of the reactor remains constant as the exhausted enzyme is removed from the reactor bottom stage while fresh biocatalyst is added at the top stage. A second advantage of dividing the reactor into compartments is the very low degree of back-mixing of the biocatalyst, and the plug-flow regime attained in the liquid phase. The use of two solid particles with different properties, particularly with different densities, can be used in a fluidized bed to achieve the in situ separation of a product of the reaction; this is a clear advantage in systems with product inhibition or when unfavorable thermodynamic equilibria limit the conversion rate for a reaction. For example, Davison and Scott (7) have proposed a system based in two different types of particles with different densities. As one type of particle, containing the biocatalysts (in this particular example, cells of *Lactobacillus delbreuckii*), remains fluidized in the bioreactor, the second type, which is heavier and contains no cells, is introduced from the top of the bioreactor and collected at the bottom. This second type of particle is selected to selectively remove the inhibitory product of the fermentation, for example, lactic acid. Another possibility is to combine both approaches, that is, to design a multistage fluidized bed working with two types of particles to achieve a selective in situ removal of the product. Van der Wielen et al. (8) have used this approach to enhance the enzymatic deacylation of benzylpenicillin, providing the desired product, 6-aminopenicillanic acid, and as a byproduct, phenylacetic acid, by means of light particles of immobilized enzymes and heavy particles of an adsorbent of the acidic by-product. Figure 4 presents a scheme of two different possibilities for the design of a two-solid-phase fluidized-bed bioreactor: semicontinous multistage pulsed flow and continuous trickle flow. In the first, the movement of the heavier solid particles downward in the reactor is obtained by periodic pulsations; in the second, the denser particles move continuously from the top to the bottom of the reactor.

MAIN ASPECTS FOR THE DESIGN AND OPERATION OF FLUIDIZED-BED BIOREACTORS

As already described, the design and operation of a fluidized-bed bioreactor has to take into account a number of aspects regarding both physical characteristics and reaction performance. These aspects include the hydrodynamics of the bioreactor (usually with three phases), characterization of the type of flow, degree of mixing for each phase and the heat transfer, mass transfer between phases, mass balancing of the species taking part in the reaction, intrinsic kinetics of the immobilized biocatalyst, diffusion transport of the species inside the solid particles, enzyme deactivation, and cell growth. Accurate knowledge concerning these points will allow complete characterization of a fluidized-bed reactor and the development of reliable mathematical models describing its performance. Next, we discuss the conceptual aspects intrinsically related to fluidized beds. For more detailed information, including mathematical formulation of the models and their

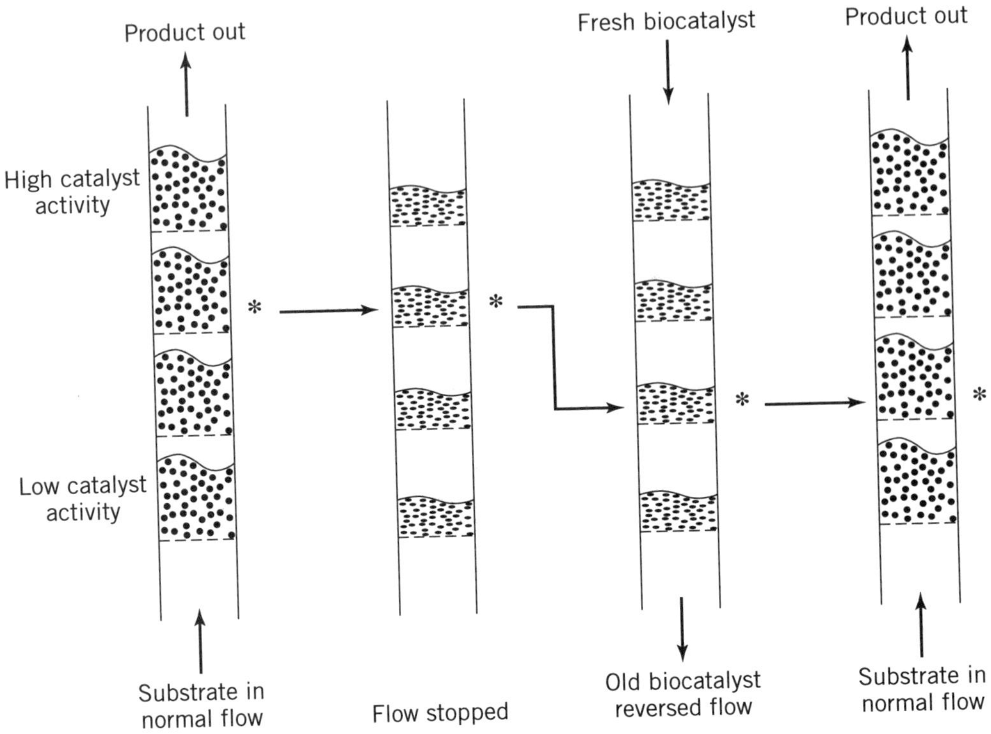

Figure 3. Operation phases of a multistage fluidized-bed reactor for a deactivating biocatalyst *Source*: From Ref. 6.

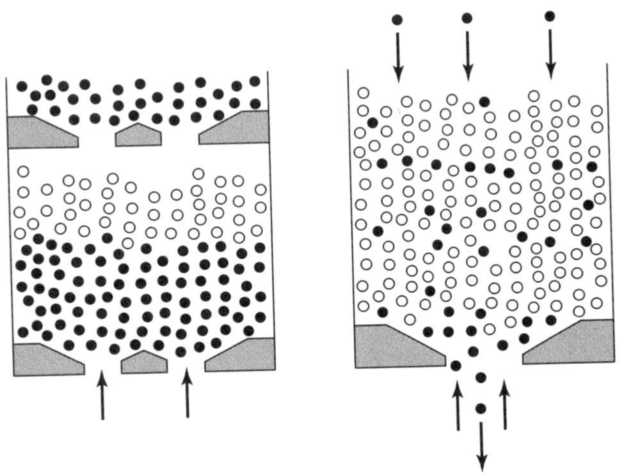

Figure 4. Schemes corresponding to two types of operation of a fluidized-bed as countercurrent adsorptive reactors, with two types of solid particles: light/small immobilized biocatalyst (○) and dense/large sorbent (●). Left: one compartment of the semicontinuous, multistage pulsed flow. Right: continuous trickle flow mode. *Source*: From Ref. 8.

application, the reader is referred to Refs. 6, 9–12 as well as general reviews in Refs. 13 and 14.

The first step in hydrodynamic characterization requires a determination of the fraction of the total volume to be occupied by the different phases (gas, liquid, solid). This is known as *phase hold-up* and can be determined using different techniques (3,15). Strictly speaking, phase hold-up may not be uniform for the whole reactor, especially with low axial dispersion. However, in systems with appreciable mixing, phase hold-up can often be considered uniform.

With respect to mixing behavior, most attention is generally paid to characterizing the liquid phase because this is, in most cases, the continuous phase in bioreactors, that is, the one in which substrates are fed and products of the reaction accumulate. For a three-phase fluidized-bed bioreactor with cocurrent up-flow circulation of liquid and gas, the type of flow pattern is very much dictated by the value and the ratio of the liquid and gas superficial velocities in the reactor (3,16). Three main regimes are possible: *(1)* the dispersed flow regime occurs at high ratios of liquid velocity to the gas velocity and is characterized by the homogeneous dispersion of small gas bubbles in the liquid; *(2)* the coalesced bubble flow regime occurs at increased gas flow rates and is characterized by the formation of bigger bubbles as a result of the coalescence of smaller ones; the coalesced bubbles have a nonuniform distribution in the liquid; *(3)* the slug-flow regime is the consequence of a further increase of gas flow rates, occurs at high ratios of gas velocity to liquid velocity, and is characterized by the formation of large gas bubbles that, in small-diameter bioreactors, tend to completely occupy their cross-sectional area. Slug formation breaks the bed continuity and causes great instability. Figure 5 gives an example of experimental data in a fluidization chart, showing these different regimes. If the gas flow rate is increased dramatically, it will eventually become the continuous phase in the bioreactor. A very detailed study on the characteristics of these dif-

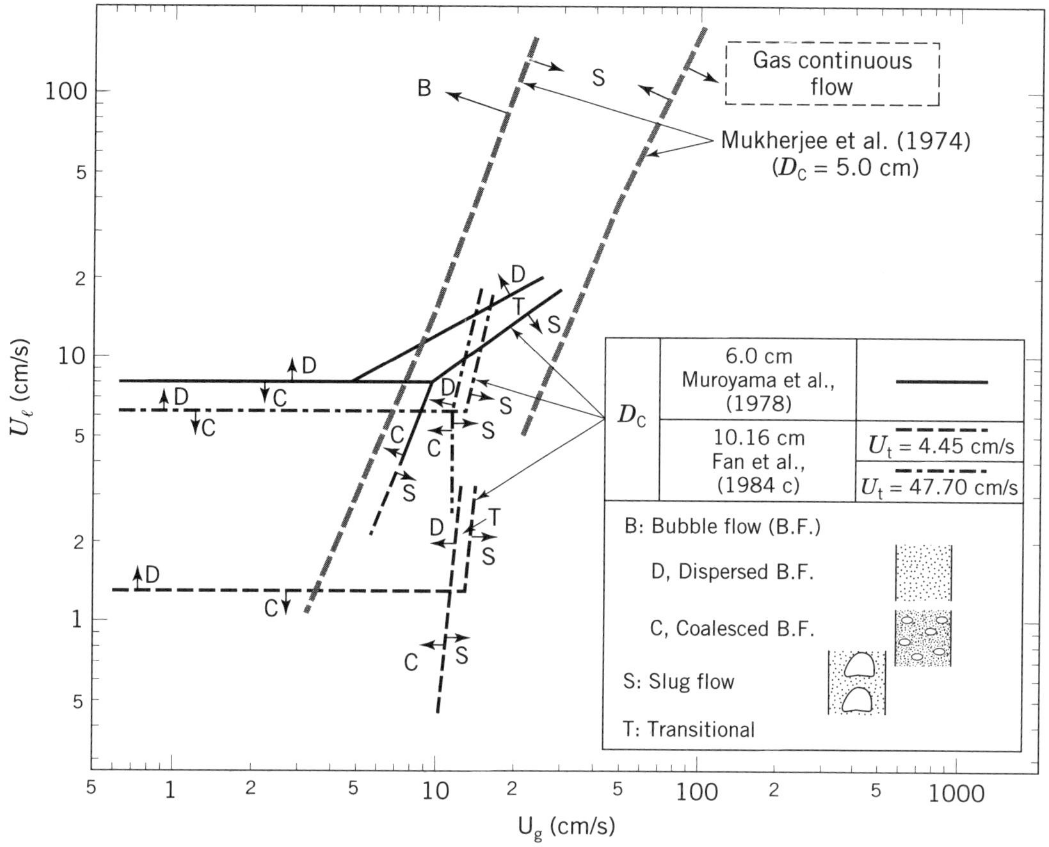

Figure 5. Flow regime diagram for the concurrent gas–liquid–solid fluidized-bed, U_g, gas superficial velocity. U_l, liquid superficial velocity. *Source*: From Ref. 3.

ferent regimes and the effect of the density and size of the solid particles on the regime transitions has been conducted by Zhang et al. (17). The liquid flow pattern in the bioreactor is directly influenced by the degree of mixing associated with these different regimes, as well as by other factors such as internal gas generation, bead size distribution, and external liquid recirculation. Basically, the flow will approach plug flow in systems with high velocity in the liquid phase. Quite often, bioreactors experience different degrees of mixing, and complete mixed flow can be observed in fluidized-bed bioreactors.

The influence of the gas phase on the liquid and solid mixing in a fluidized-bed bioreactor has been studied by Gommers et al. (18), considering two extreme situations: a reactor operating without gas and a reactor to which gas was introduced artificially from the bottom. The results showed that the gas phase greatly influences the degree of liquid mixing in the reactor, and that this effect increases sharply with the bed diameter. Another fact to consider is that in systems with gas generation associated with the reaction progress (for example, in most fermentations), the gas flow will change from the bottom to the top of the reactor, proportionally to the substrate consumption, and the axial dispersion will change with reactor height (19). An important aspect to consider in the study of the influence of the gas produced in a fluidized bed on the degree of liquid mixing is that the results will be affected by the experimental system used for gas injection, as discussed in detail by Buffière et al. (20). When liquid recirculation is used to promote fluidization because of the slow reaction rate and the long liquid-residence time required, plug flow is usually disrupted, and completely mixed flow is usually achieved.

Other factors influencing the hydrodynamic behavior of fluidized-bed bioreactors are the properties of the solid and liquid phases. In general terms, the weight and size of the solid particles will directly influence the liquid and gas flow rates required for bed fluidization. If the liquid residence time is fixed by criteria of substrate conversion, for example, then the heavier particles will require higher *L*/*D* ratios to increase liquid superficial velocity and enhance fluidization, or as an alternative, they will require high recirculation rates. In absence of recirculation, and when particles with a certain distribution in size and weight are used, solid-particle stratification is commonly found. Under stratification conditions, movement of the solid particles in the bed is very limited, and the particles are ordered by decreasing settling-velocities, from the bottom to the top of the bed. One of the consequences of this situation is that, in liquid plug-flow regimes, the particles at different reactor heights will experience different environments. For cells growing as biofilms around a solid particle, it is typical for the particles at the bottom of the bed to provide better conditions for cell growth (for example, substrate availability), and as a consequence, they will decrease their overall density and therefore migrate to the upper part of the reactor. Removal of excess biofilm may require external treatment of the particles. The operation of a stratified fluidized bed is illustrated in Figure 6, where the stratification of a bed with activated carbon as a solid support for cell growth is combined with the capacity to absorb the product of the reaction and, therefore, remove it selectively (21). The crucial effect of particle size and density distribution, as well as gas and liquid superficial velocities, on the phenomena of solid-particle stratification is discussed in detail in various recent studies (22–24). In general, solids mixing has been studied less comprehensively than liquid mixing, and it is less well understood. The possibility of using new experimental techniques (25) has enabled researchers to propose new models for the interpretation of the solids mixing, for example, trajectory length distribution (26,27), in contrast to more classical models that assume that solids mixing can be defined by means of an axial dispersion model (28).

Another important aspect to be considered from the hydrodynamic point of view is that fluidization properties depend on the difference in densities of the solid and liquid phases in the fluidized bed, and these values change during the operation time. This situation is of particular relevance for those systems where the absolute values for the liquid and solid densities are relatively close, such as for cells immobilized in natural origin polymers (alginate, agarose, etc.) or self-aggregated cells (pellets, flocs, etc.). In this case, relatively low variations in the absolute value of the solid density (usually associated with cell growth or the accumulation of CO_2 gas in the particles) and the liquid density (usually associated with substrate consumption) may cause an important percentile variation in the density difference between both phases, producing a relevant impact on the reactor hydrodynamics, as the regime changes from dispersed to coalescing (29), and on reactor stability. This problem can be of particular relevance at the start-up of a bioreactor operation, when the density values may change to a greater extent.

A second aspect to be addressed in the characterization of a fluidized-bed reactor is the definition of the flux model. As mentioned previously, most of the attention is focused on the liquid phase, and the flux model is closely connected to the hydrodynamic conditions in the reactor that will generate a given degree of internal mixing, which is somewhere between the two extreme situations of perfectly mixed or plug flow. The determination of the real liquid flux model in the bioreactor is a necessary step for the application of the mass balance equations for the species taking part in the reaction. Stimulus–response techniques are commonly used for such a purpose and are based on the introduction of an inert tracer at the reactor inlet and the analysis of the response curve obtained at the outlet, which reflects the type of flux (30). The models that describe the liquid flux in a real reactor (31) can be *(1)* an axial dispersion model, in which axial dispersion is superimposed on the liquid convective flux; *(2)* a tank-in-series model, in which the bioreactor is considered as a series of CSTR reactors of the same volume; and *(3)* a compartmented model, in which the flux model in the bioreactor is described as the combination of different ideal compartments. For fluidized-bed bioreactors, especially when fermentation gas is produced, some interesting contributions have been proposed, such as the consideration of a variable dispersion coefficient, which increases its value in proportion to the fermentation gas generated in the reactor, up to a given point (10).

In addition to hydrodynamic and mixing characteristics, another relevant aspect of a fluidized-bed bioreactor is its biocatalytic activity. Because immobilized biocatalysts are always used, the definition of the problem must simultaneously include the reaction characteristics (kinetics, stoichiometry, equilibrium) and transport characteristics, including both the transport between the liquid and the solid phase and the transport within the solid phase. When the general situation of a solid biocatalyst particle immersed in a liquid phase, as shown in Figure 7, is studied, different phenomena have to be considered. First, substrate is transported from the liquid phase to the solid ex-

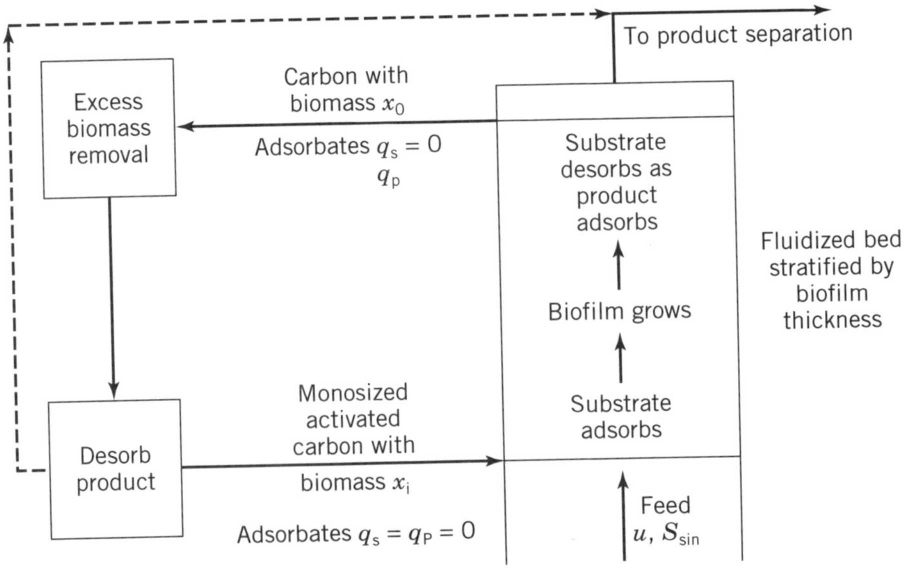

Figure 6. Operation diagram of a fluidized-bed bioreactor with simultaneous bioconversion and adsorption/desorption of substrate and product. *Source*: From Ref. 21.

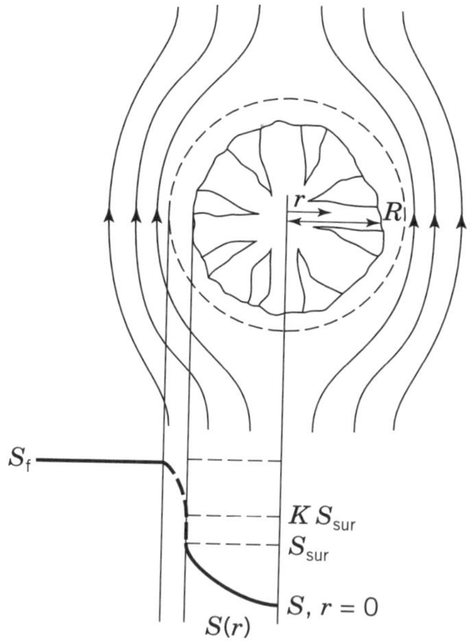

Figure 7. Inter and intraparticle mass transfer of a single porous spherical bead of radius R. Substrate concentration profiles across the stagnant liquid film and inside the solid particle, $S(r)$. S_f, concentration on the bulk liquid; S_{sur}, concentration on the solid surface; K, partition coefficient. *Source*: From Ref. 11.

ternal surface. This external mass transfer is usually modeled by means of a hypothetical liquid film that creates resistance to the transport, and it is characterized by the external mass transfer coefficient. The value of the coefficient, and thus of the external mass transfer velocity, depends on the physical properties of the liquid and the superficial velocity of the liquid around the solid; thus beds with more mixing will present better conditions for external mass transfer. When external mass transfer resistance is negligible, the concentration at the surface of the particle is the same as in the fluid phase ($S_f = S_{sur}$, according to Figure 7). An additional possibility is the occurrence of partitioning phenomena in the substrate, between the particles and the liquid, as a result of the material properties. In that case, the ratio between the substrate concentration at both sides of the solid–liquid interface will be given by the partition coefficient, K. Equally, partitioning phenomena can be determined for the product. Once in the solid particle, the substrate will diffuse inside the solid, following Fick's law, and simultaneously the reaction will take place. Therefore the corresponding equations of kinetic reaction and diffusion in the solid have to be solved simultaneously (10,11) in order to obtain the internal concentration profiles. To define this part correctly, the values for the intrinsic kinetic parameters of the reaction (that is, in absence of mass transfer limitations) and the effective diffusivities for the substrate and the product of the reaction in the solid have to be known. Moreover, the substrate conversion implies product generation, and this can affect the kinetics (for example, by product inhibition); usually the diffusion and reaction analysis is made simultaneously for both substrate and product. The overall activity of the catalytic particles will be dictated by the relative velocities of the two phenomena taking place inside them: reaction and diffusion. Systems with low diffusion rates with respect to the reaction rates will be diffusion controlled; on the other hand, systems with high diffusion rates with respect to the reaction rate will be controlled by the reaction. In the first case, the low diffusion will limit the efficiency of the particle because the reaction potential of the immobilized biocatalyst will not be fully used. Usually, this is reflected in terms of the efficiency factor, which is defined as the ratio between the actual reaction rate occurring in the system, and the reaction rate that would occur if no diffusional limitations existed, that is, when all solid particles would present a uniform concentration, equal to that of their surface. The relationship between the effectiveness factor and some moduli, such as the observable modulus, being proportional to the ratio between reaction and diffusion rates, is given in Figure 8. One of the interesting properties of such graphs is that they are very similar for different types of geometries and kinetics. Therefore they enable a direct analysis of the degree of diffusional limitations in a given type of particle in a fluidized bed and also suggest quantitative modifications to be performed to avoid such limitations, for example, changes in the particle size or diffusion conditions. In general terms, fluidized beds are interesting with regard to these aspects because since they require comparatively small-diameter particles for better fluidization, the potential diffusion limitations are reduced.

As mentioned at the beginning of this section, the combination of the different aspects we discussed allows the elaboration of mathematical models describing the behavior of fluidized-bed bioreactors. The reliability of these models depends greatly on the accuracy of the determination of the various parameters involved in the definition of the reactor and the reaction system (flux model, intrinsic kinetic parameters, external mass and heat transfer coefficients, internal effective diffusivities), and they will serve for various purposes such as conceptualization and understanding of the bioreactor itself (an important aspect required for building the model), design of similar bioreactors (especially in the case of scaling-up), simulation of the bioreactor operation at different conditions, and bioreactor control. As an example of how the output of a mathematical model can describe the internal concentration profiles in a fluidized-bed bioreactor, the data corresponding to a continuous fermenter for the production of ethanol from glucose by the bacteria *Zymomonas mobilis* immobilized in carrageenan beads (10) are given in Figure 9.

INDUSTRIAL APPLICATIONS OF FLUIDIZED-BED BIOREACTORS

Fluidized-bed bioreactors have been studied in a wide spectrum of applications as a consequence of the different advantages that they offer (14,33). Nevertheless, this interest has been translated with relatively low intensity into the industrial scale of operation. Possible reasons for this may be the greater complexity of the reactor compared with conventional types (especially regarding long-term

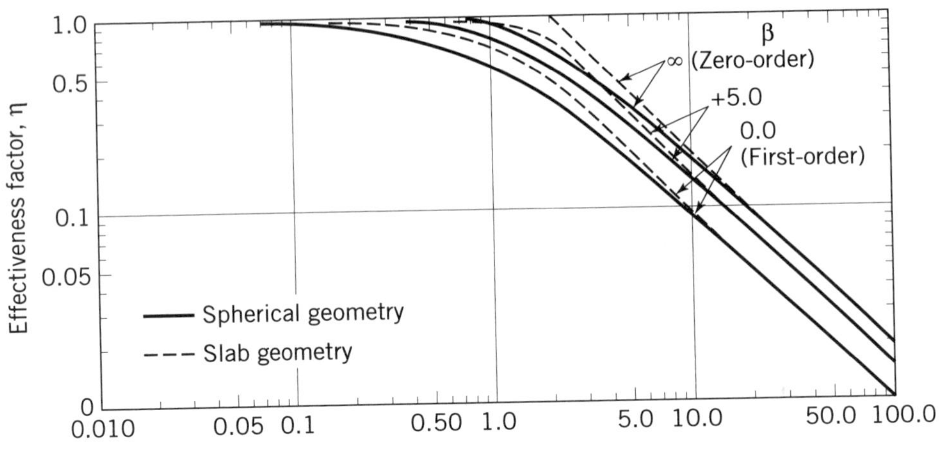

Figure 8. Variation of the effectiveness factor values for immobilized enzymes with Michaelis–Menten intrinsic kinetics with respect to the observable modulus, Φ. *Source*: From Ref. 32.

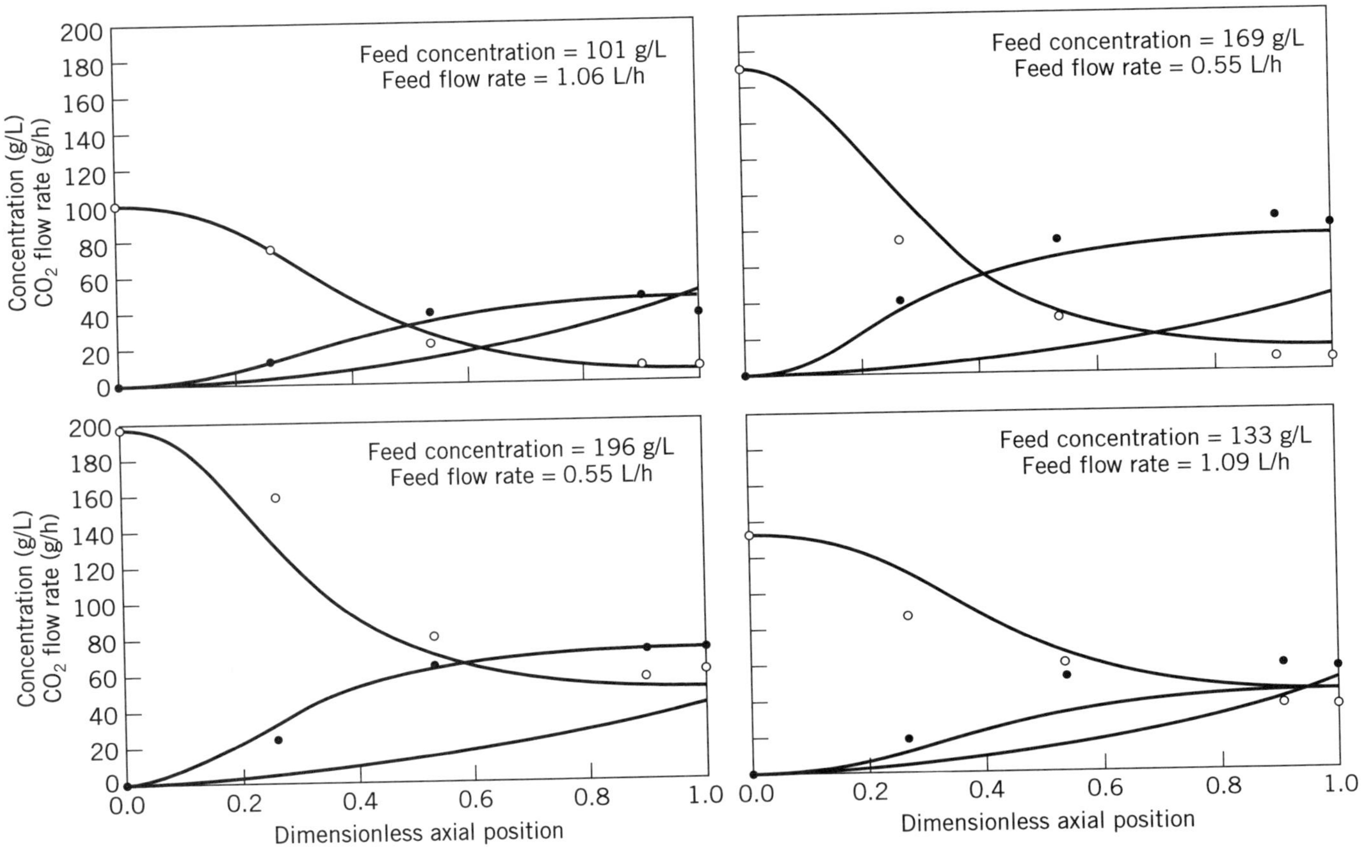

Figure 9. Results of the mathematical model for a tapered fluidized-bed producing ethanol from glucose with *Zymomonas mobilis* cells immobilized in carrageenan beads. Internal concentration profiles with the dimensionless fermenter height: experimental values (points) and values predicted by the model (solid lines). Dotted lines represent the calculated CO_2 flow rate produced by fermentation. *Source*: From Ref. 10.

operation, hydrodynamic behavior and evolution of the immobilized biocatalyst), and the time required in any industrial process to implement any change in technology. Some specific examples from different biotechnological processes illustrate the feasibility of the fluidized bed at a pilot plant and on the industrial scale. The area of wastewater treatment provides examples of high-volume operation and has made use of three-phase fluidized-bed bioreactors, as well as other related designs (such as the up-flow anaerobic sludge blanket reactor), to a significant extent. The wastewater from the baker's yeast industry, particularly from the Gist-brocades company in Delft,

Figure 10. View of an industrial anaerobic wastewater treatment plant based on the operation of fluidized-bed reactors, Gist-brocades, Delft, The Netherlands. *Source*: From Ref. 34.

Netherlands, has been treated with two anaerobic fluidized-bed reactors in series, each one 21 m high and with 390 m^3 total volume, operating with bacterial cells immobilized by attachment onto 0.25- to 0.5-mm-diameter sand particles, a support that is typical for other fluidized-bed reactors applied to anaerobic and aerobic wastewater treatment (34). The volume of the reactor employed for the disengagement of the CH_4 gas obtained was important, and the fluidized-bed volume was 215 m^3. The carbon–oxygen demand (COD) conversion of these reactors is 22 kg/(m^3 day), which is 65% of the COD in the load. A first set of these reactors started operation in 1984, and a second set in 1986. A picture of these industrial units is given in Figure 10. More recently, data from two full-scale anaerobic three-phase fluidized-bed plants have also been reported (33). A 700-m^3-total-volume reactor (500-m^3 working volume) has been installed for wastewater treatment at a sugar beet factory of United Northern Sugar Factories Co. in Clauen, Germany, using cells immobilized onto pumice and having COD conversion of 20 kg/(m^3 day). A 734-m^3-total-volume multistage reactor (650- to 700-m^3 working volume) has been installed for wastewater treatment at a yeast factory of Hamburg Co., in Hamburg, Germany, using free cells and a new principle that allows control, in each stage, of the ratio of the recirculated gas amount with respect to the rising gas flow. This enables optimization of the hydrodynamic conditions in each stage, according to the reactor loading. Nitrification of the effluent from a municipal wastewater treatment plant in order to eliminate ammonium salts has also been carried out in large-scale fluidized-bed reactors (35) with a cross-sectional area of 2.2 m^2 and a height of 5 m, using 5 tons of sand particles 0.4–0.6 mm in diameter as support particles for cell biofilm development. The reported results show the feasibility of the system, from both the technical and economic points of view.

Ethanol fermentation has been studied intensively in fluidized-bed reactors and brought up to the pilot-plant level in several cases. A pilot plant with a volume of 1 m^3, using *Saccharomyces cerevisiae* cells immobilized in carrageenan beads has been successfully operated for half a year (36). As an alternative to yeast, the bacteria *Zymomonas mobilis* can also be used. Very promising results have been reported on the operation of a series of two pilot reactors of 55 L each, under nonsterile conditions (37). The support for the immobilization in this case was a macroporous glass particle, Siran, that allowed cell attachment to the internal surface of the pores. After an initial period of sterile colonization of the beads in a conventional stirred tank, the particles, with a high *Z. mobilis* cell concentration, can be transferred to the fluidized bed, which then can be operated at high dilution rates under nonsterile conditions because all the possible contaminants in the feed (a solution of hydrolyzed B-starch) are washed out from the reactor. Complete conversion of 120 g/L sugar concentration with a 4.25-h residence time can be consistently obtained, with ethanol productivities of 18 g/(L h), calculated with the total reactor volume as a basis. A complete technical and economic evaluation of the use of fluidized-bed reactors for the production of ethanol with *Z. mobilis* has shown promising features (38).

Other examples of the use of fluidized-bed bioreactors on the pilot scale in fermentation processes include the production of alcohol-free beer with immobilized yeast in a 50-L pilot-plant fluidized-bed reactor (39) and the production of penicillin using *Penicillium chrysogenum* immobilized in urethane particles in a 160-L pilot-plant reactor (40). *Penicillium chrysogenum* cells have also been used in different studies with fluidized-bed reactors with celite as the support material (41). Also, the area of animal cell culture provides a number of examples of the application of fluidized-bed technology, taking into account that smaller production volumes are usually required in this case. The use of immobilized cells is particularly required for those cells that are anchorage dependent, and a variety of porous carriers and beads have been developed for this purpose. The production of a human anti-HIV-1 antibody using recombinant Chinese hamster ovary cells grown onto a macroporous polyethylene carrier has been carried out in an 84-L-volume pilot-scale fluidized-bed novel reactor equipped with a low shear stress internal impeller for the recirculation of the liquid (42).

BIBLIOGRAPHY

1. D. Kunii and O. Levenspiel, *Fluidization Engineering*, Robert E. Krieger, Huntington, New York, 1977.
2. L.S. Fan, *Gas–Liquid–Solid Fluidization Engineering*, Butterworths, Stoneham, 1989.
3. K. Muroyama and L.S. Fan, *AIChE J.* **31**, 1–34 (1985).
4. G. F. Andrews and J. Przezdziecki, *Biotechnol. Bioeng.* **28**, 802–810 (1986).
5. J.A.M. de Bont, J. Visser, B. Matiasson, and J. Tramper, *Physiology of Immobilized Cells*, Elsevier, Amsterdam, 1990.
6. H.J. Vos, D.J. Groen, J.M. Potters, and K.Ch.A.M. Luyben, *Biotechnol. Bioeng.* **36**, 367–376 (1990).
7. B.H. Davison and C.D. Scott, *Biotechnol. Bioeng.* **39**, 365–368 (1992).
8. L.A.M. Van der Wielen, P.J. Diepen, A.J.J. Straathof, and K.Ch.A.M. Luyben, *Ann. N.Y. Acad. Sci.* **750**, 482–490 (1995).
9. Y. Sun and S. Furusaki, *J. Ferment. Bioeng.* **69**, 102–110 (1990).
10. J.N. Petersen and B.H. Davison, *Appl. Biochem. Biotechnol.* **28/29**, 685–698 (1991).
11. T. Kiesser, G.A. Oertzen, and W. Bauer, *Chem. Eng. Technol.* **13**, 20–26 (1990).
12. N. Qureshi and I.S. Maddox, *J. Chem. Tech. Biotechnol.* **34/35**, 441–448 (1992).
13. G.F. Andrews, *Biotechnol. Gen. Eng. Rev.* **6**, 151–178 (1988).
14. F. Godia and C. Solà, *Biotechnol. Prog.* **11**, 479–497 (1995).
15. W.M. Ohkawa, A. Maezawa, and S. Uchida, *Chem. Eng. Sci.* **52**, 3941–3947 (1997).
16. C. Zheng, B. Rao, and Y. Feng, *Chem. Eng. Sci.* **43**, 2195–2200 (1988).
17. J.-P. Zhang, J.R. Grace, N. Epstein, and K.S. Lim, *Chem. Eng. Sci.* **52**, 3979–3992 (1997).
18. P.F.J. Gommers, L.P. Christoffels, J.G. Kuenen, and K.Ch.A.M. Luyben, *Appl. Microbiol. Biotechnol.* **25**, 1–7 (1986).
19. B.H. Davison, *Appl. Biochem. Biotechnol.* **20/21**, 449–460 (1989).
20. P. Buffière, Ch. Fonade, and R. Moletta, *Chem. Eng. Sci.* **53**, 617–627 (1998).
21. G.F. Andrews and J.P. Fonta, *Appl. Biochem. Biotechnol.* **20/21**, 375–390 (1989).
22. N. Funamizu and T. Takakuwa, *Chem. Eng. Sci.* **51**, 341–351 (1996).
23. L.A.M. Van der Wielen, M.H.H. Van Dam, and K.Ch.A.M. Luyben, *Chem. Eng. Sci.* **51**, 995–1008 (1996).
24. T. Matsumoto, N. Hidaka, Y. Takebayasi, and S. Morooka, *Chem. Eng. Sci.* **52**, 3961–3970 (1997).
25. J. Reese and L.S. Fan, in J. Chaouki, F. Larachi, and M.P. Dudukovuc eds., *Non-invasive Monitoring of Multiphase Flows*, Elsevier, Amsterdam, 1997.
26. J. Villermaux, *Chem. Eng. Sci.* **51**, 1939–1946 (1996).

27. K. Kiared, F. Larachi, Ch. Guy, and J. Chaouki, *Chem. Eng. Sci.* **52**, 3931–3939 (1997).
28. T.C. Bickle and M.G. Thomas, *Ind. Eng. Chem. Process Des. Dev.* **21**, 377–381 (1982).
29. P. Béjar, C. Casas, F. Gòdia, and C. Solà, *Appl. Biochem. Biotechnol.* **34/35**, 467–475 (1992).
30. O. Levenspiel, *The Chemical Reactor Omnibook*, OSU Book Stores, Corvallis, Ore., 1979.
31. D.E. Swaine and A.J. Dauglis, *Biotechnol. Prog.* **4**, 134–148 (1988).
32. J.E. Bailey and D.F. Ollis, *Biochemical Engineering Fundamentals*, McGraw-Hill, New York, 1986, pp. 213.
33. K. Schügerl, *Chem. Eng. Sci.* **52**, 3661–3668 (1997).
34. J.I. Heijnen, A. Mulder, R. Weltevrede, J. Hols, and H.L.I.M. van Leeuwen, *Water Sci. Technol.* **23**, 1427–1436 (1991).
35. P.F. Cooper and S.C. Williams, *Water Sci. Technol.* **22**, 431–442 (1990).
36. M. Nagashima, M. Azuma, S. Noguchi, K. Inuzuka, and H. Samejima, *Biotechnol. Bioeng.* **26**, 992–997 (1984).
37. D. Weuster-Botz, A. Aivasidis, and C. Wandrey, *Appl. Microb. Biotechnol.* **39**, 685–690 (1993).
38. R.M. Busche, C.D. Scott, B.H. Davison, and L.R. Lynd, *Appl. Biochem. Biotechnol.* **34/35**, 395–417 (1992).
39. A. Aivasidis, Ch. Wandrey, H.G. Eils, and M. Katzke, *Proc. 23rd. Congr. Eur. Brew. Conv.*, 569–576 (1991).
40. I. Endo, T. Nagamune, K. Tachi, and T. Kobayashi, in G. Durand, L. Bobichon, and Florent eds., *Int. Biotechnol. Symp.*, Societé Française de Microbiologie, Paris, 1988, pp. 433–441 (1988).
41. T. Keshavarz, R. Eglin, E. Walker, C. Bucke, G. Holt, A.T. Bull, and M.D. Lilly, *Biotechnol. Bioeng.* **36**, 763–770 (1990).
42. M. Reiter, G. Blüml, T. Gaida, N. Zach, F. Unterluggauer, O. Doblhoff-Dier, M. Noe, R. Plail, S. Huss, and H. Katinger, *Bio/Technology* **9**, 1100–1102 (1991).

See also BIOREACTORS, AIR-LIFT REACTORS; BIOREACTORS, CONTINUOUS STIRRED-TANK REACTORS; FERMENTATION MONITORING, DESIGN AND OPTIMIZATION; MASS TRANSFER; SCALE-UP, STIRRED TANK REACTORS.

BIOREACTORS, GAS TREATMENT

GRAHAM ANDREWS
MMBD Consulting
Gresham, Oregon
WILLIAM APEL
Idaho National Engineering and Environmental Laboratory
Idaho Falls, Idaho

KEY WORDS

Biofilters
Biophase
Bioscrubbers
Biotricking filters
Bubble-gasified bioreactors
Cometabolism
Electron acceptors
Electron acceptors
Electron donors
Nutrient addition

OUTLINE

INTRODUCTION

The most common type of bioprocess, at least in the United States, is the aerobic treatment of wastewater, in which contaminants in the aqueous phase are biodegraded with the help of oxygen transferred from the gas phase (air) to generate a metabolic product (carbon dioxide) that, being volatile, is then stripped out of the aqueous phase into the gas. The familiarity of this arrangement should not be allowed to limit our imaginations about the possibilities of bioprocessing. It is perfectly possible to construct a process in which a gas-phase contaminant (hydrogen sulfide) is biodegraded with the help of a nonvolatile oxidant (nitrate), generating a product (sulfuric acid) that remains dissolved in the aqueous phase. The bioreactor for such a process must share many of the features found in aerobic wastewater treatment, such as efficient gas–liquid contacting, cell recycling or immobilization to create a high concentration of biomass, control of temperature and pH, and addition of nutrients in order to create a comfortable environment for microbial metabolism. However, the different objective, the treatment of a gas rather than a liquid, must dictate differences in the details. This point can be illustrated by thinking of a conventional trickling filter not as a device for treating sewage, but as one for removing oxygen from the air that flows through it. It would work, but terribly inefficiently, with a fractional removal on the order of 1%.

This chapter presents a three-step study of the use of bioprocesses for the removal of undesirable components from gas streams. The existing and potential applications are described first, followed by the types of bioreactors that may be employed. A simple process analysis is then presented to illustrate how the choice and design of the bioreactor is dictated by the particular application, the type of microbial metabolism involved, the concentration and solubility of the contaminant, and so on. It should be noted that there are also a number of potential bioprocesses in which the objective is not the removal of a contaminant from a gas stream, but the bioconversion of gaseous substrates into useful products. They include the partial oxidation of methane to methanol, and the production of ethanol and acetic acid from mixtures of carbon monoxide and hydrogen. Work on these processes is not described in detail, except when doing so illustrates the common problems of bioreactors with gaseous substrates.

MICROORGANISMS AND APPLICATIONS

Due to inherent difficulties in achieving the sterile conditions necessary for prolonged maintenance of pure cultures in field-scale bioreactors, most gas treatment bioreactors contain a mixed culture or consortium of microorganisms. These consortia may be derived from a number of different inocula, from common sources such as sewage-treatment facilities, biofilms from established bioreactors used for similar applications, or soils and waters from areas contaminated with the substrate of interest. In some instances, it is possible to enrich for the desired microbial consortium directly from the bioreactor bed medium, particularly when media components such as soil, compost, peat, or bark chips are used, because these materials naturally contain a mixture of microbes with wide-ranging physiological capacities. Regardless of the inoculum source, the bioreactor must be operated under conditions that select for and maintain, over the life of the bioreactor, microorganisms with the physiological capabilities necessary to catalyze the desired bioconversion. Typical selective factors include the electron donor (i.e., the metabolizable energy source), available terminal electron acceptors, supplemental nutrients, pH, Eh, and temperature.

Regardless of the selection strategy used to achieve a microbial consortium with the desired physiological properties, the use of a consortium versus the attempted use of a pure culture of microbes has a number of practical advantages. As mentioned above, it is difficult to achieve the sterility standards necessary to maintain a pure culture of microbes in a gas treatment bioreactor under field conditions. In addition, a pure culture of microbes is often metabolically incapable of fully degrading a contaminant, so hazardous intermediates may be created and build up within the bioreactor or in effluent streams from the bioreactor. Because mixed consortia inherently tend to be composed of a number of different microbial species with a gamut of physiological capabilities, undesired intermediates that may be created by pure cultures are often fully degraded by consortia. Similarly, in many applications, the concentrations and chemical compositions of gas streams to be treated may be transient. Consortia generally have a higher probability than pure cultures of adapting to this transience. The contaminant gases and vapors that can be successfully biodegraded can be classified by their role in microbial metabolism.

Electron Donors

Microorganisms require a constant supply of metabolic energy, which is normally derived from the oxidation of either organic or inorganic compounds, with these compounds serving as electron donors to the microorganisms. Many environmentally or industrially significant gases and vapors are metabolically oxidizable by microorganisms and can serve as electron donors. Thus, many different types of gas and vapor streams can be and have been, treated microbiologically (1). Microbially oxidizable gases and vapors that have been treated in bioreactors include hydrocarbons, ketones, ammonia, xylene, alcohol, terpene, and carbon disulfide vapors, as well as gases such as methane and hydrogen sulfide (2–10). In the case of carbon-containing substrates (hydrocarbons, carbon monoxide, etc.), the metabolism of these compounds provides heterotrophic microbes not only with an electron source, but also with a source of cellular carbon. Non-carbon-containing substrates such as ammonium, while providing a metabolic electron source, must be metabolized by autotrophic microorganisms capable of obtaining their cellular carbon via the fixation of atmospheric carbon dioxide.

Cometabolism

Some substrates cannot serve as sole energy sources for microorganisms, but are nevertheless potentially bioconvertable. Their biodegradation is achieved in the presence of another compound that serves as the microorganism metabolic electron donor, a process that is termed cometabolism. Trichloroethylene (TCE) provides a well-documented example of a substrate that is oxidized cometabolically by microorganisms. Typical of cometabolic TCE degraders are the methanotrophic bacteria, aerobes that oxidize methane as their sole carbon and energy source (11). This oxidation takes place in a sequential manner, with methane first being converted to methanol, which is in turn oxidized to formaldehyde, formate, and finally, carbon dioxide. Along this oxidative pathway, the methanotrophs generate cellular energy as well as fixing methane-derived carbon into additional biomass. The first step, in which methanol is formed from methane, is catalyzed by the enzyme methane monooxygenase (MMO) (12,13). Although MMO is selective for its natural substrate, it can, under certain conditions, catalyze the oxidation of a variety of other compounds (12–14), including partially chlorinated aliphatic solvents such as TCE. Two distinct forms of MMO have been identified, membrane MMO (mMMO) and soluble (sMMO) (15,16). sMMO, expressed by methanotrophs growing under copper-limited conditions, supports much higher rates of TCE oxidation than does mMMO, which is expressed under conditions of copper sufficiency (17).

Air streams contaminated with TCE and similar chlorinated solvents are generated by industrial operations and the remediation of contaminated ground and water by air stripping. They can be treated in a bioreactor that is started up with methane as the sole carbon and energy source, while supplying the necessary oxygen and supplemental nutrients under copper-limiting conditions (6). After the methanotrophic consortium is established, the methane load to the reactor can be decreased or fed on an intermittent basis concurrently with TCE vapors. The methanotrophic microbes then will be maintained by the lessened methane feed, while catalyzing the oxidation of the TCE via their sMMO. The reduced or intermittent methane feed not only reduces the amount of methane that must be supplied, but is essential in achieving optimal TCE removal rates since methane itself is an effective competitive inhibitor of TCE oxidation. An alternate, but more expensive, strategy used to avoid this type of competitive inhibition is to supply a catabolic intermediate such as formate in order to provide energy to the methanotrophs, because formate does not compete with TCE for access to the sMMO. It should be noted that a number of other microbial oxygenases have been shown to be capable of catalyzing cometabolic TCE oxidation. These include toluene oxygenases from a number of *Pseudomonas* spp., propane monooxygenases from propanotrophs, and ammonia monooxygenase from *Nitrosomonas europea* (18–21).

Electron Acceptors

Most gas treatment bioreactors remove contaminants from air streams and oxidize them to innocuous end products (CO_2, Cl^-, etc.), using the oxygen in the air as the electron acceptor. However, there are also gas streams containing little or no oxygen where the objective is to remove a compound that can act as an alternative physiological terminal electron acceptor. This approach has been shown to work for compounds like sulfur oxides (SO_x) and nitrogen oxides (NO_x) (8,24,25). NO_x biofilters have been developed using a consortia of denitrifying microbes that have the physiological capacity, as a part of their normal metabolism, to use a variety of NO_x compounds as terminal electron acceptors, reducing them to innocuous nitrogen gas (N_2). This process generally requires low-oxygen conditions because denitrifiers preferentially use oxygen, if present, as their terminal electron acceptor. As a result, selective conditions for denitrifying microbes include a suitable carbon and energy source with a NO_x compound as a terminal electron acceptor, a lack of oxygen, and the presence of supplemental nutrients necessary for microbial growth. The potential application to stack gases raises the question of the maximum temperature at which the microbes will function. Research (25) has shown that naturally occurring thermophilic microbes selected for from compost are capable of reducing nitric oxide (NO) to nitrogen gas (N_2) at temperatures up to 60 °C.

Other processes are far more sensitive than denitrification to trace amounts of oxygen. The sulfite-reducing bacteria responsible for the removal of SO_2 are obligate anaerobes, although they may be active in anoxic microniches, for example, deep in a biofilm. Chlorinated aliphatic solvents will also accept electrons under anaerobic conditions, a degradative mechanism that works best on the completely chlorinated compounds that are not touched by the oxidative enzymes already discussed (22,23). Carbon tetrachloride, for example, will be progressively reduced to chloroform, dichloromethane, and methyl chloride. This mechanism is often categorized as cometabolism, implying that the microbes derive no metabolic benefit from it, but this remains uncertain.

BIOREACTOR TYPES

Figure 1 shows a generic gas treatment bioprocess along with the nomenclature to be used in the subsequent analysis. Gas treatment bioreactors are sometimes described as *gas-phase bioreactors*, but this term is inexact in the sense that the actual reactions happen inside a microbial cell, which is necessarily an aqueous phase. Some reactions carried out in immobilized enzyme reactors do seem to involve direct interaction between the gas and the enzyme (33), although even here the humidity remains a critical variable and the enzyme may be covered with a layer of water a few molecules thick. For our purposes, the bioreactor is divided into a gas phase, a solid phase consisting of any solid packing or biofilm support particles, and a biophase, which contains the water and the microorganisms. The fraction of the bioreactor volume occupied by each phase is called its holdup, and the three values, e_g, e_s, and e_b, must add up to 1.

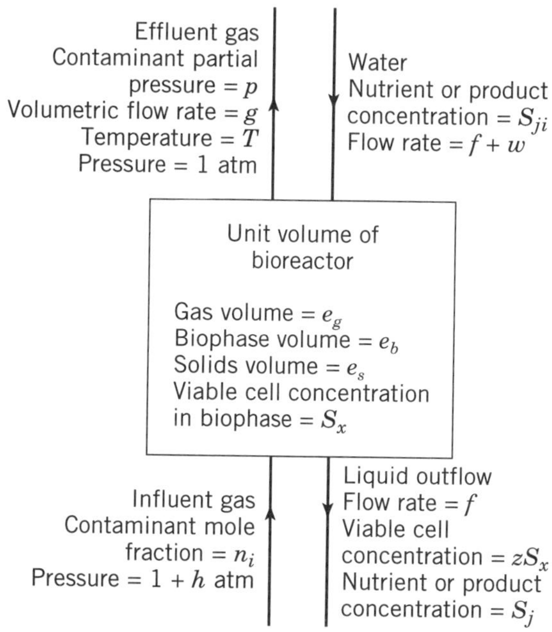

Figure 1. The generic gas-treatment bioreactor.

For convenience, we think in terms of a bioreactor of unit volume and express the gas flow rate as the volume per volume per minute (VVM) flow rate, g, at the effluent gas conditions, taken as temperature T and atmospheric pressure. Others choose to work in terms of the molar outlet gas flow rate (g/RT, where R is the gas constant) or a gas residence time, either the empty bed residence time, $1/g$, or the average time an element of gas actually spends in the reactor, e_g/g. The flow rate of the influent gas is not the same as g because its temperature may not be T, it is at a higher pressure, ($1 + h$ atmospheres, where h is the pressure drop through the reactor), and because various components (contaminant, oxygen, metabolic products, water vapor) are added to, or removed from, the gas stream as it flows through the reactor. When these effects are significant, it is essential to specify exactly what is meant by "gas flow rate."

The objective of some processes is to remove a gaseous contaminant by converting it into a soluble, nonvolatile metabolic product. Some such products, H^+ for example, can be neutralized in the bioreactor, but others, such as the Cl^- from the degradation of chlorinated organics, must be removed in the liquid if they are not to accumulate to concentrations that inhibit the microorganisms. There may therefore be a liquid outflow, f per unit bioreactor volume, and an inflow that must be greater than f to allow for the evaporation rate, w. This outflow tends to wash microorganisms out of the bioreactor, and it is usually beneficial to retain them either by cell recycle or immobilization, making the parameter z in Figure 1 less than 1.

The question is what type of bioreactor should be put in the "black box" in Figure 1. Almost every possible configuration by which a gas can be contacted with a biofilm or a suspension of microorganisms has been employed in some application, and these can be roughly divided into four categories.

Biofilters

The earliest gas-treatment bioreactors were used to control odors in air coming from slaughterhouses and rendering plants and consisted of porous pipes buried in the ground, the noxious compounds being removed by a combination of adsorption and biodegradation as the air flowed up through the soil (26). Applications of the idea have since expanded to include the removal of hazardous, volatile compounds from chemical plants, paint shops, foundries, and a variety of agricultural and food-processing operations. Although often effective, the operation of these simple soil beds depends critically on the local soil and weather conditions. Unless they are covered and/or periodically sprayed with water, they may stop working, either due to becoming waterlogged or to drying out. They have largely been replaced by custom-made biofilters, which are essentially boxes made of plastic, wood, sheet metal, concrete, or even stainless steel, containing a bed of media and a gas distribution system to ensure uniform flow of the gas through the bed.

In terms of the variables defined in Figure 1, biofilters are the small e_b, small f, and small z approach. The biophase consists of individual microorganisms attached to the media, perhaps enough to constitute a thin biofilm. The amount of moisture in this layer is critical, and is controlled by humidification of the inlet gas and/or by spraying water directly over the bed. Little or no water drips out of the bed, making it difficult to provide soluble nutrients and pH control chemicals, or to wash out nonvolatile metabolic products. Biofilters are thus best suited to applications that generate no such products and involve no pH swings, the treatment of air streams containing low levels of volatile organics being the obvious example.

The great advantage of biofilters over the early soil beds is that they allow the choice of the nature and particle size of the bed media. This media must support a high density of attached microorganisms suited to the particular application, perhaps even providing some of the nutrients needed for their growth. The ability to adsorb the contaminant is an advantage because adsorption can provide some contaminant removal during the start-up period, before a large population of well-acclimated microorganisms has developed (sometimes called "the bed-ripening period"), and during any sudden slugs of concentrated contaminant in the influent gas. Also, although the interactions between such adsorption and biodegradation are complex and poorly understood, they are generally favorable to biofilter performance. Finally, the media must be mechanically strong and resistant to disintegration, compacting, and the resulting channeling of gas flow. Soil, compost, peat, and wood-chip mixtures are all inexpensive media that have been used successfully. Materials such as activated carbon and limestone, although more expensive, may be mixed in to provide extra adsorption and pH buffering.

The particle size of the media is critical. Small particles give a huge gas–biophase interfacial area per unit bed volume (a quantity subsequently called "a"), thus eliminating concerns about mass-transfer limitations between the two phases. But beds of small particles have a higher resistance to gas flow, are more prone to waterlogging and plugging by excess biofilm, and if too light, may get blown out of the reactor altogether. A general guideline is to choose the smallest available particles that avoid the latter problems, and then ensure that the shape of the bed is optimized to give the desired contaminant removal with a reasonable pressure drop (see "Scale-Up").

If the life of the bed is found to be finite due to compaction, loss of pH buffering capacity, accumulation of salts from the evaporation of added water and so on, one solution is to employ two beds in series. Bed 1 can provide most of the contaminant removal, while bed 2 polishes the effluent, mainly by adsorption, while developing its own population of well-acclimated microorganisms. When bed 1 is exhausted it can be replaced by bed 2, which is itself replaced by a bed of fresh media. The arrangement of the media on a series of trays in the bioreactor facilitates this mode of operation.

Bubble-Gasified Reactors

Conventional bubble-aerated fermenters make poor gas treatment bioreactors because they are designed for a high rate of transfer of gaseous nutrients and products (O_2 in and CO_2 out), rather than a high fractional removal of a component from the air stream. This latter objective requires taller bioreactors with longer gas residence times, but tall reactors mean higher pressure drops due to the hydrostatic head. Nevertheless, tall, mechanically agitated bioreactors with multiple impellers have been proposed for removing H_2S from industrial gas streams (29), and bubble columns with no mechanical agitation have been demonstrated for the removal of TCE from air using a toluene-oxidizing organism (30).

These bioreactors are in many ways at the opposite extreme from biofilters. The biophase occupies most of the reactor ($e_b \rightarrow 1$ in Fig. 1), and it is dilute and well-mixed rather than dense and immobilized. The large volume of water makes it easy to add soluble nutrients, control the pH, and remove water containing nonvolatile products. However, care must be taken to not wash out the biomass, because $z = 1$ (Fig. 1) unless there are extra surfaces in the bioreactor for the attachment of a biofilm, or if a cell separation and recycle loop is added. Another major difference is that the gas–biophase interfacial area, a, is not only much smaller than in a biofilter, but is not even a constant, varying with the gas flow rate. Bubble gasified bioreactors are therefore best suited to contaminants like NO_x that are very soluble in water (thus reducing concerns about the gas–liquid mass-transfer rate), whose dissolution in water cause pH swings, and whose metabolism requires large amounts of dissolved nutrients.

Biotrickling Filters

Trickling filters for gas treatment are similar to, but usually taller than, those in use for decades for the secondary treatment of municipal and industrial wastewaters. Like biofilters, they consist of a bed of media through which the gas flows either upward or downward, but unlike biofilters, the microbial culture is recirculated continuously over the bed from a reservoir beneath it. Stone aggregate or wood chips are the traditional media, but even seashells have been used (27). Most modern filters employ ceramic or plastic media, Pall rings, Raschig rings, and so on specifically designed to make a bed with a high surface area to maximize gas transfer, and a high porosity to minimize the pressure drop and the chance of clogging or flooding (28).

These bioreactors can be thought of as intermediate between biofilters and bubble-gasified systems and are best suited to applications between those already described. If the characteristics of the media and the microorganisms are such that most of the biomass stays attached, then a trickling filter is essentially a biofilter with a higher liquid flow and a smaller gas–biophase interfacial area, a, due to the larger media. If on the other hand, most of the biomass stays suspended and the recycle ratio (RR = flow of recycled liquid/f) is high enough to keep the biophase well mixed, then it is functionally similar to a bubble-gasified bioreactor. The main differences are that the pressure drop, h, is usually much smaller, and the interfacial area, a, is a constant that is independent of the gas flow rate.

Bioscrubbers

The systems just described essentially combine two functions, stripping the contaminant from the gas into the aqueous biophase and then subjecting it to biodegradation. But what if the physicochemical conditions in the biophase resulting from the stripping are not suitable for the microbial metabolism? The removal of sulfur dioxide from stack gas is a good example because stack gases contain a small percentage of oxygen that although relatively insoluble, inhibits the bacteria responsible for the reduction of the sulfite formed by the dissolution of SO_2. One solution is the bioscrubber, which consists of an aqueous stripper in which the contaminant is transferred from the gas to water, and a separate bioreactor, in which the liquid effluent from the stripper is essentially treated as wastewater. The bioreactor should be a completely mixed, immobilized-cell type so that a large population can grow up in a well-controlled environment without continuously being recirculated through the potentially toxic environment in the stripper. Several configurations are possible, the simplest being that in which the liquid reservoir at the bottom of a packed-bed stripper is used as the bioreactor. Bioscrubbers are not considered further here because building a separate bioreactor and stripper is necessarily more expensive than combining them in a single unit, and when it is necessary, the two unit operations can be designed by conventional methods.

BIOREACTOR DESIGN

A process variable is any quantity that is under the direct control of the designer or operator of the process. For gas treatment bioreactors, the variables to be fixed by the designer include the bioreactor type, shape, and size, whereas those fixed by the operator include the addition of water, nutrients, and chemicals for pH control. The objective in fixing these variables is to satisfy a set of process requirements that specify the flow rate and composition of the gas stream, the nature and concentration of the contaminant, and the fractional removal required. The difficulty is that many designs, that is, many combinations of the process variables, will meet the process requirements for a particular case. The proper goal of engineering design is not simply to pick one of these designs at random, but to find the single design that meets the requirements at the lowest possible cost. The design of gas treatment bioreactors, like that of any other process, is an economic optimization problem, but one that is so complex that the exact solution cannot be found within the constraints of time and money normally imposed on the design process. The best solution for a particular case will certainly depend on the scale of the problem, the solubility and possible inhibitory effects of the contaminant, the type of microbial metabolism (aerobic or anaerobic, growth associated or cometabolic, etc.) needed to degrade it, and the temperature and pressure of the gas stream. Even coming close to the best solution requires a careful scale-up program in which the insights available from education and experience are used to integrate experiments at different scales with mathematical models that can extrapolate the results from one scale to the design of the equipment at the next scale. The bioreactor analysis presented in this section is intended as a guide to this program, suggesting which bioreactor types be chosen for experimental evaluation in each case, specifying the operating conditions around which the experiments should be designed, and showing how performance may be expected to vary with scale.

Because the design goals are essentially economic, the relationship of the variables shown in Figure 1 to the process costs must be kept in mind. The volume of the commercial-scale bioreactor equals the actual gas flow to be treated (a process requirement) divided by g; a large value for g translates directly into a smaller and, all other things being equal, cheaper bioreactor. The liquid flow, f should be kept as small as possible because providing clean water and disposing of any process wastewater both cost money, as do any nutrients and chemicals that must be added. The pressure drop, h, through the bioreactor may be significant, particularly on a large scale, and it determines the capital cost of the gas compressors and the energy costs for running them. Compressing the feed gas is not only expensive in itself, it heats the feed such that heat exchangers may be needed to cool it to the desired temperature.

Process Analysis

It will be assumed that the biophase is completely mixed and that the gas approximates plug flow through the bioreactor. This provides a realistic description of some bioreactors (bubble columns, some biotrickling filters), and a reasonable starting point for quantitative thinking about the others. More detailed mathematical models are available in the literature for specific types of bioreactors and metabolism (31,32). The equilibrium solubility of the contaminant in the biophase, S^*, will be described by Henry's law, $S^* = p/H$, where p = partial pressure of the contaminant in the gas and H = Henry's law constant. Although this is adequate for relatively insoluble and dilute contaminants, it is a considerable oversimplification for others. For example, the dissolution of oxides of nitrogen (NO_x: a mixture of NO and NO_2) from stack gases into water is an extremely complex process (34,35) that involves the gas-phase chemical oxidation of NO to NO_2 and a dissociation reaction that generates H^+, NO_3^-, and NO_2^- when the gas dissolves. In such cases, Henry's law can only be interpreted in the purely qualitative sense that low H means a soluble gas.

Any complete bioprocess model combines the mass conservation equations that describe the bioreactor with equations that try to approximate the complexities of microbial metabolism. Because the objective here is to compare different bioreactor configurations, it is essential to have consistent descriptions of the metabolism. In most gas treatment applications the contaminant is a major microbial nutrient, either the electron donor or the electron acceptor. Its consumption provides metabolic energy for growth and cell maintenance, so its specific consumption rate, q is given by the yield equation

$$q = \frac{\mu}{Y} + k \qquad (1)$$

where μ is the specific growth rate of cells, Y is the cell-yield coefficient, and k is the cell-maintenance coefficient.

In a well-operated bioreactor, the pH and temperature are carefully controlled and all required nutrients are provided in adequate amounts, making the specific consumption rate a function only of the concentrations of contaminant, S, and any inhibitory product, S_m, dissolved in the biophase. The exact form of this function, written $q(S, S_m)$, will not be specified because it varies among cases, but it must describe the effects on the metabolic rate of substrate limitation, product inhibition, and possibly (when n_i/H is large) substrate inhibition. It is more common to describe the specific cell growth rate, μ, by a Monod-type function of the concentration S, but this is clearly incorrect at the low concentrations found in gas treatment bioreactors, because it predicts that growth stops and substrate uptake continues when no substrate is available ($\mu = 0$ and $q = k$ when $S = 0$). In fact, when a major nutrient is exhausted, its consumption must stop, and the microorganisms go into an endogenous state in which the viable biomass declines, a phenomenon correctly approximated by requiring that $q(S, S_m) = 0$, and thus $\mu = kY$ (equation 1) when $S = 0$. Cell growth now stops at a nonzero "stationary phase" concentration, S_s, corresponding to the point where substrate uptake is just sufficient to satisfy the maintenance requirement, and defined by $q(S_s, 0) = k$.

For cometabolic processes, the contaminant is neither the main electron donor nor the acceptor; two rate equations are needed, and equation 1 contains an extra parameter whose value may be found from knowledge of the degradative pathway. See, for example, the analysis by Andrews of chloroform degradation by methanotrophic bacteria (36). The design of cometabolic bioreactors follows many of the principles described here but is too complex and case specific to be analyzed in detail.

At steady state the conservation of mass requires that the rate of growth of microorganisms in the bioreactor, $\mu e_b S_x$, equals the rate at which they flow out, fzS_x (see Fig. 1 for nomenclature). The loss of microorganisms in mist suspended in the effluent gas may be significant in some bioreactors without demisting devices but is not included explicitly in the analysis. The specific growth rate, μ, a measure of the physiological state of the microorganisms, thus equals zf/e_b, the reciprocal of the mean cell-residence time (the number of cells in the bioreactor divided by the cell outflow rate), which is a quantity under our direct control. This remarkable result, although well known for chemostats, has not been fully appreciated for gas treatment. It follows from equation 1 that the concentrations in the biophase, S and S_m, are constrained by

$$q(S, S_m) = k(1 + D) \qquad (2)$$

where $D = zf/Yke_b$ is the dimensionless dilution rate or reciprocal mean cell-residence time.

The steady-state conservation of mass for the contaminant requires that whatever is removed from the gas phase be transferred to the biophase, where it is either consumed by the microorganisms or flows out in the liquid. For a bioreactor, as opposed to a gas stripper, the latter term should be negligible, but the flows of dissolved nutrients and nonvolatile metabolic products must be included

$$\frac{g}{RT}(n_i - p) = \frac{k_1a(n_i - p)}{H \ln\frac{(n_i - HS)}{(p - HS)}}$$
$$= e_b q S_x = \pm\frac{(fS_j - (f + w)S_{ji})}{y_j} \quad (3)$$

where y_j is the amount of component j produced (+ sign) or consumed (− sign) per unit of contaminant consumed (see "Nutrient Addition and Start-Up").

The description of the driving force for mass transfer in terms of the log–mean concentration difference in the second term of this equation follows from the assumption of plug flow of the gas. The expression is not strictly correct when the pressure drop through the bioreactor is large, as in a tall bubble column, because the contaminant partial pressure in the influent is then $(1 + h)n_i$; but this complication is rarely included in practice. The essential points here are that the mass-transfer factor, k_1a, is an empirically determined parameter, that it can be based on different forms of the driving force, and that the best choice depends on the reactor type and must be clearly understood by everyone involved in a project.

Equation 3 can be rearranged to give the fractional removal of the contaminant from the gas, only the product (subscript m) being considered in the last term:

$$x = 1 - \frac{p}{n_i} = (1 - e^{-G})\left(1 - \frac{HS}{n_i}\right)$$
$$= \frac{S_x}{X}(1 + D) = FD\frac{S_m}{S'_m} \quad (4)$$

where $G = k_1aRT/gH$ is the dimensionless mass transfer coefficient; $F = YS'_m/zXy_m$ is the dimensionless product inhibition factor; $X = gn_i/ke_bRT$ is the maximum possible viable cell concentration in the biophase; and S'_m is the product concentration that completely inhibits cell growth.

Rather than solving equations 2 and 4 exactly, which would require the form of the inherent kinetic function $q(S, S_m)$, consider some limiting cases in order of decreasing liquid outflow rate. These cases are used to summarize the bioreactor performance in Figure 2, a graph that can be universal, at least within the limitations of the assumptions, because the parameters describing the contaminant (H, n_i), the reactor (f, k_1a), and the metabolism (k, S'_m) in individual cases have been incorporated into the dimensionless numbers x, D, F, and G.

- *Washout.* $D > (q(n_i/H,0)/k) - 1$: Because the dissolved concentration, S, must obviously be less than that in equilibrium with the feed gas, n_i/H, equation 2 now has no solution, meaning that all of the biomass is washed out of the reactor and there can be no contaminant removal.
- *Substrate limitation.* $D > (1 - e^{-G})/F$: Equation 4 shows that under this condition, $S \to 0$, while $S_m < S'_m$, meaning that microbial kinetics are limited by the availability of substrate rather than product inhibition and the exact value of the inhibition number, F, is irrelevant. If gas-biophase mass transfer is very rapid, the dissolved contaminant is in equilibrium with the outlet gas, giving the best possible bioreactor performance. The shape of this $G = \infty$ line in Figure 2 is simply a reflection of the relationship between $\mu(=zf/e_b)$ and $S(=p/H)$ from the inherent microbial kinetics. Mass-transfer limitation ($G < \infty$) leads to a progressive reduction in the contaminant removal, x.
- *Product inhibition.* $D < (1 - e^{-G})/F$: Now the concentration of metabolic products, S_m reaches inhibitory levels, S'_m, while dissolved contaminant is still available, meaning that metabolism is limited by product inhibition rather than substrate limitation, and the exact value of G is irrelevant. Increasingly stronger inhibition (smaller F) produces a series of curves in Figure 2 whose exact shape depends on the details of the inherent inhibition kinetics. All of these curves pass through the origin because, with no liquid outflow ($D = 0$), any soluble, nonvolatile metabolic products must eventually accumulate to inhibitory levels.

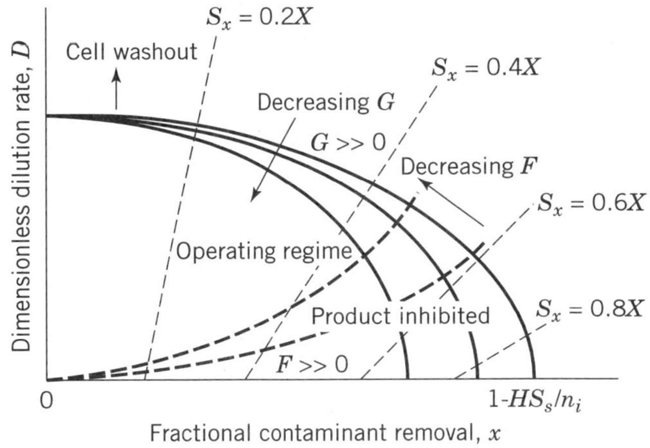

Figure 2. Operating regime for a completely mixed bioreactor.

Consequences for Process Design

Any pair of values of F and G define a regime in Figure 2 within which the bioreactor can operate. The most important result is that for any such pair of values there exists a value of D that maximizes the contaminant removal, x. If the form of the rate equation $q(S, S_m)$ were known, the pointed maximum seen in Figure 2 would be replaced by a smooth curve that defined this optimum design more accurately. In the absence of such detailed modeling the optimum must be found experimentally, the analysis providing a starting point and overall guidance. Because the gas flow rate, g, appears in the denominator of both F and G, the first step is to establish the maximum g for which the desired contaminant removal, x, falls safely within the operating regime. It is clear from equation 4 that a high removal rate (e.g., $x = 0.95$) with reasonable values of $S(\gg Hn_i)$ and $S_m(<0.5\ S'_m)$ requires both $G \sim 3$ and $FD \sim 2$, and such values are reasonable, if approximate, general design goals. Much larger values represent inefficient design because they imply excessive consumption of energy and water.

For processes that produce large amounts of very inhibitory, nonvolatile, soluble metabolic products, the problem is to wash the products out of the bioreactor without washing out the biomass. Mathematically, $y_m/S'_m \rightarrow \infty$, making it difficult to keep $FD > 2$ without D exceeding the washout criterion. These are the processes that require cell retention ($z \rightarrow 0$), either by cell recycle or by immobilization as a biofilm. For processes with more moderate values of y_m/S'_m, the analysis shows that there exists an optimum set of values of the process variables g, f, and z. Some processes, the aerobic mineralization of hydrocarbons for example, generate only innocuous (H_2O) or gaseous (CO_2) products, so $y_m/S'_m \rightarrow 0$, and the analysis suggests operating with no liquid outflow. This not only gives the best effluent-gas quality (Fig. 2 with $F = \infty$) but, because the biomass is in a maintenance condition ($\mu = 0$ when $D = 0$), it also avoids the cost of adding nutrients, such as phosphates, that are needed for microbial growth (see "Nutrient Addition and Start-Up"). In practice there are reasons to avoid this condition. Even if y_m/S'_m is too small to be observed in short-term laboratory experiments, any nonvolatile product must eventually accumulate to inhibitory levels in a continuous bioreactor with no liquid outflow. Salts in the water added to compensate for evaporation or for pH control will also accumulate in the biophase. Also, it must not be forgotten that complete mixing of the biophase is an abstraction never achieved in practice and sometimes, as in the biofilter, not even attempted. The discussion prior to equation 2 shows that any bioreactor with $D = 0$ must have a cell growth rate $\mu = 0$ as an average, but it will contain microniches where cells are dying because conditions are worse than average, and others where conditions are slightly better and cells are growing. Although the growing cells may assimilate some of the material released by the lysis of the dying cells, there must be some long-term accumulation of dead biomass, with consequent reduction in bioreactor performance. Keeping D on the of order of 1, a very low flow rate in most cases can wash out the dead biomass and ensure stable operation for the cost of a few growth nutrients.

The analysis assumed the usual case in which the cell concentration in the bioreactor is kept as high as possible, with sufficient amounts of other nutrients provided to ensure that the contaminant is the limiting nutrient. The quantity X, defined under equation 4, is then the maximum possible viable-cell concentration in the biophase, corresponding to the situation where all of the contaminant flowing into the bioreactor, gn_i/RT, is being used to satisfy the cell maintenance requirement, ke_bX. The actual concentration at any given operating condition, S_x, can be calculated from the fourth term of the equation and is shown in Figure 2 as a function of D and x. There is, however, no guarantee that this concentration corresponds to a dilute, easily mixed cell suspension. If the maintenance requirement, k, is small, and the contaminant loading gn_i is large, it may be a paste of cells too dense to be, for example, pumped around a biotrickling filter. Although it is possible to reduce S_x by restricting the amounts of growth nutrients provided, such artificial reductions in the microbial "catalyst" concentration are generally poor bioprocess design. The alternative is to choose the right type of bioreactor, recognizing that, although the mass of microorganisms in the bioreactor is fixed by the process requirements and the microbial metabolism, the amount of water is a process variable that can be controlled. The concentration, X, is an order of magnitude higher in a trickle-bed type bioreactor where $e_b \approx 0.1$, than in a bubble column where $e_b \approx 1$.

Nutrient Addition and Start-Up

In some processes, the aerobic mineralization of organics in beds of compost for example, the air provides the electron acceptor and the media provides what little growth nutrients are needed. In other processes these substances, and sometimes the electron donor, must be provided in the liquid inflow. The amounts can be estimated from the "stoichiometry" of metabolism, a procedure best illustrated by example. Ongcharit et al. (37) have shown that hydrogen sulfide can be removed from industrial gas streams by the acidophilic, autotrophic, sulfur-oxidizing, denitrifying bacterium *Thiobacillus denitrificans*. Even such esoteric metabolism can be approximated by a pseudo-chemical reaction in which all of the main nutrients and products are written in the ionic form in which they enter or leave the reactor. After performing all of the element and charge balances the result is

$$\begin{aligned} &H_2S + (1.6 - 0.2y\gamma)\,NO_3^- + yCO_2 + ypH_2PO_4^- \\ &\quad \twoheadrightarrow yCH_hN_nO_oS_sP_p + (0.8 - 0.1y(\gamma + 5n))N_2 \\ &\quad + (1 - ys)SO_4^{2-} + (0.4 + y(0.2\gamma - p - 2s))H^+ + wH_2O \end{aligned} \tag{5}$$

The first term on the right-hand side represents the dry weight of the biomass produced. $\gamma = 4 + h - 2o + 6s + 5p$ is a measure of its oxidation/reduction state and is remarkably consistent between species (36) (note that the base oxidation state for nitrogen here is N_2). The actual cell yield is y, the ratio of cells produced to H_2S consumed. For a completely mixed biophase this equals μ/q or, from equations 1 and 2,

$$y = \frac{DY}{(1 + D)} \tag{6}$$

The "stoichiometric coefficients" on the other components in equation 5 then provide the y_j values needed to calculate the concentrations of nutrients and products from the last term in equation 3. These simple calculations can provide a surprising amount of approximate information about process design and development. A feature of the example in equation 5 is the large amount of acid (H^+) produced by the metabolism. Even with the high liquid flow rate allowed by cell immobilization or recycle (29,37) its neutralization requires a high pH in the liquid inflow, so the biophase must be very well mixed if metabolism is not to be inhibited by pH gradients, suggesting some type of bubble-gasified bioreactor. If all of the NO_3^- and OH^- are provided by their sodium salts, then the salinity in the biophase would be high, suggesting a need for halophilic strains of bacteria. A reasonable alternative would be to explore the use of calcium salts, which would precipitate some of the product sulfate as gypsum, reducing inhibitory effects.

Like all the previous analyses, equation 6 applies only to steady-state operation. Gas treatment bioreactors, like any other bioprocess, are generally started with a relatively small inoculum of acclimated microorganisms, the objective of the start-up period being to increase the population to the steady-state value, S_x. This requires additional growth nutrients, and the initial feed concentration, S_{ji}, can be calculated as above, but with setting the observed cell yield, y, equal to the growth yield coefficient, Y. These concentrations can then be decreased slowly as steady state is approached. Start-up of biofilters is sometimes called ripening of the bed, and it is not always necessary to add growth nutrients, because the microorganisms capable of growth on the contaminant may obtain nutrients from the support material (compost, soil, etc.), or may grow at the expense of cells that die and lyse under conditions in the biofilter. If these sources are not sufficient to produce the biofilm thickness needed for maximum contaminant removal, some nutrient addition after steady-state contaminant removal has been reached will further improve performance. However, excessive nutrient addition must be avoided because modeling (38) suggests that, under contaminant limited conditions, a biofilm will grow much thicker than is needed. The inlet end of the biofilter may plug with useless biofilm.

Just because a steady state exists in theory, it does not follow that it can be reached by any start-up procedure. Of particular concern are those contaminants that are so concentrated (high n_i) and soluble (low H) that they will cause substrate inhibition of the microbial metabolism if their dissolved concentration in the biophase ever reaches equilibrium with the feed (n_i/H). The steady state analyzed above still exists, but $S \ll n_i/H$ only as a consequence of the metabolism of the large numbers of microorganisms present. Starting with a small inoculum, the dissolved concentration in the biophase will be much higher, metabolism will be inhibited, and the biomass may wash out. For similar reasons, batch experiments may produce the erroneous conclusion that the gas cannot be treated biologically. A proper start-up procedure may involve dilution of the feed gas and careful control of the dilution rate and nutrient addition, until a dense, well-acclimated microbial culture has developed.

Gas-Biophase Mass Transfer

Figure 2 shows that high levels of contaminant removal require high values of the dimensionless mass-transfer factor, G, which is actually the product of two dimensionless numbers: H/RT, an inverse measure of the contaminant solubility, and k_1a/g, a measure of the efficiency with which the gas flow generates mass transfer. This is not usually a difficulty for biofilters with their thin biophase and very large a but how can it be achieved for other bioreactor types? Measurements of gas–liquid mass transfer factors (28) are often given as correlations of the form

$$k_1a = AU_g^c \qquad (7)$$

The superficial gas velocity U_g is related to the flow rate g by $U_g = gL$, where L is the bioreactor height. Consider first an idealized bubble column, a hypothetical device in which all the bubbles are the same size, and doubling U_g simply doubles their number. The exponent c must then be 1, making $k_1a/g = AL$, a constant dependent upon scale but independent of gas-flow rate. In real bubble columns, with their imperfect bubble dispersion, c is closer to 0.7, but this still implies that G varies only slightly with g. The copious data available for oxygen transfer from air (28) shows that, at normal gas flows (g on the order of 1 VVM), k_1a/g is on the order of 1 at the laboratory scale, rising toward 10 in commercial-scale equipment. Assuming for the purposes of illustration that k_1 for a gas is proportional to its diffusivity in water raised to the 2/3 power (strictly valid only for relatively insoluble gases, for which most of the mass-transfer resistance is on the liquid side of the gas–biophase interface), this can be used to estimate G and the maximum possible fractional removal, $1 - e^{-G}$, (equation 4 with $S \to 0$) for contaminant gases of low, medium, and high solubility. The results shown in Table 1 suggest that bubble columns would be useless for insoluble contaminants like carbon monoxide, but worth consideration for moderately soluble contaminants like hydrogen sulfide. Most hydrocarbon vapors fall into the former category, but many alcohols and chlorinated aliphatic solvents are in the latter (30). For very soluble contaminants, such as sulfur dioxide, G tends to be unnecessarily large in bubble columns, and while acceptable removal with a moderate pressure drop could be achieved in very short, wide columns, this configuration has not been explored in practice.

The addition of a mechanical agitator can increase k_1a by roughly one order of magnitude, mainly by breaking up the gas flow into many small bubbles with a large interfacial area. Equation 7 still applies, but the A parameter becomes a function of the mechanical power input per unit reactor volume. Unfortunately, agitation also undermines the assumption that the gas moves though the reactor in plug flow. In a mechanically agitated tank, the gas bubbles are assumed to recirculate several times so that the gas phase is completely mixed; a sample of gas containing many bubbles taken from anywhere in the tank would have the same composition as the outlet gas. The effect on the driving force for mass transfer can be seen by comparing the concentration profiles shown in Figure 3a and b. Mathematically, the long-mean concentration difference used in equation 3 must be replaced by the more familiar effluent partial pressure driving force ($p/H - S$), which makes the maximum possible removal not ($1 - e^{-G}$), but $G/(1 + G)$. These two functions are virtually identical at small G, which explains why the exact definition of the driving force is not an issue for aeration (oxygen is an insoluble gas, $H/RT = 30$ at 20 °C, and its small fractional removal from the air stream in conventional fermenters is

Table 1. Performance of Bioreactors for which k_1a/g = 1–10 for O_2

			Maximum possible removal	
Contaminant gas	H/RT at 20 °C	G	Mixed gas phase $G/(1 + G)$	Plug flow gas $1 - \exp(-G)$
CO	37	0.023–0.23	0.022–0.19	0.023–0.20
H_2S	0.41	1.8–18	0.65–0.95	0.84–1
SO_2	0.0088	88–880	0.99–0.999	1

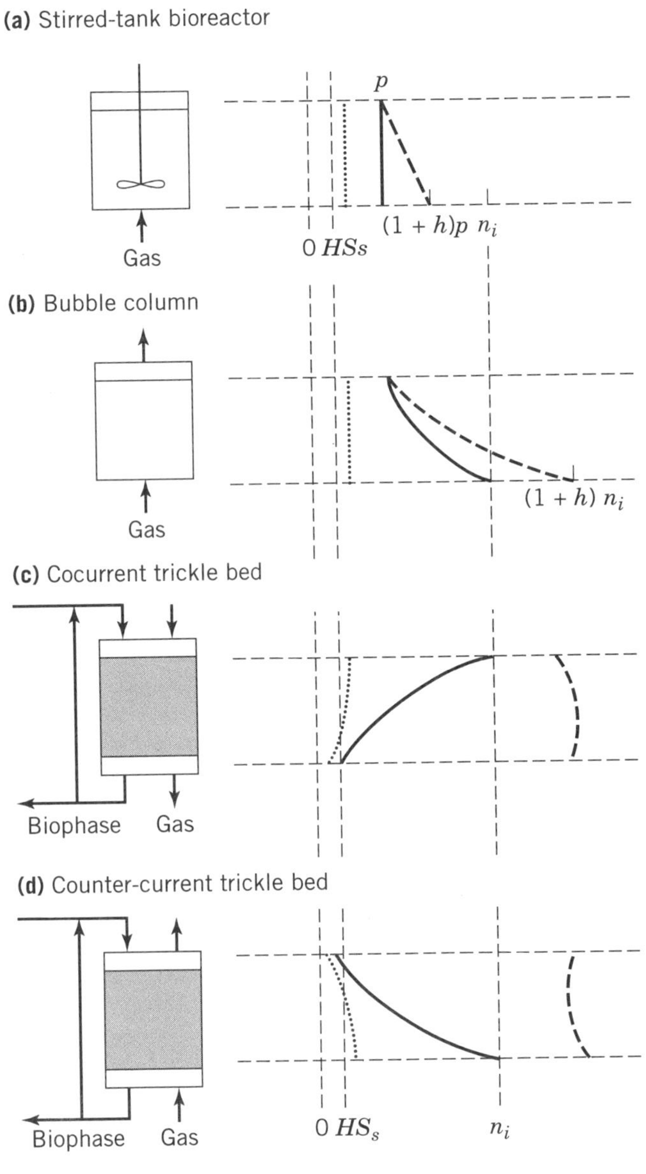

Figure 3. Concentration Profiles in Different Bioreactors. (**a**), (**b**) Partial pressure in gas phase, dot-dash line (**c**), (**d**) Viable biomass concentration, dot-dash line. Dotted line, partial pressure in biophase = HS; solid line, mole fraction in gas phase.

unimportant). However, a goal of 95% contaminant removal now requires not $G > 3$, but $G > 19$, which gives some idea of the lower effectiveness of a completely mixed gas phase. The numbers in Table 1 show that this lower effectiveness cancels out the increased k_1a leading to conclusions similar to those for bubble columns. The extra expense of mechanical agitation is rarely justified in gas treatment bioreactors, although reactors designed for a high k_1a/g with large aspect (height:diameter) ratios and multiple impellers to prevent bubble coalescence (29) have been proposed for removing hydrogen sulfide. Stirred tanks have been used successfully (39) for the microbial production of ethanol and acetic acid from carbon monoxide and hydrogen (another insoluble gas; $H/RT = 51$ at 20 °C), but the objective there was high volumetric productivity, rather than complete gas consumption. Several tanks were used in series, and they were pressurized to several atmospheres to improve gas transfer.

In contrast to bubble-gasified reactors, the gas–biophase interfacial area in biotrickling filters is almost constant, fixed by the choice of packing media and only slightly influenced by U_g and the superficial velocity at which the liquid is pumped over the bed U_1. This makes it possible to achieve the necessary value of G for contaminants of almost any solubility by appropriate choice of the packing media and the gas flow per unit volume, g. Many types of packing are available and the best choice in a particular situation is a complex trade-off between the following considerations:

- Less soluble contaminants will need the large a provided by smaller media and the small g implied by the lower permeability of such media to gas flow. For these contaminants, most of the gas–biophase mass-transfer resistance is on the biophase side of the interface, so in correlations of the form of equation 7, $c \to 0$, but A is a fairly strong function of U_1.
- More soluble contaminants can achieve the same G with the smaller a and larger g given by larger media. In this case most of the mass-transfer resistance is on the gas side of the interface, so $c > 0$ and A is less dependent on U_1.
- The conservation of mass shows that the change in concentration of a nonvolatile metabolic product in the biophase as it moves down the bed is $U_g x n_i y_m / U_1 RT$. If this number is small then the biophase approximates complete mixing. If it is large, then the concentration gradients, including that of pH, may be large enough to inhibit metabolism.

- Although a high liquid flow improves mixing and mass transfer, there is a definite upper limit of U_1 at which the bed floods. This limit is predictable (28) and is much smaller for small media.
- It is difficult to generalize about the attachment of the microorganisms to the media because this depends on both the nature of the media and the type of biomass involved. However, it is clear that a bed of small media with attached biomass may quickly become plugged if the contaminant is sufficiently concentrated and the biofilm is allowed to grow. It is reasonable to suppose that the larger the media and the higher the liquid velocity, the more cells will be suspended in the liquid rather than being attached. The z parameter in Figure 1 will be correspondingly larger.
- A biophase containing suspended and attached biomass may, if too thick or too dense, have its own mass-transfer resistance (38). While such resistance is usually to be avoided, it may be beneficial. For example, in trying to separate one electron acceptor, NO_x, from stack gases that contain another, O_2, which is preferred by the microorganisms, it should be possible to take advantage of the much higher solubility of the NO_x to create locally anaerobic environments in the biophase.

Somewhere among all of these considerations is a best choice of media for any situation. For dilute, less-soluble contaminants whose metabolism does not produce any nonvolatile metabolic products, the biotrickling filter will look more like a biofilter, with small media, little liquid flow, and attached biomass. For more soluble contaminants which do generate such products, it will tend toward the other extreme of high liquid flows and a completely mixed biophase. If there seems to be no sensible solution, the answer may be to adopt a bioscrubber. Although more expensive, it greatly simplifies the design problem by separating gas–liquid mass transfer from the microbiological considerations. It is certainly preferable to think in terms of this continuum of possibilities and to choose the one best suited to a particular application, rather than considering the different types of gas treatment bioreactors as completely separate technologies, each of which can be forced to fit any job.

Biophase Mixing and the Minimum Effluent Concentration

The single bioreactors with a completely mixed biophase analyzed here have an obvious drawback: all of the microorganisms are exposed to the same physicochemical environment, it is the worst possible environment for metabolism, with the lowest nutrient concentrations and the highest concentrations of metabolic products. One consequence appears in Figure 2. Even with no mass transfer resistance ($G = \infty$), no product inhibition ($F = \infty$), and no wastage of biomass ($D = 0$), the analysis suggests that complete contaminant removal is still impossible. With no biomass outflow there is no net cell growth, and all of the contaminant is used for maintenance metabolism (equation 2 with $\mu = D = 0$ gives $q = k$). Under these conditions, the dissolved contaminant concentration is not zero, but the stationary phase concentration, S_s, so the partial pressure in the effluent gas cannot be less than HS_s, and the fractional removal, x, can never exceed $(1 - HS_s/n_i)$, even with perfect mass transfer (equation 4 with $G = \infty$). This maximum removal also exists for completely mixed biological wastewater treatment systems, but in that case it does not matter because S_s is much less than common effluent standards. It may matter for some insoluble gas-phase contaminants because the corresponding minimum-achievable partial pressure, HS_s, may be significant, even when S_s is small. Bioreactors with a completely mixed biophase are a poor choice for such contaminants.

The simplest improvement would be two bioreactors through which the gas flows in series and between which the microbial culture is continuously recirculated. Reactor 1 would have a relatively high effluent partial pressure and dissolved contaminant concentration, providing a good environment for metabolism and cell growth. If the process variables are fixed correctly (a complete analysis is beyond the scope of this chapter) reactor 2 could run with a dissolved concentration $S < S_s$. The biomass in reactor 2 would then be in an endogenous state, but this would no longer matter because fresh, viable cells would be continuously supplied from reactor 1, and the dying cells would either flow back there to recover or be lost in the liquid effluent. The liquid inflow would go to reactor 1, where the growth nutrients are needed, and where it would dilute the concentrations of any metabolic products.

The more bioreactors in series, the better the average physicochemical environment for metabolism. Columns containing many sieve trays or bubble-cap trays have long been used for gas–liquid contacting in the chemical process industry and should be considered for the biological treatment of gases in cases where the biomass grows best in suspension and the contaminant is relatively dilute, insoluble, and a poor microbial substrate (i.e., HS_s/n_i is large). These columns are inherently countercurrent contacting devices with the gas flowing upward while the microbial culture falls down from tray to tray. An individual microorganism experiences a series of environments with ever-increasing dissolved contaminant concentration, S, until it is suddenly pumped back to the top tray where conditions are endogenous ($S < S_s$), but where it can continue metabolizing contaminant and survive long enough to start through the cycle again. Nutrient addition and pH control can be done on each tray as needed, and the liquid effluent can be taken from near the top of the column where cell viability is lowest.

A similar effect, with the gas closer to the theoretical optimum of plug flow, can be achieved in an upflow biotrickling filter with a small liquid recycle ratio. The only drawbacks are that pH control and nutrient addition can now only be done at the top of the bed, and the liquid can only be removed from the bottom, which is not necessarily the best arrangement. If all the microbial cells are suspended, the concentration profiles of viable biomass and contaminant will be as shown in Figure 3d. In real trickle beds some of the biomass remains attached, which is a great advantage over tray columns when a nonvolatile metabolic product is generated, because it allows a large liquid outflow to remove the product without the danger of washing out the biomass. (Mathematically, D is small because z is small, even though f is large.) Biotrickling filters can also be operated in the gas-downflow, cocurrent mode, which gives concentration profiles (again assuming suspended biomass) more like those shown in Figure 3c. Now, the microorganisms moving down the bed experience an ever-worsening environment for metabolism, with lower S and higher product concentration S_m. They leave the bottom of the bed in an endogenous, low-viability state, which is very suitable for the liquid outflow stream, but which may cause a significant lag phase when cells are recycled to the better environment at the top of the bed. The question of whether the cocurrent mode works better can only be answered definitively by experiments on a particular gas stream.

In biofilters the biophase is far from being completely mixed. It consists of an immobilized mass of up to xXe_b viable cells per unit bioreactor volume (less if cell growth has been limited by some nutrient other than the contaminant), and one of the goals of media selection is to provide sufficient surface area so that these cells can be distributed as a biofilm thin enough to avoid mass transfer limitation due to diffusion through the film. However, simple models (38) suggest that biofilters should share the minimum effluent partial pressure, HS_s, predicted earlier for bioreactors with a completely mixed biophase, because biofilms exposed to any lower partial pressures near the biofilter effluent would be in an endogenous state and eventually die. Experience suggests otherwise, probably because the adsorption of the contaminants on the media increases their effective concentration in the microorganisms' immediate environment. It follows that highly adsorptive support particles, such as compost or activated carbon, will work best, and that biofilters can treat very insoluble, hydrophobic contaminants if they adsorb well.

Scale-Up

First-stage scale-up experiments can be done in a reactor of any size that is convenient or available, as long as it is large enough so that the flows of gas and liquid are not significantly altered by entrance effects, channeling, or wall effects. Ideally, several different bioreactor types or packing materials are studied, but this is not always possible. The objective is to vary the gas flow, g, the liquid flow, f, nutrient addition, Sji, and so on, to determine how the specified contaminant removal can be achieved most economically. The scale-up problem is then to extrapolate the results in order to calculate the necessary volume (volume = gas flow to be treated/g) and shape (specifically the height, L) of the commercial-scale bioreactor, which must satisfy the process requirements without further experimentation immediately after start up. The best combination of process variables found in the experiments must be close to the optimum identified by the point at the right end of the operating region in Figure 2. To preserve this solution at the commercial scale, the dimensionless numbers D, F, and G that define this point must be kept the same between scales. Scaling up a gas treatment bioreactor is thus an example of Reynold's similarity principle, which states that two physical situations are similar only if the all of the relevant dimensionless numbers are identical.

Consider first the case where there is no inhibitory metabolic product and the process needs little or no liquid outflow. $F(\rightarrow\infty)$ and $D(\rightarrow 0)$ both drop from the list of relevant dimensionless numbers, and the scale-up problem reduces to finding combinations of gas flow, g, and bioreactor height, L, that keep G constant between scales. Given a mass-transfer correlation of the form of equation 7, G can be written $ARTL^c/Hg^{(1-c)}$. The problem is illustrated graphically in Figure 4, where lines of constant G are drawn on a log–log plot of g versus L. If the small-scale experiments have found different combinations of g and L that produce the specified contaminant removal, x, then this data should fall on such a line, labeled *empirical* in Figure 4, because the theory suggests that "constant x" implies "constant G." Also shown in Figure 4 is an upper limit on the gas superficial velocity, $U_g = gL$, caused by physical phenomena such as slugging of bubble columns, fluidization of the media in upflow biofilters, or flooding of the impeller in stirred tanks. This limit may be slightly different at different scales. Two extreme cases of the scale-up problem can be identified.

For a biofilter or a biotrickling filter treating sparingly soluble contaminants, the mass-transfer factor is little affected by the gas velocity, implying that $c = 0$ and making the empirical line horizontal. Several important phenomena are lumped into the parameter A, but there is no reason to expect them to vary with scale because microorganisms cannot "know" if they are in a small or large bioreactor. Consequently, performance on the large scale is defined by the same horizontal line, meaning that scale-up is done at constant g, or equivalently at constant gas-residence time ($=1/g$). Because $g = U_g/L$, constant g can be achieved either with a deep bed and a high gas velocity, the right end of the $c = 0$ line in Figure 4, or with a shallow bed and a low velocity, the left end of the line. Although the former are easier and cheaper to build, they have a much higher pressure drop though the bed, meaning higher costs for gas compression and possibly subsequent

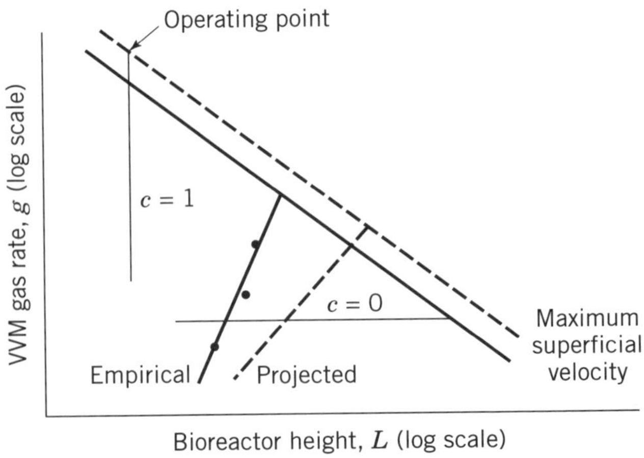

Figure 4. Dotted line, large scale; solid line, small scale.

cooling to return the gas to a temperature tolerable to the microorganisms. The latter are much cheaper to operate and work well as long as the beds are deep enough (at least several hundred times the particle size) to avoid problems of channeling and short-circuiting of the gas flow. Many biofilters consist of shallow beds of media on trays arranged in various configurations in a reactor designed to ensure an even flow of gas into the beds.

At the other extreme is the ideal bubble column described in the section "Gas Biophase Mass Transfer" for which $c = 1$. The value of G, and thus the contaminant removal, is now independent of gas flow but proportional to bioreactor height. Scale-up must therefore be done by keeping L constant but increasing the reactor width, the large-scale bioreactor operating at maximum g in order to minimize its volume (the *operating point* in Figure 4). Although this ideal bubble column is an abstraction, it illustrates a paradox that can arise in scaling up practical bubble columns (for which $c \cong 0.7$); if the desired removal can be achieved in a small-scale column, then a bubble column is probably not the best solution at the commercial scale. For very soluble contaminants, the L needed to obtain the desired fractional removal may be in the range of laboratory reactors, around 1 m. Scale-up would produce a very unusually shaped large-scale bioreactor, a similar height but several meters wide. There would be no point in making it any deeper, except to provide a safety factor in the design, and it could not be made any narrower and still meet the process requirements. For insoluble contaminants the necessary L may be several meters. The conclusion from laboratory experiments would then be that a bubble column is not a feasible bioreactor, even though the desired contaminant removal may be achievable (albeit with a high pressure drop) with the reactor heights used at the commercial scale.

All practical situations fall between these two extremes, so neither simple scale-up rule, constant g or constant L, is necessarily appropriate. The empirical line will have a positive slope, and a similar line for the same G at the larger scale can be projected from it if any variations of the parameters A and c with scale can be estimated from the published literature, experience, consultations with equipment manufacturers, and so on. The design of the large-scale bioreactor can be anywhere along this projected line in Figure 4, the exact design point again being constrained by the maximum U_g line, but essentially an economic trade-off between the large bioreactors demanded by a small g and the large pressure drops associated with a large L. The essential point is that the shape, and the size of the large-scale bioreactor are fixed by this procedure and should not be chosen based simply on guesswork or convenience.

Some bioreactor types allow design flexibility because they have variables that can be adjusted when drawing the projected line. Both A and c tend to decrease with scale in any bubble-gasified reactor due to poorer gas dispersion, but in mechanically agitated tanks, A is a function of the power input per unit volume, which can be varied to some extent. For trickle beds and biofilters, both A and the bed permeability are predictably related to packing size, allowing the packing to be chosen to give the large-scale reactor both a convenient shape and a reasonable pressure drop. Note that, whether or not such changes are made between scales, the estimates on which the projected line in Figure 4 is based should always be conservative in order to provide a factor of safety in the large-scale design.

When product inhibition is significant, the liquid outflow, f, becomes a critical variable, and all three dimensionless numbers D, F, and G must be kept constant if the physicochemical environment of the biomass (S, S_m) and the optimum bioreactor performance established in the small-scale experiments are to be extrapolated to the large scale. An exact solution is possible only when the retention of the biomass by cell immobilization or recycle (the z parameter) can be controlled, although when retention is almost complete ($z \rightarrow 0$) only the product DF (from which z cancels) is important, implying that the liquid flow, f, must be kept proportional to the gas flow, g. Biotrickling filters have an extra dimensionless number, the liquid recycle ratio, RR, which introduces the extra difficulty that increasing the bed height, while keeping f and RR constant increases the liquid superficial velocity [$U_l = Lf(1 + RR)$], thus changing the mass-transfer situation and possibly flooding the bed. The procedure just described will not work in these more complex situations (constant x no longer implies constant G), but the general considerations still apply, including the physical limits on the gas superficial velocity, the desirability of tall small-scale bioreactors to reduce uncertainties, the need for compromise between capital and energy costs, and the possibility of predicting a very inconvenient aspect ratio for the large-scale bioreactor. Judgment and calculation are needed in each case because the trivial solution to the scale-up problem—"build a bigger bioreactor with the same shape, f, and g as the small one"—is unlikely to be the best. Gas treatment bioreactors are one of the few devices whose performance tends to improve with scale, but taking advantage of this tendency in practice is not straightforward.

NOMENCLATURE

a	Gas–biophase interfacial area per unit bioreactor volume
A,c	Constants in mass-transfer correlation
D	zf/e_bYk = dimensionless liquid flow rate
e	Hold-up
f	Liquid outflow rate per unit bioreactor volume
F	YS'_m/zy_mX = dimensionless product-inhibition number
g	Volumetric gas outflow rate per unit bioreactor volume (the VVM gas rate)
G	k_1aRT/Hg = dimensionless mass-transfer-effectiveness number
h	Gauge pressure (atm) at gas inlet
H	Henry's law constant
k	Cell maintenance coefficient
k_1	Gas-to-biophase mass-transfer coefficient
L	Bioreactor height
n_i	Mole fraction of contaminant in feed gas

p Partial pressure (atm) of contaminant in effluent
q Specific consumption rate of contaminant
R Gas constant
S Dissolved concentration in biophase
S_s Stationary phase contaminant concentration (concentration at which $\mu = 0$)
S'_m Product concentration that totally inhibits metabolism
T Outlet gas temperature
U_g Superficial gas velocity at bioreactor outlet
U_1 Superficial liquid velocity in a trickle bed
X gn_i/ke_bRT = maximum possible cell concentration in biophase
x $1 - p/n_i$ = fractional removal of contaminant
w Evaporation rate per unit bioreactor volume
y Mole of nutrient consumed or product formed per unit of contaminant metabolized
Y Cell-yield coefficient
z Cell concentration in liquid effluent/cell concentration in biophase
γ Available electrons per carbon equivalent of biomass
μ Specific growth rate of biomass

Subscripts

i Inlet conditions
j Any nutrient or product
m Metabolic product
x Biomass

BIBLIOGRAPHY

1. G. Leson and A.M. Winer, *J. Air & Waste Management Assoc.* **41**, 1045–1053 (1991).
2. D.S. Hodge, V.F. Medina, R.L. Islander, and J.S. Devinny, *Environ Technol.* **12**, 655–662 (1991).
3. W.A. Apel, W.D. Kant, F.S. Colwell, B. Singleton, B.D. Lee, G.F. Andrews, A.M. Espinosa, and E.G. Johnson, *Emerging Technologies in Hazardous Waste Management IV*, American Chemical Society, Washington, D.C., 1994, pp. 142–159.
4. S. Dharmavaram, J. Casey, and C. van Lith, *Proc. Air Waste Management Association*, 93-WP-52C.01, Denver, Colo., June 13–18, 1993.
5. I. Togashi, M. Suzuki, M. Hirai, M. Shoda, and H. Kubota, *J. Ferment. Technol.*, **64**, 425–432 (1986).
6. W.A. Apel, P.R. Dugan, M.R. Wiebe, E.G. Johnson, J.H. Wolfram, and R.D. Rogers, *Emerging Technologies in Hazardous Waste Management*, vol. 3, American Chemical Society, Washington, D.C., 1993, pp. 411–428.
7. W.A. Apel, B.D. Lee, M.R. Walton, L.L. Cook, and K.B. Dinerstein, *Proceedings of the Air & Waste Management Association*, 95-TP9C.07, San Antonio, Tex., June 18–23, 1995.
8. P. Plumb, K.H. Lee, and K.L. Sublette, *Appl. Biochem. Biotechnol.* **24/25**, 785–797 (1990).
9. J. Croonenberghs, F. Varani, and P. Le Fevre, *Proceedings of the Air & Waste Management Association*, 94-RP115B.06., Cincinnati, Ohio, June 19–24, 1994.
10. Y. Yang, A.P. Tongas, and J.R. Blunk, *Proceedings of the Air & Waste Management Association*, 94-RA115A.04., 1994.
11. R. Whittenbury, K.C. Phillips, and J.E. Wilkinson, *J. Gen. Microbiol.* **61**, 205–218 (1970).
12. C.L. Haber, L.N. Allen, S. Zhao, and R.S. Hanson, *Science.* **221**, 1147–1153 (1983).
13. J. Pilkington, and H. Dalton, *Methods in Enzymology*, vol. 88, Academic, New York, 1990, pp. 181–190.
14. J. Colby, D.I. Stirling, and H. Dalton, *Biochem. J.* **165**, 395–402 (1977).
15. D. Scott, J. Brannan, and I. Higgins, *J. Gen. Microbiol.* **125**, 63–72 (1981).
16. Stanley, S. Prior, D.J. Leak, and H. Dalton, *Biotechnol. Lett.* **5**, 487–492 (1983).
17. R. Oldenhuis, R.L.J.M. Vink, D.B. Janssen, and B. Witholt, *Appl. Environ. Microbiol.* **55**, 2819–2826 (1989).
18. V. Hecht, D. Brebbermann, P. Bremer, and W.D. Deckwer, *Biotechnol. Bioeng.* **47**, 461–469 (1995).
19. B.R. Folsom, P.J. Chapman, and P.H. Pritchard, *Appl. Environ. Microbiol.* **56**, 1279–1285 (1990).
20. D. Arciero, T. Vannelli, M. Logan, and A.B. Hooper, *Biochem. Biophys. Res. Commun.* **159**, 640–643 (1989).
21. L.P. Wackett, G.A. Brusseau, S.R. Householder, and R.S. Hansen, *Appl. Environ. Microbiol.* **55**, 2960–2964 (1989).
22. M. Kastner, *Appl. Environ. Microbiol.* **57**, 2039–2046 (1991).
23. J.N. Petersen, R.S. Skeen, K.M. Amos, and B.S. Hooker, *Biotechnol. Bioeng.* **43**, 521–528 (1994).
24. J.M. Barnes, W.A. Apel, and K.B. Barrett, *J. Hazard. Mat.* **41**, 315–326 (1995).
25. W.A. Apel, J.M. Barnes, and K.B. Barrett. *Proceedings of the Air & Waste Management Association*, San Antonio, Tex., June 18–23, 1995.
26. Bohn H.L., *J. Air Pollut. Control Assoc.* **25**, 953–95 (1975).
27. M. Shields, M. Reagin, R. Gerger, C. Somerville, R. Schaubhaut, R. Campbell, and J. Hu-Primmer, in R.E. Hinchee, A. Leeson, L. Senprini, and S.K. Ong eds., Bioremediation of Chlorinated and Polycyclic Aromatic Compounds, Lewis, Ann Arbor Mich. 1994, pp. 50–65.
28. R.H. Perry and C.H. Chilton, *Chemical Engineers Handbook*, 6th ed., McGraw Hill, New York, 1988.
29. H.M. Lizama and E.N. Kaufmann, in D.S. Holmes and R.W. Smith eds., Minerals Bioprocessing II, TMS, Warrendale, Penn, 1995, pp. 241–253.
30. B.D. Ensley, Dept. of Energy Report DOE/CH-9207, 1992.
31. Z. Shareefden and B.C. Baltzis, *Chem. Eng. Sci.* **49**, 4347–4360 (1994).
32. R.M. Diks and S.P. Ottengraf, *Bioprocess Eng.* **6**, 93–99 (1991).
33. S. Lamare and M. Legoy, *Trends in Biotechnology* **11**, 413–419 (1993).
34. R.M. Counce and J.J. Perona, in N. Cheremisenoff ed., *Handbook of Heat and Mass Transfer*, vol. 2, Gulf, Houston, Tex., 1986, pp. 953–966.
35. Y.N. Lee and S.E. Schwartz, *J. Phys. Chem.* **85**, 840–848 (1981).
36. G.F. Andrews, Appl. Biochem. Biotechnol. **51/52**, 329–338 (1994).
37. C. Oncharit, Y.T. Shah, and K.L. Sublette. Chem. Eng. Sci. **45**, 2383–2389 (1990).
38. G.F. Andrews and K.S. Noah, *Biotechnol. Prog.* **11**, 498–509 (1995).
39. K.T. Klasson, M.D. Ackerson, E.C. Classen, and J.L. Gaddy, *Fuel* **72**, 1673–1678 (1993).

BIOREACTORS, MAMMALIAN CELLS.

See HYPOXIA, EFFECTS ON ANIMAL CELLS; MAMMALIAN CELL BIOREACTORS.

BIOREACTORS, PHOTO

MARIO R. TREDICI
University of Florence
Florence, Italy

KEY WORDS

Microalgae mass cultivation
Photobioreactors
Phototrophic microorganisms

OUTLINE

INTRODUCTION

Photobioreactors are bioreactors in which phototrophic microorganisms or plant cells are grown or used to carry out a photobiological reaction. In a broad sense, the open shallow basins used for microalgae cultivation could also be viewed as photobioreactors (1). In this article, however, the term *photobioreactor* is used to indicate only closed systems, that is, reactors that do not allow direct exchange of gases (e.g., CO_2, O_2) or contaminants (dust, microorganisms, etc.) between the culture and the atmosphere.

At present, commercial production of phototrophic microbial biomass is limited to two microalgae (*Dunaliella* and *Chlorella*) and one cyanobacterium (*Arthrospira*). Mass cultivation of these species is carried out in open ponds by means of a selective environment (high bicarbonate and high pH for *Arthrospira*, high salinity for *Dunaliella*), or by making use of large amounts of inoculum produced under strictly controlled conditions and taking advantage of a high growth rate (*Chlorella*). Most microalgal species cannot be maintained long enough in outdoor open-culture systems because of contamination by heterotrophic microorganisms (fungi, bacteria, protozoa) and competition by faster-growing microalgae that tend to dominate, regardless of the species used as inoculum. With the few exceptions mentioned earlier, virtually all attempts at mass cultivation of specific species of phototrophs in outdoor open ponds for prolonged periods have failed. An alternative choice is to use photobioreactors, which can provide an environment that is protected from direct fallout, are relatively safe from invasion by competing microorganisms, and in which conditions (pH, pO_2, pCO_2, temperature) can be better controlled so that dominance of the desired species is established for a sufficient period of time.

The use of photobioreactors could extend the exploitation of microbial phototrophs from the few strains commercially grown today to the more than 50,000 known species, many of which are potential sources of high-value products. Examples of such products are specialty lipids (sterols and polyunsaturated fatty acids); antifungal, antibacterial, or antiviral compounds; radioactively or isotopically labeled molecules (such as ^{13}C-labeled glucose and amino acids); vitamins (e.g., vitamin E); pigments (such as β-carotene, astaxanthin, and phycobiliproteins); pharmaceuticals (e.g., antineoplastic agents); and aquaculture feeds. The latter represent a good example of a case that could benefit in a decisive manner from the introduction of new photobioreactor concepts.

Understanding photobioreactors requires background knowledge on the fundamentals of the growth of phototrophs. The first part of this article therefore describes the principal characteristics of phototrophic microorganisms and their growth kinetics. The second part deals with photobioreactor designs, advantages, limitations, and commercial applications.

The topic of photobioreactors has also been reviewed by Lee (2), Chaumont (3), Prokop and Erickson (1), and more recently by Torzillo (4), Tredici and Chini Zittelli (5), and Pulz and Scheibenbogen (6).

PRINCIPAL CHARACTERISTICS OF PHOTOTROPHIC MICROORGANISMS

Photosynthesis and the Photosynthetic Apparatus

All organisms require a source of energy for growth to take place. Phototrophic organisms, or phototrophs, use light as an energy source. Light, or visible radiation, is the electromagnetic radiation in the 400- to 700-nm range, to which the human eye is sensitive. Because it is used by algae and plants for photosynthesis, visible radiation is also called photosynthetic active radiation (PAR).

Photosynthesis is the process by which phototrophs capture and convert the energy of photons (quanta of electromagnetic radiation, i.e., the smallest "packets" of light energy) to biochemically usable energy. It is through photosynthesis that phototrophs obtain the energy (ATP) and the reducing power (NADH or NADPH) necessary to incorporate CO_2 and to manufacture the organic molecules they require for growth.

Photosynthesis is generally associated with higher plants, yet about half the photosynthesis on earth is carried out by microorganisms. There are three major groups of phototrophic microorganisms: the photosynthetic purple and green bacteria, the cyanobacteria (also referred to by botanists as blue–green algae), and the microalgae. Photosynthetic bacteria, and cyanobacteria are prokaryotes; microalgae have a eukaryotic-cell-type organization. Photosynthetic bacteria differ from cyanobacteria and microalgae because of their photopigments and photosynthetic processes.

The ability to perform photosynthesis depends on the presence of light-absorbing pigments, the chlorophylls: large planar molecules composed of four substituted pyrrole rings with a magnesium atom coordinated to the four central nitrogen atoms. Chlorophylls absorb light primarily in the blue–violet and red regions of the spectrum and absorb weakly in the green region; this is why many phototrophs are green in color. Photosynthetic bacteria differ from cyanobacteria and microalgae in that their chlorophylls, called bacteriochlorophylls, have different molecular structure and absorption characteristics. For example, bacteriochlorophyll *b* of some purple bacteria has an in vivo maximum absorption in the infrared region between 1,020 and 1,040 nm. For purple bacteria, photosynthetic active radiation covers a much wider wavelength range than it does for microalgae and cyanobacteria, from the near ultraviolet to the long-wave infrared region (Fig. 1).

Besides the chlorophylls, other photosynthetic pigments, called accessory pigments, are also involved in the capture of light. Accessory pigments absorb light in the blue–green through yellow–orange range where chlorophylls absorb poorly and transfer the trapped energy to the chlorophylls, making photosynthesis more efficient over a broader range of wavelengths. The most widespread accessory pigments are the carotenoids, fat-soluble molecules that absorb light in the blue region of the spectrum, and the water-soluble phycobiliproteins of red algae and cyanobacteria that absorb in the green–yellow region. The accessory pigments also protect phototrophs from sunlight, which can damage the photosynthetic apparatus when it is too intense.

Cyanobacteria and microalgae also differ greatly from photosynthetic bacteria because of their photosynthetic process. Like higher plants, cyanobacteria and microalgae are oxygenic phototrophs; that is, they exhibit a type of photosynthesis that involves two photosystems (PSI and PSII), uses H_2O as electron donor, and evolves oxygen as

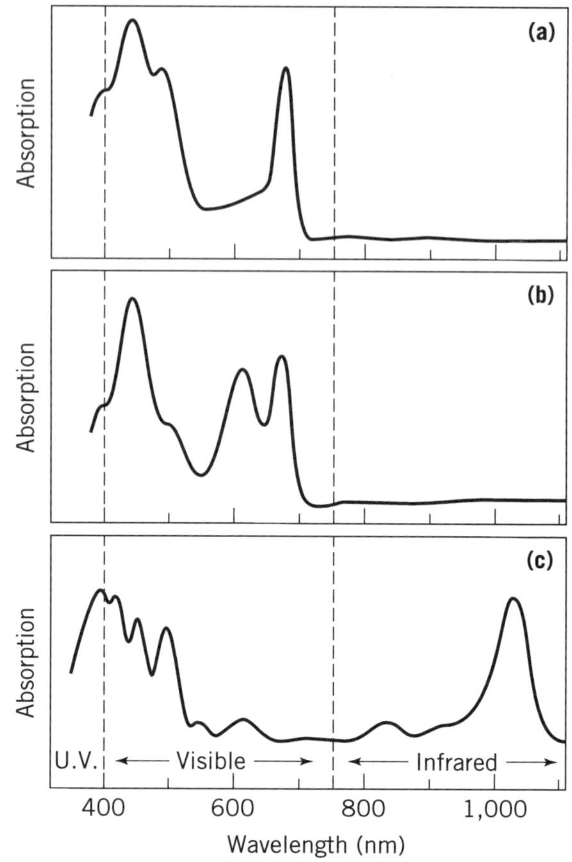

Figure 1. *In vivo* absorption spectra of (**a**) a microalga (*Scenedesmus acutus*), (**b**) a cyanobacterium (*Synechocystis* sp.), and (**c**) a purple bacterium (*Rhodopseudomonas viridis*).

a result of water photolysis. Green and purple photosynthetic bacteria lack photosystem II and cannot use water as an electron source; hence, they do not produce oxygen photosynthetically. For this reason they are called anoxygenic phototrophs. These definitions apply irrespectively of whether the organism obtains its cellular carbon principally from inorganic (autotrophically) or organic (heterotrophically) sources. Microalgae and cyanobacteria are mostly aerobic (they use oxygen as electron acceptor for respiratory metabolism) and autotrophic; purple bacteria are mainly anaerobic and heterotrophic; green bacteria are prevalently anaerobic and autotrophic. For a clear, comprehensive treatment of both plant and bacterial photosynthesis, see Hall and Rao (7).

Although this article deals exclusively with photobioreactors for the cultivation of oxygenic phototrophs (microalgae and cyanobacteria), some of the aspects covered and most of the conclusions can be extended to other phototrophic groups. For brevity, the term microalgae is used here in its botanical meaning, which includes cyanobacteria.

Response of Phototrophs to Light Quality and Quantity

The rate of photosynthesis achieved by a phototrophic cell depends on the rate of photon absorption and on the efficiency with which the absorbed photons are used. The rate of absorption and the efficiency of conversion of absorbed photons are in turn determined by the characteristics of the photosynthetic apparatus and by the spectral quality and intensity* of the incident light.

Photosynthesis as a Function of Light Quality. The wavelengths absorbed by a phototroph depend on its primary and accessory pigment content as shown by its absorption spectrum (Fig. 1). Only absorbed wavelengths are used, but not all the wavelengths absorbed are equally efficient for photosynthesis. Therefore, the spectral suitability of a light source for a phototroph is better represented by the *action spectrum* of photosynthesis (i.e., the curve obtained by plotting the *specific rate of photosynthesis [P]*† against wavelength across the photosynthetic range for the cultured organism). Although action spectra give more information than absorption spectra, the latter are more commonly reported because they are easier to determine and interpret. The curve obtained when growth, rather than photosynthesis, is plotted against wavelength is called a *growth action spectrum*.

Photosynthesis as a Function of Light Intensity. The rate of photosynthesis is not simply proportional to the rate of photon absorption, since photons may be captured by the pigments much faster than the photosynthetic apparatus can make use of them. If photosynthesis is measured at different light intensities, the specific photosynthesis rate *(P)* can be plotted as a function of light intensity *(I)* in the well-known light response curve of photosynthesis, or the *P–I* curve (Fig. 2). In the dark, where there is, of course, no photosynthetic activity, O_2 is consumed and CO_2 is released due to cellular respiration. As the light intensity gradually increases, photosynthetic O_2 evolution (and CO_2 consumption) takes place and a light intensity value, called the light compensation point or compensation irradiance (I_c), is eventually reached, at which value photosynthetic O_2 evolution balances respiratory O_2 consumption. Beyond this value, production exceeds consumption and net photosynthesis is achieved. From here on, *P* increases linearly with *I* up to a certain point, beyond which the rate of increase of *P* per increment in light intensity declines and eventually levels off. The slope of the linear portion of the curve, called alpha (α), represents the rate of photosynthesis per unit incident light or relative quantum yield (also called maximum photosynthetic efficiency or maximum quantum yield), and is a measure of the efficiency

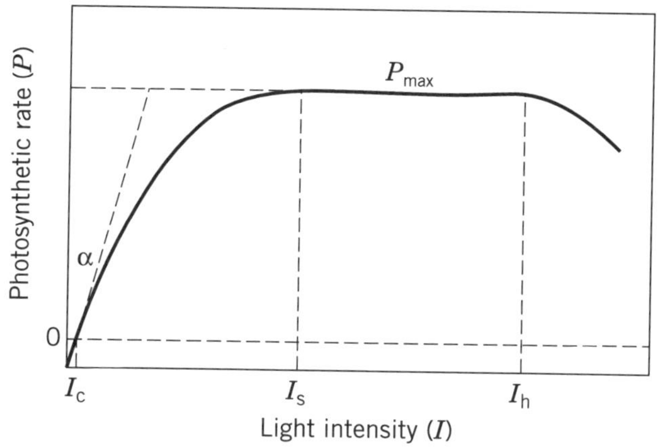

Figure 2. Light-response curve of photosynthesis (known as *P–I* curve). The intercept on the vertical axis is the measure of O_2 uptake due to dark respiration. I_c, light compensation point; I_s, light saturation intensity; I_h, light intensity value at which photoinhibition occurs. For details refer to the text.

with which the organism uses low light intensities. The relative quantum yield is expressed as moles of O_2 evolved per mole of photons of incident PAR. Plotting the absorbed light (incident light minus reflected and transmitted light) instead of incident light will give the absolute quantum yield. The reciprocal of the quantum yield (i.e., the number of photons required to evolve one molecule of oxygen) is called the quantum requirement of photosynthesis. In the range of intensities where *P* does not vary with *I*, photosynthesis is said to be light saturated and the light intensity is said to be saturating (I_s); *P* reaches its maximum value (P_{max}) in this range. If light intensity further increases, *P* may decrease again. This phenomenon involves damage to some components of the photosynthetic apparatus and is referred to as photoinhibition. At low light intensities, *P* is limited by the light reactions of photosynthesis (thermodynamic constraint). At saturating light intensities, *P* (P_{max}) is limited by the so-called dark reactions, in which the products of light reactions are used to fix CO_2 (metabolic constraint). *P–I* curves are species specific and must be determined under conditions such that all the cells receive the same photon flux and are not mutually shaded.

Hill and Bendall's widely accepted Z-scheme of oxygenic photosynthesis incorporates two photosystems (PSII and PSI), and establishes that it is necessary to move four electrons through each photosystem to evolve one molecule of O_2 and fix one molecule of CO_2. Since the transfer of each electron requires two photons (one in PSII and one in PSI), it follows that the theoretical minimal quantum requirement cannot be lower than 8 (i.e., not less than eight photons are needed to move four electrons from water to NADP through the two photosystems) and the maximal quantum efficiency cannot be higher than 0.125 (1/8).

The curve relating growth to light intensity has a pattern similar to the *P–I* curve (with growth rate increasing proportionally to light intensity when intensity is limiting and being independent of intensity when saturating values are reached), since phototrophic growth is a direct consequence of the photosynthetic rate.

*The spectral quality of light depends on the wavelength of each single photon. The energy content of a photon varies with its wavelength according to $\epsilon = hc/\lambda$, where ϵ is energy content, h is Planck's constant, c is the speed of light, and λ is the light wavelength in nm. Light intensity is given by the number of photons impinging on unit area in unit time (expressed as μmol photons $m^{-2}\ s^{-1}$). More properly the incident photon flux on a unit area in unit time is defined as photon flux density (PFD).

†Photosynthesis can be measured in terms of either CO_2 uptake or O_2 release. Rates of photosynthesis can be expressed per unit biomass, or any quantity proportional to it, to obtain the specific photosynthetic rate *(P)*. Typical units for *P* are μmol evolved O_2 (or fixed CO_2) mg Chl$a^{-1}\ h^{-1}$.

Growth of Microbial Phototrophs in Batch Culture

Phototrophic microorganisms can be grown in batch or continuous culture. A batch culture is initiated by the transfer of a small amount of cells (inoculum) into fresh growth medium; after the inoculum, nothing is added to (except light and carbon dioxide), or removed from (except oxygen), the system.

Growth Rate. Growth can be defined as the orderly increase of all the cell constituents. During this process, unicellular organisms grow to a maximum size, then divide to form two daughter cells. The daughter cells in turn repeat the process, and the number of cells doubles at regular intervals of time. The time required for a growing population to double is called the generation time, g. The rate of increase in biomass concentration is generally expressed as the specific growth rate, μ, which is related to the generation time by $\mu = \ln 2/g$. The specific growth rate, which can be defined as a proportionality constant for the autocatalytic growth reaction, is described by the following equation:

$$\mu = 1/X(dX/dt) \tag{1}$$

where μ is the specific growth rate (h^{-1}), X is the biomass concentration (g L^{-1}), t is time (h), and dX/dt is the variation in biomass concentration with respect to time (g L^{-1} h^{-1}), that is, the culture productivity per unit volume. The volumetric productivity ($dX/dt = \mu X$) of a culture in a pond or a flat reactor can be easily normalized on a per-unit-area basis by considering the culture volume (V) corresponding to the unit of illuminated surface area (A). Therefore, the areal productivity can be calculated as $\mu XV/A$, or, since V/A is the depth d of the culture, as μXd.

Growth Curve. When a phototroph grows under suitable conditions in batch culture, its growth curve follows a sigmoidal pattern showing six principal phases (Fig. 3a). Each phase is characterized by a particular value of the specific growth rate (Fig. 3b) and reflects a particular metabolic state of the cell population. After the inoculum, growth does not necessarily start right away since most cells may be viable, but not in condition to divide, especially when the parent culture is old. The interval necessary for the transferred cells to metabolically adapt to the new situation and start to grow is called the lag phase. After the lag phase, the culture enters into the acceleration phase, during which μ increases continuously. When a constant growth rate is reached, the culture is said to be in the exponential growth phase. During exponential growth, all the cells are light saturated (if a saturating light intensity is provided at the culture surface) and growth proceeds according to equation 1. The maximum μ is achieved in this phase, and in this respect phototrophs do not behave differently from chemotrophs. (*Note*: Organisms that obtain energy for growth from oxyreduction reactions involving either organic or inorganic electron donors.) The exponential growth phase of phototrophs in batch culture normally lasts for a relatively short period because the cells start to shade each other as their number increases; growth becomes light limited and μ decreases (deceleration phase). Self-shading and the fact that the energy source is instantaneously provided at the culture surface and cannot be stored in the reactor are the two fundamental factors that differentiate photobioreactors from reactors for chemotrophs. During the deceleration phase, although μ decreases continuously, the culture productivity may remain stable at maximum value because of the continuous proportional increase in cell concentration (X). Since the increase in cell biomass becomes almost linear, sometimes the initial part of this phase is called linear phase. Conditions of linear growth are typically sought in mass cultures of phototrophs, since they lead to the highest productivity. Following the linear growth phase, the cell population continues to increase, but μ decreases until it reaches zero, at which point the culture enters the stationary phase, during which the cell concentration remains constant at its maximum value. The stationary phase is followed by the death phase, in which the population of viable cells decreases.

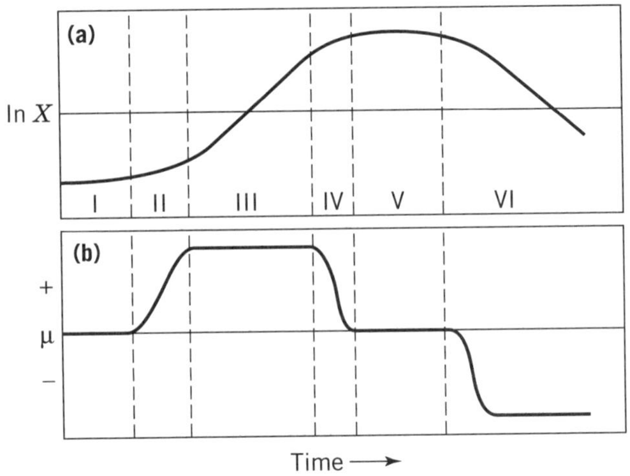

Figure 3. (**a**) Growth curve of a phototrophic microorganism in batch culture. The following six phases are shown: I, lag; II, acceleration; III, exponential; IV, deceleration; V, stationary; VI, death (after Ref. 8). (**b**) Changes in growth rate (μ) during the six typical growth phases.

To achieve continuous exponential light-saturated growth, the culture process has to be changed from batch to continuous; this is accomplished by continuously removing the newly formed cells. This condition is not adopted in photobiological process targeting production of biomass or related products, since it leads to low efficiency of conversion of the irradiance impinging on the culture surface and to low productivity (see later section).

Light distribution, Growth Rate, and Volumetric Productivity in Phototrophic Cultures

If nutrients are provided in suitable amounts and temperature is maintained at the optimum value, the productivity of phototrophic cultures becomes a function of light availability and cell concentration. In dilute cultures, self-shading is minimal and all the cells receive the same amount of light, independent of their position. It is self-evident that, at these low densities, even if the light intensity provided would permit the maximum growth rate, the culture will achieve low efficiency of light conversion and low productivity, since most of the impinging photon flux will pass through the culture unabsorbed (Fig. 4, upper curve). Mass phototrophic cultures, in order to obtain maximum productivity, must be kept dense enough to absorb all the light impinging on their surface. Under these conditions, self-shading, due to light absorption by cell pigments and scattering by the cells causes light intensity to fall off exponentially with the culture depth (Fig. 4, lower curve). In dense cultures, the light availability for each single cell is much reduced; this brings about a significant diminution of the specific growth rate, but productivity is, in any case, higher than in dilute cultures because the decrease in growth rate is more than compensated by the increase in cell concentration. For achieving maximal productivity in mass cultures it is mandatory to operate at a proper combination of culture depth and cell concentration. This concept, known as operating at the optimal population density (OPD), has been investigated by many researchers, most recently by Richmond and coworkers (9–11). In general, maximal productivity is obtained at relatively high population densities and at specific growth rates that are about one-half of the maximum. Increasing the cell concentration above the OPD is generally associated with a reduction in productivity because of the presence of a large dark zone inside the reactor, which leads to high maintenance requirements in relation to the overall

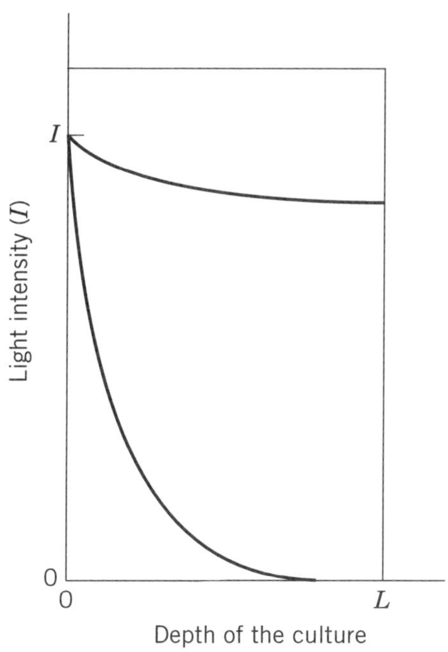

Figure 4. Light extinction in a phototrophic culture illuminated from one side with saturating light intensity. The upper curve represents a condition of low-cell concentration in which light absorption is negligible and all the cells receive almost the same light intensity. The lower curve describes a condition of high cell density in which the impinging photon flux is fully absorbed at two-thirds of the culture depth.

photosynthetic activity of the culture. Given the nonuniformity of the light environment that dominates in dense cultures, one requirement for achieving high photosynthetic efficiencies is that the individual cells be intermittently moved into and out of the illuminated zone. Although it has been shown that conventional mixing can do little to improve photosynthetic efficiency through what is known as the "flashing light effect,"* mixing is, in any case, required to provide intermittent exposure of the cells to light and darkness. A means to increase volumetric productivity in phototrophic cultures without increasing the light intensity at the culture surface is to shorten the light path by increasing the illuminated surface-to-volume ratio (*S/V* ratio) of the reactor.

Modeling Photosynthetic Growth in Photobioreactors

Mass phototrophic cultures are normally light limited, even though the light intensity striking their surface may be several times above saturation. For example, in outdoor algal cultures at midday, cells in the top layers will receive light at intensities above I_s (they will be photosaturated), or may even be photoinhibited, but a considerable portion of the culture will be photolimited or even below the compensation point (bottom layers). The rate of photosynthesis will thus be different in different regions of the culture, since it depends on the local environment experienced by the single cells, but the photosynthetic rate of the culture as a whole will be an average of all the individual rates. In the same way, while for a single cell the instantaneous specific growth rate, which depends on the instantaneous rate of light absorption and of photosynthesis, varies depending on the position of the cell, the average specific growth rate of the culture depends on the average amount of light received by the cells. It is hence a function of the culture depth (or light path) and of the cell concentration, as well as of the intensity of the incident light.

The exponential decay of the intensity of light as it penetrates into dense cultures is generally described by Lambert–Beer's law. This equation, however, accounts for the sole absorption by pigments and neglects light scattering by cells. More sophisticated models, (e.g., the radiative transfer theory developed in astrophysics), have been proposed to obtain radiant energy profiles inside reactors with a reasonable accuracy (13). Models of photosynthetic growth are based on the assumption that growth rate is closely related to the light availability inside the culture. It is thus necessary to consider both the effect of the light intensity available at each point of the reactor on growth rate and how the light environment is modified by absorption and scattering. This creates a high degree of complexity for the modeling of phototrophic cultures. A complete treatment of the kinetics and energetics of photosynthetic growth in photobioreactors has been presented by Cornet et al. (13). Accurate models for calculating local and average irradiance inside photobioreactors as functions of the design and of the impinging irradiance, and for relating

*The phenomenon, first described by Kok (12), by which the photosynthetic efficiency is increased when cells are exposed to light flashes instead of continuous light.

growth rate and productivity to average irradiance have been developed by Molina Grima and coworkers (14,15).

Light Utilization and Productivity in Phototrophic Cultures

Photosynthetic Efficiency of Outdoor Cultures. The photosynthetic efficiency of phototrophic growth is defined as the energy stored in biomass per unit of light energy absorbed (9). This parameter is of fundamental importance in mass cultures, since high photosynthetic efficiency leads to high biomass production per unit of illuminated area. As mentioned earlier, according to the current model of photosynthesis, eight photons are required to transfer four electrons from water to NADP and bring about the reduction of one molecule of CO_2 to carbohydrate in the dark reactions of photosynthesis. Since one mole of photons in the visible range contains on average about 50 kcal of free energy, and since the energy stored in carbohydrate during photosynthesis is about 114 kcal/mol of reduced CO_2, the theoretical maximum photosynthetic efficiency is approximately 28.5% [114/(8 × 50)]. The biosynthesis of the other macromolecules making up the microbial cell (lipids, protein, nucleic acids) requires additional reducing power and ATP and hence additional photons per CO_2 incorporated; this brings the quantum requirement for growing cells to 10–12, rather than 8, and reduces the maximum efficiency to about 20%. In practice, this efficiency cannot be reached with sunlight because of several losses and limitations, which include the fact that PAR represents only 43% of the total solar radiation at the earth's surface, the loss of part of the incident radiation by reflection at the culture surface, and the fact that part of the absorbed energy is invested in cell maintenance. Another major limitation, often ignored, is represented by the light saturation effect, that is, the fact that the photosynthetic apparatus of most phototrophs saturates at only 10–20% of full sunlight (about 2,000 μmol photons m^{-2} s^{-1}). According to the equation derived by Bush ($f = I_s/I\,[\ln I/I_s + 1]$) (16), which expresses the fraction (f) of the thermodynamic efficiency that can be attained as a function of the incident light intensity (I) and the saturation intensity (I_s), the quantum efficiency comes very close to the theoretical at low light intensities (<100 μmol photons m^{-2} s^{-1}). When the light intensity approaches that of full sunlight, utilization drops to less than 20% and most of the light energy absorbed by the culture is dissipated as fluorescence and heat. There are not many options available to overcome the limitation imposed by the light saturation effect. In dense phototrophic cultures, mixing partially compensates for the limitation imposed by the light saturation effect, but does not overcome the substantial inability of phototrophs to use effectively high irradiances (17). Benemann (18) discussed the alternative concept of increasing the light intensity at which photosynthesis saturates by decreasing the amount of accessory pigments in the cells. This however, is not easy to achieve and may lead to another problem: Contaminating algae with higher pigment content, because of their ability to absorb more light in the light-limited zones of the culture, will overgrow the cultured alga. Based on studies that took this direction in the 1950s, Tredici and Chini Zittelli (19) have recently proposed spatial light dilution as a means to increase the photosynthetic efficiency of outdoor algal cultures (see later section). Considering all of these limitations, it is reasonable to assume that only about 4–5% of solar energy can be converted to the chemical energy of biomass, under even the best conditions (17). For locations characterized by an annual average insolation of 4,500 kcal m^{-2} d^{-1} (e.g., southwestern United States), and assuming a caloric content of the biomass of 5.4 kcal g dry wt^{-1}, this efficiency will result in a calculated average annual productivity between 33 and 42 g m^{-2} d^{-1}.

Productivity of Outdoor Cultures of Phototrophs. The productivity of outdoor cultures is one of most debated issues connected with mass cultivation of phototrophs. Extrapolations from efficiencies of light conversion obtained in the laboratory had led to predict annual productivities as high as 140 g m^{-2} d^{-1} (20), but actual productivities are only one-tenth or less these figures, with about 40 g m^{-2} d^{-1} being considered the highest value obtainable by outdoor algal cultures under favorable conditions (17,18). A common mistake made (more or less consciously) in this respect is that the results obtained on a small scale over short periods and under optimal conditions are extrapolated to all-year-round, full commercial scale. As a matter of fact, despite statements in the early years of algal biotechnology, large algal cultures are not much more productive than terrestrial systems, and generally are much less reliable. In the author's opinion, however, productivity should not be considered as being the decisive factor in determining the viability of algal biotechnology. It is true that productivity greatly influences the economy of the process, but other factors, like sustainability and reliability, are just as, or even more, important.

Several models have been proposed to predict the productivity of outdoor cultures of phototrophs both in open ponds (21,22) and in photobioreactors (13,15).

How Productivity of Phototrophic Cultures Is Expressed. In the case of algae cultivation in open ponds, it is common to evaluate productivity as areal yield (i.e., productivity per unit of occupied land area per unit time, expressed either as g m^{-2} d^{-1} or tons ha^{-1} $year^{-1}$). This is both practical, since it makes comparison with terrestrial systems easy and immediate, and correct, since areal productivity in ponds is proportional to the amount of light entering the system per unit illuminated area. In a pond this corresponds substantially to the land surface area occupied by the system.

With photobioreactors the situation is much more complex. Except in the cases of flat horizontal systems and tubular horizontal photobioreactors with no empty space between contiguous tubes (in both these cases the areal productivity referred to the occupied land is correct and easy to interpret), expressing productivity on a per land surface area basis is misleading. For example, in some horizontal serpentine bioreactors it is difficult to decide whether and how to compute the empty space between contiguous tubes and how to account for the fact that the reactor intercepts a different proportion of the radiation impinging on the horizontal, depending on the elevation of the sun and, hence, on the hour of the day. With elevated systems, the inadequacy of the areal yield is even more evident, since the amount of radiation intercepted by a vertical or sharply inclined reactor is not related to the land surface area it occupies. Expressing the productivity of elevated systems on a per unit occupied land surface area basis (i.e., with reference to the horizontal projection of the reactor on the ground) may lead to high unrealistic figures (see later section). Productivity can also be referred to the illuminated surface area of the reactor (biomass produced per unit illuminated surface area per unit time, in g m^{-2} d^{-1}), but this parameter is also not adequate for all situations, since it can be difficult to define the correct illuminated surface area to which to refer productivity. In tubular reactors, for example, the illuminated surface area generally corresponds to half the total surface area of the reactor; using it as basis for calculation of productivity in most cases underestimates reactor performance. In some vertical reactors (e.g., large-diameter coils that receive a significant amount of radiation on their inner central part at midday, but not in the early morning nor in the evening), the illuminated surface area changes during the day with sun elevation. Another complication with vertical reactors is determined by the fact that they intercept a significant amount of diffuse radiation on the surface not directly illuminated by the sun, and this is not always taken into account.

For these reasons, photobioreactor performance, like that of fermenters, is frequently reported in terms of volumetric productivity (amount of biomass produced per unit reactor volume per unit time, commonly expressed in g L^{-1} d^{-1}). But volumetric productivity is a function of the number of photons that enter the unit reactor volume in the unit time and, as such, is dependent on the illuminated *S/V* ratio of the reactor. The higher this ratio, the higher the volumetric productivity. It is well known, for example, that small-diameter tubular reactors or thin flat systems achieve high volumetric productivities, even if they perform poorly. We should be aware of the fact that a volumetric productivity of 1 g L^{-1} d^{-1} assumes a completely different significance if obtained in a tubular reactor of 1 cm or of 5 cm diameter.

To overcome these limitations, Tredici and Chini Zittelli (19) suggested evaluating the performance of vertical photobioreactors on the basis of the photosynthetic efficiency achieved by the culture. Unfortunately this parameter is not always easy to calculate (e.g., in coiled reactors) and since it depends heavily on the irradiance (the lower the irradiance, the higher the efficiency) favors experiments carried out under low light intensities, such as in winter.

The productivity of a photobioreactor could be expressed on a per shaded surface area basis (the productivity of the reactor divided by the average ground surface area shaded by the reactor). This parameter, never suggested before, substantially reflects the efficiency of light utilization, but also takes the actual irradiance conditions into account and offers the possibility to compare very different culture systems. For example, the performance of an open pond and of a vertical coiled reactor could be compared on this basis, whereas it is rather difficult to envisage any other suitable comparison parameter for these two very different systems. The per shaded surface area parameter cannot be used alone, however, since in some cases (e.g., serpentine reactors) it does not give a correct evaluation of the efficiency of land utilization.

In conclusion, each method for expressing productivity has its own limitations, and complete evaluation of a reactor performance in terms of productivity will be achieved only by a combination of the different available methods. For example, expressing the productivity on the basis of unit shaded area, unit occupied area (including eventual empty spaces), and unit volume will allow evaluation (and comparison) of photobioreactors in terms of efficiency of light conversion, efficiency of land use, and efficiency with which the unit volume of reactor is used. This combined approach is highly recommended.

Artificial Light Sources. Solar energy is a relatively cheap energy source, but given the low photosynthetic efficiency attainable under full sunlight, and the fluctuations in supply determined by the weather, the time of the year, and the natural day/night cycle, the use of artificial light sources to grow phototrophs has been taken into serious consideration, at least for production of high-value compounds.

The use of artificial illumination offers the possibility of selecting light quality and intensity, as well as the length of the illuminated period, and thus of significantly enhancing the reliability of the production process. This, together with the higher photosynthetic efficiency attainable (a photosynthetic efficiency of 15% is not uncommon with artificial light of relatively low intensity), may compensate in many cases for the higher cost.

One of the fundamental factors guiding choice of the light source should be the efficiency of conversion of electric energy into PAR energy. This efficiency varies considerably with the type of lamp; for example, it is only about 5% in tungsten–halogen (incandescence) lamps, 10–15% for fluorescent tubes, and 20–25% for high-pressure sodium lamps. Only with xenon lamps and light-emitting diodes are very high efficiencies attained (50% and up to 80%, respectively). Considering both the efficiency of conversion of electrical energy into PAR and the efficiency of conversion of light energy into biomass, an overall efficiency of conversion of electrical energy into the energy of biomass between 0.7% and 2% has been attained by *Arthrospira* cultures grown under fluorescent lamps (13). Another aspect to consider is the spectral quality of the emitted light, which ought to match the action spectrum of the cultivated organism as closely as possible. For example, tungsten lamps have a low output in the blue region and are therefore not appropriate for oxygenic phototrophs. High-pressure sodium lamps are relatively efficient as regards electrical energy conversion but emit in a limited yellow–orange region. Practicality often prevails over efficiency considerations in the choice of an artificial light source. For example, fluorescent lamps are widely adopted, independently of the spectral quality of their emission, because they provide a relatively uniform irradiance over a large surface.

A SURVEY OF PHOTOBIOREACTORS

Classification of Photobioreactors

Photobioreactors can be classified on the basis of both design and mode of operation. In design terms, the main categories of reactors that can be envisaged are *(1)* flat or tubular; *(2)* horizontal, inclined, vertical, or spiral; and *(3)* manifold or serpentine. An operational classification of photobioreactors would include *(4)* air or pump mixed and *(5)* single-phase reactors (filled with media, with gas exchange taking place in a separate gas exchanger), or two-phase reactors (in which both gas and liquid are present and continuous gas mass transfer takes place in the reactor itself). Construction materials provide additional variation and subcategories—for example, *(6)* glass or plastic and *(7)* rigid or flexible photobioreactors.

Axenic photobioreactors are reactors operated under sterile conditions. Although a major characteristic of photobioreactors is their ability to limit contamination, it must be made clear that an effective barrier, and thus operation under truly sterile conditions, is not achieved, except in the few special designs developed expressly for that purpose. This is a major difference between photobioreactors and bioreactors for chemoheterotrophs, which typically operate under sterile conditions. Like fermenters, photobioreactors can be divided into laboratory photobioreactors used for research, pilot-plant photobioreactors used to scale-up and optimize the process, and production units. Unlike fermenters, there are only a few examples of commercial photobioreactors in operation. Some companies sell photobioreactors in a variety of sizes for use in teaching, research, and commercial applications. The emphasis in this article is on pilot-culture systems and the two commercial-scale plants realized to date. Laboratory systems, such as illuminated carboys and fermenters, immobilized cell systems, and optical-fiber reactors are not considered. Cell-immobilization techniques have been reviewed by Brouers et al. (23), whereas optical fiber reactors have been discussed by Pulz and Scheibenbogen (6).

Early Photobioreactors

The use of photobioreactors for microalgae mass culture dates back to the late 1940s and was a logical consequence of the investigations carried out with *Chlorella* on the fundamentals of photosynthesis. In those early days of algae culture, the emphasis was on continuous controlled cultivation; hence, closed reactors were given priority. Open systems were considered inappropriate to guarantee the necessary degree of control and optimization of the process.

One of the first closed systems was set up at Stanford University in California in the late 1940s and involved use of vertical tubular reactors (24). The units consisted of glass columns 1.8 m in height and 10 cm in diameter, constricted at the bottom to prevent algal settling. The system was operated turbidostatically, with medium addition being regulated by a photoelectric cell. Culture temperature was maintained constant by means of a cooling coil inserted into the reactor. Carbon dioxide–enriched air was injected at the bottom of the column to obtain mixing and pH/pCO_2 control. The system was experimented with in the cultivation of *Chlorella* under both artificial light and sunlight. Maximum volumetric productivity indoors was 0.48 g L^{-1} d^{-1}. Outdoors, productivity averaged 0.28 g L^{-1} d^{-1}, with a maximum of 0.35 g L^{-1} d^{-1}. The reason for the lower outdoor productivity was not clear.

In the early 1950s, the Arthur D. Little Company at the Massachusetts Institute of Technology, under contract with the Carnegie Institution of Washington, developed two basic types of photobioreactors: an array of 10 vertical rigid plastic tubes and a 50-m^2 horizontal flexible tubular unit (25). The vertical units, which were essentially the same as those operated at Stanford University, provided inoculum to the larger horizontal unit and were reported to operate satisfactorily, although no productivity data were actually presented. The horizontal tubular unit may be considered the first pilot plant for microalgae production. It consisted of 4-mm-thick polyethylene tubes that, once filled with algal culture up to a depth of about 8 cm, assumed a more or less elliptic shape of 1.2-m width. The tubes were joined at the ends to form a raceway 21 m in length, for a total surface area of about 56 m^2. The culture (*Chlorella*) was circulated by means of a centrifugal pump. The pumping system was not capable of generating circulation velocities over 7.5 cm s^{-1} and there was considerable settling of the culture. With the same pump and a tube width of 0.43 m, flow rates of 25 cm s^{-1} were achieved and problems of settling were eliminated. Leakage and contamination by *Chlorococcum* and protozoa were the major problems encountered in the operation of the system. Spray cooling proved inadequate for temperature control, and it was necessary to install a heat exchanger. Despite these difficulties, a 40-day run attained an average areal productivity of 9 g m^{-2} d^{-1}. The data collected with this pilot plant provided the material for an economic analysis of algae production in photobioreactors (26), which concluded that the cost of products obtained in photobioreactors was excessive in comparison to that of equivalent materials obtained using conventional methods. Cooling of the culture was found to be a major cost factor.

At about the same time in Japan, Tamiya and coworkers (27) developed a system consisting of a serpentine tubular reactor immersed in a water bath for temperature control and an external column for gas exchange. This system provided the basic design criteria for all subsequent, essentially identical, serpentine reactors (e.g., those adopted by Gudin in France and Molina Grima and coworkers in Spain) (see later section). The reactor was made up of 3-cm-diameter glass tubes joined to form a serpentine and had a total length of 33 m. The total culture volume was 40 L. The culture *(Chlorella ellipsoidea)* was circulated at a linear velocity of 15–30 cm s^{-1}. A separate column, in which a mixture of 5% CO_2 in air was injected at a rate of 1–1.5 L min^{-1}, provided mixing and gas exchange. Productivity was directly related to solar irradiance with the highest value, achieved at the maximum irradiance, of about 0.6 g L^{-1} d^{-1}. After several months of experimentation, the authors concluded that one of the main limitations was temperature control and that a significant increase of productivity could be attained through cultivation of algae exhibiting higher temperature optima and through vigorous mixing aimed at increasing growth by providing intermittent illumination. These concepts still guide most of our research.

Following these initial studies in the 1950s, little work was carried out in this field for several decades. Two exceptions were the work of Setlik in Czechoslovakia (28) and Jüttner in Germany (29). The reasons for this almost total abandonment of research activity on photobioreactors are not clear, although the successful production of *Chlorella* in open ponds in Japan, the development of microalgae-based wastewater treatment in California, and later success with open-raceway systems for outdoor production of *Spirulina* (*Arthrospira*) and *Dunaliella* perhaps explain the relative lack of interest in closed systems. The difficulties experienced by the A.D. Little Co. and Fisher's negative conclusion in his economic analysis also contributed to discouraging further studies on the application of photobioreactors.

Recent Types of Photobioreactors

Several types of photobioreactors have been designed and experimented with since the work of Pirt and coworkers sparked renewed interest in closed systems in the early 1980s (20). Most of these are small-scale systems; some are at the pilot or prepilot stage of development. Only in a very few cases has scale-up to commercial size been attempted. For convenience, these systems are described herein by categories: *(1)* axenic photobioreactors, *(2)* vertical tubular photobioreactors, *(3)* serpentine photobioreactors, *(4)* manifold photobioreactors, *(5)* helical photobioreactors, and *(6)* flat photobioreactors. Small systems and reactors for which experimentation has been conducted prevalently indoors with artificial light are not considered.

Axenic Photobioreactors Setlik (28) and later Jüttner (29) and Krüger and Eloff (30) developed glass tubular reactors for the axenic cultivation of microalgae. Jüttner's system, which substantially reproduced Setlik's design, was an all-glass, 110-L vertical flat coiled photobioreactor that could be sterilized by steam. The reactor, made of 4-cm-ID tubes, had a total length of 79 m. The algal suspension was circulated by a glass centrifugal pump at a flow rate of 20 to 60 cm s^{-1}. The temperature was regulated through a 2-m water-jacketed section. Gas exchange took place in an external exchanger in which a CO_2–air mixture was injected at a rate of 300 L h^{-1}. The system attained a maximum productivity of about 0.5 g L^{-1} d^{-1} under artificial illumination. Cell damage caused by the centrifugal pump was a problem in the case of fragile cyanobacteria; other problems encountered were foam formation and cell adhesion to the glass walls. Jüttner concluded that cell adhesion was the most discouraging feature and suggested that cultivation ought to be limited to "nonsticky" algae. Wall growth and fouling are still today major limitations in photobioreactors, although the problem is not always recognized as such because experiments generally do not last long enough for these effects to become evident.

A similar 60-L system, composed of 12 5-cm-ID glass tubes each 2 m in length, coupled by U-bends to form a vertical flattened spiral, was operated by Krüger and Eloff (30), who reported a productivity of about 0.13 g L^{-1} d^{-1} with a nontoxic species of *Mycrocystis*.

Soeder et al. (31) devised a rocking tray for the sterile cultivation of *Scenedesmus*. It consisted of a flat horizontal unit (inner dimensions 59.5 × 36 × 9 cm) made of stainless steel and covered with a glass lid. The tray was placed on a support connected to a rocking device. Temperature was controlled by circulating thermostated water through the double bottom of the unit.

Although these systems appear difficult to scale up and are rather expensive, they are mentioned here since with the exclusion of laboratory reactors they are the only systems that allow sterile cultivation of phototrophic microorganisms.

Vertical Tubular Reactors. Vertical tubular reactors (or column reactors) are simple systems in which mixing is achieved by injecting compressed air at the bottom of the column. Several vertical tubular photobioreactors were developed following Cook's first design. An example is the 30-L (2-m height, 15-cm ID) tower-type glass reactor devised by Jüttner (29). The main characteristic of this system was a wide, hollow glass finger inserted from above into the column to reduce dark space inside the reactor. Since the finger was open at the upper end, it could also be used to regulate the culture temperature. A thickness of the annular culture chamber of about 3.5 cm was found to be optimal for the growth of the light- and shear-sensitive *Microcystis aeruginosa*.

The rigid vertical columns, typically 2–2.5 m in height and 30–50 cm in diameter, that are extensively used in hatcheries to produce algal biomass for feeding the larval stages of marine bivalves and fishes must be included in this category (Fig. 5a). Most commonly, these systems are made of translucent fiberglass sheets formed into cylinders. Air is bubbled at the bottom for mixing, and either artificial illumination or natural light (or a combination of both) is provided. James and Al-Khars (32) cultivated *Chlorella* and *Nannochloropsis* in 200-L, 30-cm-diameter fiberglass cylinders of this type. Five such units were connected in series and operated continuously. Illumination was provided externally by a vertical bank of daylight fluorescent lamps. Although as a consequence of the low *S/V* ratio of the cylinders, mean volumetric productivities (0.17 g L^{-1} d^{-1} with *Chlorella* and 0.05 g L^{-1} d^{-1} with *Nannochloropsis*) were not high, exceptionally high productivities per unit occupied area were reported (up to 200 g m^{-2} land area d^{-1} for *Chlorella*). However, these values must not be taken as constituting a valid parameter for evaluation and comparison of this system, since they suffer from the limitation inherent in expressing the productivity of elevated reactors on a per occupied land area basis, as discussed earlier.

Cylinders illuminated from inside, which may be considered a variation of the annular reactor devised by Jüttner (29), have been proposed for the production of marine microalgae biomass in hatcheries (33). Although internally lit cylinders have high construction costs, they are less expensive to operate than completely filled columns. They also typically attain higher volumetric productivity and greater efficiency of light utilization since the photon flux

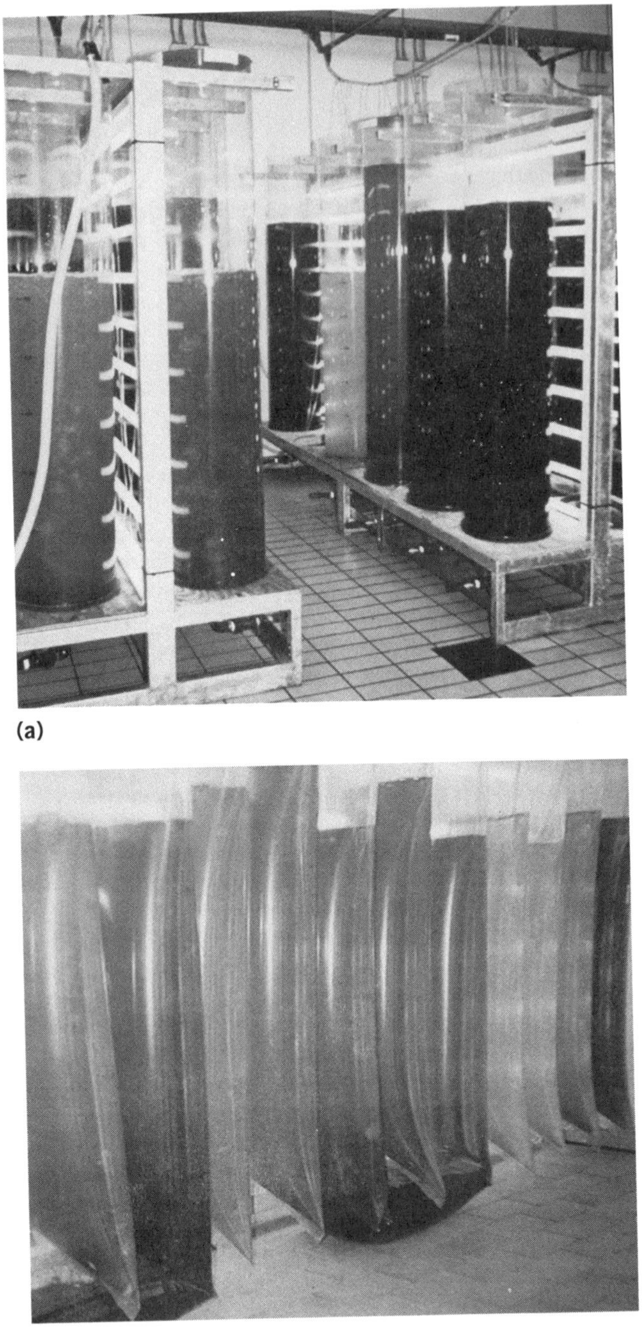

(a)

(b)

Figure 5. Vertical cylinders (**a**) and polyethylene sleeves (**b**) used in hatcheries to produce algal biomass for the larval stages of marine bivalves and fishes (photographs by G. Chini Zittelli).

provided to the culture is completely trapped by it. A patented internally illuminated annular airlift column is used by Martek Co. for specialty chemicals production from microalgae (34).

Miyamoto et al. (35) experimented with vertical tubular reactors (2.35-m height, 5-cm ID) made from the low-cost (about $2.50 per tube), mass-produced glass tubes used in the fluorescent lighting industry. Volumetric productivities of about 0.6 g L^{-1} d^{-1} were obtained with *Nostoc* and *Anacystis* in outdoor cultivation. Although these systems do not represent an innovation, they are materially useful because of their low cost and relatively high *S/V* ratio in comparison to previous reactors of this type. Using a similar system consisting of 32 glass tubes 1.5 m in height and with a 2.6-cm ID, Hu and Richmond (36) obtained a volumetric productivity of 1.6 g L^{-1} d^{-1} with *Isochrysis galbana* cultivated outdoors.

Transparent polyethylene tubing of various diameters and thicknesses is available in long rolls at very low cost. A disposable vertical reactor can be easily built by cutting a suitable length of tubing and heat-sealing one end. This bag or sleeve reactor can be suspended from a framework (Fig. 5b) or supported within a mesh frame. Such reactors are used indoors with artificial illumination (generally vertically mounted fluorescent lamps) or outdoors under sunlight. The temperature reached in the manufacturing of the polyethylene tubing is high enough to sterilize the material. These systems have a relatively short life because of biofouling but are easy and inexpensive to replace at the end of the culture run. Two-meter-long transparent polyethylene sleeves sealed at the bottom and hung on an iron structure have been used to grow *Porphyridium* and *Dunaliella* outdoors (in Israel), obtaining a significantly higher productivity compared to open ponds (37). The main drawbacks of sleeve reactors are the relatively low *S/V* ratio and the low efficiency of utilization of artificial light when it is provided from outside of the reactor.

Serpentine Photobioreactors. Serpentine photobioreactors are systems in which several straight culture tubes are connected in series by U-bends to form a flat loop that can be arranged either vertically or horizontally. Gas exchange and nutrient addition take place in a separate vessel. Circulation between the flat loop and the gas exchanger is achieved by the use of a pump or of an airlift. Several reactors of this type have been developed in the last two decades following the original design by Tamiya (27).

In the early 1980s, Gudin and coworkers (3,38), working at the Centre d'Etudes Nucléaires de Cadarache, France, devised (and since that time have further developed) a horizontal serpentine reactor the main characteristic of which is the use of a water pond to obtain temperature control (a method anticipated by Tamiya). After a first 1-m^2 version consisting of glass tubes of 3-cm diameter, a 10-m^2 and then a 100-m^2 plant were built using larger diameter flexible plastic tubes. The 100-m^2 system, experimented with from 1986 to 1989, was composed of five identical 20-m^2 units, each of which consisted of 20 tubes, 20 m in length and 6 cm in diameter, made of 0.25-mm-thickness low-density polyethylene. The total culture volume was about 7 m^3. Temperature control was attained by floating or immersing the reactor in a water bath by inflating or deflating a second layer of tubes placed below the reactor. Initially, the culture was circulated by means of a pump, but later airlift systems were adopted to limit damage to shear-sensitive cells and at the same time provide CO_2 supply and O_2 degassing. Productivities from 20 to 25 g m^{-2} d^{-1} were obtained with *Porphyridium cruentum* in a 2-month continuous run. The use of rigid rather than flexible tubes in the more recent work by Gudin's group has permitted operating a self-cleaning system consisting of two plastic balls, one with higher density than the culture medium and one lighter, which are hydraulically pushed through the system. Although flotation immersion in a water basin can provide efficient thermoregulation, the cost of such a system is prohibitive for most applications.

In 1992, Dr. Gudin founded the Heliosynthese S.A. company for development of tubular photobioreactor technology in France. In 1996, Heliosynthese changed its name to Thallia Pharmaceuticals S.A. Thallia presently makes use of serpentine reactors to produce high-value algal compounds for the nutriceutical and pharmaceutical markets. The major products under development are polyunsaturated fatty acids, carotenoids (zeaxanthin from *Porphyridium cruentum*, astaxanthin from *Haematococcus pluvialis*) and antioxidants (superoxide dismutase).

Reactors of the type developed by Gudin are also being operated by Molina Grima and coworkers in Spain (Fig. 6a) (14,15,39,40). These systems consist of a horizontal loop, made of Plexiglas® tubes (about 100 m in length and 2.6 cm in diameter), which is connected to a 3- to 3.5-m-high airlift. The tubular solar receiver is immersed in a shallow pond for temperature regulation. Dissolved oxygen, pH, and temperature probes are connected to an on-line control unit and a computer for data acquisition. The Spanish group has made important improvements to the reactor originally purchased from Heliosynthese, and with it has carried out, and is still conducting, extensive research activity with the dual aim of shedding light on basic aspects of tubular photobioreactor technology (e.g., mass transfer and light distribution), and of defining guidelines for application of closed systems to production of specialty chemicals from microalgae (e.g., polyunsaturated fatty acids).

In the early 1980s, Pirt and coworkers developed a serpentine tubular reactor at the University of London's Queen Elizabeth College (20,41). The photobioreactor consisted of a photostage formed of 52 Pyrex glass tubes (each approximately 1 m long and with a 1-cm ID) horizontally stacked. The tubes were connected to Pyrex glass U-bends by silicone-rubber tubing to form a vertical loop. The loop outlet was connected through a vertical riser to a degasser. A second tube connected the degasser to the inlet of the loop. The reactor covered an area of about 0.5 m^2. The culture suspension (about 4.6 L) was circulated by either a peristaltic pump, a rotary positive-displacement pump, or an airlift; the latter method was preferred because of the adverse effects observed with the two pumps. With a defined consortium consisting of a chlorella-type green alga and three chemotrophic bacteria, cell concentrations in excess of 20 g L^{-1} and a very high productivity of 2.2 g m^{-2} h^{-1} were achieved at a light intensity of only 38 W m^{-2} (PAR). The work carried out by Pirt and coworkers was remarkable for three reasons: It provided the first detailed analysis of the fundamental engineering parameters of a closed photobioreactor, introduced the concept of high *S/V* ratio reactors (127 m^{-1} in this case), and attained extremely high light-conversion efficiencies. It is worth noting that the about 40% reported efficiency for *Chlorella* growth corresponds to a quantum yield of about 6, as compared to the theoretical value of 8 (see earlier section). Pirt's analysis, however, did not consider the issue of the accumulation of photosynthetically generated oxygen, which represents a key limiting factor in large-scale tubular reactors (see later section), and ignored the light saturation effect, assuming that the efficiencies observed at low light intensity could be achieved in full sunlight as well. Although Pirt's serpentine reactor did deal efficiently with, and solve, many of the problems commonly encountered in closed systems, the concept of very high *S/V* ratio reactors revealed all its drawbacks when the experimental setup was scaled up to industrial level (see later section). The vertical serpentine reactor designed by Pirt was patented in 1983 (41). The patent also covers the design of a full-scale, near-horizontal manifold system according to which the culture flows from a tank through a descending conduit to the higher ends of a set of 50-m-long transparent tubes supported by a slightly inclined framework. Each tube communicates at its lowest point with its own riser, which leads up to the common reservoir. The tubes are arranged close together so as to intercept a high portion of the solar radiation impinging on the area occupied by the reactor. Both this manifold and the vertical serpentine design were scaled up to commercial size in the late 1980s by PBL (see later section).

In the mid-1980s at the Centro di Studio dei Microrganismi Autotrofi (Florence, Italy), Florenzano and coworkers started their experiments with horizontal serpentine tubular photobioreactors for the cultivation of *Arthrospira* (42). One of the first pilot reactors they set up was made of 14-cm-diameter, 0.3-mm-thick flexible polyethylene tubes. Because of problems of leakage, these collapsible sleeves were later replaced with rigid Plexiglas® tubes. The standard unit consisted of straight tubes, with a 13 cm ID and 4 mm thick, laid parallel on a white sheet and joined by PVC U-bends to form a loop 500 m in length (Fig. 6b). As it exited the reactor, the culture suspension flowed into a receiving tank, from which a diaphragm pump raised the suspension to a feeding tank placed at 3 m above ground level. The feeding tank was provided with a siphon that allowed discharge of about 340 L of culture into the photobioreactor every 4 minutes, thus sustaining a culture flow of 26 cm s^{-1}. The photobioreactor had a total volume of 8 m^3 and covered an area of 80 m^2. The productivity of *Arthrospira* in this system was 50% higher than in open ponds, due both to the better diurnal temperature profile and to the extended cultivation period. More recently, the same group developed and experimented with a 145-L two-plane serpentine reactor (43) (Fig. 6c). The system consisted of a 245-m-long loop made of 2.6-cm ID Plexiglas® tubes arranged in two planes to maximize capture of the radiation impinging on the horizontal, and thus increase areal yield. The two planes were connected to form a single loop and the culture, circulated by means of two airlifts, flowed from one plane to the other. A maximum areal productivity of 27.8 g m^{-2} d^{-1} was achieved in July. It would appear that for such a high-cost system, the loss of volumetric productivity due to shading by the upper plane would exceed the benefits deriving from improved areal yield and land saving. It must be noted, however, that the volumetric productivity achieved with this

Figure 6. Horizontal serpentine photobioreactors set up at the University of Almeria, Spain (**a**) (photograph courtesy of E. Molina Grima) and at the Centro di Studio dei Microrganismi Autotrofi of Florence, Italy (**b, c**) (photographs courtesy of G. Torzillo).

two-plane photobioreactor (1.5 g L d^{-1}) was relatively high. The main merit of this study was that it addressed fundamental engineering parameters such as the effect of the rheological behavior of the algal culture on the mixing characteristics of the system, and the energy requirement for turbulent flow, which are often ignored.

Aquaserch (Kona, Hawaii) has recently devised and patented (44) a serpentine tubular reactor made of low-density polyethylene tubes. As in several previous serpentine designs, cooling is obtained by immersion of the reactor in a pond. The system, called the Aquaserch Growth Module (AGM), has been tested with different microalgae and is presently used to grow *Haematococcus* for astaxanthin production. The AGM is claimed to be the largest commercial-scale photobioreactor ever devised and one of the least expensive.

Manifold Photobioreactors. In manifold photobioreactors, a series of parallel tubes is connected at the ends by two manifolds, one for distribution and one for collection of the culture suspension. This type of photobioreactor represents one of the simplest configurations possible. As shown by Pirt et al. (20), about 15% of the energy consumed for recycling the culture in serpentine reactors is spent in moving the culture suspension around the bends; thus, in comparison to serpentine reactors, in which the culture inverts the direction of motion at each turn, manifold systems allow a significant saving of energy.

In 1991 in Javan (Tadshikistan), Sagdiana PGT built and used a large-scale glass reactor of the manifold type (Fig. 7a) (45) for cultivation of *Chlorella*. The system, very likely the largest glass reactor ever built, comprised 10 units consisting of two banks of 28 5.7-cm-ID tubes, each 73.5 m long, for a total of about 41,000 m of tubing. The tubes were connected in parallel to two metal headers each 20 cm in diameter. The culture suspension (13 m^3 in each unit) was circulated between the two banks of tubes at a speed of 1.1 m s^{-1} by a 290-m^3 h^{-1} capacity pump. Each unit was provided with one 3.6-m-high, 2.4-m-diameter gas exchanger. Temperature control was obtained through circulation of water from a single cooling tower serving all 10 units. The reactor was operated from August to October 1991, during which time it produced 0.5 ton of *Chlorella* biomass. Instability of the production process and contamination by *Scenedesmus* were reported to be the main problems.

Tredici and coworkers, working at the University of Florence (Italy), have developed several types of near-horizontal manifold-type photobioreactors (NHTR) (Fig. 7b). Units made of rigid or flexible tubes ranging from 6 to 85 m in length and from 4 to 5 cm in diameter have been built and used to grow *Arthrospira platensis, Anabaena siamensis*, and *Nannochloropsis* sp. under the climatic conditions of central Italy. Two eight-tube reactors of this type, 20 m in length, have been installed at the University of Hawaii as part of a joint project (Fig. 7c) (46). Typically, a NHTR consists of some 10–15 flexible tubes, each 45 m long and 4 cm in diameter, connected by two manifolds; the upper manifold, about 30 cm in diameter, acts as the degasser. The tubes are placed side by side on white corrugated plastic sheeting, facing south and inclined at a slight angle to the horizontal (4–6°). The corrugated sheeting keeps the tubes well aligned at a uniform inclination, reflects light onto the culture, and also provides drainage for rain and cooling water. Temperature control during the day is obtained by water spraying; the cooling water is collected and reused. The largest unit experimented with thus far occupies an area of 30 m^2 and contains about 600 L of culture. The main characteristic of the NHTR reactors developed in Florence is the use of air injection, at rates between 0.02 and 0.1 L L^{-1} min^{-1}, to attain mixing and gas exchange. The air is injected at the lower end of each tube by means of a perforated pipe inserted into the bottom manifold. In a different configuration used more recently, a number of tubes (typically one out of four) are not gassed, and are used as return flow tubes to increase the circulation speed of the culture and to obtain better mass transfer. Carbon dioxide addition to the air stream or in the return tubes is regulated through a pH-stat system. This reactor is shown in schematic form in Figure 8. Volumetric productivities of up to 1.4 g L^{-1} d^{-1} and areal productivities of more than 28 g m^{-2} d^{-1} have been obtained with *Arthrospira platensis* grown in these manifold systems. Besides simplicity of operation and low maintenance cost (due to the absence of moving parts), this internal gas-exchange reactor has other advantages: low shear stress compared to pump mixed reactors; no need for pumping the culture to a gas exchanger; and reduced fouling and wall growth due to the scouring effect of the gas bubbles. Moreover, this system appears to have potential for unit scale-up to about 100 m^2, and is relatively low cost in comparison to other designs. A preliminary estimate for a unit of 30 m^2, built and tested at Montepaldi (near Florence), sets the cost of this type of reactor at about \$60/m^2 (M.R. Tredici and G. Chini Zittelli, unpublished material, 1997), with the following cost per square meter: land, \$1.50; site clearing, grading, and leveling, \$6; liners, \$3.50; support framework, \$14.50; corrugated sheeting, \$5.50; tubing, \$13; manifolds, connectors, and piping, \$11.50; cooling system, \$1.50. A cost analysis for a large NHTR system for biological hydrogen production was recently presented (47).

A parallel-flow tubular photobioreactor was devised and experimented with in outdoor cultivation of cyanobacteria at the Microalgal Biotechnology Laboratory of Ben Gurion University (Israel) (48). The system consisted of eight 20-m-long, 3-cm-ID polycarbonate tubes laid parallel on the ground or on a horizontal platform and connected by manifolds made of 6.3-cm-diameter black polyethylene tubing (Fig. 7d). Circulation and gas exchange were obtained by means of a 2.2-m-high, 4.6-cm-ID airlift connected to a 20-L gas separator. A manifold received the culture from the separator and distributed it among a first series of culture tubes. At the far end of these tubes, the culture suspension was collected by a second manifold and distributed among a second series of returning tubes through which the culture ran back to the airlift riser and the gas separator. Small differences in flow rates were measured between the tubes. Temperature control was provided by water spraying. A productivity of 0.55 g L^{-1} d^{-1} was obtained with *Arthrospira platensis* and *Anabaena siamensis*. The main advantages of the system over serpentine reactors are the reduction of head losses and lower oxygen

Figure 7. Manifold photobioreactors; (**a**) glass photobioreactor set up at Javan (Tadshikistan) by Sagdiana PGT (photograph by S. Markov); (**b**) NHTR photobioreactor developed at the Dipartimento di Scienze e Tecnologie Alimentari e Microbiologiche of the University of Florence, Italy (photograph by S. Pintus); (**c**) NHTR photobioreactor installed at the University of Hawaii, Honolulu, Hawaii, (photograph courtesy of O. Zaborsky); (**d**) parallel flow reactor designed at the Ben Gurion University of the Negev, Israel (photograph courtesy of A. Vonshak); (**e**) α-type photobioreactor set up at the Biotechnology Department of the National University of Singapore, Singapore (photograph courtesy of Y.K. Lee); (**f**) vertical manifold designed by Bio-Fence Ltd, Manchester, UK (photograph courtesy of Bio-Fence Ltd).

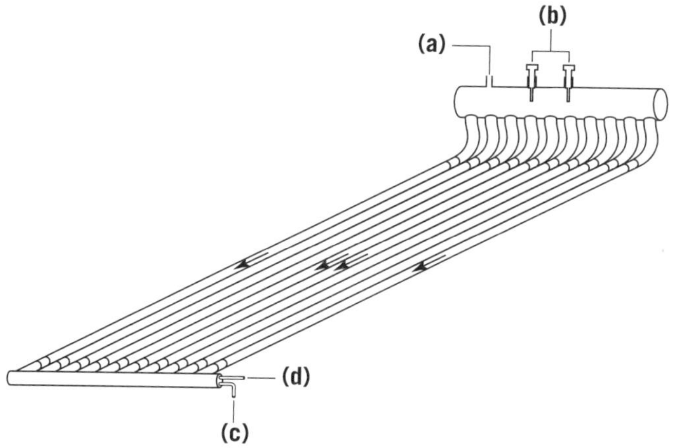

Figure 8. Scheme of the NHTR photobioreactor developed at University of Florence, Italy: (**a**), gas outlet; (**b**), electrodes and sensors; (**c**), culture outlet; (**d**), air and CO_2 injection point. Arrows indicate not-bubbled tubes used to circulate the culture suspension.

concentrations; these two factors facilitate scale up to industrial size.

A manifold elevated system called the α-type tubular photobioreactor was devised and experimented with in Singapore (Fig. 7e) (49). The reactor consisted of two sets of 10 parallel transparent PVC tubes (25 m long with a 2.5 cm ID), both placed at an angle of 25° to the horizontal but inclined in opposite directions. The algal suspension was lifted by air injection through a series of vertical PVC air-riser tubes (3.5 cm in diameter, 5 m in height) to a receiver tank from where it flowed down the first set of 25°-inclined tubes. These were connected to the base of a second series of air risers that lifted the culture to another receiver tank, from where the culture descended again through the second set of 25°-inclined tubes to the base of the first series of risers. The reactor thus looked like the Greek letter α. This elevated reactor, containing about 300 L of culture and covering a land area of about 12 m^2, was shown to be particularly well-suited to operation at low latitudes. *Chlorella pyrenoidosa* was grown for several months and reached very high cell densities (>10 g L^{-1}). The major problems were biofouling, solved by injecting air from the upper end of the inclined tubes, and foam formation at the higher cell densities. An areal productivity of 72 g m^{-2} occupied land d^{-1} was reported; as mentioned previously, the latter parameter can be misleading when referred to elevated systems.

At the University of Manchester (UK), Applied Photosynthetics Limited (APL) has recently developed a photobioreactor known as the Bio-Fence (Fig. 7f). The company offers several sizes of modular reactors of this type, designed specifically for the cultivation of the marine algal species used in aquaculture or for waste treatment with either microalgae or photosynthetic bacteria. The Bio-Fence consists of rigid transparent tubes racked together in banks and connected by manifolds in a fencelike structure. The culture suspension is circulated continuously between the photostage and a holding tank by a centrifugal pump or by an airlift. A patented scouring mechanism provides continuous cleaning of the inner walls of the tubes. The pH of the culture is controlled automatically by injection of CO_2 in the photostage. The most distinctive feature of the Bio-Fence is the stacked manifold configuration, which provides optimal gas exchange with elimination of pressure drops within the reactor. A 100-L unit requires a tube stack 5 m long by 0.75 m high. Units with a total volume of 500 L, with a tube length of 5 m, or 1,000 L volume, with a tube length of 10 m, can be assembled. The Bio-Fence is produced and marketed as units of from 50- to 1,000-L capacity that can be connected in parallel to any scale. Small steam-sterilizable systems for axenic operation are also available. In the case of systems connected in parallel, the units can be placed as close as 30 cm apart, as in the vertical system built by PBL in Spain and Pulz's alveolar plates (see later section). This arrangement will reduce volumetric productivity in case of outdoor utilization because of shading among the units but will attain two important results: high efficiency of land utilization (due to the very compact structure) and high photosynthetic efficiency (due to dilution of sunlight). The possibility of installing fluorescent tubes between the parallel units for additional lighting has also been considered. Typical cell concentrations achieved in the Bio-Fence are about 1 g L^{-1}. Productivity data are not available. The Bio-Fence is an accurately designed, well-engineered reactor. The only limitation seems to be the presence of a large dark tank that might reduce productivity in case of an excessively high ratio between darkstage and photostage volume.

Helical Photobioreactors. Helical photobioreactors consist of small-diameter, generally flexible tubing wound around an upright structure. This design is not new; it was used in the 1950s by Davis to grow *Chlorella* (50) and later adopted, in a flattened version made of glass tubes, by Setlik (28), Jüttner (29), and Krüger and Eloff (30). This coiled arrangement allows deployment of relatively long tubes on a small land area.

Helical photobioreactors called Biocoils were devised and patented in the late 1980s by Biotechna Ltd. (51). The Biotechna reactor consists of low-density polyethylene or clear food-grade PVC tubing, 2–3 cm in diameter, wound in a helix around an upright cylindrical core structure, typically 8 m in height with a core diameter of 2 m. Compressed air or peristaltic or centrifugal pumps may be used to induce culture circulation, with the culture being pumped downward through the coiled tube or upward through the tube to a header tank. Temperature control is achieved by a heat exchanger placed between the photostage and the pump. Pilot "biocoil" facilities have been installed at Luton (UK) and in Australia (52). Biotechna-Graesser A.P. Ltd. (based in the UK) marketed biocoils of different sizes (from a few liters capacity up to 10,000 L) for microalgae cultivation or for use with algae and heterotrophic bacteria in the treatment of wastewaters or toxic wastes. Small biocoil units are being operated for experimental and teaching purposes at King's College (London) (53) and at Murdoch University (Murdoch, Australia) (54).

A 120-L helical bubble reactor was devised and operated by Tredici and Chini Zittelli (19) to grow *Anabaena siamensis* and *Arthrospira platensis* outdoors (Fig. 9a). The reactor consisted of three 49-m-long transparent PVC tubes (3-cm ID, 0.5-cm wall thickness) wound around a vertical structure with an inclination of 2° to the horizontal in such a manner as to form a 1.65-m-high, 1.1-m-OD coil. The higher ends of the tubes were connected to a 20-L degasser. From the degasser, the culture flowed down through a single pipe connected to the lower ends of each tube. In this system, as in the NHTR devised by the same group, air was injected at the bottom of each tube and the bubbles flowing up the tubes provided mixing and gas exchange and prevented biofouling. Cooling was achieved by water spraying. With *A. platensis*, a mean volumetric productivity of 0.9 g L^{-1} d^{-1} and a photosynthetic efficiency of 6.6% (PAR) were achieved. The system operated with remarkable stability. The cost of a unit of this type was estimated at about $150.

A coiled, internal gas-exchange photobioreactor of about 3,000 L total volume was built in 1996 by Inalco S.p.A near Vicchio (Florence, Italy). The system consists of straight polycarbonate tubes, 5 cm in diameter, secured to an upright hexagonal frame with an inclination to the horizontal of 1.4°. The reactor is about 7 m high and occupies an area of about 40 m^2. Culture circulation is obtained by air injection at the base of the tubes. Temperature is controlled by a water-jacketed section. The reactor has been successfully experimented with in the cultivation of *Arthrospira platensis*.

Flat Photobioreactors. Flat photobioreactors and flat culture chambers have often been used to grow phototrophic microorganisms in the laboratory because they provide a simple geometry and greatly facilitate the measurement of irradiance at the culture surface. Despite their apparent simplicity, few such systems have been used for mass cultivation of algae.

The first flat culture unit devised for mass production of algae was the "rocking tray" used by Milner in 1953 (50) to grow dense *Chlorella* cultures in a thin turbulent layer. This system consisted of a 2.65-m^2, 9-cm-deep tray covered with transparent wire-reinforced plastic sheeting and supported on a shaker. At a mean culture depth of 1.7 cm, an average productivity of 8.2 g m^{-2} d^{-1} was obtained in a 20-day run. Following this experience, flat reactors were practically neglected until the 1980s.

Anderson and Eakin (55) cultivated *Porphyridium cruentum* outdoors for polysaccharide production using 3-m^2 flat plates 5–20 cm in depth (made of stainless steel and covered with glass or plastic) that were inclined to the horizontal to maximize interception of solar irradiance. About 20 g of polysaccharide m^{-2} d^{-1} were produced during the summer, when temperature control was provided.

In the mid-1980s, two groups working independently in France and in Italy introduced flat alveolar panels in algae cultivation. These systems are constructed from transparent PVC, polycarbonate, or polymethyl methacrylate sheets that are internally partitioned to form rectangular channels called alveoli. These sheets are commercially available in standard thicknesses from 4.5 to 40 mm.

Alveolar plates as culture systems for algae were first used by Ramos de Ortega and Roux (56) at the Centre d'Etudes Nucléaires de Grenoble (France). These authors experimented with 6-m-long, 0.25-m-wide, 4-cm-thick, double-layer panels made of transparent PVC for growing *Chlorella* in a greenhouse and outdoors. The plates were laid horizontally on the ground; the upper layer of channels was used for algal growth and the lower for thermoregulation. The culture suspension was circulated through a pump. A productivity of 24 g m^{-2} d^{-1} was achieved in the summer using units of 1.5-m^2 surface area. The authors stressed the fact that the double-layer plate permitted efficient thermoregulation of the culture and that significantly higher performances were obtained with the panels than with flexible sheets and semirigid tubes.

The first panel reactors built and experimented with in Florence by Tredici and coworkers were placed vertically, with the channels (and hence culture flow) running parallel to the ground, a design later adopted in Germany by Pulz. A pump was used to circulate the suspension. In 1988, the Consiglio Nazionale delle Ricerche (CNR) of Italy presented this reactor at the ITALIA 2000 exhibition, held in Moscow. The same year saw the awarding of a patent to a different design (57) in which the plate was placed vertically, but with the channels perpendicular to the ground; mixing was achieved by bubbling air at the bottom of the reactor. This latter version (Fig. 9b) has been used extensively in Florence for outdoor cultivation of cyanobacteria (5,19,58–61). The plates used in Florence are constructed from 16-mm-thick Plexiglas® alveolar sheets commonly available on the market at a cost of $30–40 m^{-2}. The main features that characterize these reactors are the high *S/V* ratio (80 m^{-1}), the vertical or tilted inclination from the horizontal of the channels, and the fact that no mechanical device is used to mix the cell suspension, circulation and degassing of the culture being attained through bubbling air at the base of each channel. Elevated "bubble column" plates of 2.2-m^2 surface area have been used to study the influence of some of the main cultural parameters, such as areal density, turbulence, oxygen tension, and orientation of the photobioreactor, on productivity and biomass composition of *Arthrospira platensis* (5). Alveolar plates, due to their high *S/V* ratio, attain high volumetric productivities (>2 g L^{-1} d^{-1}) and permit operation at high cell concentrations (4–6 g L^{-1}); these systems, however, have typically attained lower areal yields (about 24 g m^{-2} d^{-1}) compared to tubular reactors (about 28 g m^{-2} d^{-1}) (19). The lower performance achieved by cultures in plates has been attributed to the fact that these systems, unlike tubular reactors, do not achieve light dilution (unless they are placed at a high inclination with the horizontal) and thus cultures in plates suffer more strongly because of the light saturation effect and from photoinhibition (19). Alveolar plates, like other small flat systems, may be used profitably for research or in small production plants, however, due to the fact that the surface area of the cultivation unit is limited to a few square meters, they are not suitable for use in large commercial plants.

More recently, at the IGV Institut für Getreideverarbeitung (Bergholz-Rehbrücke, Germany), cultivation units based on 32-mm-thick alveolar panels were developed and

Figure 9. Biocoils and flat photobioreactors: (**a**) helical bubble-column reactor designed at the University of Florence, Italy (photograph by S. Pintus); (**b**) flat alveolar panels experimented with at the University of Florence, Italy (photograph by G. Chini Zittelli); (**c**) PBR system built by B. Braun Biotech Int., Germany (photograph courtesy of O. Pulz); (**d**) glass flat reactors experimented with at the Ben Gurion University of the Negev, Israel (photograph courtesy of A. Richmond).

operated by Pulz and coworkers (6). Typically, the plates are placed vertically with the channels arranged horizontally, just 20 cm apart, to form a compact structure called PBR (Fig. 9c). The culture, circulated by a piston pump, must reach a velocity adequate to maintain turbulent flow inside the channels. The distance to the degasser is critical in avoiding toxic concentrations of oxygen. Units from 100 to 10,000 L, made of closely packed plates, have been developed and tested both indoors and outdoors with different microalgae and cyanobacteria. Productivity levels of 1.3 g l^{-1} d^{-1}, corresponding to 28 g m^{-2} of illuminated surface d^{-1}, have been obtained with *Chlorella* under natural illumination. Among the problems that have emerged are buildup of oxygen (up to 30–40 mg L^{-1} at the outlet of the circuit) and fouling. The main characteristic of the PBR is the closely packing of the panels: 42 plates, for a total volume of 6 m^3, can be packed on 100 m^2 of land. Areal productivities of up to 130 g m^{-2} d^{-1} have been reported for such compact units, but these figures seem excessive and indeed confirm the incorrectness of referring the yield of vertical reactors to unit of occupied land. The main merit of the PBR, besides its high degree of technical development, is the achievement of significant light dilution. PBR reactors are sold commercially by B. Braun Biotech Int. in sizes from 10 to 6,000 L (6).

Richmond and coworkers (62) developed an inclined modular photobioreactor consisting of a series of flat glass chambers, 0.7 m high, 0.9 m long, and usually 2.6 cm in internal width, connected in cascade and tilted at the proper angle to maximize solar radiation capture (Fig. 9d). The culture volume varied from 6 to 50 L according to reactor width. Two mixing methods were tested: an airlift system consisting of five transparent PVC columns inserted vertically into the reactors with compressed air injected at the base of the columns, and air-bubble mixing by means of two perforated tubes running horizontally along the entire length of the reactor, one at the bottom and the other at midheight. The latter method yielded better performance. By adopting a reactor width of 2.6 cm and vigorous mixing (2.5 L air L^{-1} culture suspension min^{-1}), a record areal productivity of 49.4 g m^{-2} illuminated surface d^{-1} was achieved. When the back and side surfaces of the reactor were covered to avoid entry of diffuse light, productivity was reduced by about 20%. Nevertheless, this figure is the highest ever reported for a cyanobacterium cultivated outdoors. Using this flat modular system, the Israeli group has addressed and contributed to clarifying such fundamental aspects of photobioreactor design as optimal cell density with reference to the light path, the effect of reactor inclination, and the importance of the mode and intensity of mixing.

Commercial-Scale Photobioreactors

Only a few large commercial-scale photobioreactors have been built and operated: the PBL photobioreactors (based on the two different designs described in Pirt's patent) in southern Spain in the late 1980s and the Hidrobiologica SA manifold photobioreactor (probably inspired by previous similar designs devised in Italy by the Centro di Ricerca Agro-Industriale and Montedison) set up in the mid-1990s in northern Argentina. Despite the very different designs, both the plants suffered from similar mistakes with respect to some fundamental technical aspects and the scaling-up phase.

Photo Bioreactors Ltd. Plant. Although Pirt's measurements and extrapolations were not flawless, the quality of the basic work done at Queen Elizabeth College and the high projected productivities attracted investments and led to the creation of a company, Photo Bioreactors Ltd. (PBL UK), in 1986. A pilot plant was built and experimented with at the Reading University Research Farm (Sonning, UK). Three years later, PBL Spain (PBL SA) was founded with investments from private industries (among which the Danish Chr. Hansen Laboratory) and public Spanish sources (Instituto de Fomento de la Region de Murcia and Centro de Desarrollo Tecnico Industrial). A second pilot plant was set up in Santomera (Murcia, Spain), and finally a commercial plant for *Dunaliella* production was established in Santa Ana near Cartagena (Murcia, Spain). Full-scale activity started in May 1990, with a vertical system made of 50 rows, each of which contained 50 1.2-cm-diameter, 50-m-long polyethylene tubes (for a total tubing length of 125,000 m) (Fig. 10a). Circulation was provided by an airlift, and temperature was controlled by shading the reactors with nets or by water spraying. A near-horizontal system was built later according to the full-scale design patented by Pirt, using about 200,000 m of polyethylene tubing. Several major technical errors are apparent in these two systems: The tube diameter was too small for effective mixing considering the design (Fig. 10b), the tube material degraded quickly under sunlight, oxygen degassing was inadequate, wall growth was almost unavoidable, and control of temperature was problematic.

The very high *S/V* ratio in relation to the length of the tubes, together with the insufficient circulation and the improper management of the culture, led to poor growth of the alga, biofouling, and heavy contamination. But, as far as we know, the main reason for failure was inadequate piloting: although the pilot plants had been carefully planned and constructed with an appropriate 10-fold increase factor in the different stages, the problems encountered during the pilot stage were not taken into due consideration and had not been solved by the time commercial activity started. In September 1991, after an almost complete crash of the culture, the workers were dismissed and PBL closed without ever entering into production (63).

Hidrobiologica SA Plant. In 1996, Hidrobiologica SA, with an initial investment of about $2 million, built the largest known photobioreactor (Fig. 10c) about 15 km south of La Rioja (northern Argentina). The system consisted of 96 0.4-mm-thick polyethylene tubes, each 120 m long and 25.5 cm in diameter, laid parallel and horizontally on the ground and arranged like a manifold with feeding, connecting, and collecting channels made of concrete. The tubes, only partially filled, were maintained at very low positive pressure so that they assumed a roughly elliptical shape (about 35 cm wide and 9 cm high). The plant was divided into four modules, each of which comprised four six-tube submodules. An interspace of 0.5 m was allowed

Figure 10. Commercial-scale photobioreactors: (**a**) vertical system set up by Photo Bioreactors Ltd (PBL) in Santa Ana, Murcia, Spain (photograph courtesy of D. Hall); (**b**) a particular of the PBL vertical system as it appeared in July 1995, four years after shutdown (photograph by the author); (**c**) the Hidrobiologica SA plant built at La Rioja, Argentina (photograph by G. Chini Zittelli).

between one submodule and the next, while about 1 m separated the modules. The four modules were connected to common feeding and collecting channels. The surface occupied by the whole plant, including the gaps between tubes, was about 5,000 m^2, and the total culture volume was 600 m^3. The culture (*Arthrospira platensis*) was circulated at a flow speed between 6 and 10 cm s^{-1} by a single axial flow pump with a maximum capacity of 900 $m^3\ h^{-1}$. The pump lifted the suspension from the collecting channel, into which it was fed by the culture tubes, up to a second raised, sloped channel. The suspension flowed by gravity through this second channel, which ran parallel to the culture tubes for 120 m, to the feeding channel at the opposite end of the reactor. The feeding channel, about 42 m long, distributed the culture suspension to the inlets of the single culture tubes. During the initial period of activity, the cell concentration was typically maintained at about 1 g L^{-1} and the productivity was calculated to be around 0.2 g $L^{-1}\ d^{-1}$; the biomass produced was of excellent quality. In the long run, however, several problems became evident; the most significant were poor mixing and biofouling deriving from the low flow speed and the unequal distribution of the culture among the tubes. The limited capacity to control temperature was also a major drawback in summer. It is likely that oxygen built up to dangerous levels despite the gas outlet provided halfway along the tubes, but no data are available. The plant is at present undergoing complete restructuring; the main objectives of this effort are to subdivide it into separate units, reduce the tube diameter, and improve mixing and culture circulation. The major errors, besides poor mixing and limited temperature control, were probably the absence of accurate piloting and the fact that only one large single unit was built and operated.

DESIGN CRITERIA FOR PHOTOBIOREACTORS

The basic function of a photobioreactor is to provide a controlled environment in which to achieve optimal growth and maximum, sustainable biomass and/or product synthesis with a specific phototroph (4). This means that a photobioreactor cannot be properly designed without adequate knowledge of the growth physiology of the organism to be cultivated. Besides, since phototrophic microorganisms are highly diverse as concerns their morphology, biology, nutritional and light requirements, and resistance to stresses, a single photobioreactor design cannot adapt to all organisms and all conditions.

Design criteria for photobioreactors should aim at achieving high efficiency of light conversion and at providing the necessary reliability and stability to the cultivation process by solving the main problems encountered in photobioreactor operation, such as overheating, oxygen buildup, biofouling, contamination, and photostage material deterioration. The fundamental design criteria for photobioreactors include reactor configuration, *S/V* ratio, mixing and degassing devices, temperature regulation, orientation, and the material used to build the photostage. These aspects are briefly discussed in the following sections. Reliability, ease of operation and scale-up, and construction and operating costs also take on particular relevance in relation to commercial application of photobioreactors.

Surface-to-Volume Ratio

The surface-to-volume ratio of the photobioreactor (i.e., the ratio between the illuminated surface of the reactor and its volume) determines the amount of light that enters the system per unit volume and the light regimen to which the cell population is subjected and is consequently one of the most important issues in photobioreactor design. The higher the *S/V* ratio, the higher the cell concentration at which the reactor is operated and the volumetric productivity of the culture. The cell concentration, in turn, greatly influences medium handling and harvesting costs and, to some extent, contamination level. The temperature profile and the hydrodynamic behavior of the culture suspension are also affected by this parameter (4,5,64). For example, in small-diameter tubular systems operated outdoors, the time required to reach the optimal temperature in the morning is reduced, and in the evening the relatively low culture volume cools down more quickly to ambient night temperatures, resulting in lower respiratory losses (4). For these reasons, in recent years we have witnessed a general trend toward a reduction of the diameter of tubular reactors and of the thickness of flat panels. Yet, it must be pointed out that although high *S/V* ratio reactors may work well in the laboratory, they may become extremely inefficient when scaled up. In high *S/V* ratio photobioreactors, all the volumetric activities that depend on the input of light energy per unit volume (such as O_2 evolution, CO_2 absorption, nutrient depletion, metabolite excretion, and heat production) change at a high rate, and this could have long-term negative effects on the stability of the process. These effects are rarely observed because experiments do not last long enough for them to become evident (63). The industrial systems set up by PBL (see earlier section) provide clear examples of the kinds of problems that may be absent at laboratory scale and yet arise when high *S/V* ratio photobioreactors are adopted in large-scale plants. It is not possible to fix an optimal *S/V* ratio for large-scale photobioreactors, since this parameter depends on the organism to be grown, on the type of mixing, and on other specific conditions. However, my laboratory experience suggests that a light path between 3 and 6 cm may represent a reasonable compromise.

Oxygen Accumulation

Accumulation of photosynthetically generated oxygen is one of the main factors that limit scale-up of photobioreactors. Dissolved oxygen concentrations over 35 mg L^{-1} (equivalent to about four times air saturation at 25 °C) may be easily reached in outdoor dense algal cultures at midday. Such high oxygen levels are toxic to many phototrophs and, if coupled with prolonged exposure to full sunlight, may lead to the photooxidative death of the culture (4,9). Since volumetric oxygen production is directly related to volumetric biomass productivity (about 1 g of oxygen is evolved per gram of biomass synthesized), oxygen buildup becomes an especially serious problem in high *S/V* ratio photobioreactors operated outdoors. At maximal rates of photosynthesis, a 1-cm-diameter tubular reactor would accumulate about 8–10 mg of oxygen L^{-1} min^{-1} (64). In such systems, the maintenance of dissolved oxygen below the toxic concentration requires adoption of very short loops and this fact makes the serpentine design difficult to scale up. The situation is less dramatic in tubes of greater diameter, since the volumetric rate of oxygen evolution falls in proportion to the decrease of the *S/V* ratio, but it has been shown that even in 14-cm-diameter tubular reactors dissolved oxygen may reach levels that significantly reduce productivity (42). In serpentine reactors, the time cycle (time required for the culture to reach the degassing point), the speed of circulation, and the length of the loop must be very carefully chosen in relation to the expected volumetric productivity to prevent buildup of toxic oxygen concentrations. For example, in a tubular reactor with a 2.6-cm ID and a culture flow speed of 0.3 m s^{-1}, the time cycle should not exceed 6 min, hence, the length of the reactor must be less than 100 m (4). To increase the length of the reactor it is necessary to increase the culture circulation speed or reduce the *S/V* ratio. Internal gas-exchange systems (bubble columns) offer some advantage in this respect. Unlike serpentine reactors, in which oxygen accumulates during the loop cycle and is removed only at the degasser, bubble columns present the advantage that the whole culture is subjected simultaneously to the degassing action of air.

Mathematical models can be employed to predict the effect on oxygen concentration of various design variables, such as the length of the loop, the tube diameter, and the liquid circulation speed. A useful model for prediction of dissolved oxygen and carbon dioxide axial concentration profiles in serpentine tubular reactors has been developed and verified experimentally with *Porphyridium cruentum* in outdoor continuous culture by Camacho Rubio et al. (40).

Mixing

The type of device used to mix and circulate the culture suspension is essential in the design of a successful photobioreactor. Both the productivity of a photobioreactor and the cost of its construction and operation are influenced to a great extent by the type of device used for mixing (65). The beneficial effects that mixing has on growth and productivity of mass cultures of phototrophs are well known (9). Mixing is necessary to prevent cells from settling, to avoid thermal stratification, to distribute nutrients and break down diffusion gradients at the cell surface, to remove photosynthetically generated oxygen, and, above all, to ensure that all the cells experience alternating periods of light and darkness. It has been postulated that cells quickly brought into and out of the lighted portion of the culture could take advantage of the "flashing light effect" to increase photosynthetic efficiency. Conflicting results have been reported, however, for the light–dark fluctuations of medium frequency (i.e., in the order of seconds to minutes) that prevail in mixed-mass cultures as regards their stimulatory effect on productivity, and the true role of the mixing rate in determining algal productivity still remains to be clarified (9,64–66). With a few exceptions, significant positive effects of intense stirring on productivity of algal mass cultures have been observed only in paddle-wheel-mixed raceway ponds (9,67). Carlozzi and Torzillo (68) observed a 30–40% increase of the productivity of *A. platensis* grown in tubular reactors when the flow speed was increased from 0.18 to 0.75 m s^{-1}. Tredici and Chini Zittelli (5,60) studied the influence of the bubbling rate on productivity of *A. platensis* in inclined flat plates. At an air supply rate of 1.21 L L^{-1} min^{-1}, productivity increased by 35% over that at a rate of 0.36 L L^{-1} min^{-1}. The authors suggested that the higher bubbling rate enhanced productivity by providing a more adequate light regimen to the cells, as well as by reducing the oxygen level in the culture. Hu and Richmond (62) have recently attained astonishingly high productivities with *A. platensis* cultivated in flat plates by adopting ultrahigh cell densities (higher than 15 g L^{-1}) and providing mixing at the maximal rate permissible without causing cell damage. The choice of the intensity of mixing and of the mixing system must be dictated by the characteristics of the organism to be cultivated. Unlike *Chlorella* or the marine *Nannochloropsis*, which are endowed with strong cell walls, cultures of filamentous cyanobacteria and fragile dinoflagellates cannot be mixed by pumps that subject the cells to high shear stress and so lead to fragmentation of thricomes and to cell lysis. There is general agreement that bubble columns and airlift systems are low-shear-stress mixing systems and therefore are recommended for shear-sensitive organisms. Gudin and Chaumont (69) have reported that productivity may increase up to 75% when pumps are replaced by airlift systems. However, it is known that even air bubbling produces some stress on the cells at the level of bubble formation and breakup, bubble coalescence, and bubble-bursting at the culture surface (4). Other noticeable advantages of bubble columns and airlifts, from the point of view of industrial application, are their simplicity of operation and low maintenance costs due to the absence of moving parts. Mixing devices and power requirements for mixing have been dealt with by Weissman et al. (64) and by Torzillo (4). Torzillo has also analyzed the influence of the rheological properties of the culture on mixing and mass transfer (4). Gas–liquid mass transfer in a serpentine airlift photobioreactor has been described in detail by Camacho Rubio et al. (40).

Temperature Control

Although photobioreactors are said to provide a better process control than do ponds, the maintenance of optimal temperatures in closed reactors is not an easy task, especially when the systems are operated outdoors. Photobioreactors behave as solar collectors: Any light absorbed by the reactor walls or by the culture and not used in photosynthesis is converted into thermal energy, and may lead to an increase of the culture temperature of up to 30 °C or more above ambient temperature, causing the collapse of even rather tolerant species (59). With a few exceptions (e.g., cultivation of thermophilic species) photobioreactors operated outdoors require cooling. Among the solutions experimented with to control overheating in outdoor photobioreactors are shading, water-bath immersion, and water spraying. Shading, to be effective, requires that a large proportion of the reactor be covered for several hours a day, and this greatly reduces the irradiance at the culture surface and, consequently, the productivity (4). Cooling by water spraying has been shown to be reliable and cost-effective when used in dry climates (4,5). Immersion of the reactor in a cooling water basin is very efficient, but its cost-effectiveness is rather doubtful. It may, however, be the sole available option in humid climates, where water spraying does not provide evaporative cooling. Although it varies with the reactor design, the local climate, and the specific organism used, the need for cooling can significantly increase operating costs. In cold climates and in temperate regions during winter, heating may instead be required to maintain the optimal growth temperature.

Orientation of the Photobioreactor and Light Dilution

One of the main impediments to the efficient use of solar energy is the continuous variation of the available solar radiation throughout the day. Solar irradiance is very low in the early morning and in the evening hours and strong enough to inhibit photosynthesis at noon. Although the diurnal pattern of the solar irradiance incident on a horizontal reactor cannot be easily modified, some elevated systems can be oriented at various angles to the sun and thus offer the possibility to vary, continuously or discontinuously, the distribution of the light that strikes the reactor surface. The effect of the photobioreactor orientation on productivity of outdoor algal cultures was investigated by Lee and Low (70), who did not find the angle of inclination of the bioreactor to have any significant influence on the productivity of *Chlorella pyrenoidosa*. Because of the particular reactor design (two tubular panels were connected by hinges, erected at an angle to the ground, and placed in an east–west direction) and, above all, due to the location of the experimental facilities (about 1° latitude north), the findings of these researchers cannot be generalized. The influence of the orientation of flat photobioreactors in higher latitude regions has been studied by Tredici and Chini Zittelli (5) and by Hu et al. (11). These studies have shown that the reactor tilt angle exerts a significant effect on both productivity and photosynthetic efficiency. Systems that are sun oriented so as to intercept the maximum solar radiation on a daily basis (e.g., south-facing plates inclined at 10–30° in summer and at about 60° in winter) typically achieve the highest productivities, even if a culture placed perpendicularly to the rays of the sun at midday may suffer due to excessively high irradiance (19). Vertical plates or systems oriented so as to dilute the impinging radiation achieve lower productivities but higher photosynthetic efficiencies (19). A major objective of photobioreactor design should be to reduce the so-called light saturation effect of photosynthesis (see earlier section). One approach has been to install systems vertically (like Pulz's PBR series) or at a high inclination to the horizontal so that the reactor is at a large angle to the sun rays and solar irradiance is attenuated (diluted) at its surface (19). Another approach involves transmitting sunlight into the photobioreactor through light-diffusing prisms or optical fibers (6,71), but this option appears too complex to be effectively scaled up and applied commercially in the near future. By operating a near-horizontal flat panel and a near-horizontal tubular reactor in parallel, Tredici and Chini Zittelli (19) have shown that tubular reactors can achieve, per se, higher photosynthetic efficiencies and productivities than flat reactors because their curved surfaces achieve a certain degree of sunlight dilution and thus reduce the light saturation effect.

Materials

Materials for photobioreactors must have several properties: high transparency, high mechanical strength, lack of toxicity, resistance to weathering (durability), chemical stability, antifouling surface, and low cost. G. Chini Zittelli and M.R. Tredici (unpublished material, 1998) have carried out a preliminary investigation of the durability of some plastic materials. In Figure 11, the loss of transparency of six different tubing materials exposed to natural weathering in central Italy is shown. Polyethylene and polypropylene tubes are inexpensive, but both lose transparency very quickly and present serious limitations as regards biofouling and mechanical strength. Polycarbonate tubes are seriously weakened by alkaline solutions and become brittle and break easily after 2–3 years of exposure outdoors; they lose transparency mainly in the 300- to 500-nm range. Tubes of polymethyl methacrylate or Teflon have been shown to possess high resistance to weathering and good overall properties. PVC tubes lose transparency significantly after hydration. Some of the flexible plastic materials tested have shown limited mechanical strength that has led to trivial problems (such as leakage due to punctures inflicted by cats and wild animals) that could become rather serious in large-scale systems. Glass might be an excellent material for photobioreactors, due to its high transparency, chemical stability, and durability. Glass tubular systems, however, would require mostly field assembly and numerous connections, and this would greatly increase installation costs; furthermore, fragility may constitute a serious limitation to the use of glass tubes in large-scale installations. The tubes required for building the photostage represent a major cost component of the photobioreactor but not as high as commonly thought. Suitable long-lasting tubular material is available for about $1 per meter or about $15 per square meter of photobioreactor.

CONCLUDING REMARKS AND PROSPECTS

Photobioreactors are considered to have several major advantages over open ponds: They can *(1)* prevent or minimize contamination, *(2)* offer better control over cultural conditions (pH, pCO_2, pO_2, temperature, etc.), *(3)* prevent evaporation and reduce water use, *(4)* lower CO_2 losses due to outgassing, *(5)* permit higher cell concentrations that result in reduced operating costs, and *(6)* attain higher productivity. However, the ultimate and most important advantage is that photobioreactors permit cultivating algal species that cannot be grown in open ponds. On the other hand, certain requirements of photobioreactors, including the need for cooling, the need for strict control of oxygen accumulation and biofouling, and the need for frequent replacement of the material of the photostage, make these systems more expensive to build and operate than ponds. Thus, despite their advantages, the use of photobioreactors must be limited to production of very-high value compounds from phototrophs that cannot be cultivated in open ponds (18,52). Photobioreactors and open ponds should not be viewed as competing technologies, since it is very unlikely that photobioreactors will have a significant impact on any product or process that can be attained in large outdoor open ponds.

In recent years, significantly higher photosynthetic efficiencies and a higher degree of system reliability have been achieved due to better understanding of the growth requirements of phototrophic microorganisms in mass culture. Despite these advances, no photobioreactor has yet proven to be commercially successful, even for very-high-value products. This certainly does not depend on a lack of photobioreactor designs: There exist several valid basic concepts and myriad variations. The principle obstacle to

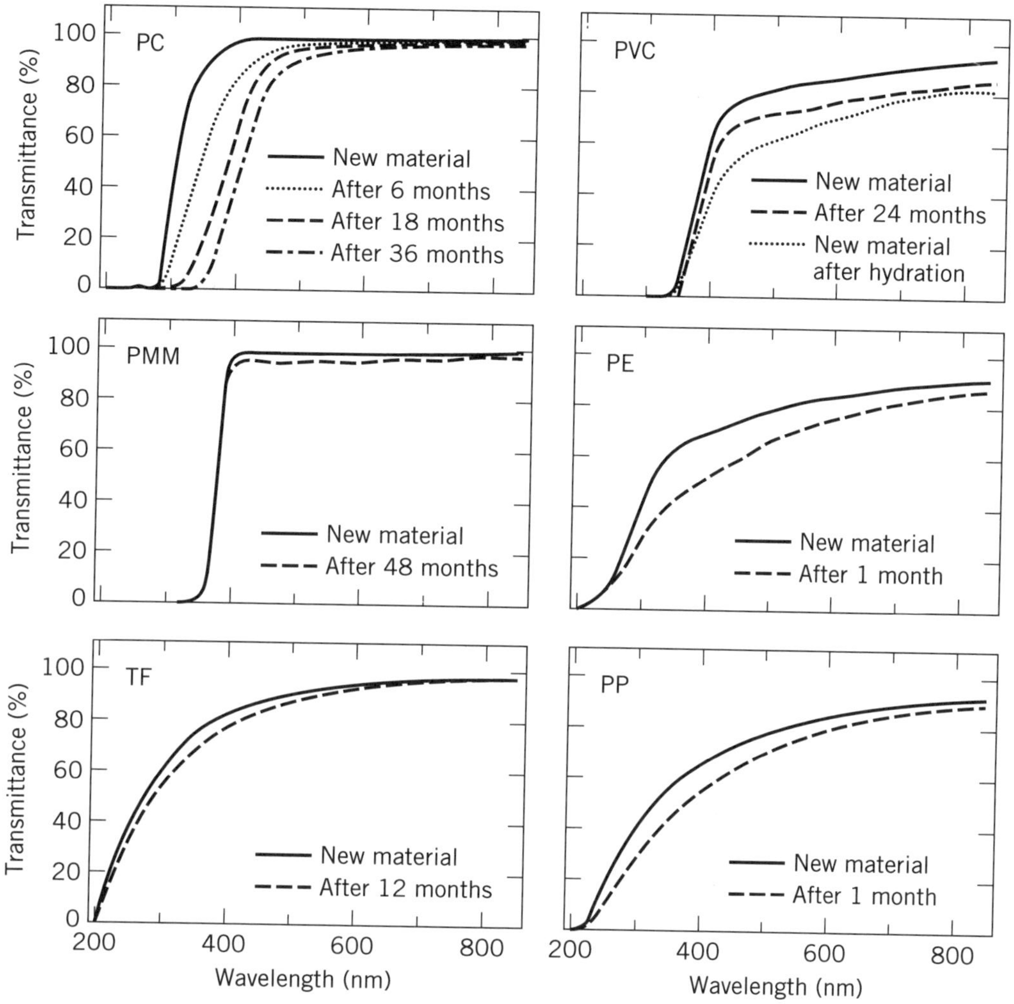

Figure 11. Loss of transmittance of different plastic tubing exposed outdoors to the climatic conditions of central Italy. PC, polycarbonate; PVC, polyvinyl chloride; PMM, polymethyl methacrylate; PE, polyethylene; TF, Teflon; PP, polypropylene. The tubing thickness was: TF, 0.15 mm; PE, 0.2 mm; PP, 0.4 mm; PC and PVC, 1 mm; PMM, 2 mm.

commercial application of this biotechnology seems instead to lie at the level of scaling-up. Transferring those processes developed at the laboratory scale to the industrial scale is never a simple proposition, and it is particularly difficult in the case of photobioreactors because of the peculiar problems of these systems. One of the major causes of the failures of photobioreactor technology at commercial scale is a lack of adequate piloting. The old rule of thumb used in the chemical industry, and still valid in fermentation technology, that each individual increase of scale should not surpass a factor of 10 (52,72) has been ignored in practice. This was especially true in the case of Hidrobiologica SA, where the full-scale plant was set up after only limited experimentation on a very small scale. At the small scale, the photobioreactor performance is usually evaluated in terms of productivity, whereas issues such as contamination, material degradation, and the reliability and sustainability of the production process, which more than the productivity determine the failure or success of the industrial-scale activity, are ignored. Only accurate piloting can shed light on the key problems and provide adequate solutions. It is not clear how much the failure of the PBL photobioreactors in Spain and of the 600-m^3 plant in Argentina depended on system design flaws or on management errors. However, in neither case had the basic technical and biological problems (mixing, oxygen buildup, temperature control, biofouling, contamination) that must be solved for successful full-scale operation been subjected to rigorous investigation at an adequately scaled pilot level. Another mistake, common to both plants, is that to keep operating costs low, the units built and operated were very large. This deprived the systems of necessary flexibility and negated the possibility of making improvements by trial and error. Under these conditions, any problem that might arise (e.g., contamination or toxic oxygen tensions) affects the whole culture and jeopardizes all of production. It is worth pointing out that both industrial plants failed in the cultivation of two phototrophs (*Dunaliella* and *Arthrospira*) that are commonly grown in the much less rigorously controlled open ponds.

The establishment of a commercial activity based on the cultivation of phototrophic microorganisms requires realizing that a photobioreactor design that is efficient at the laboratory level is not necessarily so at the level of industrial production—hence, the importance of accurate piloting. A few simple guidelines should be followed during piloting (63):

1. Experiments must last long enough to reveal the "aging" problems, including contamination, biofouling, and material deterioration, that become evident only after several months of continuous operation.
2. The units experimented with should be similar in size and type to the cultivation units envisaged for the full-scale plant.
3. The productivity of the system must be correctly evaluated and extrapolated.
4. Experiments must be carried out under the climatic and operating conditions that will prevail in the final commercial process.
5. Unless true axenic conditions are adopted, the possibility of contamination must not be underestimated.
6. Appropriate solutions to the main problems commonly encountered in photobioreactors (overheating, biofouling, oxygen accumulation, etc.) must be provided before beginning operation of the commercial plant.
7. The data obtained during the scaling-up stage must be subjected to rigorous analysis and used to reevaluate the economics of the process.

Today the efforts of many researchers, from both academia and industry, are directed toward development of new photobioreactor designs. In the majority of cases, such designs are anything but new and remain an academic exercise with little or no practical value. What we really need is a serious evaluation of the performance of photobioreactors at a relatively large scale. Through such analysis, and together with likely further improvements in engineering design, the development of new materials with suitable characteristics, and the improvement of cultivation techniques derived from a better knowledge of the physiology of phototrophs in mass culture, we can reasonably expect to achieve a degree of reliability in large-scale cultivation of phototrophic microorganisms such as to bridge the gap between laboratory experimentation and commercial application.

ACKNOWLEDGMENTS

I wish to express my sincere gratitude to my colleague Dr. Graziella Chini Zittelli, with whom I have been working on microalgae mass culture and photobioreactor development since 1989. I also wish to thank Dr. Susan Blackburn and Dr. Laura Barsanti for critically discussing the manuscript. Parts of this article have been influenced by exchange of ideas over the years with Dr. John Benemann, to whom I feel particularly indebted.

BIBLIOGRAPHY

1. A. Prokop and L.E. Erickson, in J.A. Asenjo and J.C. Merchuk eds., *Bioreactor System Design*, Marcel Dekker, New York, 1995, pp. 441–477.
2. Y.K. Lee, *Trends Biotechnol.* **4**, 186–189 (1986).
3. D. Chaumont, *J. appl. Phycol.* **5**, 593–604 (1993).
4. G. Torzillo, in A. Vonshak ed., *Spirulina platensis (Arthrospira): Physiology, Cell Biology, and Biotechnology*, Taylor & Francis, London, 1997, pp. 101–115.
5. M.R. Tredici and G. Chini Zittelli, in A. Vonshak ed., *Spirulina platensis (Arthrospira): Physiology, Cell Biology and Biotechnology*, Taylor & Francis, London, 1997, pp. 117–130.
6. O. Pulz and K. Scheibenbogen, in T. Scheper ed., *Bioprocess and Algae Reactor Technology, Apoptosis*, Springer-Verlag, Berlin, 1998, pp. 123–152.
7. D.O. Hall and K.K. Rao, *Photosynthesis 5th ed.*, Cambridge University Press, Cambridge, UK, 1995, pp. 1–194.
8. G.E. Fogg, *Algal Cultures and Phytoplankton Ecology*, The University of Wisconsin Press, Madison, Wisconsin, 1965, p. 12.
9. A. Richmond, in F.E. Round and D.J. Chapman eds., *Progress in Phycological Research*, vol. 7, Biopress, Bristol, UK, 1990, pp. 269–330.
10. Q. Hu, H. Guterman, and A. Richmond, *Biotechnol. Bioeng.* **51**, 51–60 (1996).
11. Q. Hu, D. Faiman, and A. Richmond, *J. Ferment. Bioeng.* **85**, 230–236 (1998).
12. B. Kok, in J.S. Burlew ed., *Algal Culture from Laboratory to Pilot Plant*, Carnegie Institution of Washington Publication No. 600, Carnegie Institution, Washington, D.C., 1953, pp. 63–75.
13. J.F. Cornet, C.G. Dussap, and J.B. Gros, in T. Scheper ed., *Bioprocess and Algae Reactor Technology, Apoptosis*, Springer-Verlag, Berlin, 1998, pp. 153–224.
14. F.G. Acien Fernandez, F. Garcia Camacho, J.A. Sanchez Perez, J.M. Fernandez Sevilla, and E. Molina Grima, *Biotechnol. Bioeng.* **55**, 701–714 (1997).
15. F.G. Acien Fernandez, F. Garcia Camacho, J.A. Sanchez Perez, J.M. Fernandez Sevilla, and E. Molina Grima, *Biotechnol. Bioeng.* **58**, 605–616 (1998).
16. J.S. Burlew, in J.S. Burlew ed., *Algal Culture from Laboratory to Pilot Plant*, Carnegie Institution of Washington Publication No. 600, Carnegie Institution, Washington, D.C., 1953, pp. 3–23.
17. J.C. Goldman, *Water Res.* **11**, 119–136 (1979).
18. J.R. Benemann, in R.C. Cresswell, T.A.V. Rees, and N. Shah eds., *Algal and Cyanobacterial Biotechnology*, Longman, Harlow, UK, 1989, pp. 317–337.
19. M.R. Tredici and G. Chini Zittelli, *Biotechnol. Bioeng.* **57**, 187–197 (1998).
20. S.J. Pirt, Y.K. Lee, M.R. Walach, M.W. Pirt, H.H.M. Balyuzi, and M.J. Bazin, *J. Chem. Tech. Biotechnol.* **33B**, 35–58 (1983).
21. A. Sukenik, R.S. Levy, Y. Levy, P.G. Falkowsky, and Z. Dubinsky, *J. appl. Phycol.* **3**, 191–201 (1991).
22. H. Guterman, A. Vonshak, and S. Ben-Yaakov, *Biotechnol. Bioeng.* **35**, 809–819 (1990).
23. M. Brouers, H. deJong, D.J. Shi, and D.O. Hall, in R.C. Cresswell, T.A.V. Rees, and N. Shah eds., *Algal and Cyanobacterial Biotechnology*, Longman, Harlow, UK, 1989, pp. 272–293.

24. P.M. Cook, in J. Brunel, G.W. Prescott, and L.H. Tiffany eds., *The Culturing of Algae*, Charles F. Kettering Foundation, Yellow Springs, Ohio, 1950, pp. 53–75.
25. Anonymous, in J.S. Burlew ed., *Algal Culture from Laboratory to Pilot Plant*, Carnegie Institution of Washington Publication No. 600, Carnegie Institution, Washington, D.C., 1953, pp. 235–272.
26. A.W. Fisher, in L.W. Douglas and M.L. Kastens eds., *Proc. World Symp. Appl. Solar Energy*, Stanford Research Institute, Menlo Park, Calif., 1956, pp. 243–253.
27. H. Tamiya, E. Hase, K. Shibata, A. Mituya, T. Iwamura, T. Nihei, and T. Sasa, in J.S. Burlew ed., *Algal Culture from Laboratory to Pilot Plant*, Carnegie Institution of Washington Publication No. 600, Carnegie Institution, Washington, D.C., 1953, pp. 204–232.
28. I. Setlik, J. Komarek, and B. Prokes, in J. Necas and O. Lhotsky eds., *Annual Report of the Laboratory of Experimental Algology and Department of Applied Algology for the Year 1966*, Knihtisk, Praha, Czechoslovakia, 1967, pp. 5–36.
29. F. Jüttner, *Proc. Biochem.* **17**, 2–7 (1982).
30. G.H.J. Krüger and J.N. Eloff, in J.U. Grobbelaar, C.J. Soeder, and D.F. Toerien eds., *Wastewater for Aquaculture*. University of Orange Free State Publication, Series C (no. 3), Bloemfontein, South Africa, 1981, pp. 16–23.
31. C.J. Soeder, A. Bolze, and H.D. Payer, *Br. Phycol. J.* **16**, 1–7 (1981).
32. C.M. James and A.M. Al-Khars, *Aquaculture* **87**, 381–393 (1990).
33. M.M. Helm, I. Laing, and E. Jones, *MAFF Direct. Fish. Res.* **53**, 1–7 (1979).
34. R. Radmer, P. Behrens, and K. Arnett, *Biotechnol. Bioeng.* **29**, 488–492 (1987).
35. K. Miyamoto, O. Wable, and J.R. Benemann, *Biotechnology Lett.* **10**, 703–708 (1988).
36. Q. Hu and A. Richmond, *J. appl. Phycol.* **6**, 391–396 (1994).
37. E. Cohen and S. (Malis) Arad, *Biomass* **18**, 59–67 (1989).
38. D. Chaumont, C. Thepenier, and C. Gudin, in T. Stadler, J. Morillon, M.C. Verdus, W. Karamanos, H. Morvan, D. Christiaen eds., *Algal Biotechnology*, Elsevier Applied Science, London, 1988, pp. 199–208.
39. E. Molina Grima, J.A. Sanchez Perez, F. Garcia Camacho, J.M. Fernandez Sevilla, F.G. Acien Fernandez, and J. Urda Cardona, *Appl. Microbiol. Biotechnol.* **42**, 658–663 (1995).
40. F. Camacho Rubio, F.G. Acien Fernandez, J.A. Sanchez Perez, F. Garcia Camacho, and E. Molina Grima, *Biotechnol. Bioeng.* **62**, 71–86 (1999).
41. U.K. Pat. 2,118,572 (March 17, 1983), S.J. Pirt (to Queen Elizabeth College, University of London).
42. G. Torzillo, B. Pushparaj, F. Bocci, W. Balloni, R. Materassi, and G. Florenzano, *Biomass* **11**, 61–74 (1986).
43. G. Torzillo, P. Carlozzi, B. Pushparaj, E. Montaini, and R. Materassi, *Biotechnol. Bioeng.* **42**, 891–898 (1993).
44. U.S. Pat. 5,541,056 (July 22, 1994), M.R. Huntley, P.P. Nijler, and D. Rodalje (to Aquaserch Inc.).
45. S.A. Dvorin, in P. Kretschmer, O. Pulz, and C. Gudin eds., *Proc. 1st Eur. Workshop Microalgal Biotechnology "Algology,"* Print Express K. Beyer, Potsdam, Germany, 1992, pp. 91–92.
46. J.P. Szyper, A.Y. Brandon, J.R. Benemann, M.R. Tredici, and O.R. Zaborsky, *Int. Conf. Biological Hydrogen Production*, Kona, Hawaii, June 23–26, 1997.
47. M.R. Tredici, G. Chini Zittelli, and J.R. Benemann, *Int. Conf. Biological Hydrogen Production*, Kona, Hawaii, June 23–26, 1997.
48. A. Richmond, S. Boussiba, A. Vonshak, and R. Kopel, *J. appl. Phycol.* **5**, 327–332 (1993).
49. Y.K. Lee, S.Y. Ding, C.S. Low, Y.C. Chang, W.L. Forday, and P.C. Chew, *J. appl. Phycol.* **7**, 47–51 (1995).
50. E.A. Davis, J. Dedrick, C.S. French, H.W. Milner, J. Myers, J.H.C. Smith, and H.A. Spoehr, in J.S. Burlew ed., *Algal Culture from Laboratory to Pilot Plant*, Carnegie Institution of Washington Publication No. 600, Carnegie Institution, Washington, D.C., 1953, pp. 105–153.
51. European Pat. 0,239,272 (March 6, 1987), L.F. Robinson and A.W. Morrison (to Biotechna Limited).
52. L.J. Borowitzka and M.A. Borowitzka, in R.C. Cresswell, T.A.V. Rees, and N. Shah eds., *Algal and Cyanobacterial Biotechnology*, Longman, Harlow, UK, 1989, pp. 294–316.
53. Y. Watanabe, J. de la Noüe, and D.O. Hall, *Biotechnol. Bioeng.* **47**, 261–269 (1995).
54. T. Chrismadha and M.A. Borowitzka, in P.S. Moi, Y.K. Lee, M.A. Borowitzca, and B.A. Whitton eds., *Algal Biotechnology in the Asia Pacific Region*, University of Malaya, Kuala Lumpur, 1994, pp. 122–129.
55. D.B. Anderson and D.E. Eakin, *Biotechnol. Bioeng. Symp.* **15**, 533–547 (1985).
56. A. Ramos de Ortega and J.C. Roux, *Biomass* **10**, 141–156 (1986).
57. Italian Pat. 9357-A/88 (March 11, 1988), M.R. Tredici, D. Mannelli, and R. Materassi (to Consiglio Nazionale delle Ricerche).
58. M.R. Tredici, P. Carlozzi, G. Chini Zittelli, and R. Materassi, *Biores. Technol.* **38**, 153–159 (1991).
59. M.R. Tredici and R. Materassi, *J. appl. Phycol.*, **4**, 221–231 (1992).
60. M.R. Tredici, G. Chini Zittelli, and S. Biagiolini, in P. Kretschmer, O. Pulz, and C. Gudin eds., *Proc. 1st Eur. Workshop Microalgal Biotechnology "Algology,"* Print Express K. Beyer, Potsdam, Germany, 1992, pp. 58–60.
61. B. Pushparaj, E. Pelosi, M.R. Tredici, E. Pinzani, and R. Materassi, *J. appl. Phycol.* **9**, 113–119 (1997).
62. Q. Hu and A. Richmond, *J. appl. Phycol.* **8**, 139–145 (1996).
63. M.R. Tredici and G. Chini Zittelli, in P. Kretschmer, O. Pulz, C. Gudin, and V. Semenenko eds., *Proc. 2nd Eur. Workshop Biotechnology of Microalgae*, Institut fur Getreideverarbeitung, Bergholz-Rehbrucke, Germany, 1995, pp. 21–24.
64. J.C. Weissman, R.P. Goebel, and J.R. Benemann, *Biotechnol. Bioeng.* **31**, 336–344 (1988).
65. K.L. Terry and L.P. Raymond, *Enzyme Microbiol. Technol.* **7**, 474–487 (1985).
66. J.U. Grobbelaar, *J. appl. Phycol.* **6**, 331–335 (1994).
67. E.A. Laws, K.L. Terry, J. Wickman, and M.S. Chalup, *Biotechnol. Bioeng.* **25**, 2319–2335 (1983).
68. P. Carlozzi and G. Torzillo, *Appl. Microbiol. Biotechnol.* **45**, 18–23 (1996).
69. C. Gudin and D. Chaumont, *Bioresource Technol.* **38**, 145–151 (1991).
70. Y.K. Lee and C.S. Low, *Biotechnol. Bioeng.* **38**, 995–1000 (1991).
71. K. Mori, *Biotechnol. Bioeng. Symp.* **15**, 331–345 (1985).
72. C. Solà and F. Gòdia, in J.A. Asenjo and J.C. Merchuk eds., *Bioreactor System Design*, Marcel Dekker, New York, 1995, pp. 511–552.

BIOREDUCTION

Tadashi Kometani
Toyama College of Technology
Toyama-shi, Japan

KEY WORDS

Asymmetric reduction
Chiral alcohol
Coenzyme regeneration
NAD(P)H
Resin

OUTLINE

INTRODUCTION

Bioreduction is one of the important biological transformations catalyzed by whole cells. The reducing action of fermenting baker's yeast (*Saccharomyces cerevisiae*) was first reported during the reduction of furfural to furfuryl alcohol. Since the early twentieth century, numerous bioreductions using microorganisms other than baker's yeast have been studied by both biotechnologists and organic chemists. Though the ketone, C═C double bond, azide, and nitro compound are reduced, synthetic chemists have mainly applied the microbial reduction of prochiral ketones to the synthesis of chiral alcohols in an enantiomerically pure form. Until now, a large number of references to scientific studies in this area have been reported.

During the bioreduction of carbonyl compounds, carbonyl reductases in microorganisms catalyze the stereospecific reduction using NADH, NADPH, and, sometimes, the flavine mononucleotide as hydrogen donors. The microorganisms can reduce many types of ketones and aldehydes. However, this useful bioreduction has the disadvantage of needing regeneration system of cofactors, such as NAD(P)H, in addition to the reductases in whole cells. To overcome this problem, it is important to utilize the ability of the living microorganism to regenerate NAD(P)H from $NAD(P)^+$ by their natural process of respiration (Fig. 1).

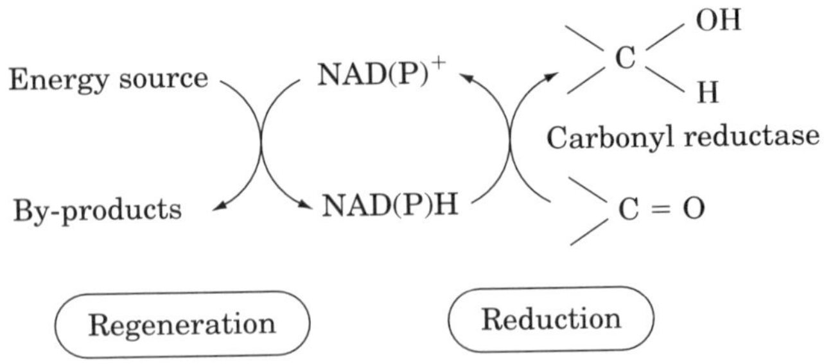

Figure 1. Outline of bioreduction.

This article describes the bioreduction of ketones with microbial cells, including reductases and nicotinamide coenzyme regeneration. An application of the bioreduction to the synthesis of an optically active alcohol is then described.

REDUCTASES OF KETONE

The most important microorganism catalyzing bioreduction is baker's yeast. The scientific studies devoted to bioreduction using baker's yeast have gradually increased since the 1950s. The transformations of various classes of synthetic organic molecules were examined. These transformations were reviewed from the viewpoint of their chemical reactions in three excellent articles published in 1990, 1991, and 1992 (1–3). In these reviews, the uses of baker's yeast in synthetic organic chemistry are fully covered.

Other microorganisms also catalyze the bioreduction of ketones. However, the reports on other microorganisms do not outnumber those on baker's yeast. The main difficulty in the widespread use of microorganisms among the organic chemists' community certainly depends on their familiarity with microbiological techniques. Therefore, good cooperation between synthetic organic chemists and microbiologists oriented toward such applications can be fruitful.

Baker's Yeast

Baker's yeast can be used by people not trained in microbiology and can be considered a useful "reagent," easily accessible in every laboratory. Fermenting baker's yeast using glucose in tap water as an energy source is generally used for highly enantioselective reactions. However, when the process is not completely enantioselective, simple modifications of the experimental conditions may influence the stereochemistry and the enantiomeric excess of the product. For instance, the use of organic media, the addition of other compounds, a change in energy source, and immobilization techniques are helpful.

Traditionally, baker's yeast can reduce variously substituted carbonyl groups to the corresponding hydroxyl compounds, the stereochemical outcome of these highly enantioselective reductions depends on the presence of dehydrogenases, which generally follow the so-called Prelog's rule. Several exceptions have been found, which can be due to the presence of competing dehydrogenases with different stereochemical requirements.

The baker's yeast–mediated reduction of acetol and ethyl acetoacetate afforded (*R*)-propylene glycol and (*S*)-ethyl 3-hydroxybutanoate, respectively. The large-scale preparations of these compounds were accomplished by using a bubble-column reactor and ethanol as the energy source in more than 98% enantiomeric excess (4). The usefulness of the yeast-mediated reduction was also shown in the diltiazem synthesis. The transformation of a racemic ketone (*RS*)-2-(4-methoxyphenyl)-1,5-benzothiazepin-3,4(2*H*,5*H*)-dione afforded only the desired (2*S*,3*S*)-2,3-dihydro-3-hydroxy-2-(4-methoxyphenyl)-1,5-benzothiazepin-4(5*H*)-one, a key intermediate in the diltiazem synthesis. The difficulty in its industrial application because of its low productivity, which is due to the low solubility of the substrate in water, was overcome by changing its physical properties. Thus, 100 g of the substrate in 1 L of reaction medium was reduced to produce the chiral intermediate with over 99% ee in 80% yield (5) (Fig. 2).

Other Microorganisms

Bioreduction using other microorganisms was systematically examined by Shimizu and Yamada using ketopantoyl lactone as the substrate (6). They found that about 10% of the microorganisms tested (yeasts, molds, bacteria, actinomycetes, and basidiomycete) reduced ketopantoyl lactone to pantoyl lactone in the presence of glucose. Their enantioselectivities varied at random and were independent of genus and species.

D-(−)-Pantoyl lactone, the γ-lactone of D-(−)-pantoic acid, is an important starting material for the synthesis of the vitamin D-(+)-pantothenic acid. The conventional synthesis involved the tedious chemical resolution of racemic pantoyl lactone. In this synthesis, the enantioselective bioreduction of ketopantoyl lactone to D-(−)-pantoyl lactone was examined. When *Candida parapsilosis* or *Rhodotorula minuta* cells were incubated with ketopantoyl lactone (more than 50 g/L) and glucose, D-(−)-pantoyl lactone was produced in excellent chemical and optical yields. In this reaction, the asymmetric reduction is catalyzed by a newly found carbonyl reductase specific to conjugated polyketones, which requires NADPH as a cofactor. Another approach is the enantioselective reduction of ketopantoic acid. *Agrobacterium* sp. was screened and incubated with ketopantoic acid and fructose to give D-(−)-pantoic acid. The acid was easily cyclized with hydrochloric acid to afford the D-(−)-pantoyl lactone (Fig. 3).

For the stereospecific reduction of 4-chloroacetoacetate, each stereoisomer of the product (*R*)- or (*S*)-4-chloro-3-hydroxybutanoate is a useful compound. Shimizu and Yamada also screened a microorganism for the asymmetric reduction of 4-chloroacetoacetate. Almost all microorganisms, such as *R. minuta* and *Paecilomyces marquandii*, reduce to (*S*)-4-chloro-3-hydroxybutanoate, which converts to (*S*)-1-benzyl-3-hydroxypyrrolidine. In the carnitine synthesis, 4-chloroacetoacetate should be reduced to (*R*)-4-chloro-3-hydroxybutanoate, which would be converted to carnitine by the chemical method. *Sporobolomyces salmonicolor*, *Micrococus luteus*, and *Cellulumonas* sp. predomi-

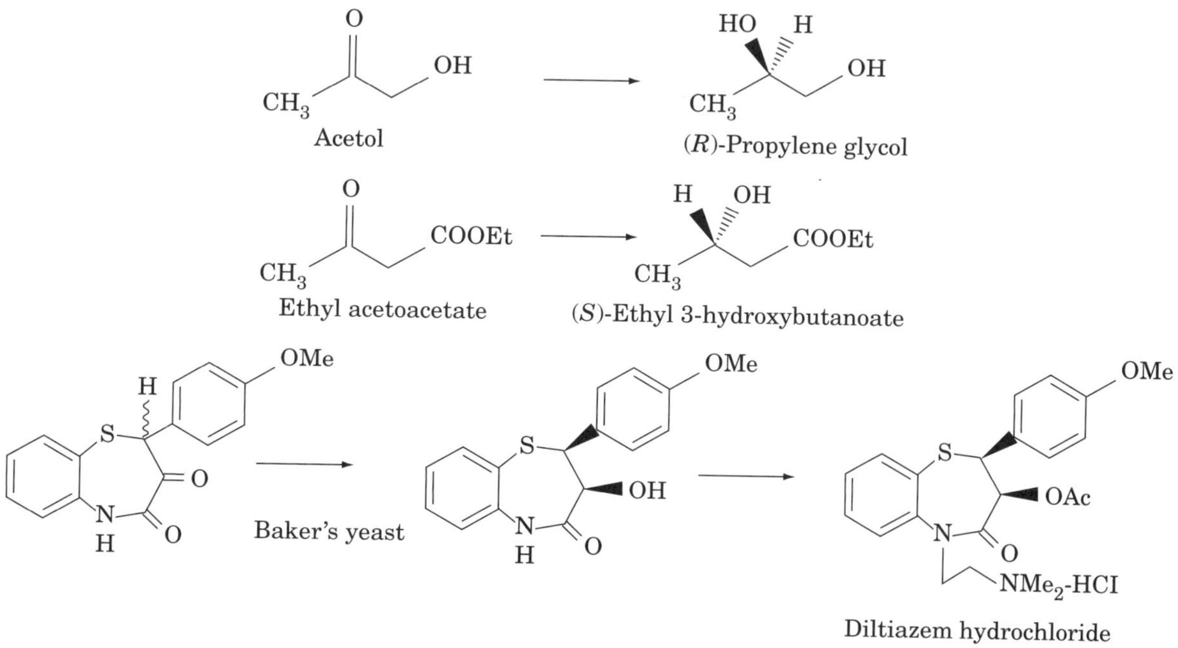

Figure 2. Production of optically pure alcohols through the baker's yeast–mediated reduction.

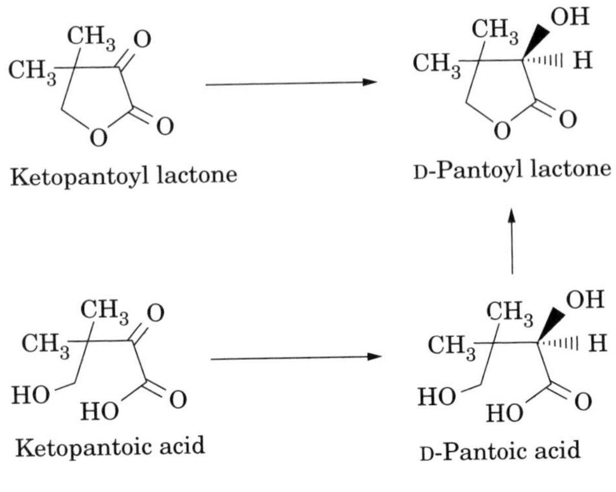

Figure 3. Production of D-pantoyl lactone.

nantly afforded the (*R*)-compound. However, they could not obtain a sufficient result for the enantiomeric excess. Therefore, *S. salmonicolor* was treated with acetone to remove (*S*)-reductase. The treated cells were used for the stereospecific reduction of ethyl 4-chloroacetoacetate to afford (*R*)-4-chloro-3-hydroxybutanoate in a water–organic solvent two-phase system. In this synthesis, NADPH must be regenerated from $NADP^+$. The regeneration was accomplished by using glucose and glucose dehydrogenase in the aqueous phase. This process can be used in the carnitine synthesis (Fig. 4).

NAD(P)H REGENERATION

During bioreduction, NAD(P)H coenzymes must be successively regenerated from the corresponding oxidized form in the cells. The system responsible for the coenzyme regeneration is coupled with the reduction of a substrate. A large amount of work has been devoted to developing and testing various types of systems for NAD(P)H regeneration.

D-Glucose Derivatives

Yamada et al. characterized the enzymes responsible for the NADPH regeneration system using glucose as the energy source (7). They examined the NADPH-forming ability of the cell-free extracts of *S. cerevisiae, Candida parapsilosis*, and *Mucor ambiguus*, with several sugars as the substrate. NADPH was not formed from D-fructose 1,6-diphosphate. This suggests that NADPH could not be produced after D-fructose 6-phosphate was phosphorylated to D-fructose 1,6-diphosphate. They found that NADPH is produced from the hexose monophosphates, D-glucose 1-phosphate, D-glucose 6-phosphate, or D-fructose 6-phosphate. Thus, with *S. cerevisiae, C. parapsilosis*, and *M. ambiguus*, the enzymes were shown to be D-glucose 6-phosphate dehydrogenase and 6-phospho-D-gluconate dehydrogenase in the hexose monophosphate pathway. By means of these systems, endogenous NADPH can be efficiently regenerated by the simple mixing of microbial cells with glucose. The reductions of various carbonyl compounds by microorganisms in synthetic organic chemistry will be mainly based on this system.

Ethanol

During the baker's yeast–mediated reduction using ethanol as the energy source, NAD(P)H is regenerated from $NAD(P)^+$ through the oxidative pathway of ethanol to carbon dioxide, and the system is also efficiently coupled with reduction of a ketone (4). The enzyme system responsible for NAD(P)H regeneration was investigated by studying the reduction of acetol and ethyl acetoacetate, catalyzed by NADH- and NADPH-dependent carbonyl reductases, respectively. NAD^+ formed during the reduction of acetol is reduced to NADH through the oxidative pathway of ethanol to acetate, catalyzed by alcohol dehydrogenase and aldehyde dehydrogenase. On the other hand, $NADP^+$ formed during the reduction of ethyl acetoacetate is reduced to NADPH through the oxidative pathway of acetate to carbon dioxide under aerobic conditions. The enzyme responsible for NADPH regeneration must be malate dehydrogenase (decarboxylating) in leaving the TCA cycle. By using the NAD(P)H regeneration system, the baker's yeast–mediated large-scale productions of (*R*)-propylene glycol from acetol and (*S*)-ethyl 3-hydroxybutanoate from ethyl acetoacetate were efficiently accomplished. Other yeasts also use ethanol as the energy source for the NAD(P)H regeneration system in the same manner (8).

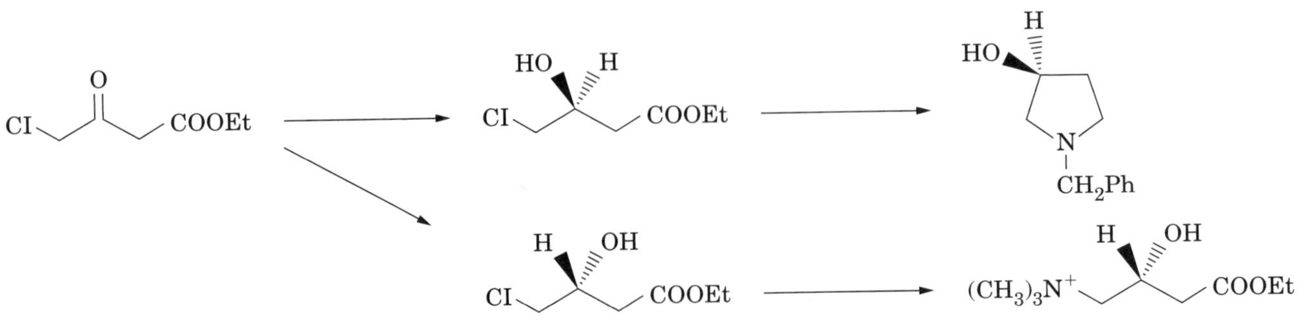

Figure 4. Production of (*S*)-1-benzyl-3-hydroxypyrrolidine and (*R*)-carnitine.

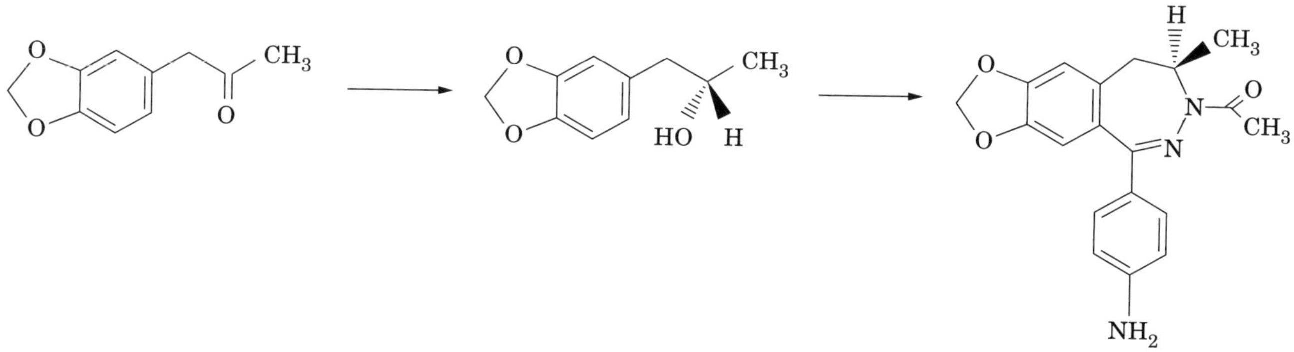

Figure 5. Production of (S)-3,4-methylenedioxyphenyl isopropanol for the synthesis of LY300164.

Methanol

Similarly, the reaction involved in the oxidation of methanol to CO_2 by methylotrophic microorganisms, methanol dehydrogenase, formaldehyde dehydrogenase, and formate dehydrogenase is considered to be a very useful system for NADH regeneration.

BIOREDUCTION OF 3,4-METHYLENEDIOXYPHENYLACETONE

The clinical potential for the (R)-isomer LY300164 has led to interest in developing syntheses that would not be limited by the inherent inefficiency of the resolution methods and would be amenable to large-scale execution. The retrosynthetic strategy required the synthesis of the benzodiazepine nucleus by intramolecular cyclization of a suitably functionalized hydrazone. This disconnection was chosen because it was anticipated that the required stereocenter could be accessed from the relatively simple alcohol. Asymmetric reduction of the corresponding ketone was considered the most desirable option for the preparation of the optically active alcohol (9,10) (Fig. 5).

Screening of Microorganisms

The chemical method for the enantioselective reduction of methyl benzyl ketones is typically nonselective or requires the stoichiometric use of costly reagents. Unlike closely related examples, however, selectivity and conversion with baker's yeast, as well as other common microorganisms, were unsatisfactory. After an extensive screening of microorganisms, the yeasts *Candida famata* and *Zygosaccharomyces rouxii* could catalyze the reduction of 3,4-methylenedioxyphenylacetone to 3,4-methylenedioxyphenyl-(S)-isopropanol with >99% enantiomeric excess in the presence of glucose. However, the productivity number using *C. famata* was too low to be practical and the microorganism was extremely sensitive to concentrations of both the ketone and the alcohol greater than 2 g/L. On the other hand, *Z. rouxii* was more tolerant to higher substrate concentrations and had a higher productivity number.

Practical Application

In this bioreduction, both the product and substrate were toxic to *Z. rouxii*, and this limited the maximum substrate concentration to only 6 g/L. It should be noted that toxicity to the substrate alone would not have been a problem, since this could be readily overcome by the slow addition of the substrate. However, the only way to overcome product toxicity was to devise a method to remove the product from the reaction as it was formed.

For in situ product removal, immiscible organic solvents are commonly used. This type of system is called a two-liquid-phase bioreactor. However, one of the major difficulties is that many organic solvents are highly toxic to microorganisms. In this reaction, polymeric hydrophobic resin (Amberlite XAD-7 from Rohm and Haas, Philadelphia, PA) was used to both supply substrate to, and remove the product from, the reaction mixture. The overall effect is that a high reaction concentration is achieved (40 g/L) while never exposing either the substrate or the product to more than 2 g/L. Thus, in many cases, a polymeric adsorbent resin will provide an attractive alternative to the use of organic solvents in whole cells.

BIBLIOGRAPHY

1. S. Servi, *Synthesis* **1990**, 1–25 (1990).
2. R. Csuk and B.I. Gläzer, *Chem. Rev.* **91**, 49–97 (1991).
3. E. Santaniello, P. Ferraboschi, P. Grisenti, and A. Manzocchi, *Chem. Rev.* **92**, 1071–1140 (1992).
4. T. Kometani, H. Yoshii, and R. Matsuno, *J. Mol. Cat. B: Enzymatic* **1**, 45–52 (1996).
5. T. Kometani, Y. Sakai, H. Matsumae, T. Shibatani, and R. Matsuno, *J. Ferment. Bioeng.* **84**, 195–199 (1997).
6. S. Shimizu and H. Yamada, *J. Synth. Org. Chem. Japan* **49**, 52–62 (1991).
7. M. Kataoka, Y. Nomura, S. Shimizu, and H. Yamada, *Biosci. Biotech. Biochem.* **56**, 820–821 (1992).
8. T. Kometani, Y. Sakai, H. Urai, H. Yoshii and R. Matsuno, *Biosci. Biotech. Biochem.* **61**, 1370–1372 (1997).
9. M.J. Zmijewski, J. Vicenzi, B.E. Landen, W. Muth, P. Marler, and B. Anderson, *Appl. Microbiol. Biotechnol.* **47**, 162–166 (1997).
10. J.F. Vicenzi, M.J. Zmijewski, M.R. Reinhard, B.E. Landen, W.L. Muth, and P.G. Marler, *Enzyme Microb. Technol.* **20**, 494–499 (1997).

BIOREMEDIATION

Ronald L. Crawford
University of Idaho
Moscow, Idaho

KEY WORDS

Air sparging
Biofiltration
Biopiles
Bioreactors
Bioremediation
Bioventing
Composting
Intrinsic remediation
Landfarming
Phytoremediation
White-rot fungi

OUTLINE

INTRODUCTION

As a result of the manufacture, use, spillage, and improper disposal of anthropogenic chemicals and their waste products during past decades, toxic chemicals have become ubiquitous contaminants of soils and waters around the globe. These compounds include everything from petroleum and synthetic organic chemicals to heavy metals and radionuclides. Governments throughout the world now require that the worst of these contamination problems be cleaned up to eliminate, insofar as possible, their hazards to human health and the environment. Traditional methods such as excavation and incineration of contaminated soil or pumping and aboveground treatment of groundwater are proving to be either too expensive, politically unpopular, or both. Thus, innovative technologies are being sought to address the needs of the environmental restoration industry. *Bioremediation* is such a technology.

Bioremediation involves the use of microorganisms, plants, or products yielded by living organisms to degrade, detoxify, or permanently sequester toxic chemicals present in environmental systems. Bioremediation can be used either ex situ or in situ to treat soil, sediment, sludge, water, or air. In ex situ bioremediation procedures, contaminated materials are first removed from the polluted site and then treated, whereas in situ techniques are used to treat contaminants in place without relocating them to a treatment facility. Bioremediation can be promoted by augmenting natural systems with exogenous biological materials or by stimulating indigenous contaminant-transforming microbial populations. Natural systems can be augmented by the introduction of bacteria that degrade the targeted contaminants or plants that sequester pollutants. Native microbial populations are stimulated by adding nutrients or modifying environmental conditions to favor the growth and activity of appropriate organisms. Although intrinsic bioremediation, or microbial processes that occur without human intervention, may ultimately be able to clean up a contaminated site, a great variety of techniques are used by the bioremediation industry to enhance and accelerate degradative processes.

BIOREMEDIATION TECHNOLOGIES

Bioreactors

Bioreactors, or vessels of various configurations (1) containing degradative microbes, are a common means of ex situ treatment of contaminated materials, typically soil that has been excavated from a polluted site or groundwater that has been pumped from a polluted aquifer. A bioreactor may contain, for example, a slurry of soil and water to which nutrients have been added, such as sources of nitrogen, phosphorus, and/or carbon that support the growth and metabolism of degradative microorganisms. Sometimes the microorganisms themselves may have to be added (40). Slurry reactors can be run either aerobically, by stirring and mixing in air or pure oxygen (2), or anoxically, with little or no mixing (3,4). Slurry phase reactor treatment cells can be designed to hold a few thousand to millions of gallons. Aeration systems for aerobic slurry reactors must be able to maintain dissolved oxygen concentrations at 2.0 mg/L or higher. Aeration is a major expense for such systems. To increase oxygen transfer rates, such systems may be sparged with pure oxygen. Often reactor designs include a series of eductors to oxygenate the mixed liquor while minimizing air emissions. Anaerobic slurry systems usually require some mixing devices to evenly distribute soil or sediment materials throughout the aqueous phase, but they spare the need and cost of transferring large amounts of oxygen to the solution phase. Aerobic slurry reactors have been used to treat materials contaminated with chemicals such as polynuclear aromatic hydrocarbons (PAHs) and chlorinated phenols (35). Anaerobic slurry reactors have been used to treat materials polluted by compounds such as chlorinated solvents, polychlorinated biphenyls (PCBs), nitrated munitions like 2,4,6-trinitrotoluene (TNT) (Fig. 1) (4), and nitrated herbicides such as dinoseb (37).

Figure 1. Pilot-scale anaerobic slurry reactor for treating TNT-contaminated soil. *Source*: University of Idaho, with permission.

Water may be treated in simple batch reactors that hold a few hundred to many thousands of liters; nutrients and microorganisms may be added as required. Fluidized beds that contain biomass-bearing carriers are popular for water treatment (5) and can be run aerobically or anaerobically, depending on the contaminants being treated (6,7). Even normally recalcitrant compounds such as aniline and nitrobenzene can be treated in such systems. Alternatively, biomass can be grown on a supporting matrix that is packed into a vessel such as a column, and the water to be treated can be passed through the packed bed to remove contaminants by degradation or absorption (8). Packed beds can also be used to treat vapor-phase contaminants such as chlorinated solvents, which are sparged through the reactor as a separate gaseous stream (9). A novel, mechanically mixed packed-bed reactor that works as a biotrickling filter is a variation on packed-bed technology for treatment of vapor-phase contaminants (43). Another variation on reactor design for water treatment is the upflow sludge blanket reactor using granules containing anaerobic bacteria acclimated to the degradation of a particular waste stream (38). Bioreactors can be used in treatment trains that vary conditions, for example, by alternating anaerobic and aerobic conditions (36), so that mixtures of contaminants or highly chlorinated compounds can be treated successfully.

Biofiltration

Biofiltration is an ex situ treatment for vapor-phase contaminants (44,45). Typically, it involves passing contaminated air through a bed of soil, peat, or compost that has been preacclimated to develop a microbial population that completely degrades the targeted pollutants (10). This type of technology works well for volatile components of petroleum, and for many years it has been used in Europe for odor control (11). Biofilters have been developed for toxic airborne industrial contaminants as diverse as styrene (46), pentane and isobutane mixtures (47), toluene (48), chlorinated benzenes (49), dimethylsulfide (50), and volatile organic compounds (VOCs) (51). For example, a biofiltration process was developed for styrene-containing industrial offgases. This process used peat as filter material. Styrene reduction after 190 days of operation was 70% to 98%. The styrene elimination capacity was 12 g m^{-3} h^{-1} (maximum, 30 g m^{-3} h^{-1}). Degradation required addition of nutrients to the peat, adjustment of the pH, and control of humidity. The optimum operation temperature was about 23 °C, but the styrene removal was still satisfactory at 12 °C (46). Composting of leaves and alfalfa was used for the treatment of toluene-contaminated air in a biofilter. During the thermophilic phase (45–55 °C), toluene biodegradation rates reached 110 g toluene m^{-3} h^{-1} at an inlet concentration of about 5 g/m^3 and a gas residence time of 90 s (48). Removal rates by biofilters for chlorinated benzenes reached 300 g solvent vapor h^{-1} m^{-3} of filter volume (49).

Air Sparging

The biodegradation of many contaminants in groundwater is limited by the shortage of electron acceptors for oxidative processes. The most effective electron acceptor for the bioremediation of many contaminants is oxygen, which is present in groundwater at only 8–10 mg/L even under optimal conditions. Air sparging is an in situ technology for introducing oxygen into contaminated aquifers (12). Air or pure oxygen is injected into the aquifer through sparging wells or infiltration galleries (13). Potential problems associated with the introduction of oxygen into groundwater include biofouling, in which injection wells become plugged with biomass, and obstruction of well screens or subsurface geologic formations by iron complexes that precipitate as a result of the oxidation of Fe(II) to Fe(III). Hydrogen peroxide, which decomposes in the presence of iron or biomass to produce oxygen (14), is sometimes substituted for air or pure oxygen. Even when oxygen is supplied by sparging, it can be used so rapidly that its supply can still be rate limiting for bioremediation processes. Thus, the bioremediation industry has developed oxygen-releasing compounds (ORCs) composed of materials such as magnesium peroxide that allow controlled, long-term release of oxygen when introduced into boreholes as a slurry (15). Most successes with air sparging, as with many other bioremediation systems, have been reported in petroleum-contaminated aquifers. The reintroduction of indigenous microorganisms isolated from a contaminated site after culturing is one approach to in situ hydrocarbon bioremediation. It is effective especially when microorganism growth is supplemented by addition of oxygen and nitrogen (52). At the Moffett Field (Sunnyvale, Calif.) groundwater test site, in situ removal of *cis*-dichloroethene (*c*-DCE) and trichloroethene (TCE) coincided with biostimulation through phenol and oxygen injection and utilization, with c-DCE removed more rapidly than TCE. Greater TCE and c-DCE removal was observed when the phenol concentration was increased. More than 90% removal of *c*-DCE and TCE was observed in a 2-m biostimulated zone (53). Thus, introduction of oxygen into contaminated aquifers can be of use even for recalcitrant compounds such as VOCs.

Bioventing

Bioventing is related to air sparging in that the objective of this in situ technology (17) is to provide oxygen to the microbial populations in the subsurface. Bioventing systems use equipment similar to that used in soil vapor extraction (SVE) processes. Air injection or extraction wells or trenches are connected by pipes to blowers or vacuum pumps which either push or pull air through the vadose zone (unsaturated soil horizons) to stimulate microbial populations capable of degrading contaminants. The objective is to provide sufficient oxygen to the subsurface to promote in situ contaminant degradation, but not so much that volatile contaminants are released undegraded to the atmosphere. Thus, bioventing is best used with contaminants that are moderately volatile, such as diesel or jet fuels or residues of aged petroleum (13,21). For bioventing to succeed, the vadose zone must be sufficiently permeable to allow air exchange at least every 24 h (16), so this process is not suitable for all sites. It may be possible in some instances to open up impermeable soils by fracturing them with hydraulic or air pressure injections (26), allowing bioventing strategies to be employed. Bioventing has produced excellent results at hundreds of sites worldwide in recent years (13,18–20).

A typical bioventing system (Fig. 2) involves soil vapor extraction followed by bioventing via vent wells located within an aquifer in areas of high pollutant contamination. Monitoring wells in both the vadose zone (unsaturated soils above the aquifer) and within the aquifer are used to monitor bioventing progress. Neutron probes may sometimes be used for soil moisture monitoring, to assure that venting does not dry out the soil so that microbial populations are inhibited. A plastic liner is usually installed over the surface of part or all of a contaminated area to prevent infiltration of air that might bypass the desired path of air to the vent wells. A catalytic incinerator to destroy extracted vapor that was not degraded by microbes in the vadose zone is often installed as a safety device, especially at sites near populated areas. If bioventing proceeds effectively, there may be no need for treatment of extracted vapors because the concentration of contaminants in the vented gases may be very low. In bioventing, as with all biological processes in soil, soil moisture must be maintained within a range of 6% to 12%, and supplemental phosphorus or nitrogen may be needed (e.g., N as ammonia provided in the gas phase) to maintain a C:N:P ratio of about 100:10:1.

Not all sites are amenable to bioventing. The Battelle Corporation has used a special site evaluation protocol to

Figure 2. Petroleum bioventing operation. *Source*: University of Idaho, with permission.

determine the suitability of a site for remediation by this technique (41). Four steps are performed sequentially for characterization of a potential bioventing site:

1. A soil-gas survey is conducted to determine if the subsurface is oxygen limited. Soil gas is analyzed for concentrations of oxygen and carbon dioxide. Oxygen is, of course, required for successful bioventing of many contaminants, such as petroleum hydrocarbons.
2. A soil-gas permeability test is conducted to see if air can be injected at rates sufficient to aerate the vadose zone. This test requires injection and withdrawal of air for measuring changes in subsurface pressures at discrete distances from the point of injection. Soils must be sufficiently permeable to be candidates for bioventing. "Tight" soils (e.g., clay-rich soils) can usually be eliminated at this stage.
3. Oxygen-limited but sufficiently permeable soils are then subjected to an in situ respiration test to measure biological contaminant degradation rates for comparison to respiration rates in similar, but uncontaminated, soils. This involves aerating the soils for 24 h and then monitoring the oxygen and carbon dioxide concentrations in the soil-gas phase after the final air injection. An oxygen utilization rate is calculated and used to estimate a contaminant degradation rate, which can in turn be used to estimate the time that will be required to biovent the site.
4. Finally, a pilot- and/or full-scale field bioventing remediation demonstration is conducted at the site. There may or may not be a need for pilot-scale testing before a full-scale process is implemented. This is determined on a site-specific basis according to the prior testing.

Landfarming

Landfarming is an aboveground ex situ process in which contaminated soils are excavated, supplemented with nutrients (primarily sources of nitrogen or phosphorus), and spread in relatively thin layers on a prepared site so that biodegradation of contaminants is promoted. These soils are aerated by periodic tilling and kept moist by irrigation, thus explaining the name of this procedure (13). Sometimes soils are spread directly on the ground at a graded site, if there is no danger that the contaminants will leach through the soil to groundwater or vaporize to dangerous levels in the atmosphere. More commonly, soils are spread on sites with preinstalled plastic liners or clay barriers and leachate collection systems to prevent any possible leaching of contaminants offsite or to groundwater, as well as vapor containment enclosures, especially if the site is located near populated areas. Petroleum is commonly treated in this manner (20). Highly toxic or biologically recalcitrant compounds generally should not be landfarmed, and the presence of heavy metals such as lead usually indicates that landfarming is not an appropriate technology for a particular soil. Some advantages and disadvantages of landfarming of petroleum-contaminated soil are shown in Table 1 (data from U.S. EPA).

The presence of vegetation can enhance the degradation of contaminants during landfarming of contaminated soil. Greenhouse experiments have shown the degradation of PAHs to be greater in the presence of plants than in their absence. PAHs were detectable in the plant tissue, but overall uptake was not significant. Enhanced microbial activity appeared to be responsible for the increased removal of PAHs. Thus, planting of vegetation is promising as an inexpensive means of speeding landfarm-based remediation of sites contaminated with petroleum (42).

Biopiles

Biopiles are basically soil-heaping operations, a variation on landfarming. The primary advantages of biopiles over landfarms are that biopiles save land space, and air emissions can be more inexpensively captured and treated (20). In this technique soil is excavated and piled into a mound within a lined treatment cell. The pile is typically covered by a plastic sheet, and an internal system of perforated pipes is usually installed to draw air in by means of vacuum pumps. Alternatively, air may be blown into the pile through the internal pumping system, or supplied from oxygen-releasing compounds (ORCs; see "Air Sparging") incorporated into the piles. Water, nutrients, and microbial inoculants are added as needed, and leachates and gaseous emissions are collected for separate treatment. In theory, biopiles can be used for most biodegradable contaminants, but the limitations of landfarming generally apply. In practice, most biopile operations have treated petroleum-contaminated soils (20).

Table 1. Advantages and Disadvantages of Landfarming for Treatment of Petroleum-Contaminated Soil

Advantages	Disadvantages
Relatively simple to design and implement	Concentration reductions >95% and constituent concentrations <0.1 ppm are very difficult to achieve
Short treatment times (usually 6 months to 2 years under optimal conditions)	May not be effective for high constituent concentrations (>50,000 ppm total petroleum hydrocarbons)
Cost competitive: \$30–60/ton of contaminated soil	Presence of significant heavy metal concentrations (>2,500 ppm) may inhibit microbial growth
Effective on organic constituents with slow biodegradation rates	Volatile constituents tend to evaporate rather than biodegrade during treatment
	Requires a large land area for treatment
	Dust and vapor generation during aeration may pose air quality concerns
	May require bottom liner if leaching from the landfarm is a concern

Source: U.S. Environmental Protection Agency (*http://www.epa.gov/swerust1/cat/landfarm.htm*).

"Biomounding" is a low-technology, low-cost variation of biopiles. It has been used to treat petroleum-contaminated soil, with the addition of bulking materials like animal manure and wood chips to promote good conditions (sufficient moisture retention, good aeration, source of nitrogen, etc.) for building up microbial populations capable of hydrocarbon degradation. This simple process usually is well accepted by local governments and the general public. Biomounding can reach fairly stringent soil cleanup standards in the range of 10 ppm as total petroleum hydrocarbons (TPH). Generally, indigenous bacteria present in petroleum-contaminated soil proliferate under biomounding conditions, so inoculation with exogenous microbes is not normally needed. By engineering a favorable environment for microorganisms in a contaminated soil, it is possible to construct biomounds that are reasonably inexpensive to manage, yet allow fairly fast breakdown of petroleum compounds in soil, without the need for continually turning of the soil as in composting. Simple passive aeration systems normally are used in biomounds. For example, a grid of horizontal piping constructed of flexible perforated drain tile, along with vertical piping (risers) of PVC, can be used in the system. Multiple layers of piping can be employed, depending on the volume of soil to be treated. Moisture and heat loss from biomounds are usually controlled by covering the mounds with plastic sheeting.

Composting

In composting, excavated soil is mixed with bulking agents such as straw, horse manure, or other agricultural residues to form compost piles, which are then managed for good aeration by periodic turning with specially designed machines. Vigorous microbial activity generates considerable heat, so the piles are termed thermophilic, operating at temperatures around 50–60 °C. Composting has been examined as a treatment for a variety of contaminants, including petroleum sludges (22), chlorophenols (23,24), and munitions residues (25). This approach is still experimental, since some questions remain as to its overall efficacy in treating some contaminants. For example, nitroaromatic compounds such as 2,4,6-trinitrotoluene (TNT) tend to be polymerized or incorporated into soil humus fractions, with little or no conversion to carbon dioxide or products that might themselves be degradable to carbon dioxide (25). The long-term stability and potential for release of toxic compounds from such uncharacterized materials in soil is not known.

In one study, a reduction in direct lethality of TNT to earthworms reflected the removal of explosive compounds by composting, as indicated by chemical analysis. However, assays by the Mutatox procedure indicated increased mutagenicity, a result that would not have been shown by chemical analysis alone, which implied that mutagenic metabolites of explosives were formed during composting and were incompletely degraded (54). In another study, [^{14}C]TNT was subjected to composting during which much of the ^{14}C becomes incorporated into the "humic" fraction of the compost where it supposedly becomes unavailable. The composted TNT resulting from this process was then fed to rats. Although only 2.3% of the total ^{14}C dose appeared in the rats' urine during the first 3 days after feeding, low levels of ^{14}C continued to be excreted in urine from compost-treated rats for more than 6 months. The total amount of ^{14}C excreted in the urine was comparable to that excreted by animals treated with pure TNT (55). Thus, it is clear that composted TNT is not irreversibly immobilized, but that TNT or its biotransformation products in compost are bioavailable. Thus, composting may not be the treatment of choice for some wastes where there is little mineralization of the target pollutant.

Despite the experimental nature of composting, it is likely to become a useful tool for treatment of some chemically polluted soils. Variations on composting appear to be possible to overcome some of these problems. For example, TNT is degraded more effectively when the compost is allowed to first undergo a fairly long anaerobic treatment phase before traditional aerobic composting (39). The critical stage of the traditional composting process is the thermophilic (hot) phase. The fate and mechanism of removal of chemicals such as pesticides and other xenobiotic compounds in composting processes are largely unknown and in need of comprehensive analysis (57–59).

The hazards of composting stem not only from chemical agents but also from microbiological agents (61,62), whose presence depends heavily on composting methods. Further research clearly is needed on the risks related to human exposure to chemical residues and microorganisms in composted wastes (56). The fate of toxic heavy metals in compost applied to land is of particular concern (60), as is the danger to the public health from pathogenic microorganisms such as *Aspergillus fumigatus*. Spores of *A. fumigatus* (the causal agent of the lung disease aspergillosis) are abundant in the outdoor air at some composting sites (61,62), indicating the importance of site location among process and design criteria for composting operations. In summary, although composting is basically a simple, affordable process, its efficacy and safety for treatment of hazardous wastes are still open to question.

Intrinsic Bioremediation

Intrinsic remediation has been defined as the combined effect of natural destruction and other nondestructive processes to reduce the mobility and mass, and therefore the associated risks, of a contaminant (27). The mechanisms that mediate intrinsic remediation of a contaminant include biodegradation by indigenous microbial populations, sorption to, or trapping within matrixes that make the compound nonbioavailable, and loss of mass by dilution or volatilization (Fig. 3).

Intrinsic remediation of petroleum hydrocarbons is well documented (28,29), and it has also been confirmed for

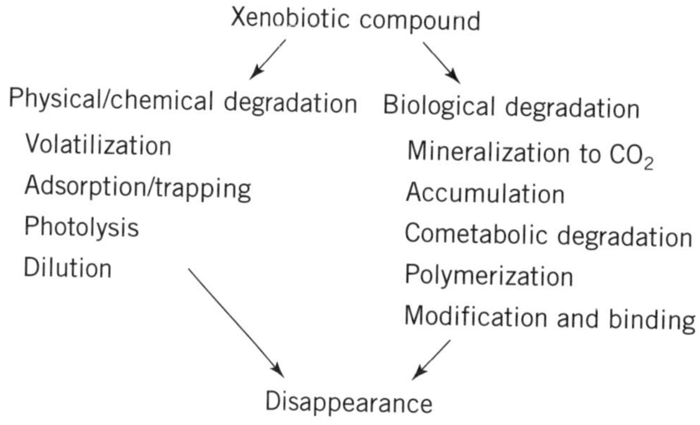

Figure 3. Intrinsic Bioremediation Processes.

chlorinated hydrocarbons, typically under anoxic environmental conditions that favor reductive dehalogenation processes (27,28). Before intrinsic processes can be accepted as a cleanup option for a particular site, regulators must agree to accept exposure and risk assessment analyses as the basis for determining cleanup end points. Therefore, in sites undergoing intrinsic remediation, the distributions of contaminants (e.g., the extent of contaminant plume migration in relation to sensitive receptors) must be thoroughly characterized, and it must be possible to confirm that pollutant remediation is proceeding at a rate sufficient to reduce risks to human health and the environment in an acceptable period. To confirm that intrinsic processes are effectively removing contaminants from soil or groundwater, several factors can be periodically measured, including the following:

- Contaminant mass and its environmental distribution
- Quantity of contaminant degraders or transformers, as indicated by microbial counts
- Temperature and pH, to ensure that they are conducive to microbial activity
- Types of electron donors and acceptors and changes in their concentrations; for example, nitrate, nitrite, metal species (especially iron), methane, ammonia, carbon dioxide, and sulfate
- Redox potential
- Other factors determined by site and contaminant

Ideally, the long-term efficacy of intrinsic remediation processes could be predicted at individual sites through modeling. It is obvious that intrinsic remediation requires specific, extensive, and long-term activities at a site and is far from a "do-nothing" option.

Phytoremediation

Phytoremediation, in which plants are used to remediate contaminated soils or waters, has been studied mainly for the removal of toxic metals but has also been proposed for remediation of some organic contaminants (30). Phytoremediation can take place through the phytoextraction of contaminants from soil or water, where contaminants are taken up by a plant and concentrated in the plant tissues (31,32). The plant materials can then be harvested, removing the sequestered contaminant from the environment. Bacteria associated with the root zones of plants may in some instances degrade organic contaminants, although more study in this emerging area of research is required to confirm which plants promote degradation of which contaminants. Plants may remove contaminants from soil by transpiration of water to the atmosphere, carrying volatile contaminants along. Plants can help in the management and stabilization of polluted sites by reducing soil erosion caused by wind and water and by reducing the infiltration of water into soil, which will in turn slow down or eliminate the leaching of mobile contaminants into groundwater (30,32). An especially promising line of phytoremediation research concerns the use of plants to remove radionuclides from soil or water (33).

After plants or trees have concentrated metals or radioactive compounds, they can be harvested and then burned, which reduces the volume of the collected material by about 95%. Concentration and volume reduction makes disposal of the ash far less expensive than disposal of the original material itself. Alternatively, useful heavy metals might be extracted from the ash as industrial products, especially if the ash contains one or a few valuable metals. It has been suggested that if a plant contained 1% by weight of a metal, then its ash would contain 20% by weight of that metal, a higher metal content than is found in most ores. Any remaining residue after metal recovery would be disposed of as a toxic material.

Phytoremediation does have limitations. It can be slow, so it is not an option for sites that represent an immediate public health or environmental threat. Plant roots also may not penetrate the soil as deeply as necessary to access the targeted contaminants. The roots of most plants generally reach depths of only 30 cm to about 10 m, although a few plants, such as certain varieties of poplar trees, may send their roots somewhat deeper. Thus, deep contamination can be inaccessible to phytoremediation. Furthermore, as living organisms, plants have specific environmental and nutritional needs. They require a rather narrow range of growth temperatures, grow well only in the presence of favorable soil qualities (texture, water-holding capacity, pH, etc.), must obtain enough water to support growth and metabolism, and usually require the presence of oxygen. Plants may also be inefficient at taking up some pollutants, allowing contaminants to reach groundwater.

Some plants proposed for use in phytoremediation (*http://www.ecological-engineering.com/phytorem.html*) include alfalfa (symbiotic with hydrocarbon-degrading bacteria); *Arabidopsis* (genetically engineered to carry a bacterial gene that transforms mercury into a gaseous state); bladder campion (accumulates zinc and copper); *Brassica juncea* (Indian mustard greens; accumulates selenium, sulfur, lead, chromium, cadmium, nickel, zinc, and copper); Buxaceae (boxwood; accumulates nickel); the Compositae family (symbiotic with bacteria; accumulates cesium and strontium); Euphorbiaceae (succulent; accumulates nickel); tomato (accumulates lead, zinc, and copper); poplar (used in the absorption of the pesticide atrazine, and solvents such as carbon tetrachloride); and *Thlaspi caerulescens* (alpine pennycress; accumulates zinc and cadmium).

White-Rot Fungi

White-rot fungi (Fig. 4) produce enzymes that degrade the plant polymer lignin and oxidize a great variety of chemicals, including many environmental pollutants (34). Because these fungi degrade such a broad range of anthropogenic compounds, they have been examined extensively as potential bioremediation agents. The peroxidases of the basidiomycete *Phanerochaete chrysosporium* are the most thoroughly investigated of the xenobiotic compound–degrading fungal enzymes. These and other oxidative fungal enzymes such as laccase are thought to be the primary catalysts responsible for initiation of degradation of many xenobiotics. For example, enzyme measurements indicated that both high and relatively stable activities of Mn-dependent peroxidase, Mn-independent peroxidase, lignin peroxidase, and laccase characterized efficient PCB degraders among a selection of white-rot fungi (63). For application in the field, white-rot fungi grown on a substrate of wood chips or other specially formulated lignocellulosic materials are mixed into a polluted soil. The soil is tilled, watered, and generally managed to encourage the growth and lignin-degrading activity of the introduced organism. As the fungus degrades the lignocellulosic carrier, it simultaneously degrades the contaminants. This technology is considered experimental, because the fungi so far examined are often inhibited by the contaminant concentrations found in some polluted soils, and even when degradation does take place, it may be incomplete. Clearly, more development work needs to be done in contaminated soil systems.

The focus of one recent study was to determine whether *P. chrysosporium* could effectively operate in an actual field sample of oil tar–contaminated soil. The soil was supplemented with [^{14}C]phenanthrene to serve as a representative polynuclear aromatic compound (PAH). This study confirmed earlier work showing that the native soil microflora could acclimate to provide full mineralization of PAHs in some contaminated soils. However, this mineralization could be enhanced by supplementation with *P. chrysosporium* (64). In another soil study, investigators examined the role of fungal oxidative enzymes in the degradation of soil contaminants by looking at gene expression in the soil matrix. Among the genes implicated in pollutant degradation were two closely related lignin peroxidase genes, whose mRNA transcripts were detected in the soil under study (65). Thus, the use of white-rot fungi appears quite promising for certain pollutants such as the PAHs (64) and chlorinated phenols (34). However, more proof of efficacy is required before this becomes a commonly used bioremediation technique. Hundreds of species of white-rot fungi have not yet been examined as bioremediation agents, so the possibility of identifying strains more suitable for fungal bioremediation is good.

LIMITATIONS OF BIOREMEDIATION

As a biological process, bioremediation has the limitations inherent in any biological system, so it may not always be

Figure 4. The white-rot fungus *Ganoderma aplanatum* (common name, Artist's conk). (Photographed by University of Idaho Photo Services.)

an appropriate technology for treating a contaminated site. For a site to be treatable by bioremediation, environmental conditions must support biological activity because extremes of pH, temperature, radioactivity, or redox potential may not be tolerated by degradative organisms. Essential nutrients such as nitrogen, phosphorus, sulfur, and trace elements must be present and available. The soil or water to be treated by bioremediation must be free of extremely high concentrations of certain toxic chemicals (e.g., the heavy metals mercury, lead, and zinc) and other antimicrobial substances. The concentrations of contaminants may be too low to provide sufficient energy for microbial growth, and cometabolism may not be an option. Contaminants may not be bioavailable, or accessible to microorganisms, but they may still be subject to remediation standards required by regulatory agencies. In complex mixtures of contaminants, the feasible bioremediation processes may be unable to remove all the components. The site to be treated, a vast, deep aquifer, for example, may not be accessible to engineers. Finally, some xenobiotic compounds simply are not biodegradable. Thus, bioremediation must be viewed as only a single tool in the environmental restoration toolbox, one that will be used frequently but not for all sites or situations.

BIBLIOGRAPHY

1. W. Admassu and R.A. Korus, in *Bioremediation: Principles and Applications*, Cambridge Univ. Press, Cambridge, UK, 1997.
2. M.D. LaGrega, P.L. Bunkingham, and J.C. Evans, *Hazardous Waste Management*, McGraw-Hill, New York, 1994.
3. S.B. Funk, D.L. Crawford, D.J. Roberts, and R.L. Crawford, *Bioremediation of Pollutants in Soil and Water*, ASTM STP 1235, American Society for Testing and Materials, Philadelphia, Pa., 1995, pp. 177–189.
4. S.B. Funk, D.L. Crawford, R.L. Crawford, G. Mead, and W. Davis-Hoover, *Appl. Biochem. Biotechnol.* **51**, 625–633 (1995).
5. E.K. Nyer, *Groundwater Treatment Technology*, Van Nostrand Reinhold, New York, 1992.
6. T.C. Voice, X. Zhao, J. Shi, and R.F. Hickey, in R.E. Hinchee, G.D. Sayles, and R.S. Skeen eds., *Biological Unit Processes for Hazardous Waste Treatment*, Battelle Press, Columbus, Ohio, 1995, pp. 29–36.
7. G.D. Sayles and M.T. Suidan, in M.A. Levin, and M.A. Gealt eds., *Biotreatment of Industrial and Hazardous Waste*, McGraw-Hill, New York, 1993.
8. R.B. King, *Practical Environmental Bioremediation*, CRC Press, Boca Raton, Fla., 1992.
9. E.H. Marsman, J.M.M. Appleman, L.G.C.M. Urlings, and B.A. Bult, in R.E. Hinchee, D.B. Anderson, and F.B. Metting, Jr. eds., *Applied Biotechnology for Site Remediation*, Lewis Publishers, Boca Raton, Fla., 1994, pp. 391–399.
10. J.M. Yudelson and P.D. Tinari, in R.E. Hinchee, G.D. Sayles, and R.S. Skeen eds., *Biological Unit Processes for Hazardous Waste Treatment*, Battelle Press, Columbus, Ohio, 1995, pp. 205–209.
11. S.P.P. Ottengraf, *Biotechnology*, **8**, 301–332 (1986).
12. R.L. Raymond, V.W. Jamison, and J.O. Hudson, API Publication 4427, American Petroleum Institute, Washington, D.C., 1975.
13. G.F. Bowlen and D.S. Kosson, in L.Y. Young, and C.E. Cerniglia eds., *Microbial Transformation and Degradation of Toxic Organic Chemicals*, Wiley-Liss, New York, 1995, pp. 515–542.
14. W.T. Frankenberger, Jr., K.D. Emerson, and D.W. Turner, *Environ. Manage.* **13**, 325–332 (1989).
15. C.-M. Kao and R.C. Borden, in R.E. Hinchee, B.C. Alleman, R.E. Hopple, and R.N. Miller eds., *Hydrocarbon Bioremediation*, Lewis Publishers, Boca Raton, Fla., 1994, pp. 262–266.

16. R.E. Hinchee and S.K. Ong, *J. Air Waste Manage. Assoc.* **42**, 1305–1312 (1992).
17. D.F. Kidd, *In Situ Treatment Technology*, Lewis Publishers, Boca Raton, Fla., 1996, pp. 245–269.
18. R.R. Dupont, *Environ. Prog.* **12**, 45–53 (1993).
19. R.E. Hinchee, D.C. Downey, and R.R. Dupont, *J. Hazard. Mater.* **27**, 315–325 (1991).
20. I.D. Bossert and G.C. Compeau, in L.Y. Young and C.E. Cerniglia eds., *Microbial Transformation and Degradation of Toxic Organic Chemicals*, Wiley-Liss, New York, 1995, pp. 77–125.
21. W.T. Frankenberger, Jr. and U. Karlson, *On-Site Bioreclamation: Applications and Investigations for Hydrocarbon and Contaminated Site Remediation*, Butterworth-Heinemann, Stoneham, Mass., 1991, pp. 239–254.
22. S.J. McMillin, J.M. Kerr, and N.R. Gray, *Proceedings of the EPA Exploration and Production Environmental Conference*, U.S. Environmental Protection Agency, Washington, D.C., March 3–7, 1993, pp. 389–401.
23. R. Valo and M. Salkinoja-Salonen, *Appl. Microbiol. Biotechnol.* **25**, 68–75 (1986).
24. J.G. Mueller, S.E. Lanz, B.O. Baltmann, and P.J. Chapman, *Environ. Sci. Technol.* **25**, 1055–1061 (1991).
25. R.T. Williams, P.S. Ziegenfuss, and W.E. Sick, *J. Ind. Microbiol.* **9**, 137–144 (1992).
26. D.L. Kaplan, in D. Kamely, A. Chakrabarty, and G.S. Omenn eds., *Biotechnology and Biodegradation*, Portfolio Publishing, Houston, Tex., 1990, pp. 155–182.
27. R.E. Hinchee, J.T. Wilson, and D.C. Downey, *Intrinsic Bioremediation*, Battelle Press, Columbus, Ohio, 1995.
28. H.S. Rifai, R.C. Borden, J.T. Wilson, and C.W. Ward, in R.E. Hinchee, J.T. Wilson, and D.C. Downey eds., *Intrinsic Bioremediation*, Battelle Press, Columbus, Ohio, 1995, pp. 1–29.
29. T.H. Wiedemeier, M.A. Swanson, J.T. Wilson, D.H. Kampbell, R.N. Miller, and J.E. Hansen, in R.E. Hinchee, J.T. Wilson, and D.C. Downey eds., *Intrinsic Bioremediation*, Battelle Press, Columbus, Ohio, 1995, pp. 31–51.
30. H. Bolton, Jr. and Y.A. Gorby, in R.E. Hinchee, J.L. Means, and D.R. Burris eds., *Bioremediation of Inorganics*, Battelle Press, Columbus, Ohio, 1995, pp. 1–16.
31. R.T. Kelly and T.F. Guerin, in R.E. Hinchee, J.L. Means, and D.R. Burris eds., *Bioremediation of Inorganics*, Battelle Press, Columbus, Ohio, 1995, pp. 25–32.
32. S.D. Cunningham, W.R. Berti, and J.W. Huang, in R.E. Hinchee, J.L. Means, and D.R. Burris eds., *Bioremediation of Inorganics*, Battelle Press, Columbus, Ohio, 1995, pp. 33–54.
33. J.E. Cornish, W.C. Goldberg, R.S. Levine, and J.R. Benemann, in R.E. Hinchee, J.L. Means, and D.R. Burris eds., *Bioremediation of Inorganics*, Battelle Press, Columbus, Ohio, 1995, pp. 55–63.
34. A. Paszczynski and R.L. Crawford, *Biotechnol. Prog.* **11**, 368–379 (1995).
35. C. Barbeau, L. Deschenes, D. Karamanev, Y. Comeau, and R. Samson, *Appl. Microbiol. Biotechnol.* **48**, 745–752 (1997).
36. B.S. Evans, C.A. Dudley, and K.T. Klasson. *Appl. Biochem. Biotechnol.* **57–58**, 885–894 (1996).
37. R.H. Kaake, D.J. Roberts, T.O. Stevens, R.L. Crawford, and D.L. Crawford, *Appl. Environ. Microbiol.* **58**, 1683–1689 (1992).
38. W.M. Wu, L. Bhatnagar, and J.G. Zeikus, *Appl. Environ. Microbiol.* **59**, 389–397 (1993).
39. J. Breitung, D. Bruns-Nagel, K. Steinbach, L. Kaminski, D. Gemsa, and E. von Low, *Appl. Microbiol. Biotechnol.* **44**, 795–800 (1996).
40. F.R. Brunsbach and W. Reineke, *Appl. Microbiol. Biotechnol.* **42**, 415–420 (1994).
41. R.E. Hinchee and S.K. Ong, *J. Air Waste Manage. Assoc.* **42**, 1305–1312 (1992).
42. A.P. Schwab and M.K. Banks, in T.A. Anderson and J.R. Coats eds., *Bioremediation through Rhizosphere Technology*, ACS Symp. Ser. 563, American Chemical Society, Washington, D.C., 1994, pp. 132–141.
43. U.S. Pat. 5,494,574 (February 27, 1996). R. Unterman, B.R. Folsom, and A.P. Togna, (to Envirogen, Inc., Lawrenceville, N.J.).
44. M.A. Deshusses, *Curr. Opin. Biotechnol.* **8**, 335–339 (1997).
45. G. Leson and A.M. Winer, *J. Air Waste Manage. Assoc.* **41**, 1045–1054 (1991).
46. M. Arnold, A. Reittu, A. von Wright, P.J. Martikainen, and M.L. Suihko, *Appl. Microbiol. Biotechnol.* **48**, 738–744 (1997).
47. J.W. Barton, K.T. Klasson, L.J. Koran, Jr., and B.H. Davison, *Biotechnol. Prog.* **13**, 814–821 (1997).
48. Y. Matteau and B. Ramsay, *Biodegradation* **8**, 135–141 (1997).
49. Y.S. Oh and R. Bartha, *Appl. Environ. Microbiol.* **60**, 2717–2722 (1994).
50. A. Pol, H.J. Op den Camp, S.G. Mees, M.A. Kersten, and C. van der Drift, *Biodegradation* **5**, 105–112 (1994).
51. G. Leson and A.M. Winer, *J. Air Waste Manage. Assoc.* **41**, 1045–1054 (1991).
52. A. Korda, P. Santas, A. Tenente, and R. Santas, *Appl. Microbiol. Biotechnol.* **48**, 677–686 (1997).
53. G.D. Hopkins, L. Semprini, and P.L. McCarty, *Appl. Environ. Microbiol.* **59**, 2277–2285 (1993).
54. S. Jarvis, V.A. McFarland, and M.E. Honeycutt, *Ecotoxicol. Environ. Saf.* **39**, 131–135 (1998).
55. W.G. Palmer, J.R. Beaman, D.M. Walters, and D.A. Creasia, *J. Toxicol. Environ. Health* **51**, 97–108 (1997).
56. I. Deportes, J.L. Benoit-Guyod, and D. Zmirou, *Sci. Total Environ.* **172**, 197–122 (1995).
57. A.M. Fogarty and O.H. Tuovinen, *Microbiol. Rev.* **55**, 225–233 (1991).
58. F.C. Michel, Jr., C.A. Reddy, and L.J. Forney, *Appl. Environ. Microbiol.* **61**, 2566–2571 (1995).
59. M. Kastner, S. Lotter, J. Heerenklage, M. Breuer-Jammali, R. Stegmann, and B. Mahro, *Appl. Microbiol. Biotechnol.* **43**, 1128–1135 (1995).
60. L. Keller and P.H. Brunner, *Ecotoxicol. Environ. Saf.* **7**, 141–150 (1983).
61. P.B. Marsh, P.D. Millner, and J.M. Kla, *Mycopathologia* **69**, 67–81 (1979).
62. P.D. Millner, P.B. Marsh, R.B. Snowden, and J.F. Parr, *Appl. Environ. Microbiol.* **34**, 765–772 (1977).
63. C. Novotny, B.R. Vyas, P. Erbanova, A. Kubatova, and V. Sasek, *Folia Microbiol. (Praha)* **42**, 136–140 (1997).
64. T.S. Brodkorb and R.L. Legge, *Appl. Environ. Microbiol.* **58**, 3117–3121 (1992).
65. R.T. Lamar, B. Schoenike, A. Vanden Wymelenberg, P. Stewart, D.M. Dietrich, and D. Cullen, *Appl. Environ. Microbiol.* **61**, 2122–2126 (1995).

See also BIOFILMS, MICROBIAL; MIXED CULTURE.

BIOSORPTION, METALS

B. VOLESKY
S. SCHIEWER
McGill University
Montreal, Canada

KEY WORDS

Aqueous environment
Biomass utilization
Biosorption process
Detoxification
Metal biosorption
Metal concentration
Metal recovery
Metal removal
Metal uptake
Passive biosorption
Wastewater treatment

OUTLINE

BIOSORPTION OF METALS

It has been known for decades that some types of microbial biomass tend to accumulate metals rather effectively. This is the case even when the biomass is dead, that is, metabolically inactive, since it is the chemical composition that is responsible for sequestering metal ions and/or metal complexes from solutions. This fact obviates the necessity to maintain special growth-supporting conditions to make the biomass sequester metals. This metal-sorption phenomenon can find enormous applications wherever metals would need to be removed from solution and concentrated.

Rather than searching among thousands of microbial species for biomass with this particular metal-sequestering feature, feasibility dictates us to look first at biomass types that are readily available and in quantities to support the potential demand. Copious quantities of seaweeds, which could easily be collected, grow in the oceans and, indeed, some of them have been identified for their especially outstanding metal-sorbing properties. Other quantities of biomass are generated as by products of large-scale industrial fermentation processes. Often posing disposal problems, some of these readily available industrial-biomass types have been found to be excellent metal sorbers.

A whole new family of metal (bio)sorbing materials appears to be on the horizon, serving as a basis for the new technology of biosorption aimed specifically at metal removal and recovery from metal-bearing industrial solutions and/or effluents.

REMOVAL AND RECOVERY OF HEAVY METALS BY BIOSORPTION

Need for Heavy-Metal-Removal Processes

By far the largest demand for particularly cost-effective metal-removal technologies from dilute solutions is in ecoapplications. Metal ions in the environment are biomagnified in the food chain and accumulated in tissues. Cd, for example, has a half-life of 10–30 years in the human body (1). Therefore, their toxic effects are especially pronounced in animals of higher trophic levels, particularly humans. From the ecotoxicological point of view, the most dangerous metals are the so-called big three: Hg, Pb, and Cd. Following well-documented and publicized disasters and problems with Hg and Pb, the large-scale use of these metals by various industries has been either eliminated or very curtailed. Technological processes making use of these metals have recently been phased out, particularly in developed countries, and replaced by environmentally safer technologies. A major step in limiting the amount of lead discharge into the environment, for instance, has been made by the elimination of leaded gasoline.

On the other hand, the use of Cd has been steadily increasing in growing applications such as in electroplating, pigments, and batteries. Cd toxicity causes kidney damage, bone disease, and cancer (1). Other common heavy metals, such as Cr, Cu, Zn, Ni, and V are less toxic, but are among the most widely used metals (after Fe and A1) (1). Cu is mainly employed in the electrical industry (e.g., wires) and brass production. The major applications of Zn are galvanization and production of brass and other alloys (1). Mine tailings and effluents from nonferrous metal industry are the major discharge sources of different combinations of dangerous heavy metals.

Therefore, it would be beneficial to devise separate pretreatment facilities for the elimination of heavy metals from wastewater, ideally at the site of those plants that discharge significant amounts of heavy metal ions with their wastewater. Such on-site treatment of waste streams contributing a load of heavy metals would be more efficient than treating large volumes of mixed wastewater in a general sewage plant. Additionally, from a policy viewpoint it seems reasonable that those who cause metal pollution should be responsible for its elimination. While industrial enterprises are presently obliged to meet certain discharge standards, drinking-water quality (Table 1) cannot be considered since such treatment would be prohibitively costly.

Natural-metal recycling in the environment makes metal recovery the only effective way of preventing heavy metals from poisoning the environment. Yet if a cost-effective way of completely eliminating heavy metals could be found, this goal could be attainable. Current research activity in the field attempts to evaluate whether biosorption may eventually provide such an effective and economical treatment-process alternative.

While biological treatment is reasonably effective in removing organic pollutants, heavy metals tend to accumulate in biological sludges (5), which are considered unsafe as fertilizer and require incineration. The accumulation of metals by the microbial biomass is a result and typical example of combined metabolism-dependent active bioaccumulation and passive biosorption of metals.

Possible Applications of Metal Biosorption

The term *biosorption* commonly refers to the passive binding of metal ions by biomass, which may even be dead. It must be distinguished from bioaccumulation, which is usually understood to be an active, metabolically mediated process occurring in living organisms.

It has been known for decades that different types of microbial biomass bind trace metal ions, achieving very high concentration factors. Black and Mitchell (6) reported a 1,000-fold increase of metal concentration in biomass of some brown marine algae as compared to the seawater. The biomass-metal-uptake phenomenon has been made use of in, for example, electron microscopy, to visualize cellular components. The focus in early studies of microbial-metal uptake has been almost exclusively on the nutritional and toxicological aspects of metal presence.

Recently, however, efforts have been made to develop the process of metal accumulation by biomass into a technique for the treatment of toxic metal-bearing industrial effluents, or for the recovery of precious elements from processing solutions, or even seawater. The process of gold sorption by *Sargassum* has been patented, and the immobilized biosorbents AlgaSORB™ and AMT-BIOCLAIM™ are being commercially developed (7).

Biosorption removal of toxic heavy metals is especially suited as a "polishing" water-treatment step because it is possible to reach drinking water quality of the treated water (initial metal concentrations, e.g., 1 to 100 mg/L, final concentration <0.01 to 0.1 mg/L), especially in packed-bed flow-through sorption column applications. To prevent unnecessarily rapid exhaustion of the sorption capacity when the metal concentrations in wastewater to be treated are high (>100 mg/L), it may be desirable to use a different pretreatment technique, such as precipitation or electrorecovery, for removal of the bulk of the metal content. However, generation of toxic sludges during pretreatment must be considered, since these represent another type of hazard and the eventual recovery of metals from them may not be feasible.

The metal-laden biosorbent can be regenerated, incinerated, or stored in landfills. The biosorption process basically serves to reduce the waste volume. Alternatively and preferably, regeneration of the biosorbent material or multiple (re)use is desirable to increase the process economy. It can be accomplished by metal desorption with, for example, acids or salt solutions. The resulting highly concentrated metal solution can be processed by other techniques such as precipitation or electro winning to remove or concentrate the metal, which could be the recovered and resold. The latter process, in particular, is aimed at recuperation of the metal. The overall achievement of the biosorption (complete adsorption + desorption cycle) process is to concentrate the metal solution, possibly by at least a factor of 100 or more. The flowchart of the biosorption process with in situ regeneration of the biosorbent is quite similar to the use of ion-exchange resins (Fig. 1).

Compared to conventional wastewater "polishing" techniques such as ion-exchange, activated carbon treatment,

Table 1. Drinking Water Standards and Discharge Limits of Heavy Metals

Metal	Drinking water standards WHO (2) (mg/L)	Discharge limits	
		Canada (3) (mg/L)	United States (4) (mg/L)
Cd	0.003[a]	1.5	0.1
Cu	2[b]	1	3.3
Cr	0.05[b]	1	2.8
Ni	0.02[a]	2	4
Pb	0.01[a]	1.5	0.6
Zn	3[c]	2	2.6

[a]Maximum acceptable concentration for health reasons.
[b]Provisional value for health reasons.
[c]Limit for aesthetic or consumer oriented reasons.

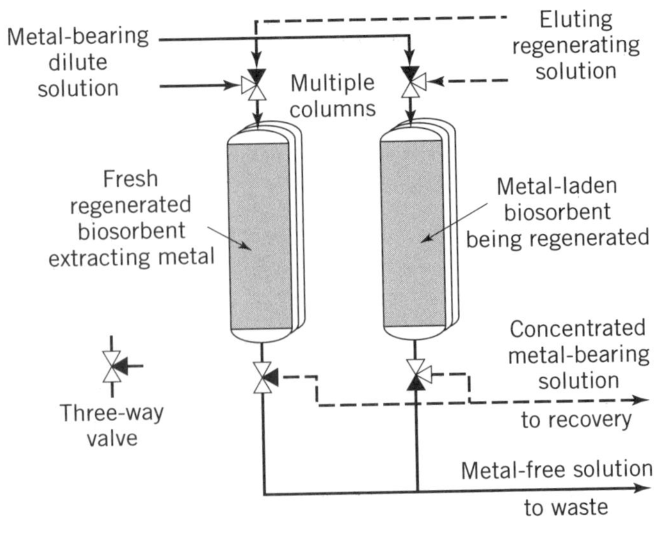

Figure 1. Schematic flowchart of a continuous-flow column biosorption process.

or membrane technologies (electrodialysis, reverse osmosis), the advantage of biosorption is not only that it can be operated under a broad range of conditions (pH, temperature), but especially that it may be economically attractive due to the cheap raw materials that can be used as biosorbents (7). These include waste products from other industries (e.g., fermentation by-products) or naturally abundant biomass (e.g., ubiquitous marine algae). Recent studies show that for certain types of seaweed biomass, for example, virtually no pretreatment may be necessary: Untreated *Sargassum* biomass was successfully applied in packed-bed column biosorption of Cu (8). This means that the sole costs of the seaweed-based biosorbent may be only those of collection and transport.

To evaluate the expected performance of biosorption for treating typical waste waters, it is desirable to reliably predict how much metal could be bound under different operating-process conditions. Computerized process simulations could effectively and economically provide this relevant information, which is essential for the engineering process design and scale-up.

Characteristics of Metal Biosorption

Biomass Types. A multitude of biomass types (9), comprising fungal biomass (10), bacterial biomass (11), algae (12–14), peat (15,16), and so on, have been studied for their biosorption of metals. It was observed that not only the species but also growth conditions such as culture medium, as well as physiological state and age of the organism, influence the biomass-sorption performance (17). In many cases, the sorption by dead material proved to be more effective than that by living organisms (18,19). Biomass of some marine brown algae has been found to be particularly effective in metal-ion binding (20,21), showing similar or better performance than industrial ion-exchange resins. Gold accumulation by the brown alga *Sargassum* can constitute up to 40% of the biomass dry weight (22). Further encouraging results with this sorbent material were obtained by Kuyucak and Volesky (22), Holan et al. (20), Leusch et al. (23), and Ramelow et al. (24).

The particular amount of metal bound in the biosorbent depends, however, not only on the chosen biosorbent but also on the type of metal ion, its concentration, and other physicochemical parameters of the solution (e.g., pH, ionic strength).

Temperature Effect. Simple physical sorption processes are generally exothermic (i.e., the equilibrium constant decreases with increasing temperature). Biosorption ion exchange among alkaline earth metals in binding to alginate was observed to be exothermic (25). However, when Cu was the replacement, the reaction exhibited a positive enthalpy change (i.e., it was endothermic—the equilibrium constant rose with temperature). The driving force in this case was attributed to a large positive entropy change, possibly caused by a larger ordering effect of Cu (than Ca) on the water molecules in the hydration sphere. This is illustrated by the following equation:

$$K = \exp\left(\frac{-\Delta G^0}{RT}\right) = \exp\left(\frac{-\Delta H^0 + T\Delta S^0}{RT}\right) = \exp\left(\frac{-\Delta H^0}{RT}\right)\exp\left(\frac{\Delta S^0}{R}\right) \quad (1)$$

A rise in K with increasing T means that $\Delta H^0 > 0$, that is, that the reaction is endothermic. But since a reaction can only proceed when $\Delta G^0 = \Delta H^0 - T\,\Delta S^0 < 0$, it follows that ΔS^0 must have been >0, that is, the system's entropy must have increased.

Similar results were obtained by Weppen and Hornburg (26) in calorimetric studies of the binding of divalent metal ions by potassium-saturated microbial biomass. The reactions were slightly endothermic and therefore driven by an increase in the entropy of the system. For most metals, the heat of reaction was constant, independent of the degree of site occupation. For Cu, however, the heat of reaction decreased with increasing degree of site occupation from 27 kJ/mol to 14 kJ/mol, which may indicate the involvement of different binding sites or the formation of different types of Cu complexes with the biomass. For other heavy metals, the heat of reaction was between about 7 and 11 kJ/mol, for light metals between about 2.5 and 6 kJ/mol (26).

Overall, however, the effect of temperature in biosorption is relatively small and perhaps only slightly more pronounced in the lower temperature range: the binding of Co by the brown alga *Ascophyllum nodosum* increased by 50–70% when the temperature was raised from 4 °C to 23 °C. With further temperature increase to 40 °C the binding increased only slightly, whereas temperatures of 60 °C or more caused a change in the texture of the sorbent and a loss in the sorption capacity. The distribution ratio (metal bound/metal in solution) for biosorption of Cd, Zn, Pb, Ni, and Cu to *Spirulina* algae increased by only about 20% when the temperature was raised from 4 °C to 55 °C. It is reasonable to expect that possible application of biosorption will be in a rather narrow temperature range (~5 to 50 °C) to avoid decomposition of biosorbent materials.

Presence of Anions (Ligands). Theoretically, the presence of ligands (at levels that do not cause precipitation) can lead to the following (27):

1. Formation of complexes that have a higher affinity to the sorbent than the free metal ions (i.e., an enhancement of sorption)
2. Formation of complexes that have a lower affinity to sorbent than the free metal ions (i.e., a reduction of sorption)
3. Interaction of anions with the biomass, changing the state of the active sites such that the binding is either enhanced or reduced

There are no indications of the third kind of interaction reported in the biosorption literature. Additionally, these are unlikely to occur in brown algal biomass since the probable binding sites (negatively charged carboxyl and sulfate groups) are not expected to interact with anions such as sulfate, nitrate, and so on.

It has been reported for sorption of Cu^{2+} and Ag^{+} to inorganic oxide surfaces that metal binding at low pH was enhanced in the presence of certain ligands. Since the ligand itself was able to bind to the oxide, and since a 1:1 stoichiometry of ligand and bound metal was noted, it was concluded that some metal complexes may bind more strongly than the respective free ions (28). Negatively charged metal–EDTA complexes were bound to oxide surfaces (probably releasing OH^{-}), with sorption decreasing with increasing pH (29).

Chen et al. (16) report that up to 50% of Cu binding by peat occurred as complexation of copper nitrate [$Cu(NO_3)_2$] with the binding sites. The remaining metal binding was due to ion exchange such that the overall charge neutrality was conserved. The data allow, however, another interpretation as well: Instead of up to half of the Cu being bound as $Cu(NO_3)_2$ and the other half as Cu^{2+}, up to 100% could have been bound as $Cu(NO_3)^{+}$. Glaus et al. (30) found that the binding of Co and uranyl (UO_2^{2+}) to humic substances could be modeled better when binding of metal–ligand (oxalate) complexes was included in the model.

In most cases of biosorption, however, metal binding tends to be reduced in the presence of ligands (exception: OH^{-}) (22,27,31). This means that the biomass apparently has less affinity for many metal–ligand complexes than for free hydrated metal ions. Therefore, the influence of ligands in solution can be understood as a competition with the biomass for binding of the metal ions.

This effect is, however, not very pronounced unless these anions show a strong complexing power for the metal ions of interest (e.g., EDTA) (24,27), which at only 1.5 times excess caused 100% inhibition of metal binding. The addition of 0.1% EDTA resulted in >90% reduction of binding for most cases of Cd, Cu, Zn, and Pb biosorption by seaweeds (also *Sargassum*) (24). A large excess (1,000 times) of the ligands glutamate, sulfate, phosphate, carbonate, or chloride resulted in the maximum reduction of, for example, Cd binding to *Rhizopus arrhizus* biomass of only 21% (27).

Since significant concentrations of strong complexing agents are rare in typical metal-bearing industrial effluents, the effect of the presence of ligands is of secondary importance.

Influence of pH. Of great importance in sorption is the pH value of the solution. It is commonly agreed that the sorption of metal cations (e.g., Cd, Cu, Zn, Pb, Ni, Mn, Al, Co) increases with increasing pH (14,20,24,32–35) as the metal ionic species become less stable in the solution. Only those metal ions that can occur as negatively charged complexes or that have a strong "b" character (i.e., tendency to form strong covalent bonds), such as Ag, Hg, or Au (e.g., as tetrachloroaureate), may show either a decrease in binding with increasing pH or may have no significant pH effect at all (24,31,33). There are three ways that the solution pH can influence metal biosorption, as described in the following.

First, the state of the active sites may be changed. When the binding groups are acidic, the availability of free sites depends on pH: At lower pH the active sites are protonated, therefore, competition between protons and metal ions for the sorption site occurs (36,37). At a low enough pH, virtually all sites become protonated and complete desorption of the bound metal ions is possible (38), which is why acid treatment is a method for metal elution and regeneration of the sorbent material. Decreasing the pH value by 2 units can in some cases result in a ~90% reduction of metal binding (14,24,31,32,38).

Second, extreme pH values, as they are employed in the regeneration (desorption) of the sorbent, may damage the structure of the biosorbent material. Microscopic observations have shown distorted cells; significant weight loss and decrease in the sorption capacity have been observed (39).

Third, the speciation of the metal in solution is pH dependent. Whereas metals in aqueous solutions occur as hydrated cations in solvation shells when the pH is low, hydroxides may form at a higher pH, especially for cations of high charge and small size (40–42). The formation of metal oxide and hydroxide complexes and precipitates is often called hydrolysis (i.e., decomposition or conversion by water) (43).

Adsorption depends not only on the attraction of the sorbate to the solid surface but also on its lyophobic behavior (44). This means that sorption increases with decreasing solubility. Since the solubility of many metal complexes in solution decreases with increasing pH, this provides an additional possible explanation as to why sorption increases with increasing pH. In the narrow pH range where the metal ions are hydrolyzed, sorption is especially enhanced, so that even a charge reversal of the surface to positive values can occur (44,45). Further possible explanations of increasing sorption with increasing pH are that hydrolyzed species have a lesser degree of hydration, that is, less energy is necessary for removal or reorientation of the hydrated water molecules upon binding. Furthermore, OH shows affinity for some sorbents (46).

With further increase of pH, the solubility of metal complexes decreases enough for precipitation to occur. Although precipitation may contribute to the overall removal of metals from solution (and therefore be desirable for metal-removal applications), it renders the study of the biosorptive binding more difficult. For scientific purposes it is therefore recommended to study biosorption at pH values where precipitation does not occur.

Presence of Other Cations. Other sorbable ions in the solution may compete with the metal ion of interest for sorption sites. The binding of this metal ion is then decreased. The amount of inhibition depends on the binding strength of the respective ions to the biomass.

For example, Mg and Na are effective in competing with Zn for binding sites of green algae [90% reduction of Zn binding at 200 mM NaCl or $Mg(NO_3)_2$], but they interfere less strongly with the binding of Pb or Cu [<50% reduction of Pb or Cu binding at 200 mM NaCl or $Mg(NO_3)_2$] (14).

The inhibition by alkali metals (K, Na) of heavy-metal-ion binding to microbial biomass was much less pronounced than the inhibition by heavy metals (Zn, Cu, Fe) of uranium or radium binding (19).

For *R. arrhizus* fungal biomass, the presence of uranium reduced the binding of Cd, Zn, and Ag more strongly than vice versa; Ag in particular was a weak competitor, but nevertheless its binding was unaffected by the presence of Na (47).

In the sorption of Cd, Cu, and Zn on formaldehyde cross-linked biomass of the brown alga *A. nodosum*, the presence of each ion reduced the binding of each other ion, with Zn being the weakest binding, and Cu the most strongest binding, of the three cations (48).

From the different literature sources it can be generally concluded that the light metals (alkaline and alkaline earth metals) bind less strongly than the heavy-metal ions or radioactive elements. Therefore, the former do not strongly interfere with the binding of the latter. There are indications that among the heavy metals, Zn binds rather weakly and its binding is therefore more strongly affected by other ions.

In order to understand why some ions bind more strongly than others, it is useful to look at the properties of different metals, to be discussed later. Such knowledge can also serve to estimate how strongly an ion may bind that has not yet been investigated. This knowledge serves as a basis to better understand the metal-binding mechanism.

Mechanism of Biosorption

The effect of any of the previously mentioned influencing factors can be best quantitatively estimated when the mechanism of biosorption for different ions is known and when a mechanism-oriented mathematical model is used for predicting the effect of these factors.

Classification of Binding Mechanisms. The metal binding in biosorption has been attributed to a number of different sequestering mechanisms, such as complexation, coordination, chelation, adsorption, ion exchange, or microprecipitation (as metal or metal complex) (49). It is a recognized fact that a combination of several mechanisms, each functioning independently, can contribute to the overall metal uptake. In the studies of biosorption conducted so far, very little attention has been paid to examination of a well-defined metal uptake (by a specific mechanism) as opposed to the overall uptake where several types of sequestration may be taking place simultaneously. Systematic understanding of the metal-uptake mechanisms and their relationships may greatly clarify the otherwise confusing broad definition of biosorption mechanisms in the literature.

For example, some mechanisms can be submechanisms of other overall mechanisms. The claim of an ion-exchange metal-binding mechanism for a specific system may not necessarily be contradictory to reports of chelation being the relevant mechanism. The overall mechanism of ion exchange can be based on the submechanism of complexation. This means that one complexed ion exchanges against another one.

A systematic presentation of the relationships between different mechanisms is compiled in Figure 2. The classification of bond types (Fig. 2, bottom) as described by Myers (50) was used, as well as information on sorption processes from Westall (51) and Pagenkopf (44).

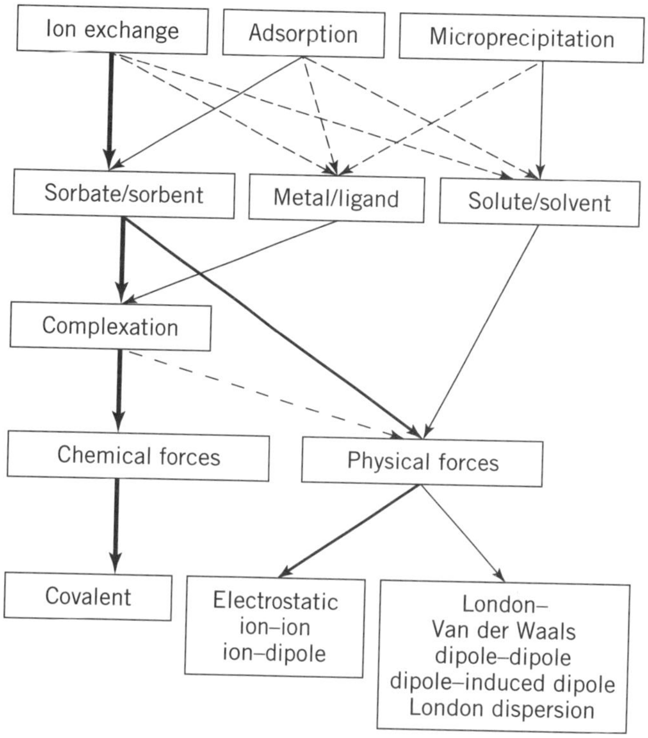

Figure 2. Metal biosorption mechanisms: bold lines, mechanisms probably important in biosorption by *Sargassum* biomass; dashed lines, biosorption binding relations of secondary importance.

In the context of this work, the term *sorption* refers to binding of a solute (usually a metal cation) to free sites that previously were not occupied by another cation. If the sites are initially occupied by another cation (regardless of the submechanism, i.e., whether covalent or electrostatic binding is involved), and if this second ion is released on the binding of the first ion, then the term *ion exchange* is used.

Microprecipitation is the deposition of electrically neutral material (metal or metal salt) at the surface of the biomass, and does not necessarily involve a bond between the biomass and the deposited layer. Microprecipitation may, however, be facilitated by initial binding of metal ions to reactive sites of the biomass, which serve as nucleation sites for further precipitation (52). This process is not limited to a monolayer (or saturation of sites): Cells can accumulate several times their dry weight in metal (53,54). Microprecipitation is based on interactions between the solute (dissolved solid) and the solvent, and occurs when the local solubility is exceeded.

Ion exchange and adsorption can be the result of three different interactions (which can act in combination): The main contribution for free metal ions (which are highly soluble in water) is usually attraction of the sorbate (metal ion) to the sorbent (biomass). In the case of binding of metal–ligand complexes, or in the case of ligand exchange (where the metal ion in solution is complexed with a ligand that has to be released when the metal ion is bound to the biosorbent [51]), interactions between ligands and metal cations must be taken into account. Additionally, hydrophobic (or, generally, lyophobic) expulsion (which is an interaction between sorbate and solution) may also play a role (44,51), especially in the adsorption of organic dipoles and large organic ions because of their low affinity to the aqueous phase (46). Among inorganic ions, the less hydrated ones are more readily accumulated at the interface (46). The hydration effect is further discussed later.

Complexation plays an important role in both metal–ligand and sorbate–sorbent interactions (51). A complex (also referred to as a coordination compound) is a polyatomic molecule that consists of one or more (then called polynuclear complex) central atoms (usually metal cations) surrounded by ligands (other atoms or groups, usually of negative or neutral charge) that are attached to it. Complexes can be neutral, positively charged, or negatively charged. The number of coordinating atoms in the ligands that are directly attached to the central atom is called the coordination number, and can be larger than the valence of the central atom. (The most common coordination numbers are 4 and 6, but 2 and 8 also occur frequently.) If one ligand is attached to the central atom through two or more coordinating atoms, then the complex is called a chelate. The bonds between the central atom and the coordinated groups can be arising from principal or auxiliary valence forces (55,56).

Binding Forces. In general, the forces between atoms or molecules can be classified into chemical and physical ones. Chemical forces extend over very short distances (0.1–0.2 nm) (50). This type of bond is rather strong: Stumm and Morgan (46) report an energy of $\gg$40 kJ/mol, Myers (50) 150–900 kJ/mol, Smith (57) for chemisorption 20–400 kJ/mol and Pagenkopf (44) 40–400 kJ/mol. Covalent bonds are formed by merging electron clouds such that a nonionic molecule is formed. These bonds are directional (characteristic bond angles and lengths) and localized (50).

Physical forces can be subdivided into electrostatic and London–van der Waals forces (50). The energy of physisorption is reported as 2–20 kJ/mol and 20–40 kJ/mol by Smith (57) and Pagenkopf (44), respectively. In the resulting bonds, the electrons stay in their original systems. Electrostatic (or coulombic) forces between ions or between ions and dipoles extend over a long range and are the strongest among the physical bonds (50), with energies $\gg$40 kJ/mol (46). The interaction is repulsive for ion charges of the same sign and attractive for unlike charges. The magnitude of the force is proportional to the charge of each ion and inversely proportional to the square of the distance between the ions.

London–van der Waals forces can be divided into three categories: dipole–dipole interactions (creating orientational energy), dipole-induced dipole interactions, and the London dispersion force (50). The first two are closely related to coulombic forces, while the last is of quantum–mechanical nature and acts over a long range, of up to about 10 nm (50). The energy of the dispersion force (8–40 kJ/mol) is larger than the one of orientational or induced dipole (or polarization) (51) energy (<8 kJ/mol) (46). An example of untypical strong (almost ionic) dipole interactions is hydrogen bonding (which may even display some covalent bond character). It occurs between molecules in which H is bound to a very electronegative atom such as O (e.g., in H_2O) (58).

The mechanisms and forces that are probably (as pointed out later) the most important ones in the case of metal binding by, for example, brown algal biomass are indicated in bold type in Figure 2, that is, complexation by chemical covalent bonds based on ion exchange (for which the driving force is the affinity of the sorbate [metal ion] to the sorbent [biomass]) supported by physical electrostatic interactions between positively charged metal ions and negatively charged biomass.

Ion Exchange versus Sorption. There have been some indications that ion exchange plays an important role in metal-ion sorption by algal biomass. Kuyucak and Volesky (59) noted that the amount of ions (K, Na, Ca, Mg) released from the marine brown alga *A. nodosum* was much more pronounced in metal (Co) bearing than in metal-free solutions. A linear correlation between Ca release and Co uptake (2:3) was found when the biomass had been previously washed with CaCl and HCl. It was concluded that ion exchange was responsible for the sorption (59). If ion exchange was responsible, an exchange ratio of 1:1 should, however, be expected for two divalent ions such as Ca and Co. Since it can be expected that some sites would have been protonated due to the pretreatment with a CaCl and HCl mixture, and since some proton release was noticed, it would have been more appropriate to relate the Co binding to the sum of Ca and H released (in milliequivalent not mole). This would have led to an exchange ratio much closer to 1:1.

Crist et al. (60) included protons in the quantitative study of ions (Ca, Mg, H) released during metal (Cu) uptake by the freshwater alga *Vaucheria* and found that the amount of divalent metal ions taken up equaled the amount of other divalent ions plus protons released (in milliequivalent terms). Treen et al. (61) reported that in column application of *Rhizopus* fungal biosorbent 2 mol of protons were released for 1 mol of uranyl sequestered.

According to Chen et al. (16), the Cu binding to peat at low metal concentrations was predominantly ion exchange (with H, Ca, Mg), while at higher concentrations, binding of copper–nitrate complexes occurred. The latter mechanism was dominant above 6 mM of Cu.

It appears reasonable that in many cases ion exchange rather than sorption to free sites is the relevant overall mechanism for the binding of metal ions in biosorption. Since the overall charge of the biomass particle has to be neutral, any binding of one cation must be accompanied by either a stoichiometric release of other cations or by the binding of anions. Since the charge of algal biomass is negative (see "Surface Charge"), it appears unlikely that free anions will be bound by the biomass. If complexes in solution occur, double exchange of cations and anions or binding of the complex may take place. Generally, sorption of neutral (metal cation + anion) complexes would be possible. However, it is unlikely to be the case at the moderate pH and metal, as well as the ligand concentrations, relevant to the biosorption process.

Desorption/Ion Exchange of the Metal from Biosorbent. Due to the weak nature of metal-binding forces, it is possible to wash the bound metal from the sorbent in a small volume of solution offering another cation (e.g., H^+, Ca^{2+}) in higher concentrations. From the process point of view, this enables the reuse of the biosorbent in multiple uptake/desorption cycles allowing for a greater process economy. Desorption can be achieved by acids (e.g., HCl, H_2SO_4) and/or salt solutions (e.g., $CaCl_2$) (38,39). The effectiveness of desorption depends in this case on the binding strength of the added cation to the biosorbent. A different approach to desorption is the addition of a complexing agent (e.g., EDTA, KSCN) (39) that binds free-metal ions and thereby lowers their concentration in solution to the extent that they are released from their binding sites on the solid. In this case the driving force is the complexation of the metal ion. The efficiency of the desorption depends on the binding strength of the desorbed ion to the complexing ligand. Nevertheless, a charge balance of the sorbent particle has to be maintained, that is, ion exchange still occurs. That means some cation (e.g., the one that was added along with the complexing ligand, often Na) has to replace the desorbed ion at the binding sites, even if its affinity toward the binding sites is lower.

Chemical (Covalent) versus Physical (Electrostatic) Binding. The proportion of protons among the released species varied with the metal taken up (25,62) and corresponded to the binding strength: More protons were released by the stronger binding ions. It could be concluded that in those cases where no protons were involved, as for most alkaline and alkaline earth metals, electrostatic attraction was the binding mechanism (63), whereas covalent binding occurred when protons were released (62).

It appears plausible that the number of protons released increases with the binding strength of the metal ion because it can more effectively compete with protons. Also, as a general rule, the force of electrostatic binding alone is weaker than binding with covalent contribution (especially since alkaline earth ions have similar radii to those of many divalent heavy metal ions, so that large differences in the binding strength are due to different amounts of covalent binding). However, it does not necessarily follow that proton release indicates covalent binding; part of the protons (depending on the relationship between pH and $p^{BH}K$) may also be bound by electrostatic attraction and therefore be easily replaceable by electrostatically bound ions.

Binding Sites. Biosorption has been attributed to different types of groups such as carboxylate, carbonyl, hydroxyl, amine, amide, imidazole, phosphate, thio, and thioether groups (33). Which binding mechanism applies in any specific case depends, of course, largely on the type of biosorbent used and on the types of binding sites it contains. Crist et al. proposed that, for example, carboxylate, sulfate, and amino groups may be responsible for metal binding by freshwater algae (62). Indications for carboxyl- and sulfate-group participation were found by titration of marine algal biomass (64). The chelating ability of polysaccharides from green freshwater alga *Chlorella* was connected to their content of uronic acids. Their carboxyl groups would be negatively charged and could bind metal ions (65). Tobin et al. (37) concluded that in *R. arrhizus* fungal biomass at pH 4, carboxylate and phosphate groups offered the primary metal-binding sites, supported by hydroxyl groups. Greene et al. (31) noticed a complete loss of available sulfhydryl groups (determined by polarimetric titration) after the binding of Au(III) to the freshwater alga *Chlorella*, thus giving evidence for the involvement of this group, but it could only account for a small amount of total gold binding. Cell modifications with succinic anhydride indicated the relevance of amine groups, since the sorption of tetrachloroaureate decreased by 50% after their modification (31).

The modification of carboxyl groups by esterification with acidic methanol resulted in a significant decrease in the binding of Cu and Al, but a slight increase in Au binding by five different algae (66). The decrease in sorption was more pronounced for Al, which as a hard metal is expected to bind preferably to hard sites, such as carboxyl groups, whereas Cu as a softer metal may bind to nitrogen- or sulfur-containing group instead. The binding of the negatively charged Au complex increases, probably due to a decrease in the negative surface charge.

For brown marine algae it is assumed that alginates play a key role in metal-ion binding (59). Fourest and Volesky (67) investigated the binding of H, Cd, and Pb by alginic acid and *Sargassum fluitans* biomass before and after modification of the carboxyl groups using acidic methanol or propylene oxide. A linear correlation between the binding capacity for Cd and the content of weak acidic groups (probably carboxylate) resulting from different degrees of active-site blocking was noted. The maximum metal-binding capacity of different algae was linearly correlated with their titer of weak acidic groups, with a stoichiometric relation between metal bound and acidic titer of 0.36 to 0.5 mol/mol. This means that about two acidic sites are necessary per bound metal ion (E. Fourest and B. Volesky, personal communication, 1997).

Surface Charge. The relevance of electrostatic attraction in biosorption depends on the types and amounts of sites present in the biomass and on whether they are ionized or occupied by protons or other ions. That, in turn, depends on the pH and on the pK_a of the respective group. Usually, a good agreement between the pK_a values of inorganic acids and their corresponding acidic groups in biomolecules exists, as evident from Table 2. Amine groups are positively charged in their protonated form and neutral

Table 2. K_a Values of Inorganic Acids and Acidic Groups in Biomolecules

Inorganic acid	pK_a	Acidic group in biomolecules	pK_a
Sulfuric acid[a]	1.9	Sulfate[b]	~1.5
Phosphoric acid[a]	2.1, 7.2	Phosphate[c]	2
Benzoic acid[a]	4.2	Carboxyl[c]	3–5
Acetic acid[a]	4.8	Carboxyl[c]	3–5
Ammonium[a]	9.2	Amine[c]	8–10
Phenol[a]	10.0	Phenolic OH[c]	10

[a]Bower and Bates (68).
[b]Crist et al. (64).
[c]Buffle (69).

when deprotonated. Carboxyl, sulfate, and phosphate groups are neutral when protonated and negatively charged when deprotonated.

If the freshwater alga *Chlorella* could be taken as an example, its isolelectric point (zero overall charge) is at pH 3 (33). At lower pH, the positive charge (possibly of amine groups) leads to an attraction of anions; at higher pH, the negative charge (probably of carboxyl groups) attracts cations and repels anions (33). Biomass of the marine brown alga *Ascophyllum nodosum* bears a negative charge whose magnitude sharply increases from pH 3 upward. It is probable that this surface charge originated from deprotonated carboxyl groups of the alginate in the cell wall (34).

Location of Bound Metal. Knowledge pertaining to the location of metal uptake in the cell can provide some indication about the relevance of binding groups if it is correlated with information about which biomolecules occur in the respective parts of the cell. Electron micrographs of bacterial (70,71), fungal (72,73), and algal (74) metal-bearing cells in the literature indicate that the cell wall plays an important role in metal-ion biosorption. While the basic biomass metal-sequestration process is usually very fast, some time-lapse electron micrography indicates slower metal penetration into cell interiors and/or building up of external metal deposits (18,31,59,63).

For a better understanding of the relevant mechanisms in metal binding by microbial biomass, more detailed and accurate knowledge about the biomolecules and binding sites offered by different types of cells is necessary.

PROPERTIES OF METAL IONS IN BIOSORPTION

Biosorption Affinity Sequence Trends

Studying metal-ion binding to alginic acid from *Laminaria digitata* Haug (75) reported that the amount of protons released into solution decreased in the order Pb > Cu > Cd > Ba > Sr > Ca > Co > Ni > Zn > Mn > Mg. This was explained by a better ability of strongly binding ions to compete with protons for organic binding sites. The affinity sequence for metal-ion binding to alginate from *L. digitata* followed a similar trend Cu > Ba > Ca > Co (76), and for binding of alkaline earth metals to polymannurate and polygulurate the binding strength decreased in the order Ba > Sr > Ca > Mg (25). According to Haug and Smidsrod, the preferential binding of the heavier ions could have been due to stereochemical effects, as larger ions might better fit a binding site with two distant active groups. The amount of proton displacement by metal ions decreased in the order Pb > Cu > Ba > Sr > Ca > Cd > Zn > Ni > Mg > Co, Mn for polyguluronic acid, and in the order Pb > Cu > Cd > Ba > Ni > Sr > Ca > Mg > Zn, Co, Mn for mannuronic acid (25). Again, the trend was concurrent with the one in binding strength.

Schweiger (77) found a different order of complex stability for binding to alginate as being Ba > Cd > Cu > Sr > Ni > Ca > Zn > Co > Mn > Mg, but still the binding of alkaline earth metals increased with their weight. For the elements Cu, Ni, Co, and Mn of the third period, an increase of binding strength with increasing atomic number was noticed. In binding to fucoidan, the affinity sequence Pb > Ba > Cd > Sr > Cu > Fe > Co > Zn > Mg > Cr > Ni > Hg > Ca was reported (78).

For *R. arrhizus* biomass, the maximum metal-binding capacity decreased in the order UO^{2+} > Cr^{3+} > Pb ~ Ag^{+} > Ba > La^{3+} > Zn > Hg > Cd > Cu > Mn (37). The metal-binding capacity increased linearly with the crystal ionic radius, and no correlation to the charge was noticed, except that the binding of alkaline metal ions was negligible (37).

The observations of higher binding (both capacity and binding constant) with increasing ionic radii were interpreted as a consequence of the binding-group conformation: Assuming that binding groups are distant from each other, large ions could not only be more easily dehydrated, but could also bind to several groups simultaneously (79). This could explain the binding strength; however, it does not explain the higher binding capacity. To the contrary, if several groups are involved in the binding of one large ion, the binding capacity of small ions, each binding to one site, could be higher.

Avery and Tobin (80) found that the amount of proton displacement decreased in the order Cu > Cd > Zn > Mn > Sr. It was assumed by these authors that proton release occurred only when the metal was covalently bound, and that the release of Ca and Mg indicated ionic binding. The ratio of (Mg + Ca) released per H^{+} released was used as an index for the relative amount of ionic/covalent binding (80). It was, however, not considered that, for example, Cu might bind covalently to a site from which Ca, which had been bound electrostatically, was released. This index should therefore not be used in a quantitative way.

Evans (81) states that the affinity sequence in alkali-metal binding to soil (Cs > Rb > K > Na > Li) reflects an increase in the hydrated radii of these ions; for divalent ions the binding strength to soil organic matter at pH 5.8 decreased in the order Hg = Pb = Cu > Cd > Zn > Ni > Co > Mn.

For freshwater algae, Crist et al. observed a trend in binding strength Cu > Sr > Zn > Mg > Na being similar to the one for proton replacement Cu > Zn > Mg > Sr > Na, and concluded that a trend from covalent to ionic bonding was responsible (62). As pointed out earlier, however, electrostatic effects could have contributed to proton binding on the one hand and, on the other hand, covalent binding to sites previously occupied by a light metal could be possible.

For binding of light metal ions to freshwater algae, the affinity sequence was Na > K > Li > Cs and Mg > Ca > Ba > Sr (63). It was concluded that if the affinity between biomass and metals was high, the binding should increase with increasing charge density, that is, with decreasing radius. If binding was weak, it would be controlled by dehydration energy and the larger ions would be preferably bound because they could be dehydrated more easily. The observed sequence, which does not show a clear preference for large or small ions, was interpreted as resulting from these opposite trends, indicating medium to strong bonds (63). It is questionable, however, whether the use of crystal ionic radii for characterizing the ionic binding of the hydrated ion is appropriate.

Leusch et al. determined the affinity sequence for untreated *Sargassum fluitans* as Pb > Cu > Mi ~ Cd > Zn

(82). Fourest and Volesky (67) observed that the binding capacity for divalent metal binding to *Sargassum* decreased in the order Pb > Cu > Cd > Zn > Ca, and related this to decreasing electronegativity. However, the binding strength, rather than the maximum binding capacity, should be influenced by the electronegativity. Possibly, the maximum binding was not reached under the conditions employed by these authors.

Conclusion. Generally, these binding-strength trends, except for the work of Crist et al. (63), follow the pattern for monovalent ions—Cs > Rb > K > Na > Li—and for divalent ions—Ba > Sr > Ca > Mg; Cu > Ni > Co > Mn; Pb > Cd > Zn. While the preceding expresses basic trends, no fixed relation between these series can be established. Whether, for example, Cu or Cd happens to be more strongly bound depends on the sorbent material.

Factors Influencing the Binding Strength

There are three factors that can determine the affinity of a metal ion (because they contribute to the free energy of binding of hydrated metal ions to biosorbent binding groups).

1. *Hydration effects:* A change in the orientation of the hydrated water molecules occurs in conjunction with electrostatic binding (69), but in such "outer-sphere" complexes both partners retain their hydration spheres (69). Complete or partial dehydration takes place for "inner-sphere" complexes resulting from covalent binding (69).

   Both types of changes in the hydration state can occur more easily if the hydrated water molecules are not held strongly by the metal ion. Williams and Hale (83) showed that a good correlation (Born equation) between the ion hydration energy and the term (z^2/r) exists. For estimating the strength of hydration, the radius of the nonhydrated cation should be used, of course. This is usually taken as the ionic crystal radius (listed in Table 3).

   Taking into account the hydration sphere, the effective hydrated radius is larger than the crystal-ion radius. In general, the larger the hydration energy of an ion (i.e., the smaller the crystal radius) the larger the effective hydrated radius (58). Consequently, the hydrated radii follow a reverse trend as the ionic crystal radii, both for the groups of alkali and alkaline earth ions (see Table 3). It is difficult to quantify the radius of the hydrated ion, but some values have been compiled (84,85).

   According to Jain (86), hydration effects can be dominating when the electric field of the sorbent is weak (i.e., weak binding). In this case, larger ions (comparing crystal radii of ions of the same charge) that exert a smaller ordering influence on their hydrated water molecules (i.e., that are less strongly hydrated) are preferably accumulated at the interface.

2. *Ionic (electrostatic) binding:* If the binding groups are negatively charged, they can attract metal cations. The interaction is stronger the higher the charge density of both biosorbent and metal ion. If a strong electric field is present, electrostatic effects may become the dominant factor such that small ions that have a higher charge density are bound more strongly (86). In the past, ionic-crystal radii have been used in estimation of the ionic binding strength, as well as in aqueous solution (69,87). However, since the cation retains its hydrated water molecules, the hydrated cation radius is more characteristic for electrostatic attraction (as opposed to hydration). According to Marcus and Kertes (85), the selectivity increases with increasing charge and decreasing hydrated radius.

3. *Covalent binding:* This type of bond involves sharing of the electrons. Therefore, the more similar the electronegativities of the metal ion and the coordinating atom of the ligand, the higher the covalent character of the bond (88). The selectivity (or binding strength) increases with increasing polarizability of the ion (i.e., softness) (85).

Conclusion. Overall, the binding strength increases with increasing ionic radius and decreasing charge if binding is weak and largely due to hydration effects, decreasing hydrated radius and increasing charge if binding is intermediately strong and due to electrostatic effects, and decreasing electronegativity difference if binding is strong and covalent.

Since the first two criteria are almost opposed to each other, the affinity sequence can be reversed with changing electric field strength of the sorbent (86), except that the ionic and hydrated radii can exhibit opposite trends. Therefore, it is possible that for the alkaline and alkaline earth metals, for example, the affinity sequences Cs > Rb > K > Na > Li and Ba > Sr > Ca > Mg > Be observed in industrial ion exchangers are due both to decreasing crystal and increasing ionic radii (85).

The Concept of Hard and Soft Ions

In order to classify the behavior of metal ions, a division into so-called hard and soft ions was introduced by Pearson (89). Hard ions are characterized by predominantly electrostatic bonds where the entropy gain in the change of the orientation of the hydrated water molecules makes an important contribution to the free energy change ΔG (69). Hard ions have a weak polarizing power and retain their electrons strongly. Williams used the charge density (z^2/r) (with z being the charge and r the cation radius) as a measure for the strength of ionic binding (90,91). Table 3 shows that the ionic binding strength of most divalent metal ions is similar and that it is higher than the one for monovalent ions. This is also illustrated in Figures 3 and 4.

The characteristics of hard ions resemble the ones of class "a" cations in the "a"–"b" categorization of metal ions. Type "a" ions form their strongest bonds with the first element of each group, "b" ions with the second or higher (92). Type "a" ions tend to form strong bonds with highly electronegative donors that are difficult to polarize (i.e., hard) (93). The difference between the electronegativity (x) of the

Table 3. Parameters Characterizing Binding Strength

Metal	z (−)	r_{cryst}[a] (Å)	r_{hyd}[b] (Å)	x[c] (−)	z^2/r_{cryst}[d] (1/Å)	z^2/r_{hyd}[e] (1/Å)	"b" type[f] (Å)	Δx[g] (−)	ionic total[h] (−)	χ[i] (1/Å)
H	1	—	2.82	2.2	—	0.35	—	1.3	0.34	1.03
Li	1	0.76	3.82	1.0	1.32	0.26	1.61	2.5	0.79	0.33
Na	1	1.02	3.58	0.9	0.98	0.28	1.51	2.6	0.82	0.34
K	1	1.51	3.31	0.8	0.66	0.30	1.51	2.7	0.84	0.36
Rb	1	1.61	3.29	0.8	0.62	0.30	1.57	2.7	0.84	0.36
Cs	1	1.74	3.29	0.7	0.57	0.30	1.27	2.8	0.86	0.35
Mg	2	0.72	4.28	1.2	5.56	0.93	2.26	2.3	0.73	1.27
Ca	2	1.00	4.12	1.0	4.00	0.97	1.85	2.5	0.79	1.23
Sr	2	1.26	4.12	1.0	3.17	0.97	2.11	2.5	0.79	1.23
Ba	2	1.42	4.04	0.9	2.82	0.99	1.84	2.6	0.82	1.21
Mn	2	0.67	4.38	1.5	5.97	0.91	3.42	2.0	0.63	1.44
Co	2	0.65	4.23	1.8	6.15	0.95	4.86	1.7	0.51	1.84
Ni	2	0.69	4.04	1.8	5.80	0.99	4.99	1.7	0.51	1.92
Cu	2	0.73	4.19	2.0	5.48	0.95	6.32	1.5	0.43	2.22
Zn	2	0.74	4.30	1.6	5.41	0.93	4.07	1.9	0.59	1.56
Cd	2	0.95	4.26	1.7	4.21	0.94	5.20	1.8	0.56	1.69
Pb	2	1.19	4.01	1.8	3.36	1.00	6.61	1.7	0.51	1.94
Al	3	0.54	4.75	1.5	16.67	1.89	3.60	2.0	0.63	3.00
Ag	1	1.15	3.14	1.9	0.87	0.29	7.22	1.6	0.47	0.62

[a]Shannon crystal radii (95).
[b]Nightingale hydrated ion radii (84).
[c]Pauling electronegativity (88).
[d]Parameter for hydration strength (91).
[e]Parameter for ionic bonding strength (91).
[f]Parameter for covalent bond ("b") character x^2 (r_{cryst} + 0.85) (96).
[g]Parameter for ionic bond character (88).
[h]Fraction of ionic bond character 1-exp($-\Delta x^2/4$) (94).
[i]Parameter for total binding strength $\xi = z^2/r_{hyd}/1\text{-exp}(-\Delta x^2/4)$.

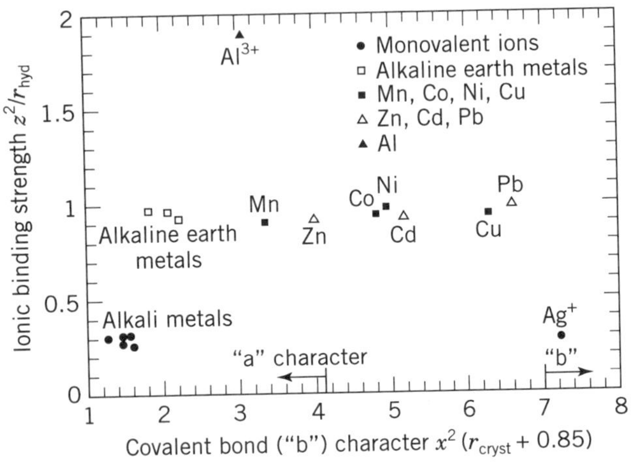

Figure 3. Ionic- and covalent-bond character for different metals.

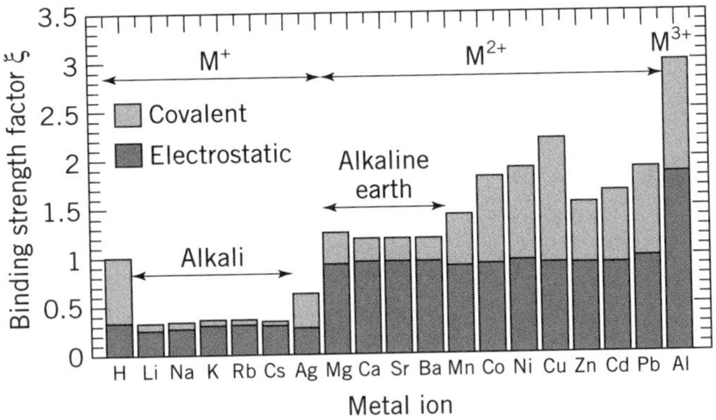

Figure 4. Binding-strength parameters for different metals.

metal ion and that of oxygen has been used as a measure for "a" (i.e., ionic bond) character, which increases with $\Delta x^2 = (^{M}x - {}^{O}x)^2$ (69). According to Pauling (94), the fraction of ionic bond character is $1 - \exp(-\Delta x^2/2)$, that is, equal contributions of covalent and ionic binding occur when $\Delta x = 1.7$. As seen in Table 3, the ionic bond character is larger than 70% for alkaline (earth) metals and about 50% for most heavy metals. Dean (88) describes a similar criterion: The ionic bond character increases with Δx and 100% ionic bonding occurs for $\Delta x > 2$ (but according to Pauling's criterion the ionic-bond character at $\Delta x = 2$ is 63%, not 100%). Table 3 shows that the Δx criterion is fulfilled for alkaline and alkaline earth ions.

Soft ions participate in more covalent bonding where the free energy is mostly enthalpic (69). The characteristics of soft ions are opposite those of hard ions, resulting in a reversed affinity sequence. The affinity sequence of soft, or class "b", ions for ligands is I > Br > Cl > F and S > N > O (97). A combined affinity sequence was described by Buffle: S > I > Br > Cl 39 N > O > F (69). Therefore, the difference between complexation constants for F and Cl, respectively, has been introduced by Turner et al. (87) to characterize softness, which increases with decreasing ΔK

$= \log({}^{F\,M}K) - \log({}^{Cl\,M}K)$. Type "a" ions are characterized by $\Delta K < 2$ and type "b" ions by $\Delta K > 2$; among the border elements, tendencies for "a" or "b" character are present for $\Delta K <$ or > 0, respectively (87). Typical hard, or class "a," ions are the alkaline and alkaline earth metals, transition elements up to class VI, and, for example, Zn (93). As opposed to the original classification of Pearson, who considered Cd a soft ion and H a typical hard ion, these two are now regarded as borderline ions (97). Table 3 and Figure 3 confirm that light metals and Mn can be classified as hard, and that most of the heavy metals may be listed as intermediate, except for Zn, which is just on the threshold to being hard, and Ag, which is soft.

For any donor (and especially for soft ones) the binding strength increases with cation softness (69). Soft cations form more stable complexes with soft donors (S, P, or As), while hard cations prefer hard donors (F or O) (89). N donors have intermediate characteristics. For O or N donor atoms, the Irving–Williams order of complex stability for divalent ions in the first series is Mn < Fe < Co < Ni < Cu > Zn (93). This can be explained by the electronegativity differences with respect to the donor atoms O or N.

Examples of hard ligands are carbonate, phosphate, sulfate, or hydroxide ions (1,94) as well as sulfate, carboxylate, alcohol, or hydroxyl groups (97). Soft ligands include sulfhydryl, thioether, and amino groups (97).

The relevant parameters for different metals are listed in Table 3. The hardness parameter (z^2/r_{hyd}) for different metal ions is plotted over the softness parameter $X^2(r_{cryst} + 0.85$ Å) in Figure 4. Since the parameter (z^2/r_{hyd}) expresses the ionic, binding strength, and the parameter $1 - \exp(-\Delta x^2/2)$ is supposed to characterize the relative contribution of ionic bonding to the total binding strength, one can obtain an indicator for the total binding strength by dividing (z^2/r_{hyd}) by $[1 - \exp(-\Delta x^2/2)]$. This parameter is called ξ; it is listed in Table 3 and plotted in Figure 4 as the total height of the bar. The lower part represents the ionic binding strength; the upper part represents the covalent binding strength. The graph should, however, be used only for illustrative purposes, rather than in a quantitative way, because the exact magnitude of the contribution of ionic bonding is disputable. According to the Δx criterion, for example, there is no covalent bonding for the light metals, while this plot, using Pauling's criterion, shows ~20% covalent binding.

The binding trends summarized in the section "Need for Heavy-Metal-Removal Processes" can now be explained: The tendency of stronger binding for the heavier elements in both the alkaline and alkaline earth metals can be explained as being due to lower charge density (z^2/r_{cryst}) (i.e., weaker hydration) of the ions themselves and higher charge density (z^2/r_{hyd}) of the hydrated ions (i.e., stronger electrostatic bonds) in the heavier elements of each series (see Table 3). Since these ions are hard, there is little contribution of covalent bonding; therefore, the electronegativity differences between them are irrelevant. The trend among the elements in the second period (Cu > Ni > Co > Mn) and the trend Pb > Cd > Zn are probably mainly due to a decrease of the covalent binding [characterized by $x^2(r_{cryst} + 0.85)$], but a decrease in the electrostatic binding (z^2/r_{hyd}) (except for the relation between Ni and Cu) and increasing hydration strength (z^2/r_{cryst}) (except for Mn) may also contribute.

Since the binding trends in biosorption by algal biomass or its constituent molecules indicate that there is significant electrostatic and covalent binding, the latest biosorption models described at the end of the next section ("Modeling") attempt to incorporate these stronger interactions and neglected hydration effects, which are weaker.

MODELING

In order to evaluate the feasibility and effectiveness of biosorption in wastewater treatment, it is essential to make predictions of the sorption performance (e.g., for facilitating process design). Therefore it is necessary to develop appropriate mathematical models of biosorption. Modeling the biosorption-binding equilibrium is a prerequisite for all further work involving batch kinetic studies and, in particular, for the continuous-flow sorption-column applications that represent the most effective configuration of the sorption-based process.

The amount of metal M (or in general, sorbate) bound per mass of sorbent is called the binding (or uptake), ${}^{M}q$. The binding is not only dependent on the sorbent material but also on the equilibrium concentration [*M*] of the sorbate in solution and on other parameters, such as pH and equilibrium concentration of other ions in solution. The relationship between the equilibrium binding and the concentration of an ion (at constant temperature) is depicted in an isotherm plot of ${}^{M}{}_{q}$ versus [M]. With increasing metal concentration in solution its binding increases from zero to the maximum (Figure 5). It is desirable for a sorbent to possess a high sorption capacity (maximum binding obtained at high equilibrium concentrations) and a high affinity for the sorbate species, which is reflected in a steep slope of the isotherm curve at low equilibrium concentrations.

Langmuir and Freundlich Isotherms

Most mathematical biosorption models used in the literature describe simple Langmuir (13,14,20,59,63,64) or Freundlich (16,35,98) sorption isotherms, where the metal-ion binding is determined as a function of the equilibrium concentration of that metal ion in the solution, without reference to pH or other ions in the same solution system.

The Freundlich isotherm (99) is originally of empirical nature, but has later been interpreted as sorption to sites with an affinity distribution; that is, it is assumed that the stronger binding sites are occupied first and that the binding strength decreases with increasing degree of site occupation. More specifically, the Freundlich isotherm is obtained when a log-normal affinity distribution [i.e., a normal distribution of log (*K*)] is assumed (57p.319;100). The Freundlich isotherm is defined by the following expression:

$$ {}^{M}q = k[\mathrm{M}]^{1/n} \tag{2}$$

where k and n are empirically determined constants, with

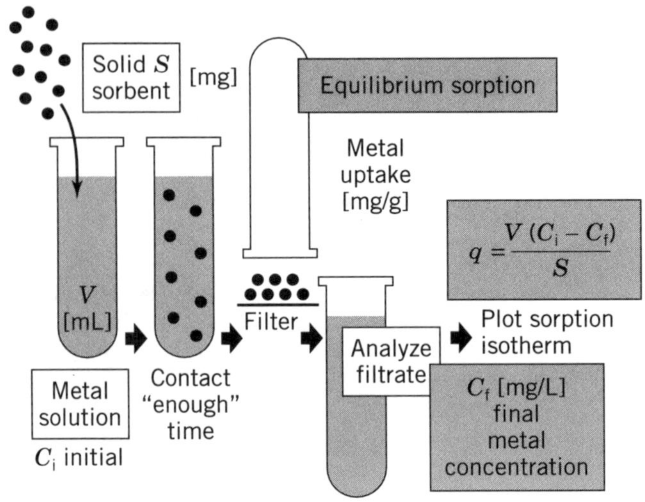

Figure 5. Experimental methodology for deriving the biosorption isotherm.

k being related to the maximum binding capacity, and n related to the affinity or binding strength (19,101).

The Langmuir isotherm (102) is based on considering sorption as a chemical phenomenon. It is assumed that the forces exerted by chemically unsaturated surface atoms (total number of binding sites B) do not extend further than the diameter of one sorbed molecule, and that therefore sorption is restricted to a monolayer. In the simplest case the following assumptions are made.

- All sorption sites are uniform (i.e., constant heat of adsorption).
- There is only one sorbate.
- One sorbate molecule reacts with one active site.
- There is no interaction between sorbed species.

The rate of adsorption is proportional to the rate constant of the forward reaction ${}^{BM}k_f$, the number of free sites ($B = {}^{t}B - {}^{M}q$), and the number of sorbate molecules hitting the surface per unit time (i.e., proportional to their concentration [M]). Under the same conditions, the rate of desorption is proportional to the rate constant of the backward reaction ${}^{BM}k_b$ and the number of occupied sites ${}^{M}q$. The equilibrium is attained when the rate of adsorption equals the rate of desorption:

$$ {}^{BM}k_f B[M] = {}^{BM}k_b {}^{M}q \quad \text{(mol/s)} \qquad (3) $$

With the equilibrium constant ${}^{BM}K$, indicating the affinity for the sorbate, defined as the ratio of adsorption and desorption rates ${}^{BM}K = {}^{BM}k_f/{}^{BM}k_b$, it follows:

$$ {}^{BM}q = \frac{{}^{t}B \cdot {}^{BM}K[M]}{1 + {}^{BM}K[M]} \quad \text{(mmol/g)} \qquad (4) $$

Both Langmuir and Freundlich isotherms have been used to describe the equilibrium of metal biosorption. While it is possible to obtain good agreement between predictions of these models and the experimentally determined metal binding, their simple hyperbole-based equations have significant limitations: They cannot describe a more complex sorption behavior often reflected by nonhyperbolic relationships and they cannot predict the effect of pH or of other ions in the solution, among other shortcomings.

pH Effect. Although the crucial role of protons in biosorption is generally known, it is usually neglected in the mathematical description of the process. The complex nature of biosorbent materials that may contain a multiplicity of binding groups makes the modeling of the pH effect in biosorption difficult. As a consequence, it has been common practice to determine a separate isotherm and a new set of parameters, for each pH value (15,103,104) or for each initial biomass saturation state (e.g., protonated or loaded with ions from the seawater). This has been necessary because the most frequently used Langmuir or Freundlich sorption models do not take into account the fact that metal-ion biosorption is largely an ion-exchange phenomenon. They do not allow the prediction of the remaining binding of those ions (e.g., protons or sodium) that were initially loaded onto the biosorbent, and neither do they incorporate the concentration of the exchanged species (e.g., protons) as a parameter, normally not allowing reliable calculated predictions of the biosorbent performance.

Some attempt has been made, assuming a simplified linear relationship between the metal uptake and pH difference, to empirically model the biosorptive binding of metal ions by a Freundlich isotherm, which was modified by a factor proportional to the pH difference (105). This approach could not represent the change of proton binding in the presence of metal ions. Though protonation reactions for two acidic sites (alga) have been considered (106), the model simplification by lumping pH and proton- and metal-binding constants into one parameter for each resulted in a standard two-site Langmuir isotherm. A rare example of a Langmuir model including protons and one or two binding sites (104,107) assumed a 1:1 stoichiometry of metal binding and proton release, although it was recognized that metal binding with a 1:2 stoichiometry probably occurred. A modified Langmuir model for competitive Cu and H binding to two sites on alginate (108) required iterative calculations of each cation uptake as a function of the binding of the other one. Another two-site model (109) based on a 2:1 proton–metal exchange constant allows direct calculation of the metal binding as a function of the concentration ratio $[H]^2/[M]$, but it does not account for free sites and it cannot easily be extended to include more species.

Competing Ions in Solution. Often, the prediction of metal-sorption equilibria is complicated by the presence of several sorbed ions, requiring the use of multicomponent isotherm equations. This is especially relevant when the sorption of one ion influences the binding of another when different ions compete for the same binding site. A variety of multicomponent sorption isotherm models is presented in a review by Yu and Neretnieks (110).

The use of multicomponent Langmuir isotherm models was demonstrated in describing the equilibrium metal biosorption (48,111,112). However, biosorption was not regarded as an ion-exchange process; the ions with which the biomass had originally been saturated were not considered.

Instead of using multicomponent isotherm models, more complex systems involving different cations and ligands can be modeled by incorporating binding constants for metal-ion–biomass complexes into a computer program for the calculation of the chemical equilibrium. Reviews of different computer programs for calculating the equilibrium speciation have been presented (44,113). These programs take acid/base reactions, coordination, and dissolution/precipitation into account. Their use is advantageous if complexation in solution is expected or if final concentrations are to be predicted from the initial state (27,47).

Ion-Exchange Constants

There have been indications that ion exchange is an important biosorption mechanism. However, this does not preclude an adequate use of simple sorption isotherms (16,59,64) instead of appropriate ion-exchange models with equilibrium constants taking into account the reversibility of the ion-exchange reactions: The ion that is displaced into the solution can compete with the sorbed metal ion for the sites (25,62,76,114).

The Langmuir equation and the ion-exchange constant for the binding of a metal ion M (for simplicity here a monovalent ion) replacing a proton H on a complexing site B are related as follows:

Langmuir: $B + M \rightleftharpoons BM$

$$^{BM}K^{*} = \frac{BM}{B[M]} \quad [B]_t = [B] + [BM] \tag{5}$$

Ion exchange: $BH + M \rightleftharpoons BM + H$

$$^{BM}K = \frac{BM[H]}{BH[M]} \quad [B]_t = [BH] + [BM] \tag{6}$$

therefore:

$$^{BM}K^{*} = {}^{BM}K/[H] \tag{7}$$

The difference between the two approaches is that the first one assumes that all sites are initially free and does not consider any reverse reaction of a displaced ion, in this case a proton, with the site. The second model assumes that all sites to which metal ions are sorbed are initially occupied, that is, the number of free sites stays constant. Crist et al. (114) compared the fit of the Langmuir sorption isotherm model and the one using ion-exchange constants. The differences between the two models were especially pronounced at low metal-ion concentrations because the reverse reaction involving the displaced ion is most noticeable if the concentration of the bound cation is low.

The ion-exchange approach is probably somewhat closer to reality than the simple Langmuir model, but it is not completely satisfying either. Although the constant number of free sites [in other words, the number of sites occupied by the bound ion is equal to the number of sites from which the exchanged ion is released so that $^{t}B = (B^{1}M) + (B^{2}M)$ can be assumed] may be a reasonable working assumption for a constant pH system, it may not hold for systems with changing pH values. According to Stumm and Morgan (46), the cation-exchange capacity increases with increasing pH above the isoelectric point. Therefore, one cannot simply model the competitive binding of metals and protons by using only a metal–proton-ion-exchange constant. It is necessary to include at least one reaction where a cation reacts with a free site.

In the modeling of industrial ion exchangers, protons have been considered (115,116). However, these synthetic resins are chemically rather simple compounds that mostly contain only one active group. Biosorbents, on the other hand, can be heterogeneous and contain different acidic groups active in metal ion binding. The $p^{BH}K$ value, the quantity of binding sites, and the metal-ion binding constant must be known for each of these sites.

Models Including Ionic Strength Effects

Ubiquitous ions such as sodium may interfere with heavy metal biosorption through the effect of ionic strength. In biosorption, no modeling of the effects of ionic strength and electrostatic attraction has been done. Several models of ionic-strength effects, however, have been published concerning metal binding to humic and fulvic acids. The physical characteristics of these molecules, which are degradation products of organic matter, are quite different from microbial biosorbents. Humic and fulvic acids are small water-soluble molecules. However, since these molecules contain carboxyl groups as one of their main functional groups they may in some aspects be comparable in particular to those biosorbent materials based on seaweeds.

Buffle (117) systematically described possible models for complexation equilibria, including site protonation, multiple sites, and secondary effects such as the influence of the electric field in polyelectrolytes on the local concentration. The simplest approach to modeling the influence of Na in the biosorption system consists of the introduction of a binding constant for Na, as done by Westall et al. (118) for the binding of H, Co, and Na to humic acids. This approach enables derivation of a multicomponent sorption isotherm model so that the binding of each species can be directly calculated. This was done for biosorbents only recently in the work of Schiewer and Volesky (119–121). Although the Na binding constant can easily be incorporated into computer programs for the calculation of chemical equilibria in aqueous solutions, such as MINEQL+ (122), the disadvantage of this approach can be that the Na binding constant is purely empirical and does not reflect the electrostatic nature of the Na binding.

Another possibility is the use of the Donnan model (123). An application of this approach for different charged polymers has been pioneered by Marinsky (124) and coworkers, who established the use of Donnan models for interpreting protonation equilibria in organic polyelectrolytes including humic and fulvic substances and some metals (125–127) and alginic acid (128). The latter contribution also involves an estimation of the water uptake (swelling) using osmotic coefficients. Unfortunately, this modeling work did not include model predictions of proton or metal binding under different conditions. More Donnan-based model work done with alginate considered competition of Cu and H for the same sites (129,130).

Güntelberg activity coefficients have been used to account for ionic-strength effects in conjunction with fulvic acid and Cu–alkaline earth metal binding competition (131). Although this method may be suitable for rather small molecules, it would not be appropriate for polyelectrolytes, such as the alginic acid found in seaweed-based biosorbents. While the isotherm sorption model did not include a term for free sites, it was noticed that increasing ionic strength (Na) or presence of Ca or Mg can reduce Cu binding.

A model that is suitable for both oligo- and polyelectrolytes was used only for modeling pH titrations and Cu binding of humic acid molecules (132). Two of the sites were assumed to be diprotic and able to bind metal ions and one was assumed to be monoprotic and not to bind metals. In this approach, the humic acid molecules were assumed to be penetrable or impenetrable spheres of two size classes (assuming that the total charge is proportional to the volume of the sphere). The advantage of this approach is its generality concerning the size of the molecule. It is particularly suited for oligoelectrolytes that show a behavior intermediate between simple ions and polyelectrolyte gels. Most biosorbent materials could be considered true polyelectrolytes. However, the semiempirical correlation used in this approach required three fitting parameters for the electrostatic effects for each of the two particle sizes. Moreover, the fitting parameters had to be determined individually for each ionic strength. This means that the effect of ionic strength could not be predicted.

A Helmholtz metal-sorption model for the charged interface (133) requires only one fitting parameter, the constant capacitance, for modeling the electrostatic effects, but not the ionic strength effects. Another model considering eight sites of two general types for light-metal binding to humic acids used relatively few fitting parameters for the prediction of a wide range of experimental conditions (134,135). Unfortunately, it also relies on an empirical relationship involving the ionic strength. A Poisson–Boltzmann model for proton binding to humic substances (136) incorporates the effect of ionic strength on the concentration factor in a realistic manner and requires only one fitting parameter for electrostatic effects (the particle radius). The assumption of small impermeable spheres or cylinders, however, does not appear applicable to the large, penetrable particles as in the preferable case for biosorbents.

Recently, Schiewer and Volesky (119–121) presented a series of equilibrium biosorption models gradually incorporating the metal-ion concentration, pH, and ionic strength. The models proposed had only one fitting parameter, instead of up to six as in some of those already mentioned. The formation of $BM_{0.5}$ complexes was assumed, and a combined Donnan–biosorption–isotherm equation was derived that allowed for direct calculation of cation binding without iterations when swelling proportional to the number of free sites was assumed. Using the parameters (number of binding sites, proton binding constant, and specific particle volume) obtained from pH titrations, it was possible to predict the effect of pH, ionic strength, and Ca concentration on Cd binding for a seaweed biosorbent.

BIOSORPTION PROCESS DESIGN

Because biosorption is just another form of a basic sorption operation, the same principles apply for the design and sizing of the equipment. The approach is always a combination consisting of basic equilibrium sorption tests (batch) and dynamic continuous-flow apparatus studies, which can become very sophisticated if empiricism is to be minimized. Invariably, numerical computer methodologies provide a powerful tool for the design of a continuous-flow (bio)sorption process.

Equilibrium Sorption Process Basis

When contacting the solid-phase biosorbent with the sorbate species (i.e., metal ions) in solution, a sorption equilibrium will become established after an appropriate length of time. When equilibrium between the species sorbed on the solid and the portion remaining in solution is established, the solids are separated and the supernatant analyzed for the residual, final, equilibrium concentration of the remaining sorbate (Fig. 6). The metal uptake by the solid sorbent is calculated (according to the formula in Fig. 6) and used for the sorption isotherm plot, which serves to compare the sorption performance of different systems. This is easily done for one-sorbate (one-metal) systems. The situation becomes more complicated when there are more than one sorbed species in multisorbate systems.

Of interest is often the case of "interference" assessment when the uptake of one key metal (e.g., toxic Cd) may be affected by the presence of a ubiquitous metal of no major interest (e.g., Fe). This situation can be still depicted by three types of three-dimensional sorption isotherm surfaces, depending on what particular metal uptake appears in the plot: (*1*) Cd uptake, (*2*) Fe uptake, and (*3*) combined (Cd and Fe) uptake. The base two axes are always the equilibrium concentrations of the two metals, respectively.

Since the experimental equilibrium (final) concentrations are the product of the experiment, there is no control

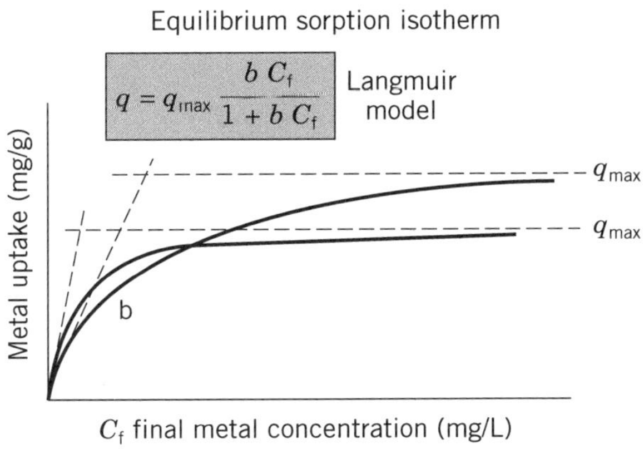

Figure 6. Biosorption-isotherm relationship curves; b is related to initial slope and indicates the sorbent "affinity."

over their values and they are generated more or less randomly. Consequently, the sorption isotherm surface needs to be fitted to the date points. Finding a suitable mathematical model capable of representing reasonably well the two-metal sorption behavior enables this and more; it can facilitate the summarizing of the effect of one metal on the uptake of the other for the desirable conclusion such as depicted in Figures 7 and 8 (137).

For many metal biosorption cases an equilibrium model considering only one type of binding site would be adequate assuming the following reaction (138):

$$M^{2+} + 2B = 2BM_{0.5} \quad H^+ + B = BH \tag{8}$$

The model provides explicit formulas for metal uptakes as functions of final concentrations of the species present in the system, even for the case of heterovalent exchange, which makes it especially suitable for use in numerical simulations of dynamic sorption systems. The model equation for the equilibrium uptake of a bivalent metal ion M1 present in a binary system with hydrogen protons can be written as follows:

$$q^*_{M1} = \frac{0.5 M_{WM1} C_t (K_{M1} C_{fM1})^{1/2}}{1 + K_H C_{fH} + (K_{M1} C_{fM1})^{1/2}} \tag{9}$$

Similarly, for a binary system containing bivalent species M1 and M2, the expressions for uptakes of each of the species can be written as follows:

$$q^*_{M1} = \frac{0.5 M_{WM1} C_t (K_{M1} C_{fM1})^{1/2}}{1 + (K_{M1} C_{fM1})^{1/2} + (K_{M2} C_{fM2})^{1/2}} \tag{10}$$

$$q^*_{M2} = \frac{0.5 M_{WM2} C_t (K_{M2} C_{fM2})^{1/2}}{1 + (K_{M1} C_{fM1})^{1/2} + (K_{M2} C_{fM2})^{1/2}} \tag{11}$$

The values of the parameters K_H, K_{M1}, and $K_{M2}$are normally determined by fitting of the model to the experimental data using, for example, the method of weighed least squares. The constant $C_t$in equations 9–11 may be obtained from pH and/or conductometric titration of protonated biomass with NaOH (67). Metal biosorption can often be viewed as a simple ion exchange; hence, the classical ion-exchange concept based on exchange-equilibrium constants and separation factors (139,140) can be applied to this case. For the ion-exchange reaction:

$$bA^{+a} + aB^{+b} = bA^{+a} + aB^{+b} \tag{12}$$

the equilibrium constant K_{AB} and the separation factor r_{AB} are defined as

$$K_{AB} = \frac{q_A^b C_{Bf}^a}{C_{Af}^b q_B^a} = \frac{y_A^b x_B^a}{x_A^b y_B^a} \cdot \frac{C_0^{a-b}}{Q^{a-b}} \tag{13a}$$

$$r_{AB} = \frac{y_a x_B}{x_a y_b} \tag{13b}$$

for the case of ideal behavior of the exchanging species in both of the phases. As can be seen from equation 13a, the total normality C_0 of the solution affects the equilibrium of a heterovalent exchange (i.e., when $a \neq b$), but it does not affect the equilibrium when the valence of the exchanging species is the same (i.e., $a = b$). Consequently, the ion-exchange isotherms of a species in a binary heterovalent system vary in shape depending on the total normality of the solution (141).

Ion-Exchange Equilibrium Relationships. The term *equilibrium sorption isotherm* is commonly used in biosorption to refer to a line connecting equilibrium metal uptakes plotted against the final concentrations of the metal. Typically, biosorption equilibrium data are obtained under the same pH. However, this is not the case in ion exchange where the sorption isotherm represents data points obtained at a constant normality of the solution. Furthermore, in compliance with the established convention, ion-exchange isotherms for binary systems usually display dimensionless uptakes as functions of dimensionless concentrations for the species of interest (Fig. 9):

$$\frac{q_M}{Q} = f\left(\frac{C_{ML}}{C_0}\right) \tag{14}$$

Connection between Equilibrium and Dynamic Sorption in Columns. Unlike the biosorption isotherm, the ion-exchange isotherm depicts an equilibrium situation that is relevant to the case of the metal biosorption in a fixed-bed adsorber in which the concentration of the species released from the biosorbent increases monotonically in the direction of the flow. In fact, the ion-exchange isotherm represents the "equilibrium line" that is, together with the "operating line," sometimes used for assessing design and performance of fixed-bed adsorbers (142,143). According to the shape of the ion-exchange isotherm, the equilibrium is classified as (*1*) favorable when the isotherm is concave, (*2*) linear when the isotherm is linear, or (*3*) unfavorable when the isotherm is convex. It is known that the favorable isotherm leads to the development of a self-sharpening dynamic sorption zone (front) in the fixed-bed adsorber, whereas the unfavorable isotherm causes its broadening (144,145). As an alternative to ion-exchange isotherms, tables of values of separation factors for the compositions

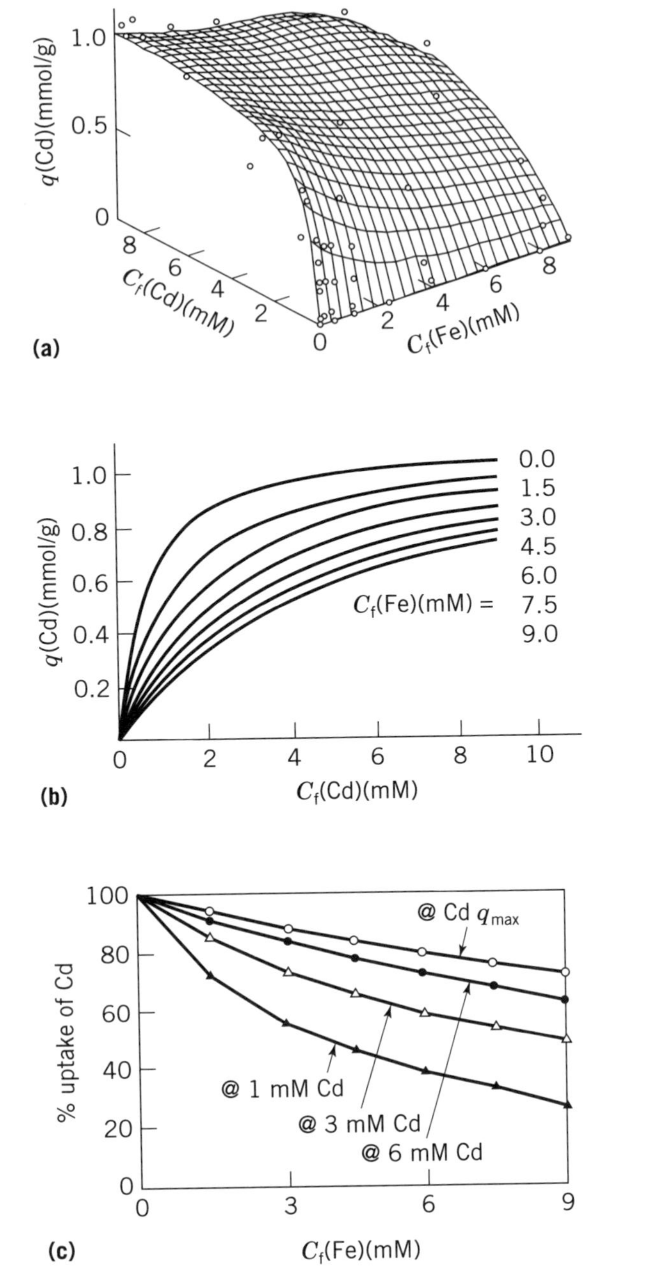

Figure 7. (**a**) A 3-dimensional sorption surface for the Cd–Fe–*Sargassum* biosorption system, cadmium uptake at pH 4.5; (**b**) the effect of Fe on the equilibrium uptake of Cd by *Sargassum* biomass at pH 4.5; (**c**) The summary of the effect of Fe presence on the Cd uptake by *Sargassum* biomass at pH 4.5. Equilibrium Cd concentrations were arbitrarily selected: ○, decreasing Cd uptake for different q_{max} at "high" C_f(Fe); ●, [Cd] = 6 mM; △, [Cd] = 3 mM; ▲, [Cd] = 1 mM.

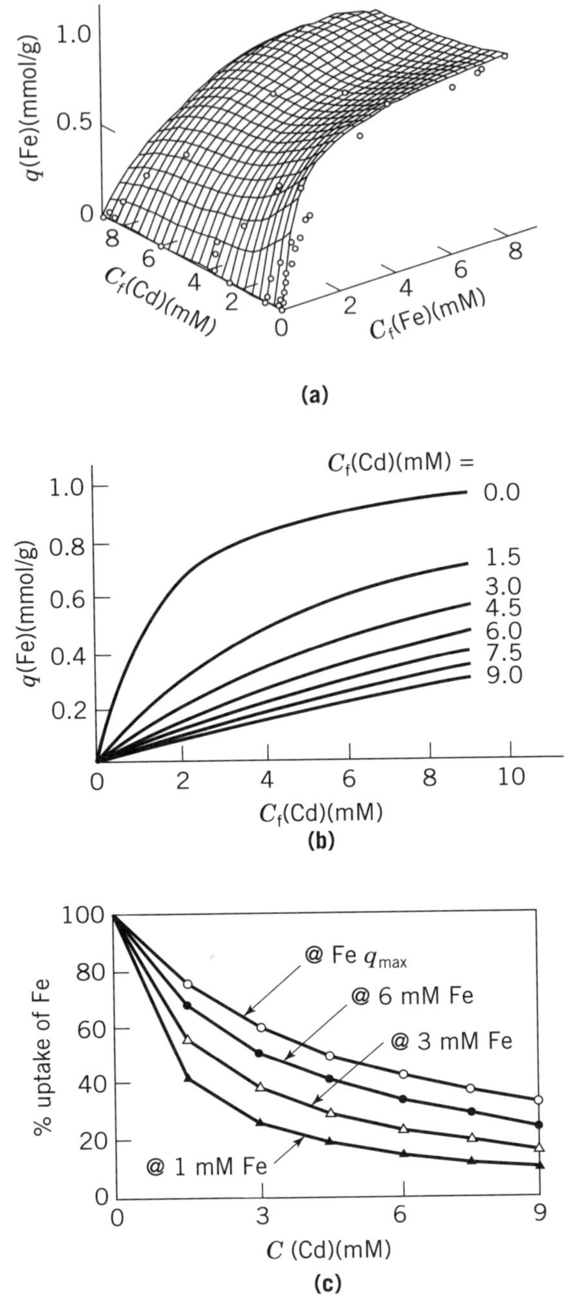

Figure 8. (**a**) A 3-dimensional sorption surface for the Cd–Fe–*Sargassum* biosorption system, iron uptake at pH 4.5. (**b**) The effect of Cd on the equilibrium uptake of Fe by *Sargassum* biomass at pH 4.5. (**c**) The summary of the effect of Cd presence on the Fe uptake by *Sargassum* biomass at pH 4.5. Equilibrium Fe concentrations were arbitrarily selected: ○, decreasing Fe uptake for different q_{max}; ●, [Fe] = 6 mM; △, [Fe] = 3 mM; ▲, [Fe] = 1 mM.

existing in an adsorber can be used to represent the equilibrium for the purpose of computer simulation of the fixed bed (146).

Continuous-Flow Fixed-Bed Column Biosorption Process

In process applications, the most effective apparatus for cyclic sorption/desorption making the most effective use of the reactor volume is a fixed-bed column that also optimizes the concentration difference known to be the driving force in sorption processes. The process of metal biosorption in a fixed-bed column is influenced by three key regimes: the sorption equilibrium, the sorption particle mass transfer, and the flow pattern through the packed bed. Combination of these three determines the overall performance of the sorption column, which is judged by its ser-

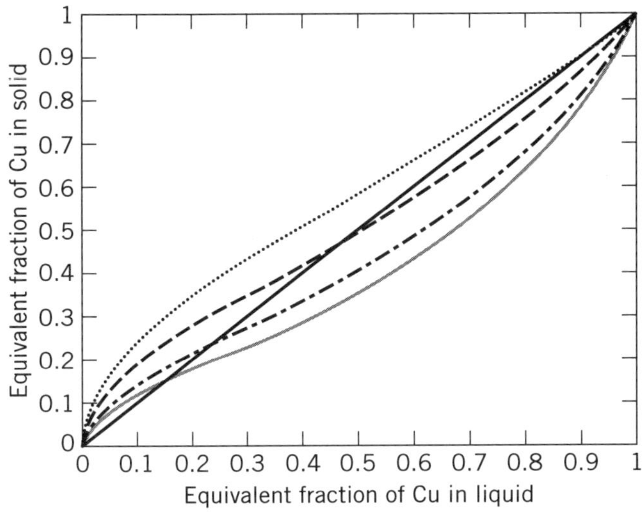

Figure 9. Calculated Cu^{2+} ion-exchange sorption isotherms for *Sargassum* biosorption system. C_o: —, 6 mequiv/L; —·—, 4 mequiv/L; — —, 2 mequiv/L; ..., 1 mequiv/L.

vice time, that is, the length of time until the sorbed species breaks through the bed to be detected at a given concentration in the column effluent (Fig. 10). At that point, the bed is considered for all practical purposes saturated, and the equipment has to be taken out of operation for some kind of regeneration (Fig. 1).

The lack of design information on biosorption in columns is due, in part, to the fact that the metal uptake by different biomaterials is still not fully understood, as well as because most of the biomasses have to be immobilized, granulated, or otherwise processed before their use in a column to provide good flow-through conditions and an acceptable pressure drop across the bed. This process condition must eventually be balanced and optimized against the maximum-sorption-performance requirement.

As opposed to activated carbon-sorption cases, metal biosorption is much closer to the heterovalent ion-exchange model because a considerable number of ions is being released from the biosorbent in exchange for the sorbed metal ion. The most universal approach seems to be that adopted by Tan and Spinner (146), who numerically solved a mixed system of partial differential, ordinary differential, and algebraic equations describing the dynamics of multicomponent ion exchange in a fixed bed. A term can be added to the differential molar balance equation to account for axial dispersion of the fluid in the packed bed. For ionic species M the molar balance is

$$\frac{C_M}{\partial z} - \frac{1}{Pe_c}\frac{\partial^2 C_M}{\partial z^2} + \frac{\partial C_M}{\partial t} + D_{gM}\frac{q_M}{t} = 0 \qquad (15)$$

The sorption rate equation can be written as

$$\frac{q_M}{t} = Sh_M(q_M^* - q_M) \qquad (16)$$

assuming a linear driving force for the sorption process and combined film and intraparticle mass transfer resistance: The equilibrium relationship for a metal species represented by equations (9) and (10) can be written schematically for a general counterion, either proton or metal, as

$$q_M^* = f(C_M, C_{CNT}) \qquad (17)$$

and the electroneutrality equation for binary systems is

$$\frac{C_{M0} - C_M}{M_{WM}} = \nu \frac{C_{CNT0} - C_{CNT}}{M_{WCNT}} \qquad (18)$$

To obtain equations 15–18, the original set of equations was transcribed with dimensionless variables using the following transformations of the variables:

$$C_M = \frac{C_{ML}}{C_0}\; q_M = \frac{q_{MS}}{Q}\; t = \frac{\tau \nu}{L_0}\; z = \frac{l}{L_0}$$

and yielding the following dimensionless groups:

$$D_{gM} = \frac{\rho_b Q}{C_0 \epsilon}\; Sh_M = \frac{K_{fM} L_0}{\nu}\; Pe_c = \frac{L_0 \nu}{D_z}$$

The initial and boundary conditions pertaining to the situations under question are specified as follows:

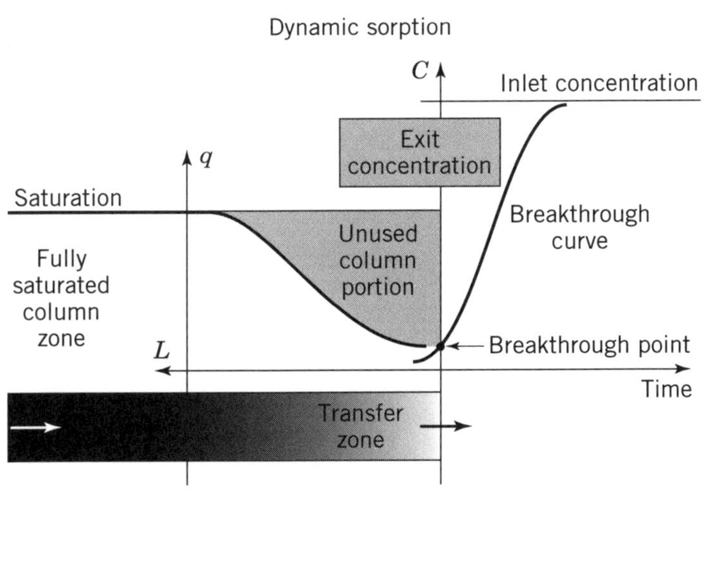

Figure 10. Biosorption column operation: the breakthrough point and the breakthrough curve for the column exit concentration.

$$t = 0 \quad l = (0, L_0) \quad C_M = 0$$

$$t > 0 \quad l = 0 \quad C_M = C_{M0} + \frac{1}{Pe_c}\frac{C_M}{z}$$

$$t > 0 \quad l = L_0 \quad \frac{C_M}{z} = 0$$

The system of equations 15–18 can be solved numerically using the Galerkin finite element method (147), combined with the implicit Euler method for time integration. Linear-basis functions would routinely be used for space discretization of both concentration C and uptake q, ensuring that the solution is mesh independent.

Fitting the Equilibrium Model and Calculation of Ion-Exchange Isotherms. Equilibrium data provide the experimental basis for the fitting of the biosorption model (148). First, the concentration of binding sites C_t can be determined from the uptake experiments; equilibrium constants for the three ions considered here (K_H, K_{Cu}, and K_{Ca}) can then be found by the least square fitting.

Once these parameters are known, the biosorption model can be used to calculate ion-exchange isotherms as follows: First, the normality of a binary mixture C_0 was fixed and the uptake Q, at the concentration of the sorbing species C_0, was calculated from equation 2 or 3. Second, a series of liquid-phase compositions was selected in such a way that the dimensionless concentration C/C_0 of one of the species covers the interval of interest. Finally, a series of dimensionless uptakes corresponding to the series of compositions was calculated using equations 2, 3, and 7.

Fitting and Use of the Flow-through Fixed-Bed Sorption Column Model. The dimensionless groups Sh, Dg, and Pe, together with the equilibrium parameters for a given binary system, represent the parameters of the fixed-bed model. However, because the equilibrium parameters can be determined from batch equilibrium studies, the values of C_0, v, L_0, and ρ_b are usually known, and Q can be calculated from equations 2 and 3, the only remaining unknown elements in the model are K_{fM}, D_z, and ϵ. The values of the three unknown elements K_{fM}, D_z, and ϵ must be determined simultaneously by fitting the fixed-bed model to an experimental breakthrough curve.

The value of an overall mass-transfer coefficient for a given species generally depends on the flow rate, the axial dispersion coefficient, and the values of the diffusivities of the species in the liquid and solid phases. The diffusion coefficients of metal ions in diluted aqueous solutions can often be obtained from the literature (149,150).

Conclusion. In preparation for the biosorption process design the sorption/desorption performance and characteristic properties (mass transfer, pressure and attrition resistance, chemical stability, etc.) of the biosorbent should be reasonably well known. Although equilibrium sorption tests yield some basic information, the sorption system behavior is invariably examined under continuous-flow conditions. The fixed-bed sorption column arrangement is usually the most effective process configuration. Its performance assessment and prediction based on both equilibrium and dynamic studies eventually leads to sizing of the equipment. Advanced methodologies used for this purpose are based on sophisticated computer-based modeling and optimization of the (bio)sorption process.

NOMENCLATURE

a,b	stoichiometric coefficients in the ion-exchange reaction equation 12
$\bar{b}$	Langmuir constant related to affinity
A,B	species in the resin phase
A,B	species in the liquid
B	biosorbent binding site
B_t	total number of binding sites on the biosorbent
BH	biosorbent–proton complex
BM	biosorbent–metal complex
C_i	initial metal concentration
C_0	normality of the solution [mequiv/L]
$C_M = x_M$	equivalent fraction of species M in liquid phase
C_{ML}	concentration of species M in liquid phase [mequiv/L]
$C_{Mf} = C_f$	equilibrium final concentration of species M in liquid phase [mequiv/L]
C_{M0}	equivalent fraction of species M in the feed to a column
C_{CNT}	equivalent fraction of counterion in liquid phase
C_{CNT0}	equivalent fraction of counterion in the feed to a column
C_t	concentration of binding sites in biomass [mequiv/L]
D_z	axial dispersion coefficient [cm^2/s]
G	free (Gibbs) energy
H or H^+	proton
H	enthalpy
$^{BM}k_f$	rate of adsorption
$^{BM}k_b$	rate of desorption
K_{AB}	ion-exchange equilibrium constant
^{BM}K	biosorbent–metal ion-exchange equilibrium constant
K_H, K_{M1}, K_{M2}	equilibrium constants of biosorption equilibrium model
K_{fM}	overall mass-transfer coefficient of species M [min^{-1}]
l	vertical distance from the top of a column
L_0	length of the column [cm]
M	metal ion in solution
M_{WM}	molecular weight of the sorbing metal
M_{WCNT}	molecular weight of the counterion
n	Freundlich sorption isotherm model constant

$q_{MS} = {}^{M}q = q$	uptake of species M [mequiv/L]
q_{max}	maximum uptake (Langmuir parameter)
$q_M = y_M$	equivalent fraction of species M in solid phase
q_M^*	dimensionless equilibrium uptake of species M at C_M
Q	equilibrium uptake of M at $C_{ML} = C_{M0}$. C_0 [mequiv/L]
r	ionic radius (subscripts: hyd = hydrated; cryst = crystalline)
r_{AB}	ion-exchange separation factor
R	gas constant
S	weight of biosorbent solids (Fig. 5) or entropy (equation 1)
t	dimensionless time
T	absolute temperature (K)
v	interstitial velocity of the fluid [cm/min]
V	volume of solution
x	electronegativity
z	ionic charge

Greek Symbols

Δ	difference (equation 1)
ϵ	column void fraction
ν	stoichiometric coefficient of the exchange reaction
ρ_b	packing density [g/L]
τ	time [min]
ξ	parameter in Table 1 $\xi = (z^2/r_{hyd})/(1 - \exp(-\Delta x^2/2))$

Dimensionless Groups

$D_{gM} = \dfrac{\rho_b Q}{C_0 \epsilon}$	solute distribution parameter
$Pe_c = \dfrac{L_0 v}{D_Z}$	column Peclet number
$Sh_M = \dfrac{K_{fM} L_0}{v}$	modified Sherwood number

BIBLIOGRAPHY

1. J.W. Moore and S. Ramamoorthy, *Heavy Metals in Natural Waters*, Springer-Verlag, New York, 1984, pp. 28–57, 77–99, 182–204.
2. WHO, *Guidelines for Drinking Water Quality*, World Health Organization, Geneva, 1993, pp. 174–181.
3. EPS, *Metal Finishing Liquid Effluent Guidelines, Report EPS 1-WP-77-5*, Water Pollution Control Directorate, Environmental Protection Service, Fisheries and Environment Canada, 1977, pp. 5958–5960.
4. UNEP, *Environmental Aspects of the Metal Finishing Industry: A Technical Guide*, United Nations Environment Programme, Industry and Environment Office, Paris, 1989, pp. 19–39, 53–57.
5. C.T. Tien and C.P. Huang, in J.P. Vernct ed., *Heavy Metals in the Environment*, Elsevier, Amsterdam, 1991, pp. 295–311.
6. W.A.P. Black and R.L. Mitchell, *J. Marine Biol. Assoc.* **30**, 575–584 (1952).
7. N. Kuyucak, in B. Volesky ed., *Biosorption of Heavy Metals*, CRC Press, Boca Raton, Fla., 1990, pp. 371–378.
8. D. Kratochvil, E. Fourest, and B. Volesky, *Biotechnol. Lett.* **17**, 777–782 (1995).
9. B. Volesky and Z.R. Holan, *Biotechnol. Prog.* **11**, 235–250 (1995).
10. A. Kapoor and T. Viraraghavan, *Biores. Technol.* **53**, 195–206 (1995).
11. H. Mann, B. Volesky ed., *Biosorption of Heavy Metals*, CRC Press, Boca Raton, Fla., 1990, pp. 93–137.
12. N. Kuyucak and B. Volesky, in B. Volesky ed., *Biosorption of Heavy Metals*, CRC Press, Boca Raton, Fla., 1990, pp. 173–198.
13. Z.R. Holan and B. Volesky, *Biotechnol. Bioeng. 43*, 1001–1009 (1994).
14. J. Ferguson and B. Bubela, *Chem. Geol.* **13**, 163–186 (1974).
15. Y.S. Ho, D.A.J. Wase, and C.F. Forster, *Wat. Res.* **29**, 1327–1332 (1995).
16. X.H. Chen, T. Gosset, and D.R. Thevenot, *Wat. Res.* **24**, 1463–1471 (1990).
17. B. Volesky, in B. Volesky ed., *Biosorption of Heavy Metals*, CRC Press, Boca Raton, Fla., 1990, pp. 3–6.
18. B. Volesky and H.A. May-Phillips, *Appl. Microbiol. Biotechnol.* **42**, 797–806 (1995).
19. M. Tsezos, in H.L. Ehrlich and C.L. Brierley eds., *Microbial Mineral Recovery*, McGraw-Hill, New York, 1990, pp. 325–340.
20. Z.R. Holan, B. Volesky, and I. Prasetyo, *Biotechnol. Bioeng.* **41**, 819–825 (1993).
21. I. Prasetyo, M. Eng. Thesis, McGill University, Montreal, Canada, 1992.
22. N. Kuyucak and B. Volesky, *Biorecovery* **1**, 189–204 (1989).
23. A. Leusch, Z.R. Holan, and B. Volesky, *J. Chem. Tech. Biotechnol.* **62**, 279–288 (1995).
24. G.J. Ramelow, D. Fralick, and Y. Zhao, *Microbios* **72**, 81–93 (1992).
25. A. Haug and O. Smidsrod, *Acta Chem. Scand.* **24**, 843–854 (1970).
26. P. Weppen and A. Hornburg, *Thermochim. Acta* **269/270**, 393–404 (1995).
27. J.M. Tobin, D.G. Cooper and R.J. Neufeld, *Biotechnol. Bioeng.* **30**, 882–886 (1987).
28. J.A. Davis and J.O. Leckie, *Environ. Sci. Technol.* **12**, 1309–1315 (1978).
29. B. Nowack, J. Luetzenkirchen, P. Behra, and L. Sigg, *Environ. Sci. Technol.* **30**, 2397–2405 (1996).
30. M.A. Glaus, W. Hummel, and L.R. Van Loon, *Environ. Sci. Technol.* **29**, 2150–2153 (1995).
31. B. Greene, M. Hosea, R. McPherson, M. Henzl, M.D. Alexander, and D.W. Darnall, *Environ. Sci. Technol.* **20**, 627–632 (1986).
32. D.W. Darnall, B. Greene, M.T. Henzl, J.M. Hosea, R.A. McPherson, J. Sneddon, and M.D. Alexander, *Environ. Sci. Technol.* **20**, 206–208 (1986).

33. B. Greene, R. McPherson, and D. Darnall, in J.W. Patterson and R. Pasino eds., *Metals Speciation, Separation, and Recovery*, Lewis, Chelsea, Mich., 1987, pp. 315–338.
34. N. Kuyucak and B. Volesky, *Biotechnol. Bioeng.* **33**, 809–814 (1989).
35. M. Tsezos and B. Volesky, *Biotechnol. Bioeng.* **23**, 583–604 (1981).
36. B. Greene, M.T. Henzl, J.M. Hosea, and D.W. Darnall, *Biotechnol. Bioeng.* **28**, 764 (1986).
37. J.M. Tobin, D.G. Cooper, and R.J. Neufeld, *Appl. Envir. Microbiol.* **47**, 821–824 (1984).
38. I. Aldor, E. Fourest, and B. Volesky, *Can. J. Chem. Eng.* **73**, 516–522 (1995).
39. N. Kuyucak and B. Volesky, *Biotechnol. Bioeng.* **33**, 815–822 (1989).
40. C.F.J. Baes and R.E. Mesmer, *The Hydrolysis of Cations*, Wiley New York, 1976, pp. 397–419.
41. F.M.M. Morel, *Principles of Aquatic Chemistry*, Wiley, New York, 1983, pp. 237–266.
42. W. Stumm and J.J. Morgan, *Aquatic Chemistry*, Wiley, New York, 1970, pp. 238–299.
43. C.F.J. Baes and R.E. Mesmer, *The Hydrolysis of Cations*, Wiley, New York, 1976, pp. 1–7.
44. G.K. Pagenkopf, *Introduction to Natural Water Chemistry*, Marcel Dekker, New York, 1978, pp. 161–167, 214–216, 220–230.
45. Y.E. Collins and G. Stotzky, *Appl. Environ. Microbiol.* **58**, 1592–1600 (1992).
46. W. Stumm and J.J. Morgan, *Aquatic Chemistry*, Wiley, New York, 1970, pp. 445–513.
47. J.M. Tobin, D.G. Cooper, and R.J. Neufeld, *Biotechnol. Bioeng.* **31**, 282–286 (1988).
48. K.H. Chong and B. Volesky, *Biotechnol. Bioeng.* **47**, 451–460 (1995).
49. B. Volesky, in B. Volesky ed., *Biosorption of Heavy Metals*, CRC Press, Boca Raton, Fla., 1990, pp. 7–43.
50. D. Myers, *Surfaces, Interfaces, Colloids: Principles and Applications*, VCH, Weinheim, Germany, 1991, pp. 39–67.
51. J.C. Westall, in W. Stumm ed., *Aquatic Surface Chemistry*, Wiley, New York, 1987, pp. 3–32.
52. I.T. Mayers and T.J. Beveridge, *Can. J. Microbiol.* **35**, 764–770 (1989).
53. L.E. Macaskie, R.M. Empson, A.K. Cheetham, C.P. Grey, and A.J. Skarnulis, *Science* **257**, 782–784 (1992).
54. L.E. Macaskie, A.C.R. Dean, A.K. Cheetham, R.J.B. Jakeman, and A.J. Skarnulis, *J. Gen. Microbiol.* **133**, 539–544 (1987).
55. R.S. Cahn and O.L. Dermer, *Introduction to Chemical Nomenclature*, Butterworths, London, 1979, pp. 17–18.
56. W.C. Fernelius, in *Chemical Nomenclature*, American Chemical Society, Washington, D.C., 1953, pp. 9–15.
57. J.M. Smith, *Chemical Engineering Kinetics*, McGraw-Hill, New York, 1981, pp. 310–322.
58. J.B. Russell, *General Chemistry*, McGraw-Hill, New York, 1980, pp. 314–316, 340–341.
59. N. Kuyucak and B. Volesky, *Biotechnol. Bioeng.* **33**, 823–831 (1989).
60. R.H. Crist, J.R. Martin, P.W. Guptill, J.M. Eslinger, and D.R. Crist, *Environ. Sci. Technol.* **24**, 337–342 (1990).
61. M.E. Treen-Sears, B. Volesky, and R.J. Neufeld, *Biotechnol. Bioeng.* **26**, 1323–1329 (1984).
62. R.H. Crist, K. Oberholser, N. Shank, and M. Nguyen, *Environ. Sci. Technol.* **15**, 1212–1217 (1981).
63. R.H. Crist, K. Oberholser, D. Schwartz, J. Marzoff, D. Ryder, and D.R. Crist, *Environ. Sci. Technol.* **22**, 755–760 (1988).
64. R.H. Crist, K. Oberholser, J. McGarrity, D.R. Crist, J.K. Johnson, and J.M. Brittsan, *Environ. Sci. Technol.* **26**, 496–502 (1992).
65. D. Kaplan, D. Christiaen, and S. Arad, *Appl. Environ. Microbiol.* **53**, 2953–2956 (1987).
66. J.L. Gardea-Torresdey, M.K. Becker-Hapak, J.M. Hosea, and D.W. Darnall, *Environ. Sci. Technol.* **24**, 1372–1378 (1990).
67. E. Fourest and B. Volesky, *Environ. Sci. Technol.* **30**, 277–282 (1996).
68. V.E. Bower and R.G. Bates, in L. Meites ed., *Handbook of Analytical Chemistry*, McGraw-Hill, New York, 1963, pp. 1.20–1.27.
69. J. Buffle, *Complexation Reactions in Aquatic Systems: An Analytical Approach*, Ellis Horwood, Chichester, U.K., 1988, pp. 59–65, 74, 284–286.
70. T.J. Beveridge and R.G.E. Murray, *J. Bacteriol.* 876–887 (1980).
71. T.J. Beveridge, in H.L. Ehrlich and D.S. Holmes eds., *Biotechnology and Bioengineering Symposium No. 16: Biotechnology for the Mining, Metal-Refining, and Fossil Fuel Processing Industries*, Wiley, New York, 1986, pp. 127–140.
72. M. Tsezos and B. Volesky, *Biotechnol. Bioeng.* **24**, 955–969 (1982).
73. M. Tsezos and B. Volesky, *Bioeng. Biotechnol.* **24**, 385–401 (1982).
74. N. Kuyucak and B. Volesky, *Biorecovery* **1**, 219–235 (1989).
75. A. Haug, *Acta Chem. Scand.* **15**, 1794–1795 (1961).
76. A. Haug and O. Smidsrod, *Acta Chem. Scand.* **19**, 341–351 (1965).
77. R.G. Schweiger, *Kolloid Z.* **196**, 47–53 (1964).
78. A.J. Paskins-Hurlburt, Y. Tanaka, and S.C. Skoryna, *Bot. Mar.* **19**, 327–328 (1976).
79. J.M. Tobin, Ph.D. Thesis, McGill University, Montreal, Canada, 1986.
80. S.V. Avery and J.M. Tobin, *Appl. Environ. Microbiol.* **59**, 2851–2856 (1993).
81. L.J. Evans, *Environ. Sci. Technol.* **23**, 1046–1056 (1989).
82. A. Leusch, Z.R. Holan, and B. Volesky, *Appl. Biochem. Biotechnol.* **61**, 231–249 (1996).
83. R.J.P. Williams and J.D. Hale, *Structure and Bonding* **1**, 249–281 (1966).
84. Y. Marcus and A.S. Kertes, *Ion Exchange and Solvent Extraction of Metal Complexes*, Wiley London, 1969, pp. 10–33.
85. Y. Marcus and A.S. Kertes, *Ion Exchange and Solvent Extraction of Metal Complexes*, Wiley London, 1969, pp. 277–298.
86. M.K. Jain and R.C. Wagner, *Introduction to Biological Membranes*, Wiley, New York, 1980, pp. 196–201.
87. D.R. Turner, M. Whitfield, and A.G. Dickson, *Geochim. Cosmochim. Acta* **45**, 855–881 (1981).
88. J.A. Dean, *Lange's Handbook of Chemistry*, 13th ed., McGraw-Hill, New York, 1985, pp. 3.11–3.12.
89. R.G. Pearson, *J. Am. Chem. Soc.* **85**, 3533–3539 (1963).
90. R.J.P. Williams, *Proc. Chem. Soc.* 20–21 (1960).
91. C.S.G. Phillips and R.J.P. Williams, *Inorganic Chemistry: Principles and Non-Metals*, Oxford University Press, New York, 1965, pp. 142–164.

92. S. Ahrland, J. Chatt, and N.R. Davis, *Quart. Rev. Chem.* **12**, 265–276 (1958).
93. C.F. Bell, *Principles and Applications of Metal Chelation*, Clarendon, Oxford, UK 1977, pp. 30–50.
94. L. Pauling, *Nature of the Chemical Bond*, Cornell University Press, Ithaca, N.Y., 1967, pp. 55–73.
95. H.T. Evans, D.R. Lide ed., *CRC Handbook of Chemistry and Physics*, CRC Press, Boca Raton, Fla., 1993, pp. 12.8–12.9.
96. E. Nieboer and W.A.E. McBryde, *Can. J. Chem.* **51**, 2512–2524 (1973).
97. E. Nieboer and D.H.S. Richardson, *Environ. Poll.* **1B**, 11–13 (1980).
98. M. Tsezos and A.A. Deutschmann, *J. Chem. Tech. Biotechnol.* **48**, 29–39 (1990).
99. H. Freundlich, *Z. Physik. Chem.* **57**, 385–470 (1907).
100. W. Stumm, *Chemistry of the Solid–Water Interface*. Wiley, New York, 1992, pp. 87–97.
101. W.J.J. Weber, in F.L. Slejko ed., *Adsorption Technology: A Step by Step Approach to Process Evaluation and Application*, Marcel Dekker, New York, 1985, pp. 9–17.
102. I. Langmuir, *J. Am. Chem. Soc.* **40**, 1361–1403 (1918).
103. H.B. Xue and L. Sigg, *Wat. Res.* **24**, 1129–1136 (1990).
104. C. Huang, C.P. Huang, and A.L. Morehart, *Wat. Res.* **25**, 1365–1375 (1991).
105. J. Votruba, in *Proceedings of IUMS Congress*, Prague, 1994, p. 283.
106. M. Gonzales-Davila, J.M. Santana-Casiano, J. Perez-Pena, and F.J. Millero, *Environ. Sci. Technol.* **29**, 288–301 (1995).
107. J.-P. Huang, C.P. Huang, and A.L. Morehart, in J.P. Vernet ed., *Heavy Metals in the Environment*, Elsevier Science, Amsterdam, 1991, pp. 329–349.
108. L.K. Jang, D. Nguyen, and G.G. Geesey, *Wat. Res.* **29**, 315–321 (1995).
109. H. Seki, A. Suzuki, and I. Kashiki, *J. Coll. Int. Sci.* **134**, 59–65 (1990).
110. J.-W. Yu and I. Neretnieks, *Ind. Eng. Chem. Res.* **29**, 220–231 (1990).
111. E.M. Trujillo, T.H. Jeffers, C. Ferguson, and H.Q. Stevenson, *Environ. Sci. Technol.* **25**, 1559–1565 (1991).
112. L.K. Jang, D. Nguyen, and G.G. Geesey, *Wat. Res.* **29**, 307–313 (1995).
113. D.K. Nordstrom and J.W. Ball, in C.J.M. Kramer and J.C. Duinker eds., *Complexation of Trace Metals in Natural Waters*, Nijhof/Junk, The Hague, The Netherlands, 1984, pp. 149–164.
114. R.H. Crist, J.R. Martin, D. Carr, J.R. Watson, H.J. Clarke, and D.R. Crist, *Environ. Sci. Technol.* **28**, 1859–1866 (1994).
115. Y. Marcus and A.S. Kertes, *Ion Exchange and Solvent Extraction of Metal Complexes*, Wiley London, 1969, pp. 345–351.
116. F. Helfferich, *Ion Exchange*, McGraw-Hill, New York, 1962, pp. 72–94.
117. J. Buffle, *Complexation Reactions in Aquatic Systems: An Analytical Approach*, Ellis Horwood, Chichester, 1988, pp. 195–303.
118. J.C. Westall, J.D. Jones, G.D. Turner, and J.M. Zachara, *Environ. Sci. Technol.* **29**, 951–959 (1995).
119. S. Schiewer and B. Volesky, *Environ. Sci. Technol.* **31**, 1863–1871 (1997).
120. S. Schiewer and B. Volesky, *Environ. Sci. Technol.* **31**, 2478–2485 (1997).
121. S. Schiewer, Ph.D. Thesis, McGill University, Montreal, Canada, 1996.
122. W.D. Schecher, *MINEOL+: A Chemical Equilibrium Program for Personal Computers, Users Manual Version 2.22.* Environmental Research Software, Inc., Hallowell, Maine, 1991.
123. F.G. Donnan, *Z. Elektroch.* **17**, 572–581 (1911).
124. J.A. Marinsky, in W. Stumm ed., *Aquatic Surface Chemistry*, Wiley, New York, 1987, pp. 49–81.
125. J.A. Marinsky and J. Ephraim, *Environ. Sci. Technol.* **20**, 349–354 (1986).
126. J. Ephraim, S. Alegret, A. Mathuthu, M. Bicking, R.L. Malcolm, and J.A. Marinsky, *Environ. Sci. Technol.* **20**, 354–366 (1986).
127. J.A. Marinsky, S. Gupta, and P. Schindler, *J. Coll. Int. Sci.* **89**, 412–426 (1982).
128. F.G. Lin and J.A. Marinsky, *React. Polymers* **19**, 27–45 (1993).
129. L.K. Jang, N. Harpt, D. Grasmick, L.N. Vuong, and G. Geesey, *J. Phys. Chem.* **94**, 482–488 (1990).
130. A. Katchalsky, R.E. Cooper, J. Upadhyay, and A. Wassermann, *J. Am. Chem. Soc.* **83**, 5198–5204 (1961).
131. S.E. Cabaniss and M.S. Shuman, *Geochim. Cosmochim. Acta* **52**, 185–193 (1988).
132. B.M. Bartschat, S.E. Cabaniss, and F.M.M. Morel, *Environ. Sci. Technol.* **26**, 284–294 (1992).
133. H.-B. Xue, W. Stumm, and L. Sigg, *Wat. Res.* **22**, 917–926 (1988).
134. E. Tipping, C.A. Backes, and M.A. Hurley, *Wat. Res.* **22**, 597–611 (1988).
135. E. Tipping, *Environ. Sci. Technol.* **27**, 520–529 (1993).
136. J.C.M. de Wit, W.H. van Riemsdijk, and L.K. Koopal, *Environ. Sci. Technol.* **27**, 2005–2014 (1993).
137. M.M. Figueira, B. Volesky, and V.S.T. Ciminelli, *Biotechnol. Bioeng.* **54**, 344–350 (1997).
138. S. Schiewer and B. Volesky, *Environ. Sci. Technol.* **29**, 3049–3058 (1995).
139. T. Vermeulen, G. Klein, and N.K. Heister, in R.H. Perry and C.H. Chilton eds., *Chemical Engineers' Handbook*, 5th ed., McGraw-Hill, New York, 1973, Ch. 16.
140. M. Shallcross, C.C. Herrmann, and B.J. McCoy, *Chem. Eng. Sci.* **43**, 279–288 (1988).
141. W.A. Selke, in J. Schubert and F.C. Nachod eds., *Ion Exchange Technology*, Academic Press, New York, 1956, pp. 53–62.
142. A.S. Michaels, *Int. Eng. Chem.* **44**, 1922–1929 (1952).
143. L. Liberti, in L. Liberti and G. Helfferich eds., *Mass Transfer and Kinetics of Ion Exchange. NATO ASI Series E. Applied Sciences No. 71*, Martinus Nijhoff, The Hague, The Netherlands, 1983, pp. 181–206.
144. F. Helfferich, *Ion Exchange*, McGraw-Hill, New York, 1962, pp. 299–319.
145. J.M. Coulson and J.F. Richardson, *Chemical Engineering*, 4th ed., Pergamon Press, New York, 1991, pp. 774–775.
146. H.K.S. Tan and I.H. Spinner, *Can. J. Chem. Eng.* **72**, 330–341 (1994).
147. C.A.J. Fletcher, *Computational Galerkin Methods*, Springer-Verlag, New York, 1984, pp. 246–276.
148. D. Kratochvil, B. Volesky, and G. Demopoulos, *Wat. Res.* **31**, 2327–2339 (1997).

BIOTRANSFORMATIONS, ENGINEERING ASPECTS

UDO KRAGL
ANDREAS LIESE
Forschungszentrum Jülich
Jülich, Germany

KEY WORDS

Chiral diol
Enzyme
GDP-mannose
Kinetics
Product inhibition
Reactor

OUTLINE

INTRODUCTION

The use of biotransformations in food, pharmaceutical, and chemical industries has increased steadily since the early 1970s. Applications range from synthesis on the milligram scale for precious oligosaccharides, to the multiton scale for bulk chemicals using either whole cells or isolated enzymes. This process has been documented in numerous reviews and textbooks (1–5). When new processes are being developed, the special properties of the biocatalyst must be considered, as well as the basic elements of chemical engineering. There are strong interactions among the individual aspects, which may be considered for process optimization. These are summarized in Figure 1.

The aim of this article is to outline the basic strategy for the development of a biocatalytic step. For some selected examples the beneficial role of (bio)chemical engineering is demonstrated. A more detailed description can be found in textbooks of (bio)chemical engineering (1,3, 6–8).

WHOLE CELLS OR ISOLATED ENZYMES?

Whole microorganisms or their components, enzymes, may be used for biotransformations. Both approaches have specific advantages and requirements, as summarized in Table 1. Commonly, the word *biotransformation* is used to describe the transformation of a substrate or precursor into a certain product in only a few steps. Fermentation processes, for example, production of amino acids from glucose and ammonia or production of (recombinant) proteins, are normally not regarded as biotransformation. Litera-

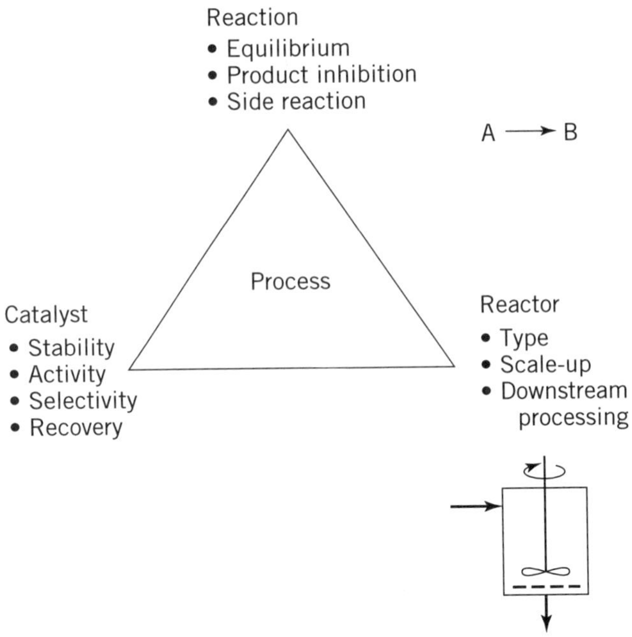

Figure 1. Elements of process development.

Table 1. Comparison of Biotransformations Using Whole Cells vs. Isolated Enzymes

Whole cells	Isolated enzymes
Large amounts of catalyst easy available	Additional step for obtaining the catalyst required
No stability problem when growing cells are used	Catalyst stability strongly influenced by reaction conditions
Limited volumetric catalyst activity	High volumetric catalyst activity possible
Multistep processes	Multistep processes difficult
Cofactor recycling solved	Cofactor recycling so far only for NADH and ATP
Side reactions may occur	Few to no side reactions
Toxicity of substrate/product may be complex	Enzyme stability may be influenced by substrate or product, but can be enhanced by choosing the appropriate reactor
Oxygen supply may control scale-up	Nonnatural substrates and reaction conditions possible
Normally no cosolvent, possibly leading to low concentrations of hydrophobic compounds	Cosolvents possible, but often reducing enzyme stability
Transport in and out of the cell may be limiting	Mass transport limitations only with immobilized enzymes

ture describing those processes can be found elsewhere in this volume or in external references (2,9–13).

Examples of industrial biotransformations using whole cells or isolated enzymes are given in Table 2. Some special information about process characteristics is given as well. When isolated enzymes are used for biotransformations, the purity must only be such that no side reactions occur. There is no need to purify the enzyme to homogeneity, as is required for biochemical characterization. On the contrary, many enzymes show reduced stability with increased purity. In some cases, the addition of inert proteins, such as bovine serum albumin, may be used as a simple way to increase the stability of the production enzyme. In others, enzyme stability may be increased by immobilization on a support. Covalent attachment fixes the conformation of the enzyme often limiting denaturation. However, the increase of stability must be balanced against the losses during the immobilization step, as well as the additional work and material costs. A method introduced quite recently for the stabilization of enzymes and their immobilization is the formation of cross-linked enzyme crystals (CLECs) (22). Reactors for immobilized enzymes are discussed elsewhere in this volume (see ENZYME, IMMOBILIZED, REACTOR). However, the engineering approach, as outlined in the following sections is the same for all forms of biocatalysts. Specific requirements and properties such as oxygen delivery, mass transport limitations, or particle size must be considered for individual cases.

CLASSIFICATION OF DIFFERENT REACTORS

There are three basic types of reactors, which have several variations (Fig. 2):

1. Stirred-tank reactors (STRs) or batch reactors
2. Continuously operated stirred-tank reactors (CSTRs)
3. Plug-flow reactors (PRFs)

Unlike the stirred-tank reactor, which is operated batchwise, the latter two are operated continuously. By knowing the main characteristics of these fundamental reactors and some of their variations, for example the cascade reactor of Figure 2, it is possible to choose the appropriate reactor for a specific application. This is especially important when dealing with a kinetically or thermodynamically limited system, as demonstrated later. In the following analyses, only the basic terms are explained. For more details the reader is referred to textbooks (6–8).

The stirred-tank reactor is operated in an instationary way (Fig. 2a). Assuming ideal mixing characteristics, the instantaneous concentration of every species is the same in every volume element. With advancing conversion the substrate concentration is decreased and the product concentration is increased. A steady state is reached only at equilibrium conversion.

In the plug-flow reactor the product concentration increases slowly over the length of the reactor (Fig. 2b). Therefore, the average reaction velocity is faster than in the continuously operated stirred-tank reactor. In each single-volume element in the reactor, the concentration is constant in the steady state. In other words, the dimension of time is exchanged with the dimension of place in comparison to the stirred-tank reactor.

The continuously operated stirred-tank reactor works under product outflow conditions, meaning that the concentrations in every volume element are the same as those at the outlet of the reactor (Fig. 2c). If the steady state is

Table 2. Examples of Industrial Biotransformations

Product	Substrate	Catalyst	Comments	References
Acrylamide	Acrylnitrile, H_2O	Nitril hydratase in whole cells of *Rhodococcus rhodochrous*	Very high substrate concentration	14
Chiral hydrophobic alcohol	Ketone	Whole cells of *Zygosaccharomyces rouxil*	XAD-resin as substrate reservoir	15
β-Tetralol derivative	β-Tetralone derivative	Whole cells of *Trichosporon capitatum*		16
Trimegestone	Corresponding ketone	Baker's yeast	In water	17
Ketodeoxynonulosonic acid	Mannose, pyruvate	Neuraminic acid aldolase	Continuously operated enzyme membrane reactor	18
Fructose	Glucose	Fructose glucose isomerase	Immobilized thermostable enzyme	4
Aspartame	L-phenylalanine-OMe,*Z*-L-asparaginic acid	Thermolysine	Precipitation of the product	19
(*S*)-Phenylethylamine	Racemic phenylethyl amine	Lipase from *Pseudomonas* spec.	Activation of enzyme by lyophilization with fatty acids	20
L-*tert*-Leucine	2-Oxo-3,3-dimethylbuturic acid, NH_4COOH	Leucine dehydrogenase, formate dehydrogenase	NADH regeneration	21

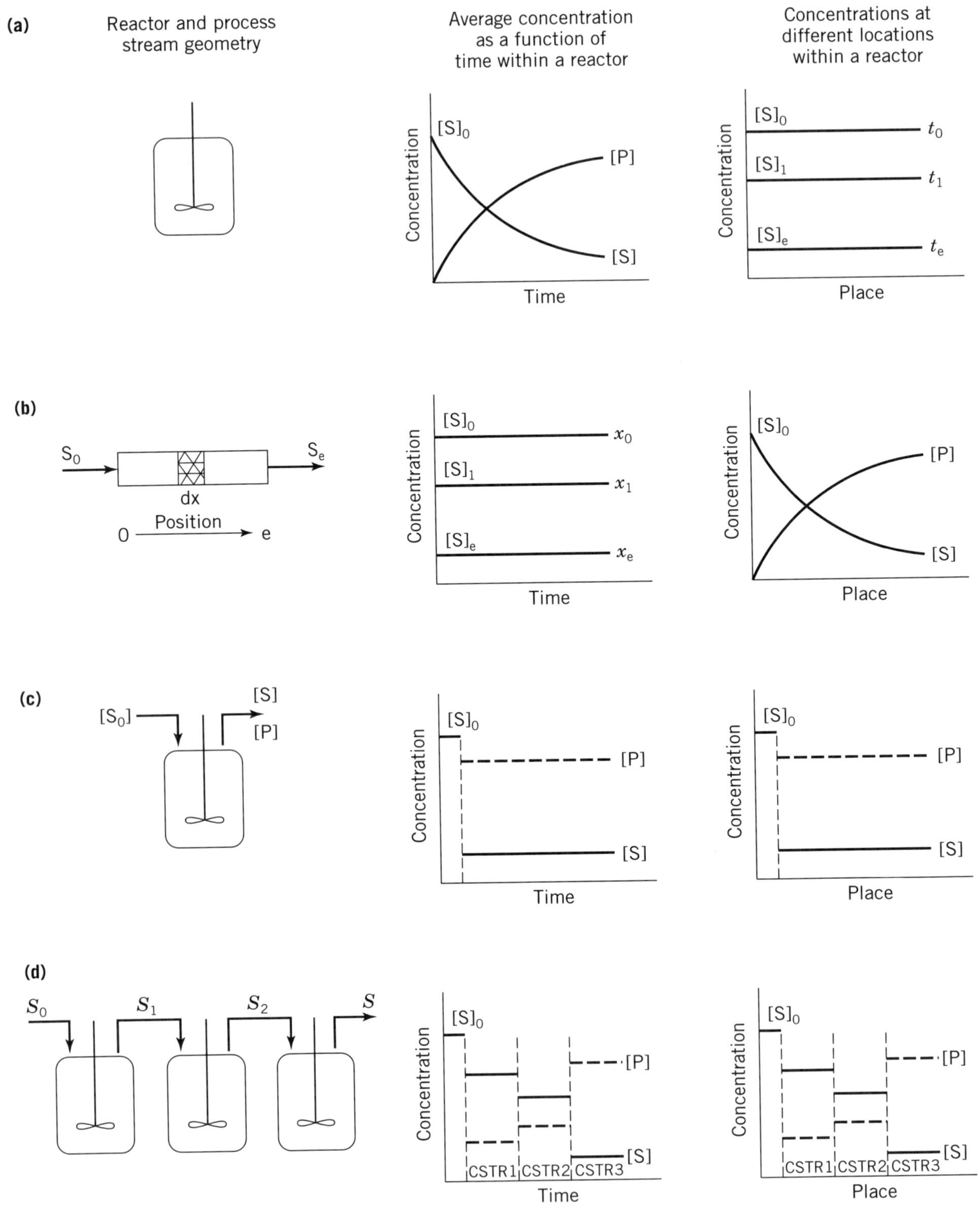

Figure 2. Characterization of different reactors. (**a**) Stirred-tank reactor. (**b**) plug-flow reactor. (**c**) continuously operated stirred-tank reactor. (**d**) Cascade.

reached, the concentrations are independent of time and place. The conversion is controlled by the catalyst concentration and the residence time, τ:

$$\tau = \frac{V}{F}$$

where $\tau[h]$ is the residence time, $V[L]$ is the total reactor volume, and F is the $[L/h]$ substrate feed rate.

One very common variation of the CSTR is the cascade of n CSTRs (Fig. 2d). With an increasing number, n, of reaction vessels, the cascade approximates the plug-flow reactor (23). The product concentration is increased stepwise from vessel to vessel.

Mass Balances

The performance of the different reactor types concerning the reaction can be simulated mathematically. This is also the verification of the kinetic model of the reaction because it should describe the course of the concentration for each compound with only a small error. The main part of the simulation model is the coupled system of differential equations of first order, which are the mass balances of all substrates and products. The change in the concentration of one compound in time and in a volume element (called the accumulation) is the sum of convection, reaction, and diffusion (equation 1).

$$\text{accumulation} = \text{convection} + \text{reaction} + \text{diffusion} \quad (1)$$

The convection term describes the change of the concentration of one compound in the reactor as the difference of the influx into the reactor and the efflux. The reaction term describes, by use of the kinetic model, the change of the concentration of one compound as result of the reaction. The reaction velocity, V_s, is the sum of the individual reaction velocities describing consumption of a substrate or formation of a product. Diffusion occurs only in the case where nonideal mixing is stated.

With regard to the chosen reactor type the mass balance can be simplified by assuming ideal mixing or ideal plug-flow behavior.

Stirred Tank Reactor

Diffusion = 0	At a defined time the concentrations are the same in every volume element.
Convection = 0	There is no influx or efflux of substrates or products to a single volume element in time.

The mass balance is simplified to

$$-\frac{d[\mathrm{S}]}{dt} = v_s \twoheadrightarrow t = [\mathrm{S}]_0 \int_0^{\text{desired conversion}} 1/v\, dx \quad (2)$$

Continuously Operated Stirred-Tank Reactor

Diffusion = 0	See previous.
Accumulation = 0	In the steady state the concentrations will not change with time.

The mass balance is simplified to:

$$0 = -\frac{d[\mathrm{S}]}{dt} = \frac{[\mathrm{S}]_0 - [\mathrm{S}]}{\tau} + v_s \quad (3)$$

Plug Flow Reactor

Diffusion = 0	Diffusion is neglected in an ideal plug-flow reactor.
Accumulation = 0	In the steady state the concentrations will not change with time.

The mass balance is the same as for the stirred-tank reactor. In the integrated form the time, t, has to be replaced by the residence time, τ.

Kinetics

The fundamental equation (equation 4) for the description of enzyme kinetics was derived by Michaelis and Menten (24).

$$v = [\mathrm{E}]\frac{A_{\max} \cdot [\mathrm{A}]}{K_m + [\mathrm{A}]} \quad (4)$$

The enzyme activity is expressed as a function of the substrate concentration for an irreversible one-substrate reaction without inhibition. As this is rarely found in nature, equation 5 gives the rate for a reversible reaction A ↔ B. In this form the product inhibition is considered as a competition of A and B for the active sites, as expressed by the two K_m values. This form might be extended to cover substrate surplus inhibition as well as multisubstrate reactions by adding additional terms to the numerator and denumerator (6,25–27).

$$v = [\mathrm{E}] \cdot \frac{\dfrac{A_{\max,A} \cdot [\mathrm{A}]}{K_{m,A}} - \dfrac{A_{\max,B} \cdot [\mathrm{B}]}{K_{m,B}}}{1 + \dfrac{[\mathrm{A}]}{K_{m,A}} + \dfrac{[\mathrm{B}]}{K_{m,B}}} \quad (5)$$

The determination of the kinetic parameters can be performed either by initial reaction velocity measurements or by batch experiments at different substrate concentration, such that substrate limitation is encountered. The kinetic parameters are determined by fitting the kinetic model to the experimental data by use of nonlinear regression; in the case of the batch experiments, numerical integration of the reaction velocity equations is required.

PROCESS DEVELOPMENT, STEP BY STEP

Figure 3 shows different phases and topics to be addressed during development and optimization of a biotransformation. The main focus is on the use of isolated enzymes, but for whole-cell processes a similar scheme must be followed. A more detailed description can be found in Ref. 6. In practice, some of the investigations listed here stepwise will be performed simultaneously. For example, first experiments should hint to possible problems from side reactions. Other factors may be considered as well.

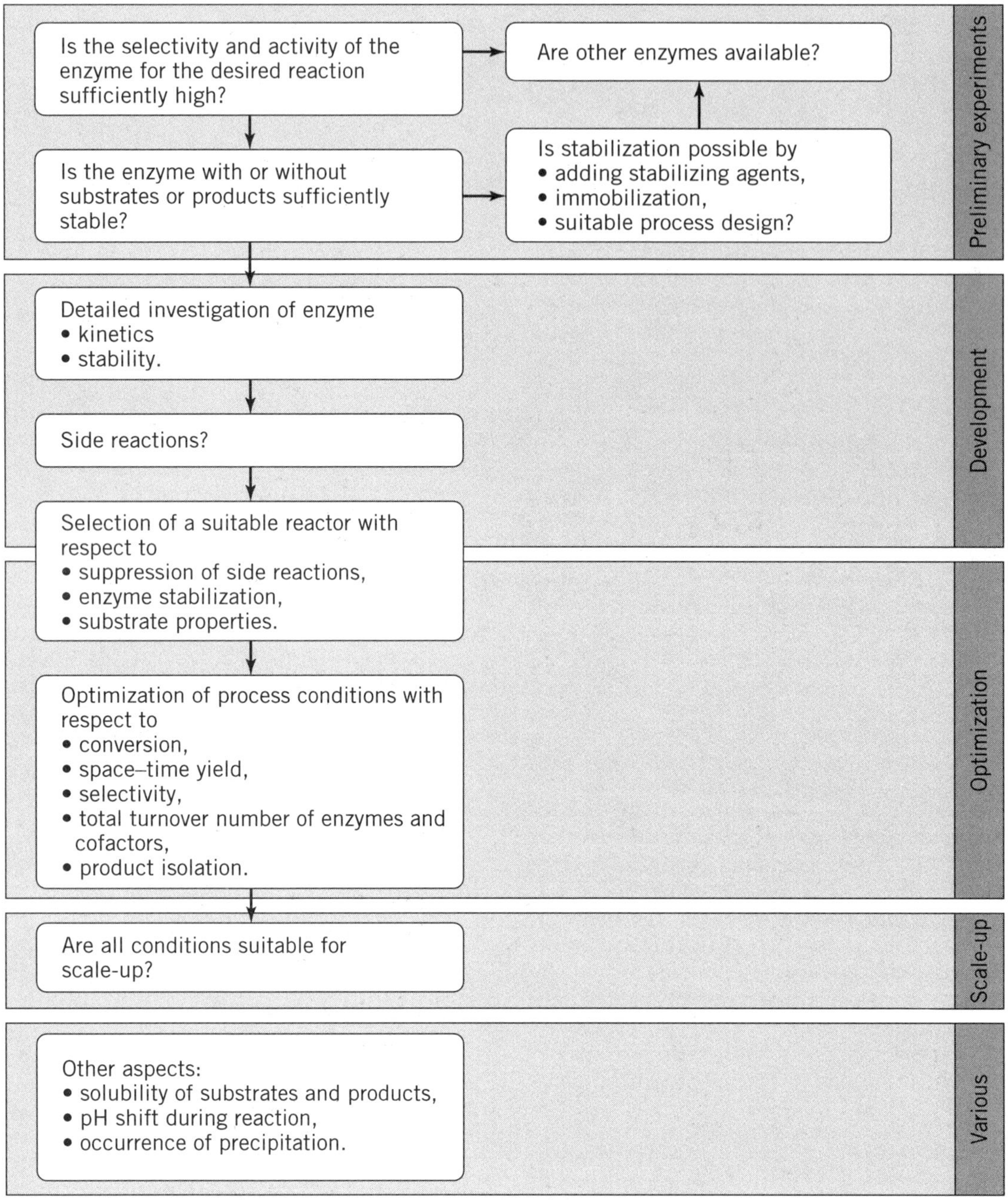

Figure 3. Process development, step by step.

- Buffer compounds may be important for enzyme stability, but must be removed during product isolation. Avoiding them may simplify product isolation and save costs.
- Any reaction exhibiting substrate surplus inhibition should not be carried out in a batch-reactor setup, since this results in longer reaction times due to the high substrate concentration at startup, which lowers the initial reaction velocity. In such a case, a continuously operated stirred-tank reactor that is operated at a high conversion in the steady state is preferred, resulting in a low substrate concentration. Another possibility is the use of a fed batch reactor, which means that the substrate is added in small portions as a function of time, also resulting in a small substrate concentration. One example of such a system is the enzymatic synthesis of L-phenylalanine by reductive amination, which can be found in the literature (25,29).
- If product inhibition occurs, either a stirred-tank reactor, a plug flow reactor, or a cascade of *n* continuously operated stirred-tank reactors should be chosen. In all these reactors the product concentration increases over time. Another possibility is the use of

a differential reactor with integrated product separation (30). These two possibilities will be discussed in more detail in the section titled "Examples".

- The biocatalyst may be recirculated in and between reactors by various means to enhance catalyst concentration. Whole cells are often concentrated by microfiltration membranes or centrifugation. Isolated enzymes are much smaller, but can be retained by an ultrafiltration membrane. This type of equipment is commercially available for the processing of volumes of a few microliters up to the cubicmeter scale. If the enzyme is immobilized on a support, its size determines the separation step. For example, penicillin amidase is immobilized on a support suitable for separation using a stainless steel mesh in a stirred tank reactor. Also, whole cells might be immobilized on a support, most often simply by adsorption, or by entrapment in alginate gel or similar material.
- The optimization process shown here and discussed later (see "Examples") is based on a kinetic model of the whole reaction system, including side reactions. This model, together with the mass balance for the reactor, can be used for simulation. By a variation of feed concentrations, catalyst concentration, and residence time, conditions can be found that result in high conversion and high space–time yield. When enzyme deactivation is known, the product-specific catalyst consumption may be calculated as well. Alternatively, an empirical optimization can be performed using a simplex method or a generic algorithm. The latter have been successfully applied to fermentation processes (31), but only tested in preliminary studies for optimization of multienzyme systems. These methods are especially powerful for optimization of processes where a direct connection between input and output values cannot be given.
- A discussion of the influence of organic solvents on the solubility of reactants and on enzyme stability can be found in literature (32–36).

Figure 4 summarizes some measures that can be taken to overcome or modulate against enzyme deactivation caused either by thermal condition or by the reaction partners. Whereas in the first case lowering of the temperature, application of stabilizers, or immobilization on a support may be helpful, the deactivation caused by the reaction partners, may be overcome with the help of the tools of reaction engineering. For the deactivation caused by substrates or products, different tools can be used. The different reactor types are discussed in "Classification of Different Reactors".

EXAMPLES

By choosing the appropriate fundamental reactor type on the basis of the individual reaction kinetics, the production of the fine chemicals as given in Table 3 is possible even in processes that exhibit a strong inhibition by one of the reaction compounds.

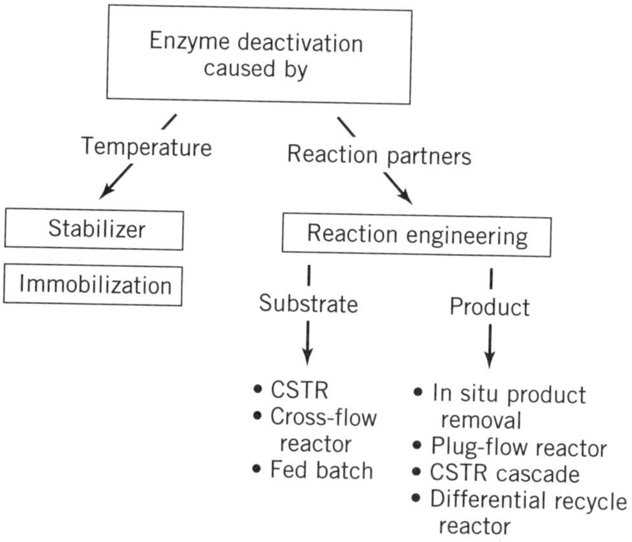

Figure 4. Tools used to overcome enzyme deactivation.

Product Inhibition

Product inhibition is the consequence of the competition of a substrate and a product for complexation in the active site of a catalyst. This kind of inhibition is not only observed with enzymes, but is also found for reactions that are catalyzed by homogeneous soluble and heterogeneous chemocatalysts (28). Whereas the reaction velocity is reduced by product bonding in the case of an irreversible reaction, the situation is more complex for reversible reactions ("principle of the microscopic reversibility"). In terms of the mathematics, product inhibition is expressed as an additional term in the denominator of the Michaelis–Menten equation (equation 6). This form of the equation describes competitive inhibition. An explicit discussion of other types of inhibition can be found in the textbooks of enzyme kinetics (6,25,26). By transformation of equation 6, equation 7 is can be obtained, which is similar in its structure to the common reaction velocity equation (equation 8), derived for chemical systems by Hougen and Watson in 1947 (28,41).

$$v = [\mathrm{E}] \cdot \frac{A_{\mathrm{max}} \cdot [\mathrm{substrate}]}{K_{\mathrm{m}}^{\mathrm{educt}} \cdot \left(1 + \dfrac{[\mathrm{product}]}{K_i^{\mathrm{product}}}\right) + [\mathrm{substrate}]} \qquad (6)$$

where v(kat/L) is reaction velocity (1 kat = 1 mol/s), [E](g/L) is enzyme concentration, [substrate](mol/L) is concentration of substrate, A_{max}(kat/g, U/mg) is maximal mass specific activity of the reaction, $K_{\mathrm{m}}^{\mathrm{substrate}}$(mol/L) is Michaelis–Menten constant of substrate, and K_i^{product}(mol/L) is competitive product inhibition.

$$v = [\mathrm{E}] \cdot \frac{\dfrac{A_{\mathrm{max}}}{K_{\mathrm{M}}^{\mathrm{substrate}}} \cdot [\mathrm{substrate}]}{1 + \dfrac{[\mathrm{substrate}]}{K_{\mathrm{M}}^{\mathrm{substrate}}} + \dfrac{[\mathrm{product}]}{K_i^{\mathrm{product}}}} \qquad (7)$$

$$v = \frac{(\text{kinetic term}) \cdot (\text{potential term})}{(\text{adsorption term})^n} \qquad (8)$$

Table 3. Examples for the Contribution of Biochemical Engineering

Product	Catalyst	Problems	Solution	Reference
(*S*)-1-Phenyl-1,2-ethanediol	Glycerol dehydrogenase	Product inhibition Chemical side reaction	Differential batch with continuous product extraction	30
GDP-mannose	GDP-Man pyrophosphorylase	Product inhibition Chemical complexation of reaction compounds	Cascade of continuously operated stirred-tank reactors	37
CMP-Neu5Ac	CMP-Neu5Ac synthetase	Substrate-surplus inhibition	Fed-batch reactor	38
Oxindol, chiral sulfoxide	Chloroperoxidase	Catalyst deactivation	H_2O_2-dosing with closed-loop control	39
Chiral lactone	Monooxygenase	Oxygen delivery	Bubble-free aeration using silicone tubing	40

where n is the number of centers, which are part of a velocity-limiting elementary reaction.

The decrease of the reaction velocity with increasing product concentration is described by the ratio of K_m to Ki. The relative reaction velocity is plotted as function of the conversion in Figure 5. Three different ratios of K_m to Ki are plotted, which relate to strong (curve A), moderate (curve B), and weak (curve C) product inhibition.

In 1985, Lee and Whitesides suggested, using the example of an oxidoreductase-catalyzed alcohol oxidation, that the feasibility of a process can be judged by the ratio of $K_m^{substrate}$ to $K_i^{product}$ (42):

- If $K_m^{substrate}$: $K^{product} \leq 1$, it is, in principle, possible that the reaction can proceed with practicable conversions (Fig. 5, curves B and C).
- If K_m^{educt}: $K_i^{product} > 1$, the reaction is unusable with regard to preparative syntheses (Figure 5: curve A).

This clearly is a simplified view. It is also important to note, that the kinetic parameters of a given reaction cannot be changed.

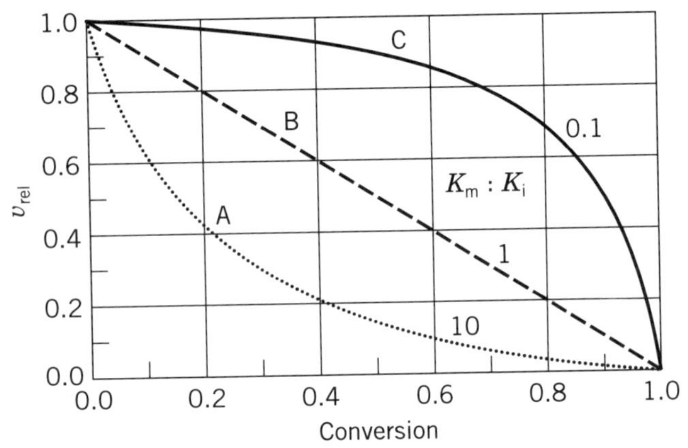

Figure 5. Influence of product inhibition on reaction velocity as function of conversion. Curve A, strong product inhibition with the ratio of Michaelis–Menten constant, K_m, and the product inhibition constant, K_i; K_m:K_i = 10. Curve B, moderate product inhibition with K_m:K_i = 1. Curve C, weak product inhibition with K_m:K_i = 0.1.

Overcoming Product Inhibition by Product Extraction

Chiral diols are important building blocks for the synthesis of chiral ligands, agrochemicals, and pharmaceutical intermediates (43–45). Using the example of the oxidative resolution of 1-phenyl-1,2-ethanediol, it is demonstrated that product inhibition can be overcome in the case of hydrophobic products. The enantioselective oxidation of the (*R*)-enantiomer to ω-hydroxyacetophenone (1-oxo-1-phenyl-2-hydroxyethane) is catalyzed by glycerol dehydrogenase (GDH) from *Enterobacter aerogenes* [EC 1.1.1.6] (46). Because GDH needs β-nicotinamide adenine dinucleotide (NAD^+) as a cofactor, a cofactor regeneration carried out by lactate dehydrogenase (LDH) [EC 1.1.1.27] was incorporated, as depicted in Figure 6. LDH reduces pyruvate to (L)-lactate enantioselectively. The cofactor regeneration must proceed faster than the racemic resolution in order not to become the limiting step. The remaining (*S*)-enantiomer of the diol is the desired product for further synthesis.

Detailed investigation of the reaction conditions and kinetics revealed strong product inhibition by the hydroxyketone, as well as by the NADH (30). This resulted in a ratio of $K_m^{substrate}$: $K_i^{product}$ = 43.2 for the ω-hydroxyacetophenone. The inhibition by NADH was not taken into account, because NADH is converted in situ to NAD^+ by LDH. Utilizing the above-determined criteria, the reaction as investigated here would never proceed efficiently. Additionally, it was observed that the hydroxyketone reacted in a chemical side reaction with NAD^+. Thus, ω-hydroxyacetophenone not only reduced the GDH activity due to its inhibition, it furthermore reduced the available NAD^+ concentration. Therefore, it is very important that the ω-hydroxyacetophenone concentration is kept as low as possible. One possibility is an integrated extraction of the inhibiting compound. Best results were obtained using hexane as the extracting solvent. In contrast to phenylethanediol, which was not extracted, up to 98% of the hydroxyketone was extracted after eight extraction steps.

The flow sheet of the two-phase membrane reactor is shown in Figure 7. The reactor consists of three different loops. The membrane reactor (loop I) integrates the stirred-reaction vessel and the membrane module (48,49). A hydrophilic ultrafiltration membrane (YM 10, cutoff 10,000 Da, Amicon GmbH, Witten, Germany) is used to retain the glycerol dehydrogenase and the lactate dehy-

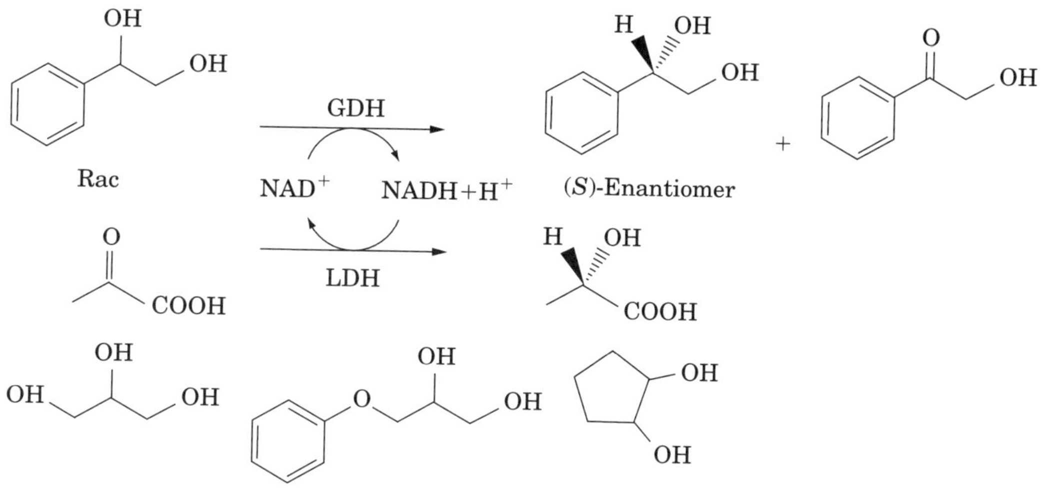

Figure 6. Enzymatic resolution of racemic 1-phenylethan-1,2-diol. GDH, glycerin dehydrogenase; LDH, lactate dehydrogenase. Relative activities: glycerol, 100%, 3-phenoxypropan-1,2-diol, 31%, 1-phenylethan-1,2-diol, 8%, *trans*-cyclopentan-1,2-diol, 3%.

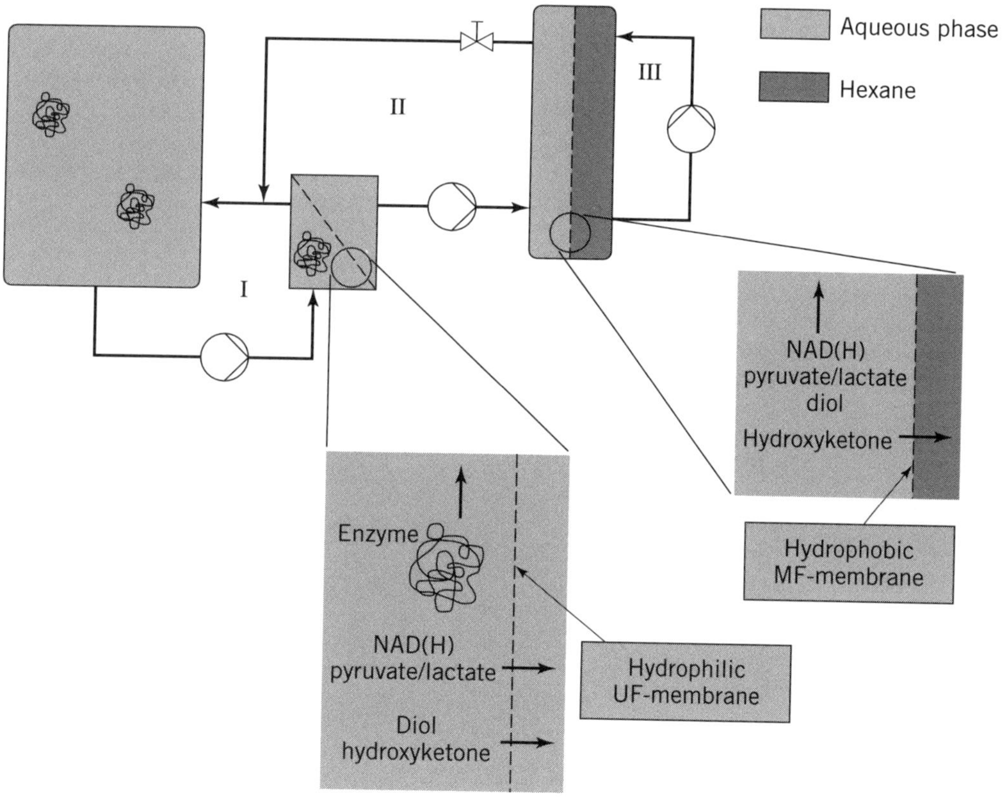

Figure 7. Flowsheet of the two-phase membrane reactor; I, membrane reactor loop; II, extraction loop; aqueous phase; III extraction loop, organic phase. For details refer to Refs 47 and 54.

drogenase. The consequence is the decoupling of the residence times of the catalysts and the reactants. Prior to the membrane reactor, an extraction module is positioned in a closed-loop arrangement (loop II). The aqueous phase is continuously extracted by means of liquid–liquid extraction using a hollow-fiber module (Liqui-Cel® 5PCM-106, Hoechst, Frankfurt, Germany) consisting of Celgard® microporous, hydrophobic hollow-fiber membranes (50). The module was circulated countercurrently by a hexane stream. The best ratio of the flow rates of the aqueous phase to the organic phase was determined to be 1:6. To reduce the amount of organic solvent used, a distillation (not shown) is incorporated in the extraction loop III. The oxidation product ω-hydroxyacetophenone is collected in the bottom of the distillation. The entire reactor behaves as a batch reactor with recirculation, meaning that only a low conversion per cycle is reached, corresponding to a small amount of oxidation product, which is directly extracted (30). The consequence is a continuous low concentration of ω-hydroxyacetophenone in the membrane reactor. This reactor represents a modification of the continuously operated bimembrane reactor that was constructed for the conversion of sparingly soluble substrates (47,51,52). In Figure 8, two experiments are compared, one with the integrated extraction and the other without. When no extraction is incorporated, an enantiomeric excess of only 58% and a conversion of 37% was reached after 170 h. The reason for this unsatisfactory result was the accumulation of the reaction product, which lowers the reaction velocity of the GDH and also reduces the concentration of the NAD^+ by the chemical side reaction. With integrated product extraction a conversion of 50% and an enantiomeric excess >99% is reached after 64 h.

This process example demonstrates that by means of reaction engineering, kinetically unfavorable reaction can still be used on a preparative scale. By application of the integrated extraction, a transition from curve A to curve C in Figure 5 is reached.

Overcoming Product Inhibition by a CSTR Cascade

D-Mannose can be found as building block of numerous glycoproteins and glycolipids, for example, in the common core structure of N-glycans (53). For the enzymatic in vitro synthesis of oligosaccharides, guanosine-5′-diphospho-α-D-mannose (GDP-Man) is used as donor for mannosyltransferases. Recombinant mannosyltransferases have been used for the synthesis of the trisaccharide Manβ1-4GlcNAcβ1-4GlcNAc (54). The enzymatic synthesis of GDP-Man starting from mannose-1-phosphate (Man-1-P) and guanosine-5′-triphosphate (GTP) using a recombinant GDP-Man pyrophosphorylase [E.C. 2.7.7.13] has been recently described (55) (Fig. 9). Both reaction products show a very strong inhibition of GDP-Man pyrophosphorylase. The inhibition by pyrophosphate can be overcome by cleaving it to phosphate using an inorganic pyrophosphatase [E.C. 3.6.11]. The product inhibition by the product GDP-Man itself can be reduced in a CSTR-cascade, as shown below. Additionally, the cofactor Mg^{2+}, required by the GDP-Man pyrophosphorylase, is complexed by the substrate GTP. Its influence on enzyme activity and the preferable use of a continuously operated reactor to establish a constant Mg^{2+} to GTP ratio, together with a complete set of kinetic data for the pyrophosphorylase, has been recently published (43). For the ratio $K_m^{\text{Man-1-P}}/K_i^{\text{GDP-MAN}}$, a value of 4.4 is obtained. According to the classification outlined in the section "The Problem of Product Inhibition", this reaction would be of only limited use. Unfortunately, the substrates and products are quite similar with respect to size, hydrophilicity, or charge, prohibiting a selective removal of the inhibiting product. However, simulations on the basis of the kinetic model and various reactor configurations revealed that a cascade of two CSTR in series reduces the required amount of enzyme by >50% (23). A two-stage cascade is a good compromise: plug-flow behavior is approximated to a large extent, and the experimental setup does not require too much effort (7). Figure 10 compares the conversion of Man-1-P as function of the product of enzyme concentration and residence time for a single CSTR and a two-stage CSTR cascade. In a two-stage cascade 98% conversion is reached for 4.5 (U/mL)min, whereas a single CSTR requires 9.0 (U/mL)min. That means that at a constant enzyme concentration the residence time can be cut in half, resulting in a doubling of the space–time yield. When the enzyme deactivation is only a function of time, by an increase of the space–time yield at

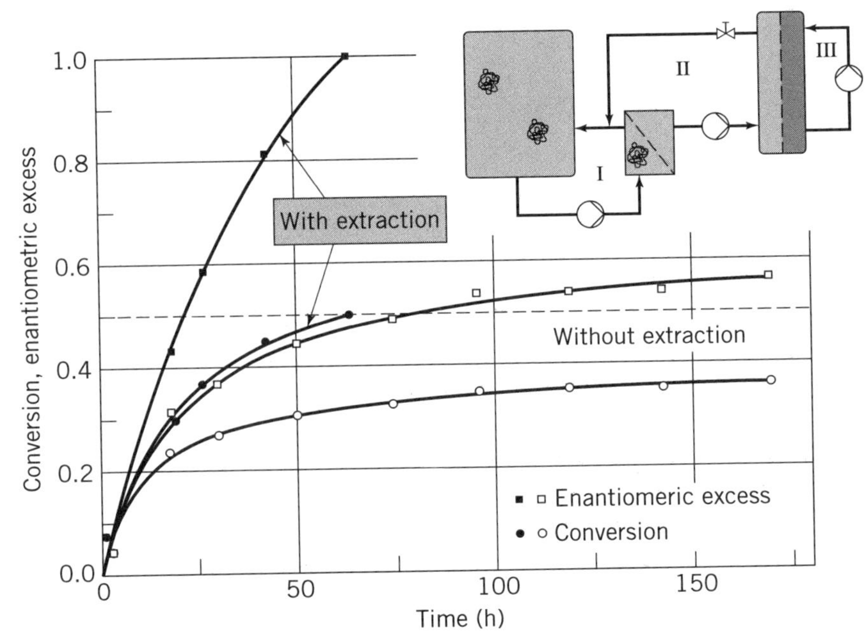

Figure 8. Increase of enantiomeric excess and conversion for the kinetic resolution of 1-phenyl-1,2-ethanediol due to product extraction. Residence time in the membrane reactor: 15 min; pressure difference: 2 bar; total volume 500 mL; 100 mM glycine/KOH buffer; pH 9; T = 25 °C; 0.054 U/mL GDH; 0.6 U/mL LDH; 200 mM ammonia; 40 mM (R,S) 1-phenyl-1,2-ethanediol; 2 mM NAD^+; 25 mM pyruvate.

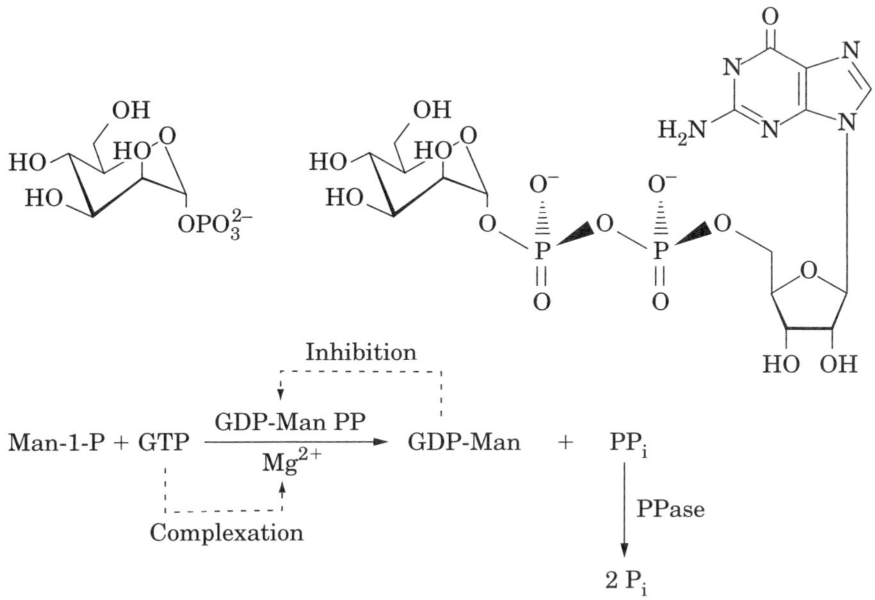

Figure 9. Enzymatic synthesis of GDP-Man starting from Man-1-phosphate and GTP; GDP-Man PP = GDP-Man pyrophosphorylase, PPase = inorganic pyrophosphate.

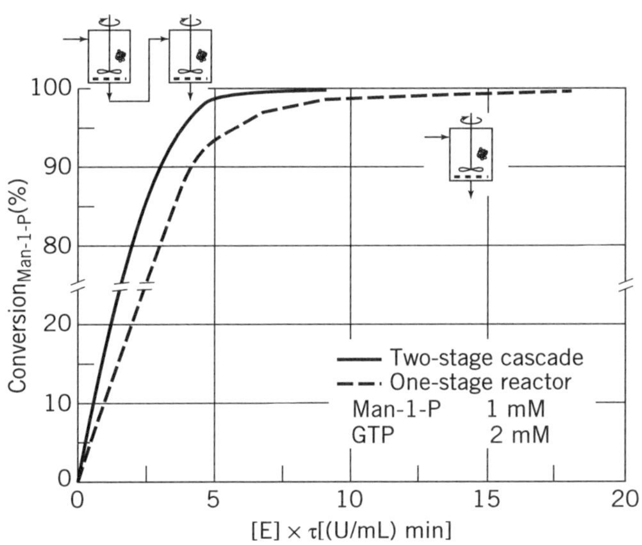

Figure 10. Simulation of GDP-Man synthesis in a one-stage CSTR (dotted line) and a two-stage CSTR cascade (solid line), Man-1-P, 1 mM; GTP, 2 mM; 25 °C; pH 8.0.

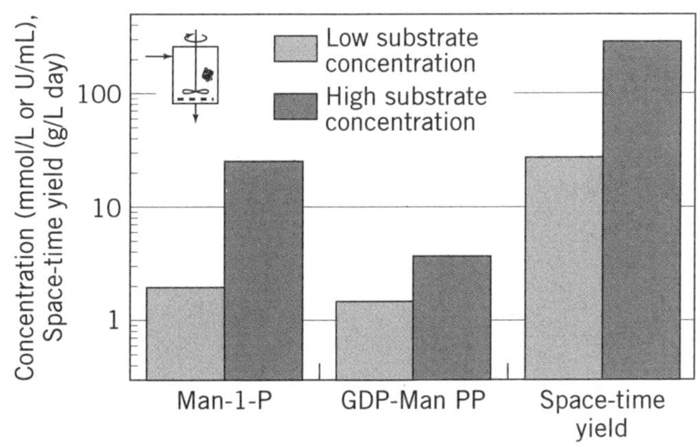

Figure 11. Improvement of space–time yield of GDP-Man by higher substrate and enzyme concentration in a one-stage CSTR; [GTP] 2.5 mmol/L or 25 mmol/L, [Mg^{2+}] 1 mmol/L or 9 mmol/L, residence time 1 h; conversion >98%. Black bars high substrate concentration, gray bars, low substrate concentration.

constant enzyme concentration, the product-specific enzyme consumption can be decreased. The experimental verification revealed good agreement between calculated and experimental values (37). At a conversion at 98%, a space–time yield of 28 g/(L day) and an enzyme consumption of 900 U/$kg_{product}$ was obtained. For a single CSTR, an enzyme consumption of 2,000 U/$kg_{product}$ was reached due to the higher enzyme concentration required to reach the desired conversion under otherwise identical conditions. The relatively low space–time yield is caused by the low substrate concentrations. Complex interactions between K_m values of the substrates and the K_i of GDP-Man require simulation to show that the space–time yield can be increased 10-fold by a corresponding increase of the substrate concentration (Fig. 11). In order to keep a high conversion at >98%, the enzyme concentration must be increased at the same time, but only 2.5-fold. This clearly demonstrates the potential of process optimization based on a model and a complete set of kinetic data.

BIBLIOGRAPHY

1. K. Faber, *Biotransformations in Organic Chemistry*, Springer-Verlag, Berlin, 1995.
2. A. Tanaka, T. Tosa, and T. Kobayashi eds., *Industrial Application of Immobilized Biocatalysts*, Marcel Decker, New York, 1993.
3. K. Drauz and H. Waldmann eds., *Enzyme Catalysis in Organic Synthesis*, VCH, Weinheim, 1995.

4. T. Godfrey and S. West eds., *Industrial Enzymology—Application of Enzymes in Industry*, 2nd ed., Macmillan, London, 1996.
5. R.A. Sheldon, *Chirotechnology: Industrial Synthesis of Optically Active Compounds*, Marcel Decker, New York, 1993.
6. M. Biselli, U. Kragl and C. Wandrey, in K. Drauz and H. Waldmann eds., *Enzyme Catalysis in Organic Synthesis*, VCH, Weinheim, 1995, pp. 89–155.
7. O. Levenspiel, *Chemical Reaction Engineering*, J Wiley, New York, 1972.
8. J.E. Bailey and D.F. Ollis, *Biochemical Engineering Fundamentals*, McGraw-Hill, New York, 1986.
9. D. Rozzell and F. Wagner, *Biocatalytic Production of Amino Acids and Derivatives*, Hanser, München, 1992.
10. W.S. Kelley, *Bio / Technology* **14**, 28–31 (1996).
11. A. Grobiollot, D.K. Boadi, D. Poncelet, and R. Neufeld, *Crit. Rev. Biotechnol.* **14**, 75–107 (1994).
12. S. Norton and J.C. Vuillemard, *Crit. Rev. Biotechnol.* **14**, 193–224 (1994).
13. J.A.M. de Bont, J. Visser, B. Mattiasson, and J. Tramper, eds., *Physiology of Immobilized Cells*, Elsevier, Amsterdam, 1990.
14. T. Nagasawa and H. Yamada, *Trends Biotechnol.* **7**, 153–158 (1985).
15. B.A. Anderson, M.M. Hansen, A.R. Harkness, C.L. Henry, J.T. Vicenzi, and M.J. Zmijewski, *J. Am. Chem. Soc.* **117**, 12358–12359 (1995).
16. J. Reddy, D. Tschaen, Y.-J. Shi, V. Pecore, L. Katz, R. Greasham, and M. Chartrain, *J. Ferment. Bioeng.* **81**, 304–309 (1986).
17. V. Crocq, C. Masson, J. Winter, C. Richard, G. Lemaitre, J. Lenay, M. Vivat, J. Buendia, and D. Prat, *Org. Process Res. Dev.* **1**, 2–13 (1982).
18. C. Salagnad, A. Gödde, B. Ernst, and U. Kragl, *Biotechnol. Progr.* **13**, 810–813 (1997).
19. V.H.M. Elferink in *Reprints of Chiral USA '96, Boston*, Spring Innovations, London, 1996, pp. 79–80.
20. F. Balkenhol, K. Ditrich, B. Hauer, and W. Ladner, *J. Prakt. Chem.* **339**, 381–384 (1997).
21. A.S. Bommarius, M. Schwarm, K. Stingl, M. Kottenhahn, K. Huthmacher, and K. Drauz, *Tetrahedron: Asymmetry* **6**, 2851–2888 (1995).
22. T. Zelinski and H. Waldmann, *Angew. Chem.* **109**, 746–748, 1997, and literature within.
23. U. Kragl, D. Vasic-Racki, and C. Wandrey, *Bioprocess Eng.* **14**, 291–297 (1996).
24. L. Michaelis and M.L. Menten, *Biochem. Z.* **49**, 333–369 (1913).
25. A. Cornish-Bowden, *Fundamentals of Enzyme Kinetics*, Butterworth, London, 1995.
26. I.H. Segel, *Enzyme Kinetics*, J Wiley, New York, 1975.
27. H. Bisswanger, *Theorie und Methoden der Enzymkinetik*, Verlag Chemie, Weinheim, 1979.
28. M. Baerns, H. Hofmann, and A. Renken, *Chemische Reaktionstechnik*, Thieme Verlag, Stuttgart, 1992.
29. E. Schmidt, E. Fiolitakis, and C. Wandrey, *Ann. N. Y. Acad. Sci.* **501**, 434–437 (1988).
30. A. Liese, M. Karutz, J. Kamphuis, C. Wandrey, and U. Kragl, *Biotechnol. Bioeng.* **51**, 544–550 (1996).
31. D. Weuster-Botz and C. Wandrey, *Process Biochem.* **30**, 563–571 (1995).
32. A.M.P. Koskinen and A.M. Klibanov eds., *Enzymatic Reactions in Organic Media*, Chapman & Hall, London, 1996.
33. P. Adlerkreutz, *Biocatal. Biotrans.* **14**, 1–30 (1996).
34. L. Kvittingen, *Tetrahedron* **50**; 8253–8274 (1994).
35. G. Carrea, G. Ottolina, and S. Riva, *Trends Biotechnol.* **13**, 63–70 (1995).
36. A.L. Gutmann and M. Shapira, *Adv. Biochem. Eng. Biotechnol.* **52**, 87–128 (1995).
37. S. Fey, L. Elling, and U. Kragl, *Carbohydr. Res.* **305**, 475–481 (1997).
38. U. Kragl, T. Klein, D. Vasic-Racki, M. Kittelmann, O. Ghisalba, and C. Wandrey, *Ann. N. Y. Acad. Sci.* **709**, 577–583 (1996).
39. K. Seelbach, M.P.J. van Deurzen, F. van Rantwijk, R.A. Sheldon, and U. Kragl, *Biotechnol. Bioeng.* **55**, 283–288 (1997).
40. S. Rissom, U. Schwarz-Linek, M. Vogel, V.I. Tishkov, and U. Kragl, *Tetrahedron: Asymm.* **8**, 2523–2526 (1997).
41. O.A. Hougen and K.M. Watson, *Chemical Process Principles*, J Wiley, New York, 1947.
42. L.G. Lee and G.M. Whitesides, *J. Am. Chem. Soc.* **107**, 6999–7008 (1985).
43. J. Hasegawa, M. Ogura, S. Tsuda, S. Maemoto, H. Kutsuki, and T. Ohashi, *Agric. Biol. Chem.* **54**, 1819–1827 (1990).
44. A. Bosetti, D. Bianchi, P. Cest, P. Golini, and S. Spezia, *J. Chem. Soc. Perkin Trans.* **1**, 2395–2398 (1992).
45. P.E. Kündig, C. Dupre, B. Bourdin, A.C. Jr., and D. Pons, *Helv. Chim. Acta* **77**, 421–428 (1994).
46. E.C. Webb, *Enzyme Nomenclature*, Academic, New York, 1992.
47. W. Kruse, W. Hummel, and U. Kragl, *Recl. Trav. Chim. Pays-Bas* **115**, 239–243 (1996).
48. M.-R. Kula and C. Wandrey, *Methods Enzymol.* **136**, 9–21 (1987).
49. R. Wichmann, C. Wandrey, A.F. Bückmann, and M.-R. Kula, *Biotechnol. Bioeng.* **23**, 2789–2802 (1981).
50. U. Daiminger, P. Plucinski, and W. Nitsch, *Chem. Ing. Tech.* **67**, 217–220 (1995).
51. U. Kragl, W. Kruse, W. Hummel, and C. Wandrey, *Biotechnol. Bioeng.* **52**, 309–319 (1996).
52. S. Kise and M. Hayashida, *J. Biotechnol.* **14**, 221–228 (1990).
53. M. Fukuda and O. Hindsgaul eds., *Molecular Glycobiology*, Oxford University Press, London, 1994.
54. S.L. Flitsch, D.M. Goodridge, B. Builbert, L. Revers, M.C. Webberly, and I.B.H. Wilson, *Bioorg. Med. Chem.* **2**, 1243–1250 (1994).
55. L. Elling, J.E. Ritter, and S. Verseck, *Glycobiology* **6**, 591–597 (1996).

BREVIBACTERIA. See CORYNEBACTERIA, BREVIBACTERIA.

C

CELL CYCLE

David R. Lloyd
Mohamed Al-Rubeai
University of Birmingham
Birmingham, England

KEY WORDS

Apoptosis
Cell cycle
Division
Flow cytometry
Growth
Productivity
Proliferation

OUTLINE

Introduction
The Cell Cycle
 Overview
 Gap Phase 1, or G_1
 Synthesis Phase, or S
 Gap Phase 2, or G_2
 Mitosis, or M: Cell Division
 Gap Phase 0, or G_0
 External (Environmental) Influence on the Cell Cycle
Methods for Describing the Cell Cycle
 Mitotic Index
 Labeled Nucleotide Uptake in S Phase
 Measuring DNA Content
 Flow Cytometry for Cell Cycle Analysis
The Importance of the Cell Cycle in Process Biotechnology
 Culture Proliferation and Growth
 Production in Relation to Cell Cycle Phase
 Cell Cycle Analysis as a Proliferation Predictor
 Cell Cycle–Related Susceptibility to Hydrodynamic Damage
 Cell Cycle and Apoptosis
Bibliography

INTRODUCTION

The cell cycle is the name given to the process by which a cell "matures," synthesizes DNA, and divides to form daughter cells. Thus the cell cycle is a fundamental process with analogous mechanisms found in all cells, from the most primitive bacterium to higher animals and plants, from the unicellular to the most complex multicellular organism. Here we describe the mammalian cell cycle, highlight its importance in biotechnology, and describe some cell cycle applications in animal cell biotechnology.

The cell cycle model (G_1-event model) currently accepted by authors of the major cell and molecular biology books (1–4) describes the cell cycle as discrete phases controlled by cell cycle proteins interacting with each other and directing cellular events including DNA synthesis and cell division. However, the validity of this model and the interpretation of many experiments that underpin the current established view of the cell cycle and cell cycle regulation have been questioned (5,6). The proposed alternative, the continuum model, suggests that the cell cycle does not exist as currently recognized but takes the form of a continuum from one cell division to the next, rather than discrete phases with discrete groups of regulatory proteins. A hypothetical initiator of DNA replication is synthesized by the cell during all phases of the cell cycle, and its concentration reaches a critical level that initiates DNA synthesis. Thus the cell can control its rate of division by regulating the amount of initiator. Nevertheless, until discussion of these competing models is concluded, with perhaps more work to determine the correctness and applicability of the continuum model, it is only reasonable to accept the weight of evidence for the established view of the cell cycle and its regulation. Thus the established model is summarized here and is the basis for the process biotechnology applications that are described.

The pathways of cell cycle regulation are, unsurprisingly perhaps, shared with another important cell process: apoptosis, or "programmed cell death," which is, in effect, the opposite of the cell proliferation cycle. Because the cell cycle and apoptosis share some common pathways, regulatory imbalance can lead to cancer in multicellular organisms (7). Regulation of the cell cycle is also crucial in embryology and ontogeny, where features such as limb formation, tissue differentiation, and growth are a function of the balance between proliferation and apoptosis. Therefore, much of the cell cycle (and apoptosis) literature is from the field of biomedical science, particularly cancer and anticancer chemotherapy research. General texts, as well as much of what is written here, provide a consensus of information describing a hypothetical typical cell and thus, by implication at least, describe what is true of all eukaryotic cells.

THE CELL CYCLE

Overview

The cell cycle and its control mechanisms are highly conserved, so most of what is described in general terms for a "typical mammalian cell" is also true for other animal cells, such as insect and plant cells, although there is the obvious exception of the cell wall and its effect on cell division. The cell cycle and its control in yeast are essentially similar to but much simpler than the analogous mechanisms in higher eukaryotes; where one protein pair exerts a particular control function in a yeast, several may be required for the same function in an animal cell. As a result of this less complex system and the fortuitous discovery of cell cycle control mutants (cdc mutants) in which the cell cycle is inhibited at specific stages, many early discoveries about cell cycle control were made using the budding yeast *Saccharomyces cerevisiae* and the fission yeast *Schizosaccharomyces pombe*. Subsequent research has shown homology between cell cycle control proteins in yeast and those in mammalian cells.

However, the bacterial cell cycle is markedly different from the eukaryotic cell cycle, and its control mechanisms are (to date) less well understood than the analogous processes in eukaryotes. Given a hypothetical typical bacterial cell, its structure and thus the function of its cell cycle are very different from those of a eukaryotic cell. Bacteria, like all cells, are enclosed in a cytoplasmic membrane, but in addition, they have a cell wall outside the cytoplasmic membrane. Within the cell there are no organelles, only free ribosomes and the nucleus, which is actually an area that contains DNA but is not segregated from the cytoplasm by a membrane. The entire genome is contained in a single, circular, double-stranded DNA chromosome; one strand is a closed circle and the other is open. There are no histone or similar DNA-associated proteins; instead there is a relatively high proportion of small, basic polyamines whose function may be to counteract the acidity of the DNA. The "joint" in the closed DNA strand, the origin of replication, is attached to a mesosome, an invagination of the cytoplasmic membrane. When DNA replication starts, the 5′ end of the open strand attaches to the cytoplasmic membrane and starts to form a second mesosome; meanwhile, the DNA unwinds in one direction from the origin of replication and splits into two single strands to form a replication fork. As the replication fork progresses along the chromosome, complementary strands of DNA are synthesized until two daughter chromosomes of double-stranded DNA are formed, each with an origin of replication attached to its mesosome. While the chromosome replicates, the cytoplasmic membrane and cell wall are synthesized in the region between the two mesosomes, enlarging the cell. After chromosome replication, a septum forms across the middle of this region, finally separating the two daughter cells. However, it takes about 40 minutes to duplicate the single chromosome; hence, to allow doubling times as low as 20 minutes, each daughter chromosome must start to replicate in exactly the same way before the parent has finished replication. Thus instead of one there will be three origins of replication operating at once. In addition, when the doubling time of a culture is less than about 40 minutes, DNA synthesis is continuous, so it may be that DNA replication and cell division cycles are independent in bacteria (8). However, despite the essentially continuous DNA synthesis and cell division shown in the fast-growing cultures likely to be used in process biotechnology, it has recently been shown (9) that slowly growing *Escherichia coli* cells (doubling time up to 5 hours) display cell cycle phases analogous to eukaryotic G_1, S, and G_2 phases, *vide infra*, followed by cell division. Furthermore it was shown that initiation of DNA replication was related to cell mass. Two genes, *seqA* and *dnaA*, were shown to be involved in the regulation of replication initiation; gene mutants exhibited decreased and increased mass, respectively, at initiation of replication. Recent work with *Caulobacter* (10) described a protein, CtrA, that has sequence homology with response regulatory proteins in other bacteria including *E. coli* and is involved in cell cycle control including cell division, and cell cycle–specific transcription. It was proposed that CtrA is part of a phosphorelay signal transduction system responsible for transcriptional-level control of bacterial cell cycle events.

The mammalian cell cycle is best described by illustration (Fig. 1a). Starting from a recently divided daughter cell, untransformed, or "normal," cells may enter gap phase 0, or G_0. This is an offshoot from the cell cycle in which cells may remain viable and nondividing but retain the potential to divide given the appropriate stimulus. A cell, depending upon its type, can remain out of the cell

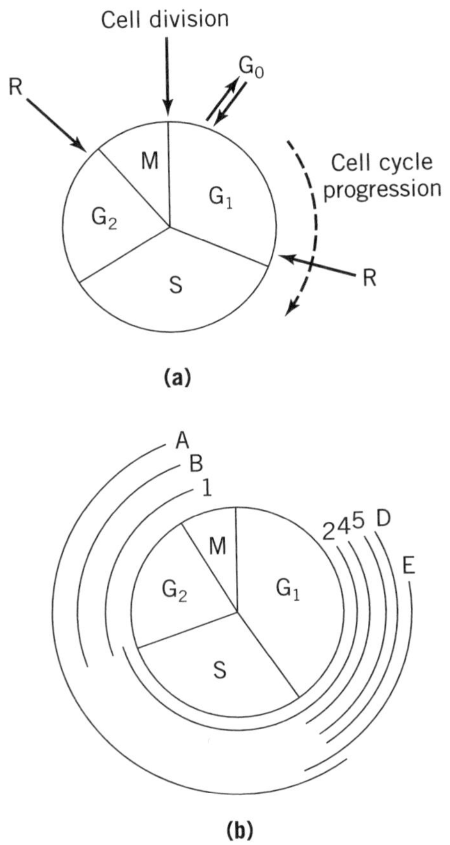

Figure 1. (**a**) A simple diagrammatic representation of the cell cycle. G_0, gap phase 0; G_1, gap phase 1; S, synthesis phase, in which the DNA and cell cytoplasmic components are duplicated; G_2, gap phase 2; M, cell division by mitosis (or meiosis); R, restriction point (check point). (**b**) A simple diagrammatic representation of cyclin and cyclin-dependent kinase activation during cell cycle traverse. A, cyclin A; B, cyclin B; D, cyclin D; E, cyclin E; 1, cyclin-dependent kinase 1 (homologous to cdc2); 2, cyclin-dependent kinase 2; 4, cyclin-dependent kinase 4; 5, cyclin-dependent kinase 5.

cycle, in G_0, for extended periods but can return to gap phase 1, or G_1, due to cell maturation after a certain time has elapsed, after a critical mass/size is reached, or as a response to an external stimulus. In contrast, cells from continuous, sometimes called *established* or *immortal*, cell lines and transformed cells (which includes most animal cells used in biotechnology) do not enter G_0 after division but restart their cell cycle after mitosis by progressing directly to G_1. During G_1, cells prepare to duplicate themselves. When a cell enters the synthesis, or S phase, it starts to precisely and accurately duplicate its genome by synthesizing DNA and the associated nuclear protein components, so that daughter cells carry a genome identical to each other and to the parent. During this time, ribosomes, mitochondria, endoplasmic reticulum, and all the other subcellular organelles are also replicated in preparation for cell division. Thus by the end of S phase, a normally diploid ($2n$) cell will be tetraploid ($4n$), containing enough DNA (and cellular components) for two cells. After S phase, the cycling cell enters the second gap, or G_2, phase. During G_2 the cell synthesizes the proteins required to direct chromosome segregation and cell division in the following mitosis, or M phase. After mitotic division the daughter cells return to G_1 at the beginning of a new cell cycle or exit the cycle to G_0. In a typical cell cycle of 24 hours' duration, G_1 would last 10 hours, S phase 9 hours, G_2 would last 4 hours, and mitosis 1 hour. However, the duration of the cell cycle varies both between cell lines and between individual cells within the same culture. Variation in cycle time is usually due either to cells exiting the cycle and arresting in G_0 or to a variation in the duration of G_1. It has been suggested that the intercellular variability of the postmitotic early G_1 period is smaller than that seen with initiation of DNA replication (presynthetic late G_1) (11). It is also postulated, by the same author, that cells that make the "yes or no" decision in the postmitotic early G_1 period about whether to proceed through the cell cycle also have the capacity to decide, in the presynthetic late G_1 period, when the cell will enter S phase. Heterogeneity in the duration of G_1 is the main reason why a synchronized culture loses synchrony with time to the extent that the culture is completely asynchronous within three or four cell cycles.

This apparently simple cell cycle is subject to complex control mechanisms. Some of the molecular pathways have been understood only in recent years and there may be others that have not yet been described. Mammalian cells periodically transcribe cell cycle genes that encode proteins directly involved in cell cycle regulation. These include the cyclins and the cyclin-dependent kinase (Cdk) families. The cyclin-dependent kinases are phosphorylating enzymes that are inactive when not associated with an appropriate cyclin; it is the associated cyclin that determines the substrate specificity of the complex. The activity of this heterodimeric complex is then regulated by phosphorylation/dephosphorylation of specific activation and/or inhibition sites, as well as by further binding with inhibitory proteins. Many of the cyclins, cyclin-dependent kinases, and related regulatory proteins are expressed or form active complexes only during particular cell cycle phases. Cdk4–cyclin D1 drives cells through mid G_1, Cdk2–cyclin E drives late G_1, Cdk2–cyclin A controls entry into S phase, and Cdk1–cyclin B drives the G_2–M transition; their phase-related activities are summarized in Figure 1b and Table 1. Others, including Cdk4 (13,14) are expressed throughout the cell cycle. The noncyclic proteins exert their cell cycle–controlling effects by association with phase-specific cyclins or cyclin-dependent kinases, association with Cdk-cyclin inhibitors, and specific amino acid phosphorylation/dephosphorylation to form active complexes in specific phases. A family of genes encoding proteins related to the retinoblastoma susceptibility gene, *RB*, and genes for transcription factors including *BYMB, E2F1*, and *JUN* are all induced in late G_1 or early S phase and have been shown to interact with cell cycle regulatory complexes. In addition, there are restriction points in the cell cycle that, once past, commit the cell to continuing the division cycle. There are also check points where, in effect, the cell ensures that the previous phase has been completed properly before proceeding or that there has been no DNA damage. To continue to mitosis before the genome is properly duplicated, for example, would be catastrophic for the cell; one if not both daughters would not be viable. It has been suggested that induction of the p53 tumor suppressor gene and its transcriptional activation of the Cdk inhibitor $p21^{WAF1/CIP1}$ with subsequent cell cycle arrest, following DNA damage, allows time for DNA repair before entering S phase (15). Activation of p53 in cells with unrepaired DNA damage may lead to apoptotic cell death.

Gap Phase 1, or G_1

G_1 was originally thought to be a gap in the cell cycle—a resting phase while the cell recovered from division before starting a new cell cycle by synthesizing more DNA—although this is now known not to be so. Rather, complex control processes are preparing the cell to synthesize the cellular machinery required for DNA synthesis in S phase. There is also a "point of no return" near the end of G_1 called the restriction point (R), which is analogous to START in yeast. Once cells pass this point they are committed to the next stages of the cell cycle. Signals for passing this point include adequate cell size and a favorable environment, including the presence of growth factors. The transition at the R point from growth-factor dependence to growth-factor independence in normal cells, as opposed to cancer or transformed cells, is likely to reflect a stringent growth regulatory mechanism that is defective in the latter cell types. Coordination between cell cycle commitment, START (or R), and cell size, involving the G_1 cyclins, has been suggested as a regulatory mechanism for cell growth in yeast. During early G_1, immediately after G_0 cells are stimulated to reenter the cell cycle, a critical step occurs involving the induction of cell cycle–specific genes including *myc, cyclin D, and fos*, and Cdk genes. It is now apparent that Cdk 2, Cdk 4, and Cdk 5, despite constitutive expression of Cdk4 thoughout the cell cycle, start to complex with cyclins D1, D2, and D3 and become active in early to mid-G_1 (possibly at a point when or soon after noncycling cells return from G_0). The level of activation of the G_1 Cdk-cyclin complexes peaks at about R; so it is possible that, by reaching a threshold activity, it is these complexes that allow cells to pass R. Cdk2 also forms active complexes

Table 1. Cyclin-Cdk Complexes, Stability, and Associated Proteins

Cyclin	Associated Cdk	Cell cycle activity	Associated proteins	Stability	Degradation
A	Cdk2 and cdc2 (Cdk1)	S and $G_2 \rightarrow M$	p107, E2F, p21, PCNA	Unstable in mitosis	Ubiquitin
B1	cdc2 (Cdk1)	$G_2 \rightarrow M$	(p21), (PCNA)	Unstable in mitosis	Ubiquitin
B2		$G_2 \rightarrow M$	(p21), (PCNA)	Unstable in mitosis	Ubiquitin
C-type	Cdk X	?	ND	ND	PEST
D1	Cdk4(2, 5, 6)	G_1	pRb, p21, p27, PCNA, p16, p15	Rapid turnover	PEST
D2	Cdk4(2, 5, 6)	G_1	pRb, p21, p27, PCNA, p16, p15	Rapid turnover	PEST
D3	Cdk4(2, 5, 6)	G_1	pRb, p21, p27, PCNA, p16, p15	Rapid turnover	PEST
E	Cdk2	G_1 and $G_1 \rightarrow S$	p107, E2F, p21, PCNA and p27 after TGFβ treatment	Rapid turnover	
F			ND	ND	
G			ND	ND	
H	Cdk7	All phases	ND	ND	ND

Source: From Ref. 12: T. Hunter and J. Pines, *Cell* **79,** 573–582 (1994), Copyright Cell Press, used with permission.
Note: ND, not determined; PCNA, proliferating cell nuclear antigen; TGFβ, transforming growth factor β; PEST, proteins that contain sequences rich in proline (P), glutamate (E), serine (S), and threonine (T).

with cyclin E starting from mid-G_1 and peaking later than the Cdk–cyclin D complexes, in other words, between R and the G_1–S phase transition.

The primary substrate of the cyclin D complexes is the retinoblastoma tumor suppresser gene product pRB. Active, unphosphorylated pRB complexes with and inhibits the transcription factor E2F. When pRB is phosphorylated and inactivated, the complex dissociates, releasing active E2F; the cell cycle is not stopped at the G_1 checkpoint, and cells are allowed to proceed to S phase (16). E2F was originally characterized as an activator of the adenovirus early region 2 transcription unit (16). It is now recognized that E2F promotes the transcription of enzymes associated with DNA synthesis (18,19). It has been proposed that the cyclin E complex also phosphorylates pRB, with similar effects (20).

Synthesis Phase, or S

The transition from G_1 to S phase is marked by the transcription (as already described) and activation of the many synthetic enzymes required for cellular duplication, including DNA polymerase α, thymidine kinase, and dihydrofolate reductase (DHFR) (21). This is followed by the exact duplication of the genetic material and replication of organelles and structural proteins. At the same time there is a marked change in regulatory proteins activated. At the onset of S phase, the activation of the G_1-phase Cdk-cyclin complexes decreases in the same temporal order as it increased, and the complexes dissociate. As Cdk2 disassociates from cyclins D and E, it complexes with cyclin A, which starts to be synthesized at about the G_1–S transition. The Cdk2–cyclin A complex retains its activity until the end of S phase. Disruption of cyclin A function inhibits DNA synthesis, suggesting that a threshold activity of the Cdk2–cyclin A complex is essential for the maintenance of S phase.

Gap Phase 2, or G_2

As with G_1, G_2 was originally thought to be a gap in the cell cycle—a resting phase while the cell recovered from DNA synthesis before dividing. This has now been shown not to be so, although G_2 remains analogous to G_1 in that, once the appropriate control signals are received, genes encoding for the proteins responsible for cell division are transcribed and ready for cell division. At the end of S phase the activity of the Cdk2–cyclin A complex decreases as the components dissociate. Cyclin A and cyclin B then form complexes with Cdk1 (also often referred to as cdc2 because of its homology with the product of *S. pombe* cell division cycle gene 2). It is these complexes, increasing in activity throughout G_2 and continuing into mitosis, that mediate activation of the cell division machinery. At the transition point between G_2 and M is a second check point that, in effect, ensures that the cells are big enough, the environment is favorable, and DNA synthesis has been fully and accurately completed before the cell continues to mitosis and division.

Mitosis, or M: Cell Division

Mitosis is the division of one normally diploid, or $2n$, somatic cell into two genetically identical (both to each other and the parent) cells. Mitosis, like the cell cycle itself, displays distinct phases that are best represented diagramatically (Fig. 2). When the cell leaves interphase at the end of G_2, it enters prophase. The nucleolus disappears (disperses), and nuclear DNA condenses into chromosomes. Following DNA synthesis in S phase, a normally diploid ($2n$) cell is now tetraploid ($4n$) and contains an identical pair of each chromosome, which in turn is organized as a pair of identical chromatids joined to each other at the centromere. During late prophase, the nuclear envelope is broken down. At the same time, the centrioles divide; one pair migrates to each cell pole with spindle fibers, which are microtubular structures, linking each pair. The transition from prophase to metaphase is marked by alignment of all the chromosome pairs on an equatorial plane, the metaphase plane, with their centromeres connected to each polar centriole pair by spindle fibers. The cell progresses into anaphase when the chromosomes separate and one chromatid from each homologous chromosome pair migrates along the spindle fibers toward each centriole. In the next phase, telophase, when the chromatids (one

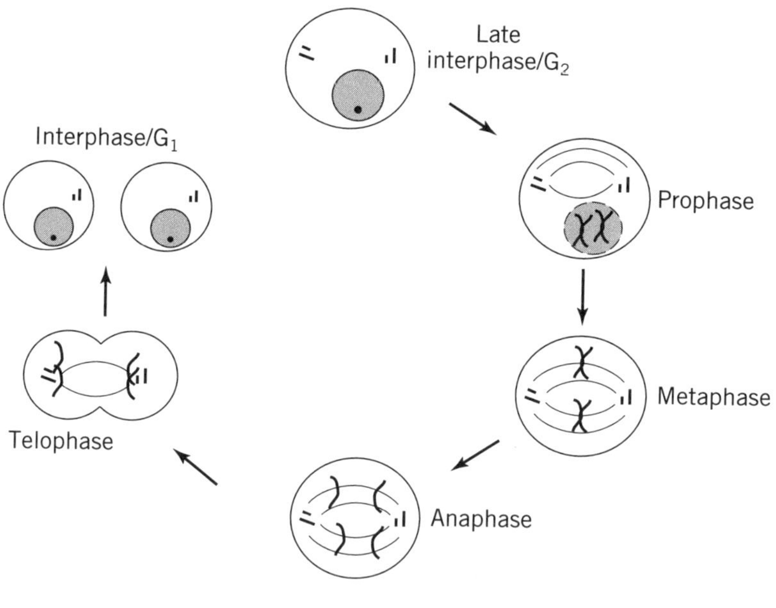

Figure 2. A simple diagrammatic representation of mitosis.

from each homologous chromosome pair) arrange at each pole, the spindle fibers depolymerize, and cytokinesis (division of cytoplasm) begins as a cytoplasmic cleavage furrow begins to form. Further changes follow: the cytoplasm cleaves completely, chromosomes uncoil, the nuclear envelope reforms around each nucleus, and the nucleoli reappear to give two genetically identical daughter cells in interphase. The final telophase changes, in which the cytoplasm divides and the cells separate, are markedly different in plant cells and in animal cells due to the constraints of the rigid cellulose cell wall. As with other cell cycle phases, Cdk-cyclin complexes perform a major regulatory function in mitosis. The activity of Cdk1–cyclin A and Cdk1–cyclin B, which started in G_2, continues to increase until the middle of mitosis, at the metaphase–anaphase boundary. At this point the activity of both complexes declines abruptly to zero, allowing the cell to complete mitosis. It is also thought that the rapid decrease in G_2 cyclin-Cdk complexes is necessary for cells to start a new cell cycle at G_1.

Meiosis. Although unlikely to be encountered in process biotechnology, it is useful to point out that germ cells may undergo an alternative form of division to form zygotes, which may also be considered to be a form of terminal differentiation, called *meiosis*. Assuming a normally diploid ($2n$) cell, initial cell division occurs exactly as in mitosis until the end of anaphase, or in this case the first meiotic anaphase. At telophase the cell may divide or partially divide and enter a brief meiotic interphase comprising two identical diploid ($2n$) daughter nuclei (if not cells). Otherwise the cell progresses directly to the second meiotic prophase. Now there is only a single copy of each chromosome, still comprising two chromatids. During the second meiotic anaphase, the chromatids separate again to each of four poles. The cell then progresses to (the second meiotic) telophase and on to interphase to give four haploid (n) daughter cells instead of the two diploid ($2n$) cells produced by mitosis.

Gap Phase 0, or G_0

The term G_0 was first used in the early 1960s when three types of cell were recognized: static cells, incapable of further division (for example, a terminally differentiated adult neurone); renewing, constantly dividing cells (for example, bone marrow cells); and conditionally renewing cells, which do not normally divide but can respond to an appropriate extraordinary stimulus (for example liver cells starting to proliferate after partial hepatectomy). G_0 is used to describe this last category of cells, which have "opted out" of the cell cycle after a mitosis because of, for example, growth factor depletion. In this circumstance, in contrast, transformed or tumor cells do not enter G_0 but continue through the cell cycle until they die through nutrient exhaustion, often by apoptosis. It has been argued that G_0 should not be used to describe cells, for example, in the stationary phase of a hybridoma, or CHO batch cultures, which are in a pseudo-steady-state of growth and death due to nutrient depletion and/or accumulation of toxic metabolites. Unlike the normal cells, they have not downregulated themselves; they have been closed down by their environment. Cells that lose their ability to respond to growth factor depletion or do not exit the cell cycle to G_0 may have a defective G_1 program, which is different from the cyclin-Cdk4-pRB phosphorylation mechanism involved in R regulation in normal cells. However, G_0 is commonly used today, some claim inappropriately, to describe any cell (containing a G_1 amount of DNA) that is not expected to proceed to S phase for any reason, including the result of any means of artificial interference with the cell cycle. Markers of G_0 include reduced RNA, reduced ribosome content, and an absence of cyclins D and E (22).

External (Environmental) Influence on the Cell Cycle

Extrinsic cell cycle regulation may be considered at two levels. At the more basic level is the effect of the physical environment and nutrient availability; the more complex level of extrinsic cell cycle control is through growth factors. The physical environment has a great effect on the cell cycle. When temperature is reduced from the optimum, the length of the cell cycle increases, and growth rates slow down. When temperature exceeds the optimum, even by a few degrees, profound changes in cell behavior occur, including the expression of heat shock proteins, and cells eventually die. Alternatively, if appropriate nutrients such as amino acids and glucose are not available in the medium, cell cycle duration and even progression will be influenced due to lack of energy and the precursors for protein and DNA synthesis.

Serum starvation, in the presence of appropriate nutrients and physical conditions, may also affect the cell cycle, but this is more likely to be caused by the absence of specific growth factors that serum provides. Most, if not all, cell lines were originally grown in medium containing animal serum, usually newborn or fetal bovine serum. The main contribution to the cell cycle and, hence, growth of a culture from these undefined and heterogeneous supplements is the supply of growth factors that promote or enhance the cell cycle. Thus, because G_0 cells reenter the cell cycle as a result of serum addition, serum is regarded as a source of positive regulatory growth factors. The presence of growth factors in the environment, their binding to specific transmembrane receptor proteins, and subsequent signal transduction effects changes in cellular processes, including the cell cycle. Expression of cyclin D, important in progression through G_1, is induced by growth factors (23). These effects are cell-type dependent because cells do not all express the same growth factor receptors. Furthermore, different cell types/lines may not only require different individual growth factors but may require a specific cocktail for optimal proliferation. Fibroblast growth, for example, requires growth factors including fibroblast growth factor (FGF), epidermal growth factor (EGF), and platelet-derived growth factor (PDGF). Other growth factors that promote cell division include insulin-like growth factor (IGF) I and II, macrophage–granulocyte colony stimulating factor (G-CSF), and endothelial cell growth factor. Each growth factor stimulates a specific receptor on the plasma membrane, which in turn activates protein kinases, many of which phosphorylate tyrosine residues on their substrates, so they are generically called tyrosine kinases. Subsequent signal transduction by phosphorylation/dephosphorylation cascades can lead to an increased rate of protein synthesis, influence the activity of DNA topoisomerase, or modify the glucose transport system—all with the similar effect of promoting cell division. Transformed cells are known to have a reduced (or no) requirement for external growth factors compared with normal cells. However, through adaptation of cell lines and extensive development of medium formulation, many cells can now be grown in protein-free medium, indicating that it is possible to bring about cell proliferation and support long-term culture in media free of (extrinsic) growth factors, hormones, and other contaminating proteins. This absence of extraneous protein is particularly important in large-scale process biotechnology for facilitating downstream processing and product recovery.

It is commonly accepted that the accumulation of toxic metabolites (ammonia and lactate, for example) inhibits culture proliferation. However, in addition to the positive growth-promoting (cell cycle–promoting) factors, there are also negative, growth-inhibiting factors, although these may be the same growth factors displaying opposite effects under different conditions. Transforming growth factor $\beta 1$ (TGF-$\beta 1$), in addition to its strong positive stimulatory effect, has been shown to produce a density-independent negative response that prolongs G_1 in the NIH-3T3 mouse embryo cell cycle and has a similar effect on mortal human fibroblasts (24).

METHODS FOR DESCRIBING THE CELL CYCLE

Mitotic Index

There are three main approaches to describing the cell cycle. The first and most historically established approach is to determine the percentage of cells in mitotic cell division. This method, called the *mitotic index*, is simple and can be applied to attached cells grown on a microscope slide or any other appropriate substratum, as well as cells in suspension. There are many fixation and staining methods described in any textbook of cytology methods, from which a protocol appropriate to animal cell biotechnology can be selected. A generic protocol is as follows. Take cells grown on a glass microscope slide (or coverslip) or prepare a smear of suspended cells on a microscope slide, using an attachment agent such as polylysine if necessary. Fix the cells with a fixative appropriate to the proposed staining method. Stain the cells with either a standard cytological or histological stain that differentiates DNA, such as Feulgen staining, or alternatively, stain with a fluorescent DNA stain such as ethidium bromide (EB), Hoechst 33258, or DAPI. Examine stained slides by conventional or fluorescence microscopy, as appropriate; count the number of nuclei in mitosis and express the result as a percentage of the total. A recent publication, investigating mitosis in a hydrodynamically stressed environment, described the use of the fluorescent DNA-intercalating stain propidium iodide (PI) with cells fixed with 3:1 ethanol:glacial acetic acid, followed by fluorescence microscopy (25). A major disadvantage to measuring the mitotic index is that, although it gives information about cell division in M phase, it can give no information about the other cell cycle phases or their associated processes. Measuring the mitotic index also exhibits disadvantages common to all manual microscopic methods, including sampling error because a small proportion of the total population is examined, and interobserver variation resulting from subjective discrimination between mitotic and interphase cells. Furthermore this slow and labor-intensive method is not appropriate for online automation or use in integrated process control.

Labeled Nucleotide Uptake in S Phase

An alternative approach for describing the cell cycle is to document the incorporation of labeled nucleotide into nuclear DNA during S phase. This method relies on $[^3H]$thymidine incorporation followed by autoradiography and can be used in one of two ways: continuous labelling, in which the cells are exposed to the label for a fixed time, usually 24 or 48 h followed by analysis at that time; or pulse–chase labeling, in which the cells are exposed to the label for a short pulse (0.5–1 h), then analyzed at regular intervals for 24–48 h. Continuous labeling gives information about the growth fraction of a culture, which is the percentage of the cells that are in the cell cycle during the labeling period. However, this information is not likely to be of great use to the process biotechnologist because most cells under optimum conditions, when viability is close to 100%, will also have a growth fraction close to 100%. A simple generic protocol for continuous labeling is as follows: [^{3}H]thymidine is added to the culture medium, and the culture continues for a specified period; cells are fixed and subjected to autoradiography; when the autoradiograph has developed, the number of radiolabeled nuclei is counted and expressed as a percentage of the total.

In contrast, pulse–chase labeling is used to determine the average duration of each cell cycle phase. A generic protocol for [^{3}H]thymidine pulse–chase labeling is as follows: [^{3}H]thymidine is added to the culture; after a short period (1–2 h), cells are removed to fresh medium without [^{3}H]thymidine; the culture is sampled regularly at times greater than the length of one cell cycle (24–48 h); each sample is subjected to autoradiography; and the number of labeled mitoses is counted as a percentage of the total. The average duration of G_2, S, M, and the total cell cycle is measured from the plot of labeled mitoses against time, then G_1 is calculated. Details of using [^{3}H]thymidine incorporation for determining growth fraction and cell cycle phase duration are summarized elsewhere (26). As with the mitotic index, any autoradiographic technique exhibits inherent disadvantages including sampling error because a small proportion of the total cell population is examined; the technique is slow, labor intensive, and not appropriate for on-line automation or use in integrated process control; and the use of radioisotopes requires additional containment and waste disposal facilities.

In recent years the thymidine analogue bromodeoxyuridine (BrdU) has been used as a safer, nonisotopic alternative to [^{3}H]thymidine for DNA labeling. BrdU can be used in exactly the same way as [^{3}H]thymidine, as either a continuous or pulse–chase label. The only difference in the protocol is that BrdU incorporation is detected by immunostaining with fluorescent anti-BrdU antibodies followed by fluorescence microscopy instead of autoradiography, so the method still suffers the disadvantages of any manual microscopic method. Alternatively, BrdU incorporation can be measured by flow cytometry, *vide infra*.

Measuring DNA Content

The third and probably easiest approach to adopt for cell cycle analysis is to measure the DNA content of each cell in a population by flow cytometry and thus determine the proportion of cells in G_0/G_1, S, and G_2/M at the time of sampling. A major and obvious advantage to this approach is that heterogeneous nonsynchronized cultures can be analyzed easily and rapidly, with the useful bonus that sub-G_1 apoptotic populations can also be identified. Other advantages include minimizing sample error by analyzing a large number of cells, and minimal interobserver variation because of the objective discrimination between G_1, G_2, and S cells. Furthermore, the method is appropriate for near on-line automation and use in integrated process monitoring or control. Although this approach is not able to resolve G_0/G_1 or G_2/M because DNA content is the same, this is of little concern in most process biotechnology applications because the (transformed) cells commonly used do not enter G_0. Furthermore the G_2/M cell subpopulation can be resolved, if necessary, by measuring the mitotic index of the population (as already described) to determine the proportion of cells in M phase.

Flow Cytometry for Cell Cycle Analysis

Flow cytometric cell cycle analysis by measuring DNA content may follow any one of many protocols. A generic protocol is as follows: fix/permeablize the cells; stain with a fluorescent DNA stain; measure stained DNA content by flow cytometry; and analyze data by simple gating or using dedicated cell cycle analysis software. Variations include pretreatment of cells; altered fixation/permablization protocols; postfixation treatment of cells; different stains, although stains such as PI or EB, which intercalate into double-stranded nucleic acid, are commonly used; and methods for data analysis. A typical flow cytometric measurement of PI-stained DNA in ethanol-fixed recombinant CHO cells and the subsequent cell cycle analysis, using software based on a polynomial S-phase algorithm with an iterative nonlinear least squares fit, is shown in Figure 3. By fitting the G_1 and G_2 peaks as Gaussian distributions, and the S-phase distribution as a Gaussian broadened distribution, the cell cycle phase distribution can be deconvoluted.

Flow cytometry can also be used with BrdU incorporation to measure the growth fraction or to determine average cell cycle phase times. For flow cytometric analysis, BrdU incorporation can be detected by either fluorescent immunostaining or its quenching effect on Hoechst DNA stains. Thus BrdU incorporation can be analyzed simultaneously with DNA content. The advantage of this method is that it can recognize arrested but non-DNA-synthesizing cells during S phase. Cell size and/or protein content can also be analyzed simultaneously to give further related information. Again, there are many variations on a theme, details of which and other flow cytometric methods for cell cycle analysis have been summarized elsewhere (27). Flow cytometry is also very useful for analyzing specific cell cycle–related proteins, following appropriate fluorescent immunostaining. For example, the absence of cyclin D or E in G_0 cells, shown by flow cytometry, has been used to distinguish G_0 from G_1 cells (22), and flow cytometric measurement of the Ki67 nuclear antigen has been shown to correlate with [^{3}H]thymidine and BrdU incorporation (28). Multiparameter flow cytometric analysis of G_1 and G_2 cyclin expression and DNA content has also been shown to be a useful tool in cell cycle analysis (29) because G_1 cells can be differentiated from G_0 cells, and cells with unusual karyotypes (for example, G_1 tetraploid) can be distinguished from G_2 diploid cells. Flow cytometry can also be applied to investigating the relationship between the cell cycle and recombinant protein expression in

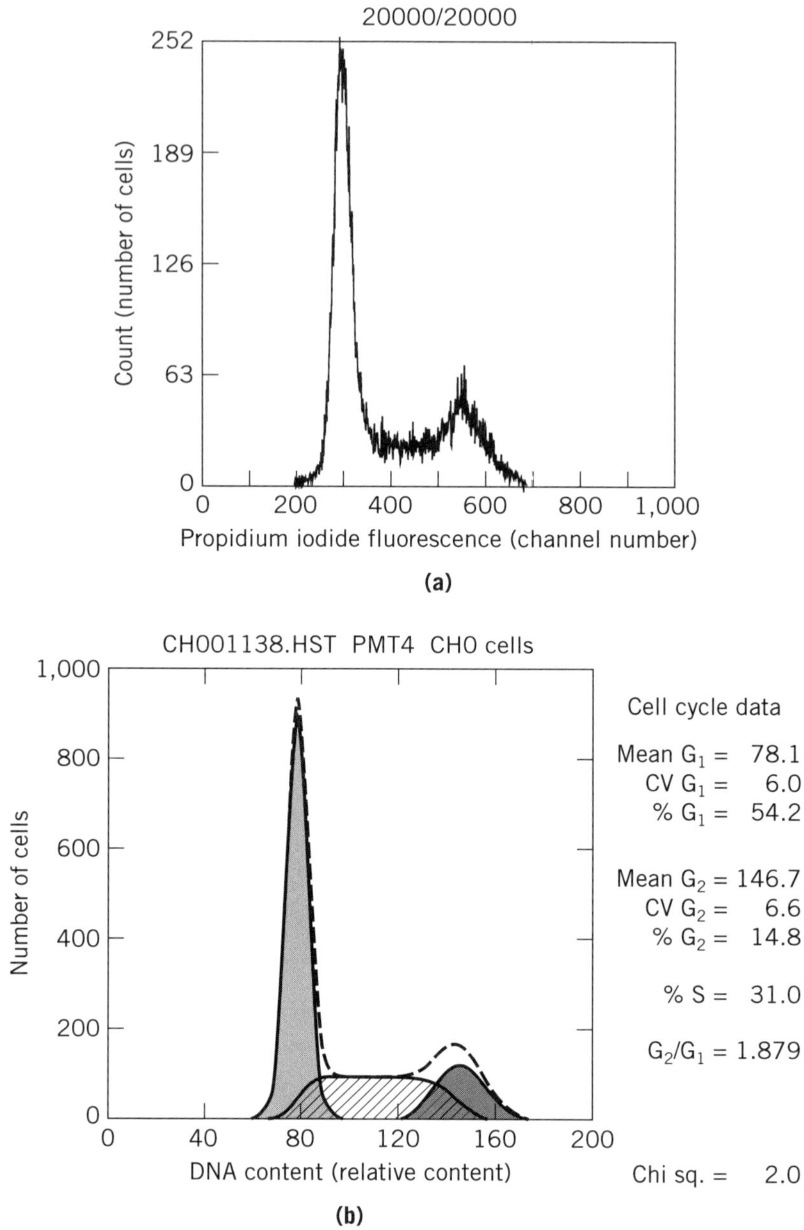

Figure 3. (**a**) Typical flow cytometric measurement of fluorescent (propidium iodide–stained) DNA content in recombinant CHO cells. (**b**) Resolution of cell cycle fractions from the fluorescent DNA content of recombinant CHO cells, using dedicated cell cycle analysis software (Multicycle®, Phoenix Flow Systems).

economically important cell lines as a means of optimizing medium and environmental conditions for successful commercial operation.

THE IMPORTANCE OF THE CELL CYCLE IN PROCESS BIOTECHNOLOGY

Culture Proliferation and Growth

The cell cycle is the means by which a culture proliferates. Cell division and cell death, both active programmed cell death (apoptosis) and passive necrosis, are responsible for the change in the number of cells present. In addition to the cell cycle control genes, the transcription of genes coding for proteins involved in nucleotide and DNA synthesis is also cyclic. However, the progress of the cell cycle is costly in nutrient utilization and energy expenditure, particularly during S phase, which is when the nucleic acid content of the cell is duplicated, but also during other phases because regulatory proteins are continuously synthesized and degraded. In theory it should be possible to arrest cells, in other words, stop them from cycling, thus preventing the "wastage" of energy and nutrients that could otherwise be directed into product synthesis. The maintenance energy model (30) states that energy (as adenosine triphosphate [ATP]) consumption is divided into that needed for cell growth (cell cycle progression, as determined by increase in cell number) and that needed for cell maintenance. Based on this model it has been determined that the maintenance energy requirement of a hybridoma is 14.4×10^{-9} mmol ATP/(cell day) at a specific growth rate (μ) of 0.66 per day (31), corresponding to 62%

of the energy (ATP) consumed. These values were confirmed by work that showed that the maintenance energy requirement of the same cells under slightly different nutrient and specific growth rate conditions was 11.4×10^{-9} and 12.0×10^{-9} mmol ATP/(cell day), again, corresponding to approximately 60% of the estimated ATP consumption rate (32). These observations agree with the value of 65%, which was determined as the proportion of energy consumption used for maintenance by mouse LS cells grown at $\mu = 0.60$ per day (33). Thus it is apparent that about two-thirds of the energy used by a culture is for its own maintenance, and this proportion is not likely to be readily altered. However the remaining third, which is used for growth (and product elaboration), may be exploited. If the growth (cell cycle progression) requirement was removed, all of the remaining resources could be directed toward product elaboration, or alternatively, cultures would require up to one-third less energy and nutrient resources for the same product yield.

Two main approaches have been used to prevent cells from cycling. The first is chemical blockade. For example, the addition of thymidine to culture medium causes cells to arrest at the G_1–S phase boundary, so cells accumulate in G_1, and specific productivity increases. However, this production advantage is short lived because the imbalanced growth state leads to culture death (usually by apoptosis) within a few days of the onset of the chemical blockade (D. C. Edwards and M. Al-Rubeai, unpublished data, 1997). The second approach is not to arrest the cell cycle but to substantially reduce its progression by reducing the culture growth rate. Using a custom-built total cell recycle bioreactor system, where the growth rate of a mouse hybridoma was substantially reduced ($\mu < 0.05\ \mu_{max}$), Flickinger et al. (34) showed that specific monoclonal antibody secretion remained constant as the specific growth rate decreased. This demonstrated that production could be uncoupled from growth (cell cycle progression), although there was no evidence that the energy saved by substantial (>95%) reduction of growth was diverted into production. However, using this method for production-scale processes may have the advantage of less medium (energy/nutrient) expenditure per unit product, although the total product yield per unit of production (from each bioreactor) would not be increased. Furthermore, questions of product fidelity, especially with respect to glycosylation, are raised when growth is limited by the stress of nutrient deprivation or limitation.

Until recently, these were the only means available for controlling the cell cycle (i.e., preventing cycle progression). In 1997, Fussenegger et al. (35) described a novel cytostatic process that enhances productivity. This was achieved by adopting a completely different approach compared with previous work. Metabolic engineering was used instead of the application of an external stress to the cells to prevent cell cycle progression. A series of transient CHO transfects was produced containing inducible dicistronic expression vectors, each of which contained a model product gene (human secretory alkaline phosphatase [hSEAP]) and a tumor-suppressor gene coding for an intracellular regulator of cell cycle progression (either p21, p27, or a mutant p53 [p53175P] that is said to have lost its proapoptotic properties while retaining its cell cycle regulation properties (36)). When the expression vector was induced, cells arrested and began to accumulate in G_1. The arrested cells showed a fourfold increase in hSEAP production compared with cells transfected with a control expression vector coding for the product only. Using stable dicistronic transfects, only p27 overexpression successfully induced cell cycle arrest (in G_1), but with it came a 10-fold increase in specific hSEAP productivity (35). Building on the success of these discistronic vectors, the tricistronic pBluescript®-based pTRIDENT family of vectors was developed (37), allowing the possibility of a single stable transfection coding not only for a product but also for cell cycle control (cell cycle arrest) and enhanced culture survival using an antiapoptosis oncogene such as *bcl-2* or $bcl\text{-}x_L$ (38–40). Subsequent work with pTRIDENT vectors containing genes encoding for hSEAP with either p21 and the CCAAT/enhancer binding protein α (which induces and stabilizes p21 (41,42)) or p27 and Bcl-x_L showed even greater benefit, with specific productivities increased 15- and 30-fold, respectively, with respect to proliferation-competent controls (43).

Production in Relation to Cell Cycle Phase

It is important not to confuse the relationship between cell cycle phase and productivity with the relationship between cell cycle arrest and productivity. The substantial increases in productivity from the metabolically engineered transfects just described are most likely to exist because the arrested cells do not need to devote resources to biomass production (35). This section addresses the situation in "simple" recombinant cells that may be grown in standard industrial processes such as batch, continuous, or draw-and-fill semicontinuous culture modes. Determining the relationship between cell cycle phase and productivity is important for developing automated on-line (or near on-line) process control strategies. These strategies will be based on models that, in addition to the basic parameters of pH, dO_2, cell density, and so forth, must incorporate information concerning cell state, including cell cycle phase and its effect on productivity.

It has been shown that specific monoclonal antibody production by a synchronized mouse–mouse hybridoma culture is maximal during G_1; furthermore, the specific activity increases when the cell cycle is arrested and cells are maintained in late G_1 (44). However, apparent specific productivity also increases when the culture declines (viability decreases); this is the result of substantial passive release of intracellular product into the culture medium from dead and dying cells. Phase dependence was also found in recombinant CHO cells transfected with an expression vector containing *dhfr* and *lacZ* genes; the former was controlled by an SV40 promoter, and the latter was controlled by the constitutive hCMV promoter (45). In this study, the nonsecreted product, β-galactosidase, was produced almost exclusively during S phase. Furthermore, cell cycle arrest in S phase, by specific inhibition of DNA polymerase, did not inhibit transcription or translation and thus had no effect on β-galactosidase expression.

However, the literature on the relationship between cell cycle and protein expression, summarized in Table 2, is

Table 2. Variation in Phase Expression of Different Products with Cell Line and Promoter

Cell line	Promoter	Product	Phase expression	Ref.
Mouse leukaemia	Endogenous	DHFR	Aphasic	46
CHO	Endogenous	DHFR	Aphasic	47
Human osteosarcoma	Endogenous	DHFR	Aphasic	47
Mouse embryo fibroblast	Endogenous	DHFR	Aphasic	47
Human prostate epithelium	Endogenous	PSAP	All except early G_1	48
Hybridoma	Endogenous	MAb	G_1/G_0 (mainly)	49
CHO	AMLP	DHFR	G_1 (mainly)	50
Hybridoma	Endogenous	MAb	G_1/S	44
Mouse plasmacytoma	Endogenous	Immunoglobulin	Late G_1/early S	51
Human lymphoid cell	Endogenous	Immunoglobulin	Late G_1/S	52
CHO	Endogenous	DHFR	G_1/S^a	22
Mouse leukemia	Endogenous	TMPS	S	46
CHO	CMV	β-Gal (*lacZ*)	S	45
CHO	SV40	tPA	S (mainly)	50
CHO	SV40	IFN-γ	S	53
Mouse L-cell	MMTV	β-Gal (*lacZ*)	S (predominantly)	54
CHO	Endogenous	DHFR	S	55
Rat adenocarcinoma	Endogenous	tPA	Shortly after S	56
Rat adenocarcinoma	Endogenous	uPA	Shortly after S	56
CHO	Endogenous	PA	G_2/M	57

Note: DHFR, dihydrofolate reductase; CHO, Chinese hamster ovary; TMPS, thymidine monophosphaste synthetase; MAb, monoclonal antibody; PSAP, prostate-specific acid phosphatase; AMLP, adenovirus major late promoter; CMV, cytomegalovirus; β-Gal (*lacZ*), β-galactosidase (product of the *Escherichia coli lacZ* gene); SV40, simian virus 40; tPA, tissuelike plasminogen activator; IFN-γ, interferon-γ; uPA, urokinase-like plasminogen activator; PA, plasminogen activator.

[a]Gene expression activates at G_1–S transition.

inconsistent. Different researchers have shown that product expression (or the maximum rate of expression) is related to different cell cycle phases. In addition, others have shown that product expression is aphasic. These observations do not necessarily contradict each other and may be a true reflection of the systems used, because a variety of cells, products, and constructs were studied. Cell cycle–dependent product expression may not be universal and could vary with cell type or cell line, the construction of the expression vector, the nature of the recombinant gene expressed, or the promoter/enhancer used to drive product expression. Other variables may include posttranscriptional regulation (mRNA level), product gene copy number, and even the culture mode, (e.g., whether the cells are attached to a substrate or are free in suspension) may be relevant. However, there are two potential flaws in studying cell cycle–related productivity: synchronizing cells using chemical means, such as thymidine block or nutrient starvation, and comparing cells of different phases from the same culture that are separated in time. The two problems often go together, for example, following a culture for a day after release from thymidine block. The difficulty with the former is that the chemical stress used (what ever its nature) may, in addition to its desired effect of cell synchrony, directly cause perturbations in productivity that could wrongfully be ascribed to the cell cycle. The difficulty with the latter is simply that if two cell populations from a culture are separated in time, they are also separated in environmental conditions; recent work (58) has shown that the effect of medium condition is dominant to any cell cycle effect on productivity. The effect of temporal separation is made more uncertain following stress synchronization; an early population may have only partially recovered from the stress, whereas a later population has fully recovered.

Cell Cycle Analysis as a Proliferation Predictor

Cells may be arrested in G_1 before the restriction point, but once past that point, they are committed to continue their cycle and divide unless prevented from doing so by catastrophe or artificial intervention. Therefore, one can presume that, once a cell enters S phase, under normal growth conditions, one single cell will in due course become two cells. This phenomenon can be used to predict culture growth. The growth-enhancing effect of peptone supplementation on hybridoma culture was predicted by cell cycle analysis 5 h after supplementation, although the effect on cell density was not apparent until 16 h later, 21 h after supplementation (59). A similar use of cell cycle analysis was made with CHO cells: the percentage of S-phase cells correlated with the specific growth rate 10.5 h later; the time difference was the average duration of S phase remaining plus the duration of G_2 and M for that population (60). Thus, the proportion of cells in S phase can be used to predict the subsequent proliferation rate of that culture.

Cell Cycle–Related Susceptibility to Hydrodynamic Damage

The effect of hydrodynamic stress, as found in stirred and sparged bioreactors, is at least in part related to cell cycle phase, and there may be a role for hydrodynamic cell damage in the induction of apoptosis (61). It has been reported that DNA synthesis (62) and the proportion of the cell population in S phase (63) were increased under conditions of hydrodynamic stress (high sparging and agitation rates). However, the cell number remained low even though the S-phase population correlated with an increase in cell number. The observed reduction in growth rate was shown to be a consequence of a decline in the proportion of G_2- and M-phase cells, with a corresponding decrease in the cell size of the population, suggesting that the larger G_2 cells were particularly susceptible to hydrodynamic damage (61). Subsequently it was shown that mitotic cells, while also being susceptible to hydrodynamic damage due to their size, were even more sensitive to hydrodynamic damage than G_2 cells, perhaps due to cytoskeletal rearrangements, alteration of plasma membrane fluidity, or morphological changes during mitosis (25).

Cell Cycle and Apoptosis

Apoptosis (active or programmed cell death) is the means by which intrinsic or extrinsic stimuli activate a series of molecular cascades, triggering events that culminate in a particular morphologically, ultrastructurally, and biochemically defined process of cell death that is, in effect, the opposite of the cell cycle. Entry of a cell into apoptosis can be restricted to a particular cell cycle phase (54–66), although this cell cycle dependence does not appear to be universal (67). Cell cycle progression and apoptosis are not only opposite in effect, they may also be alternate outcomes from a single cell control mechanism; they are inextricably linked by common components in the signal transduction pathways for both processes. In addition, circumstantial evidence from the induction of apoptosis by DNA damage or cell cycle disruption and the existence of many oncogenes that both promote the cell cycle and induce apoptosis implies that there is an equilibrium between these two antagonistic processes in all eukaryotic cells.

BIBLIOGRAPHY

1. B. Alberts, D. Bray, J. Lewis, M. Raff, K. Roberts, and J.D. Watson, *The Molecular Biology of the Cell*, 3rd ed., Garland, New York, 1994.
2. J.D. Watson, N.H. Hopkins, J.W. Roberts, J.A. Steitz, and A.M. Weiner, *Molecular Biology of the Gene*, 4th ed., Benjamin/Cummings, Melno Park, Calif., 1987.
3. H. Lodish, D. Baltimore, A. Berk, S.L. Zipursky, P. Matsudaira, and J. Darnell, *Molecular Cell Biology*, 3rd ed., Scientific American Books, New York, 1995.
4. A. Murray and T. Hunt, *The Cell Cycle: An Introduction*, Freeman, New York, 1993.
5. S. Cooper, *Nature* **280**, 17–19 (1979).
6. S. Cooper, *J. Theor. Biol.* **135**, 393–400 (1988).
7. G.I. Evan, E. Harrington, N. McCarthy, C. Gilbert, M.A. Benedict, and G. Nuñez, in N.S.B. Thomas ed., *Apoptosis and Cell Cycle Control in Cancer*, Bios Scientific, Oxford, 1996, pp. 109–129.
8. I.B. Holland, *Cell* **48**, 361–362 (1987).
9. E. Boye, T. Stokke, N. Kleckner, and K. Skarstad, *Proc. Natl. Acad. Sci.* **96**, 12206–12211 (1996).
10. K.C. Quon, G.T. Marczynski, and L. Shapiro, *Cell* **84**, 83–93 (1996).
11. R. Zetterberg, in N.S.B. Thomas ed., *Apoptosis and Cell Cycle Control in Cancer*, Bios Scientific, Oxford, 1996, pp. 17–36.
12. T. Hunter and J. Pines, *Cell* **79**, 573–582.
13. M. Matsuoka, J.Y. Kato, R.P. Fisher, D.O. Morgan, and C.J. Sherr, *Mol. Cell. Biol.* **14**, 7265–7275 (1994).
14. J.Y. Kato, M. Matsuoka, D.K. Strom, and C.J. Sherr, *Mol. Cell. Biol.* **14**, 2713–2721 (1994).
15. W.S. El-Deiry, in N.S.B. Thomas ed., *Apoptosis and Cell Cycle Control in Cancer*, Bios Scientific, Oxford, 1996, pp. 55–75.
16. R.A. Weinberg, *Cell* **81**, 323–330 (1995).
17. Kovesdi, R. Reichek, and J.R. Nevins, *Cell* **45**, 219–228 (1986).
18. E. Ogris, H. Rotheneder, I. Mudrak, A. Pichler, and E. Wintersberger, *J. Virol.* **67**, 1765–1771 (1993).
19. B.E. Pearson, H.P. Nasheuer, and T.S.F. Wang, *Mol. Cell. Biol.* **11**, 2081–2095 (1991).
20. V. Dulic, E. Lees, and S.I. Reed, *Science* **257**, 1958–1961 (1992).
21. J. Wells, P. Held, S. Illenye, and N.H. Heintz, *Mol. Cell. Biol.* **16**, 634–647 (1996).
22. Z. Darzynkiewicz, J. Gong, G. Juan, B. Ardelt, and F. Traganos, *Cytometry* **25**, 1–13 (1996).
23. C.J. Sherr, *Cell* **73**, 1059–1065 (1993).
24. D. Zhang and J.W. Jacobberger, *Cell Proliferation* **29**, 289–307 (1996).
25. F.A. Abdul Majid and M. Al-Rubeai, in M.J.T. Carrondo, B. Griffiths, and J.L.P. Moreira eds., *Animal Cell Technology from Vaccines to Genetic Medicine*, Kluwer, Dordrecht, 1997, pp. 731–736.
26. R. Baserga, in G.P. Studzinski ed., *Cell Growth and Apoptosis: A Practical Approach*, IRL Press at Oxford Univ. Press, Oxford, 1995, pp. 1–19.
27. H.A. Crissman, in G.P. Studzinski ed., *Cell Growth and Apoptosis: A Practical Approach*, IRL Press at Oxford Univ. Press, Oxford, 1995, pp. 21–43.
28. S.D. Gore, L.J. Weng, and P.J. Burke, *Exp. Hematol.* **21**, 1702–1708 (1993).
29. J. Gong, X. Li, F. Traganos, and Z. Darzynkiewicz, *Cell Proliferation* **27**, 357–371 (1994).
30. S.J. Pirt, *Principles of Microbe and Cell Cultivation*, Blackwell Scientific, Oxford, U.K., 1975.
31. W.M. Miller, C.R. Wilke, and H.W. Blanch, *Biotechnol. Bioeng.* **33**, 477–486 (1989).
32. W.M. Miller, C.R. Wilke, and H.W. Blanch, *Biotechnol. Bioeng.* **33**, 477–486 (1989).
33. D.G. Kilburn, M.D. Lilly, and F.C. Webb, *J. Cell. Sci.* **4**, 645–654 (1969).
34. M.C. Flickinger, N.K. Goebel, and M.A. Bohn, *Bioprocess Eng.* **5**, 155–164 (1990).
35. M. Fussenegger, X. Mazur, and J.E. Bailey, *Biotechnol. Bioeng.* **55**, 927–939 (1997).
36. S. Rowan, R.L. Ludwig, Y. Haupt, S. Bates, X. Lu, M. Oren, and K.H. Vousden, *EMBO J.* **15**, 827–838 (1996).
37. M. Fussenegger, X. Mazur, and J.E. Bailey, *Biotechnol. Bioeng.* **57**, 1–10 (1998).
38. Y. Itoh, H. Ueda, and E. Suzuki, *Biotechnol. Bioeng.* **48**, 118–122 (1995).
39. R.P. Singh, A.N. Emery, and M. Al-Rubeai, *Biotechnol. Bioeng.* **52**, 166–175 (1996).
40. D.C. Huang, S. Cory, and A. Strasser, *Oncogene* **14**, 405–414 (1997).
41. A.L. Gartel, M.S. Serfas, and A.L. Tyner, *Proc. Exp. Biol. Med.* **213**, 138–149 (1996).

42. N.A. Timchenko, M. Wilde, M. Nakanishi, J.R. Smith, and G.J. Darlington, *Genes Dev.* **10**, 804–815 (1996).
43. M. Fussenegger, S. Schlatter, D. Dätwyler, X. Mazur, and J.E. Bailey, *Nature Biotechnology* **16**, 468–472 (1998).
44. M. Al-Rubeai and A.N. Emery, *J. Biotechnol.* **19**, 67–86 (1990).
45. M.B. Gu, P. Todd, and D.S. Kompala, *Biotechnol. Bioeng.* **42**, 1113–1123 (1993).
46. L.H. Matherly, J.D. Schuetz, E. Westin, and I.D. Goldman, *Anal. Biochem.* **182**, 338–345 (1989).
47. J.N. Feder, Y.G. Assaraf, L.C. Seamer, and R.T. Schimke, *J. Biol. Chem.* **264**, 20583–20590 (1989).
48. F. Herz, B. Czerniak, D. Deitch, R.P. Wersto, D.A. Simmons, and L.G. Koss, *Cell Proliferation* **24**, 321–330 (1991).
49. S.J. Kromenaker and F. Srienc, *Biotechnol. Bioeng.* **38**, 665–677 (1991).
50. M. Kubbies and H. Stockinger, *Exp. Cell Res.* **188**, 267–271 (1990).
51. O. Garatun-Tjeldstø, I.F. Pryme, J.K. Weltman, and R.M. Dowben, *J. Cell Biol.* **68**, 232–239 (1976).
52. D.N. Buell and J.L. Fahey, *Science* **164**, 1524–1525 (1969).
53. V. Leelavatcharamas, A.N. Emery, and M. Al-Rubeai, in K. Funatsu, Y. Shirai, and T. Matsushita eds., *Animal Cell Technology: Basic and Applied Aspects,* Vol. 8, Kluwer, Dordrecht, 1997, pp. 55–60.
54. M.B. Gu, P. Todd, and D.S. Kompala, *Biotechnol. Bioeng.* **50**, 229–237 (1996).
55. B.D. Mariani, D.L. Slate, and R.T. Schimke, *Proc. Natl. Acad. Sci.* **78**, 4985–4989 (1981).
56. F.M. Scott, V. Sator de Serrano, and F.J. Castellino, *Exp. Cell Res.* **169**, 39–46 (1987).
57. J. Aggeler, L.N. Kapp, S.C.G. Tseng, and Z. Werb, *Exp. Cell Res.* **139**, 275–283 (1982).
58. D.R. Lloyd, V. Leelavatcharamas, A.N. Emery, and M. Al-Rubeai, *Cytotechnology*, in press.
59. M. Al-Rubeai, S. Chalder, R. Bird, and A.N. Emery, *Cytotechnology* **7**, 179–186 (1991).
60. V. Leelavatcharamas, A.N. Emery, and M. Al-Rubeai, in M. Al-Rubeai and A.N. Emery eds., *Flow Cytometry Applications in Cell Culture*, Marcel Dekker, New York, 1996, pp. 1–15.
61. M. Al-Rubeai, R.P. Singh, A.N. Emery, and Z. Zhang, *Biotechnol. Bioeng.* **46**, 88–92 (1995).
62. Lakhotia, K.D. Bauer, and E.T. Papoutsakis, *Biotechnol. Bioeng.* **40**, 978–990 (1992).
63. S. Oh, M. Al-Rubeai, A.N. Emery, and A.W. Nienow, in R.E. Spier, J.B. Grifiths, and W. Berthold eds., *Animal Cell Technology Products for Today, Prospects for Tomorrow*, Butterworth Heinmann, Oxford, 1994, pp. 405–407.
64. J. Klucar and M. Al-Rubeai, *FEBS Lett.* **400**, 127–130 (1997).
65. K.S. Sellins and J.J. Cohen, *J. Immunol.* **139,** 3199–3206 (1987).
66. S.-C. Shih and O. Stutman, *Cancer Res.* **56**, 1591–1598 (1996).
67. Y.A. Lazebnik, S. Cole, C.A. Cook, W.G. Nelson, and W.C. Earnshaw, *J. Cell Biol.* **123**, 7–22 (1993).

See also BIOENERGETICS OF MICROBIAL GROWTH; CULTURE COLLECTIONS; ENERGY METABOLISM, MICROBIAL AND ANIMAL CELLS.

CELL CYCLE, EUKARYOTES

NICHOLAS R. ABU-ABSI
FRIEDRICH SRIENC
University of Minnesota
St. Paul, Minnesota

KEY WORDS

Animal cell culture
Cell cycle
Cell cycle controls
Cell cycle models
Cell cycle regulation
Cyclin-dependent kinases
Cyclins
Hybridomas
Protein secretion

OUTLINE

INTRODUCTION

The growth of eukaryotic cells is governed by a complex array of biochemical reactions that are collectively known as the cell cycle. The cell cycle controls are intimately coupled to the internal and external states of the cell and thus ultimately determine when a cell devotes the majority of its energy to growth, DNA replication, successful cell division, or cell maintenance. A recent explosion in the amount of research devoted to cell cycle controls has revealed that the activity of enzymes known as cyclin-dependent kinases (CDKs) regulates the order of cellular events that culminate in cell division. Furthermore, these CDKs are themselves subject to strict controls that govern their expression, activation, and degradation, giving a cell the freedom to respond to a variety of situations. An overview of the findings from the numerous biological studies that have been conducted on cell cycle control is presented here. Additionally, because the cell cycle controls the most fundamental aspects of cell growth and division, it can be linked to the dynamics of many other cellular processes important to cell culture such as protein accumulation and secretion. Therefore, in order to achieve a more complete understanding of the cell cycle, a discussion of some of the investigations that have coupled the cell cycle to the dynamics of cellular process is presented. Indeed, the framework diagrammed here offers many opportunities for further research that can be used to engineer more efficient methods of cell culture.

THE BIOLOGY OF THE CELL CYCLE

The cell cycle comprises a group of complex interrelated events that exists to ensure that cells successfully divide to form two normal daughter cells. The eukaryotic cell cycle was originally divided into two stages, interphase and mitosis, as observed by light microscopy. Interphase was defined as the period between two successive rounds of cell division, whereas mitosis was identified as the period when the cell undergoes many structural changes, divides it chromosomes, and undergoes cell division. The advent of biochemical and genetic techniques later showed that interphase consists of many cellular processes necessary for the regulation of cell proliferation that can not be discerned under the light microscope. The cell cycle was then partitioned into four separate phases: G_1 (or gap 1), DNA synthesis (S phase), G_2 (gap 2), and mitosis (M phase), where G_1 refers to the gap between mitosis and DNA synthesis, and G_2 to the gap occurring after DNA synthesis and prior to mitosis.

Understanding the complex network of biochemical events that guide a cell through successful completion of cell division is of utmost importance in understanding how cells behave in response to different situations. There has been an explosion in the amount of cell cycle related research being done in the last 10 years. This is due, in part, to the fact that a wide variety cellular processes are thought to be somehow regulated by cell cycle events and proteins. Additionally, improper control of cell cycle processes in higher eukaryotes has been linked to the formation of cancerous tumors. Here we discuss the biochemical events that occur in the cell cycle and the control mechanism that maintains the temporal order of these events.

The G_1 and G_0 Phases

Upon initial inspection, it would seem that the G_1 phase of the cell cycle is a particularly uneventful time period (or gap) between the more spectacular events that occur during mitosis and DNA synthesis. Closer analysis of this interval, however, reveals that the cell is closely monitoring its environmental and internal state to determine whether the cell has sufficient resources and an intact genome to safely complete the rest of the cell cycle. In contrast to the other phases of the cell cycle, G_1 is highly variable between different cell lines and growth conditions. For example, if certain factors are not available, some cells do not progress toward S phase, but instead may enter a resting state called the G_0 phase, or quiescence. During this time, the cell undergoes a significant reduction in RNA and protein synthesis and will remain in this arrested state, with unduplicated DNA content, until the cells receive the proper stimuli.

The sequence of events that takes place during the G_1 phase has been deduced by observing how cells respond when cultured in different environments. For instance, when cultured cells are deprived of serum containing growth factors, the synthesis of new proteins is drastically reduced and a portion of newborn cells cease to advance through the cell cycle. Through observations made using time-lapse cinematography on a culture of Swiss 3T3 cells, it was determined that only those cells that exited from mitosis less than 4 h prior to growth factor deprivation were arrested in G_0, while all other cells continued to mitosis as normal (1). This observation led to the division of G_1 into two subphases. The first subphase consists of cells that are still dependent on growth factors for cell cycle progression and are referred to as G_1 postmitotic (G_{1pm}) cells, whereas cells that can proceed through the rest of the cycle are called G_1 presynthesis phase (G_{1ps}) cells. The transition point where cells pass from G_{1pm} to G_{1ps} is commonly called the restriction point in mammalian cells (2) and is analogous to "Start" in yeast. During G_{1pm}, progression through the cycle is highly dependent on growth factors and amino acids (3) that are necessary for the production of certain regulatory proteins (called cyclins) that are discussed later. It is thought that the purpose of the G_{1ps} period is to allow cells to grow, perhaps so that later in the cycle, the cell can focus its energy on other cellular processes such as DNA replication in S phase and reorganization of the cellular infrastructure during mitosis. This also explains the fact that the G_{1ps} portion of G_1 accounts for nearly all the temporal variability in the cell cycle both within a population and between cell lines (2,3). For example, after having passed the restriction point, larger cells may proceed almost immediately to the S phase, while smaller cells may linger in G_{1ps} for up to 10 h before beginning the transition to S phase (1). Additionally, although growth factors are not required to promote S phase entry once past the restriction point, the length of G_{1ps} is significantly increased when factors such as insulin and insulin-like growth factor (IGF) are lacking in the environment, presumably due to a decrease in the net protein synthesis. Under these conditions, cells undergo mitosis with a significantly reduced protein content (2), suggesting that different growth factors play an important role in growth and progression through the cell cycle.

A culture of 3T3 cells arrested in G_0 by serum deprivation can reenter the cell cycle and proceed through the restriction point only after stimulation by the addition of platelet-derived growth factor (PDGF) (1); this process is referred to as competence. Factors that initiate competence, however, vary between cell lines (4), and several other growth factor/cell-line combinations have been described (5). The biochemical effects of competency factors such as PDGF are discussed later. Time-lapse cinematography also revealed that cell cultures that lack serum do not contain cells arrested all at one point, but at numerous times throughout G_{1pm}. Additionally, when stimulated to proliferate, arrested cells require an extra 8 h in G_1 before proceeding to S phase, presumably so the cells can recover from G_0 (2).

The DNA Synthesis Phase

The S phase is the period when the cell replicates its entire genome so that each daughter cell will receive a complete, intact set of chromosomes. Due to the immense size of mammalian genomes, multiple bidirectional replication sites are necessary for the synthesis of new DNA to be accomplished in such a short time (approximately 8 h in most cell lines). In addition to chromosome duplication, the cell must also synthesize proteins such as histones for the formation of nucleosomes and the chromosomal scaffolding. The replication of DNA has been shown to be a highly ordered process. This was observed by culturing cells with a thymidine analog called bromodeoxyuridine (BrdU). Cells in the process of synthesizing DNA take up BrdU and integrate it into newly formed DNA. Cells can then be separated by their position in S phase, based on the amount of BrdU in the DNA, to determine the exact time of replication of certain genes (6–8). These studies revealed several aspects of the replication of certain genes. First, individual genes within a cell line are replicated at defined periods in the S phase, with many of the more active genes replicated earlier than others. Furthermore, newly replicated DNA is packed into nucleosomes a short distance behind the replication fork (1).

It is also very important that the cells not replicate the same chromosome more than once; the mechanism that ensures that this does not happen is known as the rereplication block (reviewed in Ref. 2). Early cell fusion experiments on human HeLa cells showed that, prior to S phase, nuclei could be induced to undergo DNA synthesis early, whereas nuclei that have already undergone a round of DNA synthesis do not undergo another round of DNA synthesis (3). Later cell fusion experiments showed that nuclei with replicated DNA can undergo another round of DNA synthesis before mitosis after the nuclear membrane has been permeablized and then repaired (4). A model involving a replication-licensing factor (RLF) was then proposed to describe these phenomena. DNA would be replicated only if RLF was bound to the chromatin to be broken down by the replication machinery during the next S phase. During mitosis, the nuclear envelope breaks down, allowing new RLF to access the chromosomes and bind to them, thus licensing the chromosomes to undergo a single round of DNA replication. Support for this model was gained when two protein complexes, called RLF-M and RLF-B, were shown to be required to modify chromatin before replication is allowed to proceed (5). These protein complexes were shown to bind to chromatin in late mitosis and be gradually removed during S phase (6,7). The rereplication block model described earlier has recently been modified to accommodate a more detailed understanding of the proteins and mechanisms involved (2). Chromosomes in G_1 must first be "primed" by the establishment of a pre-replication complex (pre-RC) containing an origin-recognition complex (ORC), an RLF complex, and other proteins. After initiation of DNA synthesis, the pre-RC is disrupted to become the post-RC, which still contains the ORC. The ORC may be some sort of "docking" factor that allows other components of the pre-RC to assemble on the DNA.

The G2 Phase and Mitosis

Cells enter the G_2 phase upon completion of DNA synthesis, and this period lasts between 2 and 4 h, culminating with the onset of chromosome condensation. During this time, duplication of the centriole is completed, and cells produce many mRNAs and proteins that will be used during mitosis. Cells may also use this period of time before mitosis to ensure that the cell is sufficiently large and is ready to undergo cell division.

Mitosis is generally the shortest phase of the cell cycle, lasting only 15 min to 1 h, depending on the cell line. However, visual inspection of a cell culture reveals that this is a very physically active period for the cell in that it goes through many structural changes in preparation for cell division. Mitosis is traditionally divided into six stages that describe the mechanical events that occur during this period: prophase, prometaphase, metaphase, anaphase, telophase, and cytokinesis. The beginning of mitosis, known as prophase, is marked by condensation of the chromosomes into compact structures that allow sister chromatids to be more easily separated. The cytoskeletal microtubules begin to disassemble, and the mitotic spindle, a weblike array of microtubules that mediates the events in mitosis, begins to form. Also during this phase, the Golgi apparatus breaks up into tiny vesicles, and pinocytosis and RNA synthesis stop. Prometaphase begins with the disruption of the nuclear envelope, followed by the attachment of the chromosomes to the microtubules that make up the mitotic spindle. The microtubules attach to protein structures on the DNA called kinetochores, which as we shall see are very important in regulating progression through mitosis. During metaphase, the mitotic spindle arranges all the chromosomes in a single plane across the middle of the cell, called the metaphase plate. After alignment of the chromosomes, the cell enters anaphase, which is when sister chromatids are pulled to opposite sides of the cell by the microtubules so that there is a complete genome on either side of the cell. Microtubules not attached to kinetochores then elongate, pushing the spindle poles further apart. Telophase then begins, resulting in the formation of two nuclei, one around each set of chromosomes on opposite sides of the cell. The cell then begins the process of cell division, called cytokinesis, when a ring of actin and myosin that has formed around the center of the cell begins to contract. Upon completion of this process, two new daughter cells are formed, each containing a complete, intact genome and cytoplasmic organelles sufficient to support further growth and proliferation.

The temporal organization of the cell cycle phases is shown in Figure 1, where the numbers and arrows represent when the corresponding cyclin–CDK complexes listed in Table 1 are active.

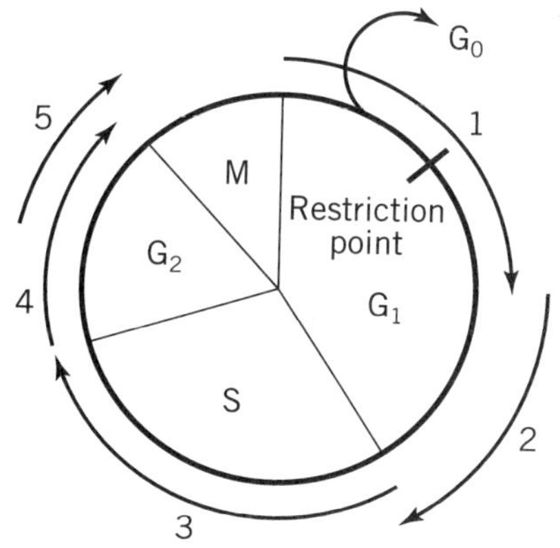

Figure 1. The temporal order of the cell cycle. Cells exit mitosis after cell division and are sensitive to the presence of growth factors until they pass the restriction point. The large block arrows depict the active period of the best-defined cyclin–CDK complexes. The numbers next to the arrows correspond with the numbers assigned to the various complexes as defined in Table 1.

CONTROL OF THE CELL CYCLE

Successful completion of cell division requires that the cell reside in an environment capable of supporting growth, and that the cell cycle proceed as normal. As discussed previously, cells require the presence of growth factors and essential amino acids to support the synthesis of new proteins. In addition, the immediate environment must not be overcrowded; daughter cells must have room to grow and subsequently divide. If any of these conditions is not met, cells enter the G_0 phase, which eventually results in cell death if environmental conditions are not improved. Similarly, if the events in the cell cycle do not occur in the proper sequence, daughter cells may not inherit the proper genetic material or organelles to properly function, and they will die. Therefore, cells must possess some means of sensing whether the environment is able to support growth. There must also be a control mechanism in place that regulates the sequence of events that make up the cell cycle. An understanding of the mechanisms that control cell growth is of interest to many scientists including oncologists and biochemical engineers. Oncologists may be able to use the knowledge of the events regulating the cell cycle to develop therapies against certain types of cancers, whereas the control network of cells in culture can be engineered to reduce the culture's dependence on serum.

Central to the mechanisms that control the events of the cell cycle are the checkpoints. Checkpoints are points in the cell cycle where certain criteria must be met before the cell cycle can proceed any further. For example, the restriction point in G_1 is a checkpoint because cells not cultured in the presence of growth factors are arrested in G_0 and not allowed to proceed through the cell cycle until growth factors are supplied. Indeed, several other checkpoints that control the order of the cell cycle phases and the events within each phase exist throughout the cell cycle. For instance, cells with unduplicated DNA do not enter mitosis, and chromosomes are replicated only once in each cycle. Additionally, cells with damaged DNA do not undergo mitosis, rather they become arrested in G_2 to allow the DNA repair machinery to correct any defects in the genetic material. Controls that mediate the order of events in mitosis also exist. For example, entry to metaphase is prevented until the chromosomes are lined up at the metaphase plate. As we will see, complexes made up of two classes of proteins are involved in regulating all these events throughout the cell cycle. These two classes of proteins are cyclins and cyclin-dependent kinases (CDKs).

Partners in Cell Cycle Regulation: Cyclins and CDKs

Cyclins were originally defined as molecules that accumulated throughout the cell cycle in marine invertebrate eggs and were then degraded during each mitosis (33). Several years later, the amino acid sequences of various cyclins were compared, and a consensus sequence of approximately 100 amino acids was found and dubbed the "cyclin box" (34). Cyclins are now defined on the basis of the presence of the cyclin box in the native sequence. Currently, 12 cyclins have been identified in mammalian cells (cyclins A–H, with two A-type, three B-type, and three D-type cyclins) (35). The cyclin box domain facilitates binding to CDKs, whose kinase activity is dependent on being bound

Table 1. Activity of the Major Cyclin–CDK Complexes

Complex (see Fig. 1)	Cyclin–CDK complex	Cell cycle phase	Function	Ref.
1	Cyclin D–CDK4	G_0, early to mid-G_1	Stimulates exit from G_0 and progression through the restriction point; phosphorylates pRb	8, 9, 10, 11–13
2	Cyclin E–CDK2	G_1–S transition	Triggers S phase entry; activates cdc25A	14, 15, 16, 17
3	Cyclin A–CDK2	S	Phosphorylates DNA synthesis machinery; inhibits E2F/DP1 transcription factor	18, 36, 37, 91
4	Cyclin A–CDK1	G_2, M	Triggers formation of mitotic spindle	22, 23, 24
5	Cyclin B–CDK1	G_2–M transition, M	Triggers entry into M; responsible for many structural changes in mitosis	18, 22, 25, 26–30
6	Cyclin H–CDK7	All cell cycle phases?	Cyclin activating kinase (CAK); phosphorylates other cyclin–CDK complexes	31, 32

Note: The numbers listed with each complex correspond with the numbers listed in Figure 1. See the text or the references for more detail on the substrates of the complexes.

to a cyclin molecule (36) and are thus defined as proteins that require binding to cyclins in order to attain activity (37). There are seven known CDKs in animal cells (named CDK1–CDK7). Different cyclins may interact with different CDKs at specific points in the cell cycle, and each of these complexes is specific for substrates with differing functions. Because several of the complexes have similar substrate specificity in vitro, the change in specificity is thought to be due not only to changes in the affinity toward different substrates, but also to localization of complexes in distinct parts of the cell by various cyclins (38). We first briefly discuss the different cyclin–CDK complexes that form throughout the cell cycle and then review the regulation of these complexes. The major cyclin–CDK complexes that are active throughout the cell cycle are listed in Table 1 along with their corresponding times of activity and their functions.

The major studies that resulted in the discovery of G_1 cyclins were performed in budding yeast (39). It was shown that three cyclin molecules, Cln1, Cln2, and Cln3, have important functions in advancing cells through the restriction point (called Start in yeast). In addition, inactivation of all three molecules is required to cause G_1 arrest of the cell cycle. By searching for factors that could complement the function of Clns in yeast, several mammalian cyclins were isolated: cyclin C, cyclin D, and cyclin E (40). Other studies revealed several other G_1 cyclins, cyclins G and F, however, neither the roles that these cyclins play nor their CDK partners are known (41). It has been shown that cyclin C mRNA synthesis reaches its peak in mid-G_1, and cyclin G mRNA synthesis starts soon after growth-factor stimulation of quiescent cells (42). Conversely, cyclins D and E are fairly well characterized and have been the focus of much attention because defects in control of these proteins have been implicated as a cause of cancer (18).

Although D-type cyclins have been shown to associate with CDK2, CDK5, and CDK6, they seem to interact primarily with CDK4 (43). Transcription of various combinations of D-type cyclins (cyclins D1, D2, and D3) is stimulated by different mitogens (growth factors), depending on the cell type, and their synthesis throughout the rest of the cell cycle is maintained by the presence of growth factors (44). Removal of growth factors halts the synthesis of D cyclins, resulting in a rapid decrease in their levels because they have very short half-lives (45). Furthermore, in cycling cells, the level of cyclin D hardly varies, and only a small peak showing its accumulation is observed. As noted previously, growth-factor deprivation has no effect on cells that have passed through the G_1 restriction point, and these cells continue to cycle until their progeny become arrested in early G_1. Additionally, microinjecting cells with cyclin D1–specific antibodies results in their cell cycle arrest only during early to mid-G_1 (8), whereas overexpression of cyclin D1 results in a reduction in the length of G_1, the size of the cell, and growth-factor dependency (46,47). These findings indicate that the primary function of D-type cyclins is to link the cell cycle machinery with the extracellular environment, which permits the cell to become committed to DNA synthesis when in a favorable environment or become arrested in G_0 under unfavorable conditions.

Cyclin E is expressed soon after cyclin D expression in G_0 cells stimulated with growth factors, and its level peaks late in G_1. It has been shown to interact with a single catalytic subunit, CDK2 (48). In contrast to D-type cyclins, cyclin E is synthesized periodically and is rapidly degraded soon after entry into S phase, thus adhering to the original definition of a cyclin (49). Similar to cyclin D, cyclin E function has been shown to be essential for progression through G_1, because microinjection of anti–cyclin E antibodies prevents cells from entering S phase, whereas its overexpression results in acceleration toward S phase (9). This acceleration is, however, balanced by an increase in the lengths of the S and G_2 phases so that the length of the entire cycle remains unchanged. Because an increased amount of cyclin E advanced progression to the S phase early, it is possible that the cyclin E–CDK2 complex is responsible for initiating the synthesis of proteins that are used during DNA synthesis. This hypothesis accounts for the increase in the length of the S and G_2 phases following increased expression of cyclin E. These findings, when taken together, suggest that the cyclin E–CDK2 complex is involved in regulating the passage from G_1 into S (41).

Upon entry into S phase, cyclin E is degraded and cyclin A takes its place as the partner of CDK2 (50). Transcription of cyclin A is begun immediately after entry into S phase, and its inhibition by microinjection of anti–cyclin A antibodies allows, at most, only 10% of the DNA to be replicated (51) and may also disrupt the temporal relationship between DNA synthesis and mitosis (10). As a cell progresses into the G_2 phase, cyclin A is no longer associated with CDK2, however, it forms a new complex with CDK1 (also known as cdc2 due to its homology with the budding yeast kinase of the same name) to regulate progression through G_2 and into mitosis. It has been shown that CDK1 forms complexes with both cyclin A and cyclin B near the G_2–M transition (52). These complexes are located in different parts of the cell because cyclin A accumulates in the nucleus, and cyclin B resides in the cytoplasm until the breakdown of the nuclear envelope during mitosis (53). Once they have completed their respective functions, both cyclins are then rapidly degraded later in mitosis.

Experiments done in fission yeast suggest that the cell monitors the presence of cyclin B, because cells undergo another round of DNA replication when cyclin B–CDK1 complexes are not present (14). This implies that the level of cyclin B–CDK1 complexes, both active and inactive, are constantly monitored, and the presence of these complexes maintains cells in G_2. Although this theory has not been investigated in other eukaryotes, it is not entirely unlikely given the high degree of conservation of cell cycle controls among eukaryotic cells.

Regulation of the Cyclin–CDK Complex Activity. Three different mechanisms allow tight control over the myriad of cyclin–CDK complexes: (*1*) cyclin transcription and degradation, (*2*) CDK phosphorylation and dephosphorylation, and (*3*) binding of the cyclin–CDK complexes to various inhibitory proteins. These three forms of cyclin–CDK regulation interact to form an intricate regulatory network that allows cells to temporally control events throughout the cell cycle and to ensure that these processes are successfully completed. These three mechanism of cyclin–CDK complex regulation are depicted in Figure 2.

Intracellular concentrations of different cyclins, and thus the activity of cyclin–CDK complexes, are regulated by synthesis and degradation. There are two different pathways by which cyclins can be degraded, depending on the type of cyclin. Cyclins D and E, which function in early G_1 and the G_1–S transition, respectively, are inherently unstable and thus have short half-lives of approximately 30 min. The instability of these cyclins has been attributed to the presence of a PEST sequence near the C-terminus of the cyclin box (54). In contrast, cyclins A and B, which function in the S, G_2, and M phases, are stable throughout interphase, even when they are not bound with CDKs. However, these cyclins are degraded during mitosis in the proteosomes through a ubiquitin-mediated pathway (55). These proteins contain a sequence of amino acids near the amino terminus, called the destruction box, that marks them for destruction. Degradation of these proteins is also regulated so that cyclin–CDK complexes that are necessary for cell cycle progression are not inactivated before their specific tasks are successfully completed. For example, it has been shown that proteolysis of cyclin A and B does not occur in *Xenopus* egg extracts until purified cyclin B–CDK1 complexes are added (19). This suggests that the proteolysis pathway is initiated by active cyclin B–CDK1 complexes. Conversely, the addition of purified cyclinA–CDK1 complexes to the extracts had the opposite effect of delaying cyclin degradation triggered by cyclin B–CDK1 kinases (20). The fact that only cyclin B–CDK1 complexes can activate proteolysis may explain why cyclin degradation does not occur until correct assembly of the mitotic spindle is achieved in metaphase (56). The exact mechanism of regulation of cyclin degradation is unknown, but it has been hypothesized that cyclin B–CDK1 activates a cyclin-specific enzyme that catalyzes the conjugation of ubiquitin to cyclins (22).

Cyclin–CDK complexes are also regulated by phosphorylation and dephosphorylation. Specifically, phosphorylation on T160/161 (depending on the CDK) is necessary for activation of the complex, whereas phosphorylation of T14 and Y15 residues inhibits cyclin–CDK activity. The T160 residue is located on a T loop, which is conserved in all CDKs. When not phosphorylated,

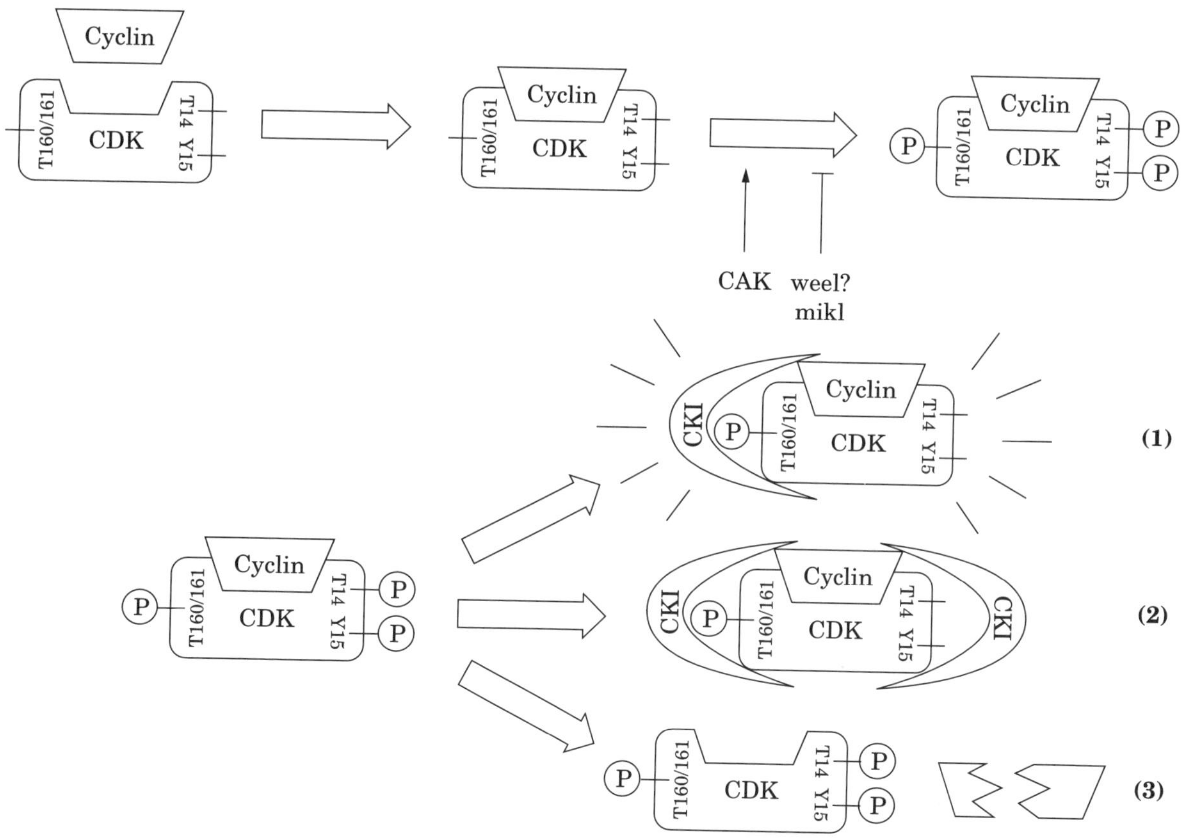

Figure 2. Control of cyclin–CDK complex activity. Cyclin–CDK activity is governed by three mechanisms: phosphorylation/dephosphorylation reactions, cyclin-dependent kinase inhibitors (CKIs), and cyclin expression and degradation. After assembly of the cyclin–CDK complex, the activity of CAK activates the complex while phosphorylation by Mik1 and Wee1 inhibits complex activity. The phosphorylated complex is then activated by the phosphatase activity of cdc25 and is denoted as complex 1 in this figure. Complex 2 has the correct residues phosphorylated, which is necessary for activation; however, it is bound by two CKIs and is inactivated. Complex 3 is shown inactivated by cyclin degradation.

this loop prevents kinase activity by covering the substrate binding domain. Phosphorylation at this site serves two purposes: it increases the affinity between the cyclin and the CDK and therefore stabilizes the complex, and it alters the conformation of the T loop, allowing access to a substrate (57).

The kinase that is responsible for phosphorylation of this residue on several CDKs is also a cyclin–CDK complex. This complex, commonly referred to as CDK-activating kinase (CAK), is made up of cyclin H, CDK 7, and a third 32-kDa protein (58). Although the level of CAK activity has not been shown to vary in a cell cycle–dependent manner, it is possible that cyclin H is regulated by its synthesis and degradation (56). The kinase responsible for activating CAK has not been identified, but it has been shown not to be an autophosphorylation process (58,59). Further studies done with this complex have shown that it may play a role in activating the RNA polymerase II C-terminus domain kinase of transcription factor IIH (60,61). This suggests that cyclin H and CDK7 may have a more diverse role than merely activating other CDKs (56).

Cyclin–CDK complexes are also subject to negative regulation through the phosphorylation of the T14 and Y15 residues. These residues are found within the ATP binding site of the enzyme. Phosphorylation of the Y15 residue inhibits kinase activity by interfering with the transfer of a phosphate group to a bound substrate (62), whereas phosphorylation of the T14 residue acts by interfering with binding of ATP (25). The regulation of phosphorylation on these residues is not yet well defined in animal cells. In addition, only the process of tyrosine phosphorylation is known in yeast because phosphorylation of the threonine residue is not necessary in this organism (35). In yeast, the tyrosine residue is phosphorylated by the Wee1 and Mik1 kinases (63). The Wee1 kinase is in turn regulated by the Nim1 kinase, which inhibits Wee1 activity. The kinase responsible for phosphorylation of the Y15 residue in animal cells has been isolated (64) and is closely related to Mik1 (35). The dephosphorylation of these residues, which is necessary for cyclin–CDK activation, is performed by the phosphatases Cdc25A, B, and C (31). Cdc25A has been shown to be activated by the cyclin E–CDK2 complex in early S phase, which then activates cyclin A–CDK2 complexes. The Cdc25C phosphatase activates cyclin B–CDK1 complexes at the transition from G_2 to mitosis (35); however, the function of cdc25B is as yet unknown.

The activity of cyclin–CDK complexes is also controlled by certain proteins known as cyclin-dependent kinase inhibitors (CKIs or CDIs) that bind to these complexes and repress their activation. Several different inhibitors have been isolated from mammalian cells that interact with either different cyclin–CDK complexes or CDKs themselves. These proteins are commonly named according to their molecular weights, such as the 21-kDa inhibitor known as p21. Other cyclin–CDK inhibitors are p27, p15, and p16. These inhibitors are intricately involved in regulating progression through the G_1 phase, and have thus been the subject of recent reviews (32,65,66).

Immunoprecipitation studies have revealed a very interesting fact: cyclin D1 is commonly found among a quaternary complex of proteins that includes either CDK4 or CDK6, p21, and proliferating-cell nuclear antigen (PCNA) (discussed later) (43). The p21 protein acts as an inhibitor by binding directly to cyclin B–CDK1, cyclin E–CDK2, cyclin A–CDK2, and cyclin D–CDK4 complexes. These cyclin–CDK complexes containing p21 are, oddly enough, still enzymatically active in vitro, which directly conflicts with studies showing that p21 is an inhibitor of CDK activity and is able to induce cell cycle arrest (67,68). It was later demonstrated that active cyclin–CDK–p21 complexes isolated from cycling cells could be inactivated by the addition of more p21 (69), suggesting that the stoichiometry of p21 binding to these complexes is very important. For example, in cycling cells, there is only enough p21 present so that each cyclin–CDK complex has a single p21 molecule and thus remains active. However, when the amount of p21 is increased or the cyclin–CDK pool depleted, the ratio of p21 to cyclin–CDK complexes increases so that more than one p21 molecule binds to each complex, resulting in inhibition of the complex's kinase activity. We explore the circumstances under which this sort of inhibition becomes important in a later section.

A second inhibitor that has been identified is called p27. p27 and p21 have been shown to have highly similar amino acid sequences near their amino terminus domains. In addition, they also share the same range of specificity for cyclin–CDK complexes. The cell cycle arrest of certain cell lines in response to some environmental signals such as contact inhibition and transforming growth factor β (TGF-β) has been shown to be due, in part, to inactivation of cyclin–CDK complexes by this inhibitor (70), and the exact mechanism of this type of arrest will be described shortly. A recent discovery has given some insight into one of the mechanisms by which this protein inhibits cyclin–CDK activity. Cyclin–CDK complexes that have not yet been activated by CAK phosphorylation can be bound by p27, which then blocks the CAK's access to the threonine residue found on the CDK subunit (71). Presumably, another mechanism exists for inhibition of active complexes.

The main function of the p27 inhibitor seems to be to prevent the G–S transition from occurring when the need arises by inhibiting the activity of cyclin E–CDK2 complexes (72). In normal cycling cells, the p27 inhibitor seems to be dispersed among the different cyclin–CDK complexes that are active during G_1, therefore it does not have a significant impact on any essential events. However, in response to stimulation by TGF-β, it has been suggested that the expression of CDK4 is downregulated, resulting in an increased amount of p27 bound to cyclin E–CDK2 complexes (73).

The p16 inhibitor was originally found associated with CDK4 in a transformed cell line (74). Unlike the p21 and p27 inhibitors, p16 was found to associate specifically with the cyclin D CDKs, specifically, CDK4 and CDK6. It competes with cyclin D for the CDK subunits and may even act to disrupt cyclin D–CDK complexes. This inhibitor is thought to play an important role in regulating growth because the p16 gene has been shown to be deleted in many tumor-cell lines (75).

Examination of TGF-β-arrested cells led to the discovery of another inhibitor, p15, because it was shown to be bound to CDK4 (76). Similar to p16, p15 binds to CDK4 and CDK6, competes with cyclin D, and may be involved in arresting cells treated with TGF-β in G_1. TGF-β stimulates the synthesis of p15 so that its level is increased approximately 30 times. This allows p15 to sequester the CDK4 complexes so that a much larger amount of the p27 inhibitor binds to and inactivates cyclin E–CDK2 complexes, resulting in G_1 arrest.

The Cyclin–CDK Complexes in Action

Now that we have discussed the events that occur in the cell cycle and defined the major regulatory proteins, we are in a position to examine the mechanisms that control cell cycle progression. Indeed, control of the activity of the cyclin–CDK complexes is at the heart of cell cycle regulation because these complexes either directly or indirectly act on cellular constituents that are capable of bringing about changes in intracellular events. This implies that the proper regulation of these cyclin–CDK complexes is essential for maintaining the temporal integrity of cell cycle events. This is accomplished by repressing the activity of the appropriate kinase through different combinations of cyclin degradation, phosphorylation/dephosphorylation of specific CDK residues, and action of CDK inhibitors. These controls not only maintain temporal stability, but also ensure that DNA synthesis is not carried out when DNA is damaged or that mitosis does not occur in the presence of damaged or misaligned chromosomes.

G_1 to S Phase Control. An important property in normal cells is their ability to sense the state of the extracellular environment, such as the presence of various growth or inhibitory factors or contact inhibition. This communication between the cell and its surroundings takes place in the G_0 and G_1 phases, and the cells respond by becoming quiescent, undergoing programmed cell death (known as apoptosis), or continuing with the cell cycle. The interaction with the environment ceases when the cell passes the restriction point, after which the cell is said to be committed to DNA synthesis. We now consider the machinery responsible for directing the path that the cell will traverse in response to signals from the environment.

Several lines of evidence indicate that the product of the retinoblastoma gene, pRb, is responsible for linking signals from the extracellular domain (77). Specifically, mutation or deletion of this gene is found in many cancerous cell lines that have lost their ability to regulate proliferation (78). Additionally, pRb accumulation was found to coincide with suppression of growth in TGF-β-induced arrest (79), whereas inhibition of pRb results in a resumption of cell cycle progression. Mammalian DNA viruses, whose main purpose is to activate quiescent cells so that they can use the cell's replication machinery to duplicate their own DNA, have been found to seek out the hypophosphorylated form of pRb (80). This led to the belief that pRb is in its active state when it is underphosphorylated, and phosphorylation of this protein results in its inactivation. This is supported by the fact that hypophosphorylated pRb is present in G_0 cells and is phosphorylated later during the G_1 phase. In its active form, pRb binds to several proteins including a heterodimer of proteins from the E2F and DP family of proteins. This heterodimer, called DRTF1/E2F (81), is a transcription factor that stimulates the transcription of numerous proteins necessary for cell cycle progression, such as CDK1, cyclin A, and a variety of proteins that are indispensable during DNA synthesis (82).

In order to further characterize G_1 events, it is necessary to find out what is responsible for inactivating pRb. A likely candidate seems to be the cyclin D–CDK4 complex. The presence of this active kinase is highly dependent on the presence of growth factors that stimulate the synthesis of cyclin D, and thus, the activity of that complex appears during the same period when pRb is inactivated. Moreover, cells that lack functional pRb do not require the presence of cyclin D–CDK4 to proceed through the cell cycle, implying that the primary function of this complex is to inactivate pRb (83). Indeed the belief that pRb is a substrate of this complex was confirmed because cyclin D–CDK4 is capable of binding to and phosphorylating pRb in vitro (84,85). The following feedback loop, involving the levels of cyclin D and the functional state of pRb, has been proposed (86). Cells that lack functional pRb have drastically reduced levels of cyclin D; however, addition of pRb to these cells results in the synthesis of cyclin D. Active kinase complexes are then allowed to form, which act by phosphorylating pRb. The hyperphosphorylated form of pRb down-regulates the levels of cyclin D during late G_1, causing the inactivation of CDK4, which in turn allows pRb to be activated once again by the action of phosphatases, and the cycle begins anew upon completion of mitosis.

Because the CDK inhibitors p21 and p27 are constitutively expressed in cycling cells, these molecules are present throughout G_1. This indicates that in order for cells to advance past the restriction point, the level of cyclin D–CDK4 complexes must increase until it exceeds the number of inhibitor molecules present so that there is at most one CKI per cyclin–CDK complex to allow kinase activation (70). This model not only satisfies the timing of cyclin D–CDK4 activation, but suggests a dual role for this complex. Its primary function is to phosphorylate pRb to promote the transcription of important genes, while its secondary role is to sequester the CKIs so that the cyclin E–CDK2 complex, which is the next kinase to be active in the cell cycle, cannot be inhibited by these molecules, thus allowing the transition to S phase to occur (32). A schematic representation of the control mechanism that regulates progression through G_1 is shown in Figure 3.

As mentioned previously, the activity of cyclin E–CDK2 complexes during the cell cycle has been narrowed down to the transition from G_1 to S. Some of the possible substrates for this complex include the so-called pocket proteins, which include pRb, p107, and p130 (82). The term *pocket protein* refers to those proteins involved in binding to certain proteins, such as transcription factors, and releasing them after activation of the pocket protein by the action of a CDK kinase. The cyclin E–CDK4 complex is localized in the nucleus where it is bound to E2F, p107, and p130 (although it has not been shown to phosphorylate these proteins) (87), suggesting that it too plays a role in regulation of transcription. The E2F subunit that interacts with cyclin E–CDK4 has been shown to differ from the E2F found associated with pRb, which hints that they have dif-

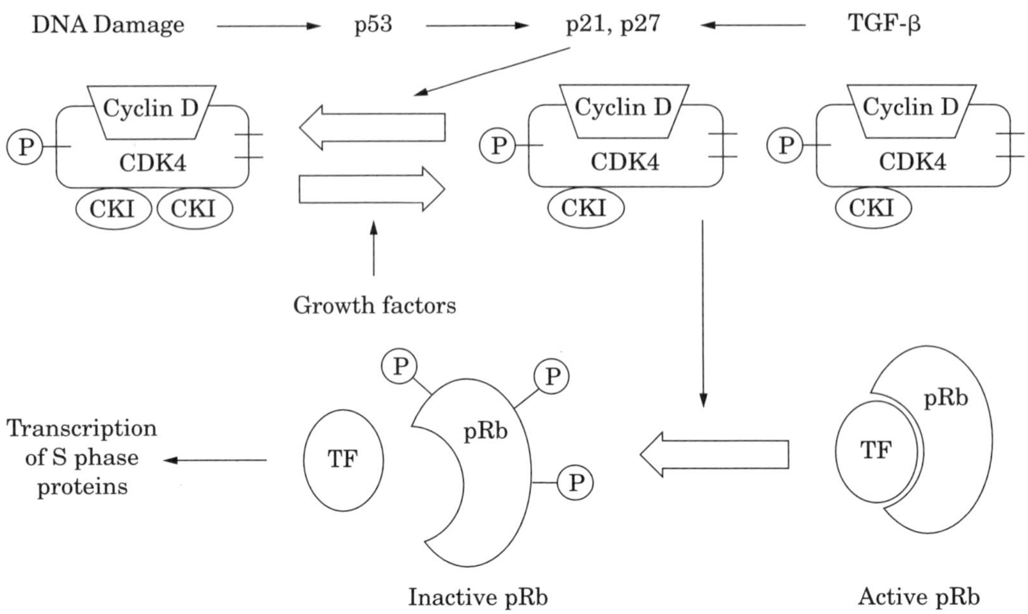

Figure 3. The G_1 control mechanism. Schematic of the control mechanism that governs the progression past the restriction point and through G_1. The presence of growth factors stimulates the expression of cyclin D to promote the formation of cyclin D–CDK4 complexes, increasing the amount of complex to a level above the inhibitory threshold of CKIs. This increase in cyclin D–CDK4 complexes results in kinase activity that then promotes the release of transcription factors. DNA damage results in the expression of p53, which increases the amount of CKIs present to arrest cells in G_1 and allow DNA repair to take place.

ferent properties (88). One confirmed substrate of cyclin E–CDK2 is the Cdc25A phosphatase that is involved in activating CDKs by cleaving the inhibitory phosphates on both a tyrosine and a threonine residue (89). The cdc25A phosphatase has been proposed to activate both cyclin E–CDK2 and cyclin A–CDK2 complexes, creating a positive feedback loop between cyclin E–CDK2 and Cdc25A (Fig. 4a). If this feedback loop does indeed exist, it has been hypothesized that the activation of Cdc25A occurs in response to the completion of some G_1 event such as the duplication of the centrosome or the accumulation of certain DNA synthesis proteins (15).

Experiments with yeast suggest that proteolysis may be involved in activating the cyclin E–CDK2 homologue that plays an important role in the G_1–S transition (11). It has been shown that a protein called Cdc34 is necessary for the G_1–S transition to take place, possibly because it is responsible for initiating the degradation of certain CKIs. Functional homologues of the Cdc34 protein have been isolated in vertebrate cells and are thought to stimulate the degradation of the p27 inhibitor (12). This, in addition to isolation of CKIs by the cyclin D–CDK4 complex, could possibly aid in beginning the onset of the synthesis phase.

In summary, a cell becomes arrested in G_0 in response to some extracellular signal that triggers a decrease in the level of cyclin D below a threshold level of CKIs. This allows p21 and p27 to bind to cyclin D–CDK4 complexes in a 2:1 ratio, resulting in an inhibition of kinase activity. Following stimulation of cyclin D synthesis due to the presence of growth factors, there is an increase in cyclin D–CDK4 kinase activity because there is no longer enough inhibitor present to inactivate all the complexes. This active CDK kinase then serves two purposes: (*1*) to inactivate pRb, allowing the release of transcription factors that induce the synthesis of other cyclins and DNA synthesis proteins, and (*2*) to isolate the CKIs to allow activation of the cyclin E–CDK2 complex. This complex then stimulates the release of additional transcription factors and initiates a positive feedback loop that triggers the onset of the DNA synthesis phase.

Once cells move beyond the restriction point, they are committed to entering the DNA synthesis phase. Cells do, however, have the power to delay the onset of later events to ensure the proper coordination of important processes. For example, failure to repair damaged DNA greatly increases the occurrence of genetic mutations, which can have disastrous consequences for the cell. Thus, an additional checkpoint exists in late G_1 phase, which ensures that the DNA is free from damage and ready to be replicated. We will now see that cells possess the means to sense when there is damage to the genetic material and then halt cell cycle progression to allow the DNA machinery time to repair any defects in the genetic material. It has been determined that the there is a single protein, p53, that is held accountable for inducing G_1 arrest in response to DNA damage (13) (see Fig. 3 for a schematic representation).

By inducing DNA damage in normal cells, it was determined that cells responded by increasing the level of p53 and the CKI p21 (90). The increase in p21 then leads to

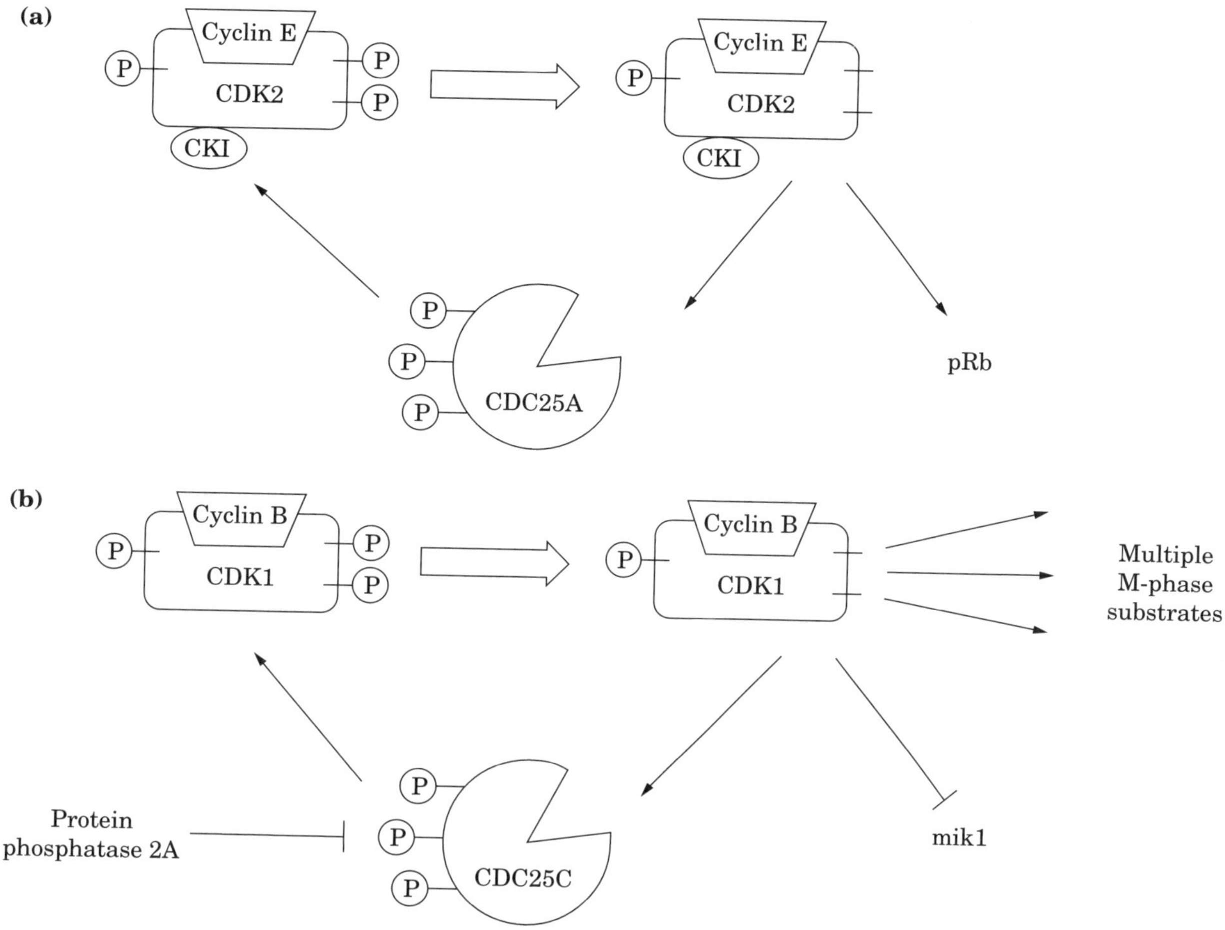

Figure 4. The S- and M-phase entry mechanisms. The positive feedback mechanisms that trigger the entry into S phase (**a**) and mitosis (**b**). The transition to S phase is triggered when an active cyclin E–CDK2 complex activates the Cdc25A phosphatase, which in turn activates more cyclin E–CDK2 complexes, causing a sharp rise in active complexes. These complexes then phosphorylate pRb to stimulate the release of transcription factors. The M-phase transition is triggered by a positive feedback mechanism between cyclin B–CDK1 complexes that is similar to that described in (**a**). In addition, the active cyclin B–CDK1 complex inactivates the mik1 kinase, which phosphorylates the complex on an inhibitory residue, thus contributing to the steep rise in cyclin B–CDK1 activity.

the inhibition of both cyclin D–CDK4 and cyclin E–CDK2 complexes to prevent entry into the S phase. Additionally, p21 also functions to oversee the proper repair of DNA by interaction with PCNA. PCNA is required for both DNA synthesis and repair. When bound with p21, PCNA is no longer able to interact with polymerase-δ, thus preventing DNA synthesis, however, the DNA repair function remains unaffected by the inhibitor (16). Therefore, p21 must inhibit both CDK activity and interaction between PCNA and polymerase-δ in order for DNA repair to be carried out before advancing to the S phase.

S Phase and G_2 Progression. Shortly after the cyclin E–CDK2 complex triggers entry into the S phase, cyclin E is degraded and CDK2 interacts with cyclin A. There is evidence that cyclin E may be degraded in a ubiquitin-mediated pathway, as in budding yeast, but this has not been confirmed (91). The cyclin A–CDK2 complex is localized in the nucleus where it associates with p107, p130, and E2F (17). Interacting with E2F allows the kinase to phosphorylate the DP1 subunit of the transcription factor complex, which inhibits its DNA binding capability (92). This suggests that when cyclin A synthesis begins in early S phase, it aids in halting the transcription of genes triggered by the E2F complex in the late G_1 phase. No other substrates of this cyclin–CDK complex have been identified except for helicase RF-A, which is one component of the DNA synthesis machinery (35).

Upon completion of DNA synthesis, CDK2 is degraded and cyclin A forms a new complex with CDK1 (93). It is not known if this complex has any function in the G_2 phase; however, its activity has been implicated in triggering initial events that occur in mitosis. During G_2, cyclin B begins associating with CDK1 in the cytoplasm and is transported to the nucleus just prior to mitosis. Mitosis begins when the cyclin B–CDK1 complex is activated, and it has been suggested that a checkpoint mechanism exists that prevents activation of this complex until duplication of the centriole has been completed (94). One model that describes this transition involves the regulation of Cdc25C activity by downregulating protein phosphatase 2A (PP2A), which controls Cdc25C activity (15). Signals could be initiated from the DNA replication machinery or from the proper organization of chromatin, which would inhibit the action of PP2A. This allows Cdc25C to become activated and dephosphorylate the inhibitory residues on CDK1, causing its activation. A positive feedback loop is then initiated (Fig. 4b), whereby active cyclin B–CDK1 complexes activate more Cdc25C and, in budding yeast, inactivate the Wee1 kinase by phosphorylizing a residue that inhibits its action (95). This mechanism brings about a sharp transition into mitosis almost immediately after a small amount of CDK1 is activated.

Another mechanism exists that allows cells with damaged DNA to become arrested in G_2. In response to DNA damage, transcription of the CHK1 kinase is initiated. This kinase targets Cdc25C and phosphorylates the serine-216 residue, which allows formation of a binding site for the 14-3-3 protein that inhibits Cdc25C activity (96). This then allows the DNA repair machinery to work because the cyclin B–CDK1 complex can not be activated by cdc25C.

Supervising the Order of Events in Mitosis. The point when the cyclin A–CDK1 complex is most active occurs just after the cell enters into mitosis, which coincides with nucleation of the centrosomes and the formation of the long microtubular extensions of the mitotic spindle. Indeed it has been shown that cyclin A–CDK1 activity has a profound affect on the nucleating ability of the centrosomes in *Xenopus* extracts (97). Once spindle formation is complete, cyclin A begins to be degraded and the activity of cyclin B–CDK1 complexes takes over. This complex is either indirectly or directly involved in regulating almost all of the structural reorganization that takes place during mitosis.

Chromosome condensation begins in prophase when cyclin B–CDK1 phosphorylates histone H1, transcription factors, and several other proteins (98). Phosphorylation of these proteins is thought to weaken the interaction between these proteins and the DNA, allowing the chromosomes to pack more tightly (21). In addition, after formation of the mitotic spindle is complete, the cyclin B–CDK1 complex reduces the nucleating ability of the centrosomes, which causes the microtubules to shorten to a length suitable for the alignment and separation of the chromosomes (35). The breakdown of the nuclear envelope, which marks the beginning of prometaphase, and the reorganization of the cytoskeleton are also brought about by phosphorylation. The nuclear lamina are present in a hypophosphorylated form in an intact nucleus, and when phosphorylated by the cyclin B–CDK1 complex, the lamina network breaks down, contributing to nuclear disassembly (99). Lamina phosphorylation itself is not sufficient to achieve disruption of the nucleus, however, and these mechanisms remain unclear (56). Caldesmon has also been shown to be a substrate of the mitotic CDK complex, and microtubular reorganization ensues after its phosphorylation. Caldesmon binds to actin and calmodulin, and its phosphorylation weakens its affinity for actin, causing the microfilaments to dissociate (23). This process causes the cells to take on a round shape that is characteristic of anchorage-dependent cells undergoing mitosis. The activity of the cyclin B–CDK1 complex is also responsible for repressing the formation of the contractile bundle until the cell enters anaphase (100). The CDK1 complex phosphorylates myosin, which prevents its association with actin filaments so the contractile bundle is unable to form. This inhibitory function of the CDK complex stays in effect until cyclin B is destroyed in the ubiquitin degradation pathway. The activity of the cyclin B–CDK1 complex is, ironically, responsible for the initiation of its own destruction (101), which stays in effect until this pathway is shut down in late G_1, possibly by the action of the cyclin E–CDK2 complex (102).

The final regulatory mechanism that exists in the cell cycle is a checkpoint that prevents the onset of anaphase until several other structural events have taken place in the cell. Some of these events include the assembly of a bipolar spindle, attachment of the chromosomes to the spindles through the kinetochore, and the migration of the chromosomes to the metaphase plate. This suggests that proper completion of these processes is somehow monitored to ensure that anaphase does not begin prematurely (24). Several experiments (recently reviewed in Ref. 26) suggest that one of two events signals that the cell is ready to undergo the chromosome separation of cell division. These events are either the attachment of the chromosomes to the mitotic spindle (27) or the presence of tension on the kinetochores due to their attachment to a bipolar spindle (28). Because all the structural reorganization processes that occur in mitosis depend on the successful completion of the other events, failure to complete any one of these processes could result in the failure of the signaling event to occur. This would then stimulate a cascade of phosphorylation events, culminating in cell cycle arrest in mitosis.

Understanding Cell Cycle Regulation through Mathematical Modeling

Understanding the complex array of biochemical events we have outlined can be difficult at times. Moreover, our knowledge of the controls responsible for cell cycle regulation is rapidly becoming more and more complicated. This complexity makes it extremely difficult to understand the mechanisms involved through logical thinking alone. Mathematical modeling has been used for the purpose of explaining the various physiological events that are commonly observed in different situations, and the events observed in *Xenopus* oocytes and yeast cells. The cell cycle controls discussed in this review have been described for animal cells. However, the majority of initial experiments on the cell cycle were carried out in yeast cells and in *Xenopus* oocyte extracts. The components that make up the control systems in these organisms are largely identical to those of animal cells, and many findings resulting from studies on these organisms have been extrapolated to apply to animal cells. An initial review of the literature would suggest that the control systems in these organisms are different; however, this is not the case—only the terminology used to represent the components differs. The cell cycle nomenclature used in these organisms has been reviewed elsewhere (35,56).

Modeling of the molecular controls that are thought to control the cell cycle in *Xenopus* oocytes (29) and yeast cells (30,103,104) has been done in order to determine the effect of changing the activity of various components of the regulatory network. We now believe that progression through the cell cycle is governed by the activity of cyclin–CDK complexes, in other words, a sharp rise in the complex's activity results in a transition to the next phase. The aim of the molecular modeling done by Novak and Tyson was to determine whether the control mechanisms presented here could be used to describe the observed behavior of cells in a variety of situations, or more simply stated, to compare theory with experiment. In order to do this, a set of equations (one for each component of the system) was generated, based on the mechanisms described, to describe the rate of change of each component as a function of time by translating the proposed control mechanisms into simple reaction rate equations. The resulting set of differential equations was then solved using numerical techniques; the activity of the cyclin–CDK complexes could be followed with time, and the sharp oscillations in complex activity could be varied. Because sets of equations commonly contain more than one solution (depending on rate constants and initial conditions used), phase-plane plots were constructed. Observation of limit-cycle-type behavior in these plots indicates an oscillatory solution that is characteristic of proposed cell cycle control (see Refs. 30 and 105 for details). This type of model is easily solvable by today's personal computers and has been successfully implemented to describe the observed behavior of a variety of yeast cell cycle mutants (103,104) and embryo and oocyte maturation (29).

IMPLICATIONS FOR ANIMAL CELL CULTURE

The use of large-scale cultures of animal cells is becoming increasingly important as the demand for their biological products grows. Animal cell cultures are used for the production of many different products of diagnostic and therapeutic value, including vaccines, monoclonal antibodies, and in some cases, the cells themselves (for example, to replace the function of damaged tissue). Determining the culture conditions that optimize the production and secretion of proteins is therefore of utmost importance in meeting these demands. Consequently, the key to designing such systems is not only determining the kinetics of growth, secretion, and proliferation, but also understanding how the cell cycle regulates certain cellular processes. The heterogeneous nature of cell populations has made it extremely difficult to obtain these data. Various mathematical models have been developed to describe cell behavior under different conditions; however, they are, for the most part, unable to account for the heterogeneous nature of cell populations and usually do not provide insight into variations introduced by the cell cycle. One model used to describe cell growth within a heterogeneous population of cells, known as population balances, has been developed (106–109). However, the implementation of this approach has been hampered by the difficulty in obtaining the single-cell parameters necessary for its use.

These problems are being tackled through the development and improvement of tools such as flow cytometry and fluorescent markers, which allow the examination of single-cell parameters within a heterogeneous population. Flow cytometry is a very sensitive tool that allows one to simultaneously obtain several single-cell parameters related to growth physiology from a large number of cells in the population. Collecting these data allows one to determine the size, DNA content (and thus cell cycle position), and protein content (when the protein is conjugated to a fluorescent marker) of individual cells in a population. Many flow cytometric studies relating the cell cycle to optimization of cell culture performance are available in the literature. Several of these analyses and their findings are outlined next.

Cell Cycle–Dependent Protein Accumulation

In the area of cell cycle–dependent protein accumulation, hybridoma cells have received the most attention because they are widely used in the production of monoclonal antibodies (MAbs). Numerous studies have suggested that lower rates of growth result in higher levels of protein production (110–112) (see also papers referenced in Ref. 113). It has been argued that because populations growing at slow rates have a larger fraction of G_1 cells, the G_1 phase must be a period of enhanced protein production and secretion. However, few studies have been done to give insight into the mechanism for this behavior. Additionally, other studies have been done that show a positive correlation between growth rate and protein production (briefly discussed in Ref. 113). These variations in response observed by different researchers may be explained by the fact that the overall growth rate of a culture depends on both cell growth and cell death. This means that the same overall growth can be observed in two cultures with different rates of cell loss and cell growth. The conclusion that can be made from the examination of these different investigations is that it is important to be aware of the different methods used to quantify growth and protein production in cell cultures because they may result in conflicting results.

Al-Rubeai and colleagues used flow cytometric results obtained from synchronous cultures and concluded that the rates of both immunoglobulin synthesis and secretion are at a maximum at the G_1–S transition (114). The synchronization of the culture was achieved by the addition of thymidine, which arrests cells prior to entry into the S phase. This type of analysis presents another problem, however, because factors used for cell cycle arrest have been shown to alter cell metabolism (115); therefore these results should be viewed with care. A more complete series of studies on rates of MAb production during the cell cycle was performed by Kromenaker and Srienc. The first study involved using flow cytometric data obtained in asynchronous cultures and applying these data to population balances to find the single-cell rates of protein accumulation in various cell cycle phases for both antibody-producing and non-antibody-producing cell lines (116). It was found that during the early G_1, producer cells with low protein content have a specific rate of accumulation of protein approximately 3–5 times greater than the specific rate observed for the remainder of G_1 cells. This indicates that the rate of protein synthesis is much greater than the rates of protein degradation and secretion in G_1. The S phase of the cell cycle was shown to a have a low specific rate of protein accumulation that remained relatively steady throughout this period. Extending this analysis to the G_2 + M phase yielded a negative specific rate of protein accumulation, indicating that protein is lost due to secretion and degradation during this period of the cell cycle. From these findings, it was concluded that the disruption of the secretion system that occurs during the division of the cell accounts for the extremely high rate of accumulation observed in early G_1 because the transport system probably has not had time to recover from its disruption.

A later report by Kromenaker and Srienc described the relation between growth rate and secretion rates during the cell cycle as determined using single-cell data and population balances (117). Lactic acid was used as a growth suppressor, and the concentrations were changed to vary the specific growth rate. This study brought several previously unconsidered factors to light. By analysis of the DNA distribution of the populations at different growth rates, it was found that the lengths of all the cell cycle phases were increased by the same factor that the specific growth rate was increased. This is in direct contrast to other studies that suggest that the increase in productivity is due to the increase in the length of the most productive phase of the cell cycle. Secondly, the application of population balances in this study revealed that the net specific rate of antibody synthesis is independent of growth rate. Additionally, the specific secretion rate of G_2 + M cells is greater than the rate in the other two phases and remains constant with decreasing growth rates. In contrast, the specific secretion rates for G_1 and S increase with decreasing growth rate. These findings are consistent with the hypothesis that the disruption of the secretion pathway during cell division results in lower secretion throughout G_1, until the Golgi apparatus has had sufficient time to reassemble. During this time, antibodies accumulate in the cell until they can be secreted during G_2 + M, before the Golgi apparatus is disrupted once again. Thus, at lower growth rates, more time is spent in G_1, so an increased secretion rate is observed because the cells have had time to recover from cell division. A subsequent study tracked the amount of cell surface antibodies as a function of DNA content (118). Because prior experiments determined that the amount of cell surface antibodies observed on the membrane is a good measure of the rate of secretion, these data were used to determine the pattern of secretion during the cell cycle. The results obtained from this method also agreed with the hypothesized regulation of secretion already described.

Understanding the regulation of foreign genes in animal cells is also of great interest because recombinant DNA is commonly used to express proteins of interest in immortal cell lines such as Chinese hamster ovary (CHO) cells. This provided the motivation for studies in which the expression of foreign genes in animal cells was tracked throughout the cell cycle. One study tracked the secretion of tissuelike plasminogen activator (tPA) in recombinant CHO cultures synchronized by fluorescence-activated cell sorting (FACS) based upon DNA content (119). This study shows that secretion of tPA reaches its maximum near late S phase, while the increase in this protein is maximum in G_1. These results are fairly similar to those obtained by Kromenaker and Srienc using different methods. Gu and colleagues also measured the expression of foreign β-galactosidase during the cell cycles in recombinant CHO cells (120) and in recombinant mouse cells (121) and hypothesize that control of foreign gene expression is promoter dependent. These particular studies, however, showed that the use of either the cytomegalovirus (CMV) promoter or the mouse mammary tumor virus (MMTV) promoter causes the production of β-galactosidase mainly in S phase.

Cell Cycle–Dependent Behavior of Culture Systems

There is a steady increase in the number of reports in the literature that look to the cell cycle to explain various phenomena such as differing growth rates between cultures and cell death due to shear stress in bioreactors. For example, the effect of different inoculation sources was examined by Al-Rubeai et al. (122). Two cultures that were inoculated with cells from either the exponential or stationary growth phase were examined using flow cytometry. The culture inoculated with cells from the exponential phase grew at a much greater rate than the other culture and therefore reached the death phase more quickly due to nutrient exhaustion and buildup of toxins. Analysis of the cell cycle, cell size, and mitochondrial activity of the cultures revealed differences among the cultures. For instance, the slower-growing culture was found to have a larger proportion of its cells in G_1, while the culture with the greater growth rate had a larger fraction of its cells in S. Because G_1 cells are generally smaller than the S-phase cells, the culture with a larger fraction of cells synthesizing new DNA uses more nutrients than the one with smaller cells. Additionally, cultures with elevated mitochondrial activity were also shown to have a greater capacity for proliferation. These findings lead the researchers to conclude that examination of these parameters can provide a means of predicting how the culture will behave and a means of explaining variations in the growth patterns of cultures.

It is also of great interest to determine how cells behave in different culture systems. For cells in continuous culture (123), it has been found that the fraction of cells in the G_1 phase decreases with increasing growth rate, while the fraction of cells in the S phase increases. As expected, the G_1 phase accounts for most of the resulting variability in the lengths of the cell cycle phases; however, the S phase is affected slightly more than the G_2 and M phases. A more detailed analysis of the cell cycle was performed in cultures containing microcarriers, and the fraction of contact-inhibited cells was determined and compared to a model (124). Microcarriers are commonly employed in cultures of anchorage-dependent cells, such as CHO cells, because these types of cells must be attached to a surface in order to proliferate. When the culture grows to the point where there is no longer any surface area available for newborn cells, contact inhibition occurs, and a portion of the cells cease to proliferate even when provided with an ample supply of nutrients. Analysis of this system yielded the distribution of cells in the cell cycle during the lag phase, exponential growth, and confluency. During the lag phase, the culture is predominantly in the G_1 and/or G_0 phases and rapidly enters the other cell cycle phases upon reaching exponential growth. As the culture reaches confluency, the culture gradually shifts back to predominantly the G_1 and/or G_0 phases; however, a portion of the cells continue cycling. These findings are similar to those of suspension cultures. Furthermore, a method was developed to determine the fraction of cells affected by contact inhibition, and the results showed good agreement with a model that had been developed.

Furthermore, when culturing animal cells in large-scale bioreactors, the cells can be injured due to the shear stress that results from agitation and sparging. There have been several studies that examined the mechanism of cell damage and sought to minimize injury to the cells through the use of various shear-protecting agents such as Pluronic F-68 and serum. However, there have been relatively few studies that examine the effect that these forces may have on cell metabolism and growth. It is evident that shear forces have an effect on the overall growth rate of the culture due to an increase in cell death. Lakhotia and coworkers (125) analyzed the rate of DNA synthesis and the cell cycle phase distributions of cultures grown in small-scale bioreactors with gas sparging and varying agitation intensities to determine the effect that these forces may have on growth kinetics of the surviving cells. Measurement of the kinetics was accomplished by measuring the uptake of BrdU and the DNA content distributions of the surviving cells with the aid of flow cytometry. They determined that, in cultures exposed to high agitation, the fraction of cells in the S phase increased by up to 45%, while the fraction of cells in G_1 decreased by up to 50%, compared to control cultures. The simultaneous analysis of the DNA synthesis rate suggests that the change in the cell cycle phase distribution is due to an increase in the DNA synthesis rate. Additionally, when cultured in the absence of shear stress, the apparent specific growth rate of the cells previously exposed to the shear forces was initially higher than and later the same as that of control cells. It was also shown that an increase in the agitation intensity resulted in a further increase of the DNA synthesis rate. By subjecting cells to extreme, short-term stresses induced by turbulent capillary flow, it was determined that larger cells are more susceptible to damage than smaller cells (126). This conclusion was made based on the preferential disappearance of S and G_2 cells when passed through the capillary. Another possible explanation for these results is that an apoptotic response triggered by the shear forces caused a decrease in the size of the cells, but verification of this hypothesis is difficult.

Engineering the Cell Cycle

Once a more complete picture of how the cell cycle regulates cell physiology and metabolism has been painted, it can possibly be manipulated in culture in such a way that cells will perform desired processes at an optimal level. For example, controlling the regulatory mechanisms of cell growth and proliferation, or controlling the rate at which cells advance through the cell cycle can lead to the enhancement of protein production, improved economic feasibility, and ease of downstream processing. Indeed there have been several studies published to date that address these issues and will undoubtedly lead to further examination of directing cellular operations through the engineering of the cell cycle in cell cultures.

Cells in culture are subjected to many different forms of stress that may ultimately lead to cell death. Cell death can be caused by external factors such as hydrodynamic forces that may cause the cell to rupture. Additionally, the response to certain stimuli such as limitation of glutamine, glucose, or serum or the buildup of toxins may result in programmed cell death, called apoptosis, which is regulated by some of the same pathways as is the cell cycle. When in culture, cells reach their maximum cell density in a few days, after which the process of programmed death is induced in the majority of the cells (127). The discovery that the product of the protooncogene *bcl-2* is able to suppress apoptosis lead to its use in cell culture to prolong the productive period of the culture (128–131). Several recent studies show that cells expressing this gene increased MAb production by 40% (129) and prolonged the viable culture time by up to 4 days by delaying the onset of apoptosis for 2 days (131). Additionally, it was determined that the increase in MAb production was equally due to both the prolonged survival of cells and the increase in protein production per cell (131). The expression of the *bcl-2* gene also allows cells to be cultured with a significant decrease in serum supplementation (130). Specifically, *bcl-2* overexpressing cells could maintain high viable cell concentrations for up to 4 days, while the control culture lasted only 1 day when supplemented with 0.5% serum. Moreover, cells cultures supplemented with 9% serum with additional 2% serum fed after 4 days sustained productivity for 10 days, while the control culture underwent apoptosis after 4 days (130). These experiments show that the cell cycle controls can be bypassed when expression of the appropriate gene is induced in culture.

A similar approach has been used in CHO cells to bypass the cell cycle controls that require the presence of growth factors for efficient proliferation. The observation that cells stimulated by growth factors possess elevated cyclin E levels led researchers to transfect the CHO cells with a cyclin E expression vector (132). The expression of this gene in cells cultured in protein-free media resulted in cell proliferation and levels of cyclin E, cell morphologies, and cell cycle phase distributions comparable to those obtained in growth-factor-stimulated cells. An approach related to the one just described employed expression of the transcription factor E2F-1 in protein-free media to bypass the growth-factor requirements for proliferation (133). Although the desired effect of eliminating growth-factor dependency of the cells was achieved in both cases, the two approaches had different effects on several cellular properties. For instance, the cells expressing cyclin E rapidly round up and move into suspension, whereas the

E2F-1 cells adhered more tightly to the surface. Additionally, the same doubling time was observed for both cell types; however, the cyclin E cells resided in G_1 for approximately 20 min longer, while the S phase was reduced by 2 h as compared with the E2F-1 cells. Finally, the expression of certain proteins varied between the two cell types. The expression of cyclin A was shown to be 2 times higher in the E2F-expressing cells, and these same cells expressed a greater number of proteins at higher levels than wild-type cells than did the cyclin E–expressing cells. These results present two different ways of circumventing growth-factor-dependent controls and show different effects on cellular processes. Additionally, these findings may give some insight into the roles that these factors play in the regulation of the cell cycle.

Another recent study involved increasing protein production in cultured cells by arresting cells in G_1 (134). This was done by transfecting CHO cells with a gene coding for either p21, p27, or p53175P controlled by a tetracycline-regulated promoter. The p53 protein used in this study is a mutant that has lost its apoptotic function, resulting in a culture that undergoes G_1 arrest rather than programmed cell death. The culture was allowed to proliferate until it reached the desired cell density, after which expression of the cell cycle arrest molecules was stimulated, thus blocking proliferation. This sort of cytostatic process has several advantages over other processes involving uncontrolled growth of cultures, including slower rates of nutrient exhaustion, higher levels of protein production, and low genetic drift.

Gene Therapy

A firm grasp of the events that control the cell cycle will also yield significant advances toward the treatment of various illnesses including cancer and genetic disorders. The connection between the cell cycle and some cancers has been reviewed elsewhere (18,65,135,136). Many genetic disorders that alter the way cells function result from either the absence of an enzyme or the presence of a flawed enzyme. This situation may be remedied by the introduction of the appropriate genes into a target cell, a procedure known as gene therapy. Although cells that have incorporated the desired gene into their genome have displayed the desired effects, this technique is limited by low rates of gene transfer (reviewed in Ref. 137). Numerous studies have attempted to uncover the factors that determine transfer efficiency, resulting in the hypothesis that gene transfer is cell cycle dependent (reviewed in Refs. 138 and 139). More specifically, the most widely used method for gene transfer is recombinant retroviruses, and genes introduced into cells in this manner are not integrated into the host's genome until mitosis. It has been postulated that nuclear envelope breakdown is necessary for the viral integration complexes to act. Based on these studies, Andreatis and Palsson suggested that transfer efficiency depends both on cell cycle and retroviral life cycle events and developed a mathematical model describing the interplay between the two (140). This model aided in the design of experiments that allow the direct measure of intracellular retroviral decay (138). Furthermore, they developed a method to monitor the cell cycle dependence of gene transfer by using the transduction of β-galactosidase as a marker. Thus it was determined that almost all transduced cells show up in the early G_1 phase, supporting the hypothesis that only mitotic cells are susceptible to gene transfer (139). These studies reveal the importance of the kinetics of both retroviral decay and cell cycle progression of the target cells in the efficiency of gene therapy.

CONCLUSIONS

An overview of the mechanisms by which cells respond to different environmental stimuli has been presented. It is evident that even after the commitment to synthesize new DNA is made, the cell possesses the ability to monitor and direct internal events. For instance, the transcription and activation of key enzymes are regulated by cell cycle controls to ensure that necessary cellular processes are carried out successfully. The existence of this type of control suggests that almost every cellular process is somehow linked to the cell cycle machinery to guarantee the proper coordination of events throughout the cell cycle. A failure to maintain the order of these events may result in cell death or a disruption of normal cell behavior. Thus, a more complete understanding of factors that have an effect on cell cycle regulation would allow one to gain insight into better directing the cellular processes bound to cell cycle control. This fact is characterized by the recent increase in the number of studies that have focused on better understanding the components involved in directing cell division and their influence on fundamental cellular processes. It seems that obtaining a better understanding of the cell cycle and understanding how the cell cycle directs fundamental components of cell growth, such as protein production and secretion, may provide the missing link that will allow researchers to not only improve cell culture, but develop treatments for many types of ailments and dysfunctions such as cancer, which are indubitably cell cycle related.

BIBLIOGRAPHY

1. R.A. Laskey, M.P. Fairman, and J.J. Blow, *Science* **246**, 609–613 (1989).
2. B. Stillman, *Science* **274**, 1659–1664 (1996).
3. P.N. Rao and R.T. Johnson, *Nature* **159**, 159–164 (1970).
4. G.H. Leno, C.S. Downes, and R.A. Laskey, *Cell* **69**, 151–158 (1992).
5. J.P.J. Chong, H.M. Mahbubani, C.Y. Khoo, and J.J. Blow, *Nature* **375**, 360–361 (1995).
6. P. Romanowski, M.A. Madine, and R.A. Laskey, *Proc. Natl. Acad. Sci. U.S.A.* **93**, 10189–10194 (1996).
7. Y. Kubota, S. Mimura, S.I. Nishimoto, H. Takisawa, and H. Nojima, *Cell* **81**, 601–609 (1995).
8. V. Baldin, J. Lukas, M.J. Marcote, M. Pagano, and G. Draetta, *Genes Dev.* **7**, 812–821 (1993).
9. M. Ohtsubo and J.M. Roberts, *Science* **259**, 1908–1912 (1993).
10. D.H. Walker and J.L. Maller, *Nature* **354**, 314–317 (1991).

11. E. Schwob, T. Bohm, M.D. Mendenhall, and K. Nasmyth, *Cell* **79**, 181–184 (1994).
12. M. Pagano, S.W. Tam, A.M. Theodoras, P. Beer-Romero, G.D. Sal, V. Chau, P.R. Yew, G.F. Draetta, and M. Rolfe, *Science* **269**, 682–685 (1993).
13. M.B. Kastan, O. Onyekwere, D. Sidransky, B. Voglestein, and R.W. Craig, *Cancer Res.* **51**, 587–597 (1991).
14. M.J. O'Conell and P. Nurse, *Curr. Opin. Cell Biol.* **6**, 867–871 (1994).
15. I. Hoffmann and E. Karsenti, *J. Cell Sci.* **Suppl. 18**, 75–79 (1994).
16. M.K.K. Shivji, S.J. Grey, U.P. Strausfeld, R.D. Wood, and J.J. Blow, *Curr. Biol.* **4**, 1062–1068 (1994).
17. M.E. Ewan, B. Faha, E. Harlow, and D.M. Livingston, *Science* **255**, 85–90 (1992).
18. C.C. Orlowski and R.W. Furlanetto, *Endocrinol. Metab. Clin. N. Am.* **25**, 491–502 (1996).
19. M.A. Felix, J.C. Labbe, M. Doree, T. Hunt, and E. Karsenti, *Nature* **346**, 379–382 (1990).
20. T. Lorca, J.C. Labbe, A. Devault, D. Fesquet, U. Strausfeld, J. Nilsson, P.A. Nygren, M. Uhlen, J.C. Cavadore, and M. Doree, *J. Cell Sci.* **102**, 55–62 (1992).
21. R. Reeves, *Curr. Opin. Cell Biol.* **4**, 413–423 (1992).
22. A. Hershko, D. Ganoth, V. Sudakin, A. Dahan, L.H. Cohen, F.C. Luka, J.V. Ruderman, and E. Eytan, *J. Biol. Chem.* **269**, 4940–4946 (1994).
23. S. Yamashiro, Y. Yamakita, R. Ishikawa, and F. Matsumura, *Nature* **344**, 675–678 (1990).
24. W. Wells, *Trends Cell Biol.* **6**, 228–234 (1996).
25. J.A. Endicott, P. Nurse, and L. Johnson, *Protein Eng.* **7**, 243–253 (1994).
26. S.J. Elledge, *Science* **274**, 664–1672 (1996).
27. X. Li and R.B. Nicklas, *Nature* **373**, 630–632 (1995).
28. C.L. Reider, R.W. Cole, A. Khodjakov, and G. Sluder, *J. Cell Biol.* **130**, 941–948 (1995).
29. B. Novak and J.J. Tyson, *J. Cell Sci.* **106**, 1153–1168 (1993).
30. B. Novak and J.J. Tyson, *J. Theor. Biol.* **165**, 101–134 (1993).
31. K. Galaktianov and D. Beach, *Cell* **67**, 1181–1194 (1991).
32. C.J. Sherr and J.M. Roberts, *Genes Dev.* **9**, 1149–1163 (1995).
33. T. Evans, E.T. Rosenthal, J. Youngblom, D. Distel, and T. Hunt, *Cell* **33**, 389–396 (1983).
34. T. Hunt, *Semin. Cell Biol.* **2**, 213–222 (1991).
35. J. Pines, *Adv. Cancer Res.* **66**, 181–212 (1995).
36. H. Kobyashi, E. Stewart, R. Poon, J.P. Ademczewski, J. Gannon, and T. Hunt, *Mol. Biol. Cell* **3**, 1279–1294 (1992).
37. J. Pines and T. Hunter, *Trends Cell Biol.* **1**, 117–121 (1991).
38. D.S. Peeper, L.S. Ewan, M.E. Ewan, M. Toebes, F.L. Hall, M. Xu, A. Zantema, A.J. van-der-Eb, and H. Piwnica-Worms, *EMBO J.* **12**, 1947–1954 (1993).
39. S.I. Reed, *Annu. Rev. Cell Biol.* **8**, 529–561 (1992).
40. C.J. Sherr, *Cell* **73**, 1059–1065 (1993).
41. X. Grana and E.P. Reddy, *Oncogene* **11**, 211–219 (1995).
42. G.F. Draetta, *Curr. Opin. Cell Biol.* **6**, 842–846 (1994).
43. Y. Xiong, H. Zhang, and D. Beach, *Cell* **71**, 504–514 (1992).
44. C.J. Sherr, *Trends Biochem. Sci.* **20**, 187–190 (1995).
45. H. Matsushime, M. Roussel, R. Ashmun, and C.J. Sherr, *Cell* **65**, 701–713 (1991).
46. D.E. Quelle, R.A. Ashmun, S.A. Shurtlcff, J. Kato, D. Bar-Sagi, M.F. Roussel, and C.J. Sherr, *Genes Dev.* **7**, 1559–1571 (1993).
47. D. Resnitzky, M. Gossen, H. Bujard, and S.I. Reed, *Mol. Cell. Biol.* **14**, 1669–1679 (1994).
48. A. Koff, A. Giordano, D. Desai, K. Yamashita, J.W. Harper, S.J. Elledge, T. Nishimoto, D.O. Morgan, B.R. Franza, and J.M. Roberts, *Science* **257**, 1689–1694 (1992).
49. V. Dulic, E. Lees, and S.I. Reed, *Science* **257**, 1958–1961 (1992).
50. L.H. Tsai, E. Harlow, and M. Meyerson, *Nature* **353**, 174–177 (1991).
51. L.H. Tsai, E. Lees, B. Faha, E. Harlow, and K. Riabowol, *Oncogene* **8**, 1593–1602 (1993).
52. G. Draetta, F. Luka, J. Westendorf, L. Brizuela, J. Ruderman, and D. Beach, *Cell* **56**, 829–838 (1989).
53. J. Pines and T. Hunter, *J. Cell Biol.* **115**, 1–17 (1991).
54. S.I. Reed, C. Wittenberg, D.J. Lew, V. Dulic, and M. Henze, *Cold Spring Harbor Symp. Quant. Biol.* **56**, 61–67 (1991).
55. M. Glotzer, A.W. Murray, and M.W. Kirschner, *Nature* **349**, 132–138 (1991).
56. J. Pines, *Biochem. J.* **308**, 697–711 (1995).
57. J. Pines, *Cancer Biol.* **5**, 305–313 (1994).
58. R.P. Fisher and D.O. Morgan, *Cell* **78**, 713–724 (1994).
59. T.P. Makela, J.P. Tassan, E.A. Nigg, S. Frutiger, G.J. Hughes, and R.A. Weinberg, *Nature* **371**, 254–257 (1994).
60. W.J. Feaver, J.Q. Svejstrup, N.L. Henry, and R.D. Kornberg, *Cell* **79**, 1103–1109 (1994).
61. R. Roy, J.P. Adamczewski, T. Seroz, W. Vermeulen, J.P. Tassan, L. Schaeffer, E.A. Nigg, J.H.J. Hoeijmakers, and J.H. Egly, *Cell* **79**, 1093–1101 (1994).
62. S. Atherton-Fessler, L.L. Parker, R.L. Geahlen, and H. Piwnica-Worms, *Mol. Cell. Biol.* **13**, 1675–1685 (1993).
63. C. Featherstone and P. Russell, *Nature* **349**, 808–811 (1991).
64. R.N. Booher, R.J. Deshaies, and M.W. Kirschner, *EMBO J.* **12**, 3417–3426 (1993).
65. T.K. MacLachlan, N. Sang, and A. Giordano, *Crit. Rev. Euk. Gene Express.* **5**, 127–156 (1995).
66. S.J. Elledge and J.W. Harper, *Curr. Opin. Cell Biol.* **6**, 847–852 (1994).
67. Y. Xiong, G.J. Hannon, H. Zhang, D. Casso, R. Kobayashi, and D. Beach, *Nature* **366**, 701–704 (1993).
68. J.W. Harper, G.W. Adami, N. Wei, K. Keyomarsi, and S.J. Elledge, *Cell* **75**, 805–816 (1993).
69. H. Zhang, G.J. Hannon, and D. Beach, *Genes Dev.* **8**, 1750–1758 (1994).
70. K. Polyak, J.-Y. Kato, M.J. Solomon, C.J. Sherr, J. Massague, J.M. Roberts, and A. Koff, *Genes Dev.* **8**, 9–22 (1994).
71. K. Polyak, M.-H. Lee, H. Erdjument-Bromage, P. Tempst, and J. Massague, *Cell* **78**, 59–66 (1994).
72. E.J. Firpo, A. Koff, M.J. Solomon, and J.M. Roberts, *Mol. Cell. Biol.* **14**, 4889–4901 (1994).
73. M.E. Ewan, H.K. Sluss, L.L. Whitehouse, and D.M. Livingston, *Cell* **74**, 1009–1020 (1993).
74. M. Serrano, G.J. Hannon, and D. Beach, *Nature* **366**, 704–707 (1993).
75. T. Nobori, K. Miura, D.J. Wu, A. Lois, K. Takabayashi, and D.A. Carson, *Nature* **368**, 753–756 (1994).
76. G.J. Hannon and D. Beach, *Nature* **371**, 257–261 (1994).
77. D.S. Peeper and R. Bernards, *FEBS Lett.* **410**, 11–16 (1997).
78. R.A. Weinberg, *Science* **254**, 1138–1146 (1991).

79. M. Laiho, J.A. DeCaprio, J.W. Ludlow, D.M. Livingston, and J. Massague, *Cell* **62**, 175–185 (1990).
80. E. Moran, *Curr. Opin. Gen. Dev.* **3**, 63–70 (1993).
81. S.P. Chellappan, S. Hiebert, M. Mudryj, J.M. Horowitz, and, J.R. Nevins, *Cell* **65**, 1053–1061 (1991).
82. E.W.-F. Lam and N.B.L. Thangue, *Curr. Opin. Cell Biol.* **6**, 859–866 (1994).
83. J. Lukas, H. Muller, J. Bartkova, D. Spitkovsky, A.A. Kjerulff, D.P. Jansen, M. Strauss, and J. Bartek, *J. Cell Biol.* **125**, 625–638 (1994).
84. M.E. Ewan, H.K. Sluss, C.J. Sherr, H. Matsushime, J. Kato, and D.M. Livingston, *Cell* **73**, 487–497 (1993).
85. S.F. Dowdy, P.W. Hinds, K. Louie, S.I. Reed, A. Arnold, and R.A. Weinberg, *Cell* **73**, 499–511 (1993).
86. H. Muller, J. Lukas, A. Schneider, P. Warthoe, J. Bartek, M. Eilers, and M. Strauss, *Proc. Natl. Acad. Sci. U.S.A.* **91**, 2945–2949 (1994).
87. E. Lees, B. Faha, V. Dulic, S.I. Reed, and E. Harlow, *Genes Dev.* **6**, 1874–1885 (1992).
88. N. La-Thangue, *Curr. Opin. Cell Biol.* **6**, 443–450 (1994).
89. I. Hoffman, G. Draetta, and E. Karsenti, *EMBO J.* **13**, 4302–4310 (1994).
90. A. Di-Leonardo, S.P. Linke, K. Clarkin, and G.M. Wahl, *Genes Dev.* **8**, 2540–2551 (1994).
91. R.W. King, R.J. Deshaies, J.M. Peters, and M.W. Kirschner, *Science* **274**, 1652–1659 (1996).
92. W. Krek, M.E. Ewan, S. Shirodkar, Z. Arany, W.G. Kaelin, and D.M. Livingston, *Cell* **78**, 161–172 (1994).
93. M. Pagano, R. Pepperkok, F. Verde, and G. Draetta, *EMBO J.* **11**, 961–971 (1992).
94. A. Bailley and M. Bornens, *Nature* **355**, 300–301 (1992).
95. T.R. Coleman and W.G. Dunphy, *Curr. Opin. Cell Biol.* **6**, 877–882 (1994).
96. Y. Sanchez, C. Wong, R.S. Thoma, R. Richman, Z. Wu, H. Piwnica-Worms, and S.J. Elledge, *Science* **277**, 1497–1501 (1997).
97. F. Verde, M. Dogterom, E. Stelzer, E. Karsenti, and S. Leibler, *J. Cell Biol.* **118**, 1097–1108 (1992).
98. E.A. Nigg, *Curr. Opin. Cell Biol.* **5**, 187–193 (1993).
99. E.A. Nigg, *Curr. Opin. Cell Biol.* **4**, 105–109 (1992).
100. Y. Yamakita, S. Yamashiro, and F. Matsumura, *J. Cell Biol.* **124**, 129–137 (1994).
101. F.C. Luka, E.K. Shibuya, C.E. Dohrmann, and J.V. Ruderman, *EMBO J.* **10**, 4311–4320 (1990).
102. J.A. Knoblich, K. Sauer, L. Jones, H. Richardson, R. Saint, and C.F. Lehner, *Cell* **77**, 107–120 (1994).
103. B. Novak and J.J. Tyson, *J. Theor. Biol.* **173**, 283–305 (1995).
104. B. Novak and J.J. Tyson, *Proc. Natl. Acad. Sci. U.S.A.* **94**, 9147–9152 (1997).
105. J.J. Tyson, B. Novak, G.M. Odell, K. Chen, and C.D. Thron, *Trends Biochem. Sci.* **21**, 89–96 (1996).
106. J.F. Collins and M.H. Richmond, *J. Gen. Microbiol.* **28**, 15–33 (1962).
107. R.J. Harvey, A.G. Marr, and P.R. Painter, *J. Bacteriol.* **93**, 605–617 (1967).
108. A.G. Fredrickson, D. Ramkrishna, and H.M. Tsuchiya, *Math. Biosci.* **1**, 327–374 (1967).
109. D. Ramkrishna, A.G. Fredrickson, and H.M. Tsuchiya, *Bull. Math. Biophys.* **30**, 319–323 (1968).
110. S. Terada, E. Suzuki, H. Ueda, and F. Makishima, *Cytokine* **8**, 889–894 (1996).
111. K. Takahashi, S. Tereda, H. Ueda, F. Makishima, and E. Suzuki, *Cytotechnology* **15**, 57–64 (1994).
112. F. Makishima, S. Terada, T. Mikami, and E. Suzuki, *Cytotechnology* **10**, 15–23 (1992).
113. M. Al-Rubeai, A.N. Emery, S. Chalder, and D.C. Jan, *Cytotechnology* **9**, 85–97 (1992).
114. M. Al-Rubeai and A.N. Emery, *J. Biotech.* **16**, 67–86 (1990).
115. N.J. Cowan and C. Milstein, *Biochem. J.* **128**, 445–454 (1972).
116. S.J. Kromenaker and F. Srienc, *Biotech. Bioeng.* **38**, 665–677 (1991).
117. S.J. Kromenaker and F. Srienc, *J. Biotech.* **34**, 13–34 (1994).
118. M. Cherlet, S.J. Kromenaker, and F. Srienc, *Biotech. Bioeng.* **47**, 535–540 (1995).
119. M. Kubbies and H. Stockinger, *Exp. Cell Res.* **188**, 267–271 (1990).
120. M.B. Gu, P. Todd, and D.S. Kompala, *Biotech. Bioeng.* **42**, 1113–1123 (1993).
121. M.B. Gu, P. Todd, and D.S. Kompala, *Biotech. Bioeng.* **50**, 229–237 (1996).
122. M. Al-Rubeai, M.S. Chalder, R. Bird, and A.N. Emery, *Cytotechnology* **7**, 179–186 (1991).
123. D.E. Martens, C.D. d. Gooijer, C.A.M. v. d. V.-d. Groot, E.C. Beuvery, and J. Tramper, *Biotech. Bioeng.* **41**, 429–439 (1993).
124. K.A. Hawboldt, T.I. Linardos, N. Kalogerakis, and L.A. Behie, *J. Biotech.* **34**, 133–147 (1994).
125. S. Lakhotia, K.D. Bauer, and E.T. Papoutsakis, *Biotech. Bioeng.* **40**, 978–990 (1992).
126. M. Al-Rubeai, R.P. Singh, A.N. Emery, and Z. Zhang, *Biotech. Bioeng.* **46**, 88–92 (1995).
127. R.P. Singh, M. Al-Rubeai, C.D. Gregory, and A.N. Emery, *Biotech. Bioeng.* **44**, 720–726 (1994).
128. E. Suzuki, S. Terada, H. Ueda, T. Fujita, T. Komatsu, S. Takayama, and J.C. Reed, *Cytotechnology* **23**, 55–59 (1997).
129. N.H. Simpson, A.E. Milner, and M. Al-Rubeai, *Biotech. Bioeng.* **54**, 1–16 (1997).
130. S. Terada, Y. Itoh, H. Ueda, and E. Suzuki, *Cytotechnology* **24**, 135–141 (1997).
131. Y. Itoh, H. Ueda, and E. Suzuki, *Biotech. Bioeng.* **48**, 118–122 (1995).
132. W.A. Renner, K.H. Le, V. Hatzimanikatis, J.E. Bailey, and H.M. Eppenberger, *Biotech. Bioeng.* **47**, 476–482 (1995).
133. K.H. Lee, A. Sburlati, W.A. Renner, and J.E. Bailey, *Biotech. Bioeng.* **50**, 273–279 (1996).
134. M. Fussenegger, X. Mazur, and J.E. Bailey, *Biotech. Bioeng.* **55**, 927–939 (1997).
135. C. Cordon-Cardo, *Am. J. Pathol.* **147**, 545–559 (1995).
136. L.H. Hartwell and M.B. Kastan, *Science* **266**, 1821–1828 (1994).
137. B. Palsson and S. Andreadis, *Exp. Hematol.* **25**, 94–102 (1997).
138. S.T. Andreadis, D. Brott, A.O. Fuller, and B.O. Palsson, *J. Virol.* **71**, 7541–7548 (1997).
139. S. Andreadis, A.O. Fuller, and B.O. Palsson, *Biotech. Bioeng.* **58**, 272–281 (1998).
140. S. Andreadis, and B.O. Palsson, *J. Theor. Biol.* **182**, 1–20 (1996).

See also BIOENERGETICS OF MICROBIAL GROWTH; CULTURE COLLECTIONS; ENERGY METABOLISM, MICROBIAL AND ANIMAL CELLS.

CELL DEATH. See APOPTOSIS; CELL CYCLE.

CELL DISRUPTION AND LYSIS

F. A. P. GARCIA
University of Coimbra
Coimbra, Portugal

KEY WORDS

Autolysis
Ball mill
Cell wall
Debris separation
Disintegration
Disruption
Homogenization
Lysis
Ultrasonics

OUTLINE

INTRODUCTION

Currently, biological products of commercial interest are excreted by producer cells into the surrounding medium, but the larger proportion of cellular products, many of which have a real or potential value, remain inside the cell (1). These intracellular products must be released into the medium before they can be recovered and purified. There are several strategies for accomplishing this: the cells can be genetically engineered to excrete the products, the cell wall can be made more permeable, cell lysis may be induced, or the cell wall can be disrupted. In preparative-scale applications the disruption of the cell wall is the most common method.

The cell wall is a rigid and complex structure that protects the cell from the environment, maintaining cell shape and providing osmotic stability, while performing biochemical functions of regulation and selective control of the exchange of nutrients and metabolites with the medium. However, microbial cell wall characteristics vary widely in composition and structure according to species; it is important to recognize the effects of these differences on the strength of the wall and their implications which disruption techniques should be adopted. Mammalian tissues, halophilic bacteria (which do not possess a true rigid cell wall), and the Mycoplasma bacteria (which lack a wall) have the lowest mechanical strength and can be easily lysed under osmotic stress in hypotonic media. The more structured and stronger cell walls in some bacteria and fungi are more difficult to disrupt, and more energy is required to overcome the toughness of the wall. Moreover, the cell age and physiological state are also important differences that contribute to the strength of a given species' cell wall.

THE CELL WALL

Chemical Composition and Structure

The diversity of cell wall composition and structure is well documented (2–5). Purified cell walls (6) can be hydrolyzed, and the different components can be identified.

Bacterial Cell Walls. Bacteria differ with respect to their ability to retain a crystal violet–iodine stain when treated with organic solvents (alcohol or acetone). Accordingly, they can be classified into two main groups: Gram positive if they retain the stain, and Gram negative if the stain is not retained. This difference reflects the diversity in cell wall composition and structure as well as differences in biochemical, physiological, or genetic features. Common to all cell walls is the element of rigidity, which is provided by a supramolecular structure known as a peptidoglycan, or murein (Fig. 1). This is a regular heteropolymer of alternating *N*-acetylglucosamine and *N*-acetylmuramic acid (the lactyl ether of *N*-acetylglucosamine) linked by β-(1–4) glycosidic bonds in linear chains of up to 65 disaccharide residues. A tetrapeptide side chain, consisting of at least three amino acids (some of the unusual D-amino acids), is attached to the carboxyl group of the *N*-acetylmuramic acid. The tetrapeptide chain of one polysaccharide chain is linked to a tetrapeptide chain of an adjacent polysaccharide chain, either directly (the terminal D-alanine of one to the *meso*-diaminopimelic acids in position 3 of the other, as in *Escherichia coli*), or indirectly through a pentaglycine peptide (the terminal D-alanine of one to the L-lysine in position 3 of the other, as in *Staphylococcus aureus*). Not all pairs of these tetrapeptide chains are involved in the cross-linking of the glycan chains; the degree of cross-linking, as well as the nature of the amino acids in positions 2 and 3 are dependent on the species.

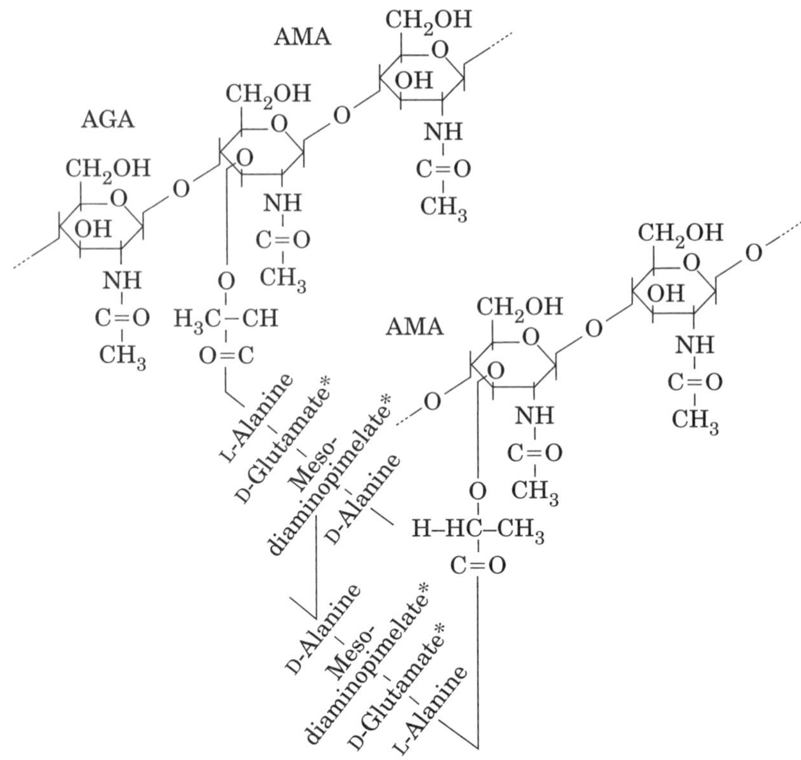

Figure 1. Structure of the building blocks of peptidoglycan in *E. coli*. The nature of the amino acids at positions 2 and 3, as well as the degree of cross-linking differ among species. AGA, *N*-acetylglucosamine; AMA, *N*-acetylmuramic acid.

Murein is the skeletal component of the cell wall in Gram-positive bacteria; it has a multilayered architecture that is 20–80 nm thick. In addition, teichoic acids (polymers of glycerol or ribitol phosphate) are associated with murein on the external side of the wall and are unique to Gram-positive cells. Teichoic acids do not contribute to wall strength but provide antigenic activity or serve as an attachment anchor. Gram-positive cell walls are devoid of significant amounts of lipids and proteins.

In Gram-negative bacteria the cell wall is thinner and more complex and stratified. Mechanical resistance is provided by an inner layer of murein, which is much thinner than in Gram-positive bacteria. This murein is covered externally by a bilayered lipid membrane, the internal leaflet being lipoprotein, containing phospholipids, and the outer leaflet being lipopolysaccharide (LPS), which has toxic properties and is able to resist the chemical aggressions of the environment.

Fungi and Yeasts. Under the electronic microscope, the walls of fungi and yeasts are seen to have a fibrillar structure that results from the presence of certain components such as cellulose or chitin. The main constituents are homo- or heteropolysaccharides, accounting for more than 80% of the wall, built up from different sugars (mainly D-glucose, D-mannose and *N*-acetylglucosamine, but also galactose, galactosamine, fucose, xylose, etc.). Usually, pairs of carbohydrates dominate the composition of the fungal walls, and a vast majority of fungi have a chitin–glucan combination, whereas the majority of yeasts have a glucan–mannan combination. Cellulose is also dominant in the walls of many lower fungi. Glucans are usually considered to be the main skeletal component in yeasts and consist of linear chains of β-(1–6)-linked D-glucose and side chains of D-glucose linked to the first ones through β-(1–3) bonds (Fig. 2a). Chitin is a polymer of *N*-acetylglucosamine and is always found in yeast cells, although in small amounts (Fig. 2c).

Lipids and proteins are also present in fungi and yeasts cell walls in different amounts. Proteins are now regarded as true structural wall components. In yeasts the wall appears stratified into two layers: the protein is associated with mannans in the form of mannoproteins that form an external layer covering an internal layer of glucans (Fig. 2b).

Algae. Cellulose is also the main skeletal component of the algal cell wall, but xylans and mannans (depending on which sugars are the most abundant) are also found. Algae differ considerably from the fungi with respect to cellulose content and crystallinity. The physical aspect is also fibrillar, as with other eukaryotes.

Effects of Upstream Conditions

The conditions that affect the synthesis of the cell wall constituents during the growth process will be reflected in the strength of the cell wall. For instance, cells of *E. coli* grown in a complex medium are less easily disrupted than cells grown in a simple medium (7). The ease of disintegration also depends on the batch growth of *E. coli*; the cells are weaker during the logarithmic phase and become stronger in the transition to the stationary phase (1,11). This corresponds to an increase in the murein thickness and cross-

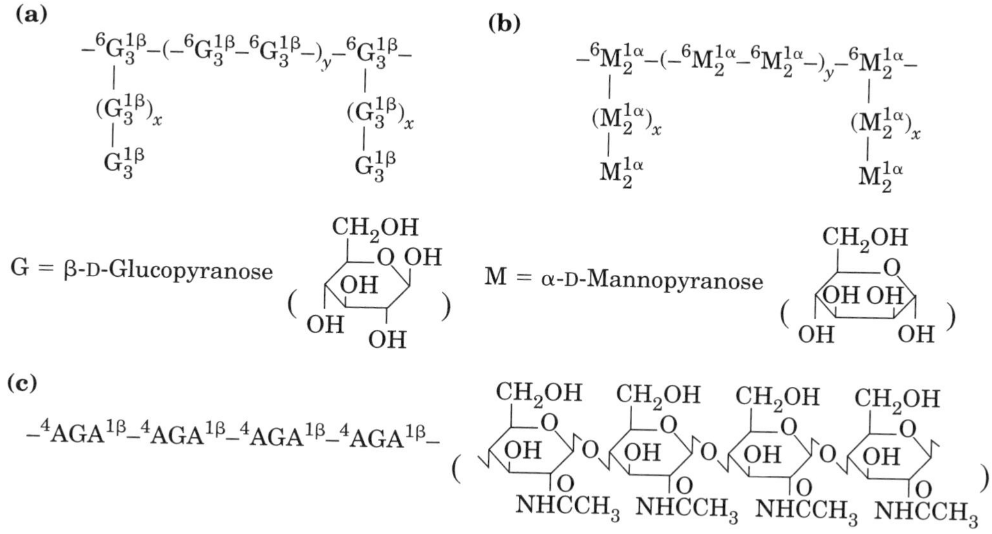

Figure 2. Structure of the building blocks of (**a**) glucan, (**b**) mannan, and (**c**) chitin in yeasts.

linking (8,10,11). A quantitative relationship has been established between mean cell strength and cell length and degree of cross-linkage (9). A higher specific rate of growth, as attained in continuous culture of *Candida utilis*, leads to a weaker cell wall, whereas a lower growth rate, as in a batch operation, allows the cells to build a stronger cell wall (12). Wall thickness and resistance to disruption in bakers' yeast increase with age, possibly due to exhaustion of the carbon source (13). Bud scars, much more frequent in a younger yeast population, may introduce local areas of lower wall strength (12,13). The compressive force required for the disruption of brewers' yeast cells was measured by using a micromanipulation technique and ranged from 40 μN for cells in the exponential phase to 90 μN for cells in the stationary phase (14).

ASSESSMENT OF THE DISRUPTION YIELD

Suitable analytical techniques are required to assess the extent of cell disruption in order to develop models for process selection, scale-up, and optimization. Different methods have been used to quantify the degree of disintegration at specified conditions; each one has its advantages and drawbacks. The number of disrupted cells must be estimated, which can be done by direct counting of nondisrupted cells or, indirectly, by determination of any intracellular metabolite whose quantitative release into the medium is proportional to the number of disrupted cells.

Direct counting of suitably diluted samples can be achieved by plating out techniques (14), microscopic counting (16,17), electronic counting (18), or using an analytical disk centrifuge (19).

Plating out techniques are time consuming, and only viable cells can be counted. However, nonviable intact cells can be present in a significant proportion and are uncountable. The situation is worsened if the cells have a tendency to agglomerate.

Microscopic counting is a faster and simpler alternative. Whole cells in suitably diluted samples of original and treated suspensions are counted in microscope hemocytometers (16,17). However, the counting process, though easier for yeasts, is tedious and difficult with very small bacterial cells, and a distinction between intact cells and cells that have been slightly damaged is generally impossible to make. Proportional counts of stained smears may be advantageous as, for instance, in the case of yeasts (20). By adopting the staining technique of Bianchi (21), it is possible to distinguish among whole intact cells, broken cells contaminated with cytoplasm, and empty debris, at 1,000× magnification. The main difficulty lies in finding a suitable staining technique for the cells concerned.

Electronic particle counters can be used to evaluate the number of whole yeast cells. The Coulter counter, for instance, was used to assess the concentration of yeast cells (18), but it is not a sensitive method for bacteria (22), and the machine tends to be fouled by cellular debris (19).

The analytical disk centrifuge was recently introduced for monitoring the disruption of *E. coli* containing recombinant inclusion bodies (19). This technique provides a size distribution in the homogenate and requires that the size distribution of the inclusion bodies be subtracted from the homogenate sample distribution by numerical or graphical deconvolution of both distributions.

Indirect counting is based on the determination of certain compounds (e.g., soluble protein, enzyme activities) that are released into the solution when the cell wall is broken: the yield of disintegration is the percentage ratio of the released material, R, to the maximum amount, R_{max}, that can theoretically be released into the solution. The most common material that has been used for this purpose is total soluble protein assayed in the sample supernatant (22–24). The "dilution method," using the Biuret method, was based on the comparison of the protein concentrations in the supernatants obtained by centrifugation of diluted and undiluted samples of homogenized suspensions (23). However, dilution interferes with protein solubility (22), because of partial denaturation. The method also fails when the protein is produced as an inclusion body (19).

Overheating, particularly during a prolonged operation to determine R_{max}, can cause thermal denaturation or enzyme deactivation.

Other bioactive products and, in particular, enzyme activities (25) released into the medium can be useful. However, the assessed value of the disruption yield under fixed conditions depends largely on the enzyme selected for analysis, its location within the cell, whether it is bound to membranous materials, and its heat or shearing lability (26), which can affect the values of both R and R_{max}. The most adequate material would be the product to be extracted and isolated, if it can be easily assayed.

Continuous monitoring of the disruption process and the performance of the equipment, particularly if wearing problems are severe, is advantageous. For instance, a relationship between the degree of cell disruption and the temperature increase of a baker's yeast suspension when passing through a mechanical homogenizer can be established. Therefore, by inserting a thermocouple in the homogenizer, continuous monitoring of its performance can be achieved (27).

METHODS OF CELL WALL DISRUPTION

Techniques for cell wall disruption have evolved over the years and are well documented in the literature (27–31). There are both mechanical and nonmechanical methods. Mechanical methods are based on shear forces that deform and rupture the cell wall. Nonmechanical methods induce the lysis of the cell by chemical, physical, or enzymatic means.

Many of these methods were developed for laboratory-scale use, with the objective of purifying biomolecules or organelles while maintaining as much integrity as possible for fundamental studies. For preparative or commercial applications, the strategy is to maximize product release, keeping the biological activity and minimizing the formation of fine cell debris, which can interfere with further downstream operations. The mechanical methods, based on solid or liquid shear, are best suited for this purpose.

High-Pressure Mechanical Homogenization

High-pressure mechanical homogenizers, typified by the APV Manton–Gaulin homogenizer (Fig. 3), have been one of the most widely used devices for the disruption of microbial cell walls in large-scale recovery of intracellular bioproducts. These devices consist of a positive displacement pump capable of generating high pressures to force the cell suspension through an adjustable discharge needle valve under high shear stresses, which disrupts the cells. The body of the valve is positioned against a valve seat by a spring-loaded rod, and the disrupting pressure is adjusted by the load that is applied, hydraulically or mechanically, to the spring. The treated suspension can be either collected or recycled for more passes through the valve.

In an earlier report (23) such a device was used to disrupt baker's yeast, and the degree of disruption was proportional to the number of passes, N, through the valve according to the model (first-order kinetics relative to N),

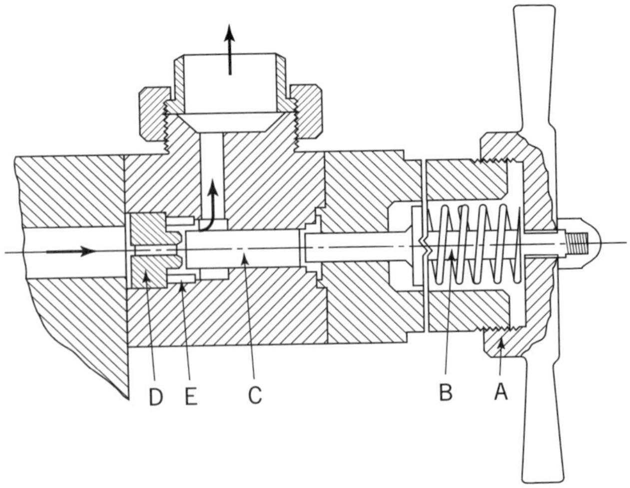

Figure 3. High-pressure homogenizer discharge valve unit. The discharge pressure is controlled by a handwheel assembly (A), which, through a spring-loaded valve rod (B) positions the valve (C) in relation to the valve seat (D). During discharge the material passes between the valve and its seat and impinges on an impact ring (F). *Source*: Reproduced by permission of the Institution of Chemical Engineers from Ref. 23.

$$\ln \frac{1}{1 - X} = kNP^{a} \qquad (1)$$

where $X = R/R_{max}$ is the fraction of disrupted cells, k is a dimensional, machine-dependent (it depends on the geometry of the needle valve) specific rate constant $[(k) = (P)^{-a} = \text{MPa}^{-a}]$; the constant a is a measure of the cell resistance to disruption (the value $a = 2.9$ was reported for baker's yeast, and 2.2 for *E. coli*) and is dependent on the type of organism and the physiologic conditions and rate of growth (12).

Besides the pressure, the type of organism and the cell growth conditions, other variables such as the temperature and the cell concentration affect the specific rate of disintegration. The process has been found to be independent of cell concentration in the range 84–168 g dry matter/L and for pressures extending from 19.6 to 53.9 MPa, although lower protein recoveries for higher cell concentrations at higher pressures were obtained (23). Further studies extended the analysis of the influence of the cell concentration up to 240 g dry matter/L, and no significant effect was detected (32). However, a theoretical analysis based on the turbulence theory of Kolmogoroff (33) showed that the amount of soluble protein release in a Manton–Gaulin homogenizer was dependent on cell concentration according to

$$R = 1 - \exp\{-[(P - P_0)/z]^{\beta}\} \qquad (2)$$

where z is a constant, P is the disrupting pressure, P_0 is a threshold pressure, and β is a parameter that depends on cell concentration. More recently, the disruption of *E. coli* in another type of homogenizer, the Microfluidizer (Microfluidics Corporation, Newton, Mass.), was correlated by (34)

$$\ln \frac{1}{1 - X} = kN^bP^a \tag{3}$$

where the exponent b was found to be a function of the cell concentration and the specific growth rate.

The rate of cell wall disruption has been shown to be highly temperature dependent, but a temperature increase when the cells are disrupted can lead to severe thermal denaturation of heat-labile products. An increase of about 2–3 °C (depending on cell concentration) can be expected for every 10 MPa increase in the backpressure of the needle valve (31,32). In fact, the suspension has to be pre-cooled in order to keep the temperature under control, and cooled immediately after leaving the homogenizer. In most cases no losses of enzyme activity or protein solubility were detected if the temperature was not allowed to rise above 30 °C (12,23).

Increased productivity can be attained with higher pressures, allowing the recovery of more product with fewer passes. However, a higher disrupting pressure demands more energy and can cause severe wearing problems in the valve seat. Erosion of the valve seat is, in fact, of great concern in using this type of equipment.

Optimization of the equipment design and operation requires a deep understanding of the mechanisms of cell disruption. Breakage of the cell wall occurs as a result of normal stresses or shear forces developed during the passage of the suspension through the needle valve. One mechanism considers the magnitude and the rate of pressure drop to be responsible for cell breakage (32). A hydrodynamic mechanism in which turbulent eddies of smaller dimensions than the cells cause the intracellular liquid to oscillate with a kinetic energy large enough to rupture the cell (33). Impingement of a high-velocity jet against a flat surface, as occurs in the needle valve when the direction of the jet is suddenly and abruptly changed, has also been considered to be a cause of cell disruption (12,35). In the case of the Manton–Gaulin–type of homogenizers, this led to the redesign of the valve geometry (35); the so-called knife-edge configuration showed the best performance, whereas the flat valve configuration had the lowest yields in the disruption of baker's yeast (Fig. 4).

These homogenizers have proved to be efficient for all types of microorganisms, with high degrees of disruption even with only one pass. Blockage has, nevertheless, remained a major concern in the use of high-pressure homogenizers for the disruption of filamentous microorganisms (36). The use of a Manton–Gaulin homogenizer resulted in a severe degradation of plasmid recovered from *E. coli* (37).

Solid-Shear Methods

Ball mills, or bead mills, originally designed for nonbiotechnological applications, have also been used to disintegrate microorganisms, particularly yeasts. They are the most suitable for scaling up and can be continuously or discontinuously operated. They consist of agitator disks mounted concentrically or excentrically on a motor-driven central shaft inside a grinding chamber (Fig. 5). During operation, the chamber is filled with small glass or stainless steel beads, which are necessary to promote an abrasive action to grind the cells. In continuous operation, a bead retainer (a sieve plate, a vibrating slot, or a special rotating disk) prevents the beads from leaving the grinding chamber. Different models of the Netzsch bead mill (Netzsch Feinmahltechnik, Germany) or the Dyno-Mill (Willy A. Bachofen AG, Basel, Switzerland) have been used.

The mechanism of disintegration is very complex and results from both a liquid shearing and a solid shearing, resulting from the impacts of the beads in a field of intense turbulence. The disruption process can be properly described by a first-order kinetics law (24,38) and, for a batch operation or continuous plug flow system, the integrated rate equation is

$$\ln \frac{R_{max}}{R_{max} - R} = kt \tag{4}$$

where R is the amount of protein released in time t, R_{max} is the maximum amount that could be released if 100% disrupted had occurred, and k is the specific velocity of disintegration (protein released).

For a continuous operation there is a certain degree of backmixing, and n tanks (CSTR) in a series model is a better representation of the hydrodynamics in the chamber. In this case a mass balance to the released protein gives (38)

$$\frac{R_{max}}{R_{max} - R} = [1 + (k\theta/n)^n] \tag{5}$$

where $\theta = V/Q$ is the mean residence time, V is the total free volume of the grinding chamber, and Q is the flow rate of the microbial suspension.

Several process parameters affect the yield and the rate of cell wall disruption. Besides the design characteristics of the mill (volume and geometry of the grinding chamber, the number of disks per unit volume of the mill, and geometry and construction materials of the disks), the rate constant is influenced by a number of process parameters such as the agitator speed, cell concentration, bead size, bead load, temperature, and flow rate of the suspension (24,38,39).

The speed of the agitator disks determines the frequency of bead collisions (40) and the intensity of shear, and, within certain limits, the specific rate of disintegration is proportional to the disk tip speed, U (38):

$$k = KU \tag{6}$$

Practical limitations arising from high energy consumption, high heat generation, erosion of the beads, and inactivation of shear-labile products require that the disk tip speeds remain between 5 and 15 ms^{-1}.

The influence of cell concentration requires a more detailed investigation because contradictory results have been reported. The disruption of baker's yeast was reported to be independent of the cell concentration in the range of 4–20% w/v on a dry matter basis (25), as expected for a first-order process. However, others (38,39) have re-

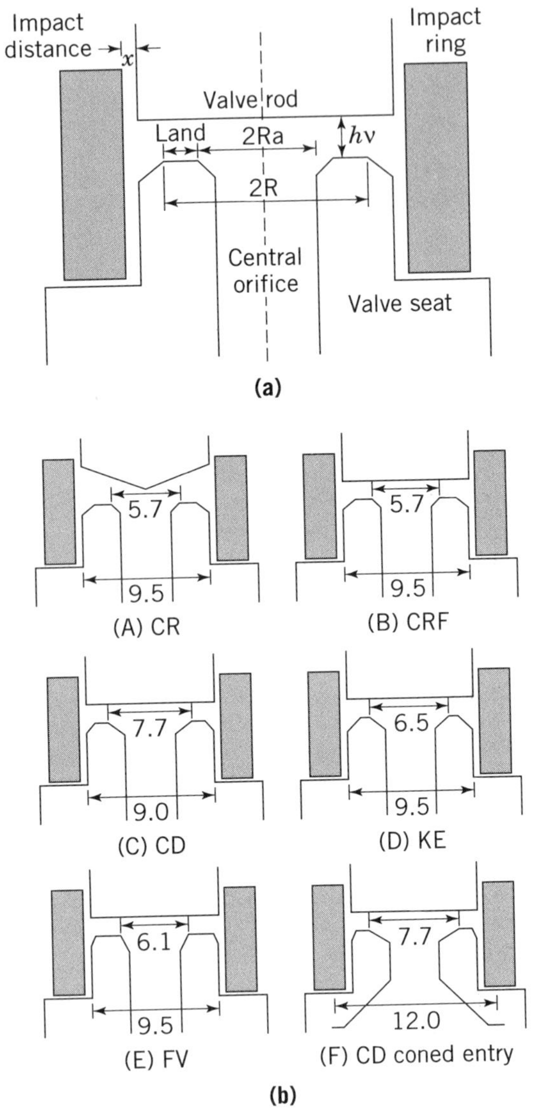

Figure 4. (**a**) Schematic diagram of a typical valve unit. (**b**) Schematic diagram of valve units used: (A) CR, cell rupture with coned valve rod; (B) CRF, cell rupture with flat valve rod; (C) CD, cell disruption; (D) KE, knife edge; (E) FV, flat valve unit; (F) CD-type unit with coned entry—dimensions in millimeters. Land distances ($R - R_o$) are for CRF, 0.7 mm; CD, 0.5 mm; KE, 0.1 mm; FV, 1.5 mm. *Source*: Reprinted by permission of Elsevier Science Inc. from Ref. 35.

ported a maximum protein release for cell concentrations of 35–40% (wet wt), and decreasing thereafter, depending on the disk materials (38), such effect being more pronounced when the agitator speed was low (39). The viscosity of the suspension increases with the cell concentration and is further increased with the progress of the disruption process, therefore increasing the rate at which the turbulence is damped out (38). However, insufficient data are generally generated or reported regarding the rheology of the suspension. The optimal cell concentration varies from case to case, and one must also consider that less heat is generated but the power consumption per unit weight of treated cells is higher for lower cell concentrations, and

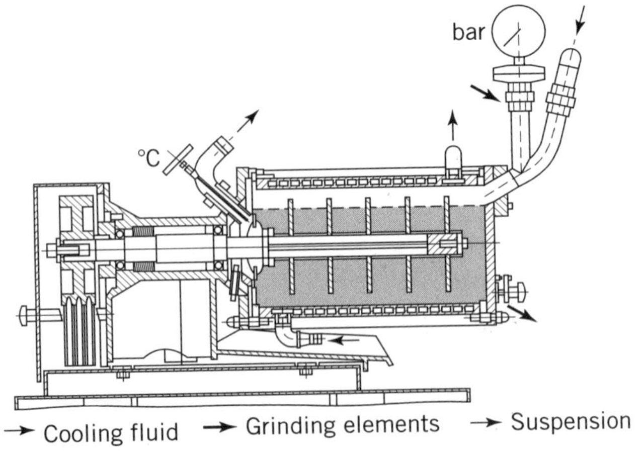

Figure 5. The Dyno-Mill. *Source*: Reprinted from a commercial brochure, by permission of Willy A. Bachofen AG, Basel/Switzerland.

that for low cell concentrations (<5% solids) increased wear is observed on the mill and beads. Hence, an optimum must be found by experiment and is generally in the range of 30–60% or, more usually, in the range of 40–50% (cell wet wt).

The bead diameters that have been reported range from 0.1 to 1.5 mm. Usually the disruption is faster with smaller beads, but the tendency of very small beads to float and difficulties in retaining them inside the grinding chamber impose practical limitations on the lower limits of size. In practice, the lower limit for batch operation of laboratory mills is 0.2 mm, whereas for production equipment working continuously the lower sizes are ≥0.4 mm. For the same bead load, larger particles correspond to fewer beads and reduced abrasive action or bead collisions. For the same bead load, there are also reports of an optimum bead size, which also has to be found by experiment and depends on the location (24,41) of the product (enzyme) inside the cells; larger beads are preferred for the recovery of enzymes located in the periplasmic region, and smaller ones for cytoplasmic material.

The rate of protein release is also enhanced by increasing the bead load (volume percentage of bead material relative to the total volume of the grinding chamber). Bead loads ranging from 30 to 90% are reported in the literature, but the packing density, depending on the bead diameter inside the chamber, is generally in the range of 80–90%. The upper limit is generally imposed by the high power consumption and difficulties in removing the heat released in the operation.

The rate constant is, obviously, dependent on the temperature that results from the balance between the heat generated during the milling process and the heat removed by a cooling liquid, brine, or a cooling liquid circulating through a cooling jacket of the grinding chamber. The heat generated in the process, if accumulated, would raise the temperature to unacceptable levels; in fact, the suspension has to be precooled and cooled immediately after leaving the chamber to avoid inactivation of heat-labile products.

The effects of the flow rate should mirror the effects of the residence time of the cells inside the grinding chamber, and assuming a first-order process, the yield after one passage should decline as the feed rate increases. However, degree of caution should be exercised because the degree of longitudinal dispersion or backmixing increases as the flow rate decreases, and the yield of disruption is not a simple function of the flow rate. On the other hand, the power consumption per unit mass of processed cells strongly decreases as the feed rate increases, and for this reason, a high flow rate should be adopted and the decline in the yield of disruption could be alleviated by recycling part of the suspension. It should be emphasized that by increasing the residence time the viscosity may be decreased because prolonged shear may cause degradation of the shear-sensitive DNA polymer (51).

In one study (41), the disintegration rate was correlated with the specific energy as determined by the following combination of parameters: small grinding beads, moderate to high cell concentrations, and/or low to moderate agitator speeds. In this case the utilization of energy input reaches its maximum. In other words, high cell concentrations are energetically favorable. This result is considered to be the basis for scale-up. For cases of low cell concentrations in the suspension, high agitator speeds, and larger beads, the disintegration rate was no longer correlated with specific energy input but was a function of the stress frequency (which is proportional to the product of the agitator speed and disintegration time), which was experimentally proved.

Ultrasonics

Ultrasound has been used on the laboratory scale for analytical and preparative extraction and purification of cell constituents of bacteria or yeasts. The mechanism of cell disruption derives from the intensive shear induced by sonicating the suspension at sound frequencies above 20 kHz. A magnetorestrictive or a piezoelectric transducer converts the alternating current of an electric oscillator into mechanical waves that are transmitted to the suspension through a metallic probe (usually made from titanium) vibrating with the same frequency as the oscillator. The sound waves create many microbubbles at various nucleation sites in the suspension, which collapse implosively during the rarefaction period of the sound waves. This phenomenon of cavitation (formation, growth, and collapse of the vapor-filled bubbles) produces intense local shock waves, and intense local shear gradients are generated that cause the cells to deform beyond the limit of elasticity and rupture. The rate of microbial cell disintegration is usually low for preparative purposes, and the inclusion of small stainless steel or glass (Ballotini) beads will help in the nucleation for cavitation as well as with a grinding action (solid shear) that will increase the efficiency of disintegration.

The kinetics of disruption (protein release) has been reported to follow a first-order law (42,43), and for a well mixed pattern inside the disruption chamber the integrated equation is

$$1 - X = \exp(-kt) \qquad (7)$$

where X is the fraction of released protein and k is the rate constant.

There are a number of variables that can markedly affect either the sample or the efficiency of the sonication process. One of these is the temperature rise. Most of the energy that is released when the bubbles collapse is absorbed by the liquid as heat. Although the value of the specific rate k increases with the temperature, the thermal energy has to be extracted to avoid temperatures that could destroy the activity of the useful bioproducts. To control the temperature within acceptable limits, the suspension should be precooled to 0–5 °C, and jacketed disruption vessels should be used with a continuous circulation of a cooling fluid. Alternatively or concomitantly, short periods of sonication can alternate with short periods of cooling. The ratio of the time of sonication to the time of cooling is known as the duty factor.

The integrity of the bioproduct can also be adversely affected if free radicals are allowed to accumulate during the intense cavitation. The addition of radical scavengers (e.g., cysteine or glutathione) or pregassing the suspension with hydrogen can alleviate this problem, particularly if a prolonged sonication is expected.

Other operating parameters affect the efficiency of the sonication process. One is the amplitude of vibrations of the sound waves (directly related to the acoustic power supplied to the suspension), which affects the rate constant according to the following relation (42):

$$k \alpha (W - W_0)^{0.9} \qquad (8)$$

where W is the acoustic power and W_0 is the cavitation threshold power. However, a linear relationship between the rate constant and the acoustic power was recently reported (43).

The viscosity of the suspension affects the hydrodynamics and the rate of energy dissipation and inhibits the phenomenon of cavitation. When the viscosity is high, the operation should be started at a lower power, and the power should be gradually increased. The initial cell concentration that affects the viscosity of the suspension was reported to have no effects on the rate constant (43).

Problems of foaming can result from the presence of surface active agents (e.g., proteins released from the disrupted cells) that lower the surface tension of the liquid medium. An intense foaming, particularly when high power is applied, contributes to cavitation "unloading" (therefore decreasing the efficiency of the process) and protein denaturation at the gas–liquid interface.

High working volumes require high acoustic power, which results in violent turbulence and in the formation of large bubbles that cause cavitation unloading. Rate constants were found to decrease linearly with increasing working volumes (43). The difficulties associated with higher working volumes (100–200 mL) can be alleviated by good design of the disruption vessel. Commonly a jacketed glass beaker with a continuous flow of cooling fluid is used; inlet and outlet tubes can be inserted for continuous operation. In continuous operation, the flow rate determines the residence time of the cells in the vessel and affects the overall yield of the process. For tougher cells the inclusion of Ballotini beads is necessary, and for further increasing the grinding action, the vessel should have internal tubing loops to circulate the suspension under the pumping energy of the sonication (27,31). The load and size of beads must in any case be optimized, but the value of the rate constant increases to a maximum and then decreases with increases in the bead diameter for the same bead load.

The shape of the probe is of particular concern. For a constant power level the amplitude of the sound waves at the probe tip is inversely proportional to the tip area. Therefore large tips dissipate the acoustic energy over a larger area, and the amplitude is smaller. However, if the tip diameter is too small the acoustic energy is dissipated over a smaller field and is less efficiently used, this is particularly important if the working volume is small because intense turbulence can lessen the power absorbed near the tip.

Manufacturing materials dictate the severity of the wearing problems that are also common with the probe tips. Titanium is a good material because of its good acoustic and mechanical properties and low toxicity to bioactive materials. However, titanium probes are expensive and difficult to machine. Pitting erosion produces deep cavities on the tip surface, particularly when Ballotini beads are included, which lessens the efficiency of sonication. The tip can usually be detached for servicing to produce a new flat surface, which is difficult with titanium, stainless or hard steels are better alternatives despite their poorer acoustic and mechanical properties.

Sonicating equipment can be found in many laboratories. In preparative applications the performance is generally not as good as with the high-pressure homogenizers or bead mills. In the case of plasmid recovery from *E. coli*, even when the disruption yield was good, sonication failed to produce intact plasmids (37). The application of this technique to large-scale purification of bioproducts is rather limited: it is difficult to transmit high acoustic power to large volumes, it is difficult to extract the large amounts of heat generated during the cavitation process, and the equipment has to be shielded to protect humans.

Other Methods

The mechanical methods are, generally, effective in disrupting tough microbial cell walls and much more efficient than the nonmechanical methods. Concomitantly, there are also several disadvantages: they are energy intensive, generate high temperatures and shear forces that can damage labile products, are nonspecific with respect to the organisms and the product to be released, and cause intense cell fragmentation. Although several attempts have been made to optimize operations and minimize these effects, process modifications that could result in a significant reduction in the overall energy requirements are still being pursued.

Nonmechanical methods are not efficient on their own, and very often they only increase the permeability of the cell envelope without breaking the wall. However, by weakening the cell wall, lytic methods can be combined with the more efficient mechanical methods, and therefore require less harsh conditions.

Among the physical methods of lysing cells, two have been reported more frequently, although they are not production oriented: osmotic shock and freeze–thawing. Osmotic stresses are generated when a cell is placed in either a hypertonic or a hypotonic medium; in the first case the cell tends to shrink (plasmolyze), and in the second case to expand. In both cases the cells can suffer potentially serious injuries. Usually the cells have osmoregulating mechanisms in addition to tough cell walls that protect them from lysis. However, if suddenly placed in a hypotonic medium (a medium of very low salt concentration), the cells suffer an osmotic shock, absorb water, and swell; eventually they reach a point of unsustainable internal pressure and burst. This bursting process is effective and upscaleable with weak animal cells. However, osmotic shocks are rather inefficient with tougher bacterial, fungal, or yeast cells, though some periplasmic enzymes and even some proteins/enzymes located near the inner side of the cytoplasmic membrane can be selectively released. Osmotic shock is frequently used to prepare membrane vesicles for studies of membrane proteins. The important operational parameter is the time to fully resuspend the cells in the hypotonic medium, which should be as short as possible to avoid the osmoadaptation of the organisms.

The release of intracellular components can also be induced by freeze–thawing the cells. The ice crystals that form and grow during the freezing step will mechanically disrupt the integrity of the inner cell membranes, which become permeable and allow several enzymes to be extracted into appropriate buffers during the thawing step. The precise mechanism for the release of proteins has not yet been elucidated; however, the formation of small transient pores (44) in the membranes is a logical explanation consistent with the facts that protein solubility is important in governing its release to the medium, the inclusion bodies remain unreleased, and large molecular weight proteins are excreted with very low yields (45). The efficiency of this method depends largely on the rates of thawing. With a large mass of frozen suspension, the rate of thawing is higher in the outside region, while the interior of the block remains frozen, allowing chemical or enzymatic (eventually, proteolytic) reactions to proceed to different degrees in different parts of the block. On the other hand, the lifetime of the transient pores (regions of discontinuity) depends on the temperature of thawing that affects the rate of membrane resealing (44) and is longer at low temperatures. This may explain the lack of reproducibility of the results in the ball mill disintegration of brewers' yeast samples that were kept frozen and separately thawed (24). The freeze–thawing process is not easily scaled up and find application mainly in laboratory-scale recovery of recombinant proteins. It can also be useful in disrupting pathogenic cells in a sealable package (7).

Chemical methods on lysis are milder and often anticipate an autolytic enzymatic activity. Organic solvents (e.g., toluene, ethanol, *n*-butanol, isobutanol, acetone, methylethyl ketone, etc.) are believed to extract, nonselectively, lipids from the cell wall or cytoplasmic membrane, therefore increasing the permeability to intracellular components. However, organic solvents might cause protein denaturation, and by diffusing into the cell, they also destroy the membranes of intracellular organelles, thus allowing destructive enzymes, such as proteases, to be liberated. These harmful effects of the organic solvents were minimized in the extraction of β-D-galactosidase by mixing yeast cells (*Kluyveromyces fragilis*) with concentrated selected solvents and, following filtration, resuspending and agitating the cells in an adequate buffer for extraction (46); 95% of the enzyme was released at the end of a 20-h extraction period. The process was claimed to be suited for scale-up, and the organic solvents were beneficial in killing the viable yeast cells, therefore avoiding the need to separate them from the lysate and the addition of antimicrobial agents.

The use of concentrated organic solvents, however, imposes a fireproof environment. Other chemicals, such as detergents, chaotropic agents (e.g., guanidine HCl or urea), EDTA, phenethyl alcohol (PEA), or alkaline solutions have been used to increase the permeability of the cell walls. All chemical methods suffer from long holding times and, thus, have long been considered unattractive for larger-scale applications. Addition of glycine in low concentrations to the medium at an appropriate pH could also increase the permeability of recombinant *E. coli* immobilized in κ-carrageenan gel beads to α-amylase.

However, the recent progress in recombinant DNA technologies has renewed interest in the development of methods to increase the permeability of the cell envelope in order to avoid a detrimental accumulation of plasmid-encoded proteins in the cytoplasm. If these proteins are allowed to accumulate, further synthesis is reduced while proteolytic degradation may cause loss of protein activity, or the plasmids may loose stability. The benefits of whole viable cell immobilization may also be lost if the foreign proteins are not excreted and the cells would require a disruption step for the product to be released and further purified. Chemical methods to recover these active recombinant products in situ, with increased selectivity concerning the intracellular location, have been investigated, with great promise. *E. coli*, a Gram-negative bacterium, has typically been used as the host organism, and many high-value recombinant proteins are secreted into its periplasmic space. Treatment of *E. coli* with guanidine HCl (a chaotropic agent) increases the permeability of the outer membrane, and periplasmically expressed proteins such as β-lactamase (47) can thus be leaked into the surrounding medium with high yield (better than 80% in 5 h), free of many intracellular contaminants. After treatment of the same cells with Triton X-100, which attacks the inner membrane, the cytoplasmic proteins could be released but with low yield and only when the cells were treated with both the guanidine HCl and Triton X-100 was the yield of intracellular protein higher than 50%. The cost of Triton X-100 is high but, nevertheless, this process was claimed to be less expensive than enzymatic methods. The release of β-lactamase was also enhanced by treating recombinant *E. coli* with EDTA and PEA at appropriate concentrations; cell viability and synthesis of the plasmid encoded β-lactamase were retained (49).

Pretreatment with guanidine HCl and Triton X-100 was also used in conjunction with a high-pressure homogenization step to disrupt recombinant *E. coli* expressing human growth hormone inclusion bodies (48). A reduced number of passes through the homogenizer at lower disrupting pressure were required to recover the protein from the inclusion bodies; due to their synergistic effects, Triton X-100 was used to reduce the concentration of guanidine HCl required to weaken the cell wall and to wash the inclusion bodies from contaminant proteins.

Pretreatment of *Candida lipolytica* with an alkaline solution resulted in a weakened cell wall and increased protein solubility, and further mechanical homogenization was thus facilitated at lower pressures than with non-treated cells (20).

Enzymic methods of cell wall hydrolysis are also becoming increasingly more attractive. The advantages associated with enzymic lysis include an increased selectivity of product release (which is desirable for further product isolation and purification), an increased rate and yield of product release, the minimization of product damage, requirements for mild conditions of pH and temperature, and no debris being produced.

More than any other methodology, enzymatic lysis requires a good understanding of the details of the chemical and physical structure of the cell walls. No single enzyme is capable of fully degrading the cell walls. As already described, peptidoglycan is the structural element in bacterial cell walls. In Gram-positive bacteria a single enzyme could lyse the cell, but a mixture of several enzymes (glycosidases [endoacetylmuramidase, *endo-* and *exo*acetylglucosaminidase], *N*-acetylmuramyl L-alanine amidase or peptidases) acting synergistically would assist in the lysis of the bacterial peptidoglycan; in Gram-negative species a pretreatment with a surface-active agent is usually required to remove the outer membrane. For yeast cells a wall lytic protease and β-(1–3)-glucanase are essential, but

Table 1. Important Microbial Cell Wall Degrading Enzymes

Organisms	Enzymes	Type of hydrolyzed linkage
Bacteria	Glycosidases	β-(1,4)-linkages between AGA and AMA residues in a peptidoglycan
	Acetylmuramoyl-L-alanine amidases	Link between *N*-acetylmuramoyl residues and L-amino acid residues in certain glycopeptides
	Peptidases	Gly-gly, ala-gly, etc., peptide bonds
Fungi, yeasts	β-(1,3)-Glucanases	Random β(1,3)-linkages in glycans
	β-(1,6)-Glucanases	Random β(1,6)-linkages in glycans
	Mannanases	(1,2)- (1,3)-, (1,6)-β-D-Mannosidic linkages
	Chitinase	*N*-Acetyl-*b*-D-glucosaminide β(1,4)-B-linkages in chitin and chitodextrins
	Proteases	
Algae	Cellulases	α(1,4)-Linkages in cellulose

the inclusion of β-(1–6)-glucanase, mannase, and chitinase assist the lytic process (Table 1). Most lytic enzyme systems are active against pretreated nonliving cells (lyophilized or heat-killed cells) or against cell walls.

The main restrictions in using multienzyme systems are the availability and high costs associated with their preparation. Different sources of lytic enzymes have been investigated. Various species of *Arthrobacter* can be induced, by whole yeast cells or cell walls, to produce yeast lytic activity (β-(1–3)-glucanase, mannase, and protease) subject to catabolite repression by glucose. Yeast lytic activity was also identified in *Cytophaga* sp. and *Oerskovia xanthineolytica*. Different strains of *Streptomyces* sp., *Pseudomonas* sp., *Staphilococcus* sp., and *Cytophaga* sp. have a wide spectrum of bacteriolytic activities. Lysozyme from hen egg white is bacteriolytic against only a few bacteria.

For the extraction of a given product from an organism, the selection of the most appropriate lytic system depends on the nature of the wall to be lysed and the nature and location of the product inside the cell. The composition of a multienzyme system determines the efficiency and the selectivity of the lytic process (50). If the product is proteinaceous, a highly proteolytic system is undesirable. Thus, if the organism to be lysed is a yeast, a high proteolytic activity should be provided in only the early part of the lytic process, when the substrate is the mannan–protein complex of the wall; the protease should then be inactivated, or a protease inhibitor should be added. However, the composition of lytic systems can be manipulated in part in the production phase, using different inducers and growth conditions. For instance, in the case of the lytic enzyme system from *Cytophaga* sp., high β-(1–3)-glucanase activities and high glucanase: protease ratios were obtained when produced in continuous culture at low dilution rates under carbon-limited conditions. The quantitative data concerning the effects of pH, temperature, concentrations of lytic enzyme, and of cells on the lysis rate should also be obtained to optimize a lytic process (50).

As with all the nonmechanical methods, extensive enzymatic lysis is a very slow process, but in conjunction with others, the rate and the yield of product extraction can be much increased. Mutants that are defective in β-glucan can be produced by mutagenesis and are, therefore, more susceptible to β-(1–3)-glucanase (commercially known as zymolyase) attack (51). Enhanced disruption of *C. utilis* was obtained by high-pressure homogenization (using a microfluidizer) when the cells were pretreated with zymolyase (51). *Bacillus cereus* pretreated with 0.5 mg cellosyl (a lytic enzyme produced by *Streptomyces coelicolor*) per gram of cells (wet wt), at pH 5.5 and 30 °C, was disrupted with 98% yield in one single passage in a Manton–Gaulin homogenizer at 7×10^7 Pa, as compared with less than 40% yield when the cells were untreated. The same enzymatic treatment facilitated disruption in a bead mill (52).

Autolysis is the process by which the cells use their own lytic capabilities and produce lytic enzymes. Autolysis, as a self-digestion process, is very slow and often requires several days, which is very inconvenient. Because of that, it is not a common method of cell disruption; however, yeasts are autolyzed on an industrial scale for the production of yeast autolysates. The process can be speeded up by drying or freeze-thawing the cells. As already mentioned, autolysis can also be induced by plasmolysis with surfactants, thiol reagents, and organic solvents (e.g., toluene or butanol) or by aging or temperature. β-Lactam antibiotics are known to cause the activation of autolytic enzymes (autolysins), and this has also been suggested to be the mechanism by which surfactants induce the autolytic process. Autolysis enzymes are important for cell growth and reproduction, but once they are deregulated, they attack bacterial peptidoglycan. A key factor is the pH, which is generally optimum around 8.0 when yeast is to be plasmolyzed by organic solvents or detergents (56).

Microbial cells differ in their potential for autolysis; often they lack the genetic information to produce lytic enzymes, in which case the genetic coding must be introduced from outside. Lytic enzyme systems that can be cloned include not only the autolysins (several peptidoglycan hydrolases) but also colicin enzymes (a class of antibiotic proteins) and phages (57).

EFFECTS OF CELL DISRUPTION ON DOWNSTREAM OPERATIONS

One of the most challenging problems in solid–liquid separation is the removal of cell debris, which includes cell wall fragments, cell membranes, and protein precipitates. Disruption and lytic treatments cause a reduction in the sedimentation properties of the cells and fine debris (<0.1 μm), leading to inefficient centrifugation and/or filtration due to changes in the surface properties, high compressibility of the solids, high viscosity of the disrupted suspension, and a small difference between the densities of the solids and the suspending medium (53). *Centrifugation* has been used, but it is unsatisfactory when the particles are smaller than 0.5 μm, particularly in viscous homogenates, therefore requiring very low flow rates (on the order of 10 times less than for whole cells). A previous flocculation can improve the centrifugation efficiency. A highly selective *flocculation* of cell debris by borax (sodium tetraborate) was improved by further addition of poly(ethylene imine) to the clarified homogenate supernatant (54). *Rotary vacuum precoat filtration* has been used for the clarification of disrupted yeasts, but *crossflow microfiltration* has the advantage of preventing cake formation and avoiding the need for expensive filter aids. The separation of cell debris is, however, the most complex problem that crossflow microfiltration has been tested against. Membrane fouling and concentration polarization are some of the difficulties that contribute to a rapid flux decline and limit the concentration of debris that can be attained. Apart from the influence of the operational parameters (pressure, feed velocity, fouling factors) on the flux and transmission, which are well documented in the literature, we have to consider that strong adsorption of proteins to cell debris may occur in the lysate, depending on the concentration, temperature, pH, or ionic strength. *Aqueous two-phase extraction* is attractive for the recovery of whole cells and debris from protein solutions but involves high costs and difficult recovery of reagents. Phase components include water-soluble polymers and electrolytes, the best-known systems being PEG-dextran and PEG-phosphate.

Separation techniques depend on many specific aspects, but the average particle size and the size distribution are crucial to all of them: the maximum amount of released protein increases with increasing micronization of the cell debris (36), but these become increasingly more difficult to separate.

Mechanical methods are generally, effective in disrupting tough microbial cell walls. The major difference among them lies in the different sizes of cell fragments that are generated; some methods disrupt the cell envelope with less overall destruction than others. This is significant because the size of the fragments has an important impact on downstream processing when they are removed from the lysate or separated from other subcellular species (37): the submicron particles are easily resuspended in the supernatant after centrifugation, hinder the fractional precipitation, and tend to foul chromatographic columns or membranes that follow (37,54).

The behavior of *E. coli* cell debris from a bead mill and high-pressure homogenizers was investigated (53). No significant differences in the yield and rate of release of protein were reported. However, the characteristics of the debris (particle size distribution, separability by centrifugation, and dead-end filtration) were quite dependent on the disruption method. The mean residence time in the bead mill did not significantly affect the particle size distribution, but the number of passes in the mechanical homogenizer broadened the debris size distribution.

The effect of operating conditions on the cell debris size distribution was investigated for bakers' yeast disrupted by high-pressure homogenization, and a two-parameter model was developed, incorporating both the operating pressure and the number of passes, to simulate the particle size distribution in the homogenate. A threshold pressure of 11.5 MPa was determined, below which no disruption of yeast cells will occur (55). Yeast cell debris particle size distribution was also reported to depend on the growth conditions (13): in batch cultivation the median diameter, d_{50}, was less in the exponential phase and greater in the stationary phase; in continuous culture the median diameter was less for higher dilution rates (younger cells).

High-pressure homogenization of of *E. coli* cells treated with a combination of guanidine HCl and Triton-X-100 produced cell debris of larger sizes than when the pretreatment was made with only guanidine HCl, which is highly desirable when the debris are to be separated by cross-flow microfiltration.

The cell suspension should also be conditioned to avoid unwanted degradation of labile products liberated from the disrupted cells. Protease inhibitors are often required and should be added to the cell slurry. Antioxidants are useful to avoid oxidation of S—S bonds. An appropriate buffer should be included to maintain the pH within the desired range. All the chemicals that are added to the system in all phases of the process (starting from the medium preparation and fermentation steps) may have effects on the product and subsequent operations; this should be taken into account.

Evolving from the recent developments in recombinant DNA technology, a large number of proteins of therapeutical importance, now in use or undergoing clinical evaluation, are produced as inclusion bodies (or "refractile" bodies) in the cytoplasm, mainly in *E. coli*, where they accumulate due to the absence of a secretory mechanism. These inclusion bodies are dense and amorphous aggregates of protein, consisting of misfolded (frequently denatured) polypeptides, often as large (0.8–1.5 μm) as a normal bacterium cell, with a relatively higher density ($\sim$1.2–1.35 kg/L) than the rest of the cell material. They require the use of chaotropic agents or detergents to be solubilized in an aqueous environment (58). The standard procedure for the purification of a protein expressed as an inclusion body begins with the isolation of the inclusion bodies, which itself begins with the disruption of the cell. The best cell disruption technique is high-pressure-homogenization. The aggregates are then separated from the soluble components and from the debris of the lysed cells, usually by centrifugation, typically using disc-stack centrifuges, at 5,000–10,000 *g* during 5–10 min; the lighter cell debris is removed along with the supernatant (58,59). Cross-flow membrane filtration has also been investigated (60). The pellets are then washed with either dilute detergents or chaotropic reagents (urea or guanidine HCl) to remove contaminants, and often reducing agents (e.g., dithiothreitol or β-mercaptoethanol) are added to disrupt incorrect disulfide bonds. The solubilized and reduced protein is then ready for subsequent purification and refolding.

BIBLIOGRAPHY

1. M.D. Lilly, *Appl. Biochem. Biotechnol.* **2**, 1–26 (1979).
2. F.C. Neidhardt, J.L. Ingraham, and M. Schaechter, *Physiology of the Bacterial Cell. A Molecular Approach*, Sinauer Associates Publishers, Sunderland, Massachussets, 1990.
3. S. Bartnicki-Garcia, *Annu. Rev. Microbiol.* **22**, 87–108 (1968).
4. R. Bonali, *Sci. Pharm. Biol. Lor.* **2**, 25–40 (1974).
5. J.L. Strominger, and J.-M. Ghuysen, *Science* **156**, 213–221 (1967).
6. E. Work, in J.R. Norris and D.W. Ribbons eds., *Methods in Microbiology*, vol 5A, Academic, London, 1971, pp. 361–418.
7. D.I.C. Wang, C.L. Cooney, A.L. Demain, P. Dunnill, A.E. Humphrey, and M.D. Lilly, *Fermentation Technology*, Wiley, New York, 1979, p. 241.
8. A.P.J. Middelberg and B.K. O'Neil, *Biotechnol. Prog.* **9**, 109–112 (1993).
9. A.P.J. Middelberg, B.K. O'Neil, and I.D.L. Bogle, *Trans. Inst. Chem. Eng.* **70**, Part C, 205–212 (1992).
10. A.P.J. Middelberg, B.K. O'Neil, I.D.L. Bogle, N.J. Gully, A.H. Rogers, and C.J. Thomas, *Trans. IChemE* **70**, Part C, 213–218 (1992).
11. T. Sauer, C.W. Robinson, and B.R. Glick, *Biotechnol. Bioeng.* **33**, 1330–1342 (1989).
12. C.R. Engler, and C.W. Robinson, *Biotechnol. Bioeng.* **23**, 765–780 (1981).
13. S.F. Siddiqi, M. Bulmer, P. Ayazi Shamlou, and N.J. Titchener-Hooker, *Bioprocess Eng.* **14**, 1–8 (1995).
14. D.A. Roberts, Z. Zhang, T.W. Young, and C.R. Thomas, *Proc. Inst. Chem. Eng. Ann. Res. Event*, University College, London, 1994, pp. 73–75.

15. J.R. Woodrow and A.V. Quirk, *Enzyme Microb. Technol.* **4**, 385–389 (1982).
16. G. Hedenskog, L. Enebo, J. Vendlova, and B. Prokes, *Biotechnol. Bioeng.* **11**, 37–51 (1969).
17. A. Heim and R. Czerski, *Bioprocess Eng.* **13**, 23–30 (1995).
18. K.E. Magnesson and L. Edebo, *Biotechnol. Bioeng.* **16**, 1273–1282 (1974).
19. A.P.J. Middelberg, B.K. O'Neil, I.D.L. Bogle, and M.A. Snoswell, *Biotechnol. Bioeng.* **38**, 363–370 (1991).
20. C.H. Lee, S.K. Tsang, R. Urakabe, and R.A. Rha, *Biotechnol. Bioeng.* **21**, 1–17 (1979).
21. D.E. Bianchi, *Stain Technol.* **40**, 79–82 (1965).
22. C.R. Englet and C.W. Robinson, *Biotechnol. Bioeng.* **21**, 1861–1869 (1979).
23. P.J. Hetherington, M. Follows, P. Dunnill, and M.D. Lilly, *Trans. Inst. Chem. Eng.* **49**, 142–148 (1971).
24. F. Marffy and M.-R. Kula, *Biotechnol. Bioeng.* **16**, 623–634 (1974).
25. H. Mogren, M. Lindblom, and G. Hedenskog, *Biotechnol. Bioeng.* **16**, 261–274 (1974).
26. M. Follows, P.J. Hetherington, P. Dunnill, and M.D. Lilly, *Biotechnol. Bioeng.* **13**, 549–560 (1971).
27. F.A.P. Garcia, in J.F. Kennedy and J.M.S. Cabral eds., *Recovery Processes for Biological Materials*, Wiley, Chichester, U.K., 1993.
28. D.E. Hughes, J.W.T. Wimpeny, and D. Lloyd, in J.R. Norris and D.W. Ribbons eds., *Methods in Microbiology*, vol 5B, Academic, London, 1971, pp. 1–54.
29. W.T. Coakley, A.J. Bater, and D. Lloyd, *Adv. Microb. Physiol.* **17**, 279–341 (1977).
30. Y. Chisti and M. Moo-young, *Enzyme Microb. Technol.* **8**, 194–204 (1986).
31. M.D. White and D. Marcus, in A. Mizrahi ed., *Downstream Processes: Equipment and Techniques*, Alan R. Liss, New York, 1988, pp. 51–96.
32. J.S.G. Brookman, *Biotechnol. Bioeng.* **16**, 371–383 (1974).
33. M.S. Doulah, T. Hammond, and J.S.G. Brookman, *Biotechnol. Bioeng.* **17**, 845–858 (1975).
34. T. Sauer, O.W. Robinson, and B.R. Glick, *Biotechnol. Bioeng.* **33**, 1330–1342 (1989).
35. E. Keshavarz Moore, M. Hoare, and P. Dunnill, *Enzyme Microb. Technol.* **12**, 764–770 (1990).
36. E. Keshavarz, J. Bonnerjea, M. Hoare, and P. Dunnill, *Enzyme Microb. Technol.* **12**, 494–498 (1990).
37. A. Carlson, M. Signs, L. Liermann, R. Boor, and K.J. Jem, *Biotechnol. Bioeng.* **48**, 303–315 (1995).
38. J. Limon-Lason, M. Hoare, C.B. Orsborn, D.J. Doyle, and P. Dunnill, *Biotechnol. Bioeng.* **21**, 745–774 (1979).
39. H. Schütte, K.H. Kroner, H. Hunstedt, and M.-R. Kula, *Microb. Technol.* **5**, 143–148 (1983).
40. Melendres, A.V., H. Unno, N. Shiragami, and H. Honda, *J. Chem. Eng. Jpn.* **25**, 354–356 (1992).
41. F. Bunge, M. Pietzsch, R. Müller, and C. Syldatk, *Chem. Eng. Sci.* **47**, 225–232 (1992).
42. M.S. Doulah, *Biotechnol. Bioeng.* **19**, 649–660 (1977).
43. R. Kuboi, H. Umakoshi, N. Takagi, and I. Komasawa, *J. Ferment. Bioeng.* **79**, 335–341 (1995).
44. B. Lee, K. McKenna, and J. Bramhall, *Biochim. Biophys. Acta* **815**, 128–134 (1985).
45. B.H. Johnson and M.H. Hecht, *Bio / Technology* **12**, 1357–1360 (1994).
46. D.M. Fenton, *Enzyme Microb. Technol.* **4**, 229–232 (1982).
47. T.J. Naglak and H.Y. Wang, *Enzyme Microb. Technol.* **12**, 603–611 (1990).
48. Bailey, S.M., P.H. Blum, and M.M. Meagher, *Biotechnol. Prog.* **11**, 533–539 (1995).
49. W. Ryan and S.J. Parulekar, *Biotechnol. Prog.* **7**, 99–110 (1991).
50. J.A. Asenjo, B.A. Andrews, J.B. Hunter, and S. LeCorre, *Process Biochem.* **20**, 158–164 (1985).
51. C.V. Baldwin and C.W. Robinson, *Biotechnol. Bioeng.* **43**, 46–56 (1994).
52. G. Vogels and M.-R. Kula, *Chem. Eng. Sci.* **47**, 123–131 (1992).
53. Agerkvist, I. and S.-O. Enfors, *Biotechnol. Bioeng.* **36**, 1083–1089 (1990).
54. A.I. Clarkson, P. Lefevre, and N.J. Titchener-Hooker, *Biotechnol. Prog.* **9**, 462–467 (1993).
55. S.F. Siddiqi, N.J. Titchener-Hooker, and P. Ayazi Shamlou, *Biotechnol. Bioeng.* **50**, 145–150 (1996).
56. K. Breddam and T. Beenfeldt, *Appl. Microbiol. Biotechnol.* **35**, 323–329 (1991).
57. R.L. Dabora and C.L. Cooney, *Adv. Biochem. Eng. Biotechnol.* **43**, 11–30 (1990).
58. A. Mukhopadhyay, *Adv. Biochem. Eng. Biotechnol.* **56**, 61–109 (1997).
59. H.H. Wong, B.K. O'Neilland, A.P.J. Middelberg, *Bioseparation* **6**, 361–372 (1997).
60. S.M. Bailey and M.M. Mragher, *Biotechnol. Bioeng.* **56**, 304–310 (1997).

See also CELL SEPARATION, CENTRIFUGATION; CELL SEPARATION, SEDIMENTATION; MEMBRANE SEPARATIONS; PROTEIN PURIFICATION, AQUEOUS LIQUID EXTRACTION.

CELL IMMOBILIZATION

ATSUO TANAKA
TAKUO KAWAMOTO
Kyoto University
Kyoto, Japan

KEY WORDS

Alginate gel method
Biocatalyst
κ-Carrageenan gel method
Carrier-binding method
Cell
Cross-linking method
Entrapment method
Immobilization

Polyacrylamide gel method
Synthetic resin prepolymer method

OUTLINE

INTRODUCTION

Immobilized cells are defined as "cells physically confined or localized in a certain defined region of space with retention of their catalytic activities, which can be used repeatedly and continuously" (1). This definition is applicable not only to cells, but also to single enzymes, multienzyme conjugating systems, and cellular organelles, that is, to all types of biocatalysts. Bioprocesses using biocatalysts have increasingly received worldwide attention because of their benefits of low environmental pollution and economical utilization of natural resources and energy. They have also contributed to a wide variety of fields. The use of immobilized biocatalysts increases the efficiency of such bioprocesses because they catalyze biochemical reactions under more stabilized states than their free counterparts; moreover, they can be used repeatedly or continuously. In particular, immobilized cells can catalyze complicated reactions and have many useful characteristics that other biocatalysts do not have, as described later. The European Federation of Biotechnology, Working Party on Applied Biocatalysis (formerly, the Working Party on Immobilized Biocatalysts) proposed that immobilized cells should be classified as viable or nonviable, where viable includes growing, nongrowing, and respiring (2). According to this proposal, terms such as treated, resting, growing, and living should not be used. However, in general, these terms are used for classifying the condition of immobilized cells. In the definition of "treated," cells are subjected to chemical or physical treatment before or after immobilization, and resting cells do not show growth during utilization, although they have the ability to grow.

HISTORY

In 1916, Nelson and Griffin fortuitously found that yeast invertase adsorbed on activated charcoal retained its catalytic ability in the hydrolysis of sucrose (3). This is said to be the first demonstration of immobilization of a biocatalyst. Although several reports were published on the immobilization of physiologically active proteins by covalent binding on various supports after this finding, immobilized biocatalysts were not used in practice until Grubhofer and Schleith immobilized several enzymes (carboxypeptidase, diastase, pepsin, and ribonuclease) on diazotized polyaminostyrene resin by covalent binding in 1953 (4). Thereafter, many studies on enzyme immobilization have been carried out, and studies on cell immobilization became to be popular in 1970s.

The first industrial application of immobilized biocatalysts was carried out by Chibata and coworkers of Tanabe Seiyaku Co., Japan, in 1969. Fungal aminoacylase immobilized on DEAE-Sephadex through ionic binding was used for stereoselective hydrolysis of *N*-acyl-D,L-amino acids to produce L-amino acids and *N*-acyl-D-amino acids (5). The first industrial application of immobilized cells was also performed by Chibata and his coworkers in 1973, producing L-aspartate from ammonium fumarate by polyacrylamide-gel-entrapped *Escherichia coli* cells containing a high level of aspartase activity.

Several other processes, such as production of L-alanine, L-malic acid, 6-aminopenicillanic acid, milk having a low lactose content, and high-fructose syrup also have been industrialized with the use of immobilized enzymes and immobilized cells, and a vast amount of literature on immobilization is available (6–15).

SIGNIFICANCE

In biological systems, such as microbial cells, plant cells, and animal cells, various chemical reactions are carried out. These chemical reactions are catalyzed by enzymes that are water-soluble globular proteins. *Compared to chemical catalysts, enzymes as catalysts have the following superior characteristics*.

1. Enzyme-catalyzed reactions take place under mild conditions, such as ordinary temperature, atmospheric pressure, and pH value around neutrality.
2. Enzymes have strict substrate specificity, stereospecificity, regiospecificity, and reaction specificity.

These facts suggest that energy-saving, resource-saving, and low-pollution processes could be designed by using biocatalysts because such processes could be carried out at relative low temperatures and at ambient pressure with little by-product formation. However, biocatalysts are, in general, liable to be deactivated under conditions such as high temperature, extremes of pH, presence of organic solvents, or even conditions suitable for their catalysis. The repeated or continuous use of biocatalysts that are required in order to construct efficient processes is another problem faced when free biocatalysts are used. Immobilization is one of the most useful methods for eliminating some of the disadvantages described above. In general, immobilization makes biocatalysts stable and easy to use repeatedly in a long-term series of batchwise reactions, or continuously in flow systems. Therefore, the immobilization of biocatalysts is attracting attention worldwide.

At present, immobilized biocatalysts have contributed to a wide variety of fields including (1) the production of useful compounds by stereospecific and/or regiospecific reactions, *(2)* the production of energy by biological processes, *(3)* the selective degradation or removal of pollutants to solve environmental problems, *(4)* analyses of various compounds with high sensitivity and high specificity, and *(5)* utilization in new types of drugs such as slow-releasing medications, in making artificial organs, and so on.

These processes require the immobilization not only of single enzymes that catalyze simple reactions, but also of multienzyme conjugating systems that mediate more complex reactions. These systems often require the regeneration system of ATP and/or coenzymes. Because cells contain metabolic systems that mediate such complicated reactions, immobilization of cells serves as immobilization of such multienzyme conjugating systems. Furthermore, in the case of immobilization of cells, the procedure of extracting enzymes from the cells is no longer required. This avoids the inactivation of enzymes during these time-consuming and expensive procedures, making it possible to use the enzymes in a stable state.

However, immobilized cells have some disadvantages that must be overcome to realize their practical applications.

1. By-products are easily synthesized because the cells contain the various metabolic systems and the many enzymes that catalyze undesirable reactions. To defeat this problem, selection of a proper strain, mutation, or the proper treatment of cells in order to lower the levels of enzymes participating in side reactions and increase the activities of desirable enzymes, or genetic improvement of a cell line is essential.
2. Cell walls and cell membranes of intact cells often hamper the permeation of substrates, products, and other reaction components into and out of cells. In this case, appropriate treatment of cells before or after immobilization is required to destroy the barriers to permeability.

If immobilized cells are kept in a growing state by a proper supply of suitable nutrients, and if their biological functions are useful, as in the case of conventional fermentation procedures, these immobilized cells (immobilized growing cells) seem very advantageous. Immobilized growing cells serve as self-proliferating and self-regenerating biocatalysts, and catalytic activities of immobilized growing cells are expected to be maintained during long-term repeated or continuous use of cells.

However, immobilized growing cells also have some disadvantages as biocatalysts, and ways to overcome these disadvantages must be studied.

1. Various nutrients and energy sources are demanded by immobilized growing cells to maintain the cells growing state(s), and yield may be lowered by the consumption of substrates and/or products as nutrients and energy sources.
2. Contamination of product is liable to occur by cells leaked from carriers.

In spite of these faults, cell immobilization is a key technology used to construct an efficient bioprocess, as immobilized cells have the attractive characteristics described earlier.

METHODS

For the application of immobilized cells, screening or breeding of cells having the desired activity and characteristics is very important. In addition, selection of the appropriate immobilization method is necessary. There is, however, no ideal universal immobilization method fitted to various types of cells at present, although a vast amount of literature on immobilization is now available.

The immobilization methods can be classified into four categories: carrier-binding, cross-linking, entrapment, and combined methods, in which combined methods are a combination of former three. Because each method has its own merits and faults (to be described later), selection should be carried out in consideration of the purpose of immobilization, characteristics of cells, type of reaction, type of reactor, and so on.

Carrier-Binding Method

The carrier-binding method is based on the binding of biocatalysts to water-insoluble carriers through covalent bonds, ionic bonds, physical adsorption, or biospecific binding. Carrier materials should have adequate function to bind biocatalysts, have sufficient mechanical strength; physical, chemical, and biological stability; and nontoxicity. Ease of shaping is also important for applying immobilized biocatalysts to various types of reactors. Furthermore, the economic feasibility should be considered. As carriers, various types of insoluble materials, such as water-insoluble polysaccharides (cellulose, dextran, agarose derivatives, etc.), proteins (gelatin, albumin, etc.), synthetic polymers (polystyrene derivatives, ion exchange resins, polyurethane, etc.), and inorganic materials (brick, sand, glass, ceramics, magnetites, etc.) have been developed and utilized as they are, or after proper modification or activation. The development of new carrier materials having desired characteristics is also important.

Any reactive components of the cell surface, for example, amino, carboxyl, thiol, hydroxyl, imidazole, or phenol groups of proteins or saccharides, can be used for covalent bonding between cells and a modified or activated carrier. The covalent-binding method is one of the most frequently used techniques for the immobilization of enzymes because of the following advantages: *(1)* enzymes do not leak or detach from carriers during utilization because of the tight binding, *(2)* enzymes can easily contact with substrates because enzymes are localized on the surface of carriers, and *(3)* an increase in stability is often observed because of the strong interaction between enzymes and carriers. However, this method is not as widely used for cells because of the toxicity of the agents needed for formation of covalent bonds, and the difficulty in finding the conditions for immobilization. In the limited number of reports on cell immobilization by this method, inorganic and organic carriers have been used as follows: Navarro and Durand (16) immobilized *Saccharomyces carlsbergensis* cells by covalent binding to porous silica beads that had been treated with aminopropyltriethoxysilane and activated with glutaraldehyde. *Serratia marcescens, Saccharomyces cerevisiae*, and *Saccharomyces amurcea* cells were also immobilized by covalent binding to borosilicate glass and zirconia ceramics with an isocyanate coupling agent (17). In the case of organic carriers, *Micrococcus luteus* cells were immobilized to the carboxyl group of agarose beads in a two-step process that avoided exposure of the cells to carbodiimide (18), and the uronic acid–forming activity of these immobilized cells was retained, even though cell viability was lost. A disadvantage of this method is that leakage of the enzymes may occur due to autolysis during the reaction.

Recently, Klein and Ziehr have summarized the immobilization of cells by adsorption to carriers (19). One of the first studies on immobilized cells (20,21) is on the adsorption of *Escherichia coli* and *Azotobacter agile* cells to Dowex-1 for the oxidation of succinic acid. This adsorption method is very simple and mild, and can be applied to the immobilization of various types of cells. However, adsorption of cells to carriers is sensitive to environmental factors such as pH, ionic strength, concentration of reactants, and temperature because the adsorption phenomenon is based primarily on van der Waals forces, ionic or hydrogen bonds, or hydrophobic interactions between the cell surface and the carrier. Therefore, cells are liable to leak from carriers.

Biospecific binding with lectins introduced on carriers was also used for immobilization of cells (22,23). Lectins that are glycoproteins bind tightly to specific carbohydrate residues on the cell surface. Concanavalin A, which can be obtained from jack beans, is one of the most useful lectins.

Cross-Linking Method

The cross-linking method utilizes bi- or multifunctional reagents, but carriers are not used. Cells are linked with each other by the reagents, and the crosslinked cells form pellets.

Glutaraldehyde is one of the most popular cross-linking reagents (24,25), and combines amino groups through a Shiff base linkage. Diisocyanates, such as toluene diisocyanate and hexamethylene diisocyanate, represent another type of cross-linking reagent, forming a urea linkage with amino groups.

E. coli having high aspertase activity was immobilized by this method with glutaraldehyde or toluene diisocyanate (26). A 34% retention of aspartase activity was obtained in the case of glutaraldehyde, but the immobilized cells were inactivated in the presence of toluene diisocyanate. So far there have been very few studies on this method, but *Bacillus coagulans* immobilized by this method is commercially available for immobilized glucose isomerase and is widely used for isomerization of glucose in the industrial field.

Entrapment Method

Entrapment methods can be classified as follows: lattice type, microcapsule type, liposome type, and membrane type.

In lattice type entrapment, biocatalysts are entrapped in gel matrices of various types of polymers. Microcapsule type is entrapment in microcapsules of semipermeable synthetic polymers. Liposome type is entrapment within a liquid membrane prepared from amphiphatic compounds. In membrane type, biocatalysts are separated from the environment as spent reaction solution by ultrafiltration membranes, microfiltration membranes, or hollow fibers.

Entrapment methods are very popular for immobilization of cells. Microbial cells, as well as plant and animal cells, can be immobilized by this method. However, this method also has disadvantages, as follows:

1. High molecular weight substrates can hardly access the cells entrapped.
2. Renewal of carriers is hardly possible.

Membrane-type entrapment can avoid these disadvantages, although it has no effects on stability of cells. The lattice-type method is most widely applied for preparing immobilized cells, and there is a great deal of literature on this method. Several representative techniques for the lattice-type method are described next.

Polyacrylamide Gel Method. Bernfeld and Wan (27) reported the entrapment of several enzymes in polyacrylamide gels in 1963, and this method was applied to the cells of the lichen *Umbilicaria pustuluta* by Mosbach and Mosbach in 1966 as reviewed by Chibata et al. (28). Various types of biocatalysts have so far been immobilized by this method. This method was used for not only the first industrial application of immobilized microbial cells as described earlier, but also for the first application of immobilized growing cells, that is, the production of L-glutamic acid by *Corynebacterium glutamicum* (29).

The procedure for the formation of gel to entrap cells is identical to that employed for the preparation of gel commonly used for electrophoresis. For the immobilization of cells by this method, acrylamide, and *N,N′*-methylenebisacrylamide (BIS) as the cross-linking reagent, are mixed with a suspension of cells and polymerized in the presence of an initiator, such as potassium persulfate, and a simulator, such as 3-dimethylaminopropionitrile (DMAPN) or *N,N,N′,N′*-tetramethylenediamine (Fig. 1).

Although this method can be used for various purposes, a major disadvantage of this method is the toxicity of the acrylamide monomer, the cross-linking reagent, the initiator, and the stimulator. In some cases, cells are damaged by free radicals during polymerization. Several analogues or derivatives of acrylamide are also useful for this method.

Table 1 summarizes some examples of cells immobilized with polyacrylamide.

Alginate Gel Method. Several natural polysaccharides, such as alginate, κ-carrageenan, and agar, are excellent gel

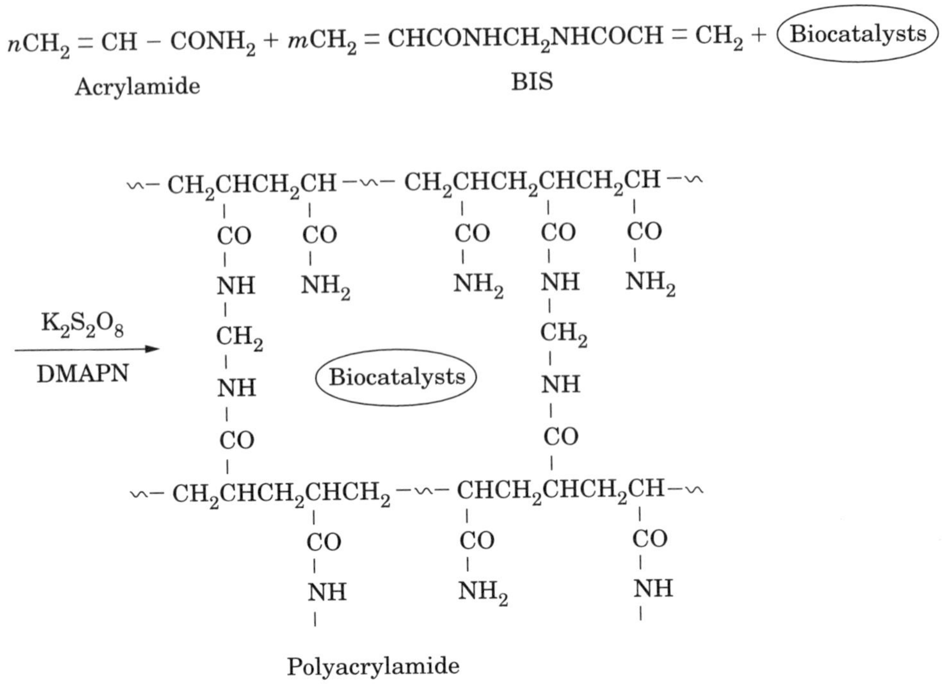

Figure 1. Immobilization with polyacrylamide.

Table 1. Examples and Applications of Cells Immobilized with Polyacrylamide

Cells	Products	Reference(s)
Acromobacter liquidum	Urocanic acid	41
Arthrobacter globiformis	Prednisolone	42
Aspergillus niger	Citric acid	43
Bacillus subtilis	α-Amylase	44
Brevibacterium ammoniagenes	L-Malic acid	45,46
Corynebacterium dismutans	L-Alanine	47
Corynebacterium glutamicum	L-Glutamic acid	29
Curvularia lunata and *Corynebacterium simplex*	Prednisolone	48
Escherichia coli	L-Aspartic acid	29,49,50
Humicola lutea	Acid protease	51
Penicillium chrysogenum	Penicillin G	52
Streptomyces clavuligerus	Cephalosporin C	53
Recombinant *Bacillus subtilis*	Proinsulin	54

materials and are widely used for the entrapment of various biocatalysts.

Alginate, which is extracted from seaweed, is a linear copolymer of D-mannuronic and L-guluronic acids (Fig. 2) and can be gelled by multivalent ions such as calcium and aluminum. For immobilization of cells, sodium alginate, which is soluble in water, is mixed with a suspension of cells and then dropped into a calcium chloride solution to form water-insoluble beads of calcium alginate gels (30).

However, calcium alginate gels are apt to disrupt by mechanical stress and are gradually solubilized in the presence of calcium-ion-trapping reagents, such as phosphate ion and citrate. Leakage of cells is then liable to occur, but free cells can be recovered from the immobilized preparation by this solubilization, which allows the characterization of the cells in gels (e.g., viability). Several multivalent metal ions, such as aluminum ion or strontium ion, can also be used instead of calcium ion. The treatment of formed calcium alginate gels with cationic polymers can improve gel resistivity against the solubilization by calcium-ion-trapping reagents (31). It is also reported that addition of polyacryl acid to sodium alginate in the entrapment by the procedure described can increase the mechanical strength of gels formed and prevent them from dissolving because of bridging polyacrylate chains with calcium ions (32).

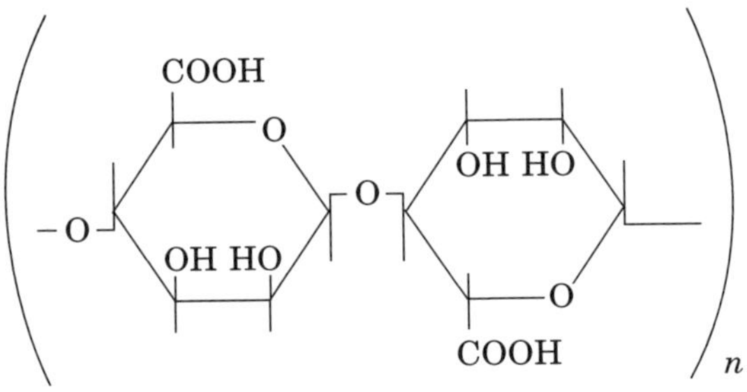

Figure 2. Structure of alginic acid (molecular weight, 50,000–200,000).

The method using alginate gel is widely applicable to immobilization of various useful biocatalysts, including very fragile ones such as plant cells and animal cells, as shown in Table 2, because of the simplicity of the procedure, easy availability of sodium alginate, and mildness of immobilization conditions. It is worthy of note that the first report on immobilization of plant cells employed this method (33).

Table 2. Examples and Applications of Cells Immobilized with Alginate

Cells	Products	Reference(s)
Arthrobacter simplex	Prednisolone	55
Aspergillus niger	Citric acid	56
	Gluconic acid	56
Curvularia lunata	Hydrocortisone	57
Corynebacterium glycinophilum	L-Serine	58
Lactobacillus lactis	Lactic acid	56
Penicillium chrysogenum	Penicillin G	59
Propionibacterium sp.	Vitamin B_{12}	60
Saccharomyces cerevisiae	Ethanol	30
Serratia marcescens	Protease	61
Streptomyces sp.	Tylosin	62
Streptomyces tendae	Nikkomycin	62
Trigonopsis variabilis	α-Keto acid	63
Zymononas mobilis	Ethanol	64
Recombinant *Bacillus subtilis*	Proinsulin	54
Recombinant *Escherichia coli*	β-Lactamase	65
Catharanthus roseus	Ajmalicine isomers	33
Digitalis lanata	Digoxin	33
Lavandula vera	Blue pigments	66
Morinda citrifolia	Anthraquinones	33
Hybridoma cell 4H11	Monoclonal IgA	67

κ-Carrageenan Gel Method. κ-Carrageenan is a readily available nontoxic natural polysaccharide, isolated from seaweed. It is composed of a unit structure of β-D-galactose sulfate and 3,6-anhydro-α-D-galactose (Fig. 3) and is widely used in the food and cosmetic industries as a gelling, thickening, and stabilizing agent.

κ-Carrageenan becomes a gel by cooling or in the presence of a gel-inducing reagent, such as ammonium ion, potassium ion, calcium ion, aluminum ion, magnesium ion, diamines, or water-miscible organic solvents (34). If the proper gel-inducing reagent is used to form a κ-carrageenan gel in the presence of cells, immobilized cells can be easily obtained.

κ-Carrageenan gels can be easily solubilized in saline or in water, and a suspension of free cells can be obtained to investigate their characteristics. On the other hand, leakage of cells easily occurs by dissolution of gel when a gel-inducing reagent is not added to the reaction mixture.

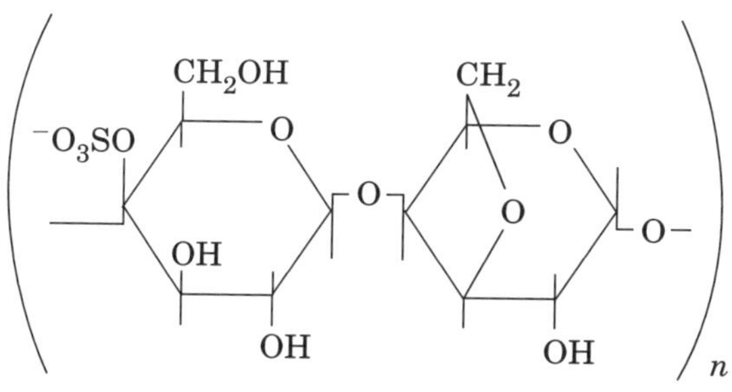

Figure 3. Structure of κ-carrageenan (molecular weight, 100,000–800,000).

In order to increase the stability of the κ-carrageenan-entrapped cells, treatment with glutaraldehyde and hexamethylenediamine after entrapment is often effective.

The conditions of immobilization by this method are also mild, and this method can be used for immobilization of various biocatalysts (Table 3). This method has replaced the polyacrylamide gel method in industrial production of L-aspartate (35) and L-malate (36) by means of immobilized microbial cells. Another advantage of this method is that various shapes of immobilized cells can be prepared (34). The typical procedures for preparing various shapes of κ-carrageenan gel-entrapped cells are as follows:

Cubic type. A suspension of cells and κ-carrageenan dissolved in physiological saline are mixed and cooled to about 10 °C to form a gel. After the gel is soaked in a cold potassium chloride solution in order to increase the gel strength, it is cut to give a cubic-type gel preparation.

Bead type. A mixture of κ-carrageenan and cells is dropped into a 0.3 M potassium chloride solution through a nozzle. Bead-type gels are obtained by this procedure.

Membrane type. A mixture of κ-carrageenan and cells is spread on a plate or sheet and soaked in a cold potassium chloride solution. Consequently, a membrane-type gel is obtained.

Synthetic Resin Prepolymer Method. With the application of a variety of bioreactions, including synthesis, transformation, degradation, or analysis of various compounds having different chemical properties under various reaction conditions, gels having controlled physicochemical properties come to be required for entrapment of biocatalysts. Selection of suitable gel materials among natural polymers for each purpose is usually difficult. For this reason, entrapment methods using synthetic resin prepoly-

Table 3. Examples and Applications of Cells Immobilized with κ-Carrageenan

Cells	Products	Reference(s)
Bacillus subtilis	α-Amylase	68
Brevibacterium flavum	L-Malic acid	36,69
Corynebacterium dismutans	L-Alanine	47
Erwinia herbicola	L-DOPA	70
Escherichia coli	L-Aspartic acid	35,71
Penicillium urticae	Patulin	72
Propionibacterium sp.	Vitamin B_{12}	60
Pseudomonas dacunhae	L-Alanine	73
Saccharomyces bayanus	Ethanol	74
Serratia marcescens	L-Alginine	75
	L-Isoleucine	76
Zymomonas mobilis	Ethanol	64
Recombinant *Escherichia coli*	Catechol-2,3-dioxygenase	77,78
Recombinant *Saccharomyces cerevisiae*	Human chorionic gonadotropin	79
Catharanthus roseus	Ajmalicine	80

mers, such as photo-cross-linkable resin prepolymers and urethane prepolymers, are promising. Synthetic resin prepolymers having desired characteristics can be synthesized chemically in advance in the absence of biocatalysts. The methods are as follows:

Photo-cross-linkable resin prepolymer method. Entrapment methods using photo-cross-linkable resin prepolymers (Fig. 4) with hydrophilic (ENT) or hydrophobic (ENTP) properties, cationic or anionic nature, and different chain lengths have been developed by Fukui and coworkers (37,38). ENT and ENTP can be prepared from hydroxyethylacrylate, isophorone diisocyanate, and poly(ethylene glycol) or poly(propylene glycol), respectively.

A mixture of the prepolymers and cells, layered (about 0.5-mm thickness) on a transparent polyester sheet is gelled by irradiation with near-ultraviolet light (wavelength rage 300–400 nm) for several minutes in the presence of a proper sensitizer, such as benzoin ethyl ether. This method has been applied for the pilot-scale continuous ethanol fermentation by immobilized growing yeast cells (39). Table 4 lists some examples of cells immobilized with photo-cross-linkable resin prepolymers and their applications. These prepolymers are available from Kansai Paint Co., Japan. Photo-cross-linkable resin prepolymers can be stored in the dark and sterilized at 120 °C.

Urethane prepolymer method. There are also various types of urethane prepolymers (PU) (Fig. 5), which have different hydrophilicities or hydrophobicities and chain length. Urethane prepolymers are synthesized from toluene diisocyanate and polyether diols composed of poly(ethylene glycol) and poly(propylene glycol) or poly(ethylene glycol) alone, although these are not commercially available. Hypol®, which has a similar structure, can be obtained from W. G. Grace Co., Boca Raton, Fla., USA.

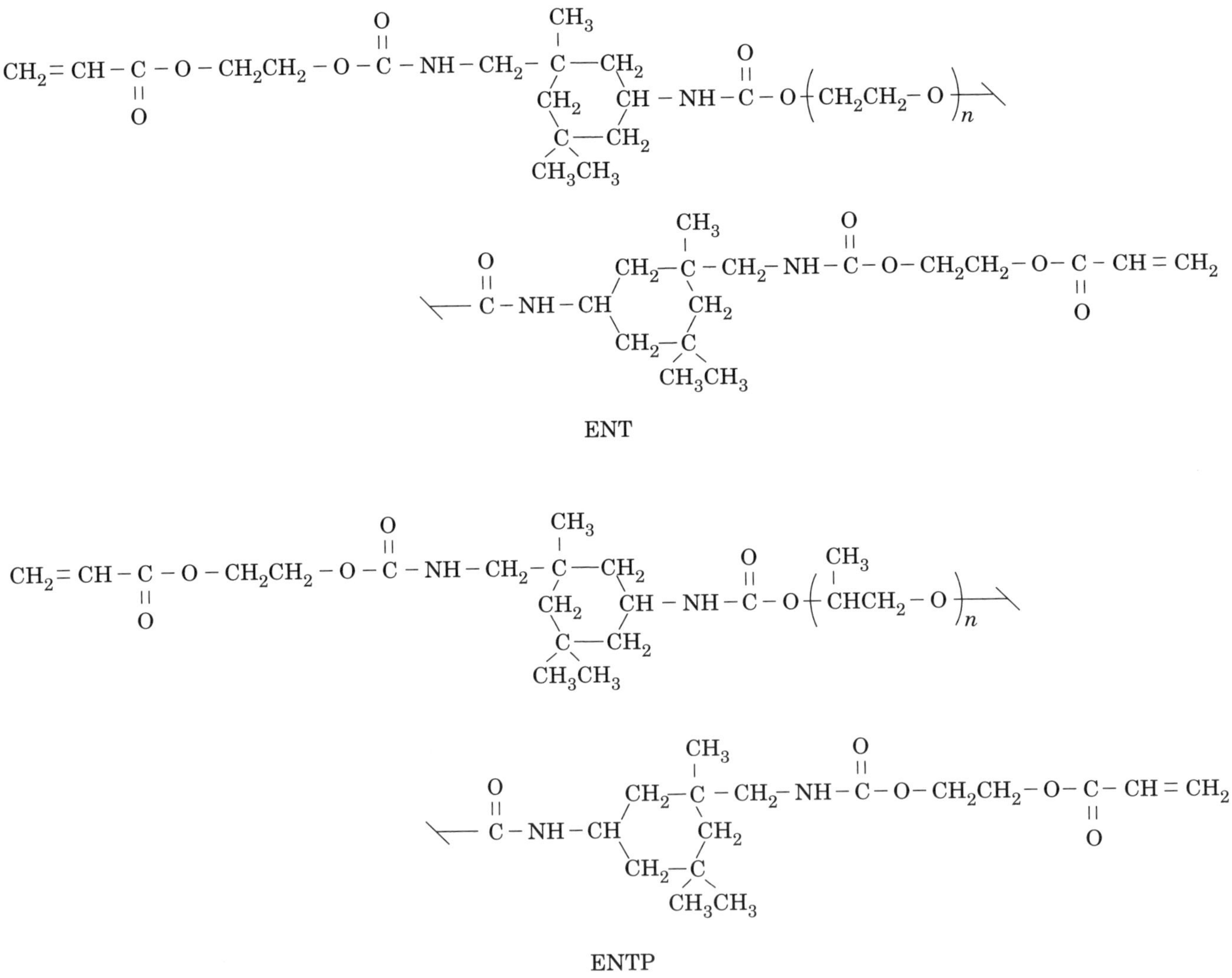

Figure 4. Structures of typical photo-cross-linkable resin prepolymers (ENT, hydrophilic; ENTP, hydrophobic).

Table 4. Examples and Applications of Cells Immobilized with Photo-Cross-Linkable Resin Prepolymers

Cells	Products	Reference(s)
Arthrobacter simplex	Prednisolone	81
Bakers' yeast	ATP	82
Brevibacterium ammoniagenes	Coenzyme A	83
Corynebacterium sp.	9α-Hydroxy-4-androstene-3,17-dione	84
Curvularia lunata	Hydrocortisone	85
Curvularia lunata and *Arthrobacter simplex*	Prednisolone	86
Enterobacter aerogenes	Adenine arabinoside	87
Nocardia rhodochrous	3-Keto-Δ^4-steroids	88
	1,4-Androstadiene-3,17-dione	89
Propionibacterium sp.	Vitamin B_{12}	60
Rhizopus stolonifer	11α-Hydroxyprosterone	90
Rhodotorula minuta var. *texensis*	*l*-Menthol	91
Saccharomyces sp.	Ethanol	39

Isocyanate functional groups at the terminals of urethane prepolymers react with each other in the presence of water to form urea bonds and liberate carbon dioxide. Therefore, gel is easily formed only by mixing the prepolymer with an aqueous suspension of cells, and entrapment of cells can be easily carried out (38,40). Some examples of cells immobilized with urethane prepolymers are listed in Table 5.

Urethane prepolymers can also be stored in a desiccator and sterilized at 120 °C, avoiding moisture.

Table 5. Examples and Applications of Cells Immobilized with Urethane Prepolymers

Cells	Products	Reference(s)
Arthrobacter simplex	Prednisolone	92
Enterobacter aerogenes	Adenine arabinoside	87
Escherichia coli	6-Aminopenicillanic acid	93
	L-Aspartate	94
Nocardia rhodochrous	3-Keto-Δ^4-steroids	88
	4-Androstene-3,17-dione	95
Propionibacterium sp.	Vitamin B_{12}	60
Rhodotorula minuta var. *texensis*	*l*-Menthol	91
Streptomyces rimosus	Oxytetracycline	96

In the cases of these synthetic prepolymer methods, the size of the network of gel matrices (which affects diffusion of substrates and products inside the gel, the mechanical strength of the gel formed, the cell-holding capacity of the gel, and the growth ability of cells inside gel matrices) can be regulated by the chain length of the prepolymers and content of the reactive functional groups.

Furthermore, a balance between hydrophilicity and hydrophobicity of gels can be optionally controlled by selecting suitable prepolymers or by mixing prepolymers that have different properties at a proper ratio. The hydrophilicity–hydrophobicity balance of the gels often has a critical effect on the bioconversion by strongly affecting the partition of reactants to the gels.

The Ionic character of the gel, which is also an important factor in bioconversion, can be introduced into the prepolymers by using ionic oligomers as the main chain for their synthesis.

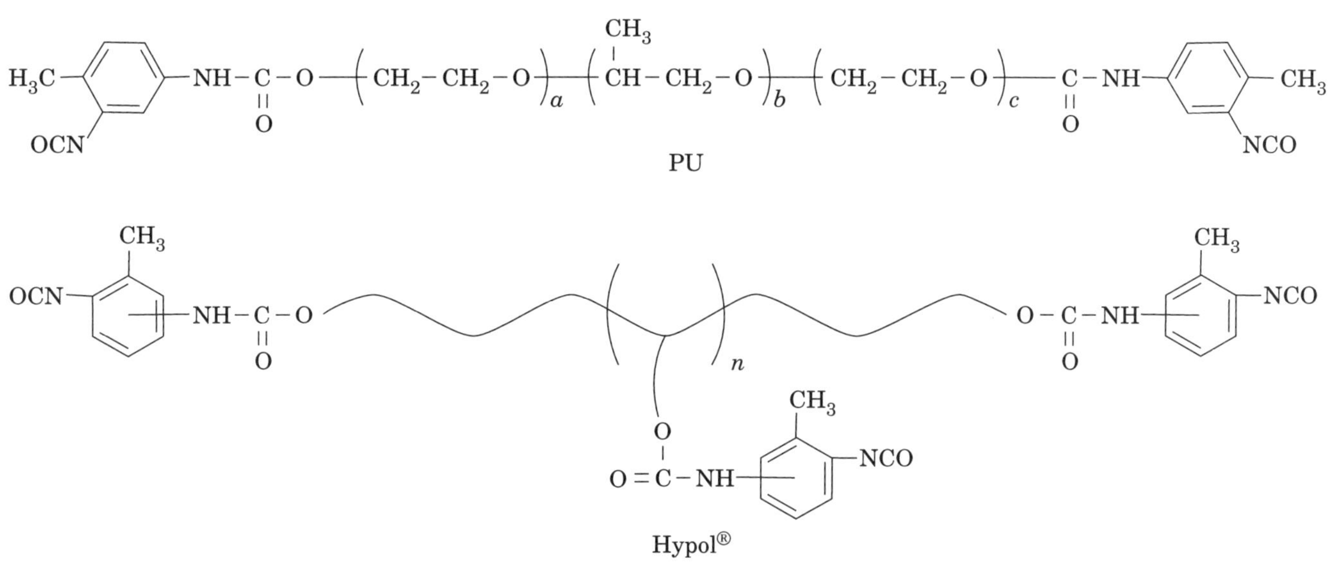

Figure 5. Structures of typical urethane prepolymers.

Other specific features and advantages of the synthetic prepolymer methods are as follows.

1. Procedures for entrapment are very simple and proceed under very mild conditions, and various types of biocatalysts can be immobilized by these methods.
2. Prepolymers do not contain monomers, which may have unfavorable effects on biocatalysts, such as cells. Therefore, the synthetic prepolymer method seems to be one of the most practically applicable immobilization methods for a wide variety of purposes.

BIBLIOGRAPHY

1. I. Chibata, *Immobilized Enzymes, Research and Development* (in Japanese), Kodansha, Tokyo, 1978.
2. M.D. Lilly, *Enzyme Microb. Technol.* **8**, 315 (1986).
3. J.M. Nelson and E.G. Griffin, *J. Am. Chem. Soc.* **38**, 1109–1115 (1916).
4. N. Grubhofer and L. Schleith, *Naturwissenschaft* **40**, 508 (1953).
5. I. Chibata, T. Tosa, T. Sato, T. Mori, and T. Matuo, in *Proceedings of the Fourth IFS. Fermentation Technology Today*, Society of Fermentation Technology, Osaka, Japan, 1972, pp. 383–389.
6. K. Mosbach, *Methods in Enzymology*, vol. 44, Academic, New York, 1976.
7. I. Chibata, *Immobilized Enzyme: Research and Development*, J Wiley, New York, 1978.
8. I. Chibata, T. Tosa, and T. Sato, *Adv. Biotechnol. Processes* **10**, 203–222 (1983).
9. I. Chibata and L.B. Wingard, Jr., *Appl. Biochem. Bioeng.* **4**, 1–349 (1983).
10. B. Mattiason, *Immobilized Cells and Organelles*, Chemical Rubber Co., Cleveland, Ohio, 1983.
11. K. Mosbach, *Methods in Enzymology*, vol. 135, Academic, New York, 1987.
12. I.A. Veliky and R.J.C. McLean, *Immobilized Biosystems, Theory and Practical Applications*, Blackie, London, 1994.
13. R.H. Wijffels, R.M. Buitelaar, C. Bucke, and J. Tramper, *Immobilized Cells: Basics and Applications*, Elsevier, Amsterdam, 1996.
14. R.G. Willaert, G.B. Baron, and L. De Backer, *Immobilized Living Cell Systems*, J Wiley, New York, 1996.
15. G.F. Bickerstaff, *Immobilization of Enzymes and Cells*, Humana Press, Totowa, N.J., 1997.
16. J.M. Navarro and G. Durand, *Eur. J. Appl. Microbiol.* **4**, 243–254 (1977).
17. R. Messing, R.A. Oppermann, and F.B. Kolot, *ACS Symp. Ser.* **106**, 13–28 (1979).
18. T.R. Jack and J.E. Zajic, *Biotechnol. Bioeng.* **19**, 631–648 (1977).
19. J. Klein and H. Ziehr, *J. Biotechnol.* **16**, 1–16 (1990).
20. T. Hattori and C. Furusaka, *J. Biochem.* **48**, 831–837 (1960).
21. T. Hattori and C. Furusaka, *J. Biochem.* **50**, 312–315 (1961).
22. C.H. Bornman and A. Zachrisson, *Plant Cell Rep.* **1**, 151–153 (1982).
23. G.S. Warren and R. Fallon, *Planta* **161**, 201 (1984).
24. F.A. Quiocho and F.M. Richards, *Proc. Natl. Acad. Sci.* **52** 833–839 (1964).
25. A. Schejter and A. Bar-Eli, *Arch. Biochem. Biophys.* **136**, 325–330 (1970).
26. I. Chibata, T. Tosa, and T. Sato, *Appl. Microbiol.* **27**, 878–885 (1974).
27. P. Bernfeld and J. Wan, *Science* **142**, 678–679 (1963).
28. I. Chibata, T. Tosa, and T. Sato, in A.L. Demain and N.A. Solomon eds., *Manual in Industrial Microbiology and Biotechnology*, ASM, Washington, D.C., 1986, pp. 217–229.
29. W. Slowinski and S.E. Charm, *Biotechnol. Bioeng.* **15**, 973–979 (1973).
30. M. Kierstan and C. Bucke, *Biotechnol. Bioeng.* **19**, 387–397 (1977).
31. I.A. Veliky and R.E. Williams, *Biotechnol. Lett.* **3**, 275–280 (1981).
32. T. Mano, S. Mitsuda, E. Kumazawa, and Y. Takeshita, *J. Ferment. Bioeng.* **73**, 486–489 (1992).
33. P. Brodelius, B. Deus, K. Mosbach, and M.H. Zenk, *FEBS Lett.* **103**, 93–97 (1979).
34. T. Tosa, T. Sato, T. Mori, K. Yamamoto, I. Takata, Y. Nishida, and I. Chibata, *Biotechnol. Bioeng.* **21**, 1697–1709 (1979).
35. T. Sato, Y. Nishida, T. Tosa, and I. Chibata, *Biochim. Biophys. Acta* **570**, 179–186 (1979).
36. I. Takata, K. Yamamoto, T. Tosa, and I. Chibata, *Enzyme Microb. Technol.* **2**, 431–437 (1980).
37. S. Fukui, A. Tanaka, T. Iida, and E. Hasegawa, *FEBS Lett.* **66**, 179–182 (1976).
38. S. Fukui, K. Sonomoto, and A. Tanaka, in K. Mosbach ed., *Methods in Enzymology*, vol. 135, Academic, New York, 1987, pp. 230–252.
39. T. Iida, in A. Tanaka, T. Tosa, and T. Kobayashi eds., *Industrial Application of Immobilized Biocatalysts*, Marcel Dekker, New York, 1993, pp. 163–182.
40. S. Fukushima, T. Nagai, K. Fujita, A. Tanaka, and S. Fukui, *Biotechnol. Bioeng.* **20**, 1465–1469 (1978).
41. K. Yamamoto, T. Sato, T. Tosa, and I. Chibata, *Biotechnol. Bioeng.* **16**, 1601–1610 (1974).
42. K.A. Koshcheenko, G.V. Sukhodolskaya, V.S. Tyurin, and G.K. Skryabin, *Eur. J. Appl. Microbiol. Biotechnol.* **12**, 161–169 (1981).
43. H. Horitsu, S. Adachi, Y. Takahashi, K. Kawai, and Y. Kawano, *Appl. Microbiol. Biotechnol.* **22**, 8–12 (1985).
44. T. Kokubu, I. Karube, and S. Suzuki, *Eur. J. Appl. Microbiol. Biotechnol.* **5**, 233–240 (1978).
45. K. Yamamoto, T. Tosa, K. Yamashita, and I. Chibata, *Eur. J. Appl. Microbiol. Biotechnol.* **3**, 169–183 (1976).
46. K. Yamamoto, T. Tosa, K. Yamashita, and I. Chibata, *Biotechnol. Bioeng.* **19**, 1101–1114 (1977).
47. J.M. Sarkar and J. Mayaudon, *Biotechnol. Lett.* **5**, 201–206 (1983).
48. K. Mosbach and P.-O. Larsson, *Biotechnol. Bioeng.* **12**, 19–27 (1970).
49. T. Tosa, T. Sato, T. Mori, and I. Chibata, *Appl. Microbiol. Biotechnol.* **27**, 886–889 (1974).
50. T. Sato, T. Mori, T. Tosa, I. Chibata, M. Furui, K. Yamashita, and A. Sumi, *Biotechnol. Bioeng.* **17**, 1797–1804 (1975).
51. P. Aleksieva, B. Tchorobanov, G. Suchodol'skaya, and K. Koschteenko, *Appl. Microbiol. Biotechnol.* **29**, 239–241 (1988).
52. Y. Morikawa, I. Karube, and S. Suzuki, *Biotechnol. Bioeng.* **21**, 261–270 (1979).

53. A. Freeman and Y. Aharonowitz, *Biotechnol. Bioeng.* **23**, 2747–2759 (1981).
54. K. Mosbach, S. Birnbaum, K. Hardy, J. Davies, and L. Bulow, *Nature*, **302**, 543–545 (1983).
55. S. Ohlson, P.-O. Larsson, and K. Mosbach, *Eur. J. Appl. Microbiol. Biotechnol.* **7**, 103–110 (1979).
56. P. Linko, in M. Moo-Young, C.W. Robinson, and C. Vezina eds., *Advances in Biotechnology*, Pergamon, Elmsford, N.Y., 1981, pp. 711–716.
57. S. Ohlson, S. Flygare, P.-O. Larsson, and K. Mosbach, *Eur. J. Appl. Microbiol. Biotechnol.* **10**, 1–9 (1980).
58. T. Tanaka, K. Yamamoto, S. Towprayoon, H. Nakajima, K. Sonomoto, K. Yokozeki, K. Kubota, and A. Tanaka, *Appl. Microbiol. Biotechnol.* **30**, 564–568 (1989).
59. W. Kurzatkowski, W. Kurytowicz, and A. Paszkiewicz, *Eur. J. Appl. Microbiol. Biotechnol.* **15**, 211–213 (1982).
60. B. Yongsmith, K. Sonomoto, A. Tanaka, and S. Fukui, *Eur. J. Appl. Microbiol. Biotechnol.* **16**, 70–74 (1982).
61. J.C. Vuillemard, S. Terre, S. Benoit, and J. Amiot, *Appl. Microbiol. Biotechnol.* **27**, 423–431 (1988).
62. M. Veelken and H. Pape, *Eur. J. Appl. Microbiol. Biotechnol.* **15**, 206–210 (1982).
63. P. Brodelius, B. Hagerdal, and K. Mosbach, *Enzyme Eng.* **5**, 383–387 (1980).
64. W. Grote, K.J. Lee, and P.L. Rogers, *Biotechnol. Lett.* **2**, 481–486 (1980).
65. G. Georgious, J.J. Chalmers, M.L. Shuler, and D.B. Wilson, *Biotechnol. Progr.* **1**, 75–79 (1985).
66. H. Nakajima, K. Sonomoto, N. Usui, F. Sato, Y. Yamada, A. Tanaka, and S. Fukui, *J. Biotechnol.* **2**, 107 (1985).
67. Y. Shirai, K. Hashimoto, H. Yamaji, and M. Takashiki, *Appl. Microbiol. Biotechnol.* **26**, 495–499 (1987).
68. P. Chevalier, and J. Noüe, *Enzyme Microb. Technol.* **9**, 53–56 (1987).
69. I. Takata, K. Yamamoto, T. Tosa, and I. Chibata, *Eur. J. Appl. Microbiol. Biotechnol.* **7**, 161–172 (1979).
70. G. Para, S. Rifai, and J. Baratti, *Biotechnol. Lett.* **6**, 703–708 (1984).
71. Y. Nishida, T. Sato, T. Tosa, and I. Chibata, *Enzyme Microb. Technol.* **6**, 85–90 (1979).
72. Y.M. Deo and G.M. Gaucher, *Biotechnol. Lett.* **5**, 125–130 (1983).
73. K. Yamamoto, T. Tosa, and I. Chibata, *Biotechnol. Bioeng.* **22**, 2045–2054 (1980).
74. M. Wada, J. Kato, and I. Chibata, *Eur. J. Appl. Microbiol. Biotechnol.* **8**, 241–247 (1979).
75. M. Fujimura, J. Kato, T. Tosa, and I. Chibata, *Appl. Microbiol. Biotechnol.* **19**, 136–139 (1984).
76. M. Wada, T. Uchida, J. Kato, and I. Chibata, *Biotechnol. Bioeng.* **22**, 1175–1188 (1980).
77. P. De Taxis du Poet, P. Dhulster, J.N. Barbotin, and D. Thomas, *J. Bacteriol.* **165**, 871–877 (1986).
78. J. Huang, P. Dhulster, J.N. Barbotin, and D. Thomas, *J. Chem. Tech. Biotechnol.* **45**, 259–269 (1989).
79. S.B. Karkare, D.H. Burke, R.C. Dean Jr., J. Lemontt, P. Souw, and K. Venkatasubramanian, *Ann. N. Y. Acad. Sci.* **469**, 91–103 (1986).
80. P. Brodelius and K. Nilsson, *FEBS Lett.* **122**, 312–316 (1980).
81. K. Sonomoto, A. Tanaka, T. Omata, T. Yamane, and S. Fukui, *Eur. J. Appl. Microbiol. Biotechnol.* **6**, 325–334 (1979).
82. M. Asada, K. Morimoto, K. Nakanishi, R. Matsuno, A. Tanaka, A. Kimura, and T. Kamikubo, *Agric. Biol. Chem.* **43**, 1773–1774 (1979).
83. M. Asada, K. Nakanishi, R. Matsuno, and T. Kamikubo, *Agric. Biol. Chem.* **46**, 1687–1688 (1982).
84. K. Sonomoto, N. Usui, A. Tanaka, and S. Fukui, *Eur. J. Appl. Microbiol. Biotechnol.* **17**, 203–210 (1983).
85. K. Sonomoto, M.M. Hoq, A. Tanaka, and S. Fukui, *J. Ferment. Technol.* **59**, 465–469 (1981).
86. T.K. Mazumder, K. Sonomoto, A. Tanaka, and S. Fukui, *Appl. Microbiol. Biotechnol.* **21**, 154–161 (1985).
87. K. Yokozeki, S. Yamanaka, T. Utagawa, K. Takinami, Y. Hirose, A. Tanaka, K. Sonomoto, and S. Fukui, *Eur. J. Appl. Microbiol. Biotechnol.* **14**, 225–231 (1982).
88. T. Omata, T. Iida, A. Tanaka, and S. Fukui, *Eur. J. Appl. Microbiol. Biotechnol.* **8**, 143–155 (1979).
89. T. Yamane, H. Nakatani, E. Sada, T. Omata, A. Tanaka, and S. Fukui, *Biotechnol. Bioeng.* **21**, 2133–2145 (1979).
90. K. Sonomoto, K. Nomura, A. Tanaka, and S. Fukui, *Eur. J. Appl. Microbiol. Biotechnol.* **16**, 57–62 (1982).
91. T. Omata, N. Iwamoto, T. Kimura, A. Tanaka, and S. Fukui, *Eur. J. Appl. Microbiol. Biotechnol.* **11**, 199–204 (1981).
92. K. Sonomoto, I.-N. Jin, A. Tanaka, and S. Fukui, *Agric. Biol. Chem.* **44**, 1119–1126 (1980).
93. J. Klein and F. Wagner, *Enzyme Eng.* **5**, 335–345 (1980).
94. M.C. Fusee, W.E. Swann, and G.J. Calton, *Appl. Environ. Microbiol.* **42**, 672–676 (1981).
95. S. Fukui, A. Ahmed, T. Omata, and A. Tanaka, *Eur. J. Appl. Microbiol. Biotechnol.* **10**, 289–301 (1980).
96. M. Ogaki, K. Sonomoto, H. Nakajima, and A. Tanaka, *Appl. Microbiol. Biotechnol.* **24**, 6–11 (1986).

CELL SEPARATION, CENTRIFUGATION

HANS AXELSSON
Alfa Laval Separation AB
Tumba, Sweden

KEY WORDS

Asepsis
Cell debris separation
Cell harvesting
Centrifugation
Containment
Scale-up
Separability
Solvent extraction
Spin test
Sterilization
Two-phase aqueous systems

OUTLINE

INTRODUCTION

The first application of centrifugal separation other than in milk processing was in the harvesting of cells in baker's yeast production. Centrifugation is still the only choice for cell separation and protein recovery in many fermentation processes.

CENTRIFUGAL SEPARATION

Centrifugal separation is a sedimentation operation accelerated by centrifugal force. Thus, a prerequisite for the separation is a difference in density between the phases. This applies to both solid–liquid and liquid–liquid separation.

The settling velocity (V_g) of a solids particle (or droplet) under the influence of gravity alone is given by Stokes' law (see "Nomenclature" for abbreviations and symbols):

$$V_g = \frac{\rho_p - \rho_f}{18 \cdot \eta} \cdot d_p^2 \cdot g \tag{1}$$

In the centrifuge field, the settling velocity becomes:

$$V_c = V_g \cdot Z \tag{2}$$

where

$$Z = \frac{\omega^2 \cdot r}{g} \tag{3}$$

is called relative centrifugal force (RCF) or G-number.

It is, however, not the RCF only that determines the performance of a given centrifugal separator. The residence time of the particle or droplet in the centrifugal force field is important, so a larger volume of the rotor will increase the possible flow rate. In the most common centrifugal separators, the disk bowl machines, a given bowl material can cope with a certain peripheral speed—the stronger the bowl material, the higher the speed. This leads to the fact that the maximal RCF is inversely proportional to the bowl diameter, but not necessarily that the lower RCF machine has a lower performance.

TYPES OF CENTRIFUGAL SEPARATORS

General

What determines the design of a centrifuge is the method by which the solids phase in the process material is handled. The first two types of centrifuges developed were applied to duties where the resulting products were in liquid form, such as cream and skim milk and yeast cream and cell-free wort. In the beginning of this century, centrifuges were applied to processes with solids, amorphous or crystalline. In the biotechnological industry, the solids phase is often recovered in a diluted suspension or is, even at high solids concentrations, a slurry with good flow properties.

Types of machines are described in the work by Rushton et al. (1). The most comprehensive description of centrifuges can be found in the book by Sokolov (2). Brochures and technical publications from sedimenting centrifuge manufacturers such as Alfa-Laval Separation AB also provide information on machine types and sizes.

In this article, centrifugal filtration will not be dealt with because of the limited use in this industry, with the exception of crystal recovery in antibiotic production. For a description of various centrifugal filters, see Rushton et al. (1).

Disk Bowl Machines

Solid Bowl. A solid bowl machine (Fig. 1a) has two liquid outlets, each with a paring disk. Other versions have

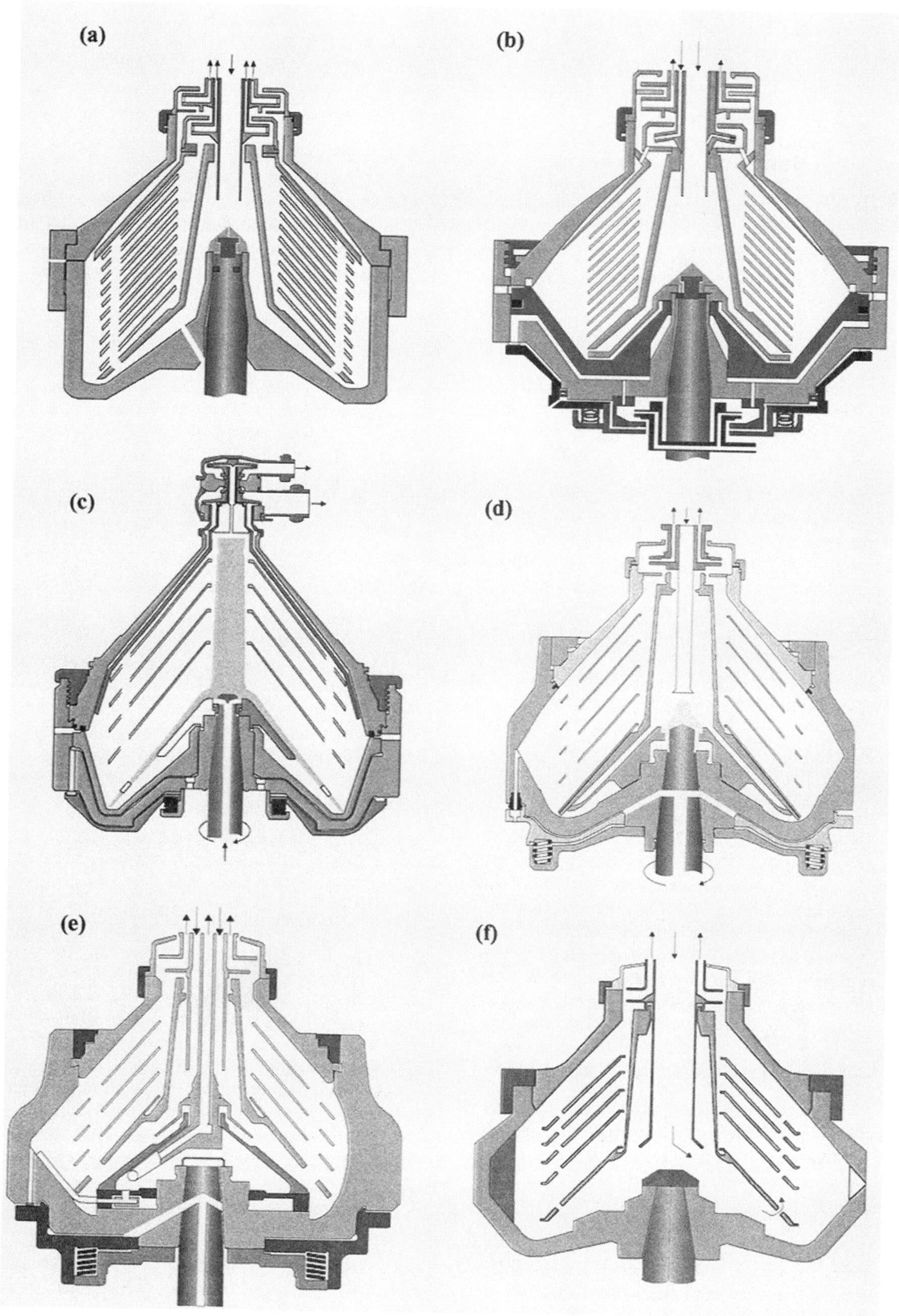

Figure 1. (**a**) Solids-retaining disk bowl, vertical cut. (**b**) Radial solids-ejecting disk bowl, top feed, vertical cut. (**c**) Radial solids-ejecting disk bowl, hermetic feed, vertical cut. (**d**) Axial solids-ejecting disk bowl, vertical cut. (**e**) Nozzle bowl with pressurized solids discharge, vertical cut. (**f**) Nozzle bowl with peripheral discharge nozzles, vertical cut.

none or one paring disk. The solids accumulate at the periphery and must be removed manually. In the separation of two nonmiscible liquids, a coalescence of the dispersed droplets takes place at an internal interface in the bowl. The positioning of this interface is very important. The nearer it is placed to the light liquid outlet, the more contaminated the light liquid will be with droplets of the heavy phase. The position is determined by a gravity disk, over which the heavy liquid flows. The gravity disk diameter is given by the relation:

$$D_{\mathrm{h}} = \sqrt{D_{\mathrm{i}}^2\left(1 - \frac{\rho_{\mathrm{fl}}}{\rho_{\mathrm{fh}}}\right) + D_{\mathrm{l}}^2 \frac{\rho_{\mathrm{fl}}}{\rho_{\mathrm{fh}}}} \qquad (4)$$

If properly designed and if the feed flow rate is high enough to prevent sedimentation in the bowl, this type of machine can be used to recover a dilute cell concentrate in the heavy liquid outlet. This is also valid for three-phase solids-ejecting machines. Solid-bowl machines are made in a large number of sizes, with bowl diameters ranging from 140 to 750 mm.

Solids-Ejecting Separators, Radial Discharge. Of all types of disk bowl machines, the solids-ejecting separators with radial discharge (Fig. 1b, c) are the most common. With an automated, periodic partial discharge of the sediment, it is possible to obtain a considerably higher concentration of the solids than in the peripheral nozzle machine (Fig. 1f). The machine is discharged in about 0.1 s by pressing a sliding bowl bottom in the bottom part of the bowl downward, thereby opening large discharge ports in the outer bowl wall. To make the discharge at the right moment, the discharge can be initiated by a self-triggering mechanism that hydraulically senses the solids level in the solids space (3). Three-phase separation units are available. The solids-ejecting machines can be equipped with hermetic seals and a hermetic feed from below in a hollow spindle (Fig. 1c). In these true hermetic machines, the split between the exiting light and heavy liquids can be made by controlling the back pressure in the outlet pipes. The heavy liquid can contain large amounts of solids (cells or proteins) if the machine is properly designed. The solids-ejecting separators are available in about 10 sizes, with bowl diameters between 180 and 900 mm. The smaller units are more often available in versions equipped for BL2-LS requirements than are larger units. A pilot-scale sterilizable separator module is seen in Figure 2. A large-scale unit is seen in Figure 3.

Solids-Ejecting Separators, Axial Discharge. When comparing axial and radial discharging separators (Fig. 1b, d), one finds that the axial machine has a different geometry. The lock ring can be placed at a smaller diameter, because no sliding bowl bottom needs to be fitted. This reduces the stresses in the winding of the lock ring. Also, the solids outlet takes place through a few small axial channels that do not induce such large stresses in the bowl shell. These two factors make it possible to increase the bowl speed considerably above that of the radially discharging machines, so that the RCF is at least doubled. The discharge system is operated by pressurized air, which forces a ring slide, equipped with valve seats, downward, opening the axial channels. To facilitate the transport of solids toward the channels, the bowl is machined to have a star-shaped inner contour. This type of machine exists at present in three sizes, with bowl diameters ranging from 500 to 900 mm. BL2-LS versions are available.

Nozzle Machines with Pressurized Discharge of Concentrate. The apparent viscosity of bacteria and yeast suspensions as well as some protein precipitates decreases with increasing velocity gradient. Therefore, they are able to flow even at high concentrations in channels or tubes from the periphery of the bowl in nozzle machines with pressurized discharge of concentrate (Fig. 1e). The tubes empty into a central chamber, where a centripetal pump, a paring tube, picks up the concentrate and pumps it out of the bowl. The nozzles are placed at the end of the concentrate tubes, just in front of the chamber. This machine type is suitable only for applications where the solids have the specific flow properties mentioned earlier and in the absence of other type of solids. In some machines, of this type of vortex chamber is placed in front of the nozzle (4). Liquid enters the chamber tangentially, creating a viscosity-dependent whirl, and leaves through the nozzle in the centre of the chamber. The viscosity effect is giving a self-regulating function so that varying solids flow rates (within limits) to the machine will result in a constant concentration of the concentrate, thereby reducing the risk of clogging the machine. One of the units of this type (Fig. 4) is available in a sterilizable, contained, and aseptic version, conforming to BL2-LS. To make this possible, the cleaning-in-place (CIP) has to be efficient. Therefore, the sterilizable machine uses the discharge system of the machine in Figure 1d during CIP. This system is also used in noncontained versions. Bowl diameters between 500 and 900 mm are available.

Nozzle Machines with Peripheral Nozzles. In nozzle machines with peripheral nozzles (Fig. 1f), nozzles with a diameter of 0.5 to 3 mm are situated at the periphery of the bowl, which has sloping walls toward the nozzles. The number of nozzles varies between 4 in small bowls and 20 in the largest bowls. It can be equipped with a recirculation device for the sediment, a separate central tube that is divided into tubes along the bowl bottom ending just in front of the nozzle. This makes it possible to increase the concentration of the solids phase without the need to decrease the nozzle size, leading to increased risk of nozzle clogging. The nozzle flow rate is proportional to number of nozzles, distance from the center of rotation, bowl speed, and square of the nozzle diameter. The peripheral nozzle machines are used for the largest flow rates; the bowl diameters are up to 1,000 mm.

Decanter Centrifuges

In the 1950s, the solid-bowl separator with scroll discharge (decanter centrifuge) (Fig. 5) was developed for large quantities of solids. It is equipped with a screw conveyor that rotates at a speed slightly higher or lower than the bowl speed.

The differential speed can be varied. Careful control of this parameter, torque control, and some newer design fea-

Figure 2. Solids-ejecting disk bowl centrifuge of pilot size for sterilizable contained installation. Type BTPX 205. The height of the separator only is 1.3 m. *Source*: Courtesy of Alfa-Laval Separation AB.

Figure 3. Solids-ejecting disk bowl centrifuge of production size. Capacity in breweries up to 75 m^3/h. Type BREW 2000. The height of the separator is 2.2 m. *Source*: Courtesy of Alfa-Laval Separation AB.

tures (e.g., the baffle disk), have recently made the decanter interesting to use also for biological sludges. However, the conventional mechanical design of these machines does not allow more than 5,000 G for a medium-diameter machine (5). To reach higher *g*-forces new design features are needed (6). With suspended main bearings, a separate gear box, and a swimming conveyor, held in place by the liquid, 10,000 G is feasible.

For three-phase decanters, the gravity disk dimensioning follows in principle equation 4. It is, however, very important to take the weir head into account, because a variation of the light-to-heavy phase ratio gives a very unstable interface position (7).

The range of bowl diameters is very wide, from 150 to 1,200 mm; the bowl length is often 4 to 5 times the diameter.

The newest addition to the decanter family is a vertical decanter equipped with a disk stack (8). The thought is to combine the best characteristics of a decanter (dry solids) and a disk bowl machine (good clarification).

Other Classical Bowl Designs

Tubular Bowl. In the tubular bowl (Fig. 6) the rotor, driven from above, consists of a long cylinder, into which liquid is pumped from below. The design shown in Figure 6 is equipped for separation of two liquid phases. Most tubular centrifuges are used for separating very valuable solids from a liquid. The liquid is pumped out by a paring disk, avoiding aerosol and foam formation. To remove the solids accumulated at the periphery, the machine must be stopped and dismantled. The bowl diameters are up to 130 mm with a solids volume of up to 6 L. In the industrial models, 20,000 G are generated. Sterilizable and contained units are available.

Multichamber Bowl. In the multichamber bowl machine (Fig. 7), the process liquid flows in the annuli between con-

Figure 4. Nozzle machine with pressurized solids discharge for yeast and bacteria separation. Capacity on *E. coli* about 3,000 L/h. Type BTUX 510. The height of the separator is 1.8 m. Sterilizable and contained installation. *Source*: Courtesy of Alfa-Laval Separation AB.

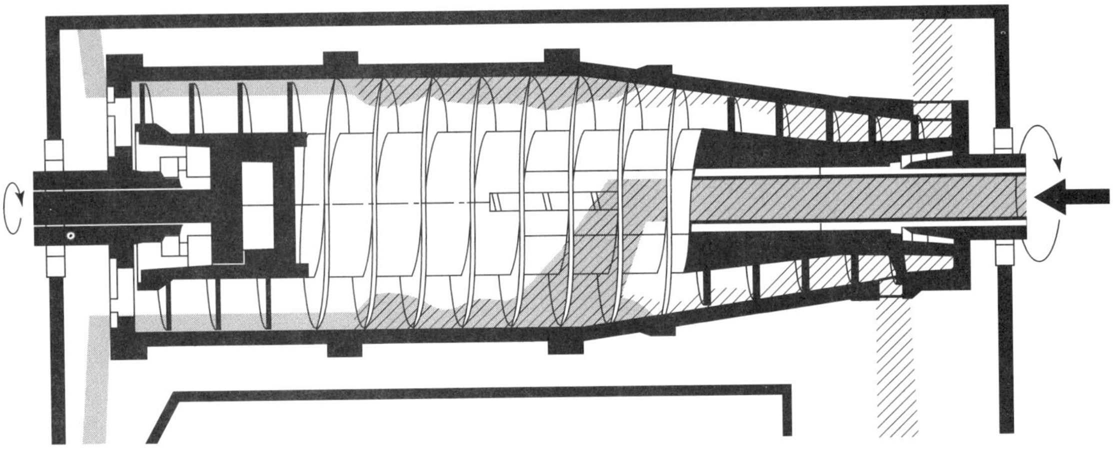

Figure 5. Decanter centrifuge, vertical cut.

centric cylinders. Having a relatively low separation capability, this type of machine is used for recovery of valuable solids, when the solids volume of the tubular bowl is too small. Some designs include cooling channels in the bowl wall, a cooling jacket around the bowl, and a cooled centripetal pump. The maximum bowl diameter is about 500 mm.

New Bowl Designs

Centritech® Machines. In the Centritech® machine (Fig. 8), a disposable separation insert in plastic is placed in a rotor. The feed liquid enters at the top of the insert, and the main liquid outflow is at the opposite end of the top of the insert. The heavy liquid phase (cell concentrate) is withdrawn intermittently through the bottom outlet. Through the principle of the inverted comma no seals between rotating and nonrotating parts are needed. The G-number is maximum 100. The mechanics are further described in Apelman et al. (9). Its only described application is separation of mammalian cells from culture fluid. Two sizes of the machine are available. The diameter of the insert is about 300 mm in the larger unit.

Inverted Chamber Bowl with Intermittently Operated Scraper. The inverted chamber bowl (Fig. 9) is a relatively new type of machine, sold under the name Powerfuge™ (10). Its bowl has its opening downward, where the centrate leaves under gravity and is collected in a chute. The feed is from the top. When the bowl is full of solids, it is decelerated. At low speed, the asymmetrically positioned scraper is engaged and removes the solids. Four sizes are available, with bowl diameters from 150 to 600 mm. They are designed for CIP and sterilization-in-place (SIP). The smallest unit can develop 20,000 G.

Summary

In Table 1, the characteristics of various separator types are found. It has to be remembered that the hydraulic capacities given by centrifuge manufacturers are the maximum possible feed flow rates that often are far above the flow rates that will give satisfactory separation results for difficult separations, such as bacteria and protein separations.

FLUID AND PARTICLE DYNAMICS

Introduction

A successful scale-up method should be based on a mathematical model that accurately describes the separation process, the hydrodynamics and particle dynamics inside the bowl. The phenomena going on inside the centrifuge bowl are, however, so complicated that no mathematical model is able to predict in absolute terms the separation efficiency as a function of machine and process media parameters. Axelsson (11) reviewed of some of the flow phenomena, which are briefly discussed in this section, along with examples of more recent findings.

Centrifuges with Conical Disks

Fluid and Particle Dynamics in the Disk Stack. The disks in the disk bowl split the flow in thin layers. The Reynold number for the flow between the disks is usually low, and the flow is laminar, resulting in an easy particle settling. The critical Reynolds number for transition to turbulent flow increases with decreasing disk spacing and increasing bowl speed (2), which is beneficial for the separation efficiency because the settling distance at a small disk spacing will be short, and the high bowl speed gives a large settling velocity. Also, a small disk spacing keeps the liquid rotating at or near the bowl speed, thus reducing the pressure drop in the disc stack.

In Figure 10 is a vertical cut of two disks in the disk stack, where the flow is driven toward the center by a radial pressure gradient. The character of the velocity profile between the disks depends on a dimensionless number, λ, which is defined as

$$\lambda = t \cdot \sqrt{\frac{\omega \cdot \sin \alpha}{\nu}} \tag{5}$$

When λ is small, the velocity profile is parabolic, but for λ values above 5, normally encountered in industrial centri-

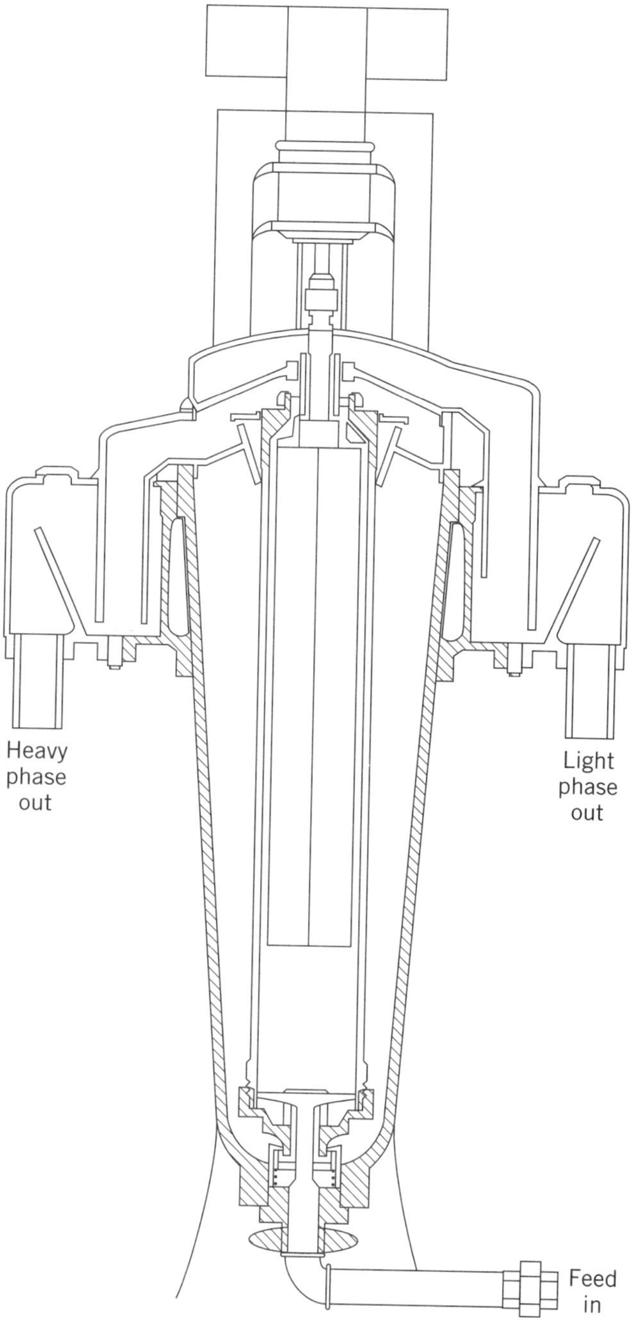

Figure 6. Tubular bowl centrifuge, vertical cut.

fuges, the radial velocity component splits up in two thin layers, Ekman layers (Fig. 10). In these layers, all the net transport of liquid toward the center takes place. Because separated particles (heavy phase) are collected at the underside of the disk, they are subjected to a strong shear force from the Ekman layer that drags the particle toward the liquid outlet. The velocity in these layers increases with increasing λ, such as with increasing bowl speed. Therefore, a higher bowl speed cannot always be fully used because of the counteracting shear force.

In between the Ekman layers is the geostrophic layer. In Figure 11 its streamlines have been visualized by dye

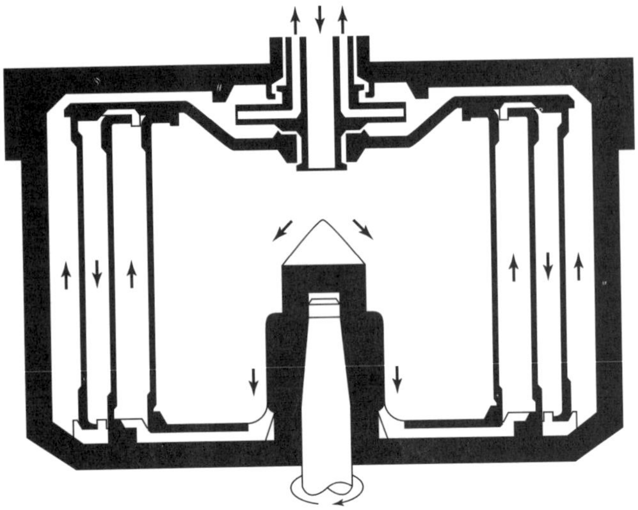

Figure 7. Multichamber bowl, vertical cut.

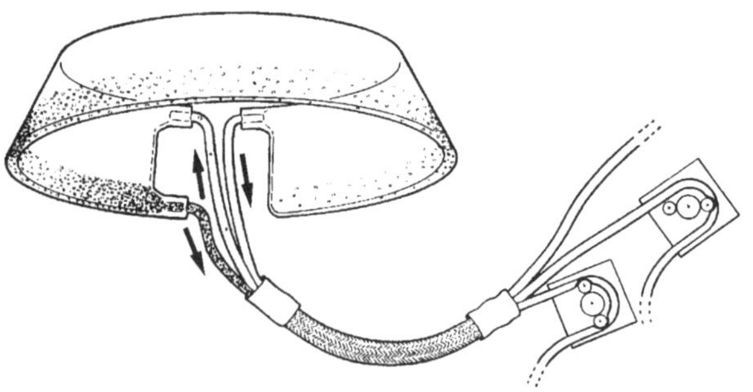

Figure 8. Centritech® Cell separation insert. *Source*: Courtesy of Sorvall (U.K.) Ltd.

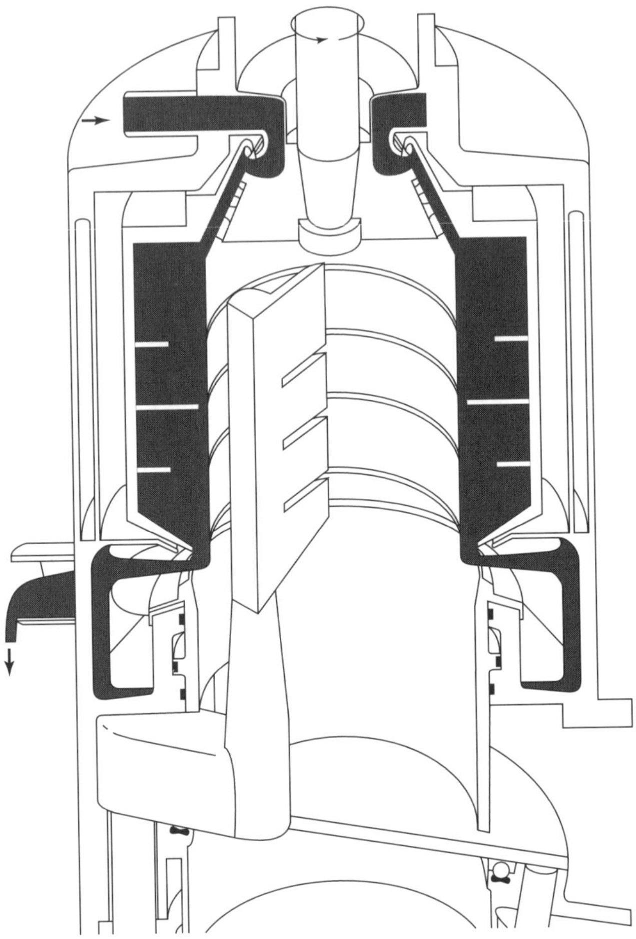

Figure 9. Vertical cut of a Powerfuge™ one-chamber bowl with scraper. Feed mode. *Source*: Courtesy of Carr Separations, Inc.

Table 1. Characteristics of Separator Types

Type	Mode of solids discharge	Feed flow rate (L/h)	Feed solids content (% by vol.)	Solids flow rate (L/h)	Σ value (m^2)	*G*-force developed	Consistency of solids
Disk solid bowl	Manual	20–100,000	<1	0	1,000–300,000	<10,000	Firm paste
Solids ejecting, radial	Intermittent	20–100,000	<25	<3,000	1,000–170,000	<14,500	Thick flowing slurry
Solids ejecting, axial	Intermittent	1,000–150,000	<15	<1,000	110,000–220,000	<15,000	Thick flowing slurry
Nozzle, pressurized discharge[a]	Continuous	1,000–180,000	4–30	>150 <40,000	69,000–180,000	<15,000	Thick flowing slurry
Peripheral nozzle	Continuous	300–250,000	2–30	>3000 <60,000	35,000–180,000	<11,000	Thin slurry
Decanter centrifuge	Continuous with scroll	20–120,000	5–50	<~50,000	400–25,000	<10,000	Thick slurry to semisolid
Tubular bowl	Manual	20–7,000	<1	0	1400–4500	<31,000	Firm paste
Multichamber bowl	Manual	100–20,000	<5	0		<9,000	Firm paste
Centritech® machines[b]	Intermittent	5–100	<1	<15	Not applicable	<100	Very thin slurry
Inverted solid bowl	Intermittent with scraper	10–6,000	1–30	<~1,000	800–~20,000	<20,000	Firm paste

[a]Only for slurry of single cells (see text).
[b]Only for mammalian cells.

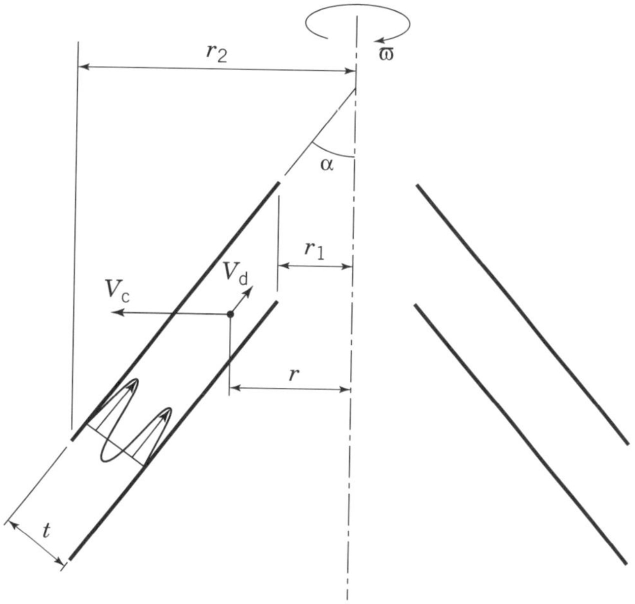

Figure 10. Flow profile between conical disks at high λ value. Particle separation.

injection, and they can be compared with the calculated streamlines in a gap between two disks with radial spacers and with a distribution hole where the liquid enters the gap (12). Particle separation has also been studied in the same geometry (13). It has been found that the drag force on the particles collected at the underside of the disk depends on the position on the disk and that some particles are easily separated, whereas others are forced toward the center. The drag force can be computed (14), and the calculations of separation efficiency including the drag force verify these findings.

There is also an accumulation of particles at the radial spacer, along which the solids slide toward the periphery (13). At the end of the spacer, there is a wake that forces the particles back toward the center until the bulk weight of the solids becomes large enough to cause an intermittent discharge of particles into the solids space outside the perimeter of the disk.

Inlet. All bowls in the machines in Figures 1 to 9, except the hermetic machine (Fig. 1c) and the Centritech® machine (Fig. 8), have an open chamber in the center of the bowl, where the nonrotating feed liquid is received. In most inlets, the liquid forms a cylindrical interface with the gas (usually air) in the receiving chamber. The nonrotating in-

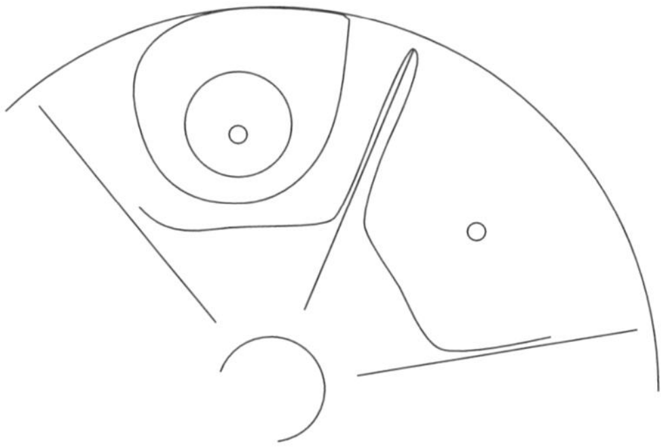

(a)

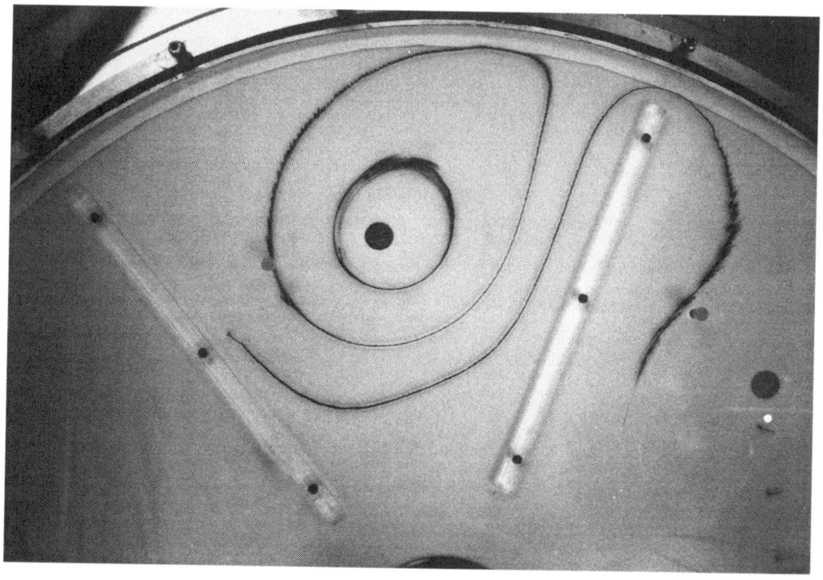

(b)

Figure 11. Comparison of calculated (**a**) and actual (**b**) flow patterns in a disk bowl.

coming liquid hitting the rotating interface results in a turbulence that entrains into the feed and splits particles that are susceptible to shear; the further out from the center the interface, the larger the turbulence. If it is not practical to use a hermetic inlet, the interface in the inlet must have as little an inner radius as possible. Also, the unavoidable energy dissipation must be distributed over a large volume, thereby diminishing the shear force. One design that allows smoother inlet conditions and a reduced air entrainment is the disk inlet (15).

A mathematical model for feed zone breakage of shear-sensitive particles has been verified for a conventional pilot-scale solids-ejecting separator (16). The extent of breakage was found to be independent of flow rate. This agrees with the finding that temperature increase in the separated liquid after the separator is also independent of flow rate.

Outlets. The liquid streams leaving the separator have picked up energy in the form of kinetic energy and heat caused by inner friction. When the liquid is decelerated outside the separator, the kinetic energy is converted to heat. The temperature increase is proportional to the square of the bowl speed and to the square of the radius at which the liquid leaves the separator. It is therefore not surprising that the discharge in a peripheral nozzle or a solids-ejecting machine causes breakage of cells because of the high kinetic energy of the fluid. A typical degree of bacterial cell breakage is 30%. The consequence of this is that if the particles to be separated are susceptible to shear and cell integrity is important, the cell concentrate should leave the centrifuge as near to the center as possible, such as in a machine according to Fig. 1c or e. It has been reported (17) that in such machines the breakage of *Escherichia coli* cells is low.

The outlet from the centrifuge is often of the paring disk type. For a good function, this outlet must be designed to avoid air entrainment and large shear forces. This is especially important for chemically sensitive substances such as proteins. A hermetic design (Fig. 1c) can be used in some applications.

Decanter Centrifuges

The objective of a decanter is usually to concentrate the solids present in concentrated suspensions. The hydrodynamics of a decanter are complicated. The conveyor adds turbulence and forces the liquid into a helical path (2). Below the top layer, vortices are formed. The transport of solids with low inner friction will influence the sedimentation process. The turbulence at the inlet zone, with possible resuspension of separated solids, and the lower solids concentration toward the liquid outlet contribute to the fact that a slender decanter will separate better than a short one with the same theoretical size (18).

THE THEORETICAL SIZE OF A CENTRIFUGAL SEPARATOR

Introduction

From the descriptions in the previous section it is evident that mathematical models describing the flow phenomena in centrifuges are very difficult to build. However, the attempts made are successful when machines of different sizes but with identical geometry are compared, for example, for scale-up purposes (see "Pilot Machine Testing").

The Σ Theory

The Generalized Σ Formula. The most-used quantity to characterize centrifuges, the Σ concept, was presented by Ambler (19,20). In its derivation he considered particles with a critical diameter (d_c) that were separated to 50%. Today, however, the most-used definition of the critical particle is the particle that is separated to 100%. This does not influence the formulas for the Σ value, but the value for the feed flow rate (Q) will be halved. Σ is defined as

$$Q_{\text{theor}} = V_g \cdot \Sigma \tag{6}$$

V_g is the Stokes settling velocity (equation 1). Σ has the general expression

$$\Sigma = \frac{V \cdot \omega^2 \cdot r_e}{g \cdot s_e} \tag{7}$$

The equation for the critical diameter becomes

$$d_c = \left(\frac{18 \cdot \eta \cdot Q_{\text{theor}}}{(\rho_p - \rho_f)} \cdot \frac{s_e}{V \cdot \omega^2 \cdot r_e}\right)^{1/2} = \left(\frac{18 \cdot \eta \cdot Q_{\text{theor}}}{\Sigma \cdot (\rho_p - \rho_f) \cdot g}\right)^{1/2} \tag{8}$$

The derivation in equation 8 is based on the following assumptions:

- Viscous drag is determining the particle movement.
- The flow in disk bowls between the disks is laminar and symmetrical.
- The liquid rotates at the same speed as the bowl.
- The particle concentration is low (no hindered settling).
- The particle always moves at its final settling velocity.
- This settling velocity (V_c) is proportional to the g-force.

Ambler uses the critical particle in his formula; therefore, his analysis is a special case of Svarovsky's grade efficiency function (21), where for each particle size the ratio of sedimented mass to the mass in the feed material is determined and integrated over all particle size to get a more realistic measure of the separation efficiency.

Σ for Disk Bowl Centrifuges. The flow system in between two conical disks is shown in Figure 10. The particle has a radial velocity component because of the centrifugal force (V_c), which increases with increasing r. It has also a drag-dependent velocity component (V_d), which can have any size or direction depending on the flow situation round the particle. In the Σ theory, it is assumed that V_d depends on uniform radial plug flow parallel to the disks. The derivation of the equation for Σ is found in the paper by Ambler (20), and the result is shown in Table 2.

Summary. In Table 2, the resulting expressions of the Σ value for four different bowl geometries in continuous centrifuges are shown. These have been obtained by inserting actual values of V, r_e, and s_e for the various geometries.

Separation Efficiency

The assumptions in the Σ theory are not fulfilled in reality, as mentioned previously. Therefore, equation 6 should include an efficiency factor, so that in practice the flow rate is reduced at the requested separation performance. The efficiency factors have been estimated to have values according to Table 3 (11). This table was compiled in 1984 from data in literature. By careful measurement of the density difference and particle dimensions, it has been shown that in one case an efficiency factor of maximum 16% was found to fit the experimental data in a pilot-size disk bowl machine (16). However, this probably also included effects of hindered settling (see "Hindered Settling").

In the absence of hindered settling, the equation describing the relationship between the feed flow rate, the Σ value, and the settling rate, will thus be

$$Q_{act} = \mu \cdot V_g \cdot \Sigma \qquad (9)$$

The KQ Formula for Disk Bowl Centrifuges

In 1961 a semi-empirical equation for the separation area was published (22). It is based on the finding that separation data fitted the relation

$$\frac{V_c}{V_g} = \left(\frac{\omega^2 \cdot r}{g}\right)^k \qquad (10)$$

with $k = 0.75$ better than the classical analysis (equations 2, 3, and 6), where $k = 1$. This fact is a consequence of the increasing shear forces in the Ekman layers (see "Fluid and Particle Dynamics in the Disk Stack"). For practical purposes, the following relation is used:

$$\mathrm{KQ} = 280 \cdot \left(\frac{n}{1000}\right)^{1.5} \cdot N \cdot \cot \alpha (r_2^{2.75} - r_1^{2.75}) \qquad (11)$$

where n = bowl speed in rpm and r_2 and r_1 are the radii of the disk in centimeters. This form of the KQ formula is dimensionally incorrect, but it is mathematically possible to make it correct, expressing the KQ value as an entity with the dimension area. In practice, the difference between the KQ and the Σ for scale-up purposes is negligible, compared to other inaccuracies.

Hindered Settling

One of the few experimental investigations into hindered settling of cells (23) shows that complex phenomena occur at high concentrations. A model with several experimental constants was developed. It showed that the settling rates drop very fast with increasing solids content, by a factor of about 2 at 20% by volume, and 15 to 20 at 50% by volume. The classical model used in the work by Clarkson et al. (16) gave similar results. In the tests by Aiba et al. (23), it was also found that cells may settle flocculated at higher concentrations. Hindered settling is one further example of the difficulty to calculate theoretically the performance of centrifuges.

SELECTION OF TYPE AND SIZE OF CENTRIFUGE

Introduction

The choice of type and size of centrifuges is a combined theoretical and practical exercise. In most cases the final choice for a new centrifuge application cannot be made without a test of the type of centrifuge preliminary selected, at between 25 and 100% of the final plant flow rate. There are at least two reasons for this. One is the elimination of the technical risk. The other is most common in

Table 2. Σ Equations for Various Separator Types

Machine type	Equation for Σ	Symbols used	References
Disk bowl	$\frac{\pi \cdot \omega^2}{g} \cdot \frac{2}{3} \cdot N \cdot (r_2^3 - r_1^3) \cdot \cot \alpha$	r_2 = max radius of disk r_1 = min. radius of disk N = number of disk α = half cone angle of disk	19,20
Decanter	$\frac{\pi \cdot \omega^2}{g} \left[L_1\left(\frac{3}{2} \cdot r_2^2 + \frac{1}{2} \cdot r_1^2\right) + L_2\left(\frac{r_2^2 + 3r_2r_1 + 4r_1^2}{4}\right) \right]$	L_1 = length of cylindrical part L_2 = length of conical part r_1 = inner radius of liquid r_2 = inner radius of bowl	20,21
Tubular bowl	$\frac{\pi \cdot \omega^2}{g} L \frac{r_2^2 - r_1^2}{\ln\left(\frac{2r_2^2}{r_2^2 + r_1^2}\right)}$	r_1 = inner radius of liquid r_2 = inner radius of bowl L = inner length of bowl	20
Multichamber bowl	$\frac{\pi \cdot \omega^2}{g} \frac{L}{3} \sum_{i=0}^{i=n} \frac{r_{2i+1}^3 - r_{2i+2}^3}{r_{2i+1} - r_{2i+2}}$	Indexes with even numbers = inner radius of chamber Indexes with odd numbers = outer radius of chamber $n + 1$ = number of chambers L = height of chambers	2

Table 3. Efficiency Factors μ for Various Separator Types (%)

Disk bowl machines	45–73
Decanter centrifuges	54–67
Tubular bowls	90–98

the pharmaceutical industry. During the preapproval inspection, those in the inspecting and license-giving institutions are concerned with the comparability between batches of product from the development phase and from the production phase (24). Because the validation procedures are costly and time consuming, it is very important to select at least the right type of centrifuge already in pilot scale. It is important that alternative process and plant designs are discussed before the centrifuge type and size are selected (Table 4).

Main Process Parameters

It is important to define at least the characteristics of the duty listed in Table 5. The *feed flow rate* is often determined by the available or necessary time to process one batch. The *separability* can be estimated on a bench scale by test tube centrifugation (spin test) (see next section). The *flow of solids* is the feed flow rate multiplied by the percent by volume of suspended solids. This solids flow rather than the suspended solids content alone influences the choice of machine type. Different machine types differ in their ability to produce high *concentration of the separated solids* (Table 1). The *required degree of separation* has a strong influence on the possible flow rate in a given machine. To increase separation efficiency of *E. coli* from 90 to 99%, the flow rate must be decreased by 30 to 35% (25). To reach 99.9%, a further reduction of about 30% is necessary. Many cell suspensions, including yeast, have the same narrow particle size distribution and the numbers are the same. With the wider distribution of cell debris, for example, the flow rate reduction percentages are much higher. *Solids rheology*, which is the reaction of the solids to pressure and shear forces, is very important for the choice of machine type. Pseudoplastic (shear thinning) solids (such as unflocculated yeast, rod-shaped bacteria, and cocci as well as some protein precipitates) will flow out of the machines seen in Figure 1c and e at 80% by volume of suspended solids, which often has positive consequences (reduced cell breakage). Shear thinning solids may have an angle of repose of close to 0 degrees (see "Important Properties of the Process Streams"), but this is not always the case, such as with flocculated or mycelium-forming bacteria. On the other hand, dilatant solids (starch or calcium carbonate) will not be discharged in concentrated form without manual handling from any machines other than those seen in Figures 1f and 5. The rheology of inclusion body concentrates is not much studied, but they are preferably discharged from the machine seen in Figure 1b. Because of the content of a valuable dissolved substance in the entrained liquid, *washing of solids* may be necessary.

Another reason for washing solids is, of course, to get rid of impurities. This is usually done by resuspension and further separation steps.

Table 4. Process and Plant Considerations

Possibility to increase particle size
Possibility to reduce viscosity
Chemical stability of product
Containment requirements
Aseptic requirements
Sterilization requirements
Mode of operation (hours/day, days/year)
Integration with unit operations upstream and downstream
Scale of pilot- and full-scale operation
Handling of solids from the centrifuge

Table 5. Main Process Parameters

Feed flow rate
Flow of solids
Required concentration of separated solids
Solids washing requirements
Separability (V_g)
Required degree of separation
Solids rheology

Test Tube Centrifugation (Spin Test)

Determination of Separability. The most important task for a spin test is the determination of separability, that is, the settling velocity. In practice, it is difficult to calculate, because often the information needed in Stokes' law is not complete. The spin test is done by spinning a sample of the feedstream for different periods of time or different G-numbers or both. For each of these spins, the supernatant is analyzed with regard to the important process parameter, for example, residual solids, turbidity and cell count, or water content in the case of light solvent. Thereafter, the time of the spin with the acceptable supernatant quality is recorded, and the necessary Σ is calculated according to equation 16.

The Spin Test as an Analytical Tool. The spin test is also a valuable tool for measuring the phase-to-volume ratios. The solids content in percent by volume from the spin test will be a very important parameter for the next step, which is to make a crude mass balance around the centrifuge. The spin test performed under different time and g-force conditions gives also an indication of solids compressibility and rheology. It is also valuable to prepare a sample of the solids phase after spinning to get a good indication of its dry solids content after the centrifugation. In cases where the fermentation substrate contains suspended solids, residues of these may be present after fermentation. A spin test will make it possible to identify the nature and amount of these, which is very important because they may obstruct the solids phase flow in nozzle and solids-ejecting machines.

Theory of the Test Tube Centrifuge. The swing-out test tube (bottle) centrifuge is shown schematically in Figure 12. In order for a particle to be separated out with 100%

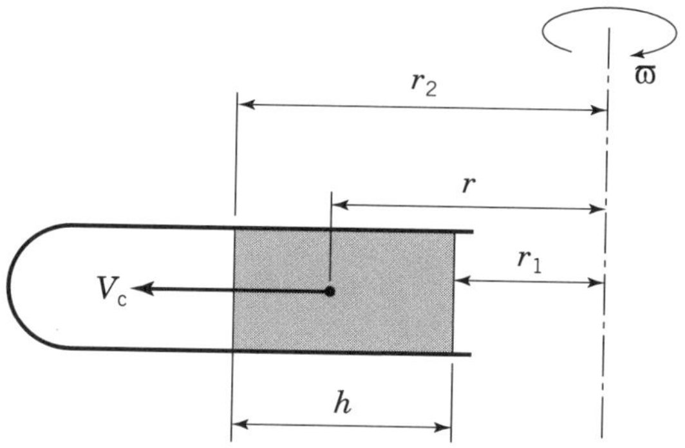

Figure 12. Separation of particles in a swing-out test tube centrifuge.

chance, it has to move from radius r_1 to radius r_2 in the spin time (T). The shaded volume with height (h) is the volume taken out for sampling of the result. The following equations describe the particle movement.

$$V_c = \frac{dr}{dt} \tag{12}$$

$$V_c = V_g \cdot \frac{r \cdot \omega^2}{g} \tag{13}$$

Insertion of equation 10 in equation 9 and rearranging and integration gives

$$\int_0^T dt = \frac{g}{V_g \cdot \omega^2} \int_{r1}^{r2} \frac{dr}{r} \tag{14}$$

$$T = \frac{g}{V_g \cdot \omega^2} \cdot {}^{e}\log \frac{r_2}{r_1} \tag{15}$$

In the test tube centrifuge, the efficiency factor μ in equation 9 is 100% because the settling process is considered ideal. Therefore, equation 6 is valid and

$$\Sigma = \frac{T \cdot Q_{\text{theor}} \cdot \omega^2}{g \cdot {}^{e}\log \frac{r_2}{r_1}} \tag{16}$$

and

$$\frac{Q_{\text{theor}}}{\Sigma} = \frac{g \cdot {}^{e}\log \frac{r_2}{r_1}}{T \cdot \omega^2} = V_g \tag{17}$$

For a test tube centrifuge with angle head, ${}^{e}\log(r_2/r_1)$ is exchanged for ${}^{e}\log(1 + D/r_1 \cdot \cos\beta)$.

Important Properties of the Process Streams

In Table 6, some solids phase data, such as *particle size* and *density difference*, have been collected. These data show that many important biological processes belong to the most demanding centrifuge applications regarding separability. The data given in Table 6 are very approximate. Cell size will vary with fermentation conditions. Cell debris size distribution is very wide and depends on the method of cell breakage. The density difference between cell and medium depends on the osmotic pressure and of the possible content of inclusion bodies. Apparent density and the density difference of inclusion bodies vary with size, so that larger ones have lower density (28).

As many as possible of the items in Table 7 should be defined before the selection of a centrifuge is made. Some of the items need further comments. Apart from Stokes' law, the *density* is important for the stress in the bowl material. *Low pH* and *high temperature* may, together with even a moderate chloride concentration, lead to difficulties with *corrosion*. In the biological process industries, steels 316 and 316L are often preferred (38). These steels have, however, such a low tensile strength that they are less suitable for high-speed centrifuges. Instead, duplex steel has been developed, it has high tensile strength coupled with a good corrosion resistance (39). The corrosive properties of process materials can be very serious, but it is in many cases possible to reduce the corrosion rate by intermittent washing with caustic solution every 4 to 5 hours of running to reestablish a passivating layer on the metal surfaces. *Erosion* may be a problem with silt-containing process water or other raw materials such as molasses. This may necessitate the use of hydrocyclones to remove these particles. The *angle of repose* (see "Main Process Parameters") is the angle a cone solid will form if allowed to settle, for example, between the nozzles in a nozzle bowl. If it is 0, or close to 0, degrees, the flow properties are extremely good. The *shear sensitivity* was mentioned in "Centrifuges with Conical Disks". The most shear-sensitive cells are mammalian cells, for which the shear forces must be minimized when the cells are used in perfusion systems.

In the Centritech® machines, the RCF is limited to 300, more than one order of magnitude lower than for other centrifuges. The cells may be so vulnerable that the residence time in the centrifugal force field must be minimized (40). Some protein precipitates are also sensitive to shear (16), including those formed in the maturation tanks in breweries. Many bacteria are sensitive, especially starter culture bacteria, where the activity of the cultures depends on intact cells. Yeast cells are the most robust cells, but they are also sheared to some extent at discharge in radial solids-ejecting centrifuges.

Improvement of Separability

According to Stokes' law, the settling velocity improves when the density difference and the particle density increase and when the viscosity decreases. Bowden (41) lists physical, chemical, and biological techniques for broth conditioning. The most-used techniques are coagulation and flocculation, but other methods that induce cell lysis can improve aggregation of cells, such as aging, pH, and change and addition of enzymes or other chemicals. In some cases it is possible to flocculate submicron suspensions, so that a decanter centrifuge can recover the precipitate (cell debris) with good efficiency and a dry cake (34).

Pilot Machine Testing

After maximizing the separability of the feedstream, the preliminary Q_{theor}/Σ is determined by spin tests. The ma-

Table 6. Size, Density Difference, and Relative Feed Flow Rate for Some Solid–Liquid Systems

Type of Solids	Particle size (μm)	Density difference (kg/m^3)	Relative throughput[a] (relative to baker's yeast)
Cell debris	0.2–0.5 (mean, bacteria)	—	0.2
	2 (median, yeast)	—	1
Streptococii	1 × 1	—	1–2
	1 (chain)	—	3–6
Clostridia	1 to 3	—	5
E. coli	1 × 2	70–85	2–6
Actinomyces spp.	10 × 10–143 × 143	1 (for large particles)	1–7
Inclusion bodies	0.7 × 0.7–1.3 × 1.3	29–154	5–8
Corynebacterium	—	—	6
Bacillus spp.	1 × 3	—	6–10
Aspergillus spp.	—	—	15
Mammalian cells	20 to 40	70	20
Flocculated material (e.g., debris)	—	—	20
Chlorella	4 × 4	—	30
Saccharomyces (alcohol yeast)	4 × 6	90	50–75
Saccharomyces (baker's yeast)	6 × 8	90	100

Source: Data from references 11, 23, 26–36.
[a]Proportional to Q_{act}/Σ.

Table 7. Important Process Liquid Properties

Chemical composition of liquid and solid phases
Densities of liquid and solid phases
Viscosities of liquid and solid phases
Particle size distribution
pH
Temperature
Vapor pressure
Solubility of one phase in the other
Corrosivity
Erosivity
Shear sensitivity
Angle of repose
Toxicity
Explosion hazard

Source: Adapted from reference 37.

chine type and size are tentatively chosen based on the preliminary mass balance and on the requested Q_{act}. With the separation efficiency μ, Q_{act}/Σ can be calculated from the equation

$$\frac{Q_{act}}{\Sigma} = \mu \cdot V_g \tag{18}$$

It is often advisable to make a test with the chosen type of machine. The purpose of such a test is to obtain a more reliable Q_{act}/Σ value, but also to assess solids flow properties, emulsification, and particle breakup tendencies and clogging risks.

Scale-Up

For all types of centrifugal separators, the Q_{act}/Σ value can be used for scale-up of the sedimentation performance if some general considerations are taken into account:

- The type of centrifuge is similar.
- The geometry is similar; for decanters, feed inlet positions, ratio of length to diameter, or beach angle, and for disk machines, cone angle, disk spacing, and number of feed holes.
- The bowl speed gives similar linear velocity of, for example, wings when treating shear-sensitive process feeds.
- The λ value (equation 5) is almost the same in the machines.
- The larger machine has no hydraulic limitations in relation to the calculated flows. The limitations could be inlet and outlet pressure drops, paring disk pumping capacities (upper and lower limit), and the upper and lower limit of solids flow.
- There are no mechanical limitations, such as power transmission and gear-box torque in decanters.

The equation to use in scale-up is

$$\left(\frac{Q_{act}}{\Sigma}\right)_1 = \left(\frac{Q_{act}}{\Sigma}\right)_2 (= \mu \cdot V_g) \tag{19}$$

where indexes 1 and 2 denote the two machines. Direct calculations of V_g and μ are thus avoided.

DESCRIPTION OF SOME APPLICATIONS

Introduction

The most important centrifuge application in biotechnology is cell harvesting, where either the liquid phase or the solid phase contains (or is) the product. Downstream of the harvesting stage the number of applications for centrifuges is growing. The operations performed are many: classification, dewatering, liquid–liquid extraction, and precipitate recovery. The applications discussed in this section are examples only; there is an infinite variation in the downstream process design.

Cell Harvesting

Introduction. As can be seen from Table 5, the throughput capacity of a given centrifuge (i.e., Q_{act}/Σ) may vary by a factor of 100 depending on the microorganism used.

Yeasts. The second application of disk bowl centrifuges developed after milk skimming is concentration of *baker's yeast*. Today, nozzle machines with pressurized discharge (Fig. 1) are very common in the industry. The flow rates are the highest in the biotechnology industry, up to 150 m^3/h of fermented broth. Cell concentrations in the fermenters have increased over the years and have now reached above 25% by volume. The baker's yeast is washed, often twice and countercurrently, to increase shelf life. The pressurized discharge of the solids concentrate makes it easy to use an inline mixer to mix the wash water with the cells. The Q_{act}/Σ is about 2×10^{-7} m/s for a 99.9% reduction of the cell count.

In *distilleries*, nozzle bowl centrifuges have been used for a long time. The centrifuge has been the prerequisite for the development of new alcohol fermentation systems (11). In the continuous fermentation from spent sulfite liquor, centrifuges are used for recycling of the yeast. This is also the case in the batchwise Melle-Boinot process, where the recovered yeast inoculates the next batch. In the continuous Biostil process, in which the distillery slop is returned to the fermenter, the centrifuge recovers the yeast from the feed to the distillation column and returns it to the fermenter.

In *breweries*, centrifuges are used in several process steps. In the separation step after fermentation, *green beer separation*, the cell count can be as high as 80×10^6 cells per mL (2% by volume), but during the emptying of the fermentation tank the solids decrease. Radial solids-ejecting machines are used, equipped with self-triggering and special outlet devices to prevent air contact. At a cell reduction of 99% and 20×10^6 cells/mL, the Q_{act}/Σ is about 2×10^{-7} m/s, increasing somewhat with higher cell count. The flow rates are up to 75 m^3/h. In this step, the same machine type as for baker's yeast could be used if the solids flow to the machine could be controlled. New instruments (42–44) measuring the dielectric constant or the inductive permittivity may be used.

For the separation step after lagering and before final filtration, *preclarification*, the cell count is much smaller, and less solids handling capacity is needed. Therefore, the bowl of the solids-ejecting machine can have the shape of the bowl in Figure 1c. The hermetic inlet is an advantage because the separation of the shear-sensitive protein particles is helped by the gentle acceleration. The bowl has only one liquid outlet. The solids are ejected at the periphery. Because the emphasis is on cell reduction, the Q_{act}/Σ is about 5×10^{-8} m/s. Flow rates are up to 45 m^3/h.

The corresponding separation step is also present in winemaking. Because the same machine unit is used also in must separation before fermentation, radial solids-ejecting machines are used.

In rDNA yeast processes, the same type of machine as for baker's yeast can be used. Such machines are available for BL2-LS containment.

Bacteria. In modern biotechnology, solids-ejecting machines, nozzle machines with pressurized discharge, and decanters are used to harvest bacteria. For microorganisms with sizes about 1 μ, axial discharging machines (Fig. 1d) are used. If the solids concentration is so high that the solids discharge frequency has to be more than 1 per 60 s, a radial discharging machine (Fig. 1b) has to be used, but this is less cost effective. The vortex nozzle machine (4) (Fig. 1e) has the disadvantage of accepting only clear substrate media. Residues of substrate may clog the concentrate tubes. If a clear substrate is used, this type of machine is as cost effective as the axial discharging machine but without its solids-handling limitation. A further possibility for clear media is the hermetic machine in Figure 1c, where the cell concentrate leaves the machine continuously and under pressure. The intermittent discharge mechanism is used for CIP.

A typical Q_{act}/Σ value for 99% recovery of *E. coli* K12 in a disk stack machine is 10^{-8} m/s. But the variation between various types of *E. coli* can be very large indeed; between 30 and 300% of the above value has been encountered by the author.

Corynebacterium, somewhat larger cells used in amino acid production, are harvested in nozzle machines with pressurized discharge, often with washing of the bacteria (repulping) and a second separation to improve the yield. A typical Q_{act}/Σ is 5 times that of *E. coli*. *Actinomyces* mycelia can be harvested by radial discharging machines. As in nozzle machines, the cells are washed, often in two steps, to improve the yield.

There are cases where repulping and reseparation are not possible because the cells break up and may damage the rest of the process. To avoid losing yield, the cells must be recovered with maximum dryness. The obvious choice of machine is then the decanter. The low Σ value of this machine type makes it impossible to use it without flocculation of cells. Polymer addition gives a separation efficiency of 98 to 99% at a Q_{act}/Σ of 3×10^{-7} m/s for a spore-free *Bacillus subtilis* broth with a solids content of 12 to 15% by volume. The dry solids (DS) content of the concentrate is 27 to 28% by weight at 3,200 G. This can be compared to solids-ejecting or nozzle machines, which give around 15% by weight for bacteria. Longer residence time in the bowl or higher RCF gives still higher DS contents. At 10,000 G and similar residence time, a decanter can produce an *E. coli* concentrate of 32 to 36% by weight at a Q_{act}/Σ comparable to that in a disk bowl centrifuge.

The high DS capability of decanters is also used to concentrate solids from nozzle machines or from ultrafiltration units, working on fermenter broths in production of amino acids and antibiotics, respectively. The improvements are yield and less costly waste disposal (e.g., drying costs).

Single-cell protein production from hydrocarbons or their simple derivatives is today a not-so-profitable business, but the technology exists. Mayer and Woernle (45) presented a flow sheet for biomass downstream processing.

Bacteria are preconcentrated in large nozzle machines to about 6% by weight DS and then flocculated. This biomass is then dewatered to about 30% by weight DS in decanters and goes after granulating to drying. The liquid phase from the decanter is clarified in high-speed separators and sent to sterilization together with the clarified liquid from the preconcentration step. From sterilization it is recycled back to the fermenter.

Mammalian Cells. It is possible to use high-speed centrifuges in mammalian cell processes (46), in which the cells are discarded after removal. Usually, the inlet shear is not large enough to release DNA into the culture fluid. Care must be taken that shear and air contact in the centrate are avoided, because some protein substances may be destroyed. The centrifugation of intact cells for perfusion must be performed at very low pressure and shear forces. Special machines have been developed recently for this purpose (9,40,47,48).

Cell Debris

In those cases where the intracellular products do not form inclusion bodies, the cell debris often has to be removed after cell breakage. The separation is difficult because of the very fine particles produced (Table 6). Moreover, the viscosity may be high. Dilution of the suspension reduces the viscosity, but the Q_{act}/Σ may still be 5 to 10% of that of the whole cell. It is necessary to use high-speed solids-ejecting machines for this step. In cases where it is possible to flocculate the debris, the Q_{act}/Σ increases by a factor of at least 10 (17). It has been reported (34) that borax-flocculated yeast cell debris can be removed to about 80% in a pilot-size decanter centrifuge at a Q_{act}/Σ of 4×10^{-4} m/s.

Inclusion Body Recovery and Washing

In cases where the substance forms inclusion bodies, it is advantageous to remove cell debris and other cell constituents before the next processing step. The inclusion body has a higher settling velocity than the cell debris, so that a classifying centrifugation will recover the inclusion bodies while leaving the cell debris in the centrifugate. There is a strong interaction between, on the one hand the homogenization method (i.e., the number of passes) and the resulting particle size distribution and, on the other hand, the ease of washing the inclusion bodies (49). There exists an optimum number of passes. The classification split is not 100% sharp, making it necessary to repeat the separation a number of times. This is done either by resuspending the recovered inclusion bodies batchwise or by a continuous simultaneous feeding of the inclusion body concentrate and water into a well-mixed flowthrough feed tank, continuing until the required purity has been reached. Axial or preferably radial discharging solids-ejecting machines are applied at a Q_{act}/Σ of 1 to 2×10^{-8} m/s. To reduce the consumption of chemicals in the next stage, the final separation is often a dewatering step. This can be performed in a high-speed decanter at a Q_{act}/Σ of 2×10^{-6} m/s.

Extraction of Cell-Free Broth

Solvent Extraction. For 50 years, liquid–liquid extraction has been used to concentrate and recover penicillin from filtered broth. This was often performed in multistage centrifugal extractors. Today, the trend for filtered broth is to use a two-stage, countercurrent system with mixers in between disk bowl machines, which are of the three-phase solids-ejecting type. Many other antibiotics need only a one-stage extraction, for example, erythromycin and tetracycline.

Two-Phase Aqueous Systems. The polyethylene glycol (PEG)/dextrane or PEG/salt systems used in protein purification are characterized by a very small density difference between the phases, from 100 down to 14 kg/m^3. The viscosity of the polymer phase could be high, perhaps 100 mPa s, and a very low interfacial tension, leading to easy splitting of drops. Therefore, it is not surprising that the Q_{act}/Σ may be rather low, from 5×10^{-8} down to 1 to 2×10^{-9} m/s. The type of equipment used for the systems with low-density difference is radial solids-ejecting machines. Decanters may be used when the density difference is larger. It is important for optimum performance that the interface between the light and heavy phase in the bowl be in the right position. Equation 4 shows the relation between the three liquid levels in the bowl. The interface should usually be placed just outside of the disk stack in the disk bowl machines.

The two-phase aqueous systems can also be used for partitioning cell debris and other impurities from a dissolved protein product (50). Datar used a PEG 4,000/potassium phosphate system. No density difference data were given. The Q_{act}/Σ was 2×10^{-7} m/s.

Precipitation

Precipitation is often used to isolate and concentrate antibiotics (e.g., rifamycin, erythromycin, and tetracycline). The suitable machine is the radial solids-ejecting disk bowl type. The methods and parameters involved in protein precipitation have been reviewed (51).

Whole-Broth Extraction

Solvent Extraction. Whole-broth extraction of penicillin was the first development of this technique (52). The solids flow is so high that decanter centrifuges must be applied. Many of the constituents of the broth (e.g., macromolecules and the mycelium) have emulsion-stabilizing properties, so de-emulsifying agents must be added (53). After extraction, the rich solvent is separated to reduce the water content before further processing. The solvent-contaminated mycelium must be further treated before disposal.

Two-Phase Aqueous Systems. The two-phase aqueous systems can be applied for extraction also in the presence of cell debris. Hustedt et al. reported on three enzyme recovery processes that use cross-current extraction (54). The phase separation is made in small solids-ejecting or nozzle disk bowl machines. Provided that the density difference between phases is large enough, extraction decanter centrifuges can also be used. Among the commercial uses of this technology on a large scale is the recovery of enzyme in a polymer–salt system, where the separation is performed in a three-phase radial solids-ejecting centrifuge.

INSTALLATION AND OPERATION

Plant Design

In modern biotechnology plants, the design philosophy must be documented as part of the validation procedure. These procedures are becoming common also in the food industry. Stringent guidelines for safe design of plants are being issued more frequently. In Europe, the draft standard for performance criteria for centrifuges in biotechnology deals with leak tightness, cleanability, and sterilizability. Centrifuge installations designed according to BL2-LS are seen in Figures 2 and 4.

Cleaning-in-Place

Automated CIP is today regarded as a quality guarantee. The validation of CIP procedures is one of the necessary protocols to obtain approval to manufacture bioproducts. An automated and logged CIP procedure is repeatable and traceable, less costly than manual cleaning, and safer for the plant personnel than manual cleaning is. Centrifuges and their peripherals are today designed with cleanability in mind. With use of modern programmable logic controllers (PLCs) for valve control, the CIP of a disk-stack centrifuge (or decanter) has been greatly simplified. A CIP cycle consists of several steps of rinsing with different cleaning liquids. The sequence could be cold water, warm (60–70 °C) 4% caustic, 3% nitric acid if necessary, hot water, and finally cold water. It is often necessary to use some type of detergent to improve the cleaning effect, mixed with the caustic or the water. In the radial solids-ejecting machines and the modern pressurized discharge nozzle machines, the discharge mechanism is activated especially during the initial stages.

Containment, Asepsis, and Explosion Protection

These three areas are related in the sense that the installation designer must define a space around the centrifuge, including tanks and piping with controlled interfaces with the environment. In the case of containment, the task is to protect process operators from harmful process substances. The National Institutes' of Health biosafety level is an attempt to classify the risks involved (55). The interface with the environment is sterile HEPA filters. This is a case of "no leaking out." The reverse case, "no leaking in," applies for asepsis and explosion protection. The barriers are formed by HEPA filters and liquid seal pots, respectively.

In both these cases, it is necessary to keep the centrifuge system under a slight, positive pressure relative to the environment, either by lowering the pressure in the room (asepsis) or increasing it in the centrifuge system. For explosion protection, this is done by filling the system with inert gas (e.g., nitrogen) and bubbling the inert gas through a water seal pot. Other measures taken with explosion hazards have been reviewed (56). Centrifuges in closed systems have to be modified in a number of ways: mechanical instead of labyrinth seals, cooling jacket around the bowl cover, and devices that can handle the pressure wave at discharge.

Sterilization

Sterilization with saturated steam at temperatures above 100 °C is by many considered to be the reliable way for avoiding contamination, and it is prescribed for BL2-LS. For the centrifuge manufacturer, this means that in addition to the modifications described in the previous section, the bowl cover must be designed as a pressure vessel. But the most complicated changes concern the installation, because the bowl internal and external surfaces, the cover, the service liquid lines, etc., must all be sterilized. To reach the desired temperature (above at least 121 °C), condensate must be drained away. Krook et al. (4) discuss the sterilizable system for a medium-sized machine with pressurized nozzle discharge in some detail. The sterilization efficiency for a radial solids-ejecting centrifuge of pilot scale was investigated (57). It was found that it was not possible to sterilize the whole system at once. By trying different combinations of open and closed valves, regimes were developed for heating different sections, and these were combined to sterilize the whole system. In the end, sterilization was validated with different chemical and biological indicators.

CENTRIFUGATION VERSUS MICROFILTRATION

After the development of sterilizable microfilters, there is available an alternative to centrifugation in separating solids and liquids. A general rule is that with increasing particle size and increasing scale of operation, centrifugation becomes more interesting than microfiltration and vice versa. It is conceivable that *E. coli* on a 200-L fermenter scale is a microfiltration application. Many factors must be considered in the choice. What is becoming more apparent is that the choice done in pilot scale becomes the most difficult one, because the scale-up and validation procedures make it easier to keep same technology on the larger scale, even if it is not the optimal choice. In many cases, a combination of a filter and a centrifuge is the best answer.

NOMENCLATURE

d_p	Particle diameter, m
D	Inner diameter of test tube, m
$D_{h,l}$	Diameter of heavy and light liquid outlet, m
D_i	Diameter of interface, m
g	Equal to 9.81 m/s^2
h	Distance between inner and outer radius in test tube, m
k	Constant in equation 10
KQ	Semi-empirical measure of centrifuge size, eq. 11

n	Bowl speed, rpm
N	Number of disks
$Q_{act,theor}$	Actual and theoretical feed flow rate, m^3/s
$r_{e,1,2}$	Equivalent, min. and max. radius, m
S_e	Equivalent settling distance, m
t	Distance between disks, m
T	Spin time, s
V	Bowl volume, m^3
$V_{g,c}$	Settling velocity at 1 g and in centrifugal field, m/s
V_d	Velocity component caused by drag forces, m/s
Z	Relative centrifugal force (RCF) or G-number
α	Half cone angle of disc
β	Angle of tube in test tube centrifuge
η	Dynamic viscosity of fluid, mPa s
λ	Dimensionless number, defined in equation 5
μ	Separation efficiency factor
ν	Kinematic viscosity of fluid, m^2/s
ρ_p	Particle density, kg/m^3
$\rho_{f,fh,fl}$	Fluid density, general, heavy liquid, light liquid, kg/m^3
ω	Velocity of rotation, rad/s
Σ	Area equivalent according to Ambler, m^2

BIBLIOGRAPHY

1. A. Rushton, A.S. Ward, R.G. Holdich, *Solid Liquid Filtration and Separation Technology*, VCH Verlagsgesellschaft mbH, Weinheim, Germany, 1996.
2. V.I. Sokolov, *Moderne Industriezentrifugen*, VEB Verlag Technik, Berlin, 1971.
3. T.A. Barker, *Filtr. Sep.* **12**, 33–36 (1975).
4. G. Krook, H. Axelsson, C. Thorsson, in A. Blazej and V. Privarova eds., *Environmental Biotechnology*, Elsevier, Amsterdam, 1991.
5. G. Bergjohann, *Chem. Tech.* **17**, 85–86 (1988).
6. N.F. Madsen, *Inst. Chem. Engrs., Symp. Ser.* No. 113, 281–299 (1989).
7. F. Litvine, *Inf. Chimie* **310**, 327–334 (1989).
8. A. Lee, *Adv. Filtr. Sep. Technol.* **10**, 248–253 (1996).
9. S. Apelman, T. Björling, *Biotech. For. Eur.* **8**, 356–358 (1991).
10. Carr Separations, Inc., brochure, Franklin, Mass., 1997.
11. H. Axelsson, in C.L. Cooney and A.E. Humphrey eds., *Comprehensive Biotechnology. vol. 2: The Principles of Biotechnology: Engineering Considerations*, Pergamon Press, Oxford, U.K., 1985, pp. 325–346.
12. T. Lagerstedt, O. Nåbo, in C. Veret ed., *Flow Visualization IV*, Hemisphere, Washington, D.C., 1986.
13. O. Nåbo, T. Lagerstedt, in R. Reznicek ed., *Flow Visualization V*, Hemisphere, Washington, D.C., 1990.
14. O. Nåbo, C.-G. Carlsson, *Proc. Filtech Conf.* **1**, 1–7 (1987).
15. L. Borgström, C.-G. Carlsson, C. Inge, T. Lagerstedt, H. Moberg, *Appl. Sci. Res.* **53**, 35–50 (1994).
16. A.I. Clarkson, M. Bulmer, N.J. Titchener-Hooker, *Bioproc. Eng.* **14**, 81–89 (1996).
17. B. Lindman, *Meded. Fac. Landbouwwet., Rijksuniv. Gent* **52**, 1383–1387 (1987).
18. B. Madsen, *Inst. Chem. Eng., Symp. Ser.* No. 113, 301–317 (1989).
19. C.M. Ambler, *Chem. Eng. Progr.* **48**, 150–158 (1952).
20. C.M. Ambler, *J. Biochem. Microbiol. Technol. Eng.* **1**, 185–205 (1959).
21. L. Svarovsky, in L. Svarovsky ed., *Solid–Liquid Separation*, Butterworth, London, 1977, pp. 124–147.
22. F.E. Sullivan, R.A. Erikson, *Ind. Eng. Chem.* **53**, 434–438 (1961).
23. S. Aiba, S. Kitai, H. Heima, *J. Gen. Appl. Microbiol.* 243–256 (1964).
24. H. Avallone, P. D'Eramo, *Pharm. Eng.* **12**, 36–39 (1992).
25. J.J. Higgins, D.J. Lewis, W.H. Daly, F.G. Mosqueira, P. Dunnill, M.D. Lilly, *Biotechnol. Bioeng.* **20**, 159–182 (1978).
26. R.V. Datar, C.-G. Rosén, in G. Stephanopoulos ed., *Biotechnology. Vol 3: Bioprocessing*, 2nd ed, VCH Verlagsgesellschaft mbH, Weinheim, Germany, 1993, pp. 469–503.
27. R. Datar, *Filtr. Sep.* **21**, 402–406 (1984).
28. G. Taylor, M. Hoare, D.R. Gray, F.A.O. Marston, *Biotechnology*, **4**, 553–557 (1986).
29. N.J. Titchener-Hooker, D. Gritsis, R. Olbrich, K. Mannweiler, S.A.M. Gardiner, N.M. Fish, M. Hoare, *Pharm. Technol. Int.* **3**, 42–48 (1991).
30. S.O. Hwang, *Biotechnol. Tech.* **10**, 157–160 (1996).
31. A.I. Clarkson, P. Lefevre, N.J. Titchener-Hooker, *Biotechnol. Prog.* **9**, 462–467 (1993).
32. S.F. Siddiqi, N.J. Titchener-Hooker, P. Ayazi Shamlou, *Biotechnol. Bioeng.* **50**, 145–150 (1996).
33. I. Agerkvist, S.-O. Enfors, *Biotechnol. Bioeng.* **36**, 1083–1089 (1990).
34. A.C. Bentham, J. Bonnerjea, C.B. Orsborn, P.N. Ward, and M. Hoare, *Biotechnol. Bioeng.* **36**, 397–401 (1990).
35. R. Datar, C.-G. Rosén, *Chem. Eng. J.* **34**, B49–B56 (1987).
36. R. Kempken, A. Preissman, W. Bertold, *Biotechnol. Bioeng.* **46**, 132–138 (1995).
37. R.J. Wakeman, *Filtr. Sep.* **32**, 337–341 (1995).
38. G.J. Curiel, G. Hauser, P. Peschel, D.A. Timperley, *Trends Food Sci. Technol.* **4**, 225–229 (1993).
39. I. Weibull, *Mater. Des.* **8**, 35–40, 82–88 (1987).
40. M. Johnson, S. Lanthier, B. Massie, G. Lefebvre, A.A. Kamen, *Biotechnol. Progr.* **12**, 855–864 (1996).
41. C. Bowden, *Chem. Eng. (Rugby)*, **415**, 50–54 (June 1985).
42. G.D. Austin, R.W.J. Watson, T. D'Amore, *Biotechnol. Bioeng.* **43**, 337–341 (1994).
43. S.A. Siano, *Biotechnol. Bioeng.* **55**, 289–304 (1997).
44. G. Siems, *Brauwelt Int.* **2**, 132–137 (1997).
45. M. Mayer, R. Woernle, *Chem.-Ing.-Tech.* **57**, 152–153 (1985).
46. M.P. Backer, L.S. Metzger, P.L. Slaber, K.K. Nevitt, and G.B. Boder, *Biotechnol. Bioeng.* **32**, 993–1000 (1988).
47. V. Jäger, in R.E. Spier, J.B. Griffiths, W. Berthold eds., *Animal Cell Technology: Developments, Processes and Products*, Butterworth-Heinemann, Oxford, U.K. 1992.
48. N. Chatzisavido, T. Björling, C. Fenge, S. Boork, E. Lindner-Olsson, S. Apelman, in T. Kobayashi, Y. Kitagawa, K. Okamura eds., *Animal Cell Technology: Basic and Applied Aspects*, Kluwer Academic Publishers, Dordrect, The Netherlands, 1993, pp. 463–468.

49. P.J. Middelberg, B.K. O'Neill, I.D.L. Bogle, *Trans. Inst. Chem. Eng.* **70**, 8–12 (1992).
50. R. Datar, C.-G. Rosén, *J. Biotech.* **3**, 207–219 (1986).
51. D.J. Bell, M. Hoare, and P. Dunnill, in A. Fiechter ed., *Advances in Biochemical Engineering / Biotechnology*, vol. 26, Springer-Verlag, Berlin, 1983, pp. 1–72.
52. H. Katinger, F. Wibbelt, and H. Scherfler, *Verfahrenstechnik* **15**, 179–182 (1981).
53. C. Szabo, *Biotechnol. Bioeng.* **40**, 247–251 (1992).
54. H. Hustedt, K-H. Kroner, and N. Papamichael, *Process. Biochem.* **23**, 129–137 (1988).
55. P. Hambleton, J. Melling, T.T. Salusbury eds., *Biosafety in Industrial Biotechnology*, Chapman & Hall, London, 1994.
56. J. Lindley, *Chem. Eng. (Rugby)* **411**, 41–44 (1985).
57. M.J. Frude, M.T. Simpson, *Proc. Biochem.* **28**, 297–303 (1993).

See also CELL DISRUPTION AND LYSIS; CELL SEPARATION, SEDIMENTATION; CENTRIFUGES, ANIMAL CELLS.

CELL SEPARATION, FLOCCULATION

C.J.-P. BOONAERT
C.C. DUPONT-GILLAIN
P.B. DENGIS
Y.F. DUFRÊNE
P.G. ROUXHET
Université Catholique de Louvain
Louvain-la-Neuve, Belgium

KEY WORDS

Activated sludge
Aggregation
Brewing yeast
Coagulation
Downstream processing
Flocculation
Hydrophobicity
Interfaces
Separation
Zeta potential

OUTLINE

INTRODUCTION

In colloid science, flocculation is defined as an aggregation of solid particles induced by addition of polymers; coagulation refers to an aggregation due to van der Waals or electrostatic attraction between particles (1). Cell aggregation was defined by Calleja (2) as the gathering together of cells to form stable, contiguous, multicellular associations under physiological conditions. The criteria of intimate cell–cell contact and of physiological conditions are not met in the case of floc formation by the addition of flocculating agents; therefore, this definition establishes a distinction between aggregation and flocculation. However, polymers formed at a cell surface or excreted into the medium may be responsible for cell–cell interactions of the same nature as those produced by polymers added to the system.

Calleja's definition of cell aggregation involves a physical movement, either active or stochastic, which is responsible for the change from a dispersed to an agglomerated state. The criterion of stability may refer to aggregation under dispersive conditions, typically in an agitated suspension (as opposed to formation of a colony on agar), of a centrifugation pellet or a sediment. The criterion of physical movement emphasizes that a bond is being formed between distinct bodies, as opposed to chains and mycelial structures. However, certain cell aggregates of great significance in biotechnological processes are not simply formed by a gathering of initially dispersed cells. In activated sludges and in cell granules formed in anaerobic digestion, a dynamic situation is obtained in which some newly formed cells remain more or less loosely attached to each other, either by cell–cell or cell–debris interactions, or because a matrix prevents their dispersion, while free cells may be captured by an existing aggregate.

From the biotechnologist's point of view, the distinction between coagulation, aggregation, flocculation, and agglomeration is not always clear. The words *aggregate, floc, agglomerate*, and *cluster* are often synonymous. They refer to an assemblage of individual cells, formed by initially free cells or by the lack of cell dispersion and occurring under physiological conditions or after addition of various agents to the medium.

Aggregated microbial cells are of great interest in biotechnology. They offer advantages, with respect to fully suspended cells in downstream processing, by facilitating cell-broth separation by sedimentation or filtration. They also allow continuous fermentation processes by keeping the cells in the bioreactor with an upflow of the broth. Further details concerning cell flocculation can be found in several books (2–4). Biofilms (5–8) are not considered explicitly here; however, as concerns the mechanisms of association, they have many features in common with flocs.

PHYSICOCHEMICAL ASPECTS OF AGGREGATION

Interactions between Particles

The essential features of the interactions between colloidal particles may be understood by considering three different aspects (9,10): (*1*) DLVO (Derjaguin, Landau, Verwey, Overbeek) theory (see "DLVO Theory") accounts for van der Waals and electrostatic forces; (*2*) when surfaces are close to molecular contact, solvation interactions (see "Solvation Interactions") may play an important role; (*3*) when mobile macromolecules are present at the interface, they may exert a particular influence at long distances (see "Influence of Macromolecules at the Interface") due to their interactions with the solvent, with the surfaces, and with one another. In the section "Charge Neutralization by Cationic Polymers", a particular mention is made of the possibility to neutralize the surface charge and thereby provoke flocculation by oppositely charged macromolecules.

DLVO Theory. According to DLVO theory (9,10), the free enthalpy of interaction, G_t, between two particles approaching each other may be accounted for by the sum of two contributions, corresponding respectively to electrostatic and van der Waals interactions

$$G_t = G_e + G_v \qquad (1)$$

The electrostatic contribution, G_e, is due to the overlap of electrical diffuse double layers of approaching particles. For two spheres of the same nature and same radius, R, suspended in a symmetrical electrolyte solution,

$$G_e = (64\pi kTRNC\kappa^{-2}T_d^2)\exp(-\kappa D) \qquad (2)$$

where k is the Boltzmann constant, T the temperature, N the Avogadro number, C the electrolyte concentration (mol/L), and D the separation between particle surfaces.

$$T_d = \tanh[ze\psi_d/4kT] \qquad (3)$$

where z is the ion charge, e is the electron charge, and ψ_d the electric potential at the inner limit of the diffuse part of the double layer. It differs from the surface potential, ψ_0, due to adsorption of ions in intimate contact with the surface, that is, in the Stern layer (chemically adsorbed or retained by localized electrostatic attraction) (10,11). For computing G_e, the value taken for ψ_d is usually the zeta potential ζ, that is, the electric potential at the plane of shear between the cell surface and the solution (11). κ, the reciprocal of the Debye length, determines the rate of the variation of the electric potential as a function of the distance to the surface, within the diffuse double layer.

$$\kappa = (1000e^2N/\epsilon kT)(2I)^{1/2} = 2.32\ 10^9(2I)^{1/2}\ (\text{at } 25\ °\text{C}) \qquad (4)$$

where I is the ionic strength of the bulk solution and ϵ the electric permittivity. At an ionic strength of 10^{-1}, 10^{-3}, and 10^{-5} M, κ^{-1} is close to 1, 10, and 100 nm, respectively. Note that the ionic strength is commonly in the range of 10^{-2} to 10^{-3} M for natural water supplies and near 10^{-1} M for microbial culture media.

When the surface potential is low ($<25/z$ mV), G_e simplifies as follows:

$$G_e \cong 2\pi R\epsilon\psi_d^2\exp(-\kappa D) = 2\pi R\sigma^2\exp(-\kappa D)/\kappa^2\epsilon \qquad (5)$$

where σ is the surface density of charge (C m^{-2}). This equation is also valid for nonsymmetrical electrolytes.

For two spheres with the same radius, R, the contribution of van der Waals interactions is given by

$$G_v = -\frac{AR}{12D} \qquad (6)$$

where A is the Hamaker constant. This is mainly determined by dispersion forces (surface–surface, surface–liquid) and thus by the polarizability of the surface-constituting molecules. For organic particles in suspension in water, it is reasonable to consider A values in the range of 0.2 to 1.6 kT (kT = 4.1×10^{-21} J mol^{-1} at room temperature). Values are close to 5 kT for mica and in the range of 2 to 50 kT for oxides depending on the atomic number of the metal ion (9,12).

Figure 1 presents the computed G_t vs D curves for spherical particles of 1-μm radius, A = 1 kT, ψ_d = −20 mV, and κ^{-1} = 10 nm. The respective influence of variation of A, ψ_d, and κ^{-1} is also illustrated. The contribution of van der Waals forces is shown in Fig. 1b (ψ_d = 0 mV). The figure shows that, even at low values of the diffuse-layer potential (Fig. 1b), a potential energy barrier appears at a distance of a few nanometers. This barrier may be very high compared with the kinetic energy associated with Brownian motion, equal to 1.5 kT. At sufficiently high ionic strength (Fig. 1c), the variations of G_e and G_v as a function of the interdistance bring about a secondary minimum that holds the particles together, although not in close contact.

The efficiency of electrolytes to provoke coagulation of colloidal particles is usually expressed by the concentration required to achieve a rapid coagulation, called critical coagulation concentration. According to the valency rule of Schulze–Hardy, this is strongly dependent upon the valency of the ion with a charge opposite to that of the surface (counterion) and the critical coagulation concentration decreases as the counterion valency increases (10). Consider a negatively charged surface; the decrease of the critical

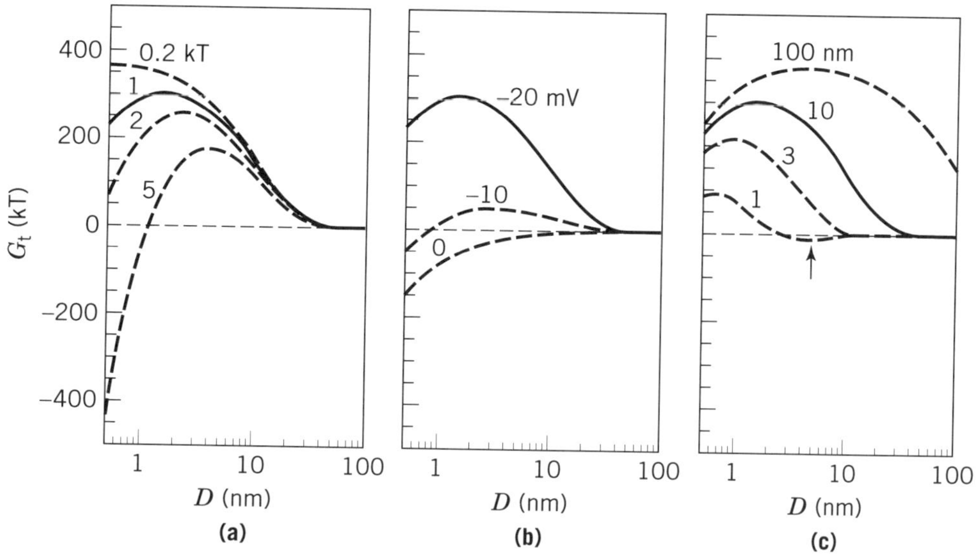

Figure 1. Variation of the free enthalpy of interaction as a function of the distance between two identical spherical particles of 1-μm radius, according to DLVO theory. Solid curve: Hamaker constant $A = 1$ kT, potential of the diffuse double layer $\psi_d = -20$ mV, $\kappa^{-1} = 10$ nm; (**a**) $\psi_d = -20$ mV; $\kappa^{-1} = 10$ nm, A varying from 0.2 to 5 kT; (**b**) $\kappa^{-1} = 10$ nm, $A = 1$ kT, ψ_d varying from 0 to -20 mV; (**c**) $A = 1$ kT, $\psi_d = -20$ mV, κ^{-1} varying from 1 to 100 nm (secondary minimum indicated by arrow).

coagulation concentration is explained by a decrease of the potential energy barrier as a result of two effects: (*1*) if the valency of the cation present in the solution is higher, its adsorption in the Stern layer rises, the potential drop in the Stern layer is larger, and ψ_d is less negative; and (*2*) as the valency of the cation increases at a given concentration, the ionic strength rises, the Debye length κ^{-1} decreases, and the potential barrier decreases (Fig. 1c). The critical coagulation concentration was found to increase in the order $Al^{3+} < Ca^{2+} < Na^+$ for *Escherichia coli* (13).

Equations 2 and 6 are restricted to situations where $D \ll R$; within this limit, the interaction energy is proportional to the particle radius. In DLVO theory, the interacting surfaces are considered as homogeneous, smooth, and rigid. This model is very rough for particles of biotechnological interest. Microbial cells may have complex shapes and bear surface appendages such as fimbriae, pili, fibrils, and flagella. As a consequence, the radius of curvature ruling the interactions may be much smaller than the overall cell radius. In particular, it is conceivable that cells become bound via appendages through a potential barrier resulting from average surface properties, as illustrated by Figure 2. It is also possible that certain zones of the surfaces of two particles are in molecular contact, and thus that solvation interactions have a significant influence, while other zones are still separated by an appreciable distance. The higher the ionic strength, the larger the possible impact of this mechanism.

Solvation Interactions. Between hydrophilic surfaces in water, a short-range repulsive force arises, which tends to prevent the removal of water from the surfaces and is commonly referred to as hydration force. This has been attributed to the structuring of water molecules at the surfaces through H-bonding with hydrophilic surface groups. However, consideration of recent experimental and theoretical results has led to the suggestion of an alternative interpretation, in which the short-range force is an entropic repulsion arising from the confinement of mobile surface groups (14). The hydration repulsion G_{rh} between two hydrophilic surfaces appears to follow the empirical equation (9)

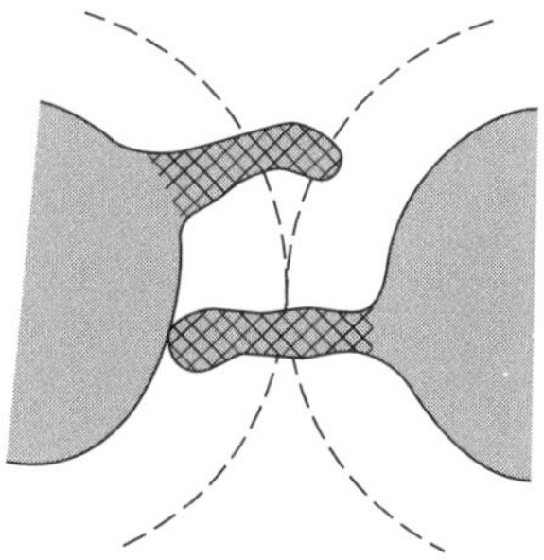

Figure 2. Schematic illustration of the penetration of a surface appendage through a potential energy barrier, the location of which is represented by dashed lines.

$$G_{rh} = +G_{rho} \exp(-D/\lambda_{rh}) \quad (7)$$

where the decay length, λ_{rh}, is of the order of 1 nm for 1:1 electrolytes, and G_{rho} (3–30 mJ m^{-2}) depends on hydration of the surface, including hydration of the cations bound to a negatively charged surface, that is, held in the Stern layer.

The effect of electrolytes on the hydration forces may thus determine whether particles will coagulate or not. The effectiveness of cations as coagulants usually increases according to the so-called lyotropic series (9,10): $Li^+ < Na^+ < K^+ < Cs^+$ for monovalent cations, $Mg^{2+} < Ca^{2+}$ for divalent cations. The decrease of ion hydration is responsible for a decrease of hydration forces between the surfaces. For some surfaces, it is also accompanied by increased ion binding at the surface (11), and thus by a larger drop of the electric potential in the Stern layer.

A hydrophobic surface cannot bind water molecules via ionic or hydrogen bonds, therefore, an attractive hydrophobic interaction arises between two hydrophobic surfaces in water. This results from the strong tendency of water molecules to associate through hydrogen bonding (9,15). To a large extent, hydrophobic bonding is due to the fact that the surface reduces the degrees of freedom offered to adjacent water molecules for forming as many hydrogen bonds as possible; as a consequence, the water structure near the surface is more ordered. Hydrophobic bonding is thus entropically determined, and the resulting association of surfaces should be viewed as the result of a thermodynamic balance rather than the effect of a force, in the common sense. As a consequence of the entropic character, the effect of hydrophobic bonding increases markedly with temperature. For two surfaces in water, the hydrophobic interaction energy, G_{ah} (ignoring DLVO and other solvation forces), is given by (9)

$$G_{ah} = -2\gamma \exp(-D/\lambda_{ah}) \quad (8)$$

where the decay length λ_{ah} is equal to 1–2 nm and γ is the surface–water interfacial energy (10–50 mJ m^{-2}). The rather long-range nature of the hydrophobic interaction accounts for the rapid coagulation of hydrophobic particles in water and the folding of proteins.

Influence of Macromolecules at the Interface. A nonionic polymer dissolved in an ideal solvent (no interaction between polymer segments and the solvent) takes the shape of an "unperturbed" random coil, characterized by the radius of gyration

$$R_g = \frac{ln^{0.5}}{\sqrt{6}} \quad (9)$$

where l is the effective segment length and n is the number of segments. R_g may thus vary from a few to about 100 nm. In a real solvent, the size of the coil may be different from R_g and is given by the Flory radius

$$R_F = \alpha R_g \quad (10)$$

In a poor solvent, the segments attract each other, the coil is more compact, and α is smaller than unity. In a good solvent, the coil swells and its Flory radius is given by

$$R_F \cong ln^{0.6} \quad (11)$$

A poor solvent may become a good solvent by adding certain solutes or changing the temperature. The temperature at which $\alpha = 1$ is known as the theta (θ) temperature, and an ideal solvent is known as theta solvent (9).

Figures 3 and 4 illustrate the different situations that may be encountered when two surfaces approach each other while carrying a nonionic polymer. Once surface separation is a few times the radius of gyration, the segments overlap, which leads to a repulsive osmotic force, that is, to the decrease of entropy associated with a closer confinement of the polymer chains between the two surfaces. This is known as steric repulsion and is responsible for steric stabilization of colloids by polymer additives (9).

As a simple case, consider a low coverage of polymer molecules (no overlap or entanglement of neighboring chains), each being grafted at one end to the surface (Fig. 3 top left; Fig. 4 right, low density). This may be relevant to microbial surfaces. In a theta solvent, if the number of chains grafted per unit area is Γ, the repulsive energy per unit area is approximately given by

$$G_s \cong 36\Gamma kT \exp(-D/R_g) \quad (12)$$

In a good solvent, the coils swell and the range of repulsion is broader. When the surface density of grafted chains is high, so that they are forced to extend away from the surface much farther than R_g, a polymer brush is obtained (Fig. 3, top right). The thickness of the polymer layer, L, no longer varies according to the power 0.5 or 0.6 of the degree of polymerization but is proportional to the polymer length. Once two brush-bearing surfaces are closer than $2L$, a repulsive energy arises (Fig. 4 right, high density). For a good solvent and for $D/2L$ in the range of 0.2 to 0.9, if s is the average distance between the grafting points, the interaction energy per unit area is approximately given by

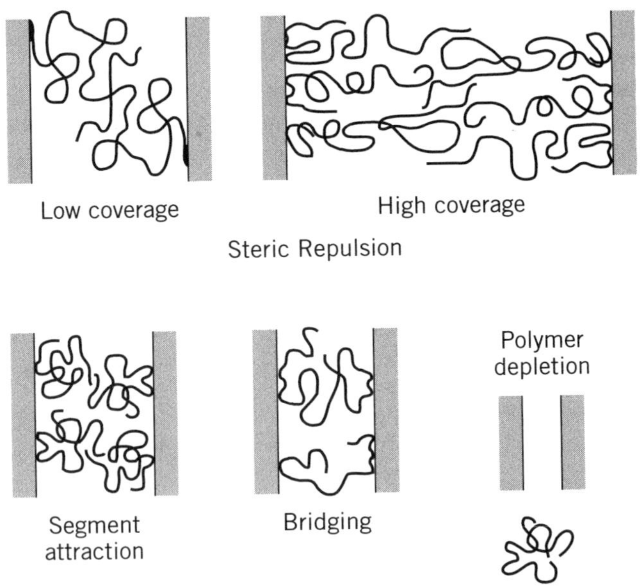

Figure 3. Illustration of steric forces generated by polymers between two surfaces (grafted or adsorbed). *Source*: Adapted from Ref. 9.

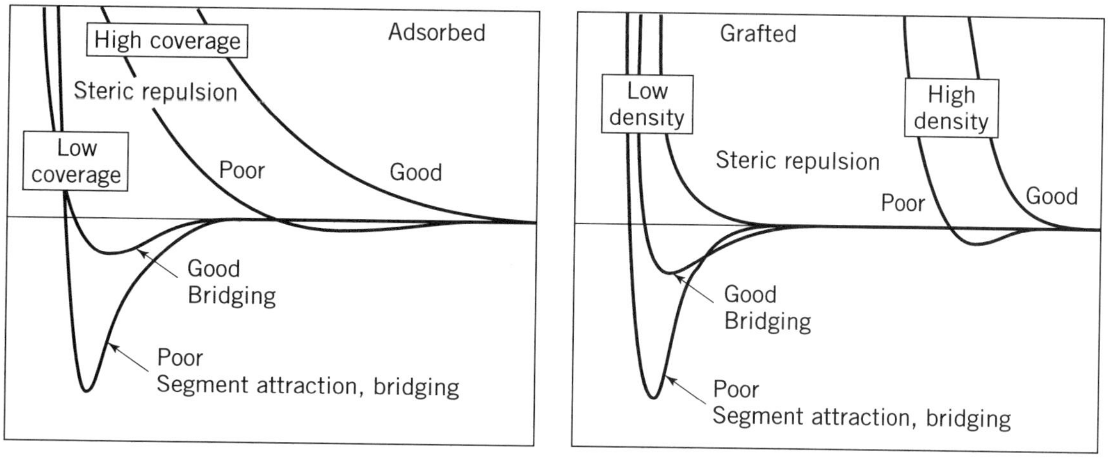

Figure 4. Variation of interaction free enthalpy as a function of the separation distance between two particles as a result of macromolecular interactions; influence of the mode of polymer retention at the surface (adsorption, grafting), polymer concentration at the surface (low, high), and solvent quality (good, poor). *Source*: Adapted from Ref. 9.

$$G_s \cong \frac{100L}{\pi s^3}\, kT \exp(-\pi D/L) \qquad (13)$$

In this case, G_s is the balance of an osmotic repulsion between the coils and the elastic energy that opposes chain stretching.

If the polymer is adsorbed (Fig. 3 left, bottom compared to top), a fraction of its segments (trains) are in contact with the surface; the other segments belong to loops and tails dangling in the solution. The polymer may adopt a conformation leading to an adsorbed thickness smaller than the radius of gyration. Things are more complicated than for grafted polymers due to the dynamics allowed by a multiplicity of weak bonds between the polymer and the surface, and to a possible variation of the adsorbed amount and the binding sites as the two surfaces approach each other. As a consequence, the variation of the interaction free energy as a function of distance is less sharp (Fig. 4, left compared with right). Moreover the force between two surfaces at a given separation distance may take a long time (hours) to reach a stationary value (16).

Attractive forces may also be mediated by adsorbed or grafted polymers (Fig. 3, bottom; Fig. 4) (9).

- In a poor solvent, segment attraction between molecules bound to opposite surfaces leads to overall attraction between the surfaces until a balance with steric repulsion is reached.
- When the degree of coverage is weak and the polymer is strongly adsorbed, the macromolecules may form bridges between the two surfaces. Under suitable conditions, bridging forces can extend over longer distances than van der Waals forces.
- When a polymer does not adsorb, bringing two surfaces at a distance smaller than R_g will push the coils out of the space between the surfaces. The polymer depletion is responsible for a difference of osmotic pressure between the bulk liquid phase and the liquid phase held between the surfaces. This effect is small compared to van der Waals and solvation interaction energies.

In a good solvent, the efficacy of a polymer to provoke flocculation is mainly determined by its molecular mass and its concentration, which determine the degree of coverage and the thickness of the adsorbed layer. An optimal flocculation concentration is found for a given molecular mass. It is influenced by pH and ionic strength, which determine the electrostatic repulsion between the surfaces. Flocs will form if the surface is partly covered and if the adsorbed molecules extend far enough from the surface to span the distance at which the electrical repulsion operates (16,17), as illustrated by Figure 5. On the left side of the figure, the molecular mass or the ionic strength are too low to allow bridging. On the right side, the ionic strength is higher (and the Debye length, lower) or the surface potential is weaker, the electric repulsion operates at short distances compared with the thickness of the adsorbed macromolecules, and bridging occurs.

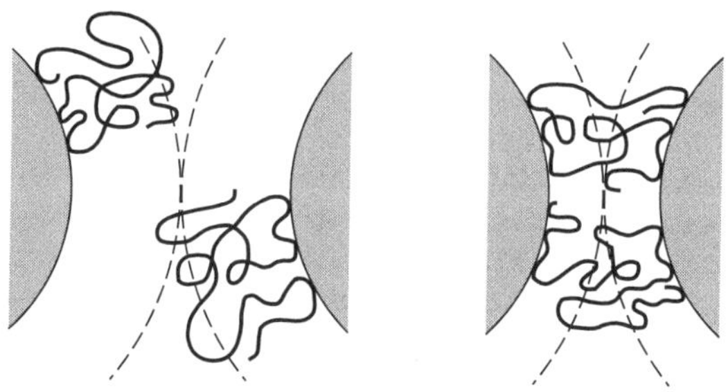

Figure 5. Illustration of the effect of ionic strength or surface potential on polymer bridging; the dashed lines represent the distance at which the electrical repulsion operates. The increase of ionic strength or decrease of surface potential from left to right allows polymer bridging. *Source*: Adapted from Ref. 18.

As polymer adsorption is slow and irreversible, flocculation may often proceed through nonequilibrium processes, which are strongly dependent on the procedure followed (manner and rate of mixing the particle suspension and the polymer solution; stirring). For instance, heterocoagulation may occur between those particles free of polymer and those that are saturated. In other situations, the surface may be partly covered by macromolecules that have not yet collapsed, thus allowing polymer bridging and producing transient flocs, the lifetime of which may vary from seconds to days, depending on the conditions (16,19).

Charge Neutralization by Cationic Polymers. As materials of biotechnological relevance are usually negatively charged, only cationic polyelectrolytes are considered here. Due to electrostatic repulsion between segments, polyelectrolytes display more or less extended conformations according to ionic strength and to their charge density, and thus to the fraction of charged segments τ. For a given molecular mass, the maximum adsorbed amount increases with τ because of the stronger polymer–surface interaction, passes through a maximum, and decreases due to increasing electrostatic repulsion between polymer segments (18,20,21). At low polymer ionicity (small τ), the maximum amount adsorbed (expressed in mass) increases as a function of molecular mass. At high ionicity, the maximum adsorbed amount and the adsorbed-layer thickness are practically independent of the molecular weight; the loop size is small and the polymer chain adopts a flat conformation on the surface. The flat conformation realizes a local neutralization of the particle surface charge, but reduces the possibility of bridging contacts with other particles.

With a cationic polyelectrolyte of high charge density, the influence of polymer molecular mass and of ionic strength can be explained by the electrostatic patch model (18,20,21) illustrated by Figure 6. The adsorbed polyelectrolyte represents patches of positive charges surrounded by the original particle surface. Particles with this mosaic surface–charge distribution may interact in such a way that positive and negative patches of different particles come into contact, causing a strong attachment. As the surface coverage increases, the particle becomes totally coated by the cationic polymer, electrostatic repulsion arises between particles, and the colloid remains in a dispersed state. The optimum flocculation concentration corresponds to a situation where the particle surface is partially covered by patches of cationic polymer. This is illustrated by Figure 7, which shows the influence of polymer concentration on flocculation of a clay, evaluated by the relative supernatant turbidity and the sedimentation time (22). The optimum flocculation concentration coincides with reversal of the zeta potential of the particles.

With a cationic polymer of low ionicity, the influence of charge density, concentration, and molecular mass is more complex, due to the combination of the charge-neutralization and bridging mechanisms. The influence of ionic strength is also complex because it affects the screening of segment/segment lateral repulsion (which decreases chain stiffness and increases loop size and adsorbed amount) and the screening of segment/surface attraction (which reduces the proportion of trains). In practice, increasing the ionic strength reduces the concentration at which the flocs initially appear and broadens the flocculation domain (18).

Rate of Aggregation

The rate of aggregation is determined by the probability of collision between particles, which depends on their concentration, and by the efficiency of a collision to make a bond between the particles. The collision may occur as a result of two types of transport processes: agitation and Brownian motion. If one considers the collision between two identical spherical particles of diameter D^*, the ratio between the probability of collision due to agitation P_o (orthokinetic conditions) and the probability of Brownian collision P_p (perikinetic conditions) is given by

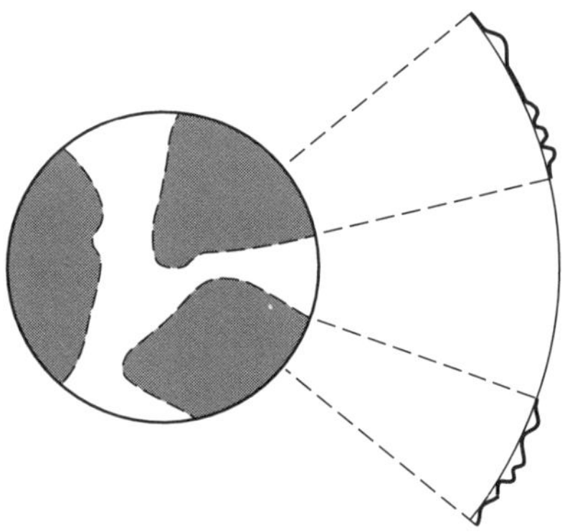

Figure 6. Patch model for a cationic polyelectrolyte (shaded area) adsorbed on a negatively charged surface. Left, view of the particle surface; right, section through a part of the surface. *Source*: Adapted from Refs. 20 and 21.

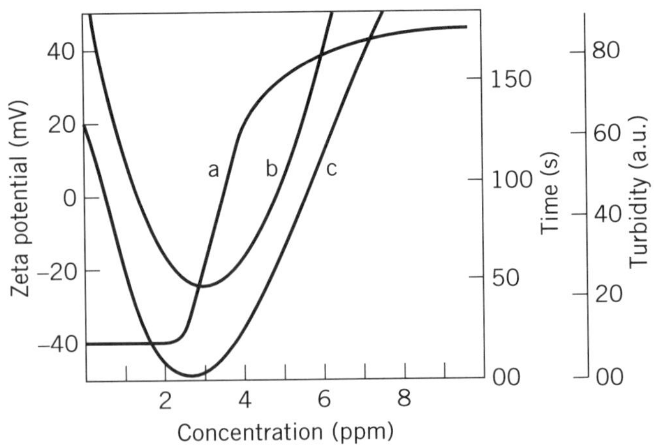

Figure 7. Flocculation of Na-montmorillonite by *N,N,N*-trimethyl aminoethyl chloride acrylate. Variation of (**a**) zeta potential, (**b**) sedimentation time, (**c**) relative supernatant turbidity, as a function of the amount of polymer introduced per unit volume. Polymer molecular weight 10^6; τ = 100%; clay concentration 0.05 g L^{-1}; NaCl concentration 1.2 mM. *Source*: Adapted from Ref. 22.

$$\frac{P_o}{P_p} = \frac{\eta D^{*3} G^*}{2kT} \tag{14}$$

where η is the dynamic viscosity, G^* the shear rate, k the Boltzmann constant, and T the temperature (23). In water at room temperature ($\eta = 1$ mPa s), with agitation typical of a magnetic stirrer ($G^* = 100\ s^{-1}$), this ratio is about 100 for particles of 2 μm (24). The rate of aggregation of microbial cells thus depends strongly on agitation due to their relatively large size and slow Brownian motion, unlike small colloidal particles. On the other hand, agitation can also break the flocs so that a stationary state will be reached, with a characteristic distribution of aggregation numbers.

The chance that a collision gives rise to a bond between the particles is determined by the variation of the free energy of particle–particle interactions as a function of the distance. Four typical situations, illustrated by Figures 1 and 4, may be encountered: *(1)* the interaction energy is attractive at all distances, and a quick coagulation occurs (low ψ_d, high ionic strength, high Hamaker constant; Fig. 1); *(2)* the interaction is attractive at short distances but is repulsive at longer distances and passes through a maximum (Fig. 1); in this case the probability that a collision leads to bond formation is the probability of passing over the energy barrier; *(3)* the interaction is repulsive at all distances and no aggregation occurs; the colloid suspension is said to be stabilized (high ψ_d, low ionic strength, low Hamaker coefficient, Fig. 1; steric repulsion, Fig. 4); *(4)* the interaction energy curve presents a well at a certain separation distance (secondary minimum, Fig. 1; bridging, segment attraction, Fig. 4), and the aggregates will be constituted by particles maintained at a certain distance from each other.

Surface Properties of Particles

The nonspecific interactions that control particle aggregation (see "Interactions between Particles") depend on the physicochemical properties of the surfaces (electrical properties, hydrophobicity, presence of macromolecules) and consequently on their surface chemical composition. Physicochemical methods that are well-established or are being developed to investigate these properties are described in the following sections.

Electrical Properties. As discussed earlier, it appears that the electrical properties of particle surfaces strongly influence coagulation by addition of salts, or influence flocculation by polyelectrolytes. The zeta potential, ζ, is usually considered to be close to the potential relevant for DLVO computations (10,11,25,26). It can be deduced from the electrophoretic mobility, μ. For spherical particles of radius R,

$$\zeta = \frac{\eta}{\epsilon} \mu B = 12.85\ \mu B \tag{15}$$

at 298 K, if ζ is given in mV and μ in $10^{-8}\ m^2\ V^{-1}\ s^{-1}$. $B = 1$ (Helmholtz–Smoluchowski equation) when $\kappa R > 100$, which is relevant for microbial cells in suspensions of common ionic strength; $B = 3/2$ (Hückel–Onsager equation) when $\kappa R < 1$. In the intermediate situations, B is a function of κR.

Equation 15 implies that: *(1)* the solid plus stagnant layer together behave electrokinetically as a solid particle, *(2)* particles are nonconducting, *(3)* particle surfaces are smooth and homogeneous, and *(4)* the viscosity, dielectric constant, and ion conductivities in the diffuse part of the double layer are equal to their bulk values (11). When the surface conductance or the bulk (or cell wall) conductance are not negligible, the relationship between electrophoretic mobility and zeta potential is more complex (27). Therefore it is recommended to report experimentally determined electrophoretic-mobility values instead of ζ potentials.

Hydrophobicity. The surface hydrophobicity of microbial cells can be assessed by the water contact angle measured on particle lawns (28). A greater cell-surface hydrophobicity is of practical concern in separation: It favors cell coagulation and cell accumulation at the liquid–gas interface. The latter effect may be used for increasing cell concentration by flotation (29).

Contact angles of liquids have been used to deduce the surface energy of solids and to predict the stability of solid–solid associations. The model is based on the balance of interfacial free energies involved in the creation of a new solid–solid interface and in the destruction of the interface of each partner with water. This approach is subject to severe limitations. *(1)* There are different theoretical frames that lead to different values of interfacial energy (30), indicating that absolute values cannot be considered individually. Nevertheless, classification of different materials according to their surface free energy is the same, whatever the approach used. For polymer materials, this classification is the same as that given by the contact angle of water. *(2)* Because interfacial energies are deduced from contact angles of liquids, the energy balance does not incorporate the influence of electrostatic interactions, which act at long distances, or the influence of ionic bridges and steric interactions. *(3)* The system is supposed to be at equilibrium and thus reversible, which is not in accordance with experimental data. *(4)* A close contact is assumed to be achieved between the surfaces, which is not reasonable in view of the roughness of most surfaces.

Other methods of evaluation of microbial-cell hydrophobicity involve adhesion at hydrophobic surfaces or aggregation at high salt concentration (28).

Surface-Chemical Composition. The surface-chemical composition of solids, including microbial cells, can be determined by X-ray photoelectron spectroscopy (XPS). The technique involves irradiation of the sample by an X-ray beam, which induces ejection of electrons. The kinetic energy of the emitted electrons is analyzed, and their binding energy determined. Each peak is then characteristic of a given element. Due to inelastic scattering of electrons in the sample, the collected information concerns only the outermost molecular layers of the surface (2–10 nm). As the position of the photoelectron peak is slightly affected by the chemical environment of the element, peak-shape analysis provides further information on chemical functions (31). The data can be converted into concentrations of model constituents: proteins, polysaccharides, and hydrocarbon-like compounds (32).

For microorganisms, correlations have been obtained between the surface-chemical composition obtained by XPS and surface properties, that is, hydrophobicity and electrical properties. Furthermore, the surface composition has been found to be related to the behavior of microorganisms at interfaces (aggregation, adhesion) (33).

Nanometer-Scale Surface Properties. The physicochemical methods just described have a poor lateral resolution, the information provided concerning only the overall surface properties of particles. Exciting new possibilities are now offered by atomic force microscopy (AFM) to investigate the molecular interactions and properties of surfaces on the nanometer scale. AFM, invented in 1986 (34), consists of measuring the forces acting between a sharp probe and the surface of a sample. It allows the nanoscale mapping of surface topography, as well as the direct measurement of mechanical properties, friction, and interaction forces (electrostatic, van der Waals, solvation, steric interactions). As opposed to more conventional surface analysis techniques, AFM can be operated in aqueous solutions, which makes it possible to study living cells in physiological conditions (35).

A very challenging goal for future research is to directly map the surface properties of living cells in aqueous conditions. The use of functionalized AFM probes, possessing different charges, surface energies, and solvation properties, should make it possible to measure: *(1)* the local surface hydrophobicity and electrical properties of individual cells, *(2)* the surface distribution of solvated macromolecules such as exopolysaccharides, and *(3)* steric interactions associated with these macromolecules. These data will provide new insight into the molecular mechanisms of cell aggregation.

BREWING YEAST: A MODEL FOR THE IMPLICATION OF SPECIFIC INTERACTIONS IN FLOCCULATION

Flocculation and Flocculence

Flocculation of brewing yeasts, thoroughly reviewed by Stratford (36), is the most extensively studied of all systems in which cell aggregates are of interest. It is illustrative of the diversity of approaches used to understand flocculation of microbial cells and the interplay between specific and nonspecific interactions. Spontaneous aggregation of yeast cells occurs near the end of fermentation and is of considerable importance in brewing. For instance, early sedimentation in lager beer production will provoke an incomplete or too slow fermentation; failure to sediment will complicate processing and may lead to alteration of taste.

In brewing technology, the flocculence of a strain means its ability or tendency to flocculate in industrial conditions. More broadly, flocculence is the ability to flocculate if all environmental conditions are favorable (37). The term *flocculability* has been proposed to designate the ability to flocculate in a given medium if hydrodynamic conditions and cell concentration are adequate (38). Beyond the terminology, it is essential to keep in mind that the occurrence of flocculation depends on *(1)* the nature of the biological material (strain; physiological state defined by growth conditions and growth phase), *(2)* the physicochemical environment (temperature, medium composition), and *(3)* the physical conditions (velocity gradient, cell concentration). Thus, the evaluation of the ability to flocculate must always be related to experimental conditions.

In order to test the ability of yeast to flocculate in a given medium, it is important to control the initial state of dispersion and to distinguish the conditions selected for the aggregation process from those allowing evaluation of the final state of dispersion. A screening test may conveniently be performed according to the following procedure (24,39). Introduce the washed cells into test tubes containing the desired medium and vigorously stir to ensure dispersion. Agitate gently for a given period to allow flocculation to take place, then let stand undisturbed to allow floc sedimentation. Finally, measure the absorbance of the supernatant, which reflects the concentration of cells remaining free. The volume of the sediment may also be a way to evaluate the final state of dispersion, as the bed volume of flocculated particles is larger than that of individual particles (40).

Yeast cell behavior in laboratory conditions does not always reproduce the behavior in industrial conditions. Among 45 strains noted as flocculent in brewing, only 20% did flocculate during a culture in laboratory conditions (shaken Erlenmeyer flasks) (41). This was attributed to agitation and aeration rather than to final pH or the requirement of particular wort constituents.

Several parameters were used to quantify yeast flocculation: initial rate of flocculation, rate of settling, number of free cells, floc size, and floc strength (38,42). It was estimated that the attraction force between flocculent cells is three orders of magnitude larger than the force calculated with the DLVO theory (43). The dynamic equilibrium between floc formation and floc breaking (i.e., splitting, which determines the average floc size, and erosion, which determines the number of single cells) was modeled. This showed the dependence of floc diameter and number of single cells on bond strength and agitation (44).

Yeast Behavior and Cell Surface Properties

There are two types of processes for beer brewing. Bottom fermentation, producing lager beer, is carried out at low temperature (near 10 °C) by *Saccharomyces carlsbergensis*; at the end of fermentation, the cells flocculate and settle. Top fermentation, typically producing ale, is carried out at higher temperature by *Saccharomyces cerevisiae*; when the cells flocculate, they do not settle but they tend to remain associated with CO_2 bubbles and rise to the top of fermented wort. The surface properties (electrophoretic mobility, water contact angle, and surface composition) of a collection of yeasts were investigated in order to examine their possible relationship with the type of strain, growth phase, flocculence, and occurrence of flocculation (45). There is no clear-cut distinction between surface properties of top and bottom strains; however, they differ on a statistical basis. On the average, top strains have more protein at their surface and are more hydrophobic (Fig. 8), which may be related to their tendency to remain associated with gas bubbles and move to the top of the wort. Bottom strains contain more phosphate. The difference between the profiles of electrophoretic mobility vs. pH (Fig. 9) may be explained by modelling the surface charge according to the surface chemical composition, as given by XPS. For the bottom strains, the electrical properties are controlled mainly by phosphate, resulting in a low isoelectric point (pH 2 or below) and an electrophoretic mobility that does not change appreciably above pH 4. For the top strains, the electrical properties are determined mainly by the balance of protonated amino and carboxylate groups in proteins, which give a higher isoelectric point (pH 4) and an electrophoretic mobility changing regularly from pH 2 to 7.

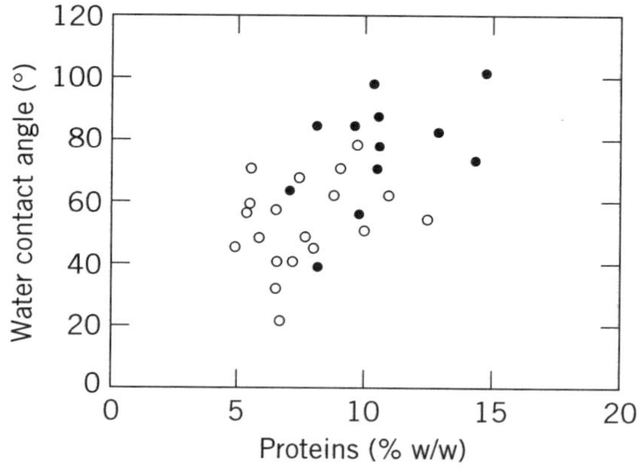

Figure 8. Water contact angles of bottom-fermenting strains (○) and top-fermenting strains (●) as a function of the cell-surface concentration (weight %) of proteins as obtained by modeling XPS data. *Source*: Adapted from Ref. 45.

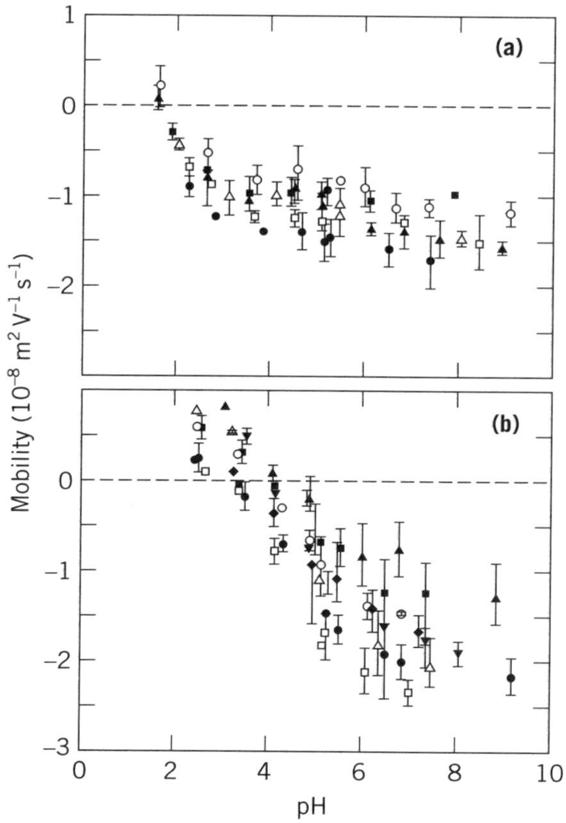

Figure 9. Electrophoretic mobility as a function of pH for bottom-fermenting strains (**a**) and top-fermenting strains (**b**) in 1 mM KNO_3 (45).

Although surface properties of brewing yeasts were statistically related to the top/bottom character, they did not show correlations with the flocculating character. No differences of surface properties were found between strains that did and did not flocculate in laboratory conditions, or between cells from the exponential and stationary growth phases, even for strains where flocculation occurred during the transition from one growth phase to the other. This is in contrast with reports (46,47) that ability to flocculate during growth coincides with an increase in cell-surface hydrophobicity correlated with the appearance of fimbriae-like structures. It was also reported (48) that flocculent cells possess a fimbriate or hairy surface, whereas nonflocculent cells are smooth.

Certain yeast strains produce aggregates with a particular geometric shape, such as cubes, lozenges, cylinders, and disks (36,49), the shape being strongly dependent on agitation. Cubic flocs, close to 1 cm in size, produced by *S. cerevisiae* used for continuous production of ethanol are shown in Figure 10. When agitation is stopped, the flocs collapse and, within a few tens of minutes give rise to a cell carpet on the bottom of the flask. This is tentatively attributed to the fact that the cells are coated by polymers that ensure a weak bridging, as schematized in Figure 11. The viscoelastic properties of the interface would be such that, upon agitation, binding prevails due to chain entanglement, with the hydrodynamic conditions controlling the floc shape; on the other hand, repulsion would prevail in resting conditions (49).

Genetic and Biochemical Aspects

According to the most widely considered mechanism of flocculation proposed by Miki et al. (50), lectinlike molecules (sugar-binding proteins that need Ca^{2+} to adopt their active conformation) recognize and bind mannans at the surface of adjacent cells (36,51), as illustrated by Figure 12(a).

Stratford and Assinder (52) distinguished two flocculation phenotypes, both based on a lectin-mediated mechanism: Flo1 phenotype strains, which included most of the laboratory strains, had *FLO1* genes and possessed mannose-specific lectins; NewFlo phenotype strains, which included most ale-brewing strains, possessed less-specific lectins, binding manno- and glucopyranoses. Masy et al. (53) found the same two phenotypes and called them MS (mannose sensitive) and GMS (glucose mannose sensitive). On the other hand, a different phenotype, called MI (mannose insensitive), representing strains that are not inhibited by sugars, including mannose and glucose, was reported (41,53).

In the last few years, much progress has been made as a result of advances in the fields of genetics and biochem-

Figure 10. Flocs produced by a strain of *S. cerevisiae*; picture taken shortly after stopping agitation (scale in cm).

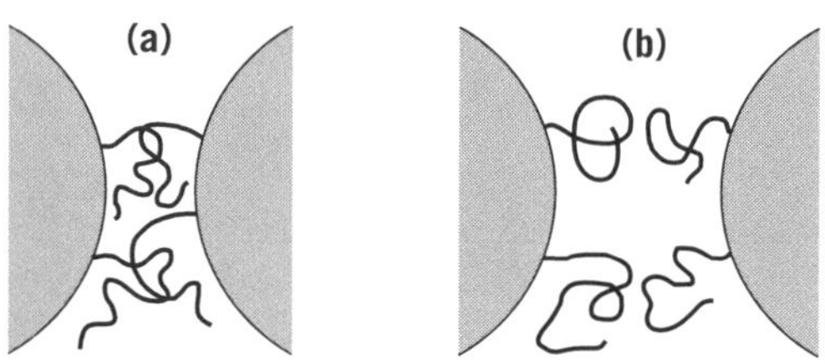

Figure 11. Schematic illustration. (**a**) Yeast surfaces weakly bound by entangled polymer chains upon agitation. (**b**) The same surfaces repelling each other in resting conditions.

istry. In a review (54), 33 genes involved in yeast aggregation have been reported; these include structural genes, regulatory genes, and genes involved in cell wall synthesis and mitochondrial functions. Four classes of strains were distinguished: *(1)* strains lacking an intact flocculation gene, *(2)* strains in which intact flocculation genes are constitutively repressed, *(3)* a strain with flocculation genes that are transcriptionally regulated, *(4)* strains with genes that encode for constitutive flocculence. For strains in which flocculation was regulated, induction was ascribed to nutrient limitation (46,55).

The best-studied dominant flocculation gene, *FLO1*, has been cloned and sequenced. Based on the genetic sequence, it was hypothesized that the Flo1 protein is present in the cell wall, and its N-terminus is exposed to the medium. The location of this protein in the cell wall was confirmed by the use of immunofluorescent probes. It was suggested that the exposed part of the Flo1 protein binds directly with neighboring cell wall mannans (54). The *FLO8* gene was also cloned (56); no homology was found with the *FLO1* gene. It was suggested that *FLO8* does not encode a cell wall protein, but is involved in the regulation of the transcription of the *FLO1* gene.

Two types of surface proteins involved in flocculation of bottom-fermenting strains were isolated from cell walls. A mannose-specific calcium-dependent agglutinin was isolated from a nonflocculent strain and from flocculent strains, both in nonflocculating and flocculating growth stages; it was released during the flocculation assay (57). A protein isolated from cell walls of flocculent cells only, called flocculin, showed homology to the protein coded by the flocculation gene *FLO1*; it was associated with fimbriae-like structures, did not show lectinlike activity, and was thought to act as a lectin ligand (58). The authors proposed a model (57) in which at least three factors determine the onset of flocculation: *(1)* release of agglutinin, *(2)* appearance or activation of flocculin, and *(3)* formation of fimbriae-like structures accompanied by an increase in cell-surface hydrophobicity.

For NewFlo phenotype *S. cerevisae* strains, there is evidence that lectins are synthesized continuously from an early stage of growth and rapidly inserted into the cell wall, but remain inactive until they are activated at flocculation onset (59).

Toward a Comprehensive View of the Role of Specific and Nonspecific Interactions

A comparative study of nine brewing strains flocculating in laboratory conditions (three bottom- and six top-fermenting strains) was performed (41,45). Flocculation most often occurred only in the stationary growth phase, and the onset of flocculation could not be related to a change of physicochemical properties (chemical composition by XPS, electrophoretic mobility, water contact angle).

Laboratory tests showed that flocculation of bottom strains requires the addition of calcium, is inhibited with mannose and sucrose, but not with galactose, and is not affected by ethanol. This is typical lectin-based behavior and the NewFlo/GMS phenotype. The following flocculation mechanism may be proposed: *(1)* there would basically be a repulsion between cell surfaces due to electrostatic or steric interactions (mobile polysaccharides); *(2)* in the stationary growth phase, lectins would appear at the cell surface or be made accessible; and *(3)* when the sugar concentration in the solution is sufficiently low, these lectins would bind sugar residues of adjacent cells across the repulsion barrier. The onset of flocculation thus seems to be triggered by the conjunction of the appearance of active lectin at the surface and of sugar attenuation, rather than by the calcium concentration (41,60). The occurrence of flocculation about 2 pH units above the isoelectric point in

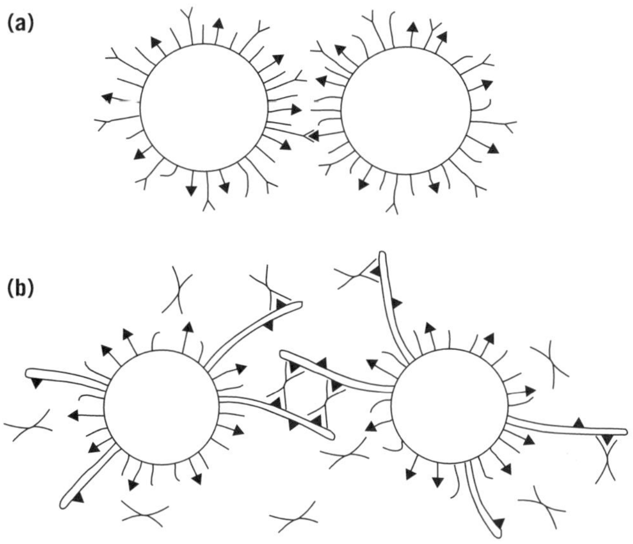

Figure 12. Schematic representation of possible cell–cell binding processes involving a lectinlike mechanism. (**a**) Direct lectin–ligand binding. (**b**) Implication of surface hairs and agglutinins (double forks).

laboratory tests, and not at the isoelectric point, indicates a major influence of pH on lectin conformation. The ionic strength has no effect on flocculation, which may be explained either by a minor role of electrostatic interactions or by a compensation of the influences of ionic strength on electrostatic interactions and lectin conformation. The penetration of the repulsion barrier will be more likely if the lectins or the lectin–ligands stick out of the surface, which is especially the case if they are located on surface hairs. Furthermore, flocculation might be favored by the release of agglutinins. The possible action of hairs and agglutinins is schematically illustrated by Figure 12b.

Flocculation of top-fermenting strains is induced by ethanol and other alcohols and does not require addition of calcium. For certain strains, there is no inhibition by mannose, sucrose, or galactose, whereas flocculation of other strains is inhibited by mannose only. The former would belong to the MI (mannose-insensitive)-flocculation phenotype, which has not been frequently reported. For this collection of top strains, there is no evidence of lectinlike interactions. Other organic solvents (methanol, *n*-propanol isopropanol, *n*-butanol, isopropanol, acetone, 1,2-propylene glycol) have the same effect as ethanol when added at a concentration provoking the same decrease of the surface tension of water. This indicates that the influence of ethanol is related to an adsorption process. The following flocculation mechanism may be proposed. There would exist a repulsion between the cells. As exponential cells, which have similar physicochemical properties as stationary cells, do not flocculate near the isoelectric point, the repulsion would be steric in nature. However, the influence of ionic strength on flocculation suggests that electrostatic interactions play a role as well. In the stationary growth phase, additional macromolecules or hairs would appear at the surface, providing binding sites for specific (lectinlike or other) or nonspecific binding. In the presence of sufficient ethanol concentration and, for the mannose-inhibited strains, sufficiently low sugar concentrations, these macromolecules or hairs would protrude from the surface and manage to bind adjacent cells across the repulsion barrier. The onset of flocculation thus seems to be triggered by the conjunction of a change of cell surface and an increase of ethanol concentration.

FLOCCULATION IN DOWNSTREAM PROCESSING

Production of fine chemicals by living cells involves a range of downstream processes, the first of which is usually the separation of cells or cell debris from the broth. This step is largely improved if flocculation can be carried out (61–63). Flocculation, as a pretreatment for the continuous centrifugal separation of whole cells, should produce flocs with high mechanical strength. In the production of microalgae, the best practical and economical way of harvesting was found to be chemical flocculation, followed by sedimentation, flotation, or filtration (64). The following sections review how flocculation can be produced in downstream processes and, to a certain extent, can be understood in the light of physicochemical theories described earlier.

Use of Colloidal Particles

S. cerevisiae forms flocs in the presence of positively charged submicrometer-sized aluminium hydroxide particles. The respective surface charges of yeast and aluminium hydroxide prevent homocoagulation and the process is a heterocoagulation of oppositely charged particles (65), which can easily be understood in the light of DLVO theory (discussed earlier).

Positively charged organic-polymer particles [poly-(styrene-divinyl benzene) with quaternary ammonium surface groups; 0.27-μm diameter] were used to obtain flocs of *S. cerevisiae* and *E. coli* whole cells or cell debris. The settling time of the cells was considerably decreased in the presence of the particles, and the centrifugation energy and power required for the removal of *E. coli* cell debris were reduced by one order of magnitude. The optimal polymer particle concentration found for the clarification of suspensions of *E. coli* cells may be related to a partial coverage of the cell surface by the polymer particles, allowing the cells to be bound together by one layer of positively charged particles; complete coverage of every cell would lead to charge reversal and overall repulsion (see "Charge Neutralization by Cationic Polymers"). An optimal polymer-particle concentration was also found for the clarification of *E. coli* cell-debris suspensions. In the recovery of several enzymes by removal of cell debris following this method, complete enzyme activity was recovered in the supernatant, indicating that the polymer particles did not adsorb or inactivate the enzymes (61).

Use of Salts

Simple electrolytes favor particle coagulation by compressing the electric double layer, as explained in the section "DLVO Theory". Metal ions of higher charge are more efficient both because they have a higher tendency to adsorb in the Stern layer and because they give a smaller Debye length. Moreover, other factors affect the efficacy of hydrolyzing metal ions, such as Al^{3+} and Fe^{3+}, which are used in water and sludge conditioning (66). Upon hydrolysis, these metal ions form positively charged polynuclear hydroxo complexes. The latter adsorb specifically and are able to reverse the surface charge. Therefore, depending on the pH and on the concentration of colloidal particles, the Al^{3+} and Fe^{3+} salts may act as coagulants only in a narrow concentration range. The concentration must be high enough to allow compression of the electric double layer or surface neutralization; however, overdosing may reverse the charge of the colloids and restabilize them. Above a certain concentration of metal ions, colloidal particles may again form flocs by enmeshing in aluminium or ferric hydroxide precipitates.

Jiang et al. (67) tested the ability of ferric and aluminium salts to coagulate algae. Hydrolized ferric species produced by oxidation of ferrous sulfate were more efficient than ferric sulfate, aluminium sulfate, or partially neutralized aluminium chloride. Floc size was larger and cell removal higher. Humic substances added to the algae suspension reduced the performances of the coagulants.

Use of Soluble Polymers

Nature and Efficiency of the Polymers. Both synthetic and biological polymers have been tested to induce cell flocculation, and results can be understood in the light of the principles presented earlier. Nonionic and anionic synthetic polymers were found to be ineffective for flocculation of *E. coli, S. cerevisiae*, and microalgae, while cationic polymers were good flocculating agents (61,64,68–70). Increasing the charge density of cationic polyacrylates and polyacrylamides enhanced flocculation (62,70). Cationic polyethyleneimine (PEI) was effective to flocculate *Chlorella ellipsoida* (68). With low molecular mass polymers, no reversal of the cell-surface charge occurred and the amount (mass) required to initiate flocculation decreased considerably as the molecular mass increased from 800 to 2,000 Da. Above 2,000 Da, the amount required to initiate flocculation did not vary much as a function of molecular mass; at high concentrations the polymer reversed the cell-surface charge; when the polymer concentration exceeded an optimal value, the flocs began to redisperse. Earlier work with silica and *E. coli* showed that an increase of pH from 4 to 6 or 9 increased the PEI concentration required for optimum flocculation, as expected from the variation of polymer ionicity. With *C. ellipsoida*, little effect of pH was observed between pH 4 and 7. The polymer concentration needed for flocculation was one order of magnitude larger for *E. coli*, compared to crystalline silica (68). With diethylaminoethylmetacrylate acrylic acid copolymers, washed cells of *E. coli* flocculated at polymer concentrations at which the isoelectric point was reached, indicating that the mechanism of flocculation was in agreement with the patch model (70).

Downstream processes are large consumers of polyacrylamide derivatives because they are economical and effective flocculating agents. However, neurotoxic and strong carcinogenic activities of acrylamide monomer were reported. Hence, there is a need to produce safe, biodegradable compounds that will minimize environmental and health risks. Several microorganisms have been reported to produce yeast-cell flocculating agents; they were effective between pH 3 and 5 and lost their effectiveness above pH 5 (71). Cultures of *Rhodococcus erythropolis* were found to flocculate *E. coli, Microcystis aeruginosa*, yeast, activated sludge, kaolin clay, and other particles. Flocculation was considerably improved by adding Ca^{2+} (72). However, the respective roles of *R. erythropolis* cells, possibly provoking heterocoagulation, and of polymers released into the suspension or at the cell surface, were not clarified. A *Paecilomyces* culture filtrate, containing a polysaccharidic compound, was shown to flocculate several microorganisms, including *Bacillus subtilis* and *E. coli*. Its effectiveness with *E. coli* did not change significantly between pH 4 and 8, and in an ionic strength range of 0.001 M to 0.1 M, except in the presence of CO_3^{2-} and Fe^{3+} ions. However, it decreased steeply below pH 4 and above pH 8, or at ionic strength of 1 M (73).

Compared to synthetic polymers, chitosan, also commercially available, possesses several advantages: monomers are not toxic, it can be heat-sterilized without degradation, and it is biodegradable. Chitosan, a partially *N*-acetylated poly (glucose-amine), is derived from chitin, the major constituent of crustacean exoskeleton (62). At acidic pH, chitosan is soluble and exists as well-extended linear chains due to repulsion of charged $-NH_3^+$ groups. At alkaline pH, the positive charges gradually disappear (the neutralization point is about pH 7.9), and chitosan tends to coil and precipitate (74).

Chitosan-induced flocculation of washed *E. coli* cells at concentrations much lower than those required to reach the isoelectric point, as opposed to the use of diethylaminoethylmetacrylate acrylic acid copolymers. This was attributed to the rigidity of chitosan, which undergoes weak deformation during adsorption and is able to form bridges between the cells (70). The optimal dose for flocculation of microalgae was about twice as low for chitosan as compared with polyacrylamides; this superiority was also attributed to the rigid backbone of chitosan (64). With washed cells of *B. subtilis* and *E. coli*, the effectiveness of chitosan and cationic polyacrylamides did not vary between pH 4 and 7. Above pH 7, the effectiveness of chitosan dropped due to a decrease of ionicity, whereas the change of polyacrylamide effectiveness was less marked. None of these compounds was effective in flocculating *Zymomonas mobilis* (62).

An optimal chitosan concentration was found near 0.2 $g\ L^{-1}$ for flocculation of *Euglena gracilis* cultures. Addition of chitosan above this concentration led to a partial cell redispersion due to repulsion of positively charged cells. With the optimal chitosan concentration, a sharp pH optimum was observed at pH 7.5, the decrease of flocculation at lower and higher pH being attributed to a lower negative charge of *Euglena* cells and to deprotonation of chitosan $-NH_3^+$ groups, respectively. At pH 7.5, flocculation did not decrease appreciably at high chitosan concentration, due to the low ionicity of the polymer at this pH (74).

Using chitosan, cell debris and nucleic acids could be separated selectively from β-D-galactosidase in *E. coli* disintegrates. For a highly charged chitosan, a pH decrease induced a reduction of the optimal flocculant dosage (63). Highly charged chitosan can achieve a more efficient removal of cell debris than less-charged chitosan. The required flocculation dosage of highly charged chitosan was relatively independent of the charge density, but increased with the molecular mass. At the required flocculation dosage, electrophoretic-mobility measurements showed that the flocs formed were not completely neutralized. It was therefore suggested that the main mechanism of flocculation was a "nonequilibrium" bridging process (75). This occurs when reconformation following polymer attachment to the surface is slow compared with polymer attachment or particle–particle collision (76). It is especially pronounced with high molecular mass polymers and leads to trapping of surface-negative sites between the bridges formed by the polycations. Highly charged and high molecular mass chitosan produced large and shear-resistant aggregates suitable for filtration, settling, and centrifugation, whereas highly charged low molecular mass chitosan gave smaller and more shear-sensitive flocs suitable only for centrifugation (75).

Influence of Medium Composition and Cell-Produced Polymers. Ionic strength can dramatically affect the flocculant effectiveness. Bilanovic et al. (64) studied the effect of salinity on microalgae flocculation with cationic polyacrylamides (1×10^7 and 2×10^7 Da) and chitosan (4×10^5 Da) at pH 7. For the two types of polycation, removal of more than 90% was obtained with freshwater cultures of *Chlorella*, whereas removal was very poor for cultures of marine species of the same genera. The efficiency of the cationic polymers to flocculate *Isochrysis galbana* dropped considerably as the ionic strength increased from 0.02 to 0.70 M, without appreciable alteration of the optimal dose. While the rise of ionic strength was expected to increase the screening of polymer–cell surface attraction, it also provoked a shrinkage of the macromolecule, as revealed by a decrease of the intrinsic viscosity.

The flocculation is also strongly influenced by the extracellular polymers (ECP) released by the cells in the solution and by the components of the culture media. The optimal dosage of cationic polymers was lower for washed *E. coli* cells than for the initial culture suspension. The interaction between anionic polymers in the medium and added cationic polymers led to an increase in flocculant requirement (69,70). On the other hand, it has been reported that the strongly acidic polymers released in solution by *Lactobacillus delbruekii* and *Lactobacillus fermentum* act as flocculating agents (strong anionic polyelectrolytes) that only need a small amount of cationic substances to be sensitized (69).

Polyacrylamides could flocculate *E. coli* and *B. subtilis* cells in defined media, but were ineffective in complex media because of precipitation. Chitosan did not precipitate below pH 7.5 in a complex medium and allowed flocculation of *E. coli* cells, while requiring twice the dose used in a defined medium; the flocculant efficiency decreased above pH 7 in both media and washed cells. On the other hand, flocculation of *B. subtilis* by chitosan in complex media occurred at pH 4 but dropped as the pH was increased from 4 to 7, and did not change with washed cells (62).

Flocculation of yeast by chitosan in a complex medium occurs simultaneously with the precipitation of chitosan with components of the medium. These can be regarded as competitive processes with respect to the availability of chitosan. The detrimental effect of precipitation (increase of dosage, contamination of biomass, removal of the fermentation product) can be minimized by a rapid addition and efficient mixing of the flocculant with fermentation media of high biomass concentration (77). Cells may also become entrapped in the precipitate of polyelectrolyte complexes, with the unexpected benefit that strong flocs are formed (78).

The effectiveness of flocculation is also dependent on the cell-surface constituents but is not simply related to the cell wall type (Gram positive or Gram negative). Polyacrylamide and chitosan were effective in flocculating *E. coli* (Gram negative) and *B. subtilis* (Gram positive) washed cells (62). Their inefficacy to flocculate washed cells of *Z. mobilis* (Gram negative) was attributed to the presence of nonionic extracellular polymers attached to the cell surface. It was reported that strongly acidic polymers adsorbed on *L. delbruekii* and *L. fermentum* surfaces induced an increase in the flocculant dosage, probably due to a high charge density and, therefore, a high polycation demand (69).

Combined Use of Flocculating Agents

The combined use of two flocculating agents has often proved to be an effective way to form flocs. Chitosan, polyethyleneimine, and Ca^{2+} were used simultaneously to perform flocculation of *E. coli* cell debris (79). This was applied in a process using a continuous centrifugal separator and allowed an at least 10-fold improvement in cell debris removal, compared with the same process without flocculation.

Sewage sludges were successfully flocculated by the addition of a cationic, followed by an anionic, polymer (80). The recovery of bacterial extracellular proteinases obtained in *B. subtilis* culture was investigated by adding bentonite (a clay with a negatively charged surface), followed by a cationic aminoamidic polyelectrolyte. The yield of recovery was more than 84% at the laboratory scale and up to 82% at the pilot-plant scale (81).

Polyelectrolytes used singly (62), or combined with an oppositely charged polymer (82), were inefficient in flocculating *Z. mobilis* in complex medium. However, effective flocculation could be performed by adding highly charged polyacrylamide or chitosan, followed by bentonite, providing high broth clarities between pH 5 and 8.5 or between pH 4 and 8, respectively. It was suggested that a combination of electrostatic interactions and bridging capability was responsible for the higher effectiveness of chitosan at low pH. Addition of bentonite after chitosan allowed avoidance of precipitation of the latter at high pH. Flocculation effectiveness was also influenced by the floc collision during settlement (82).

Use of Specific Interactions

Flocculation can be controlled by genetic factors, as discussed for brewing yeast earlier. Addition of starch to *E.*

coli suspension resulted in the formation of bacterial flocs, allowing a very efficient cell-broth separation by settling (83). Flocculation was dependent on the high expression of a cell-surface maltoporin receptor for starch: floc formation was evident with bacteria containing wild-type maltoporin, but was faster and occurred to a greater extent with strains expressing a high-affinity allele of the starch receptor or with strains grown at high levels of maltose. Because expression of the maltoporin receptor can be genetically controlled, its presence at the cell surface can be made largely independent of the variations in surface properties, thus avoiding the variability of classical flocculation. The optimal starch concentration was 1–2 mg mL^{-1} for a strain with a high-affinity allele. Decrease of flocculation at higher concentrations was attributed to saturation of binding sites. The optimal starch concentration was 5–10 mg mL^{-1} for a strain with a low-affinity allele. The optimal conditions for floc formation were pH 6–7 and a cell concentration of 2×10^9 cells mL^{-1}, which should be readily achievable in both laboratory and large-scale cultures. It was suggested that the presence of maltodextrin and that the lack of accessible α-1,4-linked chains in the starch preparation could decrease the starch effectiveness.

This starch-dependent aggregation is potentially applicable to other microorganisms if they comply with the following requirements: *(1)* the starch receptors are expressed, *(2)* the cell surface is not masked by a polysaccharide or a protein layer, and *(3)* the cell does not produce extracellular amylolytic enzymes. For harvesting bacteria, starch offers several advantages: *(1)* binding to cells is rapid; *(2)* it is nontoxic; *(3)* unlike electrically charged flocculating agents, it does not affect the pH, and its binding is effective in the pH range usually found in fermentation conditions; *(4)* it is a low-cost material; and *(5)* it is easily separable from bacteria, if necessary (83).

FLOCS AS A STATE OF PROLIFERATING BIOMASS

A biomass proliferating in the form of flocs is a system that is much more complex than flocs produced in a given downstream operation, due to the diversity of interfaces with respect to space and time. Brewing yeast, discussed earlier, illustrates the fact that a floc is the result of the integration of different processes and of their development on the time scale. In this section, even more complex systems of practical relevance are considered. Whenever possible, the presentation points out physicochemical features that provide a guide for understanding and controlling the biosystem behavior.

Continuous Ethanol Fermentation

The ability to flocculate is an important characteristic in the continuous production of ethanol, so that a high concentration of cells can be maintained within the fermentor (84). This has been used in upflow tower fermentors, which were successfully implemented at industrial scale (85). The yeast producing the particular flocs shown in Figure 10 was used in a gaslift laboratory fermentor (86). A more complete understanding of the phenomena controlling both the shape and size distribution of the flocs appeared to be a prerequisite for a full control of the highly promising fermentation process.

Continuous ethanol fermentation has also been performed with *Zymomonas mobilis*. In a laboratory upflow fermentor (volume of 0.92 L), flocs were obtained by starting-up (10 days) at a low dilution rate (0.065 h^{-1}) (87). After this period, the dilution rate was incrementally increased to 2.5 h^{-1}, allowing an ethanol productivity as high as 100 g $L^{-1}h^{-1}$. However, no attempt was made to evaluate long-term performance. A simultaneous addition of cationic polyacrylamide and titanium hydrous oxide was used to flocculate *Z. mobilis* (88). This provided an ethanol productivity of 16.4 g L^{-1} h^{-1} for at least 100 h in a gaslift tower fermentor (volume of 5 L). In similar conditions, productivity of unflocculated cells was evaluated to be about 2 g L^{-1} h^{-1}.

Mixed cultures of flocculent strains of *Z. mobilis* and *Saccharomyces* sp. gave higher ethanol production, yield, and volumetric productivity than pure cultures (84).

Activated Sludge

A common biological treatment of sewage waste is the activated sludge process, which is essentially a way to concentrate the organic matter of the waste. In its simplest form, it is a two-stage operation involving aeration followed by sedimentation. The aeration stage allows proliferation of different organisms and floc formation. An essential requirement for successful operation is the formation of settleable flocs (89). Among the physical characteristics of the flocs, the size distribution and structure are most important (90–94).

Influence of Physicochemical Factors. The floc behavior can be affected by hydrophobic interactions. Zita and Hermansson (95) isolated, by hydrophobic interaction chromatography, hydrophilic and hydrophobic bacteria from flocs coming from an aeration tank of a wastewater treatment plant. Attachment of these strains to activated sludge liquor containing flocs and wastewater was evaluated. The results showed a correlation between adhesion and surface hydrophobicity up to a threshold, above which bacterial attachment did not increase with surface hydrophobicity. This was attributed to an equilibrium between reversibly attached cells and free-living cells.

Urbain et al. (96) showed that a high internal hydrophobicity of the flocs is correlated with a decrease of the sludge volume index (SVI), which reveals an improvement of settleability. The internal hydrophobicity of the floc was determined by a salt aggregation test performed on a sonicated sludge sample mixed with increasing concentrations of $(NH_4)_2SO_4$ in phosphate buffer.

Bacterial strains of different hydrophobicity were separated by subjecting sonicated sludge to an octane–water mixture. The amount of exopolymers released by sonication of a hydrophilic strain cultured in the laboratory was greater than that of a more hydrophobic strain. This observation may suggest that the hydrophilic strains are not simply trapped in a matrix, but are also implicated in the floc organization (97).

No clear relationship has been established between the electrophoretic mobility of sludge particles and their settling properties. A strongly negative electrophoretic mobility of sludge particles was found to induce a high SVI (98,99). However, an inverse correlation (100) and the absence of correlation between electrophoretic mobility and SVI were also reported (101).

Floc stability was evaluated by gentle mixing in $CaCl_2$ and KCl solutions of increasing ionic strength (102). Floc dissociation decreased as the ionic strength rose from 50 μM to 5 mM; it decreased more slowly between 5 and 50 mM. This was explained by DLVO theory, the flocs being stronger as the diffuse double layer was more compressed. The reversibility of the process was tested by reflocculation of degraded activated sludge. Reflocculation increased between 0 and 5 mM ionic strength and remained constant up to 50 mM. Between 50 and 500 mM, floc dissociation increased and reflocculation decreased. According to the authors, at high ionic strength, the distance between particles is fairly constant and the floc structure appears to be influenced by mechanisms of unknown nature, unlike the mechanisms considered in DLVO theory. The influence of wastewater on floc stability was similar to that described for salt solutions.

Polyvalent ions can act by binding ECP and bacteria, or by binding bacteria with one another, through electrostatic interactions (90). The addition of Ca^{2+} to activated sludge induced a decrease of both the sludge volume and the residual turbidity in the supernatant after 30 min of sedimentation (103). Removal of Ca^{2+} from activated sludge by ion exchange (H^+, K^+, Na^+, Mg^{2+}) or by a chelating agent (EGTA) induced a sludge deterioration: small particles appeared in the supernatant, causing an increase of the specific resistance to filtration. On the contrary, exchange of Ca^{2+} by Cu^{2+} decreased the specific resistance to filtration and thus resulted in a more stable floc structure. This was attributed to the ability of Cu^{2+} ions to maintain the 3-dimensional structure of the exopolymers (90).

Influence of Biological Factors. Bioflocculation in activated sludge also involves extracellular polymers (ECP). Among the ECP, proteins (96,97,104,105), polysaccharides (96,97,104), lipids (104), humic acids (105), and DNA (96,97) were reported. Humic acids originate from wastewater itself, whereas other compounds arise from cell metabolism or from cell lysis. The influence of ECP on sludge settling (notably on SVI) has been widely studied. However, because there is no unified method of extraction (e.g., centrifugation with or without heating and/or sonication), care must be taken in comparing results from different studies. A general trend found in the literature is that an increasing concentration of ECP extracted from activated sludge is related to an increasing SVI (96,97,100,105); however, the absence of a relationship between ECP production and SVI has also been reported (101).

ECP were isolated after thermal extraction from sludge and their molecular mass was determined by gel filtration. A major fraction was found between 30 and 100 kDa, and an increase in its apparent molecular mass was correlated with an increase of the stirred specific volume index (SSVI). This could be due to steric interactions, allowing greater amounts of interstitial water to be retained in the sludge (106).

The presence of growing filamentous microorganisms is an important feature determining floc properties. It was suggested that they could form a backbone to which floc-forming bacteria could attach (107). This idea is supported by the presence of filamentous microorganisms in numerous activated-sludge samples (96). However, sludges have poorer settling characteristics when filaments grow out of the flocs. The outgrowth was quantified by the total extended filament length, that is, the sum of the lengths of filaments extending out of the flocs. An increase of SVI (94) or of SSVI (108) was observed as the total extended-filament length increased up to 10 m mg^{-1} of suspended solid. When the total extended-filament length exceeded this value, SVI increased dramatically and bulking phenomena were observed (101).

Microbial Aggregates in Anaerobic Wastewater Treatment

Anaerobic processes, resulting in the production of biogas, are used for the decomposition of organic compounds in wastewater; the main problem encountered is to efficiently retain the biomass in the reactor. Several reactor designs have been considered. Bacteria may be immobilized on a packing material or carrier particles; the ability of bacteria to aggregate can be used to produce granules with good settling properties. This latter is the case of the upflow anaerobic sludge blanket reactor (UASBR) (109). Bacterial aggregation ensures the conservation of the biomass inside the reactor, but also enhances the interactions between the various microorganisms needed for anaerobic digestion (110).

The formation of bacterial granules in UASBR is governed by the surface properties of the bacteria, which may themselves depend on the operational parameters. Hydrophobicity is the surface property that has been the most studied for bacteria originating from granules. Contact-angle measurements on cells found in the reactor effluent and on cells originating from granules led to the conclusion that more hydrophobic bacteria remain in the reactor in the aggregated state, whereas more hydrophilic bacteria are isolated and leave the reactor (111). It was also reported that the more hydrophobic bacteria found in UASBR had a better ability to flocculate (112). Moreover, the ECP produced by flocculent bacteria were more hydrophobic than those produced by nonflocculent bacteria. Different populations of bacteria found in UASBR are responsible for different steps of waste digestion and are classified as acidogens, acetogens, and methanogens. Most acidogens found in granules are more hydrophilic than the acetogens and methanogens; moreover, granules are formed of layers of different hydrophobicities, each layer being responsible for one step of the degradation process (113).

The surface charge of the bacteria is also recognized as a parameter governing the flocculation behavior in UASBR (109). The negative charge is attributed to carboxyl groups located on the ECP; it is indeed reduced when the cells are submitted to shear (112).

Divalent or trivalent metal ions, which can be added to the reactor, may influence granule properties by acting on cell–cell interactions, as described by DLVO theory, but also by affecting the conformation of ECP or by cross-linking the ECP to form a kind of matrix. It was shown that the size of the granules, and thereby their settleability, is significantly increased in the presence of Ca^{2+} (100 mg L^{-1}), which leads to shorter retention times in the reactor (111). When EDTA, a chelating agent, was added to the medium, the aggregation decreased but was not suppressed. It was thought that part of the Ca^{2+} was involved in irreversible bonds between the cells. The flocculation could also be induced in an anaerobic digestor by the addition of various divalent and trivalent cations, and the tendency to flocculate was related to a reduction in the negative charge of the cell surface. However, divalent and trivalent cations may also play a role through other mechanisms. Precipitates of calcium or iron may provide the bacteria with nucleation sites for granule formation. An alternative mechanism is by an increase in surface hydrophobicity provoked by complex formation between divalent cations and carboxyl or phosphate groups (112).

Other operational parameters may influence granule formation or the efficiency of the process. When the substrate is rich in protein, highly hydrophilic cells are favored, which leads to the formation of fluffy flocs rather than granules. On the contrary, a substrate rich in volatile fatty acids produces granules with a very hydrophobic surface; this may be detrimental to the process because such granules may be eliminated from the reactor by flotation (113). Lowering the surface tension of the medium by adding alkylbenzenesulfonate was shown to favor the formation of granules with a more hydrophilic surface, thus preventing their flotation (114).

CONCLUSION

Cell aggregation, which is of considerable importance in bioprocess engineering, can be understood and controlled to a certain extent in the light of established principles of colloid and surface chemistry. A major key is the interplay between electrostatic interactions and interactions involving macromolecules. In this respect, microbial cells represent a complex system that can be regarded as a solid surface, a surface grafted with macromolecules, or a surface with adsorbed macromolecules, depending on the strain and the approach followed. Furthermore, the extracellular polymers may constitute a matrix in which the cells are more or less loosely entrapped. For certain microorganisms, another important mechanism is provided by specific interactions between the cells, or between the cells and foreign substances.

In opposition with this complexity, most of the methods used so far to examine the physicochemical properties of microbial surfaces consider them to be, more or less implicitly, homogeneous and molecularly smooth, although key factors are clearly the spatial distribution of constituents and properties. Atomic-force microscopy is undoubtedly a very promising tool in this respect. In the near future, it may provide direct information, with molecular-scale spatial resolution, as regards chemical composition, physicochemical properties, and interactions of cell surfaces, in conditions relevant to practical interest. This constitutes an exciting field in which a strong synergy may develop between biotechnology and nanotechnology.

ACKNOWLEDGMENTS

We gratefully acknowledge the support of the Foundation for Training in Industrial and Agricultural Research (FRIA), the National Foundation for Scientific Research (FNRS), the Department of Education and Scientific Research (Concerted Action, Physical Chemistry of Interfaces and Biotechnology), and the Interuniversity Poles of Attraction Programme (Federal Office for Scientific, Technical and Cultural Affairs, Supramolecular Chemistry and Supramolecular Catalysis).

BIBLIOGRAPHY

1. R.J. Hunter, *Foundations of Colloid Science*, vol. 1, Clarendon Press, Oxford, 1987.
2. G.B. Calleja, *Microbial Aggregation*, CRC Press, Boca Raton, Fla., 1984.
3. K. Marshall, *Microbial Adhesion and Aggregation*, Springer-Verlag, Berlin, 1984.
4. Y.A. Attia, *Flocculation in Biotechnology and Separation Systems*, Elsevier, New York, 1987.
5. W.G. Characklis and K.C. Marshall, *Biofilms*, Wiley, New York, 1990.
6. L.F. Melo, T.R. Bott, M. Fletcher, and B. Capdeville, *Biofilms—Science and Technology*, Kluwer Academic, Dordrecht, The Netherlands, 1992.
7. J. Wimpenny, W. Nichols, D. Stickler, and H. Lappin-Scott, *Bacterial Biofilms*, Bioline, Cardiff, U.K., 1994.
8. H.M. Lappin-Scott and J.W. Costerton, *Microbial Biofilms*, University Press, Cambridge, 1995.
9. J. Israelachvili, *Intermolecular and Surface Forces*, 2nd ed., Academic, London, 1992, pp. 176–311.
10. P.C. Hiemenz and R. Rajagopalan, *Principles of Colloid and Surface Chemistry*, 3rd ed., Marcel Dekker, New York, 1997.
11. J. Lyklema, *Fundamentals of Interface and Colloid Science*, vol. 2, Academic, London, 1995, pp. 4.1.–4.135.
12. S. Nir, *Progr. Surface Sci.* **8**, 1–58 (1976).
13. L. Eriksson and C. Axberg, *Water Res.* **15**, 421–431 (1981).
14. J. Israelachvili and H. Wennerström, *Nature* **379**, 219–225 (1996).
15. J. Lyklema, *Fundamentals of Interface and Colloid Science*, vol. 1, Academic Press, London, 1991, pp. 4.48–4.51.
16. F. Lafuma, K. Wong, and B. Cabane, *J. Colloid Interface Sci.* **143**, 9–21 (1991).
17. B. Vincent, in Th. F. Tadros ed., *Solid / Liquid Dispersions*, Academic, London, 1987, pp. 149–162.
18. F. Lafuma, in J.C. Roberts ed., *Paper Chemistry*, 2nd ed., Chapman and Hall, London, 1995, pp. 44–63.
19. V. Chaplain, M.L. Janex, F. Lafuma, C. Graillat, and R.E. Audebert, *Colloid Polymer Sci.* **273**, 984–993 (1995).
20. A. Guyot, R. Audebert, R. Botet, B. Cabane, F. Lafuma, R. Jullien, E. Pefferkorn, C. Pichot, A. Revillon, and R. Varoqui, *J. Chim. Phys.* **87**, 1859–1899 (1990).

21. J. Gregory, in Th. F. Tadros ed., *Solid/Liquid Dispersions*, Academic, London, 1987, pp. 163–181.
22. G. Durand-Piana, F. Lafuma, and R. Audebert, *J. Colloid Interface Sci.* **119**, 474–480 (1987).
23. J. Th. G. Overbeek, in H.R. Kruyt ed., *Colloid Science*, Elsevier, New York, 1952, pp. 58–89, 278–301.
24. J.C. Kihn, C.L. Masy, M.M. Mestdagh, and P.G. Rouxhet, *Can. J. Microbiol.* **34**, 779–791 (1988).
25. A.M. James, in N. Mozes, P.S. Handley, H.J. Busscher, and P.G. Rouxhet eds., *Microbial Cell Surface Analysis: Structural and Physicochemical Methods*, VCH Publishers, New York, 1991, pp. 221–262.
26. R.J. Hunter, *Zeta Potential in Colloid Science—Principles and Applications*, Academic Press, London, 1981, 386 p.
27. A. van der Wal, M. Minor, W. Norde, A.J.B. Zehnder, and J. Lyklema, *J. Colloid Interface Sci.* **186**, 71–79 (1997).
28. H.C. van der Mei, M. Rosenberg, and H.J. Busscher, in N. Mozes, P.S. Handley, H.J. Busscher, and P.G. Rouxhet eds., *Microbial Cell Surface Analysis: Structural and Physicochemical Methods*, VCH Publishers, New York, 1991, pp. 263–287.
29. R. Tybussek, F. Linz, K. Schügerl, N. Mozes, A.J. Leonard, and P.G. Rouxhet, *Appl. Microbiol. Biotechnol.* **41**, 13–22 (1994).
30. M.N. Bellon-Fontaine, N. Mozes, H.C. van der Mei, J. Sjollema, O. Cerf, P.G. Rouxhet, and H.J. Busscher, *Cell Biophysics* **17**, 93–106 (1990).
31. P.G. Rouxhet and M.J. Genet, in N. Mozes, P.S. Handley, H.J. Busscher, and P.G. Rouxhet eds., *Microbial Cell Surface Analysis: Structural and Physicochemical Methods*, VCH Publishers, New York, 1991, pp. 173–220.
32. Y.F. Dufrêne and P.G. Rouxhet, *Can. J. Microbiol.* **42**, 548–556 (1996).
33. P.G. Rouxhet, N. Mozes, P.B. Dengis, Y.F. Dufrêne, P.A. Gerin, and M.J. Genet, *Colloids Surfaces B: Biointerfaces* **2**, 347–369 (1994).
34. G. Binnig, C.F. Quate, and C. Gerber, *Phys. Rev. Lett.* **56**, 930–933 (1986).
35. M. Radmacher, R.W. Tillmann, M. Fritz, and H.E. Gaub, *Science* **257**, 1900–1905 (1992).
36. M. Stratford, *Adv. Microbiol. Physiology* **33**, 1–71 (1992).
37. D.E. Amory, J-P. Dufour, and P.G. Rouxhet, *J. Inst. Brew.* **94**, 79–84 (1988).
38. E.H. van Hamersveld, M.C.M. van Loosdrecht, R.G.J.M. van der Lans, and K.Ch.A.M. Luyben, *J. Inst. Brew.* **102**, 333–342 (1996).
39. P.B. Dengis, L.R. Nélissen, and P.G. Rouxhet, *Appl. Environ. Microbiol.* **61**, 718–728 (1995).
40. N. Mozes, F. Marchal, M.P. Hermesse, J.L. Van Haecht, L. Reuliaux, A.J. Leonard, and P.G. Rouxhet, *Biotechnol. Bioeng.* **30**, 439–450 (1987).
41. P.B. Dengis and P.G. Rouxhet, *J. Inst. Brew.* **103**, 257–261 (1997).
42. M. Stratford and M.H.J. Keenan, *Yeast* **4**, 107–115 (1988).
43. E.H. van Hamersveld, M.C.M. van Loosdrecht, and K.Ch.A.M. Luyben, *Colloids Surfaces B: Biointerfaces* **2**, 165–171 (1994).
44. E.H. van Hamersveld, R.G.J.M. van der Lans, P.J.C. Caulet, and K.Ch.A.M. Luyben, *Biotechnol. Bioeng.* **57**, 330–341 (1998).
45. P.B. Dengis and P.G. Rouxhet, *Yeast* **13**, 931–943 (1997).
46. M.H. Straver, P.C. v.d. Aar, G. Smit, and J.W. Kijne, *Yeast* **9**, 527–532 (1993).
47. M.H. Straver, G. Smit, and J.W. Kijne, *Colloids Surfaces B: Biointerfaces* **2**, 173–180 (1994).
48. G.G. Stewart, I.F. Garrison, T.E. Goring, M. Meleg, P. Pipasts, and I. Russell, *Kemia-Kemi* **10**, 465–479 (1976).
49. N. Mozes, L.L. Schinckus, C. Ghommidh, J.M. Navarro, and P.G. Rouxhet, *Colloids Surfaces B: Biointerfaces* **3**, 63–74 (1994).
50. B.L.A. Miki, N.H. Poon, A.P. James, and V.L. Seligy, *J. Bacteriol.* **150**, 878–889 (1982).
51. M. Stratford, *Colloids Surfaces B: Biointerfaces* **2**, 151–158 (1994).
52. M. Stratford and S. Assinder, *Yeast* **7**, 559–574 (1991).
53. C.L. Masy, A. Henquinet, and M.M. Mestdagh, *Can. J. Microbiol.* **38**, 1298–1306 (1992).
54. A.W.R.H. Teunissen and H.Y. Steensma, *Yeast* **11**, 1001–1013 (1995).
55. A.W.R.H. Teunissen, J.A. van den Berg, and H.Y. Steensma, *Yeast* **11**, 435–446 (1995).
56. O. Kobayashi, N. Hayashi, and H. Sone, *European Brewery Convention, Proc. 25th Congress, Brussels*, Oxford University Press, London, 1995, pp. 361–368.
57. M.H. Straver, V.M. Traas, G. Smit, and J.W. Kijne, *Yeast* **10**, 1183–1193 (1994).
58. M.H. Straver, G. Smit, and J.W. Kijne, *Appl. Environ. Microbiol.* **60**, 2754–2758 (1994).
59. M. Stratford and A.T. Carter, *Yeast* **9**, 371–378 (1993).
60. Ch.L. Masy, M. Kockerols, and M.M. Mestdagh, *Can. J. Microbiol.* **37**, 295–303 (1991).
61. C.W. Kim, S.-K. Kim, C. Rha, and E. Robinson, in Y.A. Attia ed., *Flocculation in Biotechnology and Separation Systems*, Elsevier, Amsterdam, 1987, pp. 429–439.
62. J. Hughes, D.K. Ramsden, and K.C. Symes, *Biotechnol. Techniques* **4**, 55–60 (1990).
63. I. Agerkvist, L. Eriksson, and S.-O. Enfors, *Enzyme Microb. Technol.* **12**, 584–590 (1990).
64. D. Bilanovic, G. Shelef, and A. Sukenik, *Biomass* **17**, 65–76 (1988).
65. G. Kayem and P. Rouxhet, *J. Chem. Soc. Faraday Trans.* **1. 79**, 561–569 (1983).
66. W. Stumm, in W. Stumm ed., *Chemistry of the Solid Water Interface*, Wiley, New York, 1992, pp. 253–278.
67. J.-Q. Jiang, N. Graham, and C. Harward, *Water Sci. Technol.* **27**, 221–230 (1993).
68. R. Tilton, J. Murphy, and J. Dixon, *Water Res.* **6**, 155–164 (1972).
69. L. Eriksson, and A.-M. Härdin, in Y.A. Attia ed., *Flocculation in Biotechnology and Separation Systems*, Elsevier, Amsterdam, 1987, pp. 441–455.
70. A. Baran, *Colloids Surfaces* **31**, 259–264 (1988).
71. J. Nakamura, S. Miyashiro, and Y. Hirose, *Agric. Biol. Chem.* **40**, 377–383 (1976).
72. R. Kurane, K. Takeda, and T. Suzuki, *Agric. Biol. Chem.* **50**, 2301–2307 (1986).
73. H. Takagi and K. Kadowaki, *Agric. Biol. Chem.* **49**, 3151–3157 (1985).
74. P. Gualtieri, L. Barsanti, and V. Passarelli, *J. Microbiol. Methods* **8**, 327–332 (1988).

75. I. Agerkvist, *Colloids Surfaces* **69**, 173–187 (1992).
76. E. Pelssers, M. Cohen Stuart, and G. Fleer, *Colloids Surfaces* **38**, 15–25 (1989).
77. S. Weir, D.K. Ramsden, J. Hughes, and F. Le Thomas, *Biotechnol. Techniques* **7**, 199–204 (1993).
78. S. Weir, D.K. Ramsden, J. Hughes, and F. Le Thomas, *Biotechnol. Techniques* **8**, 129–132 (1994).
79. I. Persson and B. Lindman, in Y.A. Attia ed., *Flocculation in Biotechnology and Separation Systems*, Elsevier, Amsterdam, 1987, pp. 457–466.
80. UK Pat. Appl, GB 2 082 163 A (August 7, 1981), M. Hashimoto, Y. Ishii and Y. Ohi (to Kuzita Water Industries, Ltd., Osaka, Japan).
81. V. Sitkey, E. Kuzmova, R. Horenitzky, P. Michalik, and M. Minarik, *Acta Biotechnol.* **12**, 299–302 (1992).
82. J. Hughes, D. Ramsden, K. Symes, and P. Williams, *Biotechnol. Techniques* **4**, 233–236 (1990).
83. T. Ferenci and K.-S. Lee, *Biotechnol. Bioeng.* **38**, 314–318 (1991).
84. C. Abate, D. Callieri, E. Rodriguez, and O. Garro, *Appl. Microbiol. Biotechnol.* **45**, 580–583 (1996).
85. K.H. Kwok and I.G. Prince, *Enzyme Microb. Technol.* **11**, 597–603 (1989).
86. A. Fontana, C. Ghommidh, J.P. Guiraud, and J.M. Navarro, *Biotechnol. Lett.* **14**, 505–510 (1992).
87. I.G. Prince and J.P. Barford, *Biotechnol. Lett.* **4**, 525–530 (1982).
88. C. Ghommidh and J.D. Bu'Lock, *Biotechnol. Techniques* **2**, 249–252 (1988).
89. J.A. Sodell and R.J. Sevior, *J. Appl. Bacteriol.* **69**, 145–176 (1990).
90. J.H. Bruus, P.H. Nielsen, and K. Keiding, *Water Res.* **26**, 1597–1604 (1992).
91. D. Li and J.J. Ganczarczyk, *Biotechnol. Bioeng.* **35**, 57–65 (1990).
92. F. Jorand, F. Zartarian, F. Thomas, J.C. Block, J.Y. Bottero, G. Villemin, V. Urbain, and J. Manem, *Water Res.* **29**, 1639–1647 (1995).
93. B.K. Mobarry, M. Wagner, V. Urbain, B.E. Rittmann, and D.A. Stahl, *Appl. Environ. Microbiol.* **62**, 2156–2162 (1996).
94. L. Eriksson, I. Steen, and M. Tendaj, *Water Sci. Technol.* **25**, 251–265 (1992).
95. A. Zitta and M. Hermansson, *Appl. Environ. Microbiol.* **63**, 1168–1170 (1997).
96. V. Urbain, J.C. Block, and J. Manem, *Water Res.* **27**, 829–838 (1993).
97. F. Jorand, P. Guicherd, V. Urbain, J. Manem, and J.C. Block, *Water Sci. Technol.* **30**, 211–218 (1994).
98. C.F. Forster, N.J.B. Knight, and D.A.J. Wase, *Adv Biotechnol. Processes* **4**, 211–240 (1985).
99. A.E. Steiner, D.A. McLaren, and C.F. Forster, *Water Res.* **10**, 25–30 (1976).
100. Y. Magara, S. Nambu, and K. Utosawa, *Water Res.* **10**, 71–77 (1976).
101. J.B. Barber and J.N. Veenstra, *J. Water Pollut. Control Fed.* **58**, 149–156 (1986).
102. A. Zitta and M. Hermansson, *Appl. Environ. Microbiol.* **60**, 3041–3048 (1994).
103. K. Kakii, S. Kitamura, T. Shirakashi, and M. Kuriyama, *J. Ferment. Technol.* **63**, 263–270 (1985).
104. J.A.S. Goodwin and C.F. Forster, *Water Res.* **19**, 527–533 (1985).
105. L. Eriksson and B. Alm, *Water Sci. Technol.* **24**, 21–28 (1991).
106. C.F. Forster, *Water Res.* **19**, 1259–1264 (1985).
107. M. Sezgin, D. Jenkin, and D.S. Parker, *J. Water Pollut. Control Fed.* **50**, 362–381 (1978).
108. L. Eriksson and A.-M. Härdin, *Water Sci. Technol.* **16**, 55–68 (1984).
109. N. Kosaric and R. Blaszczyk, *Adv. Biochem. Eng. Biotechnol.* **42**, 27–62 (1990).
110. W. De Zeeuw and G. Lettinga, *Antonie van Leeuwenhoek* **46**, 110–112 (1980).
111. E. Mahoney, L. Varangu, W. Cairns, N. Kosaric, and R. Murray, *Water Sci. Technol.* **19**, 249–260 (1987).
112. N. Kosaric, E. Mahoney, L. Varangu, and W. Cairns, *Water Pollut. Res. J. Can.* **22**, 289–297 (1987).
113. D. Daffonchio, J. Thaveesri, and W. Verstraete, *Appl. Environ. Microbiol.* **61**, 3676–3680 (1995).
114. J. Thaveesri, D. Daffonchio, B. Liessens, P. Vandermeren, and W. Verstraete, *Appl. Environ. Microbiol.* **61**, 3681–3686 (1995).

CELL SEPARATION, SEDIMENTATION

BHAV P. SHARMA
CV Therapeutics
Palo Alto, California

KEY WORDS

Cell density
Cell separation
Cell size
Gravity
pH
Reynolds number
Sedimentation
Sedimentation velocity
Stokes' law
Viscosity

OUTLINE

INTRODUCTION

By definition, the bioprocessing industry uses biological cells as little factories to make a variety of products. Once they are presumed to have finished their product synthesis step, the cells need to be separated from the liquid medium containing them. During the course of research involving biological cells or during the isolation and purification of products derived from cells, cell separation is like the midgame of a chess match. Correct moves at this stage provide an excellent chance to obtain a product of the desired characteristics.

At least five general approaches are available for cell separation: centrifugation, cyclone separation, sedimentation, filtration, and microfiltration. In addition, cell flocculation can often be used to enhance the efficiency of these separation techniques. Sedimentation as a tool for cell separation is the focus of this article. It is also noted how centrifugation and flocculation, too, draw on sedimentation theory.

SEDIMENTATION AS A TOOL IN PRACTICE

The starting point in selecting a cell separation strategy is to locate the product of interest in a given suspension of cells. Is it the cells themselves? Is it all the cells or a specific cell type? Is the interest actually in a product inside the cells (i.e., intracellular) or in a product outside the cells (i.e., extracellular)? If the product is intracellular, cell removal may need to be achieved without significant cell damage. If the product is extracellular, the liquid phase must be separated while the solid content of the liquid phase is minimized, and the cells along with the other solids are merely the solid waste to be removed and disposed of. If the cells *are* the product, they must be separated without damage and their functionality kept intact. Sedimentation may be a useful tool for any of these situations.

Sedimentation is relevant in a variety of biotechnology/bioprocessing situations:

1. Production of therapeutic or diagnostic proteins via mammalian cells
2. Achievement of high cell concentrations via cell recycle in continuous fermentors
3. Biological waste or water treatment
4. Production of beer
5. Study of different cell types involved in immunological response
6. Harvesting of cells in microcarrier or suspension cell culture

Table 1 gives specific examples of research in which sedimentation has been used as a cell separation tool (1–16). Sedimentation is also utilized extensively in other fields such as water or solids removal in mineral recovery and processing (17).

SEDIMENTATION EQUIPMENT

In processing applications, the sedimentation step is usually carried out in a typical processing tank. Such tanks are designed to have the capabilities of a feed line, an output or an overflow line, and a line to allow the removal of the sedimented material. However, equipment designs other than a processing tank have been shown to be very effective in special situations. One such design uses an inclined, parallel-plate sedimentation device (5,18,19) shown in Figure 1, which can be used for cell recycle in continuous fermentations or cell culture. The inclined sedimentor design not only allows the fermentation or the cell culture to achieve high cell densities, it also offers an efficient means of selective cell retention.

A CelSep apparatus meant primarily for mammalian cell separations at the 1-L scale was commercially available (Wescor, Inc., Logan, Utah). Although the unit is no longer on the market, the autoclavable unit worked well and allowed gravity separations under a continuous density gradient (20).

ASPECTS OF SEDIMENTATION AS A CELL SEPARATION TOOL

1. Sedimentation as a separation tool is a mild technique. This is important in cell biology, where the functional capabilities of the isolated subpopulation of cells must not be altered by the separation step.
2. Sedimentation lends itself to aseptic handling if needed. Although other techniques may allow cell separation without contamination, sedimentation devices for cell separation are typically easier to assemble, clean, and sterilize. Hence aseptic handling is more easily performed with sedimentation than with other options.
3. From a practical viewpoint, "unaided" sedimentation can be too slow a process if the sedimentation rates are low. In industrial-scale cell separations, where microbial growth or enzymatic degradation can be a problem, long processing times are generally not acceptable. Such situations call for centrifugation, which "enhances" the sedimentation rates via the centrifugal force, because the problems of microbial growth or enzymatic degradation are less dominant for short processing times.
4. In its simplest form, sedimentation may be the lowest-cost technique for cell separation.
5. Sedimentation is a simple and a highly reliable separation technique; unlike other options, such as centrifugation; it does not require extensive training of the user.

Overall, sedimentation principles have broad applicability in cell separations and extend beyond simple gravity sedimentation. Sedimentation principles also apply to centrifugation and ultracentrifugation.

SEDIMENTATION THEORY AND APPLICATION

To consider the practical issues in cell separation problems, it is useful to understand the principles behind grav-

Table 1. Some Examples of Sedimentation as a Cell Separation Tool in Research

Cell and cell culture type	Sedimentation application	Ref.
Mammalian, suspension	Sedimentation chamber as a separation device in a perfusion system for hybridoma cell culture for antibody production	1
Mammalian, microcarrier	Sedimentation separator pipe as a separation device for microcarrier cell culture	2
Mammalian, suspension	Gravitational sedimentation used to separate cells aggregated with the help of high-frequency, low-energy ultrasonic resonance fields	3
Bacteria	Sedimentation rates of cells of different shapes and sizes used to develop a method for enumeration and quantitation of bacteria	4
Yeast, *Saccharomyces cerevisiae*	Gravitation sedimentation used to recycle yeast cells for alcohol fermentation	5
Yeast, *S. cerevisiae* and yeast–mouse/mouse hybridoma cells	Sedimentation chamber used to separate cells from a chemostat culture	6
Starch endosperm of barley, *Hordeum vulgare L.*	Endosperm protoplast isolation method utilized gravity sedimentation to get good preparations	7
Spleen B-lymphocytes	Hybridoma-forming immune spleen B-lymphocyte cells separated based on their size and density using gravity sedimentation	8
Yeast, *Candida rugosa*	Sedimentation velocity measured to compare cells grown at different temperatures	9
Bacteria (*E. coli*) and yeast (*S. cerevisiae* and baker's)	Differences in flocculation and sedimentation properties utilized to achieve selective separation of cells carrying a certain plasmic	10,11
Sperm cells	A migration–sedimentation technique used to study whether it is a useful method to separate sperm cells leading to improved fertilization	12,13
Mammalian cells, erythrocytes	Utility of the erythrocyte sedimentation rate studied as a possible indicator of clinical disease or health	14,15
Sulfate-reducing bacteria, *Desulfovibrio desulfuricans*	Gravity sedimentation utilized for cell recycle in a system for microbial reduction of sulfur dioxide in an anaerobic digester	16

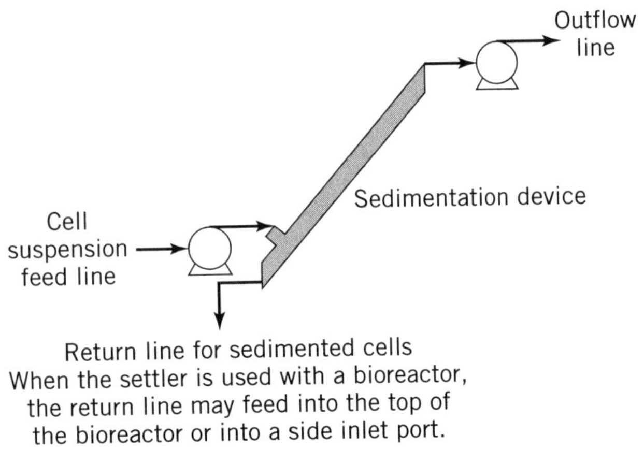

Figure 1. Conceptualized representation of an inclined, parallel-plate, cell sedimentation device used for cell recycle in a continuous fermentation or a perfusion bioreactor. When the settler is used with a bioreactor, the return line may feed into the top of the bioreactor or into a side inlet port (5,18,19).

itational sedimentation. Several factors affect the sedimentation of cells in a liquid medium:

- The *difference* between the density of a cell and that of the surrounding liquid medium
- The density of the liquid medium
- Cell size
- Cell shape
- Cell concentration
- The viscosity of the liquid medium

As a practical rule of thumb, the particle size needs to be greater than about 1 μm for gravitation sedimentation to be relevant as a separation tool. Other approaches such as centrifugation, ultracentrifugation, and membrane separation are usually needed for smaller particles.

The variables just listed are linked together by what is known as the modified *Stokes law*, named after the Ireland-born British scientist George Stokes (1819–1903), who is credited with having developed much of the general theory on the movement of a particle in a flowing fluid. Stokes was a gifted mathematician (21) who also explained the existence of fluorescence and the variation of gravity at the surface of the Earth—all during the brief span of 1845–1854. Although Stokes' law is not directly used in cell separations, it helps us understand how the various parameters affect the settling operation. For a simplified case of gravitational settling when the velocity relative to the surrounding medium is relatively modest (often expressed in chemical engineering literature via a dimensionless parameter called the Reynolds number being set at < 0.2), the Stokes' law modifications (22) suggest that

$$V_s = \frac{g(\rho_p - \rho)D_p^2}{18\mu}$$

where V_g is sedimentation rate (ft/s or m/s); g is gravitational acceleration constant (ft/s^2 or m/s^2), ρ_p is density of the particle (or cell) (s/ft^3 or kg/m^3), ρ is density of the liquid (s/ft^3 or kg/m^3), D_p is diameter of the particle (or cell) (ft or m), and μ is viscosity of the liquid [lb/(s ft) or Pa s&rsqb. Many of the factors in this equation may be beyond one's control in a given practical situation; it is not used very much for actual numerical calculation in cell separations. However, we can learn how cell sedimentation is affected under various conditions by examining this equation.

Effect of Density

In many large-scale industrial cell separation situations, the cell sedimentation velocity is too slow for sedimentation to be a practical tool. The sedimentation velocity is slow because biological cells and cell aggregates (if cell aggregates are present) tend to have a low density. The Stokes' law equation teaches us that the greater the *difference* between the densities of the cells and the liquid medium, the greater the cell sedimentation velocity. This property can be usefully exploited by a controlled increase of the liquid medium density. For illustration purposes, as shown in Figure 2, let us assume we need to separate a cell mixture containing two different cell densities in a given liquid medium. If the liquid density is increased via the addition of agents like Ficoll, fetal calf serum, or albumin, the *relative* ratio of sedimentation rates of the two cell types is increased, leading to a better separation. A refinement of this idea is to use a density gradient in the liquid medium through which the sedimentation is taking place. A technique based on this idea has been called isokinetic sedimentation (20,23,24).

Effect of Cell Size

Table 2 shows the distance traversed in 1 h by hypothetical spherical particles of specific gravity 2.0 and different cell sizes, and settling in a fluid with a specific gravity of 1.0 (25). In practice, the specific gravity of cells tends to be well below 1.2; the example merely illustrates the effect of cell size. Table 3 gives an idea of the typical sizes for the various cell types encountered in practice.

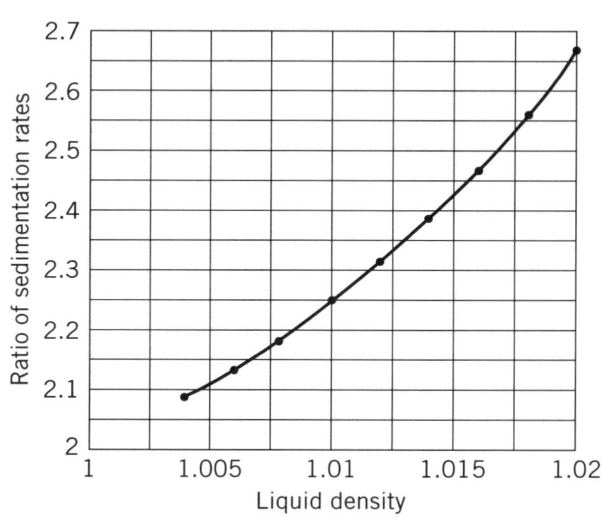

Figure 2. Effect of liquid density on relative sedimentation rate. This data assumes a hypothetical, idealized case where a mixture of two cell types, A and B (densities 1.05 and 1.10, respectively), is sedimenting in liquids of slightly differing densities.

Table 2. Effect of Spherical Particle Size on the Sedimentation Distance after 1 h in Water at 293 K

Particle radius (nm)	Distance settled
0.1	0.08 nm
1	8 nm
10	800 nm or 0.8 μm
100	80 μm
1,000	8 mm

Source: Ref. 25.

Table 3. Typical Sizes of Various Cell Types

Cell type	Size range (nm)
Bacterial virus	0.08
Bacteria	0.1–2.5
Yeast cells	2–10
Human red blood cells (erythrocytes)	6–8
Human white blood cells (leukocytes)	8–12
Filamentous fungi	10–40
Human liver cells	30
Plant cells	10–100
Paramecium, didinium (protozoa)	100–200

The sedimentation rate is proportional to the square of the diameter of the cell. Thus, any approach that leads to cell aggregation will help increase sedimentation substantially—albeit not as much as the "diameter square" relationship would seem to imply. The relative increase is less than that implied by the "square" relationship if the cell aggregate is porous and its density turns out to be less than that of the individual cells. In one study, the settling of larger-diameter activated sludge flocs was found to be proportional only to (floc diameter)$^{0.55}$ (26).

Bacterial cells are often difficult to separate owing to their small size and low density, and so flocculants such as sodium polyacrylate (27), polydimethyldiallyl ammonium chloride (28), and carboxymethyl cellulose (28) are added in some situations to achieve just this benefit. Ultrasound has been used to enhance the sedimentation of monoclonal antibody producing hybridoma cells (29) because the acoustic treatment aggregates the cells. Addition of ferromagnetic particles to a bacterial cell suspension (gluconic acid producing *Acetobacter methanolicus*) followed by an exposure to a magnetic field has been shown to enhance sedimentation (30) as a result of aggregation of cells with the magnetic particles. Two commercially products, Dynabeads® and DETACHaBEAD®, offer another way to achieve the immunomagnetic separation of cells (31). In this approach, sedimentation is enhanced by forming rosetted cells on the antibody-coated Dynabeads. The rosetted cells are washed and concentrated using a magnetic device. The positively isolated cells may then be detached for further study.

Because of the dramatic sensitivity of the sedimentation rate to the cell size, clinical researchers have attempted to determine whether the sedimentation rate of erythrocytes can be a useful indicator of patient health (19,15,32,33), results have been mixed, however.

Size differential between cells can thus also be a useful tool when one is trying to separate a specific cell type from a mixture of cells. If one were to take samples at a fixed point below the liquid surface from a container in which sedimentation of otherwise identical but differing cell sizes is under way, the initial samples would contain the largest cells. With time, smaller and smaller cells would be in the sample. This principle is useful in the separation of lymphocytes from erythrocytes where sedimentation through a dense medium in a centrifuge is one approach (32,33).

Effect of Viscosity

The sedimentation rate is inversely proportional to the viscosity of the liquid medium. This property can also lead to some practical options. Because the viscosity change with dilution is often nonlinear, one may dilute the liquid medium by 25–50% to achieve a greater decrease in viscosity, and hence, better sedimentation rate. Conversely, one may add something that increases the viscosity of the liquid medium to slow down the sedimentation of cells. An example of this is the use of 4% dextran (of average molecular weight 500,000 Da) as a viscosity-increasing agent in the study of antibody-binding rates by FACScans in immunology (34), where suspended cells must be prevented from settling.

Effect of Gravity

Cell separation via centrifugation (see CELL SEPARATION, CENTRIFUGATION) seeks to increase the sedimentation rates hundreds or thousands of times over simple sedimentation, by operating at a force much greater than the normal gravity. Because normal gravitational sedimentation is limited by the gravitational constant g, the relative efficiency of centrifugal separations has been examined by measuring the sedimentation velocity of disrupted yeast cells with and without a flocculent (35).

Effect of Chemical and Environmental Factors

It was pointed out earlier that polymers and other flocculating agents can aid sedimentation by increasing the effective cell size. In addition, the following factors can affect the behavior of particles in solution, hence can influence sedimentation rates:

- pH and temperature of the solution
- Presence of salts (e.g., of calcium, magnesium, manganese, sodium, potassium)
- Presence of sugars

Although in most cases the interaction of these parameters is complex, the question is one of how the chemical environment affects the surface charge of the cells and their tendency to form cell aggregates. The idea is that the surface interactions will control the degree of cell aggregation, and hence, the sedimentation rates. Most microbial cells usually have a negative surface charge (36,37). The system pH or temperature can affect this charge (37,38). Divalent cations, especially calcium, and certain pH ranges can promote cell aggregation, leading to better sedimentation rates (39–41). Monovalent cations and sugars can reduce the tendency of cells to aggregate, thus reducing the likely sedimentation rates (41,42). The overall effect depends on the specific cell type and the physiologic state of the cells involved.

BIBLIOGRAPHY

1. K.M.K. Lassen and C. Emborg, *FEMS Microbiol. Rev.* **14**(1), 89–91 (1994).
2. Japanese Pat 06209761 (Aug. 2, 1994), (to Toyobo Company Ltd).
3. F. Trampler, S.A. Sonderhoff, P.W.S. Pui, D.G. Kilburn, and J.M. Piret, *Bio / Technology* **12**(3), 281–284 (1994).
4. R.V. Sharma, R.T. Edwards, and R. Beckett, *Appl. Environ. Microbiol.* **59**(6), 1864–1875 (1993).
5. A.B.R.A. Maia and D.L. Nelson, *Biotechnol. Bioeng.* **41**(3), 361–369 (1993).
6. K.M.K. Lassen, P.R.E. Carcia de Cladeano, and C. Emborg, *Biotechnol. Tech.* **6**(2), 121–126 (1992).
7. I. Diaz and P. Carbonero, *Plant Cell Reprod.* **10**(12), 595–598 (1992).
8. K. Kuus-Reichel, A. Beebe, and C. Knott, *Hybridoma* **10**(4), 529–538 (1991).
9. K.Y. Lee and G. Baerwald, *Biotechnol. Lett.* **13**(8), 595–598 (1991).
10. K.L. Henry, R.H. Davis, and A.L. Taylor, *Dev. Ind. Microbiol.* **31**, 53–58 (1990).
11. K.L. Henry, C.S. Parnham, R.H. Davis, and A.L. Taylor, *Appl. Biochem. Biotechnol.* **24–25**, 651–662 (1990).
12. H. Yavetz, R. Hauser, Z.T. Homonnai, G.F. Paz, J.B. Lessing, A. Amit, and I. Yogev, *Andrologia (Germany)* **28**(1), 3–6 (1996).
13. J. Risopatron, R. Sanchez, N. Sepulveda, P. Pena, E. Villagran, and W. Miska, *Theriogenology* **46**(1), 65–73 (1996).
14. O.H. Iverson, M. Roger, H.E. Solberg, and P. Wetteland, *J. Intern. Med.* **240**(3), 133–141 (1996).
15. P. Wetteland, M. Roger, H.E. Solberg, and O.H. Iverson, *J. Intern. Med.* **240**(3), 125–131 (1996).
16. P.T. Selvaraj and K.L. Sublette, *Biotechnol. Prog.* **11**(2), 153–158 (1995).
17. D.R. Nagaraj, "Minerals Recovery and Processing, in J.I. Kroschwitz ed., *Kirk-Othmer Encyclopedia of Chemical Technology*, Wiley, New York, 1992, pp. 833–834.
18. R.H. Davis, in M.A. Hjortso and J.W. Roos eds., *Cell Adhesion*, Marcel Dekker, New York, 1995, pp. 135–185.
19. J.A. Searles, P. Todd, and D.S. Kompala, *Biotechnol. Prog.* **10**, 198–206 (1994).
20. J.R. Wells, in T.G. Pretlow II and T.P. Pretlow eds., *Cell Separation*, Academic, New York, 1982, pp. 169–189.
21. R. Porter ed., *The Biographical Dictionary of Scientists*, vol. 16, Oxford University Press, New York, 1994, pp. 647–648.
22. J.M. Coulson, J.F. Richardson, J.R. Backhurst, and J.H. Harker, *Chemical Engineering*, vol. 2, Pergamon Press, Oxford, 1978, pp. 174–175.
23. T.G. Pretlow II and T.P. Pretlow, in T.G. Pretlow II and T.P. Pretlow eds., *Cell Separation*, Academic, New York, 1982, pp. 41–60.

24. R.G. Miller, in R.H. Pain and B.J. Smith eds., *New Techniques in Biophysics and Cell Biology*, Wiley, New York, 1973, pp. 87–112.
25. A. Bleier, in J.I. Kroschwitz ed., *Kirk-Othmer Encyclopedia of Chemical Technology*, vol. 6, Wiley, New York, 1992, pp. 813–814.
26. D.-H. Li and J.J. Ganczarczyk, *Water Res.* **21**, 257–262 (1987).
27. Japanese Pat. 05003780 (Jan. 14, 1993), (to Mitsui Toatsu Chemicals, Inc.).
28. German Pat. DD298263 (Feb. 13, 1992), (to Forsch Biotechnology).
29. P.W.S. Pui, F. Trampler, S.A. Sonderhoff, M. Groeschl, D.G. Kilburn, and J.M. Piret, *Biotechnol. Prog.* **11**(2), 146–52 (1995).
30. German Pat. DD278598 (May 9, 1990), (to Karl Marx University).
31. DYNAL Dyna and DETACHaBEAD, Dynal A.S., PO Box 158, Skoyen N-0212, Oslo, Norway.
32. W. Emlen, J. Niebur, G. Flanders, and J. Rutledge, *J. Rheumatol.* **23**(6), 974–978 (1996).
32. S.L. Berger, in W.B. Jacoby and I.H. Pastan eds., *Methods in Enzymology*, vol. 58, Academic, New York, 1979, pp. 490–491.
33. M.N. Penev, P. Doukovapeneva, and K. Klinov, *Scand. J. Clin. Lab. Invest.* **56**(3), 285–288 (1996).
34. E. Martz, *J. Immunol.* **115**(1), 261–267 (1975).
35. R.L.P. Adams, in T.S. Work and R.H. Burdon eds., *Cell Culture for Biochemists*, Elsevier/North Holland, New York, 1980, pp. 70–73.
35. A.I. Clarkson, P. Lefevre, and N.J. Titchener-Hooker, *Biotechnol. Prog.* **9**(5), 462–467 (1993).
36. K. Thethi, P. Jurasz, A.J. MacDonald, A.D. Befus, S.F. Man, and M. Duszyk, *J. Biochem. Biophys. Methods* **34**(2), 137–145 (1997).
37. Y.E. Collins and G. Stotzky, *Can. J. Microbiol.* **42**(7), 621–627 (1996).
38. G.V. Sherbet and S. Jackson, *Anticancer Res.* (**1**), 129–134 (1986).
39. A.M. Porter and R.J. MacAulay, *J. Inst. Brew.* **71**, 175–179 (1965).
40. G.G. Stewart, I. Russell, and I.F. Garrison, *J. Inst. Brew.* **81**, 248–257 (1975).
41. M.A. Amri, R. Bonaly, B. Duteurtre, and M. Moll, *Eur. J. Appl. Microbiol. Biotechnol.* **7**, 235–240 (1979).
42. P.M. Jayatissa and A.H. Rose, *J. Gen. Microbiol.* **96**, 165–174 (1976).

See also CELL DISRUPTION AND LYSIS; CELL SEPARATION, CENTRIFUGATION; FILTER AIDS.

CELLULOSE CONVERSION. See ORGANIC COMPOUNDS, CELLULOSE CONVERSION.

CENTRIFUGES, ANIMAL CELLS

DANIEL D. HANLE
Kendro Laboratory Products
Newtown, Connecticut

KEY WORDS

Centrifuge
Continuous flow
Decanter
Disc stack
Fed batch
Inclined separator
Perfusion
Spin filter
Tangential flow filtration
Totally closed system

OUTLINE

INTRODUCTION

With the emergence and growth of animal cell bioprocesses during the past 15 to 20 years, the need has arisen for unit operations that gently and effectively separate animal cells at the process scale. Centrifuge-based separation systems available today are especially well suited to this task. By its nature, centrifuge separation is inherently robust and delivers consistent separation performance. Centrifuge systems now incorporate a number of design changes, including some that are quite novel, aimed at tailoring their performance to the special requirements of animal cell bioprocesses. As a result, today we have a powerful array of options available for centrifuge-based unit operations for use in both the animal cell processes under development and the existing processes that are being improved for added efficiency and productivity.

ANIMAL CELL BIOPROCESSES

As biotechnology's commercial success has grown, it has created the need for highly regulated, effective, and, most recently, efficient production processes. Such biotechnology production processes (bioprocesses) have historically drawn heavily from other process industries, such as the chemical process industry for concepts, practices, and equipment, including centrifuges. Bioprocess has therefore become, in some cases, a hybrid field where established process techniques are intermingled with laboratory methods that have been multiplied manyfold for the required process scale. Nowhere does this appear to be more the case than for the bioprocess of animal cells.

Although bacteria or yeast has been the more common choice for biological production to date, each can have such drawbacks as producing improperly formulated product or product that is unrecoverable as insoluble inclusion bodies (1). In such cases, animal cells have often been called on to overcome these limitations. Unfortunately, animal cells do not share some of the more desirable attributes of these other organisms. By comparison, animal cells are slower growing; exhibit lower protein expression levels; are susceptible to infection by a large number of adventitious agents such as viruses, prions, and so forth; and, because of their lack of a cell wall, are much less tolerant of many traditional process operations. Numerous advances in animal cell biotechnology and fermentation have enhanced the productivity of animal cells. But the bioprocess equipment has often lagged. Animal cell bioprocess has therefore become a "force fit" of techniques and equipment originally designed and developed for other purposes.

Nonetheless, the fact that animal cells are often the only practical way to produce what is required has generated a strong push by many in the industry to develop processes and equipment tailored to animal cell bioprocess. Centrifugation is no exception.

ANIMAL CELL CENTRIFUGATION

Process Applications

Centrifugation's role in animal cell bioprocess is primarily as a unit operation immediately adjacent to the bioreactor, as illustrated in Figure 1. In this role, the centrifuge is positioned to either (*1*) clarify the cell mass exiting the bioreactor, as in the case of a batch operation, or (*2*) retain the cells for further processing, as in the case of a continuous operation. These two operating modes, however, differ somewhat depending on whether the product is intracellular or extracellular. Intracellular processes, where the focus is on harvesting the product-laden cells, require that the centrifuge concentrate the intact cells from the fermentation broth. Care is taken to avoid damaging the cells during this concentration step to maximize product recovery. Any ruptured or lysed cells also release agents into the process streams that can be harmful to the desired product, especially when that product is a protein. Occasionally, this type of care is also needed when the cells are to be stored or are otherwise required to remain viable for further processing. In many of these cases, aseptic operation is essential.

Extracellular processes, on the other hand, require a different set of operating principles where the key focus is on keeping the cells viable. Any cell damage that the centrifuge causes in these processes will have a direct deleterious effect on the process productivity. Not only would the producing cells themselves be lost, but here, too, harmful intracellular agents would be released. The situation is further aggravated by the fact that the extracellular product is relatively vulnerable to any harmful agent because both are present in the bulk fluid. It is clear that centrifugation processes that avoid cellular damage are preferable in these cases, especially when operating in a continuous or perfusion process mode.

A further distinction in animal cell bioprocess that bears mentioning is that the producing cells can be either freely suspended or immobilized. The earlier discussion relates directly to suspension systems. For immobilized systems, where, for example, the cells require that they be tethered to induce growth or production, the need for centrifugation is either at the end of the process, whenever the tethered cells are released for further processing, or for clarification of debris released by the tethered cells, or both. When called for in this manner, the centrifugation considerations are similar to those for freely suspended cells.

Fed Batch versus Perfusion

Perhaps the most prevalent bioprocess mode in use today for freely suspended cells is a simple batch system using carefully constituted media. But in the past decade or so, two more involved processes, fed batch and perfusion, have gained much of the attention. The fed-batch process tends

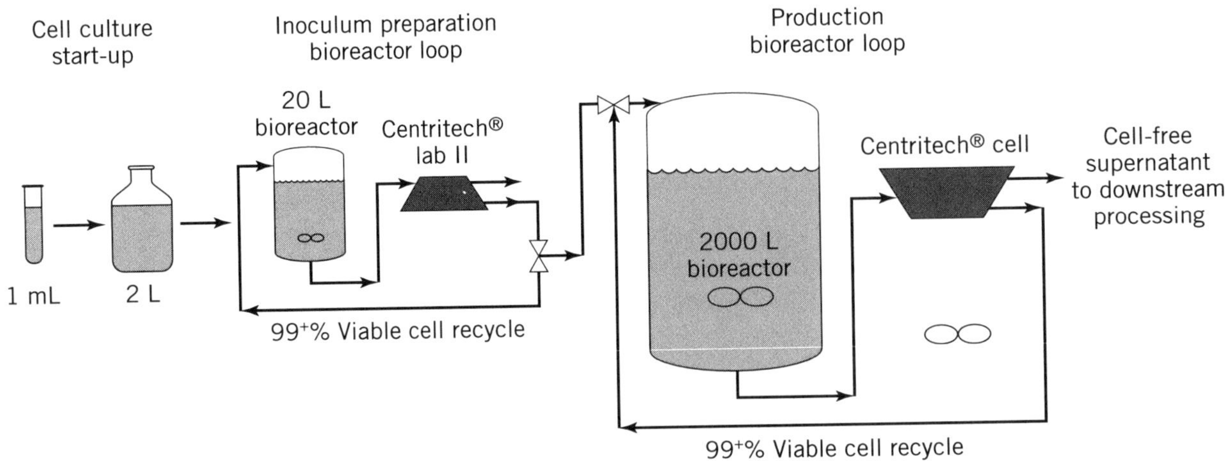

Figure 1. Centrifuge system as integral unit operation for animal cell bioprocess.

to make more efficient use of the media expense and other operating costs, whereas the perfusion process, being continuous, tends to make more efficient use of the producing cells and the physical plant investment.

The fed-batch process is currently the more prevalent of these two processes, partly because it is easier to operate in the laboratory and at the pilot and production scale. Because a batch culture is by its nature limited in its duration, process experience can be built up relatively quickly during the course of several runs. Here the need for processing the cells is confined to the end of the production run. Centrifugation may be used to concentrate the cells for intracellular product or clarify the supernatant for extracellular product. In either case, as mentioned earlier, avoiding cell damage would likely lead to important productivity gains.

Perfusion processes afford the benefits of continuous operation. To do so, however, perfusion requires an effective way to retain the cells throughout the production run with little or no cell viability loss. Centrifugation can deliver the needed performance provided it is continuous in both the liquid stream and the cell concentrate stream, can be operated aseptically, and does not damage the cells. The specific designs of centrifuges applicable to animal cell bioprocess are reviewed in the next section with reference to these important aspects. A continuous-flow centrifuge system installed in a perfusion animal cell process is shown in Figure 2.

The considerations of such processes are directly applicable to anchored cells when those cells are attached to microcarriers. Cells attached to microcarriers are, however, particularly sensitive to fluid shear forces (2–5). For example, cells that are not otherwise damaged by fluid shear are lost if the shear that is present separates them from the microcarriers and they do not reattach (6). In these cases, the bioprocess centrifuge needs to treat the process fluid especially gently.

The various roles of centrifugation in animal cell bioprocess are summarized in Table 1 for the cases discussed. Although generalizations tend to be difficult, these roles as delineated in the table are in many ways representative of the scope of possibilities.

The emergence of insect cell bioprocess has added yet another dimension to the need for gentle handling of the cells. Although insect cells have several advantages over animal cells, they tend to be even less robust than animal cells are during rough handling in the process (7–11). Coupled with the fact that insect cells typically generate intracellular product (specifically so with baculovirus systems), the centrifuge operation must generate especially low shear forces to avoid cell rupture and the resulting productivity loss.

Centrifuge Designs

Design Considerations. Centrifugation is a very robust and versatile separation technique. Because it relies primarily on density differences, centrifugation can readily account for changing process conditions by simple modulation of the intensity or duration of the *g*-field as needed during operation (12). At lower speeds and *g*-forces, the centrifuge can be used to gently remove the larger and more dense cells from the process stream; at higher speeds and *g*-forces, the centrifuge can be used to spin down ever smaller cell debris to increasingly clarify the process stream to the degree required.

The actual design of the centrifuge, however, depends on the various optimization trade-offs specific to the process application. Typical design trade-offs include high sample throughput versus instrument size, high centrifu-

Figure 2. Continuous-flow centrifuge system installed in a production-scale perfusion process.

Table 1. Centrifugation Roles in Animal Cell Bioprocess

	Suspension cultures		Immobilized cultures	
	Fed batch	Perfusion	Anchored	Microcarriers
		Extracellular product		
During the process	N/A	Retain cells with minimal cell damage	Remove/recover cells released into the process stream	Retain the microcarriers with minimal cell loss or damage
At the end of the process	Collect supernatant with minimal cell debris	Recover the final volume of the supernatant	Recover cells with minimal damage for next inoculation	Collect supernatant with minimal cell debris
		Intracellular product		
During the process	N/A	Retain most of the cells with minimal cell damage	Remove/recover cells released into the process stream	Retain most of the microcarriers with minimal cell loss or damage
At the end of the process	Collect cells with minimal cell damage	Recover the final volume of the cells	Recover cells with minimal damage for next inoculation	Collect supernatant with minimal cell debris

Note: N/A, no centrifuge role directly applicable to this case.

gal speeds versus large load volumes, versatility versus dedicated instrument specialization, batch versus continuous operation, and, for continuous operation, solids-retaining versus solids-ejecting design. Laboratory-scale applications tend toward smaller, more versatile, batch or continuous solids-retaining systems, whereas pilot- to production-scale applications tend toward high throughput, specialized, continuous solids-ejecting systems.

When applied specifically to the separation of animal cells, centrifugation typically involves the added considerations of avoiding high fluid shears and maintaining high aseptic integrity. Additional design trade-offs are therefore involved, including (*1*) increased volume throughput versus maintaining the shear forces below critical levels that can result in cell death or damage, and (*2*) providing a totally closed aseptic system versus relying on rotating seals, a source of potential contamination and heat generation as well as generation of high shear stresses.

Design Approaches. To date, the design of animal cell centrifuges has been approached from three distinct directions. The first two directions have been through engineering extensions of other centrifuge equipment. The third is a centrifuge design specifically engineered for animal cell bioprocess.

The first direction has been to take centrifuge designs for larger process scale or bacteria and yeast and adapt them for animal cells. This approach has been more or less successful depending on whether cell damage was to be avoided or continuous operation was desired. Such centrifuges tend to spin the cells down into a paste or cake, which requires some type of cell removal process such as scraping or flushing out of the cells under high shear forces.

The second direction has been to take centrifuge designs for blood component manufacture and adapt them for animal cells. Blood component manufacturers have been making commercial use of continuous-flow centrifugation for many years now. Numerous such blood separation system designs have focused on cell concentration, supernatant clarity, throughput, and other issues that are also important to animal cell bioprocess. But there are several key differences between blood component manufacture and animal cell bioprocess. Blood cells, for example, flow much more readily than typical suspended animal cells. Animal cell centrifuges therefore require some type of low shear method to nudge the animal cells out of the centrifuge in a continuous mode. Also, formed elements of blood are more tolerant of shear stresses than are most animal cells in bioprocess. Flow channel designs in the centrifuge that work for blood therefore tend to be not at all suitable for animal cells. And the usual time frame for blood separation (on the order of a few hours) is much less than that necessary for animal cell bioprocess (on the order of days or weeks).

The third direction was taken by Alfa-Laval Separation AB (Tumba, Sweden) during the 1980s when a continuous-flow centrifuge design was developed specifically for animal cell bioprocess. Combining some of the features of larger process centrifuges and those of the smaller centrifuges, this unique design focused on maintaining a continuous flow of both the clarified supernatant and the cell concentrate. Key engineering issues that had to be overcome were (*1*) gently coaxing the concentrated animal cells to smoothly exit the system without significant cell damage; (*2*) sustaining continuous operation over days and weeks; and (*3*) creating a totally closed system for inherently safe aseptic operation.

These three approaches have had the common goal to overcome the limitations of laboratory-scale and even small production–scale centrifugation. At these smaller scales, research centrifuges have often been pressed into service. Several companies provide floor-model centrifuges that can process, for example, as much as 6 L at *g*-forces more than adequate for cellular separations. Among these companies are Beckman Instruments, Inc. (Fullerton, Ca.), Kendro Laboratory Products L.P. (Norwalk, Conn.) (Sorvall® and Heraeus® brands), Hitachi Koki Co., Ltd. (Hitachi-naka City, Japan), Jouan S.A. (St. Herelain, France), and Hettich GmbH (Tüttlingen, Germany). Kendro, through its Sorvall® brand, has recently introduced a series of major software packages to their centrifuges to provide automatic process documentation for easier and more reliable process control and regulatory compliance. Other suppliers are now providing software as well. These centrifuges are designed for batch operation only. Some of these companies offer continuous-flow models, but they tend to have small throughput capacity, do not eject the cellular material, and are typically difficult to set up and clean.

Because one of the keys to successful animal cell centrifugation is the gentle handling of the cells, there have been over the years several research groups exploring the various aspects of flow-induced shear stress on different animal cell lines. Although it has been shown that cell damage in bioreactors can be attributed to bubble breakup and other gas–liquid interface phenomena (13–16), other findings indicate that cell damage can also occur at shear stress levels at lower levels of energy dissipation (17–19). Shear stress susceptibility, moreover, varies significantly among cell lines (8,20), thereby adding shear sensitivity to the list of key criteria for selecting a particular cell line for process scale-up.

Current Products. Several suppliers provide centrifuges for animal cell bioprocess that are based on proven designs for other applications. These designs typically provide high throughput and *g*-forces, clean-in-place (CIP) and steam-in-place (SIP) capability, and some method for removing the cellular material that accumulates in the centrifuge. Each design is discussed briefly in the following and then summarized in Table 2. More complete descriptions, including specific schematic depictions of how these systems operate, are available through the respective suppliers.

At the small, batch end of the spectrum, Alfa-Laval's AS-26 SP Super® Centrifuge is capable of high throughput rates in the range of 300 to 500 L/h. The AS-26, however, is strictly a batch centrifuge because it retains the solids that must later be removed by manual intervention. Also, the animal cells are removed as cell debris, so this centrifuge would be unsuitable for processes where cell viability after separation is desired.

At the pilot scale, Alfa-Laval's BTPX 205 system uses the classical disc-stack design to generate up to 12,800 $\times$ *g* and a maximum throughput of 1,200 L/h. The 205 has a solids-ejecting design that is tailored to produce very dry solids. It is therefore well suited for, but limited to, batch animal cell processes where cell viability is unimportant. Although its solids-ejecting capability would allow the 205 to be considered for use in a continuous animal cell process, its rough cell handling would result in unacceptable levels of cell damage.

At the production scale, Alfa-Laval's BTAX 510 system delivers up to 15,000 $\times$ *g* and a maximum throughput of 4,000 L/h. The 510, like the 205, uses a disc-stack, solids-ejecting design. So the 510 process applicability is, at its larger scale, similar to that for the 205.

Westfalia Separator AG's (Oelde, Germany) CFA-01 separator is also a continuous-flow system based on a disk-stack solids-ejecting design. Specific investigation (21) has demonstrated the effective use of this system for animal cell processing. This system, however, like the systems discussed earlier and the decanter designs discussed next, requires a rotating seal. To remove the frictional heat from its seal, the CFA-01 requires a sealing liquid that must be highly purified to maintain aseptic operation. These key considerations combine to significantly increase the complexity of the overall system. Later work with the West-

Table 2. Comparison of Centrifuges for Animal Cell Bioprocess

Supplier	Scale	AS 26-SP-System	Max *g*-force ($\times g$)	Max throughput (L/h)	Solids ejecting	Low shear	Closed system	Comments
Alfa-Laval	Small	AS 26-SP	20,000	600	No	No	No	Not suitable for continuous processes
Alfa-Laval	Pilot	BTPX 205	13,000	1,200	Yes	No	No	
Alfa-Laval	Prod'n	BTAX 510	15,000	4,000	Yes	No	No	
Westfalia	Pilot/Prod'n	CFA-01	300	240	Yes	No	No	
Westfalia	Pilot/Prod'n	CSA-1	5,000	50	Yes	No	No	
Alfa-Laval/ Sharples	Prod'n	Super-D-Canter®	10,000	15,000	Yes	No	No	Difficult to CIP and SIP
Carr	Pilot/Prod'n	Powerfuge® P6	20,000	60	Yes	No	No	
Kendro Heraeus®	Small	Contifuge® 17 RS	19,000	54	No	No	No	Not suitable for continuous processes
Baxter	Small	Cell Harvester®	1,000	60	No	Yes	Yes	Not suitable for continuous processes
Kendro Sorvall®	Small	Centritech® LAB	250	10	Yes	Yes	Yes	Well suited to continuous processes
Kendro Sorvall®	Pilot/Prod'n	Centritech® CELL	300	100	Yes	Yes	Yes	Well suited to continuous processes

Note: Prod'n, production.

falia CSA-1 disc-stack centrifuge (22) came to mixed conclusions and reported heating issues.

Decanter centrifuges, such as Alfa-Laval's Sharples® Super-D-Canter® systems, use a solids-ejecting design that differs from that of the disc-stack systems. The decanter design displaces the solids by pushing or scraping them out of the centrifuge using a type of helical blade arrangement. Using this design, high animal cell viabilities may be obtainable in the ejected solids. But these solids are not well suited to continuous processes where the cells must be resuspended, for example, to be introduced back into the bioreactor.

Carr Separations, Inc. (Farmington, Mass.), as a relative newcomer (the company was founded in 1987), has developed a different design specifically for bioprocess as well as for other pharmaceutical and food applications. The Powerfuge® P6 Separation System uses a new solids separation design that can generate very dry solids. This new design also allows for easier CIP than traditional disc-stack designs, an issue that can be a critical drawback for the original design, especially in the larger-scale equipment. The Powerfuge® is capable of generating up to 20,000 × g and throughput rates of 60 L/h. Recently, Carr introduced a smaller, low-end bench-top version of the Powerfuge® called the Pilot Separator®. For animal cell process, the suitability of these systems is similar to that for decanter centrifuges.

Kendro, through its Heraeus® brand, offers a pair of bench-top centrifuges designed to operate with a continuous-flow rotor system that can generate 24,000 × g and a flow rate of 54 L/h. These units are compact and readily portable. They are continuous in the liquid phase but are not designed to eject the solids. Although popular for laboratory-scale work, these units can provide only qualitative support to any planned scale-up plans during process development.

Baxter International, Inc. (Deerfield, Ill.) has modified the design of one of their blood component systems to focus on animal cell bioprocess at the laboratory and smaller production scales. Called the Cell Harvester®, this system can be used to develop centrifuge-based cell separation protocols and to prove out process concepts. Because manual intervention is needed to remove the cells, the applicability of this system to continuous processes is extremely limited. One key advantage of the Baxter system, however, is its totally closed system design. This design allows for the use of a consumable to avoid the various issues that accompany the need for providing CIP/SIP.

The third, unique design approach mentioned earlier that was taken by Alfa-Laval in the 1980s, also uses a consumable insert. These disposable inserts are provided sterile so that, upon making sterile connections to the system, the user has created a totally closed, sterile flow path and separation zone for the process fluids (23). This design approach, called Centritech® by Alfa-Laval (see Figure 3), is now available through and supported by Kendro under its Sorvall® brand. Key trade-offs of the Centritech® design are very gentle cell handling and a totally closed, inherently aseptic system versus high throughput and g-force (24).

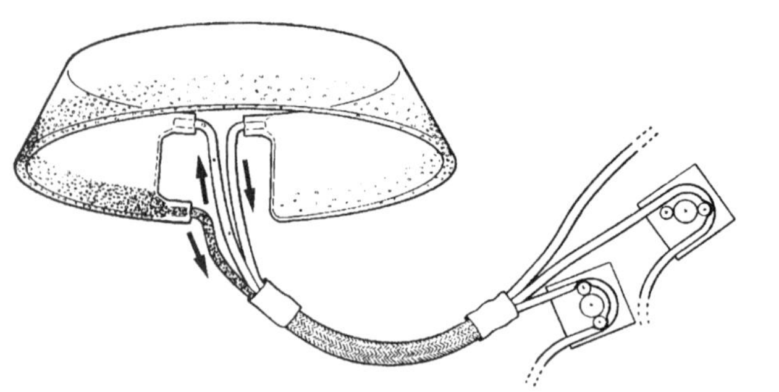

Figure 3. Sketch of the sterile, disposable insert for the Centritech® continuous-flow centrifuge system.

The Centritech® technology is embodied in both a smaller bench-top laboratory-scale system (25) and a larger floor-model pilot- to production-scale system (26). Depending on the application, however, the laboratory-scale unit has been and can be used for small production processes (27).

ALTERNATIVES TO ANIMAL CELL CENTRIFUGATION

Filtration has been the traditional choice as an alternate bioprocess unit operation to animal cell centrifugation. The most common form of filtration used to separate the cellular material from the fermentation broth is tangential flow filtration (TFF).

The foremost advantage of using TFF is that it is a technology that has been widely used to good effect (28,29). TFF does, however, have the disadvantage that, as for all filter-based systems, fouling and clogging can be a constant and often unpredictable problem if not sized properly. Filter systems also deliver performance that degrades over the course of the run, necessitating either that they be reconditioned or the run stopped to replace the filters. Even when reconditioned, filters do not usually return to their original performance, so operating them becomes a classic case of diminishing returns. Recently, issues such as documenting well-characterized products and other concerns have made the reuse of filters less desirable. In these cases, the only viable option is to stop the run, change the filters, resterilize the process, and start up again.

A further disadvantage of TFF is the inherent presence of shearing at the interface caused by the membrane. Material buildup can affect the way the system works, but a substantial shear remains across the boundary layer. Such shearing can run counter to the low shear requirements for animal cell processing and also requires that a recycle loop (with a pump) circulates the process fluid many times greater than that necessary for continuous (perfusion) operation.

Another alternative to animal cell centrifugation is the use of spin filters. Spin filters have the very attractive aspect that as a cell retention device for perfusion processes, they can be integral to the bioreactor. Thus, a single unit operation is created. Cell damage is a concern (30), but evidence suggests that with proper design and operation this issue can be minimized or even overcome (31).

A disadvantage of the spin-filter systems is that they too, like TFF, are susceptible to fouling and clogging. What can be worse, however, is that when a spin-filter system integral to the bioreactor clogs unexpectedly, the necessary maintenance typically requires that the bioreactor containment be breached. The risk of infection from such a breach can be significant. Another disadvantage is that the scaleability of spin-filter systems is limited (32). They may not be a practical unit operations choice if larger scales are anticipated during process development or once in full production.

Another device that has been used for cell retention in continuous processes is the inclined separator (33). The design of this device is mechanically simple, can be made to be relatively easy to CIP and SIP, and, because it relies on gravity (i.e., $1 \times g$), has many of the same process benefits as centrifugation. A drawback with this technology today appears to be its relative lack of throughput scaleability and, of course, that its separation efficiency is limited to that achievable at $1 \times g$ (vs. centrifugation at $100 \times g$ or more).

THE FUTURE OF ANIMAL CELL CENTRIFUGATION

Going forward, centrifugation is likely to be as commonly used for bioprocess as it is for the comparable chemical process industry. The inherent advantages of centrifugation (such as robustness and versatility) continue to fuel creative new concepts for harnessing these advantages for animal cell bioprocess. Novel centrifuge designs and the unique continuous methodologies for bioprocess will likely combine to make the centrifuge a necessary unit operation in a large number of the new animal cell bioprocesses being developed in the near future. Moreover, recent changes in the FDA regulations are conducive to the accelerated adoption of new and more efficient process methodologies. Although in the past there were often heavy disincentives to consider the type of process improvements that, for example, continuous-flow centrifugation allows, many of the regulatory-based concerns have been reduced or eliminated. As a result, we are likely to see adoption of new, higher productivity process improvements in existing processes as well as the development of new ones.

BIBLIOGRAPHY

1. G. MacMichael, *BioPharm.* **10**, 38–40 (1997).
2. M.S. Croughan and D.I.C. Wang, *Biotechnol. Bioeng.* **33**, 731–744 (1989).
3. M.S. Croughan and J.-F.P. Hamel, *Biotechnol. Bioeng.* **32**, 975–982 (1988).
4. E.T. Papoutsakis and R.S. Cherry, *Biotechnol. Bioeng.* **32**, 1001–1014 (1988).
5. R.V. Venkat and J.J. Chalmers, *Cytotechnology* **22**, 95–102 (1996).
6. M.S. Croughan, E.S. Sayre, and D.I.C. Wang, *Biotechnol. Bioeng.* **33**, 862–872 (1989).
7. J.J. Chalmers, in M.L. Shuler, H.A. Wood, R.R. Granados, and D.A. Hammer eds., *Baculovirus Expression Systems and Biopesticides*, Wiley-Liss, New York, 1995, pp. 175–204.
8. Technical Support Package, *NASA Tech. Brief* MSC-22336 (1993).
9. J. Tramper, J.B. Williams, and D. Joustra, *Enzyme Microb. Technol.* **8**, 33–36 (1986).
10. K. Trinh, M. Garcia-Briones, F. Hink, and J.J. Chalmers, *Biotechnol. Bioeng.* **43**, 37–45 (1994).
11. J.J. Chalmers, *Cytotechnology* **20**, 163–171 (1996).
12. T.C. Ford and J.M. Graham, *An Introduction to Centrifugation*, BIOS Scientific Publishers Limited, Oxford, U.K., 1991.
13. E.T. Papoutsakis, *Trends Biotechnol.* **9**, 427–437 (1991).
14. N. Kioukia, A.W. Nienow, A.N. Emery, and M. Al-Rubeai, *Trans. IChE* **70**, 143–148 (1992).
15. J.J. Chalmers, *Cytotechnology* **15**, 311–320 (1994).
16. D. Chattopadhyay, J.F. Rathman, and J.J. Chalmers, *Biotechnol. Bioeng.* **45**, 473–480 (1995).
17. M. Al-Rubeai, A.N. Emery, S. Chalder, and M.H. Goldman, *J. Biotechnol.* **31**, 161–177 (1993).
18. C. Born, Z. Zhang, M. Al-Rubeai, and C.R. Thomas, *Biotechnol. Bioeng.* **40**, 1004–1010 (1992).
19. K.T. Kunas and E.T. Papoutsakis, *Biotechnol. Bioeng.* **36**, 476–483 (1990).
20. Z. Zhang, M. Al-Rubeai, and C.R. Thomas, *Biotechnol. Tech.* **7**, 177–182 (1993).
21. V. Jager, in H. Murakami ed., *Proc. of the 4th JAACT Mtg.*, Kluwer, Dordrecht, The Netherlands, 1991.
22. R. Kempken, A. Preissmann, and W. Berthold, *Biotechnol. Bioeng.* **46**, 132–138 (1995).
23. D.D. Hanle, *Int. Biotech. Lab.* 12, (October 1996).
24. J. Hodgson, *Biotechnology* **9**, 628–629 (1991).
25. N. Chatzisavido, T. Bjorling, C. Fenge, S. Boork, E. Lindner-Olsson, and S. Apelman, in T. Kobayashi, Y. Kitagawa, and K. Okumura eds. *Animal Cell Technology: Basic and Applied Aspects*, Kluwer, Dordrecht, The Netherlands, 1993, pp. 463–468.
26. T. Bjorling, U. Malmstrom, in R.E. Spier, J.B. Griffiths, and B. Meignier eds., *Production of Biologicals from Animal Cells in Culture*, Butterworth-Heinemann, Oxford, U.K., 1991, pp. 394–399.
27. C.W. Kemp and A. Chan, *Waterside Monoclonal Antibody Mtg.*, Norfolk, Vir., 1997.
28. B. Maiorella, G. Dorin, A. Carion, and D. Harano, *Biotechnol. Bioeng.* **37**, 121–126 (1991).
29. R. vanReis, L.C. Leonard, C.C. Hsu, and S.E. Builder, *Biotechnol. Bioeng.* **38**, 413–422 (1991).
30. A.J. Brennan, J. Shevitz, and J.D. MacMillan, *Biotechnol. Tech.* **1**, 169–174 (1987).
31. D.C-H. Jan, A.N. Emery, and M. Al-Rubeai, *Biotechnol. Tech.* **7**, 351–356 (1993).
32. V.M. Yabannavar, V. Singh, and N.V. Connelly, *Biotechnol. Bioeng.* **43**, 159–164 (1994).
33. J.A. Searles, P.W. Todd, R.H. Davis, and D.S. Kompala, in R.E. Spier, J.B. Griffiths, and W. Berthold eds., *Animal Cell Technology: Products for Today, Prospects for Tomorrow*, Butterworth-Heinemann, Oxford, U.K., 1994, pp. 240–242.

See also ANTIBODY PURIFICATION; CELL SEPARATION, CENTRIFUGATION; CELL SEPARATION, SEDIMENTATION.

CEPHALOSPORINS

SHU-JEN D. CHIANG
JONATHAN BASCH
Bristol-Myers Squibb Company
Syracuse, New York

KEY WORDS

7-ACA
7-ADCA
Biosynthetic genes
Biosynthetic pathways
Cephamycins
Extraction
Fermentation
β-Lactams
Semisynthetic cephalosporins
Strain improvement

OUTLINE

INTRODUCTION

Since the discovery of penicillins in 1928 and the discovery of cephalosporin C in 1945 (1), β-lactams have become the most important class of antibiotics because of their low toxicity and their effectiveness against bacterial infection. In cephalosporins, the four-membered β-lactam ring is fused to a six-membered dihydrothiazine ring rather than to the five-membered thiazolidine ring found in penicillins (Fig. 1). The natural product, cephalosporin C, has low activity against gram-negative bacteria, but it is resistant to penicillinase hydrolysis. Many chemical derivatives of cephalosporin C have a higher and broader spectrum of antibacterial activity than cephalosporin C. After 40 years of development, cephalosporins (i.e., chemical derivatives of cephalosporin C or semisynthetic cephalosporins) have become one of the most widely used antibiotics in the world; in 1995, the world market for cephalosporins was valued at $9.3 billion at the dosage form level (2). Of this total, $6.5 billion represented products made from cephalosporin C, with the remainder derived from penicillins. In this article, we discuss the development and production of cephalosporins and their chemical derivatives.

Historical Background

Penicillin G (benzylpenicillin) was considered a miracle drug, when discovered in the 1940s, for its effective treatment of the systemic infections caused by gram-positive bacteria such as *Staphylococcus aureus*. Unfortunately, penicillin G was generally ineffective against gram-negative bacteria, and within a few years of its introduction many common gram-positive bacteria had acquired resistance to penicillin G as a result of the expression of bacterial penicillinases. At about this time, a new antibiotic, cephalosporin C, was discovered from a species of mold, *Cephalosporium acremonium*. The *C. acremonium* strain was originally isolated from seawater near a sewage outflow in Cagliari, Sardinia, by Brotzu in 1945. The culture filtrates of this fungus exhibited a broad spectrum of antibacterial activity both in vitro and in preliminary clinical studies (3). In 1948, Brotzu published his findings, which failed to attract attention. At the British Medical Research Council's suggestion, Brotzu sent his fungal culture, *C. acremonium*, which was later reclassified as *Acremonium chrysogenum* in 1971 by Gams (4), to Florey at Oxford. The work of Newton and Abraham at Oxford led to the purification of penicillin N (Fig. 1) and cephalosporin C in 1955 (5). Both compounds had activity against gram-negative bacteria. Penicillin N was as susceptible to staphylococcal penicillinase hydrolysis as was penicillin G. Cephalosporin C, however, had a high degree of resistance to hydrolysis by staphylococcal penicillinase. Further study also indicated that cephalosporin C had low toxicity in mice, but unfortunately, cephalosporin C had relatively low in vivo antibacterial activity.

In 1961, the structure of cephalosporin C was proposed by Abraham and Newton (6) and confirmed by an X-ray crystallographic analysis by Hodgkin and Maslen (7). It was found that a small quantity of the nucleus of cephalosporin C, 7-aminocephalosporanic acid (7-ACA) (Fig. 1), could be produced by gentle acid hydrolysis of cephalosporin C. Acylation of 7-ACA yielded a cephalosporin with much higher antibacterial activity than cephalosporin C against gram-positive bacteria. In 1962, Morin et al. (8) in the Lilly Research Laboratories developed a chemical method for the production of 7-ACA in kilogram quantities, and this opened the way for the extensive exploration of the synthesis of semisynthetic cephalosporins (discussed

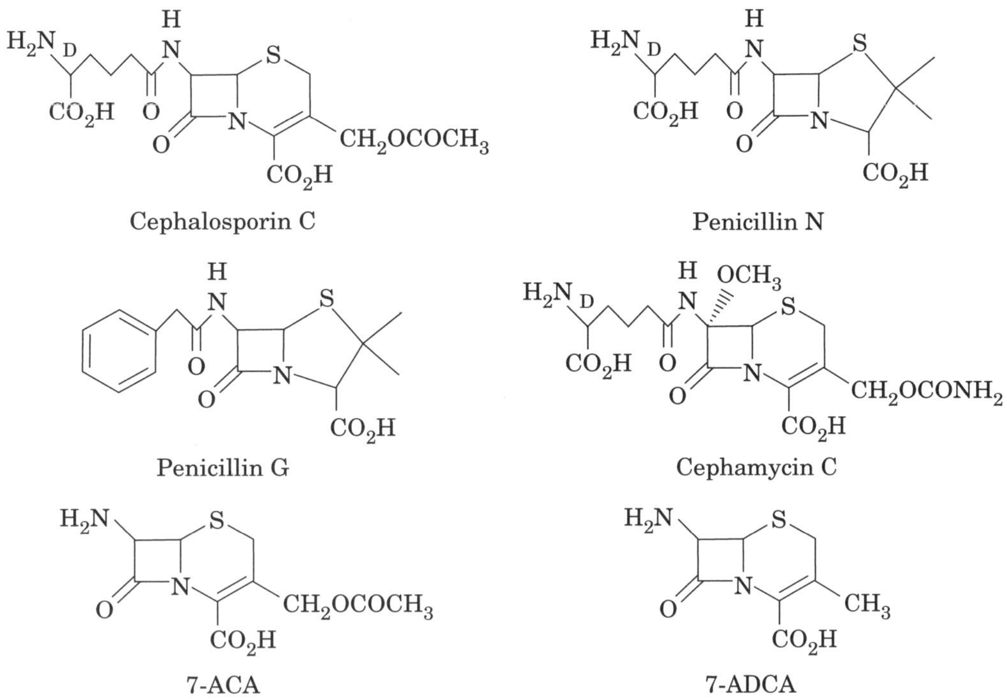

Figure 1. Chemical structure of β-lactams.

in a later section). By 1970, total synthesis of the cephalosporin double-ring system had been achieved by chemists working at Roussel Laboratories and E.R. Squibb & Sons (9).

The extensive use of semisynthetic cephalosporins resulted in the selection in nature of gram-negative bacteria containing R-plasmids that carried a gene encoding β-lactamase. This β-lactamase can hydrolyze most penicillins and all the semisynthetic cephalosporins that had been introduced in the mid 1960s. While searching for new β-lactam antibiotics that are stable in the presence of β-lactamases, culture filtrates of *Streptomyces* species yielded cephamycins (10) containing a methoxy group at the C-7α position of the cephalosporin nucleus (Fig. 1). These compounds were found to be completely resistant to the plasmid-encoded β-lactamase. Derivatives of cephamycin C will be described in "Semisynthetic Cephalosporins".

MODE OF ACTION OF β-LACTAMS

Cephalosporins and other β-lactams inhibit bacterial growth by interfering with bacterial cell wall synthesis. All bacterial cell walls have a protective layer of peptidoglycan at the outer surface of the cell membrane. The peptidoglycan layer consists of chains of alternating *N*-acetylglucosamine and *N*-acetylmuramic acid connected by peptides. These peptides cross-link adjacent peptidoglycan chains to form a net of peptidoglycan layers. In gram-positive bacteria, the cell walls are composed of peptidoglycan and teichoic acids (a polymer of glycerol phosphate or ribitol phosphate) or teichuronic acids (phosphate-free polysaccharides). In gram-negative bacteria, in addition to the peptidoglycan layer, there is a distinct outer membrane layer consisting of lipoproteins and lipopolysaccharides. During the course of bacterial cell wall synthesis, the monomers of *N*-acetylglucosamine and *N*-acetylmuramic acid peptide are linked to form the peptidoglycan chains, and three enzymes (transpeptidase, DD-carboxypeptidase, and endopeptidase) catalyze the cross-links between these peptidoglycan chains by forming peptide bridges (11). Binding of β-lactam antibiotics to the transpeptidase and the DD-carboxypeptidase enzymes, members of the penicillin-binding proteins (PBP) group, interferes with the formation of peptide bridges between peptidoglycan chains and thus inhibits the final stage of cell wall synthesis in bacteria (12).

BIOCHEMISTRY AND GENETICS OF CEPHALOSPORIN BIOSYNTHESIS

The biosynthetic pathways of cephalosporins have been well characterized (13). The current view of the overall biosynthetic scheme for cephalosporin C and cephamycin C is depicted in Figure 2. The nomenclature of genes encoding the enzymes involved in each step of the biosynthesis of cephalosporins follows the recommendation by Ingolia and Queener (14).

Biosynthetic Pathways for Cephalosporins

The biosynthesis of cephamycin C by *Streptomyces clavuligerus* and *Nocardia lactamdurans* and the biosynthesis of cephalosporin C by *A. chrysogenum* all begin with the

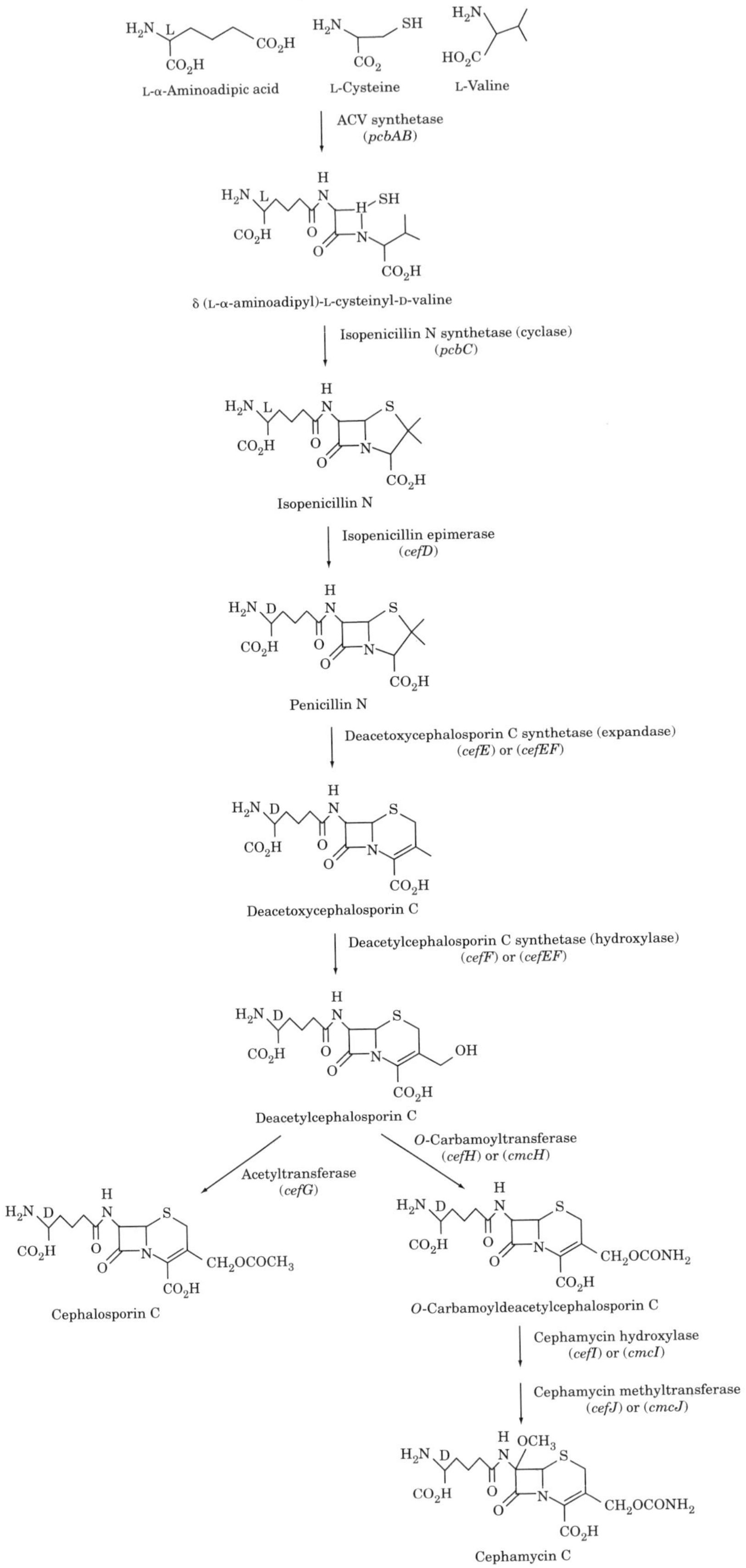

Figure 2. Biosynthetic pathways for cephalosporin C and cephamycin C.

condensation of three amino acids, L-α-aminoadipic acid, L-cysteine, and L-valine, to form the tripeptide δ-(L-α-aminoadipyl)-L-cysteinyl-D-valine (LLD-ACV) by the ACV synthetase enzyme. The racemization of valine from the L- to the D-configuration appears to take place during the synthesis of the tripeptide. The LLD-ACV tripeptide is cyclized to form the penam nucleus, isopenicillin N, by isopenicillin N synthetase (or cyclase). Both the LLD-ACV tripeptide and isopenicillin N also serve as intermediates in the biosynthesis of penicillins. The genes coding for enzymes involved in the biosynthesis of the common intermediates LLD-ACV and isopenicillin N are designated *pcb* (*p*enicillin/*c*ephalosporin *b*iosynthesis). The genes coding for the remaining enzymes involved in the biosynthesis of cephalosporins are designated *cef* (14).

The isopenicillin N epimerase (IPNE) catalyzes the conversion of the L-α-aminoadipyl side chain of isopenicillin N into the D-α-aminoadipyl side chain of penicillin N. The five-membered thiazolidine ring of penicillin N is then expanded to give the six-membered dihydrothiazine ring of deacetoxycephalosporin C by deacetoxycephalosporin C synthetase (expandase). The next enzyme in the pathway, deacetylcephalosporin synthetase (hydroxylase), catalyzes hydroxylation of deacetoxycephalosporin C at the C-3 methyl group to yield deacetylcephalosporin C. In *A. chrysogenum*, both the ring expansion and hydroxylation activities reside on the same protein, which is coded by one gene. Unlike fungi, the expandase and the hydroxylase are two distinct enzymes in *S. clavuligerus* and *N. lactamdurans* and encoded by two separate genes (13).

In *A. chrysogenum*, the last step in cephalosporin C biosynthesis, catalyzed by cephalosporin C synthetase (acetyltransferase), involves the transfer of an acetyl group from acetyl coenzyme A to the C-3 hydroxymethyl moiety of deacetylcephalosporin C.

In the cephamycin C producers *S. clavuligerus* and *N. lactamdurans*, the *O*-carbamoyltransferase catalyzes the transfer of a carbamoyl group from carbamoylphosphate to the C-3 hydroxymethyl moiety of deacetylcephalosporin C. The *O*-carbamoyldeacetylcephalosporin C is then hydroxylated at the C-7α position by cephamycin hydroxylase. The final biosynthetic step to cephamycin C (7α-methoxy-*O*-carbamoyldeacetylcephalosporin C) is the transfer of the methyl group from *S*-adenosylmethionine (SAM) to the C-7α hydroxyl moiety of 7α-hydroxyl-*O*-carbamoyldeacetylcephalosporin C by cephamycin methyltransferase. The biosynthetic pathways of other cephamycins, such as cephamycin A and cephamycin B, have not been fully investigated.

Cloning of the Cephalosporin Biosynthetic Genes

Because the DNA transformation system developed for *A. chrysogenum* is much less efficient than the bacterial systems, cloning of the cephalosporin C biosynthetic genes by complementation of blocked mutants is not feasible. However, the reverse genetics approach has been used successfully in the cloning of a number of cephalosporin C biosynthetic genes (14). This approach begins with the purification of the desired enzyme, determination of a partial amino acid sequence, and synthesis of oligonucleotide probes based on the amino acid sequence. A genomic DNA library of *A. chrysogenum* is then constructed in *Escherichia coli* and screened for clones that hybridize to the DNA probes. Sequencing of the cloned DNA will confirm that the DNA sequence matches the known amino acid sequence. The reverse genetics approach has also been used in the cloning of the cephamycin C biosynthetic genes despite the existence of highly efficient transformation systems developed for *S. clavuligerus* and *N. lactamdurans*.

In 1985, the first β-lactam biosynthetic gene, *pcbC* (encoding cyclase), was cloned from *A. chrysogenum* (15). The *A. chrysogenum pcbC* gene has no intron and encodes a polypeptide of 338 amino acids. When the open reading frame of the *A. chrysogenum pcbC* gene was expressed in an *E. coli* expression vector, the purified recombinant cyclase protein was able to convert LLD-ACV to isopencillin N with identical kinetics to the authentic protein, and it also converted ACV tripeptide analog substrates to novel β-lactam antibiotics. Recombinant cyclase is readily produced in the recombinant *E. coli* and can be used for in vitro synthesis of novel β-lactam antibiotics (16). Other *pcbC* genes have been cloned by reverse genetics or heterologous hybridization from *Aspergillus nidulans, Penicillium chrysogenum, S. clavuligerus, S. griseus, S. jumonjinensis, S. lipmanii, N. lactamdurans, Flavobacterium* sp., and *Lysobacter lactamgenus* (17). When these genes were compared at the DNA sequence and the predicted amino acid sequence levels, a high degree of similarity was found to exist among the *pcbC* genes from different sources.

The reverse genetics approach was also used to clone the *pcbAB* (encoding ACV synthetase), *cefD* (isopenicillin N epimerase), *cefE* (expandase), *cefF* (hydroxylase), *cefEF* (*A. chrysogenum* expandase–hydroxylase bifunctional enzyme), *cefG* (acetyltransferase), *cefH* (also known as *cmcH*, *O*-carbamoyltransferase), *cefI* (*cmcI*, cephamycin hydroxylase), and *cefJ* (*cmcJ*, cephamycin methyltransferase) genes (14,17–20). The only gene of the cephalosporin biosynthetic genes that has not been cloned is the *cefD* gene from *A. chrysogenum*. Most of the cephalosporin C biosynthetic genes cloned from *A. chrysogenum* do not have introns and have a high degree of similarity with their bacterial counterparts, suggesting a bacterial origin for these genes. The *cefG* gene is the exception with two introns. All the fungal genes and some of the bacterial genes encoding for the enzymes involved in cephalosporin biosynthesis have their own promoter for transcriptional control. However, the *Streptomyces cefD* and *cefE* genes are transcribed from the same promoter, with the *cefD* gene located upstream of the *cefE* gene and a 130 base pair (bp) intergenic region separating them (18). In *N. lactamdurans*, a 7-bp separation between the *cefI* and *cefJ* genes and the presence of a putative ribosome binding site upstream of the *cefI* gene suggests that the two genes are cotranscribed from the same promoter and also cotranslated (19).

Organization of the Cephalosporin Biosynthetic Genes

In 1988, Chen et al. demonstrated that a cloned 29.3-kb DNA fragment from *S. cattleya* carried the entire set of genes encoding enzymes involved in cephamycin C biosynthesis (21). No further characterization of this cloned DNA fragment was reported. Subsequent work on other bacterial cephamycin C producers also indicated that the cephamycin C biosynthetic genes are clustered in one region of the bacterial chromosome (17–20) (Fig. 3a). However, in *A. chrysogenum*, the two early cephalosporin C biosynthetic genes, *pcbAB* and *pcbC*, reside on a different chromosome from the two late biosynthetic genes, *cefEF* and *cefG*. Eight chromosomes have been detected by pulsed-field electrophoresis, an electrophoretic technique that allows separation of chromosomes in an agrose gel system. The *pcbAB* and *pcbC* genes are located on chromosome VI and transcribed in opposite directions. The *cefEF* and *cefG* genes are located on chromosome II and are also transcribed in opposite directions (17) (Fig. 3b). Because the *cefD* gene of *A. chrysogenum* has not been cloned, it is not known whether it is linked to either of the two clusters.

STRAIN IMPROVEMENT

Mutagenesis and Selection

Most of the literature describing strain improvement reports the use of the cephalosporin C producer *A. chrysogenum*. However, similar strategies could be applied to the cephamycin C producers (i.e., *Streptomyces* spp.). Improvement in the yield of cephalosporin C has been achieved primarily by increasing the productivity of the *A. chrysogenum* culture by traditional strain mutation and selection techniques. The first significant titer improvement was achieved when in 1959 mutant 8650 was isolated. This strain enabled 100 g of cephalosporin C to be purified for structure determination and is the parent of all the industrial production strains (1).

Mutagenesis, followed by high-throughput screening of the culture, has been the traditional means of strain improvement. Conidia are prepared from the mycelial culture to increase the probability of isolating mutants derived from a single nucleus. Chemical mutagens, such as *N*-methyl-*N*-nitro-*N*-nitrosoguanidine and ethyl-methanesulfonate, or *UV* light are used to induce chromosomal mutations. Growth of the mutants on selective agents can be used to increase the probability of isolating improved clones. Toxic analogs of cysteine, valine, aminoadipate, lysine, and methionine are used to select mutants that overproduce these amino acids, providing a larger precursor pool for β-lactam biosynthesis. The toxicity to *A. chrysogenum* of heavy metallic ions, such as Cu^{2+}, Mn^{2+}, and Hg^{2+}, has also been exploited for selection purposes. Because the β-lactam antibiotics complex with these ions, one can theorize that mutants resistant to inhibitory concentrations of Cu^{2+}, Mn^{2+}, or Hg^{2+} can overproduce the β-lactam nucleus to detoxify these metallic ions. Elander et al. demonstrated the isolation of superior cephalosporin-C-producing strains from mutants resistant to mercuric chloride (22).

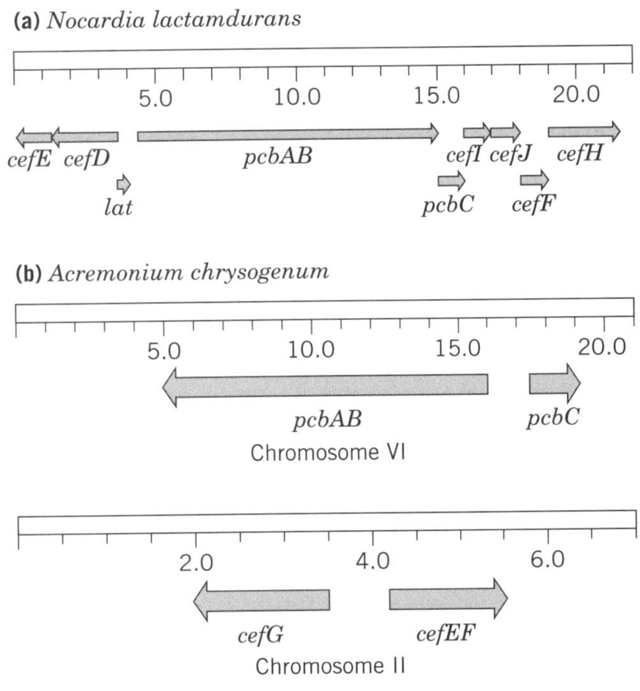

Figure 3. Organization of the cephalosporin biosynthetic genes. (**a**), Physical map of a 25-kb region of *N. lactamdurans* DNA carrying all genes related to cephamycin C biosynthesis. *lat* = gene encoding lysine-6-aminotransferase; *pcbAB* = ACV synthetase; *pcbC* = isopenicillin N synthetase (cyclase); *cefD* = isopenicillin N epimerase; *cefE* = deacetoxycephalosporin C synthetase (expandase); *cefF* = deacetylcephalosporin C synthetase (hydroxylase); *cefH* or *cmcH* = *O*-carbamoyltransferase; *cefI* or *cmcI* = cephamycin hydroxylase; *cefJ* or *cmcJ* = cephamycin methyltransferase (17–20). (**b**), Physical map of the clusters of the cephalosporin C biosynthetic genes of *A. chrysogenum*. *cefEF* = expandase/hydroxylase bifunctional enzyme; *cefG* = acetyltransferase (17). The direction of arrows indicates the transcriptional direction of a gene.

Protoplast Fusion

Protoplast fusion has been explored as a means of combining the advantageous characters of different culture lineages (e.g., high titer and fast growth). Mycelia are treated with dithiothreitol (DTT), followed by Novozym™ 234 (a multienzyme preparation from the fungus *Trichoderma harzianum* available from Novo Biolabs) in the presence of an osmotic stabilizer, usually 0.7 M NaCl. Polyethylene glycol (PEG) is then used to induce membrane fusions between protoplasts; the fusants are identified by regeneration on selective media. Auxotrophic markers have been used to select for fusants, but auxotrophic mutants often have reduced titers and are therefore not suitable for industrial strain improvement. Strains that have been transformed with positive selective markers such as phleomycin and hygromycin resistance are better candidates for protoplast fusion.

Genetic Engineering

In the past 10 years, the cloning of many of the genes involved in the biosynthetic pathway of cephalosporins has led to attempts to improve strains through genetic engineering. Enzymes that have been identified as rate-limiting in the biosynthesis of cephalosporins or their precursors are promising targets for strain improvement through genetic engineering (23). Transformation of industrial strains with additional copies of the genes coding for these rate-limiting enzymes can result in alleviation of a metabolic bottleneck and increase antibiotic production. Transformation methods for *A. chrysogenum* have been developed using protoplasts and PEG treatment.

Based on the kinetic information available for the enzymes in the cephalosporin C biosynthetic pathway in *A. chrysogenum* and the cephamycin C pathway in *S. clavuligerus*, models have been developed to predict the flux of intermediates in these pathways. The model developed for *S. clavuligerus* predicted that the amino acid precursors, in particular α-aminoadipate, were limiting. Introduction of additional copies of the lysine ϵ-aminotransferase gene *(lat)*, required for conversion of lysine to α-aminoadipate in *Streptomyces*, resulted in a fivefold increase in cephamycin C production (24).

Skatrud et al. have transformed industrial strains of *A. chrysogenum* with a recombinant plasmid containing the hygromycin resistance gene and the *cefEF* gene encoding the expandase–hydroxylase bifunctional enzyme. At the shake flask scale, a 39% increase in cephalosporin C titer was observed, with a significant reduction in the levels of penicillin N and deacetoxycephalosporin C produced. The improvement in cephalosporin C titer at the 150-L pilot plant scale was 15%. Enzyme assays indicated that the expandase–hydroxylase activity in the transformant was twofold higher than in the host strain, and Southern blot analysis revealed a single chromosomal integration at a heterologous position (25). Interestingly, when the *cefG* gene (encoding acetyltransferase) was identified, it was found to be adjacent to the *cefEF* gene and had incidentally been included on the genomic fragment that Skatrud's group had used for transformation. Chiang and Basch also reported that recombinant *A. chrysogenum* strains transformed with a plasmid containing the *cefEF* gene had a 40% reduction of deacetoxycephalosporin C present in the cephalosporin C fermentation broth as compared to the parental strain (26). This process was scaled successfully to production tanks.

CEPHALOSPORIN FERMENTATION

Physiology of Cephalosporin C Fermentation

Cephalosporin C is produced by *A. chrysogenum* as a secondary metabolite, with production of the antibiotic associated with a distinct morphological change. When no nutrient stress is placed on the culture, the organism grows as branched mycelia by extension of the hyphal tips, and little or no cephalosporin C is produced. Under glucose-limiting conditions and in the presence of DL-methionine, the mycelial tips swell and form arthrospores. The mycelial cell walls weaken and fragment, releasing the arthrospores into the media. Throughout the formation of the arthrospores, cephalosporin C is produced, with the swollen hyphae being the most productive morphology.

DL-Methionine addition significantly increases cephalosporin C production. This effect is independent of the methionine role as the sulfur donor for the formation of cysteine. The presence of DL-methionine induces small increases in the enzyme activities of expandase and cyclase and a large increase in the ACV synthetase activity (27). In the presence of high concentrations of glucose or glycerol in fermentation media, the cephalosporin C productivity decreases dramatically. The high concentration of glucose represses the production of expandase, but not of cyclase or ACV synthetase (28).

Large-Scale Fermentation

Cephalosporin C is produced industrially by fed-batch fermentation of *A. chrysogenum*. The pH is maintained between 6.0 and 7.0 in a temperature range of 24 to 28 °C. The media must supply carbon and nitrogen for growth, but must also induce the differentiation of the culture required for antibiotic production. The carbon source may be supplied as separate simple and complex carbohydrates feeds, enabling better control of growth and glucose levels. Monosaccharides, particularly glucose, promote rapid growth of the culture but reduce antibiotic titer. Simple sugars may be batched into the media or fed at higher rates earlier in the fermentation. Galactose and sucrose support slower growth than glucose but a higher rate of specific productivity. As the fermentation progresses, this sugar feed is reduced and a larger proportion of carbon is supplied as soybean or peanut oil. This limits the availability of glucose and allows the formation of arthrospores and cephalosporin C. The oil may also serve as a surfactant to reduce potential foaming of the culture.

The organic nitrogen is usually supplied by some combination of agricultural by-products, such as soybean and cottonseed meals, and supplemented with an inorganic nitrogen source such as ammonium sulfate. Ammonia gas and ammonium hydroxide can also be used to supply inorganic nitrogen and to provide pH control. Corn steep liquor is commonly used in the medium as an inexpensive source of amino acids, trace elements, and vitamins.

DL-Methionine induces the formation of arthrospores in *A. chrysogenum*, which as mentioned earlier is correlated with cephalosporin C production. Swartz et al. examined the growth and productivity of fed-batch cultures of *A. chrysogenum*. Batch addition of methionine at 3 g/L interfered with sucrose utilization of these cultures but did not affect glucose utilization. Methionine was found to be a competitive inhibitor of the invertase enzyme required for the metabolism of sucrose. Optimal productivity was achieved when sucrose and DL-methionine were fed in an exponentially ramped feed. Surprisingly, feeding of DL-methionine increased the productivity of strains that were considered methionine independent, meaning batched methionine had no effect on titer (29).

As with all fungal fermentations, dissolved oxygen is an important physical parameter. It is particularly important in the fermentation of cephalosporin C, because the cyclase and particularly the expandase enzymes in the biosynthetic pathway require oxygen for effective catalysis. If the dissolved oxygen level drops below 20%, penicillin N will accumulate and the output of cephalosporin C will be reduced.

Chemical breakdown of cephalosporin C to 2-(D-4-amino-4-carboxybutyl)-thiazole-4-carboxylic acid (compound X) (Fig. 4) results in the loss of ~40% of the cephalosporin C produced, a significant loss of the biosynthetic

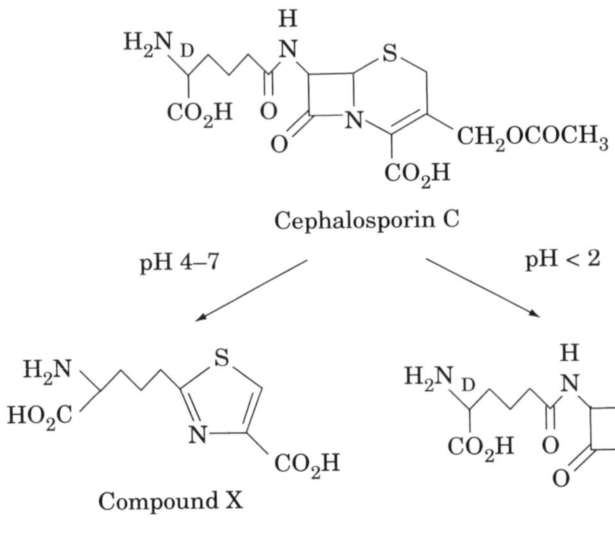

Figure 4. Chemical degradation of cephalosporin C.

potential of the organism (30). Deacetylcephalosporin C, however, is far more stable in the fermentation broth in comparison to cephalosporin C. Addition of an acetyl esterase produced by *Rhodosporidium toruloides* to the fermentation results in the accumulation of deacetylcephalosporin C and an increase in antibiotic yield of ~40% (31).

EXTRACTION OF CEPHALOSPORINS

Cephalosporin C

After the fermentation is complete, the mycelia and insoluble media components are usually removed by filtration or centrifugation. In the fermentation broth, besides cephalosporin C, there is also present small quantities of penicillin N, deacetoxycephalosporin C, and deacetylcephalosporin C. Cephalosporin C must be extracted from the fermentation broth. Under neutral and slightly acidic conditions, a temperature-dependent chemical conversion of cephalosporin C to compound X occurs in the fermentation broth (30). At a pH less than 2, cephalosporin C lactone is formed (32) (Fig. 4). To minimize this degradation, it is important that the cephalosporin C broth be processed rapidly, avoiding extreme pH conditions and keeping the temperature low.

Two main strategies are used for the purification of cephalosporin C. One strategy uses activated carbon column or nonionic resin column to adsorb cephalosporin C from fermentation broth. Because of the high selectivity of activated carbon and nonionic resin, cephalosporin C is preferentially adsorbed over penicillin N, deacetylcephalosporin C, and deacetoxycephalosporin C. A subsequent anion-exchange column gives high purity cephalosporin C product. Alternatively, a combination of anion- and cation-exchange columns arranged in series can be used to purify cephalosporin C directly from the fermentation broth (33).

Because cephalosporin C will be further derivatized to semisynthetic end products, modifications to the cephalosporin C molecule that do not affect the cephalosporin nucleus can be used to facilitate its purification. Therefore, an alternative strategy for the purification of cephalosporin C is substitution of the amine moiety on the α-aminoadipyl side chain of cephalosporin C. Some of the N-substituted derivatives, such as *N*-2,4-dichlorobenzoyl cephalosporin C and tetrabromocarboxybenzoyl cephalosporin C, can be crystallized in acidic aqueous solutions. Alternatively, salts can be formed between the N-substituted derivatives and organic base, such as dicyclohexylamine and dimethylbenzylamine, and they are solvent extractable. Several extraction steps are required to achieve an acceptable purity of the end product, although small amounts of penicillin N, deacetylcephalosporin C, and deacetoxycephalosporin C derivatives are still present (33). N-Substituted cephalosporin C derivatives can be converted to 7-ACA as effectively as cephalosporin C can.

Cephamycin C

The clarified cephamycin C fermentation broth is usually adjusted to pH 2.5 to inactivate most of the penicillin N present. Then, cephamycin C can be isolated with a combination of activated carbon adsorption and anion-exchange resin adsorption. A small amount of deacetoxycephalosporin C and the remaining penicillin N copurifies with cephamycin C. These can be removed by a subsequent silica gel column (34).

PRODUCTION OF THE CEPHALOSPORIN NUCLEUS

Cephalosporin C itself has a relatively low level of antibacterial activity. The semisynthetic cephalosporins with higher antibacterial activity, β-lactamase resistance, and desirable pharmacological properties are derived by modifications of 7-ACA or 7-aminodeacetoxycephalosporanic acid (7-ADCA) (Fig. 1).

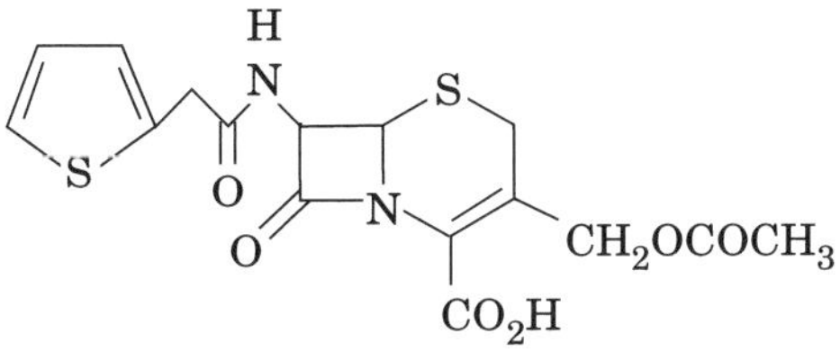

Cephalothin

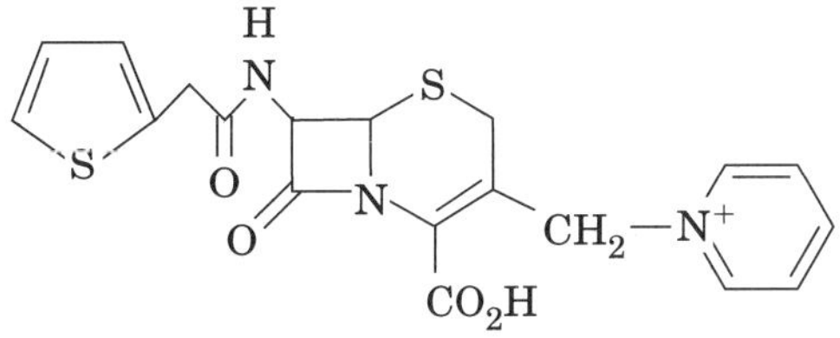

Cephaloridine

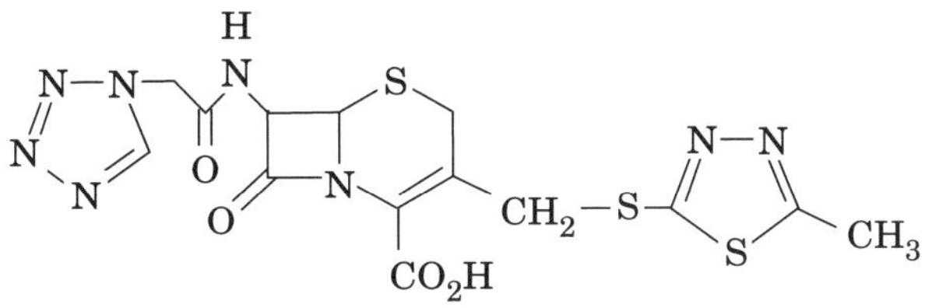

Cefazolin

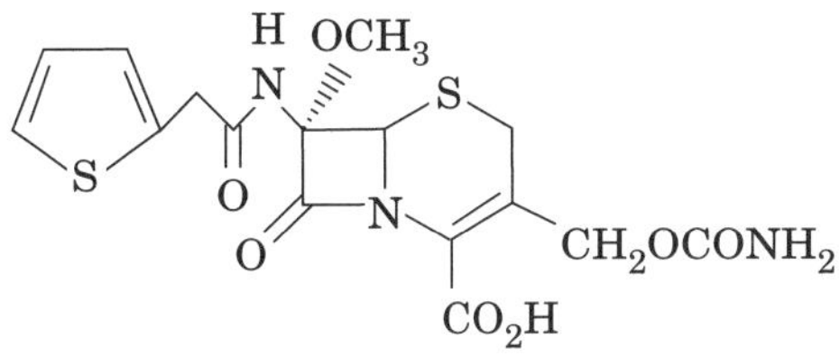

Cefoxitin

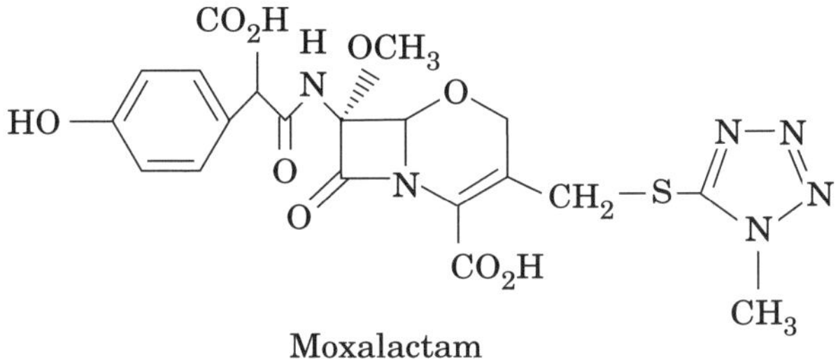

Moxalactam

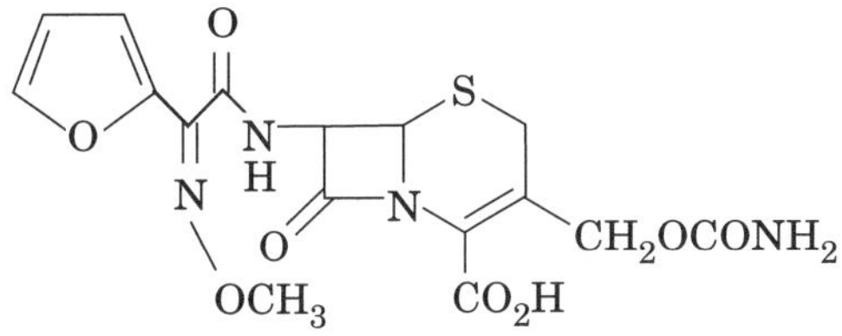

Cefuroxime

Figure 5. Chemical structure of first- and second-generation cephalosporins.

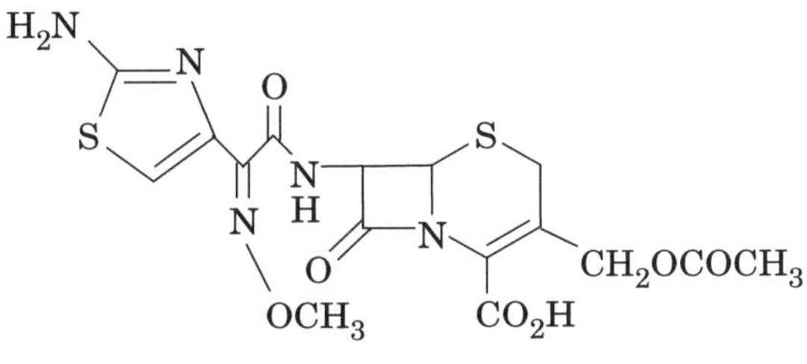

Cefotaxime

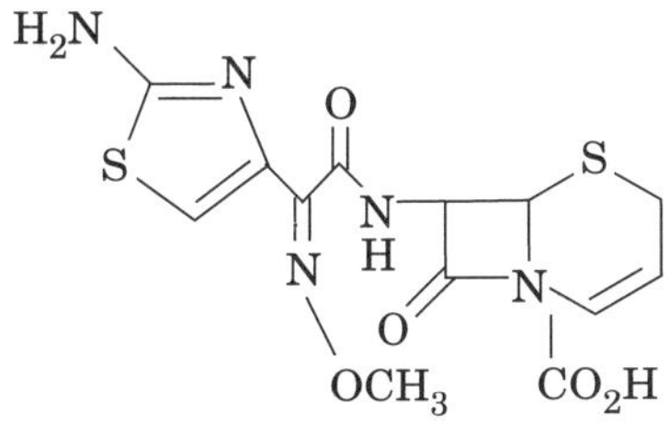

Ceftizoxime

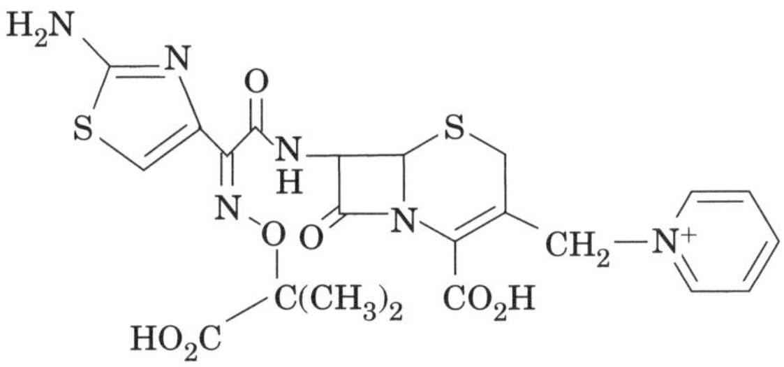

Ceftazidime

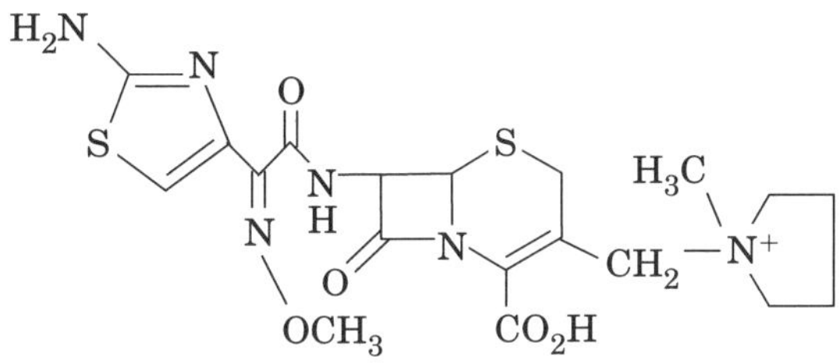

Cefepime

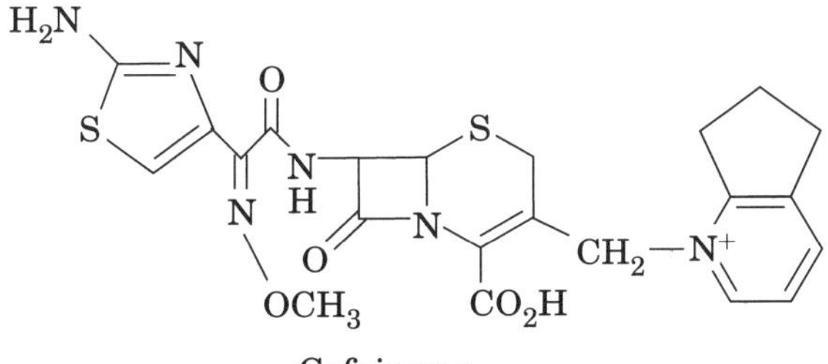

Cefpirome

Figure 6. Chemical structure of third- and fourth-generation cephalosporins.

Almost one-third of commercial cephalosporins are derived from 7-ADCA. Because of the lower cost of fermenting penicillins, 7-ADCA is most economically produced from penicillin G by ring expansion of a penicillin sulfoxide ester to yield a cephalosporin ester. The ester group is removed, followed by removal of the phenylacetyl side chain to give 7-ADCA (35). Recently, Merck and Panlabs patented a process for producing 7-ADCA. In this process, *P. chrysogenum* was transformed with the *S. clavuligerus cefE* (expandase) gene. When fed with disodium adipate, this strain produced adipoyl 6-APA, which was then converted by the recombinant expandase to adipoyl 7-ADCA in vivo. The adipoyl side chain was subsequently cleaved in vitro by a *Pseudomonas* adipoyl acylase to produce 7-ADCA (36).

Two-thirds of the commercial cephalosporins are derived from 7-ACA. 7-ACA is produced from cephalosporin C by chemical or enzymatic deacylation. In the chemical process, after protection of the amino and carboxyl groups of cephalosporin C, reaction with PCl_5 in the presence of base yields an iminochloride derivative. Addition of alcohol is used to form the imino ether, which can then be hydrolyzed to form 7-ACA (37).

Alternatively, conversion of cephalosporin C to 7-ACA can be performed enzymatically in two steps. The first step is reaction of the α-aminoadipyl group with D-amino acid oxidase to produce glutaryl-7-ACA. This reaction proceeds through a keto-7-ACA intermediate that undergoes an oxidative decarboxylation in the presence of H_2O_2. Next, a glutaryl acylase is used to remove the glutaryl side chain to produce 7-ACA. An 85% yield of this reaction can be achieved using an immobilized enzyme system (38). Acai et al. coimmobilized the D-amino acid oxidase and the glutaryl acylase enzymes to perform a single-step conversion of cephalosporin C to 7-ACA (39).

SEMISYNTHETIC CEPHALOSPORINS

The outer membrane of gram-negative bacteria is impermeable to penicillins because of their hydrophobicity. Cephalosporins, however, are less hydrophobic, and therefore can pass through the outer membrane to reach their targets. The natural product, cephalosporin C, has low antibacterial activity. Chemical modification at the C-7 position of cephalosporin C yields compounds with a much improved antibacterial activity. Since the development of an efficient method for preparation of 7-ACA, a large number of semisynthetic cephalosporins have been made in the past 30 years. The development of these compounds can be divided into four stages (40).

The first-generation semisynthetic cephalosporins introduced in the mid-1960s include cephalothin, cephaloridine, and cefazoline (Fig. 5). These cephalosporins are modified at the C-3 and C-7 positions. The antibacterial activity of these compounds is still limited because indole-positive *Proteus* species and opportunistic pathogens, such as *Pseudomonas aeruginosa*, are resistant because of the presence of β-lactamases. Since the late 1960s, an enormous effort has been devoted to the development of cephalosporins resistant to β-lactamases.

Out of this research, the second-generation cephalosporins were generated. Among this group, two types of compounds, cephamycins and 7-oximinoacyl cephalosporins, were developed. In 1971, cephamycins were discovered from actinomycetes (10). Cephamycin C has a 3-carbamoyl and a 7α-methoxy group, which confers a remarkable resistance to β-lactamases. Chemical derivation of cephamycin C led to the development of cefoxitin and moxalactam (41) (Fig. 5). The compound developed from 7-oximinoacyl cephalosporin is cefuroxime (41) (Fig. 5). Cefuroxime carries the same 3-carbamoyl-oxymethyl group as cephamycin C does.

In the search for better cephalosporins with broader antibacterial activity and resistance to different β-lactamases, the addition of a hydrophilic aminothiazole group to oximinocephalosporins produced the third-generation cephalosporins, such as cefotaxime, ceftizoxime, and ceftazidime (41) (Fig. 6). Again, the third-gener-

Table 1. U.S. Manufacturers' Trade Names for Semisynthetic Cephalosporins

Semisynthetic cephalosporins	Trade name (manufacturer)
First generation	
Cefadroxil*	Duricef (Bristol-Myers Squibb)
Cefazolin	Ancef (SmithKline Beecham), Kefzol (Lilly)
Cephalexin*	Biocef (International Ethical), Cephalexin (Biocraft, Lederle Standard Products, Novopharm USA, Warner Chilcott), Keflex (Dista), Keftab (Dista)
Cephradine*	Cephradine (Biocraft, Lederle Standard Products, Warner Chilcott)
Second generation	
Cefaclor*	Ceclor (Lilly)
Cefamandole	Mandol (Lilly)
Cefmetazole	Zefazone (Upjohn)
Cefonicid	Monocid (SmithKline Beecham)
Cefotetan	Cefotan (Zeneca)
Cefoxitin	Mefoxin (Merck & Co., Inc.)
Cefprozil*	Cefzil (Bristol-Myers Squibb)
Cefuroxime	Kefurox (Lilly), Zinacef (Glaxo Wellcome)
Cefuroxime axetil*	Ceftin (Glaxo Wellcome)
Third generation	
Cefixime*	Suprax (Lederle)
Cefoperazone	Cefobid (Roerig)
Cefotaxime	Claforan (Hoechst-Roussel)
Cefpodoxime*	Vantin (Upjohn)
Ceftazidime	Ceptaz (Glaxo Wellcome), Fortaz (Glaxo Wellcome), Tazicef (SmithKline Beecham), Tazidime (Lilly)
Ceftrizoxime	Cefizox (Fujisawa)
Ceftriaxone	Rocephin (Roche Laboratories)
Loracarbef*	Lorabid (Lilly)
Four generation	
Cefepime	Maxipime (Bristol-Myers Squibb)

Note: Orally administered cephalosporins are marked with an asterisk. All others are administered by intramuscular or intravenous injection.

ation cephalosporins have been compromised by the rapid emergence of resistant bacterial strains with plasmid-encoded β-lactamases and by their ineffectiveness against strains of *Enterobacter cloacae* and *P. aeruginosa* with constitutively expressed chromosomal β-lactamase.

By the late 1980s, the fourth-generation cephalosporins, cefepime and cefpirome (Fig. 6), had been developed. The new compounds retain the aminothiazole moiety at C-7 with an oxyimino substitution and contain a positively charged quaternary nitrogen at the C-3 position (40). Table 1 summarizes the semisynthetic cephalosporins available in the U.S. market (40,42).

CONCLUSION

Since the discovery of cephalosporin C 50 years ago, rapid progress has been made in the strain improvement and fermentation process development for the production of cephalosporin C. Genetic manipulation of *A. chrysogenum* and *S. clavuligerus* allows the potential for engineering of cephalosporin biosynthetic enzymes with broader substrate specificity for in vivo or in vitro production of new β-lactams. Because of the continuous selection of resistant bacteria carrying newly formed β-lactamases, the development of new semisynthetic cephalosporins remains a constant effort to find new compounds effective against these pathogens.

BIBLIOGRAPHY

1. S. Selwyn, in S. Selwyn ed., *The Beta-Lactam Antibiotics: Penicillins and Cephalosporins in Perspective*, Hodder & Stoughton, London, 1980, pp. 1–55.
2. M. Barber, *The Cephalosporin Business 1995–2005*, Michael Barber Assoc. Surrey, U.K., 1995.
3. H.S. Burton and E.P. Abraham, *Biochem. J.* **50**, 168 (1951).
4. W. Gams, *Cephalosporium-artige Schimmelpilze (Hyphomycetes)*, Gustav Fischer Verlag, Stuttgart, Germany, 1971.
5. G.G.F. Newton and E.P. Abraham, *Nature* **175**, 548 (1955).
6. E.P. Abraham and G.G.F. Newton, *Biochem. J.* **79**, 377–393 (1961).
7. D.C. Hodgkin and E.N. Maslen, *Biochem. J.* **79**, 393–402 (1961).
8. R.B. Morin, B.G. Jackson, E.H. Flynn, and R.W. Roeske, *J. Am. Chem. Soc.* **84**, 3400–3401 (1962).
9. S.L. Neidleman, S.C. Pan, J.A. Last, and J.E. Dolfini, *J. Med. Chem.* **13**, 386–388 (1970).
10. R. Nagarajan, L.D. Boeck, M. Gorman, R.L. Hamill, C.E. Higgens, M.M. Hoehn, W.M. Stark, and J.G. Whitney, *J. Am. Chem. Soc.* **93**, 2308–2310 (1971).
11. J. Baddiley, in M. Salton and G.D. Shockman eds., *β-Lactam Antibiotics- Mode of Action, New Developments, and Future Prospects*, Academic Press, New York, 1982, pp. 13–30.
12. J.L. Strominger, *Harvey Lect.* **64**, 171–213 (1970).
13. S.E. Jensen, *Crit. Rev. Biotechnol.* **3**, 277–301 (1986).
14. T.D. Ingolia and S.W. Queener, *Med. Res. Rev.* **9**, 245–264 (1989).
15. S.M. Samson, R. Belagaje, D.T. Blankenship, J.L. Chapman, D. Perry, P.L. Skatrud, R.M. Van Frank, E.P. Abraham, J.E. Baldwin, S.W. Queener, and T.D. Ingolia, *Nature* **318**, 191–194 (1985).
16. U.S. Pat. 5,472,853 (December 5, 1995), J.E. Baldwin (to University of Oxford).
17. J.F. Martin, S. Gutierrez, F.J. Fernendez, J. Velasco, F. Fierro, A.T. Marcos, and K. Kosalkova, *Antonie Van Leeuwenhoek* **65**, 227–243 (1994).
18. S. Kovacevic, M.B. Tobin, and J.R. Miller, *J. Bacteriol.* **172**, 3952–3958 (1990).
19. J.J.R. Coque, F.J. Enguita, J.F. Martin, and P. Liras, *J. Bacteriol.* **177**, 2230–2235 (1995).
20. W.O. Pat. Appl. 95/29253 (November 2, 1995), J.J.R. Coque, F.J. Enguita, J.L. Fuente, F.J. Llarena, P. Liras, and J.F. Martin (to Merck & Co, Inc.).
21. C.W. Chen, H.-F. Lin, C.L. Kuo, H.-L. Tsai, and J.F.-Y. Tsai, *Biotechnology* **6**, 1222–1224 (1988).
22. R.P. Elander and S.J. Chiang, in A. Prokop, R.K. Bajpai, and C.S. Ho eds., *Recombinant DNA Technology and Applications*, McGraw-Hill, New York, 1991, pp. 153–170.
23. J.J. Usher, D.W. Hughes, M.A. Lewis, and S.-J.D. Chiang, *J. Ind. Microbiol.* **10**, 157–163 (1992).
24. L.H. Malmberg, W.-S. Hu, and D.H. Sherman, *J. Bacteriol.* **175**, 6916–6924 (1993).
25. P.L. Skatrud, A.J. Tietz, T.D. Ingolia, C.A. Cantwell, D.L. Fisher, J.L. Chapman, and S.W. Queener, *Biotechnology* **7**, 477–484 (1989).
26. J. Basch and S.-J.D. Chiang, *J. Ind. Microbiol. Biotechnol.* **20**, 344–353 (1998).
27. J.-Y. Zhang, G. Banko, S. Wolfe, and A.L. Demain, *J. Ind. Microbiol.* **2**, 251–255 (1987).
28. J.-Y. Zhang, S. Wolfe, and A.L. Demain, *Curr. Microbiol.* **18**, 361–367 (1989).
29. R.W. Swartz, S.M. Vicik, A.J. Fedor, *Biotechnol. Prog.* **6**, 333–340 (1990).
30. J.J. Usher, M.A. Lewis, D.W. Hughes, and B.J. Compton, *Biotechnol. Lett.* **10**, 543–548 (1988).
31. A. Smith, in H.W. Blanch, S. Drew, and D.I.C. Wang eds., *Comprehensive Biotechnology*, vol. 3. *The Practice of Biotechnology: Current Commodity Products*, Pergamon Press, New York, 1985, pp. 168–185.
32. J. Konecny, E. Felber, and J. Gruner, *J. Antibiot.* **26**, 135–141 (1973).
33. M.E. Wildfeuer, in D. LeRoith, J. Schiloach, and T.J. Leahy eds., *Purification of Fermentation Products*, American Chemical Society, Washington D.C., 1985, pp. 155–174.
34. U.S. Pat. 3,719,563 (March 3, 1973), R.L. Hamill, C.E. Higgens, and M.M. Hoehn (to Eli Lilly & Co.).
35. R.R. Chauvette, P.A. Pennington, C.W. Ryan, R.D.G. Cooper, F.L. Jose, I.G. Wright, E.M. Heyningen, and G.W. Huffman, *J. Org. Chem.* **36**, 1259–1267 (1971).
36. U.S. Pat. 5,318,896 (June 7, 1994), M.J. Conder, P.C. McAda, and J.A. Rambosek (to Merck & Co., Inc.).
37. L.D. Hatfield, W.H.W. Lunn, B.G. Jackson, L.R. Peters, L.C. Blaszczak, J.W. Fisher, J.P. Gardner, and J.M. Dunigan, in G.I. Gregory ed., *Recent Advances in the Chemistry of β-Lactam Antibiotics*, Royal Society of Chemistry, Burlington House, London, 1981, pp. 109–124.
38. H.D. Conlon, J. Baqai, K. Baker, Y.Q. Shen, B.L. Wong, R. Noiles, and C.W. Rausch, *Biotechnol. Bioeng.* **46**, 510–513 (1995).
39. P. Acai, E. Michalkova, V. Bales, *Bioproc. Eng.* **14**, 105–109 (1996).

40. R. Periti, *J. Chemother.* **8**, 3–6 (1996).
41. C.E. Newall, *Chem. Br.* 976–980 (October 1987).
42. *Physicians' Desk Reference*, 50th ed., Medical Economics, Montvale, N.J., 1996.

See also METABOLITES, PRIMARY AND SECONDARY; SECONDARY METABOLITE PRODUCTION, ACTINOMYCETES, OTHER THAN STREPTOMYCES.

CHINESE HAMSTER OVARY CELLS, RECOMBINANT PROTEIN PRODUCTION

FLORIAN M. WURM
Swiss Federal Institute of Technology
Lausanne, Switzerland

KEY WORDS

Cell culture process
Dihydrofolate reductase
DNA integration
Gene stability
Large scale
Manufacturing
Methotrexate
Suspension cells
Transfection

OUTLINE

INTRODUCTION

Continuously propagatable mammalian cells, in general, and Chinese hamster ovary (CHO) cells, in particular, have become preferred host systems for the production of complex proteins for clinical applications. Nowadays, in pharmaoriented biotech companies, more protein species are being produced in CHO cells than in *E. coli* and in *Saccharomyces*, which were production hosts of first choice during the early years of applied DNA technology. These microbial organisms are capable of producing large amounts of recombinant protein. However, the decision to employ somewhat more complex and costly in-vitro cultures of recombinant mammalian cells has been driven by the need to obtain proteins with superior biological activity and more complex biochemical structures. Biological activity and, last but not least, the pharmacokinetic characteristics of recombinant proteins, when administered in therapy, frequently depend on a number of complex protein modifications, that is, proper folding and disulfide bridge formation, oligomerization, proteolytic processing, phosphorylation, and the addition of specific and complex carbohydrate groups. Most of these and other protein processing steps can be executed with high efficiency by mammalian cells.

CHO cells have become dominant in industry, and not only for historic reasons. Of course, they were chosen as the host substrate for the first recombinant protein from mammalian cells, human tissue plasminogen activator (Activase® tPA) (1), achieving market approval in the United States and other global markets. CHO cells are unique in many ways and exhibit a number of advantages, ranging from ease of introduction of exogenous DNA to the capacity for growth and production at very large scales. This article attempts to summarize and highlight arguments, issues, advantages, questions, and ongoing research, in a cursory way, for the industrial production of high-value proteins derived from CHO cells. In doing so, the development path followed to establish a high-yield process will be followed, from the construction and use of appropriate plasmid vectors to the planning and design of the cell culture manufacturing process. Issues raised and discussed herein are not meant to be comprehensive, yet it is hoped that the most critical points impacting development efforts will be addressed.

PRINCIPLE OF DNA TRANSFER AND AMPLIFICATION IN CHO CELLS

The following gives a brief introduction into the underlying principles for use of CHO cells as recipients of protein-encoding DNA vectors, and the most popular way to improve productivity with these cells: experimentally induced gene amplification.

A mutant cell line lacking enzyme activity for dihydrofolate reductase, developed by Urlaub and Chasin (2), can be cultivated in the presence of glycine, hypoxanthine, and thymidine (GHT). When these cells are transfected with a functional expression vector for dihydrofolate reductase (DHFR vector), cell lines that have acquired a functional DHFR gene can be selected and expanded in media lacking GHT. A second (or third or more) expression cassette for a product of interest (e.g., a protein with therapeutic value) can be included in the DHFR plasmid or cotransfected as part of one or more additional vectors. Screening a number of "clones," or more correctly, cell lines derived from colonies, will lead to the identification of cells coexpressing the product of interest.

In a second and often time-consuming phase of cell culture work, expression can be augmented by exposing recombinant cells to the drug methotrexate (MTX), a folate derivative that blocks DHFR activity completely and irreversibly. Usually, within a period of 2–3 weeks, from a population of cells a few colonies emerge (while the majority of cells die) that are resistant to the drug due to elevated expression of DHFR. It was found that these cells frequently contain chromosomally integrated plasmid sequences in a higher copy number than observed in cells prior to exposure to MTX. Treatment with stepwise elevated concentrations of MTX can be repeated several times and may result in the isolation of cells that contain dramatically increased copy numbers of the transferred genes.

The same principle for MTX-driven amplification applies to other immortalised cell lines as well (3); however, DHFR-mediated DNA transfer and methotrexate-driven amplification works best with cells that lack an endogenous DHFR gene (activity) prior to transfection.

REGULATORY ISSUES: PRODUCT SAFETY, CELL CHARACTERIZATION, IDENTITY, AND CONSISTENCY FOR PRODUCT FORMATION

CHO cells, originally derived from the ovary of a female Chinese hamster, emerged as a spontaneously immortalized line, more than 40 years ago in the hands of Kao and Puck (4,5). The cells have been the object of many studies since then and are very well characterized with respect to a variety of aspects, including karyotype, chromosome structures, gene mapping, general culture conditions, cell physiology, and media requirements. For the purpose of genetic and physiological studies on the activity of dihydrofolate reductase and its genomic alleles, in the late 1970s, Urlaub and Chasin (2) mutagenized the original cell line with the help of radiation and chemicals. The goal was to create a mutant cell line lacking functional activity of both alleles of the DHFR gene locus. The resulting cell line was called CHO-DUX (sometimes also referred to as CHO-DUKX-B11) and, although not initially intended for this purpose, proved to be an ideal substrate for a number of pioneering gene transfer experiments (6–8).

The "climate" for permission by regulatory agencies, in particular the FDA in the United States, was not favorable in the early 1980s for using "tumor" cell lines, such as immortalized CHO cells, for the production of recombinant proteins. Discussions about risks associated with the use of mammalian cells were controversial, and had been initiated more than two decades earlier (8) when a first generation of "classical" biological products (i.e., vaccines and, later, natural interferons) were developed on the basis of primary monkey kidney cells, human diploid cells, and later, transformed mammalian cells. Manufacturers of recombinant proteins for clinical applications and regulatory agencies were in agreement that it was very important to minimize eventual risks associated with the use of recombinant mammalian cell hosts. Risks were seen in tumor principles, carried by the DNA of the host and in adventitious agents (viruses, mycoplasma, etc.), which could infect the host cell lines. Also, consistency and quality of the recombinant proteins were discussed in the context of risk assessment and risk control. The result of a decade-long discussion was that very stringent manufacturing controls and monitoring procedures were enforced as a prerequisite for manufacture of proteins from such cells. A balance had to be found between the almost assured clinical benefit of some of those first recombinant products and the perceived (real or imaginary) risks associated with the unavoidable necessity to produce them in "tumor" cells.

The first product, recombinant tissue plasminogen activator (Activase® rtPA), proved to be a good candidate to achieve such a balance because the benefit, saving lives of heart-attack patients, outweighed by far anticipated risks, for example, the debated risk of exposing patients to DNA sequences that may transfer a tumor-inducing principle. However, approval was achieved only after a huge amount of data was provided to the regulatory agencies showing that (1) consistently, only a very small quantity of CHO DNA (less then 10 pg/dose recently related to less than 100 pg/dose) was present in the final product, (2) the product itself could be produced with a very high degree of reproducibility, and (3) it was produced with a purity not achieved before in any biological derivation from mammalian cell culture.

Adventitious Agents: Bacterial and Fungal Contamination

Prevention of airborne and other bacterial or fungal infections of cell culture and recovery systems can be assured, to a very high degree of confidence, by the use of a piping and vessel system that maintains absolute containment of the sterile medium fluids. The equipment used must be of a nature that allows cleaning and sterilization by clean-in-place (CIP) and sterilize-in-place (SIP) procedures (usually using high-quality stainless steel). Most cell culture processes require complex media containing amino acids, vitamins, protein hormones, and fetal bovine serum. Some of these components cannot be autoclaved. To exclude, therefore, the introduction of bacterial and fungal contamination through these raw materials, prior testing and, in addition, filtration through membranes of 0.2-μm-or even 0.1-μm-pore-size into presterilized containers is employed.

Rigorous testing cell banks (Master Seed Cell Bank, MSCB, Manufacturers Working Cell Bank, MWCB) (for review see Ref. 10), done by analyzing cells of a number of representative cryovials, which represent the bank when frozen in liquid nitrogen, assures that the production cell line itself is not contaminated with viruses, bacteria, mycoplasma, and fungi. Sterility testing of the cell line must be done in appropriate media lacking antibiotic or antifungal compounds, for obvious reasons. Virus testing is done in suitable cell systems that are validated for each of the individual virus species.

Adventitious Agents: BSE

The use of sera or other products derived from bovine sources represents a potential risk of transfer of the causative agent for bovine spongiformous encephalopathy (BSE) to patients receiving recombinant proteins.

Regulatory agencies therefore request detailed information on the origin and processing by the vendor of bovine derived products. For example, sera obtained from countries in which BSE was diagnosed, even in a small number of animals, are considered unacceptable by regulatory agencies. Companies have, in most cases anticipating these regulations, assured their supply of bovine-derived process materials from countries like New Zealand or the United States where BSE has not yet been reported. Recently, due to the high degree of concern raised in Europe, but also in the United States, very stringent process regulations have been discussed. These regulations would restrict to a very large extent, or prevent altogether, the use of bovine-derived process additives, not taking into account the processing of the materials by the vendor or the receiving company. The high degree of concern regarding BSE is based on the possibility that *(1)* a very small proteinaceous compound could be responsible for the disease, *(2)* transfer to the human population as Creutzfeld–Jacob disease has been considered possible, and *(3)* detection of the causative agent is extremely difficult, even with the most sophisticated methods available today. It appears that the complexity of the involved issues will require further discussion in the future and must occur at an international level.

Adventitious Agents: Viral Contaminants

Why were hamster cells chosen as a host for making human recombinant proteins? A strong argument, favoring nonhuman cell lines over human cell lines for the production of proteins, is the fact that certain life-threatening human viruses cannot be propagated, or grow poorly, in nonhuman cell lines. CHO cells have been found to not support the replication of pathogenic viruses such as polio, herpes, hepatitis B, HIV, measles, adenoviruses, rubella, and influenza. Thus, the risk of a viral adventitious agent, belonging to this group of viruses, involuntarily carried along with the product of interest can be considered extremely low. Wiebe et al. tested a total of 44 human pathogenic viruses for replication in CHO cells and found only seven (Reo 1,2,3; mumps; and parainfluenza 1,2,3) that were able to infect these cells (11). Exclusion of these few virus species, as well as some other viruses that can be propagated on CHO cells, such as the minute virus of the mouse (MVM), can be assured to a very high degree of certainty through testing. Tests can be performed with all materials that enter the manufacturing process and would support the viability of the virus in question. Tests are also obligatory with fluids that contain the product of interest and that represent, for example, toward the end of the cell culture production cycle, a manufactured product unit, defined as a "batch."

Sterility filtration of fluids containing a variety of raw materials, some of which may have been exposed to viruses, does not prevent the introduction of viral contaminants into the process of cell culture–based manufacturing of proteins. Only recently have membranes with pore sizes small enough to exclude passage of virus particles become available for industrial-scale operations. However, these membranes cannot be introduced into existing processes without complex consequences on regulatory issues. Testing therefore still appears to be the most efficient method to exclude viruses that may reside within the host cell line itself, or that could be introduced by biologically derived raw materials needed for the cell culture process. Specialized service companies have developed, in close collaboration with the pharmaceutical client companies, batteries of validated test procedures. These utilize cell culture systems in which supernatants or lysates of the production cells are cocultivated with the corresponding virus-sensitive substrates. Because cells in master seed banks and in working cell banks are of the highest importance for the cell culture production process, samples from these are the first to be considered for the rather expensive and time-consuming testing exercise.

Another and complementary approach to virus safety is the design of virus kill and removal steps of the recovery process for the protein of interest. These include physical and chemical principles separating (theoretical) viral contaminants from the product or inactivating them. Again, appropriate testing procedures and the demonstration of inactivation and removal of model viruses, as discussed in the important paper by Wiebe et al. (11) was considered a major provision for the achieved safety of recombinant products from hamster cells.

Product Consistency, Quality, and Purity: DNA, Endotoxin, and Protein Characterization

A very rich collection of methods has become available for the analysis of purified proteins within the last two decades. In addition, most of these methods have been recently optimized and fine-tuned to very high sensitivities and resolution. When employed as routine analytical procedures during the manufacturing process, they can assure high quality and consistency of protein products (12–17). Still, the manufacturer of a pharmaceutical protein has one major concern: will a product that has demonstrated its efficacy in extensive and expensive clinical trials remain the same in its essential characteristics when employing a defined manufacturing process for its production over many years? It seems surprising, but due to the large size and complexity of proteins under study for clinical use today, their structural features and their functional principles within the human body may not be fully understood by the manufacturer when completing successful clinical trials. This is especially true for the newer generation of pharmaceutical proteins that are larger in size and often contain multiple polypeptides and/or multiple specialized domains exhibiting secondary modifications such as glycosylaton, phosphorylation, and sialylation. Subtle changes, sometimes difficult to detect due to inherent heterogeneities of structures of protein populations, may result in loss or modification of activity and could pose risks to the patient. In order to reduce this possibility, batteries of in-process controls and tests are an inherent part of the production of clinical proteins. In the following section, sensitive and robust techniques and methods are outlined and discussed. The objective is to *(1)* prevent the occurrence of even small changes in the procedures employed to produce the product of interest and *(2)* enable the detection and exclusion from the final vial protein variants differing in a major way from the one tested in clinical trials.

Adherence to GMP Protocol for Manufacturing. The first level of protection against inadvertent changes of the product rests in effective management of the production process over time. The challenge is the same as that of any mature industry: to produce large amounts of material at competitive cost while ensuring that product consistency and quality are maintained. Obviously, manufacturing teams and their supervisors undertake every effort to reproduce the manufacturing process to the utmost detail in every production run. This is achieved by defining and describing each of the various steps in the form of detailed protocols. Almost every aspect of the procedure is documented. These documents establish, in the form of cGMP (current good manufacturing practice) protocols, the basis for the overall procedure. A "sign off" procedure by supervisors represents an integral part, assuring that the operating personnel for the manufacturing process are in fact controlling, assessing, and executing it according to the established protocols. At critical points in the overall process, the signatures are prerequisites allowing the progression of the process to the consecutive steps.

The time period of cell line and process development leading up to the establishment of the manufacturing process is very important for the definition of critical "checkpoints." During this period, knowledge about parameters and steps is acquired that can result in product changes. Once certain limits of variations of process conditions have been identified (within which no change was observed) the cGMP protocol is drafted and finalized. Specific events, defined in precise terms as part of the manufacturing protocols, can trigger a more elaborate investigation. Supervisors and managers can even order an interruption of the manufacturing process. In extreme cases, crude product batches are withheld from further processing and are discarded.

Quality Control for the Detection of Variant Product Forms. Quality control (QC) is an integral part of the manufacturing process of recombinant products. A comprehensive approach, utilizing independently validated techniques, is applied to assess the quality and identity of the product from various angles (for review see also Ref. 18). It is the goal of QC efforts to assure that products made over years of manufacturing will meet the stated specifications in terms of identity, quantity, activity, and purity.

In principle, it is no longer very difficult to produce large quantities of highly purified recombinant proteins. However, methods to produce these proteins are still considered part of a young technology, because its basis is the manipulation of genetic material of various sources in the laboratory. Those manipulations involve the creation of plasmid vectors, based on fragments of genetic information that are linked together as a first step. In a second step, the transfer of such DNA for transgene-dependent protein synthesis to a complex organism such as *E. coli*, yeast, or, as discussed here, into the chromosomes of mammalian cells cultivated in vitro occurs. A major concern has been the fidelity (amino acid sequence identity) of the final product, particularly in view of the high degree of ignorance with respect to gene transfer mechanisms in higher eucaryotes and its consequences. Also, it appears that transfected DNA may have, possibly during the transfer to the host cell, a somewhat elevated propensity for mutation (19). A discussion of issues concerning structure and fate of DNA transferred to mammalian cells will be discussed later.

Nevertheless, based on a history of experience with this technology for more than 10 years, it can be stated that this young technology is in fact a very reliable one. It has been suggested that rigorous and extensive nucleic acid based tests, most notably a complete sequence assessment of the genomically integrated DNA (or transcribed RNA from recombinant cells) should be performed (20,21). It appears that such rather complex, expensive, and time-consuming tests are not necessary. An entire session was devoted to nucleic acid considerations during a conference organized by the International Association of Biological Standardization, held in Annecy, France in 1994. Berthold specifically addressed genetic stability in CHO cells (22). As already pointed out, following the rapid development of techniques allowing manipulation and transfer of DNA molecules to various hosts, powerful and reliable new methods have emerged in analytical protein chemistry. These have increased the capacity to characterize purified protein preparations to an extremely high degree of sensitivity and resolution. These methods represent efficient tools to assess identity (including the amino acid sequence), quantity, potency, and purity of the product at the level where it is most needed—just prior to administration (delivery) to the patient. The results from QC assessments from thousands of lots of approved products are, as far as this author knows very encouraging with respect to (amino acid sequence) identity: no lot has been called into question based on a QC finding that has indicated a newly emerged mutant gene.

Still, this statement does not negate the possibility that mutants may occasionally emerge. They will, over and over again, within the large population of cells in bioreactors. A 1,000-L reactor usually contains more than 10^{12} cells. However, there is no reason to believe that mutations would occur at a frequency significantly different from the frequency of mutations occurring within the human body (1 bp change in 10^9 bp for each generation) (23). Of course, such mutations will be "scattered," and as individual entities are extremely rare. Therefore, individual and specific variants are not likely to emerge in the population of protein molecules as a sizeable subclass of identical molecules.

A telling example of the power of in-process controls and the associated biochemical assays to identify variants in a population of molecules is given in an article by Sliwkowski and coauthors (24). It describes the detection of an amino acid exchange mutant of a recombinant monoclonal antibody (anti-Her2). This variant was detected early during process-development efforts when using a methotrexate-amplified clone of CHO cells. The variant represented about 10% of the total population of antibody molecules. The origin of this mutant remains mysterious, but it seems it was the result of an early event during the development of the cell line that was used in the process.

MOLECULAR BIOLOGY OF VECTOR DESIGN, TRANSFECTION, AND IDENTIFICATION OF HIGHLY PRODUCTIVE CELL POPULATIONS

DNA Transfer to CHO Cells Using Calcium Phosphate

Graham and van der Eb (25) showed more than 20 years ago that the transfer of DNA into mammalian cells in culture can be achieved by exposing cells to microprecipitates of DNA and calcium phosphate. This simple method (termed calcium phosphate transfection) has become very popular. Typically, a glycerol shock is part of the transfection procedure. The shock is thought to promote the transport of individual DNA–calcium phosphate complexes across the cell membrane, perhaps by inducing endocytosis or phagocytosis. When optimized transfection conditions are used, a high percentage of cells take up DNA. However, in addition to the movement across the outer membrane of the cells, the transport of DNA from the cytoplasm to the nucleus seems to be a significant barrier. In 1982 Loyter et al. reported that only 1–5% of transfected cells have detectable calcium phosphate complexes within the nucleus (26,27). The stable integration of transfected DNA into chromosomes of mammalian cells was first shown by Axel and collaborators (28,29). The number of emerging colonies upon transfection by the calcium phosphate technique has been reported to be low: usually, only between 0.01% and 0.5% of the cells gave rise to recombinant colonies (30). More recent investigations and appropriate improvements of crucial physicochemical parameters of the calcium phosphate transfection methodology in several laboratories have increased transfection efficiencies for stable recombinant cells to about 5–10% (and those of transient transfections to levels of 50% or higher) of the population of cells exposed to exogenous DNA (31–33). Calcium phosphate transfection still seems to be the most frequently used method for the generation of recombinant CHO cells.

Yet, it appears that the transfer of a sufficient quantity of DNA into the nucleus and the integration of at least a few functional plasmid molecules into transcriptionally active regions of the receiving genome is by no means an efficient process. Half a million or more plasmid molecules are usually provided per cell in a typical calcium phosphate transfection, the result of which is stable gene transfer in a small percentage of the exposed cells. It has been shown that the association of the DNA with calcium phosphate protects against nuclease attack during the transport to the nucleus (26). Eventually these complexes become dissociated and nuclear endo- and exonucleases will have access to the "naked" DNA. Moreover, it has been shown that supercoiled plasmid DNA molecules will be converted into relaxed (single-stand cut), open circular, and finally linear molecules (double-strand cut) within the nucleus after 1–2 h (34). This is, of course, an essential step for integration into the linear DNA backbone of a chromosome.

The mechanism by which the DNA is transported across the nuclear membrane (nuclear pores are not large enough for diffusive transport), and finally to the site of chromosomal integration, are not yet known. On the receiving side, at least one strand of the chromosomal DNA must be opened while a linear plasmid molecule is in sufficiently close proximity. Nuclear ligases could then mediate covalent linkages. Very little is known about the mechanism and about the factors affecting site specificity for integration.

Site specificity is a very relevant factor in this context. It is believed that it has influence on the transcription rate of the integrated DNA, even if the DNA carries a powerful (viral) promotor, such as the SV40 early promoter or a CMV promoter. The current assumption is that exogenous DNA will integrate at "random" within the genome, even if the transfection cocktails contain DNA with homology to sequence segments of the receiving genome. This is an observation from numerous gene-targeting experiments in mammalian cells. It was found that only 1 in 1,000 to 1 in 10,000 events would result in targeted integration (35,36). With respect to the actual sites of integration in CHO cells, there is only limited and sporadic data available at present. Only a small proportion of recombinant cell lines should contain the transferred DNA within transcriptionally active regions of chromosomes if integration would truly occur at random subsequent to transfection. In reality, we cannot know whether DNA will end up more frequently than not in transcriptionally inactive regions of the genome. We have no opportunity to analyze cells that did not overcome the selection step. The selection step is, as we know from the low transfection efficiencies in stable transfections, a very significant barrier.

Among the emerging cell lines there is usually extensive heterogeneity in productivity for recombinant proteins. The expression levels from primary CHO clones for proteins like rtPA, a secreted version of the CD4 receptor (rCD4-truncated), and a chimeric TNFR-FC fusion protein had a range of more than two orders of magnitude (37,38), observations which this author believes can be generalized for all transfections in mammalian cells. As a consequence, the identification of high-producer cell lines is a tedious and labor-intensive exercise and requires the screening of hundreds of individual cell lines. In the author's experience, it is usually necessary to invest 2–4 months of laboratory work into this task. Only then has one enough confidence in having appropriately assessed the upper range of sustainable expression for a novel protein in CHO cells.

Nuclear enzymes such as endo- and exonucleases, but also ligases (39) and possibly recombinogenic enzymes (40,41), acting on the population of plasmid molecules within the nucleus are responsible for the modification of transfected DNA that becomes eventually integrated into the genome. This modification may not only degrade in part the plasmid DNA but is also responsible for creation of larger, that is, more extended DNA complexes. These DNA molecules seem to be created before the integration into the genome. They provide the physical basis for linking the genetic selection marker and the gene(s) of interest in those transfections that utilize separate vectors. It should be noted here that DNA transfer methods, other than the use of calcium phosphate, might deliver different quantities of DNA to the nucleus. This may have profound effects on the structure and copy number of integrated DNA molecules, as will be discussed later.

In discussing the integration loci for transfected DNA in the mammalian genome, it is interesting to note that integration sites for amplified DNA sequences, detectable by fluorescence in situ hybridization (FISH) after selection in methotrexate, did not show a preference for a particular chromosome or chromosomal region. However, in light of recent data on gene amplification mechanisms (42), it is not possible to draw conclusions from these findings toward the identification of the initial (primary) integration site(s).

Additional DNA Transfer Methods

In addition to calcium phosphate transfection (25,31), more recently electroporation (43,44), cationic lipids, and liposome mediated (45–47) transfection techniques have been developed for transfer of DNA into mammalian cells. Most of these techniques have been reported to mediate higher transfection efficiencies as compared with calcium phosphate–mediated DNA transfer. Such claims must be regarded with some caution because all DNA transfer techniques established so far suffer from high variability. Some of the variabilities observed are probably due to technical difficulties in producing reliable and consistently highly efficient carrier systems. The other factors that cause major variations in transfection efficiency are the type of cells used and the condition of the cells prior to transfection.

Because carrier principles for the transfer of DNA to cells can vary widely, one suspects that very different consequences may result within the cell, events over which the person conducting the experiment has no or little control. Just one of the possible consequences should be mentioned here: The technique applied for DNA transfer may affect the copy number average of individual clones, depending on the amount of DNA transferred to the nucleus. It has been proposed that electroporation may result in a smaller copy number of chromosomally integrated DNA than, for example, calcium phosphate–based transfection. In view of the applied selection procedures for the generation of recombinant cell "clones," very different chromosomal structures of integrated plasmid sequences may result.

Another factor to be taken into account for the genomic structure of integrated plasmid sequences is the physical form of the DNA for transfection. Because linearization of plasmid DNA is a prerequisite for integration into any chromosomal site, the structure of the plasmid prior to transfection will affect the final structure when integrated. When circular plasmids are used, as predominant in the past, linearization is dependent on cellular enzyme activity. This will happen most likely at random, and the integrated DNA structure cannot be predicted. Depending on the overall size of the plasmid molecule and the size of the DHFR gene cassette relative to the vector, a significant number of molecules would be linearized within the DHFR sequence, which jeopardizes the functionality of this plasmid. Obviously, molecules whose DHFR expression cassette are interrupted or deleted by the linearization process would be unable to transfer a functional marker gene to the host cells, and such cells would be eliminated through the selection procedure. Therefore, opening the circular plasmid molecule by restriction enzyme digestion with appropriate enzymes prior to transfection has been recommended for improved transfection efficiencies in protocols that aim at integration of plasmid DNA into chromosomal DNA.

We have seen in our laboratory that linearization is, in fact, advantageous in terms of transfection efficiency with stable transfections (32). Linear plasmid molecules are also used routinely in gene transfer experiments that attempt to transfer the desired DNA into specific (homologous) DNA sequences of the receiving genome (gene targeting). In these experiments, linearization is executed in a way that results in cutting the homologous DNA fragment approximately in half. The molecules for DNA transfer then carry DNA sequences that are homologous at their ends (48–50). We have used such an approach in our laboratory by employing (defective) retroviral DNA sequences derived from the CHO genome in expression vector cocktails. They have quite significantly improved the frequency of high-producer cell lines upon transfection (37,38).

CHO CELL ENGINEERING

Use of serum and other more or less undefined and complex media additives for growth of mammalian cells may result in problems of reproducibility in the process, and thus additional costs. To avoid a negative impact on cell growth, testing procedures have to be incorporated prior to use of these additives to identify possible variations from batch to batch. Therefore, the desire to replace some or all of these variable components in cell culture media with well-defined components having high purity and little or no variation in quality has been a priority in the industry for quite some time. Also, use of serum and animal-tissue-derived growth factors (insulin, transferrin) bears the considerable risk of transmittal of undesired adventitious agents to the production process because these additives frequently cannot be heat sterilized.

Although serum-free and growth-factor-free media may support the growth of mammalian cells, the robustness of the cell culture process, in general, and maximally achievable cell densities in particular may be reduced when using media free of serum- and/or protein growth factors. Therefore, efforts have been undertaken to modify production host cell lines prior to the transfer of expression vector DNA. The hope was to generate hosts for production that would achieve superior growth rates, and so would be more useful for scale-up and higher in productivity. Various oncogenes, cell cycle genes, and/or hormone genes (51–53) have been transferred into the genome of cells, resulting in novel and possibly superior production hosts. In view of actual discussions regarding BSE and other, so far unknown, adventitious agents, which pose the risk of entering production through process additives, the interest in such novel hosts will be very high. It needs to be seen how soon and how frequent these "super CHO" cell lines will be used in the manufacturing of novel products.

METHOTREXATE-MEDIATED VECTOR AMPLIFICATION IN DNFR-TRANSFECTED RECOMBINANT CHO CELLS

Treatment of DHFR-positive CHO cells with incremental concentrations of methotrexate has been used to select for populations and clones of populations of CHO cells with an increased (i.e., amplified) copy number of the integrated DNA within the host genome. The phenomenon of methotrexate-mediated gene amplification had been observed before the use of recombinant DNA technology, most notably in cancer patients who had been treated for extended periods of time with methotrexate (54).

With CHO cells, cell lines containing several hundreds, and sometimes a few thousand, copies of transfected DNA fragments integrated within chromosomes have been established (55,56). Most amplified cells produce more product than unamplified host cells. However, the improvement of specific productivity is highly variable when studying individual clones (57) and also varies from product to product (Wurm, unpublished observation, 1994).

Cytogenetics of Amplified CHO Populations: Long-Term Genetic Stability and Maintenance of Productivity

Subpopulations and clones of methotrexate-treated CHO cells may contain the transfected DNA sequences at very high copy numbers (55,56,58). Studies concerning genetic features of these amplified DNA sequences within the CHO genome have been ambiguous. Whether continued presence of a selective agent (i.e., methotrexate) in long-term cultures of recombinant CHO cells is required for production stability is still controversial. Whereas some studies suggest MTX is required for stable production of recombinant proteins (59), others indicate that continued culture of clonal cell lines in MTX may not be necessary (60). Cytogenetic studies using fluorescence in situ hybridization (FISH) (61) were done in the author's laboratory with clonal and nonclonal recombinant CHO cell lines. They showed that continuous exposure to MTX at the same concentration under which the cell populations were initially established promotes genetic instability at the chromosomal level (62). Cell lines selected at micromolar concentrations frequently show elongated chromosomal structures that hybridize to probes representing the transfected DNA. Some of the chromosomes, containing a large number of more or less tightly arranged bands or regions, differ dramatically from normal CHO chromosomes. This raised questions about cytogenetic stability of such aberrant chromosomes during continuous subcultivation and, in consequence, the production of recombinant proteins based on genes residing within these structures.

The first of our FISH analysis studies was done with cell lines expressing the mouse c-*myc* gene in an inducible manner. The advantage of this system is that the phenomena observed in the genetic context can be attributed solely to the status (expression level, chromosomal location, degree of amplification) of the DHFR selector gene and the inhibitor MTX. The c-*myc* gene is truly silent during regular growth conditions and could not act as a metabolic burden nor result in any selective forces. When these cells are cultivated at 37 °C, the inducible *Drosophila melanogaster* heat shock protein 70 promoter is in "off" status. Even in a cell line containing 1,000–2,000 copies (selected at 320 μM methotrexate) of the transfected DNA, no trace of transcribed c-*myc* RNA and translated c-*myc* protein was observed at 37 °C. However, high-level expression of c-*myc* RNA was induced upon shifting cells to a culture temperature of 43 °C for 1 h.

The continuous exposure of the 320-μM-MTX-selected cell line (c-*myc*) to the drug resulted in about a twofold higher copy number of plasmid DNA as compared with a population of cells that had been cultivated in the absence of any selective pressure for at least 4 weeks. A similar observation, based on a cell line with a constitutive promoter driving tPA expression, was reported by Weidle et al. (59). These authors reported a loss of about 50% in copy number when cells were grown in the absence of methotrexate for about 50 days. This decline in gene number also resulted in a similar drop in productivity. No further copy number and productivity loss was found during the subsequent observation period of up to 100 days.

Following our initial observation of a drop in copy number in cells containing the mouse c-*myc* gene, we extended our studies with this line and also analyzed the chromosomal integration patterns of transfected sequences in two additional CHO cell lines. The three cell lines studied under this program expressed one of the three following proteins: *(1)* human rtPA, *(2)* a truncated version of the human CD4 receptor, and *(3)* the murine c-*myc* protein. The three cell lines had been established by transfecting CHO-DUKX cells using standard calcium phosphate precipitation techniques. Either individual clones or pools of clones were cultivated for extended periods in media containing stepwise increasing concentrations of MTX. In each case, a single individual clone expressing the desired level of recombinant protein was expanded for long-term growth and production. FISH was performed using biotinylated probes complementary to the recombinant sequences. Three different sets of typical integration patterns (site, degree) of amplified sequences were found for lines under study.

FISH signals in the rtPA-expressing cell lines were localized, in the majority of cells, in one of the larger submetacentric chromosomes (probably either chromosome 3 or 4, or the X chromosome). The signals (a bright band of fluorescence) covered most of the length of the short arm and showed weaker fluorescing spots over the center of the chromatids in the long arm. A number of other types of integration patterns (including an elevated percentage of quasi-tetraploid cells) were found in these cells (about 25%) when they were grown in the presence of methotrexate. When methotrexate was omitted, more than 95% of metaphases had the integration described.

In cells expressing c-*myc*, continuous cultivation in media containing high concentrations of MTX was found to be associated with rearrangement and variable amplification of transfected sequences within almost half of the cells (62). Several types of integration patterns were found in these cells: *(1)* chromosomes with highly extended regions of hybridizing sequences, *(2)* cells containing transfected DNA integrated into multiple sites on an individual chromosome, *(3)* cells containing DNA integrated into multiple chromosomes, *(4)* chromosomes joined at amplified regions, *(5)* circular chromosomes, *(6)* cells with small derivative chromosomes or fragments, and *(7)* cells containing a single "normal" integration, exhibiting a bright band of hybridization in the center of the long arm of what we assume is chromosome 2. The last type *(7)* was found in about 40% of cells when grown in the presence of methotrexate. This integration type increased in frequency to 95% when cells were grown in the absence of methotrexate for 3 weeks or more.

Similar results showing quite extensive heterogeneity were observed with yet another recombinant CHO cell line (the fourth studied), producing a chimeric CD4-immunoglobulin protein. When grown in the presence of MTX, CD4-IgG-producing cells showed heterogeneity in both chromosomal location and copy number of amplified sequences coding for CD4-immunoglobulin (CD4IgG) (63).

For the cell line expressing the truncated CD4 molecule, a strong signal was found over the telomeric end of a small acrocentric chromosome (possibly a derivative of chromosome 5, 6, or 7), which covered about one-third of the total length.

In summary, when cultivating these cell lines in the absence of methotrexate, unique and characteristic integrations were found in 95–99% of metaphases. Subsequent to the first analysis, we continued to cultivate these cell lines for extended periods in the absence of methotrexate, and we occasionally performed additional FISH analysis. We found the chromosomal structures described here to be stable within the observation period (a minimum of 60 days and in one case of 160 days). Cytogenetically, these observations indicate a high degree of genetic stability of chromosomally amplified sequences in the absence of methotrexate.

Assessment by Southern Hybridization and FISH of the Chromosomal Status of Amplified DNA Sequences.

Southern hybridization (64) of genomic DNA is a useful tool to acquire insights into the molecular structure of integrated plasmid sequences. Using suitable restriction enzymes it will provide information on the integrity of the expression fragments of interest. Estimates of copy numbers of the chromosomally integrated DNA can be established when, in the same experiment, an appropriate range of quantities of plasmid DNA, restricted with the same enzyme as the one used for restriction of genomic DNA, is used. Also, Southern hybridization may indicate whether one integration locus or more than one exists for the plasmid sequences. However, conclusions on multiplicity of integration must be taken with a degree of caution. Aberrations from the expected signals can be due to postintegration rearrangement in a fraction of the cell population or in the initial cointegration of a few copies of the plasmid sequences that had been subjected to nuclease attack, resulting in the deletion of the restriction enzyme site used for the analysis.

All Southern hybridizations are based on DNA extracted from thousands of individual cells. Even if the cell lines are based on a "cloning" step, one must be aware that none of the so-called clonal cell lines are clonal in the true sense. Mammalian cells, in particular immortalized cells, have an intrinsic genomic fluidity that is most evident when studying chromosome numbers in metaphase spreads (for a review see Ref. 65).

FISH can be used to obtain insights into the extent of chromosomal amplification (not to be confused with copy number estimates) and the location of recombinant DNA within the chromosomal set of cells. In order to provide some useful information, FISH studies must be supplemented by a statistical analysis of identified integration sites (and structures observed). The time point of analysis, with respect to the total period of time (since thaw) for use of these cells, should also be taken into consideration. A reasonable value (from the invested work) may be gained from a FISH study performed shortly after cells have been thawed from a bank of cells stored in liquid nitrogen. Depending on the culture conditions (with or without methotrexate, with or without serum, etc.) and the length of time during which cells have been exposed to in vitro culture, results of FISH analysis may vary considerably. In spite of variability, in the studies discussed we were able to separate without any doubt or difficulty the identity of one recombinant cell line from a second recombinant cell line by using FISH. This was possible because we were able to identify a chromosomal marker containing hybridizing DNA that was present in a large fraction of the individual cells of the populations studied. The identifying chromosomal markers containing recombinant sequences were termed *master integrations*, which we found to be the genetically stable entities in the system.

CHOOSING THE PRODUCTION SYSTEM: ADHERENT CELLS, MICROCARRIERS, AND FREE SUSPENSION ISSUES OF ECONOMY, EXPERTISE, AND TIME FRAME TO CLINIC AND MARKET

Process scientists and their managers must eventually decide which type of production system to choose for the product in question. The choice of a production system is driven by the anticipated scale of operation for the manufacturing process, as well as by other issues. A number of very successful recombinant products from mammalian cells, such as EPO and GM-CSF, have been administered worldwide to a large number of patients. For these peptide hormones, the applied doses for treatment of patients are relatively small (about 100 μg/patient). Another successful protein therapeutic, recombinant tissue plasminogen activator (hu-rtPA, Activase®) requires a dose of 100 mg/patient, a 1,000-fold difference in dose as compared with erythropoeitin. The number of patients treated per annum with the two categories of products is about the same. It is obvious that such a difference in the annual product requirement would drive very different manufacturing decisions.

In the case of EPO and GM-CSF, a process based on roller bottles appears to be sufficient to supply the market with product. Using roller bottles, a "simple" and robust process can be established with minimal initial investment in hardware.

The Roller Bottle to Roller Bottle Process

Cells thawed from a cell bank can be expanded by subcultivation into a fixed number of roller bottles. This approach is based on cells growing attached to the inner surface of the roller bottle. Standard 2-L roller bottles are usually filled with 300–500 ml of medium. The remaining volume (air) provides the necessary oxygen while the bottles are slowly rolled (at about 1 rpm) in an incubator at 37 °C. The caps on the bottles are closed during the period of incubation. A sufficiently large number of roller bottles containing a healthy cell population represents the starting point for several production cycles. Part of the cell mass in the roller bottles can be immediately used for reproduction of cells in order to initiate the production cycle following the one already initiated. In other words, the bulk of the cell mass will serve as a seed culture to initiate the first production cycle. The remainder will provide, within one or two subcultivations, the basis for the subsequent production cycle.

Media for production are usually rich in nutrient content in order to maintain viability and productivity of the cells for a minimum of 5–7 days. Subsequently, the medium with the secreted product will be harvested, leaving the adherent cells inside the bottle. Sometimes, a refeeding with fresh medium for a second production cycle is possible, on the condition that the product of the first and the second harvest will be similar in composition and quality. Since a standard 2-L roller bottle will have about 300 ml of medium for harvest, 1,000 rollerbottles can provide product from 300 L or 600 L of supernatant, from one or two harvests, respectively. If such a campaign lasts all year long, a manufacturing process based on this principle will deliver 15,000 L (30,000 L) of cell-free culture medium containing the product of interest. Product concentrations in the 10- to 100-mg/L range are possible, resulting in 150 to 1,500 of unpurified product. Of course it must be realized that such a process is very labor intensive, requires the repeated use of trypsin for detachment of cells, and carries with it a considerable risk of contamination with adventitious agents through the handling personnel. However, because subcultivation in plastic cell-culture vessels represents a standard procedure in all laboratories, the scale-up from the lab-scale procedure to the pilot or manufacturing scale is a convenient approach for obtaining considerable (but not large) quantities of recombinant proteins.

The Bioreactor to Roller Bottle Process

The ability to cultivate CHO cells both in suspension and, subsequently, in an attached form allows the roller bottle production process to be streamlined. The seed culture for the roller bottle production phase can be generated in spinners or in bioreactors. The advantage of such a process would be that fewer interactions in terms of subcultivation are needed to generate sufficient cell mass. Being less labor intensive, it also reduces the risk of contamination by adventitious agents of the feed culture.

The Stirred Tank Bioreactor Process

Kilogram-scale production of recombinant proteins from CHO cells was first achieved in stirred bioreactors at very large scale (1,000 L and 10,000 L) with cells synthesizing while floating in free suspension (66). While such adaptation to suspension growth had been, in the early years of CHO technology, a somewhat tedious process, today multiple factors seem to have made this easier. For one, cell culture media have been developed that support the growth of cells in suspension better than those based on formulations such as DMEM and Ham's F12. Also, a number of purposeful genetic modifications (transfection of hormone genes or cell cycle genes, selection of CHO populations, or extended growth of CHO cells in media with little or no fetal calf serum and other modifications) appear to have generated CHO phenotypes that more readily support the transition from adherent growth to suspension growth.

In a bioreactor-based process, the scale-up to very large volumes can occur rather fast. It is usually executed by subcultivations, that is, by diluting the entire volume of one bioreactor into 5–20 volumes of fresh medium, maintained prewarmed in the next larger reactor. Within 10–15 days, therefore, a suspension culture at the 50-L scale could inoculate a 10,000-L reactor. It is major goal of process development work to carefully optimize media for the very last step—the production phase. Such a medium must support good growth in order to achieve the highest cell density possible, and it also must provide the nutritional basis and physiological balance needed to maintain viability and productivity for extended periods. While the termination (i.e., harvest) of such a culture is driven mainly by issues of plant capacity and volumetric productivity, the other important issue to consider is the quality of the derived product. The continuously changing composition of the culture medium during the production phase can affect the quality of earlier synthesized product through degradative activities, such as sialidase activity and proteolytic activity.

BATCH, EXTENDED-BATCH, AND PERFUSION PROCESSES

The batch process is considered a very simple, robust production process for stirred bioreactors. In the step preceding production, cells in a (somewhat smaller) bioreactor are cultivated to maximal cell density and are then transferred, still suspended in the exhausted growth medium, to the production reactor. After addition of fresh medium and eventual adjustments in control of certain process parameters (O_2 supply, pH, temperature, etc.) the production process begins. It ends at a predetermined point in time when synthesis of recombinant protein seizes or slows down due to exhaustion of nutritional components in the medium and/or accumulation of toxic end products of cellular metabolism. For CHO cells the production phase will usually last, depending on the proteins' susceptibility to degradative enzymes and a number of other process-related factors, between 5 and 14 days. The advantage of such a process is obvious. Providing that the inoculating cell mass can be made reproducible and with little variation in biological quality, the resulting production process will show a high degree of reproducibility with respect to cell growth, viability, quantity, and quality of the product being synthesized.

The issue of reproducibility of process parameters and of achievable product quantity and quality is of highest significance, of course. It provides confidence-generating data for the long-term future. Another important point is the obligation to present such data to regulatory agencies that will ultimately give permission to use the process for the generation of the product. For investigational new drug applications (IND) and process licence applications (PLA), rather specific requirements must be met with respect to the minimal number of product batches analyzed. For INDs no less than three product runs are recommended to be presented. Shorter batch processes (5–7 days) have the advantage of generating more data within a given time frame. This can be a very important cost factor because the time period necessary to acquire and evaluate necessary data from the new process will eventually affect the overall time necessary for entering the market.

On the other hand, there are viable arguments that could sway process-development decisions in another direction. The option to prolong the synthetic activities of cells in the production vessel would capitalize on the process investment that allows generation of the necessary cell mass for the production vessel. Therefore, more and more extended-batch or perfused-batch cultures are used. With a longer-lasting production phase in cell culture, feeding additional medium components becomes necessary. Obviously, extending the process for a considerable period of time, (for example, from 7 days to 13 days) makes sense only when the return for this investment in labor and in occupation of the production facility results in a sufficiently high increase in product concentration within the vessel. There are various ways in which medium and medium components can be added to a culture that was initiated a few days earlier in the same tank. This could be done by feeding (batchwise) highly concentrated mixtures of essential amino acids and other medium components, thus not significantly affecting the volume in the tank. Alternatively, a culture may be started in the production vessel at half or so of the working volume, then standard concentration medium could be pumped slowly and continuously, or batchwise, into the tank until the final working volume has been reached. The choice between the modes, or combinations thereof, is in the hands of the process development scientist, who has to carefully evaluate advantages and disadvantages. No matter what principle will be used for extending the production phase, the overall result will always be a tank that contains, in one batch, the entire protein population for subsequent recovery and purification.

A very different philosophy for manufacturing is represented by continuously perfused production processes. Here, the goal is to achieve the highest cell concentrations possible within small tanks. In this process up-front investment for manufacturing equipment is reduced, and product quantities can be quite high. The much improved knowledge base in technology, and in physiology of CHO cells in culture, has made this more-complicated approach to manufacturing attractive. Perfused cultures can be maintained for many weeks and months, with product harvests occurring repeatedly throughout that period. A protein of high interest to the pharmaceutical industry for decades, Factor VIII, is reliably being manufactured using perfusion technology with baby hamster kidney cells, a "cousin" of Chinese hamster ovary cells. While the production of Cognate® Factor VIII (67) has pioneered the use of perfusion technology for recombinant proteins with mammalian cells in a non-CHO cells, other products from CHO cells have recently been approved as well using this technology (68).

Finally, due to the fact that many monoclonal antibodies have been produced at the laboratory scale in perfusion systems for many years by using hollow-fiber technology, the transfer of this technology to the production scale for pharmaceutical manufacturing has been an obvious conclusion for quite a while (69,70). Recently, a diagnostic monoclonal antibody for imaging in patients with prostate cancer (ProstaScint™) was the first FDA-approved product for injection into humans derived from the hollow-fiber perfusion process (71). The application of this and other perfusion technologies, using immobilization of CHO cells on macroporous carriers to produce recombinant antibodies or other proteins, is likely to be yet another option for future manufacturing processes (72).

CONCLUSIONS

The technology of using mammalian cells, and in particular CHO cells, for recombinant protein production is, in spite of significant success stories, still in its infancy. Much has to be done to establish production processes in a more straitforward way and to make them more productive. A data explosion is occurring with DNA sequences from microorganisms, plants, and animals, including the human genome. Subsequently, a tremendous need is arising for more detailed studies and, frequently, the urge to use the translation products of these sequences in many applications. Also, the possibility of designing new proteins, based on concepts provided by nature, will even increase the need for flexible and robust synthetic machinery. Mammalian cells have been used as production hosts in some of the first pioneering experiments for the design of novel proteins (73,74). However, to deliver significant quantities of the desired proteins to analytical chemists, cell biologists, clinical researchers, and so on, takes far too long and is still a very demanding task. Some of the bottlenecks in the knowledge base have been addressed in this article. It is hoped that research efforts will solve some of them in the near future, resulting in more efficient, faster, and less expensive synthesis of recombinant proteins by mammalian cells. It appears that mammalian cells inherently have, due to their large size, the capacity to take up large quantities of foreign DNA without any negative effects, the responsiveness to perform a large number of synthetic tasks, and the capability to meet the challenges of the future.

BIBLIOGRAPHY

1. D. Pennica, W.E. Holmes, W.J. Kohr, R.N. Harkins, G.A. Vehar, C.A. Ward, W.F. Bennett, E. Yelverton, P.H. Seeburg, H.L. Heyneker, D.V. Goeddel, and D. Collen, *Nature* **301**, 214–221 (1983).
2. G. Urlaub and L.A. Chasin, *Proc. Natl. Acad. Sci. USA* **77**, 4216–4220 (1980).
3. M.B. Hendricks, C.A. Luchette, M.J. Banker, *Biotechnology* **7**, 1271–1274 (1989).
4. F.-T. Kao, and T.T. Puck, *Proc. Natl. Acad. Sci. USA* **60**, 1275–1281 (1968).
5. T.T. Puck, in M.M. Gottesman ed., *Molecular Cell Genetics*, Wiley, New York, 1985, pp. 37–64.
6. G. Ringold, B. Dieckmann, and F. Lee, *J. Mol. Appl. Genetics* **1**, 165–175 (1981).
7. R.J. Kaufman and P. Sharp, *J. Mol. Biol.* **159**, 601–621 (1982).
8. R.J. Kaufman, L.C. Wasley, A.J. Spiliotes, S.D. Gossels, S.A. Latt, G.R. Larsen, and R.M. Kay, *Mol. Cell. Biol.* **5**, 1750–1759 (1985).
9. L. Hayflick, *Science* **276**, 336–337 (1997).
10. M.E. Wiebe, and L.H. May, in A.S. Lubiniecki ed., *Large Scale Mammalian Cell Culture Technology*, Marcel Dekker, New York, 1990, pp. 147–160.
11. M.E. Wiebe, F. Becker, R. Lazar, L. May, B. Casto, M. Semense, C. Fautz, R. Garnick, C. Miller, G. Masover, D. Bergman, and A.S. Lubiniecki, in R.E. Spier, J.B. Griffiths, J. Stephenne, and P.J. Crooy eds., *Advances in Animal Cell Biology and Technology for Bioprocesses*. 1987, pp. 68–71.
12. J.V. O'Connor, R.G. Keck, R.J. Harris, and M.J. Field, in: F. Brown, and A.S. Lubiniecki eds., *Genetic Stability and Recombinant Product Consistency*, Karger, Basch, 1994, pp. 165–173.
13. R.L. Garnick, *Develop. Biol. Standard*, **76**, 117–130 (1992).
14. H.S. Lu, L.B. Tsai, W.C. Kenney, and P-H. Lai, *Biochem. Biophys. Res. Comm.* **156**, 807–813 (1988).
15. R.C. Chloupek, R.J. Harris, C.K. Leonard, R.G. Keck, B.A. Keyt, M.W. Spellman, A.J.S. Jones, and W.S. Hancock, *J. Chromatog.* **463**: 375–396 (1989).
16. J.V. O'Connor, *Biologicals* **21**, 111–117 (1993).
17. R.C. Stephenson, and S. Clarke, *J. Biol. Chem.* **264**, 6164–6170 (1989).
18. V. Anicetti, and W.S. Hancock, in R. Harrison ed., *Protein Purification Process Engineering*, Marcel Dekker, Inc. New York, 1994, pp. 25–56.
19. M. Calos, J.S. Lebkowski, and M.R. Botchan, *Proc. Natl. Acad. Sci. USA*, **80**, 3015–3019 (1983).
20. F. Galibert, *Biologicals* **18**, 221–224 (1990).
21. F. Galibert, in F. Brown and A.S. Lubiniecki eds., *Genetic Stability and Recombinant Product Consistency*, Karger, Basel, 1994, pp. 27–30.
22. W. Berthold, in F. Brown and A.S. Lubiniecki eds., *Genetic Stability and Recombinant Product Consistency*. Karger, Basel, 1994, pp 67–79.
23. J.W. Drake, *Annu. Rev. Genet.* **25**, 125–146 (1991).
24. R.J. Harris, A.A. Murnane, S.L. Utter, K.L. Wagner, E.T. Cox, G.D. Polastri, J.C. Helder, and M.B. Sliwskowski, *Bio / Technology* **11**, 1293–1297 (1993).
25. F.L. Graham, and A.J. van der Eb, *Virology* **52**, 456–467 (1973).
26. A. Loyter, G.A. Scangos, and F.H. Ruddle, *Proc. Natl. Acad. Sci. USA* **79**, 422–426 (1982).
27. A. Loyter, G. Scangos, D. Juricek, D. Keene, and F. Ruddle, *Exp. Cell Res.* **139**, 223–234 (1982).
28. M. Wigler, S. Silverstein, L.-S. Lee, A. Pellicer, Y.-C. Cheng, and R. Axel, *Cell* **11**, 223–232 (1977).
29. A. Pellicer, M. Wigler, and R. Axel, *Cell* **14**, 133–141 (1978).
30. D.M. Robins, S. Ripley, A. Henderson, and R. Axel, *Cell* **23**, 29–39 (1981).
31. C. Chen, and H. Okayama, *Mol. Cell. Biol.* **7** 2745–2752 (1987).
32. M. Jordan, A. Schallhorn, and F.M. Wurm, *Nucl. Acid. Res.* **24**, 596–601 (1996).
33. M. Jordan, C. Köhne, and F.M. Wurm in M.J. Carrondo, B. Griffiths and J.P. Moreira eds., *Animal Cell Technology—from Vaccines to Genetic Medicine*, Kluwer Academic, Dordrecht, 1997, pp. 47–50.
34. G.K. Finn, B.W. Kurz, R.Z. Cheng, and R.J. Shmookler *Mol. Cell. Biol.* **9**, 4009–4017 (1989).
35. S.L. Mansour, K.R. Thomas, and M.R. Capecchi, *Nature* **336**, 348–352 (1988).
36. H. Zheng, and J.H. Wilson, *Nature* **344**, 170–173 (1990).
37. F.M. Wurm, A. Johnson, T. Ryll, C. Koehne, H. Scherthan, F. Glaab, Y.S. Lie, C.J. Petropoulos, and W.R. Arathoon, *Ann. of the Acad. Sci.* **782**, 70–78 (1996).
38. F.M. Wurm, A. Johnson, Y.S. Lie, M.T. Etcheverry and K.P. Anderson, in R.E. Spier, J.B. Griffiths, and C. MacDonald eds., *Animal Cell Technology: Developments, Processes and Products*, Butterworth-Heinemann, Oxford, 1992, pp. 35–41.
39. M. Perucho, D. Hanahan, and M. Wigler, *Cell*, **22**, 309–317 (1980).
40. G.K. Finn, B.W. Kurz, R.Z. Cheng, and R.J. Shmookler Reis, *Mol. Cell. Biol.* **9**, 4009–4017 (1989).
41. K.R. Folger, E.A. Wong, G. Wahl, M.R. Capecchi, *Mol. Cell. Biol.* **2**, 1372–1387 (1982).
42. J. Ruiz, and G.M. Wahl, *Mol. Cell. Biol.* **10**, 3056–3066 (1990).
43. G. Chu, H. Hayakawa, and P. Berg, *Nucl. Acids Res.* **15**, 1311–1326 (1987).
44. J. Barsoum, *DNA Cell Biol.* **9**, 293–300 (1990).
45. P.L. Felgner, and G.M. Ringold, *Nature* **337**, 387–388 (1989).
46. J.-P. Behr, B. Demeneix, J.-P. Loeffler, and J. Perez-Mutul, *Proc. Natl. Acad. Sci. USA* **86**, 6983–6986 (1989).
47. S.R. Muller, P.D. Sullivan, D.O. Clegg, and S.C. Feinstein, *DNA Cell Biol.* **9**, 221–229 (1990).
48. M. Jasin, and P. Berg, *Genes Develop.* **2**, 1353–1363 (1988).
49. O. Smithies, R.G. Gregg, S.S. Boggs, M.A. Koralewski, and R.S. Kucherlapati, *Nature* **317**, 230–234 (1985).
50. H. Zheng, and J.H. Wilson, *Nature* **344**, 170–173 (1990).

51. S. Shirahata, K. Teruya, K. Seki, T. Mori, H. Ohashi, H. Tachibana, and H. Murakami, in R.E. Spier, J.B. Griffiths, and C. MacDonald eds., *Animal Cell Technology: Developments, Processes and Products*, Butterworth-Heinemann, Oxford, 1992, pp. 54–59.
52. K.H. Lee, A. Guerini-Sburlati, W.A. Renner, and J.E. Bailey in M.J.T. Carrondo, et al. eds., *Animal Cell Technology from Vaccines to Genetic Medicine*, Kluwer Academic, Dordrecht, 1997, pp. 621–623.
53. P. Gray, S.C.O. Pak, S.M. Hunt, M.W. Bridges, N.A. Sunstrom, C. Bailey and M.J. Sleigh, *Proc. 10th Intern. Biotechnology Symposium*, Sydney, Australia, 1996.
54. J.L. Biedler and B.A. Spengler, *J. Natl. Cancer Inst.* **57**, 683–689 (1976).
55. R.J. Kaufman and P.A. Sharp, *J. Mol. Biol.* **159**, 601–621 (1982).
56. F.M. Wurm, K.A. Gwinn, and R.E. Kingston, *Proc. Natl. Acad. Sci. USA* **83**, 5414–5418 (1986).
57. G. Zettlmeissl, M. Wirth, H.J. Hauser, and H.A. Küpper, *Gene* **73**, 419–426 (1988).
58. J.H. Nunberg, R.J. Kaufman, R.T. Schimke, G. Urlaub, and L.A. Chasin, *Proc. Natl. Acad. Sci. USA* **75**, 5553–5556 (1978).
59. U.H. Weidle, P. Buckel, and J. Wienberg, *Gene* **66**, 193–203 (1988).
60. J.A. Vitek, *Neoplasma* **34**, 665–670 (1987).
61. D. Pinkel, T. Straume, and J.W. Gray, *Proc. Natl. Acad. Sci. USA* **83**: 2934–2938 (1986).
62. M.G. Pallavicini, P.S. DeTeresa, C. Rosette, J.W. Gray, and F.M. Wurm, *Mol. Cell. Biol.* **10.1**, 401–404 (1990).
63. F.M. Wurm, in H.J. Hauser and R. Wagner eds., *Genetic Manipulation of Mammalian Cells*, Walter de Gruyter, Berlin, pp. 87–120.
64. F.M. Ausubel, R. Brent, R.E. Kingston, D.D. Moore, J.G. Seidman, J.A. Smith, and K. Struhl, *Current Protocols in Molecular Biology*, Wiley, New York, 1988.
65. T.C. Hsu, *Int. Rev. Cytol.* **12**, 69–121 (1961).
66. J.R. Birch and W.R. Arathoon, in A.S. Lubiniecki ed., *Large Scale Mammalian Cell Culture Technology*, Marcel Dekker, New York, 1990, pp. 251–270.
67. B.G.D. Bödecker, R. Newcomb, P. Yuan, A. Braufman, and W. Kelsey, in R.E. Spier, J.B. Griffiths, and W. Berthold eds., *Animal Cell Technology: Products of Today, Prospects for Tomorrow*, Butterworth-Heinemann, Oxford, U.K., pp. 580–590.
68. Annual Report 1995, ARES-SERONO: Follicle Stimulating Hormone.
69. R.J. van Wedel, in S. Seaver, ed., *Commercial Production of Monoclonal Antibodies*, Marcel Dekker, New York, 1987, pp. 159–173.
70. J.M. Davies, C.M. Lavender, K.J. Bowes, J.A.J. Hanak, B.S. Combridge, and S.L. Kingsland, in E.C. Beuvery, J.B. Griffiths, and W.P. Zeijlemaker eds., *Animal Cell Technology: Developments towards the 21st Century*, Kluwer Academic, Dordrecht, 1995, pp. 149–153.
71. S. Heimbuch, Press Release of Cellex Biosciences: Cellex Biosciences perfusion equipment used for 1st injectable product produced in hollow fiber licensed by FDA, October 30, 1996.
72. A. Preissmann, R. Bux, P. Schorn, W. Noe, in E.C. Beuvery, J.B. Griffiths, and W.P. Zeijlemaker eds., *Animal Cell Technology: Developments towards the 21st Century*, Kluwer Academic, 1995, pp. 841–845.
73. R.A. Byrn, J. Mordenti, C. Lucas, D. Smith, S.A. Marsters, J.S. Johnson, S.M. Chamow, F.M. Wurm, T. Gregory, J.E. Groopman, and D.J. Capon, *Nature* **344**, 667–670 (1990).
74. W.F. Bennett, N. Paoni, D. Botstein, A.J.S. Jones, B. Keyt, L. Presta, F.M. Wurm, and M. Zoller, *J. Biol. Chem.* **266**, 5191–5201 (1991).

See also EXPRESSION SYSTEMS, *E. COLI*; EXPRESSION SYSTEMS, MAMMALIAN CELLS; INSECT CELL CULTURE, PROTEIN EXPRESSION; INSECT CELLS AND LARVAE, GENE EXPRESSION SYSTEMS; PICHIA, OPTIMIZATION OF PROTEIN EXPRESSION.

CHOLESTEROL OXIDASE

KAZUO AISAKA
Kyowa Hakko Kogyo Co., Ltd.
Tokyo, Japan

KEY WORDS

Brevibacterium sterolicum
4-Cholesten-3-one
Cholesterol
Cholesterol oxidase (EC 1.1.3.6)
3β-Hydroxysteroids
Steroid δ-isomerase (EC 5.3.3.1)

OUTLINE

INTRODUCTION

Cholesterol oxidase is the enzyme that catalyzes the oxidation of the 3β-hydroxy group of various kinds of steroids. The oxidation is the important step in the catabolism of the sterols. This catalytic property plays an important role in various bioprocesses

The catalytic and structural properties of cholesterol oxidases from microbial origins are described. Further, some applications are described, including cholesterol determination in medical diagnosis, reduction of cholesterol content in the food industry, and pest control (based on its insecticidal activity) in the agriculture industry.

CHOLESTEROL

Cholesterol (cholest-5-en-3β-ol) is principal sterol of higher animals. Cholesterol is practically insoluble in water, but it is soluble in aqueous solutions of bile salts. It is found in all body tissues, especially the brain and spinal cord, and in animal fats or oils.

Cholesterol is considered to be a key intermediate in the biosyntheses of fecal sterols, bile acids, and steroid hormones in animals. Turfitt first demonstrated the formation of 4-cholesten-3-one from cholesterol by *Proactiomyces erythropolis* (*Nocardia erythropolis*) and showed that this oxidation of the 3β-hydroxy group to the keto group of cholesterol was the initial step in the aerobic catabolism of the sterol (1). On the other hand, in the intestinal tract of humans, cholesterol can be anaerobically metabolized by bacterial microflora to coprostanol (65%) and coprostanone (10%) and to a lesser extent to cholestanone and epicoprostanol (2) (Fig. 1).

CHOLESTEROL OXIDASE

Cholesterol oxidase (EC 1.1.3.6) catalyzes the oxidation by dioxygen of the 3β-hydroxyl group of cholesterol and of related steroids (other 3β-hydroxy steroids) to a keto group and catalyzes the successive isomerization of the δ5 double bond to the conjugated position: δ4 (Fig. 2). Thus, in many cholesterol oxidases, both the oxidase and isomerase activities are catalyzed by the single enzyme protein, although the two kinds of enzymes, cholesterol dehydrogenase and δ5-3-ketosteroid isomerase (EC 5.3.3.1) are present in some microorganisms containing *Pseudomonas testosteroni* and *P. putida* (3).

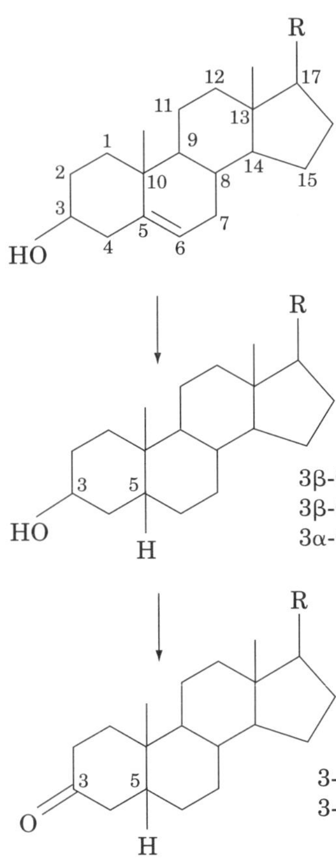
3β-OH, 5-en—Cholesterol
3β-OH, 5β-H—Coprostanol
3β-OH, 5α-H—Cholestanol
3α-OH, 5α-H—Epicoprostanol
3-Keto, 5β-H—Coprostanone
3-Keto, 5α-H—Cholestanone

Figure 1. Anaerobic metabolism of cholesterol.

The oxidation is estimated by the following methods.

1. Conjugated enone formation by the increase in UV absorption
2. Monitoring oxygen uptake
3. Hydrogen peroxide production by the oxidative coupling of 4-aminoantipyrine with phenol by peroxidase to give a quinoneimine dye absorbing at 500 nm

Generally, the enzyme is a flavin-dependent oxidase containing one mole of tightly bound flavin adenine dinucleotide (FAD) per mole of protein (4). However, cholesterol oxidases from *Nocardia rhodochrous, Streptomyces griseocarneus*, and *Streptomyces violascens* lack FAD or any other cofactor.

The enzyme has been isolated from a number of different soil bacteria that are capable of using cholesterol as the sole source of carbon and energy. They include Gram-positive bacteria such as *Arthrobacter simplex, Brevibacterium sterolicum, Corynebacterium sterolicum*, and *Rhodococcus equi*; Gram-negative bacteria such as *Pseudomonas* spp., actinomycetes such as *Actinomyces lavendulae, N. erythropolis, N. rhodochrous, Nocardia* spp., *Streptomyces griseocarneus*, and *Streptomyces violascens*; and fungi such as *Schizophyllum commune* (5). There are also two types of enzymes, secreted and membrane bound, according to their localization.

Commercial preparations of cholesterol oxidase can be obtained from the following suppliers: *Streptomyces* spp. (15 units/mg) from Toyobo Co.; *B. sterolicum* ATCC21387 (13 units/mg) from Kyowa Hakko Kogyo Co.; *Pseudomonas* spp. (3 units/mg) from Wako Pure Chemical Industries; *R. erythropolis* (25 units/mg) from Boehringer Mannheim; and *Cellulomonas* spp. (20–60 units/mg) and *Schizophyllum commune* (5 units/mg) from Sigma Chemicals.

The enzymes from *B. sterolicum* (6), *Schizophyllum commune* (7), and *Arthrobacter simplex* (8) are purified, and their enzymatic properties have been examined.

This enzyme can oxidize 3β-hydroxy groups in cholesterols, the biosynthetic precursors such as desmosterol, 17 ketosteroids such as dehydroepiandrosterone, and plant sterols such as stigmasterol, β-sitosterol, and campesterol. Of these, the length of the C17 side chain is especially an important factor in determining the reaction rate. On the other hand, 3α-hydroxysteroids, estrogens, and cholecalciferol do not serve as substrates (9). Further, the relative activities are also dependent on the origin of the enzymes.

Murooka et al. cloned the cholesterol oxidase gene from a *Streptomyces* sp. SA-COO and demonstrated high productivity of the extracellular cholesterol oxidase by using the *Streptomyces lividans* host-vector system (10). Ohta et al. cloned the cholesterol oxidase gene from *B. sterolicum* (11). The gene was also overexpressed into *Streptomyces lividans* (12). Both genes show identities of 64% in nucle-

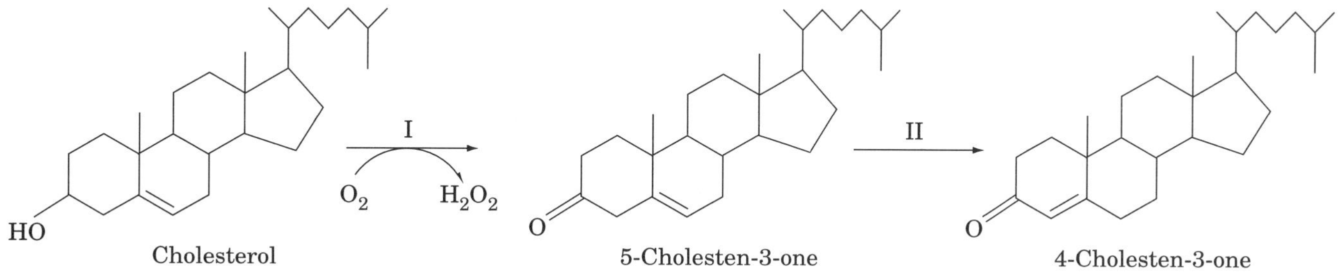

Figure 2. Scheme for the cholesterol oxidase reaction. I, oxidation step; II, isomerization step.

otide sequence level and 58% in amino acid sequence level, respectively. The gene analysis shows that the mature enzyme consists of 503 or 507 amino acids and is preceded by a 42 or 45 amino acid signal sequence for *Streptomyces* sp. or *B. sterolicum*. There are at least six highly conserved regions between the two enzymes, which may play structurally or functionally important roles such as substrate binding or catalysis. The FAD-binding domain in a number of FAD-dependent enzymes contains a pattern of three glycine residues, Gly-X-Gly-X-X-Gly, common to NAD and FAD-binding proteins. This pattern of residues is also found near the N-terminal of the sequences of cholesterol oxidases from *Streptomyces* sp. and *B. sterolicum*.

The crystal structure of the enzyme from *B. sterolicum* has been determined using the method of isomorphous replacement and refined to 1.8-Å resolution (13). The structure is made up of two domains: an FAD-binding domain and a steroid-binding domain. The FAD-binding domain consists of three noncontinuous segments of sequence, including the N-terminal and the C-terminal. The overall topology of this domain is very similar to other FAD-binding proteins. The steroid-binding domain consists of two noncontinuous segments of sequence and contains the roof of the active-site cavity. The active site lies at the interface of the two domains. The FAD lies on the floor of the cavity. The steroid substrate is buried within the internal cavity. The hydroxyl group of the steroid is hydrogen bonded to the flavin ring system of the FAD. The active-site cavity contains one charged residue, Glu361. This residue is appropriately positioned to act as the proton acceptor in the proposed mechanism for the isomerization step.

FAD-dependent cholesterol oxidase shares significant structural homology with another flavoenzyme, glucose oxidase, suggesting that it might also be a member of the glucose-methanol-choline (GMC) oxidoreductase family. A histidine residue is completely conserved in the known sequences of GMC oxidoreductases and may act as the general base in the oxidation reaction.

APPLICATION OF CHOLESTEROL OXIDASE

Diagnosis

Cholesterol oxidase is used in the determination of serum cholesterol concentration for the assessment of arteriosclerosis and other lipid disorders and of the risk of thrombosis.

Richmond (14) and Flegg (15) first showed the possibility of using cholesterol oxidase from microorganisms as a basis for a specific assay for serum cholesterol. In the procedure, 4-cholesten-3-one produced was assayed by its UV absorption after solvent extraction of the reaction mixture. However, their procedures were tedious and time consuming. Allain et al. developed the excellent method in which the hydrogen peroxide produced during the cholesterol oxidase reaction was oxidatively coupled with 4-aminoantipyrine and phenol by peroxidase to form a quinoneimine dye (16). Figure 3 shows the scheme for the enzymatic determination of total cholesterol. First, cholesterol esters are able to be converted to free cholesterol by the action of cholesterol esterase. Then, the liberated cholesterol can be determined by cholesterol oxidase and peroxidase. This enzymatic method is simple and specific under extremely mild conditions. The results with serum samples agreed very well with those obtained by chemical procedures based on the Liebermann-Burchard reaction. This assay system is marketed as a serum cholesterol determination kit; worldwide annual sales amount to $100 to 200 million.

Food Industry

Dietary cholesterol has been implicated as one of the causes of coronary artery disease. Because of this, reducing dietary cholesterol is said to cause a gradual reduction in serum cholesterol and regression of atherosclerotic plaque. The National Cholesterol Education Program coordinating board recommends a daily cholesterol intake in the American diet of 300 mg or less. As a result, reduction of cholesterol in food items relatively rich in cholesterol (dairy foods, eggs) has become a challenging problem.

Some organisms have been shown to degrade cholesterol, as described earlier, but only a few do so without accumulating any steroid intermediates. Watanabe et al. described the degradation of cholesterol in egg yolk and lard by crude cholesterol oxidase in *R. equi* no. 23 and the bacterial cell suspension (17). *R. equi* no. 23 degraded cholesterol with little or no accumulation of steroid intermediates such as 4-cholesten-3-one and 1,4-androstadiene-3,17-dione, which are recognized as key intermediates in a comprehensive pathway of cholesterol degradation.

A biotechnological approach to reducing the cholesterol content of milk involves the genetic engineering of starter culture bacteria (lactococci, lactobacilli, streptococci), leading to the expression of phenotypic traits that are related to cholesterol metabolism. These cultures are then used in

$$\text{Cholesterol-ester} + H_2O \longrightarrow \text{Cholesterol} + \text{Fatty acid} \quad \textbf{(1)}$$

$$\text{Cholesterol} + O_2 \longrightarrow \text{4-Cholesten-3-one} + H_2O_2 \quad \textbf{(2)}$$

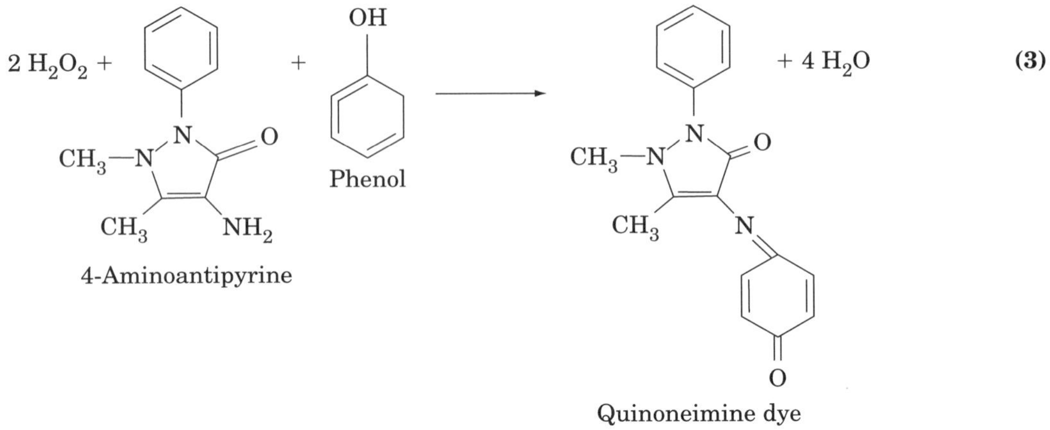

Figure 3. Scheme for the enzymatic determination of total cholesterol. Reactions 1, 2, and 3 are catalyzed by cholesterol esterase, cholesterol oxidase, and peroxidase, respectively.

the production of fermented dairy foods (yogurt and cheeses).

Somkuti et al. have investigated the possibility of introducing the cholesterol oxidase gene from *Streptomyces* sp. SA-COO into a common dairy starter culture bacteria such as *Streptococcus thermophilus* (18) and *Lactobacillus casai* (19). The biosynthesis of a functional cholesterol oxidase in *Streptococcus thermophilus* was confirmed by TLC analysis of reaction products after incubation of sonically disrupted cells with cholesterol. However, unlike *Streptomyces lividans, Streptococcus thermophilus* did not secrete cholesterol oxidase into the culture medium.

At present, the use of cholesterol-degrading enzymes in the food industry remains unrealistic because of the lack of toxicological data on enzyme extracts and the intermediate products of cholesterol degradation.

Agriculture

The discovery of novel insecticidal proteins is vital for pest control strategies using transgenic crops. Filtrates from several thousand microbial fermentations were assayed for insecticidal activity against crop-damaging insects. Two *Streptomyces*-culture filtrates killed boll weevil larvae in feeding studies, and the bioactive component was identified as cholesterol oxidase (20). The enzyme was active in a dose-dependent manner, and the LC50 against neonate boll weevil larvae was 20.9 μg/mL after 6 days in diet-incorporation bioassay. *Bacillus thuringensis* subsp. Kurstaki protein has an LC50 of 1 μg/mL against tobacco budworm and 37 μg/mL against European corn borer in similar diet-incorporation bioassay. Thus, cholesterol oxidase is active against the boll weevil in the same concentration range that *B. thuringiensis* protein is active against lepidopterans.

Morphological changes induced by ingestion of cholesterol oxidase suggest that the enzyme has a direct effect on the midgut tissue of boll weevil larvae. Thus, cholesterol oxidase disrupted the midgut epithelium at low doses and lysed the midgut cells at higher doses.

The cholesterol oxidase gene from *Streptomyces* spp. was expressed in tobacco and tomato, and successful control of boll weevil was achieved by the Monsanto company (21).

BIBLIOGRAPHY

1. G.E. Turfitt, *Biochem. J.* **38**, 492–496 (1942).
2. I.A. MacDonald, V.D. Bokkenheuser, J. Winter, A.M. McLernon, and E.H. Mosbach, *J. Lipid. Res.* **24**, 675–681 (1983).
3. P. Talalay, *Ann. Rev. Biochem.* **34**, 347–380 (1965).
4. T. Uwajima, H. Yagi, and O. Terada, *Agric. Biol. Chem.* **38**, 1149–1156 (1974).
5. *Enzyme Handbook*, Springer-Verlag, Berlin, 1995, EC 1.1.3.6.
6. T. Uwajima, H. Yagi, S. Nakamura, and O. Terada, *Agric. Biol. Chem.* **37**, 2345–2350 (1973).
7. M. Fukuyama and Y. Miyake, *J. Biochem.* **85**, 1183–1193 (1979).
8. W.-H. Liu, M.-H. Meng, and K.-S. Chen, *Agric. Biol. Chem.* **52**, 413–418 (1988).
9. A.G. Smith and C.J.W. Brooks, *J. Steroid Biochem.* **7**, 705–713 (1976).
10. Y. Murooka, T. Ishizaki, O. Nimi, and N. Maekawa, *Appl. Environ. Microbiol.* **52**, 1382–1385 (1986).
11. T. Ohta, K. Fujishiro, K. Yamaguchi, Y. Tamura, K. Aisaka, T. Uwajima, M. Hasegawa, *Gene* **103**, 93–96 (1991).
12. T. Ohta, K. Fujishiro, K. Yamaguchi, T. Uwajima, K. Aisaka, and M. Hasegawa, *Biosci. Biotech. Biochem.* **56**, 1786–1791 (1992).
13. A. Vrielink, L.F. Lloyd, and D.M. Blow, *J. Mol. Biol.* **219**, 533–554 (1991).
14. W. Richmond, *Scan. J. Clin. Lab. Invest.* **29**, suppl. 126, abs. no. 3.25 (1972).
15. H.M. Flegg, *Ann. Clin. Biochem.* **10**, 79–84 (1973).

16. C.C. Allain, L.S. Poon, C.S.G. Chan, W. Richmond, and P.C. Fu, *Clin. Chem.* **20**, 470–475 (1974).
17. K. Watanabe, H. Aihara, N. Tachi, and R. Nakamura, *J. Appl. Bacteriol.* **62**, 151–155 (1987).
18. G.A. Somkuti, D.K.Y. Solaiman, T.L. Johnson, and D.H. Steinberg, *Biotech. Appl. Biochem.* **13**, 238–245 (1991).
19. G.A. Somkuti, D.K.Y. Solaiman, and D.H. Steinberg, *Appl. Microbiol. Biotechnol.* **37**, 330–334 (1992).
20. J.P. Purcell, J.T. Greenplate, M.G. Jennings, J.S. Ryerse, J.C. Pershing, S.R. Sims, M.J. Prinsen, D.R. Corbin, M. Tran, R.D. Sammons, and R.J. Stonard, *Biochem. Biophys. Res. Commun.* **196**, 1406–1413 (1993).
21. D.R. Corbin, J.T. Greenplate, E.Y. Wong, and J.P. Purcell, *Appl. Environ. Microbiol.* **60**, 4239–4244 (1994).

CHROMATOGRAPHY, COMPUTER-AIDED DESIGN

ALOIS JUNGBAUER
OLIVER KALTENBRUNNER
Institute of Applied Microbiology
Vienna, Austria

KEY WORDS

Biochromatography
Computer simulation
Neural network
Optimization
Productivity
Simplex optimization
Throughput

OUTLINE

INTRODUCTION

Currently, chromatographic purification processes are optimized empirically. Separation media, running conditions, and sequence of purification steps are selected according to the expertise of the responsible scientists.

In biotechnology, computer-aided design is less far advanced than in certain areas of chemical engineering. The lack of expert systems is the reason for the traditional optimization, which is based on trial and error. Mathematical models for description of the chromatographic separation process have been around since the early 1940s (1–4). The common use of computers in the separation laboratory, the need of rapid process development, and the theoretical knowledge accelerated the development of expert systems and software tools. They are considered to be beneficial for design and optimization of purification steps.

The area of computer-aided chromatography design includes the design of experiments, the appropriate interpretation of experimental results (by the aid of mathematical algorithms), the prediction of separation conditions, and the rescaling of unit operations (5).

Four levels of complexity have to be considered for optimization of separation steps. The first level is the general process economics. This level will be influenced by the second level: flow sheeting. The third level is the optimization of particular unit operations. The fourth level is the sectioning of unit operations into their thermodynamical and molecular considerations by the aid of molecular simulation.

Simulated moving bed, countercurrent chromatography, and annular chromatography are not been considered in this chapter, because these modes of chromatography have not been used for bioprocessing to large extent.

BASIC CONSIDERATIONS FOR COMPUTER-AIDED DESIGN OF CHROMATOGRAPHY

Terminology

The word *model* denotes a system of algebraic or differential equations, which are held to be a representation of a physical system. From this general point of view, any relationship between quantities is a model, and the principles of modeling are the principles of chemical engineering science. *Process models*, and here particularly chromatography, can be counted to a chemical process. They are established by application of the three fundamental conservation principles: conservation of mass, momentum, and energy. Because liquid chromatography is carried out under isothermal conditions, the conservation of energy can be neglected in general. The contribution of liberated or consumed thermal energy during adsorption and desorption is negligibly small. The conservation of mass and momentum and the virtual breakdown of a chromatography system to a small unit, where the adsorption process takes place, are the fundamentals for modeling of chromatography (6).

In principle, there is no difference between *computer models* and other models. Many models are so complex that they can be solved only numerically. In these cases, computers can be used for reasons of convenience, independent of the complexity of the model applied. Further considerations about modeling are given by Vajda and Valkó (7).

We usually assume that the input, initial conditions, and parameters of a model determine the response (result). The *parameters*, which are the different coefficients in mathematical equations, must be evaluated. This process is called parameter estimation. Once a model is established, the performance of a system can be predicted by varying the input and initial conditions. The prediction will yield a correct answer (response) as long as the predictions are carried out within a certain range of validity.

The words *prediction* and *simulation* describe, in this context of modeling, a similar process. The forecast of process conditions, using process parameters acquired at a small scale, is called prediction. Frequently parameters cannot be estimated because of the lack of a determination method or accessibility. Thus, a hypothetical value is used to solve the set of mathematical equations. This process is called simulation.

Depending on the task, some of the *operational variables* may be kept constant or may not be taken into consideration. The most important variables are listed in Table 1. The residence time of a product in a column is a critical factor, when the product undergoes a conformational change or a reaction in the column. Alteration of the residence time in a column can completely change the result of the purification. Reversed-phase chromatography (RPC) and hydrophobic interaction chromatography (HIC) are more susceptible to residence time changes than ion-exchange chromatography or affinity chromatography are.

The *response values* depend on the strategy and the stage where the chromatographic purification is placed. Frequently used response values are:

- *Productivity* (*P*) is defined as

$$P = \frac{Q_R \cdot C_0 \cdot V_F}{V_t \cdot t_C} \qquad (1)$$

  where Q_R is the recovery ratio, C_0 is the sample concentration, V_F is the sample volume, V_t the column volume, and t_C the cycle time, which includes all time periods necessary to start the next cycle (process time). The purity has to be defined before optimization. The productivity in the aforementioned case does not consider the lifetime of a column.
- *Yield* or *recovery* is simply the percentage of recovered material and material present in the feed. Yield is not easily accessable by mathematical models. In most cases, product loss is caused by a very complex event. Unspecific (irreversible) adsorption to stationary phases, aggregation, and denaturation together with enzymatic degradation are responsible for incomplete recovery. These effects are also dependent on age and history of the column. These events are not covered by chromatography models.
- *Dynamic capacity* is the capacity at a certain flow in contrast to the capacity determined by batch experiments. BTC experiments at 5% breakthrough are usually used to determine the dynamic capacity. Originally, the analytical solution of an idealized model, known as the Thomas solution (8), has been used for approximation and prediction of breakthrough curve (BTC). Under conditions applied in preparative chromatography, the BTC deviates significantly from the ideal model. Recently, models have been published that take effects of kinetics under overloaded conditions into consideration (9).
- To minimize the number of purification steps, the highest *purity factor* should be achieved at each step. The regulations for production of pharmaceuticals by recombinant DNA technologies require very high purity standards. The standards can only be met when a purification power of each step is used to the optimal extent. Other factors, such as design of a plant, quality of the materials and buffers, operator training, etc., must also be obeyed. The optimization of the

Table 1. Important Variables for Optimization of Preparative Chromatography

Column size/shape
Height (Z)
Diameter
Height-to-diameter ratio
Height-to-particle diameter
Flow rate (u)
Flow rate during loading, during elution
During regeneration
Residence time (Z/u)
Physical conditions of packing
Packing quality
Packing density
Mobile phase
Modifier composition
Rate of change of mobile phase
Modifier composition (gradient shape)
Loading
Volume
Concentration
Flow rate
Stationary phase
Ligand type
Ligand density
Type of support
Porosity, tortuosity
Bead diameter
Beads size distribution (monodispers)
Temperature
Scarcely a process in biotechnology, which uses temperature gradients for elution

chromatographic process is only one piece of the puzzle.

- A high *concentration factor* leads to a smaller volume for the next step, which has to be handled and the product becomes more stable. Also, an upper limit in the concentration can be of consideration, because the solubility of the product is exceeded. The concentration factor (CF) can be defined as:

$$CF = \frac{C_R \cdot V_R}{C_0 \cdot V_F} \quad (2)$$

where C_0 is the initial (feed) concentration, V_F is the feed volume, C_R is the recovered concentration, and V_R is the recovered volume.

- The *process time* may be limited by the stability of the product. Economical considerations also come into play, such as shift time and batches, which have to be produced by a certain time period. The process time includes the optimization of the gradient, flow rate, etc. Process time has to be optimized with regard to total process time. Cleaning, regeneration, and, if necessary, repacking have to be included.

General Criteria of Mathematical Models

Today, broadening of scientific and technological knowledge is based on three cornerstones: theory, experiments, and computer simulation. Reliable computer simulation needs an appropriate mathematical model. Without a valid mathematical model that reflects the real physical situation, the most sophisticated computer algorithm will fail. A model will never be as good as the physical reality, because the complexity has to be reduced to a certain formalism. A fictional model of the complexity of the reality would be as slow and laborious as the reality.

Vajada and Valkó (7) compiled a list of criteria for establishing a model in chemical engineering. These rules also apply for the chromatography of proteins.

- First, scientific curiosity is a strong motivation for establishing a model. A priori curious-appearing models may be of considerable value, when they are integrated with the appropriate objective. This requires the innovative selection of parameters and design of experiments to determine those parameters.
- The second reason for using models is the reduction information and indirect measurements. The most common information reduction in chromatography peak is the approximation of a peak by its first and second moment, instead of using the information from the entire peak curve. Using these information reduction processes, the appropriate model (10–14) must be selected to describe the peak area, width, and the asymmetric shape.
- The ability to design and optimize a separation scheme quickly and effectively is also a motivation for developing a model. To minimize process time and determine optimal column size and optimal gradient shape, loading and sequence of the different process steps (equilibration, loading, washing, elution, and regeneration) are necessary.
- The design of optimal particle properties is also an important goal for manufacturers of resins and for engineers selecting a resin. Modeling of packing provides a directive on the pressure drop and performance loss caused by irregularities (15).
- Control and simulation for validation purposes will become an important issue in modeling of chromatography. The ability to calculate acceptance criteria for scale-up or scale-down will allow rapid, accurate validation of tasks, such as virus clearance. The correct column dimensions, gradient shapes, limits of operation parameters, and rapid diagnosis of separation deficiencies are also added benefits. Scale-up of the production of a biologic is often considered a major change of the process, requiring additional clinical trials that are time consuming and costly. However, if the appropriate scale-up process can be performed with the generation of the same product exhibiting identical qualities, no additional clinical trial would be required. Computer-assisted optimization evaluates the influence of the process change. It would save significant manufacturing costs if the altered scheme were not considered a major change.

The existing models for simulating chromatography are based on different core assumptions. Depending on these assumptions, different levels of complexity are handled. At each level of complexity, different simplifications are made in order to make the mathematical treatment accessible.

According to van Deemter et al. (16) the following assumptions are made:

- The equilibrium concentrations in the two phases are proportional (linear chromatography).
- The exchange process is thermodynamically reversible, that is, the equilibrium between particle and fluid is immediate; the mass transfer coefficient is infinitely high; and longitudinal diffusion and other processes with similar effects can also be ignored. The latter processes are considered ideal chromatography.

From these assumptions one can categorize chromatography into four cases (Table 2). Although the cases 1, 2, and 3 are of great value for understanding chromatography, only case 4 reflects the real situation in preparative chromatography.

Table 2. Modes of Complexity in Chromatography Modeling

	Ideal chromatography	Nonideal chromatography
Linear isotherm	1	2
Nonlinear isotherm	3	4

Source: van Deetmer et al. (Ref. 16).

Differences in Analytical Chromatography

Besides the size and the shape of the separation column, the difference between analytical and preparative chromatography is the intended use (Table 3). In analytical chromatography, the components are separated and identified according to their retention times. In some cases, a component-specific detector is used for the identification. Generally, a small volume (mass) is applied onto the column. Therefore, the separation is carried out under linear isothermal conditions. Frequently, the chromatography can be described as case (Table 2). Preparative chromatography is intended for isolation of a compound. In contrast to analytical chromatography, the separation is carried out in the nonlinear regime. Peak-broadening effects resulting from the nonlinearity of the isotherm must be taken into consideration. The sample is applied as a high concentrated solution or in a large volume as a very dilute solution. The Dirac δ function can no longer be used for describing the sample volume and concentration. The Dirac δ function is a hypothetical approximation for an infinitesimal small volume containing one mass unit.

The whole system is designed so that extra-column effects cannot be neglected. When the performance of a chromatographic separation should be predicted from small scale to large scale, those effects must not be neglected. Special emphasis is put on this problem, because the successful application of computer-assisted optimization is strongly dependent on the extrapolation from small scale to large scale. The extra-column effects can dominate a preparative separation in small scale. In the mathematical description of an analytical column, various mass–transfer problems can be handled by using fine adsorbent particles at a high flow rate. However in a large-scale column, fine particles introduce hydrodynamical problems such as high pressure drop, channeling, and fingering of flow. Consequently, in large chromatography, larger particles are used and mass–transfer resistencies are more pronounced.

Objectives for Computer-Assisted Optimization

The general objective for computer-assisted optimization of chromatography is the reduction of the number of experiments required to identify an optimum. The optimization is accomplished by an iterative process. Frequently, optimal running conditions must be identified. Optimal running conditions can be defined as best buffer for loading, washing, and elution while maintaining highest resolution. The same can be valid for the best throughput (productivity) under highest resolution (17). The latter require chemical parameters (i.e., buffer composition) physical and engineering parameters (i.e., flow rate, height equivalent to theoretical plate). In other words, selectivity, efficiency, and productivity (throughput) must be optimized. Depending on the task, one must rate selectivity, efficiency, and throughput in different ways. Selectivity is mainly achieved by the chromatography sorbent and buffer composition and to some extent by engineering parameters. Efficiency depends on the physical characteristics of the sorbent, the overall setup of the chromatography system, and the chemistry of the mobile phases used. Throughput is influenced by the dynamic capacity of the sorbent, which depends on the composition of the sorbent and the buffers used. All three important characteristics of preparative chromatography are interconnected. Changing one parameter influences the others. Traditionally, after selection of a particular sorbent, one can identify the appropriate separation conditions by seeking the appropriate buffer conditions. Then the throughput is maximized by overloading and increasing the flow rate (18).

Another very important objective in preparative chromatography is the extrapolation from small-scale data to the performance of large scale (scale-up) and the extrapolation from a large scale to small scale (scale-down). When a process is translated from a laboratory scale to a pilot and industrial scale, very often the linear gradient elution must be changed to stepwise elution, because this elution mode is more simple (19,20).

After optimization, the single steps in the sequence of steps have to be optimized, and thus flow sheeting comes into play. In flow sheeting, the performance of the different steps is interconnected, and a strong reciprocal action is observed. However, multiple interconnected experiments must be carried out. The more steps and performance characteristics that are simulated, the fewer experiments needed and thus the faster an optimal separation process is identified.

What are the prerequisites for computer-assisted optimization of preparative chromatography? Because experiments have to be performed in order to have access to parameters for simulation, the experiments should be carried out at the smallest applicable scale. This is not only a heuristic for computer-assisted optimization, but is also valid for all kind of optimization. Whenever possible, the experiments should be carried out with equipment that allows a high degree of automation (21). Electronic data acquisition of a chromatogram is essential for mathematical treatment. The predicted data should be valid for a broad range of flow rates, column sizes, and loading conditions. The main parameters influencing the performance (including the extra-column parameters) should be known. Once the extra-column peak broadening can be discriminated from the parameters of the column, the simulation is fed with this information (22). During scale-up, the contribution of extra columns and column band broadening

Table 3. Differences between Analytical and Preparative Chromatography Other than the Intended Use of the Technique

Analytical	Preparative
Identification by retention or selective detection	Separation coupled with collection of separated proteins
High efficiency	High capacity
Small particles	Large particles
Small loading of columns	Overloading of columns
Linear adsorption isotherms	Nonlinear adsorption isotherms
Sample volume negligible	Sample volume not negligible, (Dirac δ function not applicable)
	Extracolumn space not negligible

changes. Observing these facts improves the accuracy of the prediction of chromatography from small scale to large scale.

Objectives for Computer-Aided Design

The objective in computer-aided design is the proper selection and order of chromatography steps in a purification process. Based on a primary knowledge on the properties (e.g., isoelectric point, molecular weight, surface hydrophobicity, biospecificity) of the protein of interest and its impurities, expert systems can assist in setting up a prospective purification design.

In parallel, these systems have to consider some basic rules for the separation sequence consisting of capture, purification, and polishing (23):

- Choose separation steps based on different properties.
- Separate the largest impurity first.
- Choose steps that use differences in the properties of the product and the impurities best.
- Use a high-resolution step as soon as possible.
- Set the most laborious step last.

After characterizing the chosen steps experimentally, economic considerations concerning productivity, capital, or production costs must be evaluated. Using computer-aided design routines, several characteristics of the product (amount required in a calculation period, purity required, recovery), resin (cost per liter, lifetime), labor constraints (labor costs, number of operators required), and production constraints (column dimensions available, costs of waste disposal, costs of overheads) can be evaluated to find the optimal scale of an individual purification step.

MATHEMATICAL DESCRIPTION OF THE SEPARATION PROCESS

Chromatographic theories have been applied for description of band broadening and retention of a solute while migrating through the column. In liquid chromatography, the column geometry can be described by the radius (r) and the length (z) of the column. Independently of the model applied, some basic assumptions are made in preparative chromatography of biomolecules:

- The column is considered to be homogeneous in radial direction. The applied protein is distributed homogeneously over the cross-sectional area of the column. As a consequence, the r dimension can be disregarded, and the effective parameter for the column geometry is z.
- The mobile phase is incompressible, and the mobile phase velocity is constant along the column.
- There is no influence of thermal effects. The adsorption process is considered to be isothermal, although this is not consistent with regard to rigorous thermodynamic treatment. The temperature is constant during the experiment, independent of position and time.
- Mass exchanges between mobile and stationary phase are infinitely fast, and the two phases are permanently in composition equilibrium. The need of this assumption is circumvented by kinetic models.
- The ratio between the mobile and stationary phase volume is constant during the experiment, independent of position and time. It is also independent of the mobile phase concentrations of protein and salt.

Depending on the type of chromatography and the intended use, these basic assumptions can heavily deviate from the real conditions.

Plate Models

Using these models, the column is divided into identical theoretical plates, and for each of these plates a mass balance is established. Originally, a linear isotherm was assumed, and in each plate mobile and stationary phase concentrations were in equilibrium. These models introduce the concept of the height equivalent of a theoretical plate (HETP). For a column of length (L) with N theoretical plates, HETP is

$$\mathrm{HETP} = \frac{L}{N} \tag{3}$$

Comparison of HETP of different columns or different conditions is a well-accepted measure for the performance of a column. HETP is not a constant value for a sorbent species, it depends on the column length (24), packing quality, flow rate, solute used for determination, and calculation method (25).

Martin and Synge Plate Model. This model (2) considers the column as a series of stages, and an infinitesimal fraction of the mobile phase volume passes from one stage to the next (Fig. 1). Because an infinitesimal small fraction of volume is transferred from one stage to the next, this model is denominated a continuous plate model. The ratios between mobile and stationary phase volume in each vessel are identical and constant. The mass balance for the jth vessel can be written as

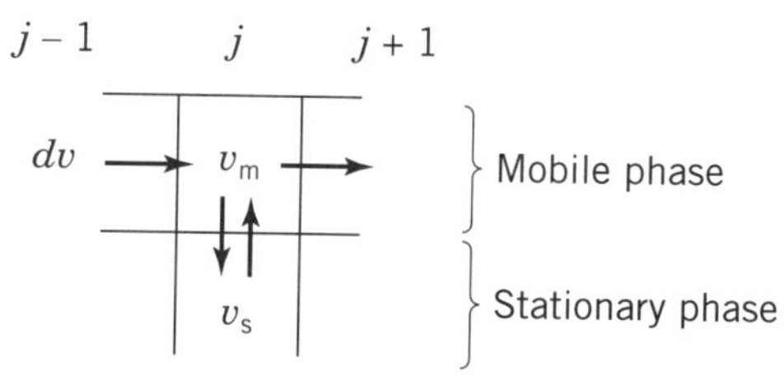

Figure 1. Schematic diagram of the distribution in a Martin and Synge plate model. The infinitesimal small fraction of the mobile phase volume (dv) enters stage j while the identical volume fraction leaves the stage. The distribution between the mobile and stationary phases occurs immediately and infinitely fast.

$$C_{\mathrm{m,j-1}} \cdot dv = C_{\mathrm{m,j}} \cdot dv + v_{\mathrm{m}} \cdot dC_{\mathrm{m,j}} + v_{\mathrm{s}} \cdot dC_{\mathrm{s,j}} \quad (4)$$

when dv denotes the movement of the mobile phase volume, v_{m} and v_{s} are the mobile and stationary phase volume of a stage and C is the concentration of the species under consideration. Martin and Synge introduced one mass unit into the first stage. Solving equation 4 with respect to a linear isotherm and sample only in the first stage gives a Poisson distribution.

$$C_{\mathrm{N}} = C_1^0 \cdot \exp\left(-\frac{v}{v_{\mathrm{m}} + K \cdot v_{\mathrm{s}}}\right) \cdot \frac{\left(\frac{v}{v_{\mathrm{m}} + K \cdot v_{\mathrm{s}}}\right)^{N-1}}{(N-1)!} \quad (5)$$

which can be approximated by a Gaussian distribution for high plate numbers. In equation 5, C_{N} is the concentration of the Nth stage, C_1^0 is the initial concentration of the first stage, and v is the volume of mobile phase that has left stage N from the start.

Craig Plate Model. Analogously to the Martin and Synge model, this model (26) divides the column into a series of discrete stages (plates), and each stage contains the same volume of stationary and mobile phase. But in contrast to the previous model, the flow is not continuous. It is assumed that the whole amount of mobile phase moves to the next stage after equilibrium is attained (Fig. 2). Then again, the distribution between stationary and mobile phase takes place immediately. Hence, the Craig model is a discontinuous plate model. The mass balance for the jth plate can be written as

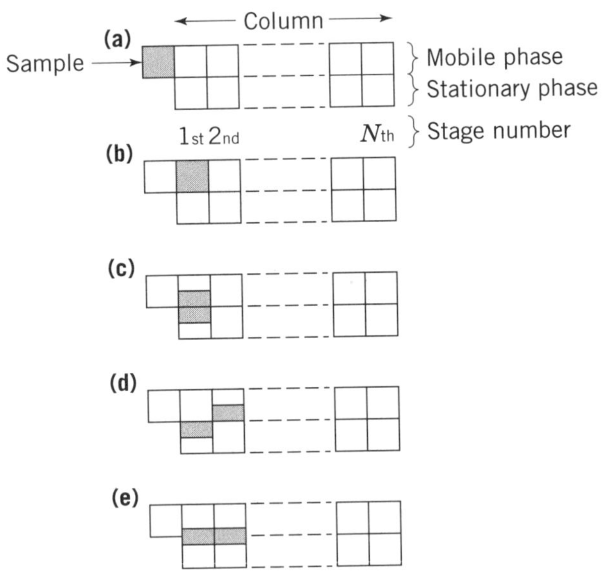

Figure 2. Schematic diagram of a Craig machine for N stages and a constant distribution coefficient. (**a**) A sample segment is arriving at the first stage. (**b**) The segment arrives in the mobile phase of the first stage. (**c**) The sample is distributed between mobile and stationary phases according to the distribution coefficient. (**d**) The mobile phase volumes are moving one stage further. (**e**) Mobile and stationary phases are distributed again.

$$v_{\mathrm{m}} \cdot C_{\mathrm{m},j,\mathrm{t}} + v_{\mathrm{s}} \cdot C_{\mathrm{s},j,\mathrm{t}} = v_{\mathrm{m}} \cdot C_{\mathrm{m},j-1,t-1} + v_{\mathrm{s}} \cdot C_{\mathrm{s},j-1} \quad (6)$$

The indexes t and $t-1$ denote for the actual and the previous time increment. For a linear isotherm and sample only in the first plate, this model gives a binomial distribution that also can be approximated by a Gaussian distribution for large plate numbers.

In the Craig model, peak dispersion is smaller than in the Martin and Synge model (27). This difference can be related by

$$N_{\mathrm{C}} = N_{\mathrm{M,S}} \cdot \frac{k_0'}{1 + k_0'} \quad (7)$$

when N_{C} is the number of plates of the Craig model, $N_{\mathrm{M,S}}$ is the number of plates of the Martin and Synge model, and k_0' is the retention factor

$$k_0' = \frac{t_{\mathrm{R}} - t_0}{t_0} \quad (8)$$

at infinite dilution. Here, t_{R} is the retention time at the conditions of interest and t_0 is the void time, that is, the retention time assuming nonbinding conditions.

Other Plate Models. The plate models described in this section are extended to kinetic models by introduction of mixing in the stages. Hence, the equilibrium in the stages is no longer instantaneous. The finite time for adsorption and transport kinetics between the phases are compensated by mixing kinetics within the stages. The sectional model is an extension of the Craig model (28). At the end of each time, increment the whole content of the liquid phase of each stage is transferred to the next stage. Therefore, the flow of the mobile phase is discontinuous.

The tanks-in-series model is an extension of the Martin and Synge model (29). The flow of the mobile phase is continuous, and the column is considered as an array of well-mixed cells.

Rate Theory

Although the plate theory was originally developed for linear chromatography with the intention to describe the nonideal effects, this theory was, in the early works, focused on ideal nonlinear chromatography.

The Ideal Model. The mass balance equation was derived independently by Wicke (30) and Wilson (1). In their mass balance equations, axial dispersion was disregarded. DeVault (4) introduced the influence of the pore volume, and Weiss (3) discussed Wilson's approach for several types of adsorption isotherms. The mass balance equation of this ideal model is

$$\frac{\partial C_{\mathrm{m},i}}{\partial t} + H \cdot \frac{\partial C_{\mathrm{s},i}}{\partial t} + u \cdot \frac{\partial C_{\mathrm{m},i}}{\partial z} = 0 \quad (9)$$

Here, C_{m} and C_{s} denote for the protein concentration of species i in the mobile and stationary phase, respectively. H is the phase ratio, t is time, z is column length and u is the mobile phase velocity. This model is of interest only for nonlinear isotherms. In the case of a linear isotherm and a column of infinite efficiency, the sample pulse travels through the column without any change (31).

Equilibrium-Dispersive Model. According to Van Deemter et al. (16), Giddings (13), and Haarhoff and van der Linde (32), equation 9 can be replaced by

$$\frac{\partial C_{m,i}}{\partial t} + H \cdot \frac{\partial C_{s,i}}{\partial t} + u \cdot \frac{\partial C_{m,i}}{\partial z} = D_{a,i} \cdot \frac{\partial^2 C_{m,i}}{\partial z^2} \quad (10)$$

when the mass transfer is not infinitely fast. Here, $D_{a,t}$ is the apparent dispersion coefficient that lumps together axial dispersion and nonequilibrium from adsorption and desorption.

$$D_{a,i} = \frac{\mathrm{HETP} \cdot u}{2} \quad (11)$$

It is assumed that the mobile and stationary phases are instantaneously in equilibrium. Assuming a column of infinite length, a Dirac $\delta(z)$ pulse as an injection profile, and a linear isotherm, the solution of this equation is a Gaussian distribution (33).

Lumped Kinetic Model. In this model, the dispersion coefficient D_L accounts only for the axial dispersion and is not lumped to the influence of nonequilibrium effects.

$$\frac{\partial C_{m,i}}{\partial t} + H \cdot \frac{\partial C_{s,i}}{\partial t} + u \cdot \frac{\partial C_{m,i}}{\partial z} = D_{L,i} \cdot \frac{\partial^2 C_{m,i}}{\partial z^2} \quad (12)$$

Depending on the assumptions, different models can be distinguished. If the influence of the mass transfer resistance is negligible and only the kinetics of adsorption and desorption are taken into account, the model is called reaction-dispersive. For a linear isotherm, the kinetic equation is

$$\frac{\partial C_s}{\partial t} = k_a \cdot C_m - k_d \cdot C_s \quad (13)$$

when k_a and k_d are the rate constants of adsorption and desorption, respectively. When a Langmuir-type isotherm is assumed, the Langmuir kinetic model is obtained (8).

$$\frac{\partial C_s}{\partial t} = k_a \cdot (C_{s,\max} - C_s)C_m - k_d \cdot C_s \quad (14)$$

If we assume the adsorption-desorption kinetics as infinitely fast but not the mass transfer kinetics, the model is called transport-dispersive. Then, the kinetic equation for C_s is

$$\frac{\partial C_s}{\partial t} = k_m \cdot (C_m - C_s) \quad (15)$$

when k_m is the apparent mass transfer coefficient.

Assuming a Dirac $\delta(z)$ pulse as an injection profile, a linear isotherm, and that mass transfer kinetics in the transport-dispersive model and adsorption-desorption kinetics in the reaction-dispersive model are fast, the solution of these equations is a Gaussian distribution (16). The general rate model consists of two partial differential equations for each component, one for the mobile phase in the interparticle area of the column and one for the fluid in the intraparticle area. It was used by Kucera (34) for linear isotherms and by Lee et al. (35) and Yu and Wang (36) for nonlinear isotherms.

For numerical solutions of these equations several different approaches can be used. The techniques generally applied are finite-difference and finite-element methods. The direct calculation of numerical solutions of the mass balance equations by finite-difference models leads to stability problems of the calculations (37). A method applied extensively replaces the axial dispersion by introducing a numerical dispersion. In this case, a solution for the ideal model is looked for that gives a first-order error equivalent to the axial dispersion. The numerical dispersion can be adjusted by changing the space increment (z) or time increment (τ) of the iteration. The basis for all calculations is a modified form of equation 10

$$\frac{\partial C_m}{\partial z} + \frac{\partial\left[\frac{1}{u} \cdot (C_m + H \cdot C_s)\right]}{\partial t} = \frac{D_a}{u} \cdot \frac{\partial^2 C_m}{\partial z^2} \quad (16)$$

which is identical with

$$\frac{\partial C_m}{\partial z} + \frac{\partial G(C_m)}{\partial t} = \frac{D_a}{u} \cdot \frac{\partial^2 C_m}{\partial z^2} \quad (17)$$

For the ideal model this equation simplifies to

$$\frac{\partial C_m}{\partial z} + \frac{\partial G(C_m)}{\partial t} = 0 \quad (18)$$

Replacing the first term by the backward finite difference and the second by the forward finite difference (backward-forward differences), this equation for the $j + 1$ at the position n is

$$\frac{C^{j}_{m,n} - C^{j}_{m,n-1}}{h} + \frac{G^{j+1}_{n} - G^{j}_{n}}{\tau} = 0 \quad (19)$$

This calculation scheme is identical to the Craig model if the ratio of the calculation intervals h/τ equals u. This Craig model was widely used by several authors (38–42). In contrast to a forward-backward scheme (43), which is put to use for isocratic conditions (44,45), it can be applied to calculate gradient elution chromatography (Fig. 3).

More exact calculations with respect to the estimation of the axial dispersion effects are possible using finite-element methods. In chromatography, the method of orthogonal collocation on fixed elements is preferred to the method of moving elements (36,46,47). Arve and Liapis (48–50) call this approach the finite bath method instead of finite element. All these methods are limited to isocratic

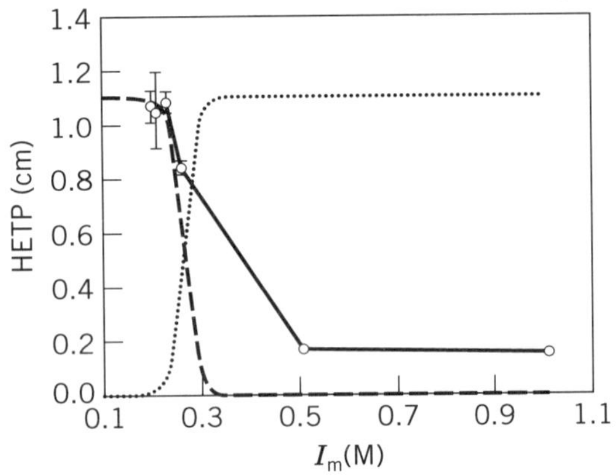

Figure 3. Dependence of HETP on I_m measured by BSA pulses applied to an anion exchanger (Q-HyperD). The column was equilibrated with elution buffers containing various amounts of NaCl. u = 3,006 cm/h; i.d. × h = 5 mm × 5.2 cm. The *dashed line* is a calculation of HETP values according to a Craig model. The *dotted line* occurs when using forward-backward calculation of finite differences.

conditions by means of numerical stability of the calculations (31).

Recently, a fast algorithm based on the assumption of elution as a collocation of point masses migrating with a velocity dependent on the local concentration and the flow rate was presented (51). This model gives comparable results to a Craig model, but the computation time is two orders of magnitude smaller.

Stochastic Theory

A statistical approach developed by Giddings and Eyring (52) has been used for a stochastic model to describe migration and broadening of a sample in a chromatographic column. They considered the migration of a solute in a column as a Poisson distribution process and ignored the axial dispersion. They obtained following probability distribution:

$$P = \frac{k_a \cdot k_d \cdot t_0}{2 \cdot \sqrt{\pi} \cdot (t - t_0)^{0.75}} \cdot \exp[-(\sqrt{k_d \cdot (t - t_0)} - \sqrt{k_a \cdot t_0})^2] \tag{20}$$

This model was extended to more complex situations accounting for axial dispersion and mass transfer resistances (53). For an increasing number of entries of a molecule into the stationary phase and a linear adsorption isotherm, this probability distribution tends to a Gaussian distribution.

Extra-Column Broadening

To be able to describe the behavior of a protein sample in different scales of chromatographic systems, the influence of the extra-column space on the band broadening must be distinguished from the broadening occurring in the column. A comprehensive description of the extra-column effects on gas chromatography is given by Sternberg (54). Most of the explanations are also valid for liquid chromatography (55).

Contributions to Extra-Column Broadening. The extra-column effects can be subdivided into dispersion in tubes, contributions of dead volumes, finite detector volume, and dynamic behavior of transducers and electronics. According to Taylor (56), tubes introduce a symmetrical Gaussian-type broadening (σ_t^2) to the sample. The band spreads under the combined action of molecular diffusion and the parabolic axial flow velocity profile. This broadening can be described by

$$\sigma_t^2 = \frac{t_R \cdot r^2}{24 \cdot D_M} \tag{21}$$

where t_R is the retention time, r is the radius of the tube, and D_M is the molecular diffusion coefficient. Dead volumes create a type of washout kinetic (57). Therefore, they introduce an exponential contribution to the band broadening. This exponential contribution (τ_{dead}) is a function of the dead volume (V_{dead}) and the flow rate (F)

$$\tau_{dead} = \frac{V_{dead}}{F} \tag{22}$$

It was also shown by Sternberg (54) that the finite sensing volume of a detector introduces an additional symmetrical, rectangular broadening. This broadening (σ_d) depends on the sensitive detector volume (V_d) and the flow rate. According to Cram and Glenn (57) it can be described by

$$\sigma_d^2 = \frac{V_d^2}{12 \cdot F^2} \tag{23}$$

An exponential contribution to the band broadening is introduced by the finite response rate of the electronical amplifier and recorder. This contribution (τ_{el}) is independent of the flow rate (57,58).

Mathematical Treatment. Assuming additivity of the variances, the total extra-column broadening can be written as

$$\sigma_{ex}^2 = \sigma_s^2 + \sigma_t^2 + \sigma_d^2 + \tau_{dead}^2 + \tau_{el}^2 \tag{24}$$

where σ_{ex}^2 is the lumped extra column dispersion and σ_s^2 is the variance of the initial theoretical injection profile. Assuming that in liquid chromatography the variances introduced by the electronics and the detector volume are negligibly small compared to the other variances, equation 24 simplifies to

$$\sigma_{ex}^2 = \sigma_s^2 + \sigma_t^2 + \tau_{dead}^2 \tag{25}$$

Lumping together the F-independent variances (σ_A^2) and the F-dependent variances (σ_B^2), the extra-column broadening (σ_{ex}^2) is (59)

$$\sigma_{ex}^2 = \sigma_A^2 + \frac{\sigma_B^2}{F^2} \tag{26}$$

An equation for the prediction of the extra-column contribution to the total broadening can be derived based on the definition of the plate number N of a system

$$N = \frac{V_P^2}{\sigma_{c.i.}^2} \tag{27}$$

If the total broadening (σ_{total}^2) is considered as the sum of the column internal broadening ($\sigma_{c.i.}^2$) and the extra-column broadening (σ_{ex}^2)

$$\sigma_{total}^2 = \sigma_{c.i.}^2 + \sigma_{ex}^2 \tag{28}$$

and N is replaced according to equation 3, the following relationship can be obtained

$$\sigma_{total}^2 = \frac{\text{HETP}}{L} \cdot V_R^2 + \sigma_{ex}^2 \tag{29}$$

This equation was derived for analytical chromatography by Huber and Rizzi (60) and has also been applied for preparative chromatography (22). Using this relationship it is possible to exhibit a quantitative picture of the contribution of extra-column broadening on the total broadening (Fig. 4).

ACADEMIC AND COMMERCIAL AVAILABLE COMPUTER PROGRAMS

Computer-aided tools have been developed to different levels of complexity. Programs have been written for prediction of peak profiles, breakthrough curves, gradient shapes, and optimization of resolution (61). On the other hand, expert systems have been created for development of chromatographic separations (62), by using the prediction of peak profiles as tools for optimization.

Expert Systems

An expert system applies heuristic rules for the solution of a problem by using a knowledge base and an interference engine. The knowledge base is a collection of rules and

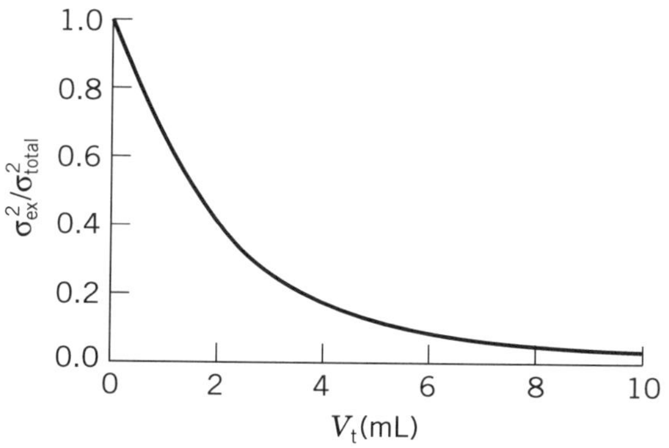

Figure 4. Contribution of extra-column band broadening to total broadening as a function of total column volume (V_t) (22). Original data from experiments using a 0.5-cm i.d. column.

facts gathered from practical experience. The interference engine applies domain knowledge to draw conclusions for the problem under consideration (Fig. 5).

Although the usefulness of expert systems was shown in the late 1960s for chemical applications with the expert system DENDRAL (64), expert systems for chromatography were not introduced until the early 1980s. The expert system ECAT (Expert Chromatographic Assistance Team) (65) provides information for column packing and geometry, and known analyte properties determine the type of chromatographic mode (reversed phase, ion exchange, etc.). Optimizations based on the linear solvent strength model (LSS), assuming a linear relationship between the log k' and the mobile phase modifier concentration, Φ, (66) are offered. A linear relationship between the log k' and the solvent strength of the mobile phase modifier is assumed

$$\log k' = \log k_0' - S\Phi \tag{30}$$

where k' is the capacity factor (normalized retention time in respect to the dead time) defined as

$$k' = \frac{t_R - t_0}{t_0} \tag{31}$$

and φ is the volume fraction of the mobile phase modifier. S reflects the dependency of k' on φ, and k_0' is the capacity factor without the mobile phase modifier.

The expert system expert system for chromatography (ESC) (67) was generated for method development in gas chromatography and HPLC. A set of about 500 chromatograms from the literature constitutes the basis for optimization. Peaks can be simulated, and mobile and stationary phases can be selected. For the peak identification, a curve-fitting tool based on the exponential modified Gaussian function is implemented. Fell et al. (68) developed an expert system to optimize the mobile phase composition in RPC. Therefore, they applied simplex optimization and iterative regression techniques based on experimentally determined retention times.

The European project ESCA (Expert Systems for Chemical Analysis) provided three complex expert systems for method development in chromatography. The first deals with method ruggedness (69); the second, with method (70), and the third, with method development from initial method selection to optimization (71).

The expert system HPLC Doctor (72) and a similar system developed by Tsuji and Jenkins (73) promise help when specific problems with HPLC arise.

Numerous expert systems were developed for RPC (74–76).

A comprehensive description of expert systems in chromatography is given by Hamoir and Massart (62). Presently, for preparative chromatography expert systems are not available. The starting material is very complex, and the purification is generally not completed in one cycle. Expert systems on downstream processing of fermentation products have been developed by Leser and Asenjo (63). These systems comprise optimal flow sheeting of the possible unit operation suited for recovery of bioproducts.

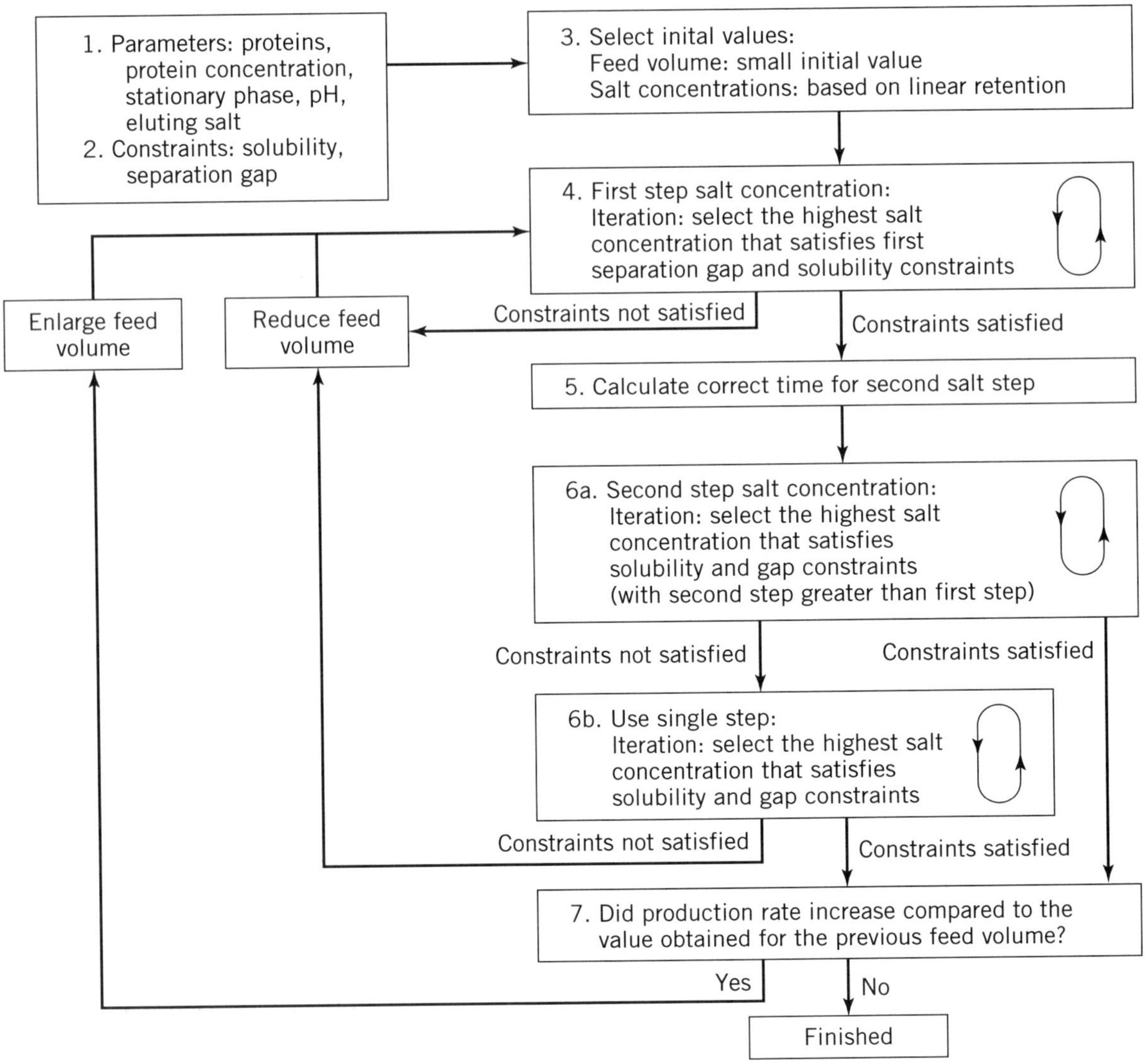

Figure 5. Architecture of the expert system and the links with external databases and algorithms. *Source*: Leser and Asenjo (63).

Leser et al. built an expert system for selecting the sequence of processes for downstream processes, using a commercially available shell (77). The knowledge base contains around 600 logical rules, methods, and structures that emulate the expert's reasoning in selecting downstream processes. The creation of the rationale for the selection of a sequence of high-resolution purification steps must be based on heuristic rules because the existing models for chromatographic separations are not adequate for solving questions of selecting processes, they relate more to the mechanisms of interaction between proteins and matrixes. The heuristic rules are similar to those used to guide the selection of processes for multicomponent mixtures in chemical engineering (Table 4). The rationale for selection of high-resolution purification operations uses the theoretical concept of separation coefficients (23,80).

Simulation Programs

Since the early 1980s, several computer programs for simulating chromatographic behavior have been developed. Most of the programs are only of academic applicability because the experimental accessibility of model parameters is very cumbersome.

Smit et al. (81,82) describe a discrete mathematical model for nonlinear, nonideal chromatography. The nonlinearity can be represented by several different isotherms (linear, Langmuir, Freundlich, sigmoidal). The nonideality is represented by a lumped parameter reflecting axial molecular diffusion, Eddy diffusion, and different sample shapes. No dispersion occurs in the stationary phase. The conservation of mass is used to set up a mass balance. The resulting differential equations are solved by a fourth order Runge-Kutta method.

Phillips et al. (83) use a probabilistic approach for the simulation of a chromatographic separation process, called discrete event simulation. All molecules are executed with the same adsorption–desorption mechanism. Each molecule undergoes a certain number of events (adsorption and desorption). The program can only simulate isocratic conditions and was originally developed for gas chromatog-

Table 4. Classification of Heuristics

Type	Applications and examples
Method heuristics	Specify choice between different unit operations.
	Describe conditions for preference of microfiltration over centrifugation.
Design heuristics	Specify order or sequence of process steps.
	Perform low-resolution step before high-resolution step.
	Use large particles at the beginning of a purification sequences.
Species heuristics	Specify based on the properties of the components.
	Choose separation method based on the greatest difference in properties of product and impurities.
	This rule tells us to use always affinity chromatography.
	There are certain other limitations.
Composition heuristics	Specify feed composition based on separation costs.
	This rule tells us that affinity chromatography may be too expensive.

Source: Wheelright (78), Nadgir and Liu (79).

raphy. Nonlinear isotherms are also taken into account by applying a molecular density map. The probability of adsorption is related to the accessible (free) binding sites.

Whitley et al. (84) established a very advanced computer program for simulation of peak profiles named VERSEC-LC. The calculation is based on an highly evolved rate model, where the column is virtually divided into different stages (vessels). This model treats the column as a series of N equal volume stages (equivalent to plates or series of vessels or transfer units). The mobile phase flows through these stages at a given rate. Each stage is divided into three fractions: the bulk solution, pore solution, and the solid phase. A gradient between the bulk solute concentration and pore solute concentration results from the combined effects of film and pore diffusion mass transfer resistances. Equilibrium assumed between the solute in the pore solution and the solute on the solid phase is described by nonlinear isotherms. The program can simulate gradients, different loading concentrations, and volumes using different pore structures and diffusion coefficients.

Several computer programs (e.g. Dry-Lab®) use the linear solvent strength theory (LSS) for the calculation of peak positions (27,85). A column is calibrated by two runs, and the optimal gradient shape can be calculated by using a resolution map. The program is mainly suited for RPC, because ion-exchange chromatography (IEC) does not follow the LSS theory (86). The program is intended for optimization of analytical chromatography.

Gallant et al. developed a computer model based on the stochastic mass action (SMA) formalism. SMA is a further improvement of the stoichiometric displacement model (87). The SMA formalism represents the chromatographic adsorption process as a stoichiometric exchange of mobile phase protein and bound counterions. Furthermore, a fraction of counterions is shielded by the adsorbed protein. The accessible sites are corrected by a so-called shielding factor, which is proportional to the stationary phase concentration of protein. For acquisition of the parameters, describing the protein adsorption at different salt concentrations, the LSS formalism is applied, although this formalism is not completely valid for IEC (86). Prediction of protein zone migration and broadening is carried out by a single lumped dispersion model (which is described by equation 10). Gallant et al. (88) propose an optimization routine using simulation of chromatography by the aforementioned models. The optimization runs through different stages. Initially, modeling parameters have to be acquired, and protein constraints (e.g., solubility, resolution) have to be defined. These constraints meet actual requirements of large-scale protein production (Fig. 6).

The program Simulus uses a similar approach as described by Whitley et al. (84), and diffusional resistances and a surface reaction resistance were included. The fundamentals for this program were published by Cowan et al. (29). Furthermore, they have integrated moving grids for calculation of the peak profiles of Wiblin et al. (89). The program is marketed by BIOSEP, AEA Technology (Harwell, U.K.).

ProSys™ Simulation, distributed by BioSepra Inc. (Marlborough, Mass.) is a computer package using the simple Craig simulation (5,90). The computer package is connected to hardware that allows automated determination of the parameters required for optimization of preparative chromatography (ProSys™ Workstation). Currently, it is the only commercially available integrated package of parameter acquisition and simulation software. For IEC, distribution coefficient values as functions of salt concentrations are predicted by linear gradient experiments as proposed by Yamamoto et al. (19). The calculation of simulation parameters can be carried out semiautomatically, assisted by the software. Validation of the simulated peak profiles can easily be accomplished because the software does not discriminate between experimental and simulated runs. So, comparison and superposition are simple. The software package implemented in the ProSys™ Workstation also contains a tool for economical evaluation of chromatographic unit operations (ProSys™ Scale-Up).

All the aforementioned computer programs do not take into account the extra-column band spreading. This mechanism may become very important when it comes to scale-down and parameter acquisition at very small scale (22).

The authors of this chapter have developed a computer program that is an extended version of the simulation program used in the commercial system ProSys. This extended version allows the prediction of peak profiles using a big variety of different adsorption isotherms (90). Furthermore, the influence of extra-column inlet and outlet can be simulated. The modifications of gradient profiles by the mixing device is also part of the simulation (91). Rescaling and influence of the setup can be virtually studied. The simulation program can also be used for validation of chromatographic conditions.

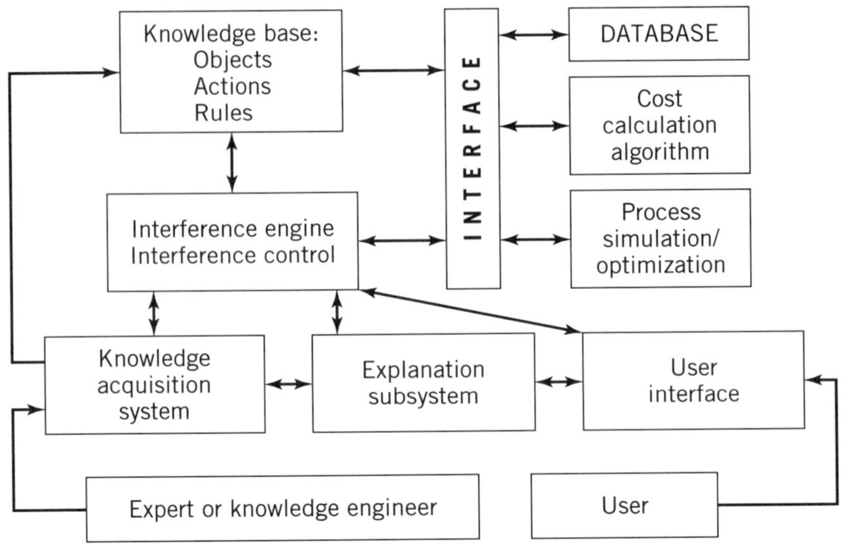

Figure 6. Iterative optimization scheme for step-gradient chromatography, according to Gallant et al. (20). *Source*: Reproduced by kind permission of the authors.

Artificial Neural Networks

Artificial neural networks have been recently applied for chromatographic peak classification (92). Neural networks are specialized computer systems having processing capabilities similar to the human brain. They are distinct from conventional computer systems, because they process entire patterns, rather than successive individual items of data. They are trained by example to carry out their tasks, instead of being explicitly programmed. The processing patterns and learning capabilities displayed by human brains are difficult to mirror using conventional computer programming. Nevertheless, a neural network is, ultimately, a particular kind of computer program, far removed in its theory or implementation from any biological system.

Modern digital computers built with traditional designs have a fundamental limitation, the so-called von Neumann bottleneck. Traditional computation requires a problem to be broken down to a set of operations that are performed in serial fashion, that is, one instruction at a time. Typically, each instruction must be completed before the next instruction is executed. Artificial neural networks represent a fundamentally different approach to computation. They are explicitly designed to mimic the basic organizational features of biological nervous systems: parallel, distributed processing. Artificial neural networks consist of a large number of simple interconnected processing elements, where the processing elements are simplified models of neurons and the interconnections between the processing elements are simplified models of the synapses. The processing of information in such networks therefore occurs in parallel and is distributed throughout each unit composing the network.

Neural networks are particularly appropriate for a defined class of tasks, which may be characterized as follows:

- Pattern recognition, rather than sequential data processing
- tasks demanding nonlinear capabilities in producing their classification, which can be a problem for other techniques
- technically ill-defined tasks, even though the decision may be obvious
- tasks with sufficient examples of the target data available to allow a learning system to be considered

Peak-shape classification is a pattern recognition task that is pertinent to chromatography. The automated determination of the peak profiles in a chromatogram gives an objective answer when two chromatograms are identical or different. It may be a valuable contribution to validation for a purification sequence or validation of scale-up.

APPLICATIONS AND LIMITATIONS OF MODELING

For a conventional software, exact definition of the conditions leading to the optimum is necessary. Mathematical algorithms (explicit equations or differential equations) must be available to calculate the optimum. The major obstacle in preparative chromatography is the complexity of the variables, the problem to estimate the parameters, and the response value. It may be also possible that an optimum is not found within the investigated range of variables.

In Figure 7a and b, a system with a distinct optimum is described. Independently from the starting point (initial pair of variables 1 and 2), an optimum will be found. Systems, where the variable and the response value are related by a peak function or where the variable consists of several variables lumped together, produce a surface with a unique maximum of the response value. The lumped variables or reciprocal action with other variables should be carefully investigated. Minor changes in the experimental conditions may cause a big variation in the optimal response value.

The optimum obtained in a system shown in Figure 7c and d depends on the starting point. Several optima and even local minima are present. Such a system can be characterized by carrying out an optimization strategy with starting points over the whole area in small increments of the considered variables and comparing the reached min-

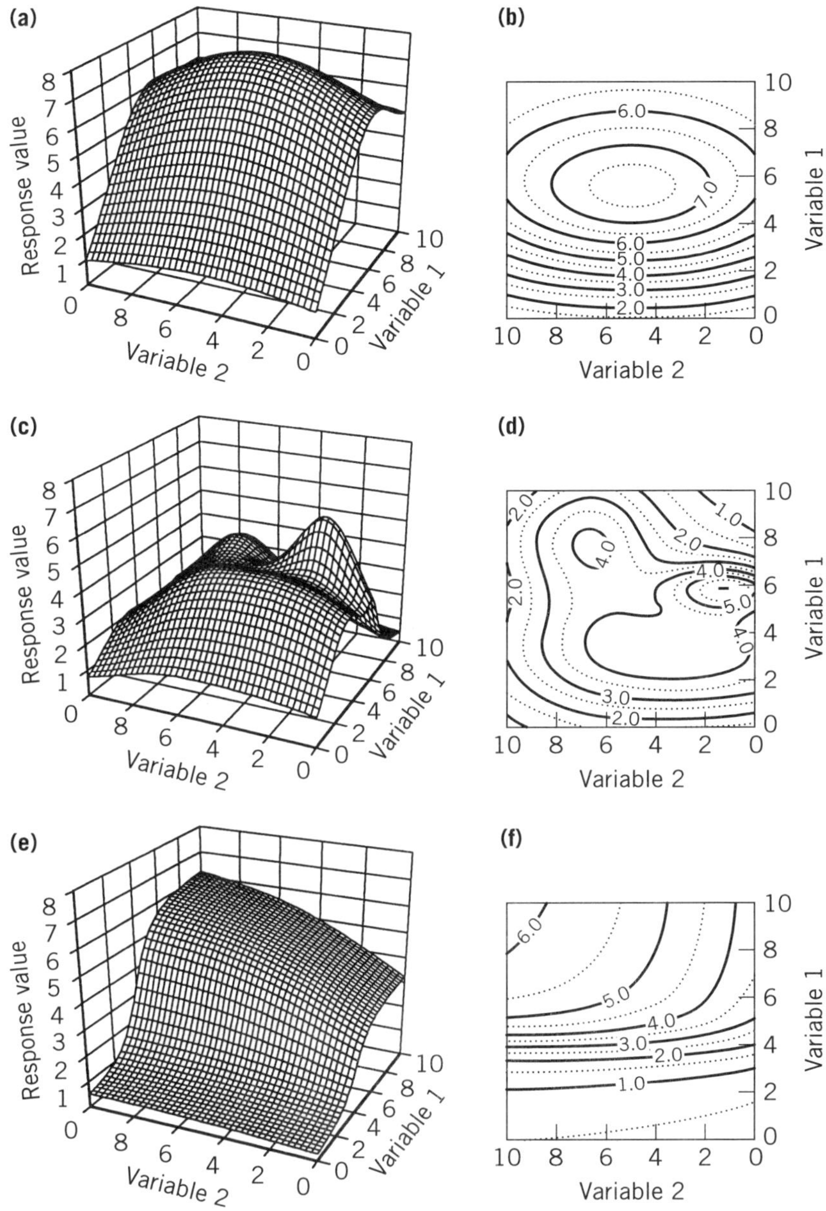

Figure 7. (**a** to **d**) Evaluation of optimum with two variables and one response value. (**e** and **f**) An optimal response value is not found in the considered range. In chromatography, the variable can be, for example, flow rate or loading, and the response value can be, for example, resolution, productivity, or purity.

ima and maxima. Figure 7e and f shows a system where the optimum cannot be reached within the allowed range of variables. This situation is often observed, when the productivity of a chromatographic separation is optimized in respect to flow rate and loading. When the separation is not sensitive to resolution, the system is hypothetically only constrained by the range of the variables. The performance of chromatography is influenced by far more variables than two, but it is very complicated to plot a function with more than two variables in a nonconfusing manner. One way is the overlay of contour plots, as shown in Figure 7b, d, and f and in Figure 8.

Let us assume that a chromatographic separation depends on two variables, such as flow rate and loading. The hypothetical task is to find the optimal response value, such as productivity. Using experimental and simulated results, the optimal response values are searched. A certain purity is presumed. Then the productivity and the purity are plotted versus flow rate and loading. Both contour plots are superimposed. The area, where the presumed purity is reached, indicates the optimal range of operational variables. This approach was used by Mao et al. (93).

Sometimes the conditions are constrained by pragmatic issues, such as availability of equipment (commercially available columns with a certain diameter, costs of a buffer) or by the physicochemical characteristics of the product. Many proteins have limited stability and solubility. There are conditions where the protein may be inactivated. These conditions are not easily transferred into mathematical equations. Sadana (94) summarized possible causes for inactivation of biological macromolecules during separation. During the adsorption and desorption process, proteins may undergo a conformational change. Proteins can associate or they can precipitate on the surface (95–

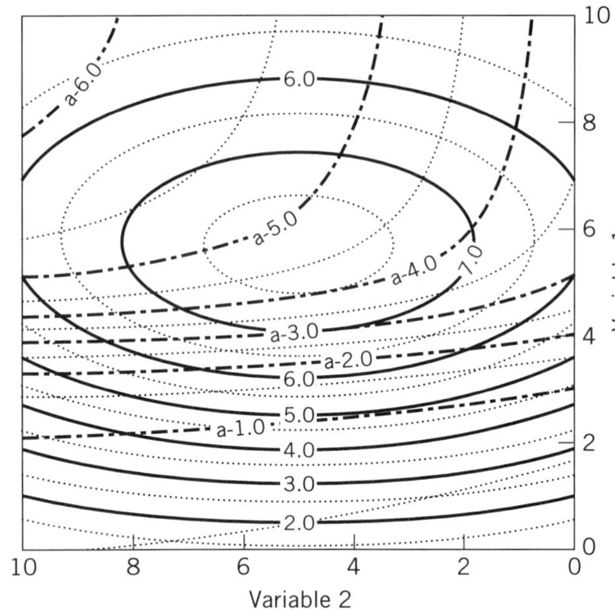

Figure 8. Evaluation of an optimal operational range, when two different response values are related to the two variables. A hypothetical system represented by the graphs **a** and **b** and **e** and **f** in Figure 7 are used to describe the strategy. The contour plots in Figure 7**b** and **f** are superimposed. The intersecting plane formed by the contours with the lower limits of the response values (e.g., a.5.5 and 7.0) yield the optimal range.

97). Self-association was modeled by Lemque et al. (98). They could derive quantitative information on cooperativity effects from the adsorption isotherms. From their model, they showed that protein–protein interaction exists in the adsorbed state and led to an increase in protein self-adsorption. Considering a crude protein sample from a clarified fermentation broth, many proteins and other biological macromolecules are present in this solution. Changing the concentration or pH will affect the self-association. On the other hand, one can also expect the formation of hetero di- and oligomeric protein complexes. Protein-DNA complexes can also occur. The capacity and the lifetime of a column are affected by the aforementioned aggregation phenomenon. This short discussion of protein interaction shows a very important aspect in optimization of chromatography, and complications arise when they have to be translated into a mathematical formulation. One possible way to circumvent these obstacles is the standardization of the sample as far as possible.

Conventional Optimization Methods

Conventionally, resolution has been optimized, because the major activities in computer-assisted optimizations have been carried out in analytical applications. Though some rules may also be applied for optimization of preparative chromatography, in analytical chromatography resolution (R_S) is defined as:

$$R_S = \frac{2 \cdot (t_2 - t_2)}{w_1 + w_2} \tag{32}$$

where t_2 and t_1 are the retention time of the separated substance, and w_1 and w_2 are the peak width at the base. A resolution map is the response value of the optimization. A resolution higher than 1 should be achieved throughout the whole chromatogram. A high resolution is not desired, because it increases the analysis time. The optimization of the running conditions (mobile-phase modifier composition, gradient shape, flow rate, etc.) can be carried out by factorial design, simplex optimization, and others.

Simplex Optimization. Although there are many variants of the simplex procedure (99), they are all based on the basic procedure of Spendley et al. (100). A simplex is defined as a geometric figure having one more point (vertex) than the number of variables being optimized. Thus, for two variables, a simplex is a triangle and for three variables the simplex is a tetrahedron. Although it is difficult to visualize a simplex for more than three variables, the mathematics does not become significantly more complex and the procedure is easily handled by manual or digital computation. Figure 9 shows a two-variable (dimension) simplex as it moves, with constant step sizes, across a response surface. The optimization proceeds by rejection of the vertex, which has the worst experimental response, and by reflecting its coordinates through the midpoint of the hyperplane.

Unfortunately, this original, fixed-step size procedure suffers from several severe limitations. In particular, progress across the response surface is made at a constant rate because the step size is fixed; this fixed size also means that the optimum may not be precisely located because the simplex will be forced (in two dimensions at

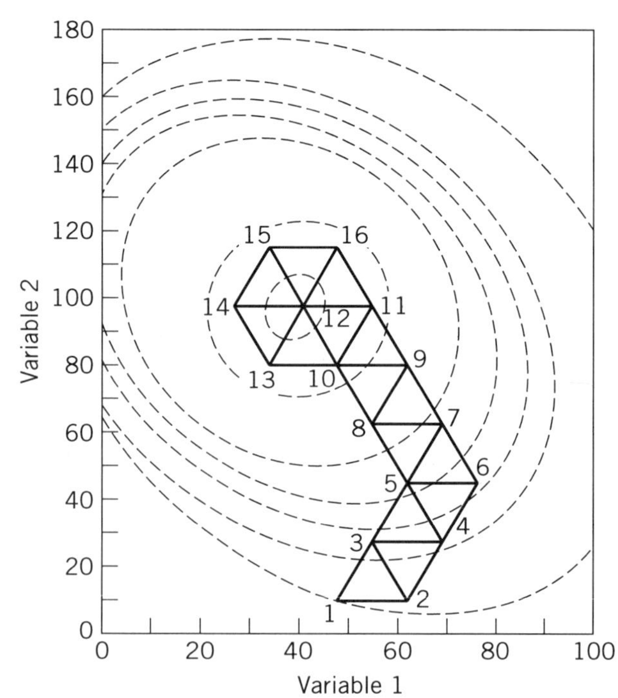

Figure 9. Fixed-step-size simplex optimization of two variables. Initial simplex is 123 and the optimum region is at vertex 10, 11, 12. *Source*: Reproduced with kind permission of Berridge (101).

least) to circle around it. Additionally, the fixed-step size simplex is prone to failure on a surface ridge, in that it will become stranded and not make progress toward the optimum. These limitations have been largely overcome in the modified procedure of Nelder and Mead (102). They introduced two new operations to the fixed-step size procedure, expansion and contraction. The latter operations allow the simplexes to expand and thus accelerate toward the optimum region. When they have approximately located the optimum region, the simplexes contract and reduce the search region until the optimum is precisely located. Figure 10 shows a variable-step size simplex on the same response surface as in Figure 9.

Berridge (101) reviewed the chromatography response functions (CRF) used in analytical chromatography. These functions can be also applied to preparative chromatography (103). The functions successfully used to date for simplex optimization are in general based on sums of terms, usually reflecting resolution and analysis time. The CRF are derived for the separation of a small number of components from a complex sample. A few response functions will be discussed as examples. A comprehensive treatment of the CRF used in analytical chromatography would exceed the scope of this chapter.

For isocratic separations, Berridge (104) developed a function to provide greater flexibility in meeting the needs of fully automated optimization:

$$\mathrm{CRF} = \sum R_{\mathrm{i}} + n^{\mathrm{a}} - |t_{\mathrm{m}} - t_{\mathrm{n}}| - c(t_0 - t_1) \quad (33)$$

where R_i is the actual resolution, n is the number of detected peaks, t_m is the desired retention time, t_n is the retention time of the last peak, t_0 is the dead time, t_1 is the retention time of the first elute peak, and a is a selectable weighting parameter.

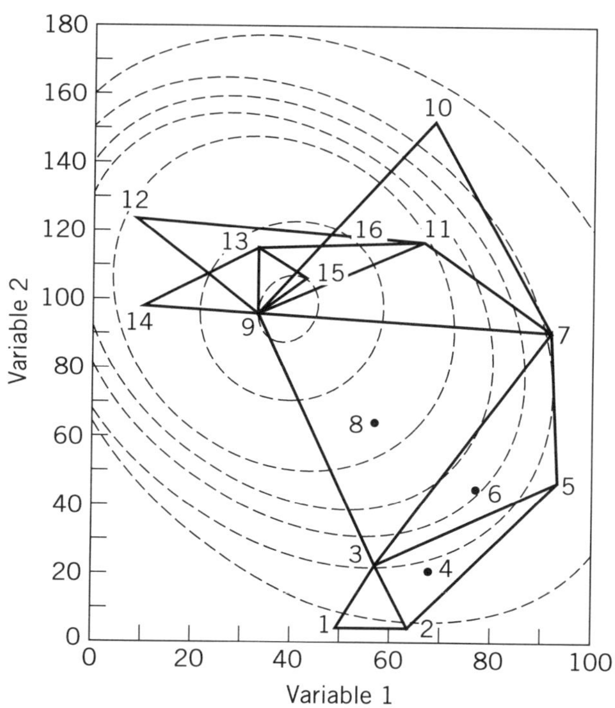

Figure 10. Variable-step simplex. *Source*: Reproduced by kind permission from Berridge (101).

To overcome some of the anomalies generated by using a multicriterion CRF in simplex optimization, Wright et al. (105) simplified the CRF (equation 33) to

$$\mathrm{CFR} = \sum R_{\mathrm{i}} + n - (t_{\mathrm{m}} - t_{\mathrm{n}}) \quad (34)$$

with $t_m - t_n > 1$. The time constraint for the first eluted peak is deleted and that for the last peak is considered only if the difference between the retention time and the target retention time is greater than 1 min. The term for the number of peaks (n) is not raised to a power to ensure that when the term is included, it has a significant influence on the CRF value.

Isocronal Analysis. Atamna and Grushka (106) describe the concept of optimization by isochronal analysis, formerly known as the time normalization for HPLC. Briefly, in this mode of optimization, two experimental conditions are changed simultaneously. One is altered to improve resolution, whereas the other is changed to maintain the analysis time constant. The constant analysis time means:

$$t_{\mathrm{R1}} = t_{\mathrm{R2}} \quad (35)$$

where the subscripts 1 and 2 indicate two different experimental conditions. From equation 36, the following conditions can be derived:

$$\frac{Z_1(1 + k_1')}{u_1} = \frac{Z_2(1 + k_2')}{u_2} \quad (36)$$

where Z is the column length, u is the linear velocity, and k' is the capacity factor. The need to change two experimental conditions simultaneously is clearly seen from equation 36. The most practical pairs of experimental conditions that can be exploited in isochronal analysis are mobile phase velocity and composition, mobile phase velocity and column temperature, or mobile phase composition and column temperature. When the residence time of a product is of concern, this type of optimization may have some value also for preparative chromatography.

Flow Sheeting

Flow sheeting in bioprocessing is a very complex task. A process for purification of a protein consists of several steps. Usually there is a link between the performance of each step and the performances of the previous step and the subsequent steps. The complexity of the problem is obvious when considering a tree structure, assuming a three-step purification with only two possibilities at each step (Fig. 11). Even in this simplified case we end up with eight different purification schemes. Additionally, an economical evaluation of the individual arrangement is hard because the steps of one sequence are influencing each other.

The use of expert systems is suggested for the rational selection of purification processes starting from different

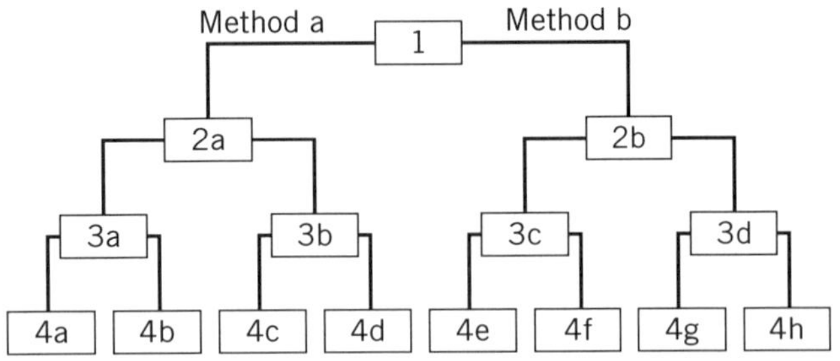

Figure 11. Tree structure of an idealized three-step purification. Each step offers only two possibilities.

microbial sources, such as harvesting, preconditioning, concentrating, purification, and polishing (77,107).

Artificial neural networks are suited to improve the quality of expert systems for flow sheeting. The numerical certainty for defined steps can be fine tuned by feedback with existing results. A successful application has not yet been shown today. However, the large number of software tools for artificial neural networks will promote this application.

The Biopro Designer is a computer program allowing the economical evaluation of each separation step. As in real life, chromatography is limited to other separation steps, such as ultrafiltration, precipitation, etc.

BIBLIOGRAPHY

1. J.N. Wilson, *J. Am. Chem. Soc.* **62**, 1583–1591 (1940).
2. A.J.P. Martin and R.L.M. Synge, *Biochem. J.* **35**, 1358–1368 (1941).
3. J. Weiss, *J. Chem. Soc.* 297–303 (1943).
4. D. DeVault, *J. Amer. Chem. Soc.* **65**, 532–540 (1943).
5. A. Jungbauer and O. Kaltenbrunner, *Biotechnol. Bioeng.* **52**, 223–239 (1996).
6. E.N. Lightfoot, R.J. Sanches-Palma, and D.O. Edwards, in H.M. Schoen ed., *New Chemical Engineering Separation Technique*, Interscience, New York, 1962, pp. 99–181.
7. S. Vajada and P. Valkó, in I. Pallai and G.E. Veress eds., *Studies in Computer-Aided Modeling, Design and Operation. Part B: Systems*, Elsevier, Amsterdam, 1992, pp. 28–95.
8. H.C. Thomas, *J. Am. Chem. Soc.* **66**, 1664–1666 (1944).
9. R.D. Whitley, R. Wachter, F. Liu, and N.-H.L. Wang, *J. Chromatogr.* **465**, 137–156 (1989).
10. E. Grushka, *Anal. Chem.* **44**, 1733–1738 (1972).
11. W.W. Yau, *Anal. Chem.* **49**, 395–398 (1977).
12. J.P. Foley and J.G. Dorsey, *Anal. Chem.* **55**, 730–737 (1983).
13. J.C. Giddings, *Dynamics in Chromatography. Part I. Principles and Theory*, Dekker, New York, 1965.
14. J.L. Wade, A.F. Bergold, and P.W. Carr, *Anal. Chem.* **59**, 1286–1295 (1987).
15. P.L. Spedding and R.M. Spencer, *Comput. Chem. Eng.* **19**, 43 (1995).
16. J.J. van Deemter, F.J. Zuiderweg, and A. Klinkenberg, *Chem. Eng. Sci.* **5**, 271–289 (1956).
17. J.C. Janson and P. Hedman, *Biotechnol. Prog.* **3**, 9 (1987).
18. J.H. Knox and H.M. Pyper, *J. Chromatogr.* **363**, 1 (1986).
19. S. Yamamoto, K. Nakanishi, and R. Matsuno, *Ion-Exchange Chromatography of Proteins*, Dekker, New York, 1988.
20. S.R. Gallant, A. Kundu, and S.M. Cramer, *Biotechnol. Bioeng.* **47**, 355–372 (1995).
21. D. Berger and D. Gillard, *J. Chromatogr.* **210**, 33 (1981).
22. O. Kaltenbrunner, A. Jungbauer, and S. Yamamoto, *J. Chromatogr., A* **760**, 47–53 (1997).
23. J.A. Asenjo, in JA. Asenjo ed., *Separation Processes in Biotechnology*, vol. 9, Dekker, New York, 1990, pp. 3–16.
24. R.M. Nicoud and M. Perrut, in G. Ganetsos and P.E. Barker eds., *Preparative and Production Scale Chromatography*, Dekker, New York, 1993, pp. 47–77.
25. H.P. Lettner, O. Kaltenbrunner, and A. Jungbauer, *J. Chromatogr. Sci.* **33**, 451–457 (1995).
26. L.C. Craig, *J. Biol. Chem.* **155**, 519–534 (1944).
27. L.R. Snyder, J.W. Dolan, and G.B. Cox, *J. Chromatogr.* **483**, 63–84 (1989).
28. Q.M. Mao and M.T.W. Hearn, *Biotechnol. Bioeng.* **52**, 204–222 (1996).
29. G.H. Cowan, in A.E. Rodrigues, M.D. LeVan, and D. Tondeur eds., *Adsorption: Science and Technology, NATO ASI Series E.* vol. 158, Kluwer, Dordrecht, The Netherlands, 1989, pp. 517–537.
30. E. Wicke, *Kolloid-Z.* **86**, 295–313 (1939).
31. G. Guiochon, S. Golshan-Shirazi, and A.M. Katti, *Fundamentals of Preparative and Nonlinear Chromatography*, Academic Press, Boston, 1994.
32. P.C. Haarhoff and H.J. van der Linde, *Anal. Chem.* **38**, 573–582 (1966).
33. L. Lapidus and N.R. Amundson, *J. Phys. Chem.* **56**, 984–988 (1952).
34. E. Kucera, *J. Chromatogr.* **19**, 237–248 (1965).
35. C.K. Lee, Q. Yu, S.U. Kim, and N.-H.L. Wang, *J. Chromatogr.* **484**, 29–59 (1989).
36. Q. Yu and N.-H.L. Wang, *Comput. Chem. Eng.* **13**, 915–926 (1989).
37. B. Lin, Z. Ma, S. Golshan-Shirazi, and G. Guiochon, *J. Chromatogr.* **475**, 1–11 (1989).
38. J.E. Eble, R.L. Grob, P.E. Antle, and L.R. Snyder, *J. Chromatogr.* **384**, 25–44 (1987).
39. M. Czok and G. Guiochon, *Comput. Chem. Eng.* **14**, 1435–1443 (1990).
40. M. Czok and G. Guiochon, *Anal. Chem.* **62**, 189–200 (1990).
41. M.Z. El Fallah and G. Guiochon, *Anal. Chem.* **63**, 859–867 (1991).
42. A. Felinger and G. Guiochon, *J. Chromatogr., A* **658**, 511–515 (1994).
43. M.Z. El Fallah and G. Guiochon, *Anal. Chem.* **63**, 2244–2252 (1991).
44. P. Rouchon, P. Valentin, M. Schoenauer, C. Vidal-Madjar, and G. Guiochon, *J. Phys. Chem.* **89**, 2076–2082 (1985).
45. P. Rouchon, M. Schoenauer, P. Valentin, and G. Guiochon, *Sep. Sci. Technol.* **22**, 1793–1833 (1987).
46. A.I. Liapis and D.W.T. Rippin, *Chem. Eng. Sci.* **33**, 593–600 (1978).
47. T. Gu, *Mathematical Modeling and Scale-Up of Liquid Chromatography*, Springer, Heidelberg, 1995.
48. B.H. Arve and A.I. Liapis, *AIChE J.* **33**, 179–193 (1987).
49. B.H. Arve and A.I. Liapis, *Biotechnol. Bioeng.* **30**, 638–649 (1987).

50. B.H. Arve and A.I. Liapis, *Biotechnol. Bioeng.* **31**, 240–249 (1988).
51. S.E. Asplund and R.K. Edvinson, *Comput. Chem. Eng.* **20**, 507–516 (1996).
52. J.C. Giddings and H. Eyring, *J. Phys. Chem.* **59**, 416–421 (1955).
53. F. Dondi and M. Remelli, *J. Phys. Chem.* **90**, 1885–1891 (1986).
54. J.C. Sternberg, in J.C. Giddings and R.A. Keller eds., *Advances in Chromatography*, vol. 2, Dekker, New York, 1966, pp. 205–270.
55. J.Å. Jönsson, in J.Å. Jönsson ed., *Chromatographic Theory and Basic Principles*, Dekker, New York, 1987, pp. 27–157.
56. G. Taylor, *Proc. R. Soc. London* **219**, 186–203 (1953).
57. S.P. Cram and T.H. Glenn, *J. Chromatogr.* **112**, 329–341 (1975).
58. H. Poppe, *Anal. Chim. Acta* **114**, 59–70 (1980).
59. J.L. Rocca, J.W. Higgins, and R.G. Brownlee, *J. Chromatogr. Sci.* **23**, 106–113 (1985).
60. J.F.K. Huber and A. Rizzi, *J. Chromatogr.* **384**, 337–348 (1987).
61. J.L. Glaich and L.R. Snyder, *Computer-Assisted Method Development for High-Performance Liquid Chromatography*, Elsevier, Amsterdam, 1990.
62. T. Hamoir and D.L. Massart, in J.C. Giddings and R.A. Keller eds., *Advances in Chromatography*, vol. 33, Dekker, New York, 1993, pp. 97–145.
63. E.W. Leser and J.A. Asenjo, *J. Chromatogr.* **584**, 43–57 (1992).
64. J. Lederberg, G.L. Sutherland, B.G. Buchanan, E.A. Feigenbaum, A.V. Robertson, A.M. Duffield, and C. Djerassi, *J. Am. Chem. Soc.* **91**, 2973–2976 (1969).
65. R. Bach, J.F. Karnicky, J. Excoffier, and S.R. Abbot, in R.T. Pierce and B.A. Hohne eds., *ACS Symp. Ser. No. 306*, American Chemical Society, Washington D.C., 1986, p. 278.
66. L.R. Snyder, *Principles of Adsorption Chromatography*, Dekker, New York, 1968.
67. L. Peichang and H. Hongxin, *J. Chromatogr.* **452**, 175–189 (1988).
68. A.F. Fell, T.P. Bridge, and M.H. Williams, *J. Pharm. Biomed. Anal.* **6**, 555 (1988).
69. P.J. Schoenmakers, N. Dunand, A. Cleland, G. Musch, and T. Blaffert, *Chromatographia* **26**, 37–44 (1988).
70. M. Mulholland, J.A. van Leeuwen, and B. Vandeginste, *Anal. Chim. Acta* **223**, 183–192 (1989).
71. T. Hamoir, M. de Smet, H. Pyrins, P. Conti, N. vanden Driessche, D.L. Massart, F. Maris, H. Hindriks, and P.J. Schoenmakers, *J. Chromatogr.* **589**, 31–43 (1992).
72. S.A. Borman, *Anal. Chem.* **58**, 1192A (1986).
73. K. Tsuji and K.M. Jenkins, *J. Chromatogr.* **485**, 297–309 (1989).
74. H. Yuzhu, A. Peeters, G. Musch, and D.L. Massart, *Anal. Chim. Acta* **223**, 1–17 (1989).
75. K. Valkó, G. Szabó, J. Röhricht, K. Jemnitz, and F. Darvas, *J. Chromatogr.* **485**, 349–363 (1989).
76. R.M. Smith and C.M. Burr, *J. Chromatogr.* **485**, 325–340 (1989).
77. E.W. Leser, M.E. Lienqueo, and J.A. Asenjo, *Ann. N.Y. Acad. Sci.* **782**, 441–455 (1996).
78. S. Wheelwright, *Protein Purification: Design and Scale Up of Downstream Processing*, Hanser Publishing, Munich, 1991.
79. V.M. Nadgir and Y.A. Liu, *AIChE J.* **29**, 926–934 (1993).
80. E.W. Leser and J.A. Asenjo, *Ann. N.Y. Acad. Sci.* **721**, 337–347 (1994).
81. J.C. Smit, H.C. Smit, and E.M. De Jager, *Anal. Chim. Acta* **122**, 1–26 (1980).
82. J.C. Smit, H.C. Smit, and E.M. De Jager, *Anal. Chim. Acta* **122**, 151–169 (1980).
83. J.B. Phillips, N.A. Wright, and M.F. Burke, *Sep. Sci. Technol.* **16**, 861–884 (1981).
84. R.D. Whitley, J.A. Berninger, N. Rouhana, and N.-H.L. Wang, *Biotechnol. Prog.* **7**, 544–553 (1991).
85. J.W. Dolan, D.C. Lommen, and L.R. Snyder, *J. Chromatogr.* **485**, 91–112 (1989).
86. M.A. Quarry, R.L. Grob, and L.R. Snyder, *Anal. Chem.* **58**, 907–917 (1986).
87. W. Kopaciewicz, M.A. Rounds, J. Fausnaugh, and F.E. Regnier, *J. Chromatogr.* **266**, 3–21 (1983).
88. S.R. Gallant, A. Kundu, and S.M. Cramer, *J. Chromatogr., A* **702**, 125–142 (1995).
89. D.J. Wiblin, S.D. Roe, and R.G. Myhill, *J. Chromatogr., A* **702**, 81–87 (1995).
90. O. Kaltenbrunner and A. Jungbauer, *J. Chromatogr., A* **734**, 183–194 (1996).
91. O. Kaltenbrunner and A. Jungbauer, *J. Chromatogr., A* **769**, 37–48 (1997).
92. R.C. Rowe, V.J. Mulley, J.C. Hughes, I.T. Nabney, and R.M. Debenham, *LC-GC Int.* **7**, 36–42 (1994).
93. Q.M. Mao, I.G. Prince, and M.T.W. Hearn, *J. Chromatogr.* **646**, 81–89 (1993).
94. A. Sadana, *BioSeparation* **3**, 145–165 (1992).
95. R.H. Ingraham, S.Y.M. Lau, A.K. Taneja, and R.S. Hodges, *J. Chromatogr.* **327**, 77–92 (1985).
96. S.-L. Wu, A. Figueroa, and B.L. Karger, *J. Chromatogr.* **371**, 3–27 (1986).
97. J. Fausnaugh and F.E. Regnier, *J. Chromatogr.* **359**, 131–146 (1986).
98. R. Lemque, A. Jaulmes, B. Sébille, V.-M.C., and P. Cysewski, *J. Chromatogr.* **599**, 255–265 (1992).
99. F.H. Walters, L.R. Parker Jr., S.L. Morgan, and S.N. Deming, *Sequential Simplex Optimization*, CRC Press, Boca Raton, Fla., 1991.
100. W. Spendley, G.R. Hext, and F.R. Himsworth, *Technometrics* **4**, 441–461 (1962).
101. J.C. Berridge, *Chemom. Intell. Lab. Sys.* **5**, 195–207 (1989).
102. N.A. Nelder and R. Mead, *Computer J.* **7**, 308–314 (1965).
103. L.S. Jennings, K.L. Teo, F.Y. Wang, and Q. Yu, *Comput. Chem. Eng.* **19**, 567–573 (1995).
104. J.C. Berridge, *J. Chromatogr.* **244**, 1–14 (1982).
105. A.G. Wright, A.F. Fell, and J.C. Berridge, *J. Chromatogr.* **458**, 335–353 (1988).
106. I. Atamna and E. Grushka, *J. Chromatogr.* **355**, 41 (1986).
107. G.J. Prokopakis and J.A. Asenjo, in J.A. Asenjo ed., *Separation Processes in Biotechnology*, vol. 9, Dekker, New York, 1990, pp. 571–601.

See also ADSORBENTS, INORGANIC; ADSORPTION, EXTENDED BED; ADSORPTION, PROTEIN, BATCH; CHROMATOGRAPHY, RADIAL FLOW; PILOT PLANTS: DESIGN AND OPERATION.

CHROMATOGRAPHY, HYDROPHOBIC INTERACTION

Per Kårsnäs
Percreative
Göteborg, Sweden

KEY WORDS

Chromatography
HIC
Hydrophobic interaction
Protein purification
Protein separation

OUTLINE

INTRODUCTION

Hydrophobic interaction chromatography (HIC) is an important chromatography technique used for the separation of biomolecules. The interaction is based on physicochemical characteristics of the molecules other than the charge, as in ion-exchange chromatography (IEC), or the size, as in gel filtration. Although its importance is increasing in industrial applications, the start has been slow. The technique is certainly very powerful when the optimal conditions for a certain separation are determined, but to achieve this has often been far from easy. One reason is that the mechanism of the interaction is complex, depending on several properties of the separation medium as well as of the separated substances, and not yet fully understood. There are a number of hypotheses that approach the phenomenon of hydrophobic interaction from different points of view, but the practitioner will still find that there is no simple theory that will always facilitate the optimization of a separation, as is the case when using other chromatography techniques. Hydrophobic interaction is favored by high salt concentration and is thus a logical purification step after salt precipitation or IEC, where the environment has a high ionic strength.

HIC has been used as a group separation method, rapidly separating what retards on the column from what is not interacting, and as a powerful high resolution technique with gradients of descending salt concentration.

In addition to what is usually designated as HIC, there are several related chromatography techniques based entirely or partly on hydrophobic interaction. Reversed-phase chromatography (RPC) is the most important, working with strong hydrophobic interaction that is often achieved by high substitution with long aliphatic chains. The principal difference between HIC and RPC is methodological in nature. In HIC, the interaction is increased by salt in order to retard the sample on the column and, basically, a return to native conditions will elute the compounds, whereas in RPC the sample is retarded on the column and eluted by addition of organic solvents. The methods based on what are described as salt-mediated and thiophilic interactions gain in importance, whether they should be considered as HIC techniques is perhaps a matter of semantics. Moreover, the binding in affinity chromatography and to dye–ligand matrixes often have significant elements of hydrophobic interaction.

HYDROPHOBIC INTERACTION

Basic Theory

There are several approaches to theoretically elucidate the nature of hydrophobic interaction, ranging from thermodynamics over surface tension to van der Waals forces (1,2). From different angles they all try to explain the observed influence on hydrophobic interaction by ions with salting-out effect, chaotropic ions, type of salt, temperature, pH, and additives. No single theory, however, manages to fully cover the mechanisms of interaction. In HIC the added complexity, contributed by matrixes, spacers, ligands, and the effects of buffer environment on different species of biomolecules, implicates that, apart from very general recommendations, the design of a HIC procedure still has to be founded on careful practical and empirical studies.

Salting Out and Hydrophobic Interaction

There is definitely a relationship between the salting out process used for the precipitation of proteins and hydrophobic interaction as used in chromatography. The Hofmeister series list anions and cations according to decreasing ability to precipitate and increasing ability to dissolve proteins, and the combinations (salts) from the beginning of the list are well known to promote hydrophobic interaction (3–6) (Fig. 1). Probably precipitation by salting-out and HIC are based on the same phenomenon, the principal difference being that in precipitation the hydrophobic interaction has to take place between molecules of the same kind or others that happen to be present in the solution, whereas in chromatography a well-defined counterpart is hydrophobic enough to bind some of the passing compounds, preferably at lower salt concentration than is needed for precipitation.

Thermodynamic Theory

The main and very elegant theory of hydrophobic interaction (7) refers directly to the Gibbs free energy equation (equation 1), which describes the change in energy when a chemical system is transferred from one state to another:

$$\Delta G = \Delta H - T\Delta S \qquad (1)$$

where ΔG is the change in Gibbs free energy, ΔH is the change in enthalpy or heat content, T is the temperature in Kelvin, and ΔS is the change in entropy or disorder of the system. If ΔG is negative, the second state has a lower energy and the system will adapt to it, provided that there are no energy barriers. In hydrophobic interaction, the entropy change is of greatest importance (8). If hydrophobic molecules such as aliphatic carbon chains are immersed in water, the monolayer of water molecules in contact with a carbon chain will be in higher order than those in the bulk of the water. If two or more hydrophobic structures come together, the surface that has to be covered by ordered water molecules decreases, some water molecules join the less-ordered bulk water, and the entire system gains in entropy and thus, according to equation 1, decreases its free energy (Fig. 2). This state will be favorable for energetic reasons and thus promoted. If some of the hydrophobic molecules are attached to a chromatographic matrix as ligands, the setup for HIC is obvious (Fig. 3).

The phenomenon of increased interaction in the presence of salting out ions is explained by a higher gain in entropy when water is transferred from the surface of a nonpolar molecule to the bulk of water. Whether this is achieved because the surface water in the presence of salt is initially more ordered or the bulk water less ordered or both is not quite clear.

Equation 1 reveals that at a higher temperature an increase in entropy will give a higher energy gain because the entropy factor of the equation will be multiplied by a higher number. Consequently, if the binding to a medium has been shown to increase with the temperature, this has been taken as an indication that the interaction is really hydrophobic.

Surface Tension, van der Waals Forces

Another theory (3) suggests that the surface tension of the water surrounding a nonpolar structure and the tendency to minimize it are responsible for the interaction. If salt is added to the water, it has been shown that the surface tension increases. As a consequence, the energy gain in minimizing the surface exposed to the hydrophobic molecules is larger and thus the interaction is stronger. Some salts that influence the surface tension are listed in order of decreasing effect in Fig. 1. Although the salts that increase the surface tension most are among those that should promote hydrophobic interaction according to the Hofmeister series, the order is not quite the same.

A third theory suggests that van der Waals forces (4) are responsible for the hydrophobic interaction in HIC. This is supported by the fact that these forces should be increased as the order of water increases in the presence of salt. Probably the theories mentioned here all describe one part of the complicated interaction, and a combination of them is more relevant to describe the phenomenon of hydrophobic interaction.

(a)

Anions: Increasing "salting-out" effect ←

PO_4^{3-}; SO_4^{2-}; CH_3 $COOH^-$; CI^-; Br^-; NO_3^-; CIO_4^-; I^-; SCN^-

(b)

Cations: Increasing chaotropic effect →

NH_4; Rb^+; K^+; Na^+; Cs^+; Li^+; Mg^{2+}; Ca^{2+}; Ba^{2+}

(c)

Salts promoting the molal surface tension of water:

$Na_2SO_4 > K_2SO_4 > (NH_4)_2SO_4 > Na_2HPO_4 > LiCI > KSCN$

Figure 1. The Hofmeister series of anions (**a**) and cations (**b**). The *arrows* indicate increasing salting-out effect and chaotropic effect, respectively. Salts with high ability to promote hydrophobic interaction are combinations of ions found to the left in the series. The same salts are found early in the series of salts listed in order of decreasing ability to promote the surface tension of water (**c**). This indicates the relationship between hydrophobic interaction and surface tension. Which salt to chose depends on the solubility and the production cost of the salt as well as the cost of preventing pollution of the environment by waste buffers.

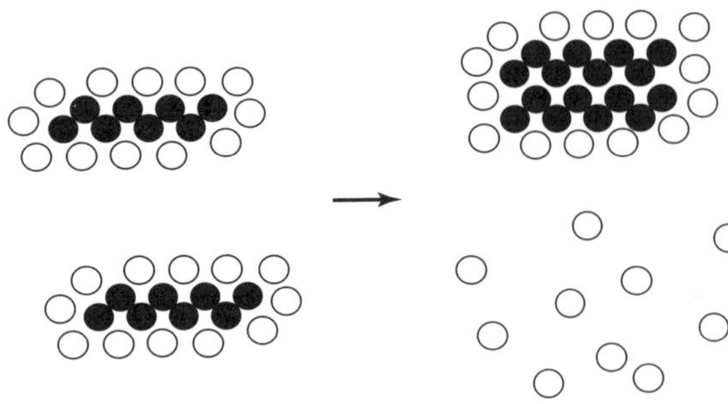

Figure 2. Hydrophobic interaction. Two aliphatic carbon chains (*black circles*) are immersed in water. Only the water molecules (*white circles*) with high order in the monolayer adjacent to the nonpolar molecules are indicated. When two chains are in contact, the surface area that is covered with ordered water molecules is decreased and some water is in a state of less order in the bulk of the water. The free energy of the system decreases, and thus the binding together of the two molecules is favored.

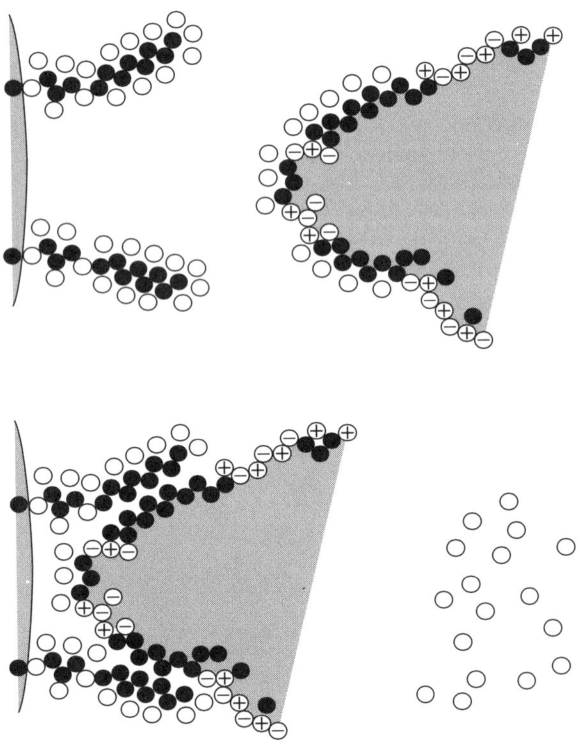

Figure 3. Hydrophobic interaction used for chromatography. Two octyl chains are coupled to a chromatographic medium. An energy gain, equivalent to that described in Figure 1, is achieved by the interaction of the octyl chains with nonpolar moieties on a protein. The carbon atoms of the hydrophobic parts of the protein are indicated with *black circles*, the oxygen atoms of the water with *white circles*, and charged atoms or groups of the protein with *white circles* containing plus or minus signs. The *hatched area* indicates the interior part of the protein.

HYDROPHOBIC INTERACTION CHROMATOGRAPHY

Development of the Technique

As early as 1948 Tiselius presented what he called Adsorption separation by salting out (9), but not until the beginning of the 1970s were reports on the synthesis of media published. Some researchers were working with a mixed mode because of their medias' content of charges (10–13), but charge-free hydrophobic media were also synthesized (5,14,15), and the first commercially available products were presented some years later (16). At present, a wide range of products, including HIC-media for HPLC, are available but phenyl, octyl, butyl, and alkyl ligands are still by far the most common (17,18).

Factors Influencing HIC

In addition to the environmental parameters, including ionic strength, temperature, and pH that influence the hydrophobic interaction as such, there are several factors that have to be controlled to use the hydrophobic interaction for chromatographic purposes. The chromatographic medium is constructed by coupling hydrophobic ligands to a chromatographic matrix, often by use of a short spacer to enhance the sterical possibilities of interaction with large molecules. The type of ligand and the ligand density affect the characteristics of the medium. The choice of buffer ions has been found to change the interaction pattern more profoundly than generally anticipated. The optional use of organic solvents and detergents have to be taken into consideration. The molecules themselves have certain characteristics and will often change chromatographic interaction behavior with the buffer salts in an individual manner (19). Finally, the choice of dimensions of the chromatographic column and the mode of elution will depend on the size range of biomolecules that are subject to processing and the desired throughput of the separation step. One approach, to describe the complex interaction of HIC from a practical point of view, has been to characterize a HIC medium by the interaction pattern of a significant set of protein molecules and to evaluate the result by the use of principal component analysis (20). As described in later sections, the media can be classified by this approach, and the importance of different characteristics of the proteins can be correlated to the interaction of different media.

The Matrix

Agarose-based matrixes with varying degrees of cross-linking are most commonly used for HIC, but organic resins and silica-based media are also used, especially in HPLC applications. All common chromatographic matrixes, such as cross-linked agarose (21,22), organic polymers (23), and silica (24), can be derivatized with hydrophobic groups to give HIC media. Normally, the influence of the by-design hydrophilic chromatographic matrix will just add slightly, if at all, to the performance of the medium, but it is a well-known fact that unsubstituted agaroses intended for gel filtration will expose what can only be characterized as hydrophobic interaction at extremely high salt concentrations (25–27). This allows for using HIC even for rather hydrophobic proteins that would stick irreversibly to conventional hydrophobic gels. It has to be noted, though, that under normal salt concentrations (e.g., those used in gel filtration) no hydrophobic interaction will be taking place because the gels were successfully designed to be as hydrophilic as possible.

The Ligands

Alkyl, butyl, phenyl, and octyl groups are the most common ligands in commercially available HIC media (Fig. 4). They are listed in what is commonly considered as in order of increasing hydrophobicity, but as mentioned earlier there is no linear or single dimension relationship between their interaction with different species or sets of molecules. It is obvious that the aromatic part of the phenyl group can bind with π-to-π binding as well thus adding to the complexity of the interaction. Other ligands of interest include mercapto derivates, which have yet another spectrum of interaction. They have been claimed to work partly with a special thiophilic interaction, but there is some doubt whether their content of sulfur is really responsible for their interaction pattern or whether they are just another type of hydrophobic medium among the others and working by a combination of the same mechanisms as other hydrophobic media.

The Coupling Reactions

Glycidyl ether is the coupling agent used most often to immobilize hydrophobic ligands (18), but in principle all the common coupling procedures can be used to attach the ligands to the chromatographic matrix. An extensive review of coupling methods is presented in the work by Carlsson et al. (28). The coupling procedure will almost inevitably result in the ligands being attached via an often very short spacer, which is an advantage because conformational restrictions when large protein molecules bind are decreased. On the other hand, the choice of coupling chemistry will always influence the performance of the chromatographic medium by introducing hydrophobic or charged moieties, and the selectivity and the capacity may be very different even if the coupled ligand happens to be the same.

Binding Capacity

The binding capacity of an HIC medium depends partly on the ligand density (11,21,29,30), but it is important to remember that any figures given will always be related to the specific protein that was used to determine the capacity and to the conditions under which the determination was performed. The capacity can be determined as total capacity, with the protein exposed to the medium without flow, or as dynamic capacity, where the linear flow and the residence time of the molecules in the column will influence the result. There may well be a difference of magnitude of 10 to 20 between the capacities for different proteins, a fact that has to be considered when choosing a medium from a catalog where the dynamic capacity is often determined using serum proteins like albumin.

The capacity that can be used is dependent on the contamination picture as well and whether the main part of the impurities are more or less hydrophobic than the desired substance, so in practice it is necessary to determine the dynamic capacity in every specific application; this should be seen as a normal part of an optimization process.

Much work is being done to optimize the dynamic capacity of media, both by suppliers and by users at research sites, but the question is whether the most convenient and uncostly way of increasing the capacity of a column is to increase the amount of gel!

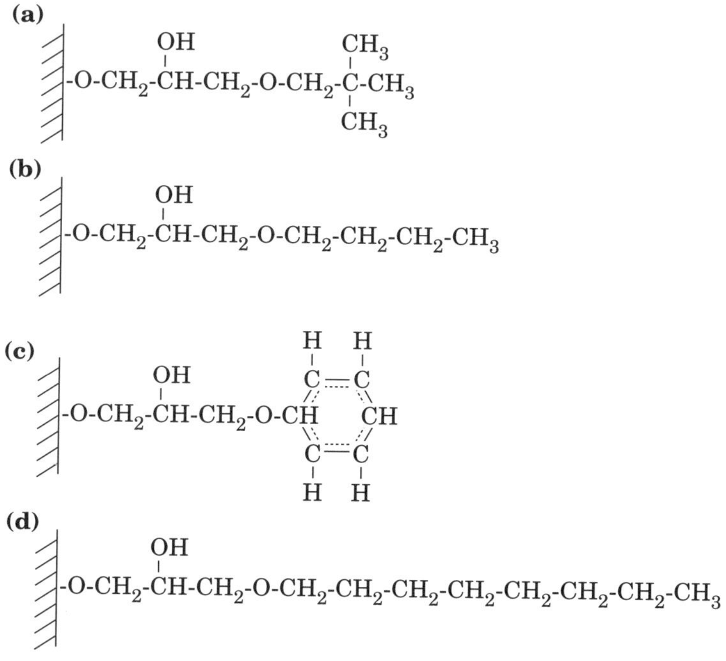

Figure 4. Common hydrophobic ligands. Alkyl (**a**), butyl (**b**) phenyl (**c**), and octyl (**d**) ligands coupled to a chromatographic matrix by glycidyl ether.

HIC and RPC and Column Design

HIC and RPC are both based on hydrophobic interaction, and the borderline between the techniques is more defined by difference in methodology and the type of sample than by the interaction as such.

In RPC the hydrophobicity of the media is higher because the ligands are more hydrophobic and the ligand density is higher. When the sample is run through an RPC column at conditions in the range of physiological pH and salt concentrations, most compounds are retarded on the column. Elution has to be achieved by using increasing gradients of organic solvents. Commonly used solvents are, in order of increasing elution ability, methanol, isopropyl alcohol, and acetonitrile.

In contrast, in HIC very little of the sample is retarded at physiological conditions. The ionic strength of the sample has to be increased to promote attraction, and the elution is normally done by decreasing salt gradients. If some components are still attached to the matrix in weak buffer, a mild organic solvent such as ethylene glycol will be used to wash them out.

As a consequence, HIC is more often used for protein separation where the task often is to retain the tertiary conformation and the biological activity of the separated molecules, whereas RPC is often used for smaller molecules, such as peptides, which lack tertiary structure and subsequently will withstand the elution conditions without denaturation.

Historically, RPC was developed as an HPLC technique to separate small molecules using mainly silica and organic polymer-based matrixes, and HIC was developed as a technique for the separation of proteins, as an alternative to IEC and gel filtration and using the common agarose matrixes as carriers of the hydrophobic ligands. The difference in molecular size range of the molecules to be separated leads to several important differences in the methodologies of the two techniques (31). The differences in behavior between small and large molecules are important in all attraction techniques, such as IEC, HIC, and RPC. But because of the restrictions to expose large molecules to organic solvents, the borderline between separation of small and large molecules happens to coincide more or less with the borderline between the two hydrophobic techniques. But there are other technical reasons behind the difference of methodology based on molecular size (31).

Small molecules interact with the matrix by single point attachment. A molecule is bound to the gel for a certain fraction of the time and is free to move with the buffer the rest of the time. Another way to express the same fact is that at a certain time one fraction of the molecules attach to the column while the rest are moving with the flow. Even if the equilibrium is shifted far toward binding, the macroscopic effect will be that small substances move down the column at some rate in a broad range of salt concentrations, although the speed may be rather low. In a mixture of substances, the rates of movement down the column will be different, as described by the individual equilibrium constants, and a separation will take place. To allow for the molecules to separate, a long path length is needed, and thus long columns and keeping the separating zones sharp are essential for high effectivity. This fact has led to a hunt for high plate numbers, a feature that can be achieved by using small matrix particles. This in turn will cause a high backpressure and the need to work under HPLC conditions. The full utilization of the selectivity of the column will be achieved by elution under isocratic conditions or by shallow gradients. The resulting extremely resolving techniques have been very successful and impressive and have been used for peptide fingerprinting and other types of qualified analysis of small organic molecules.

Proteins, on the other hand, interact with the matrix by multipoint attachment. Even if at some salt concentration, some of the interacting parts of the large molecules may be unbound, there are often several points of interaction left that keep the protein attached to the column. Not until a certain elution condition is reached, such as higher ionic strength in IEC or lower salt concentration in HIC, where all binding sites of the protein are simultaneously unbound, will the protein move down the column. The range between when the protein is not moving at all and moving at full speed is very narrow compared to when small molecules are separated. Isocratic elution is not possible because at constant buffer concentration some proteins will move without retardation and some will hardly move at all. A gradient is needed to elute the proteins in order. Because the resolution is not very dependent on the effectivity of the column, an HIC step where proteins are separated should be performed in short columns, and the linear flow should be kept low by increasing the diameter of the column. In large scale, the use of short and wide columns in protein separation based on IEC and HIC opens up the possibility to run with a high total flow and throughput and still keep the linear flow low enough to allow for the adsorption and desorption to take place without being too far from equilibrium.

CLASSIFYING HIC MEDIA BY MULTIVARIATE ANALYSIS

The performance of HIC media can be monitored by multivariate analysis, using a set or probe of protein molecules with different hydrophobic properties. The interactions of the molecule set with the HIC media could be correlated to molecular weight, isoelectric point, fraction of hydrophobic amino acids, surface charge, and several hydrophobicity indexes (20) without the need for detailed knowledge about the parameters or the mechanisms involved. Instead, the different combinations of media and chromatography conditions can be fingerprinted by their performance in practice and easily compared. As well as artificial mixtures of proteins, serums with 15 to 20 of the proteins identified have been used as probes to determine adsorption patterns (32).

Inversely, biomolecules can be classified by their hydrophobic characteristics as measured by their interactions with a standard set of different hydrophobic media. Molecules with similar adsorption patterns over the media should have similarities on a molecular level.

The results hint that in the tested buffer system, there are at least two main mechanisms that have an impact on HIC (20,33). One mode of interaction was typical on highly hydrophobic media, for example, with an octyl chain as ligand, where the binding seems to be entirely correlated to the fraction of hydrophobic amino acids and to the hydrophobicity index of the protein molecule. An explanation may be that a higher content of hydrophobic amino acids automatically leads to more hydrophobic amino acids being exposed on the surface of the protein. But because the fraction of exposed hydrophobic areas of proteins is always in the vicinity of 40 to 50% (34,35), another possible explanation takes into consideration that proteins are not rigid bodies with rigid surfaces but that the conformation should be considered as a statistical mean around which the molecule must have the property to breath or change the microconformation, thus allowing a long hydrophobic chain to come in contact with hydrophobic structures that statistically are not always exposed at the surface.

An interesting finding is that the binding to the octyl medium is highly correlated with the molecular weight as well. This phenomenon is logical, though, taking into consideration that larger protein molecules generally have to have a higher fraction of hydrophobic amino acids to keep the tertiary structure.

The other mode of action does not correlate with the content of hydrophobic amino acids of the proteins at all, but instead correlates inversely with the surface charge of the molecules as found in biochemical handbooks. This interaction was displayed on weakly hydrophobic matrixes with short alkyl chains but also with some thiophilic media. An explanation is that charged groups will mask the type of hydrophobic interaction taking place on these gels under the tested conditions. Probably this interaction will be influenced significantly by altering pH.

Very commonly used HIC media, based on phenyl groups and butyl groups as ligands, were shown to work with a mixed mode, that is, combination of the two pure mechanisms. This fact adds to the explanation of the difficulty, especially using these HIC media, to find simple rules after which the separation should be optimized.

It should also be noted that the performance of a HIC medium will always be the result of the combination of ligand, spacer, and matrix. Phenyl gels from different manufacturers have quite different patterns of interaction if tested by a set of proteins as described earlier. This has to be remembered when, after the optimization in laboratory scale, alternative media are tested before scaling-up of a HIC procedure. This optimization is only valid for the medium used. There is thus a risk that competing media, which might be advantageous, will be tested under non-optimal conditions and rejected under false premises.

THE CHROMATOGRAPHY CONDITIONS

Ionic Strength

A high ionic strength is needed by definition to attach biomolecules to the column. Very often, 1.5 to 1.7 M ammonium sulphate is used as initial buffer concentration without further trials to optimize. Another common approach is to equilibrate the column in weak tris-buffer to preserve the pH and increase the ionic strength by the addition of sodium chloride to a concentration of up to 3 M.

It has to be noted, though, that ammonium sulphate, at a certain concentration, is between three and four times as effective in promoting the hydrophobic interaction compared to sodium chloride, so if the substances in question need a stronger interaction to be retarded on the column than can be achieved by almost saturated sodium chloride, ammonium sulphate is a natural choice. Using ammonium sulphate extends the range of HIC to be used for the separation of less hydrophobic molecules.

In production-scale operations, the use of ammonium sulphate has been questioned because of the effect the nitrogen of used buffers may have on the environment by increasing the overnutrition. The choice between overnutrition and possible toxic effects of substitute buffers such as sodium sulphate may not be easy, and all the factors, from the conveniency of the separation procedure to the cost of purifying the sewage, will have to be considered.

Buffer Ions

Recently, it has been shown that the choice of buffer ions greatly influences the function of an HIC medium (32,36). The overall behavior of a buffer-ligand-spacer-matrix system could be switched in a wide range, such as from typical phenyl-like to typical butyl-like or vice versa, by just changing the buffer ions. Optimization work by testing different buffers might then be more beneficial than usually anticipated but, as with pH, the task still includes much work based on trial and error.

pH

It has been shown that hydrophobic interaction between proteins and an octyl gel generally increases with lowered pH and most often decreases with high pH, but the pH dependence in this case is not very profound from pH 6 to pH 8.5. For some proteins, though, the interaction increases with an increase in pH (21). One reason for the different results might be that sodium sulphate was used above pH 8.5 and ammonium sulphate buffer was used below pH 8.5. According to Oscarsson and Kårsnäs (32), the adsorption behavior of serum proteins on octyl gel is quite different in the two environments. The reason for this complexity again has certainly to do with the multimechanism nature of HIC, with irregularities implied by different kinds of interactions between buffer ions and protein surfaces (18) and thus, secondarily, interaction with the HIC matrix.

In practice, changes in pH to get a desired separation can be very powerful because it may alter the entire adsorption spectrum, but, because of the trial-and-error nature of the task, pH is not the primary parameter to use for an optimization.

Temperature

High temperature promotes the binding, and HIC is preferably run at room temperature if the compounds to be separated can withstand this condition without losing activity (4,7,37,38). In room temperature, slightly lower buffer concentrations can be used than if the separation is made in a cold room. Elution of hardly bound substances may also be facilitated by performing the elution step at cold room temperature. In practice, however, the increase of temperature at the reequilibration of the column in room temperature inevitably produces air bubbles that will destroy the packing of the column, so if a two-temperature scheme is chosen, repacking of the column will most certainly have to be included in the separation cycle.

Flow, Residence Time

Hydrophobic interaction shows signs of being a slower process compared to other interactive chromatography techniques. The desorption of compounds from the column is often run under nonequilibrium conditions, resulting in long tailing peaks that seriously damage the resolution.

It is also generally found that the adsorption of molecules sometimes seems to be a two-step reaction, where the initial binding is weaker and more readily broken under the elution conditions. After some time the bound molecules seem to adapt themselves to the column, possibly involving conformational changes (39). More harsh conditions may have to be introduced for the elution, and there is a risk that the recovery will decrease. On the other hand, there are reports showing that an elution pattern can be quite reproducible even when the bound substances have been immobilized on the column for a long time (21), and HIC is generally considered a high-recovery technique (40).

Elution

Generally, elution of substances bound to an HIC column can be achieved by altering one of the parameters that promote binding. It is possible to elute by increasing pH, if the adsorption was performed at low pH, or by lowering the temperature, although the effect of this may be too small. But most commonly, elution is performed by a descending salt gradient or, in large scale, by stepwise lowering of the salt concentration (5,14).

If substances remain on the column at very low buffer concentration, an organic additive such as ethylene glycol or a detergent can be used for the elution. The ethylene glycol will alter the water structure and weaken the hydrophobic interaction, and the detergent will act by competing with and displacing the protein on the chromatographic medium.

REGENERATION AND CLEANING-IN-PLACE

For process economy, the regeneration of the HIC column is essential. Fouling of the column and irreversible binding of contaminants to the column are common obstacles in HIC. There are three degrees of regeneration: removal of strongly bound or precipitated proteins, removal of hydrophobic substances like lipids and lipoproteins, and sanitation, which means removal or deactivation of microorganisms. Preferably, the regeneration is performed as a cleaning-in-place (CIP) procedure without costly and cumbersome repacking of the column. All washing steps should be performed with reversed flow to give the contaminants the shortest distance out of the column and avoid secondary interaction with the rest of the column. First, all conditions that will decrease possible hydrophobic interaction can be used for the regeneration. Detergents that will compete with the attached substances and ethanol and isopropyl alcohol, buffers normally used in RPC, are obvious choices NaOH, 0.5 to 1 M, has been shown to almost universally sanitize the column and has a good cleaning ability as well. It is important to note that washing with narrow zones of the washing solution is often more effective in removing the column contaminants than exposing the medium for a constant concentration of the washing agent for an extended time. It is not unusual when cleaning HIC media that material will be washed out both in the beginning and at the end of a washing zone where the concentration is increasing and decreasing, respectively. Probably, a lot of equilibria are present, including hydrophobic interaction and interactions involving charges, and if one type of binding is minimized another may be enhanced and keep the contaminants on the column. By wobbling the concentration, the differences of reaction velocities are probably utilized, with elution as the result.

OPTIMIZATION PROCEDURES

Although all the parameters mentioned above can be used to optimize a hydrophobic interaction chromatography step, for many of them the effect of a change is not easy to predict because it will also be dependent on the individual properties of the proteins subject to separation. Thus, a pH change may alter the entire elution pattern or just influence the binding of a single component. Other parameters that may change the entire interaction picture are the buffer ions and the chosen chromatography medium.

Choice of Medium

Ideally, it is preferable that the desired components attach to the medium under moderate salt concentrations (e.g., up to 1 M ammonium sulphate), and a screening of the most common commercially available media is always advisable. This screening is easily performed in small scale and should include a test of whether elution with a high recovery is possible with water or weak buffer. If a lower salt concentration is sufficient on a certain gel, this is a preferable choice for economical and environmental reasons provided that the resolution obtained is still acceptable.

Determination of the Starting Buffer Concentration

An HIC run is normally performed by equilibrating the column in a starting buffer, applying the sample, and eluting with a stepwise descending ionic strength or a descending continuous gradient.

Optimizing the starting concentration, or the concentration at which the proteins are adsorbed to the column, is important to really utilize the HIC step (Fig. 5). In laboratory scale and analytical experiments, where the resolution can be achieved by gradient elution, the ionic strength should be high enough to retard the desired component and to let it elute somewhere in the middle of the

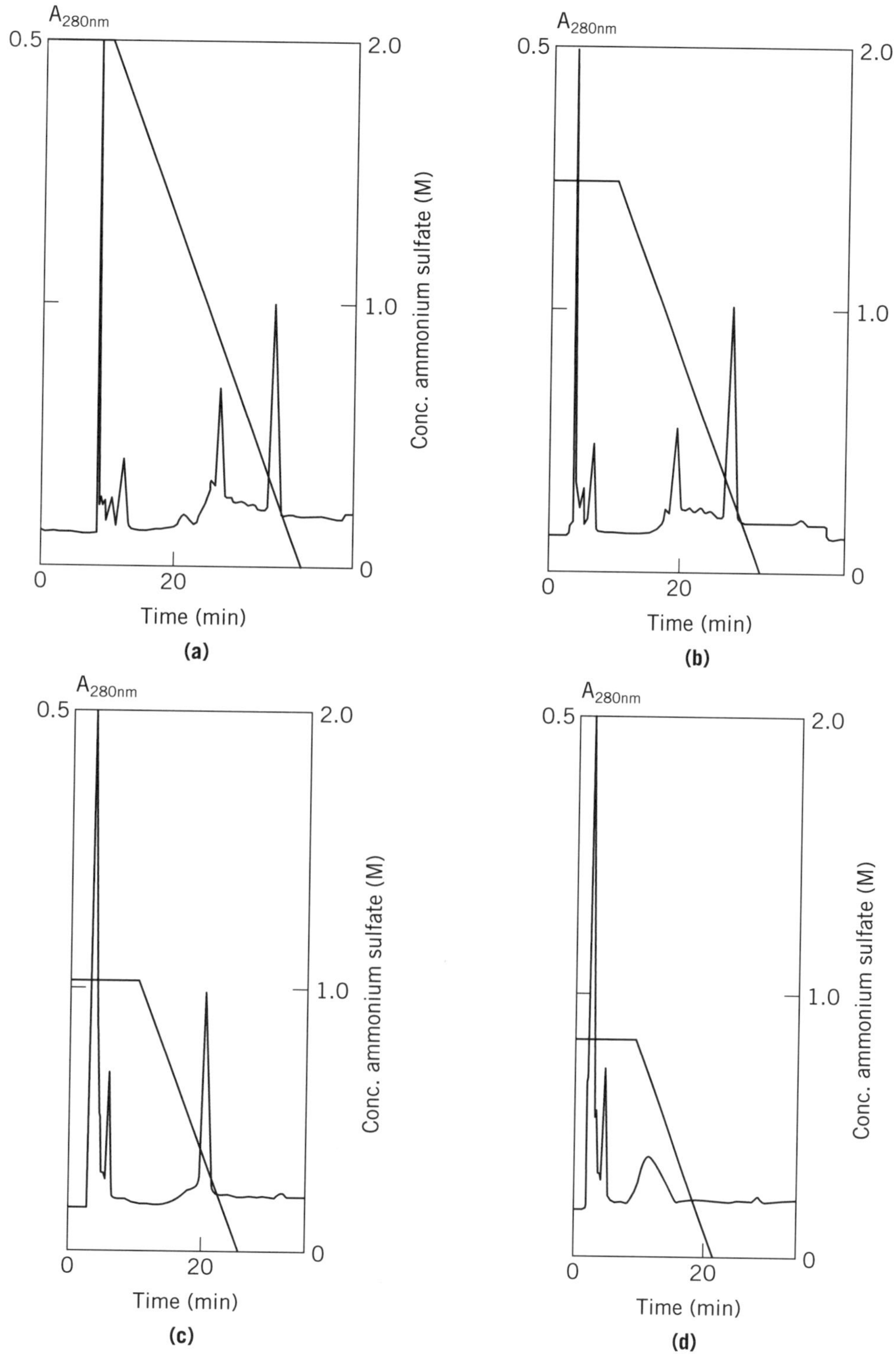

Figure 5. The effect of starting concentration. 100 μL anti-CEA MAB ($-$IgG1) from mouse ascites fluid, in 0.8 M $(NH_4)_2SO_4$ (AS), applied to an Alkyl Superose HR 5/5 column with a flow rate of 2.5 cm/min. The different concentrations of AS in the initial buffer were mixed in 0.1 M sodium phosphate, pH 7. (**a**) The sample is applied in 2 M AS. Both albumin and IgG are accumulated on the column. (**b**) The sample is applied in 1.5 M AS. Less albumin is bound, which indicates a weaker interaction and less binding capacity. (**c**) In 1 M AS only IgG is binding, and (**d**) in 0.8 M AS the IgG is still retarded but eluted in a broad peak. *Source*: Figure reprinted from ref. 17 by kind permission of Pharmacia Biotech, Uppsala, Sweden.

descending gradient. In this way, the combination of the capacity for the component of interest and the resolution will be favorable. In large-scale runs, well-defined and reproducible gradients are more difficult to achieve, so the elution is preferably done by stepwise decrease of the ionic strength. In some cases, it will work beautifully to chose an ionic strength that is just enough to retard the substance of interest, but in other cases, the peak containing the desired substance will broaden unacceptably at this condition, and the salt concentration will have to be increased (Fig. 5).

Gradient Volume

A gradient volume between 5- and 10-column volumes will give enough resolution in most cases. By increasing the gradient volume, the resolution will increase, but the components will be eluted in a larger volume.

Selectivity of the On and Off Reactions

It is often found that the reaction of binding to the column is rather rapid and selective, whereas the elution of a compound from the column is slower. As a result of this, tailing of zones at elution is very often experienced and the resolution of substances is jeopardized. A way of optimizing a method fully, utilizing the selectivity of the adsorption, is to use the following scheme:

1. Equilibrate the column with a starting buffer that has a lower concentration than is needed to retard the desired substance, adjust the sample accordingly, and run it through the column.
2. Elute everything that was retarded on the column with water or low salt. Check for activity.
3. Add some salt to the buffer and the sample and repeat steps 1 to 3 the desired substance is eluted from the column. To speed up the procedure, a prerun with a gradient starting at high salt concentration will hint at a proper initial concentration when step 1 is run the first time.

An optimized separation scheme includes as its first step running the column in a salt concentration just below the one where the desired substance was found to be retarded. All impurities that bind harder than the desired substance attach to the column and should be eluted. The next step is to bind the desired substance using a slightly higher salt concentration. If this is done carefully, very few contaminants will bind, and subsequently the desired substance can be eluted in high yield by a steep descending gradient, if any contaminating substances were retarded as well, or by a step of very diluted buffer. The described method is advantageous not only because the selectivity of the interaction is fully utilized but also because of the simplicity to recondition sample and buffers by just adding salt. Moreover, the capacity of the column is utilized to bind mostly the compound of interest and very little contaminants. A drawback may be that the capacity of the column for the substance of interest might be lower be-

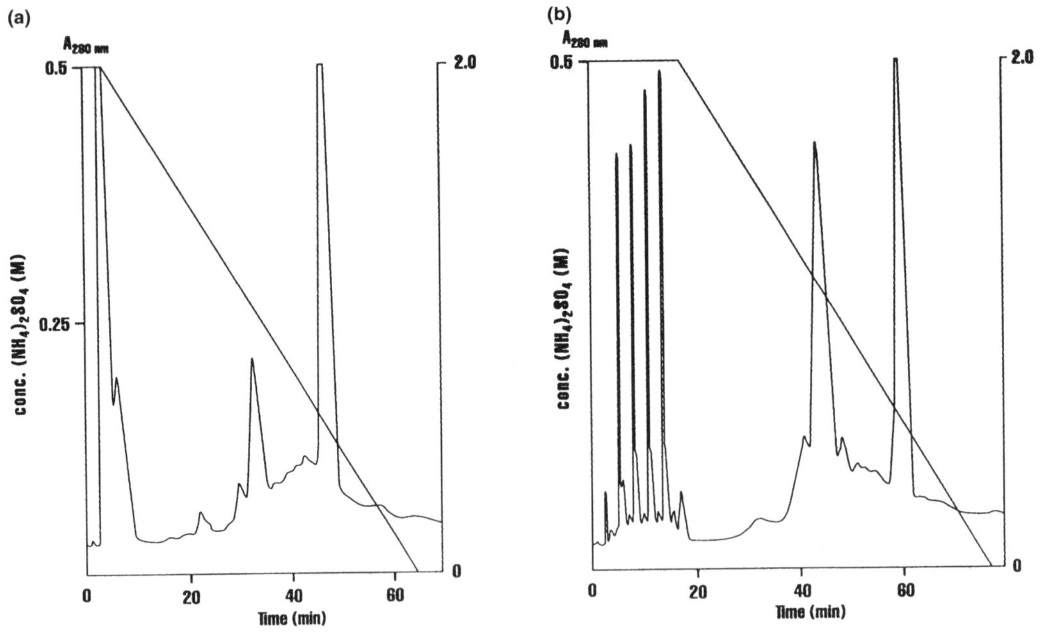

Figure 6. The effect of the suggested method of injecting the sample in diluted aliquots intermissioned by zones of concentrated buffer. In (**a**) 500 μL anti-CEA MAB (−IgG1) from mouse ascites fluid, in 0.9 M $(NH_4)_2SO_4$ (AS), was injected on a column equilibrated in 2 M AS in 0.1 M sodium phosphate, pH 7. In (**b**) the same sample was injected in 100 μL aliquots. 1.3 mL of 2 M AS were injected between every injection of sample. The capacity of the column was fully utilized without the risk of precipitating the sample in tubings and adapters. *Source*: Figure reprinted from ref. 17 by kind permission of Pharmacia Biotech, Uppsala, Sweden.

cause the binding is taking place near the elution concentration.

Equilibration of the Column and Sample Application

There is always a risk that the high salt concentration needed to attach the sample to the column will cause precipitation in tubings, pumps, columns, and detectors. This may be prevented by the following scheme: The column should be equilibrated with the proper buffer concentration but the sample with slightly lower salt concentration. This is easily accomplished by dilution of the sample. Instead of injecting the whole sample volume at one time, the injection should be done in aliquots divided by zones of the equilibration buffer. If the volume of the aliquots is kept less than 10% of the column volume and the volume of the intermediate zones with full buffer concentration exceeds 10% of the column volume, the sample condition will adapt to that of the equilibrated column. In this way, the precipitation in wrong places is avoided and the interaction with the column is still taking place in the correct environment (Fig. 6).

Scaling-Up

To get a similar result in large scale as in laboratory scale, it is important to keep some parameters constant. The column length, the linear flow, and the ratio between gradient and column volumes belong to this category. A constant column length leads to the fact that the desired capacity will be proportional to the cross-sectional area of the column and actually decide it, because the same proportion of the column length should be occupied by the sample in both scales.

Because of the relationship between the volumetric flow, the linear flow, and the cross-sectional area of the column on one side and the gradient-to-column-volume ratio on the other side, the gradient volume will be decided when the diameter of the column is set. In principle, a gradient elution in process scale will then take the same time as in laboratory scale. In practice, a decrease in resolution caused by a more ineffective solvent distribution system at the inlet of a process scale column will sometimes have to be compensated for by a lower flow.

APPLICATIONS

HIC has been used for the purification of serum proteins (41,42), nuclear proteins (43), membrane-bound proteins (44), and viruses and cells (45), and recombinant proteins (46) and insulin receptors (47) have been successfully prepared. Interesting applications include the use of HIC media as biochemical reactors by coupling enzymes by hydrophobic interaction (48) and a method for the exchange of detergents bound to membrane proteins (18).

For further applications and more discussion about large-scale processing using HIC, the works by Pharmacia Biotech and Eriksson (17,18) are recommended.

BIBLIOGRAPHY

1. C. Tanford, *The Hydrophobic Effect: Formation of Micelles and Biological Membranes*, Wiley, New York, 1973.
2. T.E. Crighton, *Proteins, Structures and Molecular properties*, W.H. Freeman, New York, 1984.
3. W. Melander, C. Horwath, *Arch. Biochem. Biophys.* **183**, 200–215 (1977).
4. R. Srinivasan, E. Ruckenstein, *Sep. Purif. Methods* **9**, 267–370 (1980).
5. J. Porath, L. Sundberg, N. Fornstedt, I. Olsson, *Nature* **245**, 465–466 (1973).
6. J. Porath, *J. Chromatogr.* **376**, 331–341 (1986).
7. S. Hjertén, *J. Chromatogr.* **87**, 325–331 (1973).
8. S. Lewin, *Displacement of Water and Its Control of Biochemical Reactions*, Academic Press, New York (1974).
9. A. Tiselius, *Ark. Kemi* **26B**, 1–5 (1948).
10. B.H.J. Hofstee, *Anal. Biochem.* **52**, 430–448 (1973).
11. S. Shaltiel, Z. Er-el, *Proc. Natl. Acad. Sci. USA* **70**, 778–781 (1973).
12. G. Halperin, M. Breitenbach, M. Tauber-Finkelstein, S. Shaltiel, *J. Chromatogr.* **215**, 211–228 (1981).
13. M. Wilchek, T. Miron, *Biochem. Biophys. Res. Commun.* **72**, 108–113 (1976).
14. S. Hjertén, J. Rosengren, and S. Påhlman, *J. Chromatogr.* **101**, 281–288 (1974).
15. S. Påhlman, J. Rosengren, and S. Hjertén, *J. Chromatogr.* **131**, 99–108 (1977).
16. J.-C. Jansson, T. Låås, in E. Roger ed., *Chromatography of Synthetic and Biological Macromolecules*, Ellis Horwood, Chichester, U.K., 1978.
17. Pharmacia Biotech, *Hydrophobic Interaction Chromatography. Principles and Methods*, Uppsala (1992).
18. K.-O. Eriksson, in J.-C. Jansson, L. Rydén ed., *Protein Purification: Principles, High Resolution Methods, and Applications*, VCH Publishers, Cambridge, U.K. 1989, pp. 207–226.
19. J.L. Fausnaugh, F.E. Reigner, *J. Chromatogr.* **359**, 131 (1986).
20. P. Kårsnäs, T. Lindblom, *J. Chromatogr.* **599**, 131–136 (1992).
21. S. Hjertén, K. Yao, K.-O. Eriksson, B. Johansson, *J. Chromatogr.* **359**, 99–109 (1974).
22. F. Maisano, M. Belew, J. Porath, *J. Chromatogr.* **321**, 305–317 (1985).
23. Y. Kato, T. Kitamura, T. Hashimoto, *J. Chromatogr.* **360**, 260 (1986).
24. J.L. Fausnaugh, E. Pfannkoch, S. Gupta, F.E. Reigner, *Anal. Biochem.* **137**, 464 (1984).
25. F. von der Haar, *Biochim. Biophys. Res. Commun.* **70**, 1009–1013 (1976).
26. K. Adachi, *Biochim. Biophys. Acta* **912**, 139 (1987).
27. J. Porath, *Nature* **196**, 47 (1962).
28. J. Carlsson, J.-C. Jansson, and M. Sparrman, in J.-C. Jansson, L. Rydén ed., *Protein Purification: Principles, High Resolution Methods, and Applications*, VCH Publishers, Cambridge U.K., 1989, pp. 275–329.
29. H.P. Jennissen, I.M.G. Heilmeyer, *Biochemistry* **14**, 754–760 (1975).
30. J. Rosengren, S. Påhlman, M. Glad, S. Hjertén, *Biochim. Biophys. Acta* **412**, 51–61 (1975).
31. B. Ekström, G. Jacobson, *Anal. Biochem* **142**, 134–139 (1984).
32. S. Oscarsson, P. Kårsnäs, *J. Chromatogr* **803**, 83–93 (1988).
33. W. Melander, Z. El Rassi, C. Horvath, *J. Chromatogr.* **496**, 3–27 (1989).
34. S. Miller, J. Janin, A.M. Lesk, C. Chotia, *J. Mol. Biol.* **196**, 641 (1987).
35. B. Lee, F.M. Richards, *J. Mol. Biol.* **55**, 379–400 (1971).
36. T. Arakawa, L.O. Narhi, *Biotechnol. Appl. Biochem.* **13**, 151–172 (1991).

37. V.A. Parsegian, B.W. Ninham, *Biophys. J.* **10**, 664–674 (1970).
38. H.P. Jennissen, *J. Chromatogr.* **159**, 71–83 (1978).
39. H.P. Jennissen, *J. Colloid. Interface Sci.* **111**, 570 (1986).
40. R.K. Scopes, *Protein Purification, Principles and Practice*, Springer, New York, 1994.
41. J.-C. Jansson, T. Låås, in E. Roger ed., *Chromatography of Synthetic and Biological Macromolecules*, Ellis Horwood, Chichester, U.K., 1978.
42. Z. Hrkal, J. Rejnkova, *J. Chromatogr.* **242**, 385–388 (1982).
43. D.E. Comings, A.G. Miguel, H.H. Lesser, *Biochim. Biophys. Acta* **563**, 253–260 (1979).
44. R.D. McNair, A.J. Kenny, *Biochem. J.* **179**, 379–395 (1979).
45. S. Hjertén, in D. Glick ed., *Methods of Biochemical Analysis*, Wiley, New York, 1981, pp. 89–108.
46. M. Belew, M. Yafang, L. Bin, J. Berglöf, J.-C. Jansson, *Bioseparation* **1**, 397–408.
47. L. Kuehn, H. Meyer, H. Reinauer, *Proc. 2nd. Int. Insulin Symp.* 243–250 (1980).
48. K.D. Caldwell, R. Axén, J. Porath, *Biotech. Bioeng.* **17**, 613–616 (1975).

CHROMATOGRAPHY, ION EXCHANGE

STEVEN M. CRAMER
VENKATESH NATARAJAN
Rensselaer Polytechnic Institute
Troy, New York

KEY WORDS

Ion exchange
Models
Nonlinear chromatography
Preparative chromatography

OUTLINE

INTRODUCTION

Preparative ion exchange chromatography is a technique for separating biomolecules based primarily on their electrostatic interactions with charged stationary phase materials. This technique is the most widely employed chromatographic separation technique in the biotechnology industry today. It is used for both capture and high-resolution separations. The last decade has seen dramatic advances in the state of the art of chromatographic bioprocessing with ion exchange materials. This article presents recent advances in resin materials, modes of operation, and modeling.

BACKGROUND

Ion exchange chromatography (IEC) is the most widely employed chromatographic separation technique in the biotechnology industry today. It is based on the principle of electrostatic interactions between charged surfaces. In IEC, one employs a charged stationary phase material to separate oppositely charged solutes. Cation exchange employs a negatively charged surface to separate positively charged solutes, and anion exchange employs a positively charged surface to separate negatively charged solutes. The binding of solutes to ion exchangers may be viewed as an exchange process in which the solutes compete for binding with salt counterions and other charged components in the mobile phase.

This article focus on preparative protein separations in ion exchange systems, beginning with a brief discussion of various factors that affect protein retention (e.g., pH, ionic strength, mobile phase modifiers). A detailed treatment of the state of the art of stationary phase materials is then presented. A description of different modes of preparative IEC is given, along with separation examples from the literature. Finally, the modeling of preparative ion exchange is addressed.

Factors Affecting the Ion Exchange of Proteins

pH of the Mobile Phase. Proteins are ampholytes, possessing both negative and positive surface charges. In fact, the zwitterionic nature of proteins is what makes ion exchange such a powerful tool for these separations. The charges on the proteins arise because of the presence of acidic and basic amino acid residues on the protein surface as well as the amino and carboxyl termini of the protein.

Thus, the net charge on a protein is a strong function of the pH of the solution. Under acidic conditions, the basic side chains are charged, under basic conditions, the acidic residues will be charged. At the isoelectric point (pI), the net charge on the protein will be zero. In general, when the pH of the mobile phase is greater than the pI of the protein, the protein will have a net negative charge and will be retained on an anion exchanger. On the other hand, when the pH of the mobile phase is less than the pI of the protein, the protein will have a net positive charge and will be retained on a cation exchanger. It turns out that the net charge of the protein is not the sole determinant of its retention (1). The distribution of charges on the surface of the protein is also important. The protein may have patches of positively or negatively charged residues that may interact with the stationary phase even when its net charge is zero. Furthermore, the microenvironment pH often can differ slightly in the resin and bulk solutions. Thus, a protein may be retained even at its pI. In addition, it is important to consider protein solubility constraints in the development of an ion exchange separation, since proteins are sparingly soluble near their isoelectric point. In practice, to prevent protein denaturation, the operating pH in IEC is limited to the range 2–9.

The choice of buffer to establish the operating pH is of paramount interest in ion-exchange chromatography. It is desirable to employ a buffering agent that has the same charge as the stationary phase material or is neutral. This precludes the binding of these ions onto the stationary phase and thus prevents pH shifts within the column. Hence, it is desirable to use cationic buffers (e.g., alkylamines, Tris, aminoethyl alcohol) in anion exchange chromatography and anionic buffers (e.g., phosphate, acetate, citrate, barbiturate) in cation exchange chromatography.

Ionic Strength. The chromatographic retention of proteins is modulated by the ionic strength of the mobile phase. In cation exchange systems, the relative strength of cations is as follows (2): $Ba^{2+} > Pb^{2+} > Sr^{2+} > Ca^{2+} > Ni^{2+} > Cd^{2+} > Cu^{2+} > Co^{2+} > Zn^{2+} > Mg^{2+} > Ti^{2+} > Ag^{+} > Cs^{+} > Rb^{+} > K^{+} > NH_4^{+} > Na^{+} > H^{+} > Li^{+}$. In anion exchange, the relative strength of anions is citrate $>$ sulfate $>$ oxalate $> I^{-} >$ nitrate $> CrO_4^{2-} > Br^{-} > SCN^{-} > Cl^{-} >$ formate $>$ acetate $> OH^{-}$.

Thus, Br^{-} will bind stronger than Cl^{-} in an anion exchange column. In cation exchange, an equivalent solution of KCl would elute a protein faster than NaCl. Continuous and step gradients of ionic strength are often employed to effect protein separations. While protein retention is generally reduced as the ionic strength is increased, at very high ionic strength (>1 M), hydrophobic interactions can become pronounced.

Mobile Phase Modifiers. Addition of modifiers to the mobile phase can often influence the selectivity of a given separation. Recently, Gagnon *et al.* (3) reported an enhancement of the selectivity of proteins on the addition of poly(ethylene glycol) to the mobile phase. Other additives include chelating agents such as EDTA (to protect enzymes against metal impurities), reducing agents (e.g., dithiothreitol) to retard oxidation, and surfactants and chaotropic agents (e.g., urea) to solubilize membrane proteins (4).

Temperature. There are conflicting reports in the literature on the effect of temperature on resolution and band broadening. Increasing the temperature may reduce the viscosity of the mobile phase and increase the kinetics of the adsorption process. This may result in more efficient separations. However, changes in the temperature may also induce conformational changes in the protein. This may denature the protein or expose more hydrophobic patches to the stationary phase, enhancing nonspecific interactions and causing dramatic increases in protein retention (5). In general, the practice in industry is to employ either room temperature or low temperatures (4 °C) to minimize protease activity and to ensure protein stability.

STATIONARY-PHASE MATERIALS

The stationary phase is of paramount importance in IEC. Ideally, the stationary phase should be rigid enough to withstand high flow rates, provide sufficient selectivity and capacity for the problem at hand, and not denature the protein of interest. Furthermore, it should be stable over a wide range of operating conditions (pH, solvents, etc.) and amenable to rigorous cleaning-in-place with acid or alkali. The choice of the manufacturer is also important (6). The manufacturer should be able to ensure a regular supply with little lot-to-lot variation. Finally, the resin should not leach materials into the process stream.

Selectivity in ion exchange materials is affected by the ligand density, spacer arm chemistry, and the backbone resin material. Unfortunately, state-of-the-art chromatographic modeling cannot a priori predict the selectivity of different resin materials for various classes of separation problem. Thus, selection of an ion exchange resin is problem specific and requires the screening of a large variety of materials for a given separation. In bioprocessing applications, IEC is used for both capture and high-resolution separations. Capture applications demand that the resin possess sufficient dynamic capacity and be rigid enough to allow operations at high flow rates. On the other hand, selectivity is paramount for high-resolution separations.

Particle Size

While reversed-phase chromatography often employs materials of relatively small particle diameter for preparative applications, stationary-phase materials of 30- to 200-μm diameter are typically employed for preparative IEC. Higher resolutions can be achieved with smaller particles at the cost of increased backpressure. Finally, the size of the particle can also influence the mode of packing (7).

Ion Exchange Functional Groups

The stationary phase of an ion exchanger consists of a base matrix that is functionalized with basic (anion exchange) or acidic (cation exchange) groups. The charge on the stationary phase is pH dependent. This pH dependency allows one to categorize ion exchange stationary phases into strong and weak exchangers. Strong ion exchangers retain their charge over a very wide range of pH, whereas weak ones do so over a narrower range of pH. Thus, the capacity of weak ion exchangers is a function of the pH. Typically, strong exchangers have functional groups with very low pK_a (sulfopropyl in cation exchange) or very high pK_a (quaternary amino–type anion exchangers). Table 1 lists some of the functional groups commonly employed in ion exchange stationary phases and their corresponding pK_a values (8,9).

Capacity and Porosity

The capacity of the ion exchanger can be measured as both the total number of charges per unit stationary-phase volume or the mass of protein adsorbed per unit volume. When one is determining protein-binding capacity, it is important to ensure that the activity of the adsorbed protein is not compromised. The capacity expressed in terms of adsorbed protein is the effective capacity of the column. It is dependent on both the specific protein and the operating conditions (pH, ionic strength). It is also dependent on the pore size and the porosity of the particles. If the particle is pellicular (i.e., not porous), the surface area is relatively low, resulting in reduced capacity. On the other hand, if the particle is highly porous, the capacity increases, but the resistance to mass transfer also increases. Pores that are too small may be inaccessible to the proteins. The pore size distribution is usually unavailable, and a nominal pore size is supplied by the manufacturer.

Conventional ion exchange media have nominal pore sizes between 100 and 1,500 Å. At times, the pore size is also specified as the molecular weight of the largest globular protein that can be accommodated in the material.

Novel and Traditional Matrices

Table 2 lists representative resins available in the market today along with their manufacturers and trade names. As seen in the table, the resins can be roughly divided into five categories. Polysaccharide gels have been in existence since the 1950s (10). These matrices are inert toward proteins, can be easily derivatized, have very high capacities, and are stable over a wide range of pH values. Polysaccharide gels based on weakly cross-linked agarose, dextran, or cellulose have weak mechanical properties and are limited to low-pressure regimes. In contrast, Sepharose® Fast Flow ion exchangers (Pharmacia) are composed of a higher degree of cross-linked agarose, which increases their mechanical stability. High-performance polysaccharide materials (Pharmacia) are also available in relatively small particle diameter (34 μm) and have been shown to be particularly useful for difficult high resolution separations. These polysaccharide-based resins are currently the most widely employed chromatographic materials in the biopharmaceutical industry.

Although silica-based ion exchangers have good mechanical properties, they are not as widely used for bioprocessing because their instability under alkaline conditions precludes alkaline cleaning. Furthermore, surface silanol and siloxane groups can have adverse effects on proteins (11). To overcome this problem, silica supports have been coated with a polymeric layer (12).

Resins based on porous particles of zirconium oxide have been synthesized and characterized for biochromatography (13). Zirconia possesses all the mechanical advantages offered by silica. In addition, it has good thermal and chemical stability. Unfortunately, this backbone material tends to bind proteins, which necessitates coating with a biocompatible layer. Clausen and Carr (14) recently employed a phosphonate analog of EDTA for the modification of zirconia supports to provide a biocompatible stationary phase. As described shortly (see "Dynamic Capacity"), a potential advantage of zirconium is its relatively high density, which has been shown to have potential for expanded-bed applications (15).

The polymer-based matrices (16) have good mechanical properties and are stable over a wide pH range. Furthermore, they have a relatively high loading capacity because of their macroporous structure. The hydrophobic polymer matrices are typically coated with a hydrophilic polymer to reduce nonspecific interactions. Polymeric supports are being used increasingly in the preparation of ion exchangers for bioprocessing (17). In addition, many of the novel materials described below are based on polymeric supports.

The penultimate category of particulate ion exchangers in Table 2 is entitled "Novel" media. Each of the stationary phases in this category possesses unique features that distinguish it from more traditional materials. The POROS® supports are synthesized by a suspension polymerization process. These chromatographic materials consist of a rigid polystyrene–divinylbenzene base matrix covered with a polyhydroxyl layer that can be further derivatized to produce various functional groups (18). This type of resin belongs to the class of particles termed "perfusive" (19,20). Such particles have a bimodal pore distribution containing macropores (6,000–8,000 Å) that transect the particle and interconnecting "diffusive" micropores (500–1,500 Å). The macropores facilitate mass transport, while the diffusive pores increase the available surface area. Their permeability has been shown to be independent of

Table 1. Functional Groups Used in IEC

Name	Abbreviation	PK_a
Anion exchangers		
Diethyl aminoethyl	DEAE	9–9.5
Quaternary aminoethyl	Q	9.5
Dimethyl aminoethyl	DMAE	9
Trimethyl aminoethyl	TMAE	>13
Cation exchangers		
Carboxymethyl	CM	3.5–4
Phosphate	Phospho	<2 & 6
Sulfonate	S	2
Sulfoethyl	SE	2
Sulfopropyl	SP, SO_3^-	<1

Source: Compiled from Refs. 8 and 9 and from manufacturers' data.

Table 2. Representative Commercial Ion Exchange Media

Trade name	Manufacturer	Function groups	Particle sizes (μm)	Matrix
		Polysaccharide gels		
Bio-Gel A	Bio-Rad	DEAE, CM	80–150	Cross-linked agarose
Sepharose CL-6B	Pharmacia LKB	DEAE, CM	45–165	Agarose
Sephacel	Pharmacia LKB	DEAE	40–160	Cross-linked cellulose
Express-Ion	Whatman	DEAE, Q, CM, SE	60–130	Microgranular cellulose
Sephadex	Pharmacia LKB	DEAE, Q CM, SP	40–125	Cross-linked dextran
Sepharose Fast Flow	Pharmacia LKB	Q, DEAE, CM, SP	45–165	Cross-linked agarose
Sepharose High Performance	Pharmacia LKB	Q, SP	24–44	Cross-linked agarose
Ultrogel	BioSepra	Hydroxy-apatite	60–180	Cross-linked agarose
		Inorganic matrix		
Accell	Waters	CM, QMA	40	Polymer-coated silica
TSKgel 3SW	TosoHaas	CM, DEAE	5, 10	Silica
Spherodex	BioSepra	DEAE, SP, CM	40–300	Silica–dextran
Spherosil	BioSepra	DEAE, QMA	40–300	Polymer-grafted silica
Zephyr	Sepracore	DEAE, S	20	Silica–dextran
		Synthetic polymers		
TSKgel 5PW	TosoHaas	SP, DEAE	30	Methacrylate copolymer
Mono Q/S	Pharmacia LKB	Q, S	10	Cross-linked polystyrene–divinylbenzene
Macro-Prep	Bio-Rad	S, Q, CM, DEAE	50	Methacrylate copolymer
Protein-Pak	Waters	DEAE, Q, SP, CM	8, 15, 40	Rigid polymeric gel
Trisacryl	BioSepra	Q, DEAE, CM, SP	40–160	Macroporous synthetic polymer
Trisacryl Plus	BioSepra	DEAE, SP	40–160	Base-resistant polymer
Toyopearl	TosoHaas	DEAE, CM, SP	35, 65, 100	Methacrylate
Super Q	TosoHaas	Q	35, 65, 100	Synthetic polymer
		Novel media		
POROS	Perseptive	Q, S	10, 20	Coated polystyerene
SOURCE	Pharmacia LKB	Q, S	15, 30	Cross-linked polystyrene–divinylbenzene
Hyper D	BioSepra	Q, S	10, 20, 35, 60	Polystyrene shell and hydrogel
Fractogel EMD (tentacle resins)	Merck	TMAE, DEAE, DMAE, SP, CM	20–40	Vinylic grafted polymer
STREAMLINE	Pharmacia	SP, DEAE	200	Macroporous cross-linked agarose with a quartz core
		Membrane ion exchangers		
MemSep	Millipore	DEAE, Q, CM		Cellulose
QuickDisk	Knauer	DEAE, Q		Poly(glycidylmethacrylate)
Sartobind	Sartorius	Q, S		Synthetic copolymer

the particle size (21). The solute breakthrough curves for the POROS media have been shown to be relatively unaffected by the flow rate in contrast to conventional ion exchangers (20). Gerstner et al. (22) have demonstrated that displacement chromatography of proteins can be carried out on these particles at very high linear velocities. The SOURCE™ materials have also been shown to exhibit chromatographic behavior that is independent of flow rate. In addition, the SOURCE™ resins have relatively high efficiency owing to their monodisperse particle size distribution. Recently, the development of agarose-based perfusive materials has also been reported (23).

Hyper-D® materials are composite stationary phases composed of a rigid macroporous support filled with a weakly cross-linked ion exchange gel. The rigid particle provides the necessary mechanical strength for high flow rate operations, while the soft gel provides high binding capacity. The equilibrium uptake of model proteins and the mass transfer rates in Q-Hyper-D® has been studied by Fernandez and Carta (24,25). These materials have a very high binding capacity even at elevated flow rates. In fact, it has recently been shown (26) that large-particle-diameter Hyper-D® materials (200 μm) also exhibit high binding capacities even at elevated linear velocities. It has been proposed that the mass transport mechanism in these materials is surface diffusion due to the high concentrations obtained in the gel (25).

The "tentacular" stationary phases comprise another novel sorbent (27). These chromatographic materials have polyelectrolyte chains (typically 5–50 monomers long) grafted onto porous matrices. In contradistinction to the supports mentioned thus far, the charge moieties in tentacular phases are attached to the matrix via relatively long, flexible spacer arms. This results in reduced nonspecific interactions between the proteins and the base matrix. Furthermore, since the ligands are able to move relatively freely in space, they can explore the steric distribution of charges on a protein in a more efficient manner. This can potentially result in higher-resolution separations. Tentacular ion exchangers have been employed to differentiate between native and deamidated recombinant human DNAse which differ in a single surface charge (Fig. 1) (28).

Dynamic Capacity

The dynamic capacity of the column is the effective capacity determined from the breakthrough curve (i.e., when the effluent concentration reaches 10–50% of the inlet protein concentration). Table 3 lists the dynamic capacities of several anion exchange media obtained at low linear velocities; these capacities are seen to range from 30 to 165 mg/mL. The cellulosic anion exchangers such as Sepharose® Fast Flow have high dynamic capacities under these conditions. Importantly, dynamic capacities can be strongly affected by the flow rate. In Figure 2, which compares the flow rate dependence of the dynamic capacities of five Q exchangers, the dynamic capacities of the tentacular (Frac-

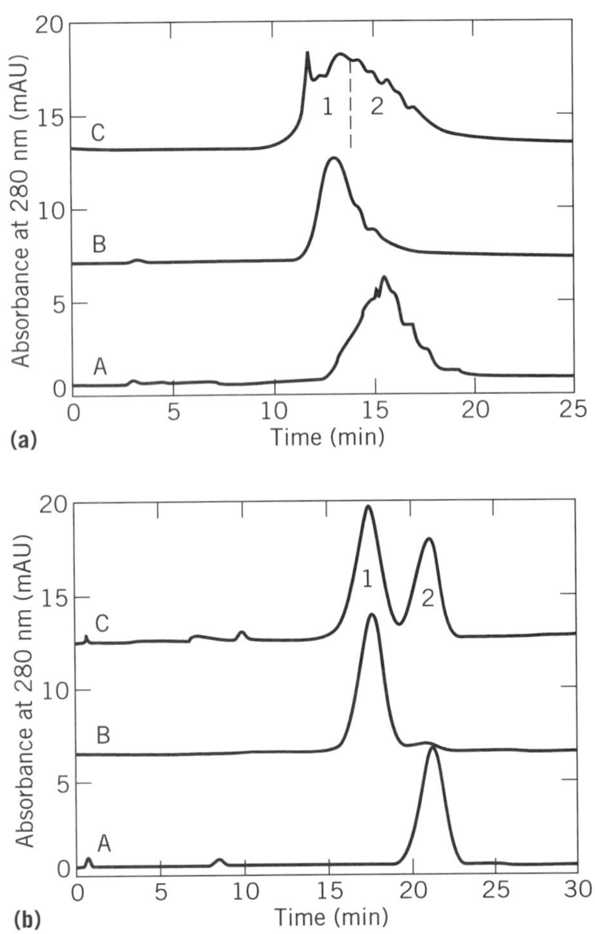

Figure 1. (**a**) Strong cation exchange chromatography on a sulfopropyl column (TSK SP 5PW, 75 mm × 7.5 mm i.d.) of three samples: A, rhDNAse; B, its deamidated variant; and C, a mixture of the two variants. (**b**) Cation exchange chromatography of three samples: A, rhDNAse; B, its deamidated variant; and C, a mixture of the two variants on a tentacle ion exchanger (LiChrosphere SO_3^- 50 mm × 4.6 mm i.d.). The two sets of chromatograms were obtained under the same chromatographic conditions. (*Source*: Reproduced with permission from Ref. 28.)

Table 3. Ionic Capacities of Commonly Used Anion Exchangers

Ion exchanger	BSA sorption capacity (mg/mL)[a]
QMA Accell	111
DEAE Bio Gel A	38
DEAE Fractogel 650	32
DEAE Sephacel	131
DEAE Sepharose CL 6B	164
DEAE Sepharose Fast Flow	88
DEAE Spherodex	98
DEAE Trisacryl	86
DEAE Fractogel EMD	100[b]
Q Fractogel EMD	55
Q POROS II	40
Q Sepharose Fast Flow	115
Q Super	127
Q Hyper D	125
Express-Ion Q	55

Source: From Ref. 8 and manufacturers' data.
[a]Determined for BSA in 50 mM Tris–HCl buffer (pH 8.6) except as noted.
[b]Determined for BSA in 50 mM Tris–HCl buffer (pH 8.3).

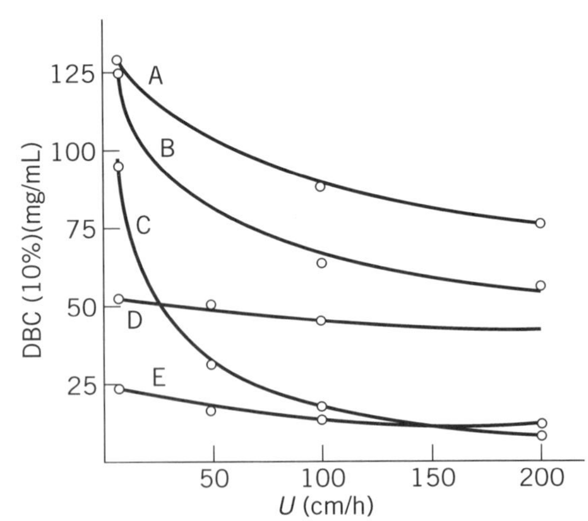

Figure 2. Variation of dynamic binding capacity (DBC) as a function of the flow rate (U) for Q anion exchangers. Determination of DBC were performed using breakthrough curves (10%) with a column of 1 cm i.d. containing 4 mL of sorbent. Protein solution was bovine serum albumin at 5 mg/mL in 50 mM Tris–HCl buffer (pH 8.6): A, Q HyperD F; B, Super Q; C, Q Sepharose Fast Flow; D, Q Fractogel EMD; E, Q POROS. *Source*: Reproduced with permission from Ref. 8.

togel EMD) and perfusive (POROS®) resins are relatively low at low flow rates, but their dependence on the flow rate is rather weak. On the other hand, the dynamic capacity of the Sepharose® FF resin exhibits a stronger dependence on the flow rate. The Hyper D® and the Super materials exhibit a moderate flow rate dependence. However, the Hyper D® material has a capacity substantially higher than those of the other materials at elevated flow rates. An important caveat to this discussion on dynamic capacity is that it is often desirable to use low operating pressures in a bioprocessing environment. This practice may affect the linear velocities that can be employed under actual operating conditions and may make performance at elevated velocities less important.

An important recent advance in the state of the art of chromatographic bioprocessing is the development of expanded-bed systems, in which the bed expands during the loading phase, enabling it to directly process unclarified feedstock. This can potentially obviate the need for membrane or centrifugal processing prior to the ion exchange capture step. A schematic of the basic principles of expanded bed adsorption is shown in Figure 3. The Streamline adsorbent, which has a mean particle size of approximately 200 μm, is a specially designed stationary phase for expanded-bed chromatography. It is stable over a wide pH range but unstable in solutions containing oxidizing agents. Accordingly, in situ sterilization may be a problem with these sorbents. Streamline forms a stable expanded bed without the need for external magnetic fields. Its capacities are lower than those of typical commercial adsorbents such as Sepharose FF and Spherodex (29,30). This may be due to the presence of the quartz core, which provides the necessary particle density but reduces the available surface area. The performance of this sorbent has been shown to be similar in packed-bed and expanded-bed modes (29), indicating that the expanded-bed mode behaves like a packed bed with increased voidage.

In addition to the commercially available Streamline material, several other expanded-bed supports are in development. Flickinger and Carr (15) are exploring the potential of zirconia stationary phases for expanded-bed applications. These materials have the desirable property of having sufficient density without further modification. Gilchrist et al. (30) synthesized a polysaccharide–ceramic composite sorbent for use in expanded-bed chromatography. Thommes et al. (31) derivatized Bioran (controlled pore glass) with sulfonic acid residues. These particles compared well with Streamline media in initial evaluations.

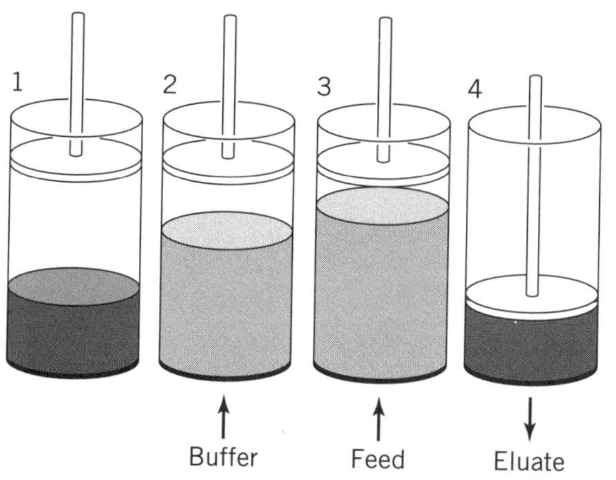

Figure 3. Principles of expanded-bed operation: **1**, sedimentation of adsorbent, **2**, upward flow of equilibration buffer resulting in a stable expanded bed, **3** application of unclarified feed by upward flow, and **4**, downward flow of the eluent to elute the bound protein of interest. *Source*: Adapted from Pharmacia catalog.

Membrane Adsorbers

In contrast to all the chromatographic supports discussed thus far, an alternative approach to ion exchange is the use of charged membrane systems. Mass transfer in membrane exchangers is typically dominated by convection, which enables these systems to be operated at relatively fast linear velocities resulting in high throughputs. The typical configuration of membrane ion exchangers is the "pancake" configuration, which has a large width-to-length aspect ratio that provides higher surface areas and lower backpressures. The principal disadvantage of these configurations is the difficulty of achieving a uniform inlet flow distribution and collection of the eluate with minimal backmixing.

Monoclonal antibodies and human recombinant antithrombin III were subjected to preparative purification on Sepharose® FF columns and Sartobind units (32). The membrane exchanger yielded higher throughputs (>13-fold in the case of monoclonal antibodies) and comparable or higher recoveries. However, higher concentration factors were achieved on the Sepharose columns. An extensive review of the state of the art of membrane chromatography can be found in Ref. 33.

Regeneration, Cleaning-in-Place, and Sanitization

A critical issue in ion exchange bioprocessing is the ability to clean the column. This includes column regeneration (between each separation cycle), cleaning-in-place operations (after every 1–10 cycles), and sanitization (between batches). Any commercial ion exchange material must be stable enough to withstand high pH as well as oxidizing conditions.

Column regeneration and cleaning-in-place operations are important for establishing long-term use of ion exchange columns. The conditions are usually determined empirically and are strongly affected by the feed composition. Typical conditions include high salt (1 M NaCl) and/or 1 M NaOH. A complete list of regeneration protocols can be found in Ref. 34. The monograph by Sofer and Nystrom (35) offers a more detailed discussion of column maintenance procedures.

In situ sanitization of the column is necessary to prevent microbial contamination between batches. Formalin, peracetic acid, and sodium hypochlorite are effective chemical disinfectants. Alkaline solutions of ethanol are effective for spore destruction. Clearly, the stability of the resin must be kept in mind when the sanitization conditions are chosen. A recommended validation protocol for the in situ sanitization of an ion exchanger is outlined in Ref. 34.

Relative Efficacy of Resin Materials

In process chromatography, it is important to distinguish between initial capture (recovery) and high-resolution or polishing steps. For product recovery, the capacity and production rate are paramount. In addition, it is often necessary to process large volumes of relatively dilute feedstock, which often contain particulates. It is precisely this challenge that has catalyzed the development of many of the "novel" materials described above. The advantage of perfusive-type materials (e.g., POROS®) and membrane chromatography is that they can be operated at high linear velocities. However, these materials often have relatively low capacities. While the problem can be overcome by carrying out many column cycles for a given batch, this approach has not been widely adopted by the industry. Composite stationary phases (e.g., Hyper-D®) possess high binding capacities and can be operated at reasonably high flow rates without a drastic loss of dynamic capacity or increased backpressure. This makes their use for capture operations potentially very attractive. Expanded-bed sorbents are particularly attractive for capture operations because they combine clarification, concentration, and capture into a single step. It turns out that the stability of expanded beds is quite sensitive to such factors as the solids content, pH, ionic strength, and viscosity of the feed (36). Since the effect of these factors has not been fully elucidated, the choice of the mobile phase conditions in expanded-bed applications will be highly problem specific. Nevertheless, expanded-bed chromatography has generated tremendous interest from the industry.

As mentioned in the beginning of this section, the effect of stationary phase chemistry on selectivity is still unpredictable, and the selection of an ion exchange resin is problem specific. Recently, Levison et al. (37) evaluated the functional and physical performance of a range of ion exchange media based on polysaccharide and polymeric or composite materials. Although the data suggest significant differences in the performance of various media for a common set of criteria, the selection of an ion exchanger remains problem specific. Thus, until the state of the art of chromatographic modeling is further improved, the biochromatographic engineer is left with the daunting task of screening a large variety of materials for a given separation. Fortunately, there are several automated chromatographic workstations that can readily screen a large number of resins in a relatively short period of time. Once the resin with the optimal selectivity has been identified, the task remains to develop an efficient, high-throughput chromatographic process. This requires the selection of the appropriate mode of operation.

MODES OF COLUMN OPERATION

Preparative ion exchange can be carried out by means of a variety of resin materials, column configurations, and modes of operation. While the selection and development of resin materials often receives significant attention, the choice of the proper mode of operation is often given less attention. In fact, the selection of the optimal mode of operation for a given bioseparation problem can have a profound effect on the production rate, yield, and purity for that separation. To date, linear gradient or step gradient modes have been used to carry out most industrial chromatographic bioprocessing. This section briefly describes these gradient modes and presents examples from the recent literature. In addition, the displacement and chromatofocusing modes of operation are discussed.

Step Gradient

Step gradient is often employed for initial capture or recovery operations where the primary objective is to recover and to concentrate the bioproduct of interest. This operation is usually carried out in short, large-diameter columns packed with relatively large-particle diameter stationary-phase materials (>90 μm). This mode of chromatography consists of first loading the column with a large volume of low ionic strength feed solution, followed by a wash step, and finally an elution step based on an elevated salt and/or modified pH eluent. This type of chromatography can be characterized as essentially an on/off operation. The most important stationary-phase characteristic in preparative step gradient chromatography is the dynamic capacity of the resin. Thus, these separations do not require very efficient columns and can be readily carried out in short "pancake"-type columns. These capture steps can be carried out by means of conventional packed-bed columns or with "expanded-bed"-type configurations. As described earlier, it is important to screen a wide variety of ion exchange resins to identify a material that has sufficient selectivity and capacity for a given separation. Several examples of step gradient ion exchange chromatography are now presented.

The use of an expanded bed in the pilot-scale recovery of the anticoagulant recombinant Annexin V from unclarified *E. coli* homogenate has been reported (36). The optimal binding and elution conditions for Annexin V were determined in a 1.6 cm × 20 cm packed bed of Streamline DEAE adsorbent using clarified homogenate. Then 1.7 L of unclarified homogenate was applied to the laboratory-scale Streamline 50 column (5 cm × 100 cm, Streamline DEAE adsorbent) under the mobile phase conditions established with the small packed bed. The protein was eluted using a 250-mM step gradient of NaCl. The elution was carried out in the sedimented mode at a mobile phase velocity of 100 cm/h. The results were very similar to those obtained with the small packed bed. The presence of cell debris did not seem to affect the process. The biomass dry weight and the viscosity of the homogenate were 3.6% and 10 mPa/s, respectively.

Finally, the process was scaled up to a pilot-scale expanded bed: Streamline 200 (20 cm × 95 cm, Streamline DEAE adsorbent). The adsorbent volume, feed volume, and flow rate were scaled up by a factor of 16 to account for the increased diameter of the column, resulting in a feed load of 26.5 L. The rest of the conditions were kept the same. Figure 4 shows the chromatogram from the pilot-scale recovery of the protein. Yields greater than 95% were obtained with both the Streamline 50 and Streamline 200 columns. The expanded bed was cleaned after each run using 0.5 M NaOH with 1 M NaCl followed by distilled

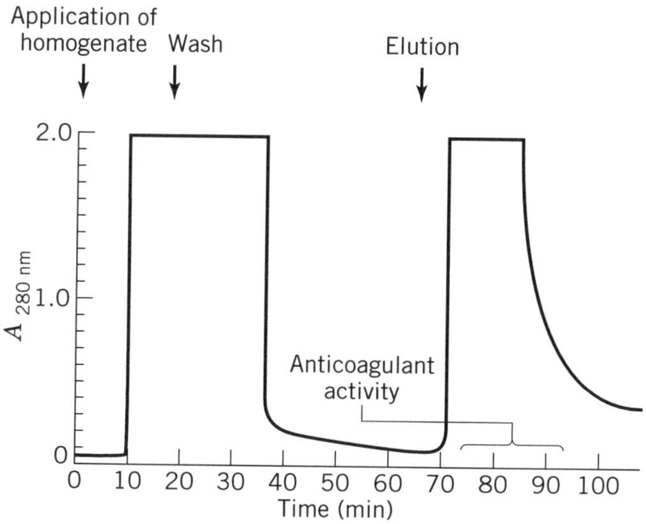

Figure 4. Chromatogram from pilot-scale, expanded-bed recovery of recombinant Annexin V from unclarified *E. coli* homogenate: feed, 26.5 L of unclarified *E. coli* homogenate in 30 mM ammonium acetate (pH 5.5); column, Streamline 200 with Streamline DEAE adsorbent; elution with 30 mM ammonium acetate (pH 5.5) containing 250 mM NaCl; flow rate: upward flow at 300 cm/h during application and wash and downward flow at 100 cm/h during elution. *Source*: Reproduced with permission from Ref. 36.

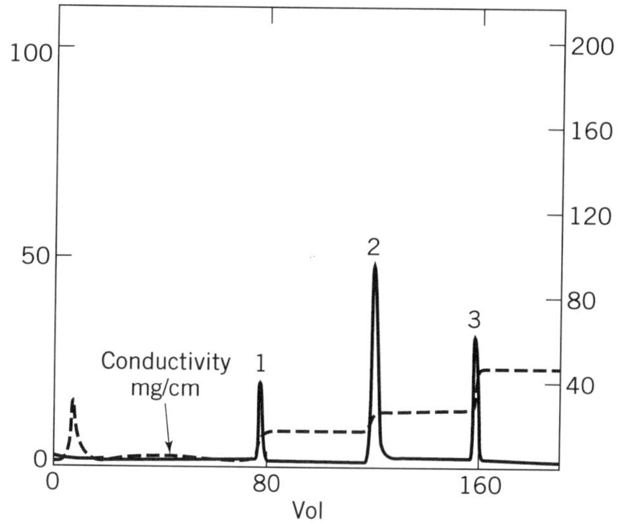

Figure 5. Purification of *Pseudomonas aeruginosa* exotoxin A on Q Sepharose HP (10 cm × 10 cm i.d.): buffer, 20-mM Tris, 1-mM EDTA (pH 7.4). Gradient program, first step salt concentration, 0.15 M NaCl (5 column volumes); second step salt concentration, 0.25 M NaCl (5 column volumes); final step salt concentration, 0.5 M NaCl (5 column volumes). *Source*: Reproduced with permission from Ref. 38.

water, 30% isopropanol, and 25% acetic acid. This example illustrates the potential of expanded-bed chromatography for ion exchange step gradient chromatography.

The use of step gradient in the production and purification of *Pseudomonas aeruginosa* exotoxin A (PE) from *E. coli* has been reported (38). The initial recovery of this protein from the fermentation broth involved cell lysis and centrifugation. Purification was achieved in three steps: weak anion exchange followed by hydrophobic interaction and a strong anion exchange step. The weak anion exchange step was carried out by means of a DEAE Sepharose® Fast Flow column in the flowthrough mode to remove DNA and pyrogens. In flowthrough column operations the impurities are retained on the column, but the product is not. Thus, this type of operation is quite simple to operate and can result in a rapid reduction of impurities from the feedstock. This flowthrough step resulted in a 10-fold reduction in DNA and a protein purity of approximately 25%. In fact, it is quite common to add a flowthrough anion exchange step in a process for DNA clearance.

The final anion exchange operation was carried out on a Q Sepharose high-performance (10 cm × 10 cm) column using multiple step gradients (Fig. 5). The protein of interest was eluted with the second step, resulting in the purification of 0.3 g of protein at a purity of greater than 99%. Since the column loading was relatively low, multiple cycles were necessary to produce sufficient amount of material. The ion exchange columns were cleaned and regenerated between runs with 1 M NaOH and 3 M NaCl. This example illustrates how step gradient chromatography can sometimes be employed for high-resolution separations for systems exhibiting sufficient separation factors. However, as shown in this example, these operations are often limited by relatively low column loading.

Step gradient has many advantages, namely, ease of operation, no need for complicated gradient pumping apparatus, and concentration of the bioproduct. When the separation problem is more difficult, however, a higher-resolution mode of operation is often required.

Linear Gradient

For high-resolution separations, linear gradient chromatography is often employed. This operation is usually carried out in large-diameter columns using a range of column lengths and particle diameters. It turns out that depending on the difficulty of the separation, it may be necessary to employ longer columns and/or materials of smaller particle diameter for these separations. This mode of chromatography is typically carried out by first loading the column with a relatively large volume of low-ionic-strength feed solution, followed by a continuous linear gradient of increasing ionic strength. This type of chromatography cannot be characterized as an on/off operation and is thus dependent on the length of the column and the column efficiency. The most important stationary-phase attribute in preparative linear gradient chromatography is the selectivity of the resin for the given separation. Accordingly, it is again essential that a wide variety of resin materials be screened to identify the material with the highest selectivity. In addition to the selectivity, the dynamic capacity of the stationary phase is also important for preparative linear gradient separations and must be considered when selecting a resin. The major advantage of linear gradient is its ability to resolve difficult separation problems. Several examples of linear gradient ion exchange chromatography are now presented.

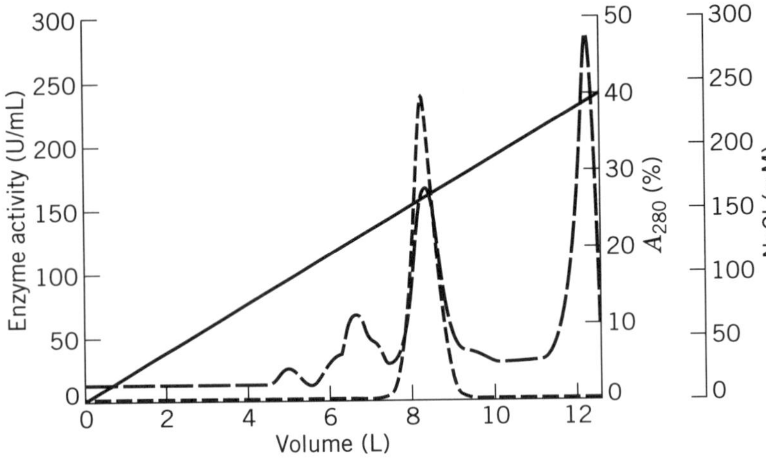

Figure 6. Cation exchange of L-asparaginase on a CM-Sepharose column (24 cm × 11.3 cm i.d.): buffer, 5 mM succinic acid–NaOH (pH 6); feed, dilute solution of cell extract with a conductivity of less than 2 mS; flow rate, 24 cm/h; gradient, 0–300 mM NaCl in 16 L of buffer. Curves: solid, NaCl; dotted, enzyme, dashed, protein. *Source*: Reproduced with permission from Ref. 39.

The use of linear gradients in the purification of L-asparaginase from *Vibrio (Wolinella) succinogenes* has been reported (39). The purification of this enzyme was achieved through a cation exchange step followed by an anion exchange polishing step. Approximately 13 g of protein (184,000 units) was loaded onto a 11.3 cm × 24 cm CM Sepharose cation exchanger at a mobile phase velocity of 24 cm/h. Upon elution of the enzyme by means of a linear gradient of 0–300 mM NaCl, the chromatogram shown in Figure 6 was obtained. This gradient resulted in the recovery of 91% of the enzyme and a 28-fold purification. An anion exchange polishing step followed. Approximately 0.4 g of protein (167,000 units) was loaded onto a 5 cm × 10.2 cm Q Sepharose® FF column at a mobile phase velocity of 15 cm/h. A linear gradient of 0–200 mM NaCl was used to elute the enzyme. This separation scheme resulted in an 87% recovery and a 47-fold purification of the enzyme. These separations required quite low mobile-phase velocities. In addition, the high-resolution step was limited to a column loading of 2g/L, which is typical of high-resolution linear gradient separations.

Often, the purification scheme for a protein involves both step and linear gradient operations. The purification of factor VIII (FVIII) and von Willebrand factor (vWF) from human plasma (40) illustrates the use of both step and linear gradients in the purification protocol. The FVIII–vWF complex is a complicated system in which the two proteins are highly glycosylated, have low pIs, and exhibit strong binding even under conditions of relatively high ionic strength. Furthermore, FVIII is prone to proteolytic degradation without its carrier protein vWF. The naturally adhesive vWF exhibits strong nonspecific interactions with organic, hydrophobic supports.

A wide variety of stationary-phase materials were evaluated for this separation. Silica-based supports could not be used in the purification scheme because they tended to activate FVIII. Supports such as cellulose and poly(styrene–divinylbenzene) were unable to effect the separation owing to the strong nonspecific interactions of vWF. Columns packed with traditional media such as Toyopearl® DEAE, MonoQ, MonoP, and TSK® DEAE 5PW were subject to fouling. On the other hand, tentacular supports gave good results for this separation, possibly because their extended tentacular arms prevent the protein from interacting with the base matrix.

The purification scheme for this complex involved three anion exchange steps and two virus inactivation steps. All the anion exchange steps involved the use of tentacular supports. Weak Fractogel® EMD-DEAE supports were used in the step gradient mode for the first two steps, and strong Fractogel® TMAE in the linear gradient mode was employed for the final purification. The FVIII–vWF complex was strongly retained on the DEAE support, enabling the use of step gradient chromatography to remove a variety of contaminating proteins, solvents, detergents, and stabilizers (employed during viral inactivation steps).

The FVIII–vWF complex was dissociated and separated in the final step by means of a linear gradient. Calcium ions were added to the buffer to cause the dissociation of the complex. Figure 7, which illustrates the separation achieved with the strong tentacular exchanger, indicates that this difficult separation was complicated by the heterogeneity of the protein glycosylation and incomplete dissociation. This example illustrates how linear gradient ion exchange with highly selective resins can be employed for

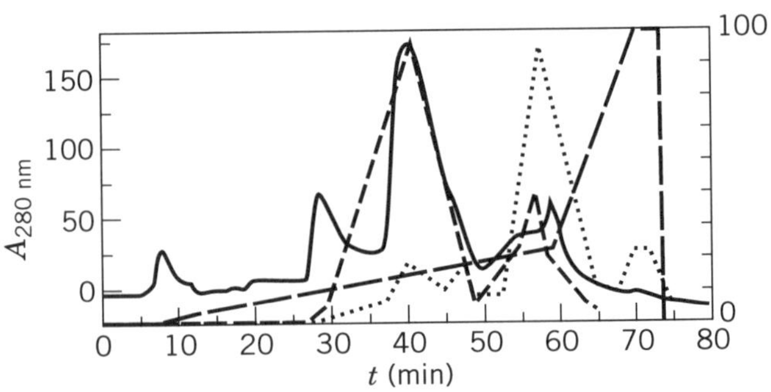

Figure 7. Dissociation and separation of FVIII and vWF from their complex in the presence of 50 mM calcium: buffer, 20 mM Tris–HCl (pH 7.4): feed, 300 units of FVIII in buffer; column, Fractogel® EMD TMAE (20–40 μm, 120 mm × 16 mm i.d.); flow rate, 3 mL/min; gradient, 0–200 mM NaCl. Curves: short dashes, vWF activity; dots, FVIII activity; long dashes, % B (IM NaCl) *Source*: Reproduced with permission from Ref. 40.

very difficult separations. This process resulted in a purification factor of 8,000 with an overall yield of approximately 20%.

According to the conventional wisdom, the more difficult the separation, the lower the column loading that can generally be processed by linear gradient. Thus, linear gradient is often limited to lower column loadings. In addition, linear gradient requires more complicated gradient pumping apparatus. In contrast, displacement chromatography is able to carry out high-resolution separations at very high column loadings, using the simple pumping systems employed for step gradient chromatography.

Displacement

The displacement mode of chromatography, first identified in the 1940s by Tiselius (41), combines the high throughput and concentrating power of step gradient chromatography with the high resolving power of linear gradient chromatography. Although displacement chromatography has been shown to hold significant promise for high-resolution preparative separations, a major impediment to the implementation of displacement chromatography has been the lack of suitable displacer compounds. Recent advances in this field have generated significant interest in displacement technology.

Operationally, displacement chromatography is performed in a manner similar to step gradient chromatography, where the column is subjected to sequential step changes in the inlet conditions. The column is initially equilibrated with a carrier buffer of relatively low ionic strength. The feed mixture is then introduced into the column, followed by a constant infusion of the displacer solution. The displacer is selected such that it has a higher affinity for the stationary phase than any of the feed components. Under appropriate conditions, the displacer induces the feed components to develop into adjacent "square-wave" zones of highly concentrated pure material. After the breakthrough of the displacer from the column effluent, the column is regenerated and reequilibrated with the carrier buffer.

The separation of oligonucleotides from their failure sequences presents a significant challenge to chromatographic engineers. The failure sequences are typically very closely related to the main product. Furthermore, there is considerable conformational heterogeneity of both the product and the failure sequences. Displacement chromatography has been successfully employed for resolving oligonucleotide mixtures characterized by low separation factors. For example, 1.2 g of a 24-mer phosphorothioate oligonucleotide was purified on a POROS® HQ/M column using dextran sulfate as the displacer (42) (Fig. 8). The separation yielded 70% of the product at a purity of about 96%. Methanol was added to the mobile phase to eliminate hydrophobic interactions between the oligonucleotide and the stationary phase and to improve the recovery. The column was regenerated with 3 M NaCl in 0.5 M NaOH. On a semipreparative scale, the entire separation (equilibration + loading + displacement + regeneration) could be accomplished in approximately 32 mins.

An important recent advance in the state of the art of displacement chromatography has been the discovery that low molecular weight displacers can be successfully employed for protein purification in ion exchange systems (43). Pentaerythritol-based dendritic polyelectrolytes ranging in molecular weight from 480 to 5,100 Da have been shown to be effective displacers for protein purification (44). In addition, protected amino acids (45), antibiotics (46), and a variety of anionic displacers have been developed (47–49). Low molecular weight displacers have been successfully employed for very difficult protein separations (50). These novel displacers have significant operational advantages over large polyelectrolyte displacers. First and foremost, if there is any overlap between the displacer and the protein of interest, the low molecular weight materials can be readily separated from the purified protein during postdisplacement downstream processing involving size-based purification methods (e.g., size exclusion chromatography, ultrafiltration).

The relatively low cost of synthesizing low molecular weight displacers can be expected to significantly improve the economics of displacement chromatography. Furthermore, the salt-dependent adsorption behavior of these low molecular weight displacers greatly facilitates column regeneration. In addition, these displacers can be employed in "selective" displacement chromatography, where the product(s) of interest can be selectively displaced while the low-affinity impurities can be desorbed in the induced salt gradient ahead of the displacement train and the high-affinity impurities either retained or desorbed in the displacer zone (51).

Figure 9 shows an example of a selective displacement separation: a displacement experiment was carried out using a feed mixture of ribonuclease B, bovine heart cytochrome *c*, horse heart cytochrome *c*, and lysozyme with *N*-α-benzoylarginine ethyl ester (BAEE) as the displacer. As seen in the figure, the low-affinity protein ribonuclease B eluted ahead of the displacement train in the induced salt gradient, whereas the two cytochrome *c*'s were well resolved in the displacement train. Thus, in spite of the extreme closeness in their adsorption behavior, displacement chromatography resulted in excellent separation of the two cytochrome-*c* proteins with both high yields and high purity. The highest-affinity protein, lysozyme, remained on the column under the conditions of this experiment. This experiment demonstrates that selective displacement chromatography can result in baseline separation of the product of interest from the low- and the high-affinity impurities, while resolving the target bioproduct from very similar affinity impurities in an isotachic displacement train. Recent results have indicated that displacement chromatography may have significant utility for extremely difficult bioseparations such as protein variants (52) and glycoforms.

Chromatofocusing

In chromatofocusing retained, internal pH gradients are employed to effect the separation of protein mixtures. These pH gradients are produced internally by running a buffer of a particular pH through an ion exchanger equilibrated at a different pH. The focusing effects produced by these pH gradients were originally studied by Sluyterman

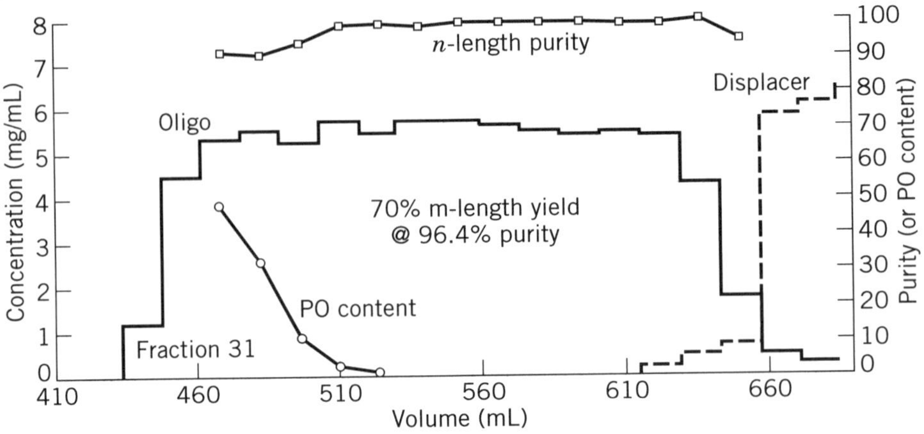

Figure 8. Preparative-scale purification of a 24-mer phosphorothioate oligonucleotide: feed, 1.2 g in 100 mM NaCl, 10 mM NaOH, 5% methanol; displacer, 800 mL of 7 mg/mL dextran sulfate; column, POROS HQ/M (20 μm, 100 mm × 50 mm i.d.); flow rate, 50 mL/min; fraction size, 14 mL. *Source*: Reproduced with permission from Ref. 42.

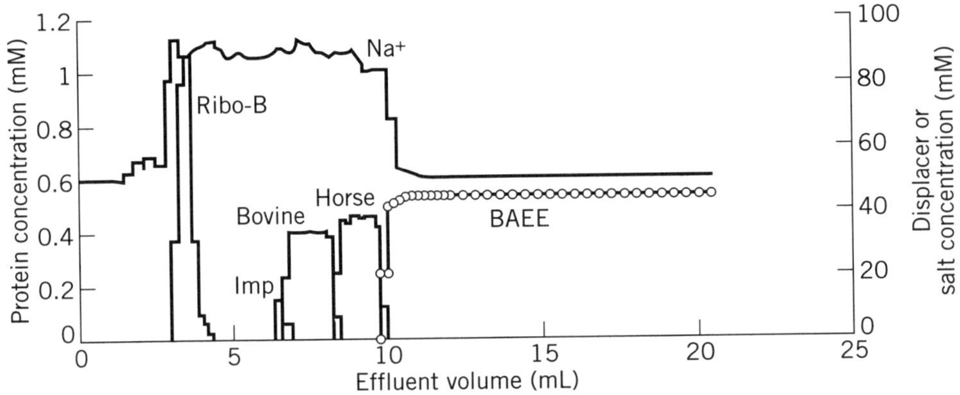

Figure 9. Displacement separation of a four-component protein mixture using BAEE as a displacer-column, 100 mm × 5 mm i.d. strong cation exchanger (8 μm); buffer, 50 mM sodium phosphate (pH 6.0); feed, 1.6 mL of 0.38 mM ribonuclease B, 0.45 mM horse heart cytochrome *c*, 0.37 mM bovine heart cytochrome *c*, and 0.34 mM lysozyme in buffer; displacer concentration, 45 mM; flow rate, 0.2 mL/min; fraction size, 200 μL. *Source*: Reproduced with permission from Ref. 51.

and coworkers (53,54). Typically, polyampholyte buffers are employed in this mode of chromatography. These buffers are not only expensive but also contain components that can associate with proteins (55). However, Hearn and Lyttle (56) and Frey et al. (57) demonstrated the efficacy of employing multicomponent mixtures of common buffer systems to produce internal pH gradients in ion exchange columns. Recently, Strong and Frey (58) demonstrated the chromatofocusing of model proteins under trace- and volume-overloaded feed conditions using mixtures of common buffers. The ability to carry out chromatofocusing with inexpensive, commonly employed buffers offers new opportunities for this high-resolution technology.

MODELS

Methods development for chromatographic bioprocessing typically involves an evaluation of the linear chromatographic behavior followed by empirical scale-up. Mathematical models are not widely employed for methods development and optimization of chromatography in the biopharmaceutical industry. Thus a significant amount of time is needed for methods development, as well as for the implementation of nonoptimal processes. As a result, there is increasing interest in the use of mathematical models for the development of efficient preparative ion exchange processes.

The classical view of protein retention in ion exchange columns involves the stoichiometric exchange of proteins and mobile-phase counterions (1,59–61). This is referred to as the stoichiometric displacement model (SDM). In this model, the retention of proteins on ion exchange stationary phases is described as follows (for a 1:1 salt, e.g., NaCl)

$$C_{\mathrm{pb}} + \nu Q_{\mathrm{m}} \Leftrightarrow Q_{\mathrm{p}} + \nu C_{\mathrm{m}} + \nu C_{\mathrm{b}} \tag{1}$$

where p is the protein molecule, ν is the ratio of the characteristic charge of the protein to the valence of the counterion, b is the co-ion (the ion with the same charge as the stationary phase) associated with the protein, m is the mobile-phase counterion (ion with the opposite charge of the stationary phase), C denotes the concentrations in the mobile phase, and Q denotes the concentrations on the stationary phase. The characteristic charge of the protein is a statistical average of the electrostatic interactions of the protein with the stationary phase as it travels down the column.

Some of the counterions remain bound to the protein on adsorption, while the rest make way for the interaction of the protein with the stationary phase. Drager and Regnier (60) argued that the equilibrium between the protein and the co-ion can be neglected. This results in the simplified equilibrium reaction

$$C_p + \nu Q_m \Leftrightarrow \nu C_m + Q_b \quad (2)$$

The SDM predicts that the retention of the protein under isocratic, linear conditions (low solute concentrations) is related to the counterion concentration as follows:

$$\log(k) = -\nu\log(C_m) + \text{const} \quad (3)$$

where k is the retention factor under isocratic linear conditions, and C_m is the concentration of the counterion in the mobile phase. This relationship has been experimentally demonstrated (60,62,63) under conditions usually employed in the IEC of proteins. Equation 3 forms the basis for the quantification of the effect of the ionic strength on the selectivity of linear separations.

Linear Chromatography

In linear chromatography, the various solutes migrate down the column independent of each other. The retention is characterized by a linear isotherm and an absence of intersolute competition for the adsorption sites. The plate model (64) is used extensively to characterize the efficiencies and resolutions obtained in these systems. In the plate model, all the nonidealities of the column are lumped into one parameter: N, the number of plates [or, alternatively, in height equivalent to one theoretical plate, HETP = (length of column)/N]. The HETP is widely used for the comparison of different sorbents and the validation of packing procedures. It is usually determined from the effluent profiles of isocratic pulse injections of a tracer. For a gaussian effluent peak, the HETP can be determined from the following equation

$$\text{HETP} = \left(\frac{w_h}{t_r}\right)^2 \frac{L}{5.545} \quad (4)$$

where w_h is the peak width at half-height, t_r is the retention time of the peak, and L is the length of the column. In ion exchange systems, the HETP measurement is affected by the salt in the background. It is recommended that pulse injections be made in a relatively high salt background to reduce Donnan exclusion (65).

The HETP is also a function of the flow rate. For conventional, macroporous sorbents, the plot of HETP versus flow rate (termed a Van Deemter or Knox plot) exhibits a minimum, indicating the presence of an optimum flow rate for operation. The HETP can be viewed as a sum of all the mass transfer, kinetic, and diffusional resistances in the column (66,67).

Yamamoto and coworkers have employed a continuous-flow plate model to characterize and scale up linear gradient and step gradient operations (68). Graphical techniques have been developed for scaling up linear gradient elution in linear chromatography. In this approach, the ionic strength at which the peak will elute in a large column is obtained directly from a plot of GH (gradient slope times the stationary phase volume) versus I_r (ionic strength at the elution time of the peak), which is established with the use of a relatively small column (69). The dimensionless retention time and the peak width in the large column can then be determined from analytical expressions (68). These methods have been applied to the scale-up of β-galactosidase purification in a 30-L column based on linear gradient data obtained with a 23-mL column (69).

Although the foregoing approach appears to work well for linear gradient scale-up of single solutes and multicomponent separations with baseline resolution, an alternative method (70) is required for general multicomponent linear separations. The resolution R_s is related to the operating conditions and the column dimensions as follows (70):

$$R_s \propto \left(\frac{L}{G\ \text{HETP}}\right)^{1/2} = \left(\frac{D_m I_a L}{GHud_p^2}\right)^{1/2} \quad (5)$$

where D_m is the molecular diffusivity, u is the mobile-phase velocity, d_p is the particle size, L is the length of the column, G is the gradient slope times the column dead volume, and I_a is a constant, introduced to dedimensionalize the right-hand side.

The HETP in equation 5 is the HETP in linear gradient elution. An algorithm to estimate the HETP from a set of linear gradient experiments at various flow rates has been developed (71). Thus, for scale-up,

$$\left(\frac{L}{G\ \text{HETP}}\right)_{\text{small column}} = \left(\frac{L}{G\ \text{HETP}}\right)_{\text{large column}} \quad (6)$$

While these approaches have proved useful for the scale-up of linear chromatography, the production rate in these ion exchange systems can be dramatically enhanced by operating these columns under nonlinear conditions. To model various modes of operation under nonlinear (high loading) conditions, an alternative approach must be employed.

Nonlinear Chromatography

To study the effects of nonlinear chromatography, one needs a suitable isotherm that accurately describes nonlinear protein adsorption. Guiochon et al. (72) have employed the Langmuir isotherm extensively to study the nonlinear chromatographic behavior of both small molecules and macromolecules. Berninger et al. (73) have developed the VERSE-LC model, which solves the complete mass transfer equations in conjunction with the Langmuir or modulator-modified Langmuir isotherms to describe nonlinear chromatography. While the Langmuir isotherm has many advantages, it is inadequate for describing protein adsorption in ion exchange systems. Protein adsorption is multipointed in nature (74) and is affected by the presence of mobile-phase modifiers (75–77). The Langmuir isotherm is unable to accurately describe these features.

The multivalent ion exchange model (75,78) incorporates both the multipointedness of the adsorption process and the effect of the mobile phase modifier on the process. It is derived from the SDM in conjunction with the law of mass action. Thus the multivalent ion exchange formalism predicts an isotherm of the form

$$Q_\mathrm{p} = K_\mathrm{eq} C_\mathrm{p} \left(\frac{\bar{Q}_1}{C_1} \right)^{\nu_\mathrm{p}} \quad (7)$$

where $\bar{Q}_1$ is the number of sites available for the adsorption of proteins, C_1 is the mobile-phase concentration of the counterions, K_eq is the equilibrium constant for the exchange process, and the subscript p denotes the protein.

The effect of the size of the protein molecules and lateral interactions on the adsorption process can be considerable (76,79–81). Furthermore, the adsorption of a protein molecule will result in the steric shielding of a number of stationary phase sites underneath the molecule. Brooks and Cramer (76) have defined a steric factor, σ, to account for such steric hindrance and have used the postulate that the steric hindrance will be proportional to the concentration of proteins on the surface (79). A steric mass action (SMA) model was developed that essentially employs equation 7 in concert with the following electroneutrality condition for single component equilibrium (76):

$$\Lambda = \bar{Q}_1 + (\nu_\mathrm{p} + \sigma_\mathrm{p}) Q_\mathrm{p} \quad (8)$$

The parameters for the SMA model (K_eq, ν, and σ) can be readily obtained using standard chromatographic techniques. The equilibrium constant K_eq and the characteristic charge (for a monovalent counterion) ν are obtained from either isocratic or linear gradient elution chromatography (62). The steric factor can be obtained from nonlinear frontal chromatography. Knowledge of the bed capacity and the protein parameters allows the generation of nonlinear isotherms at any salt concentration.

The SMA formalism can be extended to multicomponent mixtures. In this case, the electroneutrality condition becomes

$$\Lambda = \bar{Q}_1 + \sum_{i=2}^{N} (\nu_i + \sigma_i) K_{\mathrm{eq},i} \left(\frac{\bar{Q}_1}{C_1} \right)^{\nu_i} C_i \quad (9)$$

where N is the total number of components (including the mobile phase counterion) in the system. Equations 7 and 9 then describe multicomponent nonlinear protein equilibria at various salt concentrations.

The SMA model accurately predicts a variety of nonlinear ion exchange behavior in different modes of chromatography. Two distinct approaches have been employed to date. The first involves the development of graphical techniques based on the method of characteristics for the prediction of single solute peaks in nonlinear isocratic and gradient chromatography (77,82). The second approach has employed the SMA formalism in concert with appropriate mass transport equations to describe multicomponent separations. As seen in Figures 10 and 11, the SMA model can accurately predict separations in multistep gradient and linear gradient separations, respectively. In addition, the model can predict sample displacement effects, which can have a profound effect on linear gradient separations under high loading conditions (Fig. 11). This model has also been shown to accurately predict the behavior of complex displacement systems (83). Furthermore, the SMA formalism has been used along with iterative optimization techniques to develop optimal step (84) and linear gradient (85) separations. Finally, the SMA model has been employed to describe the dynamic affinity of low molecular weight displacers in displacement systems, resulting in a simple approach to methods development in displacement chromatography (86).

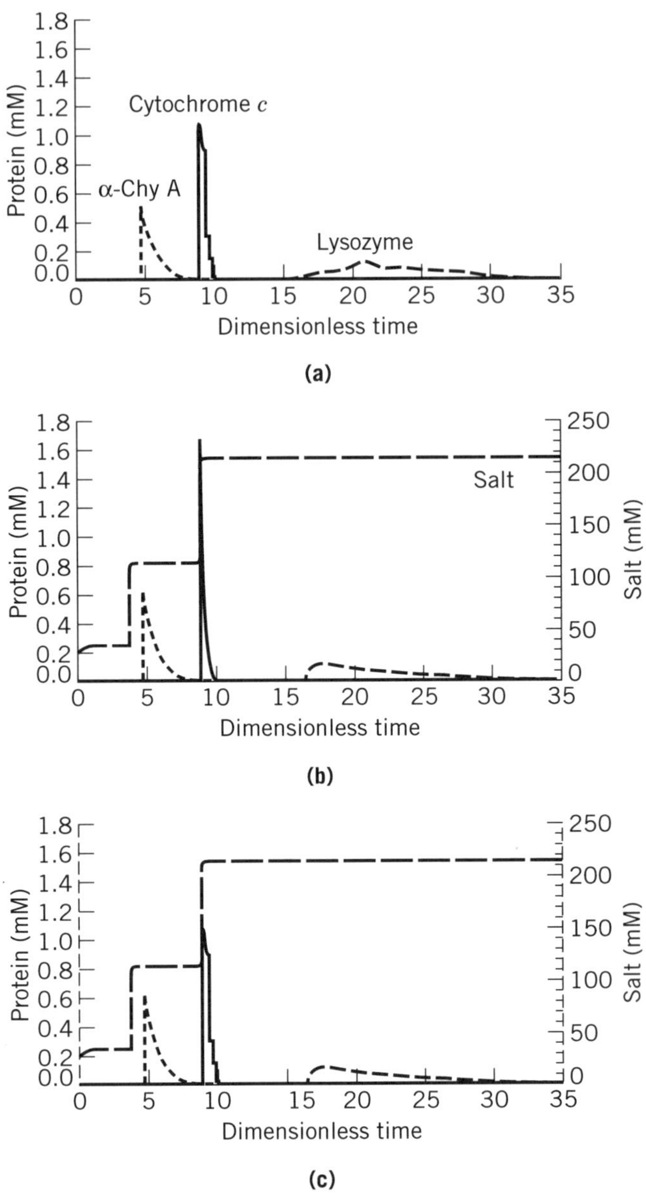

Figure 10. Comparison of step gradient chromatograms: (**a**) experimental, (**b**) simulated (using SMA), and (**c**) simulated (using SMA with fraction collection. Feed 2.8 column dead volumes of 0.19 mM α-chymotrypsinogen A, 0.19 mM cytochrome c, and 0.21 mM lysozyme in 30 mM sodium phosphate (pH 6). Gradient program: first step salt concentration, 112 mM, second step salt concentration, 213 mM; delay of 5.1 column dead volumes; flow rate, 0.4 mL/min; column, Waters SP (8 μm, 50 mm × 5 mm i.d.); fraction size, 147 μL (before 11 column dead volumes) and 413 μL (after 11 column dead volumes). *Source*: Reproduced with permission from Ref. 84.

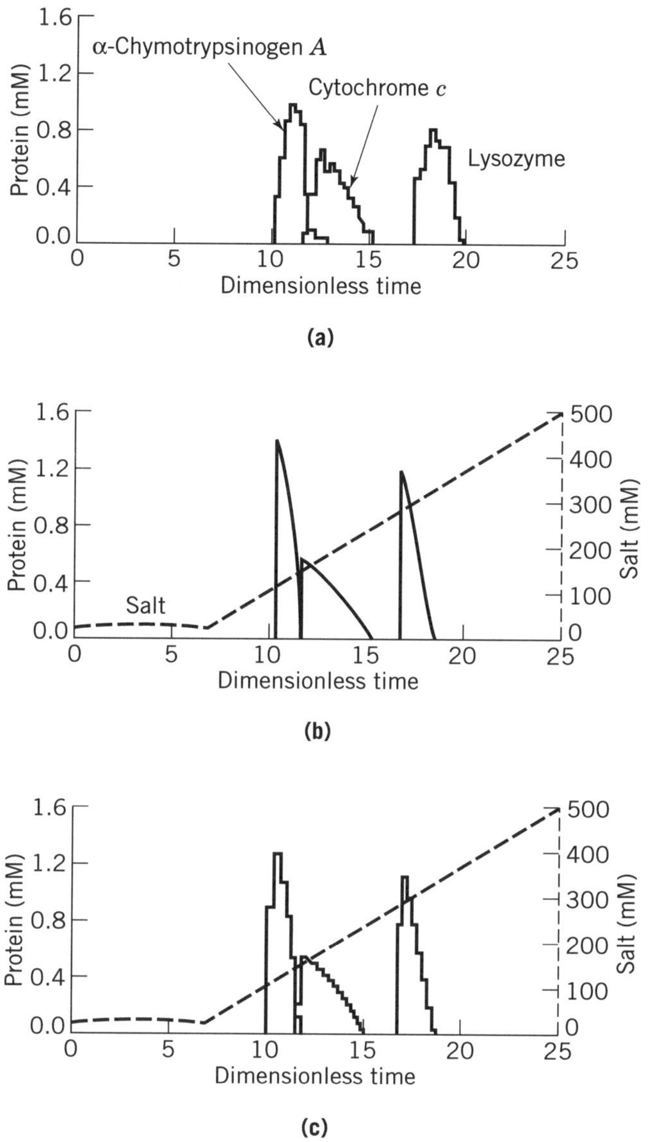

Figure 11. Comparison of linear gradient chromatograms: (**a**) experimental, (**b**) simulated (using SMA), and (**c**) simulated (using SMA with fraction collection). Feed, 6 column dead volumes of 0.2 mM α-chymotrypsinogen A, 0.2 mM cytochrome *c* and 0.2 mM lysozyme in 30 mM sodium phosphate (pH 6); gradient slope, 25 mM Na^+ per column dead volume; flow rate, 0.5 mL/min; column: strong cation exchanger (15 μm, 54 mm × 5 mm i.d.); fraction size, 200 μL. *Source*: Reproduced with permission from Ref. 85.

Nonstoichiometric Models

Though the SMA model is capable of accurately describing the behavior of nonlinear ion exchange chromatographic systems, it assumes that the individual charges on the protein molecule interact with the discrete charges on the ion exchange surface. In reality, the retention in ion exchange is more complex and is primarily due to the interaction of the *electrostatic fields* of the protein molecules and the chromatographic surface. Haggerty and Lenhoff (87) demonstrated a correlation between the retention of various proteins with their surface potential. Stahlberg and coworkers have developed the electrostatic interaction model (EIM), which treats the retention process as a coulombic interaction between two charged flat plates (88,89). Roth and Lenhoff (90) employ van der Waals forces in concert with electrostatic interactions to quantify the adsorption of proteins onto charged stationary phases. Recently, a mechanistic model has been developed that incorporates stationary-phase properties such as the surface charge density and short-range interaction energies to predict retention on various stationary phases (91). The drawback with the nonstoichiometric approaches reported thus far is that they are limited to describing retention in the linear mode of chromatography.

CONCLUSIONS

The last decade has seen dramatic advances in the state of the art of chromatographic bioprocessing with ion exchange materials. This article has presented recent advances in resin materials, modes of operation, and mathematical modeling. The need to process large volumes of relatively dilute feedstock has catalyzed the development of an array of "novel" chromatographic materials. While many novel materials have been developed, the effect of stationary-phase chemistry on selectivity is still essentially unknown, and thus the selection of an ion exchange resin requires significant column screening. Once a resin with sufficient selectivity has been identified, the task remains to develop an efficient, high-throughput chromatographic process. This requires the selection of an appropriate mode of operation. In fact, the selection of the optimal mode of operation for a given bioseparation problem can have a profound effect on the production rate, yield, and purity for that ion exchange step. Finally, the development of mathematical models capable of predicting scale-up and nonlinear protein chromatography now make it possible to employ modeling for bioprocess methods development.

NOMENCLATURE

C	mobile phase concentration (mM)
d_p	Particle size (cm)
D_m	Molecular diffusivity (cm^2/s)
G	Gradient slope times column dead volume (mM)
H	Column phase ratio (dimensionless)
I_r	Ionic strength at peak elution
k	Retention factor (dimensionless)
K_{eq}	Equilibrium constant (dimensionless)
L	Length of column (cm)
Q	Stationary phase concentration (mM)
$\bar{Q}_1$	Concentration of sites available for protein adsorption (mM)
R_s	Resolution (dimensionless)
t_r	Retention time (min)
u	linear velocity (cm/min)
w_h	Peak width at half-height (min)

Subscripts

i solute index ($i = 1$ for salt counterion)
p Protein
m mobile-phase counterion

Greek

β Phase ratio (dimensionless)
ν Characteristic charge (dimensionless)
σ Steric factor (dimensionless)
Λ Bed capacity (mM)

BIBLIOGRAPHY

1. W. Kopaciewicz, M.A. Rounds, J. Fausnaugh, and F.E. Regnier, *J. Chromatogr.* **266**, 3–21 (1983).
2. Pharmacia, *FPLC Ion Exchange and Chromatofocusing*, Pharmacia, Uppsala, Sweden, 1985.
3. P. Gagnon, B. Godfrey, and D. Ladd, *J. Chromatogr. A* **743**, 51–55 (1996).
4. Dj. Josic, W. Hofmann, and W. Reutter, *J. Chromatogr.* **371**, 43–54 (1986).
5. F. Fang, M-I. Aguilar, and M.T.W. Hearn, *J. Chromatogr. A* **729**, 49–66 (1996).
6. W.F. Prouty, in Cs. Horvath and L.S. Ettre eds. *Chromatography in Biotechnology*, American Chemical Society, Washington D.C., 1993, pp. 43–58.
7. R. Henry and M.V. Piserchio, in C.T. Mant and R.S. Hodges eds., *High-Performance Liquid Chromatography of Peptides and Proteins: Separation, Analysis, and Conformation*, CRC Press, London, 1991, pp. 537–549.
8. E. Boschetti, *J. Chromatogr. A* **658**, 207–236 (1994).
9. E. Karlsson, L. Ryden, and J. Brewer, in J.-C. Janson and L. Ryden eds. *Protein Purification Principles, High Resolution Methods, and Applications*, VCH, New York, 1989, pp. 107–148.
10. H.A. Sober and E.A. Peterson, *J. Am. Chem. Soc.* **76**, 1711–1712 (1954).
11. S.H. Chang, R. Noel, and F.E. Regnier, *Anal. Chem.* **48**, 1839–1845 (1976).
12. A.J. Alpert and F.E. Regnier, *J. Chromatogr.* 185, 375–392 (1979).
13. P.W. Carr, J.A. Blackwell, T.P. Weber, W.A. Schafer, and M.P. Rigney, in Cs. Horvath and L.S. Ettre eds., *Chromatography in Biotechnology*, American Chemical Society, Washington D.C., 1993, pp. 146–164.
14. A.M. Clausen and P.W. Carr, *Anal. Chem.* **70**, 378–385 (1998).
15. M.C. Flickinger and P.W. Carr, presentation at Recovery of Biological Products VII, San Diego, CA 1994.
16. M.A. Rounds, W.D. Rounds, and F.E. Regnier, *J. Chromatogr.* **397**, 25–38 (1987).
17. C.K. Ratnayeke and F.E. Regnier, *J. Chromatogr. A* **743**, 15–23 (1996).
18. J.A. Gerstner, T. Londo, T. Hunt, J. Morris, P. Pedroso, and R. Hamilton, *CHEMTECH* **25**, 27–32 (1995).
19. N.B. Afeyan, N.F. Gordon, I. Mazsaroff, L. Varady, S.P. Fulton, Y.B. Yang, and F.E. Regnier, *J. Chromatogr.* **519**, 1–29 (1990).
20. N.B. Afeyan, S.P. Fulton, and F.E. Regnier, *J. Chromatogr.* **544**, 267–279 (1991).
21. J.F. Pfeiffer, J.C. Chen, and J.T. Hsu, *AIChE J.* **42**, 932–939 (1996).
22. J.A. Gerstner, J. Morris, T. Hunt, R. Hamilton, and N.B. Afeyan, *J. Chromatogr. A* **695**, 195–204 (1995).
23. P-E. Gustavsson and P-O. Larsson, *J. Chromatogr. A* **734**, 231–240 (1996).
24. M.A. Fernandez and G. Carta, *J. Chromatogr. A* **746**, 169–183 (1996).
25. M.A. Fernandez and G. Carta, *J. Chromatogr. A* **746**, 185–198 (1996).
26. J. Coffman, presentation at ISPPP 96, Boston, Mass., 1996.
27. W. Muller, *J. Chromatogr.* **510**, 133–140 (1990).
28. J. Cacia, C.P. Quan, M. Vasser, M.B. Sliwkowski, and J. Frenz, *J. Chromatogr.* **634**, 229–239 (1993).
29. Y.K. Chang and H.A. Chase, in D.L. Pyle ed. *Separations for Biotechnology*, Vol. 3, Royal Society of Chemistry, Cambridge, 1994, pp. 106–112.
30. G.R. Gilchrist, M.T. Burns, and A. Lyddiatt, in D.L. Pyle ed. *Separations for Biotechnology*, Vol. 3, Royal Society of Chemistry, Cambridge, 1994, pp. 186–192.
31. J. Thommes, M. Halfar, S. Lenz, and M-R. Kula, *Biotechnol. Bioeng.* **45**, 205–211 (1995).
32. D. Lutkemeyer, M. Bretschneider, H. Buntemeyer, and J. Lehmann, *J. Chromatogr.* 639, 57–66, (1993).
33. J. Thommes and M.-R. Kula, *Biotechnol. Prog.* **11**, 357–367 (1995).
34. A. Jungbauer and E. Boschetti, *J. Chromatogr. B* **662**, 143–179 (1994).
35. G.K. Sofer and L.-E. Nystrom, *Process Chromatography: A Guide to Validation*, Academic Press, London, 1991, pp. 5–30.
36. A.-K. Barnfield Frej, R. Hjorth, and A. Hammarstrom, *Biotechnol. Bioeng.* **44**, 922–929 (1994).
37. P.R. Levison, C. Mumford, M. Streater, A. Brandt-Nielsen, N.D. Pathirana, and S.E. Badger, *J. Chromatogr. A* **760**, 151–158 (1997).
38. A. Tsai, M. Gallo, T. Petterson, and J. Shiloach, *Bioprocess Eng.* **12**, 115–118, (1995).
39. R.J. Hinton, R.F. Sherwood, and J.R. Ramsay, in D.L. Pyle ed. *Separations for Biotechnology*, vol. 3, Royal Society of Chemistry, Cambridge, 1994, pp. 207–213.
40. Dj. Josic, H. Schwinn, M. Stadler, and A. Strancar, *J. Chromatogr. B* **662**, 181–190 (1994).
41. A. Tiselius, *Ark. Kemi Mineral. Geol.* **16A**, 1 (1943).
42. J.A. Gerstner, P. Pedroso, J. Morris, and B.J. Bergot, *Nucleic Acids Res.* **23**, 2292–2299 (1995).
43. U.S. Pat. 5,478,924 (December 26, 1995), S.M. Cramer, J.A. Moore, A. Kundu, Y. Li, and G. Jayaraman (to S.M. Cramer).
44. G. Jayaraman, Y.F. Li, J.A. Moore, and S.M. Cramer, *J. Chromatogr. A* **702**, 143–155 (1995).
45. A. Kundu, S. Vunnum, G. Jayaraman, and S.M. Cramer, *Biotechnol. Bioeng.* **48**, 452–460 (1995).
46. A. Kundu, S. Vunnum, and S.M. Cramer, *J. Chromatogr. A* **707**, 57–67 (1995).
47. S.D. Gadam and S.M. Cramer, *Chromatographia* **39**, 409–418 (1994).
48. J.A. Gerstner and S.M. Cramer, *Biopharm* **5**, 42–45 (1992).
49. S.C.D. Jen and N.G. Pinto, *J. Chromatogr.* 519, 87–98, (1990).
50. A. Kundu and S.M. Cramer, *Anal. Biochem.* **248**, 111–116 (1997).
51. A. Kundu, K.A. Barnthouse, and S.M. Cramer, *Biotechnol. Bioeng.* **56**, 119–129 (1997).

52. K.A. Barnthouse, W. Trompeter, R. Jones, P. Inampudi, R. Rupp, and S.M. Cramer, *J. Biotechnol.* (in press).
53. L.A. Sluyterman and O. Elgersma, *J. Chromatogr.* **150**, 17 (1978).
54. L.A. Sluyterman and J. Wijdenes, *J. Chromatogr.* **150**, 31 (1978).
55. J.H. Scott, K.L. Kelner, and H.P. Pollard, *Anal. Biochem.* **149**, 163 (1985).
56. M.T.W. Hearn and D.J. Lyttle, *J. Chromatogr.* **218**, 483–495 (1981).
57. D.D. Frey, A. Barnes, and J. Strong, *AIChE J.* **41**, 1171–1183 (1995).
58. J.C. Strong and D.D. Frey, *J. Chromatogr. A* **769**, 129–143 (1997).
59. N.K. Boardman and S.M. Partridge, *Biochem. J.* **59**, 543–552 (1955).
60. R.R. Drager and F.E. Regnier, *J. Chromatogr.* **359**, 147–155 (1986).
61. M.A. Rounds and F.E. Regnier, *J. Chromatogr.* **283**, 37–45 (1984).
62. S.D. Gadam, G. Jayaraman, and S.M. Cramer, *J. Chromatogr.* **630**, 37–52 (1993).
63. M.T.W. Hearn, in C.T. Mant and R.S. Hodges eds., *High-Performance Liquid Chromatography of Peptides and Proteins: Separation, Analysis, and Conformation*, CRC Press, London, 1991, pp. 105–122.
64. A.J.P. Martin and R.L.M. Synge, *Biochem. J.* **35**, 1358–1368 (1941).
65. H.P. Lettner, O. Kaltenbrunner, and A. Jungbauer, *J. Chromatogr. Sci.* **33**, 451–457 (1995).
66. J.J. van Deemter, F.J. Zuiderweg, and A. Klinkenberg, *Chem. Eng. Sci.* **5**, 271–289 (1956).
67. Cs. Horvath and H.J. Lin, *J. Chromatogr.* **149**, 43–70 (1978).
68. S. Yamamoto, K. Nakanishi, R. Matsuno, and T. Kamikubo, *Biotechnol. Bioeng.* **25**, 1465–1483 (1983).
69. S. Yamamoto, M. Nomura, and Y. Sano, *AIChE J.* **33**, 1426–1434 (1987).
70. S. Yamamoto, M. Nomura, and Y. Sano, *J. Chromatogr.* **409**, 101–110 (1987).
71. S. Yamamoto, *Biotechnol. Bioeng.* **48**, 444–451 (1995).
72. G. Guiochon, S.G. Shirazi, and A.M. Katti, *Fundamentals of Preparative and Nonlinear Chromatography*, Academic, New York, 1994, pp. 245–286.
73. J.A. Berninger, R.D. Whitley, X. Zhang and N.-H.L. Wang, *Comput. Chem. Eng.* **15**, 749–768 (1991).
74. H.P. Jennissen, *Biochemistry* **15**, 5683–5692 (1976).
75. A. Velayudhan and Cs. Horvath, *J. Chromatogr.* **443**, 13–29 (1988).
76. C.A. Brooks and S.M. Cramer, *AIChE J.* **38**, 1969–1978 (1992).
77. S.R. Gallant, A. Kundu, and S.M. Cramer, *J. Chromatogr. A* **702**, 125–142 (1995).
78. P. Cysewski, A. Jaulmes, R. Lemque, B. Sebille, C. Vidal-Madjar, and G. Jilge, *J. Chromatogr.* **548**, 61–79 (1991).
79. A. Velayudhan, *Studies in Nonlinear Chromatography*, Ph.D. dissertation, Yale University, New Haven, Conn., 1990.
80. Y.-L. Li, and N.G. Pinto, *J. Chromatogr. A* **658**, 445–457 (1994).
81. R.D. Whitley, R. Wachter, F. Liu, and N.-H.L. Wang, *J. Chromatogr.* **465**, 137–156 (1989).
82. S.R. Gallant, S. Vunnum, and S.M. Cramer, *AIChE J.* **42**, 2511–2520 (1996).
83. S.D. Gadam, S.R. Gallant, and S.M. Cramer, *AIChE J.* **41**, 1676–1686 (1995).
84. S.R. Gallant, A. Kundu, and S.M. Cramer, *Biotechnol. Bioeng.* **47**, 355–372 (1995).
85. S.R. Gallant, S. Vunnum, and S.M. Cramer, *J. Chromatogr. A* **725**, 295–314 (1996).
86. C.A. Brooks and S.M. Cramer, *Chem. Eng. Sci.* **51**, 3847–3860 (1996).
87. L. Haggerty and A.M. Lenhoff, *J. Phys. Chem.* **95**, 1472–1477 (1991).
88. J. Stahlberg, B. Jonsson, and Cs. Horvath, *Anal. Chem.* **63**, 1867–1874 (1991).
89. J. Stahlberg, B. Jonsson, and Cs. Horvath, *Anal. Chem.* **64**, 3118–3124 (1992).
90. C. M. Roth and A.M. Lenhoff, *Langmuir* **9**, 962–972 (1993).
91. C.M. Roth, K.K. Unger, and A.M. Lenhoff, *J. Chromatogr. A* **726**, 45–56 (1996).

See also ADSORBENTS, INORGANIC; ADSORPTION, EXPANDED BED; ADSORPTION, PROTEIN, BATCH; ADSORPTION, PROTEINS WITH SYNTHETIC MATERIALS; CHROMATOGRAPHY, SIZE EXCLUSION; MEMBRANE CHROMATOGRAPHY.

CHROMATOGRAPHY, RADIAL FLOW

TINGYUE GU
Ohio University
Athens, Ohio

KEY WORDS

Chromatography
Protein
Purification
Radial
Scale-up
Separation

OUTLINE

INTRODUCTION

The production of a modern biotechnology product, typically a recombinant protein, requires a multistage downstream process. Such a process usually centers on two or more liquid chromatography steps. As production scales escalate, the columns used become larger and more expensive. It is not uncommon to have bed volumes reaching hundreds of liters at industrial scales.

Radial flow columns were first used for gas–solid catalytic reactions in large packed beds. They were designed to increase gas flow rate and reduce pressure by increasing the cross-sectional flow area. Radial flow chromatography (RFC) columns first entered the commercial biotechnology market in the mid-1980s (1). RFC was intended as an alternative to the conventional axial flow chromatography (AFC) for preparative- and large-scale applications. In a radial flow chromatography column (Fig. 1), the mobile phase flows in the radial direction, not in the axial direction. The mobile enters from the outside tube and merges into the center tube (Fig. 2). In comparison to a slim AFC column, an RFC column provides a relatively larger flow area and a shorter flow path. It allows a higher volumetric flow rate with a lower bed pressure. The effect is equivalent to using a short pancakelike AFC column (Fig. 3). Pancakelike AFC columns are quite common in industrial applications. They are available from most major commercial column vendors. In scale-up to accommodate large feed loads, it is impractical to increase the column height of AFC columns by too much because excessive bed pressure drop will result. This is especially troublesome when soft gels are used. Thus, pancakelike AFC columns are used.

This article is devoted to the discussion of the applications of RFC columns in bioseparations and RFC modeling and scale-up issues. Applications examples will be provided based on existing literature. A general rate model will be presented for the modeling of RFC. Experimental and theoretical comparisons between AFC and RFC will be discussed.

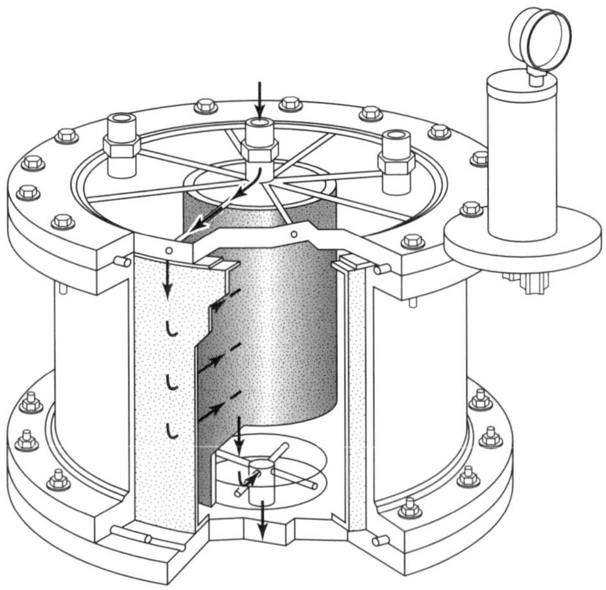

Figure 1. Diagram of an acrylic Superflo® column. *Source*: Courtesy of Sepragen Corp.

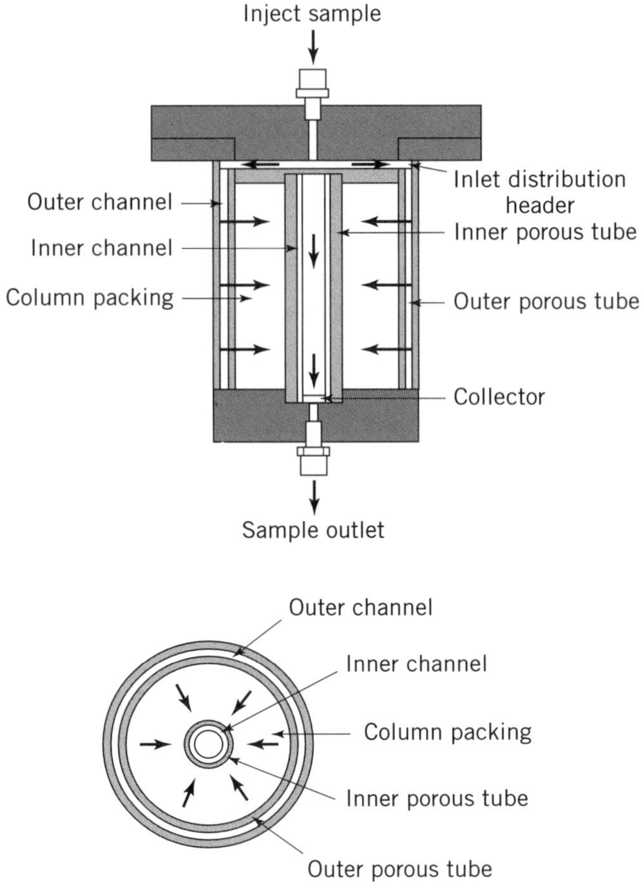

Figure 2. Anatomy of a Superflo® column. *Source*: Courtesy of Sepragen Corp.

RADIAL FLOW COLUMN CONFIGURATIONS

In 1947, Hopf (2) described a radial chromatography device. The device had a feed pipe in the center. The outward liquid flow in the radial direction was forced by the centrifugal force when the device was rotated. He called this device a chromatofuge. Such a device was obviously too complex and expensive for large-scale industrial application, and thus it was not adopted in biotechnology.

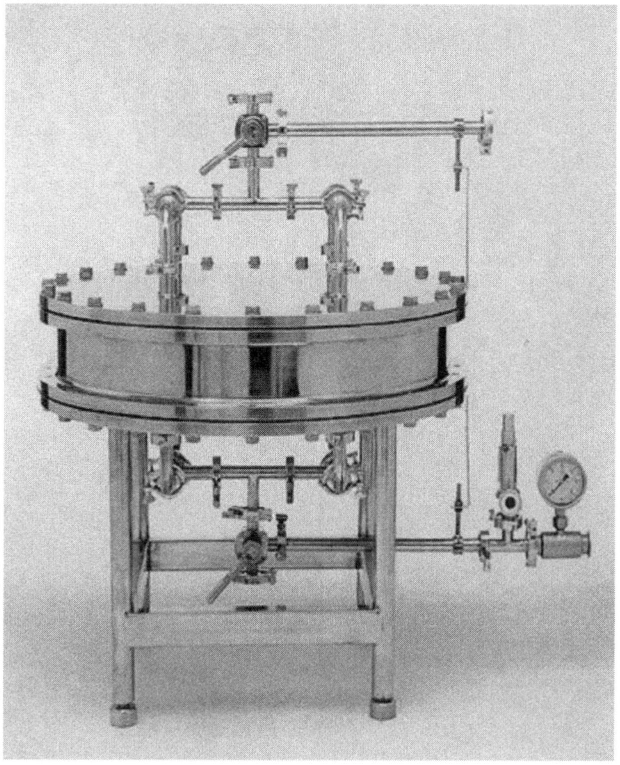

Figure 3. A large-scale pancakelike axial flow column. Courtesy of Pharmacia Biotech, Inc.

To date, two commercial companies have marketed RFC columns. The first one is CUNO, Inc., in Connecticut. They marketed the Zetaffinity series of preparative-scale radial flow cartridges, which looked like a spiral-wound filtration cartridge (Fig. 4). Such a design obviously was rooted in the fact the CUNO is a major manufacturer of industrial filtration systems. The packing for CUNO's RFC cartridges was based on modified cellulose (3). These cartridges were used for affinity chromatography, hence the trade name Zetaffinity. CUNO discontinued its Zetaffinity product line in 1991.

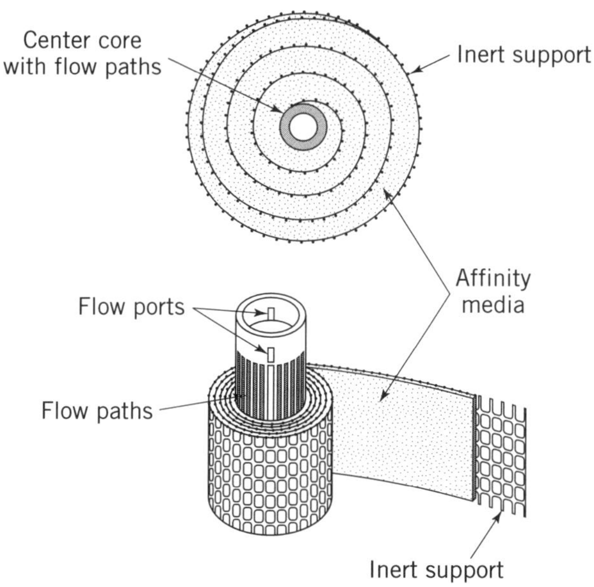

Figure 4. Structure of a Zetaffinity® cartridge.

The second vendor and manufacturer of RFC columns is Sepragen Corporation in Hayward, California. Their RFC product line carries the trade name Superflo®. Sepragen markets unpacked RFC columns ranging from 50 mL to 200 L. Figure 5 shows several Superflo columns with a similar diameter but different column heights. Superflo columns come in stainless steel, acrylic, polycarbonate, and polyethylene. They can sustain a pH range of 2 to 12 and a maximum pressure of 50 psi. They are used by major biopharmaceutical companies at production scales. Sepragen's president owns a U.S. patent (4) on the design of unpacked RFC columns. Figure 1 shows an acrylic RFC column with inward radial flow. The feedstream enters from a center input port at the top. It is then distributed through several flow channels to the outer shell and subsequently enters the packing media in the radial direction toward a center collection tube. The effluent exits the tube through an outlet port at the bottom. The two ports next to the center input port at the top in Figure 1 are packing ports used during media packing. A bubble trap with a pressure gauge sits on the edge of the column's top lid. Figure 2 is an anatomical view of a Superflo column. Superflo columns usually use inward flow instead of outward flow because it is more difficult to distribute outward flow without increasing flow distortion caused by gravity. The other reason for not using outward flow is because inward flow provides slightly sharper peaks than outward flow based on computer simulation (5). Outward flow is usually used for the packing or regeneration of a RFC column.

PACKING PROCEDURES FOR RFC COLUMNS

Because of their structure, CUNO's Zetaffinity cartridges do not require special packing procedures. Sepragen's Superflo columns are packed by first displacing air using outward flow with a buffer solution. Packing slurry is then pumped into the column through the two packing ports on the column top surface (Fig. 1). Excess buffer is squeezed out of the column through the input port on the column top surface. After that, the column is ready for use.

PRESSURE DROPS OF RFC COLUMNS

Because of a short flow path and a low pressure drop, both CUNO's and Sepragen's RFC devices exhibit a highly linear relationship between bed pressure drop versus flow rate. Figure 6 shows a linear relationship between bed pressure and flow rate for an 800-mL Zetaffinity cartridge studied by Huang et al. (3).

Figure 7 shows the pressure drops for a Superflo 20 L column and a Superflo 50 L column packed with several different soft-gel chromatographic media (6). It indicates that the pressure drops are quite low at relatively high flow rates. At lower flow rates, the pressure drop curve is linear. This behavior is similar to that of short pancakelike AFC columns.

Figure 5. Superflo columns. *Source*: Courtesy of Sepragen Corp.

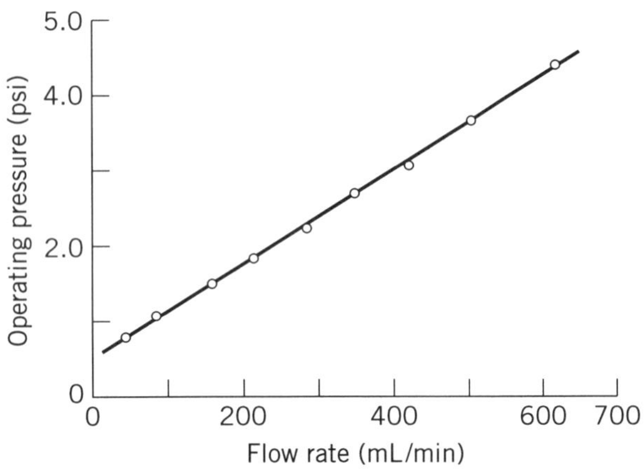

Figure 6. Pressure drop vs. flow rate for an 800-mL Zetaffinity cartridge.

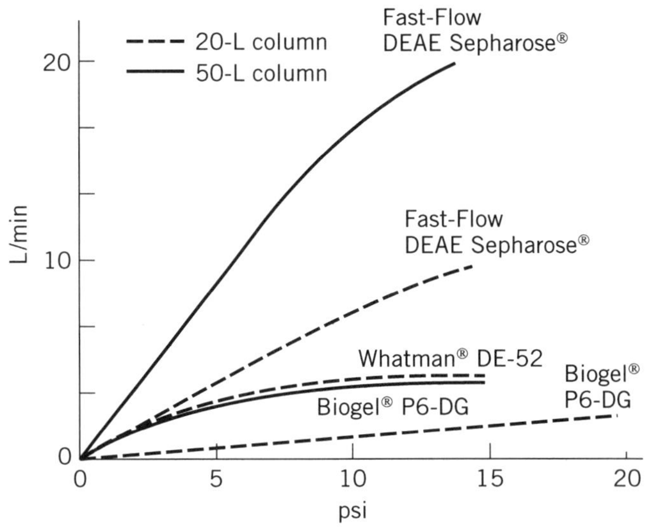

Figure 7. Pressure drop vs. flow rate for 20-L and 50-L Superflo® columns. *Source*: Courtesy of Sepragen Corp.

COMPARISON OF RADIAL FLOW AND AXIAL FLOW COLUMNS

Saxena and Weil (7) did an experimental case study of the comparison of RFC and AFC. They used a 100-mL axial flow glass column (Econocolumn®) with 2.5-cm i.d. from BioRad Laboratories (Richmond, Calif.) and a Superflo-100 RFC column from Sepragen with a bed volume of 100 mL. Both columns were packed with QAE cellulose with the following procedures. The Superflo column was packed using a 25% slurry (with 0.5 M NaCl) that was pumped in through the two packing ports at a flow rate of 30 mL/min. The final QAE cellulose density after packing was 6 mL per dry gram. The Superflo column was packed in about 20 min. The AFC column was packed using 50% slurry added from the top of the column. After the liquid was drained and the bed was settled, additional slurry was added. This process was repeated until the same amount of QAE cellulose was packed into the AFC column. The two columns were used to separate two 10-mL ascites fluid samples. The samples were pretreated by dialyzing the ascites fluid with 10 mM phosphate buffer of pH 8.0 for 2 days with three changes. Precipitates and debris were then removed by centrifugation. Stepwise salt gradients were used. Figure 8 shows the comparison between the 100-mL

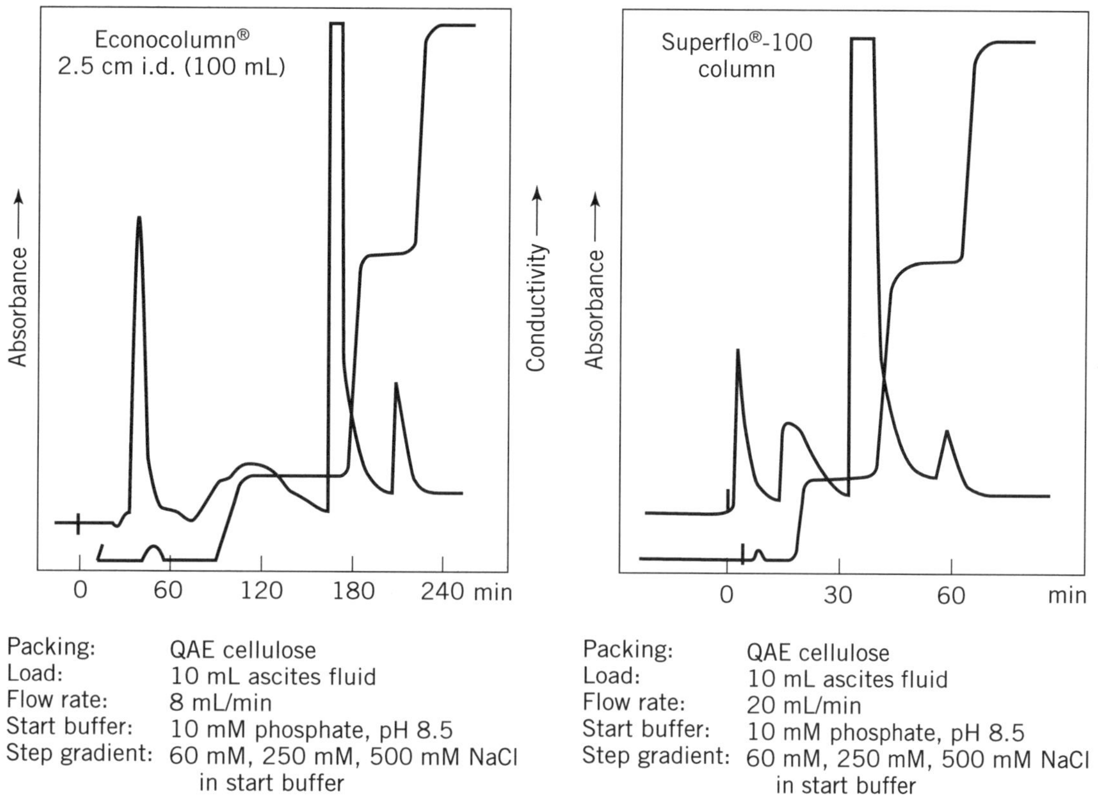

Figure 8. Comparison of AFC and RFC columns. *Source*: Courtesy of Sepragen Corp.

AFC column and the 100-mL Superflo-100 RFC column. From Figure 8, it can seen that the RFC column achieved similar separation results with much less time. In this case, the two columns were equivalent only in bed volume. A stricter comparison should use a pancakelike, short AFC column with its bed height about the same as the packing's thickness in radial direction for the RFC column, and both columns should have equal bed volume. The case study discussed next is close to satisfying these conditions.

Tharakan and Belizaire (8,9) used a 50-mL RFC column with a bed height of 0.95 cm and packing thickness of 3.0 cm in the radial direction and a 50-mL AFC column with a bed height of 2.8 cm. Both columns were packed with Sepharose CL2B resin-containing monoclonal antibody from Pharmacia (Piscataway, N.J.). They studied the purification of a protein called factor IX. Their experimental results indicated that the two columns gave similar purification results. This was expected because the AFC column was pancakelike. Tharakan and Belizaire (8) also packed the same column with S-200 Sephacryl size-exclusion gel from Sigma Chemical Co. (St. Louis). The protein band was more diffused using RFC than AFC.

Lane et al. (10) compared the performance of a Superflo-100 RFC column and a 6.6 × 4.4 cm i.d. AFC column for the separation of egg-white proteins. Both columns had a nominal bed volume of 100 mL. They were packed with the same media for comparison. Two anion-exchange cellulose media were tested, Whatman DE52 and QA52 from Whatman Specialty Products Division (Maidstone, U.K.). Egg whites were first separated from fresh hen eggs and then treated with a buffer. After treatment with a cell debris remover, egg-white suspensions were filtered using filter paper. The samples for chromatography had a protein concentration of 14 mg/mL. Sample load volume was 40 mL. After sample loading, the column was washed with a buffer. Elution was carried out using a linear gradient of 0 to 0.5 M NaCl in 0.025 M tris/HCl buffer at pH 7.5. Various flow rates were tested, ranging from 5 to 50 mL/min for the AFC column and 5 to 150 mL/min for the RFC column. Figures 9 and 10 represent typical results obtained by Lane et al. (10). They indicate that the AFC column gave slightly sharper peaks and faster elution times for both DE52 and QA52 media at a flow rate of 25 mL/min.

PROS AND CONS OF RADIAL FLOW COLUMNS

RFC columns provide a short flow path and a large cross-sectional area. This has the same effect as short pancakelike AFC columns. However, RFC columns occupy considerably less floor space. Both RFC and pancakelike AFC columns face flow distribution problems. According to Sepragen, its Superflo columns have better flow distribution than typical large pancakelike columns.

Compared to long AFC columns, RFC columns produce smaller pressure drops, thus enabling larger volumetric flow rates. If soft gels are used as separation media, the low pressure drop of RFC columns helps prevent bed compression (11,12). RFC is especially suitable for affinity chromatography using soft gels. In affinity chromato-

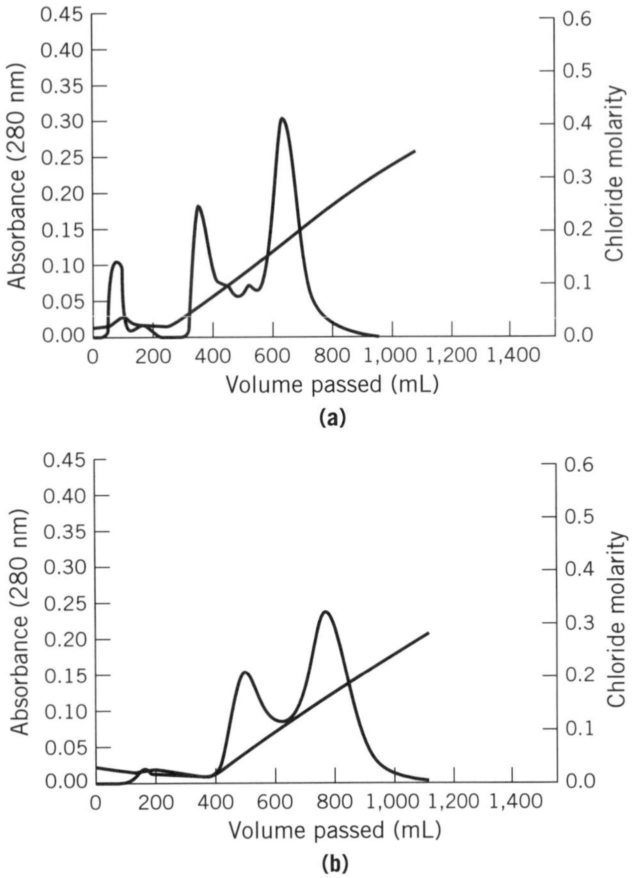

Figure 9. Comparison between a 100-mL AFC column (a) and a 100-mL RFC column (b) packed with DE52 in the separation of egg-white proteins.

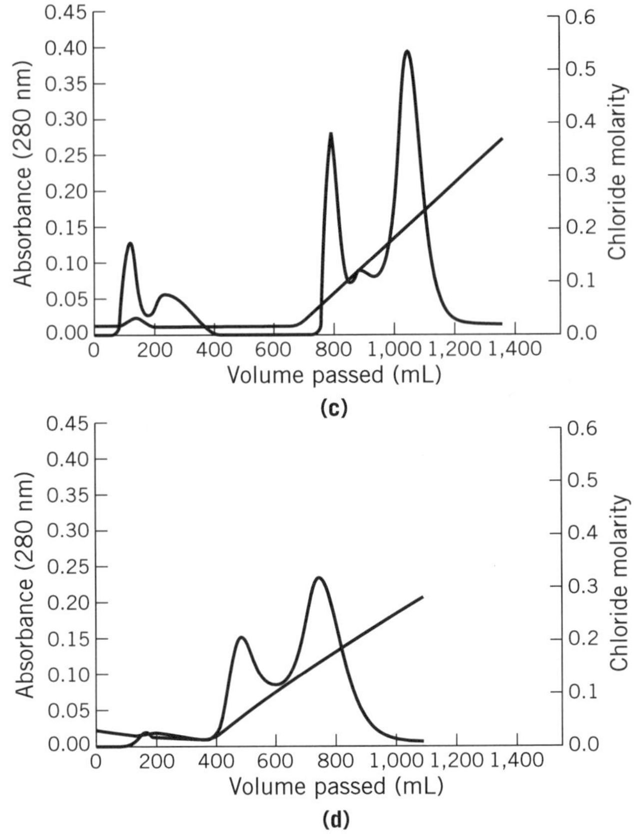

Figure 10. Same as Figure 9, but the columns were packed with QA52.

graphic operations, dilute feeds are typically used. Because of the extremely high affinity between the product and the gel matrix, very high flow rates are permitted without sacrificing column resolution. RFC is also a good choice for strong reversed phase, hydrophobic interaction and ion-exchange media.

Scale-up of RFC columns are relatively more straightforward because it is usually done by increasing the column height. To a certain extent, this does not seems to increase flow distortion problems in practice. If an AFC column is scaled up by increasing diameter, the flow distribution behaves quite differently from a small AFC column for which plug flow is easy to achieve. Thus, predicting actual performance of a large pancakelike AFC column from a small one is difficult.

The disadvantage of the RFC column is primarily its limited resolution caused by a short flow path. If the flow path is increased to a large extent, there will be flow distribution problems in the radial direction because of gravity. High resolution for more demanding separations can only come from using AFC columns with sufficiently large column length. This is precisely the reason why RFC has no use in analytical HPLC. RFC is not suitable for separations in which solute-stationary interactions are weak. For example, RFC is not a suitable choice for size-exclusion chromatography (SEC), because SEC depends strongly on the length of flow path for its resolution. RFC's short flow length cannot meet the demand. It is also expected that RFC is not suitable for other chromatography with weak solute-stationary phase interactions.

If mechanically strong packing materials, such as silica-based particles, are used, a longer column flow path can be used because the bed can sustain a much higher pressure. If a relatively high resolution is desired, RFC would be at a disadvantage compared to AFC since RFC is limited by its short flow path.

APPLICATION EXAMPLES

Applications Using Zetaffinity Cartridges

Huang et al. (3) studied several Zetaffinity cartridges from CUNO that contained modified cellulose-based affinity media. Table 1 lists the dimensions of three Zetaffinity cartridges tested by Huang et al. Figure 11 shows a chromatogram for the removal of proteases from human plasma using an 800-mL Zetaffinity cartridge with a flow rate of 100 mL/min (3). The cartridge contained modified cellulose with para-aminobenzamidine (PAB) as ligand. A 70% efficiency of protease removal was achieved in a single pass. The treated plasma was expected to have a three-fold increase in stability. The same type of 800-mL Zetaf-

Table 1. Dimensions of CUNO Radial Flow Cartridges Tested by Huang et al. (3)

Parameter	Small	Medium	Large
Nominal size	250 mL	800 mL	3200 mL
Bed volume	210 mL	810 mL	3020 mL
Outer diameter	7.0 cm	12.7 cm	12.7 cm
Inner diameter	0.6 cm	0.9 cm	0.9 cm
Height	6.4 cm	6.4 cm	23.8 cm

finity cartridge loaded with 1,260 mg PAB ligand was also used to purify crude trypsin purchased from Sigma Chemical. The data in Figure 12 were obtained by Huang et al. (3) using a flow rate of 295 mL/min.

Planques et al. (13) used a 250-mL Zetaffinity cartridge from CUNO packed with modified cellulose as chromatography media. They first chemically treated the cartridge to couple the media with L-lysine. This affinity chromatography media was able to bind with a human plasminogen protein. After centrifugation and microfiltration, human plasma was diluted with a buffer and then applied to the cartridge at a flow rate of 20 mL/min. After washing and elution, a 85% recovery yield and an increase of 110-fold in specific activity were achieved.

Applications Examples Using Sepragen's Superflo Columns

Akoum et al. (14) used a Superflo-400 RFC column packed with histidyl-Sepharose gel for the purification of myxalin, a glycopeptide with anticoagulant property. The feed for the column was obtained from a fermentation broth with the microorganism *Myxococcus xanthus*. Before the feed was applied to the column, it was clarified and concentrated using centrifugation, microfiltration, and reverse osmosis. A relative short processing time was achieved.

Strætkvern et al. (15) used a 60-mL, 2.2-cm i.d. AFC column, a Superflo-250 column (250 mL in bed volume), and a 2,500-mL AFC column for the separation of DNase from the extracts of cod pyloric cacca. The packing medium was Q-Sepharose Fast Flow anion exchange gel from Pharmacia. Column dimensions and operating conditions are listed in Table 2. Table 3 is a summary of their experimental results. Their results indicate that the RFC column required much less time and achieved higher productivity. The superior performance in this case should not be inter-

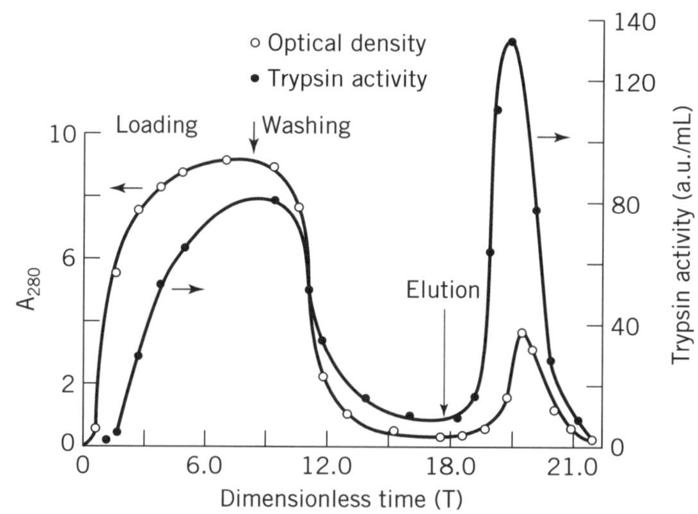

Figure 11. Protease removal from human plasma using a Zetaffinity® cartridge. a.u., affinity unit.

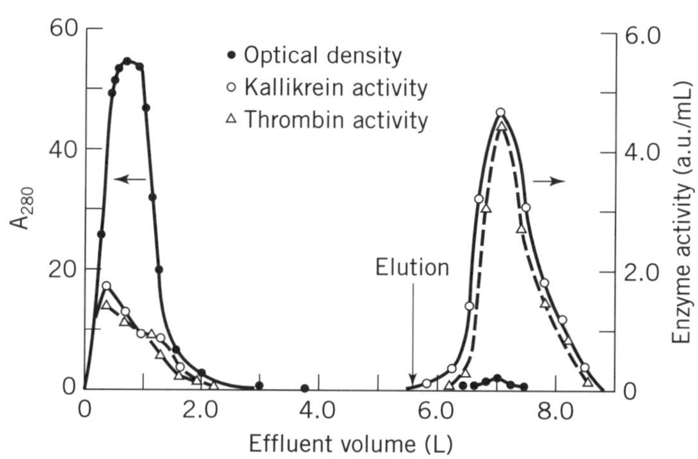

Figure 12. Trypsin purification using a Zetaffinity® cartridge. a.u., affinity unit.

Table 2. Columns Used by Strætkvern et al. (15)

Parameter	Small column	Medium column	Large column
Column type	Axial flow	Radial flow	Axial flow
Column volume (mL)	60	250	2,500
Cross-sectional flow area (cm^2)	3.8	120 (outer)	154
Volumetric flow rate (L/h)	0.6	17.4	20.9
Sample volume (L)	0.033	0.135	1.32
Protein concentration (mg/mL)	2.6	2.6	2.4
Scale-up factor	1	4	40

Table 3. Purification Results Obtained by Strætkvern et al. (15)

Parameter	Small column	Medium column	Large column
Elution volume (L)	0.20	0.87	3.57
Protein (mg)	10	44	350
Total activity (units $\times 10^{-6}$)	20	78	783
Specific activity (units $\times 10^{-6}$ mg^{-1})	2.00	1.77	1.35
Yield (%)	100	107	76
Purification (fold)	20	17	13
Cycle time (h)	1.3	0.25	1.4
Productivity (units $\times 10^{-6}$ $h^{-1} \cdot mL^{-1}$ gel)	0.256	1.25	0.135
Productivity (mg $\cdot h^{-1} \cdot mL^{-1}$ gel)	0.13	0.70	0.1

preted as a fixed rule here because the AFC columns they used were not equivalent pancakelike columns.

Weaver et al. (16) used a 10-L Superflo RFC column packed with Q-Sepharose Fast Flow medium from Pharmacia for the separation of uridine phosphorylase from total crude extracts of *Escherichia coli.* After fermentation, 375 L of broth was concentrated to 10 L. It was then washed with 50 L of 20 mM K_3PO_4 buffer containing 1 mM $MgCl_2$. After adding 20 μg lysozyme per milliliter of broth, 100 mg of DNase, and 100 mg of RNase, the cells were homogenized using a bead mill. The supernatant was diluted to 30 L and then applied to the 10-L RFC column. The feed was recirculated back to the column at a flow rate of 1.3 L/min for 3 h and then discharged. The column was washed with three different buffer solutions to remove bound lipid and hydrophobic proteins. Elution was then carried out using a buffer containing 0.225 mM NaCl. The collected effluent was 50 L. The final product after dialysis had a purity of 85% and a recovery yield of 82%.

McCartney (17) used two Superflo-100 RFC columns packed with S-Sepharose FF (Pharmacia) ion-exchange media in tandem for the purification of an undisclosed recombinant protein from *E. coli.* The system was able to process 4 L feed in less than 2 h with a concentration factor of 64 times.

Saxena et al. (12) applied a Superflo-1500 column (1,500 mL in bed volume) packed with immobilized Protein-A Sepharose in the purification of an antimelanoma IgG2A antibody from ascites fluids. Sample was loaded with a flow rate of 104 mL/min. After loading, the column was washed with a buffer using a flow rate of 170 mL/min. Elution was carried out with a flow rate of 92 mL/min. An actual recovery of 3.1 g of the antibody with a product purity greater than 97% was achieved in 3.5 h. A Superflo-1500 packed with riboflavin immobilized on a Sepharose-4B matrix via an epoxy linkage was used by the same researchers to obtain a crude preparation of a riboflavin-binding protein (12). The entire run was carried out using a flow rate of 350 mL/min. They also used a Superflo-200 column packed with antiricin B-chain antibody immobilized on cross-link agarose (12). The flow rate used was 45 mL/min. A 100% yield was achieved with a product amount of 2.1 g and a purity of 100%.

MATHEMATICAL MODELING OF RADIAL FLOW CHROMATOGRAPHY

In 1950, Lapidus and Amundson (18) proposed a simplified theoretical model for RFC. Their model ignores radial diffusion in the bulk-fluid phase and intraparticle diffusion. It is similar to that used by Rachinskii (19). Inchin and Rachinskii (20) subsequently included molecular diffusion in the bulk-fluid phase. Lee et al. (21) proposed several single component rate models for the comparison of statistical moments for RFC and AFC. They included radial dispersion, intraparticle diffusion, and external mass transfer effects. Kalinichev and Zolotarev (22) performed an analytical study on moments for single component RFC in which they treated the radial dispersion coefficient as a variable.

A rate model for nonlinear single component RFC was solved numerically by Lee (23) by using the finite difference and orthogonal collocation methods. His model considered radial dispersion, intraparticle diffusion, external mass transfer, and nonlinear isotherms. It used averaged radial dispersion and mass transfer coefficients instead of treating them as variables. A nonlinear model of this kind of complexity has no analytical solution and must be solved numerically.

Realistic modeling of RFC should treat the radial dispersion and external mass transfer coefficients as variables rather than constants because the linear flow velocity (v) in the RFC column changes continuously along the radial coordinate of the column. Without this distinctive important feature the curvature in the flow path is lost, and thus the column can be imaginatively cut and spread out to become exactly like a pancakelike AFC column.

General Rate Model for Multicomponent RFC

Gu et al. (24,25) presented a general rate model for RFC in which radial dispersion and external mass transfer coefficients are treated as variables rather than constants. The model was solved numerically using finite element and the collocation methods for discretizations of the bulk-fluid phase and the particle phase governing partial differential equations (PDEs), respectively. Figure 13 is an anatomy of a RFC column for the purpose of modeling. The following basic assumptions are made in order to formulate a general rate model for RFC.

1. The column is isothermal.
2. The porous particles in the bed can be treated as spherical and uniform in diameter.
3. The concentration gradients in the axial direction are negligible. This means that the maldistribution of radial flow is ignored.
4. The fluid inside particle macropores is stagnant, that is, there is no convective flow inside macropores.
5. An instantaneous local equilibrium exists between the macropore surfaces and the stagnant fluid in the macropores.
6. The film mass transfer theory can be used to describe the interfacial mass transfer between the bulk-fluid and particle phases.
7. The diffusional and mass transfer coefficients are constant and independent of the mixing effects of the components involved.

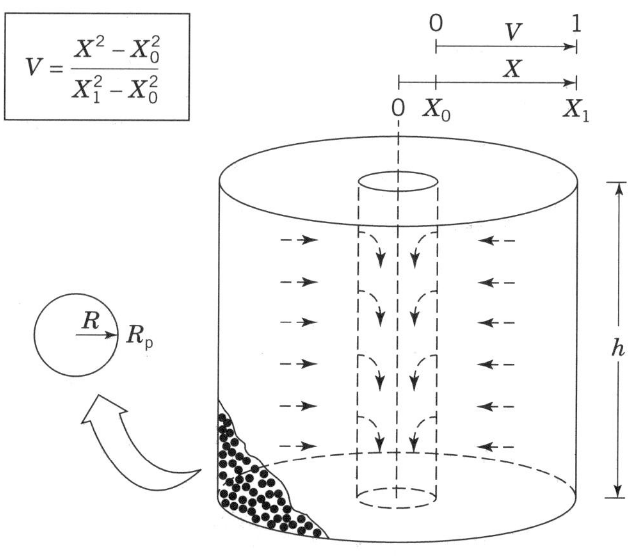

Figure 13. Anatomy of an inward-flow RFC column.

Based on these basic assumptions, equations 1 and 2 are formulated from the differential mass balance for each component in the bulk-fluid and particle phases. In equation 1, the plus sign represents outward flow, and $-v$ is inward flow.

$$-\frac{\partial}{\partial X}\left(D_{bi}X\frac{\partial C_{bi}}{\partial X}\right) \pm v\frac{\partial C_{bi}}{\partial X} + \frac{\partial C_{bi}}{\partial t} + \frac{3k_i(1-\epsilon_b)}{\epsilon_b R_p}(C_{bi} - C_{pi,R=R_p}) = 0 \quad (1)$$

$$(1-\epsilon_p)\frac{\partial C_{pi}^*}{\partial t} + \epsilon_p\frac{\partial C_{pi}}{\partial t} - \epsilon_p D_{pi}\left[\frac{1}{R^2}\frac{\partial}{\partial R}\left(R^2\frac{\partial C_{pi}}{\partial R}\right)\right] = 0 \quad (2)$$

in which C_{pi}^* is related to C_{pi} via isotherm (24).

The initial conditions for the PDE system are at

$$t = 0,\ C_{bi} = C_{bi}(0, X) \quad (3)$$

and

$$C_{pi} = C_{pi} = C_{pi}(0, R, X) \quad (4)$$

The boundary conditions are at the inlet X position

$$\partial C_{bi}/\partial X = (v/D_{bi})[C_{bi} - C_{fi}(t)] \quad (5)$$

and at the outlet X position

$$\partial C_{bi}/\partial X = 0 \quad (6)$$

Equations 1 and 2 can be written in dimensionless forms as follows.

$$-\frac{\partial}{\partial V}\left(\frac{\alpha}{Pe_i}\frac{\partial c_{bi}}{\partial V}\right) \pm \frac{\partial c_{bi}}{\partial V} + \frac{\partial c_{bi}}{\partial \tau} + \chi_i(c_{bi} - c_{pi,r=1}) = 0 \quad (7)$$

$$\frac{\partial}{\partial \tau}[(1-\epsilon_p)c_{pi}^* + \epsilon_p c_{pi}] - \eta_i\left[\frac{1}{r^2}\frac{\partial}{\partial r}\left(r^2\frac{\partial c_{bi}}{\partial r}\right)\right] = 0 \quad (8)$$

In equations 1 to 7, the dimensionless variable, $V = (X^2 - X_0^2)/(X_1^2 - X_0^2) \in [0, 1]$, is based on the local volume averaging method [24]. $\alpha = 2\sqrt{V + V_0}(\sqrt{1 + V_0} - \sqrt{V_0})$ is a function of V.

The dimensionless initial conditions are at

$$\tau = 0,\ c_{bi} = c_{bi}(0, V) \quad (9)$$

and

$$c_{pi} = c_{pi}(0, r, V) \quad (10)$$

The dimensionless boundary conditions are

$$\partial c_{bi}/\partial V = Pe_i[c_{bi} - C_{fi}(\tau)/C_{0i}] \quad (11)$$

At the inlet V position, for frontal adsorption, $C_{fi}(\tau)/C_{0i} = 1$. For elution,

$$C_{fi}(\tau)/C_{0i} = \begin{cases} 1 & 0 < \tau < \tau_{imp} \\ 0 & \text{otherwise} \end{cases}$$

After the introduction of a sample in the form of a rectangular pulse, it component I is displaced, $C_{fi}(\tau)/C_{0i} = 0$. If component I is a displacer, $C_{fi}(\tau)/C_{0i} = 1$. At the outlet V position, $\partial c_{bi}/\partial V = 0$. For the particle phase governing equation, the boundary conditions are at r = 0

$$\partial c_{pi}/\partial r = 0 \tag{12}$$

and at r = 1

$$\partial c_{pi}/\partial r = Bi_i(c_{bi} - c_{pi,r=1}) \tag{13}$$

Note that all the dimensionless concentrations are based on C_{0i}, that is, the maximum of the feed profile $C_{fi}(\tau)$ for each component.

The radial dispersion coefficient (D_{bi}) depends on the linear velocity (v). In liquid chromatography, it can be assumed (5,22,25) that $D_{bi} \propto v$. Thus, $Pe_i = v\,(X_1 - X_0)/D_{bi}$ can be considered constant in liquid RFC. The variation of Bi_i values observes the following relationship:

$$Bi_i \propto k_i \propto v^{1/3} \propto (1/X)^{1/3} \propto (V + V_0)^{-1/6} \tag{14}$$

If $Bi_{i,V=1}$ values are known, Bi_i values anywhere else can be obtained from equation 15.

$$Bi_{i,V} = [(1 + V_0)/(V + V_0)]^{1/6} Bi_{i,V=1} \tag{15}$$

χ_i can be calculated from Bi_i using its definition $\chi_i = 3Bi_i\eta_i(1 - \epsilon_b)/\epsilon_b$.

Numerical Solution

The PDE system of the governing equations is first discretized. The finite element and orthogonal collocation methods are used to discretize the bulk-fluid phase and the particle phase governing equations, respectively. The resulting ordinary differential equation (ODE) system is then solved using the public domain ODE solver called DVODE written by Brown et al. (26). The compiled Fortran program is free to academic researchers. Both DOS and Windows 95 versions are available. Information on the software is available at *http://www.ent.ohiou.edu/~guting/CHROM/*.

A study of the effects of treating D_{bi} and k_i as variables compared to treating them as constants was carried out by Gu et al. (5,25). The comparison between RFC and AFC was also studied through computer simulation. Figure 14 shows that outward flow gives slightly sharper concentration profile (5). The figure also shows that RFC gives similar concentration profiles as AFC with equivalent physical parameters. These theoretical results support the experimental results obtained by Tharakan and Belizaire (8).

SCALE-UP OF RFC COLUMNS

One of the advantages of RFC columns is the relative ease for scale-up. In Sepragen's Superflo column series, increasing the column height is a rather safe way to accommodate an increase of sample size to a certain degree. However, one must keep in mind that maldistribution in radial flow may deteriorate slightly. If the bed thickness in the radial direction is increased, bed pressure usually increases proportionally. The performance prediction will be less reliable if the column diameter is increased. Figure 15 shows that a 15-fold increase of sample load and bed column produced very similar performances when DEAE cellulose is used to separate an ascites fluid (27,28). Figure 16 is an example with a 50-fold increase of sample load and bed column (27). These two examples are very successful examples.

The mathematical model introduced above can be used to help the scale-up process. Before a column is bought, its performance can be predicted using computer simulation. Isotherm data must be obtained experimentally or be supplied by the vendor of the packing material. Mass transfer parameters can be estimated by using existing correlations (24).

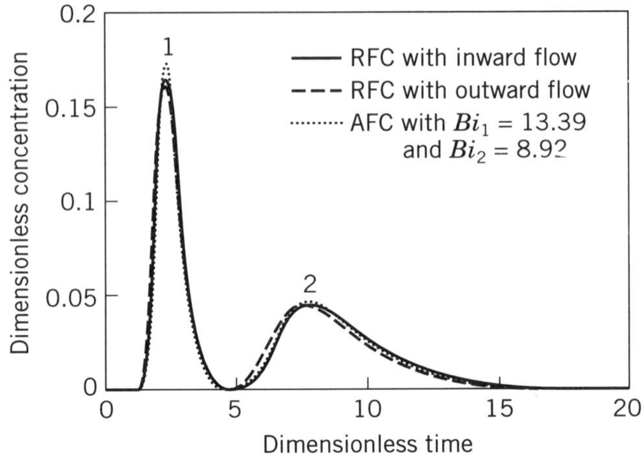

Figure 14. Simulated comparison for inward- and outward-flow RFC and AFC.

CONCLUSIONS

RFC columns have a short flow path and a large flow area, resulting in a small bed pressure. They are especially suitable when soft-gels are used. Because of its limited resolution, RFC should be used in chromatographic separations involving strong solute–stationary phase interactions, such as affinity chromatography, strong anion or cation exchange, strong reversed phase, and strong hydrophobic interaction chromatography. RFC is not suitable for SEC and other chromatography with weak solute–stationary phase interactions. RFC is an attractive alternative to AFC in preparative- and large-scale separations. Both experimental and theoretical modeling indicate that to a large extent an RFC column behaves much like a pancake-style AFC column with the same packing volume and with its bed height about the same as the packing's radial thickness in RFC column. However, RFC has a much

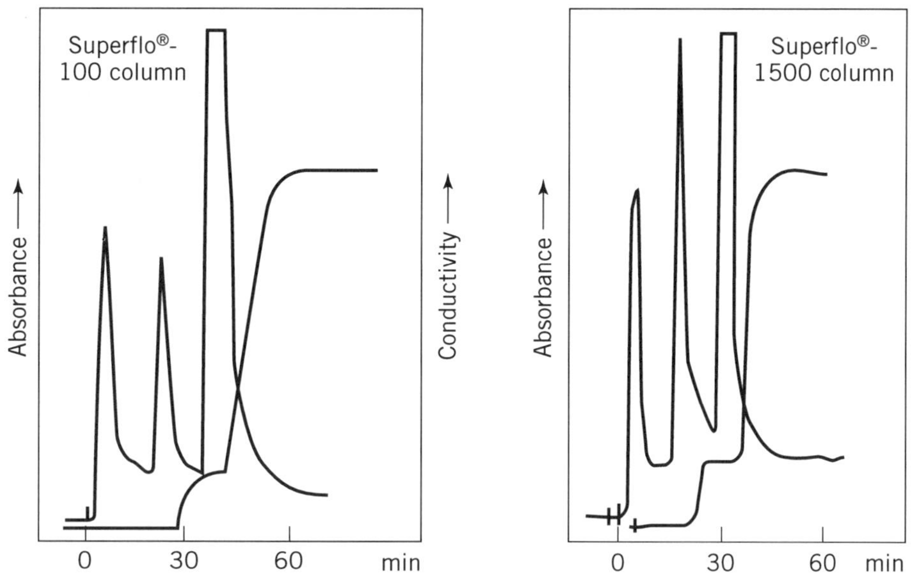

Packing:	DEAE cellulose
Load:	10 mL ascites fluid
Flow rate:	10 mL/min
Start buffer:	10 mM phosphate, pH 8.5
Step gradient:	60 mM, 500 mM NaCl in start buffer

Packing:	DEAE cellulose
Load:	150 mL ascites fluid
Flowrate:	150 mL/min
Start buffer:	10 mM phosphate, pH 8.5
Step gradient:	60 mM, 500 mM NaCl in start buffer

Figure 15. A 15-fold scale-up example using Superflo® columns.

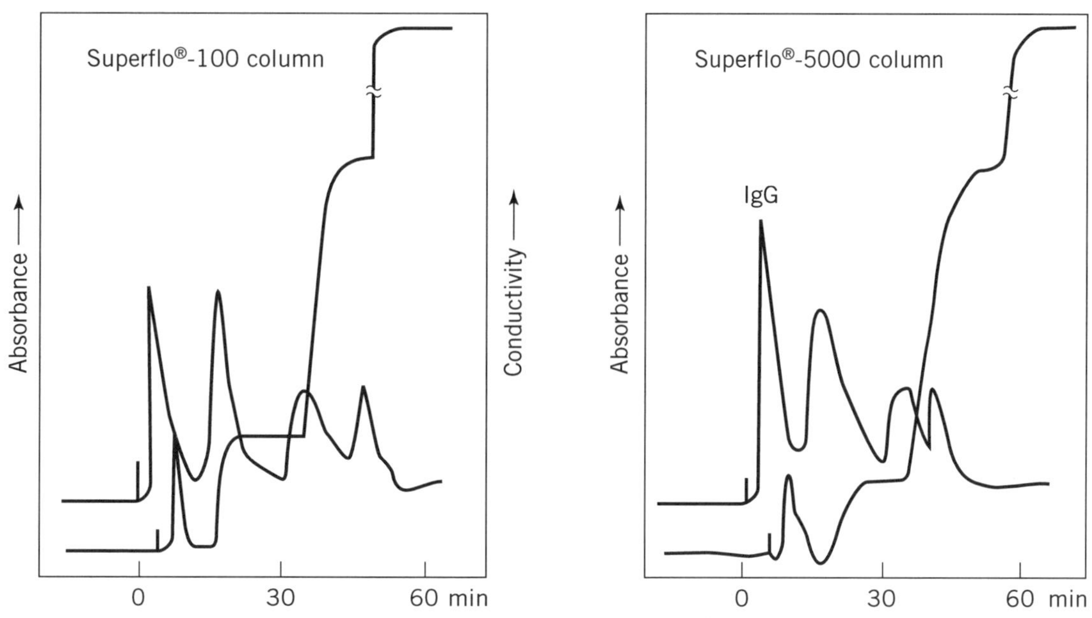

Packing:	DEAE cellulose
Load:	10 mL cell culture fluid (murine IgG)
Flow rate:	10 mL/min
Start buffer:	10 mM phosphate, pH 8.5
Step gradient:	60 mM, 250 mM, 700 mM NaCl in start buffer

Packing:	DEAE cellulose
Load:	500 mL cell culture fluid (murine IgG)
Flow rate:	500 mL/min
Start buffer:	10 mM phosphate, pH 8.5
Step gradient:	60 mM, 250 mM, 700 mM NaCl in start buffer

Figure 16. A 15-fold scale-up example using Superflo® columns.

smaller footprint, and it seems to handle flow distribution better than very wide pancakelike columns.

NOMENCLATURE

Bi_i	Biot number of mass transfer for component i, $k_i R_p(\epsilon_p D_{pi})$
C_{0i}	Concentration used for nondimensionalization, $\max\{C_{fi}(t)\}$
C_{bi}	Bulk-fluid phase concentration of component i
C_{fi}	Feed concentration profile of component i, a time-dependent variable
C_{pi}	Concentration of component i in the stagnant fluid phase inside particle macropores
C^*_{pi}	Concentration of component i in the solid phase of particle (based on unit volume of particle skeleton)
c_{bi}	$= C_{bi}/C_{0i}$
c_{pi}	$= C_{pi}/C_{0i}$
c^*_{pi}	$= C^*_{pi}/C_{0i}$
D_{bi}	Axial or radial dispersion coefficient of component i
D_{pi}	Effective diffusivity of component i, porosity not included
k_i	Film mass transfer coefficient of component i
Pe_i	Peclet number of radial dispersion for component i, $v(X_1 - X_0)/D_{bi}$
R	Radial coordinate for particle
R_p	Particle radius
r	$= R/R_p$
t	Dimensional time ($t = 0$ is the moment a sample enters a column)
v	Interstitial velocity
V_0	Dimensionless constant corresponding to X_0
V	Dimensionless volumetric coordinate
X	Coordinate in the radial direction for a RFC column.

Greek Letters

α	$= 2\sqrt{V + V_0}(\sqrt{1 + V_0} - \sqrt{V_0})$ for RFC
ϵ_b	Bed void volume fraction
ϵ_p	Particle porosity
η_i	Dimensionless constant, $\epsilon_p D_{pi} L/(R_p^2 v)$
χ_i	Dimensionless constant for component i, $3Bi_i\eta_i(1 - \epsilon_b)/\epsilon_b$
τ	Dimensionless time, vt/L
τ_{imp}	Dimensionless time duration for a rectangular pulse of the sample

ACKNOWLEDGMENT

The author wishes to thank Mr. Sanjeev Sexena of Sepragen Corp., Hayward, Calif., for providing information on Sepragen's Superflo® radial flow columns.

BIBLIOGRAPHY

1. D. McCormick, *Biotechnology* **6**, 158–160 (1988).
2. P. Hopf, *Ind. Eng. Chem.* **39**, 938–940 (1947).
3. S.H. Huang, S. Roy, K.C. Hou, and G.T. Tsao, *Biotechnol. Prog.* **4**, 159–165 (1988).
4. U.S. Pat. 4,627,918 (Dec. 9, 1986), V. Saxena (to Sepragen Corporation).
5. T. Gu, G.-J. Tsai, and G.T. Tsao, in A. Fiechter ed., *Advances in Biochemical Engineering / Biotechnology*, Springer-Verlag, Berlin, 1993, pp. 73–95.
6. V. Saxena and M. Dunn, *Biotechnology* **7**, 250–255 (1982).
7. V. Saxena and A.E. Weil, *BioChromatography* **2**, 90–97 (1987).
8. J.P. Tharakan and M. Belizaire, *J. Liq. Chromatogr.* **18**, 39–49 (1995).
9. J.P. Tharakan and M. Belizaire, *J. Chromatogr.* **702**, 191–196 (1995).
10. L. Lane, M.L. Koscielny, P.R. Levison, D.W. Toome, and E.T. Butts, *Bioseparation* **1**, 141–147 (1990).
11. P. Ernst, *Aust. J. Biotechnol.* **1**, 22–26 (1987).
12. V. Saxena, A.E. Weil, R.T. Kawahata, W.C. McGregor, and M. Chandler, *Am. Lab.* **19**, 112–120 (1987).
13. Y. Planques, H. Pora, and F.D. Menozzi, *J. Chromatogr.* **539**, 531–533 (1991).
14. A. Akoum, F. Devichi, M. Kalyanpur, J.P. Neff, M.A. Vijayalakshmi, and M. Sigot, *Process Biochem.* 55–59 (April 1989).
15. K.O. Strætkvern, A.J. Raae, K. Folkvord, B.A. Næss, I.M. Aasen, *Bioseparation* **2**, 81–93 (1991).
16. K. Weaver, D. Chen, L. Walton, L. Elwell, and P. Ray, *BioPharm* 25–29 (July/August 1990).
17. J.E. McCartney, *BioTechniques* **11**, 648–649 (1991).
18. L. Lapidus and N.R. Amundson, *J. Phys. Colloid. Chem.* **54**, 821–825 (1950).
19. V.V. Rachinskii, *J. Chromatogr.* **33**, 234–242 (1968).
20. P.A. Inchin and V.V. Rachinskii, *Russ. J. Phys. Chem.* **47**, 1331–1333 (1977).
21. W.-C. Lee, S.H. Huang, and G.T. Tsao, *AlChE J.* **34**, 2083–2087 (1988).
22. A.I. Kalinichev and P.P. Zolotarev, *Russ. J. Phys. Chem.* **51**, 871–873 (1977).
23. W.-C. Lee, Ph.D. Thesis, Purdue University, West Lafayette, Ind., 1989.
24. T. Gu, *Mathematical Modeling and Scale-Up of Liquid Chromatography*, Springer-Verlag, Berlin, 1995, pp. 102–106, 116.
25. T. Gu, G.-J. Tsai, and G.T. Tsao, *Chem. Eng. Sci.* **46**, 1279–1288 (1991).
26. P.N. Brown, G.D. Byrne, and A.C. Hindmarsh, *Sci. Stat. Comput.* **10**, 1038–1051 (1989).
27. V. Saxena and A.E. Weil, *BioChromatography* **2**, 90–97 (1987).
28. V. Saxena, K. Subramanian, S. Saxena, and M. Dunn, *BioPharm* 46–49 (March 1989).

See also CHROMATOGRAPHY, COMPUTER-AIDED DESIGN; CHROMATOGRAPHY, ION EXCHANGE; CHROMATOGRAPHY, SIZE EXCLUSION.